Introduction to Organic Chemistry

FOURTH EDITION

ANDREW STREITWIESER
UNIVERSITY OF CALIFORNIA, BERKELEY

CLAYTON H. HEATHCOCK
UNIVERSITY OF CALIFORNIA, BERKELEY

EDWARD M. KOSOWER
TEL AVIV UNIVERSITY

MACMILLAN PUBLISHING COMPANY
NEW YORK
MAXWELL MACMILLAN CANADA
TORONTO
MAXWELL MACMILLAN INTERNATIONAL
NEW YORK OXFORD SINGAPORE SYDNEY

Editor: Paul F. Corey
Production Supervisor: Elisabeth H. Belfer
Production Manager: Pamela Kennedy Oborski
Text and Cover Designer: Sheree L. Goodman
Cover photograph: © Richard Megna/Fundamental Photographs
Photo Researcher: Chris Migdol

This book was set in Times Roman and Goudy Old Style by York Graphic Services, Inc., printed and bound by Von Hoffmann Press. The cover was printed by Von Hoffmann Press.

Printed in the United States of America

Macmillan Publishing Company
866 Third Avenue, New York, New York, 10022

Macmillan Publishing Company is part of the Maxwell Communication Group of Companies.

Maxwell Macmillan Canada, Inc.
1200 Eglinton Avenue East
Suite 200
Don Mills, Ontario M3C 3N1

Library of Congress Cataloging-in-Publication Data

Streitwieser, Andrew (date)
Introduction to organic chemistry.—4th ed. / Andrew Streitwieser, Clayton H. Heathcock, Edward M. Kosower.
p. cm.
Includes index.
ISBN 0-02-418170-6
1. Chemistry, Organic. I. Heathcock, Clayton H. II. Kosower, Edward M. III. Title.
QD251.2.S76 1992 91-33039
547—dc20 CIP

Printing: 1 2 3 4 5 6 7 8 Year: 2 3 4 5 6 7 8 9 0 1

Symbols for Amino Acids

Ala	alanine
Arg	arginine
Asn	asparagine
Asp	aspartic acid
Cys	cysteine
Gln	glutamine
Glu	glutamic acid
Gly	glycine
His	histidine
Ile	isoleucine
Leu	leucine
Lys	lysine
Met	methionine
Phe	phenylalanine
Pro	proline
Ser	serine
Thr	threonine
Trp	tryptophan
Tyr	tyrosine
Val	valine

Greek Alphabet

Lower Case	Capital	Name
α	A	alpha
β	B	beta
γ	Γ	gamma
δ	Δ	delta
ϵ	E	epsilon
ζ	Z	zeta
η	H	eta
θ	Θ	theta
ι	I	iota
κ	K	kappa
λ	Λ	lambda
μ	M	mu
ν	N	nu
ξ	Ξ	xi
o	O	omicron
π	Π	pi
ρ	P	rho
σ	Σ	sigma
τ	T	tau
υ	Υ	upsilon
ϕ	Φ	phi
χ	X	chi
ψ	Ψ	psi
ω	Ω	omega

Chemical Abbreviations

Ac	acetyl, $CH_3C(=O)—$
Boc	*t*-butoxycarbonyl, $(CH_3)_3COC(=O)—$
n-Bu	*n*-butyl, $CH_3CH_2CH_2CH_2—$
t-Bu	*t*-butyl, $(CH_3)_3C—$
Cbz	benzyloxycarbonyl, $C_6H_5CH_2OC(=O)—$
DCC	dicyclohexylcarbodiimide, $C_6H_{11}N{=}C{=}NC_6H_{11}$
DIBAL	diisobutylaluminum hydride, $[(CH_3)_2CHCH_2]_2AlH$
diglyme	bis(2-methoxyethyl) ether, $(CH_3OCH_2CH_2)_2O$
DMF	dimethylformamide, $(CH_3)_2NCHO$
DMSO	dimethyl sulfoxide, $(CH_3)_2SO$
DNP	2,4-dinitrophenyl, $2,4—(O_2N)_2C_6H_3—$
Et	ethyl, $CH_3CH_2—$
glyme	1,2-dimethoxyethane, $CH_3OCH_2CH_2OCH_3$
HMPT	hexamethylphosphoric triamide, $[(CH_3)_2N]_3PO$
LAH	lithium aluminum hydride, $LiAlH_4$
LDA	lithium diisopropylamide, $LiN[CH(CH_3)_2]_2$
Me	methyl, $CH_3—$
PPA	polyphosphoric acid
THF	tetrahydrofuran, $CH_2CH_2CH_2CH_2O$ (ring)
TMS	tetramethylsilane, $(CH_3)_4Si$
Ts	*p*-toluenesulfonyl, *p*-$CH_3C_6H_4SO_2—$

Equilibria and Free Energy

$$A \rightleftharpoons B \qquad K = \frac{[B]}{[A]}$$

$$\Delta G^\circ = -RT \ln K$$

K	Percent B	Percent A	ΔG°, kcal mole^{-1} (25°)
0.0001	0.01	99.99	+5.46
0.001	0.1	99.9	+4.09
0.01	0.99	99.0	+2.73
0.1	9.1	90.9	+1.36
0.33	25	75	+0.65
1	50	50	0
3	75	25	−0.65
10	90.9	9.1	−1.36
100	99.0	0.99	−2.73
1000	99.9	0.1	−4.09
10000	99.99	0.01	−5.46

Introduction to Organic Chemistry

PREFACE

Organic chemistry is alive and continuously being transformed by new concepts, new technologies, and new reactions. Any modern textbook has to provide the fundamentals in a sound and clear manner, and to give a glimpse of the frontier. This new edition was motivated by the need to expand the coverage in a biochemical/biological direction, and to introduce certain significant new developments in approach. Organic chemistry is so broad and encompasses so much that those of us concerned with teaching the subject to new generations of students keep searching for the best way. So it is that textbooks spawn new editions. We are gratified by the response to previous editions of our textbook and are indebted to numerous teachers and students of organic chemistry for suggested improvements to make our text still better. The new edition continues this trend with the addition of a third author. The color introduced in the third edition has been recognized as a significant aid to study and in emphasizing the location of the site of reaction. Experienced organic chemists view structures as a unified *gestalt* and see quickly where structural change has occurred in a reaction. The inexperienced student needs to search for such changes. The use of color helps by ''spotlighting'' where the action is. We have also used color to indicate the essential stoichiometry of many reactions, showing where specific atoms go in a manner resembling the use of isotopic labels. That is, we have tried to use color as a pedagogic tool. A ''photoessay'' insert shows that the sparing use of multiple colors allows one to combine good pedagogy with the aesthetic aspects of organic chemistry.

Many small changes and additions have been made to improve presentation, to introduce newer material, or to bring certain facts up to date. More obvious changes include some reorganization. There are now two chapters on the chemistry of aldehydes and ketones (Chapters 14 and 15). The topic of molecular recognition has emerged as a significant part of modern organic chemistry, and its appearance in limited number of contexts is reflected in the new Chapter 33. The discussion on nuclear magnetic resonance has been reorganized to begin with ^{13}C spectra (Chapter 13). Chapter 19 has been expanded to include much more on the biological roles for derivatives of carboxylic acids.

Frontier orbital concepts have been retained to a considerable extent. However, resonance theory, with its symbolism rooted in conventional structures, continues to be a sound way to leach much organic chemistry. The molecular orbital theory has been written as a separate chapter (Chapter 22) that can be skipped by those who prefer a conventional treatment.

The most striking change throughout the book is the addition of ''Highlights,'' which present the essence of some of the preceding material in a way that should remind the student of what he or she has read. Immediately rereading the relevant sections should be an effective way to study. We have also introduced ''Asides'' at the ends of chapters to present particular topics in a different context that should intrigue students.

We are pleased that Professor Judith G. Koch has prepared a Student Study Guide that includes answers to most of the exercises and problems as well as numerous supplemental problems. This supplement also contains useful outlines of the essential points of each chapter, tips for effective study strategies, and an extensive appendix of reactions keyed by page to the text.

We have retained many of the attributes of the past editions that we regard as strengths: explanations of physical properties and phenomena, use of specific examples for reactions with experimental details, descriptive chemistry of important inorganic reagents, and up-to-date values of important properties such as acidities and bond dissociation energies. The stereoscopic projections emphasize the three-dimensional nature of organic structures. We only hope that with the growing capabilities of computer graphics, stereo viewers will become increasingly available to readers.

As in previous editions, we have many people to thank for their comments, corrections, and suggestions. We are particularly indebted to

Paul A. Bartlett
University of California, Berkeley

Dr. K. C. Bass
The City University, London

Janet Carlson
Macalester College

James A. Deyrup
University of Florida

Francois N. Diedrich
University of California, Los Angeles

Leonard W. Fine
Columbia University

James P. Hagen
University of Nebraska, Omaha

Dr. Lionel S. Hart
Bristol University

Dr. Ann Jarvie
Aston University

Judith G. Koch
Ithaca College

Ronald G. Lawler
Brown University

Terrence C. Morrill
Rochester Institute of Technology

Stanley M. Roberts
University of Exeter

Paul F. Schatz
University of Wisconsin–Madison

Peter G. Schultz
University of California, Berkeley

Alan Shusterman
Reed College

Marc S. Silver
Amherst College

H. Stephen Stoker
Weber State College

James K. Wood
University of Nebraska, Omaha

A. S.
C. H. H.
E. M. K.

CONTENTS

A NOTE TO THE STUDENT

Before you begin your adventure in organic chemistry it is perhaps appropriate for you to take a few minutes to plan your journey. The first chapter of this book provides a succinct history of the development of chemical science up to the beginnings of organic chemistry in the middle of the last century together with a table indicating the broad scope of organic chemistry. Immediately following is a brief review of the important concepts of orbitals and chemical bonds. Although Chapter 2 is a review of topics you learned in your general chemistry course, it is essential that you be familiar with this material before proceeding with your study. Therefore, take an hour or so to go over this chapter and work the problems, even if your instructor does not specifically repeat the review material in lecture. The Lewis structures and simple hybrid orbitals and bond concepts reviewed in this chapter are fundamental to the chemistry that follows.

Chapters 3 and 4 are intended to introduce you to the two important aspects of organic chemistry—structures and reactions. Although some general chemistry courses will have covered the subject matter of these two chapters, many will not. Again, you should be thoroughly acquainted with the material in Chapters 3 and 4 before going further.

In Chapter 5 you will encounter the simplest organic compounds, those made up solely of carbon and hydrogen. This chapter also introduces two basic principles—thermodynamics and conformations. In Chapter 6 you will find the first detailed study of an organic reaction, free radical halogenation, and you will be able to put into practice the general ideas of reaction mechanisms and thermodynamics presented in Chapters 4 and 5. Chapter 7 will introduce you to a fascinating topic—stereochemistry. This special aspect of molecular structure is of fundamental importance to organic chemistry and to biochemistry. Although you may find thinking in three dimensions difficult at first, practice pays off. Once you can freely visualize organic compounds as three-dimensional objects just like the familiar objects of everyday life, you will discover that organic chemistry is suddenly ‘‘much easier’’ than you thought.

Chapter 8 introduces two new groups of organic compounds, those containing halogen atoms and those with carbon-metal bonds. New structures are involved, and your repertoire of reactions will start to grow. The displacement reaction, which is treated in Chapter

9, is one of the most important reactions in organic chemistry. It is vital that you grasp the mechanism of this reaction because you will find that the same relationships of structure and reactivity occur over and over again. By acquiring an early understanding of the principles of the displacement reaction, you can avoid mindlessly memorizing dozens of reactions; you will be able to recognize that many "new" reactions are only different versions of reactions you already understand.

As you progress in your study of organic chemistry, you will discover that the science is conveniently organized in terms of functional groups—the parts of organic compounds other than the carbon-carbon and carbon-hydrogen single bonds that are common to all organic structures. The next group of three chapters systematically treats the chemistry of alcohols and ethers (Chapter 10), alkenes (Chapter 11), alkynes and nitriles (Chapter 12). In each of these chapters the topical sequence is similar.

1. The functional group itself—its characteristic geometry and its effect on the geometry of the hydrocarbon part of the molecule containing it.
2. How compounds of the class are named.
3. The common physical properties of the class of compounds.
4. The chemical reactions that are characteristic of the functional group, to which the bulk of the chapter is devoted.

In most organic reactions one functional group is typically transformed into another. Thus, you will find that an organic reaction can usually be thought of both as an attribute of a given class of compounds and as a method of preparation of another class of compounds. In this book you will find that reactions are generally introduced as a property of a class of compounds. However, each functional group chapter also contains a section on preparative methods. In general, the emphasis in these preparation sections is on the practical aspects of the reactions rather than on the mechanistic aspects.

With the chemistry of several key functional groups and a number of structures and reactions under our belts, we now shift gears and turn in Chapter 13 from chemical reactivity back to chemical structure. In this chapter you will have your first encounter with spectroscopy, which is the principal way modern organic chemists find out how the atoms of a molecule are joined together. In all, four sections of the book are devoted to various forms of spectroscopy—Chapter 13 (nuclear magnetic resonance spectroscopy), Chapter 17 (infrared spectroscopy), Section 22.6 (ultraviolet spectroscopy), and Chapter 34 (mass spectroscopy). Each kind of spectroscopy gives us different pieces of the molecular jigsaw puzzle, but nuclear magnetic resonance spectroscopy is the most important. In subsequent chapters on functional groups the discussion of physical properties includes spectroscopy. After Chapter 13 on nuclear magnetic resonance spectroscopy, Chapter 14 on the important carbonyl functional group in aldehydes and ketones, and Chapter 15 on the especially important enol derivatives of aldehydes and ketones, we come to a convenient break in the sequence.

Chapter 16 is rather different from the other chapters in the book in that it is essentially a review of the organic chemistry learned up to that point. In some ways, learning organic chemistry is like learning a language. The simple reactions and mechanistic principles are like the vocabulary of the language. As in learning a language, you must first learn the vocabulary. However, if you only know the words of a language, you will not be able to compose a poem, or even rent a hotel room with hot and cold running water. It is necessary also to learn how the words are put together to make sentences, the grammar and syntax of the language. In organic chemistry we learn to put several simple reactions together to achieve an overall transformation that cannot be accomplished by any single reaction. Chapter 16 will give you an opportunity to practice multistep synthesis using the reactions you have learned. This chapter is followed by Chapter 17 on infrared spectroscopy and two more chapters or carbonyl-containing functional groups, the carboxylic acids and their derivatives (Chapters 18 and 19).

In the first half of the book, you will have considered the typical chemical properties of molecules having a single functional group. In Chapter 20 you will discover that compounds having two functional groups can have properties that are very different from those of compounds having only one of the groups. This study of "conjugated systems" is fundamental to the understanding of aromatic chemistry (Chapters 21 and 23) and provides a convenient way of introducing some new concepts in molecular orbital theory (Chapter 22), a theory that is especially important with conjugated and "aromatic" compounds and related reactions and has had a major impact on modern organic chemistry.

The remainder of the book contains a good deal of chemistry of such "polyfunctional" compounds. For example, Chapter 27 treats the special chemistry of compounds with two oxygen-containing functional groups. Chapters 28 and 29 cover the chemistry of two important families of polyfunctional "natural products"—carbohydrates and amino acids. Chapter 30 discusses aromatic halogen- and oxygen-containing compounds. Through Chapter 23 your study of organic chemistry will have dealt with compounds made up mainly of carbon, hydrogen, oxygen, and the halogens. In Chapters 24–26, you will encounter organic compounds of nitrogen, phosphorus, sulfur, and silicon. Of these compounds, the amines (Chapter 24) are the most important, but the other nitrogen functions (Chapter 25) play a significant role, especially in aromatic chemistry. Although your instructor may choose to omit Chapter 26 (sulfur, phosphorus, and silicon compounds), if you are bound for a career in medicine or in the health sciences, you will want at least to read through this chapter.

Chapters 30 and 31 deal with further aspects of aromatic chemistry. The astute student will recognize that there are virtually no new concepts in these two chapters; rather, they serve to add flesh to the bones of the subject. However, it is interesting flesh, and the future chemical engineer or physician will find in these chapters many hints of things to come. Chapter 32 covers some aspects of the chemistry of heterocyclic compounds, which constitute a major fraction of all organic compounds. Chapter 33 introduces the new idea of molecular recognition, and shows how the simple chemical principles on which it is based have enormous consequences in all branches of chemistry and biology.

The final chapters of the book are optional reading. Chapter 34 introduces mass spectroscopy and Chapter 35 is an introduction to the literature of chemistry. Although you may not need to use the chemical literature at this point in your career, many of you will need this knowledge later. Chapter 35 will give you a start at the appropriate time. Finally, Chapter 36 is a collection of brief essays to give the interested student a glimpse of some of the exciting areas of modern research.

It is also appropriate at this point to mention several tools we have provided to assist you in learning organic chemistry. The first is the "Highlight," of which there are generally three or four per chapter. Some chapters have a greater number and some have a smaller number, but in all cases the purpose is the same, to extract the essence of some (not all!) of the preceding material. You should be able to read the "Highlights" and say to yourself that you know and understand that material. If not, you should go back and read those sections of the chapter again. Another feature is the "boxed sections." These sections, which are found at various points within each chapter, contain several types of information. Some give more detailed information on the topic immediately preceding. Others contain specific reaction conditions for a reaction that has been used for an example. Still others convey information of interest about specific compounds, often inorganic compounds that are employed as reagents in organic chemistry. These indented sections are set apart so that they may be skipped over by the student who is just reviewing the important principles of the chapter. Our rule of thumb at Berkeley and at Tel Aviv is that the material in these sections is for enrichment and that students are not held responsible for it on examinations. You should ask your instructor about the policy in your course.

A third invaluable tool is the exercises and problems in each chapter, the latter usually preceded by an Aside that should intrigue you and enhance your understanding of the material. The exercises, which occur in most sections, are often cast in the form of ''drill'' to provide you with immediate practice in using new principles or reactions that have just been introduced. For many of you, these exercises will seem ridiculously easy, as you will be asked to write out an equation you have just learned. However, they are an important part of the learning process. Everyone has had the experience of ''daydreaming'' while reading merrily along. It is possible to read several pages and be totally unconscious of what you have read. The exercises force you to pause periodically and check to see that you have really been assimilating what you have been reading.

The problems at the end of each chapter also contain some drill questions, but the parts of a single question may draw from many different sections of the chapter. Thus, these questions provide for a second check on your retention of the various reactions and principles you have studied. There are also ''thought questions'' that ask you to take several reactions or principles and put them together to solve a problem or in some cases to extend your knowledge and discover something yourself. To derive the greatest benefit from the exercises and problems, you will want the Student Study Guide, which contains worked-out answers to all of the exercises and problems as well as a key-word index and study hints for each chapter. In addition, the study guide contains supplemental problems, with answers. It is important that you give any problem a good try before looking up the answer. It is human nature to quit worrying about a problem as soon as the answer is known, and it is also true that we learn more from a problem we have labored over than from one we haven't given much thought to.

One teaching device we have used requires special mention—the three-dimensional stereoscopic projections that are distributed throughout the book. These computer-generated images are designed so that you may see the figure in three dimensions. With a suitable viewer each eye can focus independently on one of the two images of such a projection, and there is an illusion of depth to the resulting picture one sees. It is actually possible, albeit a bit more difficult, to see a stereo image without a special viewer. To do this, hold the page about 20 inches from your eyes and focus on a point behind the book in such a way that the two images merge. Generally, the merged image will suddenly seem three dimensional. One caution: with this method the right eye will sometimes focus on the left image and vice versa. The result is the perception of the mirror image, a concern when precise stereochemistry is important, as in Chapter 7. The stereo pictures should be considered as an adjunct to and not a substitute for molecular models. We urge you to acquire and regularly use a set of molecular models; most bookstores stock one or more relatively inexpensive student sets as learning aids for organic chemistry courses. Organic structures are three-dimensional structures, and practice with models is essential to give proper meaning to the structural symbols commonly recorded on two-dimensional pages.

With these general suggestions in mind, it remains only for us to wish you luck as you set out upon your journey through organic chemistry. All three of us look back with fond remembrance upon our own discovery of this fascinating subject; we hope that you will find it as rewarding.

Andrew Streitwieser
Clayton H. Heathcock
Edward M. Kosower

CHAPTER 1

INTRODUCTION

CHEMISTRY emerged as a coherent science only in the seventeenth century, but its roots extend back into antiquity. The first deliberate chemical changes were probably produced by paleolithic man in the use of fire to warm his body and roast his food. Being a curious and a resourceful creature, man observed and exploited other natural phenomena as well. By neolithic times he had discovered such arts as smelting, glass making, the dyeing of textiles, and the manufacture of beer, wine, butter, and cheese. However, the nature of matter and its transformations were not systematically discussed in a theoretical sense until the period of the Greek philosophers, beginning in about 600 B.C. One popular theory that emerged during this period saw all matter as being made up of the four ''elemental'' substances: fire, earth, air, and water. For a time, the atomist school, of which Democritus (ca. 460–370 B.C.) was the chief spokesman, gained popularity. The atomists hypothesized all matter to be made up of a finite number of different kinds of particles called atoms. Although the atomists promulgated their views, as in the poem *De rerum natura* (On the nature of things) by Lucretius (95–55 B.C.), for several centuries, the theory was rejected as speculative by the highly respected philosophers Plato and Aristotle. The advent of stoicism and the subsequent rise of popular religious movements in the Western world kept the idea of fundamental particles in limbo for almost two millennia.

Around the time of Christ, craftsmen intent on simulating gold and silver led the Greek philosophers of Alexandria to the idea of changing (or ''transmuting'') base metals such as lead and iron into gold and silver. Alchemy was first practiced in a serious sense by the Greeks, quickly spread to other cultures, and continued as a lively discipline throughout the world for over a thousand years. The alchemical period has often been put down as a ''dark age'' of science. However, there is nothing inherently wrong with the notion that one metal may be transformable into another. Chemistry is, in fact, based upon changes in matter. The alchemists had no way of recognizing the elemental nature of the metals with which they dealt.

Although the alchemists were uniformly unsuccessful in their quest for perfecting metals with ''the philosopher's stone,'' they contributed a great deal to chemical technology. Not only did they develop numerous processes for the production of relatively pure compounds but they also invented tools and apparatus—beakers, flasks, funnels, mortars, crucibles—many of which persist in similar form to the present day. Perhaps the most important invention of the alchemists was the still. The important technique of distillation was probably discovered by an early Greek alchemist, Dioscorides, after he noticed condensate on the lid of a vessel in which some mercury was being heated. It was only a short step from this observation to the realization that this technique could be used to separate volatile substances from nonvolatile animal and vegetable matter. The still was quite inefficient in its infancy but its design improved steadily. By 1300 fractionation was being practiced, and alcoholic distillates of fairly high alcohol concentration could be prepared. The production of whiskey and brandy became an established industry in short order.

The invention and development of the still had an interesting consequence in another area—medicine. Through the Middle Ages, medicine was practiced as a mystical blend of magic and folklore. It had long been noticed that certain animal and plant substances seemed to possess curative powers (''folk'' remedies from less developed parts of the world are still studied for possible new therapeutic compounds). With the use of the still, it became possible to concentrate the ''essence'' of various natural materials. The use of alcohol (*aqua vitae*, water of life) and other distilled materials quickly became a widespread practice. Over a period of several hundred years, all manner of natural substances were distilled by physicians and their associates. In the course of the search, a number of relatively pure organic compounds were isolated, such as acetic acid from vinegar and formic acid from ants.

During the period before 1600, the tools for handling matter were being developed and numerous relatively pure chemical substances were discovered. There was no advance at all in the theory of matter. During the seventeenth and eighteenth centuries, chemistry was born as a science in Europe. The first objects of serious investigation were gases. During the period of ''pneumatic chemistry'' (Hales used a ''pneumatic'' trough to collect gases) from 1650 to 1750, theory explained combustion by the gain or loss of a substance called ''phlogiston.'' Almost every kind of matter was supposed to contain a certain amount of phlogiston, and fire caused transfer of the phlogiston to another body.

The phlogiston theory was a step away from the purely qualitative notions of chemistry that had existed since the time of the Greek philosophers toward quantitation of chemical change. Individual scientists like Boyle, Hales, Cavendish, Priestley, and Scheele made important breakthroughs in chemistry, but Antoine Laurent Lavoisier (1743–1794) laid the real foundation for modern chemistry. It was Lavoisier who first realized that the gain in weight that occurs during combustion and calcination of metals is due to the combination of the substance being heated with a component of the air. During the 1770s Lavoisier carried out the quantitative experiments with Priestley's ''dephlogisticated air'' that subsequently led him to recognize its elemental nature; the name oxygen was first used in a memoir dated September 5, 1775. The characterization of oxygen as an elemental substance allowed quantitative studies of combustion and led rapidly to the notion of combining weights. By 1789, Lavoisier had assembled a Table of the Elements containing 33 substances, most of which appear in the modern periodic table.

In this formative stage in the development of the science of chemistry, the substances derived from the animal and vegetable worlds were largely ignored. These materials were recognized as being different—more complex—than the compounds of the atmosphere or those compounds derived from the mineral kingdom. Lavoisier himself noted that ''organic compounds,'' as they came to be known, differed from the inorganic compounds in that they all seemed to be composed of carbon and hydrogen and occasionally nitrogen or phosphorus. For a time it was thought that organic compounds did not obey the new law of definite proportions, and that a "vital force," present only in living organisms, was responsible for the production of organic compounds.

The idea of vitalism survived into the middle of the nineteenth century. In 1828 Friedrich Wöhler, while a medical student at the University of Heidelberg, reported that, upon treating lead cyanate with ammonium hydroxide, he obtained urea. Since urea was a well-known organic compound, having been isolated from human urine by Roulle in about 1780, Wöhler had succeeded in preparing an organic compound in the laboratory for the first time. In a letter to Jöns Jakob Berzelius, a leading Swedish chemist, Wöhler proclaimed, "I must tell you that I can make urea without the use of kidneys, either man or dog. Ammonium cyanate is urea." Although the synthesis of urea was recognized as an important accomplishment by the leading chemists of the day, the concept of vitalism did not die quickly. It was not until the synthetic work of Kolbe in the 1840s and Berthelot in the 1850s that the demise of vitalism was complete.

At this time, chemists recognized that it was not the vital force that imparted uniqueness to organic chemistry but rather the simple fact that organic compounds are all compounds of carbon. This definition—**organic chemistry is the chemistry of carbon compounds**—has persisted. During the eighteenth and nineteenth centuries, analytical methods were also being perfected. With the advent of these methods, particularly the technique of combustion analysis, organic chemistry began to take on new dimensions. For the first time accurate formulas were available for fairly complicated organic compounds. There ensued a confusing period, which lasted from about 1800 until about 1850, during which various theories were advanced in an attempt to explain such complexities as isomerism (the existence of two compounds with the same formula) and substitution (the replacement of one element by another in a complex organic formula). Organic chemistry began to emerge from this chaotic period in 1852 when Edward Frankland advanced the concept of valence. In 1858, Friedrich August Kekulé and Archibald Scott Couper, working independently, introduced a simple, but exceedingly important, concept. Making use of the new structural formulas shown, which had come into vogue since 1850, Kekulé and Couper proposed that the carbon atom is always tetravalent and that carbon atoms have the ability to link to each other.

$$\left.\begin{array}{l}H\\H\\H\end{array}\right\}N \qquad \left.\begin{array}{l}H\\ \\H\end{array}\right\}O \qquad \left.\begin{array}{c}C_2H_5\\ \\H\end{array}\right\}O \qquad \left.\begin{array}{c}C_2H_5\\ \\C_2H_5\end{array}\right\}O$$

ammonia water alcohol ether

A third event ushered organic chemistry into its modern period. Stanislao Cannizzaro in 1858 showed that Avogadro's hypothesis, formulated in 1811, allowed the determination of accurate molecular weights for organic compounds. With this last piece of the structural puzzle available, it was possible to think in terms of molecular structure and the chemical bond. Kekulé introduced the idea of a bond between atoms and depicted it with his "sausage formulas" in the first edition of his textbook in 1861.

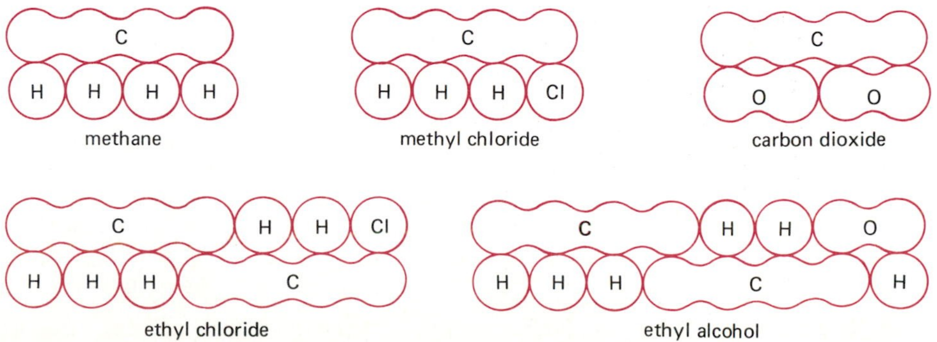

In the century since Kekulé, organic chemistry has matured as a scientific discipline in its own right. Well over 95% of all known chemical compounds are compounds of carbon. Over one half of present-day chemists classify themselves as organic chemists. The organic chemical industry plays a major role in world economy. For example, some commercial herbicides (Roundup™, from the Monsanto Chemical Company, Glean™ from the Du Pont Company, and paraquat from ICI Ltd.) had 1989 sales of more than $1 billion, and are currently significant factors in increasing agricultural yields and therefore the availability of food, both in industrialized and third-world countries. Finally, because organic chemicals are literally the "stuff of life," the significant advances in unraveling the nature of life are discoveries in organic chemistry. The scope of the subject encompasses almost every aspect of modern life and basic science, as shown in the list of topics and compounds given in Table 1.1.

Why study organic chemistry? There are different answers to this question depending upon who you are. It may be that you will devote your life to a career in organic chemistry per se, although if this is the case, you probably do not know it yet. Or you may plan to specialize in some other area of chemistry and want a knowledge of organic chemistry as an adjunct to your specialty area. You may become a chemical engineer; if so, organic chemistry will be an important part of your life, since most of the industrial processes you will encounter will be organic reactions. If you are headed for medicine or nursing, you should be aware of the fact that worldwide sales of pharmaceutical products amounted to over $150 billion in 1990 and that chemotherapy is one of the major techniques in modern medicine. You may be going into biochemistry, molecular biology, biotechnology, or some other life science to which organic chemistry is essential. Biochemistry is an exten-

TABLE 1.1 Scope of Organic Chemical Knowledge

Technical/Engineering	Physical	Biochemical/Biological	Organic
polymers	thermodynamics	proteins (polypeptides)	amino acids
Nylon (polyamide)	energetics	polynucleic acids (RNA, DNA)	nucleotides
Dacron™ (polyester)	structure	carbohydrates	saccharides
Lucite™ (polyester)	dynamics	lipids	esters, cholesterol
Orlon™ (polyacrylonitrile)	spectroscopy	hormones	estrone, adrenaline
Delrin™ (polyformaldehyde)	NMR, UV–vis, IR	vitamins (A, B_1, B_2, B_6, C,	retinol, thiamine
polyethylene	fluorescence	D, E, K)	riboflavin
polypropylene	photoelectron	neurotransmitters	pyridoxal
polystyrene	(chemiluminescence)	hemoglobin (oxygen)	nicotinamide
Teflon™ (polytetrafluoroethylene)	conductivity	chlorophyll	tachysterol
cellulose (polycellobiose)	superconductivity	enzymes, catalytic antibodies	tocopherol
rubber (polyisoprene)	stereochemistry	ionophores	naphthoquinone
wool (polyamide)	molecular mechanics		acetylcholine
small molecules	quantum mechanics		glutamate
fuel (isooctane)	separations (phase)		γ-aminobutyrate
pigments, dyes (indigo)	preferences)		dopamine
solvents (methyl ethyl ketone)	molecular graphics		porphyrins
drugs (penicillin, acetylsalicylate)	strain		organometallics
pesticides (DDT, parathion)	photochemistry		catalysts
herbicides (Roundup™)	lasers		crowns, cryptates
antiflame agents (polybromoorganics)	solvent effects		cyclopropane
flavors (sweet, Aspartame™)	micelles		cubane
odorants (pentanethiols, geraniol)	scanning tunnel		ethanol, methanol
refrigerants (fluorocarbons)	microscopy		chlorodifluoromethane
detergents			ion complexes
fertilizer (urea)			

sion of organic chemistry to the materials and transformations in living organisms; the molecules of molecular biology are organic molecules.

Even if you do not have any of the foregoing reasons to study organic chemistry, there is a purely intellectual justification. The previous arguments are vocational motivations. But organic chemistry also provides a fascinating area of natural philosophy for the student who wants to obtain a broad liberal arts education. If you approach the subject in the proper frame of mind, you will find it to be an extremely stimulating intellectual pursuit. The subject has a highly logical structure. As you will see, organic chemists make much use of symbolic logic, the logical principle of analogy, and deductive reasoning. In fact, it has been intimated that medical school admission boards value organic chemistry courses as much for the test in logical thinking that they provide as for their factual content. Some of these aspects of organic chemistry are inherent in Table 1.1, which illustrates the scope of organic chemical knowledge. Of twelve technologies classified as "emerging" as the basis of large-scale industry in 1990, eight utilize or depend on the application of organic chemistry. These are advanced materials, biotechnology, digital imaging technology, high density data storage, medical devices and diagnostics, optoelectronics, sensor technology, and superconducting materials. The highest critical temperature for an organic superconductor is 29 K, well below the 125 K for a thallium-based ceramic high temperature superconductor; this offers a clear challenge to the organic chemist.

Finally, organic chemistry has a unique content as an art form. The building up of complex molecular architecture by appropriate choice of a sequential combination of reactions provides syntheses that are described as "elegant" and "beautiful." The design of an experiment in reaction mechanism can be similarly imaginative. Such elegantly conceived experiments can evoke that delightful feeling of pleasure that one obtains from the appreciation of human creativity—but only in the mind of the knowledgeable spectator. These unique works of art can only be appreciated by those who know some organic reactions and have tried to design some simple syntheses and experiments themselves, such as those suggested in problem sets throughout this textbook. Because of the breadth and depth of the subject, organic chemistry is a subject in which intuition (a combination of imagination, knowledge, and experience) can be powerfully utilized to design, prepare, and study new molecules. Only one who has played chess can feel that special pleasure of following a game between Grand Masters.

CHAPTER 2

ELECTRONIC STRUCTURE AND BONDING

2.1 Periodic Table

The periodic table of the elements was developed well over 100 years ago. At that time it was an empirically derived arrangement based on the chemical and physical properties of the known elements. The table now embraces almost 110 elements. Compounds of carbon, organic compounds, are known that contain virtually all of the elements except some of the noble gases. However, only a small number of elements is important in an introductory study of organic chemistry. In the condensed form of the periodic table shown in Table 2.1, the names of the most important elements are emphasized with color; these are C, H, N, O, S, Mg, Cl, Br, and I. Secondary but still important elements are shown in italics: Li, B, F, and P.

TABLE 2.1 Abbreviated Periodic Table

first period:	**H**							He
second period:	*Li*	Be	*B*	**C**	**N**	**O**	*F*	Ne
third period:	Na	**Mg**	Al	Si	*P*	**S**	**Cl** **Br** **I**	Ar

2.2 Lewis Structures

The "noble" gases, He, Ne, Ar, Kr, Xe, and Ra, are almost inert chemically. Acquisition of this inert character is the feature that dominates much of the chemistry of the other elements. The noble gases have characteristic numbers of electrons, 2 for helium, 10 for neon (2 + 8), 18 for argon (2 + 8 + 8), and so on. They are described as having "filled shells" or, for neon and argon, as having filled outer **octets.** Other elements can achieve such stable electronic configurations by gaining or losing electrons.

The energy required to remove an electron is known as the **ionization potential, IP.**

$$\text{M} \longrightarrow \text{M}^+ + \text{e}^- \qquad \Delta H^\circ = \text{IP}$$

For elements at the far left of the periodic table, loss of an electron produces the electronic configuration of the next lower noble gas.

$$\underset{\text{3 electrons}}{\text{Li}} \xrightarrow{-\text{e}^-} \underset{\substack{\text{2 electrons}\\\text{(same as helium)}}}{\text{Li}^+}$$

$$\underset{\text{11 electrons}}{\text{Na}} \xrightarrow{-\text{e}^-} \underset{\substack{\text{10 electrons}\\\text{(same as neon)}}}{\text{Na}^+}$$

Such elements have relatively low ionization potentials and are described as being **electropositive.**

The energy liberated when an electron is acquired is called the **electron affinity, EA.**

$$\text{X} + \text{e}^- \longrightarrow \text{X}^- \qquad -\Delta H^\circ = \text{EA}$$

The electron affinity of an atom is also the ionization potential of the corresponding anion. Elements at the far right of the periodic table readily acquire electrons to produce the stable electronic configuration of the next higher noble gas.

$$\underset{\text{9 electrons}}{\text{F}} + \text{e}^- \longrightarrow \underset{\substack{\text{10 electrons}\\\text{(same as neon)}}}{\text{F}^-}$$

$$\underset{\text{17 electrons}}{\text{Cl}} + \text{e}^- \longrightarrow \underset{\substack{\text{18 electrons}\\\text{(same as argon)}}}{\text{Cl}^-}$$

Such elements have relatively high electron affinities and are described as being **electronegative.**

The electrons outside the shell of the next lower noble gas are the **valence electrons** and are the only ones normally included in symbols. The above ionization processes are then symbolized as follows.

$$\text{Li}\cdot \longrightarrow \text{Li}^+ + \text{e}^- \qquad \text{IP} = 123.6 \text{ kcal mole}^{-1},^* \ 517.1 \text{ kJ mole}^{-1}$$

$$\text{Na}\cdot \longrightarrow \text{Na}^+ + \text{e}^- \qquad \text{IP} = 118.0 \text{ kcal mole}^{-1},\ 493.7 \text{ kJ mole}^{-1}$$

$$:\ddot{\underset{\cdot\cdot}{\text{F}}}\cdot + \text{e}^- \longrightarrow :\ddot{\underset{\cdot\cdot}{\text{F}}}:^- \qquad \text{EA} = 78.3 \text{ kcal mole}^{-1},\ 327.6 \text{ kJ mole}^{-1}$$

$$:\ddot{\underset{\cdot\cdot}{\text{Cl}}}\cdot + \text{e}^- \longrightarrow :\ddot{\underset{\cdot\cdot}{\text{Cl}}}:^- \qquad \text{EA} = 83.3 \text{ kcal mole}^{-1},\ 348.5 \text{ kJ mole}^{-1}$$

Electropositive elements such as the alkali metals tend to lose electrons to electronegative elements such as the halogens to form pairs of ions, which are thus connected by ionic

---

*Another system of units that is being adopted in many parts of the world is known as SI (*Système International d'Unités*). In this international system of units, the unit of energy is the joule, J; 1 cal = 4.184 J. The six fundamental units in SI are length = meters (m), mass = kilograms (kg), time = seconds (s), electrical current = amperes (A), temperature = degrees Kelvin (K), and luminosity = candelas (cd). These units are modified by 10^3 (kilo-), 10^6 (mega-), 10^9 (giga-), 10^{12} (tera-), 10^{-3} (milli-), 10^{-6} (micro-), 10^{-9} (nano-), 10^{-12} (pico-), 10^{-15} (femto-), 10^{-18} (atto-). Many traditional units are still in common use among chemists. Examples are calories (cal), kilocalories (kcal), and centimeters (cm). In this text we will generally use such traditional units but will often include the SI units as well for comparison.

bonding. Typical examples are lithium chloride ($Li^+ \ :\ddot{\underset{..}{Cl}}:^-$) and sodium fluoride ($Na^+ \ :\ddot{\underset{..}{F}}:^-$).

For elements in the middle of the periodic table, too much energy is required to gain or lose a sufficient number of electrons to form similar octet ions. Compare the energy required to generate a triply positive boron or quadruply positive carbon with the energies required to form Li^+ or Na^+.

$$\cdot\dot{B}\cdot \longrightarrow 3\,e^- + B^{3+} \qquad IP = 870.4 \text{ kcal mole}^{-1}, 3642 \text{ kJ mole}^{-1}$$

$$\cdot\dot{\underset{\cdot}{C}}\cdot \longrightarrow 4\,e^- + C^{4+} \qquad IP = 1480.7 \text{ kcal mole}^{-1}, 6195 \text{ kJ mole}^{-1}$$

Consequently, such elements tend to complete their electron octets by sharing electrons, as in the following examples.

H H:C:H H	H H:N: H	H:Ö:H	H H:C:F: H
methane	ammonia	water	methyl fluoride

Such bonds are described as **covalent bonds.**

The symbols used to describe the foregoing examples are called **Lewis structures.** Such structures not only provide simple and convenient representations of ions and compounds but are also valuable in providing an accurate accounting for electrons. They form an important basis for predicting relative stabilities. Lewis structures are important in the study of organic chemistry and the student should be able to write them with facility. The following general rules are useful for deriving suitable structures.

1. *All valence electrons are shown.* The total number of such electrons is equal to the sum of the numbers contributed by each atom, modified by addition or subtraction of the number of any ionic charges. Some examples are worked out in Table 2.2.

TABLE 2.2 Valence Electrons

Species	Atomic Contributions	−	Cation Charges	+	Anion Charges	=	Total Valence Electrons
CH_4	4(C) + 4 × 1 (H) = 8	−	0	+	0	=	8
NH_3	5(N) + 3 × 1 (H) = 8	−	0	+	0	=	8
H_2O	6(O) + 2 × 1 (H) = 8	−	0	+	0	=	8
H_3O^+	6(O) + 3 × 1 (H) = 9	−	1	+	0	=	8
HO^-	6(O) + 1 (H) = 7	−	0	+	1	=	8
BF_3	3(B) + 3 × 7 (F) = 24	−	0	+	0	=	24
NO_2^-	5(N) + 2 × 6 (O) = 17	−	0	+	1	=	18
CO_3^{2-}	4(C) + 3 × 6 (O) = 22	−	0	+	2	=	24

2. *Each element should, to the greatest extent possible, have a complete octet.* Exceptions are hydrogen, which has a duet shell, and elements beyond the first row. For example, elements such as sulfur and phosphorus have been considered to accommodate more than eight valence electrons (''expand their octets'') in certain circumstances.

Correct Structure	*Incorrect Structures*		
:Ö::C::Ö:	:Ö:C::Ö:	:Ö:C:Ö:	:Ö::C:Ö:
H:C̈l:		:H:C̈l:	
:N:::N:	:N̈::N:	:N̈:N̈:	

3. ***Formal charges*** *are assigned by dividing each bonding pair of electrons equally between the bonded atoms.* The number of electrons ''belonging'' in this way to each atom is compared with the neutral atom and appropriate positive or negative charges are assigned. Lone pairs ''belong'' to a single atom.

```
     H             H              :Ö:⁻
  H:N⁺:H         H:C:Ö:⁻      ⁻:Ö:S²⁺:Ö:⁻
     H             H              :Ö:⁻
```

ammonium ion methoxide ion sulfate ion

In NH_4^+, each electron pair is divided between N and H. This gives one electron for each H, the same as a hydrogen atom. The N has a total of 4, one less than atomic nitrogen; hence, the formal charge of +1 is associated with N. This procedure assigns the entire positive charge of $(NH_4)^+$ to the nitrogen; in practice, the electrons are spread over the entire molecule. In particular, this procedure does not take account of differences in electronegativity. Nitrogen, for example, is more electronegative than hydrogen and will tend to attract electrons to it. Thus, most of the positive charge is actually associated with the hydrogens in NH_4^+. However, this method of assigning formal charges does keep strict account of the total numbers of electrons and charges present and, when used with care, it helps to interpret chemistry. For example, the formal charge of −1 assigned to the oxygen of methoxide ion helps to explain why this ion is a strong base that readily adds a proton to the oxygen.

The example of sulfate ion is more complex. Some students tend to write this ion as ⁻:Ö:Ö:S̈:Ö:Ö:⁻, an arrangement that has the proper number of valence electrons and a less complex formal structure assignment. However, sulfate ion is known experimentally to have each oxygen bound to sulfur in an equivalent manner.

When a species has an incomplete octet, it is usually unstable or highly reactive. Examples are methyl cation and methyl radical.

```
   H          H
 H:C:H      H:C:H
   +          ·
```

methyl cation methyl radical

Multiple bonds are written in a straightforward manner, with two pairs of electrons for a double bond, and three pairs for a triple bond.

```
H.    .H
 :C::C:        H:C:::C:H        ⁻:C:::N:
H˙    ˙H
```

ethylene acetylene cyanide ion

A further simplifying convention is to replace each electron-pair bond by a line. For convenience, electron pairs not involved in bond formation are frequently omitted, unless needed to call attention to a particular property of the molecule.

$$\mathrm{H-\overset{H}{\underset{H}{N^{+}}}-H} \qquad \mathrm{H-\overset{H}{\underset{H}{C}}-O^{-}} \qquad \mathrm{{}^{-}O-\overset{O^{-}}{\underset{O_{-}}{S^{2+}}}-O^{-}} \qquad \mathrm{H-\overset{H}{\underset{+}{C}}-H}$$

$$\mathrm{H-\overset{H}{\underset{\bullet}{C}}-H} \qquad \mathrm{\overset{H}{\underset{H}{}}C{=}C\overset{H}{\underset{H}{}}} \qquad \mathrm{H-C{\equiv}C-H} \qquad \mathrm{{}^{-}C{\equiv}N}$$

In these symbolic representations, the lone-pair electrons are understood to be present and their presence is signified by appropriate formal charges. This is another reason for assigning formal charges properly. *The use of such symbols is widespread in organic chemistry, and practice in reading and writing these electronic representations cannot be overemphasized*. The simplified symbols correspond to the notational system proposed by Kekulé and Couper in 1858 (Chapter 1). Such symbols, in which each electron-pair bond is represented by a line and the lone-pair electrons are omitted, are frequently called **Kekulé structures.**

The use of a ''coordinate covalence'' bond is sometimes convenient. In this convention, an arrow represents a two-electron bond in which both electrons are considered to ''belong'' to the donor atom for the bookkeeping purpose of assigning formal charges.

$$\mathrm{H-O-\overset{+}{N}\overset{\displaystyle /\!/ O}{\underset{\displaystyle \backslash O_{-}}{}}} \qquad \text{or} \qquad \mathrm{H-O-N\overset{\displaystyle /\!/ O}{\underset{\displaystyle \searrow O}{}}}$$

This type of symbolism finds most use in representing ligands in inorganic complexes and will rarely be used in this text.

EXERCISE 2.1 Rewrite the following Kekulé structures as Lewis structures, including all of the valence electrons.

(a) chloride ion, Cl^{-}
(b) water, H—O—H
(c) hydroxide ion, $H—O^{-}$
(d) hypochlorite ion, $Cl—O^{-}$
(e) ammonia, $\mathrm{H-\underset{H}{N}-H}$
(f) cyanogen fluoride, F—C≡N
(g) nitric oxide, NO
(h) hydronium ion, H_3O^+
(i) sodium cation, Na^+
(j) helium, He
(k) hydrogen peroxide, H—O—O—H
(l) carbon dioxide, O=C=O

Highlight 2.1

All chemical bonding results from the coulombic attraction of negative and positive charges. We approximate the complex interactions in molecules with the help of simplifying concepts. The chemical bond is one such concept. Atoms form bonds by losing electrons to, gaining electrons from, or sharing electrons with neighboring atoms. Each atom thus achieves a stable noble gas arrangement of electrons. This arrangement is an octet for first and second row (second and third period) elements.

2.3 Geometric Structure

One of the really important achievements of modern experimental physics was the development of methods for the determination of crystal structures by x-ray diffraction. Other approaches to the precise determination of molecular structures include the techniques of electron diffraction and microwave spectroscopy. These experimental approaches have yielded a wealth of detailed structures at the molecular level. For example, H_2O is known to have a structure with a bent H—O—H angle of 104.5° and an oxygen-hydrogen bond distance of 0.96 Å [1 Å ≡ 1 × 10^{-8} cm ≡ 100 pm (picometers, SI units)].

H—O—H (0.96 Å, 104.5°)

It should be emphasized that water is not a rigid molecule with the atoms fixed in this geometry. The atoms are constantly in motion, even at a temperature of absolute zero. This motion is conveniently described in terms of the bending and stretching of bonds, processes detected by infrared and Raman spectroscopies. At any instant of time, the actual O—H distance may vary from 0.96 Å by several hundredths of an Ångstrom around the average distance. Similarly, bond angles are constantly changing, and the value given is an average value.

An important result has emerged from these many structural studies. Specific bonds retain a remarkably constant geometry from one compound to another. For example, the oxygen-hydrogen bond distance is almost always 0.96–0.97 Å.

Compound	O—H Bond Distance, Å
HO—H, water	0.96
HOO—H, hydrogen peroxide	0.97
H_2NO—H, hydroxylamine	0.97
CH_3O—H, methyl alcohol	0.96

In fact, the consistent bond distances allow us to treat the oxygen-hydrogen bond as an individual unit in different compounds.

Lewis structures can be useful in the interpretation of bond distances. For example, the nitrogen-oxygen bond distance is longer in hydroxylammonium ion than in nitronium ion.

$H_3\overset{+}{N}$——OH (1.45 Å) — hydroxylammonium ion

O=$\overset{+}{N}$=O ≡ :Ö::$\overset{+}{N}$::Ö: (1.15 Å) — nitronium ion

> The hydroxylammonium ion should not be confused with ammonium hydroxide, NH_4^+ OH^-. The hydroxylammonium ion is an ammonium ion with one hydrogen replaced by an OH group. In the ammonium ion there are four nitrogen-hydrogen bonds; in the hydroxylammonium ion, there are three nitrogen-hydrogen bonds and one nitrogen-oxygen bond.

In $HONH_3^+$, one electron pair binds the nitrogen to the oxygen, and the compound is said to have a nitrogen-oxygen **single bond.** In NO_2^+, the Lewis structure shows that each nitrogen-oxygen bond involves two pairs of electrons and is therefore said to be a **double bond.** At the equilibrium bond distance, the electrostatic forces between the negative electrons and positive nuclei are balanced. For the nitrogen-oxygen single bond, two bonding electrons are involved, and the balance of attraction and repulsion occurs at an internuclear distance of 1.45 Å. For the nitrogen-oxygen double bond, four electrons produce a greater net electrostatic attraction to the nuclei. The increased internuclear repulsion required to balance this additional attraction occurs at the shorter distance of 1.15 Å.

EXERCISE 2.2 On the basis of the Lewis structure you wrote for nitric oxide in Exercise 2.1, what nitrogen-oxygen bond distance would you expect?

EXERCISE 2.3 In the following comparisons of bonds, determine which bond is the shorter:

(a) CO in H—C(=O)—O—H (b) NO in O=N—O—H

(c) CO in H_3C—O—H or H_2C=O

2.4 Resonance Structures

In some cases, it is not possible to describe the electronic structure of a species adequately with a single Lewis structure. An example is nitryl chloride, NO_2Cl.

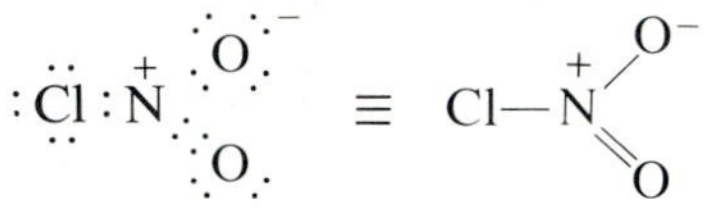

$$:\ddot{Cl}:\overset{+}{N}(\ddot{O}:^-)(\ddot{O}) \equiv Cl—\overset{+}{N}(—O^-)(=O)$$

nitryl chloride

The Lewis structure shown has one nitrogen-oxygen single bond and one nitrogen-oxygen double bond. However, it has been determined experimentally that both nitrogen-oxygen bonds are equivalent. Furthermore, the nitrogen-oxygen bond distance of 1.21 Å is intermediate between the nitrogen-oxygen single- and double-bond distances described in the previous section. Actually, two alternative structures, which differ only in the positions of electrons, may be written for nitryl chloride.

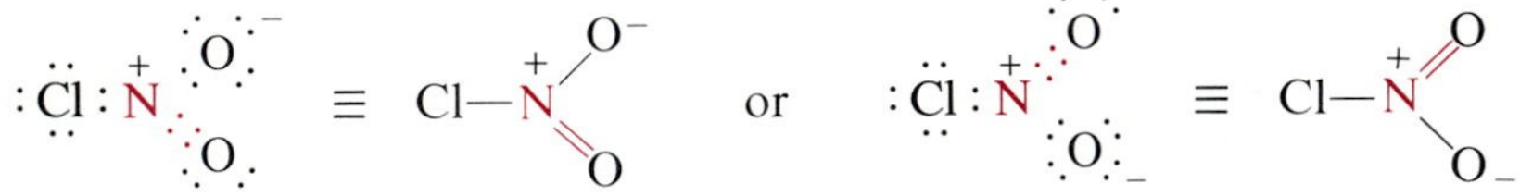

$$:\ddot{Cl}:\overset{+}{N}(\ddot{O}:^-)(\ddot{O}) \equiv Cl—\overset{+}{N}(—O^-)(=O) \quad \text{or} \quad :\ddot{Cl}:\overset{+}{N}(\ddot{O})(\ddot{O}:^-) \equiv Cl—\overset{+}{N}(=O)(—O^-)$$

The actual electronic structure of NO_2Cl is a *composite* or *weighted average* of the two Lewis structures. The alternative structures are called **resonance structures,** and the molecule is said to be a **resonance hybrid.** Resonance structures may be defined as alternative representations of the electronic configuration of a particular set of nuclei.

It is important to recognize that nitryl chloride *has only one geometric structure:* that in which the two nitrogen-oxygen bonds are equivalent. It is *not* Cl—$\overset{+}{N}$(=O)(—O⁻) half of the time and Cl—$\overset{+}{N}$(—O⁻)(=O) the other half. It is a hybrid in the same sense that a mule is a hybrid of a horse and a donkey. Resonance structures are necessary only because of inadequacies in our simplified system for describing bonding and electron distribution in molecules. When one conventional Lewis structure does not adequately describe what we know to be the actual structure of a species, we use two or more structures (resonance structures) for the species and bear in mind that the species has some characteristics of each structure.

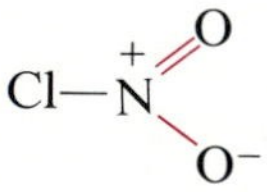

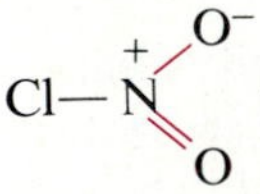

Resonance structures are written with a double-headed arrow, and the resonance hybrid is frequently written with dotted lines to represent partial bonds. Even in such cases, the individual Lewis structures provide an accurate accounting of the electrons and are to be preferred to dotted-line formulas. In the case of nitryl chloride, the Lewis structures indicate that the nitrogen-oxygen bond is halfway between single and double, and we expect an intermediate bond distance. Because each nitrogen-oxygen bond is single in one resonance structure and double in the other, the nitrogen-oxygen bond in the resonance hybrid is said to have a **bond order** of $1\frac{1}{2}$.

Another species that is not adequately described by a single structure is formate ion, HCO_2^-. Two of the resonance structures of the formate ion hybrid are

These structures give the two carbon-oxygen bonds a bond order of $1\frac{1}{2}$. Accordingly, the carbon-oxygen bond distance of 1.26 Å is intermediate between the carbon-oxygen double-bond distance of 1.20 Å in $H_2C{=}O$ and the carbon-oxygen single-bond distance of 1.43 Å in HO—CH_3.

Carbonate ion, CO_3^{2-}, is somewhat more complicated in that the hybrid includes three equivalent resonance structures. These resonance structures correspond to three equivalent carbon-oxygen bonds, each having a bond order of $1\frac{1}{3}$. Because the carbon-oxygen bonds in carbonate ion (order $1\frac{1}{3}$) have more single-bond character than those in formate ion (order $1\frac{1}{2}$), they are slightly longer (1.28 Å).

EXERCISE 2.4 One resonance structure for ozone, O_3, is $^-O{-}O^+{=}O$. Write two Lewis resonance structures showing all valence electrons and compare the expected oxygen-oxygen bond length with that of hydrogen peroxide (Exercise 2.1).

In each of the foregoing examples, the important resonance structures are equivalent. Other resonance structures can be written for each of these cases. For nitryl chloride a third resonance structure can be written as

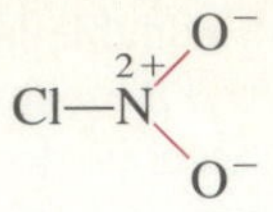

$$\mathrm{Cl{-}\overset{2+}{N}(O^-)_2}$$

This resonance structure has a high degree of charge separation and may therefore be thought to be less stable than the two other structures above and would therefore be considered *to contribute less to the resonance hybrid.* This structure does, however, reflect the greater electronegativity of O compared to N. That is, oxygen in a nitrogen-oxygen bond attracts the electrons toward itself because of its higher nuclear charge. Note that to the extent that this resonance structure contributes, the bond order of the nitrogen-oxygen bond will be less than $1\frac{1}{2}$.

Similarly, another resonance structure for formate ion is

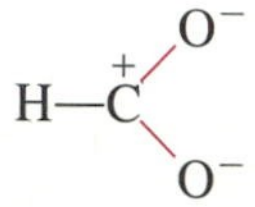

$$\mathrm{H{-}\overset{+}{C}(O^-)_2}$$

Again, any contribution of this structure would reduce the bond order of the carbon-oxygen bond below $1\frac{1}{2}$.

EXERCISE 2.5 Show that the third resonance structures for nitryl chloride and formate ion are proper Lewis structures by indicating all valence electrons. Write another resonance structure for carbonate ion that has only single carbon-oxygen bonds.

In some cases, a species is best described by two or more resonance structures that are not energetically equivalent. One such species is protonated formaldehyde, $(H_2COH)^+$. Formaldehyde itself may be represented by a Lewis structure in which there are two carbon-hydrogen single bonds and a carbon-oxygen double bond (Exercise 2.3).

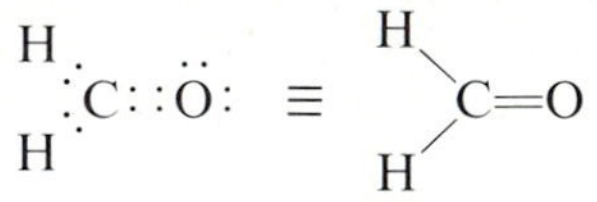

$$\mathrm{H_2C{::}\ddot{O}{:}} \equiv \mathrm{H_2C{=}O}$$

formaldehyde

In protonated formaldehyde, an additional oxygen-hydrogen single bond is present. Two Lewis structures may be written for $(H_2COH)^+$.

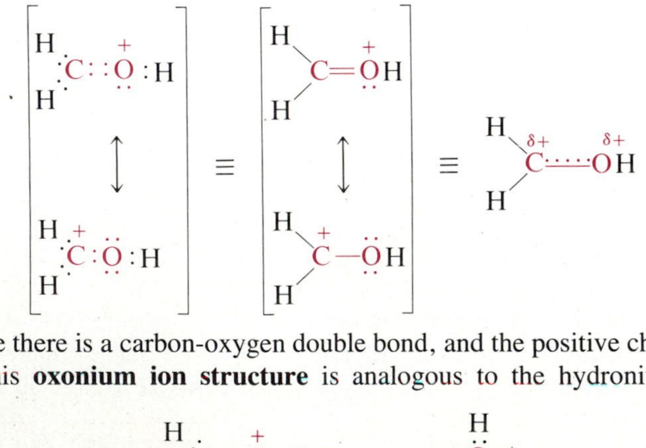

$$\left[\begin{array}{c}\mathrm{H_2C{::}\overset{+}{O}{:}H}\\ \updownarrow \\ \mathrm{H_2\overset{+}{C}{:}\ddot{O}{:}H}\end{array}\right] \equiv \left[\begin{array}{c}\mathrm{H_2C{=}\overset{+}{O}H}\\ \updownarrow \\ \mathrm{H_2\overset{+}{C}{-}\ddot{O}H}\end{array}\right] \equiv \mathrm{H_2\overset{\delta+}{C}{\cdots\cdots}\overset{\delta+}{O}H}$$

In one structure there is a carbon-oxygen double bond, and the positive charge is assigned to oxygen. This **oxonium ion structure** is analogous to the hydronium ion, H_3O^+.

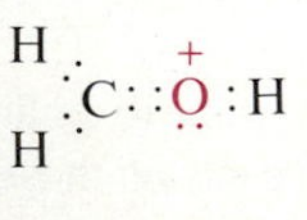

$$\mathrm{H_2C{::}\overset{+}{O}{:}H} \qquad \mathrm{H{:}\ddot{O}{:}^{+}}\ (\text{with H above and below O})$$

oxonium ion structure 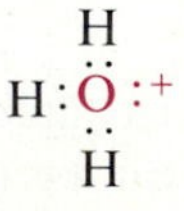　　　hydronium ion

In the alternative structure there is a carbon-oxygen single bond, and the positive charge is assigned to carbon. This **carbocation structure** is analogous to the methyl cation, CH_3^+.

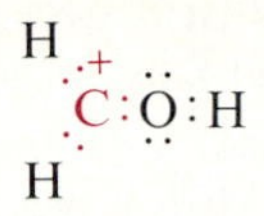

carbocation structure 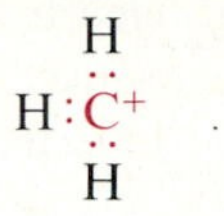   methyl cation

Which structure more adequately represents protonated formaldehyde? The carbon-oxygen bond length in $(H_2COH)^+$ is 1.27 Å, which is closer to the normal carbon-oxygen double-bond length of 1.20 Å than to the normal carbon-oxygen single-bond length of 1.43 Å. On this basis, we conclude that $(H_2COH)^+$ is more nearly described by the oxonium ion structure than by the carbocation structure. However, the carbon-oxygen bond length is significantly longer than a normal double bond, and calculations show that there is a substantial partial positive charge on carbon. We shall see in Chapters 14 and 15 that much of the chemistry of $(H_2COH)^+$ is best explained by the contribution of the less important carbocation structure.

Again, let us reiterate that *neither oxonium nor carbocation structure provides a totally accurate description of* $(H_2COH)^+$. The structure of the ion is a resonance hybrid of the oxonium and carbocation structures and it has some of the characteristics of each. The carbon-oxygen bond order is something between 1 and 2. The positive charge is spread over both atoms. Because oxygen is more electronegative than carbon, the positive charge would normally be on carbon. However, in this structure, carbon does not have an electron octet. In order for carbon to fill its octet, the positive charge must be borne by the more electronegative oxygen. Thus, neither structure is ideal and the actual electronic structure is a compromise between the tendency of electrons to complete octets and their tendency to be attracted to more electronegative atoms.

An extreme example is trifluoromethyl cation, CF_3^+. It has been calculated that the carbon-fluorine bond length in this ion is 1.27 Å, much less than the normal carbon-fluorine single-bond length of 1.38 Å. Thus, the fluoronium ion structure is a substantial contributor to the resonance hybrid, even though some of the positive charge must thus be assigned to fluorine, the most electronegative of all the elements.

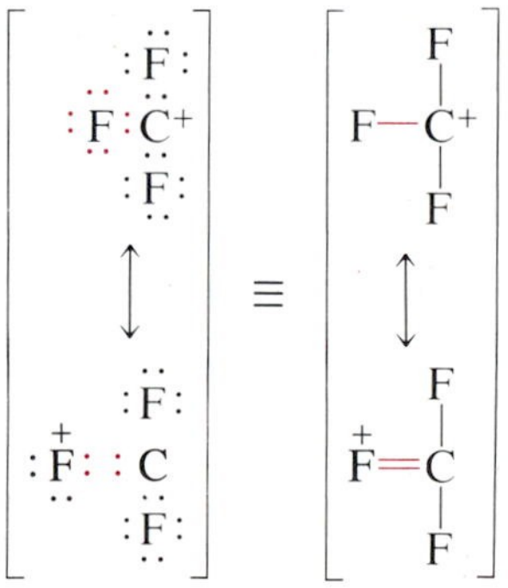

Another interesting example is provided by protonated methyleneimine, $(H_2CNH_2)^+$. The carbon-nitrogen bond length of 1.29 Å is almost exactly the same as the carbon-nitrogen double-bond length in methyleneimine itself (1.27 Å) and is much less than the normal carbon-nitrogen single-bond length of 1.47 Å.

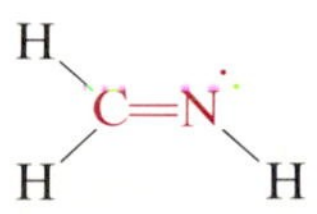

methyleneimine
C=N length = 1.27 Å

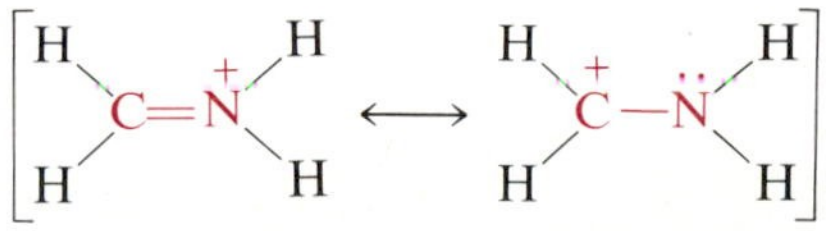

protonated methyleneimine
C—N length = 1.29 Å

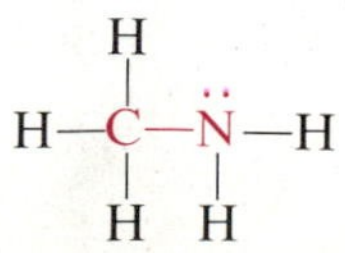

methylamine
C—N length = 1.47 Å

In this case, the immonium ion structure dominates the hybrid even more than the oxonium ion structure does in the case of $(H_2COH)^+$ because the difference in electronegativity between carbon and nitrogen is less than that between carbon and oxygen.

In summary, let us set out some empirical rules for assessing the relative importance of the resonance structures of molecules and ions.

1. Resonance structures involve *no* change in the positions of nuclei; only electron distribution is involved.
2. Structures in which all first-row (second-period) atoms have filled octets are generally important; however, resulting formal charges and electronegativity differences can make appropriate nonoctet structures comparably important.

More Important *Less Important*

$$\left[\mathrm{H_2C{=}\overset{+}{N}H_2} \longleftrightarrow \mathrm{H_2\overset{+}{C}{-}\ddot{N}H_2}\right]$$

$$\left[\mathrm{H_2\overset{+}{C}{-}\ddot{F}:} \longleftrightarrow \mathrm{H_2C{=}\ddot{F}:^{+}}\right]$$

Comparably Important

$$\left[\mathrm{H_2C{=}\overset{+}{O}:{-}H} \longleftrightarrow \mathrm{H_2\overset{+}{C}{-}\ddot{O}:{-}H}\right]$$

3. The more important structures are those involving a minimum of charge separation, particularly among atoms of comparable electronegativity. Structures with negative charges assigned to electronegative atoms may also be important.

More Important *Less Important*

$$\left[\mathrm{H{-}\ddot{O}{-}C({=}O){-}H} \longleftrightarrow \mathrm{H{-}\overset{+}{O}{=}C(O^{-}){-}H}\right]$$

$$\left[\mathrm{H{-}\ddot{N}(H){-}C{\equiv}N:} \longleftrightarrow \mathrm{H{-}\overset{+}{N}(H){=}C{=}\ddot{N}:^{-}}\right]$$

$$\left[\mathrm{:\ddot{F}{-}BF_2} \longleftrightarrow \mathrm{:\overset{+}{F}{=}\overset{-}{B}F_2}\right]$$

In many cases we shall find the charge-separated structure to be useful in interpreting some chemical reactions.

In some cases, Lewis structures with complete octets cannot be written without charge separation. In such alternative structures, the more important structure is again that in which the negative charge is borne by the more electronegative element and the positive charge by the more electropositive element.

More Important *Less Important*

$$\left[\mathrm{H{-}C(H){=}\overset{+}{N}{=}\overset{-}{\ddot{N}}:} \longleftrightarrow \mathrm{H{-}\overset{-}{\ddot{C}}(H){-}\overset{+}{N}{\equiv}N:}\right]$$

diazomethane

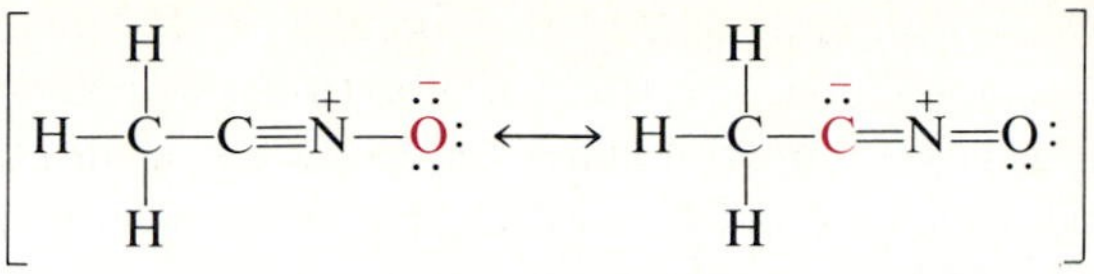

$$\left[H-\underset{H}{\overset{H}{C}}-C\equiv\overset{+}{N}-\overset{-}{\ddot{O}}: \longleftrightarrow H-\underset{H}{\overset{H}{C}}-\overset{\cdot\cdot}{\overset{-}{C}}=\overset{+}{N}=\ddot{O}: \right]$$

acetonitrile oxide

Elements beyond the second period form structures with an apparent expansion of their octets. Examples are sulfur hexafluoride, SF_6, and gaseous phosphorus pentachloride, PCl_5. The structures of such compounds are probably best considered in terms of "no-bond" contributions, by writing resonance forms in which there are no electrons between the two atoms of an individual bond. Solid PCl_5 in the crystal has the structure $PCl_4^+\ PCl_6^-$ with the bond distances shown:

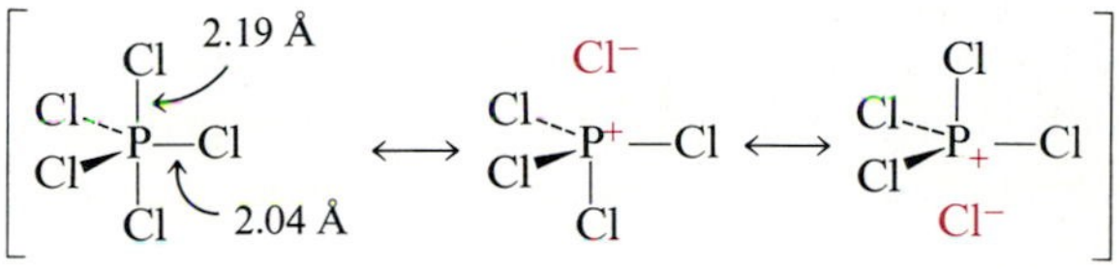

$$\left[PCl_5 \ (\text{P–Cl } 2.19\ \text{Å},\ 2.04\ \text{Å}) \longleftrightarrow Cl^-\ P^+Cl_4 \longleftrightarrow P^+Cl_4\ Cl^- \right]$$

EXERCISE 2.6 Write a Lewis structure for PCl_6^- in which each atom has a complete octet. What is the bond order of a phosphorus-chlorine bond in this ion? Indicate how the bond distance is rationalized by the phosphorus-chlorine bond order.

In the same manner, some compounds of these elements are often written as resonance hybrids with expanded octet resonance structures. An example is sulfuric acid.

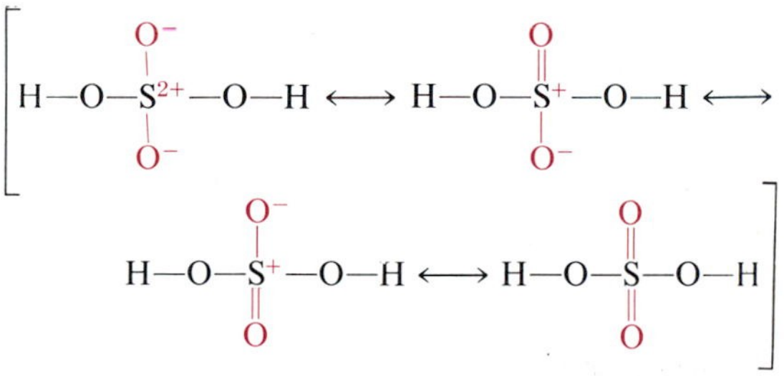

$$\left[H-O-\underset{O^-}{\overset{O^-}{S^{2+}}}-O-H \longleftrightarrow H-O-\underset{O^-}{\overset{O}{\overset{\|}{S^+}}}-O-H \longleftrightarrow H-O-\underset{\underset{O}{\|}}{\overset{O^-}{S^+}}-O-H \longleftrightarrow H-O-\underset{\underset{O}{\|}}{\overset{O}{\overset{\|}{S}}}-O-H \right]$$

The normal Lewis octet structure at the far left has a formal charge of +2 on sulfur. Sulfuric acid is known to be a strong acid, and the high formal charge on sulfur helps to explain the ease of loss of a proton.

EXERCISE 2.7 In this section we have made use of the concept of bond order and the generalization that the higher the bond order (degree of bonding), the shorter (and stronger) the bond. An example is the application to nitric acid and nitrate ion. Write two equivalent resonance structures for nitric acid and three for nitrate ion. Rank these three different types of nitrogen-oxygen bonds in order of length.

Highlight 2.2

The structures of molecules are determined by physical or chemical experiment. We express the structure in terms of one or more resonance structures that in sum reflect the electron and bond distribution in the molecule.

2.5 Atomic Orbitals

Careful use of Lewis structures and the related straight-line structural shorthand is clearly important in developing an understanding of the physical and chemical properties of molecules. But such structures are themselves only symbolic representations of electronic structures. In the real world, electrons do not stand still in octets. A more complete understanding of the chemical bond and the structure of molecules requires a discussion of the modern theory of electronic structure in terms of wave functions or orbitals. Unfortunately, this theory involves new and unfamiliar concepts that do not relate to human experience. Atomic and molecular orbitals are usually covered in depth in courses on physical chemistry, but the qualitative aspects are so important to understanding modern organic chemistry that a brief survey of some results of quantum mechanics is highly desirable at this point. In the next few sections, we shall review those aspects of atomic and molecular orbital theory that are particularly important in the study of organic chemistry.

As mentioned in Section 2.1, the periodic table of the elements was first developed in a purely empirical fashion. The known chemical elements were arranged into groups and rows on the basis of similarities in their chemical and physical properties. The "periodicity" of the table first became understandable with the early development of electronic theory. The early theory was based on the Bohr model, presenting an atom as a miniature solar system in which electrons are pictured as revolving in fixed orbits around a nucleus much as planets revolve about the sun.

In the context of the quantum mechanics introduced more than a half century ago, this analogy was shown to be seriously deficient in an extremely important respect. A basic tenet of quantum mechanics is the **Heisenberg uncertainty principle,** which states that it is not possible to determine simultaneously both the precise position and momentum of an electron. In other words, the laws of nature are such that we cannot determine an exact trajectory for an electron. The best we can do is to describe a probability distribution that gives the probability of finding an electron in any region around a nucleus. The mathematical description that leads to this probability distribution has the same form as that which describes a wave. Thus, we may use the mathematics and concepts of wave motion to describe electron distributions. Consequently, it is common to refer to the motion of an electron around a nucleus as a "wave motion" or in terms of a "wave function." This does not mean that the electron actually bobs up and down like a cork in a stormy sea. It is only a convenient phrase that helps to characterize the mathematical equations that describe the electron probability distribution.

In quantum mechanics, an **atomic orbital** is defined as a one-electron wave function, ψ. For each point in space there is associated a number whose square is proportional to the probability of finding an electron at that point. Such a probability function corresponds to the more useful concept of an **electron density** distribution. The mathematical function that describes an atomic orbital has all of the properties associated with waves; hence, it is called a **wave function.** It has a numerical magnitude (its amplitude), which can be either positive or negative (corresponding to a wave crest or a wave trough, respectively), and nodes. A **node** is the region where a crest and a trough meet. For the three-dimensional waves characteristic of electronic motion, the nodes are two-dimensional surfaces at which $\psi = 0$. Consequently, atomic orbitals may be characterized by their corresponding nodes as given by quantum numbers (Table 2.3).

TABLE 2.3 Atomic Quantum Numbers

Quantum Number	Symbol	Possible Values	Relationship to Nodes
principal	n	$1, 2, 3, \ldots$	one more than the total number of nodes[a]
azimuthal or angular momentum	l	$0, 1, \ldots, n-1$	number of nonspherical nodes
magnetic	m	$-l, \ldots, 0, \ldots +l$	character (planes or cones) and orientation of nonspherical nodes
spin	—	$-\frac{1}{2}, +\frac{1}{2}$	none

[a] Because atomic orbitals are exponential functions, they have very small values at distances far from the nucleus but never reach zero. The extra node could therefore be taken at infinity, but such a node could never be represented in conventional symbols. If the node "at infinity" is included, the principal quantum number is the same as the total number of nodes.

If one recognizes the relationship between quantum numbers and the number and character of the nodes in a wave, it is clear why quantum numbers are integers; that is, it is meaningless to talk of a fraction of a node. In labeling a particular atomic orbital, the principal, azimuthal, and magnetic quantum numbers are specified. The three quantum numbers are expressed in the order nlm, but in a particular manner. The principal quantum number n is given as the appropriate integer. The azimuthal number l is expressed in code, where $0 = s$, $1 = p$, $2 = d$, and $3 = f$. In spatial descriptions m is not given explicitly, but is implied in a subscript code that defines the orientation of the orbital.

Examples

- $1s$ no nodes. This wave function is a spherically symmetric function whose numerical value decreases exponentially from the nucleus.
- $2s$ one spherical node
- $2p_x$ one node, the yz-plane
- $2p_y$ one node, the xz-plane
- $2p_z$ one node, the xy-plane

These orbitals are the most important for the organic compounds we will study. They are usually represented symbolically as in Figure 2.1. The plus and minus signs in Figure 2.1 have no relationship to electric charge. They are simply the arithmetic signs associated with the wave function, much as a positive sign for an ocean wave is a wave crest and a negative sign denotes a wave trough. We shall see in the next section that the positive and negative signs determine how two or more wave functions combine when they interact.

In the symbolic representations given in Figure 2.1, the solid line represents the angular part of the wave function and defines a three-dimensional closed surface. A useful approximation is to regard the surface as a locus of points of constant value of ψ such that some given, but arbitrary, proportion of the total electron density is contained within the surface. For example, the value of ψ may be selected so that the resulting surface will enclose 80%, 90%, 95%, and so on of the electron density. The dotted lines in Figure 2.1 represent nodal surfaces. Remember that at the nodal surface the value of the wave function is zero. These nodes are a sphere for the $2s$-orbital and a plane for the $2p$-orbital.

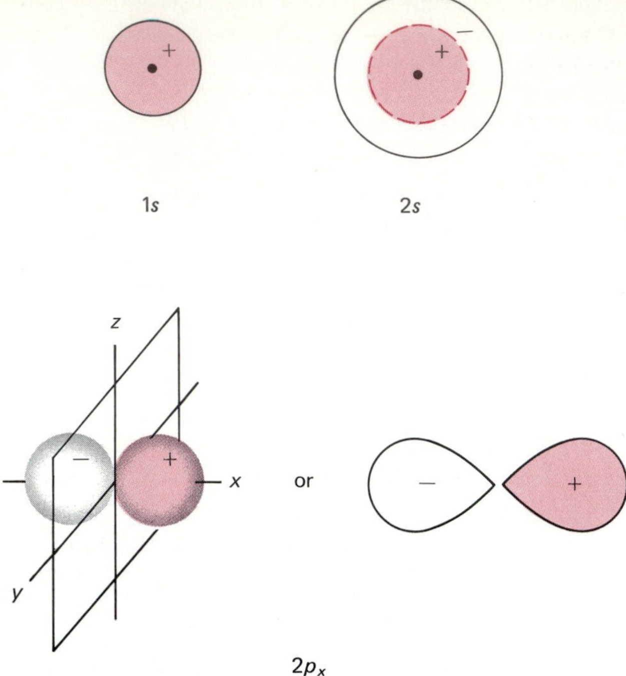

FIGURE 2.1 Symbolic representation of some atomic orbitals.

The strange shape of the *p*-orbital is determined by the central attractive force of the nucleus and the constraint of a planar node. The representation of 2*p*-orbitals in Figure 2.2 gives a better feeling for the three-dimensional shape and shows the orientations for the three orthogonal p_x-, p_y-, and p_z-orbitals.

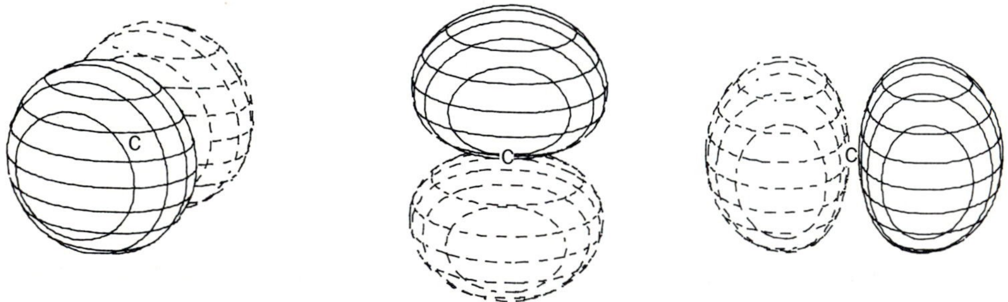

FIGURE 2.2 Perspective diagrams of $2p_x$-, $2p_y$-, and $2p_z$-orbitals.

2.6 Electronic Structure of Atoms

In atoms, the Pauli principle requires that no two electrons can have identical quantum numbers. Three quantum numbers characterize an atomic orbital. Electrons have a fourth quantum number associated with the characteristics of spin. This quantum number may have a value of either $+\frac{1}{2}$ or $-\frac{1}{2}$. Consequently, each atomic orbital may have associated with it no more than two electrons, and these two electrons must have "opposite spin."

In general, the more nodes a wave function has, the higher is its energy. In atoms that have more than one electron, the energies of atomic orbitals increase in the order $1s < 2s < 2p < 3s < 3p$, and so on. If we begin with an appropriate nucleus, putting one electron in the lowest energy atomic orbital, $1s$, produces a hydrogen atom. If a second electron has a spin opposite to that of the first electron, it can also be put into a $1s$-orbital, leading to a helium atom with two electrons. These two electrons "fill" the $1s$-shell, and helium has the filled-shell configuration characteristic of noble gases. Thus, hydrogen and helium constitute the "first period" of the periodic table. The third electron of lithium must be put into a higher energy atomic orbital, $2s$. If the spin of the fourth electron of beryllium is opposite to that of the third electron, it can also be put into the $2s$-orbital. The $2s$-orbital is now also filled, and the additional electrons of the first row (second period) elements must go into $2p$ atomic orbitals. The $2p_x$-, $2p_y$-, and $2p_z$-orbitals may each accept two electrons, giving a total of six for the p-set. Consequently, eight electrons fill the $n = 2$ shell and again give the stable filled-shell electronic configuration characteristic of the noble gases.

The process of filling successive atomic orbital levels with electron pairs is used to build up the entire periodic table. The atomic configurations of the first ten elements are summarized in Table 2.4. Each filled principal quantum shell corresponds to a stable noble gas. Other elements chemically react in such a way as to achieve the stability associated with filled orbital shells. One way of achieving this higher stability is by combining atomic orbitals into molecular orbitals, as discussed in the next section.

TABLE 2.4 Electronic Configurations of Some Elements

First Period		Second Period	
H	$1s$	Li	$1s^2 2s$
He	$1s^2$	Be	$1s^2 2s^2$
		B	$1s^2 2s^2 2p$
		C	$1s^2 2s^2 2p^2$
		N	$1s^2 2s^2 2p^3$
		O	$1s^2 2s^2 2p^4$
		F	$1s^2 2s^2 2p^5$
		Ne	$1s^2 2s^2 2p^6$

2.7 Bonds and Overlap

One of the useful concepts derived from treating atomic orbitals as wave functions is that two such orbitals may overlap to form a bond. The combination of two waves having the same sign is *reinforcing* (Figure 2.3). This is true for light waves, sound waves, or the waves of an ocean. It is also true for the combination of two electron waves or wave functions having the same sign.

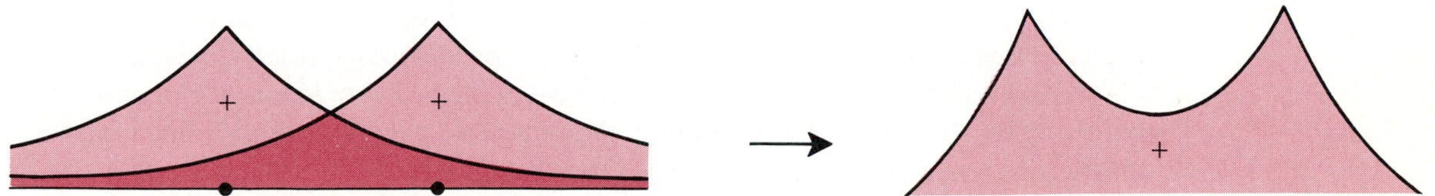

FIGURE 2.3 Two interacting waves or wave functions of the same sign add or **reinforce.**

The increased magnitude of the wave function between the atoms corresponds to higher electron density in this region. Electrons are attracted electrostatically to both nuclei, and the net effect of increased electron density between the nuclei counterbalances the internuclear repulsion. The result is a **covalent bond.** An example is the combination of two 1*s* atomic orbitals to give a new wave function. This wave function now encompasses both nuclei and is therefore called a **molecular orbital.** Figure 2.4 shows a symbolic representation of a molecular orbital (b) formed by the overlap of the two atomic orbitals shown in (a). Figure 2.4c is a contour diagram that depicts the value of the wave function in such a covalent bond as a function of distance from the nuclei, which are symbolized by the two heavy dots. The diagram represents a plane passing through the nuclei; each contour line connects points having the same value of the wave function in the same way that a contour line on a map connects points having the same altitude relative to sea level. In the example shown, all of the contours are positive. Figure 2.4d is a perspective view of a contour surface of such an orbital for a given contour value. Each bond that we have heretofore symbolized by a shared electron pair or by a straight line may now be interpreted as a "two-center" molecular orbital (an orbital encompassing two nuclei). Each such two-center molecular orbital contains two electrons of opposite spin.

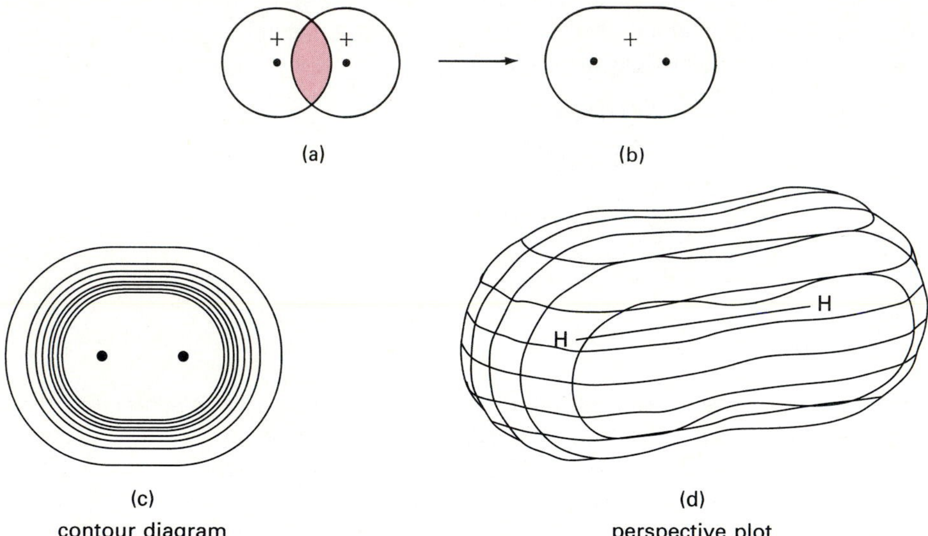

FIGURE 2.4 The combination of two H 1*s* orbitals to form an H_2 molecular orbital.

When two waves of opposite sign interact, they interfere or cancel each other. It is this characteristic of waves that can produce regions of darkness in the interaction of two light beams or regions of silence from the combination of two sound waves. At the point of interference the wave function has the value of zero; that is, interference of waves creates a node. The same pattern holds for electron waves. The interaction of two orbitals of opposite sign produces a node between the nuclei, as illustrated in Figure 2.5. Because there is no electron density at a node, the net effect is a reduced electron density between the nuclei that does not compensate for the nuclear repulsion; the result is a higher energy or lower stability than that associated with the two orbitals before interaction. A molecular orbital of this type is called **antibonding.** A perspective plot of an antibonding molecular orbital is shown in Figure 2.6.

The interaction of two orbitals can be positive or reinforcing to give a bonding molecular orbital, or the combination can be negative or interfering to give an antibonding molecular orbital. The bonding combination corresponds to a decrease in energy (greater stability); the antibonding combination corresponds to an increase in energy (lower stability). Two atomic orbitals give rise to two molecular orbitals. We can then place the two paired electrons of opposite spin, one from each atom, into the bonding molecular orbital.

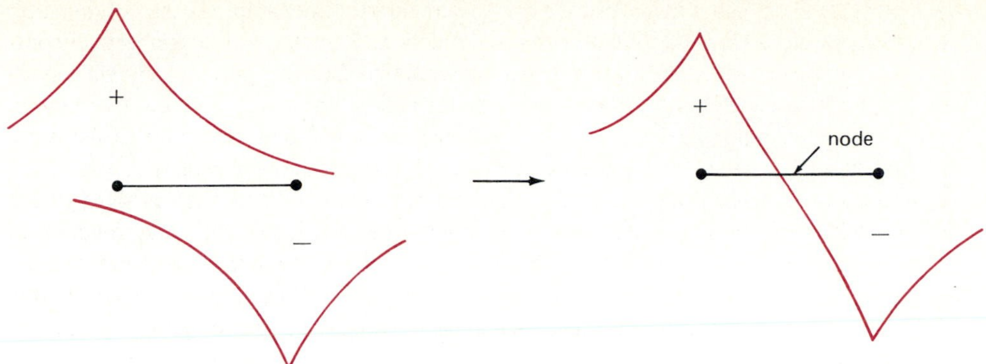

FIGURE 2.5 Interaction of two waves of opposite sign gives subtraction or **interference.**

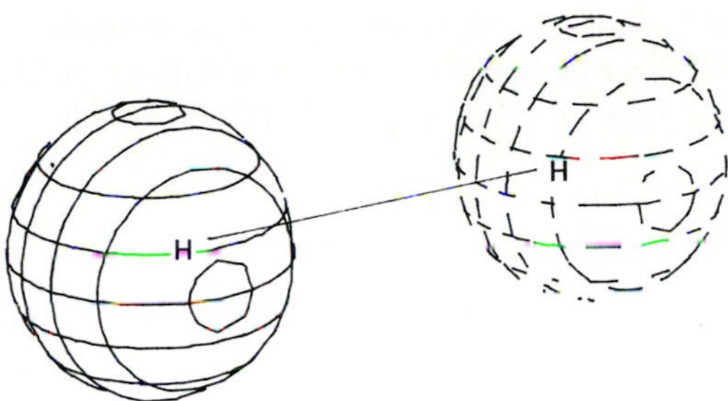

FIGURE 2.6 A two-center antibonding molecular orbital. The dashed lines indicate a wave function of negative sign.

We will not refer to antibonding molecular orbitals in most of the discussion of organic compounds, but we will need them to understand the excited states treated in Chapter 22.

The energy relationships of two combining orbitals are summarized in Figure 2.7. Note how the energies of the two starting orbitals separate or spread apart when they interact to form the two molecular orbitals. The amount of the separation depends on the degree to which the orbitals occupy the same space or **overlap.** A slight overlap gives two molecular orbitals that differ little in energy; a large overlap results in strong energy separation

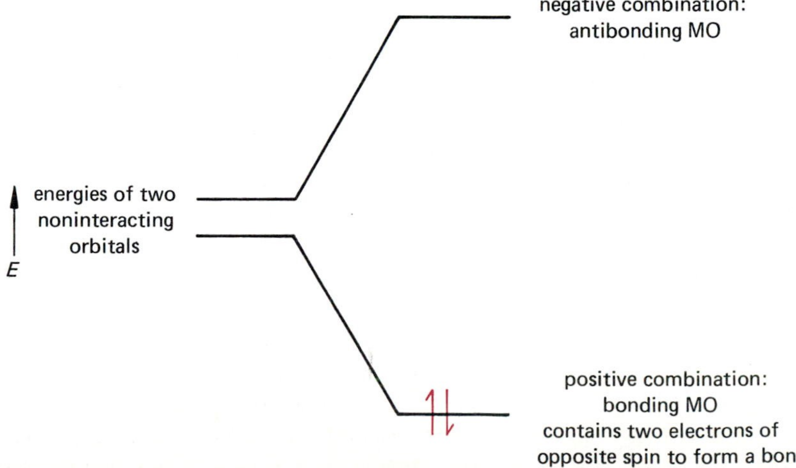

FIGURE 2.7 Energy relationships of combining orbitals.

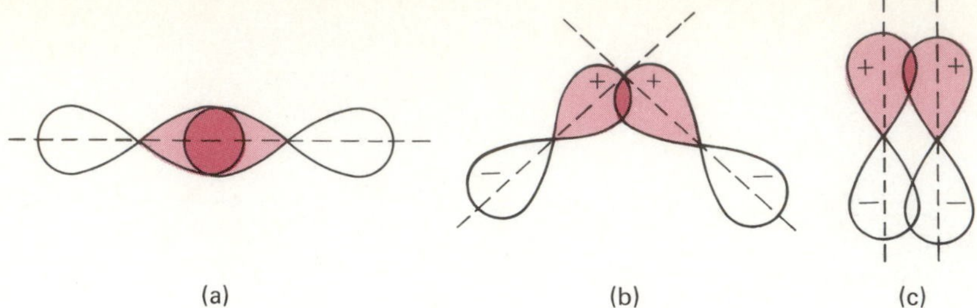

FIGURE 2.8 Illustrating the overlap of two *p*-orbitals (a) along the internuclear axis and (b and c) off the internuclear axis. The bonding molecular orbital resulting from the overlap shown in (a) has lower energy than the other two.

such as that shown in Figure 2.7. For axially symmetric orbitals, such as *p*-orbitals, the greatest overlap occurs when the orbitals are allowed to interact along the nuclear axis, that is, to form straight bonds. We shall see later that if orbitals are so constrained that overlap is not along the internuclear axis, then the resulting ''bent bonds'' (Figure 2.8b and c) are weaker than equivalent straight bonds (Figure 2.8a).

2.8 Hybrid Orbitals and Bonds

When more than two valence electrons on the same atom are involved in bonding, the individual bonds are not generally describable in terms of overlap of simple atomic orbitals as in the foregoing example. Consider the molecule BeH_2 as an example. Spectroscopic measurements show that the two beryllium-hydrogen bonds are of equal length. These two bonds clearly cannot be described adequately by using the beryllium 2*s*-orbital for one bond and a 2*p*-orbital for the other. These two orbitals have different spatial extensions and different energies and would be expected to give different bonds. The bonding can be explained if we construct *two equivalent* **hybrid** *orbitals* by combining the 2*s*- and a 2*p*-orbital. This is done mathematically by taking the sum and the difference of the two orbitals, as in Figure 2.9. This example shows how the mathematical signs of wave functions enter into arithmetic operations.

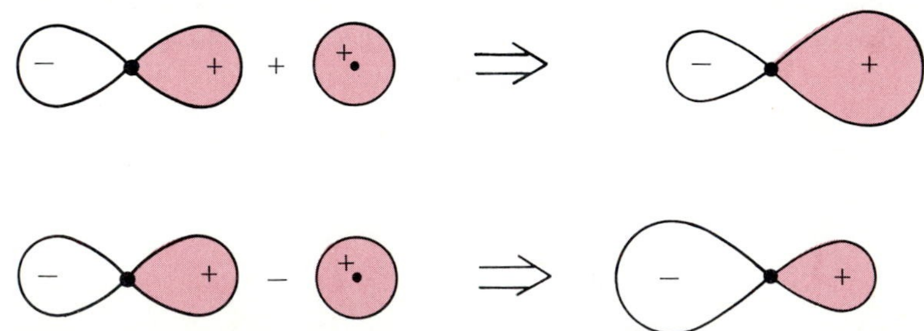

FIGURE 2.9 Mathematical combination of *s*- and *p*-orbitals to form two *sp*-hybrid orbitals.

The two orbitals that result from this operation are designated *sp*-hybrid orbitals because they are each constructed from equal amounts of an *s*- and a *p*-orbital. The *sp*-hybrid orbital is shown in contour form in Figure 2.10a and as the three-dimensional perspective plot in Figure 2.10b. The two hybrid orbitals are each well suited to form a bond by overlapping with a hydrogen 1*s*-orbital. They are equivalent and are directed opposite each other. Furthermore, the two lobes of a hybrid orbital are unequal in ''size''—the larger lobe can overlap well with another orbital. That is, overlap at the large lobe can occur readily in a straight line to produce stronger bonding.

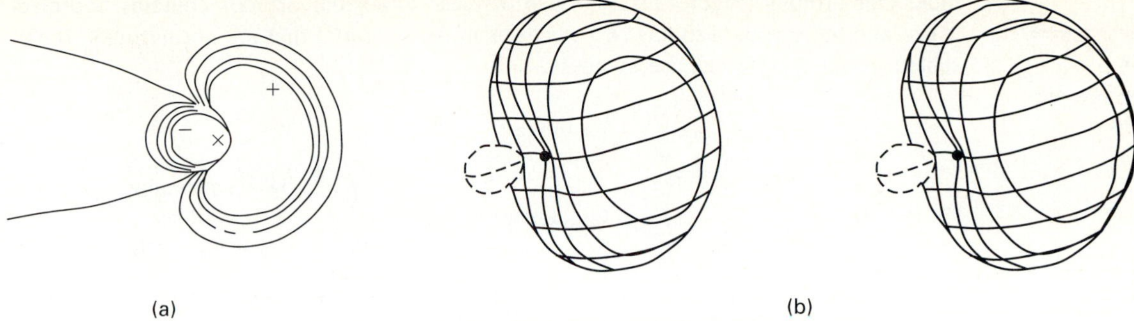

FIGURE 2.10 Contour and perspective plots of an *sp*-hybrid orbital.

Why does beryllium form bonds in this manner rather than by overlap of the simple atomic orbitals? The answer is simply that stronger bonds and a more stable structure result when the system H—Be—H is linear and the two bonds are of equal length. In this manner *the two electron pairs involved in the bonds are directed as far apart from each other as possible*. This principle is a useful guide for predicting the geometry of a molecule in which several groups are bonded to a central atom. In general, the bonding may be described by constructing as many hybrid orbitals from the simple *s*- and *p*-atomic orbitals as are needed to accommodate all of the valence electrons associated with the central atom. In the BeH_2 example, we used one *s*- and one *p*-orbital and constructed two equivalent hybrids. Each such hybrid is described as 50% *s* and 50% *p*. In constructing such combinations, we must again obey the *rule of conservation of orbitals*. We must end up with as many orbitals as we started with. The beryllium atom has, of course, two remaining *p*-orbitals that are not occupied by electrons in the molecule BeH_2.

As a further example, consider a species in which three groups are to be bonded to a central atom. From an *s*-orbital and two *p*-orbitals—for example, a p_x- and p_y-orbital—we may construct three equivalent sp^2-hybrids. Each such hybrid is $\frac{1}{3}$ *s* and $\frac{2}{3}$ *p*. The three equivalent hybrids lie in the *xy*-plane (the same plane defined by the two *p*-orbitals) and are directed 120° from each other (Figure 2.11).

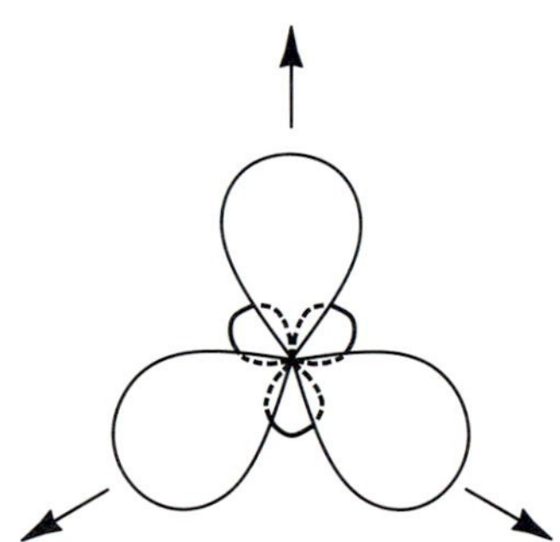

FIGURE 2.11 Three sp^2-hybridized orbitals.

Methyl cation, CH_3^+, is an example of such a species. It is planar and the three carbon-hydrogen bonds are equal in length. It may be regarded as being derived from overlap of three equivalent carbon sp^2-orbitals with hydrogen 1*s*-orbitals. Each bond may be represented as C_{sp^2}—H_{1s}. The remaining carbon *p*-orbital is perpendicular to the molecular plane and contains no electrons. In this conceptual process of orbital construction, we have used the sequence of combining three atomic orbitals to form three hybrid orbitals (Figure 2.11) that are allowed to overlap (Figure 2.12a) to form three two-center molecular orbitals (Figure 2.12b). Each of these molecular orbitals contains two electrons, and the carbon also has two electrons in its $1s$ orbital that are not normally represented in our simple valence symbols.

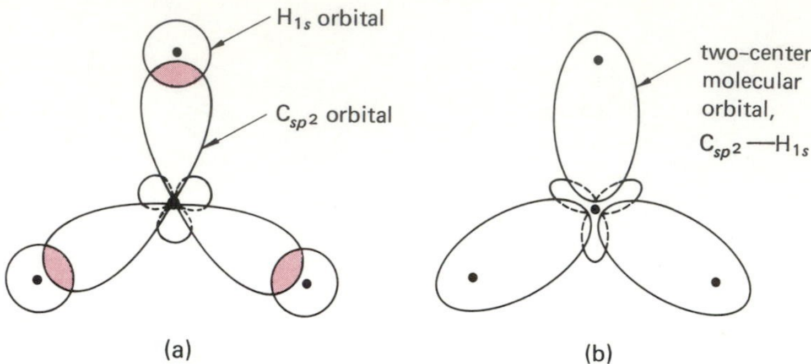

FIGURE 2.12 Construction of the electronic structure of CH_3^+.

Finally, from an *s*-orbital and three *p*-orbitals we may derive four sp^3-hybrids directed to the corners of a tetrahedron with an interorbital angle of 109.5°, the tetrahedral angle. Each such hybrid orbital is 25% *s* and 75% *p*. A three-dimensional perspective plot of one sp^3-hybrid orbital is shown in Figure 2.13. Note that the "small lobe" is much larger than in the *sp*-hybrid depicted in Figure 2.10b.

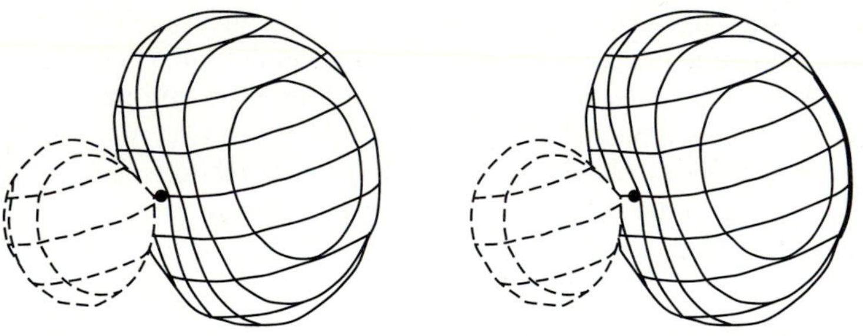

FIGURE 2.13 Three-dimensional perspective view of an sp^3-hybrid orbital.

The tetrahedral structure of methane, CH_4, is illustrated in stereo plot in Figure 2.14a or by the perspective model in Figure 2.14b. Each bond between C and H may be described as a C_{sp^3}—H_{1s} bond. Each such bond is derived by the interaction of a C_{sp^3}-hybrid orbital with a hydrogen $1s$, as in Figure 2.14c, to produce the resulting two-center molecular orbital shown in Figure 2.14d. The actual wave function—the mathematical form of the molecular orbitals—for which Figure 2.14d is only a symbolic representation is shown in contour form in Figure 2.14e.

The hybrid orbitals considered thus far are equivalent, but it is not necessary that all orbitals on an atom be equivalent when the molecule lacks symmetry. It is possible to have a hybrid orbital that is, for example, 23% *s* and 77% *p*. In NH_3, for example, the H—N—H angle of 107.1° does not correspond to any simple hybrid. Recall that sp^3-hybridization corresponds to a bond angle of 109.5°, sp^2-hybridization corresponds to 120°, and that pure *p*-orbitals form a 90° angle. In addition to its three nitrogen-hydrogen bonds, ammonia also has a nonbonding pair of electrons on the nitrogen. These electrons are in an orbital that has more *s*-character than a simple sp^3-orbital. Consequently, the

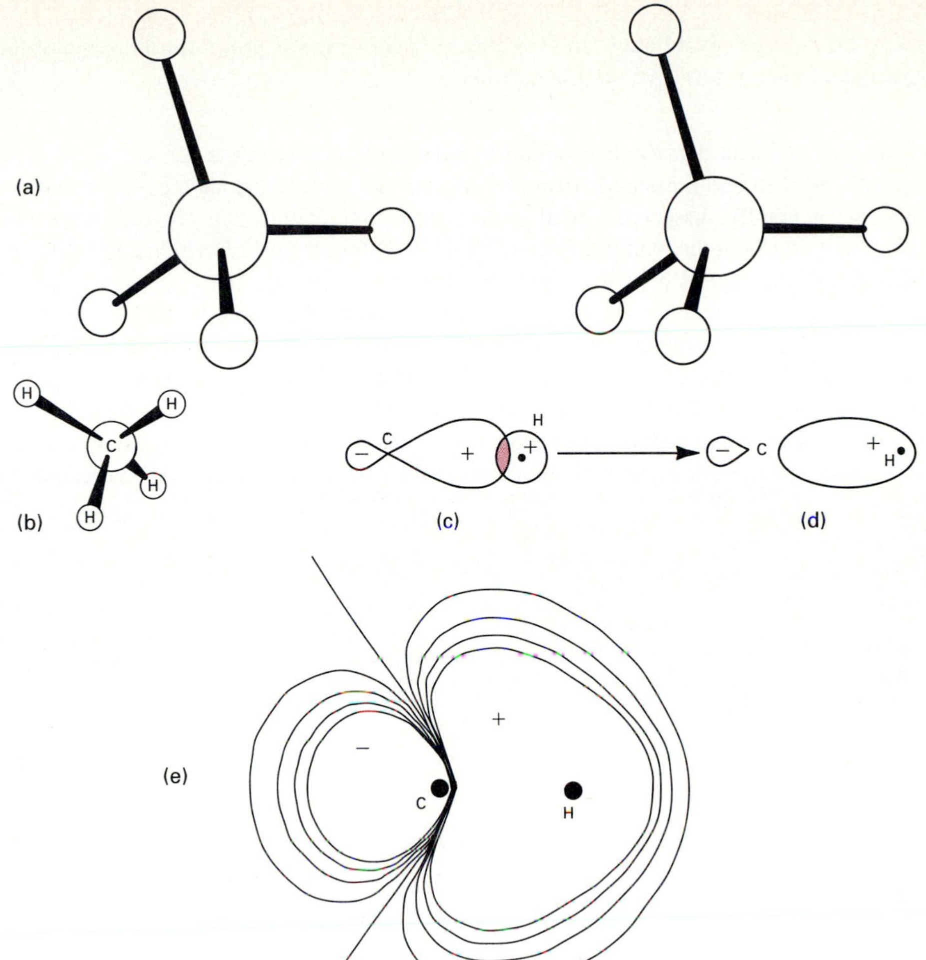

FIGURE 2.14 Methane, CH_4, and its C_{sp^3}—H_{1s} bond.

three hybrid orbitals that overlap with the three hydrogen atoms contain less *s*-character than a sp^3-orbital (actually, these orbitals are each approximately 23% *s* and 77% *p*).

Electrons in *s*-orbitals have lower energy than electrons in *p*-orbitals. Therefore, bonds with more *s*-character tend to be stronger. However, an electron pair in a bond is affected by two nuclei, whereas nonbonding electrons are attracted only by a single nucleus. Hence, *s*-character is more important for lone-pair electrons than for bonding electron pairs. In the division of the available *s*-orbital among bonds and lone pairs, the lone pairs generally receive a higher proportion.

This type of result is general. In water, for example, the H—O—H angle is 104.5° and each oxygen-hydrogen bond clearly involves an oxygen hybrid with more *p*-character than in ammonia. The two oxygen lone pairs require a large fraction of the available oxygen 2*s*-orbital. In HF, the hydrogen-fluorine bond is an almost pure F_{2p}—H_{1s} bond, and the fluorine 2*s*-orbital is used almost entirely for the three hybrid orbitals that contain the lone pairs. Nevertheless, despite these complexities, it is frequently convenient and sufficient to regard the two-center molecular orbitals that comprise electron-pair covalent bonds to be composed *approximately* of simple hybrids: *sp*, sp^2, sp^3, and so on.

EXERCISE 2.8 The bond angle in H_2S is 93.3°. How would you describe the sulfur-hydrogen bond in terms of simple hybrids?

The total electron density distribution in a molecule is real in the sense that it can, in principle, be seen and measured. However, in order to understand such electron distributions, we generally dissect the total system into components that we can work with conceptually through the manipulation of symbols. Our concepts of orbitals, hybrids, and bonds should be regarded in this light. In principle, there are many possible ways of dissecting a total molecular electron density distribution into smaller and smaller parts. The traditional way is merely one such method, but one with historical roots, having a grammar and language of its own. It is also a system that can be represented by simple symbols and serves as a powerful and widespread method for correlating and predicting a wide range of chemical results. The success of this symbolism and language has promoted its extension to the chemical problems associated with biochemistry and molecular biology.

Highlight 2.3

Electron distributions as seen in the combinations of atomic orbitals (s, p_x, p_y, p_z) into hybrid atomic orbitals (sp, sp^2, sp^3) help us to understand the three-dimensional structures of molecules.

ASIDE

Gilbert N. Lewis (1875–1946) was one of the preeminent physical chemists of the twentieth century; his ideas on covalent bonding and acid-base behavior have had an immense influence on chemistry. Our current view of these ideas is that atoms form bonds by losing electrons to, gaining electrons from, or sharing electrons with neighboring atoms. Each atom thus reaches a stable octet (noble gas) arrangement of electrons except for hydrogen, which acquires a duet. At this point the idea should be familiar to you.

As we learned in the introduction, our ideas about how to express the formulas of organic compounds are only about 130 years old. Modern theory has given effective ways of understanding the bonds. The simplest organic hydrocarbon, methane, may one day be the major source of organic chemicals after the supplies of readily accessible petroleum are exhausted. We must understand the molecule on several levels in order to understand its transformations into useful materials. The first step in understanding a compound is to know its formula. Deeper understanding comes with a knowledge of its electronic structure as given by molecular orbitals and symbolized by orbital structures and resonance diagrams.

Practice in reading and writing these formulas and electronic representations cannot be overemphasized.

PROBLEMS

1. Write a valid Lewis structure for each of the following inorganic compounds.
   a. bisulfate ion, HSO_4^-
   b. amide ion, NH_2^-

c. nitrite ion, NO_2^- (arranged ONO)
d. dinitrogen trioxide, N_2O_3 (arranged ONONO)
e. nitrous oxide, N_2O (arranged NNO)
f. hydroxylamine anion, $(ONH_2)^-$
g. nitronium ion, NO_2^+ (arranged ONO)
h. cyanamide, H_2NCN
i. nitrosonium ion, NO^+
j. hydrazoic acid, HN_3 (arranged HNNN)
k. azide ion, N_3^- (arranged NNN)
l. carbonate ion, CO_3^{2-}
m. cyanic acid OCNH

2. Write out the Lewis structures and corresponding Kekulé structures for each of the following organic compounds.

a. ethyl cation, $CH_3CH_2^+$
b. ethyl radical, $CH_3CH_2^{\cdot}$
c. ethyl anion, $CH_3CH_2^-$
d. methylacetylene, $CH_3C{\equiv}CH$
e. methyl ethyl ketone, $CH_3COCH_2CH_3$
f. dimethyl ether, CH_3OCH_3
g. methylamine, CH_3NH_2
h. methylammonium cation, $CH_3NH_3^+$
i. methoxide ion, CH_3O^-
j. methyloxonium ion, $CH_3OH_2^+$
k. vinyl chloride, $CH_2{=}CHCl$
l. formyl cation, HCO^+ (arranged HCO)

3. For each of the following compounds, describe each bond in terms of its component atomic orbitals.

a. ethane, CH_3CH_3
b. ethyl anion, $CH_3CH_2^-$
c. ethyl cation, $CH_3CH_2^+$
d. methylborane, CH_3BH_2
e. methylberyllium hydride, CH_3BeH
f. methanol, CH_3OH

4. Which of the following pairs of Kekulé structures do *not* constitute resonance structures?

a. $CH_3C({=}O){-}O^-$ and $CH_3C(O^-){=}O$

b. $CH_3C({=}O){-}OH$ and $CH_3C(OH){=}O$

c. $CH_3C({=}O)CH_3$ and $CH_3C(OH){=}CH_2$

d. $^+CH_2{-}CH{=}CH_2$ and $CH_2{=}CH{-}\overset{+}{C}H_2$

e. $CH_2{=}CH{-}CH({=}O)$ and $^+CH_2{-}CH{=}CH(O^-)$

f. $CH_3CH{=}CHCH_3$ and $CH_3CH_2CH{=}CH_2$

g. $CH_2{=}C{=}CH_2$ and $CH_3C{\equiv}CH$

h. $H{-}C{\equiv}\overset{+}{N}H$ and $H{-}\overset{+}{C}{=}NH$

i. $CH_3{-}\overset{+}{N}({=}O){-}O^-$ and $CH_3{-}\overset{+}{N}(O^-){=}O$

j. $CH_3CH(OH)CH_3$ and $CH_3CH_2CH_2OH$

k. $CH_3N{=}C{=}O$ and $CH_3O{-}C{\equiv}N$

l. $CH_3N{=}C{=}O$ and $CH_3\overset{+}{N}{=}C{-}O^-$

m. $^-O{-}\overset{+}{O}{=}O$ and $O{=}\overset{+}{O}{-}O^-$

n. $\dot{N}{=}O$ and $^-N{=}O\cdot^+$

o. HOCN and HNCO

5. For each of the following resonance hybrids, rank the contributing structures in order of their relative importance.

a. $H{-}C({=}\ddot{O}{:}){-}\ddot{N}H_2 \longleftrightarrow H{-}C({-}\ddot{O}{:}^-){=}\overset{+}{N}H_2$

b. $H{-}C({=}\ddot{O}{:}){-}\ddot{C}H_2^- \longleftrightarrow H{-}C({-}\ddot{O}{:}^-){=}CH_2$

c. $H{-}C({=}\ddot{O}{:}){-}\ddot{N}H^- \longleftrightarrow H{-}C({-}\ddot{O}{:}^-){=}\ddot{N}H$

d. $CH_2{=}CH{-}CH({=}\ddot{O}{:}) \longleftrightarrow {}^+CH_2{-}CH{=}CH({-}\ddot{O}{:}^-) \longleftrightarrow {}^-\ddot{C}H_2{-}CH{=}CH({-}O{:}^+)$

e. (benzene, C_6H_6, Kekulé structure) $\longleftrightarrow$ (benzene, C_6H_6, alternate Kekulé structure)

f. $H{-}C({=}\overset{+}{O}H){-}\ddot{O}H \longleftrightarrow H{-}C({-}\ddot{O}H){=}\ddot{O}H^+$

g. $CH_3{-}\overset{+}{N}{\equiv}C{-}O^- \longleftrightarrow CH_3{-}{}^-N{-}C{\equiv}O^+$

h. $^-O{-}\overset{+}{O}{=}O \longleftrightarrow O{=}O{=}O$

i. $\dot{N}{=}O \longleftrightarrow {}^-N{=}O\cdot^+$

j. $HONC \longleftrightarrow HO\overset{+}{N}{\equiv}C^-$

k. $CH_2{=}C{=}O \longleftrightarrow \overset{-}{C}H_2{-}C{\equiv}O^+$

6. Arrange the following compounds in order of increasing C—O bond length. Explain your answer.
   a. dimethyl ether, CH_3OCH_3
   b. acetone (dimethyl ketone), CH_3COCH_3
   c. acetate ion, $CH_3CO_2^-$
   d. carbon monoxide, CO

7. Arrange the following compounds in order of increasing C—N bond length. Explain your answer.
   a. acetonitrile, CH_3CN
   b. acetone oxime, $(CH_3)_2C{=}NOH$
   c. dihydrogen cyanide ion, $HCNH^+$
   d. methylamine, CH_3NH_2

8. Use the information given on page 7 to show that the reaction $Na\cdot + Cl\cdot \rightarrow Na^+ + Cl^-$ is endothermic in the gas phase. Sodium chloride has m.p. 801 °C and boils at 1465 °C. The vapor consists of ion pairs, $Na^+\ Cl^-$, held together by electrostatic attraction at a bond distance of 2.36 Å. A proton and an electron at a distance of 1 Å have an electrostatic attraction of 330 kcal mole^{-1}. What is the electrostatic energy of the proton and electron at the bond distance of sodium chloride? As a model for $Na^+\ Cl^-$, is this value enough to make the reaction $Na\cdot + Cl\cdot \rightarrow Na^+\ Cl^-$ exothermic?

CHAPTER 3

ORGANIC STRUCTURES

3.1 Introduction

By the middle of the nineteenth century, organic chemistry was being actively explored in Europe. Many new compounds had been isolated from natural sources, and chemists found more and more transformations of one organic compound into another. However, the fundamental concept of **structure** had not yet evolved. The atomic theory of matter was well established, and it was clear that, by and large, inorganic compounds could be characterized by simple formulas, such as NH_3, P_2O_5, HNO_3, or H_2SO_4.* The satisfying theory of valency was well on its way to providing an organized view of the multitude of known chemical substances. However, some vexing problems remained, largely with the "organic" compounds. For example, "marsh gas" (methane) was shown to have the formula CH_4, which agreed with the quadrivalence of carbon in inorganic compounds such as CS_2. However, "olefiant gas" (ethylene) was found to have the formula C_2H_4, in which the apparent valence of carbon was two. To make matters worse, a gaseous hydrocarbon (acetylene) discovered by Edmund Davy in 1836 was found to have the formula C_2H_2 in which carbon had the ridiculous valence of one!

A second difficult problem was presented by the phenomenon of **isomerism** (the existence of two or more compounds having the same molecular formula). Isomerism had first been discovered in the 1820s by Liebig and Wöhler when they found that silver cyanate (AgNCO) and silver fulminate (AgONC) have the same atomic composition. Joseph Louis Gay-Lussac made the then-revolutionary suggestion that the two compounds dif-

*Actually, the formulas in use in the early nineteenth century were not standardized, since different chemists were prone to use different atomic weights for various elements. For example, the German chemist Justus von Liebig used C = 6, O = 8, while the French chemist Jean Baptiste André Dumas used C = 6, O = 16. In Sweden, Berzelius used atomic weights that agree with the present-day values, but he also used double volume formulas. In this section, we use only the modern formulas.

fered "in the way the elements were combined together." From this point on, it was clear that a formula alone was not adequate to characterize a compound uniquely.

The third important event that paved the way for the formulation of chemical structures was the discovery of **substitution reactions** (replacement of an atom or group of atoms in a compound by another atom or group of atoms). Although substitution reactions had been carried out by Faraday, Gay-Lussac, Liebig, and Wöhler, it was first actively investigated by Dumas and his protégé Auguste Laurent in the 1830s and 1840s. One of the first substitution reactions studied was halogenation (Chapter 6). For example, it was found that chlorine combines with many organic compounds by replacing hydrogen atoms one-for-one, as in the conversion of methane to chloroform.

$$CH_4 + 3\ Cl_2 \longrightarrow CHCl_3 + 3\ HCl$$

However, it was also found that in some compounds all of the hydrogens did not seem to be equivalent. For example, acetic acid was found to undergo substitution of only three of its four hydrogens, no matter how vigorous the reaction conditions. The product, trichloroacetic acid, had properties that differed only quantitatively from those of acetic acid itself.

$$HC_2H_3O_2 + 3\ Cl_2 \longrightarrow HC_2Cl_3O_2 + 3\ HCl$$

Thus, it became increasingly clear that it must matter how a given hydrogen is joined to the remainder of the molecule.

At the time, a considerable controversy raged over the work of Laurent and Dumas. In fact, strange as it may now seem, the very existence of substitution was not even accepted by Berzelius, since it appeared to be leading toward nullification of the "dualistic theory," of which he was the architect and principal advocate. The controversy was not solely the result of inflexibility on the part of Berzelius, since Dumas was prone to make rather rash extensions of his experiments, such as his claim that even carbon atoms were susceptible to substitution by halogens. The argument eventually led Wöhler, who was something of a practical joker, to perpetrate an amusing hoax. In an article published in the German journal *Liebigs Annalen* under the pen name S. C. H. Windler, Wöhler described work on the reaction of chlorine with manganese acetate, $Mn(C_2H_3O_2)_2$. He reported that the hydrogen was first replaced by chlorine as expected. However, he reported that on longer reaction the oxygen and manganese were also replaced and that finally even the carbon was replaced by chlorine. There was obtained a substance which analyzed for 100% chlorine, but which had all the physical and chemical properties of manganese acetate!

Gradually, it came to be realized that the atoms in a molecule are "hooked together" in a certain way and that this assembly or structure is a more accurate way of describing a substance than a simple molecular formula. As shown on page 3, the first attempts to depict structures were fairly clumsy. However, these initial attempts to depict the manner of connecting the atoms in a compound soon gave way to a standard formalism in which a line was used to symbolize a point of connection, or "bond," between two atoms. With this simple concept came the realization that "valency," which worked so well for inorganic compounds, could be easily extended to organic compounds by the expedient of using "double" and "triple" bonds, as in ethylene and acetylene.

$$\mathrm{H_2C{=}CH_2} \qquad \mathrm{H{-}C{\equiv}C{-}H}$$

It was much later that the idea of the electron-pair (covalent) bond was introduced, and the lines of the nineteenth century acquired a physical significance as shown in the Lewis "electron dot" structures we reviewed in Chapter 2. In modern orbital terms each of these lines represents a two-center molecular orbital "localized" on a pair of atoms and derived from the overlap of two atomic or hybrid orbitals. These **structural formulas** are often further abbreviated for convenience by omitting the lines. The resulting expressions are called **condensed formulas.**

$$\text{H—C(H)(H)—H} \equiv CH_4 \qquad \text{H—C(H)(H)—}\ddot{\underset{..}{\text{Cl}}}\text{:} \equiv CH_3Cl$$

methane   methyl chloride

$$\text{H—C(H)(H)—}\ddot{\underset{..}{\text{O}}}\text{—H} \equiv \text{H—C(H)(H)—O—H} \equiv CH_3OH$$

methyl alcohol

Among chemical entities, carbon-carbon bonds are ubiquitous and responsible for the wide variety of organic compounds. Although some other atoms can bond to themselves to form short or long chains, carbon is unique in the extent and versatility of its catenation (chain formation; Lt., *catena*, a chain). Such carbon-carbon bonds are treated in the same way as others, as shown by the following examples.

$$\text{H—C(H)(H)—C(H)(H)—H} \equiv CH_3CH_3 \qquad \text{H—C(H)(H)—C(H)(H)—Cl} \equiv CH_3CH_2Cl$$

ethane   ethyl chloride

$$\text{H—C(H)(H)—C(H)(H)—O—H} \equiv CH_3CH_2OH \qquad \text{H—C(H)(H)—C(H)(H)—C(H)(H)—H} \equiv CH_3CH_2CH_3$$

ethyl alcohol   propane

These compounds involve carbon-hydrogen bonds that are all approximately C_{sp^3}—H_{1s}. Correspondingly, all of these carbon-hydrogen bonds are about the same length, 1.10 Å. Similarly, all of the carbon-carbon bonds in these compounds are approximately C_{sp^3}—C_{sp^3}, and these bond lengths are all about the same, 1.54 Å.

Compounds with multiple bonds can also be represented by condensed formulas.

$$\text{(H)(H)C=C(H)(H)} \equiv CH_2\text{=}CH_2 \qquad \text{H—C≡C—H} \equiv \text{HC≡CH}$$

ethylene   acetylene

The carbon-hydrogen bonds in ethylene are approximately C_{sp^2}—H_{1s} and are slightly shorter than the C_{sp^3}—H_{1s} bonds in ethane. Similarly, the carbon-hydrogen bonds in acetylene are approximately C_{sp}—H_{1s} and are shorter still. The double and triple carbon-carbon bonds in ethylene and acetylene are also shorter and stronger than the single carbon-carbon bond in ethane and are discussed in detail in subsequent chapters (Sections 11.1 and 12.1).

Further examples in which several different types of bonds are involved are

$$\mathrm{H{-}\underset{H}{\overset{H}{C}}{-}\overset{O}{\overset{\|}{C}}{-}\underset{H}{\overset{H}{C}}{-}H} \equiv \mathrm{CH_3COCH_3} \qquad \mathrm{H{-}\underset{Cl}{\overset{Cl}{C}}{-}Cl} \equiv \mathrm{CHCl_3}$$

acetone chloroform

$$\mathrm{O{=}C{=}O} \equiv \mathrm{CO_2} \qquad \mathrm{H{-}\underset{H}{\overset{H}{C}}{-}\overset{O}{\overset{\|}{C}}{-}O{-}H} \equiv \mathrm{CH_3COOH}$$

carbon dioxide acetic acid

In our subsequent discussion of organic structures, we shall make frequent use of these simple bonding concepts and symbols. We shall find them to be common and powerful devices for understanding physical properties and reactions. Organic structures are generally so large and complex that it is essential to have such systematic methods for dissecting the whole molecule into component parts and individual bonds.

EXERCISE 3.1 Write structural and condensed formulas for two compounds having the formula C_4H_{10} and three compounds having the formula C_5H_{12}.

3.2 Functional Groups

One consequence of the Laurent-Dumas investigations of substitution during the 1830s and 1840s was the evolution of the "type theory." It was gradually recognized that organic compounds can be classified into a certain rather small number of "types," each of which shows similar chemical properties. The type theory was greatly extended in the 1840s and 1850s, principally by August Wilhelm von Hofmann at the Royal College of Chemistry and Alexander William Williamson at University College, both in London. Hofmann studied the relationship between ammonia and the group of organic compounds known as amines, and showed that the amines can be prepared from ammonia by the successive replacement of one, two, or three hydrogens of ammonia by organic groups. Furthermore, the amines all show the characteristic basic properties of ammonia. Accordingly, Hofmann suggested that amines are organic molecules of the "ammonia type."

$$\left.\begin{matrix}\mathrm{H}\\\mathrm{H}\\\mathrm{H}\end{matrix}\right\}\mathrm{N} \qquad \left.\begin{matrix}\mathrm{C_2H_5}\\\mathrm{H}\\\mathrm{H}\end{matrix}\right\}\mathrm{N} \qquad \left.\begin{matrix}\mathrm{C_2H_5}\\\mathrm{C_2H_5}\\\mathrm{H}\end{matrix}\right\}\mathrm{N} \qquad \left.\begin{matrix}\mathrm{C_2H_5}\\\mathrm{C_2H_5}\\\mathrm{C_2H_5}\end{matrix}\right\}\mathrm{N}$$

ammonia ethylamine diethylamine triethylamine

At about the same time as Hofmann was investigating amines, Williamson carried out research on the relationship of water, alcohols, and ethers and eventually realized that, just as there is an ammonia type of organic compound, alcohols and ethers belong to a "water type."

$$\left.\begin{matrix}\mathrm{H}\\\mathrm{H}\end{matrix}\right\}\mathrm{O} \qquad \left.\begin{matrix}\mathrm{C_2H_5}\\\mathrm{H}\end{matrix}\right\}\mathrm{O} \qquad \left.\begin{matrix}\mathrm{C_2H_5}\\\mathrm{C_2H_5}\end{matrix}\right\}\mathrm{O}$$

water ethyl alcohol ethyl ether

Chemists were quick to realize the power of this new idea of grouping the rapidly expanding plethora of organic molecules into a relatively small number of types characterized by similar chemical properties. For example, the OH group in water reacts avidly with potassium to form potassium hydroxide and molecular hydrogen.

$$H_2O + K \longrightarrow K^+ OH^- + \tfrac{1}{2} H_2$$

Under the type theory, it is no surprise to find that the same reaction occurs with methyl alcohol.

$$CH_3OH + K \longrightarrow CH_3O^- \ K^+ + \tfrac{1}{2}H_2$$

Furthermore, a large number of different alcohols are known. Each consists of an OH group attached to a carbon framework, and all show this same reaction. Because of this constancy in the chemical properties of the OH group, it is unnecessary to study in detail the reactions of each of these many alcohols. Instead, it suffices to study alcohols as a class of organic compounds characterized by the chemical properties of the hydroxy group. This is a fortunate situation, for it gives organic chemistry a logical and systematic structure. There are a number of atoms or groups of atoms that show a relative constancy of properties when attached to different carbon chains. In modern terminology, such groups of atoms are called **functional groups.** The OH group is an example of a functional group. In this book, we will organize our systematic study of organic chemistry according to the chemistry of the important functional groups.

The simplest organic compounds are those that have no functional groups. These compounds consist only of carbon and hydrogen and are molecules in which carbons are joined to each other only by single bonds. These **saturated hydrocarbons** (''saturated'' means having no double or triple bonds) may be noncyclic (the **alkanes**) or cyclic (the **cycloalkanes**). They form the framework to which functional groups may be attached. The symbol R is often used to denote an alkyl group, the simplest being the methyl group, CH_3. Since the simplest alkane is methane, CH_4, we see that with this symbolism the alkane class may be represented by RH. We shall see in Chapter 6 that alkanes undergo only a limited number of reactions—precisely because they have no functional groups.

Some Alkanes

CH_4 $\quad$ CH_3CH_3 $\quad$ $CH_3CH_2CH_3$ $\quad$ $CH_3CH_2CH_2CH_3$

Some Cycloalkanes

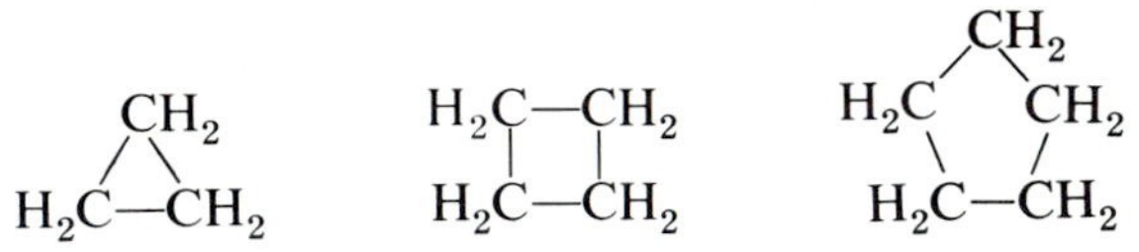

```
   CH2          H2C—CH2          CH2
   / \            |   |        /    \
H2C—CH2         H2C—CH2     H2C      CH2
                               \    /
                               H2C—CH2
```

In a similar manner, all hydrocarbons containing one or more carbon-carbon double bonds form a logical class, the **alkenes.** The hydrocarbons having a carbon-carbon triple bond form a third structurally similar set, the **alkynes.** The double bond of an alkene and the triple bond of an alkyne are functional groups.

Some Alkenes

```
                 CH3       H
                    \     /
H2C=CH2              C=C            H2C—CH
                    /     \          |   ||
                   H       CH3      H2C—CH
```

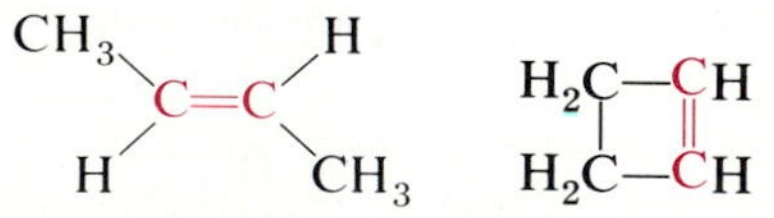

Some Alkynes

$CH_3—C{\equiv}C—H$ $\quad$ $CH_3CH_2CH_2—C{\equiv}C—CH_3$ $\quad$ $H—C{\equiv}C—H$

We will find a number of reactions characteristic of carbon-carbon multiple bonds that are not shared by single bonds.

Organic compounds that contain carbon-oxygen single bonds are classed as **alcohols** or **ethers,** depending on whether or not the oxygen is also bonded to a hydrogen.

Some Alcohols

$$CH_3{-}OH \qquad CH_3{-}C(CH_3)_2{-}OH \qquad \text{cyclopentanol: } H_2C{-}CH_2{-}CH(OH){-}CH_2{-}CH_2 \text{ (ring)}$$

Some Ethers

$$CH_3{-}O{-}CH_3 \qquad CH_3{-}O{-}CH(CH_3)_2 \qquad \text{tetrahydrofuran: } H_2C{-}CH_2{-}CH_2{-}O{-}CH_2 \text{ (ring)}$$

The carbon-oxygen double bond, the carbonyl group, is found in **aldehydes** and **ketones.** When the carbonyl group is bonded to an OH group, it becomes a carboxy group. Compounds containing this functional group are called **carboxylic acids.**

Some Aldehydes

$$H_2C{=}O \qquad CH_3\overset{O}{\overset{\|}{C}}H \qquad CH_3CH_2\overset{O}{\overset{\|}{C}}H$$

Some Ketones

$$CH_3\overset{O}{\overset{\|}{C}}CH_3 \qquad CH_3\underset{}{\overset{CH_3}{\overset{|}{C}}}H{-}\overset{O}{\overset{\|}{C}}CH_3 \qquad \text{cyclobutanone: } H_2C{-}C({=}O){-}CH_2{-}CH_2 \text{ (ring)}$$

Some Carboxylic Acids

$$H\overset{O}{\overset{\|}{C}}OH \qquad CH_3\overset{O}{\overset{\|}{C}}OH \qquad \text{cyclopropanecarboxylic acid: } H_2C{-}CH_2{-}CHCOOH \text{ (ring)}$$

Table 3.1 lists a number of the important functional groups. The structures and names of these groups should be committed to memory. They form an essential part of the language of organic chemistry. In our subsequent studies we will develop the chemistry of the individual functional groups in terms of structural and electronic theory, nomenclature (names), physical properties, formation from other functional groups, and the characteristic reactions that produce other groups. Although in molecules occurring in biological systems nitriles and nitro compounds are rare, acyl halides are replaced by acyl thioesters, and there seems to be only one extremely important organometallic compound, vitamin B_{12}, almost all of the functional groups are important, except for boranes and sulfones.

Interconversions of functional groups constitute a large proportion of organic chemistry. After the individual groups have been studied, the effect of one group on another can be considered, for the organic chemistry of compounds with more than one functional group is not always simple. Groups affect each other, sometimes in complex ways. One of

TABLE 3.1

Class	General Structure	Characteristic Functional Group	Example
alkanes	$R{-}H$	none	CH_4
alkenes	$\begin{matrix} R & & R_2 \\ & C{=}C & \\ R_1 & & R_3 \end{matrix}$	$>C{=}C<$	$CH_3CH{=}CH_2$
aromatic ring	$\begin{matrix} & R_2 & R_3 & \\ & C{-}C & & \\ R_1{-}C & & & C{-}R_4 \\ & C{=}C & & \\ & R_6 & R_5 & \end{matrix}$	$\begin{matrix} & C{-}C & \\ {-}C & & C{-} \\ & C{=}C & \end{matrix}$	$\begin{matrix} & CH{-}CH & \\ CH_3{-}C & & CH \\ & CH{=}CH & \end{matrix}$
alkynes	$R{-}C{\equiv}C{-}R'$	$-C{\equiv}C-$	$CH_3{-}C{\equiv}C{-}CH_3$
alkyl halides	RF, RCl, RBr, RI	—F, —Cl, —Br, —I	CH_3Cl
alcohols	$R{-}OH$	—OH	CH_3OH
ethers	$R{-}O{-}R'$	—O—	CH_3OCH_3
amines			
primary amines	$R{-}NH_2$	$-NH_2$	CH_3NH_2
secondary amines	$R{-}NH{-}R'$	$>N{-}H$	CH_3NHCH_3
tertiary amines	$R{-}\underset{\underset{R''}{\mid}}{N}{-}R'$	$>N{-}$	$CH_3\underset{\underset{CH_3}{\mid}}{N}CH_3$
thiols	$R{-}SH$	—SH	CH_3SH
sulfides	$R{-}S{-}R'$	—S—	CH_3SCH_3
disulfides	$R{-}S{-}S{-}R'$	—S—S—	CH_3SSCH_3
boranes	R_3B	$-\underset{\mid}{B}-$	$(CH_3)_3B$
organometallic	RM, R_2M, R_3M	—M	CH_3Li, $(CH_3)_2Mg$, $(CH_3)_3Al$
aldehydes	$R{-}\overset{\overset{O}{\|}}{C}{-}H$	$-\overset{\overset{O}{\|}}{C}{-}H$	$CH_3{-}\overset{\overset{O}{\|}}{C}{-}H$
ketones	$R{-}\overset{\overset{O}{\|}}{C}{-}R'$	$-\overset{\overset{O}{\|}}{C}-$	$CH_3{-}\overset{\overset{O}{\|}}{C}{-}CH_3$
imines	$R{-}\overset{\overset{N-R'}{\|}}{C}{-}R''$	$-\overset{\overset{N-}{\|}}{C}-$	$CH_3{-}\overset{\overset{N-CH_3}{\|}}{C}{-}H$
carboxylic acids	$R{-}\overset{\overset{O}{\|}}{C}{-}OH$	$-\overset{\overset{O}{\|}}{C}{-}OH$	$H{-}\overset{\overset{O}{\|}}{C}{-}OH$

TABLE 3.1 (continued)

Class	General Structure	Characteristic Functional Group	Example
esters	$R-\overset{O}{\overset{\Vert}{C}}-OR'$	$-\overset{O}{\overset{\Vert}{C}}-O-$	$CH_3-\overset{O}{\overset{\Vert}{C}}-OCH_3$
amides	$R-\overset{O}{\overset{\Vert}{C}}-NR'_2$	$-\overset{O}{\overset{\Vert}{C}}-N\langle$	$CH_3-\overset{O}{\overset{\Vert}{C}}-NH_2$
acyl halides	$R-\overset{O}{\overset{\Vert}{C}}-X$	$-\overset{O}{\overset{\Vert}{C}}-X$	$CH_3-\overset{O}{\overset{\Vert}{C}}-Cl$
nitriles	$R-C\equiv N$	$-C\equiv N$	$CH_3C\equiv N$
nitro compounds	$R-NO_2$	$-NO_2$	CH_3-NO_2
sulfones	$R-SO_2-R'$	$-SO_2-$	$CH_3-SO_2-CH_3$
sulfonic acids	$R-SO_2-OH$	$-SO_2-OH$	CH_3-SO_2-OH

the reasons for studying the theory of organic chemistry is that the mutual interactions of functional groups can be understood. A series of books called *Chemistry of Functional Groups* included 80 volumes at the end of 1991 (see Chapter 35).

The aromatic ring in Table 3.1 is written with three carbon-carbon double bonds. Nevertheless, we shall see later (Chapters 21 and 23) that compounds containing this ring system differ substantially in their chemistry from the alkenes. Compounds containing this ring system are known collectively as **aromatic compounds.** Compounds with no aromatic ring are known as **aliphatic compounds.**

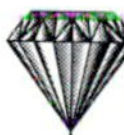

Highlight 3.1

Functional groups are combinations of atoms that undergo chemical reactions in a consistent way from molecule to molecule.

EXERCISE 3.2 Using R = ethyl, write structural and condensed formulas for one example each of an alkene, an alkyne, an alcohol, an ether, an aldehyde, a ketone, a carboxylic acid, an amine, and a nitrile.

3.3 The Shapes of Molecules

The Kekulé-Couper concept of bonds and the powerful organizational theory of types, or functional groups, catalyzed an explosive period in the development of the new science of organic chemistry. One important aspect of molecular structure remained unrecognized. The German chemist Johannes Wislicenus first conceived of the idea of molecular shape. He studied pairs of isomeric organic compounds and concluded in 1873 that "If molecules can be structurally identical and yet possess dissimilar properties, this can be explained only on the ground that the difference is due to a different arrangement of the atoms in space." In 1874, the Dutch chemist Jacobus Hendricus van't Hoff and the French chemist Joseph Achille Le Bel independently suggested that the four valences of a carbon atom are oriented in space in a tetrahedral manner. That is, the four hydrogens of methane can be viewed as occupying the four corners of a regular tetrahedron (Figure 2.14).

Molecular shape is a fundamental concept in organic chemistry. Molecules are three-dimensional, and the spatial relations between different parts of a molecule can be very important in determining interactions which can affect chemical and physical properties. Consequently, the student must cultivate the ability to think of organic molecules as ''objects'' having a definite shape. Since this is a difficult task for most people, especially with complex molecules, various aids to visualization are employed. One such technique, which we shall use in this book, is stereoscopic projection.

Other tools that are indispensable for visualizing spatial relations in molecules are **molecular models.** Since bonds in organic compounds are formed from hybrid orbitals which approximate simple *sp*-, sp^2-, and sp^3-hybrids, and bond angles and distances are relatively constant from one molecule to another, it is possible to use simple objects as models for various atoms. These objects can be joined together in the same manner as ''Tinkertoys'' to produce models of molecules. There are model sets, some that are expensive precision tools used primarily for research purposes, and several that are relatively inexpensive and designed for student use in the study of organic chemistry. The nature of some sets is summarized below; some models of a molecule are illustrated in Figure 3.1.

Dreiding stereomodels are skeletal models constructed from welded stainless steel tubing. The bond lengths and angles are precisely proportional to the average molecular dimensions. They are relatively expensive and are widely used by professional chemists for research purposes.

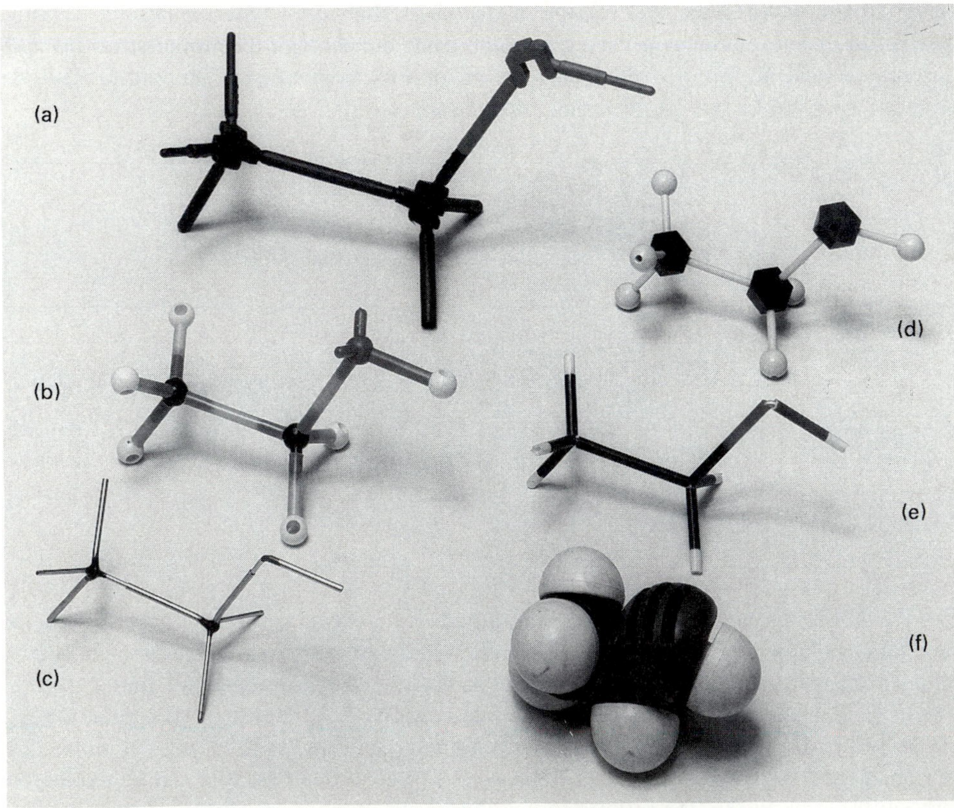

FIGURE 3.1 Some molecular model representations of ethyl alcohol, CH_3CH_2OH. Models used are (a) Darling, (b) Theta, (c) Dreiding, (d) HGS Maruzen, (e) Framework, and (f) CPK™.

The Theta Molecular Model Set marketed by John Wiley & Sons uses small plastic nuclei like ''jacks'' and plastic tubing. The Prentice-Hall Framework molecular models are similar, but use metal nuclei. The HGS-Maruzen models use plastic atoms with holes drilled to accommodate bonds. The Darling models also use plastic parts that are fitted together. Models such as these are relatively inexpensive and are indispensable as an aid in visualizing three-dimensional aspects of organic structures and reactions.

Corey-Pauling-Koltun (CPK™) molecular models are an example of space-filling models. The models are constructed from an acrylic polyester plastic and have sizes proportional to the covalent and atomic radii of the atoms. They are held together by connectors made of a hard rubber-like elastomer. They are fairly expensive and are mainly used by professional chemists for constructing models where a knowledge of molecular shape and intramolecular interactions is important.

A new modeling tool called *molecular graphics* is coming into use, both on personal computers of the IBM and Macintosh types and on more sophisticated workstations like the Silicon Graphics IRIS. Both 2D formulas and 3D models can be generated on the computer screen. The 3D models can be turned, brought together (''docked''), and manipulated in some ways like physical models.

EXERCISE 3.3 Using your molecular model set, construct models of some of the organic molecules previously illustrated in this chapter.

3.4 The Determination of Organic Structures

In previous sections, we have reviewed some basic concepts of electronic structure and bonding and have introduced the subject of organic structures and functional groups. In subsequent chapters, we shall take up the structures and chemical reactions of various classes of organic compounds and examine them in detail. At this point, an additional question must be addressed before we embark upon our systematic study of organic chemistry. How does the chemist know the structure of a compound? The question is an important one, and it is encountered over and over again by researchers in the field. In fact, the rate of development of organic chemistry as a science has been intimately related to our ability to *determine structure*.

Mentally transport yourself back a hundred years—imagine that you are a nineteenth-century scientist and have laboriously purified an organic substance from some source. How do you determine its structure? It is possible to measure various physical properties such as boiling point, melting point, density, and refractive index. A catalog of these properties constitutes the **characterization** of a compound. Just as no two people have identical fingerprints, no two compounds have all physical properties in common. It is relatively easy to assemble a list of physical properties for the compound and to decide that it is different from other previously isolated substances. Your new compound also undergoes various chemical reactions, and from its behavior with various reagents you may be able to decide what kinds of functional groups are present. For example, suppose that your material reacts with sodium to produce hydrogen gas. Since simple alcohols such as methyl alcohol are known to undergo this reaction, you can make the deduction that your new compound contains the functional group OH. But still, from all these data, how do you write a molecular structure for the compound? This problem challenged chemists for over a hundred years.

The first major breakthrough came with the development of methods of elemental analysis. The first attempts to determine the elemental compositions of organic compounds were made by Lavoisier in the late eighteenth century in connection with his pioneering work on the reactions of oxygen. Lavoisier examined the combustion products from various compounds and could deduce which elements were present in the substance burned. For example, combustion of methane gives carbon dioxide and water. Hence, methane must be built up from carbon and hydrogen in some way.

$$\text{methane} + O_2 \longrightarrow CO_2 + H_2O$$

Although Lavoisier attempted to quantitate combustion as a method to determine accurate formulas for organic compounds, the apparatus at his disposal was not sufficiently accurate for him to obtain precise formulas. The next significant advance came in 1831 when Liebig developed the Lavoisier method into a precise quantitative technique. For the first time, it was possible to determine accurate empirical formulas for organic compounds. In connection with methods for the determination of molecular weights, it was then possible to determine molecular formulas.

The method of combustion analysis, as developed by Liebig, is conceptually very simple. A weighed quantity of the sample to be analyzed is burned in the presence of red-hot copper oxide, which is reduced to metallic copper. The sample is swept through the combustion tube with pure oxygen gas, which reoxidizes the copper to copper oxide.

$$\underset{\text{naphthalene}}{C_{10}H_{14}} + 27\ CuO \longrightarrow 10\ CO_2 + 7\ H_2O + 27\ Cu$$

$$2\ Cu + O_2 \longrightarrow 2\ CuO$$

The combustion products are swept through a calcium chloride tube, which absorbs the water formed, and then through a tube containing aqueous potassium hydroxide, which absorbs the carbon dioxide produced. The two tubes are weighed before and after combustion to determine the weights of water and carbon dioxide produced. From the weights of the two products, the weight of sample burned, and the atomic weights of carbon and hydrogen, it is possible to compute an empirical formula for the substance burned.

$$\text{weight of H in sample} = \text{weight of } H_2O \times \frac{2.016}{18.016}$$

$$\text{weight of C in sample} = \text{weight of } CO_2 \times \frac{12.01}{44.01}$$

$$\%\ \text{H in sample} = \frac{\text{weight of H}}{\text{weight of sample}} \times 100$$

$$\%\ \text{C in sample} = \frac{\text{weight of C}}{\text{weight of sample}} \times 100$$

If the percentages of carbon and hydrogen do not add up to 100 and no other element has been detected by qualitative tests, the deficiency is taken as the percentage of oxygen.

As an example, consider the analysis of propyl alcohol, C_3H_8O. An ideal analysis on a 0.500-g sample would give 0.600 g of H_2O and 1.099 g of CO_2. The calculations proceed as follows.

$$\text{weight of H in sample} = 0.600\ \text{g} \times \frac{2.016}{18.016} = 0.067\ \text{g}$$

$$\text{weight of C in sample} = 1.099\ \text{g} \times \frac{12.01}{44.01} = 0.300\ \text{g}$$

$$\%\ \text{H in sample} = \frac{0.067}{0.500\ \text{g}} \times 100 = 13.4$$

$$\%\ \text{C in sample} = \frac{0.300}{0.500} \times 100 = 60.0$$

The percentages of hydrogen and carbon add up to 73.4%, and the remaining 26.6% is taken as the percentage of oxygen in the sample. In actual practice, the analytical values are usually accurate to ± 0.3%.

EXERCISE 3.4 How much CO_2 and H_2O are produced by the combustion of 3.74 mg of $C_6H_{12}O$?

From the elemental analysis of a compound, one can easily calculate its **empirical formula,** which expresses the ratio of the elements present. In the present case, for example, the analysis tells us that 100 g of propyl alcohol contains 60.0 g of carbon, 13.4 g of hydrogen, and 26.6 g of oxygen. Dividing each of these weights by the appropriate atomic weights gives us the number of moles of each element in 100 g of sample.

$$\frac{60.0}{12.01} = 5.00 \text{ moles of carbon}$$

$$\frac{13.4}{1.008} = 13.29 \text{ moles of hydrogen}$$

$$\frac{26.6}{16} = 1.66 \text{ moles of oxygen}$$

This calculation gives us an empirical formula of $C_{5.00}H_{13.29}O_{1.66}$. However, because the atoms in a molecule must be present in whole numbers, the initially derived formula must be normalized. If we divide each of the factors derived above by the smallest, we have

$$C_{5.00/1.66}H_{13.29/1.66}O_{1.66/1.66} = C_{3.01}H_{8.01}O_{1.00}$$

Thus, the empirical formula of propyl alcohol is calculated from its elemental analysis to be C_3H_8O. The molecular formula expresses the total number of each atom present and is the same as the empirical formula or some multiple of it. For example, if the molecular formula of propyl alcohol were $C_6H_{16}O_2$, the percentages of carbon, hydrogen, and oxygen would be the same. (Actually, because of the rules of valence, $C_6H_{16}O_2$ is an impossible formula, as a little trial and error will readily reveal.)

EXERCISE 3.5 Combustion of 0.250 g of di-*n*-butyl ether gives 0.677 g of CO_2 and 0.311 g of H_2O. Calculate the empirical formula of di-*n*-butyl ether.

The Lavoisier-Liebig method of analysis provided a tremendous boost to the development of organic chemistry but required relatively large amounts of sample, on the order of 0.25–0.50 g. In 1911 Fritz Pregl introduced a technique of microanalysis that allows combustion analysis to be carried out with only 3–4 mg of sample. Elements such as N, S, Cl, Br, I, and P are determined on a micro scale by other analytical methods that we shall not detail. Highly accurate molecular formulas can now be determined with a few micrograms of substance by the technique of high-resolution mass spectrometry (Chapter 34).

From the molecular formula, the next step is to derive a molecular structure. How are the atoms bonded to one another? For our present example of C_3H_8O, which of the following structures corresponds to propyl alcohol?

```
   H  H  H            H  OH  H            H     H  H
   |  |  |            |  |   |            |     |  |
H—C—C—C—OH        H—C—C——C—H        H—C—O—C—C—H
   |  |  |            |  |   |            |     |  |
   H  H  H            H  H   H            H     H  H
```

As indicated previously, some insight may be gained by a consideration of the gross chemical properties of the material. For example, if we know that the compound reacts with sodium to liberate hydrogen, this would indicate the presence of the OH functional group and would eliminate the third possible structure above. In a similar manner, an examination of other chemical properties could lead to the elimination of other candidate structures. For example, we shall see in later chapters that the OH functional group has slightly different chemical properties when it is bonded to a carbon having two hydrogens than when it is bonded to a carbon having only one hydrogen. Thus, a careful consideration of the properties of a substance could eventually lead to a structure consistent with all the data. The modern chemist relies heavily on spectroscopic methods for the determination of structure. As we shall see in subsequent chapters, spectroscopy is the experimental evaluation of the manner in which a substance interacts with electromagnetic radiation. Thus, by examining the various spectra of a material, the chemist is evaluating some physical properties of the substance. In fact, knowledge of these particular physical properties is a very powerful aid to structural analysis. In our study of organic chemistry we shall consider various kinds of spectroscopy: nuclear magnetic resonance (Chapter 13), infrared spectroscopy (Chapter 17), mass spectrometry (Chapter 34), and ultraviolet spectroscopy (Chapter 22).

EXERCISE 3.6 Elemental analysis of an organic compound reveals it to have the empirical formula C_3H_6O. Determination of its molecular weight gives a value of 57 ± 2. These data, together with the rules of valence, are compatible with eight structures. Write these eight structures. What two structures are possible if the infrared spectrum of the unknown compound reveals the presence of a carbon-oxygen double bond?

3.5 *n*-Alkanes, the Simplest Organic Compounds

The straight-chain alkanes constitute a family of hydrocarbons in which a chain of CH_2 groups is terminated at both ends by a hydrogen. They have the general formula $H(CH_2)_nH$ or C_nH_{2n+2}. Such a family of compounds, which differ from each other by the number of CH_2 groups in the chain, is called a **homologous series.** The individual members of the family are known as **homologs** of one another. Straight-chain alkanes are called **normal alkanes,** or simply ***n*-alkanes,** to distinguish them from the branched alkanes, which we shall study later.

Alkanes are sometimes called **saturated** hydrocarbons. This term means that the carbon skeleton is "saturated" with hydrogen. That is, in addition to its bonds to other carbons, each carbon bonds to enough hydrogens to give a maximum covalence of 4. The general formula for an *n*-alkane is C_nH_{2n+2} because there are two hydrogens bonded to each carbon atom of the chain plus one additional hydrogen bonded to each end of the chain. In saturated hydrocarbons, there are only single bonds. Later, we shall study **unsaturated** hydrocarbons, compounds that contain double and triple carbon-carbon bonds. The normal alkanes are named according to the number of carbon atoms in the chain (Table 3.2).

These names derive from the generic name alkane with the **alk**- stem replaced by a stem characteristic of the number of carbons in the chain. The first four members of this series, methane, ethane, propane, and butane, are names that came into widespread use before any attempts were made to systematize the names of organic compounds. The remaining names derive quite obviously from Greek numbers; compare **pent**agon, **oct**al, **dec**imal, and so on. The student should memorize the names of the *n*-alkanes up through dodecane and know the logical procedure for developing names for larger compounds.

TABLE 3.2

n	Name	Formula
1	methane	CH_4
2	ethane	CH_3CH_3
3	propane	$CH_3CH_2CH_3$
4	butane	$CH_3CH_2CH_2CH_3$
5	pentane	$CH_3(CH_2)_3CH_3$
6	hexane	$CH_3(CH_2)_4CH_3$
7	heptane	$CH_3(CH_2)_5CH_3$
8	octane	$CH_3(CH_2)_6CH_3$
9	nonane	$CH_3(CH_2)_7CH_3$
10	decane	$CH_3(CH_2)_8CH_3$
11	undecane	$CH_3(CH_2)_9CH_3$
12	dodecane	$CH_3(CH_2)_{10}CH_3$
13	tridecane	$CH_3(CH_2)_{11}CH_3$
14	tetradecane	$CH_3(CH_2)_{12}CH_3$
15	pentadecane	$CH_3(CH_2)_{13}CH_3$
20	eicosane	$CH_3(CH_2)_{18}CH_3$
21	heneicosane	$CH_3(CH_2)_{19}CH_3$
22	docosane	$CH_3(CH_2)_{20}CH_3$
30	triacontane	$CH_3(CH_2)_{28}CH_3$
40	tetracontane	$CH_3(CH_2)_{38}CH_3$

A "group" is a portion of a molecule in which a collection of atoms is considered together as a unit. For purposes of naming more complicated compounds, it is necessary to have names for such groups. A group name is derived by replacing the **-ane** of the corresponding alkane name by the suffix **-yl.**

Alkane	*Group*	*Sample Molecule*
CH_4 methane	$CH_3—$ methyl group	$CH_3—OH$ methyl alcohol
CH_3CH_3 ethane	$CH_3CH_2—$ ethyl group	$CH_3CH_2—Cl$ ethyl chloride
$CH_3CH_2CH_3$ propane	$CH_3CH_2CH_2—$ propyl group	$CH_3CH_2CH_2—Br$ propyl bromide

EXERCISE 3.7 Write the structures of the pentyl group and of pentyl iodide.

Highlight 3.2

Chains of carbon atoms form families of organic structures. A straight chain of carbons combined with hydrogen forms the alkane family C_nH_{2n+2} or $H(CH_2)_nH$ in which the first member is methane CH_4. The longest alkane in current use is linear polymethylene (usually called polyethylene) with $n > 1000$.

3.6 Systematic Nomenclature

There is only one compound corresponding to each of the formulas CH_4, C_2H_6, and C_3H_8. There are two **isomeric** compounds having the formula C_4H_{10}. **Isomers** are defined as compounds that have identical formulas but differ in the nature or sequence of bonding of their atoms or in the arrangement of their atoms in space. One of the C_4H_{10} isomers is *n*-butane, discussed previously. The other is isobutane.

$$CH_3CH_2CH_2CH_3 \qquad CH_3{-}\underset{\displaystyle H}{\overset{\displaystyle CH_3}{C}}{-}CH_3$$

n-butane   isobutane

In general, isomers have different physical and chemical properties. For example, with the two C_4H_{10} compounds, isobutane has the lower melting point and boiling point. The lower boiling point reflects the branched-chain structure of isobutane, which provides less effective contact area for van der Waals attraction (Chapter 5).

Interconversion of the two butane isomers requires breaking bonds. Since carbon-carbon bonds have bond strengths of about 80 kcal mole^{-1}, these isomers are completely stable under normal conditions. Interconversion requires very high temperatures or special catalysts.

There are three C_5H_{12} isomers.

$$CH_3CH_2CH_2CH_2CH_3 \qquad CH_3CH_2\overset{\displaystyle CH_3}{C}HCH_3 \qquad CH_3{-}\underset{\displaystyle CH_3}{\overset{\displaystyle CH_3}{C}}{-}CH_3$$

pentane   isopentane   neopentane

The prefix **iso-** serves to name one of these isomers, the one with the greatest structural similarity to isobutane. The third isomer was named neopentane, the prefix **neo-** being derived from the Greek word for "new." With more carbons the number of possible isomers increases rapidly. As shown in Table 3.3, there are 5 possible hexanes, 9 heptanes, and 75 decanes, and with larger alkanes the number of possible isomers becomes astronomic. It is obviously not possible to coin a set of unique prefixes like iso- and neo- to name all of the dodecane isomers, or, for that matter, even the heptane isomers. Clearly, it is essential that a method of systematic nomenclature be devised, so that each different compound may be assigned an unambiguous name.

TABLE 3.3 Number of Isomers of C_nH_{2n+2}

n	Number of Isomers
4	2
5	3
6	5
7	9
8	18
9	35
10	75
12	355
15	4,347
20	366,319

This problem was first addressed by an International Congress of Chemists, held in Geneva in 1892. The forty chemists in attendance at this meeting adopted a set of systematic rules for naming organic compounds. For the most part, the Geneva rules were based on a suggestion originally made by Laurent, whereby hydrocarbons are the base compounds, with other classes of compounds being treated as substitution products. The International Congresses evolved into an organization now known as the International Union of Pure and Applied Chemistry (IUPAC), which maintains several committees whose job it is to see that the rules for systematic nomenclature are continually updated.

The IUPAC system of alkane nomenclature is based on the simple fundamental principle of considering all compounds to be **derivatives of the longest single carbon chain** present in the compound. Appendages to this chain are designated by appropriate prefixes. The chain is then numbered from one end to the other. The end chosen as number 1 is that which gives the *smaller number (a locant) at the first point of difference*.

$$\begin{array}{l} \quad CH_3 \\ \quad | \\ CH_3CHCH_2CH_2CH_3 \\ 1 \quad 2 \quad 3 \quad 4 \quad 5 \end{array} \qquad \begin{array}{l} \qquad\quad CH_3 \\ \qquad\quad | \\ CH_3CH_2CHCH_2CH_2CH_3 \\ 1 \quad 2 \quad 3 \quad 4 \quad 5 \quad 6 \end{array}$$

2-methylpentane
(not 4-methylpentane)

3-methylhexane
(not 4-methylhexane)

EXERCISE 3.8 Write the structure of 3-ethylhexane. Why is "2-ethylhexane" not an acceptable name under the IUPAC system?

When there are two or more identical appendages, the modifying prefixes di-, tri-, tetra-, penta-, hexa-, and so on are used, but every appendage group still gets its own number.

$$\begin{array}{l} \quad CH_3 \\ \quad | \\ CH_3CCH_2CH_2CH_3 \\ \quad | \\ \quad CH_3 \end{array} \qquad \begin{array}{l} \quad CH_3 \qquad\quad CH_3 \\ \quad | \qquad\qquad\quad | \\ CH_3C-CH_2-CCH_3 \\ \quad | \qquad\qquad\quad | \\ \quad CH_3 \qquad\quad CH_3 \end{array}$$

2,2-dimethylpentane
(not 2-dimethylpentane)

2,2,4,4-tetramethylpentane

EXERCISE 3.9 (a) Write the structure of 2,3,4-trimethyloctane. (b) Why is 1,4-dimethylbutane an incorrect name?

When two or more appendage locants are employed, the longest chain is numbered from the end that produces the lowest series of locants. When comparing one series of locants with another, that series is lower which contains the *lower number at the first point of difference*.

$$\begin{array}{l} \qquad\qquad\quad CH_3 \\ \qquad\qquad\quad | \\ CH_3CH_2CHCHCH_3 \\ \qquad\quad | \\ \qquad\quad CH_3 \end{array} \qquad \begin{array}{l} \quad CH_3 \qquad\qquad\quad CH_3 \\ \quad | \qquad\qquad\qquad\quad | \\ CH_3CHCH_2CH_2CHCHCH_2CH_3 \\ \qquad\qquad\qquad\quad | \\ \qquad\qquad\qquad\quad CH_3 \end{array}$$

2,3-dimethylpentane
(not 3,4-dimethylpentane)

2,5,6-trimethyloctane
(not 3,4,7-trimethyloctane)

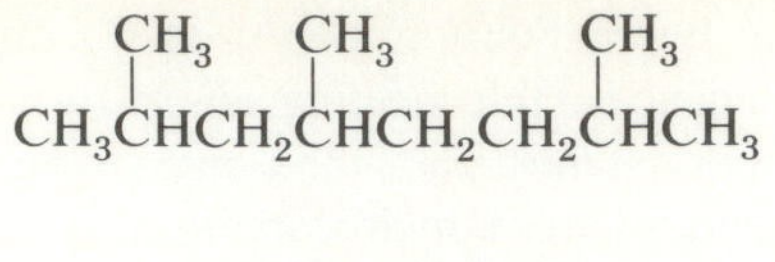

$$\begin{array}{c} CH_3 \quad CH_3 \qquad\quad CH_3 \\ CH_3CHCH_2CHCH_2CH_2CHCH_3 \end{array}$$

2,4,7-trimethyloctane
(not 2,5,7-trimethyloctane)

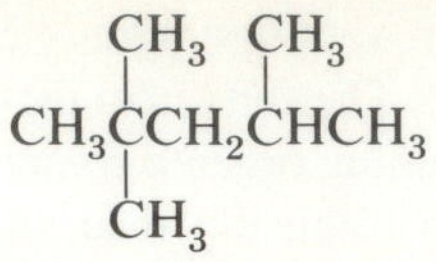

$$\begin{array}{c} CH_3\ CH_3 \\ CH_3CCH_2CHCH_3 \\ CH_3 \end{array}$$

2,2,4-trimethylpentane
(not 2,4,4-trimethylpentane)

EXERCISE 3.10 Two possible names for an alkane are 2,3,5-trimethylhexane and 2,4,5-trimethylhexane. Which is correct?

Groups derived from the terminal position of a *n*-alkane are named as discussed in Section 3.5. Several other common groups have special names that must be memorized by the student.

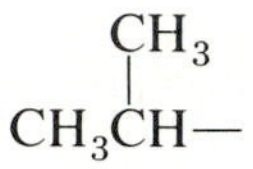

$$\begin{array}{c} CH_3 \\ CH_3CH- \end{array}$$
isopropyl

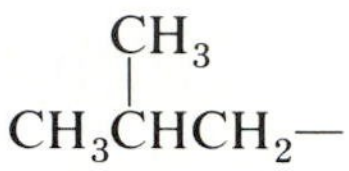

$$\begin{array}{c} CH_3 \\ CH_3CHCH_2- \end{array}$$
isobutyl

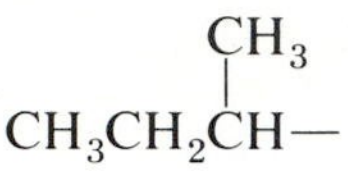

$$\begin{array}{c} CH_3 \\ CH_3CH_2CH- \end{array}$$
sec-butyl

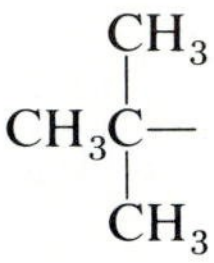

$$\begin{array}{c} CH_3 \\ CH_3C- \\ CH_3 \end{array}$$
tert-butyl
or *t*-butyl

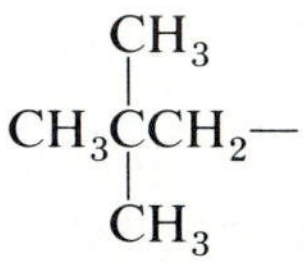

$$\begin{array}{c} CH_3 \\ CH_3CCH_2- \\ CH_3 \end{array}$$
neopentyl

A more complex appendage group is named as a derivative of the **longest carbon chain in the group** starting from the carbon that is attached to the principal chain. The description of the appendage is distinguished from that of the principal chain by enclosing it in parentheses.

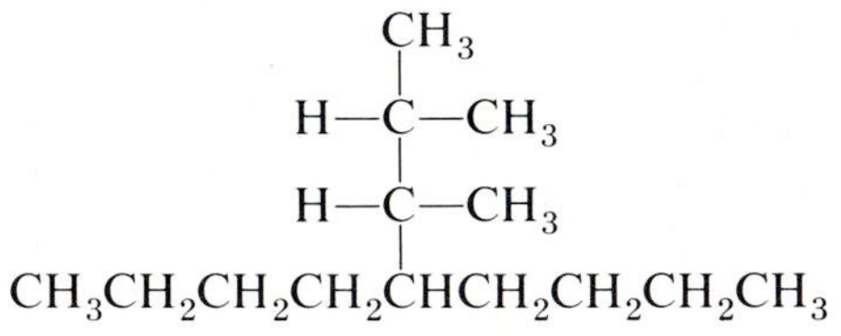

$$\begin{array}{c} CH_3 \\ H-C-CH_3 \\ H-C-CH_3 \\ CH_3CH_2CH_2CH_2CHCH_2CH_2CH_2CH_3 \end{array}$$

5-(1,2-dimethylpropyl)nonane

When two or more appendages of different nature are present, they are cited as prefixes in alphabetical order. Prefixes specifying the number of identical appendages (di-, tri-, tetra-, and so on) and hyphenated prefixes (*tert*- or *t*-, *sec*-) are ignored in alphabetizing except when part of a complex substituent. The prefixes cyclo-, iso-, and neo- count as a part of the group name for purposes of alphabetizing.

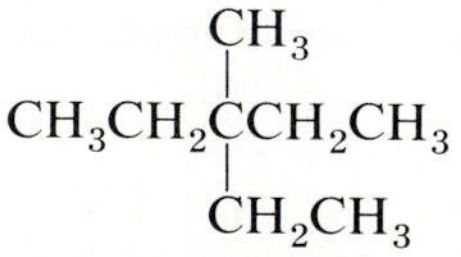

$$\begin{array}{c} CH_3 \\ CH_3CH_2CCH_2CH_3 \\ CH_2CH_3 \end{array}$$

3-ethyl-3-methylpentane

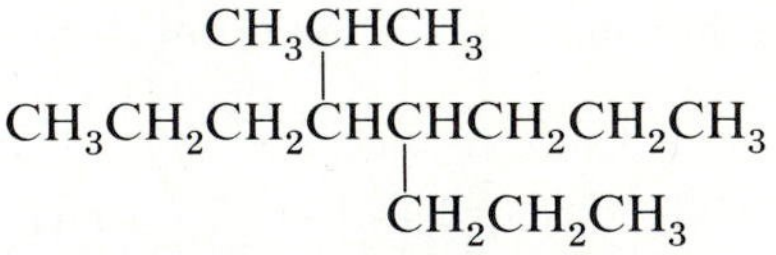

$$\begin{array}{c} CH_3CHCH_3 \\ CH_3CH_2CH_2CHCHCH_2CH_2CH_3 \\ CH_2CH_2CH_3 \end{array}$$

4-isopropyl-5-propyloctane

When chains of equal length compete for selection as the main chain for purposes of numbering, that chain is selected which has the greatest number of appendages attached to it.

$$
\begin{array}{c}
CH_2CH_3 \\
| \\
CH_3{-}C{-}CH_3 \\
| \\
CH_2 \\
| \\
CH_3CH_2CH_2CH_2CH_2CHCH_2CHCH_2CH_3 \\
| \\
CH_2CH_3
\end{array}
$$

5-(2-ethylbutyl)-3,3-dimethyldecane
[not 5-(2,2-dimethylbutyl)-3-ethyldecane]

When two or more appendages are in equivalent positions, the lower number is assigned to the one that is cited first in the name (that is, the one that comes first in the alphabetic listing).

$$
\begin{array}{c}
CH_3CH_2 \quad CH_3 \\
| \qquad\quad | \\
CH_3CH_2CHCH_2CHCH_2CH_3
\end{array}
\qquad
\begin{array}{c}
CH_3CHCH_3 \\
| \\
CH_3CH_2CH_2CHCHCH_2CH_2CH_3 \\
| \\
CH_2CH_3
\end{array}
$$

3-ethyl-5-methylheptane
(not 5-ethyl-3-methylheptane)

4-ethyl-5-isopropyloctane
(not 4-isopropyl-5-ethyloctane)

Note that the direction of numbering of the main chain may already have been decided by application of a higher priority rule.

$$
\begin{array}{c}
CH_3 \qquad CH_2CH_3 \\
| \qquad\qquad | \\
CH_3CH_2CCH_2CH_2CHCH_2CH_3 \\
| \\
CH_3
\end{array}
\qquad
\begin{array}{c}
CH_3 \qquad\qquad CH_2CH_3 \\
| \qquad\qquad\qquad | \\
CH_3CHCH_2CH_2CH_2CHCH_2CH_3
\end{array}
$$

6-ethyl-3,3-dimethyloctane
[3,3,6 is lower than 3,6,6 at the first point of difference]

6-ethyl-2-methyloctane
(2,6 is lower than 3,7)

EXERCISE 3.11 Which three of the following names are incorrect according to the IUPAC rules: (a) 2-methyl-4-ethylheptane (b) 3-ethyl-2,2-dimethylhexane (c) 2-isopropyl-3-methylhexane (d) 4-isopropyl-3-ethylheptane? What is the correct name in each case?

The complete IUPAC rules actually allow a choice regarding the order in which the appendage groups may be cited. One may cite the appendages alphabetically, as above, or in order of increasing complexity. In this book, we shall adhere to the alphabetic order in citing appendage prefixes. The alphabetic order is also used by *Chemical Abstracts* for indexing purposes.

The prefix halo- is used as a generic expression for the halogens, which are treated in the same manner as alkyl appendages for purposes of nomenclature. The individual halogen prefixes are fluoro-, chloro-, bromo-, and iodo-.

$$
\begin{array}{c}
Cl \qquad CH_3 \\
| \qquad\quad | \\
CH_3CHCH_2CHCH_3
\end{array}
\qquad
\begin{array}{c}
CH_2Cl \\
| \\
CH_3CH_2CHCH_2CH_3
\end{array}
$$

2-chloro-4-methylpentane

3-(chloromethyl) pentane

$ClCH_2CH_2CH_2CH_2Br$
1-bromo-4-chlorobutane

$BrCH_2CHFCH(CH_3)CH_2I$
1-bromo-2-fluoro-4-iodo-3-methylbutane

$BrCH_2CH_2CHBrCH_2Br$
1,2,4-tribromobutane

EXERCISE 3.12 Why is the name 4-chloro-2-methylpentane incorrect? Write the structure of 2,6-dichloro-4-ethylheptane. If the two chlorine substituents of this compound are replaced by iodines, what is the correct IUPAC name of the resulting compound?

As with many other classes of organic compounds, haloalkanes can be named with systematic names, such as the foregoing, or with common names (names that evolved before attempts were made to systematize nomenclature). Simple haloalkanes are often named as though they were salts of alkyl groups—as alkyl halides.

CH_3F
methyl fluoride

$(CH_3)_2CHBr$
isopropyl bromide

$CH_3C(CH_3)_2CH_2I$
neopentyl iodide

$CH_3CH(CH_3)CH_2Cl$
isobutyl chloride

Nomenclature is an essential element of organic chemistry for several reasons. First, it is our basic tool for communicating about the subject. It is not always convenient to draw a structure on every bottle we want to label or for every compound we want to talk about. Therefore, it is important that we be able to assign a name to every compound, and it is essential that the name we use correspond to only one compound. This is the most fundamental rule of nomenclature. If more than one structure can be written that corresponds to a name, or if the name is so ambiguous that no unique structure can be written, then the name is incorrect.

However, another important use of nomenclature is in searching the chemical literature for the physical and chemical properties of various compounds. We shall consider the chemical literature in Chapter 35; however, a brief mention at this point is appropriate to the subject of nomenclature. New chemical information is made public for the first time as scientific **papers** that appear in various chemical magazines called **journals.** Examples are the *Journal of the American Chemical Society*, the *Journal of Organic Chemistry*, and *Chemische Berichte*. Hundreds of such journals are published, mostly on a monthly or twice-monthly basis. The back issues of these basic journals contain all of the accumulated knowledge of the science and are known collectively as the **chemical literature.** To facilitate the retrieval of information from this mass of data, the American Chemical Society publishes a reference journal known as *Chemical Abstracts*. This journal is published twice monthly and contains short abstracts of all of the chemical papers published in the basic journals. At the end of each year, an extensively cross-referenced index is published. At the end of each 5-year period (10-year period before 1957), a cumulative index covering that period is published.

In order to search the chemical literature for a compound *by name* it is necessary to know the correct name of any given structure. For such a name search to be successful, it follows that every structure must correspond to *only one name*. The IUPAC system is one example of an attempt to construct such an unambiguous system of nomenclature. The actual rules are quite extensive and allow for all sorts of special situations. More complete versions of the rules may be found in standard reference works such as the *CRC Handbook of Chemistry and Physics*. Unfortunately, the rules as they have been formulated are not totally unambiguous, and all of the special situations that may arise have not been anticipated. Moreover, in some situations the rule makers have been permissive rather than compulsory, as in allowing the retention of certain common names and in allowing a choice in deciding which appendage is cited first in a name. Therefore, one finds that the same compound may be named several different ways in different important reference works. In fact, in the most important reference source, *Chemical Abstracts*, one will find that the name used for a compound has often been changed over the years. Indeed, many of the names currently in use by *Chemical Abstracts* bear little resemblance to those derivable under the IUPAC rules. Fortunately, the necessity of using compound names only for searching the literature is partially alleviated by the existence of formula indices. A welcome development of the "computer age" is the ability to conduct literature searches by computer, using a graphical formula for the structure of a compound, rather than a name.

Nevertheless, if the student is to learn organic chemistry effectively, it is essential that he or she thoroughly learn the simplified IUPAC system that we have presented in this Section, and which may be summarized as follows.

1. Find the longest carbon chain in the compound.
2. Name each appendage group that is attached to this principal chain.
3. Alphabetize the appendage groups.
4. Number the principal chain from one end in such a way that the smaller number is used at the first point of difference in the two possible series of locants.
5. Name the parent alkane using the set of names summarized in Table 3.2 and assign to each appendage group a number signifying its point of attachment to the principal chain.

We shall see that this basic system is used in naming all other classes of organic compounds. For naming substances that have functional groups, the foregoing system is simply modified by the use of appropriate prefixes or suffixes. For example, the characteristic suffix denoting the functional group OH is **-ol** and that for the group NH_2 is **-amine.** Thus, alcohols and amines are named as follows.

$$CH_3CH_2\overset{\displaystyle CH_3}{\overset{|}{C}}HCH_2CH_2OH \qquad CH_3CH_2CH_2CH_2CH_2NH_2$$

3-methyl-1-pentanol   1-pentanamine

Complete details on the modifications which are necessary to adapt the basic system to a given class of compound will be given when we discuss each class.

Highlight 3.3

Systematic numbering of the carbons in the longest chain of a molecule allows us to locate recognizable groups (methyl, ethyl, chloro-, etc.) in an unequivocal way. The names (nomenclature) are important in discussion, in recording information about the compound for the literature, and for organizing the systematic study of organic compounds.

ASIDE

Auguste Laurent (1808–1853) was a modest, creative chemist who began his career as assistant to Dumas. His suggestion that hydrocarbons be used as the basis for the names of organic compounds was developed into the Geneva (IUPAC) system of nomenclature. The longest single carbon chain is the basis for the names; atoms attached to the main chain are named as substituent groups.

Justus Liebig (1803–1875) and then Fritz Pregl (1869–1930) improved the analytical method for organic compounds introduced by Antoine L. Lavoisier (1743–1794) to the point that empirical formulas could be derived by elemental analysis of relatively small samples. Liebig introduced formal laboratory instruction in organic chemistry. There were no hoods in the laboratory, and unpleasant-smelling materials were made in the open, with everyone leaving the room if a flask had broken. Our laboratories have improved enormously, but nevertheless we still analyze organic compounds by combustion. This method alone, however, is generally insufficient.

The local office of the Environmental Protection Agency has come to you with two samples labeled as 2-methylpentane and 3-methylhexane. Will a combustion analysis suffice to distinguish these compounds? (Note that acceptable errors in such analyses are generally $\pm 0.3\%$.) What additional information, if any, would be required? Would the empirical formulas provide an unambiguous structure for each compound? If not, write all the possibilities and name them.

PROBLEMS

1. Write condensed formulas and IUPAC names for the eight isomers having the molecular formula $C_5H_{11}Cl$.

2. Write condensed formulas for seven isomers having the molecular formula $C_4H_{10}O$.

3. Write condensed formulas for the nine isomers having the molecular formula C_7H_{16}. Give the IUPAC name for each isomer.

4. How many monobromo derivatives may be formed from the nine alkanes in problem 3?

5. Each of the following molecules contains one principal functional group. Locate and name the group, and classify the molecule for each case.

a. CH, CH_2, CH, CH_2, CH, CH_2, CH, CH (bicyclic ring)

b. CH, CH_2, CH_2, O, CH_2, CH_2, CH (bicyclic ring)

c. CH_2—CH_2 / CH_2—CHOH (ring)

d. CH_3—$C(CH_3)_2$—$C(CH_3)_2$—Cl

e. CH_3—$C(CH_3)_2$—CH_2—C(=O)—OH

f. C=O, H_2C, CH_2, H_2C—CH_2 (ring)

g. H, C(=O)H, C, H_2C, CH_2, H_2C—CH_2 (ring)

h. S, H_2C, S, H_2C—CH_2 (ring)

i. $CH_3CH(SH)CH_3$

j. 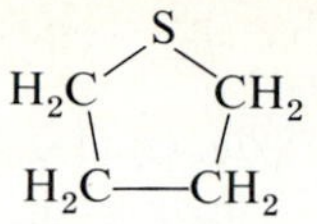 (ring: S, H_2C, CH_2, H_2C—CH_2; thiolane)

$$\begin{array}{c} \text{S} \\ \text{H}_2\text{C} \quad\quad \text{CH}_2 \\ \text{H}_2\text{C}\text{—}\text{CH}_2 \end{array}$$

k. 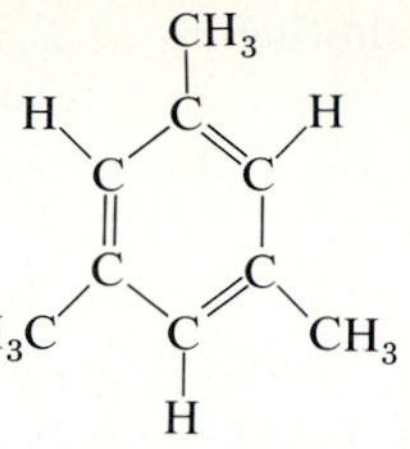 H, CH_3, H, H_3C, CH_3, H substituted benzene ring (C=C alternating)

l. 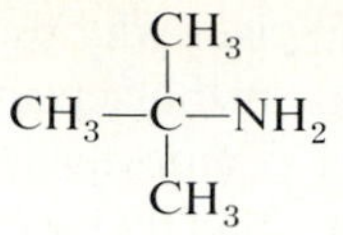 $CH_3—\overset{CH_3}{\underset{CH_3}{C}}—NH_2$

m. 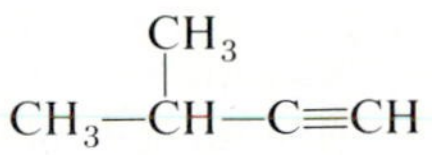 $CH_3—\overset{CH_3}{CH}—C{\equiv}CH$

n. $CH_3CH_2CH_2CH_2Li$

6. Write the structural formula corresponding to each of the following common names.

a. neopentane
b. isobutane
c. *t*-butyl bromide
d. isobutyl iodide
e. isopropyl chloride
f. *sec*-butyl bromide

7. Draw a Kekulé structure of a compound representing each of the following.

a. a six-carbon alkene
b. an eight-carbon ketone containing a six-membered ring
c. an aromatic carboxylic acid
d. a four-carbon amide (i) with one C—N bond, (ii) with two C—N bonds

8. Write the structural formula corresponding to each of the following IUPAC names.

a. 3,4,5-trimethyl-4-propyloctane
b. 3-fluoro-3-ethylhexane
c. 6-(3-methylbutyl)undecane
d. 4-*t*-butylheptane
e. 2-methylheptadecane
f. 4-(1-chloroethyl)-3,3-dimethylheptane
g. 6,6-dimethyl-5-(1,2,2-trimethylpropyl)dodecane
h. 5,5-diethyl-2-methylheptane

9. Give the IUPAC name for each of the following compounds.

a. $(CH_3)_2CHCH_2CH_2CH(CH_3)_2$

b. 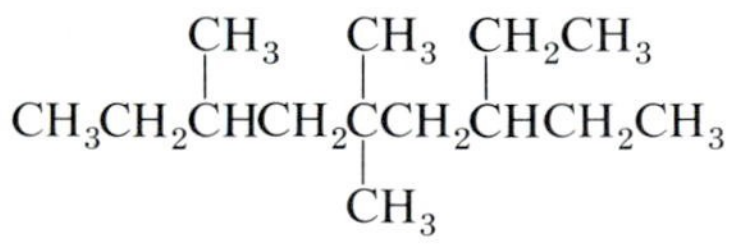 $CH_3CH_2\overset{CH_3}{C}HCH_2\overset{CH_3}{\underset{CH_3}{C}}CH_2\overset{CH_2CH_3}{C}HCH_2CH_3$

c. $CH_3CH_2CH_2\overset{CH_3CHCH_2CH_3}{C}HCH_2CH_3$

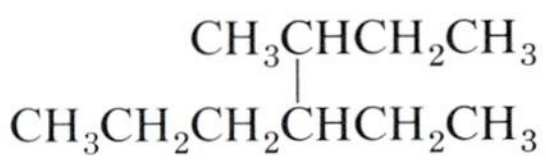

d. 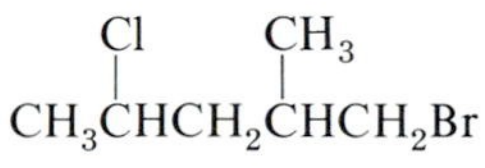 $CH_3\overset{Cl}{C}HCH_2\overset{CH_3}{C}HCH_2Br$

e. 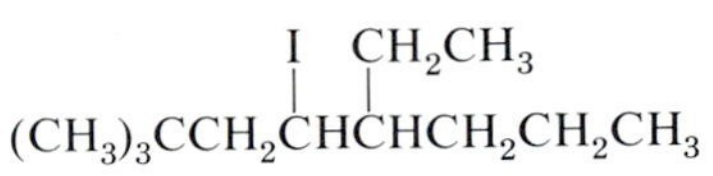 $(CH_3)_3CCH_2\overset{I}{C}H\overset{CH_2CH_3}{C}HCH_2CH_2CH_3$

f. 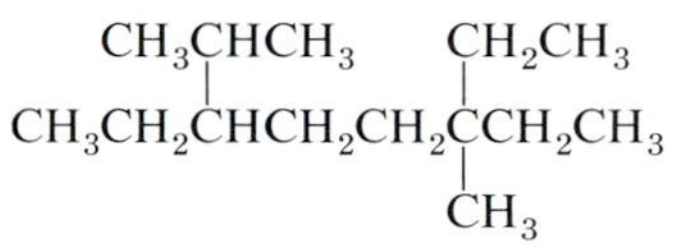 $CH_3CH_2\overset{CH_3CHCH_3}{C}HCH_2CH_2\overset{CH_2CH_3}{\underset{CH_3}{C}}CH_2CH_3$

g. 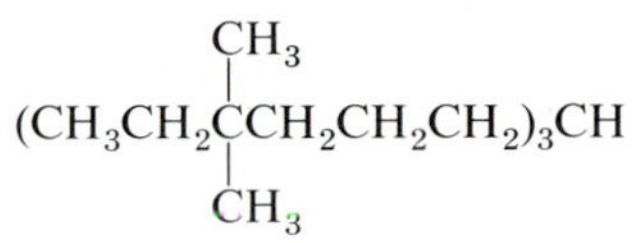 $(CH_3CH_2\overset{CH_3}{\underset{CH_3}{C}}CH_2CH_2CH_2)_3CH$

h. $(CH_3CH_2)_4C$

i. 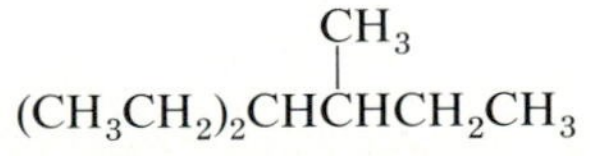 $(CH_3CH_2)_2CH\overset{CH_3}{C}HCH_2CH_3$

j. 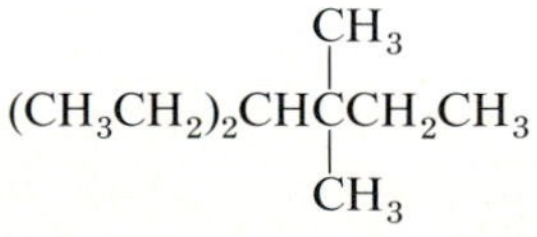 $(CH_3CH_2)_2CH\overset{CH_3}{\underset{CH_3}{C}}CH_2CH_3$

10. Explain why each of the following names is incorrect.

a. methylheptane
b. 4-methylhexane
c. 3-propylhexane
d. 3-isopropyl-5,5-dimethyloctane
e. 3-methyl-4-chlorohexane
f. 2,2-dimethyl-3-ethylpentane
g. 3,5,6,7-tetramethylnonane
h. 2-dimethylpropane

11. The empirical formula C_4H_8 may correspond to many molecular formulas: C_8H_{16}, $C_{12}H_{24}$, etc. Explain why the empirical formula C_5H_{12} can only correspond to one molecular formula.

12. Convince yourself by trial and error that the molecular weight of an organic compound containing only carbon, hydrogen, and oxygen will always be an even number and that the molecular weight for compounds having an odd number of nitrogens will be an odd number.

13. Calculate the elemental composition of each of the following compounds.

a. 3,3-dimethylheptane
b. ethyl acetate (fingernail polish remover), $C_4H_8O_2$
c. nitromethane (racing-engine fuel), CH_3NO_2
d. trinitrotoluene (TNT), $C_7H_5N_3O_6$

14. A series of compounds was analyzed by the Liebig method. In each case, the specified amount of sample was burned and the indicated quantity of CO_2 and H_2O was obtained. Calculate the empirical formula for each unknown compound.

a. 5.72 mg of sample gave 15.73 mg of CO_2 and 6.38 mg of H_2O.
b. 3.81 mg of sample gave 12.3 mg of CO_2 and 3.90 mg of H_2O.
c. 2.58 mg of sample gave 5.87 mg of CO_2 and 2.40 mg of H_2O.
d. 3.41 mg of sample gave 4.99 mg of CO_2 and 2.05 mg of H_2O.

15. From the analytical values for each compound, derive its empirical formula.

a. hexanol: 70.4% C, 13.9% H
b. benzene: 92.1% C, 7.9% H
c. pyrrole: 71.6% C, 7.5% H, 20.9% N
d. morphine: 71.6% C, 6.7% H, 4.9% N
e. quinine: 74.1% C, 7.5% H, 8.6% N
f. DDT: 47.4% C, 2.6% H, 50.0% Cl
g. vinyl chloride: 38.3% C, 4.8% H, 56.8% Cl
h. thyroxine: 23.4% C, 1.4% H, 65.3% I, 1.8% N

16. In each of the following examples, qualitative analysis shows the presence of no elements other than C, H, and O. Calculate the empirical formula for each case.

a. Combustion of 0.0132 g of camphor gave 0.0382 g of CO_2 and 0.0126 g of H_2O.
b. Combustion of 1.56 mg of the sex-attractant of the common honey bee (*Apis mellifera*) gave 3.73 mg of CO_2 and 1.22 mg of H_2O.
c. Benzo[*a*]pyrene is a potent carcinogenic compound that has been detected in tobacco smoke. Combustion of 2.16 mg gave 7.50 mg of CO_2 and 0.92 mg of H_2O.

17. Halogen may be determined by burning the sample under conditions such that the halogen is converted into halide ion, which is then determined by titration with standard silver nitrate solution. Carbon and hydrogen are determined by conversion into CO_2 and H_2O as usual. Calculate the empirical formula for the following case: 2.03 mg of sample gave 4.44 mg of CO_2 and 0.91 mg of H_2O; a separate 5.31 mg sample gave a chloride solution that required 4.80 mL of 0.0110 N $AgNO_3$.

CHAPTER 4

ORGANIC REACTIONS

4.1 Introduction

The two principal components of organic chemistry are **structure** and **reactions.** Each of these components has experimental and theoretical aspects, and they are interrelated. *Structures* are built with bond angles and bond distances obtained from the interpretation of experimentally obtained rotational spectra and x-ray or electron diffraction patterns. In Chapter 2 we reviewed the symbolism used to represent such *structures*—structural formulas and Lewis structures—and their expression in terms of atomic, hybrid, and molecular orbitals.

In this chapter we introduce some concepts concerning reactions. Many reactions are known in organic chemistry that allow us to convert one structure to another. For these conversions we must distinguish between **equilibrium** and **rate.** Equilibrium refers to the relative amounts of reactants and products expected by thermodynamics, *if a suitable pathway exists between them.* A simple example is the state of glucose in the presence of oxygen. These two reagents can exist together for indefinite periods without change, but if the sugar is ignited it will burn to produce the equilibrium products CO_2 and H_2O. Alternatively, the same result is accomplished in living organisms by a series of catalysts (enzymes) that cause this oxidation to occur in a sequence of controlled steps.

Reactants can reach equilibrium over a range of rates from immeasurably slow to exceedingly fast. The rate at which equilibrium is reached depends on the reaction and on the structures of the reactants. Consequently, we will be much concerned with the effect of structural change on reactivity. We will also find that many reactions are characteristic of individual functional groups and form much of the chemistry of functional groups. Finally, we will also be concerned with the pathway, or *mechanism*, by which reactants find their way to products.

4.2 An Example of an Organic Reaction: Equilibria

Although methyl chloride is a gas at room temperature, it is sufficiently soluble in water to give a solution of about 0.1 M concentration. If the solution also contains hydroxide ion, reaction occurs to form methyl alcohol and chloride ion.

$$\underset{}{HO^-} + \underset{\text{methyl chloride}}{CH_3Cl} \rightleftharpoons \underset{\text{methyl alcohol}}{CH_3OH} + Cl^- \qquad (4\text{-}1)$$

At equilibrium all four compounds are present, but the equilibrium constant K is such an exceedingly large number that the amount of methyl chloride present in the equilibrium mixture is vanishingly small.

$$K = \frac{[CH_3OH][Cl^-]}{[CH_3Cl][OH^-]} = 10^{16}$$

If the reaction started with 0.1 M CH_3Cl and 0.2 M NaOH, at the end of the reaction we would have a solution of 0.1 M CH_3OH, 0.1 M NaCl, 0.1 M NaOH, and 10^{-17} M CH_3Cl. That is, 1 mL of such a solution would contain only a few thousand molecules of CH_3Cl.

Such a reaction is said to **go to completion.** In practice a reaction may be considered to go to completion if the final equilibrium mixture contains less than about 0.1% of reactant.

The reaction of methyl chloride and hydroxide ion may also be characterized by the **Gibbs standard free energy change at equilibrium, $\Delta G°$.**

$$\Delta G° = -RT \ln K$$

For reaction (4-1), $\Delta G°$ is -22 kcal mole^{-1}, a rather large value. We may speak of this reaction as having a large **driving force.** The driving force for a reaction is a qualitative description of an equilibrium property and is related to the overall free energy change. For comparison, $\Delta G°$ for a reaction that proceeds to 99.9% completion at equilibrium is -4.1 kcal mole^{-1} at 25 °C. As pointed out in the footnote on page 7, 1 kcal mole^{-1} = 4.184 J mole^{-1}. The student should use this conversion when necessary.

There is an important difference between ΔG and $\Delta G°$. The free energy of a given system is ΔG. The free energy of that system with the components in their standard states is $\Delta G°$. For gases, the standard state is generally the corresponding ideal gas at 1 atm pressure. For solutions, the standard state is normally chosen to be the ideal 1 M solution. The standard free energy of a system $\Delta G°$ is defined as the free energy ΔG of an ideal solution in which each reactant and product is present in a concentration of 1 M. For such a system, $\Delta G = \Delta G°$. When reaction has reached equilibrium, the free energy of the system $\Delta G = 0$. The concentrations of the components at this point are given as

$$\Delta G = \Delta G° + RT \ln K$$

where

$$K = \frac{[\text{products}]_{eq}}{[\text{reactants}]_{eq}}$$

Hence, the standard free energy is given by

$$\Delta G° = -RT \ln K$$

Note also that the units of K are determined by the standard states.

The Gibbs standard free energy may be dissected into **enthalpy** and **entropy** components, $\Delta H°$ and $\Delta S°$, respectively.

$$\Delta G° = \Delta H° - T\Delta S°$$

Enthalpy is the heat of reaction and is generally associated with bonding. If stronger bonds are formed in a reaction, $\Delta H°$ is negative and the reaction is **exothermic.** A reaction with positive $\Delta H°$ is **endothermic.** Entropy may be thought of simply as freedom of motion. The more a molecule or portion of a molecule is restricted in motion, the smaller is the entropy. Both the formation of stronger bonds and greater freedom of motion can contribute to a favorable driving force for reaction (negative $\Delta G°$).

For reaction (4-1), $\Delta H° = -18$ kcal mole^{-1} and $\Delta S° = +13$ eu (ΔS is usually expressed as entropy units, eu, which mean cal deg^{-1} mole^{-1}). The driving force in this case comes mostly from bond energy changes; a carbon-oxygen bond is stronger than a carbon-chlorine bond. The formation of stronger bonds is usually an important component of the driving force of a reaction.

In the vapor phase, where intermolecular interactions are negligible, the strength of the internal bonds in a molecule is especially important in determining its stability. In solutions, however, one must also consider the intermolecular interactions with solvent molecules **(solvation).** Solvent interactions that involve varying degrees of ionic and covalent bonding are particularly important for ions. They provide the main driving force for breaking up the stable crystal lattices when ionic substances dissolve. Although solvation of an ion provides bonding stabilization, which is reflected in $\Delta H°$, it is partially offset by a decrease in entropy, $\Delta S°$. The association of solvent molecules with an ion restricts the freedom of motion of these molecules. In the present case the entropy of reaction $\Delta S°$ is positive because chloride ion is less strongly solvated than hydroxide ion. That is, the solvent molecules are less restricted after reaction than before. Since $\Delta S°$ is positive, the quantity ($-T\Delta S°$) contributes a negative value to $\Delta G°$ and provides an additional driving force for reaction to occur.

EXERCISE 4.1 Using the basic thermodynamic relationships given in this section, calculate $\Delta G°$ and the equilibrium constant K for a reaction that has $\Delta H° = -10$ kcal mole^{-1} and $\Delta S° = -22$ eu (a) if the reaction is carried out at 27 °C (300 K or Kelvin) and (b) if the reaction temperature is 227 °C.

4.3 Reaction Kinetics

Because it has such a large driving force, it seems remarkable that the reaction of methyl chloride with hydroxide ion is relatively slow. For example, a 0.05 M solution of methyl chloride in 0.1 M aqueous sodium hydroxide will have reacted only to the extent of about 10% after 2 days at room temperature. It is an important principle in all reactions that favorable thermodynamics is not enough; a suitable reaction pathway is essential.

Reactions generally involve an **energy barrier** that must be surmounted in going from reactants to products. This barrier exists because molecules generally repel each other as they are brought together and reactants must be forced close to each other to cause the bond changes involved in reaction. The resulting energy barrier is called the **activation energy** or the **enthalpy of activation** and is symbolized by $\Delta H^{\ddagger}$. In the reaction of methyl chloride with hydroxide ion, $\Delta H^{\ddagger}$ is about 25 kcal mole^{-1}. This appears to be a rather formidable hurdle when one realizes that the average kinetic energy of molecules at room temperature is only about 0.6 kcal mole^{-1}. However, this latter number is only an average. Molecules are continually colliding with each other (picosecond time scale) and

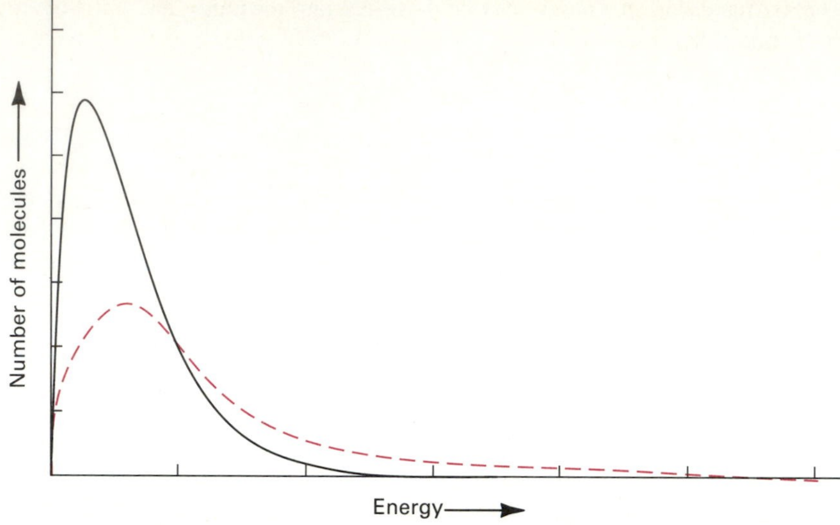

FIGURE 4.1 Boltzmann distribution functions for molecules at 300 and 600 K. Dashed line shows the function for the higher temperature.

exchanging kinetic energy. At any given instant some molecules have less than this average energy, some have more, and a few even have very large energies—like 25 kcal $mole^{-1}$.

The relative number of molecules with any given energy is given by the Boltzmann distribution function, shown schematically in Figure 4.1. Most of the molecules have an energy close to the average energy represented by the large hump. Only the minute fraction of molecules in the far end of the asymptotic tail have sufficient energy to overcome the barrier to reaction. Thus, only a very small number of the collisions between methyl chloride and hydroxide ion are energetic enough to result in reaction.

At a higher temperature, the average kinetic energy of the molecules is greater, and the entire distribution function is shifted to higher energies, as shown by the dashed curve in Figure 4.1. The fraction of molecules with kinetic energy sufficient for reaction is larger, and the rate of reaction is correspondingly larger. For example, the reaction of methyl chloride with hydroxide ion is 25 times faster at 50 °C than it is at 25 °C. A useful rule of thumb for many organic reactions is that a 10° change in temperature causes a two- to threefold change in rate of reaction.

Highlight 4.1

The thermodynamic relationship between reactants and products determines the **equilibrium position** and the free energy $\Delta G°$ of a reaction. The height of the barrier between the reactants and the products determines the **rate** of the reaction.

The rate of a chemical reaction depends not only on the fraction of molecules that have sufficient energy for reaction but also on their concentration because this determines the probability of an encounter that could lead to reaction. Reaction rates are directly proportional to the concentrations of the reactants, and the proportionality constant is called a **rate constant, *k*.** The reaction of methyl chloride with hydroxide ion is an example of a **second-order reaction,** since the rate depends on two concentrations.

$$\text{rate} = k[CH_3Cl][OH^-] \qquad (4\text{-}2)$$

"Rate" involves a change in concentration of something per unit time, usually expressed as moles per liter per second, M sec^{-1}. In equation (4-2), therefore, k must have units of M^{-1} sec^{-1}.

$$(M^{-1}\ sec^{-1})(M)(M) = M\ sec^{-1}$$

The "something" whose concentration is changing is either a reactant or a product.

$$\text{rate} = -\frac{\Delta[CH_3Cl]}{\Delta t} = -\frac{\Delta[OH^-]}{\Delta t} = \frac{\Delta[CH_3OH]}{\Delta t} = \frac{\Delta[Cl^-]}{\Delta t} \tag{4-3}$$

The minus signs for the reactants indicate that their concentrations decrease with increasing time. All of the changes shown are equal by stoichiometry.

In the language of calculus, this rate equation becomes

$$-\frac{d[CH_3Cl]}{dt} = -\frac{d[OH^-]}{dt} = \frac{d[CH_3OH]}{dt} = \frac{d[Cl^-]}{dt} = k[CH_3Cl][OH^-]$$

The actual value of k at 25 °C is 6×10^{-6} M^{-1} sec^{-1}.

The reaction of methyl chloride with hydroxide ion may be compared with its reaction with water. The reaction with water is an example of a **first-order reaction.** The reaction involves water molecules, but, because water is the solvent, it is present in large excess. Therefore the concentration of water remains effectively the same, even after all of the methyl chloride has reacted. Since the concentration of water appears not to change during the reaction, it does not appear in the kinetic expression; thus the rate of reaction depends only on the concentration of methyl chloride.

$$\text{rate} = k[CH_3Cl] \tag{4-4}$$

Because only one concentration is involved, equation (4-4) is the equation of a first-order reaction. The rate is a change in concentration per unit time, for example, moles per liter per second, M sec^{-1}. Therefore, k has the units of sec^{-1}. For methyl chloride at 25 °C, $k = 3 \times 10^{-10}$ sec^{-1}.

The reaction of methyl chloride with water is experimentally a first-order reaction because the concentration of one reactant (water) does not change significantly during the reaction. A second-order reaction becomes effectively a first-order reaction if the concentration of one component is much greater than the other. For example, in the reaction of a solution of 0.01 M CH_3Cl with 1 M NaOH the concentration of hydroxide ion changes from 1 M to 0.99 M during the reaction. That is, its concentration remains essentially constant and the reaction of methyl chloride under these conditions appears to be a first-order reaction with a rate constant of 6×10^{-6} sec^{-1}. Such a reaction is also called a **pseudo first-order reaction.**

The student should be careful not to confuse "rate" and "rate constant." The rate constant k for a reaction is simply a numerical measure of how fast a reaction can occur if the reactants are brought together. It relates the actual concentrations of the reactants to the rate of reaction, which is the "through-put" or "flux" of the reaction.

EXERCISE 4.2 Using the rate constant $k = 6 \times 10^{-6}$ M^{-1} sec^{-1} for the reaction of methyl chloride with hydroxide ion, calculate the initial rate of reaction for each of the following initial reactant concentrations.
(a) 0.1 M CH_3Cl and 1.0 M OH^- (b) 0.1 M CH_3Cl and 0.1 M OH^-
(c) 0.01 M CH_3Cl and 0.01 M OH^-
Compare each of these initial rates with the rate of reaction when 90% of the methyl chloride has reacted.

Highlight 4.2

The reaction of a substance, A, shows a rate of disappearance of A, $-d[A]/dt = k[A]$. That is, the disappearance of A has a **first-order** dependence given by the **rate constant** k only on the concentration of A. The reaction of A and B shows a rate of disappearance of A, $-d[A]/dt = k[A][B]$. That is, the disappearance of A has a **second-order** dependence given by the **rate constant** k on the concentrations of A and B. If B is the solvent, its concentration will change very little during the reaction, and the rate of disappearance of A will be given by $-d[A]/dt = k'[A]$. That is, the disappearance of A has a **pseudo first-order** dependence given by the **rate constant** k' essentially on the concentration of A.

4.4 Reaction Profiles and Mechanism

In the reaction of methyl chloride and hydroxide ion, atoms must move around and bonds must change in order that the products methyl alcohol and chloride ion can be produced. One of the important concepts in organic chemistry involves the consideration of the structure of the system as reaction proceeds. Each configuration of the atoms that occurs during the process of changing from reactants to products has an energy associated with that configuration. Since reaction generally involves bringing the reactants close together and breaking bonds, these structures generally have higher energy than the isolated reactants. That is, as the reactants approach each other and start to undergo the molecular changes that will eventually result in products, the potential energy of the reacting system increases. As the reaction encounter continues, the potential energy continues to increase until the system reaches a structure of **maximum energy.** Thereafter the changes that result in the final products continue, but the structures represent lower and lower energy until the products are fully formed.

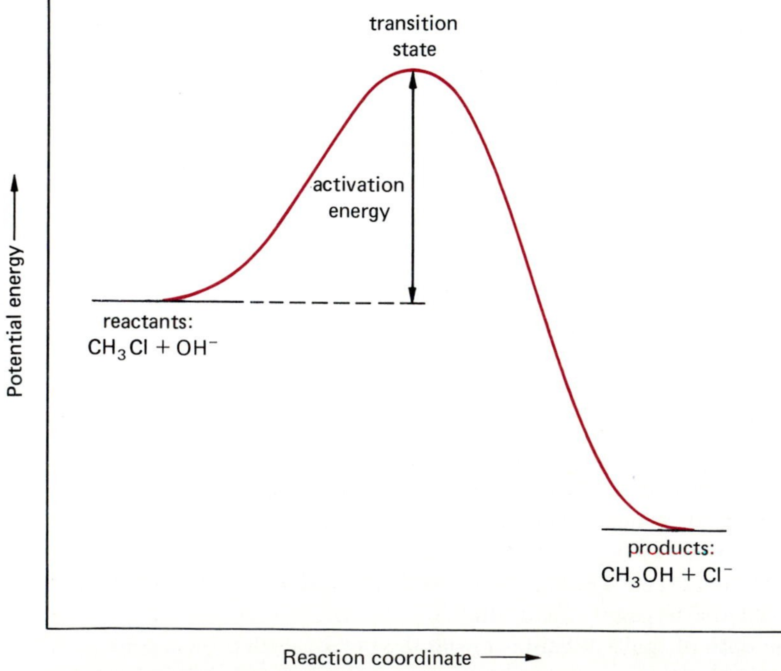

FIGURE 4.2 A reaction profile for the reaction $CH_3Cl + OH^- \rightarrow CH_3OH + Cl^-$.

The difference in the energy of the isolated reactants and the maximum energy structure that the system passes through on the path to products is the **activation energy** of the reaction. This maximum energy corresponds to a definite structure called the **transition state.** The measure of the progress of reaction from reactants to products is the **reaction coordinate.** This coordinate is usually not specified in detail because the qualitative concept is usually sufficient, but in our reaction, for example, it could be represented by the carbon-oxygen bond length or the carbon-chlorine bond length as the reaction progresses, or by the net electronic charge on chlorine. Whatever measure is used, the general **reaction profile** is given by Figure 4.2. In this figure, the energy shown is the potential energy. This quantity is related to but is not identical with $\Delta H°$. Similarly, the difference in potential energy between reactants and products contributes to $\Delta G°$ for the reaction. The magnitude of this difference determines the *position* of equilibrium. The magnitude of the activation energy determines the *rate* at which equilibrium is established.

The energy quantities involved in reactions are given more precise definitions in the **theory of absolute rates.** In this theory the transition state is characterized by the dynamic properties: free energy, enthalpy, and entropy. The rate constant for reaction is related to the Gibbs free energy difference between the transition state, sometimes called an **activated complex,** and the reactant state by the equation

$$k = \nu^{\ddagger} e^{-\Delta G^{\ddagger}/RT} = \nu^{\ddagger} e^{-\Delta H^{\ddagger}/RT} e^{\Delta S^{\ddagger}/R}$$

The proportionality constant, $\nu^{\ddagger}$ is a kind of frequency. Its magnitude is 6.2×10^{12} sec^{-1} at 25 °C, a magnitude comparable to ordinary vibration frequencies. In fact, the reaction process can be described as one of the modes of vibration of the activated complex. The activated complex or transition state is thus a specially unstable type of molecule and is only a transient phase in the course of reaction.

Knowing the structure of the transition state is an important aspect of understanding a reaction. If we can estimate its energy, we can predict the reaction rate, at least roughly. For example, a transition state in which several bonds are broken is likely to have a high energy and be associated with a slow reaction. Furthermore, and most important, from the structure of the transition state we can often evaluate how a given change in structure will change the rate. Unfortunately, we cannot directly observe a transition state—we cannot take its spectrum or determine its structure by x-ray diffraction. Instead, we must infer its structure indirectly.

The reaction of hydroxide ion with methyl chloride is an example of a general reaction known as an S_N2 reaction, which we will study in Chapter 8. The structure of the transition state will be analyzed at that time. We will find, for example, that the S_N2 reaction probably involves a single step with one transition state, as suggested in Figure 4.2. Many other reactions, however, involve more than one step; for example, two reactants can combine to form a compound that is not the final product, but an **intermediate.** The corresponding reaction profile such as that in Figure 4.3 is not uncommon. Such a reaction involves one or more intermediates, and each intermediate is flanked by transition states. Reaction intermediates correspond to energy minima on the reaction coordinate diagram. They may be sufficiently stable that they can be isolated and stored in bottles, or they can have such fleeting existence that their presence must be inferred from subtle observations in cleverly designed experiments.

The **reaction mechanism** is a sequential description of each transition state and intermediate in a total reaction. The overall rate of reaction is determined approximately by the transition state of highest energy in the sequence, so that this structure has particular importance. The step involving this transition state is often called the **rate-determining** step.

The reaction profile shown in Figure 4.3 corresponds to the equations

$$\text{reactants} \underset{k_{-1}}{\overset{k_1}{\rightleftharpoons}} \text{intermediate} \underset{k_{-2}}{\overset{k_2}{\rightleftharpoons}} \text{products}$$

The relative energies in this figure correspond to rate constants having the relationships

$$k_{-1} > k_1 > k_2 > k_{-2}$$

Remember that the lower the energy barrier, the larger the rate constant. Because it is the highest point between reactants and products, the second transition state in Figure 4.3 corresponds to the rate-determining step.

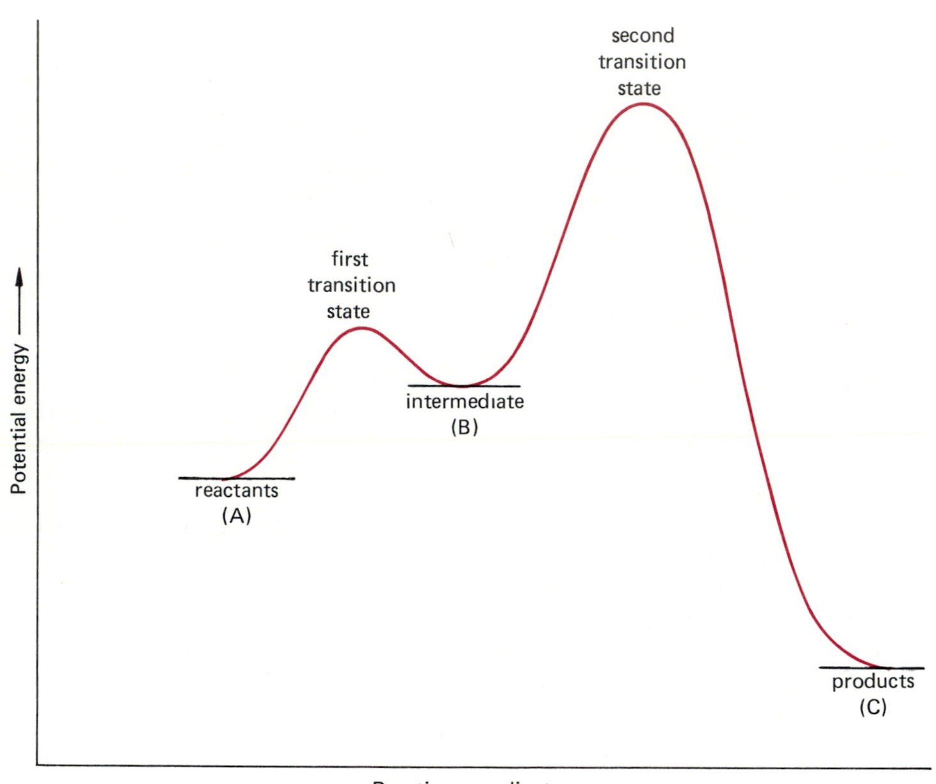

FIGURE 4.3 Profile of a complex reaction that involves an intermediate.

Again, it must be emphasized that rate constant and rate are not the same. For example, in Figure 4.3 the first step has a greater rate constant than the second step (i.e., $k_1 > k_2$). It is also clear from the graph that since B is less stable than A, the equilibrium between A and B is such that the concentration of B at any given time is much less than the concentration of A, or [A] > [B]. Thus, the first step in the reaction has a faster *rate* than the second ($k_1[A] > k_2[B]$).

As a further example of this point, consider the reaction coordinate diagram for a two-step reaction shown in Figure 4.4. Again, the highest energy transition state occurs in the second step, which is therefore the rate-determining step. In this example $k_2 > k_1$. However, since $[A] \gg [B]$, the rate of the first step is still faster than the rate of the second.

An example of a reaction that involves intermediates is the hydrolysis of ethyl acetate with aqueous sodium hydroxide.

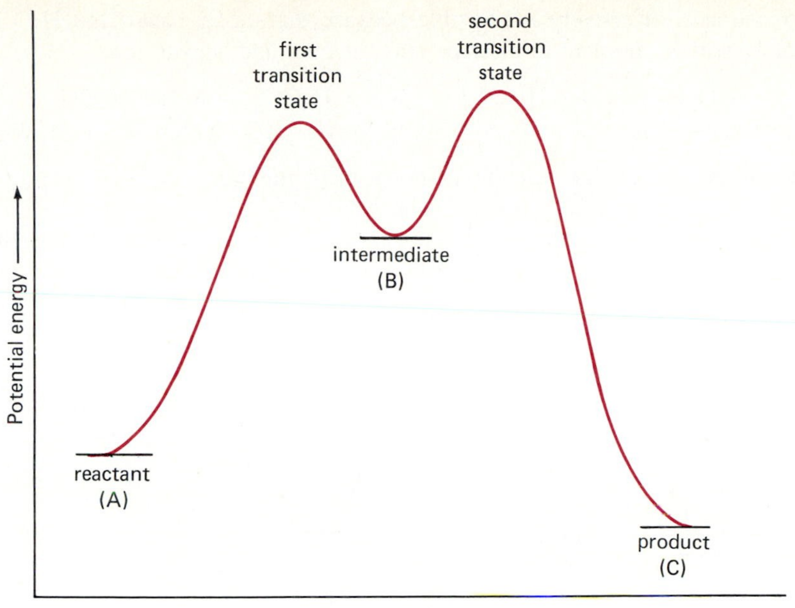

FIGURE 4.4 Profile of another two-step reaction.

$$\underset{\text{ethyl acetate}}{CH_3\overset{\overset{O}{\|}}{C}OCH_2CH_3} + OH^- \xrightarrow[H_2O]{25^\circ} \underset{\text{acetate ion}}{CH_3CO_2^-} + \underset{\text{ethyl alcohol}}{CH_3CH_2OH}$$

> This example illustrates the way in which organic reactions are typically written. The arrow shows the direction of the reaction and implies that the equilibrium lies far to the right. Reaction conditions such as solvent, temperature, and any catalysts used are written with the arrow as shown. Abbreviations are often used in this convention. An example is the use of the symbol Δ for heat. If the reaction mixture shown was heated or refluxed in order to speed reaction, we could represent the reaction as
>
> $$CH_3\overset{\overset{O}{\|}}{C}OCH_2CH_3 + OH^- \xrightarrow[H_2O]{\Delta} CH_3CO_2^- + CH_3CH_2OH$$

The rate expression for the hydrolysis reaction is

$$\text{rate} = 0.1\,[OH^-][CH_3COOCH_2CH_3]$$

The second-order rate constant 0.1 M^{-1} sec^{-1} is relatively large and corresponds to a rather fast reaction. As we shall learn in Chapter 19, the mechanism of this reaction appears to be

$$CH_3\overset{\overset{O}{\|}}{C}OCH_2CH_3 + OH^- \rightleftharpoons CH_3\underset{\underset{OH}{|}}{\overset{\overset{O^-}{|}}{C}}{-}OCH_2CH_3$$

$$CH_3\underset{\underset{OH}{|}}{\overset{\overset{O^-}{|}}{C}}{-}OCH_2CH_3 \longrightarrow CH_3COOH + CH_3CH_2O^-$$

The reaction is effectively irreversible because the strong base $CH_3CH_2O^-$ reacts immediately with acetic acid to produce ethyl alcohol and acetate ion.

$$\underset{\text{acetic acid}}{CH_3COOH} + CH_3CH_2O^- \xrightarrow{\text{fast}} CH_3CO_2^- + CH_3CH_2OH$$

The reaction profile for this reaction is shown in Figure 4.5.

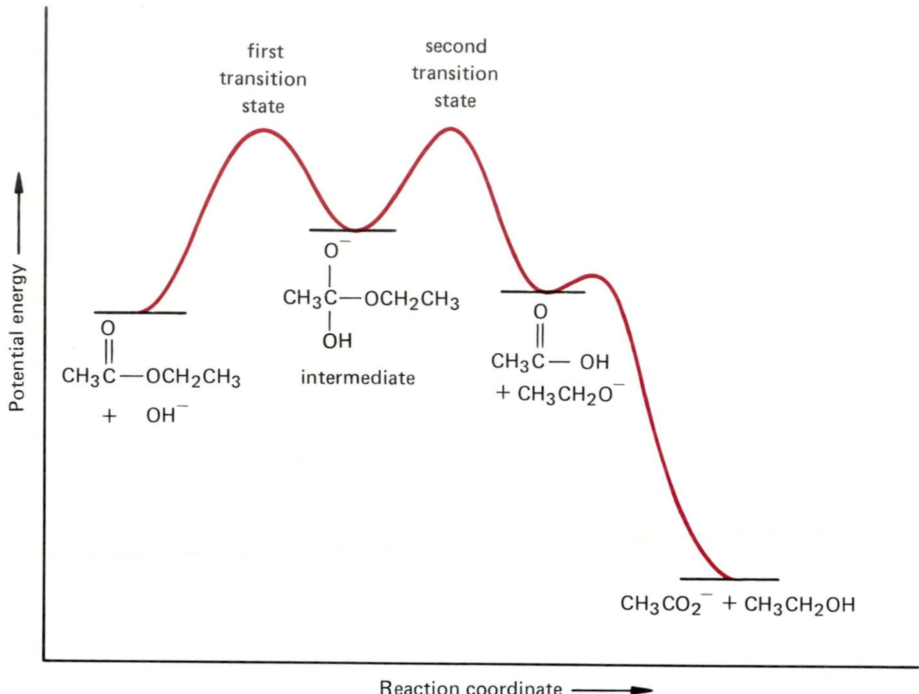

FIGURE 4.5 Reaction profile for hydrolysis of an ester.

Reaction profiles such as those in Figures 4.2–4.5 can be useful because they are a visualization of a rather complex set of events in a diagrammatic way that is easy to follow. But such diagrams have only a qualitative significance. The reason is that the rate constant for a given reaction step depends on the Gibbs free energy difference between the reactants and transition state, $\Delta G^{\ddagger}$, a thermodynamic quantity that has significance only for a statistical collection of species. It is not defined for the distorted structures that result as the reactant molecule is changing to the product molecule. The potential energy is defined for individual molecules and distorted structures, but is only one component of the free energy. Nevertheless, it is generally (but not always) true that a larger potential energy increase between reactants and transition state is associated with a larger $\Delta G^{\ddagger}$ and a smaller rate constant. Moreover, for these same reasons the term "rate-determining step" has only approximate significance. In multistep reactions the actual rate of reaction may be a complex function of time and reactant concentrations that does not lend itself to a simple pictorial representation.

The elucidation of reaction mechanisms is a fascinating branch of organic chemistry. In our study of organic chemistry, we will deal frequently with reaction mechanisms because they help enormously in our classification and understanding of the vast array of known organic reactions. In some important cases, such as the hydrolysis of ethyl acetate, we will also study some of the experimental evidence from which the reaction mechanisms and transition state structures have been deduced.

EXERCISE 4.3 Construct a reaction coordinate diagram for a reaction $A \rightleftharpoons B \rightleftharpoons C$ in which the relative stabilities of the three species are $C > A > B$ and for which the relative order of the four rate constants is $k_2 > k_{-1} > k_1 \gg k_{-2}$. Which is the rate-determining step in your diagram? What is the relative order of the four free energies of activation, $\Delta G^{\ddagger}$? Of the two forward steps, which has the faster rate? Is there any way you can adjust the relative levels within the foregoing specified conditions so that the rate of the other step becomes faster?

Highlight 4.3

A potential energy versus reaction coordinate diagram shows the reaction profile in terms of reactants, intermediates, and products. It is a pictorial way of expressing the energetics of the mechanism of a reaction.

4.5 Acidity and Basicity

Some of the most important reactions in chemistry are those involving the properties of acidity and basicity. **Acidity** refers to the loss of a proton.

$$HY \rightleftharpoons H^+ + Y^-$$

Such reactions can be measured for the gas phase and are invariably highly endothermic. For example,

$$HCl(g) \rightleftharpoons H^+(g) + Cl^-(g) \qquad \Delta H^\circ = 328.8 \text{ kcal mole}^{-1}$$

This reaction is extremely endothermic because a bond is broken and charges are separated; both of these processes require energy. In aqueous solution acidity is defined in terms of a dissociation equilibrium involving solvated species.

$$HCl(aq) \rightleftharpoons H^+(aq) + Cl^-(aq)$$

The species $H^+(aq)$ is often written as H_3O^+ for convenience. However, in actuality much more extensive solvation occurs through hydrogen bonds to the oxygens of other water molecules.

H_2O---H ⟍ $\overset{+}{O}$ ⟋ H---- OH_2, with O–H ⋯ OH_2 below

Similarly, the Cl^- ion is solvated by hydrogen bonds to water molecules.

Cl^- hydrogen-bonded to four water molecules: H–O–H---Cl^-----H–O–H, with H–O–H above and below

Hydrogen bonds will be discussed in greater detail in Section 10.3. They result from the electrostatic attraction between lone-pair electrons, such as those on oxygen and chlorine, and hydrogens that are bound to an electronegative element. Such hydrogens are good "acceptors" for these electrostatic bonds because of the partial positive charge they bear.

Each ion has many of these bonds to solvent molecules; all of the bonds together are called the solvation of the ion. The bonds to the solvating molecules are sufficiently strong in the aggregate to compensate for the energy that must be supplied to break the hydrogen-chlorine bond and the electrostatic energy required to separate negative from positive charges.

Acidity measurements have been made in many different solvents. Nevertheless, water is by far the most common solvent and is assumed to be the reference solvent if none is specified. *The acidity of an acid HA in water is defined as the equilibrium constant for the reaction*

$$\text{HA(aq)} \overset{K}{\rightleftharpoons} \text{H}^+\text{(aq)} + \text{A}^-\text{(aq)}$$

$$K = \frac{[\text{H}^+][\text{A}^-]}{[\text{HA}]}$$

The equilibrium constants for acid dissociations are normally symbolized as K_a. Note that the concentration of water does not appear in the expression and that K_a normally has units of M or moles liter^{-1}.

Acidity equilibrium constants vary over a wide range. Acids with $K_a > 1$ are referred to as strong acids; acids with $K_a < 10^{-4}$ are weak acids, and many compounds are very weak acids for which K_a is exceedingly small. For example, methane is a very weak acid with $K_a \cong 10^{-50}$. This number is so small (it corresponds to having approximately one pair of dissociated ions in a whole universe of solution) that it is known only approximately and must be measured indirectly.

Acidity equilibrium constants are usually expressed as an exponent of 10 in order to accommodate this large range of possible values. The pK_a is defined as the negative exponent of 10, or as

$$\text{p}K_a = -\log K_a$$

The pK_a values for some common acids are summarized in Table 4.1. A more extensive list is given in Appendix IV. The term ''acidity'' is also used in a qualitative sense to refer to acidic character relative to water. Solutions of acids that are substantially more acidic than water have significant hydrogen ion concentrations. That is, their aqueous solutions have pH values less than 7. Recall that the pH is defined as the negative logarithm of the hydrogen-ion concentration.

$$\text{pH} = -\log\,[\text{H}^+]$$

Neutral water has $[\text{H}^+] = 1.0 \times 10^{-7}$ M or pH = 7.0. Solutions with pH < 7 are ''acidic'' and have a distinctive sharp taste at the tip of the tongue. The strong acids HI, HBr, HCl, HNO_3, and H_2SO_4 are commonly referred to as ''mineral acids.'' Their solutions have the low pH values characteristic of solutions that are ''acidic.'' Acetic acid is a weaker acid, but its aqueous solutions also have pH values that are definitely ''acidic.'' Alcohols have about the same acidity as water itself and are therefore not ''acidic'' in this sense.

EXERCISE 4.4 Calculate the pH of a solution of 1 mole of each of the following acids in 1 L of water.

(a) HI (b) HCl (c) HF (d) acetic acid (e) H_2S

Table 4.1 gives second and third dissociation constants where appropriate. Thus, the second dissociation constant of HSO_4^- with pK_a = 1.99 is about as large as the first

TABLE 4.1 Acidities of Common Acids at 25°C

Name	Formula	pK_a
acetic acid	CH_3COOH	4.76
ammonium ion	NH_4^+	9.24
hydriodic acid	HI	−5.2
hydrobromic acid	HBr	−4.7
hydrochloric acid	HCl	−2.2
hydrocyanic acid	HCN	9.22
hydrofluoric acid	HF	3.18
hydrogen selenide	H_2Se	3.71
hydrogen sulfide	H_2S	6.97
methyl alcohol	CH_3OH	15.5
nitric acid	HNO_3	−1.3
nitrous acid	HNO_2	3.23
phosphoric acid	H_3PO_4	2.15 (7.20, 12.38)[a]
phenol	C_6H_5OH	10.00
sulfuric acid	H_2SO_4	−5.2 (1.99)[a]
water	H_2O	15.74

[a] Values in parentheses are the constants for dissociation of the second and third protons from phosphoric acid and the second proton from sulfuric acid.

dissociation constant of H_3PO_4, $pK_a = 2.15$. The acidity of $H_2PO_4^-$, $pK_a = 7.20$, is comparable to that of H_2S, $pK_a = 6.95$, and both pK_as are comparable to the pH of neutral water, 7. The significance of this point is emphasized from the following analysis. When an acid is exactly 50% dissociated, the remaining acid concentration [HA] is equal to the concentration of the conjugate anion [A^-]. For such a case, the hydrogen-ion concentration of the solution is numerically equal to the acidity constant, or pH = pK_a.

$$K = \frac{[A^-][H^+]}{[HA]}$$

$$pK = -\log K = -\log [H^+] = pH$$

For aqueous solutions pK_a values in the range of about 2–12 are known fairly accurately. The acidity constants of stronger acids ($pK_a < 2$) are known with less precision because such acids are extensively dissociated in aqueous solution.

EXERCISE 4.5 Determine the concentrations of H^+, A^- and undissociated acid, HA, for a 1 M solution of an acid having a $pK_a = -2$.

We have discussed acidity in terms of the process of acid HA giving up a proton to the solvent water. But the process is an equilibrium, the reverse of which involves the reaction of the conjugate anion A^- to accept a proton from the solvent. We could speak equally of the acidity of HA or of the **basicity** of A^-.

$$\underset{\text{conjugate acid}}{HA} \rightleftharpoons H^+ + \underset{\text{conjugate base}}{A^-}$$

The stronger HA is as an acid, the weaker A^- is as a base. Conversely, a strong base has a weak conjugate acid. Because of these relationships, basicity is generally characterized in terms of the acidity of the conjugate acid.

EXERCISE 4.6 For each of the acids listed in Table 4.1 write the conjugate base and give an evaluation of the basicity considering hydroxide ion to be a relatively strong base.

Knowledge of the relationship between structure and acidity is extremely important in understanding chemistry. We shall refer frequently throughout our study of organic chemistry to the acidities of different functional groups and the manner in which these acidities change as the structure is changed. We shall find that the mechanisms of many organic reactions involve intermediates functioning in one way or another as acids or bases. For comparisons of the acidities of two or more compounds, some generalizations are useful. Removing a proton from a molecule involves breaking a bond to hydrogen and leaving a negative charge on the rest of the molecule. The weaker the bond to hydrogen in HX, the easier it is to remove the proton and the greater the acidity of HX. As we move from element to element down the periodic table, bond strengths diminish because valence atomic orbitals become larger and more diffuse and their overlap to a hydrogen 1*s*-orbital is less effective. This factor can have a dominating effect on acidity. Note the acidity of the compounds in the series $H_2O < H_2S < H_2Se$ and $HF < HCl < HBr < HI$.

The easier it is to place a negative charge on the anion, the greater the acidity of the acid from which it is derived. An anion can be formed easily or is stable if there is an electronegative atom (an atom that attracts negative charge) or if the negative charge can be redistributed over several atoms. Along any given row of the periodic table, electronegativity increases as we move to the right. This is the dominating factor in such acidity comparisons as $HF > H_2O$ and $HCl > H_2S$. We will learn many examples of the redistribution of charge principle as we develop the chemistry of various functional groups in subsequent chapters. The presence of a formal positive charge within the anion results in a dramatic increase in acidity because of the electrostatic attraction between positive and negative charges. Some examples are

$$\mathrm{HO{-}S^{2+}(O^-)_2{-}OH} \quad > \quad \mathrm{HO{-}\ddot{S}^{+}(O^-){-}OH}$$

sulfuric acid   sulfurous acid

$$\mathrm{O{=}N^{+}(O^-){-}OH} \quad > \quad \mathrm{O{=}\ddot{N}{-}OH}$$

nitric acid   nitrous acid

A simple corollary is that a nearby negative charge will reduce acidity because of charge repulsion in the resulting anion. Thus the second dissociation constants are generally smaller than the first. Examples of such acidity orders are

$$\mathrm{HO{-}S^{2+}(O^-)_2{-}OH} \quad > \quad \mathrm{HO{-}S^{2+}(O^-)_2{-}O^-}$$

$$\begin{array}{c}OH\\ | \\ ^{-}O{-}P^{+}{-}OH\\ | \\ OH\end{array} > \begin{array}{c}O^{-}\\ | \\ ^{-}O{-}P^{+}{-}OH\\ | \\ OH\end{array} > \begin{array}{c}O^{-}\\ | \\ ^{-}O{-}P^{+}{-}OH\\ | \\ O_{-}\end{array}$$

Another important effect in many kinds of anions is resonance stabilization. Recall from page 13 that the electronic structure of formate ion cannot be represented accurately by a single Lewis structure but includes two equivalent structures:

$$H{-}C(=O){-}O^{-} \longleftrightarrow H{-}C(-O^{-})=O$$

The negative charge is divided between the two equivalent oxygens. Accordingly, each oxygen is less basic than an oxygen with a full charge such as hydroxide ion or methoxide ion. As a result, it is easier to generate the anion and, equivalently, the conjugate acid is more acidic. Thus, the carboxylic acids are more acidic than alcohols. This an important point and we shall discuss it again in greater detail in Chapter 18. The fundamental principle involved will also enter into some reactions of acetate ion to be studied in Chapter 9.

Some organic compounds are related chemically to corresponding inorganic acids. For example, methanesulfonic acid, CH_3SO_3H, is related to sulfuric acid, $HOSO_3H$, and is also a strong acid. Ethylamine, $CH_3CH_2NH_2$, is related to ammonia, NH_3, and has comparable base strength. Some important generalizations are

1. The *lower* Y is located in a given column of the periodic table the *greater* the acidity of Y—H.
2. The *farther to the right* Y is along a given row of the periodic table the *greater* the acidity of Y—H.
3. Positive charges within the anion increase acidity; negative charges decrease acidity.
4. Resonance stabilization of the anion increases acidity. This effect will be discussed more in later chapters.

To summarize, acidity and basicity are important chemical properties. Many organic reactions involve proton-transfer equilibria, and we need to appreciate the effect of structural change on such equilibria.

Highlight 4.4

The position of the equilibrium for dissociation of a proton from the substance H—Y to form H^+ and Y^- **is** the **acidity** of HY. Dissociation is favored by solvents that can interact with and stabilize the ions. Dissociation is also favored by molecular features that stabilize the anion, either with electronegative atoms or by distributing the negative charge over more than one atom.

ASIDE

Remember the Chinese adage that the longest journey begins with a single step. The most complex reaction sequence can be broken down into a step-by-step description of the reaction pathway, usually involving only a small number of bonds. We use a potential energy versus reaction coordinate diagram to describe both the rate and equilibrium properties of many reactions; these diagrams help us to see the individual steps. For example, the reaction of hydroxide ion with methyl chloride proceeds in one step, from the initial reactants through a transition state to the final products. This reaction is not very reversible, since the final state is much lower in energy than the initial state.

PROBLEMS

1. Consider the following reaction sequence in which B is an intermediate. Sketch energy profiles for each of the possible relationships among rate constants shown.

$$A \underset{k_2}{\overset{k_1}{\rightleftharpoons}} B$$

$$B \xrightarrow{k_3} C$$

The back-reaction from C is negligible.

a. k_1 and k_2 large; k_3 small.
b. k_1 large; k_2 and k_3 large, but $k_2 > k_3$.
c. k_1 and k_3 large; k_2 small.
d. k_1 small; k_2 and k_3 large, but $k_3 > k_2$.

Identify the rate-determining transition state for each of the four cases.

2. Consider the hypothetical two-step reaction

$$A \underset{k_2}{\overset{k_1}{\rightleftharpoons}} B \underset{k_4}{\overset{k_3}{\rightleftharpoons}} C$$

that is described by the following energy profile.

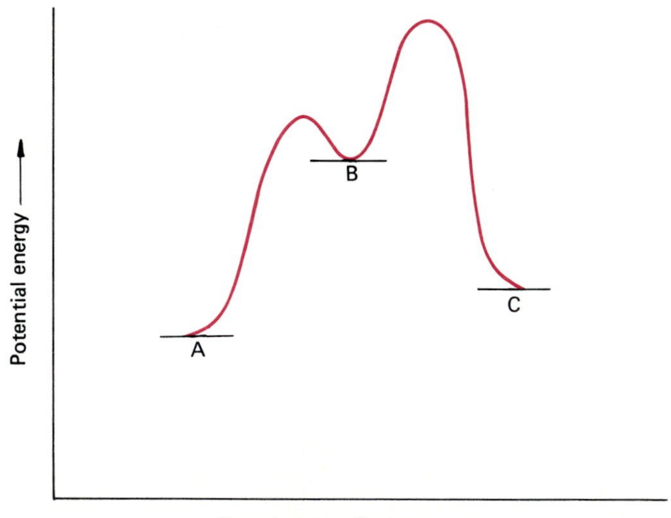

a. Is the overall reaction (A → C) exothermic or endothermic?
b. Label the transition states. Which transition state is rate-determining?
c. What is the correct order of magnitude of rate constants?

i. $k_1 > k_2 > k_3 > k_4$
ii. $k_2 > k_3 > k_1 > k_4$
iii. $k_4 > k_1 > k_3 > k_2$
iv. $k_3 > k_2 > k_4 > k_1$

d. Which is the thermodynamically most stable compound?
e. Which is the thermodynamically least stable compound?

3. *t*-Butyl chloride reacts with hydroxide ion according to the following equation.

$$\underset{t\text{-butyl chloride}}{CH_3-\overset{\overset{CH_3}{|}}{\underset{\underset{CH_3}{|}}{C}}-Cl} + OH^- \overset{K}{\rightleftharpoons} \underset{t\text{-butyl alcohol}}{CH_3-\overset{\overset{CH_3}{|}}{\underset{\underset{CH_3}{|}}{C}}-OH} + Cl^-$$

We will learn in Chapter 9 that the reaction is believed to proceed by the following mechanism.

$$(1)\ CH_3-\overset{\overset{CH_3}{|}}{\underset{\underset{CH_3}{|}}{C}}-Cl \underset{k_2}{\overset{k_1}{\rightleftharpoons}} CH_3-\overset{\overset{CH_3}{|}}{\underset{\underset{CH_3}{|}}{C^+}} + Cl^-$$

$$(2)\ CH_3-\overset{\overset{CH_3}{|}}{\underset{\underset{CH_3}{|}}{C^+}} + OH^- \underset{k_4}{\overset{k_3}{\rightleftharpoons}} CH_3-\overset{\overset{CH_3}{|}}{\underset{\underset{CH_3}{|}}{C}}-OH$$

The order of rate constants is $k_3 > k_2 > k_1 \gg k_4$.

a. Construct a reaction coordinate diagram for the reaction.
b. Is the first step exothermic or endothermic?
c. Is the overall reaction exothermic or endothermic?
d. Does the first or second step govern the rate of disappearance of *t*-butyl chloride?

4. In this chapter we discuss the hydrolysis of methyl chloride in aqueous solution. Consider the same reaction in the gas phase at 25 °C:

$$CH_3Cl + H_2O \rightleftharpoons CH_3OH + HCl$$

a. $\Delta H° = 7.3$ kcal mole^{-1}; $\Delta S° = 0.3$ eu. Calculate $\Delta G°$.
b. Calculate the equilibrium constant.
c. Can this reaction be said to ''go to completion'' in the direction shown?

5. Consider the equilibrium between butane and ethane plus ethylene.

$$C_4H_{10} \rightleftharpoons C_2H_6 + C_2H_4$$

a. At 25 °C $\Delta H° = 22.2$ kcal mole^{-1} and $\Delta S° = 33.5$ eu. What is $\Delta G°$? On which side does the equilibrium lie?
b. Calculate $\Delta G°$ at 800 K (527 °C) and determine the position of equilibrium.
c. How does the relative effect of $\Delta H°$ and $\Delta S°$ change with temperature? (Actually, $\Delta H°$ and $\Delta S°$ change somewhat with temperature, but the effect is not large enough to change the qualitative result.)

6. Hydride ion, H^-, is known in the form of salts such as sodium hydride, NaH. When sodium hydride is added to water, the hydride ion is converted completely into hydrogen. What does this say about H_2 as an acid relative to water? What are the products of the reaction? What would happen if acetic acid (Table 4.1) were used in place of water?

7. Ammonia is a weak acid with a pK_a estimated as 33. The conjugate anion NH_2^-, amide ion, is available as alkali metal salts such as sodium amide, $NaNH_2$. Calculate the pH of a solution prepared by adding 0.1 mole of sodium amide to 1 L of water. Does this pH differ appreciably from that of a solution prepared from 0.1 mole of NaOH in 1 L of water? How does the pK_a of phosphine, PH_3, compare with that of ammonia?

8. Consider the reaction of A + B as a second-order reaction for which rate = k[A][B]. If A and B start off with equal concentrations, how has the rate changed at 50% reaction?

9. Explain the following acidity orders using the principles summarized on page 69.
- **a.** $H_2SO_4 > HSO_4^-$
- **b.** Nitric acid ($HONO_2$) > nitrous acid (HONO)
- **c.** $H_2Te > H_2Se$
- **d.** Nitrous acid > hydroxylamine (H_2NOH)
- **e.** $H_2S > PH_3$
- **f.** Perchloric acid ($HOClO_3$) > hypochlorous acid (HOCl)
- **g.** Oxalic acid ($\mathrm{HO{-}\overset{\overset{\textstyle O}{\|}}{C}{-}\overset{\overset{\textstyle O}{\|}}{C}{-}OH}$) > hydrogen oxalate ion ($\mathrm{HO{-}\overset{\overset{\textstyle O}{\|}}{C}{-}CO_2^-}$)
- **h.** A sulfinic acid (RSO_2H) > a sulfenic acid (RSOH)

10. **a.** At room temperature what change in free energy in units of kcal mole^{-1} will change an equilibrium constant by a factor of 10? By a factor of 100? The change in free energy for a factor of 10 is a handy number to remember.
- **b.** These numbers can be converted to equivalent ΔH and ΔS values. Consider ΔG for the factor of 10 change in equilibrium constant. What is the equivalent value for ΔH in kcal mole^{-1} if $\Delta S = 0$; what is the equivalent value for ΔS in eu (cal deg^{-1} mole^{-1}) if $\Delta H = 0$?

11. During the course of a reaction, the concentration of reactants decreases; hence, the rate of reaction is reduced.
- **a.** In the example of 0.05 M methyl chloride and 0.10 M OH^- discussed on page 57, what is the rate of reaction at the start of the reaction, using the rate constant given on page 59?
- **b.** Using this rate, determine the time required for 10% reaction. What are the concentrations of reactants after 10% reaction? What is the rate of reaction at this point? Using this rate, determine how long it takes for the second 10% of reaction to occur.
- **c.** Repeat the calculation to estimate the time for 50% completion of the reaction.

12. The equilibrium reaction in the gas phase of ethylene and HCl to give ethyl chloride, $C_2H_4 + HCl = C_2H_5Cl$, has a favorable enthalpy, $\Delta H° = -15.5$ kcal mole^{-1}, but an unfavorable entropy, $\Delta S° = -31.3$ eu.
- **a.** Why is the entropy negative?
- **b.** What is $\Delta G°$ at room temperature (25 °C)?
- **c.** If the reaction mixture started with 1 atm pressure each of HCl and C_2H_4, what pressure of each is left at equilibrium?
- **d.** For the system to be at equilibrium with all three components present in equal amounts, what total pressure is required? Incidentally, in this system a mixture of pure, dry HCl and C_2H_4 will not react at room temperature. Establishment of the equilibrium requires a suitable catalyst.

13. **a.** The reaction of methyl chloride with water was described as a first-order reaction because the concentration of water does not change during the reaction. If the reaction with water is exactly analogous to the reaction with hydroxide ion, we should write the kinetic equation as

$$\text{rate} = k_2[CH_3Cl][H_2O]$$

What is the value of $[H_2O]$ in this expression?
- **b.** Because $[H_2O]$ remains constant, this case is an example of pseudo first-order kinetics. For the expression

$$\text{rate} = k_1[CH_3Cl]$$

we found that $k_1 = 3 \times 10^{-10}$ sec^{-1}. Using the value of $[H_2O]$ found above, derive k_2. How does the value of k_2 for the reaction of methyl chloride with water compare with that for reaction with hydroxide ion?

c. For a first-order reaction, the time for half of the remaining reactant to react—the half-life—is given by

$$t_{1/2} = 0.693/k$$

From the value of k_1, calculate the half-life in years of an aqueous solution of methyl chloride.

CHAPTER 5

ALKANES

5.1 *n*-Alkanes: Physical Properties

Table 5.1 lists the boiling points, melting points, and densities of some *n*-alkanes. These properties vary in a regular manner. Note that the alkanes from methane through butane are gases at room temperature, pentane boils just above room temperature, and the remaining alkanes show regular increases in boiling point with each additional methylene unit. This regularity of physical properties stems from a regularity of structure. In all of the alkanes the bonds to carbon are nearly tetrahedral and the carbon-hydrogen bond lengths are all essentially constant at 1.095 ± 0.01 Å. Similarly, the carbon-carbon bonds are uniformly 1.54 ± 0.01 Å in length.

The boiling point of a substance is defined as the temperature at which its vapor pressure is equal to the external pressure, usually 760 mm. The vapor pressure of a compound is inversely related to the energy of attraction between the molecules. If the intermolecular attractive force is weak, little energy must be supplied in order for vaporization to occur and the compound has a high vapor pressure. If the intermolecular attractive force is large, more energy must be supplied to cause vaporization and the compound has a low vapor pressure. Interactions between neutral molecules generally result from **van der Waals forces,** dipole-dipole electrostatic attraction, and hydrogen bonding. For hydrocarbons, only the van der Waals interaction is important. This force of attraction results from an electron correlation effect also called the **London force** or **dispersion force.**

Although we normally think of atoms and molecules in terms of smeared-out electron-density distributions, it should be emphasized that this is a time-average picture. At any given instant the electrons are positioned as far from each other as possible, and these positions change from one instant to the next. Consider the simplified models shown in Figure 5.1. The system of charges on the left has a small net attraction that binds the two molecules together. In the system on the right, the electrons have all moved but there is

TABLE 5.1 Physical Properties of *n*-Alkanes

Hydrocarbon	Boiling Point, °C	Melting Point, °C	Density[a] d^{20}
methane	−161.7	−182.5	
ethane	−88.6	−183.3	
propane	−42.1	−187.7	0.5005
butane	−0.5	−138.3	0.5787
pentane	36.1	−129.8	0.5572
hexane	68.7	−95.3	0.6603
heptane	98.4	−90.6	0.6837
octane	125.7	−56.8	0.7026
nonane	150.8	−53.5	0.7177
decane	174.0	−29.7	0.7299
undecane	195.8	−25.6	0.7402
dodecane	216.3	−9.6	0.7487
tridecane	235.4	−5.5	0.7564
tetradecane	253.7	5.9	0.7628
pentadecane	270.6	10	0.7685
eicosane	343	36.8	0.7886
triacontane	449.7	65.8	0.8097
polyethylene			0.965

[a] Note that densities vary with temperature; d^{20} refers to the density in grams per milliliter at 20°C.

still net attraction. The motion of the electrons is mutually correlated to produce net attraction at all times. This attractive force is sensitive to distance and varies as $1/r^6$. It is significant only for molecules close to each other—but not too close. As molecules get too close, the electron charge clouds overlap appreciably and electron repulsion dominates.

Van der Waals attraction depends on the approximate "area" of contact of two molecules—the greater this area, the greater is the attractive force. Each additional methylene unit provides an additional area of contact that increases the total attractive force and gives rise to a greater boiling point. The energy of attraction per methylene group is approximately 1–1.5 kcal mole^{-1}. Van der Waals forces are even greater in solids, and there is also a progressive change in melting point with increasing chain length, as shown in Table 5.1.

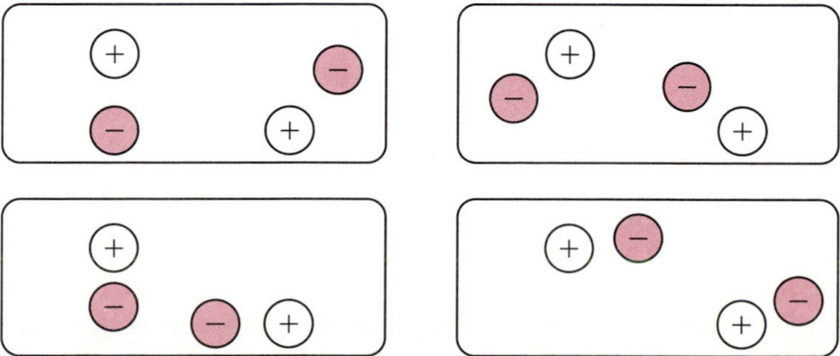

FIGURE 5.1 Van der Waals attraction. At small distances, the dipoles induced by the effect of one molecule on another lead to net electrostatic attraction at every instant.

Highlight 5.1

Van der Waals attraction between two molecules arises from the correlated motion of the electrons, requires close contact, and amounts to 1–1.5 kcal $mole^{-1}$ for a CH_2 group.

Because of the tetrahedral nature of carbon, alkane chains tend to have a zigzag geometry. For example, one of the geometric arrangements adopted by butane is shown in stereo-plot form in Figure 5.2. This type of zigzag arrangement of butane and pentane is symbolized in Figure 5.3.

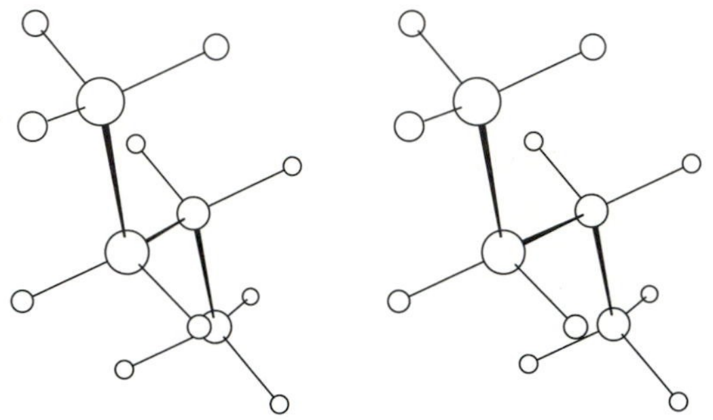

FIGURE 5.2 Stereo representation of one conformation of butane.

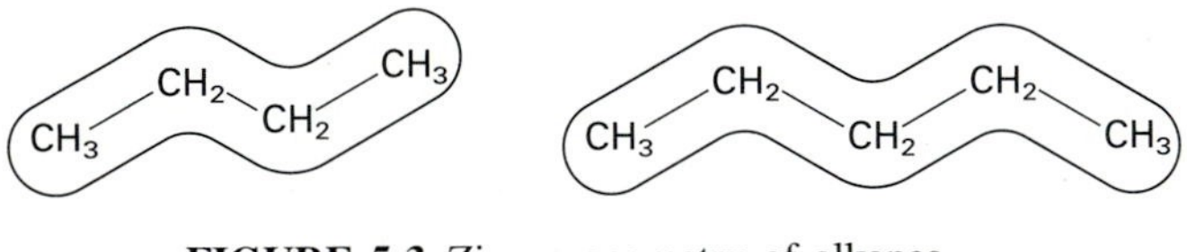

FIGURE 5.3 Zigzag geometry of alkanes.

5.2 *n*-Alkanes: Barriers to Rotation

Experiments of various sorts have shown that the methyl groups of ethane are free to **rotate** with respect to each other about the carbon-carbon single bond. Two extreme structures are possible, the **eclipsed** and **staggered** forms (Figure 5.4). In the eclipsed form, the carbon-hydrogen bonds of one methyl group are ''lined up'' with those of the other. In the staggered form, each carbon-hydrogen bond of one methyl group bisects a H—C—H angle of the other methyl group, when the molecule is viewed along the carbon-carbon bond axis. The eclipsed form of ethane is 3 kcal $mole^{-1}$ higher in energy than the more stable staggered form.

Structures that differ only by rotation about one or more bonds are defined as **conformations** of a compound. In order to represent the three-dimensional character of such conformations, two useful systems are commonly employed. In Figure 5.5 the eclipsed and staggered conformations of ethane are depicted as ''sawhorse'' structures. In this kind of representation a dashed bond (---) projects *away* from the viewer, a heavy wedge bond (▬) projects *toward* the viewer, and a normal bond (—) lies *in the plane* of the page.

Another useful representation is the **Newman projection.** Newman projections for eclipsed and staggered ethane are shown in Figure 5.6. In a Newman projection the carbon-carbon bond is being viewed end on. The nearer carbon is represented by a point. The three other bonds attached to that carbon radiate as three lines from the point. The

(a) eclipsed structure of ethane

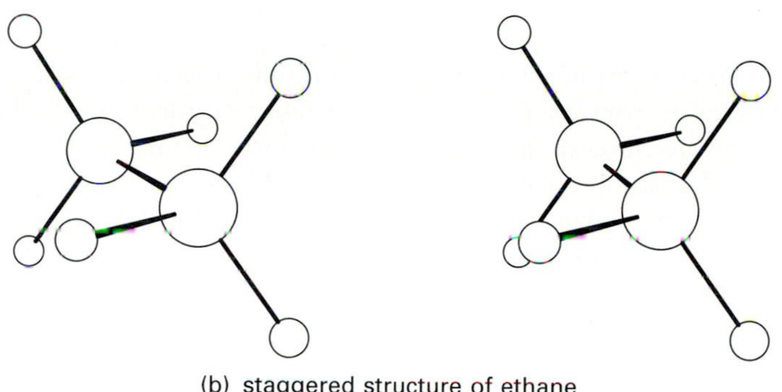

(b) staggered structure of ethane

FIGURE 5.4 Stereo plots illustrating the eclipsed and staggered structures of ethane.

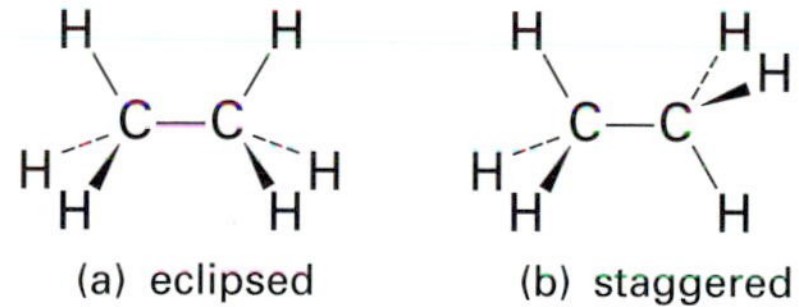

FIGURE 5.5 Sawhorse structures illustrating the eclipsed and staggered conformations of ethane.

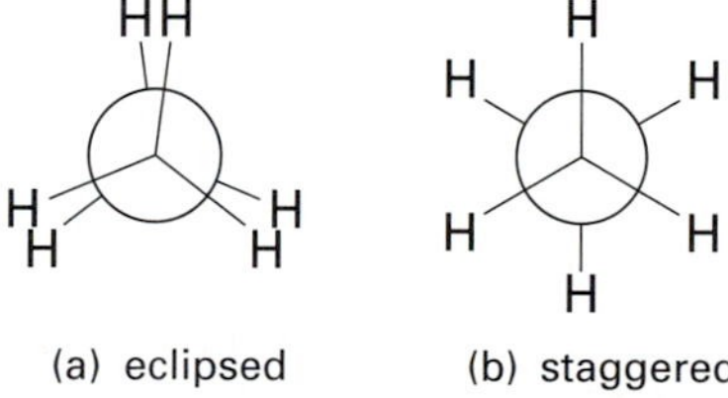

FIGURE 5.6 Newman projections illustrating the eclipsed and staggered conformations of ethane.

farther carbon is represented by a circle with its bonds radiating from the edge of the circle. These projections show that a rotation of 60° about the C—C axis converts the staggered form to the eclipsed structure. As the rotation is continued another 60°, a new staggered conformation is produced, which is identical with the first staggered conformation. A plot of potential energy versus degree of rotation for one complete 360° rotation about the carbon-carbon bond in ethane is shown in Figure 5.7.

The diagram shows that, in rotating about the carbon-carbon bond, there is a 3.0 kcal mole^{-1} energy barrier in passing from one staggered conformation to another. The insta-

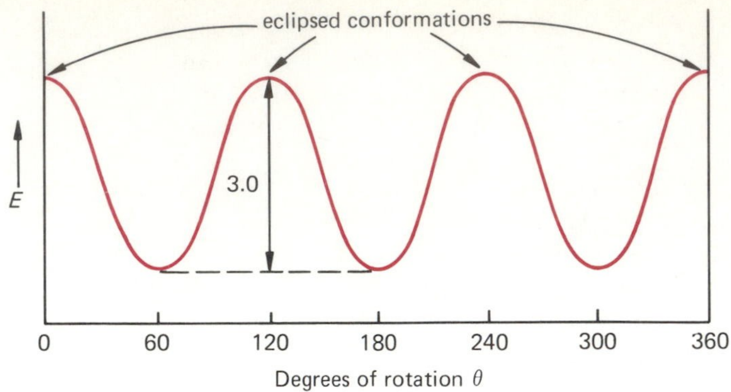

FIGURE 5.7 Potential energy of ethane as a function of degree of rotation about the carbon-carbon bond.

bility of the eclipsed form of ethane appears to result from repulsion of some of the hydrogen orbitals. The hydrogen orbitals on one methyl group are rather far from those on the other, but they are closer in the eclipsed conformation than in the staggered one. The internuclear H-H distance in staggered ethane is 2.55 Å, whereas it is only 2.29 Å in the eclipsed form. Orbital overlap between hydrogens on adjacent atoms is antibonding or repulsive. At these distances the magnitude of this effect is small, only about 1 kcal mole^{-1} per pair of hydrogens, but this small repulsive energy has *staggering* structural consequences, controlling the conformations of a vast number of organic compounds. As we shall see, staggered conformations are the common arrangements for alkane chains and cycloalkane rings.

In propane the barrier to rotation now involves the interaction of one C—CH_3 bond with a carbon-hydrogen bond, as well as two C—H : C—H interactions. Consequently, the rotational barrier is slightly higher in energy at 3.4 kcal mole^{-1}. The most stable conformation is again the staggered one, as illustrated in Figure 5.8.

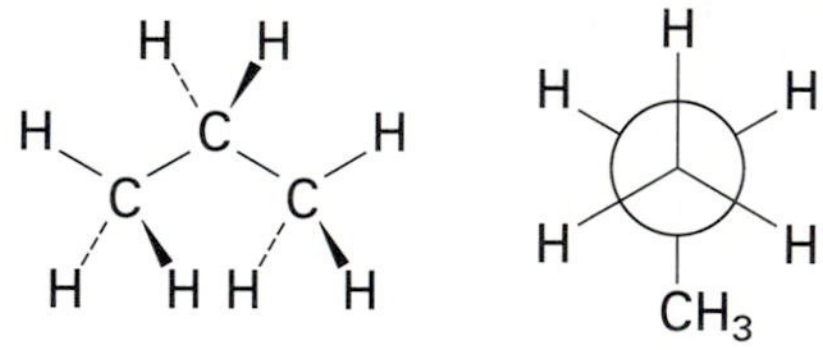

FIGURE 5.8 Most stable conformation of propane.

A potential energy plot for rotation about the C-2—C-3 bond in butane is shown in Figure 5.9. The conformations at the various unique maxima and minima are depicted in Figure 5.10, both as Newman projections and as stereo plots. Note that there are two different kinds of staggered conformation: **anti,** in which the two methyl groups attached to carbons 2 and 3 are farthest apart, and **gauche,** in which these two methyl groups are adjacent. These two kinds of staggered conformation have different energies. The anti conformation is more stable than the gauche by 0.9 kcal mole^{-1}. At room temperature butane is a mixture of 72% anti and 28% gauche conformations.

If these two structures could be isolated, they would have different physical properties such as density, spectra, and melting points. However, the energy barrier separating them is rather small, only 3.8 kcal mole^{-1}. A barrier of such magnitude is far too small to permit isolation of the separate anti and gauche conformations at normal temperatures. In order to separate these two species, one would have to slow the conversion by working at very low temperatures, below approximately −230 °C.

Also note that there are two distinct eclipsed conformations. One of these maxima is reached in rotation from the anti to a gauche conformation. In this conformation, there are

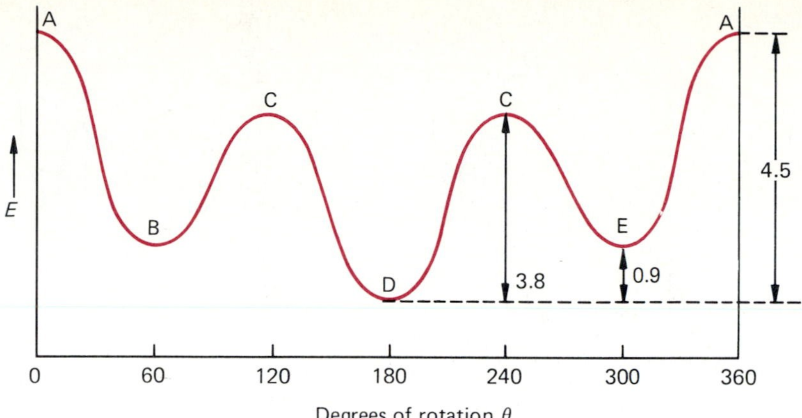

FIGURE 5.9 Potential energy of butane as function of degree of rotation about the C-2—C-3 bond.

two $CH_3:H$ and one $H:H$ eclipsed interactions. This eclipsed conformation is 3.8 kcal mole^{-1} less stable than the anti conformation. The other eclipsed conformation, through which the molecule passes in the course of rotation from one gauche conformation to the other, has one $CH_3:CH_3$ and two $H:H$ eclipsed interactions. Its energy is about 4.5 kcal mole^{-1} above that of the anti conformation.

Finally, note that the gauche conformations labeled B and E, although energetically equivalent, are not really the same. They are actually mirror images of one another. They are the same only in the sense that your right and left hands are the same. We shall return to this phenomenon in Chapter 7.

The same principles apply to larger alkanes. In general, the most stable structure of a given compound is the completely staggered one with all alkyl groups having an anti relationship to one another. However, keep in mind that gauche conformations are only slightly less stable, and there will always be a sizable fraction of the molecules with one or more gauche conformations. The two conformations of pentane given here are shown in stereo-plot form in Figure 5.11.

When writing the structures of such compounds, it is usually not convenient to depict the full geometry as is done in the preceding illustrations. It is important to recognize that a given structure may be written in many ways. For example, all of the following formulas represent pentane.

$CH_3{-}CH_2{-}CH_2{-}CH_2{-}CH_3$

$CH_3{-}CH_2{-}CH_2$
$\qquad\qquad\quad |$
$\qquad\qquad CH_2{-}CH_3$

$CH_3{-}CH_2$
$\qquad\quad |$
$\qquad CH_2$
$\qquad\quad |$
$\qquad CH_2{-}CH_3$

$\quad CH_2 \quad\quad CH_2$
$CH_3 \quad\quad CH_2 \quad\quad CH_3$

$\quad CH_2$
$CH_3 \quad\quad CH_2$
$\qquad CH_2$
$\qquad\quad CH_3$

EXERCISE 5.1 Using molecular models, compare the gauche and anti conformations of butane. Estimate the distance between different pairs of hydrogens. Which pair of hydrogens is closest together? Compare this separation with that of the eclipsed hydrogens in ethane. Does this comparison suggest a principal reason for the higher energy of the gauche conformation?

(a) eclipsed (A)

H_3C CH_3

H H H H

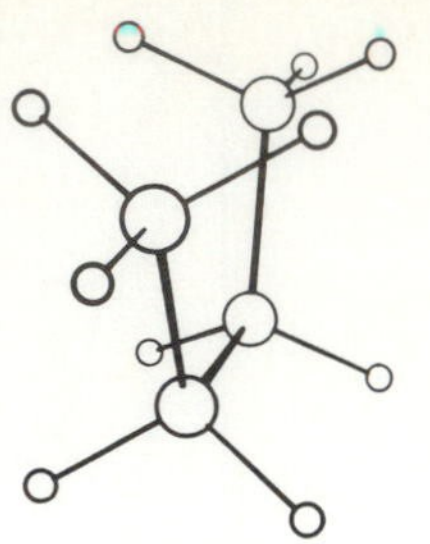

(b) *gauche* (B)

CH_3 H CH_3 H H H

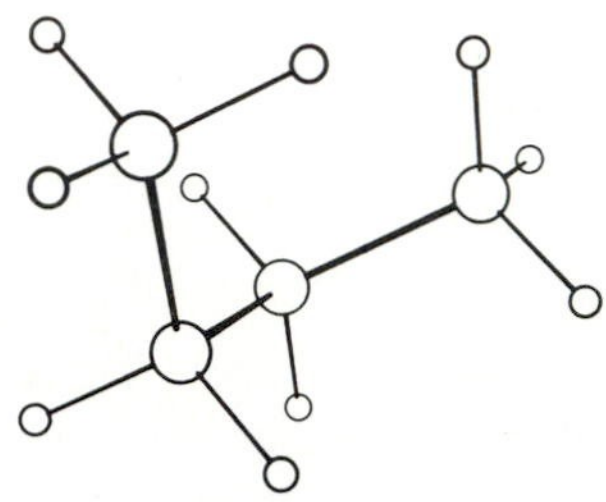

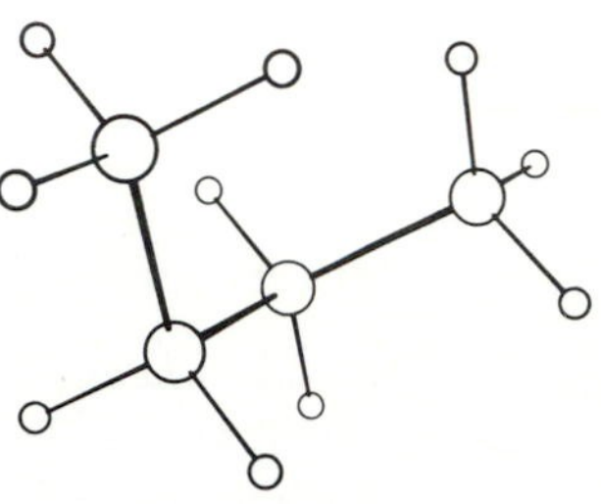

(c) eclipsed (C)

H CH_3 H H CH_3 H

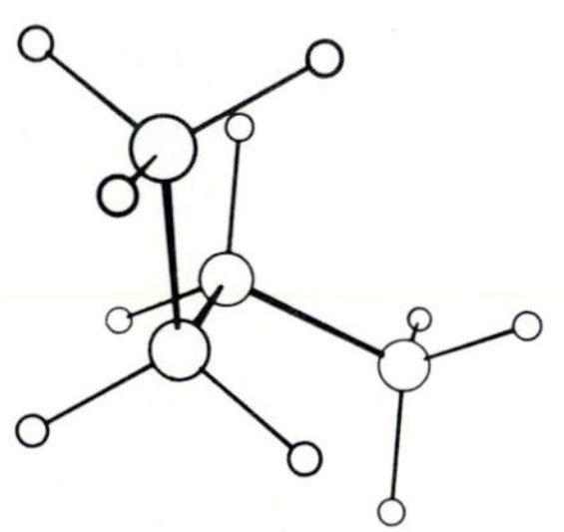

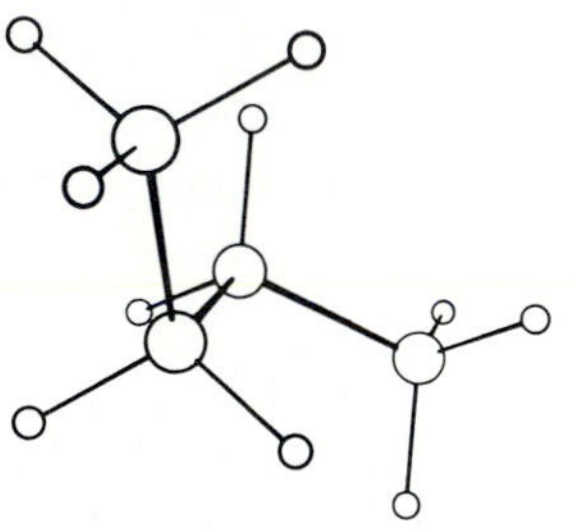

(d) *anti* (D)

CH_3 H H H H CH_3

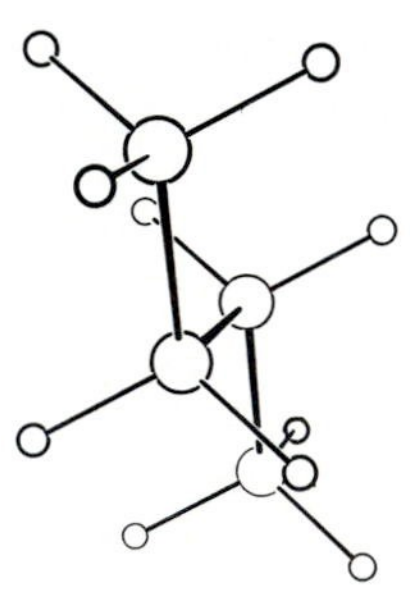

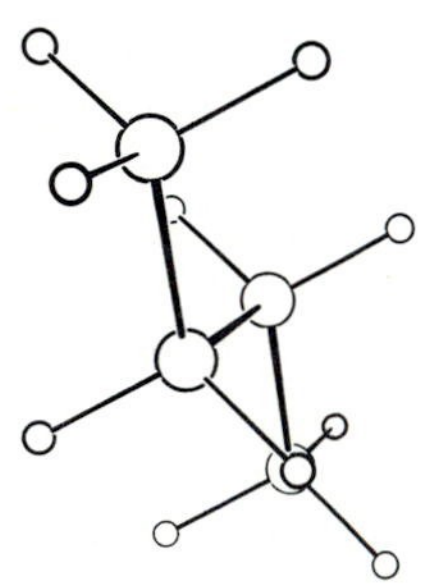

(e) *gauche* (E)

CH_3 CH_3 H H H H

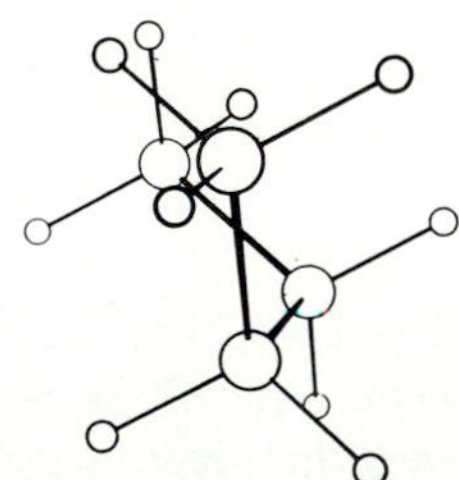

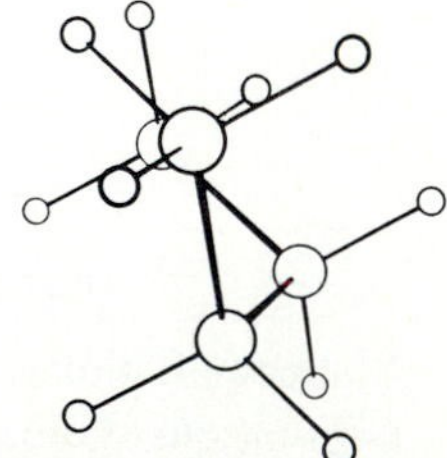

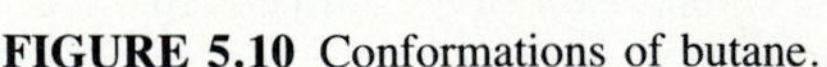

FIGURE 5.10 Conformations of butane.

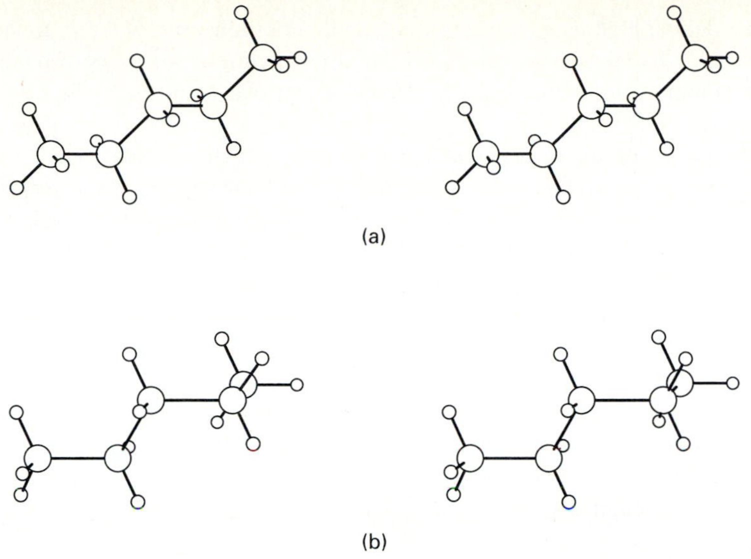

(a)

(b)

FIGURE 5.11 Stereo representations of two pentane conformations: (a) anti-anti; (b) anti-gauche.

Highlight 5.2

Groups bonded to one another by a single bond can usually rotate around that bond; each arrangement represents a conformation of the molecule. It is useful to focus on the most stable and the most extreme forms. Thus, in ethane, CH_3CH_3, the **most stable conformations** are **staggered,** with the C—H bonds of one CH_3 group located in between two of the C—H bonds of the other CH_3 group. The **least stable conformations** are those in which the C—H bonds of one CH_3 group **eclipse** the C—H bonds of the second CH_3 group.

5.3 Branched-Chain Alkanes

The branched-chain alkanes also exist as mixtures of rapidly interconverting staggered conformations. For example, the two staggered conformations of 2-methylbutane for rotation about the C-2—C-3 bond, $(CH_3)_2CH—CH_2CH_3$, are shown in Figure 5.12. The rotational barrier separating these two conformations is about 5 kcal mole^{-1}. At room temperature 2-methylbutane exists as a mixture of the two conformations, 90% of the one with only one gauche interaction and 10% of the one with two gauche interactions.

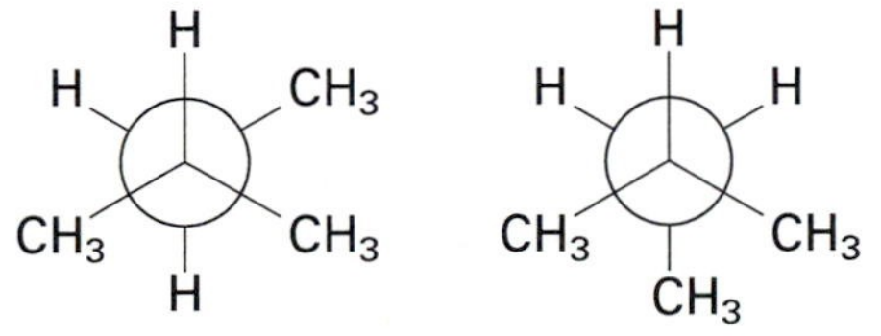

H, H, CH_3, CH_3, CH_3, H | H, H, H, CH_3, CH_3, CH_3

FIGURE 5.12 Staggered conformations of 2-methylbutane.

According to the definition of conformations given in Section 5.2, different conformations of a compound are isomers since they have the same formula but differ in the arrangement of the atoms in space. However, it is convenient to distinguish such isomers, which rapidly interconvert at ordinary temperatures, from other kinds of isomers that interconvert only at high temperatures or not at all. Consequently, we refer to these easily interconvertible spatial isomers as **conformational isomers,** or as **conformations,** or even as **conformers** to distinguish them from **structural isomers,** such as butane and 2-methylpropane.

Physical properties for some branched hydrocarbons are summarized in Table 5.2. Branched-chain hydrocarbons are more compact than their straight-chain isomers. Thus, there is less surface area per unit mass, and van der Waals interactions between molecules are weaker. For this reason, branched hydrocarbons tend to have lower boiling points and melting points than their straight-chain isomers. However, when a molecule has sufficient symmetry, it forms a crystal lattice *more* easily and therefore has a higher melting point but a relatively low boiling point. For an example of such a case, compare the melting points of pentane, 2-methylbutane and 2,2-dimethylpropane. An extreme example is given by 2,2,3,3-tetramethylbutane, which boils only a few degrees above its melting point. Hydrocarbons having a high degree of symmetry or ''ball-like'' character tend to sublime rather than boil. On heating they pass directly from the solid to vapor state without passing through the intermediate liquid state.

This result can be cast into the language of entropy in a straightforward way. In the crystal, molecules are locked in and have greatly restricted movement. In the liquid phase, molecules have enhanced freedom of movement. Consequently, the entropy of melting is a positive quantity whose magnitude is a measure of this increased freedom of movement. The entropy of melting of pentane is +14 eu, whereas that of 2,2-dimethylpropane is only +3 eu. Both molecules have increased freedom of translational motion in the liquid hydrocarbon. In addition, pentane has a floppy chain with many rotational degrees of freedom so that the liquid hydrocarbon is a mixture of many staggered conformational isomers having relatively high entropy. Rotation about the carbon-carbon bonds in 2,2-dimethylpropane, however, always gives back the same structure, and it has a lower entropy. Note that 2-methylbutane has an intermediate value of the entropy of melting of +11 eu.

TABLE 5.2 Physical Properties of Some Branched Alkanes

	Boiling Point, °C	Melting Point, °C	Density d^{20}
2-methylpropane	−11.7	−159.4	0.5572
2-methylbutane	29.9	−159.9	0.6196
2,2-dimethylpropane	9.4	−16.8	0.5904
2-methylpentane	60.3	−153.6	0.6532
3-methylpentane	63.3		0.6644
2,2-dimethylbutane	49.7	−100.0	0.6492
2,3-dimethylbutane	58.0	−128.4	0.6616
2,2,3,3-tetramethylbutane	106.3	100.6	0.6568

EXERCISE 5.2 Write Newman projections for two different staggered conformations of 2,3-dimethylbutane for rotation about the C-2—C-3 bond.

EXERCISE 5.3 Draw a potential energy diagram (see Figure 5.9) for rotation about the C-2—C-3 bond of 2,3-dimethylbutane.

For relatively simple organic structures, such as the examples discussed so far in this chapter, it is possible to use condensed structural formulas, in which each carbon and hydrogen is explicitly shown. However, for more complex molecules, these formulas are rather awkward, and an even more highly abbreviated representation is used. In these **line structures** an alkane chain is represented by a zigzag line. The ends and each "bend" in the line represent a carbon atom. Hydrogens are not shown; each carbon is understood to have enough hydrogens to satisfy its tetravalency. Thus, pentane and octane are depicted as

$\equiv$ $CH_3CH_2CH_2CH_2CH_3$

pentane

$\equiv$ $CH_3CH_2CH_2CH_2CH_2CH_2CH_2CH_3$

octane

Branched alkanes are represented in an analogous fashion, using a branched line. Remember that at a branch point, either one or no hydrogens are attached, depending on the number of carbons that are also attached to that position.

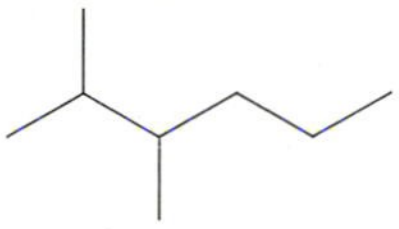

2,3-dimethylhexane  2,2-dimethylheptane 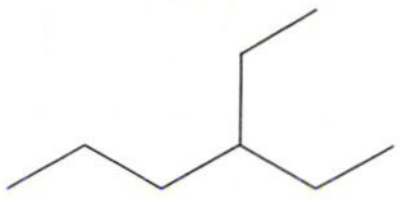 3-ethylhexane

Substituent groups such as halogens are shown as

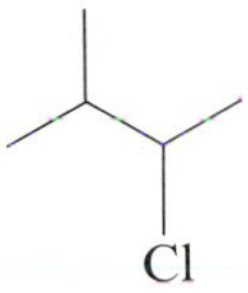

Cl Cl

Cl

2-chloro-3-methylbutane 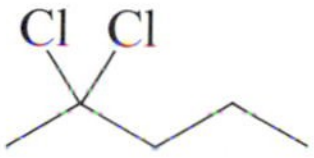 2,2-dichloropentane

EXERCISE 5.4 Write the simple line structures corresponding to the following alkanes and compare with complete structures.

(a) 2-methylnonane
(b) 2,3,3-trimethylpentane
(c) 3-ethyl-4-methylnonane
(d) 3-chloromethyl-3-methylhexane

5.4 Cycloalkanes

Carbon chains can also form rings. Because there are no ends to the carbon chain in a cyclic alkane, the general formula is $(CH_2)_n$ or C_nH_{2n}. Like straight-chain alkanes, they are saturated hydrocarbons. They are named according to the number of carbons in the ring with the prefix **cyclo**-.

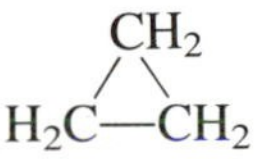

CH_2 / $H_2C—CH_2$ $H_2C—CH_2$ / $H_2C—CH_2$ CH_2 / H_2C CH_2 / $H_2C—CH_2$

cyclopropane 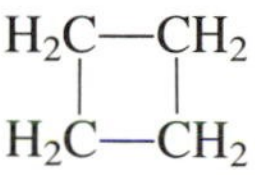 cyclobutane 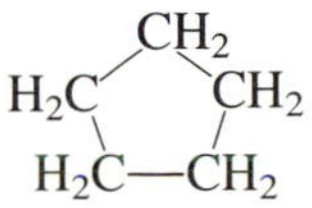 cyclopentane

The physical properties of some cycloalkanes are summarized in Table 5.3. Note that their symmetry and more restricted rotations result in higher melting points and boiling points than comparable *n*-alkanes.

TABLE 5.3 Physical Properties of Some Cycloalkanes

	Boiling Point, °C	Melting Point, °C	Density d^{20}
cyclopropane	−32.7	−127.6	
cyclobutane	12.5	−50.0	
cyclopentane	49.3	−93.9	0.7457
cyclohexane	80.7	6.6	0.7786
cycloheptane	118.5	−12.0	0.8098
cyclooctane	150.0	14.3	0.8349

Because of symmetry, there is only one monosubstituted cycloalkane, and a number to designate the position of the appendage is not necessary.

CH_3 — CH (ring: H_2C, CH_2, H_2C, CH_2, CH_2)

methylcyclohexane

CH_2CH_3 — CH (ring: H_2C, CH_2, H_2C—CH_2)

ethylcyclopentane

When there is more than one substituent, numbers are required. One substituent is always given the number 1, and the other is given the next lowest possible number.

H_3C, CH_3 — C (ring: H_2C, CH_2, H_2C, $CHCH_2CH_2CH_3$, CH_2)

1,1-dimethyl-3-propylcyclohexane

In more complex compounds, the cycloalkyl radical may be named as a prefix.

H_2C—CH_2 (ring) CH — $CH_3CH_2CHCH_2CH_3$

3-cyclopropylpentane

CH_2 (ring: H_2C, CH_2) CH — $CH_3CHCHCH_3$ — CH_3

2-cyclobutyl-3-methylbutane

Cycloalkanes are usually symbolized by simple geometric figures in which a carbon atom with its appropriate number of attached hydrogens is understood to be present at each apex. Thus, the four smallest cycloalkanes are depicted as

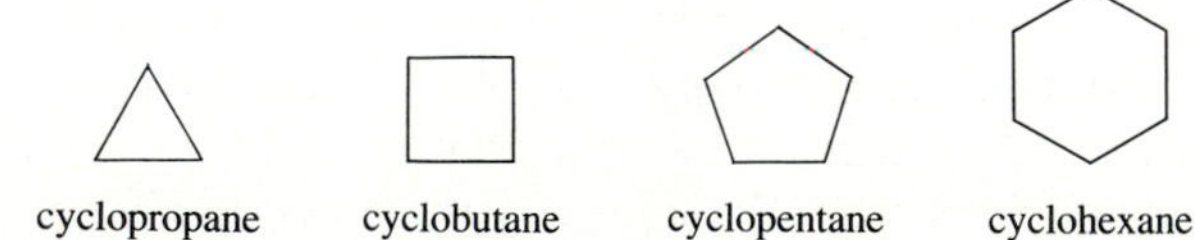

cyclopropane cyclobutane cyclopentane cyclohexane

Simple substituted cycloalkanes are depicted by the appropriate geometric figure, with attached substituent groups.

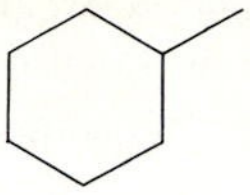

methylcyclohexane 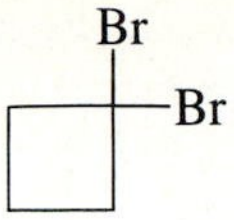 1,1-dibromocyclobutane

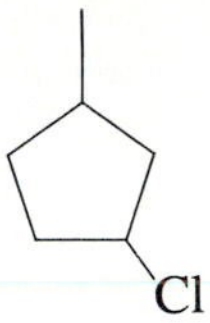

1-chloro-3-methylcyclopentane 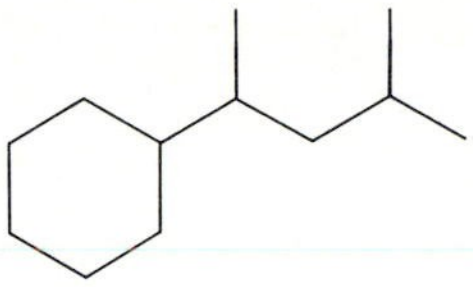 2-cyclohexyl-4-methylpentane

The alkanes and cycloalkanes are the parent structures in the general class of aliphatic compounds. Most of the chemistry of cycloalkanes is similar to that of the alkanes. There are some differences in stability and in their conformations, which will be discussed in Sections 5.6 and 7.7, respectively.

EXERCISE 5.5 Using simple geometric figures and line structures, depict the following compounds. Compare your structures with complete structural representations.

(a) 1,1,3-trimethylcyclohexane
(b) 3-cyclopentylpentane
(c) 1-chloro-4-chloromethylcyclohexane
(d) 1,1,2,2-tetramethylcyclopropane

5.5 Heats of Formation

The **heat of formation** of a compound from its elements in their standard states is a thermodynamic property with considerable use in organic chemistry. This quantity, symbolized ΔH_f°, is defined as the enthalpy of the reaction of elements in their standard states to form the compound. The standard state of each element is generally the most stable state of that element at 25 °C and 1 atm pressure. The standard state of carbon is taken as the graphite form, whereas those of hydrogen and oxygen are H_2 and O_2 gases, respectively. By definition, ΔH_f° for an element in its standard state is zero. The standard heat of formation of butane is -30.36 ± 0.16 kcal mole^{-1} and that of 2-methylpropane is -32.41 ± 0.13 kcal mole^{-1}.

$$4\ \text{C (graphite)} + 5\ \text{H}_2(\text{g}) = n\text{-C}_4\text{H}_{10}(\text{g}) \qquad \Delta H^\circ \equiv \Delta H_f^\circ(n\text{-C}_4\text{H}_{10}) = -30.36 \text{ kcal mole}^{-1}$$

$$4\ \text{C (graphite)} + 5\ \text{H}_2(\text{g}) = (\text{CH}_3)_3\text{CH}(\text{g}) \qquad \Delta H^\circ \equiv \Delta H_f^\circ(i\text{-C}_4\text{H}_{10}) = -32.41 \text{ kcal mole}^{-1}$$

We shall see in Chapter 6 how these hypothetical enthalpies of reaction are determined. For now, suffice it to say that they *can* be determined, although indirect methods are required. The ΔH_f° of a compound may either be negative, as in the two foregoing examples, or positive. A negative ΔH_f° means that heat would be liberated if the compound could be prepared directly by combination of its elements. That is, butane and 2-methylpropane are both more stable (have lower enthalpy) than four carbon atoms and five hydrogen molecules in their standard states. The heats of formation of the butane isomers reveal that 2-methylpropane is more stable than butane by 2.05 kcal mole^{-1}. Thus, in the following hypothetical equilibrium, 2-methylpropane would predominate.

$$\text{CH}_3\text{CH}_2\text{CH}_2\text{CH}_3 \overset{K}{\rightleftharpoons} \text{CH}_3\overset{\text{CH}_3}{\overset{|}{\text{C}}}\text{HCH}_3 \qquad \Delta H^\circ = -2.05 \text{ kcal mole}^{-1}$$

The heats of formation of these two hydrocarbons are depicted graphically in Figure 5.13. In using energy diagrams such as these, remember that down represents less energy and greater stability ("downhill in energy"), whereas up represents higher energy and lower stability.

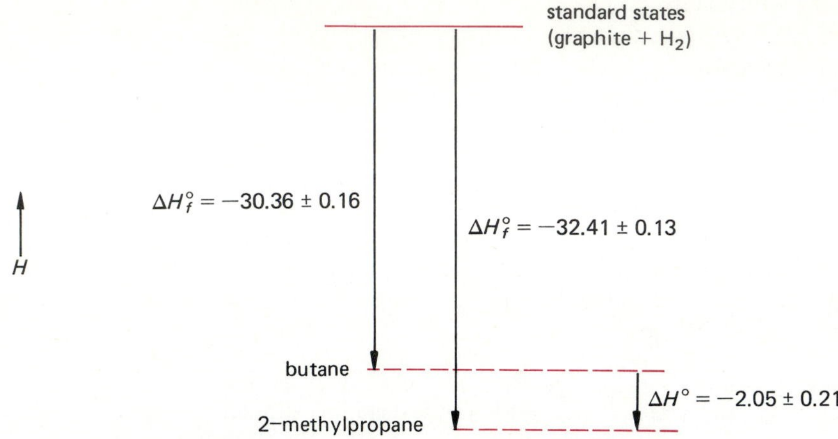

FIGURE 5.13 The heats of formation of butane and 2-methylpropane, illustrating the use of ΔH_f° values to compute ΔH° for a simple reaction.

TABLE 5.4 Some Heats of Formation

Compound	Heat of Formation at 25°C ΔH_f°, kcal mole^{-1}
CH_4	−17.9
CH_3CH_3	−20.2
$CH_3CH_2CH_3$	−24.8
$CH_3CH_2CH_2CH_3$	−30.4
$(CH_3)_3CH$	−32.4
$CH_3CH_2CH_2CH_2CH_3$	−35.1
$(CH_3)_2CHCH_2CH_3$	−36.9
$(CH_3)_4C$	−40.3
CO	−26.4
CO_2	−94.1
H_2O(g)	−57.8
H_2O(1)	−68.3
H_2	0
O_2	0
C (graphite)	0

Some values for heats of formation are listed in Table 5.4. A more complete list is given in Appendix I. These ΔH_f° values are useful for estimating the energetics of possible reactions, providing that a pathway or reaction mechanism is accessible. For example, the hydrogenation of butane to ethane is exothermic by 10 kcal mole^{-1}.

$$\begin{array}{llll} & C_4H_{10} + H_2 = & 2\,C_2H_6 & \\ \Delta H_f^\circ: & -30.4 + 0 & 2 \times -20.2 & \Delta H^\circ = -10.0 \text{ kcal mole}^{-1} \end{array}$$

If a suitable catalyst or reaction pathway could be found, this reaction would proceed toward the right. However, no such catalyst or pathway is known for the reaction at ordinary temperatures. The reaction remains hypothetical even though, if realized, it would be exothermic. This example illustrates the difference between thermodynamics and kinetics. A given reaction may have favorable thermodynamics but will occur only if a pathway with a sufficiently low activation barrier can be found. Because of the importance of pathways, our studies of organic reactions will also often include discussions of reaction mechanism. In biochemical reactions, proteins called enzymes have groups arranged in a way to create pathways for reaction, often by mechanisms similar to those that we will discuss for organic reactions.

EXERCISE 5.6 Using the heats of formation given in Table 5.4, construct a diagram analogous to Figure 5.13 showing the relative energies of pentane, 2-methylbutane, and 2,2-dimethylpropane. Note that each new branch provides 2–3 kcal $mole^{-1}$ in stabilization.

The hydrogenation of ethylene to ethane is also highly exothermic.

$$\begin{array}{lll} C_2H_4 + H_2 = C_2H_6 & & \\ 12.5 + 0 \quad -20.2 & & \Delta H^\circ = -32.7 \text{ kcal mole}^{-1} \end{array}$$

In this case a number of catalysts are known that provide reaction pathways, and this reaction is an important general reaction of alkenes (Section 11.6).

One important limitation on the use of heats of formation is that equilibria are determined by free energy rather than by enthalpy alone.

$$\Delta G^\circ = -RT \ln K = \Delta H^\circ - T\Delta S^\circ$$

That is, an entropy change plays a large role in determining an equilibrium constant. For example, in the equilibrium between *n*-butane and 2-methylpropane discussed previously, ΔH° is -2.05 kcal $mole^{-1}$, but ΔG° is only -0.89 kcal $mole^{-1}$, corresponding to an equilibrium constant at 25 °C of 4.5. Since entropy is a measure of freedom of motion, the largest entropy changes result from a difference in numbers of molecules on the two sides of an equilibrium. The magnitude of this effect depends on physical state (gas, liquid, and so on), molecular weight, and temperature. For a gas at ordinary temperature and pressure, a difference of one molecule on the two sides of an equilibrium (for example, $A \rightleftharpoons B + C$) corresponds to about 30–40 eu, which is equivalent to 9–12 kcal $mole^{-1}$ in enthalpy at room temperature. At higher temperatures any entropy change has a still greater effect. For example, at 25 °C the conversion of butane to one molecule of ethane and one of ethylene is highly endothermic.

$$\begin{array}{rccc} & C_4H_{10} & = \; C_2H_6 & + \; C_2H_4 \\ \Delta H_f^\circ: & -30.4 & -20.2 & 12.5 \end{array}$$

$$\begin{aligned} \Delta H^\circ &= \Delta H_f^\circ(\text{products} = -20.2 + 12.5) - \Delta H_f^\circ(\text{reactants} = -30.4) \\ &= +22.6 \text{ kcal mole}^{-1} \end{aligned}$$

Even though this reaction involves one molecule going to two, the resulting ΔS° of 33 eu still leaves a positive free energy change at room temperature: $\Delta G^\circ = +12.7$ kcal $mole^{-1}$. At 500 °C, although the equilibrium is still highly endothermic in enthalpy, the positive entropy change gives a ΔG° of -3.8 kcal $mole^{-1}$. The equilibrium now favors the products. As we shall see in Section 6.2, this reaction is involved in the refining of petroleum (''cracking''). However, the thermodynamics of the molecules demand that the reaction be carried out at high temperature, as the foregoing simple calculations show.

EXERCISE 5.7 Calculate $\Delta H°$ for the reaction of hydrogen with 2-methylbutane to give ethane and propane.

Highlight 5.3

By definition, the heats of formation of elements in their standard states are zero. A negative heat of formation corresponds to the evolution of heat in a hypothetical reaction that produces a compound from its elements. Comparing the heats of formation of two compounds tells us about their relative stabilities and about structural effects on those stabilities. However, favorable thermodynamics in a reaction does not always imply favorable kinetics because the barrier to reaction may be high.

5.6 Cycloalkanes: Ring Strain

Ring strain is an energy effect that can be seen clearly in the heats of formation of the cycloalkanes. In alkanes each CH_2 group contributes about -5 kcal mole^{-1} to ΔH_f° of a molecule. That is, the heats of formation of compounds differing by only one CH_2 differ by a regular increment of about 5 kcal mole^{-1}.

	ΔH_f°, *kcal mole*$^{-1}$
$4\,C + 5\,H_2 \longrightarrow n\text{-}C_4H_{10}$	-30.4
$5\,C + 6\,H_2 \longrightarrow n\text{-}C_5H_{12}$	-35.1
$6\,C + 7\,H_2 \longrightarrow n\text{-}C_6H_{14}$	-39.9

Since cycloalkanes have the empirical formula $(CH_2)_n$, one can obtain the ΔH_f° for each CH_2 group by simply dividing ΔH_f° for the molecule by n. The heats of formation for a number of cycloalkanes are tabulated in Table 5.5. Examination of the table shows that most of these cycloalkanes have less negative values of $\Delta H_f^\circ/n$ than the alkane value of about -5 kcal mole^{-1}. That is, many cycloalkanes have a higher energy content per CH_2

TABLE 5.5 ΔH_f° of Cycloalkanes, $(CH_2)_n$

n	Cycloalkane	ΔH_f°, kcal mole^{-1}	$\Delta H_f^\circ/n$, kcal mole^{-1} per CH_2 group	Total Strain Energy, kcal mole^{-1}
2	ethylene	+12.5	+6.2	22
3	cyclopropane	+12.7	+4.2	27
4	cyclobutane	+6.8	+1.7	26
5	cyclopentane	−18.4	−3.7	6
6	cyclohexane	−29.5	−4.9	(0)
7	cycloheptane	−28.2	−4.0	6
8	cyclooctane	−29.7	−3.7	10
9	cyclononane	−31.7	−3.5	13
10	cyclodecane	−36.9	−3.7	12
11	cycloundecane	−42.9	−3.9	11
12	cyclododecane	−55.0	−4.6	4
13	cyclotridecane	−58.9	−4.5	5
14	cyclotetradecane	−57.1	−4.1	12
15	cyclopentadecane	−72.0	−4.8	2
16	cyclohexadecane	−76.9	−4.8	2

group than a typical acyclic alkane. This excess energy is called **ring strain.** The total excess energy of a cycloalkane is simply the excess energy per CH_2 multiplied by the number of CH_2 groups in the particular cycloalkane.

Cyclohexane shows essentially no ring strain; its CH_2 groups have essentially the same ΔH_f° as those of normal alkanes. For the purpose of computing the ring strain of a particular cycloalkane, cyclohexane is considered to be strain-free; it is the standard for comparison. For cyclohexane $\Delta H_f^\circ = -29.5$ kcal mole^{-1} and $\Delta H_f^\circ/n = -29.5/6 = -4.92$ kcal mole^{-1}. This value is taken as ΔH_f° for a "strainless" CH_2 group. For example, ΔH_f° for a hypothetical "strainless" cyclopentane would be 5×-4.92 kcal mole^{-1} $= -24.6$ kcal mole^{-1}. Hence, the strain energy of cyclopentane $= (-18.4) - (-24.6) = +6.2$ kcal mole^{-1}. In other words, cyclopentane is 6 kcal mole^{-1} less stable than it would be if each CH_2 group were in some hypothetical strain-free state.

EXERCISE 5.8 Using the data in Table 5.5, verify that the total strain energy of cyclooctane is 10 kcal mole^{-1}.

The inherent ring strain of a given molecule results from three distinct contributions: bond strain ("bent bonds"), eclipsing of adjacent pairs of carbon-hydrogen bonds, or transannular nonbonded interaction (the "bumping together" of two hydrogens that are bonded to atoms "across the ring" from one another).

A bond is strongest when it is formed by the overlap of two atomic orbitals along the internuclear bond axis. The strength of the bond is reduced if overlap of the constituent orbitals is not along the bond axis.

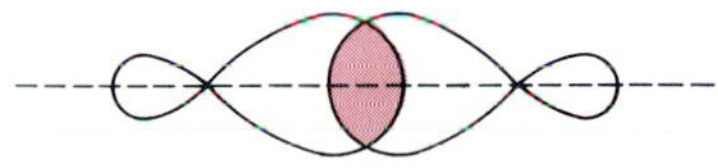

stronger,
more efficient overlap

weaker,
less efficient overlap

The structure of cyclopropane is shown in Figure 5.14. For purely geometric reasons the internuclear C—C—C angle in cyclopropane is 60°. The natural bond angle for $C_{sp}3$-orbitals overlapping linearly would be 109.5°. Even with pure *p*-orbitals the natural bond angle cannot be less than 90°. In practice, the carbon-carbon bonds in cyclopropane do have more *p*-character than normal sp^3. As a result, the orbitals form bent bonds (Figure 5.15). Consequently, the carbon-carbon bonds in cyclopropane are weaker than those in normal alkanes. This reduced bond strength shows up as a ring strain in the ΔH_f°.

To compensate for the fact that extra *p*-character is used in the carbon-carbon bonds of cyclopropane, extra *s*-character is used for the carbon-hydrogen bonds. Consequently, these bonds are somewhat shorter and stronger than alkyl carbon-hydrogen bonds, and the H—C—H bond angle is greater than tetrahedral. Another factor that contributes to the ring strain in cyclopropane is the eclipsing of the carbon-hydrogen bonds. Recall that the eclipsed conformation of ethane is 3.0 kcal mole^{-1} less stable than the staggered conformation; each pair of eclipsed hydrogens raises the energy by 1.0 kcal mole^{-1} (Section 5.2). In cyclopropane there are six pairs of eclipsed hydrogens, which could contribute a maximum of 6 kcal mole^{-1} to the energy of the molecule. However, the eclipsed hydrogens are farther apart in cyclopropane than they are in ethane because of the small C—C—C angle in the former. Therefore, the actual magnitude of the eclipsing interaction is somewhat less than the maximum of 6 kcal mole^{-1}.

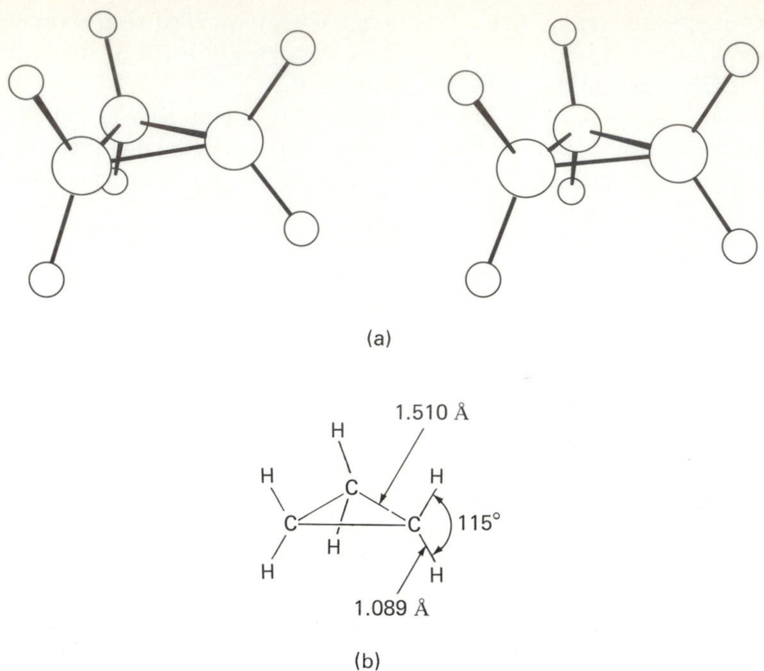

FIGURE 5.14 Cyclopropane: (a) stereo representation; (b) geometric structure.

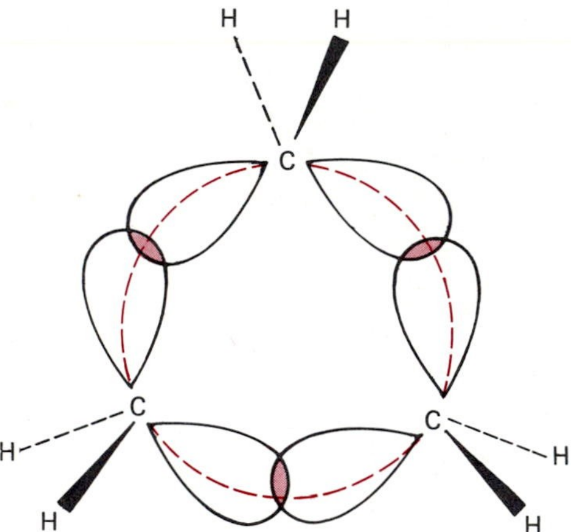

FIGURE 5.15 Orbital structure of cyclopropane ring showing bent-bond strain.

In cyclobutane the internuclear angles of 90° are not as small as in cyclopropane. The carbon-carbon bonds are not as bent, and there is less strain per bond. However, there are four strained bonds rather than three, and there are eight pairs of eclipsed hydrogens rather than six. Also the eclipsing in a planar cyclobutane would be more important than in cyclopropane because the hydrogens are closer. The result is that the total ring strain in the two compounds is about the same.

Since three points define a plane, the carbon framework of cyclopropane must have a planar structure. However, cyclobutane can exist in a nonplanar conformation. Spectroscopic studies show that cyclobutane and many of its derivatives do have nonplanar structures in which one methylene group is bent at an angle of about 25° from the plane of the other three ring carbons. In this structure, shown in Figure 5.16, some increase in bond-angle strain is compensated by the reduction in the eclipsed-hydrogen interactions.

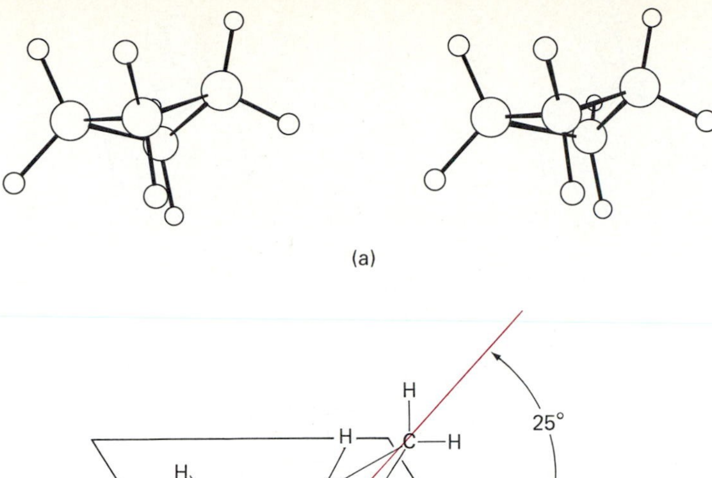

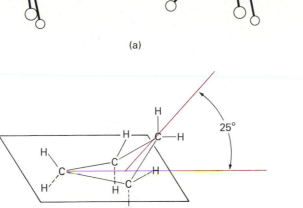

FIGURE 5.16 Bent cyclobutane: (a) stereo representation; (b) illustrating the angle of bend.

A planar pentagonal ring structure for cyclopentane would have C—C—C angles of 108°, a value so close to the normal tetrahedral angle of 109.5° that no important strain effect would be expected. However, all of the hydrogens are completely eclipsed in such a structure and it would have about 10 kcal mole^{-1} of strain energy.

The molecule finds it energetically worthwhile to twist somewhat from a planar conformation. The actual structure has the "envelope" shape shown in Figure 5.17. The additional bond-angle strain involved in this structure is more than compensated by the reduction in eclipsed hydrogens. The out-of-plane methylene group is approximately staggered with respect to its neighbors.

Cyclohexane is the most important of the carbocycles; its structural unit is widespread in compounds of natural origin. Its importance no doubt stems from the fact that it can adopt a conformation that is essentially strain-free. This structure, shown in Figure 5.18, is known as a **chair conformation.** In this structure the bond angles are all close to tetrahedral, and all pairs of hydrogens are completely staggered with respect to each other. The latter point can easily be seen by looking down each carbon-carbon bond in turn to produce the Newman projections shown in Figure 5.19. Cyclohexane has neither bond-angle strain nor eclipsed-hydrogen strain.

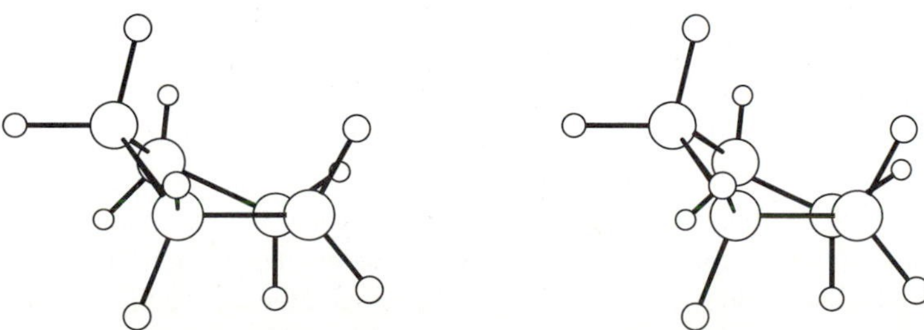

FIGURE 5.17 Stereostructure of cyclopentane.

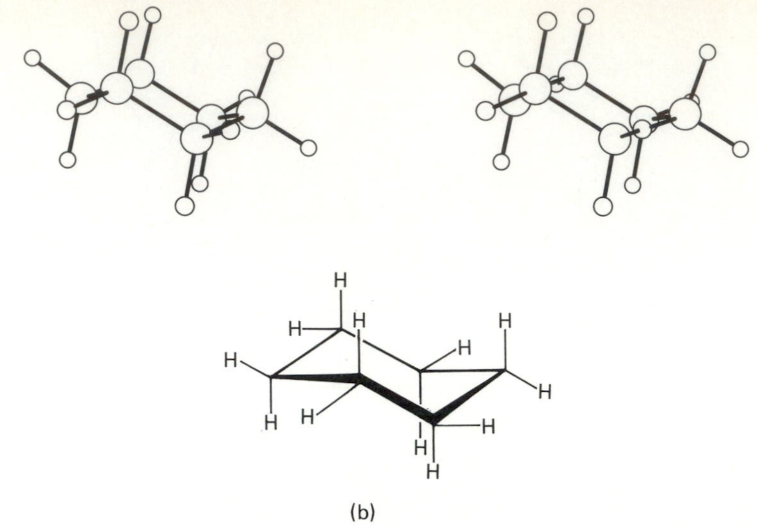

(b)

FIGURE 5.18 Chair conformation of cyclohexane: (a) stereo representation; (b) conventional perspective drawing.

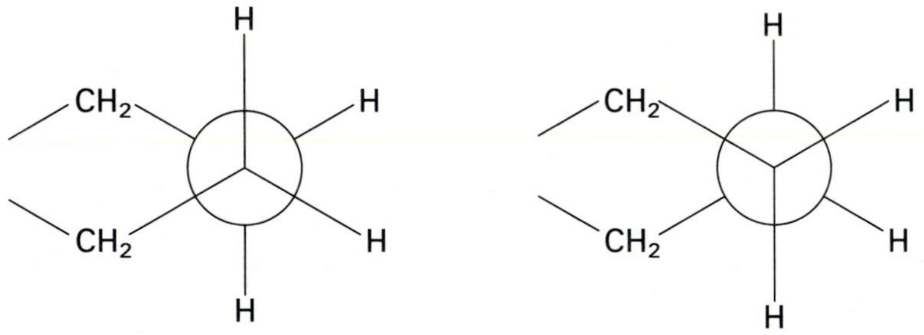

FIGURE 5.19 Newman projections of carbon-carbon bonds in cyclohexane.

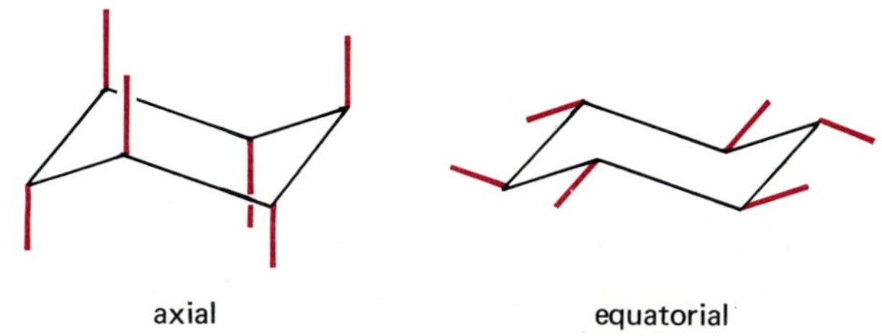

axial

equatorial

FIGURE 5.20 Cyclohexane bonds.

The chair conformation has two distinct types of hydrogens. These different hydrogens correspond to two sets of exocyclic bonds, the **axial** and **equatorial** bonds shown in Figure 5.20.

The chair conformation of cyclohexane is so important that the student should learn to draw it legibly. Notice should be taken of the sets of parallel lines in the structure shown in Figure 5.21. The molecular axis shown in Figure 5.21 is a threefold axis; rotation by 120° about this axis leaves the molecule unchanged.

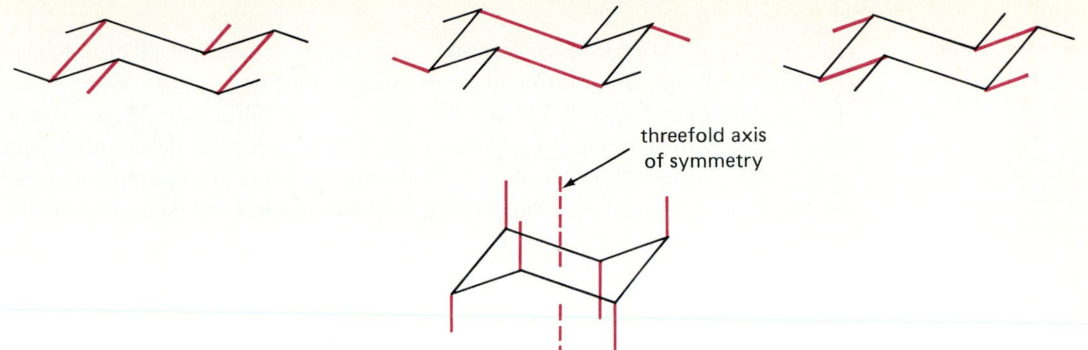

FIGURE 5.21 Construction of chair conformations.

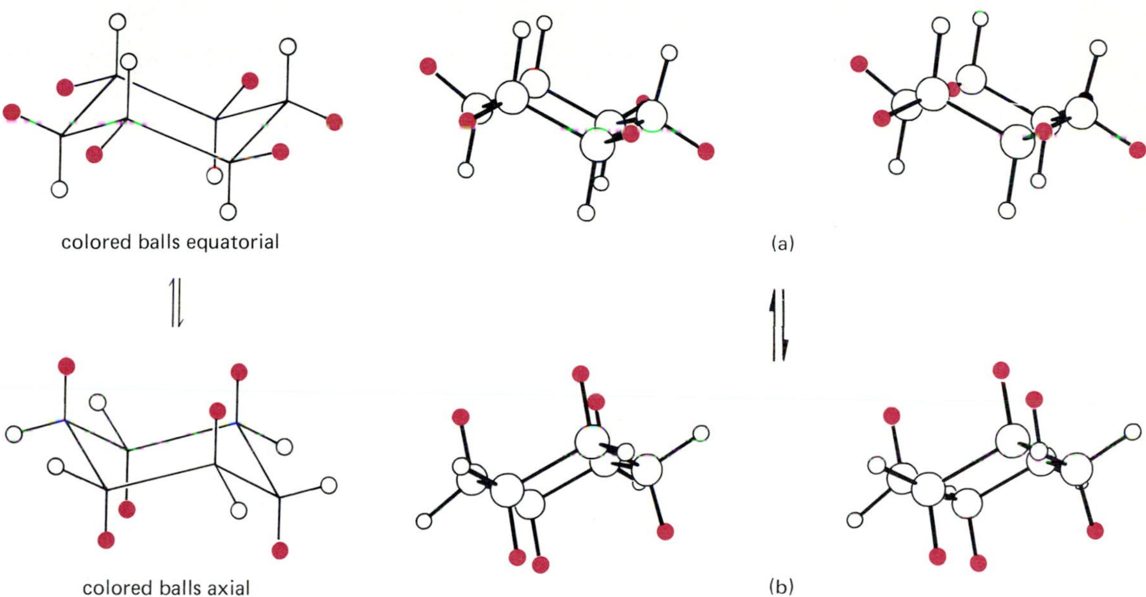

FIGURE 5.22 Two chair conformations of cyclohexane: (a) colored balls equatorial; (b) colored balls axial. Left: normal projection; right: stereo.

Cyclohexane is a dynamic structure. A concerted rotation about the carbon-carbon bonds changes one chair conformation to another in which the axial and equatorial bonds have changed places. This change is shown in Figure 5.22, in which two sets of bonds are marked by open and filled circles. The conversion of cyclohexane from one chair form to another is a conformational change that involves only rotation about carbon-carbon bonds. The process has an energy barrier of 10.8 kcal $mole^{-1}$.

The larger cycloalkanes are less important, and we will not dwell on them. In general, the medium-ring C_7–C_{12} cycloalkanes have conformations in which some form of hydrogen-hydrogen repulsion is inescapable. When the carbocyclic ring is sufficiently large, the effect of size on the energetics of ring formation is no longer significant. This point is reached by about C_{15}. Segments of such rings behave much as long, linear alkanes; in such large rings there are generally a number of possible conformations in which the hydrogens are sufficiently separated from each other in staggered arrangements.

EXERCISE 5.9 Using a molecular model set, construct a model of cyclohexane. Place the model in a chair conformation (reference to the stereoscopic projection of cyclohexane in Figure 5.22 may be useful in this regard), and identify the axial and equatorial hydrogens. Mark the end of one of the carbon-hydrogen bonds. Experiment with "flipping" the model from one chair conformation to the other, noting that the marked hydrogen is in an axial position in one conformation and in an equatorial position in the other.

EXERCISE 5.10 Practice drawing a chair conformation of cyclohexane, paying careful attention to the sets of parallel lines as shown in Figure 5.22. Compare the drawing you have made to the molecular model you constructed in Exercise 5.9. Draw a chair conformation of chlorocyclohexane in which the substituent is axial and one in which the substituent is equatorial. Do not include the hydrogens. Show your drawings to your professor or teaching assistant and see if he or she can tell which is which.

Highlight 5.4

Cyclohexane exists in a strain-free chair conformation. A substituent can replace hydrogen in the equatorial position or in the axial position. A line drawn through the center of the cyclohexane in a direction perpendicular to the average plane of the molecule is called an axis. Bonds parallel to the axis are called **axial** and bonds almost perpendicular to the axis or almost parallel to the average plane are called **equatorial.**

5.7 Occurrence of Alkanes

Alkanes are widespread as natural products. On earth, they are primarily the product of living processes, but one theory holds that there is a large quantity of methane of prebiotic origin trapped in the earth's crust. It is presumed that the enormous quantities of methane present on Jupiter are also abiogenic. A search for prebiotic methane down to a depth of 6.8 km at Siljan, Sweden was unsuccessful. Methane is produced by the anaerobic bacterial decomposition of vegetable matter under water. Because it was first isolated in marshes, it was long called "marsh gas." It is also an important constituent of the gas produced in some sewage-disposal processes. Methane also occurs in the atmosphere of coal mines, where it is called "fire damp" because of the explosive nature of methane-air mixtures.

Natural gas is a mixture of gaseous hydrocarbons and consists primarily of methane and ethane, along with small amounts of propane. Natural gas production in the world in 1985 was the equivalent of 10.6 billion barrels of oil. The smaller alkanes are also by-products of petroleum refining operations. For example, propane is the major constituent of liquefied petroleum gas (LPG), a domestic fuel used in mobile homes, among other places.

Petroleum itself is a complex mixture of hydrocarbons, mostly alkanes and cycloalkanes. It is the end result of the decomposition of animal and vegetable matter that has been buried in the earth's crust for long periods of time. The hydrocarbon mixture collects as a viscous black liquid in underground pockets, whence it is obtained by drilling wells. The resulting crude oil is refined by distillation into useful fuels and lubricants. Crude oil has a very broad boiling range. The more volatile constituents are propane, which is used as LPG, and butane, which is used as a chemical raw material. Light petroleum ether consists of pentanes and hexanes and boils at 30–60 °C. Ligroin is a mixture of heptanes and boils at 60–90 °C. These relatively volatile mixtures are often used as solvents, both in industry and in chemical laboratories. The most important petroleum distillates are gasoline and heating oils.

Fractional distillation of a typical crude oil yields the following fractions.

	Boiling Range, °C
natural gas (C_1 to C_4)	below 20
petroleum ether (C_5 to C_6)	30–60
ligroin or light naphtha (C_7)	60–90
straight-run gasoline (C_6 to C_{12})	85–200
kerosene (C_{12} to C_{15})	200–300
heating fuel oils (C_{15} to C_{18})	300–400
lubricating oil, greases, paraffin wax, asphalt (C_{16} to C_{24})	over 400

In 1985 total world production of crude oil was 21.5 billion barrels (bbl). About 45% of this total was converted into gasoline, 31% into heating oils, and the remainder into a variety of products, including kerosene, jet fuel, and petrochemical feedstocks.

One of the major problems facing mankind is energy. As consumption increases, the supply of nonrenewable fossil fuels obviously decreases. A large fraction of the world's crude oil reserve is under the control of a group of Middle Eastern and South American countries that have banded together and formed an organization known as the Organization of Petroleum Exporting Countries (OPEC). In 1974, as an aftermath of the armed conflict between Israel and several Arab countries, OPEC severely limited production and curtailed crude oil exports, in an action known as the "Arab oil embargo." Subsequently, prices of crude oil have increased dramatically. In the industrialized Western world, there has been a strong emphasis on conservation, with the result that demand for petroleum products leveled off in about 1978. At the same time, the dramatic increases in the price of crude oil resulted in intensified exploration. The proven reserves of crude oil and natural gas have actually increased dramatically since 1978, the total for the whole world amounting to 1800 billion bbl in 1985. However, it is clear that one day the known reserves will begin to decline and we must eventually turn to alternative sources of fuel. This need is further emphasized by the instability in world oil markets caused by Iraq's seizure of Kuwait in 1990, and the consequent unavailability of a significant portion of the world oil supply. Iraq was expelled from Kuwait by a U.S.-led coalition in early 1991 but not before setting afire half of the oil wells in that country and dumping sufficient oil to pollute seriously the Persian Gulf.

Another problem connected to the use of fossil fuels is the rate of carbon dioxide emission into the atmosphere. One model suggests that the current estimated rate of 6×10^9 tons of CO_2 will contribute to an appreciable warming of the earth by the "greenhouse effect" due to the absorption of infrared radiation by CO_2. International discussion of the idea of limiting CO_2 emissions has begun. A related problem is that of SO_2 emission (ca 50×10^6 tons per year) the cause of "acid rain" and a danger to the source of atmospheric oxygen, the plants. The major source of sulfur dioxide emission is the burning of fossil fuels. Simple chemical procedures such as adding $CaCO_3$ directly to the coal can diminish SO_2 emission but create the problem of disposing of $CaSO_4$.

Although other sources of energy will undoubtedly replace the fossil fuels as energy sources, there will still be a need for the fossil fuels as a source of **carbon.** At present, petroleum and coal hydrocarbons are the basic raw materials of much of the chemical industry. As the reserves become depleted, it is essential that we develop new sources of carbon raw materials to augment and eventually replace petroleum and natural gas.

An obvious source of additional petroleum is the vegetable matter from which it derives in the first place. However, the natural production of petroleum by the decomposition of vegetation requires eons of time. Some current research is directed toward developing ways to speed up this process, since vegetation may be grown relatively quickly and is therefore replaceable. Renewable ''biomass'' will be used in connection with more efficient techniques for conversion to fuel, food, and electricity.

Hydrocarbons also result from some inorganic reactions. Examples are the production of methane by the hydrolysis of beryllium carbide or aluminum carbide.

$$Be_2C + 4\ H_2O \longrightarrow CH_4 + 2\ Be(OH)_2$$

$$Al_4C_3 + 12\ H_2O \longrightarrow 3\ CH_4 + 4\ Al(OH)_3$$

Methane and ethane are odorless, but many of the higher hydrocarbons have distinctive odors. Typical functional groups are not needed for the interaction of odorous compounds (odorants or olfactants) with the proteins that function as olfactory receptors. These proteins in the olfactory nerve membrane combine with odorants and cause changes that send a signal to the brain giving information about the odorant. Hydrocarbons can effect the changes that initiate the nerve signals we sense as odor. Alkanes are among the compounds that can function as **pheromones,** chemicals used in nature as signals in communication among members of the same species. For example, 2-methylheptadecane and 17,21-dimethylheptatriacontane are the sex-attractants for the tiger moth and the tsetse fly, respectively. Control of insect populations might be achieved through interfering with their ability to receive or send chemical signals. Such a method of insect control would be environmentally and ecologically safe.

EXERCISE 5.11 Write a line structure for the tsetse fly sex-attractant. (Refer to Table 3.2.).

ASIDE

Pliny (Roman, 23–79) refers to crude oil as *naphtha,* a Greek word derived from the Babylonian *naptu,* to blaze. In Babylon, the tarry residues of crude oil (bitumen) found on the surface were used as a binder and water-proofing for bricks and clay. Little did the technicians of the ancient civilizations suspect the chemical riches contained within. In the hot desert climate of the present Middle East, many of the *n*-alkanes and cycloalkanes would have evaporated soon after emerging from the ground. Name those compounds and draw their structures. Which of the hydrocarbons in Tables 5.1–5.3 would be found in solid form?

PROBLEMS

1. Give the IUPAC name for each of the following compounds.

a.

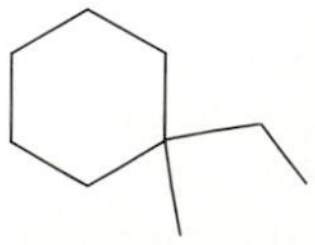

b.

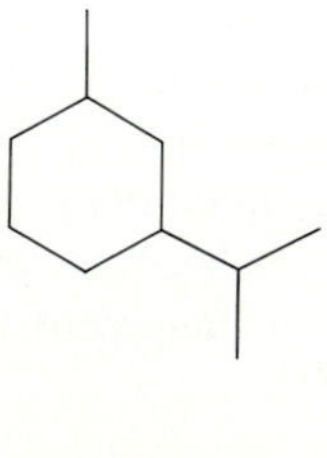

c.

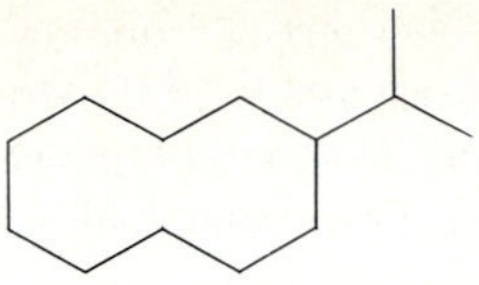

d.

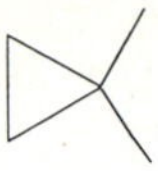

e.

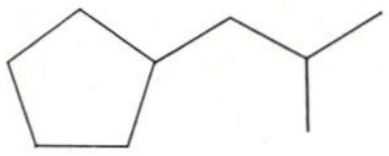

f.

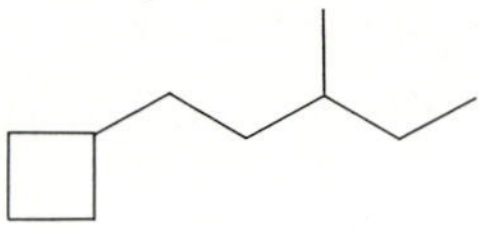

g. 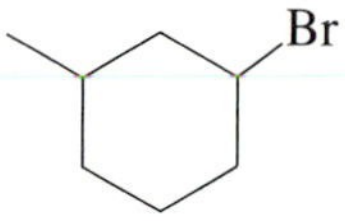 Br

h. 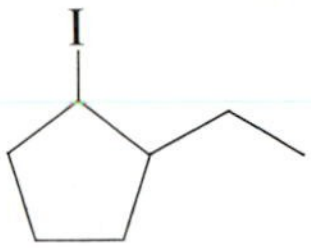 I

2. Using line structures and geometric figures, depict the following compounds.
   a. 3-ethyl-3-cyclopentylhexane
   b. 1,1-dichloro-3-ethylcyclohexane
   c. isopropylcyclohexane
   d. 3,3-dichloropentane
   e. cyclohexylcyclohexane
   f. 2-methyl-3-isopropylheptane

3. Draw and name all of the isomers of C_6H_{12} that contain a four-membered ring.

4. Write the structures of the nine possible heptane isomers and assign IUPAC names. The b.p. of heptane is 101 °C. By referring to the boiling points of the isomeric hexanes in Table 5.2, estimate the boiling points of the heptane isomers. Check your answers by looking up these compounds in the *Handbook of Chemistry and Physics* or *Lange's Handbook of Chemistry*. Not all of these hydrocarbons are listed in these handbooks. Browse through your library and see if you can find their properties in other reference works.

5. Write Newman projections showing the possible staggered conformations about the C-2—C-3 bond of pentane. Which ones correspond to the two stereo projections shown in Figure 5.11?

6. Examine a molecular model of adamantane. Give a rough estimate of its boiling point. Would you expect the melting point to be far below the boiling point? Look up its melting point and boiling point in a handbook.

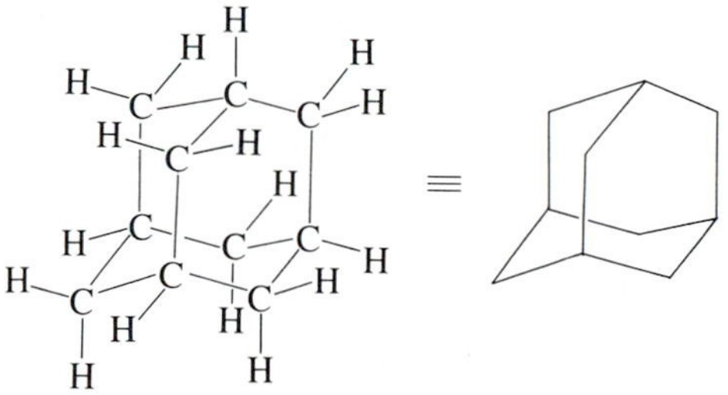

≡

7. With a set of molecular models find each of the four staggered conformations of pentane. Sketch each of these structures using dashed bonds and wedges as appropriate. Try to rank these conformations in order of increased energy (remember that a gauche conformation is less stable than an anti conformation).

8. For each of the following compounds, construct a potential energy diagram for rotation about the C-2—C-3 bond. For each unique energy maximum or minimum, illustrate the structure with a Newman projection.
   a. 2-methylbutane
   b. 2,2-dimethylbutane
   c. 2,2,3,3-tetramethylbutane

9. Draw a Newman projection for the C-1—C-2 bond of methylcyclohexane with the

methyl group in the equatorial position (see Figure 5.19). Note that the methyl group is anti with respect to C-3 of the ring. Now draw a Newman projection for the conformation having an axial methyl group. What is the relationship of the methyl group and the C-3 ring carbon? By analogy to the conformational analysis of butane, which conformation of methylcyclohexane do you think is more stable?

10. Using the heats of formation in Appendix I, calculate $\Delta H°$ for each of the following reactions.

(a) $CH_2{=}CH_2$ (ethylene) + H_2 $\longrightarrow$ CH_3CH_3
(b) $CH_2{=}CH_2$ + HCl $\longrightarrow$ CH_3CH_2Cl
(c) $CH_2{=}CH_2$ + H_2O $\longrightarrow$ CH_3CH_2OH (ethanol)

What do your calculations tell you about the equilibrium constant for each reaction? What do they tell you about the rates of the three reactions?

11. Write Newman projections and "sawhorse" structures for the three possible staggered conformations of 2,3-dimethylbutane. Note that two of these conformations are equivalent. The two different types of conformation differ in enthalpy by 0.9 kcal $mole^{-1}$. Which has the lower energy? Assume that $\Delta G° = \Delta H°$ and calculate the equilibrium composition at room temperature and compare your answer with that given for butane on page 7.

12. Ethylcyclohexane and cyclooctane are isomers. Using the heats of formation of the two cycloalkanes (Appendix I) and assuming $\Delta H° = \Delta G°$, calculate the equilibrium constant for the equilibrium

$$\text{cyclooctane} \underset{}{\overset{K}{\rightleftharpoons}} \text{ethylcyclohexane}$$

cyclooctane ethylcyclohexane

The actual equilibrium constant at 25 °C is 6.7×10^8. Explain.

13. 2-Methylpropane is thermodynamically more stable than butane. Which has the lower boiling point? Is there any relationship between thermodynamic stability and boiling point? Would you expect such a relationship between thermodynamic stability and melting point?

14. We found in Section 5.5 that at 25 °C the equilibrium butane $\rightleftharpoons$ 2-methylpropane has $\Delta H° = -2.05$ kcal $mole^{-1}$; however, $\Delta G°$ is only -0.89 kcal $mole^{-1}$. Calculate the entropy change for the reaction and explain the direction of the effect. Calculate the equilibrium constant at 25 °C.

15. The definition of free energy of formation, $\Delta G_f°$, is analogous to that of the enthalpy of formation, $\Delta H_f°$. Some values of $\Delta G_f°$ (25 °C, gas) are pentane, -2.00; 2-methylbutane, 3.54; 2,2-dimethylpropane, -3.64 kcal $mole^{-1}$. Calculate the composition of the equilibrium mixture of these three isomers at 25 °C.

16. Careful inspection of the heats of formation in Table 5.4 will show regular increments per CH_2 group in a homologous series. In fact, $\Delta H_f°$ increments can be associated with the groups

CH_3, $>CH_2$, $-\overset{|}{\underset{|}{C}}H$ and $-\overset{|}{\underset{|}{C}}-$

Determine average values for these groups from Table 5.4, or, with a small computer, calculate the values that give the best least squares fit to the experimental data. Use your results to estimate ΔH_f° for hexane, 2-methylpentane, 3-methylpentane, 2,2-dimethylbutane, and 2,3-dimethylbutane. Compare with the experimental values in Appendix I. You can see how far your "group equivalents" will go by comparing your calculated value with the experimental ΔH_f° for nonane of 54.7 kcal mole^{-1}. Your value should agree to within several tenths of a kilocalorie per mole. However, compare your calculated value for 2,2,4,4-tetramethylpentane with the experimental ΔH_f° of -57.8 kcal mole^{-1}. Why is there a discrepancy? (*Hint*: Look at a molecular model. Will steric interferences or strain increase or decrease ΔH_f°?)

CHAPTER 6

REACTIONS OF ALKANES

6.1 Bond-Dissociation Energies

Heat is kinetic energy. When a substance is heated, this kinetic energy increases the motion of atoms and molecules. When a molecule is heated, much of the added energy goes into translational motion, and the molecules move about faster. However, some of the energy absorbed appears as increased rotation of the molecule around given axes and as vibration of the atoms within the molecule. Vibrational energy levels are quantized, like the electronic energy levels of atoms and molecules. Figure 6.1 is a schematic diagram of the potential energy of a diatomic molecule as a function of the bond distance. The diagram shows how the vibrational quantum levels change with bond stretching. At room temperature, only the lowest quantum state is significantly populated.

Remember that even at absolute zero the atoms are still vibrating. If the atoms were at rest, we would know both their position and momentum exactly—in violation of the Heisenberg uncertainty principle. This lowest vibrational quantum state has an energy ϵ_0 above the potential minimum. The quantity ϵ_0 is called the zero-point energy of the vibration.

As heat is applied, higher vibrational states are increasingly populated. The higher the vibrational quantum state, the greater the bond distance. When sufficient energy D, as shown in Figure 6.1, has been absorbed, the bond breaks. The distance between the nuclei increases without restraint and two separated atoms result.

$$\text{A—B} \longrightarrow \text{A}\cdot + \cdot\text{B}$$

For polyatomic molecules diagrams such as Figure 6.1 are more complicated because they are multidimensional. Instead of the energy quantity D we refer instead to the enthalpy of a bond-dissociation reaction, $\Delta H°$. The enthalpy value for a bond-dissociation reaction is generally called the **bond-dissociation energy** and is given the special symbol $DH°$.

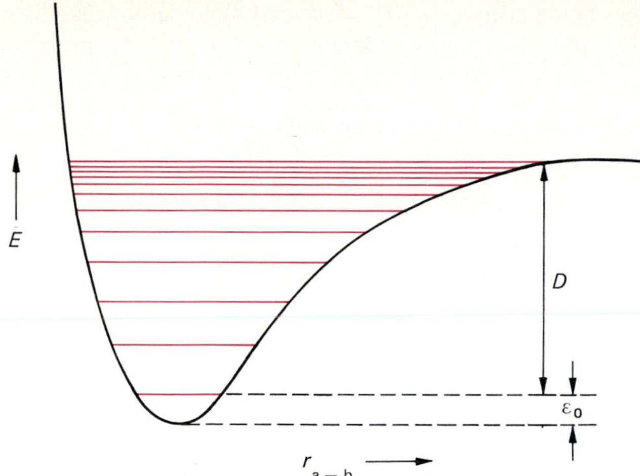

FIGURE 6.1 Schematic diagram of the potential function for a diatomic molecule. Horizontal lines represent the various vibrational energy levels.

When methane is heated to a high temperature dissociation occurs to give a hydrogen atom and methyl radical.

$$H{-}CH_3 \longrightarrow H\cdot + \cdot CH_3$$

For this reaction $DH°$ is 105 kcal mole^{-1}.

Bond-dissociation energies are often measured at rather high temperatures. Because enthalpies generally vary somewhat with temperature, they are usually extrapolated back to room temperature for convenience. That is, $DH°$ values refer to 25 °C. Bond-dissociation reactions are difficult to measure accurately and $DH°$ values are rarely known more accurately than about ±1 kcal mole^{-1}.

A $DH°$ of 105 kcal mole^{-1} represents a rather strong bond. Temperatures of the order of 1000 °C are required for dissociation of methane to occur at an appreciable rate. A carbon-hydrogen bond in ethane has a slightly lower $DH°$ (101 kcal mole^{-1}), but $DH°$ for the carbon-carbon bond is only 90 kcal mole^{-1}. Consequently, when ethane is heated, C—C fission occurs more rapidly than does C—H fission.

$$H_3C{-}CH_3 \longrightarrow 2\ CH_3\cdot$$

This reaction occurs at about 700 °C.

In general, pyrolysis of a compound results in fission of the weakest bond. The products are **free radicals.**

Free radicals contain an odd number of electrons. The Lewis structure for methyl radical is

$$\begin{array}{c} H \\ H\!:\!\overset{..}{\underset{..}{C}}\cdot \\ H \end{array}$$

Alkyl radicals can exist only at low concentrations at ordinary temperatures. Nevertheless, many such radicals have been "seen" by various spectroscopic methods. For example, methyl radical has been shown by spectroscopic measurements to be essentially flat—all four atoms lie in the same plane. Free radicals are important intermediates in many organic reactions, and we shall encounter them in this context later in this chapter.

The bond-dissociation energies, $DH°$, of several hydrocarbons are listed in Table 6.1. A more extensive table for a wider variety of compounds is given in Appendix II.

TABLE 6.1 Bond-Dissociation Energies for Some Alkanes[a]

Compound	$DH°$, kcal mole^{-1}	Compound	$DH°$, kcal mole^{-1}
CH_3—H	105	CH_3—CH_3	90
C_2H_5—H	101	C_2H_5—CH_3	89
$CH_3CH_2CH_2$—H	101	C_3H_7—CH_3	90
$(CH_3)_2CHCH_2$—H	101	C_2H_5—C_2H_5	88
$(CH_3)_2CH$—H	98	$(CH_3)_2CH$—CH_3	89
$(CH_3)_3C$—H	96	$(CH_3)_3C$—CH_3	87

[a]The bond dissociated is shown as a bond.

Note that $DH°$ depends on the character of the radical products. The $DH°$ for dissociation of a terminal C—H of an alkane is always about 101 kcal mole^{-1}. The product of this bond cleavage is called a **primary** alkyl radical. It is a species with **one** R substituent attached to the radical carbon, with the structure $RCH_2\cdot$, where R is any alkyl group. A methyl radical is a superprimary radical.

$$CH_3CH_3 \longrightarrow CH_3CH_2\cdot + H\cdot \qquad DH° = 101 \text{ kcal mole}^{-1}$$

When an interior carbon-hydrogen bond of a linear alkane is broken, the product $R_2CH\cdot$ is a **secondary** alkyl radical with **two** R substituents. Such bonds have a lower $DH°$ of 98 kcal mole^{-1}.

$$CH_3CH_2CH_3 \longrightarrow CH_3\dot{C}HCH_3 + H\cdot \qquad DH° = 98 \text{ kcal mole}^{-1}$$

A carbon-hydrogen bond at a branch point is the weakest type of carbon-hydrogen bond. Such bonds have $DH°$ of about 96 kcal mole^{-1}. The product is a **tertiary** alkyl radical, $R_3C\cdot$ with **three** R substituents.

$$CH_3\overset{\overset{\displaystyle CH_3}{|}}{C}HCH_3 \longrightarrow CH_3\underset{\bullet}{\overset{\overset{\displaystyle CH_3}{|}}{C}}CH_3 + H\cdot \qquad DH° = 96 \text{ kcal mole}^{-1}$$

The relative stability of alkyl radicals depends on the number of alkyl groups attached to the radical carbon; alkyl radicals have the order of stability

$$\text{tertiary} > \text{secondary} > \text{primary} > \text{methyl}$$

The same principle applies to carbon-carbon bonds. The strength of this bond also depends on the relative stabilities of the radical products.

$$CH_3CH_3 \longrightarrow 2\,CH_3\cdot \qquad DH° = 90 \text{ kcal mole}^{-1}$$

$$CH_3\underset{\underset{\displaystyle CH_3}{|}}{\overset{\overset{\displaystyle CH_3}{|}}{C}}CH_3 \longrightarrow CH_3\underset{\underset{\displaystyle CH_3}{|}}{\overset{\overset{\displaystyle CH_3}{|}}{C}}\cdot + CH_3\cdot \qquad DH° = 87 \text{ kcal mole}^{-1}$$

Carbon-carbon bond-dissociation energies are also tabulated in Table 6.1.

Consider fission of the two types of carbon-hydrogen bonds in 2-methylpropane (Table 6.1). In order to break one of the terminal carbon-hydrogen bonds, 101 kcal mole^{-1} of energy must be absorbed. In order to break the carbon-hydrogen bond at the branch point, only 96 kcal mole^{-1} is required.

$$\begin{array}{l} CH_3CH(CH_3)CH_3 \longrightarrow CH_3CH(CH_3)CH_2\cdot + H\cdot \qquad DH° = 101 \text{ kcal mole}^{-1} \\ CH_3CH(CH_3)CH_3 \longrightarrow CH_3C(CH_3)_2\cdot + H\cdot \qquad DH° = 96 \text{ kcal mole}^{-1} \end{array}$$

The starting point is the same for the two reactions; one of the products (H $\cdot$) is the same in each reaction. Thus, the difference in DH°s for these reactions is a direct measure of the difference in stability of the two alkyl radicals. *The t-butyl radical is more stable than the isobutyl radical by 5 kcal mole^{-1}* (Figure 6.2).

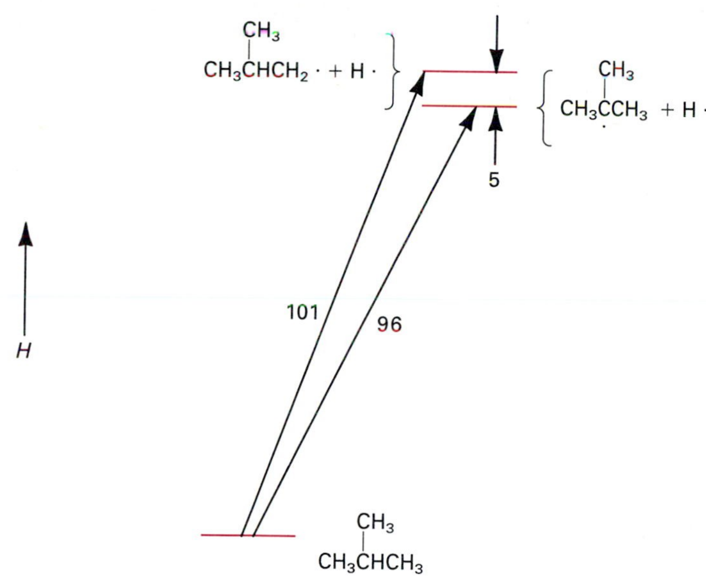

FIGURE 6.2 The DH°s for the two carbon-hydrogen bonds in 2-methylpropane.

These results can also be expressed in terms of heats of formation. The heat of formation of a radical is derived from the DH° of the reactants and the DH° of the bond-breaking reaction; for example,

$$H_2 \longrightarrow 2H\cdot \qquad \Delta H° = DH° = 104 \text{ kcal mole}^{-1}$$

$$\Delta H_f°(H\cdot) = [DH° + H_f(H_2)] \div 2 = 52 \text{ kcal mole}^{-1}$$

$$\Delta H_f°[(CH_3)_2CHCH_2\cdot] = \Delta H_f°[(CH_3)_3CH] + DH°[(CH_3)_2CHCH_2{-}H] - \Delta H_f°(H\cdot)$$

$$= -32.4 + 101 - 52 = 17 \text{ kcal mole}^{-1}$$

Heats of formation of some radicals are summarized in Table 6.2. A more complete list is given in Appendix I.

TABLE 6.2 Heats of Formation of Some Radicals

Compound	ΔH_f°, kcal mole^{-1} at 25 °C	Compound	ΔH_f°, kcal mole^{-1} at 25 °C
$H\cdot$	52	$(CH_3)_3C\cdot$	12
$CH_3\cdot$	35	$CH_3CH_2CH_2CH_2\cdot$	19
$C_2H_5\cdot$	29	$(CH_3)_2CHCH_2\cdot$	17
$CH_3CH_2CH_2\cdot$	24	$CH_3CH_2\dot{C}HCH_3$	15
$(CH_3)_2CH\cdot$	21		

EXERCISE 6.1 For each of the reactions discussed in this section, calculate the bond-dissociation energies using the DH° values in Table 6.2 and Appendix I.

Highlight 6.1

Bond dissociation energies of C—C and C—H bonds vary with the number of alkyl substituents attached to carbon; the greater the number of alkyl groups, the lower the dissociation energy.

6.2 Pyrolysis of Alkanes: Cracking

The process of breaking up a molecule by heat is called **pyrolysis** (Gk., *pyros*, fire; *lysis*, a loosening). When alkanes are pyrolyzed, the carbon-carbon bonds cleave to produce smaller alkyl radicals. With higher alkanes, the cleavage occurs randomly along the chain.

$$CH_3CH_2CH_2CH_2CH_3 \longrightarrow CH_3\cdot + CH_3CH_2CH_2CH_2\cdot$$

$$CH_3CH_2CH_2CH_2CH_3 \longrightarrow CH_3CH_2\cdot + CH_3CH_2CH_2\cdot$$

One possible reaction that these radicals may undergo is recombination to form an alkane. A mixture of different alkanes is produced.

$$CH_3\cdot + CH_3\cdot \longrightarrow CH_3CH_3$$
$$CH_3\cdot + CH_3CH_2\cdot \longrightarrow CH_3CH_2CH_3$$
$$CH_3CH_2\cdot + CH_3CH_2CH_2CH_2\cdot \longrightarrow CH_3CH_2CH_2CH_2CH_2CH_3$$

Another reaction that occurs is disproportionation. In this process one radical transfers a hydrogen atom to another radical to produce an alkane plus an alkene.

$$CH_3CH_2\cdot + CH_3CH_2CH_2\cdot \longrightarrow CH_3CH_3 + CH_3CH{=}CH_2$$

The net result of pyrolysis is the conversion of a large alkane to a mixture of smaller alkanes and alkenes. This reaction is not a useful one in the organic laboratory, where the aim is generally to produce a single pure compound in high yield. However, thermal cracking of hydrocarbons has been an important industrial process. As it comes from the ground, crude oil varies widely in composition depending on its source. For example, fractional distillation of a typical "light oil" affords 35% gasoline, 15% kerosene, and only a trace of asphalt, the balance being mainly high-boiling heating and lubricating oils. On the other hand, a typical "heavy oil" affords only 10% gasoline, 10% kerosene, and 50% asphalt. In order to decrease the average molecular weight of heavy oils and increase the production of the desirable more volatile fractions, "cracking" techniques such as thermal cracking have been used. Modern cracking methods employ various catalysts, mainly alumina and silica, that catalyze degradation of the large hydrocarbons into smaller ones at temperatures lower than those used in thermal cracking. Catalytic cracking probably involves cationic rather than free-radical intermediates.

EXERCISE 6.2 2-Methylbutane has three different carbon-carbon bonds that can break in the initial step of a thermal cracking process. Give the disproportionation products expected from the radicals produced in each case. From a consideration of bond-dissociation energies determine which products predominate.

6.3 Halogenation of Alkanes

When a mixture of methane and chlorine is heated to about 120 °C or irradiated with light of a suitable wavelength, a highly exothermic reaction occurs.

$$CH_4 + Cl_2 \xrightarrow[\text{or light}]{\text{heat}} \underset{\text{methyl chloride}}{CH_3Cl} + HCl \qquad \Delta H° = -24.7 \text{ kcal mole}^{-1}$$

This reaction is a significant industrial process for preparing methyl chloride. It has limited usefulness as a laboratory preparation because the reaction does not stop with the introduction of a single chlorine. As the concentration of methyl chloride builds up, it undergoes chlorination in competition with methane.

$$CH_3Cl + Cl_2 \longrightarrow \underset{\text{methylene chloride}}{CH_2Cl_2} + HCl$$

$$CH_2Cl_2 + Cl_2 \longrightarrow \underset{\text{chloroform}}{CHCl_3} + HCl$$

$$CHCl_3 + Cl_2 \longrightarrow \underset{\text{carbon tetrachloride}}{CCl_4} + HCl$$

The actual product of the reaction of methane and chlorine is a mixture of methyl chloride (b.p. −24.2 °C), methylene chloride (CH_2Cl_2, b.p. 40.2 °C), chloroform ($CHCl_3$, b.p. 61.2 °C), and carbon tetrachloride (CCl_4, b.p. 76.8 °C). The composition of the mixture depends on the relative amounts of starting materials used and the reaction conditions. In this case it is easy to separate the products by fractional distillation because of the difference in boiling points.

A good deal of experimental evidence is in accord with the following mechanism for the chlorination of methane. The reaction begins with the **homolysis** of a chlorine molecule to two chlorine atoms (equation 6-1).

$$Cl_2 \xrightarrow[\text{or } h\nu]{\Delta} 2\,Cl\cdot$$

When a bond breaks in such a way that each fragment retains one electron of the bond, the process is called **homolytic cleavage** or **homolysis.** The bond dissociation energy usually refers to homolysis of the bond.

$$A{:}B \longrightarrow A\cdot + B\cdot$$

When one fragment retains both electrons, the process is termed **heterolytic cleavage** or **heterolysis.**

$$A{:}B \longrightarrow A^+ + B{:}^-$$

Since molecular chlorine has a rather low bond-dissociation energy ($DH^\circ = 58$ kcal mole^{-1}), chlorine atoms may be produced by light of relatively long wavelength (a *photochemical* process) or by heating to moderate temperatures (equation 6-1). Once chlorine atoms are present, a set of reactions called a **chain reaction** commences. A chlorine atom reacts with a methane molecule to give a methyl radical and HCl (equation 6-2). The methyl radical then reacts with a chlorine molecule to give methyl chloride and a chlorine atom (equation 6-3).

$$Cl\cdot + CH_4 \rightleftharpoons CH_3\cdot + HCl \qquad \Delta H^\circ = +2 \text{ kcal mole}^{-1} \qquad (6\text{-}2)$$

$$CH_3\cdot + Cl_2 \longrightarrow CH_3Cl + Cl\cdot \qquad \Delta H^\circ = -26.7 \text{ kcal mole}^{-1} \qquad (6\text{-}3)$$

The chlorine atom produced in equation (6-3) can react with another methane molecule to continue the chain. Reaction (6-1) is called the **initiation** step, and reactions (6-2) and (6-3) are called the **propagation** steps.

In principle, only one chlorine molecule need homolyze in order to convert many moles of methane and chlorine to methyl chloride and HCl. In practice, the chain process only goes through, on the average, about 10,000 cycles before it is terminated. Termination occurs whenever two radicals happen to collide, for example, equations (6-4) and (6-5).

$$CH_3\cdot + Cl\cdot \longrightarrow CH_3Cl \qquad (6\text{-}4)$$

$$CH_3\cdot + CH_3\cdot \longrightarrow CH_3CH_3 \qquad (6\text{-}5)$$

Another possible termination step involves the collision of two chlorine atoms—reverse of the initiation step (equation 6-1). However, when two chlorine atoms collide to form Cl_2, the resulting molecule has as vibrational energy all of the kinetic energy of translation of the two atoms. This energy is always in excess of the bond energy, and the two atoms simply separate again. Only if collision occurs in the presence of a third body or on the wall of the reaction vessel to remove some of this energy does the chlorine molecule formed stay intact.

$$Cl\cdot + Cl\cdot + M \longrightarrow Cl_2 + M$$

This fundamental principle holds for all radical recombinations; a third body is required to remove the excess energy.

Other reactions that may (and probably do) occur are unproductive and do not terminate the chain reaction.

$$CH_4 + CH_3\cdot \longrightarrow CH_3\cdot + CH_4$$

$$Cl_2 + Cl\cdot \longrightarrow Cl\cdot + Cl_2$$

Let us look at each of the foregoing propagation steps in some detail. Reaction (6-2) is slightly endothermic and reversible, but it has a low activation energy of only about 4 kcal mole^{-1}. The reaction, which begins with an attack by Cl· on hydrogen, may be considered in further detail.

$$\text{Cl}\cdot + \text{H—CH}_3 \longrightarrow [\text{Cl}\cdots\text{H}\cdots\text{CH}_3]^{\ddagger} \longrightarrow \text{Cl—H} + \cdot\text{CH}_3$$

The H—Cl and H—CH_3 bonds have similar strength [the $DH°$s are 103 and 105 kcal mole^{-1}, respectively (Appendix II)]. As chlorine and hydrogen come closer, a chlorine-hydrogen bond begins to form. As the Cl—H bond becomes stronger, the hydrogen-carbon bond becomes weaker and breaks. The product methyl radical appears to be planar (Figure 6.3). Methyl radical can be described to a good approximation in terms of three C_{sp^2}—H_{1s} bonds with the odd electron contained in the remaining C_{2p}-orbital. At the transition state the methyl group has started to flatten out from its original tetrahedral structure.

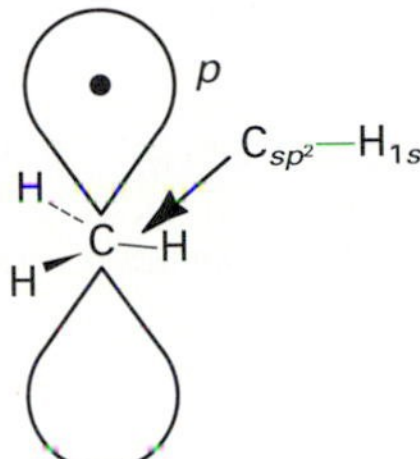

FIGURE 6.3 Methyl radical.

For the reverse process, HCl + CH_3· → CH_4 + Cl·, the same mechanism applies in reverse. The carbon of the methyl radical attacks the hydrogen of HCl from the side opposite to the hydrogen-chlorine bond and a carbon-hydrogen bond begins to form. As the forming carbon-hydrogen bond distance decreases and the bond strength increases, the remaining carbon-hydrogen bonds begin to bend back toward their tetrahedral geometry in CH_4. At the same time, the hydrogen-chlorine bond distance increases.

$$\text{H}_3\text{C—H} \quad \cdot\text{Cl} \rightleftharpoons \left[\text{H}_3\text{C}\cdots\text{H}\cdots\text{Cl}\right]^{\ddagger} \rightleftharpoons \cdot\text{CH}_3 \quad \text{H—Cl}$$

A stereo representation of the transition state is shown in Figure 6.4.

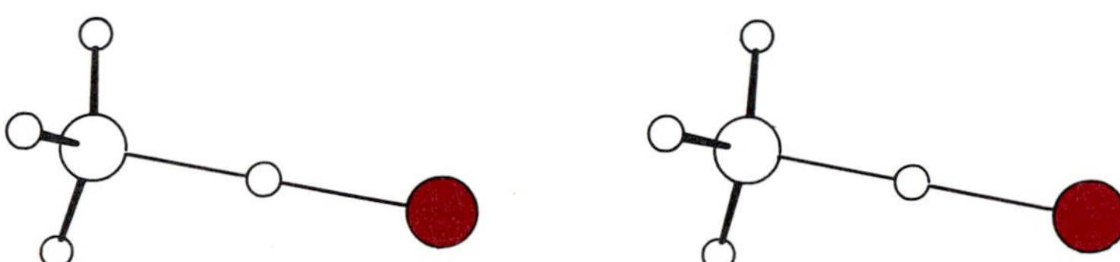

FIGURE 6.4 Stereo representation of the transition state for the reaction CH_4 + Cl· ⇌ CH_3· + HCl.

> The structure of the transition state is the same for both directions by the **principle of microscopic reversibility.** That is, the reverse reaction from products to reactants must have the same reaction mechanism as the forward reaction. If it did not, we could, in principle, set up a perpetual motion machine in violation of the second law of thermodynamics.

An equivalent description may be given in orbital terms. As the chlorine orbital containing one electron overlaps with the hydrogen 1*s*-orbital, electron repulsion causes a decrease in the overlap of the H_{1s}-orbital with the C_{sp^3}-orbital, and the carbon-hydrogen bond begins to lengthen and become weaker. As this carbon-hydrogen bond gets weaker, it has less demand for *s*-orbital character, and the carbon *s*-orbital is used more for bonding to the other carbon-hydrogen bonds. Rehybridization occurs progressively from sp^3 toward sp^2. The carbon begins to flatten out, and the remaining carbon-hydrogen bonds become somewhat shorter and stronger. The structure of the transition state is depicted in terms of component atomic orbitals in Figure 6.5. A reaction coordinate diagram for reaction (6-2) is shown in Figure 6.6.

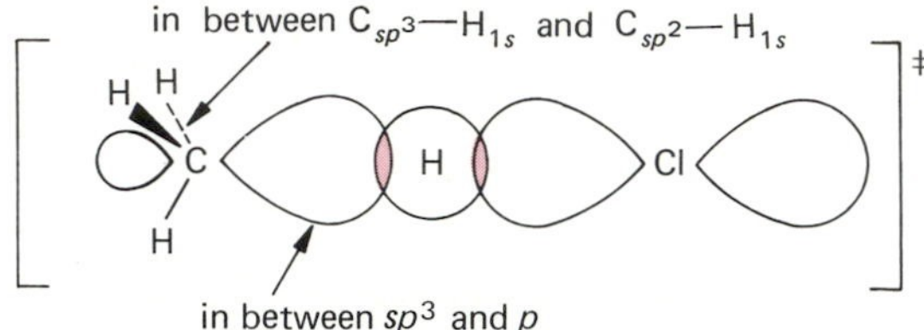

FIGURE 6.5 Orbital description of the transition state for the reaction $CH_4 + Cl\cdot \rightleftharpoons CH_3\cdot + HCl$.

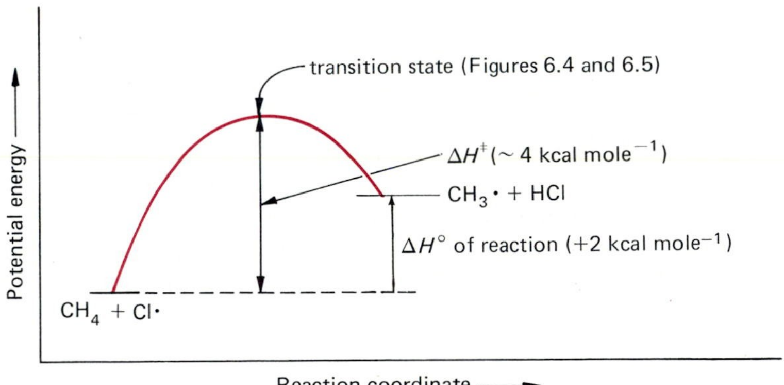

FIGURE 6.6 Reaction profile for the reaction $CH_4 + Cl\cdot \rightleftharpoons CH_3\cdot + HCl$.

Reaction (6-3) has a small activation energy of about 2 kcal mole^{-1}. This reaction is rapid and highly exothermic. The reverse, reaction of Cl· with methyl chloride to form chlorine and methyl radical, is highly endothermic and has a correspondingly high activation energy of $26.7 + 2 \cong 29$ kcal mole^{-1}. Consequently, the overall forward reaction is effectively irreversible. A reaction coordinate diagram for this step is shown in Figure 6.7. The transition state for this reaction is one in which the carbon-chlorine bond is partly formed and the chlorine-chlorine bond is partly broken.

$$[CH_3\cdot\cdots\cdots Cl\cdots Cl]^{\ddagger}$$

A stereo representation of this transition state is given in Figure 6.8.

The overall $\Delta H°$ of the net chlorination reaction may be obtained by summing equations (6-2) and (6-3).

		$\Delta H°$, *kcal mole*$^{-1}$
(6-2):	$CH_4 + \cancel{Cl}\cdot \longrightarrow \cancel{CH_3}\cdot + HCl$	+2
(6-3):	$\cancel{CH_3}\cdot + Cl_2 \longrightarrow CH_3Cl + \cancel{Cl}\cdot$	−26.7
	$CH_4 + Cl_2 \longrightarrow CH_3Cl + HCl$	−24.7

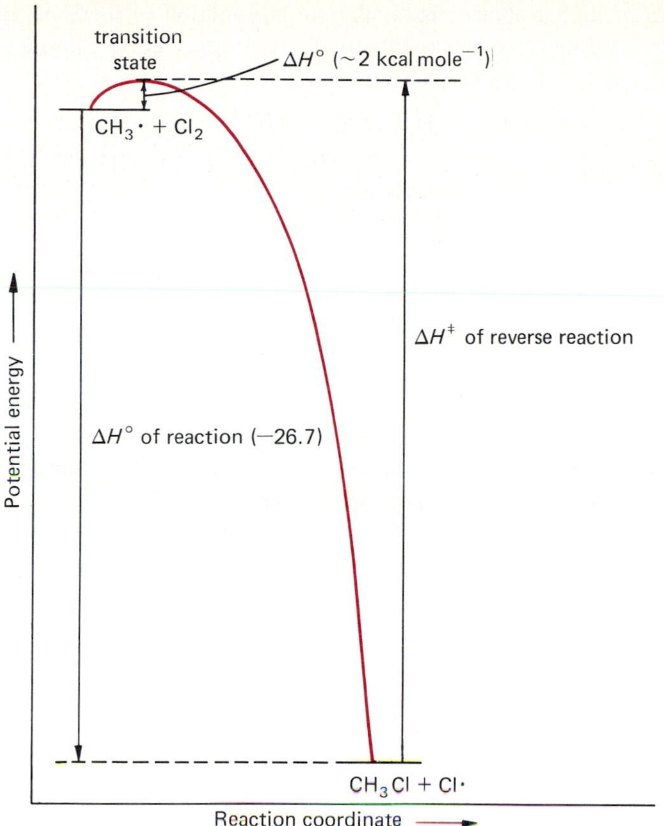

FIGURE 6.7 Reaction profile for the reaction $CH_3\cdot + Cl_2 \rightleftharpoons CH_3Cl + Cl\cdot$.

FIGURE 6.8 Stereo representation of the transition state for the reaction $CH_3\cdot + Cl_2 \rightleftharpoons CH_3Cl + Cl\cdot$.

Note that $\Delta H°$ for the initiation step is *not* added to the $\Delta H°$ values for the propagation steps in deriving $\Delta H°$ for the overall reaction. If one does this, one is actually calculating $\Delta H°$ for another reaction.

	$\Delta H°$, *kcal mole*$^{-1}$
$Cl_2 \longrightarrow 2\,Cl\cdot$	+58
$CH_4 + \cancel{Cl\cdot} \longrightarrow \cancel{CH_3\cdot} + HCl$	+ 2
$\cancel{CH_3\cdot} + Cl_2 \longrightarrow CH_3Cl + \cancel{Cl\cdot}$	−26.7
$Cl_2 + CH_4 + Cl_2 \longrightarrow CH_3Cl + HCl + 2\,Cl\cdot$	+32.3

This equation is just the sum of the overall chlorination reaction and the chlorine homolysis.

This is often a point of confusion because the student reasons that heat had to be put in to initiate the reaction. However, the question is not how much heat is applied, but what is $\Delta H°$, the *heat of the reaction*?

Chlorination of higher alkanes is similar to chlorination of methane except that the product mixtures are more complex. Ethane gives not only ethyl chloride, but also 1,1-dichloroethane and 1,2-dichloroethane.

$$CH_3CH_3 + Cl_2 \longrightarrow CH_3CH_2Cl + HCl$$

$$CH_3CH_2Cl + Cl_2 \longrightarrow \underset{\text{1,1-dichloroethane}}{CH_3CHCl_2} + \underset{\text{1,2-dichloroethane}}{ClCH_2CH_2Cl} + HCl$$

EXERCISE 6.3 Write equations showing the initiation, propagation, and termination steps for the monochlorination of ethane. Compute the $\Delta H°$ for each step and the overall $\Delta H°$ of the reaction. Write equations for the chain propagation steps involved in the formation of the two dichloroethanes.

With propane, two monochloro products may be formed. Both *n*-propyl chloride and isopropyl chloride are formed, but not in equal amounts.

$$CH_3CH_2CH_3 + Cl_2 \longrightarrow \underset{\substack{(43\%) \\ n\text{-propyl chloride}}}{CH_3CH_2CH_2Cl} + \underset{\substack{(57\%) \\ \text{isopropyl chloride}}}{CH_3CHClCH_3}$$

In carbon tetrachloride solution at 25 °C, the two isomers are produced in the relative amounts 43:57. Further reaction gives a mixture of the four possible dichloropropanes.

Let us examine the monochlorination of propane in greater detail. Recall that $DH°$ for the secondary hydrogen in propane is about 3 kcal mole^{-1} lower than $DH°$ for the primary hydrogen (Table 6.1). We might anticipate, then, that the secondary hydrogen would be removed by a chlorine atom more easily than a primary hydrogen. However, there are six primary hydrogens that may be replaced, whereas there are only two secondary hydrogens. The **relative reactivity** per hydrogen is then

$$\frac{\text{secondary}}{\text{primary}} = \frac{57/2}{43/6} = \frac{4}{1}$$

A similar trend is noticed in the monochlorination of 2-methylpropane, which gives 36% *t*-butyl chloride and 64% isobutyl chloride.

$$(CH_3)_3CH + Cl_2 \longrightarrow \underset{\substack{(36\%) \\ t\text{-butyl chloride}}}{(CH_3)_3CCl} + \underset{\substack{(64\%) \\ \text{isobutyl chloride}}}{(CH_3)_2CHCH_2Cl}$$

The relative reactivity of tertiary and primary hydrogens on a per-hydrogen basis is

$$\frac{\text{tertiary}}{\text{primary}} = \frac{36/1}{64/9} = \frac{5.1}{1}$$

Thus, the relative rates of reaction of different hydrogens with $Cl\cdot$ are just as we expect on the basis of $DH°$ for the various hydrogens

$$\text{tertiary} > \text{secondary} > \text{primary}$$

However, the degree of preference is relatively low. That is, there is less difference between the activation energies for the various reactions than there is between the heats of reaction (Figure 6.9).

For example, in chlorination of propane, the $Cl\cdot$ can abstract a hydrogen from the methyl group or from the methylene group. In the former case $\Delta H°$ is -2 kcal mole^{-1}, and in the latter it is -5 kcal mole^{-1}. Thus, the difference in heats of reaction $\Delta H° =$

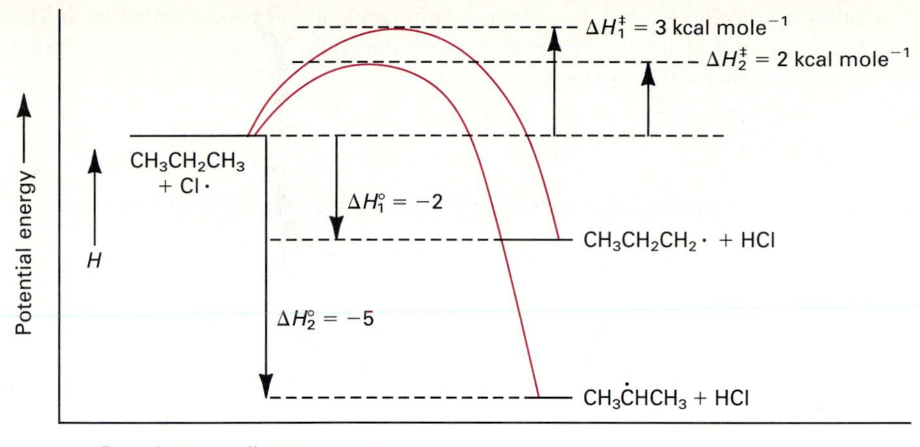

FIGURE 6.9 Reaction profiles for the reaction of Cl· with C_3H_8.

3 kcal mole^{-1}. However, $\Delta H^\ddagger$ for abstraction of a CH_2 hydrogen is 2 kcal mole^{-1}, whereas $\Delta H^\ddagger$ for abstraction of a hydrogen from a CH_3 group is 3 kcal mole^{-1}; the difference in activation energies $\Delta\Delta H^\ddagger$ = 1 kcal mole^{-1}. This result seems reasonable when one realizes that in the transition state the alkyl free radical is not yet fully formed. The greater stability of a secondary free radical than a primary free radical will be reflected in the two transition states. However, that effect will be muted in the transition state to the extent that carbon has not achieved complete free radical character.

With more complicated alkanes, chlorination mixtures are hopelessly complex. Hence, chlorination of alkanes is not a good general reaction for preparing alkyl chlorides. There is only one type of compound for which chlorination has practical utility in laboratory preparations. When all hydrogens are equivalent, there is only one possible monochloro product. In such cases the desired product can generally be separated from hydrocarbon and di- and higher chlorinated species by fractional distillation. Two examples are the chlorination of cyclohexane and 2,2-dimethylpropane.

$$C_6H_{12} + Cl_2 \longrightarrow C_6H_{11}Cl + HCl$$

cyclohexyl chloride

$$CH_3-C(CH_3)_2-CH_3 + Cl_2 \longrightarrow CH_3-C(CH_3)_2-CH_2Cl + HCl$$

neopentyl chloride

The handling of gaseous chlorine in the laboratory is frequently inconvenient and such chlorinations are often done with sulfuryl chloride, SO_2Cl_2, instead.

> Sulfuryl chloride is a colorless liquid, b.p. 69 °C, produced by reaction of Cl_2 and SO_2. It fumes in moist air because it reacts rapidly with water according to the equation
>
> $$SO_2Cl_2 + 2\,H_2O \longrightarrow 2\,HCl + H_2SO_4$$

When sulfuryl chloride is used as a chlorinating agent, a compound called an **initiator** must be used to provide the free radicals that start the chain reaction. Peroxides are often used for this purpose because the oxygen-oxygen bond is weak and readily broken at relatively low temperatures (see Appendix II).

$$\text{ROOR} \longrightarrow 2\ \text{RO}\cdot$$

The chlorination of cyclohexane by sulfuryl chloride provides a typical example.

$$\text{C}_6\text{H}_{12} + SO_2Cl_2 \longrightarrow \text{C}_6\text{H}_{11}\text{Cl} + HCl + SO_2$$

A mixture of 1.8 mole of cyclohexane, 0.6 mole of sulfuryl chloride, and 0.001 mole of benzoyl peroxide, $(C_6H_5COO)_2$, is refluxed for 1.5 hr. Fractional distillation gives 89% of chlorocyclohexane, b.p. 143 °C, and 11% of a mixture of dichlorocyclohexanes.

EXERCISE 6.4 Write the structures expected from the monochlorination of 2-methylbutane. Using relative reactivities of 1:4:5 for replacement of primary, secondary, and tertiary hydrogens, determine the percent of each of the monochloro compounds expected in the product mixture.

EXERCISE 6.5 Each of the following hydrocarbons contains no double or triple bonds and reacts with chlorine to give a single monochloride. Deduce the structure of each hydrocarbon and its chloride.

(a) C_8H_{18} (b) C_8H_{16} (c) C_8H_8

Although the mechanism for chlorination might apply to other halogenations, the actual reactions show important differences. The overall enthalpies of halogenation of methane by various halogens are summarized in Table 6.3.

TABLE 6.3
$CH_4 + X_2 \rightleftharpoons CH_3X + HX$

X	ΔH°, kcal mole^{-1}
F	-102.8
Cl	-24.7
Br	-7.3
I	$+12.7$

The reaction with fluorine is so highly exothermic that controlled fluorination is difficult to accomplish. The energy liberated is sufficient to break most bonds. The hydrogen-fluorine bond is so strong ($DH^\circ = 136$ kcal mole^{-1}) that the following reaction is endothermic by only 7 kcal mole^{-1}.

$$CH_4 + F_2 \longrightarrow CH_3\cdot + F\cdot + HF \qquad \Delta H^\circ = +7 \text{ kcal mole}^{-1}$$

Consequently, when methane and fluorine are mixed, fluorine atoms formed by thermal dissociation can initiate chain reactions. The heat liberated by this reaction causes a rapid rise in temperature, and more bonds break to form radicals, which initiate more chain reactions. A radical chain reaction that is highly exothermic and produces radicals faster than they are destroyed results in an explosion. Organofluorine compounds are important because they frequently have unique and desirable properties. However, they are generally *not* made by direct fluorination, and this reaction is not a usual laboratory preparation.

> 3-Methylnonane can be fluorinated by F_2 diluted with N_2 in good yield at a tertiary position in a 1 : 1 $CFCl_3$-$CHCl_3$ mixture at −78 °C

Iodination is at the opposite extreme. As shown in Table 6.3, the reaction of methane with iodine is endothermic. In fact, methyl iodide reacts with HI to generate CH_4 and I_2. Iodine atoms are relatively unreactive. The reaction with methane is so endothermic that no significant reaction occurs at ordinary temperatures.

$$CH_4 + I\cdot \longrightarrow HI + CH_3\cdot \qquad \Delta H° = +34 \text{ kcal mole}^{-1}$$

Any iodine atoms produced ultimately dimerize to reform I_2.

The bromination of methane is less exothermic than is chlorination. Of the two chain propagation steps only one is relatively exothermic.

$$CH_4 + Br\cdot \longrightarrow CH_3\cdot + HBr \qquad \Delta H° = +18 \text{ kcal mole}^{-1}$$
$$CH_3\cdot + Br_2 \longrightarrow CH_3Br + Br\cdot \qquad \Delta H° = -25 \text{ kcal mole}^{-1}$$

Consequently, bromination is much slower than chlorination. It is instructive to examine the bromination of methane from a mechanistic standpoint. The two propagation steps are plotted in reaction-coordinate form in Figures 6.10 and 6.11.

In its reactions with other alkanes, bromine is a much more *selective* reagent than chlorine. For example, bromination of propane at 330 °C in the vapor phase gives 92% isopropyl bromide and only 8% *n*-propyl bromide.

$$CH_3CH_2CH_3 + Br_2 \longrightarrow \underset{\substack{(8\%) \\ n\text{-propyl bromide}}}{CH_3CH_2CH_2Br} + \underset{\substack{(92\%) \\ \text{isopropyl bromide}}}{(CH_3)_2CHBr}$$

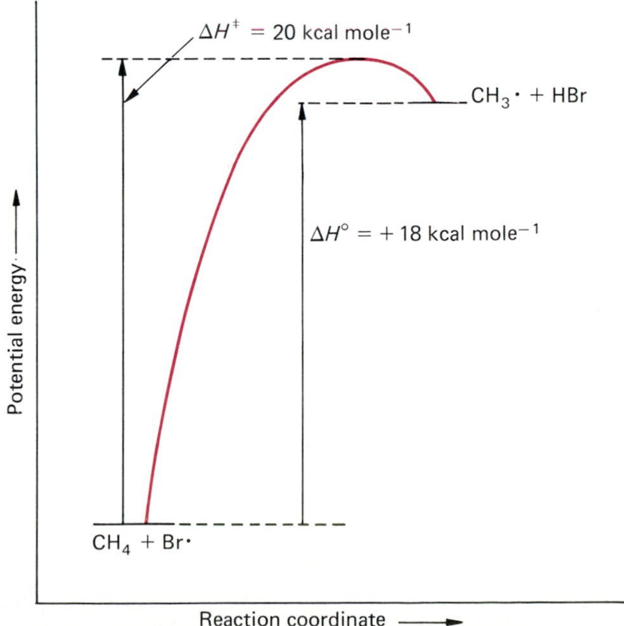

FIGURE 6.10 Reaction profile for the reaction $CH_4 + Br\cdot \rightleftharpoons CH_3\cdot + HBr$.

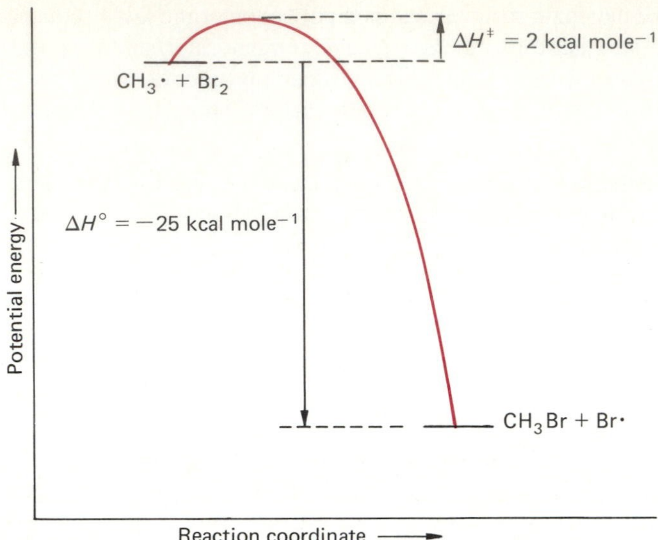

FIGURE 6.11 Reaction profile for the reaction $CH_3\cdot + Br_2 \rightleftharpoons CH_3Br + Br\cdot$.

The hydrogen abstraction steps for formation of the two isomers are

$$CH_3CH_2CH_3 + Br\cdot \longrightarrow CH_3CH_2CH_2\cdot + HBr \qquad \Delta H^\circ = +13 \text{ kcal mole}^{-1}$$

$$CH_3CH_2CH_3 + Br\cdot \longrightarrow CH_3\dot{C}HCH_3 + HBr \qquad \Delta H^\circ = +10 \text{ kcal mole}^{-1}$$

The two reactions are plotted in reaction coordinate form in Figure 6.12.

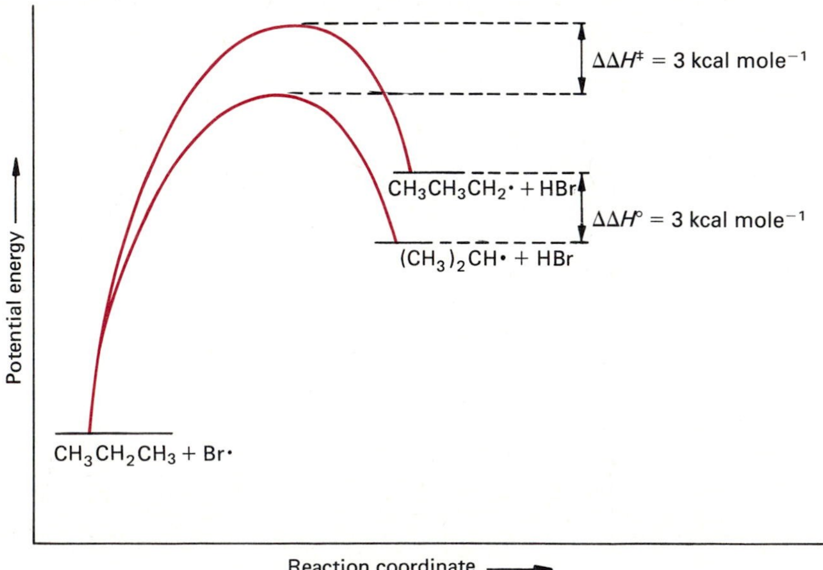

FIGURE 6.12 Reaction profiles for the reaction of $Br\cdot$ with C_3H_8.

The rates of reaction of a bromine atom with the two types of hydrogen in propane are given by

$$\text{rate } (1^\circ) = k_{1^\circ}[CH_3CH_2CH_3][Br\cdot]$$
$$\text{rate } (2^\circ) = k_{2^\circ}[CH_3CH_2CH_3][Br\cdot]$$

The ratio of the products formed is simply the ratio of the two rate constants.

$$\frac{\text{rate }(2^\circ)}{\text{rate }(1^\circ)} = \frac{k_{2^\circ}}{k_{1^\circ}}$$

For two similar reactions such as these, the ratio of the rate constants is related in an exponential manner to the two activation energies. The reaction with the larger activation energy has the smaller rate constant. In the chlorination of propane $\Delta\Delta H^{\ddagger}$ is only 1 kcal mole^{-1}, and consequently chlorination is relatively nonselective. For bromination $\Delta\Delta H^{\ddagger}$ is 3 kcal mole^{-1}, and hence bromination gives a greater *ratio* of secondary to primary products.

EXERCISE 6.6 From the difference in activation energies of 3 kcal mole^{-1} calculate the relative reactivity of secondary and primary hydrogens for vapor phase bromination at 330 °C.

The selectivity of bromine relative to chlorine is even more apparent when there are tertiary hydrogens in the alkane. For example, 2,2,3-trimethylbutane undergoes bromination to give more than 96% of the tertiary bromide, even though the alkane has only one tertiary hydrogen and fifteen primary hydrogens.

$$(CH_3)_3CCH(CH_3)_2 + Br_2 \xrightarrow[CCl_4]{h\nu} \underset{(>96\%)}{(CH_3)_3CCBr(CH_3)_2}$$

Thus, bromination is a somewhat more useful process for preparative purposes than chlorination. However, when there is only one tertiary hydrogen and many secondary hydrogens in a molecule, complex mixtures will still be produced.

EXERCISE 6.7 Using the data in Appendices I and II calculate the heats of reaction for

$$(CH_3)_3CH + X_2 \longrightarrow (CH_3)_3CX + HX$$

for $X_2 = F_2$, Cl_2, Br_2, and I_2.

Highlight 6.2

Chlorine atom ($Cl\cdot$) reacts with methane (CH_4) to form hydrogen chloride (HCl) and a methyl radical ($CH_3\cdot$). The hydrogen atom is transferred from carbon to chlorine via a transition state $[Cl{-}{-}{-}H{-}{-}{-}CH_3]^{\bullet} \equiv [Cl{-}H\ \cdot CH_3 \rightleftharpoons Cl\cdot\ H{-}CH_3]$.

Highlight 6.3

Along the reaction coordinate for the transfer of one hydrogen from methane to a chlorine atom, the arrangement of the bonds to the other hydrogens changes from tetrahedral in methane to planar in the methyl radical ($CH_3\cdot$). The bonding orbitals change from $C_{sp^3}{-}H_{1s}$ to $C_{sp^2}{-}H_{1s}$.

Highlight 6.4

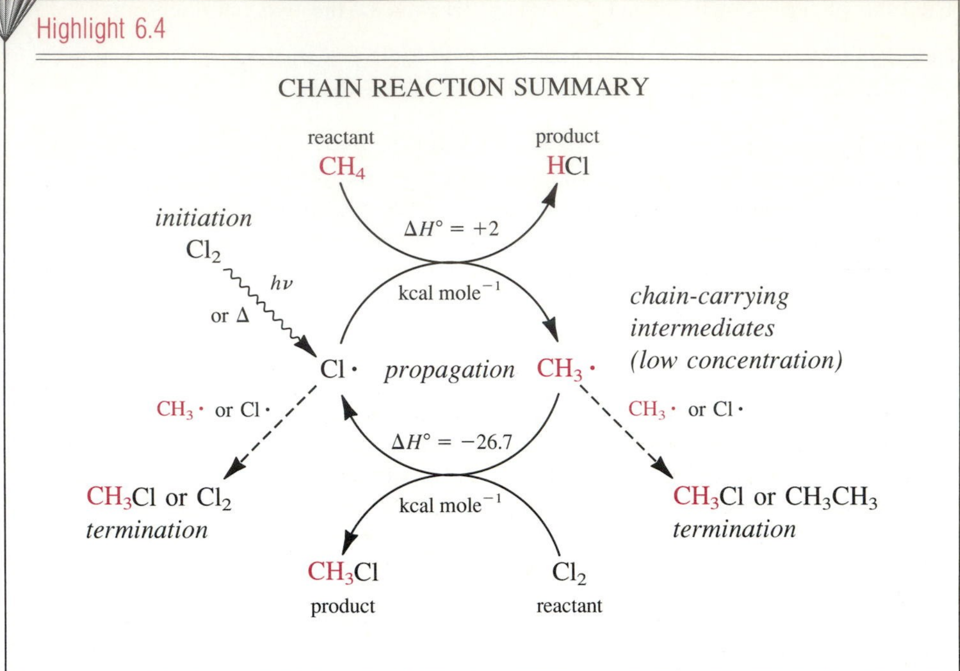

6.4 Combustion of Alkanes

Combustion of hydrocarbons is one of the most important organic reactions. The burning of huge amounts of natural gas, gasoline, and fuel oil involves mostly the combustion of alkanes. However, this combustion is an atypical organic reaction in two respects. First, mixtures of alkanes are normally the "reactants" in this reaction. Second, the desired product is heat and not particular compounds. Indeed, the chemical products are frequently undesirable, and their sheer mass creates significant problems of disposal. The equation for complete combustion of an alkane is simple.

$$C_nH_{2n+2} + \left(\frac{3n + 1}{2}\right)O_2 \longrightarrow n\,CO_2 + (n + 1)\,H_2O$$

However, many combustion processes, such as the burning of gasoline in an internal combustion engine, are incomplete. In an automobile, 1 gal of gasoline produces more than 1 lb of carbon monoxide. There are many other products resulting from incomplete combustion. Among these other products are aldehydes (RCHO), compounds that contribute significantly to the problem of smog, a mixture of fog and smoke common in urban areas around the world.

The mechanism by which alkanes react with oxygen is a complex one that has not been worked out in detail. There are many partially oxidized intermediates. Radical chain reactions are certainly involved. An especially important reaction is the combination of alkyl radicals with oxygen to give alkylperoxy radicals, which abstract hydrogen from an alkane to give intermediate alkyl hydroperoxides.

$$R\cdot + O_2 \longrightarrow ROO\cdot$$
$$ROO\cdot + RH \longrightarrow ROOH + R\cdot$$

Alkyl hydroperoxides contain a weak oxygen-oxygen bond ($DH^\circ \cong 44$ kcal mole^{-1}), which breaks readily at elevated temperatures to produce more radicals.

$$ROOH \longrightarrow RO\cdot + HO\cdot$$

Thus combustion is another example of a radical-multiplying reaction (a **branching** chain reaction), which leads to explosions under proper conditions.

When such an explosion occurs in the reaction chamber of an internal combustion engine, the piston is driven forward with a violent, rather than a gentle, stroke. Such premature explosions cause the phenomenon known as "knocking." The tendency of a fuel to cause engine knock depends markedly on the nature of the hydrocarbons used. In general, branching of an alkane chain tends to inhibit knocking. The knocking characteristic of a fuel is expressed quantitatively by an "octane number." On this arbitrary scale, heptane is given a value of 0 and 2,2,4-trimethylpentane ("isooctane") is assigned the value of 100. An octane number of 86, typical of a medium-grade "standard" or "regular" gasoline, has a knocking characteristic equivalent to that of a mixture of 86% 2,2,4-trimethylpentane and 14% heptane. The octane rating may be upgraded by the addition of small amounts of tetraethyllead $(C_2H_5)_4Pb$, which is called an "antiknock" agent. Its function is to control the concentration of free radicals and prevent the premature explosions that are characteristic of knocking. "Catalytic converters" decrease smog by destroying precursors in burned fuel, but are inactivated by lead-containing combustion products. This has reduced the use of leaded fuel, with the added benefit that less lead is delivered to the atmosphere. Lead in certain forms is toxic and exposure should be restricted. Thus, for environmental protection reasons, other ways are used for increasing the octane rating of gasoline. Various "reforming" processes cause straight-chain alkanes to be rearranged to their branched-chain isomers and to cycloalkanes, both of which have intrinsically higher octane ratings. However, the additional processing requires the expenditure of additional energy and entails additional losses of a limited raw material. For this reason, there is currently an active program under way to find other compounds that will improve the knocking characteristics of gasoline without damaging smog-control devices. Two leading candidates are *t*-butyl alcohol and methyl *t*-butyl ether (Chapter 10). The ether has an octane rating of 110.5.

The **heat of combustion** is defined as the enthalpy of the complete oxidation. The heat of combustion of a pure alkane can be measured experimentally with high precision (±0.02%) and constitutes an important thermochemical quantity. For example, the heat of combustion of butane is $\Delta H° = -634.82 \pm 0.15$ kcal mole^{-1}, whereas that of 2-methylpropane is $\Delta H° = -632.77 \pm 0.11$ kcal mole^{-1}. The general equation for combustion of these two isomers is the same.

$$C_4H_{10} + 6\tfrac{1}{2}\,O_2 \longrightarrow 4\,CO_2 + 5\,H_2O$$

Heats of combustion are often expressed in terms of $H_2O(l)$ as one product. To avoid confusion with other energy terms in this text we have given values for the heats of combustion with $H_2O(g)$ as one product. This difference is the heat of vaporization of water, 10.52 kcal mole^{-1}.

A direct comparison of these two heats of combustion shows that the branched hydrocarbon is 2.0 kcal mole^{-1} more stable than the straight-chain hydrocarbon at room temperature (Figure 6.13). The products, carbon dioxide and water, are more stable than the reactants. Because the products have a lower energy content, energy is released as heat—the heat of combustion. The less stable the reactants, the more heat is evolved. Since butane has a heat of combustion of greater magnitude than 2-methylpropane, butane must have a higher energy content and is less stable thermodynamically than 2-methylpropane.

It is from accurate heats of combustions that heats of formation (Section 5.5) have been determined. Figure 6.14 illustrates the relationship between heats of combustion and heats of formation for *n*-butane and 2-methylpropane. Note that the conversion of the heats of combustion to heats of formation requires only the heats of combustion of graphite and of hydrogen.

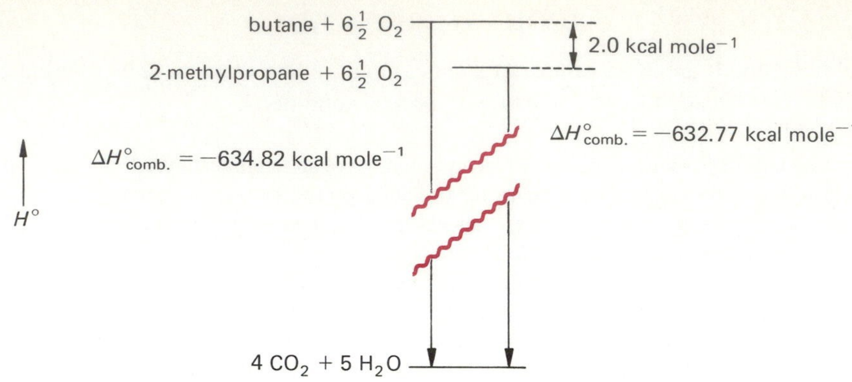

FIGURE 6.13 Illustrating the heats of combustion of butane and 2-methylpropane.

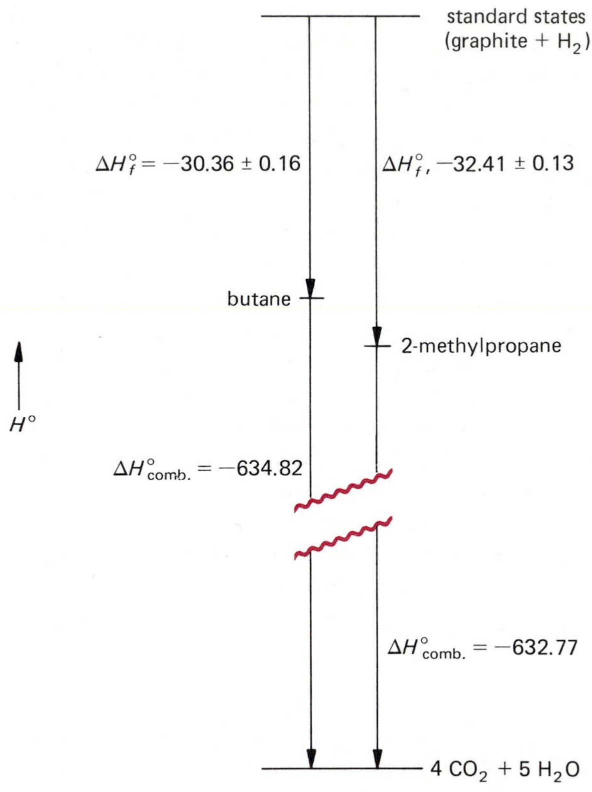

FIGURE 6.14 The relationship between the heats of formation and combustion of butane and 2-methylpropane.

EXERCISE 6.8 The heat of combustion of graphite (per carbon) and hydrogen (H_2) are -94.05 and -57.80 kcal mole^{-1}, respectively. From the heats of combustion of butane and 2-methylpropane shown in Figure 6.14, calculate ΔH_f° for both hydrocarbons and compare with the values in Figure 6.14.

EXERCISE 6.9 From heats of formation given in Appendix I, calculate the heats of combustion of pentane, 2-methylbutane, and 2,2-dimethylpropane.

Appendix I includes heats of formation for a number of free atoms. With these values we can calculate **heats of atomization,** the enthalpy required to dissociate a compound into all of its constituent atoms. For example, the heat of atomization of methane is 397 kcal mole^{-1}.

$$\begin{array}{llll} & CH_4 = & C & + 4\,H \\ \Delta H_f^\circ: & -17.9 & 170.9 + (4 \times 52.1) & \quad \Delta H^\circ = 397.2 \text{ kcal mole}^{-1} \end{array}$$

> Note that ΔH_f° of atomic carbon is much higher than ΔH_f° of C(graphite), carbon bound as graphite, which is defined as the standard state. Graphite has strong carbon-carbon bonds that must be broken to obtain carbon atoms. The heat of formation of carbon atoms is actually the heat of atomization of graphite per carbon.

This reaction requires breaking four carbon-hydrogen bonds. Hence, we can consider each bond to have an **average bond energy** of 397/4 = 99 kcal mole^{-1}. Note that this number differs from the bond-dissociation energy of methane (DH° = 105 kcal mole^{-1}), which is the energy required to break only one carbon-hydrogen bond in methane.

A similar calculation for ethane gives 674.6 kcal mole^{-1} as the heat of atomization required to break six carbon-hydrogen bonds and one carbon-carbon bond. If we assume that the average carbon-hydrogen bond energy in ethane is the same as it is in methane, we obtain 675 − (6 × 99) = 81 kcal mole^{-1}, a number that we could call the average bond energy of the carbon-carbon bond in ethane. If the same technique is applied to propane, we find a carbon-carbon bond energy similar to that in ethane.

In practice, data for a large number of compounds have been used to derive best overall values for such average bond energies. A table of such values is given in Appendix III. With this table one can calculate heats of atomization that are accurate to a few kilocalories per mole. The use of such a table is important for determining the approximate energy content of molecules whose heats of formation have not been determined experimentally or are too unstable to be isolated. Note that the results are only approximations. Butane and 2-methylpropane, for example, have the same numbers of carbon-carbon and carbon-hydrogen bonds. Such an approximate calculation using average bond energies results in identical heats of atomization; however, accurate heats of combustion show that butane and 2-methylpropane differ in energy content by 2.0 kcal mole^{-1}. Nevertheless, such approximate values will be found to have important uses in our study of organic chemistry.

EXERCISE 6.10 Use the average bond energies given in Appendix III to estimate the heat of reaction for chlorination of methane. Compare your answer with that obtained by using heats of formation of reactants and products.

EXERCISE 6.11 From Appendix I calculate the heats of atomization of butane and 2-methylpropane. Compare these values with the approximate one obtained from the use of average bond energies (Appendix III).

PROBLEMS

1. **a.** What products are expected from thermal cracking of pentane?
   b. Write reaction mechanisms leading to each product.

c. From heats of formation calculate the enthalpy of each of the net reactions involved.

2. a. Using the appropriate $DH°$ values from Appendix II, calculate $\Delta H°$ for each of the following reactions.

$$Br_2 \rightarrow 2\ Br\cdot$$
$$CH_3CH_3 + Br\cdot \rightarrow CH_3CH_2\cdot + HBr$$
$$CH_3CH_2\cdot + Br_2 \rightarrow CH_3CH_2Br + Br\cdot$$

b. What is the overall $\Delta H°$ for bromination of ethane?

$$CH_3CH_3 + Br_2 \rightarrow CH_3CH_2Br + HBr$$

c. How does this value compare with that obtained using heats of formation?

3. In the course of the bromination of ethane (problem 2), both bromine atoms and ethyl radicals will be present but not in equal amounts. Which is present in larger quantity? Explain.

4. The reaction of the unusual hydrocarbon spiropentane with chlorine and light is one of the best ways of preparing chlorospiropentane.

H_2C CH_2 C H_2C CH_2 $\xrightarrow[h\nu]{Cl_2}$ H_2C $CHCl$ C H_2C CH_2

spiropentane chlorospiropentane

a. Explain why chlorination is such a useful preparative method in this case.
b. Write the reaction mechanism.

5. For each of the following compounds, write the structures of all of the possible monochlorination products and predict the relative amounts in which they will be produced.

a. butane
b. 2,3-dimethylbutane
c. 2,2,4-trimethylpentane
d. 2,2,3-trimethylbutane
e. pentane

6. Answer problem 5 for bromination, using the relative reactivities of carbon-hydrogen bonds toward bromine atoms at 40 °C: primary, 1; secondary, 220; tertiary, 19,000.

7. In the chlorination of ethane, the observed reaction is

$$CH_3CH_3 + Cl_2 \longrightarrow CH_3CH_2Cl + HCl$$

An alternative reaction that might have occurred is

$$CH_3CH_3 + Cl_2 \longrightarrow 2\ CH_3Cl$$

a. Calculate $\Delta H°$ for each reaction.
b. Propose a radical chain mechanism by which the alternative reaction might occur. Calculate $\Delta H°$ for each of the propagation steps.
c. Suggest a reason why the alternative reaction does not occur.

8. From the average bond energies in Appendix III calculate the heat of atomization of cyclopropane. Compare with the value derived from heats of formation in Appendix I. Do average bond energies take ring strain into account? Does this result suggest how average bond energies can be used to estimate ring strain? The unusual molecule cubane, C_8H_8 (see Exercise 6.5) has $\Delta H_f° = 148.7$ kcal mole^{-1}. Estimate its strain energy by comparing its heat of atomization with that calculated from average bond energies.

9. From the heats of formation given in Appendix I calculate the heat of combustion of cyclopropane and cyclohexane. For combustion of an equal weight under the same conditions, which is the better fuel?

10. Fluorine reacts readily with alkanes.

a. What are the energetics of the reaction of octane with fluorine? Why does fluorine react so rapidly?

b. Why is fluorine diluted with nitrogen to carry out reactions with alkanes?

c. Why is the tertiary hydrogen the preferred site of attack in 2-methylnonane or in *t*-butylcyclohexane?

11. Chlorine fluoride, ClF, is a colorless gas (b.p. −101 °C) which behaves chemically as a reactive halogen. In principle, it can react with methane in two alternative ways

$$CH_4 + ClF \begin{array}{l} \rightarrow CH_3F + HCl \\ \rightarrow CH_3Cl + HF \end{array}$$

a. From ΔH_f° (ClF) = −12.2 kcal mole^{-1} and other data in Appendix I, calculate ΔH° for both reactions.

b. ΔS° for both reactions is expected to be approximately the same; therefore, which set of products is expected to predominate at equilibrium?

c. The reaction mechanism involves radical chain reactions similar to the reactions with chlorine except that the reaction of methyl radical with ClF can take two possible courses.

$$CH_3\cdot + ClF \begin{array}{l} \rightarrow CH_3Cl + F\cdot \\ \rightarrow CH_3F + Cl\cdot \end{array}$$

What is the difference in ΔH° for these two reactions? To the extent that the difference in activation energies reflects the difference in ΔH°, which reaction is expected to be the faster?

12. Chlorofluorocarbons such as CCl_3F (CFC-11) are excellent, stable, nontoxic refrigerants and foaming agents (for example, for polyurethane foam). Yet, in 1990, the nations of the world agreed to ban most of the agents by the year 2000. Their stability and relatively low molecular weights allow the CFCs to reach the upper atmosphere. There, short wavelength ultraviolet light from the sun dissociates the chlorofluorocarbons and bromofluorocarbons to chlorine atoms and bromine atoms. The halogen atoms participate in a chain reaction involving the ozone that absorbs much of the short wavelength ultraviolet light, which would otherwise reach the earth's surface (and us). Using the data in Appendices I and III, calculate the ΔH° for each step of the chain reaction.

$$(1)\ Cl\cdot + O_3 \rightarrow ClO\cdot + O_2$$
$$(2)\ O\cdot + ClO\cdot \rightarrow O_2 + Cl\cdot$$

What is the overall result of these two steps? Why is the presence of ozone in the upper atmosphere important?

13. Nitromethane, CH_3NO_2, is prepared by reaction of methane with nitric acid in the gas phase at temperatures over 400 °C. Appendix I includes values for some nitrogen compounds. Calculate ΔH° (298 K) for the equilibrium

$$CH_4 + HNO_3 \rightleftharpoons CH_3NO_2 + H_2O$$

It may seem strange that such an exothermic reaction requires such a high temperature. The actual reaction steps are believed to be

$$CH_4 + \cdot NO_2 \rightleftharpoons CH_3\cdot + HNO_2$$

$$HNO_2 + HNO_3 \rightleftharpoons 2\ NO_2\cdot + H_2O$$

$$CH_3\cdot + NO_2\cdot \rightleftharpoons CH_3NO_2$$

Calculate $\Delta H°$ for each step. Which step is expected to be the slow step that requires the high temperature? The reaction is initiated by traces of oxygen or radicals to produce some $NO_2\cdot$ radicals, which start the reaction. Note that this reaction sequence involves a radical chain propagation. Nitrogen dioxide is a rather stable radical, and its concentration in the reaction mixture is relatively high. It reacts rapidly with the methyl radicals and keeps these radicals at a very low concentration so that alternative free radical chain reactions are kept to minor importance; that is, nitrogen dioxide scavenges the methyl radicals. List several possible reactions of methyl radicals with nitric acid to produce methyl alcohol or nitromethane directly and show that these reactions are exothermic.

Nitrogen dioxide is a resonance-stabilized radical in which the odd electron can be placed on both oxygens and nitrogen. Write Lewis structures to demonstrate this point. In view of these structures it may seem surprising that methyl radical reacts with the nitrogen of NO_2. Actually, reaction at oxygen also occurs to give methyl nitrite, CH_3ONO. Compare $\Delta H°$ for this reaction with that for production of nitromethane:

$$CH_3NO_2 \longleftarrow CH_3\cdot + NO_2\cdot \longrightarrow CH_3ONO$$

Methyl nitrite is unstable under the reaction conditions and gives other products. The entire reaction is complex, and the discussion has treated only the most important of the many reactions that actually occur in this system.

14. In the chlorination of methane, the propagation steps (equations 6-2 and 6-3) do not have the same activation energies, as shown by Figures 6.6 and 6.7. However, by varying the relative concentrations of the methane and chlorine, either step can be made to be the slow step. Explain. Which step is rate-determining when $[CH_4] = [Cl_2]$? Under these conditions of relative concentrations, which free radical will be present in higher concentration, $CH_3\cdot$ or $Cl\cdot$? Show that the nature of the principal termination step depends on which propagation step is faster.

15. The description of rotational barriers in the alkanes (Section 5.2) was actually oversimplified, since, like all motion, the torsional motion around a carbon-carbon bond is quantized and a molecule will exist in one or more torsional-rotational states. Construct a more accurate version of Figure 5.9 to show this quantization. Assume that the zero-point energy level is about 0.5 kcal $mole^{-1}$ above the potential minimum and that the torsional quantum levels are separated by about 1.0 kcal $mole^{-1}$ increments. Which quantum level corresponds to continuous, uninhibited rotation about the bond?

CHAPTER 7

STEREOISOMERISM

7.1 Chirality and Enantiomers

The two objects depicted in Figure 7.1 appear to be identical in all respects. For every edge, face, or angle on one there is a corresponding edge, face, or angle on the other. And yet the two objects are not superimposable upon each other and are therefore *different objects*. They are related to one another as an object is related to its mirror image.

Another pair of familiar objects that are related to each other in this way are your right and left hands. They are (to a first approximation) identical in all respects. Yet your right hand will fit into a right glove and not into a left glove. The general property of ''handedness'' is called **chirality.** An object that is not superimposable upon its mirror image is **chiral.** If an object and its mirror image can be made to coincide in space, then they are said to be **achiral.**

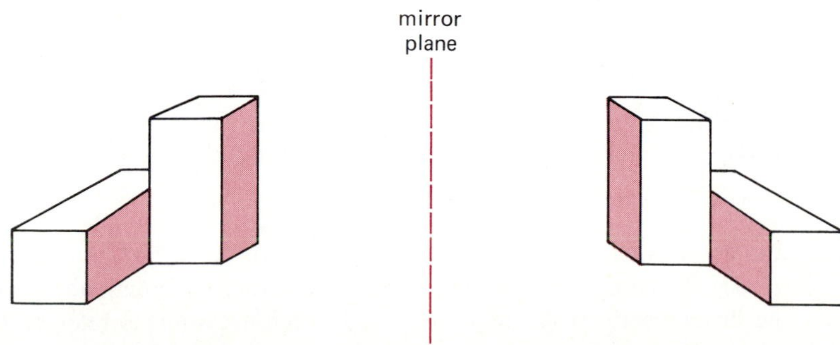

FIGURE 7.1 Two nonsuperimposable mirror-image objects.

EXERCISE 7.1 Which of the following familiar objects are chiral?

(a) an American football (b) an egg (c) a corkscrew
(d) a golf club (e) screwdriver (f) a person
(g) a hand mitten (h) a spiral staircase (i) a pencil
(j) a pocket computer (k) a pair of scissors (l) a screw-cap bottle
(m) a Greek vase (n) a portrait

Careful inspection of the structure of 2-iodobutane reveals that the molecule is chiral. There are actually two isomeric 2-iodobutanes, which are nonsuperimposable mirror images (Figure 7.2). Two compounds that differ in handedness in this way are called **enantiomers** and are said to have an **enantiomeric** relationship to each other. In order to convert one of the enantiomeric 2-iodobutanes into the other, it is necessary either to break and reform bonds or to distort the molecule through a planar geometry. Either process requires substantial energy; consequently, there is a rather large energy barrier to interconversion of enantiomers of this type. The two enantiomeric forms of 2-iodobutane may be kept in separate flasks or bottles for an indefinite period of time; under normal conditions they show no tendency to interconvert. This type of isomerism is called stereoisomerism. **Stereoisomers** are compounds that have the same sequence of covalent bonds and differ in the relative disposition of their atoms in space.

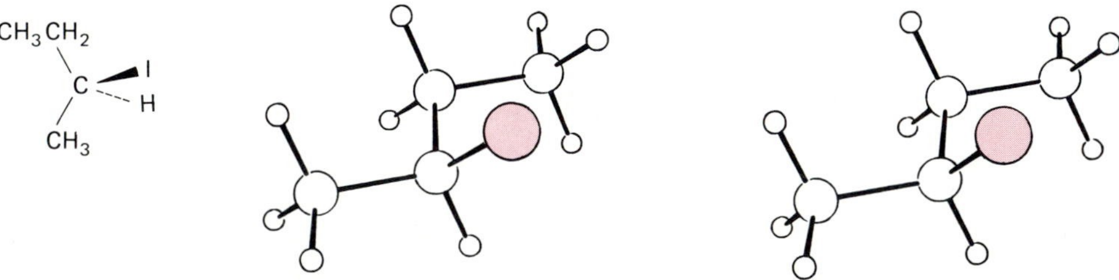

FIGURE 7.2 The mirror-image relationship of the two 2-iodobutanes.

2-Iodobutane owes its chirality to C-2, which has four different groups attached to it (C_2H_5, CH_3, I, H). If the positions of any pair of these four different substituents are exchanged, the enantiomeric structure is produced. Such a carbon is referred to as a **stereocenter.**

> From the time of Le Bel and van't Hoff, an atom with four different substituents was called an "asymmetric atom." It is true that most compounds with one such atom are truly asymmetric in the sense that they lack symmetry. However, as we shall see, it is possible for molecules that have atoms with four different substituents also to have various symmetry elements, including planes of symmetry. Use of the term "asymmetric atom" in such a case is confusing. The term *stereocenter* avoids such confusion.

Notice that 1,1-dichloroethane, which has three different groups attached to C-1 (CH_3, Cl, H), is achiral; it is superimposable upon its mirror image (Figure 7.3). When a compound has one stereocenter, its molecules are always chiral. However, a stereocenter is not a necessary condition for chirality, as we shall soon see. Also, as we shall see in Section 7.5, a molecule may still be achiral if it contains more than one stereocenter.

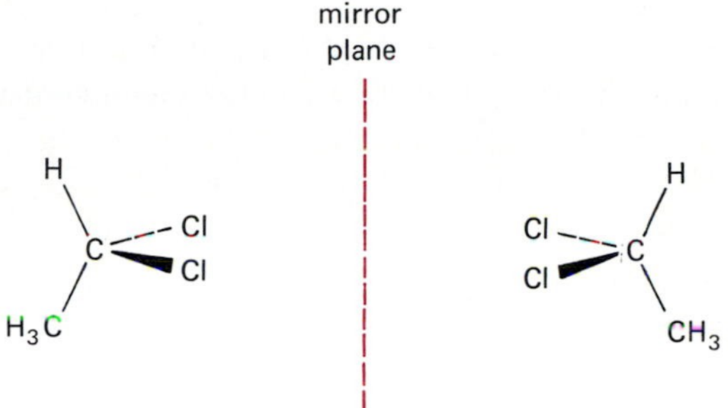

FIGURE 7.3 The achirality of 1,1-dichloroethane.

EXERCISE 7.2 Which of the following compounds has a stereocenter?

(a) 3-chloropentane (b) 3-methylhexane
(c) 1-bromo-1-chloroethane (d) bromocyclobutane

Two of the conformational isomers of butane (Section 5.2) also have an enantiomeric relationship to each other (Figure 7.4). However, in this case, the two enantiomers may interconvert simply by rotation about the central carbon-carbon bond.

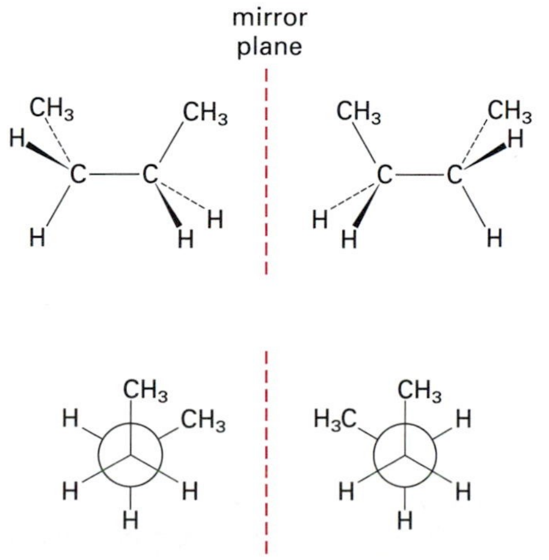

FIGURE 7.4 The enantiomeric relationship between the two gauche conformations of butane.

Since rotational barriers are generally rather small, enantiomers such as these interconvert rapidly at room temperature. The individual enantiomers could be obtained in a pure state only by working at exceedingly low temperatures, on the order of −230 °C (page 78). Note that the enantiomeric gauche butanes do not have stereocenters. Instead, these molecules are chiral because of the asymmetric disposition of the groups around the central C—C bond.

EXERCISE 7.3 Construct molecular models of the two enantiomeric forms of gauche butane. Confirm that the two enantiomers are not superimposable and that rotation about the C-2—C-3 bond interconverts them.

7.2 Physical Properties of Enantiomers: Optical Activity

Most of the physical properties of the two enantiomeric 2-iodobutanes are identical. They have identical melting points, boiling points, solubilities in common solvents, densities, refractive indices, and spectra. However, they differ in one important respect—the way in which they interact with **polarized light.**

Light may be treated as a wave motion of changing electric and magnetic fields that are at right angles to each other. When an electron interacts with light, it oscillates at the frequency of the light in the direction of the electric field and in phase with it. In normal light, the electric field vectors of the light waves are oriented in all possible planes. **Plane-polarized** light is light in which the electric field vectors of all the light waves lie in the same plane, the plane of polarization.

normal light　　　　plane-polarized light

Plane-polarized light may be produced by passing normal light through a polarizer, such as a polaroid sheet or a device known as a Nicol prism.

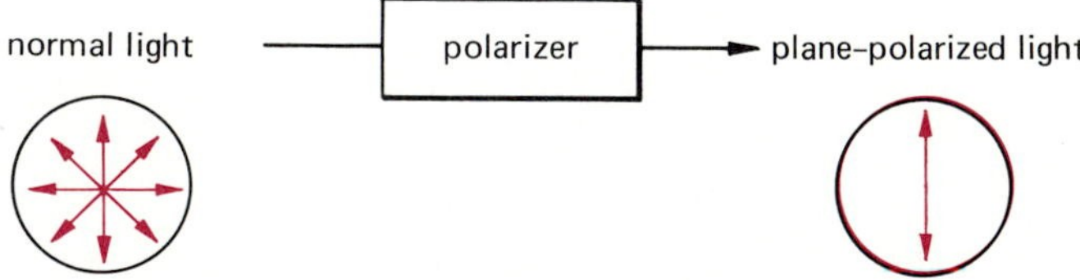

In a molecule, an electron is not free to oscillate equally in all directions. That is, its polarizability is "anisotropic," which means different in different directions. When electrons in molecules oscillate in response to plane-polarized light, they generally tend, because of their anisotropic polarizability, to oscillate out of the plane of polarization. Because of its interaction with the oscillating electrons, the light has its electric and magnetic fields changed. Thus, when plane-polarized light interacts with a molecule, the plane of polarization rotates slightly. However, in a large collection of achiral molecules, for any orientation of a molecule that changes the plane of polarization of the light, there is apt to be another molecule with a mirror-image orientation having the opposite effect. Consequently, when a beam of plane-polarized light is passed through such a compound, it emerges with the plane of polarization unchanged.

plane-polarized light
emergent light

However, for chiral molecules such as one of the enantiomeric 2-iodobutanes, the stereoisomer with mirror-image orientations are lacking, and the plane of polarization of the light is usually measurably altered in its passage through the sample. Such compounds are said to be **optically active.** If a compound causes the plane of polarization to rotate in a clockwise (positive) direction as measured by the detector facing the beam, it is said to be **dextrorotatory.** If it causes the plane to rotate in a counterclockwise (negative) direction, it is called **levorotatory.** The amount by which the plane is rotated is expressed as the angle of rotation α and by the appropriate sign, which shows whether rotation is in the dextro (+) or levo (−) sense.

Rotations are measured with a device called a **polarimeter.** Since the degree of rotation depends on the wavelength of the light used, monochromatic light (light having a single wavelength) is necessary. Common polarimeters use the sodium D line (5890 Å). The monochromatic light is first passed through the polarizer (usually a Nicol prism), from which it emerges polarized in one plane. The plane-polarized light is then passed through a tube that contains the sample, either as a liquid or dissolved in some achiral solvent. It emerges from the sample with the plane of polarization rotated in either the plus or minus direction by some amount. The light beam then passes through a second Nicol prism, which is mounted on a circular marked dial (the analyzer). The analyzer is rotated by an amount sufficient to allow the light beam to pass through at maximum intensity. Readings are compared with and without the sample tube to obtain the rotation value. Precision polarimeters using the sodium yellow line (D line) or the mercury green line are generally precise to about ±0.01°. Modern spectropolarimeters use photocells in place of the human eye and can give even more precise data over a wide spectral region. A schematic representation of a polarimeter is shown in Figure 7.5.

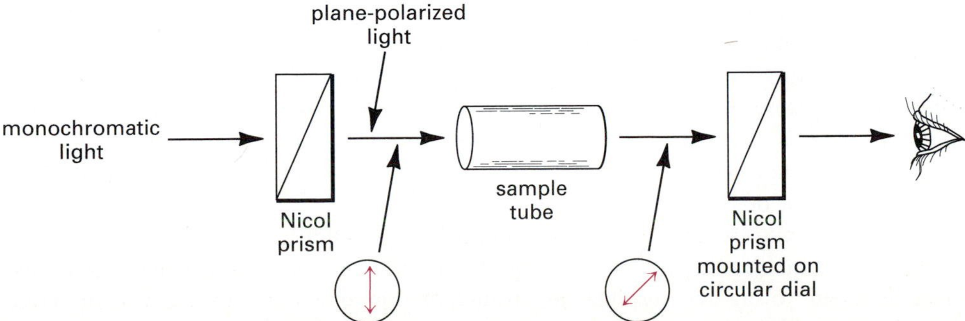

FIGURE 7.5 Polarimeter schematic.

The student may easily experience the phenomenon of optical rotation by performing a simple experiment. Take two pairs of Polaroid sunglasses and line them up, one in front of the other. Look through one lens of each pair of glasses at a bright light. Now rotate one of the lenses. When the glasses are parallel, the maximum amount of light is transmitted. When they are oriented at right angles to each other, no light is transmitted. What you have constructed is a simple polarimeter. The first pair of glasses corresponds to the polarizer and the second to the analyzer. Now dissolve several tablespoons of table sugar (sucrose, an optically active compound) in a small glass of water and place the glass between the two sunglasses. Again rotate one pair of glasses and note that the orientation for maximum and minimum transmission of light is now different. It is easier to observe the change at the point of minimum transmission.

The observed angle of rotation α is proportional to the number of optically active molecules in the path of the light beam. Therefore, α is proportional to the length of the sample tube and to the concentration of the solution being observed. The **specific rotation** $[\alpha]$ is obtained by dividing α by the concentration, c (expressed in g mL^{-1} solution) and by the length of the cell, l (expressed in decimeters). The wavelength of light used is given as a subscript, and the temperature at which the measurement was made is given as a superscript.

$$[\alpha]_D^t = \frac{\alpha}{l \cdot c} \quad \text{(for solutions)}$$

The decimeter is used as the unit of length simply because a 1-dm (10-cm) tube is a common length for measurements of rotation. For a pure liquid the definition of c (g mL^{-1}) is simply the density of the compound, d.

$$[\alpha]_D^t = \frac{\alpha}{l \cdot d} \quad \text{(for liquids)}$$

When the temperature is not given, the rotation is assumed to be that at room temperature, generally 25 °C.

Actually, it is not possible to determine whether the rotation is (+) or (−) from a single measurement. Is a reading of 60° to be interpreted as +60° or −300°? The sign may be determined by measuring the rotation at different sample concentrations. For example, if a 1 M sample gives a reading of 60° on the polarimeter, this may be either +60° or −300°. For a 1.1 M sample, the values would be either +66° or −330°, which are easily distinguished.

As was mentioned earlier, enantiomers differ from one another in the manner in which they interact with plane-polarized light. In fact, two enantiomers cause the plane of polarization to rotate by exactly the same amount, but in *opposite* directions. For example, one of the two enantiomeric 2-iodobutanes has $[\alpha]_D^{24°} = +15.9$, and the other has $[\alpha]_D^{24°} = -15.9$. This information still does not tell us which enantiomer is which. *There is no simple relationship between the sign of α and the absolute stereostructure of a molecule.* For example, (+)-lactic acid, (+)-$CH_3CH(OH)COOH$, is convertible to its sodium salt, (−) sodium lactate, (−)-$CH_3CH(OH)COO^-\ Na^+$, by reaction with NaOH, and recovered by addition of HCl.

Absolute stereostructure can be determined by x-ray diffraction using a technique known as anomalous dispersion. Although the technique is too sophisticated to discuss here, suffice it to say that absolute stereostructures for some optically active compounds have been established in this way. Once the absolute stereostructures for a few optically active compounds are known, other molecular configurations may be determined by correlating them chemically with the compounds of known structure. We shall show how this is done in later sections. By these methods, the structures of (+)- and (−)-2-iodobutane are known to be

CH_3CH_2, CH_3, C, H, I

(+)-2-iodobutane
$[\alpha]_D^{24} = +15.9$

CH_2CH_3, CH_3, C, H, I

(+)-2-iodobutane
$[\alpha]_D^{24°} = +15.9$

EXERCISE 7.4 The specific rotation of sucrose is +66. Assuming that 5 tablespoons weighs 60 g and that a small glass 5 cm in diameter holds 300 mL of solution, calculate how much rotation should have been observed in the experiment described on page 128.

Highlight 7.1

A molecule and its nonidentical mirror image are a **pair of enantiomers.** One will rotate the plane of polarized light to the right (clockwise); it is **dextrorotatory,** defined as (**+**). The second will rotate the plane of polarized light to the left (counterclockwise); it is **levorotatory,** defined as (**−**).

7.3 Nomenclature of Enantiomers: The *R*,*S* Convention

Suppose we have one bottle containing only one of the two enantiomeric 2-iodobutanes and another bottle containing only the other enantiomer. What labels do we attach to the two bottles? We cannot simply label each bottle ''2-iodobutane'' because they contain different compounds. We can label the bottles ''(+)-2-iodobutane'' and ''(−)-2-iodobutane.'' By this we mean: ''this bottle contains the 2-iodobutane that rotates the plane of polarized light in the dextro sense'' and ''this bottle contains the 2-iodobutane that rotates the plane of polarized light in the levo sense.'' Since it has also been determined which absolute stereostructure corresponds to (+)-2-iodobutane, these labels are sufficient to define unambiguously which compounds are in the bottles.

However, if a chemist were to encounter a bottle labeled (+)-2-iodobutane, chances are that he or she would not know which of the two possible absolute configurations correspond to dextrorotation. For this reason, it is highly desirable to have a system whereby the precise arrangement of the atoms may be specified in the name of the compound. A system of nomenclature for this purpose was first introduced by R. S. Cahn and Sir Christopher Ingold of University College, London, in 1951. The proposal was subsequently modified in collaboration with Vlado Prelog, of the Swiss Federal Institute of Technology and has since been adopted for this purpose by the IUPAC. The Cahn-Ingold-Prelog system is also called the *R*,*S* convention, or the ''sequence rule.''

The application of the sequence rule to naming enantiomers that owe their chirality to one or more stereocenters is straightforward and involves the following simple steps.

1. Identify the four different substituents attached to the stereocenter. Assign to each of the four substituents a priority 1, 2, 3, or 4 using the sequence rule, such that $1 > 2 > 3 > 4$.

2. Orient the molecule in space so that one may look down the bond from the stereocenter to the substituent with lowest priority, 4.

When one looks along that bond, one will see the stereocenter with the three attached substituents 1, 2, and 3 radiating from it like the spokes of a wheel. Trace a path from 1 to 2 to 3. If the path describes a clockwise motion, then the stereocenter is called *R* (L., *rectus*, right). If the path describes a counterclockwise motion, then the stereocenter is called *S* (L., *sinister*, left).

Stereo representations of *R* and *S* structures are shown in Figure 7.6.

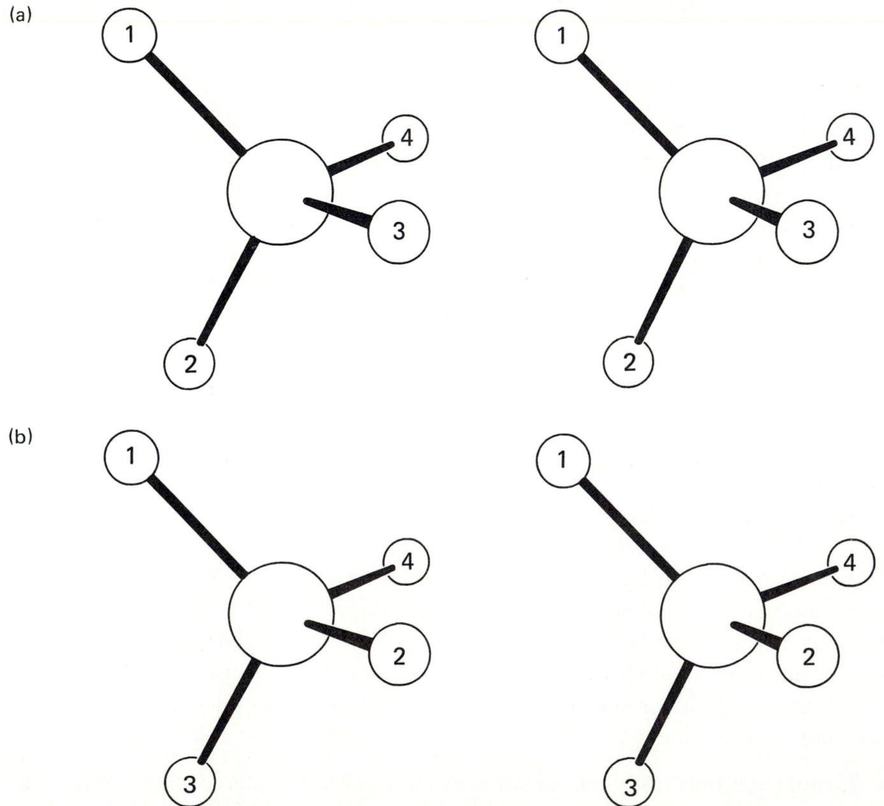

FIGURE 7.6 Stereo diagrams of a stereocenter, illustrating the arrangement of 1, 2, 3, and 4 priority groups for assignment of configurations as (a) *S* and (b) *R*.

The actual sequence rule is the method whereby the four substituents are assigned priorities 1, 2, 3, and 4 so that the symbols *R* and *S* may be assigned. There are a number of parts to the sequence rule, but we need only consider four aspects of it.

1. If the four atoms directly attached to the stereocenter are different, **higher atomic number takes precedence over lower.** For example, in 1-bromo-1-chloroethane, the four atoms attached to the stereocenter are ranked Br > Cl > C > H.

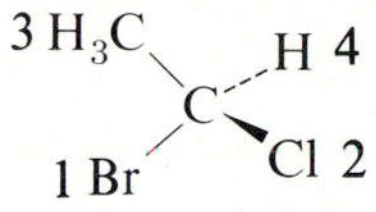

3 H_3C, H 4, Cl 2, 1 Br (around C)

(*S*)-1-bromo-1-chloroethane

Note that, in applying this aspect of the sequence rule, only the atoms attached *directly to the stereocenter* are considered. For example, in 1-chloro-1-fluoropropane, the four *groups* attached to C-1 are ranked Cl > F > CH_2CH_3 > H. For stereocenters that have three groups and a lone pair, the lone pair has the lowest priority (i.e., the lone pair is treated as having an atomic number of zero).

EXERCISE 7.5 Write a three-dimensional structure for (*R*)-1-chloro-1-fluoropropane.

2. In cases where two of the attached atoms are isotopes of each other, **higher atomic mass has priority over lower.** In 1-deuterio-1-fluoroethane, the four groups are therefore ranked F > C > D > H.

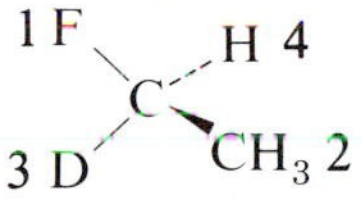

1 F, H 4, CH_3 2, 3 D (around C)

(*R*)-1-deuterio-1-fluoroethane

3. For many chiral compounds, two of the atoms directly attached to the stereocenter will be the same. In this case, work outward concurrently along the two chains atom by atom until a point of difference is reached. **The priorities are then assigned at that first point of difference,** using the same considerations of atomic number and atomic mass. In 2-iodobutane, the iodine is assigned priority 1 and the hydrogen is assigned priority 4. The two remaining groups are CH_2—**C**H_3 and —CH_2—**H**. The first point of difference is at the two carbons attached to the stereocenter. The group —CH_2—CH_3 takes priority over —CH_2—H because carbon has a higher atomic number than hydrogen. Thus, we may assign *R* and *S* configurations to the two enantiomers as

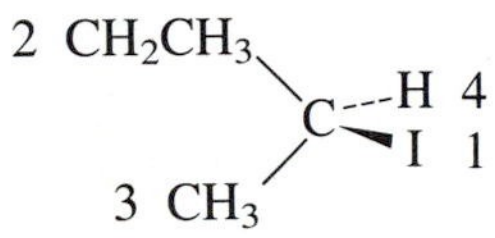

2 CH_2CH_3, H 4, I 1, 3 CH_3 (around C) | CH_2CH_3 2, 4 H, 1 I, CH_3 3 (around C)

(*S*)-2-iodobutane 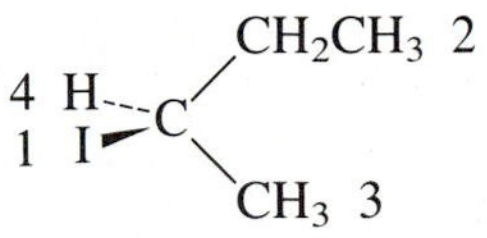 | (*R*)-2-iodobutane

For 4-chloro-2-methyloctane, the four atoms attached to the stereocenter are Cl (1), H (4), C, and C. In order to rank the isobutyl and *n*-butyl groups, we work along the chains until we reach the second carbon from the stereocenter before we reach a point of difference. The group —CH_2**C**$H(CH_3)_2$ takes priority over —CH_2**C**$H_2CH_2CH_3$ because at the first point of difference the carbon in isobutyl has *two other carbons* attached to it, whereas the analogous carbon in *n*-butyl has only *one other attached carbon*.

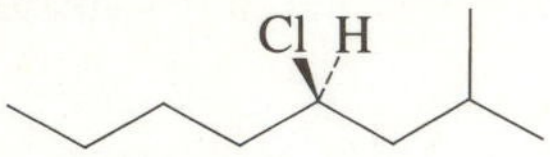

Cl H

EXERCISE 7.6 Write the structure of (*S*)-3-chloro-2,6-dimethylheptane.

Note that the configuration *R* or *S* is assigned by considering only the *first* point of difference as one works out the two chains, even if other differences exist further out the chain. For example, in 1,3-dichloro-4-methylpentane, the priority order of the four groups attached to the stereocenter is Cl > —$CH(CH_3)_2$ > —CH_2CH_2Cl > H. In this case, the isopropyl and 2-chloroethyl groups are ranked on the basis of the fact that the isopropyl group has two carbons and one hydrogen bonded to the indicated carbon, whereas the 2-chloroethyl group has one carbon and two hydrogens attached to the analogous carbon.

EXERCISE 7.7 Assign *R* and *S* configurations to the following compounds.

(a) 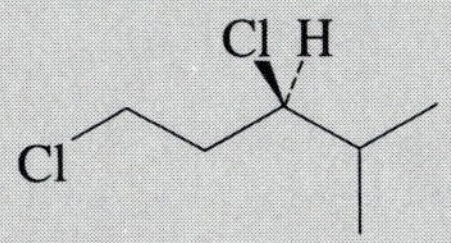 Cl ... Cl H

(b) 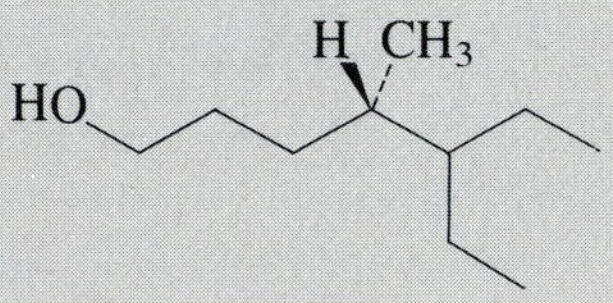 HO ... H CH_3

4. **Double and triple bonds are treated by assuming that each such bonded atom is duplicated or triplicated.** For example, a carbon that is doubly bonded to oxygen is considered to be bonded to *two oxygens*.

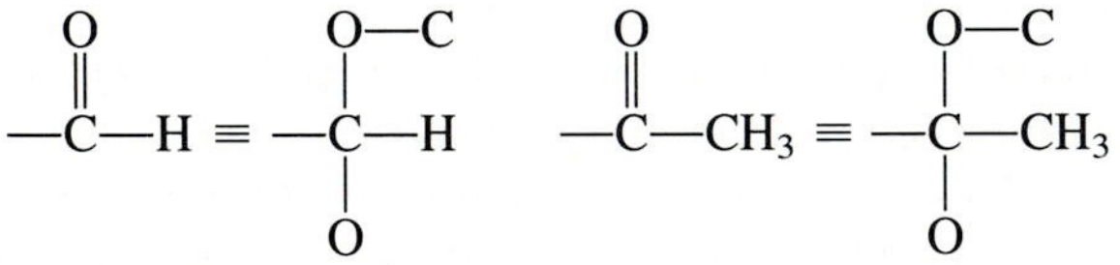

$$-\overset{\overset{\displaystyle O}{\|}}{C}-H \equiv -\overset{\overset{\displaystyle O-C}{|}}{\underset{\underset{\displaystyle O}{|}}{C}}-H \qquad -\overset{\overset{\displaystyle O}{\|}}{C}-CH_3 \equiv -\overset{\overset{\displaystyle O-C}{|}}{\underset{\underset{\displaystyle O}{|}}{C}}-CH_3$$

EXERCISE 7.8 What is the absolute configuration of the following dihydroxy aldehyde (glyceraldehyde)?

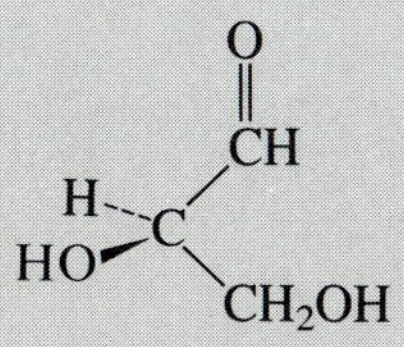

O ‖ CH — C (H, HO) — CH_2OH

For a carbon-carbon double bond, *both* doubly-bonded carbons are considered to be duplicated.

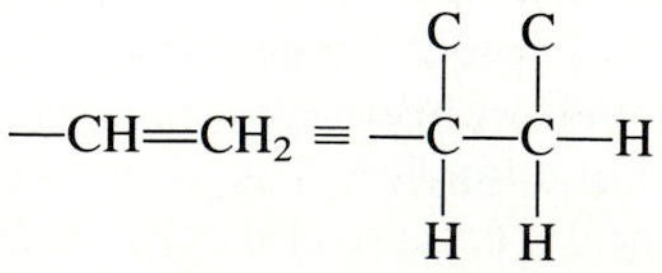

$$-CH{=}CH_2 \equiv -\overset{\overset{\displaystyle C}{|}}{\underset{\underset{\displaystyle H}{|}}{C}}-\overset{\overset{\displaystyle C}{|}}{\underset{\underset{\displaystyle H}{|}}{C}}-H$$

Thus, the group —CH=CH_2 takes priority over the group —$CH(CH_3)_2$. In this example, the first carbon of each group is considered to be bonded to two carbons and one hydrogen, so that a priority cannot be assigned on the basis of this position. However, C-2 of the —CH=CH_2 group is considered to be bonded to one carbon and two hydrogens, while the analogous carbon in the —$CH(CH_3)_2$ group is considered to be bonded to three hydrogens.

```
    C   C             CH3   H
    |   |             |     |
  —C—C—H  >  —C———C—H
    |   |             |     |
    H   H             H     H
```

EXERCISE 7.9 What are the absolute configurations of the following compounds?

```
                                              O
                                              ‖
      CH3CH=CH         H                     HC         H
(a)            \   ,'              (b)         \   ,'
                C◀OH                              C◀C≡N
               /                                 /
      (CH3)2CH                                CH3
```

It is important to remember that the *R* and *S* prefixes discussed in this section are nomenclature devices. They specify the absolute configuration of individual molecules. The terms (+) and (−) refer to the experimental property of optical activity.

Highlight 7.2

The **R,S convention** is a system of nomenclature that is used to unambiguously specify the **absolute configuration** of the four groups attached to a stereocenter. In using the **sequence rule,** you assign the highest priority (lowest number) to the atom that has the highest atomic number and the lowest priority to the atom that has the lowest atomic number. When two or more of the attached groups are bonded to the stereocenter by atoms of the same atomic number, you continue out the respective chains atom by atom until a point of difference is reached and then assign the higher priority to the chain that has the higher atomic number at this point of difference. Multiple bonds are counted as branches (C=O is taken as carbon connected to two oxygens). When the four groups attached to the stereocenter have been assigned priorities (1, 2, 3, and 4), you view the stereocenter along the bond from it to group 4. If the arc traced by the remaining three groups 1–2–3 is clockwise, the absolute configuration at the stereocenter is **R**; if the arc 1–2–3 is counterclockwise, the absolute configuration at the stereocenter is **S**. The convention is used to assign a stereochemical name, *R* or *S*, to each stereocenter in a molecule that has two or more stereocenters. There is no general relationship between the *R* or *S* designation and the sign of rotation of the plane of polarized light.

7.4 Racemates

When equal amounts of enantiomeric molecules are present together, the sample is said to be **racemic.** Such a mixture is also called a **racemate.** Since a racemate contains equal numbers of dextrorotating and levorotating molecules, the net optical rotation is zero. A racemate is often specified by prefixing the name of the compound with the symbol (±), for example, (±)-2-iodobutane.

The physical properties of a racemate are not necessarily the same as those of the pure enantiomers. A sample composed solely of right-handed molecules experiences different intermolecular interactions than a sample composed of equal numbers of right- and left-handed molecules. (In order to verify this in a simple way, use your right hand to shake hands with another person. The interaction is clearly different depending on whether the other person extends his right or his left hand.)

A racemate may crystallize in several ways. In some cases, separate crystals of the (+) and (−) forms result. In this case, the crystalline racemate is a mechanical mixture of two different crystalline compounds and is called a **racemic mixture.** The melting-pôint diagram for such a mixture is like that for any other mixture of two compounds (Figure 7.7). The eutectic point in such a case is always at the 50:50 point. Addition of a little of either pure enantiomer will cause the melting point of the mixture to increase. The racemate may also crystallize as a **racemic compound.** In this case only one type of crystal is formed, and it contains equal numbers of (+) and (−) molecules. The racemic compound acts as though it were a separate compound; its melting point is a peak on the phase diagram. However, the racemic compound may melt either higher or lower than the pure enantiomers. Addition of a small amount of either pure enantiomer causes a melting-point depression (Figure 7.8).

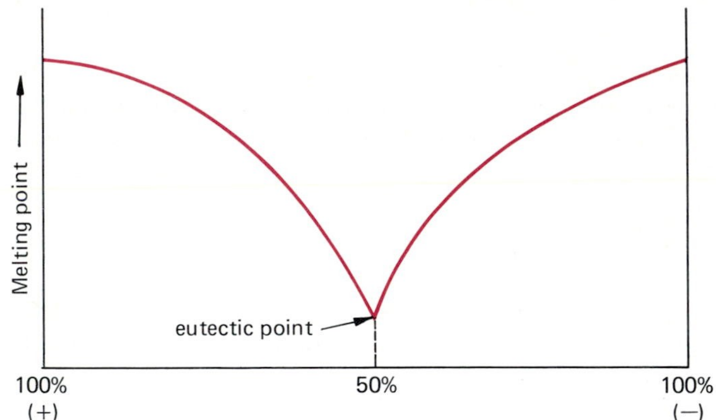

FIGURE 7.7 Melting point diagram for a racemic mixture.

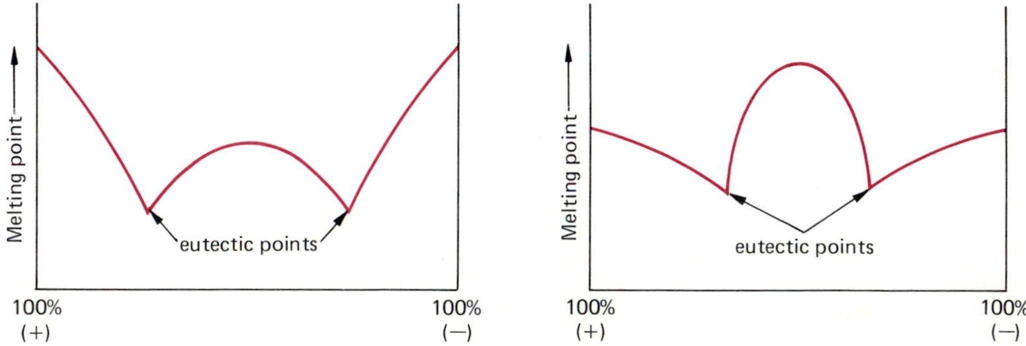

FIGURE 7.8 Representative melting point diagrams for racemic compounds.

Because of these differential intermolecular interactions, racemates frequently differ from the pure enantiomers in other physical properties. Differences have been observed in density, refractive index, and in various spectra.

The process whereby a pure enantiomer is converted into a racemic mixture is called **racemization.** Racemization may be accomplished in a trivial sense by simply mixing equal amounts of two pure enantiomers. Racemization may also result from chemical interconversion; we shall see many examples of this in future chapters. We have already encountered one racemization process in Section 7.1, the interconversion of the two enantiomeric gauche forms of butane by rotation about the central carbon-carbon bond.

Highlight 7.3

A racemic mixture is formed by physically mixing equal amounts of two enantiomers or by a physical or chemical process that equilibrates the two.

7.5 Compounds Containing More Than One Stereocenter: Diastereomers

If a molecule has more than one stereocenter, the number of possible stereoisomers is correspondingly larger. Consider 2-chloro-3-iodobutane as an example. There are four isomers, which are depicted in Figure 7.9. Of the four stereoisomeric 2-chloro-3-iodobutanes, two pairs bear an enantiomeric relationship to one another. The 2*R*,3*R* and 2*S*,3*S* compounds are one enantiomeric pair, and the 2*R*,3*S* and 2*S*,3*R* compounds are another enantiomeric pair. As with the other enantiomeric pairs previously discussed, the 2*R*,3*R* and 2*S*,3*S* compounds have identical boiling points, melting points, densities, solubilities, and spectra. They cause the plane of polarized light to rotate to the same degree but in opposite directions; one is dextrorotatory, and the other is levorotatory. A similar correspondence in physical properties is observed for the 2*R*,3*S* and 2*S*,3*R* compounds.

Compounds that are stereoisomers of one another, but are not enantiomers, are called **diastereomers** and are said to have a **diastereomeric** relationship. The stereoisomeric relationships for a compound having two unlike stereocenters are summarized in schematic form in Figure 7.10.

In general, the maximum number of possible stereoisomers for a compound having n stereocenters is given by 2^n. Thus, for a compound with one stereocenter, there are $2^1 = 2$ stereoisomers. For a compound with two stereocenters, there can be as many as $2^2 = 4$ stereoisomers. In some cases, there are fewer than the maximum number of possible

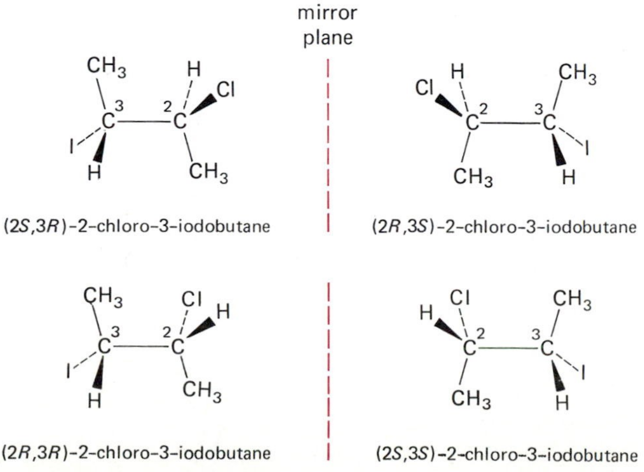

FIGURE 7.9 Stereoisomers of 2-chloro-3-iodobutane.

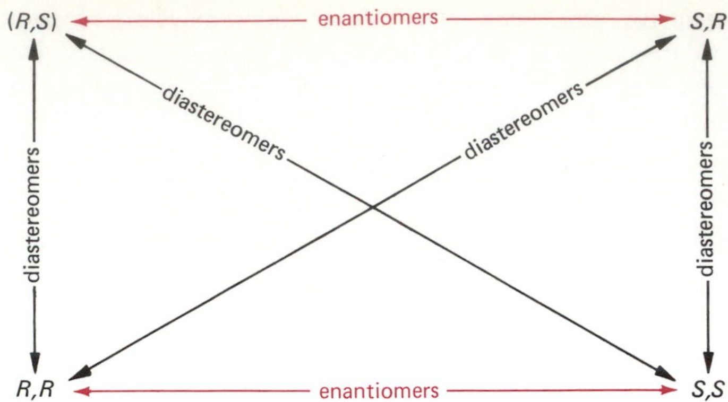

FIGURE 7.10 Stereoisomeric relationships for a compound having two unlike stereocenters.

stereoisomers. As an example, consider 2,3-dichlorobutane. The 2*R*,3*R* and 2*S*,3*S* compounds are enantiomers of one another.

mirror plane

(2*S*,3*S*)-2,3-dichlorobutane (2*R*,3*R*)-2,3-dichlorobutane

However, careful inspection reveals that the 2*R*,3*S* and 2*S*,3*R* compounds are actually the *same* compound (mentally perform a 180° rotation of the entire molecule about the axis of the C-2—C-3 bond).

mirror plane

(2*S*,3*R*)-2,3-dichlorobutane (meso)

Since this isomer of 2,3-dichlorobutane is achiral, it is not optically active. Such a compound, which has stereocenters yet is achiral, is called a **meso** compound. It is important not to confuse meso compounds with racemates, which are actually equimolar mixtures of two enantiomers. Both show no optical activity, but a meso compound is a single achiral substance, whereas a racemate is a 50 mole % mixture of two chiral substances.

Meso compounds may be recognized by looking for a plane or a point of symmetry *within a molecule* that has stereocenters. When such an element of symmetry exists, the maximum number of possible stereoisomers is less than 2^n. In one of the eclipsed conformations of *meso*-2,3-dichlorobutane the plane of symmetry is clearly obvious.

plane of symmetry

A point of symmetry may be seen in one of the staggered conformations.

center of symmetry

An object has a point of symmetry (or is centrosymmetric) when the identical environment is encountered at the same distance in both directions along any line through a given point.

EXERCISE 7.10 The tsetse fly sex-attractant (page 96) has the 17*R*,21*S* configuration. Is it a meso compound, or does it show optical activity?

In writing more complex structures than we have considered heretofore in this chapter, it is often convenient to employ the line structures that were introduced in Chapter 5 (page 83). In such structures, an emboldened line represents a part of the molecule that projects *forward* from the general plane of the molecule and a dashed line represents a part of the molecule that projects *away* from the general plane of the molecule. Remember that hydrogens are not shown in these simplified drawings, even if attached to a stereocenter.

Cl

(*S*)-3-chloroheptane

Cl

(4*R*,5*S*)-5-chloro-4-ethyl-
2,8-dimethylnonane

EXERCISE 7.11 Write line structures for (*R*)-2-chloro-5-methylhexane and (2*R*,4*S*)-2,4-dichloropentane. The latter compound is meso. Identify an internal symmetry element in the drawing you have made.

7.6 Stereoisomeric Relationships in Cyclic Compounds

Chlorocyclopropane and 1,1-dichlorocyclopropane are achiral molecules, as can be readily deduced from the fact that each has an internal plane of symmetry. For both molecules, the symmetry plane includes C-1 and its two substituents and bisects the C-2—C-3 bond.

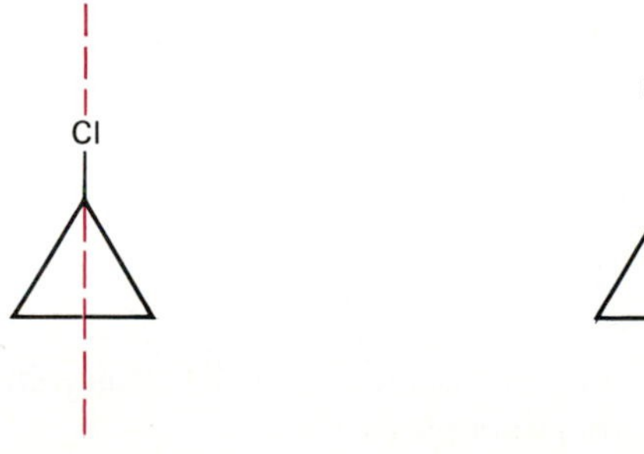

For 1,2-dichlorocyclopropane there are three stereoisomers. In one of the isomers, the two chloro substituents are on the same side of the plane of the ring. This isomer is called *cis*-1,2-dichlorocyclopropane (L., *cis*, on this side). This isomer also contains a symmetry plane, and is therefore achiral (even though it has two stereocenters).

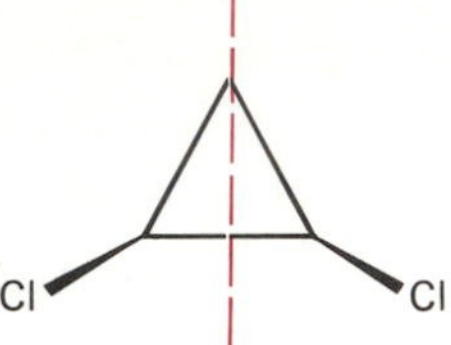

In the other two stereoisomeric 1,2-dichlorocyclopropanes, the two chloro substituents are attached to opposite sides of the cyclopropane ring. These isomers are called trans (L., *trans*, across). Neither of these isomers have an internal symmetry plane. Indeed, as the following structures show, the two *trans*-1,2-dichlorocyclopropanes are enantiomers.

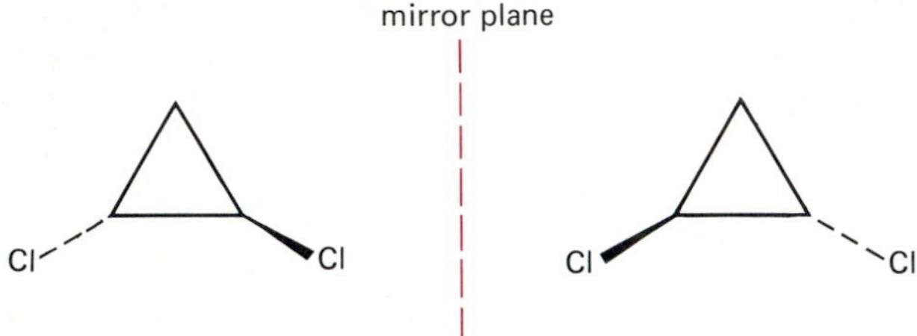

The set of stereoisomeric 1,2-dichlorocyclopropanes is completely analogous to the set of stereoisomeric 2,3-dichlorobutanes we discussed in Section 7.5. The cis isomer has the *R,S* configuration and is a meso compound. It has a diastereomeric relationship to each of the trans isomers, which have *R,R* and *S,S* configurations, respectively.

EXERCISE 7.12 Construct molecular models of the three stereoisomeric 1,2-dichlorocyclopropanes. Verify that the two trans isomers have a mirror-image relationship. Find the internal symmetry plane in the cis isomer. Assign *R* and *S* configurations to the two stereocenters in each of the three isomers you have constructed.

Chlorocyclohexane and 1,1-dichlorocyclohexane are analogous to their cyclopropane counterparts. Both have symmetry planes, which include C-1 and its two substituents. In these molecules, the internal symmetry planes also pass through C-4 and its two substituents. The symmetry planes are clearly evident in ''flat'' projection structures of the two compounds.

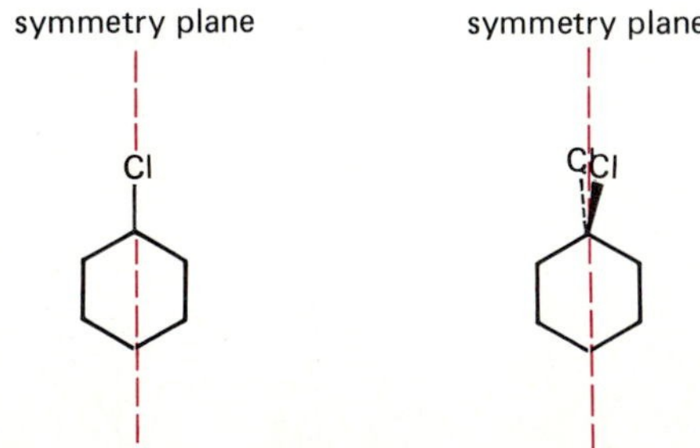

In the chair perspective structure of 1,1-dichlorocyclohexane, the symmetry plane may also be seen, but with somewhat more difficulty.

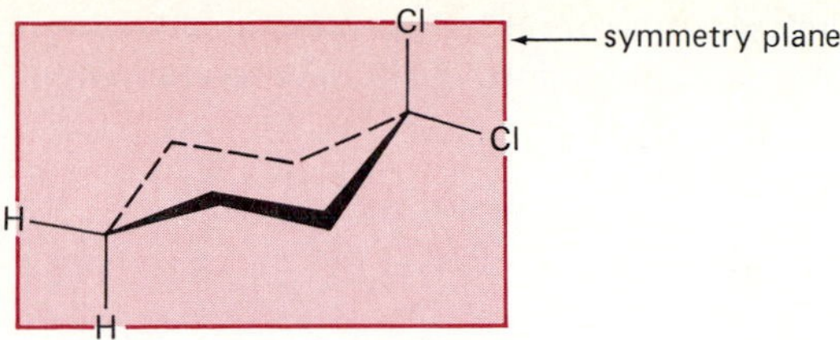

1,1-dichlorocyclohexane

Chlorocyclohexane can exist in two different chair conformations since the chloro group may be either axial or equatorial. As shown by the following structural drawings, both conformations have an internal symmetry plane, and are therefore achiral.

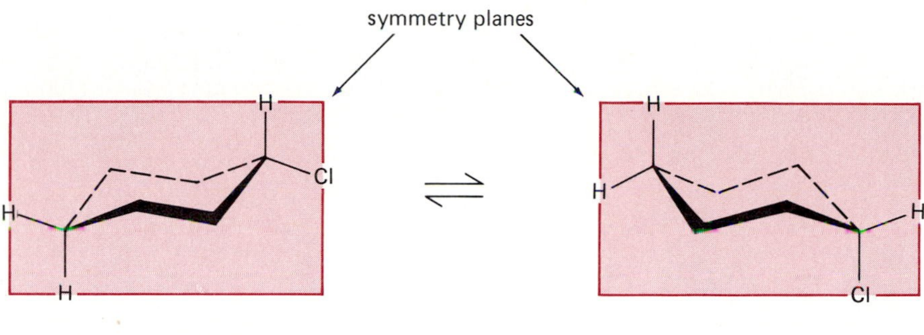

equatorial 1-chlorocyclohexane

axial 1-chlorocyclohexane

The two chair conformations of 1-chlorocyclohexane actually have a diastereomeric relationship, since they are stereoisomers that do not have a mirror-image relationship. However, as we saw in Section 5.6, the energy barrier for the conversion of one cyclohexane chair conformation to another is on the order of 10 kcal mole^{-1}. Thus, at normal temperatures, they are in rapid equilibrium. In fact, the situation is analogous to that seen with butane (Sections 5.2 and 7.1), where the two enantiomeric gauche conformations are in rapid equilibrium.

For 1,4-dichlorocyclohexane, there is a cis and a trans isomer. Both have an internal symmetry plane and are thus achiral.

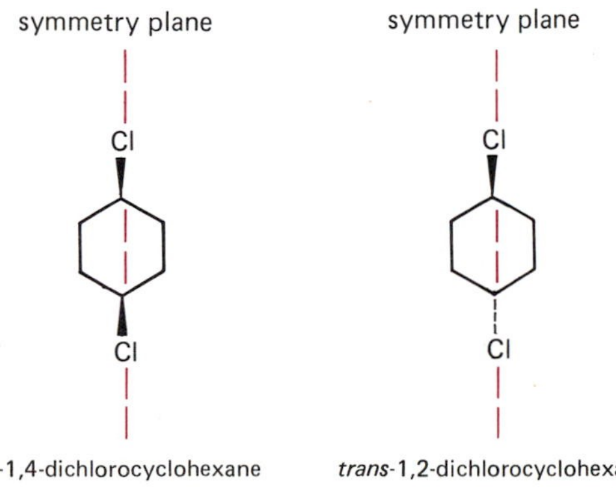

cis-1,4-dichlorocyclohexane   *trans*-1,2-dichlorocyclohexane

EXERCISE 7.13 Construct molecular models of *cis*- and *trans*-1,4-dichlorocyclohexane. Note that the trans isomer has two *different* chair conformations. Identify the symmetry plane in each of these isomers. What relationship do the two chair conformations of the cis isomer have?

The 1,3-dichlorocyclohexane stereoisomers comprise an interesting case. A plane of symmetry is apparent in the cis isomer. Since this isomer has stereocenters, *cis*-1,3-dichlorocyclohexane is a meso compound.

symmetry plane

(1*R*,3*S*)-1,3-dichlorocyclohexane

However, no symmetry plane exists in the trans isomer. Thus, as is the case with *trans*-1,2-dichlorocyclopropane, there are two enantiomeric *trans*-1,3-dichlorocyclohexanes.

mirror plane

(1*R*,3*R*)-1,3-dichlorocyclohexane (1*S*,3*S*)-1,3-dichlorocyclohexane

Note that with the absolute configuration of each stereocenter specified it is not necessary to specify that either is a trans isomer.

EXERCISE 7.14 Construct molecular models of the two enantiomeric *trans*-1,3-dichlorocyclohexanes. Convince yourself that there are no conformations in which the two isomers are superimposable.

7.7 Conformations of Substituted Cyclohexanes

As we saw in the preceding section, a monosubstituted cyclohexane can exist in two different conformations, since the substituent can be either axial or equatorial. The two chair conformations of methylcyclohexane are shown in Figure 7.11. It is instructive to look at Newman projections of the ring C-1—C-2 bond of methylcyclohexane (Figure 7.12).

In the equatorial conformation, the methyl group is anti to the C-3 CH_2 group of the ring, whereas in the axial conformation these groups have a gauche relationship. The

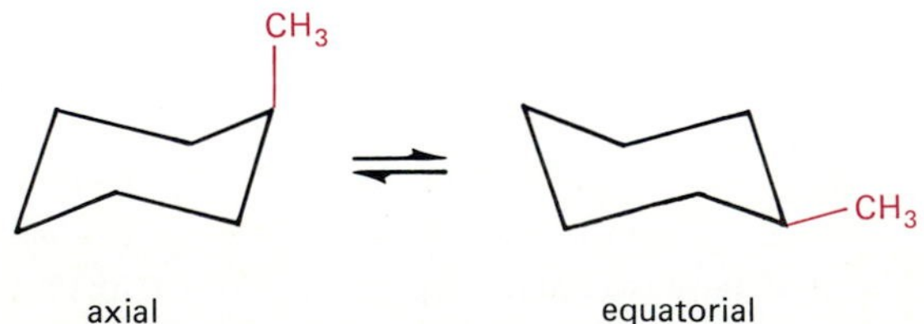

FIGURE 7.11 Chair conformations of methylcyclohexane.

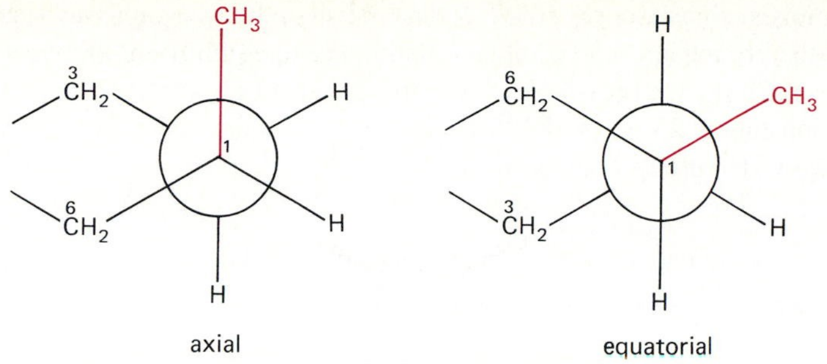

FIGURE 7.12 Newman projections of methylcyclohexane.

interaction of a methyl hydrogen with the axial hydrogen of the C-3 CH_2 group is much like the interaction of the corresponding hydrogens in a gauche conformation of butane (Section 5.2). Recall that this interaction in butane causes an enthalpy increase of 0.9 kcal mole^{-1}. There are two such interactions in axial methylcyclohexane—between the methyl group and the C-3 CH_2 and the C-5 CH_2, as shown in Figure 7.12. The axial conformation of methylcyclohexane is, as expected, about 1.8 kcal mole^{-1} less stable than the equatorial conformation. This difference in energy for the two conformations may be approximated as $\Delta G°$ and transformed into an equilibrium constant for the equilibrium between the two isomers at 25 °C:

$$\Delta G° = -RT \ln K$$
$$(-1.8 \text{ kcal mole}^{-1}) = -(1.987 \times 10^{-3} \text{ kcal mole}^{-1} \text{ deg}^{-1})(298 \text{ deg}) \ln K$$
$$K = 21$$

Thus, at 25 °C, methylcyclohexane exists as an equilibrium mixture of the two conformations, with 95% of the molecules having the equatorial-methyl structure and 5% having the axial-methyl structure. Because of interaction of axial groups with the other axial hydrogens on the same side of the ring, axial conformations of substituted cyclohexanes are generally less stable than the corresponding equatorial conformations. Actual energy differences for various substituents, expressed as $\Delta G°$ values, are summarized in Table 7.1.

TABLE 7.1 Conformational Energies for Monosubstituted Cyclohexanes

Group	$-\Delta G°$ (axial $\rightleftharpoons$ equatorial), kcal mole^{-1} (25°C)
F	0.25
Cl	0.5
Br	0.5
I	0.45
OH	1.0
CH_3	1.7
CH_2CH_3	1.8
$C{\equiv}CH$	0.41
$CH(CH_3)_2$	2.1
$C(CH_3)_3$	~5–6
C_6H_5	3.1
COOH	1.4
CN	0.2

Bulky groups such as isopropyl and *t*-butyl have such strong interactions in the axial position that the proportion of axial conformation in the equilibrium mixture is small. For example, the $\Delta G°$ of 3.1 kcal mole^{-1} for the group C_6H_5 (phenyl) corresponds to an equilibrium constant at 25 °C of 189; for phenylcyclohexane 995 out of every 1000 molecules have the C_6H_5 group equatorial.

EXERCISE 7.15 Calculate the equilibrium constants for the axial $\rightleftharpoons$ equatorial equilibria in isopropylcyclohexane and *t*-butylcyclohexane.

Like other substituted cyclohexanes, 1,1-dimethylcyclohexane can exist in two chair conformations. In each conformation, one methyl group is axial and one is equatorial.

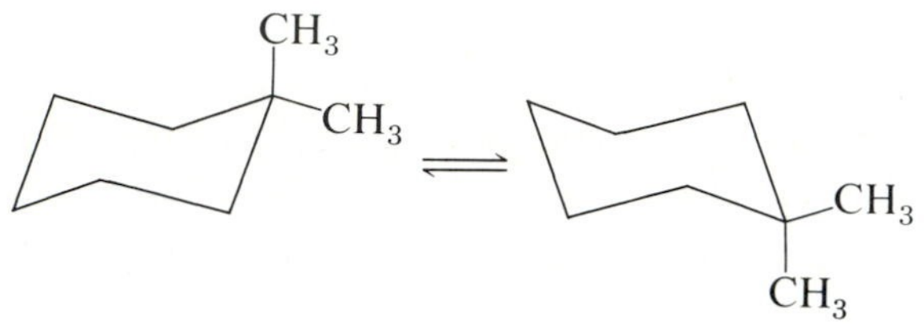

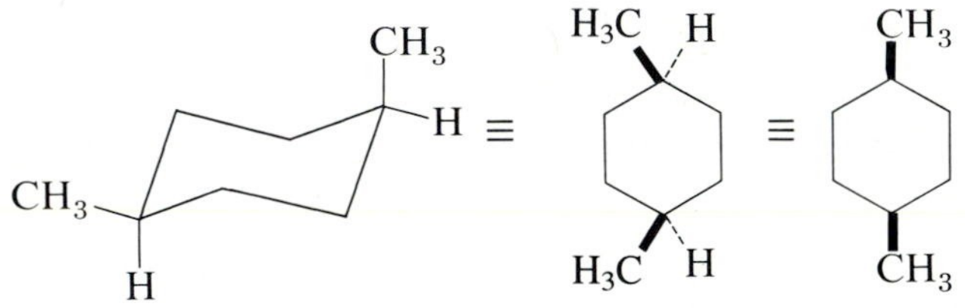

cis-1,4-dimethylcyclohexane

Of course, the $\Delta G°$ for this equilibrium is zero, and the corresponding equilibrium constant is 1.

The lower energy of equatorial substituents compared to axial is also seen in the disubstituted compounds. For example, *trans*-1,4-dimethylcyclohexane ($\Delta H_f^\circ =$ -44.1 kcal mole^{-1}) is more stable than the cis isomer ($\Delta H_f^\circ = -42.2$ kcal mole^{-1}) by 1.9 kcal mole^{-1}. In the trans isomer both methyl groups can be accommodated in equatorial positions, whereas in the cis hydrocarbon one methyl must be axial.

Both *cis*- and *trans*-1,4-dimethylcyclohexane can exist in two chair conformations. For the cis isomer, the two conformations are of equal energy, since each has one axial substituent and one equatorial substituent.

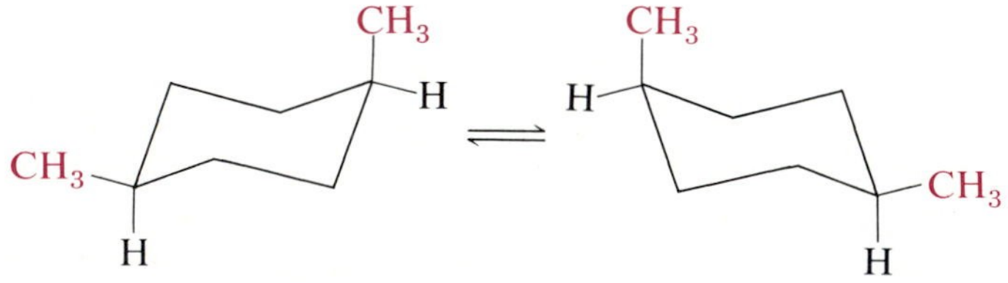

cis-1,4-dimethylcyclohexane, $\Delta G° = 0$

For the trans isomer, one conformation has both substituents axial and the other has both substituents equatorial. The diequatorial conformation predominates greatly at equilibrium ($\Delta G° \cong 3.4$ kcal mole^{-1}).

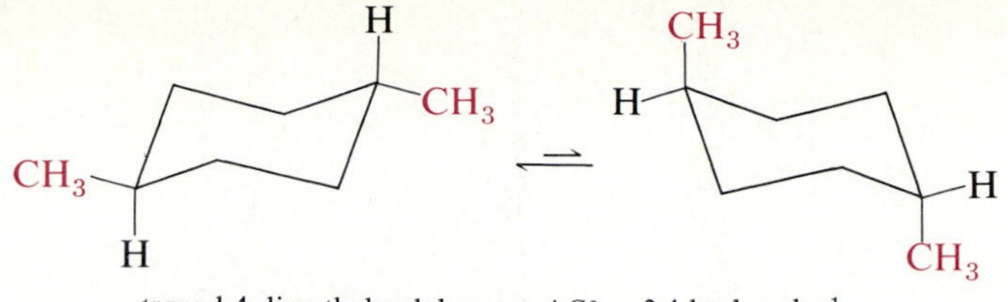

H CH3 CH3 H H CH3 H CH3

trans-1,4-dimethylcyclohexane, $\Delta G° \cong 3.4$ kcal mole^{-1}

For 1,3-dimethylcyclohexane the cis isomer is more stable than the trans. In *cis*-1,3-dimethylcyclohexane both methyls can be equatorial, whereas one methyl must be axial in the trans isomer.

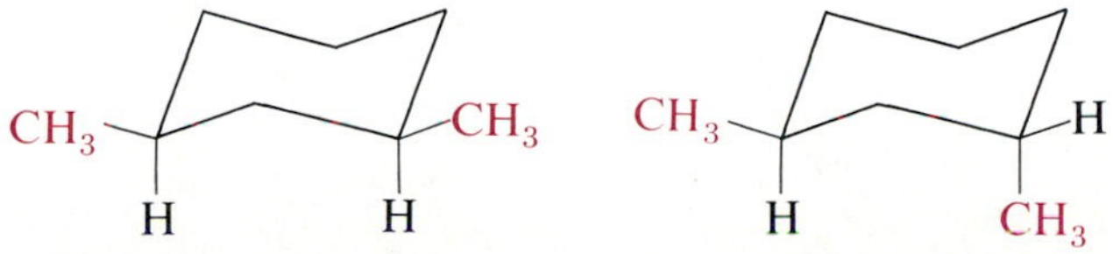

CH3 CH3 H H CH3 H H CH3

cis-1,3-dimethylcyclohexane
both methyls equatorial

trans-1,3-dimethylcyclohexane
one methyl equatorial

The *t*-butyl group is so bulky that it effectively demands an equatorial position. Indeed, an axial *t*-butyl group represents so strained a structure that it is not possible to measure its concentration directly. Consequently, the $\Delta G°$ value in Table 7.1 for the difference between axial and equatorial *t*-butyl groups is only a rough estimate. In *cis*-1-*t*-butyl-4-methylcyclohexane, for example, the conformation with axial methyl and equatorial *t*-butyl groups dominates completely.

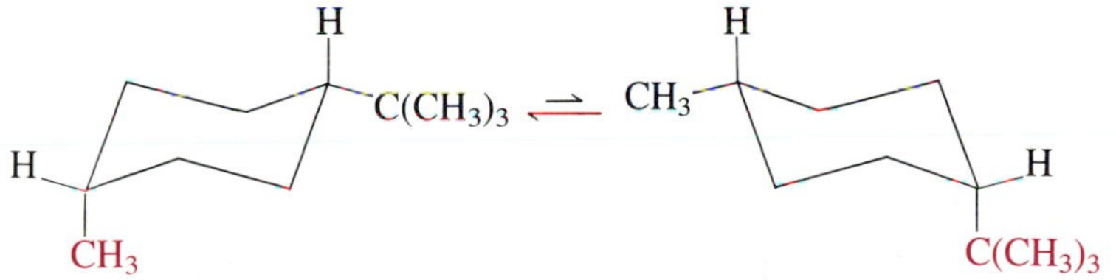

H H C(CH3)3 CH3 H H CH3 C(CH3)3

EXERCISE 7.16 Using the data in Table 7.1, estimate the equilibrium constant for the foregoing equilibrium.

When excessive strain is involved, a distortion of the cyclohexane ring occurs. For example, phenyl and *t*-butyl are both rather bulky groups. A crystal structure analysis of a compound that has a *cis*-1-*t*-butyl-4-phenylcyclohexane structure shows that the ring has been stretched out somewhat but still has essentially a chair conformation with axial phenyl and equatorial *t*-butyl groups.

In *trans*-1,3-di-*t*-butylcyclohexane a chair cyclohexane ring would require one *t*-butyl group to be axial as in Figure 7.13. Actually, in this compound the cyclohexane ring is twisted in order to avoid placing the *t*-butyl group in an axial position. This new conformation of cyclohexane is related to the hypothetical **boat** conformation shown in Figure 7.14. In this conformation, however, two of the hydrogens are so close together that a slight further twisting occurs to give the so-called "twist-form" or "skew-boat" structure as shown in Figure 7.15. This skew-boat form occurs in several compounds containing bulky groups but is not an important conformation for cyclohexane itself. In the skew-boat conformation several hydrogens are partially eclipsed. The structure has a strain energy of about 5 kcal mole^{-1} relative to chair cyclohexane.

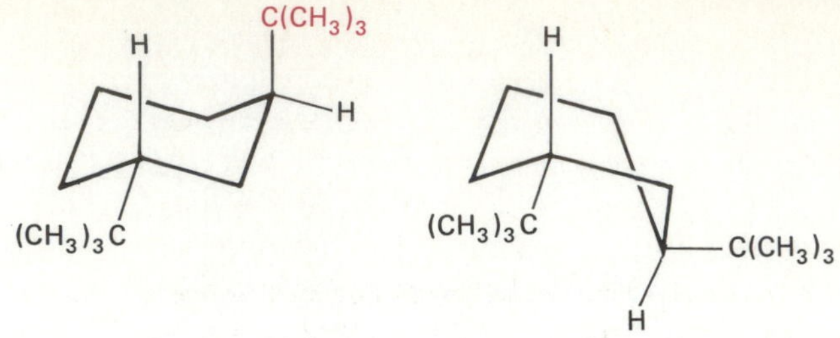

FIGURE 7.13 Conformations of *trans*-1,3-di-*t*-butylcyclohexane.

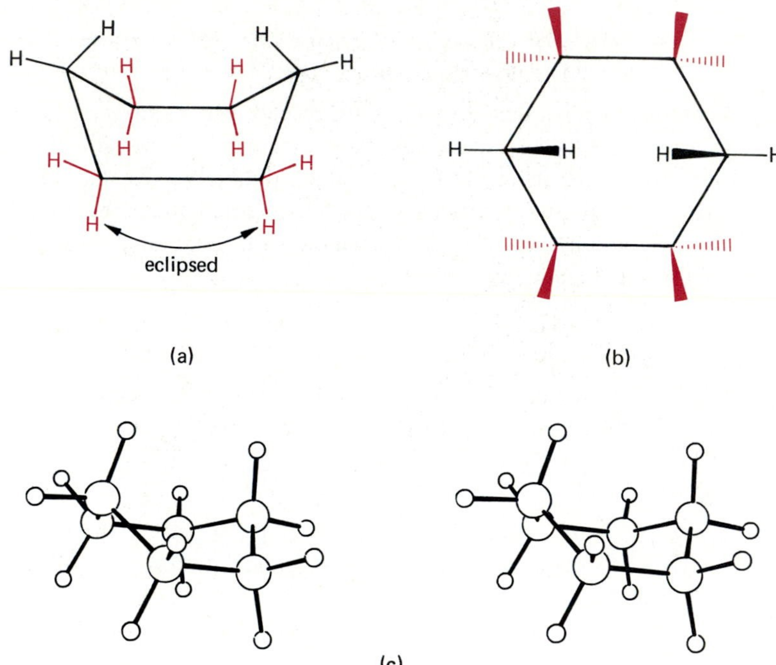

FIGURE 7.14 Boat conformation of cyclohexane: (a) side view; (b) top view; (c) stereo view.

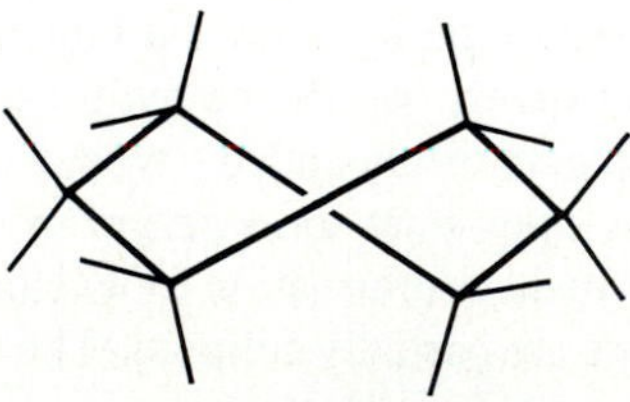

FIGURE 7.15 Skew-boat conformation of cyclohexane. The hydrogens are omitted for clarity.

EXERCISE 7.17 Construct a molecular model of cyclohexane and place it in the ideal boat conformation depicted in Figure 7.14. Note that four of the ring carbons (C-2, C-3, C-5, and C-6) lie in a common plane, and that C-1 and C-4 are tipped above this plane. Verify that this conformation has four eclipsed H:H interactions. Note that one of the C-1 hydrogens is very close to one of the C-4 hydrogens. With a ruler, measure the distance between these hydrogens and compare with the distance between a pair of eclipsed hydrogens. Now manipulate your model so that its carbon skeleton matches the representation of the twist-boat conformation shown in Figure 7.15. Note that in this conformation, all H:H interactions are minimized.

Highlight 7.4

The **most stable** conformation of cyclohexane is the **chair**; a **boat** conformation is **destabilized** by four eclipsed hydrogen-hydrogen interactions plus that between the opposed 1,4-hydrogens or about 6.5 kcal $mole^{-1}$. A **twist-boat** conformation is destabilized by about 5 kcal $mole^{-1}$. If a bulky group would be axial in a chair conformation, the twist-boat conformation may be preferred to give the least unstable arrangement, as in the case of *trans*-1,3-di-*t*-butylcyclohexane.

7.8 Chemical Reactions and Stereoisomerism

When a chemical reaction involves only achiral reactants, solvents, and reagents, the reaction products must be either achiral or, if chiral, they must be racemates. As an example, consider the monochlorination of butane. After the monochlorobutane fraction has been isolated, it is found to be a mixture of 1-chlorobutane and 2-chlorobutane. The 1-chlorobutane is, of course, achiral. The 2-chlorobutane formed in the reaction is a racemate, an equimolar mixture of (*R*)-2-chlorobutane and (*S*)-2-chlorobutane.

$$CH_3CH_2CH_2CH_3 + Cl_2 \xrightarrow{h\nu} CH_3CH_2CH_2CH_2Cl + \left\{ \begin{array}{c} CH_3CH_2\text{—}\overset{Cl}{\underset{H}{C}}\text{—}CH_3 \\ \\ CH_3CH_2\text{—}\overset{H}{\underset{Cl}{C}}\text{—}CH_3 \end{array} \right.$$

Moreover, recall that the reactive intermediate in the reaction leading to 2-chlorobutane is the *sec*-butyl free radical, which is approximately planar. Being planar, it is achiral and may react with Cl_2 on either of its faces. Reaction on one face yields (*R*)-2-chlorobutane, and reaction on the other side yields (*S*)-2-chlorobutane. The foregoing reaction illustrates an interesting point of stereochemistry. Because of its planar geometry, the *sec*-butyl free radical is achiral. However, reaction on its two faces gives rise to enantiometric products. The two faces are said to have an **enantiotopic** relationship. The sequence rule is used to assign names to the enantiotopic faces in the following way. First, assign priorities 1, 2, and 3 to the three groups attached to the carbon of the free radical. Second, orient the planar molecule so that you are looking down on one of the two faces. Third, trace the arc

1–2–3; if this arc is clockwise, you are looking at the ***re*** face; if it is counterclockwise, you are looking at the ***si*** face.

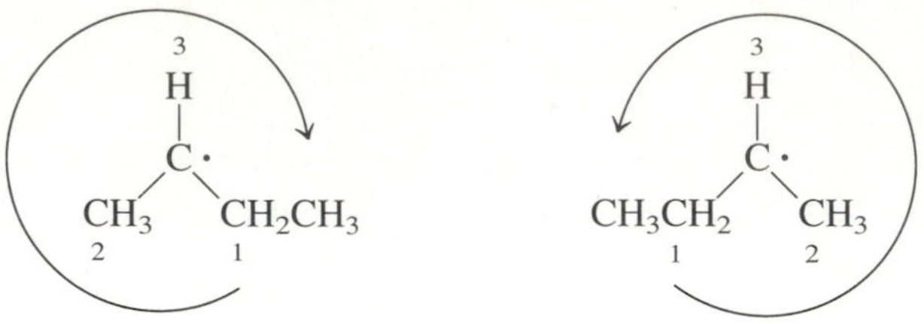

the *re* face

the *si* face

Note that not all planar species have enantiotopic faces. For example, consider the isopropyl free radical, in which two of the groups attached to the free-radical carbon are the same. In this case, reaction of Cl_2 with the two faces gives the same achiral product, 2-chloropropane. In this case, the faces are said to be **homotopic.** Since reaction is equally probable on the two faces, a racemate results. Consequently, any reaction that involves an achiral intermediate will give racemic products.

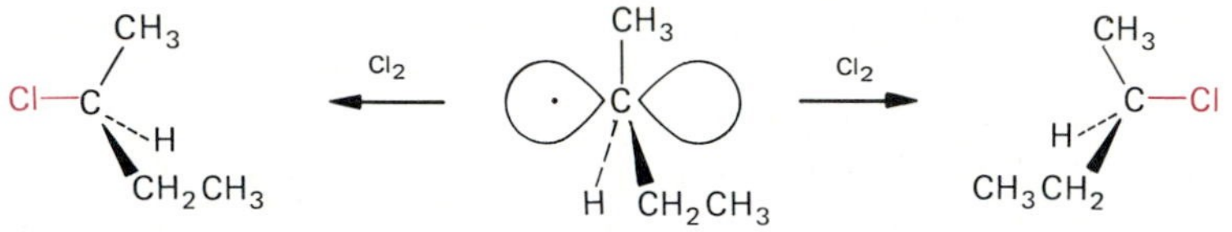

(*S*)-2-chlorobutane

(*R*)-2-chlorobutane

This result may also be discussed in terms of the relative rates of two competing reactions, *sec*-butyl free radical reacting with Cl_2 to give either (*R*)- or (*S*)-2-chlorobutane. The transition states for the two reactions are depicted as follows.

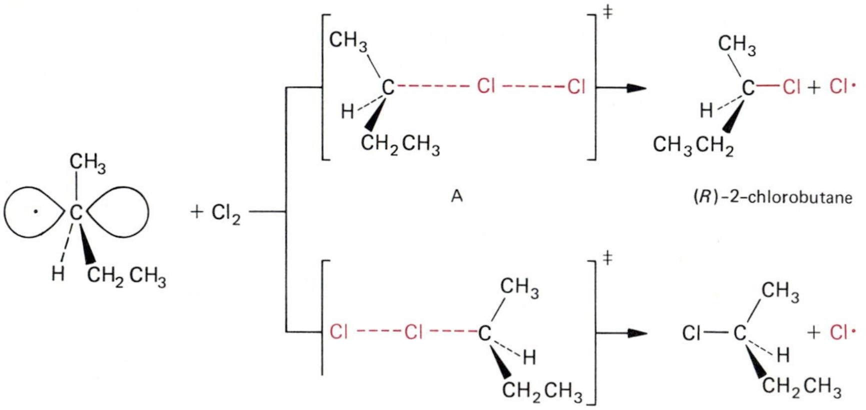

A

(*R*)-2-chlorobutane

B

(*S*)-2-chlorobutane

Notice that transition state A leading to the *R* enantiomer and transition state B leading to the *S* enantiomer are themselves enantiomeric. Because they are enantiomeric, they have identical physical properties, including bond angles, bond lengths, and free energies of formation. Since the two competing reactions begin at the same place and pass through transition states of equal energy, they have identical activation energies, and a 50:50 mixture of (*R*)- and (*S*)-2-chlorobutane results (Figure 7.16).

In Sections 4.2 and 4.3, we briefly considered the reaction of methyl chloride with hydroxide ion, in which the hydroxy group replaces the chloro group. Analogous reactions occur with other alkyl halides. For example, the enantiomeric 2-iodobutanes undergo a similar reaction. As we shall see in the next Chapter, an **inversion of absolute**

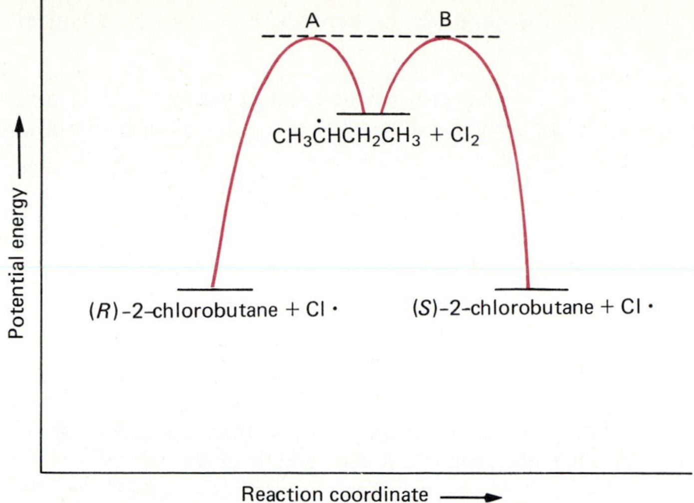

FIGURE 7.16 An achiral intermediate gives enantiomeric transition states with equal activation energies.

configuration takes place in this reaction, so that (*R*)-2-iodobutane gives (*S*)-2-butanol and (*S*)-2-iodobutane gives (*R*)-2-butanol.

(*R*)-2-iodobutane + OH^- $\longrightarrow$ (*S*)-2-butanol + I^-

(*S*)-2-iodobutane + OH^- $\longrightarrow$ (*R*)-2-butanol + I^-

In Chapter 9, we shall study this kind of reaction in detail. For now, let us only note that the **principle of enantiomeric transition states** that we have just learned demands that the two reactions take place with precisely the same rate. This may be readily appreciated by inspection of Figure 7.17. Since (*R*)- and (*S*)-2-iodobutane are enantiomeric, they have equal free energies of formation. Similarly, the transition states leading from each

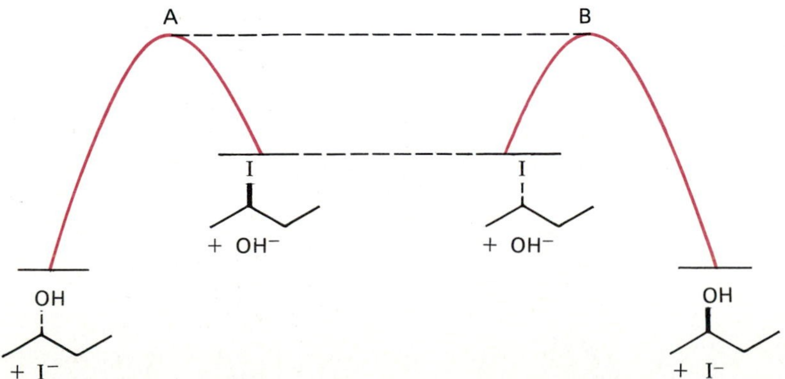

FIGURE 7.17 Enantiomers react with achiral reagents with equal rates.

enantiomer to the respective alcohols (A and B) bear an enantiomeric relationship. Consequently, the energies of activation for the two reactions must be identical.

> Catalytically active proteins such as enzymes are chiral molecules and the transition states of the reactions catalyzed are chiral and different in energy for enantiomeric reactants.

ASIDE

An Elamite cemetery at Susa in Khuzistan (part of present-day Iran) contained large numbers of copper mirrors, and elaborate bronze mirrors are known of Chinese and Celtic origin. Inexpensive modern silver-coated glass mirrors, or even imaginary mirrors, can be used just as effectively to illustrate mirror-image relationships. Construct two models of bromochlorofluoromethane as mirror images. Show that they are not superimposable. Which model is *R* and which is *S*?

PROBLEMS

1. Calculate $[\alpha]_D$ for each of the following compounds.

a. A 1 M solution of 2-chloropentane in chloroform in a 10-cm cell gives an observed α of $+3.64°$.

b. A solution containing 0.96 g of 2-bromooctane in 10 mL of ether gives an observed α of $-1.80°$ in a 5-cm cell.

2. How many stereoisomers may exist for each of the following compounds?

a. 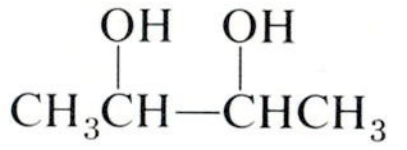 $CH_3\overset{OH}{\overset{|}{C}}H-\overset{OH}{\overset{|}{C}}HCH_3$ **b.** $CH_3CH_2\overset{OH}{\overset{|}{C}}H-\overset{OH}{\overset{|}{C}}HCH_3$

c. $CH_3\overset{Cl}{\overset{|}{C}}H-\overset{OH}{\overset{|}{C}}HCH_3$ **d.** 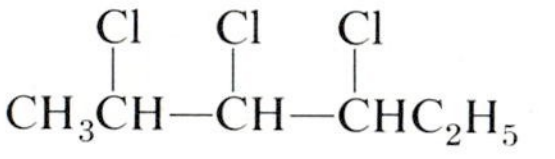 $CH_3\overset{Cl}{\overset{|}{C}}H-\overset{Cl}{\overset{|}{C}}H-\overset{Cl}{\overset{|}{C}}HC_2H_5$

e. 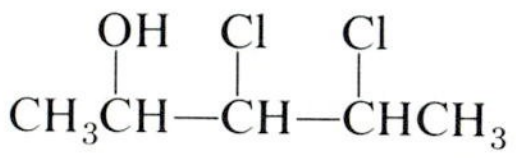 $CH_3\overset{OH}{\overset{|}{C}}H-\overset{Cl}{\overset{|}{C}}H-\overset{Cl}{\overset{|}{C}}HCH_3$ **f.** 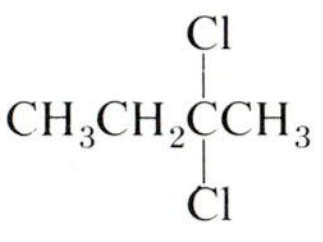 $CH_3CH_2\underset{Cl}{\underset{|}{\overset{Cl}{\overset{|}{C}}}}CH_3$

3. For parts **a–e** of problem 2, show which pairs of stereoisomers are enantiomeric and which pairs are diastereomeric. Assign *R* or *S* to each stereocenter.

4. For each of the following compounds assign *R* or *S* to each stereocenter.

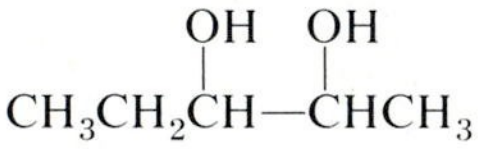

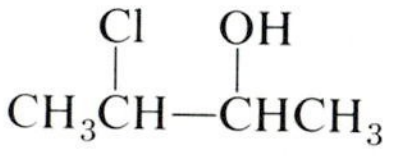

a. 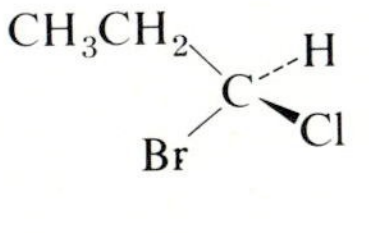 C bearing CH_3CH_2, H (dashed), Cl (wedge), Br

b. CH_3, Cl (dashed), H (wedge) on C—C bearing H (dashed), CH_3 (wedge), Cl

c. C bearing H, F (dashed), Br (wedge), Cl

d. C bearing CH_3, H (dashed), OH (wedge), D

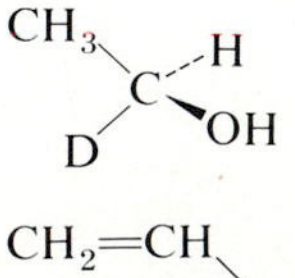

e. 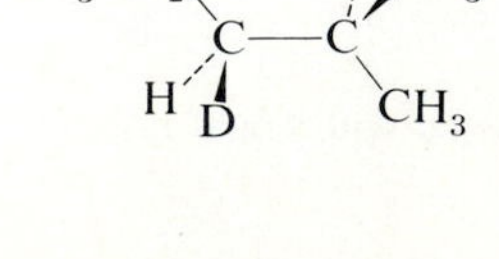 CH_3CH_2, H (dashed), D (wedge) on C—C bearing Br (dashed), CH_3 (wedge), CH_3

f. 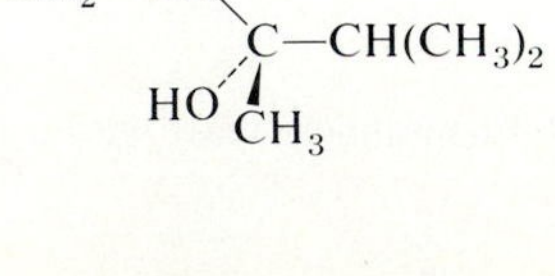 $CH_2{=}CH$, HO (dashed), CH_3 (wedge) on C—$CH(CH_3)_2$

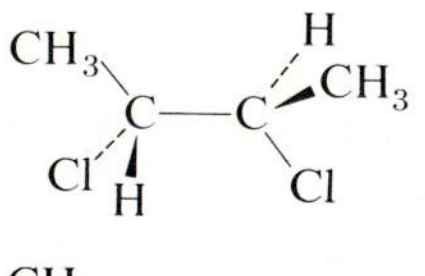

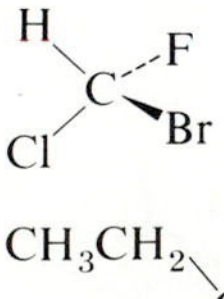

g. 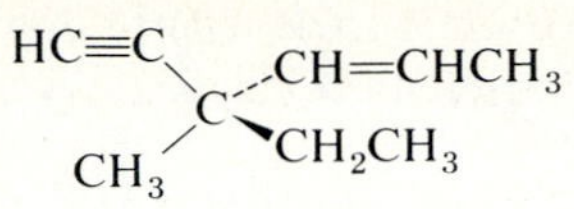

h. 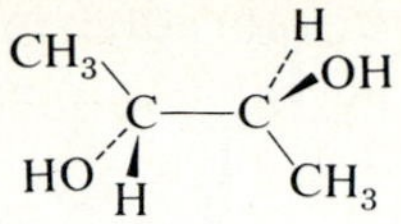

5. For each of the following compounds, assign *R* or *S* to each stereocenter.

a. 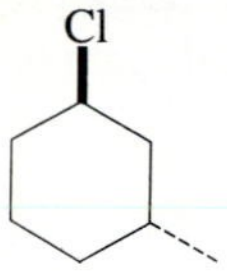

b. 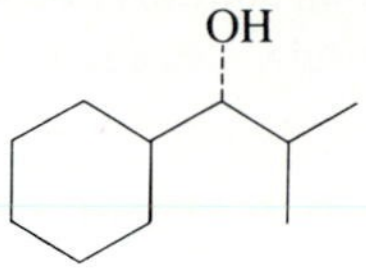

c. 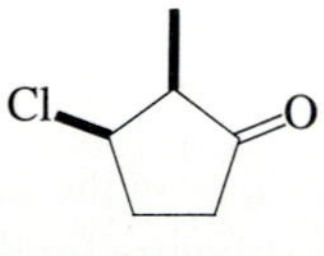

d. 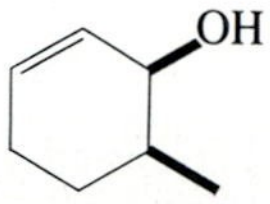

6. Write structures for each of the following compounds.
a. (1*R*,3*R*)-1,3-dichlorocyclohexane
b. (3*S*,4*R*)-3-chloro-4-methylhexane
c. the meso isomer of 1,3-dimethylcyclopentane
d. an optically active isomer of 1,2-dimethylcyclobutane

7. Write structures for all of the stereoisomeric 2,4-dichloro-3-methylpentanes. Identify the two meso forms.

8. Given the Newman projection

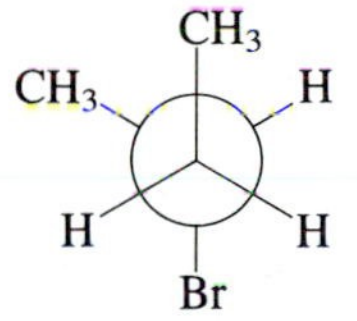

Is this structure *R* or *S*? Determine whether each of the following structural symbols is equivalent to the above Newman projection or to its enantiomer.

a. 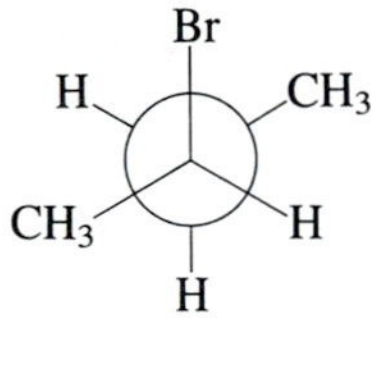

b. 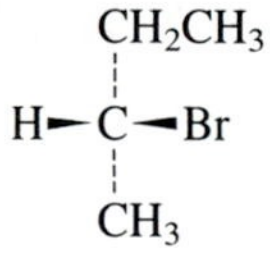

c. 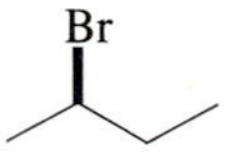

d. 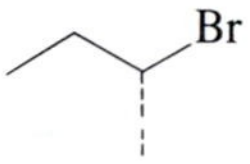

9. How many stereoisomers exist for 1,3-dichloro-2,4-dimethylcyclobutane? Write all the structures. Which are chiral and which are achiral? Identify all planes and points of symmetry in the various structures.

10. Assign *R* or *S* to each stereocenter in the following compounds.

a. 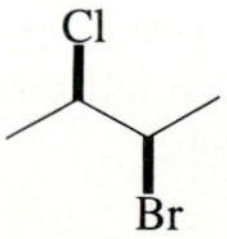

b. 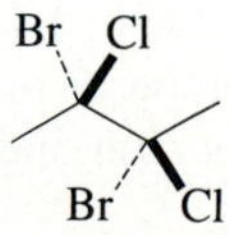

c. 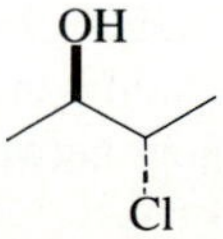

d. 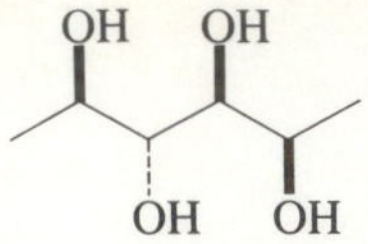 e. 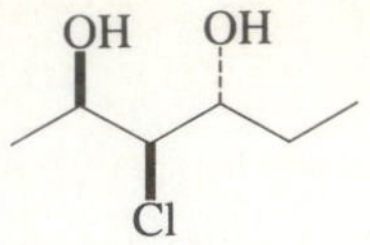 f.

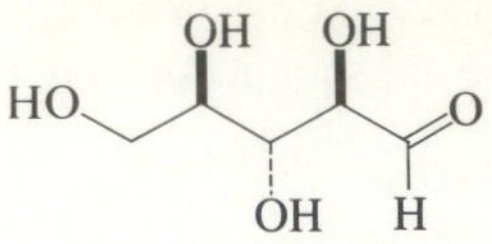

11. A fifth sub-rule of the sequence rule (Section 7.3) provides that when two substituents are identical except in absolute sense of chirality, *R* configuration takes precedence over *S* configuration. Using this sub-rule, assign configurational labels to the stereocenters in the following compounds.

a. 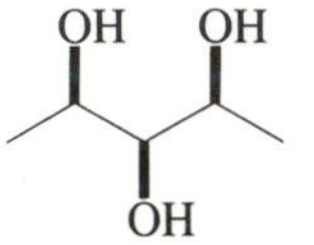 b. 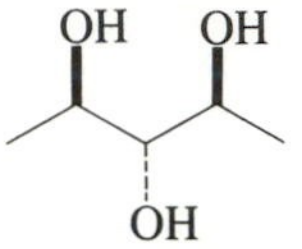 c.

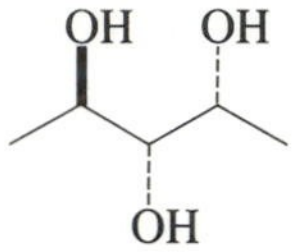

12. Draw a chair perspective structure for *cis*-1,2-dimethylcyclohexane. Is the molecule in the conformation you have drawn chiral or achiral? Now draw the structure for the other chair conformation. What relationship do the two conformations have?

13. Using the conformational energies in Table 7.1, account for the large difference in energies between the following as cyclohexane substituents.

a. $C(CH_3)_3$ and $CH(CH_3)_3$ **b.** CH_2CH_3 and C≡CH

14. From the data in Table 7.1, calculate the percentage of molecules having the substituent in the equatorial position for each of the following compounds.

a. 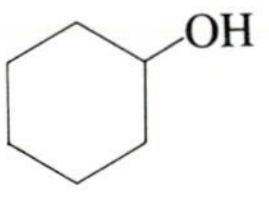 b. 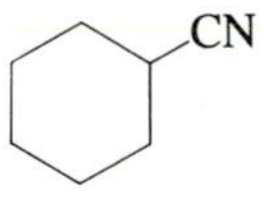 c.

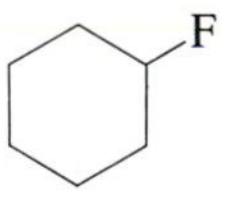

15. Using the conformational energies in Table 7.1 as a guide, estimate the difference in energy between the diaxial and diequatorial conformers of *trans*-1,4-dichlorocyclohexane. Experimentally, it is found that the diequatorial conformer predominates in the equilibrium, as expected. However, for *trans*-1,2-dichlorocyclohexane, the diaxial conformer is found to predominate at equilibrium. Explain.

16. Draw Newman projections for the C-1—C-2 bond, the C-1—C-6 bond, and the C-2—C-3 bonds of *cis*- and *trans*-1,2-dimethylcyclohexane (for the trans isomer, use the diequatorial conformation). Identify all gauche interactions involving the methyl groups. Estimate the difference in energy in the two isomers. How does your estimate compare with the difference in energy calculated from the heats of formation given in Appendix I?

17. Make perspective structural drawings of the two chair conformations of *cis*-1,3-dimethylcyclohexane. Assuming that the only important interactions are the *gauche*-butane interactions involving the methyl groups, estimate the difference in energy in the two isomers. The actual difference is not known accurately, but it has been estimated to be greater than 5 kcal mole^{-1}. Can you offer an explanation for the difference between this value and the value you estimated?

18. (*S*)-1-Chloro-2-methylbutane has been shown to have (+) rotation. Among the products of light-initiated chlorination are (−)-1,4-dichloro-2-methylbutane and (±)-1,2-dichloro-2-methylbutane.

a. Write out the absolute configuration of the (−)-1,4-dichloro-2-methylbutane produced by the reaction and assign the proper *R* or *S* label. What relationship does this example show between sign of rotation and configuration?

b. What does the fact that the 1,2-dichloro-2-methylbutane produced is totally racemic indicate about the reaction mechanism and the nature of the intermediates?

19. Consider the chlorination of (*S*)-2-fluorobutane. The monochlorination fraction of the reaction product contains 1-chloro-2-fluorobutane, 2-chloro-2-fluorobutane, 2-chloro-3-fluorobutane, and 1-chloro-3-fluorobutane.

a. The 1-chloro-2-fluorobutane constitutes 1% of the monochloro product. What is the absolute configuration of this material (*R* or *S*)?

b. The 1-chloro-3-fluorobutane constitutes 26% of the monochloro product. What is its absolute configuration?

c. The 2-chloro-2-fluorobutane fraction amounts to 31% of the monochloro product. This material is found to be racemic. How do you explain this result?

d. Careful examination of the 2-chloro-3-fluorobutane product (obtained in 40% yield) reveals that it is actually a mixture consisting of 16% of the 2*S*,3*S* diastereomer and 24% of the 2*R*,3*S* diastereomer. Can you offer an explanation for the fact that these two isomers are not produced in equal amounts? [*Hint*: Construct a reaction coordinate diagram analogous to Figure 7.16.]

20. If the apposed 1,4-hydrogen atoms in a boat conformer of cyclohexane are replaced by a CH_2 group, how would the stability of the system be affected relative to the boat? How would the stability of the twist-boat conformer be affected? The bicyclo[2.2.1]heptane (for naming the system, see Section 20.5) is present in many naturally occurring molecules (''natural products'').

CHAPTER 8

ALKYL HALIDES AND ORGANOMETALLIC COMPOUNDS

8.1 Structure of Alkyl Halides

Alkyl halides are the first group of organic compounds we will study in which there is a functional group. The halogen group can be converted into several other functional groups, and some of these reactions are among the most important in organic chemistry.

The carbon atoms in alkyl halides are essentially tetrahedral. The carbon-halogen bond may be regarded as resulting from overlap of a C_{sp^3} orbital with a hybrid orbital from the halogen. Molecular orbital calculations suggest that the hybrid halogen orbital is mostly p, with only a small amount of s-character. In methyl fluoride, for example, the hybrid orbital from fluorine in the carbon-fluorine bond is calculated to be about 15% s and 85% p. The reason for the relatively small amount of s-character is that the halogen has three lone pairs and most of the X_{2s} atomic orbital is used to bind these lone-pair electrons. Only a small amount of s-orbital is available for the orbital bonded to carbon. Note that the hybridization of the halogen must be computed; it is not amenable to experimental tests with currently available methods.

The carbon-halogen bond lengths of the methyl halides are shown in Table 8.1. The size of the halogen atom increases as we go down the periodic table. The fluorine atom is

TABLE 8.1 Bond Lengths of Methyl Halides

Compound	r_{C-X}, Å
CH_3F	1.39
CH_3Cl	1.78
CH_3Br	1.93
CH_3I	2.14

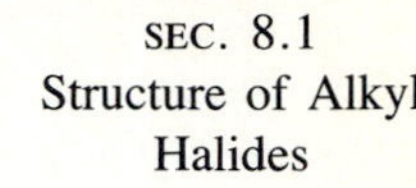

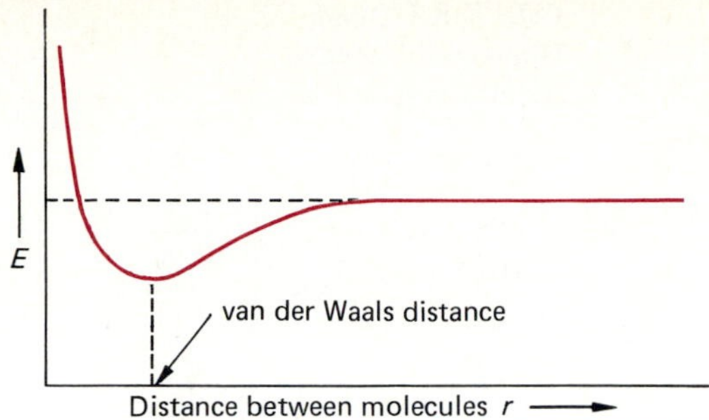

FIGURE 8.1 Van der Waals forces.

somewhat larger than hydrogen, but smaller than carbon: compare the C—F bond distance of 1.39 Å with C—C, 1.54 Å, and C—H, 1.10 Å. The higher halogens are all substantially larger than carbon.

The *van der Waals radius* of a group is the effective size of the group. As two molecules approach each other, the van der Waals attractive force (Section 5.1) increases to a maximum, then decreases and becomes repulsive (Figure 8.1). The van der Waals radius is defined as one half the distance between two equivalent atoms at the point of the energy minimum. It is an equilibrium distance and is usually evaluated from the structures of molecular crystals. Van der Waals radii for several atoms and groups are summarized in Table 8.2. The van der Waals radius of bromine (1.95 Å) is about the same as that for a methyl group (2.0 Å).

TABLE 8.2 Van der Waals Radii, Å

H 1.2	N 1.5	O 1.4	F 1.35
CH_2 2.0	P 1.9	S 1.85	Cl 1.8
CH_3 2.0			Br 1.95
			I 2.15

Although the carbon-halogen bonds in alkyl halides are covalent, they have a polar character because halogens attract electrons better (have greater electronegativity) than carbon. The "center of gravity" of the electron density in the carbon-halogen region does not coincide with the center of nuclear positive charge. This imbalance results in a **dipole moment** μ, which is expressed as the product of the charge q and the distance of separation d: $\mu = q \times d$. The distance involved has direction; hence, dipole moments are vectors. In the case of methyl halides this vector is directed along the carbon-halogen bond and is usually symbolized as

$$\overset{+\!\!\!\longrightarrow}{CH_3{-}X}$$

The direction of the arrow is from positive to negative charge. The magnitudes of the dipole-moment vectors for methyl halides are summarized in Table 8.3. Since the charge involved is on the order of 10^{-10} esu and the distance is on the order of 10^{-8} cm, dipole moments are on the order of 10^{-18} esu cm. This unit is named the Debye, abbreviated D,

TABLE 8.3 Dipole Moments of Methyl Halides (Vapor Phase)

Compound	μ, D
CH_3F	1.82
CH_3Cl	1.94
CH_3Br	1.79
CH_3I	1.64

after Peter Debye who discovered this molecular property. Thus, if a positive charge and a negative charge of 10^{-10} esu are separated by 10^{-8} cm, the system has a dipole moment of 1 Debye or 1 D.

EXERCISE 8.1 The electron charge is 4.8×10^{-10} esu. Using the bond distances in Table 8.1 calculate the fraction of an electronic charge at the ends of the carbon-halogen bond that would give rise to the dipole moments in Table 8.3. Note that fluorine is more electronegative than chlorine but gives rise to a smaller carbon-halogen dipole moment because of its shorter bond distance.

8.2 Physical Properties of Alkyl Halides

The lower molecular weight *n*-alkyl halides are gases at room temperature. Starting with *n*-butyl fluoride, *n*-propyl chloride, ethyl bromide, and methyl iodide, the alkyl halides are liquids at room temperature. This result comes mostly from the increasing effective "size" of the halogens as we proceed down the periodic table. We saw in the previous section how this changing size is reflected in increasing carbon-halogen bond distances along the series from fluorine to iodine. Increasing size carries with it an increase in the effective "area of contact" at the van der Waals radius that produces van der Waals attraction.

However, size does more than increase this area of contact. We learned on page 74 that van der Waals attraction results from the mutual correlation of electronic motions. The movement of one electron describes a changing electric field. The ability of a second electron to respond to such a changing field is measured by its **polarizability.** The smaller and "tighter" the atom, the lower the polarizability of its electrons and the lower the van der Waals attraction for a given area of contact. Consequently, along the series F, Cl, Br, I the polarizability increases. Furthermore, lone-pair electrons are generally held more loosely than bonding electrons and can be more polarizable. Although bromine has a van der Waals radius similar to that of a methyl group, it has much higher polarizability, and an alkyl bromide, RBr, has a much higher boiling point than that of the corresponding RCH_3. We have emphasized the role of van der Waals attractions in boiling points, but it should be mentioned that molecular weight also plays a role because of the effect of mass on kinetic energy. Along a given homologous series, however, the van der Waals force depends on the overall size of the molecule which, in turn, parallels the molecular weight.

The tightly held electrons and consequent low polarizability of fluorine results in the unique and distinctive properties of fluorocarbons, compounds composed entirely of carbon and fluorine. The boiling points of fluorocarbons are much closer to those of related hydrocarbons than might have been expected from the difference in molecular weights or size; for example, C_2H_6 has b.p. −89 °C; C_2F_6 has b.p. −79 °C.

Increasing the alkyl chain also causes a normal progressive increase in boiling point. As with alkanes themselves, branched systems have lower boiling points than isomeric linear systems. Some boiling point data are summarized in Table 8.4 and in Figure 8.2.

TABLE 8.4 Boiling Points of Alkyl Halides (RX)

	Boiling Point, °C				
R	X = H	F	Cl	Br	I
CH_3—	−161.7	−78.4	−24.2	3.6	42.4
CH_3CH_2—	−88.6	−37.7	12.3	38.4	72.3
$CH_3(CH_2)_2$—	−42.1	−2.5	46.6	71.0	102.5
$CH_3(CH_2)_3$—	−0.5	32.5	78.4	101.6	130.5
$CH_3(CH_2)_4$—	36.1	62.8	107.8	129.6	157.
$(CH_3)_2CH$—	−42.1	−9.4	34.8	59.4	89.5
$(CH_3)_2CHCH_2$—	−11.7		68.8		
$CH_3CH_2CH(CH_3)$—	−0.5		68.3	91.2	120.
$(CH_3)_3C$—	−11.8		50.7	73.1	dec.

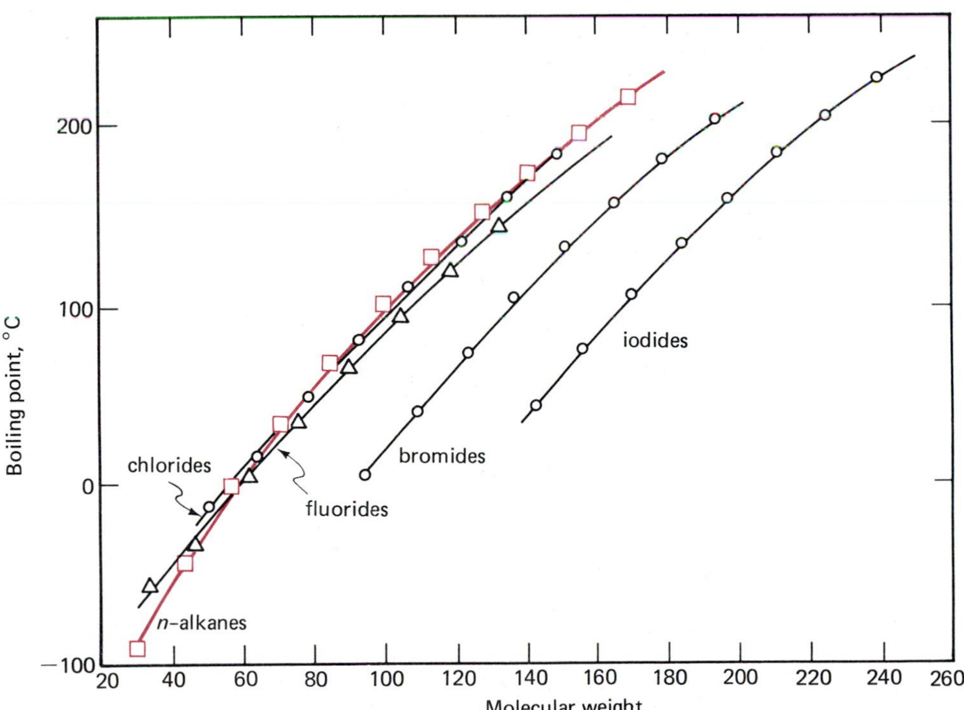

FIGURE 8.2 Boiling points of *n*-alkanes and *n*-alkyl halides.

Alkyl halides are insoluble in water but soluble in most organic solvents. They vary greatly in stability. Monofluoroalkanes are difficult to keep pure; on distillation they tend to lose HF to form alkenes. Chlorides are relatively stable and generally can be purified by distillation. However, higher molecular weight tertiary alkyl chlorides tend to lose HCl on heating and must be handled more carefully. Indeed, this property holds for most tertiary alkyl halides. Note in Table 8.4 that *t*-butyl iodide decomposes on attempted distillation at atmospheric pressure.

Chloroform slowly decomposes on exposure to light. This tendency is diminished by the presence of small amounts of alcohol. Commercially available chloroform has about 0.5% alcohol added as a stabilizer. Alkyl bromides and iodides are also light sensitive. Upon exposure to light they slowly liberate the free halogen and turn brown or violet, respectively. Thus, these halides are generally stored in opaque vessels or brown bottles and should generally be redistilled before use.

EXERCISE 8.2 (a) Which has the higher melting point, *n*-butyl bromide or *t*-butyl bromide? Explain. Compare your answer with melting points found in a handbook. (b) From the generalizations and data provided in this section and in Chapter 5, estimate the boiling points of $CH_3CH_2CH(CH_3)CH_2CH_2Cl$ and $CH_3CH_2CH_2CH(C_2H_5)CH_2Cl$. Look up these boiling points in a handbook.

8.3 Conformations of Alkyl Halides

Barriers to rotation about carbon-carbon bonds bearing halogens are comparable to those in hydrocarbons, and these compounds also prefer staggered conformations. Some rotation barriers are summarized in Table 8.5. Note that there is no simple relationship between the barrier and the size of the halogen. One reason is that as the size of a halogen increases, its bond length to carbon also increases. The same principle operates in the halogenated cyclohexanes; recall that the halogens show a relatively small and almost constant preference for the equatorial conformation (Table 7.1). In alkyl halides where rotation involves eclipsing C—H with C—H and C—H with C—X, the barriers are about 3.2–3.7 kcal mole^{-1}. Even for hexafluoroethane, where rotation involves eclipsing three pairs of carbon-fluorine bonds, the barrier is only 3.9 kcal mole^{-1}. However, rotation of one carbon-chlorine bond past another is more difficult; the barrier in hexachloroethane is 10.8 kcal mole^{-1}.

TABLE 8.5 Barriers to Rotation in Alkyl Halides

Compound	Rotation Barrier, kcal mole^{-1}
CH_3—CH_2F	3.3
CH_3—CF_3	3.25
CF_3—CF_3	3.9
CH_3—CH_2Cl	3.7
CCl_3—CCl_3	10.8
CH_3—CH_2Br	3.7
CH_3—CH_2I	3.2

1,2-Dichloroethane, like butane, exists in two conformations, gauche and anti.

Cl, H, Cl, H, H, H

gauche

Cl, H, H, H, H, Cl

anti

One conformation is converted to another by rotating a carbon-chlorine bond past a carbon-hydrogen bond. It is not necessary to rotate C—Cl past C—Cl. Accordingly, the barrier between these conformations is only about 3.2 kcal mole^{-1}, not much different than for rotation in ethyl chloride.

We saw in Chapter 5 that the gauche and anti conformations of butane differ in energy by about 0.9 kcal mole^{-1}. We might expect, therefore, that the two analogous conformations of 1,2-dichloroethane will also have different energies. In fact, in the vapor phase, the anti conformation is more stable by 1.2 kcal mole^{-1}. Remarkably, however, the energy difference in the pure liquid is about zero!

How do we account for this interesting observation? One explanation involves two opposing factors, dipole repulsion and van der Waals attraction. Each carbon-chlorine bond has an associated dipole moment. The electrostatic repulsion for two dipoles oriented in the anti conformation is lower than that for two dipoles oriented in the gauche conformation.

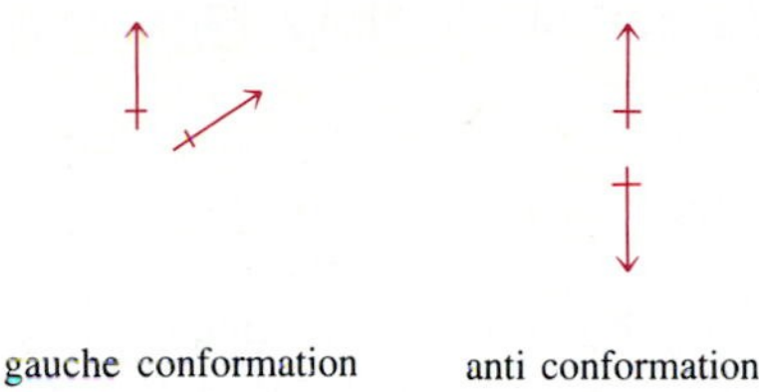

gauche conformation anti conformation

The other factor involved is van der Waals attraction. Two chlorines separated by little more than the sum of their van der Waals radii attract each other in exactly the same manner as two neighboring alkanes (Section 5.2). Such van der Waals attraction is especially important for the large halogen atoms because the lone-pair electrons are spread through a relatively large volume and respond easily to changing neighboring charge fields. That is, such electrons have relatively high polarizability.

The net result for the gauche and anti conformations is a balance. Van der Waals attraction favors the gauche conformation, but dipole-dipole repulsion favors the anti conformation. In the vapor phase the dipole effect dominates, and the anti structure is more stable. In the liquid phase dipole-dipole interaction is reduced by the proximity of other polar or polarizable molecules. The two effects now just cancel.

8.4 Some Uses of Halogenated Hydrocarbons

The simple alkyl halides and polyhaloalkanes are readily available and are used extensively as solvents. Chlorides are most important because of the low cost of chlorine relative to bromine and iodine. In fact, chlorine is one of the basic raw materials of the chemical industry. In 1988 the United States produced 11,330,000 tons of chlorine, most of which was used to produce chlorinated hydrocarbons.

The polychloromethanes are produced industrially by the chlorination of methane. Carbon tetrachloride has been used extensively in drycleaning establishments. However, it must be handled with care because it is an accumulative poison that causes liver damage. Consequently, its use in drycleaning has declined. Chloroform was once used as an anesthetic, but its use for this purpose has now been abandoned because it is toxic and a suspected carcinogen. The polyhalogenated compound $CF_3CHClBr$, "Halothane," has found important use as an inhalative general anesthetic because it is effective and relatively nontoxic. However, a rare and sometimes fatal genetic sensitivity to halothane is known.

Anesthesia may involve several different effects on the nervous system, including a decrease in the activity of a protein, the sodium channel. The membranes of the cells in the nervous system contain proteins called "ion channels" surrounded by lipid (see Chapter 18) that admit or expel sodium or potassium ions. Controlled movement of the ions is the way in which the nerve cells conduct the signals that allow us to move, breathe, eat, and remember the facts of organic chemistry. Local anesthetics work mainly by temporarily blocking the sodium channel. General anesthetics have activity related to solubility in the lipid of the membrane, and may affect the conformational transitions in the sodium channel necessary for the motion of the sodium. Anesthetics can be chemically inert; for example, xenon has anesthetic action. It is remarkable that the effective concentration of an anesthetic is species-independent. The same partial pressure of anesthetic functions as well in man as in a goldfish, reflecting a similarity in the mechanism of nerve conduction. The application of anesthetics needs to be carefully monitored. Lethal concentrations are typically only about double the useful anesthetic concentration.

A number of partially fluorinated alkanes are widely marketed for use as cooling fluids in refrigeration systems and as aerosol propellants. These compounds are often known by their trade names.

Compound	Trade Name	Systematic Name
$CFCl_3$	Freon 11	trichlorofluoromethane
CF_2Cl_2	Freon 12	dichlorodifluoromethane
CF_3Cl	Freon 13	chlorotrifluoromethane
CF_4	Freon 14	tetrafluoromethane

Photolysis of chlorofluorocarbons (CFCs) yields Cl· atoms that catalytically convert ozone, O_3, to oxygen, O_2. The "ozone layer" in the upper atmosphere absorbs dangerous ultraviolet light from the sun (see problem 12, Chapter 6). To lower the risk to humans, the production of CFCs will be decreased in many countries and the use of many will be phased out altogether by the year 2000. The skill of organic chemists will be needed to provide economical substitutes for these extremely useful materials.

Ethyl chloride is used as a local anesthetic. It is a gas at ambient temperature (b.p. 12 °C) and is kept in pressurized containers. When it is sprayed onto the skin, rapid vaporization occurs. The heat required to cause vaporization is drawn from the local surroundings, in this case the skin, and the resultant cooling decreases the ability of nerve endings to produce sensory signals.

Another significant use of chlorinated hydrocarbons is as pesticides, a general term that includes fungicides, herbicides, insecticides, fumigants, and rodenticides. There are three main types of pesticides in use: carbamates, organophosphorus compounds, and chlorinated hydrocarbons. The use of such compounds for the control of disease-bearing pests has increased sharply during the past decades. However, certain halogenated pesticides persist in the environment and are not used in some countries. An appropriate environmental residence time must now be designed into a biologically active molecule by the organic chemist.

The best known pesticide is DDT, which has been used extensively since 1939.

$$Cl-C_6H_4-CH(CCl_3)-C_6H_4-Cl$$

1,1,1-trichloro-2,2-bis(*p*-chlorophenyl)ethane
DDT

DDT is effective against many organisms, but its most spectacular success has been in control of the *Anopheles* mosquito, which transmits malaria. Malaria has been a scourge of mankind for centuries. According to the World Health Organization, malaria is still the chief cause of human death in the world, aside from natural causes. The disease acquired its name in ancient Rome (L. *mala*, bad; *aria*, air), where it was believed to be a result of the bad air in the city. It is actually caused by a parasite of the *Plasmodium* family which infects and ruptures erythrocytes in the bloodstream. The organism has a complex life cycle requiring both vertebrate and invertebrate hosts. Humans are infected by sporozoites of the organism that are injected into the bloodstream by the bite of an infected mosquito.

Although malaria can be treated, the most effective method of controlling it is to eliminate the insect vector that is essential for its transmission. DDT was especially effective for this purpose, and malaria has been essentially eliminated from large areas of the world through its use. It has been estimated that because of the efficacy of DDT in checking malaria and other mosquito-borne diseases (yellow fever, encephalitis), more than 75 million human deaths have been averted. A striking example is Sri Lanka (the island of Ceylon). In 1934–35, there were 1.5 million cases of malaria resulting in 80,000 deaths. After an intensive mosquito abatement program using DDT, malaria effectively disappeared and there were only 17 cases reported in 1963. When the use of DDT was discontinued in Sri Lanka, malaria rebounded, and there were over 600,000 cases reported in 1968 and the first quarter of 1969. In the absence of control measures for the mosquito or prophylactic measures for infected individuals, an estimated 100 million cases existed worldwide in 1989, leading to as many as a million deaths.

In spite of its obvious value in combatting diseases such as malaria, DDT has been abused. It is a ''hard'' insecticide, in that its residues accumulate in the environment. Although it is not especially toxic to mammals (the fatal human dose is 500 mg kg^{-1} of body weight, about 35 g for a 150 lb person), it is concentrated by lower organisms such as plankton and accumulates in the fatty tissues of fish and birds. The toxicity of DDT was first noted in 1949 by the Fish and Wildlife Service, but indiscriminate use as an agricultural pesticide for the control of crop-destroying pests continued to grow. In 1962, following the publication of *Silent Spring* by the late biologist Rachel Carson, an intensive campaign against the use of pesticides such as DDT commenced. In 1972 its use as an agricultural pesticide in the United States was banned by the Environmental Protection Agency.

Many species of mosquitoes have become resistant to DDT and other insecticides. *Plasmodia* have also become resistant to the best drugs for the treatment of malaria, quinine and chloroquine. Research on new chemical agents against *Plasmodia* is active (mefloquine, artemisinin). New types of pesticides that are species-specific and biodegradable and will not accumulate in the environment may eventually be found.

Alkyl halides are important as reagents. Many reactions are known for transforming the halogens to other functional groups. For industrial reactions, chlorides are used almost exclusively because of the high cost of bromine and iodine. For laboratory uses, where cost is not as great a consideration, bromides are used preferentially because alkyl bromides are generally more reactive than alkyl chlorides. Methyl iodide is a commonly used laboratory reagent because it is the only methyl halide that is liquid at room temperature. The preparation of alkyl halides from alcohols and alkenes will be discussed in Chapters 10 and 11, respectively.

Another important use of alkyl halides is in the preparation of organometallic compounds. We will next turn our attention to this group.

8.5 Nomenclature of Organometallic Compounds

Organometallic compounds are substances in which an organic group is bonded directly to a metal, R—M. They are named by prefixing the name of the metal with the appropriate organic group name. The names are written as one word.

$(CH_3)_3C$ Li *t*-butyllithium	$(CH_3CH_2)_2$ Mg diethylmagnesium	$(CH_3)_3$ Al trimethylaluminum
$(CH_3CH_2CH_2)_2$ Cd dipropylcadmium	$(CH_3CH_2)_2$ Zn diethylzinc	$(CH_3)_2$ Hg dimethylmercury
CH_3 Cu methylcopper	$(Ch_3)_4$ Si tetramethylsilicon	$(CH_3CH_2)_4$ Pb tetraethyllead

Compounds of boron, tin, and silicon are also named as derivatives of the simple hydrides: borane, BH_3; stannane, SnH_4; and silane, SiH_4. These compounds are indexed by *Chemical Abstracts* in this manner.

$(CH_3CH_2)_3B$ triethylborane	$(CH_3CH_2)_4Sn$ tetraethylstannane	$(CH_3)_3SiCH_2CH_3$ ethyltrimethylsilane

In some organometallic compounds, the valences of the metal are not all utilized in bonding to carbon but include bonds to inorganic atoms as well. Such compounds are named as organic derivatives of the corresponding inorganic salt.

CH_3CH_2MgBr ethylmagnesium bromide	CH_3HgCl methylmercuric chloride	$CH_3CH_2AlCl_2$ ethylaluminum dichloride

EXERCISE 8.3 Provide an acceptable name for each of the following compounds.
(a) $(CH_3)_2CHMgCl$ (b) $(CH_3)_4Si$ (c) $CH_3CH_2CH_2CH_2Li$ (d) $(CH_3)_4Pb$
(e) $(CH_3)_2Sn(CH_2CH_3)_2$

8.6 Structures of Organometallic Compounds

Since metals are **electropositive** elements, carbon-metal bonds can have a high degree of **ionic character.** That is, dipolar resonance structures are often important contributors to the structures of such compounds.

$$\left[R{-}M \longleftrightarrow R^- \; M^+\right] \equiv \overset{\delta-}{R}{-}\overset{\delta+}{M}$$

The degree of covalency of such carbon-metal bonds depends markedly on the metal and is related to the metal's **electronegativity.** Electronegativity is a measure of the ability of an element to attract or to hold onto electrons. Linus Pauling has established a semiquantitative scale in which each element is assigned an electronegativity value. On this scale a larger number signifies a greater affinity for electrons. When two elements of differing electronegativity are bonded, the bond is polar with the ''center of gravity'' of electron density in the bond closer to the more electronegative element. The greater the difference in electronegativity, the more polar is the bond. Pauling electronegativities for some of the elements are listed in Table 8.6.

Note that, in Table 8.6, carbon is assigned an electronegativity of 2.5, a value that is approximately midway between the extremes. Fluorine, the most electronegative element, has a value of 4.1, and the alkali metals, the most electropositive elements, have values of 1 or less. Thus, the bond in lithium fluoride is virtually completely ionic. That is, the two bonding electrons are associated almost solely with F^-, and the lithium-fluorine bond is almost totally electrostatic. We have already discussed the polarity of the carbon-fluorine bond, which arises from the difference in electronegativity between carbon (2.5) and

TABLE 8.6 Electronegativity Values for Some Elements

IA	IIA	IB	IIB	IIIA	IVA	VA	VIA	VIIB
H 2.2								
Li 1.0	Be 1.5			B 2.0	C 2.5	N 3.1	O 3.5	F 4.1
Na 0.9	Mg 1.2			Al 1.5	Si 1.7	P 2.1	S 2.4	Cl 2.8
K 0.8	Ca 1.0	Cu 1.9	Zn 1.6	Ga 1.8	Ge 2.0	As 2.2	Se 2.5	Br 2.7
		Ag 1.9	Cd 1.7	In 1.5	Sn 1.7	Sb 1.8	Te 2.0	I 2.2
Cs 0.7			Hg 1.9	Tl 2.0	Pb 1.6	Bi 1.7		

fluorine (4.1). The electron density in the carbon-fluorine bond is shifted toward the fluorine.

Because lithium is so much less electronegative than carbon, we expect the electrons in the carbon-lithium bond to be polarized toward carbon. The concept of relative electronegativity is thus useful for understanding the relative polarities of different bonds. But the argument is oversimplified because the electronic structure of bonds depends not just on relative electronegativities but also on the effectiveness of orbital overlap. The valence orbitals of the alkali metals are highly diffuse and their overlap with other orbitals is generally ineffective. For example, the difference between the electronegativities of lithium and carbon, 1.5, is the same as that between carbon and fluorine. Both bonds are polar but the lithium-carbon bond is almost as ionic as the lithium-fluorine bond, whereas the carbon-fluorine bond is a strong covalent bond. Organometallic compounds of the alkali metals may be regarded simply as essentially ionic salts of the alkali metal cation and an organic anion called a **carbanion.**

The ionic character of alkyllithium, alkylsodium, and alkylpotassium compounds is manifest in their structures. Methylpotassium displays almost classic salt-like behavior. Its crystal structure is similar to that of NaCl; each potassium ion is symmetrically surrounded by six methyl anions and vice versa. Alkyllithium compounds generally form aggregates of tetramers, hexamers, and so on, that are usually soluble in organic solvents, even hydrocarbons. Those aggregates that are ionic clusters with a hydrocarbon-like exterior have appreciable volatility.

With the less electropositive metals, such as Be, Mg, B, and Al, the carbon-metal bonds are fairly polar but still partially covalent. The metals in these compounds generally do not have an inert gas configuration. This electron deficiency has several structural consequences. For example, **three-center two-electron bonds** are encountered commonly. An example is trimethylaluminum, which exists as a dimer in which two aluminum atoms are bridged by two methyl groups (Figure 8.3a). There is attraction arising from the ionic character of the atoms; however, a three-center two-electron bond formed by mutual overlap of an orbital from each aluminum with an sp^3-orbital of carbon is a useful way to describe the structure (Figure 8.3b). In this way both aluminum atoms achieve an octet electronic configuration. Structural symbols such as Figure 8.3a are in common use but can be confusing because they do not correspond to a Lewis structure in which each line represents two electrons. An alternative symbolism has been suggested, shown in Figure 8.3c, in which ⅄ represents a three-center two-electron bond.

Beryllium compounds are frequently highly toxic and are not important in organic chemistry. Magnesium compounds are particularly important. Dialkylmagnesium com-

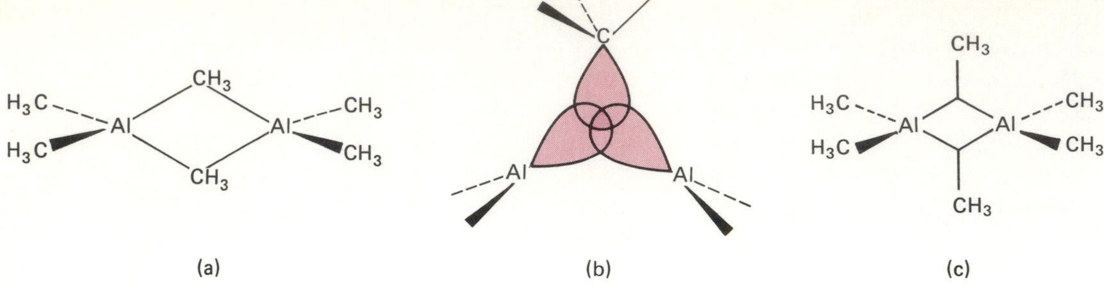

FIGURE 8.3 (a) Trimethylaluminum dimer. (b) Three orbitals overlap to give a molecular orbital containing two electrons. (c) Structure showing the ⅄ symbol for a three-center two-electron bond.

FIGURE 8.4 (a) Dialkylmagnesium polymer. Ionic contributions are shown in a second resonance structure. (b) Dialkylmagnesium coordinated to two ether molecules. Note the use of arrows to show the coordinative ''donor'' bonds. Ionic contributions are indicated in a resonance form.

pounds exist as polymeric structures in the solid phase or in hydrocarbon solvents; an ionic contribution to the structure is also illustrated (Figure 8.4a). Note how bridging three-center two-electron bonds lead to electron octets around each magnesium in the polymer. These compounds are usually soluble in ether solvents because the monomeric dialkylmagnesium compounds are coordinated to two ether oxygens (Figure 8.4b). In these structures the oxygen lone pairs are more effective than three-center two-electron bonds in completing the octets of the magnesium. Such interaction is usually represented with a coordinative bond or as an ionic structure to reflect the electrostatic character of the bonds as shown in Figure 8.4b.

EXERCISE 8.4 Write a Lewis structure for dimethylmagnesium coordinated to two molecules of dimethyl ether, $(CH_3)_2O$. Include all electrons and formal charges and compare with the structure using dative bonds.

Alkylmagnesium halides, also known as Grignard reagents, are important organometallic compounds that have many uses in synthesis. In dilute ether solution (about 0.1 M) Grignard reagents exist as monomers in which the magnesium is coordinated to two solvent molecules; their structures are similar to that of dialkylmagnesium (Figure 8.4b). However, in more concentrated solution (0.5–1 M) the principal species is a dimer in which two magnesium atoms are bridged by two bromines (Figure 8.5). In this structure each magnesium acquires its octet by additional coordination with one bromine from the other RMgBr and an ether oxygen. The bridging bonds in this structure make effective use of the additional lone-pair electrons available on the halogen. Note that the bridging bromines are *not* involved in three-center two-electron bonds because each magnesium-bromine bond involves a separate pair of electrons.

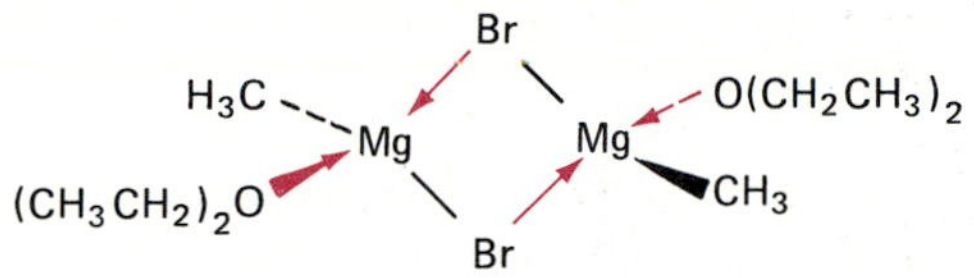

Br
H_3C Mg Mg $O(CH_2CH_3)_2$
$(CH_3CH_2)_2O$ CH_3
Br

FIGURE 8.5 Methylmagnesium bromide dimer in ether.

EXERCISE 8.5 Write Lewis and ionic structures for methylmagnesium bromide coordinated to diethyl ether, $(CH_3CH_2)_2O$. Include all electrons and formal charges and compare with the structure using dative bonds.

Because of the high ionic character of these carbon-metal bonds, they generally react avidly with water and other compounds containing relatively acidic hydrogens. The reaction is essentially an acid-base reaction with the proton being transferred to the carbanion, a strong base. Methane ($pK_a \cong 50$), for example, is a much weaker acid than water, and its conjugate base, methide ion, is a much stronger base than hydroxide ion. We will learn more about such weak acids in Chapter 12.

$$H_3C:^- \ Li^+ + \underset{pK_a = 15.7}{H_2O} \rightleftharpoons \underset{pK_a \cong 50}{H_3C{-}H} + Li^+ \ OH^-$$

In compounds of metals in groups IV and V (Si, Ge, Sn, Pb, Sb, Bi), there are sufficient electrons for the metals to engage in normal covalent bonding. Furthermore, since these metals tend to be much nearer to carbon in electronegativity than metals of groups I–III, the carbon-metal bonds are not very polar. These organometallic compounds tend to resemble conventional organic compounds in their properties. For example, tetramethylsilane and tetramethyllead are similar in structure to neopentane. The metal in each case has tetrahedral geometry (sp^3-hybridization) and makes covalent bonds to the four methyl groups.

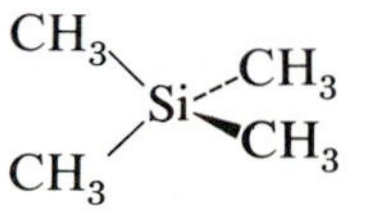

$Si(CH_3)_4$ — tetramethylsilicon (tetramethylsilane) b.p. 26.5°C

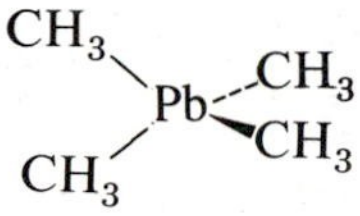

$Pb(CH_3)_4$ — tetramethyllead b.p. 110°C

Trimethylantimony and trimethylbismuth have pyramidal structures in which the metal has an unshared electron pair, similar to that in ammonia (page 26).

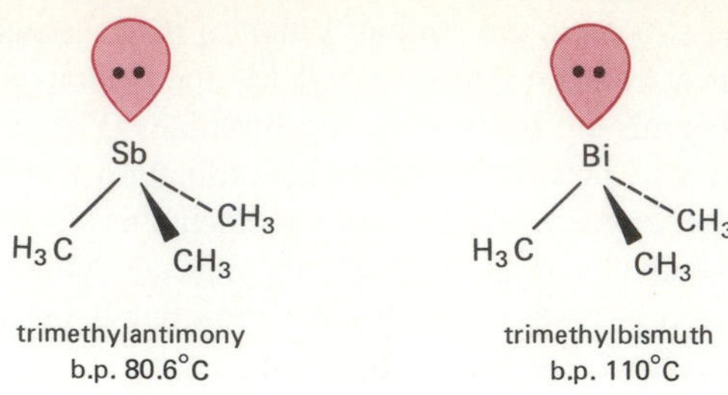

Highlight 8.1

The structures of organometallic compounds depend on the difference in electronegativity between carbon and the metal. Large differences result in substantial ionic character. Such compounds behave as metal salts of carbanions, $R^- M^+$ (derivatives of Li, Na, K, Cs, Mg, Ca). The small differences in electronegativity between carbon and elements in groups IV and V (Si, Ge, Sn, Pb, Sb, Bi) lead to compounds with covalent bonds and structures similar to those of compounds bonded to carbon or nitrogen.

8.7 Physical Properties of Organometallic Compounds

The melting points and boiling points of some simple organometallic compounds are summarized in Table 8.7. If sufficient caution is exercised, many organometallics can be prepared and handled in the same manner as other organic compounds. However, as we have noted, many types of organometallics react vigorously with water or other protic compounds; they also react with oxygen. Consequently, care must be taken in performing such operations as distillation and recrystallization.

Many of the organometallic compounds in Table 8.7 react with water, but are soluble in various inert aprotic organic solvents. Typical solvents are ethers and alkanes. Because

TABLE 8.7 Physical Properties of Organometallic Compounds

Compound	Melting Point, °C	Boiling Point, °C
CH_3CH_2Li	95	subl. 95 (aggregated)
$(CH_3)_2Mg$	240	(probably polymeric)
$(CH_3)_3Al$	0	130
CH_3AlCl_2	73	97–100 (100 torr)
$(CH_3)_2Cd$	−4.5	106
$(CH_3)_2Hg$	—	96
CH_3CH_2HgCl	193	subl. 40
$(CH_3)_3Ga$	−19	56
$(CH_3)_3Te$	38.5	147
$(CH_3)_4Si$	—	26.5
$(CH_3)_4Ge$	−88	43
$(CH_3)_4Sn$	−55	78
$(CH_3)_4Pb$	−27.5	110

of their extreme reactivity, a number of the organometallic compounds used in organic syntheses are normally not purified but are prepared and used as solutions in inert organic solvents without isolation.

8.8 Preparation of Organometallic Compounds

A. Reaction of an Alkyl Halide with a Metal

Reaction of the metal with an alkyl halide is most generally used for the laboratory preparation of organolithium and organomagnesium compounds. The reaction is normally carried out by treating the metal with an ether or hydrocarbon solution of the alkyl halide.

$$(CH_3)_3CCl + Mg \xrightarrow{\text{ether}} (CH_3)_3CMgCl$$

A solution of 227 g of *t*-butyl chloride in 1300 mL of dry ether is stirred in contact with 61 g of magnesium turnings for 6–8 hr. A cloudy gray solution is obtained that is approximately 2 M in *t*-butylmagnesium chloride. This solution is used directly for further reactions.

This type of reaction is an example of a heterogeneous reaction—a reaction that occurs at the interface between two different phases. The alkyl halide in solution must react with magnesium on the surface of solid magnesium. In such reactions the surface area and its character are important in allowing maximum contact between the elemental magnesium and alkyl halide. In the preparation of Grignard reagents, the magnesium is usually in the form of metal shavings or lathe "turnings." The reaction mechanism consists of the following steps.

$$R{-}X + {-}Mg{-}Mg{-}Mg{-} \longrightarrow {-}Mg{-}Mg^{+}{-}Mg{-} \;\; (\text{with } R\dot{\bar{}}{-}X \text{ above})$$

$$R\dot{\bar{}}{-}X \longrightarrow R\cdot + X^{-}$$

$$\overset{R\dot{\bar{}}-X}{{-}Mg{-}Mg^{+}{-}Mg{-}} \longrightarrow \overset{R\cdot \quad X^{-}}{{-}Mg{-}Mg^{+}{-}Mg{-}} \longrightarrow \boxed{\overset{R:^{-} \quad X^{-}}{{-}Mg^{+}{-}Mg^{+}{-}Mg{-}}}$$

$$\boxed{\overset{R:^{-} \quad X^{-}}{{-}Mg^{+}{-}Mg^{+}{-}Mg{-}}} \longrightarrow \overset{R:^{-} \quad X^{-}}{{-}Mg^{2+}{-}Mg{-}Mg{-}} \longrightarrow R:^{-}Mg^{2+}X^{-} \longleftrightarrow R{-}Mg{-}X$$

$$+$$

reactive region of surface

$$\begin{matrix} Mg- & & -Mg-Mg- \\ | & | & | \quad | \\ Mg- & Mg- & Mg-Mg \end{matrix}$$

Reaction of RX at the magnesium surface produces an alkyl radical and a Mg—X species probably still associated with the metal surface. The resulting free radical, R·, then reacts with the ·MgX to produce the Grignard reagent, RMgX. The principal side reactions involve alternative reactions of organic radicals, mostly dimerization and disproportionation (page 104).

$$CH_3CH_2CH_2CH_2\cdot + CH_3CH_2CH_2CH_2\cdot \longrightarrow CH_3CH_2CH_2CH_2CH_2CH_2CH_2CH_3$$

$$CH_3CH_2CH_2CH_2\cdot + CH_3CH_2CH_2CH_2\cdot \longrightarrow CH_3CH_2CH_2CH_3 + CH_3CH_2CH{=}CH_2$$

However, for simple alkyl halides the yields of alkylmagnesium halide are high—frequently above 90%. The reaction works well with chlorides, bromides, and iodides. Reaction of alkyl chlorides is frequently somewhat sluggish and iodides are generally expensive. Hence, alkyl bromides are common laboratory reagents in Grignard syntheses.

The mechanism by which the alkyl-halogen bond is cleaved into radicals is interesting. Since the size of the alkyl group or its degree of branching does not strongly affect the rate of the reaction, the halogen of the halide must approach the metal. The electron-rich metal then transfers an electron to the alkyl halide, forming a highly unstable species in which the electron is attached to the carbon-halogen bond. The unstable species is called a **radical anion** since it has an odd number of electrons and a negative charge. The radical anion decomposes rapidly to a halide ion and an alkyl radical. The alkyl radical does not leave the surface but combines with MgX to form RMgX, the alkylmagnesium halide. In certain cases, one enantiomer of an alkyl halide yields the corresponding enantiomeric alkylmagnesium halide, showing that an intermediate radical had no chance to become planar or react to give the opposite enantiomer. The surface which has lost the magnesium atom becomes more reactive because the other magnesium atoms now have unsatsified links.

A suitable solvent is essential for formation of the Grignard reagent because of the necessity for solvating the magnesium, as discussed in Section 8.6. The reaction is commonly carried out by adding an ether solution of the alkyl halide to magnesium turnings stirred in ether. The reaction is exothermic, particularly with bromides and iodides, and a reflux condenser is provided for returning the boiling ether to the flask. Anhydrous conditions must be maintained throughout the reaction, since Grignard reagents react rapidly with traces of moisture.

Alkyllithium compounds are prepared in the same manner, using lithium instead of magnesium.

$$CH_3CH_2CH_2CH_2Br + 2\,Li \xrightarrow{\text{ether}} CH_3CH_2CH_2CH_2Li + LiBr$$

A solution of 68.5 g of *n*-butyl bromide in 300 mL of dry ether is added slowly to 8.6 g of lithium wire. The mixture is stirred at −10 °C for about 1 hr, during which time all of the lithium dissolves. The resulting ether solution of *n*-butyllithium is stored under nitrogen in a well-stoppered flask.

The reaction of alkyl halides with lithium is probably similar to that with magnesium; that is, alkyl radicals are probably produced at the surface of the lithium metal and react further with the metal surface.

Although many other organometallic compounds can be prepared by direct reaction of the metal with an alkyl halide (for example, C_2H_5ZnI, CH_3HgCl), a more convenient and general laboratory method is metal exchange with an organolithium or organomagnesium compound (next section).

Ultrasound, sound waves beyond the range of normal human hearing or 20,000 Hz, can be used to promote the formation of organolithium, organozinc, and organocopper compounds. ''Sonochemical'' syntheses can sometimes be carried out at lower reaction temperatures, are faster, and give higher yields. The ultrasonic waves produce cavitation, disruption of the surface due to rapid growth (low pressure) and collapse (high pressure) of tiny bubbles. Remember that sound waves represent the passage of alternating high and low pressure waves through any medium. The laboratory apparatus for generating ultrasonic waves in reactions is inexpensive and readily available.

EXERCISE 8.6 Write the structure and name of each organometallic compound derived from reaction of (a) methyl iodide, (b) cyclohexyl bromide, and (c) 2-chloro-2-methylbutane with magnesium and with lithium.

B. Reaction of Organometallic Compounds with Salts

One of the most useful methods for preparing organometallic compounds in the laboratory is the exchange reaction of one organometallic with a salt to give a new organometallic and a new salt.

$$RM + M'X \rightleftharpoons RM' + MX \qquad (8\text{-}1)$$

This is an equilibrium process, and the equilibrium constant is dominated by the relative electronegativity (or electropositivity) of the two metals. In general, the halide ion is more electronegative than carbon and tends to be associated with the cation of the more electropositive metal. Recall that the more electropositive elements are those that are low and to the left in the periodic table. For example, Grignard reagents react readily with cadmium chloride to give organocadmium compounds and magnesium chloride. Magnesium is to the left of cadmium in the periodic table and is more electropositive [electronegativities are 1.2 and 1.7, respectively (Table 8.6)].

$$2\,CH_3CH_2MgCl + CdCl_2 \xrightarrow{\text{ether}} (CH_3CH_2)_2Cd + 2\,MgCl_2$$

In a similar way, Grignard reagents may be used to prepare tetraalkylsilanes and dialkylmercury compounds.

$$4\,CH_3MgCl + SiCl_4 \longrightarrow (CH_3)_4Si + 4\,MgCl_2$$

$$2\,CH_3CH_2\overset{\overset{\large CH_3}{|}}{C}HMgCl + HgCl_2 \longrightarrow (CH_3CH_2\overset{\overset{\large CH_3}{|}}{C}H)_2Hg + 2\,MgCl_2$$

EXERCISE 8.7 From inspection of a periodic table of the elements, and Table 8.6, predict the direction of each of the following equilibria:

(a) $2\,(CH_3)_3Al + 3\,ZnCl_2 \rightleftharpoons 3\,(CH_3)_2Zn + 2\,AlCl_3$
(b) $2\,(CH_3)_2Hg + SiCl_4 \rightleftharpoons (CH_3)_4Si + 2\,HgCl_2$
(c) $(CH_3)_2Mg + CaBr_2 \rightleftharpoons (CH_3)_2Ca + MgBr_2$

Highlight 8.2

Alkyl halides react with many metals to yield organometallic compounds. Metals that are very electropositive (large difference in electronegativity between the metal and carbon) can react directly. Exchange reactions in which a metal halide reacts with an organometallic compound can produce most other types of organometallic compounds.

8.9 Reactions of Organometallic Compounds

A. Hydrolysis

Organometallic compounds in which the metal has an electronegativity value of about 1.7 or less (Table 8.6) react with water to give the hydrocarbon and a metal hydroxide. The more electropositive the metal is, the faster is the hydrolysis. As mentioned earlier, alkyllithium, alkylmagnesium, and alkylaluminum compounds react vigorously, even violently, with water.

$$CH_3Li + H_2O \longrightarrow CH_4 + LiOH$$
$$CH_3CH_2MgBr + H_2O \longrightarrow CH_3CH_3 + HOMgBr$$
$$(CH_3)_3Al + 3\ H_2O \longrightarrow 3\ CH_4 + Al(OH)_3$$

Such compounds react similarly with other hydroxylic compounds, such as alcohols and carboxylic acids.

$$CH_3{-}\underset{CH_3}{\overset{CH_3}{\underset{|}{\overset{|}{C}}}}{-}Li + CH_3OH \longrightarrow CH_3\overset{CH_3}{\overset{|}{C}}HCH_3 + CH_3O^- \ Li^+$$

They also react with other compounds having relatively acidic hydrogens, such as thiols and amines.

Since the product of hydrolysis is an alkane, hydrolysis is not a very useful preparative reaction. However, it is important to recognize the limitation that this ready hydrolysis puts on the use of such organometallic compounds for other purposes. For example, it is not possible to prepare a Grignard reagent from an alkyl halide that also has an acidic hydrogen in the molecule, such as $ClCH_2CH_2CH_2OH$.

One important use for such hydrolysis reactions is **specific deuteriation.** When one carries out the hydrolysis with heavy water, deuterium oxide, the product is an alkane containing a deuterium at the position formerly occupied by the metal.

$$CH_3CH_2\underset{MgBr}{\overset{CH_3}{\underset{|}{\overset{|}{C}}}}CH_3 + D_2O \longrightarrow CH_3CH_2\underset{D}{\overset{CH_3}{\underset{|}{\overset{|}{C}}}}CH_3 + DOMgBr$$

2-methylbutane-*2-d*

The nomenclature used for isotopically labeled compounds is illustrated in this example. The use of *-d* for deuterium and *-t* for tritium is common, although more generally the isotope is specified by the atomic symbol and a prefix superscript giving the atomic mass of the isotope; for example, our labeled compound may also be named as 2-methylbutane-*2-^{2}H*. Finally, deuterium may be specified by a prefix as in 2-deuterio-2-methylbutane.

Heavy water is readily available, and this reaction is an excellent way of making hydrocarbons ''labeled'' with deuterium in a specific position. After reaction, the magnesium salts are removed, and the ether solution of the labeled hydrocarbon is dried and distilled. In subsequent chapters we will see several examples of the use of labeled compounds in studies of reaction mechanisms.

Alkylzinc and alkylcadmium compounds also react with protic materials, but their reactions are not so vigorous. Compounds of silicon, tin, mercury, and lead are unaffected by water, but in acidic solution they also undergo hydrolysis.

$$(C_2H_5)_4Si + 4\ HCl \xrightarrow{H_2O} 4\ CH_3CH_3 + SiCl_4$$

EXERCISE 8.8 Suggest how each of the following deuterated compounds can be prepared from a hydrocarbon.

(a) cyclohexane-*d*
(b) 2-methylpropane-2-*d*
(c) 1-deuterio-2,2-dimethylpropane

B. Reaction with Halogens

Most organometallic reagents react vigorously with chlorine and bromine.

$$RM + Cl_2 \longrightarrow RCl + M^+ \ Cl^-$$

The reaction is not preparatively useful because the product is an alkyl halide and organometallic compounds are frequently derived from alkyl halides.

C. Reaction with Oxygen

Organic compounds of many metals react rapidly with oxygen. Some are so reactive that they spontaneously inflame in air, often with spectacular consequences.

> Alkylboranes burn with a brilliant green flame. In his graduate student days, one of the authors was briefly immersed in a sea of such green fire. Only the wearing of safety glasses and rapid reflexes of a lab partner with a fire extinguisher allowed the current textbook to come to fruition.

Because of this reactivity, it is common to carry out organometallic reactions under an inert atmosphere such as nitrogen or argon. Oxidation is a side reaction of Grignard reactions run in the presence of air. In refluxing diethyl ether, a common solvent for Grignard syntheses, the ether vapor forms a suitable ''blanket'' of inert atmosphere.

These few reactions do little more than hint at the great versatility and usefulness of organometallic compounds in organic synthesis but we will encounter many more examples as we study the chemistry of individual functional groups.

Highlight 8.3

Organometallic compounds from electropositive metals are carbanionic in character and react with protons, proton donors, and oxidizing agents. Reaction of $R^- M^+$ with a D^+ source is an excellent way of preparing specifically deuteriated hydrocarbons, RD.

ASIDE

The English chemist Edward Frankland (1825–1899) added water to the tube used for the first preparation of dimethylzinc in 1849. A greenish blue flame several feet long shot out of the tube, and an abominable odor filled the laboratory (remember, there were no hoods). The German chemist Bunsen thought that the smell was that of cacodyl (tetramethyldiarsine, $(CH_3)_4As_2$) and that Frankland would die from its toxic effects, but the residues on the laboratory equipment proved to be zinc. Write the overall equation for the formation of dimethylzinc from methyl iodide and zinc. Write a possible mechanism for the reaction showing any important intermediates involved. What do you think should happen if water is added to dimethylzinc? Write equations for the expected reaction of dimethylzinc with water.

PROBLEMS

1. Write structures and IUPAC names for all structural isomers corresponding to each of the following formulas.

a. C_3H_7Cl **b.** C_4H_9Br **c.** $C_5H_{11}I$

d. $C_5H_{14}Sn$ **e.** $C_4H_{10}Mg$ **f.** C_4H_9Li

2. Write Newman projections for the conformations of 1,1,2-trichloroethane. Two of these are the same, and the third is different. The two types of conformation differ in energy by 2.6 kcal $mole^{-1}$ in the vapor phase. Which is the more stable? This energy difference is reduced to 0.2 kcal $mole^{-1}$ in the liquid. Explain why. Interconversion of the two similar conformations requires about 2 kcal $mole^{-1}$, but conversion of either to the third structure requires about 5 kcal $mole^{-1}$. Explain why these two rotation barriers differ.

3. Show how the following conversions may be accomplished.

(a) $(CH_3)_3CCH(CH_3)_2 \longrightarrow (CH_3)_3CCD(CH_3)_2$

(b) $CH_3CH_2CHClCH_3 \longrightarrow \left(CH_3CH_2\overset{CH_3}{\overset{|}{C}H}\right)_4 Sn$

(c) $(CH_3)_3CCH_2Cl \longrightarrow (CH_3)_3CCH_2Br$

(d) $CH_3CH_2Cl \longrightarrow (CH_3CH_2)_2Cd$

(e) $(CH_3)_3CCl \longrightarrow [(CH_3)_3C]_2Hg$

4. Predict whether the equilibrium constant will be greater than or less than unity for each of the following reactions.

(a) $2\,(CH_3)_3Al + 3\,CdCl_2 \rightleftharpoons 3\,(CH_3)_2Cd + 2\,AlCl_3$
(b) $(CH_3)_2Hg + ZnCl_2 \rightleftharpoons (CH_3)_2Zn + HgCl_2$
(c) $2\,(CH_3)_2Mg + SiCl_4 \rightleftharpoons (CH_3)_4Si + 2\,MgCl_2$
(d) $CH_3Li + HCl \rightleftharpoons CH_4 + LiCl$
(e) $(CH_3)_2Zn + 2\,LiCl \rightleftharpoons 2\,CH_3Li + ZnCl_2$

5. a. Dimethylberyllium has a polymeric structure analogous to that described for dimethylmagnesium in Figure 8.4a. By contrast, di-*t*-butylberyllium is monomeric. Suggest a reason for the difference and predict the geometry of di-*t*-butylberyllium.

b. Dimethylmercury is monomeric. Predict its geometry. The measured dipole moments of $(CH_3)_2Hg$ and $(C_2H_5)_2Hg$ are $\mu = 0.0$ D. Explain.

6. Trimethylborane reacts with methyllithium to give a product having the formula $C_4H_{12}BLi$. Propose a structure for this substance. What is the hybridization of boron? Describe the geometry of the species.

7. Write the structures of the eight compounds having the formula $C_5H_{11}D$. Which two of these isomers can be prepared in good yield from alkanes using reactions you have learned in the last two chapters? Explain why the other six isomers cannot be prepared in this way.

8. In the formation of Grignard reagents from magnesium and alkyl halides, the most frequent side reactions are dimerization and disproportionation, as in the following example

Cyclohexyl–Cl $\xrightarrow[\text{ether}]{\text{Mg}}$ Cyclohexyl–MgCl + Cyclohexyl–cyclohexyl + Cyclohexane + Cyclohexene

Propose a mechanism for these side reactions.

CHAPTER 9

NUCLEOPHILIC SUBSTITUTION

9.1 The Displacement Reaction

The replacement of the halogen in an alkyl halide by another group is one of the most important reactions in organic chemistry. In Section 4.2, we took a brief look at one such reaction, the reaction of methyl chloride with hydroxide ion.

$$HO^- + CH_3Cl \xrightarrow{H_2O} CH_3OH + Cl^-$$

We learned that this reaction is effectively irreversible and is a rather slow reaction at room temperature. By raising the temperature the reaction rate is increased.

Another example is the reaction of ethyl bromide with potassium iodide in acetone solution.

$$CH_3CH_2Br + K^+ I^- \xrightarrow{\text{acetone}} CH_3CH_2I + KBr\downarrow$$

In this case the reaction is reversible; that is, the equilibrium constant is not a large number. Nevertheless, the reaction proceeds virtually to completion because potassium iodide is soluble in acetone and potassium bromide is not.

The foregoing reactions are but two of a large number of closely related reactions that have in common the replacement of one Lewis base by another. Recall that a Lewis base has a lone pair of electrons available for bonding. We shall first look at reactions in which the leaving group is a halide ion but we shall see later that these reactions are not limited to alkyl halides. Many Lewis bases qualify as suitable incoming groups in such reactions. Some additional examples are summarized in Table 9.1.

Many of the examples given in Table 9.1 involve anions as incoming groups. Some of these anions are rather strong bases; that is, they are the conjugate bases of weak acids. Examples are hydroxide ion, HO^-, and ethoxide ion, $C_2H_5O^-$. Others, such as cyanide ion, CN^-, azide ion, N_3^-, and acetate ion, $CH_3CO_2^-$, are the conjugate bases of moderately strong acids and are weaker bases than hydroxide ion. Halide ions are the weak

TABLE 9.1 Some Displacement Reactions with Ethyl Bromide

Attacking Reagent		Product	
Formula	Name	Formula	Name
HO^-	hydroxide ion	C_2H_5OH	ethyl alcohol
$C_2H_5O^-$	ethoxide ion	$CH_3CH_2OCH_2CH_3$	diethyl ether
HS^-	hydrosulfide ion	CH_3CH_2SH	ethanethiol
SCN^-	thiocyanate ion	CH_3CH_2SCN	ethyl thiocyanate
CN^-	cyanide ion	CH_3CH_2CN	ethyl cyanide propionitrile
N_3^-	azide ion	$CH_3CH_2N_3$	ethyl azide
NH_3	ammonia	$CH_3CH_2NH_3^+\ Br^-$	ethylammonium bromide
H_2O	water	$CH_3CH_2OH_2^+\ Br^-$	ethyloxonium bromide
$CH_3CO_2^-$	acetate ion	$CH_3CO_2C_2H_5$	ethyl acetate
NO_3^-	nitrate ion	$CH_3CH_2ONO_2$	ethyl nitrate
$P(CH_3)_3$	trimethylphosphine	$C_2H_5P(CH_3)_3^+\ Br^-$	ethyltrimethyl-phosphonium bromide
$N(C_2H_5)_3$	triethylamine	$(C_2H_5)_4N^+\ Br^-$	tetraethylammonium bromide
$S(C_2H_5)_2$	diethyl sulfide	$(C_2H_5)_3S^+\ Br^-$	triethylsulfonium bromide

conjugate bases of rather strong mineral acids, the hydrohalic acids, such as HCl and HI. When used in displacement reactions, such bases are referred to as **nucleophilic reagents** or simply as **nucleophiles** (L., *nucleus,* kernel; Gr., *philos,* loving; hence ''nucleus loving''). The reaction is then also called a **nucleophilic displacement reaction.**

A number of the nucleophiles listed in Table 9.1 are neutral molecules. The reaction of trimethylphosphine with ethyl bromide may be written

$$(CH_3)_3\,P\!:\, + \underset{CH_3}{CH_2}\!-\!Br \longrightarrow (CH_3)_3\overset{+}{P}CH_2CH_3\ Br^-$$

The phosphorus in the phosphine has a lone pair of electrons that bonds to the methylene group in the product. Thus, trimethylphosphine is a neutral nucleophilic reagent.

EXERCISE 9.1 Give the structure and name of the product of the displacement reaction of each of the nucleophiles in Table 9.1 with methyl iodide. For each case, identify the lone-pair electrons that bond to carbon in the product.

Highlight 9.1

Neutral or anionic Lewis bases are nucleophilic and can displace the same or other Lewis bases from carbon in an organic molecule, RX. Typical nucleophiles are $RO:^-$, $RS:^-$, $R_3N:$, and $R_3P:$. The displaced Lewis base is called a leaving group. Halide ions, $X:^-$, are typical leaving groups.

9.2 Mechanism of the Displacement Reaction

Two important tools in studying the mechanism of a reaction are the **reaction kinetics** and **stereochemistry.** The reaction kinetics can reveal the *composition* of the rate-determining transition state, and stereochemical studies with chiral molecules help to determine its *structure*. We will demonstrate the application of these tools to the displacement reaction of alkyl halides with acetate ion. The reaction of methyl iodide with lithium acetate in methanol solution produces methyl acetate and is effectively irreversible. The reactants and products are all soluble in the alcohol; hence, this is a *homogeneous* reaction.

$$CH_3CO_2^- + CH_3{-}I \longrightarrow CH_3\overset{\overset{\displaystyle O}{\|}}{C}{-}OCH_3 + I^-$$

The condensed formula given for acetate ion, $CH_3CO_2^-$, is a shorthand for the two resonance structures that show that the negative charge in acetate ion is divided between the two oxygens (pages 13 and 69):

$$CH_3CO_2^- \equiv \left[CH_3\overset{\overset{\displaystyle O}{\|}}{C}{-}O^- \longleftrightarrow CH_3\overset{\overset{\displaystyle O^-}{|}}{C}{=}O \right]$$

The rate of the reaction can be determined by following the rate of disappearance of reactants or the rate of formation of products. It is proportional to the product of the concentrations of the two reactants.

$$\text{rate} = -\frac{d[CH_3I]}{dt} = -\frac{d[CH_3CO_2^-]}{dt} = \frac{d[CH_3CO_2CH_3]}{dt} = \frac{d[I^-]}{dt}$$
$$= k[CH_3I][CH_3CO_2^-]$$

> This equation is expressed in the symbolism of calculus. The expression
>
> $$-\frac{d[CH_3I]}{dt}$$
>
> means simply the rate with which the concentration of CH_3I changes with time. The negative sign indicates that the concentration of CH_3I decreases as time increases.

The concentrations of CH_3I, $CH_3CO_2^-$, $CH_3CO_2CH_3$ and LiI can be determined at different times during the reaction by chemical or spectroscopic analysis. As the reaction proceeds, the concentrations of the reactants become reduced and the rate of reaction decreases. For example, at 60 °C, with the two reactants each present in an initial concentration of 0.1 M, the reaction is 50% complete in 20 hr but only 90% complete after 180 hr. Furthermore, the reaction has an activation energy, $\Delta H^\ddagger$, of 24 kcal mole^{-1}; it is about 80 times faster at 60 °C than it is at 25 °C. This activation energy is considerably higher than those of the free radical reactions we studied in Chapter 6.

When the rate of a chemical reaction depends on the concentration of two species, as in this case, it is said to display **second-order kinetics.** Both components are present in the rate-determining transition state. This suggests a **bimolecular** mechanism, one in which one molecule of each reactant collide and react. The relatively high activation energy shows that only a minute fraction of such collisions actually result in reaction—those involving reactant molecules with sufficient kinetic energy.

The **molecularity** of a reaction is defined as the number of reactant molecules involved in the rate-determining transition state. It is sometimes, but not always, equal to the kinetic order of the reaction. For example, consider a reaction of the type

$$A + B \underset{k_{-1}\ \text{fast}}{\overset{k_1}{\rightleftharpoons}} C \xrightarrow[\text{slow}]{k_2} D$$

The rate of appearance of the product D is given by the rate law

$$\text{rate of formation of D} = \frac{d[D]}{dt} = \frac{k_1 k_2}{k_{-1}}[A][B] = k'[A][B]$$

which shows second-order kinetics. However, in the slow step (k_2), the rate-determining step, only one molecule, species C, is involved. Hence, the reaction is unimolecular.

The **stereochemistry of reaction** refers to the relative configurations of the leaving group and entering group bonded to a stereocenter in the reactant and product, respectively. This stereochemistry cannot be determined for methyl iodide because it is not chiral. But many such reactions have been carried out with optically active compounds in which the leaving group has been bonded to the stereocenter. These experiments have shown that in such displacement reactions every replacement of the leaving group by the incoming group occurs with *inversion of configuration*. For example, the reaction of (*R*)-2-bromobutane with acetate ion produces (*S*)-2-butyl acetate.

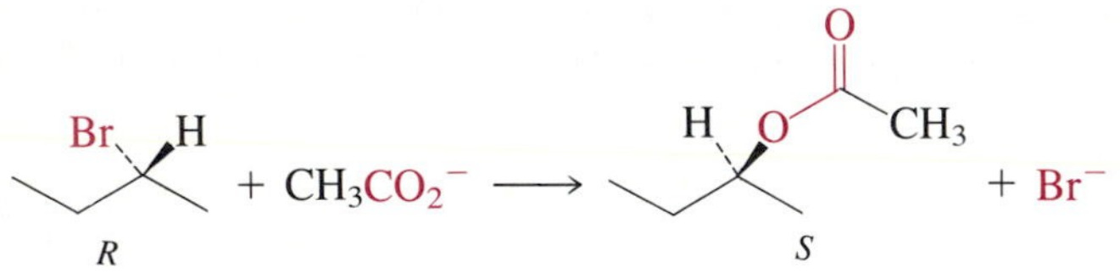

Br H (*R*) + $CH_3CO_2^-$ ⟶ H O (C=O) CH_3 (*S*) + Br^-

This example poses the problem of relating the configurations of the reactant and product. Remember that there is no necessary relation between the sign of rotation and the structural configuration. We will not discuss the specific example shown here, but in subsequent studies of different functional groups we will encounter many similar examples where such configurations have been interrelated. One such example is given at the end of this chapter.

In most cases, the inversion of configuration that occurs in the S_N2 mechanism converts a reactant with *R* configuration into a product with *S* configuration, or vice versa. This is because it is common for both the leaving group and incoming nucleophile to have sequence rule priority number 1. However, you should be cautious and remember that this relationship will not hold in all cases, as is shown by the following example.

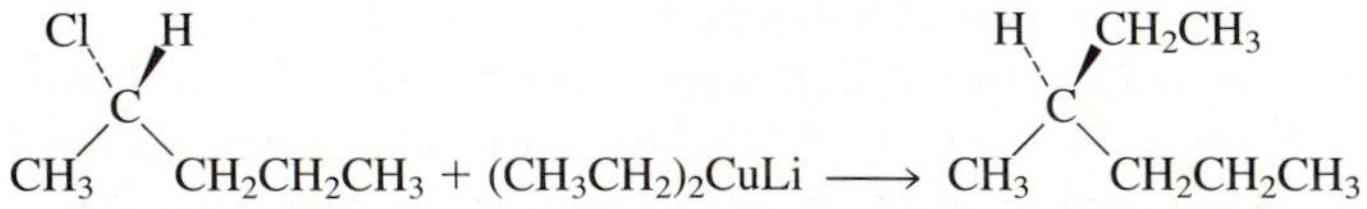

Cl, H, C, CH_3, $CH_2CH_2CH_3$ + $(CH_3CH_2)_2CuLi$ ⟶ H, CH_2CH_3, C, CH_3, $CH_2CH_2CH_3$

(*S*)-2-chloropentane ⟶ (*S*)-3-methylhexane

The nucleophilic reagent in this substitution reaction is lithium diethylcopper, a member of a class of organometallic compounds called cuprates; we will encounter them in Section 18.7.D. Note that in this example, the leaving group has sequence rule priority number 1, but the incoming nucleophile has sequence rule priority number 2, so both the reactant and product have the same stereochemical name, *S*.

EXERCISE 9.2 For each of the following displacement reactions write the structure of the product with inversion of configuration.

(a) (S)-$CH_3CH_2CH_2CHDCl + CH_3CO_2^-$

(b) 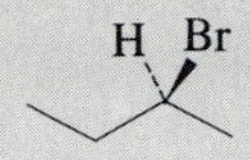 [structure: H, Br on a carbon chain] $+ P(CH_3)_3$

(c) 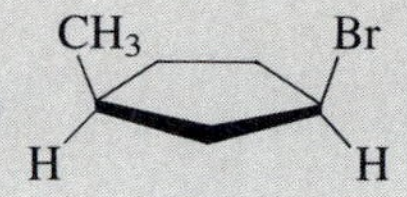 [structure: cyclohexane ring with CH_3, H, Br, H] $+ CN^-$

Consideration of the second-order reaction kinetics and the stereochemistry of the reaction leads to a mechanism in which the entering nucleophilic reagent attacks the carbon atom at the rear of the bond to the leaving group. As the carbon forms a progressively stronger bond to the attacking group, the bond to the leaving group is being weakened. During this change, the other three bonds to the central carbon progressively flatten out and end up on the other side of the carbon in a manner similar to the spokes of an umbrella inverting in a windstorm.

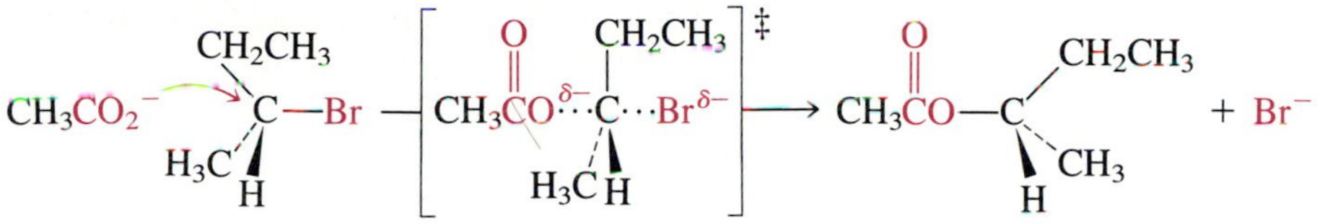

$$CH_3CO_2^- + (CH_2CH_3)(H_3C)(H)C\text{—}Br \longrightarrow \left[CH_3\overset{O}{\overset{\|}{C}}O^{\delta-}\cdots C(CH_2CH_3)(H_3C)(H)\cdots Br^{\delta-}\right]^{\ddagger} \longrightarrow CH_3\overset{O}{\overset{\|}{C}}O\text{—}C(CH_2CH_3)(CH_3)(H) + Br^-$$

The foregoing equation illustrates the use of a curved arrow as a symbol for the flow of electrons in the course of the reaction. This symbolism finds much use in organic chemistry. An electron pair is thought of as originating at the end of an arrow and flowing in the direction of the point. In this case, a pair of electrons belonging to acetate ion flows toward the bromine-bearing carbon and eventually forms a new carbon-oxygen bond. Simultaneously, the pair of electrons comprising the carbon-bromine bond flows away from carbon and eventually ends up as a fourth lone pair on the product bromide ion. The transition state is shown in brackets. The dotted lines indicate a partially formed or partially broken bond. The symbols δ^- indicate that the negative charge is spread over both entering and leaving groups in the transition state. Using curved arrows, the entire reaction can be symbolized as follows with the transition state being implied.

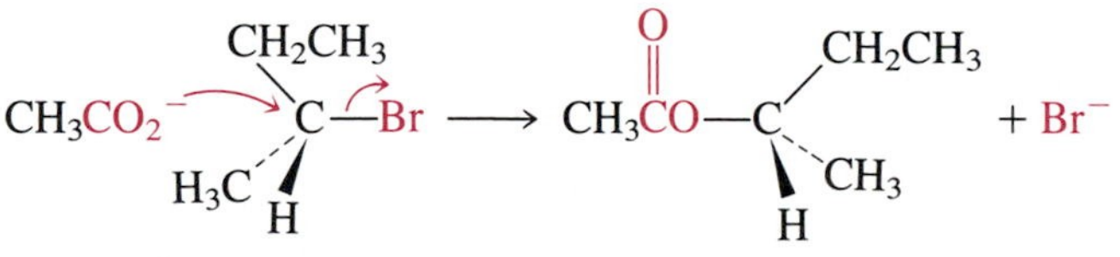

$$CH_3CO_2^- + (CH_2CH_3)(H_3C)(H)C\text{—}Br \longrightarrow CH_3\overset{O}{\overset{\|}{C}}O\text{—}C(CH_2CH_3)(CH_3)(H) + Br^-$$

EXERCISE 9.3 For each of the reactions in Exercise 9.2, write Lewis structures showing the electrons around the reacting carbon and use curved arrows to show the electron flow corresponding to the displacement reaction.

During the course of the reaction, the reacting system has greater potential energy than either the reactants or the products. The two weak bonds to the entering and leaving groups are weaker together than the single bond in either the reactant or the product. Hence, energy is required in order for reaction to occur. The necessary potential energy is supplied by the conversion of kinetic energy. Only the minute fraction of reactants that have sufficient kinetic energy can react. Furthermore, even if the colliding reactants have sufficient kinetic energy, they must have the proper orientation or they will simply bounce apart. The transition state is not a discrete molecule that can be isolated and studied. In fact, the whole act of displacement occurs in the space of about 10^{-12} sec, the period of a single vibration, so the system has the transition state geometry for only a fleeting moment.

Transition states can be examined by ultrashort laser light pulses on the time scale of 20–30 fs (10^{-15} sec) for such reactions as the dissociation of ICN into I and CN radicals.

This reaction mechanism may be generalized. The general reaction is called an **S_N2** reaction for **substitution, nucleophilic, bimolecular.** The geometry of the transition state appears to be that in which incoming and leaving groups are both weakly bonded to carbon in a linear fashion and in which the three remaining bonds to carbon lie in a plane perpendicular to the two weak bonds. The mechanism for reaction of an entering group Y^- and a leaving group X^- is shown in Figure 9.1, where the structure of the reacting system is illustrated at several points along the reaction coordinate. At point (b) the C—X bond has started to lengthen and the central carbon has started to flatten out. At the transition state, point (c), the central carbon is approximately flat and both bonds to the leaving and entering groups are long. Point (d) occurs on the final road to products (e); the central carbon has bent, the C—Y bond is approaching normal length, and the leaving group X is receding. The structures at points (a) through (e) are represented in stereo form in Figure 9.2.

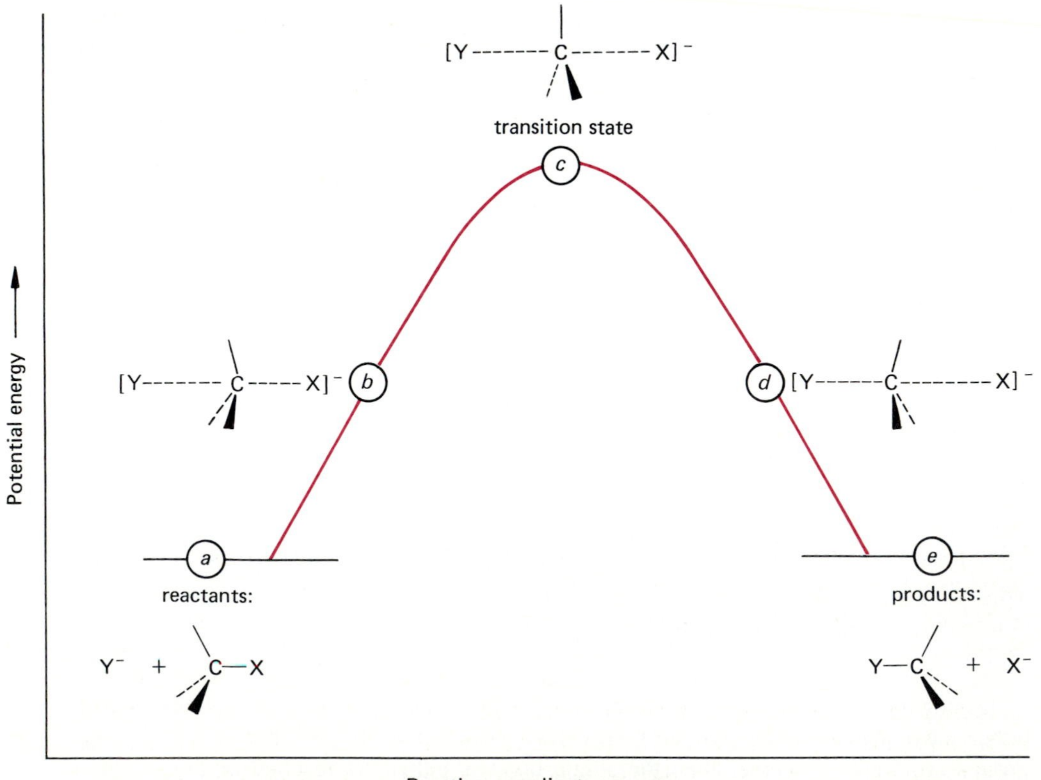

FIGURE 9.1 Reaction mechanism profile for a displacement reaction by Y^- on RX.

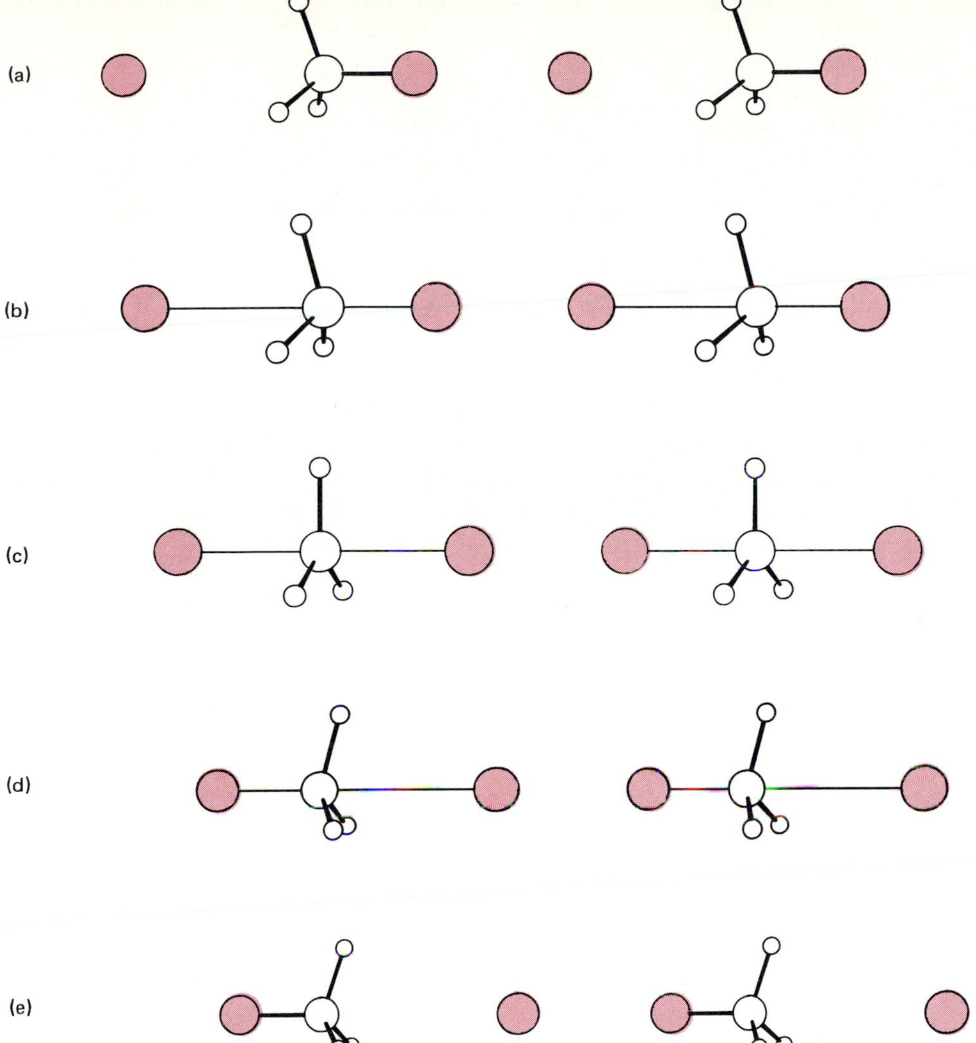

FIGURE 9.2 The structure of the reaction system at points (a)–(e) in Figure 9.1.

The reaction can also be represented by changes in orbitals. Both reactant and product are tetrahedral and the central carbon has sp^3-hybrid orbitals. As reaction proceeds, the orbital to the leaving halide acquires greater *p*-character and the three remaining bonds to the central carbon develop greater *s*-character. In the transition state structure, the weak bonds to X and Y may be considered to derive from overlap of their orbitals with the two lobes of a *p*-orbital on the central carbon. The other three bonds to this carbon are formed from sp^2-hybrid orbitals, as shown in Figure 9.3.

Table 9.1 shows that the mechanism label S_N2 covers a wide variety of specific reactions. All of these reactions are bimolecular and occur with inversion of configuration at the reacting carbon. In such reactions we need to be concerned not only with the position of equilibrium but with rate. The reactions can range from incredibly fast at low temperatures to exceedingly slow even at higher temperatures, depending on the nature of the alkyl group, the nucleophile, the solvent, and the leaving group. We shall discuss each of these variables in turn. The important principles that enter into evaluating the effects of these variables recur frequently in organic chemistry and therefore warrant careful study at

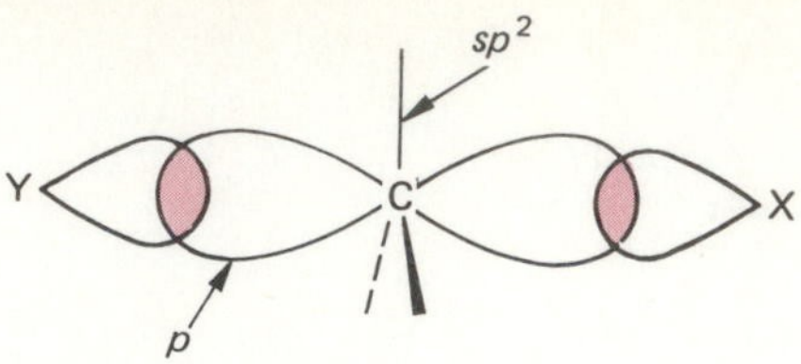

FIGURE 9.3 Orbital formulation of the transition state of a displacement reaction.

this time. Moreover, we shall see how a knowledge of transition state structure helps to rationalize a large body of diverse facts.

EXERCISE 9.4 Displacement reactions can be carried out in which the entering and leaving groups are the same. Consider the reaction of *cis*-1-iodo-4-methylcyclohexane with iodide ion in acetone. What is the product of this displacement reaction? What is the composition of the mixture when equilibrium is reached (note Table 7.1, page 141)?

Highlight 9.2

In the reaction of a nucleophile (Nu) with an alkyl derivative (RX) to give RNu and X, if the stereochemical outcome of attack of Nu at the carbon of RX is inversion of configuration and the reaction shows second-order kinetics (dependence on the concentrations of Nu and RX), the reaction has an S_N2 mechanism.

Highlight 9.3

The bonds to the carbon undergoing nucleophilic substitution are arranged in a tetrahedral fashion, with sp^3-orbitals from the central carbon to the attached groups. As the carbon forms a bond to the incoming nucleophile with a *p*-type orbital, the bond to the leaving group also changes from sp^3- to a *p*-type orbital. The orbitals that make the bonds to the other three groups, which remain bonded to the carbon, change during the course of the substitution to sp^2-type. At the transition state, the other three groups are disposed approximately in a planar fashion.

9.3 Effect of Alkyl Structure on Displacement Reactions

A large variety of alkyl halides undergo substitution by the S_N2 mechanism. The ease of reaction depends markedly upon the structure of the alkyl group to which the halogen is attached. Reactivities vary widely and in a consistent manner. Branching of the chain at the carbon where substitution occurs (the α-carbon) has a significant effect on the rate of reaction. Relative rates of S_N2 reactions for methyl, ethyl, isopropyl, and *t*-butyl halides are approximately as shown in Table 9.2.

This use of the Greek alphabet is widespread in organic chemistry, and it is important to learn the first few letters, at least through delta (the entire Greek alphabet is given inside the front cover of this book). Many of the letters, small and capital, have evolved standard meanings in the mathematical and physical sciences (for example, the number π). In organic chemistry, the lowercase letters are used more frequently than the capital letters.

TABLE 9.2 Effect of Branching at the α-Carbon on the Rate of S_N2 Reactions

Alkyl Halide	Relative Rate
CH_3—	30
CH_3CH_2—	1
$(CH_3)_2CH$—	0.02
$(CH_3)_3C$—	~0

These effects on reaction rate are interpreted with the concept of **steric hindrance** to approach of the attacking nucleophile. The halide is on the front side of a methyl halide. The rear side of a methyl group is relatively open to such attack, and methyl compounds are generally quite reactive in displacement reactions. As the hydrogens of the methyl group are replaced successively by methyl groups, the space to the rear of the leaving group becomes more encumbered. It becomes more difficult for the attacking group to approach closely enough to the rear of the C—X bond for reaction to occur, and the rate of reaction diminishes (Figure 9.4).

A similar effect may be seen in branching at the β-carbon. Some typical relative rates are shown in Table 9.3. The reduction in rate with branching is also attributable to steric hindrance. In one conformation, the rear of a *n*-propyl carbon is seriously blocked (Figure 9.5a), but in two other conformations the situation is no worse than for ethyl (Figure 9.5b). Consequently, *n*-propyl halides undergo S_N2 displacement only slightly less readily than do ethyl halides.

For the isobutyl group, it is possible to rotate both of the β-methyl groups out of the way of the attacking group, but the resulting conformation is highly congested and has relatively high energy (Figure 9.6). Accordingly, isobutyl halides are much less reactive than either ethyl or *n*-propyl compounds.

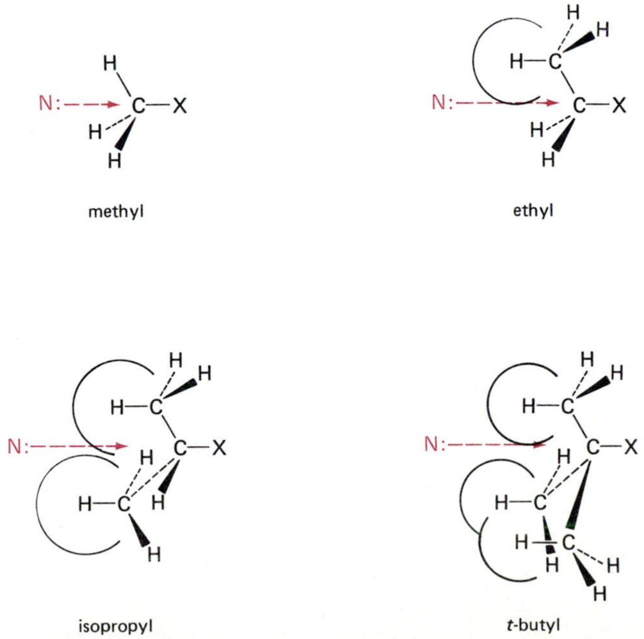

methyl

ethyl

isopropyl

t-butyl

FIGURE 9.4 Effect of α-branching on S_N2 reactions.

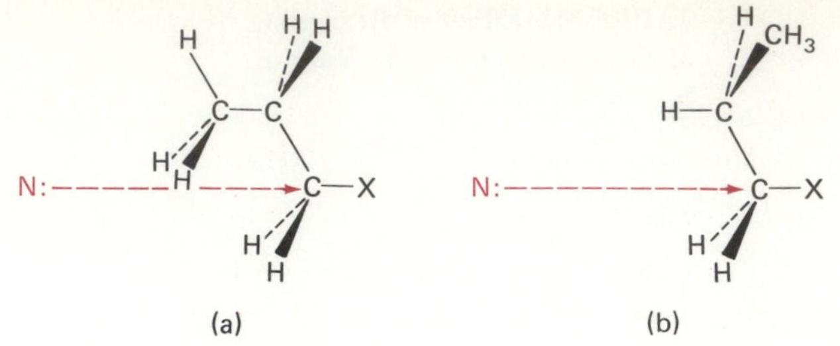

FIGURE 9.5 S_N2 attack at two conformations of *n*-propyl compounds.

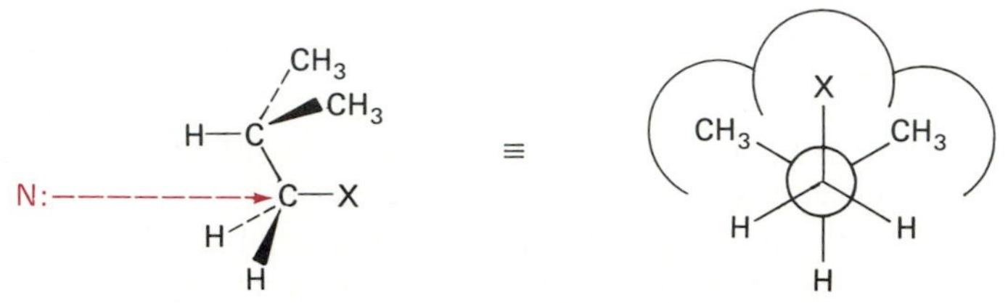

FIGURE 9.6 S_N2 reaction at isobutyl systems.

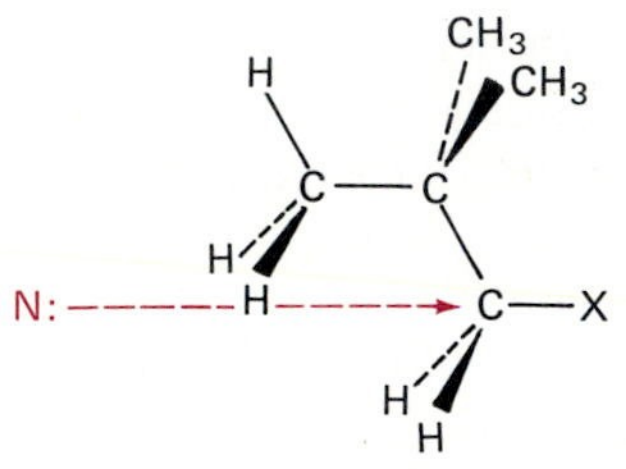

FIGURE 9.7 S_N2 reaction at neopentyl compounds.

Neopentyl halides are particularly interesting because there is no conformation in which a blocking methyl group can be avoided (Figure 9.7). Neopentyl halides are unreactive in S_N2 reactions except under very drastic conditions.

Substitution of sites more remote than the β-carbon have little or no effect on the ease of S_N2 reactions. For example, *n*-butyl and *n*-pentyl halides react at essentially the same rate as *n*-propyl halides.

The type of steric interaction we have discussed here forces groups to bend away from each other. Such deformation often forces orbitals to overlap in a noncolinear fashion. Recall that the resulting bent bonds are weaker than the corresponding straight bonds (Section 5.6).

TABLE 9.3 Effect of Branching at the β-Carbon on the Rate of S_N2 Reactions

Alkyl Halide	Relative Rate
$CH_3CH_2—$	1
$CH_3CH_2CH_2—$	0.4
$(CH_3)_2CHCH_2—$	0.03
$(CH_3)_3CCH_2—$	0.00001

In summary, the effect of the structure of the alkyl group on the rate of S_N2 reaction is apparent in two ways.

1. Branching at the α-carbon hinders reaction: rate order is methyl > primary > secondary ≫ tertiary.
2. Branching at the β-carbon hinders reaction: neopentyl compounds are particularly unreactive.

Displacements that proceed by the S_N2 mechanism are most successful with primary compounds having no branches at the β-carbon. Yields are poor to fair with secondary halides and with primary halides having branches at C-2. Neopentyl systems undergo the reaction only under very drastic conditions, and tertiary halides rarely react by this mechanism. When the rate of the S_N2 reaction is slowed down by these structural effects, alternative side reactions begin to compete. With tertiary halides, and to an important degree with secondary and highly branched primary halides, the side reactions tend to dominate. The most important of these side reactions is *elimination* of hydrogen halide to form alkenes. This reaction will be discussed in Sections 9.7 and 11.5.

EXERCISE 9.5 For each of the following pairs of isomeric alkyl halides indicate which isomer is the more reactive in S_N2 reactions.

(a) I I

(b) Br Br

(c) Cl Cl

(d) Cl Cl

Highlight 9.4

Approach to the central carbon in a displacement reaction by a nucleophile is made more difficult by groups that block access to the backside of the leaving group. The steric hindrance by these groups, usually at the positions α or β to the leaving group, increases the energy required to reach the transition state and decreases the rate of the substitution reaction.

9.4 Effect of Nucleophile Structure on Displacement Reaction

Halide ions, acetate ion, and many other nucleophiles can react at only one atom in the system either as bases or as nucleophiles. Similarly, sulfate ion is straightforward in its reaction with either H^+ or CH_3I in an S_N2 reaction. Both types of reaction occur at an oxygen.

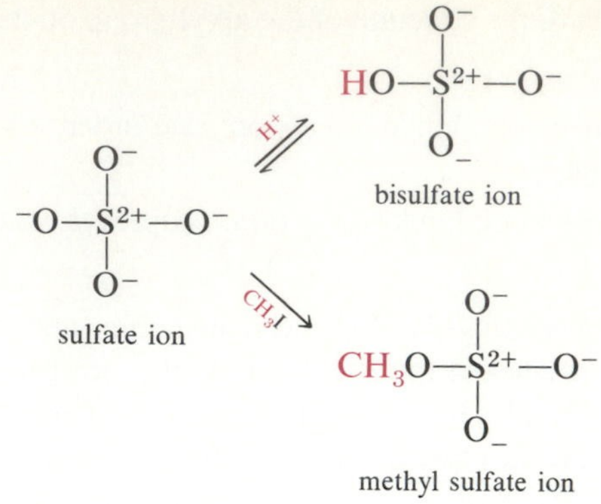

$$^{-}O-\overset{\overset{O^-}{|}}{\underset{\underset{O^-}{|}}{S^{2+}}}-O^- \quad \xrightleftharpoons{H^+} \quad HO-\overset{\overset{O^-}{|}}{\underset{\underset{O_-}{|}}{S^{2+}}}-O^- \text{ (bisulfate ion)}$$

sulfate ion

$$\xrightarrow{CH_3I} \quad CH_3O-\overset{\overset{O^-}{|}}{\underset{\underset{O_-}{|}}{S^{2+}}}-O^-$$

methyl sulfate ion

Sulfite ion, however, behaves quite differently. In hydroxylic solvents it reacts with a proton mainly on oxygen to form bisulfite ion and with methyl iodide on sulfur to form the methanesulfonate ion.

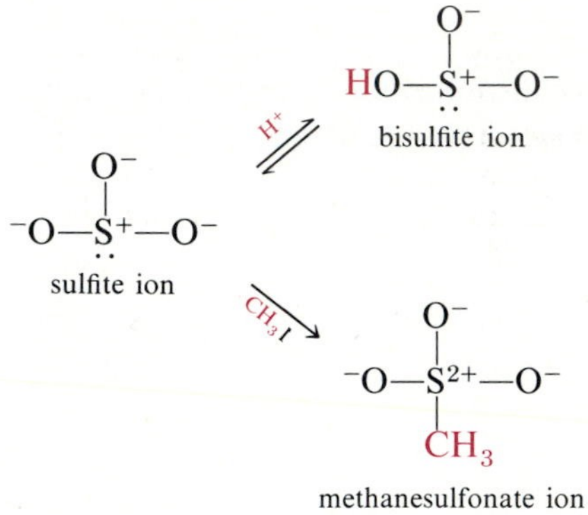

$$^{-}O-\overset{\overset{O^-}{|}}{\underset{\cdot\cdot}{S^{+}}}-O^- \quad \xrightleftharpoons{H^+} \quad HO-\overset{\overset{O^-}{|}}{\underset{\cdot\cdot}{S^{+}}}-O^- \text{ (bisulfite ion)}$$

sulfite ion

$$\xrightarrow{CH_3I} \quad ^{-}O-\overset{\overset{O^-}{|}}{\underset{\underset{CH_3}{|}}{S^{2+}}}-O^-$$

methanesulfonate ion

The oxygen in sulfite ion is the more basic atom and prefers to attack H^+, but the oxygens are also solvated by hydrogen bonding. The lone pair on sulfur is the more nucleophilic center and it has preference in the S_N2 transition state. Sulfate ion has no lone pair on sulfur, and both reactions have no alternative but to occur at the oxygen. Thiosulfate ion is a simple sulfur analog of sulfate. This ion reacts with methyl iodide exclusively on sulfur, even though there are three oxygens and only one sulfur.

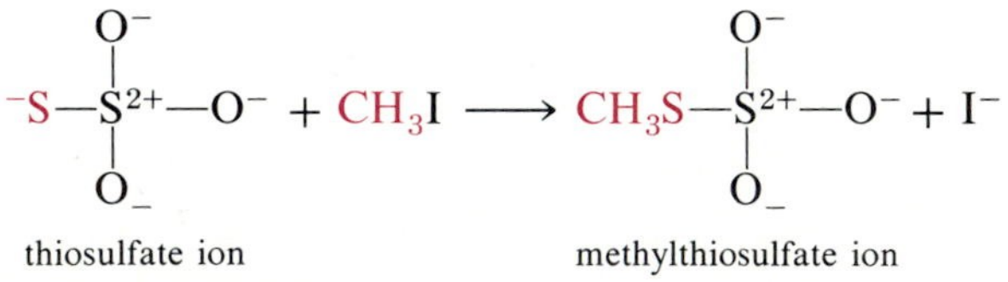

$$^{-}S-\overset{\overset{O^-}{|}}{\underset{\underset{O_-}{|}}{S^{2+}}}-O^- + CH_3I \longrightarrow CH_3S-\overset{\overset{O^-}{|}}{\underset{\underset{O_-}{|}}{S^{2+}}}-O^- + I^-$$

thiosulfate ion methylthiosulfate ion

Finally, there are some nucleophiles that show measurable nucleophilic properties at two different atoms. Nitrite ion is an example. The ion undergoes protonation exclusively on oxygen to give nitrous acid.

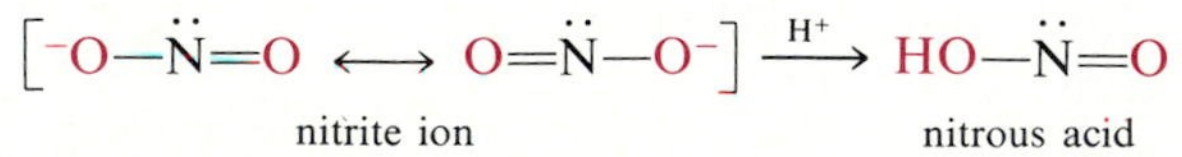

$$\left[{}^{-}O-\ddot{N}=O \longleftrightarrow O=\ddot{N}-O^-\right] \xrightarrow{H^+} HO-\ddot{N}=O$$

nitrite ion nitrous acid

However, the reaction of nitrite ion with methyl iodide gives both methyl nitrite and nitromethane.

$$^{-}O—\ddot{N}=O + CH_3I \longrightarrow CH_3O—\ddot{N}=O \text{ (methyl nitrite)} + \left[CH_3—\overset{+}{N}(=O)O^- \longleftrightarrow CH_3—\overset{+}{N}(O^-)=O \right] \text{ (nitromethane)}$$

In this case both nitrogen and oxygen are first-row elements and have comparable nucleophilicities. The ratio of the products actually depends on the reaction conditions. Another example is the reaction of methyl iodide with cyanide ion. In addition to methyl cyanide (acetonitrile, see Chapter 12), the major product, small amounts of methyl isocyanide are also produced.

Hydrogen cyanide, a dangerously toxic substance, smells like almonds to many people, but a genetic defect in the olfactory receptor prevents 16–30% from detecting its odor. Those who have handled sodium or potassium cyanide and are not aware of the characteristic smell must assume that they are not sensitive and must work with special care. Isocyanides are relatively nontoxic and have such a nauseating and offensive smell that they have been considered as warning agents.

$$:\bar{C}{\equiv}N: + CH_3I \longrightarrow CH_3—C{\equiv}N: \text{ (major, methyl cyanide)} + CH_3—\overset{+}{N}{\equiv}\overset{-}{C}: \text{ (minor, methyl isocyanide)}$$

Anions such as these, which can react at two different positions, are called ambident (L., *ambo,* both; *dentis,* tooth), "two-fanged" nucleophiles.

EXERCISE 9.6 The sodium salt of methylsulfinic acid reacts with methyl iodide in methanol to give a mixture of two isomeric products. What are the structures of these two products?

$$CH_3\ddot{S}(=O)—OH \xrightarrow{NaOH} \xrightarrow{CH_3I} \mathbf{A}\ (C_2H_6SO_2) + \mathbf{B}\ (C_2H_6SO_2)$$

9.5 Nucleophilicity and Solvent Effects

A. Solvent Properties

The solvent plays an important role in determining the effect of a displacement reaction. For example, the reaction of acetate ion with methyl iodide that we examined earlier is over ten million times faster in dimethylformamide than in methanol.

$(CH_3)_2NCHO$
dimethylformamide (DMF)

In this section we will study some of the factors that give rise to such large rate variations.

Nucleophilic displacement reactions involve ions either as nucleophiles or as products. Ions interact well with solvent molecules possessing ''polar'' groups, groups having bonds in which there is partial charge separation. Sometimes these solvents have high dipole moments. The interaction of the partial charges in different molecules is an additional electrostatic energy of attraction, so the boiling points of these solvents are higher than those of less polar materials.

A solvent is considered polar if it interacts well with polar molecules. Solvent polarity is an experimental property that can be measured in a number of ways. A widely used parameter is the dielectric constant.

The dielectric constant D of a substance is the factor by which an electrostatic interaction in a vacuum is reduced by a medium of the substance. Coulomb's law for the electrostatic interaction of two charges, q_1 and q_2, separated by a distance r in a medium of dielectric constant D is given as

$$E = \frac{q_1 q_2}{Dr}$$

As the dielectric constant of a solvent is reduced, electrostatic interaction between ions in the solvent increases. They form ion pairs and, for a sufficiently low D, insoluble crystals.

$$M^+ + X^- \rightleftharpoons \underset{\text{ion pair}}{M^+\ X^-} \rightleftharpoons \underset{\text{crystal}}{(M^+X^-)}$$

Ion pairs are relatively unreactive in S_N2 reactions.

Polar solvents orient around ions such that their dipole moments provide electrostatic attraction to both anions and cations.

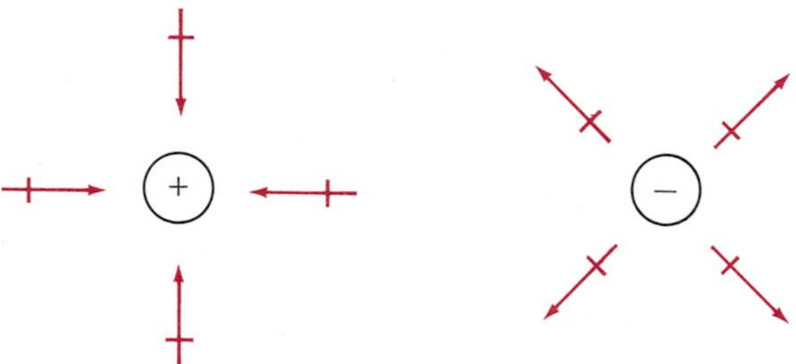

This attraction of ions and dipoles is one of the microscopic mechanisms that leads to the macroscopic property of a high dielectric constant.

> Another way of measuring solvent polarity is with the solvent polarity parameters, Z or E_T, described in Section 22.6.D.

The properties of compounds that are important as solvents in displacement reactions are summarized in Table 9.4. It is convenient to consider them in two categories: *hydroxylic* (water, alcohols, acetic acid) and *polar aprotic* (acetonitrile, dimethylformamide, dimethyl sulfoxide, acetone). Hydroxylic solvents have hydroxy groups that can hydrogen bond with anions. Polar aprotic solvents are polar solvents that have no hydroxy groups. The nonpolar solvent, *n*-hexane, is included for comparison. Note that 1,2-ethanediol with two functional groups has a relatively high boiling point and viscosity, since the groups in different molecules interact with one another.

TABLE 9.4 Solvent Properties

Solvent	m.p., °C	b.p., °C	d^a	vis.[b] (η), cp	d.m.[c] (μ), D	diel.[d] (D)	r.i.[e] (n_D^{20})
water	0.0	100.0	1.000^4	1.00	1.8	78.5	1.3330
1,2-ethanediol	−13.0	198	1.11	26.1^{15}	2.0	37.7	1.4318
methanol	−97.8	65	0.791	0.551	2.87	32.6	1.3286
ethanol	−117	78.3	0.789	1.08^{25}	1.7	24.3	1.3611
acetic acid	16.6	118.5	1.05	1.31	1–1.5	6.2	1.3721
acetonitrile	−43.8	81.6	0.782	0.375^{15}	3.5	37.5	1.3442
dimethyl sulfoxide (DMSO)	18.5	189	1.10^{25}	2.00^{25}	3.9	48.9	1.4783
dimethylformamide (DMF)	−61	153	0.949	0.924	3.8	36.7	1.4269^{25}
acetone	−95.3	56.2	0.790	0.337^{15}	2.7	20.7	1.3588
hexamethylphosphoric acid triamide (HMPA)	7.2	235	1.03	3.47	5.5	29.6	1.4582
n-hexane	−95	68.8	0.659	0.313	0.0	1.9	1.3749

[a]Density in g cm^{-3} at 20 °C except as noted by a superscript to the density value. [b]Viscosity in centipoises (cp) at 20 °C. [c]Dipole moment in Debyes. [d]Dielectric constant at 25 °C. [e]Refractive index at the sodium D line at 20 °C.

The pronounced effects of solvent on the rate of reaction with different nucleophiles are exemplified by the data in Table 9.5. This table gives the free energies of activation for reaction of many of the nucleophiles in Table 9.5 with methyl iodide at 25 °C in dimethylformamide, a representative polar aprotic solvent, and in methanol, a representative hydroxylic solvent. Although the two solvents have about the same dielectric constant, the reaction rate constants are quite different. (Recall that the transition state free energy is related to the rate constant; the lower the energy the greater the rate constant). The relative reactivity of a nucleophile is referred to as its *nucleophilicity*. The representative data in Table 9.5 show that nucleophilicity (the tendency to react with carbon) is not the same as basicity (the tendency to react with a proton) and that nucleophilicity is sensitive to solvent.

TABLE 9.5 Free Energies of Activation for Reaction of Nucleophiles with Methyl Iodide at 25°C in Methanol and in Dimethylformamide (DMF)

Nucleophile	DMF	CH_3OH
CN^-	14.0	21.8
$CH_3CO_2^-$	15.7	25.1
NO_2^-	16.8	22.5
N_3^-	16.8	23.0
Cl^-	16.9	25.0
Br^-	17.3	23.0
SCN^-	19.0	22.0
I^-	20.9	18.0
$(CH_3)_2S$	21.8	23.6

To put the numbers in Table 9.5 in practical perspective, note that a value of $\Delta G^\ddagger$ of 22.5 kcal mole^{-1} corresponds to a reaction having a half-life of approximately 1 hr with a reagent concentration of 1 M. Recall also that a change of 1.4 kcal mole^{-1} corresponds approximately to a factor of ten in rate.

B. Polar Aprotic Solvents

Many of the anions listed in Table 9.5 are good nucleophiles in polar aprotic solvents. This includes nucleophiles that are somewhat basic (cyanide ion, acetate ion) as well as the conjugate bases of strong acids (halide ions). The reactivity depends primarily on the standard free energy—the thermodynamic driving force—of reaction. That is, in these solvents the more exothermic the reaction, the more readily it occurs. Figure 9.8 is a general reaction profile for these reactions.

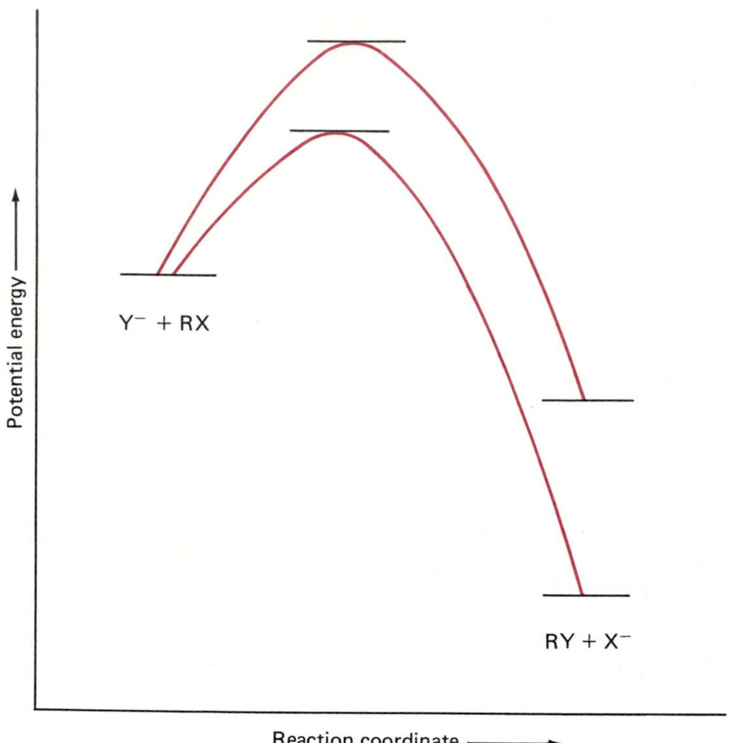

FIGURE 9.8 S_N2 reactions in polar aprotic solvents: the greater the driving force, the faster the reaction.

As in the case of the free-radical reactions we have already studied, one important component of the driving force is the bond strength of the new bond formed compared to that of the carbon-halogen bond being broken. Thus, groups in which the nucleophilic center is a first row atom (O, N) tend to be more reactive. Among halide ions, the reactivity order is $Cl^- > Br^- > I^-$. The nucleophile forms a bond to the carbon at the **reaction center,** the atom at which the bonding changes occur, with an electron pair. In a sense, the nucleophile is donating its electrons to the reaction center, and its ability to hold the electrons influences the ease of reaction. Thus, the relative electron affinity of the nucleophile compared to the leaving group is also important. The more readily a nucleophile can give up an electron (the lower its electron affinity), the greater is its nucleophilicity and the more facile is its reaction. Since the ability of the halides to give up electrons increases in the order $Cl^- < Br^- < I^-$, the rate differences among the halides as nucleophiles are less than suggested on the basis of bond strengths alone. Note that, in general, any factor that stabilizes a nucleophile anion would be expected to reduce the rate of reaction.

C. Hydroxylic Solvents

Nucleophiles are generally much less reactive in methanol than in dimethylformamide, often by many powers of ten, as a glance at Table 9.5 would show. The reason for this difference is that hydroxylic solvents interact with an ion by hydrogen bonding. Hydrogen bonding is especially important for small anions with concentrated charges. The nucleophile must lose at least one solvent molecule in order to form the bond to carbon in the transition state. The charge is more dispersed in the transition state, decreasing the strength of hydrogen bonding and the stabilization relative to the initial state. The difference between the initial state and the transition state is larger for reactions in hydroxylic solvents than in polar aprotic solvents. Thus, the energy barrier to reaction is larger and the rate constants are smaller.

$$\mathrm{Br:^{-}\text{---}H{-}O{-}R} + CH_3I \longrightarrow \left[Br^{\delta-}\cdots\overset{H}{\underset{H\ \ H}{C}}\cdots I^{\delta-} \right]^{\ddagger} \longrightarrow$$

hydrogen-bond solvation more important for smaller atoms (C, N, O, F)

much larger ion hydrogen-bond solvation less important

The effect is particularly important for first-row ions—ions with negative charge on N, O, and F. This effect leads to large reactivity changes between hydroxylic and polar aprotic solvents. Acetate ion, which is highly nucleophilic in dimethylformamide, is much less reactive in methanol. The reactivity order for halide ions is the reverse of that in dimethylformamide; in methanol the order is $I^- > Br^- > Cl^-$. The smaller chloride ion is more hydrogen bonded in methanol than are the larger halide ions. The rate is affected more by this solvation than by the greater carbon-chlorine bond strength compared to carbon-bromine and carbon-iodine. In general, nucleophiles with second row and larger atoms are less hydrogen bonded in hydroxylic solvents and are relatively more nucleophilic compared to polar aprotic solvents. Compare, for example, the relative nucleophilicities of thiocyanate ion (SCN^-) in methanol and in dimethylformamide (Table 9.5).

> Iodide ion is less reactive in dimethylformamide than in methanol. The large iodide ion fits poorly into the hydrogen-bonded liquid structure of methanol.

EXERCISE 9.7 Table 9.5 lists nucleophiles in order of decreasing reactivity in dimethylformamide. Rewrite the table in order of decreasing reactivity in methanol.

Despite the generally lesser reactivity in hydroxylic solvents, methanol and ethanol are important solvents for carrying out displacement reactions because they are inexpensive, relatively inert, and dissolve many organic substrates and inorganic salts. Sometimes some water is added to increase the solubility of the inorganic salt used as the displacing agent.

$$CH_3O^- + CH_3CH_2CH_2CH_2Br \xrightarrow{CH_3OH} CH_3CH_2CH_2CH_2OCH_3 + Br^-$$

> *n*-Butyl bromide is refluxed with sodium methoxide in methanol for $\frac{1}{2}$ hr. Water is added, and the organic layer is separated, dried, and distilled to give methyl *n*-butyl ether.

$$\mathrm{Na^{+}SCN^{-} + CH_3\underset{}{\overset{Br}{C}}HCH_3 \xrightarrow[H_2O]{C_2H_5OH} CH_3\overset{SCN}{C}HCH_3 + Na^{+}Br^{-}}$$

Isopropyl bromide and sodium thiocyanate (NaSCN) are refluxed in 90% aqueous ethanol for 6 hr. The precipitated sodium bromide is filtered. The filtrate is diluted with water and extracted with ether. Distillation gives isopropyl thiocyanate, $(CH_3)_2CHSCN$, in 76–79% yield.

In summarizing the effect of structure on nucleophilicity the following generalizations are useful. Nucleophiles with second row and larger atoms at the nucleophilic center tend to have useful reactivity. Nucleophiles with first row atoms at the nucleophilic center tend to be relatively more nucleophilic in polar aprotic solvents but less so in hydroxylic solvents. When the attacking atom remains the same, stronger bases are also more nucleophilic. For example, methoxide ion is a stronger base than acetate ion and is generally also more reactive in S_N2 reactions.

EXERCISE 9.8 The average $\Delta G^{\ddagger}$ in Table 9.5 is 20 kcal mole^{-1}, which corresponds to $\log k = -2$ or $k = 1 \times 10^{-2}\ M^{-1}\ sec^{-1}$. Consider the reaction of 0.1 M methyl iodide with 1 M nucleophile having this rate constant. How long would it take to achieve 99.9% reaction (10 half-lives)? Note that under these conditions the nucleophile concentration changes only from 1 M to 0.9 M and the reaction becomes pseudo-first order.

Highlight 9.5

Solvent molecules interact with reactant molecules through hydrogen bonds, electrostatic interactions, and van der Waals interactions. The sum of these interactions is called solvent polarity. Solvent polarity is modeled by several parameters, among which is *D*, the dielectric constant.

Highlight 9.6

The charge distribution often changes in reactant molecules on proceeding from the initial state to the transition state. High solvent polarity will lower the free energy of a charged transition state relative to an uncharged initial state, thus accelerating the reaction.

9.6 Leaving Groups

Alkyl chlorides, bromides, and iodides all react with nucleophiles by the S_N2 mechanism. The ease of reaction is dependent on the nature of the leaving group, alkyl iodides reacting most rapidly and alkyl chlorides most slowly. Alkyl fluorides are essentially unreactive by the S_N2 mechanism. Since chlorine is much cheaper than bromine, alkyl chlorides are the least expensive alkyl halides. However, for laboratory uses where only small amounts of material are involved, alkyl bromides are commonly used because they are 50–100 times more reactive than the corresponding chlorides. Iodides are somewhat more reactive than bromides but are quite a bit more expensive, and this slightly increased reactivity does not justify their additional cost. In industrial processes, where massive amounts of materials are involved and cost is a prime consideration, alkyl chlorides are used almost exclusively.

The S_N2 reaction is not restricted to alkyl halides. Any group that is the conjugate base of a strong acid can act as a leaving group. An example is bisulfate ion, HSO_4^-, which is the conjugate base of sulfuric acid, pK_a -5. Dimethyl sulfate is an inexpensive commercial compound, and reacts readily by the S_N2 mechanism. The leaving group is the methyl sulfate ion, which is similar in its base strength to bisulfate ion.

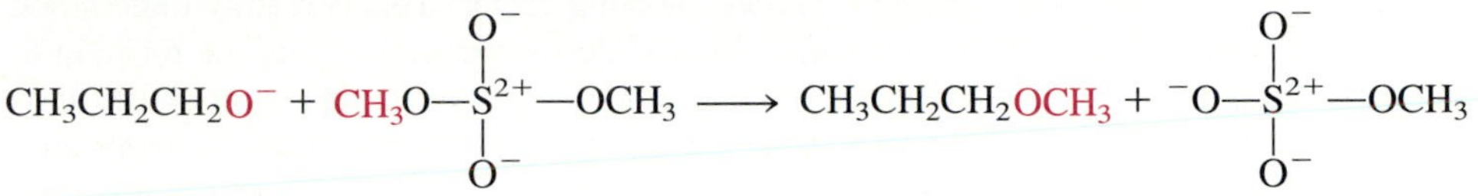

$$CH_3CH_2CH_2O^- + CH_3O{-}\overset{O^-}{\underset{O^-}{S^{2+}}}{-}OCH_3 \longrightarrow CH_3CH_2CH_2OCH_3 + {}^-O{-}\overset{O^-}{\underset{O^-}{S^{2+}}}{-}OCH_3$$

methyl sulfate ion

The chief disadvantage of dimethyl sulfate is its toxicity. It is water soluble and reacts readily with the nucleophilic groups in body tissues and fluids. Although dimethyl sulfate is the only sulfate in common use, alkyl sulfonates are often employed. Sulfonic acids, RSO_2OH, are similar to sulfuric acid in acidity, and the sulfonate ion, RSO_3^-, is an excellent leaving group. Alkyl benzenesulfonates, alkyl *p*-toluenesulfonates, and alkyl methanesulfonates are extremely useful substrates for S_N2 reactions. These compounds are readily prepared from alcohols as described in Sections 10.6.B and 26.5).

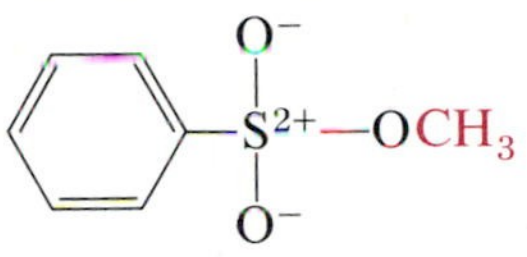

$$C_6H_5{-}\overset{O^-}{\underset{O^-}{S^{2+}}}{-}OCH_3 \qquad CH_3{-}C_6H_4{-}\overset{O^-}{\underset{O^-}{S^{2+}}}{-}OCH_2CH_3 \qquad CH_3{-}\overset{O^-}{\underset{O^-}{S^{2+}}}{-}OCH_3$$

methyl benzenesulfonate 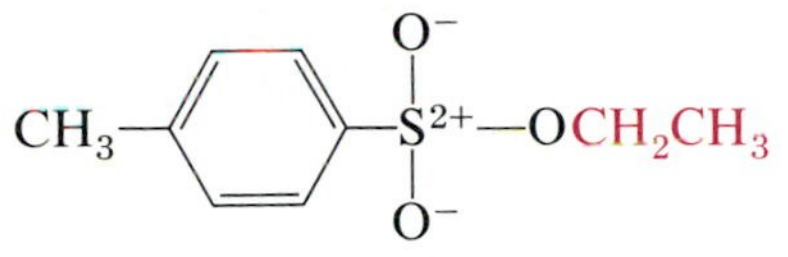 | ethyl *p*-toluenesulfonate 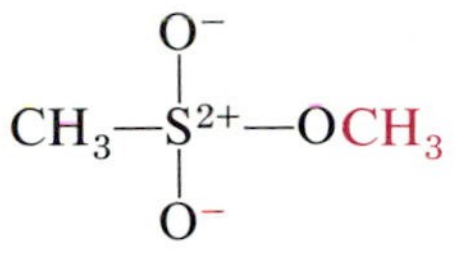 | methyl methanesulfonate

One instructive example of the use of such sulfonates in studying reaction mechanisms is shown by the following sequence of reactions involving chiral compounds and compounds enriched with the oxygen isotope ^{18}O. (**1.** OH → OAc) The butyl alcohol enriched with ^{18}O as shown can be converted to the acetate without loss of the oxygen isotope and shows that the carbon-oxygen bond has not been broken in the process. Under these conditions the optically active (+)-alcohol is converted to acetate that (coincidentally) also has a (+) optical rotation. Since the carbon-oxygen bond remains intact during this reaction, whatever the configuration the (+)-alcohol has around the stereocenter must be the same configuration in the (+)-acetate. (**2.** OH → OSO_2R) Similarly, the formation of the sulfonate ester as shown also proceeds with retention of the configuration around the stereocenter. (**3.** OSO_2R → OAc) When the sulfonate is treated with acetate ion in a displacement reaction, the acetate formed has lost the ^{18}O label; that is, the old carbon-

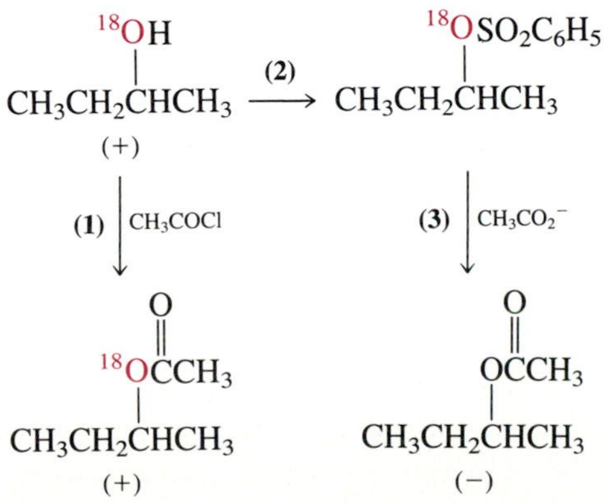

$$\underset{(+)}{CH_3CH_2\overset{^{18}OH}{C}HCH_3} \xrightarrow{(2)} CH_3CH_2\overset{^{18}OSO_2C_6H_5}{C}HCH_3$$

$$(1)\ CH_3COCl \downarrow \qquad\qquad (3)\ CH_3CO_2^- \downarrow$$

$$\underset{(+)}{CH_3CH_2\overset{^{18}OCOCH_3}{C}HCH_3} \qquad \underset{(-)}{CH_3CH_2\overset{OCOCH_3}{C}HCH_3}$$

oxygen bond was broken in order to form the new bond to an oxygen of the acetate. This reaction produces acetate having an optical rotation opposite that of acetate produced directly from the alcohol. The displacement reaction must have proceeded with *complete inversion of configuration*. Many experiments of this type were used to form the generalization, discussed in Section 9.2, that S_N2 reactions occur with inversion of configuration.

The facility with which a group can function as a leaving group in an S_N2 reaction can be gauged approximately by its basicity. If a group is a weak base (that is, the conjugate base of a strong acid), it will generally be a "good" leaving group. This is readily understood on recalling that the leaving group L gains electron density in going from the reactant to transition state.

$$N:^- + \rangle C{-}L \longrightarrow \left[\overset{\delta-}{N}\cdots C \cdots \overset{\delta-}{L} \right]^{\ddagger} \longrightarrow N{-}C\langle + :L^-$$

The more this electron density or negative charge is stabilized, the lower is the energy of the transition state and the faster is the reaction. The degree to which a group can accommodate a negative charge is also related to its affinity for a proton, its basicity. The acids HCl, HBr, HI, and H_2SO_4 are all strong acids because the anions Cl^-, Br^-, I^-, and HSO_4^- are stable anions. These anions are also good leaving groups in S_N2 reactions.

Hydrocyanic acid (HCN) is a weak acid (pK_a = 10), and the displacement of cyanide is never observed.

$$N:^- + RCN \not\longrightarrow R{-}N + CN^-$$

Hydrazoic acid (HN_3) and acetic acid (CH_3CO_2H) are also weak acids (pK_as of 5.8 and 4.8, respectively). Correspondingly, azide ion and acetate ion are poor leaving groups.

$$N:^- + R{-}N_3 \not\longrightarrow RN + N_3^-$$

$$N:^- + R{-}O\overset{O}{\overset{\|}{C}}CH_3 \not\longrightarrow RN + CH_3CO_2^-$$

The reason that alkyl fluorides are ineffective substrates in S_N2 reactions is related to the relatively low acidity of HF (pK_a = 3).

EXERCISE 9.9 Write out the complete displacement reactions of trimethylamine, $(CH_3)_3N$, with (a) $CH_3SO_2OCH_3$ and (b) $(CH_3)_3S^+$. Use curved arrows to show the electronic changes involved.

EXERCISE 9.10 Draw reaction coordinate diagrams for the displacement of bromide from ethyl bromide by methylamine in a low polarity solvent and in a high polarity solvent. Predict how the rate would change for the change in solvent polarity.

9.7 Elimination Reactions

One of the side reactions that occurs in varying degree in displacement reactions is the elimination of the elements of HX to produce an alkene.

$$B:^- \ H{-}\overset{|}{\underset{|}{C}}{-}\overset{|}{\underset{|}{C}}{-}X \longrightarrow BH + \rangle C{=}C\langle + X^-$$

Under appropriate conditions, this reaction can be the principal reaction and becomes a method for preparing alkenes. Accordingly, it is discussed in more detail in Section 11.5.A. For the present, it suffices to know that this reaction occurs by attack of a base on a hydrogen with concomitant formation of a carbon-carbon double bond and breaking of the carbon-halogen bond to form halide ion.

Mechanistically, the reaction is classified as **bimolecular elimination,** or **E2.** Since attack on a proton is involved, it is the basicity, rather than the nucleophilicity, of the Lewis base that is important. Strongly basic species such as alkoxide or hydroxide ions favor elimination; highly nucleophilic species such as second- and third-row elements favor substitution. For example, the S_N2 and E2 reactions of 3-bromo-2-methylbutane with chloride ion in acetone occur at about equal rates.

+ Cl^- $\xrightarrow{\text{acetone}}$ + +

Br — Cl

50% — 50%

With acetate ion, however, elimination is about 8 times faster than substitution.

+ $CH_3CO_2^-$ $\xrightarrow{\text{acetone}}$ + +

Br — O

11% — 89%

The structure of the substrate compound is also important in determining the substitution/elimination ratio. Straight-chain primary compounds show little tendency toward elimination because the alternative S_N2 reaction is relatively rapid. For example, ethyl bromide reacts with N_3^-, Cl^-, or $CH_3CO_2^-$ in acetone to give only substitution products. Even the strong base sodium ethoxide in ethanol gives virtually none of the elimination product. However, with more highly branched compounds, the S_N2 reaction is slower, and attack at hydrogen can compete more favorably. Consequently, larger amounts of the elimination product are obtained. Some data are presented in Table 9.6 for the reactions of various alkyl halides with acetate ion in acetone.

TABLE 9.6 Substitution and Elimination with Acetate Ion

$$\text{R—Br} + CH_3CO_2^- \xrightarrow{\text{acetone}} \text{R—O}\overset{\text{O}}{\overset{\|}{\text{C}}}CH_3 + Br^-$$

RBr	Percent Substitution	Percent Elimination
CH_3CH_2Br	100	0
$(CH_3)_2CHBr$	100	0
$(CH_3)_2CHCHBrCH_3$	11	89
$(CH_3)_3CBr$	0	100

Note the resulting generalizations that elimination by-products are quite minor with simple primary halides, but with branching at either the α- or β-carbon, the alkene elimination products become increasingly important. The behavior of tertiary halides is more complex. Since they undergo S_N2 reactions so very slowly, one would expect that tertiary halides would give complete elimination, even with weak bases. However, tertiary halides can undergo substitution by another mechanism (next section). Consequently, the elimination/substitution for tertiary halides is highly dependent on reaction conditions. In general, they give mainly the elimination products, especially under conditions that favor the bimolecular mechanism (high concentrations of strong base).

The distinction between *nucleophilicity* and *basicity* helps us to understand some of the chemistry of organo-alkali metal compounds. We would normally expect that carbanions generated as the alkali metal salts of hydrocarbons (Section 8.6) would be excellent nucleophiles. But they are also very strong bases. Alkylsodium and -potassium compounds do react with straight-chain halides to give higher alkanes—this is probably the reaction that occurs when alkyl halides are allowed to react with sodium or potassium metals.

$$2\ CH_3CH_2CH_2Br + 2\ Na \longrightarrow CH_3CH_2CH_2CH_2CH_2CH_3 + 2\ NaBr$$

The initially formed carbanion salt displaces halide from a molecule of alkyl halide that has not yet reacted.

$$CH_3CH_2CH_2Br + 2\ Na \longrightarrow CH_3CH_2CH_2{:}^{-}\ Na^{+} + NaBr$$
$$CH_3CH_2CH_2{:}^{-} + CH_3CH_2CH_2Br \longrightarrow CH_3CH_2CH_2CH_2CH_2CH_3 + Br^{-}$$

The reaction is limited in its usefulness. Branched-chain systems generally give largely or entirely elimination.

EXERCISE 9.11 In each of the following pairs of reactions which gives more of the elimination product?

(a) I^- + ([2-chlorobutane] or [2-chloro-3-methylbutane])

(b) ($CH_3CO_2^-$ or $CH_3CH_2O^-$) + [2-chlorobutane]

(c) [propyl]$^-$ Na^+ + ([1-chloropropane] or [2-chloropropane])

(d) ([2-chloropropane] or [2-chloro-2-methylpropane]) + CN^-

9.8 S_N1 Reactions: Carbocations

Ethyl bromide reacts rapidly with ethoxide ion in refluxing ethanol (78 °C); reaction is complete after a few minutes. If ethyl bromide is refluxed in ethanol not containing any added sodium ethoxide, S_N2 displacement still occurs, but the reaction is exceedingly slow. After refluxing for 4 days, the reaction is only 50% complete. This reactivity difference is due to the fact that the negatively charged ethoxide ion is much more nucleophilic than the neutral ethanol molecule. On an equal concentration basis, ethoxide ion is more than 10,000 times more reactive than ethanol itself. In general, anions are much more basic and nucleophilic than their conjugate acids.

$$CH_3O^- > CH_3OH$$
$$CH_3S^- > CH_3SH$$
$$CH_3CO_2^- > CH_3COOH$$
$$SO_3^{2-} > HSO_3^-$$

We saw earlier that tertiary alkyl halides rarely react by the S_N2 mechanism. Yet *t*-butyl bromide reacts quite rapidly in pure ethanol; in refluxing ethanol the half-life for reaction is only a few minutes! Various observations show that this reaction does not proceed by an S_N2 displacement even though the principal product is ethyl *t*-butyl ether, $(CH_3)_3COCH_2CH_3$. For example, the addition of ethoxide ion to an ethanol solution of ethyl bromide causes a large increase in the rate of reaction. The rate law for the reaction of ethyl bromide is

$$\text{rate} = k_1[C_2H_5Br] + k_2[C_2H_5Br][C_2H_5O^-]$$

The first term represents the reaction of ethyl bromide with neutral ethanol; because ethanol is the solvent and its concentration remains virtually unchanged, its concentration does not appear in the rate equation. The second term represents the reaction of ethyl bromide with ethoxide ion. For ethyl bromide, as we saw above, the reaction with ethoxide is much faster than the reaction with ethanol, that is, $k_2 \gg k_1$. Since k_1 is so small relative to k_2, the rate expression is approximately

$$\text{rate} = k_2[C_2H_5Br][C_2H_5O^-]$$

The rate of reaction of *t*-butyl bromide (*t*-BuBr) is given by a similar equation.

$$\text{rate} = k_1[t\text{-BuBr}] + k_2[t\text{-BuBr}][C_2H_5O^-]$$

Here, however, the second term is unimportant; $k_1 \gg k_2$. Addition of sodium ethoxide has no effect on the rate of reaction. Therefore, in this case the rate is effectively

$$\text{rate} = k_1[t\text{-BuBr}]$$

> This is true only for small concentrations of sodium ethoxide. High concentrations of strong base lead to formation of alkene by a competing elimination reaction (Section 9.6).

This change in kinetic behavior is indicative of a change in mechanism. With the tertiary alkyl halide, the rear of the molecule is effectively blocked and the S_N2 mechanism cannot operate. However, a competing mechanism is possible. A great deal of experimental work over the past several decades has established that this mechanism involves two steps. In the first step, ionization of the carbon-bromine bond occurs and an intermediate **carbocation** is produced. The carbocation then reacts rapidly with solvent or whatever nucleophiles are around.

$$\text{(1)}\quad CH_3C(CH_3)_2Br \underset{}{\overset{\text{slow}}{\rightleftharpoons}} CH_3C^+(CH_3)_2 + Br^-$$

$$\text{(2)}\quad CH_3C^+(CH_3)_2 + C_2H_5OH \xrightarrow{\text{fast}} CH_3C(CH_3)_2\text{—}\overset{+}{O}(H)CH_2CH_3 \longrightarrow CH_3C(CH_3)_2OCH_2CH_3 + H^+$$

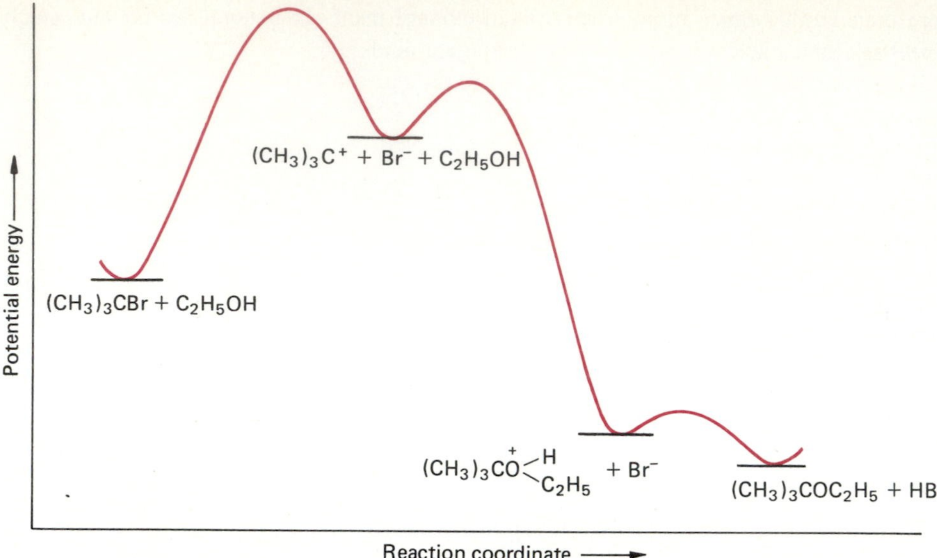

FIGURE 9.9 Reaction coordinate diagram for the solvolysis of *t*-butyl bromide.

When a compound reacts with the solvent, as is the case here, the process is called a **solvolysis reaction.** This solvolysis reaction is classified mechanistically as **unimolecular nucleophilic substitution** or S_N1 because only one species is involved in the rate-limiting step. A reaction coordinate diagram for the reaction is shown in Figure 9.9. In the intermediate carbocation, the central carbon has only a sextet of electrons (compare with the Lewis structure of methyl cation, Section 2.2). In orbital terms, the ion is best described in terms of a central sp^2-hybridized carbon with an empty *p*-orbital (Figure 9.10).

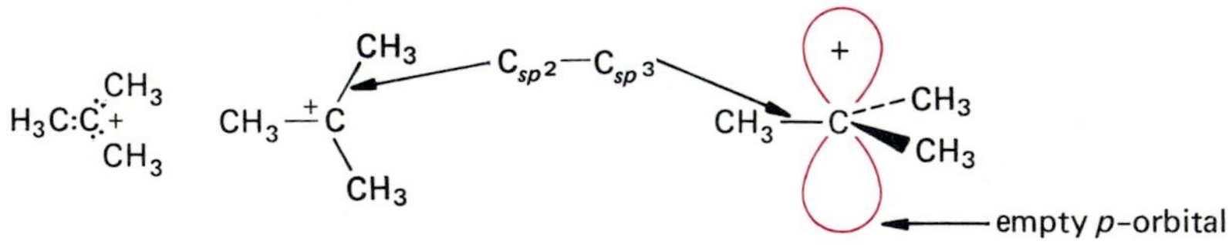

FIGURE 9.10 The *t*-butyl cation.

> We have already encountered *carbanions* (Section 8.6). The terms *carbocation* and *carbanion* are generic names for organic cations R^+ and anions R^-, respectively. For many years the term *carbonium ion* was used as a generic name for organic cations having an electron-deficient carbon with a sextet of electrons. The term carbonium ion has been considered to be nonsystematic because comparable oxonium ions (for example, H_3O^+) and ammonium ions (NH_4^+) have electron octets; the term *carbenium ion* has been suggested instead, but is not yet in common use, although it seems to be gaining acceptance. To avoid confusion, we will frequently use the term carbocation as a generic name for organic cations with positive formal charge on carbon, but specific ions will be referred to as alkyl cations (methyl cation, *t*-butyl cation, etc.).

Note that alkyl cations are planar; the central carbon and its three attached atoms all lie in one plane. The structure is favored because it places the three groups as far apart from one another as possible and allows the carbocation to achieve the strongest bonds and lowest energy. This electronic structure can also be understood on the basis that electrons in a 2*s*-orbital are more stable than electrons in a 2*p*-orbital. By using the *s*-orbital to form three sp^2-hybrid orbitals, it is involved most effectively in bonding; the remaining $2p$-orbital is left vacant.

It is important to point out that cations are highly reactive reaction intermediates. They have only a short, but finite, lifetime in solution. Under normal conditions they cannot be observed directly in the reaction mixture because they react almost as soon as they are produced. However, carbocations vary widely in stability with structure. The enthalpies for ionization of various alkyl halides *in the gas phase* are summarized in Table 9.7. Note that ethyl cation is more stable than methyl cation, but n-propyl and n-butyl cations are of similar stability. These systems may be described as *primary carbocations,* RCH_2^+. Isopropyl cation is an example of a *secondary carbocation,* R_2CH^+, and t-butyl cation is a *tertiary carbocation.*

CH_3^+	methyl cation
$CH_3CH_2^+$	ethyl cation, a *primary carbocation*
$(CH_3)_2CH^+$	isopropyl cation, a *secondary carbocation*
$(CH_3)_3C^+$	t-butyl cation, a *tertiary carbocation*

TABLE 9.7 Enthalpy for Ionization of Alkyl Chlorides in the Gas Phase

$$RCl \longrightarrow R^+ + Cl^-$$

R	$\Delta H°$, kcal mole^{-1}
CH_3	229
CH_3CH_2	193
$CH_3CH_2CH_2$	196
$CH_3CH_2CH_2CH_2$	196
$(CH_3)_2CH$	172
$(CH_3)_3C$	157

Table 9.7 shows that the relative stabilities of various cations are

$$\text{tertiary} > \text{secondary} > \text{primary} > \text{methyl}$$

Note that the difference in the ionization energy between methyl chloride and t-butyl chloride is 72 kcal mole^{-1} in the gas phase. In solution, the ions are solvated and ionization is facilitated. Consequently, the $\Delta H°$ for ionization is much lower than given in Table 9.7. Thus, the energy required to ionize t-butyl chloride in hydroxylic solvents is low enough that reaction proceeds at easily observable rates. Tertiary alkyl cations are rather common intermediates in organic reactions. Secondary alkyl cations are considerably less stable and much more difficult to produce, but they do occur as intermediates in some reactions. Primary alkyl cations are so much less stable that they virtually never occur as reaction intermediates in solution. These generalizations may be summarized as follows.

Alkyl Group	Occurrence of Carbocation Intermediates in Solution	Occurrence of S_N2 Displacement Reactions
tertiary	common	rare
secondary	sometimes	sometimes
primary	rare	common
methyl	never	common

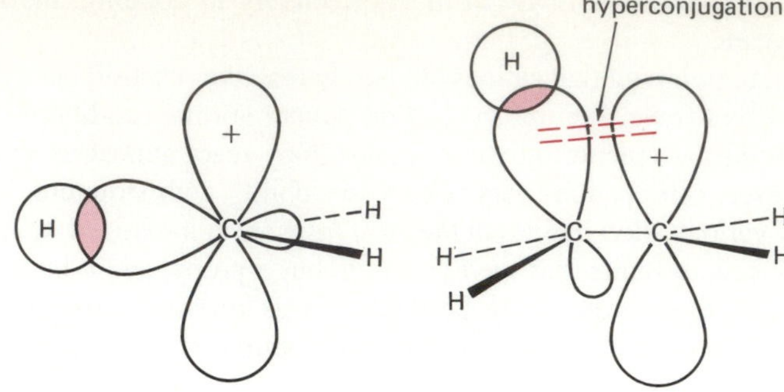

FIGURE 9.11 Overlap of a carbon-hydrogen bond orbital with the empty *p*-orbital of a carbocation.

The order of stabilities of carbocations is in large part to be attributed to the greater polarizability of alkyl groups compared to hydrogen. Another factor is an interesting aspect of overlapping orbitals. Consider the methyl cation, CH_3^+. The carbon-hydrogen bonds in methyl cation lie in the nodal plane of the vacant $2p_z$-orbital and hence cannot overlap with it. In the ethyl cation, $CH_3CH_2^+$, there can be some overlap between this empty orbital and one of the bonds of the methyl group (Figure 9.11). This type of overlap is readily shown by quantum mechanical calculations. It has the effect of stabilizing the ion because electron density from an adjacent bond can ''spill over'' into the empty orbital. This results in spreading the positive charge over a larger volume. We shall see frequently that ions with concentrated charge are less stable and more reactive than ions in which the charge is spread over a greater volume. As more alkyl groups are attached to the cationic carbon, it becomes even more stable.

The interaction of a bond orbital with a *p*-orbital as shown in Figure 9.11 is referred to as *hyperconjugation*. Its relationship to *conjugation* will become apparent when we get to Chapter 20.

The relative stabilities of carbocations can be understood on the basis of simple electrostatics. Electrons are attracted to nuclei and are repelled by other electrons. The electrons in a bond repel each other, but they are prevented from getting too far apart because of their attraction to the two nuclei in the bonded atoms. When there is an adjacent atom with a vacant orbital available for overlap and a positive charge, the original bonding pair of electrons can reduce their mutual repulsion by getting farther apart; they can do this and still maintain the stability of being associated with a positive nucleus. Electron repulsion is decreased in such a carbocation and it is convenient to describe the result in terms of a spreading out of positive charge.

The first step of the S_N1 reaction is an ionization process that involves separation of charge.

$$R{-}X \longrightarrow R^+ \ X^-$$

Accordingly, the reaction is highly sensitive to solvent effects. The reaction is rapid in aqueous and hydroxylic solvents and slow in nonpolar solvents.

The solvolysis reaction of tertiary alkyl halides is only in part an S_N1 reaction. The intermediate carbocation can react rapidly in several alternative ways. Two of these reaction paths are

1. Reaction with any nucleophiles present (the S_N1 process)
2. Elimination of a proton (the E1 process).

These reactions are illustrated by the behavior of *t*-butyl bromide in ethanol at 25 °C. The solvolysis product, ethyl *t*-butyl ether, is produced along with a significant amount of the elimination product, an alkene.

$$(CH_3)_3CBr \xrightarrow[C_2H_5OH]{slow} (CH_3)_3C^+ \begin{cases} \xrightarrow{fast} (CH_3)_3COCH_2CH_3 \quad (81\%) \\ \xrightarrow{fast} (CH_3)_2C{=}CH_2 \quad (19\%) \end{cases}$$

In a mixed solvent, the carbocation can react with both components in addition to eliminating a proton. For example, in a mixture of 80% ethanol and 20% water, *t*-butyl bromide gives three products.

$$(CH_3)_3CBr \xrightarrow[H_2O \; 25°]{C_2H_5OH} (CH_3)_3C^+ \begin{cases} \longrightarrow (CH_3)_3COCH_2CH_3 \quad (29\%) \\ \xrightarrow{H_2O} (CH_3)_3COH \quad (58\%) \\ \xrightarrow{-H^+} (CH_3)_2C{=}CH_2 \quad (13\%) \end{cases}$$

Because product mixtures are so frequently obtained, solvolysis reactions are generally not important in synthesis. Such reactions have been studied in great detail over the past several decades, but primarily for the purpose of evaluating the properties and relative stabilities of carbocations.

EXERCISE 9.12 When cyclopentyl chloride is refluxed with a solution of sodium thiocyanate in methanol, the principal product is cyclopentyl thiocyanate. Under the same conditions, however, 1-chloro-1-methylcyclopentane gives principally 1-methoxy-1-methylcyclopentane. Explain.

Highlight 9.7

Tertiary alkyl halides in a sufficiently polar solvent readily form *t*-alkyl cations that can undergo either substitution by a nucleophile (S_N1) or attack by base to eliminate a proton (E1).

9.9 Ring Systems

Cyclic systems provide some special insights in substitution reactions of alkyl halides. For example, the reaction of amines with alkyl halides is a typical S_N2 reaction (Section 24.6.A).

$$RCH_2NH_2 + R'CH_2Br \longrightarrow RCH_2\overset{+}{N}H_2CH_2R' + Br^-$$

When both functional groups are present in the same molecule, the reaction is an **intramolecular** S_N2 reaction, which creates a ring. Ring formation necessarily has been initiated at the transition state; hence, the relative energies of transition states depend on the ring size. Relative rates for reactions of ω-bromoalkylamines, $Br(CH_2)_{n-1}NH_2$, are given in Table 9.8.

> Omega, ω, is the last letter of the Greek alphabet and its use conveniently designates substituents at the end of the alkyl chain.

TABLE 9.8 Relative Rates of Cyclization of Aminoalkyl Halides

$$Br(CH_2)_{n-1}NH_2 \longrightarrow (CH_2)_{n-1}NH_2^+ \ Br^-$$

n (Ring Size)	Relative Rate
3	0.1
4	0.002
5	100
6	1.7
7	0.03
10	10^{-8}
12	10^{-5}
14	3×10^{-4}
15	3×10^{-4}
17	6×10^{-4}

1-Amino-4-bromobutane, which gives a five-membered ring, is the most reactive compound, followed by 1-amino-5-bromopentane, which gives a six-membered ring. We see that the stability of the ring is not the only factor—the probability that the ends can get together is also important. The three-membered ring, for example, has high energy, but the functional groups are so close together that ring formation is relatively probable. The pattern shown in Table 9.8 is common. The general order of ring formation in intramolecular S_N2 reactions is $5 > 6 > 3$ with other rings being formed much more slowly.

Large rings are relatively unstrained, but the groups are so far apart that their probability of getting together for reaction is low. Indeed, it becomes more probable for reaction of the amino group of one molecule to displace a bromide belonging to another molecule. That is, **intermolecular** S_N2 reaction is an important side reaction for such cases unless conditions of high dilution are used. At ordinary concentrations the reaction product is a polymer chain.

> Note that an **intramolecular** S_N2 reaction will have a rate law that is *first order* in the halide. Other characteristics of a displacement reaction, such as inversion of configuration at the reaction center, will still be seen.

Displacement reactions on cyclic compounds show large effects of ring size. Table 9.9 summarizes some relative rates of reaction of alkyl bromides with lithium iodide in acetone.

Halocyclopropanes are relatively unreactive in S_N2 reactions. At the transition state of an S_N2 reaction, the central carbon has sp^2-hybridization in which the normal bond angle is 120°. The imposition of such an increased bond angle on a cyclopropyl ring would result in additional bond angle strain (Figure 9.12). The same effect is apparent in the slow reactions of cyclobutyl systems, but since the bond angle strain is lower, the effect is not as great.

TABLE 9.9 Relative Rates of Reaction of Alkyl Bromides with Lithium Iodide in Acetone

Alkyl Group	Relative Rate
Isopropyl	1.0
Cyclopropyl	<0.0001[a]
Cyclobutyl	0.008
Cyclopentyl	1.6
Cyclohexyl	0.01
Cycloheptyl	1.0
Cyclooctyl	0.2

[a] Approximate upper limit; no reaction was detected.

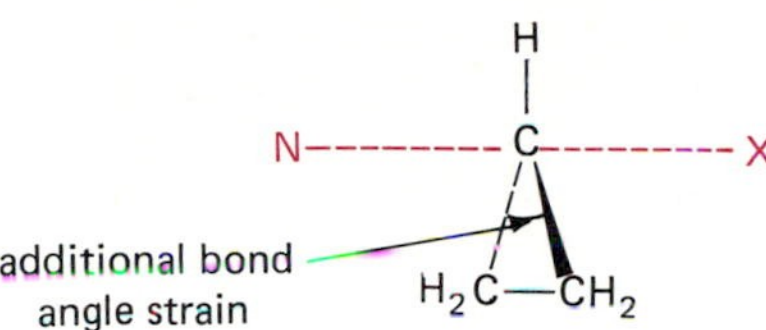

FIGURE 9.12 Transition state for S_N2 reaction on a cyclopropyl halide.

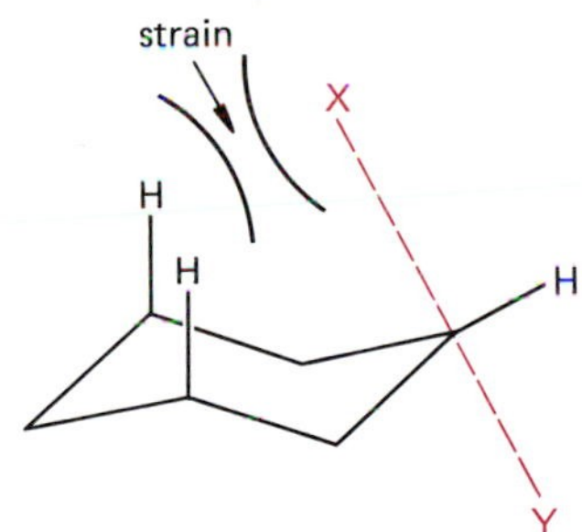

FIGURE 9.13 Transition state for S_N2 reaction with cyclohexyl compounds.

Cyclopentyl compounds undergo S_N2 reactions at rates comparable to open-chain systems. The reactions usually proceed in good yield.

Cyclohexyl compounds react rather slowly in S_N2 reactions. Ring strain does not appear to be an important factor for five- and six-membered rings, but a novel kind of strain involving axial hydrogens does appear to be significant for cyclohexyl systems (Figure 9.13). Because the displacement reaction has a reduced rate, elimination reactions to alkenes are frequently important side reactions with cyclohexyl compounds and often dominate. For larger rings, the reactions are roughly comparable to open-chain systems.

EXERCISE 9.13 In each of the following pairs, which is more reactive in S_N2 reactions?

(a) cyclobutyl bromide, cyclopentyl bromide
(b) cyclohexyl iodide, 3,3-dimethylcyclohexyl iodide
(c) cyclopropyl chloride, cyclopropylmethyl chloride

1. Show how each of the following compounds can be prepared from an alkyl halide.

a. $CH_3CH_2CH_2SH$ **b.** $(CH_3)_2CHCH_2CH_2CN$
c. $CH_3CH_2CH_2OCH_3$ **d.** $CH_3CH_2CH_2OH$
e. CH_3ONO_2 **f.** $CH_3CH_2CH_2CH_2N_3$

2. Predict the principal product of each of the following reactions:

a. 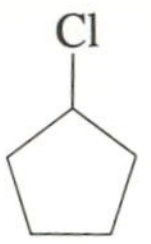 (chlorocyclopentane) $+ K^+ I^- \xrightarrow{\text{acetone}}$

b.  (1-chloro-1-methylcyclopentane) $+ K^+ I^- \xrightarrow{\text{acetone}}$

c. $CH_3I + Na^+ SCN^- \xrightarrow[\Delta]{CH_3OH}$

d. $(CH_3CH_2)_3CCl + Na^+ OCH_3^- \xrightarrow[\Delta]{CH_3OH}$

e. 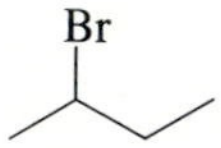 (2-bromobutane) $+ Na^+ CN^- \xrightarrow{\text{DMF}}$

f. $(H_3C)_2NCH_2CH_2CH_2CH_2Br \xrightarrow{\text{DMF}}$

g. $(CH_3)_3CCl + Na \xrightarrow{\text{ether}}$

h. $(CH_3)_2CHCH_2CH_2Br + CH_3CO_2^- Na^+ \xrightarrow[\Delta]{C_2H_5OH}$

i. $CH_3CH_2CH_2I + Na^+ NO_2^- \xrightarrow{\text{DMSO}}$

j. $(CH_3)_2CHCH_2Cl + Na^+ N_3^- \xrightarrow[\Delta]{CH_3OH}$

3. (*R*)-2-Butanol, labeled with ^{18}O, is subjected to the following sequence of reactions.

$$CH_3CH_2\overset{^{18}OH}{\overset{|}{C}}HCH_3 \xrightarrow{CH_3SO_2Cl} CH_3CH_2\overset{^{18}OSO_2CH_3}{\overset{|}{C}}HCH_3 \xrightarrow[\substack{H_2O \\ \text{dioxane}}]{OH^-} CH_3CH_2\overset{OH}{\overset{|}{C}}HCH_3 + CH_3SO_2{-}^{18}O^-$$

What is the absolute configuration of the product?

4. 2-Bromo-, 2-chloro-, and 2-iodo-2-methylbutanes react at different rates with pure methanol but produce the same mixture of 2-methoxy-2-methylbutane and alkenes as products. Explain these results in terms of the reaction mechanism.

5. Explain each of the following observations.

a. (*S*)-3-Bromo-3-methylhexane reacts in aqueous acetone to give racemic 3-methyl-3-hexanol.

$$CH_3CH_2CH_2\overset{Br}{\underset{CH_3}{\overset{|}{\underset{|}{C}}}}CH_2CH_3 \xrightarrow[\text{acetone}]{H_2O} (\pm)CH_3CH_2CH_2\overset{OH}{\underset{CH_3}{\overset{|}{\underset{|}{C}}}}CH_2CH_3$$

b. (*R*)-2-Bromo-2,4-dimethylhexane reacts in aqueous acetone to give optically active 2,4-dimethyl-2-hexanol.

6. For each of the following pairs of reactions, predict which one is faster and explain why.

a. $(CH_3)_2CHCH_2Cl + N_3^- \xrightarrow{C_2H_5OH} (CH_3)_2CHCH_2N_3 + Cl^-$

$(CH_3)_2CHCH_2I + N_3^- \xrightarrow{C_2H_5OH} (CH_3)_2CHCH_2N_3 + I^-$

b. $(CH_3)_3CBr \xrightarrow[\Delta]{H_2O} (CH_3)_3COH + HBr$

$(CH_3)_2CHBr \xrightarrow[\Delta]{H_2O} (CH_3)_2CHOH + HBr$

c. $CH_3CH_2CH(CH_3)CH_2Br + SH^- \longrightarrow CH_3CH_2CH(CH_3)CH_2SH + Br^-$

$CH_3CH_2CH_2CH_2Br + SH^- \longrightarrow CH_3CH_2CH_2CH_2SH + Br^-$

d. $CH_3CH_2Br + CN^- \xrightarrow[\text{(solvent)}]{CH_3OH} CH_3CH_2CN + Br^-$

$CH_3CH_2Br + CN^- \xrightarrow[\text{(solvent)}]{(CH_3)_2NCHO} CH_3CH_2CN + Br^-$

e. $(CH_3)_2CHBr + NH_3 \xrightarrow{CH_3OH} (CH_3)_2CH\overset{+}{N}H_3\ Br^-$

$CH_3CH_2CH_2Br + NH_3 \xrightarrow{CH_3OH} CH_3CH_2CH_2\overset{+}{N}H_3\ Br^-$

f. $CH_3I + Na^+OH^- \xrightarrow{H_2O} CH_3OH + Na^+\ I^-$

$CH_3I + Na^+\ {}^-O_2CCH_3 \xrightarrow{H_2O} CH_3O_2CCH_3 + Na^+\ I^-$

g. $CH_3Br + (CH_3)_3N \longrightarrow (CH_3)_4N^+\ Br^-$

$CH_3Br + (CH_3)_3P \longrightarrow (CH_3)_4P^+\ Br^-$

h. $SCN^- + CH_3CH_2Br \xrightarrow{\text{aq. } C_2H_5OH} CH_3CH_2SCN + Br^-$

$SCN^- + CH_3CH_2Br \xrightarrow{\text{aq. } C_2H_5OH} CH_3CH_2NCS + Br^-$

i. $N_3^- + (CH_3)_2CHCH_2Br \xrightarrow{C_2H_5OH} (CH_3)_2CHCH_2N_3 + Br^-$

$C_6H_5S^- + (CH_3)_2CHCH_2Br \xrightarrow{C_2H_5OH} (CH_3)_2CHCH_2SC_6H_5 + Br^-$

[*Note:* $pK_a(HN_3) \cong pK_a(C_6H_5SH)$.]

j. $CH_3CO_2^-$ + chlorocyclobutane (Cl on cyclobutane ring) ⟶ CH_3COO–cyclobutane + Cl^-

$CH_3CO_2^-$ + chlorocyclopentane (Cl on cyclopentane ring) ⟶ CH_3COO–cyclopentane + Cl^-

k. $^-OCH_2CH_2Cl \longrightarrow$ CH_2—CH_2 (three-membered ring bridged by O) $+ Cl^-$

$^-OCH_2CH_2CH_2Cl \longrightarrow$ O—CH_2 / CH_2—CH_2 (four-membered ring) $+ Cl^-$

7. Of the following nucleophilic substitution reactions, which ones will probably occur and which will probably not occur or be very slow. Explain.

a. $CH_3CN + I^- \longrightarrow CH_3I + CN^-$
b. $CH_3F + Cl^- \longrightarrow CH_3Cl + F^-$
c. $(CH_3)_3COH + NH_2^- \longrightarrow (CH_3)_3CNH_2 + OH^-$
d. $CH_3OSO_2OCH_3 + Cl^- \longrightarrow CH_3Cl + CH_3OSO_3^-$
e. $CH_3NH_2 + I^- \longrightarrow CH_3I + NH_2^-$
f. $CH_3CH_2I + N_3^- \longrightarrow CH_3CH_2N_3 + I^-$
g. $CH_3CH_2OH + F^- \longrightarrow CH_3CH_2F + OH^-$
h. cyclopropyl bromide + $N(CH_3)_3 \longrightarrow C_3H_5N(CH_3)_3^+ Br^-$

8. Of the following statements, which are true for nucleophilic substitutions occurring by the S_N2 mechanism?

a. Tertiary alkyl halides react faster than secondary.
b. The absolute configuration of the product is opposite to that of the reactant when an optically active substrate is used.
c. The reaction shows first-order kinetics.
d. The rate of reaction depends markedly on the nucleophilicity of the attacking nucleophile.
e. The probable mechanism involves only one step.
f. Carbocations are intermediates.
g. The rate of reaction is proportional to the concentration of the attacking nucleophile.
h. The rate of reaction depends on the nature of the leaving group.

9. Answer problem 8 for nucleophilic substitutions occurring by the S_N1 mechanism.

10. Give a specific example of two related reactions having different rates for which each of the following is the principal reason for the relative reactivities.

a. The less basic leaving group is more reactive.
b. A nitrogen anion is hydrogen bonded more strongly than a sulfur anion.
c. Tertiary carbocations are more stable than secondary carbocations.
d. Steric hindrance.
e. Protic solvents can form hydrogen bonds.

11. Consider the reaction of isopropyl iodide with various nucleophiles. For each pair, predict which will give the larger substitution/elimination ratio.

a. SCN^- or OCN^-
b. I^- or Cl^-
c. $N(CH_3)_3$ or $P(CH_3)_3$
d. CH_3S^- or CH_3O^-

12. When a solution of *cis*-1-*t*-butyl-4-chlorocyclohexane in ethanol is refluxed for several hours, the major product is found to be *trans*-1-*t*-butyl-4-ethoxycyclohexane. However, if the solution is also made 2.0 M in sodium ethoxide, the major product after the same treatment is found to be 4-*t*-butylcyclohexene. Explain.

13. The reaction of methyl bromide with methylamine to give dimethylammonium bromide is a typical S_N2 reaction that shows second-order kinetics.

$$CH_3Br + CH_3NH_2 \longrightarrow (CH_3)_2NH_2^+ Br^-$$

However, the analogous cyclization of 4-bromobutylamine shows first-order kinetics. Explain.

$$H_2C(CH_2Br)-CH_2-CH_2-CH_2NH_2 \longrightarrow \text{ring } H_2C{-}CH_2,\ H_2C,\ CH_2,\ \overset{+}{N}H_2 \ \ Br^-$$

The foregoing intramolecular displacement reaction is a useful method for making cyclic amines. However, a competing side reaction is the intermolecular displacement

$$Br(CH_2)_4NH_2 + Br(CH_2)_4NH_2 \longrightarrow Br(CH_2)_4\overset{H}{\underset{H}{N^+}}(CH_2)_4NH_2\ Br^-$$

Suggest the experimental conditions to minimize this side reaction.

14. Chloroacetic acid is more acidic than acetic acid by 1.9 p*K* units. Chloroacetate ion is 10 times less reactive than acetate ion in an S_N2 reaction with methyl iodide in methanol. What do these data suggest concerning the amount of negative charge left on the carboxylate group in the transition state?

15. Optically active 3-bromo-2-butanol is treated with KOH in methanol to obtain an optically inactive product having the formula C_4H_8O. What is the structure of this material?

16. Suggest an explanation for each of the following observations.

a. Compound A reacts faster by the S_N2 mechanism than compound B.

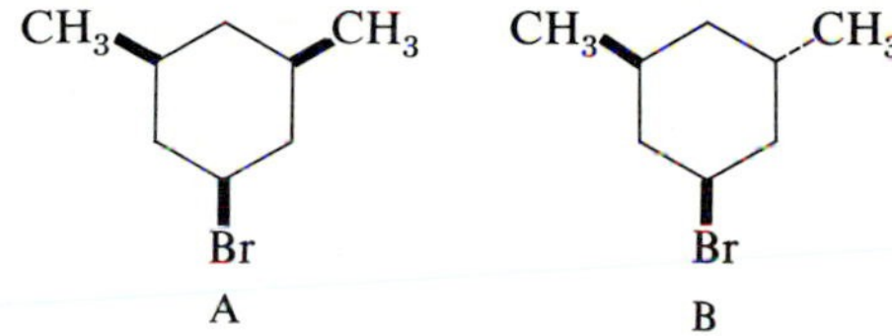

b. Compound C reacts faster by the S_N1 mechanism than compound D.

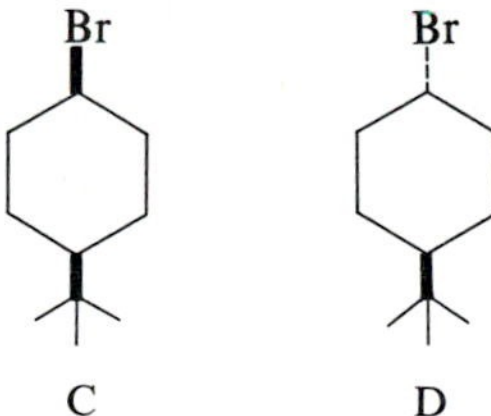

c. *t*-Butyl chloride reacts in methanol solution at 25 °C 8 times faster than it does in ethanol solution.

17. Appendix I gives some heats of formation for anions in the gas phase. Use these data to compute the $\Delta H°$ for those S_N2 reactions listed in Table 9.5 for which sufficient data are available. Do these heats of reaction correlate with the reaction rates in either DMF or methanol? What do the results imply?

CHAPTER 10

ALCOHOLS AND ETHERS

10.1 Introduction: Structure

Alcohols are compounds in which an alkyl group replaces one of the hydrogens of water. They are organic compounds that contain the functional group OH. As we shall see, this functional group dominates the chemistry of alcohols. **Ethers** are analogs of water in which both hydrogens are replaced by alkyl groups.

$CH_3—OH$ — an alcohol (methyl alcohol)

$CH_3—O—CH_3$ — an ether (dimethyl ether)

Methyl alcohol has been found by microwave spectroscopy to have the following geometry.

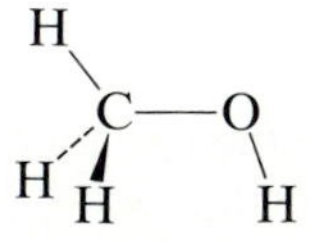

Bond Lengths, Å		Bond Angles, deg	
C—H	1.10	H—C—H	109
O—H	0.96	H—C—O	110
C—O	1.43	C—O—H	108.9

The hybridization of carbon is approximately sp^3, as shown by the H—C—H and H—C—O bond angles of 109° and 110°. The hybridization of oxygen may also be described as approximately sp^3. Oxygen makes one bond to carbon and one to hydrogen. The oxygen-hydrogen bond distance is the same as the oxygen-hydrogen bond distance in water. The molecule exists in a conformation that has the oxygen-hydrogen bond staggered between two carbon-hydrogen bonds, and the barrier to rotation about the carbon-oxygen bond is 1.1 kcal mole^{-1}. It is frequently useful to consider the oxygen lone-pair electrons to occupy orbitals that are each approximately sp^3; such lone-pair orbitals are each staggered between two adjacent carbon-hydrogen bonds.

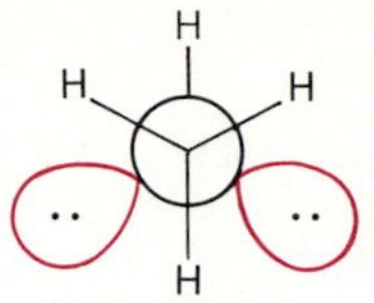

The simple alcohols are important industrial materials. They are also used extensively as laboratory reagents and as solvents. The most important representative of the class is undoubtedly ethyl alcohol, which has been known as an intoxicant since the dawn of civilization. Indeed, it has been suggested that the discovery of alcohol fermentation and the physiological effects of its product provided a major incentive for agriculture and the *start* of civilization! Alcohols can be prepared readily from many other classes of compounds and can, in turn, be transformed into many others. For this reason, alcohols play a key role as synthetic intermediates.

10.2 Nomenclature of Alcohols

Like most other classes of organic compounds, alcohols can be named in several ways. Common names are useful only for the simpler members of a class. However, common names are widely used in colloquial conversation and in the scientific literature. In order to communicate freely, the student must know common names. Since the systematic IUPAC names are often used for indexing the scientific literature, the student must be thoroughly familiar with systematic names in order to retrieve data from the literature.

The common names of alcohols are derived by combining the name of the alkyl group with the word alcohol. The names are written as two words.

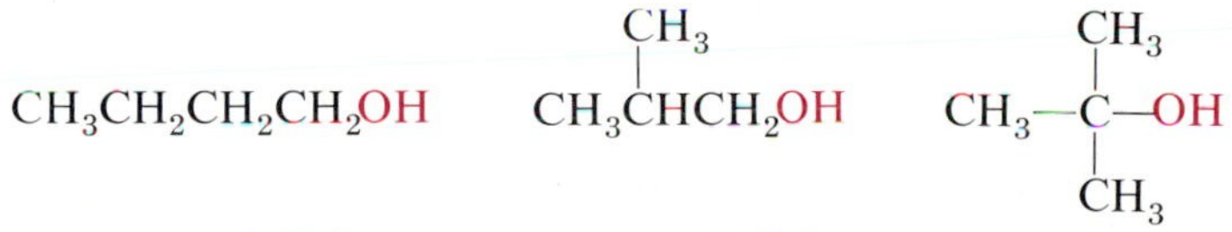

$CH_3CH_2CH_2CH_2OH$ — *n*-butyl alcohol

$CH_3CH(CH_3)CH_2OH$ — isobutyl alcohol

$CH_3C(CH_3)_2OH$ — *t*-butyl alcohol

In common names the position of an additional substituent is indicated by letters of the Greek alphabet rather than by numbers.

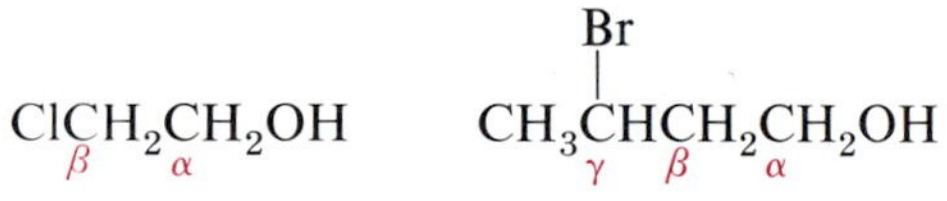

$\underset{\beta}{Cl\text{CH}_2}\underset{\alpha}{\text{CH}_2}OH$ — β-chloroethyl alcohol

$CH_3\underset{\gamma}{CH(Br)}\underset{\beta}{CH_2}\underset{\alpha}{CH_2}OH$ — γ-bromobutyl alcohol

EXERCISE 10.1 Give common names for each of the following alcohols.

(a) 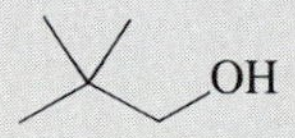 [skeletal structure: 2,2-dimethylpropyl–OH]   (b) 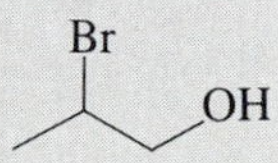 [skeletal structure with Br, OH]   (c) 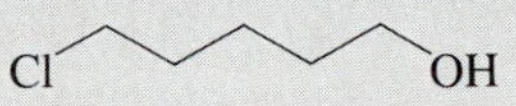 Cl〰〰OH

In the IUPAC system of nomenclature alcohols are named by replacing the **-e** of the corresponding alkane name by the suffix **-ol,** that is, as **alkanols.**

CH_3OH — methanol

CH_3CH_2OH — ethanol

The alkan- stem corresponds to the longest carbon chain in the molecule *that contains the OH group*. The chain is numbered so that the OH group gets the smaller of two possible numbers.

$CH_3CH_2CH_2CH_2OH$
1-butanol
or butan-1-ol

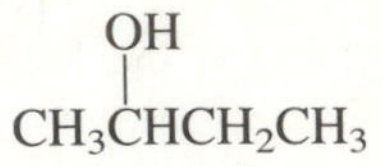

$CH_3CH(OH)CH_2CH_3$
2-butanol
or butan-2-ol

Substituents are appended as prefixes and are numbered according to the numbering system established by the position of the OH group. Names are written as one word with no spaces.

$CH_3CHClCH_2OH$
2-chloro-1-propanol

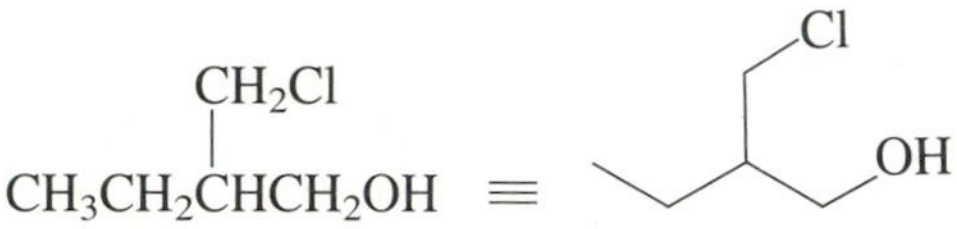

$CH_3CH_2CH(CH_2Cl)CH_2OH$
2-(chloromethyl)-1-butanol

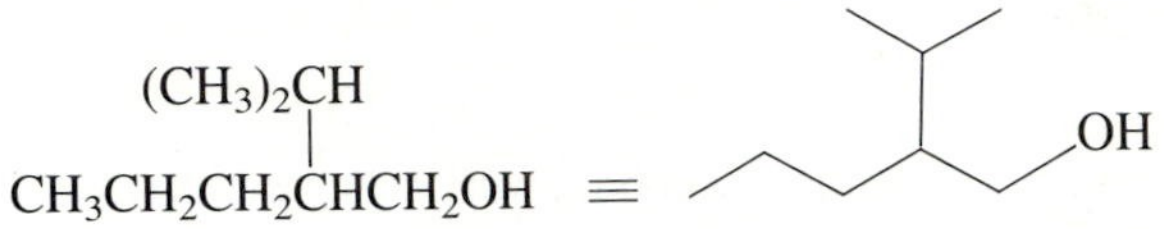

$CH_3CH_2CH_2CH(CH(CH_3)_2)CH_2OH$
2-isopropyl-1-pentanol
(not 3-methyl-2-propyl-1-butanol; pentanol has a longer chain)

1-methylcyclopentanol

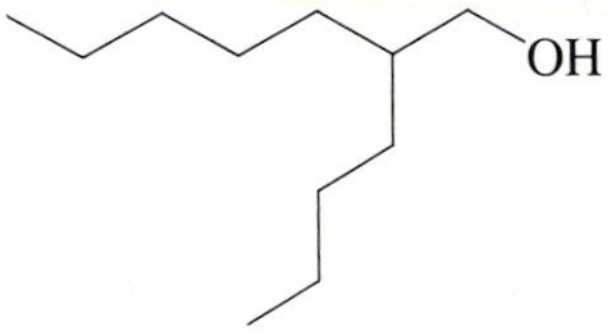

2-butyl-1-heptanol

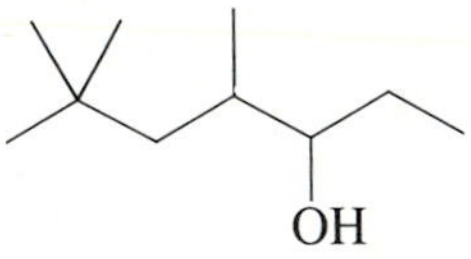

4,6,6-trimethyl-3-heptanol

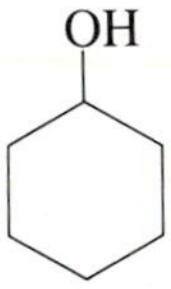

cyclohexanol

The general rule in the IUPAC system of nomenclature is that a functional group named as a suffix becomes a parent system that dominates the numbering scheme. Prefix groups are considered to be substituents or appendages to the parent compound.

Some kinds of alcohols are too difficult or cumbersome to name as alkanols. For such compounds it is preferable to use the appropriate hydroxyalkyl name as a prefix.

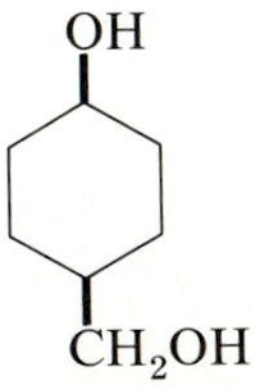

OH / CH_2OH

cis-4-(hydroxymethyl)cyclohexanol

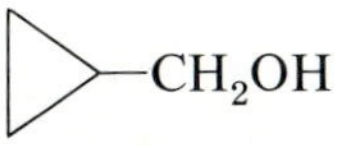

$-CH_2OH$

(hydroxymethyl)cyclopropane

EXERCISE 10.2 Write structures corresponding to each of the following names.

(a) 2,4-dimethyl-3-pentanol
(b) 2-(2-bromoethyl)-3-methyl-1-pentanol
(c) *trans*-4-methylcyclohexanol

EXERCISE 10.3 Explain why each of the following names is incorrect.

(a) 2-isopropyl-1-butanol
(b) 2-ethyl-4-butanol
(c) 2,2-dichloro-5-(hydroxymethyl)heptane

EXERCISE 10.4 Name each of the following compounds by the IUPAC system.

(a) OH

(b) $(CH_3)_3C\overset{CH_3}{\underset{CH_3}{C}}OH$

(c) OH Cl

(d) Cl OH

10.3 Physical Properties of Alcohols

The lower molecular weight alcohols are liquids with characteristic odors and sharp tastes. One striking feature is their relatively high boiling points (Table 10.1). The OH group is roughly equivalent to a methyl group in size and polarizability, but alcohols have much higher boiling points than an alkane of corresponding size. For example, compare ethanol (mol. wt. 46, b.p. 78.5 °C) and propane (mol. wt. 44, b.p. −42 °C). A plot of boiling point versus molecular weight for straight-chain alcohols and alkanes is shown in Figure 10.1.

TABLE 10.1 Physical Properties of Alcohols

Compound	Common Name	IUPAC Name	Melting Point, °C	Boiling Point, °C	Density d^{20}	Solubility, g/100 mL H_2O
CH_3OH	methyl alcohol	methanol	−97.8	65.0	0.7914	∞
CH_3CH_2OH	ethyl alcohol	ethanol	−114.7	78.5	0.7893	∞
$CH_3CH_2CH_2OH$	*n*-propyl alcohol	1-propanol	−126.5	97.4	0.8035	∞
$CH_3CHOHCH_3$	isopropyl alcohol	2-propanol	−89.5	82.4	0.7855	∞
$CH_3CH_2CH_2CH_2OH$	*n*-butyl alcohol	1-butanol	−89.5	117.3	0.8098	8.0
$CH_3CH_2CHOHCH_3$	*sec*-butyl alcohol	2-butanol	−114.7	99.5	0.8063	12.5
$(CH_3)_2CHCH_2OH$	isobutyl alcohol	2-methyl-1-propanol		107.9	0.8021	11.1
$(CH_3)_3COH$	*tert*-butyl alcohol	2-methyl-2-propanol	25.5	82.2	0.7887	∞
$CH_3(CH_2)_4OH$	*n*-pentyl alcohol	1-pentanol	−79	138	0.8144	2.2
$CH_3CH_2CH_2CHOHCH_3$	—	2-pentanol		119.3	0.809	4.9
$CH_3CH_2CHOHCH_2CH_3$	—	3-pentanol		115.6	0.815	5.6
$(CH_3)_3CCH_2OH$	neopentyl alcohol	2,2-dimethyl-1-propanol	53	114	0.812	∞
$CH_3(CH_2)_5OH$	*n*-hexyl alcohol	1-hexanol	−46.7	158	0.8136	0.7

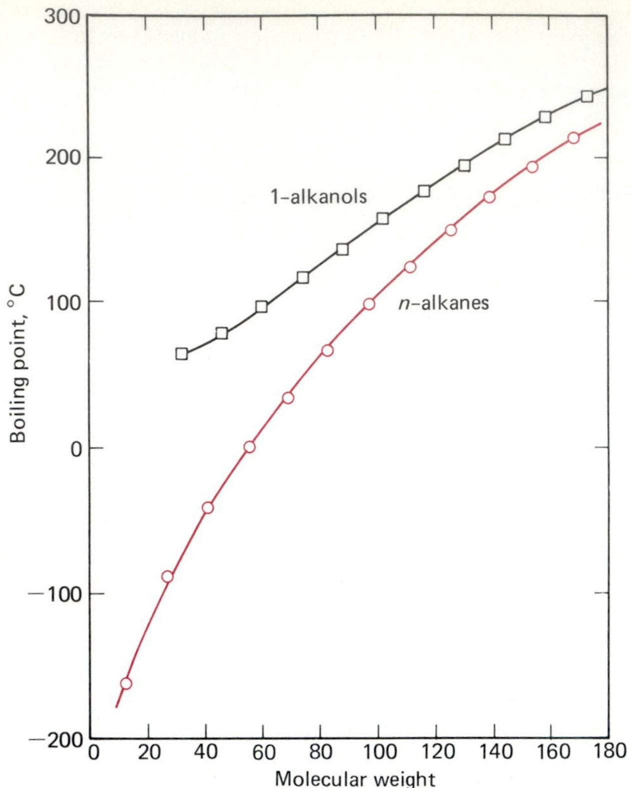

FIGURE 10.1 Boiling points of 1-alkanols as a function of molecular weight.

The abnormally high boiling points of alcohols are the result of a special type of dipolar association that these molecules experience in the liquid phase. Because oxygen is more electronegative than carbon or hydrogen, both the carbon-oxygen and the oxygen-hydrogen bonds are polar. These polar bonds cause alcohols to have substantial dipole moments. However, dipole moments of alcohols are no greater than those of corresponding chlorides.

CH_3OH	CH_3CH_2OH
$\mu = 1.71$ D	$\mu = 1.70$ D
CH_3Cl	CH_3CH_2Cl
$\mu = 1.94$ D	$\mu = 2.04$ D

Molecules with substantial dipole moments should show enhanced intermolecular interaction in the liquid phase because of electrostatic interaction between the dipoles. However, we have just seen that the dipole moments of alcohols are actually smaller than those of alkyl chlorides. In fact, alkyl chlorides differ very little in boiling point from alkanes of corresponding molecular weight (see Figure 8.2). Consequently, it would seem that dipolar attraction is not the cause of the elevated boiling points of alcohols. Or is it? The magnitudes of the individual dipole moments are not the only important factor. How closely the negative and positive ends of the dipoles can approach one another is also important.

By Coulomb's law two opposite charges attract each other with an energy proportional to $1/r$, where r is the distance between the charges. The electrostatic energy of two dipoles is related to $1/r^3$ and therefore falls off sharply with distance. In alkyl halides the negative end of the dipoles is out at the lone-pair electrons, but the positive end is in the C—X bond close to carbon. Because of the van der Waals size of carbon, the positive and negative ends of adjacent dipoles cannot get close together and the electrostatic energy of dipole-dipole attraction is relatively small.

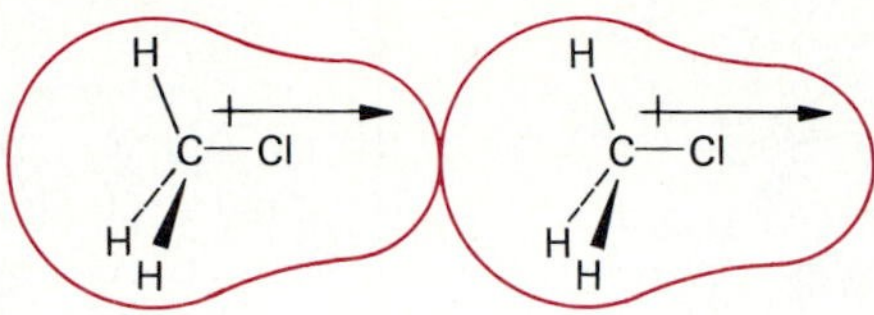

Consequently, such dipole association does not have much of an effect on the energy required to separate alkyl halide molecules.

For alcohols the negative end of the dipole is out at the oxygen lone pairs, and the positive end is close to the small hydrogen. For hydrogen atoms bonded to electronegative elements dipole-dipole interaction is uniquely important and is called a **hydrogen bond.**

In alcohols, the hydrogen bond is O—H---O. The proximity of approach is shown by bond-distance data. The oxygen-hydrogen bond length in alcohols is 0.96 Å. The distance between the hydrogen and the second oxygen H---O is 1.69–1.79 Å, almost twice as large. In fact, this distance is sufficiently small that some hydrogen bonds may have a small amount of covalent or shared electron character. The overall O—H---O distance is often quoted to characterize a hydrogen bond, which for alcohols is 2.65–2.75 Å. In condensed phases, alcohols are associated via a chain of hydrogen bonds.

A three-dimensional view of liquid methanol showing the hydrogen-bonding network is given in Figure 10.2. The figure is of a random cube of the liquid, showing only the C—O—H parts of the various molecules. Some of the oxygen atoms are colored in the figure. The six at the top of the cube are the oxygens of a "chain" of methanol molecules; the molecules are held in the chain by hydrogen bonds. At the bottom of the cube may be seen a cyclic tetramer of methanol molecules.

The oxygen-hydrogen bond in alcohols is stronger than most carbon-hydrogen bonds. It has a bond dissociation energy of 103 kcal $mole^{-1}$. The hydrogen bond is far weaker—only about 5 kcal $mole^{-1}$. Nevertheless, this additional heat term in the heat of vaporization results in relatively high boiling points. A bond strength of 5 kcal $mole^{-1}$ does not sound like much; however, when there are many such bonds, as in polyhydroxy com-

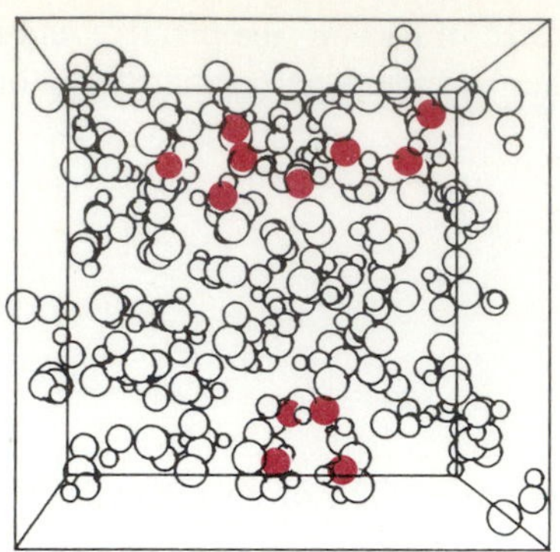
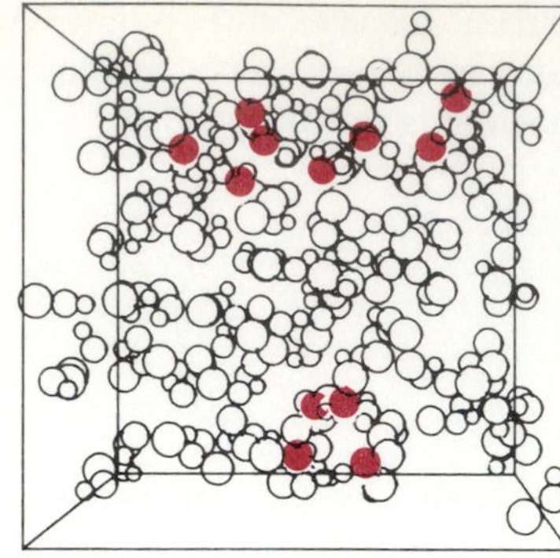

FIGURE 10.2 Three-dimensional view (stereo plot) of a random cube of liquid methanol, showing the hydrogen-bonded networks. The large circles represent methyl groups, the small circles hydrogen, and the intermediate circles oxygen. Colored oxygens describe a chain and a cycle of molecules. [Courtesy of W. L. Jorgensen, Department of Chemistry, Purdue University.]

pounds such as carbohydrates, the total strength is sufficient to hold up 100-meter-tall redwood trees.

Since alcohols and water both contain the OH group, they have many properties in common. We should emphasize that water is a remarkable substance. Its boiling point of 100 °C is exceedingly high for a compound with a molecular weight of only 18. The extensive hydrogen-bond networks in liquid water make it a highly polar liquid with a dielectric constant of 78.5 at 25 °C. In aqueous solution the interaction between ions is relatively small. Therefore, water is a good solvent for ionic compounds. The lower alcohols also have relatively high dielectric constants (Table 10.2) and high solvent polarities (Table 9.4). As the carbon chain gets longer, the importance of the OH group is reduced relative to that of the alkyl group and the dielectric constant approaches the alkane value of about 2.

TABLE 10.2 Dielectric Constants of Alcohols

Compound	Dielectric Constant
H_2O	78.5
CH_3OH	32.6
CH_3CH_2OH	24.3
$CH_3(CH_2)_3OH$	17.1
$CH_3(CH_2)_4OH$	13.9
$CH_3(CH_2)_{11}OH$	6.5

Because of their high polarities, methanol and ethanol are reasonably good solvents for salt-like compounds. Since they are also good solvents for organic compounds, they are used frequently for organic reactions such as S_N2 displacement reactions (Chapter 9).

The OH group of alcohols can participate in the hydrogen-bond network of water. The smaller alcohols are completely soluble in water. As the hydrocarbon chain of an alcohol gets larger, the compound begins to look more like an alkane. To make room for the hydrocarbon chain, hydrogen bonds in water form a tighter, ''ice-like'' structure around the chain. The entropy cost of the extra organization is not compensated for by enthalpy gained through the van der Waals interaction of the hydrocarbon chains with the ''ice-like'' structure plus hydrogen bonding to the alcohol OH. Thus, solubility decreases as the hydrocarbon chain gets larger. Roughly, the change occurs for an alcohol containing four carbons to one oxygen. Above this ratio, alcohols tend to have little solubility in water. This guideline is only approximate because the shape of the hydrocarbon portion is also important. *t*-Butyl alcohol is much more soluble than *n*-butyl alcohol because the *t*-butyl group is more compact and requires less room, and thus less structure formation than with straight-chain alkanols. A similar pattern is seen with the branched pentyl alcohols.

Highlight 10.1

Alcohols are compounds with an alkyl group bonded to a hydroxy group, R—OH. Water, H—O—H, can hydrogen-bond in four directions to form a three-dimensional network. Water has a high-melting (for its molecular size) solid form, m.p. 0 °C, and is a high-boiling liquid, b.p. 100 °C. Alcohols form mainly chains and cycles of hydrogen bonds. Solid methanol melts and boils at lower temperatures than water.

10.4 Acidity of Alcohols: Inductive Effects

One of the important properties of water is its self-ionization.

$$H_2O + H_2O \overset{K_w}{\rightleftharpoons} H_3O^+ + OH^-$$

In pure water, the concentrations of H_3O^+ and OH^- are very low, only 10^{-7} mole L^{-1}. The ion product or self-dissociation constant K_w is defined as

$$K_w = [H_3O^+][OH^-] = 1.0 \times 10^{-14} \text{ mole}^2 \text{ L}^{-2} \text{ (or M}^2)$$

Remember that this is not a normal equilibrium constant, which includes the concentrations of reactants and products. For water, the concentration is 55.5 moles L^{-1}. The equilibrium constant for dissociation is therefore

$$K = \frac{(10^{-7})(10^{-7})}{(55.5)(55.5)} = 3.25 \times 10^{-18}$$

Note that K is unitless. The relationship between K and K_w is

$$K_w = K \times (55.5 \text{ M})^2$$

Alcohols also undergo self-dissociation, but to a much smaller extent than water.

$$2\,CH_3OH \rightleftharpoons CH_3OH_2^+ + CH_3O^-$$

For methanol the ion product $K_{CH_3OH} = 1.2 \times 10^{-17}$ M^2, and for larger alcohols the value is even smaller.

$$K_{CH_3OH} = [CH_3OH_2^+][CH_3O^-] = 1.2 \times 10^{-17} \text{ M}^2$$

The reduced value comes in large part from the lower dielectric constant of alcohols—it takes greater energy to separate charges. But the relative acidities and basicities of alcohols are also important.

The acidity of an alcohol in water is defined in the usual way.

$$ROH + H_2O \overset{K_a}{\rightleftharpoons} H_3O^+ + RO^-$$

The acid dissociation constant K_a is defined as

$$K_a = \frac{[H_3O^+][RO^-]}{[ROH]}$$

Since these equilibria refer to dilute solutions in water, the concentration of water is generally omitted in the expression for an equilibrium constant and K_a has units of mole L^{-1}, or molarity, M. Recall that the acid dissociation constant is generally such a small number that it is usually more convenient to refer to the negative logarithm (pK_a).

$$pK_a = -\log K_a$$

Values of pK_a for some alcohols are listed in Table 10.3. For comparison, the pK_a values of some common inorganic acids are also given. Note that the K_a for water is obtained by dividing K_w by the concentration of water, 55.5 moles L^{-1}. This change is necessary to put all of the ionizations on the same scale and in the same units. Recall that the ion product of water, K_w, has units of moles^2 L^{-2} or M^2, whereas K_a values are given in units of moles L^{-1} or M.

TABLE 10.3 pK_a Values for Alcohols and Some Acids

Compound	pK_a	Compound	pK_a
H_2O	15.7	HCl	−2.2
CH_3OH	15.5	H_2SO_4	−5
C_2H_5OH	15.9	H_3PO_4	2.15
$(CH_3)_3COH$	≅18	HF	3.18
$ClCH_2CH_2OH$	14.3	H_2S	6.97
CF_3CH_2OH	12.4	HOCl	7.53
C_6H_5OH	10.0	H_2O_2	11.64
CH_3COOH	4.8		

Methanol and ethanol are about as acidic as water itself. The larger alcohols are less acidic. Alcohols (and water) are generally much less acidic than other compounds commonly regarded as acids. ''Strong'' acids, such as HI, HBr, HCl, and H_2SO_4, have negative pK_a values. Such compounds are completely dissociated in water and are 10^{15} to 10^{20} more acidic than alcohols. Typical ''weak'' acids such as acetic acid (to be discussed in Chapter 18), HF, H_2S, and HOCl are still 10^7 to 10^{10} stronger than alcohols. For a brief review of acidity, see Section 4.5.

The weakly acidic character of alcohols is seen primarily in their reactions with very strong bases. Alcohols, like water, react with alkali metals to liberate hydrogen and form the corresponding metal alkoxide.

$$CH_3OH + Na \longrightarrow CH_3O^- + Na^+ + \tfrac{1}{2}H_2$$

The reaction of sodium with an alcohol tends to be less vigorous than that with water. In fact, isopropyl alcohol is often used to decompose scraps of sodium in the laboratory because its reaction is relatively slow and moderate. When sodium reacts with water, the reaction is so rapid that the heat produced cannot be dissipated quickly enough; the evolved hydrogen catches fire, and an explosion results. Tertiary alcohols react so sluggishly with sodium that potassium must often be used to convert such an alcohol to the alkoxide.

Potassium has a relatively low melting point, 64 °C. In laboratory use it is sometimes converted to a finely divided state in order to render it more reactive. Solid pieces of potassium are added to benzene (b.p. 80 °C), and the mixture is heated to the boiling point of the benzene. The potassium melts, and the mixture is allowed to cool with vigorous stirring. The molten potassium solidifies as small particles of "potassium sand." The alcohol is added to this mixture with stirring and generally reacts readily because of the large surface area of the potassium sand. This procedure is especially useful with tertiary alcohols. Disposing of scraps of the more reactive potassium can lead to fire, as one of the authors found to his dismay.

Another reagent commonly used to prepare the sodium salts of alcohols is sodium hydride, NaH. This compound is a nonvolatile, insoluble salt, Na^+H^-, and reacts readily with acidic hydrogens.

$$CH_3OH + Na^+ H^- \longrightarrow CH_3O^- + Na^+ + H_2$$

The reaction may be regarded as a combination of hydride ion with a proton.

$$H{:}^- + H^+ \longrightarrow H_2$$

The sodium salts of primary alcohols are common reagents in organic chemistry.

$Na^+ \ ^-OCH_3$ sodium methoxide $Na^+ \ ^-OCH_2CH_3$ sodium ethoxide

Because sodium and sodium hydride react so sluggishly with tertiary alcohols, the corresponding potassium salts are more commonly used as reagents.

$$K^+ \ ^-O{-}C(CH_3)_3$$

potassium *t*-butoxide

Note in Table 10.3 that 2-chloroethanol is significantly more acidic than ethanol. This difference in acidity is best understood in terms of the electrostatic interaction of the C—Cl dipole with the negative charge of the alkoxide ion.

$$Cl{-}CH_2{-}CH_2{-}O^-$$

The negative charge on oxygen is closer to the positive end of the dipole than it is to the negative end. Thus, electrostatic attraction exceeds repulsion, and the result is a net stabilization of the anion. Stabilization of the anion increases its ease of formation and the conjugate acid, 2-chloroethanol, is more acidic than ethanol itself (Figure 10.3).

This effect is generally called an **inductive field effect** or, more simply, an **inductive effect.** The magnitude of the effect falls off as the distance between the dipolar group and the charged group is increased. The effect increases with the number of dipolar groups. Note the relatively large effect of the three fluorines in 2,2,2-trifluoroethanol. Halogen groups are said to be **electron attracting** and to stabilize anions. The effect is present in inorganic systems as well: HOCl is a stronger acid than HOH. Conversely, alkyl groups are generally considered to be somewhat **electron donating** and, therefore, to weaken acids. We will make use of inductive effects frequently in our subsequent discussions of the effects of structure on chemical reactivity.

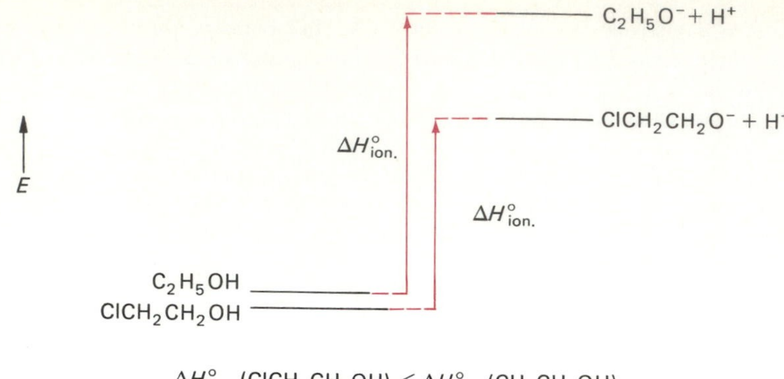

$$\Delta H^\circ_{ion.}(ClCH_2CH_2OH) < \Delta H^\circ_{ion.}(CH_3CH_2OH)$$

FIGURE 10.3 The effect of a dipolar substituent on the ionization energy of an alcohol. The stabilizing effect of the substituent is greater in the anion (charge-dipole interaction) than in the alcohol (dipole-dipole interaction).

Like water, alcohols are not only acids, but also bases. For example, when gaseous hydrogen chloride is passed into an alcohol it protonates the oxygen just as it does the oxygen of water.

$$H_2O + HCl \longrightarrow H_3O^+ + Cl^-$$

$$ROH + HCl \rightleftharpoons \underset{\text{alkyloxonium chloride}}{ROH_2^+\ Cl^-} \rightleftharpoons ROH_2^+ + Cl^-$$

The initially formed species in such a protonation is an **ion pair** (two ions in close juxtaposition). In water and the smaller alcohols, the dielectric constant is sufficiently high that the initially formed ion pairs largely dissociate to free ions. However, as the dielectric constant becomes smaller, too much work is required to separate the ion pairs, and the oxonium chloride remains largely associated.

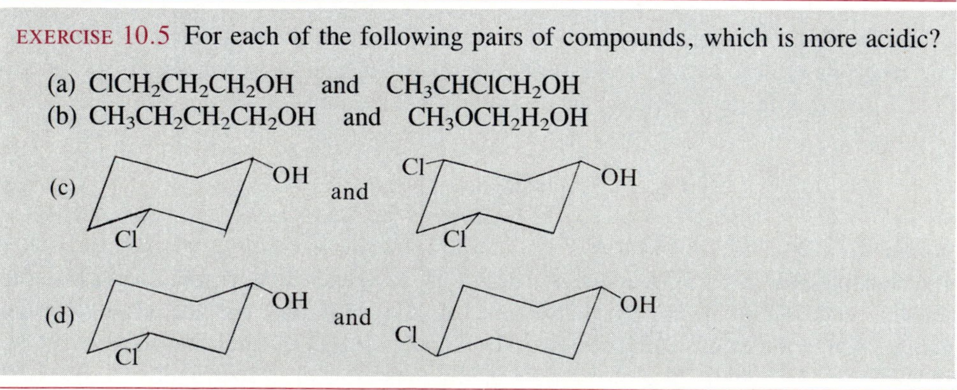

EXERCISE 10.5 For each of the following pairs of compounds, which is more acidic?

(a) $ClCH_2CH_2CH_2OH$ and $CH_3CHClCH_2OH$
(b) $CH_3CH_2CH_2CH_2OH$ and $CH_3OCH_2H_2OH$
(c)

(d)

Highlight 10.2

Alcohols are weak acids. For the simplest alcohols, methanol and ethanol, the pK_a in water is similar to that of water itself. p$K_a = -\log K_a$, $K_a = [H_3O^+][RO^-]/[ROH]$. The p$K_a$ values for alcohols depend on the nature of R, the acidity increasing for groups that attract electrons. Metal salts, $M^+\ OR^-$, can be prepared and are useful strong bases in organic syntheses.

10.5 Preparation of Alcohols

The hydroxy group plays a central role in organic chemistry. Alcohols can be prepared from many other classes of compounds, and they also serve as important starting materials for the synthesis of other materials. In this section, we will consider the preparation of alcohols from alkyl halides. In later sections, as we study the chemical reactions of other functional groups, we shall learn many other important ways of preparing alcohols.

Hydrolysis of alkyl halides in aqueous solvents may occur by either the S_N1 or S_N2 mechanism. With some halides, elimination is a major side reaction (Chapter 8). The hydrolysis of most primary halides occurs by the S_N2 path and is sufficiently clean that this reaction is a good preparative method.

$$CH_3CH_2CH_2CH_2Cl + OH^- \xrightarrow[100°C]{H_2O} CH_3CH_2CH_2CH_2OH + Cl^-$$

The reaction can be carried out in refluxing aqueous sodium hydroxide, especially with lower molecular weight halides. Even though the reaction mixtures are two-phase because of the low solubility of alkyl halides in water, reaction takes place. If the alcohol is water soluble, the end of the hydrolysis is marked by the formation of a homogeneous solution. Alternatively, the reaction mixture can be made homogeneous by the addition of an inert organic cosolvent such as dioxane, and the end of the reaction detected by pH change or chromatography.

```
      O
 H2C     CH2
  |       |
 H2C     CH2
      O
```

dioxane

> Dioxane is a colorless liquid, b.p. 100 °C, and is completely miscible with water. It is relatively inert to many reagents and is frequently used in mixtures with water to increase the solubility of organic compounds.

Remember that with secondary alkyl halides and primary halides with a β-branch, elimination is an important side reaction and may be the principal reaction.

$$(CH_3)_2CHCH_2Br \xrightarrow{OH^-} (CH_3)_2C{=}CH_2$$

An alternative procedure that avoids the use of strong base employs acetate ion ($CH_3CO_2^-$) as the nucleophilic reagent (Section 9.2). Since acetate is much less basic than hydroxide (the pK_a of acetic acid, CH_3COOH, is 4.8, whereas that of water is 15.7), the E2 mechanism is suppressed, and alkene formation is minimized.

$$CH_3CH_2CH_2CH_2Br + CH_3\overset{\overset{O}{\|}}{C}O^-\ K^+ \xrightarrow[100°C]{DMF} CH_3CH_2CH_2CH_2O\overset{\overset{O}{\|}}{C}CH_3 + K^+\ Br^-$$

The product, an ester, can be readily hydrolyzed to the desired alcohol (Section 18.6).

$$CH_3CH_2CH_2CH_2O\overset{\overset{O}{\|}}{C}CH_3 \xrightarrow[H_2O]{OH^-} CH_3CH_2CH_2CH_2OH + CH_3CO_2^-$$

Since displacement by acetate ion proceeds by the S_N2 mechanism, the alcohol product has an absolute configuration opposite that of the starting halide.

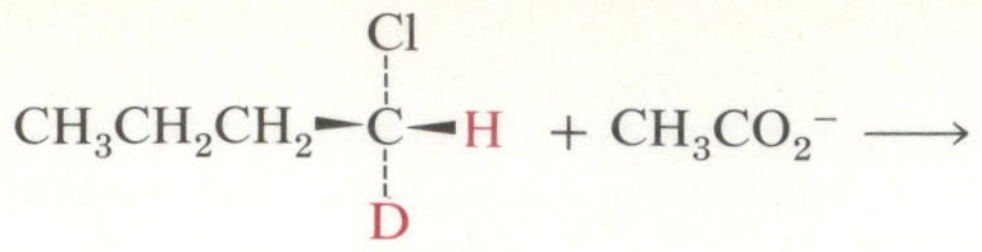

$$CH_3CH_2CH_2-\underset{D}{\overset{Cl}{C}}-H + CH_3CO_2^- \longrightarrow$$

(*R*)-1-chloro-1-deuteriobutane
(*R*)-chlorobutane-*1-d*

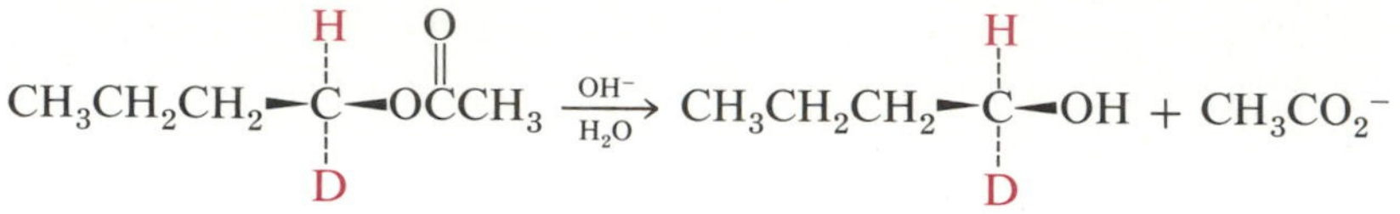

$$CH_3CH_2CH_2-\underset{D}{\overset{H}{C}}-O\overset{O}{\overset{\|}{C}}CH_3 \xrightarrow[H_2O]{OH^-} CH_3CH_2CH_2-\underset{D}{\overset{H}{C}}-OH + CH_3CO_2^-$$

(S)-butyl-*1-d* acetate (S)-butanol-*1-d*

Implicit in the foregoing statement is the assumption that the newly formed carbon-oxygen bond is not broken in the ester hydrolysis step. As we shall see in Section 19.6, the normal mechanism for this reaction involves cleavage of the other carbon-oxygen bond, as indicated below.

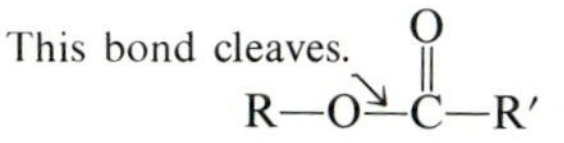

This bond cleaves.

$$R-O-\overset{O}{\overset{\|}{C}}-R'$$

That is, ester hydrolysis *does not proceed by the S_N2 mechanism,* even when the ester group is attached to a primary carbon.

Tertiary halides also undergo hydrolysis, but this reaction occurs by the S_N1 rather than the S_N2 mechanism. The reaction is best carried out by shaking the halide with aqueous sodium carbonate. The carbonate neutralizes the acid formed by hydrolysis but avoids the high concentration of hydroxide ion that encourages elimination. Unimolecular elimination is more difficult to avoid. However, this side reaction may be minimized by using highly aqueous solvents and by operating at low temperature. Nevertheless, hydrolysis of tertiary halides involves carbocation intermediates, and such intermediates have several modes of reaction available besides reaction with water. As discussed in Section 10.6.B, rearrangements frequently occur.

The conversions of alkyl halides to alcohols discussed in this section are, by and large, *not important synthetic laboratory processes*. This is not due solely to deficiencies in the methods (although there obviously are some) but also to the practical fact that the halides are commonly obtained from alcohols in the first place. Hydrolysis of sulfonate esters is also a perfectly good reaction, but they are invariably prepared from alcohols. Hydrolysis of halides is an important industrial reaction for those halides obtained commercially by direct halogenation of hydrocarbons. More important laboratory syntheses of alcohols will be discussed in subsequent chapters as reactions of other functional groups.

EXERCISE 10.6 Write equations showing how each of the following alcohols can be prepared from an alkyl halide.

(a) (*R*)-1-butanol-*1-^{2}H* (b) 4-methyl-1-pentanol
(c) 1-methylcyclohexanol

Methanol at one time was prepared commercially by the dry distillation of wood and once had the commercial name of "wood alcohol." It is now prepared on a large scale by catalytic hydrogenation of carbon monoxide. Methanol is toxic; ingestion of small amounts causes nausea and blindness, and death can result from ingestion of 100 mL or less. It is an important industrial solvent and reagent and is also used to denature ethanol, i.e. make the material unfit for human consumption. About 7 billion pounds of methanol were produced in 1988 in the United States.

Ethanol is prepared for consumption in beverages by fermentation of sugars, but industrial alcohol is prepared by other routes such as the hydration of ethylene (Section 11.6.C). Ethanol is toxic in large quantities but is a normal intermediate in metabolism and, unlike methanol, is metabolized by normal enzymatic body processes. It is an important solvent and reagent in industrial processes and in such use is often "denatured" by addition of toxic and unappetizing diluents. Alcohol for consumption is heavily taxed, but denatured alcohol is not (for example, the 1991 price of 100% ethanol was \$2.12–2.15 per gal; the federal distilled spirits tax was \$21.00 per gal).

10.6 Reactions of Alcohols

A. Reactions of Alkoxides with Alkyl Halides

The alkali alkoxides, produced by reaction of alcohols with alkali metals, are important reagents as bases in nonaqueous media and as nucleophilic reagents. An example of the latter use is the reaction of potassium *t*-butoxide and methyl iodide to form *t*-butyl methyl ether.

$$(C_2H_5)_3COH + K \longrightarrow \tfrac{1}{2}\,H_2 + (C_2H_5)_3CO^-\ K^+$$

$$(C_2H_5)_3CO^-\ K^+ + CH_3I \longrightarrow (C_2H_5)_3COCH_3 + KI$$

This is a typical example of a reaction that proceeds by the S_N2 mechanism. The reaction is a good method for preparing ethers from primary halides that have no β-substituents. Other kinds of halides give more or less elimination.

$$(C_2H_5)_3CCl + CH_3O^-\ K^+ \longrightarrow \underset{\text{100\% elimination}}{(C_2H_5)_2C{=}CHCH_3} + KCl + CH_3OH$$

If elimination is dominant, the reaction may be used as a method for the synthesis of alkenes. Potassium *t*-butoxide is frequently used as a reagent for dehydrohalogenation because of its high basicity and because it is moderately soluble in nonpolar organic solvents such as benzene (C_6H_6).

EXERCISE 10.7 Which of the following ethers could be prepared from an alcohol and an alkyl halide?

(a) $CH_3\underset{}{\overset{OCH_3}{\overset{|}{C}}}HCH_2CH_3$ (b) $(CH_3)_3COCH(CH_3)_2$ (c) $CH_3CH_2OCH_2C(CH_3)_3$

B. Conversion of Alcohols into Alkyl Halides

Sodium bromide is slightly soluble in ethanol. Such a solution can be refluxed indefinitely with no reaction; ethyl bromide and sodium hydroxide are *not* formed. Indeed, we have just seen that the reverse reaction, *hydrolysis* of ethyl bromide, can be carried out readily.

$$C_2H_5OH + Br^- \overset{K}{\rightleftharpoons} C_2H_5Br + OH^-$$

For this system, the thermodynamics is such that equilibrium lies far to the left (Section 4.2).

$$K = \frac{[C_2H_5Br][OH^-]}{[C_2H_5OH][Br^-]} \cong 10^{-19}$$

Consequently, the rate of reaction of ethanol with bromide ion is totally negligible compared to the reverse reaction. (Remember that $K = k_1/k_{-1}$; for the foregoing reaction, k_{-1} must be greater than k_1 by a factor of 10^{19}!)

However, if some sulfuric acid is added to our solution of sodium bromide in ethanol, reaction does occur (of course, the same result is obtained if a mixture of ethanol and hydrobromic acid is used from the start).

$$C_2H_5OH + HBr \longrightarrow C_2H_5Br + H_2O$$

Why this dramatic difference? For one thing, the equilibrium constant for the reaction of ethanol and hydrogen bromide to give ethyl bromide and water must obviously be favorable, or net reaction would not be observed. Indeed, it has been estimated that the equilibrium constant for reaction of ethanol and hydrogen bromide is approximately 10^2.

$$K = \frac{[C_2H_5Br][H_2O]}{[C_2H_5OH][HBr]} \cong 10^2$$

This large change in equilibrium constant (more than 20 orders of magnitude) results from both an increase in rate constant of the forward reaction (k_1) and a decrease in the rate constant for the reverse reaction (k_{-1}). The rate constant for the reaction of ethyl bromide with water (k_{-1}) is less than that for reaction with hydroxide ion, because H_2O is a much weaker nucleophile than OH^-. In the forward reaction, the leaving group is now H_2O, which is a weaker base, and hence a better leaving group, than OH^-. The displacement reaction in this case actually occurs on the intermediate **alkyloxonium salt** which is formed when the strong acid protonates the oxygen of ethanol.

$$C_2H_5OH + HBr \rightleftharpoons C_2H_5OH_2^+ + Br^-$$

$$Br^- + CH_3CH_2\text{—}\overset{+}{O}H_2 \longrightarrow Br\text{—}CH_2CH_3 + OH_2$$

Protonation converts the substrate from one with a very poor leaving group (hydroxide ion) to one with a better leaving group (water).

The foregoing reaction is only another example of a nucleophilic displacement. For primary alcohols, the reaction proceeds by the S_N2 mechanism and can be a useful method for preparing primary alkyl halides. The reaction is carried out by refluxing the alcohol with a mixture of concentrated sulfuric acid and either sodium bromide or hydrobromic acid.

A mixture of 71 mL of 48% hydrobromic acid, 30.5 mL of concentrated sulfuric acid, and 37 g of *n*-butyl alcohol is refluxed for 2 hr. The product is separated, washed, and distilled to yield 50 g (95%) of *n*-butyl bromide.

For the preparation of primary alkyl chlorides, more vigorous conditions are required because chloride ion is a poorer nucleophile than bromide. A mixture of concentrated hydrochloric acid and zinc chloride, the so-called Lucas reagent, is frequently used. Zinc chloride is a powerful Lewis acid that serves the same purpose as does a proton in coordinating with the hydroxy oxygen.

$$CH_3CH_2CH_2CH_2CH_2OH + HCl \xrightarrow[150°C]{ZnCl_2} CH_3CH_2CH_2CH_2CH_2Cl$$

(76–78%)

A mixture of 371 g of *n*-butyl alcohol, 1.363 kg of anhydrous zinc chloride, and 864 mL of concentrated hydrochloric acid is heated in an oil bath that is maintained at 150 °C. The product *n*-butyl chloride, 352–361 g (76–78%), is distilled, b.p. 78.4 °C, from the reaction mixture over a 1-hr period.

Secondary alcohols react more readily than primary alcohols under these conditions. For example, the foregoing HCl-$ZnCl_2$ procedure may also be applied for the conversion of 2-butanol to 2-chlorobutane. In this case, maintaining the oil bath at only 140 °C is required, and the yield of alkyl halide, b.p. 68.2 °C, is 85–88%.

$$\text{CH}_3\text{CH(OH)CH}_2\text{CH}_3 + \text{HCl} \xrightarrow[140^\circ\text{C}]{\text{ZnCl}_2} \text{CH}_3\text{CH(Cl)CH}_2\text{CH}_3 \quad 85\text{–}88\%$$

Tertiary alcohols react the most rapidly of all. For example, *t*-butyl alcohol is converted into *t*-butyl chloride simply by shaking with hydrochloric acid at room temperature.

$$(CH_3)_3COH + HCl \xrightarrow[25^\circ C]{H_2O} (CH_3)_3CCl$$

A mixture of 74 g of *t*-butyl alcohol and 247 mL of concentrated hydrochloric acid is shaken in a separatory funnel at room temperature. After 15–20 min, the upper layer is drawn off and washed with dilute sodium bicarbonate solution until neutral. The yield of *t*-butyl chloride is 72–82 g (78–88%).

There is ample evidence that, unlike primary alcohols, secondary and tertiary alcohols react with hydrohalic acids by the S_N1 mechanism; that is, the reactions involve carbocation intermediates. One strong piece of evidence in this regard is the observation of skeletal rearrangements, a phenomenon that we shall consider in detail in the next section.

EXERCISE 10.8 Write a mechanism for reaction of *t*-butyl alcohol with HBr. Be careful to show each step (there are three).

In summary, OH^- is too basic to function as an effective leaving group in nucleophilic substitution reactions, either by the S_N1 or S_N2 mechanism. Protic acids can catalyze the substitution process by protonating the OH group of the alcohol, converting it to a much better leaving group, H_2O. The ensuing substitution reaction takes place by the S_N2 mechanism in primary systems and by the S_N1 mechanism in secondary and tertiary systems.

Recall that basicity towards a proton source is inversely related to reactivity as a leaving group, showing that strength of bonding to hydrogen and carbon are often parallel (Section 9.5).

There are other ways in which the poor leaving group OH^- may be converted into a better one. Thionyl chloride, $SOCl_2$, is a colorless liquid, b.p. 79 °C, that reacts rapidly with water to give sulfur dioxide and HCl.

$$SOCl_2 + H_2O \longrightarrow 2\ HCl + SO_2$$

Thionyl chloride also reacts readily with primary and secondary alcohols, giving the corresponding alkyl chloride, sulfur dioxide, and HCl. An organic base is usually added to catalyze the substitution reaction and to neutralize the acid that is produced.

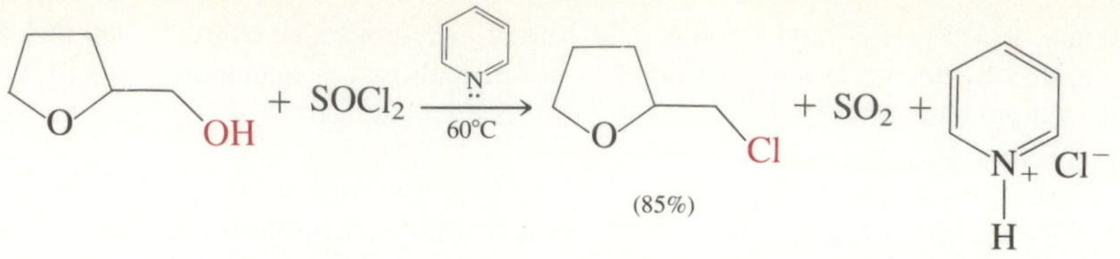

OH + $SOCl_2$ → (pyridine, 60°C) Cl (85%) + SO_2 + pyridinium Cl^-

Pyridine is a heterocyclic amine, b.p. 115 °C. Like the nitrogen of ammonia, the pyridine nitrogen has a nonbonded electron pair that may accept a proton. Pyridine is a weak base (approximately 10^6 less basic than ammonia), and is soluble in water. Because of these properties, it is frequently used as an acid scavenger and mildly basic catalyst in organic reactions. Heterocyclics are ring compounds including at least one atom other than carbon (Section 10.11).

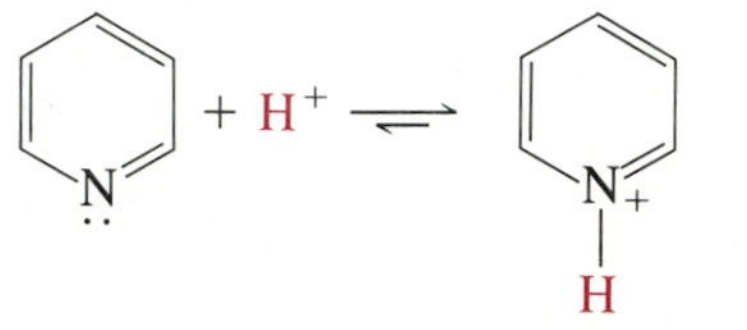

pyridine + H^+ ⇌ pyridinium ion

In the reaction of an alcohol with thionyl chloride, the first step is formation of a **chlorosulfite ester.** In this intermediate, the hydroxy group of the alcohol has effectively been converted into a better leaving group. Displacement by a nucleophile produces SO_2 and Cl^-, species with stabilities that are reflected in a lower transition state energy.

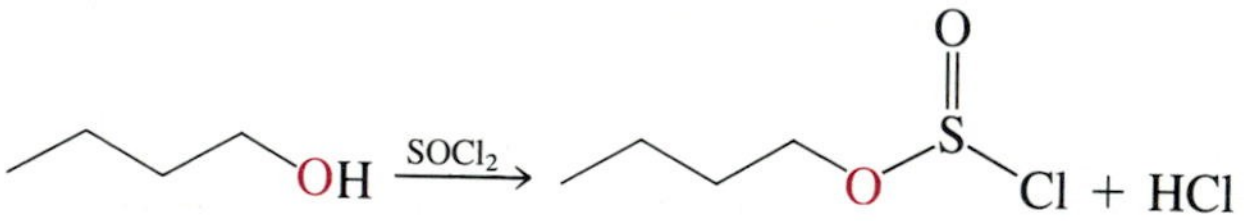

OH $\xrightarrow{SOCl_2}$ *n*-butyl chlorosulfite + HCl

Cl^- → Cl + O=S=O + Cl^-

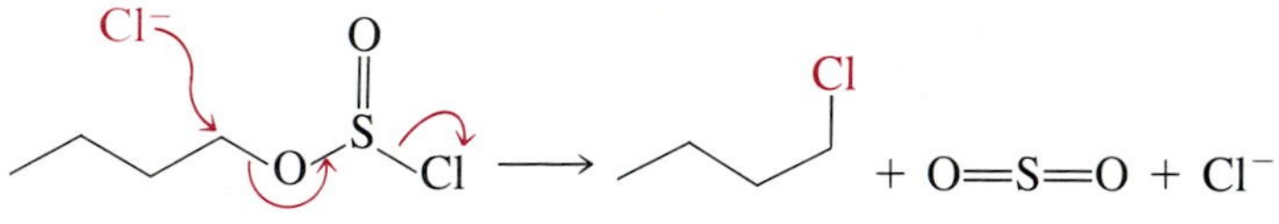

EXERCISE 10.9 On the basis of the foregoing mechanism, explain how pyridine promotes the reaction of an alcohol with $SOCl_2$.

Another reagent that can be used to activate a hydroxy group so that it can function as a leaving group is phosphorus tribromide, PBr_3. This substance is a dense, colorless liquid, b. p. 173 °C, that is prepared by the direct reaction of phosphorus with bromine. It reacts with water to give phosphorous acid and HBr.

$$PBr_3 + 3\ H_2O \longrightarrow 3\ HBr + H_3PO_3$$

The reaction of PBr_3 with three mole-equivalents of an alcohol produces first a phosphite ester, which may be isolated if the reaction is conducted at low temperature and the reaction time is minimized.

$$3\ (CH_3)_2CHCH_2CH_2OH + PBr_3 \longrightarrow [(CH_3)_2CHCH_2CH_2O]_3P + 3\ HBr$$

isoamyl alcohol triisoamyl phosphite

However, the mixture of alcohol and PBr_3 is usually allowed to react further, until the initially formed phosphite has reacted with the HBr to give the alkyl bromide and phosphorous acid.

$$[(CH_3)_2CHCH_2CH_2O]_3P + 3\ HBr \longrightarrow 3\ (CH_3)_2CHCH_2CH_2Br + H_3PO_3$$

The substitution reaction occurs by the S_N2 mechanism, with bromide being the attacking nucleophile and the protonated phosphite being the leaving group.

$$Br^- + (CH_3)_2CHCH_2\overset{+}{O}(H){-}P(OR)_2 \longrightarrow (CH_3)_2CHCH_2Br + (RO)_2POH$$

A mixture of isobutyl alcohol and PBr_3 is maintained at 0 °C for 4 hr. The product is washed, dried, and distilled to give 60% of isobutyl bromide.

The reaction also works well with most secondary alcohols, although with such compounds, the reaction temperature must be no higher than 0 °C in order to avoid carbocation rearrangements.

Thionyl chloride and phosphorus tribromide are commercially available and are common laboratory chemicals. Phosphorus triiodide is a red solid that decomposes on heating. For the preparation of alkyl iodides, the PI_3 is usually generated *in situ* by heating red phosphorus, iodine, and the appropriate alcohol.

$$3\ CH_3OH + \tfrac{3}{2}\ I_2 + P \longrightarrow 3\ CH_3I + H_3PO_3$$

Iodine is added over a period of several hours to a refluxing mixture of red phosphorus and methanol. Methyl iodide, b.p. 42.4 °C, is distilled, washed, dried, and redistilled: yield 94%.

EXERCISE 10.10 Write balanced equations showing how the following alkyl halides may be prepared from the corresponding alcohols. In each case carefully write out each step in the reaction mechanism, showing all reaction products, both organic and inorganic.

(a) $CH_3CH_2OCH_2CH_2Br$ (b) 2-chloropentane (c) 1-iodobutane

The foregoing inorganic reagents ($SOCl_2$, PBr_3, PI_3) are mainly useful for the conversion of primary and secondary alcohols to the corresponding halides. The major advantages of these reagents, compared to the hydrohalic acids, is that milder reaction conditions are required and carbocation rearrangements (Section 10.6.C) are less important. The ultimate way to minimize carbocation formation, and thus avoid the problems that are usually associated with reactions involving these intermediates, is to convert the alcohol first to the **sulfonate ester,** which is then allowed to react with halide ion under conditions favorable for the S_N2 mechanism.

Common reagents for the formation of sulfonate esters are benzenesulfonyl chloride, $C_6H_5SO_2Cl$, and *p*-toluenesulfonyl chloride, $CH_3C_6H_4SO_2Cl$ (Section 26.5). These compounds react with primary and secondary alcohols, generally in the presence of a tertiary amine such as pyridine, to produce sulfonate esters.

Cl⁀⁀⁀OH + C_6H_5–SO_2Cl + C_5H_5N ⟶

benzenesulfonyl chloride   pyridine

Cl⁀⁀⁀OSO$_2$–C_6H_5 + $C_5H_5\overset{+}{N}H$

4-chlorobutyl benzenesulfonate

We saw in Chapter 9 that sulfonate ions are good leaving groups in substitution reactions, since the sulfonate anion is the conjugate base of a strong acid (notice the similarity of benzenesulfonic acid to sulfuric acid). Such esters react with halide ion in inert solvents to give alkyl halides and the sulfonate ion.

$Na^+ I^-$ + Cl⁀⁀⁀OSO$_2$–C_6H_5 $\xrightarrow{\text{acetone}}$

Cl⁀⁀⁀I + C_6H_5–SO_3^- Na^+

Note that the benzenesulfonate ion is displaced selectively in this reaction, even though displacement of chloride ion might also have occurred. The benzenesulfonate ion is about 100 times more reactive as a leaving group than chloride.

Other common reagents for preparation of halides are lithium chloride in dimethylformamide or ethanol and sodium bromide in dimethylformamide (DMF) or dimethyl sulfoxide (DMSO).

OSO$_2$–C_6H_4–CH_3 $\xrightarrow[\text{DMSO, 25°C}]{\text{NaBr}}$ Br

3-pentyl *p*-toluenesulfonate   (85%) 3-bromopentane

Of course, the preceeding method is useful only with primary and secondary alcohols. With tertiary alcohols, the S_N2 displacement reaction is so slow that side reactions (mainly elimination) dominate.

EXERCISE 10.11 What product is expected when each of the following alcohols is converted into the corresponding alkyl bromide by the sulfonate displacement method?

(a) (*S*)-2-pentanol   (b) *cis*-4-methylcyclohexanol

C. Carbocation Rearrangements

In the preceding section, we saw that the hydroxy group of an alcohol may be replaced by a halide nucleophile if it is converted into a better leaving group by reaction with a proton, an inorganic halide such as $SOCl_2$, or a sulfonyl chloride such as benzenesulfonyl chloride. Many of the examples given were primary alcohols, which react by the S_N2 mechanism. In these cases, the reactions are relatively uncomplicated by side reactions. Other examples given were for tertiary alcohols, which react by the S_N1 mechanism. Again, the transformation is relatively simple and the reaction conditions are mild; one simply treats the tertiary alcohol with either the aqueous hydrohalic acid or with the anhydrous HX at low temperature. Yields of alkyl halide are generally good, and there is little complication by side reactions. With secondary alcohols, the situation is often more complicated. With the sulfonate method, the actual substitution reaction can be carried out under conditions that are favorable for the S_N2 process, and the situation is relatively straightforward. However, with HBr, HCl-$ZnCl_2$ (''Lucas reagent''), or PBr_3, the substitution reaction usually occurs by the S_N1 process, by way of the intermediate carbocation.

$$R{-}OH + HBr \rightleftharpoons R{-}\overset{+}{O}H_2 + Br^-$$

$$R{-}\overset{+}{O}H_2 \rightleftharpoons R^+ + H_2O$$

$$R^+ + Br^- \longrightarrow R{-}Br$$

One important drawback of carbocation reactions is the alternative reaction pathways that are available. We have already discussed one such reaction, E1 elimination. The electron-deficient carbocation center tends to attract electron density from adjacent bonds, and these bonds become weaker. One result is the ready loss of a proton to a basic solvent molecule. In some systems, another important side reaction can occur—rearrangement. The hydrogen attached by the weakened bond *along with its bonding electrons* can move to the cationic center, thus generating a new carbocation.

Note that in this process the positive charge moves to the carbon to which the hydrogen was originally attached. Such rearrangements are especially important when the new carbocation is more stable than the old, but the reaction can occur even when the two carbocations have comparable stability. Such reactions are common for secondary carbocations but almost never involve primary or tertiary carbocations. For primary systems, the alternative S_N2 process is generally so favorable that the relatively energetic primary carbocation is never produced. Rearrangements are generally unimportant in tertiary systems because it is usually the case that any rearrangement would produce a *less stable* carbocation.

An example of a process involving carbocation rearrangement is seen in the reaction of 3-methyl-2-butanol with HBr; the sole product is 2-bromo-2-methylbutane. In this case, the intermediate secondary carbocation rearranges to the more stable tertiary carbocation much more rapidly than it reacts with bromide ion.

$$\mathrm{CH_3\overset{H}{\underset{CH_3}{C}}-CHOHCH_3 \overset{H^+}{\rightleftharpoons} CH_3\overset{H}{\underset{CH_3}{C}}-\overset{\overset{+}{O}H_2}{C}HCH_3 \xrightleftharpoons{-H_2O}}$$

$$\mathrm{CH_3\overset{H}{\underset{CH_3}{C}}-\overset{+}{C}HCH_3 \rightleftharpoons CH_3\overset{+}{\underset{CH_3}{C}}-\overset{H}{C}HCH_3 \xrightarrow{HBr} (CH_3)_2\overset{Br}{C}CH_2CH_3}$$

secondary carbocation — tertiary carbocation (more stable) — actual product

Another interesting example is seen in the reactions of 2-pentanol and 3-pentanol with hot hydrobromic acid; both alcohols give mixtures of 2- and 3-bromopentane.

$$\mathrm{CH_3CH_2CH_2CHOHCH_3 \xrightarrow{HBr} \underset{(86\%)}{CH_3CH_2CH_2CHBrCH_3} + \underset{(14\%)}{CH_3CH_2CHBrCH_2CH_3}}$$

$$\mathrm{CH_3CH_2CHOHCH_2CH_3 \xrightarrow{HBr} \underset{(20\%)}{CH_3CH_2CH_2CHBrCH_3} + \underset{(80\%)}{CH_3CH_2CHBrCH_2CH_3}}$$

The mechanism whereby these two products are formed is

$$\mathrm{CH_3CH_2CH_2\underset{^+OH_2}{C}HCH_3 \rightleftharpoons CH_3CH_2CH_2\overset{+}{C}HCH_3 \xrightarrow{Br^-} CH_3CH_2CH_2\underset{Br}{C}HCH_3}$$

$$\mathrm{CH_3CH_2C\overset{H}{H}\overset{+}{C}HCH_3 \rightleftharpoons CH_3CH_2\overset{+}{C}H\overset{H}{C}HCH_3}$$

$$\mathrm{CH_3CH_2\underset{^+OH_2}{C}HCH_2CH_3 \rightleftharpoons CH_3CH_2\overset{+}{C}HCH_2CH_3 \xrightarrow{Br^-} CH_3CH_2\underset{Br}{C}HCH_2CH_3}$$

Note that equilibration of the carbocations is not complete. In this case, the rate of reaction of the carbocation with bromide ion is comparable to the rate of rearrangement.

EXERCISE 10.12 Suggest a reason why 3-methyl-2-butanol gives completely the rearranged bromide, while rearrangement is not complete in the 2-pentanol case.

It is important to remember that these rearrangements generally do not occur in primary systems. For example, the reaction of 1-butanol with HBr gives a good yield of 1-bromobutane; reaction occurs by the S_N2 mechanism, and carbocations are not involved.

$$\mathrm{CH_3CH_2\overset{H}{C}HCH_2-\overset{+}{O}H_2 \xrightarrow{Br^-} CH_3CH_2CH_2CH_2-Br}$$

$$\not\downarrow$$

$$\mathrm{CH_3CH_2\overset{+}{C}H\overset{H}{C}H_2}$$

Even isobutyl alcohol on heating with HBr and H_2SO_4 gives mainly isobutyl bromide, although formation of some *t*-butyl bromide *does* occur via the relatively stable *t*-butyl cation.

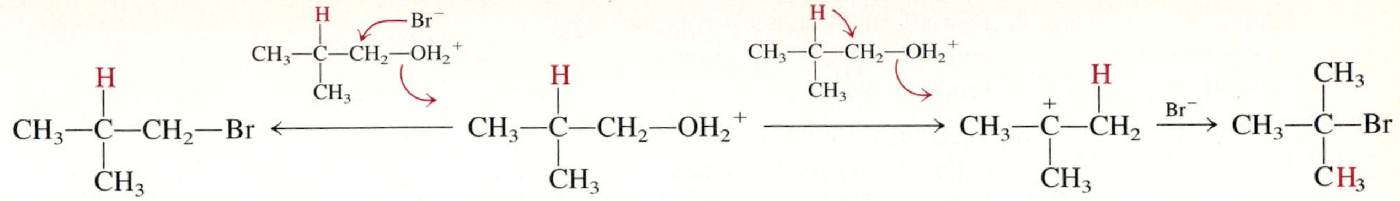

$$CH_3CH(CH_3)CH_2Br \xleftarrow{Br^-} CH_3CH(CH_3)CH_2OH_2^+ \xrightarrow{\text{H shift}} CH_3\overset{+}{C}(CH_3)CH_3 \xrightarrow{Br^-} CH_3C(CH_3)_2Br$$

However, primary alcohols that have a **quaternary carbon** (a carbon attached to four other carbons) next to the alcohol carbon react with complete rearrangement.

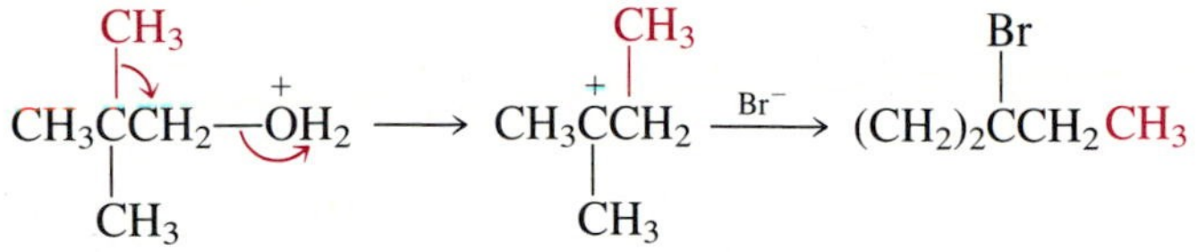

$$CH_3C(CH_3)_2CH_2\overset{+}{O}H_2 \longrightarrow CH_3\overset{+}{C}(CH_3)CH_2CH_3 \xrightarrow{Br^-} (CH_3)_2CBrCH_2CH_3$$

Recall that S_N2 reactions of compounds such as this (neopentyl-type compounds) are very slow (Section 9.3). Consequently the alternative rearrangement reaction is able to compete and becomes the dominating reaction. This example shows also that alkyl groups can migrate as well as hydrogen.

Alcohols with a cycloalkyl group next to the alcohol carbon frequently undergo rearrangement with resultant **ring expansion.** Ring expansion is particularly prone to occur when there is a decrease in ring strain.

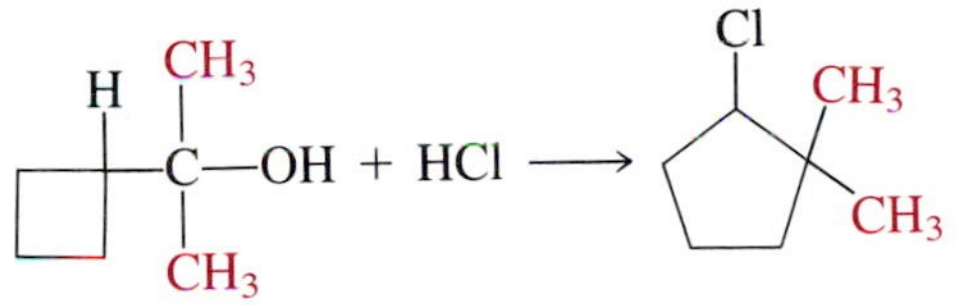

$$\text{cyclobutyl}-C(CH_3)_2OH + HCl \longrightarrow \text{1-chloro-2,2-dimethylcyclopentane}$$

This reaction involves formation of an intermediate carbocation, which undergoes rearrangement of one of the ring bonds to give a cyclopentyl cation.

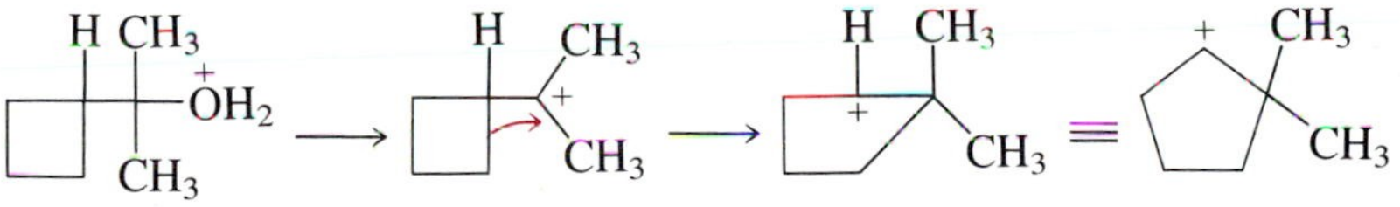

$$\text{cyclobutyl}-C(CH_3)_2\overset{+}{O}H_2 \longrightarrow \text{cyclobutyl}-\overset{+}{C}(CH_3)_2 \longrightarrow \text{2,2-dimethylcyclopentyl cation} \equiv \text{2,2-dimethylcyclopentyl cation}$$

The driving force for this rearrangement is relief of ring strain—the cyclobutyl system has about 26 kcal $mole^{-1}$ of strain, while the cyclopentyl cation has only about 6 kcal $mole^{-1}$ (see Table 5.5). For this reason, ring expansion occurs even though it involves rearrangement of a tertiary to a secondary carbocation.

EXERCISE 10.13 Make a molecular model of 2-cyclobutyl-2-propanol. Remove the hydroxy group; the vacant valence on the propane chain represents the carbocation center. Now disconnect one of the C-1—C-2 bonds of the cyclobutane ring and reattach it to the carbocation center. Note that the product of these operations is a model of the 2,2-dimethylcyclopentyl cation. Repeat these operations until you are thoroughly at ease with the bond reorganization that occurs in the ring expansion.

In the last two sections, we have considered a number of reactions that may be used for accomplishing the important conversion ROH → RX, and we have various complications to watch out for. The best overall methods may be summarized as follows.

1. Primary alcohols with no β-branching
   chloride: $SOCl_2$ + pyridine (generally better than $ZnCl_2$-HCl)
   bromide: PBr_3 or HBr-H_2SO_4
   iodide: P + I_2

2. Primary alcohols with β-branching
   chloride: $SOCl_2$ + pyridine
   bromide: PBr_3
3. Secondary alcohols
   chloride: $SOCl_2$ + pyridine
   bromide: PBr_3 (low temperature, less than 0 °C)
   (*The two-step sequence* alcohol → sulfonate ester → halide *gives a product of higher purity*.)
4. Tertiary alcohols
   chloride: HCl at 0 °C
   bromide: HBr at 0 °C

EXERCISE 10.14 Suggest a method for preparation of each of the following alkyl halides from an alcohol.

(a) 3-chloro-3-ethylpentane
(b) 1-chloropentane
(c) 3-(bromomethyl)pentane
(d) (*S*)-2-bromooctane
(e) *trans*-1-chloro-3-methylcyclohexane

Highlight 10.3

Activation of the OH group in alcohols is accomplished by converting the OH to a better leaving group. Lewis acids (H^+, Zn^{2+}) form ROH_2^+ or $ROHZn^{2+}$ in which the leaving groups are H_2O or $ZnOH^+$. RCl or RBr are formed in this way. Certain halides ($SOCl_2$, $C_6H_5SO_2Cl$, PCl_3, PBr_3, PI_3) undergo reactions in which the O of the ROH displaces a halide to form ROSOCl, $ROSO_2C_6H_5$, or $(RO)_3P$. In each case, the R—O bond is more easily broken in subsequent reactions. The intermediates can react further to form R—Cl, R—Br, or R—I.

Highlight 10.4

The nature of the R group in ROH strongly influences the mechanism of the conversion to RX after activation. Primary alkyl derivatives undergo S_N2 substitution unless there are two β substituents that sterically prevent approach to the backside of the central carbon, and a rearranged RX results, as in the case of neopentyl alcohol. Tertiary alkyl derivatives undergo S_N1 substitution rapidly to form *t*-alkyl—Cl or *t*-alkyl—Br. See Highlight 10.5 for secondary alkyl derivatives.

Highlight 10.5

Secondary alkyl derivatives from the activation of ROH can undergo S_N2 or S_N1 substitution to form R—Cl or R—Br. Rearrangement to an isomeric secondary or a tertiary alkyl derivative reflects carbocation formation and S_N1 reaction. A tertiary alkyl derivative can also rearrange at the stage of the tertiary carbocation to an isomeric carbocation if the driving force is great enough. Relief of angle strain in a small ring provides such driving force and leads to rearrangement.

D. Dehydration of Alcohols: Formation of Ethers and Alkenes

When a primary alcohol is treated with sulfuric acid alone, the product is an **alkylsulfuric acid.** The reaction is an equilibrium process—the product alkylsulfuric acid is readily hydrolyzed by excess water. If the transformation is carried out with excess sulfuric acid at 0 °C, the reaction proceeds to completion and the alkylsulfuric acid may be isolated.

$$C_2H_5OH + \underset{\text{(excess)}}{H_2SO_4} \xrightarrow{0°C} C_2H_5OSO_2OH + H_2O$$

However, if ethanol is *heated* with concentrated sulfuric acid, ethyl ether is produced in high yield.

$$2\ C_2H_5OH \xrightarrow[140°C]{H_2SO_4} C_2H_5OC_2H_5 + H_2O$$

The detailed mechanism of the reaction under these conditions is not known. One possibility is that the alkyloxonium ion formed by reaction of alcohol with sulfuric acid is attacked by another alcohol molecule instead.

$$RCH_2{-}\underset{H}{O} + \underset{R}{CH_2}{-}\overset{+}{O}H_2 \longrightarrow RCH_2{-}\underset{H}{\overset{+}{O}}{-}CH_2R + OH_2$$

Alternatively the alkylsulfuric acid may be an intermediate.

$$RCH_2{-}\underset{H}{O} + \underset{R}{CH_2}{-}OSO_2OH \longrightarrow RCH_2{-}\underset{H}{\overset{+}{O}}{-}CH_2R + HSO_4^-$$

In the case of primary alcohols, the reaction is an acceptable way of preparing symmetrical ethers.

n-Butyl alcohol and concentrated sulfuric acid are refluxed with provision to remove water as it is formed, either with a suitable trap or by a fractionating column. The reaction mixture is maintained at 130–140 °C. The reaction mixture is allowed to cool and, after washing and drying, is distilled to give di-*n*-butyl ether.

At still higher temperature, elimination occurs to give the alkene.

$$C_2H_5OH \xrightarrow[170°C]{H_2SO_4} CH_2{=}CH_2 + H_2O$$

Secondary and tertiary alcohols generally give only the elimination products.

$$\text{cyclohexanol} \xrightarrow[165\text{–}200°C]{H_3PO_4} \text{cyclohexene}$$

cyclohexanol   cyclohexene (79–84%)

A mixture of 1 kg of technical grade cyclohexanol and 200 g of 85% phosphoric acid is heated at 165–170 °C for 4–5 hr and at 200 °C for 30 min. The upper layer is separated, dried, and distilled to obtain 660–690 g (79–84%) of cyclohexene, b.p. 81–83 °C.

Tertiary alcohols eliminate water readily on heating with even traces of acid. If elimination is to be avoided, the alcohols should be distilled at low temperature (vacuum distillation) or in apparatus that has been rinsed with ammonia.

EXERCISE 10.15 Explain why acid-catalyzed dehydration of an alcohol is not a desirable method for the preparation of 1-butene.

E. Oxidation of Alcohols: Formation of Aldehydes and Ketones

Primary and secondary alcohols can be oxidized to carbonyl compounds.

$$RCH_2OH \xrightarrow[-2\,H]{[O]} R{-}\overset{\overset{\displaystyle O}{\|}}{C}{-}H$$

an aldehyde

$$R{-}\underset{}{\overset{\overset{\displaystyle OH}{|}}{C}H}{-}R \xrightarrow[-2\,H]{[O]} R{-}\overset{\overset{\displaystyle O}{\|}}{C}{-}R$$

a ketone

Many procedures are available for accomplishing these transformations, but the most common general oxidizing agent is some form of chromium(VI), which becomes reduced to chromium(III).

Chromium trioxide, CrO_3, also known as chromic anhydride, forms red, deliquescent crystals. It is soluble in water and in sulfuric acid.

Sodium or potassium dichromate forms orange aqueous solutions; under basic conditions the yellow chromate salt is produced.

$$\underset{\text{yellow}}{2\,CrO_4^{2-}} + 2\,H^+ \rightleftharpoons \underset{\text{orange}}{Cr_2O_7^{2-}} + H_2O$$

Pyridinium chlorochromate (PCC) is produced by the reaction of equimolar amounts of pyridine (page 220), chromium trioxide, and HCl. It is a yellow-orange, crystalline material, m.p. 205 °C, and is soluble in organic solvents such as methylene chloride and chloroform.

$$\text{pyridine} + CrO_3 + HCl \longrightarrow \text{pyridinium (N}^+\text{–H)}\ ClCrO_3^-$$

pyridine

pyridinium chlorochromate (PCC)

Primary alcohols give aldehydes on warming with sodium dichromate and aqueous sulfuric acid. However, in aqueous solution, aldehydes are also readily oxidized to give carboxylic acids (Section 14.8.A). Thus, this method is only successful for the synthesis of an aldehyde if it is of sufficiently low molecular weight that it may be distilled from solution as formed. For example, *n*-butyl alcohol can be oxidized in this way to give butyraldehyde in 50% yield.

$$CH_3CH_2CH_2CH_2OH + H_2Cr_2O_7 \longrightarrow CH_3CH_2CH_2CHO$$

(50%)
butyraldehyde

In practice, only aldehydes that boil considerably below 100 °C can be conveniently prepared in this manner (since it would not be possible to distill a higher boiling aldehyde from an aqueous reaction mixture). This effectively limits the method to the production of a few simple aldehydes, and it is not an important synthetic method.

A virtue of pyridinium chlorochromate (PCC) is that it is soluble in organic solvents. Therefore, it may be used for the oxidation of primary alcohols to aldehydes under non-aqueous conditions. An example is seen in the preparation of oct-2-yn-1-al from oct-2-yn-1-ol.

$$\text{oct-2-yn-1-ol (OH)} \xrightarrow[CH_2Cl_2]{PCC} \text{oct-2-yn-1-al (CHO)}\ (84\%)$$

oct-2-yn-l-ol (84%) oct-2-yn-l-al

Since ketones are more stable to general oxidation conditions than aldehydes, chromic acid oxidations are more important for secondary alcohols. In one common procedure a 20% excess of sodium dichromate is added to an aqueous mixture of the alcohol and a stoichiometric amount of acid.

$$\text{4-ethylcyclohexanol (H, OH; } CH_2CH_3) \xrightarrow[H_2SO_4]{Na_2Cr_2O_7} \text{4-ethylcyclohexanone (O; } CH_2CH_3)\ (90\%)$$

4-ethylcyclohexanol (90%) 4-ethylcyclohexanone

An especially convenient oxidizing agent is Jones reagent, a solution of chromic acid in 8N sulfuric acid. The secondary alcohol in acetone solution is ''titrated'' with the reagent with stirring at 15–20 °C. Oxidation is rapid and efficient. The green chromium salts separate from the reaction mixture as a heavy sludge; the supernatant liquid consists mainly of an acetone solution of the product ketone.

$$\text{cyclooctanol (OH, H)} \xrightarrow[\substack{H_2SO_4 \\ \text{acetone} \\ 25^\circ C}]{H_2CrO_4} \text{cyclooctanone (=O)}\ (95\%)$$

cyclooctanol (95%) cyclooctanone

Chromium(VI) oxidations are known to proceed by way of a chromate ester of the alcohol. If the alcohol has one or more hydrogens attached to the carbinol position, a base-catalyzed elimination occurs, yielding the aldehyde or ketone and a chromium(IV) species. The overall effect of these two consecutive reactions is oxidation of the alcohol and reduction of the chromium.

$$(1)\ R_2CHOH + H_2CrO_4 \rightleftharpoons R_2CH{-}O{-}Cr(=O)_2{-}OH + H_2O$$

$$(2)\ B{:}\ \ R_2C(H){-}O{-}Cr(=O)_2{-}OH \longrightarrow R_2C{=}O + HCrO_3^- + BH^+$$

The chromium(IV) produced in the elimination undergoes rapid reaction with a molecule containing chromium(VI) to produce two chromium(V) species. These chromium(V)-containing molecules function further as two-electron oxidants to produce more aldehyde or ketone and two chromium(III) species, the ultimate oxidation state of the chromium in such reactions.

Under conditions such as these, tertiary alcohols do not generally react, although under proper conditions the chromate ester can be isolated.

$$2\,(CH_3)_3COH + H_2CrO_4 \rightleftharpoons (CH_3)_3CO{-}\overset{\overset{\displaystyle O}{\|}}{\underset{\underset{\displaystyle O}{\|}}{Cr}}{-}OC(CH_3)_3 + 2\,H_2O$$

Since there is no proton on the carbon with the OH group, the elimination noted above cannot occur in the case of a tertiary alcohol, and such esters are stable. If the chromate ester is treated with excess water, simple hydrolysis occurs with regeneration of the tertiary alcohol and chromic acid.

Balancing Oxidation-Reduction Reactions. Organic redox reactions may be balanced by the method of half-cells often taught in beginning chemistry courses for inorganic redox reactions. To use this method, first write an equation for the substance being oxidized, showing the hydrogens that are lost in the process as *protons*. Add enough *electrons* to the side of the equation showing the protons so that the equation is *balanced for charge*. For oxidation of an alcohol, two hydrogens are lost, so we must add two electrons to the right side of the equation. (In other words, oxidation of an alcohol is a ''two-electron'' oxidation.)

$$RCH_2OH = RCHO + 2\,H^+ + 2\,e^-$$

Next, write an equation showing the change that the oxidizing agent undergoes. One side of this equation will contain the oxidizing agent and the other will contain its ultimate reduced form. In the case of chromium(VI), the reduced form is chromium(III). Add enough electrons to balance this equation.

$$3\,e^- + Cr^{6+} = Cr^{3+}$$

Now add the two half-reactions together, multiplying by the appropriate factor so that the electrons cancel.

$$\begin{array}{c} 3 \times [RCH_2OH = RCHO + 2\,H^+ + 2\,e^-] \\ \underline{2 \times [3\,e^- + Cr^{6+} = Cr^{3+}]\qquad\qquad} \\ 3\,RCH_2OH + 2\,Cr^{6+} = 3\,RCHO + 6\,H^+ + 2\,Cr^{3+} \end{array}$$

As shown by the foregoing procedure, 3 moles of alcohol are oxidized by 2 moles of chromium(VI) reagent.

EXERCISE 10.16 How many grams of sodium dichromate are required to convert 70 g of cyclohexanol to cyclohexanone?

10.7 Nomenclature of Ethers

The common names of ethers are derived by naming the two alkyl groups and adding the word ether.

$CH_3OCH_2CH_3$ — ethyl methyl ether

$CH_3OC(CH_3)_3$ — *t*-butyl methyl ether

In symmetrical ethers the prefix di- is used. Although the prefix is often omitted, it should be included to avoid confusion.

$CH_3CH_2OCH_2CH_3$ — diethyl ether (or ethyl ether)

$(CH_3)_2CHOCH(CH_3)_2$ — diisopropyl ether (or isopropyl ether)

In the IUPAC system ethers are named as alkoxyalkanes. The larger alkyl group is chosen as the stem.

$CH_3CH_2OCH(CH_3)CH_2CH_2CH_3$ — 2-ethoxypentane

$CH_3C(CH_3)_2CH_2OCH_3$ — 1-methoxy-2,2-dimethylpropane

$CH_3OCH_2CH_2OCH_3$ — 1,2-dimethoxyethane

EXERCISE 10.17 Write common and IUPAC names for the following ethers.

(a) (b) (c)

10.8 Physical Properties of Ethers

The physical properties of some ethers are listed in Table 10.4. Note that dimethyl ether is a gas at room temperature and that diethyl ether has a boiling point only about 10 °C above normal room temperature. Diethyl ether is an important solvent and has a characteristic odor. It was once used as an anesthetic, but it has been largely replaced for this purpose by other compounds.

TABLE 10.4 Physical Properties of Ethers

Compound	Name	Melting Point, °C	Boiling Point, °C
CH_3OCH_3	dimethyl ether	−138.5	−23
$CH_3OCH_2CH_3$	ethyl methyl ether		10.8
$(CH_3CH_2)_2O$	diethyl ether	−116.62	34.5
$CH_3CH_2OCH_2CH_2CH_3$	ethyl propyl ether	−79	63.6
$(CH_3CH_2CH_2)_2O$	dipropyl ether	−122	91
$(CH_3CH(CH_3))_2O$	diisopropyl ether	−86	68
$(CH_3CH_2CH_2CH_2)_2O$	dibutyl ether	−95	142

The rule of thumb that compounds having no more than four carbons per oxygen are water soluble holds for ethers as well as for alcohols. Dimethyl ether is completely miscible with water. The solubility of diethyl ether in water is about 10 g per 100 g of H_2O at 25 °C. Tetrahydrofuran (b.p. 67 °C) is another important solvent. This cyclic ether, commonly abbreviated THF, has essentially the same molecular weight as diethyl ether, but it is much more soluble in water.

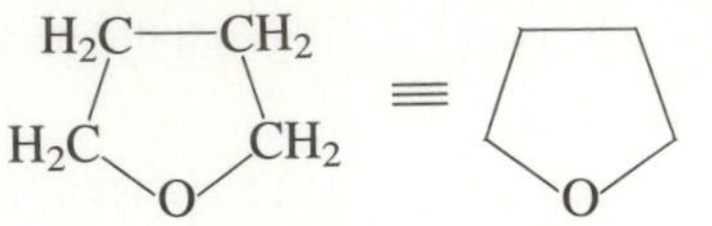

tetrahydrofuran (THF)
(b.p. 67°C; miscible with H_2O in all proportions at 25°C)

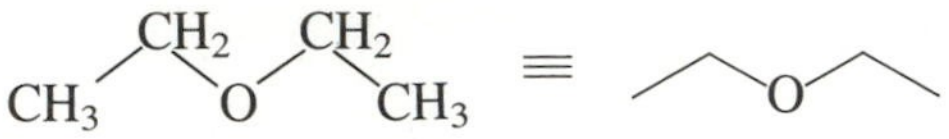

diethyl ether
(b.p. 34.5°C; 10 g dissolves in 100 g of H_2O at 25°C)

Because of its cyclic structure, THF has lone-pair electrons that are more accessible for hydrogen bonding than those of diethyl ether. The ''floppy'' ethyl groups in the acyclic compound interfere with hydrogen bonding and cause the water solubility of diethyl ether to be lower. Also note that the cyclic compound has a significantly higher boiling point; its more compact structure allows for more efficient van der Waals attraction between molecules.

As we shall see, ethers are fairly inert to many reagents. Because of their unreactivity, they are not generally important as chemical reagents. However, their general lack of reactivity, combined with their favorable solvent properties, makes ethers useful solvents for many other reactions. Several ethers that are important as solvents are dioxane (page 215), methyl *t*-butyl ether (MBE), 1,2-dimethoxyethane (glyme), and bis-β-methoxyethyl ether (diglyme).

$CH_3OCH_2CH_2OCH_3$	$CH_3OCH_2CH_2OCH_2CH_2OCH_3$	$(CH_3)_3COCH_3$
1,2-dimethoxyethane (glyme, DME)	bis-β-methoxyethyl ether (diglyme)	methyl *t*-butyl ether (MBE)

10.9 Preparation of Ethers

Ethers may be prepared by the **Williamson ether synthesis** or by the reaction of alcohols with sulfuric acid. The Williamson ether synthesis is simply an S_N2 reaction of a primary alkyl halide or sulfonate ester in which an alkoxide ion is the nucleophile. The alkoxide may be derived from a primary or secondary alcohol, but the substrate must be primary and have no β-branches.

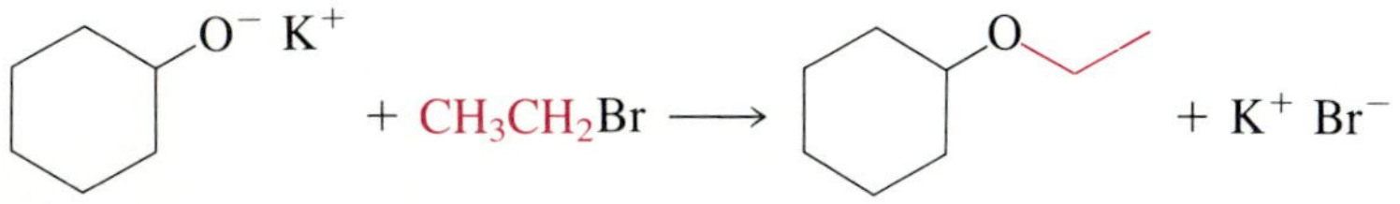

$$C_6H_{11}O^- K^+ + CH_3CH_2Br \longrightarrow C_6H_{11}OCH_2CH_3 + K^+ Br^-$$

Methyl and ethyl substrates may be used with tertiary alkoxides. Other halides and sulfonates give too much elimination (Section 11.5).

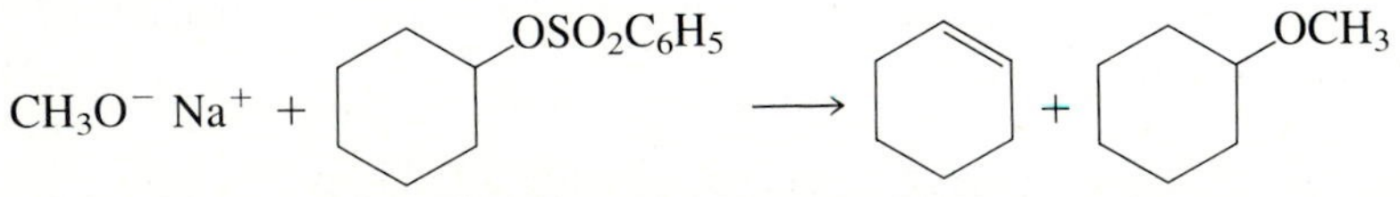

$$CH_3O^- Na^+ + C_6H_{11}OSO_2C_6H_5 \longrightarrow \text{cyclohexene (85\%)} + C_6H_{11}OCH_3 \text{ (5\%)}$$

Since the reaction is a classic example of an S_N2 reaction, one must keep in mind the principles of that mechanism (Chapter 9) when planning an ether synthesis by this route.

EXERCISE 10.18 Methyl neopentyl ether can be prepared readily from sodium neopentoxide and methyl benzenesulfonate, but not from sodium methoxide and neopentyl benzenesulfonate. Write equations for these two reactions and explain why one combination works and the other does not.

The alcohol-sulfuric acid reaction, which was discussed in Section 10.6.D, is most often used for the conversion of simple primary alcohols into symmetrical ethers.

$$ClCH_2CH_2OH \xrightarrow[\Delta]{H_2SO_4} \underset{\substack{(75\%) \\ \text{bis-}\beta\text{-chloroethyl ether}}}{ClCH_2CH_2OCH_2CH_2Cl}$$

Secondary and tertiary alcohols undergo predominant dehydration when subjected to these conditions. Occasionally some of the symmetrical ether is formed as a by-product in the case of secondary alcohols.

$$CH_3\overset{\overset{CH_3}{|}}{C}HOH \xrightarrow{H_2SO_4} \underset{\text{major product}}{CH_3CH{=}CH_2} + \underset{\text{by-product}}{CH_3\overset{\overset{CH_3}{|}}{C}H{-}O{-}\overset{\overset{CH_3}{|}}{C}HCH_3}$$

The method is generally useless for the preparation of unsymmetrical ethers because complex mixtures are formed.

$$ROH + R'OH \xrightarrow{H^+} ROR + ROR' + R'OR'$$

An exception is the case in which one alcohol is tertiary and the other alcohol is primary or secondary. Since tertiary carbocations form under mild conditions, this method is generally a satisfactory synthetic method.

$$CH_3\underset{\underset{CH_3}{|}}{\overset{\overset{CH_3}{|}}{C}}OH + CH_3\overset{\overset{CH_3}{|}}{C}HOH \xrightarrow[\substack{H_2O \\ 100°C}]{NaHSO_4} \underset{(82\%)}{CH_3\underset{\underset{CH_3}{|}}{\overset{\overset{CH_3}{|}}{C}}{-}O{-}\overset{\overset{CH_3}{|}}{C}HCH_3}$$

EXERCISE 10.19 Show how the following ethers may be prepared.

(a) 1-methoxy-1-methylcyclohexane (b) di-*n*-butyl ether

10.10 Reactions of Ethers

Ethers are relatively inert to most reagents. They are stable to base and to most reducing agents. They are also stable to dilute acid but do react with hot concentrated acids. Strong HBr or HI causes cleavage.

$$CH_3CH_2OCH_2CH_3 + 2\,HBr \xrightarrow{\Delta} 2\,CH_3CH_2Br + H_2O$$

The mechanism of this reaction is S_N2 and involves displacement by bromide ion on the protonated ether. The alcohol produced reacts further with HBr to yield more alkyl bromide.

$$(1)\ CH_3CH_2OCH_2CH_3 + HBr \rightleftharpoons CH_3CH_2\overset{H}{\underset{+}{O}}CH_2CH_3 + Br^-$$

$$(2)\ Br^-: + \underset{CH_3}{CH_2}-\overset{+}{O}(H)CH_2CH_3 \longrightarrow CH_3CH_2Br + CH_3CH_2OH$$

$$(3)\ CH_3CH_2OH + HBr \rightleftharpoons CH_3CH_2\overset{+}{O}H_2\ Br^-$$

$$(4)\ Br^-: + \underset{CH_3}{CH_2}-\overset{+}{O}H_2 \longrightarrow CH_3CH_2Br + H_2O$$

With tertiary ethers, carbocations are involved and the reactions tend to be much more complex. Heating such ethers with strong acid generally leads to elimination as the major reaction.

$$CH_3\overset{CH_3}{\underset{CH_3}{C}}-OCH_3 \xrightarrow[\Delta]{H_2SO_4} CH_3\overset{CH_3}{C}=CH_2 + CH_3OH$$

t-butyl methyl ether   isobutylene

The reaction is not generally useful for preparations unless one of the alkyl groups of the ether is a small tertiary group. In such a case the alkene formed upon elimination evaporates as it is produced.

EXERCISE 10.20 When methyl neopentyl ether is treated with anhydrous hydrogen bromide, an alcohol and an alkyl bromide are produced. What are they?

One of the most important reactions of ethers is an undesirable one—the reaction with atmospheric oxygen to form hydroperoxides and peroxides (**autoxidation**).

$$CH_3\underset{CH_3}{CH}-O-\underset{CH_3}{CH}CH_3 + O_2 \longrightarrow CH_3\overset{OOH}{\underset{CH_3}{C}}-O-\underset{CH_3}{CH}CH_3$$

a hydroperoxide

Autoxidation occurs by a free radical mechanism.

$$(1)\ RO-\overset{H}{\underset{|}{C}}- + R'\cdot \longrightarrow RO-\overset{\bullet}{\underset{|}{C}}- + R'H$$

$$(2)\ RO-\overset{\bullet}{\underset{|}{C}}- + O_2 \longrightarrow RO-\overset{OO\cdot}{\underset{|}{C}}-$$

$$(3)\ \mathrm{RO{-}\overset{\overset{\displaystyle OO\cdot}{|}}{\underset{|}{C}}{-}} + \mathrm{RO{-}\overset{\overset{\displaystyle H}{|}}{\underset{|}{C}}{-}} \longrightarrow \mathrm{RO{-}\overset{\overset{\displaystyle OOH}{|}}{\underset{|}{C}}{-}} + \mathrm{RO{-}\overset{\cdot}{\underset{|}{C}}{-}}$$

or

$$\mathrm{RO{-}\overset{\overset{\displaystyle OO\cdot}{|}}{\underset{|}{C}}{-}} + \mathrm{RO{-}\overset{\cdot}{\underset{|}{C}}{-}} \longrightarrow \mathrm{RO{-}\overset{|}{\underset{|}{C}}{-}OO{-}\overset{|}{\underset{|}{C}}{-}OR}$$

a peroxide

Ethers of almost any type that have been exposed to the atmosphere for any length of time invariably contain peroxides. Isopropyl ether is especially treacherous in this regard, but ethyl ether and tetrahydrofuran are also dangerous. One virtue of using methyl *t*-butyl ether (MBE) as a laboratory solvent is the fact that it is not as prone to autoxidation as ethyl ether and THF. Peroxides and hydroperoxides are hazardous because they decompose violently at elevated temperatures, and serious explosions may result. When an ether that contains peroxides is distilled, the less volatile peroxides concentrate in the residue. At the end of the distillation, the temperature increases, and the residual peroxides may explode. For this reason, ethers should never be evaporated to dryness, unless care has been taken to exclude peroxides rigorously.

The fact that methyl *t*-butyl ether does not easily form peroxides makes it practical for large scale use. This ether has a much higher octane number than straight-run gasoline (110.5 versus 70). For some time, it has been used for increasing the octane rating in the production of unleaded gasoline. It has the advantage of being a ''clean'' octane booster, in contrast to tetraethyllead, which is converted to potentially dangerous lead oxides upon combustion. The amount of methyl *t*-butyl ether produced in the United States during 1988 was 4.7 billion pounds, sufficient to treat only a small fraction of the gasoline produced in the United States.

A simple test for peroxides is to shake a small volume of the ether with aqueous KI solution. If peroxides are present, they oxidize I^- to I_2. The iodine complexes with excess iodide to form triiodide I_3^-. The characteristic brown color of I_3^- is diagnostic for the presence of peroxides. The dark blue color of a starch-polyiodide complex is produced in the presence of starch. Contaminated ether may be purified by shaking with aqueous ferrous sulfate to reduce the peroxides.

> One of the authors has had an impressive demonstration of the violence of a peroxide explosion. In a laboratory adjacent to his office, a laboratory technician was engaged in purifying about 2 L of old THF, later found to contain substantial amounts of peroxides. The material exploded, virtually demolishing the laboratory, and moving the wall several inches toward the author's desk. Several large bookshelves were knocked over and emptied their contents onto his desk and chair. Only by good fortune was no one injured.

10.11 Cyclic Ethers

A. Epoxides: Oxiranes

Heterocyclic compounds are cyclic structures in which one or more ring atoms are **hetero atoms,** a hetero atom being an element other than carbon. The three-membered ring containing oxygen is the first class of heterocyclic compounds whose chemistry we shall consider. Because they are readily prepared from alkenes (Section 11.6), they are commonly named as ''olefin oxides.'' Hence, the parent member is often called ethylene oxide, although the formal IUPAC nomenclature is oxirane. Oxiranes are also called

epoxides. Substituents on the oxirane ring require a numbering system. The general rule for heterocyclic rings is that the hetero atom gets the number 1.

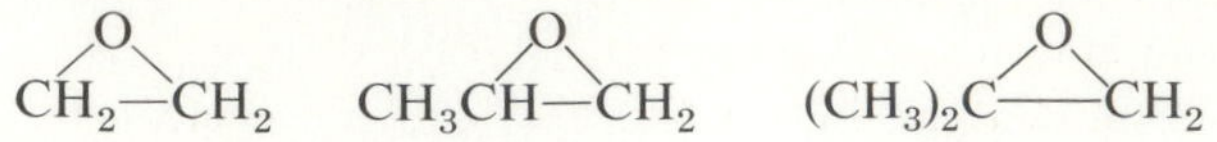

ethylene oxide
oxirane

propylene oxide
2-methyloxirane

isobutylene oxide
2,2-dimethyloxirane

Ethylene oxide is a significant article of commerce and is prepared industrially by the catalyzed air oxidation of ethylene. About 2.7 million tons were produced in the United States in 1988.

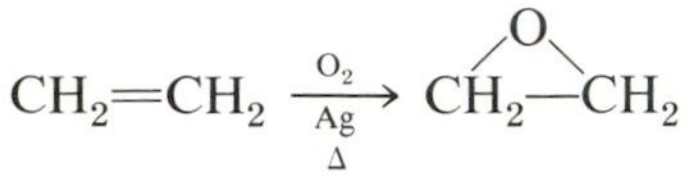

$$CH_2{=}CH_2 \xrightarrow[Ag,\ \Delta]{O_2} \text{ethylene oxide}$$

Several general laboratory preparations are available. One reaction is an internal S_N2 displacement reaction starting with a β-halo alcohol.

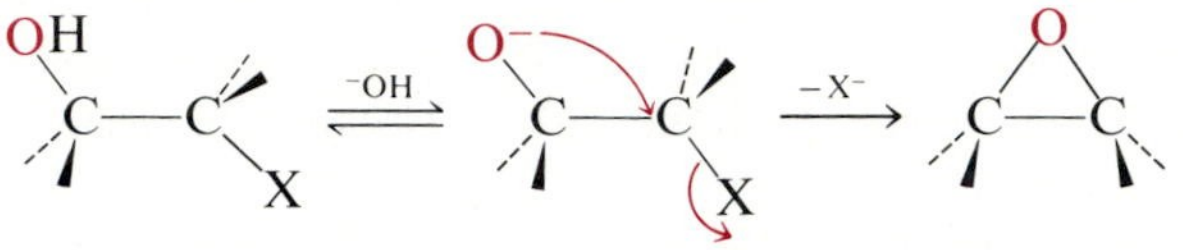

As shown by the foregoing equation, the intramolecular displacement of a halide ion by a neighboring alkoxide group requires that the reacting groups have a specific relationship such that the alkoxide can attack the rear of the carbon-chlorine bond. As we shall see in Section 11.6.B, the required stereoisomer is readily available by the reaction of hypochlorous acid (HOCl) with an alkene.

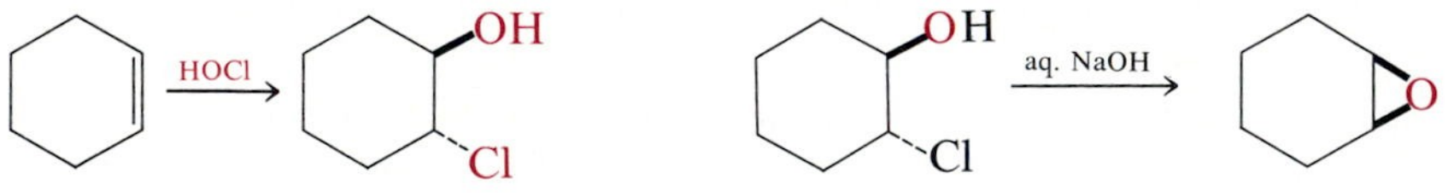

HOCl
(70–73%)

aq. NaOH
(70–73%)
cyclohexene oxide

EXERCISE 10.21 (2*R*,3*R*)-3-Chloro-2-butanol is treated with KOH in ethanol. Is the product *cis*- or *trans*-2,3-dimethyloxirane?

Like other ethers, oxiranes undergo carbon-oxygen bond cleavage under acidic conditions. However, because of the large ring strain associated with the three-membered ring (about 25 kcal mole^{-1}, see Section 5.6), they are much more reactive than normal ethers.

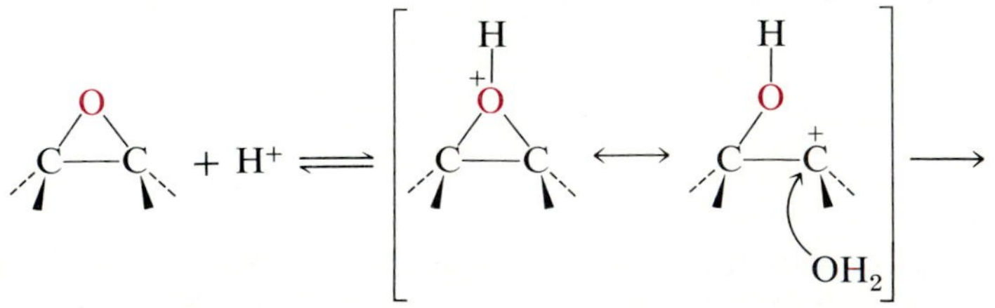

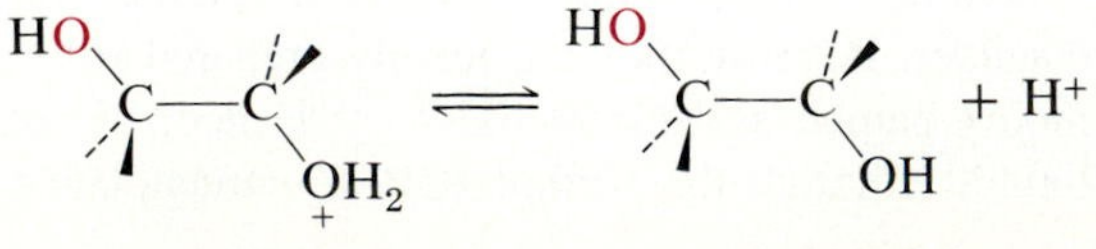

The product is a 1,2-diol. As a class, 1,2-diols are called ''glycols.'' They are important industrial compounds. The simplest 1,2-diol, ''ethylene glycol,'' is commonly used as antifreeze for automobile radiators.

As shown by the foregoing mechanism, ring opening of an epoxide is another example of the S_N2 mechanism (Chapter 9) in which the attacking nucleophile is H_2O and the leaving group is ROH. The ''leaving group'' is, of course, part of the molecule and doesn't leave completely.

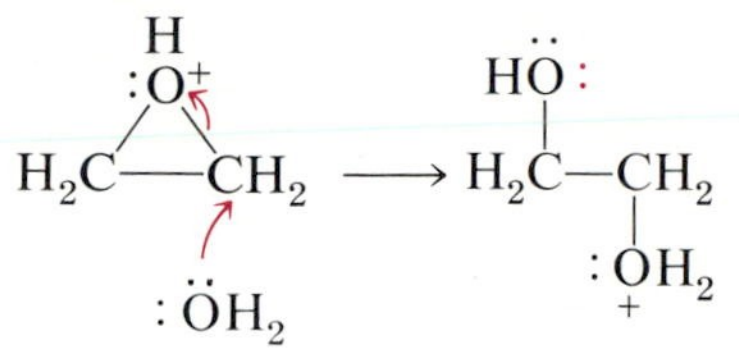

The stereochemical outcome of the reaction is inversion at the reaction center. Thus, cyclohexene oxide reacts with aqueous acid to give exclusively *trans*-cyclohexane-1,2-diol.

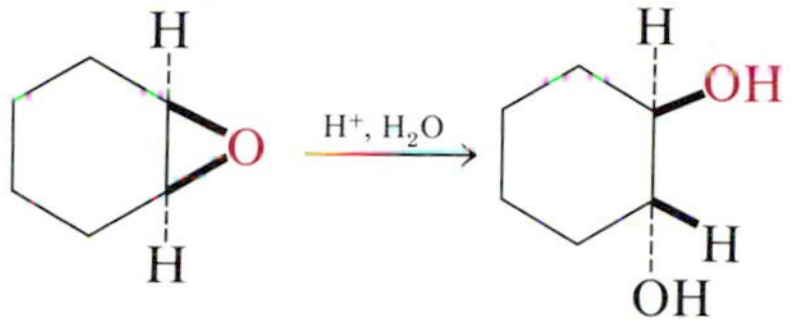

The oxirane ring is so prone to undergo ring-opening reactions that it even reacts with aqueous base.

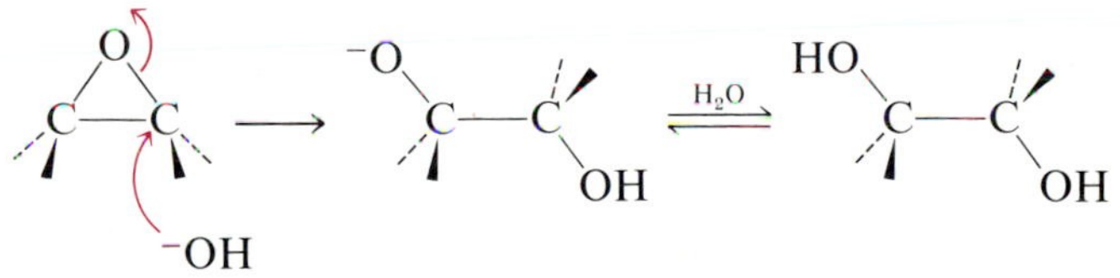

The base-catalyzed opening of an epoxide occurs by the S_N2 mechanism with an alkoxide ion as the leaving group and has no counterpart with normal ethers. It occurs with epoxides only because the relief of ring strain provides a potent driving force for reaction.

Ring strain is distributed over all of the ring bonds. Perhaps half of the total ring strain is relieved at the transition state, which is lowered by that amount compared to a similar reaction with an open-chain ether. A decrease of 12.5 kcal $mole^{-1}$ would raise the rate by 9 powers of 10.

The two ring-opening reactions have different orientational preferences. The acid-catalyzed process is essentially a carbocation reaction, and reaction tends to occur at the ring carbon that corresponds to the more stable carbocation. That is, reaction with solvent occurs at the more highly substituted ring carbon. Like other reactions that occur by the S_N2 mechanism, the base-catalyzed reaction is subject to steric influences. Reaction occurs at the less hindered carbon. The difference is exemplified by the reactions of 2,2-dimethyloxirane in methanol under acidic and basic conditions.

Acidic conditions

$$(CH_3)_2C\overset{O}{\frown}CH_2 \xrightleftharpoons{H^+} \left[(CH_3)_2C\overset{\overset{H}{O^+}}{\frown}CH_2 \longleftrightarrow (CH_3)_2\overset{+}{C}\overset{\overset{H}{O}}{\frown}CH_2 \right] \xrightarrow{CH_3OH}$$

(tertiary carbocation)

$$CH_3-\underset{CH_3\overset{+}{\underset{H}{O}}:}{\overset{CH_3}{C}}-CH_2OH \xrightleftharpoons{-H^+} (CH_3)_2\underset{OCH_3}{C}CH_2OH$$

Basic conditions

$$(CH_3)_2C\overset{O}{\frown}CH_2 + CH_3O^- \longrightarrow \left[(CH_3)_2C\overset{O}{\frown}\underset{OCH_3}{CH_2} \right]^{-\ddagger} \longrightarrow$$

S_N2 reaction at less hindered primary position

$$CH_3-\underset{CH_3}{\overset{O^-}{C}}-CH_2OCH_3 \xrightleftharpoons{H^+} (CH_3)_2\overset{OH}{C}CH_2OCH_3$$

The product from the reaction of ethylene oxide with hydroxide ion is an alkoxide. Under the proper conditions this species can react with more ethylene oxide.

$$HO^- \xrightarrow{\text{oxirane}} HOCH_2CH_2O^- \xrightarrow{\text{oxirane}} HOCH_2CH_2OCH_2CH_2O^- \longrightarrow \text{etc.}$$

The final products are ether alcohols, which are called diethylene glycol, triethylene glycol, and so on. These diols are methylated to produce a series of polyethers that have useful solvent properties. These materials are known by the trivial names **glyme, diglyme, triglyme,** and so on.

$$\underset{\text{diethylene glycol}}{HOCH_2CH_2OCH_2CH_2OH} \longrightarrow \underset{\text{diethylene glycol dimethyl ether (diglyme)}}{CH_3OCH_2CH_2OCH_2CH_2OCH_3}$$

EXERCISE 10.22 Write the equation for the reaction of *trans*-2,3-dimethyloxirane with sulfuric acid in methanol. What is the stereostructure of the product?

An important variation of the ring opening of epoxides with bases is the reaction with organometallic reagents. For example, the sodium salts of 1-alkynes react readily with oxirane itself to give alk-3-yn-1-ols.

$$\underset{\text{1-butyne}}{CH_3CH_2C\equiv CH} \xrightarrow[NH_3]{NaNH_2} CH_3CH_2C\equiv C^-\ Na^+ \xrightarrow{H_2C\overset{O}{\frown}CH_2} \xrightarrow{H} \underset{\text{(80\%) hex-3-yn-1-ol}}{CH_3CH_2C\equiv CCH_2CH_2OH}$$

The ring opening is an S_N2 reaction in which a carbanion is the nucleophile. This reaction does not occur with ordinary ethers because the RO^- group is normally a poor leaving group; indeed, diethyl ether and tetrahydrofuran are common solvents for these reactions.

A similar reaction occurs with Grignard reagents. The reaction of an organometallic reagent with oxirane is a useful synthetic transformation because *the carbon chain is extended by two carbons in one step*.

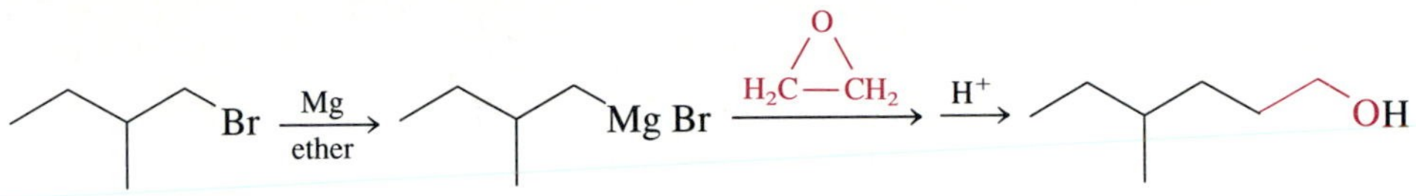

Br $\xrightarrow[\text{ether}]{\text{Mg}}$ Mg Br $\xrightarrow{H_2C—CH_2 \text{ (O)}}$ $\xrightarrow{H^+}$ OH

1-bromo-2-methylbutane

(68%)
4-methyl-1-hexanol

EXERCISE 10.23 Show how isobutyl alcohol can be converted into 1-bromo-4-methylpentane.

B. Higher Cyclic Ethers

One of the most important of the cyclic ethers is tetrahydrofuran, THF, a material we have encountered frequently as a solvent.

O

tetrahydrofuran
THF

Like many other ethers, tetrahydrofuran reacts with oxygen to form dangerous peroxides. The five-membered ring is stable to ring opening, but ring opening can be accomplished under conditions that cause ether-cleavage reactions of open-chain compounds.

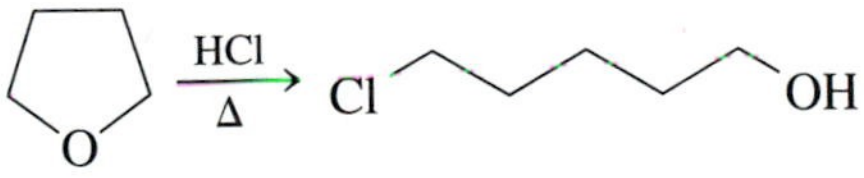

O $\xrightarrow[\Delta]{\text{HCl}}$ Cl OH

A group of large-ring polyethers that have attracted a good deal of recent attention is the **crown ethers.** The compounds are cyclic polymers of ethylene glycol, $(OCH_2CH_2)_n$, and are named in the form x-crown-y, where x is the total number of atoms in the ring and y is the number of oxygens. An example is 18-crown-6, the cyclic hexamer of ethylene glycol.

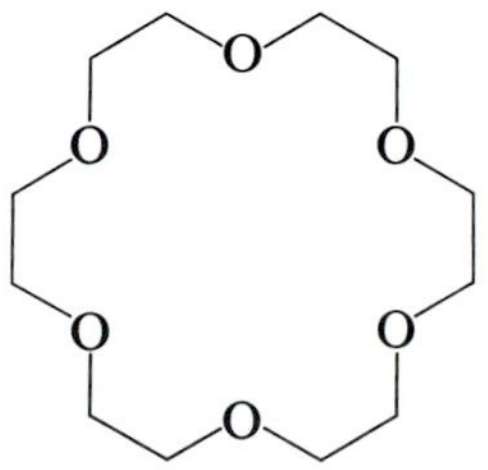

O O O O O O

18-crown-6

EXERCISE 10.24 Write the structure of 15-crown-5.

The crown ethers are important for their ability to solvate cations strongly. The six oxygens in 18-crown-6 are ideally situated to solvate a potassium cation, just as water molecules would normally do. The interior of the complex contains the oxygen-solvated cation, but the exterior has hydrocarbon properties. As a result, the complexed ion is soluble in nonpolar organic solvents. For example, the complex of 18-crown-6 and potassium permanganate is soluble in benzene.

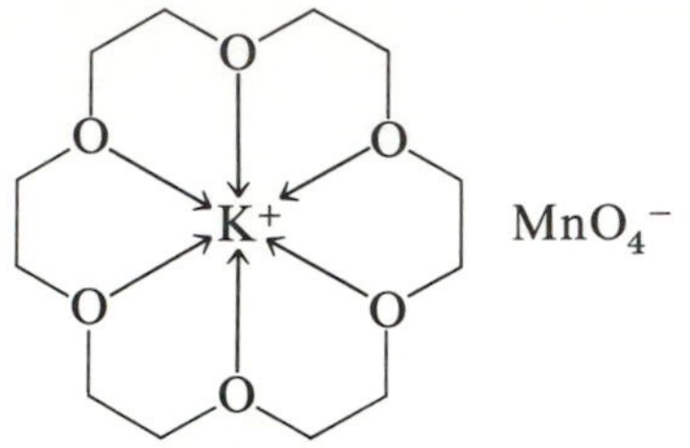

MnO_4^-

18-crown-6–$KMnO_4$ complex
soluble in benzene

18-Crown-6 may be prepared by treating a mixture of triethylene glycol and the corresponding dichloride with aqueous KOH.

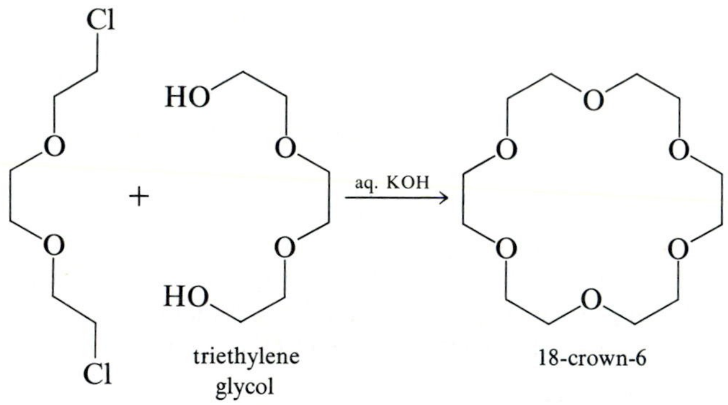

aq. KOH

triethylene glycol

18-crown-6

The reaction mechanism is S_N2 and involves two successive displacements with chloride ion being the leaving group. Even though a large ring is being formed, the reaction need not be carried out at high dilution. After the initial alkylation, the potassium cation apparently acts as a "template" to bring the two reacting ends of the long chain close together for rapid reaction.

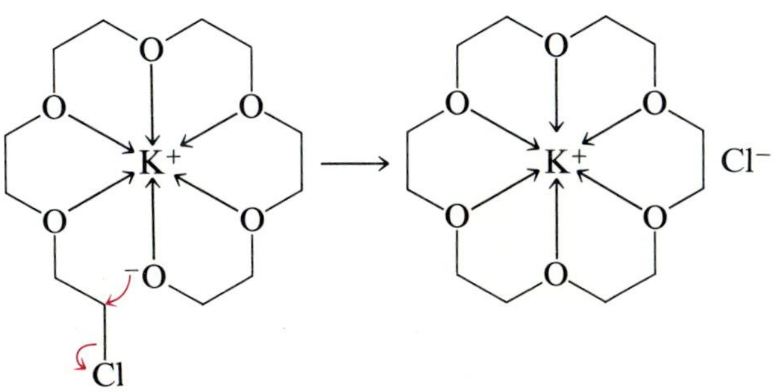

Cl^-

In the last 25 years, organic chemists have actively investigated the chemistry of synthetic crown ethers. Hundreds of different examples have been synthesized and their properties studied. A relationship has been discovered between the structure of a crown ether and its ability to complex various cations. For example, 18-crown-6 shows a high affinity for K^+, 15-crown-5 for Na^+, and 12-crown-4 for Li^+. Measurement of molecular models of these three molecules reveals that the "cavity size," is in each case a good match for the ionic diameter of the cation most strongly bound by the molecule.

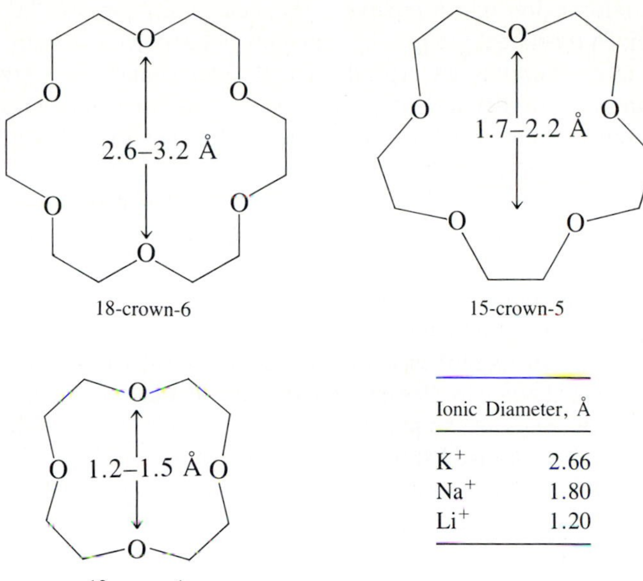

18-crown-6

15-crown-5

12-crown-4

	Ionic Diameter, Å
K^+	2.66
Na^+	1.80
Li^+	1.20

A number of naturally occurring cyclic compounds are now known with oxygens or nitrogens in the ring that coordinate with metal cations. Some of these compounds are involved in the transport of ions across biological membranes. An example is nonactin, an antibiotic that functions by transporting potassium ions out of cells. Cell rupture eventually occurs. Nonactin favors complexation with potassium over sodium by a factor of ten; four tetrahydrofuran and four carbonyl oxygens solvate the potassium. A related noncyclic antibiotic, monensin, widely used as a feed additive for animals, complexes with sodium ten times better than potassium via six ether oxygens. The utility of open-chain polyethers such as glyme and diglyme as reaction solvents depends in part on their ability to complex cations.

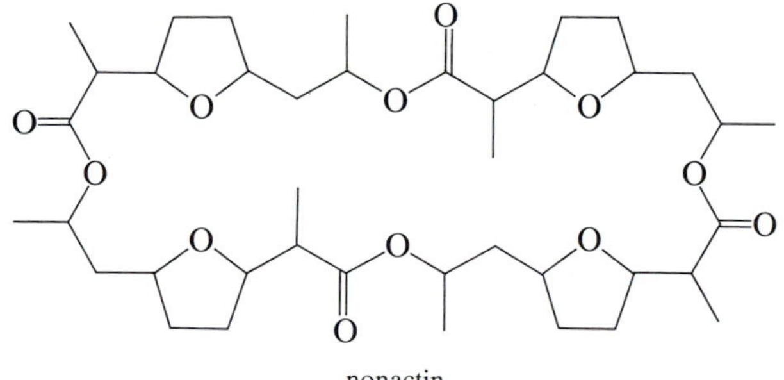

nonactin

Highlight 10.6

An ether oxygen acts as a Lewis base toward metal ions, making ethers effective and chemically inert solvents for the preparation of organometallic compounds such as RMgX (Section 8.6). Crown ethers are constructed so that the interaction of the oxygens is optimized to take the place of water oxygens around a metal ion. This is **molecular recognition,** a subject taken up in Chapter 33.

10.12 Multistep Synthesis

We have seen in our study so far that organic compounds may conveniently be classified according to **functional groups,** and that such a classification aids in organizing a massive amount of information into a relatively few categories. One of the characteristic properties of a given functional group is its "qualitative chemical reactivity." By this we mean the *kinds* of reaction that are typical for molecules containing a given functional group. For example, two chemical properties of the class of compounds that we call alcohols are the replacement of the hydroxy group by halogen to give alkyl halides, and the reaction with oxidizing agents to give aldehydes and ketones. Of course, every chemical reaction is at the same time a characteristic property of one functional group and a **method of preparation** of another. For example, the reaction of a secondary alcohol with chromium(VI) is a property of secondary alcohols and a method of preparation of ketones.

Because of the ability of carbon to form chains and rings, and because of the different functional groups that exist, the number of possible organic structures that may exist is virtually limitless. Indeed, more than *ten million* different substances, most of them organic compounds, were known by 1990. Of course, only a relatively small number (about 30,000) of these compounds can be purchased from companies that sell chemicals. Most of the rest have been **synthesized** from other compounds.

One of the powerful aspects of organic chemistry is that the chemical reactions characteristic of various classes of compounds may be carried out in sequence to synthesize almost any desired structure, *whether or not that compound has ever existed before!* As a simple example, suppose that we wish to have a sample of the ketone 2-butanone, and that the only chemicals we have are alkanes and inorganic reagents. We saw in Section 10.6 that ketones are produced by the oxidation of secondary alcohols. Thus, if we had the secondary alcohol 2-butanol, we might oxidize it to prepare 2-butanone.

$$\underset{\text{2-butanone}}{CH_3\overset{\overset{\displaystyle O}{\|}}{C}CH_2CH_3} \xleftarrow{[O]} \underset{\text{2-butanol}}{CH_3\overset{\overset{\displaystyle OH}{|}}{C}HCH_2CH_3}$$

Now our problem changes to how to obtain 2-butanol. We have also learned, in Sections 9.1 and 10.5, that alcohols may be prepared by the hydrolysis of alkyl halides. Thus, we might hydrolyze a 2-halobutane as a method of preparing 2-butanol.

$$CH_3\overset{\overset{\displaystyle OH}{|}}{C}HCH_2CH_3 \xleftarrow{H_2O} CH_3\overset{\overset{\displaystyle X}{|}}{C}HCH_2CH_3$$

But where do we obtain the 2-halobutane? We saw in Chapter 6 that alkyl halides may be prepared by the free-radical halogenation of alkanes. Thus, a possible method of preparation of 2-halobutanes is by halogenation of butane itself.

$$CH_3\overset{\overset{\displaystyle X}{|}}{C}HCH_2CH_3 \xleftarrow{X_2} CH_3CH_2CH_2CH_3$$

Since our storeroom contains butane, the foregoing **multistep synthesis** plan represents a possible way for us to obtain 2-butanone.

Of course, at this point, the plan has been sketched out in only rough form. To evaluate whether the plan is really workable, it is necessary to consider the details of the transformations we have proposed. The last step in our proposed multistep synthesis presents no problem, since the oxidation of secondary alcohols by chromium(VI) reagents is a reaction that is not subject to any real limitations.

$$\mathrm{CH_3\underset{\displaystyle |}{\overset{\displaystyle OH}{C}}HCH_2CH_3 \xrightarrow{PCC} CH_3\overset{\displaystyle O}{\overset{\|}{C}}CH_2CH_3}$$

However, the proposed hydrolysis of a 2-halobutane to 2-butanol may be tricky. We saw in Sections 9.6 and 10.5 that hydrolysis of secondary alkyl halides is accompanied by more or less elimination. In many cases, this competing elimination reaction is the main reaction.

$$\mathrm{CH_3\overset{\displaystyle X}{\overset{|}{C}}HCH_2CH_3 \xrightarrow[H_2O]{OH^-} CH_2{=}CHCH_2CH_3 + CH_3CH{=}CHCH_2}$$

However, we also learned that, by using the less basic nucleophile sodium acetate, elimination may be minimized. The initial product of this substitution is an ester, which can be hydrolyzed to obtain the alcohol.

$$\mathrm{CH_3\overset{\displaystyle X}{\overset{|}{C}}HCH_2CH_3 \xrightarrow{CH_3CO_2^-} CH_3\overset{\displaystyle O\overset{\displaystyle O}{\overset{\|}{C}}CH_3}{\overset{|}{C}}HCH_2CH_3 \xrightarrow[H_2O]{OH^-} CH_3\overset{\displaystyle OH}{\overset{|}{C}}HCH_2CH_3}$$

Thus, 2-butanol is a viable intermediate, *if* we can obtain a 2-halobutane. Here we must consider the fine points of the free-radical halogenation process. We saw in Section 6.3 that chlorination is a rather indiscriminate reaction; the relative reactivity of secondary and primary carbon-hydrogen bonds is only 4:1. Since butane contains four secondary carbon-hydrogen bonds and six primary carbon-hydrogen bonds, we can expect that chlorination of butane would give something on the order of a 16:6 ratio of 2-chlorobutane and 1-chlorobutane. Thus, the maximum yield of 2-chlorobutane would be about 70%. In addition, we would be faced with the problem of *separating* the two isomers.

$$\mathrm{CH_3CH_2CH_2CH_3 \xrightarrow[h\nu]{Cl_2} \underset{\cong 73\%}{CH_3CH_2\overset{\displaystyle Cl}{\overset{|}{C}}HCH_3} + \underset{\cong 27\%}{CH_3CH_2CH_2CH_2Cl}}$$

A better solution would be to use bromine, rather than chlorine, as the halogenating agent. We saw in Section 6.3 that bromine is much more selective than chlorine, and we could expect the vapor phase bromination to give almost completely 2-bromobutane.

$$\mathrm{CH_3CH_2CH_2CH_3 \xrightarrow[h\nu]{Br_2} \underset{>95\%}{CH_3\overset{\displaystyle Br}{\overset{|}{C}}HCH_2CH_3} + \underset{<5\%}{CH_3CH_2CH_2CH_2Br}}$$

Thus, the multistep synthesis of 2-butanone that we have sketched out holds up to closer scrutiny. The full plan, complete with reagents, is summarized as follows.

$$\mathrm{CH_3CH_2CH_2CH_3 \xrightarrow[h\nu]{Br_2} CH_3\overset{\displaystyle Br}{\overset{|}{C}}HCH_2CH_3 \xrightarrow{CH_3CO_2^-}}$$

$$\mathrm{CH_3\overset{\displaystyle O\overset{\displaystyle O}{\overset{\|}{C}}CH_3}{\overset{|}{C}}HCH_2CH_3 \xrightarrow[H_2O]{NaOH} CH_3\overset{\displaystyle OH}{\overset{|}{C}}HCH_2CH_3 \xrightarrow{PCC} CH_3\overset{\displaystyle O}{\overset{\|}{C}}CH_2CH_3}$$

Of course, this is a highly simplified and somewhat artificial example of a multistep synthesis. Actually, 2-butanone *is* available from chemical supply houses; in fact, it would be found in any well-stocked storeroom. Furthermore, it is not likely that a chemist would actually carry out a free-radical bromination reaction as part of a multistep synthesis, since this kind of reaction is relatively inconvenient for normal laboratory operation. However, the principle is illustrative of how real-life multistep syntheses are planned. Note that we began by *working backwards* from our desired final product toward the simpler starting materials we had in our storeroom. At first we sketched out the plan qualitatively. After we had a rough plan, we thought about the details of each step and made suitable modifications until we arrived at a plan that appeared practical.

EXERCISE 10.25 Plan a synthesis of each of the following compounds, starting with alkanes and employing any inorganic reagents necessary.

(a) [cyclohexanone structure] (b) CH_3CH_2CN (c) [cyclopentene structure]

ASIDE

The fine black powder, usually antimony trisulfide, used to emphasize the eyelids of the women of ancient Egypt was called in Arabic *al-koh'l,* related to the Hebrew word for stain, *kakhal*. By extension, a fine powder produced by sublimation was called an alcohol. Sublimation produced the essence or ''spirit'' of a substance, which then became known as an alcohol. The spirit of wine was the alcohol of wine. Eventually the substance ethanol was called alcohol, the prototype of the compound class alcohols. In the introduction (Chapter 1) we noted that ethanol was referred to as *aqua vitae,* water of life, anticipating its chemical relationship to water.

Give the names and structures for ten alcohols. What compounds are formed if we replace the O—H hydrogen with an alkyl group? Give a chemical method for executing this conversion.

PROBLEMS

1. Give the structure corresponding to each of the following common names.
a. isobutyl methyl ether
b. neopentyl alcohol
c. tetrahydrofuran (THF)
d. dioxane
e. 18-crown-6
f. bis-β-bromopropyl ether
g. *sec*-butyl alcohol
h. isobutyl alcohol

2. Give the structure corresponding to each of the following IUPAC names.
a. 3-ethoxy-2-methylhexane
b. 4-methyl-2-pentanol
c. 4-*t*-butyl-3-methoxyheptane
d. (1*S*,3*R*)-3-methylcyclohexanol
e. (2*R*,3*R*)-dimethyloxirane
f. 1,6-hexanediol

3. Give the IUPAC name corresponding to each of the following structures.

a. 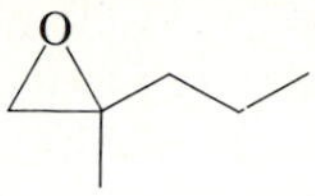 O

b. 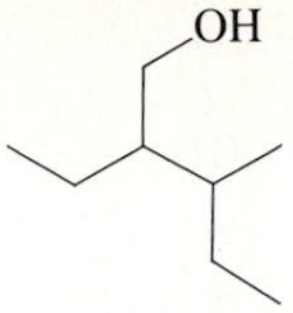 OH

c. 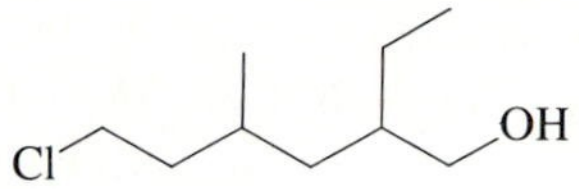 Cl OH

d. 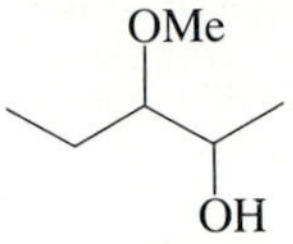 OMe OH

e. 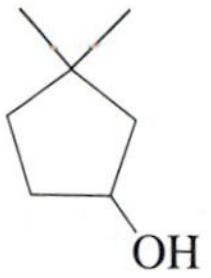 OH

f. 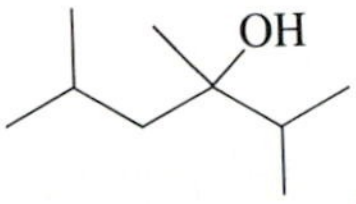 OH

g. 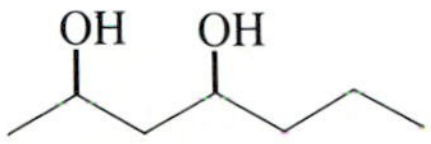 OH OH

h. 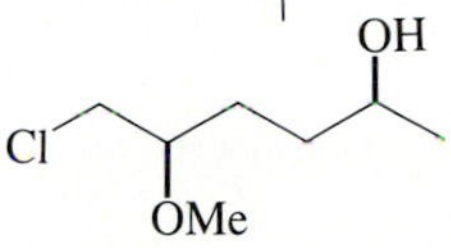 Cl OH OMe

4. Explain why each of the following is not a correct name.

a. 4-hexanol
b. 2-hydroxy-3-methylhexane
c. 3-(hydroxymethyl)-1-hexanol
d. 2-isopropyl-1-butanol

5. How many isomeric ethers correspond to the molecular formula $C_5H_{12}O$? Give common and IUPAC names for each structure. Which of these ethers is capable of optical activity? Write out the structures of the two mirror images and show that they are not superimposable.

6. There are 17 isomeric alcohols of the formula $C_6H_{13}OH$. Write out the structure and give the IUPAC name of each one. Identify the primary, secondary, and tertiary alcohols.

7. a. In a popular handbook the compound $CH_3CH_2C(CH_3)_2OCH_3$ is listed as ether, *sec*-butylmethyl, 2-methyl-. Is this name consistent with any approved nomenclature you have studied? Give correct common and IUPAC names.

b. Note the resemblance in shape of this ether to 3,3-dimethylpentane. Compare the boiling points of the two compounds as given in a handbook.

c. 2,2-Dimethyl-3-pentanol and *t*-butyl isopropyl ether are isomers that are not listed in common handbooks. How would you expect their boiling points to compare?

8. Give the principal product(s) from each of the following reactions.

a. $CH_3CH_2CH_2OH \xrightarrow[130°C]{H_2SO_4}$

b. $CH_3CH_2CH_2O^- + (CH_3)_3CCl$

c. $(CH_3)_3CO^- + CH_3CH_2CH_2Br$

d. $CH_3CH_2CH_2OH + K_2Cr_2O_7 + H_2SO_4 \xrightarrow{50°C}$

e. 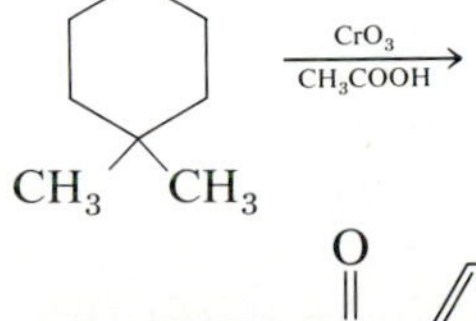 4,4-dimethylcyclohexanol (H, OH; CH_3, CH_3) $\xrightarrow[CH_3COOH]{CrO_3}$

f. 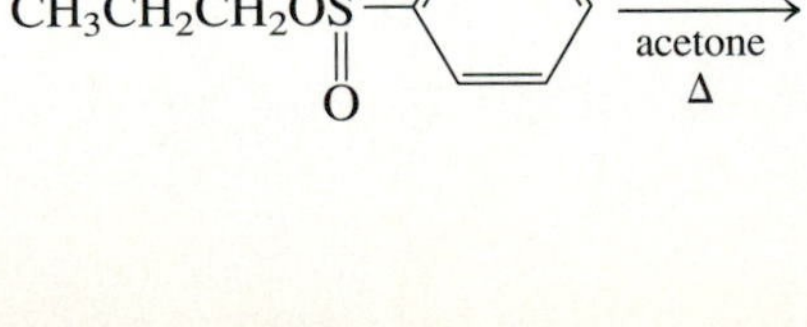 $CH_3CH_2CH_2OS(=O)_2-C_6H_5 \xrightarrow[\text{acetone}, \Delta]{NaI}$

g. $CH_3CH_2CH_2OH \xrightarrow[H_2SO_4,\ \Delta]{HBr}$

h. $(CH_3CH_2)_3CCH_2OH \xrightarrow[H_2SO_4,\ \Delta]{HBr}$

$\quad\quad\quad\quad\quad OH$

i. (S)-$CH_3CHCH_2CH_3 + SOCl_2 \xrightarrow{\text{pyridine}}$

j. $(CH_3CH_2)_2CHOH + KH \longrightarrow$

k. $(CH_3CH_2CH_2)_2O + HI$ (excess) $\xrightarrow{\Delta}$

l. $CH_3CH_2CHCH_2CH_3 + HBr$ (excess) $\xrightarrow{\Delta}$

$\quad\quad\quad\quad\quad OCH_3$

9. Give the reagents and conditions for the best conversions of alcohol to alkyl halide as shown.

a. $CH_3CH_2CH_2CH_2OH \longrightarrow CH_3CH_2CH_2CH_2Cl$

$\quad\quad\quad\quad\quad CH_3 \quad\quad\quad\quad\quad\quad\quad\quad\quad\quad CH_3$

b. $CH_3CH_2CHCH_2OH \longrightarrow CH_3CH_2CHCH_2Cl$

$\quad\quad\quad\quad\quad CH_3 \quad\quad\quad\quad\quad\quad\quad\quad\quad\quad CH_3$

c. $CH_3CH_2CHCH_2OH \longrightarrow CH_3CH_2C{-}CH_3$

$\quad\quad\quad\quad\quad\quad\quad\quad\quad\quad\quad\quad\quad\quad\quad\quad\quad\quad Cl$

d. $CH_3CH_2CH_2CHOHCH_3 \longrightarrow CH_3CH_2CH_2CHICH_3$

e. CH_2OH, H $\longrightarrow$ CH_2Br, H

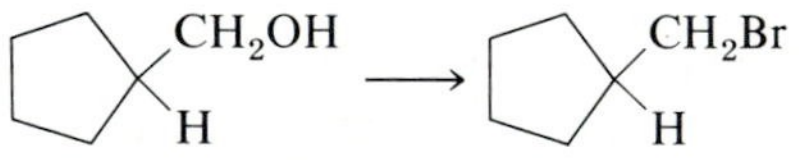

f. HO CH_3 $\longrightarrow$ Cl CH_3

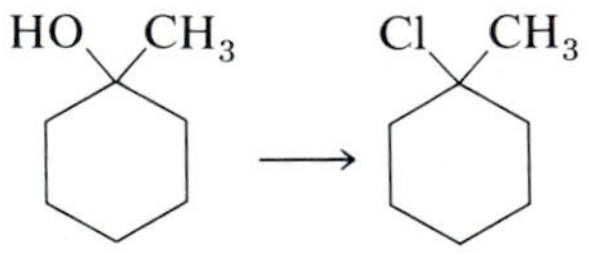

10. Show how to accomplish each of the following conversions in a practical manner.

a. $(CH_3)_3CCH_2OH \longrightarrow CH_3CH_2CCl(CH_3)_2$
b. $(CH_3)_2CHCHOHCH_3 \longrightarrow (CH_3)_2CHCHBrCH_3$
c. $(CH_3CH_2)_3COH \longrightarrow (CH_3CH_2)_2C{=}CHCH_3$
d. $CH_3CH_2CH_2CH_2OH \longrightarrow CH_3CH_2CH_2CH_2CN$
e. $(CH_3)_3CCl + CH_3I \longrightarrow (CH_3)_3COCH_3$
f. $(CH_3)_2CHCH_2OH \longrightarrow (CH_3)_2CHCH_2SCH_3$
g. $(CH_3)_3COH \longrightarrow (CH_3)_3CCH_2CH_2OH$

11. Outline multistep syntheses of the following compounds, starting with alkanes and using any necessary inorganic reagents.

a. methanol
b. methyl cyclohexyl ether
c. propene, $CH_3CH{=}CH_2$
d. ethanethiol, CH_3CH_2SH

12. A naive graduate student attempted the preparation of $CH_3CH_2CDBrCH_3$ from $CH_3CH_2CDOHCH_3$ by heating the deuterio alcohol with HBr and H_2SO_4. He obtained a product having the correct boiling point, but a careful examination of the spectral properties by his research director showed that the product was a mixture of $CH_3CHDCHBrCH_3$ and $CH_3CH_2CDBrCH_3$. What happened?

13. Explain why 2-cyclopropyl-2-propanol reacts with HCl to give 2-chloro-2-cyclopropylpropane instead of 1-chloro-2,2-dimethylcyclobutane.

14. Write balanced equations for the following reactions.

a. $(CH_3)_2CHOH + KMnO_4 + H^+ \longrightarrow (CH_3)_2C{=}O + MnO_2 + K^+ + H_2O$

b. $ClCH_2CH_2CH_2OH + HNO_3 \longrightarrow ClCH_2CH_2CO_2H + NO_2 + H_2O$

c. 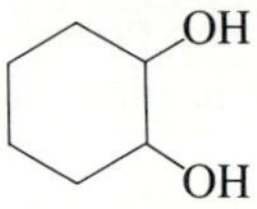 (cyclohexane-1,2-diol: ring with OH, OH) $+ K_2Cr_2O_7 + H^+ \longrightarrow HO\overset{O}{\overset{\|}{C}}(CH_2)_4\overset{O}{\overset{\|}{C}}OH + Cr^{3+} + K^+ + H_2O$

15. a. Although isobutyl alcohol reacts with HBr and H_2SO_4 to give isobutyl bromide, with little rearrangement, 3-methyl-2-butanol reacts on heating with conc. HBr to give only the rearranged product, 2-bromo-2-methylbutane. Explain this difference using the reaction mechanisms involved.

b. Unsymmetrical ethers are generally not prepared by heating two alcohols with sulfuric acid. Why not? Yet, when *t*-butyl alcohol is heated in methanol containing sulfuric acid, a good yield of *t*-butyl methyl ether results. Explain this result by means of the reaction mechanism.

c. Prolonged reaction of ethyl ether with HI gives ethyl iodide. Write out the reaction mechanism.

16. *n*-Butyl ethyl ether is cleaved by hot conc. HBr to give both ethyl bromide and *n*-butyl bromide. However, *t*-butyl ethyl ether is cleaved readily by cold conc. HBr to give almost completely *t*-butyl bromide and ethyl alcohol. Write the reaction mechanisms of both reactions, showing all intermediates, and explain briefly how the reaction mechanisms relate to these experimental observations.

17. Triethyloxonium cation can be prepared as a crystalline tetrafluoroborate salt, $(C_2H_5)_3O^+\ BF_4^-$, which is appreciably soluble in methylene chloride. Write the Lewis structure of this salt. The compound is a reactive ethylating reagent and reacts with alcohols, for example, to yield ethers.

$$ROH + (CH_3CH_2)_3O^+\ BF_4^- \longrightarrow ROC_2H_5 + (C_2H_5)_2O + HBF_4$$

Write out the mechanism of the reaction. Why is the reagent so reactive in this process?

18. 1-Butanol-*1-d*, $CH_3CH_2CH_2CHDOH$, has a relatively small optical activity, $[\alpha]_D = 0.5°$, due solely to a difference between hydrogen isotopes. The (−) enantiomer has been shown to have the *R* configuration. On treatment with thionyl chloride and pyridine, (−)-1-butanol-*1-d* gives (+)-1-chlorobutane-*1-d*. What is the configuration of the chloride? Draw perspective diagrams of the two compounds. Show how (*R*)-$CH_3CH_2CH_2CHDOH$ may be converted to (*S*)-$CH_3CH_2CH_2CHDOCH_3$.

19. Provide a mechanistic rationalization for the following reaction course.

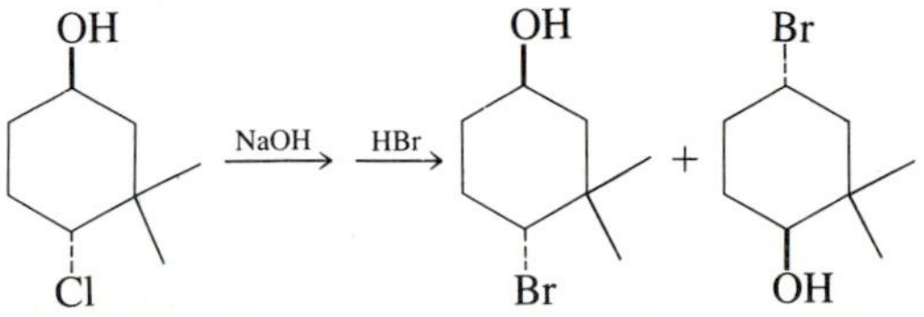

(cyclohexane with OH, Cl and *tert*-butyl substituents) $\xrightarrow{NaOH}$ $\xrightarrow{HBr}$ (cyclohexane with OH, Br and *tert*-butyl) + (cyclohexane with Br, OH and *tert*-butyl)

20. Optically active (2*R*,3*S*)-3-chloro-2-butanol is allowed to react with sodium hydroxide in ethanol to give an optically active oxirane, which is treated with potassium hydroxide in water to obtain 2,3-butanediol. What is the stereostructure of the diol? What can you say about its optical rotation?

21. Optically active 5-chloro-2-hexanol is allowed to react with KOH in methanol. The product, $C_6H_{12}O$, is found to have $[\alpha]_D = 0$. What can you say about the stereochemistry of the reactant?

22. 18-Crown-6 is a useful catalyst for some reactions of potassium salts. For example, KF reacts with 1-bromooctane to give 1-fluorooctane much faster in the presence of 10 mole % 18-crown-6 than in its absence. However, the crown ether is ineffective as a catalyst if NaF is used. Explain.

23. Treatment of 1,2,2-trimethylcycloheptanol with sulfuric acid gives a mixture of 1,7,7-trimethylcycloheptene, 1-*t*-butylcyclohexene, and 1-isopropenyl-1-methylcyclohexane. Write a mechanism that accounts for the formation of these products.

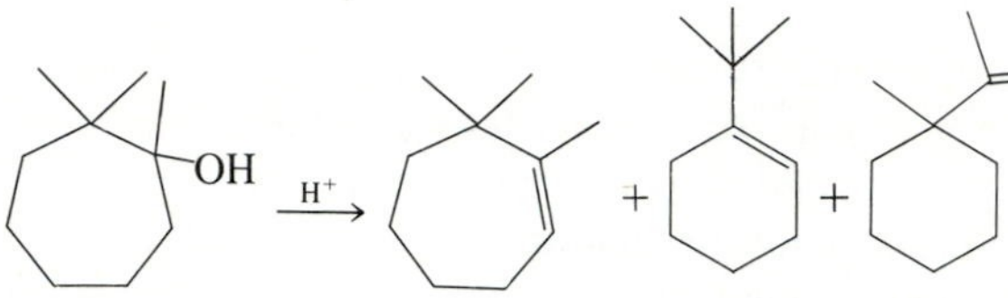

OH $\xrightarrow{H^+}$ + +

24. The thermodynamics of reactions in a solution are often quite different from those in the gas phase because of the importance of solvation energies. However, thermodynamic data for solvents other than water are sparse, whereas many heats of formation are now available for the gas phase. If one considers reactions in which the number of ions or ion pairs remains the same, the relative gas-phase values can be instructive. Compare $\Delta H°$ for the following two reactions in the gas phase.

$$CH_3CH_2OH + Br^- \rightleftharpoons CH_3CH_2Br + OH^-$$
$$CH_3CH_2OH + HBr \rightleftharpoons CH_3CH_2Br + H_2O$$

25. Propose a mechanism for the following reaction.

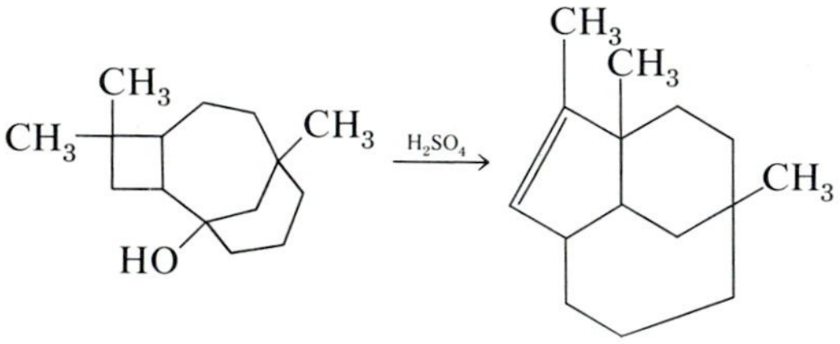

CH_3 CH_3 CH_3 HO $\xrightarrow{H_2SO_4}$ CH_3 CH_3 CH_3

26. Diisopropyl ether was left in a loosely stoppered bottle in a chemistry stockroom for some years. A clerk moved the bottle when looking for another material, whereupon the contents violently exploded, injuring the clerk and starting a serious fire. What was the chemistry underlying the accident? What precautions should have been taken?

CHAPTER 11

ALKENES

An alkene is a hydrocarbon with a carbon-carbon double bond. The double bond is a stronger bond than a single bond, yet paradoxically the carbon-carbon double bond is much more reactive than a carbon-carbon single bond. Unlike alkanes, which generally show rather nonspecific reactions, the double bond is the site of many specific reactions and is a functional group.

11.1 Electronic Structure

The geometric structure of ethylene, the simplest alkene, is well known from spectroscopic and diffraction experiments and is shown in Figure 11.1. The entire molecule is planar. In the Lewis structure of ethylene, the double bond is characterized as a region with two pairs of electrons.

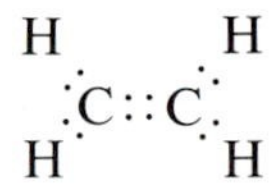

```
H       H
 :C::C:
H       H
```

In molecular orbital descriptions, we need one orbital for each pair of electrons. Hence, we need two orbitals between the carbons to accommodate the electrons. The orbital model of ethylene starts with the two sp^2-hybrids from each carbon to the hydrogens. A third sp^2-hybrid on each carbon is used to form a C_{sp^2}—C_{sp^2} single bond. So far,

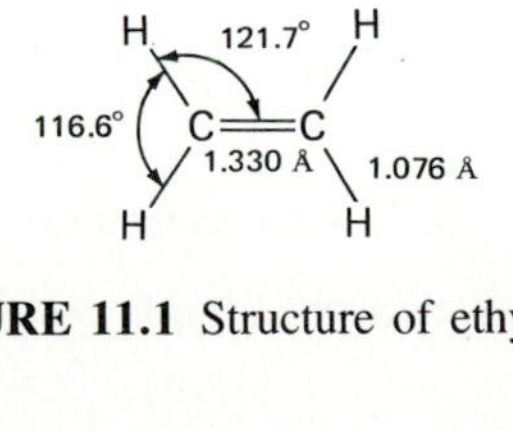

FIGURE 11.1 Structure of ethylene.

we have used the *s*-orbital and two of the *p*-orbitals of each carbon—each carbon has a *p*-orbital "left over." These *p*-orbitals are perpendicular to the plane of the six atoms. They are parallel to each other and have regions of overlap above and below the molecular plane. This type of bond, in which there are two bonding regions above and below a nodal plane, is called a π-bond. This notation is used in order to distinguish it from the type of bond formed by overlap of two carbon sp^2-orbitals. Such a bond has no node and is called a σ-bond (Figure 11.2). This orbital picture of ethylene is shown in Figure 11.3.

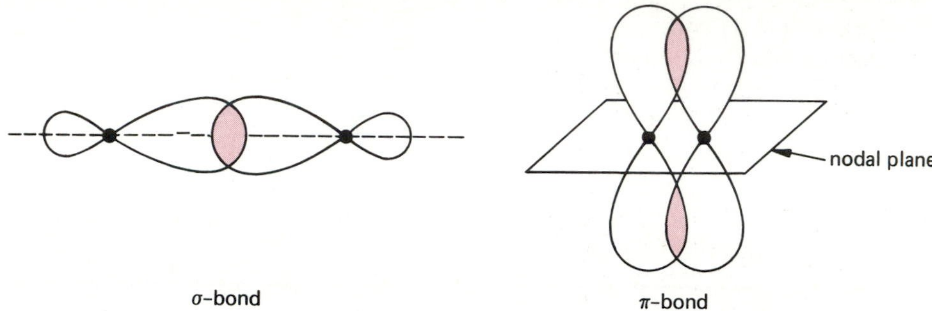

FIGURE 11.2 σ- and π-bonds.

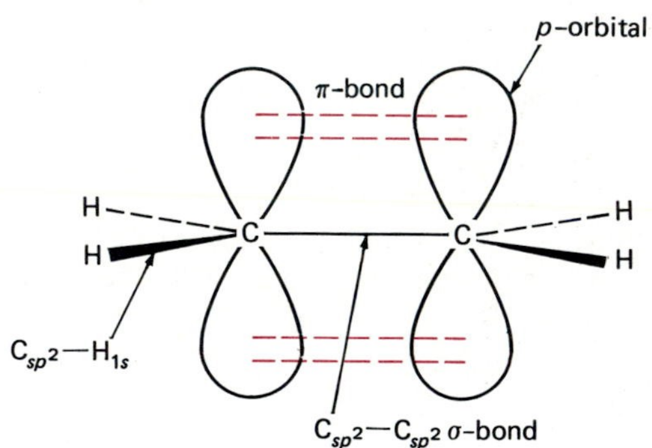

FIGURE 11.3 σ-π-bond model of ethylene.

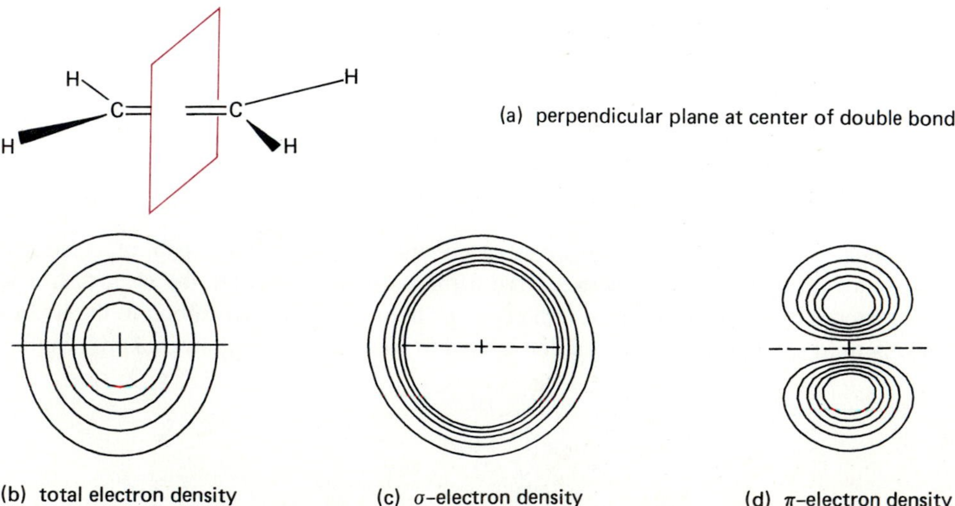

FIGURE 11.4 Electron-density distribution in center of ethylene double bond.

An interesting view of the electron density in the ethylene double bond is seen in Figure 11.4. Each contour diagram represents the electron density of a slice cut through the midpoint of the carbon-carbon bond (see plane in Figure 11.4a). Each line in such a contour diagram represents a constant definite value of the electron density. For example, the innermost oval in Figure 11.4b represents a rather high value of the electron density, and the outermost oval has a rather low value. The σ-electron density is almost cylindrically symmetric but not quite. The outermost contours are oval rather than circular because of electron repulsion by the π-electrons. The π-electron density is much less than the σ-density and decreases to zero in the midpoint plane (Figure 11.4d). However, the total electron density in Figure 11.4b has a smooth oval character. This total electron distribution does not have the appearance one might expect from the simple representation in Figure 11.3.

The noncylindrically symmetric distribution of electron density about the carbon-carbon π-bond axis has an important consequence; there is a barrier to rotation about this axis. For example, two dideuterioethylenes are known and can be distinguished by their different spectroscopic properties.

two different dideuterioethylenes

Interconversion of these isomers takes place only at high temperatures (≅500 °C) and has an activation energy of 65 kcal mole^{-1}. The transition state for the reaction has a half-twisted structure in which the p-orbitals have zero overlap. This structure is represented in Figure 11.5.

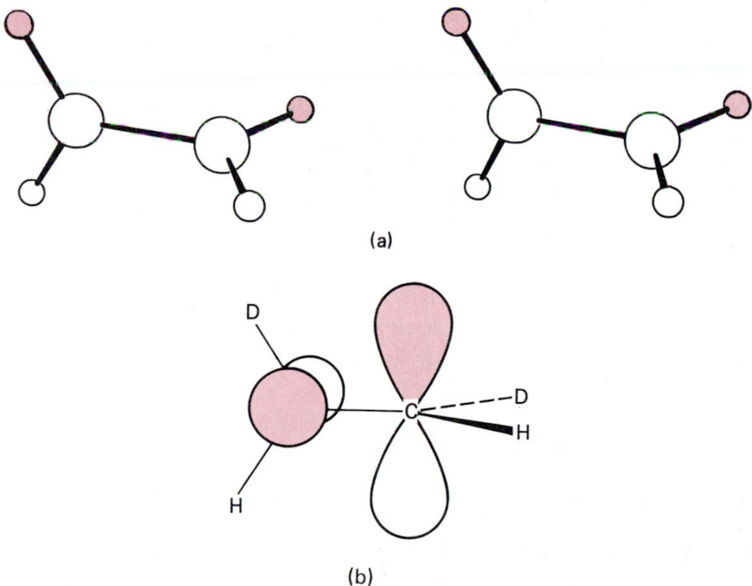

FIGURE 11.5 The half-twisted transition state for interconversions of dideuterioethylenes. (a) The structure is shown in the stereo plot. (b) The orbital representation shows one electron in each of the noninteracting π-orbitals, which are at right angles to each other.

The two forms of dideuterioethylene represent another case of stereoisomerism (Chapter 7). Recall that stereoisomers are compounds having the same sequence of bonded atoms but differing in the orientation of the atoms in space. In older books, one often finds alkene isomers such as these classified as geometric isomers. However, since they are stereoisomers that *do not* have a mirror image relationship to one another, they are actually diastereomers.

The two dideuterioethylene stereoisomers may be interconverted by rotation about a bond, just as the anti and gauche stereoisomers of butane.

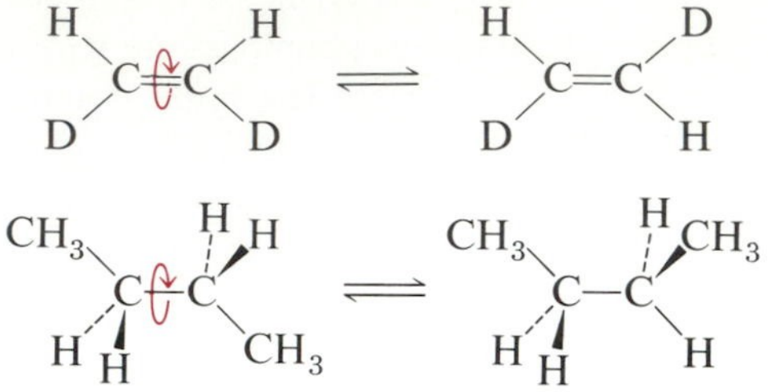

$$\mathrm{HDC{=}CDH} \rightleftharpoons \mathrm{HDC{=}CHD}$$

$$\mathrm{CH_3CH_2{-}CH_2CH_3} \rightleftharpoons \mathrm{CH_3CH_2{-}CH_2CH_3}$$

However, in the first case the barrier to interconversion (the rotational barrier) is 65 kcal mole^{-1}, and in the second it is only 3.3 kcal mole^{-1}. The dideuterioethylene stereoisomers can be obtained separately, and each isomer is perfectly stable at normal temperatures. On the other hand, the anti and gauche butane stereoisomers interconvert easily at temperatures above about −230 °C. This difference in the ease of interconversion has resulted in the two types of stereoisomers having different names. Stereoisomers that can be easily interconverted by rotation about a bond are called **conformational isomers.** Stereoisomers that are not easily interconverted are called **configurational isomers.** The dideuterioethylenes are two such configurational isomers.

EXERCISE 11.1 How many stereoisomers exist for each of the following compounds? Write their structures.

(a) 1,4-dideuterio-1,3-butadiene, $CHD{=}CH{-}CH{=}CHD$
(b) 3-chloro-1-butene, $CH_2{=}CHCHClCH_3$
(c) 1,3-dichloro-1-butene, $CHCl{=}CHCHClCH_3$

11.2 Nomenclature

Historically, hydrocarbons with a double bond were known as **olefins.** This rather strange class name comes from the French *olefiant,* based on the Latin words *oleum,* an oil, and *ficare,* to make. It arose because derivatives of such compounds often had an oily appearance.

As with other classes of organic compounds, two systems of nomenclature are used: common and systematic. In the common system, which is only used for fairly simple compounds, the final **-ane** of the alkane name is replaced by **-ylene.**

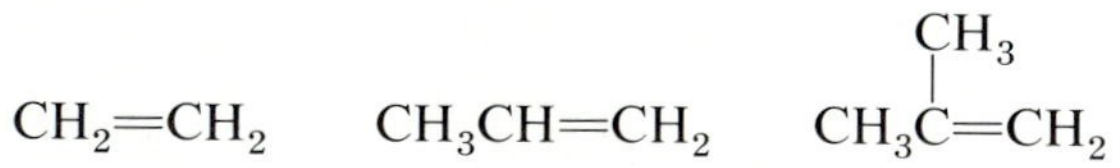

$CH_2{=}CH_2$ ethylene — $CH_3CH{=}CH_2$ propylene — $CH_3C(CH_3){=}CH_2$ isobutylene

A few simple molecules are named as derivatives of ethylene.

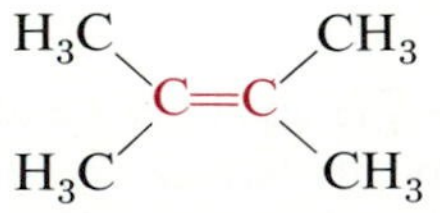

$(CH_3)_2C{=}C(CH_3)_2$ tetramethylethylene — $Cl_2C{=}CHCl$ trichloroethylene

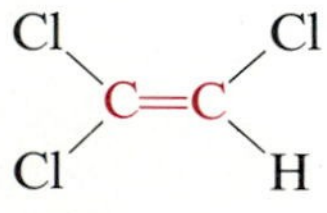

Configurational isomers are distinguished by the use of the prefixes *cis*- (L., on this side) and *trans*- (L., across), as with disubstituted cycloalkanes (page 138).

Br, Br / C=C / H, H	H, Br / C=C / Br, H
cis-dibromoethylene	*trans*-dibromoethylene

Some monosubstituted ethylenes are named as radical combinations in which the $CH_2{=}CH{-}$ group is called **vinyl.**

$CH_2{=}CHCl$	$CH_2{=}CHOCH_3$
vinyl chloride	methyl vinyl ether

In the IUPAC system alkenes are named as derivative of a parent alkane. The **alk-** stem specifies the number of carbons in the chain and the **-ene** suffix specifies a double bond. A number is used to indicate the position of the double bond along the chain. Since the double bond joins one carbon to a carbon with the next higher number, only one number need be given. Finally, a prefix *cis*- or *trans*- is included where necessary.

$CH_2{=}CH_2$	$CH_3CH{=}CH_2$	$CH_3CH_2CH{=}CH_2$
ethene	propene	1-butene

trans-2-butene

cis-2-pentene

cyclohexene
(no number necessary)

Substituent groups are included as prefixes with appropriate numbers to specify position. Since the **-ene** stem is a suffix, it dominates the numbering. That is, the parent alkene chain is named first, including *cis*- or *trans*- where necessary, and then the substituents are appended as prefixes.

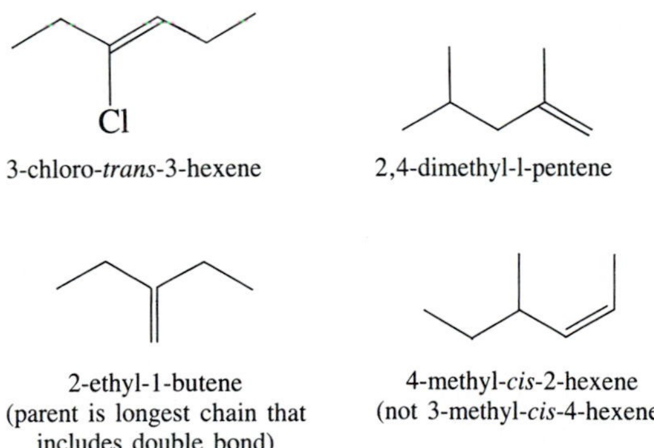

3-chloro-*trans*-3-hexene

2,4-dimethyl-1-pentene

2-ethyl-1-butene
(parent is longest chain that includes double bond)

4-methyl-*cis*-2-hexene
(not 3-methyl-*cis*-4-hexene)

EXERCISE 11.2 Name the following alkenes.

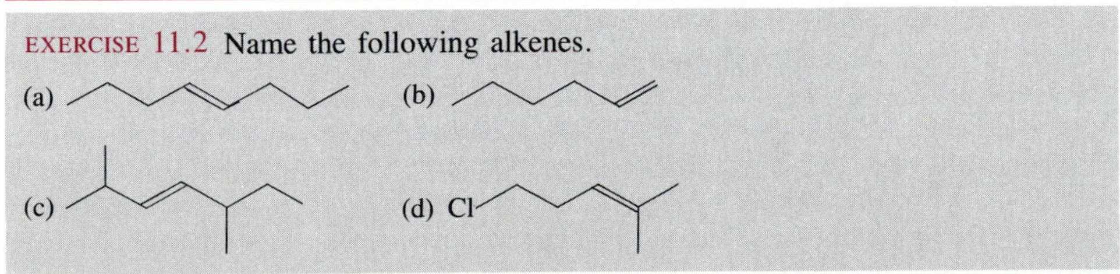

The cis-trans system for naming configurational isomers frequently leads to confusion. The Chemical Abstracts Service has proposed an unambiguous system. In this system, the two groups attached to each end of the double bond are assigned priority numbers as is done in naming enantiomers by the *R,S* system (Chapter 7). When the two groups of higher priority number are on the *same side* of the molecule, the compound is the **Z** isomer (Ger., *zusammen,* together). When the two groups of highest priority are on *opposite sides* of the molecule, the compound is the **E** form (Ger., *entgegen,* opposite).

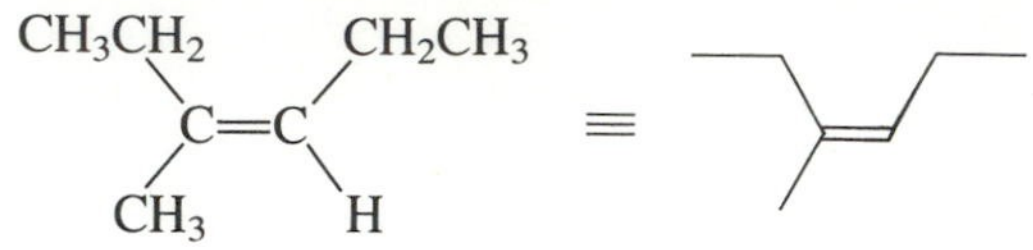

CH_3CH_2 CH_2CH_3 C=C CH_3 H ≡

(*Z*)-3-methyl-3-hexene

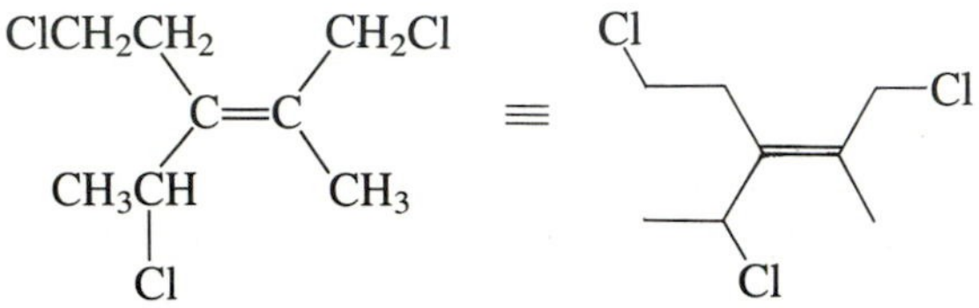

$ClCH_2CH_2$ CH_2Cl C=C $CH_3CH(Cl)$ CH_3 ≡ Cl Cl Cl

(*E*)-1,4-dichloro-3-(2-chloroethyl)-2-methyl-2-pentene

The *E,Z* system for specifying stereoisomerism in alkenes has been sanctioned by the IUPAC, although this body still recognizes the use of cis and trans in cases where there is no ambiguity. In normal use, most alkenes are named by the IUPAC system (using either cis-trans or *E,Z* to specify stereostructure), although several common compounds (ethylene, propylene, isobutylene) are invariably called by their common or trivial names.

EXERCISE 11.3 Name the following alkenes using the *E,Z* system to specify stereostructure.

(a) 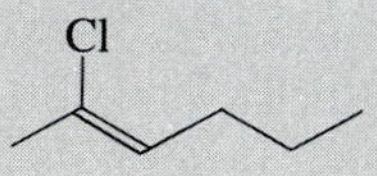 Cl (b) 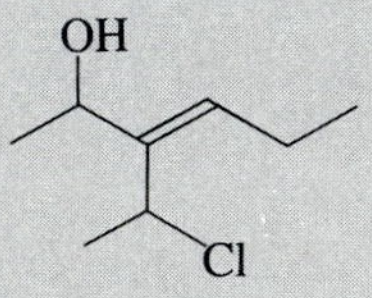 OH Cl

(c) 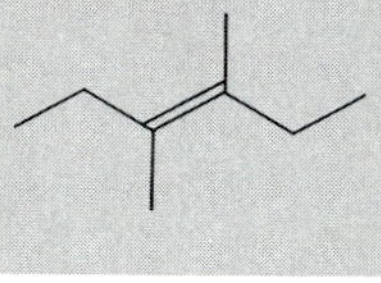 (d) 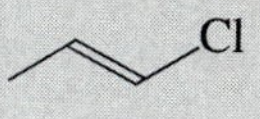 Cl

In compounds that have both a double bond and a hydroxy group, two suffixes are used. The -ol suffix has priority, and the longest chain *containing both the double bond and the OH group* is numbered in such a way as to give the OH group the smaller number.

OH OH

but-3-en-1-ol
(not but-1-en-4-ol)

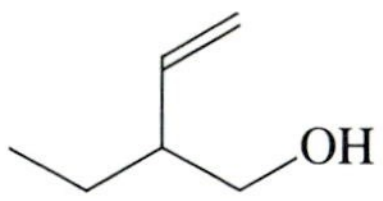

2-ethylbut-3-en-1-ol

[Note that the final -e of the alkene suffix -ene is dropped when it is followed by another suffix.]

EXERCISE 11.4 Explain why the following names are incorrect.

(a) 2-(chloromethyl)-2-pentene (b) (*E*)-2-methyl-2-hexene
(c) *trans*-pent-2-en-4-ol

11.3 Physical Properties

Physical properties of some alkenes are summarized in Table 11.1. These properties are similar to those of the corresponding alkanes, as shown by the boiling point plot in Figure 11.6. The smaller alkenes are gases at room temperature. Starting with the C_5 compounds, the alkenes are volatile liquids. Isomeric alkenes have similar boiling points, and mixtures can be separated only by careful fractional distillation with efficient columns. 1-Alkenes tend to boil a few degrees lower than internal alkenes and can be separated readily by careful fractionation.

TABLE 11.1 Physical Properties of Alkenes

Name	Structure	Boiling Point, °C	Density d^{20}
ethylene	$CH_2{=}CH_2$	−103.7	
propene	$CH_3CH{=}CH_2$	−47.4	0.5193
1-butene	$CH_3CH_2CH{=}CH_2$	−6.3	0.5951
cis-2-butene	$CH_3C(H){=}C(H)CH_3$ (both H on top)	3.7	0.6213
trans-2-butene	$CH_3C(H){=}C(H)CH_3$ (H above first C, H below second C)	0.9	0.6042
2-methyl-1-propene	$(CH_3)_2C{=}CH_2$	−6.9	0.5942
1-pentene	$CH_3CH_2CH_2CH{=}CH_2$	30.0	0.6405
cis-2-pentene	$CH_3CH_2C(H){=}C(H)CH_3$ (both H on top)	36.9	0.6556
trans-2-pentene	$CH_3CH_2C(H){=}C(H)CH_3$ (H above first C, H below second C)	36.4	0.6482
2-methyl-2-butene	$(CH_3)_2C{=}CHCH_3$	38.6	0.6623

The two carbon orbitals making up the =C—C bond in an alkyl-substituted double bond have different amounts of *s*-character. Such a bond may be approximated as $C_{sp^2}{-}C_{sp^3}$. The resulting change in electron density distribution gives such bonds an effective dipole moment with the negative end of the dipole toward the sp^2-hybridized carbon.

$$\overset{\leftarrow\!\!+}{\text{C—}}\text{C=C}$$

These dipole moments are small for hydrocarbons, but still permit a distinction between cis and trans isomers. For example, *cis*-2-butene has a small dipole moment, whereas *trans*-2-butene has a resultant dipole moment of zero because of its symmetry.

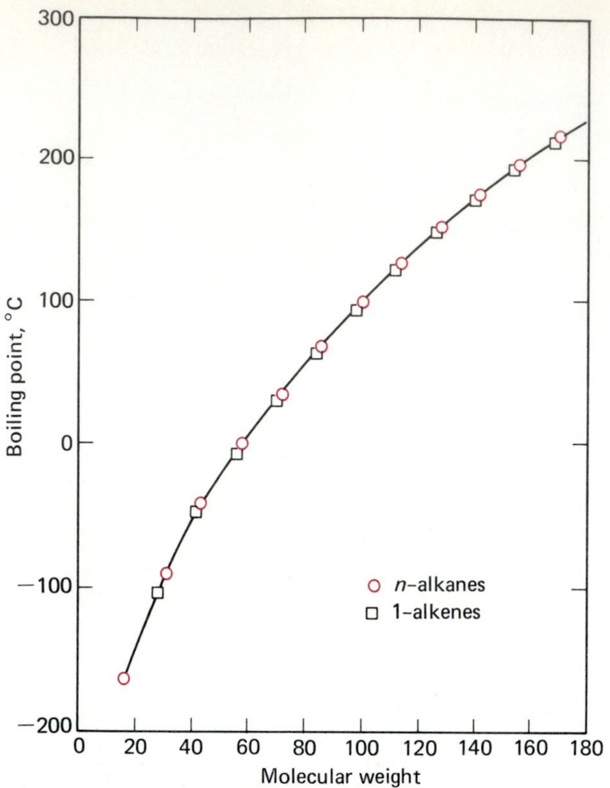

FIGURE 11.6 Boiling point relationships of alkenes and alkanes.

vector sum

cis-$H_3C-CH=CH-CH_3$ (both CH_3 on the same side) $\quad \mu = 0.33$ D

trans-$H_3C-CH=CH-CH_3$ $\quad \mu = 0$ D

With substituents such as halogens, the dipole moment differences are greater.

vector sum

cis-$ClCH=CHCl$ $\quad \mu = 2.95$ D

trans-$ClCH=CHCl$ $\quad \mu = 0$ D

11.4 Relative Stabilities of Alkenes: Heats of Formation

Heats of formation have been evaluated for a number of alkenes. Examination of these values shows that trans alkenes are generally more stable than the isomeric cis alkenes by about 1 kcal mole^{-1}. (Remember that a *more negative* heat of formation ΔH_f° corresponds to a *more stable* compound.)

$$\underset{\Delta H_f^\circ:\ -1.9}{\text{cis-}H_3C\text{–}CH{=}CH\text{–}CH_3} \rightleftharpoons \underset{-3.0}{\text{trans-}H_3C\text{–}CH{=}CH\text{–}CH_3} \qquad \Delta H^\circ = -1.1 \text{ kcal mole}^{-1}$$

$$\underset{\Delta H_f^\circ:\ -7.0}{\text{cis-}H_3C\text{–}CH{=}CH\text{–}CH_2CH_3} \rightleftharpoons \underset{-7.9}{\text{trans-}H_3C\text{–}CH{=}CH\text{–}CH_2CH_3} \qquad \Delta H^\circ = -0.9 \text{ kcal mole}^{-1}$$

$$\underset{\Delta H_f^\circ:\ -13.7}{\text{cis-}H_3C\text{–}CH{=}CH\text{–}CH(CH_3)_2} \rightleftharpoons \underset{-14.7}{\text{trans-}H_3C\text{–}CH{=}CH\text{–}CH(CH_3)_2} \qquad \Delta H^\circ = -1.0 \text{ kcal mole}^{-1}$$

The distance between the adjacent methyl groups in *cis*-2-butene is about 3 Å. Since the sum of the van der Waals radii for two methyl groups is 4 Å, the hydrogens in these two groups are sufficiently close that there is a net repulsion not present in the trans compound (Figure 11.7). This effect of repulsion for sterically congested systems is another form of **steric hindrance** (Section 9.3).

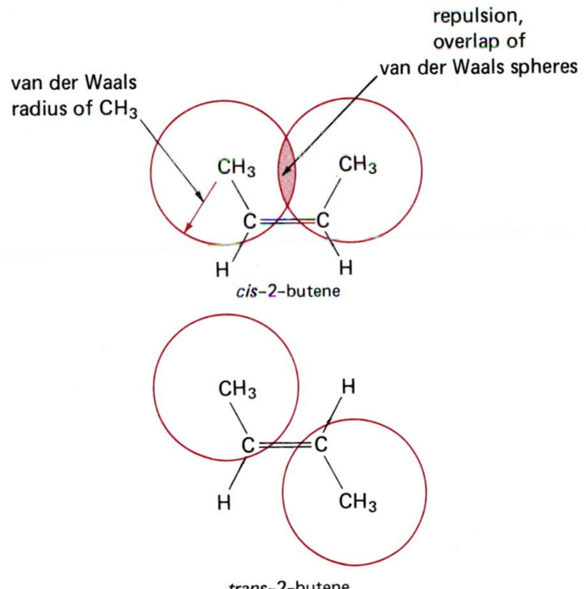

FIGURE 11.7 Steric hindrance in *cis*-2-butene, relative to *trans*-2-butene.

Monosubstituted ethylenes are 2–3 kcal mole^{-1} less stable than disubstituted ethylenes. The examples in the following table may be compared with the corresponding isomers listed above.

Compound	ΔH_f°, kcal mole^{-1}
$CH_3CH_2CH{=}CH_2$	−0.2
$CH_3CH_2CH_2CH{=}CH_2$	−5.3
$(CH_3)_2CHCH_2CH{=}CH_2$	−12.3

The stabilizing effect of substituents on the double bond continues with additional substituents, although the incremental effect is reduced because of cis interactions. For example, 2-methyl-2-butene, with $\Delta H_f^\circ = -10.1$ kcal mole^{-1}, is the most stable five-carbon alkene. Similarly, 2,3-dimethyl-2-butene is the most stable six-carbon alkene.

These results are most simply interpreted on the basis of relative bond strengths. A C_{sp^2}—H bond is a stronger bond than a C_{sp^3}—H bond. If this were the only important factor, the least substituted alkenes would be the most stable. However, a C_{sp^2}—C_{sp^3} bond is also stronger than a C_{sp^3}—C_{sp^3} bond, and it seems that putting more *s*-character in a carbon-carbon bond has a greater effect than in a carbon-hydrogen bond. This hybridization effect may also be observed in the bond lengths because bond lengths are inversely related to bond strengths. We noted earlier that the carbon-hydrogen bond in ethylene is about 0.01 Å shorter than the carbon-hydrogen bond in ethane. The carbon-carbon single bond in propene (1.505 Å) is 0.03 Å shorter than the carbon-carbon bond in propane. The difference $r_{C_{sp^3}—C_{sp^3}} - r_{C_{sp^2}—C_{sp^3}}$ is greater than $r_{C_{sp^3}-H} - r_{C_{sp^2}-H}$.

The difference in stability of various alkene isomers is only a few kilocalories per mole, but this makes an important difference in equilibria. From the thermodynamic equation

$$\Delta G^\circ = -RT \ln K$$

a free-energy difference of 1.4 kcal mole^{-1} at room temperature corresponds to an equilibrium constant of 10 (Table 11.2). Consequently, in an equilibrium mixture of alkenes, the more highly substituted isomers predominate. For example, the equilibrium composition of the butenes is

$$\underset{(74\%)}{\text{H}_3\text{C(H)C=C(H)CH}_3} \rightleftharpoons \underset{(23\%)}{\text{H}_3\text{C(H)C=C(H)CH}_3} \rightleftharpoons \underset{(3\%)}{CH_3CH_2CH{=}CH_2}$$

We will find that such equilibria can be established and that the relative stabilities of alkene isomers affect the product composition in some synthetic methods.

TABLE 11.2 Equilibrium Concentrations as a Function of ΔG° at 25°C: A ⇌ B

ΔG°, kcal mole^{-1}	Percent A	Percent B
−5	0.02	99.98
−2	3.3	96.7
−1	15.6	84.4
−0.5	30.1	69.9
0	50.0	50.0
+0.5	69.9	30.1
+1	84.4	15.6
+2	96.7	3.3
+5	99.98	0.02

EXERCISE 11.5 Using the heats of formation for *cis*- and *trans*-2-pentene given on page 257, calculate the composition of an equilibrium mixture of these two isomers (ignore 1-pentene and assume $\Delta S^\circ = 0$) at 25 °C and at 300 °C.

In principle, the double bond in a cyclic alkene can be either cis or trans. In a cis cycloalkene two carbons that are attached cis to each other on the double bond are also part of the ring. Several examples are shown below.

cyclohexene 1-methylcycloheptene 1,2-dimethylcyclobutene

In a trans cycloalkene the two ring carbons are trans with respect to the double bond. In practice, the trans isomers are too strained to exist at room temperature for three- through seven-membered rings. The smallest cycloalkene for which an isolable trans isomer is known is cyclooctene. Even in this case there is significant strain—*cis*-cyclooctene is 9.1 kcal mole^{-1} more stable than the trans isomer.

cis-cyclooctene *trans*-cyclooctene

EXERCISE 11.6 Calculate the composition at 25 °C for an equilibrium mixture of *cis*- and *trans*-cyclooctene. Construct molecular models of the stereoisomeric cyclooctenes. Note that it is much more difficult to construct a model of the trans isomer than of the cis. Demonstrate that the trans isomer is chiral, whereas the cis isomer is not.

Highlight 11.1

The structure and stability of alkenes reflect (a) the two-fold character of the double bond, sp^2-sp^2 (σ-bond) + p-p (π-bond), (b) bonding to hydrogen (sp^2-s) or alkyl substituents (sp^2-sp^3), and (c) steric interference between alkyl groups on the cis or *Z* side of the double bonds.

11.5 Preparation

The important preparations of alkenes that we have discussed thus far have all been elimination reactions—dehydrohalogenation of alkyl halides and dehydration of alcohols. Other important procedures for creating the carbon-carbon double bond will be discussed in subsequent chapters.

A. Dehydrohalogenation of Alkyl Halides

Dehydrohalogenation was discussed previously (Section 9.7), but primarily as a complication of nucleophilic substitution. Elimination can often be made the principal reaction by using a suitably strong base. One common reagent used for this purpose is potassium hydroxide in refluxing ethanol. This solution is really a solution primarily of potassium ethoxide in ethanol because of the equilibrium

$$OH^- + C_2H_5OH \rightleftharpoons C_2H_5O^- + H_2O$$

EXERCISE 11.7 Assuming that ethanol and water have equal dissociation constants in ethanol solution, estimate the hydroxide/ethoxide ratio in a solution made by dissolving 1 mole of KOH in enough ethanol to make 1 L of solution.

In such a solution, the elimination reaction shows second order kinetics, first order in alkyl halide and first order in base. As we have seen (Section 9.7), eliminations that occur under such conditions are classified mechanistically as **E2,** for bimolecular elimination. Later in this section, we shall consider some of the considerable amount of evidence upon which the E2 mechanism is based.

For secondary and tertiary alkyl halides, treatment with alcoholic potassium hydroxide gives the elimination products in good yield.

$$CH_3CH(I)CH_3 \xrightarrow[EtOH]{EtOK} CH_3CH{=}CH_2 \quad (94\%)$$

For primary halides, especially for those with no β-branches, nucleophilic substitution by the S_N2 mechanism is so facile that it dominates, and ethers are the principal products.

$$n\text{-}C_5H_{11}Br \xrightarrow[EtOH,\ 55°C]{EtONa} \underset{(12\%)}{CH_3CH_2CHCH{=}CH_2} + \underset{(88\%)}{n\text{-}C_5H_{11}OC_2H_5}$$

Bimolecular elimination almost invariably gives a mixture of the possible alkene products. The mixture usually reflects the thermodynamic stabilities of the isomeric alkenes; the most stable isomers tend to predominate.

$$CH_3CH_2CH_2CHBrCH_3 \xrightarrow[EtOH,\ \Delta]{EtOK}$$

$$\underset{(41\%\ trans)\ (14\%\ cis)}{CH_3CH_2CH{=}CHCH_3} + \underset{(25\%)}{CH_3CH_2CH_2CH{=}CH_2} + \underset{(20\%)}{CH_3CH_2CH_2CH(OEt)CH_3}$$

In this example note that more trans isomer is produced than cis isomer and that the total amount of 2-pentenes is greater than that of 1-pentene. However, the more basic and bulkier reagent potassium *t*-butoxide in the less polar solvent *t*-butyl alcohol tends to give more of the terminal alkene. A further increase in the fraction of terminal alkene is obtained by the use of the potassium alkoxide derived from 3-ethyl-3-pentanol.

$$(CH_3)_2CBrCH_2CH_3 \longrightarrow (CH_3)_2C{=}CHCH_3 + CH_2{=}C(CH_3)CH_2CH_3$$

	2-methyl-2-butene	2-methyl-1-butene
EtOK/EtOH	71%	29%
$KOC(CH_3)_3/HOC(CH_3)_3$	28%	72%
$KOC(C_2H_5)_3/HOC(C_2H_5)_3$	11%	89%

Potassium *t*-butoxide gives good yields of elimination product even with straight-chain primary halides.

$$n\text{-}C_{18}H_{37}Br \xrightarrow[HOC(CH_3)_3,\ 40°C]{KOC(CH_3)_3} \underset{(88\%)}{n\text{-}C_{16}H_{33}CH{=}CH_2} + \underset{(12\%)}{n\text{-}C_{18}H_{37}OC(CH_3)_3}$$

These various effects of structure are best rationalized in the context of reaction mechanism. A correct mechanism needs to explain these facts as well as several other generalizations that can be made about such eliminations.

1. The rate of bimolecular elimination reactions depends on the concentration of both the alkyl halide and the base.
2. The rate of reaction depends on the nature of the leaving group. The general reactivity order for alkyl halides is iodides > bromides > chlorides.
3. The reaction has a high primary hydrogen isotope effect; that is, reactions in which carbon-deuterium bonds are broken proceed more slowly than those involving the corresponding carbon-hydrogen bonds.
4. The reaction is **stereospecific.** The leaving hydrogen must generally have a conformation that is **anti** to that of the leaving halide.

The first generalization has an obvious consequence for any proposed mechanism. Both the base and the alkyl halide must be involved in the rate-determining step. In fact, it is this observation that gives the mechanism its name—E2 or ''elimination, bimolecular.''

EXERCISE 11.8 We saw in Chapter 8 that elimination is often a complicating side reaction in nucleophilic substitution reactions. For a primary alkyl halide, how would you expect the substitution/elimination ratio to vary with nucleophile (base) concentration?

The second generalization also has an obvious consequence: The bond to the leaving halide must be partially broken in the transition state. Bonds that are broken more easily lead to a lower energy transition state and a faster reaction rate.

The third generalization establishes that the bond to the leaving hydrogen is also partially broken in the transition state. To a first approximation, it effectively takes more energy to break a carbon-deuterium bond than it does to break a carbon-hydrogen bond.

EXERCISE 11.9 Compare the substitution/elimination ratio for the reactions of ethyl bromide and 2,2,2-trideuterioethyl bromide (CD_3CH_2Br) with potassium *t*-butoxide in *t*-butyl alcohol?

To be more precise, isotope effects originate in the nature of the vibrational energy levels of bonds. These quantum states were discussed previously in connection with bond dissociation energies of alkanes (Section 6.1), and we will encounter them again in studying infrared spectra (Section 14.2). For the present purpose, consider the two carbon-hydrogen bond motions in Figure 11.8. In the strong carbon-hydrogen bond the potential energy is very sensitive to the value of the carbon-hydrogen bond distance. Thus, in order to accommodate the Heisenberg uncertainty principle, the lowest vibrational energy state is relatively high above the potential minimum. This difference is called the **zero point energy,** ϵ_0, and is relatively large. For the carbon-hydrogen bond in alkanes ϵ_0 is about 4 kcal $mole^{-1}$. For a weaker bond, as shown in Figure 11.8b, a given uncertainty in the position of the hydrogen corresponds to a smaller change in potential energy. Hence, ϵ_0 is smaller.

A deuterium atom is twice as heavy as hydrogen because its nucleus has a neutron as well as a proton. Because of its greater mass, the same momentum corresponds to a slower velocity. A given uncertainty in momentum corresponds to a smaller uncertainty in position compared to the hydrogen case. Accordingly, the zero point energy for C—D is less than that for C—H, as shown in Figure 11.8.

In a reaction in which a carbon-hydrogen bond is broken, the bond is weaker in the transition state. In the change from reactant to transition state, the carbon-hydrogen bond has lost some zero point energy. In the corresponding case of a carbon-deuterium bond the loss in zero point energy is lower because the heavier isotope had less zero point energy to begin with. As a result, as shown in Figure 11.9, the activation energy for breaking a carbon-deuterium bond is greater than that for a carbon-hydrogen bond. The difference in reaction rates can be substantial. For hydrogen isotopes, the effect on ϵ_0 is approximately the square root of the ratio of masses. If ϵ_0(H) is 4 kcal mole^{-1}, the corresponding ϵ_0(D) is 2.8 kcal mole^{-1}. If all of this zero point energy were lost in the transition state, the difference in activation energies would correspond to a reaction rate difference of a factor of about 9. In practice, primary isotope effects for E2 reactions have been observed commonly in the range of 4–8.

$$CD_3CHBrCD_3 \xrightarrow{EtO} CD_2{=}CHCD_3$$

$$\frac{k_{CH_3CHBrCH_3}}{k_{CD_3CHBrCD_3}} = 6.7$$

If no bond to hydrogen is broken at the transition state, isotope effects are generally rather small. For example, in substitution reactions occurring by the S_N2 mechanism, deuterium compounds react at virtually the same rates as the corresponding hydrogen compounds.

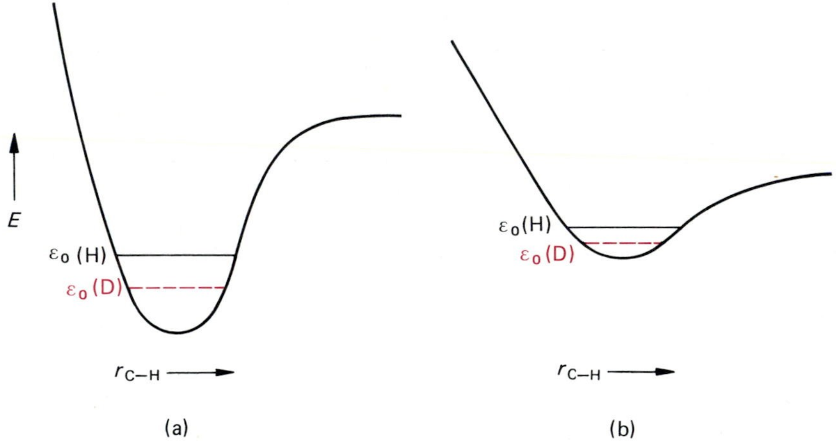

FIGURE 11.8 Potential energies for stretching motion of (a) a strong and (b) a weak carbon-hydrogen bond.

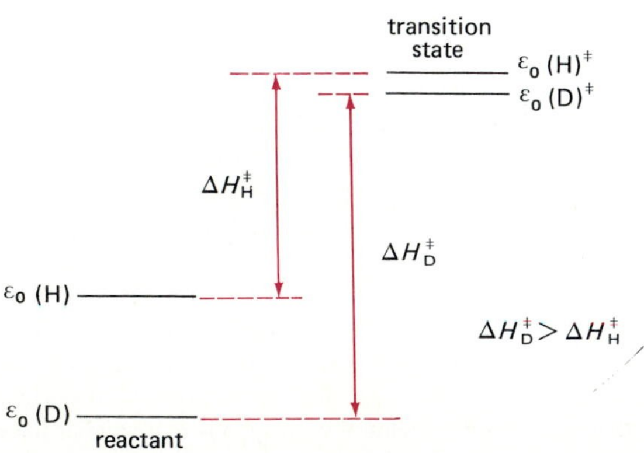

FIGURE 11.9 The effect on the activation energy of loss of zero point energy in a reaction.

The fourth generalization has been established by many examples. A particularly striking case is the behavior of the two diastereomeric 1,2-dibromo-1,2-diphenylethanes upon treatment with base. (The phenyl group, C_6H_5, is derived from the benzene ring, and is analogous to an alkyl group. We shall consider the chemistry of benzene and other molecules containing the phenyl group in Chapter 21. In the current example, the phenyl group plays no role other than that of a stereochemical marker.)

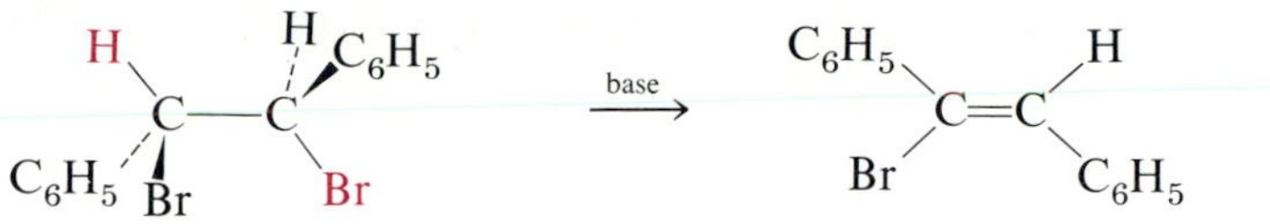

(1*R*,2*R*)-1,2-dibromo-1,2-diphenylethane $\xrightarrow{\text{base}}$ (*Z*)-1-bromo-1,2-diphenylethene

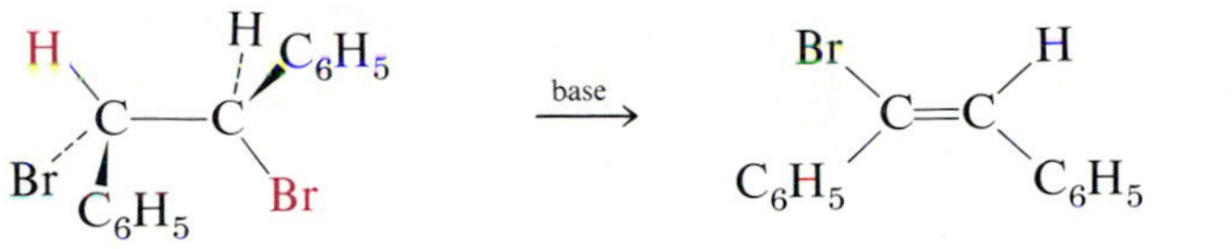

meso-1,2-dibromo-1,2-diphenylethane $\xrightarrow{\text{base}}$ (*E*)-1-bromo-1,2-diphenylethene

The two stereoisomers react by the E2 mechanism to give *totally different* products. From a knowledge of the configurations of the two reactant dibromides and the product that each gives, it may be deduced that each isomer reacts in a conformation that has H anti to Br. The staggered conformations of the two compounds are shown in Newman-projection form in Figure 11.10. Note that removal of H and Br from opposite sides of the molecule results in conversion of the 1*R*,2*R* stereoisomer into the alkene having the two phenyl groups trans.

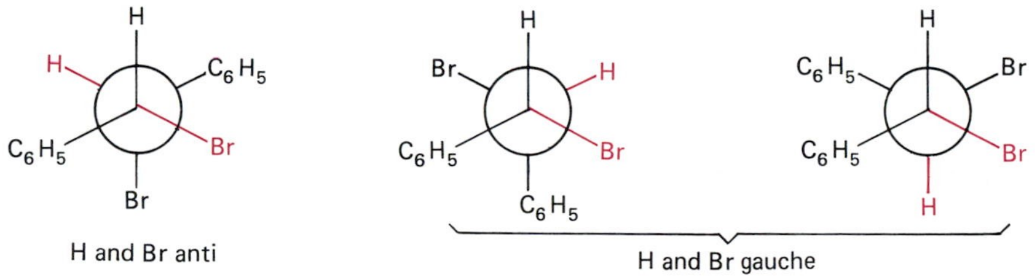

H and Br anti

H and Br gauche

(a) staggered conformation of (1*R*,2*R*)-1,2-dibromo- 1,2-diphenylethane

H and Br anti

H and Br gauche

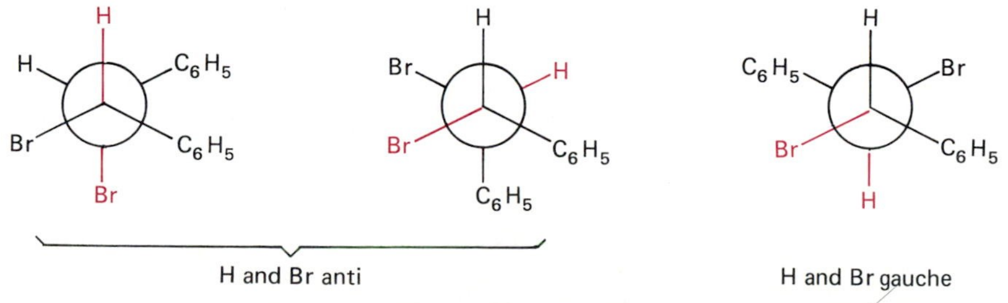

(b) staggered conformations of *meso*-1,2 dibromo-1,2-diphenylethane

FIGURE 11.10 Newman projections illustrating the staggered conformations of the stereoisomeric 1,2-dibromo-1,2-diphenylethanes.

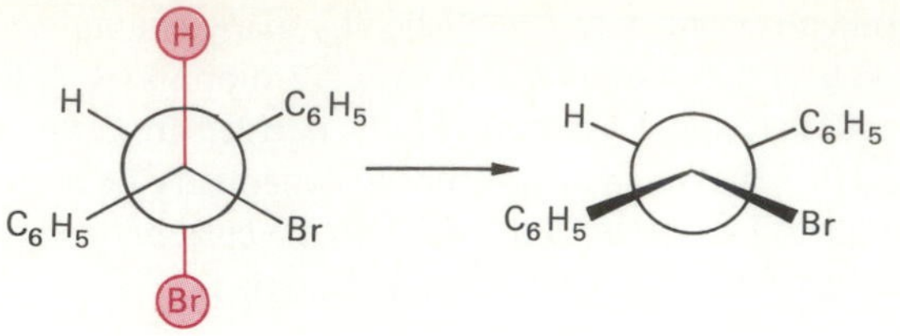

Similarly, anti elimination in the meso stereoisomer gives the alkene having the phenyl groups cis.

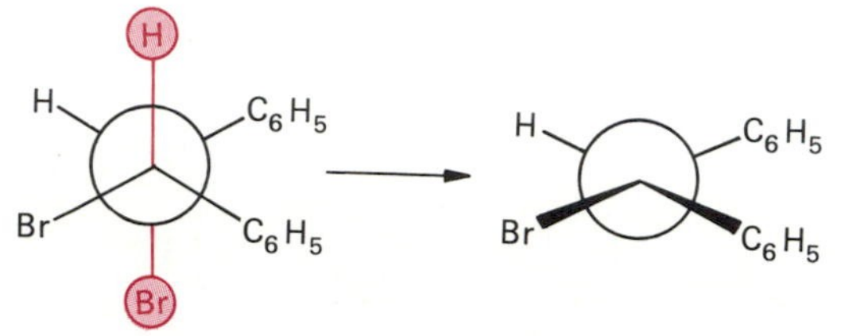

This example demonstrates that anti elimination can give either a cis or a trans alkene, depending on the structure of the starting halide.

The strict requirement for anti elimination may also be seen in the reactions of cyclic alkyl halides. For example, *cis*-1-bromo-4-*t*-butylcyclohexane reacts rapidly with sodium ethoxide in ethanol to give 4-*t*-butylcyclohexene. However, the trans isomer undergoes elimination extremely slowly.

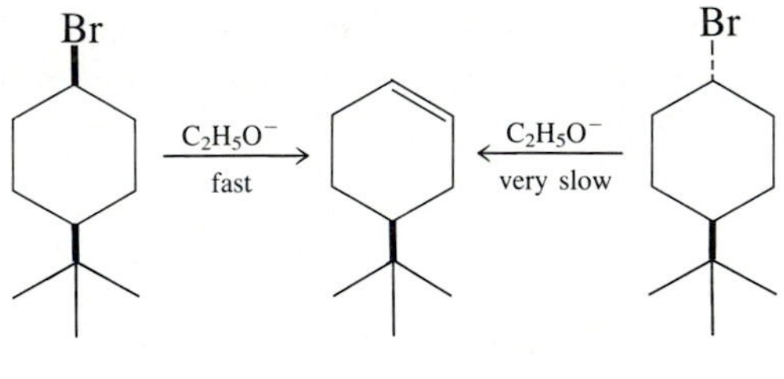

In this example the *t*-butyl group is so sterically demanding that it determines which chair conformation is adopted by each molecule. *cis*-1-Bromo-4-*t*-butylcyclohexane exists almost entirely in the conformation shown, in which bromine is axial. To determine the relative energy of the axial-*t*-butyl, equatorial-bromo conformation, the energies in Table 7.1 may be used.

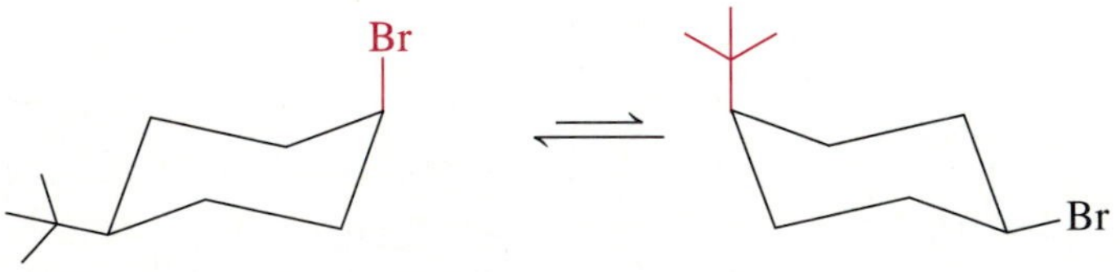

relative energy: $0 + 0.5 = 0.5$ kcal mole^{-1} $\sim 5 + 0 = \sim 5$ kcal mole^{-1}

$\Delta G° \cong 4.5$ kcal mole^{-1}

The corresponding equilibrium constant, $K \cong 10^{-3}$; that is, *cis*-1-bromo-4-*t*-butylcyclohexane exists at least 99.9% in the axial-bromo conformation.

Applying the same approach to the trans isomer reveals that the conformation with axial bromo is only about 0.01% of the total.

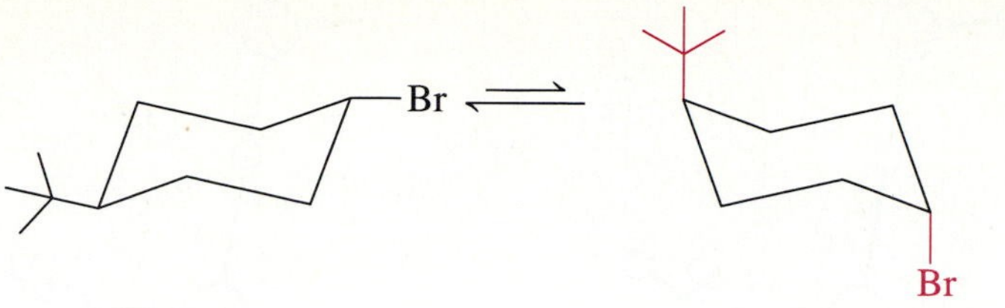

relative energy: $0 + 0 = 0$ $\qquad$ $\sim 5 + 0.5 = \sim 5.5$ kcal mole^{-1}

$\Delta G^\circ \cong 5.5$ kcal mole^{-1}

Thus, only in the cis isomer is there any significant number of molecules in which the bromine has an anti relationship to a hydrogen.

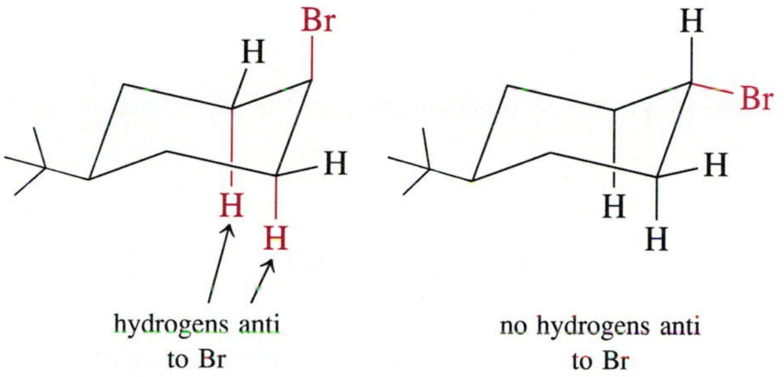

hydrogens anti to Br $\qquad$ no hydrogens anti to Br

EXERCISE 11.10 Construct a molecular model of bromocyclohexane with the bromine in an axial position. Verify that the trans hydrogens at C-2 and C-6 are anti and coplanar with the carbon-bromine bond. Now flip the cyclohexane ring into the other chair conformation and note the relationship of the carbon-bromine and carbon-hydrogen bonds.

The mechanism that results from putting these facts together is one in which the attacking base removes a proton concurrently with loss of halide ion from the other side of the molecule. As the carbon-hydrogen bond lengthens and weakens, the carbon-halogen bond also lengthens and weakens. The remaining four groups on the two carbons start to move into coplanarity, and a carbon-carbon double bond starts to form. We may represent this reaction as follows:

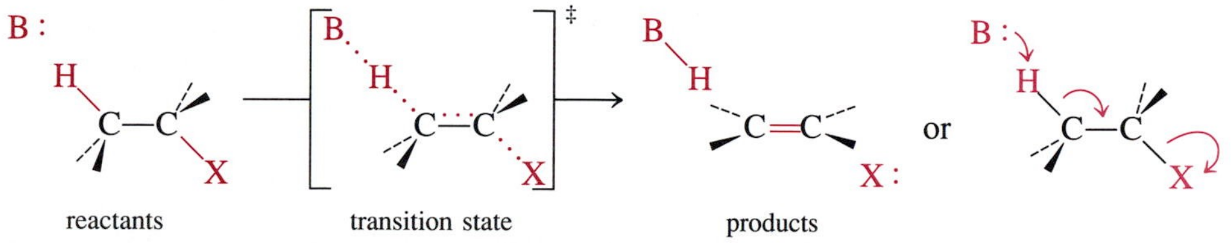

reactants $\qquad$ transition state $\qquad$ products

Recall that each curved arrow represents the movement of one pair of electrons. The proton has been displaced in a typical acid-base proton-transfer reaction, and the leaving halide has been displaced by a pair of electrons at the rear of the carbon-halogen bond. The transition state for the E2 reaction is represented in stereo form in Figure 11.11.

In orbital terms the bonds on the alkyl halide all involve C_{sp^3}-orbitals. In the product alkene some orbitals are C_{sp^2} and others are C_p. At the transition state the orbitals have an intermediate character, as illustrated in Figure 11.12. Because π-bonding is significant in the transition state, those effects that stabilize double bonds, such as added substituents, also stabilize the transition state. However, inasmuch as π-overlap is only partial, these stabilizing effects are not as important as they are in the product alkenes.

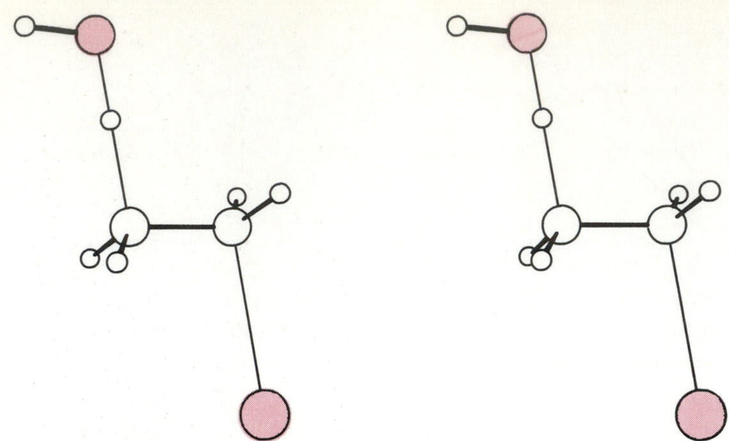

FIGURE 11.11 Stereo representation of the E2 transition state.

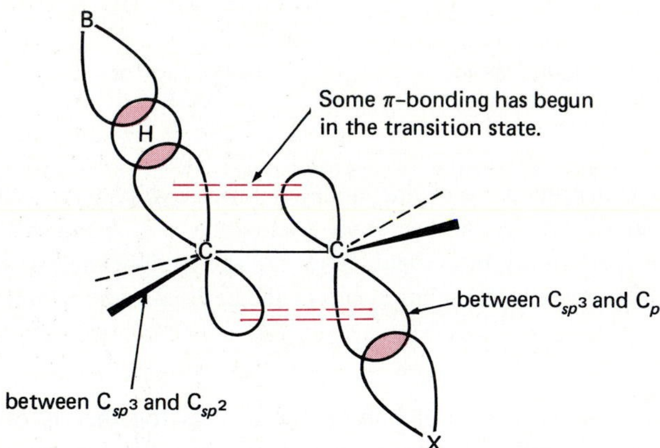

FIGURE 11.12 Orbital representations of the E2 transition state.

It is clear that for π-bonding to be significant in the transition state, the carbon-hydrogen and carbon-halogen bonds to be broken must lie in the same plane, but we may well ask why this requires anti elimination. The answer is simply that in the corresponding syn elimination, groups eclipse one another; in addition the incoming base and leaving halide repel one another. The transition state is then a structure of higher energy.

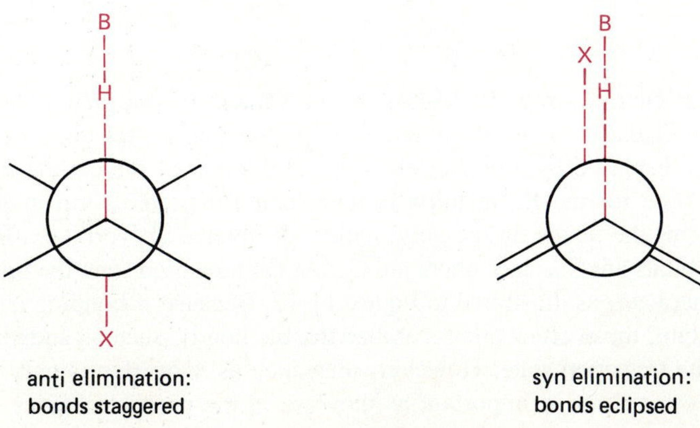

Finally, primary hydrogens are less sterically hindered and more open to attack than secondary or tertiary hydrogens. Refer back to the reaction of 2-bromopentane with potassium ethoxide in ethanol (page 260). 1-Pentene is formed in 25% yield, although at equilibrium it would constitute only about 3% of a mixture of pentene isomers. With the larger bases the disparity in the amount of the 1-alkene is even greater (compare the examples on page 260).

EXERCISE 11.11 Treatment of *meso*-2,3-dibromobutane [the 2*R*,3*S* diastereomer] with KOH in ethanol gives a mixture of 3-bromo-1-butene and one of the isomeric 2-bromo-2-butenes. What is the stereostructure of the latter product? Use your molecular models in working out the answer to this exercise.

Highlight 11.2

The zero-point energy of a carbon-deuterium bond is less than that of a carbon-hydrogen bond. More energy is therefore required to break a C—D bond than a C—H bond. Reactions that involve the cleavage of a C—D bond in the rate-determining step are slower than analogous reactions that involve the cleavage of a C—H bond, a difference called the **hydrogen isotope effect.** The mechanisms of reactions that involve a C—H bond cleavage are often probed by the measurement of such isotope effects.

Highlight 11.3

Base-induced elimination of HX from alkyl halides is usually E2, meaning that the elimination involves two molecules, the base and the alkyl halide, in the rate-determining step. The kinetics of such reactions is usually first order in the base and first order in the alkyl halide. The geometry of the transition state is usually anti-coplanar, meaning that four atoms of the reacting system H—C—C—X lie in a plane and that the C—H and C—X partial bonds are anti, rather than eclipsed. The rate of the reaction depends on the leaving group ability of X, the order for halides being I > Br > Cl.

B. Dehydration of Alcohols

We saw in Section 10.6.D that alcohols undergo dehydration when heated with strong acids. In the case of ethyl alcohol, the reaction requires concentrated sulfuric acid at 170 °C. At lower temperatures (140 °C) diethyl ether is the major reaction product.

$$2\ CH_3CH_2OH \xrightarrow[140°C]{H_2SO_4} CH_3CH_2OCH_2CH_3 + H_2O$$

$$CH_3CH_2OH \xrightarrow[170°C]{H_2SO_4} CH_2{=}CH_2 + H_2O$$

n-Propyl alcohol may be similarly dehydrated to propene.

$$CH_3CH_2CH_2OH \xrightarrow[170°C]{H_2SO_4} CH_3CH{=}CH_2 + H_2O$$

With primary alcohols larger than propyl, mixtures of alkenes result.

$$CH_3(CH_2)_6CH_2OH \xrightarrow[\Delta]{H_3PO_4} \underbrace{CH_3(CH_2)_4CH{=}CHCH_3 + CH_3(CH_2)_3CH{=}CHCH_2CH_3}_{\text{major products}}$$

Essentially no 1-octene is produced in this reaction. The problem in this case is that the initially formed 1-octene undergoes acid-catalyzed isomerization to give the more stable internal octene isomers.

$$CH_3(CH_2)_6CH_2OH \longrightarrow CH_3(CH_2)_4CH_2CH{=}CH_2 \underset{-H^+}{\overset{+H^+}{\rightleftharpoons}}$$

$$CH_3(CH_2)_4CH_2\overset{+}{C}HCH_3 \underset{+H^+}{\overset{-H^+}{\rightleftharpoons}} CH_3(CH_2)_4CH{=}CHCH_3 \rightleftharpoons \text{etc.}$$

This type of isomerization is common under the conditions of acid-catalyzed dehydration. The intermediate secondary carbocations can also rearrange (see page 224). (Note that isomerization of the double bond in propene does not lead to a new product.) As a result, acid-catalyzed dehydration is not a generally useful procedure for the conversion of primary alcohols to alkenes.

EXERCISE 11.12 Suppose you had a sample of $CD_3CH_2CH_2OH$ (3,3,3-trideuterio-1-propanol) and wished to convert it into $CD_3CH{=}CH_2$ (3,3,3-trideuteriopropene). Could dehydration with hot sulfuric acid be used? Explain.

In contrast, secondary and tertiary alcohols are more easily dehydrated.

$$\text{cyclohexanol} \xrightarrow[130^\circ C]{95\%\ H_2SO_4} \text{cyclohexene} + H_2O$$

cyclohexanol — cyclohexene (83%)

$$CH_3CH_2C(OH)(CH_3)CH_3 \xrightarrow[90^\circ C]{46\%\ H_2SO_4} CH_3CH{=}C(CH_3)_2 + H_2O$$

2-methyl-2-butanol — 2-methyl-2-butene (84%)

$$CH_3CH_2CH_2CH(OH)CH_3 \xrightarrow[95^\circ C]{62\%\ H_2SO_4} CH_3CH_2CH{=}CHCH_3 + H_2O$$

2-pentanol — *cis*- and *trans*-2-pentene (65–80%)

> 2-Pentanol, 214 mL, is heated with a mixture of 200 mL of sulfuric acid and 200 mL of water, and the alkene produced is distilled as formed. The distillate is washed, dried, and redistilled; yield, 65–80%. This product is mostly a mixture of *cis*- and *trans*-2-pentenes. The small amount of 1-pentene also present can be removed by careful fractional distillation.

In these cases dehydration probably occurs by the E1 mechanism, by way of intermediate carbocations.

(1) 2-pentanol (OH) + H^+ $\rightleftharpoons$ protonated 2-pentanol ($^+OH_2$)

(2) protonated 2-pentanol ($^+OH_2$) $\rightleftharpoons$ 2-pentyl cation (+) + H_2O

(3) 2-pentyl cation (+) $\rightleftharpoons$ *trans*-2-pentene + *cis*-2-pentene + H^+

Acid-catalyzed dehydration is especially suitable for relatively simple alcohols. In more complex cases, rearrangements may occur.

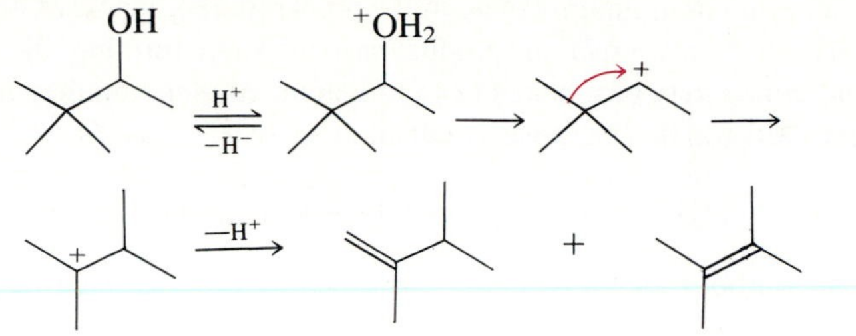

OH $\underset{-H^+}{\overset{H^+}{\rightleftharpoons}}$ $^+OH_2$ ⟶ + ⟶

$\xrightarrow{-H^+}$ +

2,3-dimethyl-1-butene (minor product)

2,3-dimethyl-2-butene (major product)

EXERCISE 11.13 What major product would you expect from the acid-catalyzed dehydration of 3-methylcyclohexanol?

A simple and effective procedure for dehydration of many alcohols, including primary alcohols, involves passing the vapors over alumina at 350–400 °C.

Alumina is a common name for aluminum oxide, Al_2O_3. It occurs naturally in crystalline form as ruby, sapphire, and corundum. Commercial alumina for laboratory use is a white powder, of which types of many different reactivities are available. It is highly insoluble in water and in organic solvents and has an extremely high melting point (over 2000 °C). It is used as an adsorbent in liquid chromatography, as a catalyst for some reactions, and as a catalyst support in other cases.

Like many aluminum salts, alumina is a Lewis acid. The dehydration reaction probably occurs by some version of the E1 mechanism on the alumina surface.

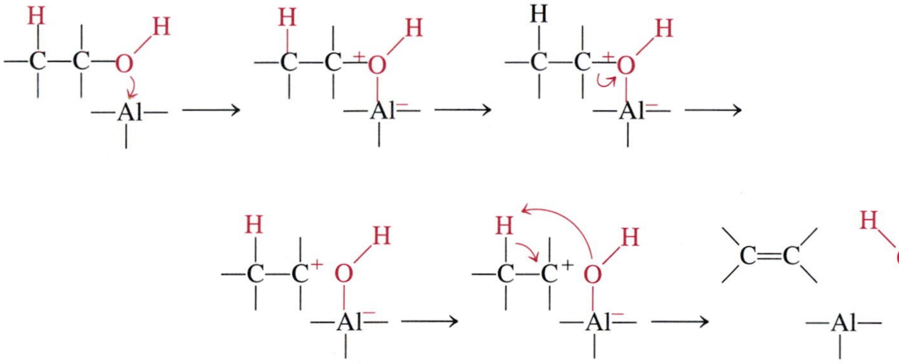

Accordingly, isomerization of olefins and rearrangements are common. These reactions can be suppressed by first treating the alumina with a base such as ammonia.

EXERCISE 11.14 1-Cyclobutylethanol reacts with 60% aqueous sulfuric acid to give 1-methylcyclopentene. Write a mechanism, showing each step, for this rearrangement reaction. If the rearrangement step is not clear, construct a molecular model of the 1-cyclobutylethyl cation and perform the following operation: break the C-1—C-2 bond of the cyclobutane ring and attach C-2 to the former cationic carbon. What is the structure of the product?

C. Industrial Preparation of Alkenes

Ethylene is an important item of commerce. It is used in large quantities for the manufacture of polyethylene and as an intermediate in the preparation of a host of other chemicals. It is obtained primarily as a cracking product in petroleum refining (Section 6.2). Although any hydrocarbon may be cracked to yield mainly ethylene, in the United States the primary material used for this purpose is ethane.

$$CH_3CH_3 \xrightarrow[\text{1 atm}]{\text{700–900°C}} CH_2{=}CH_2 + H_2$$

When higher hydrocarbons are submitted to the cracking process, significant amounts of propene are produced.

$$\underset{}{n\text{-}C_6H_{14}} \xrightarrow{\text{700–900°C}} \underset{(15\%)}{CH_4} + \underset{(40\%)}{CH_2{=}CH_2} + \underset{(20\%)}{CH_3CH{=}CH_2} + \underset{(25\%)}{\text{other}}$$

A large amount of the propene produced in this country goes into the manufacture of polypropylene. Other important industrial alkenes are the butenes and 1,3-butadiene.

The 1989 U.S. industrial production of various alkenes is summarized in Table 11.3.

TABLE 11.3 1989 U.S. Production of Alkenes

Compound	Formula	Tons
Ethylene	$CH_2{=}CH_2$	17,475,000
Propylene	$CH_3CH{=}CH_2$	10,115,000
Styrene	$C_6H_5CH{=}CH_2$	4,065,000
Vinyl chloride	$CH_2{=}CHCl$	4,810,000
Butadiene	$CH_2{=}CHCH{=}CH_2$	1,545,000
Acrylonitrile	$CH_2{=}CHCN$	1,304,000
Vinyl acetate	$CH_2{=}CHOCOCH_3$	1,235,000

11.6 Reactions

In Appendix III, ''Average Bond Energies,'' we find that the value for C=C is 146 kcal mole^{-1}. This value is 63 kcal mole^{-1} higher than the normal C—C strength of 83 kcal mole^{-1}. The difference is reminiscent of the 65 kcal mole^{-1} required for rotation about the double bond in ethylene (Section 11.1) and may be considered roughly as the bond strength of the second or π-bond in ethylene. That is, the ''second'' bond of a double bond is substantially weaker than the first.

The reaction of this ''weak'' π-bond with a normal single bond to produce a molecule containing two new single bonds is generally a thermodynamically favorable process. For example, in gas phase reactions at 25 °C the reactions of ethylene with a number of inorganic compounds are exothermic processes.

$$\begin{array}{llll} CH_2{=}CH_2 + H_2 \longrightarrow CH_3CH_3 & \Delta H^\circ & = & -32.7 \text{ kcal mole}^{-1} \\ CH_2{=}CH_2 + Cl_2 \longrightarrow CH_2ClCH_2Cl & \Delta H^\circ & = & -43.2 \text{ kcal mole}^{-1} \\ CH_2{=}CH_2 + HBr \longrightarrow CH_3CH_2Br & \Delta H^\circ & = & -19.0 \text{ kcal mole}^{-1} \\ CH_2{=}CH_2 + H_2O \longrightarrow CH_3CH_2OH & \Delta H^\circ & = & -10.9 \text{ kcal mole}^{-1} \end{array}$$

Not only do such *additions across a double bond* have favorable thermodynamics; many also have energetically accessible pathways or reaction mechanisms. Such additions form an important part of the chemistry of alkenes.

A. Catalytic Hydrogenation

Even though the reaction is highly exothermic ($\Delta H^\circ = -32.7$ kcal mole^{-1}), ethylene does not react with hydrogen at an appreciable rate without an appropriate catalyst. The following "four-center" mechanism, which is conceptually very simple, is apparently not energetically accessible for the reaction of ethylene with hydrogen.

$$\begin{matrix} H \\ | \\ H \end{matrix} \quad \begin{matrix} CH_2 \\ \| \\ CH_2 \end{matrix} \longrightarrow \left[\begin{matrix} H\cdots CH_2 \\ \vdots \quad \vdots \\ H\cdots CH_2 \end{matrix} \right]^{\ddagger} \longrightarrow \begin{matrix} H-CH_2 \\ | \\ H-CH_2 \end{matrix}$$

cyclic four-membered transition state

Such four-center mechanisms are rare because cyclic four-membered transition states, such as that shown, have unusually high energy. The high activation energy corresponds to an impractically slow reaction rate. The relatively high energies of such four-center transition states can be explained by molecular orbital concepts, and further discussion is deferred to Section 22.4.

The hydrogenation reaction does take place readily on the surface of some metals, particularly platinum, palladium, and nickel. These metals are known to coordinate with double bonds and to absorb hydrogen. The detailed reaction mechanism is complex and involves various types of metal-carbon bonds. A schematic representation that is suitable for our purposes is approximated as follows.

H_2 ⇌ H H $\xrightleftharpoons{C_2H_4}$ $CH_2{=}CH_2$ H H → $CH_2{-}CH_2$ (H H)

(catalyst surface)

Platinum is usually used as the black oxide known as "Adams' catalyst." The oxide is prepared by fusion of chloroplatinic acid, $H_2PtCl_6{\cdot}6H_2O$, hygroscopic red-brown crystals, or ammonium chloroplatinate, $(NH_4)_2PtCl_6$, yellow crystals, with sodium nitrate. It reacts readily with hydrogen gas even at low pressures to form a finely divided platinum metal catalyst. This transformation is usually accomplished by stirring with a suitably inert solvent, such as ethanol or acetic acid. The alkene is then added, and when the solution is stirred with the suspension of platinum under an atmosphere of hydrogen, hydrogen gas is absorbed rapidly. The hydrogen is usually contained in a gas buret so that the amount absorbed can be measured. The resulting mixture is filtered and the product is isolated from the filtrate. Only small amounts of platinum catalyst are required, but the filter paper residues are normally saved for recovery of the platinum, a rare and expensive material.

Palladium is usually used as a commercial preparation in which the finely divided metal is supported on an inert surface, frequently charcoal (Pd/C) or barium sulfate ($Pd/BaSO_4$). Alkenes are normally hydrogenated in ethanol solution by stirring with the supported palladium catalyst at room temperature under an atmosphere of hydrogen.

Nickel is usually used in a finely divided state called "Raney nickel." The catalyst is prepared by allowing nickel-aluminum alloy to react with aqueous sodium hydroxide. The aluminum dissolves and leaves the nickel as a finely divided suspension. Typical hydrogenations are conducted at moderately high pressures of hydrogen (~1000 psi).

Other hydrogenation catalysts that are used for specific purposes are rhodium, ruthenium, and iridium, but platinum, palladium, and nickel in their various forms are the most common. They are subject to "poisoning" by some compounds, notably sulfur-containing compounds such as thiols and sulfides. These compounds bind firmly to the catalyst surface and destroy its catalytic activity.

Since the two hydrogens are added to the double bond from the surface of the metal, they are normally both added *to the same face of the double bond*. This type of addition is referred to as **syn** (Gk., *syn*, together). The other stereochemical possibility, addition of two "pieces" of a reagent to opposite faces of a double bond is called **anti** addition (Gk., *anti*, opposite). In the case of catalytic hydrogenation the alkene molecule is adsorbed to the catalyst surface with one face of the double bond coordinated to the surface; the two hydrogens are both added to this face.

$$(CH_3)(COOH)C{=}C(CH_3)(COOH) \xrightarrow[Pd]{H_2} HOOC(CH_3)(H)C{-}C(H)(CH_3)COOH \equiv HOOC(CH_3)(H)C{-}C(H)(CH_3)COOH$$

Although syn addition is the general rule, anti addition is sometimes observed. For example, hydrogenation of 1,2-dimethylcyclohexene over palladium gives *trans*-1,2-dimethylcyclohexane as the major product.

$$\text{1,2-dimethylcyclohexene} \xrightarrow[Pd/Al_2O_3,\ CH_3CO_2H]{H_2} \underset{(73\%)}{\textit{trans}\text{-1,2-dimethylcyclohexane}} + \underset{(27\%)}{\textit{cis}\text{-1,2-dimethylcyclohexane}}$$

Anti addition in this example results because double bond isomerization occurs more rapidly than hydrogenation. The isomerized alkene, 1,6-dimethylcyclohexene, reacts with hydrogen from both its faces, giving a mixture of the cis and trans products.

$$\text{1,2-dimethylcyclohexene} \overset{Pd}{\rightleftharpoons} \text{1,6-dimethylcyclohexene} \xrightarrow[Pd]{H_2} \text{products}$$

Palladium is particularly prone to catalyze double bond isomerization. Platinum, rhodium, or iridium should be used if such isomerization is a possible problem.

EXERCISE 11.15 Hydrogenation of optically active 3,7-dimethyl-1-octene using platinum as catalyst gives optically active 2,6-dimethyloctane. If Pd/C is used as a catalyst, the 2,6-dimethyloctane produced is racemic. Explain.

B. Addition of Halogens

An important general reaction of double bonds is the addition of chlorine and bromine.

$$>C{=}C< + X_2 \longrightarrow X{-}\overset{|}{\underset{|}{C}}{-}\overset{|}{\underset{|}{C}}{-}X \qquad X = Cl, Br$$

This reaction is rapid and can be regarded as a nucleophilic displacement reaction on a halogen. The alkene is the nucleophile and halide ion is the leaving group.

$$>C{=}C< + Br{-}Br \longrightarrow Br{-}C{-}C^{+} + Br^{-}$$

The resulting cation reacts with halide ion to give the observed product.

$$Br{-}\overset{|}{\underset{|}{C}}{-}C^{+}< + Br^{-} \longrightarrow Br{-}\overset{|}{\underset{|}{C}}{-}\overset{|}{\underset{|}{C}}{-}Br$$

The intermediate cation contains an electron-deficient cationic carbon and a halogen atom with nonbonding electron pairs. Consequently there is a tendency for overlap to produce a **cyclic halonium ion** as in Figure 11.13.

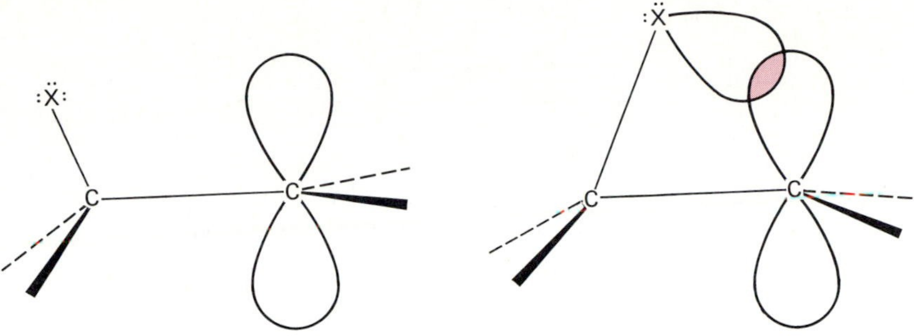

FIGURE 11.13 Formation of cyclic halonium ion.

The cyclic halonium ion may be written in Lewis form as

$$\overset{\cdot\cdot}{\underset{\cdot\cdot}{X}}{}^{+}$$
$$\diagdown\!\!\!\!\!\; C-C \;\!\!\!\!\!\diagup$$

The advantage in terms of energy in forming such a structure is primarily the formation of an additional covalent bond. Furthermore, all of the atoms now have an octet electronic configuration. However, a price is paid for these gains. The bond angles in the three-membered ring structure are bent far from the desired tetrahedral geometry, and the positive charge is localized on the more electronegative halogen atom rather than on carbon.

In practice, the tendency of such a cation to exist in the cyclic form depends on the stability of the "open" carbocation. The intermediate formed from the addition of bromine to ethylene is best described as a symmetrical bromonium ion with relatively strong carbon-bromine bonds. The alternative open form would be a highly unstable primary carbocation.

$$\overset{Br^{+}}{CH_2\!-\!CH_2} \quad \text{better than} \quad Br\!-\!CH_2\!-\!\overset{+}{C}H_2$$

The ion formed by addition of bromine to isobutylene is better described as a tertiary carbocation with a long and weak bond to bromine.

$$CH_3-\overset{+}{\underset{CH_3}{C}}\cdots\overset{Br}{CH_2}$$

Cations such as these may be described in terms of three resonance structures: The actual ion is a composite or hybrid of the three structures A, B, and C.

$$\underset{A}{-\overset{+}{C}-C-\ (\cdot X \text{ bridged})} \longleftrightarrow \underset{B}{-C-C-\ (\overset{+}{X} \text{ bridged})} \longleftrightarrow \underset{C}{-C-C^{+}-\ (X\cdot \text{ bridged})}$$

If both A and C correspond to unstable carbocations, then structure B is a more important contributor to the actual structure of the ion. If either A or C corresponds to a relatively stable carbocation, then that structure contributes more and the ion has substantial carbocation character without as much halonium ion character.

The crystal structure of a stable yellow bromonium salt derived from bromine and the alkene adamantylidene adamantane shows that the three-membered ring has carbon-bromine bonds of somewhat different lengths, 2.116 Å and 2.194 Å, about 10% longer than a usual carbon-bromine bond.

Tribromide ion is analogous to, but less stable than, the triiodide ion (Section 10.10)

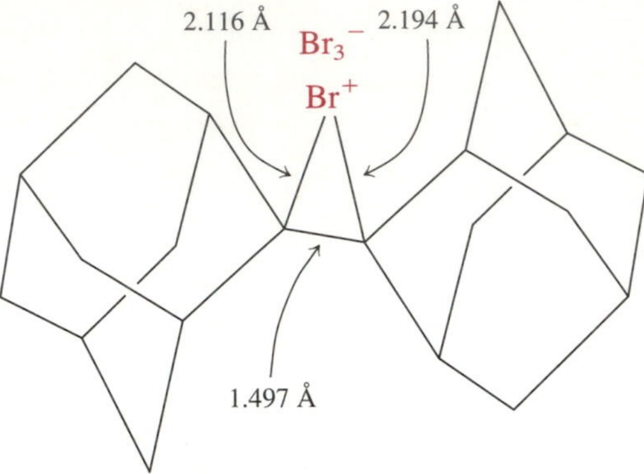

bromonium tribromide from adamantylidene adamantane

The cyclic halonium ion intermediate has an important effect on the *stereochemistry* of halogen additions. The reaction of halide ion with the cyclic ion is a nucleophilic displacement reaction. Since the nucleophile Br^- must approach carbon to the rear of the leaving group, the net result is anti addition of Br_2 to the double bond.

nucleophile $Br:^-$ — Br^+ leaving group

As a result, addition of halogens to cycloalkenes having cis double bonds gives the trans product.

Cyclopentene $\xrightarrow[\text{CCl}_4,\ -5°\text{C}]{\text{Br}_2}$ *trans*-1,2-dibromocyclopentane

In additions to cyclohexene the initial product is the diaxial dibromide, which immediately undergoes chair-chair conformational interconversion to give the more stable diequatorial conformer.

trans-1,2-dibromocyclohexane

Note the conformation of cyclohexene that is indicated in this equation. The double bond and the four atoms attached to it lie in one plane, as shown by the stereoscopic representation in Figure 11.14. The remaining two ring carbons lie above and below this plane in order to stagger the hydrogens. In the conventional symbol used in the foregoing equation, we are viewing the molecule *in the plane of the double bond*. We see the carbon-carbon double bond in front, with C-4 and C-5 above and below this plane, respectively. The hydrogens are usually omitted for clarity, although in the present example they are included to emphasize the stereochemistry of the addition reaction.

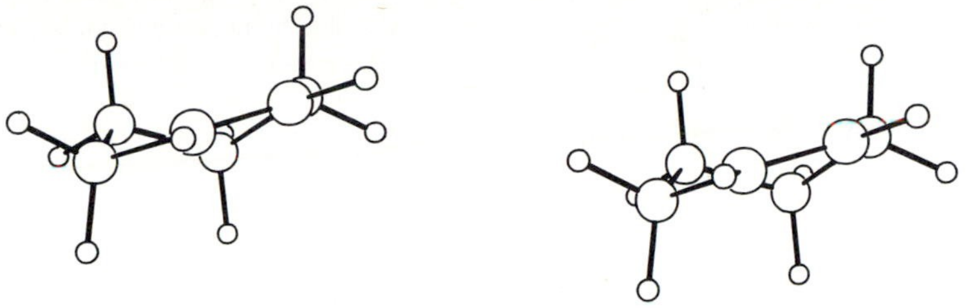

FIGURE 11.14 Stereoscopic representation of cyclohexene. The preference for diaxial addition in cyclohexenes is related mechanistically to the preference for diaxial elimination (Section 11.5.A).

EXERCISE 11.16 Construct a molecular model of *trans*-1,2-dibromocyclohexane with the two bromine atoms in axial positions. Note that there is an anti-coplanar arrangement of the four-atom array Br—C—C—Br. Imagine that one bromine uses one of its nonbonding electron pairs to attack the other C—Br from the rear and displace bromide ion. What is the product of this internal S_N2 process? Now change your *trans*-1,2-dibromocyclohexane model to the conformation with the two bromines in equatorial positions. Can a similar S_N2 process occur in this conformation? Keeping in mind the principle of microscopic reversibility (page 107), what does this exercise tell you about the stereochemistry of the addition of bromine to cyclohexene?

When reaction of an alkene with bromine is carried out in an inert solvent such as carbon tetrachloride, the only nucleophilic reagent available for reaction with the intermediate cation is bromide ion. In hydroxylic solvents, the solvent itself is nucleophilic and can react in competition with the bromide ion.

$$C_6H_5CH{=}CH_2 \xrightarrow[\text{Br}_2\text{ in CH}_3\text{OH}]{\text{dilute solution of}} \underset{\text{minor}}{C_6H_5CHBrCH_2Br} + \underset{\text{major}}{C_6H_5\overset{\overset{\displaystyle OCH_3}{|}}{C}HCH_2Br}$$

The relative amounts of dibromide and bromo ether produced depend on the concentration. For dilute solutions the product is almost exclusively the bromo ether.

Similarly, aqueous or aqueous alkaline solutions of chlorine or bromine produce the corresponding halo alcohols.

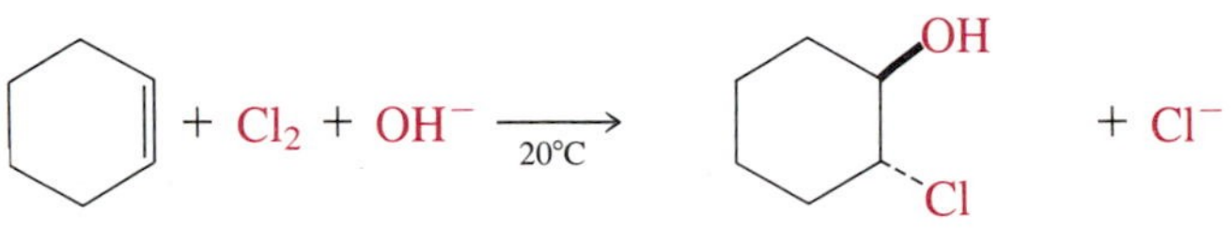

$$\text{cyclohexene} + Cl_2 + OH^- \xrightarrow[20°C]{} \text{(OH, Cl trans)} + Cl^-$$

trans-2-chlorocyclohexanol

A solution of chlorine in aqueous sodium hydroxide cooled with ice is added in portions to cyclohexene keeping the temperature at 15-20 °C. The mixture is saturated with salt and steam distilled. The distillate is saturated with salt and extracted with ether. Distillation of the dried ether solution gives 70–73% yield of product.

Solutions of chlorine and bromine in water are in equilibrium with the corresponding hypohalous acids.

$$X_2 + H_2O \rightleftharpoons H^+ + X^- + HOX$$

A saturated solution of chlorine in water at 25 °C is 0.09 M and is one-third converted to chloride ion and hypochlorous acid. In saturated bromine water (0.2 M) 0.5% of the bromine is converted to hypobromous acid. Thus, although the formation of halo alcohol may be formally regarded as an addition of HO—X across the double bond, the mechanism probably involves a reaction of an intermediate halonium ion with water or hydroxide ion.

The halo alcohols that result from these additions provide a useful route to epoxides (Section 10.11).

OH
Cl
$+ OH^- \longrightarrow$ O $+ H_2O + Cl^-$

cyclohexene oxide

EXERCISE 11.17 Make a conformational drawing showing the chair conformation of *trans*-2-chlorocyclohexanol that is the *initial product* of the reaction of hydroxide ion with the intermediate halonium ion. What is the geometric relationship of the carbon-oxygen bond to the carbon-chlorine bond? What is the stable conformation of this molecule? Now write a mechanism for the conversion of *trans*-2-chlorocyclohexanol into cyclohexene oxide, taking care to indicate any conformational changes that must occur as separate steps.

The 1,2-dihalides produced by the addition of halogens to alkenes are called **vicinal** dihalides (L., *vicinus,* near). They have many chemical properties in common with simple alkyl halides. For example, 1,2-dibromoethane readily enters into nucleophilic displacement reactions.

$$BrCH_2CH_2Br + 2\ CN^- \longrightarrow NCCH_2CH_2CN + 2\ Br^-$$

As with the simple monohalides, nucleophilic displacement is usually accompanied by some elimination, particularly when one or both of the halogens is attached to a secondary carbon. With strong bases, elimination is the principal reaction.

$$\underset{\text{1,2-dichlorobutane}}{CH_3CH_2\overset{Cl}{\overset{|}{C}}HCH_2Cl} + CH_3O^- \longrightarrow \underset{\text{1-chloro-1-butene}}{CH_3CH_2CH{=}CHCl} + \underset{\text{2-chloro-1-butene}}{CH_3CH_2\overset{Cl}{\overset{|}{C}}{=}CH_2}$$

Such dehydrohalogenations are not generally useful ways to prepare haloalkenes because both isomers are usually produced.

EXERCISE 11.18 What products are expected to result from each of the following reactions? Show the complete stereostructure of the product where relevant.

(a) 1-methylcyclohexene + bromine
(b) *cis*-2-butene + chlorine + aqueous NaOH at 10 °C
(c) the product of (b) + aqueous NaOH at 100 °C

Highlight 11.4

Halogens are **electrophilic** reagents (Gk., *electron,* amber [which acquires negative charge by rubbing] and Gk., *philos,* loving). The halogens Br_2 and Cl_2 react with alkenes in two stages. The first step produces a cyclic halonium ion, a positively charged three-membered ring, and a halide ion. The halide ion then reacts on the backside of a carbon of the three-membered ring, forming a dihalide with a trans arrangement of the halide groups, X—C—C—X.

C. Addition of HX and Water

The region above and below a double bond is electron-rich because of the π-bond. Consequently, alkenes have a tendency to act as Lewis bases and react with electrophilic reagents. An example is the reaction of 2-methylpropene with HCl. In the first step the double bond reacts with a proton to give a carbocation intermediate, which combines with chloride ion to give *t*-butyl chloride.

$$(CH_3)_2C{=}CH_2 + H{-}Cl \longrightarrow (CH_3)_2\overset{+}{C}{-}CH_3 + Cl^- \longrightarrow (CH_3)_3C{-}Cl$$

Some further examples of this reaction are

4-methyl-1-pentene + HBr ⟶ 2-bromo-4-methylpentane

1-methylcyclohexene + HCl ⟶ 1-chloro-1-methylcyclohexane

Gaseous hydrogen chloride is bubbled slowly into a solution of 1-methylcyclohexene in methylene chloride at 0 °C. The end of the reaction is marked by the appearance of HCl fumes over the surface of the reaction mixture. The methylene chloride solution is washed with aqueous sodium bicarbonate, dried, and evaporated to obtain 1-chloro-1-methylcyclohexane.

With unsymmetrical alkenes the initial protonation occurs so as to afford the *more stable carbocation*. Since alkyl substituents stabilize carbocations, the proton adds to the less substituted carbon of the double bond.

$$(CH_3)_2C{=}CH_2 + HCl \longrightarrow (CH_3)_2\overset{+}{C}{-}CH_3 \quad \text{not} \quad CH_3{-}CH(CH_3){-}\overset{+}{C}H_2$$

This generalization is commonly referred to as **Markovnikov's rule.** It was formulated by Markovnikov long before the foregoing mechanistic interpretation was developed to explain it.

EXERCISE 11.19 Refer back to the examples given for the addition of HCl and HBr to unsymmetrical alkenes in this section. Note the operation of Markovnikov's rule. Write a reaction mechanism for each example, showing the intermediate carbocation in each case.

If two intermediate carbocations of comparable stability can be formed, a mixture of products results.

$$(CH_3CH_2)_2C{=}C(CH_3)_2 + HCl \longrightarrow (CH_3CH_2)_2CH{-}\overset{+}{C}(CH_3)_2 + (CH_3CH_2)_2\overset{+}{C}{-}CH(CH_3)_2$$

3-ethyl-2-methyl-2-pentene

$$\longrightarrow (CH_3CH_2)_2CH{-}CCl(CH_3)_2 + (CH_3CH_2)_2CCl{-}CH(CH_3)_2$$

2-chloro-3-ethyl-2-methylpentane 3-chloro-3-ethyl-2-methylpentane

The addition of HX to a double bond is a significant reaction because of what it reveals about the general chemistry of alkenes, but it is not an important method for preparing the simpler alkyl halides, which are prepared either by halogenation of alkanes or from the corresponding alcohols.

EXERCISE 11.20 Addition of HBr to *trans*-2-pentene gives a mixture of 2-bromopentane and 3-bromopentane. Write a mechanism, showing all intermediates, and explain this result.

The **hydration** of alkenes is an important industrial method for the manufacture of alcohols. Hydration is usually accomplished by passing the alkene into a mixture of sulfuric acid and water. For example, gaseous isobutylene is absorbed in 60–65% aqueous sulfuric acid. The intermediate formed is undoubtedly the *t*-butyl cation, which reacts with water to give *t*-butyl alcohol.

$$(CH_3)_2C{=}CH_2 + H^+ \rightleftharpoons (CH_3)_3C^+ \xrightarrow{H_2O} (CH_3)_3C{-}\overset{+}{O}H_2 \rightleftharpoons (CH_3)_3COH + H^+$$

The reaction is the reverse of acid-catalyzed dehydration. Low temperatures and aqueous solution favor formation of the alcohol, whereas high temperatures and distillation of the alkene as it is formed shift the equilibrium toward the alkene. Under more vigorous conditions dimeric and polymeric products are produced.

Ethylene is also absorbed by sulfuric acid, but in this case 98% H_2SO_4 is required. The product is ethylsulfuric acid, which is hydrolyzed to ethyl alcohol in a separate step. This reaction may involve ethyl cation as an intermediate and is a rare example of the involvement of primary alkyl cations in solution.

$$H_2SO_4 + CH_2{=}CH_2 \longrightarrow CH_3CH_2^+ + HOSO_3^- \longrightarrow CH_3CH_2OSO_2OH \xrightarrow{H_2O} CH_3CH_2OH + H_2SO_4$$

Although direct hydration is an important industrial process, it is seldom used as a laboratory procedure. Yields of alcohol are highly variable and depend strongly on the exact reaction conditions. A much more reliable procedure for small-scale work involves the use of mercuric ion, Hg^{2+}, as an electrophile. If an alkene is treated with an aqueous solution of mercuric acetate or mercuric perchlorate, a hydroxyalkylmercuric salt is formed.

$$\mathrm{CH_3CH{=}CHCH_3 + Hg(OAc)_2 \xrightarrow[25°C]{H_2O\text{-}THF} CH_3\underset{\underset{HgOAc}{|}}{CH}\overset{\overset{OH}{|}}{C}HCH_3}$$

The mechanism of this reaction involves initial ionization of the mercuric acetate to provide the acetoxymercuric ion, $AcOHg^+$. This species reacts with the carbon-carbon double bond, in much the same manner as a proton. The carbocation so formed reacts with water to give the hydroxyalkylmercuric salt.

(1) $\mathrm{Hg(OAc)_2 \rightleftharpoons AcO^- + AcOHg^+}$

(2) $\mathrm{AcOHg^+ + \diagdown C{=}C \diagup \rightleftharpoons AcOHg{-}\overset{|}{\underset{|}{C}}{-}C^+ \diagdown}$

(3) $\mathrm{AcOHg{-}\overset{|}{\underset{|}{C}}{-}C^+ + H_2O \rightleftharpoons AcOHg{-}\overset{|}{\underset{|}{C}}{-}\overset{|}{\underset{|}{C}}{-}OH + H^+}$

Mercuric acetate in methanol or ethanol readily yields the corresponding alkoxyalkylmercuric acetate.

$$\mathrm{CH_3CH{=}CH_2 + Hg(OAc)_2 \xrightarrow[25°C]{CH_3OH} CH_3\overset{\overset{OCH_3}{|}}{C}HCH_2HgOAc}$$

These compounds are readily reduced with sodium borohydride, which replaces the carbon-mercury bond by C—H with liberation of free mercury. The intermediate organomercury compounds need not be isolated. The net result of mercuration in alcohol or water, followed by sodium borohydride reduction, is addition of alcohol or water to the alkene. The reduction is an excellent method for the synthesis of alcohols and ethers. **Addition follows the Markovnikov rule,** Hg^{2+} (and ultimately, hydrogen) becoming attached to the less substituted carbon.

$$\underset{\text{1-hexene}}{n\text{-}\mathrm{C_4H_9CH{=}CH_2}} \xrightarrow[\text{aq. THF}]{\mathrm{Hg(OAc)_2}} \xrightarrow[\mathrm{NaOH}]{\mathrm{NaBH_4}} \underset{\substack{(96\%)\\ \text{2-hexanol}}}{n\text{-}\mathrm{C_4H_9CHOHCH_3}} + \mathrm{Hg}$$

> 1-Hexene is added with stirring to an equivalent amount of mercuric acetate in 1:1 water-THF. After stirring for 10 min at 25 °C, aqueous NaOH is added, followed by a 0.5 M solution of $NaBH_4$ in 3 M NaOH. The organic layer is separated, dried and distilled to yield 2-hexanol.

The sodium borohydride used in the foregoing reaction is an important reagent in organic chemistry.

> Sodium borohydride, $NaBH_4$, is a salt containing the borohydride ion, BH_4^-. This anion has a tetrahedral structure and can be regarded as being derived from BH_3 and hydride ion, H^-. Sodium borohydride is a white powder and dissolves in water to form stable solutions at basic pH. In acidic medium, $NaBH_4$ reacts rapidly to form hydrogen and sodium borate. Sodium borohydride is soluble in methanol and ethanol, but decomposes slowly in these solvents. It is appreciably soluble in diglyme (5.5 g per 100 g of solvent), but is almost insoluble in glyme or tetrahydrofuran.

The reaction combination of mercuration and reduction is a useful laboratory alternative to acid-catalyzed hydration of olefins. Of course, it cannot compete in cost with sulfuric acid for large-scale commercial production.

EXERCISE 11.21 What products are expected from each of the following reactions?

(a) 1-methylcyclohexene + $Hg(OAc)_2$ in methanol, followed by $NaBH_4$
(b) 2-methyl-2-pentene + $Hg(OAc)_2$ in water-THF, followed by $NaBH_4$

D. Hydroboration

Although the reaction of alkenes with diborane was discovered less than fifty years ago, it has become one of the most important reactions in the repertoire of the synthetic chemist.

Diborane B_2H_6 is a colorless, toxic gas that is spontaneously flammable in air. It is usually prepared by the reaction of sodium borohydride with boron trifluoride.

$$3\ NaBH_4 + 4\ BF_3 \longrightarrow 2\ B_2H_6 + 3\ NaBF_4$$

Borane itself, BH_3, is not known. In this compound boron has a sextet of electrons and is a Lewis acid. In ethers such as tetrahydrofuran or diglyme, common solvents for hydroboration reactions, diborane dissolves readily to form an ether-monomer complex, $R_2O{:}BH_3$. Commercially, borane is often sold as the THF complex.

Diborane has an unusual bridged structure because it is an electron-deficient compound. The 12 valence electrons are too few to provide enough normal two-electron bonds for an ethane-like structure with six boron-hydrogen bonds. In the actual structure four hydrogens and the two borons define a plane with four two-electron boron-hydrogen bonds.

H
H B B H
H H
H

diborane

The other two hydrogens lie above and below this plane and involve three-center two-electron bonds symbolized by ⅄ (see pages 161–162).

Boron trifluoride is a colorless gas, b.p. −100 °C, and is available commercially in cylinders. The compound has a planar structure. The Lewis structure shows that there are only six electrons around boron.

:F̤—B—F̤:
|
:F̤:

The tendency of boron to combine with an electron pair to form an octet is augmented by the electron-attracting character of the attached fluorines. Boron trifluoride is a strong Lewis acid. It reacts avidly with water to form a hydrate, $F_3\overset{-}{B}—\overset{+}{O}H_2$, which is itself a strong acid but slowly hydrolyzes in water to form boric acid and HF. In fact, BF_3 has a strong affinity generally for oxygen, nitrogen, and fluorine. With HF it forms fluoroboric acid, HBF_4, a strong acid in aqueous solution. With ethyl ether it forms the complex $(C_2H_5O){:}BF_3$, boron trifluoride etherate, which can be formulated as $(C_2H_5)_2\overset{+}{O}—\overset{-}{B}F_3$. This compound is a distillable liquid, b.p. 126 °C, and is water-white when pure. We will encounter it often as a useful Lewis acid catalyst.

The boron-hydrogen bond adds rapidly and quantitatively to many multiple bonds including carbon-carbon double bonds. With simple alkenes the product is a trialkylborane.

$$3\ CH_3CH_2CH{=}CH_2 + BH_3 \xrightarrow{THF} (CH_3CH_2CH_2CH_2)_3B$$

tri-*n*-butylborane

The addition appears to be dominated by steric considerations. The boron generally becomes attached to the less substituted and less sterically congested carbon. With highly substituted or hindered alkenes, addition may stop at the mono- or dialkylborane stage. The reaction appears to involve initial coordination of BH_3 with the π-electrons of the double bond followed by formation of the carbon-hydrogen bond.

H—BH_2 ... C⋯C ⟶ H BH_2 C—C

In cases where stereochemistry may be defined, exclusive syn addition is observed.

CH_3 $\xrightarrow{B_2H_6}$ H, CH_3, H, BR_2

The alkylborane products are generally not isolated but are converted by subsequent reactions directly into desired products. The most important general reaction of alkylboranes is that with alkaline hydrogen peroxide.

CH_3, BR_2 $\xrightarrow[OH^-]{H_2O_2}$ CH_3, OH

trans-2-methylcyclopentanol

Three separate processes are involved in the oxidation of alkylboranes to alcohols. In the first step, hydroperoxide anion adds to the electron-deficient boron atom.

$$R_3B + {}^-OOH \longrightarrow R_3B^-{-}O{-}OH$$

The resulting intermediate rearranges with loss of hydroxide ion. The driving force for the rearrangement is liberation of the stable anion OH^- and formation of the strong boron-oxygen bond.

$$R_3B^-{-}O{-}OH \longrightarrow R_2B{-}OR + OH^-$$

The boron-oxygen bond is much stronger than a boron-carbon bond because of overlap of an oxygen *p*-orbital with its lone pair of electrons and the empty *p*-orbital on boron.

empty ⟶ •• ⟵ full
---B — O —

The oxygen lone pair becomes polarized toward boron in the sense indicated by the resonance structures

$$>B-\ddot{O}- \longleftrightarrow >\overset{-}{B}=\overset{+}{O}-$$

Hydrogen peroxide is available as aqueous solutions ranging from 3 to 30% in concentration. The compound is thermodynamically unstable with respect to water and oxygen, and high-strength solutions are explosively hazardous. The 3% solution is used medicinally as a topical antiseptic, but the 30% solution is commonly used in the organic laboratory. Even with the 30% reagent, experiments should be carried out behind safety shields and the material should be kept out of contact with skin and eyes. The pure anhydrous liquid, b.p. 150.2 °C, is stable in inert containers and in the absence of catalytic impurities.

The migration of an alkyl group, and its bonding electron pair, is analogous to the rearrangements of carbocations that we considered in Section 9.8.

$$-\underset{R}{C}-\overset{+}{C}- \longrightarrow -\overset{+}{C}-\underset{R}{C}-$$

$$-\underset{R}{\overset{-}{B}}-O-OH \longrightarrow -B-\underset{R}{O} + OH^-$$

The reaction of boranes with alkaline hydrogen peroxide is rapid and exothermic. The product R_2BOR reacts further by the same process to give a trialkyl borate ester.

$$R_2BOR + 2\ OOH^- \longrightarrow B(OR)_3 + 2\ OH^-$$

The borate ester is then hydrolyzed under the reaction conditions to the alcohol and sodium borate.

$$B(OR)_3 + OH^- + 3H_2O \longrightarrow 3\ ROH + B(OH)_4^-$$

EXERCISE 11.22 Write all of the steps in a reaction mechanism for the alkaline hydrolysis of trimethyl borate.

The net result of hydroboration and oxidation-hydrolysis is **anti-Markovnikov hydration** of a double bond. The reaction is a relatively simple and convenient laboratory procedure and has become an important synthetic reaction in organic chemistry.

$$\text{1-methylcyclopentene} \xrightarrow{B_2H_6} \xrightarrow[OH^-]{H_2O_2} \text{trans-2-methylcyclopentanol (85\%)}$$

1-methylcyclopentene

(85%)
trans-2-methylcyclopentanol

Diborane prepared by reaction of sodium borohydride and boron trifluoride etherate in diglyme is swept by a stream of nitrogen into a solution of 1-methylcyclopentene in THF at 0 °C. The reaction is completed by addition of aqueous sodium hydroxide followed slowly by 30% hydrogen peroxide. After stirring for an additional period the layers are separated, the aqueous phase is extracted with ether and the combined organic layers are dried and distilled to give 85% of *trans*-2-methylcyclopentanol.

EXERCISE 11.23 Write the structure, paying attention to stereochemistry where pertinent, for the alcohol produced by hydroboration-oxidation of each of the following alkenes.

(a) 4-methyl-1-pentene (b) (*E*)-3-methyl-2-pentene
(c) 1,2-dideuteriocyclohexene

E. Oxidation

Alkenes are oxidized readily by potassium permanganate, $KMnO_4$, but the products depend on the reaction conditions.

Potassium permanganate forms dark purple crystals that dissolve in water to give intense purple solutions. In permanganate anion, MnO_4^-, manganese has an oxidation state of +7. As an oxidizing agent in basic solution, manganese is reduced to manganese dioxide, MnO_2, an insoluble brown compound that is frequently difficult to filter because it tends to form colloidal suspensions. Treatment with SO_2 at this point forms the soluble $MnSO_4$. In acid solution reduction of permanganate to Mn^{2+} occurs. The half-reactions for these two reactions are as follows.

$$3\,e^- + MnO_4^- + 2\,H_2O \longrightarrow MnO_2 + 4\,OH^-$$

$$5\,e^- + MnO_4^- + 8\,H^+ \longrightarrow Mn^{2+} + 4\,H_2O$$

In acid solution potassium permanganate is a strong reagent that attacks organic compounds almost indiscriminately. It will even oxidize HCl to Cl_2. Hence, in organic reactions potassium permanganate is almost always used in neutral or alkaline solutions in which MnO_2 is produced.

Cold dilute potassium permanganate reacts with double bonds to give vicinal diols, which are commonly called glycols.

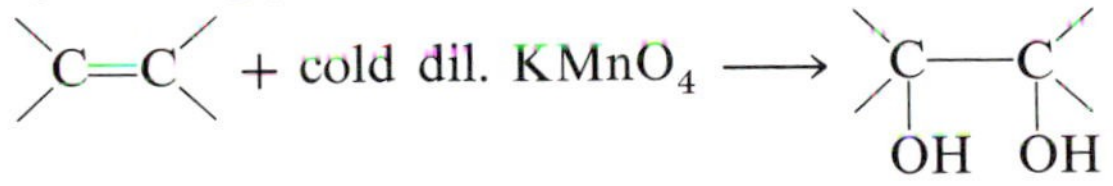

$$\rangle C{=}C\langle \;+\; \text{cold dil. } KMnO_4 \longrightarrow -\underset{OH}{C}-\underset{OH}{C}-$$

a glycol

Reaction conditions need to be carefully controlled. Yields are variable and often low. The reaction occurs with syn addition and is thought to involve an intermediate cyclic manganate ester that is rapidly hydrolyzed.

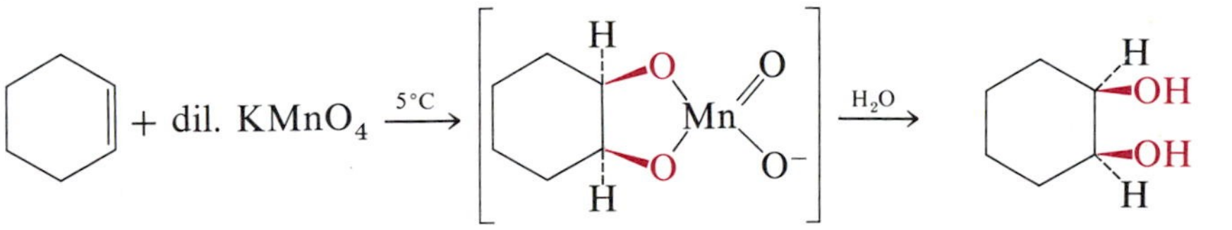

cyclohexene + dil. $KMnO_4$ $\xrightarrow{5^\circ C}$ [cyclic manganate ester, H, O, Mn, O, O⁻, H] $\xrightarrow{H_2O}$ product (H, OH, OH, H)

(30–33%)
cis-1,2-cyclohexanediol

The same overall reaction can be accomplished with osmium tetroxide, a reagent that forms isolable cyclic adducts with alkenes.

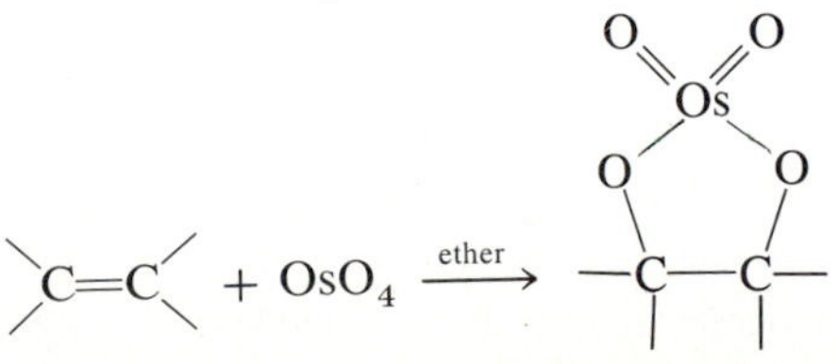

$$\rangle C{=}C\langle \;+\; OsO_4 \xrightarrow{\text{ether}} \text{cyclic osmate ester (O, O, Os, O, O, } -C-C-)$$

black precipitate

Osmium tetroxide, OsO_4, forms colorless or yellow crystals that are soluble in water and in organic solvents. The compound sublimes readily and is highly toxic. It is an expensive reagent (greater than $10 per gram). Because of its cost, it is normally used only in relatively small-scale preparations. It is supplied commercially in small sealed tubes.

The cis diol can be isolated from the alkene-OsO_4 adduct by treatment with H_2S. A more convenient (and less expensive) procedure involves the combination of hydrogen peroxide with a *catalytic* amount of osmium tetroxide. The alkene-OsO_4 adduct is formed as usual, but is converted by the peroxide directly to the cis diol. In this procedure, osmium tetroxide is constantly regenerated, so that only a small amount need be used.

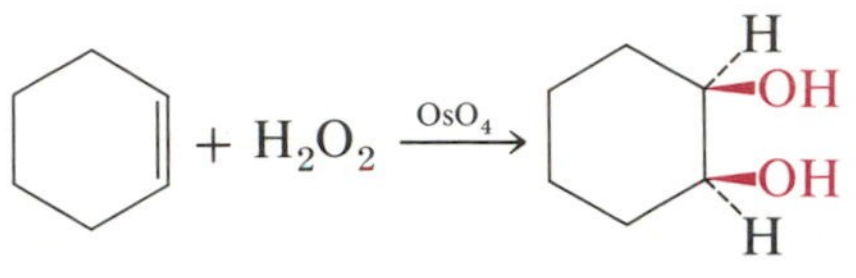

$$\text{cyclohexene} + H_2O_2 \xrightarrow{OsO_4} \textit{cis}\text{-1,2-cyclohexanediol}$$

(58%)

Oxidative cleavage of the double bond can be accomplished by reaction of an alkene with ozone.

Ozone, O_3, is an important constituent of the upper atmosphere, where it is produced by action of solar ultraviolet radiation on atmospheric oxygen. Ozone, in turn, absorbs in the ultraviolet region of the spectrum and provides an important screen that limits the amount of this radiation that reaches the earth's surface. The ozone layer is decreased as a result of photolytic reactions involving chlorofluorocarbons (Section 8.4). Ozone is thermodynamically unstable with respect to oxygen.

$$O_3 \longrightarrow \tfrac{3}{2}\,O_2 \qquad \Delta H° = -34 \text{ kcal mole}^{-1}$$

Ozone is produced in the laboratory with an ''ozonator,'' a special apparatus in which an electrodeless discharge is induced in dry air passing through an alternating electric field. Ozone concentrations as high as 4% in air can be produced. The gas has the characteristic odor usually associated with electric arcs.

Reactions of alkenes with ozone are normally carried out by passing ozone-containing air through a solution of the alkene in an inert solvent at a low temperature (usually −78 °C, the temperature of a dry ice–isopropyl alcohol cooling bath). Reaction is rapid; completion of reaction is determined by testing the effluent gas for ozone. Aqueous potassium iodide reacts with ozone to give iodine which complexes with iodide ion to produce the intensely brown color of triiodide. Suitable solvents for ozonizations include methylene chloride, ethanol, and ethyl acetate. The first-formed addition product, which is called a molozonide, rearranges rapidly, even at low temperatures, to the ozonide structure.

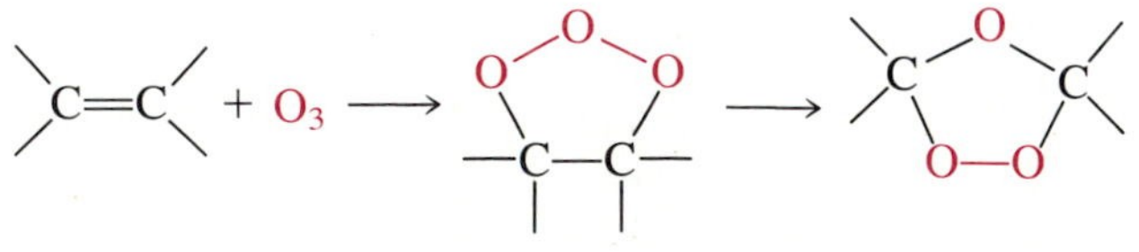

$$\text{C=C} + O_3 \longrightarrow \text{molozonide} \longrightarrow \text{ozonide}$$

molozonide ozonide

In some cases polymers are obtained. Some ozonides, especially the polymers, decompose with explosive violence on heating; hence, the ozonides are generally not isolated but are decomposed directly to desired products.

Hydrolysis with water occurs readily to give carbonyl compounds and hydrogen peroxide.

$$\text{ozonide} + H_2O \longrightarrow \rangle C{=}O + O{=}C\langle + H_2O_2$$

Aldehydes are oxidized by hydrogen peroxide to carboxylic acids. Hence, reducing agents are often used in decomposing the ozonides. Such agents include zinc dust-acetic acid and H_2-Pd/C.

$$\text{6-methyl-1-heptene} \xrightarrow[CH_2Cl_2,\ -78°C]{O_3} \xrightarrow[CH_3CO_2H]{Zn} \text{5-methylhexanal}$$

A solution of 6-methyl-1-heptene in methylene chloride at −78 °C is treated with ozone and is then added to a stirring mixture of zinc dust and 50% aqueous acetic acid. The mixture is refluxed for 1 hr and extracted with ether. The last traces of peroxides are removed from the ether with aqueous potassium iodide and the washed and dried solution is distilled to give 5-methylhexanal, b.p. 144 °C.

Treatment of the ozonide with sodium borohydride gives a mixture of the corresponding alcohols. In the case of a cyclic alkene, the ring is cleaved at the stage of the ozonide, and an open chain diol is produced by hydrolysis and reduction.

$$\text{cyclohexene} \xrightarrow[CH_2Cl_2,\ 0°C]{O_3} \xrightarrow{NaBH_4} HO(CH_2)_6OH$$

(63%)
1,6-hexanediol

EXERCISE 11.24 What products are produced by each of the following reactions. Indicate stereostructure where pertinent.

(a) 1-methylcyclohexene + O_3, followed by zinc-acetic acid
(b) *cis*-2-butene + 10 mole % OsO_4 + H_2O_2
(c) cyclopentene + aqueous $KMnO_4$ at 5 °C

Alkenes may also be oxidized by peroxycarboxylic acids. The product is an oxirane (Section 10.11.A).

$$\text{trans-3-hexene} + CH_3\overset{O}{\overset{\|}{C}}OOH \xrightarrow{CH_2Cl_2} \text{trans-2,3-diethyloxirane} + CH_3CO_2H$$

trans-3-hexene — peroxyacetic acid — *trans*-2,3-diethyloxirane

Peroxycarboxylic acids are related to hydrogen peroxide, in that they have an oxygen-oxygen single bond.

$$R-\overset{O}{\overset{\|}{C}}-O-OH$$

Like H_2O_2, peroxycarboxylic acids are oxidizing agents and are often used for that purpose. The formal name in *Chemical Abstracts* for the peroxy analog of benzoic acid (sometimes the per- analog, perbenzoic acid or peroxybenzoic acid) is benzenecarboperoxoic acid. Peroxycarboxylic acids are generally unstable and must either be stored at low temperatures or prepared as needed. The once widely used 3-chloroperoxybenzoic acid, a crystalline solid, has proven unreliable and is no longer available commercially.

$$CH_3(CH_2)_3CH{=}CH_2 + \text{3-ClC}_6H_4CO_3H \xrightarrow{CHCl_3} CH_3(CH_2)_3\underset{\text{O}}{CH{-}CH_2} + \text{3-ClC}_6H_4COOH$$

3-chloroperoxybenzoic acid; (60%) 2-butyloxirane; 3-chlorobenzoic acid

A useful substitute is magnesium monoperoxyphthalate, a water-soluble, shock-insensitive reagent that is currently sold as the hexahydrate, which is equivalent to monoperoxyphthalic acid. The reagent can be used in a suitable aqueous organic solvent to convert 2-methyl-2-butene to the corresponding oxirane. If reactant solubility is too low, as in the case of limonene, a two-phase system can be used together with a phase-transfer catalyst (Section 23.5.B), enabling the perphthalic acid to be transferred into the organic phase.

2-methyl-2-butene + magnesium monoperoxyphthalate ($\frac{1}{2}Mg^{2+}$) $\xrightarrow[\text{5 hr, 40 °C}]{\text{i-PrOH : }H_2O}$ (98%) trimethyloxirane + magnesium phthalate monoacid

limonene + magnesium monoperoxyphthalate ($\frac{1}{2}Mg^{2+}$) $\xrightarrow[\text{4.5 hr, 50 °C}]{CHCl_3{-}H_2O \text{ (two phases)}}$ (80%) limonene [oxide] + magnesium phthalate monoacid

$(C_8H_{15})_3N^+CH_3Cl^-$ *phase-transfer agent*

The mechanism of the oxidation of alkenes by peroxycarboxylic acids is such that *both new carbon-oxygen bonds are formed at one time*. Thus, the oxidation is stereospecific, and may be considered as an addition reaction in which the two new bonds (which happen both to be to the same oxygen) are formed in a syn manner.

Recall that oxiranes may also be synthesized by addition of aqueous halogen to an alkene, followed by base-catalyzed cyclization (page 275).

1-methylcyclohexene $+ Br_2 \xrightarrow{H_2O}$ trans-2-bromo-1-methylcyclohexanol $\xrightarrow[C_2H_5OH]{C_2H_5O^-}$ 1-methylcyclohexene oxide

---

EXERCISE 11.25 We saw in Section 10.11.A that oxiranes undergo ring-opening upon treatment with aqueous acid. What is the stereostructure of the product resulting from application of the following reaction sequence to *trans*-2-butene? (1) peroxyacetic acid, CH_2Cl_2; (2) 10% H_2SO_4, 100 °C.

F. Addition of Carbenes and Carbenoids

Carbenes are reactive intermediates that have the general formula R_2C:, in which carbon has only a sextet of electrons. Although carbenes are neutral species, the electron-deficient carbon is still ''hungry'' for electrons, and hence carbenes behave as **electrophiles.** One way in which carbenes may be generated is by the reaction of chloroform with a strong base. Because of the strong electron-attracting effect of the three chlorines, chloroform is rather acidic ($pK_a \cong 25$). Thus, upon treatment of chloroform with hydroxide or alkoxide ion, the trichloromethyl anion is formed in a small equilibrium concentration. If the reaction is carried out in water, the $Cl_3C:^-$ is mainly protonated to regenerate chloroform. However, in less acidic solvents such as *t*-butyl alcohol, the trichloromethyl anion lives long enough to lose a chloride ion to give dichlorocarbene.

$$\mathrm{CHCl_3 + RO^- \rightleftharpoons ROH + \underset{\text{trichloromethyl anion}}{Cl_3C:^-} \longrightarrow Cl^- + \underset{\text{dichlorocarbene}}{Cl_2C:}}$$

If dichlorocarbene is generated in the presence of an alkene, it adds to the double bond to give a 1,1-dichlorocyclopropane.

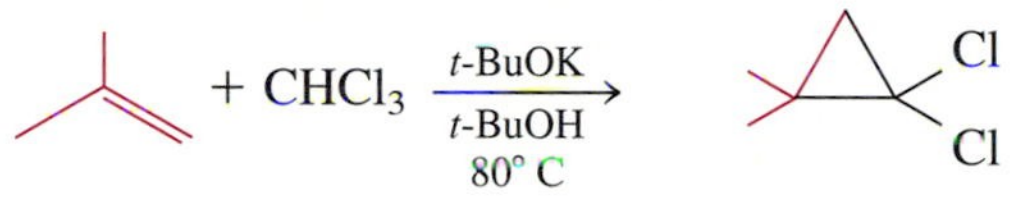

1,1-dichloro-2,2-dimethyl-cyclopropane

The electronic structure of the intermediate dichlorocarbene has a pair of electrons in an orbital that has hybridization of approximately sp^2. In addition, the carbene carbon also has a vacant *p*-orbital (Figure 11.15). Thus, a carbene has both a carbanion lone pair and a carbocation vacancy on a single carbon.

As in the epoxidation of alkenes, the addition of dichlorocarbene is stereospecific—syn addition of the two new carbon-carbon bonds is observed.

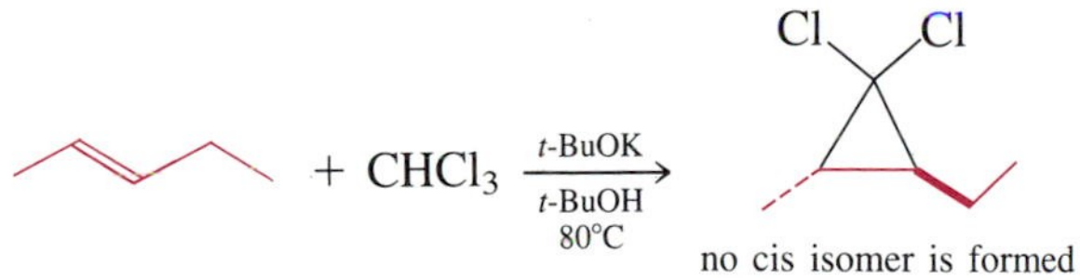

no cis isomer is formed

This reaction can also be applied to bromoform, $CHBr_3$, to yield the corresponding dibromocyclopropanes.

Carbene itself, $:CH_2$, can be produced by photolysis or pyrolysis of a compound known as diazomethane, CH_2N_2. However, a better way to add $:CH_2$ to an alkene utilizes

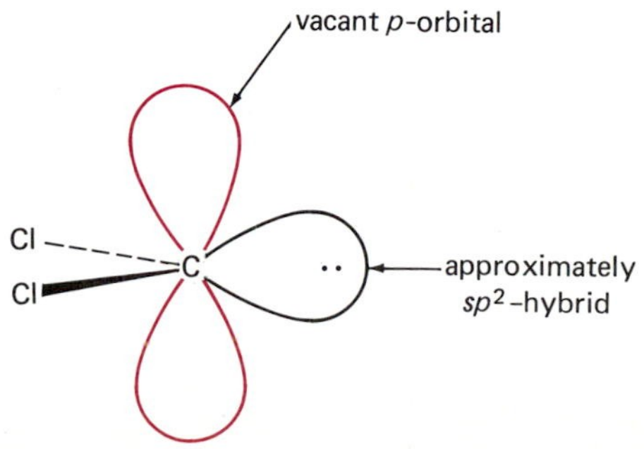

FIGURE 11.15 Electronic structure of dichlorocarbene.

a mixture of diiodomethane and zinc dust (Simmons-Smith reaction). In practice, the zinc dust is usually activated by alloying it with a small amount of copper or by ultrasonication (Section 8.8.A).

$$\text{1-heptene} + CH_2I_2 \xrightarrow[\text{ether}]{\text{Zn(Cu)}} \text{pentylcyclopropane (30\%)}$$

The reaction is applicable to many kinds of double bonds. Yields are generally only fair (30–70%), but the products are often difficult to prepare by alternative routes. The method appears to involve the formation of an iodomethylzinc organometallic species (Section 8.8.A). It is proposed that this species, which is sometimes termed a **carbenoid** because it behaves in some ways like a true carbene, adds to the double bond to give an adduct, which then eliminates zinc iodide.

(1) $CH_2I_2 + Zn \longrightarrow ICH_2ZnI$

(2) $>C{=}C< + ICH_2ZnI \longrightarrow ICH_2{-}\overset{|}{\underset{|}{C}}{-}\overset{|}{\underset{|}{C}}{-}ZnI$

(3) $ICH_2{-}\overset{|}{\underset{|}{C}}{-}\overset{|}{\underset{|}{C}}{-}ZnI \longrightarrow$ (cyclopropane with H, H on the new CH_2)

EXERCISE 11.26 What is the stereostructure of the product of each of the following reactions.

(a) (*E*)-3-methyl-3-hexene + $CHCl_3$ + potassium *t*-butoxide
(b) *trans*-2-pentene + CH_2I_2 + Zn(Cu) in ether

G. Free Radical Additions

The early literature of organic chemistry contained considerable disagreement on the mode of addition of HBr to terminal olefins. In some cases Markovnikov's rule appeared to hold; in other cases it did not. Often two chemists would add HBr to the same alkene and obtain contradictory results.

$$RCH{=}CH_2 + HBr \longrightarrow RCH_2CH_2Br \quad \text{or} \quad R\overset{Br}{\overset{|}{C}}HCH_3$$

In the 1930s this apparent dilemma was resolved when it was discovered that HBr (but *not* HCl or HI) can add to alkenes by two different mechanisms. Pure materials and pure solvents encourage addition by the electrophilic mechanism discussed in Section 11.6.C, which leads to normal Markovnikov addition.

$$RCH{=}CH_2 + HBr \xrightarrow{\text{(ionic mechanism)}} R\overset{Br}{\overset{|}{C}}HCH_3$$

Impure materials, oxygen, and some other additives were found to promote ''abnormal'' addition by a mechanism involving free radical intermediates.

$$RCH{=}CH_2 + HBr \xrightarrow{\text{(free radical mechanism)}} RCH_2CH_2Br$$

The free radical mechanism starts with an initiation step that results in oxidation of HBr to bromine atoms, sometimes with hydroperoxides present in ether solvents.

$$ROOH + 2\,HBr \longrightarrow ROH + HOH + Br\cdot$$

The bromine atom then adds to the alkene to give a free radical that continues the chain by abstracting hydrogen from a molecule of HBr. Both of the propagation steps are exothermic and have low activation energies.

$$RCH{=}CH_2 + Br\cdot \longrightarrow R\dot{C}HCH_2Br \qquad \Delta H^\circ = -7 \text{ kcal mole}^{-1}$$
$$R\dot{C}HCH_2Br + HBr \longrightarrow RCH_2CH_2Br + Br\cdot \qquad \Delta H^\circ = -10 \text{ kcal mole}^{-1}$$

Note that the bromine atom adds to the alkene in such a way as to give the more highly substituted (more stable) free radical. The overall outcome is thus **anti-Markovnikov** orientation. This abnormal addition or ''peroxide effect'' is a useful reaction with HBr, but is not significant with HCl or HI. The carbon-iodine bond is so weak that the addition of iodine atoms to double bonds is endothermic. It becomes exothermic only at elevated temperatures.

$$RCH{=}CH_2 + I\cdot \longrightarrow R\dot{C}HCH_2I \qquad \Delta H^\circ = +8 \text{ kcal mole}^{-1}$$
$$R\dot{C}HCH_2I + HI \longrightarrow RCH_2CH_2I + I\cdot \qquad \Delta H^\circ = -27 \text{ kcal mole}^{-1}$$

The hydrogen-chlorine bond is so strong that the second step in the sequence is endothermic and slow.

$$RCH{=}CH_2 + Cl\cdot \longrightarrow R\dot{C}HCH_2Cl \qquad \Delta H^\circ = -19 \text{ kcal mole}^{-1}$$
$$R\dot{C}HCH_2Cl + HCl \longrightarrow RCH_2CH_2Cl + Cl\cdot \qquad \Delta H^\circ = +5 \text{ kcal mole}^{-1}$$

Free radical chain reactions work best when both propagation steps are exothermic. An endothermic step corresponds to a slow and reversible reaction that breaks the chain.

Other compounds that have appropriate bond strengths can add to double bonds under free radical conditions. Examples include chlorine, bromine, hydrogen sulfide, thiols, and polyhaloalkanes.

$$CH_3CH_2CH{=}CH_2 + H_2S \xrightarrow[0^\circ C]{h\nu} \underset{(68\%)}{CH_3CH_2CH_2CH_2SH} + \underset{(12\%)}{(CH_3CH_2CH_2CH_2)_2S}$$

$$(CH_3)_2C{=}CH_2 + CH_3CH_2SH \xrightarrow[O_2]{100^\circ C} \underset{\substack{(94\%)\\ \text{ethyl isobutyl sulfide}}}{(CH_3)_2CHCH_2SCH_2CH_3}$$

Carbon tetrachloride and carbon tetrabromide react readily with alkenes and free radical initiators to give 1:1 adducts. The propagation steps are addition of the trihalomethyl radical to C=C and reaction of the free radical so produced with CX_4 to regenerate the trihalomethyl radical.

$$RCH{=}CH_2 + \cdot CX_3 \longrightarrow R\dot{C}HCH_2CX_3$$

$$R\dot{C}HCH_2CX_3 + CX_4 \longrightarrow R\overset{\displaystyle X}{\overset{|}{C}}HCH_2CX_3 + \cdot CX_3$$

Again, the overall result is Markovnikov addition; the trihalomethyl group ends up bonded to the carbon bearing the greater number of hydrogens.

$$CH_3(CH_2)_5CH{=}CH_2 + CCl_4 \xrightarrow[\Delta]{\text{peroxide}} \underset{\substack{(75\%)\\ \text{1,1,1,3-tetrachlorononane}}}{CH_3(CH_2)_5CHClCH_2CCl_3}$$

$$CH_3(CH_2)_5CH{=}CH_2 + CBr_4 \xrightarrow[75^\circ]{h\nu} \underset{\substack{(88\%)\\ \text{1,1,1,3-tetrabromononane}}}{CH_3(CH_2)_5CHBrCH_2CBr_3}$$

At this point it is instructive to review the meanings of the terms "Markovnikov addition" and "anti-Markovnikov addition." When first formulated, the Markovnikov rule simply stated that "in the addition of HX to an alkene, the H goes to the carbon already having the greater number of hydrogens."

Markovnikov addition

$$CH_3{-}CH{=}CH_2 + HX \longrightarrow CH_3{-}\overset{\overset{\displaystyle X}{|}}{C}H{-}\overset{\overset{\displaystyle H}{|}}{C}H_2$$

Anti-Markovnikov addition

$$CH_3{-}CH{=}CH_2 + HX \longrightarrow CH_3{-}\overset{\overset{\displaystyle H}{|}}{C}H{-}\overset{\overset{\displaystyle X}{|}}{C}H_2$$

We now recognize that electrophilic additions obey the Markovnikov rule simply because H^+ is the reagent that adds to the double bond first and *it always adds to the end that produces the more stable intermediate carbocation.*

$$CH_3CH{=}CH_2 + HX \longrightarrow CH_3\overset{+}{C}H{-}\overset{\overset{\displaystyle H}{|}}{C}H_2 \qquad (\text{not } CH_3\overset{\overset{\displaystyle H}{|}}{C}H{-}\overset{+}{C}H_2)$$

Free radical additions "disobey" the Markovnikov rule because in this case it is X· that adds first and this species *adds so as to produce the more stable intermediate free radical.*

$$CH_3CH{=}CH_2 + X\cdot \longrightarrow CH_3\dot{C}H{-}\overset{\overset{\displaystyle X}{|}}{C}H_2 \qquad (\text{not } CH_3\overset{\overset{\displaystyle X}{|}}{C}H{-}CH_2\cdot)$$

EXERCISE 11.27 Write the two propagation steps for addition of methanethiol, CH_3SH, to ethylene. Estimate $\Delta H°$ for each of these steps, using the following bond energies (all in kcal $mole^{-1}$): C=C, 150; C—C, 83; C—S, 65; S—H, 83; C—H, 101.

Highlight 11.5

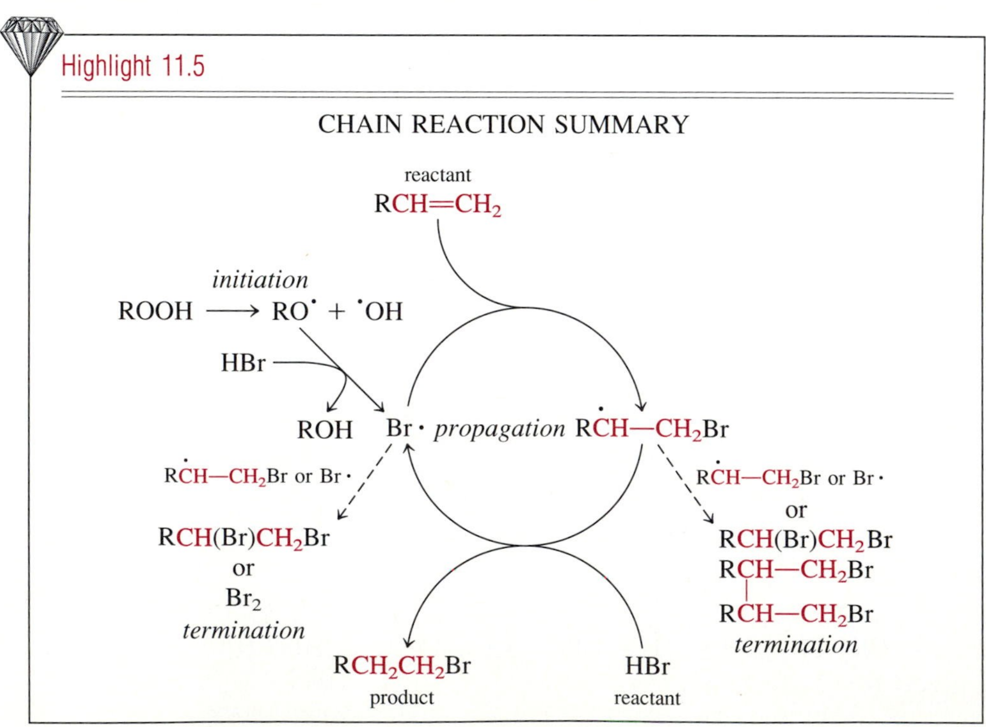

EXERCISE 11.28 Write chain reaction summaries for radical additions of methanethiol, HBr and CCl_4 to 1-butene.

Highlight 11.6

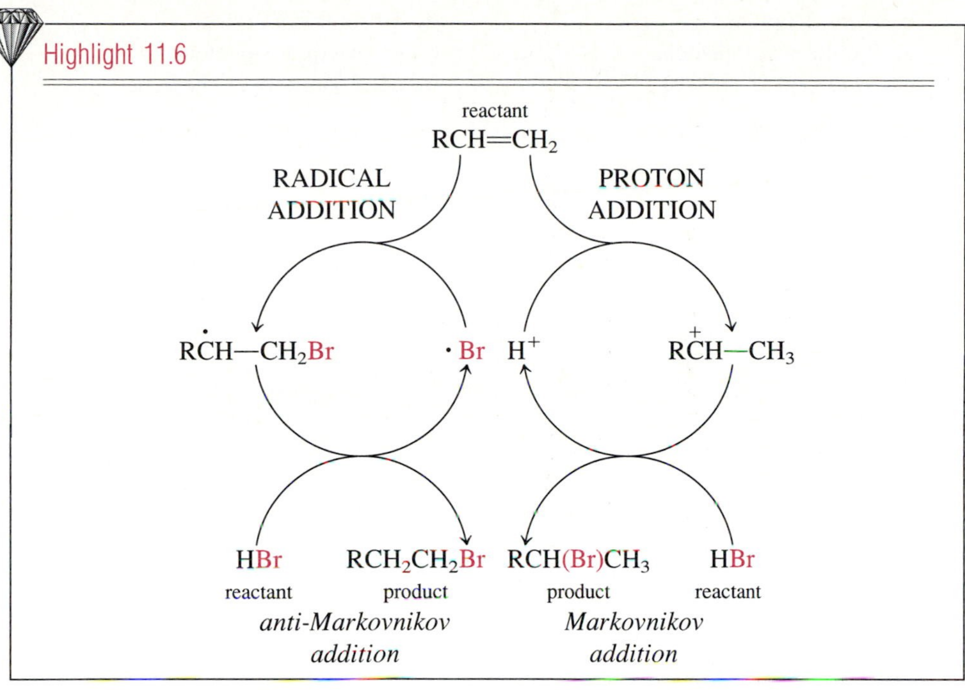

ASIDE

Olefin, the old name for alkenes, is derived from the French, *olefiant,* having made oil, because ethylene, obtained by Dutch chemists from ethanol and concentrated sulfuric acid, reacted with chlorine to give a heavy oil. The word olefin is still sometimes used. Find all six instances of the word in Chapter 11. Give another name for ethylene, and write an equation for its formation from ethanol and sulfuric acid. Under what conditions could another product be formed in this reaction? What is the heavy oil obtained on reaction with chlorine? Replace one hydrogen of ethylene with deuterium: how many isomers are there of deuterioethylene? Replace two hydrogens with deuterium; how many isomers are there of dideuterioethylene?

PROBLEMS

1. **a.** Give the structure and IUPAC name of each of the isomeric pentenes. Which ones are stereoisomers? Which ones are capable of optical activity?
   b. Answer part (a) for the methylcyclopentenes.

2. Write the structure corresponding to each of the following names.
   a. *trans*-3,4-dimethyl-2-pentene
   b. 4-methyl-3-penten-1-ol
   c. *cis*-3-ethyl-2-hexene
   d. vinyl fluoride
   e. 1-bromocyclohexene
   f. (*R*)-3-methylcyclohexene
   g. (*Z*)-2-bromo-2-pentene
   h. (*E*)-3-methyl-2-hexene

i. 3-methyl-*cis*-cyclooctene
j. 3-methyl-*trans*-cyclooctene
k. *trans*-3,4-dimethylcyclobutene
l. 2,3-dimethyl-2-butene

3. Explain why each of the following names is incorrect.

a. 2-methylcyclopentene
b. 2-methyl-*cis*-3-pentene
c. *trans*-1-butene
d. 1-bromoisobutylene
e. 2-chlorocyclopentene
f. (*E*)-3-ethyl-3-pentene
g. *trans*-pent-2-en-4-ol
h. (*Z*)-3-isopropyl-3-heptene

4. Name each of the following compounds. For compounds able to form stereoisomers, use the *E*,*Z* and *R*,*S* nomenclature.

a.

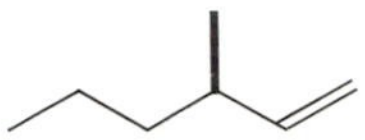

b.

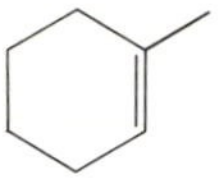

c.

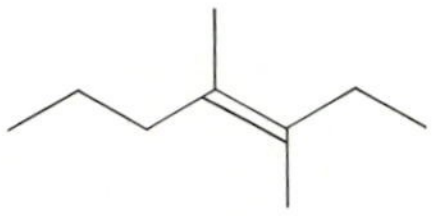

d.

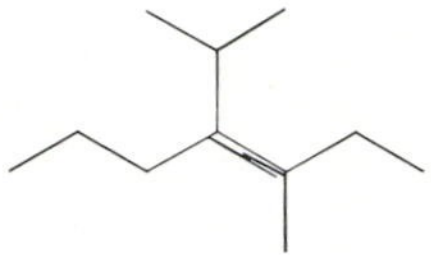

e. 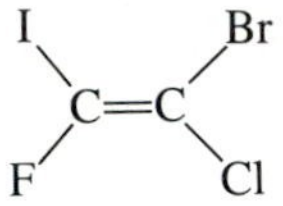 I, F, Br, Cl; C=C

f.  Cl

g. 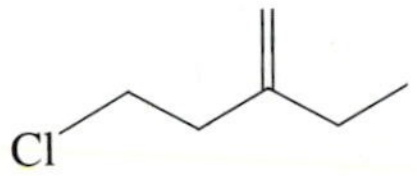 Cl

h. 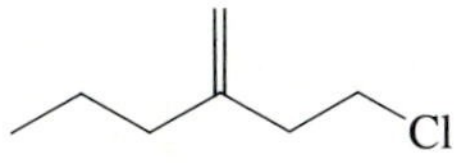 Cl

i. Cl

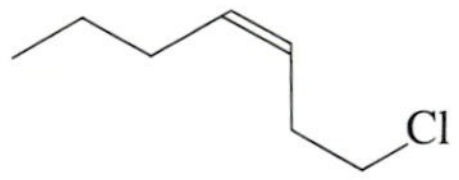

j.  HO

k. 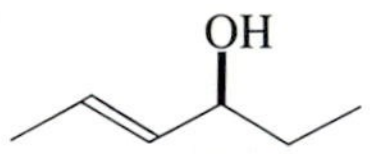 OH

l.

5. Give the alkyl halide and reaction conditions needed to yield each of the following as the primary alkene product. For **d** there are two possible alkyl halides to use as starting material. Give both and be specific about stereoisomers where appropriate.

a. $(CH_3)_2CHCH{=}CHCH_3$
b. $(CH_3)_2CHCH_2CH{=}CH_2$
c. 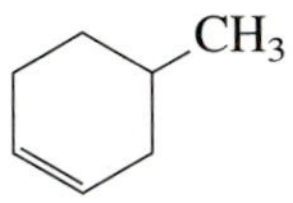 CH_3
d. 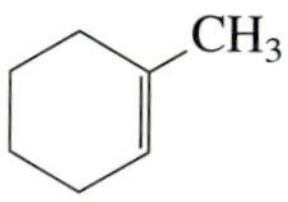 CH_3

6. Give the structure of the principal organic product(s) produced from 3-ethyl-2-pentene under each of the following reaction conditions.

a. H_2/Pd-C
b. H_2O, Br_2
c. Cl_2 in CCl_4 at 0 °C
d. cold dilute $KMnO_4$
e. (i) B_2H_6; (ii) NaOH-H_2O_2
f. (i) aq. $Hg(OAc_4)_2$; (ii) $NaBH_4$
g. (i) O_3; (ii) Zn dust, aq. CH_3COOH
h. HBr, free-radical inhibitor
i. HBr, peroxides
j. Br_2, dilute solution in CH_3OH
k. peroxybenzoic acid in chloroform
l. $CHBr_3$, *t*-BuOK, *t*-BuOH
m. CH_2I_2, Zn(Cu), ether

7. Apply each of the reaction conditions in problem 6 to *cis*- and *trans*-3-hexene. For which reactions are the same products obtained from both stereoisomers? For which reactions do the products differ and how do they differ?

8. Apply each of the reaction conditions in problem 6 to cyclohexene. Specify stereochemistry where pertinent.

9. What is the principal organic product of each of the following reaction conditions? Specify stereochemistry where appropriate.

a. 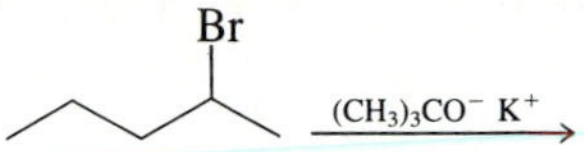 Br $\xrightarrow{(CH_3)_3CO^-\ K^+}$

b. 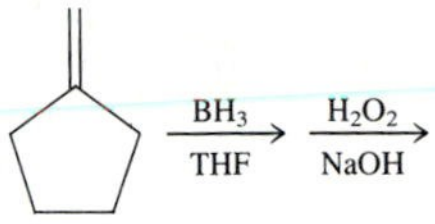 $\xrightarrow[THF]{BH_3}$ $\xrightarrow[NaOH]{H_2O_2}$

c. 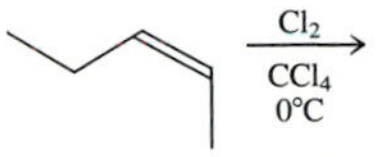 $\xrightarrow[CCl_4,\ 0°C]{Cl_2}$

d. 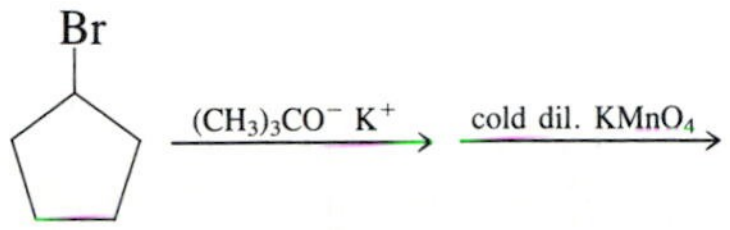 Br $\xrightarrow{(CH_3)_3CO^-\ K^+}$ $\xrightarrow{\text{cold dil. } KMnO_4}$

e. 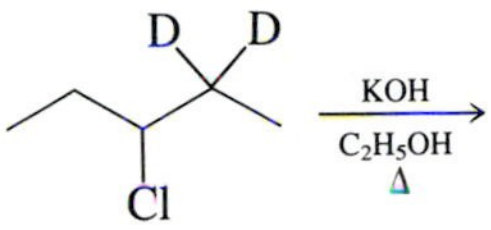 D D Cl $\xrightarrow[C_2H_5OH,\ \Delta]{KOH}$

f. 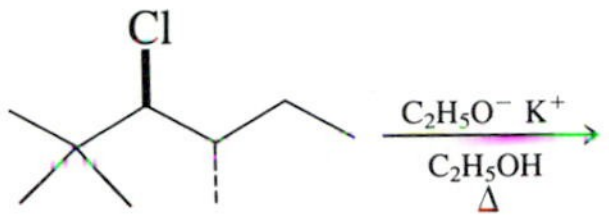 Cl $\xrightarrow[C_2H_5OH,\ \Delta]{C_2H_5O^-\ K^+}$

g. 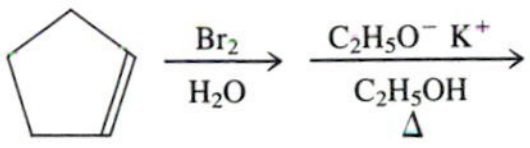 $\xrightarrow[H_2O]{Br_2}$ $\xrightarrow[C_2H_5OH,\ \Delta]{C_2H_5O^-\ K^+}$

h. $\xrightarrow[CH_3OH,\ -60°C]{O_3}$ $\xrightarrow{\text{Zn-HOAc}}$

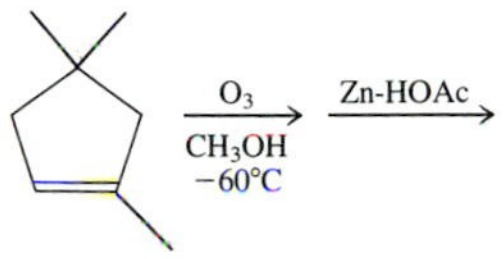

i. 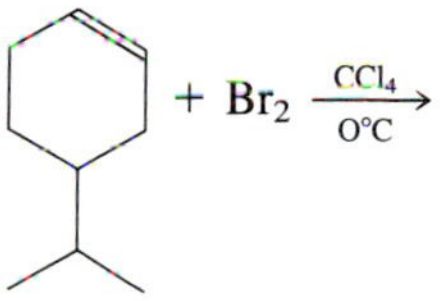 + Br_2 $\xrightarrow[0°C]{CCl_4}$

10. Show how one may accomplish each of the following transformations in a practical manner.

a. 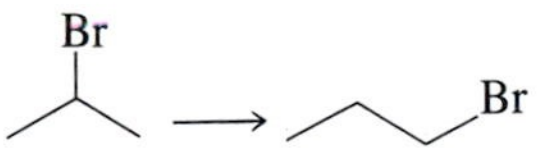 Br → Br

b. 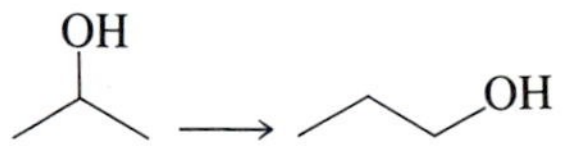 OH → OH

c. 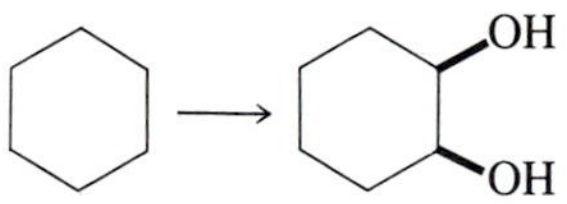 → OH OH

d. 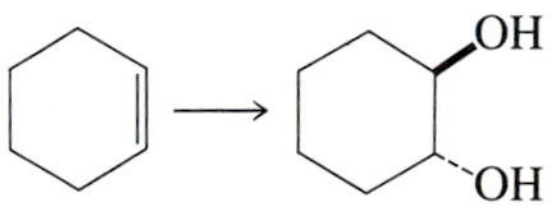 → OH OH

e. 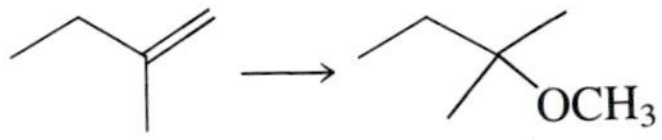 → OCH_3

f. 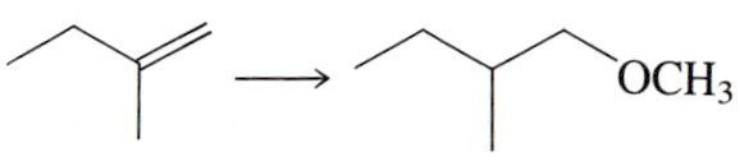 → OCH_3

g. 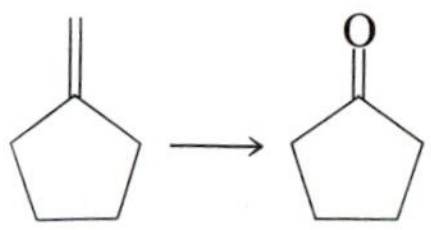 → O

11. Show how one may accomplish each of the following transformations in a practical manner.

a. 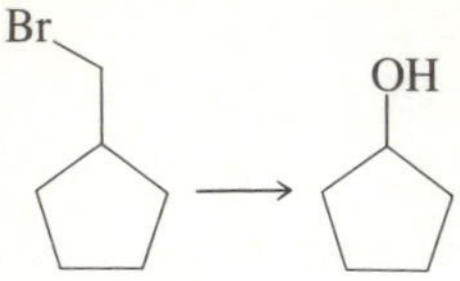 (Bromomethyl)cyclopentane → cyclopentanol structures

b. 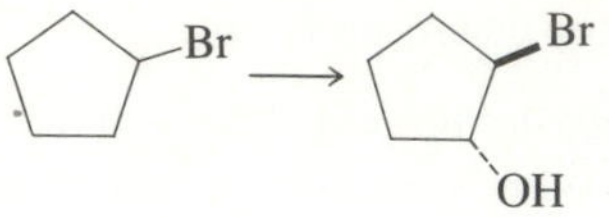 Bromocyclopentane → trans-2-bromocyclopentanol structures (Br, OH)

c. 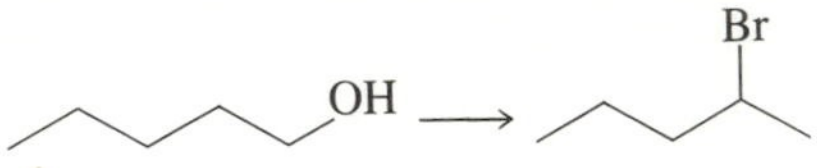 OH → Br

d. 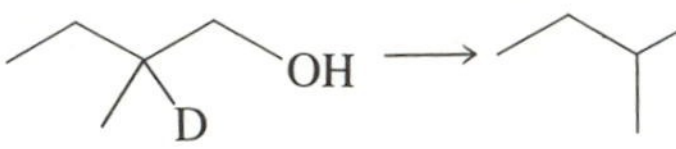 D, OH →

e. 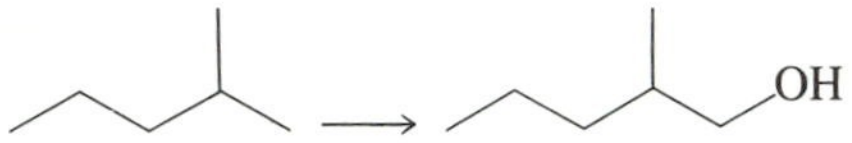 → OH

f. 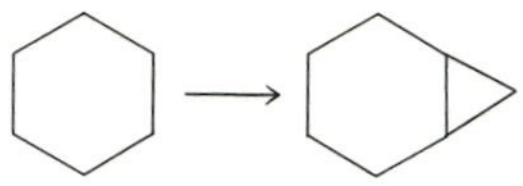 →

g. 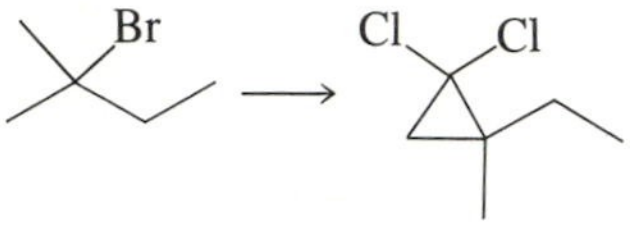 Br → Cl, Cl

12. Starting with alcohols, outline multistep syntheses of each of the following compounds.

a. 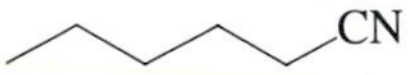 CN

b. 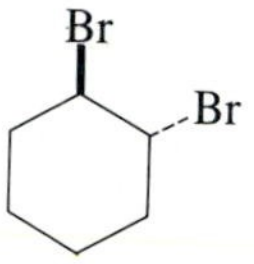 Br, Br

c.  HO OH

d. 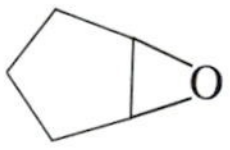 O

13. The potential function for rotation of the methyl group in propylene is approximately that of a threefold barrier with a barrier height of 2.0 kcal mole^{-1}. The most stable conformation is A in which a methyl hydrogen is eclipsed with the double bond. The least stable conformation is B in which H-2 is eclipsed to a methyl hydrogen. Plot the energy of the system as a function of a 360° rotation of the methyl group. Identify the points along this plot that correspond to conformations A and B.

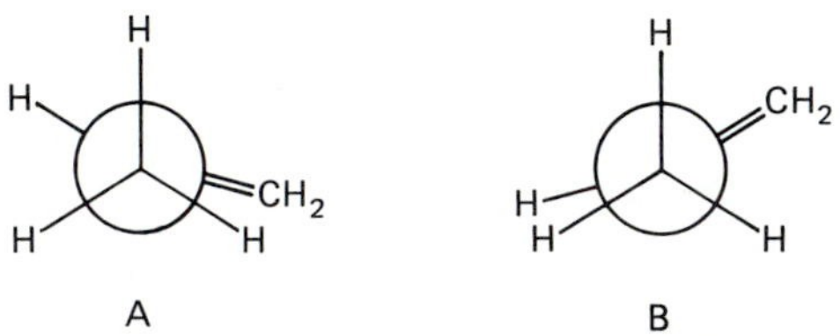

14. Although the difference in energy between cis and trans alkenes is generally about 1 kcal mole^{-1}, for 4,4-dimethyl-2-pentene the cis isomer is 3.8 kcal mole^{-1} less stable than the trans isomer. Explain.

15. In the acid-catalyzed dehydration of 6-methyl-1,6-heptanediol, it is easy to find conditions that give smooth loss of one molecule of water to yield 6-methyl-5-hepten-1-ol. Explain.

16. Explain why the disubstituted olefin 2,4,4-trimethyl-1-pentene predominates in the acid-catalyzed equilibrium with its trisubstituted isomer 2,4,4-trimethyl-2-pentene.

17. Compare the product of addition of bromine to *cis*-1,2-dideuterioethylene and to *trans*-1,2-dideuterioethylene. On treatment with base each of the dibromides gives predominantly a single different dideuteriovinyl bromide. Show the structure in each case. (Remember: HBr is eliminated faster than DBr.)

18. When isopropyl bromide is treated with sodium ethoxide in ethanol, propene and ethyl isopropyl ether are formed in a 3:1 ratio. If the hexadeuterioisopropyl bromide, $CD_3CHBrCD_3$, is used, $CD_3CH{=}CD_2$ and $(CD_3)_2CHOC_2H_5$ are formed in a ratio of 1:2. Explain.

19. The heat of hydrogenation, ΔH°_{hydrog}, is defined as the enthalpy of the reaction of an alkene with hydrogen to give the alkane.

$$CH_2{=}CH_2 + H_2 \longrightarrow CH_3CH_3 \qquad \Delta H^\circ_{hydrog.} = -32.7 \text{ kcal mole}^{-1}$$

From the heats of formation given in Appendix I calculate heats of hydrogenation for a number of simple alkenes. Note that all monoalkyl ethylenes have about the same ΔH°_{hydrog} which is more positive than that for ethylene. Explain. How would you expect ΔH°_{hydrog} to compare for isomeric cis and trans olefins?

20. Reaction of either 1-butene or 2-butene with HCl gives the same product, 2-chlorobutane, via the same carbocation, 2-butyl cation. Yet, the reaction of 1-butene is faster than that of 2-butene. Explain why, using simple energy diagrams. Using this explanation, predict which is more reactive, *cis*-2-butene or *trans*-2-butene.

21. Optically active 1-chloro-3-methylcyclopentane was treated with potassium *t*-butoxide in *t*-butyl alcohol. Two isomeric alkenes were obtained. The major product was optically active whereas the minor product was optically inactive. What are the structures of the major and minor products?

22. Propose a mechanism for each of the following reactions.

a. 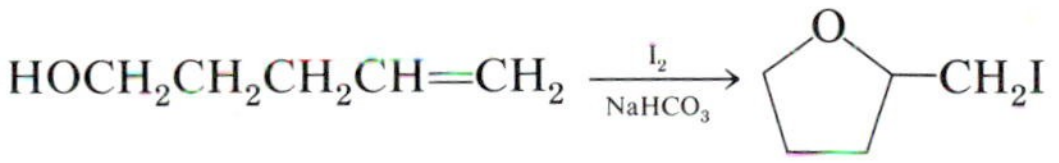 $HOCH_2CH_2CH_2CH{=}CH_2 \xrightarrow[NaHCO_3]{I_2}$ (tetrahydrofuran ring, O)–CH_2I

b. $CH_3CH_2C(CH_3){=}CH_2 + INCO \longrightarrow CH_3CH_2C(CH_3)(NCO)CH_2I$

c. $\overset{O}{CH_2{-}CH}{-}CD_2Cl + CH_3O^- \xrightarrow[\Delta]{CH_3OH} CH_3OCH_2{-}\overset{O}{CH{-}CD_2}$

23. 3,3,5-Trimethyl-1-methylenecyclohexane reacts with 3-chloroperoxybenzoic acid to give mainly the diastereomeric epoxide shown. However, when the same alkene is treated first with Br_2 in water and then with alcoholic base, the other diastereomer is the major product.

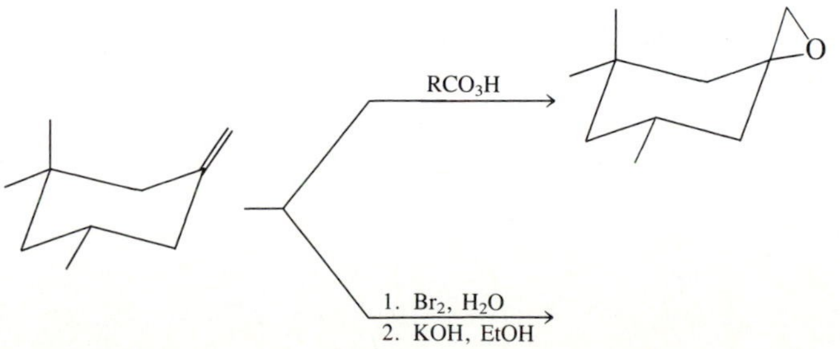

RCO$_3$H

1. Br_2, H_2O
2. KOH, EtOH

Draw the structure of the second diastereomer and explain the results.

24. Compound H, $C_{11}H_{24}O$, reacts with PBr_3 in ether at 0 °C to give I, $C_{11}H_{23}Br$. Treatment of I with potassium ethoxide in ethanol gives a mixture of J (major) and K (minor), both $C_{11}H_{22}$. Each of these compounds was treated first with ozone in $CHCl_3$ at 0 °C, and then with $NaBH_4$. In each case a mixture of 3-methyl-1-butanol and 4-methyl-1-pentanol was produced. What are compounds H through K?

25. Treatment of $C_7H_{15}Br$ with strong base gave an alkene mixture that was shown by careful gas chromatographic analysis and separation to consist of three alkenes, C, D, and E, each having the formula C_7H_{14}. Catalytic hydrogenation of each alkene gave 2-methylhexane. Reaction of C with B_2H_6 in ether, followed by H_2O_2 and OH^- gave mostly an alcohol, F. Similar reaction of D or E gave approximately equal amounts of F and an isomeric alcohol G. What structural assignments can be made for C through G on the basis of these observations? What structural element is left undetermined by these data alone?

26. The propagation steps for the radical addition of HY to propylene are as follows.

a. $CH_3CH{=}CH_2 + Y\cdot \longrightarrow CH_3\dot{C}HCH_2Y$

b. $CH_3\dot{C}HCH_2Y + HY \longrightarrow CH_3CH_2CH_2Y + Y\cdot$

$CH_3CH{=}CH_2 + HY \longrightarrow CH_3CH_2CH_2Y$

The following table gives $\Delta H°$ values for steps (a) and (b) and for the net reaction with a number of reagents of the type HY in the gas phase. For which reagents is such a radical chain mechanism plausible?

Y:	F	Cl	Br	I	HS	HO	H_2N	CH_3	$(CH_3)_3C$
$\Delta H_a°$	−45	−19	−7	+8	+9	−30	−20	−24	−20
$\Delta H_b°$	+38	+5	−10	−27	−7	+21	9	+7	−2
$\Delta H_{net}°$	−7	−14	−17	−18	−16	−8	−11	−17	−22

27. Consider a proposed free radical chain addition of HCN to $CH_3CH{=}CH_2$ to give *n*-propyl cyanide, $CH_3CH_2CH_2CN$. Use data in Appendix I and the following $DH°$ values.

H—CN	120 kcal $mole^{-1}$
$CH_3CH(\text{—}H)CH_2CN$	98 kcal $mole^{-1}$

a. Determine $\Delta H°$ for the net reaction,

$$CH_3CH{=}CH_2 + HCN \rightleftharpoons CH_3CH_2CH_2CN$$

b. Write the two chain-propagation steps for the proposed reaction and calculate $\Delta H°$ for each.

c. Is the proposed reaction feasible? Explain.

d. Draw a chain reaction diagram for the reaction.

CHAPTER 12

ALKYNES AND NITRILES

12.1 Electronic Structure of the Triple Bond

Acetylene is known experimentally to have a linear structure. The C≡C distance of 1.20 Å is the shortest carbon-carbon bond length known. The carbon-hydrogen bond length of 1.06 Å is shorter than that in ethylene (1.08 Å) or in ethane (1.10 Å) (Figure 12.1). These structural details are readily interpreted by an extension of the σ-π electronic structure of double bonds. In acetylene the σ-framework consists of C_{sp}-hybrid orbitals as indicated in Figure 12.2.

In Section 11.1 we saw that sp^2-s σ-bonds are shorter than are sp^3-s σ-bonds. The trend also holds for the sp-s bonds in acetylene. The effect of the amount of s-character in the

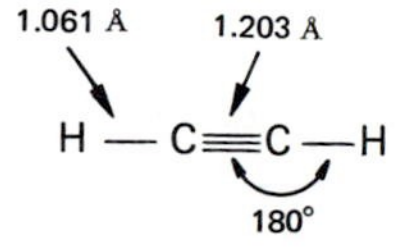

1.061 Å
1.203 Å
H — C≡C — H
180°

FIGURE 12.1 Structure of acetylene.

H_{1s}—C_{sp}
C_{sp}—C_{sp}
C_{sp}—H_{1s}
H
C
C
H

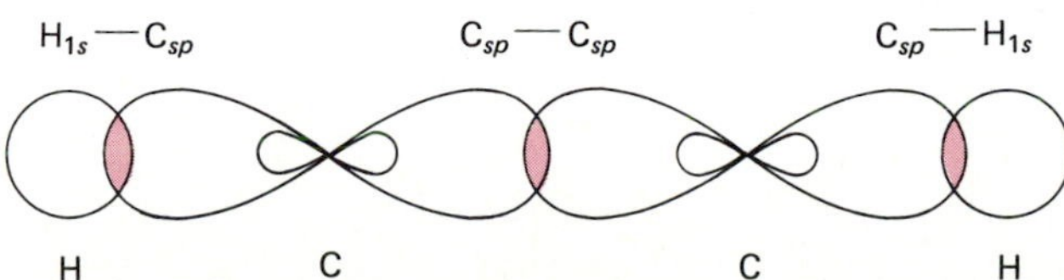

FIGURE 12.2 σ-electronic framework of acetylene.

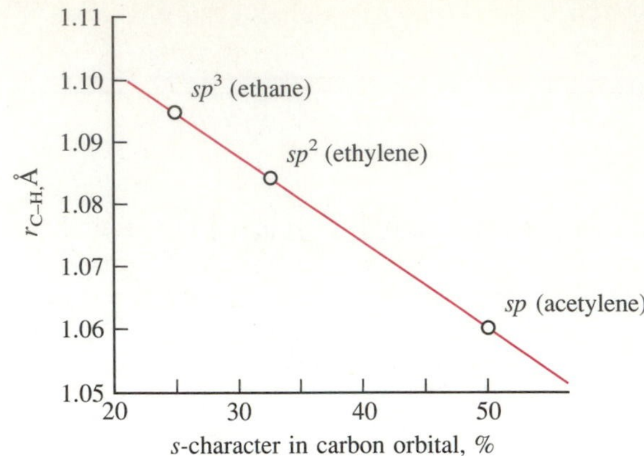

FIGURE 12.3 Relationship between carbon-hydrogen bond distance and the approximate amount of *s*-character in carbon orbital.

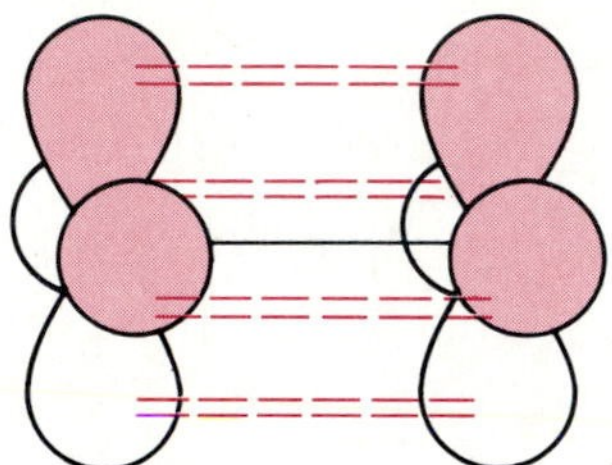

FIGURE 12.4 π-systems of acetylene.

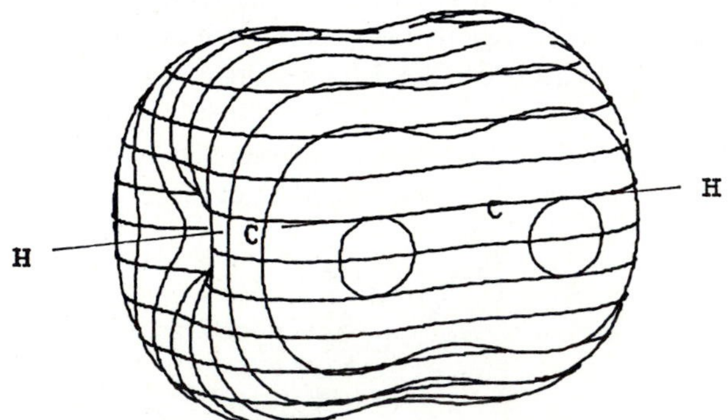

FIGURE 12.5 π-electron density in the acetylene triple bond.

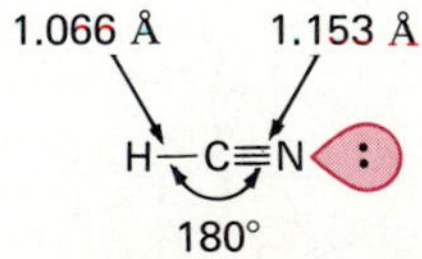

FIGURE 12.6 Structure of hydrogen cyanide.

carbon-hydrogen bond distance is shown graphically in Figure 12.3. Superimposed on the σ-electrons are two orthogonal π-electron systems as shown in Figure 12.4.

The symbolic representations in Figure 12.4 are actually misleading because the electrons in two orthogonal p-orbitals form a cylindrically symmetrical torus or doughnut-like electron density distribution. A perspective view of the total π-electron density is seen in Figure 12.5.

The structure of acetylene presented in Figures 12.1–5 is similar to that of the inorganic compound hydrogen cyanide (Figure 12.6). In the latter compound, the triple bond may be viewed as one $C_{sp}—N_{sp}$ and two $C_p—N_p$ bonds. The remaining electron pair on the nitrogen occupies the other N_{sp}-hybrid orbital.

In fact, a close examination of the nuclear and electronic structures of acetylene and hydrogen cyanide reveals that they differ in only two aspects:

1. In hydrogen cyanide, the two electrons in the *sp*-orbital not bonded to carbon are nonbonding, while in acetylene they form a σ-bond to a hydrogen.
2. In hydrogen cyanide, there are seven protons and seven neutrons in the nucleus of the nitrogen, whereas in acetylene the nucleus of the corresponding carbon contains only six protons and six neutrons.

When two compounds have equivalent electronic distributions, and only differ in their nuclei, they are said to be **isoelectronic.** In such cases, it is usually found that their chemical properties are qualitatively similar. We shall see that parallels do exist between the reactions of acetylene and hydrogen cyanide. Similar analogies are found in the reactions of alkyl substituted acetylenes, and the organic derivatives of hydrogen cyanide, which are called **nitriles.**

$H—C{\equiv}C—CH_3$	$:N{\equiv}C—CH_3$
a substituted acetylene	a nitrile

Unlike ethane (page 96), acetylene has a characteristic aromatic odor. Unlike highly toxic hydrogen cyanide, which has an almond odor for most people (Section 9.4), a genetic defect in detection of acetylene has never been reported. The interaction of these similarly shaped molecules with olfactory receptor(s) must be different, so that our perception of their odors differs.

Highlight 12.1

The carbon-carbon triple bond is much shorter than C=C or C—C because the carbons are held together by six bonding electrons instead of four (for C=C) or two (for C—C). In addition, the single bonds to an sp-hybridized carbon ($\equiv$C—H or $\equiv$C—C) are also shorter and stronger than analogous bonds to sp^2- or sp^3-hybridized carbons. This is because *s*-electrons spend more time near the nucleus and are therefore held more tightly than *p*-electrons. In addition, an *sp*-hybridized carbon is effectively more electronegative than one that is sp^2 or sp^3.

12.2 Nomenclature of Alkynes and Nitriles

The simple alkynes are readily named in the common system as derivatives of acetylene itself.

$CH_3C{\equiv}CH$	$CH_3C{\equiv}CCH_2CH_3$	$F_3CC{\equiv}CH$
methylacetylene	ethylmethylacetylene	trifluoromethylacetylene

In the IUPAC system the compounds are named as alkynes in which the final **-ane** of the parent alkane is replaced by the suffix **-yne.** The position of the triple bond is indicated by a number when necessary.

$CH_3C{\equiv}CH$ — propyne

$(CH_3)_2CHC{\equiv}CH$ — 3-methyl-1-butyne

$CH_3CH_2CH_2CH(CH_2CH_2CH_3)C{\equiv}CCH_3$ — 4-propyl-2-heptyne

If both -yne and -ol endings are used, the -ol is last and determines the numbering sequence.

$HC{\equiv}CCH_2CH_2OH$ — 3-butyn-1-ol (not but-4-ol-1-yne)

$HOCH_2C{\equiv}CCH_2OH$ — 2-butyn-1,4-diol

When both a double and triple bond are present, the hydrocarbon is named an **alkenyne** with numbers as low as possible given to the multiple bonds. In case of a choice, the *double bond gets the lower number*.

$CH_3CH{=}CHC{\equiv}CH$ — 3-penten-1-yne (not 2-penten-4-yne)

$HC{\equiv}CCH_2CH{=}CH_2$ — 1-penten-4-yne (not 4-penten-1-yne)

In complex structures the alkynyl group is used as a modifying prefix.

(cyclopentyl)—$C{\equiv}CH$

ethynylcyclopentane

Nitriles are named in the IUPAC system by adding the suffix **nitrile** to the name of the alkane corresponding to the longest carbon chain in the molecule (*including the nitrile carbon*).

CH_3CN — ethanenitrile (acetonitrile)

$CH_3CH(CH_3)CH_2CN$ — 3-methylbutanenitrile

$BrCH_2CH_2CH_2CH_2CH_2CN$ — 6-bromohexanenitrile

The simplest nitrile, CH_3CN, is usually referred to by the common name of acetonitrile.

EXERCISE 12.1 Write the structures and name all ten isomeric pentynols (do not forget stereoisomerism).

EXERCISE 12.2 Write structures corresponding to each of the following names.

(a) dimethylacetylene (b) (*S*)-pent-3-yn-2-ol (c) 2-chlorobutanenitrile

12.3 Physical Properties

The physical properties of alkynes are similar to those of the corresponding alkenes. The lower members are gases with boiling points somewhat higher than the corresponding alkenes. Terminal alkynes have lower boiling points than isomeric internal alkynes (Table 12.1) and can be separated by careful fractional distillation.

The CH_3—C bond in propyne is formed by overlap of a C_{sp^3}-hybrid orbital from the methyl carbon with a C_{sp}-hybrid from the acetylenic carbon. The bond is C_{sp^3}—C_{sp}. Since the orbital with more *s*-character is more electronegative than the other, the electron

TABLE 12.1 Physical Properties of Alkynes

Compound	Boiling Point, °C	Melting Point, °C	d^{20}
ethyne (acetylene)	−84.0[a]	−81.5[b]	
propyne	−23.2	−102.7	
1-butyne	8.1	−122.5	
2-butyne	27	−32.3	
1-pentyne	39.3	−90.0	
2-pentyne	55.5	−101	
1-hexyne	71	−132	0.7152
2-hexyne	84	−88	0.7317
3-hexyne	81	−105	0.7231

[a] Sublimation temperature. [b] Under pressure.

density in the resulting bond is not symmetrical. The unsymmetrical electron distribution results in a dipole moment larger than that observed for an alkene, but still relatively small.

$$\underset{\mu = 0.80\ \text{D}}{CH_3CH_2C{\equiv}CH} \qquad \underset{\mu = 0.30\ \text{D}}{CH_3CH_2CH{=}CH_2} \qquad \underset{\mu = 0\ \text{D}}{CH_3C{\equiv}CCH_3}$$

Symmetrically disubstituted acetylenes, of course, have no net dipole moment.

In contrast, the boiling points for nitriles are markedly higher than those of analogous acetylenes. The difference may be readily seen in Figure 12.7, which shows a plot of molecular weight versus boiling point for 1-alkynes and straight-chain nitriles. Note that the nitrile line lies approximately 60 °C above the alkyne line.

In contrast to the relatively small dipole moments of acetylenes, those of nitriles are substantial, and result from the greater electronegativity of nitrogen relative to carbon, as well as the lone electron pair at the nitrogen end of the molecule.

$$\underset{\mu = 3.30\ \text{D}}{CH_3\overset{\longmapsto}{C{\equiv}N}:} \qquad \underset{\mu = 3.39\ \text{D}}{CH_3CH_2\overset{\longmapsto}{C{\equiv}N}:}$$

It is the dipole moments of nitriles that are responsible for their abnormally high boiling points. In order for a nitrile molecule to be brought into the vapor state, it is necessary that rather strong polar intermolecular forces be disrupted.

$$\begin{array}{c} R{-}C{\equiv}N \\ \longmapsto \\ \\ \longleftarrow\!\!\!\!+ \\ :N{\equiv}C{-}R \end{array}$$

Highlight 12.2

The electronegativity of nitrogen is much greater than that of carbon, leading to an asymmetric charge distribution in the C≡N bond, and a substantial dipole moment μ of 3.3 D. The compact size of the C≡N group allows close approach of these groups to one another. The large dipole-dipole attraction at short distances leads to relatively high boiling points for liquid nitriles.

12.4 Acidity of Alkynes

The hydrogens in terminal alkynes are relatively acidic. Acetylene itself has a pK_a of about 25. It is a far weaker acid than water (pK_a 15.7) or the alcohols (pK_a 16–19), but it is much more acidic than ammonia ($pK_a \cong 34$). A solution of sodium amide in liquid

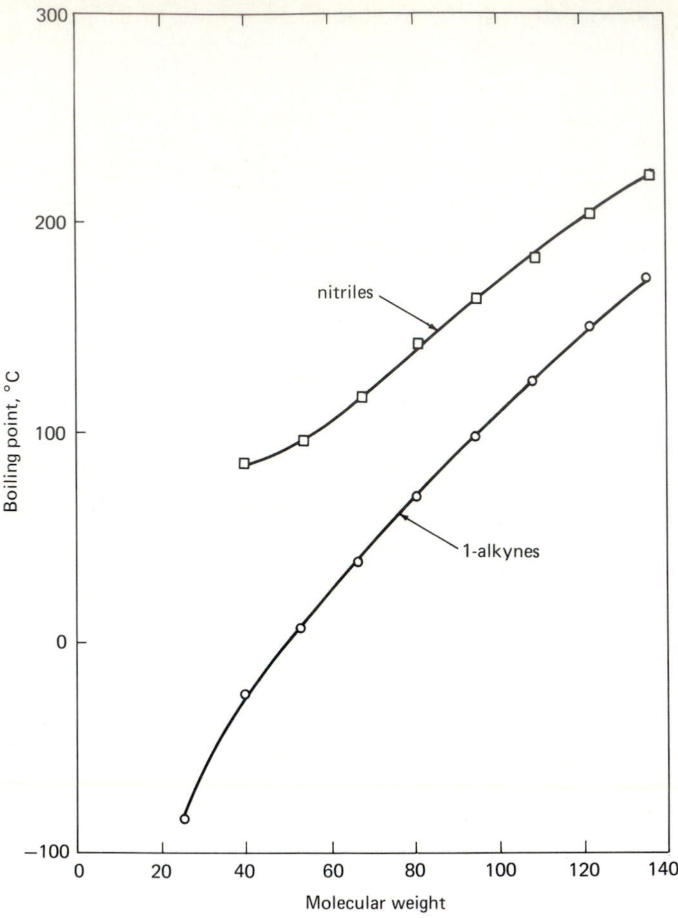

FIGURE 12.7 Boiling points of 1-alkynes and nitriles.

ammonia readily converts acetylene and other terminal alkynes into the corresponding carbanions.

$$RC{\equiv}CH + NH_2^- \rightleftharpoons RC{\equiv}C^- + NH_3$$

This reaction does not occur with alkenes or alkanes. Ethylene has a pK_a of about 44 and methane has a pK_a of about 50.

$$CH_2{=}CH_2 + NH_2^- \rightleftharpoons CH_2{=}CH^- + NH_3$$

$$CH_4 + NH_2^- \rightleftharpoons CH_3^- + NH_3$$

EXERCISE 12.3 Using the pK_a values given in the preceding paragraph, estimate the equilibrium constants for the reactions of sodium amide with acetylene, ethylene, and methane.

From the foregoing pK_as we see that there is a vast difference in the stability of the carbanions $RC{\equiv}C^-$, $CH_2{=}CH^-$, and CH_3^-. This difference may readily be explained in terms of the character of the orbital occupied by the lone-pair electrons in the three anions. Methyl anion has a pyramidal structure with the lone-pair electrons in an orbital that is approximately sp^3 ($\frac{1}{4}$ s and $\frac{3}{4}$ p). In vinyl anion the lone-pair electrons are in an sp^2-orbital ($\frac{1}{3}$ s and $\frac{2}{3}$ p). In acetylide ion the lone pair is in an sp-orbital ($\frac{1}{2}$ s and $\frac{1}{2}$ p).

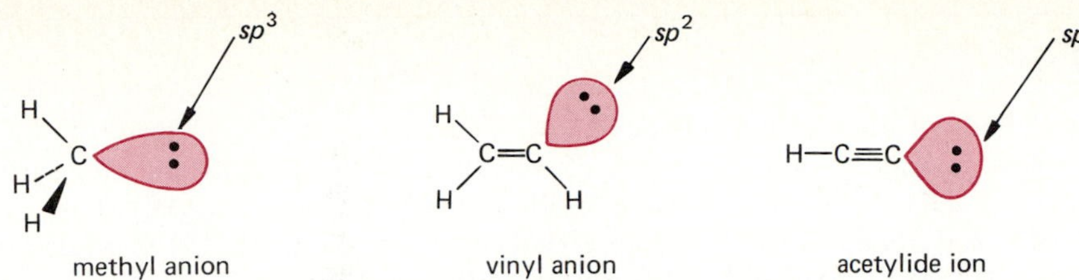

Electrons in *s*-orbitals are held, on the average, closer to the nucleus than they are in *p*-orbitals. This increased electrostatic attraction means that *s*-electrons have lower energy and greater stability than *p*-electrons. In general, *the greater the amount of* s-*character in a hybrid orbital containing a pair of electrons, the less basic is that pair of electrons, and the more acidic is the corresponding conjugate acid.*

base strength ↑			acid strength ↓
	$CH_3:^-$	CH_4	
	$CH_2{=}CH:^-$	$CH_2{=}CH_2$	
	$HC{\equiv}C:^-$	$HC{\equiv}CH$	

Of course, the foregoing argument applies to hydrogen cyanide as well. In this case, the conjugate base, $N{\equiv}C^-$, is further stabilized by the presence of the electronegative nitrogen. Consequently, HCN is sufficiently acidic (pK_a 9.2) that it is converted to its salt with hydroxide ion in water.

$$HCN + OH^- \rightleftharpoons CN^- + H_2O$$

EXERCISE 12.4 Estimate the equilibrium constant for reaction of 1 M HCN with 1 M NaOH in water.

Alkynes are also quantitatively deprotonated by alkyllithium compounds, which may be viewed as the conjugate bases of alkanes (Section 8.9).

$$CH_3(CH_2)_3C{\equiv}CH + n\text{-}C_4H_9Li \longrightarrow CH_3(CH_2)_3C{\equiv}CLi + n\text{-}C_4H_{10}$$

The foregoing transformation is simply an acid-base reaction, with 1-hexyne being the acid and *n*-butyllithium being the base. Since the alkyne is a much stronger acid than the alkane (by over 20 p*K* units!), equilibrium lies essentially completely to the right.

Terminal alkynes give insoluble salts with a number of heavy metal cations such as Ag^+ and Cu^+. The alkyne can be regenerated from the salt, and the overall process serves as a method for purifying terminal alkynes. However, many of these salts are touch-sensitive, exploding when dry, and should always be kept moist.

$$CH_3(CH_2)_3C{\equiv}CH + AgNO_3 \longrightarrow CH_3(CH_2)_3C{\equiv}CAg + HNO_3$$

Impure 1-hexyne is dissolved in 95% ethanol and aqueous silver nitrate is added. The resulting white precipitate is filtered and washed with alcohol. On refluxing with sodium cyanide solution, the alkyne is regenerated and distilled.

Cyanide ion regenerates the alkyne by converting silver cation to the stable complex $Ag(CN)_2^-$.

EXERCISE 12.5 Since 1-alkynes have substantially lower boiling points than internal alkynes, mixtures can be separated by careful fractional distillation. In practice, however, complete separation of such a mixture is difficult. Suggest a simple way in which the last traces of 1-pentyne might be removed from a sample of 2-pentyne.

Highlight 12.3

The stability of the acetylide anion, RC≡C$^-$, is due to the 50% *s*-character in the *sp*-orbital, making it relatively easy to convert the acid RC≡CH to the anion with amide ion, NH_2^-. Terminal alkynes readily form salts such as RC≡C$^-$ Cu$^+$ and RC≡C$^-$ Ag$^+$. Some of these metal acetylides are explosive when dry. Cyanide ion, C≡N$^-$, is further stabilized by the electronegative nitrogen atom.

12.5 Preparation of Alkynes and Nitriles

A. Acetylene

Acetylene itself is formed from the reaction of the inorganic compound calcium carbide with water.

$$CaC_2 + 2\,H_2O \longrightarrow Ca(OH)_2 + HC{\equiv}CH$$

Calcium carbide is a high-melting (m.p. 2300 °C) gray solid prepared by heating calcium oxide (lime) and carbon (coke) in an electric furnace.

$$CaO + 3\,C \longrightarrow CaC_2 + CO$$

This method was once an important industrial process for the manufacture of acetylene. However, the method has now been replaced by a pyrolytic process in which methane has contact with a high temperature zone for a short time.

$$2\,CH_4 \xrightarrow[0.01\text{–}0.1\text{ sec}]{1500\,^\circ C} HC{\equiv}CH + 3\,H_2$$

This reaction is endothermic at ordinary temperatures, but is thermodynamically favored at high temperatures.

At room temperature acetylene is thermodynamically unstable with respect to its elements, as shown by its large positive heat of formation ($\Delta H_f^\circ = +54.3$ kcal mole^{-1} at 25 °C).

$$C_2H_2 = 2\,C(s) + H_2 \qquad \Delta H^\circ = -54.3 \text{ kcal mole}^{-1}$$

This instability causes certain problems in the handling and storage of the material. When under pressure or in the presence of copper, it can convert to carbon and hydrogen with explosive violence. Although acetylene gas can be condensed readily (b.p. −84 °C), the liquid is similarly unstable. Since the gas is extremely soluble in acetone, commercial cylinders of acetylene contain pieces of pumice saturated with acetone. When the cylinder is filled, the acetylene mostly dissolves, giving a relatively stable solution. Acetylene is also appreciably soluble in water. A saturated aqueous solution at 25 °C and 1 atm pressure has a concentration of 0.05 M (0.13 g C_2H_2 per 100 mL).

B. Nucleophilic Substitution Reactions

We saw in Section 9.1 that cyanide ion is an effective nucleophile and readily displaces halide ion from primary alkyl halides. In fact, the reaction represents a general method of preparation of nitriles.

$$\text{CH}_3\text{CH}_2\text{CH}_2\text{CH}_2\text{Cl} + \text{NaCN} \xrightarrow[140\text{–}150\,^\circ\text{C}]{\text{DMSO}} \text{CH}_3\text{CH}_2\text{CH}_2\text{CH}_2\text{CN}$$

(93%)
pentanenitrile

A solution of 92.5 g of 1-chlorobutane and 53 g of sodium cyanide in 250 mL of dimethyl sulfoxide (DMSO) is heated at 140–150 °C for 15 min. The reaction mixture is diluted with water and extracted with ether to obtain pentanenitrile in 93% yield.

Since the acetylide anion is isoelectronic with cyanide, it is no surprise to find that it is also highly nucleophilic and that it participates readily in displacement reactions that occur by the S_N2 mechanism.

$$RC{\equiv}C{:}^- + R'X \longrightarrow RC{\equiv}CR' + X^-$$

This reaction is a useful general method for the preparation of certain types of alkynes. The reaction may be carried out in liquid ammonia solution or in an ether such as tetrahydrofuran (THF). The acetylide anion is formed with sodium amide or with *n*-butyllithium.

Liquid ammonia is available commercially in cylinders. Although the compound boils at −33 °C it has a relatively high heat of vaporization, due to extensive hydrogen bonding in the liquid. Because of this high heat of vaporization, boiling is a relatively slow process at room temperature. When using liquid ammonia, the material is kept in a normal reaction flask, which is equipped with a type of trap or condenser containing dry ice (−78 °C). The liquid ammonia in the flask refluxes gently and condenses on the dry ice condenser. The terminal alkyne is added to a solution of sodium amide in ammonia. After it has been converted into its salt, the alkyl halide is added. The mixture is stirred for a few hours, and water is then added. The hydrocarbon is separated from the aqueous ammonia layer and purified.

A variety of alkynes may be made using this method. Acetylene itself may be alkylated either once to make a terminal alkyne or twice to make an internal alkyne.

$$HC{\equiv}CH + NaNH_2 \xrightarrow[-33\,°C]{\text{liq. } NH_3} HC{\equiv}C^-Na^+ \xrightarrow{n\text{-}C_4H_9Br} \underset{\substack{(89\%)\\ \text{1-hexyne}}}{CH_3(CH_2)_3C{\equiv}CH}$$

$$HC{\equiv}CH \xrightarrow[\substack{\text{liq. } NH_3\\ -33\,°C}]{2\,NaNH_2} \xrightarrow{2\,n\text{-}C_3H_7Br} \underset{\substack{(60\text{–}66\%)\\ \text{4-octyne}}}{CH_3CH_2CH_2C{\equiv}CCH_2CH_2CH_3}$$

In Section 9.7, we saw that elimination by the E2 mechanism competes with nucleophilic substitution, and that this side reaction is particularly competitive when the alkyl halide is secondary or tertiary. Furthermore, we saw that the elimination/substitution ratio is a function of the **basicity** of the nucleophile. Since acetylide ions are highly basic, competing elimination is a common side reaction. The products of such an elimination reaction are an alkene (from the alkyl halide) and an alkyne (from the acetylide ion).

$$RC{\equiv}C{:}^- + H{-}CH_2{-}\overset{Br}{\overset{|}{C}}HR \longrightarrow RC{\equiv}C{-}H + CH_2{=}CHR + Br^-$$

In practice, the alkylation of acetylene or another terminal alkyne is only a good method for the synthesis of alkynes when applied to primary halides that do not have branches close to the reaction center. With secondary halides, and even with primary halides that have branches close to the reaction center, elimination is usually the major reaction.

$$CH_3(CH_2)_3C{\equiv}C^- Li^+ + CH_3\underset{}{\overset{Br}{\overset{|}{C}}}HCH_3 \xrightarrow[25°C]{HMPT} \begin{cases} CH_3CH_2CH_2CH_2C{\equiv}CCH(CH_3)_2 \\ (6\%) \\ + \\ \{CH_3CH_2CH_2CH_2C{\equiv}CH + CH_2{=}CHCH_3\} \\ (85\%) \end{cases}$$

$$CH_3(CH_2)_3C{\equiv}C^- Li^+ + CH_3\overset{CH_3}{\overset{|}{C}}HCH_2Br \xrightarrow[25°C]{HMPT} \begin{cases} CH_3CH_2CH_2CH_2C{\equiv}CCH_2CH(CH_3)_2 \\ (32\%) \\ + \\ \{CH_3CH_2CH_2CH_2C{\equiv}CH + (CH_3)_2C{=}CH_2\} \\ (68\%) \end{cases}$$

Highlight 12.4

Alkylation of acetylene is an excellent way to prepare other alkynes. Treatment of the sodium or lithium salt with a primary alkyl halide gives a terminal alkyne. This product can be deprotonated again and the resulting sodium or lithium acetylide alkylated with either the same alkyl halide or a different one to obtain symmetrical or unsymmetrical internal alkynes.

$$HC{\equiv}CH \longrightarrow HC{\equiv}C^- M^+ \xrightarrow{CH_3CH_2CH_2CH_2Br} HC{\equiv}CCH_2CH_2CH_2CH_3$$

$$\downarrow$$

$$CH_3CH_2CH_2CH_2C{\equiv}CCH_2CH_2CH_2CH_3 \xleftarrow{CH_3CH_2CH_2CH_2Br} M^+ \ {}^-C{\equiv}CCH_2CH_2CH_2CH_3$$

$$\downarrow CH_3I$$

$$CH_3C{\equiv}CCH_2CH_2CH_2CH_3$$

This synthetic method is an important one because it involves the making of new carbon-carbon bonds and permits us to build up more complicated molecules from simple ones. It is limited by the fact that the mechanism of the alkylation is S_N2, so one can only use primary alkyl halides.

EXERCISE 12.6 The nucleophilic substitution of alkyl halides by cyanide ion is not nearly as subject to competing elimination as is acetylene alkylation. For example, isobutyl chloride and isopropyl chloride may both be converted into the corresponding nitriles in excellent yield by treatment with sodium cyanide in DMSO. Explain.

EXERCISE 12.7 Outline syntheses of each of the following compounds from alkyl halides.

(a) 1-octyne (b) 2-octyne (c) 4-methylpentanenitrile

C. Elimination Reactions

A triple bond can be introduced into a molecule by elimination of two molecules of HX from either a **geminal** (L., *geminus,* twin) or a **vicinal** (page 276) dihalide.

$$—CBr_2—CH_2— \xrightarrow{-2\ HBr} —C{\equiv}C—$$

a *gem*-dibromide

$$—CHBr—CHBr— \xrightarrow{-2\ HBr} —C{\equiv}C—$$

a *vic*-dibromide

Both kinds of dehydrohalogenation are known. The reactions proceed in stages, with the second molecule of HX being removed with greater difficulty than the first.

$$\left.\begin{matrix} —CX_2CH_2— \\ \text{or} \\ —CHXCHX— \end{matrix}\right\} \xrightarrow{\text{faster}} —CX{=}CH— \xrightarrow{\text{slower}} —C{\equiv}C—$$

Typical reaction conditions involve the use of molten KOH, solid KOH moistened with alcohol, or concentrated alcoholic KOH solutions at temperatures of 150–200 °C. In practice, these conditions are so drastic that the method is only useful for the preparation of certain kinds of alkynes. Under these highly basic conditions the triple bond can **migrate** along a chain. Since disubstituted alkynes are thermodynamically more stable than terminal alkynes, the triple bond will migrate from the end of a chain to an internal position. For example, treatment of 1-butyne with hot alcoholic KOH affords 2-butyne.

$$CH_3CH_2C{\equiv}CH \underset{\Delta}{\overset{\text{alc. KOH}}{\rightleftharpoons}} CH_3C{\equiv}CCH_3$$

EXERCISE 12.8 Using the heats of formation in Appendix I, estimate the equilibrium constant for the foregoing reaction at 100 °C.

In the case of longer chains, complex mixtures result, since there are generally several isomeric internal alkynes of comparable stability. For example, application of the foregoing conditions to 1-hexyne gives an equimolar mixture of 2-hexyne and 3-hexyne. Consequently, the dehydrohalogenation of a dihalide by KOH or NaOH is *not* a generally useful procedure for the preparation of alkynes.

Sodium amide is an effective strong base that is particularly useful for the preparation of 1-alkynes.

Sodium amide, $NaNH_2$, is a white solid prepared by the reaction of sodium with ammonia. Sodium actually dissolves in liquid ammonia to give a blue solution of "solvated electrons" and sodium cations.

$$Na \xrightarrow{NH_3} Na^+ + e^-(NH_3)$$

These solutions are relatively stable. However, in the presence of small amounts of ferric ion, a rapid reaction takes place with the liberation of hydrogen.

$$e^- + NH_3 \xrightarrow{Fe^{3+}} NH_2^- + \tfrac{1}{2} H_2$$

The overall equation for the ferric ion-catalyzed reaction of sodium with ammonia is as follows:

$$2\ Na + 2\ NH_3 \xrightarrow{Fe^{3+}} 2\ NaNH_2 + H_2$$

In liquid ammonia sodium amide is a strong base, just as sodium hydroxide is in water.

Since NH_3 is much less acidic than water, sodium amide reacts quantitatively with water. Solutions of $NaNH_2$ in NH_3 readily absorb moisture from the atmosphere.

$$NH_2^- + H_2O \longrightarrow NH_3 + OH^-$$

In the organic laboratory, sodium amide is generally prepared and used as a solution in liquid ammonia. However, for some applications, suspensions of solid $NaNH_2$ in an inert medium such as benzene or mineral oil are used.

In one classical procedure for the dehydrohalogenation of a dihalide, a suspension of sodium amide in mineral oil is heated to 150–165 °C. The dihalide is added slowly and a vigorous reaction ensues. Ammonia is evolved, and the sodium salt of the alkyne is formed. After cooling, the hydrocarbon is liberated by the addition of water.

$$CH_3(CH_2)_3CH_2CHCl_2 + NaNH_2 \xrightarrow{160°C} CH_3(CH_2)_3C{\equiv}C^- \, Na^+ + NH_3$$

$$CH_3(CH_2)_3C{\equiv}C^-Na^+ + H_2O \longrightarrow CH_3(CH_2)_3C{\equiv}CH + Na^+ \, OH^-$$

Since the reaction product is the salt of an alkyne, this method is useful for preparing terminal alkynes even when migration of the triple bond is possible.

$$n\text{-}C_{14}H_{29}CHBrCH_2Br \xrightarrow{NaNH_2} \underset{(65\%)}{n\text{-}C_{14}H_{29}C{\equiv}CH}$$

In fact, internal alkynes can even be isomerized to terminal alkynes by the use of sodium amide or alkali metal salts of amines. A particularly useful reagent for this type of reaction is potassium 3-aminopropylamide (''KAPA''), which is used in 1,3-diaminopropane as solvent. This reagent causes rapid reaction even at room temperature.

$$\underset{\text{2-hexyne}}{CH_3(CH_2)_2C{\equiv}CCH_3} \xrightarrow[\substack{H_2N(CH_2)_3NH_2 \\ \text{(1,3-propanediamine)} \\ 20°C}]{\substack{H_2NCH_2CH_2CH_2NH^- \, K^+ \\ \text{(KAPA)}}} CH_3(CH_2)_3C{\equiv}C^- \, K^+ \xrightarrow{H_2O} \underset{\substack{\text{1-hexyne} \\ (100\%)}}{CH_3(CH_2)_3C{\equiv}CH}$$

EXERCISE 12.9 Suggest a way in which 2-pentene can be converted into
(a) 2-pentyne and (b) 1-pentyne.

EXERCISE 12.10 With 1-bromobutane as the only source of carbon, show how each of the following compounds can be synthesized.
(a) 1-butyne (b) 2-butyne (c) 3-octyne (d) 1-octyne

12.6 Reactions of Alkynes and Nitriles

Many reactions of the triple bond of alkynes are analogous to comparable reactions of alkenes. However, just as a double bond is weaker than two single bonds, a triple bond is weaker still than three single bonds. This comparison is apparent in the average bond energies tabulated in Table 12.2. As a result the carbon-carbon triple bond enters into some reactions not generally seen with alkenes.

TABLE 12.2 Average Bond Energies of C—C Bonds

Bond	Average Bond Energy, kcal mole^{-1}
C—C	83
C=C	146
C≡C	200

A. Reduction

Hydrogenation of an alkyne to an alkane occurs readily with the same general catalysts that are used for the reduction of alkenes.

$$R{-}C{\equiv}C{-}R' + 2\,H_2 \xrightarrow{\text{Pt, Pd, Ni}} R{-}CH_2CH_2{-}R'$$

The first step in the reduction is a more exothermic reaction than is the second.

$$HC{\equiv}CH + H_2 \longrightarrow CH_2{=}CH_2 \qquad \Delta H° = -41.9 \text{ kcal mole}^{-1}$$

$$CH_2{=}CH_2 + H_2 \longrightarrow CH_3CH_3 \qquad \Delta H° = -32.7 \text{ kcal mole}^{-1}$$

The second reaction is so facile that, with many catalysts, it is not possible to stop the reduction at the alkene stage. However, with palladium or nickel, alkynes undergo hydrogenation extremely readily—faster than any other functional group. By taking advantage of this catalytic effect, one may accomplish the **partial hydrogenation** of an alkyne to an alkene. In practice, specially deactivated or "poisoned" catalysts are usually used. An effective catalyst for this purpose is palladium metal that has been deposited in a finely divided state on solid $BaSO_4$ and then treated with quinoline (the actual poison).

The function of the poison is to moderate the catalyst's activity to a point where triple bonds are still reduced at a reasonable rate but double bonds react only slowly. One can then readily stop the reduction after absorption of 1 mole of hydrogen and isolate the alkene in excellent yield.

Quinoline is a heterocyclic amine and is discussed in Section 32.7.

N

quinoline

It is isolated industrially from coal tar, but the commercial material contains trace amounts of sulfur compounds that are difficult to remove. Divalent sulfur compounds are such exceedingly powerful catalyst poisons that they completely inhibit the catalytic activity. For this reason, only pure synthetic quinoline may be used for this purpose.

For the reduction, 10 g of alkyne in 75–100 mL of methanol is treated with 0.2 g of 5% palladium on barium sulfate and 5–6 drops of pure, synthetic quinoline is added. The reaction mixture is stirred under an atmosphere of hydrogen until hydrogen absorption ceases.

As in the hydrogenation of alkenes, the hydrogenation of acetylenes is a syn process. That is, the disubstituted alkene product has predominantly the cis stereostructure. In fact, alkyne hydrogenation is even more stereoselective than alkene hydrogenation, although small amounts (5–10%) of the trans isomer are often formed.

$$C_2H_5C{\equiv}CC_2H_5 + H_2 \xrightarrow[\text{quinoline}]{\text{Pd/BaSO}_4} \begin{array}{ccc} C_2H_5 & & C_2H_5 \\ & C{=}C & \\ H & & H \end{array}$$

Reduction of triple bonds can also be accomplished by treating the alkyne with sodium in liquid ammonia at −33 °C. This reduction produces exclusively the trans alkene.

$$C_4H_9C{\equiv}CC_4H_9 \xrightarrow[\text{liq. NH}_3]{\text{Na}} \xrightarrow{\text{NH}_4\text{OH}} \begin{array}{ccc} C_4H_9 & & H \\ & C{=}C & \\ H & & C_4H_9 \end{array}$$

(80–90%)
trans-5-decene

The mechanism of this reaction involves the reduction of the triple bond by two electrons from sodium atoms. The first electron goes into an antibonding π-orbital to give a radical anion. This strongly basic species is protonated by ammonia to give a vinyl radical, which is reduced by another electron to give a vinyl anion. Final protonation of the vinyl anion by ammonia (acting as an acid) yields the trans alkene and amide ion.

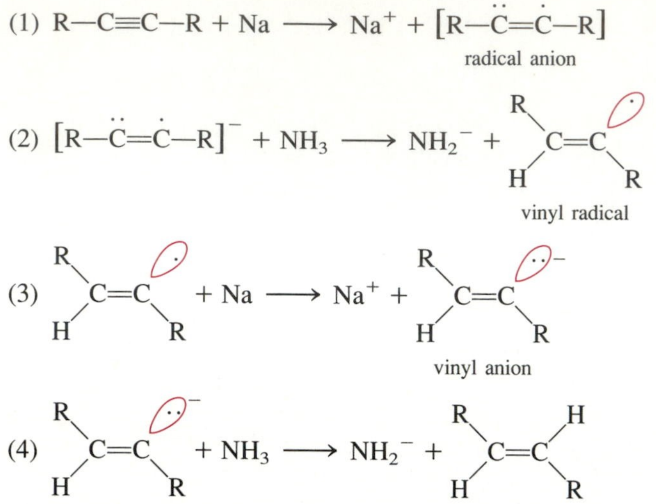

$$(1)\ R{-}C{\equiv}C{-}R + Na \longrightarrow Na^+ + [R{-}\ddot{C}{=}\dot{C}{-}R]$$

radical anion

$$(2)\ [R{-}\ddot{C}{=}\dot{C}{-}R]^- + NH_3 \longrightarrow NH_2^- + \text{(R)(H)C=C(R)}^{\cdot}$$

vinyl radical

$$(3)\ \text{(R)(H)C=C(R)}^{\cdot} + Na \longrightarrow Na^+ + \text{(R)(H)C=C(R)}^{:-}$$

vinyl anion

$$(4)\ \text{(R)(H)C=C(R)}^{:-} + NH_3 \longrightarrow NH_2^- + \text{(R)(H)C=C(H)(R)}$$

The stereochemistry of the final product is probably established in the reduction of the vinyl radical (step 3). The two vinyl radicals with the R groups trans or cis interconvert rapidly, but the trans form is preferred because of nonbonded interactions in the cis form. Since reduction of the two vinyl radicals probably proceeds at comparable rates and the trans form is present in much greater amount, the vinyl anion formed is mostly trans. The vinyl anion interconverts between cis and trans forms only relatively slowly and appears to protonate before it has a chance to isomerize.

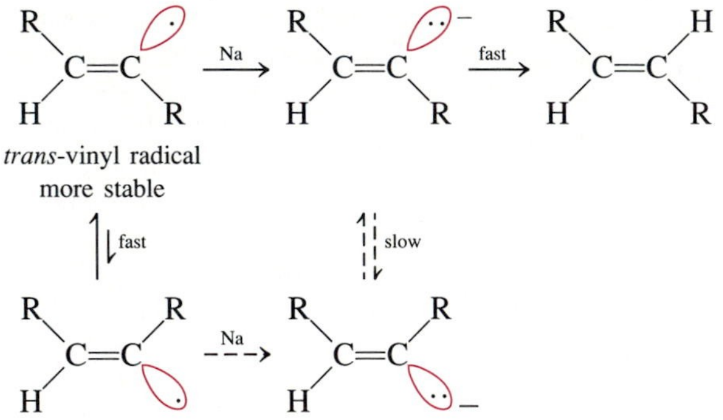

trans-vinyl radical
more stable

Na → ; fast → ; fast ⇅ ; slow ⇅ ; Na ⇢

cis-vinyl radical
less stable

Note that we have used the term **vinyl** in two different senses. It refers to the common name for the specific organic function, $-CH{=}CH_2$ (for example, vinyl chloride, $CH_2{=}CHCl$) but it is also used generically to refer to substitution at a carbon that is part of an alkene double bond (for example, $CH_3CCl{=}CH_2$, a vinyl or vinylic chloride).

Simple alkenes are not reduced by sodium in liquid ammonia, so it is easy to perform the partial reduction of an alkyne to an alkene by this method. It is important not to confuse a solution of Na in liquid NH_3 (which is actually a solution containing Na^+ ions and solvated electrons, e^-) with a solution of $NaNH_2$ in liquid NH_3 (which is a solution containing Na^+ ions and NH_2^- ions). The former solution reduces alkynes. The latter solution does not reduce alkynes, but does deprotonate terminal alkynes.

Disubstituted alkynes may also be reduced to trans alkenes by lithium aluminum hydride, $LiAlH_4$.

$$CH_3CH_2C{\equiv}CCH_2CH_2CH_3 \xrightarrow[\substack{\text{THF} \\ \text{diglyme} \\ 138°\text{C}}]{LiAlH_4} \underset{\substack{(96\%) \\ \textit{trans}\text{-3-hexene}}}{\overset{CH_3CH_2\qquad H}{\underset{H\qquad CH_2CH_2CH_3}{C{=}C}}}$$

3-hexyne

A solution of 85 mmol of $LiAlH_4$ in 50 mL each of THF and diglyme is heated while distilling off solvent until the internal temperature reaches 138 °C. 3-Hexyne (50 mmol) is added, and the solution is refluxed for 4.5 hr. Distillation of the product gives 96% yield of *trans*-3-hexene, contaminated with 4% of the cis isomer.

Lithium aluminum hydride, $LiAlH_4$, is a white, salt-like compound that is prepared by the reaction of lithium hydride with aluminum chloride.

$$4\ LiH + AlCl_3 \longrightarrow LiAlH_4 + 3\ LiCl$$

It is easily soluble in ethers such as diethyl ether, tetrahydrofuran (THF), glyme, and diglyme. It has a clear structural relationship to sodium borohydride, $NaBH_4$ (page 279), and is a salt of the cation Li^+ and the anion AlH_4^-. It reacts avidly with traces of moisture to liberate hydrogen.

$$LiAlH_4 + 4\ H_2O \longrightarrow LiOH + Al(OH)_3 + 4\ H_2$$

All hydroxylic compounds (alcohols, carboxylic acids, and so on) react similarly. The dry crystalline powder must be used with care. It produces dust particles that are highly irritating to mucous membranes. It may also inflame spontaneously while being crushed with a mortar and pestle and explodes violently when heated to about 120 °C.

By means of these several reactions it is possible to construct larger chains from smaller ones and to prepare either cis or trans alkenes with little contamination from the other. For example, acetylene can be converted into propyne, which can, in turn, be alkylated with *n*-propyl bromide to give 2-hexyne. This disubstituted acetylene can be reduced by sodium in ammonia to yield *trans*-2-hexene, or with hydrogen in the presence of a poisoned palladium catalyst to produce the cis isomer.

$$HC{\equiv}CH \xrightarrow{NaNH_2} \xrightarrow{CH_3I} HC{\equiv}CCH_3 \xrightarrow{NaNH_2} \xrightarrow{n\text{-}C_3H_7Br} n\text{-}C_3H_7C{\equiv}CCH_3$$

$$n\text{-}C_3H_7C{\equiv}CCH_3 \xrightarrow[\text{liq. } NH_3]{Na} \underset{\textit{trans}\text{-2-hexene}}{\overset{n\text{-}C_3H_7\qquad H}{\underset{H\qquad CH_3}{C{=}C}}}$$

$$n\text{-}C_3H_7C{\equiv}CCH_3 \xrightarrow[\substack{Pd/BaSO_4 \\ \text{quinoline}}]{H_2} \underset{\textit{cis}\text{-2-hexene}}{\overset{n\text{-}C_3H_7\qquad CH_3}{\underset{H\qquad H}{C{=}C}}}$$

We begin to see the sensitivity and power of organic syntheses and we have only barely scratched the surface of the many and varied reactions known and used in the organic laboratory.

EXERCISE 12.11 Using 1-chlorobutane as the only source of carbon, show how the following compounds may be prepared.

(a) *cis*-3-octene (b) *trans*-3-octene (c) 1-octyne

The carbon-nitrogen triple bond in nitriles is also readily reduced. However, in this case it is the completely reduced product that is usually desired. Catalytic hydrogenation can be employed, as shown in the following example.

$$\text{CH}_3\text{CH}_2\text{CH}_2\text{CH}_2\text{CN} + \text{H}_2 \xrightarrow[\text{C}_2\text{H}_5\text{OH, HCl}]{\text{PtO}_2} \text{CH}_3\text{CH}_2\text{CH}_2\text{CH}_2\text{CH}_2\overset{+}{\text{N}}\text{H}_3\ \text{Cl}^-$$

(95%)

The product of this reaction is an example of an amine, an organic analog of ammonia. Since the hydrogenation is carried out in the presence of slightly more than one equivalent of hydrochloric acid, the amine is isolated in the form of its ammonium salt. We will discuss the chemistry of amines in detail in Chapter 24.

Nitriles may also be reduced to amines by lithium aluminum hydride. Because the carbon-nitrogen triple bond is more polar than the carbon-carbon triple bond, the reduction can be accomplished under considerably milder conditions.

$$\text{CH}_3\text{CH}_2\text{CD}_2\text{CN} + \text{LiAlH}_4 \xrightarrow[70^\circ\text{C}]{\text{diglyme}} \xrightarrow{\text{ROH}} \text{CH}_3\text{CH}_2\text{CD}_2\text{CH}_2\text{NH}_2$$

(86%)

2,2-Dideuteriobutanenitrile (71 g) is added slowly to a solution of 38 g of lithium aluminum hydride in 500 mL of diglyme heated at 70 °C. The solution is cooled and treated with 2-butoxyethanol to destroy excess $LiAlH_4$. Distillation of the product affords 65 g (86%) of 3,3-dideuteriobutanamine.

EXERCISE 12.12 Starting with 1-chlorobutane, and using any necessary inorganic reagents, show how 1-pentanamine, $CH_3CH_2CH_2CH_2CH_2NH_2$, can be prepared.

B. Electrophilic Additions

The triple bond reacts with HCl and HBr in much the same manner as does the double bond. The addition goes in stages, and Markovnikov's rule is followed.

$$\text{RC}{\equiv}\text{CH} \xrightarrow{\text{HCl}} \text{RC(Cl)}{=}\text{CH}_2 \xrightarrow{\text{HCl}} \text{RCCl}_2\text{CH}_3$$

Alkynes are only about 10^{-2} to 10^{-3} as reactive as comparable alkenes. It is sometimes possible to stop the reaction of an alkyne with HX at the monoadduct stage, since the presence of a halogen in the initial product reduces its reactivity. However, yields of 1:1 adducts are usually poor.

$$\text{CH}_3(\text{CH}_2)_3\text{C}{\equiv}\text{CH} + \text{HBr} \xrightarrow[\text{FeBr}_3,\ 15^\circ\text{C}]{\text{inhibitor}} \text{CH}_3(\text{CH}_2)_3\text{C(Br)}{=}\text{CH}_2$$

(40%)
2-bromo-1-hexene

Although addition across a triple bond is a more exothermic process than comparable addition across a double bond, alkynes are generally less reactive than alkenes toward electrophilic reagents. This apparent anomaly is rationalized by comparison of the intermediate carbocations produced from alkynes and alkenes.

$$\mathrm{RC{\equiv}CH + H^+ \longrightarrow R\overset{+}{C}{=}CH_2}$$

a vinyl cation

$$\mathrm{RCH{=}CH_2 + H^+ \longrightarrow R\overset{+}{C}H{-}CH_3}$$

an alkyl cation

The carbocation produced from the alkyne is a vinyl cation, $\mathrm{R\overset{+}{C}{=}CH_2}$, whose electronic structure is shown in Figure 12.8. This type of carbocation is substantially less stable than an ordinary alkyl cation such as $\mathrm{R\overset{+}{C}HCH_3}$, since the vacant *p*-orbital belongs to an *sp*-hybridized carbon rather than to an sp^2-hybridized carbon. Since a carbon that has *sp*-hybridization is more electronegative than one having sp^2-hybridization, it is less tolerant of the positive charge. We shall return to the subject of vinyl cations in Section 12.7.

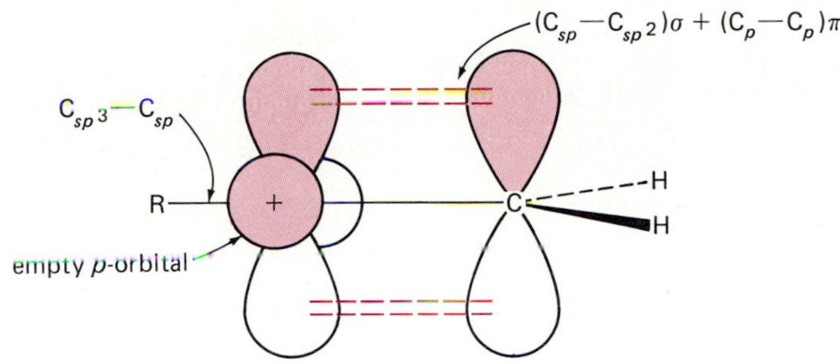

FIGURE 12.8 Electronic structure of a vinyl cation, $\mathrm{R\overset{+}{C}{=}H_2}$.

Once formed, the vinyl cation reacts with whatever nucleophiles are present. For example, the overall reaction with HCl involves initial formation of the vinyl cation, followed by its reaction with chloride ion.

$$\text{(1)}\quad \mathrm{RC{\equiv}CH + H^+ \longrightarrow R\overset{+}{C}{=}CH_2}$$

$$\text{(2)}\quad \mathrm{R\overset{+}{C}{=}CH_2 + Cl^- \longrightarrow R\overset{Cl}{\overset{|}{C}}{=}CH_2}$$

Addition of a second HX to the initially formed vinyl halide gives a product in which the two halogens are attached to the same carbon, a *gem*-dihaloalkane.

$$\mathrm{R\overset{Cl}{\overset{|}{C}}{=}CH_2 + HCl \longrightarrow R\underset{\underset{Cl}{|}}{\overset{\overset{Cl}{|}}{C}}CH_3}$$

a *gem*-dichloroalkane

In reactions with aqueous sulfuric acid the intermediate carbocation reacts with water to produce an intermediate vinyl alcohol, RC(OH)=CHR. This reaction is poor with sulfuric acid alone, but is catalyzed by mercuric salts. Vinyl alcohols are unstable and rearrange immediately under the reaction conditions to give ketones (Section 14.6).

$$\text{4-hexyne} \xrightarrow[\mathrm{H_2O,\ HOAc}]{\mathrm{H_2SO_4,\ HgSO_4}} \text{4-octanone (89\%)}$$

4-hexyne

4-octanone
(89%)

As shown in this example, hydration of *symmetrical* internal alkynes can be a good method for the preparation of ketones. This is only true if the alkyne is symmetrical, since for an unsymmetrical internal alkyne, there are two possible secondary vinyl carbocations that can be formed. Since these will have comparable stabilities, they will both be formed, and a mixture of ketones will result.

EXERCISE 12.13 Consider the hydration of 2-pentyne. Write a complete mechanism for the reaction, assuming that the electrophile that initiates carbocation formation is H^+. Note that H^+ may be added either to C-2 or to C-3. Are the two resulting carbocations different? Can you think of any reason why they should be very different in stability?

EXERCISE 12.14 Hydration of *terminal* alkynes is a general method for the preparation of methyl ketones:

$$\mathrm{RC{\equiv}CH} \xrightarrow[\mathrm{H_2O,\ HOAc}]{\mathrm{H_2SO_4,\ HgSO_4}} \mathrm{R\overset{O}{\overset{\|}{C}}CH_3}$$

a methyl ketone

Explain the observed selectivity with a reaction mechanism.

The carbon-nitrogen triple bond also undergoes acid-catalyzed hydration. Since they have nonbonding electron pairs on nitrogen, nitriles are Lewis bases in the same way that ammonia is a Lewis base. However, the nitrile nonbonding pair is in a N_{sp}-orbital, and it is therefore *not* very basic; the pK_a of protonated acetonitrile is -10.1.

$$\mathrm{CH_3C{\equiv}\overset{+}{N}{-}H \rightleftharpoons CH_3C{\equiv}N{:} + H^+}$$

Nevertheless, in acidic medium, the nitrile nitrogen is reversibly protonated. The protonated form is much more receptive to nucleophiles than the neutral nitrile and adds water slowly. The initial product rearranges to give an amide.

$$\mathrm{RC{\equiv}N{:} + H^+ \rightleftharpoons \left[RC{\equiv}\overset{+}{N}H \longleftrightarrow R\overset{+}{C}{=}\ddot{N}H\right] \xrightarrow{H_2O} R\overset{\overset{+}{O}H_2}{\overset{|}{C}}{=}NH \xrightarrow{-H^+} R\overset{OH}{\overset{|}{C}}{=}NH \rightleftharpoons R\overset{O}{\overset{\|}{C}}NH_2}$$

Typical catalysts are aqueous hydrochloric acid and concentrated sulfuric acid.

$$\mathrm{C_6H_5CH_2CN \xrightarrow[\mathrm{H_2O,\ 50^\circ C}]{HCl} C_6H_5CH_2\overset{O}{\overset{\|}{C}}NH_2}$$

(82–86%)

A mixture of 200 g of phenylacetonitrile and 800 mL of concentrated hydrochloric acid is stirred at 50 °C for 20–30 min. At the end of this time, ice-cold water is added and the crystalline phenylacetamide is isolated by filtration. The yield is 190–200 g (82–86%).

EXERCISE 12.15 Show how 1-chlorobutane can be converted into the following compounds.

(a) [pentanamide: line structure with C=O and NH_2] (b) [2-methyl-3-heptanone-type line structure with C=O]

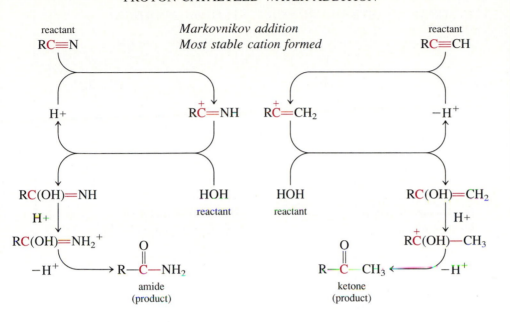

Highlight 12.5

PROTON-CATALYZED WATER ADDITION

Markovnikov addition
Most stable cation formed

reactant $RC{\equiv}N$ → (H^+) → $R\overset{+}{C}{=}NH$ → (HOH reactant) → $RC(OH){=}NH$ → (H^+) → $RC(OH){=}NH_2^+$ → ($-H^+$) → $R{-}C({=}O){-}NH_2$ amide (product)

reactant $RC{\equiv}CH$ → (H^+) → $R\overset{+}{C}{=}CH_2$ → (HOH reactant) → $RC(OH){=}CH_2$ → (H^+) → $R\overset{+}{C}(OH){-}CH_3$ → ($-H^+$) → $R{-}C({=}O){-}CH_3$ ketone (product)

C. Nucleophilic Additions

Unlike simple alkenes, alkynes undergo nucleophilic addition reactions. For example, acetylene reacts with alkoxides in alcoholic solution to yield vinyl ethers. The reaction usually requires conditions of high temperature and pressure.

$$HC{\equiv}CH + RO^- \xrightarrow[\text{150°C, pressure}]{ROH} ROCH{=}CH^- \xrightarrow{ROH} ROCH{=}CH_2 + RO^-$$

The reaction of a stable alkoxide ion to produce a less stable and more basic vinyl anion may seem surprising. However, the addition also involves the formation of a strong carbon-oxygen bond at the expense of the relatively weak "third bond" of a triple bond. The net effect of stronger bonding is more than enough to compensate for the creation of a stronger base. The intermediate vinyl anion is immediately protonated by the alcohol solvent to regenerate the alkoxide ion.

As might be expected, the more polar carbon-nitrogen triple bond in nitriles is more susceptible to nucleophilic addition. A common procedure is the base-catalyzed hydration, which is a good method for the synthesis of amides from nitriles.

$$CH_3CH_2CH_2CH_2CN + KOH \xrightarrow[80°C]{t\text{-BuOH}} CH_3CH_2CH_2CH_2C({=}O)NH_2$$

(84%)

To a solution of 5 g of pentanenitrile in 50 mL of *t*-butyl alcohol is added 10 g of powdered KOH. The mixture is refluxed for 20 min, poured into aqueous sodium chloride solution, and extracted to obtain 5.1 g (84%) of pentanamide.

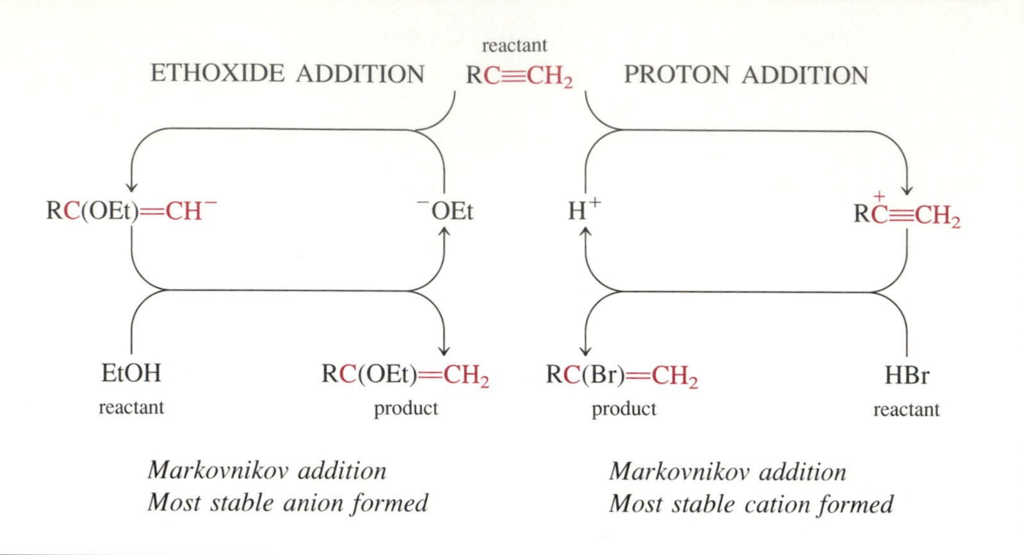

D. Hydroboration

Hydroboration of alkynes is a useful laboratory process for the synthesis of several types of compounds. Diborane reacts with alkynes at 0 °C to produce the intermediate trivinylborane. The reaction is generally useful for terminal alkynes. As with alkenes, the boron adds to the terminal carbon. The reaction is also useful with symmetrical disubstituted alkynes. Unsymmetrical disubstituted alkynes generally give a mixture of products. The net reaction is syn addition of H—BR_2 to the triple bond.

$$3\ RC{\equiv}CR + BH_3 \longrightarrow (RCH{=}CR)_3B$$

$$3\ RC{\equiv}CH + BH_3 \longrightarrow (RCH{=}CH)_3B$$

The resulting vinylboranes, like alkylboranes (Section 11.6.D), enter into a number of reactions, the most useful of which is oxidative cleavage with alkaline hydrogen peroxide. The initial product is a vinyl alcohol, which rearranges quantitatively to the corresponding aldehyde or ketone (Section 14.6.A).

$$CH_3(CH_2)_3C{\equiv}CH \xrightarrow[THF]{BH_3} \xrightarrow[NaOH]{H_2O_2} [CH_3(CH_2)_3CH{=}CHOH] \longrightarrow CH_3(CH_2)_4CHO$$

The overall effect of hydroboration-oxidation is that of hydration of the triple bond. Note that with terminal alkynes the aldehyde is formed (anti-Markovnikov hydration), whereas with direct H_2SO_4-$HgSO_4$ hydration the ketone is produced.

EXERCISE 12.16 What are the principal organic products of the reaction of borane in THF followed by alkaline hydrogen peroxide with (a) 1-butyne, (b) 2-butyne, and (c) 2-pentyne?

12.7 Vinyl Halides

We saw in the last section that alkenyl halides may be prepared by addition of 1 mole of hydrogen halide to an alkyne, often with the aid of a mild Lewis-acid catalyst.

$$CH_3C{\equiv}CH + HCl \xrightarrow{CuCl} CH_3CCl{=}CH_2$$

Partial dehydrohalogenation of a vicinal dihalide generally gives a mixture of the possible haloalkenes.

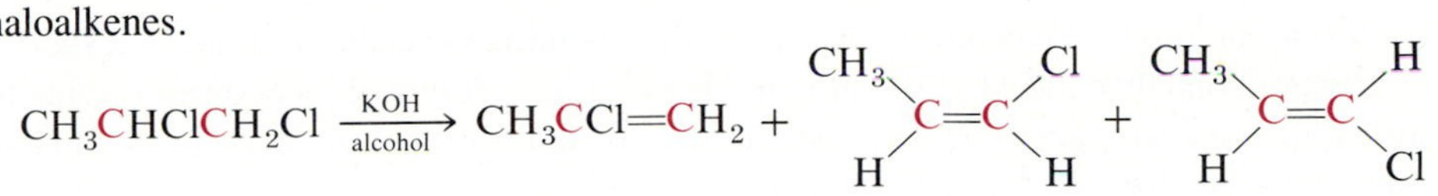

$$CH_3CHClCH_2Cl \xrightarrow[\text{alcohol}]{KOH} CH_3CCl{=}CH_2 + \underset{\text{(CH}_3\text{, H)}}{C{=}C}\text{(Cl, H)} + \underset{\text{(CH}_3\text{, H)}}{C{=}C}\text{(H, Cl)}$$

However, the *vic*-dichlorides obtained from symmetrical olefins give good yields of single products.

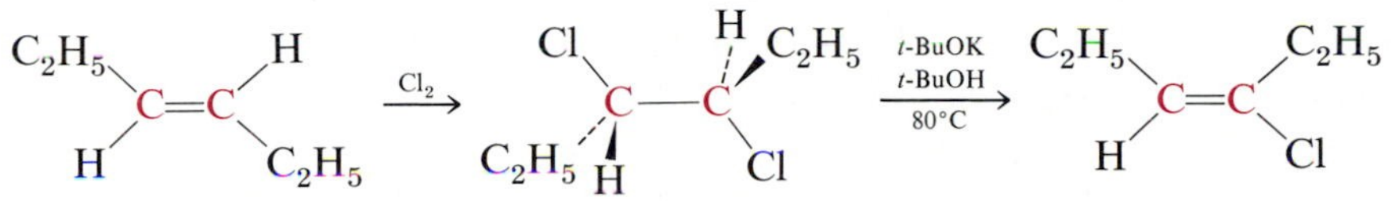

$$\underset{\text{(C}_2\text{H}_5\text{, H)}}{C{=}C}\text{(H, C}_2\text{H}_5\text{)} \xrightarrow{Cl_2} \text{Cl(C}_2\text{H}_5\text{)(H)C—C(H)(C}_2\text{H}_5\text{)Cl} \xrightarrow[80°C]{t\text{-BuOK}, \; t\text{-BuOH}} \underset{\text{(C}_2\text{H}_5\text{, H)}}{C{=}C}\text{(C}_2\text{H}_5\text{, Cl)}$$

(*E*)-3-chloro-3-hexene

Haloalkenes in which the halogen is attached directly on the double bond have exceptionally low reactivity in nucleophilic substitution reactions, either by the S_N1 or S_N2 mechanism. For example, 1-chloropropene is inert to potassium iodide in acetone under conditions where *n*-propyl chloride undergoes rapid substitution.

$$CH_3CH_2CH_2Cl + I^- \longrightarrow CH_3CH_2CH_2I + Cl^- \qquad \text{fairly fast}$$

$$CH_3CH{=}CHCl + I^- \longrightarrow \text{no reaction}$$

Simple alkenyl halides do not readily form carbocations. For such halides, reaction by the S_N1 mechanism is exceedingly slow. Consequently, other reactions, such as addition to the double bond, occur instead. The relative difficulty of ionizing a vinylic carbon-chlorine bond is shown by the following gas phase enthalpies.

	$\Delta H°$, *kcal mole*$^{-1}$
$CH_3Cl \longrightarrow CH_3^+ + Cl^-$	228
$CH_3CH_2Cl \longrightarrow CH_3CH_2^+ + Cl^-$	193
$CH_2{=}CHCl \longrightarrow CH_2{=}CH^+ + Cl^-$	225
$(CH_3)_2CHCl \longrightarrow (CH_3)_2CH^+ + Cl^-$	172

The difference between the energy required to form a vinyl cation and that needed for a simple primary carbocation is comparable to the difference between primary and secondary carbocations. Recall that secondary carbocations are common intermediates in many reactions but that simple primary carbocations are virtually unknown in solution. Primary vinyl cations are similarly unknown in S_N1 reactions; however, secondary vinyl cations of the type $R\overset{+}{C}{=}CH_2$ have been detected under special conditions. They are not important in most organic reactions of the simple vinyl halides.

This lack of reactivity is explained most simply as an increased difficulty in removing an atom with its pair of electrons from a bond to a vinyl orbital with its higher *s*-character than from a simple primary sp^3-orbital. The increased strength of the vinyl-halogen bond compared to the ethyl-halogen bond is manifest also in the relative bond lengths and bond-dissociation energies as shown in the following examples.

	r_{C-X}, Å	$DH°$, kcal mole^{-1}
$CH_3CH_2{-}Cl$	1.78	83
$CH_2{=}CH{-}Cl$	1.72	92
$CH_3CH_2{-}Br$	1.94	71
$CH_2{=}CH{-}Br$	1.89	80

The sp^2-carbon orbital involved in the vinyl halide bond is expected to produce a shorter and stronger bond than the ethyl sp^3-orbital. However, an additional component leading to a still shorter and stronger bond is π-overlap between the π-orbital of the double bond and a lone-pair orbital of the halogen, as depicted in Figure 12.9.

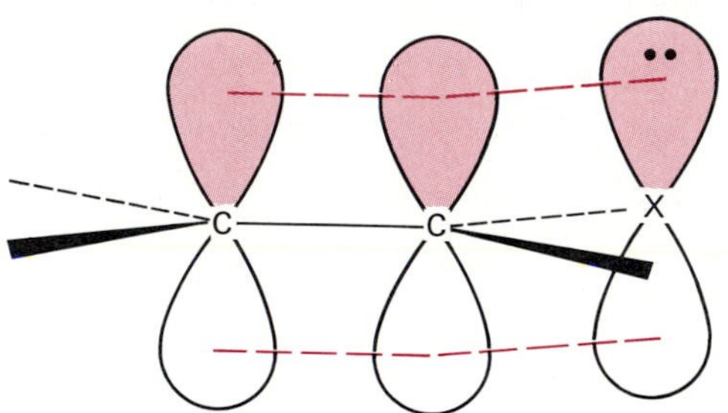

FIGURE 12.9 π-orbital overlap in a vinyl halide.

Such π-overlap can also be represented by resonance involving Lewis structures.

$$\left[CH_2{=}CH{-}\ddot{\underset{\cdot\cdot}{X}}\!: \longleftrightarrow \bar{C}H_2{-}CH{=}\overset{+}{\underset{\cdot\cdot}{X}}\!:\right]$$

As a result of such overlap the carbon-halogen bond in a vinyl halide has partial double bond character. The amount of double bond character (that is, the contribution of resonance structure $^-CH_2{-}CH{=}X^+$) is relatively small, but has a significant effect on reactivity.

EXERCISE 12.17 Using your molecular models, construct a model of (3*R*,4*S*)-3,4-dichlorohexane. Convince yourself that if elimination of HCl occurs in an anti fashion, the product will be (*E*)-3-chloro-3-hexene, regardless of which chlorine is lost. Suggest a way in which 3-hexyne can be converted into (*Z*)-3-chloro-3-hexene (three steps are necessary).

ASIDE

Acetylene was named by the French chemist Berthelot on the basis of a supposed relationship to acetyl ($C_2H_3 - H = C_2H_2$, even though it was known that acetyl was C_2H_3O), in analogy to the derivation of the name of ethylene from ethyl (C_2H_5 −

$H = C_2H_4$). The word nitrile can be traced via nitrogen and the Greek *nitron* to the Hebrew *neter* and the ancient Egyptian *ntrj* for natron or cleaning material. Give the formal IUPAC names for acetylene and acetonitrile. Suppose we wanted an acetylene derivative that would distill at a temperature close to the boiling point of diethyl ether, as, for example, in a special fuel mixture. Pick an appropriate acetylene derivative and show how it could be synthesized from acetylene itself.

PROBLEMS

1. Write the structure corresponding to each of the following common names.
a. methylisopropylacetylene
b. acetonitrile
c. vinylacetylene
d. di-*t*-butylacetylene
e. vinyl bromide
f. isobutylacetylene
g. methyl ethynyl ether

2. Write the structure corresponding to each of the following IUPAC names.
a. cyclodecyne
b. (*R*)-3-methyl-1-pentyne
c. 1-ethynylcyclohexanol
d. 1-methoxyethyne
e. 3-methoxy-1-pentyne
f. pent-2-yn-1-ol
g. (*S*)-2-chlorobutanenitrile
h. pent-1-en-3-yne
i. 2,2,5,5-tetramethyl-3-hexyne
j. 4-methyl-2-pentyne

3. Give an acceptable name for each of the following structures.
a. $BrCH_2CH_2CH_2CH_2C{\equiv}CH$
b. $(CH_3)_3CC{\equiv}CCH_2CH_3$
c. CH_3CH_2CN
d. (cyclopropyl)$-C{\equiv}CH$
e. $CH_2{=}CH{-}C(OH)(H){-}C{\equiv}CH$
f. $HOCH_2C{\equiv}CCH_2OH$
g. $CH_2{=}CHCH_2CH_2C{\equiv}CH$
h. $(CH_3)(Cl)C{=}C(H)(CH_3)$

4. Give the principal reaction product(s) for the reaction of 1-butyne with each of the following reagents. If no reaction is expected, so indicate.
a. *n*-butyllithium, THF
b. H_2/Pt
c. Na, liquid NH_3
d. aq. NaCl
e. aq. $AgNO_3$
f. (i) B_2H_6; (ii) H_2O_2, NaOH
g. H_2SO_4, H_2O, $HgSO_4$
h. H_2 (1 mole), Pd($BaSO_4$)-quinoline

5. Answer problem 4 for 2-butyne.

6. Give the principal reaction product for reaction of butanenitrile with each of the following reagents.
a. H_2, Pd/C, ethanol, HCl
b. conc. HCl, 50 °C
c. KOH, *t*-butyl alcohol, reflux
d. (i) $LiAlH_4$, ether; (ii) water

7. Using propyne or sodium cyanide as the only sources of carbon, devise practical syntheses for the following compounds.
a. $(CH_3)_2CHBr$
b. CH_3COCH_3
c. CH_3CH_2CHO
d. $CH_3(CH_2)_4CH_3$

e. 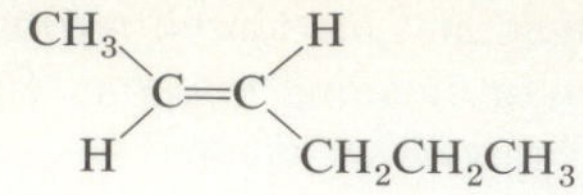 $\underset{H}{\overset{CH_3}{}}\!\!C{=}C\!\!\underset{CH_2CH_2CH_3}{\overset{H}{}}$ f. $CH_3CCl{=}CH_2$

g. $CH_3CH_2CH_2CN$ h. $CH_3CH_2CH_2CH_2NH_2$

8. Show how each of the following conversions may be accomplished in good yield. In each case, use only the indicated starting material as a source of carbon.

a. $CH_3CH_2CH_2CH{=}CH_2 \longrightarrow CH_3CH_2CH_2C{\equiv}CH$
b. $CH_3CH_2CH_2Br \longrightarrow CH_3CH_2CH_2C{\equiv}CCH_3$
c. $HC{\equiv}CH \longrightarrow CH_3CH_2CH_2CH_2OH$
d. $HC{\equiv}CH \longrightarrow CH_3CH_2\overset{OCH_3}{\overset{|}{C}}HCH_2CH_2CH_3$
e. 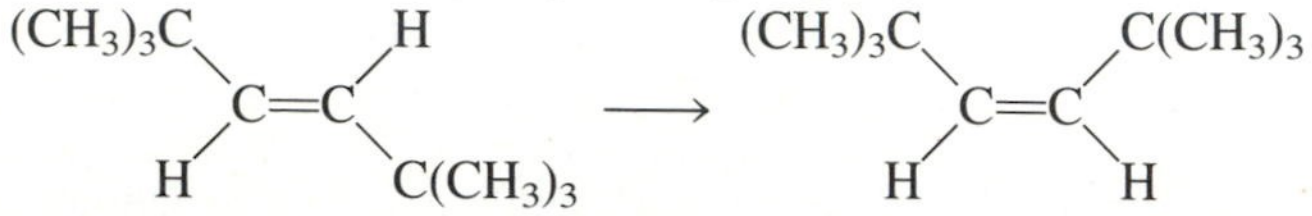 $\underset{H}{\overset{(CH_3)_3C}{}}\!\!C{=}C\!\!\underset{C(CH_3)_3}{\overset{H}{}} \longrightarrow \underset{H}{\overset{(CH_3)_3C}{}}\!\!C{=}C\!\!\underset{H}{\overset{C(CH_3)_3}{}}$
f. $CH_3CH_2CH_2OH \longrightarrow CH_3COCH_3$
g. $CH_3CH_2C{\equiv}CH \longrightarrow \underset{H}{\overset{CH_3CH_2}{}}\!\!C{=}C\!\!\underset{H}{\overset{D}{}}$
h. $HC{\equiv}CH \longrightarrow CH_3CH_2CH_2CH_3$

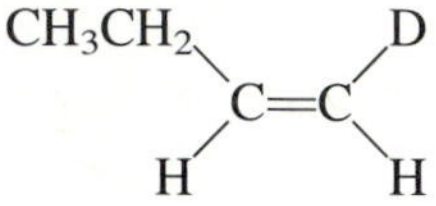

9. Assume that you are shipwrecked on a desert island and that you find the ship pharmacist's trunk, which turns out to be well-stocked with inorganic chemicals and also contains bottles of ethyl iodide, isopropyl bromide, *n*-butyl chloride, DMSO, diethyl ether, and *t*-butyl alcohol. How might these materials be used to prepare the following compounds, and thereby start a chemical industry in your new homeland?

a. 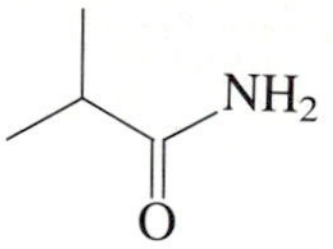 NH_2 / O b. 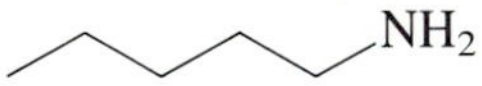 NH_2

c. 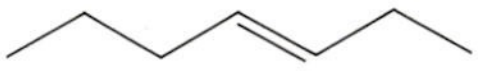 d. 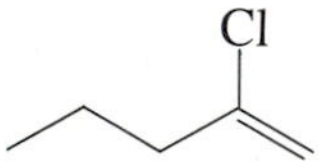 Cl

10. Show how each of the following conversions may be accomplished in good yield. In addition to the indicated starting material, other organic compounds may be used as necessary.

a. $HC{\equiv}CH \longrightarrow CH_3CH_2\overset{O}{\overset{\|}{C}}CH_3$
b. $CH_3OCH_2CH_2C{\equiv}CH \longrightarrow CH_3OCH_2CH_2CH_2CH_2CN$
c. $HC{\equiv}CH \longrightarrow CH_3CH_2C{\equiv}CCH_2CH_2CH_3$
d. $(CH_3)_2\overset{OH}{\overset{|}{C}}CH_2CH_3 \longrightarrow (CH_3)_2CHC{\equiv}CCH_3$

11. From 3-methyl-1-butanol, acetylene, and any required straight-chain primary alcohols, derive a practical synthesis for 2-methylheptadecane, the sex attractant for the Tiger moth (page 96).

12. Muscalure, *cis*-9-tricosene, is the sex-attractant insect pheromone of the common housefly. Give a practical synthesis of this compound from acetylene and straight-chain alcohols.

13. The hydration of alkynes is catalyzed by Hg^{2+}. Write a mechanism for the hydration that accounts for this catalysis. (*Hint:* The mercuric ion adds to the triple bond to give a mercuricarbocation).

14. The pK_as of ethane, ethylene, and acetylene are approximately 50, 44, and 25, respectively.

a. If each hydrocarbon is treated with sodium amide in liquid ammonia and the resulting solution then treated with methyl iodide, different results are obtained. There is no reaction with ethane or ethylene, but acetylene gives a good yield of propyne. Explain this observation using the pK_as.

b. The bond-dissociation energies, *DH*°, for the carbon-hydrogen bonds are ethane, 101; ethylene, 110; acetylene, 132 kcal mole^{-1}. The carbon-hydrogen bonds are progressively harder to break along this series, yet the compounds are increasingly acidic. Explain this apparent paradox.

15. Alkenyl chlorides react with a solution of sodium in liquid ammonia to replace the Cl by H with *retention of configuration*. That is, reduction of (*E*)-2-chloro-2-pentene gives (*Z*)-2-pentene. The reaction may be regarded as a reduction with solvated electrons to produce a vinyl anion which is protonated to give the observed product.

$$\mathrm{>C{=}C<^{Cl}} \xrightarrow{2\,e^-} \mathrm{>C{=}C^-<} + \mathrm{Cl^-} \xrightarrow{NH_3} \mathrm{>C{=}C<^{H}} + \mathrm{NH_2^-}$$

a. What does the stereochemistry of the reaction reveal concerning the geometrical structure and configurational stability of the intermediate vinyl anion?

b. This reaction is used in a sequence to invert the configuration of internal olefins. Show how *trans*-3-hexene may be converted to *cis*-3-hexene by use of this reaction as the final step.

16. The reaction of (*Z*)-1,5-dibromo-1-pentene with ethanolic $NaOC_2H_5$ can give principally (*Z*)-1-bromo-5-ethoxy-1-pentene, 5-ethoxy-1-pentyne, or 2,5-diethoxy-1-pentene, depending on the reaction conditions. Explain. Why is 1,5-diethoxy-1-pentene not a principal product under any of these conditions?

17. Compound A has the formula C_8H_{12} and is optically active. It reacts with hydrogen in the presence of platinum metal to give B, which has the formula C_8H_{18} and is optically inactive. Careful hydrogenation of A using H_2 and a poisoned palladium catalyst gives C, which has the formula C_8H_{14} and is optically active. Compound A reacts with sodium in ammonia to give D, which also has the formula C_8H_{14} and is optically inactive. What are compounds A through D?

18. Compound E has the formula C_7H_{12}. It reacts with dry HCl at −20 °C to give F, $C_7H_{13}Cl$. Compound F reacts with potassium *t*-butoxide in *t*-butyl alcohol to give a small amount of E and mainly G, which as the formula C_7H_{12}. Ozonization of G gives cyclohexanone and formaldehyde. What are compounds E through G?

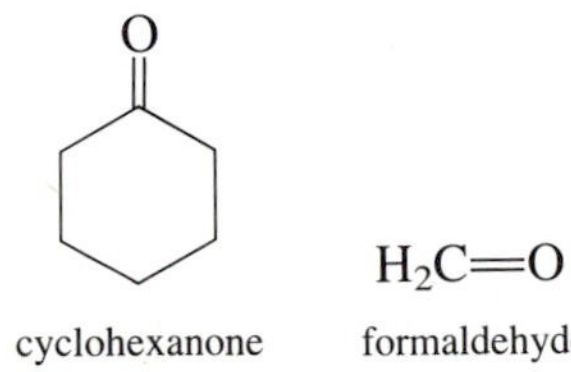

$H_2C{=}O$

cyclohexanone formaldehyde

19. Appendix II, "Bond-Dissociation Energies," gives values of *DH*° for the carbon-halogen bonds in $CH_2{=}CH{-}Cl$ and $CH_2{=}CH{-}Br$. Compare with the corresponding values for ethyl halides and explain any difference. Estimate *DH*° for vinyl fluoride and vinyl iodide.

CHAPTER 13

NUCLEAR MAGNETIC RESONANCE SPECTROSCOPY

13.1 Structure Determination

Structure determination is one of the fundamental operations in chemistry. How does the chemist determine the structure of a compound? In order to answer this question, let us consider a simple, hypothetical experiment. Imagine that we carry out a reaction between propane (C_3H_8) and chlorine, both of which are gases at room temperature. After the reaction is completed, we obtain a liquid product. We distill this liquid and obtain two main fractions, one boiling at 36 °C and one at 47 °C. These two liquids are obviously not the reactants, which are both gases. Therefore, they must be reaction products. What are they?

As a first step, we might perform an elemental analysis (Section 3.4) and determine their empirical formulas. When we do this, we find that they both have the formula C_3H_7Cl. We conclude that a reaction has occurred in which a hydrogen has been replaced by a chlorine and that two isomeric products have been produced in the reaction. Since there are only two types of hydrogen in propane, we can write structures for the two products. One is 1-chloropropane and the other is 2-chloropropane. Therefore, the reaction that has occurred is substitution of hydrogen by chlorine and the two possible substitution products have been formed.

$$CH_3CH_2CH_3 + Cl_2 \longrightarrow CH_3CH_2CH_2Cl + CH_3\overset{\overset{\displaystyle Cl}{|}}{C}HCH_3$$

But which is which? Is the product that boils at 36 °C 1-chloropropane or 2-chloropropane?

One way to answer this question is to look up the boiling points of 1-chloropropane and 2-chloropropane in a handbook. But suppose for a moment that the two compounds have never been prepared before and their boiling points are not known. [Remember, this is a *hypothetical* case.] Another way to answer our question would be to convert the two isomers into compounds that *are* known. For example, suppose we have samples of 1-propanol and 2-propanol and that we know which is which. We can treat each of our two C_3H_7Cl isomers with aqueous KOH. Under these conditions, each alkyl halide is converted into the structurally analogous alcohol, by the S_N2 reaction. The isomer that boils at 47 °C gives 1-propanol and the isomer that boils at 36 °C gives 2-propanol. (In addition, both isomers give some propene.) Because we know that the OH group in 1-propanol is attached to the end of the propane chain, it follows that the C_3H_7Cl isomer with b.p. 47 °C also has its chlorine attached to the end of the chain. In this simple example, we have assigned structures to the two isomers on the basis of their conversion to products of known structure.

$$\underset{\substack{\text{1-chloropropane} \\ \text{b.p. 47°C}}}{CH_3CH_2CH_2Cl} + OH^- \longrightarrow \underset{\text{1-propanol}}{CH_3CH_2CH_2OH} + CH_3CH{=}CH_2$$

$$\underset{\substack{\text{2-chloropropane} \\ \text{b.p. 36°C}}}{CH_3\overset{\overset{\large Cl}{|}}{C}HCH_3} + OH^- \longrightarrow \underset{\text{2-propanol}}{CH_3\overset{\overset{\large OH}{|}}{C}HCH_3} + CH_3CH{=}CH_2$$

A more direct method of structure determination involves the careful examination of certain physical properties of the compound of unknown structure. The most useful properties for this purpose are spectra. Spectroscopy is a powerful tool for structure determination. There are many different types of spectroscopy. In the following section we shall have a brief introduction to spectroscopy generally, and then we shall take up one specific type of spectroscopy, nuclear magnetic resonance spectroscopy, in detail.

13.2 Introduction to Spectroscopy

Molecules, and parts of molecules, are constantly in **motion.** The entire molecule rotates, the bonds vibrate, and even the electrons move—albeit so rapidly that our measurements probe only electron density distributions. It is a fundamental law of nature that each of these kinds of motion is **quantized.** That is, the molecule can exist only in distinct states (**quantum states**) that correspond to discrete energy contents. Each state is characterized by one or more quantum numbers.

> The quantum nature of molecules conflicts with our day-to-day experience, which leads us to assume that there is a continuous range of velocities and energies. For example, we are used to the idea that we can drive our automobile at 55, 56, . . . , 60 mph, or even at $57\frac{1}{3}$ mph. However, even though it is not intuitive, the student must get used to the idea that on the molecular level, velocities and energies *are* restricted to certain specific values. It is just as though there were a law of nature that restricted the velocity of a Volkswagen only to the speeds 40 mph, 48 mph, 55 mph, and 61 mph.

The energy difference, ΔE, between two quantum states is related to a light frequency ν by Planck's constant h (Figure 13.1).

$$\Delta E = h\nu \tag{13-1}$$

Spectroscopy is an experimental process in which the energy differences between allowed states of a system are measured by determining the frequencies of the corresponding light absorbed.

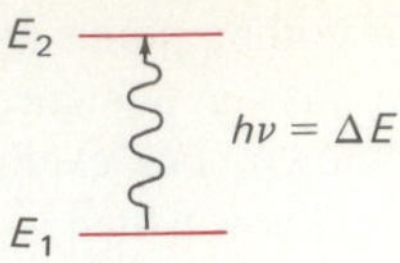

FIGURE 13.1 Light of frequency ν corresponds to an energy difference ΔE between states corresponding to energies E_1 and E_2.

The energy difference between the different quantum states depends on the type of motion involved. The wavelength of light required to bring about a transition is different for the different types of motion. That is, each type of motion corresponds to the absorption of light in a different part of the electromagnetic spectrum. Because the detection methods for each of the wavelengths required are so vastly different, different instrumentation is required for each spectral region. For example, the energy differences between molecular rotational states are rather small, on the order of 1 cal mole^{-1}. Light having this energy has a wavelength of about 3 cm and is called microwave radiation. The energy spacings of molecular rotational states depend on bond distances and angles and the atomic masses of the bonded atoms (moments of inertia). Hence, **microwave spectroscopy** is a powerful tool for precise structure determination. However, the technique must be applied in the vapor phase and it is restricted to rather simple molecules. Although it is an important technique in the hands of a specialist, it is not commonly used by organic chemists.

Energy differences between different states of bond vibration are of the order 1–10 kcal mole^{-1} and correspond to light having wavelengths of 30–3 μ (1 $\mu \equiv 10^{-6}$ m). This is the **infrared** region of the spectrum. Infrared spectrometers are relatively inexpensive and easy to use, and infrared spectroscopy is an important technique in organic chemistry. It is used mainly to determine which functional groups are present in a compound. We will study its use in more detail in Chapter 17.

Different electronic states of organic compounds correspond to energies in the visible (4000–7500 Å; 1 Å$=10^{-8}$ cm; 70–40 kcal mole^{-1}) and ultraviolet (1000–4000 Å; 300–70 kcal mole^{-1}) regions of the electromagnetic spectrum. Spectrometers for this region are also common, and ultraviolet-visible spectroscopy is an important technique in organic chemistry, especially for conjugated systems. Such compounds will be discussed in detail later, and our study of this spectroscopy will be deferred to Chapter 22.

Highlight 13.1

Spectroscopy involves the measurement of the energy difference between two quantum states of a molecule. The wavelength of the light used for a spectroscopic experiment depends on the difference in energy between the states; if the difference is great, we must use light of relatively high energy (short wavelength), and if the difference in energy is small, we use light of relatively low energy (long wavelength). Thus, in Figure 2.7 we see that absorption of a photon of considerable energy is required to promote an electron from a bonding to an antibonding orbital, a process studied by **ultraviolet** or **visible spectroscopy** (see Chapter 22 for a more extensive discussion). In Figure 11.8 we see curves that describe the quantum levels for stretching or vibration of a bond. Since the energy difference between vibrational quantum states is smaller, we may study this phenomenon with **infrared spectroscopy.** Because groups of atoms usually have characteristic bond vibrations that are relatively unperturbed by the remainder of a molecule, infrared spectroscopy is a useful tool to detect some functional groups. The energy differences between the quantum states associated with rotations about bonds are even smaller than for vibrational states, and bond rotations are studied with **microwave spectroscopy.**

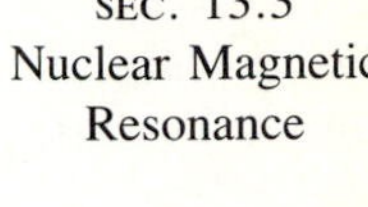

13.3 Nuclear Magnetic Resonance

Nuclear magnetic resonance (NMR) spectroscopy has only been important in organic chemistry since the mid-1950s, yet in this relatively brief time it has taken its place as our most important spectroscopic tool. In NMR spectroscopy, a solution of the sample is placed in the instrument—actually the sample tube is fitted precisely between the poles of a powerful magnet—and the spectrum is recorded. A typical example is the NMR spectrum of 1-chloropropane shown in Figure 13.2. Some appreciation of the usefulness of this technique can be sensed by comparing the spectrum of the isomeric compound, 2-chloropropane, shown in Figure 13.3. In this chapter we will develop the rules for interpreting such spectra. We will find it rather simple, for example, to *deduce* the structures of 1-chloropropane and 2-chloropropane from their NMR spectra. It is possible to treat the NMR spectrometer as a ''magic box'' and simply memorize a few rules that

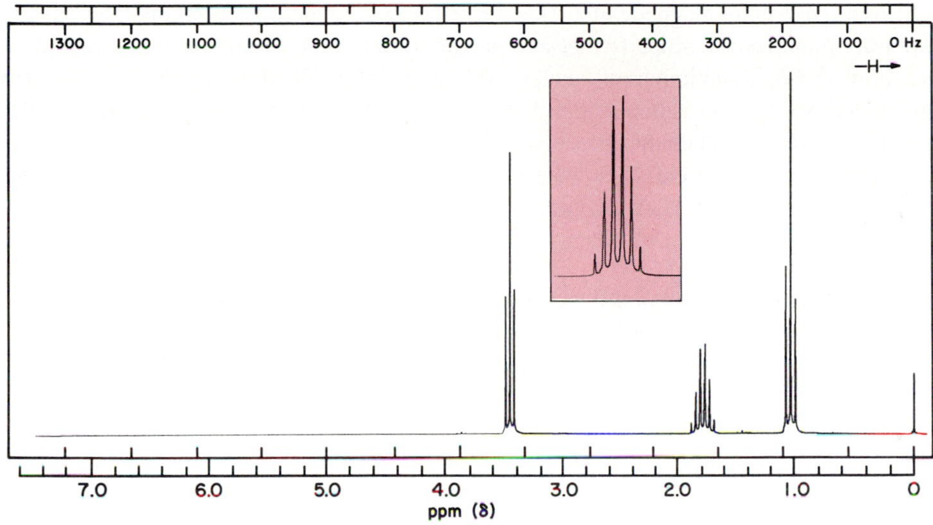

FIGURE 13.2 NMR spectrum of 1-chloropropane, $CH_3CH_2CH_2Cl$.

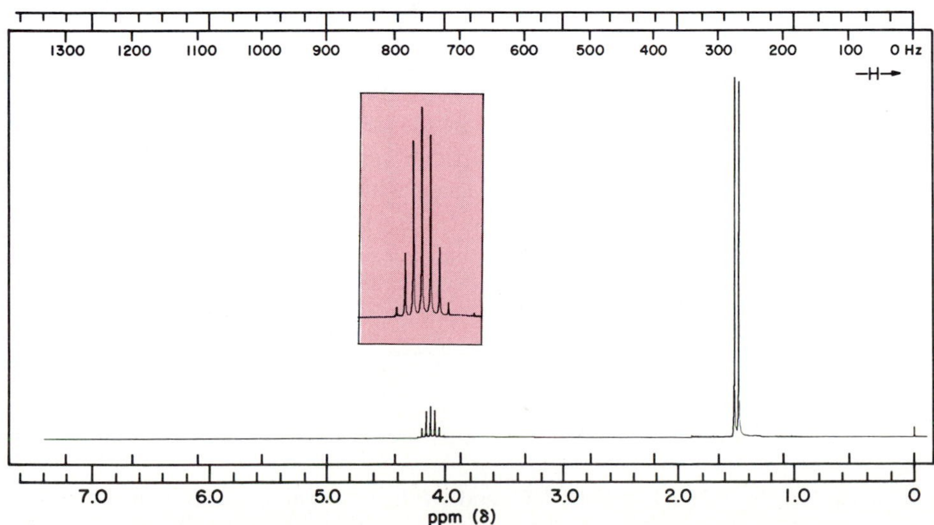

FIGURE 13.3 NMR spectrum of 2-chloropropane, $CH_3CHClCH_3$.

suffice for deducing the structure of a compound from its spectrum. In this chapter we will also go into some of the theoretical background of NMR spectroscopy to show why the rules take the form they do.

Nuclear magnetic resonance spectroscopy differs from the spectroscopic techniques discussed in the previous section in that the states being examined have different energies in a magnetic field. That is, the molecules are placed in a powerful magnetic field to create a difference in energy between two states, which are then detected by absorption of light of the appropriate energy. In the absence of the magnetic field, these different states all have nearly the same energy.

The motion involved in NMR spectroscopy is that of **nuclear spin.** The nuclei of many atoms behave as though they are spinning on an axis. Since they are positively charged, such nuclei must obey the physical laws of spinning charged particles. A moving charge, positive or negative, is associated with a magnetic field. Consequently, the spinning nuclei behave as tiny bar magnets; that is, such nuclei have **magnetic moments.** In field-free space these magnetic moments are oriented in random fashion, but they have the important quantization property that in a magnetic field only certain discrete orientations are allowed. For some important nuclei, ^{1}H (but not ^{2}H), ^{13}C (but not ^{12}C), and ^{19}F (the only common fluorine isotope), the nuclear spin can have only two alternative values associated with the quantum numbers, $+\frac{1}{2}$ ($=\alpha$) and $-\frac{1}{2}$ ($=\beta$). When these nuclei are placed in a magnetic field, their magnetic moments tend either to align with the field (corresponding to α-spin) or against the field (corresponding to β-spin) (Figure 13.4).

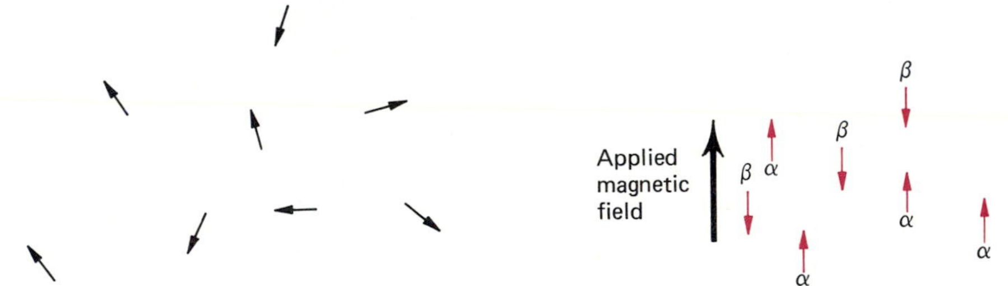

FIGURE 13.4 Orientation of nuclear magnetic moments.

In an applied field, a magnetic moment tends to align with the field (for example, a compass needle in the earth's magnetic field). A magnet aligned against the magnetic field is in a higher energy state than one aligned with the field. For ^{1}H, ^{13}C, and ^{19}F the β-spin state (magnetic moment aligned against the field) corresponds to a higher energy state than the α-spin state. If the system is irradiated with light of the proper frequency or wavelength, a nucleus with α-spin can absorb a light quantum and be converted to the higher-energy β-spin state, a process colloquially described as "flipping the spin" (Figure 13.5).

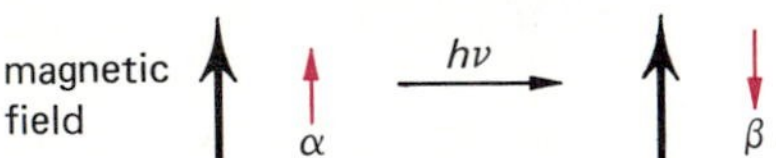

FIGURE 13.5 Absorption of light of proper frequency changes the nuclear spin state.

To recapitulate, the nuclei of ^{1}H, ^{13}C, and ^{19}F have spinning nuclei with spins of $\pm\frac{1}{2}$. Because of the restrictions imposed by quantization, only two orientations are permitted for these nuclei in a magnetic field:

(+$\frac{1}{2}$) **α-spin** (nuclear magnetic moment aligned *with* the applied magnetic field, lower energy).

(−$\frac{1}{2}$) **β-spin** (nuclear magnetic moment aligned *against* the applied magnetic field, higher energy).

Many nuclei have no spin. All even-even nuclei (those having an even number of protons and an even number of neutrons) are in this class. In this important class, which includes ^{12}C and ^{16}O, individual pairs of protons and neutrons have opposed spins so that the net spin of the nucleus as a whole is zero. Other nuclei, such as ^{14}N, have three or more possible spin states in a magnetic field. We will not consider such cases, but will restrict our attention primarily to those nuclei that have spin of $\pm\frac{1}{2}$: ^{1}H, ^{13}C, and ^{19}F. The energy difference between the two states is given by the relationship

$$\nu = \frac{\gamma \mathbf{H}}{2\pi} \qquad (13\text{-}2)$$

in which **H** is the magnetic field strength *at the nucleus* and γ is the magnetogyric ratio of the nucleus. This quantity is the ratio of the angular momentum (from the rotating nuclear mass) and the magnetic moment (from the rotating nuclear charge) and is characteristic and different for each nucleus. That is, the energy difference between the α- and β-spin states in a magnetic field is proportional to the strength of the magnetic field with a proportionality constant that is characteristic of the nucleus (Figure 13.6). In other words, the difference in energy between the α- and β-spin states is greater the greater the strength of the magnetic field; when there is no field, the two spin states have the same energy.

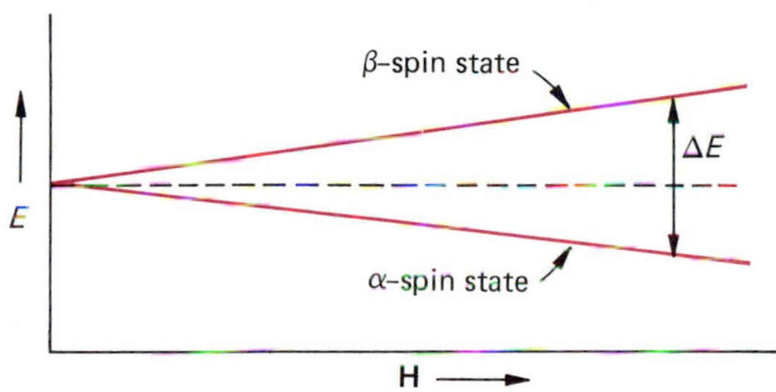

FIGURE 13.6 The energy difference ΔE between the α- and β-spin states is a function of the magnetic field at the nucleus.

For ^{1}H, γ has the value 2.6753×10^4 radians sec^{-1} $gauss^{-1}$. When **H** is given in units of gauss, the frequency ν is given in units of cycles per second or Hertz, Hz. This energy unit is used commonly in NMR, and it may be converted to the more familiar units of cal $mole^{-1}$ by the following equation.

$$E\ (\text{cal mole}^{-1}) = 9.54 \times 10^{-11}\ \nu\ (\text{Hz}) \qquad (13\text{-}3)$$

According to equation (13-2), the energy differences involved are proportional to the magnetic field and are exceedingly small. For example, for an isolated hydrogen nucleus in a magnetic field of 42,276 gauss (the field strength of the magnet used to obtain the spectra shown in Figures 13.2 and 13.3), the energy difference between α- and β-spin states is given by $\Delta E = (26{,}753)(42{,}276)/2\pi = 180 \times 10^6$ Hz = 180 MHz (megaHertz). From equation (13-3) this energy value is equivalent to only 0.0171 cal $mole^{-1}$ (not kcal!). The frequency of 180 MHz corresponds to a wavelength of about 170 cm and is in the radio region of the electromagnetic spectrum. A field of 42,000 gauss is a rather strong magnetic field but one that is readily accessible with modern superconducting magnets. NMR spectrometers with field strengths of 20,000–80,000 gauss are now relatively common, and commercial instruments are also available with field strengths in which the proton ‘‘flip’’ corresponds to 400, 500, and even 600 MHz.

Highlight 13.2

Certain nuclei (^{1}H, ^{13}C, ^{19}F, and ^{31}P) that are important in organic chemistry have nuclear spins, α- ($+\frac{1}{2}$) [magnetic moment aligned **with** applied magnetic field] and β- ($-\frac{1}{2}$) [magnetic moment aligned **against** applied magnetic field]. The NMR spectrum is a measurement of the absorption of radiofrequency waves that convert a nucleus in an α state to one in a β state.

13.4 Chemical Shift

If NMR spectroscopy related only to free protons floating in a magnetic field, we would hardly expect to find that thousands of NMR spectrometers now operate in laboratories throughout the world. It is when we look at magnetic phenomena in bonds to protons in molecules that we find why NMR is such an invaluable asset to the organic chemist. A proton in a molecule is surrounded by a cloud of electronic charge. In a magnetic field these electrons move in such a way that their motion induces their own small magnetic field, characterized by a magnetic moment that is **opposed** to the field applied by the external magnet. Consequently, the *net* magnetic field at the hydrogen nucleus is slightly *less* than the applied field (Figure 13.7).

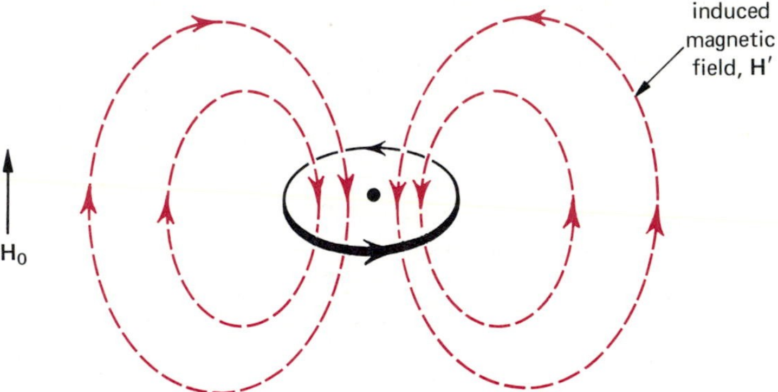

FIGURE 13.7 An external magnetic field induces an electron flow in an electron cloud that, in turn, induces a magnetic field. At the nucleus the induced field opposes the external field.

A frequent source of confusion is the difference between electron flow and electrical current. By a convention established before the discovery of electrons, current flows from anode (+) to cathode (−), which turns out to be exactly the opposite of the actual movement of electrons. The figures in this book (e.g., Figure 13.7) represent the actual flow of electrons and are the reverse of the direction of a positive electrical current.

The magnetic field **H** experienced by the nucleus is therefore

$$\mathbf{H} = \mathbf{H}_0 - \mathbf{H}' \qquad (13\text{-}4)$$

where $\mathbf{H}_0$ is the applied field and $\mathbf{H}'$ is the induced field. Because the nucleus experiences a smaller magnetic field than that applied externally, it is said to be **shielded.** This particular type of shielding is called **diamagnetic shielding.** If we are irradiating the proton with radio waves of exactly 180 MHz frequency, the change of $\alpha \rightarrow \beta$ energy states of "spin flipping" requires a field of 42,276 gauss *at the proton*. However, since the nucleus is shielded by electrons, *the applied field* must be made somewhat higher than 42,276 gauss in order for the field at the nucleus to have the **resonance** value of 42,276 gauss, the field strength that corresponds to the radio frequency required to produce "spin flipping." Protons in different electronic environments experience different amounts of

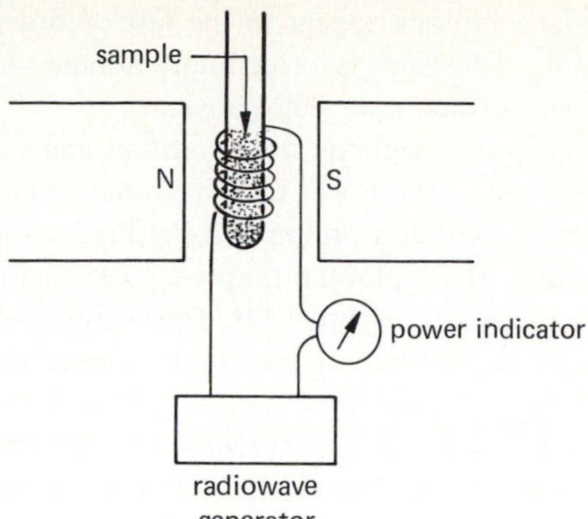

FIGURE 13.8 Schematic of an NMR spectrometer.

shielding, and the resonance absorption of light energy will occur at different values for the applied field or irradiating light frequency. These changes are referred to as **chemical shifts.**

A nuclear magnetic resonance spectrometer is arranged schematically as shown in Figure 13.8. A liquid sample or solution contained in a narrow glass tube is put between the poles of the powerful magnet. The magnetic field creates the two energy states for various hydrogen nuclei in the sample. The sample is irradiated with radio waves from a simple coil. In one mode of operation we fix the radio frequency at, say, 180 MHz. We then vary the magnetic field, and as the field at each kind of proton reaches the **resonance** value, energy is absorbed from the radio waves as the nuclear spins "flip," and this absorption is measured and recorded on a graph.

For example, 1,2,2-trichloropropane, $CH_3CCl_2CH_2Cl$, gives the spectrum shown in Figure 13.9. This spectrum consists of two sharp peaks corresponding to the methylene group and the methyl group. The CH_2 group is attached to chlorine, an electronegative element, which withdraws electrons from carbon and hydrogen. Since there is less electron density around the methylene protons, the diamagnetic shielding of these protons is less than it is for the methyl protons. The induced magnetic field is therefore lower at CH_2 than at CH_3, and the applied field must be increased less in order to achieve resonance.

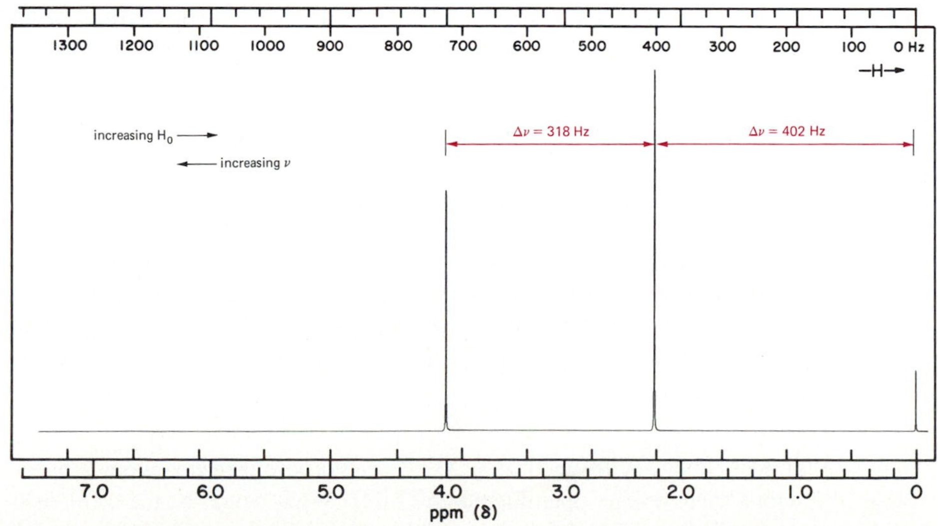

FIGURE 13.9 NMR spectrum of $CH_3CCl_2CH_2Cl$ at 180 MHz.

Consequently, the methylene protons appear to the left or **downfield** compared to the methyl protons. Note that the difference is exceedingly minute—about 0.012 gauss compared to a total applied field of about 42,000 gauss.

Alternatively, we can keep the magnetic field constant and vary the frequency of the radio electromagnetic irradiation. The lower the electronic shielding of the nucleus, the higher the effective magnetic field at a proton and the higher the frequency required to reach the resonance condition. If we plot the frequency increasing from right to left, the resulting spectrum looks exactly like Figure 13.9. The methylene hydrogens now appear at higher frequency than the methyl hydrogens, the frequency difference being 318 Hz. Frequency differences can be measured more precisely than differences in magnetic field strength. Consequently, the difference in peaks is always given in frequency units, regardless of the specific mode of operation of the NMR spectrometer. In practice, most spectrometers today operate at constant field and vary the frequency. Nevertheless, the kind of language in common use is illustrated by the statement that in Figure 13.9 the methylene group appears *downfield* with a frequency difference of 318 Hz. Note again that this is a small difference between large numbers; if the methyl group **resonates** at 180,000,000 Hz, the methylene is at 180,000,318 Hz! Since these *absolute numbers* are difficult to reproduce, in practice we compare differences relative to a standard.

The standard compound used for most proton NMR spectra is tetramethylsilane, $(CH_3)_4Si$, commonly abbreviated **TMS,** a volatile liquid, b.p. 26.5 °C (Sections 8.5, 8.6). This useful compound is inert to most reagents and is soluble in most organic liquids. A small amount has been added to our sample of 1,2,2-trichloropropane, and it gives rise to the peak at the far right in Figure 13.9. All of the hydrogens in TMS are equivalent and give rise to the single sharp line 402 Hz upfield from the methyl of the trichloropropane. Furthermore, silicon is electropositive relative to carbon and tends to donate electron density to the methyl groups, thereby increasing their shielding. The relatively high shielding of the protons in TMS causes them to resonate upfield from most other protons commonly encountered in organic compounds.

When the spectrum is recorded with a spectrometer operating at 70,460 gauss, the resonance frequency of hydrogen is 300 MHz. The larger magnetic field induces a larger electron current, which causes a larger diamagnetic shielding at the nucleus. The difference in diamagnetic shielding is proportionally larger and the peaks spread apart, as shown in Figure 13.10. The chemical shift of the methyl group is now 670 Hz downfield from TMS instead of 402 Hz. Because different NMR instruments are in common use, it is convenient to define a unitless measure that is independent of field strength. The unit used is δ. It is simply the ratio of the chemical shift of the resonance in question, in Hertz, to the total radio frequency used. Since the resulting number is small, it is multiplied by 10^6 so as to be convenient to handle. Thus, δ has the units of parts per million (ppm) and represents a chemical shift downfield (higher frequency) from TMS.

$$\delta_i = \frac{\nu_i - \nu_{\text{TMS}}}{\nu_0} \times 10^6 \text{ ppm} \qquad (13\text{-}5)$$

In equation (13-5) δ_i is the chemical shift of proton i, ν_i is the resonance frequency of that proton, ν_{TMS} is the resonance frequency of TMS, and ν_0 is the operating frequency of the instrument. Thus, for $CH_3CCl_2CH_2Cl$, $\delta(CH_3) = 402/180$ (or 669/300) = 2.23 ppm and $\delta(CH_2) = 720/180$ (or 1200/300) = 4.00 ppm. If a resonance is upfield from TMS, its δ value has a negative sign.

EXERCISE 13.1 Sketch the NMR spectrum of 1,2,2-trichloropropane as it would appear if measured with a spectrometer operating at 250 MHz. What would be the difference in Hz between the TMS and CH_3 resonances? Between the CH_3 and CH_2 resonances?

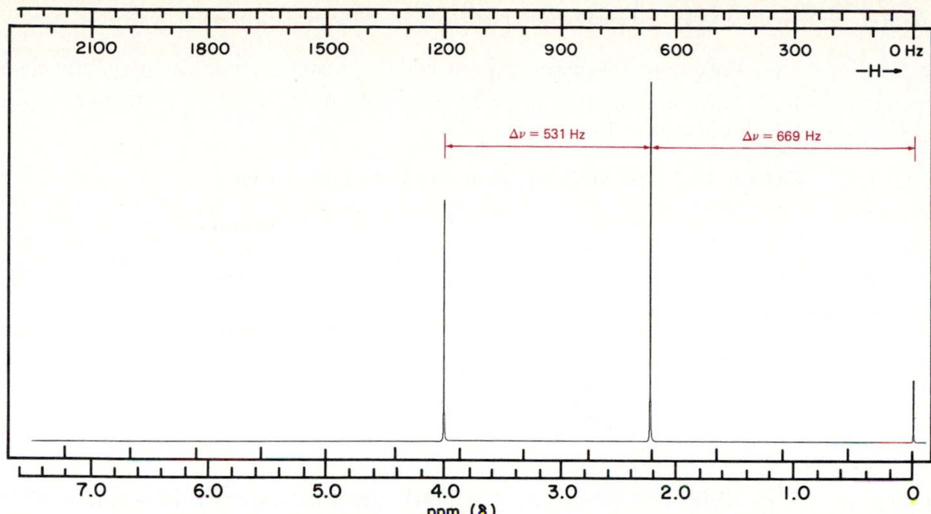

FIGURE 13.10 NMR spectrum of $CH_3CCl_2CH_2Cl$ at 300 MHz.

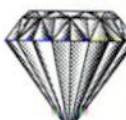

Highlight 13.3

The **chemical shift** δ_H or δ_C of a resonance signal in ^1H or ^{13}C NMR is measured with respect to a standard, usually tetramethylsilane, $(CH_3)_4Si$, with δ_H or $\delta_C = 0.00$. The magnetic field at the nucleus is the sum of the externally **applied** magnetic field and the magnetic field **induced** by the applied magnetic field in electrons surrounding the nucleus. The resultant magnetic field at the nucleus determines the radiofrequency required to "flip the spin" and is a sensitive characteristic of the electronic, and therefore the chemical, environment of the nucleus.

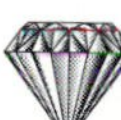

Highlight 13.4

The stronger the field at the nucleus, the higher the frequency needed to "flip the spin." Compare the following figure with Figure 13.6.

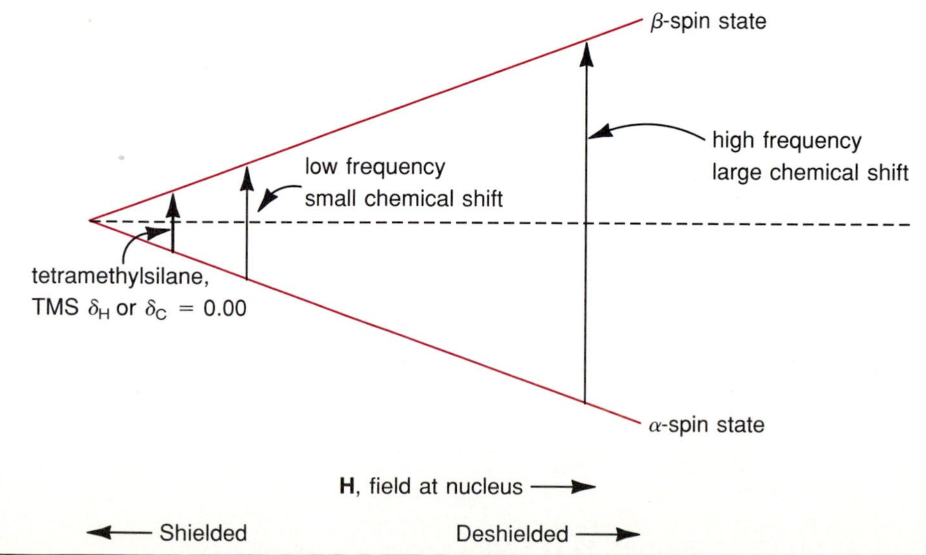

13.5 Carbon NMR Spectroscopy

Up until now, we only have discussed proton NMR. However, nuclear magnetic resonance experiments may be done with any element whose nuclei have a net magnetic spin. A few examples are given in Table 13.1.

TABLE 13.1 The Magnetic Properties of Some Nuclei

Isotope	Natural Abundance, %	Spin States	Resonance Frequency at 42,276 Gauss, MHz
^{1}H	99.88	$\pm\frac{1}{2}$	180
^{13}C	1.1	$\pm\frac{1}{2}$	45.3
^{19}F	100	$\pm\frac{1}{2}$	169.2
^{31}P	100	$\pm\frac{1}{2}$	73.2

In addition to ^{1}H NMR, ^{19}F NMR and ^{13}C NMR are used extensively. Carbon NMR (CMR) is of particular value in organic chemistry. While proton NMR allows us to "see" the protons attached to the carbon framework of an organic compound, CMR allows us to see the carbons themselves. Therefore, CMR is a perfect complement to proton NMR in solving structural problems.

There is one important difference between proton NMR and carbon NMR. In proton NMR, we observe the most abundant isotope, ^{1}H. For carbon, the most abundant isotope, ^{12}C, has an even-even nucleus that has no net nuclear spin or magnetic moment. Thus we must observe the isotope ^{13}C, which has a natural abundance of only about 1%. The low abundance of ^{13}C means that we must use a larger sample than is normally required for proton NMR. Whereas an excellent proton NMR spectrum can usually be obtained with less than 0.1 mg of sample, something on the order of 1–5 mg is required to obtain a CMR spectrum of comparable quality.

Figure 13.11 shows the CMR spectrum of 2-butanol in deuteriochloroform, $CDCl_3$. The spectrum was measured with a spectrometer operating at a field strength of 58,717 gauss, which corresponds to a frequency of 62.5 MHz for ^{13}C. Although 2-butanol has only four different carbons, the spectrum contains many absorption lines. This complication results from a phenomenon called "spin-spin splitting", or "coupling." We will

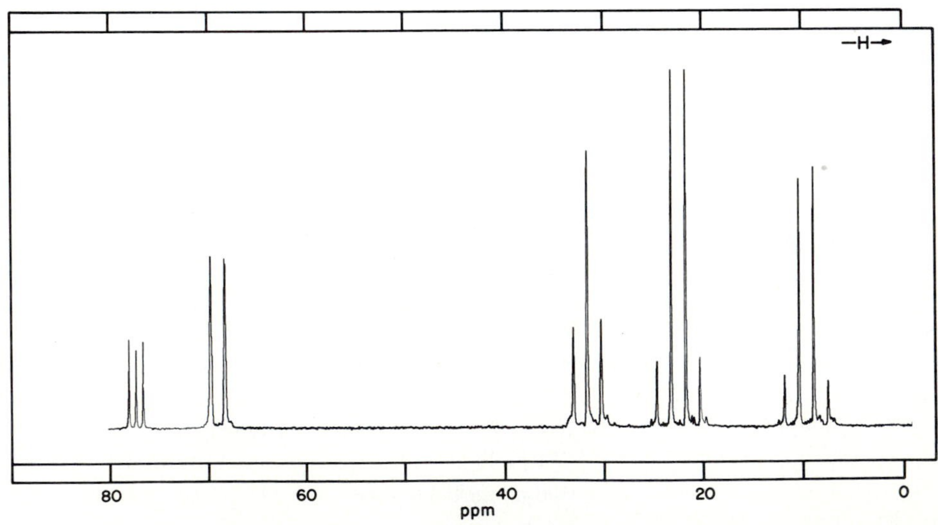

FIGURE 13.11 CMR spectrum of 2-butanol.

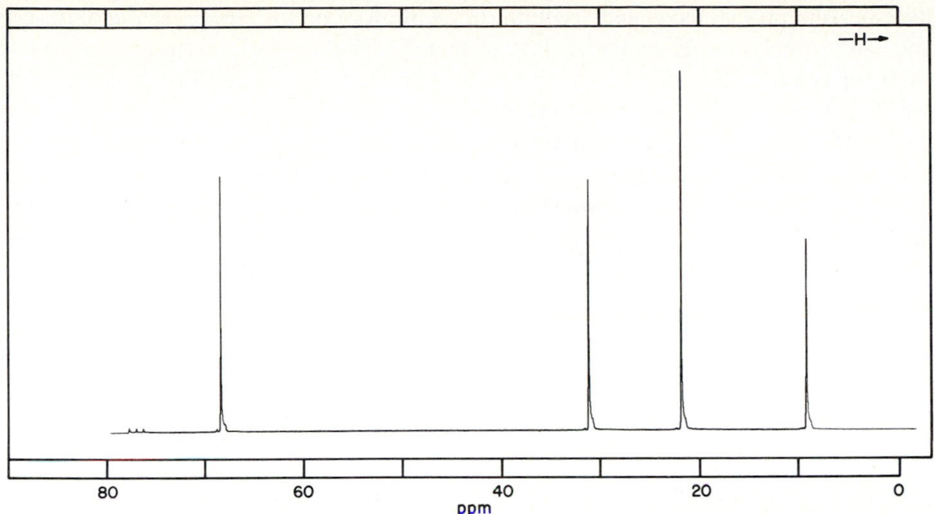

FIGURE 13.12 Proton-decoupled CMR spectrum of 2-butanol.

explain the phenomenon in detail in Section 13.7. For the present, just accept the fact that the absorption peak of a carbon is complicated by nearby hydrogens. Under appropriate experimental conditions, called "off-resonance decoupling," the carbon resonance is affected only by the hydrogens directly bonded to it. If a carbon has three attached hydrogens (CH_3), its CMR resonance appears as a quartet of lines with relative intensities of 1:3:3:1. If there are two attached hydrogens (CH_2), the carbon resonance has the form of a triplet of lines with relative intensities of 1:2:1. A carbon with only one attached hydrogen (CH) appears as a doublet of equally intense lines. Finally, a carbon with no attached hydrogens is unsplit. A simplifying aspect of CMR is that, because of its low abundance, it is highly unlikely that two ^{13}C nuclei will be bonded together, so we do not have to worry about splitting caused by other carbons.

In the spectrum in Figure 13.11 note that we see four signals corresponding to the four carbon atoms. The 1:1:1 triplet at $\delta = 76$ ppm is due to the carbon of $CDCl_3$, the solvent. Deuterium, ^2H, has three spin states, -1, 0, and $+1$, and splits a coupled resonance into a 1:1:1 triplet. The four carbon atoms in 2-butanol appear as a doublet, triplet, and two quartets due to splitting by their attached hydrogens (one, two, and three, respectively). The electronegative oxygen causes C-2 to resonate downfield from the other carbons, as in proton NMR. The chemical shifts in CMR are much greater than we will find in proton NMR; in this example, they range from $\delta = 10.2$ ppm for C-4 to $\delta = 69.0$ for C-2. Tetramethylsilane (TMS) serves as the reference compound from which shifts are measured in *both* ^{13}C and ^1H spectra.

The spectrum in Figure 13.12 is also of 2-butanol, but in this case the spectrum was measured while simultaneously applying a strong radio frequency field of 250 MHz. The irradiating field is not a sharp signal, but has sufficient bandwidth to cover the resonance frequencies of all of the protons in the molecule, which resonate at about 250 MHz at 58,727 gauss. Since the hydrogen nuclei are being constantly excited, they do not spend sufficient time in either the α- or β-spin state to couple with the ^{13}C nuclei. That is, on the NMR time scale each hydrogen is in an average or effectively constant state, and the result is that no coupling is observed. This process is called **decoupling,** and the spectrum in Figure 13.12 is said to be **proton-decoupled.** Note that this type of decoupling is different from "off-resonance decoupling." Each carbon nucleus now appears as a sharp singlet and the entire spectrum is greatly simplified.

For structure work, it is convenient to obtain both types of spectra. For complex molecules the proton-decoupled spectrum often allows one to ''see'' each carbon resonance and to measure its chemical shift accurately. This type of spectrum is especially useful for *counting the number of different carbons in a molecule*. The proton-coupled spectrum then allows the analyst to determine the number of hydrogens attached to each carbon. By using these data together with the proton NMR spectrum, even complex structures may be solved.

Because of their simplicity, proton-decoupled CMR spectra are particularly suited for detecting symmetry in fairly complicated molecules. For example, Figures 13.13 and 13.14 are the proton-decoupled spectra of the two diastereomers of 1,3,5-trimethylcyclohexane.

1-*cis*,3-*cis*,5-trimethylcyclohexane 1-*cis*,3-*trans*,5-trimethylcyclohexane

Note that the CMR spectrum of one of the isomers shows only three resonances. This must be the isomer in which all methyl groups are cis. Since all of the methyl groups occupy equatorial positions, there is only one kind of CH_3 resonance. Likewise, there are only one kind of CH_2 and only one kind of CH. In the other diastereomer the two equatorial methyls are equivalent, but the axial one gives rise to a different resonance. Similarly, this isomer has two different kinds of CH_2 resonance and two different kinds of CH resonance. Note that the six resonances of this molecule occur in three sets with the relative intensities being 2:1 in each set.

EXERCISE 13.2 A compound known to be one of the stereoisomeric 1,5-dichloro-2,4-dimethyl-3-pentanols is found to have a four-signal CMR spectrum. Write the structures of the four stereoisomers and indicate which are eliminated by the CMR spectrum.

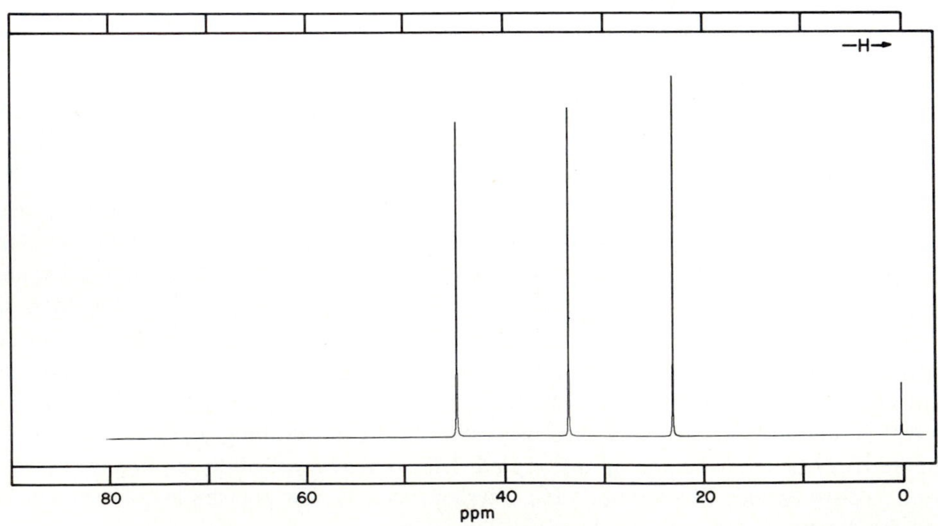

FIGURE 13.13 Proton-decoupled CMR spectrum of 1-*cis*,3-*cis*,5-trimethylcyclohexane.

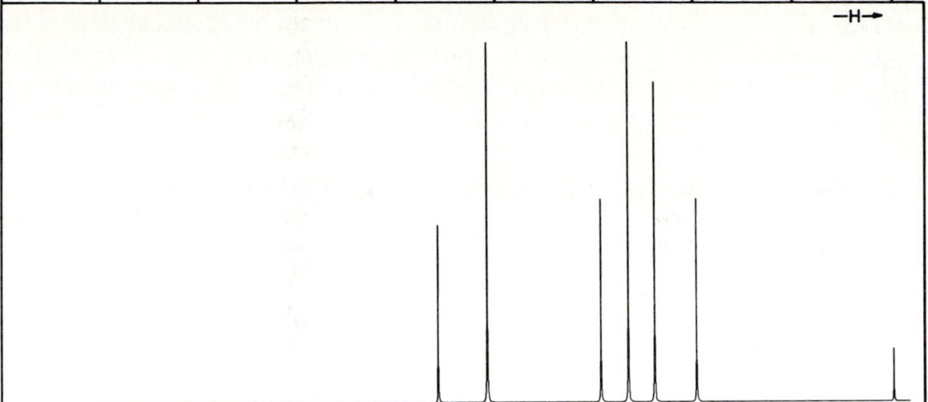

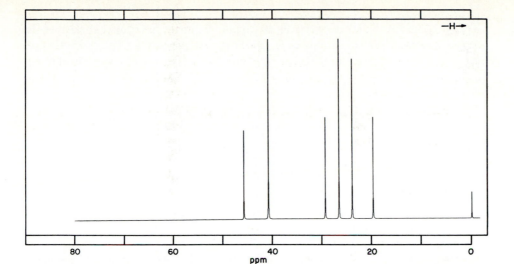

FIGURE 13.14 Proton-decoupled CMR spectrum of 1-*cis*-3-*trans*-5-trimethylcyclohexane.

As we will see in proton NMR spectra, the resonance frequencies in CMR spectra are influenced by the nature and number of substituents attached to the carbon under observation. However, simple alkyl substitution has a much greater effect on CMR resonance than it does on proton NMR resonances. Notice the chemical shifts of the simple alkanes tabulated in the first four lines of Table 13.2. The chemical shift of methane is −2.1 ppm (that is, CH_4 resonates 2.1 ppm *upfield* from the reference compound TMS). The chemical shift of the two equivalent CH_3 carbons in ethane is 5.9 ppm. Thus, replacement of a hydrogen by a methyl group causes a *downfield* shift of 8.0 ppm. Further replacement of hydrogens by methyl groups, as in the central carbons of propane and isobutane, results in further downfield shifts of 10.2 and 9.1 ppm. This regular substituent effect is called an "α-effect" and is usually taken to have the value +9 ppm. This empirical value carries over to other classes of compounds as well, and can be of great use in *predicting* the approximate resonance positions of carbons in compounds for which data are not available.

TABLE 13.2 CMR Chemical Shifts for Alkanes

	C-1	C-2	C-3	C-4	C-5	C-6	Other
methane	−2.1						
ethane	5.9	5.9					
propane	15.6	16.1	15.6				
isobutane	24.3	25.2	24.3				
neopentane	31.5	27.9	31.5				
n-butane	13.2	25.0	25.0	13.2			
2-methylbutane	22.0	29.9	31.8	11.5			
2,3-dimethylbutane	19.3	34.1	34.1	19.3			
2,2,3-trimethylbutane	27.2	32.9	38.1	15.9			
n-pentane	13.7	22.6	34.5	22.6	13.7		
2-methylpentane	22.5	27.8	41.8	20.7	14.1		
3-methylpentane	11.3	29.3	36.7	29.3	11.3		18.6[a]
3,3-dimethylpentane	6.8	25.1	36.1	25.1	6.8		4.4[a]
n-hexane	13.9	22.9	32.0	32.0	22.9	13.9	

[a] C-3 methyl carbon(s).

EXERCISE 13.3 Methanol resonates at 49.3 ppm. Estimate the chemical shift of the alcohol carbon in ethanol, 2-propanol, and 2-methyl-2-propanol. Compare your estimates with the values given in Table 13.3. Note that your estimates are only approximate.

Another difference between CMR and proton NMR chemical shifts is the rather large effect that results from substitution at the adjacent carbon and at the position one carbon removed from that under observation. For example, compare the chemical shifts of the methyl resonances in ethane (5.9 ppm) and propane (15.6 ppm). There is a downfield shift of 9.7 ppm as a result of replacing a hydrogen *on an adjacent carbon* by a methyl group. This "β-effect" may also be seen in the shift of the methyl resonance of isobutane (24.3 ppm). Thus, replacement of a second hydrogen by a methyl group results in a further downfield shift of 8.7 ppm. The β-effect is usually taken to have the value +9.5 ppm.

Finally, examination of the chemical shifts tabulated in Table 13.3 reveals that there is a γ-effect that results in an *upfield* shift of 2–3 ppm. It may be seen clearly in the C-1 resonance of propane and butane and the C-4 resonance of isopentane. The γ-effect is usually taken to have the value −2.5 ppm.

Let us see how these empirical substituent effects might be useful. Suppose we want to predict the chemical shifts of the various carbon resonances in the CMR spectrum of 2-methylhexane. We may do this by adding the appropriate substituent corrections (α-, β-, and γ-effects) to the known chemical shifts of the various carbons of *n*-hexane (Table 13.2). The calculation is illustrated below.

Estimation of CMR Chemical Shifts for 2-Methylhexane

Carbon	Chemical Shift in *n*-Hexane	Correction	Estimated Chemical Shift in 2-Methylhexane	Actual Chemical Shift in 2-Methylhexane
1	13.9	β, +9.5	23.4	22.4
2	22.9	α, +9.0	31.9	28.1
3	32.0	β, +9.5	41.5	38.9
4	32.0	γ, −2.5	29.5	29.7
5	22.9	none	22.9	23.0
6	13.9	none	13.9	13.6

Although such calculations are admittedly crude, they are useful in assigning observed resonances to the proper carbons, and in some cases they may be useful in assigning a structure to a compound from its CMR spectrum.

EXERCISE 13.4 Predict the resonance frequencies of the seven carbons in 3-methylhexane.

Table 13.3 summarizes the CMR chemical shifts for some simple alcohols. By comparing these data with those given for alkanes in Table 13.2, it may be seen that the hydroxy group causes a large downfield shift of the carbon to which it is attached. This OH α-effect is generally taken to be +48 ppm. The β- and γ-effects are about the same as for a methyl group, +9.5 ppm and −2.5 ppm, respectively.

Carbon NMR is also a useful tool for the analysis of alkenes. Chemical shifts for some simple alkenes are collected in Table 13.4. Comparison of this table with Table 13.2 shows that sp^2-hybridized carbons resonate 90–120 ppm downfield from sp^3-hybridized

TABLE 13.3 CMR Chemical Shifts for Some Alcohols

	C-1	C-2	C-3	C-4	C-5
methanol	49.3				
ethanol	57.3	17.9			
1-propanol	63.9	26.1	10.3		
2-propanol	25.4	63.7	25.4		
1-butanol	61.7	35.3	19.4	13.9	
2-butanol	22.9	69.0	32.3	10.2	
2-methyl-1-propanol	69.2	31.1	19.2		
2-methyl-2-propanol	31.6	68.7	31.6		
1-pentanol	62.1	32.8	28.5	22.9	14.1
2-pentanol	23.6	67.3	41.9	19.4	14.3
3-pentanol	10.1	30.0	74.1	30.0	10.1

carbons having the same degree of substitution. Substituent effects are about the same as for alkanes, with one significant exception. The γ-effect is highly dependent on the steric relationship between the carbon being observed and the substituent. For example, the CH_3 resonances in *cis*- and *trans*-2-butene occur at 11.4 and 16.8 ppm, respectively. In essence, each methyl in the cis isomer is exerting a γ-effect of -5.4 on the other.

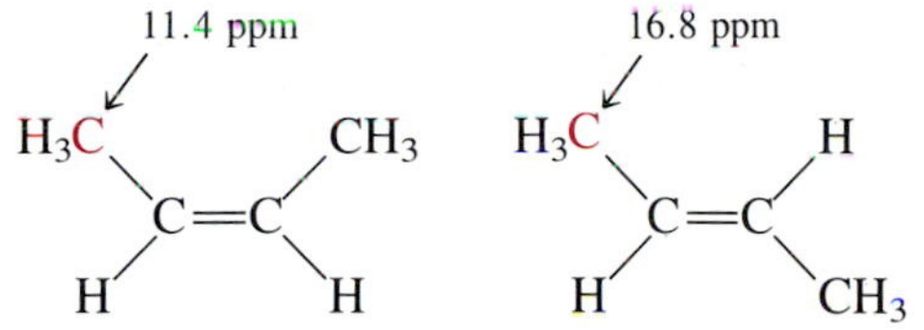

The appearance of alkene CMR spectra is shown in Figures 13.15 (1-octene) and 13.16 (*trans*-4-octene). Note that *trans*-4-octene shows only four resonances because of the symmetry of the molecule. In both spectra a small 1:1:1 triplet may be seen at about 75 ppm. This triplet is caused by $CDCl_3$, which is a common solvent for CMR measurements.

TABLE 13.4 CMR Chemical Shifts for Alkenes

	C-1	C-2	C-3	C-4	C-5	C-6
propene	115.4	135.7	18.7			
1-butene	112.8	140.2	23.8	9.3		
cis-2-butene	11.4	124.2	124.2	11.4		
trans-2-butene	16.8	125.4	125.4	16.8		
1-pentene	114.3	138.9	36.0	22.1	13.6	
cis-2-pentene	12.0	122.8	132.4	20.3	13.8	
trans-2-pentene	17.3	123.6	133.2	25.8	13.6	
1-hexene	114.1	139.1	33.6	31.2	22.2	13.9
cis-2-hexene	12.3	123.7	130.6	29.3	23.0	13.5
trans-2-hexene	17.5	124.7	131.5	35.1	23.1	13.4
cis-3-hexene	14.3	20.6	131.0	131.0	20.6	14.3
trans-3-hexene	13.9	25.8	131.2	131.2	25.8	13.9

Proton NMR and CMR are the preferred techniques for detecting the presence of a carbon-carbon double bond in a molecule and for deducing something about its nature. The low-field resonances of vinyl hydrogens in the proton NMR spectrum ($\delta = 5$ ppm) and of olefinic carbons in the CMR spectrum ($\delta = 110$–140 ppm) are used to deduce the presence of a double bond. The magnitudes of the proton NMR coupling constants (Sec-

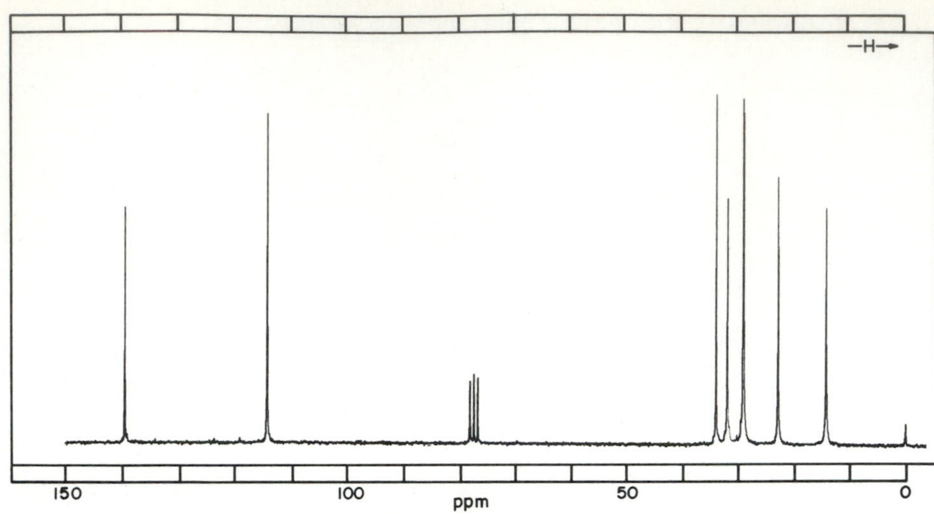

FIGURE 13.15 Proton-decoupled CMR spectrum of 1-octene.

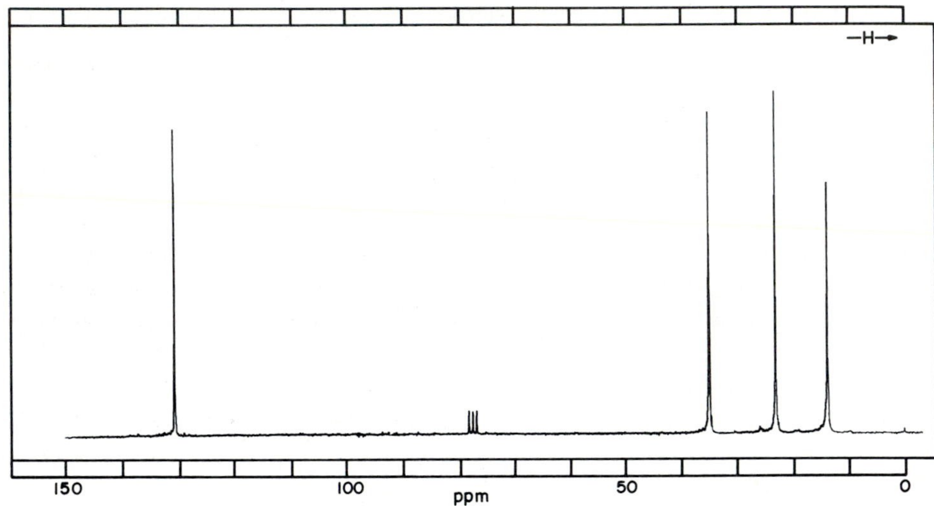

FIGURE 13.16 Proton-decoupled CMR spectrum of *trans*-4-octene.

tion 13.9) and of the γ-effects in the CMR spectra are diagnostic of the stereostructure of the alkene.

EXERCISE 13.5 Explain how CMR spectroscopy can be used to distinguish between (*E*)- and (*Z*)-3-methyl-2-pentene.

EXERCISE 13.6 Sketch the proton-decoupled CMR spectrum expected for cycloheptene.

13.6 Relative Peak Areas

We saw in previous sections that we can obtain a valuable piece of information from an NMR spectrum—the number of magnetically different carbons and hydrogens in the compound. The amount of energy absorbed at each resonance frequency is proportional to

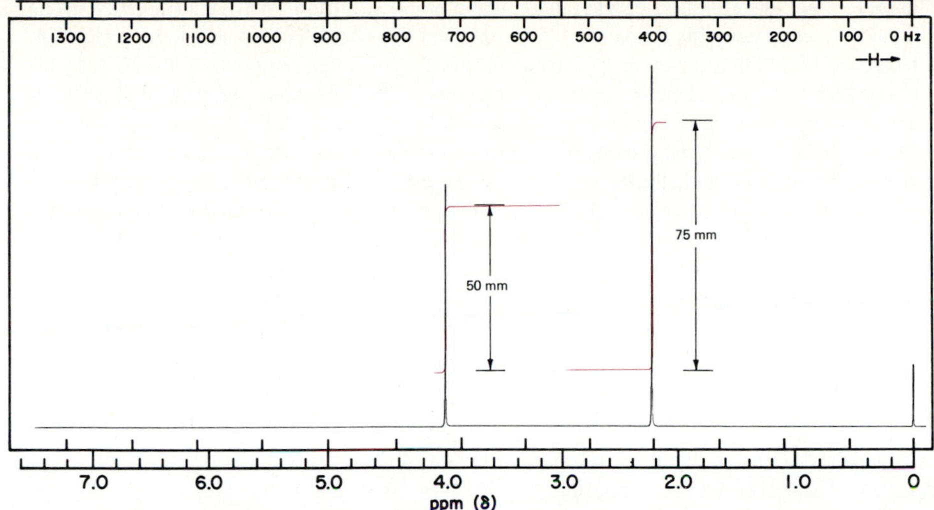

FIGURE 13.17 Integrated intensities superimposed on the NMR spectrum are proportional to the relative numbers of hydrogens.

the number of nuclei that are absorbing energy at that frequency. By measuring the **areas** of each of the resonance lines, we may determine the relative number of each different kind of hydrogen. In practice, this is accomplished with an electronic integrator. After the NMR spectrum has been recorded, the instrument is switched to an ''integrator mode'' of operation and the spectrum is recorded again. The recorder output in this mode of operation is illustrated in Figure 13.17. The integral line for each of the two peaks in the spectrum of 1,2,2-trichloropropane is shown superimposed on the appropriate peak. The ratio of the heights of the two integral lines is equal to the ratio of the number of protons giving rise to the two peaks, in this case 3:2. In the remaining sample spectra in this book, integral lines are shown in a separate color. Measurement of the heights of the separate ''steps'' on the integral line will give the relative numbers of hydrogens that are responsible for the corresponding resonances. In some cases, where it might be difficult to measure small steps on the integral line, the relative values of the integral are also printed below the appropriate resonances. Some modern instruments also print the chemical shift for every line and a value for the integrated area for selected regions. Note that it is more difficult to obtain correct integrations for ^{13}C spectra because of the ''slow'' relaxation of the ^{13}C nuclei after spin flip (see next paragraph) and other reasons that go beyond the scope of this chapter.

We should note that the NMR experiment does not measure all of the protons, just those that have α-spin and absorb energy in ''flipping'' to β-spin. The difference in the population of α- and β-spins is rather small. We learned above that the energy difference between the proton α- and β-spin states in the magnetic field of a 180-MHz NMR instrument is only 0.018 cal mole^{-1}. At equilibrium, the population difference is given by the Boltzmann distribution as

$$\frac{N_\alpha}{N_\beta} = e^{0.017/RT} = 1.00003$$

When we now turn on the applied radio frequency field, we excite the slight excess of α nuclei to β. Of course, those nuclei with β-spin can also *give up* energy to the applied field, and **relax** back to the α-spin state. If there were no other mechanism for converting β back to α, we would quickly have exactly equal populations in both spin states and would no longer observe a net absorption of energy; there would be no spectrum. With a sufficiently strong radio frequency field this can generally be done and the system is then said to be **saturated.** However, in normal operation, hydrogens in the β-spin state continually relax back to the α-state because of local fields, and equilibrium imbalance is maintained. These local fields are associated with other spinning nuclei. That is, even in the absence of the applied radio frequency field, individual protons convert from one state to another quite readily because, in moving about the liquid, they experience the magnetic fields of other nearby nuclear magnetic moments. Occasionally, such moving and changing fields happen to have the resonance value and energy interchange can occur, resulting in spin flipping. In the NMR experiment the net result of all this activity is the conversion of our measuring radio waves into heat within the sample. In normal operation the distribution of α- and β-spins remains close to the equilibrium value and the amount of energy absorbed is proportional to the number of protons.

EXERCISE 13.7 What are the relative peak areas in the different hydrogens in the NMR spectra of the following compounds?

(a) 1,1,3-trichloro-2,2-dimethylpropane
(b) 1,1,3,3-tetramethylcyclobutane
(c) 1,2-dimethoxyethane (glyme)
(d) (*E*)-1-chloropropene

13.7 Spin-Spin Splitting

We have seen that the NMR spectrum of 1,2,2-trichloropropane shows two sharp peaks that are easy to interpret. From the fact that there are two peaks, we deduce that the compound has two types of hydrogen that are magnetically nonequivalent, and, from the relative areas of the two peaks, we conclude that they are present in a ratio of 3:2. Let us now consider the NMR spectrum of a related trihaloalkane, 1,1,2-tribromo-3,3-dimethylbutane, $(CH_3)_3CCHBrCHBr_2$ (Figure 13.18).

We recognize the small peak at $\delta = 0.0$ as that of TMS added as a standard to define the zero on our scale. The large peak at $\delta = 1.2$ ppm comes from the nine equivalent

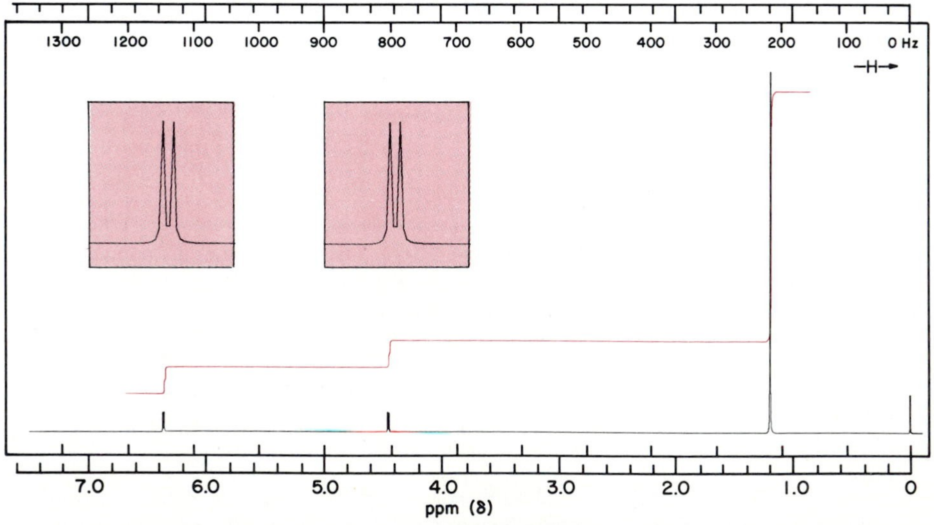

FIGURE 13.18 NMR spectrum of $(CH_3)_3CCHBrCHBr_2$. The scale expansion shows the doublets more clearly.

methyl protons. The other two protons are responsible for the downfield resonances; the downfield shifts are explained by their proximity to one and two electronegative bromines, respectively. But these two resonances are not single lines, as are the CH_3 and CH_2 resonances in 1,2,2-trichloropropane. Instead, each resonance appears as a **doublet.** This ''splitting'' of peaks is common in NMR spectra. The phenomenon has its origin in the magnetic field associated with each individual spinning proton. These small magnetic fields affect the total magnetic field experienced by another proton. For convenience we will label these hydrogens as H_a and H_b.

$$\begin{array}{c} \quad\ \ \mathrm{Br}\ \ \mathrm{Br} \\ \mathrm{Br{-}C{-}C{-}C(CH_3)_3} \\ \quad\ \ \mathrm{H}_a\ \ \mathrm{H}_b \end{array}$$

In the applied magnetic field of the NMR spectrometer we would expect H_a normally to show up as a single peak. However, the magnetic field associated with the spin of the nearby proton, H_b, contributes to the effective field experienced by H_a This is not a through-space effect of the magnetic field associated with the spinning nucleus H_b but results instead from interaction between each H nucleus and electron spins. That is, the spin of H_b is relayed to H_a by way of shared electrons. For most cases, if H_b has α-spin, the effect is as if the total magnetic field at H_a were slightly greater than that provided by the NMR instrument's applied field alone. Consequently, a slightly higher frequency is required to achieve resonance than in the absence of H_b, and we find a slight ''downfield'' shift (Figure 13.19). But only half of the H_b nuclei have α-spin. The rest have β-spin, in which the opposite effect results. For these molecules the effect is as if the effective magnetic field at H_a were slightly *weaker* than that given by the applied field alone. The

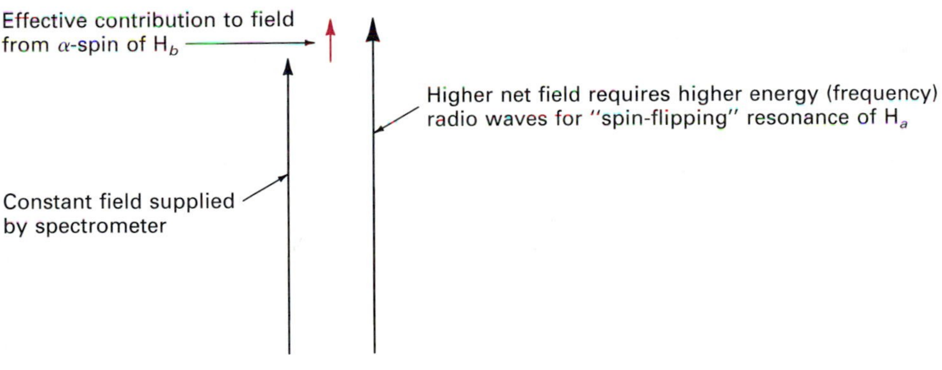

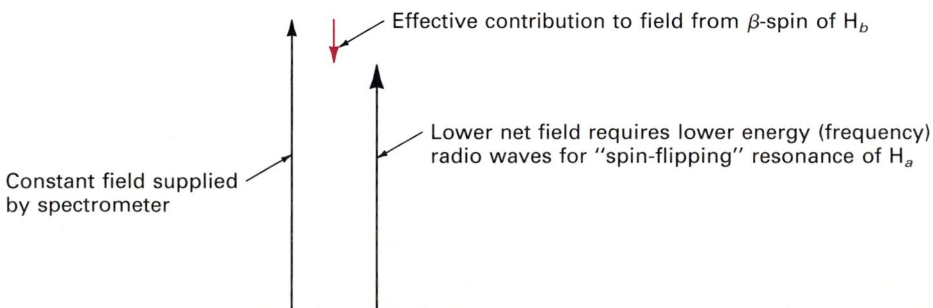

FIGURE 13.19 Source of spin-spin splitting.

NMR spectrometer must then provide a slightly smaller frequency in order to achieve the "spin-flipping" resonance condition with H_a. Now the result is an upfield shift (Figure 13.19). In such a case, H_a and H_b are said to be "coupled."

Let us recapitulate the conditions of the experiment. We apply a constant magnetic field in the NMR instrument and irradiate our sample with a high frequency radio signal. As we slowly decrease the frequency, we reach a point where the energy of the irradiating radio waves matches the effective magnetic field at the half of the H_a protons that are in molecules where the H_b protons have α-spin. At this frequency, H_a protons of α-spin absorb photons and "flip" to β-spins. Motion in the liquid sample provides a mechanism for the β-spins to change to α with the excess energy given up as heat. The absorption of radio waves is recorded by the NMR instrument as a "peak." As the frequency of the radio waves is decreased still more, the resonance condition is destroyed and the recorder pen returns to the base line. At a still lower frequency, we reach a point where the other half of the H_a protons absorb radio energy. These H_a protons are in molecules where H_b has β-spin; in this case the coupling effect subtracts from the applied field, and a lower frequency must therefore be applied to achieve resonance.

These relationships are summarized in Figure 13.20. Note that only one kind of hydrogen is in resonance at any given point in the spectrum. Both lines in the low-field doublet in Figure 13.20 correspond to transitions of H_a. If there were no coupling to H_b, the resonance position of H_a would be at the point marked δ_a. However, since H_b *is* there, and since its spin *does* affect the magnetic field experienced by H_a, we see two lines, one at slightly lower field (higher frequency) for the one half of the molecules having H_b nuclei with α-spin and one at slightly higher field (lower frequency) for the one half of the molecules having H_b nuclei with β-spin. At these frequencies H_b, with its different chemical shift, is not in resonance even though its presence is "felt" by H_a, and it produces two resonance positions instead of one. As the frequency is decreased still further, the nuclei eventually come into resonance. However, now we must reckon with the effect of α- or β-spin of H_a on the effective magnetic field experienced by H_b. The effect is an exact reciprocity—the effect of H_a on H_b is exactly the same as the effect of H_b on H_a. Consequently, the splitting of the H_b peaks has the same magnitude as that of the H_a peaks. The spacing between the peaks in a resonance multiplet is called the **coupling constant** between the two protons. The coupling constant is conventionally labeled J, and is given in units of cycles per second or Hertz. For the case under discussion, $J_{ab} = 1.6$ Hz.

In the foregoing example, we note that there is no coupling to the nine methyl protons, which appear as a sharp singlet. Because the coupling effects of proton spin are relayed by way of shared electrons, the effect is attenuated rapidly with the number of bonds and is usually quite small if more than two atoms intervene between the protons. Thus, the methyl protons do indeed couple to the other two protons in our example, but the magnitude of each such J is so small as to be unobservable with normal spectrometers. As a further example, in the spectrum of 1,2,2-trichloropropane, which was discussed in Section 13.4, the protons on C-1 and C-3 do not noticeably split each other.

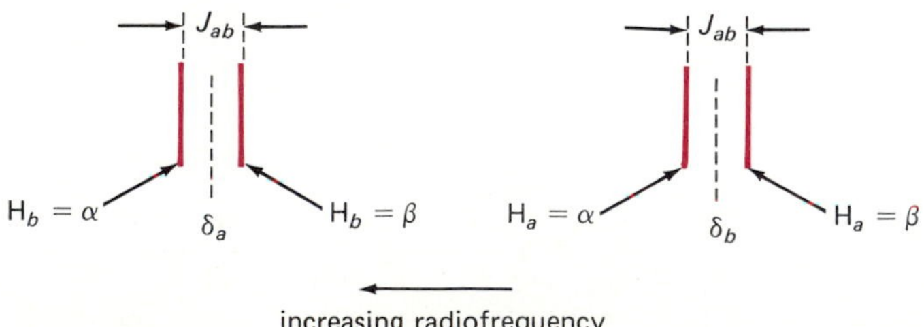

FIGURE 13.20 J_{ab} causes equal spin-spin splitting in H_a and H_b.

$$\begin{array}{c} H_a\ H_b \\ |\quad | \\ C\!-\!C \end{array}$$ spin-spin splitting normally observed

$$\begin{array}{c} H_a\qquad H_b \\ |\qquad\quad | \\ C\!-\!C\!-\!C \end{array}$$ spin-spin splitting normally not observed

Now let us consider a slightly more complex spectrum, that of 1,1,2-trichloroethane. In this compound, there are also two types of hydrogen, H_a and H_b.

$$\begin{array}{c} \quad Cl\ \ H_b \\ \quad |\quad\ \ | \\ Cl\!-\!C\!-\!C\!-\!Cl \\ \quad |\quad\ \ | \\ \quad H_a\ \ H_b \end{array}$$

The spectrum of 1,1,2-trichloroethane (Figure 13.21) shows two resonances, a triplet of relative area 1 centered at $\delta = 5.8$ ppm and a doublet of relative area 2 centered at $\delta = 3.9$ ppm. The triplet is associated with H_a, which is more deshielded because it is bonded to a carbon that also has two chlorines. The two equivalent H_b hydrogens are less deshielded because their carbon has only one attached chlorine.

The spectrum of 1,1,2-trichloroethane is shown in diagrammatic form in Figure 13.22. If there were no spin-spin coupling, the resonance positions of H_a and H_b would be at the points marked δ_a and δ_b, respectively. However, in any given molecule, the two H_b nuclei may have their spins as $\alpha\alpha$, $\alpha\beta$, $\beta\alpha$, or $\beta\beta$. If both H_b nuclei have α-spin, the effect of coupling is to augment the applied field and a higher frequency is required for H_a to be in resonance. Thus, H_a in molecules with this arrangement of H_b spins comes into resonance at slightly higher frequency than if there were no coupling to H_b. If the H_b spins are $\alpha\beta$ or $\beta\alpha$, there is no effect on the field experienced by H_a because the opposed spins of the two H_b nuclei cancel. If both H_b nuclei have β-spin, the magnetic field associated with their coupling subtracts from $\mathbf{H}_0$ and a lower frequency is necessary to achieve resonance at H_a. Statistically, the probability of the two H_b nuclei being $\alpha\alpha$, $\alpha\beta$, $\beta\alpha$, and $\beta\beta$ is 1:1:1:1. Since $\alpha\beta$ and $\beta\alpha$ are energetically equivalent with the effects of the two different nuclear spins canceling each other, the H_a resonance appears as three lines with relative intensities

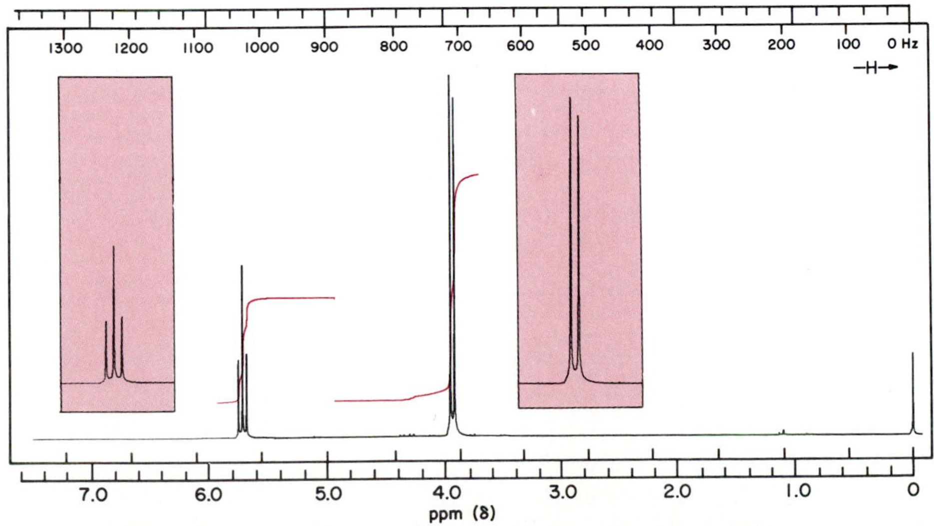

FIGURE 13.21 NMR spectrum of $CHCl_2CH_2Cl$.

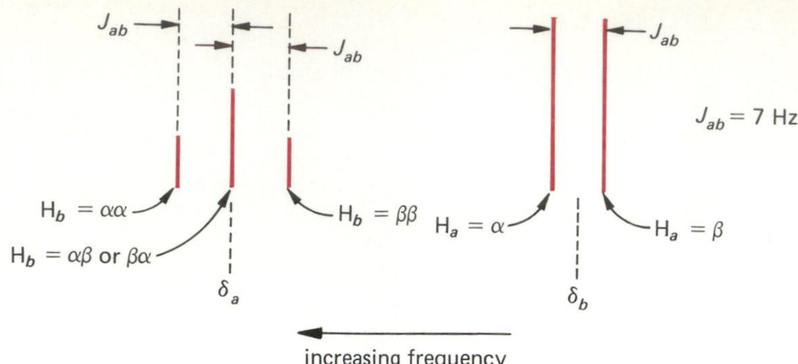

FIGURE 13.22 Spin-spin splitting analysis of 1,1,2-trichloroethane.

of 1:2:1. At the resonance frequency of the two H_b hydrogens we find two lines because H_a can have either α- or β-spin. In this case the coupling constant J has the value 7 Hz.

Extension to different numbers of equivalent neighboring hydrogens is straightforward. Three hydrogens, as in a methyl group, can have the possible spin states $\alpha\alpha\alpha$, $\alpha\alpha\beta$, $\alpha\beta\alpha$, $\beta\alpha\alpha$; $\alpha\beta\beta$, $\beta\alpha\beta$, $\beta\beta\alpha$, and $\beta\beta\beta$. An adjacent proton would therefore give four peaks with area ratios of approximately 1:3:3:1. These states can be counted simply with the coefficients of the polynomial expansion, 1: $n/1$: $n(n-1)/2!$: $n(n-1)(n-2)/3!$, etc., and are summarized in Table 13.5. In general, n neighboring equivalent hydrogens cause splitting into $n+1$ peaks.

TABLE 13.5 Number of Peaks and Area Ratios for NMR Multiplets

Number of Equivalent Adjacent Hydrogens	Total Number of Peaks	Area Ratios
0	1	1
1	2	1:1
2	3	1:2:1
3	4	1:3:3:1
4	5	1:4:6:4:1
5	6	1:5:10:10:5:1
6	7	1:6:15:20:15:6:1

EXERCISE 13.8 Sketch the NMR spectra of the following compounds. Show the relative resonance positions (not the precise values of δ) and the expected appearance of each resonance multiplet. Finally, indicate the expected *relative heights* of the various peaks.

(a) (*E*)-1-chloro-3,3-dimethyl-1-butene
(b) 1,1-dimethoxyethane
(c) diethyl ether
(d) 1-methoxy-3,3-dimethylbutane

EXERCISE 13.9 Sketch the NMR spectra that you would expect for 1-chloropropane and 2-chloropropane. Compare your sketches with Figures 13.1 and 13.2.

Since J has its origin in the magnetic spin of the proton, its magnitude is not dependent on the strength of the applied magnetic field. That is, the same J value applies for spectra determined at 60 MHz, 180 MHz, 300 MHz, and so on. Because J is field-independent, and the resonance position of a multiplet does vary with field strength, spectra of the same

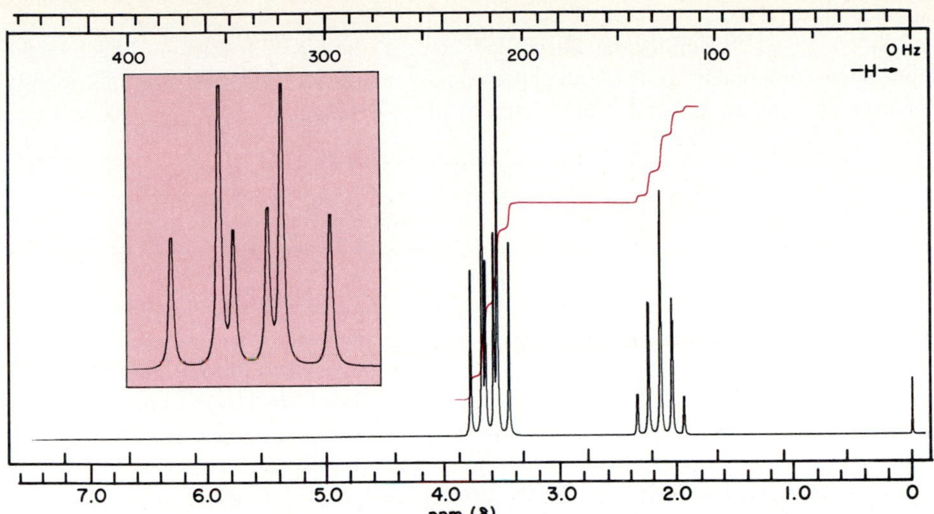

FIGURE 13.23 60-MHz NMR spectrum of $BrCH_2CH_2CH_2Cl$.

compound that are measured with different instruments can sometimes have rather different appearance. An example is seen in the 60-MHz and 180-MHz spectra of 1-bromo-3-chloropropane (Figures 13.23 and 13.24). In this compound, the protons attached to C-3 are slightly more deshielded than those attached to C-1, because chlorine is more electronegative than bromine. Thus, in the 180-MHz spectrum (Figure 13.24), we see two downfield triplets ($\delta = 3.67$ and 3.54, respectively). In this spectrum, the two triplets are readily apparent, since the magnitudes of the coupling constants ($\cong$7 Hz) are considerably less than the difference in resonance frequencies ($661 - 637 = 24$ Hz). However, in the 60-MHz spectrum (Figure 13.23), the two coupling constants have the same values, but the difference in resonance frequencies is only 8 Hz. Therefore, the two triplets *overlap*, making the spectrum rather difficult to decipher at first examination. Note that, even though CH_2Br and CH_2Cl are not magnetically equivalent, the two *J*s are accidentally equal. Therefore, the center CH_2 appears as a 1:4:6:4:1 quintet with $\delta = 2.15$ ppm.

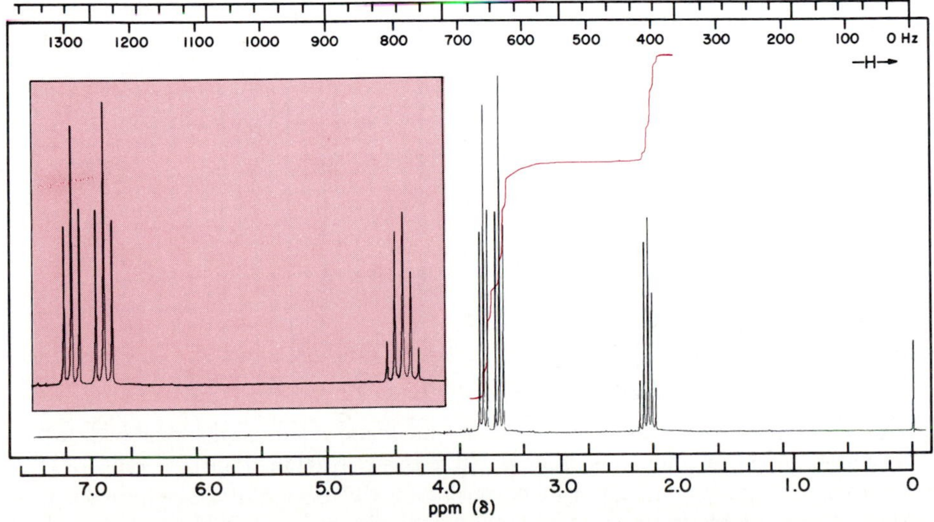

FIGURE 13.24 180-MHz NMR spectrum of $BrCH_2CH_2CH_2Cl$.

EXERCISE 13.10 Sketch the expected appearance of the NMR spectrum of each of the following compounds. Indicate the approximate resonance positions, the multiplicity of each resonance, and the relative area of the resonance.

(a) 1,3-dichloropropane (b) 1,2-dimethoxypropane

Highlight 13.5

Spin-spin splitting is the "fine structure" seen in the NMR signal for one nucleus that is caused by the magnetic states of nearby nuclei. The situation is illustrated for the case of 1,1-dichloroethane, CH_3CHCl_2, which has two kinds of hydrogen. The resonance caused by the H attached to C-1 is at lower field because of the inductive effect of the two attached chlorines. However, even though there is only one H nucleus at C-1, its resonance appears in the proton NMR spectrum as four lines, in a ratio of 1:3:3:1. This splitting is caused by the three Hs attached to C-2. In any given molecule, each of these three Hs can have either α spin (magnetic moment aligned with the external field) or β spin (magnetic moment aligned against the external field). There are eight possible arrangements (2^3) for the three C-1 Hs.

	$\alpha\alpha\alpha$	$\alpha\alpha\beta$	$\alpha\beta\alpha$	$\beta\alpha\alpha$	$\alpha\beta\beta$	$\beta\alpha\beta$	$\beta\beta\alpha$	$\beta\beta\beta$
Net spin:	3α	α	α	α	β	β	β	3β

That is, in one eighth of the compounds the net magnetic field effect on the H of the $CHCl_2$ group is that from the net spin of 3α (meaning that in those molecules all three of the CH_3 hydrogens have their spins aligned with the external field). In three eighths of the molecules, the net effect is that of a single α; in another three eighths, the net effect is a single β; and in the remaining one eighth, the net effect is 3β. Thus, the actual NMR resonance of the proton of the $CHCl_2$ looks like this.

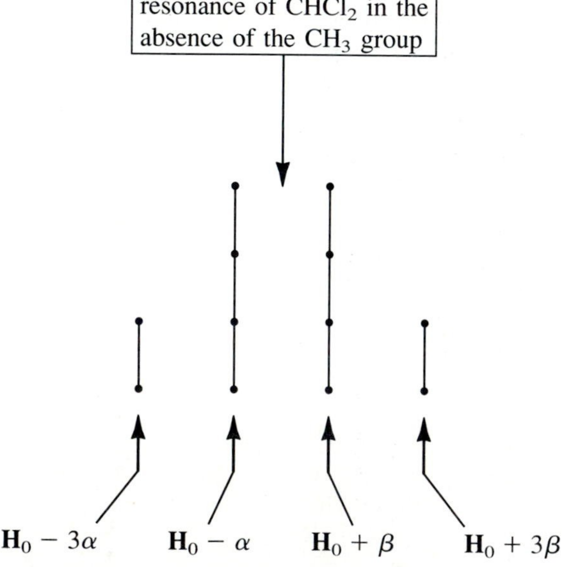

The three Hs of the CH_3 group (C-2) are equivalent and cannot be distinguished. If the hydrogen at C-1 were not there, they would resonate together at a single value of $\mathbf{H_0}$. This resonance position is considerably upfield of the position of the $CHCl_2$ hydrogen, because there are no electronegative atoms directly attached to C-2. However, the H at C-1 *is* there, and it can have either α or β spin. Therefore, half the molecules will give a resonance for the CH_3 hydrogens that is slightly downfield and half will give a resonance for these hydrogens that is slightly upfield of the unperturbed resonance position. Thus, the actual resonance for the CH_3 group looks like this.

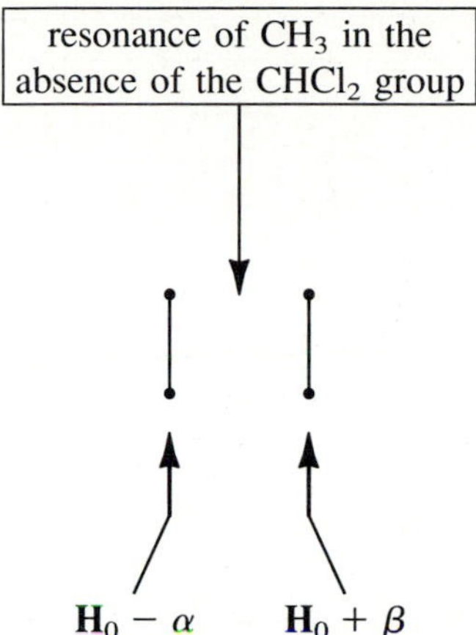

Finally, we must remember that the *total* integrated intensity of the CH_3 doublet is three times that of the $CHCl_2$ quartet because the doublet comes from the resonances of three protons and the quartet comes from the resonances of only one proton.

13.8 More Complex Splitting

Our simple splitting rules, with peak numbers and intensities that follow the polynomial coefficients as given in Table 13.5, apply for cases where splitting is small compared to the difference in chemical shift between the neighboring hydrogens; that is, $J \ll \Delta\nu$. As $\Delta\nu$ is reduced (as the peaks for two nonequivalent hydrogens approach each other), the inner peaks increase in intensity and the outer ones diminish. In practice, such perturbations are almost always apparent. This effect may be seen in a comparison of the spectra in Figures 13.18, 13.21, 13.23, and 13.24.

If $\Delta\nu$ is too small compared to J, the simple rules do not apply at all. Such spectra are often rather complex and require a detailed analysis beyond the scope of this book. One especially simple case, however, is the extreme one for which $\Delta\nu = 0$. Such hydrogens are *magnetically equivalent and do not split each other*. In the cases discussed previously, for example, a methylene group was treated as a unit—because the two hydrogens are magnetically equivalent, they have no effect on each other. This effect is a direct and exact outcome of the quantum mechanics of magnetic resonance and the detailed reason is not important for our purposes.

The following simple explanation may be helpful. Because the NMR experiment does not distinguish magnetically equivalent hydrogens, different spin properties cannot be assigned to individual protons. For example, in a methylene group in which the protons have α- and β-spin, we cannot assign one spin to one proton and the other spin to the remaining proton. Instead, the $\alpha\beta$-spin property belongs to the methylene protons *as a unit*. Since we cannot assign individual equivalent protons, it follows that we should not observe the splitting or J-coupling normally associated with such assignments.

The effect of the relative magnitudes of J and $\Delta\nu$ is illustrated in Figure 13.25 for a two-proton system, H_a—C—C—H_b. The first spectrum, for $\Delta\nu \gg J$, is close to the "first-order" analysis that we outlined above. As $\Delta\nu$ and J become of comparable magnitude, the inner peaks increase in intensity and the outer ones fade until, in the limit where $\Delta\nu = 0$, the two inner peaks have merged and the outer peaks have vanished.

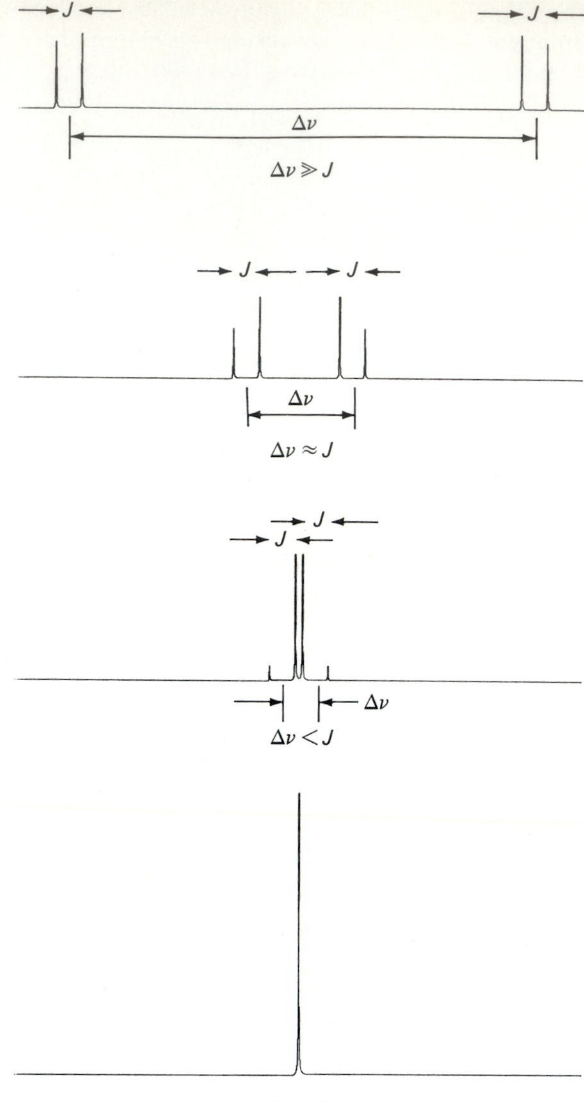

FIGURE 13.25 NMR spectra of the system H_a—C—C—H_b for various values of J and $\Delta\nu$.

The spectra of 1-methoxybutane in Figures 13.26 and 13.27 illustrate the complications of small $\Delta\nu/J$ ratios. In the 180-MHz spectrum of this ether (Figure 13.26), the C-4 methyl resonance at $\delta = 0.95$ ppm is only barely discernible as a triplet, and the resonances of the C-2 and C-3 methylene hydrogens appear as an ill-defined lump centered at about $\delta = 1.5$ ppm. Furthermore, the methoxy singlet at $\delta = 3.32$ ppm overlaps and partially obscures the C-1 triplet. However, the 300-MHz spectrum of this same compound (Figure 13.27) is readily interpretable in terms of the simple rules we have discussed.

Splitting can become more complex if coupling occurs to more than one type of hydrogen. Consider the case of H_a coupled to two different hydrogens, H_b and H_c, with $J_{ab} > J_{ac}$.

$$\begin{array}{ccccc} H_b & & H_a & & H_c \\ | & & | & & | \\ C & — & C & — & C \end{array}$$

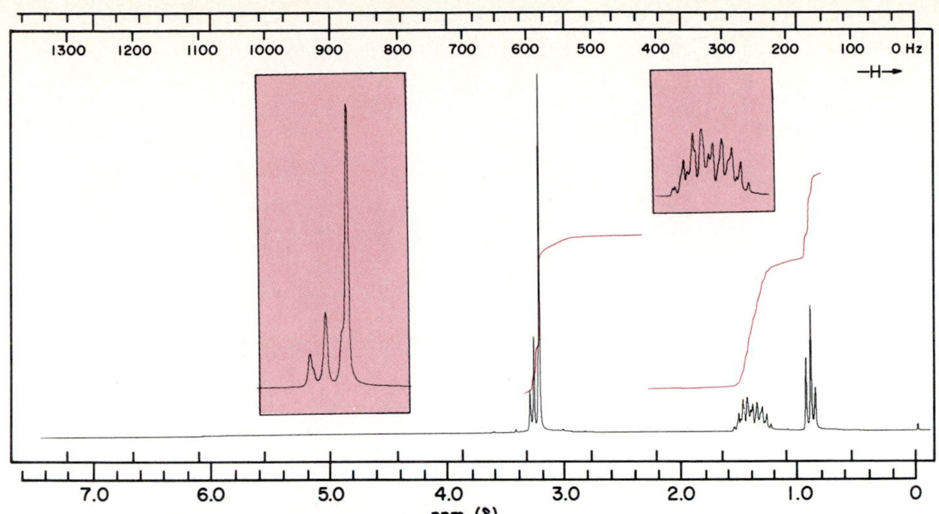

FIGURE 13.26 NMR spectrum of 1-methoxybutane (180 MHz).

In this case, H_a is split into a doublet by H_b, and each of the lines of the doublet is split into a further doublet to give a total of four lines, as shown in Figure 13.28. The four lines will have approximately the same intensity and correspond to transitions of H_a when the H_b and H_c nuclei have the following spin states.

Line	Spin of H_b	Spin of H_c
1	α	α
2	α	β
3	β	α
4	β	β

In this example, $J_{ab} > J_{ac}$. This means that the effect of nucleus H_b on H_a is greater than the effect of nucleus H_c on H_a. Thus, line 2 appears at higher frequency than the reso-

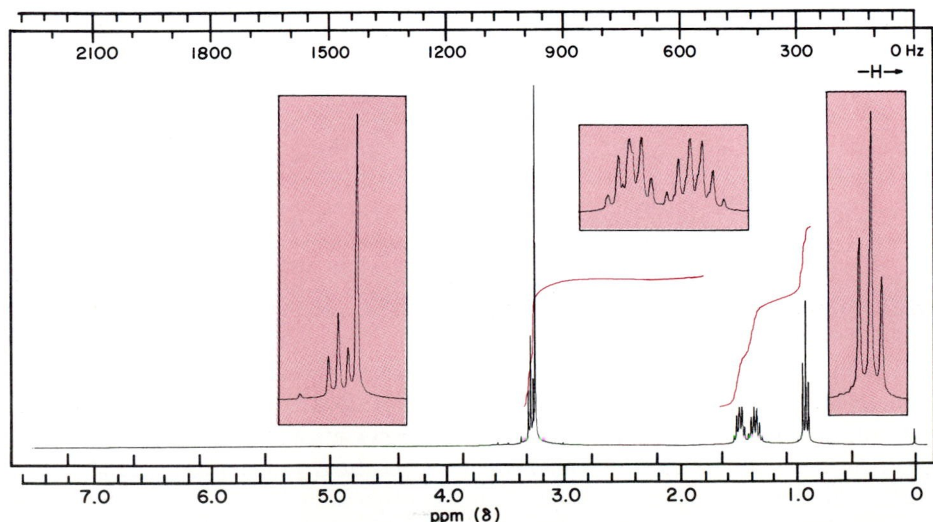

FIGURE 13.27 NMR spectrum of 1-methoxybutane (300 MHz).

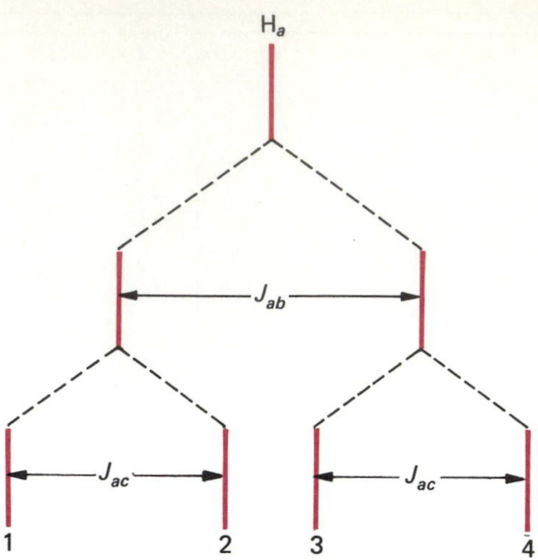

FIGURE 13.28 Effect of two coupling constants, $J_{ab} > J_{ac}$, on the resonance frequency of H_a.

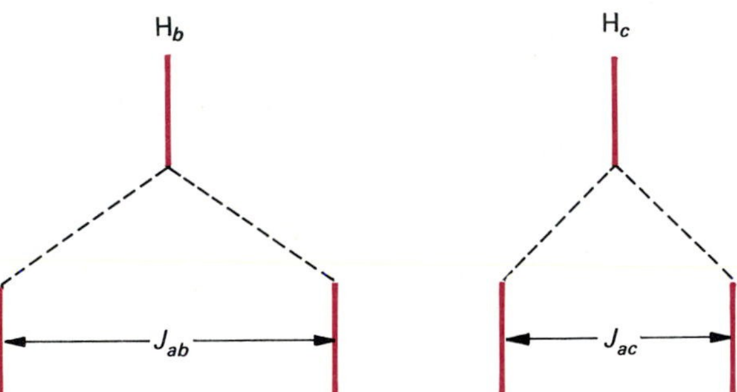

FIGURE 13.29 The H_b and H_c resonances of the system H_b—H_a—H_c when $J_{ab} > J_{ac}$.

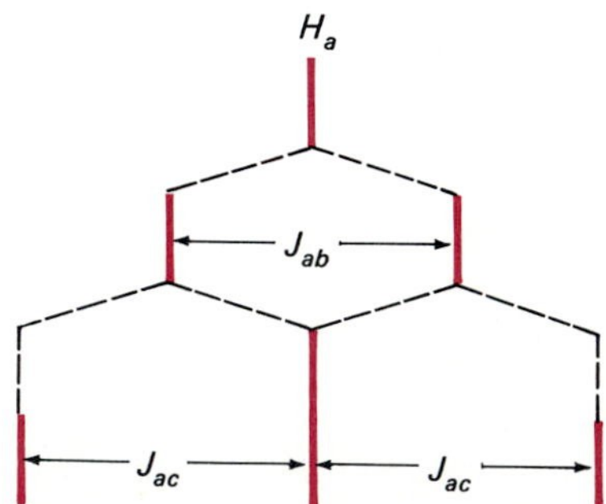

FIGURE 13.30 Effect of equal coupling constants, $J_{ab} = J_{ac}$, on the resonance frequency of H_a.

nance position of H_a in the absence of H_b and H_c because the α- and β-spins of the H_b and H_c nuclei do not cancel. Similarly, line 3, which corresponds to a transition of H_a when H_b is β and H_c is α, is at slightly lower frequency than the resonance position of H_a in the absence of the other nuclei. The remainder of the spectrum will show H_b and H_c each as doublets due to their respective couplings to H_a (Figure 13.29).

Note what happens if $J_{ab} = J_{ac}$. This will occur if H_b and H_c are magnetically equivalent (that is, if they have the same chemical shift) or if the two *J*s accidentally have the

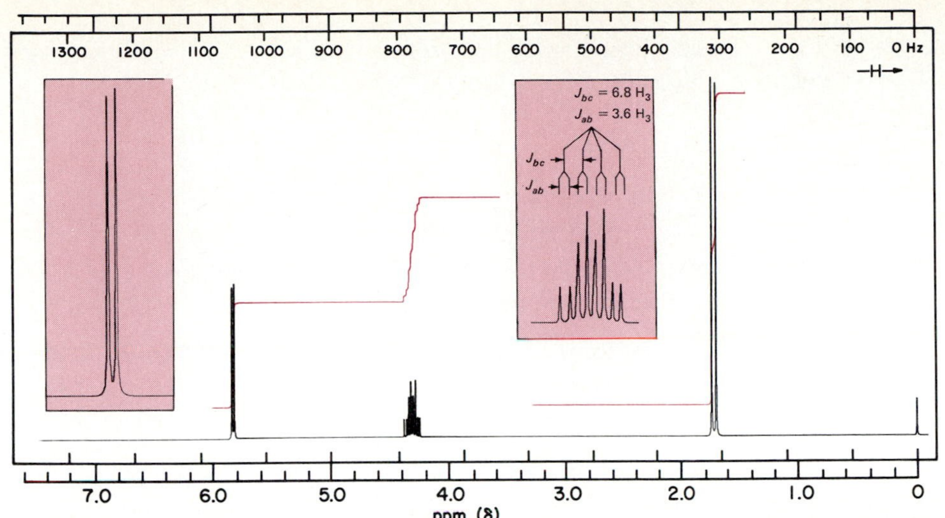

FIGURE 13.31 NMR spectrum of 1,1,2-trichloropropane, $CH_3CHClCHCl_2$. The center multiplet ($\delta = 4.30$ ppm) is expanded in the insert.

same value. In such a case the two inner lines of the quartet occur at the same point and appear as a single line of double the intensity. The net result is a triplet with intensity ratios of 1:2:1 (Figure 13.30).

The spectrum of 1-bromo-3-chloropropane (Figure 13.24) is a relevant example. The CH_2Cl hydrogens and the CH_2Br hydrogens are not magnetically equivalent, and therefore they do not have the same chemical shift. Each is adjacent to two hydrogens (the center CH_2 group) and each appears as a 1:2:1 triplet. Even though CH_2Cl and CH_2Br are not magnetically equivalent, the two *J*s are accidentally equal. Therefore, the center CH_2 still appears as a 1:4:6:4:1 quintet with $\delta = 2.15$ ppm.

Figure 13.31 shows the spectrum of 1,1,2-trichloropropane, a compound in which there are two different coupling constants. There are three different types of hydrogen, which we may label H_a, H_b, and H_c.

```
    Cl  Cl
    |   |
Cl—C—C—CH3(c)
    |   |
    Ha  Hb
```

The H_a hydrogen is most deshielded and appears as a low-field (high-frequency) resonance with relative area of unity. It is a doublet due to coupling with H_b, and the separation between the two lines, J_{ab}, is 3.6 Hz. The three equivalent CH_3 hydrogens, H_c, are least deshielded and appear as a high-field (low-frequency) resonance of area 3. They are also coupled to H_b and appear as a doublet. In this case the separation between the lines, J_{bc}, is 6.8 Hz. The two coupling constants in this case are unequal. The resonance for H_b is in between those of H_a and H_c, and it has a relative area of unity. Because of the two unequal *J*s, it appears as a "doubled quartet" and may be analyzed as shown on the insert in Figure 13.31. The chemical shift for H_b is $\delta = 4.30$ ppm, the midpoint of the multiplet.

Finally, remember that J is independent of applied field, whereas the normal shielding by electrons $\Delta\nu$ results from an induced field and is proportional to the applied field. One of the incentives for seeking larger magnetic fields for NMR instruments is the spreading of $\Delta\nu$ relative to J. At a sufficiently high field strength, all spectra reduce to the simple "first-order" type ($\Delta\nu \gg J$) we have treated.

EXERCISE 13.11 Consider the following three-spin system, with resonance frequencies $\delta_a = 6.0$ ppm, $\delta_b = 4.0$ ppm, $\delta_c = 7.0$ ppm

```
  Hb Ha Hc
  |  |  |
—C—C—C—
  |  |  |
  Hb
```

Sketch the expected NMR spectrum for each of the following situations. [The use of graph paper will greatly facilitate working this exercise.]

(a) $J_{ab} = 6$ Hz, $J_{ac} = 4$ Hz, $J_{bc} = 0$ (b) $J_{ab} = 6$ Hz, $J_{ac} = 3$ Hz, $J_{bc} = 0$
(c) $J_{ab} = 4$ Hz, $J_{ac} = 6$ Hz, $J_{bc} = 0$ (d) $J_{ab} = 3$ Hz, $J_{ac} = 6$ Hz, $J_{bc} = 0$

13.9 The Effect of Conformation on Coupling Constants

Although spin-spin coupling works through the bonding electrons, the magnitude of J is related to the relative orientation of the coupled nuclei in space. For example, the J value between adjacent axial hydrogens in cyclohexanes is 10–13 Hz, whereas J between axial and equatorial or between two equatorial hydrogens is 3–5 Hz.

The dihedral angle (the solid angle between the C—C—H_a and C—C—H_b planes) between two axial hydrogens is 180°, whereas the axial-equatorial and equatorial-equatorial dihedral angles are both 60° (Figure 13.32). The relation between J and the dihedral angle between hydrogens has been established theoretically and may be depicted in a familiar graphic form known as the **Karplus curve** (Figure 13.33). The coupling constant between two hydrogens reaches a minimum when the dihedral angle between them is 90°. The curve can be expressed analytically by two equations.

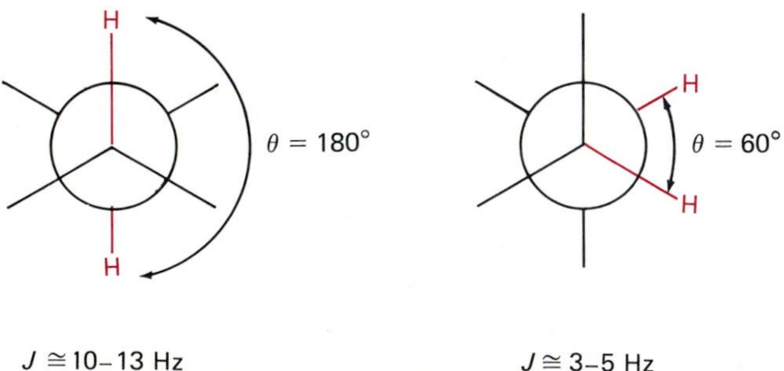

FIGURE 13.32 J_{HH} and conformation. The angle θ is the dihedral angle between the carbon-hydrogen bonds.

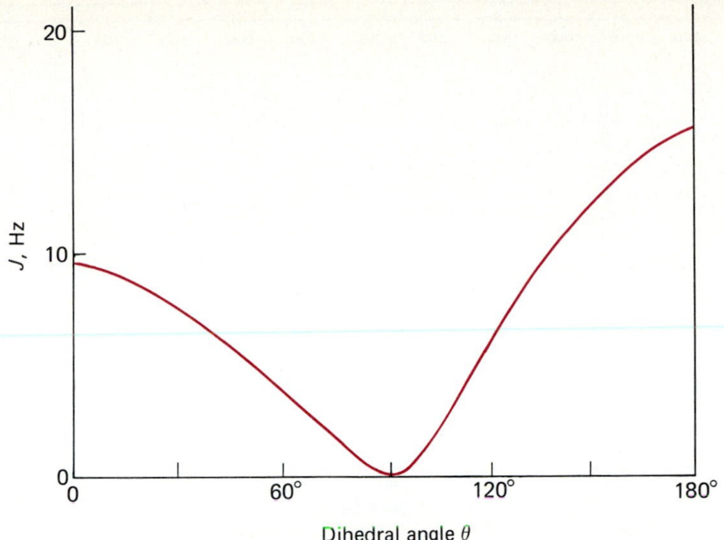

FIGURE 13.33 Karplus curve.

$$J_{HH'} = 10 \cos^2 \theta \qquad (0° \leq \theta \leq 90°)$$
$$J_{HH'} = 16 \cos^2 \theta \qquad (90° \leq \theta \leq 180°)$$

EXERCISE 13.12 Use graph paper and your calculator to plot the magnitude of J_{HH} as a function of dihedral angle, from $\theta = 0°$ to 180°.

The Karplus relationship also applies to hydrogens attached to the carbon-carbon double bond. Figures 13.34 and 13.35 show the NMR spectra of (*E*)- and (*Z*)-3-chloropropenenitrile. Note that the coupling constant for the trans hydrogens, in which the

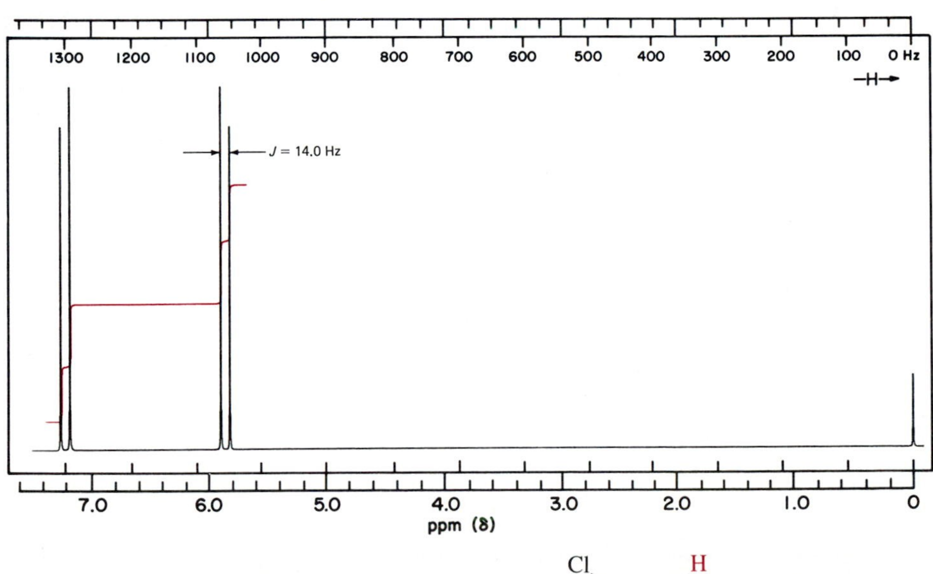

FIGURE 13.34 NMR spectrum of (Cl)(H)C=C(H)(CN)

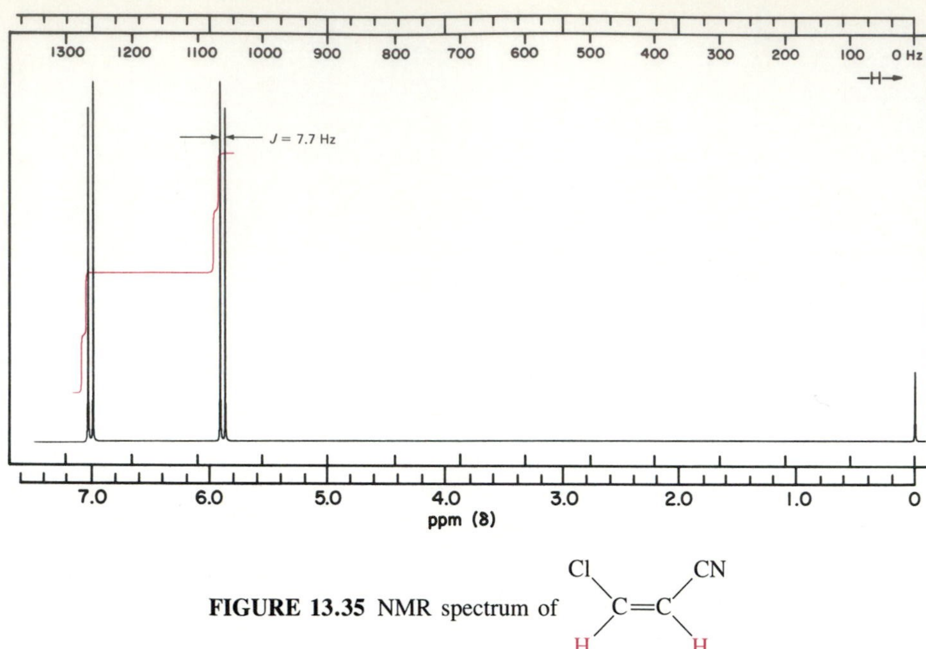

FIGURE 13.35 NMR spectrum of Cl(H)C=C(H)CN

dihedral angle is 180°, is 14.0 Hz, while that for the cis hydrogens, with a dihedral angle of 0°, is 7.7 Hz.

As shown by the foregoing example, the value of the vicinal coupling constant is an excellent criterion for stereostructure of alkenes. In general, J is greater for trans hydrogens than for cis; J_{trans} is usually in the range 11–19 Hz and J_{cis} is in the range 5–14 Hz. Although the two ranges overlap, J_{trans} is invariably greater than J_{cis} for a pair of isomeric cis and trans alkenes. Thus, when one has only one isomer of an alkene, it is possible confidently to assign stereochemistry if the vicinal coupling constant is quite small (<10 Hz) or quite large (>14 Hz). If the coupling constant is in the overlap range (10–14 Hz), it is necessary to have both isomers before a reliable assignment of structure can be made.

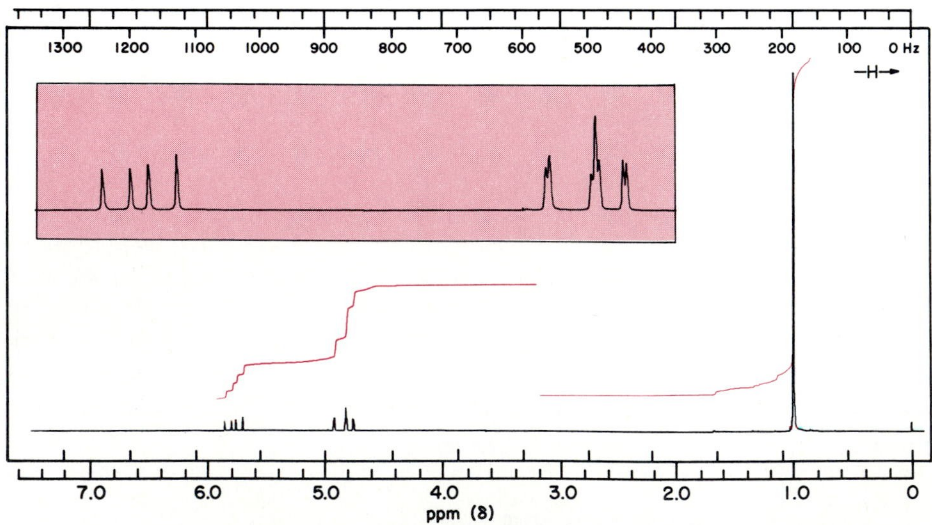

FIGURE 13.36 NMR spectrum of 3,3-dimethyl-1-butene, $(CH_3)_3CCH{=}CH_2$.

In a monosubstituted alkene, $H_2C{=}CHX$, all three vinyl hydrogens are nonequivalent. In such a case, we see a fully coupled three-spin system, with $J_{ab} \neq J_{ac} \neq J_{bc}$. Such an example is shown in Figure 13.36. In this case, $J_{trans} = 18.0$ Hz and $J_{cis} = 10.0$ Hz. The third coupling constant, between the two nonequivalent C-1 hydrogens, is 2.0 Hz. This value is a typical value for the geminal coupling constant for two alkene hydrogens.

In summary, the *magnitude* of J is related to the spatial orientation of two coupled hydrogens relative to one another. With the aid of the Karplus relationship, useful structural information may be garnered from an examination of these absolute values of J. The effect is most easily seen in conformationally locked cyclohexanes and in alkenes. In cyclohexanes, $J_{axial\text{-}axial}$ is generally large, in the range 10–13 Hz, with $J_{axial\text{-}equatorial}$ and $J_{equatorial\text{-}equatorial}$ being smaller, in the range 3–5 Hz. In alkenes, J_{trans} is large (11–19 Hz) and J_{cis} is small (5–14 Hz).

EXERCISE 13.13 What is the expected appearance of the resonance of the C-1 hydrogen in *trans*-4-*t*-butyl-1-methoxycyclohexane?

13.10 Remote Shielding of Multiple Bonds: Magnetic Anisotropy

In the previous section, we discussed the effect of stereochemistry on the magnitude of J and illustrated our discussion with the NMR spectra of several alkenes (Figures 13.34–13.36). We saw that, because of the dependence of J on the dihedral angle of the system H—C—C—H, NMR is an excellent tool for assigning stereochemistry to disubstituted alkenes. However, the NMR spectra of alkenes are also distinguished by the fact that the hydrogens attached to the carbon-carbon double bond (vinyl hydrogens) are found to resonate far downfield of normal alkane resonances, in the region $\delta = 4.5$–5.5 ppm. The actual values vary from about $\delta = 4.7$ ppm when the vinyl hydrogen is at the end of a chain to about $\delta = 5.3$ ppm when it is at some other position along the chain.

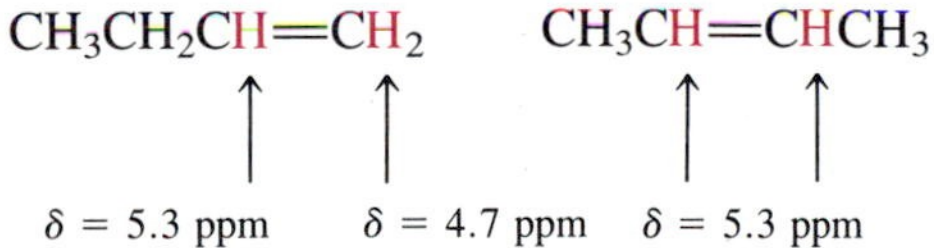

These values may appear to be rather far downfield for a C—H function. It is true that the increased *s*-character of the carbon orbital makes the carbon effectively more electronegative, but the observed change is too large to be a simple electronegativity effect.

The effect has its origin in the induced motion of bond electrons, just as discussed earlier in the diamagnetic shielding (Section 13.4) by electrons around the hydrogen nucleus. In a double bond the π-electrons are more polarizable than σ-electrons and are freer to move in response to a magnetic field. Such an electron cloud is said to be magnetically **anisotropic.** Thus when the molecule is oriented so that the plane of the double bond is perpendicular to the applied magnetic field, the π-electrons tend to circulate about the direction of the applied field. As shown in Figure 13.37, the circular motion of the π-electrons produces an induced magnetic field opposed to the applied field *at the middle of the double bond*. Out by the vinyl hydrogens, the magnetic lines of force are in the same direction as the applied field. Hence, a higher frequency is necessary for resonance at the total field at the hydrogen nucleus. For double bonds with orientations other than that shown, the effect will be smaller and the actual effect will be the average for all orientations because the tumbling and rotation of molecules is so rapid that only an "time-averaged state" is observed.

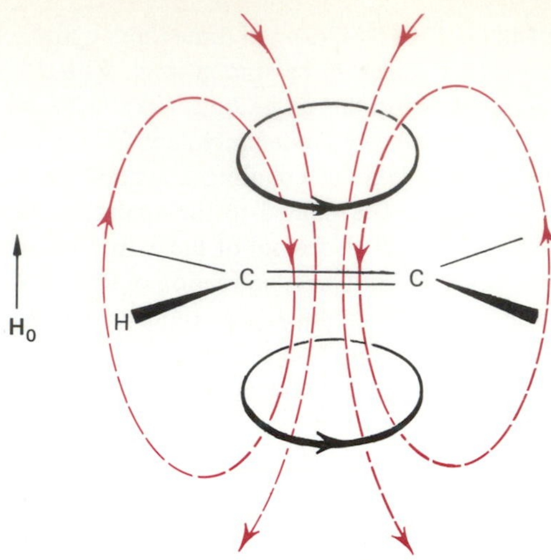

FIGURE 13.37 Induced motion of π-electrons of a double bond in a magnetic field; the induced field has the same direction as the applied field at a vinyl proton. Recall our convention of representing circulating electrons rather than positive current.

It should be emphasized that vinyl hydrogens are still subject to the normal diamagnetic shielding effects of the electron clouds that envelope them. The net effect is still a chemical shift far upfield from a really unshielded proton. However, the effect of the **π-electron circulation** partially opposes the normal effect of local electrons in such a way that resonance occurs downfield from that observed for a saturated C—H by a significant amount. The resulting chemical shift of vinyl hydrogens provides an important analytical method for establishing both their presence and number.

The applied magnetic field also induces electron currents in carbon-carbon and carbon-hydrogen single bonds, and it is these induced currents rather than electronegativity effects that give rise to the characteristic pattern of chemical shifts for differently substituted alkyl hydrogens.

$$\delta_{(-\overset{|}{\underset{|}{C}}-H)} > \delta_{(>CH_2)} > \delta_{(-CH_3)}$$

The NMR resonance for a hydrogen attached to the carbon *next to a double bond* is shifted downfield by about 0.8 ppm. This effect is due partly to the weak inductive effect of the sp^2-hybridized carbon of the double bond and partly to the anisotropy of the π-system of the double bond.

In contrast to alkene protons, the hydrogens attached to carbon-carbon triple bonds resonate at rather high field—$\delta = 2$ ppm. In fact, the alkyne C—H resonates at approximately the same place as does a comparable alkane proton. The observed position is due to a deshielding effect of the electronegative triple bond superimposed on the effect of the magnetic anisotropy of the triple bond π-electrons. Recall that a triple bond has a cylindrically symmetric sheath of π-electrons. The electrons in this torus can circulate in a magnetic field, just as the electrons in the π-system of the carbon-carbon double bond. As shown in Figure 13.38, the alkyne π-electrons are most free to circulate *around the symmetry axis of the triple bond*. When the molecule finds itself in an orientation that has the triple bond aligned with the external magnetic field, this induced electronic motion in turn induces a local field (dashed lines in Figure 13.38). At the acetylenic proton the induced field opposes the applied field. Thus, a lower frequency is required to bring this

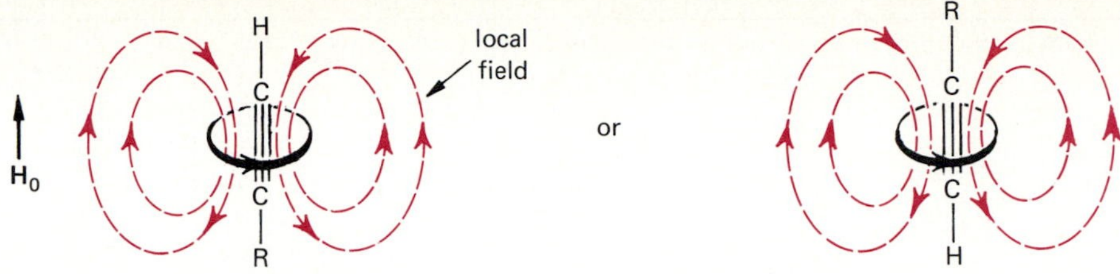

FIGURE 13.38 Shielding of acetylenic protons by a triple bond in parallel orientation to the applied field.

proton into resonance. In this case, the result of the induced field is additional shielding of the alkyne proton.

Actually, the diamagnetic shielding of acetylenic protons is a result of two factors. When the molecule is aligned perpendicular to $\mathbf{H}_0$, the acetylenic proton is deshielded, just as in the case of alkene hydrogens (Figure 13.39). This deshielding component is smaller than the shielding component diagrammed in Figure 13.38, because the electrons are not as free to move in this direction. When averaged over all possible orientations, the effect is a net diamagnetic shielding.

In the alkene case, the long-range effect of the anisotropic C═C π-electrons has the effect of deshielding the vinyl hydrogens. It therefore *augments* the inductive effect of the somewhat electronegative sp^2-hybridized alkene carbon. In the alkyne case, the long-range effect of the C≡C π-electrons *diminishes* the normal inductive effect. As a result of these two opposing effects, the resonance positions of acetylenic protons are similar to those of alkane protons. Similar effects operate on hydrogens bound to carbons adjacent to triple bonds, causing them to resonate about 1 ppm downfield from the corresponding alkane position.

The NMR spectra of 3,3-dimethyl-1-butyne and 1-hexyne are shown in Figures 13.40 and 13.41. Note that in the spectrum of 1-hexyne, the alkyne proton ($\delta = 1.7$ ppm) appears as a triplet with $J = 2.5$ Hz. This small splitting is the result of **long-range coupling** through the triple bond. Although coupling through more than three bonds is not usually observed in alkanes, small four-bond and five-bond coupling is frequently seen in alkenes and alkynes. In these cases, the coupling occurs through the π-electrons of the multiple bond.

EXERCISE 13.14 An unknown hydrocarbon is found to have the formula C_6H_{10}. Its NMR spectrum shows a three-proton triplet ($J = 6.5$ Hz) at 0.9 ppm, a three-proton doublet ($J = 6.6$ Hz) at 1.1 ppm, a two-proton multiplet at 1.5 ppm, a one-proton doublet ($J = 2.3$ Hz) at 1.8 ppm, and a one-proton multiplet at 2.3 ppm. Suggest a structure for the compound.

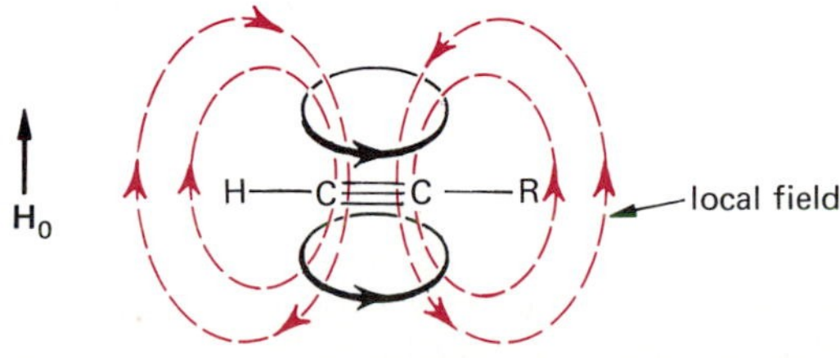

FIGURE 13.39 Shielding of acetylenic protons by a triple bond perpendicular to the applied field.

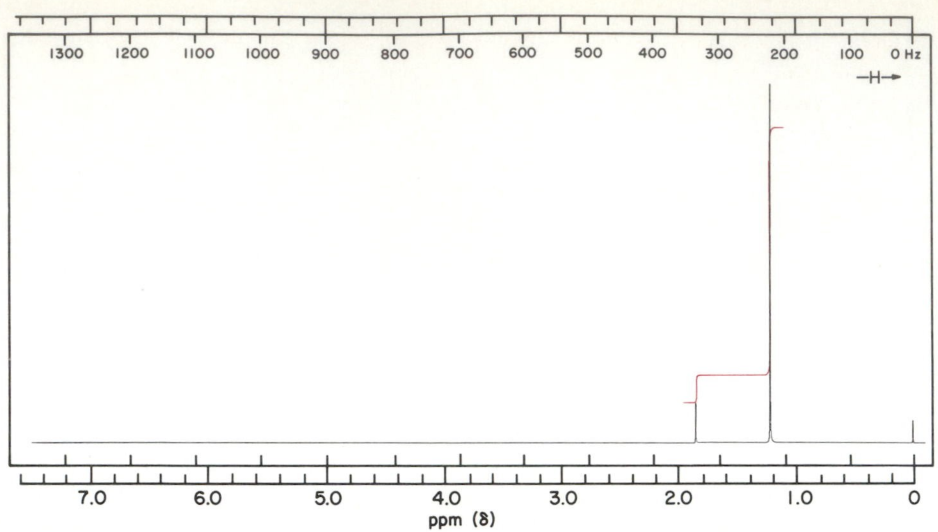

FIGURE 13.40 NMR spectrum of 3,3-dimethyl-1-butyne.

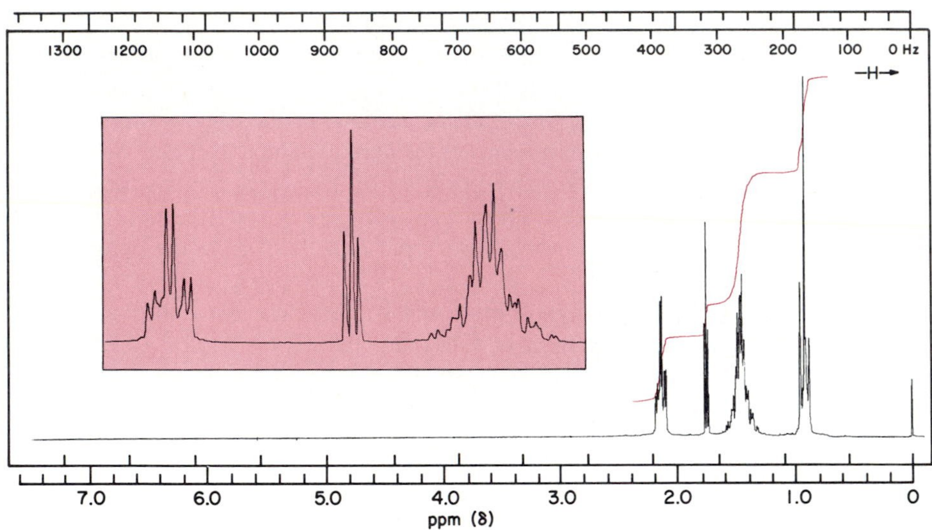

FIGURE 13.41 NMR spectrum of 1-hexyne.

13.11 Dynamical Systems

In our discussion of the NMR spectrum of 1,2,2-trichloropropane (Section 13.4) it seemed quite natural to expect both hydrogens in the methylene group to absorb in the same place because they appear to be equivalent. However, if we examine the structure in more detail, this equivalence is not so apparent. The compound actually exists as an equilibrium mixture of three conformations, symbolized by the following Newman projections.

A: Cl (top), Cl, Cl, H_a, H_b, CH_3 (bottom)

B: Cl (top), CH_3, Cl, H_a, H_b, Cl (bottom)

C: Cl (top), Cl, CH_3, H_a, H_b, Cl (bottom)

In structure A hydrogens H_a and H_b are clearly equivalent, but this is not the case in B and C. For example, in conformation B H_b is flanked by two chlorines and would be expected to be deshielded relative to H_a. Why, then, do we not see two or more peaks for these hydrogens?

The answer comes from the Heisenberg uncertainty principle of quantum mechanics. One expression of this principle is

$$\Delta E\,\Delta t \cong h/2\pi \qquad \text{or} \qquad \Delta\nu\,\Delta t \cong 1/2\pi \quad (\text{since } \Delta E = h\Delta\nu)$$

where $\Delta\nu$ and Δt are the uncertainties in energy and time in units of Hertz and seconds, respectively. That is, we cannot know precisely both the energy and the lifetime of a given state. The longer-lived the state, the more precisely can its energy content be evaluated. In NMR, as in the case of 1,2,2-trichloropropane, suppose that $\delta(H_a)$ and $\delta(H_b)$ differ by 1 ppm. This amount in a 180-MHz instrument corresponds to an energy difference of 180 Hz or 18×10^{-9} cal mole^{-1}, an exceedingly small energy quantity. In order to measure this small difference for H_a and H_b as separate states, they would have to have lifetimes in each conformation of at least

$$\Delta t \cong 1/(2\pi\,\Delta E) = 1/(2\pi \cdot 180) = 0.00088 \text{ sec}$$

But with an energy barrier of only 3–4 kcal mole^{-1} between one conformer and another, the average lifetime of a given conformation is only about 10^{-10} to 10^{-11} sec! (See Section 5.2.) In other words, the lifetime of a given methylene hydrogen in the magnetic environment of a given conformation is too short to permit us to distinguish it from the other methylene hydrogen. The "state" measured in a NMR spectrometer is a weighted average of all of the rotational conformations. The energy differences measured in NMR are so small that one frequently refers to the "NMR time scale," a time period ranging from milliseconds to seconds.

To summarize, a consequence of the Heisenberg uncertainty principle and the small energy changes characteristic of NMR spectroscopy is that two hydrogen states that are interconvertible but have separate lifetimes of more than about 1 sec can be seen as two sharp peaks whose separation can be measured accurately. If the lifetimes are less than about 1 msec, they can be seen only as a combined single sharp peak; that is, on the "NMR time scale" the two hydrogens are **magnetically equivalent.** If the lifetimes are in an intermediate region, a broad peak results.

The foregoing discussion shows why the NMR sample tube is spun rapidly between the magnet faces. It is difficult to prevent slight changes in a magnetic field at different places. In a tube placed between the pole faces of even high-quality magnets, different protons would experience slightly different fields at different points. The result would be a rather broad NMR signal. Rapid spinning of the tube causes all of the protons to experience the same average field *on the NMR time scale*. It is also clear why the normal tumbling and rotation of molecules results in our seeing only a time-averaged spectrum, since such molecular motions are exceedingly rapid relative to the NMR time scale. This same principle is clearly involved in the procedure for obtaining proton-decoupled CMR spectra (page 333).

The rates of reactions that have half-lives of the order of minutes or hours can generally be determined without difficulty. Faster reactions, with half-lives of the order of seconds, can frequently be measured with careful work. Nuclear magnetic resonance spectroscopy provides an excellent extension for determining the rates of reactions having half-lives in the NMR time scale range from about 0.001 to 1 sec. The chair-chair interconversion of cyclohexane provides an example of the type of techniques used. At room temperature cyclohexane gives a sharp singlet. The rate of chair-chair interconversion is so fast that NMR measures only the average state of the axial and equatorial hydrogens. However, at

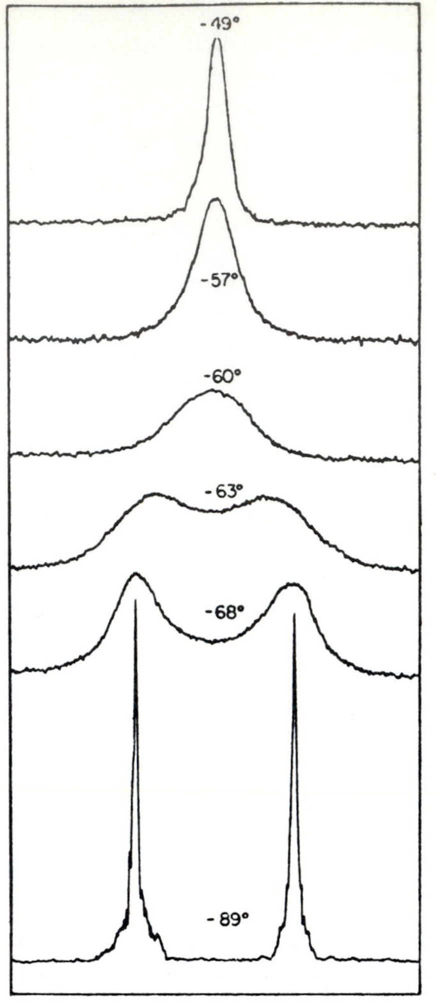

FIGURE 13.42 NMR spectrum of cyclohexane-d_{11} at different temperatures. [Reproduced with permission from F. A. Bovey, *Nuclear Magnetic Resonance Spectroscopy,* Academic Press, New York, 1969.]

sufficiently low temperature, < -70 °C, the rate of interconversion is so slow that the molecular state measured by NMR is a single conformation.

In one chair conformation, all equatorial hydrogens are equivalent but different from the axial hydrogens. The two sets of hydrogens have different chemical shifts and give rise to two broad bands separated by $\delta = 0.5$ ppm with $\delta_{\text{equatorial}} > \delta_{\text{axial}}$. The bands are broad because of J-splittings between the two sets of protons. The NMR spectrum of cyclohexane-d_{11} is a simpler case to interpret because the J-coupling between the proton nucleus and a deuteron is rather small (on the order of 15% of the corresponding H-H coupling), and each cyclohexane molecule now has only a single proton. The NMR spectrum is reproduced in Figure 13.42 as a function of temperature. At the lowest temperature (-89 °C) half of the deuteriated cyclohexane molecules have their lone proton in an axial position and the other half have the proton equatorial. Interconversion of the two isomers is slow, and since the chemical shifts differ, we see two sharp singlets. At the highest temperature (-49 °C) the ring interconversions are rapid, and the NMR spectrometer ''sees'' only a time-average position, a singlet with δ midway between δ_{axial} and $\delta_{\text{equatorial}}$. At intermediate temperatures, the rate of interconversion of the conformations is comparable to the frequency difference between the states, and a broad signal results. The results can be analyzed completely to give rate constants as a function of temperature and an enthalpy of activation, $\Delta H^{\ddagger}$, of 10.8 kcal mole^{-1}. This value is relatively high compared to other conformational interchanges we have studied. The transition state involves a partially planar cyclohexane that has both bond angle strain and eclipsed hydrogen strain.

EXERCISE 13.15 1,3,5-Trioxane exists in a chair conformation similar to that of cyclohexane.

O
H_2C CH_2
O O
C
H_2

What is the expected appearance of its NMR spectrum at room temperature and −100 °C?

13.12 Chemical Exchange of Hydrogens Bonded to Oxygen

Alcohols and ethers often have hydrogens attached to a carbon bearing an electronegative atom (oxygen). Thus, we expect to find resonance of such protons downfield from normal alkane protons, as in the case of alkyl halides. Indeed, we have already seen an example of such a spectrum (Figures 13.26 and 13.27). A further example is seen in the spectrum of diisopropyl ether (Figure 13.43).

EXERCISE 13.16 Using the simplified prescription that you learned in Section 13.7 for predicting the multiplicity of NMR resonances, predict the number of peaks and relative intensities expected for the C—H resonance in diisopropyl ether. Assuming J = 6 Hz, plot the expected appearance of the multiplet on graph paper. Compare your predicted spectrum with Figure 13.43.

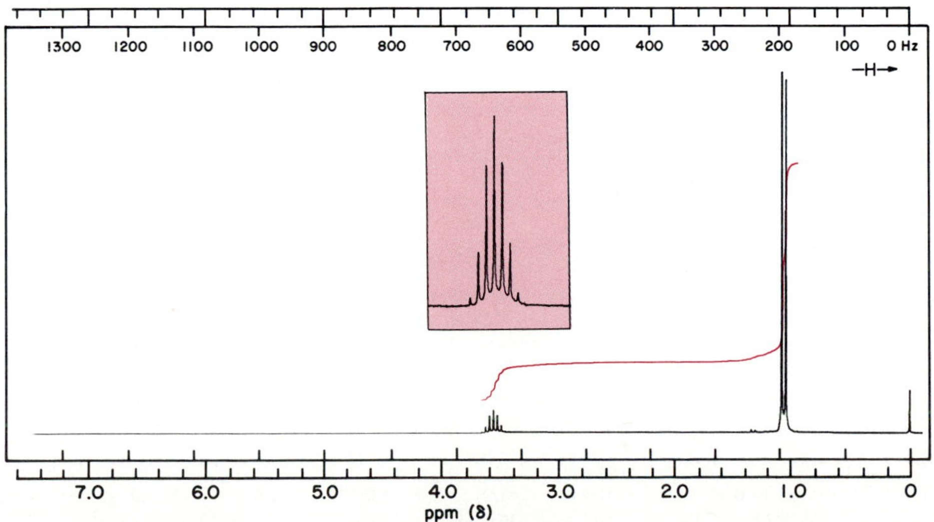

FIGURE 13.43 NMR spectrum of diisopropyl ether, $(CH_3)_2CHOCH(CH_3)_2$.

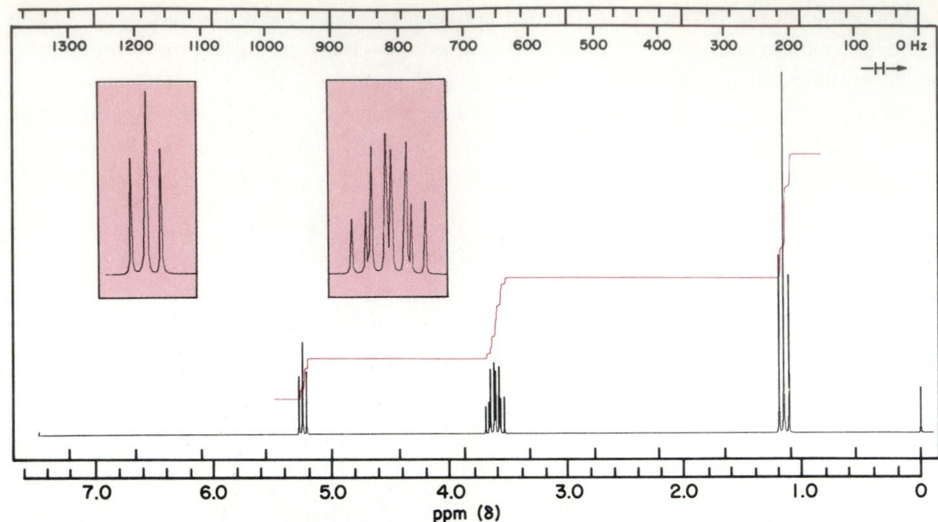

FIGURE 13.44 NMR spectrum of pure ethyl alcohol.

The hydroxy proton itself shows more complex behavior. In dilute solutions of rigorously purified alcohols, the hydroxy proton shows normal splitting by adjacent C—H protons. Under these conditions, the proton exchange caused by autoprotolysis is sufficiently slow that a given proton is associated with a given oxygen on the NMR time scale. However, the protons are still hydrogen bonded, and this leads to deshielding. The spectrum of pure ethyl alcohol in Figure 13.44 is illustrative. Note that the H—C—C—H and the H—C—O—H couplings have different magnitudes. The CH_2 resonance appears as a complex multiplet rather than a simple quartet.

Traces of acid or base cause the resonance of the hydroxy proton to collapse to a sharp singlet (see Figure 13.45). In such cases proton exchange is rapid on the NMR time scale, and the "state" observed is that of a proton in a weighted average of a number of environments. The situation may be thought of in the following way. Imagine that the proton jumps rapidly from one oxygen to another during the time that the NMR measure-

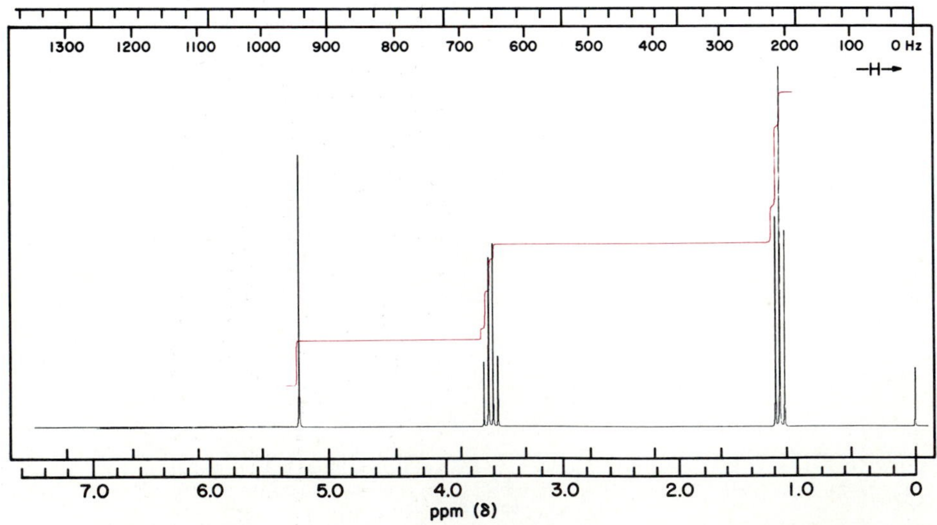

FIGURE 13.45 NMR spectrum of ethyl alcohol containing 1% formic acid.

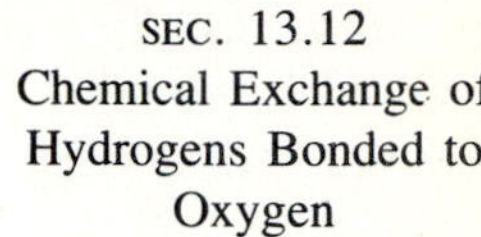

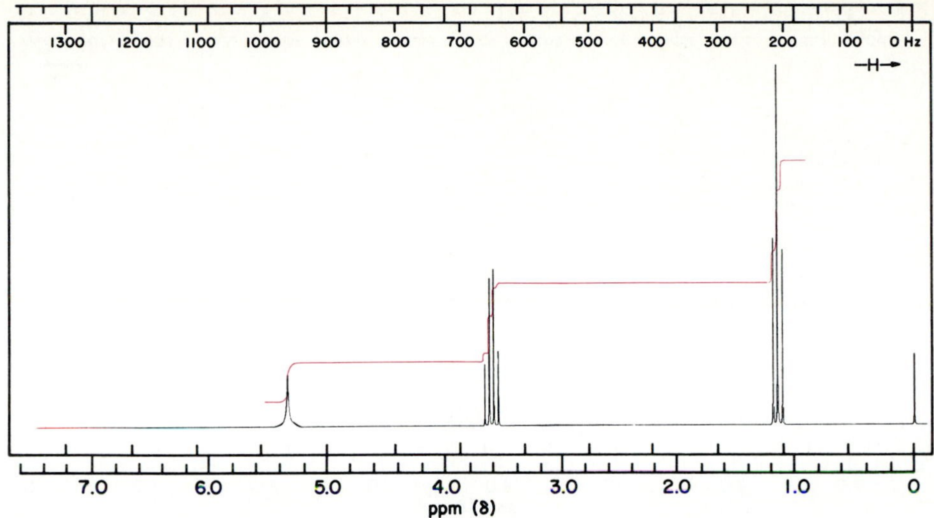

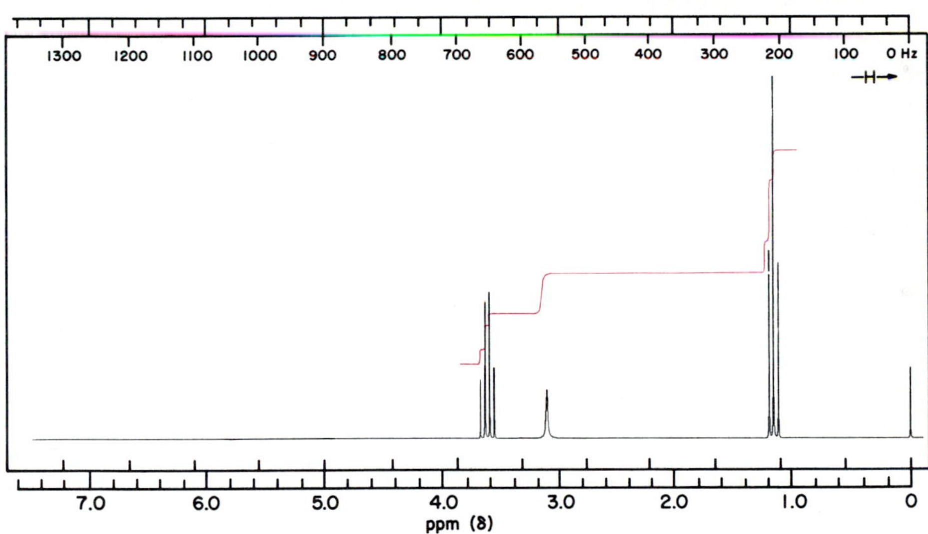

FIGURE 13.46 NMR spectra of ethyl alcohol in carbon tetrachloride. The ethyl alcohol concentration for the top spectrum is 1.0 M; for the bottom 0.1 M.

ment is carried out. In its first environment, the spins of the adjacent CH_2 may be $\alpha\alpha$, in the second they may be $\beta\beta$, in the third $\beta\alpha$, in the fourth $\alpha\beta$ and so on. Thus, the average environment will be one in which the effects of the spins of CH_2 cancel, and no spin-spin splitting is observed.

When an alcohol is diluted by an inert solvent, its hydroxy proton resonance shifts to higher field because hydrogen bonding becomes less important (Figure 13.46). In very dilute solutions the hydroxy proton may resonate as high as 0.5 ppm. However, because of a combination of hydrogen bonding and some exchange, the hydroxy proton is often observed as a broad featureless peak at a position varying from 2 to 4.5 ppm. The exact appearance and position depend on the solvent, purity, temperature, and structure. One simple diagnosis for an OH group is the addition of D_2O to the NMR solution. Rapid exchange replaces the OH groups by OD and the NMR signal for OH vanishes or becomes less intense (Figure 13.47).

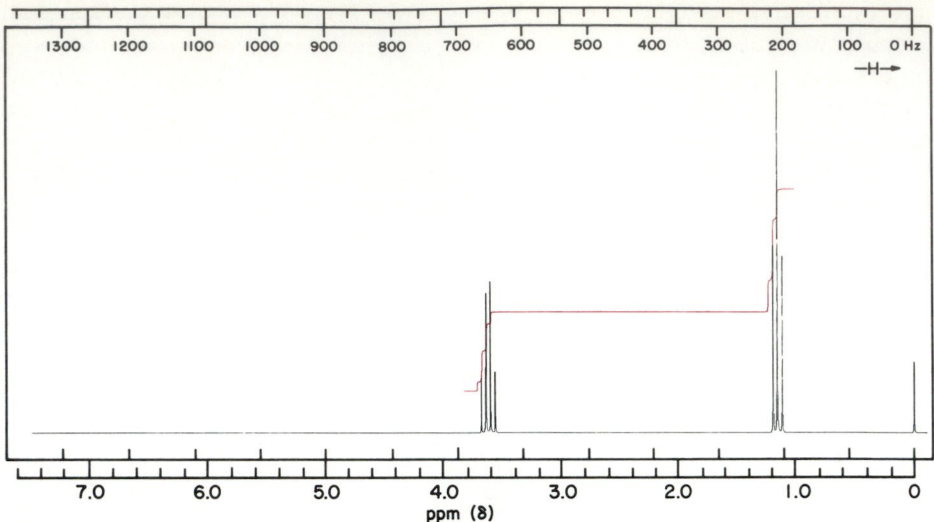

FIGURE 13.47 NMR spectrum of ethyl alcohol in carbon tetrachloride. The solution was shaken with D_2O and the layers separated before the spectrum was measured.

EXERCISE 13.17 The CH_2 resonance of ethyl alcohol is a double quartet with coupling constants of $J = 5$ and 7 Hz. Using graph paper, plot the expected appearance of such a double quartet. Remember that a simple doublet has relative intensities of 1:1 and a simple quartet has relative intensities of 1:3:3:1. What would be the appearance of the CH_2 resonance if the two *J*s were equal?

EXERCISE 13.18 Plot the expected appearance of the NMR spectrum of pure isopropyl alcohol. Assume that the coupling constants are 5 and 7 Hz, as they are in ethyl alcohol.

13.13 Solving Spectral Problems

We can now apply our knowledge to interpret the spectra of the propyl chlorides introduced in Section 13.2. In the NMR spectrum of *n*-propyl chloride (Figure 13.2), for example, the methyl group is clearly distinguished as the group furthest upfield ($\delta = 1.2$), split into a triplet by its neighboring methylene group. The chlorine-bearing methylene group is furthest downfield ($\delta = 3.6$), also split into a triplet by its neighboring methylene group. The center methylene group is expected to be split into a quartet by the adjacent methyl and into a triplet by the adjacent methylene. If these two interactions had different *J* values, we would indeed see a total of twelve lines under sufficient resolution. However, because the two *J* values are approximately the same (note that the CH_3 quartet and the CH_2Cl triplet have approximately equal splittings), the splitting in the middle CH_2 group is that expected for five magnetically equivalent hydrogens, namely six peaks.

The spectrum of 2-chloropropane in Figure 13.3 is simpler. The two methyl groups are equivalent and appear as a pair of superimposed doublets because of coupling to the C-2 hydrogen. They have a combined area six times that of the downfield resonance of the single C-2 hydrogen. The downfield position of $\delta = 4.05$ ppm results from the deshielding effect of the neighboring electronegative chlorine. This peak is split into seven peaks by the six adjacent methyl hydrogens. The theoretical relative intensities of such a septet are 1:6:15:20:15:6:1 (Table 13.5). It is difficult to see the two weak outer lines.

Proton and carbon NMR spectroscopy are now the primary methods whereby the structures of organic compounds are determined. Spectra are deduced from a consideration of several important spectral features. From the CMR spectrum, we can usually determine the number of carbons in the molecule. In addition, from the general resonance positions, we can often deduce the presence of functional groups. Furthermore, when we know the number of carbons from some other piece of data (such as elemental analysis), the appearance of fewer than the expected number of lines in the CMR spectrum often allows us to infer molecular symmetry.

A simple example will illustrate the use of CMR spectroscopy in solving a structure problem. Let us assume that we subject 3-hexanol to treatment with boiling sulfuric acid and isolate a mixture of isomeric alkenes.

OH

$\xrightarrow{H_2SO_4}$ + + +

The isomers are separated by preparative gas chromatography and the CMR spectra are measured. The data are as follows:

isomer A: 12.3, 13.5, 23.0, 29.3, 123.7, 130.6
isomer B: 13.4, 17.5, 23.1, 35.1, 124.7, 131.5
isomer C: 14.3, 20.6, 131.0
isomer D: 13.9, 25.8, 131.2

We can immediately deduce that isomers C and D are the two 3-hexenes, from the fact that each shows only three resonances. In these spectra, the resonances at about 131 ppm are clearly the double bond carbons and the resonances at about 14 ppm are probably the methyl carbons. However, note that the C-2 carbon resonances differ by about 5 ppm. This is exactly what we expect for the γ-effect in a pair of cis-trans isomers; we can confidently deduce that isomer C (with the upfield C-2) must be the cis isomer. Similar arguments can be used to deduce that isomer A is *cis*-2-hexene and isomer B is *trans*-2-hexene. Note in isomer A, two of the resonances are about 5 ppm upfield of their counterparts in isomer B (35.1 compared to 29.3, 17.5 compared to 12.3).

It is important to reiterate that it is not necessary to memorize a lot of chemical shift information to use CMR spectroscopy. It is only necessary to remember that double-bonded carbons resonate far downfield and that carbons bonded to halogen or oxygen resonate moderately far downfield. When we know enough about the possible structure of the unknown sample, as in the example just given, it is often possible to determine the structure completely from the CMR spectrum. However, in other cases we only derive partial information—pertaining to the carbon skeleton and the functional groups that are present.

In proton NMR we look at the resonance positions (chemical shifts) and spin-spin splitting patterns for structural information. Again, it is not recommended that the student try to memorize a lot of precise chemical shift information. Rather, the practicing chemist usually thinks of the NMR spectral chart as a sort of football field. We remember that alkanes tend to resonate around the 10–20 yard line on the right side, while protons attached to carbons that also bear an electronegative group (Cl, Br, OH) and vinyl protons resonate near mid-field. It is also useful to remember that inductive deshielding effects are cumulative—when there are more halogens or oxygens, the C—H resonance can move over mid-field into the other end of the field. If more precise chemical shift information is desirable, the practicing chemist refers to compilations of data.

The most useful aspect of proton NMR is undoubtedly spin-spin splitting. From this complication, we are often able to deduce the relative placements of most or all of the hydrogens in the molecule. It is important to remember that, when we look at a multiplet, we are not evaluating the hydrogens that actually give rise to the resonance, but rather *the neighboring hydrogens*. For example, the student should learn to associate a triplet with a proton (or protons) that has two neighboring hydrogens. A few common multiplets, with the types of structures that give rise to them, are shown in Figure 13.48.

The way in which proton NMR data are frequently presented and a structural problem is solved is shown in the following example. Assume that an unknown compound is found to have the formula $C_4H_7Cl_3$ and the following proton NMR spectrum.

δ, ppm: 0.9 (t, 3H); 1.7 (m, 2H); 4.3 (m, 1H); 5.8 (d, 1H)

In this shorthand the δ value in ppm is given for the center of a multiplet. The number of peaks in the multiplet is indicated by the code: s = singlet, d = doublet, t = triplet. Quartet and quintet are obvious, but it may not always be possible to resolve all of these peaks, and such multiple peak groups are frequently recorded as m = multiplet. Finally, the number of hydrogens represented by each multiplet as determined from the integral line is indicated.

To solve the problem shown, we generate hypotheses of structural units from the information given about δ values and put the units together with the help of the splitting information. The three-proton triplet with δ = 0.9 ppm clearly corresponds to a methyl

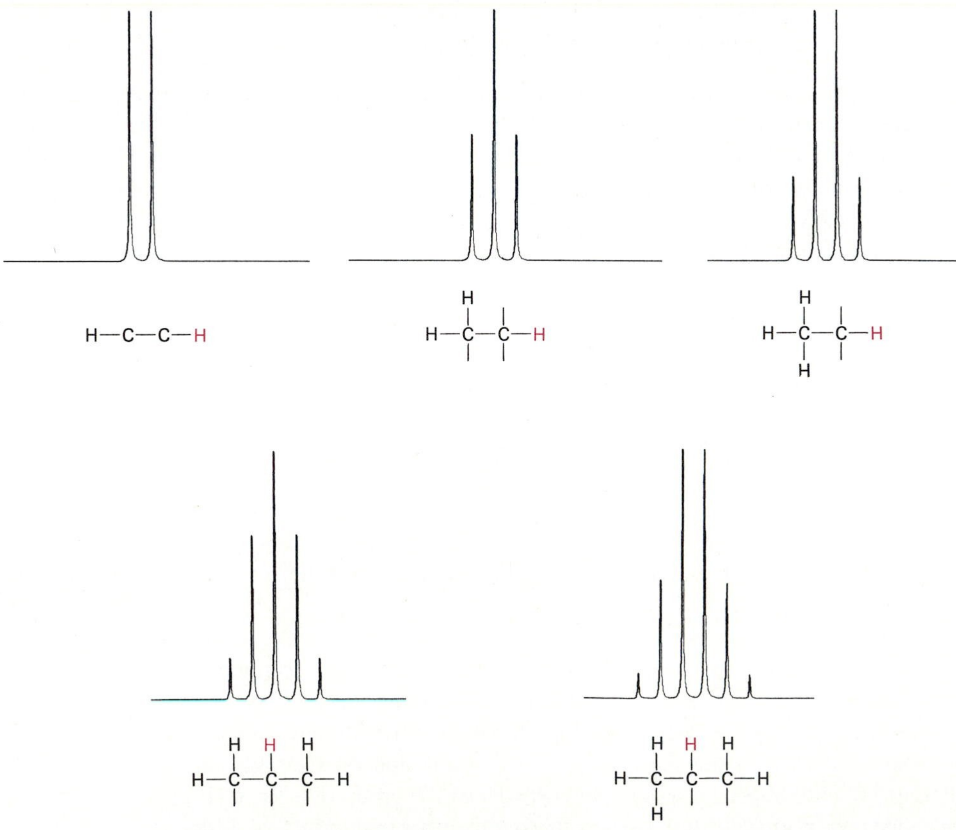

FIGURE 13.48 Common NMR multiplets.

group; both the number of hydrogens and the upfield resonance position fit this hypothesis. The one-proton doublet at $\delta = 5.8$ ppm is so far downfield it must correspond to $CHCl_2$. Its multiplicity tells us that it has only one neighboring H. The one-proton multiplet with $\delta = 4.3$ ppm is probably H—C—Cl. We are left with a two-proton multiplet at rather high field ($\delta = 1.7$ ppm); this must be from a CH_2 group. Our structural units are therefore CH_3, CH_2, CHCl, and $CHCl_2$. Since the methyl group is a triplet, it must be attached to the CH_2 group. Since $CHCl_2$ is a doublet, it must be attached to CHCl. The entire structure then becomes

$$CH_3—CH_2—CHCl—CHCl_2$$

the compound is 1,1,2-trichlorobutane. The CH_2 and CHCl protons give rise to complex multiplets because of the unequal coupling constants to their adjacent neighbors (see Figure 13.31).

The foregoing example illustrates the power of NMR spectroscopy in deducing structure and exemplifies the general approach to be used. In working actual problems look first for methyl groups—they are common in organic structures and are frequently readily recognizable since they are generally the furthest upfield and often occur as sharp peaks even when the compound contains several types of methylene groups and sharp multiplets in the 1–2 ppm region. Downfield peaks next help to identify protons on double bonds or near electronegative functions. The molecular formula as given by combustion analysis tells us whether multiple bonds are likely to be present and which electronegative elements to account for. When these resonances are identified and their splittings are analyzed, many structural problems are essentially solved.

EXERCISE 13.19 The proton NMR spectra of several $C_5H_{10}Br_2$ isomers are summarized below. Deduce the structure of each compound.

(a) δ, 1.0 (s, 6H); 3.4 (s, 4H)
(b) δ, 1.0 (t, 6H); 2.4 (quart, 4H)
(c) δ, 0.9 (d, 6H); 1.5 (m, 1H); 1.85 (t, 2H); 5.3 (t, 1H)
(d) δ, 1.0 (s, 9H); 5.3 (s, 1H)
(e) δ, 1.0 (d, 6H); 1.75 (m, 1H); 3.95 (d, 2H); 4.7 (quart, 1H)
(f) δ, 1.3 (m, 2H); 1.85 (m, 4H); 3.35 (t, 4H)

ASIDE

Three alcohols of similar boiling point (80 ± 3 °C) were brought to an NMR spectroscopist. He sniffed cautiously (not a recommended procedure for unknown materials!) and described their odors as those of vodka, rubbing alcohol, and camphor. He found that he could easily identify them on the basis of their ^1H and ^{13}C NMR spectra. Do you agree? Name the alcohols, draw their formulas, and sketch both the ^1H and ^{13}C NMR spectra, giving all possible information.

PROBLEMS

1. Fill in the blank spaces in the following statement.

In the NMR spectrum of ethyl bromide, the methyl hydrogens have $\delta = 1.7$ ppm, the methylene hydrogens have $\delta = 3.3$ ppm, and J = 7 Hz. The number of peaks given by the methyl hydrogens is ____ with the approximate area ratio ____. These peaks are separated from one another by ____ Hz. The number of peaks given by the methylene hydrogens is ____ with the approximate area ratio ____. These peaks are separated by ____ Hz. The total area of the methyl peaks compared to the methylene peaks is in the ratio ____. Of these two groups of peaks, the ____ peaks are farther downfield. The chemical shift difference between these peaks of 1.6 ppm corresponds in a 180-MHz instrument to ____ Hz and in a 250-MHz instrument to ____ Hz.

2. Sketch the expected NMR spectra of the following compounds. Be sure to represent the approximate δ expected for each group of peaks, the relative areas, and the splittings.

a. $CH_2BrCH_2CH_2CH_2Br$ **b.** $CH_2ClCH(CH_3)CH_2Cl$ **c.** $CH_3CBr(CH_3)CH_2Br$

3. The NMR spectra for some isomers having the formula $C_4H_{10}O$ are summarized as follows. Deduce the structure of each compound.

a. δ, 0.95 (t, 3H); 1.52 (sextet, 2H); 3.30 (s, 3H); 3.40 (t, 2H)
b. δ, 1.15 (s, 1H); 1.28 (s, 9H)
c. δ, 1.20 (t, 3H); 3.45 (quartet, 2H)
d. δ, 0.90 (d, 6H); 1.78 (m, 1H); 2.45 (t, 1H); 3.30 (t, 2H)
e. δ, 1.13 (d, 6H); 3.30 (s, 3H); 3.65 (septet, 1H)
f. δ, 0.95 (t, 3H); 1.50 (m, 4H); 2.20 (t, 1H); 3.70 (dt, 2H)
g. δ, 0.92 (t, 3H); 1.18 (d, 3H); 1.45 (m, 2H); 1.80 (d, 1H); 3.75 (m, 1H)

4. Free radical chlorination of propane using 1 mole of C_3H_8 and 2 moles of Cl_2 gives a complex mixture of chlorination products. By careful fractional distillation of the product mixture, one may isolate four dichloropropanes, A, B, C, and D. From the NMR spectra of the four isomers, deduce their structures.

isomer A: (b.p. 69 °C) δ 2.4 (s, 6H)
isomer B: (b.p. 88 °C) δ 1.2 (t, 3H), 1.9 (quintet, 2H), 5.8 (t, 1H)
isomer C: (b.p. 96 °C) δ 1.4 (d, 3H), 3.8 (d, 2H), 4.3 (sextet, 1H)
isomer D: (b.p. 120 °C) δ 2.2 (quintet, 2H), 3.7 (t, 4H)

5. There are nine possible isomers (not counting stereoisomers) of $C_4H_8Br_2$. Two of them have the following NMR spectra. Deduce the structures of each and indicate the logic used in your assignment.

a. δ 1.7 (d, 6H), 4.4 (quart, 2H)
b. δ 1.7 (d, 3H), 2.3 (quart, 2H), 3.5 (t, 2H), 4.2 (m, 1H)

6. A compound having the formula $C_5H_{10}Br_2$ has a proton-coupled CMR spectrum consisting of a doublet, a triplet, and a quartet. What is its structure?

7. Free radical chlorination of (*R*)-2-chlorobutane gives a mixture of isomeric products that is subjected to careful fractional distillation. Five different dichlorobutanes are obtained. For each compound, the optical rotation and the CMR spectrum are measured. The following results are obtained.

isomers E, F: optically active; CMR, 4 resonances
isomer G: optically active; CMR, 2 resonances
isomer H: optically inactive; CMR, 4 resonances
isomer I: optically inactive; CMR, 2 resonances

From these data, make whatever structural assignments are possible.

8. Deduce the structure corresponding to each of the following NMR spectra.

a. C_2H_5I

b. $C_2H_3Cl_3$

c. $C_5H_{12}O$

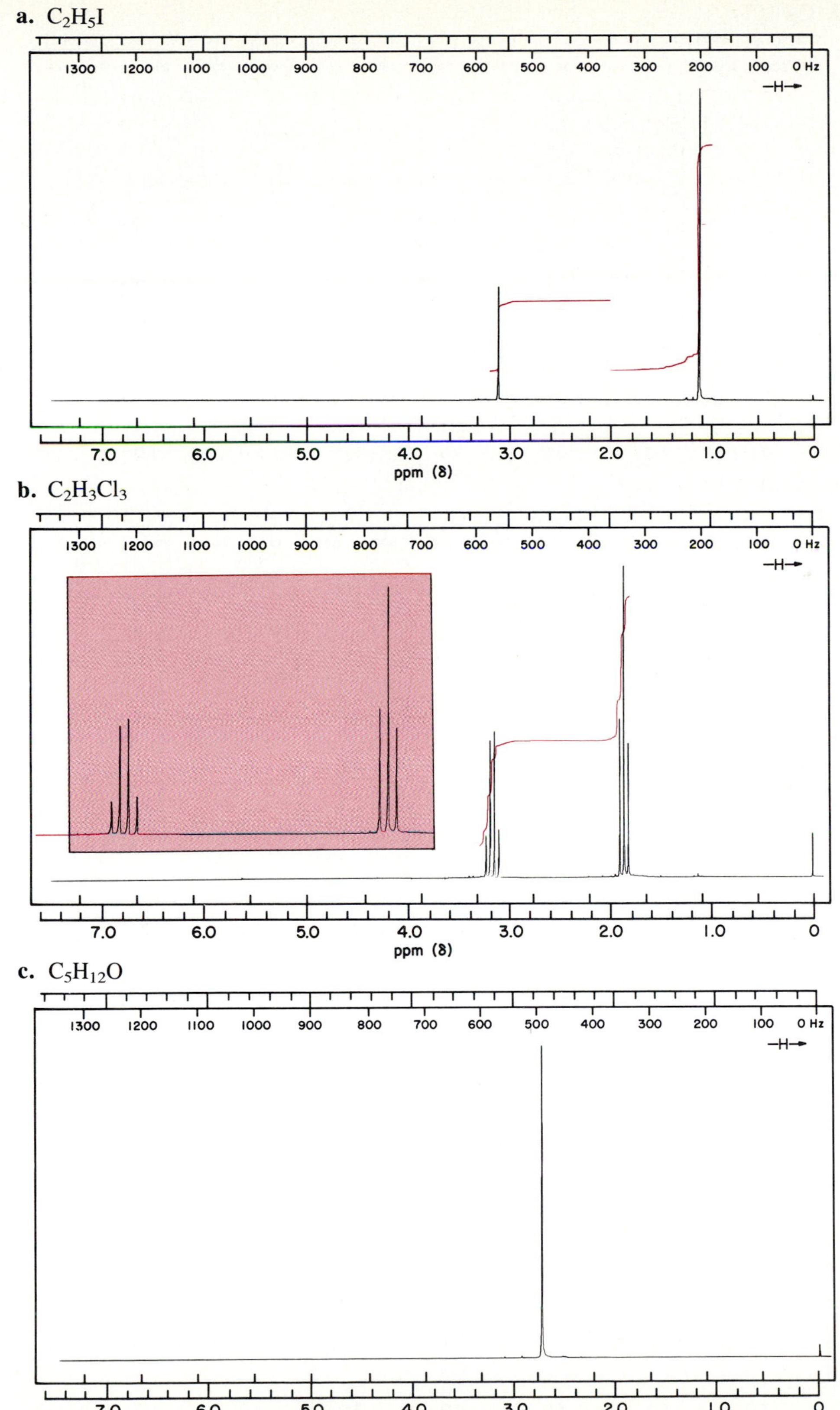

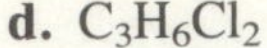

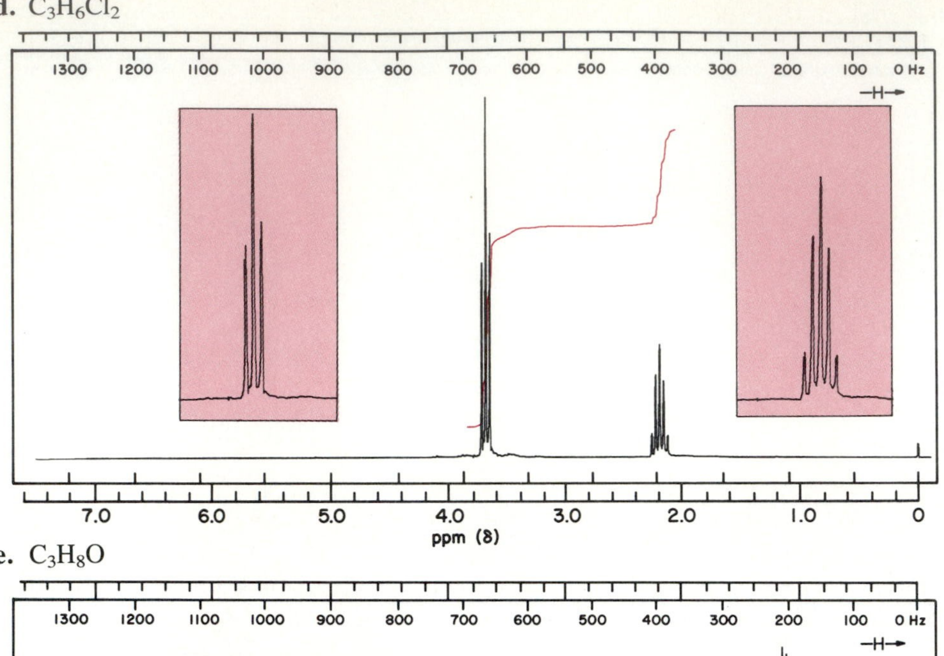
1300 1200 1100 1000 900 800 700 600 500 400 300 200 100 0 Hz
7.0 6.0 5.0 4.0 3.0 2.0 1.0 0
ppm (δ)

e. C_3H_8O

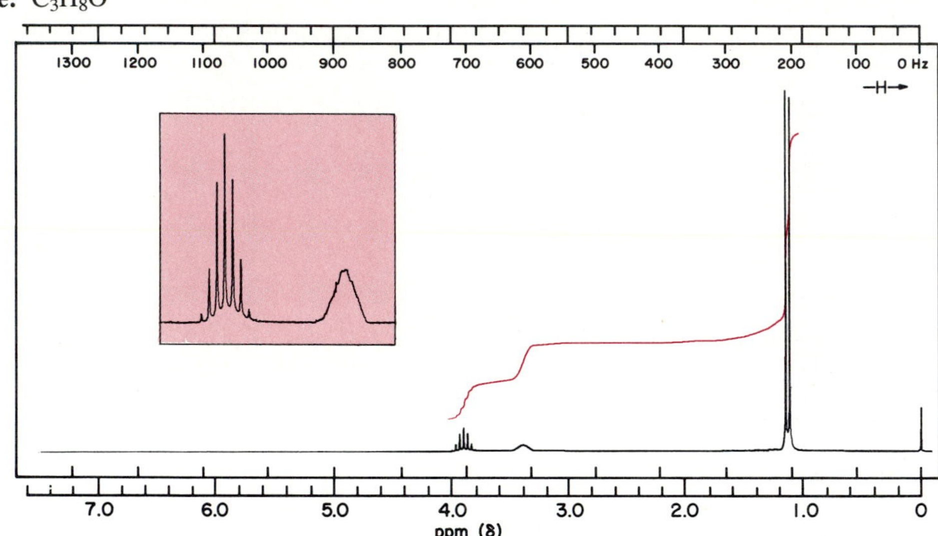
1300 1200 1100 1000 900 800 700 600 500 400 300 200 100 0 Hz
7.0 6.0 5.0 4.0 3.0 2.0 1.0 0
ppm (δ)

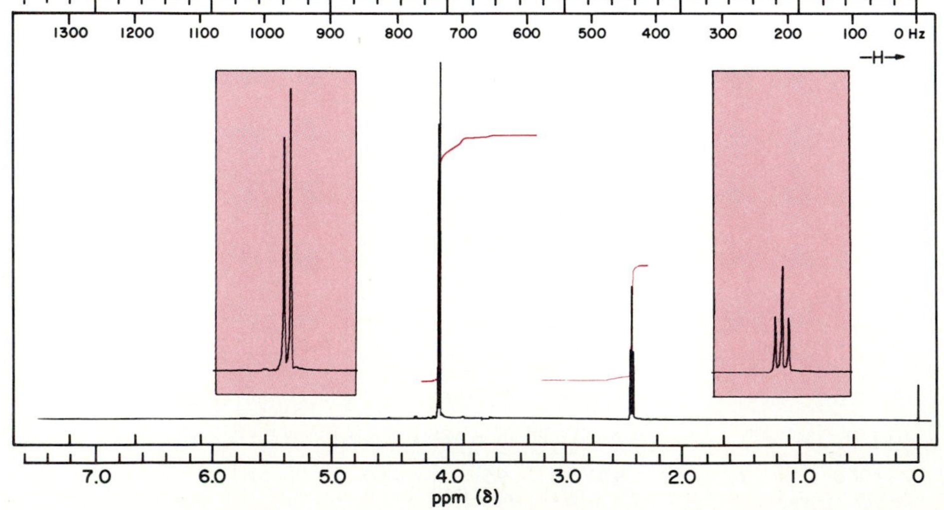
1300 1200 1100 1000 900 800 700 600 500 400 300 200 100 0 Hz
7.0 6.0 5.0 4.0 3.0 2.0 1.0 0
ppm (δ)

9. While in the process of writing this chapter, the authors ordered a sample of 1,2,2-trichloropropane from a chemical supplier in order to obtain its proton NMR and CMR spectra. The proton-coupled and -decoupled CMR spectra of the commercial sample were determined first and are reproduced below. The bottle was obviously mislabeled. What is the actual structure of this compound?

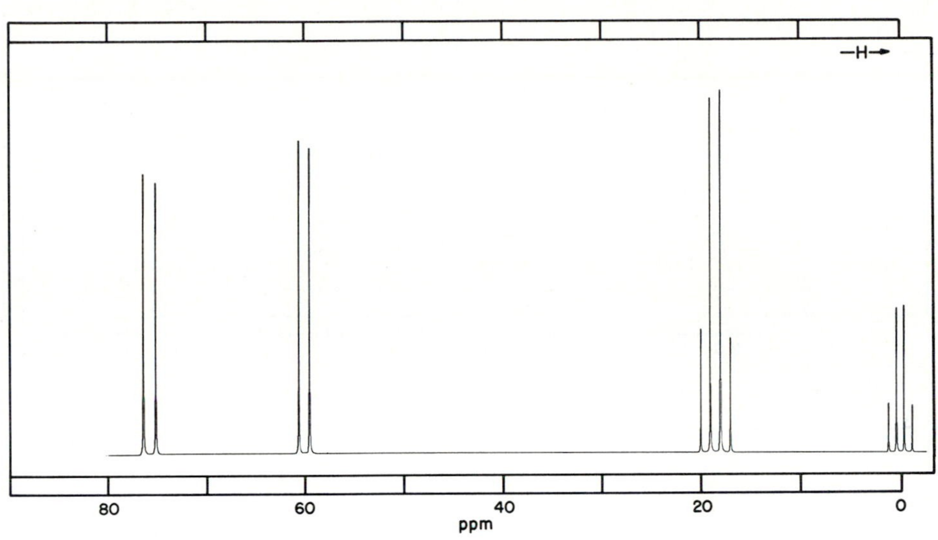

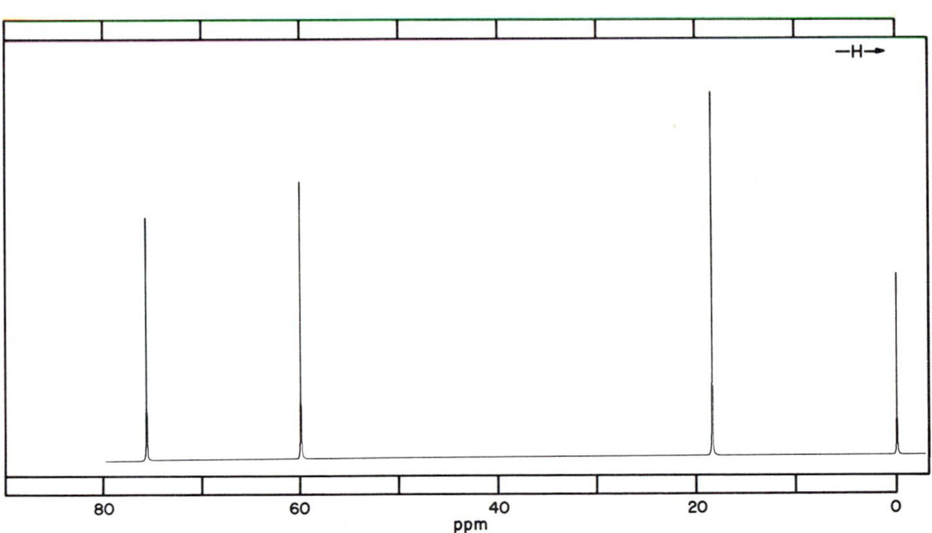

10. The proton NMR spectrum of chloroform shows a single intense peak at $\delta =$ 7.27 ppm. However, careful examination shows a small peak -104.5 Hz above and below the main peak. These peaks are associated with $^{13}CHCl_3$. Explain. What CMR spectrum would you expect for $^{13}CHCl_3$?

11. Deduce the structure of each of the following compounds from the NMR and proton-decoupled CMR spectra.

a. C_4H_8O

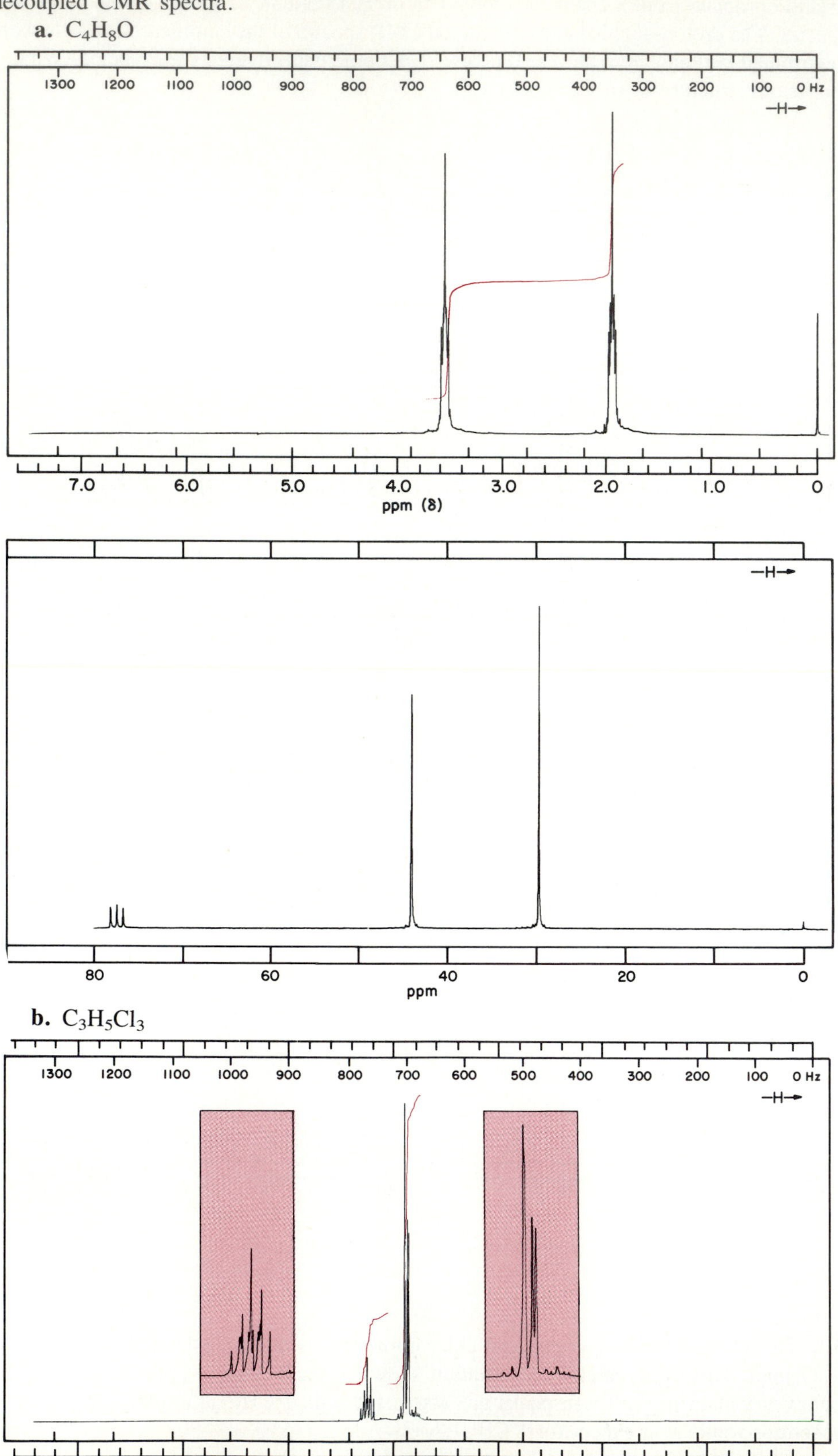

b. $C_3H_5Cl_3$

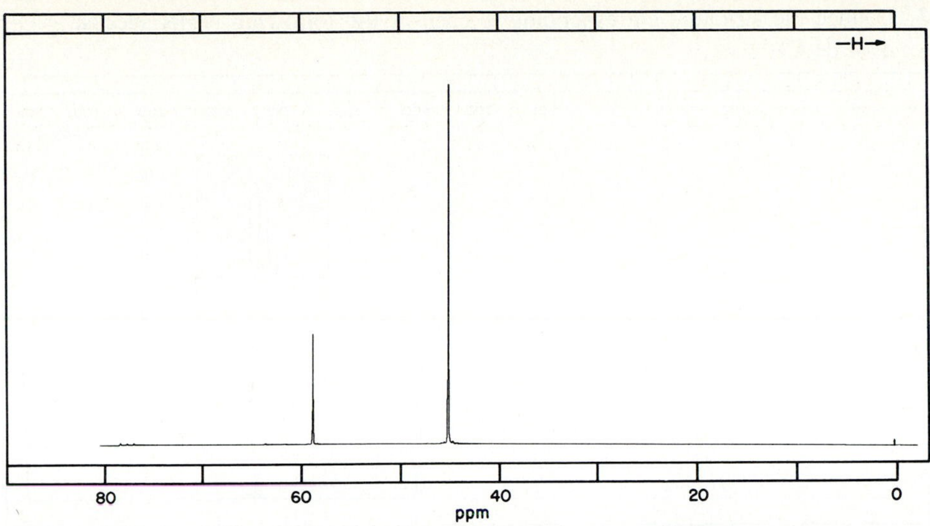
—H→
80
60
40
20
0
ppm

c. C_4H_8O

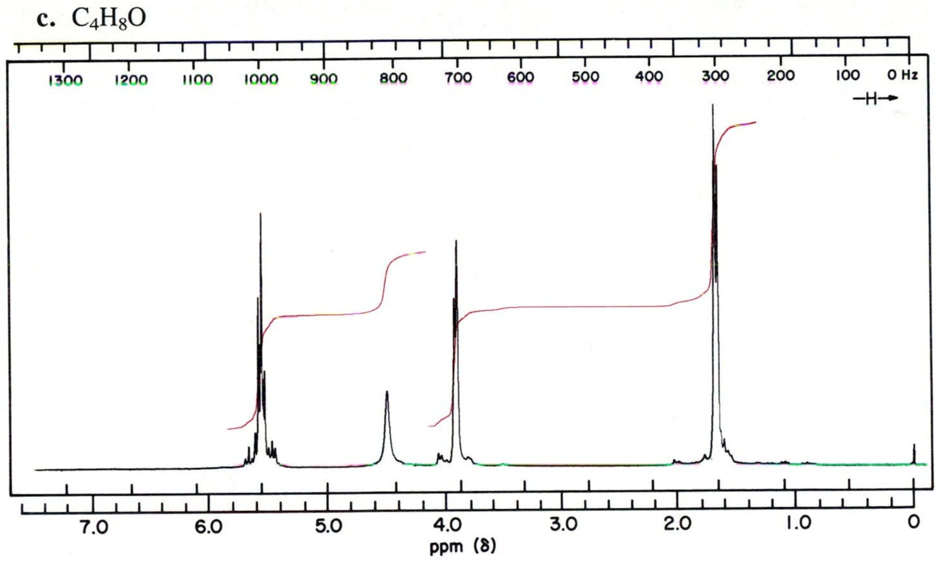
1300
1200
1100
1000
900
800
700
600
500
400
300
200
100
0 Hz
—H→
7.0
6.0
5.0
4.0
3.0
2.0
1.0
0
ppm (δ)

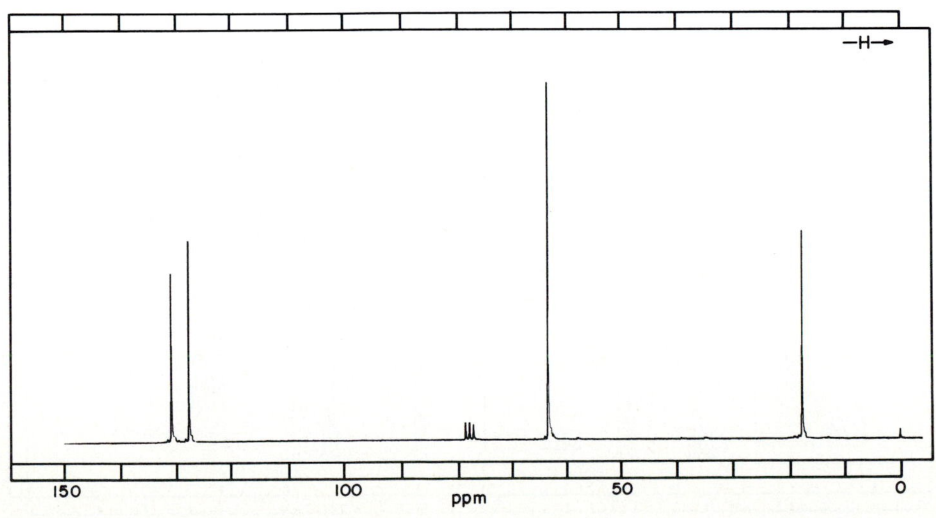
—H→
150
100
50
0
ppm

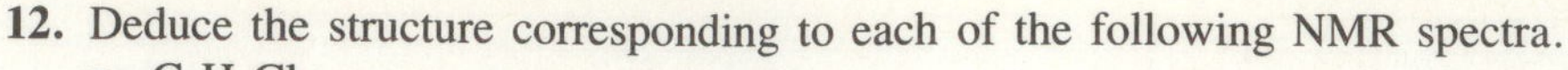

12. Deduce the structure corresponding to each of the following NMR spectra.

a. C_4H_9Cl

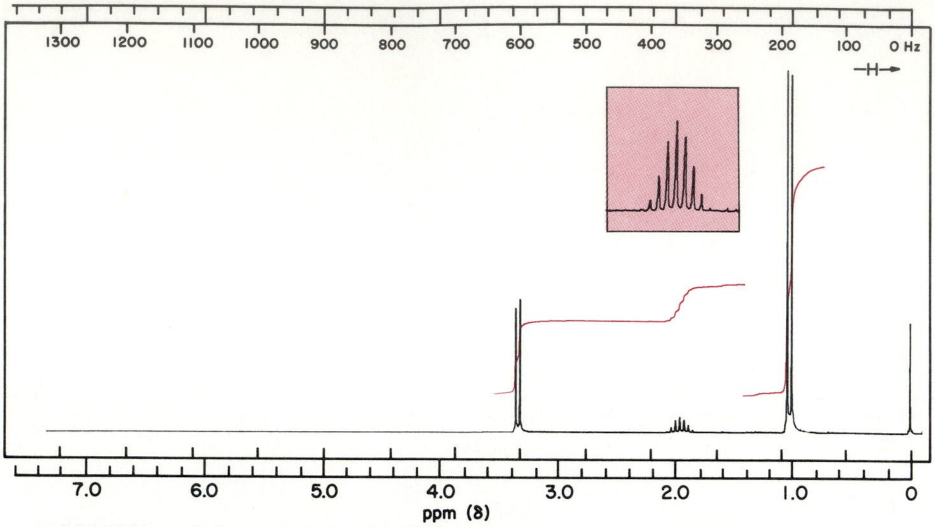

b. C_4H_9I

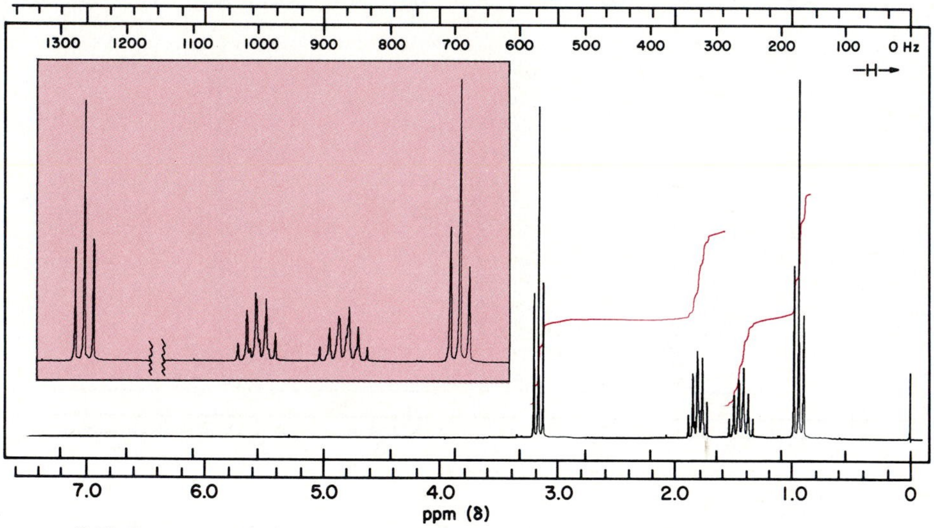

c. C_4H_7Br

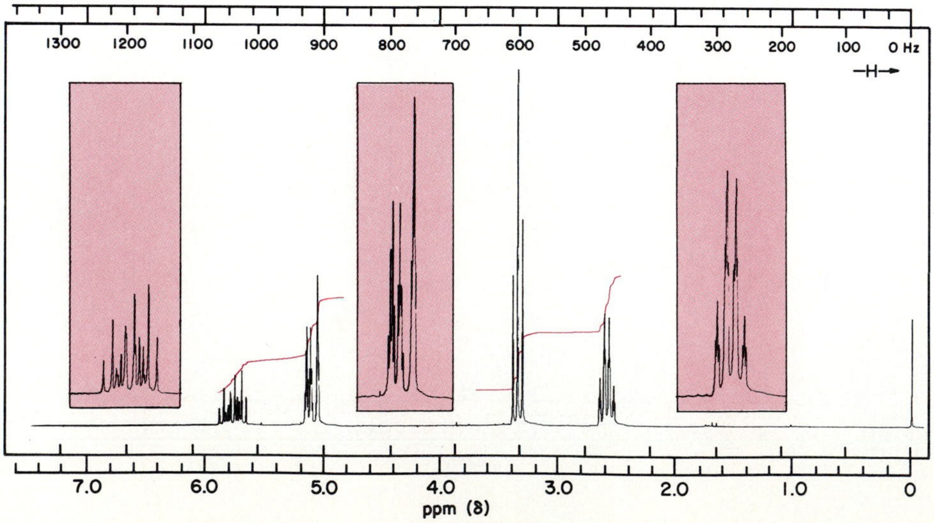

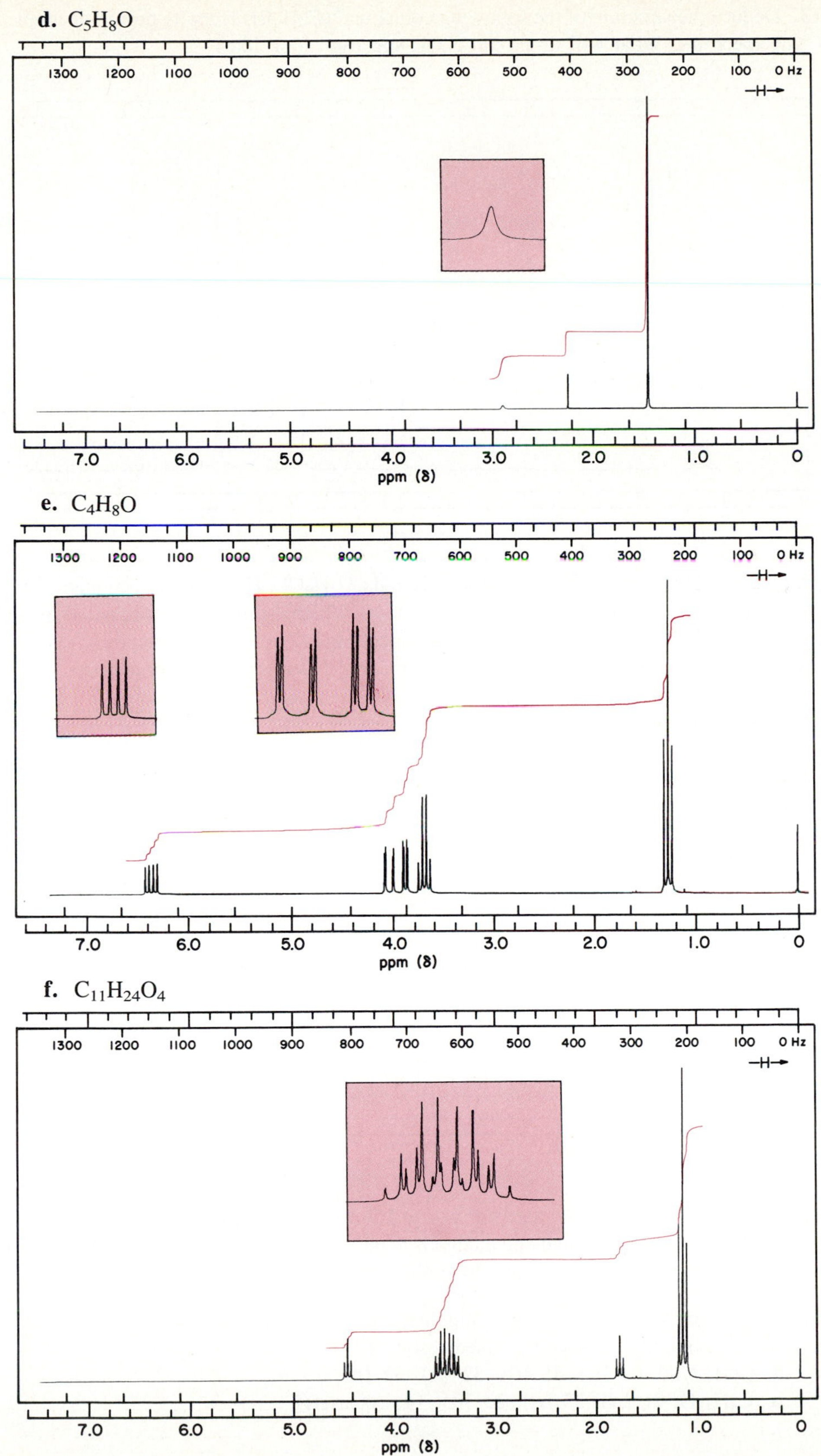
d. C5H8O
1300 1200 1100 1000 900 800 700 600 500 400 300 200 100 0 Hz
7.0 6.0 5.0 4.0 3.0 2.0 1.0 0
ppm (δ)
e. C4H8O
1300 1200 1100 1000 900 800 700 600 500 400 300 200 100 0 Hz
7.0 6.0 5.0 4.0 3.0 2.0 1.0 0
ppm (δ)
f. C11H24O4
1300 1200 1100 1000 900 800 700 600 500 400 300 200 100 0 Hz
7.0 6.0 5.0 4.0 3.0 2.0 1.0 0
ppm (δ)

13. Deduce the structure of the following compound (C_4H_9Br) from its proton-decoupled CMR spectrum. What will its proton NMR spectrum look like?

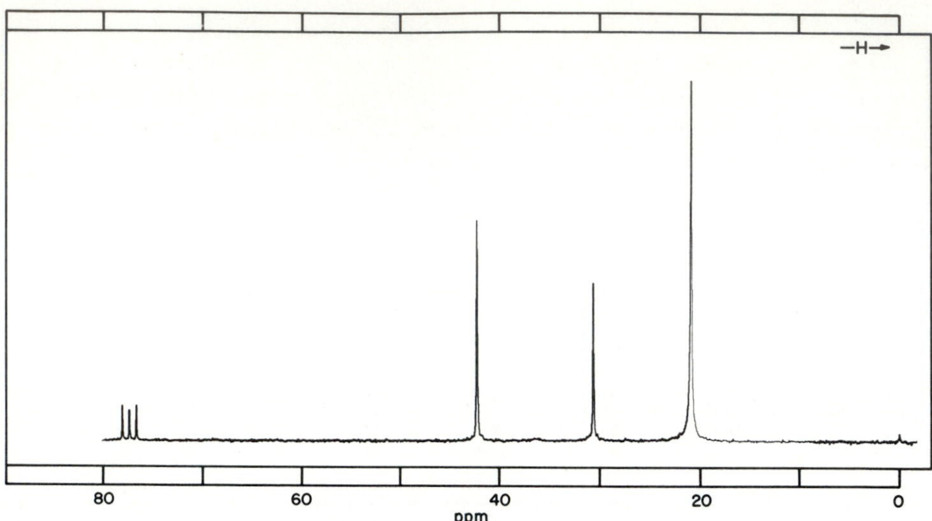

14. There are four compounds with the formula C_4H_9Cl. The proton-decoupled CMR spectrum of one of the isomers is shown below. Which isomers are eliminated by the spectrum? Which isomers might give such a spectrum? Describe how the proton-coupled CMR spectrum can be used to decide which C_4H_9Cl isomer the compound is.

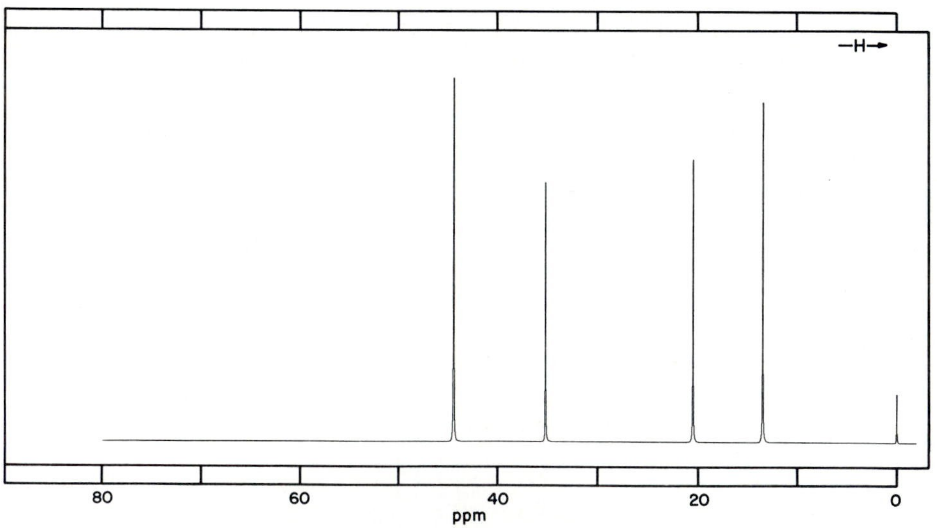

15. Explain why the CMR spectrum of *cis*-1,2-dimethylcyclohexane shows only one methyl resonance, even though one methyl is axial and the other is equatorial. How would you expect the spectrum to change on cooling below −70 °C?

16. Identify each of the following compounds from the CMR data presented [resonance position (multiplicity of proton-coupled resonance)].

a. $C_5H_{11}ClO$; 62.2(t), 45.4(t), 32.9(t), 32.1(t), 23.7(t)

b. $C_8H_{18}O$; 71.2(t), 33.1(t), 20.3(t), 14.6(quartet)

c. $C_6H_{14}O$; 75.1(d), 35.3(s), 25.8(quartet), 18.2(quartet); the resonance at 25.8 ppm is much more intense than the others

d. $C_6H_{14}O$; 65.5(d), 49.2(t), 25.1(d), 24.3(quartet), 22.7(quartet)

e. $C_6H_{12}O_2$; 71.1(d), 33.9(t)

17. 1-Chloro-3-methylcyclopentane was treated with potassium *t*-butoxide in *t*-butyl alcohol. Two isomeric alkenes were formed. The CMR spectrum of the major product has only four resonances and that of the minor product has six. What are the structures of the two products?

18. 3-Bromo-2,3-dimethylpentane was treated with KOH in refluxing ethanol. The resulting alkene mixture was separated by preparative gas chromatography into four fractions. The principal product was found by proton NMR to be 2,3-dimethyl-2-pentene (no vinyl hydrogens). A second product was found to be 2-ethyl-3-methyl-1-butene (two vinyl hydrogens). The other two products have the following CMR chemical shifts.

isomer J: 12.4, 17.8, 20.6, 28.5, 117.9, 141.0
isomer K: 13.0, 13.1, 21.6, 37.4, 116.2, 141.5

What are the structures of these two isomers?

19. Three compounds, all having the formula C_6H_{10}, are found to have the following CMR spectra. Propose structures for the three compounds.

isomer L: 12.9, 124.9, 125.3
isomer M: 17.6, 125.8, 132.3
isomer N: 13.0, 18.0, 123.1, 127.4, 128.3, 130.2

20. The CMR spectrum of *n*-butanol, $CH_3CH_2CH_2CH_2OH$, shows four resonances with chemical shifts of 61.7, 35.3, 19.4, and 13.9 ppm. Using these data and the chemical shifts of butane (Table 13.2), calculate α-, β-, and γ-effects for the OH group. Chlorination of *n*-butanol gives several products, one of which shows the CMR resonances 62.0, 29.7, 30.0, and 45.4 ppm. What is the structure of this chlorination product?

21. Triply-bonded carbons have CMR resonances in the range $\delta = 60–80$ ppm. For example, in 1-butyne, the C-1 resonance is at 67.3 ppm and the C-2 resonance is at 85.0 ppm. These resonance positions are in between those of the corresponding carbons in butane (C-1 = 13.2 ppm, C-2 = 25.0 ppm) and 1-butene (C-1 = 112.8 ppm, C-2 = 140.2 ppm). Explain.

22. Treatment of *trans*-3-methylcyclohexanol with concentrated sulfuric acid gave a mixture of three alkenes, having the following CMR spectra.

isomer O: 23.8, 24.4, 26.7, 31.5, 122.3, 134.3
isomer P: 22.4, 26.1, 30.9, 32.6, 126.6, 134.0
isomer Q: 23.0, 26.0, 29.6, 32.2, 34.9, 127.1

[Note that, in each case, there is an accidental coincidence of a pair of resonances.] In the proton NMR spectra, isomer O shows one vinyl proton and isomers P and Q show two each. Assign structures to the three isomers.

23. The NMR spectrum of difluoromethane measured as a dilute solution in carbon tetrachloride is shown below. Provide an interpretation of the spectrum. (*Hint*: Recall that the fluorine nucleus, ^{19}F, also has spin of $\pm\frac{1}{2}$.)

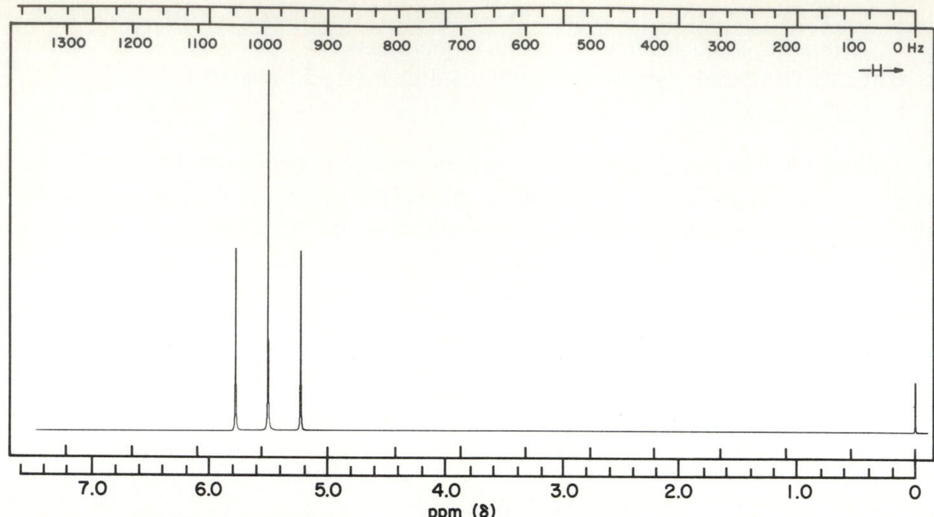

24. The proton-decoupled CMR spectrum of ethyl fluoride consists of two doublets with chemical shifts of 79.3 and 13.6 ppm. Explain.

25. Several simple compounds containing only ^{13}C at one or more positions are available commercially. These compounds may be used to synthesize more complex molecules in which one or more positions are *labeled* with ^{13}C. Such materials are useful for elucidating reaction mechanisms. In one such experiment, a sample of 2-pentanol containing only ^{13}C at C-2 and C-3 was prepared and treated with hot H_2SO_4. From the reaction mixture, *trans*-2-pentene was isolated.

- **a.** What is the expected appearance of the CMR spectrum of the dilabeled 2-pentanol? (Remember that the natural abundance of ^{13}C is only 1%, so the synthetic material has 1% ^{13}C at C-1 and C-4 and 100% at C-2 and C-3.)
- **b.** The isolated *trans*-2-pentene has a CMR spectrum that consists of doublets at 25.8 ppm (J = 44 Hz) and 123.6 ppm (J = 70 Hz); at 133.2 ppm there are two superimposed doublets (J = 44 Hz and 70 Hz). Explain.

26. The radio waves used to irradiate the NMR sample are absorbed in converting α-spin states to β and are converted ultimately to heat. To see how much heat is involved, calculate approximately the temperature increase produced in 1 mL of an NMR solution containing 0.01 mole of protons in a 180-MHz NMR instrument. For the purpose of this calculation consider that the entire excess population of α-spins is converted to β and that the heat capacity of the solution is 1 cal deg^{-1} mL^{-1}.

CHAPTER 14

ALDEHYDES AND KETONES

14.1 Structure

Aldehydes and ketones are compounds containing the **carbonyl group,** C═O. When two alkyl groups are attached to the carbonyl, the compound is a **ketone.** When two hydrogens, or one hydrogen and one alkyl group, are attached to the carbonyl, the compound is an **aldehyde.**

:O: ∷ R:C:R′	O ‖ R—C—R′	RCOR′
Lewis structure	Kekulé structure	Condensed structure

The structure of formaldehyde, the simplest member of the class, is depicted below, along with its experimental bond lengths and bond angles.

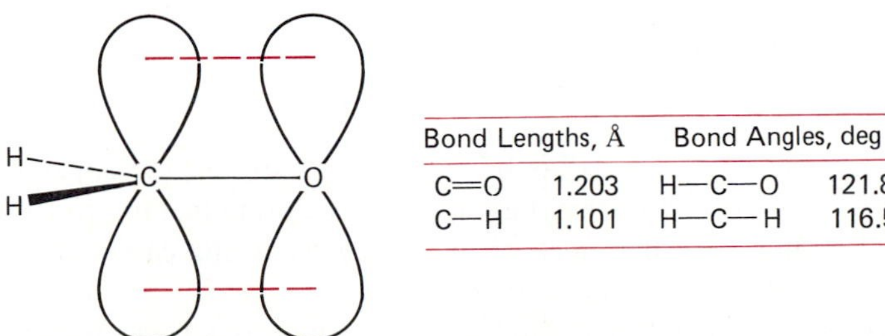

Bond Lengths, Å		Bond Angles, deg	
C═O	1.203	H—C—O	121.8
C—H	1.101	H—C—H	116.5

The carbon atom is approximately sp^2-hybridized and forms σ-bonds to two hydrogens and one oxygen. The molecule is planar and the H—C—O and H—C—H angles are close to 120°, the idealized sp^2-angles. The remaining carbon p-orbital overlaps with the oxygen p-orbital, giving rise to a π-bond between these atoms. The oxygen atom also has two nonbonding electron pairs (the lone pairs) that occupy the remaining orbitals. A stereo

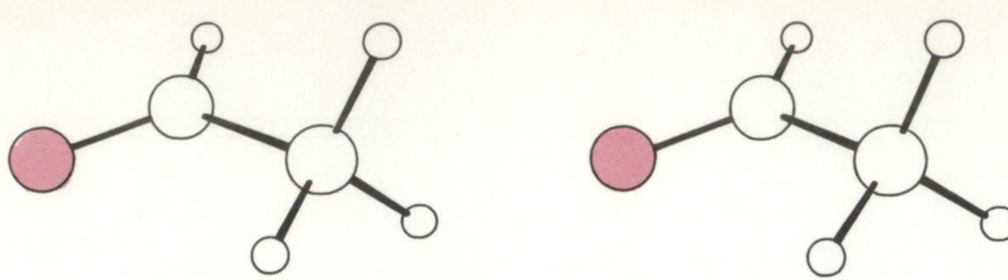

FIGURE 14.1 Stereo representation of acetaldehyde.

representation of acetaldehyde is shown in Figure 14.1. Note the planarity of the carbonyl group. Also note that one carbon-hydrogen bond of the methyl group is eclipsed with the C=O and that the carbonyl C—H is staggered with respect to the other two carbon-hydrogen bonds.

Oxygen is more electronegative than carbon and attracts the bonding electrons more strongly; that is, the higher nuclear charge on oxygen provides a greater attractive force than carbon. Accordingly, the carbon-oxygen bond is polarized in the direction $C^+—O^-$. This effect is especially pronounced for the π-electrons and can be represented by the following resonance structures for formaldehyde.

$$\left[\begin{matrix} H \\ \quad :C::\ddot{O}. \\ H \end{matrix} \longleftrightarrow \begin{matrix} H \\ \quad :\overset{+}{C}:\ddot{O}:^- \\ H \end{matrix} \right] \quad \text{or} \quad \left[\begin{matrix} H \\ \quad C=O \\ H \end{matrix} \longleftrightarrow \begin{matrix} H \\ \quad \overset{+}{C}—\overset{-}{O} \\ H \end{matrix} \right]$$

The actual structure is a composite of the normal octet structure, $CH_2=O$, and the polarized structure, $^+CH_2—O^-$, which corresponds to a carbocation oxide. The composite structure may be represented with dotted line symbolism that shows the partial charges in carbon and oxygen and the partial single-bond character of the carbon-oxygen bond.

$$\begin{matrix} H \\ \quad \overset{\delta+}{C}\overset{\delta-}{=\!=}O \\ H \end{matrix}$$

One physical consequence of this bond polarity is that carbonyl compounds generally have rather high dipole moments. The experimental dipole moments of formaldehyde and acetone are 2.27 D and 2.85 D, respectively.

$$\begin{matrix} O \\ \| \\ C \\ H \quad H \end{matrix} \; \updownarrow \qquad\qquad \begin{matrix} O \\ \| \\ C \\ CH_3 \quad CH_3 \end{matrix} \; \updownarrow$$

formaldehyde
dipole moment, 2.27 D

acetone
dipole moment, 2.85 D

The chemical consequences of this bond polarity will become apparent during our discussions of the reactions of carbonyl groups. We shall find that the positive carbon can react with nucleophiles and that much of the chemistry of the carbonyl function corresponds to that of a relatively stable carbocation.

The lone-pair electrons in the carbonyl oxygen have weakly basic properties. In acidic solution acetone acts as a Lewis base and is protonated to a small but significant extent.

$$CH_3—\overset{\overset{:O:}{\|}}{C}—CH_3 + H^+ \rightleftharpoons CH_3—\overset{\overset{:\overset{+}{O}H}{\|}}{C}—CH_3$$

In fact, acetone is a much weaker Lewis base than is water. An acid strength corresponding to 82% sulfuric acid is required to give 50% protonation of acetone. This corresponds to an approximate pK_a for the conjugate acid of acetone of −7.2 (the approximate pK_a of H_3O^+ is −1.7). Even though the carbonyl group has only weakly basic properties, we shall find that this basicity plays an important role in the chemistry of aldehydes, ketones, and related compounds.

14.2 Nomenclature

Traditionally, aldehyde names were derived from the name of the corresponding carboxylic acids (Section 18.2) by dropping the suffix **ic** (or **-oic**) and adding in its place the suffix **-aldehyde.** These common names are still widely used for simpler aldehydes.

$$\underset{\text{formic acid}}{H{-}\overset{\overset{\large O}{\|}}{C}{-}OH} \qquad \underset{\text{formaldehyde}}{H{-}\overset{\overset{\large O}{\|}}{C}{-}H}$$

$$\underset{\text{acetic acid}}{CH_3{-}\overset{\overset{\large O}{\|}}{C}{-}OH} \qquad \underset{\text{acetaldehyde}}{CH_3{-}\overset{\overset{\large O}{\|}}{C}{-}H}$$

$$\underset{\text{benzoic acid}}{C_6H_5{-}\overset{\overset{\large O}{\|}}{C}{-}OH} \qquad \underset{\text{benzaldehyde}}{C_6H_5{-}\overset{\overset{\large O}{\|}}{C}{-}H}$$

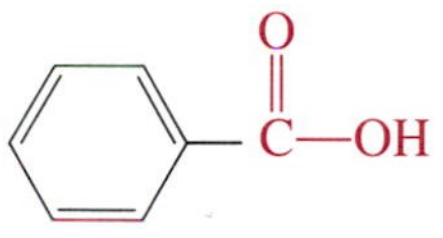

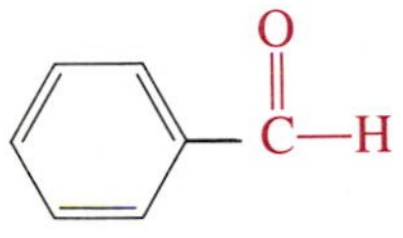

> Because of their characteristic acidic properties, which rendered them relatively easy to purify and characterize, the carboxylic acids were among the first well-known organic compounds. As they were found, they were given names stemming from their natural sources. Thus, formic (L., *formica,* ant); acetic (L., *acetum,* vinegar); benzoic (from gum benzoin, *Styrax benzoin*). As the "type theory" developed, and functional group interconversions were discovered, the carboxylic acid names were translated, with appropriate modification, to related compounds.

In the common names of aldehydes, appendage groups are traditionally designated by the appropriate Greek letters as prefixes. The chain is labeled by using α, β, γ, and so on, beginning with the carbon *next to the carbonyl group.*

$$\underset{\alpha\text{-chloropropionaldehyde}}{CH_3\overset{\overset{\large Cl}{|}}{C}HCHO} \qquad \underset{\beta\text{-bromobutyraldehyde}}{CH_3\overset{\overset{\large Br}{|}}{C}HCH_2CHO}$$

The common names of ketones are derived by prefixing the word **ketone** by the names of the two alkyl radical groups; the separate parts are separate words. Dimethyl ketone has the additional trivial name acetone, which is universally used.

$$\underset{\substack{\text{dimethyl ketone}\\ \text{(acetone)}}}{CH_3\overset{\overset{\large O}{\|}}{C}CH_3} \qquad \underset{\text{ethyl methyl ketone}}{CH_3\overset{\overset{\large O}{\|}}{C}CH_2CH_3} \qquad \underset{\text{diisopropyl ketone}}{(CH_3)_2CH\overset{\overset{\large O}{\|}}{C}CH(CH_3)_2}$$

EXERCISE 14.1 Write structures for each of the following common names.

(a) isopropyl *n*-propyl ketone
(b) isobutyl neopentyl ketone
(c) γ-methoxybutyraldehyde (butyric acid is the four-carbon carboxylic acid; its name derives from L., *butyrum,* butter).
(d) *t*-butyl cyclohexyl ketone

In the IUPAC system aldehyde names are derived from the name of the alkane of the same carbon number. The final **-e** of the alkane is replaced by the suffix **-al.** Since the carbonyl group is necessarily at the end of a chain, it is not necessary to designate its position by a number, but as a suffix group it determines the direction in which the chain is numbered; the carbonyl carbon in an aldehyde is *always C-1.* [Note that the carbonyl carbon is C-1 in the IUPAC system, but it is given no designation in the common system.]

$CH_3CH_2CH{=}O$
propanal

$CH_3CH(CH_3)CH_2CH_2CH_2CH{=}O$
5-methylhexanal

More complicated aldehydes may be named using the suffix **-carbaldehyde.** When it is necessary to name a compound based on another functional group, the aldehyde grouping is designated with the prefix **formyl-.**

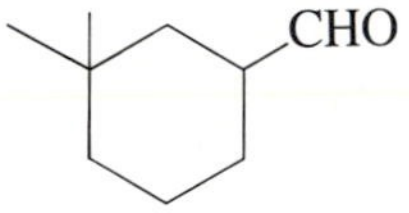

3,3-dimethylcyclohexanecarbaldehyde

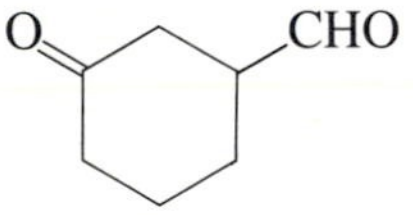

3-formylcyclohexanone

The IUPAC names of ketones are derived from the name of corresponding alkane by replacing the final *-e* by *-one*. In acyclic ketones it is necessary to prefix the name by a number indicating which carbon along the longest chain is the carbonyl carbon. The longest chain containing the carbonyl group is numbered from the end that gives the carbonyl carbon the lower number. In cyclic ketones it is understood that the carbonyl carbon is number 1.

$CH_3CH_2CH(CH_2CH_3)CH_2C({=}O)CH_2CH_3$
5-ethyl-3-heptanone

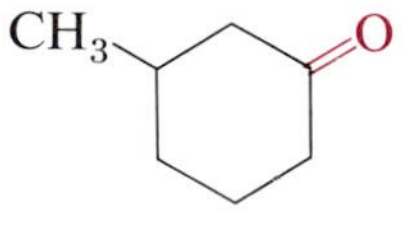

3-methylcyclohexanone

Occasionally it is necessary to name a molecule containing a carbonyl group as a derivative of a more important function. In such a case, the prefix **oxo**- is used, along with a number, to indicate the position and nature of the group. One such example is shown below.

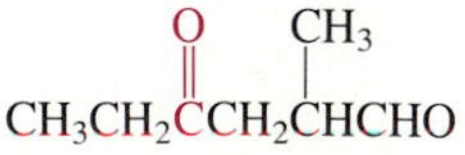

$CH_3CH_2C({=}O)CH_2CH(CH_3)CHO$
2-methyl-4-oxohexanal

It is generally desirable that the common and IUPAC nomenclature systems not be mixed. Ambiguity can result because counting by Greek letters in the common system starts from the carbon next to the carbonyl group, whereas the numbers in the IUPAC system always include the carbonyl group.

$ClCH_2CH_2CHO$

correct: β-chloropropionaldehyde or 3-chloropropanal

incorrect: β-chloropropanal

allowed but not recommended: 3-chloropropionaldehyde

EXERCISE 14.2 Give IUPAC names for each of the compounds in Exercise 14.1.

14.3 Physical Properties

Physical data for a number of aldehydes and ketones are collected in Tables 14.1 and 14.2. As in other homologous series, there is a smooth increase in boiling point with increasing molecular weight. Aldehydes and ketones boil higher than alkanes of comparable molecular weights. This boiling-point elevation results from the attraction between the carbonyl dipoles.

$$R_2C{=}O \quad R_2C{=}O$$

The discrepancy is largest with the simplest aldehyde, formaldehyde (mol. wt. 30, b.p. −21 °C), which boils 68° higher than ethane (mol. wt. 30, b.p. −89 °C). With higher members of the series, as the polar functional group becomes a smaller and smaller fraction of the molecule, the boiling point tends to come closer and closer to that of a corresponding alkane (see 2-decanone, mol. wt. 156, b.p. 210 °C; *n*-undecane, mol. wt. 155, b.p. 196 °C).

TABLE 14.1 Physical Properties of Some Aldehydes

Compound	Structure	Molecular Weight	Boiling Point, °C	Melting Point, °C
formaldehyde	$HCHO$	30	−21	−92
acetaldehyde	CH_3CHO	44	21	−121
propionaldehyde	CH_3CH_2CHO	58	49	−81
butyraldehyde	$CH_3(CH_2)_2CHO$	72	76	−99
pentanal	$CH_3(CH_2)_3CHO$	86	103	−92
hexanal	$CH_3(CH_2)_4CHO$	100	128	−56
heptanal	$CH_3(CH_2)_5CHO$	114	153	−43
octanal	$CH_3(CH_2)_6CHO$	128	171	
nonanal	$CH_3(CH_2)_7CHO$	142	192	
decanal	$CH_3(CH_2)_8CHO$	156	209	−5

TABLE 14.2 Physical Properties of Some Ketones

Compound	Structure	Molecular Weight	Boiling Point, °C	Melting Point, °C	H_2O Solubility, wt. % (25°C)
acetone	CH_3COCH_3	58	56	−95	∞
2-butanone	$CH_3CH_2COCH_3$	72	80	−86	25.6
2-pentanone	$CH_3(CH_2)_2COCH_3$	86	102	−78	5.5
3-pentanone	$CH_3CH_2COCH_2CH_3$	86	102	−40	4.8
2-hexanone	$CH_3(CH_2)_3COCH_3$	100	128	−57	1.6
3-hexanone	$CH_3(CH_2)_2COCH_2CH_3$	100	125		1.5
2-heptanone	$CH_3(CH_2)_4COCH_3$	114	151	−36	0.4
2-octanone	$CH_3(CH_2)_5COCH_3$	128	173	−16	
2-nonanone	$CH_3(CH_2)_6COCH_3$	142	195	−7	
2-decanone	$CH_3(CH_2)_7COCH_3$	156	210	14	

14.4 Nuclear Magnetic Resonance Spectra

Nuclear magnetic resonance spectroscopy is an important technique for identifying aldehydes. The hydrogen attached to the carbonyl carbon gives rise to a characteristic band at very low field, usually around $\delta = 9.5$ ppm. In a magnetic field the circulating π-electrons produce an induced field that effectively deshields the aldehyde proton (Figure 14.2). That is, the induced field adds to the applied field in such a way that a smaller applied field is required to achieve resonance. The same phenomenon was discussed earlier with alkenes and accounts for the substantial downfield shift of vinyl protons (Section 13.10). In aldehydes the effect is greater and, in addition, the positive character of the carbonyl carbon produces a further downfield shift. The net result is a relatively large downfield resonance position for the aldehyde proton. Few other kinds of protons appear in this region; thus a peak at $\delta = 9.5$ is strongly indicative of the presence of a CHO function.

The same induced field that causes deshielding of a proton attached directly to a carbonyl group also produces significant deshielding of protons somewhat farther away at the α-carbon atoms. Typical chemical shifts for these protons are summarized in Table 14.3.

The spectra of acetaldehyde and 3-methyl-2-butanone shown in Figures 14.3 and 14.4 are characteristic. Note that the vicinal coupling constant in acetaldehyde is quite small, only 3 Hz.

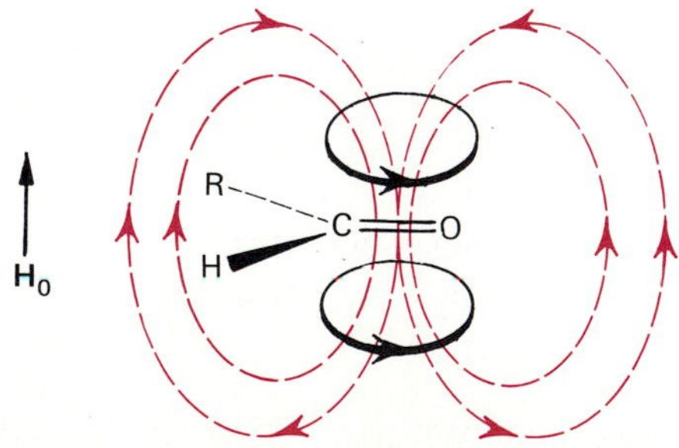

FIGURE 14.2 The diamagnetic anisotropy of the carbonyl group.

TABLE 14.3 Chemical Shifts for Aldehyde and Ketone Hydrogens

Hydrogen	Approximate Chemical Shift δ, ppm
$R-C(=O)-\mathbf{H}$	9.5
$R-C(=O)-CH_3$	2.0
$R-C(=O)-CH_2R$	2.2
$R-C(=O)-CHR_2$	2.4

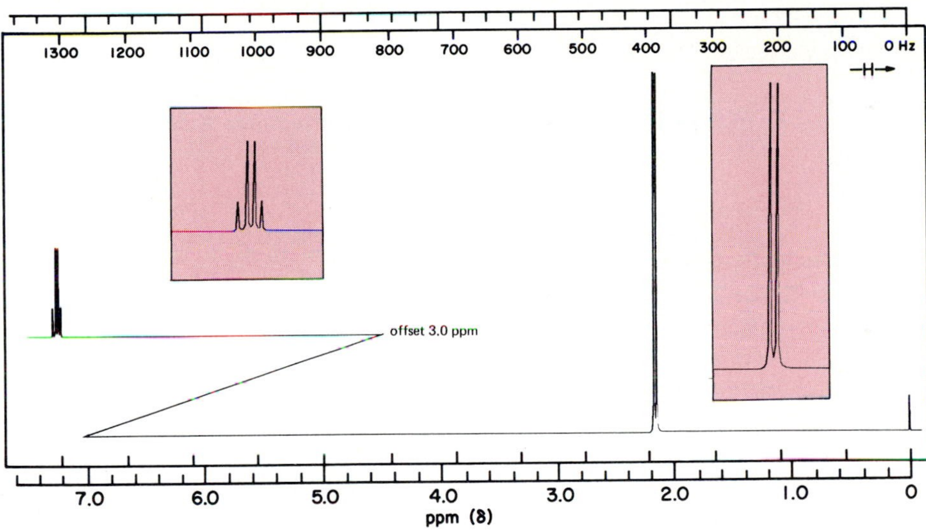

FIGURE 14.3 NMR spectrum of acetaldehyde, CH_3CHO.

The CMR chemical shifts for a few simple aldehydes and ketones are collected in Table 14.4. Note the extreme downfield shift of the carbonyl resonance—about 200 ppm from TMS. Carbonyl groups resonate further downfield than any other type of carbon. There are two main reasons for this effect. First, there is the fact that sp^2-hybridized carbons resonate downfield from sp^3-hybridized carbons. However, comparison of the shifts in Table 14.4 with those given for alkenes in Table 13.4 shows that aldehyde carbons resonate about 70 ppm downfield from alkene carbons of comparable substitution (compare propanal with 1-butene and butanal with 1-pentene). The reason for this additional shift is the polar nature of the carbonyl bond, which *deshields* the carbon nucleus even more than a double bond. Because the carbonyl carbon has a long relaxation time, carbonyl resonances are normally much weaker than the resonances of other carbons, as shown in the CMR spectrum of 2-octanone (Figure 14.5).

EXERCISE 14.3 Give the origin of the signal at $\delta = 76$ ppm in Figure 14.5.

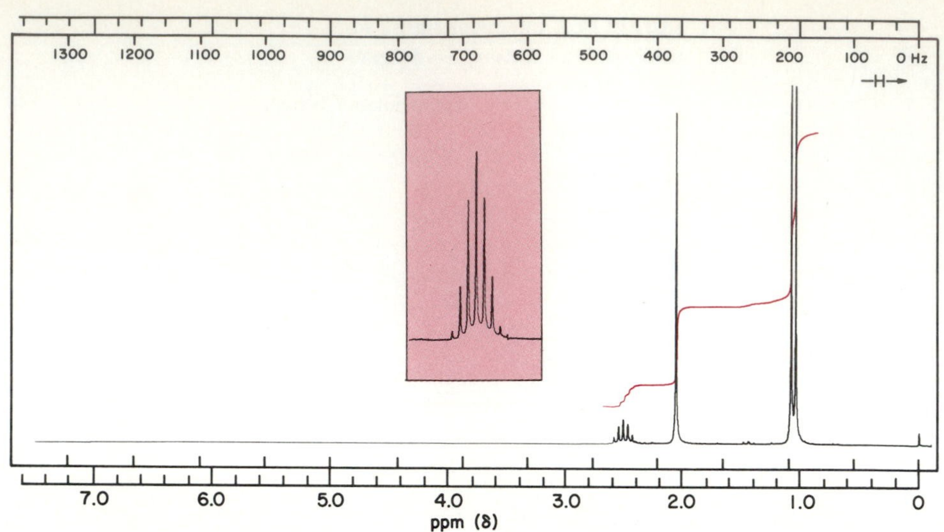

FIGURE 14.4 NMR spectrum of 3-methyl-2-butanone, $CH_3COCH(CH_3)_2$.

TABLE 14.4 CMR Chemical Shifts of Aldehydes and Ketones

	C-1	C-2	C-3	C-4	C-5
ethanal	199.6	31.2			
propanal	201.8	36.7	5.2		
butanal	201.6	45.7	15.7	13.3	
propanone	30.2	205.1	30.2		
butanone	28.8	206.3	36.4	7.6	
2-pentanone	29.3	206.6	45.2	17.5	13.5
3-pentanone	7.3	35.3	209.3	35.3	7.3

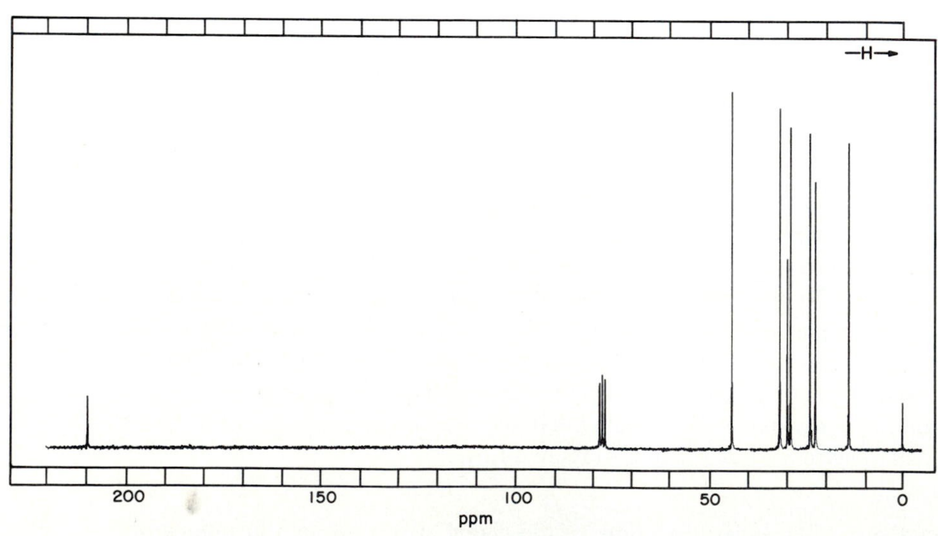

FIGURE 14.5 Proton-decoupled CMR spectrum of 2-octanone.

On page 339 we discussed the "relaxation" of excited spin states to lower energy spin states. The dominant relaxation mechanisms are different for different nuclei. For protons a major mechanism involves motion of the molecule relative to other molecules in the sample. Although this mechanism also contributes to relaxation of carbon nuclei, a more important one involves interaction with hydrogens that are *bonded directly to the nucleus undergoing relaxation*. As a consequence, carbons that have no attached hydrogens have long relaxation times. The relationship of relaxation times to the NMR spectrum is described in Section 13.6. This is true for the carbonyl carbons of ketones (but not of aldehydes) and for other carbons that are fully substituted by alkyl groups ("quaternary carbons").

EXERCISE 14.4 Write the structure for a compound having the formula $C_5H_{10}O$, that shows four signals in its CMR spectrum and has a triplet with $\delta = 9.7$ ppm and $J = 3$ Hz in its NMR spectrum.

14.5 Synthesis of Aldehydes and Ketones

The carbonyl group in aldehydes and ketones is one of the most important functional groups. In this section we shall review several reactions that are good methods for the synthesis of aldehydes and ketones.

We saw in Section 10.6.E that aldehydes and ketones may be obtained by the oxidation of primary and secondary alcohols, respectively.

$$\text{R—CH}_2\text{OH} \xrightarrow{[O]} \text{R—C(=O)—H}$$

$$\text{R—CH(OH)—R}' \xrightarrow{[O]} \text{R—C(=O)—R}'$$

In the latter case the product is not easily oxidized further, so there is no special problem in controlling the reaction to obtain the ketone in good yield. Although many oxidants have been used, the most commonly employed ones are chromium(VI) compounds.

$$\text{menthol} + Na_2Cr_2O_7 \xrightarrow[H_2SO_4]{H_2O} \text{menthone (84\%)}$$

A mixture of 120 g of $Na_2Cr_2O_7$, 100 g of conc. H_2SO_4, and 600 mL of water is prepared. To this solution is added 90 g of menthol (2-isopropyl-5-methylcyclohexanol). Heat is evolved, the temperature of the mixture rising to 55 °C. As soon as the reaction is complete, the oily product layer is removed by ether extraction and distilled to obtain 75 g (84%) of menthone.

In the case of primary alcohols, this simple picture is complicated by the fact that the product aldehyde may be further oxidized to a carboxylic acid. In most cases, the primary alcohol undergoes oxidation *more rapidly* than the corresponding aldehyde. However, in aqueous solution the product aldehyde forms a **hydrate,** which is oxidized even more rapidly than the primary alcohol. Aldehyde hydrates are discussed thoroughly in Section 14.7.A.

$$R{-}CH_2OH \xrightarrow[\text{fast}]{[O]} R{-}\overset{O}{\overset{\|}{C}}{-}H \quad \begin{matrix} \xrightarrow[\text{slow}]{[O]} R{-}\overset{O}{\overset{\|}{C}}{-}OH \\ \underset{H_2O}{\rightleftharpoons} R{-}\overset{OH}{\overset{|}{C}H}{-}OH \xrightarrow[\text{fast}]{[O]} R{-}\overset{O}{\overset{\|}{C}}{-}OH \end{matrix}$$

We mentioned previously (Section 10.6.E) that the preparation of aldehydes by oxidation of primary alcohols with aqueous chromic acid is limited to compounds of low molecular weight that can be removed from the reaction mixture by distillation as they are formed. However, in nonhydroxylic solvents, selective oxidation can be accomplished, and several oxidants that can be used in organic solvents have been developed for this purpose. One such oxidant is the complex formed between chromium trioxide, pyridine, and HCl (pyridinium chlorochromate, PCC, page 228). This material is soluble in chloroform and dichloromethane and can therefore be used to oxidize primary alcohols to aldehydes under anhydrous conditions.

$$n\text{-}C_9H_{19}CH_2OH + C_5H_5\overset{+}{N}H\ CrO_3Cl^- \xrightarrow[\substack{25^\circ C \\ 1\ hr}]{CH_2Cl_2} n\text{-}C_9H_{19}CHO$$

(92%)

Aldehydes and ketones may also be prepared by oxidative cleavage of carbon-carbon double bonds. A particularly useful reagent for this purpose is ozone (Section 11.6.E). Hydrolysis of the ozonide, usually under reductive conditions, results in the production of two carbonyl compounds.

$$\xrightarrow[-78^\circ C]{O_3,\ CH_2Cl_2} \xrightarrow[HOAc]{Zn} \text{—CHO} + CH_2O$$

(62%)
5-methylhexanal

In our study of the carbon-carbon triple bond (Section 12.6), we found that alkynes undergo hydration to yield unstable vinyl alcohols that immediately rearrange to the corresponding ketones. The hydration reaction is usually catalyzed by mercuric ion and sulfuric acid.

$$R{-}C{\equiv}C{-}R + H_2O \xrightarrow[H_2SO_4]{Hg^{2+}} \left[R{-}\overset{OH}{\overset{|}{C}}{=}CH{-}R \right] \rightleftharpoons R{-}\overset{O}{\overset{\|}{C}}{-}CH_2R$$

The reaction is generally useful as a preparative method only when the alkyne is terminal, in which case a methyl alkyl ketone is always formed, or in cases where the molecule is symmetrical.

$$\text{HO} \xrightarrow[H_2O,\ MeOH]{HgO,\ H_2SO_4} \text{HO}\ \overset{O}{\|}$$

(84%)

$$\xrightarrow[\substack{i\text{-PrOH, } H_2O \\ 100^\circ C}]{HgSO_4,\ H_2SO_4} \overset{O}{\|}$$

(80%)

Since the direct addition of water to a terminal alkyne always occurs in such a way that the hydroxy group becomes attached to the carbon bearing the alkyl group, the only alkyne that will yield an aldehyde upon hydration is acetylene itself. Indirect hydration of the triple bond, by the hydroboration route, yields the opposite result—terminal alkynes yield aldehydes (Section 12.6.D).

$$n\text{-}C_5H_{11}C{\equiv}CH \xrightarrow{B_2H_6} \xrightarrow[OH^-]{H_2O_2} n\text{-}C_5H_{11}CH_2CHO$$

EXERCISE 14.5 For each of the following carbonyl compounds, write equations showing how the compound could be prepared by oxidation of an alcohol, ozonization of an alkene, and (if possible) by hydration of an alkyne. Which of the preparations that you have written would give an unacceptable mixture of isomers?

(a) pentanal (b) 3-hexanone (c) 3-heptanone (d) 2-methylbutanal

14.6 Addition of Oxygen and Nitrogen Nucleophiles

A. Carbonyl Hydrates: *gem*-Diols

Aldehydes and ketones react with water to give an equilibrium concentration of the hydrate, a *gem*-diol.

$$\gt C{=}O + H_2O \overset{K}{\rightleftharpoons} \gt C(OH)_2$$

The reaction illustrates many of the important principles of reactions of carbonyl groups. The equilibrium constant for hydration, K, is sensitive to the nature of the carbonyl group. In aqueous solution, the equilibrium constant is

$$K = \frac{[C(OH)_2]}{[C{=}O][H_2O]}$$

The equilibrium constant has a value of about 18 for formaldehyde, roughly 0.01 for other aldehydes such as propionaldehyde, and about 10^{-5} for ketones. Thus a solution of formaldehyde in water is almost all $CH_2(OH)_2$. Aqueous solutions of other aldehydes contain comparable amounts of the hydrated and nonhydrated forms, and ketones are present almost wholly in their carbonyl forms.

$$CH_2{=}O + H_2O \overset{K}{\rightleftharpoons} HOCH_2OH \qquad K \cong 18$$

$$CH_3CH_2CH{=}O + H_2O \overset{K}{\rightleftharpoons} CH_3CH_2{-}CH(OH)_2 \qquad K \cong 10^{-2}$$

$$CH_3COCH_3 + H_2O \overset{K}{\rightleftharpoons} CH_3{-}C(OH)_2{-}CH_3 \qquad K \cong 4 \times 10^{-5}$$

EXERCISE 14.6 Using the approximate values given above for K, estimate the ratio of hydrated and nonhydrated forms of formaldehyde, acetaldehyde, and acetone in aqueous solutions nominally 1 M in carbonyl compound.

The relative tendency for various carbonyl compounds to undergo hydration can be explained with concepts that come from our knowledge of carbocation chemistry. Two important resonance structures can be written for a carbonyl group. The resonance forms shown in Chapter 2 (pages 13–15) should be reviewed.

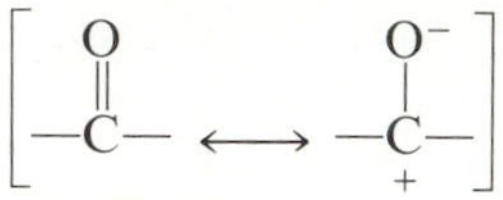

$$\left[-\overset{\overset{\displaystyle O}{\|}}{C}- \longleftrightarrow -\overset{\overset{\displaystyle O^-}{|}}{\underset{+}{C}}- \right]$$

The structure with the double bond is important because in it all atoms have complete octets. However, the other structure contributes significantly because of the electronegativity difference between C and O. This dipolar structure has the character of a carbocation. Recall that the order of carbocation stability is secondary > primary > methyl.

The dipolar resonance structure of formaldehyde is analogous to a methyl cation, that for propionaldehyde is analogous to a primary carbocation, and that for a ketone is analogous to a secondary carbocation.

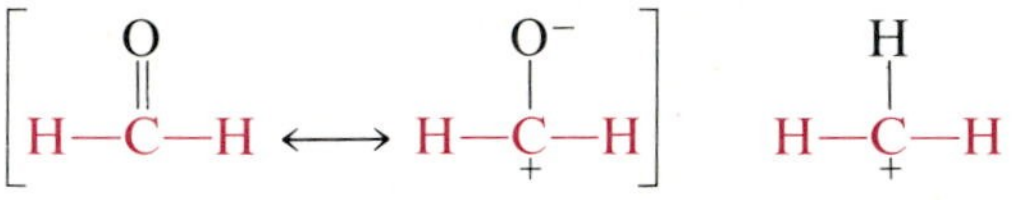

$$\left[H-\overset{\overset{\displaystyle O}{\|}}{C}-H \longleftrightarrow H-\overset{\overset{\displaystyle O^-}{|}}{\underset{+}{C}}-H \right] \qquad H-\overset{\overset{\displaystyle H}{|}}{\underset{+}{C}}-H$$

formaldehyde methyl cation

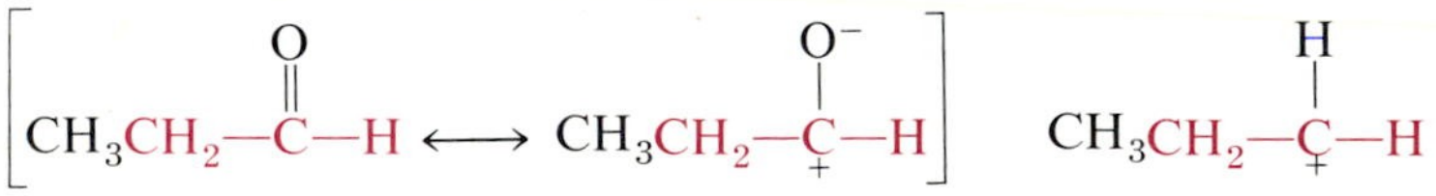

$$\left[CH_3CH_2-\overset{\overset{\displaystyle O}{\|}}{C}-H \longleftrightarrow CH_3CH_2-\overset{\overset{\displaystyle O^-}{|}}{\underset{+}{C}}-H \right] \qquad CH_3CH_2-\overset{\overset{\displaystyle H}{|}}{\underset{+}{C}}-H$$

propionaldehyde n-propyl cation

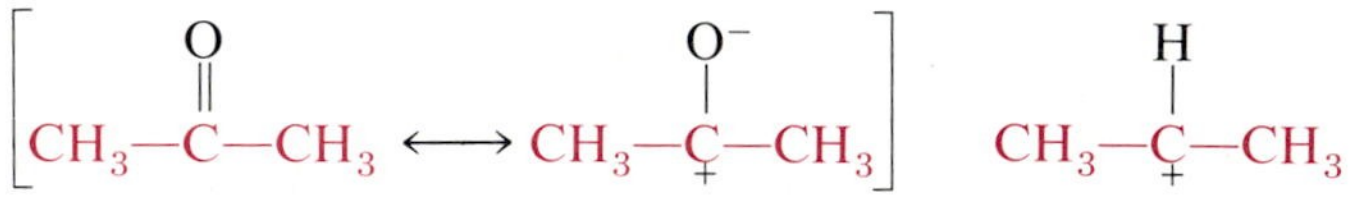

$$\left[CH_3-\overset{\overset{\displaystyle O}{\|}}{C}-CH_3 \longleftrightarrow CH_3-\overset{\overset{\displaystyle O^-}{|}}{\underset{+}{C}}-CH_3 \right] \qquad CH_3-\overset{\overset{\displaystyle H}{|}}{\underset{+}{C}}-CH_3$$

acetone isopropyl cation

Just as isopropyl cation is more stable than n-propyl cation, acetone is more stable than propionaldehyde owing to the extra stabilization imparted by the dipolar resonance structure.

We can see this effect by an examination of heats of formation of isomeric aldehydes and ketones. Some of these comparisons are summarized in Table 14.5. The data show that ketones are about 7 kcal mole^{-1} more stable than the isomeric aldehydes.

Consideration of the carbocation character of a carbonyl group has other corollaries as well. Consider the effect of a nearby polar substituent such as a chlorine, as in chloroacetaldehyde. The C—Cl dipole acts to destabilize the carbocation resonance structure. Hence, this structure contributes less to the overall resonance hybrid and the resonance hybrid is less stable as a result.

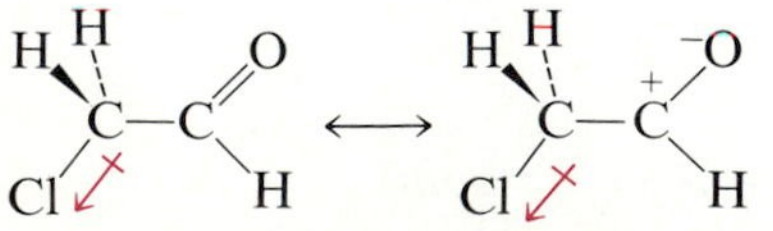

$$ClCH_2-CH{=}O \longleftrightarrow ClCH_2-\overset{+}{C}H-O^-$$

contributes less

TABLE 14.5 Heats of Formation of Some Aldehydes and Ketones

Aldehydes	ΔH_f°, kcal mole^{-1} 25°C, gas	Ketones	ΔH_f°, kcal mole^{-1} 25°C, gas
HCHO	−26.0		
CH_3CHO	−39.7		
CH_3CH_2CHO	−45.5	CH_3COCH_3	−51.9
$CH_3CH_2CH_2CHO$	−48.9	$CH_3CH_2COCH_3$	−57.0
$CH_3CH_2CH_2CH_2CHO$	−54.5	$CH_3CH_2CH_2COCH_3$	−61.8
		$CH_3CH_2COCH_2CH_3$	−61.8

No comparable effect operates on the corresponding hydrate. Thus, the chlorine substituent destabilizes the nonhydrated form, relative to the hydrated form, and as a result the aldehyde is more hydrated at equilibrium. For trichloroacetaldehyde, "chloral," the equilibrium constant is about 500. This compound exists almost wholly as the hydrate.

Chloral hydrate, $CCl_3CH(OH)_2$, is a crystalline solid, m.p. 57 °C, having a distinctive odor. The biologically active agent for its useful hypnotic effect is 2,2,2-trichloroethanol, produced by enzymatic reduction. The agent, CCl_3CH_2OH, decreases the rate of enzymatic oxidation of ethanol, and thereby enhances the lifetime and effect of ethanol in the body. The doubly depressant combination of ethanol and chloral hydrate has been illegally used as "knockout drops." A hypnotic drug is one that induces drowsiness. Chronic abuse of chloral hydrate is known.

The equilibrium between a carbonyl compound and its hydrate can also be described as the resultant of two rate constants.

$$CH_3\overset{O}{\overset{\|}{C}}CH_3 + H_2O \underset{k_{-1}}{\overset{k_1}{\rightleftharpoons}} CH_3\underset{OH}{\underset{|}{\overset{OH}{\overset{|}{C}}}}CH_3$$

The equilibrium constant is given by the ratio, $K = k_1/k_{-1}$. In the case of a ketone, such as acetone, the amount of hydrate present is so small that its rate of formation cannot be determined directly. The rate constant k_1 can be determined indirectly by an isotope exchange reaction. Water consists mostly of $H_2{}^{16}O$, but it also contains 0.20% of the heavy oxygen isotope, $H_2{}^{18}O$. Water enriched in the heavy isotope is available, and the rate of hydration can be followed by measuring the rate of incorporation of ^{18}O into acetone. The ^{18}O content of the ketone can be determined by mass spectrometry (Chapter 34).

$$CH_3\overset{^{16}O}{\overset{\|}{C}}CH_3 + H_2{}^{18}O \underset{k_{-1}}{\overset{k_1}{\rightleftharpoons}} CH_3\overset{^{18}O}{\overset{\|}{C}}CH_3 + H_2{}^{16}O$$

This exchange reaction is slow in pure water, but is much faster in the presence of small amounts of either acid or base. In the uncatalyzed reaction a molecule of water attacks the electron-deficient carbonyl carbon to produce an intermediate that undergoes rapid proton exchange to give the *gem*-diol.

$$H_2O \quad {}^{\delta+}\underset{CH_3}{\underset{|}{\overset{CH_3}{\overset{|}{C}}}}{=}O^{\delta-} \rightleftharpoons H_2\overset{+}{O}{-}\underset{CH_3}{\underset{|}{\overset{CH_3}{\overset{|}{C}}}}{-}O^- \rightleftharpoons HO{-}\underset{CH_3}{\underset{|}{\overset{CH_3}{\overset{|}{C}}}}{-}OH$$

The *gem*-diol decomposes to give back the ketone by an exact reversal of this sequence. If one of the oxygens in the diol is heavy oxygen, the dehydration process has an equal probability of expelling the labeled oxygen in the leaving water or of retaining it in the ketone. However, water is a rather weakly basic reagent, and the carbonyl carbon is only partially positive. Furthermore, the product of this addition has a charge-separated structure, as shown in the foregoing equation. Consequently, the direct attack by water on the carbonyl carbon is a slow process.

Hydroxide ion is a much more basic reagent and its reaction with a carbonyl group is much faster than that of water.

$$HO^- + (CH_3)_2C{=}O \rightleftharpoons HO{-}C(CH_3)_2{-}O^- \underset{HO^-}{\overset{H_2O}{\rightleftharpoons}} HO{-}C(CH_3)_2{-}OH$$

Note that the presence of hydroxide ion does not affect the *position* of equilibrium in the hydration reaction. Hydroxide ion catalyzes the reverse reaction exactly as much as the forward reaction.

In the acid-catalyzed reaction the ketone oxygen is first protonated in a rapid equilibrium process.

$$(CH_3)_2C{=}O + H^+ \rightleftharpoons \left[(CH_3)_2C{=}OH^+ \longleftrightarrow (CH_3)_2\overset{+}{C}{-}OH\right]$$

In the protonated compound the carbonyl carbon has more positive charge than in the neutral ketone. One resonance structure is that of a hydroxycarbocation.

> A carbonyl group has dipolar character because oxygen is more electronegative than carbon and has greater attraction for electrons than carbon. The π-bond of the carbonyl group is relatively polarizable, and the electron density in this bond is displaced toward oxygen. In a protonated carbonyl group the oxonium ion oxygen is even more electronegative, and the electron density is displaced even more toward oxygen, leaving an effectively more positive carbon.

The hydroxycarbocation reacts rapidly with water to give a protonated form of the ketone hydrate.

$$H_2O + \left[(CH_3)_2\overset{+}{C}{-}OH \longleftrightarrow (CH_3)_2C{=}\overset{+}{O}H\right] \rightleftharpoons H_2\overset{+}{O}{-}C(CH_3)_2{-}OH \rightleftharpoons HO{-}C(CH_3)_2{-}OH + H^+$$

Note that this reaction is closely analogous to the S_N1 reaction involving carbocations (Section 9.8).

$$H_2O + (CH_3)_2\overset{+}{C}{-}CH_3 \rightleftharpoons H_2\overset{+}{O}{-}C(CH_3)_2{-}CH_3 \rightleftharpoons HO{-}C(CH_3)_2{-}CH_3 + H^+$$

Furthermore, the reverse reaction, dehydration of the ketone hydrate, is analogous to unimolecular elimination of an alcohol (E1 reaction).

The two-step scheme (Section 10.6.E) for chromic acid oxidation indicates (1) that the alcohol forms an ester of chromic acid, H_2CrO_4 called an alkyl chromate and (2) that a bimolecular elimination forms the carbonyl derivative and a reduced chromium species. The scheme makes clear why the aldehyde hydrate (the *gem*-diol described in Section 14.6.A) is so easily converted into the carboxylic acid, by way of chromate ester formation and E2 elimination.

$$CH_3CH{=}O \underset{}{\overset{H^+}{\rightleftharpoons}} CH_3CH{=}OH^+ \overset{H_2O}{\rightleftharpoons} CH_3CH(OH)(\overset{+}{O}H_2)$$

$$CH_3CH(OH)(\overset{+}{O}H_2) \longrightarrow CH_3CH(OH)_2 \xrightarrow{CrO_3} CH_3CH(OH)(O{-}CrO_3H)$$

$$H_2O + CH_3CH(OH)(O{-}CrO_3H) \longrightarrow CH_3C(=O){-}OH + H_3O^+ + CrO_3H^-$$

EXERCISE 14.7 The NMR spectrum of a solution of chloroacetaldehyde in tetradeuteriomethanol, CD_3OD, shows no signals in the region $\delta = 9$–10 ppm. Explain.

EXERCISE 14.8 Write out the steps for the oxidation of the following aldehydes with chromium trioxide.

(a) formaldehyde
(b) acetaldehyde
(c) dichloroacetaldehyde
(d) trifluoroacetaldehyde

Highlight 14.1

The acid-catalyzed reaction of acetone with water has a close analogy to the S_N1 reaction of *t*-butyl bromide with ethanol.

$$(CH_3)_2C{=}O \overset{H^+}{\rightleftharpoons} \left[(CH_3)_2C{=}OH^+ \longleftrightarrow (CH_3)_2\overset{+}{C}{-}OH\right] \overset{H_2O}{\rightleftharpoons} (CH_3)_2C(OH)(\overset{+}{O}H_2) \xrightarrow{-H^+} (CH_3)_2C(OH)_2$$

$$(CH_3)_3C{-}Br \xrightarrow{-Br^-} (CH_3)_2\overset{+}{C}{-}CH_3 \xrightarrow{C_2H_5OH} (CH_3)_3C{-}\overset{+}{O}(H){-}C_2H_5 \longrightarrow (CH_3)_3C{-}O{-}C_2H_5$$

B. Acetals

The equilibrium between carbonyl compounds and water to form *gem*-diols occurs in certain synthetic reactions, in which the generally unstable diols may undergo elimination of water by the E1 or E2 mechanism. The analogous reaction of aldehydes and ketones with alcohols has considerable utility in forming isolable products. The addition of 1 mole of an alcohol to the carbonyl group of an aldehyde or ketone yields a **hemiacetal.**

$$CH_3CH_2CHO + CH_3OH \rightleftharpoons \underset{\text{a hemiacetal}}{CH_3CH_2\overset{\overset{\displaystyle OH}{|}}{C}H{-}OCH_3}$$

$$CH_3\overset{\overset{\displaystyle O}{\|}}{C}CH_3 + CH_3OH \rightleftharpoons \underset{\text{a hemiacetal}}{CH_3\overset{\overset{\displaystyle OH}{|}}{\underset{\underset{\displaystyle OCH_3}{|}}{C}}CH_3}$$

Addition of 2 moles of an alcohol, with the consequent formation of 1 mole of water, yields an **acetal.**

$$CH_3CH_2CHO + 2\ CH_3OH \rightleftharpoons \underset{\text{an acetal}}{CH_3CH_2\overset{\overset{\displaystyle OCH_3}{|}}{\underset{\underset{\displaystyle OCH_3}{|}}{C}}H} + H_2O$$

$$CH_3\overset{\overset{\displaystyle O}{\|}}{C}CH_3 + 2\ CH_3OH \rightleftharpoons \underset{\text{an acetal}}{CH_3\overset{\overset{\displaystyle OCH_3}{|}}{\underset{\underset{\displaystyle OCH_3}{|}}{C}}CH_3} + H_2O$$

> The 1,1- or *n,n*-dialkoxy compounds derived from either aldehydes or ketones are both called acetals. Those derived from ketones were formerly called ketals. The monoalkoxy-monohydroxy derivatives are hemiacetals. Those derived from ketones were formerly hemiketals.

Formation of the hemiacetal is directly analogous to addition of water and is also subject to both acid and base catalysis. As with hydration, aldehydes give more of the addition product at equilibrium than do ketones.

Acid-catalyzed hemiacetal formation

$$(1)\ CH_3\overset{\overset{\displaystyle O}{\|}}{C}H + H^+ \rightleftharpoons CH_3\overset{\overset{\displaystyle {}^+OH}{\|}}{C}H$$

$$(2)\ CH_3\overset{\overset{\displaystyle {}^+OH}{\|}}{C}H + CH_3OH \rightleftharpoons CH_3\overset{\overset{\displaystyle OH}{|}}{C}H\underset{\underset{\displaystyle H}{|}}{\overset{+}{O}}CH_3$$

$$(3)\ CH_3\overset{\overset{\displaystyle OH}{|}}{C}H\underset{\underset{\displaystyle H}{|}}{\overset{+}{O}}CH_3 \rightleftharpoons CH_3\overset{\overset{\displaystyle OH}{|}}{C}HOCH_3 + H^+$$

$$\text{(1)}\ CH_3\overset{\overset{O}{\|}}{C}H + CH_3O^- \rightleftharpoons CH_3\overset{\overset{O^-}{|}}{C}HOCH_3$$

$$\text{(2)}\ CH_3\overset{\overset{O^-}{|}}{C}HOCH_3 + CH_3OH \rightleftharpoons CH_3\overset{\overset{OH}{|}}{C}HOCH_3 + CH_3O^-$$

Like *gem*-diols, simple hemiacetals are generally not sufficiently stable for isolation.

Acetals are formed by way of the intermediate hemiacetal. *However, replacement of the OH group by OR is only brought about by acid catalysis.*

Acid-catalyzed acetal formation

$$\text{(1)}\ CH_3\overset{\overset{OH}{|}}{C}HOCH_3 + H^+ \rightleftharpoons CH_3\overset{\overset{^+OH_2}{|}}{C}HOCH_3$$

$$\text{(2)}\ CH_3\overset{\overset{^+OH_2}{|}}{C}HOCH_3 \rightleftharpoons \left[CH_3\underset{\underset{H}{|}}{\overset{+}{C}}{-}OCH_3 \longleftrightarrow CH_3\underset{\underset{H}{|}}{C}{=}\overset{+}{O}CH_3\right] + H_2O$$

$$\text{(3)}\ \left[CH_3\underset{\underset{H}{|}}{\overset{+}{C}}{-}OCH_3 \longleftrightarrow CH_3\underset{\underset{H}{|}}{C}{=}\overset{+}{O}CH_3\right] + CH_3OH \rightleftharpoons CH_3\underset{\underset{H}{|}}{\overset{\overset{H\overset{+}{O}CH_3}{|}}{C}}{-}OCH_3$$

$$\text{(4)}\ CH_3\underset{\underset{H}{|}}{\overset{\overset{H\overset{+}{O}CH_3}{|}}{C}}{-}OCH_3 \rightleftharpoons CH_3\underset{\underset{H}{|}}{\overset{\overset{OCH_3}{|}}{C}}{-}OCH_3 + H^+$$

The net equilibrium that occurs when an aldehyde or ketone is treated with an alcohol and an acid catalyst is as follows.

$$CH_3\overset{\overset{O}{\|}}{C}CH_3 + 2\,CH_3OH \overset{H^+}{\rightleftharpoons} CH_3\underset{\underset{OCH_3}{|}}{\overset{\overset{OCH_3}{|}}{C}}CH_3 + H_2O$$

For simple aldehydes the overall equilibrium constant is favorable, and the acetal may be prepared simply by treating the aldehyde with two equivalents of alcohol and an acid catalyst.

$$CH_3CHO + 2\,CH_3CH_2OH \overset{H^+}{\rightleftharpoons} CH_3CH(OCH_2CH_3)_2 + H_2O$$

A mixture of 1305 mL of ethanol (21.7 moles), 500 g of acetaldehyde (11.4 moles) and 200 g of anhydrous $CaCl_2$ is placed in a 4-L bottle and kept at 25 °C for 1–2 days. At the end of this time the upper layer is washed with water and distilled to yield 790–815 g of 1,1-diethoxyethane, b.p. 101–103 °C. Note that $CaCl_2$ serves as a catalyst by hydrolyzing to give a small amount of HCl as well as displacing the equilibrium by forming the hydrate.

With larger aldehydes and with ketones, the equilibrium constant for acetal formation is generally unfavorable, more so for ketones than for aldehydes. For this reason the reaction is usually carried out with the alcohol as solvent in order to drive the equilibrium to the right. With aldehydes this usually allows the acetal to be produced in good yield.

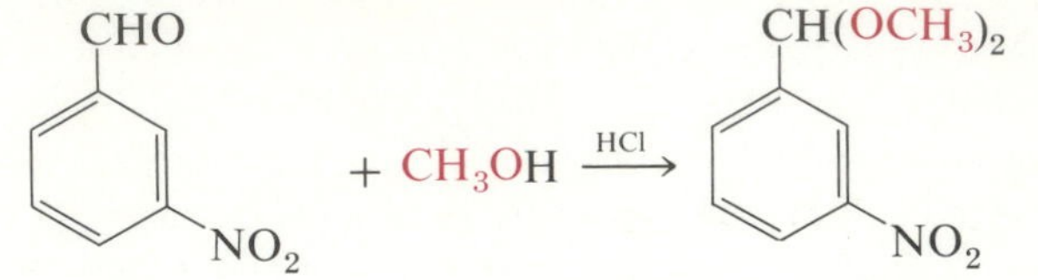

CHO, NO_2 + CH_3OH $\xrightarrow{HCl}$ $CH(OCH_3)_2$, NO_2

3-nitrobenzaldehyde

(76–85%)
3-nitrobenzaldehyde dimethyl acetal

For ketones the equilibrium lies even further to the left, and special techniques are used to remove water as it is formed and thus drive the equilibrium to the right.

The equilibria for acetal formation provide an illustration of the role of entropy in equilibria. In the formation of an acetal, three reactant molecules combine to form two product molecules. The resulting loss of the freedom of motion of one molecule corresponds to a negative entropy change. As in the formation of *gem*-diols from aldehydes (page 389), the formation of acetals from aldehydes is so exothermic that the equilibrium lies far to the right despite the unfavorable entropy change. However, in the case of ketones, the equilibrium constant for acetal formation is unfavorable, both as a result of ΔH and ΔS. This unfavorable entropy effect is avoided by the use of a 1,2- or 1,3-diol to form a cyclic acetal.

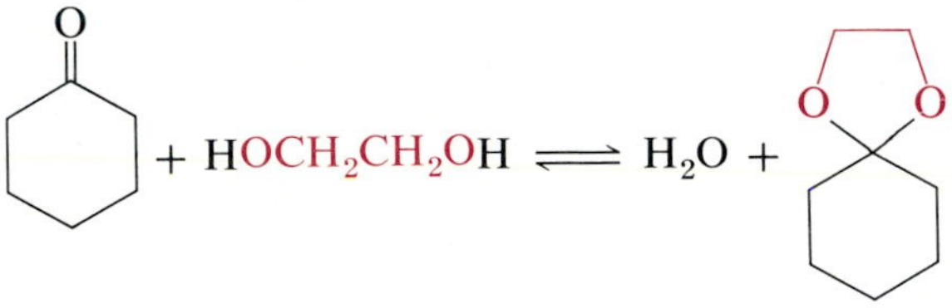

O + $HOCH_2CH_2OH$ $\rightleftharpoons$ H_2O + O O

(80%)

Note in this case that two reactant molecules produce two product molecules. The overall entropy change is approximately zero, resulting in less unfavorable equilibria for ketones. However, for most cases, $\Delta G°$ is still positive, and it is necessary to drive the reaction to completion by removing the water produced.

Acetals are generally stable to basic conditions and are hydrolyzed back to carbonyl compounds in acidic solution. They play an important role in carbohydrate chemistry (Chapter 28). They are also used to **protect** a carbonyl group during the execution of a synthetic scheme. The following synthesis of 4-heptynal from 3-bromopropanal illustrates the use of an acetal as a **protecting group.**

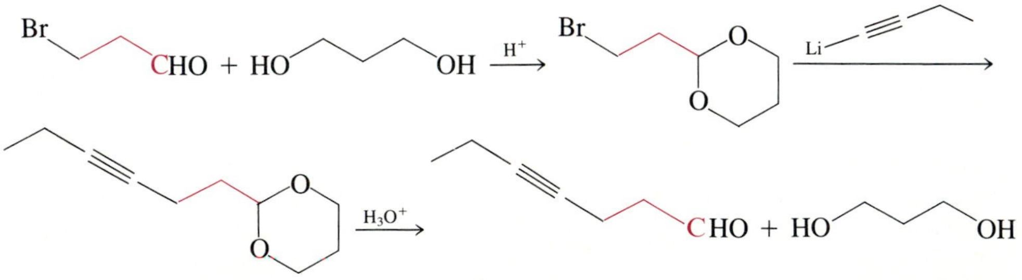

Br CHO + HO OH $\xrightarrow{H^+}$ Br O O $\xrightarrow{Li}$

O O $\xrightarrow{H_3O^+}$ CHO + HO OH

In this case it is desired to replace Br by the 1-butynyl group. As we saw in Section 12.5.B, alkynyllithium compounds undergo ready alkylation by primary alkyl bromides. However, we shall see in Section 14.8 that organolithium reagents also react readily with the carbonyl group. Thus, direct replacement of bromine in the bromoaldehyde is not possible. Therefore the aldehyde is temporarily "protected" by conversion to the acetal, which is an ether and does not react with the organolithium reagent. After the displacement reaction has been carried out, the aldehyde functional group is regenerated by treatment of the acetal with aqueous acid. Excess water in the regeneration step shifts the equilibrium back to the side of the aldehyde and the diol.

A group of compounds that are related to acetals are the **enol ethers,** which are produced by the nucleophilic addition of alcohols to alkynes (Section 12.6.C). Enols (**ene** + alcoh**ol**) are hydroxyalkenes; these are important intermediates that we will discuss further. Enol ethers are alkoxyalkenes.

$$HC{\equiv}CH + C_2H_5OH \xrightarrow{C_2H_5O^-} \underset{\text{ethyl vinyl ether}}{C_2H_5OCH{=}CH_2}$$

Like other ethers, enol ethers are stable to basic conditions and to basic reagents such as organolithium reagents. However, under acidic conditions they undergo rapid hydrolysis to give the aldehyde or ketone and alcohol.

$$C_2H_5OCH{=}CH_2 + H_2O \xrightarrow{H^+} C_2H_5OH + CH_3CHO$$

The mechanism of this ready hydrolysis starts with the addition of a proton to the carbon-carbon double bond. The resulting cation is completely analogous in structure to a protonated aldehyde or ketone. From this point on, the hydrolysis mechanism involves the same type of intermediates as are involved in the formation and hydrolysis of acetals.

$$(1)\ C_2H_5O{-}CH{=}CH_2 + H^+ \rightleftharpoons [C_2H_5O{-}\overset{+}{C}H{-}CH_3 \longleftrightarrow C_2H_5\overset{+}{O}{=}CH{-}CH_3]$$

$$(2)\ [C_2H_5O{-}\overset{+}{C}H{-}CH_3 \longleftrightarrow C_2H_5\overset{+}{O}{=}CH{-}CH_3] + H_2O \rightleftharpoons C_2H_5O{-}\overset{\overset{+}{O}H_2}{\overset{|}{C}H}{-}CH_3$$

$$(3)\ C_2H_5O{-}\overset{\overset{+}{O}H_2}{\overset{|}{C}H}{-}CH_3 \rightleftharpoons C_2H_5O{-}\overset{OH}{\overset{|}{C}H}{-}CH_3 + H^+$$

$$(4)\ C_2H_5O{-}\overset{OH}{\overset{|}{C}H}{-}CH_3 + H^+ \rightleftharpoons C_2H_5\underset{H}{\underset{|}{\overset{+}{O}}}{-}\overset{OH}{\overset{|}{C}H}CH_3$$

$$(5)\ C_2H_5\underset{H}{\underset{|}{\overset{+}{O}}}{-}\overset{OH}{\overset{|}{C}H}CH_3 \rightleftharpoons C_2H_5OH + [HO{-}\overset{+}{C}HCH_3 \longleftrightarrow H\overset{+}{O}{=}CHCH_3]$$

$$(6)\ [HO{-}\overset{+}{C}HCH_3 \longleftrightarrow H\overset{+}{O}{=}CHCH_3] \rightleftharpoons H\overset{O}{\overset{\|}{C}}CH_3 + H^+$$

EXERCISE 14.9 Write the steps in the mechanism for acid-catalyzed hydration of propene (Section 11.6.C) and compare with the steps in the foregoing mechanism for hydrolysis of ethyl vinyl ether. Note that there is a parallel between the steps of the alkene hydration mechanism and the first three steps of the mechanism for enol ether hydrolysis. Can you offer an explanation for the observation that enol ether hydrolysis is *much* more facile than is alkene hydration?

EXERCISE 14.10 Propanal reacts with 2-methyl-1,3-propanediol in the presence of acid to give two isomeric acetals. Write the structures of these two isomers and indicate which is expected to be the major product.

Acetals are ethers and will form dangerous peroxides on exposure to air. Appropriate precaution should be taken when heating acetals that have had long exposure to oxygen.

The tendency of aldehydes to form acetals shows up in the formation of cyclic or polymeric acetals that are derived from three to many molecules of aldehyde. Formaldehyde itself is a gas that is available commercially as a 37% aqueous solution called formalin, or as a solid polymer, paraformaldehyde. For use in syntheses, formaldehyde is normally obtained by heating the dry polymer.

The linear polymer, HO—(CH_2—O)$_n$—H, forms the basis of several commercial plastics such as Delrin and Celcon. In these cases the terminal OH groups are ''capped'' with alkyl or ester groups to prevent depolymerization or ''unzipping'' upon heating.

Formaldehyde also forms a cyclic trimer, trioxane, a solid having m.p. 64 °C, which can be sublimed unchanged. Paraldehyde, a cyclic trimer derived from acetaldehyde, is a liquid, b.p. 128 °C, that regenerates acetaldehyde on heating with a trace of acid.

trioxane paraldehyde

Acetaldehyde also forms a cyclic tetramer, metaldehyde, a solid that sublimes readily. Other low molecular weight aldehydes form cyclic trimers related to paraldehyde. This kind of behavior is not shown by ketones.

EXERCISE 14.11 Careful purification of paraldehyde yields two isomeric substances. The CMR spectrum of one isomer has only two signals, whereas that of the other isomer shows four signals. What are the structures of these two isomers?

Highlight 14.2

The addition of ROH to the C=O group in aldehydes and ketones yields **hemiacetals,** >C(OH)OR, in a base- or acid-catalyzed reaction. Only acid can form the stabilized carbocation, $>\overset{+}{C}OR$, which is able to react with another ROH to yield the **acetal,** >C(OR)$_2$. If the alcohol is a 1,2- or 1,3-diol, a **cyclic acetal** is formed.

C. Imines and Related Compounds

Ammonia reacts with aldehydes and ketones to form compounds called **imines.** These compounds contain the functional group C=N, which may be considered to be the nitrogen analog of a carbonyl group.

$$-\overset{O}{\overset{\|}{C}}- + NH_3 \rightleftharpoons \left[-\overset{OH}{\underset{NH_2}{\overset{|}{\underset{|}{C}}}}- \right] \rightleftharpoons H_2O + \underset{\text{an imine}}{-\overset{NH}{\overset{\|}{C}}-}$$

Imines derived from ammonia are not important class of compounds. They hydrolyze rapidly in water to generate carbonyl compounds.

However, the substituted imines that are produced from the reactions of aldehydes and ketones with primary amines (amines having only one alkyl group attached to nitrogen) are much more stable. These compounds are sometimes referred to as **Schiff bases.**

$$RCHO + R'NH_2 \rightleftharpoons RCH{=}NR' + H_2O$$

substituted imine, a Schiff base

> Imines and immonium ions play an important role in many biochemical reactions and physiological processes. Imine intermediates are formed from carbonyl compounds and the amino group of enzymes (Chapter 24). Ready formation and hydrolysis via the protonated form makes them especially apt for their niche in biological evolution. An imine derivative of the aldehyde retinal is the light-absorbing group in vision (Section 22.6.H).

Even though N-substituted imines are more stable than their N—H relatives, they are still fairly reactive compounds. They readily undergo hydrolysis back to the amine and carbonyl compound and are often prone to polymerization. However, if either the carbon or the nitrogen is substituted by a phenyl group, the resulting imine is generally rather stable.

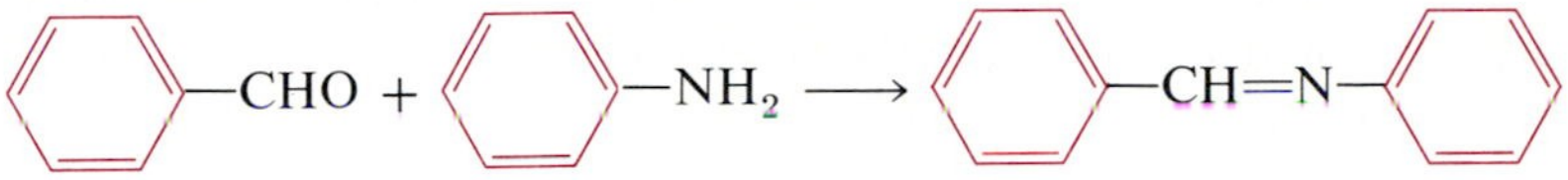

$$C_6H_5\text{—}CHO + C_6H_5\text{—}NH_2 \longrightarrow C_6H_5\text{—}CH{=}N\text{—}C_6H_5$$

(84–87%)

Imines prepared from aliphatic aldehydes and ketones and aliphatic amines are more reactive than phenyl analogs and are somewhat more difficult to prepare. Since imine formation in this case is not as favorable as when there is a phenyl group attached to the carbon-nitrogen double bond, it is usually necessary to drive the reaction to completion by removal of water from the reaction mixture as it is formed, as in the formation of acetals from ketones. An example of such a case is the condensation of cyclohexanone with *t*-butylamine.

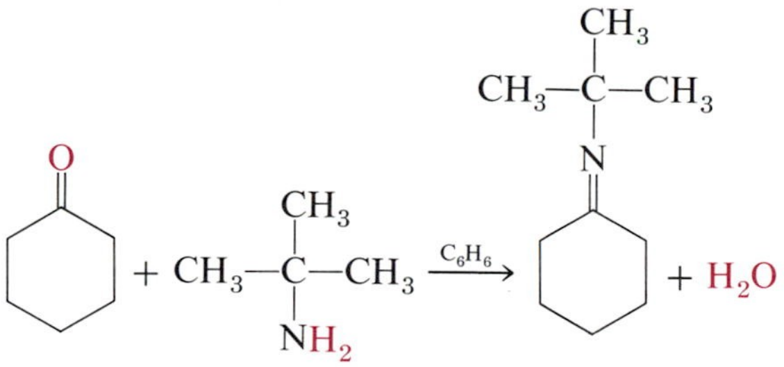

$$\text{cyclohexanone} + CH_3\text{—}\underset{NH_2}{\overset{CH_3}{C}}\text{—}CH_3 \xrightarrow{C_6H_6} \text{cyclohexylidene}{=}N\text{—}C(CH_3)_3 + H_2O$$

t-butylamine (85%)

Aldehydes and ketones also react with ammonia derivatives to give analogous adducts. Common reagents of this class are hydroxylamine (H_2NOH), hydrazine (H_2NNH_2), and phenylhydrazine ($H_2NNHC_6H_5$). Examples of such reactions follow. Unlike imines, the products of these reactions are generally quite stable.

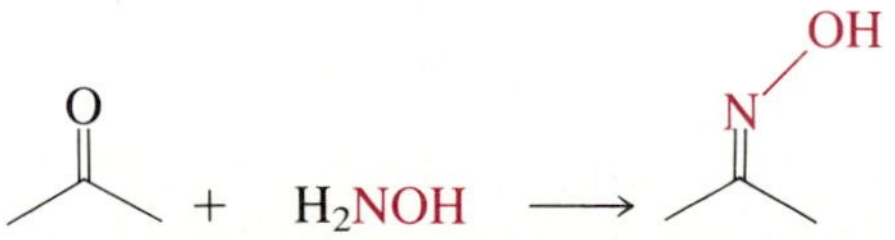

$$(CH_3)_2C{=}O + H_2NOH \longrightarrow (CH_3)_2C{=}N\text{—}OH$$

hydroxylamine an oxime

cyclopentanone + H_2NNH_2 ⟶ cyclopentanone hydrazone

hydrazine — a hydrazone

$CH_3CH_2CH_2CHO$ + $C_6H_5NHNH_2$ ⟶ $CH_3CH_2CH_2CH{=}N{-}NH{-}C_6H_5$

phenylhydrazine — a phenylhydrazone

EXERCISE 14.12 The reaction of acetaldehyde with hydroxylamine gives a product that shows *two* CH_3 doublets in its NMR spectrum. Explain.

The reactions of carbonyl compounds with substituted ammonia compounds are generally catalyzed by mild acid. The mechanism is directly analogous to the reactions discussed previously with water and alcohols.

$$\text{(1)}\quad \gt C{=}O + R{-}\ddot{N}H_2 \rightleftharpoons {-}\underset{|}{\overset{^{-}O}{\overset{|}{C}}}{-}\overset{+}{N}H_2R$$

$$\text{(2)}\quad {-}\underset{|}{\overset{^{-}O}{\overset{|}{C}}}{-}\overset{+}{N}H_2R \rightleftharpoons {-}\underset{|}{\overset{OH}{\overset{|}{C}}}{-}NHR$$

$$\text{(3)}\quad {-}\underset{|}{\overset{OH}{\overset{|}{C}}}{-}NHR + H^+ \rightleftharpoons {-}\underset{|}{\overset{^{+}OH_2}{\overset{|}{C}}}{-}NHR$$

$$\text{(4)}\quad {-}\underset{|}{\overset{^{+}OH_2}{\overset{|}{C}}}{-}NHR \rightleftharpoons \gt C{=}\overset{+}{N}HR + H_2O$$

$$\text{(5)}\quad \gt C{=}\overset{+}{N}HR \rightleftharpoons \gt C{=}NR + H^+$$

The first step is a nucleophilic addition to the carbonyl group. Rapid proton transfer gives the product of net addition of RNH_2 to C=O, a **hemiaminal,** also sometimes called a carbinolamine. This substance is generally so reactive that it cannot normally be isolated. A second acid-catalyzed reaction occurs in which water is eliminated from the hemiaminal. The resulting product is the imine, oxime, or hydrazone, and so on.

In the foregoing mechanism, steps (1)–(3) are rapid equilibria. The rate-limiting step is generally step (4), the elimination of water from the protonated hemiaminal. The overall reaction obeys the following rate law.

$$\text{rate} = k[\text{ketone}][H^+][RNH_2]$$

Highlight 14.3

Nitrogen nucleophiles, XNH_2, add to C=O bonds, usually under mild acid catalysis. The intermediate **hemiaminal** cannot usually be isolated but undergoes acid-catalyzed ionization to a cationic intermediate that loses a proton to give a stable **imine** derivative.

$$\text{HNHX} + \text{>C=O} \xrightarrow{H^+} \text{>C(OH)(NHX)} \xrightarrow{H^+} \text{>C=N}^+\text{(H)X} \xrightarrow{-H^+} \text{>C=NX}$$

hemiaminal — imine derivative

If X = OH, the imine derivative is called an **oxime;** if it is NH_2, the derivative is called a **hydrazone;** and if it is NHC_6H_5, the derivative is called a **phenylhydrazone.**

EXERCISE 14.13 Assuming that steps (1)–(3) in the mechanism on page 400 are rapid equilibria, characterized by equilibrium constants K_1, K_2, and K_3, and that step (4) is the rate-limiting step, characterized by rate constant k_1, derive the expected rate law. What rate law would be observed if the first step was rate limiting?

Although the reaction is catalyzed by acid at moderate pH, at higher acid concentration the rate actually diminishes with increasing acid concentration because the nitrogen base is itself protonated by acid. Therefore the concentration of free nucleophile is inversely related to the acid concentration. In solutions having high acid concentrations (low pH) the concentration of unprotonated nitrogen base is so low that step (1) becomes rate limiting. At moderate acid concentrations enough free nitrogen base is available that step (1) is a rapid equilibrium, yet enough acid is also available to catalyze step (4). For this reason the reaction is often carried out in the presence of a buffer such as sodium acetate. In some cases, particularly in the formation of simple imines, the reaction proceeds satisfactorily without acid catalysis.

14.7 Addition of Carbon Nucleophiles

A. Addition of Organometallic Reagents: Synthesis of Alcohols

One of the most useful techniques in organic chemistry for building up more complex molecules from simple ones involves the reaction of Grignard reagents and organolithium compounds with carbonyl compounds. We have seen previously (Section 8.6) that the carbon-metal bonds in these highly reactive organometallic compounds are polarized in the sense $C^- M^+$. The negative carbon (carbanion) of Grignard reagents reacts readily and rapidly with the positive carbon of the carbonyl group of aldehydes and ketones, the final products being alcohols. Some examples illustrating the synthesis of primary, secondary and tertiary alcohols, respectively, show the scope of this important reaction.

$$\text{CH}_3\text{CH}_2\text{Br} \xrightarrow[\text{ether}]{\text{Mg}} \xrightarrow{\text{(CH}_3)_2\text{CHCH}_2\text{COCH}_3} \xrightarrow{H_3O^+} \text{(CH}_3)_2\text{CHCH}_2\text{C(OH)(CH}_3)\text{CH}_2\text{CH}_3$$

3,5-dimethyl-3-hexanol

$$\text{C}_6\text{H}_{11}\text{Cl} \xrightarrow[\text{ether}]{\text{Mg}} \xrightarrow{\text{CH}_2\text{O}} \xrightarrow{\text{H}_3\text{O}^+} \text{C}_6\text{H}_{11}\text{CH}_2\text{OH}$$

(64–69%)
cyclohexylmethanol

> The Grignard reagent is prepared from 26.7 g of magnesium turnings and 118.5 g of cyclohexyl chloride in 450 mL of dry ether. In a separate flask 50 g of dry paraformaldehyde (page 402) is heated to 180–200 °C, and the formaldehyde formed by depolymerization is carried by a stream of nitrogen gas into the solution of the Grignard reagent. When the reaction is complete, ice and dilute sulfuric acid are added, and the mixture is steam distilled. The distillate is extracted with ether and distilled at reduced pressure to yield 72.5–78.5 g of cyclohexylmethanol.

$$(\text{CH}_3)_2\text{CHBr} \xrightarrow[\text{ether}]{\text{Mg}} \xrightarrow{\text{CH}_3\text{CHO}} \xrightarrow{\text{H}_3\text{O}^+} (\text{CH}_3)_2\text{CHCH(OH)CH}_3$$

(53–54%)
3-methyl-2-butanol

> A solution of 600 g of isopropyl bromide in ether is slowly added to a mixture of 146 g of dry magnesium turnings in ether. The Grignard solution is then cooled to −5 °C, and a solution of 200 g of acetaldehyde in ether is added. Ice and dilute sulfuric acid are added, and the mixture is extracted with ether. The dried extract is distilled to give 210–215 g of 3-methyl-2-butanol, b.p. 110–111.5 °C.

These examples show that the reaction of organomagnesium halides with carbonyl compounds is useful for preparation of primary, secondary, and tertiary alcohols. Reaction of a Grignard reagent with formaldehyde gives a primary alcohol, other aldehydes yield secondary alcohols, and ketones lead to tertiary alcohols. Note that secondary and tertiary alcohols may generally be prepared by more than one combination of Grignard reagent and carbonyl component.

$$\text{CH}_3\text{MgBr} + \text{CH}_3\text{CH}_2\overset{\text{O}}{\overset{\|}{\text{C}}}\text{CH}_2\text{CH}_2\text{CH}_3 \longrightarrow \text{CH}_3\text{CH}_2\underset{\text{CH}_3}{\underset{|}{\overset{\text{OH}}{\overset{|}{\text{C}}}}}\text{CH}_2\text{CH}_2\text{CH}_3$$

$$\text{CH}_3\text{CH}_2\text{MgBr} + \text{CH}_3\overset{\text{O}}{\overset{\|}{\text{C}}}\text{CH}_2\text{CH}_2\text{CH}_3 \longrightarrow \text{CH}_3\text{CH}_2\underset{\text{CH}_3}{\underset{|}{\overset{\text{OH}}{\overset{|}{\text{C}}}}}\text{CH}_2\text{CH}_2\text{CH}_3$$

$$\text{CH}_3\text{CH}_2\text{CH}_2\text{MgBr} + \text{CH}_3\overset{\text{O}}{\overset{\|}{\text{C}}}\text{CH}_2\text{CH}_3 \longrightarrow \text{CH}_3\text{CH}_2\underset{\text{CH}_3}{\underset{|}{\overset{\text{OH}}{\overset{|}{\text{C}}}}}\text{CH}_2\text{CH}_2\text{CH}_3$$

The particular combination used is governed by such practical matters as cost and availability of reagents and ease of handling reactants.

For the preparation of many alcohols the Grignard reaction is a simple and straightforward process. However, side reactions are often important and can dominate in sterically congested cases in which the normal addition reaction is retarded. One such side reaction is enolization.

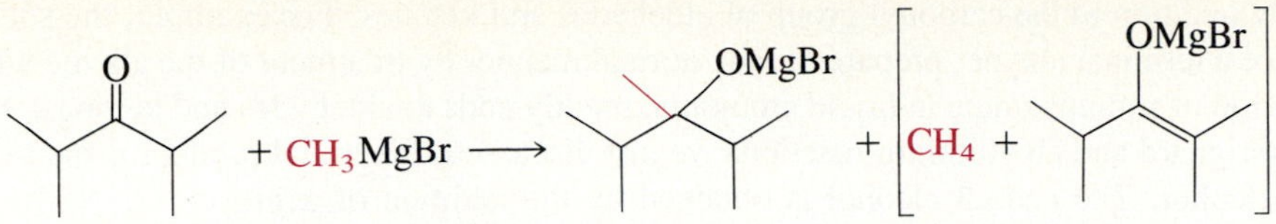

+ CH_3MgBr ⟶ OMgBr + [CH_4 + OMgBr]

In this reaction, the organomagnesium reagent acts as a *base,* rather than as a nucleophile, and abstracts an α-proton from the ketone to give the enolate ion. When water is added during normal work-up, the enolate is hydrolyzed to give back the starting ketone. The formation of enolates is discussed in Chapter 15.

Another side reaction is important in cases in which the ketone has two large alkyl groups, which sterically hinder the close approach of the organometallic R group to the carbonyl carbon. If the Grignard reagent has a β-hydrogen, the initial complex of the magnesium and the carbonyl oxygen can donate the equivalent of a hydride ion to the carbonyl carbon. In this reaction the carbonyl group is *reduced* and an alkene is formed.

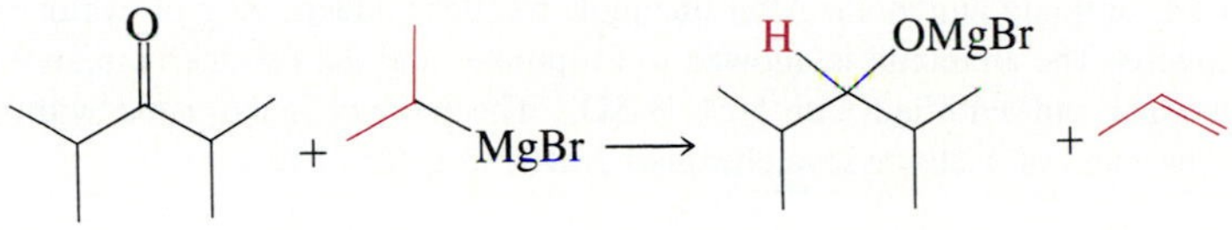

+ MgBr ⟶ H OMgBr +

The initial product is the magnesium alkoxide, which is hydrolyzed to the secondary alcohol upon workup. The reaction can be formulated as shown.

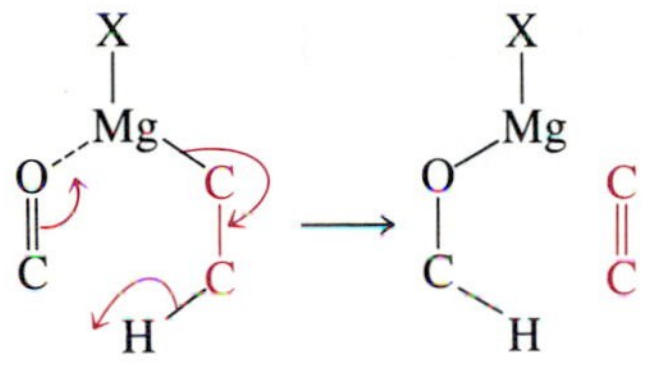

In situations such as the foregoing, alkyllithium reagents are especially useful because they are more reactive than Grignard reagents and can be used at low temperature where the alternative reduction and enolization reactions are less important. A spectacular example of a hindered system prepared by an organolithium reaction is

$$(t\text{-Bu})_2C{=}O + t\text{-BuLi} \xrightarrow[-78^\circ C]{\text{ether}} \xrightarrow{H_3O^+} (t\text{-Bu})_3COH$$

di-*t*-butyl ketone; *t*-butyllithium; (81%) 3-*t*-butyl-2,2,4,4-tetramethyl-3-pentanol

EXERCISE 14.14 Show how 1-bromobutane can be converted into each of the following compounds.

(a) 1-pentanol (b) 3-heptanol
(c) 2-methyl-2-hexanol (d) 1-hexanol (see Section 10.11.A)

EXERCISE 14.15 Outline three different combinations of alkyl halide and ketone that can be used to prepare each of the following tertiary alcohols.

(a) 3-methyl-3-nonanol (b) 3,4,5-trimethyl-3-heptanol
(c) 4-ethyl-2-methyl-4-octanol

The organometallic derivatives of alkynes (Sections 12.4 and 12.5.B) also undergo ready addition to the carbonyl group of aldehydes and ketones. For example, the sodium salt of a terminal alkyne, prepared in the normal manner by treatment of the alkyne with a solution of sodium amide in liquid ammonia, readily adds to aldehydes and ketones. As in the Grignard and alkyllithium reactions we just discussed, the initial product is the salt of the alcohol. The neutral alcohol is obtained by the addition of acid.

$$\text{cyclohexanone} + HC{\equiv}C^- \ Na^+ \xrightarrow[-33°C]{NH_3} \xrightarrow{H_3O^+} \text{1-ethynylcyclohexanol}$$

(65–75%)
1-ethynylcyclohexanol

A stream of dry acetylene is passed into the sodium amide solution prepared from 23 g of sodium in 1 L of liquid ammonia. After one mole has been added, 98 g of cyclohexanone is added dropwise. The ammonia is allowed to evaporate, and the residue is treated with 400 mL of ice water and acidified with 50% H_2SO_4. The product is extracted with ether and distilled. The yield of 1-ethynylcyclohexanol is 81–93 g (65–75%).

EXERCISE 14.16 Show the principal products from reaction of 1-ethynylcyclohexanol with the following reagents.

(a) H_2SO_4, H_2O, $HgSO_4$ (b) H_2, poisoned Pd catalyst
(c) (i) BH_3, THF; (ii) NaOH, H_2O_2

Highlight 14.4

Organometallic compounds add irreversibly to aldehyde and ketone carbonyl groups to form alkoxide salts. Protonation with water yields the corresponding alcohol.

$$R^- M^+ \ \ \ce{C=O} \longrightarrow R-C-O^- M^+ \xrightarrow{H_2O} R-C-OH$$

The addition reaction is favored by the highly nucleophilic character of R^- and by complexation of the developing charge on oxygen by M^+. If the aldehyde used is formaldehyde ($CH_2{=}O$), the resulting alcohol is primary, if any other aldehyde is used, the alcohol product is secondary, and if a ketone is used the product is a tertiary alcohol.

B. Addition of HCN

Because of the analogy that exists between the chemical reactivity of acetylide and cyanide ions (Chapter 12), we might expect that cyanide would also add to aldehydes and ketones. Such addition does occur, but the equilibrium constant for the reaction is often unfavorable, and net reaction is generally not observed.

$$(CH_3)_2C{=}O + CN^- \rightleftharpoons (CH_3)_2C(O^-)CN$$

The relatively unfavorable equilibrium constant in this reaction is primarily the result of replacing the salt of HCN ($pK_a = 9.2$) by the salt of an alcohol ($pK_a \cong 16$).

Although the addition of $Na^+ \, CN^-$ to ketones is a generally poor reaction, the addition of HCN itself usually proceeds in good yield. For example, acetone reacts readily with HCN to give the 1:1 adduct, which is called a **cyanohydrin.**

$$(CH_3)_2C{=}O + HCN \rightleftharpoons (CH_3)_2C(OH)CN$$

acetone cyanohydrin

Although most acids will add to the carbonyl group to some extent, usually the adducts are not stable. For example, with HCl the equilibrium lies far to the left and α-chloro alcohols cannot be isolated.

$$(CH_3)_2C{=}O + HCl \rightleftharpoons (CH_3)_2C(OH)Cl$$

Furthermore, if such an α-chloro alcohol is produced by some other process, it immediately decomposes to give HCl and the corresponding aldehyde or ketone.

The relative stabilities of 1-chloro alcohols and cyanohydrins can be appreciated by comparing bond energies.

$$RCH{=}O + HY \rightleftharpoons RCH(Y){-}OH$$

$$\Delta H^\circ = E(C{=}O) + E(H{-}Y) - E(O{-}H) - E(C{-}O) - E(C{-}Y)$$

The differences for ΔH° from one Y group to another are in the comparisons of $E(H{-}Y) - E(C{-}Y)$. To evaluate these bond-strength differences, compare the DH° values of H—Y and CH_3—Y for Y = Cl and CN.

$$Y = Cl: \quad DH^\circ(H{-}Cl) - DH^\circ(CH_3{-}Cl) = 103 - 85 = 18 \text{ kcal mole}^{-1}$$
$$Y = CN: \quad DH^\circ(H{-}CN) - DH^\circ(CH_3{-}CN) = 124 - 122 = 2 \text{ kcal mole}^{-1}$$

The difference in bond strengths between H—Cl and C—Cl is much greater than between H—CN and C—CN; hence, formation of the cyanohydrin has a more favorable energy change than formation of a 1-chloroalkanol.

For aldehydes and most aliphatic ketones the equilibrium favors the adduct. For some aliphatic ketones the equilibrium constant is small, and the reaction is not a useful one. The reaction is a typical nucleophilic addition with the attacking nucleophile being CN^-. Addition is therefore catalyzed by base, which increases the cyanide concentration. The reaction can be carried out using liquid hydrogen cyanide (b.p. 26 °C) as the solvent.

$$ClCH_2CHO + HCN \longrightarrow ClCH_2CH(OH)CN$$

(95%)

Because of the high toxicity of HCN, procedures such as the foregoing are seldom used. A more common procedure is to generate the HCN *in situ* by the addition of HCl or H_2SO_4 to a mixture of the carbonyl compound and sodium or potassium cyanide.

$$CH_2O + KCN \xrightarrow[H_2O]{H_2SO_4} HOCH_2CN$$

(76–80%)

A mixture of 130 g of potassium cyanide in 250 mL of H_2O and 170 mL of 37% formaldehyde solution (formalin) is prepared. To this solution is added a mixture of 57 mL of conc. H_2SO_4 and 173 mL of H_2O. The product, hydroxyacetonitrile, is obtained by exhaustive extraction with ether, and weighs 87–91 g (76–80%).

Under strongly basic conditions cyanohydrin formation is reversed. The equilibrium is shifted by transformation of HCN into its conjugate base.

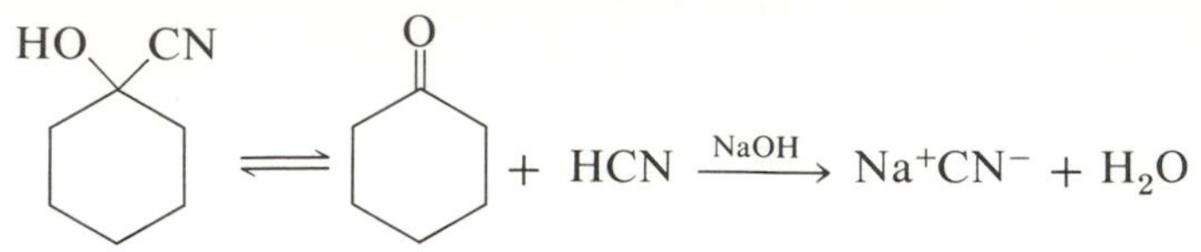

HO CN (cyclohexane) ⇌ O (cyclohexanone) + HCN $\xrightarrow{NaOH}$ Na^+CN^- + H_2O

EXERCISE 14.17 By combining the chemistry you have learned in this section with reactions you have learned previously, show how to convert acetone into each of the following compounds.

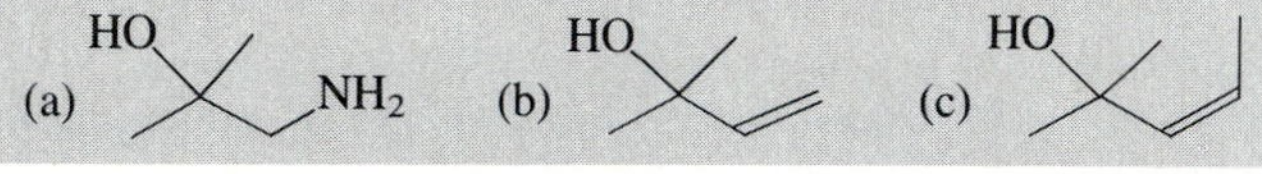

(a) HO, NH_2 (b) HO (c) HO

C. Diastereomeric Transition States

In Section 7.8, we saw that reactions that involve achiral reactants and media but that produce a new stereocenter, must give racemic products. This truth stems from the fact that, under such circumstances, the transition states leading to the two stereoisomeric products are enantiomeric. Since the competing reactions start from the same point and pass through enantiomeric transition states that must necessarily have identical energies, their energies of activation must be equal. A further example is seen in the addition of methylmagnesium bromide to pentanal, which gives racemic 2-hexanol.

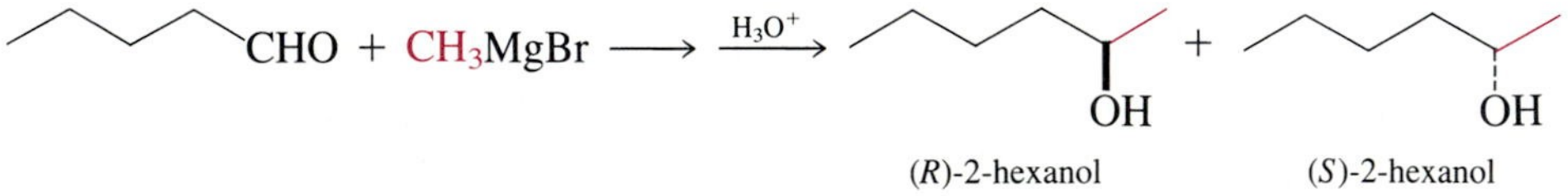

CHO + CH_3MgBr ⟶ $\xrightarrow{H_3O^+}$ OH + OH

(*R*)-2-hexanol (*S*)-2-hexanol

When one of the reactants is chiral, and a new stereocenter is created in the reaction, the two products bear a diastereomeric relationship to one another. In such a case, the two competing transition states are also diastereomeric, and therefore have *different* energies. For example, in addition of methyllithium to 2-methylcyclopentanone, the two diastereomeric products are produced in unequal amounts.

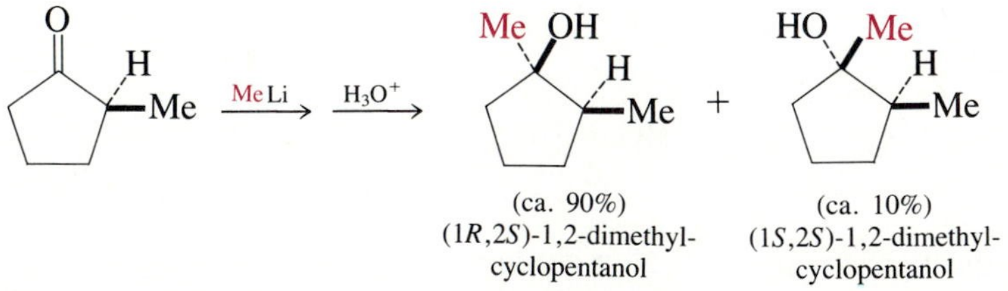

O H Me $\xrightarrow{MeLi}$ $\xrightarrow{H_3O^+}$ Me OH H Me + HO Me H Me

(ca. 90%)
(1*R*,2*S*)-1,2-dimethyl-cyclopentanol

(ca. 10%)
(1*S*,2*S*)-1,2-dimethyl-cyclopentanol

It is intuitively clear that the two isomers should be produced in unequal amount in this example. To form the 1*R*,2*S* compound, the organometallic reagent may approach the face of the cyclopentanone C=O from the side of the ring *opposite the C-2 methyl substituent*. However, in order to form the 1*S*,2*S* compound, it must attack *from the same* *side of the ring as the C-2 methyl group*. There will obviously be more steric repulsion in the latter case than in the former.

It is perhaps *not* so obvious that the foregoing situation holds even for additions to carbonyl groups that are in noncyclic structures. For example, consider the addition of methylmagnesium bromide to the chiral aldehyde 2-cyclohexylpropanal. Again, a new stereocenter is produced in the reaction, and the two products have a diastereomeric relationship. In this case, just as in the 2-methylcyclopentanone case, the two diastereomers are formed in unequal amounts.

CHO $\xrightarrow{CH_3MgBr}$ $\xrightarrow{H_3O^+}$ OH + OH

(65%) (2*R*,3*R*)-3-cyclohexyl-2-butanol (35%) (2*S*,3*R*)-3-cyclohexyl-2-butanol

The relationship that holds in both the cyclic and noncyclic cases is illustrated in Figure 14.6. Since the two competing reactions start at the same point, and since they proceed through diastereomeric transition states (which have different energies), it follows that the two reactions *must* have different activation energies, and therefore the two diastereomers must be formed in unequal amounts.

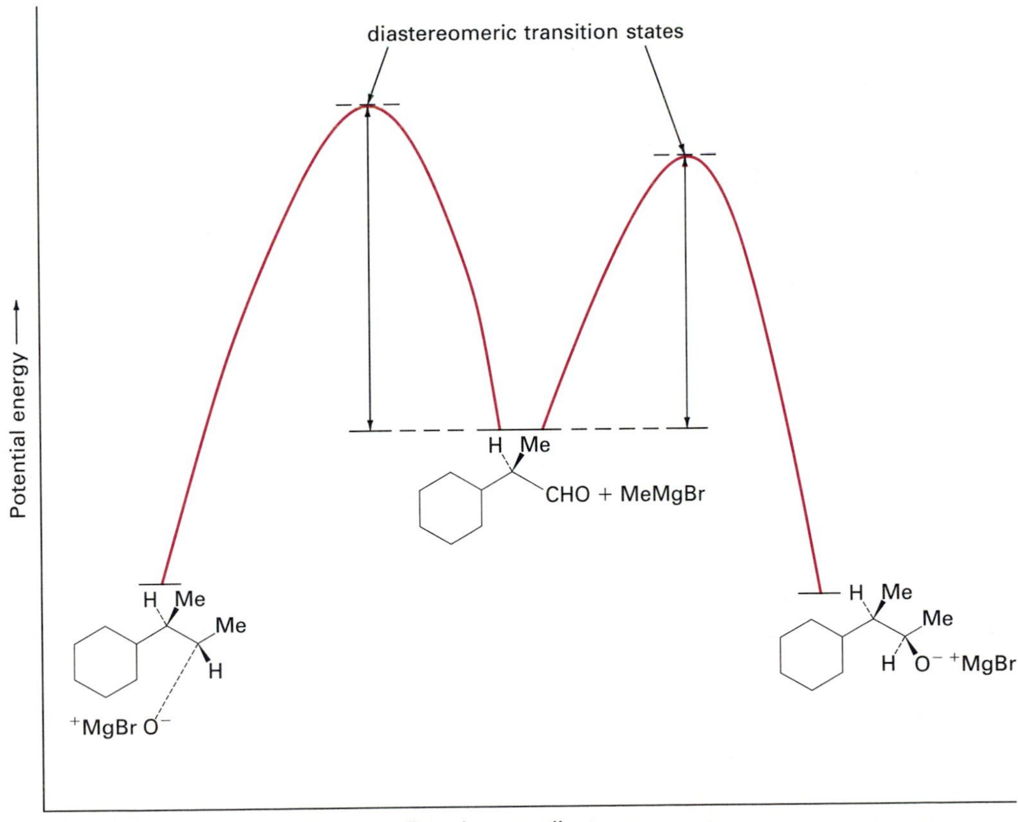

FIGURE 14.6 Nucleophilic addition to a chiral aldehyde gives diastereomeric products in unequal amounts.

While the foregoing analysis assures us that the two diastereomers must be formed in unequal amounts, it does not help us to decide which will be the major product. In the 2-methylcyclopentanone case, it is relatively easy to predict that the reagent will prefer to attack the face of the C=O from the less-hindered side of the cyclopentanone ring. In the acyclic case, it is not obvious which face is more reactive, since the molecule is free to rotate about the C-1—C-2 bond.

Be careful not to fall into the trap of neglecting this free rotation. For example, some students will look at the following depiction of the reaction and assume that the 2*R*,3*R* isomer predominates because "the methyl group is on top and the hydrogen is on the bottom."

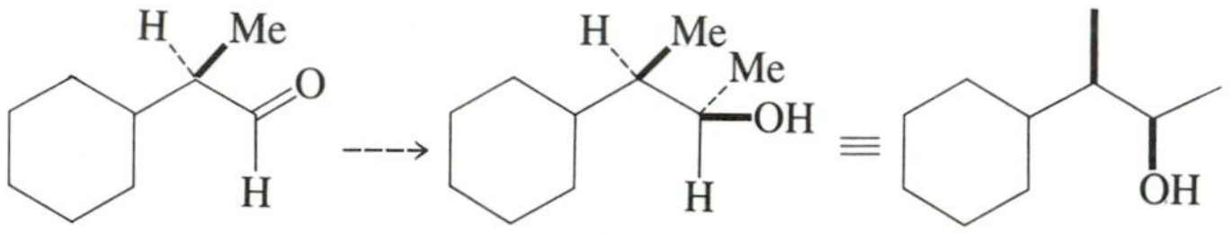

This is a completely falacious argument, since we could just as well have written the structure of (*R*)-2-cyclohexylpropanal in the following way. Application of the foregoing argument to this depiction of the molecule would lead one to expect the reagent to attack the top face of the C=O, since "hydrogen is on top and cyclohexyl is on the bottom."

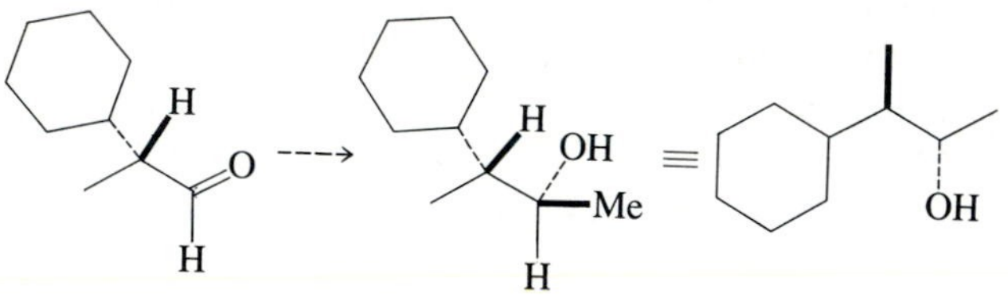

This way of looking at the question leads to *exactly the opposite prediction*.

Several useful models have been developed for predicting the relative stereochemistry of the major and minor products in reactions such as these. However, they are beyond the scope of an introductory textbook.

EXERCISE 14.18 Predict the product, including stereochemistry, of the acid-catalyzed bromination of 3-methylcyclobutanone.

D. The Wittig Reaction

Alkyl halides react with triphenylphosphine, $(C_6H_5)_3P$, by the S_N2 mechanism to give crystalline phosphonium salts (Note that the corresponding nitrogen derivatives are ammonium salts).

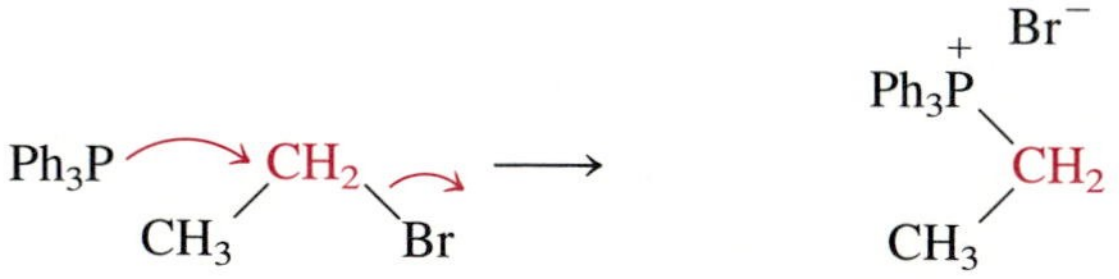

$$Ph_3P + CH_3CH_2Br \longrightarrow Ph_3\overset{+}{P}CH_2CH_3 \ Br^-$$

ethyltriphenylphosphonium bromide

Phosphine, PH_3, is the phosphorus analog of ammonia (Section 26.6). It is a poisonous gas and is usually spontaneously flammable because of the presence of impurities. Triphenylphosphine, Ph_3P, is a commercially available crystalline solid, m.p. 80 °C. Note that the abbreviation Ph— is used to represent the phenyl group, C_6H_5, a group derived from benzene, a ring system for which the name **phene** (Gk., *phainein,* to shine, a name applied by

Laurent because Michael Faraday first isolated benzene from *illuminating* gas) survives in many derivatives (Chapter 21). Triphenylphosphine is insoluble in water, but is soluble in most organic solvents.

Since phosphines are good nucleophiles and weak bases, competing elimination is not as important here as in other bimolecular substitutions (Section 9.7). Consequently, most primary and secondary alkyl halides give good yields of phosphonium salts.

$$\text{C}_6\text{H}_{11}\text{I} + Ph_3P \longrightarrow \text{C}_6\text{H}_{11}\overset{+}{P}Ph_3\ I^-$$

cyclohexyltriphenylphosphonium iodide

The alkyl proton adjacent to the positive phosphorus is moderately acidic ($pK_a \cong 35$) and can be removed by strong bases such as *n*-butyllithium or sodium hydride to give a neutral phosphorus compound called an **ylide** or **phosphorane.**

$$Ph_3\overset{+}{P}CH_2CH_2CH_3\ Br^- + n\text{-BuLi} \xrightarrow{\text{ether}} \left[Ph_3P{=}CHCH_2CH_3 \longleftrightarrow Ph_3\overset{+}{P}\overset{-}{C}HCH_2CH_3\right] + C_4H_{10}$$

The formula $Ph_3P{=}CHCH_2CH_3$ implies an expansion of the phosphorus octet and orbital overlap with phosphorus 3*d*-atomic orbitals. Detailed quantum-mechanical studies show that the actual participation of such *d*-orbitals in bonding in phosphorus ylides is minor. Instead, the dipolar structure is stabilized by polarization of the electrons around phosphorus. Such polarization can be represented in terms of an induced dipole on phosphorus.

$$Ph_3\overset{+}{P}{-}\overset{-}{C}HR$$

Nevertheless, we will frequently use the simple pentacoordinate formula for convenience.

Ylides react rapidly with aldehydes and ketones, even at −80 °C, to give neutral products called **oxaphosphetanes.** The carbon-carbon and phosphorus-oxygen bonds may form simultaneously or the mechanism may involve initial nucleophilic addition of the ylide carbon to the carbonyl group, giving a dipolar intermediate called a **betaine** (pronounced "bay-ta-ene"), which then reacts further to give the oxaphosphetane. At −80 °C the oxaphosphetane is stable in solution. Upon warming the solution to 0 °C, it decomposes to give an alkene and triphenylphosphine oxide. The overall process, illustrated by the following equations for the reaction of acetone and the ylide derived from ethyl bromide, is called the **Wittig reaction,** in honor of the German organic chemist Georg Wittig, who developed the process and subsequently received a share of the 1979 Nobel Prize in chemistry for the achievement.

$$Ph_3P{=}CHCH_3 + (CH_3)_2C{=}O \xrightarrow{-78°C} \left[(CH_3)_2\underset{\overset{|}{O^-}}{C}{-}\underset{\overset{|}{\overset{+}{P}Ph_3}}{C}HCH_3\right] \longrightarrow$$

a **betaine**

$$\left[\text{O—PPh}_3 \text{ ring: } (CH_3)_2C{-}CHCH_3\right] \xrightarrow{0°C} (CH_3)_2C{=}CHCH_3 + Ph_3P{=}O$$

an oxaphosphetane

triphenylphosphine oxide

Betaine is $(CH_3)_3N^+ CH_2COO^-$, first isolated from beets, *Beta vulgaris,* in 1866, hence the name. The name of the plant family, an important food source since prehistoric times, may be related to *beta,* the second letter of the Greek alphabet derived originally from the Semitic *bet',* or house.

The Wittig reaction is an exceedingly useful method for the synthesis of alkenes. Although a mixture of cis and trans isomers usually results, *only a single positional isomer is produced.* Consider, as an example, the synthesis of methylenecyclohexane. Dehydration of 1-methylcyclohexanol gives mainly 1-methylcyclohexene, since this isomer is more stable.

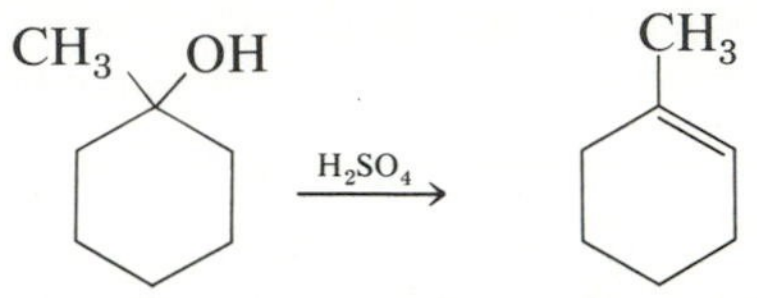

CH_3 OH $\xrightarrow{H_2SO_4}$ CH_3

1-methylcyclohexene

The less stable isomer may be readily prepared from cyclohexanone by the Wittig reaction.

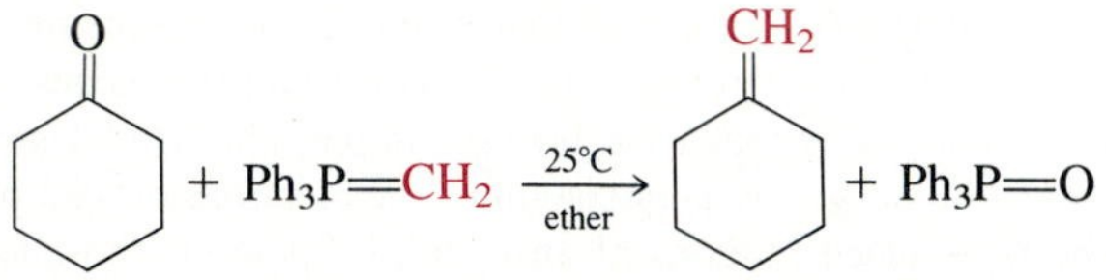

O + $Ph_3P{=}CH_2$ $\xrightarrow[\text{ether}]{25°C}$ CH_2 + $Ph_3P{=}O$

EXERCISE 14.19 Starting with triphenylphosphine and any desired alkyl halides, aldehydes, and ketones, show how the following alkenes can be prepared.

(a) 2-methyl-2-pentene (b) 2,2-dimethyl-3-heptene (c) 3-methyl-2-hexene

14.8 Oxidation and Reduction

A. Oxidation of Aldehydes and Ketones

Aldehydes are oxidized to carboxylic acids with ease. Oxidizing agents that have been used are Ag_2O, H_2O_2, $KMnO_4$, CrO_3, and peroxy acids.

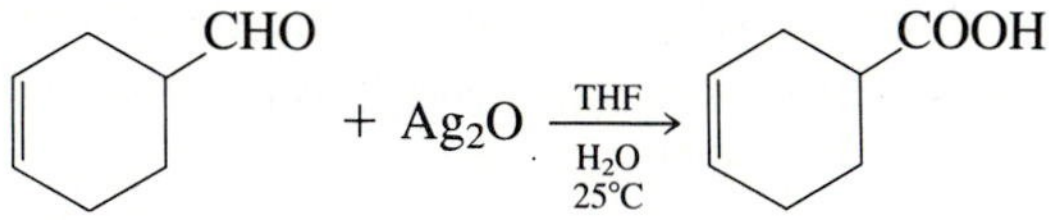

CHO + Ag_2O $\xrightarrow[\substack{H_2O \\ 25°C}]{THF}$ COOH

(97%)

CHO + O OOH ⟶ COOH

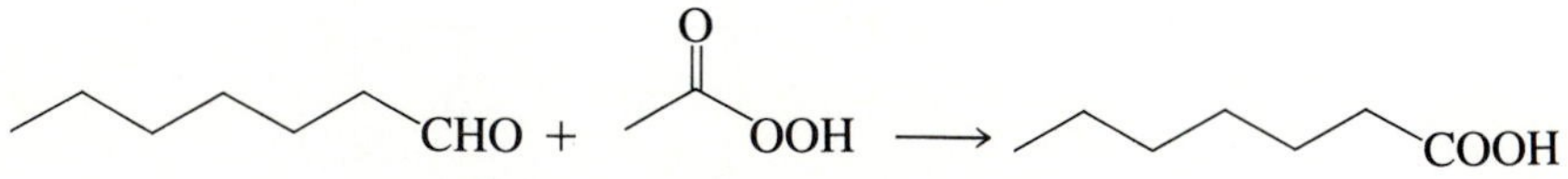

peroxyacetic acid

(88%)

This oxidation is so facile that even atmospheric oxygen will bring it about. Most aldehyde samples that have been stored for some time before use are found to be contaminated with variable amounts of the corresponding carboxylic acid. In the case of oxidation by air (**autoxidation**) the initial oxidation product is a **peroxycarboxylic acid** (page 285). The peroxycarboxylic acid reacts with another molecule of aldehyde to give two carboxylic acid molecules.

$$\text{RCHO} + \text{O}_2 \longrightarrow \text{RC(=O)OOH}$$

a peroxycarboxylic acid

$$\text{RC(=O)OOH} + \text{RCHO} \longrightarrow 2\,\text{RC(=O)OH}$$

The initial oxidation (to the peroxycarboxylic acid stage) is a free radical chain process, probably via small amounts of *gem*-diol or a similar adduct. The adducts have a hydrogen at the carbon next to the two hydroxy groups, and thus should resemble an ether in sensitivity toward free-radical attack (see Section 10.10). The second stage oxidation of the aldehyde by the initially formed peroxycarboxylic acid, is an example of the **Baeyer-Villiger oxidation.** The probable course of this oxidation follows.

$$\text{R—C(=O)—H} + \text{R—C(=O)—OOH} \rightleftharpoons \text{R—CH(OH)—O—O—C(=O)—R}$$

$$\text{R—CH(:OH)—O—O—C(=O)—R} \rightleftharpoons \text{R—C(=}\overset{+}{\text{O}}\text{H)—OH} + {}^{-}\text{O—C(=O)—R}$$

$$\text{R—C(=}\overset{+}{\text{O}}\text{H)—OH} + {}^{-}\text{O—C(=O)—R} \longrightarrow 2\,\text{R—C(=O)—OH}$$

In contrast to aldehydes, ketones are oxidized only under rather special conditions. The Baeyer-Villiger oxidation is one reaction in which a ketone undergoes oxidation. In this case the product is an ester.

$$\text{CH}_3\text{CH}_2\text{C(=O)CH}_2\text{CH}_3 \xrightarrow[\text{CH}_2\text{Cl}_2]{\text{CF}_3\text{CO}_3\text{H}} \text{CH}_3\text{CH}_2\text{C(=O)OCH}_2\text{CH}_3$$

(78%)

The mechanism of the reaction is believed to be similar to that outlined for the oxidation of an aldehyde, except that the migrating group is an alkyl group rather than a hydrogen.

$$\text{C}_2\text{H}_5\text{—C(:OH)(C}_2\text{H}_5\text{)—O—O—C(=O)—CF}_3 \longrightarrow \text{C}_2\text{H}_5\text{—C(=}\overset{+}{\text{O}}\text{H)—OC}_2\text{H}_5 + \text{CF}_3\text{CO}_2^- \longrightarrow$$

$$\text{C}_2\text{H}_5\text{C(=O)OC}_2\text{H}_5 + \text{CF}_3\text{COOH}$$

The overall result of the Baeyer-Villiger oxidation is insertion of an oxygen atom into one of the bonds to the carbonyl group; consequently, the ketone is converted into an ester. The reaction is a preparatively useful one for the oxidation of certain ketones to esters. Cyclic ketones give cyclic esters, which are called lactones (Section 27.5.C).

$$\text{cyclohexanone} + CH_3CO_3H \xrightarrow[\substack{40^\circ C \\ 6.5\ hr}]{CH_3COOEt} \text{caprolactone}\ (90\%)$$

Symmetrical ketones can give only one product in the Baeyer-Villiger oxidation. However, unsymmetrical ketones can give two oxidation products, and this is sometimes observed. When the two alkyl groups differ substantially, a clear selectivity can often be observed. The approximate order of decreasing ease of migration (the migratory aptitude) for various groups is hydrogen > tertiary alkyl > secondary alkyl > phenyl > primary alkyl > methyl. The following examples illustrate this selectivity.

$$\xrightarrow[CH_2Cl_2]{PhCO_3H}\quad (67\%)$$

$$\xrightarrow[\substack{BF_3 \\ ether}]{H_2O_2}\quad (62\%)$$

Although ketones can be oxidized by other reagents, oxidative cleavage is seldom a useful preparative method. The conditions required for the oxidation of most ketones are sufficiently vigorous that complex mixtures result. The chief exception to this generalization is in symmetrical cyclic ketones, where the reaction can be useful. The oxidation of cyclohexanone by nitric acid is catalyzed by vanadium pentoxide. The product, adipic acid, is an important industrial chemical because it is one of the constituents of nylon 66 (Section 36.4.E).

$$\xrightarrow[V_2O_5]{HNO_3} HOOC(CH_2)_4COOH$$

adipic acid

EXERCISE 14.20 Give the principal expected product of Baeyer-Villiger oxidation of the following compounds.

(a) $C_6H_5COCH_3$
(b) cyclopentanone
(c) 3,3-dimethyl-2-butanone
(d) ethyl cyclohexyl ketone

B. Metal Hydride Reduction

Aldehydes and ketones are easily reduced to the corresponding primary and secondary alcohols, respectively.

$$R{-}CHO \xrightarrow{[H]} R{-}CH_2OH$$

$$R{-}\overset{O}{\overset{\|}{C}}{-}R' \xrightarrow{[H]} R{-}\overset{OH}{\overset{|}{C}H}{-}R'$$

Many different reducing agents can be used. For laboratory applications the complex metal hydrides are particularly effective. Lithium aluminum hydride ($LiAlH_4$, page 311) is a powerful reducing agent that has been used for this purpose. Reductions are normally carried out by adding an ether solution of the aldehyde or ketone to an ether solution of $LiAlH_4$. Reduction is rapid even at −78 °C (dry ice temperature). At the end of the reaction the alcohol is present as a mixture of lithium and aluminum salts and is liberated by hydrolysis.

$$\text{cyclobutanone} + LiAlH_4 \xrightarrow[30°C,\ 1\ hr]{ether} \xrightarrow[H_2O]{H_2SO_4} \text{cyclobutanol}$$

(90%)
cyclobutanol

The reagent also reduces many other oxygen- and nitrogen-containing functional groups, as illustrated in Table 14.6. The chief drawbacks of the reagent are its cost, which renders it useful only for fairly small-scale laboratory applications, and the hazards involved in handling it.

TABLE 14.6 Functional Groups Reduced by $LiAlH_4$

Functional Group	Product	Moles of $LiAlH_4$ Required
RCHO	RCH_2OH	0.25
$R_2C{=}O$	R_2CHOH	0.25
RCOOR′	$RCH_2OH + R'OH$	0.5
RCOOH	RCH_2OH	0.75
$RC(=O)NH_2$	RCH_2NH_2	1
$RC{\equiv}N$	RCH_2NH_2	0.5
RNO_2	RNH_2	1.5
RCl(Br, I)	RH	0.25

Sodium borohydride, $NaBH_4$ (page 279), offers certain advantages. This hydride is much less reactive than $LiAlH_4$ and is consequently more selective. Of the functional groups in Table 14.6 that are reduced by $LiAlH_4$, only aldehydes and ketones are reduced at a reasonable rate by $NaBH_4$. The reagent is moderately stable in aqueous and in alcoholic solution, especially at basic pHs. The following example illustrates the selectivity that may be achieved with the reagent.

$$\xrightarrow[CH_3OH]{NaBH_4}$$

(Steroid 3-ketone with Br, H and $COOCH_3$ side chain → corresponding 3-alcohol, HO, Br, H, $COOCH_3$)

The carbonyl group is also reduced rapidly and quantitatively by diborane in ether or THF (page 280). The initial product is the ester of boric acid and an alcohol, a trialkyl borate. This material is rapidly hydrolyzed upon treatment with water.

$$(CH_3)_2CHCHO + B_2H_6 \xrightarrow{THF} \underset{(100\%)}{[(CH_3)_2CHCH_2O]_3B} \xrightarrow{H_2O} \underset{(100\%)}{(CH_3)_2CHCH_2OH}$$

EXERCISE 14.21 Suggest reaction sequences for carrying out the following transformations.

(a) 1-octyne to 2-octanol (b) cyclopentanone to 1,5-pentanediol

C. Catalytic Hydrogenation

Aldehydes and ketones may also be reduced to alcohols by hydrogen gas in the presence of a metal catalyst (catalytic hydrogenation; Section 11.6.A). The chief advantages of this method are that it is relatively simple to accomplish and usually affords a quantitative yield of product because no complicated work-up procedure is required. However, it suffers from the disadvantages that many of the catalysts used (Pd, Pt, Ru, Rh) are relatively expensive and that other functional groups (C=C, C≡C, NO_2, C≡N) also react.

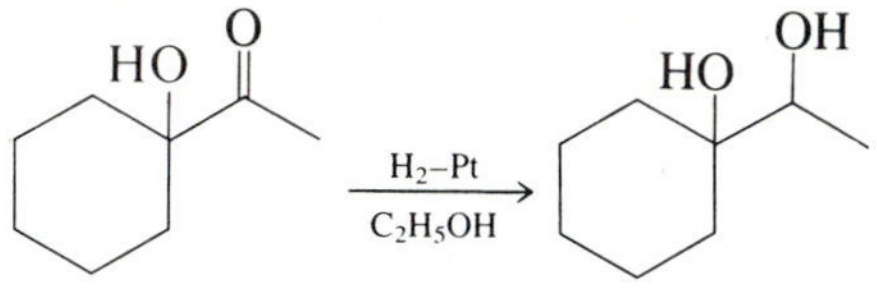

HO O → (H_2–Pt, C_2H_5OH) → HO OH

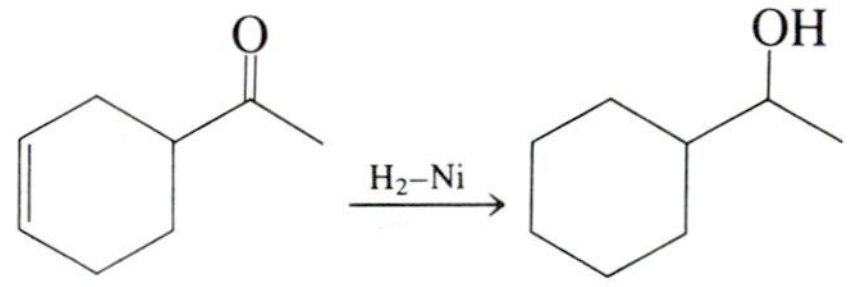

O → (H_2–Ni) → OH

(96%)

D. Deoxygenation Reactions

Several methods are known whereby the oxygen of an aldehyde or ketone is replaced by two hydrogens; this process is known as **deoxygenation.**

$$R{-}\overset{\overset{\displaystyle O}{\|}}{C}{-}R' \longrightarrow R{-}CH_2{-}R'$$

One such method is the **Wolff-Kishner reduction.** The first step in the reaction sequence is the formation of the hydrazone by addition of hydrazine and elimination of water. Base attacks the somewhat acidic NH_2 group (hydrogen α to an electronegative double bond) to form an ambident anion that can react with a proton donor to produce a diazene $R_2CH{-}N{=}NH$. The acidic diazene ($pK_a \cong 23$) reacts with base forming an anion that loses nitrogen to give the hydrocarbon. The reaction is normally carried out by heating the ketone with hydrazine hydrate (Section 14.6.C) and sodium hydroxide in diethylene glycol, $HOCH_2CH_2OCH_2CH_2OH$, which has a b.p. of 245 °C. Alternatively, the reduction may be carried out in the polar aprotic solvent DMSO at 100 °C. The hydrazone forms, and water distills out of the mixture. On refluxing, nitrogen is evolved, and the product is isolated.

$$\text{cyclononanone} \xrightarrow[\text{(HOCH}_2\text{CH}_2)_2\text{O, NaOH, }\Delta]{\text{H}_2\text{NNH}_2} [\text{cyclononylidene}{=}\text{N—NH}_2] \xrightarrow{-\text{N}_2} \text{cyclononane (47\%)}$$

cyclononanone — cyclononane

The decomposition of the diazene involves an intermediate anion. This diazene anion loses nitrogen to produce the alkyl anion. The carbanion is exceedingly basic and rapidly abstracts a proton from the solvent. Alkyl anions are rarely encountered in reactions because of the low acidity of hydrocarbons. In the present case they are formed only because the nitrogen also produced is an extremely stable molecule and its production provides the driving force for the reaction.

$$(1)\ R_2C{=}N\ddot{N}H_2 + B^- \rightleftharpoons BH + [R_2C{=}N{-}\ddot{N}H^- \longleftrightarrow R_2\ddot{C}^-{-}N{=}NH]$$

$$(2)\ [R_2C{=}N{-}\ddot{N}H^- \longleftrightarrow R_2\ddot{C}^-{-}N{=}NH] + BH \rightleftharpoons R_2C(H){-}N{=}NH + B^-$$

$$(3)\ R_2CH{-}N{=}\ddot{N}H + B^- \rightleftharpoons R_2CH{-}N{=}\ddot{N}^- + BH$$

$$(4)\ R_2CH{-}N{=}\ddot{N}^- \xrightarrow{\text{slow}} R_2\ddot{C}H^- + N_2$$

$$(5)\ R_2\ddot{C}H^- + BH \xrightarrow{\text{fast}} R_2CH_2 + B^-$$

Methyldiazene, $CH_3N{=}NH$, and phenyldiazene, $C_6H_5N{=}NH$, can be generated in neutral solution by decarboxylation (Section 27.7.C) of $CH_3N{=}NCOOH$ and $C_6H_5N{=}NCOOH$. Note that the IUPAC name for the nitrogen (Fr., *azote*) analog of an alkene is diazene, from **diaza** + alk**ene**. Monosubstituted diazenes have lifetimes from minutes to hours in dilute solution but are reactive toward oxygen. Diazene, HN=NH, has the widely used common name diimide; it is a reducing agent and can be produced *in situ* from HOOCN=NCOOH. It reacts with symmetrical double bonds to give cis addition of hydrogen.

An alternative procedure for the direct reduction of a carbonyl group to a methylene group involves refluxing the aldehyde or ketone with amalgamated zinc and hydrochloric acid (**Clemmensen reduction**).

Amalgamated zinc is zinc with a surface layer of mercury. It is prepared by treating zinc with an aqueous solution of a mercuric salt. Since zinc is higher on the electromotive force scale than mercury, it reduces mercuric ions to mercury.

$$Zn + Hg^{2+} \longrightarrow Zn^{2+} + Hg$$

The mercury then alloys with the surface of zinc to produce an amalgam.

Reduction of the carbonyl compound occurs on the surface of the zinc, and, like many heterogeneous reactions, this reaction does not have a simple mechanism.

The Clemmensen reduction is suitable for compounds that can withstand treatment with hot acid. Many ketones are reduced in satisfactory yields.

$$C_6H_5COCH_3 \xrightarrow[\text{HCl, }\Delta]{\text{Zn(Hg)}} C_6H_5CH_2CH_3$$

acetophenone — ethylbenzene (80%)

EXERCISE 14.22 Which method is preferable for deoxygenation of each of the following aldehydes or ketones? If neither the Wolff-Kishner nor the Clemmensen method is expected to be suitable, so indicate.

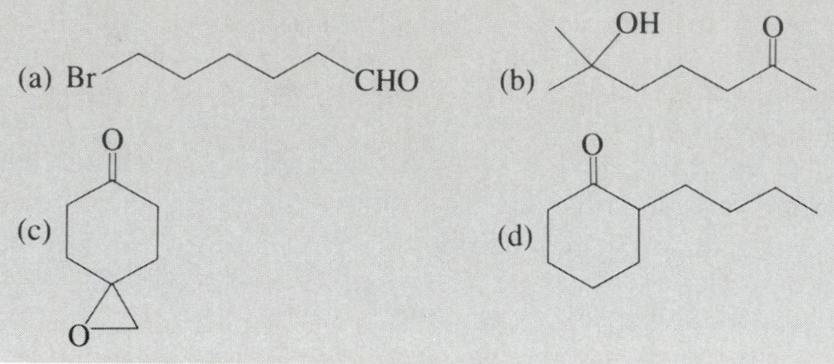

(a) Br～～CHO (b) (c) (d)

E. Enzymatic Reductions

Enzymes are proteins that catalyze reactions with one or more compounds called substrates. We will discuss proteins further in Chapter 29. For now we must know that they are chiral and exist in only one enantiomeric form, so it is not surprising that enzymes catalyze reactions of the substrates via diastereomeric transition states. Many enzymes are commercially available and are finding a place in organic synthesis, especially for the preparation of pure enantiomers or one of two possible diastereomers. A class of enzymes called alcohol dehydrogenases includes the enzyme from yeast that produces ethanol from acetaldehyde. The stereochemical outcome of a stereospecific reaction is characterized by the e.e., *enantiomeric excess,* defined by the difference in the amounts of the two enantiomers formed. Zero e.e. would mean that the reaction had no stereospecificity. In the examples given the e.e.'s are 85–98%.

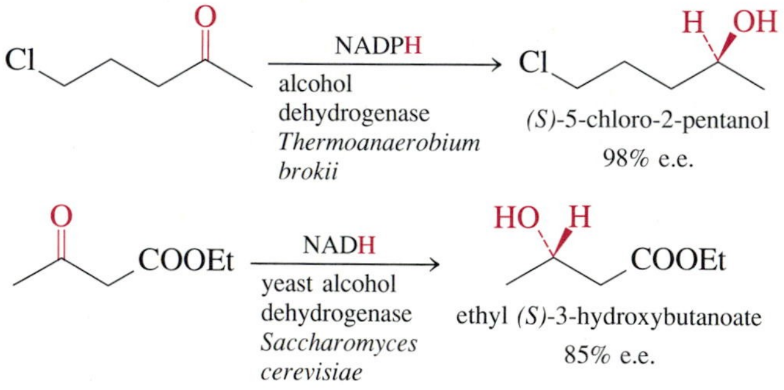

NADPH, alcohol dehydrogenase *Thermoanaerobium brokii* → (*S*)-5-chloro-2-pentanol, 98% e.e.

NADH, yeast alcohol dehydrogenase *Saccharomyces cerevisiae* → ethyl (*S*)-3-hydroxybutanoate, 85% e.e.

The ketones in the foregoing examples have enantiotopic faces (page 145), so addition of hydride can give two enantiomers. In both cases, the enzyme causes preferential delivery of hydride to the *re* face to give the *S* enantiomer.

Biological receptors are generally proteins and display a preference for one enantiomeric or diastereomeric ligand; in fact, the inactive isomer sometimes blocks the site occupied by the active isomer, thereby decreasing its potency. Thus, stereospecific reactions are important in the synthesis of drugs and other biologically active materials.

ASIDE

Liebig gave the name aldehyde to the compound derived from oxidation of ethanol by chromic acid, based on Latin, as al(cohol) dehyd(rogenatus). The French chemist Bussy named acetone as the substantive feminine form of acetic acid from which it could be prepared by pyrolysis of the barium salt. Ketone was given for "acetones in general" by L. Gmelin. These classical syntheses of aldehydes and ketones are still used. What functional group transformations have you learned so far for the preparation of aldehydes and ketones? What are the special precautions needed to successfully prepare aldehydes? In the mechanism for oxidation of aldehydes by chromic acid, you will note a step involving water addition to the carbonyl group. Give other examples of water addition to aldehydes and write the mechanism for the reaction. What is the product if methanol replaces the water?

PROBLEMS

1. Provide common names for the following ketones and aldehydes.

a. 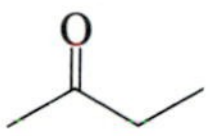 b. 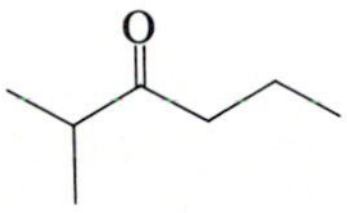 c. 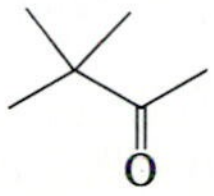 d.

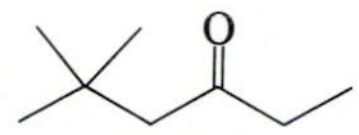

e. 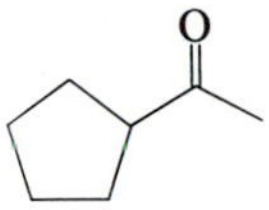 f. 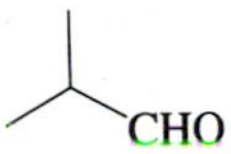 CHO g. 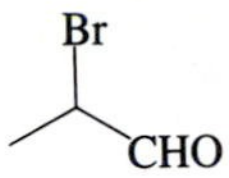 Br CHO h. 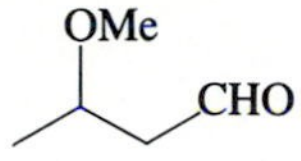 OMe CHO

i. 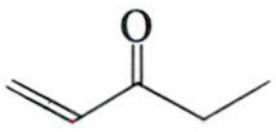 j.

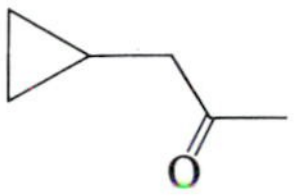

2. Provide IUPAC names for the compounds in problem 1.

3. Write the structure corresponding to each of the following names.

a. methyl isobutyl ketone
b. propionaldehyde diethyl acetal
c. β-chlorobutyraldehyde
d. 2,2-dimethylcyclopentanone
e. cyclododecanone
f. formaldehyde phenylhydrazone
g. cyclohexanone oxime
h. acetone hydrazone

4. What is the product of the reaction of 4,4-dimethylcyclohexanone with each of the following reagents?

a. peroxyacetic acid
b. (i) $LiAlH_4$ in ether; (ii) H_2O
c. $NaBH_4$ in ethanol
d. (i) $CH_3C{\equiv}C^-\ Na^+$ in liq. NH_3; (ii) H_2O
e. KCN and aqueous sulfuric acid
f. H_2NOH + sodium acetate in acetic acid
g. 0.5 mole of H_2NNH_2 + sodium acetate in acetic acid
h. $H_2NNH_2 + HOCH_2CH_2OCH_2CH_2O^-\ Na^+$, 200 °C
i. zinc amalgam + hot conc. hydrochloric acid
j. $Ph_3P{=}CHCH_2CH_3$

5. Answer problem 4 for 2-methylpentanal. In this case, some of the reactions will not work. Give products for those reactions that are expected to be successful and indicate those that are expected to fail.

6. Show how each of the following compounds can be prepared from alkyl halides and alcohols containing four or fewer carbons.

a. 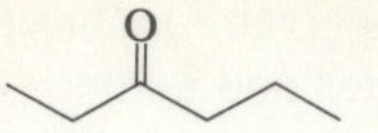 O

b.

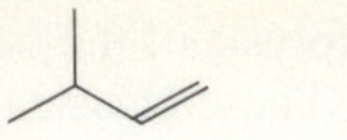

c. 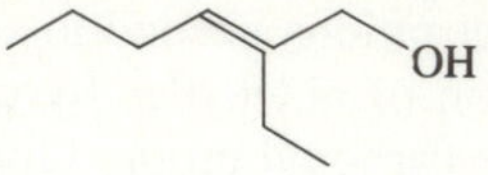 OH

d.

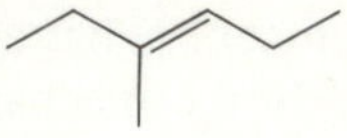

e. 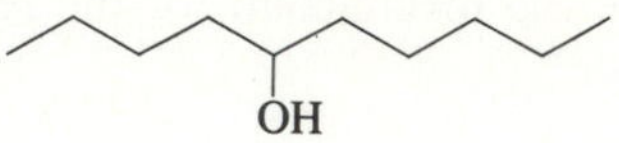 OH

f. 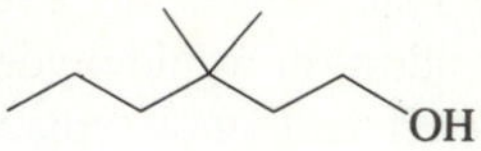 OH

g. 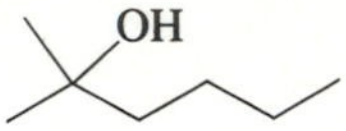 OH

h. 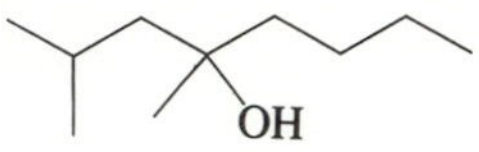 OH

7. Show how each of the following compounds can be prepared from pentanal. Any other reagents, organic or inorganic, may be used.

a.

b. 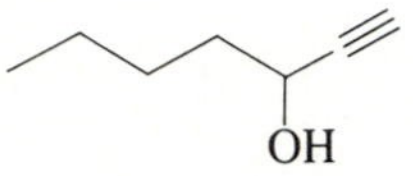 OH

c. 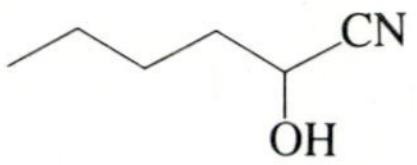 CN OH

d.  COOH

e.

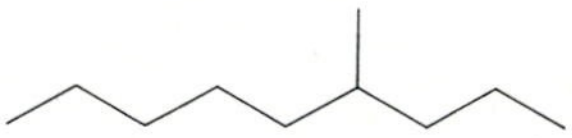

f. 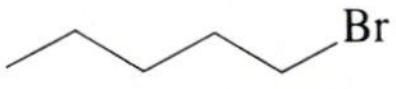 Br

g. 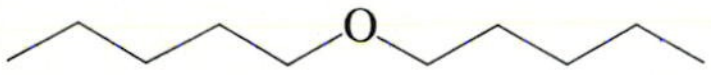 O

h.

8. Show how one can accomplish each of the following conversions in a practical manner using any necessary organic or inorganic reagents.

a. 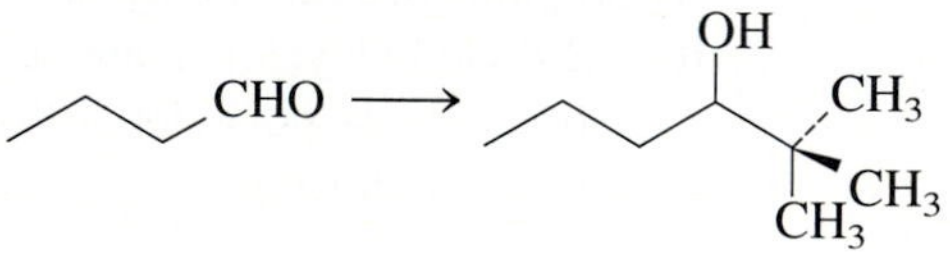 CHO ⟶ OH CH_3 CH_3 CH_3

b. 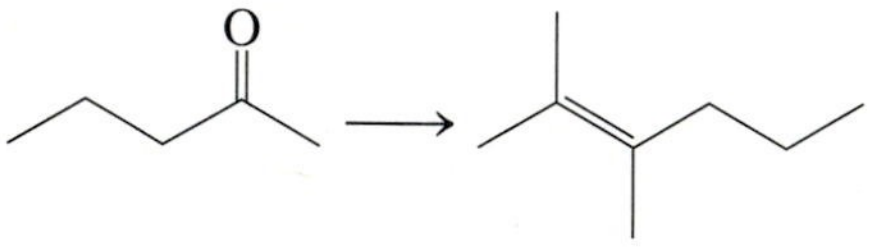 O ⟶

c. 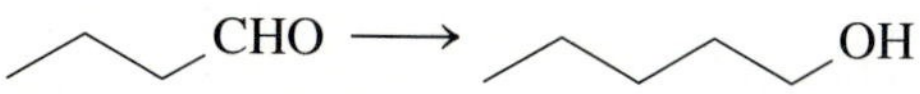 CHO ⟶ OH

d. 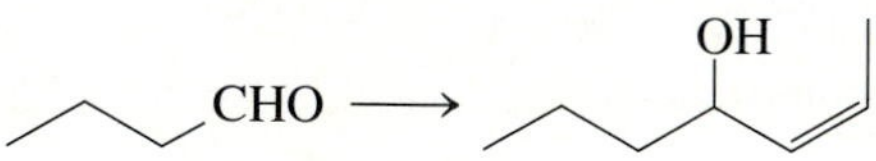 CHO ⟶ OH

e. 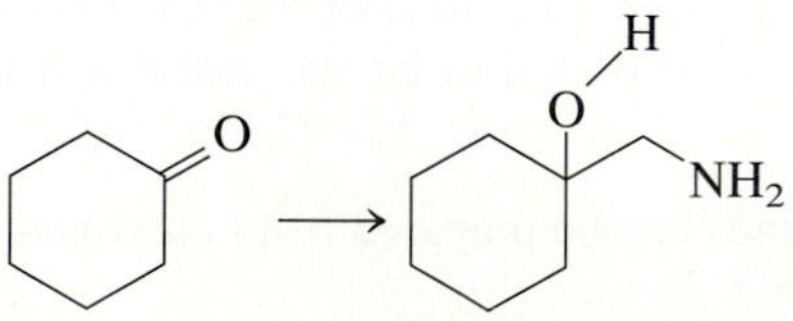 O ⟶ H O NH_2

9. Show how each of the following compounds can be prepared from aldehydes, ketones, and alkyl halides containing four carbons or less using the Grignard synthesis as a key step in each case.

a. 1-bromopentane **b.** 3-hexanone
c. 4-propyl-3-heptene **d.** 2,5-dimethyl-3-ethyl-3-hexanol

10. Show how each of the following isotopically labeled compounds can be synthesized. You may use any necessary organic starting materials, as long as they are unlabelled at the start of your synthesis. You may use any necessary inorganic reagents, labelled or otherwise.

a. 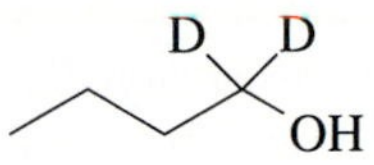 D D OH

b. 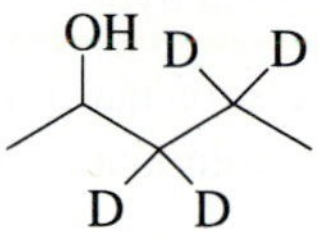 OH D D D D

c. 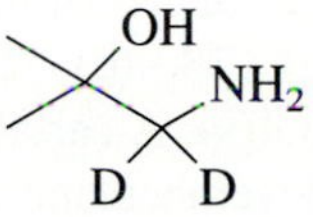 OH NH_2 D D

d. D OH

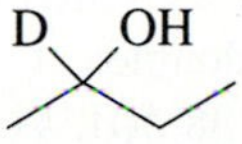

11. Explain why each of the following syntheses cannot be accomplished in the specified manner.

a. 2,2,5,5-tetramethyl-3-hexene by the Wittig reaction.
b. 3-methyl-2-hexanol by Clemmensen reduction of 4-methyl-5-hydroxy-3-hexanone

12. Outline a sequence of reactions that can be used to convert hept-6-en-2-one into 7-hydroxyheptan-2-one. [*Caution*: Remember that ketones are reduced by diborane.]

13. Acetaldehyde reacts with (*R*)-1,2-propanediol to give *two* isomeric acetals. What are the structures of these two compounds?

14. Propose a mechanism for the following reaction.

O O + Br_2 $\xrightarrow{\text{trace } H^+}$ O O Br

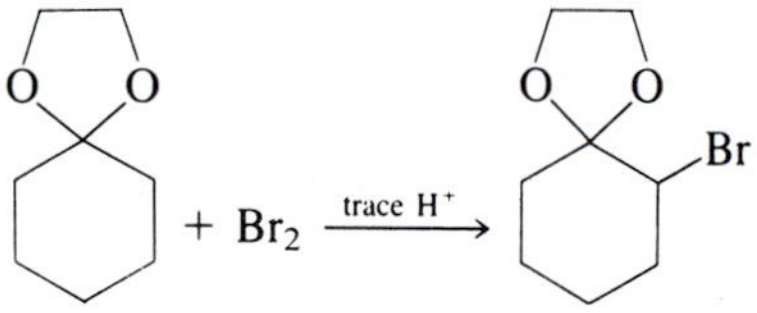

15. 1,1-Diethoxyethane hydrolyzes readily to acetaldehyde and ethanol in water containing some sulfuric acid. Write a complete reaction mechanism for this transformation including each significant intermediate and reaction step.

16. Is the equilibrium constant for the following equilibrium greater than, less than, or equal to unity? Explain briefly.

$$CF_3CHO + CH_3CH(OCH_3)_2 \rightleftharpoons CH_3CHO + CF_3CH(OCH_3)_2$$

17. When 4-hydroxybutanal is dissolved in methanol containing HCl, the following reaction occurs.

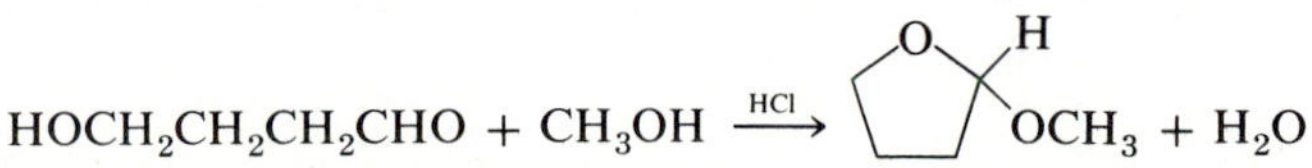

$$HOCH_2CH_2CH_2CHO + CH_3OH \xrightarrow{HCl}$$ O H OCH_3 + H_2O

a. What type of compound is the product?
b. Propose a mechanism for this reaction.

Actually, 4-hydroxybutanal exists in solution mostly in the cyclic form.

$$HOCH_2CH_2CH_2CHO \rightleftharpoons$$ (cyclic hemiacetal: five-membered ring with O, H, OH)

c. What type of compound is this product?
d. Propose a mechanism for the equilibrium.

18. Write a plausible reaction mechanism for the trimerization of acetaldehyde to paraldehyde with a trace of acid. How does this mechanism compare to the acid-catalyzed depolymerization of paraldehyde?

19. Undecanal is a sex attractant for the greater wax moth (*Galleria mellonella*). Show how to synthesize this compound efficiently from (a) 1-decanol and (b) 1-dodecanol.

20. *trans*-3-Isopropyl-6-methylcycloheptanone has the following proton-coupled CMR spectrum (letters in parentheses after each chemical shift indicate multiplicity of the resonance: s = singlet, d = doublet, t = triplet, q = quartet): 18.5(q), 18.9(q), 23.9(q), 32.0(d), 32.5(t), 34.9(d), 38.5(t), 41.9(d), 47.0(t), 51.6(t), and 213.4(s). A certain reaction provided samples of the cis and trans isomers of 3,6-dimethylcycloheptanone were obtained. However, it was not known which isomer had which stereostructure. The isomers have the following CMR spectra.

isomer J: 20.9(q), 29.2(d), 33.9(t), 50.8(t), 212.1(s)
isomer K: 23.7(q), 31.1(d), 37.9(t), 51.7(t), 212.1(s)

Which isomer has the cis structure, and which the trans?

21. Unsymmetrical alkynes normally undergo hydration to give a mixture of the two possible ketones. For example, hydration of 2-heptyne gives both 2-heptanone and 3-heptanone. However, treatment of hept-5-yn-2-one with aqueous sulfuric and a small amount of mercuric sulfate gives *only* heptan-2,5-dione. Suggest a mechanism that might account for this behavior.

22. Ethyl vinyl ether, $CH_2{=}CHOCH_2CH_3$, reacts with *n*-butanol and a trace of sulfuric acid in ether to give 1-*n*-butoxy-1-ethoxyethane. Treatment of this product with aqueous sulfuric acid affords a mixture of ethanol, *n*-butanol, and acetaldehyde. Propose a mechanism for each reaction. How might this chemistry be used to advantage in accomplishing the following conversion?

HO–(Br-substituted chain)–C(=O)– ⟶ HO–(Br-substituted chain)–CH(OH)–

23. Treatment of an aldehyde or ketone with a mixture of ammonium chloride and sodium cyanide yields an α-amino nitrile **(Strecker synthesis).**

$$(CH_3)_2CHCH_2CHO + NH_4Cl + NaCN \longrightarrow (CH_3)_2CHCH_2CH(NH_2)CN$$

an α-amino nitrile

Propose a mechanism for the reaction. What product is expected when 2,6-heptanedione is subjected to the Strecker synthesis?

CHAPTER 15

ALDEHYDES AND KETONES: ENOLS

15.1 Enolization

The carbonyl group increases greatly the reactivity of the hydrogens on the carbons α to the carbonyl. The resonance structures written for the carbonyl group make the origin of the activation clear. The positive charge on the carbonyl carbon makes it easier to remove a proton from the α-carbon. Base-catalyzed reaction is illustrated with hydroxide ion.

$$\left[>C{=}O \longleftrightarrow >\overset{+}{C}{-}\overset{-}{O} \right]$$

carbonyl group

$$\left[H_2C{-}C{=}O \longleftrightarrow H_2C{-}\overset{+}{C}{-}\overset{-}{O} \right] \xrightarrow{-H_2O}$$

HO⁻

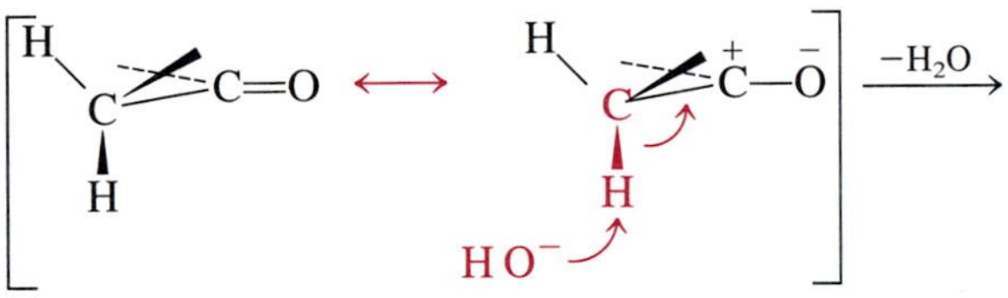

carbonyl group with α-CH_2 attacked by OH^-

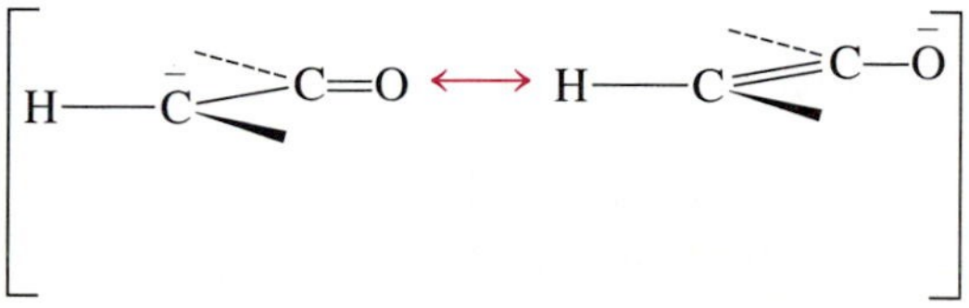

$$\left[H{-}\overset{-}{C}{-}C{=}O \longleftrightarrow H{-}C{=}C{-}\overset{-}{O} \right]$$

enolate anion

The α-proton may be lost through attack by solvent after protonation, even though the basicity of the carbonyl group is relatively low.

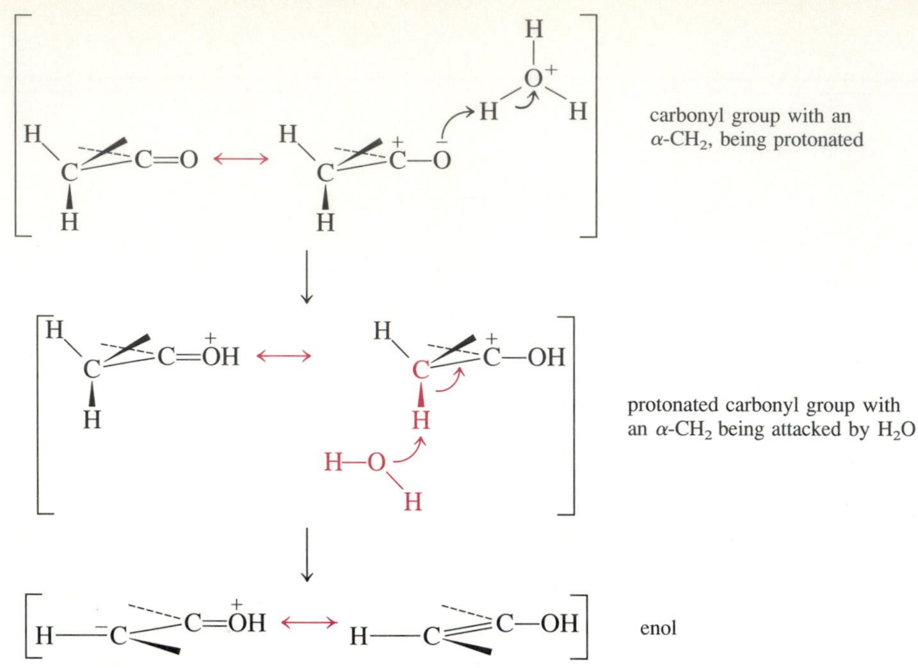

The products of proton loss by base attack are enolate ions (alk**ene** + alcoh**olate**) and those of proton loss from carbon after protonation on oxygen are enols (hydroxyalkenes, alk**ene** + alcoh**ol**), both of which are central intermediates in many organic syntheses.

A. Keto-Enol Equilibria

Aldehydes and ketones exist in solution as equilibrium mixtures of two isomeric forms, the keto form and the **enol** form. For simple aliphatic ketones there is very little of the enol form present at equilibrium, as shown by the following examples.

	$K = \frac{[\text{enol form}]}{[\text{keto form}]}$
$CH_3{-}\overset{O}{\overset{\|}{C}}H \rightleftharpoons CH_2{=}CHOH$	2×10^{-5}
$CH_3{-}\overset{O}{\overset{\|}{C}}{-}CH_3 \rightleftharpoons CH_2{=}\overset{OH}{\overset{\|}{C}}{-}CH_3$	1.5×10^{-7}
cyclohexanone $\rightleftharpoons$ cyclohexenol (OH)	5×10^{-5}

Enols are important in many reactions, even though the percentage of enol isomer at equilibrium is quite small. As we shall soon see, many reactions of aldehydes and ketones occur by way of the unstable enol form.

As outlined in the introduction, enolization is subject to both acid and base catalysis. In aqueous solutions the base is hydroxide ion. The base forms a bond to a proton α to the carbonyl group as the bond between the proton and the α-carbon breaks to give an anion that is called an **enolate** ion. The enolate ion may be protonated on carbon, which regenerates the keto form, or on oxygen, which yields the enol form.

$$(1)\quad HO:^- + H{-}CH_2{-}\overset{\overset{O}{\|}}{C}CH_3 \rightleftharpoons \left[CH_2{=}\overset{\overset{O^-}{|}}{C}CH_3 \longleftrightarrow {}^-\ddot{C}H_2{-}\overset{\overset{O}{\|}}{C}CH_3\right] + H_2O$$

$$(2)\quad \left[CH_2{=}\overset{\overset{O^-}{|}}{C}CH_3 \longleftrightarrow {}^-\ddot{C}H_2{-}\overset{\overset{O}{\|}}{C}CH_3\right] + H_2O \rightleftharpoons CH_2{=}\overset{\overset{OH}{|}}{C}CH_3 + OH^-$$

Note that the first step in base-catalyzed enolization is formally analogous to the E2 mechanism. The "leaving group" may be considered to be the π-bond electron pair.

The first step of base-catalyzed enolization is simply an acid-base reaction, with the ketone acting as a protic acid. The pK_a for acetone is approximately 19. Although this makes it an extremely weak acid compared to such familiar acids as HCl (pK_a −2.2), HF (pK_a +3), acetic acid (pK_a +5), or water (pK_a +15.7), we must remember that acidity is relative. In all of these acids, the acidic proton is bonded to an electronegative hetero atom (Cl, F, O), whereas acetone is a **carbon acid.** If we compare acetone to ethane (pK_a estimated to be approximately +50), we see that the carbon-hydrogen bonds in acetone are enormously more acidic than those in an alkane.

$$CH_3CH_3 \rightleftharpoons CH_3CH_2^- + H^+ \qquad K_a \cong 10^{-50}\ M$$

The reason for this enhanced acidity is apparent from a consideration of the conjugate bases produced by ionization of the two carbon acids. The anion produced from ethane has its negative charge localized on carbon. Since carbon is a fairly electropositive element, a carbanion is a high-energy species and the ionization that produces it is highly endothermic. On the other hand, the enolate ion produced by ionization of acetone is not a localized carbanion but a resonance hybrid of two structures.

$$CH_3\overset{\overset{O}{\|}}{C}CH_3 \rightleftharpoons H^+ + \left[\ddot{C}H_2{-}\overset{\overset{O}{\|}}{C}CH_3 \longleftrightarrow CH_2{=}\overset{\overset{O^-}{|}}{C}CH_3\right] \qquad K_a \cong 10^{-19}\ M$$

In one of the resonance structures, the negative charge is borne by carbon, as in the ethyl anion. In the other the negative charge is on the more electronegative oxygen. Although the resonance hybrid has the character of both structures, it is more like the structure in which the negative charge is on oxygen. It is important to point out once again that the two structures connected by the double-headed arrow are *not isomers, but resonance structures*. The anion derived from acetone is neither one nor the other of the two indicated structures; it has the character of both. An alternative symbol that gives a somewhat more accurate picture of the electronic distribution is

$$CH_3{-}C\genfrac{}{}{0pt}{}{\diagup\!\!\!\diagup O^{\delta-}}{\diagdown\!\!\!\diagdown CH_2^{\,\delta-}}$$

wherein we see that the negative charge is divided between the carbon and the oxygen. When this anion reacts with water, it can undergo protonation either on carbon, in which case the keto form results, or on oxygen, in which case the enol form is produced. The rate-limiting step for base-catalyzed enolization is usually the deprotonation step.

EXERCISE 15.1 What is the kinetic rate law for base-catalyzed enolization? (Remember that, for most aldehydes and ketones, enolization is an *endothermic* process. In order to actually measure the rate of enolization, you would have to devise some indirect method to determine that the reaction had actually occurred.)

In neutral solution enolization is much slower than it is in basic medium. At pH 7 the principal base present is H_2O. Since H_2O is a much weaker base than OH^-, proton transfer from an aldehyde or ketone is not as rapid, and consequently enolization is slower.

However, below pH 7, the rate of enolization is proportional to the acid concentration. In acidic solution some of the weakly basic carbonyl groups are protonated. The protonated aldehyde or ketone loses a proton from carbon with much greater ease, even to such a weak base as H_2O. A carbonyl group is not very basic, and only a small amount of the protonated structure is present at equilibrium. The presence of the positive charge, however, greatly increases the rate of proton loss from carbon to solvent.

$$(1)\quad H{-}CH_2{-}\overset{O}{\overset{\|}{C}}CH_3 + H_3O^+ \rightleftharpoons \left[H{-}CH_2{-}\overset{\overset{+}{O}H}{\overset{\|}{C}}CH_3 \longleftrightarrow H{-}CH_2{-}\overset{OH}{\overset{|}{\underset{+}{C}}}CH_3 \right] + H_2O$$

$$(2)\quad \left[H{-}CH_2{-}\overset{\overset{+}{O}H}{\overset{\|}{C}}CH_3 \longleftrightarrow H{-}CH_2{-}\overset{OH}{\overset{|}{\underset{+}{C}}}CH_3 \right] + H_2O \rightleftharpoons CH_2{=}\overset{OH}{\overset{|}{C}}CH_3 + H_3O^+$$

In fact, deprotonation of the protonated ketone (step 2) is analogous to the E1 reaction, deprotonation of a carbocation.

$$H_2O \;\; H{-}CH_2{-}\underset{+}{C}(CH_3)_2 \longrightarrow H_3O^+ + CH_2{=}C(CH_3)_2$$

In acid-catalyzed enolization the first step is a rapid equilibrium. Loss of a proton from carbon (step 2) is slower and is rate-determining.

Highlight 15.1

We pointed out in Highlight 14.1 (page 393) that the acid-catalyzed reaction of acetone with water has a close analogy to the S_N1 reaction of *t*-butyl bromide with ethanol. We now see that acid-catalyzed enolization of acetone is analogous to the E1 mechanism for elimination of *t*-butyl bromide to give 2-methylpropene.

$$(CH_3)_2C{=}O \overset{H^+}{\rightleftharpoons} \left[(CH_3)_2C{=}OH^+ \updownarrow (CH_3)_2\overset{+}{C}{-}OH \right] \overset{H_2O}{\rightleftharpoons} (CH_3)_2C(OH)(\overset{+}{O}H_2) \xrightarrow{-H^+} (CH_3)_2C(OH)_2$$

$$\left[(CH_3)_2C{=}OH^+ \leftrightarrow (CH_3)_2\overset{+}{C}{-}OH \right] \xrightarrow{E1} CH_2{=}C(CH_3){-}OH$$

$$(CH_3)_3C{-}Br \xrightarrow{-Br^-} (CH_3)_2\overset{+}{C}{-}CH_3 \xrightarrow{C_2H_5OH} (CH_3)_3C{-}\overset{+}{O}(H){-}C_2H_5 \xrightarrow{-H^+} (CH_3)_3C{-}O{-}C_2H_5$$

$$(CH_3)_2\overset{+}{C}{-}CH_3 \xrightarrow{E1} (CH_3)_2C{=}CH_2$$

Let us summarize. In aqueous solution aldehydes and ketones are in equilibrium with their corresponding enol forms. Interconversion of the enol and keto forms is catalyzed by either acid or base. At any given moment the vast majority of molecules are present as the more stable keto form. However, as we shall see, the small amount of enol form present is involved as an important *intermediate* in many of the reactions of aldehydes and ketones.

One way in which the intermediate enols and enolates can be detected is by **deuterium exchange.** If one dissolves a ketone in D_2O containing DCl or Na^+ OD^-, all of the α-hydrogens are exchanged for deuterium.

$$CH_3CH_2\overset{O}{\overset{\|}{C}}CH_2CH_3 + D_2O \xrightarrow[OD^-]{D^+ \text{ or}} CH_3CD_2\overset{O}{\overset{\|}{C}}CD_2CH_3 + H_2O$$

3-pentanone 3-pentanone-*2,2,4,4-d*$_4$

The amount of deuterium incorporation at equilibrium is related to the initial concentrations of the ketone and D_2O. In dilute solution the D_2O is present in large excess, and replacement of the α-hydrogens by deuterium is essentially complete.

EXERCISE 15.2 For a 1 M solution of 3-pentanone in D_2O in the presence of a catalytic amount of DCl, calculate how many deuterium atoms will be present in each 3-pentanone molecule (on the average) at equilibrium. (Remember that there is a statistical factor; each 3-pentanone molecule starts with four exchangeable Hs, whereas each D_2O molecule starts with two exchangeable Ds. Also remember that the molarity of pure, liquid D_2O is 55.)

The *rate* of deuterium incorporation is proportional to the concentration of ketone and the catalyst, either D^+ or OD^-.

$$\text{rate}_{ex} = k[\text{ketone}][D^+] \qquad \text{or} \qquad k'[\text{ketone}][OD^-]$$

Such exchange reactions may be applied even when the aldehyde or ketone is not very soluble in water. Shaking such a compound with NaOD or DCl in D_2O for several hours results in virtually complete exchange.

$$\xrightarrow[D_2O]{NaOD}$$

2-methylcyclohexanone 2-methylcyclohexanone-*2,6,6-d*$_3$

Since the number of deuteriums is easily determined by mass spectrometry or by NMR, this reaction is a useful technique for *counting* the number of α-hydrogens in an aldehyde or ketone.

EXERCISE 15.3 How many hydrogens are replaced by deuterium when each of the following compounds is treated with NaOD in D_2O?

(a) 2,2,4-trimethyl-3-pentanone (b) 2-ethylbutanal
(c) 3-methylcyclopentanone (d) *trans*-2-pentene

Highlight 15.2

The conversion of an aldehyde or ketone to its **enol form** is catalyzed by either H^+ or OH^-. The double bond of an enol is "electron-rich" because of the contribution of a resonance form that has carbanion character. Enols can react at the nucleophilic carbon with electron-poor species like H^+, giving the O-protonated carbonyl compound.

$$\left[\begin{array}{c} \text{O—H} \\ \text{>C=C<} \end{array} \longleftrightarrow \begin{array}{c} \text{O}^+\text{H} \\ \text{>}\bar{\text{C}}\text{—C<} \end{array} \right] \xrightarrow{H^+} \begin{array}{c} \text{H} \quad \overset{+}{\text{O}}\text{—H} \\ \text{—C—C<} \end{array}$$

enol form of aldehyde or ketone — O-protonated aldehyde or ketone

B. Enolate Ions

From the pK_a of acetone, +19, it is clear that in aqueous solution where the strongest base is OH^- the amount of enolate ion present is small.

$$\underset{\text{p}K_a\ 19}{CH_3\overset{O}{\overset{\|}{C}}CH_3} + OH^- \rightleftharpoons CH_2{=}\overset{O^-}{\overset{|}{C}}CH_3 + \underset{\text{p}K_a\ 15.7}{H_2O}$$

$$K = \frac{[\text{enolate}][H^+]}{[\text{acetone}]} \cdot \frac{[H_2O]}{[OH^-][H^+]} = \frac{10^{-19}}{10^{-15.7}} \cong 10^{-3}\ M$$

However, if a ketone is treated with a much stronger base, it can be converted completely into the corresponding enolate ion.

$$\underset{\text{p}K_a\ 19}{CH_3\overset{O}{\overset{\|}{C}}CH_3} + (i\text{-}C_3H_7)_2N^-\ Li^+ \xrightarrow{\text{THF}} CH_3\overset{O^-\ Li^+}{\overset{|}{C}}{=}CH_2 + \underset{\text{p}K_a\ \cong\ 40}{(i\text{-}C_3H_7)_2NH}$$

$$K = \frac{[\text{enolate}]}{[\text{acetone}]} \times \frac{[(i\text{-}C_3H_7)_2NH]}{[(i\text{-}C_3H_7)_2N^-\ Li^+]} = \frac{10^{-19}}{10^{-40}} \cong 10^{21}$$

Lithium diisopropylamide, $(i\text{-}C_3H_7)_2N^-\ Li^+$, is commonly abbreviated as LDA. It is prepared by treating a solution of diisopropylamine in ether, THF, or 1,2-dimethoxyethane with *n*-butyllithium.

$$(i\text{-}C_3H_7)_2NH + n\text{-BuLi} \longrightarrow (i\text{-}C_3H_7)_2N^-\ Li^+ + C_4H_{10}$$

Since diisopropylamine has p$K_a \cong 40$, whereas *n*-butyllithium may be regarded as the conjugate base of butane (p$K_a \cong 50$), proton transfer is virtually quantitative. Because of its highly basic nature, and because it is relatively soluble in common ether solvents, LDA is widely used for converting ketones quantitatively into their corresponding lithium enolates.

$$\text{cyclohexanone} \xrightarrow[\text{THF, } -78°\text{C}]{\text{LDA}} \text{cyclohexenolate } O^-\ Li^+ + \text{diisopropylamine (HN)}$$

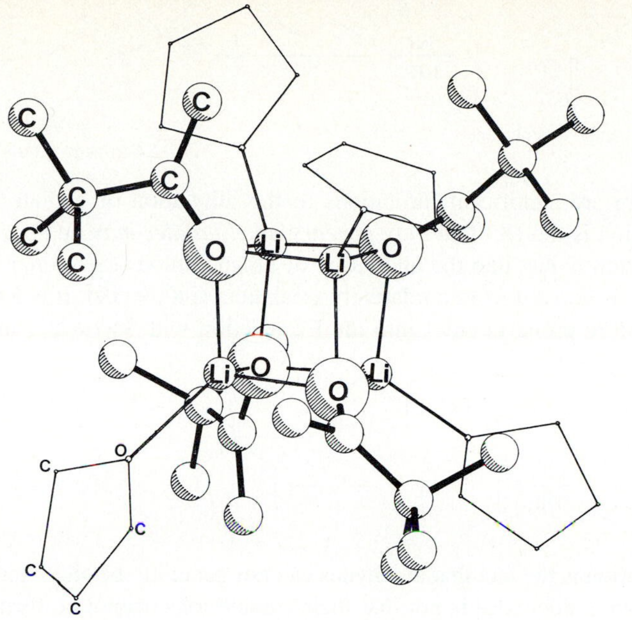

FIGURE 15.1 Structure of the crystalline lithium enolate of 3,3-dimethyl-2-butanone. The hydrogens are omitted. Each lithium is at the center of a tetrahedron of oxygen atoms, three from neighboring enolate ions and one from a tetrahydrofuran molecule. The four tetrahydrofuran molecules are shown as simple pentagons. [Courtesy of D. Seebach and J. Dunitz, Eidgenössische Technische Hochschule, Zürich, Switzerland.]

Solutions of enolate ions may be prepared by the use of strong bases such as LDA. They are quite stable if air and moisture are rigorously excluded. Spectroscopic measurements have shown that lithium enolates exist as tetrameric aggregates in ether solvents. In fact, crystalline lithium enolates have been isolated and found by x-ray analysis to have dimeric and tetrameric structures. One such example is the lithium enolate of 3,3-dimethyl-2-butanone (Figure 15.1). In this structure, note that each lithium cation is surrounded by four oxygens, three from neighboring enolates and one from a tetrahydrofuran molecule.

Enolate ions are useful nucleophiles and are readily alkylated by primary alkyl halides.

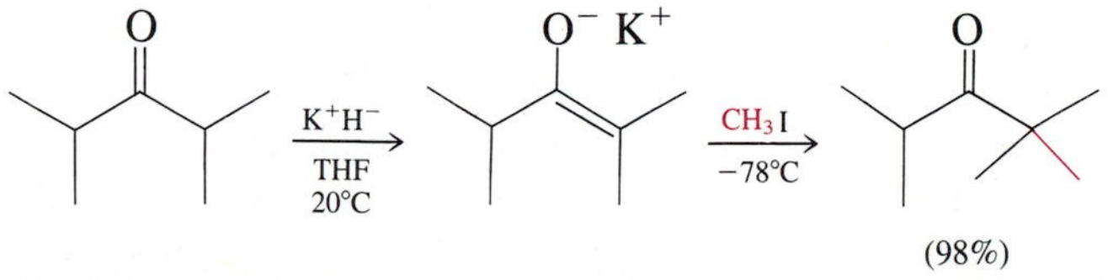

O → (K^+H^-, THF, 20°C) → O^- K^+ → (CH_3I, −78°C) → O

(98%)

Potassium hydride, KH, is commercially available as a grey microcrystalline material dispersed in white mineral oil. Potassium hydride is insoluble in hydrocarbons, ethers, ammonia, and amines. For use, the mineral oil is washed away with an appropriate solvent and the hydride is used as a slurry. It is much more reactive than sodium hydride (page 213). It converts ketones to the potassium enolates in minutes at room temperature.

The alkylation of ketone enolates is a significant synthetic method, since it allows one to construct larger molecules from small, readily available ketones and alkyl halides.

2,2-dimethyl-3-pentanone NaH / THF → 1-iodobutane → (57%) 2,2,4-trimethyl-3-octanone

However, there are significant limitations to the alkylation of enolate ions. The most serious limitation is the fact that *only primary alkyl halides may be used*. In this regard, enolate alkylation is just like the alkylation of alkyne anions (Section 12.5.B). Because the enolate ion is derived from a relatively weak acid ($pK_a \cong 19$), it is a relatively strong base and therefore prone to enter into the E2 reaction with secondary and tertiary alkyl halides.

cyclohexanone —LDA, THF→ $O^- Li^+$ enolate —2-bromopropane→ cyclohexanone + propene + $Li^+ Br^-$

Another limitation is the fact that aldehydes can not generally be alkylated in this manner. The problem with aldehydes is not that their enolate ions cannot be formed, but that the aldehyde carbonyl is itself so reactive to enolate ions that it reacts with the enolate as fast as it is formed. We shall return to a consideration of this reaction in Section 15.2.

Enolate ions are ambident anions (Section 9.4). Just as they may undergo protonation on either carbon or oxygen, they may also react with other electrophilic species at either of these two centers. An example that illustrates this ambident character is the reaction with chlorotrimethylsilane.

ketone —LDA, THF, −78°C→ $O^- Li^+$ enolate —Me_3SiCl→ $OSiMe_3$ (85%) + $Li^+ Cl^-$

Chlorotrimethylsilane, $(CH_3)_3SiCl$, is a clear liquid, b.p. 57 °C, that has a number of uses in organic synthesis. The chemistry of organosilicon compounds is discussed in Sections 26.9–11

Whether or not the oxygen or the carbon of an enolate ion is the site of reaction with an electrophile is determined by a number of factors. One important factor, which is illustrated by the examples just given, is the nature of the electrophile. In general, alkylation of an enolate always occurs on carbon. However, some electrophiles, including silyl halides, react predominantly on oxygen. In the latter case it is the great strength of the silicon-oxygen bond that governs the course of reaction. In Appendix III, we see that the average bond energy for the silicon-oxygen bond is approximately 108 kcal $mole^{-1}$, whereas that for the silicon-carbon bond is only 72 kcal $mole^{-1}$. This large difference in the heats of reaction for the two competing processes partially appears at the transition state. On the other hand, the carbon-carbon and carbon-oxygen single bonds have comparable bond energies, 83 and 86 kcal $mole^{-1}$, respectively. In the absence of a strong bias for the formation of C—C or C—O, alkylation of an enolate tends to produce the more stable product. We have seen in the previous section that the keto form is much more stable than the enol form.

EXERCISE 15.4 From the pK_a of acetone, 19, and the equilibrium constant for formation of the enol form (page 422), derive the pK_a of the enol. How does this value compare with the pK_a values for alcohols?

EXERCISE 15.5 Show how the following compounds may be synthesized from cyclohexanone.

(a) [1-(triethylsilyloxy)cyclohexene, $OSiEt_3$] (b) [2-methylcyclohexanone]

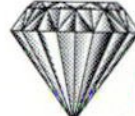

Highlight 15.3

The conversion of an aldehyde or ketone to an **enolate ion** is brought about by base, which takes a proton from an α position. The double bond of an enolate ion is even more electron-rich than the double bond of an enol, due to the contribution of the resonance form with negative charge on carbon. Enolates react with some electrophiles (for example, H^+) at carbon, forming the aldehyde or ketone. Other electrophiles, notably silicon halides, react at the oxygen atom, giving enol ethers.

$Cl—Si(CH_3)_3$; H^+

enolate of aldehyde or ketone

reaction at O ; reaction at C

$O—Si(CH_3)_3$

silyl enol ether ; aldehyde or ketone

C. Racemization

When (*R*)-3-phenyl-2-butanone is dissolved in aqueous ethanol that contains NaOH or HCl, the optical rotation of the solution gradually drops to zero. Reisolation from the reaction mixture yields a racemate, an equimolar mixture of the *R* and *S* enantiomers. Such a process, in which an enantiomerically homogeneous compound changes into a racemate, is called **racemization.** In this case, the rate of racemization is proportional to the concentration of the ketone and also to the concentration of NaOH or HCl. Clearly, racemization occurs by way of the intermediate enol form in which the former stereocenter is planar and hence achiral.

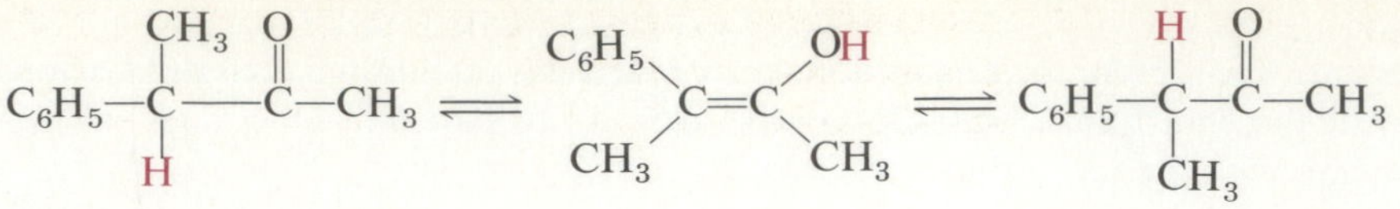

$$\mathrm{C_6H_5{-}\underset{H}{\overset{CH_3}{C}}{-}\overset{O}{\overset{\|}{C}}{-}CH_3} \rightleftharpoons \mathrm{\overset{C_6H_5}{\underset{CH_3}{}}C{=}C\overset{OH}{\underset{CH_3}{}}} \rightleftharpoons \mathrm{C_6H_5{-}\underset{CH_3}{\overset{H}{C}}{-}\overset{O}{\overset{\|}{C}}{-}CH_3}$$

(*R*)-3-phenyl-2-butanone (achiral) (*S*)-3-phenyl-2-butanone

Since racemization involves the formation of the enol form, the rate of racemization is exactly equal to the rate of enolization.

$$\text{rate} = k[\text{ketone}][\text{H}^+] \quad \text{or} \quad k'[\text{ketone}][\text{OH}^-]$$

Furthermore, the rate of racemization is equal to the rate of deuterium incorporation because both reactions involve the same intermediate enol.

Note that racemization of an optically active ketone will occur *only* when the stereocenter is α to the carbonyl group. If the aldehyde or ketone is chiral because some other carbon is a stereocenter, the enol form is also chiral, and enolization does not result in racemization.

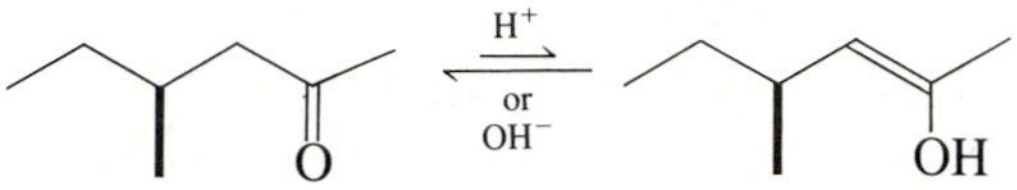

$\xrightleftharpoons[\text{or OH}^-]{\text{H}^+}$

(*S*)-4-methyl-2-hexanone (still chiral)

EXERCISE 15.6 Which of the following compounds will racemize in basic solution (sodium ethoxide in ethanol)?

(a) (*R*)-2-methylbutanal
(b) (*S*)-3-methylcyclohexanone
(c) (*S*)-3-methyl-2-heptanone
(d) (*S*)-2-methyl-1-pentanol

D. Halogenation

Aldehydes and ketones undergo acid- and base-catalyzed halogenation. The reaction occurs with chlorine, bromine, and iodine.

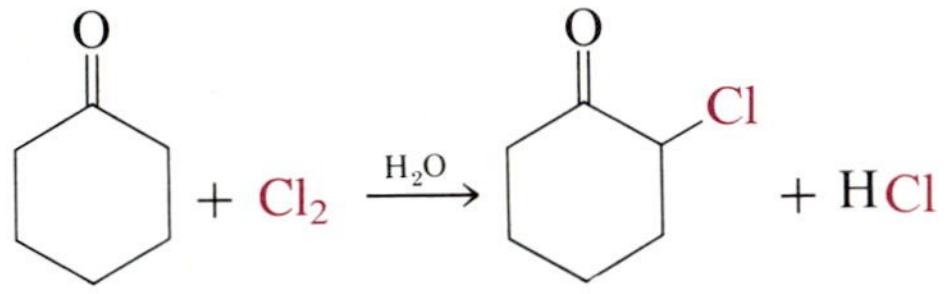

$+ \ Cl_2 \xrightarrow{H_2O}$ Cl $+ \ HCl$

(61–66%)

$+ \ Br_2 \xrightarrow{CH_3OH}$ Br $+ \ HBr$

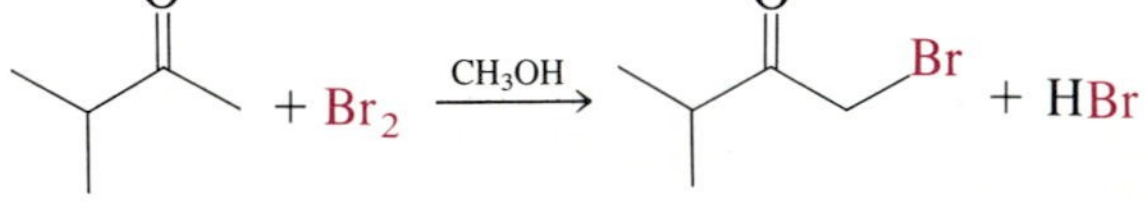

(70%)

Halogenation is often carried out without added catalyst. In this case the reaction is **autocatalytic** (catalyzed by one of the reaction products). There usually is no apparent reaction when bromine and the ketone are first mixed. However, there is a slow uncatalyzed reaction that produces some HBr. As HBr is produced in the reaction, it catalyzes the bromination, producing more HBr, which makes the reaction proceed even faster. In practice, autocatalytic reactions often have an **induction period** (a time when no apparent reaction is occurring), after which a rapid and vigorous reaction sets in and is soon over.

In base-catalyzed halogenation, a full equivalent of base must be used because an equivalent of HX is formed in the reaction.

The actual species that reacts with the halogen is the enol form of the aldehyde or ketone; the purpose of the acid or base is simply to catalyze enolization. The first step in the acid-catalyzed halogenation is simply the normal electrophilic reaction of halogen with a double bond. The probable mechanism of the acid-catalyzed reaction is as follows.

$$(1)\quad \mathrm{-\overset{H}{\underset{|}{\overset{|}{C}}}-\overset{O}{\overset{\|}{C}}-} + \mathrm{H^+} \underset{}{\overset{\text{fast}}{\rightleftharpoons}} \mathrm{-\overset{H}{\underset{|}{\overset{|}{C}}}-\overset{\overset{+}{O}H}{\overset{\|}{C}}-}$$

$$(2)\quad \mathrm{-\overset{H}{\underset{|}{\overset{|}{C}}}-\overset{\overset{+}{O}H}{\overset{\|}{C}}-} \overset{\text{slow}}{\rightleftharpoons} \mathrm{\rangle C{=}C\langle^{OH}} + \mathrm{H^+}$$

$$(3)\quad \mathrm{\rangle C{=}C\langle^{OH}} + \mathrm{X_2} \overset{\text{fast}}{\rightleftharpoons} \mathrm{-\overset{X}{\underset{|}{\overset{|}{C}}}-\overset{\overset{+}{O}H}{\overset{\|}{C}}-} + \mathrm{X^-}$$

$$(4)\quad \mathrm{-\overset{X}{\underset{|}{\overset{|}{C}}}-\overset{\overset{+}{O}H}{\overset{\|}{C}}-} \overset{\text{fast}}{\rightleftharpoons} \mathrm{-\overset{X}{\underset{|}{\overset{|}{C}}}-\overset{O}{\overset{\|}{C}}-} + \mathrm{H^+}$$

There is considerable evidence for this mechanism. First, the rate of halogenation depends *only* upon the concentration of the ketone and the acid; it is independent of halogen concentration. Second, chlorination and bromination occur at the same rate. Finally, acid-catalyzed halogenation occurs at the same rate as does acid-catalyzed exchange of an α-proton for deuterium, strongly suggesting that these two reactions proceed by way of the same intermediate.

In the base-catalyzed reaction, the enolate ion is the probable intermediate.

$$(1)\quad \mathrm{-\overset{H}{\underset{|}{\overset{|}{C}}}-\overset{O}{\overset{\|}{C}}-} + \mathrm{OH^-} \rightleftharpoons \mathrm{\rangle C{=}C\langle^{O^-}} + \mathrm{H_2O}$$

$$(2)\quad \left[\mathrm{\rangle C{=}C\langle^{O^-}} \longleftrightarrow \mathrm{\rangle \overset{-}{C}{-}C\langle^{\!\!/\!\!/ O}}\right] + \mathrm{X_2} \rightleftharpoons \mathrm{-\overset{X}{\underset{|}{\overset{|}{C}}}-\overset{O}{\overset{\|}{C}}-} + \mathrm{X^-}$$

The acid- and base-catalyzed halogenation reactions differ in several important aspects. In acid-catalyzed halogenation each successive halogenation step is normally slower than the previous one. Therefore it is usually possible to prepare a monohalo ketone in good yield by carrying out the halogenation under conditions of acid catalysis using one equivalent of halogen, as shown by the examples on page 430. In the base-catalyzed reaction, each successive halogenation step is faster than the previous one, since the electron-attracting halogens increase the acidity of halogenated ketones; consequently base-catalyzed halogenation is not a generally useful method for preparation of a monohalo ketone.

EXERCISE 15.7 Show how acetone can be converted into each of the following compounds,

(a) O, Br (b) O

Highlight 15.4

When a ketone that has one stereocenter α to its carbonyl group enolizes, chirality is lost because the intermediate enol has a plane of symmetry. The enol may react from either side of the α-carbon with an electrophile (H^+, Br_2) to yield a racemic derivative.

15.2 The Aldol Addition Reaction

When acetaldehyde is treated with aqueous sodium hydroxide solution, 3-hydroxybutanal is formed in 50% yield. When this reaction was first discovered, in the early nineteenth century, the product was known as ''aldol,'' from ''*ald*ehyde-alcoh*ol*.''

$$2\,CH_3CHO \xrightarrow[5\,°C,\ 4\text{–}5\ hr]{OH^-,\ H_2O} CH_3\overset{\overset{OH}{|}}{C}HCH_2CHO$$

(50%)
aldol

The reaction is a general one for aldehydes that have a hydrogen α to the carbonyl group. Consequently, the term ''aldol'' addition reaction has come to be a generic term for the addition of the enol or enolate of one aldehyde or ketone to the carbonyl group of another. Under more vigorous conditions (base concentration, temperature), elimination of the β-hydroxy group occurs and an α,β-unsaturated aldehyde is produced.

$$2\,CH_3CH_2CH_2CHO \xrightarrow[80\text{–}100\,°C,\ 3\ hr]{1\,M\ NaOH,\ H_2O} \left[CH_3CH_2CH_2\overset{\overset{OH}{|}}{C}H\underset{\underset{CH_3}{|}}{\underset{CH_2}{|}}{C}HCHO \right] \xrightarrow{-H_2O}$$

$$CH_3CH_2CH_2CH{=}\overset{\overset{CHO}{|}}{C}CH_2CH_3$$

(86%)
2-ethylhex-2-enal

Mechanistically, the addition reaction is simply the nucleophilic addition of an enolate ion to the carbonyl group of another, nonionized molecule. Thus, in one reaction, both of the important chemical properties of the carbonyl group are expressed. The mechanism for the dimerization of acetaldehyde consists of the following three simple steps.

$$(1)\ CH_3CHO + OH^- \rightleftharpoons [CH_2{=}CHO^- \longleftrightarrow \bar{C}H_2{-}CH{=}O] + H_2O$$

$$(2)\ CH_3\overset{\overset{O}{\|}}{C}H + {}^-CH_2{-}CH{=}O \rightleftharpoons CH_3\overset{\overset{O^-}{|}}{C}HCH_2CHO$$

$$\text{(3)}\quad CH_3\overset{O^-}{\overset{|}{C}}HCH_2CHO + H_2O \rightleftharpoons CH_3\overset{OH}{\overset{|}{C}}HCH_2CHO + OH^-$$

The slow, rate-limiting step is usually the addition step. The dehydration of the β-hydroxy aldehyde or ketone mechanism to produce an α, β-unsaturated carbonyl compound involves the enolate ion of the aldol product.

$$\text{(4)}\quad CH_3\overset{OH}{\overset{|}{C}}HCH_2CHO + OH^- \rightleftharpoons CH_3\overset{OH}{\overset{|}{C}}H{-}\overset{-}{C}HCHO + H_2O$$

$$\text{(5)}\quad CH_3\overset{OH}{\overset{|}{C}}H{-}\overset{-}{C}HCHO \rightleftharpoons CH_3CH{=}CHCHO + OH^-$$

When a mixture of two different aldehydes is treated with base, four aldol products are possible.

$$RCH_2CHO + R'CH_2CHO \longrightarrow \begin{cases} RCH_2\overset{OH}{\overset{|}{C}}H{-}\overset{R}{\overset{|}{C}}HCHO \\ RCH_2\overset{OH}{\overset{|}{C}}H{-}\overset{R'}{\overset{|}{C}}HCHO \\ R'CH_2\overset{OH}{\overset{|}{C}}H{-}\overset{R}{\overset{|}{C}}HCHO \\ R'CH_2\overset{OH}{\overset{|}{C}}H{-}\overset{R'}{\overset{|}{C}}HCHO \end{cases}$$

In practice, such a complex mixture is the usual result. However, when one of the aldehydes cannot form an enolate ion or when one has an unusually unreactive carbonyl group, such a "mixed aldol reaction" is often a practical synthetic alternative.

$$\text{furfural (2-furyl-CHO)} + CH_3CH_2CHO \xrightarrow{NaOH,\ H_2O} \text{2-furyl-}CH{=}C(CHO)CH_3 \quad (72\%)$$

furfural

(72%)

Although ketones also undergo aldol addition, the reaction in this case often requires rather special conditions. The overall reaction is an equilibrium process, and it appears that the equilibrium constant in most ketone aldol reactions is unfavorable. For example, acetone and its aldol product (4-hydroxy-4-methyl-2-pentanone) are in rapid equilibrium in the presence of base catalysts. The amount of the aldol product in the equilibrium is only a few percent. However, if the product is removed from the basic catalyst as it is formed, the conversion can be accomplished in 80% yield.

$$CH_3COCH_3 \underset{}{\overset{BaO}{\rightleftharpoons}} CH_3COCH_2C(OH)(CH_3)_2$$

4-hydroxy-4-methyl-2-pentanone

Intramolecular aldol addition of diketones is an important method for the synthesis of cyclic compounds. In this case, the dehydrated product is usually desired, and such reactions are carried out under conditions of base concentration and temperature that lead to the α, β-unsaturated ketone.

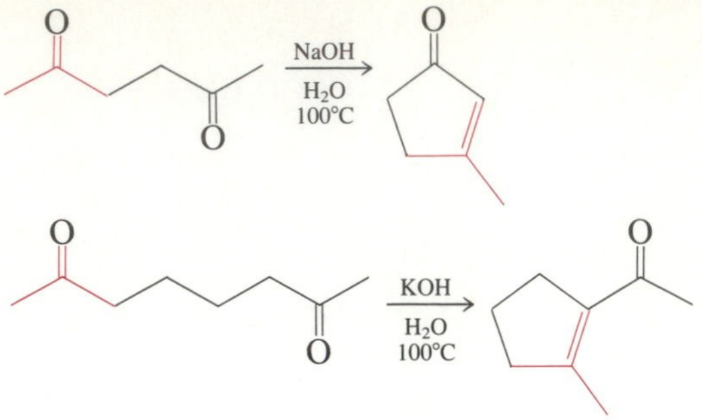

NaOH, H_2O, 100°C

KOH, H_2O, 100°C

Although intramolecular aldol addition is exceedingly useful for the synthesis of cyclic compounds, the method has an important limitation in that it is only generally applicable for the synthesis of five- six-, and seven-membered ring systems.

The intermolecular aldol addition of one ketone to another is slightly *endothermic*. In part this is due to the fact that the bonds formed in the product are about the same strength as those broken in the reactants and in part it is due to the unfavorable entropy resulting from combining two molecules into one. In the intramolecular version of the reaction, the entropy of reaction is not unfavorable, since one reactant molecule gives one product molecule. Consequently, intramolecular aldol additions of one ketone to another are generally successful.

However, the $\Delta G°$ of reaction is not very negative in most cases, and as a result, the addition reaction is easily reversible. Cyclopropane and cyclobutane rings are not produced by intramolecular aldol addition because it is necessary to create approximately 25 kcal mole^{-1} of ring strain in the formation of these rings. Rings larger than seven members are not formed, both because of ring strain in the product (unfavorable $\Delta H°$, see Table 5.5) and the improbability of bringing the two ends of the chain into proximity (unfavorable $\Delta S°$).

Since ketones undergo self-addition much more slowly than aldehydes, mixed aldol additions between a ketone and a nonenolizable aldehyde are usually clean. In order to ensure the formation of a 1:1 adduct, an excess of ketone is often used, and the reaction is carried out under fairly mild conditions.

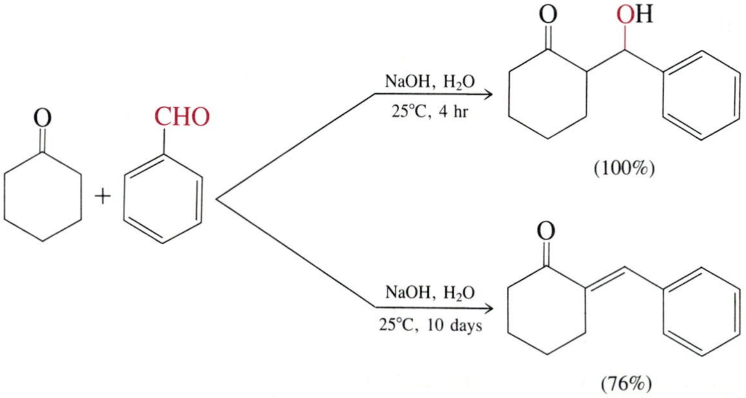

CHO

NaOH, H_2O, 25°C, 4 hr → (100%)

NaOH, H_2O, 25°C, 10 days → (76%)

EXERCISE 15.8 Write the structure of the diketone that would give each of the following cyclic α,β-unsaturated ketones upon treatment with KOH in ethanol.

(a)

(b)

The aldol reactions discussed up to this point are carried out with a base such as hydroxide ion or ethoxide ion in a protic solvent such as water or ethanol. Since aldehydes and ketones are a good deal less acidic than these solvents, the enolate ions are formed reversibly and in only small amounts (see Section 15.1.B). However, the reaction may also be carried out in another manner. Recall that the strong base lithium diisopropylamide (LDA) converts a ketone completely into the corresponding enolate ion (see page 426). Since all of the ketone is consumed in this essentially irreversible reaction, self-addition of the ketone to itself does not occur. However, if an aldehyde is subsequently added to the cold enolate solution, a rapid and efficient mixed aldol reaction occurs. The initial product is the lithium salt of the β-hydroxy ketone. Work-up with water affords the aldol.

$$CH_3\overset{\overset{O}{\|}}{C}CH_3 \xrightarrow[\text{THF, }-78^\circ\text{C}]{\text{LDA}} CH_3\overset{\overset{O^-Li^+}{|}}{C}{=}CH_2 \xrightarrow{CH_3CH_2CHO}$$

$$CH_3\overset{\overset{O}{\|}}{C}CH_2\overset{\overset{O^-Li^+}{|}}{C}HCH_2CH_3 \xrightarrow[\text{(work-up)}]{H_2O} CH_3\overset{\overset{O}{\|}}{C}CH_2\overset{\overset{OH}{|}}{C}HCH_2CH_3$$

(85%)
4-hydroxy-2-hexanone

With unsymmetrical ketones, the strong base LDA always removes a proton from the *least sterically hindered position*; that is, from the position having the larger number of α-hydrogens. Thus, in such a case the mixed aldol reaction occurs specifically at one of the two α-positions.

$$CH_3\overset{\overset{O}{\|}}{C}CH_2CH_3 \xrightarrow[\text{THF, }-78^\circ\text{C}]{\text{LDA}} \xrightarrow{CH_3CHO} \xrightarrow{H_2O} CH_3\overset{\overset{OH}{|}}{C}HCH_2\overset{\overset{O}{\|}}{C}CH_2CH_3$$

(75%)
5-hydroxy-3-hexanone

This method of preforming the lithium enolate and subsequently adding the carbonyl acceptor compound is the best way of carrying out a mixed aldol condensation.

EXERCISE 15.9 Show how each of the following compounds can be prepared from simple aldehydes and ketones utilizing the aldol addition reaction.

(a) CHO
(b) O
(c) CHO OH
(d) O OH

Highlight 15.5

The carbonyl group of an un-ionized aldehyde or ketone is a good acceptor of nucleophiles (that is, is **electrophilic**), as shown by the fact that it reacts with even weak nucleophiles like H_2O, NH_2X, and HCN. At the same time, hydrogens attached to the carbon α to C=O are sufficiently acidic to be removed by base to give the enolate ion, which has a very **nucleophilic** double bond carbon.

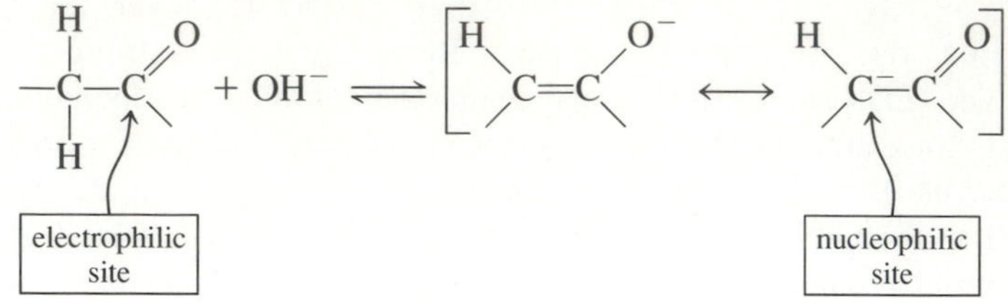

electrophilic site

nucleophilic site

Thus, in basic solution, aldehydes and ketones undergo a self-reaction between the nucleophilic ionized form and the electrophilic un-ionized form. The product, called an **aldolate ion,** reacts with a proton from the solvent to give an **aldol.** This product readily loses a molecule of water to give an α,β-unsaturated aldehyde or ketone.

ROH

aldolate ion

$-H_2O$

an "aldol"

an α,β-unsaturated aldehyde or ketone

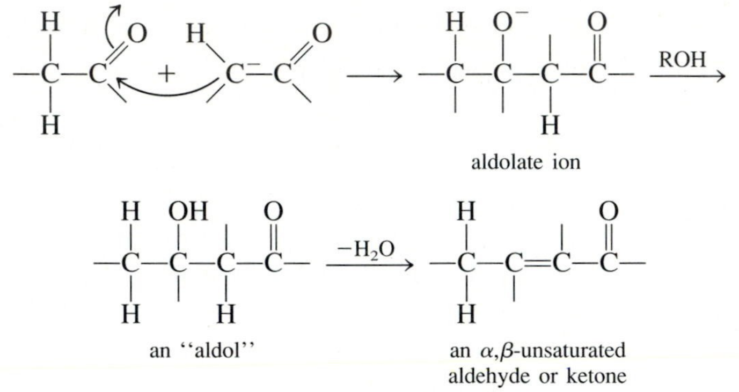

PROBLEMS

1. Name the following materials.

a. 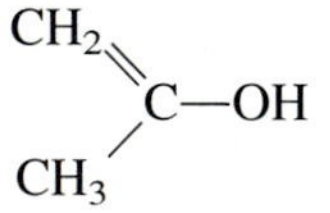 $CH_2=C(OH)CH_3$

b. $CH_3C(=OH^+)CH_3$

c. OH

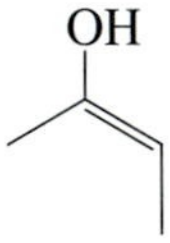

d. 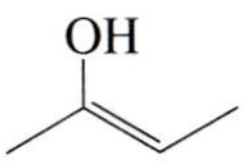 OH

e. 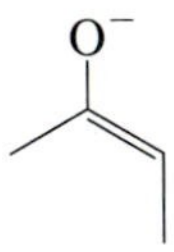 O^-

f. 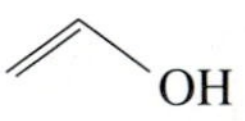 OH

g. 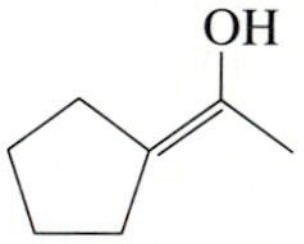 OH

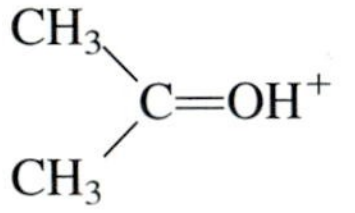

h. 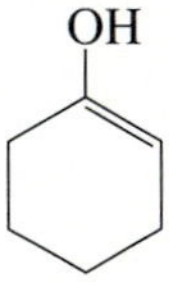 OH

i. 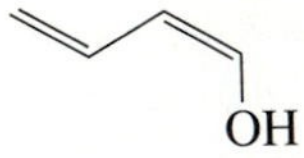 OH

j. 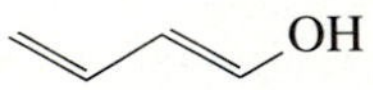 OH

2. Provide IUPAC names for at least one isomer of each compound in problem 1.

3. Write structures corresponding to each of the following names.
   a. 1-hydroxy-1-butene
   b. 2-hydroxy-1-butene
   c. lithium propenolate
   d. magnesium cyclohexenolate
   e. lithium diisopropylamide
   f. 1-cyclopropylethenol

4. What is the product of the reaction of 4,4-dimethylcyclohexanone with each of the following reagents?
   a. (i) lithium diisopropylamide (LDA) in THF; (ii) CH_3CH_2Br
   b. bromine in acetic acid
   c. [benzene ring]—CHO + aqueous NaOH
   d. NaOD in D_2O at 25 °C

5. Answer problem 4 for 2-methylpentanal. In this case, some of the reactions will not work. Give products for those reactions that are expected to be successful and indicate those that are expected to fail.

6. Explain why each of the following syntheses cannot be accomplished in the specified manner.
   a. 2,2-dimethyl-3-hydroxycyclobutanone by an intramolecular aldol addition
   b. 2-ethyl-3-hydroxybutanal by a mixed aldol addition reaction

7. In reactions of protonated acetone, why does reaction with a base always occur at carbon rather than at the positive oxygen?

8. If the acid-catalyzed bromination of bromoacetone is analyzed immediately after the bromine has all reacted, the major product present is 1,1-dibromoacetone. If the reaction mixture is allowed to stir at room temperature for several hours and then analyzed, the sole product is found to be 1,3-dibromoacetone. Explain these observations with a reasonable mechanism for each step.

9. Compound A, $C_7H_{16}O$, reacts with sodium dichromate in aqueous H_2SO_4 to give B, $C_7H_{14}O$. When B is treated with NaOD at 25 °C for several hours, analysis shows that it has incorporated two deuteriums. Compound B is not oxidized by Ag_2O. What are A and B?

10. Compound C has the formula $C_{12}H_{20}$ and is optically active. It reacts with H_2 and Pt to give two isomers, D_1 and D_2, $C_{12}H_{22}$. Ozonolysis of C gives only E, $C_6H_{10}O$, which is optically active. Compound E reacts with hydroxylamine to give F, $C_6H_{11}NO$. When E is treated with DCl in D_2O for several hours and then analyzed by mass spectrometry, it is found to have a molecular weight of 101. The NMR spectrum of E shows that it has only one methyl group, which appears as a doublet with J =6.5 Hz. What are compounds C through F?

11. An unknown compound, G, has the formula C_6H_8O and shows a singlet methyl in its NMR spectrum. When treated with hydrogen gas and palladium it absorbs 1 equivalent of H_2 to give a product, H, with the formula $C_6H_{10}O$. The infrared spectrum of compound H shows that it has a carbonyl group. Compound H reacts with NaOD in D_2O to give a product shown by mass spectrometry to have the formula $C_6H_7D_3O$. Compound H reacts with peroxyacetic acid, CH_3CO_3H, to give I, which has the formula $C_6H_{10}O_2$. The NMR spectrum of I contains one and only one absorption due to a CH_3 group, a doublet with J = 8 Hz at δ =1.2 ppm. Propose structures for compounds G, H, and I.

12. Cyclopropanones undergo a reaction that is not common for other ketones—cleavage of one of the carbon-to-carbonyl bonds. Thus, cyclopropanone itself reacts with NaOH in water to give sodium propionate. Propose a mechanism for this unusual reaction. Predict the product that will result when 2-methylcyclopropanone is treated in the same manner. Treatment of 2-bromo-3-pentanone with NaOH in water yields, after acidification of the reaction mixture, 2-methylbutanoic acid. Propose a mechanism for this reaction (the **Favorskii reaction**).

13. Each of the following compounds may be obtained by an intramolecular aldol condensation. Show the precursor in each case.

a. 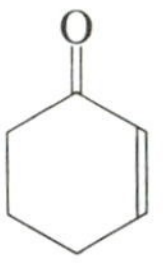 b.

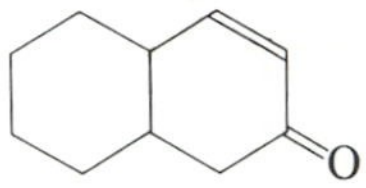

c. 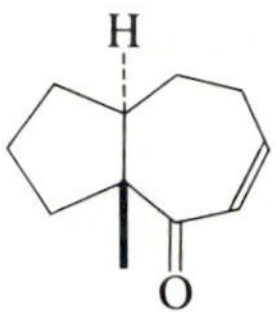 d.

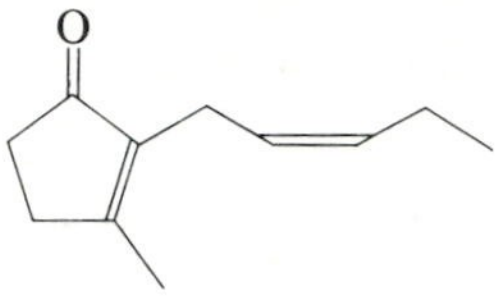

14. Bromination of 7-cholestanone gives 6-bromo-7-cholestanone with $J_{5H,6H} = 2.8$ Hz. This initial product slowly converts to the more stable 6-bromo stereoisomer that has $J_{5H,6H} = 11.8$ Hz. Assign stereostructures to the two isomers and explain their relative stabilities.

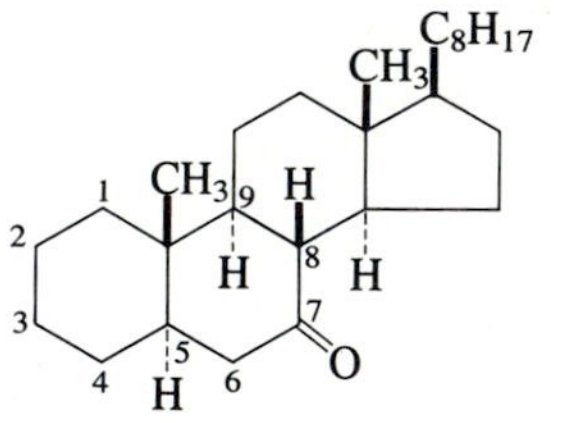

7-cholestanone

15. When a mixture of an aldehyde or ketone and an α-halo ester is treated with a strong base, an α, β-epoxy ester is obtained.

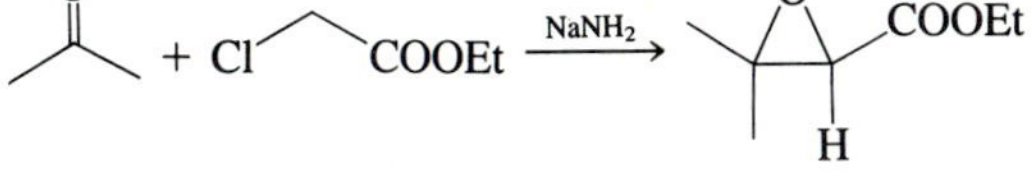

Propose a mechanism for this reaction (**Darzen's condensation**).

CHAPTER 16

ORGANIC SYNTHESIS

16.1 Introduction

Organic synthesis is the preparation of a desired organic compound from a commercially available material, usually by some multistep procedure. It is an important element of organic chemistry and the cornerstone upon which the organic chemical industry is based. A scientist who wishes to study the physical, chemical, or physiological properties of a compound obviously must have a sample of it. Since relatively few organic compounds are commercially available from chemical suppliers, the scientist often must synthesize the desired material. In Section 10.12 we had a brief look at multistep synthesis. In this chapter we shall go into the topic in somewhat greater detail using the chemistry previously learned.

16.2 Considerations in Synthesis Design

The goal in any synthesis is to obtain a pure sample of the desired product by the most efficient and convenient procedure possible. For this reason one usually strives to use reactions that can reliably be expected to give only a single product and avoids reactions that will give a mixture of products. It is also important to plan a synthesis that entails the fewest possible steps. This is necessary both in terms of the amount of time consumed in an overly long route and in the ultimate yield that may be realized. A ten-step synthesis averaging 80% yield per step will give an overall yield of only 10.7%.

In planning a synthesis, three interrelated factors are involved.

1. Construction of the Proper Carbon Skeleton. In a sense, reactions that result in formation of a new carbon-carbon bond are the most important reactions in organic chemistry because these reactions allow us to build up more complicated structures. A brief summary of the carbon-carbon bond-forming reactions that we have encountered follows.

(a) Reaction of primary alkyl halides with cyanide ion (Section 9.1).

$$RCH_2X + CN^- \longrightarrow RCH_2CN$$

(b) Reaction of primary alkyl halides with acetylide ions (Section 12.5.B).

$$RCH_2X + {}^-C{\equiv}CR' \longrightarrow RCH_2C{\equiv}CR'$$

(c) Addition of HCN to aldehydes and ketones (Section 14.7.B).

$$R_2C{=}O + HCN \longrightarrow R_2\overset{\displaystyle OH}{\overset{|}{C}}CN$$

(d) Addition of acetylide ions to aldehydes and ketones (Section 14.7.A).

$$R_2C{=}O + {}^-C{\equiv}CR' \longrightarrow \xrightarrow{H^+} R_2\overset{\displaystyle OH}{\overset{|}{C}}C{\equiv}CR'$$

(e) Reaction of Grignard reagents with oxirane (Section 10.11.A).

$$RMgX + H_2C\overset{O}{\frown}CH_2 \longrightarrow \xrightarrow{H^+} RCH_2CH_2OH$$

(f) Reactions of Grignard reagents and alkyllithium compounds with aldehydes and ketones (Section 14.7.A).

$$RMgX + R'CHO \longrightarrow \xrightarrow{H^+} R{-}\overset{\displaystyle OH}{\overset{|}{C}}H{-}R'$$

$$RMgX + R'_2C{=}O \longrightarrow \xrightarrow{H^+} R{-}\underset{\displaystyle R'}{\underset{|}{\overset{\displaystyle OH}{\overset{|}{C}}}}{-}R'$$

$$RMgX + CH_2O \longrightarrow \xrightarrow{H^+} RCH_2OH$$

(g) Alkylation of enolate ions with primary alkyl halides (Section 15.1.B).

$$R_2CH\overset{\displaystyle O}{\overset{\|}{C}}R' \xrightarrow{base} R_2C{=}\overset{\displaystyle O^-}{\overset{|}{C}}R' \xrightarrow{R''CH_2X} R_2\underset{\displaystyle CH_2R''}{\underset{|}{C}}\overset{\displaystyle O}{\overset{\|}{C}}R'$$

(h) The Wittig reaction (Section 14.7.D).

$$R_2C{=}O + R'CH{=}P(C_6H_5)_3 \longrightarrow R_2C{=}CHR'$$

(i) The aldol addition reaction (Section 15.2).

$$2\ RCH_2CHO \longrightarrow RCH_2\overset{\displaystyle HO}{\overset{|}{C}}H\overset{\displaystyle R}{\overset{|}{C}}HCHO \longrightarrow RCH_2CH{=}\overset{\displaystyle R}{\overset{|}{C}}CHO$$

$$R\overset{\displaystyle O}{\overset{\|}{C}}CH_3 \xrightarrow{LDA} R\overset{\displaystyle O^-}{\overset{|}{C}}{=}CH_2 \xrightarrow{R'CHO} R\overset{\displaystyle O}{\overset{\|}{C}}CH_2\overset{\displaystyle OH}{\overset{|}{C}}HR'$$

A more complete summary of carbon-carbon bond-forming reactions is given in Appendix II of the Study Guide.

Highlight 16.1

The carbon-carbon bond forming reactions described above may be summarized in a compact way as follows. The experimental conditions for the individual reaction types vary greatly.

carbon nucleophile	+	carbon electrophile	⟶	carbon-carbon bonded products *(Partial list)*
$N{\equiv}C^-$ cyanide				
$R{-}C{\equiv}C^-$ acetylide		C=O carbonyl		nitriles
$>C^-$ alkide		O / C—C oxirane		cyanohydrins
				alcohols
				alkenes
$O{=}C{-}C^-$ enolate		$CH_2{-}X$ primary alkyl halide		alkynes
				alkenals
$Ph_3P^+{-}C^-$ ylide				

2. Placement of Desired Functional Groups in Their Proper Place. This aspect of a synthesis involves the introduction, removal, or interconversion of functional groups. We have encountered a great many such reactions. Rather than summarize them all, we shall only give a few illustrative examples.

(a) Introduction of a functional group.

$$R{-}CH(R'){-}R'' + Br_2 \xrightarrow{h\nu} R{-}CBr(R'){-}R''$$

(b) Removal of a functional group.

$$R{-}\overset{O}{\overset{\|}{C}}{-}R' + H_2NNH_2 \xrightarrow[(HOCH_2CH_2)_2O,\ \Delta]{NaOH} RCH_2R'$$

$$-\overset{H}{\underset{|}{\overset{|}{C}}}-\overset{OH}{\underset{|}{\overset{|}{C}}}- \xrightarrow[(-H_2O)]{H^+} >C{=}C< \xrightarrow[Pd]{H_2} -\overset{H}{\underset{|}{\overset{|}{C}}}-\overset{H}{\underset{|}{\overset{|}{C}}}-$$

(c) Interconversion of functional groups.

$$-\overset{\text{OH}}{\overset{|}{\text{CH}}}- \underset{\text{reduction}}{\overset{\text{oxidation}}{\rightleftharpoons}} -\overset{\text{O}}{\overset{\|}{\text{C}}}-$$

$$-CH_2Br \underset{PBr_3}{\overset{OH^-}{\rightleftharpoons}} -CH_2OH$$

Insofar as possible, one should choose reactions for building up the carbon skeleton so that the least amount of subsequent functional group manipulation is necessary. A complete summary of functional group interconversions is included in Appendix II of the Study Guide.

Highlight 16.2

Functional groups are the keys to organic synthesis. Such groups are introduced into or removed from organic compounds. Functional groups can be converted into other functional groups, used to link molecular fragments to form a larger molecule, or to create two smaller molecules from a large molecule.

3. Control of Stereochemistry Where Relevant. When more than one stereoisomer of the desired product is possible, it is necessary to design a synthesis that will yield only that isomer. In such cases it is important to use reactions that are stereoselective—that is, reactions yielding largely or completely one stereoisomer when two or more might result. The stereoselective reactions we have encountered are summarized as follows.

(a) S_N2 displacement reactions of secondary halides (Chapter 9).

$$\text{R(H)(R')C—X} + \text{Y}^- \longrightarrow \text{Y—C(R)(R')(H)} + \text{X}^-$$

(b) Catalytic hydrogenation of alkynes (Section 12.6.A).

$$\text{R—C}\equiv\text{C—R}' \xrightarrow[\text{cat.}]{H_2} \text{(H)(R)C=C(H)(R')}$$

(c) Metal ammonia reduction of alkynes (Section 12.6.A).

$$\text{R—C}\equiv\text{C—R}' \xrightarrow[\text{NH}_3]{\text{Na}} \text{(H)(R)C=C(R')(H)}$$

(d) Oxidation of alkenes with osmium tetroxide (Section 11.6.E).

$$\text{cyclohexene} + H_2O_2 \xrightarrow{OsO_4} \text{cis-1,2-cyclohexanediol (H, OH; H, OH)}$$

(e) Addition of halogens to alkenes (Section 11.6.B).

+ Br_2 ⟶

(f) Bimolecular elimination of alkyl halides (Section 11.5.A)

t-BuO⁻

(g) Hydroboration (Section 11.6.D).

B_2H_6 ⟶ H_2O_2, OH^- ⟶

(h) Epoxidation of alkenes (Section 11.6.E).

RCO_3H

Br_2, H_2O

KOH, C_2H_5OH

(i) Ring opening of epoxides (Section 10.11.A).

H^+, H_2O

(j) Cyclopropanation (Section 11.6.F).

Zn(Cu), CH_2I_2

(k) Enzymatic reduction (Section 14.8.E)

NADPH
alcohol dehydrogenase
Thermoanaerobium brokii

(*S*)-6-methyl-5-hepten-2-ol
99% e.e.

Sometimes it is not possible to control a synthesis so that only the desired stereoisomer is produced because a method that would accomplish that goal is lacking. In such a case the next best solution is to prepare the mixture of isomers and separate the desired isomer.

At this point in our study of organic synthesis we shall only touch on this subject of stereoselectivity, but the topic will come up frequently in our subsequent discussions. In 1990, there were 1327 totally synthetic drugs, of which 528 are chiral, with various numbers of stereocenters. Only 61 drugs were sold as single enantiomers, the remainder as racemates. However, the U.S. Food and Drug Administration has let it be known that it will in future favor the introduction of single enantiomers of new drugs in cases for which pharmacological data indicate a difference in biological behavior. Stereoselectivity in synthesis or efficiency in separation of pure enantiomers thus has economic as well as medicinal and scientific importance (see Section 14.8.E).

EXERCISE 16.1 Consider the methods available for the general conversion of RY to RZ in which Y and Z are the seven groups given below. Set up a 7 × 7 matrix with Y down the side and Z across the top. Mark with a minus sign those conversions for which no *general* reaction sequence is presently known to you. Mark with a 1 those interconversions of functional groups that can usually be accomplished by a simple reaction process we have studied. Finally, mark with a + those interconversions that can be accomplished in a sequence of two or more reaction steps. Y and Z: H, Br, OH, CH_2OH, CHO, $COCH_3$, CN.

Highlight 16.3

The mechanism of a reaction determines its stereospecificity and the stereochemistry of the product. A bimolecular displacement reaction, S_N2, on an alkyl halide leads to inversion of configuration at a stereocenter. Reduction of an alkyne with hydrogen over an appropriate metal catalyst forms a cis or *Z* alkene; reduction with sodium in liquid NH_3 yields the trans or *E* alkene.

16.3 Planning a Synthesis

In planning a synthesis, one starts from the product and works backward. Remember that the goal is to connect the desired product with some commercially available starting material by a series of reactions each of which will, insofar as possible, give a single product in high yield. For this reason, the practicing chemist usually acquires a fairly good working knowledge of what types of starting materials are available from chemical suppliers. Of course, if the chemist is not sure whether or not a possible starting material is available, he or she checks for its availability in the catalogs of various suppliers. A good rule of thumb is that most monofunctional aliphatic compounds containing five or fewer carbons may be purchased. A great many others are also available, but this type of information is acquired only by experience. For the purpose of learning how to design a synthesis, we will assume that the only available starting materials are those monofunctional compounds containing five or fewer carbons.

For relatively simple synthetic problems, one may decide how to synthesize the desired product from what appear to be accessible materials. These "accessible materials" must be made from still simpler materials. Working backward from the final product is continued until we arrive at possible starting materials that are known to be available. The process of analyzing synthetic problems in a reverse manner is called **retrosynthesis** and the record of such an analysis has been called a **synthetic tree.**

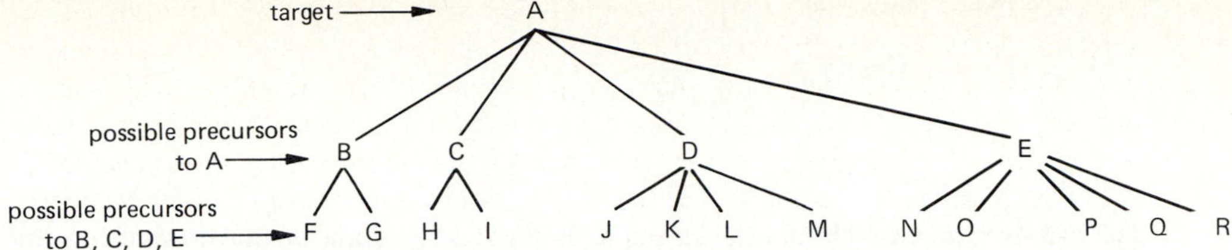

For more complex problems the synthetic tree soon becomes unwieldy and current research is being directed toward the application of computers to synthetic design. In most cases the practicing chemist solves such problems by a combination of logical analysis and intuition. The synthetic tree is built until the chemist recognizes, by insight or intuition, a complete path from one of the possible intermediates to an available starting material.

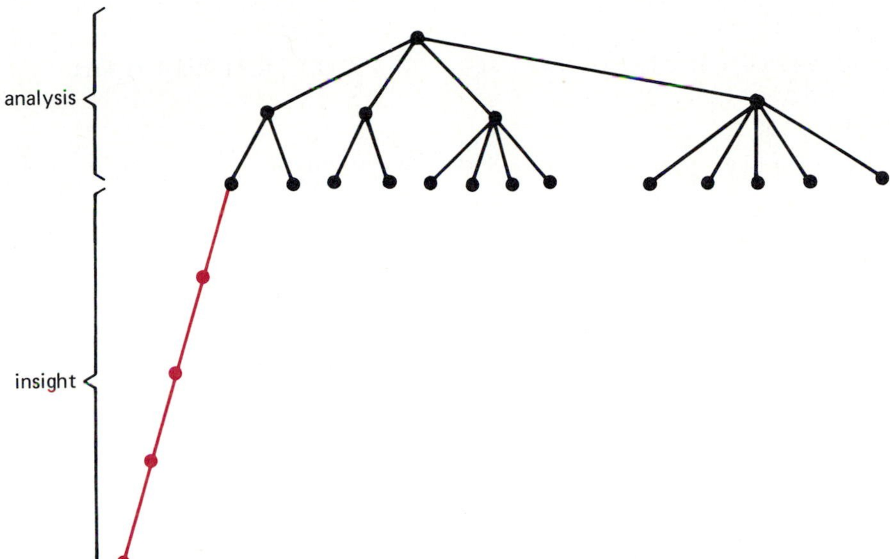

The importance of insight and intuition, relative to analytical reasoning, should not be underestimated for science in general and synthetic design in particular. Nevertheless, insight and intuition cannot function in the absence of a body of facts—in the present case, a thorough knowledge of organic reactions.

The best way to illustrate synthesis design is to demonstrate with a few specific simple examples.

Example 16.1 Plan a synthesis of 4-methyl-1-pentanol.

$$\begin{array}{c} CH_3 \\ | \\ CH_3CHCH_2CH_2CH_2OH \end{array}$$

Since the product contains six carbons, we must build up the skeleton from a simpler starting material. In principle, there are a number of ways in which this can be done.

$$\begin{array}{l} C \\ | \\ \textbf{1.}\ C—C—C—C—C^* \end{array}$$

$$\begin{array}{l} C \\ | \\ \textbf{2.}\ C—C—C—C—C^* \end{array}$$

```
        C
        |
3. C—C—C—C—C*

        C
        |
4. C—C—C—C—C*
```

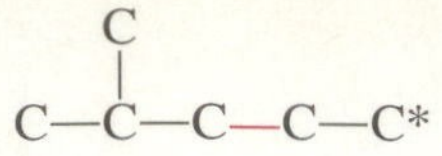

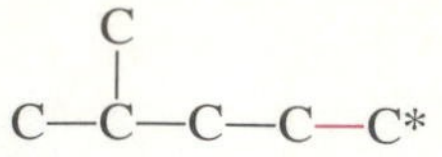

The carbon marked with an asterisk indicates the desired point of functionality, in this case, a hydroxy group. Of the various approaches, the fourth is most attractive, since the carbon-carbon bond to be created is adjacent to the functional group in the final product. In planning a synthesis, *it is generally most productive to look first at combinations in which the functional group is close to the carbon-carbon bonds that will be formed in the synthesis*. We immediately recognize that 4-methyl-1-pentanol is readily available by reaction of the Grignard reagent derived from 1-chloro-3-methylbutane with formaldehyde.

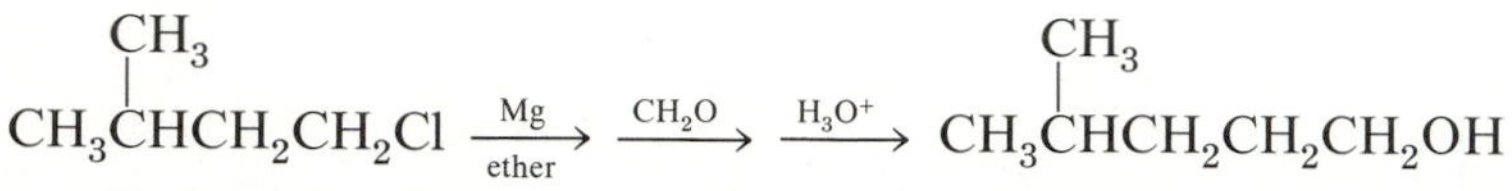

$$\underset{\displaystyle CH_3CHCH_2CH_2Cl}{\overset{CH_3}{|}} \xrightarrow[\text{ether}]{Mg} \xrightarrow{CH_2O} \xrightarrow{H_3O^+} \underset{\displaystyle CH_3CHCH_2CH_2CH_2OH}{\overset{CH_3}{|}}$$

EXERCISE 16.2 Plan a synthesis of 4-methylpentanenitrile, $(CH_3)_2CHCH_2CH_2CN$.

Example 16.2 Plan a synthesis of 4-methyl-1-hexanol.

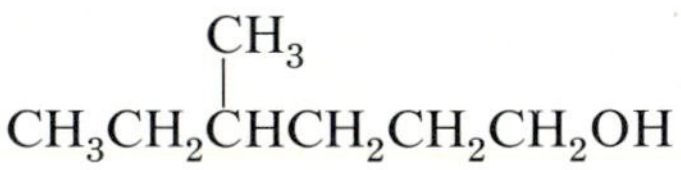

$$\underset{\displaystyle CH_3CH_2CHCH_2CH_2CH_2OH}{\overset{CH_3}{|}}$$

In this example, the desired product contains seven carbons. Analysis of the carbon skeleton gives the following combinations.

```
           C
           |
1. C—C—C—C—C—C*

           C
           |
2. C—C—C—C—C—C*

           C
           |
3. C—C—C—C—C—C*

           C
           |
4. C—C—C—C—C—C*

           C
           |
5. C—C—C—C—C—C*

           C
           |
6. C—C—C—C—C—C*
```

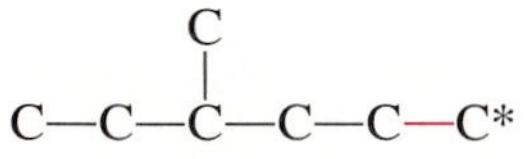

In approach number 6 the functional group is nearest the bond to be formed. We might consider solving this synthetic problem in the same way we solved Example 16.1, by application of the Grignard synthesis to 1-chloro-3-methylpentane.

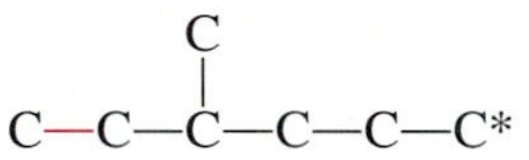

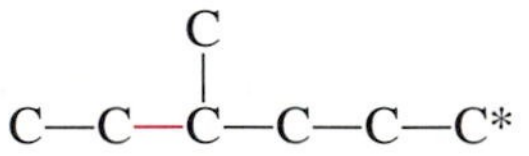

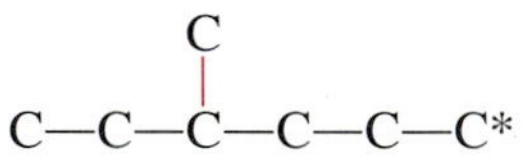

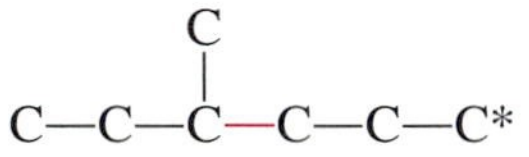

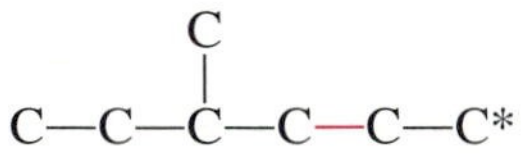

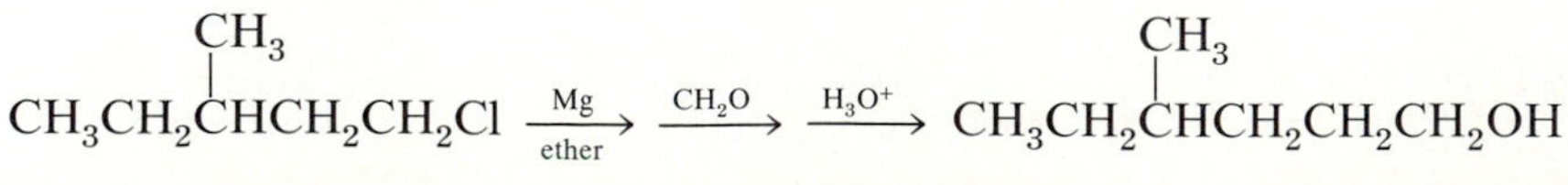

$$\underset{\displaystyle CH_3CH_2CHCH_2CH_2Cl}{\overset{CH_3}{|}} \xrightarrow[\text{ether}]{Mg} \xrightarrow{CH_2O} \xrightarrow{H_3O^+} \underset{\displaystyle CH_3CH_2CHCH_2CH_2CH_2OH}{\overset{CH_3}{|}}$$

However, this starting material contains six carbons and, by our ground rules, is not readily available. Thus we would have to synthesize it. Of course, this is easily accomplished as follows.

$$CH_3CH_2\underset{}{\overset{CH_3}{\overset{|}{C}}}HCH_2Cl \xrightarrow[\text{ether}]{\text{Mg}} \xrightarrow{CH_2O} \xrightarrow{H_3O^+}$$

$$CH_3CH_2\overset{CH_3}{\overset{|}{C}}HCH_2CH_2OH \xrightarrow[\text{pyridine}]{SOCl_2} CH_3CH_2\overset{CH_3}{\overset{|}{C}}HCH_2CH_2Cl$$

The overall synthesis of 4-methyl-1-hexanol would therefore require three separate operations beginning with 1-chloro-2-methylbutane.

Although this route will provide the desired product, note that approach number 5 could afford it in only one operation, since this combination employs a five-carbon starting material. If we search through our repertoire of reactions for one in which the unit CH_2CH_2OH may be joined to another molecule, we recall that Grignard reagents react with ethylene oxide to produce exactly this result. Thus we can prepare the Grignard reagent of 1-chloro-2-methylbutane and allow it to react with ethylene oxide to obtain 4-methyl-1-hexanol.

$$CH_3CH_2\overset{CH_3}{\overset{|}{C}}HCH_2Cl \xrightarrow[\text{ether}]{\text{Mg}} \xrightarrow{\overset{O}{\triangle}\; CH_2-CH_2} \xrightarrow{H_3O^+} CH_3CH_2\overset{CH_3}{\overset{|}{C}}HCH_2CH_2CH_2OH$$

This example illustrates a second rule that should be observed when planning a synthesis. *It is more productive to add a large fragment in a single reaction than to add several smaller fragments sequentially*. One should therefore examine all of the various carbon-carbon bonds that might be formed, even though they are not immediately adjacent to the desired functional group. In some cases a reaction may exist that allows the formation of such a remote bond.

This example illustrates another concept of synthesis design, that of **synthetic equivalents,** or **synthons.** A synthon is defined as a part of a molecule that is related to some other structural unit by a reliable reaction or sequence of reactions. In the foregoing example, the molecule oxirane (ethylene oxide) is a synthon for the structural unit $\mathbf{CH_2CH_2OH}$. That is, the reaction of a Grignard reagent (RMgX) with oxirane to give RCH_2CH_2OH is a general process that can be relied upon.

EXERCISE 16.3 What synthons exist for each of the following structural units?

(a) $-COCH_3$ (b) $-CH_2NH_2$ (c) $-CH_2OH$ (d) $-CH{=}CHCH_3$

Example 16.3 Plan a synthesis of 3-methyl-1-pentene.

$$CH_3CH_2\overset{CH_3}{\overset{|}{C}}HCH{=}CH_2$$

First, we might note that $\mathbf{HC{\equiv}CH}$ is a synthon for the structural unit $\mathbf{CH{=}CH_2}$. Thus, at first glance, partial hydrogenation of 3-methyl-1-pentyne is attractive, since the necessary alkyne appears to be available by alkylation of acetylene with 2-bromobutane.

$$CH_3CH_2\overset{Br}{\overset{|}{C}}HCH_3 + Na\,C{\equiv}CH \longrightarrow CH_3CH_2\overset{CH_3}{\overset{|}{C}}H\,C{\equiv}CH \longrightarrow CH_3CH_2\overset{CH_3}{\overset{|}{C}}H\,CH{=}CH_2$$

However, recall that acetylide ions cause *elimination* of secondary alkyl halides (Section 12.5.B).

$$\underset{\text{Br}}{CH_3CH_2\overset{|}{C}HCH_3} + \text{Na}\,C{\equiv}CH \longrightarrow CH_3CH_2CH{=}CH_2 + \text{H}\,C{\equiv}CH + \text{NaBr}$$

(plus other isomers)

Therefore, this direct two-step synthesis of 3-methyl-1-pentene is not applicable. This example illustrates an important aspect of synthesis design. After writing out a possible route to a desired product, *carefully review the chemistry that would be involved to see if there are any subtle structural features that render the route inoperable for the specific carbon skeleton under consideration.*

The functional group in our desired product is a carbon-carbon double bond. We should now review the other methods whereby this functional group can be introduced into a molecule.

(a) Dehydration of an alcohol (Sections 10.6.D and 11.5.B)

$$R\overset{OH}{\overset{|}{C}}HCH_3 \xrightarrow{H^+} RCH{=}CH_2 + H_2O$$

(b) Elimination of an alkyl halide (Section 11.5.A).

$$RCH_2CH_2X + R'O^- \longrightarrow RCH{=}CH_2 + X^- + R'OH$$

(c) Wittig reaction (Section 14.7.D).

$$RCHO + (C_6H_5)_3P{=}CH_2 \longrightarrow RCH{=}CH_2 + (C_6H_5)_3P{=}O$$

For the problem under consideration, alcohol dehydration is not suitable, since carbocation rearrangements could result in a mixture of products (Section 10.6.C). Base-catalyzed elimination of an alkyl halide is a possibility. The required halide would be 1-chloro-3-methylpentane, which could be prepared as discussed in Example 16.2. This would provide a three-step synthesis of the desired alkene.

$$CH_3CH_2\overset{CH_3}{\overset{|}{C}}HCH_2Cl \longrightarrow CH_3CH_2\overset{CH_3}{\overset{|}{C}}HCH_2CH_2OH \longrightarrow$$

$$CH_3CH_2\overset{CH_3}{\overset{|}{C}}HCH_2CH_2Cl \longrightarrow CH_3CH_2\overset{CH_3}{\overset{|}{C}}HCH{=}CH_2$$

The final method for creation of a carbon-carbon double bond is the Wittig reaction. In this case the reaction of 2-methylbutanal with methylenetriphenylphosphorane would yield the desired alkene in one step.

$$CH_3CH_2\overset{CH_3}{\overset{|}{C}}HCHO \longrightarrow CH_3CH_2\overset{CH_3}{\overset{|}{C}}HCH{=}CH_2$$

This is clearly a more efficient synthesis of 3-methyl-1-pentene than either of the two routes considered heretofore, since only one step is involved. Even so, the practicing chemist might not choose this method, for practical reasons. The required reagent is prepared by reaction of methyltriphenylphosphonium bromide with a strong base such as *n*-butyllithium.

$$(C_6H_5)_3\overset{+}{P}{-}CH_3\ Br^- \xrightarrow{n\text{-}C_4H_9Li} (C_6H_5)_3P{=}CH_2$$

Both of these reactants are rather expensive. Furthermore, the molecular weight of the phosphonium salt is 357. Thus, to carry out the reaction on a 1-mole scale (which would provide a maximum of 82 g of alkene), one must employ 357 g of phosphonium salt. Consequently, for preparation of a large quantity of 3-methyl-1-pentene, the chemist would probably use the longer Grignard route rather than the one-step Wittig reaction. For preparation of only a small amount of material, where cost is of less importance, one might well use the Wittig procedure.

EXERCISE 16.4 Plan a synthesis of 1-heptene.

Example 16.4 Plan a synthesis of *cis*-2-methyl-5-decene.

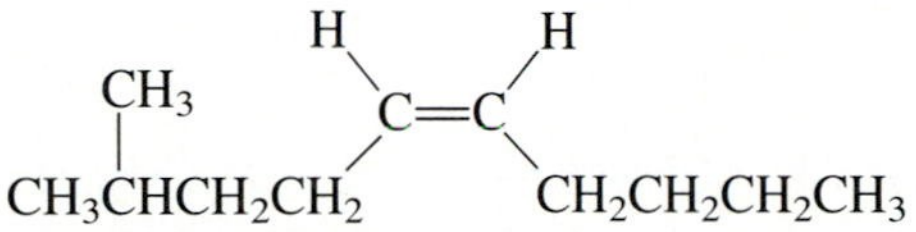

$$\underset{\displaystyle CH_3CHCH_2CH_2}{\overset{CH_3}{|}}\!\!\!\backslash\; C{=}C \;/\; CH_2CH_2CH_2CH_3 \quad (\text{H and H cis})$$

Here stereochemistry is important because only the cis isomer is desired. This consideration dominates the planning, since we have at this point only one method available for the stereospecific production of a cis alkene, partial catalytic hydrogenation of an internal alkyne.

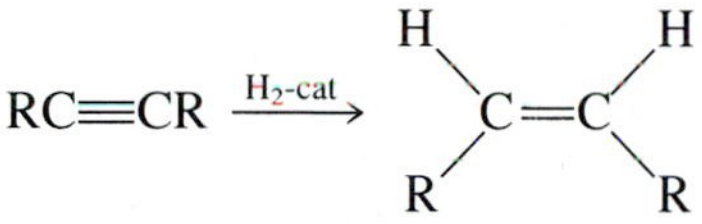

$$RC{\equiv}CR \xrightarrow{H_2\text{-cat}} \underset{R\quad\;\; R}{\overset{H\quad\;\; H}{C{=}C}}$$

Furthermore, we recognize that the acetylene group is a synthon for internal alkynes of many types, since acetylene can be successively alkylated with two different alkyl halides (page 305).

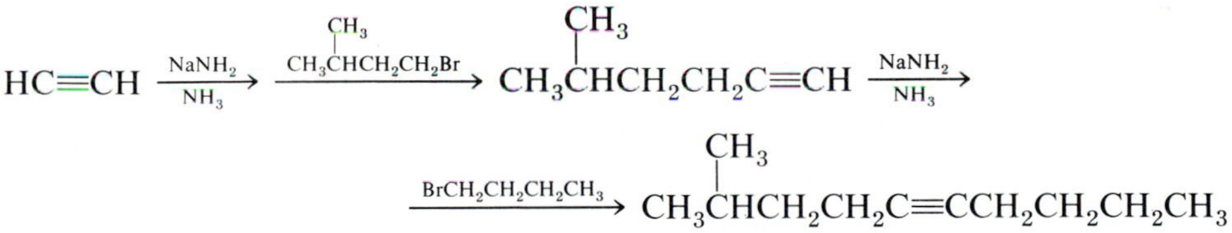

$$HC{\equiv}CH \xrightarrow[NH_3]{NaNH_2} \xrightarrow{CH_3CH(CH_3)CH_2CH_2Br} CH_3CH(CH_3)CH_2CH_2C{\equiv}CH \xrightarrow[NH_3]{NaNH_2}$$

$$\xrightarrow{BrCH_2CH_2CH_2CH_3} CH_3CH(CH_3)CH_2CH_2C{\equiv}CCH_2CH_2CH_2CH_3$$

EXERCISE 16.5 Plan a synthesis of non-3-yn-1-ol, $CH_3(CH_2)_4C{\equiv}C(CH_2)_2OH$. [*Hint*: look for two synthons.]

Example 16.5 Plan a synthesis of 2-ethylhexanol.

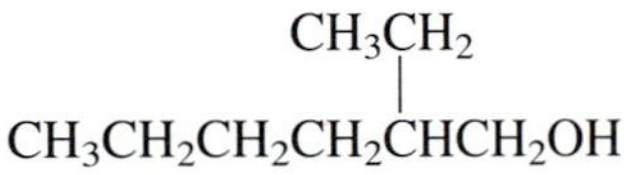

$$CH_3CH_2CH_2CH_2\overset{CH_3CH_2}{\overset{|}{C}}HCH_2OH$$

To simplify the problem, consider only the combinations involving five carbons or smaller fragments.

1. 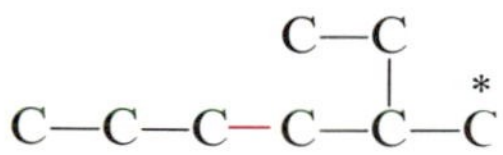 C—C—C—C—C(—C—C)—C*

2. 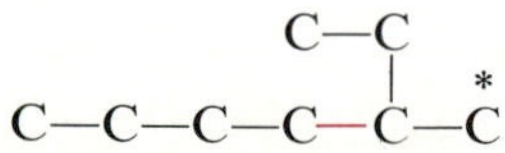 C—C—C—C—C(—C—C)—C*

Combination number 1 is not practical, since we know no simple methods for making a bond so far from a functional group. However, the second possible approach has an interesting feature—it requires joining two similar four-carbon units *at the position adjacent to the position where functionality appears in the desired final product*. Thus, a simple solution to the synthetic problem would be in hand if we had a way to join two $CH_3CH_2CH_2CH_2X$ synthons, provided CH_2X can be converted readily into CH_2OH. We can accomplish this end with the aldol addition reaction (Section 15.2).

$$2\ CH_3CH_2CH_2CHO \xrightarrow[\Delta]{KOH,\ H_2O} CH_3CH_2CH_2CH{=}\underset{}{C}(CH_2CH_3)CHO$$

With this realization, we recognize a very efficient synthesis of 2-ethylhexanol, since catalytic hydrogenation will saturate both double bonds.

$$CH_3CH_2CH_2CH{=}C(CH_2CH_3)CHO \xrightarrow{H_2\text{-}Pt} CH_3CH_2CH_2CH_2CH(CH_2CH_3)CH_2OH$$

This example illustrates yet another aspect of synthesis design. Look for ''hidden functionality.'' In this case it is important to notice that although the carbon-carbon double bond does not appear in the ultimate product, the most effective synthesis proceeds through an intermediate containing this functional group. This example also demonstrates the economy that results when a synthetic target can be dissected into two identical synthons.

EXERCISE 16.6 Plan a synthesis of 2,9-dimethyldecan-5-ol.

Example 16.6 Plan a synthesis of 4-methylnonane.

$$CH_3CH_2CH_2CH_2CH_2CH(CH_3)CH_2CH_2CH_3$$

Here we have a compound with no functional group. Since it has ten carbons, we would like to assemble it from two five-carbon building blocks. Thus the questions are how to accomplish the following combination and how to get rid of any functional groups that might be present in our intermediates.

$$C{-}C{-}C{-}C{-}C{-}C(C){-}C{-}C{-}C$$

First, let us review the methods at our disposal for removal of a functional group. We have learned three such methods—hydrogenation of a multiple bond, removal of the oxygen from an aldehyde or ketone, and removal of halogen by formation and hydrolysis of a Grignard reagent.

(a) Hydrogenation of alkenes or alkynes (Sections 11.6.A and 12.6.A).

$$\gt C{=}C\lt + H_2 \xrightarrow{cat.} {-}CH{-}CH{-}$$

$${-}C{\equiv}C{-} + 2\ H_2 \xrightarrow{cat.} {-}CH_2{-}CH_2{-}$$

(b) Deoxygenation of aldehydes and ketones (Section 14.8.D).

$$-\overset{\overset{\Large O}{\|}}{C}- \xrightarrow[\text{Wolff–Kishner reduction}]{\text{Clemmensen or}} -\underset{H}{\overset{H}{\underset{|}{\overset{|}{C}}}}-$$

(c) Hydrolysis of organometallic compounds (Section 8.9.A).

$$-\overset{|}{\underset{|}{C}}-Cl \xrightarrow[\text{ether}]{Mg} -\overset{|}{\underset{|}{C}}-MgCl \xrightarrow{H_2O} -\overset{|}{\underset{|}{C}}-H$$

There are a number of ways by which we might use these reactions in a synthesis of 4-methylnonane. For example, we could make the Grignard reagent from 1-chloropentane, add it to 2-pentanone, and dehydrate the resulting tertiary alcohol. A mixture of $C_{10}H_{20}$ isomers is expected to result, but this is of no consequence, since all of the isomers will give 4-methylnonane upon hydrogenation.

$$CH_3CH_2CH_2CH_2CH_2Cl \xrightarrow[\text{ether}]{Mg} \xrightarrow{CH_3\overset{O}{\overset{\|}{C}}CH_2CH_2CH_3} CH_3CH_2CH_2CH_2CH_2\underset{OH}{\underset{|}{\overset{CH_3}{\overset{|}{C}}}}CH_2CH_2CH_3 \xrightarrow[\Delta]{H_2SO_4}$$

$$\left\{\begin{array}{c} CH_3CH_2CH_2CH_2CH{=}\overset{CH_3}{\overset{|}{C}}CH_2CH_2CH_3 \\ \text{(cis and trans)} \\ + \\ CH_3CH_2CH_2CH_2CH_2\overset{CH_3}{\overset{|}{C}}{=}CHCH_2CH_3 \\ \text{(cis and trans)} \\ + \\ CH_3CH_2CH_2CH_2CH_2\overset{CH_2}{\overset{\|}{C}}CH_2CH_2CH_3 \end{array}\right\} \xrightarrow[\text{Pd/C}]{H_2} CH_3CH_2CH_2CH_2CH_2\overset{CH_3}{\overset{|}{C}}HCH_2CH_2CH_3$$

EXERCISE 16.7 Suppose that the compound needed in the foregoing example were actually 4-(deuteriomethyl)nonane. Plan an efficient synthesis for this labeled compound.

16.4 Protecting Groups

In the design of a synthesis that one wishes to carry out, we find often that the reagent that produces a desired transformation on one functional group will also react with some other functional group present in the same molecule. The reagent may still be used by temporarily **protecting** one of the functional groups by changing it into another functional group that is unreactive to the reagent in question. For example, suppose it is desirable to carry out the following conversion.

4-oxocyclohexanecarbaldehyde (O=, CHO) $\longrightarrow$ 4-hydroxycyclohexanecarbaldehyde (HO, CHO)

The method for reducing a ketone to a secondary alcohol is reduction with sodium borohydride or lithium aluminum hydride (Section 14.8.B). However, there is a problem, since the aldehyde would react with either of these reagents *more rapidly than the ketone*. The problem can be circumvented by first transforming the aldehyde into an acetal. In principle, the ketone can also be converted into a acetal. However, the same factors that cause the aldehyde carbonyl to be reduced more rapidly than the ketone carbonyl also cause it to be transformed more readily into the corresponding acetal.

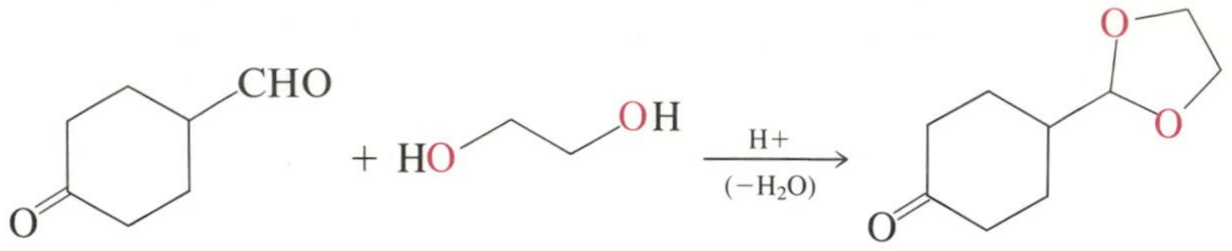

CHO + HO⌒⌒OH $\xrightarrow[(-H_2O)]{H+}$ O

Since an acetal is an ether and ethers do not react with either $NaBH_4$ or $LiAlH_4$, the ketone reduction may now be carried out. After the reduction, the acetal is hydrolyzed by treatment with acid and excess water in order to regenerate the aldehyde function.

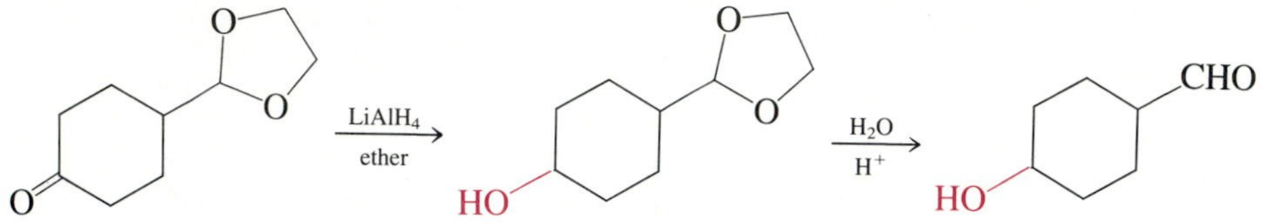

$\xrightarrow[\text{ether}]{LiAlH_4}$ HO $\xrightarrow[H^+]{H_2O}$ HO CHO

In this example the acetal group is a protecting group. It is a good protecting group for aldehydes because it is easily introduced and removed and, since it is an ether, is stable to many reagents. In a similar manner, acetals are sometimes useful as protecting groups for the ketone carbonyl.

EXERCISE 16.8 Suggest a strategy for conversion of 5-bromo-2-pentanone into 7-hydroxy-2-heptanone.

A method often used for the protection of primary alcohols is to convert them into *t*-butyl ethers. For example, suppose it is desired to convert 3-bromopropanol into 3-deuterio-1-propanol. Hydrolysis of a Grignard reagent with D_2O is often an effective method for introducing deuterium into a molecule. However, in this case the hydroxy function would interfere (Section 8.8.A). What is needed is a protecting group for OH. The group must be easily installed, it must be stable to the conditions of formation and reaction of Grignard reagents, and it must be conveniently removed. Primary and secondary alcohols may be transformed into *t*-butyl ethers by treatment with *t*-butyl alcohol and acid. Isobutylene and acid may also be used to produce *t*-butyl ethers.

$$BrCH_2CH_2CH_2OH + CH_3\overset{\displaystyle CH_3}{\overset{|}{C}}{=}CH_2 \xrightarrow{H_2SO_4} BrCH_2CH_2CH_2O\overset{\displaystyle CH_3}{\overset{|}{\underset{\underset{\displaystyle CH_3}{|}}{C}}}{-}CH_3$$

The hydroxy group has now been protected by conversion into an ether function, which is stable to Grignard reagents. The Grignard reagent is prepared and hydrolyzed with D_2O, and the product is hydrolyzed with aqueous sulfuric acid to obtain the deuterated propanol.

$$\mathrm{BrCH_2CH_2CH_2OC(CH_3)_2{-}CH_3 \xrightarrow[\text{ether}]{Mg} BrMgCH_2CH_2CH_2OC(CH_3)_2{-}CH_3 \xrightarrow{D_2O}}$$

$$\mathrm{DCH_2CH_2CH_2OC(CH_3)_2{-}CH_3 \xrightarrow[H_2O]{H_2SO_4} \underset{\text{3-deuterio-1-propanol}}{DCH_2CH_2CH_2OH} + CH_3C(CH_3){=}CH_2}$$

Another protecting group commonly used for alcohols is the trimethylsilyl ether, which is formed by treating the alcohol with trimethylchlorosilane and an organic base such as triethylamine.

$$\mathrm{{-}\overset{|}{\underset{|}{C}}{-}OH + (CH_3)_3SiCl \xrightarrow{(C_2H_5)_3N} {-}\overset{|}{\underset{|}{C}}{-}O{-}Si(CH_3)_3}$$

The silyl ether grouping is stable to most neutral and basic conditions. Upon treatment with mild aqueous acid, the alcohol is regenerated.

$$\mathrm{{-}\overset{|}{\underset{|}{C}}{-}O{-}Si(CH_3)_3 \xrightarrow[25^\circ C]{H_3O^+} {-}\overset{|}{\underset{|}{C}}{-}OH + (CH_3)_3SiOH}$$

EXERCISE 16.9 Suggest a strategy for the following conversion.

HO–cyclohexyl–CHO ⟶ HO–cyclohexyl–C(=O)CH₂CH₂CH₂CH₃

16.5 Industrial Syntheses

Organic compounds are synthesized for two fundamentally different kinds of reasons. On the one hand, we may need a specific compound in order to study its properties or to use it for further research purposes. For such purposes relatively small amounts generally suffice, and cost is not an important criterion—within limits. On the other hand, a compound may have commercial significance, and for such purposes economic factors take on a vital importance. The cost of a medicinal used in small quantity where no other product will work is clearly of a different magnitude than that of a structural polymer that must compete with wood and steel.

Most of the reactions and syntheses we have studied are useful for understanding the chemistry of different kinds of functional groups and for the laboratory preparation of various compounds. Few of these reactions, however, are suitable for the industrial preparation of compounds, some of which are produced on the scale of millions of pounds a year. An important distinction is the following: the reactions and laboratory preparations that we study have generality. These methods, with minor modifications, are suitable for the preparation of whole classes of compounds. On the other hand, many industrial preparations are specific. They apply for making one and only one compound. Many such reactions are gas-phase catalytic processes with the precise catalysts and reaction conditions carefully worked out. An important advantage of such processes is that they are continuous rather than batch processes. Ideally, in a continuous process the reactants are fed into one end of a chemical plant and the product comes out at the other end without intermediate stops, ready for marketing or for the next step. In practice, this ideal is rarely achieved. Catalysts lose their efficiency with time and need to be replaced. By-products build up and need to be cleaned out.

Much research in the chemical industry is devoted to discovering new products with useful properties. But much other research is devoted to existing products, improving processes to obtain these products cheaper and more efficiently, and, in many cases, purer. Research in process development is a fascinating area of its own that requires special talents of creativity, patience, and chemical knowledge. Close attention in process development is also given to the environment and to energy conservation. A suitable industrial process must involve a minimum of waste products that require disposal.

Hightlight 16.4

Two retrosynthetic analyses for 1-(5-methyl-1-hexyl)-1-cyclopentanol are shown below. The symbol ⟹ signifies a reverse synthetic step and is sometimes called a **transform.** The main transforms are *disconnection* or cleavage of C—C bonds, functional group interchange **(FGI),** and protection/deprotection of functional groups that might be altered by particular reactions.

C—C(C)—C—C—C—C(HO)(cyclopentyl) ⟹ C—C(C)—C—C—C—C—Br + O=cyclopentane

retrosynthetic cut [disconnection site]

cyclopentanone (available)

functional group transformation (left) ⇓

C—C(C)—C—C—C≡C—C(HO)(cyclopentyl)

retrosynthetic cut [disconnection site] ⇓

C—C(C)—C—C—C≡C + O=cyclopentane

cyclopentanone (available)

retrosynthetic cut [disconnection site] ⇓

C—C(C)—C—C—Br + C≡C

isopentyl bromide (available) acetylene (available)

functional group transformation (right) ⇓

C—C(C)—C—C—C—C—OH

retrosynthetic cut [disconnection site] ⇓

C—C(C)—C—C—Br + oxirane (C—C with O)

isopentyl bromide (available) oxirane (available)

EXERCISE 16.10 Write in detail the steps for an actual synthesis according to the retrosynthetic analyses shown in Highlight 16.4.

EXERCISE 16.11 Write in detail the steps for an actual synthesis according to the retrosynthetic analyses shown in Highlight 16.4 for the isomeric target molecule shown below.

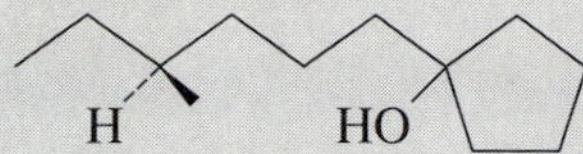

(4*S*)-1-(4-methyl-1-hexyl)-1-cyclopentanol

Which synthetic pathway would be preferred in this case and why? Identify the stereocenter in the molecule. Plan a synthesis that will deliver the enantiomerically pure target molecule. *Hint:* The naturally occurring alcohol (*S*)-(+)-2-methyl-1-butanol (''primary active amyl alcohol'') is commercially available.

EXERCISE 16.12 Identify the synthons for the pathways shown in Highlight 16.4.

ASIDE

''Science is fun, but it is not entertainment''—twentieth-century aphorism. Building a molecule is similar to constructing any complex object such as an airplane, a house, or a microcomputer chip. It is a blend of science, technique, and art that requires many different skills. The plane designers must not exceed the performance characteristics of the wing materials (either aluminum or a composite of carbon-fiber-reinforced polymer) and must finally have an aircraft that will carry passengers economically to their destinations. The molecular designer must plan ahead (often by working backward) so as to achieve his aims in a reasonable period of time at reasonable cost.

Show how to synthesize 2-pentanol from one- and two-carbon compounds. Write down the reasons why a particular reaction was chosen. List as many alternatives for a given reaction step as you can think of.

PROBLEMS

1. Plan a synthesis for each of the following compounds from starting materials containing five or fewer carbons.

a. 1-hexanol
b. 2-hexanol
c. 3-hexanol
d. 1-heptyne
e. *trans*-3-heptene
f. 5-methyl-5-nonanol
g. 2,6-dimethyl-2-heptanol
h. 1-cyclopentyl-2-methylpropene

2. Plan a synthesis for each of the following compounds from starting materials containing five or fewer carbons. For these compounds, the most efficient synthesis will involve removal of a functional group from an intermediate.

a. 2-methylnonane
b. cyclopentylcyclopentane
c. 2,9-dimethyldecane
d. 2,3,6-trimethylheptane
e. 4-methyl-1-heptene
f. 5-(3-methylbutyl)-2,8-dimethylnonane

3. Plan a synthesis for each of the following compounds from monofunctional starting materials containing five or fewer carbons. In each case, the final product may be racemic; it is not necessary to plan the synthesis using optically active reactants. Relative stereochemistry is important.

a. *trans*-2-hexene
b. *trans*-2-butylcyclopentanol
c. *cis*-4-octene
d. (3*R*,4*R*)-hexane-3,4-diol
e. (2*R*,3*S*)-3-methyl-2-heptanol
f. OH; CH_2CH_3; C=C; H; H
g. *trans*-1,2-diethylcyclopropane

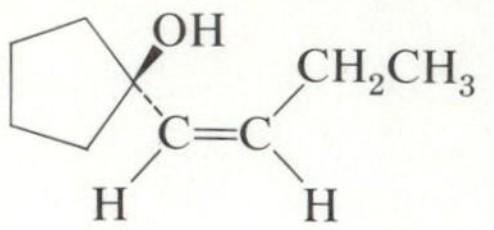

4. Show how each of the following *optically active* compounds may be prepared from optically active starting materials containing five or fewer carbons.

a. (*S*)-4-methyl-1-hexanol
b. (*R*)-2-methylpentanenitrile
c. (*R*)-4,4-dimethyl-2-pentanol
d. (*S*)-3-deuterio-4-octyne

5. Show how one may carry out each of the following conversions in good yield. In each case a protecting group will be necessary.

a. $BrCH_2CH_2CH_2CH_2CHO \rightarrow CH_2{=}CHCH_2CH_2CHO$
b. 4-bromo-1-butanol → 1,5,9-nonanetriol
c. $BrCH_2CH_2CHO \rightarrow NCCH_2CH_2CHO$

6. Plan a synthesis for each of the following compounds from starting materials containing five or fewer carbons. Difunctional starting materials may be used if necessary.

a. *cis*-2,2-dimethyl-3-octene
b. $CH_3CH_2CH_2CH_2\overset{\overset{\large O}{\|}}{C}CH_2CH_2CHO$
c. 5-hydroxy-2-hexanone
d. 4-ethyl-2-methyl-2,4-octadiene
e. 2-methyloct-2-en-5-yn-4-ol
f. 3-ethylhept-6-en-4-yn-3-ol
g. 6-methylheptan-1,4-diol

CHAPTER 17

INFRARED SPECTROSCOPY

17.1 The Electromagnetic Spectrum

There are many different forms of radiant energy that display wave properties. Examples are x-rays, visible light, infrared radiation and radio waves. These apparently different types of radiation are known collectively as **electromagnetic radiation.** They are all considered as waves that travel at a constant velocity (the "speed of light," $c = 3.0 \times 10^{10}$ cm sec^{-1}) and differ in wavelength or frequency. The **electromagnetic spectrum** is diagrammed in Figure 17.1 along with the wavelength in centimeters of its various regions. The divisions between regions are arbitrary and are established in practice by the different instrumentation required to produce and record electromagnetic radiation in the different regions. As pointed out in Section 13.2, compounds can absorb radiant energy in various regions of the electromagnetic spectrum and thereby become excited from their ground state to a more energetic state. **Spectroscopy** is a technique whereby we measure the amount of radiation a substance absorbs at various wavelengths. From the spectrum of a compound we can often obtain useful information about the structure of the compound.

The relationship between the wavelength and frequency of radiation is given by the relationship

$$\nu = \frac{c}{\lambda}$$

where λ = wavelength in centimeters, ν = frequency in Hertz (Hz), and c = the velocity of light (2.998×10^{10} cm sec^{-1}). The relationship between energy and frequency is

$$\epsilon = h\nu$$

where ϵ = the energy of a photon and h = Planck's constant (6.6242×10^{-27} erg sec), or

$$E = Nh\nu$$

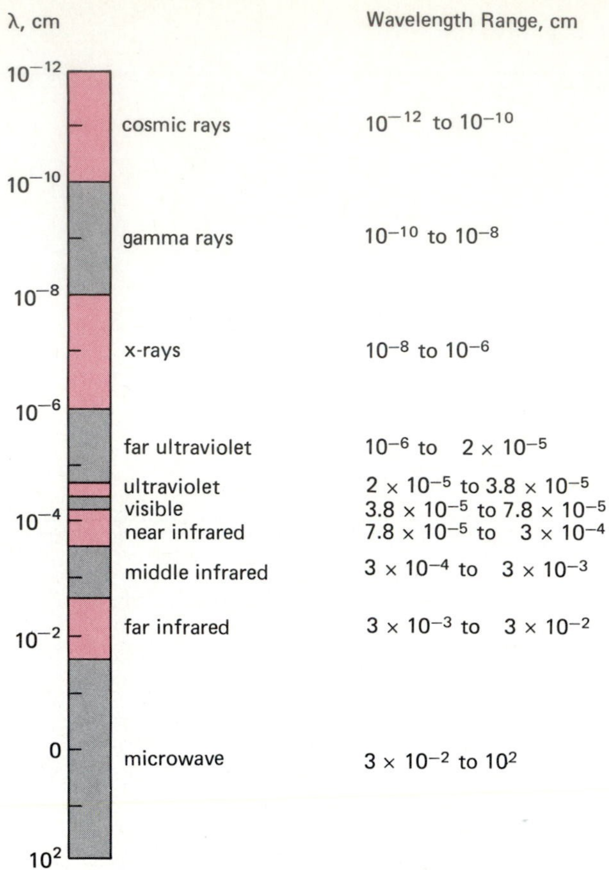

FIGURE 17.1 The electromagnetic spectrum.

where E = the energy of an Avogadro number N of photons ($E = \epsilon \times 6.023 \times 10^{23}$). Thus, when a compound absorbs radiation of a given wavelength, each molecule absorbs an amount of energy ϵ, and each mole of the compound absorbs an amount of energy E. In organic chemistry energy is traditionally expressed in units of kilocalories per mole.

$$E\,(\text{kcal mole}^{-1}) = \frac{2.857 \times 10^{-3}}{\lambda\ (\text{cm})}$$

In this chapter, we are concerned with absorption of light in the infrared region of the electromagnetic spectrum. Wavelengths in this region are traditionally expressed in microns (μ), where $10^4\ \mu = 1$ cm. More commonly, another unit of measurement, the wavenumber, is used to describe infrared spectra. The wavenumber $\tilde{\nu}$ is defined as the number of waves per centimeter and is expressed in units of cm^{-1} (reciprocal centimeters).

$$\tilde{\nu} = \frac{1}{\lambda}$$

By definition, the infrared region is split up into three parts (Figure 17.2): the near infrared ($\lambda = 0.78$–$3.0\ \mu$; $\tilde{\nu} = 12{,}820$–$3333\ \text{cm}^{-1}$), the middle infrared ($\lambda = 3$–30μ; $\tilde{\nu} = 3333$–$333\ \text{cm}^{-1}$), and the far infrared ($\lambda = 30$–$300\ \mu$; $\tilde{\nu} = 333$–$33\ \text{cm}^{-1}$). The near infrared region corresponds to energies in the range 37–10 kcal mole^{-1}. Since there are few absorptions of organic molecules in this range, it is seldom used for spectroscopic purposes. Radiation in the middle infrared region has $E = 10$–1 kcal mole^{-1}, which corresponds to the differences commonly encountered between vibrational states. Spec-

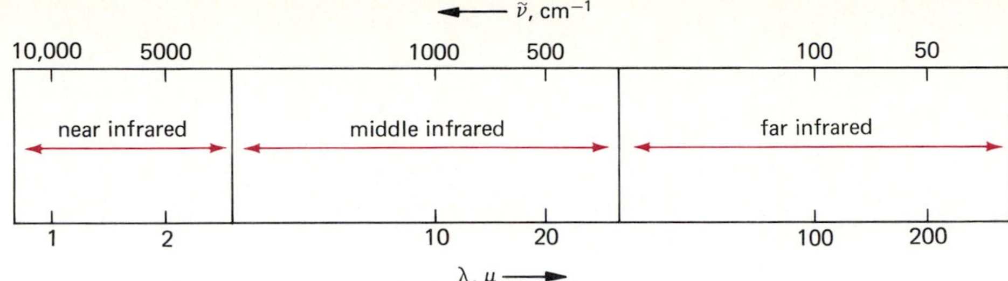

FIGURE 17.2 Regions of the infrared spectrum. Notice that the scales are logarithmic.

troscopy in this region is useful to the organic chemist. The far infrared region has $E = 1.0\text{–}0.1$ kcal mole^{-1}. This region is little used for organic spectroscopy, again because few useful absorptions occur here.

An infrared spectrometer can be dispersive or interferometric. A dispersive instrument can be either single-beam or double-beam. In a single-beam spectrophotometer, light from the radiation source (usually an oxide-coated ceramic rod that is heated electrically to about 1500 °C) is focused and passed through the sample, contained in a special cell. After passing through the sample, the emergent light beam is dispersed by a monochromator (either a prism or a diffraction grating) into its component wavelengths. The spectrum is scanned by slowly rotating the prism or grating. A double-beam spectrophotometer operates on a similar principle except that the original light is split into two beams, one of which passes through the sample while the other passes through a reference cell. The instrument records the difference in intensity of these two beams. This type of instrument is especially useful when spectra are to be measured in solution. In such a case the reference cell contains pure solvent. Thus, if the solvent absorbs weakly in a given region of the spectrum, its absorption may be "canceled out." Since glass absorbs strongly in the useful infrared region, it cannot be used for the optical parts of a spectrophotometer. The prism and sample cell walls are usually fabricated from large NaCl or KBr crystals. Other materials useful in infrared measurements include germanium, silver chloride or bromide, and the highly toxic KRS-5, thallium bromoiodide.

An interferometric instrument splits the light from a high-temperature source into two beams, half going to a fixed mirror and half to a moving mirror. The reflected beams recombine at the beam-splitter. For a given wavelength there will be either constructive interference (more light energy) or destructive interference (less light energy). The complex mixture of modulated frequencies is passed through the sample in which some of the light is absorbed and then to a detector from which the signal is digitized to produce an interferogram. A helium-neon laser provides a precise reference wavelength with which the interferogram can be fast Fourier-transformed into an infrared spectrum. Fourier-transform infrared (FTIR) spectrometers are faster and more versatile than dispersive instruments, yet cost only somewhat more. The digitized spectrum can be manipulated in many ways, such as subtracting an initial spectrum from a subsequent one to obtain the spectrum of a substance formed in a reaction. Under favorable conditions, a good spectrum can be obtained in 10 sec. Using a technique known as ATR, attenuated total internal reflectance, one can measure an adequate spectrum of a mono- or bilayer of an organic compound.

17.2 Molecular Vibration

As discussed previously (Section 6.1), atoms in a molecule do not maintain fixed positions with respect to each other, but actually vibrate back and forth about an average value of the interatomic distance. This vibrational motion is quantized, as shown in the accom-

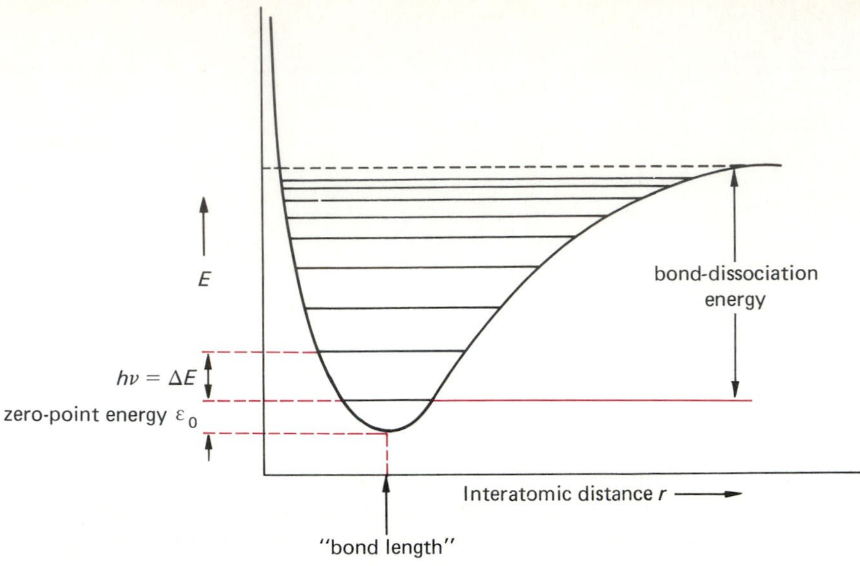

FIGURE 17.3 Vibrational levels for a vibrating bond.

panying familiar diagram for a diatomic molecule (Figure 17.3). At room temperature most of the molecules in a given sample are in the lowest vibrational state. Absorption of light of the appropriate energy allows the molecule to become "excited" to the second vibrational level. In this level the amplitude of the molecular vibration is greater. In general, such absorption of an infrared light quantum can occur only if the dipole moment of the molecule is different in the two vibrational levels. The variation of the dipole moment with the change in interatomic distance during the vibration corresponds to an oscillating electric field that can interact with the oscillating electric field associated with electromagnetic radiation. The requirement that absorption of a vibrational quantum be accompanied by a change in dipole moment is known as a **selection rule.** Such a vibrational transition is said to be **infrared-active.** Furthermore, the greater the change in dipole moment, the more intense is the absorption. Vibrational transitions that do not result in a change of dipole moment of the molecule are not observed directly and are referred to as infrared-inactive transitions. Thus, carbon monoxide and iodine chloride absorb infrared light, but hydrogen, nitrogen, chlorine, and other symmetrical diatomics do not.

C≡O, I—Cl	H_2, N_2, Cl_2
absorb in infrared	*do not absorb in infrared*

For more complex molecules, there are more possible vibrations. A nonlinear molecule containing n atoms has $3n - 6$ possible fundamental vibrational modes. Polyatomic molecules exhibit two distinct types of molecular vibration, stretching and bending. Vibrations of bonds involving hydrogens are especially significant because atoms of low mass tend to do a lot of moving compared to atoms of higher mass. The stretching and bending motions in a methylene group are diagrammed in Figure 17.4.

For polyatomic molecules of the size of typical organic compounds, the possible number of infrared absorption bands becomes very large. For example, pentane has 45 possible infrared absorption bands, decane has 90, and triacontane has 270. Many of these vibrations occur at the same frequency (that is, some vibrations are "degenerate"), and not all of the possible bands are generally seen as independent absorptions. However, additional bands, usually of low intensity, may occur as overtones (at approximately $\frac{1}{2}$, $\frac{1}{3}$, $\frac{1}{4}$, . . . , and so on, the wavelength of the fundamental mode).

Stretching Vibrations

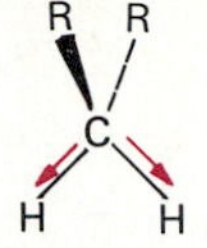

(a) symmetric

(b) asymmetric

Bending Vibrations

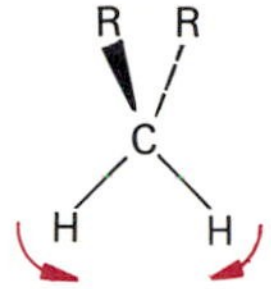

(a) scissoring (in-plane)

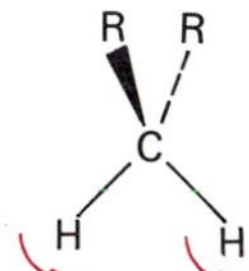

(b) rocking (in-plane)

(c) wagging (out-of-plane)

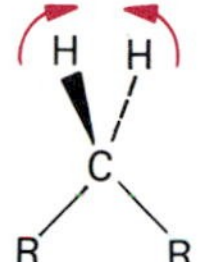

(d) twisting (out-of-plane)

FIGURE 17.4 Some vibrational modes for the methylene group.

> Overtones can arise in two ways. If a molecule in the lowest or first vibrational state is excited to the third vibrational level, the energy required is almost twice that required for excitation to the second vibrational level. It is not exactly twice as much because the higher levels tend to lie closer together than the lower levels (see Figure 17.3).
>
> Another type of overtone, commonly referred to as a "combination band," occurs when a single photon has precisely the correct energy to excite two vibrations at once. For this to happen, the energy of the combination band must be the exact sum of the two independent absorptions.

As a result, the infrared spectrum of an organic compound is usually rather complex.

The spectrum of octane; shown in Figure 17.5, illustrates several features of an infrared spectrum. Note that the wavelength is plotted against the percent transmittance of the sample. An absorption band is therefore represented by a "trough" in the curve; zero transmittance corresponds to 100% absorption of light of that wavelength.

The curve in Figure 17.5 is a spectrum of pure octane. The spectrum was measured with a Perkin-Elmer Model 735 Spectrometer using a cell 0.016 mm in length. Only four major absorption bands are apparent, at 2925, 1465, 1380, and 720 cm^{-1}. These four bands correspond to the C—H stretching vibrations, the CH_2 and CH_3 scissoring mode, the CH_3 rocking mode, and the CH_2 rocking mode, respectively. Figure 17.6 is a spectrum of the same sample measured in a cell 0.20 mm long. Since the amount of light absorbed is proportional to the number of molecules encountered by the beam as it passes through the sample, the longer cell allows absorption bands of low intensity to be observed. Many more bands can now be seen, especially in the region from 700 to 1300 cm^{-1}.

Because of its complexity, the spectrum cannot be analyzed completely. However, a peak-by-peak correspondence in the infrared spectra of two different samples is an excellent criterion of identity, as a comparison of the octane spectrum in Figure 17.6 with the heptane spectrum in Figure 17.7 readily shows. That is, the IR spectrum of heptane is similar to, but differs in significant respects from, that of octane.

Molecular vibrations are actually rather complex. Generally, all of the atoms in a molecule contribute to a vibration. Fortunately, however, some molecular vibrations can be treated by considering the motion of a few atoms relative to one another and ignoring

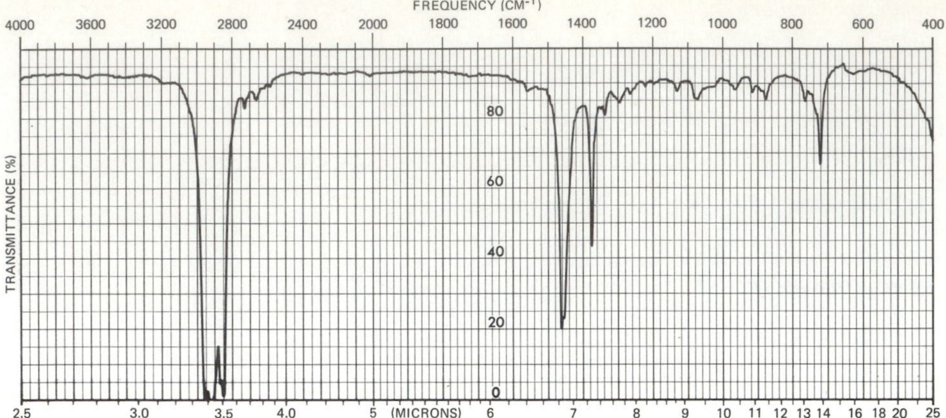

FIGURE 17.5 Infrared spectrum of octane, 0.016-mm cell.

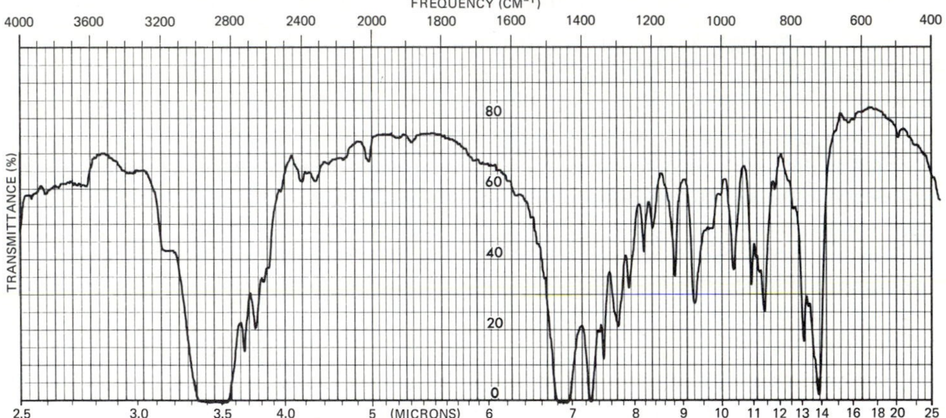

FIGURE 17.6 Infrared spectrum of octane, 0.2-mm cell.

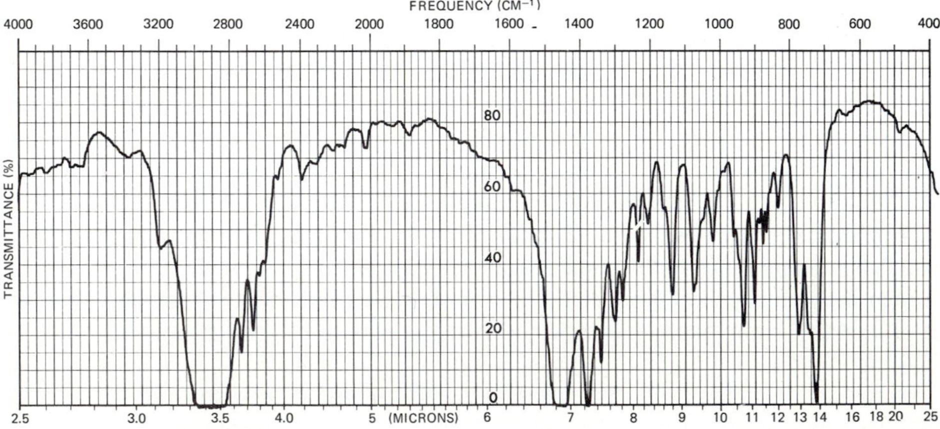

FIGURE 17.7 Infrared spectrum of heptane, 0.2-mm cell.

the rest of the atoms in the molecule. For example, it is convenient to refer to the vibration of individual bonds. To a useful approximation (the harmonic oscillator approximation), the vibration frequency of a bond is related to the masses of the vibrating atoms and the force constant, *f*, of the vibrating bond by the following equation.

$$\tilde{\nu} = \frac{1}{2\pi c}\sqrt{\frac{f(m_1 + m_2)}{m_1 m_2}}$$

In this relationship $\tilde{\nu}$ = vibrational frequency in cm^{-1} (wavenumber), c = velocity of light in cm sec^{-1}, m_1 = mass of atom 1 in g, m_2 = mass of atom 2 in g, and f = force constant in dyne cm^{-1} (g sec^{-2}). The equation corresponds to a simple Hooke's law model of two units coupled by a spring in which the force constant is the restoring force provided by the spring.

The larger the force constant, the higher the vibration frequency and the greater the energy spacings between vibrational quantum levels. The force constants for single, double, and triple bonds are approximately 5×10^5, 10×10^5, and 15×10^5 dynes cm^{-1}, respectively.

> Recall that 1 dyne is the force required to accelerate a 1 g mass 1 cm sec^{-2}. Therefore 1 dyne = 1 g cm sec^{-2}. The units of f, the force constant, are thus g sec^{-2}.

Force constants provide another measure of bond strength and generally are roughly proportional to bond-dissociation energies. On the other hand, vibration frequencies relate inversely to the masses of the vibrating atoms. Bonds to hydrogen occur at relatively high frequencies compared to bonds between heavier atoms—a light weight on a spring oscillates faster than a heavy weight.

In spite of its gross assumptions, the Hooke's law approximation is useful because it helps us to identify the general region in which a vibration will occur. For example, we may easily estimate the $^{12}C—^{1}H$ stretching frequency by

$$\tilde{\nu} = \frac{1}{2\pi\, 2.998 \times 10^{10}\text{ cm sec}^{-1}}\sqrt{\frac{5 \times 10^5\text{ g sec}^{-2}\left(\frac{12}{6.023} + \frac{1}{6.023}\right) \times 10^{-23}\text{ g}}{\left(\frac{12}{6.023} \times 10^{-23}\text{ g}\right)\left(\frac{1}{6.023} \times 10^{-23}\text{ g}\right)}}$$

$$\tilde{\nu} = 3032\text{ cm}^{-1}$$

The actual range for C—H absorptions is 2850–3000 cm^{-1}.

The regions of the infrared spectrum where various bond stretching vibrations are observed depend primarily on whether the bonds are single, double, or triple or bonds to hydrogen. These regions are summarized in Table 17.1.

TABLE 17.1 Approximate Values for Infrared Absorptions

Bond	General Absorption Region, cm^{-1}
C—C, C—N, C—O	800–1300
C=C, C=N, C=O	1500–1900
C≡C, C≡N	2000–2300
C—H, N—H, O—H	2850–3650

EXERCISE 17.1 Using the Hooke's law approximation, estimate $\tilde{\nu}$ for each of the following stretching vibrations.

(a) O—H (b) O—D (c) C=C (d) C≡C (e) C—O
(f) C=O (g) C≡N (h) C—F

Highlight 17.1

A *change* in dipole moment with the vibration is required for infrared light absorption. The oscillating field associated with the atomic motion can interact with the electric field of the photon.

$$\left[\rangle C{=}O \longleftrightarrow \rangle \overset{+}{C}{-}\overset{-}{O} \right] \longrightarrow \left[\rangle C{=\!=}O \longleftrightarrow \rangle \overset{+}{C}{-\!-}\overset{-}{O} \right]$$

$$C^{\delta+}{-}Cl^{\delta-} \longrightarrow C^{\delta+}{-\!-}Cl^{\delta-}$$

vibrational level, $v = 0$
dipole moment, μ

vibrational level, $v = 1$
dipole moment, $\mu + \delta\mu$

17.3 Characteristic Group Vibrations

As was pointed out in the previous section, the infrared spectrum of a polyatomic molecule is so complex that it is usually inconvenient to analyze it completely. However, valuable information can nevertheless be gleaned from the infrared spectrum of an organic compound. Consider the infrared spectra of 1-octene and 1-octadecene shown in Figures 17.8 and 17.9. Aside from the C—H stretching and bending vibrations at 2925, 1450, and 1370 cm^{-1}, we see several distinctive bands that do not appear in the spectra of typical alkanes (compare Figure 17.5). These new bands occur in the following general positions: 3080 cm^{-1}, 1640 cm^{-1}, 995 cm^{-1}, and 915 cm^{-1}. The Hooke's law approximation tells us that the band in the 3080 cm^{-1} region is the C—H stretch of the olefinic carbon-hydrogen bonds, and the 1640 cm^{-1} band is the carbon-carbon double bond stretching vibration. Other theoretical considerations suggest that the 995 cm^{-1} and 915 cm^{-1} bands are from out-of-plane bending of the olefinic C—H bonds. The weak band near 1820 cm^{-1} is an overtone of the fundamental band at 915 cm^{-1}.

The absorption bands mentioned in the foregoing discussion are characteristic of compounds containing a carbon-carbon double bond and may be used to determine unsaturation in an organic compound. Organic chemists use infrared spectroscopy in this semiempirical way. Most of the common functional groups give rise to characteristic absorption

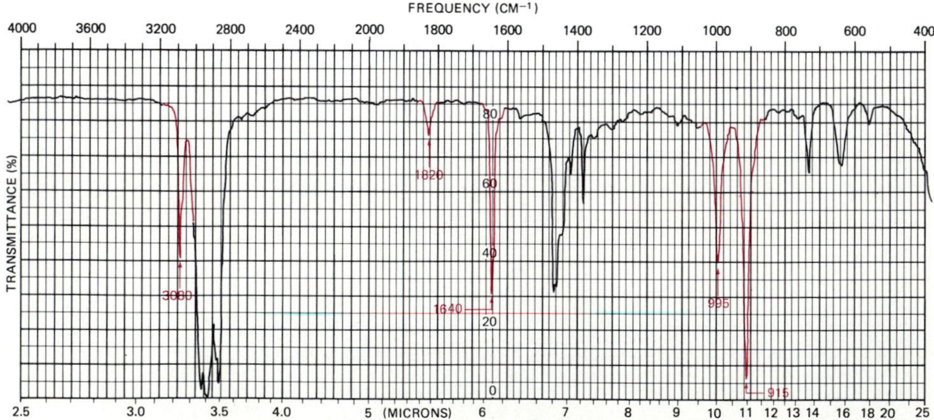

FIGURE 17.8 Infrared spectrum of 1-octene.

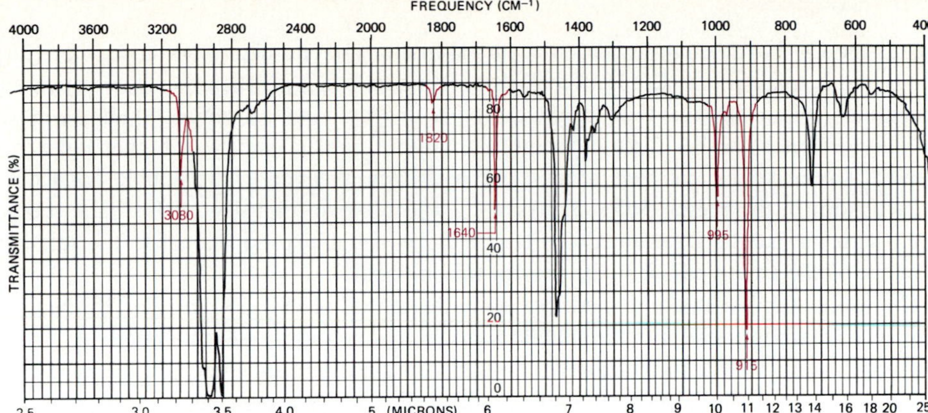

FIGURE 17.9 Infrared spectrum of 1-octadecene.

bands in defined regions of the infrared range. The chemist uses the presence or absence of a band in that region of the infrared spectrum as a diagnosis for the presence or absence of the corresponding functional group in the compound. One example will illustrate this point. The spectrum of 4-bromo-1-butene is shown in Figure 17.10. The spectrum is a fairly complex one, with a number of bands not characteristic of simple alkanes. The highlighted bands in Figure 17.10 are all due to vibrational transitions associated with the double bond. The occurrence of these bands in the spectrum is taken as evidence for the presence of such a functional group in the molecule.

In the next few sections, we shall consider the characteristic group vibrations for various classes of compounds we have encountered.

17.4 Alkanes

As we saw previously, the major bands that appear in the infrared spectra of alkanes are those due to C—H stretching in the 2850–3000 cm^{-1} region, those due to CH_2 and CH_3 scissoring in the 1450–1470 cm^{-1} region, the band due to CH_3 rocking at about 1370–1380 cm^{-1}, and the CH_2 rocking bands at 720–725 cm^{-1}. These bands are of essentially no diagnostic value because most alkanes contain all of these groupings.

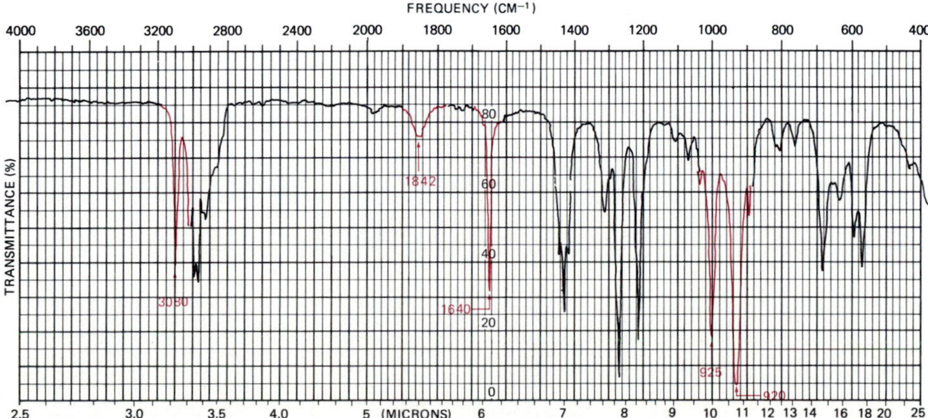

FIGURE 17.10 Infrared spectrum of 4-bromo-1-butene, $CH_2{=}CHCH_2CH_2Br$.

17.5 Alkenes

The alkene C—H stretching vibration occurs at higher wavenumber (shorter wavelength) than that due to an alkane C—H. Recall that alkene carbon-hydrogen bonds have greater *s*-character and are stronger than alkane carbon-hydrogen bonds. Stronger bonds are more difficult to stretch (higher force constant) and require greater energy or higher light frequency. Thus alkenes that have at least one hydrogen attached to the double bond normally absorb in the region 3050–3150 cm^{-1}. The relative intensity of this band, compared with the band for saturated C—H stretch, is roughly proportional to the relative numbers of the two types of hydrogens in the molecule.

The alkene C=C stretching mode occurs in the region 1645–1670 cm^{-1}. This band is most intense when there is only one alkyl group attached to the double bond. As more alkyl groups are added, the intensity of the absorption diminishes because the vibration now results in a smaller change of dipole moment. For trisubstituted, tetrasubstituted, and relatively symmetrical disubstituted alkenes, the C—C stretching bond is often of such low intensity that it is not observable.

Alkene C—H out-of-plane bending vibrations occur between 650 and 1000 cm^{-1}. For terminal alkenes, these vibrations are particularly intense and appear between 890 and 990 cm^{-1}

EXERCISE 17.2 The infrared spectrum for an unknown hydrocarbon is reproduced in Figure 17.11. What structural information can be gleaned from the highlighted absorption bands in this spectrum?

EXERCISE 17.3 Explain why *trans*-4-octene shows no infrared absorption for its carbon-carbon double bond.

17.6 Alkynes and Nitriles

Terminal alkynes show a sharp C—H stretching band at 3300–3320 cm^{-1} and an intense C—H bending mode at 600–700 cm^{-1}. The C≡C stretch for terminal alkynes appears as a sharp absorption of moderate intensity at 2100–2140 cm^{-1} (Figure 17.12). For internal alkynes the C≡C stretch is a weak band occurring at 2200–2260 cm^{-1}. In hydrocarbons, there is little change in dipole moment when C≡C is stretched; consequently, this band is

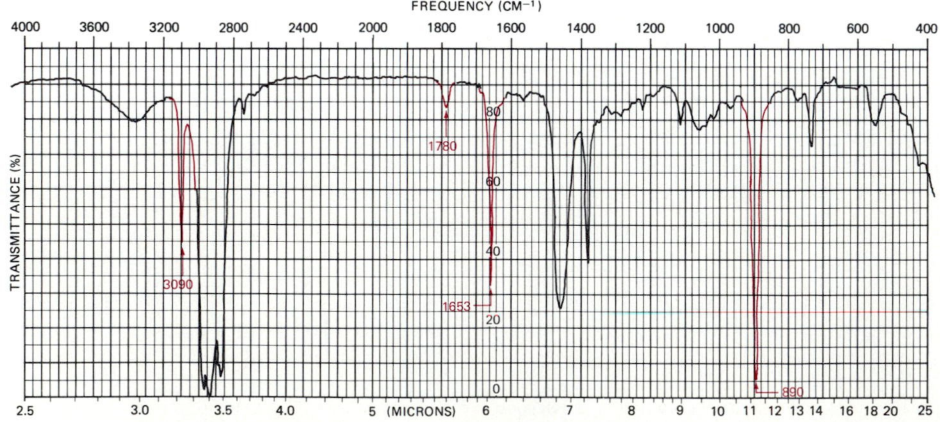

FIGURE 17.11 Infrared spectrum of an unknown hydrocarbon.

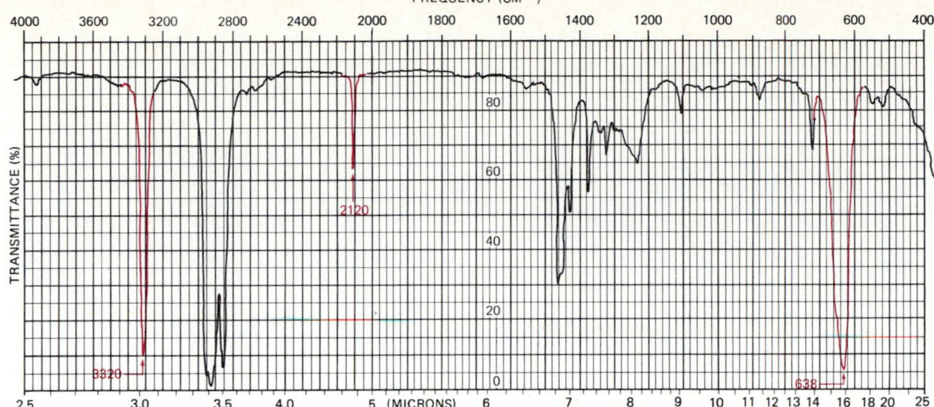

FIGURE 17.12 Infrared spectrum of 1-octyne.

so weak that it is usually not observed. The C≡N stretch occurs in the 2210–2260 cm^{-1} region.

EXERCISE 17.4 1-Octyne is treated with butyllithium and the resulting 1-lithio-1-octyne is hydrolyzed with D_2O. Using the simple Hooke's law relationship, predict the position of the C—D stretch in the resulting product.

17.7 Alkyl Halides

The characteristic absorption of alkyl halides is the band due to the C—X stretch. Typical positions for these bands are 1000–1350 cm^{-1} for C—F, 750–850 cm^{-1} for C—Cl, 500–680 cm^{-1} for C—Br, and 200–500 cm^{-1} for C—I. None of these absorptions is particularly useful for diagnosis.

17.8 Alcohols and Ethers

Alcohols and ethers have a characteristic infrared absorption associated with the C—O stretch (1050–1200 cm^{-1}). Since these bands occur in a region of the spectrum where there are usually many other bands, they are not of great diagnostic value. However, the O—H stretch of alcohols, which occurs in the 3200–3600 cm^{-1} region, is of more use. The infrared spectrum of *t*-butyl alcohol shown in Figure 17.13 is illustrative. Note the very intense O—H stretch, which is centered at about 3360 cm^{-1}.

In Figure 17.14a–c are plotted the dispersive spectra of the O—H and C—H regions of *t*-butyl alcohol dissolved in carbon tetrachloride. (Carbon tetrachloride is a frequently used solvent for infrared studies because it is relatively inert chemically and is "transparent" to infrared light in most of the useful spectral regions.) Notice that in the first spectrum the 3440 cm^{-1} O—H absorption is accompanied by a sharp peak at 3620 cm^{-1}. In more dilute solutions, the 3620 cm^{-1} band is more intense relative to the 3440 cm^{-1} band. These two bands are both due to O—H stretch. The band at shorter wavelength (higher energy) is due to the stretching mode of the "free" hydroxy group. The stretching mode of hydrogen-bonded or "associated" O—H occurs at lower energy.

free hydroxy	[O—H]	3620–3640 cm^{-1}
associated hydroxy	[O—H···O]	3250–3450 cm^{-1}

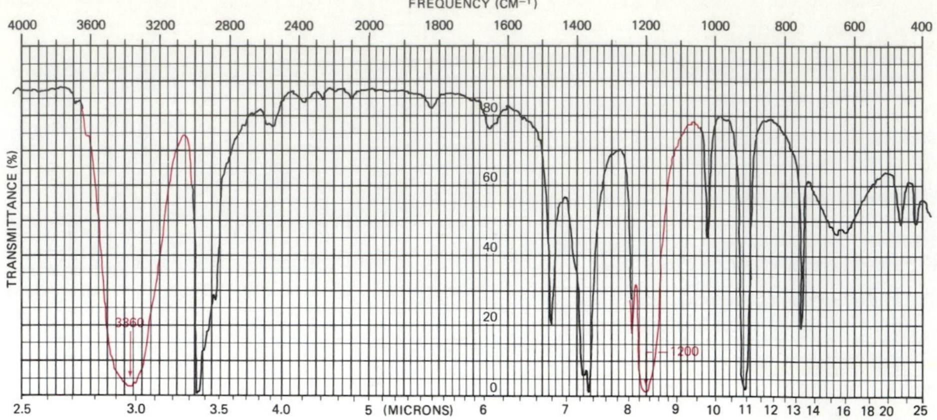

FIGURE 17.13 Infrared spectrum of 2-methyl-2-propanol (*t*-butyl alcohol).

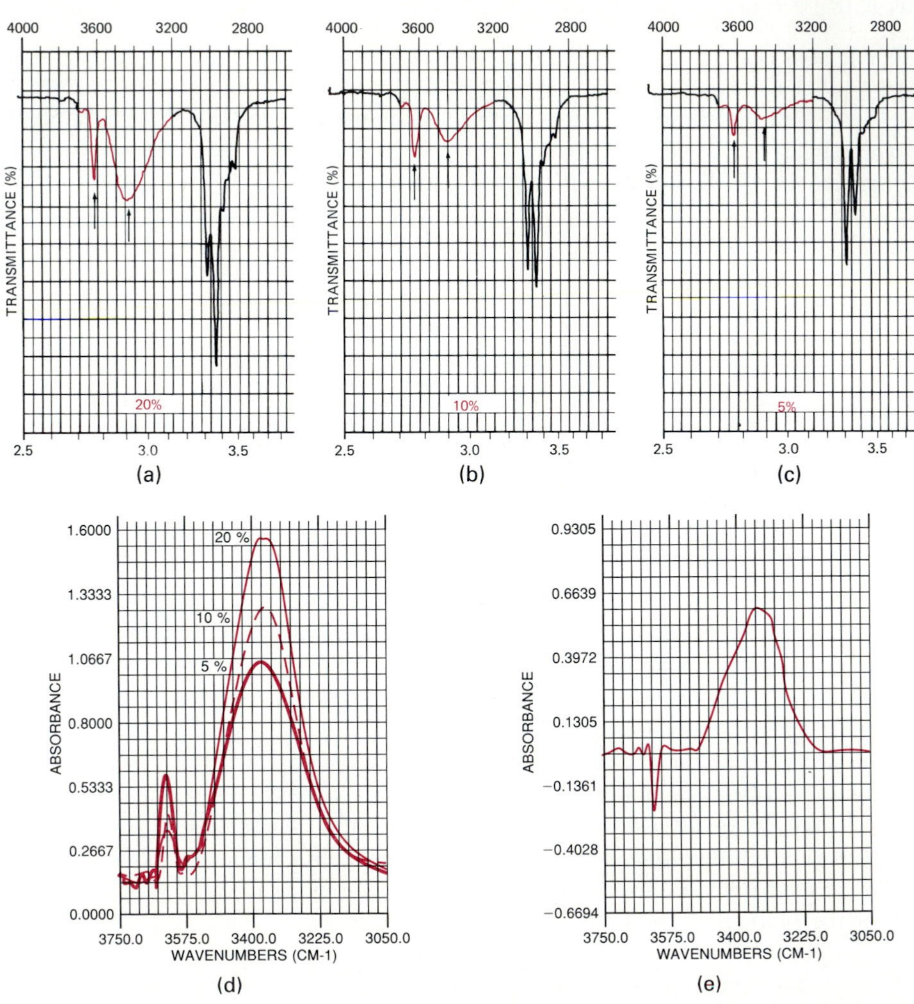

FIGURE 17.14 (a–c) Dispersive infrared spectra of three solutions of *t*-butyl alcohol in carbon tetrachloride. (d, e) A set of superimposed spectra of the O—H stretching region of the same three solutions of *t*-butyl alcohol normalized to the same alcohol concentration taken with a Fourier transform infrared spectrometer is also shown along with a difference spectrum for the highest and lowest concentration.

As the solution is made progressively more dilute, it is more likely that a molecule will exist in an unassociated state. Consequently, the relative intensity of the "bonded" O—H absorption is less in dilute solutions. A set of superimposed spectra of the O—H stretching region of the same three solutions of *t*-butyl alcohol normalized to the same alcohol concentration taken with a Fourier transform infrared spectrometer is also shown along with a difference spectrum for the highest and lowest concentration (Figure 17.14d, e). The "associated" hydrogen-bonded absorption is more intense in more concentrated solution, and the "free" hydroxy group absorption is less intense. The hydroxy group must be partially associated with at least one other *t*-butyl alcohol even in 5% solution. The position (wavelength), the shape, and the intensity of infrared absorption bands are thus important.

17.9 Aldehydes and Ketones

The characteristic infrared absorption for aldehydes and ketones is the band due to the C=O stretching vibration. Since the carbonyl group is highly polar, stretching of this bond results in a relatively large change in dipole moment. Consequently the carbonyl stretching band is an intense spectral feature. Because of its intensity, and also because it occurs in a region of the infrared spectrum commonly devoid of other absorptions, the carbonyl stretch is a reliable method for deducing the presence of such a functional group in a compound.

For simple saturated aldehydes the band occurs at about 1725 cm^{-1}. For saturated acyclic ketones the band occurs at about 1715 cm^{-1}. The distinctive nature of the C=O stretch is apparent in the spectrum of 2-heptanone shown in Figure 17.15. Since the carbonyl stretch is such an intense absorption, it often gives rise to a noticeable overtone in the 3400–3500 cm^{-1} region. In 2-heptanone the carbonyl overtone occurs at 3440 cm^{-1} and may be seen in Figure 17.15. One must be cautious not to mistake this overtone for an OH absorption.

In cyclic ketones the exact stretching frequency depends on the size of the ring containing the carbonyl carbon. The magnitude of the effect is shown in Table 17.2. The observed relationship between C=O stretching frequency and C—C=O angle has its origin in a "coupling" of the C=O stretch with that of the carbon-carbon bonds. For example, consider the hypothetical molecule depicted in Figure 17.16a in which the C—C=O angle is 90°. When the carbon-oxygen double bond is stretched, the carbon and oxygen move apart from one another. As shown in Figure 17.16b, this motion may occur without seriously affecting the carbon-carbon bond length. However, the situation is different in

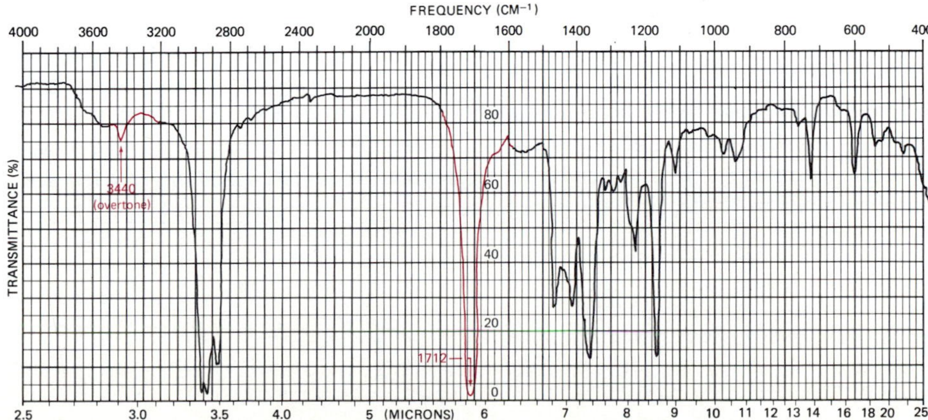

FIGURE 17.15 Infrared spectrum of 2-heptanone.

TABLE 17.2 Carbonyl Stretching Frequencies

Type of Ketone	Frequency, cm^{-1}
normal open chain	1715
cyclic	
three-membered	1850
four-membered	1780
five-membered	1745
six-membered	1715
seven-membered	1705

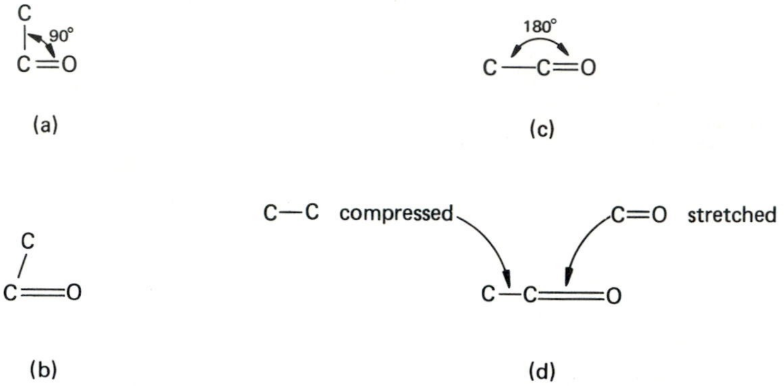

FIGURE 17.16 Illustrating the coupling of C=O and C—C stretching in ketones.

the hypothetical ketone shown in Figure 17.16c, where the C—C=O angle is 180°. When the carbon-oxygen double bond is stretched in this case, the carbon-carbon bond must be simultaneously shortened (Figure 17.16d). In a way, the carbon-carbon bond acts as a kind of ''ballast'' that resists stretching of the adjacent carbon-oxygen double bond. It is obvious from these simple extremes that the magnitude of coupling between the two bonds will relate to the C—C=O angle. Cyclohexanone and normal acyclic ketones have C—C=O angles of about 112°, which is close to the normal angle for sp^2-hybridization (120°). Smaller ring size is associated with smaller internal bond angles and hence a larger C—C=O angle. The larger the angle, the more the coupling and the greater the energy required to stretch the carbonyl bond.

The characteristic stretching frequencies of cyclic carbonyl groups apply to polycyclic systems as well. For example, the carbonyl group of camphor is part of a five-membered ring and the stretching frequency of 1740 cm^{-1} is characteristic of a cyclopentanone. The fact that the CO group in the bicyclic camphor is also part of a six-membered ring is not relevant—it is the character of ring strain that shows up in the spectrum. However, since ring-strain effects do depend somewhat on the detailed structure, variations of a few cm^{-1} from the values in Table 17.2 are to be expected.

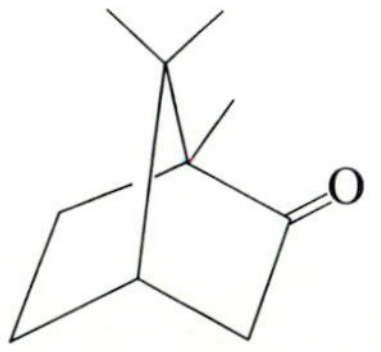

$\nu(CO) = 1740\ cm^{-1}$
camphor

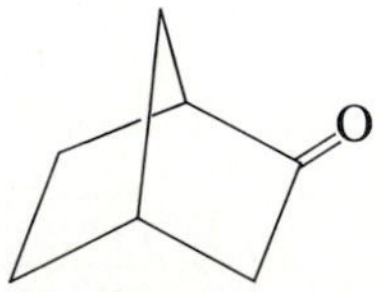

$\nu(CO) = 1751\ cm^{-1}$
norbornanone

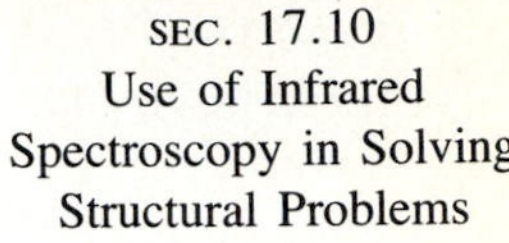

The following bicyclic ketone is an apparent exception. Its C═O stretching band at 1731 cm^{-1} is far higher than expected for a cyclohexanone ring. A second look at the structure of this bicyclic ketone reveals, however, that its six-membered ring is that of a boat conformation rather than a chair.

ν(CO) = 1731 cm^{-1}
bicyclo[2.2.2]octan-2-one

EXERCISE 17.5 The carbonyl stretching frequency of ketene, $H_2C{=}C{=}O$, is 2150 cm^{-1}. Is this value consistent with the argument advanced to account for the magnitude of C═O stretching frequencies?

17.10 Use of Infrared Spectroscopy in Solving Structural Problems

In this chapter we have considered the infrared spectra of the classes of organic compounds taken up so far in this book. We have seen that the infrared spectra of organic compounds are so complex that it is impractical to analyze a spectrum completely and assign each absorption to a given vibration. However, for each functional group there are characteristic absorptions that may be used empirically as a diagnosis for that particular functional group. Table 17.3 summarizes the infrared characteristics of alkanes, alkenes, alkynes, alkyl halides, alcohols, ethers, aldehydes, and ketones. The most useful bands are those printed in boldface type. As we consider other classes of compounds, we shall point out their characteristic infrared absorption bands.

TABLE 17.3 Principal Infrared Absorption

Class	Frequency, cm^{-1}	Intensity[a]	Assignment
1. Alkanes	2850–3000	s	C—H stretch
	1450–1470	s	CH_2 and CH_3 bend
	1370–1380	s	
	720–725	m	
2. Alkenes			
(a) $RCH{=}CH_2$	**3080–3140**	m	═C—H stretch
	1800–1860	m	overtone
	1645	m	C═C stretch
	990	s	C—H out-of-plane bend
	910	s	
(b) $R_2C{=}CH_2$	**3080–3140**	m	═C—H stretch
	1750–1800	m	overtone
	1650	m	C═C stretch
	890	s	C—H out-of-plane bend
(c) *cis*-RCH═CHR	3020	w	═C—H stretch
	1660	w	C═C stretch
	675–725	m	C—H out-of-plane bend

TABLE 17.3 (continued)

Class	Frequency, cm^{-1}	Intensity[a]	Assignment
(d) *trans*-RCH=CHR	3020	w	=C—H stretch
	1675	vw	C=C stretch
	970	s	C—H out-of-plane bend
(e) R_2C=CHR	3020	w	=C—H stretch
	1670	w	C=C stretch
	790–840	s	C—H out-of-plane bend
(f) R_2C=CR_2	1670	vw	C=C stretch
3. Alkynes			
(a) RC≡CH	**3300**	s	≡C—H stretch
	2100–2140	m	C≡C stretch
	600–700	s	≡C—H bend
(b) RC≡CR	2190–2260	vw	C≡C stretch
4. Alkyl Halides			
(a) R—F	1000–1350	vs	C—F stretch
(b) R—Cl	750–850	s	C—Cl stretch
(c) R—Br	500–680	s	C—Br stretch
(d) R—I	200–500	s	C—I stretch
5. Alcohols			
(a) RCH_2OH	**3600**	var	free O—H stretch
	3400	s	bonded O—H stretch
	1050	s	C—O stretch
(b) R_2CHOH	**3600**	var	free O—H stretch
	3400	s	bonded O—H stretch
	1150	s	C—O stretch
(c) R_3COH	**3600**	var	free O—H stretch
	3400	s	bonded O—H stretch
	1200	s	C—O stretch
6. Ethers	1070–1150	s	C—O stretch
7. Aldehydes	**1725**	s	C=O stretch
	2720, 2820	m	C—H stretch
8. Ketones			
(a) acyclic	**1715**	s	C=O stretch
(b) three-membered	**1850**	s	C=O stretch
(c) four-membered	**1780**	s	C=O stretch
(d) five-membered	**1745**	s	C=O stretch
(e) six-membered	**1715**	s	C=O stretch
(f) seven-membered	**1705**	s	C=O stretch

[a]vs = very strong, s = strong, m = medium, w = weak, vw = very weak, var = variable.

It is important that the student recognize that it is not necessary to memorize all of the absorption ranges in Table 17.3. However, it *is* useful to commit a few numbers to memory. For the time being, it is sufficient if you remember that the O—H stretch is in the region 3400–3600 cm^{-1}, that the C=O stretch is an intense absorption in the region around 1700 cm^{-1}, that terminal alkynes are characterized by a relatively sharp and distinct C—H stretch at about 3300 cm^{-1}, and that the C=C stretch is at around 1650 cm^{-1}. Further values can be obtained by reference to tables of data, such as Table 17.3.

In modern practice, infrared spectroscopy is used as an adjunct to NMR and CMR spectroscopy—mainly for the identification of the functional groups in a molecule. For example, suppose that an unknown compound shows NMR resonances at δ 2.42 (quartet) and 1.05 (triplet) and has a strong band in its IR spectrum at 1710 cm^{-1}. The NMR splitting pattern tells us immediately that the compound has an ethyl group, CH_3CH_2, and the chemical shift of the quartet tells us that this group is attached to a relatively electronegative atom. The 1710 cm^{-1} band in the IR spectrum tells us that the compound has a carbonyl group. Putting these two facts together, we conclude that the unknown compound is 3-pentanone.

As CMR spectrometers have become more available, the importance of IR spectroscopy has diminished somewhat, since CMR is actually superior to IR for the identification of many functions. For example, the characteristic CMR resonances of alkenes (δ 110–140) are observed whether or not the double bond bears hydrogens. Indeed, the exact resonance position is highly characteristic of the substitution pattern of the double bond. In a similar manner, the carbonyl carbons of aldehydes and ketones give rise to characteristic CMR resonances in unique positions (δ 200–220). However, CMR spectroscopy has the drawback that relatively large amounts of sample are necessary (on the order of 1–5 mg for a normal spectrum), whereas IR spectra can be measured with much less than a milligram of material.

EXERCISE 17.6 A compound with the formula $C_8H_{14}O$ has a strong infrared absorption at 1710 cm^{-1}. The NMR spectrum shows a sharp, six-proton singlet at δ 1.0 ppm and a four-proton multiplet centered at δ 2.3 ppm. The other four protons are seen as a complex multiplet centered at δ 1.6 ppm. What structural information can be gained from this data?

Highlight 17.2

Chemical compounds absorb infrared light over wavelength ranges that are characteristic for individual groups and their **external** and **internal** chemical environments. An example of the former is the O—H bond for which the absorption due to stretching is at 3600 cm^{-1}, and at 3400 cm^{-1} for a group associated with another O—H group, symbolized as O—H $\cdots$ O. The 3400 cm^{-1} band is broader because there are various conformational forms of the associated compound. Recall that the position (wavelength), the shape, and the intensity of infrared absorption bands are important. An example of the influence of the internal environment is that of the C=O group of cyclic ketones, with stretching absorptions at 1705 cm^{-1} (seven-membered ring), 1715 cm^{-1} (six-membered ring), to 1745 cm^{-1} (five-membered ring), 1780 cm^{-1} (four-membered ring), and 1850 cm^{-1} (three-membered ring). These values are to be compared with 1715 cm^{-1} for an open-chain ketone.

ASIDE

The rise in the atmospheric carbon dioxide concentration and the very strong infrared absorption of carbon dioxide at about 2350 cm^{-1} might be the critical factors in causing a rise in the temperature of our planet, the "greenhouse effect." The temperature rise has not been definitively established, but the indications are suggestive enough to create some pressure for reducing the emissions of carbon dioxide. Fortunately, this absorption does not overlap very much with most of the important and characteristic infrared absorption bands of organic compounds. Compare the infrared spectra expected for ethane, ethanol, and acetaldehyde. Which bands would you use for analysis of a mixture of the three compounds?

PROBLEMS

1. Identify the functional groups in each compound from the following infrared spectra. Note that the spectra in **f** and **g** were obtained using a different instrument than that used for the spectra in **a–e.** This is a common situation. Although the spectra obtained with the two instruments have different appearances, the characteristic absorptions are independent of the spectrometer used.

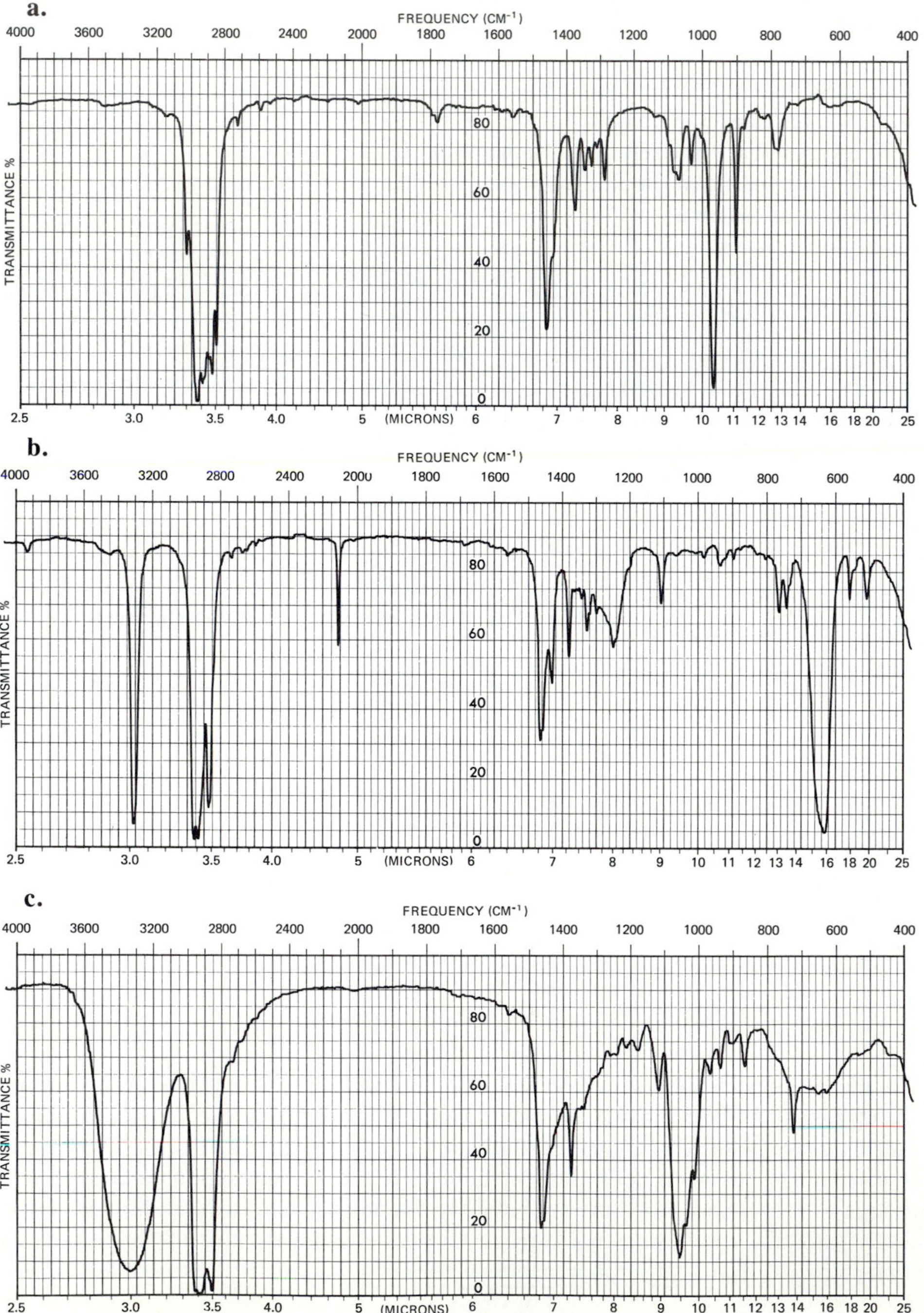

d.

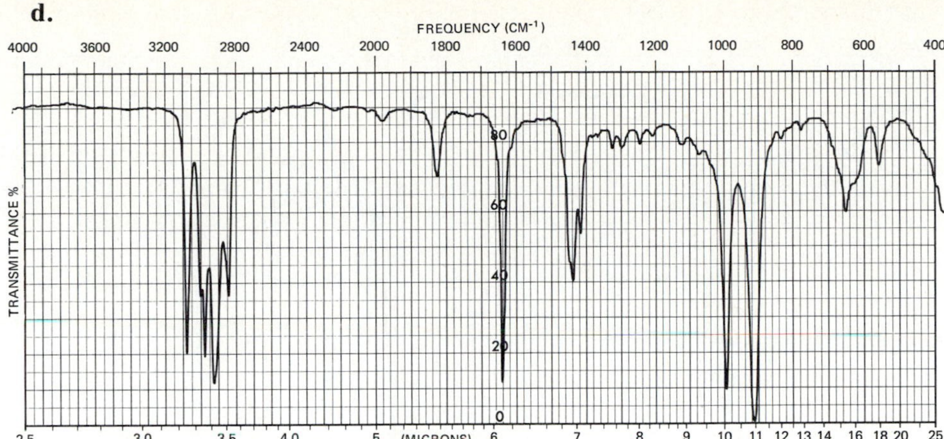
FREQUENCY (CM^{-1})
4000 3600 3200 2800 2400 2000 1800 1600 1400 1200 1000 800 600 400
TRANSMITTANCE %
80 60 40 20 0
2.5 3.0 3.5 4.0 5 (MICRONS) 6 7 8 9 10 11 12 13 14 16 18 20 25

e.

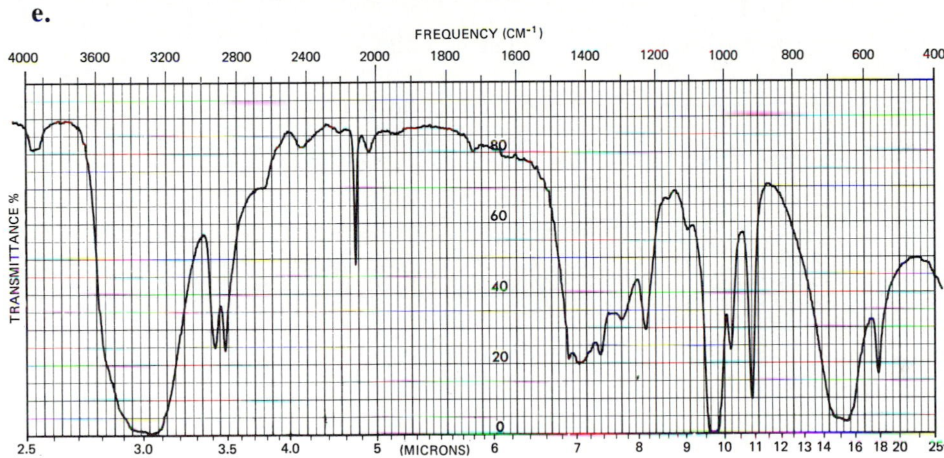
FREQUENCY (CM^{-1})
4000 3600 3200 2800 2400 2000 1800 1600 1400 1200 1000 800 600 400
TRANSMITTANCE %
80 60 40 20 0
2.5 3.0 3.5 4.0 5 (MICRONS) 6 7 8 9 10 11 12 13 14 16 18 20 25

f.

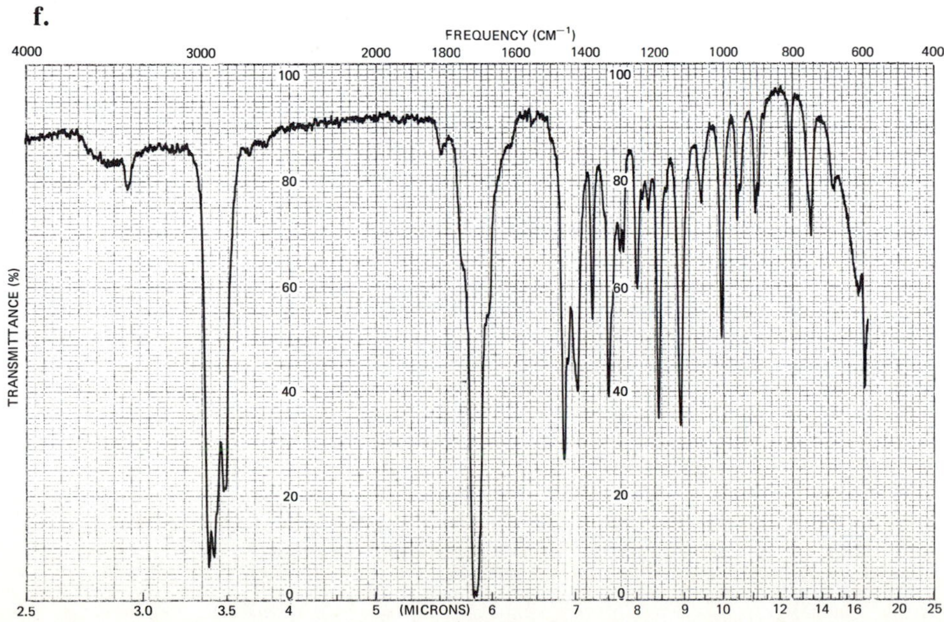
FREQUENCY (CM^{-1})
4000 3000 2000 1800 1600 1400 1200 1000 800 600 400
TRANSMITTANCE (%)
100 80 60 40 20 0
2.5 3.0 3.5 4 5 (MICRONS) 6 7 8 9 10 12 14 16 20 25

g.

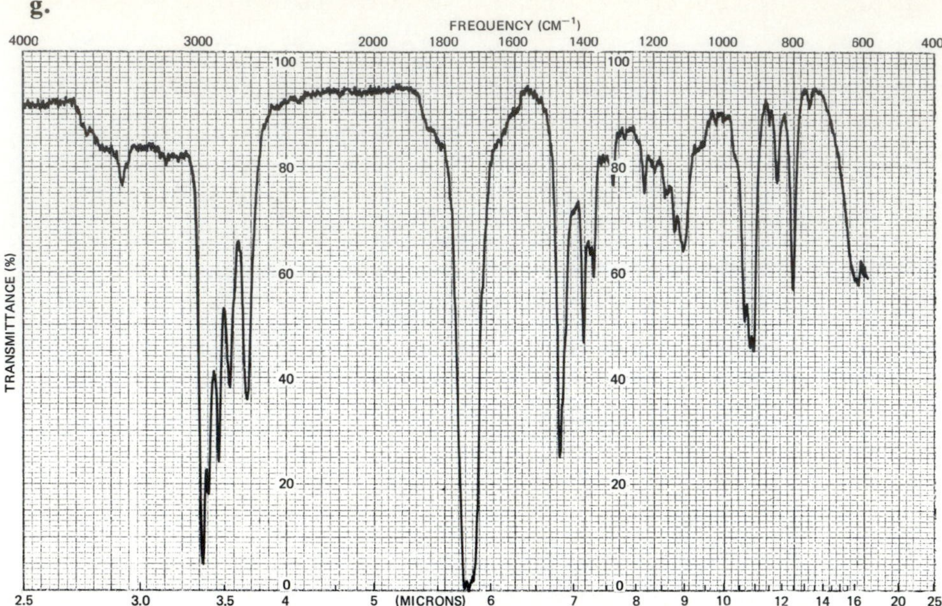

2. Identify the following compound from its IR and NMR spectra.

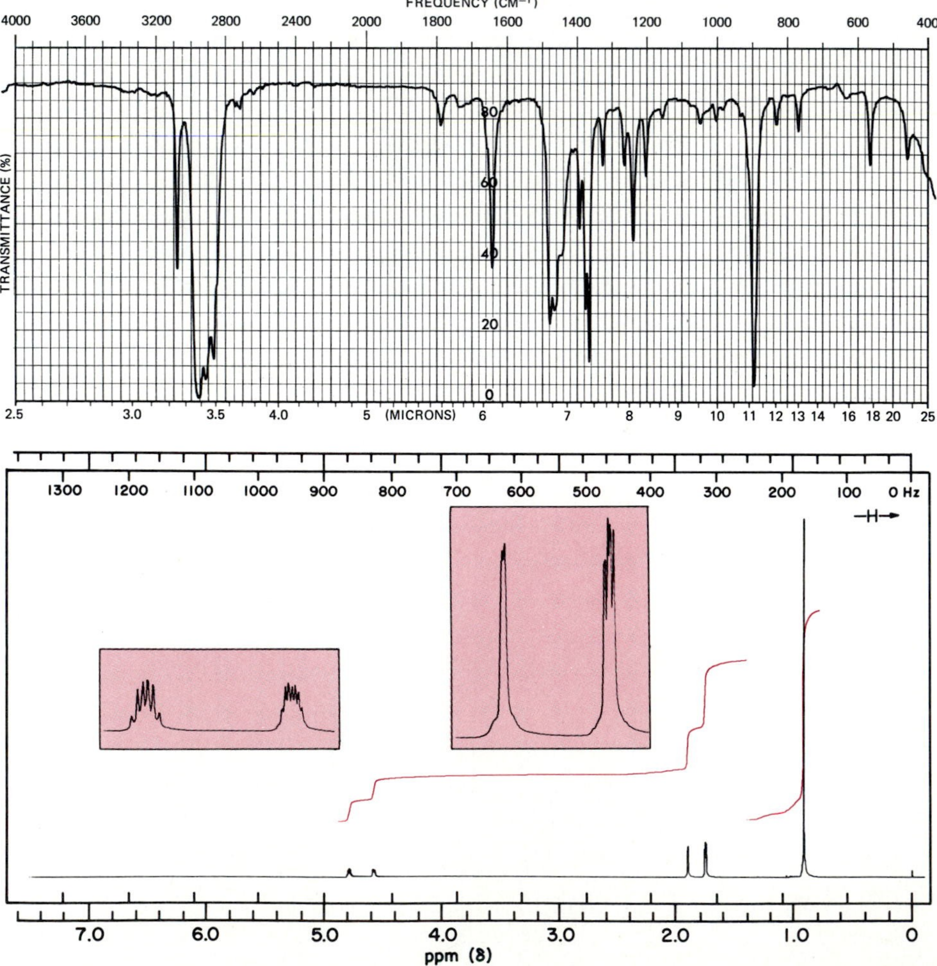

3. Identify the following compound from its IR and NMR spectra.

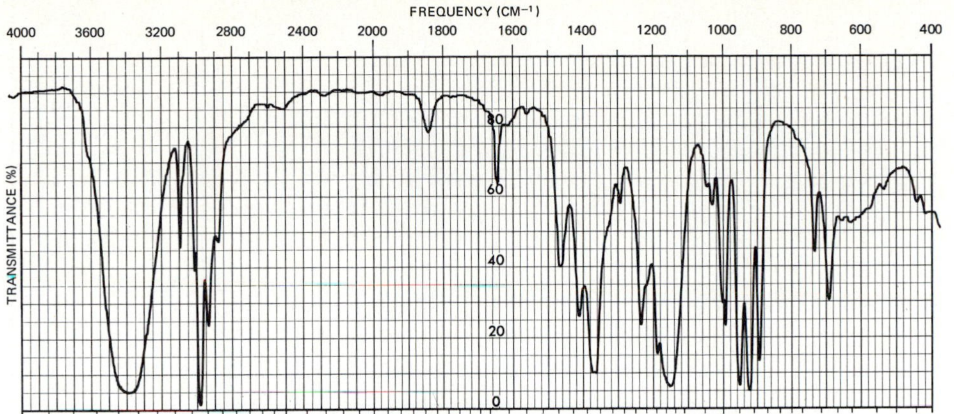

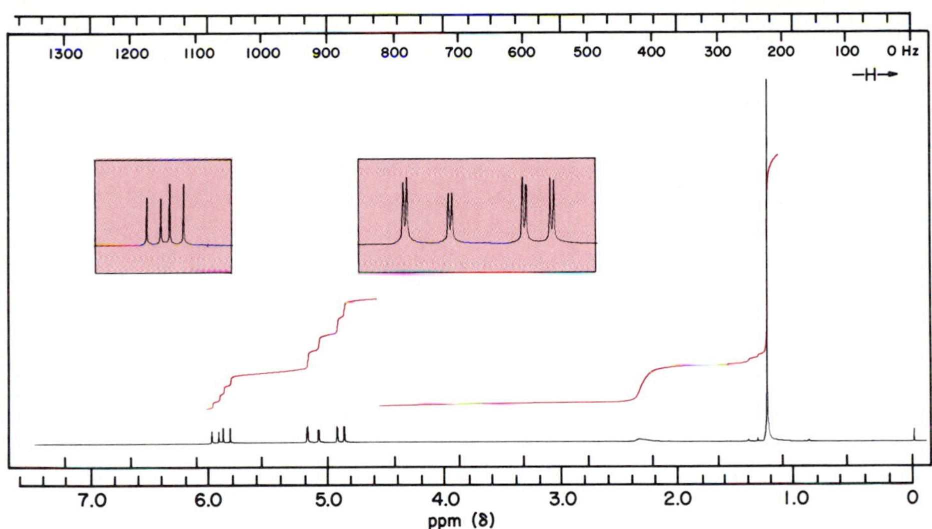

4. Identify the following compound from its IR and NMR spectra.

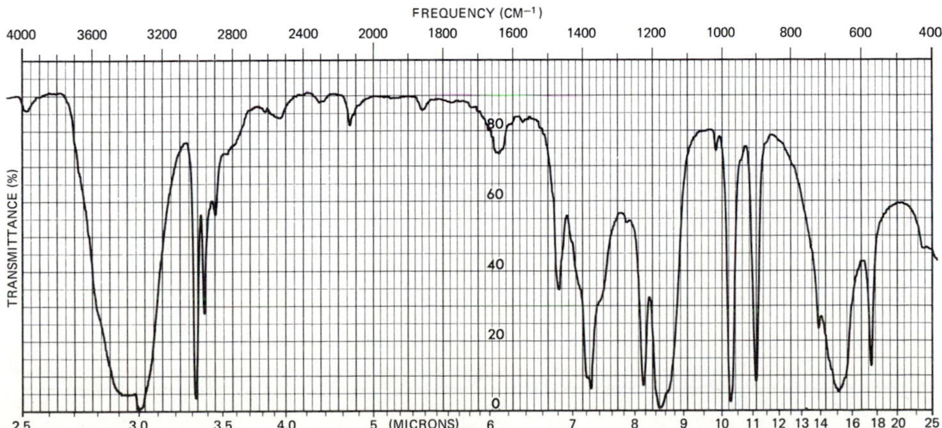

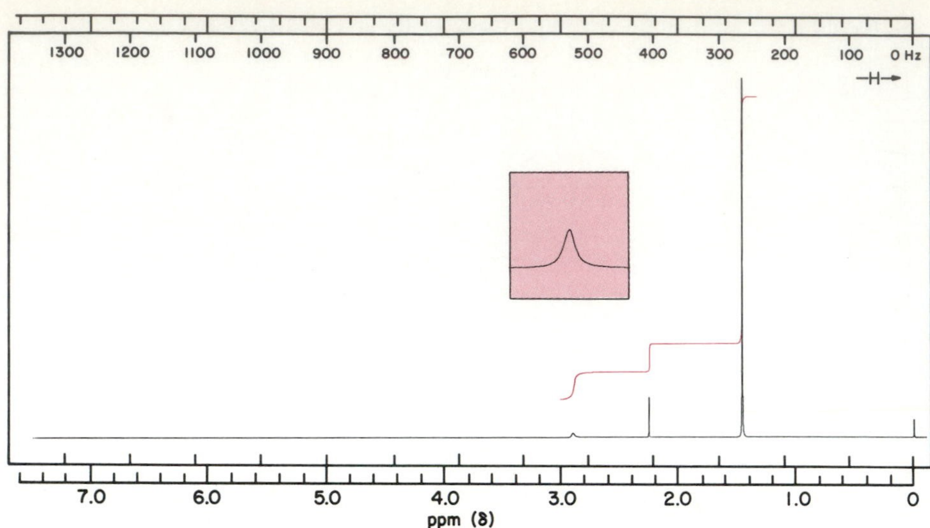

5. Identify the following compound from its IR and NMR spectra.

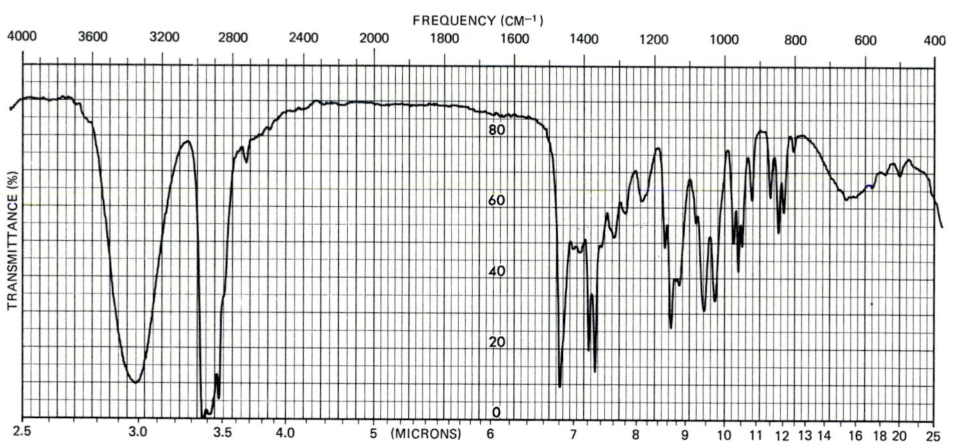

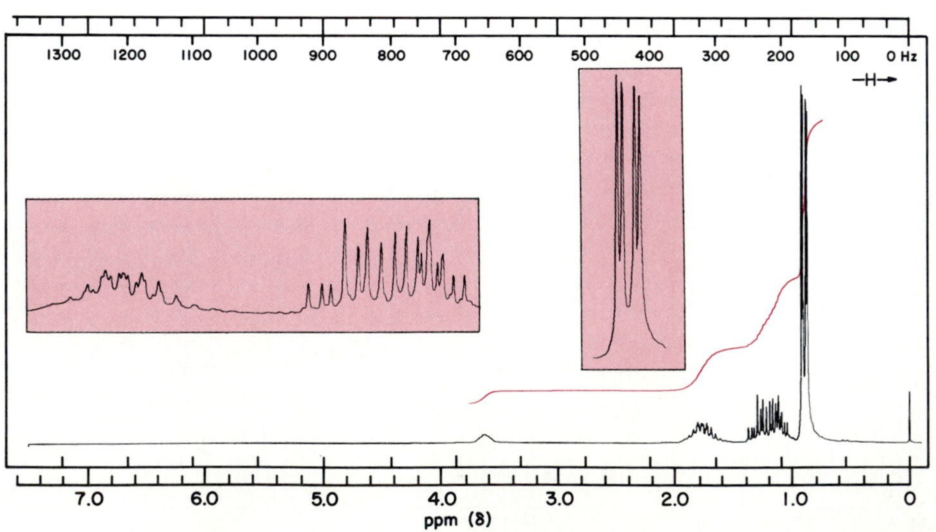

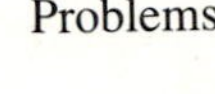

6. A compound gives the following IR spectrum. Its CMR spectrum has absorptions at 7.1, 8.5, 26.4, 30.7, 76.8, 212.7 ppm. Suggest a structure for the compound.

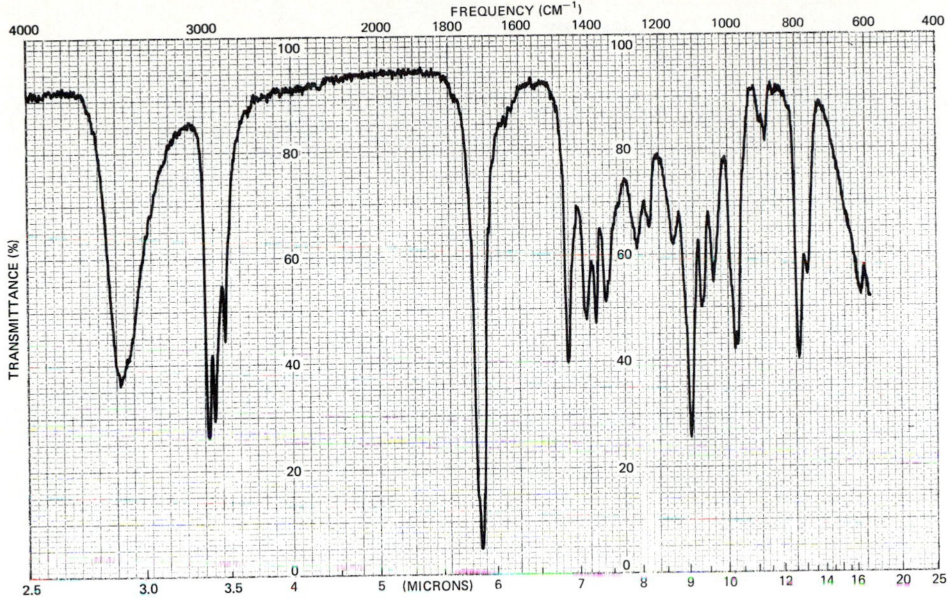

7. Identify the following isomeric compounds from their IR and CMR spectra.

a. CMR: 7.9, 25.5, 33.5, 72.6 ppm.

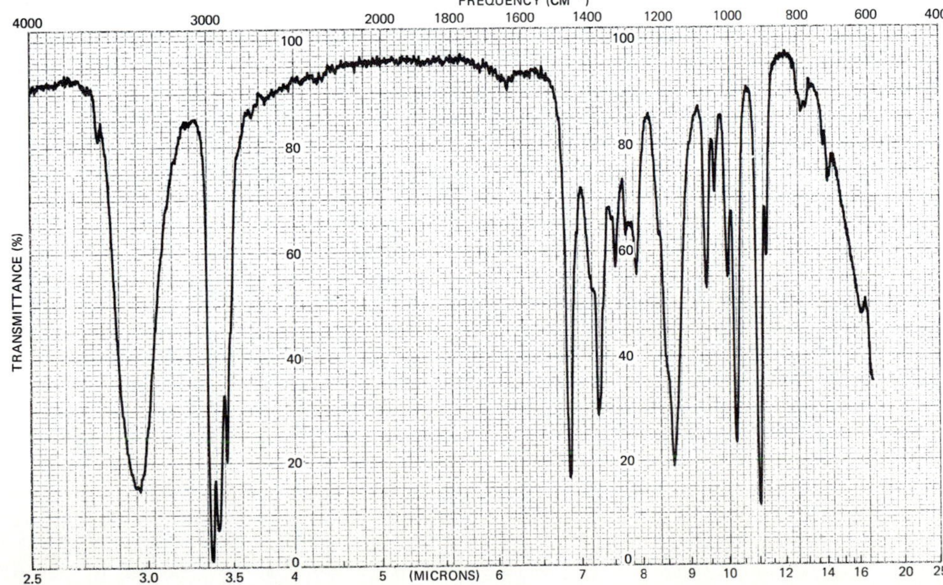

b. CMR: 22.7, 23.5, 24.3, 25.1, 49.2, 65.5 ppm.

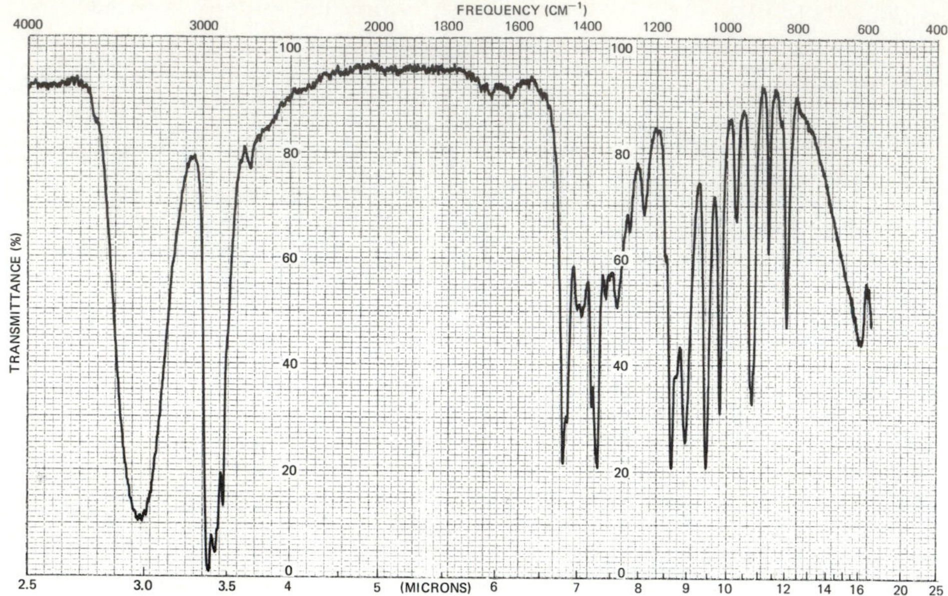

8. a. For the heavier methyl halides one infrared frequency can be treated to an excellent approximation as a C—X stretching vibration. The position of this band is CH_3Cl, 732 cm^{-1}; CH_3Br, 611 cm^{-1}; CH_3I, 533 cm^{-1}. Find the corresponding C—X force constants and determine whether they are proportional to the corresponding $DH°$ (CH_3—X) values (see Appendix II).

b. Astatine, element no. 85, is a halogen with no stable isotopes. The longest-lived isotope is ^{210}At with a half-life of 8.3 hr. At what value of $\tilde{\nu}$ would you expect to find the CH_3—At stretching band of methyl astatide?

c. Methanethiol has a corresponding vibration (C—S stretch) at 705 cm^{-1}. Use your results from (a) to calculate the corresponding $DH°$ value. The experimental value for $DH°$ (CH_3—SH) is about 76 kcal $mole^{-1}$.

9. Identify the following compound from its IR and proton-decoupled CMR spectra.

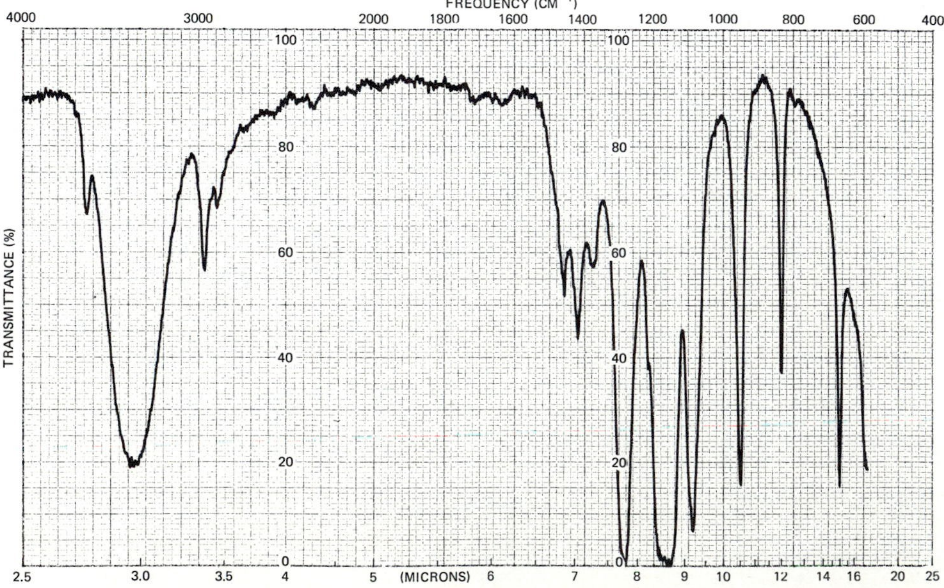

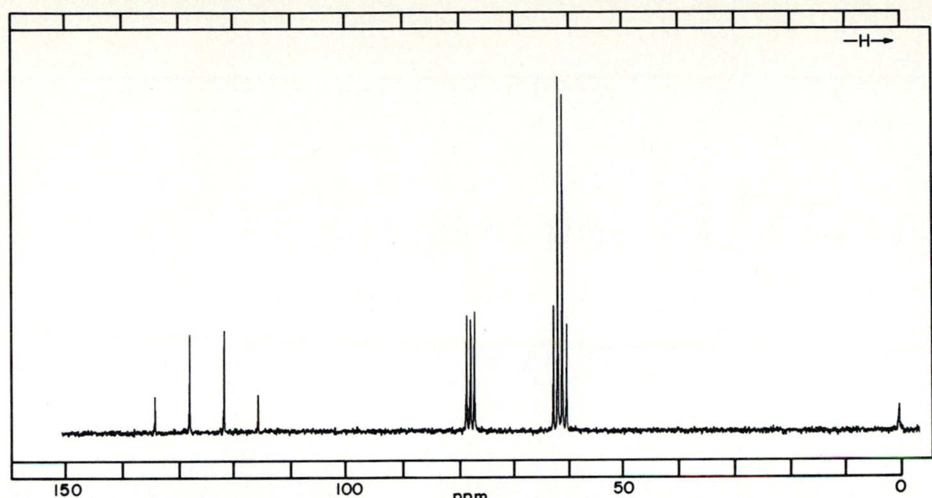

10. Dialkyl peroxides, ROOR, have an absorption in the region 820–1000 cm^{-1}, but this band is extremely weak and difficult to detect. Explain. Using the Hooke's law approximation, find the force constant for an O—O stretch of 900 cm^{-1}. How does it compare with the normal single bond f of 5×10^5? Explain.

11. Give a plausible explanation for the species responsible for the two absorption bands around 3500 cm^{-1} observed for *t*-butyl alcohol in carbon tetrachloride solution. (*Hint*: Consider the hydrogen bond to have a ΔH_f of 5 kcal $mole^{-1}$)

CHAPTER 18

CARBOXYLIC ACIDS

18.1 Structure

Carboxylic acids are distinguished by the functional group COOH, referred to as the **carboxy group.** Four ways of depicting the carboxy group are shown below.

$$\underset{\text{Lewis structure}}{\mathrm{R{:}\overset{\overset{:O:}{::}}{C}{:}\ddot{\underset{\cdot\cdot}{O}}{:}H}} \qquad \underset{\text{Kekulé structure}}{\mathrm{R{-}\overset{\overset{O}{\|}}{C}{-}O{-}H}} \qquad \underbrace{\mathrm{RCOOH}\quad \mathrm{RCO_2H}}_{\text{condensed structures}}$$

The R group can be either an organic group or a hydrogen.

The carbon atom in a carboxy group uses three hybrid orbitals to bond to the oxygen of the OH group, to the carboxy oxygen, and to hydrogen or an organic group. These three orbitals are approximately sp^2-hybrids that lie in one plane. The remaining p-orbital on the carbon forms a π-bond to a p-orbital on the carboxy oxygen. There are two distinct carbon-oxygen bond distances, corresponding to C=O and C—O. The bond angles and bond lengths of formic acid, as determined by microwave spectroscopy, are shown in Figure 18.1. Note that the bond angles around the carboxy carbon are only approximately those expected for sp^2-hybridization. The array HCOO is planar, and the hydroxy hydrogen lies outside of this plane.

In the solid and liquid phases, as well as in the vapor phase at moderately high pressure, carboxylic acids exist largely in a dimeric form in which there is mutual hydrogen bonding from the OH of one molecule to the carbonyl oxygen of another.

$$2\,\mathrm{RCOOH} \rightleftharpoons \mathrm{R{-}C}\begin{matrix}\nearrow \mathrm{O\cdots H{-}O} \searrow \\ \searrow \mathrm{O{-}H\cdots O} \nearrow\end{matrix}\mathrm{C{-}R}$$

	Bond Lengths, Å		Bond Angles, deg
C=O	1.202	H—C=O	124.1
C—O	1.343	O—C=O	124.9
C—H	1.097	H—C—O	111.0
O—H	0.972	H—O—C	106.3

FIGURE 18.1 Structure of formic acid.

For formic acid in the vapor phase, $\Delta H°$ for the dimerization has been determined to be -14 kcal mole^{-1}.

18.2 Nomenclature

Two systems of nomenclature are currently in use for carboxylic acids, and the student should be acquainted with both. Since many of the simpler acids are naturally occurring and were discovered early in the history of organic chemistry, they have well-entrenched common names. At the 1892 Geneva Congress it was agreed to derive the name of a carboxylic acid systematically from that of the normal alkane having the same number of carbon atoms by dropping the ending **-e** and adding the suffix **-oic acid.** The common and IUPAC names for the first ten straight-chain acids, as well as other selected examples, are given in Table 18.1. The name used by *Chemical Abstracts* in indexing is printed in bold type.

Past caproic acid, the even-numbered carboxylic acids are the more important because only the even-numbered acids occur in nature. Carboxylic acids are biosynthesized by the combination of two-carbon units related to acetic acid. **Biosynthesis** is the synthesis from small precursor molecules in reactions catalyzed by the multiple enzyme systems found in living organisms. Because they are built up from acetic acid, most of the naturally occurring acids have an even number of carbon atoms in the chain.

When naming a substituted carboxylic acid in the IUPAC system, the longest carbon chain *containing the carboxy group* is numbered from 1 to *n*, beginning with the carboxy

TABLE 18.1 Nomenclature of Carboxylic Acids

Compound	Common Name	IUPAC Name
HCOOH	**formic acid**	methanoic acid
CH_3COOH	**acetic acid**	ethanoic acid
CH_3CH_2COOH	propionic acid	**propanoic acid**
$CH_3(CH_2)_2COOH$	butyric acid	**butanoic acid**
$CH_3(CH_2)_3COOH$	valeric acid	**pentanoic acid**
$CH_3(CH_2)_4COOH$	caproic acid	**hexanoic acid**
$CH_3(CH_2)_5COOH$	enanthic acid	**heptanoic acid**
$CH_3(CH_2)_6COOH$	caprylic acid	**octanoic acid**
$CH_3(CH_2)_7COOH$	pelargonic acid	**nonanoic acid**
$CH_3(CH_2)_8COOH$	capric acid	**decanoic acid**
$CH_3(CH_2)_{10}COOH$	lauric acid	**dodecanoic acid**
$CH_3(CH_2)_{12}COOH$	myristic acid	**tetradecanoic acid**
$CH_3(CH_2)_{14}COOH$	palmitic acid	**hexadecanoic acid**
$CH_3(CH_2)_{16}COOH$	stearic acid	**octadecanoic acid**

carbon. The name of this parent straight-chain carboxylic acid is then prefixed by the names of the various substituents.

$$\underset{6\quad 5\quad 4\quad 3\quad 2\quad 1}{CH_3\overset{CH_3}{\overset{|}{C}}HCH_2CH_2CH_2\overset{OH}{\overset{|}{C}}HCOOH}$$

2-hydroxy-6-methylheptanoic acid

$$\underset{6\quad 5\quad 4\quad 3\quad 2\quad 1}{CH_3(CH_2)_{11}\overset{CH_3}{\overset{|}{C}}HCH_2CH_2CH_2\overset{CH_3CH_2CH_2}{\overset{|}{C}}HCOOH}$$

6-methyl-2-propyloctadecanoic acid

When using common names, the chain is labeled α, β, γ, δ, and so on, *beginning with the carbon adjacent to the carboxy carbon* (see table of Greek letters inside the front cover).

As in the case of aldehydes and ketones, it is desirable not to mix IUPAC and common nomenclature (page 382).

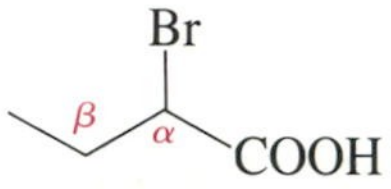

Br, β, α, COOH

α-bromobutyric acid
2-bromobutanoic acid
(not 2-bromobutyric acid)

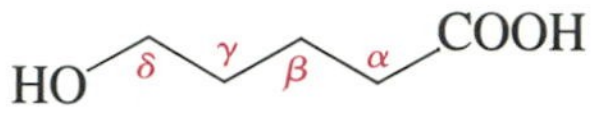

HO, δ, γ, β, α, COOH

δ-hydroxyvaleric acid
5-hydroxypentanoic acid
(not δ-hydroxypentanoic acid)

It is not always possible or convenient to name a carboxylic acid in the foregoing way. This is the case with cyclic acids.

COOH

cyclopentanecarboxylic acid

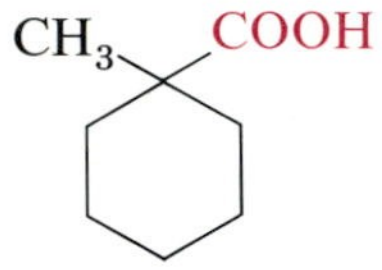

CH_3, COOH

1-methylcyclohexanecarboxylic acid

In rare cases it may be necessary to name a compound containing a carboxy group as a derivative of some other function. In such a case the COOH group is designated by the prefix **carboxy-**.

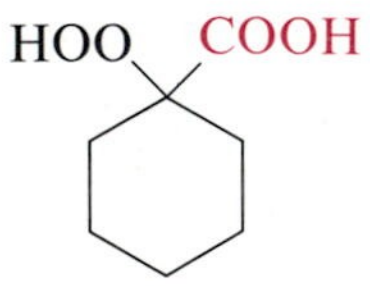

HOO, COOH

1-carboxycyclohexyl hydroperoxide

EXERCISE 18.1 What is the IUPAC name of each of the following carboxylic acids?

(a) $ClCH_2CH_2CH_2CH_2COOH$ (b) $(CH_3)_3CCH_2CH_2CH_2COOH$

(c) $CH_3CH_2\overset{CH_3}{\overset{|}{C}}HCH_2\overset{I}{\overset{|}{C}}HCOOH$ (d) $CH_2{=}CHCH{=}CHCOOH$

18.3 Physical Properties

Table 18.2 lists the melting point, boiling point, and water solubility of a number of straight-chain carboxylic acids. The boiling points of carboxylic acids are higher than expected for their molecular weights because of hydrogen bonding. The lower molecular weight acids are liquids at room temperature. The first four acids are fully miscible with water in all proportions. As the chain length is increased, the water solubility steadily decreases.

TABLE 18.2 Physical Properties of Carboxylic Acids

Acid	Melting Point, °C	Boiling Point, °C (760 mm)	Solubility in H_2O g/100 ml, 20°C
formic	8.4	101	∞
acetic	16.6	118	∞
propanoic	−21	141	∞
butanoic	−5	164	∞
pentanoic	−34	186	4.97
hexanoic	−3	205	0.968
heptanoic	−8	223	0.244
octanoic	17	239	0.068
nonanoic	15	255	0.026
decanoic	32	270	0.015
dodecanoic	44	299	0.0055
tetradecanoic	54	251 (100 mm)	0.0020
hexadecanoic	63	267 (100 mm)	0.00072
octadecanoic	72	—	0.00029

EXERCISE 18.2 Make a graph of the properties presented in Table 18.2 versus molecular weight. Use your graph to estimate the boiling point and water solubility of undecanoic acid. (The actual values are 280 °C and 0.0093 g/100 mL.)

18.4 Acidity

A. Ionization

We saw in Section 4.5 that compounds containing the carboxy group are weakly acidic. In fact, it is this property from which the class derives its name. When acetic acid is dissolved in water, the equilibrium shown in equation (18-1) exists.

$$CH_3-\overset{\overset{\displaystyle O}{\|}}{C}-OH + H_2O \rightleftharpoons CH_3-\overset{\overset{\displaystyle O}{\|}}{C}-O^- + H_3O^+ \qquad (18\text{-}1)$$

The equilibrium constant for this reaction, denoted as K_a or the ''acid dissociation constant,'' has the magnitude

$$K_a = \frac{[CH_3CO_2^-][H^+]}{[CH_3COOH]} = 1.8 \times 10^{-5}\ M \qquad (18\text{-}2)$$

Remember that the concentration of H_2O, which remains essentially invariant for dilute solutions (at 55.5 M), is not carried in the denominator of the expression for K_a. More correctly, equation (18-2) is an expression for the equilibrium

$$CH_3COOH \rightleftharpoons CH_3CO_2^- + H^+$$

The exact equilibrium expression for equation (18-1) is

$$K = \frac{[CH_3CO_2^-][H_3O^+]}{[CH_3COOH][H_2O]} = 3.25 \times 10^{-7}$$

It follows that $K_a = [H_2O] \times K = 1.8 \times 10^{-5}$ M.

The corresponding $pK_a = -\log(1.8 \times 10^{-5}) = 5 - \log 1.8 = 4.74$.

Dissociation constants of this magnitude put the carboxylic acids in the class of relatively weak acids. For example, a 0.1 M aqueous solution of acetic acid is only 1.3% dissociated into ions. Strong acids, such as HCl and H_2SO_4, are completely dissociated in dilute aqueous solution. Nevertheless, the carboxylic acids are distinctly acidic—their aqueous solutions have the characteristic sour taste of hydronium ion. Although the carboxylic acids are weak acids compared with mineral acids, they are much stronger than alcohols. Recall that ethanol has a dissociation constant of about 10^{-16}; ethanol is only 10^{-11} as strong an acid as acetic acid.

The question immediately arises, "Why is acetic acid more acidic than ethanol?" The answer lies mostly in the relative stability of the negative charge of the anion. In ethoxide ion the negative charge is concentrated on a single oxygen atom; ethoxide ion is basic because this concentrated charge provides strong attraction for a proton. In acetate ion, however, the charge on the carboxy group is divided between two oxygens. Acetate ion is not adequately represented by a single Lewis structure. A second and equivalent structure may be written that differs only in the position of electrons. A third structure with full negative charges on each oxygen balanced by a positive charge on carbon may also be written, and reflects a charge distribution indicated by quantum mechanical calculations.

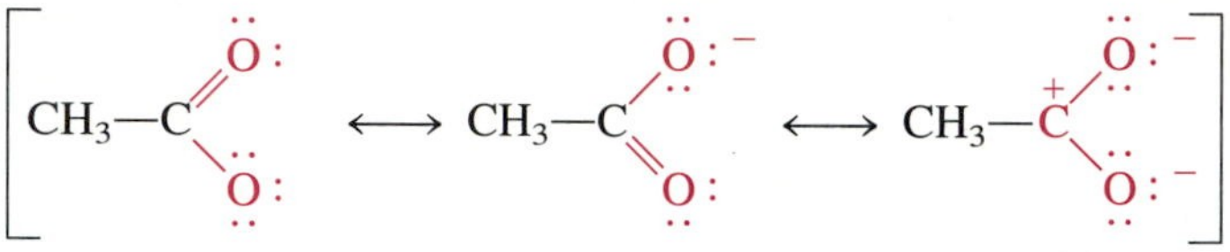

$$\left[CH_3{-}C(=\ddot{O}:)(-\ddot{O}:^-) \longleftrightarrow CH_3{-}C(-\ddot{O}:^-)(=\ddot{O}:) \longleftrightarrow CH_3{-}\overset{+}{C}(-\ddot{O}:^-)(-\ddot{O}:^-) \right]$$

Acetate ion is described as a resonance hybrid of these three principal structures (Section 2.4). The hybrid structure may also be written as

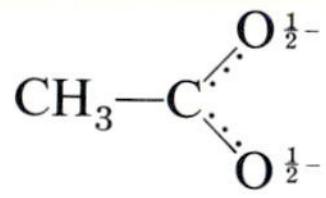

$$CH_3{-}C(\cdots O^{\frac{1}{2}-})(\cdots O^{\frac{1}{2}-})$$

For simplicity, the symbol shows half of a negative charge on each oxygen. The attraction for a proton is therefore reduced.

Another way of describing this phenomenon is wholly in terms of energy. The energies required to remove a proton from ethanol and acetic acid in the dilute gas phase are

$$CH_3CH_2OH \rightleftharpoons CH_3CH_2O^- + H^+ \qquad \Delta H^\circ = 378 \text{ kcal mole}^{-1}$$

$$CH_3COOH \rightleftharpoons CH_3CO_2^- + H^+ \qquad \Delta H^\circ = 349 \text{ kcal mole}^{-1}$$

It takes a large amount of energy to separate charges because of their Coulombic attraction. The energy required for dissociation of acetic acid is substantially less than that for ethanol because in acetate ion the negative charge is attracted by two oxygen nuclei, but since it is spread over a larger volume of space, there is less electron-electron repulsion. This energy effect in acetate ion is sometimes referred to as **resonance stabilization** or as a **resonance energy** (Chapter 20).

An alternative description can be given in terms of the overlap of atomic orbitals. Each oxygen contributes a *p*-orbital that can overlap in π-bonding to a *p*-orbital of the carboxy carbon, as illustrated in Figure 18.2. The resulting three-center π-molecular orbital system has four electrons with excess electron density on the two oxygens. Such multicenter π-molecular orbital systems are characteristic of conjugated systems and are discussed in more detail in Chapter 22.

In either representation, the two carbon-oxygen bonds in a carboxylate ion are equivalent. The carbon-oxygen bond length is 1.26 Å, which is between the bond length of 1.20 Å for C=O and 1.34 Å for C—O in formic acid (see Figure 18.1).

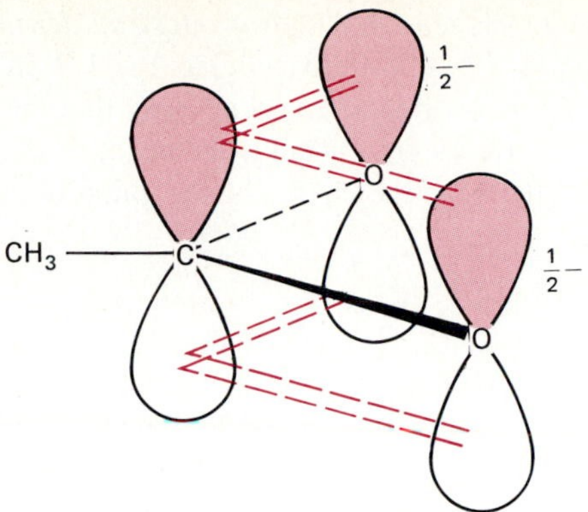

FIGURE 18.2 π-orbital interactions in acetate ion.

EXERCISE 18.3 Formic acid has a dissociation constant, $K_a = 1.77 \times 10^{-4}$. What is its p$K_a$? Calculate the approximate concentration of formate ions in a solution nominally 0.1 M in formic acid.

B. Inductive Effects

In Section 10.4 we found that electronegative groups whose bonds to carbon are highly polar have important effects on the acidity of alcohols. We saw that this effect could be interpreted in terms of the electrostatic interaction of a bond dipole with the negative charge on oxygen in the alkoxide ion. Substituent groups also affect the acidity of carboxylic acids. Because carboxylic acids are more acidic than alcohols, it is easier to determine their dissociation constants. Consequently, a wealth of such quantitative acidity data is available. Some of these results are summarized in Table 18.3.

TABLE 18.3 Acidity of Some Substituted Acetic Acids

Acid	K_a, M	pK_a
CH_3COOH	1.8×10^{-5}	4.74
FCH_2COOH	2.6×10^{-3}	2.59
F_3CCOOH	0.59	0.23
$ClCH_2COOH$	1.4×10^{-3}	2.86
$Cl_2CHCOOH$	5.5×10^{-2}	1.26
Cl_3CCOOH	0.23	0.64
$BrCH_2COOH$	1.3×10^{-3}	2.90
ICH_2COOH	6.7×10^{-4}	3.18
$HOCH_2COOH$	1.5×10^{-4}	3.83
CH_3OCH_2COOH	2.9×10^{-4}	3.54
$CH_2{=}CHCH_2COOH$	4.5×10^{-5}	4.35
$HC{\equiv}CCH_2COOH$	4.8×10^{-4}	3.32
CH_3CH_2COOH	1.3×10^{-5}	4.87
$NCCH_2COOH$	3.4×10^{-3}	2.46

Highly electronegative atoms tend to withdraw electron density from carbon and have a marked acid-strengthening effect. Chloroacetic acid is 1.9 pK_a units more acidic than acetic acid. The C—Cl dipole is oriented in such a way that the positive end is closer to the negative charge on the carboxy group than is the negative end. Electrostatic attraction exceeds the repulsion and the negative charge of the anion is therefore stabilized by the chlorine.

$$\text{Cl}\diagdown\text{CH}_2\text{—CO}_2^-$$

We used the same explanation to interpret the effect of a chlorine substituent on the acidity of ethanol (page 213).

$$\text{Cl}\diagdown\text{CH}_2\text{—CH}_2\text{O}^-$$

The acid-strengthening effect, in this case of 1.6 pK_a units, is similar to that in chloroacetic acid. In both anions the negative charge is two atoms away from the carbon-chlorine bond.

Carbon-carbon double and triple bonds have a significant electron-attracting effect that is reflected in the enhanced acidity of vinylacetic and ethynylacetic acids. An sp^2-hybridized carbon orbital with its greater s-character is effectively more electronegative than an sp^3-orbital. Recall that alkenes and alkynes have small but significant dipole moments (Sections 11.3 and 12.3).

The higher alkanoic acids are somewhat less acidic than acetic acid. Alkyl groups show a small but significant electron-donating inductive effect in appropriate systems in solution.

The inductive effect of remote substituents falls off dramatically with increased distance from the charged center. This effect is expected because electrostatic interactions between charges are inversely proportional to the distance between them. An example is seen in the acidity constants for butanoic acid and its three monochloro derivatives (Table 18.4). Beyond a few methylene groups, the effect becomes negligible.

TABLE 18.4 Acidity of Butanoic Acids

Acid	K_a, M	pK_a
2-chlorobutanoic acid	139×10^{-5}	2.86
3-chlorobutanoic acid	8.9×10^{-5}	4.05
4-chlorobutanoic acid	3.0×10^{-5}	4.52
butanoic acid	1.5×10^{-5}	4.82

EXERCISE 18.4 Plot the pK_as of chloroacetic acid and trichloroacetic acid versus the pK_as of the corresponding fluoroacetic acids. Note that the pK_a of acetic acid falls on the line. What is the predicted pK_a of difluoroacetic acid?

C. Salt Formation

There is one more aspect of acidity of carboxylic acids that we should consider. In Section 18.3 we saw that carboxylic acids of more than five carbons are essentially insoluble in water. Beyond this point, the polar portion of the molecule (COOH) becomes less important than the nonpolar hydrocarbon tail (R). Consider the reaction of a carboxylic acid such as dodecanoic acid with hydroxide ion.

$$CH_3(CH_2)_{10}COOH + OH^- \overset{K}{\rightleftharpoons} H_2O + CH_3(CH_2)_{10}CO_2^- \tag{18-3}$$

The equilibrium constant for reaction (18-3) can be derived as follows.

$$K_a = \frac{[CH_3(CH_2)_{10}CO_2^-][H^+]}{[CH_3(CH_2)_{10}COOH]} = 1.3 \times 10^{-5}\text{ M} \tag{18-4}$$

$$K_w = [H^+][OH^-] = 10^{-14}\text{ M}^2 \tag{18-5}$$

Rearranging (18-5), we have

$$[H^+] = \frac{10^{-14}}{[OH^-]}\text{ M} \tag{18-6}$$

Substituting (18-6) into (18-4) and expanding, we have

$$K = \frac{[CH_3(CH_2)_{10}CO_2^-]}{[CH_3(CH_2)_{10}COOH][OH^-]} = 1.3 \times 10^9\text{ M}^{-1} \tag{18-7}$$

Equation (18-7) is merely the equilibrium expression for reaction (18-3). The large value of K shows that the reaction proceeds to completion; dodecanoic acid is converted by aqueous sodium hydroxide completely into the salt, sodium dodecanoate. Note that the anions of carboxylic acids are named by dropping **-ic** from the name of the parent acid and adding the suffix **-ate.** Although dodecanoic acid is a neutral molecule, sodium dodecanoate is a salt. Dissolution of this salt gives an anion and a cation, which can be solvated by water. It is not surprising that the solubility of sodium dodecanoate (1.2 g per 100 mL) is much greater than that of dodecanoic acid itself (0.0055 g per 100 mL).

EXERCISE 18.5 Equation (18-7) can be used to calculate the ratio of ionized and non-ionized dodecanoic acid at a given pH, by inserting the proper value for $[OH^-]$. Calculate this ratio for pH = 2, 4, 6, and 8.

D. Soaps

The sodium and potassium salts of long-chain carboxylic acids (''fatty acids'') are obtained by the reaction of natural fats with sodium or potassium hydroxide. These salts, referred to as soaps (Tartar to early Teutonic, *saipon*), have the interesting and useful ability to solubilize nonpolar organic substances. This phenomenon can easily be understood if one considers the structure of such a salt, potassium octadecanoate or more commonly, potassium stearate.

$$CH_3CH_2CH_2CH_2CH_2CH_2CH_2CH_2CH_2CH_2CH_2CH_2CH_2CH_2CH_2CO_2^-\ K^+$$

The molecule has a polar ionic region and a large nonpolar hydrocarbon region or ''tail.'' Molecules that incorporate both possibilities are called by optimists **amphiphilic** (Gk. *amphi*, both; *philos*, love) or by pessimists **amphipathic** (Gk. *amphi*, both; pathos, suffering). In aqueous solution the tail interferes with the hydrogen-bonded structure of water and leads to increased order (structure) in the surrounding water. The resulting decrease in entropy is not compensated for by the van der Waals attraction between the structured water and a single hydrocarbon chain, leading to association of two (or more) hydrocarbon tails. This arrangement diminishes the loss of water hydrogen bonding, and the amount of structure formation, shown schematically in Figure 18.3. Molecules that are expelled from the water structure associate with one another owing to this **hydrophobic effect.** The stability of cell membranes and the existence of life itself depend upon the hydrophobic effect.

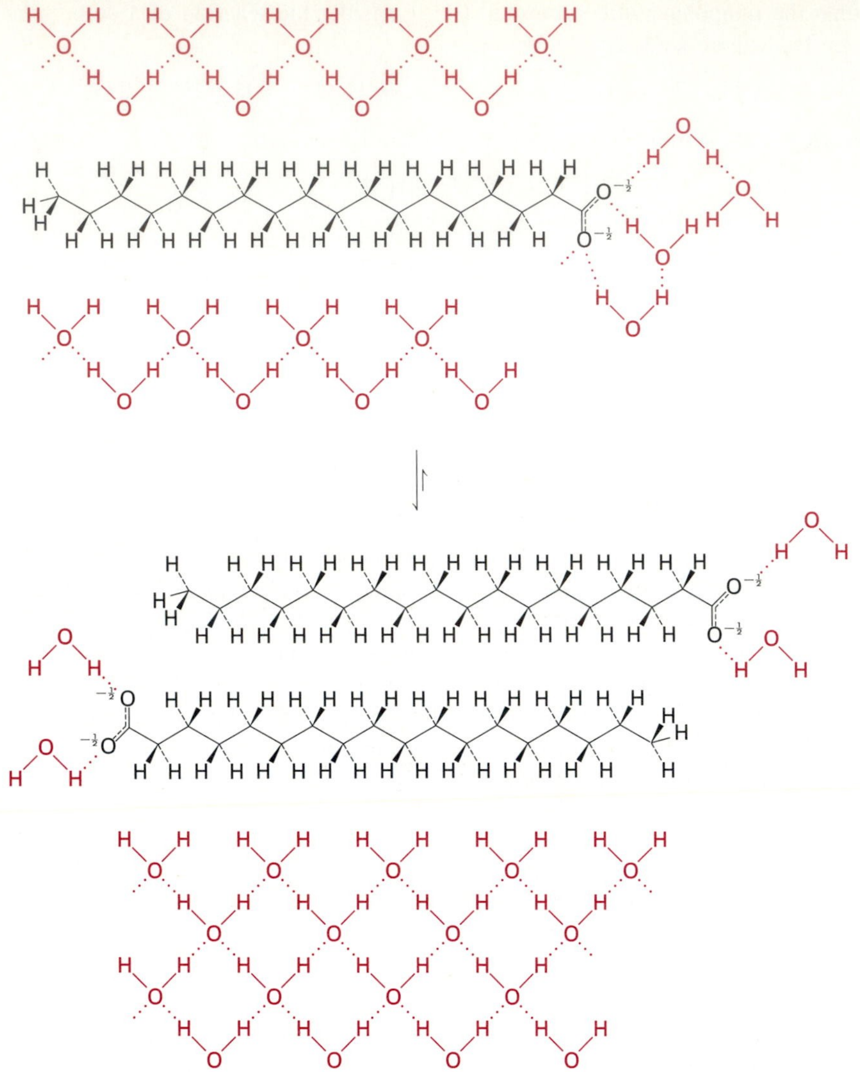

FIGURE 18.3 Structure forming dimerization of potassium stearate.

This simple and important idea has other consequences. The minimization of the loss of hydrogen bonding and gain in water structure causes a number of carboxylate ions to cluster together so that many hydrocarbon tails are close to each other. Some energy is gained through the attractive van der Waals forces expected for normal hydrocarbons. The surface of the sphere-like cluster of amphiphilic molecules is then occupied by the highly polar CO_2^- groups. These polar groups face the medium, where they may be solvated by H_2O or paired with a cation. The resulting spherical structure, called a **micelle,** is depicted in cross section in Figure 18.4. The wavy lines in the figure represent the long hydrocarbon chains of the salt molecules. Simple geometry implies that the outer part of the micelle must be exposed to water, just as the smallest carboxylic acids are soluble in water. Micelles can have many shapes and sizes, can be made from both charged and neutral amphiphiles, and can form at a concentration of amphiphile called the **critical micelle concentration** or **c.m.c.,** typically 10^{-4} to 10^{-3} M.

Organic material such as butter or motor oil that is not normally soluble in water may ''dissolve'' in the hydrocarbon interior of a micelle. The overall process of soap solubilization is diagrammed schematically in Figure 18.5.

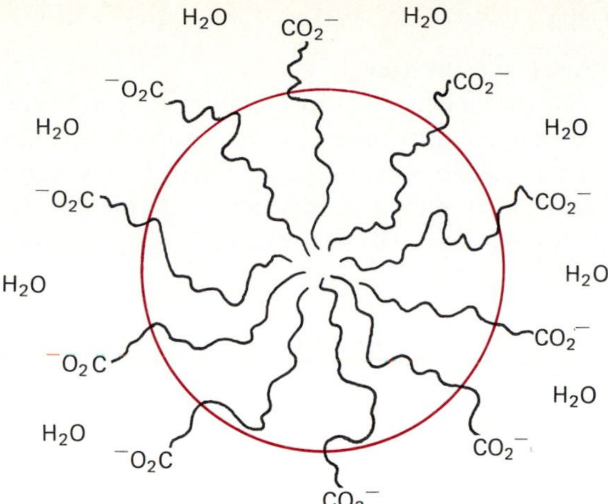

FIGURE 18.4 Cross section of a micelle.

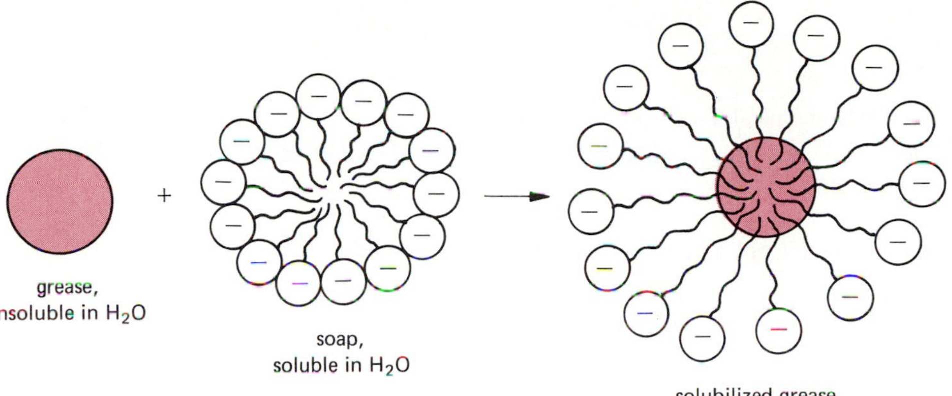

FIGURE 18.5 Schematic diagram of soap solubilization.

Certain bacteria can metabolize soaps. This degradation is most rapid when there are no branches in the hydrocarbon chain of the soap molecule. Since the naturally occurring fatty acids are all unbranched compounds, soaps derived from natural fats are said to be **biodegradable.** Before 1933 all cleaning materials were soaps. In that year the first synthetic detergents were marketed. Detergents have the useful property of not forming the hard "scum" that often results from the use of a soap with hard water. This scum is actually made of the insoluble magnesium and calcium salts of fatty acids. The first detergents were alkylbenzenesulfonates. Like soaps, they had a large nonpolar hydrocarbon tail and a polar end.

$$R-C_6H_4-SO_3^-\ K^+$$

R = branched alkyl chain

However, being branched compounds, these early detergents were not rapidly biodegradable. Since the materials could not be completely metabolized by the bacteria that operate in sewage treatment plants, they were passed into natural waterways with the treated sewage and often reappeared as foam or suds on the surface of lakes and rivers. After an intensive research project, the detergent industry in 1965 introduced **linear alkanesulfonate detergents** (Section 26.5.B).

$$CH_3(CH_2)_nCH_2SO_3^-\ K^+$$

Since the new detergents are straight-chain compounds, they can be metabolized by bacteria.

18.5 Spectroscopy

A. Nuclear Magnetic Resonance

The resonance positions for various types of hydrogens in carboxylic acids are summarized in Table 18.5. Hydrogens attached to C-2 of a carboxylic acid resonate at roughly the same place as do the analogous hydrogens in aldehydes and ketones. The very low-field resonance of the carboxy proton is associated with the dimeric hydrogen-bonded structure discussed in Section 18.1. The spectrum of 2-methylpropanoic acid is shown in Figure 18.6.

TABLE 18.5 Chemical Shifts of Carboxylic Acid Hydrogens

Type of Hydrogen	Chemical Shift, δ, ppm
$C\mathbf{H}_3COOH$	2.0
$RC\mathbf{H}_2COOH$	2.36
$R_2C\mathbf{H}COOH$	2.52
$RCOO\mathbf{H}$	about 10–13

The CMR chemical shifts of carboxylic acids are similar to those seen with aldehydes (Table 14.4), except that the carbonyl carbon itself resonates at much lower field. Representative data are summarized in Table 18.6.

TABLE 18.6 CMR Chemical Shifts of Carboxylic Acids

	C-1	C-2	C-3	C-4	C-5
formic acid	166.3				
acetic acid	177.2	21.1			
propanoic acid	180.4	27.8	9.0		
butanoic acid	179.6	36.3	18.5	13.4	
pentanoic acid	179.7	34.1	27.0	22.0	13.5

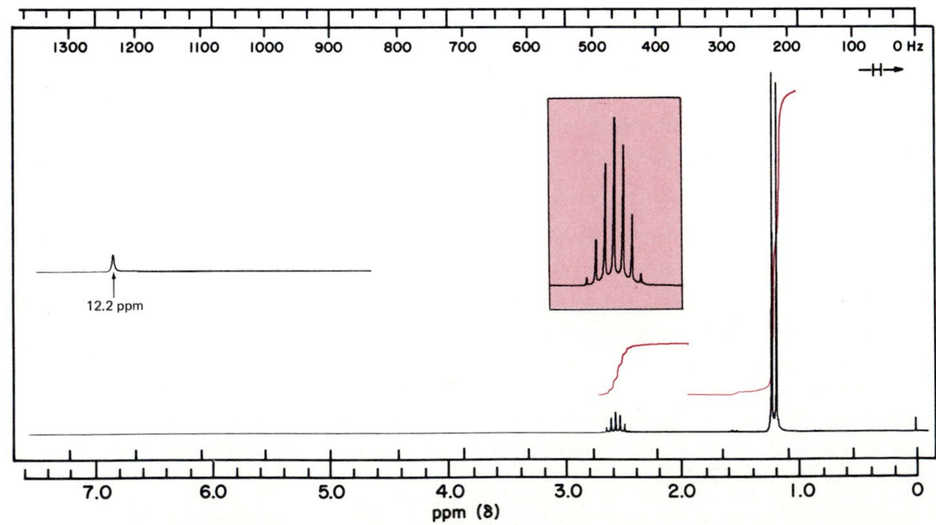

FIGURE 18.6 NMR spectrum of 2-methylpropanoic acid, $(CH_3)_2CHCOOH$.

EXERCISE 18.6 An unknown carboxylic acid shows NMR peaks at δ 12.5 (1H, s), 1.25 (6H, s), 0.95 (3H, t, J = 6) and 1.50 (2H, q, J = 6) ppm. The CMR spectrum shows signals at δ 9.5, 24.8, 33.7, 43.0, and 182.2 ppm. What is the structure of the compound?

B. Infrared

One of the characteristic absorptions of acids is the C═O stretch, which occurs in the region 1710–1760 cm^{-1}. The exact position and appearance of this absorption depends on the physical state in which the measurement is made. In pure liquids or in the solid state the C═O stretch occurs as a broad band at about 1710 cm^{-1}. In dilute solution (CCl_4, $CHCl_3$) the absorption band is narrower and is seen at about 1760 cm^{-1}. The O—H stretch occurs as a broad, relatively intense absorption that usually extends from 2400 to 3400 cm^{-1}. The spectrum of hexanoic acid in Figure 18.7 illustrates these bands.

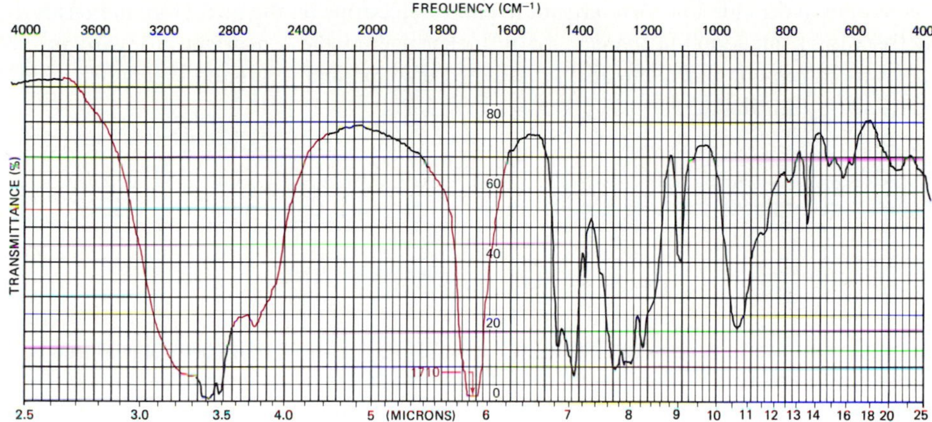

FIGURE 18.7 Infrared spectrum of hexanoic acid.

Highlight 18.1

Carboxylic acids are weak acids with the anion stabilized by the presence of two electronegative oxygens. The resonance forms make clear (a) the difference in acidity of ROH and RCOOH, (b) the effect of electronegative groups on the acidity of RCOOH, and (c) the oxygen nucleophilicity of the COO^- group.

$$\left[CH_3{-}C({=}\ddot{O}{:})(\ddot{O}{:}^-) \longleftrightarrow CH_3{-}C(\ddot{O}{:}^-)({=}\ddot{O}{:}) \longleftrightarrow CH_3{-}\overset{+}{C}(\ddot{O}{:}^-)(\ddot{O}{:}^-) \right] \equiv CH_3{-}C(O)(O)^-$$

$$Cl{-}CH_2{-}C(O)(O)^-$$

$$CH_3{-}C({=}\ddot{O}{:})(\ddot{O}{:}^-) + CH_3{-}I \longrightarrow CH_3{-}C({=}\ddot{O}{:}){-}\ddot{O}{-}CH_3$$

18.6 Synthesis

A. Hydrolysis of Nitriles

We have seen that nitriles undergo acid-catalyzed and base-catalyzed hydration (Sections 12.6.B,C). Under more drastic conditions, the product of such a hydration, called an **amide,** undergoes hydrolysis to give a mole of carboxylic acid and a mole of ammonia.

$$\underset{\text{nitrile}}{RC{\equiv}N} \xrightarrow{H_2O} \underset{\text{amide}}{R\overset{O}{\overset{\|}{C}}NH_2} \xrightarrow{H_2O} \underset{\text{carboxylic acid}}{RCOOH} + NH_3$$

For the preparation of a carboxylic acid from a nitrile, it is not necessary to isolate the intermediate amide. Either acidic or basic hydrolysis can be employed. In acidic medium, the ammonia produced is protonated as ammonium ion. Under basic conditions, gaseous ammonia is evolved as the hydrolysis reaction proceeds. Hydrolysis can be accelerated in basic solution by hydrogen peroxide, since the hydroperoxide ion, ^-OOH, is more nucleophilic than hydroxide ion. A nonbonding pair of electrons on the next atom increases the reactivity of a nucleophilic center, an influence called the "α-effect." Hydrolysis of nitriles to carboxylic acids can also be effected enzymatically in good yields under mild conditions by nitrilases.

$$\underset{\text{phenylacetonitrile}}{C_6H_5CH_2CN} + 2\,H_2O \xrightarrow[\substack{100^\circ C \\ 3\text{ hr}}]{H_2SO_4} \underset{\text{phenylacetic acid}}{C_6H_5CH_2COOH} + NH_4^+\,HSO_4^-$$

A mixture of 1150 mL of water, 840 mL of conc. H_2SO_4, and 700 g of phenylacetonitrile is heated at reflux for 3 hr. Phenylacetic acid (630 g, 78% yield) is obtained when the reaction mixture is poured into cold water.

$$\underset{\text{undecanenitrile}}{CH_3(CH_2)_9CN} + 2\,H_2O \xrightarrow[\substack{C_2H_5OH \\ \text{reflux} \\ 77\text{ hr}}]{KOH} \xrightarrow{HCl} \underset{\text{undecanoic acid}}{CH_3(CH_2)_9COOH} + NH_3$$

A mixture of 27 g of undecanenitrile, 200 g of 20% ethanolic KOH, and 50 mL of water is refluxed for 77 hr, during which time ammonia is evolved. The solvent is evaporated, and the residue is treated with conc. HCl. After washing with water, 24 g of undecanoic acid (80%), m.p. 29° C, is obtained.

As we saw in Chapters 9 and 12, nitriles are conveniently prepared from primary alkyl halides by treatment with cyanide ion. Thus, cyanide ion is a synthon for the carboxy group.

$$\underset{\text{1-bromobutane}}{CH_3CH_2CH_2CH_2Br} + NaCN \xrightarrow{DMSO} \underset{\substack{(92\%) \\ \text{pentanenitrile}}}{CH_3CH_2CH_2CH_2CN} \xrightarrow[\substack{H_2O \\ \Delta}]{KOH}$$

$$CH_3CH_2CH_2CH_2CO_2^-\,K^+ \xrightarrow{H^+} \underset{\substack{(90\%) \\ \text{pentanoic acid}}}{CH_3CH_2CH_2CH_2COOH}$$

The mechanism of the hydrolysis reaction will be discussed in Section 18.6.

EXERCISE 18.7 Which of the following carboxylic acids can be prepared from an alkyl halide by formation and hydrolysis of a nitrile?

(a) 2,2-dimethylpropanoic acid
(b) 2-methylpentanoic acid
(c) 1-methylcyclohexanecarboxylic acid
(d) 7-methyloctanoic acid

B. Carbonation of Organometallic Reagents

Alkyl halides may also be converted into carboxylic acids by formation of an organometallic reagent, which is then allowed to react with carbon dioxide. The product is the salt of a carboxylic acid. Treatment of this salt with aqueous mineral acid liberates the free acid. Grignard reagents are commonly used (Section 8.8).

Cl → (Mg, ether) → MgCl → (CO_2) → CO_2^- ^+MgCl → (H_2O, H_2SO_4) → COOH

Carbon dioxide gas is bubbled through a solution of *sec*-butylmagnesium chloride (prepared from 46 g of 2-chlorobutane and 13.4 g of magnesium in 400 mL of ether) at −10 °C. When CO_2 is no longer absorbed, the mixture is hydrolyzed with 25% aqueous H_2SO_4. Distillation of the crude product gives 40 g (80%) of 2-methylbutanoic acid.

A similar reaction occurs between carbon dioxide and organolithium compounds.

Li (phenyllithium) + CO_2 → CO_2^- Li^+ → (H_3O^+) → COOH (benzoic acid)

The Grignard method of converting an alkyl halide to the corresponding carboxylic acid must be used when the alkyl halide is unreactive in S_N2 reactions. With primary and unbranched secondary alkyl halides the nitrile displacement method discussed in the previous section may also be used.

EXERCISE 18.8 Show how the following conversions can be accomplished.

(a) cyclohexanone → 1-methylcyclohexanecarboxylic acid
(b) 2-methylbutane → 2,2-dimethylbutanoic acid
(c) 2-methyl-1-propanol → 3-methylbutanoic acid

C. Oxidation of Primary Alcohols or Aldehydes

The third generally useful method for preparing carboxylic acids involves oxidation of aldehydes (obtained in the aldol reaction, Section 15.2) or primary alcohols (obtained, for example, by hydroboration of terminal alkenes; Section 11.6.D). A useful oxidizing agent for this purpose is potassium permanganate.

$$\text{2-ethyl-1-hexanol (OH)} + KMnO_4 \xrightarrow[25°C]{H_2O,\ NaOH} \xrightarrow{H_2SO_4} \text{2-ethylhexanoic acid (COOH)}$$

$$\text{2-ethylhexanal (CHO)} \xrightarrow[(78\%)]{\text{same as above}} \text{2-ethylhexanoic acid}$$

2-ethyl-1-hexanol 2-ethylhexanal 2-ethylhexanoic acid

The initial product in oxidation of a primary alcohol is the corresponding aldehyde. However, with aqueous permanganate the aldehyde undergoes subsequent oxidation more rapidly than the primary alcohol, so it is normally not observed in the reaction mixture.

A mild and selective reagent for the oxidation of aldehydes to carboxylic acids is silver oxide suspended in aqueous base. Although the method usually affords the desired acid in good yield, it is expensive to carry out on a large scale owing to the cost of silver oxide, unless one reclaims and recycles the silver metal.

$$\underset{CH_3CH_2}{\overset{H}{>}}C{=}C\underset{CH_3}{\overset{CHO}{<}} + Ag_2O \xrightarrow[NaOH]{H_2O} \xrightarrow{HCl} \underset{(95\text{–}98\%)}{\underset{CH_3CH_2}{\overset{H}{>}}C{=}C\underset{CH_3}{\overset{COOH}{<}}} + Ag$$

The foregoing example illustrates the virtue of a mild oxidant, since the double bond is sensitive to stronger oxidizing agents.

Silver oxide is a brown solid that has only slight solubility in water. It is usually prepared as needed by mixing a solution of silver nitrate with sodium hydroxide. The precipitate may be filtered, washed with water until the pH is lower than 11, and used as an aqueous suspension.

EXERCISE 18.9 Write the balanced equation for oxidation of ethanol to acetic acid by $KMnO_4$ (see page 283).

18.7 Reactions

A. Reactions Involving the OH Bond

We have already seen one important reaction of carboxylic acids involving the OH bond—the reaction with bases to give salts.

$$CH_3CH_2CH_2COOH + Na^+\ OH^- \longrightarrow CH_3CH_2CH_2CO_2^-\ Na^+ + H_2O$$

The resulting carboxylate salts enter into the S_N2 reaction with alkyl halides to give carboxylic acid **esters.**

$$\underset{\text{1-bromobutane}}{CH_3CH_2CH_2CH_2Br} + CH_3CO_2^-\ Na^+ \xrightarrow[95°C]{DMF} \underset{\substack{(95\text{–}98\%)\\ \text{butyl acetate}}}{CH_3CH_2CH_2CH_2O\overset{\overset{O}{\|}}{C}CH_3}$$

Of course, dehydrohalogenation is an important competing reaction, especially with tertiary halides (Sections 9.7, 11.5.A). However, because carboxylate ions are rather weak bases, even secondary halides may often be converted into the corresponding esters with little competing elimination.

Another important reaction involving the OH bond is the reaction of carboxylic acids with diazomethane. The products of this reaction are the methyl ester and nitrogen.

$$RCOOH + CH_2N_2 \longrightarrow RCOOCH_3 + N_2$$

> Diazomethane is a yellow gas boiling at about 0 °C. It is highly toxic and, under certain conditions, explosive. Diazomethane is another example of a compound for which multiple Kekule' or Lewis structures can be written. The molecule is considered to be a resonance hybrid of the following forms.
>
> $$[\,:\overset{-}{C}H_2—\overset{+}{N}{\equiv}N: \longleftrightarrow CH_2{=}\overset{+}{N}{=}\ddot{N}:^-]$$

The reaction of diazomethane with carboxylic acids probably involves the following steps.

$$(1)\ R—\overset{O}{\overset{\|}{C}}—O—H + :\overset{-}{C}H_2—\overset{+}{N}{\equiv}N: \longrightarrow R—\overset{O}{\overset{\|}{C}}—\ddot{\underset{..}{O}}:^- + CH_3—\overset{+}{N}{\equiv}N:$$

$$(2)\ R—\overset{O}{\overset{\|}{C}}—\ddot{\underset{..}{O}}:^- + CH_3—\overset{+}{N}{\equiv}N: \longrightarrow R—\overset{O}{\overset{\|}{C}}—OCH_3 + :N{\equiv}N:$$

The first step is a simple acid-base reaction; the moderately acidic carboxylic acid transfers a proton to the basic carbon atom of diazomethane. The pair of ions thus formed immediately reacts by the S_N2 mechanism; carboxylate ion is the entering nucleophile and nitrogen is the leaving group.

Because of the toxicity and danger of explosion, diazomethane reactions are rarely carried out on a large scale. However, because of the convenience of the procedure (yields are usually quantitative and the only by-product is a gas), it is frequently used for the small-scale conversion of an acid into its methyl ester, especially when the acid is a relatively precious one.

CH_3 … COOH $+ CH_2N_2 \xrightarrow[25°C]{ether}$ CH_3 … $COOCH_3$ $+ N_2$

(100%)

EXERCISE 18.10 Write the structure of the product of the reaction of (*R*)-1-deuterio-1-iodobutane with sodium propanoate.

B. Reactions Involving the Hydrocarbon Side Chain

Carboxylic acids undergo the normal reactions of alkanes, as modified by the presence of the carboxy group, in the hydrocarbon chain of the molecule. For example, butanoic acid undergoes combustion and free radical chlorination.

$$CH_3CH_2CH_2COOH \xrightarrow{O_2} CO_2 + H_2O$$

$$CH_3CH_2CH_2COOH \xrightarrow[\text{heat or light}]{Cl_2} \text{mixture of mono- and polychlorobutanoic acids}$$

Since these reactions are not selective for any particular position along the chain they generally have no preparative utility.

The indiscriminate nature of such free radical reactions is demonstrated by the light-initiated chlorination of butanoic acid in CCl_4 at 25 °C.

$$CH_3CH_2CH_2COOH \xrightarrow[h\nu]{Cl_2} \begin{cases} ClCH_2CH_2CH_2COOH \quad (31\%) \\ CH_3CH(Cl)CH_2COOH \quad (64\%) \\ CH_3CH_2CH(Cl)COOH \quad (5\%) \end{cases}$$

One useful reaction sequence involving the aliphatic chain is the conversion of carboxylic acids to α-bromoacyl bromides with phosphorus tribromide and bromine.

$$3\ CH_3CH_2CH_2COOH + PBr_3 + 3\ Br_2 \longrightarrow 3\ CH_3CH_2CH(Br)C(=O)Br + H_3PO_3 + 3\ HBr$$

Note that the reaction is *positionally selective*—only the hydrogen at C-2 is replaced. The reaction is not a free radical halogenation process. The overall result, α-bromination, is accomplished by a sequence of steps. The key step involves the reaction of bromine with the **enol form** of the corresponding **acyl bromide.** Phosphorus tribromide facilitates the reaction by reacting with the carboxylic acid to yield the acyl bromide (Section 18.7.D), which undergoes enolization more readily than the acid itself.

$$(1)\quad 3\ RCH_2COOH + PBr_3 \longrightarrow 3\ RCH_2C(=O)Br + H_3PO_3$$

$$(2)\quad \underset{\text{keto form}}{RCH_2C(=O)Br} \overset{H^+}{\rightleftharpoons} \underset{\text{enol form}}{RCH{=}C(OH)Br}$$

$$(3)\quad RCH{=}C(OH)Br + Br_2 \longrightarrow RCH(Br)C(=O)Br + HBr$$

The reaction is analogous to the acid-catalyzed bromination of ketones (Section 15.1.D). Only a catalytic amount of PBr_3 is needed because the product acyl bromide enters into the following equilibrium with the starting acid.

$$RCH(Br)C(=O)Br + RCH_2C(=O)OH \rightleftharpoons RCH(Br)C(=O)OH + RCH_2C(=O)Br$$

Under these conditions the product that is isolated is the α-bromo carboxylic acid rather than the α-bromo acyl bromide.

$$CH_3CH_2CH_2COOH + Br_2 \xrightarrow{PBr_3} CH_3CH_2\underset{\substack{| \\ Br}}{C}HCOOH + HBr$$

(82%)

Since phosphorus reacts rapidly with bromine to give PBr_3, the reaction is often carried out by simply heating the carboxylic acid with a mixture of phosphorus and bromine. In the bromination reaction one normally begins with a carboxylic acid and ends up with the α-bromo carboxylic acid, as illustrated in the last example. However, it is important to remember that the crucial reaction, introduction of the bromine into the molecule, is actually a reaction of the intermediate acyl bromide.

EXERCISE 18.11 Show how hex-2-enoic acid ($CH_3CH_2CH_2CH{=}CHCOOH$) can be synthesized starting with 1-chloropentane.

C. Formation of Esters

Carboxylic acids react readily with alcohols in the presence of catalytic amounts of mineral acids to yield compounds called **esters** (Chapter 19). The process is called **esterification.**

$$CH_3COOH + C_2H_5OH \overset{H^+}{\rightleftharpoons} CH_3COOC_2H_5 + H_2O \qquad (18\text{-}8)$$

Unlike most of the reactions we have encountered, this one has an equilibrium constant of relatively low magnitude. The experimental equilibrium constant for the reaction of acetic acid with ethanol is

$$K_{eq} = \frac{[CH_3COOC_2H_5][H_2O]}{[CH_3COOH][C_2H_5OH]} = 3.38$$

As in any equilibrium process, the reaction may be driven in one direction by controlling the concentration of either the reactants or products (Le Châtelier's principle). For reaction (18-8) the equilibrium constant tells us that an equimolar mixture of acetic acid and ethanol will eventually reach equilibrium to give a mixture containing 0.35 mole each of acetic acid and ethanol and 0.65 mole each of ethyl acetate and water. Of course, the same equilibrium mixture is obtained if one starts with equimolar quantities of ethyl acetate and water.

If we increase the concentration of either reactant relative to the other, the reaction will be driven to the right and the equilibrium mixture will contain proportionately more ethyl acetate and water. Table 18.7 shows the equilibrium compositions that will be achieved starting with various mixtures of acetic acid and ethanol.

TABLE 18.7 Equilibrium Compositions

	CH_3COOH +	C_2H_5OH $\rightleftharpoons$	$CH_3COOC_2H_5$ +	H_2O
at start	1	1	0	0
at equilibrium	0.35	0.35	0.65	0.65
at start	1	10	0	0
at equilibrium	0.03	9.03	0.97	0.97
at start	1	100	0	0
at equilibrium	0.007	99.007	0.993	0.993

Similar results will obviously be obtained by increasing the acetic acid concentration rather than the ethanol concentration. In a practical situation, when one wants to prepare an ester, it is desirable to obtain the maximum yield of pure product. It is often done as suggested in the preceding paragraph—by using a large excess of one of the reactants that, for economic reasons, is usually the less expensive of the two.

The mechanism of the acid-catalyzed esterification reaction has been studied thoroughly. All of the experimental facts are consistent with a mechanism involving the following steps (illustrated for acetic acid and methanol).

$$\text{(1)}\quad CH_3-\overset{O}{\overset{\|}{C}}-OH + H^+ \rightleftharpoons CH_3-\overset{OH^+}{\overset{\|}{C}}-OH$$

$$\text{(2)}\quad CH_3-\overset{OH^+}{\overset{\|}{C}}-OH + CH_3\ddot{O}H \rightleftharpoons CH_3-\overset{OH}{\underset{H\overset{+}{O}CH_3}{C}}-OH$$

$$\text{(3)}\quad CH_3-\overset{OH}{\underset{H\overset{+}{O}CH_3}{C}}-OH \rightleftharpoons CH_3-\overset{OH}{\underset{OCH_3}{C}}-OH + H^+$$

$$\text{(4)}\quad CH_3-\overset{OH}{\underset{OCH_3}{C}}-OH + H^+ \rightleftharpoons CH_3-\overset{OH}{\underset{OCH_3}{C}}-OH_2^+$$

$$\text{(5)}\quad CH_3-\overset{\ddot{O}H}{\underset{OCH_3}{C}}-OH_2^+ \rightleftharpoons CH_3-\overset{\ddot{O}H^+}{\overset{\|}{C}}-OCH_3 + H_2O$$

$$\text{(6)}\quad CH_3-\overset{OH^+}{\overset{\|}{C}}-OCH_3 \rightleftharpoons CH_3-\overset{O}{\overset{\|}{C}}-OCH_3 + H^+$$

Steps 1, 3, 4, and 6 are rapid proton-transfer steps—simple acid-base reactions. Although we show ''bare'' protons in each case, they are actually solvated by some Lewis base, which may be methanol, water, or any of the other oxygenated species present. In steps 2 and 5, carbon-oxygen bonds are formed or broken. These steps have higher activation energies than the proton-transfer steps.

The foregoing mechanism is an important one in organic chemistry. As mentioned previously, it is based on a large amount of experimental data. Two of the more definitive experiments involved the use of ^{18}O-labeled materials. The first of these interesting experiments demonstrated that the oxygen-carbonyl bond is broken during the esterification process. Benzoic acid was treated in the presence of HCl with methanol enriched in ^{18}O. The water produced in the reaction was isolated and shown to be normal $H_2{}^{16}O$.

$$C_6H_5\overset{O}{\overset{\|}{C}}-OH + CH_3{}^{18}OH \overset{H^+}{\rightleftharpoons} C_6H_5\overset{O}{\overset{\|}{C}}-{}^{18}OCH_3 + H_2{}^{16}O$$

This experiment rules out mechanisms such as the following, in which the oxygen in the water produced comes from the alcohol.

$$C_6H_5\overset{O}{\overset{\|}{C}}-\ddot{O}H + CH_3-OH_2^+ \not\longrightarrow C_6H_5\overset{O}{\overset{\|}{C}}-\underset{H}{\overset{+}{O}}CH_3 + H_2O$$

The second important labeling experiment showed that a symmetrical **intermediate** intervenes in the process. Ethyl benzoate enriched in ^{18}O in the carbonyl oxygen was hydrolyzed with HCl and normal water. The reaction was stopped short of completion, and the recovered ethyl benzoate was analyzed. It was found that exchange of ^{18}O in the ester by ^{16}O had occurred. Although hydrolysis occurs approximately five times faster than exchange, this experiment demonstrates that an intermediate is formed that can go on to give acid or reverse to give exchanged ester. Mechanisms such as the following, which is analogous to the S_N2 displacement in saturated systems, are ruled out.

$$\left\{ R-\overset{\overset{O}{\|}}{C}-\underset{\underset{H}{|}}{\overset{+}{O}}R + H_2O \longrightarrow \left[\overset{\delta+}{H_2O}\cdots\underset{\underset{R}{|}}{\overset{\overset{O}{\|}}{C}}\cdots\underset{H}{\overset{\delta+}{O}R} \right]^{\ddagger} \longrightarrow H_2\overset{+}{O}-\overset{\overset{O}{\|}}{C}-R + ROH \right\} \quad \text{not observed}$$

Note that the accepted mechanism involves simply an acid-catalyzed addition of an alcohol to the carbonyl group and is analogous to the similar reactions with aldehydes and ketones to form intermediate hemiacetals (Section 14.6.B). Also note that the overall process of esterification amounts to no more than **nucleophilic substitution**—the nucleophile OR substitutes for the nucleophile OH.

EXERCISE 18.12 In the mechanism for acid-catalyzed esterification (page 501), step 1 is a rapid equilibrium and step 2 is usually the rate-determining step. Derive the rate law for acid-catalyzed esterification as a function of K, the equilibrium constant for step 1, and k, the rate constant for the forward reaction of step 2.

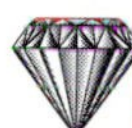

Highlight 18.2

Acid-catalyzed esterification is one of the fundamental processes in organic chemistry. The mechanism of the reaction consists of a series of six steps. Four of the steps (1, 3, 4, and 6) involve addition or removal of a proton from oxygen, and two (2 and 5) involve formation or breakage of a C—O bond.

1. $R-C(=O)OH + H^+ \rightleftharpoons R-C(=\overset{+}{O}H)OH$

2. $R-C(=\overset{+}{O}H)OH + ROH \rightleftharpoons R-C(OH)_2-\overset{+}{O}(H)R$

3. $R-C(OH)_2-\overset{+}{O}(H)R \rightleftharpoons R-C(OH)_2-OR + H^+$

4. $R-C(OH)_2-OR + H^+ \rightleftharpoons R-C(OH)(\overset{+}{O}H_2)-OR$

5. $R-C(OH)(\overset{+}{O}H_2)-OR \rightleftharpoons R-C(OR)=\overset{+}{O}H + H_2O$

6. $R-C(OR)=\overset{+}{O}H \rightleftharpoons R-C(OR)=O + H^+$

The overall process, $RCO_2H + ROH \rightleftharpoons RCO_2R + H_2O$, has an equilibrium constant, K, of about unity. The reverse process, starting with step 6 and going backward to step 1, is the mechanism for acid-catalyzed hydrolysis of an ester. This basic mechanism is one of the most important processes in biochemistry.

D. Formation of Acyl Halides

Carboxylic acids react with thionyl chloride, phosphorus pentachloride, and phosphorus tribromide in the same way that alcohols do (Section 10.6.B). The products are **acyl halides** (Chapter 19).

$$CH_3CH_2CH_2CH_2COOH + SOCl_2 \longrightarrow CH_3CH_2CH_2CH_2COCl + HCl + SO_2$$

pentanoic acid → pentanoyl chloride

$$(CH_3)_2CHCOOH + PCl_5 \longrightarrow (CH_3)_2CHCOCl + POCl_3 + HCl$$

2-methylpropanoic acid → 2-methylpropanoyl chloride

$$3\ CH_3CH_2CH_2COOH + PBr_3 \longrightarrow 3\ CH_3CH_2CH_2COBr + H_3PO_3$$

butanoic acid → butanoyl bromide

Mechanistically, these reactions are related to the acid-catalyzed esterification mechanism. In each case, the inorganic chloride reacts with the carboxylic acid to give an intermediate in which the OH group has been converted into a relatively good leaving group. The ensuing reaction is nucleophilic substitution, which occurs by the ''addition-elimination'' mechanism. For example, in the formation of acyl chlorides with $SOCl_2$ the leaving group in the nucleophilic substitution process is the chlorosulfite ion.

$$(1)\ CH_3\overset{O}{\overset{\|}{C}}OH + SOCl_2 \longrightarrow CH_3\overset{O}{\overset{\|}{C}}{-}O{-}\overset{O}{\overset{\|}{S}}Cl + HCl$$

$$(2)\ CH_3\overset{O}{\overset{\|}{C}}{-}O{-}\overset{O}{\overset{\|}{S}}Cl + Cl^- \longrightarrow CH_3\overset{O^-}{\overset{|}{\underset{Cl}{\underset{|}{C}}}}{-}O{-}\overset{O}{\overset{\|}{S}}Cl$$

$$(3)\ CH_3\overset{O^-}{\overset{|}{\underset{Cl}{\underset{|}{C}}}}{-}O{-}\overset{O}{\overset{\|}{S}}Cl \longrightarrow CH_3\overset{O}{\overset{\|}{C}}Cl + {}^-O\overset{O}{\overset{\|}{S}}Cl$$

We shall see in the next chapter that acyl halides are important intermediates for the conversion of carboxylic acids into other derivatives.

E. Reaction with Ammonia: Formation of Amides

As in aldehydes and ketones, the carbonyl group in carboxylic acids is polarized. That is, the bonding electrons have higher density in the neighborhood of the oxygen than at the carbon.

$$\left[R{-}\overset{O}{\overset{\|}{C}}{-}OH \longleftrightarrow R{-}\overset{O^-}{\overset{|}{\underset{+}{C}}}{-}OH\right]$$

With aldehydes and ketones, we found that, in addition to acid-catalyzed nucleophilic addition, base-catalyzed addition is also observed. With carboxylic acids base-catalyzed nucleophilic additions are rare, and with good reason. Consider the reaction of acetic acid with sodium methoxide. Since methoxide ion is a strong base (the pK_a of methanol is about 16) and acetic acid is a moderately strong acid (pK_a about 5), the following simple acid-base equilibrium lies strongly to the right ($K \cong 10^{11}$) and is established very rapidly.

$$CH_3{-}\overset{\overset{\displaystyle O}{\|}}{C}{-}OH + CH_3O^- \xrightleftharpoons{K} CH_3{-}\overset{\overset{\displaystyle O}{\|}}{C}{-}O^- + CH_3OH$$

In other words, the acetic acid is converted immediately and quantitatively into acetate ion and the methoxide into methanol. Even in the presence of excess methoxide ion no further reaction occurs, since the acetate carbonyl is less electrophilic. That is, nucleophilic addition to the carbonyl would require that two anions combine to give a dianion. Since like charges repel one another, this reaction is unlikely.

$$CH_3{-}\overset{\overset{\displaystyle O}{\|}}{C}{-}O^- + CH_3O^- \not\longrightarrow CH_3{-}\underset{\underset{\displaystyle O^-}{|}}{\overset{\overset{\displaystyle O^-}{|}}{C}}{-}OCH_3$$

Even so, several base-catalyzed nucleophilic additions of carboxylic acids are known. As we shall see, each involves rather special conditions.

The most common reaction of this type is the reaction of carboxylic acids with ammonia or amines to give amides. When ammonia is bubbled through butanoic acid at 185 °C, butanamide is obtained in 85% yield. The reaction involves two stages. At room temperature, or even below, butanoic acid reacts with the weak base ammonia to give the salt ammonium butanoate.

$$CH_3CH_2CH_2COOH + NH_3 \xrightarrow{25\,°C} CH_3CH_2CH_2CO_2^-\ NH_4^+$$

This salt is perfectly stable at normal temperatures. However, pyrolysis of the salt results in the elimination of water and formation of the amide.

$$\underset{\text{ammonium butanoate}}{CH_3CH_2CH_2CO_2^-\ NH_4^+} \xrightarrow{185\,°C} \underset{\text{butanamide}}{CH_3CH_2CH_2\overset{\overset{\displaystyle O}{\|}}{C}NH_2} + H_2O$$

The reaction occurs only because ammonium butanoate, being the salt of a weak acid and a weak base, is in equilibrium with a significant amount of ammonia and butanoic acid. The actual dehydration step is probably the result of nucleophilic addition of ammonia to the carbonyl group of butanoic acid itself, which might be activated through hydrogen bonding to the carbonyl by another butanoic acid.

$$R{-}\overset{\overset{\displaystyle O}{\|}}{C}{-}O^-\ NH_4^+ \rightleftharpoons R{-}\overset{\overset{\displaystyle O}{\|}}{C}{-}OH + NH_3$$

$$R{-}\overset{\overset{\displaystyle O}{\|}}{C}{-}OH + :NH_3 \rightleftharpoons R{-}\underset{\underset{\displaystyle NH_3^+}{|}}{\overset{\overset{\displaystyle O^-}{|}}{C}}{-}OH$$

$$R{-}\underset{\underset{\displaystyle NH_3^+}{|}}{\overset{\overset{\displaystyle O^-}{|}}{C}}{-}OH \rightleftharpoons R{-}\underset{\underset{\displaystyle NH_2}{|}}{\overset{\overset{\displaystyle O^-}{|}}{C}}{-}OH_2^+$$

$$R{-}\underset{NH_2}{\overset{O^-}{C}}{-}OH_2^+ \rightleftharpoons R{-}\overset{O}{\overset{\|}{C}}{-}NH_2 + H_2O$$

EXERCISE 18.13 From the pK_as of butanoic acid and ammonium ion, 4.82 and 9.24, respectively, calculate K for the following equilibrium.

$$CH_3CH_2CH_2CO_2^- + NH_4^+ \rightleftharpoons CH_3CH_2CH_2COOH + NH_3$$

F. Reduction of the Carboxy Group

Another nucleophilic addition to the carboxylate group that is of some interest is in the reduction of carboxylic acids by lithium aluminum hydride.

$$3,5\text{-}(CH_3O)_2C_6H_3{-}COOH + LiAlH_4 \xrightarrow{\text{ether}} \xrightarrow[H_2O]{H_2SO_4} 3,5\text{-}(CH_3O)_2C_6H_3{-}CH_2OH$$

(93%)

The first step in this reaction is an acid-base reaction, giving the lithium salt of the acid, hydrogen gas, and aluminum hydride.

$$RCOOH + LiAlH_4 \longrightarrow RCOOLi + H_2 + AlH_3$$

The lithium carboxylate is then reduced further, eventually to the salt of the corresponding primary alcohol. Tetrahydroaluminate ion, AlH_4^-, is so reactive that it reduces even a carboxylate ion.

$$RCO_2^- + AlH_4^- \longrightarrow R{-}\underset{H}{\overset{O^-}{C}}{-}O^- + AlH_3$$

The reaction is undoubtedly assisted by the Lewis-acid character of aluminum salts, which reduce the effective negative charge on oxygen. The remaining steps in the reduction are still more complex, but undoubtedly also involve lithium and aluminum salts. For example, further reaction of the bis-alkoxide dianion could involve expulsion of O^{2-} as an aluminum oxide with formation of an intermediate aldehyde. The aldehyde is then rapidly reduced to the alcohol with lithium aluminum hydride.

EXERCISE 18.14 Show how pentanenitrile might be converted into 1-pentanol.

G. One-Carbon Degradation of Carboxylic Acids

Carboxylic acids undergo several reactions in which the carboxy group is replaced by halogen.

$$RCOOH \longrightarrow RX$$

Such reactions, in which carbons are lost from a molecule, are called "degradations."

In the **Hunsdiecker reaction** the silver salt of a carboxylic acid, prepared by treating the acid with silver oxide, is treated with a halogen. Bromine is the usual reagent, but iodine may also be used. Carbon dioxide is evolved and the corresponding alkyl halide is obtained, usually in fair to good yield.

$$\mathrm{MeOC(=O)CH_2CH_2CH_2CH_2CO_2^-Ag^+} \xrightarrow[\mathrm{CCl_4}]{\mathrm{Br_2}} \mathrm{MeOC(=O)CH_2CH_2CH_2CH_2Br} + \mathrm{AgBr} + \mathrm{CO_2}$$

(65–68%)

The reaction appears to proceed by a free radical path and may be formulated as follows.

$$(1)\ \mathrm{R{-}\overset{O}{\overset{\|}{C}}{-}O^-\ Ag^+} + \mathrm{Br_2} \longrightarrow \mathrm{R{-}\overset{O}{\overset{\|}{C}}{-}OBr} + \mathrm{AgBr}$$

$$(2)\ \mathrm{R{-}\overset{O}{\overset{\|}{C}}{-}OBr} \longrightarrow \mathrm{R{-}\overset{O}{\overset{\|}{C}}{-}O\cdot} + \mathrm{Br\cdot}$$

initiation step

$$(3)\ \mathrm{R{-}\overset{O}{\overset{\|}{C}}{-}O\cdot} \longrightarrow \mathrm{R\cdot} + \mathrm{CO_2}$$

$$(4)\ \mathrm{R\cdot} + \mathrm{R{-}\overset{O}{\overset{\|}{C}}{-}OBr} \longrightarrow \mathrm{RBr} + \mathrm{R{-}\overset{O}{\overset{\|}{C}}{-}O\cdot}$$

(3) and (4): propagation steps

In a useful modification of the Hunsdiecker reaction, the carboxylic acid is treated with mercuric oxide and bromine.

$$2\ \text{(cyclopropyl)}\mathrm{CH{-}COOH} + \mathrm{HgO} + 2\ \mathrm{Br_2} \longrightarrow 2\ \text{(cyclopropyl)}\mathrm{CH{-}Br} + \mathrm{HgBr_2} + 2\ \mathrm{CO_2} + \mathrm{H_2O}$$

(41–46%)

In the **Kochi reaction** the carboxylic acid is treated with lead tetraacetate and lithium chloride; the product is an alkyl chloride.

$$\text{(cyclobutyl)}\mathrm{CH{-}COOH} + \mathrm{Pb(O\overset{O}{\overset{\|}{C}}CH_3)_4} + \mathrm{LiCl} \longrightarrow \text{(cyclobutyl)}\mathrm{CH{-}Cl} + \mathrm{CO_2} + \mathrm{LiPb(O\overset{O}{\overset{\|}{C}}CH_3)_3} + \mathrm{CH_3COOH}$$

(100%)

The Hunsdiecker and Kochi reactions complement each other, the former giving best results with primary alkyl carboxylic acids and the latter being preferred for secondary and tertiary alkyl carboxylic acids.

EXERCISE 18.15 Show how 2,2-dimethylpentanoic acid could be converted into 2-methyl-1-pentene.

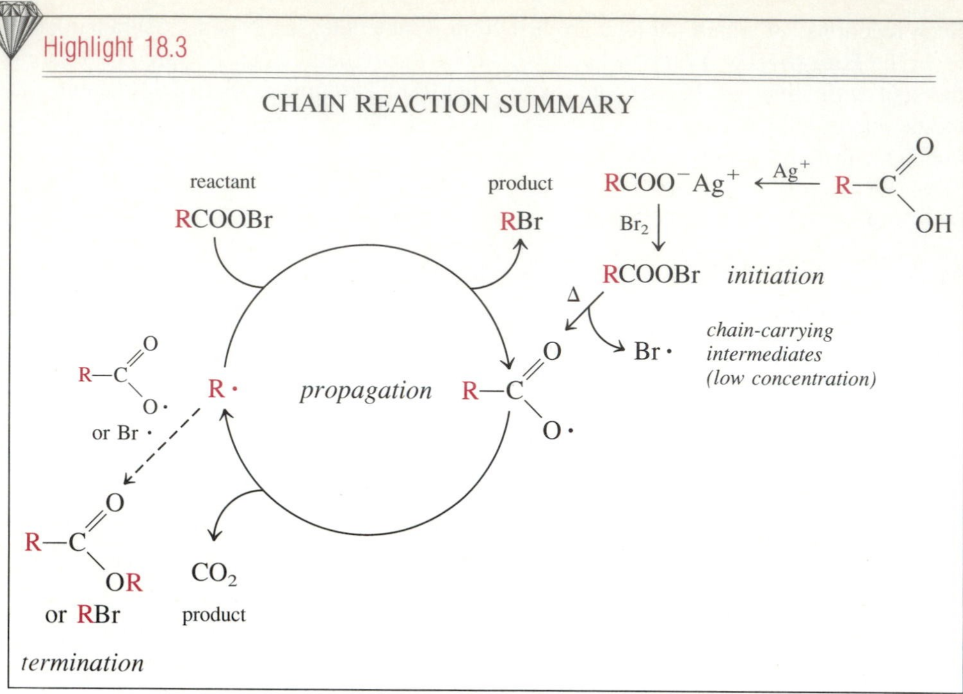

18.8 Natural Occurrence of Carboxylic Acids

Carboxylic acids are widespread in nature, both as such and in the form of esters. Partly because they are easily isolated as salts, they were among the earliest known organic compounds.

Formic acid was first discovered in 1670 by the distillation of ants. Its name comes from the Latin word for ant, *formica*. Formic acid is partially responsible for the irritation resulting from the sting of an insect, the red ant, and a plant, the stinging nettle.

Acetic acid is a product of fermentation. The characteristic taste of sour wine is due to the oxidation of ethanol to acetic acid. Vinegar is a dilute solution of acetic acid. Although the acid has been known in the form of vinegar since antiquity, it was first isolated in pure form by Stahl in 1700. Pure acetic acid is known as glacial acetic acid. This term arises from the relatively high melting point of the compound (17 °C, 63 °F). In earlier times, when buildings were not heated as they are now, pure acetic acid was commonly observed to be a solid at ''room temperature.'' Acetic acid is also one of the products of pyrolysis of wood (destructive distillation).

Butyric acid is responsible for the sharp odor of rancid butter. It was first isolated from this source. Caproic acid also has a penetrating unpleasant odor described as ''goat-like.'' Indeed, its name, as well as those of caprylic acid and capric acid, is derived from the Latin word for goat, *caper*. These acids and their esters are widespread in nature.

Juvenile hormone and juvabione are examples of carboxylic acids that occur in nature in the form of their methyl esters.

O COOCH$_3$

juvenile hormone

H$_3$C H H O COOCH$_3$

juvabione

> Failure of the growth of the European bug *Pyrrhocoris apteris* on American paper towels rather than Whatman filter paper led to the discovery of juvabione, a compound present in the balsam fir tree, *Abies balsamea*. Juvabione arrests the bugs in the fifth larval stage, and may serve in the tree as defense against certain insects.

These compounds are associated with the pupal development of various insects; an excess of agent arrests further development. Such compounds offer some promise as insect-control agents. Since they are highly species-specific and leave no residues, they have obvious advantages over other commonly used pesticides. However, juvenile hormones and pheromones are relatively expensive. Their successful use depends on stability in a field formulation and wide dispersion at times when insects are present. Although several have shown promise in field trials, further research is needed on commercially viable syntheses, stable dispersion vehicles, and choice of appropriate targets such as the cotton pest, the pink bollworm *Pectinophora gossypiella*.

ASIDE

The Tatar or Teutonic barbarians apparently invented soap, which became common in Europe only in the Middle Ages. Animal fat or olive oil heated with potassium hydroxide yields soft soap; sodium hydroxide gives hard soap. Supposing the soaps made from animal fat were to be derived from a saturated C_{18} acid, what would happen if one equivalent of acetic acid were added to a 0.01 M soap solution? How about chloroacetic acid? Or trichloroacetic acid? What would you observe if only one-tenth of an equivalent of trichloroacetic acid were used? Can you explain the appearance of the solution under these conditions? Why are different effects obtained for the three acids?

PROBLEMS

1. Give the IUPAC name for each of the following compounds.

a. $CH_3CH_2CH(CH_3)CH_2COOH$ **b.** $(CH_3)_3CCOOH$ **c.** $BrCH_2CH_2CH_2COOH$

d. ICH_2COOH **e.** $CH_3CH_2CH(OH)COOH$ **f.** $CH_3(CH_2)_8COOH$

g. cyclobutane–COOH **h.** $CH_3CH(OCH_3)CH_2COOH$ **i.** $(CH_3)_2CHCH(CH_3)CH_2COOH$

j. $NCCH_2COOH$

2. Write the correct structure for each of the following names.

a. β-chlorobutyric acid **b.** hexanoic acid
c. γ-methoxyvaleric acid **d.** cyclopentanecarboxylic acid
e. 3-bromopropanoic acid **f.** *cis*-2-pentenoic acid
g. (E,R)-4-hydroxypent-2-enoic acid **h.** α-chloro-β-bromopropionic acid

3. Give the products in each of the following reactions of cyclohexanecarboxylic acid.

a. $LiAlH_4$ in ether, then dilute hydrochloric acid
b. P + Br_2, heat, then water
c. diazomethane in ether
d. isopropyl alcohol (excess) and a trace of sulfuric acid
e. ammonia, 200 °C
f. methylamine (CH_3NH_2), 200 °C
g. $SOCl_2$
h. PBr_3
i. Pt/H_2, room temperature
j. dilute aqueous sodium hydroxide at room temperature
k. silver hydroxide, followed by bromine in carbon tetrachloride
l. $Pb(OAc)_4$ + LiCl

4. Two general methods for converting alkyl halides to carboxylic acids are displacement by cyanide ion, followed by hydrolysis, and conversion to the Grignard reagent, followed by carbonation with carbon dioxide. Which method is superior for each of the following transformations? In which cases would a protecting group facilitate the conversion? Explain why.

a. $(CH_3)_3CCl \rightarrow (CH_3)_3CCOOH$
b. $BrCH_2CH_2CH_2Br \rightarrow HOOCCH_2CH_2CH_2COOH$
c. $CH_3COCH_2CH_2CH_2Br \rightarrow CH_3COCH_2CH_2CH_2COOH$
d. $(CH_3)_3CCH_2Br \rightarrow (CH_3)_3CCH_2COOH$
e. $CH_3CH_2CH_2CH_2Br \rightarrow CH_3CH_2CH_2CH_2COOH$
f. $HOCH_2CH_2CH_2CH_2Br \rightarrow HOCH_2CH_2CH_2CH_2COOH$

5. Show how neopentane can be converted into each of the following compounds.

a. $(CH_3)_3CCH_2COOH$
b. $(CH_3)_3CCHBrCOOH$
c. $(CH_3)_3CCH_2CH_2OH$
d. $(CH_3)_3CCH_2COCl$

6. Show how butanal can be converted into each of the following compounds.

a. $CH_3CH_2CH_2COOH$
b. $CH_3CH_2CH_2CH{=}C(COOH)CH_2CH_3$
c. $CH_3CH_2CH_2CH(OH)COOH$
d. $CH_3CH_2CH_2CH_2COOH$

7. Show how 3-methylbutanoic acid can be converted into each of the following compounds.

a. $(CH_3)_2CHCH_2CH_2OH$
b. $(CH_3)_2CHCH_2CH_2COOH$
c. $(CH_3)_2CHCH_2CH_2CHBrCH_2Br$
d. $(CH_3)_2C{=}CHCOOCH_3$
e. $(CH_3)_2CHCH_2C(=O)CH_3$
f. $(CH_3)_2CHCH_2C(=O)NH_2$
g. $(CH_3)_2CHCH_2Br$

8. Show how each of the following transformations can be accomplished in a practical manner.

a. (cyclohexane)$=CH_2 \longrightarrow$ (cyclohexane)$-CH_2COOH$

b. $(CH_3)_3CCH{=}CH_2 \longrightarrow (CH_3)_3CCOOH$

c. $CH_3COCH_2CH_2CH_2C(CH_3)_2Br \longrightarrow CH_3COCH_2CH_2CH_2C(CH_3)_2COOH$

d. $CH_3CH_2COOH \longrightarrow CH_3CH_2CH_2COOH$
e. $CH_3CH_2CH_2COOH \longrightarrow CH_3CH_2COOH$
f. $CH_3CH_2CH_2COOH \longrightarrow CH_3CH_2CH_2N_3$

9. The dissociation constant of acetic acid is 1.8×10^{-5} M. Calculate the percent dissociation when the following amounts of acetic acid are made up to 1 L with water at 25 °C.

a. 0.1 mole **b.** 0.01 mole **c.** 0.001 mole

10. The following dissociation constants are given. Calculate the corresponding pK_a values.

a. $(CH_3)_2CHCH_2CH_2COOH$: $K_a = 1.4 \times 10^{-5}$ M

b. 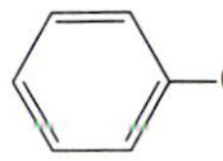 C_6H_5–COOH: $K_a = 6.3 \times 10^{-5}$ M

c. $Cl_2CHCOOH$: $K_a = 5.5 \times 10^{-2}$ M

d. Cl_3CCOOH: $K_a = 0.23$ M

e. $CH_3CONHCH_2COOH$: $K_a = 2.1 \times 10^{-4}$ M

11. In each of the following pairs, which is the stronger *base*? Explain briefly.

a. $CH_3CH_2O^-$; $CH_3CO_2^-$
b. $ClCH_2CH_2CO_2^-$; $CH_3CH_2CH_2CO_2^-$
c. $ClCH_2CH_2CO_2^-$; $CH_3CHClCO_2^-$
d. $FCH_2CO_2^-$; $F_2CHCO_2^-$
e. $HC{\equiv}CCH_2CO_2^-$; $CH_3CH_2CH_2CO_2^-$
f. Cl^-, $CH_3CO_2^-$

12. The two carboxy groups in 3-chlorohexanedioic acid are not equivalent and have different dissociation constants. Which carboxy group is the more acidic?

13. From the progression of acidity constants for chlorobutanoic acids in Table 18.4 and the pK_a of 3-cyanobutanoic acid, 4.44, estimate the pK_a of 2-cyanobutanoic acid.

14. When propanoic acid is refluxed with some sulfuric acid in water enriched with $H_2{}^{18}O$, ^{18}O gradually appears in the carboxylic acid group. Write the mechanism for this reaction, showing each intermediate in the reaction pathway.

15. When 5-hydroxyhexanoic acid is treated with a trace of sulfuric acid in benzene solution, the following reaction occurs.

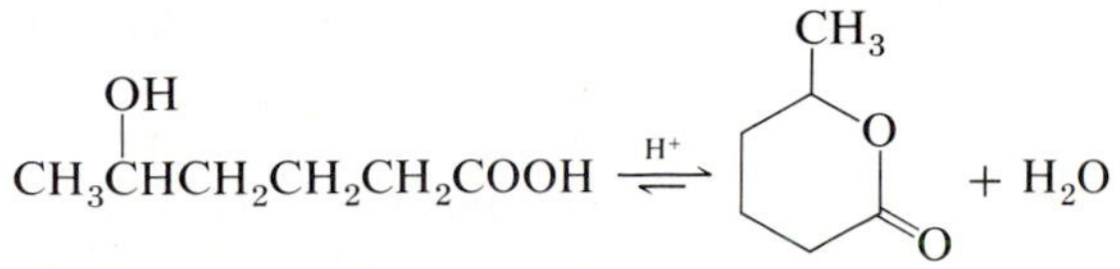

$$CH_3CH(OH)CH_2CH_2CH_2COOH \xrightleftharpoons{H^+} \text{(6-methyltetrahydropyran-2-one, ring with } CH_3\text{, O, =O)} + H_2O$$

a. Propose a mechanism for the reaction.
b. The equilibrium constant for this process is much larger than that normally observed for an esterification reaction. Explain.

16. On refluxing with D_2O containing a strong acid, propanoic acid is slowly converted to CH_3CD_2COOD. Write a plausible mechanism for this reaction.

17. Values of heats of formation, ΔH_f°, for the ideal gas state at 25 °C are given in the table that follows for several compounds. Calculate ΔH° for the following equilibrium in the gas phase.

$$CH_3COOH + C_2H_5OH \rightleftharpoons CH_3COOC_2H_5 + H_2O$$

In the liquid phase, ΔH° for this equilibrium is -0.9 kcal mole^{-1}. Why is there such a difference between the two values?

Compound	ΔH_f° at 25°C, kcal mole^{-1}
CH_3COOH	-103.3
C_2H_5OH	-56.2
$CH_3COOC_2H_5$	-106.3
H_2O	-57.8

18. The following reaction is exothermic in the gas phase.

$$CH_3CO_2^- + ClCH_2COOH \rightleftharpoons CH_3COOH + ClCH_2CO_2^- \qquad \Delta H^\circ = -13 \text{ kcal mole}^{-1}$$

a. Explain briefly why this reaction is exothermic.

b. Perform a simple calculation to determine whether the electrostatic interaction of a C—Cl dipole with a carboxylate anion has the proper magnitude to account for this energy difference. For this purpose treat $ClCH_2CO_2^-$ as having the following structure in which the CCCl plane is perpendicular to the OCO plane.

in which θ is the angle between the dipole and the charge. For a charge q of one electron, $\mu = 1$ D, $r = 1$ Å, and $\theta = 0°$, the energy E is 69 kcal mole^{-1}. Consider the effect of the chlorine to be that of a point dipole of 1.9 D. The electrostatic energy for a point dipole and a charge is given by

$$E = \frac{q\mu \cos\theta}{r^2}$$

19. What is the mechanism by which hydrocarbons (grease or those from an oil spill) are solubilized with amphiphilic (or amphipathic) agents? How could you convert stearic acid into an amphiphilic agent? If each agent molecule occupies the same surface area as stearic acid (25 Å^2), how much stearic acid derivative would be needed in principle to solubilize 50 mL of oil? 500 mL of oil? 500,000 L of oil? The last would correspond to the amount released in a substantial oil spill. (Assume that the oil particles must be spheres with a radius of 10 μ and that a single surface monolayer of agent is needed for solubilization.) Does the amount of water in which the oil is found make any difference? Why? Let us suppose that the 50 mL of oil were floating on 500 mL of water in a 1 L beaker. How much stearic acid derivative would be needed? And if the oil were on the water in a filled Olympic swimming pool (assume 50 m length, 10 m width and a 3 m depth)?

CHAPTER 19

DERIVATIVES OF CARBOXYLIC ACIDS

19.1 Structure

Functional group derivatives of carboxylic acids, **RCOX**, are compounds that are transformed into carboxylic acids, **RCOOH**, by simple hydrolysis.

Esters, RCOOR, are the most common such derivatives. The hydroxy group in RCOOH is replaced by an alkoxy group.

$$R{-}\overset{\overset{\large O}{\|}}{C}{-}OCH_3$$

an ester

Amides, $RCONH_2$, are compounds in which the hydroxy group is replaced by an amino group. The nitrogen of the amino group may bear zero, one, or two alkyl groups.

$$R{-}\overset{\overset{\large O}{\|}}{C}{-}NH_2 \qquad R{-}\overset{\overset{\large O}{\|}}{C}{-}N\begin{matrix}\diagup H\\ \diagdown CH_3\end{matrix} \qquad R{-}\overset{\overset{\large O}{\|}}{C}{-}N\begin{matrix}\diagup CH_3\\ \diagdown CH_3\end{matrix}$$

amides

Acyl halides, RCOX, are derivatives in which the carboxy OH is replaced by a halogen atom; acyl chlorides and acyl bromides are the most commonly encountered acyl halides.

$$R{-}\overset{\overset{\large O}{\|}}{C}{-}Cl \qquad R{-}\overset{\overset{\large O}{\|}}{C}{-}Br$$

an acyl chloride   an acyl bromide

Acid anhydrides, $(RCO)_2O$, are molecules in which one molecule of water has been removed from two molecules of a carboxylic acid. The only acyclic anhydride of general importance is acetic anhydride.

$$CH_3-\overset{\overset{O}{\|}}{C}-O-\overset{\overset{O}{\|}}{C}-CH_3$$

acetic anhydride

In a formal sense, nitriles are functional derivatives of carboxylic acids because they may be hydrolyzed to carboxylic acids (Section 18.6.A). The chemistry of nitriles has been discussed in Chapter 12.

$$R-CN$$

a nitrile

The simplest ester, methyl formate, may be considered a derivative of formic acid in which the OH group is replaced by the OCH_3 group. Correspondingly, the molecular geometry of methyl formate is similar to that of formic acid. Experimental bond lengths and bond angles, determined by microwave spectroscopy, are given in Figure 19.1.

Note that the C_{sp^2}—O σ-bond is considerably shorter than the C_{sp^3}—O σ-bond. Two factors are apparently important in accounting for this bond shortening. In Section 11.1 we saw that because of the difference in "length" of various hybrid orbitals, C_{sp^3}—C_{sp^3} σ-bonds are longer than C_{sp^3}—C_{sp^2} σ-bonds. This factor is probably also important in methyl formate. Another factor involves the dipolar resonance form (19-1) as a contributor to the structure of methyl formate.

$$\left[H-C\begin{matrix} \nearrow\!\!\!\!= \ddot{O}: \\ \searrow \ddot{\underset{..}{O}}-CH_3 \end{matrix} \longleftrightarrow H-C\begin{matrix} \diagup :\ddot{O}:^- \\ \diagdown\!\!\!\!= \overset{+}{\underset{..}{O}}-CH_3 \end{matrix} \right] \quad (19\text{-}1)$$

To the extent that this form contributes to the actual structure of the molecule, the C_{sp^2}—O σ-bond will be shorter because it has some double bond character. This latter factor is especially important in amides.

Microwave measurements on formamide indicate the structure shown in Figure 19.2. The entire molecule is planar. Note that the two hydrogens attached to nitrogen are distinguishable. The barrier to rotation about the carbon-nitrogen bond has been measured experimentally and is found to be 18 kcal mole^{-1}. A high degree of **double-bond character** in this bond, as indicated in the dipolar resonance form (19-2), has been invoked to explain this relatively high rotational barrier. Because of the high barrier to rotation about the carbon-nitrogen bond, amides have a relatively rigid structure.

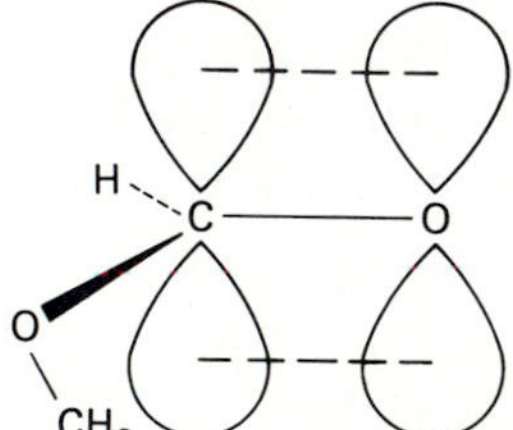

Bond Lengths, Å		Bond Angles, deg	
C=O	1.200	H—C=O	124.95
C(=O)—O	1.334	O—C=O	125.87
C(H$_3$)—O	1.437	H—C—O	109.18
C(=O)—H	1.101	CH$_3$—O—C	114.78

FIGURE 19.1 Structure of methyl formate.

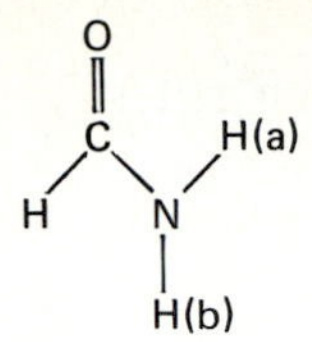

Bond Lengths, Å		Bond Angles, deg	
C=O	1.193	H—C=O	122.97
C—N	1.376	H—C—N	113.23
C—H	1.102	N—C=O	123.80
N—H(a)	1.014	C—N—H(a)	117.15
N—H(b)	1.002	C—N—H(b)	120.62
		H—N—H	118.88

FIGURE 19.2 Structure of formamide.

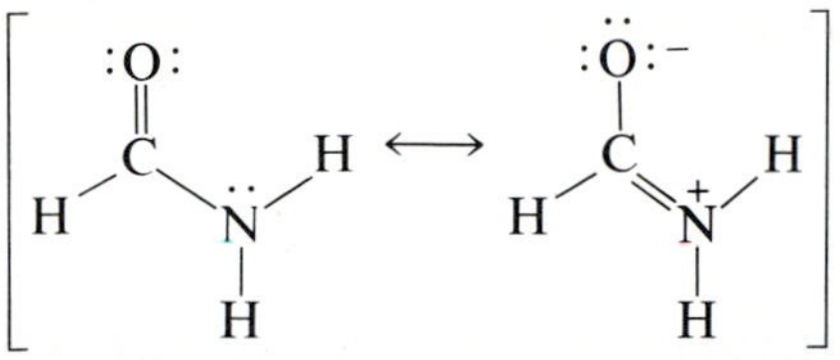

(19-2)

Since the simplest acyl chloride, formyl chloride, is not stable at temperatures above −60 °C, its structural parameters have not been measured. However, the bond lengths and bond angles have been determined for acetyl chloride (Figure 19.3). The carbon-chlorine bond is not appreciably shorter than the analogous bond in methyl chloride (1.784 Å), suggesting that dipolar resonance structures are not particularly important in the case of acyl halides.

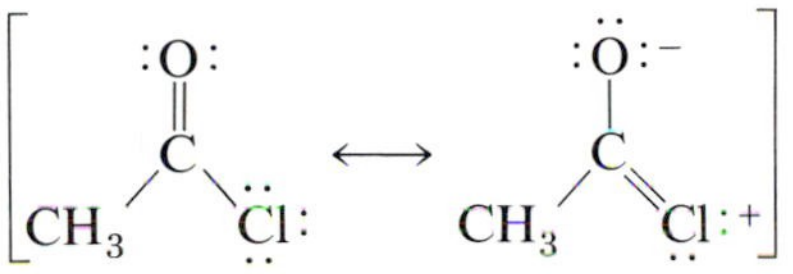

EXERCISE 19.1 For methylamine, dimethyl ether, and methyl fluoride, the CH_3—X distances are 1.47 Å, 1.42 Å, and 1.38 Å, respectively. For acetamide, methyl acetate, and acetyl fluoride, the analogous distances are 1.36 Å, 1.36 Å, and 1.37 Å, respectively. Use the differences in C—X distances to evaluate the relative importance of dipolar resonance structures in the three carboxylic acid derivatives. Is there a correlation between the importance of the dipolar resonance structure and the relative basicity of F^-, OH^-, and NH_2^-?

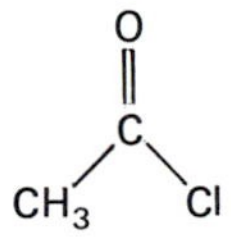

Bond Lengths, Å		Bond Angles, deg	
C=O	1.192	C—C=O	127.08
C—C	1.499	C—C—Cl	112.66
C—Cl	1.789	O=C—Cl	120.26
C—H	1.083		

FIGURE 19.3 Structure of acetyl chloride.

19.2 Nomenclature

Esters are named in the following way. The first word of the name is the stem name of the alkyl group attached to oxygen. The second word of the name is the name of the parent acid with the suffix -**ic** replaced by -**ate.** This nomenclature applies for both common and IUPAC names of acids.

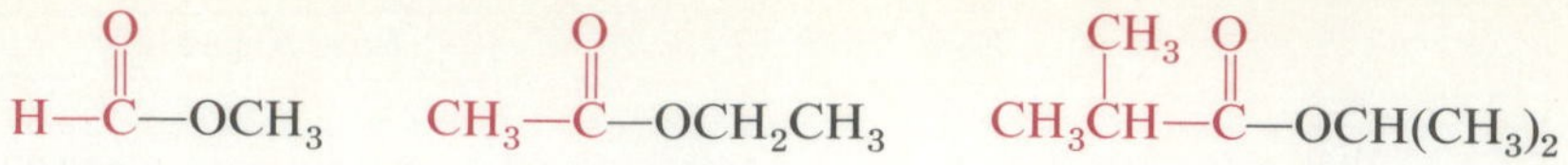

$$H{-}\overset{\overset{\displaystyle O}{\|}}{C}{-}OCH_3 \qquad CH_3{-}\overset{\overset{\displaystyle O}{\|}}{C}{-}OCH_2CH_3 \qquad CH_3\overset{\overset{\displaystyle CH_3}{|}}{C}H{-}\overset{\overset{\displaystyle O}{\|}}{C}{-}OCH(CH_3)_2$$

methyl formate — ethyl acetate — isopropyl isobutyrate / isopropyl 2-methylpropanoate

Amides are named by dropping the suffix -**ic** or -**oic** from the name of the parent acid and adding the suffix -**amide.**

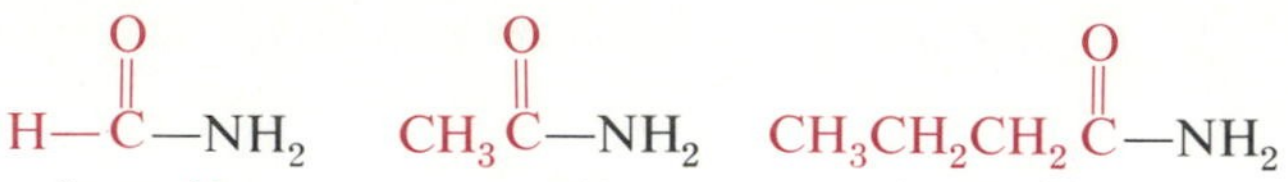

$$H{-}\overset{\overset{\displaystyle O}{\|}}{C}{-}NH_2 \qquad CH_3\overset{\overset{\displaystyle O}{\|}}{C}{-}NH_2 \qquad CH_3CH_2CH_2\overset{\overset{\displaystyle O}{\|}}{C}{-}NH_2$$

formamide — acetamide — butanamide

A substituted nitrogen is indicated by prefixing the name of the simple amide by N-, followed by the name of the substituent group.

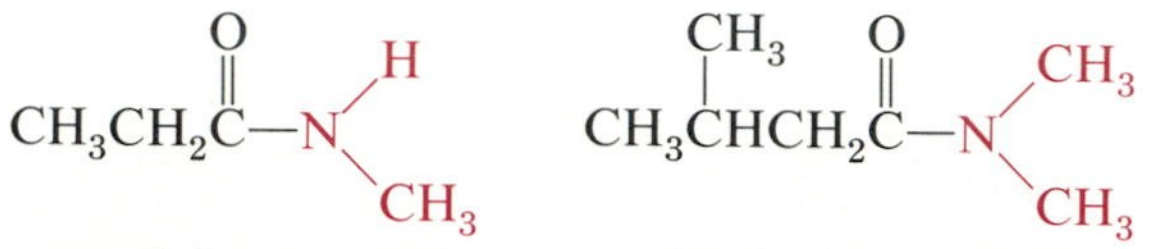

$$CH_3CH_2\overset{\overset{\displaystyle O}{\|}}{C}{-}N\begin{matrix}\diagup H\\ \diagdown CH_3\end{matrix} \qquad CH_3\overset{\overset{\displaystyle CH_3}{|}}{C}HCH_2\overset{\overset{\displaystyle O}{\|}}{C}{-}N\begin{matrix}\diagup CH_3\\ \diagdown CH_3\end{matrix}$$

N-methylpropanamide — N,N,3-trimethylbutanamide

Acyl halides are named in a similar manner. In this case the suffix -**ic** is replaced by the suffix -**yl,** and the halide name is added as a second word. (Note that for acyl halides, the "o" of the ending **oic** is retained.)

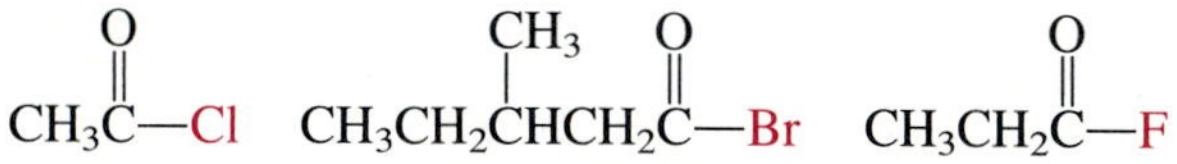

$$CH_3\overset{\overset{\displaystyle O}{\|}}{C}{-}Cl \qquad CH_3CH_2\overset{\overset{\displaystyle CH_3}{|}}{C}HCH_2\overset{\overset{\displaystyle O}{\|}}{C}{-}Br \qquad CH_3CH_2\overset{\overset{\displaystyle O}{\|}}{C}{-}F$$

acetylchloride — 3-methylpentanoyl bromide — propanoyl fluoride

Anhydrides are named by adding **anhydride** to the name of the corresponding carboxylic acid.

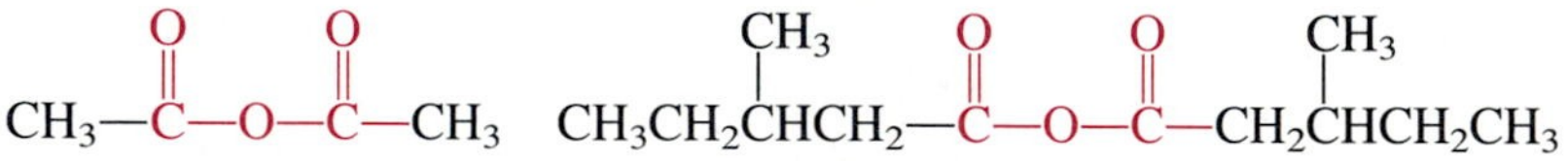

$$CH_3{-}\overset{\overset{\displaystyle O}{\|}}{C}{-}O{-}\overset{\overset{\displaystyle O}{\|}}{C}{-}CH_3 \qquad CH_3CH_2\overset{\overset{\displaystyle CH_3}{|}}{C}HCH_2{-}\overset{\overset{\displaystyle O}{\|}}{C}{-}O{-}\overset{\overset{\displaystyle O}{\|}}{C}{-}CH_2\overset{\overset{\displaystyle CH_3}{|}}{C}HCH_2CH_3$$

acetic anhydride — 3-methylpentanoic anhydride

For mixed anhydrides, the parent name of each acid is given, followed by the word **anhydride.**

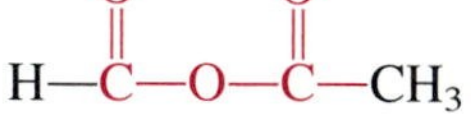

$$H{-}\overset{\overset{\displaystyle O}{\|}}{C}{-}O{-}\overset{\overset{\displaystyle O}{\|}}{C}{-}CH_3$$

acetic formic anhydride

For functional derivatives of carboxylic acids that are named as alkanecarboxylic acids (page 483) the suffix -**carboxylic** acid is replaced by -**carboxamide** or -**carbonyl halide.**

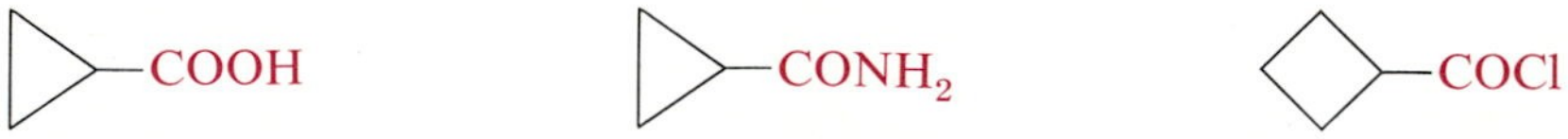

▷—COOH — ▷—$CONH_2$ — ◇—COCl

cyclopropanecarboxylic acid — cyclopropanecarboxamide — cyclobutanecarbonyl chloride

Occasionally it is necessary to name an acid derivative function as a derivative of some other functional stem. The group names for the various radicals as prefixes are given in Table 19.1.

TABLE 19.1 Functional Group Names

Radical	Group Name as Prefix
$—COOCH_3$	methoxycarbonyl
$—COOCH_2CH_3$	ethoxycarbonyl
$—CONH_2$	carbamoyl
—COCl	chloroformyl
—COBr	bromoformyl
—CN	cyano

EXERCISE 19.2 What are the IUPAC names for the methyl ester, amide, acyl chloride, and anhydride corresponding to each of the following carboxylic acids?

(a) $CH_3CH_2CH_2CH_2CH_2COOH$ (b) $(CH_3)_2CHCH_2CH_2COOH$
(c) $CH_2{=}CHCH_2CH_2COOH$ (d) $BrCH_2CH_2COOH$

What are the IUPAC names for the following ester and amide?

(e) $(CH_3)_2CHCH_2CH_2COOCH_2CH(CH_3)_2$ (f) $(CH_3)_3CCH_2CH_2CON(CH_2CH_3)_2$

19.3 Physical Properties

In Figure 19.4 the boiling points of straight-chain acids, methyl esters, and acyl chlorides are plotted against molecular weight. For comparison the boiling point curve for *n*-alkanes is also given. It can readily be seen that esters and acyl halides have approximately the boiling points expected for hydrocarbons of the same molecular weight. This correspondence indicates that the main attractive forces in the condensed phase for these compounds are the relatively weak van der Waals forces. On the other hand, carboxylic acids boil much higher than hydrocarbons of equivalent weight, a consequence of strong intermolecular hydrogen bonds (page 482). Enough energy must be supplied to overcome the normal van der Waals attractive forces, and additional energy must be supplied to overcome the "extra" polar attractive forces. The result is a higher boiling point.

All methyl and ethyl esters and acyl chlorides for the straight-chain acids lower than tetradecanoic are liquids at room temperature. Simple anhydrides above nonanoic anhydride are solid at room temperature.

The melting points and boiling points of acetamide, N-methylacetamide, and N,N-dimethylacetamide are tabulated in Table 19.2. Note that acetamide boils 215 °C higher than a comparable alkane. Dimethylacetamide still boils 95 °C higher than a comparable

TABLE 19.2 Physical Properties for Acetamide Derivatives

	Molecular Weight	Melting Point, °C	Boiling Point, °C
$CH_3—C(=O)—NH_2$	59	82	221
$CH_3—C(=O)—NHCH_3$	73	28	204
$CH_3—C(=O)—N(CH_3)_2$	87	−20	165

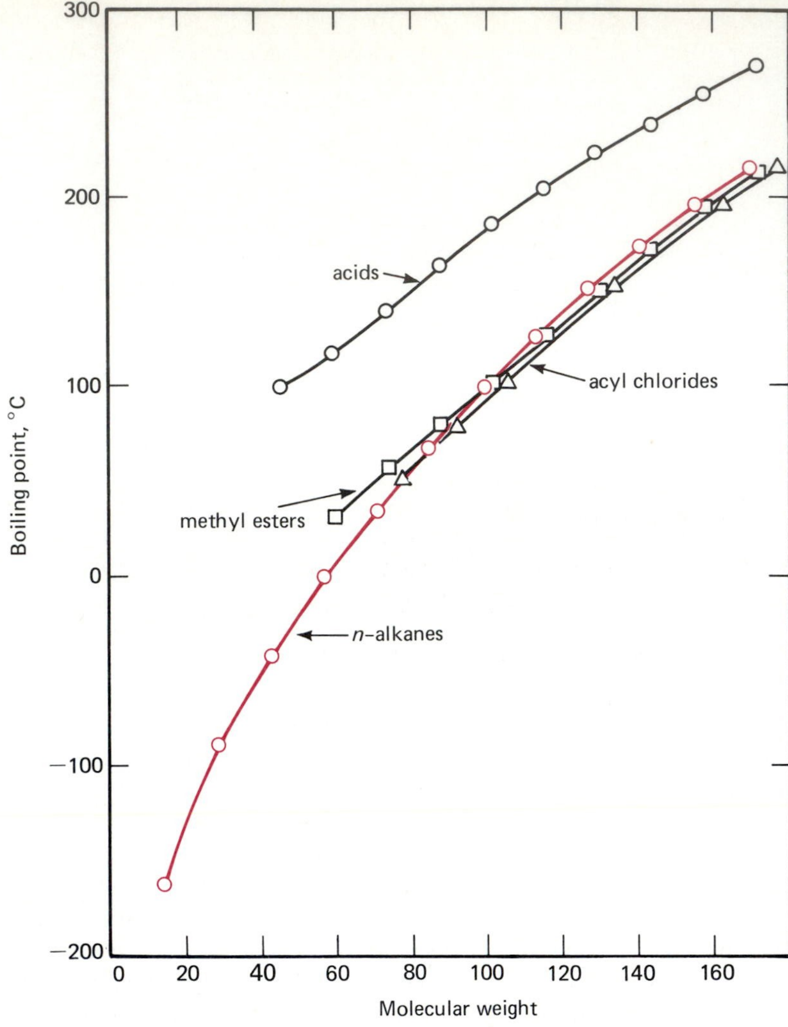

FIGURE 19.4 Boiling points of various compounds.

alkane, but it has a boiling point almost exactly the same as that for an acid nitrile of comparable weight. A similar downward trend is seen in the melting points of the three compounds.

The explanation of these interesting trends lies in the phenomenon of hydrogen bonding. Acetamide, with two hydrogens attached to nitrogen, is extensively hydrogen bonded in both the solid and liquid phases. In N-methylacetamide there is only one N—H, and therefore the hydrogen bonding is less extensive. However, an extended linear combination of hydrogen bonds gives rise to a very high dielectric constant ($D = 176$). Finally, dimethylacetamide cannot engage in hydrogen bonding at all, since it has no hydrogens attached to nitrogen.

Esters, amides, acyl halides, and anhydrides are generally soluble in common organic solvents (ether, chloroform, benzene, and so on). Dimethylformamide and dimethylacetamide are miscible with water in all proportions. Because of their polar, aprotic nature, these compounds are excellent solvents. Ethyl acetate, which is only sparingly soluble in water, is also a common solvent. Because of its excellent solvent properties, ethyl acetate is a common constituent of many brands of paint remover and is also used as a fingernail polish remover. It may easily be recognized by its characteristic ''fruity'' odor.

Esters usually have pleasant odors, which vary markedly with even small changes in structure. Examples are amyl acetate: banana, ethyl butyrate: pineapple, heptyl isobutyrate: cyclamen, heptyl caproate: bruised green leaves. Esters are important components of the fragrance of fruits and flowers and are widely used in perfumes.

Highlight 19.1

The physical properties of carboxylic acids, RCOOH, and carboxylic acid derivatives, RCOX, are dominated by the possibilities for intermolecular interaction through hydrogen bonds. RCOOH readily forms cyclic dimers. RCONHR′ forms long chains that lead to a high dielectric constant for the liquid. RCOOR′, RCOOCOR′, and RCOCl are relatively volatile.

19.4 Spectroscopy

A. Nuclear Magnetic Resonance

Protons in the vicinity of the carbonyl group have similar NMR resonance positions regardless of the exact nature of the compound. Typical values are summarized in Table 19.3. A typical example is the spectrum of methyl propanoate, shown in Figure 19.5.

TABLE 19.3 Chemical Shifts in Compounds of the Type R—Y

	Chemical Shift in δ		
Y	$\mathbf{CH_3}$Y	R$\mathbf{CH_2}$Y	$\mathbf{CH_3}CH_2$Y
—CHO	2.20	2.40	1.08
—COOH	2.10	2.36	1.16
—$COOCH_3$	2.03	2.13	1.12
—COCl	2.67		
—$CONH_2$	2.08	2.23	1.13
—CN	2.00	2.28	1.14

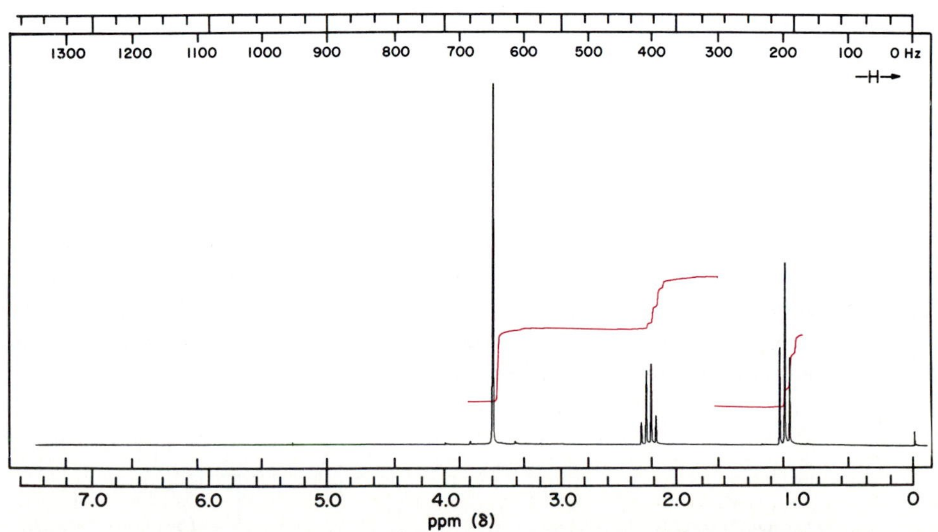

FIGURE 19.5 NMR spectrum of methyl propanoate, $CH_3CH_2COOCH_3$.

We saw in Section 14.4 that the CMR resonance position of the carbonyl carbon in aldehydes and ketones is in the region δ190–200 ppm. The carbonyl resonances of carboxylic acids and of acid derivatives also appear at relatively low field, as shown in Table 19.4. However, in the case of RCOX compounds, the carbonyl resonances are in the range δ 168–178 ppm. Thus, CMR spectroscopy is an excellent tool for distinguishing among certain types of carbonyl-containing compounds.

TABLE 19.4 CMR Chemical Shifts in Carbonyl Derivatives, RCOX

X	δ, ppm
CH_3	205.1
H	199.6
OH	177.3
OCH_2CH_3	169.5
$N(CH_3)_2$	169.6
Cl	168.6

B. Infrared

In Chapter 17 we saw that the characteristic absorption of aldehydes and ketones is the C═O stretch that occurs in the 1710–1825 cm^{-1} region. Other compounds containing the carbonyl group also absorb in this general region. The exact position of the absorption depends on the nature of the functional group. Typical values are listed in Table 19.5.

TABLE 19.5 Carbonyl Stretching Bands of Carboxylic Acid Derivatives in Solution

Functional Group	C═O Stretch, cm^{-1}
—C(═O)—OR	1735
—C(═O)—Cl	1800
—C(═O)—O—C(═O)—	1820 and 1760 (two peaks)
—C(═O)—NR_2	1650–1690

Amides that have one or two hydrogens on nitrogen show a characteristic N—H stretch. For compounds of the type $RCONH_2$ the N—H absorption occurs as two peaks at 3400 and 3500 cm^{-1}. For RCONHR compounds the N—H stretch comes at 3440 cm^{-1}.

Recall that the exact stretching frequency for cyclic ketones is related to ring size. This same effect is seen with cyclic esters, which are called **lactones** (Section 27.5.C), and with cyclic amides, which are called **lactams.**

$CH_3C(═O)OCH_3$
1736 cm^{-1}

1735 cm^{-1}

1770 cm^{-1}

1800 cm^{-1}

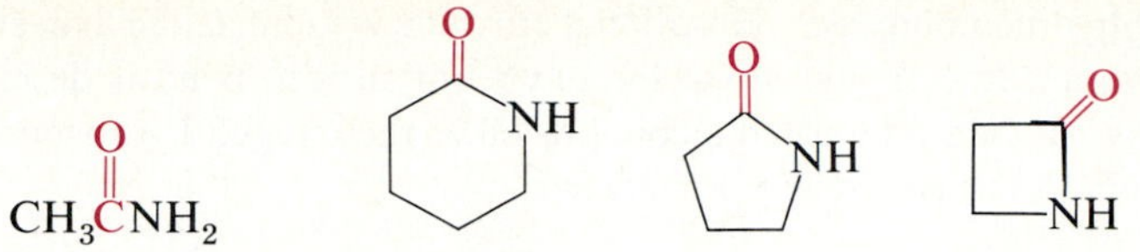

$CH_3\overset{O}{\overset{\|}{C}}NH_2$ 1680 cm^{-1} — (six-membered lactam, NH) 1670 cm^{-1} — (five-membered lactam, NH) 1700 cm^{-1} — (four-membered lactam, NH) 1745 cm^{-1}

EXERCISE 19.3 Give a plausible structure for an unknown compound that has the following spectral properties.
IR: 1740 cm^{-1}
NMR: δ 0.9 (3H, t, J = 7), 1.2–1.9 (6H, m), 1.25 (6H, d, J = 7), 2.3 (2H, t, J = 7), 5.0 (1H, septet, J = 7) ppm
CMR: δ 14.3, 20.2, 23.4, 25.5, 32.2, 33.9, 65.6, 172.0 ppm (the resonance at δ 20.2 ppm is approximately twice as strong as the other resonances in that region of the spectrum.)

19.5 Basicity of the Carbonyl Oxygen

As in the cases of aldehydes, ketones (Section 14.7), and carboxylic acids (Section 18.7.C), the carbonyl oxygen of carboxylic acid derivatives has basic properties. The conjugate acid, an oxonium salt, plays an important role as an intermediate in acid-catalyzed reactions of all of these types of compounds. The actual basicity of the lone-pair electrons of the carbonyl oxygen depends markedly on the nature of the group attached to the carbonyl. This basicity is expressed quantitatively in terms of the acidity or pK_a of the conjugate acid.

$$R\overset{^{+}OH}{\overset{\|}{C}}Y \rightleftharpoons R\overset{O}{\overset{\|}{C}}Y + H^+$$

$$K_a = \frac{[\mathrm{RCOY}][\mathrm{H}^+]}{\left[R\overset{^{+}OH}{\overset{\|}{C}}Y\right]}$$

$$\mathrm{p}K_a = -\log K_a$$

Some pK_a values are summarized in Table 19.6. Most protonated carbonyl compounds are strong acids, stronger than H_3O^+ and comparable in acidity to sulfuric acid (pK_a = −5.2). That is, the carbonyl compounds themselves are weak bases in water, in a class with bisulfate ion. Some of the individual structural effects warrant comment.

Alcohols are generally a little weaker as bases than water itself, and ethers are weaker bases still. These variations are probably the result of solvation differences. The fewer the number of protons on an oxonium oxygen, the less the amount of solvation stabilization by hydrogen bonds to water. If the acid structure is less stable for whatever reason, the conjugate base has lower basicity.

The carbonyl oxygen of aldehydes and ketones is less basic than an alcohol or ether oxygen by several powers of ten. The lone-pair electrons of the carbonyl oxygen may be considered to be approximately sp^2 in character. These electrons have greater σ-character than the lone pairs of alcohol oxygens. Hence, the lone-pair electrons of carbonyl oxygens are more tightly held. As a result, the carbonyl group as a conjugate base is more stable, and the corresponding protonated carbonyl is more acidic. This system is analogous to the corresponding hydrocarbon cases. Recall that ethylene is more acidic than ethane (Section 12.4). In a protonated carbonyl group the oxygen-hydrogen bond is described approximately as O_{sp^2}—H. In a protonated alcohol or ether the oxygen-hydrogen bond involves an oxygen orbital that has greater *p*-character.

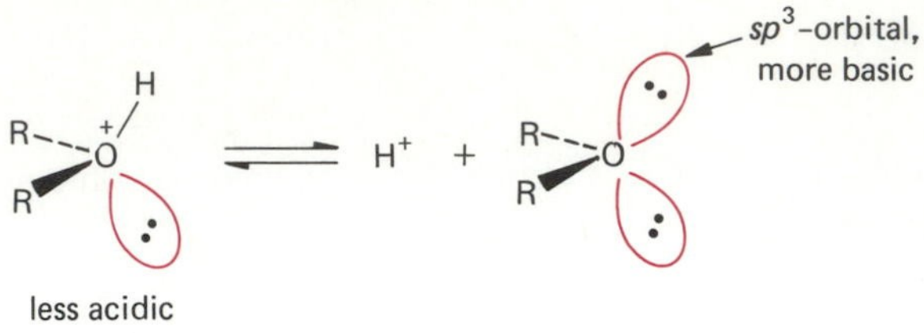

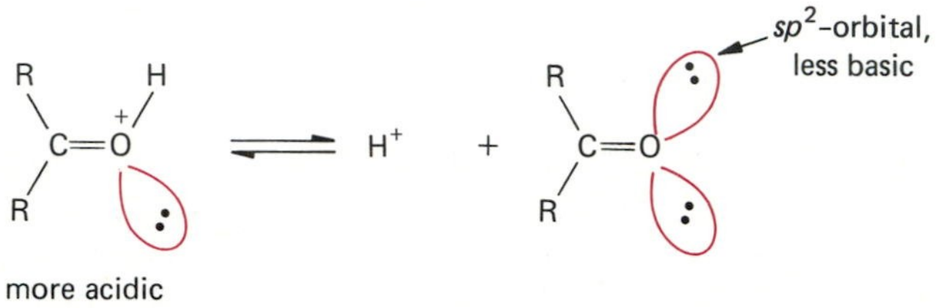

The structure of protonated carboxylic acids and esters is shown in Table 19.6 with the proton on the carbonyl oxygen rather than on the OR oxygen, despite the argument just presented that carbonyl oxygens are generally less basic than singly bonded oxygens. This result is a manifestation of conjugation. If the carbonyl group and hydroxy or alkoxy group are separated by one or more carbons (for example, in hydroxyacetone), we would anticipate the carbonyl group to be the less basic. That is, in an acidic medium the hydroxy or alkoxy group is protonated to a greater degree than is the carbonyl group.

TABLE 19.6 Acidities of Protonated Compounds

Compound	Conjugate Acid	pK_a of Conjugate Acid
CH_3CONH_2	$CH_3C(=\overset{+}{O}H)NH_2$	0.0
H_2O	H_3O^+	−1.7
CH_3OH	$CH_3\overset{+}{O}H_2$	−2.2
$(CH_3CH_2)_2O$	$(CH_3CH_2)_2\overset{+}{O}H$	−3.6
CH_3COOH	$CH_3C(=\overset{+}{O}H)OH$	−6
$CH_3COOC_2H_5$	$CH_3C(=\overset{+}{O}H)OC_2H_5$	−6.5
CH_3COCH_3	$CH_3C(=\overset{+}{O}H)CH_3$	−7.2
CH_3CHO	$CH_3C(=\overset{+}{O}H)H$	~−8
CH_3COCl	$CH_3C(=\overset{+}{O}H)Cl$	~−9
CH_3CN	$CH_3C\equiv\overset{+}{N}H$	−10.1

$$\underset{\text{hydroxyacetone}}{CH_3\overset{O}{\overset{\|}{C}}CH_2OH} + H^+ \rightleftharpoons \underset{\text{acetonyloxonium ion (more)}}{CH_3\overset{O}{\overset{\|}{C}}CH_2\overset{+}{O}H_2} \quad \text{or} \quad \underset{\text{hydroxyacetonium ion (less)}}{CH_3\overset{\overset{+}{O}H}{\overset{\|}{C}}CH_2OH}$$

> Note that both protonated isomers are more acidic than their monofunctional counterparts: acetonyloxonium ion is more acidic than propyloxonium ion, $CH_3CH_2CH_2OH_2^+$, and hydroxyacetonium ion is more acidic than acetonium ion. In both cases the substituent has an electron-attracting inductive effect that results in an increase in acidity, just as in the case of substituted acetic acids (Section 18.4.B).

When the OH or OR group is attached directly to the carbonyl group, some of the electron density on the singly bonded oxygen can be accommodated by the electron-attracting carbonyl group, as symbolized by the following resonance structures.

$$\left[R{-}\overset{O}{\overset{\|}{C}}{-}OR' \longleftrightarrow R{-}\overset{O^-}{\overset{|}{C}}{}^{+}{-}OR' \longleftrightarrow R{-}\overset{O^-}{\overset{|}{C}}{=}\overset{+}{O}R'\right]$$

The actual electronic structure of the carboxylic acid or ester group may be represented as

$$R{-}\overset{O^{\delta-}}{\overset{\vdots}{\overset{\delta+}{C}}}{\cdots}\overset{\delta+}{O}R'$$

The partial positive charge or oxonium character of the OR group makes this oxygen less basic than an ether oxygen. The partial negative charge on the carbonyl oxygen makes it more basic than a ketone oxygen. This argument does not mean that the alternative protonated compound cannot exist. It does say that this protonated compound, an acyloxonium ion, is much more acidic than a simple oxonium ion.

$$\underset{\text{acyloxonium ion}}{R{-}\overset{O}{\overset{\|}{C}}{-}\overset{+}{O}H_2} \quad \text{more acidic than} \quad \underset{\text{alkyloxonium ion}}{RCH_2{-}\overset{+}{O}H_2}$$

On the other hand, the carbonyl-protonated carboxylic grouping is stabilized by conjugation; the positive charge is distributed between the two hydroxy groups, much as the negative charge is distributed in acetate ion (Section 18.4.A).

$$\left[R{-}\overset{{}^+OH}{\overset{\|}{C}}{-}OH \longleftrightarrow R{-}\overset{OH}{\overset{|}{C}}{=}\underset{+}{O}H \longleftrightarrow R{-}\overset{OH}{\overset{|}{\underset{+}{C}}}{-}OH \longleftrightarrow R{-}\overset{O^-}{\overset{|}{C}}{}^{+}{-}NH_2\right]$$

These same considerations apply to an even greater extent in the case of amides. Ammonia is much more basic than water: $pK_a(NH_4^+) = 9.2$; $pK_a(H_3O^+) = -1.7$. The nitrogen in an amide is far less basic than that in ammonia because of the important contribution of the dipolar resonance structure.

$$\left[R{-}\overset{O}{\overset{\|}{C}}{-}NH_2 \longleftrightarrow R{-}\overset{O^-}{\overset{|}{C}}{}^{+}{-}NH_2 \longleftrightarrow R{-}\overset{O}{\overset{|}{C}}{=}\overset{+}{N}H_2\right] \equiv R{-}\overset{O^{\delta-}}{\overset{\vdots}{\overset{\delta+}{C}}}{\cdots}\overset{\delta+}{N}H_2$$

That is, the nitrogen in an amide already has some of the character of an ammonium ion. If this nitrogen becomes protonated, the conjugation stabilization of the amide is lost.

$$R-\overset{\overset{\displaystyle O}{\|}}{C}-\overset{+}{N}H_3$$

(no lone pair for conjugation with carbonyl group)

However, the O-protonated amide is greatly stabilized by resonance.

$$\left[R-\overset{\overset{\displaystyle ^{+}OH}{\|}}{C}-NH_2 \longleftrightarrow R-\underset{+}{\overset{\overset{\displaystyle OH}{|}}{C}}-NH_2 \longleftrightarrow R-\overset{\overset{\displaystyle OH}{|}}{C}=\underset{+}{N}H_2 \right]$$

In fact, as shown in Table 19.6, the O-protonated amide is almost 100 times less acidic than H_3O^+.

EXERCISE 19.4 In Exercise 19.1 we gave the lengths of a number of C—X bonds in compounds of the types R—XCH_3 and RCO—XCH_3. How do these bond lengths correlate with the carbonyl basicity of amides, esters, and acyl halides as given in Table 19.6?

19.6 Hydrolysis: Nucleophilic Addition-Elimination

The most characteristic reaction of the functional derivatives of carboxylic acids is **hydrolysis,** the reaction with water to give the carboxylic acid itself.

$$R-\overset{\overset{\displaystyle O}{\|}}{C}-X + H_2O \longrightarrow R-\overset{\overset{\displaystyle O}{\|}}{C}-OH + HX$$

X = halogen, OR, NR_2, OOCR

Esters react with water to yield the corresponding carboxylic acid and alcohol.

$$CH_3\overset{\overset{\displaystyle O}{\|}}{C}-OCH_3 + H_2O \rightleftharpoons CH_3\overset{\overset{\displaystyle O}{\|}}{C}-OH + CH_3OH$$

The reaction is generally slow, but is strongly catalyzed by acid. The acid-catalyzed hydrolysis of esters of primary and secondary alcohols is simply the reverse of the acid-catalyzed esterification reaction (Section 18.7.C). The reaction is an equilibrium process, but can be driven practically to completion by using a large excess of water (page 499). The probable mechanism involves the following steps.

$$(1)\ CH_3-\overset{\overset{\displaystyle O}{\|}}{C}-OCH_3 + H^+ \rightleftharpoons CH_3-\overset{\overset{\displaystyle ^{+}OH}{\|}}{C}-OCH_3$$

$$(2)\ CH_3-\overset{\overset{\displaystyle ^{+}OH}{\|}}{C}-OCH_3 + H_2O \rightleftharpoons CH_3-\underset{\underset{\displaystyle _{+}OH_2}{|}}{\overset{\overset{\displaystyle OH}{|}}{C}}-OCH_3$$

$$(3)\ CH_3-\underset{\underset{\displaystyle _{+}OH_2}{|}}{\overset{\overset{\displaystyle OH}{|}}{C}}-OCH_3 \rightleftharpoons CH_3-\underset{\underset{\displaystyle OH}{|}}{\overset{\overset{\displaystyle OH}{|}}{C}}-OCH_3 + H^+$$

$$(4)\ CH_3-\overset{OH}{\underset{OH}{C}}-OCH_3 + H^+ \rightleftharpoons CH_3-\overset{OH}{\underset{OH}{C}}-\overset{H}{\underset{+}{O}}CH_3$$

$$(5)\ CH_3-\overset{OH}{\underset{OH}{C}}-\overset{H}{\underset{+}{O}}CH_3 \rightleftharpoons CH_3-\overset{^+OH}{\overset{\|}{C}}-OH + CH_3OH$$

$$(6)\ CH_3-\overset{^+OH}{\overset{\|}{C}}-OH \rightleftharpoons CH_3-\overset{O}{\overset{\|}{C}}-OH + H^+$$

Let us examine the role of the acid catalyst in the preceding reaction. In neutral water the preponderant nucleophile present is water. Even though the carbonyl double bond is polarized, water is not a sufficiently strong nucleophile to add to it at a reasonable rate. Furthermore, addition of water to methyl acetate would produce an intermediate bearing both a positive and a negative charge. Since charge separation requires electrostatic energy, this type of addition is exceptionally slow.

$$CH_3-\overset{O}{\overset{\|}{C}}-OCH_3 + H_2O \rightleftharpoons CH_3-\overset{O^-}{\underset{^+OH_2}{C}}-OCH_3$$

In the presence of strong acids, the ester may be protonated (Section 19.5).

$$CH_3-\overset{O}{\overset{\|}{C}}-OCH_3 + H^+ \rightleftharpoons \left[CH_3-\overset{^+OH}{\overset{\|}{C}}-OCH_3 \longleftrightarrow CH_3-\overset{OH}{C}=\overset{+}{O}CH_3 \longleftrightarrow CH_3-\overset{OH}{\underset{+}{C}}-OCH_3\right]$$

resonance-stabilized cation

Since there is a very large excess of water molecules, and since the ester carbonyl is actually less basic than water (Table 19.6), only a small percentage of the ester is protonated at moderate acid concentration. However, the carbonyl carbon in the protonated species is much more electrophilic and reacts much faster with the weak nucleophile water than does the unprotonated ester. Furthermore, addition now involves no charge separation.

$$CH_3-\overset{OH^+}{\overset{\|}{C}}-OCH_3 + H_2\ddot{O}: \rightleftharpoons CH_3-\overset{OH}{\underset{^+OH_2}{C}}-OCH_3$$

In some cases acid-catalyzed ester hydrolysis involves cleavage of the *alkyl-oxygen* bond rather than the *acyl-oxygen* bond. Such is the case with *t*-butyl acetate (19-3).

$$CH_3\overset{O}{\overset{\|}{C}}-O-\overset{CH_3}{\underset{CH_3}{C}}CH_3 + H_2O \xrightleftharpoons{H^+} CH_3COOH + CH_3\overset{CH_3}{\underset{CH_3}{C}}-OH \qquad (19\text{-}3)$$

Although the products are the same in both cases, the different mechanisms may be demonstrated by the labeling experiments (19-4) and (19-5).

$$CH_3\overset{O}{\overset{\|}{C}}-{}^{18}OCH_3 + H_2O \longrightarrow CH_3\overset{O}{\overset{\|}{C}}-OH + CH_3{}^{18}OH \qquad (19\text{-}4)$$

$$CH_3\overset{O}{\overset{\|}{C}}{-}^{18}O{-}\underset{CH_3}{\overset{CH_3}{C}}CH_3 + H_2O \longrightarrow CH_3\overset{O}{\overset{\|}{C}}{-}^{18}OH + CH_3\underset{CH_3}{\overset{CH_3}{C}}{-}OH \qquad (19\text{-}5)$$

The probable mechanism for this reaction involves the following steps.

$$(1)\ CH_3\overset{O}{\overset{\|}{C}}{-}O{-}\underset{CH_3}{\overset{CH_3}{C}}CH_3 + H^+ \rightleftharpoons CH_3\overset{^+OH}{\overset{\|}{C}}{-}O{-}\underset{CH_3}{\overset{CH_3}{C}}CH_3$$

$$(2)\ CH_3\overset{^+OH}{\overset{\|}{C}}{-}O{-}\underset{CH_3}{\overset{CH_3}{C}}CH_3 \rightleftharpoons CH_3\overset{OH}{\overset{|}{C}}{=}O + CH_3\underset{CH_3}{\overset{CH_3}{C^+}}$$

$$(3)\ CH_3\underset{CH_3}{\overset{CH_3}{C^+}} + H_2O \rightleftharpoons CH_3\underset{CH_3}{\overset{CH_3}{C}}{-}\overset{+}{O}H_2$$

$$(4)\ CH_3\underset{CH_3}{\overset{CH_3}{C}}{-}\overset{+}{O}H_2 \rightleftharpoons CH_3\underset{CH_3}{\overset{CH_3}{C}}{-}OH + H^+$$

It is reasonable that *t*-butyl acetate should react by this mechanism whereas methyl acetate does not. The reaction is simply an acid-catalyzed S_N1 process in which the nucleophile water replaces the nucleophile acetic acid. The reactive intermediate is the *t*-butyl cation. Esters of other alcohols that give rise to relatively stable carbocations also undergo hydrolysis by this mechanism.

Ester hydrolysis is also strongly catalyzed by hydroxide ion. Since the carboxylic acid product neutralizes one equivalent of hydroxide, it is actually necessary to employ stoichiometric amount of base. That is, hydroxide ion is actually a reagent instead of just a catalyst. The products are the salt of the carboxylic acid and the corresponding alcohol.

$$CH_3(CH_2)_8CH{=}\overset{CH_3}{\overset{|}{C}}COOCH_3 + KOH \xrightarrow[\substack{H_2O\\ \Delta}]{C_2H_5OH}$$

$$CH_3(CH_2)_8CH{=}\overset{CH_3}{\overset{|}{C}}CO_2^-\ K^+ \xrightarrow{H_2SO_4} \underset{(68\text{–}83\%)}{CH_3(CH_2)_8CH{=}\overset{CH_3}{\overset{|}{C}}COOH}$$

To a solution of 20 g of methyl (*E*)-2-methyl-2-dodecenoate in 100 mL of 95% aqueous ethanol is added 8.8 g of potassium hydroxide. The solution is refluxed for 1.5 hr, concentrated to 40 mL, and acidified by the addition of 5 N sulfuric acid. The product is isolated by extraction with petroleum ether. After removal of the petroleum ether, the crude product is distilled to yield 18 g of pure acid, m.p. 29.5–32.5 °C.

The probable mechanism for base-catalyzed hydrolysis is illustrated below with methyl acetate.

$$\text{(1)}\ CH_3\overset{\overset{\displaystyle O}{\|}}{C}{-}OCH_3 + OH^- \rightleftharpoons CH_3\overset{\overset{\displaystyle O^-}{|}}{\underset{\underset{\displaystyle OH}{|}}{C}}{-}OCH_3$$

$$\text{(2)}\ CH_3\overset{\overset{\displaystyle O^-}{|}}{\underset{\underset{\displaystyle OH}{|}}{C}}{-}OCH_3 \rightleftharpoons CH_3\overset{\overset{\displaystyle O}{\|}}{C}{-}OH + CH_3O^-$$

$$\text{(3)}\ CH_3\overset{\overset{\displaystyle O}{\|}}{C}{-}OH + CH_3O^- \longrightarrow CH_3\overset{\overset{\displaystyle O}{\|}}{C}{-}O^- + CH_3OH$$

In basic aqueous solution two nucleophiles are present, H_2O and OH^-. As we have seen, H_2O is a poor nucleophile and therefore reacts slowly with the carbonyl carbon. On the other hand, OH^- is a much stronger nucleophile and adds more rapidly to the carbonyl carbon.

After addition has taken place, elimination of a nucleophile from the tetrahedral intermediate can occur. Elimination of hydroxide ion merely reverses the initial addition step. However, elimination of methoxide ion gives a new species—acetic acid. Since acetic acid is a weak acid and methoxide ion is a strong base, a rapid acid-base reaction then occurs, yielding acetate ion and methanol. Because of the great difference in acidity between acetic acid ($pK_a \cong 5$) and methanol ($pK_a \cong 16$), this last step is essentially irreversible (K for the last reaction $\cong 10^{11}$). Thus basic hydrolysis of esters differs from acid-catalyzed hydrolysis in that the equilibrium constant for the overall reaction is very large, and it is sufficient to use only one equivalent of water in order for the reaction to proceed to completion.

It is interesting to compare the carbon-oxygen bond-forming step in the acid-catalyzed and base-catalyzed mechanisms. In the former case the "weak" nucleophile H_2O adds to the "strongly" electrophilic bond $C{=}OH^+$.

$$CH_3\overset{\overset{\displaystyle \overset{+}{O}H}{\|}}{C}{-}OCH_3 + \ddot{O}H_2 \rightleftharpoons CH_3\overset{\overset{\displaystyle OH}{|}}{\underset{\underset{\displaystyle \underset{+}{O}H_2}{|}}{C}}{-}OCH_3$$

In the latter case the "strong" nucleophile OH^- adds to the "weakly" electrophilic bond $C{=}O$.

$$CH_3\overset{\overset{\displaystyle O}{\|}}{C}{-}OCH_3 + :\ddot{O}H^- \rightleftharpoons CH_3\overset{\overset{\displaystyle O^-}{|}}{\underset{\underset{\displaystyle OH}{|}}{C}}{-}OCH_3$$

Either case is better than the case where the "weak" nucleophile H_2O adds to the "weakly" electrophilic bond $C{=}O$.

The most rapid hydrolysis would involve *both* acid and base catalysis.

$$CH_3\overset{\overset{\displaystyle \overset{+}{O}H}{\|}}{C}{-}OCH_3 + OH^- \longrightarrow CH_3\overset{\overset{\displaystyle OH}{|}}{\underset{\underset{\displaystyle OH}{|}}{C}}{-}OCH_3$$

In aqueous solution this mechanism is not observed for a simple reason. In acidic solution, where the concentration of $C{=}OH^+$ species is appreciable, the concentration of OH^- is very small. In basic solution, where the concentration of OH^- is appreciable, the concentration of $C{=}OH^+$ is low.

Amides undergo hydrolysis to 1 mole of carboxylic acid and 1 mole of ammonia or amine. The reaction is catalyzed by acid or base. The hydrolysis of amides is more difficult than the hydrolysis of esters, and more vigorous conditions are normally required.

$$C_6H_5\text{—}\underset{}{CH}(CH_2CH_3)CONH_2 \xrightarrow[H_2O]{H_2SO_4} C_6H_5\text{—}CH(CH_2CH_3)COOH$$

(88–90%)

> A mixture of 600 g of 2-phenylbutanamide, 1 L of water, and 400 mL of conc. sulfuric acid is refluxed for 2 hr. After cooling the mixture and diluting with 1 L of water, the oily organic layer is separated and distilled to yield 530–554 g of 2-phenylbutanoic acid, m.p. 42 °C.

Both the acid- and base-catalyzed reactions are essentially irreversible. In the former case ammonium ion is produced; in the latter case a carboxylate ion is formed.

$$RC(=O)\text{—}NH_2 + H_2O \overset{H^+}{\rightleftharpoons} RC(=O)\text{—}OH + NH_3 \xrightarrow{H^+} NH_4^+$$

$$RC(=O)\text{—}NH_2 + H_2O \overset{OH^-}{\rightleftharpoons} RC(=O)\text{—}OH + NH_3 \xrightarrow{OH^-} RCO_2^-$$

Acyl halides and anhydrides undergo hydrolysis with ease. Acetyl chloride reacts with water to give acetic acid and hydrogen chloride, whereas acetic anhydride gives two equivalents of acetic acid.

$$CH_3C(=O)\text{—}Cl + H_2O \longrightarrow CH_3C(=O)\text{—}OH + HCl$$

$$CH_3C(=O)\text{—}O\text{—}C(=O)CH_3 + H_2O \longrightarrow 2\ CH_3C(=O)\text{—}OH$$

As with esters and amides, hydrolysis of acyl halides and anhydrides is subject to acid or base catalysis. However, both acyl halides and anhydrides react much more rapidly than esters, and uncatalyzed hydrolysis occurs readily, provided the reaction mixture is homogeneous. Since most acyl halides and anhydrides are only sparingly soluble in water, hydrolysis often appears to be slow. However, if a solvent is used that dissolves both water and the organic reactant, hydrolysis is rapid.

$$n\text{-}C_9H_{19}C(=O)\text{—}Cl + H_2O \xrightarrow[25°C]{\text{dioxane}} n\text{-}C_9H_{19}COOH + HCl$$

In hydrolysis of an ester, acyl halide, or anhydride the overall result is replacement of the nucleophile HO^- for one of the nucleophiles RO^-, X^-, or RCO_2^-. Thus hydrolyses of these compounds, as well as some of the reactions of carboxylic acids themselves (Section 18.7.D–F), are but further examples of **nucleophilic substitution** (Chapter 9). However, as we have seen in this section, the mechanism for these nucleophilic substitution reactions is different from the substitution mechanisms we have previously encountered (S_N1, S_N2). At this point it is instructive to examine the three distinct ways in which the bond-breaking and bond-making operations of a nucleophilic substitution process may be timed.

1. Bond breaking may occur first, followed in a subsequent step by bond making.

$$\text{R—Y} \rightleftharpoons \text{R}^+ + \text{Y}^-$$
$$\text{R}^+ + :\text{N}^- \longrightarrow \text{R—N}$$

This sequence of steps is involved in the S_N1 process for substitution in alkyl halides (Section 9.8). In the case of carboxylic acid derivatives, the intermediate carbocation is called an **acylium ion.**

$$\text{R—}\overset{\overset{\text{O}}{\|}}{\text{C}}\text{—Y} \rightleftharpoons :\text{Y}^- + \left[\text{R—}\overset{+}{\text{C}}\text{=}\ddot{\text{O}}: \longleftrightarrow \text{R—C}\equiv\overset{+}{\text{O}}:\right]$$

acylium ion

$$\text{R—}\overset{+}{\text{C}}\text{=}\ddot{\text{O}}: + :\text{N}^- \longrightarrow \text{R—}\overset{\overset{\text{O}}{\|}}{\text{C}}\text{—N}$$

Only a few reactions of carboxylic acids and their derivatives occur by this mechanism.

2. Bond breaking and bond making may occur more or less synchronously.

$$\text{R—Y} + :\text{N}^- \longrightarrow \left[\overset{\delta-}{\text{N}}\cdots\text{R}\cdots\overset{\delta-}{\text{Y}}\right]^{\ddagger} \longrightarrow \text{N—R} + :\text{Y}^-$$

This is the familiar S_N2 mechanism (Sections 9.1-9.5). The synchronous mechanism is rare in the chemistry of carboxylic acid derivatives. It has been suggested by a few workers that some reactions of acyl halides may occur by this path, but actual evidence for the mechanism is sparse.

3. Bond making may occur first, followed in a subsequent step by bond breaking. This mechanism is not possible in the case of simple alkyl halides, since it would require the intervention of a pentacoordinate carbon. However, the mechanism is the most common one in the chemistry of carboxylic acids and their derivatives.

$$\text{R—}\overset{\overset{\text{O}}{\|}}{\text{C}}\text{—Y} + :\text{N}^- \rightleftharpoons \text{R—}\underset{\underset{\text{N}}{|}}{\overset{\overset{\text{O}^-}{|}}{\text{C}}}\text{—Y}$$

$$\text{R—}\underset{\underset{\text{N}}{|}}{\overset{\overset{\text{O}^-}{|}}{\text{C}}}\text{—Y} \rightleftharpoons \text{R—}\overset{\overset{\text{O}}{\|}}{\text{C}}\text{—N} + :\text{Y}^-$$

In this case *addition* to the carbonyl group occurs first, giving an intermediate in which the former carbonyl carbon now has sp^3-hybridization. This intermediate then decomposes by ejection of a nucleophile, restoring the carbonyl group. Almost all nucleophilic substitution reactions of carboxylic acids and their derivatives occur by this pathway, the so-called **nucleophilic addition-elimination** mechanism.

EXERCISE 19.5 Write the steps involved in the acid- and base-promoted hydrolyses of acetyl azide, CH_3CON_3.

19.7 Other Nucleophilic Substitution Reactions

The nucleophilic substitution mechanism discussed in the previous section is general. Carboxylic acids and their derivatives react with nucleophiles other than water in the same manner. In this section we shall consider some of these other reactions.

A. Reaction with Alcohols

Acyl halides react with alcohols to yield esters and the corresponding hydrohalic acid.

$$\underset{\text{acetyl chloride}}{CH_3\overset{O}{\overset{\|}{C}}-Cl} + \underset{\text{isobutyl alcohol}}{CH_3\overset{CH_3}{\overset{|}{C}}HCH_2OH} \longrightarrow \underset{\text{isobutyl acetate}}{CH_3\overset{O}{\overset{\|}{C}}-OCH_2\overset{CH_3}{\overset{|}{C}}HCH_3} + HCl$$

Such reactions are usually carried out in the presence of a weak base that serves to neutralize the HX formed in the reaction.

$$CH_3\underset{CH_3}{\overset{CH_3}{C}}-OH + CH_3\overset{O}{\overset{\|}{C}}-Cl + \underset{\text{N,N-dimethylaniline}}{C_6H_5N(CH_3)_2} \xrightarrow{\text{ether}} \underset{(63\text{–}68\%)}{CH_3\underset{CH_3}{\overset{CH_3}{C}}-O-\overset{O}{\overset{\|}{C}}CH_3} + C_6H_5\overset{+}{N}H(CH_3)_2\ Cl^-$$

To a refluxing solution of 114 g of *t*-butyl alcohol and 202 g of N,N-dimethylaniline in 200 mL of dry ether is added dropwise 124 g of acetyl chloride. After addition of all the acyl chloride, the mixture is cooled in an ice bath, and the solid N,N-dimethylaniline hydrochloride is removed by filtration. The ether layer is extracted with aqueous sulfuric acid to remove excess amine and worked up to yield 110–119 g of *t*-butyl acetate, b.p. 95–98 °C.

Other bases commonly used for this purpose are triethylamine and pyridine (page 220).

Since acyl halides are readily available from the corresponding carboxylic acids (Section 18.7.D), the following sequence is often used for the preparation of esters.

$$RCOOH \xrightarrow[PCl_3]{SOCl_2 \text{ or}} R\overset{O}{\overset{\|}{C}}-Cl \xrightarrow{R'OH} R\overset{O}{\overset{\|}{C}}-OR'$$

Anhydrides also react readily with alcohols. The product is 1 mole of ester and 1 mole of the carboxylic acid corresponding to the anhydride used.

$$\underset{\text{cholesterol}}{\text{HO–(steroid)}} + \underset{\text{acetic anhydride}}{CH_3\overset{O}{\overset{\|}{C}}O\overset{O}{\overset{\|}{C}}CH_3} \longrightarrow \underset{\text{cholesteryl acetate}}{CH_3\overset{O}{\overset{\|}{C}}O\text{–(steroid)}} + CH_3COOH$$

A mixture of 5 g of cholesterol and 7.5 mL of acetic anhydride is boiled for 1 hr. The mixture is cooled and filtered to yield 5 g of cholesteryl acetate, m.p. 114–115 °C.

The mechanism of the reaction is the same as that for the reaction of an alcohol with an acyl halide; the leaving group in this case is the carboxylate anion rather than a halide ion. This reaction is an important method for the preparation of acetates, since acetic anhydride is an inexpensive reagent and the reaction is convenient to carry out.

Esters undergo reaction with alcohols to give a new ester and a new alcohol. The reaction is catalyzed by either acid or base and is called **transesterification.**

$$CH_3\overset{O}{\overset{\|}{C}}-OCH_2CH_3 + CH_3OH \xrightleftharpoons{H^+ \text{ or } OCH_3^-} CH_3\overset{O}{\overset{\|}{C}}-OCH_3 + C_2H_5OH$$

The mechanism for the transesterification process involves steps identical to those given for acid-catalyzed and base-catalyzed ester hydrolysis, with one significant exception. In base-catalyzed transesterification, step 3 of the mechanism on page 525 cannot occur because the free carboxylic acid is never formed. Thus, base-catalyzed transesterification is subject to the same equilibrium conditions that apply to the acid-catalyzed reaction. Some of the steps of transesterification and hydrolysis are shown below.

$$CH_3-\overset{O}{\overset{\|}{C}}-OCH_2CH_3 \xrightarrow{H^+} CH_3-\overset{^+OH}{\overset{\|}{C}}-OCH_2CH_3 \;(N)$$

$$\xrightarrow{N = H_2O} CH_3-\underset{^+OH_2}{\overset{OH}{C}}-OCH_2CH_3 \xrightarrow{-C_2H_5OH} CH_3-\overset{O}{\overset{\|}{C}}-OH$$

$$\xrightarrow{N = CH_3OH} CH_3-\underset{^+OCH_3(H)}{\overset{OH}{C}}-OCH_2CH_3 \xrightarrow{-C_2H_5OH} CH_3-\overset{O}{\overset{\|}{C}}-OCH_3$$

$$CH_3-\overset{O}{\overset{\|}{C}}-OCH_2CH_3 \longrightarrow CH_3-\overset{O}{\overset{\|}{C}}-OCH_2CH_3 \;(N^-)$$

$$\xrightarrow{N = ^-OCH_3} CH_3-\underset{OCH_3}{\overset{O^-}{C}}-OCH_2CH_3 \xrightarrow{-C_2H_5O^-} CH_3-\overset{O}{\overset{\|}{C}}-OCH_3$$

$$\xrightarrow{N = ^-OH} CH_3-\underset{OH}{\overset{O^-}{C}}-OCH_2CH_3 \xrightarrow{-C_2H_5O^-} CH_3-\overset{O}{\overset{\|}{C}}-O^-$$

EXERCISE 19.6 What products are produced when a solution of 1.0 mole of cholesteryl acetate and 0.05 mole of sodium methoxide in 1 L of methanol is refluxed for 30 min? Write a stepwise mechanism for the reaction.

B. Reaction with Amines and Ammonia

Acyl halides react with ammonia or amines that have at least one hydrogen bound to nitrogen to give amides. Since one equivalent of hydrogen chloride is formed in the reaction, two equivalents of ammonia or the amine must be used.

$$CH_3CH_2\overset{O}{\overset{\|}{C}}{-}Cl + 2\ HNR_2 \longrightarrow CH_3CH_2\overset{O}{\overset{\|}{C}}{-}NR_2 + R_2\overset{+}{N}H_2\ Cl^-$$

(R = H or alkyl)

The reaction of ammonia and amines with anhydrides follows a similar course; the products are 1 mole of amide and 1 mole of carboxylic acid. Since the liberated acid reacts to form a salt with the ammonia or the amine, it is necessary to employ an excess of that reactant.

$$2\ CH_3CH_2NH_2 + CH_3\overset{O}{\overset{\|}{C}}{-}O{-}\overset{O}{\overset{\|}{C}}CH_3 \longrightarrow CH_3CH_2NH{-}\overset{O}{\overset{\|}{C}}CH_3 + CH_3CH_2NH_3^+\ CH_3CO_2^-$$

As in the analogous reaction of amines with acyl halides, one may carry out the reaction in the presence of one equivalent of tertiary amine.

$$C_6H_{11}NH_2 + CH_3\overset{O}{\overset{\|}{C}}O\overset{O}{\overset{\|}{C}}CH_3 + C_5H_5N \longrightarrow C_6H_{11}NH{-}\overset{O}{\overset{\|}{C}}CH_3 + C_5H_5\overset{+}{N}{-}H\ \ CH_3CO_2^-$$

Esters also react with ammonia and amines to yield the corresponding amide and the alcohol of the ester. This synthetic path is useful in cases where the corresponding acyl halide or anhydride is unstable or not easily available. An interesting example of such a case is

$$CH_3\overset{OH}{\overset{|}{C}}HCOOC_2H_5 + NH_3 \xrightarrow[24\ hr]{25°C} CH_3\overset{OH}{\overset{|}{C}}H{-}\overset{O}{\overset{\|}{C}}NH_2 + C_2H_5OH$$

(70–74%)

In this case the acyl halide method for preparing the amide may not be used, since the molecule contains an OH group, which will react rapidly with an acyl halide.

EXERCISE 19.7 Treatment of 2-hydroxypropanoic acid (lactic acid) with thionyl chloride gives a product having the formula $C_6H_8O_4$. Propose a structure for this material.

C. Reaction of Acyl Halides and Anhydrides with Carboxylic Acids and Carboxylate Salts. Synthesis of Anhydrides

A mixture of an acid anhydride and a carboxylic acid undergoes equilibration when heated.

$$R\overset{O}{\overset{\|}{C}}O\overset{O}{\overset{\|}{C}}R + 2\ R'COOH \overset{\Delta}{\rightleftharpoons} R'\overset{O}{\overset{\|}{C}}O\overset{O}{\overset{\|}{C}}R' + 2\ RCOOH$$

The reaction is preparatively useful when the anhydride is acetic anhydride. In this case, acetic acid can be removed by distillation as it is formed because it is the most volatile component in the equilibrium mixture.

$$CH_3\overset{O}{\overset{\|}{C}}O\overset{O}{\overset{\|}{C}}CH_3 + 2\ C_6H_5\overset{O}{\overset{\|}{C}}OH \longrightarrow C_6H_5\overset{O}{\overset{\|}{C}}O\overset{O}{\overset{\|}{C}}C_6H_5 + 2\ CH_3\overset{O}{\overset{\|}{C}}OH$$

b.p. 142°C — benzoic acid — (72–74%) benzoic anhydride — b.p. 117°C

The only carboxylic acid derivatives that undergo a useful reaction with carboxylate salts are acyl halides. The product is an anhydride.

$$\underset{\text{heptanoyl chloride}}{C_6H_{13}\overset{O}{\overset{\|}{C}}Cl} + \underset{\text{sodium heptanoate}}{C_6H_{13}\overset{O}{\overset{\|}{C}}O^- \ Na^+} \xrightarrow{H_2O} \underset{\substack{(60\%)\\ \text{heptanoic anhydride}}}{\left(C_6H_{13}\overset{O}{\overset{\|}{C}}\right)_2O} + Na^+ \ Cl^-$$

D. Reaction with Organometallic Compounds

Acyl halides react with various organometallic reagents to give ketones. Since ketones are also reactive towards many of the reagents, it is best to use less reactive agents when the target compound is the ketone. Reaction with a Grignard reagent can give ketones if the reaction is carried out at low temperature and the Grignard reagent solution is added to the acyl halide to avoid further reaction of the ketone to form tertiary alcohol. Anhydrous ferric chloride is often added as a catalyst.

$$\text{1-methylcyclopentane-}\overset{O}{\overset{\|}{C}}Cl + CH_3MgI \xrightarrow[-15°C]{\text{ether}} \underset{(71\%)}{\text{1-methylcyclopentyl-}\overset{O}{\overset{\|}{C}}CH_3}$$

$$CH_3\overset{O}{\overset{\|}{C}}Cl + CH_3CH_2CH_2CH_2MgCl \xrightarrow[\substack{FeCl_3\\ -70°C}]{\text{ether}} \underset{(72\%)}{CH_3\overset{O}{\overset{\|}{C}}CH_2CH_2CH_2CH_3}$$

Ferric chloride may serve as a Lewis acid, complexing with the oxygen of the acyl chloride. Many organometallic reactions are now thought to proceed by **single electron transfer (SET).** The carbanion of the Grignard transfers an electron to the acyl chloride–ferric chloride complex to yield a stabilized radical anion and a highly reactive alkyl radical. The radical then combines with the radical anion to form the "tetrahedral intermediate" familiar from our discussion of hydrolysis. The intermediate decomposes to yield the ketone, magnesium halide and the catalyst ferric chloride.

$$R\overset{O}{\overset{\|}{C}}Cl \xrightarrow[\substack{FeCl_3\\ THF,\ -78\ °C}]{R'MgX} R\overset{\overset{\delta-}{FeCl_3}\cdots\overset{\delta+}{O}}{\overset{\|}{C}}Cl \quad R'^{-}MgX^+ \xrightarrow{SET} R\overset{FeCl_3\cdots{}^-O}{\overset{|}{\dot{C}}}Cl \quad R'\cdot \ MgX^+ \longrightarrow$$

$$R\overset{FeCl_3\cdots{}^-O}{\overset{|}{C}}(R')Cl \quad MgX^+ \longrightarrow R\overset{\overset{\delta-}{FeCl_3}\cdots\overset{\delta+}{O}}{\overset{\|}{C}}R' + ClMgX \longrightarrow R\overset{O}{\overset{\|}{C}}R'$$

(See Sections 8.6 and 8.8.) Rapid reaction at low temperatures is consistent with an SET mechanism.

If excess Grignard reagent is used the product ketone reacts further, giving a tertiary alcohol.

$$R-\overset{O}{\overset{\|}{C}}Cl \xrightarrow{R'MgX} R-\overset{O}{\overset{\|}{C}}-R' \xrightarrow{R'MgX} R-\underset{R'}{\underset{|}{\overset{O^- \, ^+MgX}{\overset{|}{C}}}}-R' \xrightarrow{H_2O} R-\underset{R'}{\underset{|}{\overset{OH}{\overset{|}{C}}}}-R'$$

Some of the less reactive organometallic compounds still react rapidly with acyl halides, but react with ketones only sluggishly or not at all. Such is the case with lithium organocuprates, which are obtained by treating an organolithium compound with cuprous iodide.

$$2\,CH_3Li + CuI \xrightarrow{\text{ether}} LiCu(CH_3)_2 + LiI$$

The cuprate reacts rapidly with acyl halides and aldehydes, slowly with ketones, and not at all with esters, amides, and nitriles.

$$CH_3\overset{CH_3}{\overset{|}{C}}HCH_2\overset{O}{\overset{\|}{C}}Cl + \left(CH_3\overset{CH_3}{\overset{|}{C}}=CH\right)_2CuLi \xrightarrow[-5°C]{\text{ether}} CH_3\overset{CH_3}{\overset{|}{C}}HCH_2\overset{O}{\overset{\|}{C}}CH=C(CH_3)_2$$

3-methylbutanoyl chloride

(70%)
2,6-dimethylhept-2-en-4-one

Since esters are generally less reactive than ketones, the preparation of ketones by nucleophilic substitution on an ester is usually unsatisfactory; in most cases the only isolable product is the tertiary alcohol. In fact, the reaction of esters with two equivalents of a Grignard reagent or an alkyllithium is probably the best general method for the synthesis of tertiary alcohols in which two of the alkyl groups are equivalent.

$$2\,C_6H_5MgBr + CH_3\overset{O}{\overset{\|}{C}}OC_2H_5 \longrightarrow \xrightarrow{H_2O} C_6H_5-\underset{CH_3}{\underset{|}{\overset{OH}{\overset{|}{C}}}}-C_6H_5$$

1,1-diphenylethanol

$$C_6H_{11}COOCH_3 + 2\,CH_3Li \longrightarrow \xrightarrow{H_3O^+} C_6H_{11}C(CH_3)_2OH$$

methyl cyclohexanecarboxylate

2-cyclohexyl-2-propanol

In this reaction, the initially formed ketone reacts with the second equivalent of Grignard reagent to give the tertiary alcohol. If an ester of formic acid is used, the product is a secondary alcohol.

$$2\,CH_3CH_2CH_2CH_2MgBr + H\overset{O}{\overset{\|}{C}}OC_2H_5 \xrightarrow{\text{ether}} \xrightarrow{H_3O^+} CH_3CH_2CH_2CH_2\overset{OH}{\overset{|}{C}}HCH_2CH_2CH_2CH_3$$

(85%)
5-nonanol

Carbonate esters yield tertiary alcohols in which all three of the carbinol alkyl groups come from the organometallic reagent.

$$3\,C_2H_5MgBr + CH_3O\overset{O}{\overset{\|}{C}}OCH_3 \longrightarrow \xrightarrow{H_3O^+} C_2H_5-\underset{C_2H_5}{\overset{OH}{C}}-C_2H_5$$

dimethyl carbonate

(85%)
3-ethyl-3-pentanol

Carbonic acid diesters, dialkyl carbonates, are generally prepared by the reaction of alcohols with phosgene, $COCl_2$.

$$COCl_2 + 2\,CH_3CH_2OH \longrightarrow CH_3CH_2O\overset{O}{\overset{\|}{C}}OCH_2CH_3 + 2\,HCl$$

Phosgene is a highly toxic colorless gas, b.p. 7.6 °C, having a distinctive odor. It was used in World War I as a war gas. This compound is the diacid chloride of carbonic acid and reacts accordingly. It is hydrolyzed by water to give carbon dioxide and HCl.

$$COCl_2 + 2\,H_2O \longrightarrow 2\,HCl + [CO(OH)_2] \longrightarrow CO_2 + H_2O$$

Dimethyl carbonate is a commercially available, colorless liquid, b.p. 90 °C.

EXERCISE 19.8 Show how 2-chlorobutane can be converted into each of the following compounds.

(a) 3-methyl-2-pentanone (b) 3,4,5-trimethyl-4-heptanol
(c) 3,5-dimethyl-4-heptanol

Highlight 19.2

The chemical properties of carboxylic acid derivatives, RCOX, are dominated by the reactivity of the carbonyl group in addition-elimination reactions. The charge distributions differ from one X to another. Write out all of the important resonance forms for each derivative in each step.

$$\left[CH_3-C(=\ddot{O}:)-X \longleftrightarrow CH_3-C(-\ddot{O}:^-)=X^+ \longleftrightarrow CH_3-\overset{+}{C}(-\ddot{O}:^-)-X\right]$$

$$CH_3-C(=\ddot{O}:)-X + Nu \longrightarrow CH_3-\underset{X}{\overset{:\ddot{O}:^-}{C}}-Nu^+ \xrightarrow{\pm H^+} CH_3-\underset{X}{\overset{:\ddot{O}H}{C}}-OH,\ NH_2,\ NHR$$

$X = Cl,\ ORH^+,\ OCOR,\ OR,\ NH_2$

$$CH_3-\underset{X}{\overset{:\ddot{O}H}{C}}-OH,\ NH_2,\ NHR \xrightarrow{-X^-} CH_3-\overset{:OH^+}{\overset{\|}{C}}-OH,\ NH_2,\ NHR$$

$$CH_3-\overset{:O:}{\overset{\|}{C}}-OH,\ NH_2,\ NHR$$

19.8 Reduction

Acyl halides may be reduced to aldehydes or to primary alcohols.

$$R{-}\overset{O}{\overset{\|}{C}}Cl \xrightarrow{[H]} R{-}\overset{O}{\overset{\|}{C}}H \xrightarrow{[H]} RCH_2OH$$

The selective reduction of an acyl halide is one of the most useful ways of preparing aldehydes. Such selective reduction is possible because acyl halides are generally more reactive than the product aldehydes. One procedure for accomplishing the selective reduction is catalytic hydrogenation; the method is called a **Rosenmund reduction.** The acyl halide is hydrogenated in the presence of a catalyst such as palladium deposited on barium sulfate. As in the reduction of alkynes to alkenes, a "regulator" or "catalyst poison" (page 309) is frequently added in order to moderate the effectiveness of the catalyst and thereby inhibit subsequent reduction of the product aldehyde.

$$CH_3O\overset{O}{\overset{\|}{C}}CH_2CH_2\overset{O}{\overset{\|}{C}}Cl + H_2 \xrightarrow[\substack{\text{quinoline}\\ \text{sulfur}\\ \text{xylene, 110°C}}]{Pd/BaSO_4} CH_3O\overset{O}{\overset{\|}{C}}CH_2CH_2\overset{O}{\overset{\|}{C}}H \quad (65\%)$$

Another reagent that has found use for the selective reduction of an acyl halide to an aldehyde is lithium tri-*t*-butoxyaluminohydride, $LiAl(t\text{-}C_4H_9O)_3H$.

$$\text{(4-methyl-3-methoxybenzoyl chloride)} + Li(t\text{-}C_4H_9O)_3Al\,H \xrightarrow{\text{ether}} \xrightarrow{H_2O} \text{(4-methyl-3-methoxybenzaldehyde, CHO; CH}_3\text{; OCH}_3\text{)} \quad (60\%)$$

In general, acyl halides are reduced more rapidly than aldehydes. However, lithium aluminum hydride is so extremely reactive that a selective reduction is difficult to accomplish. The tri-*t*-butoxy derivative, prepared by treating the hydride with three equivalents of *t*-butyl alcohol in ether, is less reactive and therefore more selective.

$$LiAlH_4 + 3\ t\text{-}C_4H_9OH \longrightarrow Li(t\text{-}C_4H_9O)_3AlH + 3\ H_2$$

If excess reducing agent is used, the product aldehyde is reduced further to a primary alcohol.

$$(1)\quad R{-}\overset{O}{\overset{\|}{C}}{-}Cl + Li(t\text{-}C_4H_9O)_3Al\,H \xrightarrow{\text{faster}} R{-}\underset{H}{\overset{O^-}{C}}{-}Cl + Li^+ + (t\text{-}C_4H_9O)_3Al$$

$$(2)\quad R{-}\underset{H}{\overset{O^-}{C}}{-}Cl \rightleftharpoons R{-}\overset{O}{\overset{\|}{C}}{-}H + Cl^-$$

$$(3)\quad R{-}\overset{O}{\overset{\|}{C}}{-}H + Li(t\text{-}C_4H_9O)_3AlH \xrightarrow{\text{slower}} RCH_2O^- + Li^+ + (t\text{-}C_4H_9O)_3Al$$

With the more reactive lithium aluminum hydride, it is difficult to achieve selectivity.

Since esters are generally less reactive than aldehydes, they cannot be selectively reduced to the aldehyde stage. However, the reduction of an ester to a primary alcohol is an important preparative method. The most generally used reducing agents are lithium aluminum hydride and lithium borohydride.

$$C_6H_5COC_2H_5 + LiAlH_4 \longrightarrow \xrightarrow{H_2O} C_6H_5CH_2OH + C_2H_5OH$$

(90%)

$$CH_3(CH_2)_{14}COOCH_2CH_2CH_2CH_3 + LiBH_4 \xrightarrow[\Delta]{THF} \xrightarrow{H_2O} CH_3(CH_2)_{14}CH_2OH + CH_3CH_2CH_2CH_2OH$$

(95%)

Lithium borohydride $LiBH_4$ is a hygroscopic white solid, m.p. 284 °C. It is prepared by the reaction of sodium borohydride (page 279) with lithium chloride in ethanol.

$$NaBH_4 + LiCl \longrightarrow LiBH_4 + NaCl$$

It is a more reactive reducing agent than sodium borohydride, but is less reactive than lithium aluminum hydride. It is much more soluble in ether (4 g per 100 mL) than is sodium borohydride.

Esters are also reduced by sodium in ethanol (the **Bouveault-Blanc reaction).** Before the discovery of lithium aluminum hydride, this was the most common laboratory method for reducing esters, and it is still an important method for large-scale preparations where reagent cost is a concern.

$$n\text{-}C_{11}H_{23}COOC_2H_5 + Na \xrightarrow[\Delta]{C_2H_5OH} n\text{-}C_{11}H_{23}CH_2OH + C_2H_5OH$$

ethyl dodecanoate — (65–75%) 1-dodecanol

The reaction mechanism is not known in detail, but undoubtedly involves electron transfer from sodium to the carbonyl group as a first step. The reaction is *not* a reduction by hydrogen liberated from the reaction of sodium with ethanol.

Reduction of amides having at least one hydrogen on nitrogen with lithium aluminum hydride in ether or THF provides the corresponding primary or secondary amines.

$$C_6H_5OCH_2CNH_2 + LiAlH_4 \xrightarrow[\Delta]{ether} \xrightarrow{H_2O} C_6H_5OCH_2CH_2NH_2$$

(80%)

$$CH_3(CH_2)_{10}CNHCH_3 + LiAlH_4 \xrightarrow[\Delta]{ether} \xrightarrow{H_2O} CH_3(CH_2)_{10}CH_2NHCH_3$$

(81–95%)

A solution of 38 g of lithium aluminum hydride in 1800 mL of dry ether is placed in a 5-L three-necked flask equipped with a condenser and a mechanical stirrer. The solution is gently refluxed while 160 g of N-methyldodecanamide is slowly added over a period of 3 hr. The mixture is refluxed an additional 2 hr, then stirred overnight. The reaction mixture is worked up by the addition of 82 mL of water. After filtration to remove the solid aluminum and lithium salts, the ether is evaporated, and the residue is distilled to yield 121–142 g of N-methyldodecylamine.

The first step in this reduction is probably reaction of the strongly basic $LiAlH_4$ with the weakly acidic NH bond, giving the lithium salt of the amide.

$$R{-}\overset{O}{\overset{\|}{C}}{-}NH_2 + LiAlH_4 \longrightarrow R{-}\overset{O}{\overset{\|}{C}}{-}NH^- \, Li^+ + AlH_3 + H_2$$

Aluminum hydride may then add to the carbonyl group.

$$R{-}\overset{O}{\overset{\|}{C}}{-}NH^- \, Li^+ + AlH_3 \longrightarrow R{-}\overset{O-AlH_2}{\overset{|}{\underset{H}{\underset{|}{C}}}}{-}NH^- \, Li^+$$

This tetrahedral intermediate may decompose by elimination of H_2AlO^-, which is a reasonably good leaving group (recall that aluminum is an amphoteric metal; H_3AlO_3 is a protonic acid).

$$R{-}\overset{OAlH_2}{\overset{|}{\underset{H}{\underset{|}{C}}}}{-}NH^- \, Li^+ \longrightarrow R{-}\underset{H}{\underset{|}{C}}{=}NH + H_2AlO^- \, Li^+$$

The resulting imine is now reduced by another hydride (from AlH_4^- or H_2AlO^-).

$$R{-}\overset{NH}{\overset{\|}{C}}H + AlH_4^- \text{ (or } H_2AlO^-) \longrightarrow R{-}\overset{NH-Al\langle}{\overset{|}{\underset{H}{\underset{|}{C}}}}{-}H$$

Upon work-up with water, the N—Al bond is hydrolyzed to liberate the amine.

$$RCH_2{-}NH{-}Al\langle + H_2O \longrightarrow RCH_2NH_2 + HO{-}Al\langle$$

N,N-Dialkylamides may also be reduced to amines by lithium aluminum hydride.

$$\underset{\text{N,N-diethylacetamide}}{CH_3\overset{O}{\overset{\|}{C}}N(CH_2CH_3)_2} + LiAlH_4 \xrightarrow[\text{reflux}]{\text{ether}} \xrightarrow{H_2O} \underset{\substack{(50\%) \\ \text{triethylamine}}}{CH_3CH_2N(CH_2CH_3)_2}$$

With disubstituted amides the reduction can generally be controlled so that the aldehyde may be obtained. This occurs when the initial tetrahedral intermediate is sufficiently stable so that it survives until all of the hydride has been consumed. If one wishes to prepare an aldehyde in this manner, it is necessary to keep the amide in excess by slowly adding the reducing agent to it. Several modified hydrides have been used for this purpose. One reagent that is particularly useful with simple dimethylamides is lithium triethoxyaluminohydride, prepared *in situ* by the reaction of 3 moles of ethanol with 1 molar equivalent of lithium aluminum hydride.

$$LiAlH_4 + 3\,C_2H_5OH \xrightarrow{\text{ether}} LiAlH(OC_2H_5)_3 + 3\,H_2$$

$$(CH_3)_3C\overset{O}{\overset{\|}{C}}N(CH_3)_2 + LiAlH(OC_2H_5)_3 \xrightarrow{\text{ether}} \xrightarrow{H_2O} \underset{\substack{(88\%) \\ \text{2,2-dimethylpropanal} \\ \text{pivalaldehyde}}}{(CH_3)_3C\overset{O}{\overset{\|}{C}}H}$$

EXERCISE 19.9 Show how each of the following transformations can be accomplished.

(a) $CH_3CH_2CH_2COCl \rightarrow CH_3CH_2CH_2CHO$
(b) $CH_3CH_2CH_2COOCH_3 \rightarrow CH_3CH_2CH_2CH_2OH$
(c) $CH_3CH_2CH_2CONH_2 \rightarrow CH_3CH_2CH_2CH_2NH_2$
(d) $CH_3CH_2CH_2CON(CH_3)_2 \rightarrow CH_3CH_2CH_2CHO$

19.9 Acidity of the α-Protons

Like the α-protons of aldehydes and ketones (Section 15.1), protons adjacent to the carbonyl group in carboxylic acid derivatives are weakly acidic. Table 19.7 lists the pK_a values for some representative compounds. Recall that the main reason for the acidity of aldehydes and ketones relative to the alkanes is the fact that the resulting anion is resonance stabilized. In fact, the resonance contributor with the negative charge on oxygen (the enolate ion) is the more important structure because of the greater electronegativity of oxygen. This stabilization of the anion greatly reduces the $\Delta H°$ for the ionization process and is responsible for the fact that acetone is approximately 30 powers of ten more acidic than ethane.

$$CH_3CH_3 \rightleftharpoons H^+ + CH_3CH_2^-$$

charge localized on carbon

$$CH_3\overset{\overset{O}{\|}}{C}CH_3 \rightleftharpoons H^+ + \left[CH_3\overset{\overset{O}{\|}}{C}CH_2^- \longleftrightarrow CH_3\overset{\overset{O^-}{|}}{C}{=}CH_2\right]$$

charge delocalized—mainly on oxygen

The anions obtained upon deprotonation of carboxylic acid derivatives are also stabilized by delocalization of the negative charge onto the carbonyl oxygen. Consequently these compounds also act as weak acids; representative pK_as are summarized in Table 19.7.

TABLE 19.7 Acidity of Carboxylic Acid Derivatives

Compound	pK_a of **H**
$\mathbf{CH_3}\overset{\overset{O}{\|}}{C}Cl$	~16
$\mathbf{CH_3}\overset{\overset{O}{\|}}{C}H$	17
$\mathbf{CH_3}\overset{\overset{O}{\|}}{C}CH_3$	19
$\mathbf{CH_3}\overset{\overset{O}{\|}}{C}OCH_3$	25
$\mathbf{CH_3}CN$	25
$\mathbf{CH_3}\overset{\overset{O}{\|}}{C}N(CH_3)_2$	~30
$\mathbf{CH_3}CH_3$	~50

The α-proton acidity manifests itself in the chemistry of carboxylic acid derivatives in several ways. For example, if ethyl acetate is dissolved in deuterioethanol that contains a catalytic amount of sodium ethoxide, exchange of the α-protons by deuterium occurs.

$$CH_3\overset{O}{\overset{\|}{C}}OC_2H_5 + C_2H_5OD \xrightleftharpoons{C_2H_5O^-} CD_3\overset{O}{\overset{\|}{C}}OC_2H_5 + C_2H_5OH$$

Ethyl acetate is a much weaker acid than ethanol (pK_a 15.9), so that the foregoing equilibrium is established relatively slowly. For example, a solution of ethyl acetate in deuterioethanol containing 0.1 M sodium ethoxide is only 50% exchanged after 2 weeks at 25 °C. For this reason, if one wishes to exchange the α-protons of an ester, it is necessary to reflux the solution for several hours.

If an ester is treated with a sufficiently strong base, it can be completely converted to the corresponding anion. For example, t-butyl acetate reacts with lithium isopropylcyclohexylamide in THF at −78 °C to give t-butyl lithioacetate, which can be isolated as a white crystalline solid.

$$CH_3\overset{O}{\overset{\|}{C}}OC(CH_3)_3 + \text{lithium isopropylcyclohexylamide} \xrightarrow[-78°C]{THF} CH_2{=}C(O^-\,Li^+)OC(CH_3)_3 + \text{isopropylcyclohexylamine}$$

p$K_a \cong 25$ | lithium isopropylcyclohexylamide | t-butyl lithioacetate | p$K_a \cong 40$

Such ester anions are strong bases and are also good nucleophiles. They undergo reactions similar to those of the corresponding enolate ions derived from ketones. Examples are S_N2 displacements with primary alkyl halides and additions to aldehyde and ketone carbonyl groups.

methyl hexanoate $\xrightarrow[\text{THF, } -78\,°C]{LDA}$ lithium enolate (O⁻ Li⁺, OMe) + CH_3—I $\xrightarrow{-Li^+\,I^-}$

(83%)
methyl 2-methylhexanoate

CH_3 C(=O) $OC(CH_3)_3$ (t-butyl acetate) $\xrightarrow[\text{THF, } -78\,°C]{LDA}$ $CH_2{=}C(O^-\,Li^+)OC(CH_3)_3$ + cyclopentanone ⟶

$Li^+\ ^-O$–cyclopentyl–$CH_2C(=O)OC(CH_3)_3$ $\xrightarrow[-Li^+\,X^-]{H_2O(HX)}$ HO–cyclopentyl–$CH_2C(=O)OC(CH_3)_3$

(93%)
t-butyl
1-hydroxycyclopentyl)acetate

A related reaction results when an aldehyde or ketone is treated with an α-halo ester and zinc in an inert solvent.

$$\text{cyclohexanone} + \text{Br}CH_2COOC_2H_5 + \text{Zn} \xrightarrow{\text{toluene}} \xrightarrow{H_3O^+} \text{1-hydroxycyclohexyl}-CH_2COOC_2H_5$$

(70%)
ethyl (1-hydroxycyclohexyl)acetate

This reaction, which is known as the **Reformatsky reaction,** involves the formation of an intermediate organozinc compound. The intermediate has been isolated and examined spectroscopically. It crystallizes as a dimer, with each zinc partially bonded to both an oxygen and a carbon.

$$\begin{array}{c} \text{Br} \\ | \\ O\rightarrow Zn - CH_2 \\ CH_3OC \qquad\qquad COCH_3 \\ CH_2 - Zn \leftarrow O \\ | \\ \text{Br} \end{array}$$

The Reformatsky reaction is a convenient alternative to the reaction of an aldehyde or ketone with the preformed lithium enolate of an ester.

Ester anions also undergo another reaction that is an important synthetic procedure, the **Claisen condensation.** In this reaction the ester anion condenses with an nonionized ester molecule to give a β-keto ester.

$$2\,CH_3\overset{O}{\overset{\|}{C}}OC_2H_5 \xrightarrow[C_2H_5OH]{NaOC_2H_5} \xrightarrow{H_3O^+} CH_3\overset{O}{\overset{\|}{C}}CH_2\overset{O}{\overset{\|}{C}}OC_2H_5$$

(75%)

> A mixture of 1.2 moles of ethyl acetate and 0.2 mole of alcohol-free sodium ethoxide is heated at 78 °C for 8 hr. The mixture is then cooled to 10 °C, and 36 g of 33% aqueous acetic acid is slowly added. The aqueous layer is washed with ether, and the combined organic layers are dried and distilled to give ethyl acetoacetate, b.p. 78–80 °C (16 torr), in 75–76% yield.

The product in this example, ethyl 3-oxobutanoate, is known by the trivial name ethyl acetoacetate, or simply acetoacetic ester. For this reason the self-condensation of esters is sometimes called the **acetoacetic ester condensation,** even when other esters are involved.

The reaction is mechanistically similar to the aldol addition reaction (Section 15.2) in that the conjugate base of the ester is a reactive intermediate.

$$(1)\quad CH_3\overset{O}{\overset{\|}{C}}OC_2H_5 + C_2H_5O^- \rightleftharpoons {}^-CH_2\overset{O}{\overset{\|}{C}}OC_2H_5 + C_2H_5OH$$

$$(2)\quad {}^-CH_2\overset{O}{\overset{\|}{C}}OC_2H_5 + CH_3\overset{O}{\overset{\|}{C}}OC_2H_5 \rightleftharpoons CH_3\underset{OC_2H_5}{\underset{|}{\overset{O^-}{\overset{|}{C}}}}CH_2\overset{O}{\overset{\|}{C}}OC_2H_5$$

$$(3)\quad CH_3\underset{OC_2H_5}{\underset{|}{\overset{O^-}{\overset{|}{C}}}}CH_2\overset{O}{\overset{\|}{C}}OC_2H_5 \rightleftharpoons CH_3\overset{O}{\overset{\|}{C}}CH_2\overset{O}{\overset{\|}{C}}OC_2H_5 + C_2H_5O^-$$

$$\text{(4) } CH_3\overset{O}{\overset{\|}{C}}CH_2\overset{O}{\overset{\|}{C}}OC_2H_5 + C_2H_5O^- \rightleftharpoons CH_3\overset{O^-}{\overset{|}{C}}{=}CH\overset{O}{\overset{\|}{C}}OC_2H_5 + C_2H_5OH$$

As we shall see in Section 27.7.B, 1,3-dicarbonyl compounds are fairly strong carbon acids; the pK_a of acetoacetic ester itself is about 11. Thus the equilibrium for the last step in this mechanism lies far to the right. Since the pK_a of ethanol is about 16, K for step 4 is about 10^5. This final, essentially irreversible, step provides the driving force for the Claisen condensation. A dramatic illustration is provided by attempted Claisen condensation with ethyl 2-methylpropanoate.

$$2\ (CH_3)_2CH\overset{O}{\overset{\|}{C}}OC_2H_5 \xrightarrow[C_2H_5OH]{NaOC_2H_5} \!\!/\!\!/\ (CH_3)_2CH\overset{O}{\overset{\|}{C}}-\underset{CH_3}{\overset{CH_3}{C}}-\overset{O}{\overset{\|}{C}}OC_2H_5$$

In this case there are no protons in the normally acidic position between the two carbonyl groups; hence, the final step in the mechanism cannot occur. The overall equilibrium constant for steps 1 through 3 in the mechanism is apparently too small for condensation to be observed in the absence of step 4. The most acidic proton available in the hypothetical product is the proton at C-4, which is a normal proton α to a ketone carbonyl; its pK_a is therefore about 20. If a much stronger base is used to catalyze the reaction, this proton can be removed and reaction can now be observed.

$$(CH_3)_2CHCOOC_2H_5 + (C_6H_5)_3C^-\ Na^+ \longrightarrow \xrightarrow[\text{(work-up)}]{H_3O^+} (CH_3)_2CH\overset{O}{\overset{\|}{C}}-\underset{CH_3}{\overset{CH_3}{C}}-COOC_2H_5$$

(45–60%)

The base in the foregoing example, sodium triphenylmethide, is much more basic than ethoxide ion. It is the conjugate base of the weak carbon acid triphenylmethane (pK_a = 31.5), and it will be discussed in detail in Section 21.5.D. Since it is such a strong base, K for the last step in the Claisen condensation is large (10^{11}), even though a normal ketone is being deprotonated.

$$\underset{pK_a \cong 20}{(CH_3)_2CH\overset{O}{\overset{\|}{C}}-\underset{CH_3}{\overset{CH_3}{C}}-COOC_2H_5} + (C_6H_5)_3C^- \overset{K}{\rightleftharpoons} (CH_3)_2C{=}\overset{O^-}{\overset{|}{C}}-\underset{CH_3}{\overset{CH_3}{C}}-COOC_2H_5 + \underset{pK_a = 31.5}{(C_6H_5)_3CH}$$

Mixed Claisen condensations between two esters are successful when one of the esters has no α-hydrogens, as shown by the following examples.

$$CH_3\overset{O}{\overset{\|}{C}}OC_2H_5 + H\overset{O}{\overset{\|}{C}}OC_2H_5 \xrightarrow[C_2H_5OH]{NaOC_2H_5} \xrightarrow{H_3O^+} H\overset{O}{\overset{\|}{C}}CH_2\overset{O}{\overset{\|}{C}}OC_2H_5$$

(79%)
ethyl formylacetate

$$C_6H_5COC_2H_5(\text{C=O}) + CH_3COC_2H_5(\text{C=O}) \xrightarrow[C_2H_5OH]{NaOC_2H_5} \xrightarrow{H^+} C_6H_5C(=O)CH_2C(=O)OC_2H_5$$

(55–70%)

The overall result of a mixed Claisen condensation is given by the general equation

$$RCOOC_2H_5 + H{-}\underset{R''}{\overset{R'}{C}}{-}COOC_2H_5 \longrightarrow R\overset{O}{\overset{\|}{C}}{-}\underset{R''}{\overset{R'}{C}}{-}COOC_2H_5$$

Best results are obtained when R has no α-hydrogens and either R′ or R″ is hydrogen.

A Claisen-like condensation is used extensively in biological systems to build up alkanoic acid chains. Biochemical evolution has created a special source for the nucleophilic component of the reaction. Instead of a normal ester, the nucleophile source is a derivative of the thioester, acetyl coA. Acetyl coA is converted into a carboxylic acid derivative, malonyl coA. As in β-keto esters, the CH_2 between two electron-attracting groups is acidic, and readily yields the anion at the active site of the condensing enzyme. The carbanion is the nucleophile that attacks the carbonyl group of a second thioester. The larger and less electronegative sulfur atom in the thioester is less effective in stabilizing the positive charge on the carbonyl carbon than the oxygen of the normal ester. The carbonyl group is then more reactive and is more readily attacked by a nucleophile. The point at which the condensation product loses carbon dioxide by decarboxylation to form acetoacetyl coA is uncertain, but is probably after the loss of coA from the condensation intermediate. The decarboxylation step is analogous to enolate formation in the usual Claisen condensation and drives the reaction in the direction of C—C bond formation. Condensation with another anion derived from malonyl coA extends the chain a further two atoms, and continuation of the process eventually leads to the long-chain alkanoic acids that are common in nature. The prevalence of even-numbered alkanoic acids in biological systems is thus understandable.

An enzyme cofactor or coenzyme is a molecule that participates in the enzymatic reaction but is regenerated either at the end of the catalyzed reaction or in a separate enzymatic reaction. In this case the cofactor, coenzyme A, is a thiol (Chapter 26) and forms a thioester analogous to the ester derived from an alkanol. A multienzyme complex catalyzes the condensation sequence. Carboxylation is promoted by a biotin-dependent enzyme complex, acetyl coA-carboxylase, through a reaction of carboxybiotin with the anion of acetyl coA. The key condensation step is catalyzed by β-ketoacyl synthase. The remaining steps in alkanoic acid synthesis involve reduction of the carbonyl group to an alcohol, dehydration to an alkene, and either release of an alkenoic acid or reduction of the alkene to an alkane. We have seen chemical conversions similar to each biochemical step (Section 16.2).

carboxybiotin

coenzyme A

$CH_3CO{-}coA$ (acetyl coA) ⟶ $HOOCCH_2CO{-}coA$ (malonyl coA) ⟶ $HOOCCH^{-}CO{-}coA$ $\xrightarrow{CH_3CO{-}coA}$

$CH_3{-}C(O^-)({-}coA){-}CHCO{-}coA$ with $HOOC{-}$ on CH ⟶ $CH_3{-}C(=O){-}CH(^-OOC)CO{-}coA$ ⟶ $CH_3{-}C(O^-){=}CHCO{-}coA$ ⟶

$CH_3CO{-}CH_2CO{-}coA$ (acetoacetyl coA) $\xrightarrow{\text{1. carbonyl reduction; 2. dehydration; 3. alkene reduction}}$ $CH_3CH_2{-}CH_2CO{-}coA$ (butanoyl coA) $\xrightarrow{HOOCCH^{-}CO{-}coA}$

$CH_3CH_2CH_2{-}C(O^-)({-}coA){-}CHCO{-}coA$ with $HOOC{-}$ on CH ⟶ ⟶ $CH_3CH_2CH_2CH_2{-}CH_2CO{-}coA$ (hexanoyl coA) ⟶ ⟶

C_{14}, C_{16}, C_{18} etc.

Degradation of alkanoic acids is chemically the reverse of synthesis except that malonyl coA is not involved; acetyl coA is formed directly by ''thiolysis'' of a β-ketoacyl coA. The reactions are catalyzed in eukaryotes (Gk., *eu-*, good, and *karyon,* kernel, meaning ''true nucleus'' and thus, cells containing nuclei) by a set of mitochondrial enzymes different from those involved in synthesis.

EXERCISE 19.10 Show how ethyl acetate can be converted into each of the following compounds.

(a) ethyl hexanoate
(b) ethyl 3-hydroxyhexanoate
(c) ethyl 3-hydroxy-3-methylbutanoate
(d) ethyl 2,2,2-trideuterioacetate
(e) 3-methylpentane-1,3-diol
(f) $CH_3CH_2OOCCH_2CHO$

Highlight 19.3

The carbonyl group reacts with carbanions in addition-elimination reactions, including the Claisen condensation, the Reformatsky reaction, and the Grignard reaction. Write out all of the important resonance forms for each derivative in each step.

$$\left[CH_3{-}C(=\ddot{O}{:}){-}X \longleftrightarrow CH_3{-}C(\ddot{O}{:}^-){=}X^+ \longleftrightarrow CH_3{-}\overset{+}{C}(\ddot{O}{:}^-){-}X\right]$$

$$CH_3{-}C(=\ddot{O}{:}){-}X + Nu^- \longrightarrow CH_3{-}C(\ddot{O}{:}^-)(X){-}Nu \xrightarrow{-X^-} CH_3{-}C(={:}O{:}){-}CH_2COOR,\ CH_2R$$

X = Cl, OR Nu = $^-CH_2COOR$, $M^+\ ^-CH_2R$

19.10 Reactions of Amides That Occur on Nitrogen

The nitrogen-hydrogen bonds of amides are also acidic. In fact, since the negative charge in the resulting amidate anion is shared between oxygen and nitrogen rather than oxygen and carbon, they are substantially more acidic than ketones; acetamide has $pK_a \cong 15$.

$$R{-}\overset{\overset{\displaystyle O}{\|}}{C}{-}NH_2 \rightleftharpoons H^+ + \left[R{-}\overset{\overset{\displaystyle O}{\|}}{C}{-}NH^- \longleftrightarrow R{-}\overset{\overset{\displaystyle O^-}{|}}{C}{=}NH \longleftrightarrow R{-}\underset{+}{\overset{\overset{\displaystyle O^-}{|}}{C}}{-}NH^-\right]$$

an **amidate** anion

The lability of the amide N—H is reflected in many of the reactions of amides. One of these is dehydration. A primary amide can be converted into the corresponding nitrile by treatment with an efficient dehydrating agent such as P_2O_5, $POCl_3$, $SOCl_2$, or acetic anhydride.

$$\text{2,3-dimethylhexanamide} \xrightarrow[\text{benzene, } 80^\circ\text{C}]{SOCl_2} \text{2,3-dimethylhexanenitrile (90\%)}$$

2,3-dimethylhexanamide — 2,3-dimethylhexanenitrile (90%)

EXERCISE 19.11 Consider the four compounds ethyl bromide, propanoic acid, propanamide, and propanenitrile. Make a 4 × 4 matrix with the four compounds on each side of the matrix. There are twelve intersections of the matrix representing conversion of one compound to another (ignore the four that connect a compound with itself). We have learned reactions that correspond to eight of these twelve possibilities. Review these eight reactions.

19.11 Complex Derivatives

Complex derivatives of alkanoic acids abound in nature. The most biologically active is the eicosanoid group (C_{20} alkanoic acid derivatives) which includes prostaglandins, prostacyclin, thromboxanes, and leukotrienes. Some of these are physiologically active in micromolar to nanomolar concentrations. On the other extreme are pharmacologically inactive compounds such as waxes, which serve as structural elements and to control water permeability. In between are the lipids, which are both metabolically active and structurally important to cell membranes, as well as those that are stored and consumed as fats (Gk., *lipos,* fat; fat, Old English, *faett*). Although the terms lipid and fat are roughly equivalent, lipid has been used more generally and fat more specifically for glyceryl esters of alkanoic acids (Section 19.11.B).

Lipid is a term that has been used to describe the group of natural substances that are soluble in hydrocarbons and insoluble in water. It includes fats, waxes, phosphoglycerides, natural hydrocarbons, and so on. Most biochemists reserve the term lipid for natural compounds that yield fatty acids upon hydrolysis.

A. Waxes

Waxes are naturally occurring esters of long-chain carboxylic acids (C_{16} or greater) with long-chain alcohols (C_{16} or greater). They are low-melting solids that have a characteristic "waxy" feel. We present three examples. **Spermaceti** is a wax that separates from the oil of the sperm whale on cooling. It is mainly cetyl palmitate (cetyl alcohol is the common name for 1-hexadecanol) and melts at 42–47 °C.

$$C_{15}H_{31}\overset{\overset{\large O}{\|}}{C}OC_{16}H_{33}$$

spermaceti
cetyl palmitate
1-hexadecyl hexadecanoate

Beeswax is the material from which bees build honeycomb cells and also the origin of the name "wax" which has been generalized to include materials with similar properties. It melts at 60–82 °C and is a mixture of esters. Hydrolysis yields mainly the C_{26} and C_{28} straight-chain carboxylic acids and the C_{30} and C_{32} straight-chain primary alcohols.

$$C_{25\text{-}27}H_{51\text{-}55}\overset{\overset{\large O}{\|}}{C}OC_{30\text{-}32}H_{61\text{-}65}$$

beeswax

Carnauba wax occurs as the coating on Brazilian palm leaves. It has a high melting point (80–87 °C) and is impervious to water. It is widely used as an ingredient in automobile and floor polish. It is a mixture of esters of the C_{24} and C_{28} carboxylic acids and the C_{32} and C_{34} straight-chain primary alcohols. Other components are present in smaller amounts.

EXERCISE 19.12 Develop a molecular theory for the properties of waxes, using your knowledge of dipole-dipole attractions, the molecular geometry of esters, and the magnitude of van der Waals forces.

B. Fats

Fats are naturally occurring esters of long-chain carboxylic acids and the triol glycerol (propane-1,2,3-triol). They are also called **glycerides.**

CH_2OH	CH_2OCOR
$CHOH$	$CHOCOR'$
CH_2OH	CH_2OCOR''
glycerol propane-1,2,3-triol	a fat or glyceride

Hydrolysis of fats yields glycerol and the component carboxylic acids. Over 500 different acids have been detected in natural fats, the majority by gas chromatography. The straight-chain carboxylic acids that may be obtained from fats are frequently called **fat acids** or **fatty acids.** Fatty acids may be saturated or unsaturated. The most common saturated fatty acids are lauric acid, myristic acid, palmitic acid, and stearic acid.

$CH_3(CH_2)_{10}COOH$	$CH_3(CH_2)_{12}COOH$	$CH_3(CH_2)_{14}COOH$	$CH_3(CH_2)_{16}COOH$
lauric acid dodecanoic acid	myristic acid tetradecanoic acid	palmitic acid hexadecanoic acid	stearic acid octadecanoic acid

The most important unsaturated fatty acids have 18 carbon atoms, with one or more double bonds. Examples are oleic acid, linoleic acid, and linolenic acid.

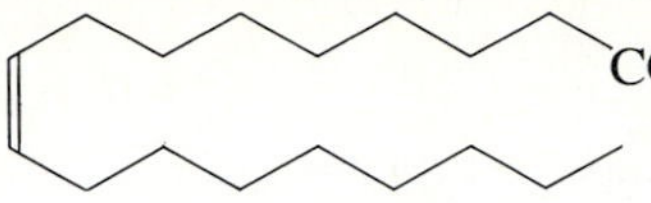

COOH

oleic acid
(Z)-9-octadecenoic acid

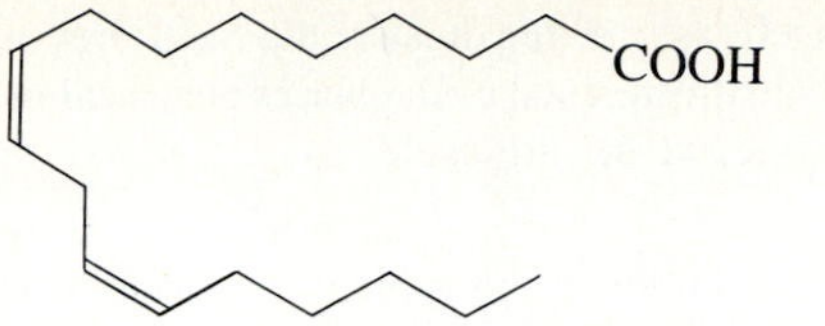

COOH

linoleic acid
(Z,Z)-9,12-octadecadienoic acid

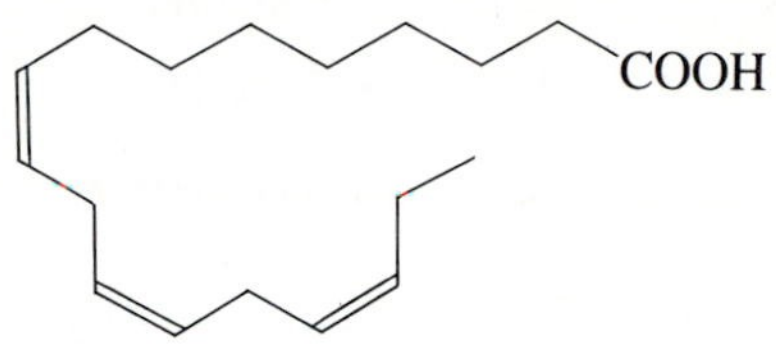

COOH

linolenic acid
(Z,Z,Z)-9,12,15-octadecatrienoic acid

Almost invariably, the double bonds have the cis or Z configuration. A significant exception is eleostearic acid, which has one cis and two trans double bonds.

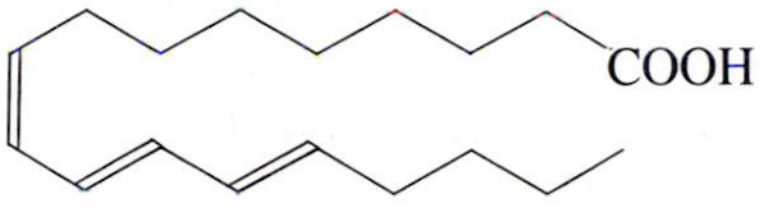

COOH

eleostearic acid
(*Z*,*E*,*E*)-9,11,13-octadecatrienoic acid

Natural fats are generally complex mixtures of triesters of glycerol (triglycerides). The acyl groups are sometimes described briefly by the number of carbon atoms followed by the number of double bonds, as 18:1 for oleic acid. Human liver fat contains 14:0, 16:0, 16:1, 18:0, 18:1, 18:2, and 20:4 acyl groups as well as a few others in smaller amount. In general, the distribution of the various acyl groups among the three possible positions varies with the tissue origin of the fat in humans. The acyl composition of fats from different species is different, and subject to some genetic manipulation to change the nature of the fat in a commercially successful organism. Hydrolysis of the common palm oil yields 1–3% myristic acid, 34–43% palmitic acid, 3–6% stearic acid, 38–40% oleic acid, and 5–11% linoleic acid. Some natural fats yield large amounts of a single fatty acid on hydrolysis; tung oil yields 2–6% stearic acid, 4–16% oleic acid, 1–10% linoleic acid, and 74–91% eleostearic acid. Cottonseed oil contains sterculic acid, a cyclopropene derivative, which must be removed before the oil can be used in feed for chickens or cows. The reactive cyclopropene ring apparently inhibits certain enzymatic transformations involving alkanoic acids, resulting in off-color eggs or butter with a sticky feel and little economic value.

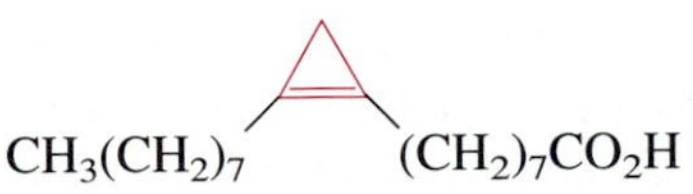

$CH_3(CH_2)_7$ $(CH_2)_7CO_2H$

sterculic acid

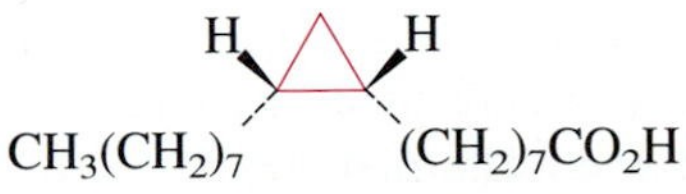

H H

$CH_3(CH_2)_7$ $(CH_2)_7CO_2H$

cis-9,10-dihydrosterculic acid

Certain physical properties of alkanoic acid derivatives are profoundly affected by interference with the extensive van der Waals interaction between hydrocarbon chains (see Figure 18.3). A cis double bond causes the chain to be bent, and the melting point of methyl esters, for example, to be considerably lowered due to decreased interchain interaction. The **disorder** in the interchain arrangements has been increased. The magnitude of the disordering effect varies with the position of the double bond in the chain. A *cis*-cyclopropane derivative such as *cis*-dihydrosterculic acid has a similar, but smaller disordering effect. *cis*-Dihydrosterculic acid replaces oleic acid in the membrane lipids of *Escherichia coli* at certain stages of the life cycle.

Fats undergo the typical reactions of esters. An important commercial reaction of fats is alkaline hydrolysis. The product fatty acid salts are used as soaps (Section 18.4.D).

$$\begin{array}{l} CH_2OCOR \\ | \\ CH_2OCOR \\ | \\ CH_2OCOR \end{array} + 3\ NaOH \xrightarrow{H_2O} \begin{array}{l} CH_2OH \\ | \\ CHOH \\ | \\ CH_2OH \end{array} + 3\ RCO_2^-\ Na^+$$

The alkaline hydrolysis of an ester is often referred to as **saponification,** from this process.

The melting point of a fat depends on the amount of unsaturation in the fatty acids. Fats with a preponderance of unsaturated fatty acids have melting points below room temperature and are called oils. Fats with little unsaturation are solid at normal temperatures. For the manufacture of soaps and for certain food uses, solid fats are preferable to oils. The melting point of a natural fat may be increased by hydrogenation. Industrially, the process is called **hardening.** Vegetable oils, such as cottonseed and peanut, are often hardened to the consistency of lard by partial hydrogenation. The increase in van der Waals interactions and the decrease in the disorder produced by the chains with cis double bonds leads to an increase in melting point.

A variety of unsaturated glyceryl esters occur in nature. Ricinoleic acid, the 12-hydroxy derivative of oleic acid, is the major component of the oil from the castor bean. Unsaturated octadecanoic acids such as linoleic acid and linolenic acid are found in linseed oil, which is used as a drying oil. On exposure to air, free radical chain reactions are initiated by reaction with oxygen. Further reaction leads eventually to a tough transparent polymer. The bonds between the monomeric units are called **cross-links.** Oil-based paint is a combination of pigment suspended in a solution of a drying oil and solvent. Varnish contains such drying oils; its conversion to a protective film on wood involves the formation of a tough waterproof film by oxygen-promoted free radical cross-linking.

EXERCISE 19.13 Develop a molecular theory for the properties of hydrogenated (hardened) food oils using your knowledge of dipole-dipole attractions, the molecular geometry of lipid esters, and the magnitude of van der Waals forces.

C. Eicosanoids

Arachidonic acid is a C_{20} tetraenoic acid derived from phospholipids by the action of a specific enzyme, phospholipase A_2, in a reaction promoted by various chemical and mechanical stimuli. The reaction is rate-limiting for synthesis of prostaglandins and thromboxanes. The first step in the formation of biologically active **eicosanoids** is the reaction of arachidonic acid with oxygen in the presence of an enzyme, cycloxygenase. The anti-inflammatory action of aspirin involves formation of an acetate ester on a serine OH group of the cyclooxygenase, inhibiting further formation of prostaglandins. (Serine is an amino acid that is described in Chapter 29.) The initial product PGG_2, a hydroperoxy-peroxide, is converted to PGH_2, a hydroxy-peroxide, at a different site on the same enzyme. Both cyclic peroxides have half-lives of 5 min at pH 7.5 and 37 °C.

COOH $\xrightarrow[\text{cyclooxygenase}]{O_2,}$

arachidonic acid
(Z,Z,Z,Z)-5,8,11,14-eicosatetraenoic acid

O—O COOH $\xrightarrow{O_2}$

O—O COOH $\xrightarrow{\text{peroxidase}}$

OOH

prostaglandin G_2, PGG_2

O—O COOH

OH

prostaglandin H_2, PGH_2

There are a number of interesting chemical features associated with these enzymatic reactions. First, a carbon-carbon bond is produced as a result of the oxygen addition reaction. Second, the formation of the peroxide bond must be stepwise since the substituents on the cyclopentane ring are trans to one another rather than cis, as expected from a concerted addition. Third, the conversion of the hydroperoxide to the hydroxy group resembles common protective reactions in biological systems, so that free radical chain reactions (see Section 10.10) that lead to oxidation and chain degradation might be avoided.

In blood platelets, TXA_2 synthase converts PGH_2 into thromboxane TXA_2. Another enzyme, PGI_2 synthase, forms prostacyclin, PGI_2, in arterial walls. The platelet-compound TXA_2 promotes aggregation whereas PGI_2 inhibits platelet aggregation at concentrations between 1–10 nM. The TXA_2 is a strained acetal (note the four-membered ring) and has a half-life of 30 sec at pH 7.5 and 37 °C. Both TXA_2 and PGI_2 formation resemble Baeyer-Villiger oxidations (Section 14.8.A) in which the enzymatically controlled product is determined by which oxygen becomes positive. The reaction mechanisms are shown without the groups on the enzyme that participate in creating and directing the reactive intermediates.

> Formation of a blood clot is a complex process that is initiated by platelet (thrombocyte) adhesion to the site of injury. Platelet aggregation, stimulated by TXA_2 at the site, reinforces the plug, which must then reinforced by an insoluble protein, fibrin. The fibrin is generated from a precursor, fibrinogen, by an enzyme, thrombin, which itself results from a cascade of proteolytic processes.

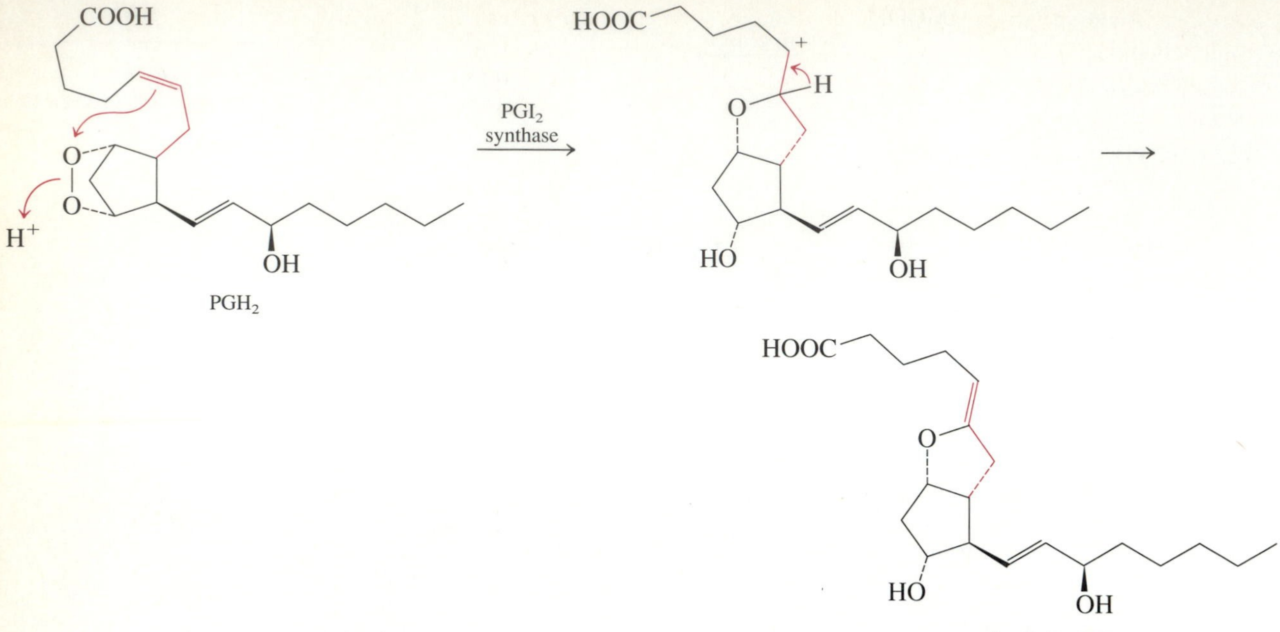

COOH

H^+

PGH$_2$

PGI$_2$ synthase

HOOC

H

O

HO

OH

HOOC

O

HO

OH

prostaglandin I$_2$, PGI$_2$

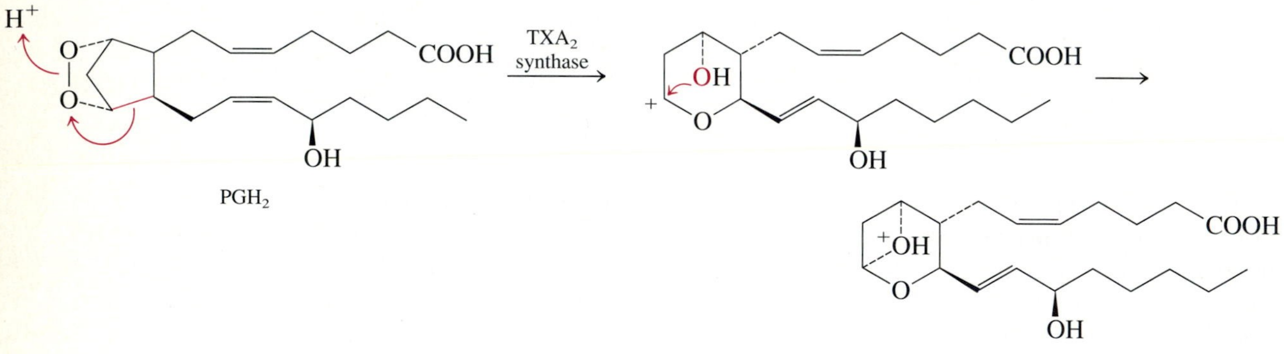

H^+

COOH

OH

PGH$_2$

TXA$_2$ synthase

OH

COOH

O

OH

COOH

$^+$OH

O

OH

thromboxane A$_2$, TXA$_2$

Another important reaction of arachidonic acid with oxygen is catalyzed by 5-lipoxygenase and forms 5-HPETE. The latter is converted into the 5,6-epoxide, leukotriene A, by an enzyme.

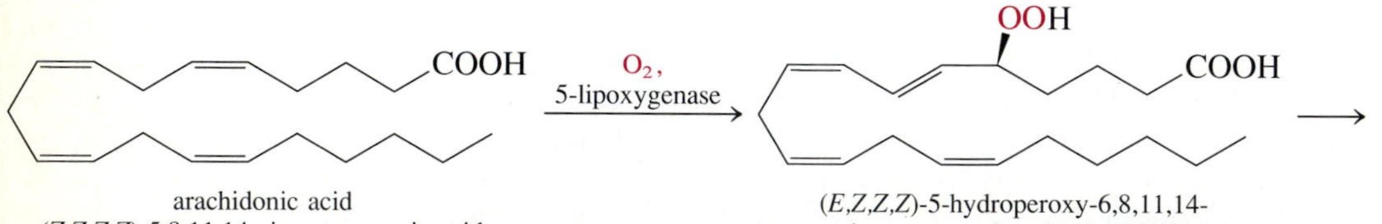

COOH

arachidonic acid
(Z,Z,Z,Z)-5,8,11,14-eicosatetraenoic acid

O_2, 5-lipoxygenase

OOH

COOH

(*E*,*Z*,*Z*,*Z*)-5-hydroperoxy-6,8,11,14-eicosatetraenoic acid, 5-HPETE

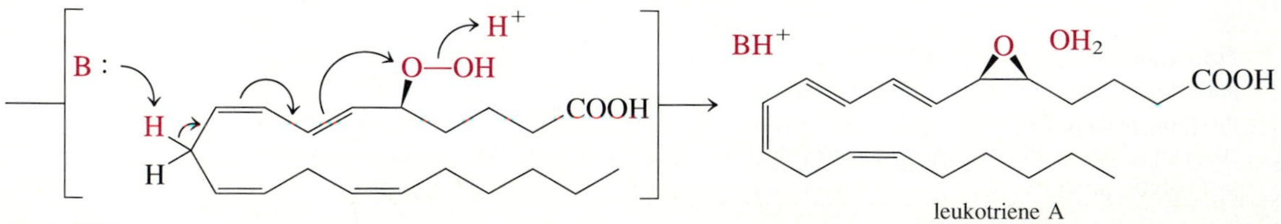

B :

H

H

H^+

O—OH

COOH

BH^+

O

OH_2

COOH

leukotriene A

Leukotriene A reacts with the tripeptide thiol, glutathione, to yield leukotriene C. Cleavage of an amide bond leads to leukotriene D, one of the ''slow-reacting substances'' implicated in the muscle spasms that accompany asthma attacks. (Glutathione itself is an important tripeptide that will be taken up in Chapter 29.)

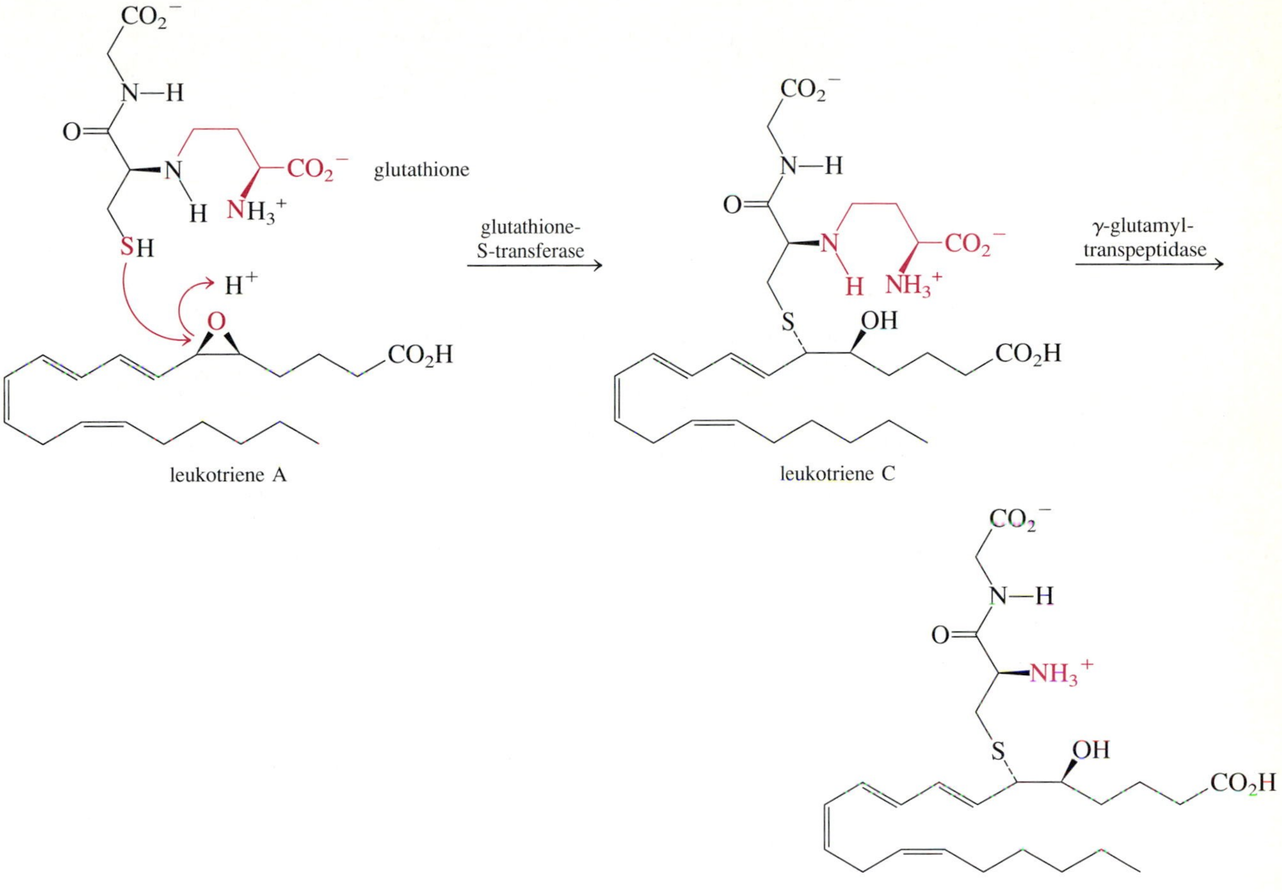

glutathione

glutathione-S-transferase →

γ-glutamyl-transpeptidase →

leukotriene A

leukotriene C

leukotriene D

Designing agents to control some of the biological responses mediated by these compounds requires a thorough understanding of the chemistry, the biochemistry and of the mechanisms by which the transformations occur.

EXERCISE 19.14 Arachidonic acid is an **essential** fatty acid, and must be biosynthesized from linoleic acid derived from food through the addition of malonyl coenzyme A to the coenzyme A ester followed by desaturation. To test the idea that a closely related compound would inhibit enzymes for the conversion of arachidonate into biologically active materials, we require a homologue of arachidonic acid with one additional CH_2 group. Devise a synthesis beginning with arachidonic acid for the C_{21} acid.

D. Phospholipids

Phosphoglycerides are lipids in which glycerol is esterified to two fatty acids and to phosphoric acid; they are generally called **phospholipids.** Such monophosphate esters are called phosphatidic acids. Most phosphatidic acids contain one saturated and one unsaturated fatty acid. Because different acids are esterified at the C-1 and C-3 hydroxy groups of glycerol, phosphatidic acids are chiral; the absolute configuration at the C-2 ester link is R.

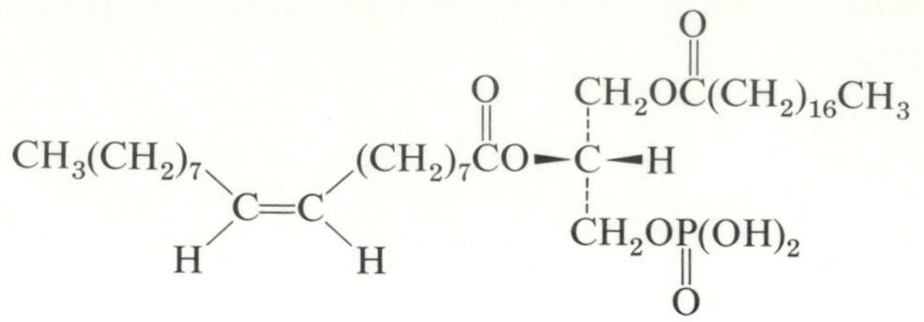

a phosphatidic acid

Free phosphatidic acids are rare in nature. Usually the phosphoric acid moiety is esterified to a second alcohol component. The phosphatidyl derivatives thus formed are major components of biological cell membranes. Quantitatively, the most important are phosphatidylethanolamine (PE) and phosphatidylcholine (PC).

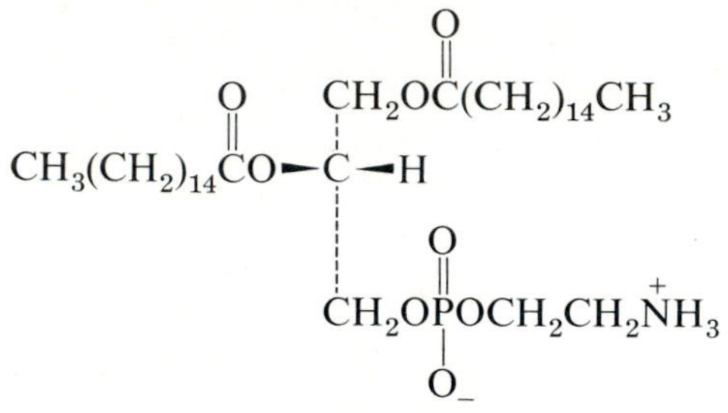

phosphatidylethanolamine

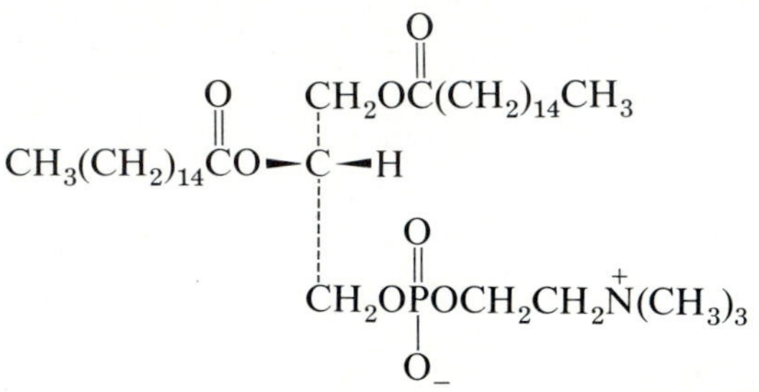

phosphatidylcholine

Other phospholipids include phosphatidylserine (PS) and phosphatidylinositol (PI). In rat liver cell membranes, the composition is PC, 46%; PE, 25%; PI, 7%; and PS, 4%. Most of the remaining phospholipid is sphingomyelin, a class of compounds derived from sphingosine rather than glycerol. The hydrolysis of PI diphosphate to inositol triphosphate and a diacylglycerol by phospholipase C is closely linked to a major class of neurotransmitter actions, making the process a potential target for neuroactive drugs.

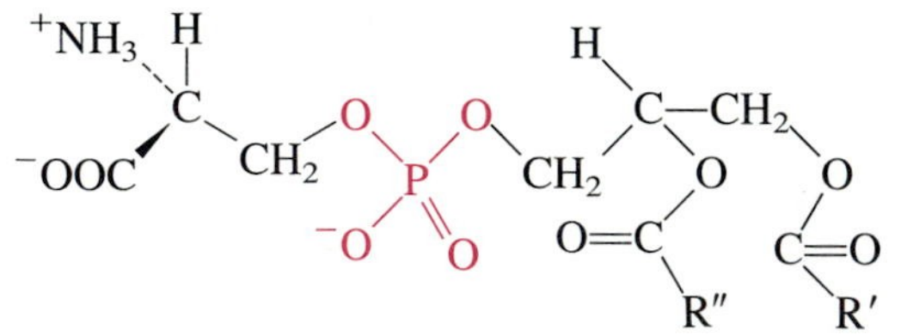

phosphatidylserine
R′ saturated, R″ unsaturated

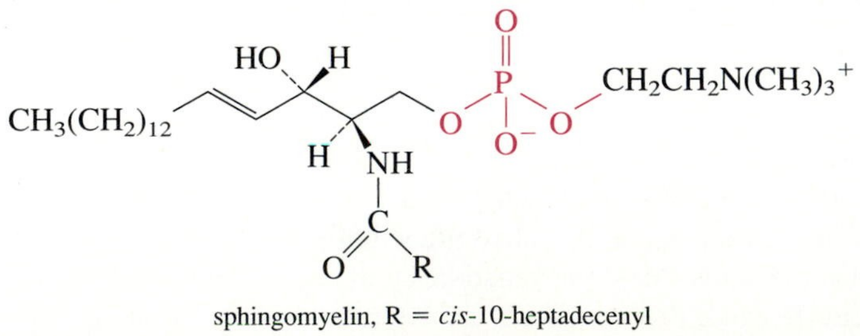

sphingomyelin, R = *cis*-10-heptadecenyl

phosphatidylinositol 4,5-diphosphate
R′ saturated, R″ unsaturated

Phosphoglycerides are amphiphilic molecules, with two long nonpolar ''tails'' and a small, highly polar ''head'' (Figure 19.6). These molecules form aggregated structures more easily than soaps and synthetic detergents, as reflected in lower c.m.c. values.(Section 18.4.D) Palmitoyl lysophosphatidylcholine (lyso means that the 2-acyl group has been removed by hydrolysis) has a c.m.c of less than 10^{-5} M and dipalmitoylphosphatidylcholine a c.m.c of 5×10^{-10} M. The c.m.c. decreases to 10^{-3} M for diheptanoylphosphatidylcholine, showing that the shorter the chain, the less the interference with water-water interactions. Judging from the ubiquitous occurrence of phospholipids in cell membranes, the stability of such structures cannot be doubted. Among the molecular arrangements formed by phospholipids to minimize the disruption of water structure by

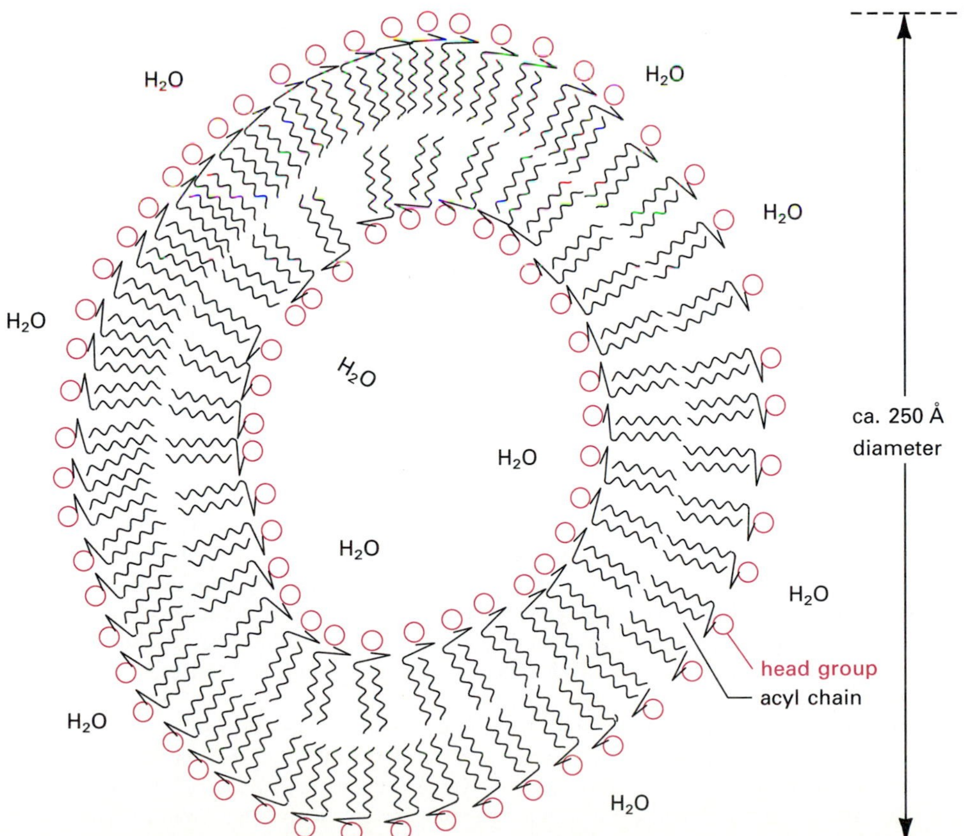

FIGURE 19.6 Schematic formula for a small (250 Å) phospholipid vesicle.

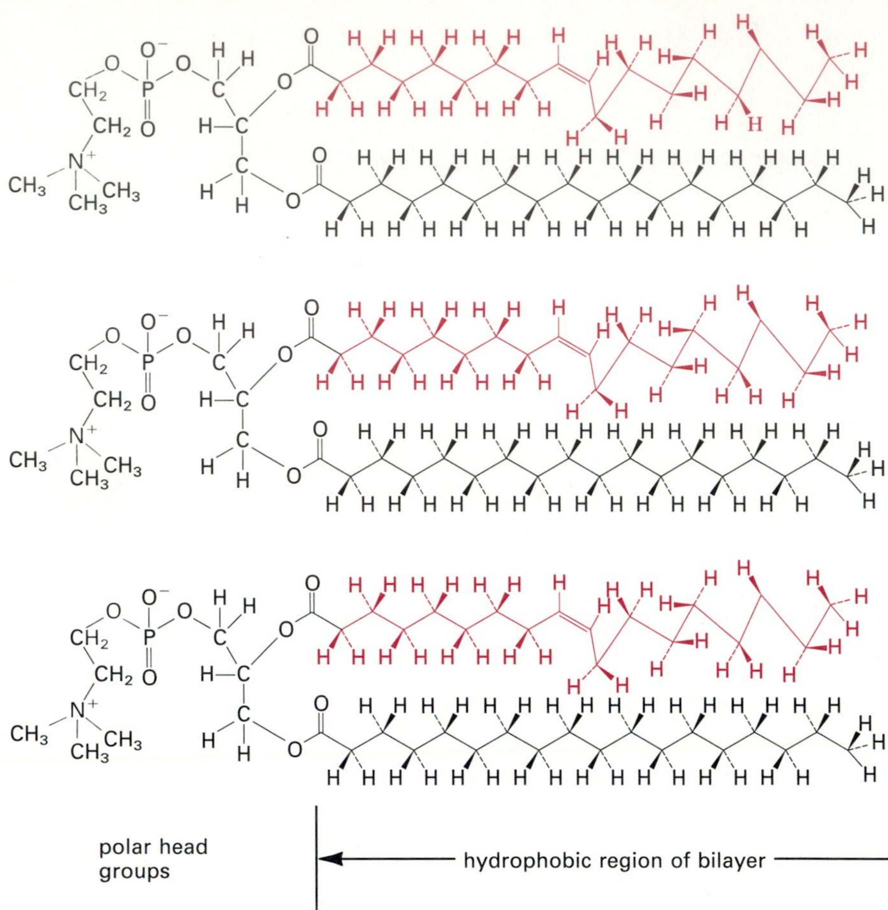

FIGURE 19.7 Cross section of a phospholipid bilayer. Note the disorder present on the side having the unsaturated acyl chains.

the nonpolar tails are **bilayers.** An unsaturated chain is usually attached to the middle position of the glycerol. It has a less regular structure than a saturated chain; the disorder introduced into the hydrophobic region of the bilayer (compare the left- and right-hand sides of Figure 19.7) makes the chains more mobile or more "fluid." Cell membranes contain proteins that move in order to function biologically; their activity is influenced by membrane "fluidity."

Exposure to ultrasound (sonication) of phospholipids mixed with cholesterol and related compounds produces vesicles or liposomes that enclose the components of the source solution. These vesicles (liposomes) have been considered as drug delivery vehicles, with stability enhanced by the inclusion of head groups that have been linked to one another. Geometry requires that the number of molecules on the inner leaflet of the bilayer be smaller than the number in the outer leaflet, a difference that is most important for the stability of small vesicles.

Phospholipids and other amphiphiles form monolayers on water. If these are prepared in a device called a Langmuir-Blodgett trough, they may be studied as if the monolayer were a two-dimensional gas. Compression of a monolayer until the pressure rises leads to a measure of the area occupied by a single molecule. The monolayers can be transferred to plates by careful dipping and have become popular for the formulation of prospective microscopic optical devices.

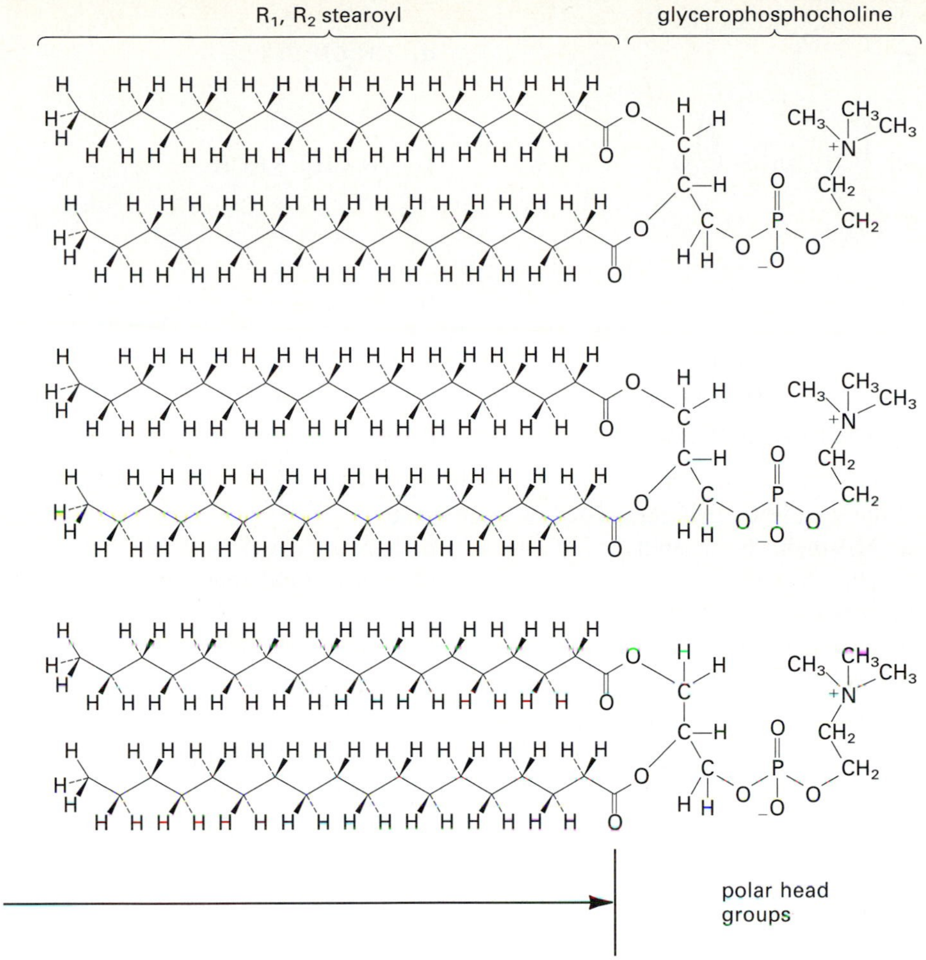

R_1, R_2 stearoyl

glycerophosphocholine

polar head groups

ASIDE

Bond (''marry'') an acyl group to a hydroxy group and one has created the carboxylic acid family. Combining the acyl group with other groups produces other families of carboxylic acid derivatives. The bond between two people has a strength that often reflects the interaction between them. Carboxylic acids were not what Shakespeare had in mind when he had Juliet say to Romeo, ''Good night, good night! parting is such sweet sorrow,'' but it reminds us that strong measures are required to break the bond between the acyl and hydroxy groups. List four different types of carboxylic acid derivatives in order of how strongly the groups interact. Give at least one synthesis for each type and one reaction in which each type participates.

PROBLEMS

1. Write the IUPAC name for each compound.

a. $(CH_3CH_2)_2CHCH_2COOCH_2CH_2CH_3$ **b.** 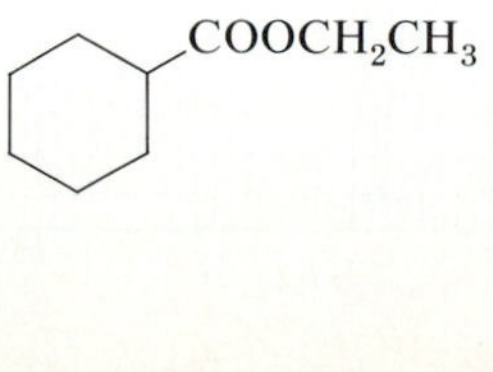 (cyclohexane ring bearing) $COOCH_2CH_3$

c. 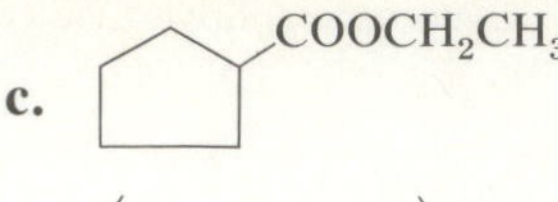

d. $CH_3CH_2\overset{\overset{O}{\|}}{C}Cl$

e. $\left(CH_3CH_2CH_2\overset{\overset{O}{\|}}{C}\right)_2O$

f. $CH_3CH_2\overset{\overset{O}{\|}}{C}NHCH_3$

g. (cyclohexyl)$NH\overset{\overset{O}{\|}}{C}CH(CH_3)_2$

h. $CH_3CH_2\overset{\overset{Cl}{|}}{C}HCOOCH_3$

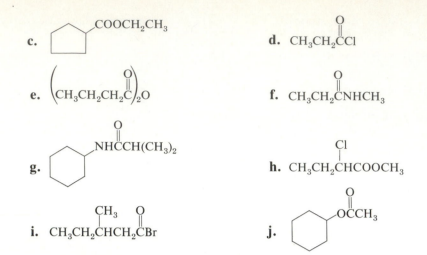

i. $CH_3CH_2\overset{\overset{CH_3}{|}}{C}HCH_2\overset{\overset{O}{\|}}{C}Br$

j. (cyclohexyl)$O\overset{\overset{O}{\|}}{C}CH_3$

2. Write a structure that corresponds to each name.

a. N,3-diethylhexanamide
b. N,N-dimethylformamide
c. ethyl butanoate
d. methyl 3-chloropropanoate
e. propanoic anhydride
f. acetic formic anhydride
g. cyclohexanecarboxamide
h. butanoyl bromide
i. cyclobutyl formate
j. 3-formylhexanoic acid
k. ethyl acetoacetate
l. N-bromoacetamide

3. In a dilute solution of acetic acid in 0.1 M aqueous HCl, what percentage of the acetic acid is present as $CH_3CO_2^-$ and what percentage is present as $CH_3C(OH)_2^+$?

4. The pK_a of $CH_3\overset{\overset{O}{\|}}{C}OH_2^+$ may be estimated to be approximately −12. Calculate the ratio of $CH_3\overset{\overset{OH^+}{\|}}{C}OH$ to $CH_3\overset{\overset{O}{\|}}{C}OH_2^+$ in an acidic solution of acetic acid.

5. A neutral compound, $C_7H_{13}O_2Br$, shows bands in its infrared spectrum at 1740 cm^{-1} and 2850–2950 cm^{-1} but none above 3000 cm^{-1}. The NMR spectrum shows the following pattern: δ 1.0 ppm (3H, t), 1.3 ppm (6H, d), 2.1 ppm (2H, m), 4.2 ppm (1H, t) and 4.6 ppm (1H, m). The CMR spectrum has a resonance at δ 168 ppm. Deduce the structure and indicate the origin of the foregoing spectral features.

6. Identify the compound having the following IR and NMR spectra and the formula $C_6H_{11}BrO_2$.

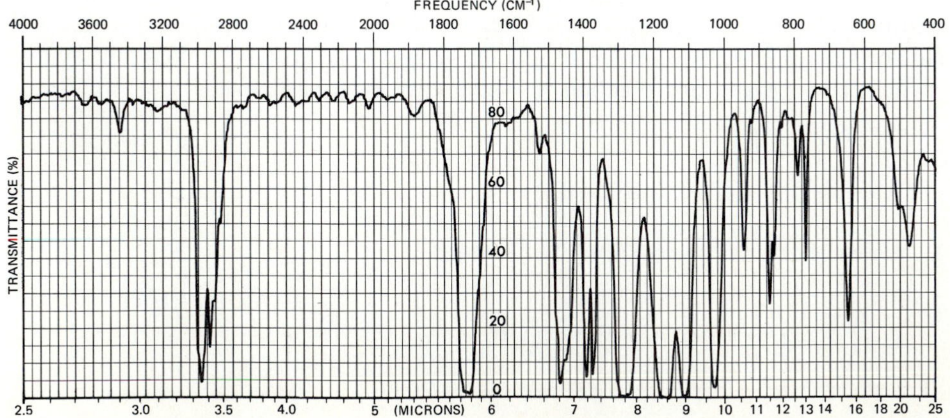

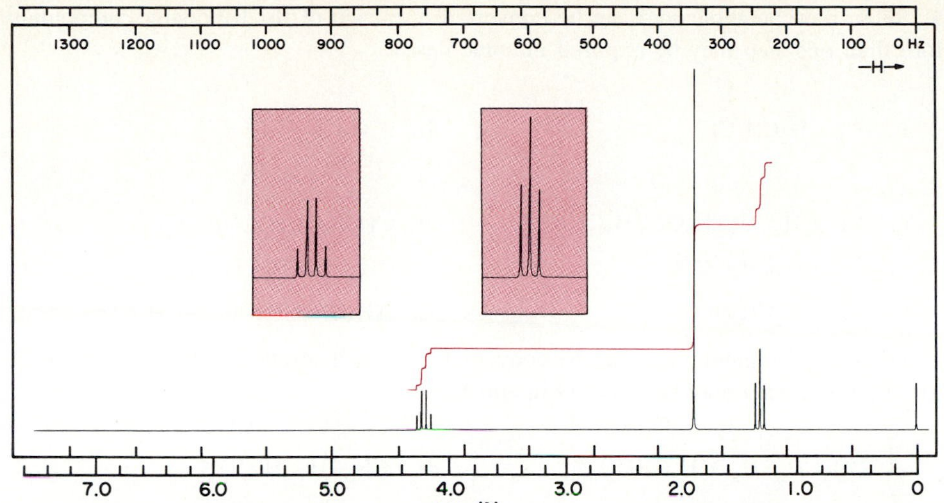

7. What are the organic products of each of the following reactions or sequences of reactions?

a. $CH_3CH_2CH_2OH + CH_3\overset{O}{\overset{\|}{C}}Cl \longrightarrow \xrightarrow{500°C}$

b. (cyclohexane bearing H and OH on the same carbon) $\xrightarrow[NaOH]{CS_2} \xrightarrow{CH_3I} \xrightarrow{200°C}$

c. (cyclopentane with CH_3 and OH on adjacent carbons) $\xrightarrow[NaOH]{CS_2} \xrightarrow{CH_3I} \xrightarrow{200°C}$

d. $CH_3CH_2CH_2Cl \xrightarrow[\substack{ethanol \\ 75°C}]{NaCN} \xrightarrow[\substack{H_2O \\ 100°C}]{NaOH}$

e. C_6H_5—$\overset{O}{\overset{\|}{C}}OH \xrightarrow{SOCl_2} \xrightarrow{NH_3} \xrightarrow{LiAlH_4}$

f. C_6H_5—$\overset{O}{\overset{\|}{C}}NH_2 \xrightarrow[\Delta]{P_2O_5}$

8. What are the organic products of each of the following reactions or sequences of reactions?

a. $(CH_3)_3\overset{O}{\overset{\|}{C}}NH_2 \xrightarrow[D_2O]{0.1\ M\ NaOD}$
25°C, 5 min

b. (steroid with $CH_3\overset{O}{\overset{\|}{C}}O$— on ring A and side chain ending in $COOCH_3$) $\xrightarrow[CH_3OH]{CH_3O^-Na^+}$

c. (bicyclic ring system with angular methyl, ring double bond, and $COOCH_3$) $\xrightarrow[ether]{LiAlH_4} \xrightarrow[25°C]{H_3O^+}$

d. $CH_3CH_2COOCH_2CH_3 \xrightarrow[EtOH]{NaOEt} \xrightarrow{H_3O^+}$

e. $BrCH_2COOCH_3 + CH_3CH_2CHO + Zn \xrightarrow{benzene}$

f. $CH_3CH_2COOCH_2CH_3 \xrightarrow[THF,\ -78°C]{LDA} \xrightarrow{CH_3CH_2CHO} \xrightarrow{H_2O}$

9. Show how butanoic acid can be converted into each of the following compounds. More than one step may be required in some cases.

a. $CH_3CH_2CH_2C(=O)Cl$ **b.** $(CH_3CH_2CH_2C(=O))_2O$

c. $CH_3CH_2CH_2CH_2OC(=O)CH_2CH_2CH_3$ **d.** $CH_3CH_2CH_2C(=O)N(CH_3)_2$

e. $CH_3CH_2CH_2C\equiv N$ **f.** $CH_3CH_2CH_2CH_2NH_2$

g. $CH_3CH_2CH_2Br$ **h.** $CH_3CH_2CH_2CH_2N(CH_2CH_3)_2$

10. Show how butanoic acid can be converted into each of the following compounds. More than one step may be required in some cases.

a. $CH_3CH_2CH_2C(=O)CH_2CH(CH_3)CH_3$ **b.** $CH_3CH_2CH_2C(CH_3)_2OH$

c. $CH_3CH_2CH_2CHO$ **d.** $CH_3CH_2CH_2CH_2OH$

e. $CH_3CH_2CH_2CH(OH)CH(CH_2CH_3)CH_2OH$ **f.** $CH_3CH_2CH_2CH(OH)CH(CH_2CH_3)COOC_2H_5$

g. $CH_3CH_2CH_2COCH(CH_2CH_3)COOC_2H_5$

11. Predict the product of each of the following reactions. Rationalize your predictions.

a. $H_2N{-}C(=O){-}Cl + CH_3O^- \longrightarrow$ **b.** $CH_3O{-}C(=O){-}Cl + H_2N^- \longrightarrow$

12. How may each of the following compounds be prepared from monofunctional compounds containing five or fewer carbons.

a. $(CH_3)_2CHCH_2CH_2CH_2COOCH_3$ **b.** $CH_3CH_2CH_2CH(OH)CH_2COOCH_2CH_3$

c. $(CH_3)_3CC(=O)CH_2COOCH_2CH_3$ **d.** 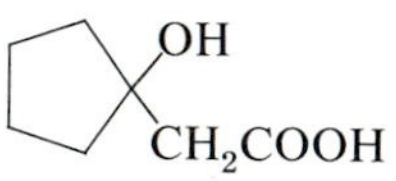 1-hydroxycyclopentyl with CH_2COOH (cyclopentane ring bearing OH and CH_2COOH on the same carbon)

13. N-Methylacetamide reacts with triethyloxonium tetrafluoroborate to give a salt, $C_5H_{12}NO^+\ BF_4^-$. When this salt is treated with sodium bicarbonate, a compound $C_5H_{11}NO$ is produced.

$$CH_3C(=O)NHCH_3 + (C_2H_5)_3O^+\ BF_4^- \longrightarrow C_5H_{12}NO^+\ BF_4^- \xrightarrow{NaHCO_3} C_5H_{11}NO$$

a. What are the structures of the two compounds?
b. Rationalize the formation of the salt in mechanistic terms.
c. Predict the product that will be obtained if the salt is treated with aqueous acid.

14. Write a mechanism that explains the following reaction (the Ritter reaction).

$$CH_3C(CH_3)_2{-}OH + CH_3CN \xrightarrow[\Delta]{H_2O{-}H_2SO_4} CH_3C(CH_3)_2{-}NHC(=O)CH_3$$

What product is expected when 2-methyl-2,4-pentanediol is treated with acetonitrile and aqueous sulfuric acid?

15. Define the eicosanoid group of compounds. Draw structures for at least three typical eicosanoids.

16. Orthoesters are compounds that have three alkoxy groups attached to the same carbon, for example, ethyl orthoacetate, $CH_3C(OC_2H_5)_3$. When ethyl orthoacetate is treated with dilute aqueous acid, ethyl acetate is obtained. Explain with a mechanism.

$$CH_3C(OC_2H_5)_3 + H_2O \xrightarrow{H^+} CH_3COOC_2H_5 + 2\ C_2H_5OH$$

17. Treatment of diethyl adipate, $C_2H_5OOC(CH_2)_4COOC_2H_5$, with sodium ethoxide in ethanol, followed by neutralization with aqueous acid, yields a compound having the formula $C_8H_{12}O_3$. Propose a structure for this product, and write a mechanism that accounts for its formation. This reaction is a general reaction of certain diesters (the Dieckmann condensation) and will be discussed in Section 27.6.C.

18. A solution of methyl cyclohexyl ketone in chloroform is treated with peroxybenzoic acid for 16 hr at 25 °C. The reaction mixture is worked up to obtain A, which has infrared absorption at 1740 cm^{-1}. The NMR spectrum of A shows a sharp three-proton singlet at $\delta = 2.0$ ppm and a one-proton multiplet at $\delta = 4.8$ ppm. Elemental analysis of A shows that it has the formula $C_8H_{14}O_2$. What is A and how is it formed?

19. (*R*)-2-Butanol, $[\alpha]_D = -13.5°$, reacts with methanesulfonyl chloride to give a methanesulfonate. Treatment of the methanesulfonate with aqueous sodium hydroxide affords 2-butanol having $[\alpha]_D = +13.5°$.

a. From this result, what conclusions may you draw regarding the mechanism of the hydrolysis?

b. How may (*S*)-2-octanol be converted into (*R*)-2-methoxyoctane?

20. Dimethyl carbonate is an ester of carbonic acid (page 533). Dimethyl carbonate is dissolved in isopropyl alcohol containing a small amount of sodium isopropoxide.

$$CH_3O\text{–}C(=O)\text{–}OCH_3 + (CH_3)_2CHOH \xrightarrow[\text{reflux}]{(CH_3)_2CHO^-\ Na^+}$$

The solution is heated at reflux for several days and is occasionally monitored by gas chromatography. This analysis revealed the slow appearance of an intermediate (A), which slowly went away and was replaced by another compound, B. The analyses were plotted against reaction time and the following graph was obtained:

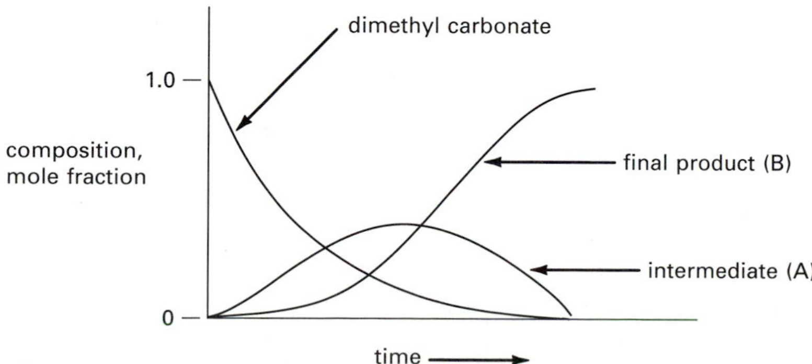

Explain these observations. What are A and B? Write a mechanism showing each step in the transformation of dimethyl carbonate into these two products.

21. Arachidonic acid is converted into PGH_2 by oxygen through a set of reactions promoted by the enzyme cyclooxygenase. Write out the steps of the conversion. What would happen if all of the CH_2 groups next to double bonds were replaced by the corresponding deuterated group, CD_2? What would happen if the double bonds, CH═CH, were replaced by CD═CD?

22. Write out the steps for the conversion of PGH_2 to TXA_2 and PGI_2. Would there be any consequences of deuterium replacements as noted in problem 21?

23. Write out the steps for the formation of leukotriene A. Would there be any consequences of deuterium replacements as noted in problem 21?

24. Write out the steps for the formation of leukotriene D. Would there by any consequences of deuterium replacements as noted in problem 21?

25. The model phospholipid bilayer shown in Figure 19.7 sums up some of the characteristics of the membranes of biological cells. Describe the structure and how the unsaturated acyl chains affect the behavior of the bilayer. What would you conclude from the fact that the membranes of organisms that live in hot springs at temperatures up to 89 °C contain only saturated lipids?

CHAPTER 20

CONJUGATION

20.1 Allylic Systems

A. Allylic Cations

When 2-buten-1-ol is treated with hydrogen bromide at 0 °C, a 3:1 mixture of 1-bromo-2-butene and 3-bromo-1-butene is produced. A comparable mixture is produced when 3-buten-2-ol is treated with HBr under the same conditions.

$$\left.\begin{array}{l} CH_3CH{=}CHCH_2OH \\ CH_3CHOHCH{=}CH_2 \end{array}\right\} \xrightarrow[0\,°C]{HBr} \underset{3:1}{CH_3CH{=}CHCH_2Br + CH_3CHBrCH{=}CH_2}$$

Similarly, when 1-chloro-3-methyl-2-butene is hydrolyzed in water containing silver oxide at room temperature, a mixture of alcohols consisting of 15% of 3-methyl-2-buten-1-ol and 85% of 2-methyl-3-buten-2-ol is produced. Essentially the same mixture is obtained by the reaction of 3-chloro-3-methyl-1-butene with silver oxide in water.

$$\left.\begin{array}{l} (CH_3)_2C{=}CHCH_2Cl \\ (CH_3)_2CClCH{=}CH_2 \end{array}\right\} \xrightarrow[H_2O,\ 25\,°C]{Ag_2O} \underset{(15\%)}{(CH_3)_2C{=}CHCH_2OH} + \underset{(85\%)}{(CH_3)_2C(OH)CH{=}CH_2} + AgCl$$

> Silver cation catalyzes the formation of carbocations from alkyl halides. The Ag^+ tends to coordinate with the leaving halide group; that is, it provides a potent "pull" that contributes to the driving force of the reaction.
>
> $$RX + Ag^+ \longrightarrow \underset{\text{(transition state)}}{[R\cdots X\cdots Ag]^{+}} \longrightarrow R^+ + AgX\downarrow$$

Furthermore, silver chloride, bromide, and iodide are highly insoluble salts and remove the halide ion from further equilibration reactions. The net reaction of an alkyl halide with silver oxide is generally written as follows:

$$2\,RX + Ag_2O + H_2O \longrightarrow 2\,ROH + 2\,AgX\downarrow$$

However, since these reactions often involve carbocation intermediates, other reactions such as rearrangements and eliminations frequently occur. Because of these reaction possibilities and the high cost of the reagent, the reaction of silver oxide with alkyl halides has little preparative significance and is used mainly to study the properties of carbocation intermediates.

These observations are explained by the formation of an intermediate cation in which the positive charge is delocalized over two carbons.

$$CH_3CH{=}CHCH_2\overset{+}{O}H_2 \text{ and } CH_3CH(\overset{+}{O}H_2)CH{=}CH_2 \xrightarrow{-H_2O} [CH_3CH{=}CH\overset{+}{C}H_2 \longleftrightarrow CH_3\overset{+}{C}HCH{=}CH_2] \xrightarrow{Br^-}$$

$$CH_3CH{=}CHCH_2Br + CH_3CH(Br)CH{=}CH_2$$

$$(CH_3)_2C{=}CHCH_2Cl \text{ and } (CH_3)_2C(Cl)CH{=}CH_2 \xrightarrow{-Cl^-} [(CH_3)_2C{=}CH\overset{+}{C}H_2 \longleftrightarrow (CH_3)_2\overset{+}{C}CH{=}CH_2] \xrightarrow{H_2O}$$

$$(CH_3)_2C{=}CHCH_2\overset{+}{O}H_2 + (CH_3)_2C(\overset{+}{O}H_2)CH{=}CH_2$$

The intermediate carbocations in the foregoing reactions are described as resonance hybrids of two important structures. The simplest such cation is the 2-propen-1-yl or **allyl cation.**

$$[CH_2{=}CH{-}\overset{+}{C}H_2 \longleftrightarrow \overset{+}{C}H_2{-}CH{=}CH_2] \equiv \overset{\frac{1}{2}+}{C}H_2\cdots CH\cdots\overset{\frac{1}{2}+}{C}H_2$$

allyl cation

Various symbols may be used to describe the hybrid electronic structure of the allyl cation. The two resonance structures in brackets show that the positive charge in the cation is divided equally between the two indicated positions. The alternative structure with dotted bonds shows that each carbon-carbon bond has an order of $1\frac{1}{2}$ and that each end carbon has $\frac{1}{2}$ of a positive charge.

The grouping $CH_2{=}CHCH_2{-}$ is called the **allyl group,** just as $CH_3CH_2{-}$ is called the ethyl group. The group name is used in naming many compounds containing the allyl group.

$CH_2{=}CHCH_2Cl$	$CH_2{=}CHCH_2OH$	$CH_2{=}CHCH_2OC({=}O)CH_3$
allyl chloride	allyl alcohol	allyl acetate

Substituents may be indicated by the use of Greek letters.

$$\underset{\gamma\quad\ \ \beta\ \ \alpha}{CH_2{=}CH\overset{\overset{\displaystyle CH_3}{|}}{C}HCl}$$

α-methylallyl chloride
3-chloro-1-butene

$$\begin{matrix} CH_3 \diagdown & & \\ & C{=}CHCH_2OH & \\ CH_3 \diagup & \gamma \quad \beta \quad \alpha & \end{matrix}$$

γ,γ-dimethylallyl alcohol
3-methyl-2-buten-1-ol

One important structural feature of allyl cation is that all of the atoms lie in one plane in such a way that the empty *p*-orbital of the carbocation can overlap with the π-orbital of the double bond. The electron density in the double bond is shared in the manner indicated by the following symbols.

$$[CH_2{=}CH{-}\overset{+}{C}H_2 \longleftrightarrow \overset{+}{C}H_2{-}CH{=}CH_2]$$

It is important to recall that resonance structures are used to symbolize alternative configurations of electron density. The relative positions of the nuclei remain precisely the same in all resonance structures. The symbol $CH_2{=}CH{-}CH_2^+$ would normally indicate a carbon-carbon double bond having a short distance and a carbon-carbon single bond having a longer distance. However, the alternative structure $^+CH_2{-}CH{=}CH_2$ contributes equally to the actual structure symbolized by dotted lines as $(CH_2{\cdots}CH{\cdots}CH_2)^+$. Allyl cation has two equivalent carbon-carbon bonds of equal length. The parent ion has been detected in the gas phase and is inferred as an intermediate in some solution reactions. The structure has not been determined experimentally, but sophisticated quantum mechanical calculations show the carbon-carbon bond length to be between those for single and double bonds. Similarly, the two terminal carbon atoms share the positive charge equally.

A stereo representation of the planar structure of allyl cation and the corresponding orbital description are shown in Figure 20.1. The two electrons in the π-orbital are in a molecular orbital extending over all three carbon atoms (Section 22.1).

Since the positive charge is spread over a larger volume, we expect allyl cation to be more stable than a simple primary alkyl cation. This expectation is confirmed by a comparison of gas phase enthalpies of ionization of alkyl chlorides.

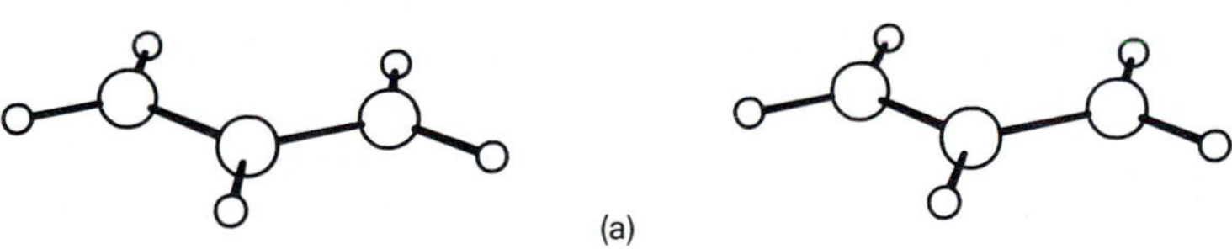

(a)

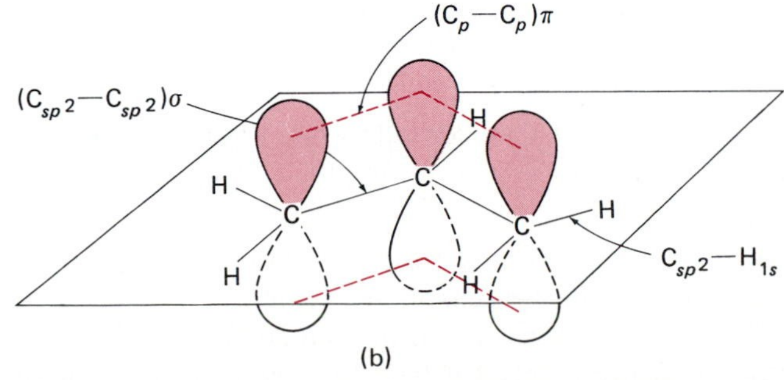

(b)

FIGURE 20.1 Allyl cation: (a) stereo structure; (b) orbital description.

$$RCl \longrightarrow R^+ + Cl^-$$

R	$\Delta H°$, kcal mole^{-1}
CH_3CH_2Cl	193
$CH_2{=}CHCH_2Cl$	171
$(CH_3)_2CHCl$	172

In fact, *allyl cation is roughly comparable in relative stability to a secondary alkyl cation.* Similarly, methylallyl cation has the positive charge spread between a secondary and primary position, and the net stabilization is comparable to that of a tertiary alkyl cation.

$$[CH_3\overset{+}{C}H{-}CH{=}CH_2 \longleftrightarrow CH_3CH{=}CH{-}\overset{+}{C}H_2] \equiv CH_3\overset{\delta+}{C}H\text{---}CH\text{---}\overset{\delta+}{C}H_2$$

When an allylic cation reacts with a nucleophilic reagent, it can react at either positive center and a mixture of products is generally produced. As a result, reactions that proceed by way of allyl cations often appear to give "rearranged" products. Such reactions are called **allylic rearrangements.** For example, in the first reaction we encountered in this chapter, 2-buten-1-ol reacts with HBr to give 1-bromo-2-butene, a "normal" product. However, the reaction also gives 3-bromo-1-butene, a product of allylic rearrangement. Allylic rearrangement can be observed even with the parent system labeled at one carbon with radioactive ^{14}C.

$$CH_2{=}CH^{14}CH_2OH \xrightarrow{SOCl_2} CH_2{=}CH^{14}CH_2Cl + {}^{14}CH_2{=}CHCH_2Cl$$

(ratio depends on reaction conditions)

EXERCISE 20.1 Write the structures for all of the isomeric products expected from the reaction of (*R*)-2-chloro-(*E*)-3-hexene with silver oxide in aqueous dioxane at 25 °C.

B. S_N2 Reactions

In addition to forming carbocations relatively easily, allylic halides and alcohols also undergo substitution by the S_N2 mechanism more readily than analogous saturated systems. For example, allyl bromide undergoes bimolecular substitution about 40 times more rapidly than does ethyl bromide. This enhanced reactivity results from stabilization of the transition state for substitution by the double bond with consequent lowering of the activation energy for the reaction (Figure 20.2). Recall (Figure 9.3) that the S_N2 transition state can be viewed as a *p*-orbital on the central carbon that simultaneously overlaps with orbitals from the entering and leaving nucleophiles. In an allylic system this *p*-orbital also participates in π-overlap with the adjacent double bond. Thus, even though allylic systems are prone to react by the S_N1 mechanism, because they form carbocations easily, it is possible by a careful choice of reaction conditions to cause them to react without allylic rearrangement by way of the S_N2 mechanism.

$$\underset{\text{but-2-en-1-ol}}{CH_3CH{=}CHCH_2OH} \xrightarrow[\text{pyridine}]{C_6H_5SO_2Cl} CH_3CH{=}CHCH_2OSO_2C_6H_5 \xrightarrow{Cl^-} \underset{\text{1-chloro-2-butene}}{CH_3CH{=}CHCH_2Cl}$$

In the foregoing example, the allylic hydroxy group is first converted into a good leaving group by reaction with benzenesulfonyl chloride in pyridine (page 222). This leaving group is then displaced by treatment with sodium chloride. Since substitution occurs by the S_N2 mechanism, only 1-chloro-2-butene is obtained. If substitution is accomplished by treatment of the alcohol with HCl, a mixture of 1-chloro-2-butene and 3-chloro-1-butene is obtained, analogously to the reaction with HBr shown on page 559.

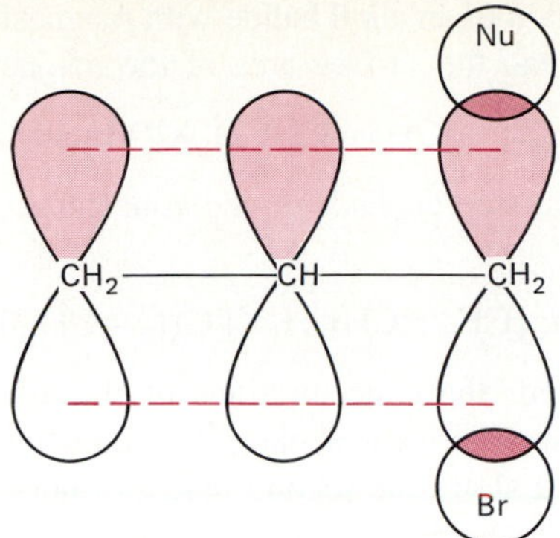

FIGURE 20.2 Transition state of an S_N2 reaction of allyl bromide with a nucleophile, Nu, showing interaction of the reacting center with the double bond.

The reactivity of allylic compounds in displacement reactions is sufficiently high that they even react with Grignard reagents.

$$CH_2{=}CHCH_2Br + \text{(cyclopentyl)}{-}MgBr \longrightarrow \text{(cyclopentyl)}{-}CH_2CH{=}CH_2 + MgBr_2$$

(70%)

Cyclopentylmagnesium bromide is prepared from 745 g of bromocyclopentane and 125 g of magnesium in 3 L of anhydrous ether. The mixture is refluxed while 605 g of allyl bromide is added slowly. After 2 hr, cold 6 N HCl is added and the ether layer is separated, washed, dried, and distilled to obtain 70% of allylcyclopentane, b.p. 121–125 °C.

The reaction of Grignard reagents with allyl bromide is a good method for preparing 1-alkenes. The corresponding reaction with normal saturated alkyl halides does not occur.

EXERCISE 20.2 Show how 4,4-dimethyl-1-pentene and 4-methyl-1-hexene can be synthesized from allyl chloride.

C. Allylic Anions

The allyl Grignard reagent may be prepared by treating an allyl halide with magnesium.

$$CH_2{=}CHCH_2X + Mg \xrightarrow{\text{ether}} CH_2{=}CHCH_2MgX$$

However, since allyl halides react with Grignard reagents to give 1-alkenes, care must be taken in preparing the allyl Grignard reagent. If the allyl halide concentration is too great, a large amount of the coupling product, 1,5-hexadiene or "biallyl," is formed.

$$CH_2{=}CHCH_2MgX + CH_2{=}CHCH_2X \longrightarrow \underset{\substack{\text{1,5-hexadiene} \\ \text{biallyl}}}{CH_2{=}CHCH_2CH_2CH{=}CH_2} + MgX_2$$

The allyl Grignard reagent can be prepared in good yield by minimizing the further S_N2 reaction of the Grignard reagent with allyl halide. This result may be accomplished by slow addition of a dilute solution of the allyl halide in ether to a large excess of a vigorously stirred suspension of magnesium. This technique is an example of the **dilution**

principle. The rate of reaction of an alkyl halide with magnesium depends on the concentration of the alkyl halide and the surface area of the magnesium.

$$\text{rate}_1 = k_1[\text{CH}_2\text{=CHCH}_2\text{X}][\text{Mg surface}]$$

The rate of the displacement step depends on the concentrations of allyl halide and Grignard reagent.

$$\text{rate}_2 = k_2[\text{CH}_2\text{=CHCH}_2\text{X}][\text{CH}_2\text{=CHCH}_2\text{MgX}]$$

When the solution is diluted, the concentrations of the allyl halide and the Grignard reagent decrease, but the surface area of the magnesium remains unchanged. Dilution retards both reactions, but it slows the second reaction more than the first. The surface area can be increased by sonochemical activation. Recall the mechanism for forming the Grignard reagent by electron transfer from the magnesium metal surface (see Section 8.8.A).

Isomeric allylic halides give Grignard reagents with indistinguishable properties. For example, the Grignard reagents derived from 3-bromo-1-butene and 1-bromo-2-butene give identical mixtures of alkenes upon hydrolysis.

$$\text{CH}_3\text{CHBrCH=CH}_2 \xrightarrow{\text{Mg}} \text{C}_4\text{H}_7\text{MgBr} \xleftarrow{\text{Mg}} \text{CH}_3\text{CH=CHCH}_2\text{Br}$$

$$\downarrow \text{aq H}^+$$

$$\underset{(57\%)}{\text{CH}_3\text{CH}_2\text{CH=CH}_2} + \underset{\substack{(27\%\text{ cis})\\(16\%\text{ trans})}}{\text{CH}_3\text{CH=CHCH}_3}$$

This result is understood in terms of rapid isomerization of the allylic Grignard reagent.

$$\text{CH}_3\text{CH=CHCH}_2\text{MgBr} \overset{\text{fast}}{\rightleftharpoons} \text{CH}_3\underset{\substack{|\\ \text{MgBr}}}{\text{C}}\text{HCH=CH}_2$$

Although Grignard reagents apparently have substantial carbon-magnesium covalent bonding, we have seen that in many reactions they behave as carbanion salts, $[\text{R—MgX} \longleftrightarrow \bar{\text{R}}\ \overset{+}{\text{M}}\text{gX}]$. Similarly, allylic Grignard reagents have a high degree of ionic character involving the magnesium cation salt of an allylic anion.

$$[\text{CH}_2\text{=CH—CH}_2\text{—MgX} \longleftrightarrow \text{CH}_2\text{=CH—}\bar{\text{C}}\text{H}_2\ \overset{+}{\text{M}}\text{gX} \longleftrightarrow {}^-\text{CH}_2\text{—CH=CH}_2\ \overset{+}{\text{M}}\text{gX}]$$

The negative charge is spread between two **conjugated carbons.** *A conjugated system is one having alternating single and double bonds.* This spreading of ionic character facilitates rearrangement of the magnesium. To summarize, there are two isomeric allylic Grignard reagents derived from 1-bromo-2-butene and 3-bromo-1-butene. The two isomers are in *rapid equilibrium* by allylic rearrangement of the magnesium. In the reaction of either isomer with an acid, protonation may occur at either of the carbons where there is negative charge. The net result on hydrolysis is that a mixture of product alkenes, in the present case 1-butene and 2-butene, is produced.

EXERCISE 20.3 What products are expected from treatment of 3-chloro-1-pentene with magnesium in ether followed by carbon dioxide, and then dilute sulfuric acid?

The spreading of charge in an allylic anion is a stabilizing mechanism; such anions are more stable than simple unconjugated anions. The allylic hydrogen of propene, for example, is substantially more acidic than any hydrogen in ethane or propane.

$$CH_2{=}CHCH_3 + B{:}^- \longrightarrow BH + [CH_2{=}CH{-}\ddot{C}H_2 \longleftrightarrow \ddot{C}H_2{-}CH{=}CH_2]$$

more acidic — delocalized allyl anion

$$CH_3CH_3 + B{:}^- \longrightarrow BH + CH_3\ddot{C}H_2$$

less acidic — localized ethyl anion

We have encountered similar kinds of conjugated or resonance-stabilized anions before: carboxylate ion (Section 18.4.A), enolate ion (Section 15.1.B), amidate ion (Section 19.10), even nitrite ion (Section 9.4).

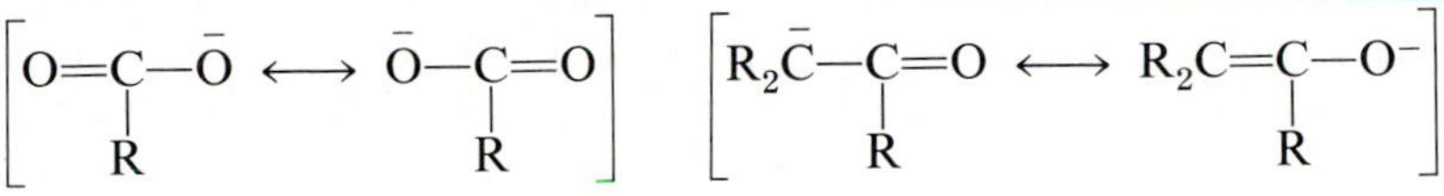

$$\left[O{=}\underset{R}{\underset{|}{C}}{-}\bar{O} \longleftrightarrow \bar{O}{-}\underset{R}{\underset{|}{C}}{=}O\right] \qquad \left[R_2\bar{C}{-}\underset{R}{\underset{|}{C}}{=}O \longleftrightarrow R_2C{=}\underset{R}{\underset{|}{C}}{-}O^-\right]$$

carboxylate ion — enolate ion

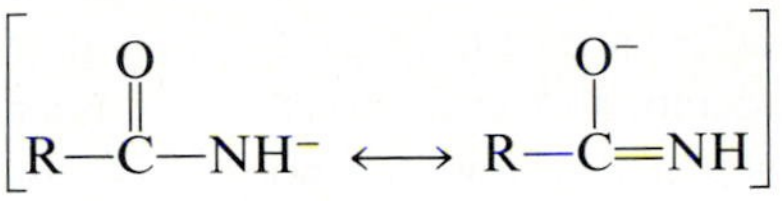

$$\left[R{-}\overset{O}{\overset{\|}{C}}{-}NH^- \longleftrightarrow R{-}\overset{O^-}{\overset{|}{C}}{=}NH\right] \qquad [O{=}N{-}O^- \longleftrightarrow {}^-O{-}N{=}O]$$

amidate ion — nitrite ion

All such ions are described in the same way. The resonance structures provide a way of representing a rather complex electronic distribution by means of the formal symbolism of Lewis structures. The actual structure is a composite of the resonance structures.

The mathematics of linear combinations is especially useful for describing this situation. In this approach, the wave function of a molecule Ψ is represented as a linear combination of simpler wave functions ψ.

$$\Psi = a\psi_a + b\psi_b + c\psi_c + \cdots$$

For many molecules a single structure provides a satisfactory representation. For example, methane is well represented by a single Lewis structure ψ_a, and

$$\Psi = \psi_a$$

Other possible Lewis structures, such as

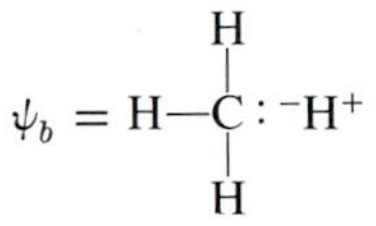

$$\psi_b = H{-}\overset{H}{\overset{|}{\underset{H}{\underset{|}{C}}}}{:}^-H^+$$

are so unlikely that their contribution to the linear combination is negligible; that is, b for methane is a very small number.

In allyl cation and anion, carboxylate ion, and nitrite ion, two resonance structures are equivalent and must have equal coefficients in the linear combination

$$\Psi(CH_2{\overset{...}{-}}CH{\overset{...}{-}}CH_2)^+ = a\psi_a(CH_2{=}CH{-}\overset{+}{C}H_2) + b\psi_b(\overset{+}{C}H_2{-}CH{=}CH_2)$$

where $a = b$.

In enolate ions the corresponding two structures are not equivalent. The structure with negative charge on the electronegative oxygen is more stable than that with the charge on carbon. Hence, these two structures enter into the linear combination with unequal coefficients.

$$\Psi(CH_2{\overset{...}{-}}CH{\overset{...}{-}}O)^- = a\psi_a(CH_2{=}CH{-}O^-) + b\psi_b(\bar{C}H_2{-}CH{=}O)$$

where $a > b$. We say that ψ_a contributes more than ψ_b; hence, the amount of negative charge on oxygen in Ψ, the resonance hybrid, is greater than that on carbon.

Resonance structures whose coefficients are very small contribute so little to the actual structure of the resonance hybrid that their contribution is generally neglected. It is this important property that allows us to represent even complex organic molecules by what is really a rather simple symbolism.

We have already encountered a number of neutral functions that have electronic structures closely related to that of allyl anion. In all such systems a lone pair of electrons is conjugated with a multiple bond.

$$\left[R-\overset{\overset{:O:}{\|}}{C}-\ddot{O}H \longleftrightarrow R-\overset{\overset{:\ddot{O}:^-}{|}}{C}=\overset{+}{\ddot{O}}H \right] \qquad \left[R-\overset{\overset{:O:}{\|}}{C}-\ddot{N}H_2 \longleftrightarrow R-\overset{\overset{:\ddot{O}:^-}{|}}{C}=\overset{+}{N}H_2 \right]$$

carboxylic acid — amide

These cases involve nonequivalent resonance structures. The dipolar structures involve charge separation and are often less stable than the normal Lewis structures. Hence such dipolar structures contribute less to the overall resonance hybrids. But they do contribute, and we have seen (Section 19.1) how consideration of such structures is essential to the understanding of the chemistry of such functional groups; the normal Lewis structures alone provide an inadequate description of the actual electronic structures of such groups.

EXERCISE 20.4 Amidines are to amides as imines are to ketones:

$$R-C\begin{matrix} \diagup\!\!\!\diagup NH \\ \diagdown NH_2 \end{matrix}$$

an amidine

Write the equation for reaction of an amidine with a proton. Compare the basicity of amidines and amides.

Highlight 20.1

Reaction of an allylic derivative, $CH_2{=}CH{-}CH_2{-}X$ under ionizing conditions (S_N1) gives an allylic carbocation that can react at either end. The reactive carbocation intermediate is delocalized.

$$CH_2{=}CH{-}CH_2{-}X \xrightarrow{-X^-} [CH_2{=}CH{-}CH_2^+ \longleftrightarrow {}^+CH_2{-}CH{=}CH_2] \longrightarrow$$

(Nu attacks either end of the allyl cation)

allyl cation

$$CH_2{=}CH{-}CH_2Nu \quad or \quad NuCH_2{-}CH{=}CH_2$$

Reaction of an allylic derivative, $CH_2{=}CH{-}CH_2{-}X$ with a nucleophile under conditions not favorable to ionization (S_N2) results in displacement of X and bonding of the nucleophile at the same position. The transition state is stabilized by delocalization, but reaction occurs at a specific position.

$$CH_2{=}CH{-}\underset{}{\overset{X}{\overset{|}{C}}}H_2 \xrightarrow{Nu^-} CH_2{=}CH{-}\underset{Nu^{\frac{1}{2}-}}{\overset{X^{\frac{1}{2}-}}{CH_2}} \xrightarrow{-X^-} CH_2{=}CH{-}\underset{Nu}{\underset{|}{CH_2}}$$

$$\underset{Nu^-}{\overset{X^-}{CH_2{=}CH{-}\overset{+}{C}H_2}} \longleftrightarrow \underset{Nu^-}{\overset{X^-}{\overset{+}{C}H_2{-}CH{=}CH_2}}$$

transition state
resonance forms

X

$CH_2{-}CH{-}CH_2$

Nu

orbital form
(see Figure 9.11 for hyperconjugation)

An allylic anion (for example, one associated with magnesium in a Grignard reagent) can react at either end. The reactive intermediate is delocalized, but the Grignard reagent is a mixture of two ion pairs in rapid equilibrium.

$$CH_2{=}CH{-}CH_2{-}X \xrightarrow{Mg}$$

magnesium monobromide allyl anion
ion pair A

$$\left[\underset{^+MgX}{CH_2{=}CH{-}CH_2^-} \longleftrightarrow \underset{^+MgX}{^-CH_2{-}CH{=}CH_2}\right]$$

$\rightleftharpoons$

$$\left[\underset{^+MgX}{CH_2{=}CH{-}CH_2^-} \longleftrightarrow \underset{^+MgX}{^-CH_2{-}CH{=}CH_2}\right] \xrightarrow[D_2O]{D^+} CH_2{=}CH{-}CH_2D \quad \textit{or} \quad DCH_2{-}CH{=}CH_2$$

magnesium monobromide allyl anion
ion pair B

D. Allylic Radicals

Allylic radicals are also stabilized by resonance.

$$[CH_2{=}CH{-}\dot{C}H_2 \longleftrightarrow \dot{C}H_2{-}CH{=}CH_2]$$

The odd-electron character is shared by two carbons, and this radical is more stable than a simple alkyl radical. This increased stability is reflected in the relatively low bond-dissociation energy of bonds conjugated to a double bond.

	$DH°$, kcal mole^{-1}
$CH_3CH_2{-}H$	101
$CH_2{=}CHCH_2{-}H$	87

Advantage can be taken of the low bond-dissociation energy of allylic carbon-hydrogen bonds in free radical halogenation—but only under special circumstances because of the alternative reaction path of addition to the double bond. One method for accomplishing **allylic bromination** is with the reagent N-bromosuccinimide. This material is available commercially and is prepared by bromination of the cyclic imide of succinic acid (Section 27.6.C).

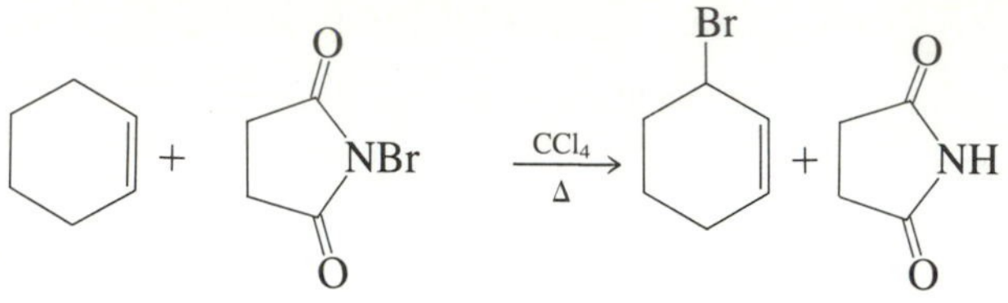

$$\text{cyclohexene} + \text{N-bromosuccinimide} \xrightarrow[\Delta]{CCl_4} \text{3-bromocyclohexene} + \text{succinimide}$$

N-bromosuccinimide   succinimide

The reaction is normally carried out in carbon tetrachloride, in which both the reactant N-bromosuccinimide and the product succinimide are insoluble. Reaction occurs in part on the surface of the N-bromosuccinimide, although the active reagent appears to be bromine formed in dilute solution from the reaction of traces of acid and moisture with the bromo imide.

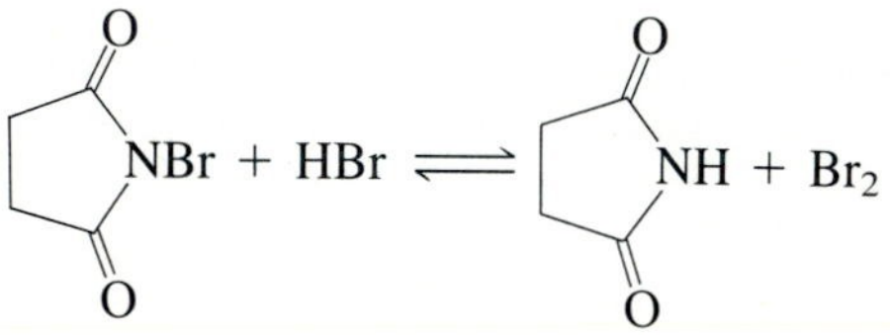

$$\text{(succinimide)NBr} + HBr \rightleftharpoons \text{(succinimide)NH} + Br_2$$

The bromine is then involved in free radical chain bromination of the allylic hydrogen. Under these conditions of high dilution no significant addition of bromine to the double bond occurs.

One of the reasons for using a nonpolar solvent such as CCl_4 in this reaction is that the normal addition of Br_2 to a double bond is an ionic reaction (see Section 11.6.B).

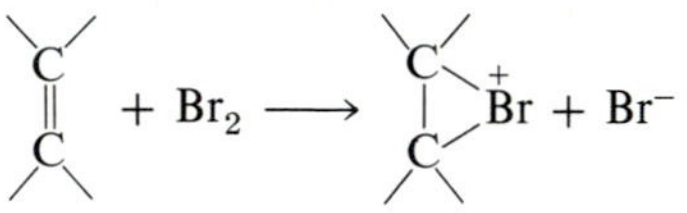

$$>C{=}C< + Br_2 \longrightarrow \text{(bromonium ion)}\ \overset{+}{Br} + Br^-$$

In the absence of a suitable solvent to solvate these ions, one or more excess bromine molecules are required for this role. Recall that the crystal structure of an isolated bromonium tribromide derivative has been determined (page 274, Section 11.6.B).

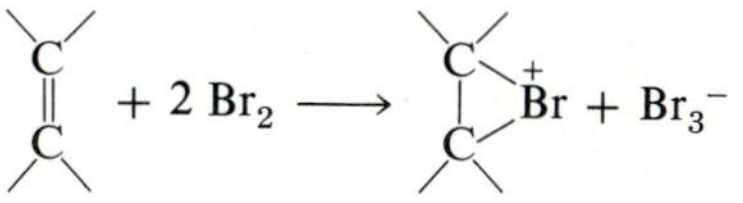

$$>C{=}C< + 2\,Br_2 \longrightarrow \text{(bromonium ion)}\ \overset{+}{Br} + Br_3^-$$

Thus the reaction kinetics has a relatively high order in bromine and the ionic addition has a low reaction rate when bromine is kept in low concentration.

A free radical initiator or light is often used to promote the reaction. Because the reaction intermediate is a resonance-stabilized radical, two products can be obtained in unsymmetrical cases.

$$Br_2 \xrightarrow[\text{or } (C_6H_5COO^-)_2]{h\nu} 2\ Br\cdot$$

$$Br\cdot + H{-}CHR{-}CH{=}CHR' \longrightarrow$$

$$[Br{-}{-}{-}H{-}{-}{-}\dot{C}HR{-}CH{=}CHR' \longleftrightarrow Br{-}{-}{-}H{-}{-}{-}CHR{=}CH{-}\dot{C}HR'] \longrightarrow$$

transition state

$$Br{-}H + [\dot{C}HR{-}CH{=}CHR \longleftrightarrow CHR{=}CH{-}\dot{C}HR]$$

allylic radical

$$[R\dot{C}H{-}CH{=}CHR' \longleftrightarrow RCH{=}CH{-}\dot{C}HR'] \xrightarrow{Br_2} R\underset{}{\overset{Br}{C}}H{-}CH{=}CHR' + RCH{=}CH{-}\overset{Br}{C}HR'$$

Allyl chloride is prepared commercially in large quantity by the direct free radical chlorination of propylene at high temperature. At higher temperatures the normal addition of chlorine atom to the double bond becomes unfavorable for entropic reasons, and hydrogen abstraction is the principal reaction. In the addition reaction two species become one and freedom of motion is lost, whereas the entropy change in hydrogen abstraction is small. Entropy considerations are especially important at high temperatures.

$$CH_3CH{=}CH_2 + Cl_2 \xrightarrow{500^\circ C} ClCH_2CH{=}CH_2 + HCl$$

$$Cl_2 \rightleftharpoons 2\ Cl\cdot$$

$$CH_2{=}CHCH_3 + Cl\cdot \longrightarrow HCl + CH_2{=}CH\dot{C}H_2$$

$$CH_2{=}CH\dot{C}H_2 + Cl_2 \longrightarrow Cl\cdot + CH_2{=}CHCH_2Cl$$

Allyl alcohol is prepared from the chloride by hydrolysis.

EXERCISE 20.5 Consider two reactions A and B having $\Delta\Delta H^{\ddagger} = 10$ kcal mole^{-1} and $\Delta\Delta S^{\ddagger} = 20$ e.u. Calculate the relative rate constants, k_A/k_B, at 300 K and at 700 K. What is the relevance of this exercise to the reactions discussed in this section?

20.2 Dienes

A. Structure and Stability

Double bonds separated by one or more carbon atoms react more or less independently. The heats of hydrogenation of such double bonds are essentially those of independent units. For example, ΔH° for the reaction of 1,5-hexadiene with hydrogen is exothermic by 60 kcal mole^{-1}, exactly twice that for the reaction of 1-hexene with hydrogen. (Recall that the heat of hydrogenation of an alkene is ΔH° for the reaction alkene + $H_2 \rightarrow$ alkane.)

$$CH_2{=}CHCH_2CH_2CH{=}CH_2 + 2\ H_2 \longrightarrow CH_3CH_2CH_2CH_2CH_2CH_3 \qquad \Delta H^\circ_{hydrog} = -60 \text{ kcal mole}^{-1}$$

$$CH_3CH_2CH_2CH_2CH{=}CH_2 + H_2 \longrightarrow CH_3CH_2CH_2CH_2CH_2CH_3 \qquad \Delta H^\circ_{hydrog} = -30 \text{ kcal mole}^{-1}$$

Heats of hydrogenation for other alkenes and dienes are included in Table 20.1.

TABLE 20.1 Heats of Hydrogenation

	ΔH°_{hydrog}, kcal mole^{-1}
$CH_3CH_2CH{=}CH_2$	−30.2
$CH_3CH_2CH_2CH{=}CH_2$	−29.8
$CH_3CH_2CH_2CH_2CH{=}CH_2$	−30.0
$CH_2{=}CH{-}CH{=}CH_2$	−56.5
$CH_2{=}CHCH_2CH{=}CH_2$	−60.4
$CH_2{=}CHCH_2CH_2CH{=}CH_2$	−60.0

Note that 1,3-butadiene is a significant exception to the preceding generalization. Its hydrogenation is about 4 kcal mole^{-1} less exothermic than for the other two dienes. This compound is an example of a **conjugated diene,** a diene in which the two double bonds are separated by a single bond. Dienes in which two or more single bonds separate the double bonds are called **unconjugated dienes.** The double bonds in unconjugated dienes are called **isolated double bonds.**

Conjugated dienes are significantly more stable than would be expected for a compound with completely independent double bonds. This relatively small but significant difference is attributed to two factors, which are shown in Figure 20.3.

First, the two double-bond distances are essentially normal, but the C-2—C-3 single bond is shorter than the 1.54 Å distance normally associated with carbon-carbon single bonds.

$$CH_2 \overset{1.34\ \text{Å}}{=\!=} CH \overset{1.48\ \text{Å}}{-\!-} CH =\!= CH_2$$

This decreased bond length results in part from the increased *s*-character of the carbon orbitals comprising this bond; the single bond between the double bonds may be described approximately as C_{sp^2}—C_{sp^2}. This shorter bond is somewhat stronger than carbon-carbon bonds having less *s*-character. Second, the p_z-orbitals on C-2 and C-3 can also overlap to give some double-bond character to the C-2—C-3 single bond. This factor also contributes some additional stability to the conjugated double-bond system. However, this overlap is much less than those between C-1 and C-2 and between C-3 and C-4 carbons because of the greater distance between the C-2 and C-3 *p*-orbitals.

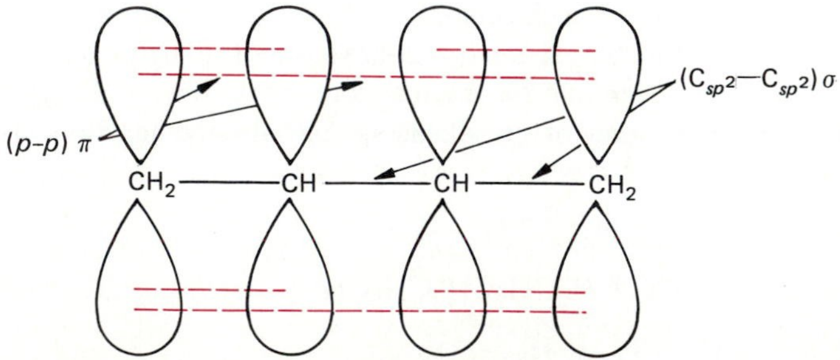

FIGURE 20.3 Structure of 1,3-butadiene.

B. Addition Reactions

The conjugated character of 1,3-dienes is shown in two-step addition reactions. Such additions are almost invariably initiated at the end of a chain of conjugation to produce a resonance-stabilized allylic intermediate rather than a nonconjugated intermediate.

$$CH_2{=}CH{-}CH{=}CH_2 + HBr \not\rightarrow \overset{+}{C}H_2{-}\underset{\displaystyle H}{\underset{|}{C}H}{-}CH{=}CH_2$$

$$CH_2{=}CH{-}CH{=}CH_2 + HBr \longrightarrow \left[\underset{\displaystyle H}{\underset{|}{C}H_2}{-}\overset{+}{C}H{-}CH{=}CH_2 \longleftrightarrow \underset{\displaystyle H}{\underset{|}{C}H_2}{-}CH{=}CH{-}\overset{+}{C}H_2\right] + Br^-$$

This intermediate reacts in a second step to give a mixture of products characteristic of the intermediate allylic system.

$$[CH_3\overset{+}{C}H{-}CH{=}CH_2 \longleftrightarrow CH_3CH{=}CH{-}\overset{+}{C}H_2] + Br^- \longrightarrow \underset{\substack{(80\%)\\ \alpha\text{-methylallyl}\\ \text{bromide}}}{CH_3\overset{\displaystyle Br}{\overset{|}{C}H}CH{=}CH_2} + \underset{\substack{(20\%)\\ \text{crotyl bromide}}}{CH_3CH{=}CHCH_2Br}$$

A further example is seen in the addition of bromine to 1,3-butadiene.

$$CH_2{=}CH{-}CH{=}CH_2 + Br_2 \xrightarrow[-15^\circ C]{\text{hexane}} \left[\begin{array}{c} CH_2{=}CH{-}\overset{+}{C}HCH_2Br \\ \updownarrow \\ \overset{+}{C}H_2{-}CH{=}CHCH_2Br \end{array}\right] Br^- \longrightarrow$$

$$\underset{46\%}{BrCH_2CH{=}CHCH_2Br} + \underset{54\%}{CH_2{=}CHCHBrCH_2Br}$$

In this case, the allylic carbocation is sufficiently stabilized and may not be converted to a cyclic bromonium ion (see Highlight 20.2 and Section 11.6.B).

When the mixture of dibromides in the foregoing example is warmed to 60 °C, the composition changes to 90% (*E*)-1,4-dibromo-2-butene and 10% 3,4-dibromo-1-butene. The major isomer of this mixture is easy to isolate in a pure state because it is a solid, m.p. 54 °C, and crystallizes readily.

$$\left.\begin{array}{lr} BrCH_2CH{=}CHCH_2Br & (46\%) \\ + & \\ CH_2{=}CHCHBrCH_2Br & (54\%) \end{array}\right\} \xrightarrow{60^\circ C} \underset{(90\%)}{\begin{array}{ccc} BrCH_2 & & H \\ & C{=}C & \\ H & & CH_2Br \end{array}}$$

Thus, (*E*)-1,4-dibromo-2-butene is the most stable product, but it is formed at a rate comparable to the rate of formation of the other isomer.

This example illustrates an important concept in organic chemistry, **kinetic versus thermodynamic control.** In the addition reaction, the product composition is determined by the relative rates of reaction of the nucleophilic reagent at the two positions of positive charge. These relative reactivities need not, and generally do not, reflect the relative thermodynamic stabilities of the products. In the present case, the reaction of the intermediate carbocation with bromide ion occurs approximately equally at both cationic centers; the reaction shows little selectivity. However, 1,4-dibromo-2-butene has a disubstituted double bond and is somewhat more stable than 3,4-dibromo-1-butene, which has a

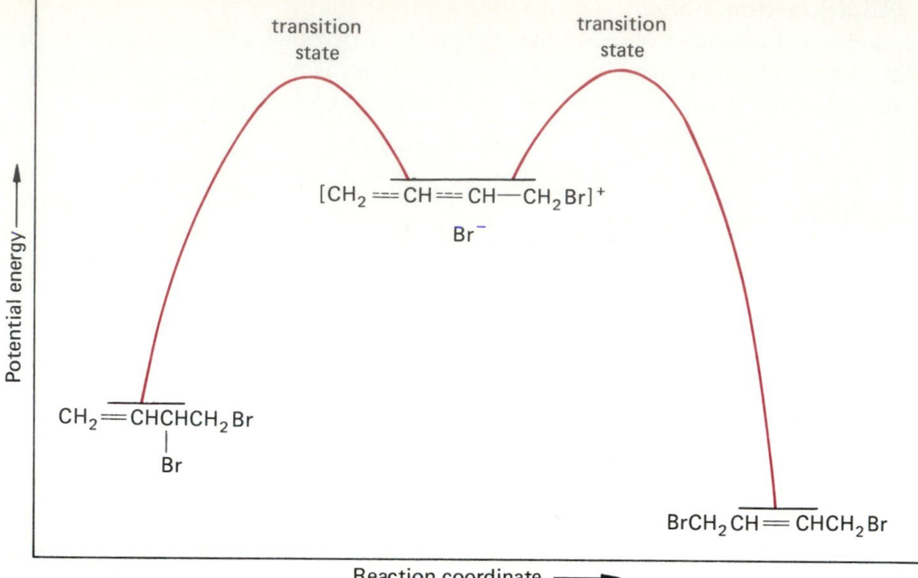

FIGURE 20.4 Kinetic and thermodynamic effects in the formation of dibromobutenes.

monosubstituted double bond (Section 11.4). Under conditions where the dibromides can react to reform the carbocation, the more stable isomer predominates. Such a process provides a mechanism for establishing equilibrium.

The diagram in Figure 20.4 shows the energy relationships between the products and the intermediate. The two alternative transition states derived from the intermediate carbocation have comparable energies and give the alternative products at approximately equal rates. Actually, the rate of formation of 3,4-dibromo-1-butene, which involves reaction at the more positive secondary carbocation, is a little faster than the formation of the 1,4-dibromo isomer. However, the 1,4-isomer is the more stable; it reforms the carbocation less readily than the 3,4-isomer. Hence, at equilibrium, some of the 3,4-isomer is converted to 1,4-isomer, and the latter predominates.

The contrast between kinetic and thermodynamic control is important and will be encountered from time to time in our further study of organic chemistry. Another example is found in the reaction products of butadiene with hydrogen bromide. 3-Bromo-1-butene (α-methylallyl bromide) is the dominant product of the addition reaction. However, the equilibrium mixture consists of only 15% of α-methylallyl bromide and 85% of 1-bromo-2-butene (crotyl bromide). Once again, equilibrium favors the more highly substituted double bond. On prolonged reaction or treatment with strong Lewis acids such as ferric bromide, the equilibrium mixture is produced (Figure 20.5).

EXERCISE 20.6 The reaction of 1,3-butadiene with aqueous bromine is expected to afford three isomeric bromobutenols (see page 275). Give the structures of these three products and predict the principal product of the reaction of each isomer with sodium hydroxide (page 237).

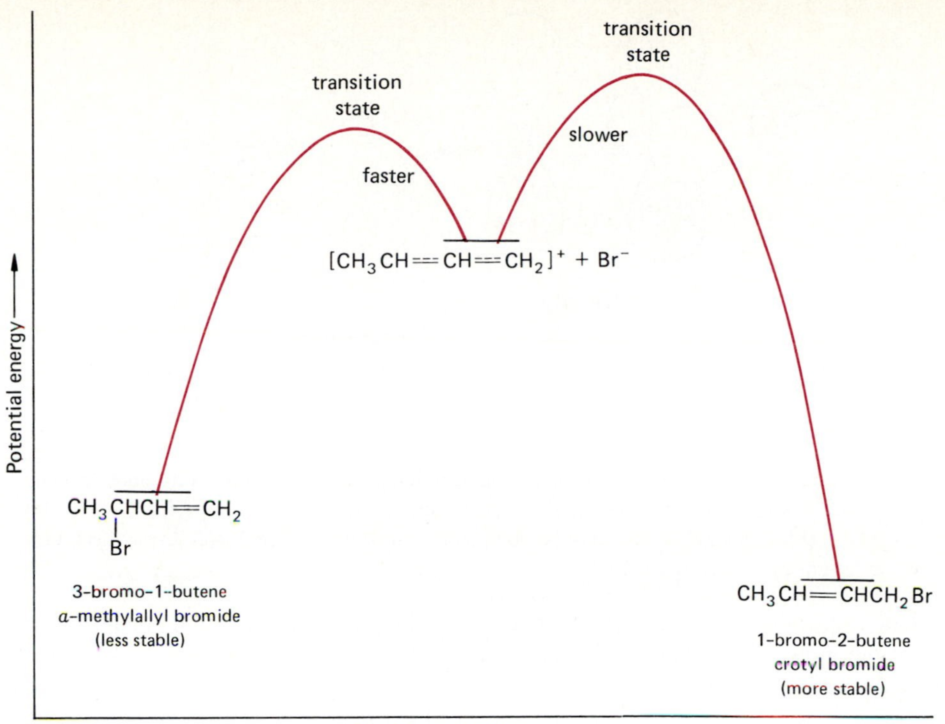

FIGURE 20.5 Kinetic and thermodynamic effects in the formation of bromobutenes.

Highlight 20.2

The addition of bromine to 1,3-butadiene (a 1,3-diene) first forms a bromomethyl allylic cation bromide ion pair. The low polarity (in this case, low solvating power for ions) prevents dissociation of the ion pair. Cyclization of the allylic cation to a bromonium bromide is energetically disfavored. The initial attack of bromide ion on the cation takes place about equally at the end carbon to give the 1,4-adduct, or on the second carbon to yield the 1,2-adduct.

$$CH_2{=}CH{-}CH{=}CH_2 + Br_2 \xrightarrow[-15\ ^\circ C]{\text{hexane}}$$

$$\left[\overset{+}{C}H_2{-}CH{=}CH{-}\underset{Br}{CH_2}\ \ Br^- \longleftrightarrow CH_2{=}CH{-}\overset{+}{C}H{-}\underset{Br}{CH_2}\ \ Br^-\right] \rightleftharpoons CH_2{=}CH{-}CH{-}CH_2\ (\text{bridged } Br^+)\ \ Br^-$$

60 °C, slow (↑ from 1,4-adduct); 46% ↙ −15 °C ↘ 54%; 60 °C, fast (↑ from 1,2-adduct)

$$\underset{Br}{CH_2}{-}CH{=}CH{-}\underset{Br}{CH_2} \qquad CH_2{=}CH{-}\underset{Br}{CH}{-}\underset{Br}{CH_2}$$

On warming, the 1,2-dibromide reforms the allylic cation bromide ion pair more rapidly than the 1,4-dibromide. The balance of these rate processes at 60 °C leads to a mixture containing 90% 1,4-dibromide.

C. 1,2-Dienes: Allenes

1,2-Propadiene, $CH_2{=}C{=}CH_2$, has the trivial name ''allene.'' Both double bonds in this hydrocarbon are especially short; the bond distance of 1.31 Å is between those for the double bond in ethylene, 1.34 Å, and the triple bond in acetylene, 1.20 Å. The electronic structure can be represented in terms of two double-bond systems at right angles, as in

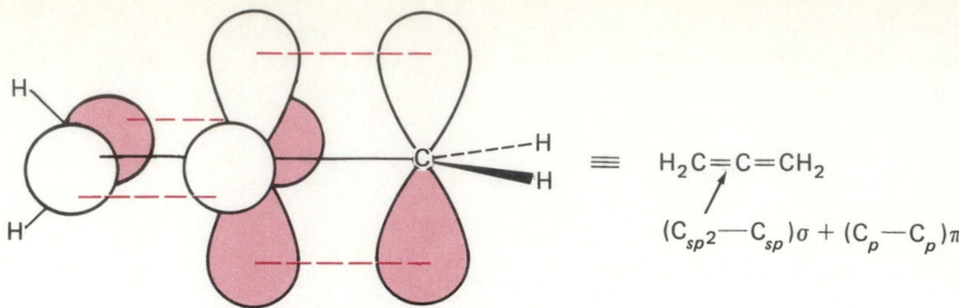

FIGURE 20.6 Orbital structure of allene.

Figure 20.6. Note that the central carbon is *sp*-hybridized. The additional *s*-character in these carbon-carbon double bonds accounts for the rather short length.

> A strong electron-withdrawing substituent on acetylene decreases the bond distance further. The shortest known acetylene bond, 1.145 Å, occurs in the iodonium salt, HC≡C—I$^+$—Ph $CF_3SO_3^-$ (Ph = phenyl). This may be compared with the value of 1.153 Å for the C≡N bond length given in Figure 12.6.

One especially interesting feature of allenes, which results from the nonplanar character of the molecule, is that suitably substituted allenes are chiral and can be obtained as optically active enantiomers. For example, 2,3-pentadiene has no plane of symmetry. Its mirror images are not superimposable, and this hydrocarbon is capable of existence as a pair of enantiomers.

$CH_3(H)C{=}C{=}C(H)(CH_3)$	$H(H_3C)C{=}C{=}C(CH_3)(H)$ ≡ $CH_3(H)C{=}C{=}C(CH_3)(H)$

mirror images of penta-2,3-diene

A stereo representation of such an allene is given in Figure 20.7.

Chiral allenes are examples of compounds that are chiral but do not have a stereocenter. Instead, such allenes have an asymmetric distribution of groups around the C—C—C axis, which can be labeled a **stereoaxis.** The two enantiomers of such a chiral allene can be named using a straightforward extension of the R,S convention (Section 7.3). To make such an assignment, the molecule is viewed along the stereoaxis. The two substituents on the nearer carbon are given priorities of 1 and 2, according to the sequence rule. The two substituents on the farther carbon are given priorities of 3 and 4, on the same basis. The four ranked substituents, 1, 2, 3, and 4, describe the four corners of a distorted

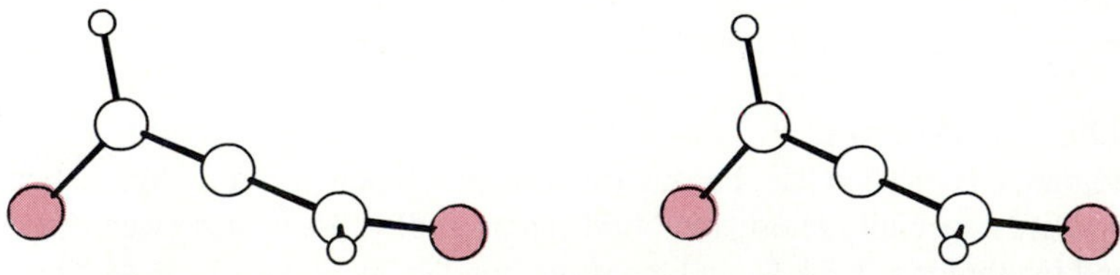

FIGURE 20.7 Stereo representation of a 1,3-disubstituted allene.

tetrahedron. The 1,2,3 face of this tetrahedron is viewed, with apex 4 to the rear. If the arc 1,2,3 is clockwise, the molecule has the R configuration; if it is counterclockwise, the configuration is S.

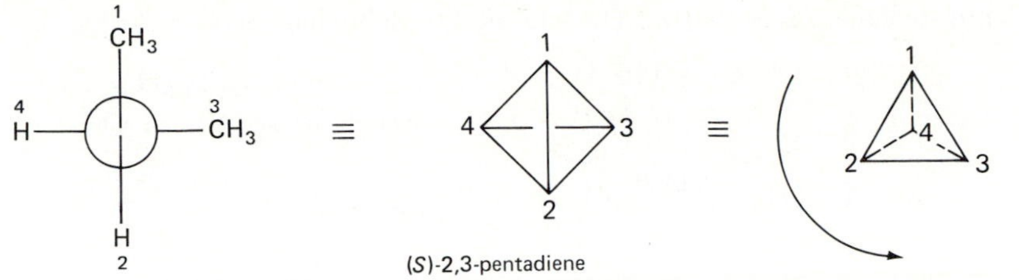

(S)-2,3-pentadiene

EXERCISE 20.7 In applying the foregoing procedure, it does not matter from which direction the stereoaxis is viewed. Make a model of (*S*)-2,3-pentadiene and convince yourself that this is true. Make a model of one of the stereoisomers of 1-chloro-1,2-butadiene. Which enantiomer have you prepared?

Molecules with **cumulated** double bonds, double bonds on successive carbon atoms as in allene, do not constitute an important class of compounds. They are generally difficult to prepare and can frequently be isomerized to more stable dienes. Allene, for example, with $\Delta H_f^\circ = 45.9$ kcal mole^{-1} is 1.6 kcal mole^{-1} less stable than propyne with $\Delta H_f^\circ =$ 44.3 kcal mole^{-1}; and 1,2-butadiene (methylallene, $CH_3CH{=}C{=}CH_2$, $\Delta H_f^\circ =$ 38.3 kcal mole^{-1}), though slightly more stable than 1-butyne ($\Delta H_f^\circ = 39.5$ kcal mole^{-1}), is almost 4 kcal mole^{-1} less stable than 2-butyne ($\Delta H_f^\circ = 34.7$ kcal mole^{-1}) and almost 13 kcal mole^{-1} less stable than 1,3-butadiene ($\Delta H_f^\circ = 26.1$ kcal mole^{-1}). Note that the heats of formation are positive. (See Section 5.5)

EXERCISE 20.8 Use molecular models to show that 1,3-dichloro-1,2-propadiene is chiral and that 1,4-dichloro-1,2,3-butatriene is not. Which enantiomer of 1,3-dichloro-1,2-propadiene have you constructed?

D. Preparation of Dienes

Many dienes can be prepared in much the same way as monoenes except that two functional groups are involved. Some examples are shown below.

$$(CH_3)_2C(OH){-}C(OH)(CH_3)_2 \xrightarrow[420\text{–}470^\circ C]{Al_2O_3} CH_2{=}C(CH_3){-}C(CH_3){=}CH_2$$

(79–85%)
2,3-dimethyl-1,3-butadiene

$$CH_3COOCH_2CH_2CH_2CH_2CH_2OOCCH_3 \xrightarrow{575^\circ C} 2\ CH_3COOH + CH_2{=}CHCH_2CH{=}CH_2$$

(63–71%)
1,4-pentadiene

Many other synthetic methods are known, but most require difunctional compounds (to be studied later) as starting materials.

Allylic halides are useful for preparing both conjugated and unconjugated dienes.

Displacement by a vinylmagnesium halide on an allyl halide is another route to 1,4-dienes (see Section 20.1.B).

$$CH_2{=}CHCH_2Cl + CH_2{=}CHMgBr \xrightarrow{\text{ether}} CH_2{=}CHCH_2CH{=}CH_2 + MgBrCl$$

The Wittig reaction (Section 14.7.D) with allylphosphoranes gives 1,3-dienes.

$$CH_2{=}CHCH_2Cl + P(C_6H_5)_3 \longrightarrow CH_2{=}CHCH_2\overset{+}{P}(C_6H_5)_3\ Cl^-$$

allyltriphenylphosphonium chloride

$$CH_2{=}CHCH_2\overset{+}{P}(C_6H_5)_3 \xrightarrow{LiOC_2H_5} [CH_2{=}CH\overset{-}{C}H\overset{+}{P}(C_6H_5)_3] \xrightarrow{C_6H_5CHO}$$

$$CH_2{=}CHCH{=}CHC_6H_5 + (C_6H_5)_3PO$$

1-phenyl-1,3-butadiene

Another use of allylic intermediates is exemplified in one preparation of 2-methyl-1,3-butadiene (isoprene).

$$CH_3COCH_3 + HC{\equiv}CH \xrightarrow{KOH} (CH_3)_2\underset{\displaystyle\,}{\overset{\displaystyle OH}{\overset{|}{C}}}C{\equiv}CH \xrightarrow{Pd,\ H_2}$$

2-methyl-3-butyn-2-ol

$$(CH_3)_2\underset{\underset{\displaystyle OH}{|}}{C}CH{=}CH_2 \xrightarrow[\Delta]{Al_2O_3} CH_2{=}\overset{\overset{\displaystyle CH_3}{|}}{C}{-}CH{=}CH_2$$

2-methyl-3-buten-2-ol — isoprene

EXERCISE 20.9 Show how 3-chlorocyclohexene can be converted into (a) 3-vinylcyclohexene and (b) 3-isopropylidenecyclohexene. Note the nomenclature used to name a compound with an *exocyclic* double bond, a double bond attached to a ring.

3-isopropylidenecyclohexene

20.3 Unsaturated Carbonyl Compounds

A. Unsaturated Aldehydes and Ketones

Compounds having both a carbonyl group and a double bond are known as unsaturated aldehydes or ketones. As with dienes, the two centers of unsaturation can be conjugated or unconjugated and the conjugated isomers are generally more stable.

$$\underset{\text{unconjugated isomer}}{CH_2{=}CHCH_2CHO} \longrightarrow \underset{\text{conjugated isomer}}{CH_3CH{=}CHCHO} \qquad \Delta H° = -6 \text{ kcal mole}^{-1}$$

$$\underset{\text{unconjugated isomer}}{CH_2{=}CHCH_2CH{=}CH_2} \longrightarrow \underset{\text{conjugated isomer}}{CH_3CH{=}CHCH{=}CH_2} \qquad \Delta H° = -7 \text{ kcal mole}^{-1}$$

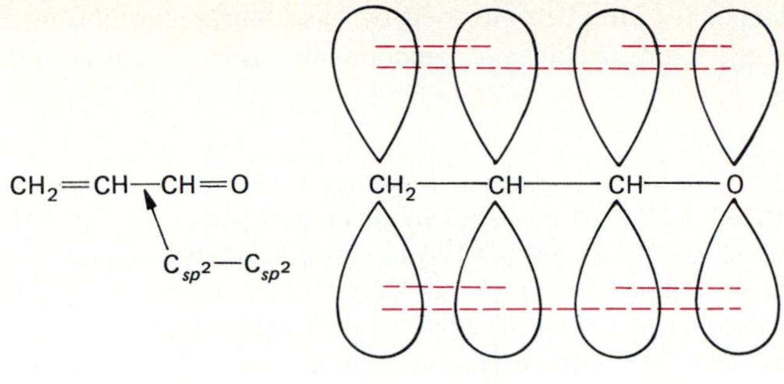

$$CH_2{=}CH{-}CH{=}O \quad (C_{sp^2}{-}C_{sp^2})$$

FIGURE 20.8 Orbital structure of propenal (acrolein).

The stabilizing effect of conjugation in unsaturated carbonyl compounds is of the same order of magnitude as that in the corresponding dienes. It is explained in the same manner in terms of the stabilizing effect of the central $C_{sp^2}{-}C_{sp^2}$ bond and by the overlap of p-orbitals to give π-bonding (Figure 20.8).

Because of the greater stability of the conjugated unsaturated aldehydes and ketones, an isolated double bond will tend to **move into conjugation** if a suitable pathway is available. This migration of double bonds is especially facile for double bonds that are β,γ to the carbonyl group because both acid- and base-catalyzed reactions produce an intermediate conjugated enol.

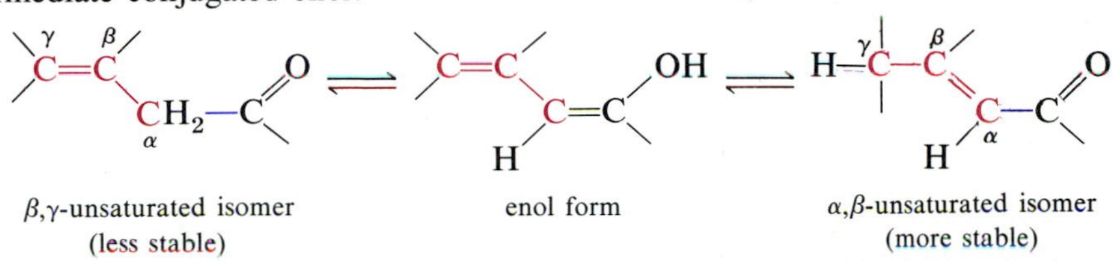

$$\underset{\substack{\beta,\gamma\text{-unsaturated isomer}\\ \text{(less stable)}}}{\overset{\gamma}{C}{=}\overset{\beta}{C}{-}\underset{\alpha}{CH_2}{-}C{=}O} \rightleftharpoons \underset{\text{enol form}}{C{=}C{-}C(H){=}C{-}OH} \rightleftharpoons \underset{\substack{\alpha,\beta\text{-unsaturated isomer}\\ \text{(more stable)}}}{H{-}\overset{\gamma}{C}{-}\overset{\beta}{C}{=}\underset{\alpha}{C}(H){-}C{=}O}$$

For example, the methylene hydrogens in 3-butenal (vinylacetaldehyde) are appreciably acidic because they are α to the carbonyl group and give rise to a resonance-stabilized enolate ion. However, in this enolate ion a further resonance structure can be written that shows that negative charge is also spread to the γ-carbon.

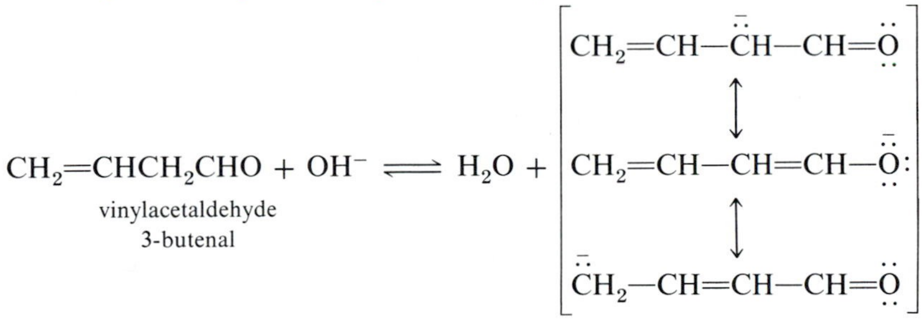

$$\underset{\substack{\text{vinylacetaldehyde}\\ \text{3-butenal}}}{CH_2{=}CHCH_2CHO} + OH^- \rightleftharpoons H_2O + \left[\begin{array}{c} CH_2{=}CH{-}\ddot{C}H^-{-}CH{=}\ddot{O} \\ \updownarrow \\ CH_2{=}CH{-}CH{=}CH{-}\ddot{O}:^- \\ \updownarrow \\ \ddot{C}H_2^-{-}CH{=}CH{-}CH{=}\ddot{O} \end{array}\right]$$

This delocalized enolate anion now has three positions that can be protonated by water. Protonation on oxygen gives the enol form, protonation of the α-carbon regenerates vinylacetaldehyde, and protonation at the γ-carbon generates crotonaldehyde.

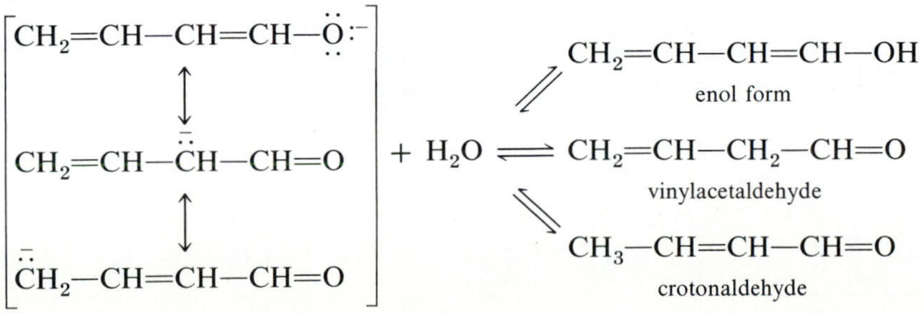

$$\left[\begin{array}{c} CH_2{=}CH{-}CH{=}CH{-}\ddot{O}:^- \\ \updownarrow \\ CH_2{=}CH{-}\ddot{C}H^-{-}CH{=}O \\ \updownarrow \\ \ddot{C}H_2^-{-}CH{=}CH{-}CH{=}O \end{array}\right] + H_2O \rightleftharpoons \begin{cases} \underset{\text{enol form}}{CH_2{=}CH{-}CH{=}CH{-}OH} \\ \underset{\text{vinylacetaldehyde}}{CH_2{=}CH{-}CH_2{-}CH{=}O} \\ \underset{\text{crotonaldehyde}}{CH_3{-}CH{=}CH{-}CH{=}O} \end{cases}$$

All of these compounds are interconverted by base, but at equilibrium the most stable isomer, the conjugated crotonaldehyde, predominates to the extent of greater than 99.9% (Figure 20.9).

Both (*Z*)- and (*E*)-trimethylsilyl ethers of 1-hydroxy-1,3-butadiene are known (see Sections 15.1.B and 16.4), and can be converted by dilute acid hydrolysis (10^{-3} M HCl in 90% CH_3CN-H_2O at 32 °C) into (*Z*)- and (*E*)-1-hydroxy-1,3-butadienes, respectively. Under these conditions, the enols could be kept for 2–3 days before slowly changing to mixtures of 3-butenal (vinylacetaldehyde) and (*E*)-2-butenal (crotonaldehyde), which were initially different from each enol. The NMR coupling constant $J_{2\text{-}3}$ is 10–11 Hz which suggests that the central single bond has a trans arrangement of the double bonds as in 1,3-butadiene (see Section 13.9). Conformers that have groups on the same side of a *single bond* (s) are s-cis; if the groups are on opposite sides, the conformer is s-trans. The barrier to interconversion of the s-cis and s-trans conformers in 1,3-butadiene is somewhat higher than for the rotation of ethane.

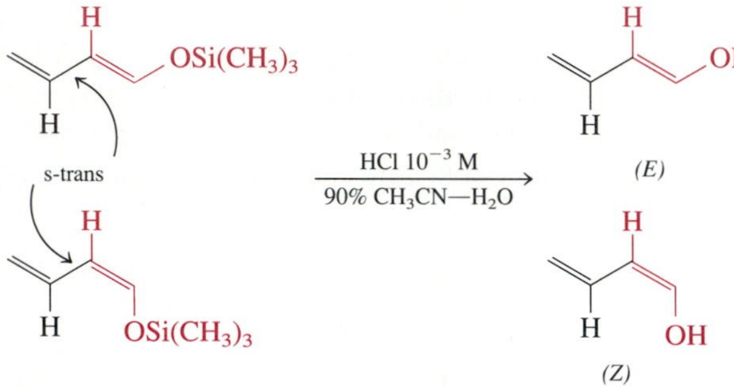

H OSi(CH$_3$)$_3$ H s-trans H H OSi(CH$_3$)$_3$ — HCl 10^{-3} M, 90% CH_3CN—H_2O → H OH H (*E*) H H OH (*Z*)

The rate constant for acid-catalyzed ketonization to the aldehyde mixture is a factor of 1500 less for the *Z* isomer and 900 less for the *E* isomer than for the transformation of ethenol (vinyl alcohol) to acetaldehyde. Recall that each factor of 10 in rate corresponds to a transition state energy difference of 1.4 kcal mole^{-1}. Conjugation retards the rate of conversion of the enol to the carbonyl form, raising $\Delta\Delta G^{\ddagger}$ by about 4 kcal mole^{-1}, the same as the decrease in the $\Delta\Delta H°$ for hydrogenation (Section 21.2.A).

One way in which these interconversions can be demonstrated is by deuterium exchange. The enolate ion derived from crotonaldehyde or vinylacetaldehyde can react with D_2O at either of the carbons that bear the negative charge.

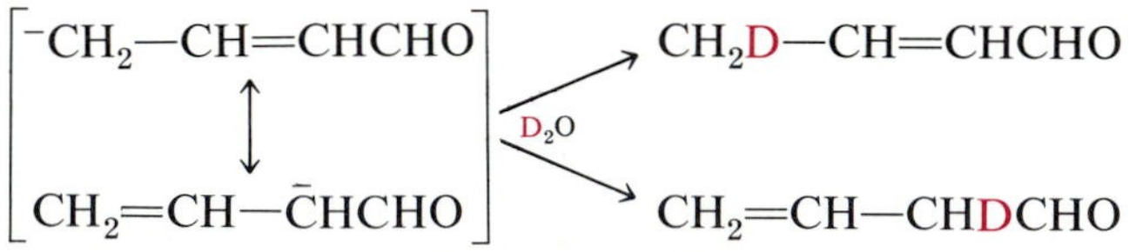

$$\left[\begin{array}{c} ^{-}CH_2{-}CH{=}CHCHO \\ \updownarrow \\ CH_2{=}CH{-}\bar{C}HCHO \end{array}\right] \xrightarrow{D_2O} \begin{array}{c} CH_2D{-}CH{=}CHCHO \\ CH_2{=}CH{-}CHDCHO \end{array}$$

Repeated reaction with base to reform the enolate ion and reaction with D_2O eventually produces the tetradeuterio compound $CD_3CH{=}CDCHO$.

EXERCISE 20.10 What is the structure of the deuterated product formed by base-catalyzed exchange of each of the following unsaturated ketones with excess D_2O?

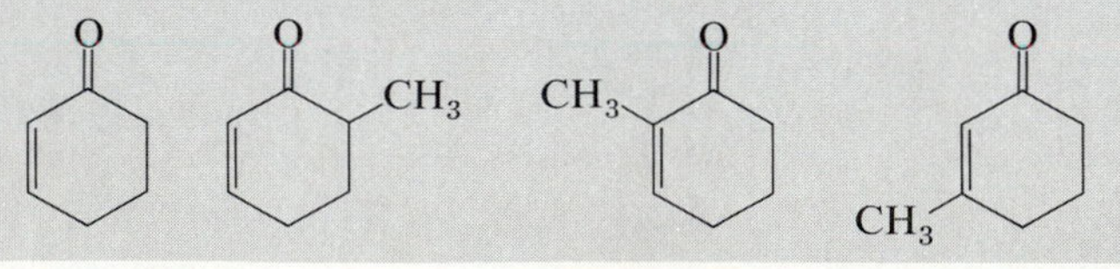

O O CH$_3$ CH$_3$ O O CH$_3$

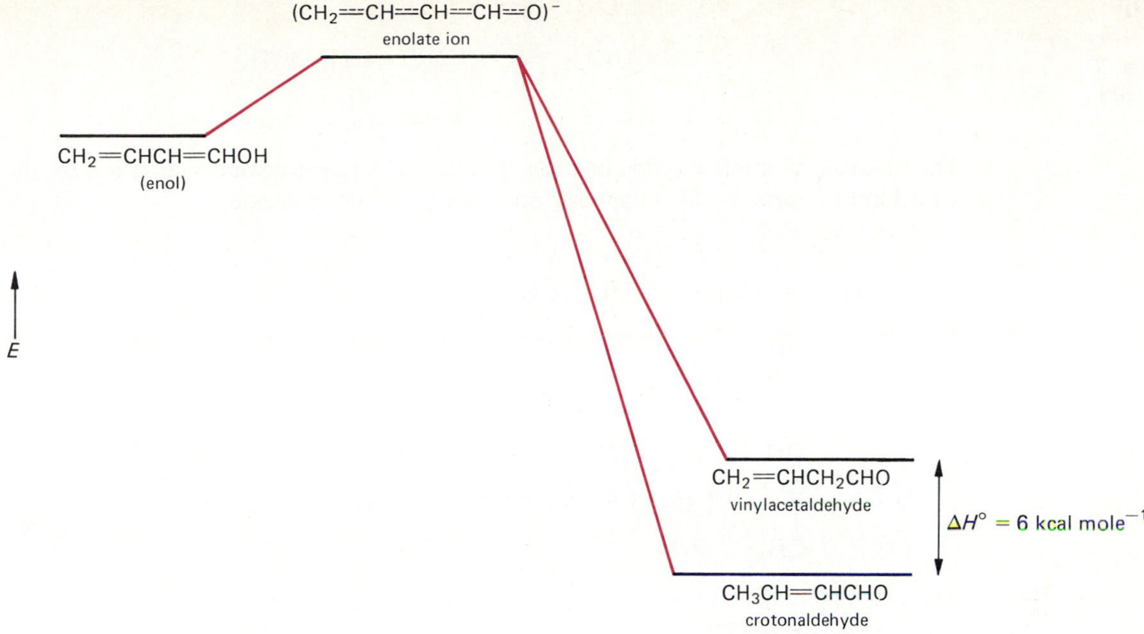

FIGURE 20.9 Some energy relationships of an unsaturated aldehyde.

The acid-catalyzed interconversions involve intermediate enols in exact analogy to the similar reactions of simple aldehydes and ketones (Section 15.1.A). Rapid and reversible protonation occurs at the carbonyl oxygen to form a hydroxycarbocation, which can lose a proton from carbon to form an enol. This enol is also a diene and can reprotonate at either the α- or the γ-carbon to produce the β,γ- or α,β-unsaturated carbonyl derivative, respectively.

$$CH_2{=}CH{-}CH_2{-}\overset{\overset{O}{\|}}{C}H \overset{H^+}{\rightleftharpoons} \left[CH_2{=}CH{-}CH_2{-}\overset{\overset{\overset{+}{O}H}{\|}}{C}H \longleftrightarrow CH_2{=}CH{-}CH_2{-}\underset{+}{\overset{\overset{OH}{|}}{C}}H \right] \overset{-H^+}{\rightleftharpoons} \underset{\text{a dienol}}{CH_2{=}CH{-}CH{=}\overset{\overset{OH}{|}}{C}H} \overset{H^+}{\rightleftharpoons}$$

$$\left[CH_3\overset{+}{C}H{-}CH{=}\overset{\overset{OH}{|}}{C}H \longleftrightarrow CH_3CH{=}CH{-}\underset{+}{\overset{\overset{OH}{|}}{C}}H \longleftrightarrow CH_3CH{=}CH{-}\overset{\overset{\overset{+}{O}H}{\|}}{C}H \right] \underset{H^+}{\overset{-H^+}{\rightleftharpoons}} CH_3CH{=}CHCHO$$

α,β-Unsaturated aldehydes and ketones are often obtained by dehydration of the β-hydroxy aldehydes and ketones that are produced in the aldol addition reaction (Section 15.2). Under certain circumstances, the treatment of an aldehyde under basic conditions leads directly to the α,β-unsaturated aldehyde. Such products from aldehydes or ketones may also be obtained under acidic conditions. For example, the acid-catalyzed self-reaction of acetone produces 4-methylpent-3-en-2-one, commonly called "mesityl oxide."

$$2\ CH_3COCH_3 \xrightarrow{H_2SO_4} CH_3\overset{\overset{\large CH_3}{|}}{C}{=}CHCOCH_3$$

mesityl oxide
4-methylpent-3-en-2-one

The mechanism of this reaction involves a number of straightforward steps. To start, the enol form of acetone adds to another protonated acetone molecule.

$$(1)\quad CH_3\overset{\overset{O}{\|}}{C}CH_3 + H^+ \rightleftharpoons CH_3\overset{\overset{^+OH}{\|}}{C}CH_3$$

$$(2)\quad CH_3{-}\overset{\overset{^+OH}{\|}}{C}CH_3 \rightleftharpoons H^+ + CH_2{=}\overset{\overset{OH}{|}}{C}CH_3$$

$$(3)\quad CH_3{-}\underset{\underset{CH_3}{|}}{\overset{\overset{^+OH}{\|}}{C}} + CH_2{=}\overset{\overset{:OH}{|}}{C}CH_3 \rightleftharpoons CH_3\underset{\underset{CH_3}{|}}{\overset{\overset{OH}{|}}{C}}{-}CH_2{-}\overset{\overset{^+OH}{\|}}{C}CH_3$$

The resulting oxonium ion can lose a proton from oxygen to give 4-hydroxy-4-methyl-2-pentanone (page 433) or from carbon to give the enol form 4-methylpent-2-en-2,4-diol.

$$(4)\quad CH_3\underset{\underset{CH_3}{|}}{\overset{\overset{OH}{|}}{C}}{-}CH_2{-}\overset{\overset{^+OH}{\|}}{C}CH_3 \rightleftharpoons CH_3\underset{\underset{CH_3}{|}}{\overset{\overset{OH}{|}}{C}}{-}CH_2{-}\overset{\overset{O}{\|}}{C}CH_3 + H^+$$

$$(4')\quad CH_3\underset{\underset{CH_3}{|}}{\overset{\overset{OH}{|}}{C}}{-}CH_2{-}\overset{\overset{^+OH}{\|}}{C}CH_3 \rightleftharpoons CH_3\underset{\underset{CH_3}{|}}{\overset{\overset{OH}{|}}{C}}{-}CH{=}\overset{\overset{OH}{|}}{C}CH_3 + H^+$$

The latter species is an enol form of a ketone and is unstable relative to the ketone; it is present in only low concentration. Protonation on the tertiary hydroxy gives an oxonium ion that readily eliminates water to form a new cation. The cation is resonance stabilized with the positive charge spread over oxygen and two carbons. The oxonium ion structure is the more important structure because all atoms have octet configurations.

$$(5)\quad CH_3\underset{\underset{CH_3}{|}}{\overset{\overset{OH}{|}}{C}}{-}CH{=}\overset{\overset{OH}{|}}{C}CH_3 + H^+ \rightleftharpoons CH_3\underset{\underset{CH_3}{|}}{\overset{\overset{^+OH_2}{|}}{C}}{-}CH{=}\overset{\overset{OH}{|}}{C}CH_3$$

$$(6)\quad CH_3\underset{\underset{CH_3}{|}}{\overset{\overset{^+OH_2}{|}}{C}}{-}CH{=}\overset{\overset{OH}{|}}{C}CH_3 \rightleftharpoons$$

$$H_2O + \left[CH_3\underset{\underset{CH_3}{|}}{\overset{+}{C}}{-}CH{=}\overset{\overset{OH}{|}}{C}CH_3 \longleftrightarrow CH_3\underset{\underset{CH_3}{|}}{C}{=}CH{-}\underset{+}{\overset{\overset{OH}{|}}{C}}CH_3 \longleftrightarrow CH_3\underset{\underset{CH_3}{|}}{C}{=}CH{-}\overset{\overset{^+OH}{\|}}{C}CH_3\right]$$

Loss of the proton from oxygen gives the product mesityl oxide.

$$(7)\ CH_3\underset{\displaystyle CH_3}{\underset{|}{C}}{=}CH{-}\overset{\displaystyle {}^{+}OH}{\overset{\|}{C}}CH_3 \rightleftharpoons CH_3\underset{\displaystyle CH_3}{\underset{|}{C}}{=}CH{-}\overset{\displaystyle O}{\overset{\|}{C}}CH_3 + H^+$$

It should be emphasized that in this reaction sequence, simple alkyl cations are not involved. *Every organic cationic intermediate either is an oxonium ion or has an oxonium ion resonance form.*

EXERCISE 20.11 Write a mechanism for the acid-catalyzed reaction of cyclohexanone with itself to give 2-cyclohexylidenecyclohexanone.

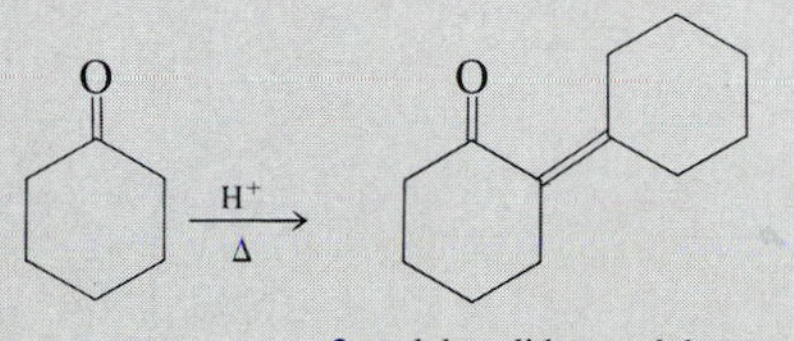

$\xrightarrow[\Delta]{H^+}$

2-cyclohexylidenecyclohexanone

Recall from Section 15.2 that mixed aldol reactions are successful ways of preparing α,β-unsaturated ketones in cases where one carbonyl compound cannot form an enol or enolate (because it has no α-protons) and/or has an especially reactive carbonyl group. These conditions are met by aromatic aldehydes, since the aldehyde function is generally more reactive than a ketone carbonyl and aromatic aldehydes have no α-protons. Reactions of aromatic ketones with aromatic aldehydes generally give good yields of enones, which are usually nicely crystalline substances. Such enones are sometimes referred to as "chalcones." Aromatic compounds are derived from benzene and will be discussed in Chapter 21.

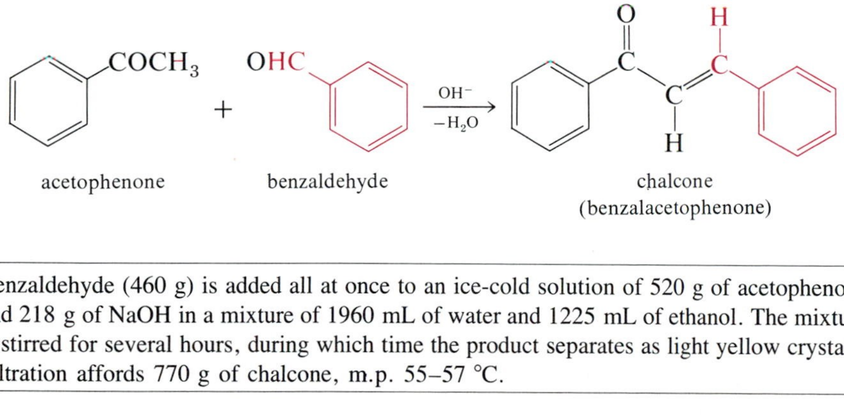

$C_6H_5COCH_3 + OHCC_6H_5 \xrightarrow[-H_2O]{OH^-}$ $C_6H_5C(=O)CH{=}CHC_6H_5$

acetophenone — benzaldehyde — chalcone (benzalacetophenone)

Benzaldehyde (460 g) is added all at once to an ice-cold solution of 520 g of acetophenone and 218 g of NaOH in a mixture of 1960 mL of water and 1225 mL of ethanol. The mixture is stirred for several hours, during which time the product separates as light yellow crystals. Filtration affords 770 g of chalcone, m.p. 55–57 °C.

The simplest conjugated unsaturated carbonyl compound is propenal, $CH_2{=}CHCHO$, commonly known as acrolein. This compound is a liquid, b.p. 53 °C, having a powerful, pungent odor. It may be prepared by a special reaction in which the readily available triol, glycerol, is heated with sulfuric acid or potassium acid sulfate.

$$HOCH_2CHOHCH_2OH \xrightarrow[200\text{–}230\,°C]{KHSO_4} CH_2{=}CHCHO$$

glycerol, 1,2,3-propanetriol — acrolein, propenal

Most of us are familiar with the odor of acrolein because a similar dehydration occurs thermally when fats burn or decompose on a hot surface. Recall that fats are esters of glycerol (Section 19.11.B).

α,β-Unsaturated aldehydes and ketones are also available by oxidation of the corresponding unsaturated alcohols, which in turn are frequently available by Grignard syntheses. The oxidation requires mild conditions in order not to oxidize the double bond. One reagent that is specific for allylic alcohols is a specially active form of manganese dioxide, obtained by treatment of manganese sulfate with base and potassium permanganate. Retinol, a polyene allylic alcohol is oxidized to retinal, a polyenal.

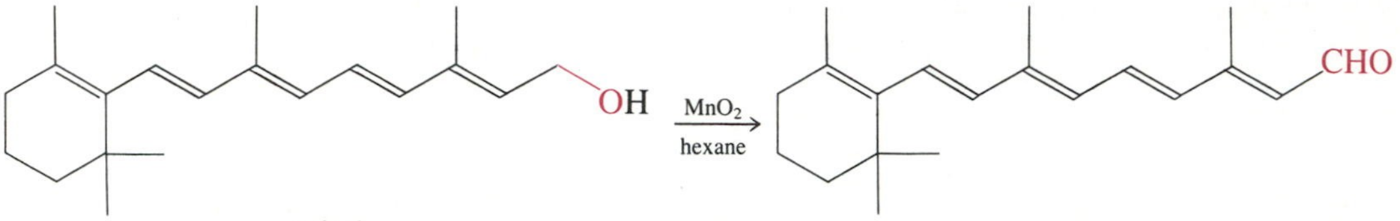

OH $\xrightarrow[\text{hexane}]{MnO_2}$ CHO

retinol
(vitamin A_1)

(80%)
retinal

α,β-Unsaturated aldehydes and ketones undergo many of the reactions expected separately for the double bond and carbonyl functions. The C=O group forms normal derivatives such as oximes, phenylhydrazones, and so on (Section 14.6.C).

$$CH_3CH{=}CH\overset{\overset{\displaystyle O}{\|}}{C}CH_3 + H_2NOH \longrightarrow CH_3CH{=}CH\overset{\overset{\displaystyle NOH}{\|}}{C}CH_3$$

an oxime

The aldehyde group is oxidized under mild conditions to a carboxylic acid (Section 14.8.A).

$$CH_2{=}CHCHO + Ag_2O \longrightarrow CH_2{=}CHCOOH + 2\ Ag$$

Bromine can be added to the double bond (Section 11.6.B)

$$C_6H_5CH{=}CHCOCH_3 + Br_2 \xrightarrow[10\text{–}20^\circ C]{CCl_4} C_6H_5CHBrCHBrCOCH_3$$

(52–57%)

However, some reactions are unique to the conjugated system. Additions can occur to the ends of the conjugated system or to either one of the double bonds, just as in the case of conjugated dienes (Section 20.2.B). Additions that occur to a single double bond are called **1,2-additions** or **normal additions.** Additions that occur to the ends of the conjugated system are called **1,4-additions** or **conjugate additions.**

1,2-additions

$$C{=}C{-}C{=}O + X{-}Y \longrightarrow \overset{X}{\overset{|}{C}}{-}\overset{Y}{\overset{|}{C}}{-}C{=}O \quad \text{or} \quad C{=}C{-}\overset{X}{\overset{|}{C}}{-}\overset{Y}{\overset{|}{O}}$$

1,4-additions

$$C{=}C{-}C{=}O + X{-}Y \longrightarrow \overset{X}{\overset{|}{C}}{-}C{=}C{-}\overset{Y}{\overset{|}{O}}$$

Do not be confused by the terms *1,2-addition* and *1,4-addition*. The numbers do not refer to the carbon numbers in any given compound. The terms mean that the addition is to the 1 and 2 positions or the 1 and 4 positions of a conjugated system. For example, the addition of Br_2 to the double bond in pent-3-en-2-one is an example of a 1,2-addition.

$$\underset{5\quad 4}{CH_3CH}{=}\underset{3}{CH}{-}\underset{2\ 1}{\overset{O}{\overset{\|}{C}}CH_3} + Br_2 \longrightarrow CH_3\overset{Br}{\overset{|}{C}}H{-}\overset{Br}{\overset{|}{C}}H{-}\overset{O}{\overset{\|}{C}}CH_3$$

Cyanide ion, which normally adds to the carbonyl bond in aldehydes and ketones (Section 14.7.B), frequently adds instead to a carbon-carbon double bond that is conjugated with a carbonyl group.

$$C_6H_5CH{=}CH{-}\overset{O}{\overset{\|}{C}}C_6H_5 + KCN \xrightarrow{CH_3CO_2H} C_6H_5\overset{CN}{\overset{|}{C}}HCH_2\overset{O}{\overset{\|}{C}}C_6H_5$$

(93–96%)

The reaction appears to be 1,2-addition to the double bond. However, the mechanism actually involves 1,4-addition of HCN to the conjugated system. The initial product of the 1,4-addition is an enol, which isomerizes to the more stable keto form (Section 15.1.A).

$$(1)\ C_6H_5CH{=}CH{-}\overset{O}{\overset{\|}{C}}C_6H_5 + H^+ \rightleftharpoons C_6H_5CH{=}CH{-}\overset{^+OH}{\overset{\|}{C}}C_6H_5$$

$$(2)\ C_6H_5CH{=}CH{-}\overset{^+OH}{\overset{\|}{C}}C_6H_5 + CN^- \rightleftharpoons C_6H_5\overset{CN}{\overset{|}{C}}H{-}CH{=}\overset{OH}{\overset{|}{C}}C_6H_5$$

$$(3)\ C_6H_5\overset{CN}{\overset{|}{C}}H{-}CH{=}\overset{OH}{\overset{|}{C}}C_6H_5 + H^+ \rightleftharpoons C_6H_5\overset{CN}{\overset{|}{C}}H{-}CH_2{-}\overset{^+OH}{\overset{\|}{C}}C_6H_5$$

$$(4)\ C_6H_5\overset{CN}{\overset{|}{C}}H{-}CH_2{-}\overset{^+OH}{\overset{\|}{C}}C_6H_5 \rightleftharpoons C_6H_5\overset{CN}{\overset{|}{C}}HCH_2\overset{O}{\overset{\|}{C}}C_6H_5 + H^+$$

A particularly effective method for accomplishing the 1,4-addition of HCN to α,β-unsaturated ketones employs triethylaluminum as a catalyst. The procedure gives high yields of the conjugate adduct even when the enone is highly substituted at the β-position.

H + HCN $\xrightarrow{(C_2H_5)_3Al}$ H, CN

(85%)

EXERCISE 20.12 Suggest a method for the synthesis of 2,2-dimethyl-4-oxopentanoic acid, starting with acetone.

O, COOH

2,2-dimethyl-4-oxopentanoic acid

Organometallic compounds can add either 1,2 or 1,4. Grignard reagents show variable behavior depending on the structure of the conjugated system. The most important factor in determining whether the addition is 1,2 or 1,4 seems to be steric hindrance. All α,β-unsaturated aldehydes and many α,β-unsaturated ketones undergo normal 1,2-addition to the carbonyl group.

$$CH_3CH{=}CHCHO + C_2H_5MgBr \longrightarrow \xrightarrow{H_3O^+} CH_3CH{=}CHCH(OH)C_2H_5$$

(70%)
trans-4-hexen-3-ol

$$CH_3CH{=}CHC(=O)CH_3 + CH_3MgBr \longrightarrow \xrightarrow{H_3O^+} CH_3CH{=}CHC(CH_3)_2OH$$

(80%)
2-methyl-*trans*-3-penten-2-ol

However, other α,β-unsaturated ketones give substantial amounts of the 1,4-adduct as well.

$$C_6H_5CH{=}CH{-}\overset{O}{\overset{\|}{C}}CH_3 + C_2H_5MgBr \longrightarrow$$

$$\left\{ C_6H_5CH{=}CH{-}C(O^-\,{}^+MgBr)(C_2H_5)CH_3 + C_6H_5CH(C_2H_5){-}CH{=}C(O^-\,{}^+MgBr)CH_3 \right\} \xrightarrow{H_3O^+}$$

$$C_6H_5CH{=}CH{-}C(OH)(C_2H_5)CH_3 + C_6H_5CH(C_2H_5){-}CH_2{-}\overset{O}{\overset{\|}{C}}CH_3$$

(37%) 1,2-adduct; (57%) 1,4-adduct

In some cases the 1,4-adduct is the preponderant product. Conjugate addition is particularly likely in reactions of α,β-unsaturated ketones that have large substituents attached to the C=O group.

$$C_6H_5CH{=}CH\overset{O}{\overset{\|}{C}}C_6H_5 \xrightarrow{C_6H_5MgBr} \xrightarrow{H_3O^+} (C_6H_5)_2CHCH_2\overset{O}{\overset{\|}{C}}C_6H_5$$

(96%)
1,3,3-triphenyl-1-propanone

Organolithium compounds show a much greater tendency to engage in 1,2-addition.

$$C_6H_5CH{=}CHCC_6H_5 \;(\text{with } C{=}O) \xrightarrow{C_6H_5Li} \xrightarrow{H_3O^+} C_6H_5CH{=}CHC(OH)(C_6H_5)_2$$

(75%)

When one wants to maximize 1,2-addition, it is customary to utilize the organolithium reagent.

On the other hand, 1,4-addition can be achieved by using lithium dialkylcuprates, which are readily prepared from the corresponding alkyllithium reagent and cuprous iodide (Section 8.8).

$$2\ CH_3Li + CuI \longrightarrow (CH_3)_2CuLi + LiI$$

These reagents add to α,β-unsaturated ketones exclusively in a 1,4-fashion.

$$CH_3C(CH_3){=}CHC(O)CH_3 + (CH_2{=}CH)_2CuLi \longrightarrow \xrightarrow{H_3O^+} CH_2{=}CHC(CH_3)_2CH_2C(O)CH_3$$

(72%)

The same result may often be achieved by forming the dialkylcuprate *in situ* from the corresponding Grignard reagent. In practice, it is only necessary to use a catalytic amount of cuprous bromide or iodide.

$$\text{2-cyclohexenone} + CH_3C(MgBr){=}CH_2 \xrightarrow{5\text{ mole \% CuI}} \xrightarrow{H_3O^+} \text{3-(}CH_2{=}C(CH_3)\text{)cyclohexanone}$$

(68%)

The mechanisms of the 1,4-additions of organometallic reagents to α,β-unsaturated carbonyl compounds are not fully understood, but it is clear that the metal must have higher oxidation states that are relatively accessible in order for 1,4-addition to occur. The most useful reagents for 1,4-addition are the alkyl cuprates, and copper salts are most often used to catalyze 1,4-additions of Grignard reagents. Dialkyl cuprates, nominally R_2CuLi, are known to exist in solution as dimeric forms, $(R_2CuLi)_2$, and probably have structures such as that shown below, with the lithium cations being further solvated by ether molecules. In this structure the carbon-copper bonds are viewed as being covalent, with the copper bearing a formal negative charge. Two of these R_2Cu^- units combine with two lithium cations to form the dimeric cuprate structure. Copper, being a transition metal, has a number of d-electrons that are not involved in bonding and can act as a nucleophile by adding to the electrophilic double bond. In the initial adduct, copper has the formal oxidation state of +3. In order to return to its more preferred oxidation state, **reductive elimination** occurs, giving RCu with formation of a carbon-carbon bond between the R group formerly attached to copper and the carbon that was formerly the β-carbon of the enone system. The product of the reductive elimination has been shown to be a lithium enolate. The RCu forms the highly insoluble polymeric product $(RCu)_n$.

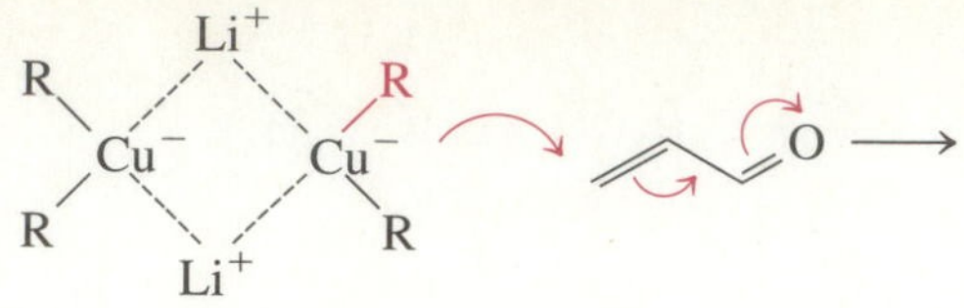

dialkyl cuprate, dimeric structure

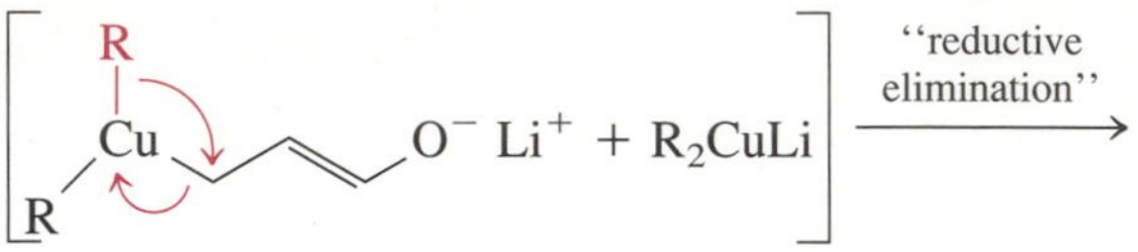

"reductive elimination"

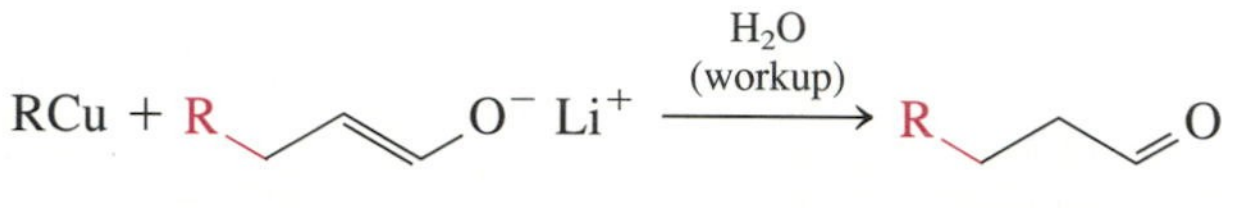

$O^- Li^+ + R_2CuLi$

$$RCu + R\text{–CH=CH–}O^- Li^+ \xrightarrow[\text{(workup)}]{H_2O} R\text{–CH}_2\text{CH}_2\text{CH=O}$$

lithium enolate

1,4-addition product

The 1,4-additions that are observed with Grignard reagents result from the catalytic effect of traces of transition metals, particularly copper, iron, and manganese, in the commercially available magnesium. If highly purified magnesium is used to prepare the Grignard reagent, 1,4-addition reactions are not observed.

EXERCISE 20.13 Write equations showing how hept-4-en-3-one can be converted into (a) 3-ethylhept-4-en-3-ol and (b) 5-ethylheptan-3-one.

Reduction of α,β-unsaturated carbonyl compounds can also involve either the carbon-carbon or the carbon-oxygen double bond. Lithium aluminum hydride reduction of most such compounds usually gives high yields of the products of simple carbonyl reduction.

$$CH_3CH{=}CHCHO + LiAlH_4 \xrightarrow{\text{ether}} \xrightarrow{H_2O} CH_3CH{=}CHCH_2OH$$

(82%)

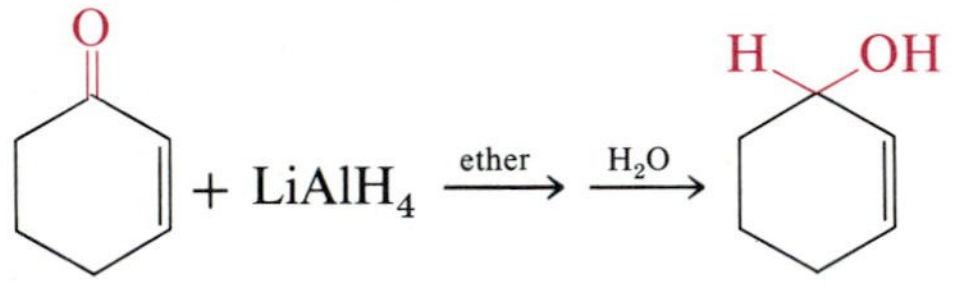

$$\text{cyclohex-2-enone} + LiAlH_4 \xrightarrow{\text{ether}} \xrightarrow{H_2O} \text{cyclohex-2-enol}$$

(97%)

In contrast, sodium borohydride in ethanol often gives substantial amounts of the 1,4-addition product.

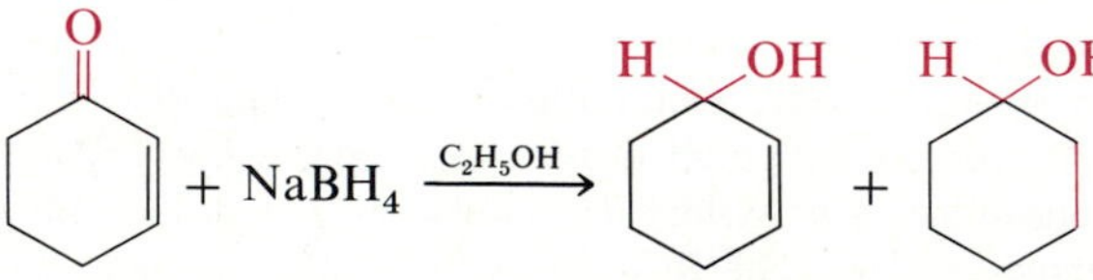

$$\text{cyclohex-2-enone} + NaBH_4 \xrightarrow{C_2H_5OH} \text{cyclohex-2-enol} + \text{cyclohexanol}$$

(59%) (41%)

The fully reduced product in the preceding example arises from the following pathway, which begins with the conjugate addition of hydride to the enone system.

(1) cyclohexenone $+ BH_4^- \longrightarrow$ [enolate $\longleftrightarrow$ carbanion] $+ BH_3$

(2) [enolate $\longleftrightarrow$ carbanion] $\xrightarrow{C_2H_5OH}$ cyclohexanone $+ C_2H_5O^-$

(3) cyclohexanone $+ BH_4^- \longrightarrow$ cyclohexyl alkoxide $+ BH_3$

(4) cyclohexyl alkoxide $+ C_2H_5OH \rightleftharpoons$ cyclohexanol $+ C_2H_5O^-$

The double bond of such a conjugated system can generally be reduced cleanly by either of two procedures, catalytic hydrogenation or lithium-ammonia reduction.

3-methylcyclohex-2-enone $+ H_2 \xrightarrow{Pd\text{-}C}$ 3-methylcyclohexanone (100%)

$+ Li \xrightarrow[-33°C]{NH_3} \xrightarrow{H_3O^+}$ (95%)

The mechanism of the latter reduction is similar to that seen earlier in the reduction of alkynes to alkenes (Section 12.6.A). The first step involves addition of an electron to the conjugated system, giving a resonance-stabilized radical anion. This radical anion protonates on carbon giving a radical, which is reduced by another electron with the formation of an enolate ion.

$$(1)\ \gt C{=}C{-}C{=}O + Li \longrightarrow \left[-\overset{-\cdot\cdot}{C}{-}C{=}C{-}\ddot{O}\cdot \longleftrightarrow -\dot{C}{-}C{=}C{-}\ddot{O}{:}^{-} \longleftrightarrow -C{=}C{-}\overset{\cdot\cdot}{\underset{-}{C}}{-}\dot{O}{:} \longleftrightarrow -C{=}C{-}\dot{C}{-}\ddot{O}{:}^{-} \right] + Li^{+}$$

$$(2)\ \left[\gt C{\cdots}C{\cdots}C{\cdots}O \right]^{\dot{-}} + NH_3 \longrightarrow -\overset{H}{C}{-}C{=}C{-}\dot{O}{:} + NH_2^{-}$$

$$(3)\ -\overset{H}{C}{-}C{=}C{-}\dot{O}{:} + Li \longrightarrow \left[-\overset{H}{C}{-}C{=}C{-}\ddot{O}{:}^{-} \longleftrightarrow -\overset{H}{C}{-}\overset{\cdot\cdot}{\underset{-}{C}}{-}C{=}O \right] + Li^{+}$$

Under the conditions of the reaction, the enolate ion is stable. Its reduction potential is too high for it to accept another electron and be reduced further, and it is not basic enough to be protonated by such a weak acid as ammonia. Upon workup with water, the enolate ion is protonated to give the ketone.

$$\left[\gt CH{\cdots}C{\cdots}C{\cdots}O \right]^{-} + H_3O^{+} \longrightarrow \gt CH{-}CH{-}C{=}O + H_2O$$

The partial reduction of a conjugated enone in this reaction is a particularly impressive example of the special properties of conjugated systems, since isolated carbon-carbon double bonds are not reduced by lithium in ammonia and isolated carbon-oxygen double bonds are reduced by the reagent.

EXERCISE 20.14 Write the equations showing the reaction of 4-methylpent-3-en-2-one with each of the following reagents.

(a) n-C_4H_9Li (b) n-C_4H_9MgBr, CuBr (c) H_2/Pd
(d) Li-NH_3 (e) HCN, $(C_2H_5)_3Al$ (f) Br_2-CCl_4

Highlight 20.3

The connection of two double bonds is called "**conjugation**" (L., *conjugere,* joining together in marriage). The properties of the conjugated molecule are different from molecules in which the double bonds are not joined together.

Thermodynamically, conjugated molecules are generally more stable than their unconjugated isomers.

$$CH_2{=}CHCH_2CHO \longrightarrow CH_3CH{=}CHCHO \qquad \Delta H^\circ = -6\ \text{kcal mole}^{-1}$$

Chemically, conjugated systems often react in ways involving both double bonds. For electrophilic addition of Br_2, see Highlight 20.2. Nucleophilic addition to conjugated double bonds is enhanced by polarization as in α,β-enones. The contribution of the polar resonance form is promoted by association with H^+ (addition of HCN) or M^+ (reaction of organometallic derivatives).

$$\left[CH_3{-}CH{=}CH{-}C(H){=}O \longleftrightarrow CH_3{-}\overset{+}{C}H{-}CH{=}C(H){-}O^- \right]$$

$$\left[CH_3{-}CH{=}CH{-}C(H){=}\overset{+}{O}{-}H \longleftrightarrow CH_3{-}\overset{+}{C}H{-}CH{=}C(H){-}O{-}H \right]$$

Conjugated intermediates form more easily than similar unconjugated compounds; the acidity of the methyl group in crotonaldehyde is orders of magnitude greater than the methyl group in butanal.

$$CH_3{-}CH{=}CH{-}C(H){=}O \longrightarrow {}^-CH_2{-}CH{=}CH{-}C(H){=}O$$ *stabilized by conjugation with CH=CHC=O*

$$CH_3{-}CH_2{-}CH_2{-}C(H){=}O \longrightarrow CH_3{-}CH_2{-}\overset{-}{C}H{-}C(H){=}O$$ *stabilized by conjugation with C=O*

$$CH_3{-}CH_2{-}CH_2{-}C(H){=}O \longrightarrow {}^-CH_2{-}CH_2{-}CH_2{-}C(H){=}O$$ *not formed, not stabilized by conjugation*

B. Unsaturated Carboxylic Acids and Derivatives

Both conjugated and unconjugated unsaturated carboxylic acids and acid derivatives are known. As with other multiply unsaturated systems, conjugation provides added stabilization, but the magnitude of this stabilization is rather small, substantially smaller than for dienes or unsaturated aldehydes and ketones. The heats of formation of isomeric ethyl pentenoates summarized in Table 20.2 demonstrate this point. In other words, the carboxylic function is less effective in conjugation than is a simple carbonyl group.

One way of rationalizing this behavior is to consider that the carbonyl group in a carboxylic function is already involved in conjugation to an atom with a pair of electrons to donate—such as the oxygen of OH or OR or the nitrogen of NH_2. Such a carbonyl group is less able to conjugate with another group. This situation is representative of a **cross-conjugated** system as illustrated by the π-overlap of p-orbitals in Figure 20.10. Much qualitative evidence, as well as theoretical considerations, shows that cross conjugation is less effective than linear conjugation in stabilizing a molecule.

As a result of the reduced effectiveness of conjugation in unsaturated acid functions, in some compounds the unconjugated isomer may be the more stable. For example,

TABLE 20.2 Heats of Formation of Ethyl Pentenoates

Isomer	ΔH_f°, kcal mole^{-1}
$CH_2{=}CHCH_2CH_2COOC_2H_5$	-92.1 ± 0.6
$CH_3(H)C{=}C(H)CH_2COOC_2H_5$ (trans)	-92.6 ± 0.9
$CH_3(H)C{=}C(H)CH_2COOC_2H_5$ (cis)	-93.2 ± 0.7
$CH_3CH_2(H)C{=}C(H)COOC_2H_5$ (trans)	-94.3 ± 0.7
$CH_3CH_2(H)C{=}C(H)COOC_2H_5$ (cis)	-94.2 ± 0.9

4-methylpent-3-enoic acid, with a trisubstituted double bond, is more stable than 4-methylpent-2-enoic acid, which has a conjugated π-system but has only a disubstituted double bond.

$$\underset{\text{4-methylpent-3-enoic acid}}{CH_3\overset{CH_3}{\overset{|}{C}}{=}CHCH_2COOH} \rightleftharpoons \underset{\text{4-methylpent-2-enoic acid}}{CH_3\overset{CH_3}{\overset{|}{C}}HCH{=}CHCOOH}$$

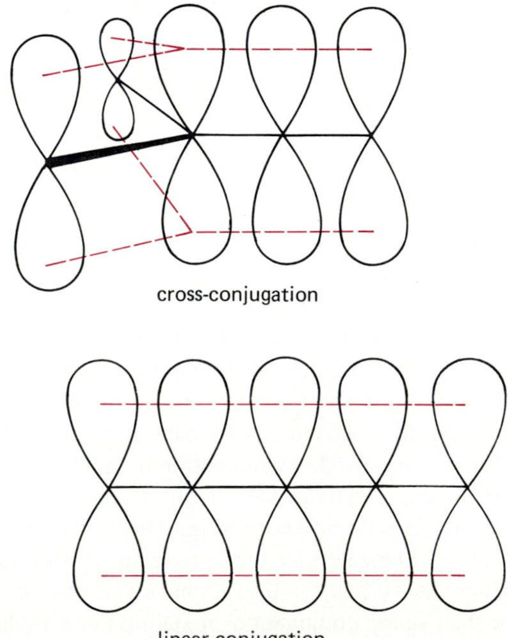

FIGURE 20.10 Orbital diagrams of linear conjugation and cross-conjugation.

Unsaturated carboxylic acids and their derivatives may be prepared by many of the same routes appropriate for the saturated analogs. An example is the sequence RX $\rightarrow$ RCOOH, in which R contains a double bond.

$$CH_2{=}CHCH_2Cl \xrightarrow{CuCN} CH_2{=}CHCH_2CN \xrightarrow[\Delta]{aq.\ HCl} CH_2{=}CHCH_2COOH$$

(52–62%)

allyl chloride → vinylacetic acid, but-3-enoic acid

α,β-Unsaturated acids and derivatives are also available by elimination of HX from α-halo acids and esters.

$$CH_3(CH_2)_9CH(CH_3)COOH \xrightarrow[Br_2]{PBr_3} CH_3(CH_2)_9CBr(CH_3)COBr \xrightarrow{CH_3OH} CH_3(CH_2)_9CBr(CH_3)COOCH_3$$

2-methyldodecanoic acid

$$CH_3(CH_2)_9CBr(CH_3)COBr \xrightarrow{(CH_3)_3COK,\ (CH_3)_3COH} CH_3(CH_2)_9C({=}CH_2)COOC(CH_3)_3 \xrightarrow{alc\ KOH} \xrightarrow{H^+} CH_3(CH_2)_9C({=}CH_2)COOH$$

2-decylpropenoic acid

$$CH_3(CH_2)_9CBr(CH_3)COOCH_3 \xrightarrow{quinoline,\ 160\text{–}170^\circ C} CH_3(CH_2)_8CH{=}C(CH_3)COOCH_3 \xrightarrow{alc\ KOH} \xrightarrow{H^+} CH_3(CH_2)_8CH{=}C(CH_3)COOH$$

(70–85%) ... (68–83%)

2-methyldodec-2-enoic acid

Note in the foregoing examples that the direction of elimination can sometimes be controlled by the choice of basic reagent used. Although both of these eliminations occur by the E2 mechanism, recall that the bulky reagent potassium *t*-butoxide tends to abstract primary hydrogens (Section 11.5.A).

Aromatic aldehydes may be converted into α,β-unsaturated acids by the **Perkin reaction,** in which the aldehyde is heated with an acid anhydride and the corresponding carboxylate salt. Acetic anhydride and sodium acetate are most commonly used.

$$o\text{-}ClC_6H_4CHO \xrightarrow[180^\circ C]{(CH_3CO)_2O,\ CH_3COONa} o\text{-}ClC_6H_4CH{=}CHCOOH$$

(72%)
o-chlorocinnamic acid

The reaction is a typical base-catalyzed condensation in which the enolate ion of an acid anhydride is an intermediate.

The Perkin reaction appears to proceed by way of the following interesting mechanism.

$$\text{(1)}\ CH_3COCCH_3(\text{O O}) + AcO^- \rightleftharpoons :\overset{-}{C}H_2COCCH_3 + AcOH$$

$$\text{(2)}\ ArCH{=}O + :\overset{-}{C}H_2COAc \rightleftharpoons ArCH(O^-)-CH_2COAc$$

(3) Ar—CH(O⁻)—CH₂—C(=O)—O—C(=O)CH₃ ⇌ cyclic intermediate: Ar—CH—O—C(CH₃)(O⁻)—O—C(=O)—CH₂ (ring)

(4) cyclic intermediate ⇌ $ArCH(OAc)CH_2CO_2^-$

$$\text{(5)}\ ArCH(OAc)CH_2CO_2^- + Ac_2O \rightleftharpoons ArCH(OAc)CH_2COAc + AcO^-$$

$$\text{(6)}\ ArCH(OAc)CH_2COAc + AcO^- \rightleftharpoons ArCH(OAc)\overset{-}{C}HCOAc + AcOH$$

$$\text{(7)}\ ArCH(OAc)-\overset{-}{C}HCOAc \rightleftharpoons ArCH{=}CHCOAc + AcO^-$$

$$\text{(8)}\ ArCH{=}CHCOAc + AcOH \rightleftharpoons ArCH{=}CHCOOH + Ac_2O$$

The simplest unsaturated carboxylic acid is propenoic acid, $CH_2{=}CHCOOH$, commonly known as acrylic acid, a liquid having b.p. 141.6 °C. The corresponding nitrile, $CH_2{=}CHCN$, acrylonitrile, is an important industrial material that is made in large quantity for use in synthetic fibers and polymers; its 1989 production in the United States was 1,304,000 tons. Acrylonitrile is also a liquid, b.p. 78.5 °C. It was once prepared industrially by addition of HCN to acetylene.

$$HC{\equiv}CH + HCN \xrightarrow[NH_4Cl]{CuCl} CH_2{=}CHCN$$

It is now prepared by a cheaper process that involves the catalytic oxidation of propene in the presence of ammonia.

$$2\ CH_3CH{=}CH_2 + 3\ O_2 + 2\ NH_3 \xrightarrow[\text{catalyst}]{450°C} 2\ CH_2{=}CHCN + 6\ H_2O$$

The methyl ester of α-methylacrylic acid is also an important industrial product. It is prepared from acetone by the following sequence.

$$CH_3COCH_3 + HCN \longrightarrow (CH_3)_2C(OH)CN \xrightarrow{H_2SO_4} CH_2{=}C(CH_3)CONH_2 \xrightarrow[H_2SO_4]{CH_3OH} CH_2{=}C(CH_3)COOCH_3$$

C. Ketenes

The compound $CH_2{=}C{=}O$ is known as ketene and is the carbonyl analog of allene. It is a toxic gas, b.p. −48 °C, and is prepared by the pyrolysis of acetone at high temperatures.

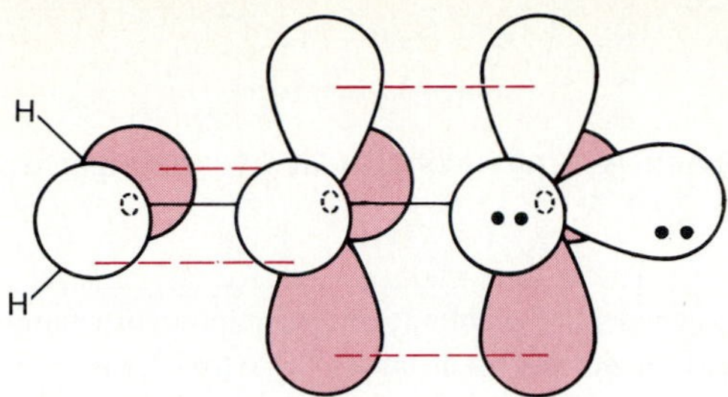

FIGURE 20.11 Orbital structure of ketene.

$$CH_3COCH_3 \xrightarrow{700\,°C} CH_4 + CH_2{=}C{=}O$$

The reaction mechanism appears to involve free radical chain decomposition.

Substituted ketenes are prepared by treatment of acyl halides with triethylamine or by treatment of α-halo acyl halides with zinc.

$$\text{(cyclohexane-H, COCl)} \xrightarrow[\text{ether}]{(C_2H_5)_3N} \text{(cyclohexylidene)}{=}C{=}O + (C_2H_5)_3NH^+\ Cl^-$$

(32%)
pentamethyleneketene

$$CH_3{-}\underset{CH_3}{\overset{Br}{C}}{-}COBr \xrightarrow{Zn} (CH_3)_2C{=}C{=}O + ZnBr_2$$

(46–54%)
dimethylketene

Ketenes react as "super anhydrides." With water they give carboxylic acids, and with alcohols they give esters. One commercial synthesis of acetic anhydride involves the combination of acetic acid with ketene.

$$CH_3COOH + CH_2{=}C{=}O \longrightarrow (CH_3CO)_2O$$

The world production of acetic anhydride in 1989, mostly by this reaction, was 1,710,000 tons.

The ketene group is extremely reactive and is a relatively unimportant functional group. Like allenes, ketenes have two π-systems at right angles. The two double bonds are cumulated and *not* conjugated (Figure 20.11). An acetal of ketene is used to generate vinyl alcohol (Section 15.1.B)

EXERCISE 20.15 In its addition to ketene ethanol probably acts as a base, whereas acetic acid probably adds by an acid-catalyzed mechanism. Write reasonable mechanisms for both processes.

20.4 Higher Conjugated Systems

Many compounds having more than two conjugated double bonds are known. In such systems each double bond alternates with a single bond to allow extensive π-overlap of p-orbitals. One example is *trans*-1,3,5-hexatriene, a liquid, b.p. 79 °C.

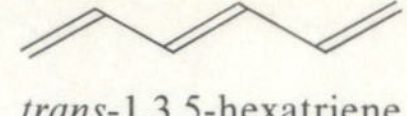

trans-1,3,5-hexatriene

Another is retinol (vitamin A_1), an alcohol with five conjugated double bonds that we encountered in Section 20.3.A.

Despite its vital role as an essential vitamin, overconsumption of vitamin A is dangerous. The recommended intake is 1 mg per day for an adult. Acute toxicity is observed after eating polar bear liver, which contains up to 12 mg of retinol per gram. A normal portion (100 g) could provide 1200 mg, more than twice the toxic dose for an adult human. By 1984, more than 600 cases of hypervitaminosis A had been reported in the literature. Lipid soluble vitamin E protects against the hypervitaminosis of A; vitamin E functions as a free radical trap. Autoxidation within cell membranes may underlie the toxic effects. Recall that cell membranes contain phospholipids with unsaturated acyl groups (Section 19.11.D).

Despite the stabilization that such highly conjugated compounds derive from their extensive π-electronic systems, they are generally more reactive, not less reactive, than their nonconjugated isomers. The reason for this apparent paradox is simply that the intermediate radicals or ions are even more stabilized by conjugation. For example, 1,3,5-hexatriene reacts rapidly with acids, bromine, free radicals, and other reagents. The addition of a proton to the terminal carbon atom gives a pentadienyl cation that is highly stabilized by resonance. That is, in this carbocation the positive charge is distributed among three carbons.

$$CH_2{=}CH{-}CH{=}CH{-}CH{=}CH_2 + H^+ \longrightarrow$$

$$\left[\begin{array}{c} CH_2{=}CH{-}CH{=}CH{-}\overset{+}{C}H{-}CH_3 \\ \updownarrow \\ CH_2{=}CH{-}\overset{+}{C}H{-}CH{=}CH{-}CH_3 \\ \updownarrow \\ \overset{+}{C}H_2{-}CH{=}CH{-}CH{=}CH{-}CH_3 \end{array}\right] \equiv \overset{\delta+}{C}H_2{\cdots}CH{\cdots}\overset{\delta+}{C}H{\cdots}CH{\cdots}\overset{\delta+}{C}H{-}CH_3$$

The resonance stabilization of such carbocation, carbanion, and free radical intermediates and of the transition states leading to them is much greater than the stabilization afforded by π-overlap in the starting polyenes. As a result, such polyenes are highly reactive. Exposure to air or light is often sufficient to initiate free radical chain polymerization.

EXERCISE 20.16 Write the structures of all of the products expected from the addition of 1 mole of bromine to 1,3,5-hexatriene at low temperature. What principal product is expected if this reaction mixture is allowed to stand?

20.5 The Diels-Alder Reaction

Conjugated dienes undergo a **cycloaddition reaction** with certain multiple bonds to form cyclohexenes and related compounds. This reaction, which is an important synthetic method, is called the **Diels-Alder** reaction, after Otto Diels and Kurt Alder, two German chemists who received the 1950 Nobel Prize for its discovery. The simplest Diels-Alder reaction is the reaction of 1,3-butadiene and ethylene to yield cyclohexene.

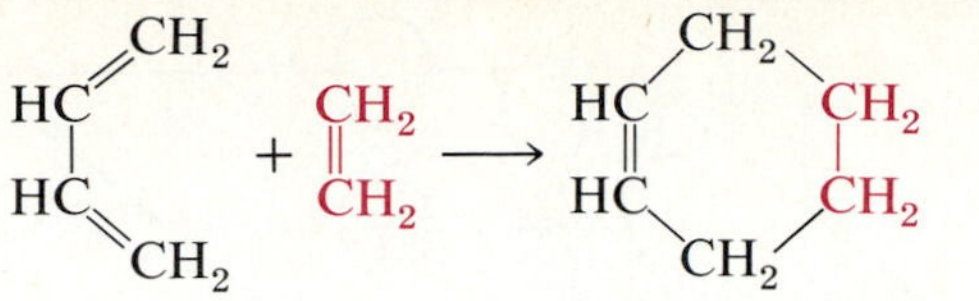

Although this simple example is known, it is very slow and only occurs under pressure with heating. However, the Diels-Alder reaction is facilitated by the presence of electron-donating groups on the diene component and by the presence of electron-attracting groups on the monoene component, often referred to as the "dienophile."

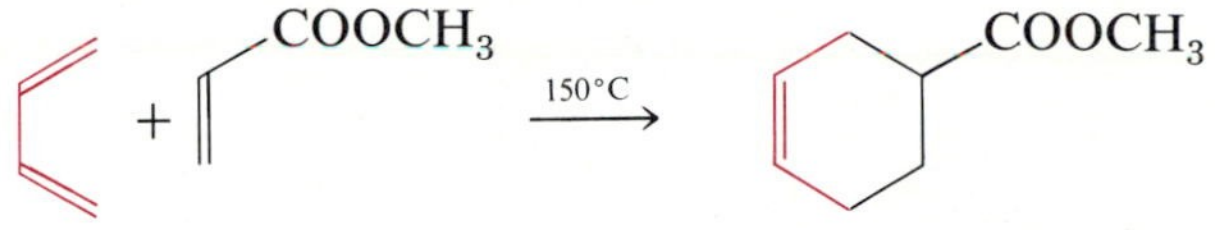

diene dienophile methyl 4-cyclohexenecarboxylate

The reaction is a particularly versatile method for the preparation of cyclohexane derivatives of varied sorts.

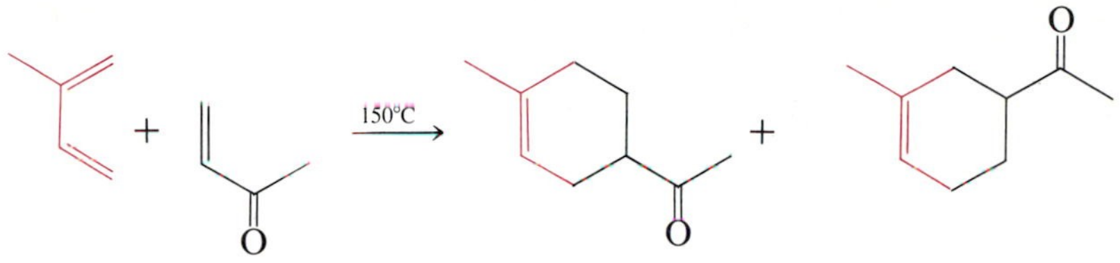

isoprene methyl vinyl ketone (70%) methyl 4-methyl-3-cyclohexenyl ketone (30%) methyl 3-methyl-3-cyclohexenyl ketone

The reaction has wide scope because multiple bonds other than C=C may be used

acrolein

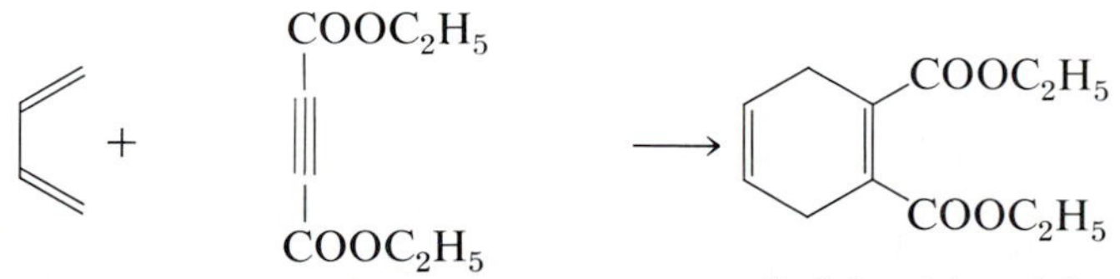

diethyl acetylenedicarboxylate diethyl cyclohexa-1,4-diene-1,2-dicarboxylate

The Diels-Alder reaction is also known as a thermal cycloaddition reaction. The mechanism of the reaction involves σ-overlap of the π-orbitals of the two unsaturated systems, as illustrated in Figure 20.12. The Diels-Alder reaction involves specifically four π-electrons on one system and two on another and is therefore referred to as a [4 + 2] cycloaddition reaction. A remarkable fact is that although such [4 + 2] reactions are common and general, analogous [2 + 2] and [4 + 4] thermal cycloadditions are less common and involve different reaction mechanisms.

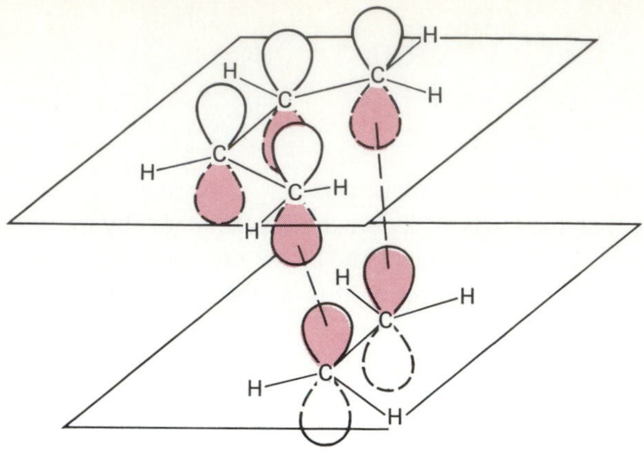

FIGURE 20.12 Transition state of a Diels-Alder reaction.

$$\begin{matrix} CH_2 \\ \| \\ CH_2 \end{matrix} + \begin{matrix} CH_2 \\ \| \\ CH_2 \end{matrix} \xrightarrow[\Delta]{\not\;} \square$$

2 + 2

4 + 4

The requirement of *six* electrons in the cyclic transition state in Figure 20.12 is now well understood and is closely related to the stability of the benzene ring. Accordingly, this topic is discussed further in the next chapter (Section 21.7).

The transition state of the Diels-Alder reaction depicted in Figure 20.12 has stereochemical consequences. The following examples illustrate that the reaction is a syn addition with respect to both the diene and the dienophile.

$$\text{H}, COOCH_3 / \text{H}, COOCH_3 \xrightarrow{\Delta} \text{H}, COOCH_3 / \text{H}, COOCH_3$$

$$CH_3, \text{H}, \text{H}, CH_3 + COOCH_3\!-\!C{\equiv}C\!-\!COOCH_3 \xrightarrow{\Delta} \text{H}, CH_3, COOCH_3, COOCH_3, \text{H}, CH_3$$

When both the diene and dienophile are suitably substituted, a further stereochemical feature arises because the reactants may approach each other in two distinct orientations. The substituent on the dienophile may be directed away from the diene (*exo* approach) or toward the diene (*endo* approach), resulting in two stereoisomeric products.

Exo approach

Endo approach

Diels-Alder reactions often show a preference for *endo* approach, but *endo*/*exo* ratios depend strongly on reaction conditions (such as temperature and solvent polarity) and on the exact structures of diene and dienophile.

When cyclic dienes are used in the Diels-Alder reaction, bicyclic adducts result. An especially important cyclic diene is cyclopentadiene.

cyclopentadiene + methyl acrylate ($COOCH_3$) $\xrightarrow{75\text{–}80°C}$ methyl bicyclo[2.2.1]-hept-5-ene-*endo*-2-carboxylate (79–91%)

Polycyclic compounds are an important group of organic structures. The number of rings (cycles) in such a compound is determined by the minimum number of ring bonds that must be broken to obtain an acyclic compound. For example, in a bicyclic compound, if one of the ring bonds is broken, a monocyclic compound results. If one of the ring bonds of this monocyclic compound is broken, an acyclic product is produced. Bicyclic structures have the following structural features in common.

1. Two bridgehead atoms.
2. Three arms connecting the two bridgehead atoms.

Bicyclic compounds are named as derivatives of the alkane corresponding to the total number of carbons in both ring skeletons. The prefix **bicyclo-** indicates that the compound has a bicyclic structure. The numbers of carbons in each of the three connecting arms are given in brackets.

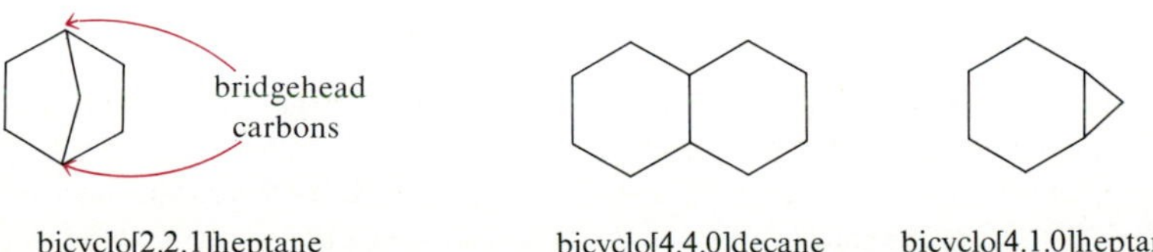

bicyclo[2.2.1]heptane　　bicyclo[4.4.0]decane　　bicyclo[4.1.0]heptane

The resulting compound names are written as all one word, with no spaces or hyphens.

The numbering system used to assign substituents starts at a bridgehead position, proceeds along the *longest* arm to the other bridgehead position, and continues along the next longest arm. The bridgehead position chosen to start the numbering is that which gives the lower substituent number.

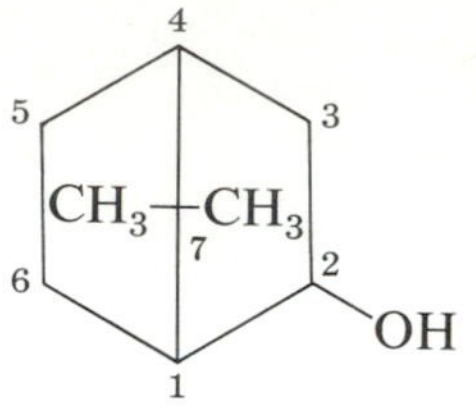

7,7-dimethylbicyclo[2.2.1]heptan-2-ol

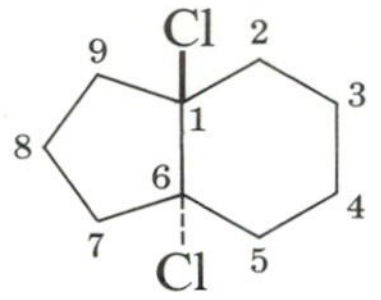

trans-1,6-dichlorobicyclo[4.3.0]nonane

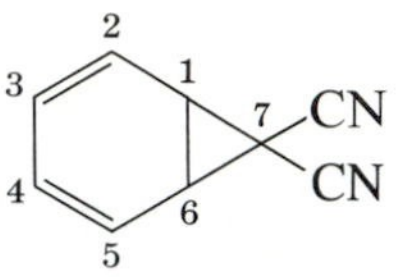

7,7-dicyanobicyclo[4.1.0]hepta-2,4-diene

EXERCISE 20.17 Name the following bicyclic compounds.

(a)

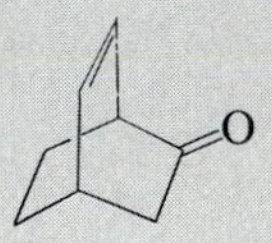

(b)

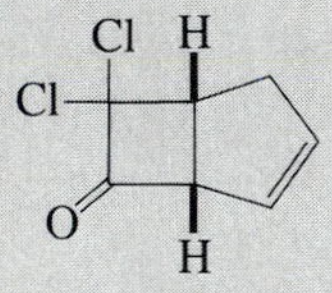

In many bicyclic structures the stereochemistry at the bridgehead position is established by steric constraints. There is only a single bicyclo[2.2.1]heptane, for example, that in which the methylene bridge is joined cis at the 1,4-positions of a boat cyclohexane. The corresponding compound with a trans attachment is too strained to exist (Figure 20.13). In other systems both cis and trans ring fusions can occur and are specified appropriately in the nomenclature. Some bicyclic systems have a further aspect of stereochemistry that must be noted. For example, the Diels-Alder reaction between cyclopentadiene and methyl acrylate, discussed on page 597, could have given two stereoisomeric products, one in which the methoxycarbonyl group is cis to the two-carbon bridge (called the **endo** isomer) or one in which this group is cis to the one-carbon bridge (called the **exo** isomer). Diels-Alder reactions of cyclopentadiene generally produce the endo isomer as the major product.

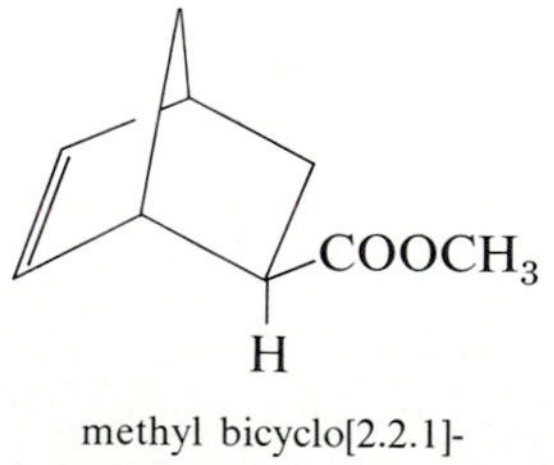

methyl bicyclo[2.2.1]-hept-5-ene-*exo*-2-carboxylate

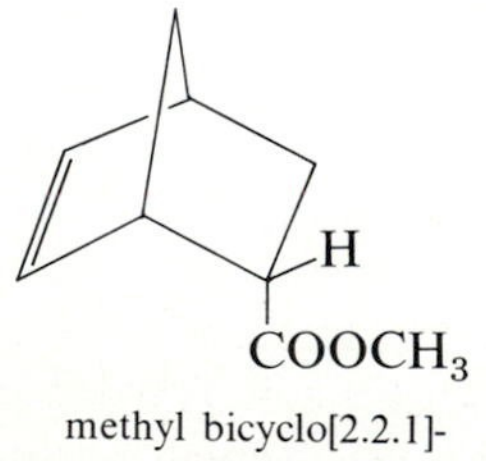

methyl bicyclo[2.2.1]-hept-5-ene-*endo*-2-carboxylate

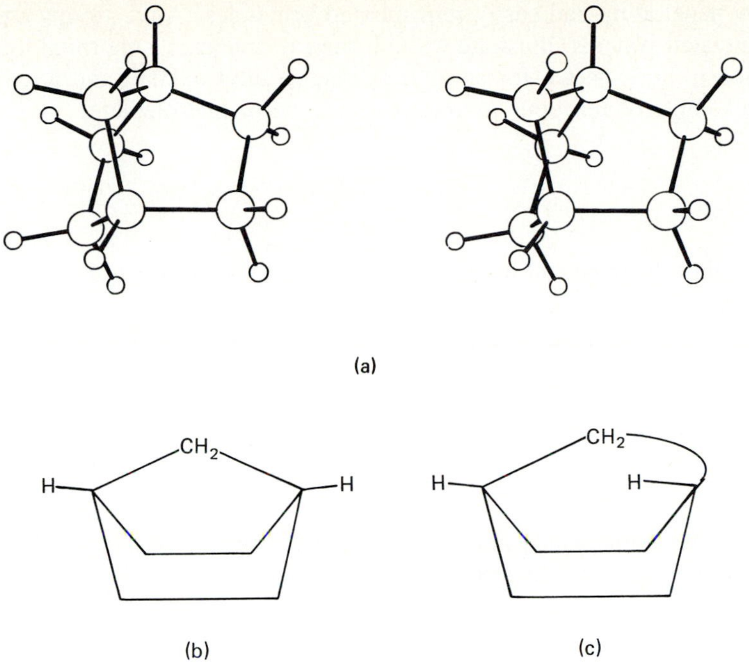

FIGURE 20.13 Bicyclo[2.2.1]heptane (norbornane): (a) stereo view; (b) conventional perspective drawing; (c) hypothetical trans isomer—too strained to exist.

Cyclopentadiene is a low-boiling hydrocarbon, b.p. 46 °C, available commercially as a dimer that can be readily cracked thermally. The dimer boils at 170 °C. When the free monomer has been prepared by slow distillation of the dimer, it must be used immediately as it re-dimerizes on standing. The dimerization reaction is a Diels-Alder reaction in which one molecule acts as the diene and another takes the role of the dienophile. The endo dimer is produced.

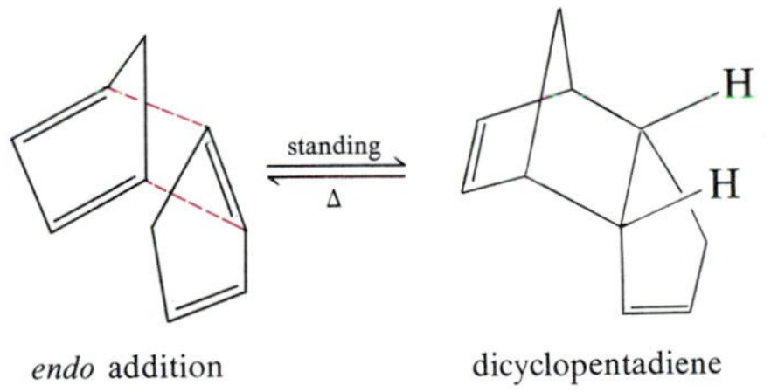

EXERCISE 20.18 Write the equations for the Diels-Alder reactions of cyclopentadiene with (a) vinyl acetate, (b) acrylic acid, (c) dimethyl acetylenedicarboxylate, $CH_3OOCC{\equiv}CCOOCH_3$. What is the name of each product?

ASIDE

As pointed out in Chapter 25, the name allyl is derived from the Latin name for onion and garlic, plants that yield pungent and unforgettable aromas. The connection of reactive allylic intermediates with conjugation is a more recent development, characterized by penetrating and easily remembered principles. Allyl groups are the simplest conjugated systems. Illustrate with resonance structures and orbital formulas the conjugation to be expected for an allyl cation, an allyl radical, and an allyl anion. Give an example of reactions involving each type of intermediate.

PROBLEMS

1. Draw all of the important resonance structures for each of the following allylic type carbocations.

a. $CH_2{=}CH{-}CH{=}CH{-}CH_2^+$ ≡

b. $CH_2{=}CH{-}C(CH_2^+){=}CH_2$ ≡ 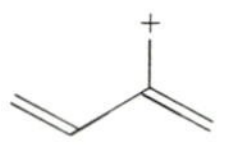   **c.**

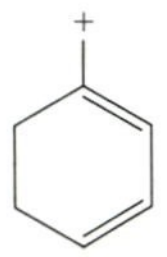

2. For each of the following pairs of allylic resonance structures, indicate which contributes more to the resonance hybrid.

a. $[^+CH_2{-}CH{=}CHCH_3 \longleftrightarrow CH_2{=}CH{-}\overset{+}{C}HCH_3]$

b. $[CF_3\overset{-}{C}H{-}CH{=}CHCH_3 \longleftrightarrow CF_3CH{=}CH{-}\overset{-}{C}HCH_3]$

c. $[CF_3\overset{+}{C}H{-}CH{=}CHCH_3 \longleftrightarrow CF_3CH{=}CH{-}\overset{+}{C}HCH_3]$

d. $[CH_2{=}CH{-}O^- \longleftrightarrow {}^-CH_2{-}CH{=}O]$

e. $\left[CH_3\overset{O}{\overset{\|}{C}}{-}NH_2 \longleftrightarrow CH_3\overset{O^-}{\overset{|}{C}}{=}\overset{+}{N}H_2\right]$

f. $[CH_2{=}CH{-}OCH_3 \longleftrightarrow {}^-CH_2{-}CH{=}\overset{+}{O}CH_3]$

g. 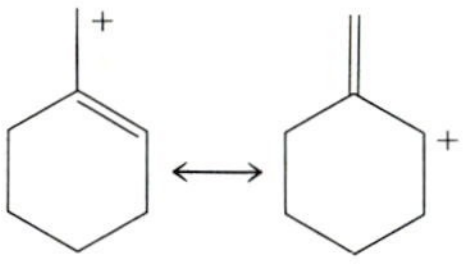 ⟷

3. Illustrate the use of allylic halides and Grignard reagents in preparation of the following alkenes.

a. $CH_3CH_2CH_2CH{=}CH_2$
b. $C_6H_5CH_2CH{=}CH_2$
c. $CH_2{=}CHCH_2CH{=}CH_2$
d. $(CH_3)_3CCH_2CH{=}CH_2$

4. What is the principal product of reaction of each of the following alcohols with activated MnO_2?

a. $CH_3CH{=}CHCH_2OH$
b. $CH_3CH_2CH_2CH_2OH$
c. $HOCH_2CH_2CH{=}CHCH_2OH$
d. 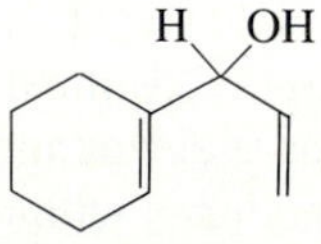 H OH

e. 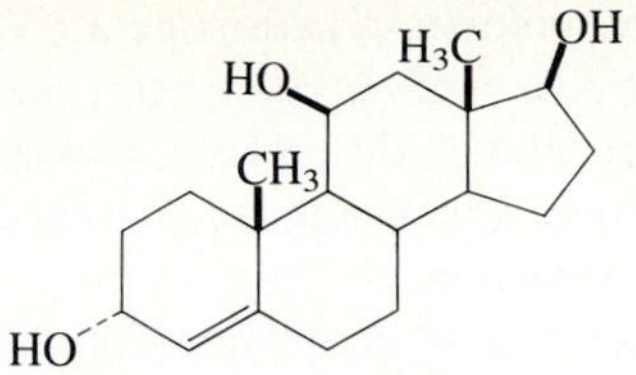 [steroid structure: HO, H_3C, OH, CH_3, HO]

f. $CH_3C{\equiv}CCHOHCH_3$

5. The reaction of 1-octene with N-bromosuccinimide in carbon tetrachloride with a small amount of benzoyl peroxide, $(C_6H_5COO)_2$, gives a mixture of 17% 3-bromo-1-octene, 44% *trans*-1-bromo-2-octene, and 39% *cis*-1-bromo-2-octene. Account for these products with a reaction mechanism showing all significant intermediates.

6. In the reaction of 1,3-cyclopentadiene with hydrogen chloride at 0 °C, no significant amount of 4-chlorocyclopentene is produced. Explain.

7. When 1,3-butadiene is allowed to react with hydrogen chloride in acetic acid at room temperature, there is produced a mixture of 22% 1-chloro-2-butene and 78% 3-chloro-1-butene. On treatment with ferric chloride or on prolonged treatment with hydrogen chloride, this mixture is converted to 75% 1-chloro-2-butene and 25% 3-chloro-1-butene. Explain.

8. Show how each of the following conversions can be accomplished.

a. 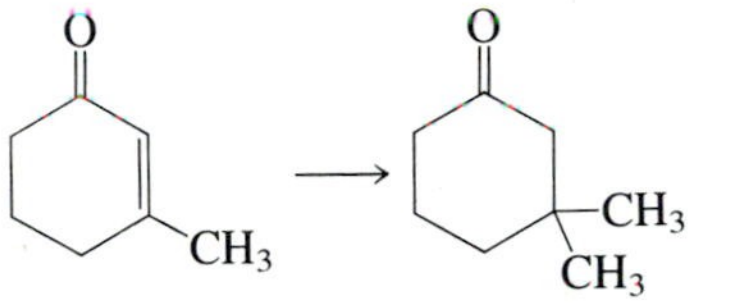 [3-methylcyclohex-2-enone, CH_3] ⟶ [3,3-dimethylcyclohexanone, CH_3, CH_3]

b. 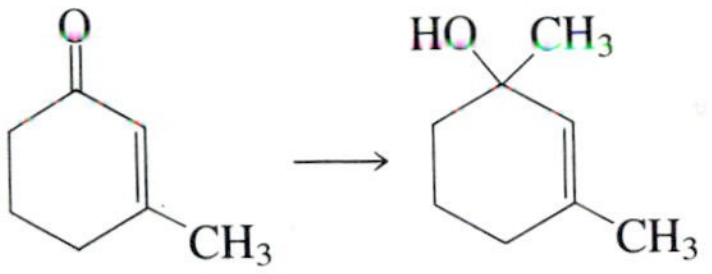 [3-methylcyclohex-2-enone, CH_3] ⟶ [1,3-dimethylcyclohex-2-en-1-ol, HO, CH_3, CH_3]

c. 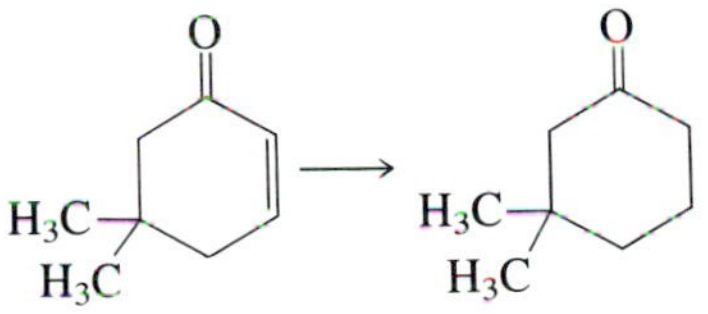 [4,4-dimethylcyclohex-2-enone, H_3C, H_3C] ⟶ [4,4-dimethylcyclohexanone, H_3C, H_3C]

d. 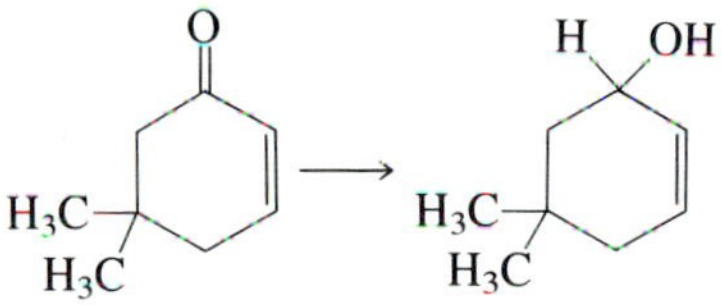 [4,4-dimethylcyclohex-2-enone, H_3C, H_3C] ⟶ [4,4-dimethylcyclohex-2-en-1-ol, H, OH, H_3C, H_3C]

e. 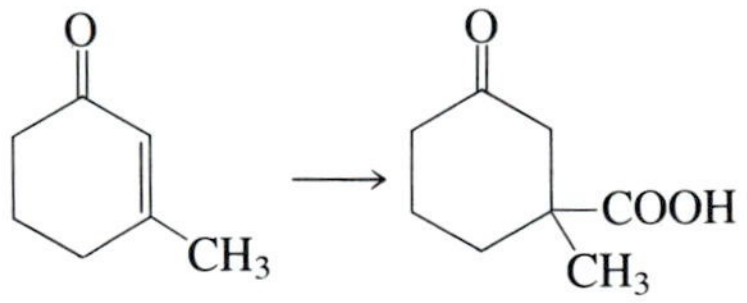 [3-methylcyclohex-2-enone, CH_3] ⟶ [3-methyl-3-carboxycyclohexanone, COOH, CH_3]

9. a. A common procedure for measuring ^{14}C is to obtain it in the form of carbon dioxide, which is passed into aqueous barium hydroxide and precipitated as barium carbonate. The white product is dried and pressed into a pellet, which is counted with a Geiger counter. Given $Ba^{14}CO_3$ as starting material, present a practical synthesis of $CH_2 = CH^{14}CH_2OH$. How would you show that allylic rearrangement occurs when this labeled allyl alcohol reacts with thionyl chloride (page 562)?

b. Allylic rearrangement can also be demonstrated with deuterium labeling. Present a practical synthesis of $CH_2{=}CHCD_2OH$. What would the NMR spectrum look like? What product would you expect on treatment with thionyl chloride? What is the expected NMR spectrum of this product?

10. On heating 2-buten-1-ol with dilute sulfuric acid, a mixture of three structurally different isomeric ethers of the type $(C_4H_7)_2O$ is produced. Give the structures of these ethers [do not count cis-trans or R-S isomers] and write a plausible reaction mechanism for their formation.

11. a. When 1-pentyne is treated with 4 N alcoholic potassium hydroxide at 175 °C, it is converted slowly into an equilibrium mixture of 1.3% 1-pentyne, 95.2% 2-pentyne, and 3.5% 1,2-pentadiene. Calculate $\Delta G°$ differences between these isomers for the equilibrium composition. Write a reasonable reaction mechanism showing all intermediates in the equilibrium reaction.

b. Sodium amide in liquid ammonia is a stronger base system than alcoholic KOH, yet prolonged treatment of 1-pentyne by $NaNH_2$ in liquid ammonia leads to recovery of the 1-pentyne essentially unchanged. Explain.

12. 5-Methylcyclopent-2-en-1-one reacts with refluxing aqueous sodium hydroxide to give 2-methylcyclopent-2-en-1-one. Explain with a mechanism.

13. The gas phase enthalpy of ionization ($\Delta H°$ for $RCl \rightarrow R^+ + Cl^-$) for *trans*-1-chloro-2-butene is about 161 kcal mole^{-1} and is significantly lower than that for 3-chloro-2-methyl-1-propene, 169 kcal mole^{-1}. Explain, using resonance structures where appropriate.

14. What diene and dienophile produce the following Diels-Alder adducts?

a. 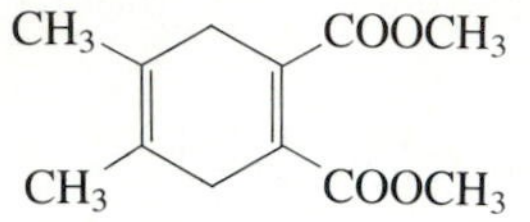 CH_3, CH_3, $COOCH_3$, $COOCH_3$ **b.** 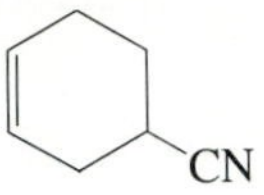 CN **c.** 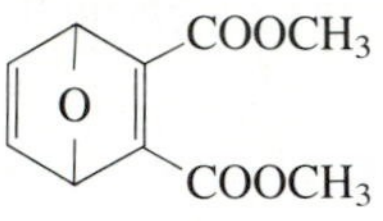 O, $COOCH_3$, $COOCH_3$

d. 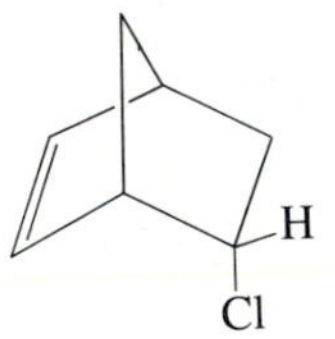 H, Cl **e.**  $COCH_3$ **f.**

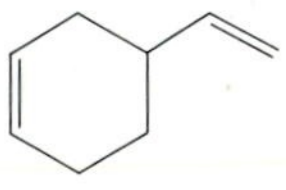

g. 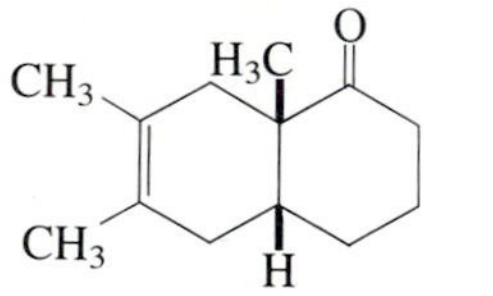 CH_3, CH_3, H_3C, O, H **h.** H, O, H, O

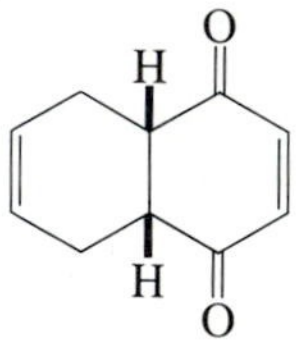

15. Allylic chlorination can be accomplished by the use of *t*-butyl hypochlorite, a reagent prepared by passing chlorine into an alkaline solution of *t*-butyl alcohol.

a. Write a plausible mechanism for this reaction.

b. An example of the use of $(CH_3)_3COCl$ in allylic chlorination is

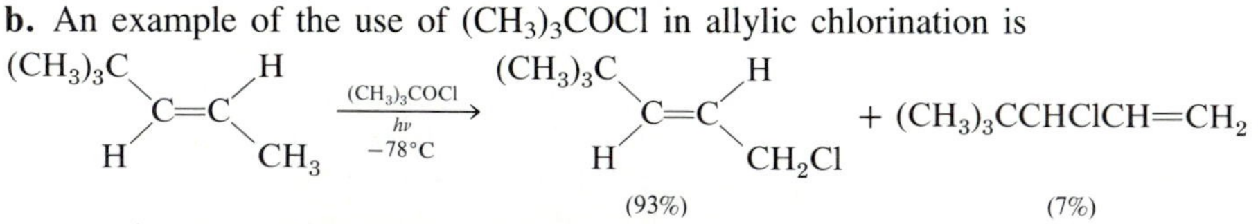

$$\underset{\text{(trans)}}{(CH_3)_3C(H)C{=}C(H)CH_3} \xrightarrow[\;h\nu,\ -78°C\;]{(CH_3)_3COCl} \underset{(93\%)}{(CH_3)_3C(H)C{=}C(H)CH_2Cl} + \underset{(7\%)}{(CH_3)_3CCHClCH{=}CH_2}$$

Write a reasonable reaction mechanism.

c. In the example in (b), note that none of the cis isomer is obtained. If we start with the cis olefin, the reaction takes the following course.

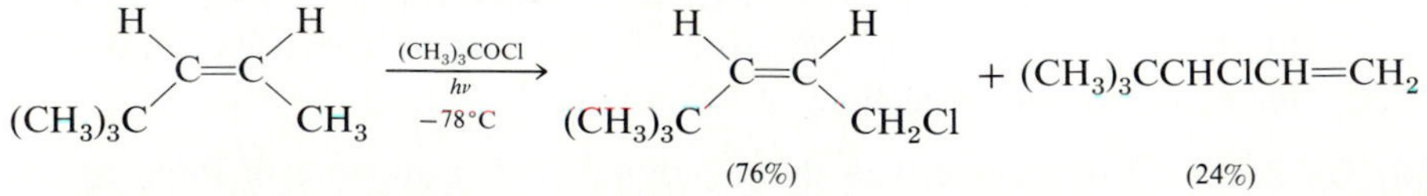

$$\underset{\text{(cis)}}{(CH_3)_3C(H)C{=}C(H)CH_3} \xrightarrow[\;h\nu,\ -78°C\;]{(CH_3)_3COCl} \underset{(76\%)}{(CH_3)_3C(H)C{=}C(H)CH_2Cl} + \underset{(24\%)}{(CH_3)_3CCHClCH{=}CH_2}$$

What does this experiment reveal concerning the configurational stability around the carbon-carbon bond in an allyl radical, at least at −78 °C?

d. Why was it necessary to do the experiment with both the cis and trans olefins? Could the conclusion in (c) have been derived from the results of part (b) alone?

e. The rotation barrier in allyl radical is estimated to be about 10 kcal $mole^{-1}$. Explain why such a barrier exists.

16. Reaction of 2,3-dimethyl-1,3-butadiene with Cl_2 in carbon tetrachloride in the dark at -20 °C gives 45% of the expected product, 1,4-dichloro-2,3-dimethyl-2-butene, in addition to 54% of A and 1% of B, both of which are determined to have the formula C_6H_9Cl. The NMR spectrum of compound A shows singlets at $\delta = 1.90$ ppm (3H) and $\delta = 4.20$ ppm (2H) and four peaks at $\delta = 6.06, 6.19, 6.22, 6.30$ ppm (4H).
The NMR spectrum of compound B shows singlets at $\delta = 1.78$ ppm (3H), 1.85 ppm (3H), and 6.20 ppm (1H), and two peaks at $\delta = 5.08, 5.00$ ppm (2H).
Deduce the structures of A and B and write a plausible mechanism for their formation.

17. The NMR spectrum at -90 °C of the Grignard reagent prepared from 4-bromo-2-methyl-2-butene shows the following signals: $\delta = 0.6$ ppm, doublet (2H); $\delta = 1.6$ ppm, doublet (6H); $\delta = 5.6$ ppm, triplet (1H). Which structure best fits this NMR spectrum? On warming the solution to room temperature, the doublet at $\delta = 1.6$ ppm first broadens and then becomes a sharp singlet. How do you interpret this behavior?

18. Reaction of 3-methyl-1,2-butadiene with Cl_2 under free radical conditions gives 3-chloro-2-methyl-1,3-butadiene. Write a reasonable reaction mechanism.

19. One convenient preparation of acrolein involves the treatment of glycerol, $CH_2OHCHOHCH_2OH$, with sulfuric acid. Write a plausible reaction mechanism.

20. Consider the following equilibria between α,β- and β,γ-isomers. For R = CH_3 the equilibrium constant is about unity. For R = $(CH_3)_3C$, however, the equilibrium is displaced substantially in the α,β-direction. Explain this result. The use of molecular models may be helpful.

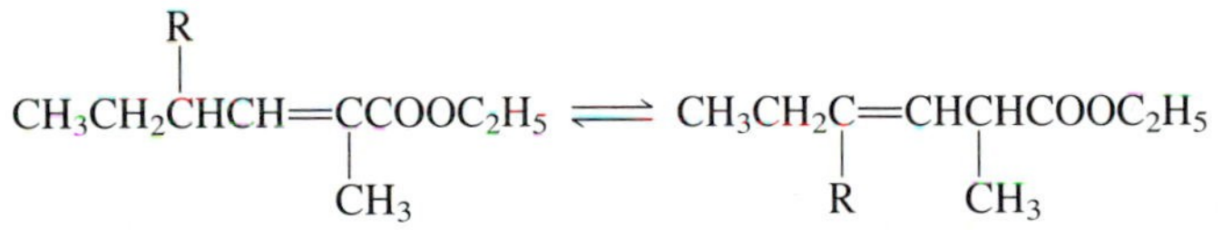

$$CH_3CH_2\underset{}{\overset{R}{\overset{|}{C}}}HCH{=}\underset{\underset{CH_3}{|}}{C}COOC_2H_5 \rightleftharpoons CH_3CH_2\underset{\underset{R}{|}}{C}{=}CH\underset{\underset{CH_3}{|}}{C}HCOOC_2H_5$$

R = CH_3:	(45%)	(55%)
R = $C(CH_3)_3$:	(86%)	(14%)

21. Cinnamaldehyde reacts with manganese dioxide and cyanide ion in methanol to yield a product with a pleasant odor [Do not smell until cyanide has been removed!]. Write the product of the reaction, sketch the ^{1}H and ^{13}C NMR spectra and discuss possible side reactions.

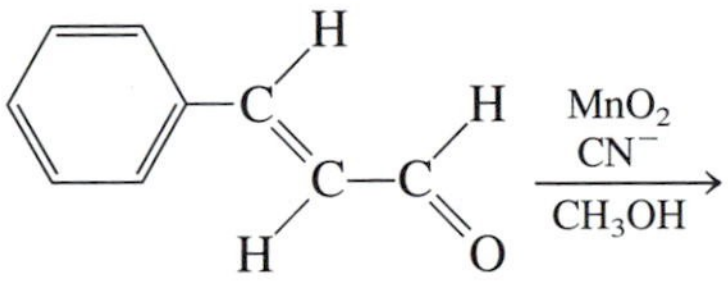

$$C_6H_5CH{=}CHCHO \xrightarrow[CH_3OH]{MnO_2,\ CN^-}$$

cinnamaldehyde

CHAPTER 21

BENZENE AND THE AROMATIC RING

21.1 Benzene

A. The Benzene Enigma

The hydrocarbon now known as benzene was first isolated by Michael Faraday in 1825 from an oily condensate that deposited from illuminating gas. Faraday determined that it has equal numbers of carbons and hydrogens and named the new compound ''carbureted hydrogen.'' In 1834 Mitscherlich found that the same hydrocarbon may be produced by pyrolysis with lime of benzoic acid, which had been isolated from gum benzoin. By vapor density measurements, Mitscherlich established the molecular formula to be C_6H_6. He named the compound benzin, but other influential chemists protested that this name implied a relationship to alkaloids such as quinine. Finally, the German name benzol, based on the German *öl,* oil, was adopted. In France and England, the name **benzene** was adopted, to avoid confusion with the typical alcohol ending.

Early in the history of benzene, Laurent proposed the name pheno as we noted in Section 14.7.D. Although the name never gained acceptance, it persists in **phenyl,** the name of the C_6H_5 group.

Other preparations of benzene followed these early discoveries, and it was soon recognized that benzene is the parent hydrocarbon of a whole family of organic compounds. The physical properties of benzene (b.p. 80.1 °C, m.p. 5.5 °C) are consistent with its molecular formula of C_6H_6. For example, cyclohexane, C_6H_{12}, has b.p. 80.7 °C and m.p. 6.5 °C. A six-carbon saturated alkane would have the formula C_6H_{14}. Therefore, benzene must have four double bonds and/or rings. Yet, it does not exhibit the high reactivity of typical polyenes. In fact, it is remarkably inert to many reagents. For example, it does not react with aqueous potassium permanganate or with bromine water. It does not even react with cold concentrated sulfuric acid. It is stable to air and tolerates free radical initiators.

It may be used as a solvent for Grignard reagents and alkyllithium compounds. All of these properties are totally inconsistent with such C_6H_6 structures as the following.

$$CH_2{=}C{=}\overset{H}{\overset{|}{C}}{-}\overset{H}{\overset{|}{C}}{=}C{=}CH_2 \qquad \text{or} \qquad CH_3C{\equiv}C{-}C{\equiv}CCH_3$$

The fact that benzene has a formula that suggests a polyene structure but does not behave at all like other polyenes was a dilemma for nineteenth century chemists. Furthermore, new compounds were continually being discovered that were structurally related to benzene. It was clear that there is something fundamentally different about benzene and its derivatives. As a group, the benzene-like compounds were called **aromatic** compounds because many of them have characteristic aromas.

The Kekulé theory of valence, first proposed in 1859, allowed acceptable structures to be written for aliphatic compounds such as ethane and ethylene, but at first it did not appear to be applicable to aromatic compounds. In 1865, Kekulé suggested a regular hexagon structure for benzene with a hydrogen attached at each corner of a hexagonal array of carbons.

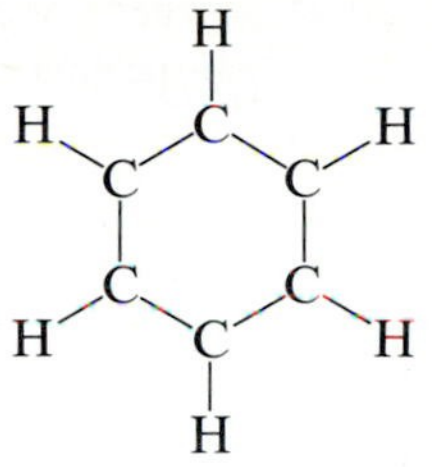

However, this structure violates the tetravalence of carbon inherent in his theory. He later modified his structure to treat benzene as an equilibrating mixture of cyclohexatrienes. However, this structure does not account for the nonolefinic character of benzene.

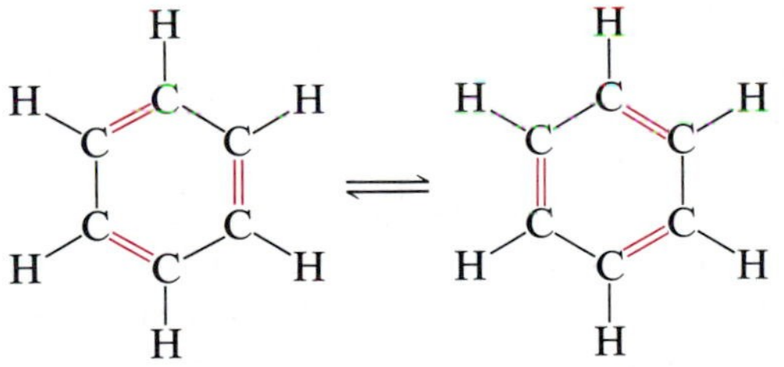

cyclohexatriene

Other attempts by nineteenth century chemists to explain the benzene problem only emphasize the frustrations inherent in the limited theory of the day. One such example was Armstrong's centroid formula in which the fourth valence of each carbon is directed toward the center of the ring.

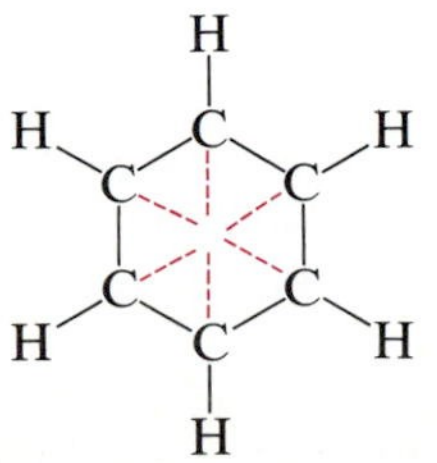

Ladenburg, in 1879, proposed an interesting structure that would solve the problem of why benzene displays no polyene properties. In the Ladenburg proposal, benzene was treated as a tetracyclic compound with no double bonds.

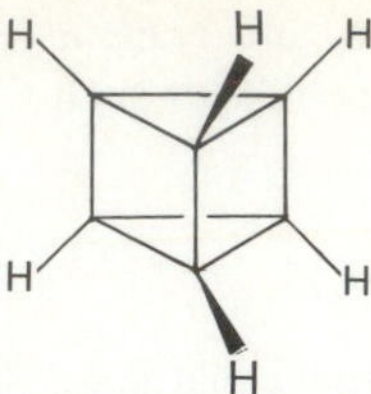

EXERCISE 21.1 The nature of a polycyclic compound as bicyclic, tricyclic, etc., is determined by the number of cuts or bond cleavages required to convert the polycyclic framework to an acyclic chain of atoms. Using this definition, show that the Ladenburg structure is tetracyclic.

Ladenburg's representation of benzene, although not the structure of benzene, is a perfectly valid structure for an organic compound. It has come to be known as ''Ladenburg benzene'' or ''prismane.'' After considerable effort, prismane was finally synthesized in 1973 by organic chemists at Columbia University. Upon heating to 90 °C, it isomerizes to benzene.

EXERCISE 21.2 Compare a model of prismane with that of benzene. Consider the various isomers of prismane having two different substituents, A and B. Are any of these isomers chiral? Are any of the corresponding disubstituted benzene isomers chiral? How do your answers provide a method by which nineteenth century chemists could have distinguished between the Kekulé and Ladenburg representations of benzene?

Only with the advent of modern wave mechanics did the structure of benzene take its place within a unified electronic theory. The x-ray crystal structure of benzene shows that the compound does indeed have a regular hexagonal structure as Kekulé had originally suggested. The carbon-carbon bond distance of 1.40 Å is intermediate between those for a single bond (1.54 Å) and a double bond (1.33 Å). In a regular hexagon the bond angles are all 120°, and this suggests the involvement of sp^2-hybrid orbitals. We can now recognize the ''fourth valence,'' which was so difficult for nineteenth century chemists to explain, as being π-bonds from *p*-orbitals *extending equally around the ring,* as in Figure 21.1.

In resonance language, we may depict benzene by two equivalent resonance structures.

Note the important difference in meaning between this formulation and that of equilibrating cyclohexatrienes. Cyclohexatriene would have alternating single and double bonds, and the chemical equilibrium between the two alternative structures requires the movement of nuclei.

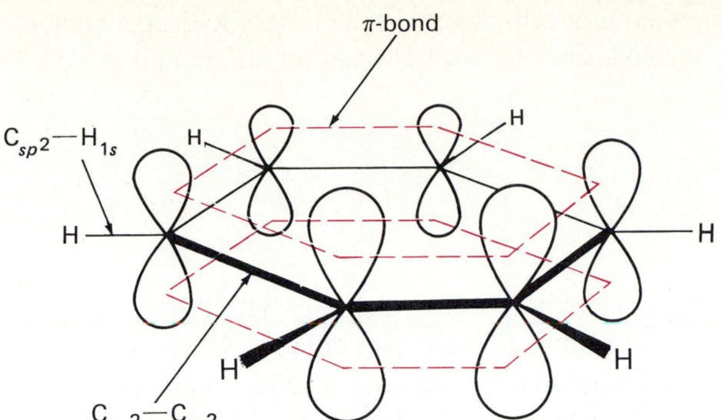

FIGURE 21.1 Orbital structure of benzene.

In the resonance structures the carbon-carbon distances remain the same. The resulting resonance hybrid may be written with dotted lines to indicate the partial double-bond character of the benzene bonds.

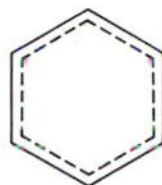

Highlight 21.1

The stability of benzene is its most salient characteristic. Two resonance forms express the equivalent lengths of all double bonds in benzene as does a *p*-orbital formula.

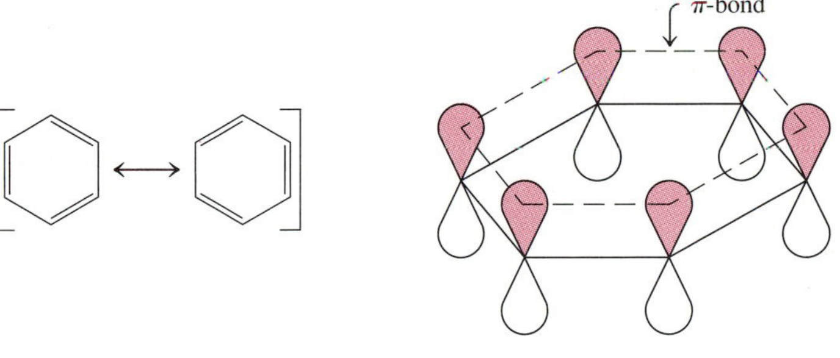

B. Resonance Energy of Benzene

From an examination of the heat of hydrogenation of benzene it is possible to estimate how much more stable benzene is compared to a hypothetical "cyclohexatriene." This imaginary quantity is called the **resonance energy** of benzene. The heat of hydrogenation of the double bond in cyclohexene is -28.4 kcal mole^{-1}. That for one double bond in 1,3-cyclohexadiene is -26.5 kcal mole^{-1}.

$$+ H_2 \longrightarrow \qquad \Delta H^\circ = -28.4 \text{ kcal mole}^{-1}$$

$$+ H_2 \longrightarrow \qquad \Delta H^\circ = -26.5 \text{ kcal mole}^{-1}$$

By a simple extrapolation, we might expect the heat of hydrogenation of one of the double bonds in a 1,3,5-cyclohexatriene with alternating single and double bonds to be about -24.5 kcal mole^{-1}.

$$+ H_2 \longrightarrow \qquad \Delta H° \cong -24.5 \text{ kcal mole}^{-1}(?)$$

Benzene can in fact be hydrogenated, but only with difficulty. It hydrogenates slowly under conditions where simple alkenes react rapidly. When hydrogenation does occur, it generally goes all the way and cyclohexane results. The heat of hydrogenation for the complete reduction of benzene to cyclohexane is -49.3 kcal mole^{-1}.

$$+ 3\,H_2 \longrightarrow \qquad \Delta H° = -49.3 \text{ kcal mole}^{-1}$$

Since the heat of hydrogenation of 1,3-cyclohexadiene to cyclohexane is -54.9 kcal mole^{-1}, the heat of hydrogenation of benzene to 1,3-cyclohexadiene is $-49.3 - (-54.9) = +5.6$ kcal mole^{-1}; the process is actually endothermic!

$$+ H_2 \longrightarrow \qquad \Delta H° = +5.6 \text{ kcal mole}^{-1}$$

These energy relationships are shown graphically in Figure 21.2.

By comparison with the actual heat of hydrogenation of one bond in benzene, we find that benzene is about 30 kcal mole^{-1} more stable than it would be if it had the cyclohexatriene structure. This stabilization energy defines the resonance energy of benzene; that is, the resonance energy is the difference in energy between the real benzene and that of a

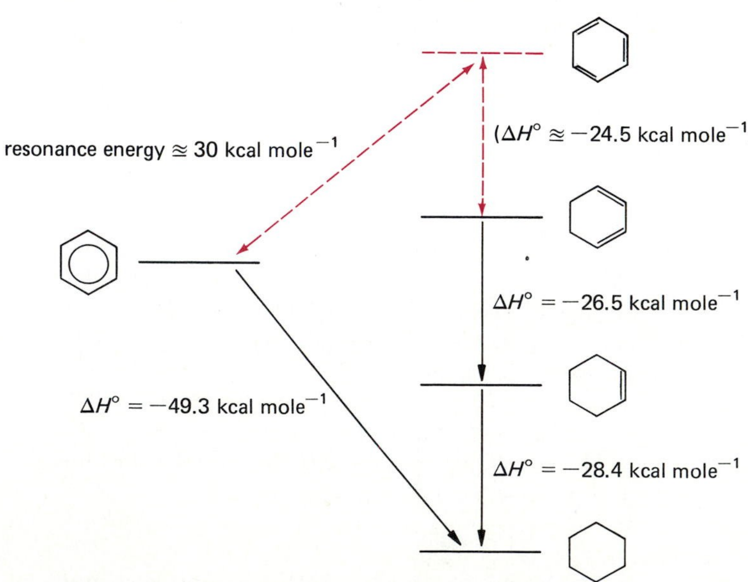

FIGURE 21.2 Estimation of the resonance energy of benzene.

single principal Lewis resonance structure. Other derivations of this quantity give somewhat different values; one commonly used number is 36 kcal $mole^{-1}$.

Actually, the true resonance energy of benzene should not be referred to a cyclohexatriene with alternating bonds of different lengths. Rather, it should be referred to a hypothetical model having the geometry of benzene but with π-overlap allowed only between alternating bonds. Such a structure requires the deformation of cyclohexatriene—stretching the double bonds and compressing the single bonds

distortion energy ⟶ "resonance energy" or delocalization energy ⟶

Various estimates have been made of this distortion energy, but one estimated value of about 30 kcal $mole^{-1}$ appears to be reasonable. This would make the actual resonance energy of benzene about 60 kcal $mole^{-1}$. To distinguish between these different energy quantities, this number of about 60 kcal $mole^{-1}$ is referred to as a **delocalization energy** because it is the energy liberated when electrons are allowed to *delocalize* or *relax* from a hypothetical compound, with the benzene geometry but with the electrons constrained to alternating single and double bonds, to the electronic structure of benzene itself. The value of about 30 kcal $mole^{-1}$, derived above from heats of hydrogenation, is referred to as the **empirical resonance energy.**

Furthermore, this use of the term *resonance* should not be confused with the resonance that occurs in, for example, NMR (nuclear magnetic resonance) in which resonance refers to a matching of the frequency of an irradiating electromagnetic beam with the energy difference between two nuclear spin states in a magnetic field. However, both kinds of resonance are in fact related to the resonance phenomena of vibrations that allowed Joshua's horn to bring down the walls of Jericho.

The benzene ring can also conjugate with other π-electron groups and provide additional stabilization. Comparison of some heats of hydrogenation shows that the benzene ring is less effective in this regard than a double bond.

$$CH_3CH_2CH{=}CH_2 + H_2 \longrightarrow CH_3CH_2CH_2CH_3 \qquad \Delta H^\circ = -30.2 \text{ kcal mole}^{-1}$$

$$CH_2{=}CH{-}CH{=}CH_2 + H_2 \longrightarrow CH_2{=}CHCH_2CH_3 \qquad \Delta H^\circ = -26.3 \text{ kcal mole}^{-1}$$

$$C_6H_5{-}CH{=}CH_2 + H_2 \longrightarrow C_6H_5{-}CH_2CH_3 \qquad \Delta H^\circ = -28.2 \text{ kcal mole}^{-1}$$

The resonance energy of benzene associated with its cyclic π-electronic system of six electrons gives benzene a special character known as *aromatic stability*. Examples of other aromatic systems, which are not based on the benzene ring, are discussed later in this chapter. As has already been stated, this use of the term *aromatic* has nothing whatsoever to do with smell. Although the term was first used to describe a class of compounds that had strong odors, it is now recognized that the special property setting benzene and related compounds apart from aliphatic compounds is its resonance energy and cyclic conjugation. Thus, the term aromatic has come to be associated in organic chemistry with the general class of compounds possessing cyclic π-electron systems that have this special stabilizing electronic character.

Many such "aromatic" compounds are known. They include derivatives of benzene in which one or more groups are attached to the ring. Examples are toluene and benzoic acid.

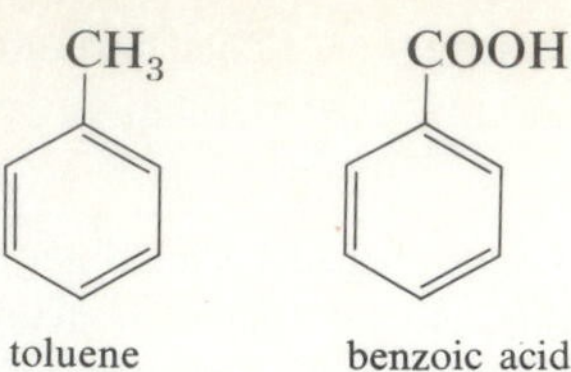

toluene benzoic acid

There are polycyclic benzenoid compounds in which two or more benzene rings are fused together; examples are naphthalene and coronene.

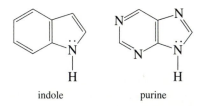

naphthalene coronene

There are also aromatic heterocyclic compounds, compounds in which one or more atoms other than carbon participate in the cyclic conjugated ring. Examples are pyridine, furan, thiophene, and pyrrole.

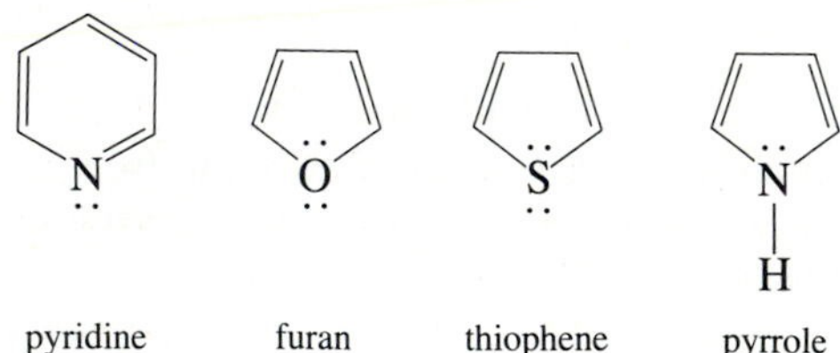

pyridine furan thiophene pyrrole

Polycyclic examples made up of one benzene ring and one heterocyclic ring, or of two or more heterocyclic rings are also known. Examples are indole and purine.

N
H
N
N
N
N
H

indole purine

All of these cases involve cyclic systems of six π-electrons.

EXERCISE 21.3 Draw Lewis structures for one of the resonance structures each of toluene, pyridine, and pyrrole. Compare these structures with an orbital diagram showing the cyclic conjugated system of overlapping π-orbitals. Pay special attention to the difference between the ring systems of pyridine and pyrrole: compare the number of p-orbitals in the cyclic π-system with the number of π-electrons for each compound.

Highlight 21.2

Derivation of the empirical resonance energy of benzene: Found − expected heat of hydrogenation = +5.6 − (−24.5) = 30 kcal $mole^{-1}$

C. Symbols for the Benzene Ring

The symbolism used for the benzene ring deserves further comment. We have discussed the electronic structure of benzene in terms of its cyclic system of orbitals and of resonance structures with reference to hypothetical formulations of cyclohexatriene. We have used symbolic representations of cyclic orbitals and various symbols based on hexagons. These symbols are all in common use in various contexts and may be summarized as follows.

The orbital diagram in Figure 21.1 is especially useful for understanding the high stability of the benzene ring, but it is too complex and cumbersome a symbolism for normal use. The hexagon with an inscribed circle is a simple and commonly used representation of the aromatic π-system and is especially useful for the representation of aromatic structures.

benzene   benzoic acid   toluene

1,2,4-trimethylbenzene   phenylacetylene   1-methylnaphthalene

However, this symbol has an important disadvantage in not allowing an accurate accounting of electrons; that is, it does not correspond to a Lewis structure. In all of our other structural representations, a bond symbolized by a straight line corresponds to two electrons. No such simple correspondence applies to the inscribed circle; for example, the circle in benzene corresponds to six π-electrons, whereas the two circles in naphthalene correspond to a total of ten π-electrons.

EXERCISE 21.4 Three resonance structures of the Kekulé type can be drawn for naphthalene. For each structure determine the number of π-electrons involved in each six-membered ring. Draw an orbital structure of naphthalene and compare the total number of p-orbitals in the π-system with the total number of π-electrons.

The alternating-double-bonds symbol does allow a simple and accurate accounting of electrons and does correspond to a Lewis structure.

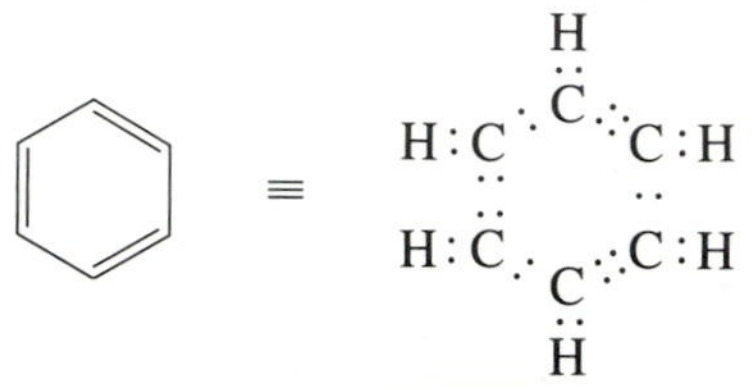

This symbol is used frequently to represent the benzene ring, and the student must be wary not to read this symbol as that of cyclohexatriene—that is, as a cyclic polyene. Generally, this symbol is used as a shorthand for a resonance hybrid of Kekulé structures.

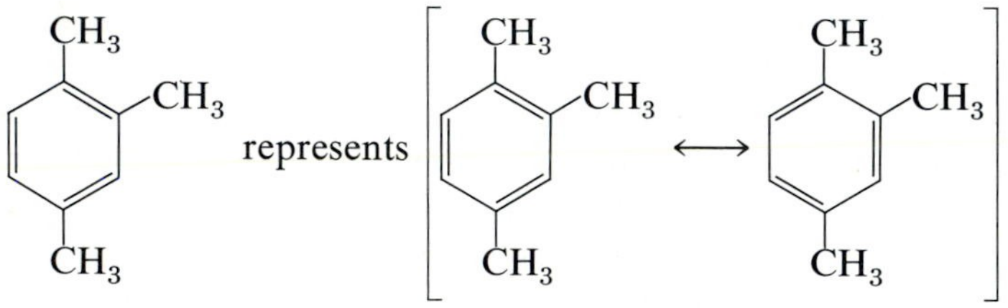

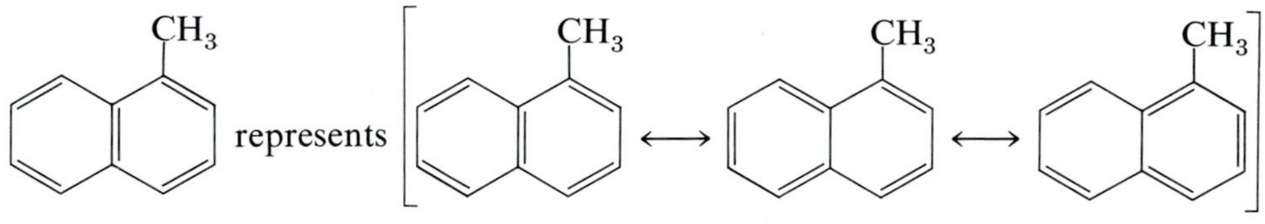

This is the symbol for benzene rings that we will generally use throughout this textbook when it is desirable to call attention to the aromatic ring itself. We shall see later in this chapter and in Chapter 22 how this symbol lends itself readily to following the mechanisms of reactions at the benzene ring.

In many reactions, the bonding electrons of the benzene ring do not take active part. In such cases, it is often convenient to use the symbol **Ph** for the phenyl group, just as we use Me, Et, *i*-Pr, and *t*-Bu to stand for simple alkyl groups. The following examples illustrate this convention.

$PhCH_3$	$PhCOOH$	$PhCl$
toluene	benzoic acid	chlorobenzene

D. Formation of Benzene

The high stability of the benzene ring is further demonstrated by reactions that produce this ring system. Cyclohexane rings can be **dehydrogenated** with suitable reagents or catalysts.

$$\text{cyclohexane} \xrightarrow[\Delta]{\text{Pd or Pt}} \text{benzene} + 3\ H_2$$

$$\text{cyclohexane} + 3\ S \xrightarrow{\Delta} \text{benzene} + 3\ H_2S$$

Dehydrogenation with cyclization can be accomplished from aliphatic hydrocarbons.

$$CH_3(CH_2)_4CH_3 \xrightarrow[450\text{–}550°C]{Cr_2O_3} \text{benzene}\ (20\%)$$

$$CH_3(CH_2)_5CH_3 \xrightarrow[450\text{–}550°C]{Cr_2O_3} C_6H_5CH_3\ (36\%)$$

Such reactions form the basis of the **hydroforming** process of petroleum refining. Gasoline fractions are heated with platinum catalysts (**platforming**) to produce mixtures of aromatic hydrocarbons by cyclization and dehydrogenation of aliphatic hydrocarbons. Most of the benzene used commercially comes from petroleum. United States production in 1989 was 5.84 million tons. Benzene itself is an important starting material for the preparation of many other compounds. Many of these compounds result from electrophilic aromatic substitution, an important reaction that will be discussed in Chapter 23.

21.2 Substituted Benzenes: Nomenclature

Benzene derivatives are named in a systematic manner by combining the substituent prefix with the word benzene. The names are written as one word with no spaces. Since benzene has sixfold symmetry, there is only one monosubstituted benzene for each substituent, and no position number is necessary.

NO_2	$C(CH_3)_3$	Br
nitrobenzene	*t*-butylbenzene	bromobenzene

A number of monosubstituted benzene derivatives have special names that are in such common use that they have IUPAC sanction. We shall refer to the following 12 compounds by their IUPAC-approved common names; thus the student should commit them to memory at this time. Before 1978 *Chemical Abstracts* also used these names for indexing. Beginning with the 1978 indices, however, a new system of nomenclature was introduced. The *Chemical Abstracts* names currently in use are shown in brackets. Although a scientist must be aware of these indexing names, they are not in widespread use for other purposes.

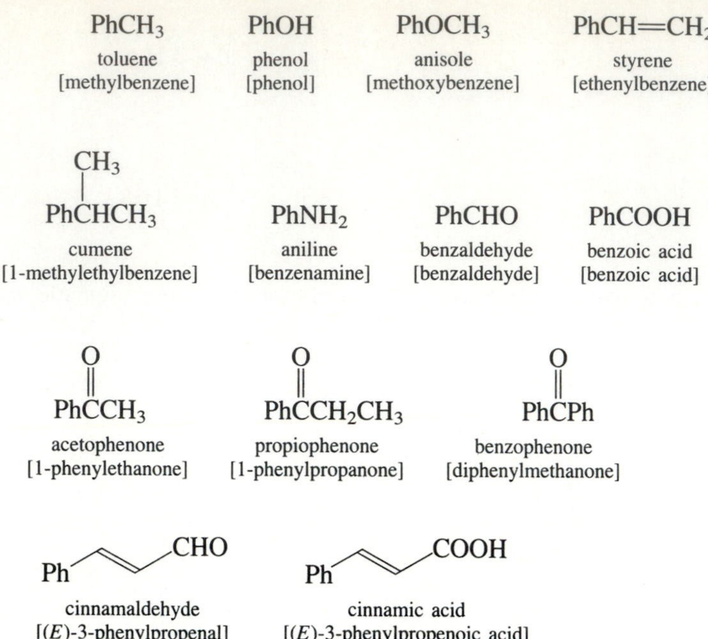

When there are two or more substituents, some specification of position is required. The numbering system is straightforward.

1 2 3 4 5 6

For disubstituted benzene derivatives, the three possible isomers are named using the Greek prefixes ***ortho-***, ***meta-***, and ***para-*** (often shortened to ***o-***, ***m-***, and ***p-***).

X Y X Y X Y

ortho- or *o-* *meta-* or *m-* *para-* or *p-*

The following examples illustrate the use of these prefixes.

Cl Cl Cl Br NO_2 I

ortho-dichlorobenzene
o-dichlorobenzene

meta-bromochlorobenzene
m-bromochlorobenzene

para-iodonitrobenzene
p-iodonitrobenzene

Note that the substituent prefixes are ordered alphabetically. When one of the substituents corresponds to a monosubstituted benzene that has a special name, the disubstituted compound is named as a derivative of that parent.

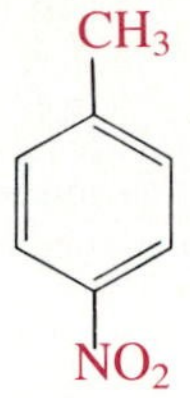

p-nitrotoluene 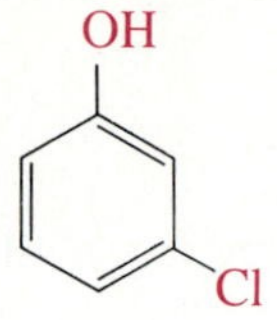   *m*-chlorophenol 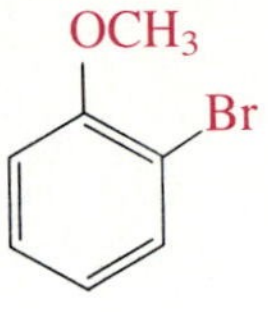   *o*-bromoanisole

However, if a compound has two or more identical substituents that would normally generate one of the foregoing special names, then the compound is named as a derivative of benzene. That is, it is a general rule that identical substituents in a molecule are treated equally for nomenclature purposes.

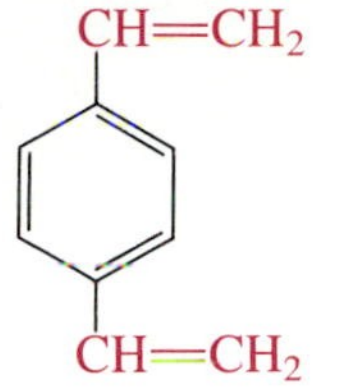

p-divinylbenzene (not *p*-vinylstyrene)   1,2,3-trimethylbenzene (not 3-methyl-*o*-xylene)

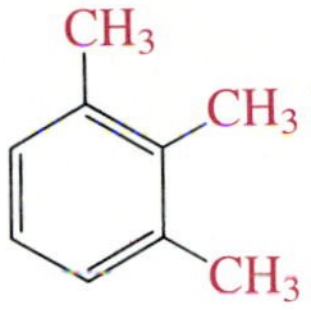

For polysubstituted benzenes, the numbering system should be used.

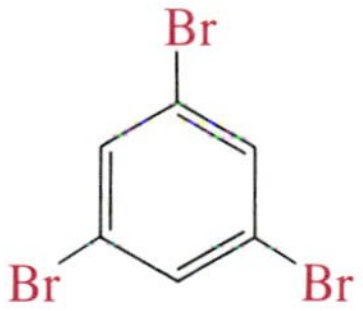

1,3,5-tribromobenzene 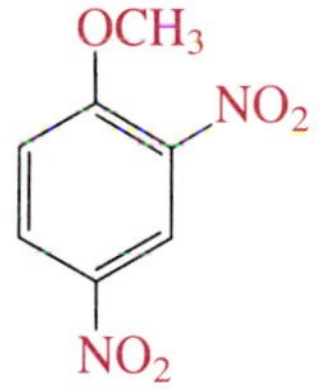   2,4-dinitroanisole 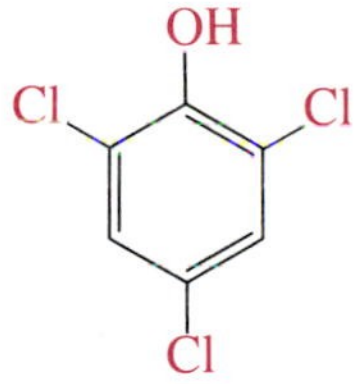   2,4,6-trichlorophenol

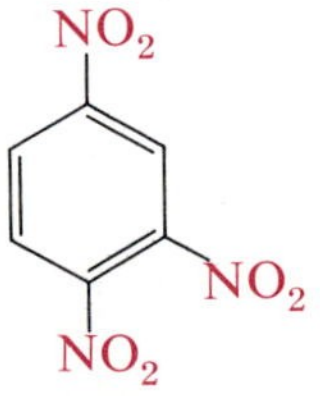

1,2,4-trinitrobenzene (not 1,3,4-trinitrobenzene; the lower numbers are used)   2-bromo-6-nitrotoluene (prefixes are alphabetic)

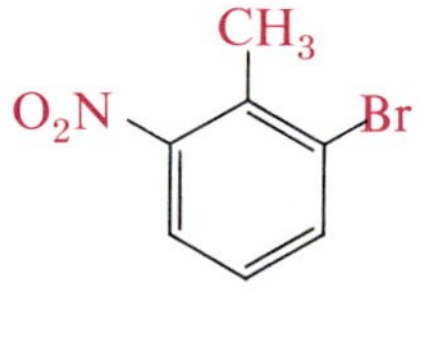

Some di- and polysubstituted benzenes have common or trivial names that are widely used and should be learned by the student. Some of these special names follow; others will be brought up in subsequent chapters dealing with the chemistry of such compounds.

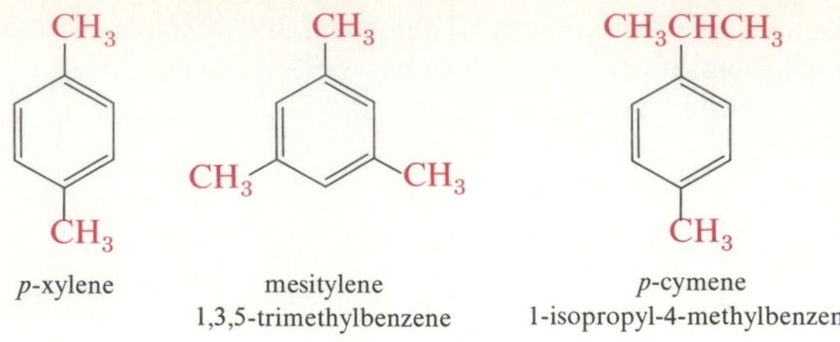

p-xylene | mesitylene 1,3,5-trimethylbenzene | *p*-cymene 1-isopropyl-4-methylbenzene

o-toluic acid *o*-methylbenzoic acid | *m*-toluic acid *m*-methylbenzoic acid | *p*-toluic acid *p*-methylbenzoic acid

Aromatic hydrocarbons have the generic name of **arene.** Accordingly, for many purposes the general abbreviation Ar, for aryl, is used just as R is used for alkyl; thus, the symbol ArR refers to arylalkanes. We have already learned that for benzene itself the term **phenyl-** is used. Examples of names employing this prefix are

$PhC{\equiv}CH$

trans-2-phenylcyclohexanol | 3-phenylpropanal | phenylacetylene

Similarly, derivatives of toluene, the xylenes, and mesitylene, where the additional substituent is attached to the ring, may be named by using the prefixes **tolyl-, xylyl-,** and **mesityl-.**

2-methyl-3-*o*-tolylbutane | 3,4-xylylacetic acid | 3-mesitylpropanoic acid

Certain other group names are used for derivatives of these hydrocarbons when the substituent is attached to a side chain. Note in some of these cases how Greek letters are used to define the side-chain position relative to the benzene ring. The prefix **benzyl-** for the phenylmethyl group is especially important.

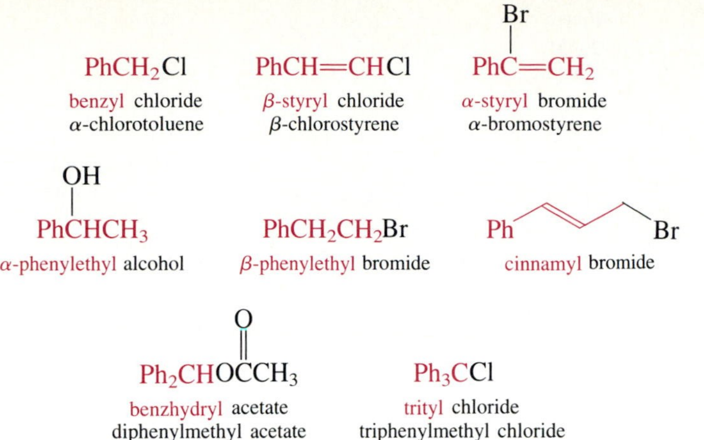

EXERCISE 21.5 Write the structures and names of the twelve methyl and polymethylbenzenes and the six (monochlorophenyl)propanoic acids.

21.3 NMR and CMR Spectra

The NMR spectrum of benzene is shown in Figure 21.3. Since the six hydrogens are equivalent, the spectrum consists of a single line. The unusual feature of the spectrum is the position of the singlet, $\delta = 7.27$ ppm. Recall that δ for olefinic protons is generally about 5 ppm (Section 13.10).

The downfield shift of the benzene hydrogens results from the cyclic nature of the π-electrons of the aromatic ring. This cyclic electronic system can be likened to a circular wire, which in a magnetic field produces a current around the ring. This current is exactly analogous to the current induced in the π-electrons of a double bond (Figure 21.4), except that in benzene this electron current extends around the ring rather than being localized in

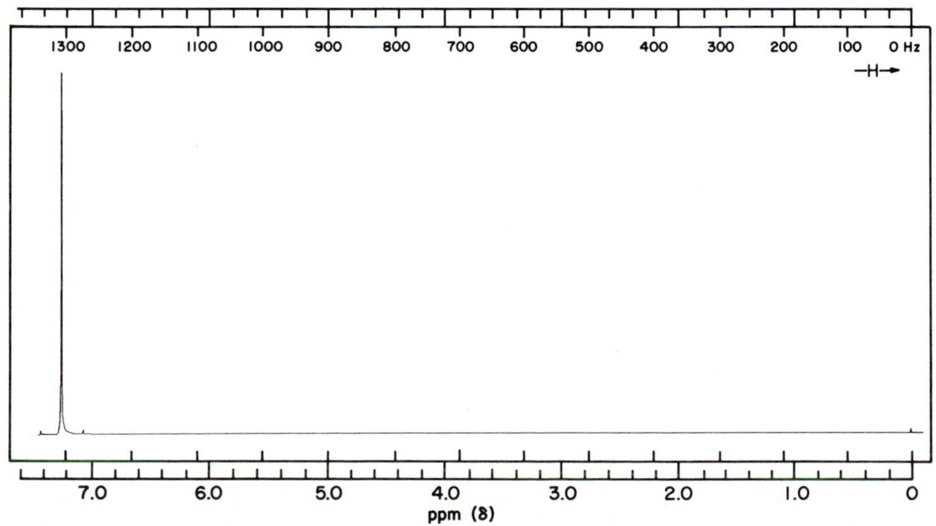

FIGURE 21.3 NMR spectrum of benzene.

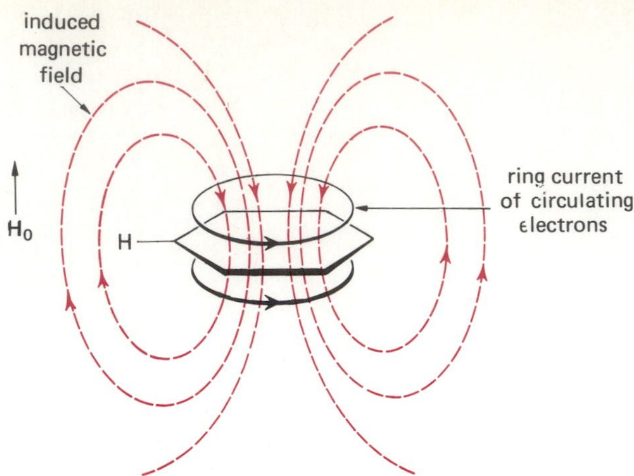

FIGURE 21.4 Effect of ring current in benzene π-system increases effective magnetic field at the proton.

one double bond. The resulting **ring current** has an induced magnetic field that *adds* to the externally applied field at the protons, just as in the related case of olefinic hydrogens (Figure 13.37). Since a smaller applied field is required to achieve resonance at the nucleus of the proton, the net result is a downfield shift.

> Recall that the ring current shown in Figure 21.4 refers to *circulating electrons* rather than the *positive current* to which the ''right hand rule'' applies as usually taught in physics courses.

The effect is greater for a benzene ring than for a simple alkene, in part because the benzene ring has six π-electrons in the cycle. The proton resonance of substituted benzene rings occurs generally in the region of $\delta = 7$–8, a region in which few other kinds of proton resonances occur. NMR peaks in this region are diagnostic for aromatic protons. Substituents have a normal type of effect: electronegative substituents generally cause a downfield shift, and electron-donating groups usually produce an upfield shift.

	p-$CH_3C_6H_4CH_3$	C_6H_6	p-BrC_6H_4Br	p-$CCl_3C_6H_4CCl_3$
δ for aromatic H, ppm:	6.95	7.27	7.34	7.98

In some monosubstituted benzenes all five benzenoid hydrogens have approximately the same chemical shift, and the aromatic protons appear as a single group of peaks. This is usually the case when the substituent is an alkyl group or some other group having approximately the same electronegativity as carbon. An example is toluene, the NMR spectrum of which is shown in Figure 21.5.

Also note in Figure 21.5 that the methyl group attached to the benzene ring resonates at $\delta = 2.32$ ppm, about 1.4 ppm downfield from the resonance position of a methyl group in an alkane. The main cause of this downfield shift is the diamagnetic anisotropy or ring current of the aromatic ring. The effect is not as great with the methyl group as it is for hydrogens directly attached to the ring because the methyl hydrogens are farther from the circulating electrons than the benzenoid hydrogens. The effect is comparable to that found for allylic protons in alkenes (Section 13.10).

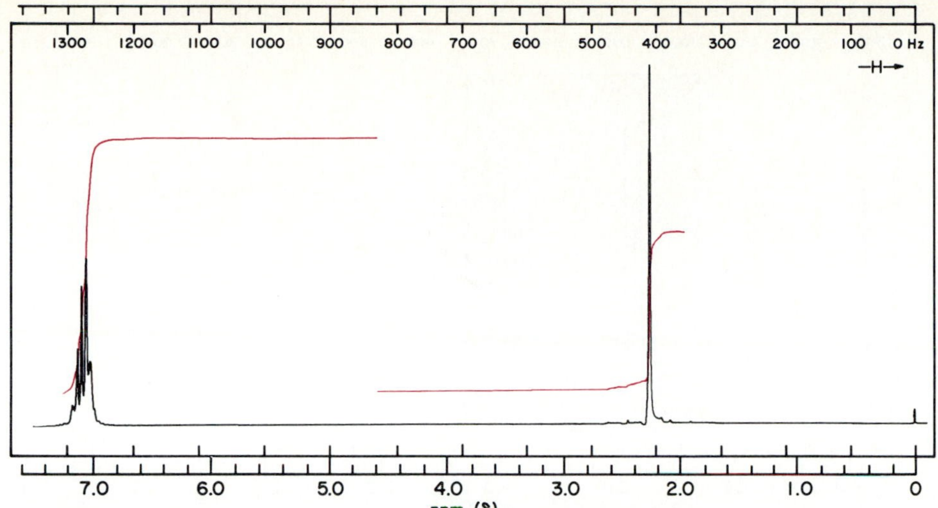

FIGURE 21.5 NMR spectrum of toluene, $PhCH_3$.

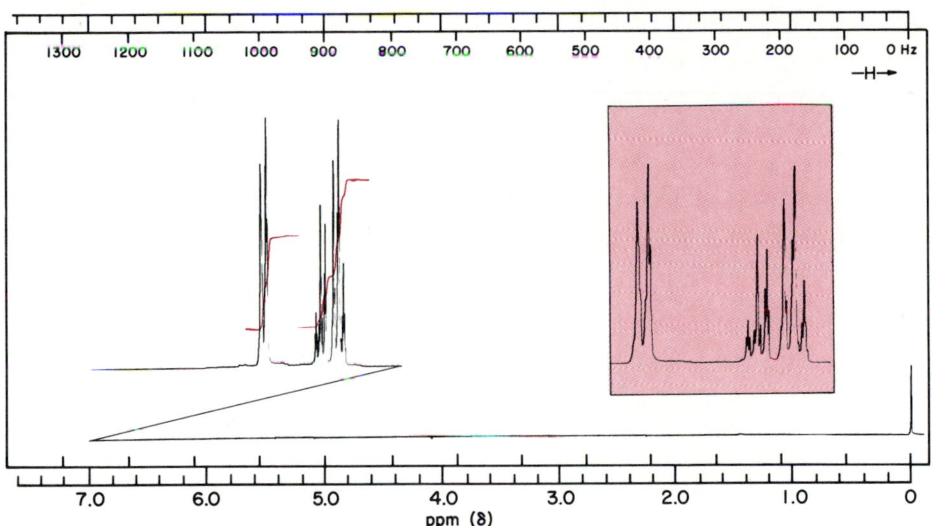

FIGURE 21.6 NMR spectrum of nitrobenzene, $PhNO_2$.

When the substituent in a monosubstituted benzene is sufficiently electronegative or electropositive relative to carbon, the *ortho, meta,* and *para* hydrogens have significantly different chemical shifts and the NMR spectrum becomes more complex. Such a spectrum is shown by nitrobenzene (Figure 21.6).

The spectra of disubstituted benzenes can sometimes be rather complex. In *p*-dichlorobenzene the four benzenoid hydrogens are equivalent, and the NMR spectrum is a sharp singlet at $\delta = 7.22$ ppm. On the other hand, there are 24 lines in the NMR spectrum of *o*-dichlorobenzene (Figure 21.7). Some of these signals are of low intensity, and others are so close together that they appear merged if instrument resolution is inadequate. The analysis of such complex splitting patterns is beyond the scope of this text, but the student should know that such spectra can be analyzed and interpreted by experts to give structural information.

When the two substituents are of different electronegativity, the NMR spectra are sometimes sufficiently simple to be interpretable by a "first-order" approximation. An

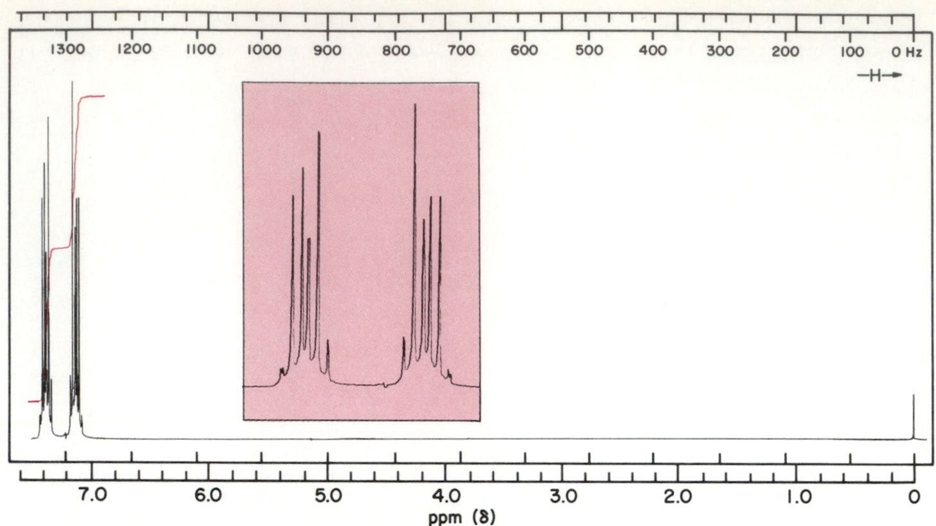

FIGURE 21.7 NMR spectrum of *o*-dichlorobenzene, *o*-$Cl_2C_6H_4$.

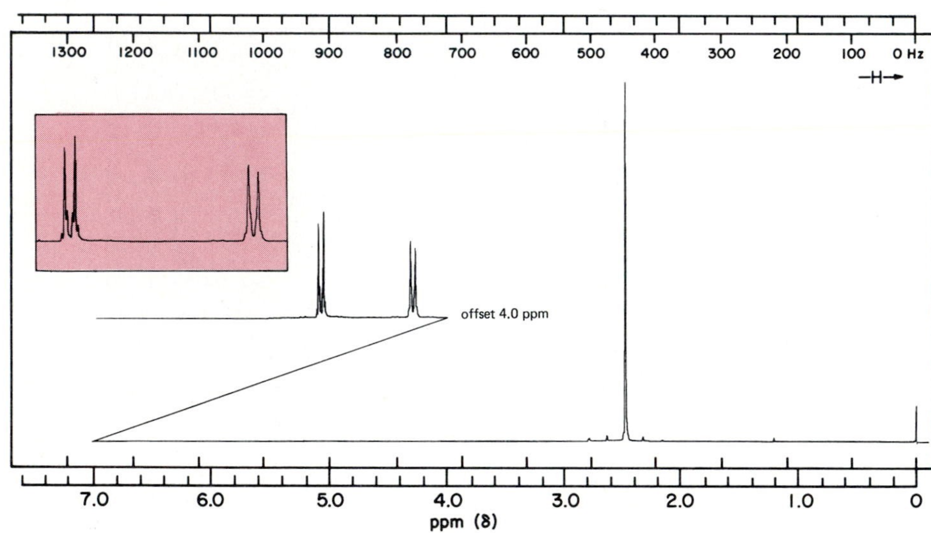

FIGURE 21.8 NMR spectrum of *p*-nitrotoluene, *p*-$NO_2C_6H_4CH_3$.

example is the spectrum of *p*-nitrotoluene, shown in Figure 21.8. To a first approximation, the benzenoid region in *p*-nitrotoluene may be regarded as a pair of doublets arising from coupling between the hydrogens on C-2 and C-3 with $J = 8$ Hz. Each doublet has an intensity of 2, relative to 3 for the methyl group because the hydrogens at C-2 and C-6 are equivalent and the hydrogens at C-3 and C-5 are equivalent.

CH_3

H 6 2 H

$J = 8$ Hz

H 5 3 H

NO_2

The CMR chemical shifts for benzene and several simple alkyl benzenes are collected in Table 21.1. The data point out an important difference between NMR and CMR spectra. *Diamagnetic anisotropy effects are of only minor importance in determining carbon chemical shifts.* For comparison, the chemical shifts of two alkenes analogous to two of the aromatic compounds listed in Table 21.1 are shown below.

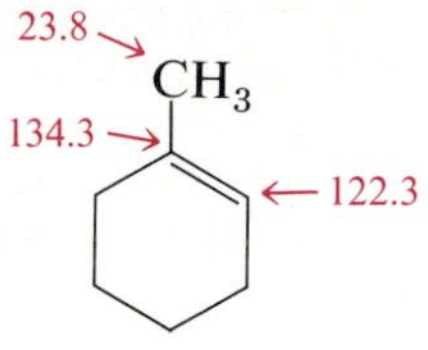

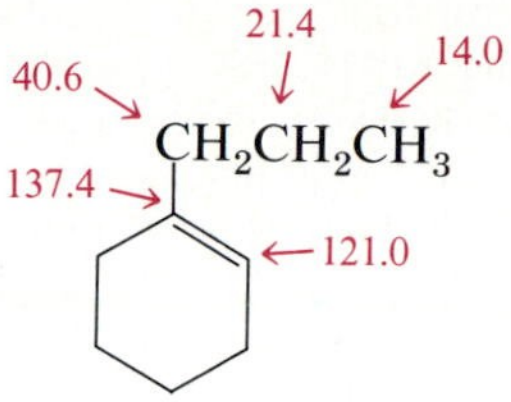

Thus, the carbon chemical shifts of aromatic compounds for both the sp^2-carbons of the ring itself and the sp^3-carbons bonded to the ring are similar to those of a comparable alkene.

TABLE 21.1 CMR Spectra of Some Aromatic Compounds

	Aromatic Resonances					Side-Chain Resonances		
Compound	C-1	C-2	C-3	C-4	C-5	C-1′	C-2′	C-3′
benzene	128.7							
toluene	137.8	129.3	128.5	125.6		21.3		
o-xylene	136.4		129.9	126.1		19.6		
m-xylene	137.5	130.1		126.4	128.3	21.3		
p-xylene	134.5	129.1				20.9		
mesitylene	137.6	127.4				21.2		
ethylbenzene	144.1	128.1	128.5	125.9		29.3	16.8	
n-propylbenzene	142.5	128.7	128.4	125.9		38.5	25.2	14.0

However, CMR spectroscopy can be useful in determining the substitution pattern on a benzene ring. Note, for example, that *o*-, *m*-, and *p*-xylene have CMR spectra consisting of four, five, and three signals, respectively.

EXERCISE 21.6 Sketch the expected NMR and CMR spectra of (a) 1-chloro-3-nitrobenzene and (b) 1-chloro-4-nitrobenzene.

21.4 Dipole Moments in Benzene Derivatives

Methyl chloride has a dipole moment of 1.94 D in the gas phase. The experimental measurement of the dipole moment gives only its magnitude and not its direction. Nevertheless, there is no doubt that the dipole moment in methyl chloride is oriented from carbon to chlorine.

$$\underset{\longmapsto}{CH_3{-}Cl}$$

1.94 D

This orientation agrees with quantum-mechanical calculations and with spectroscopic interpretations of related compounds.

ESSAY 1 Scanning Tunnel Microscopy of Liquid Crystals

Shiny flakes of graphite, composed of stacked planes of benzene-like rings (see Chapter 31), prefer to slide over one another rather than stick, and thus provide lubrication between two surfaces with this sliding mechanism. An especially pure form, freshly cleaved highly oriented pyrolytic graphite, has a regular planar surface suitable for the new atomic microcopies, techniques such as STM, scanning tunnel microscopy, and AFM, atomic force microscopy (see Chapter 33). In STM, a needle-like tip is brought to within 5–10 Å of a conducting surface. A small electrical current will flow between the tip and the conducting surface by a process called tunneling. The presence of a molecule on the surface changes the current. Such current changes are recorded over the whole surface by moving the tip in a scanning motion with sensitive piezoelectric tubes. The resulting current map is converted by a computer into images of the surface or molecules sitting on the surface; with careful interpretation, these images provide direct and very local (on the Ångstrom scale) information about atoms and molecules. The images are especially good for seeing the relationships between molecules and surfaces.

Many digital displays rely on materials that form liquid crystals. Simply stated, crystals are ordered in three dimensions, liquids are not ordered in any dimension, and liquid crystals are ordered in either one dimension (nematic, Gk., *nema,* thread) or two dimensions (smectic, Gk., *smektikos,* cleanse, i.e., soap-like). The order and the property of optical transparency can be altered by small electric fields. Digital numbers appear in a watch when selected areas have their molecular order, and therefore their light transmission, changed by application of a localized electrical field.

A typical liquid crystal material is 4-cyano-4′-octylbiphenyl (8CB), a molecule with a strong dipole at one end and a long alkyl chain at the other. The arrangement of the molecules on the graphite surface is shown in an idealized way (opposite), against a background of hexagons for the carbons of the graphite. Note how the spacing of the methylene groups of the alkyl chain matches the center-to-center distance of the benzenoid rings of the graphite substrate. In another picture, false color is used to make the visualization of the STM image easier, the dark blue regions in the center showing the less-conducting octyl chains, and the light blue being the biphenyl rings. The cyano groups are shown in orange. We see that the molecules form an ordered pattern on the graphite surface, that the rows of molecules change their relative positions every four molecules, that the cyano groups of one row are sandwiched between the cyano groups of the neighboring rows, and that the alkyl chains lie next to one another. Such STM results can eventually be applied to creating new digital display devices.

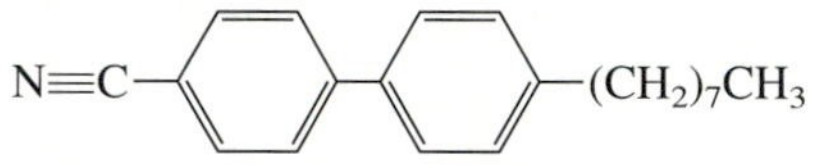

N≡C–(biphenyl)–$(CH_2)_7CH_3$

4-cyano-4′-octylbiphenyl

ESSAY 2 Borscht Color Has Prickly Relatives

"High" cuisine does not often include certain common vegetables. Social position notwithstanding, the admirably colored beet, *Beta vulgaris,* lends its deep red color to a famous Russian soup, borscht, through its pigment, betanin. The red pigment belongs to a general class named betacyanins. Betacyanins and the related betaxanthins, both having a nitrogen-containing polymethine chromophore, are called betalains. Without the benzene ring, the conjugation is decreased and the absorption shifts to shorter wavelengths to yield the yellow color of the betaxanthins, a prominent example being the cactus flower pigment, indicaxanthin. (Gk. *xanthos,* yellow). The elegant and bright colors of the betalain family are often due to mixtures; the early stages of the cactus fruit, *Opuntia,* contain only indicaxanthin, whereas the red flowers of the *Bougainvillea* bush contain a 9:1 mixture of betacyanins and betaxanthins.

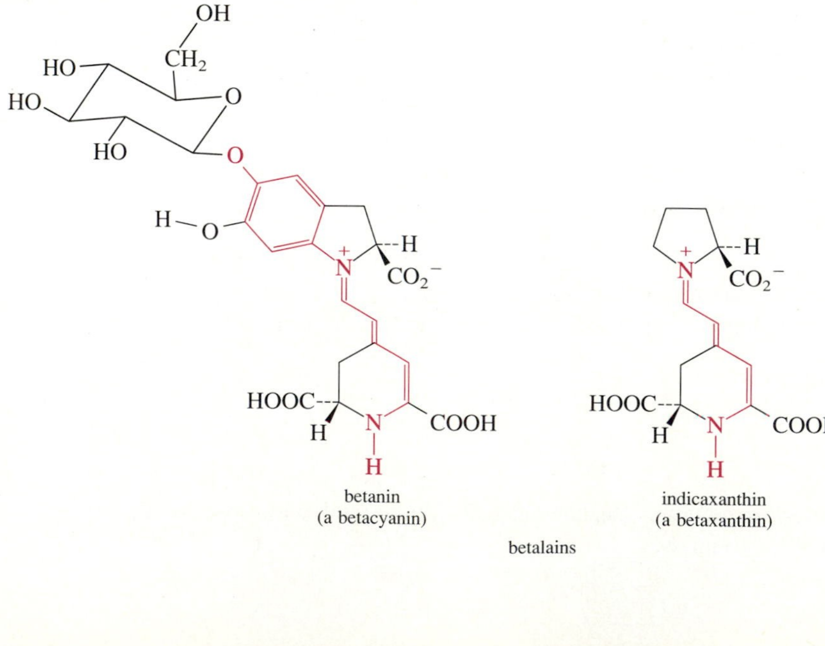

OH, CH_2, HO, HO, O, HO, O, H–O, N+, H, CO_2^-, HOOC, H, N, COOH, H

betanin
(a betacyanin)

N+, H, CO_2^-, HOOC, H, N, COOH, H

indicaxanthin
(a betaxanthin)

betalains

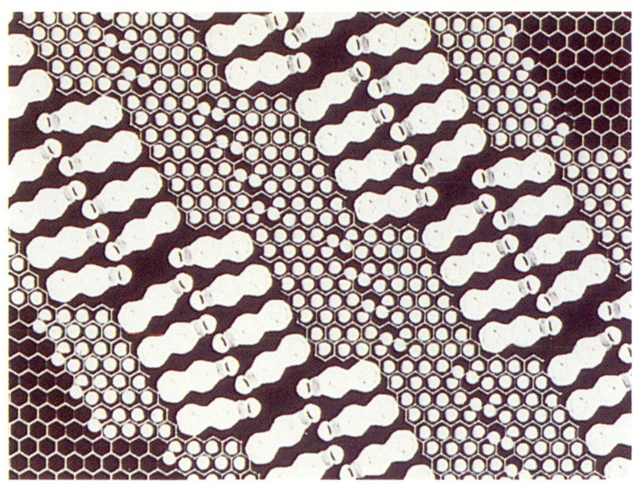

Dr. D. P. E. Smith

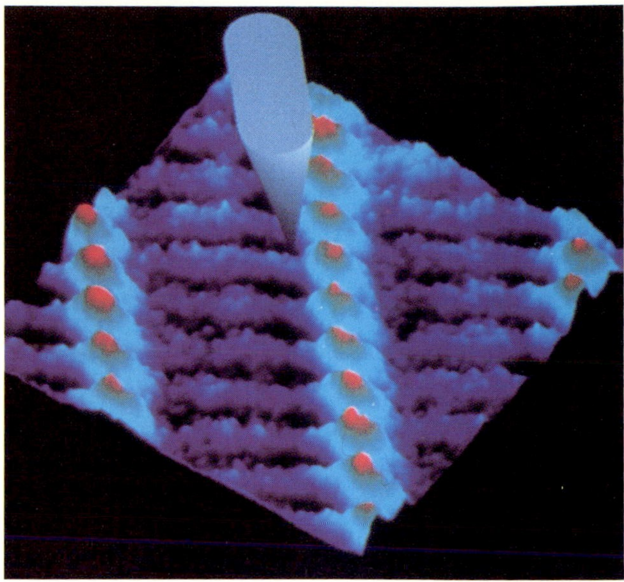

Dr. J. Frommer (IBM)

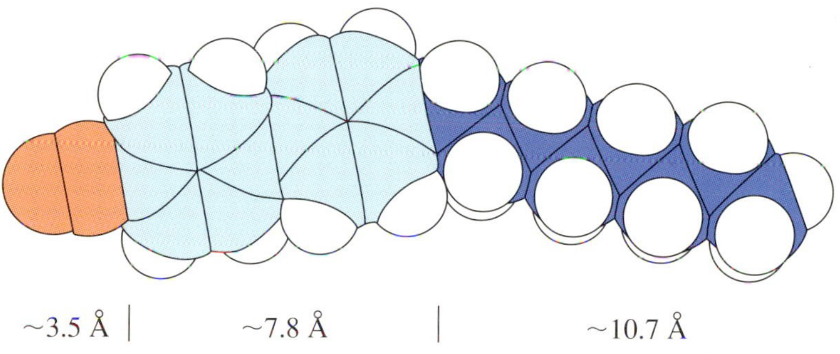

octylcyanobiphenyl (8CB)

Prof. André Dreiding, Universität Zurich

Prof. André Dreiding, Universität Zurich

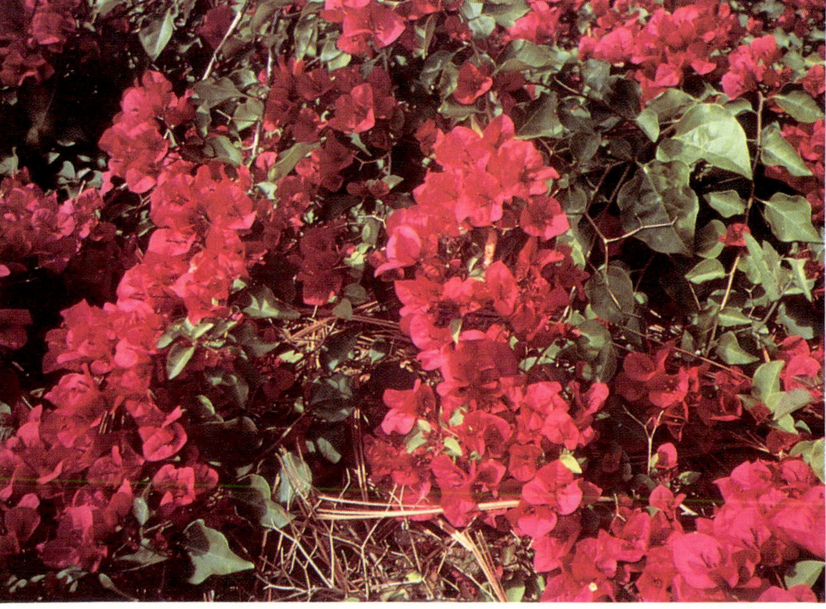

ESSAY 3 Color Attracts

Color is the spice of the natural world; the rainbows of the sky and of nature have been analyzed, duplicated by organic chemists, and cast into the world to delight us with their spectral beauty. To know the physical and chemical basis of color enhances our appreciation of its attractiveness. How marvelous that canthaxanthin, the β-carotene derivative that is the red coloring matter of the male scarlet tanager, *Piranga olivacea,* is needed in the feed of Scotch fish farm salmon to yield the familiar attractive pink flesh. The males in the fall season and the female tanagers always are yellow-green, the pigment having been reduced to the corresponding dihydroxy β-carotene, isozeaxanthin.

[H] ⇌ [O]

canthaxanthin
4,4′-diketo-β-carotene

isozeaxanthin
4,4′-dihydroxy-β-carotene

ESSAY 4 A Measure of Solvent Polarity Is Color

The pyridinium phenol betaine 30 (an internal ion-pair or *zwitterion*) absorbs light in a way that is very sensitive to solvent polarity. The charge-transfer transition of betaine 30 [$E_1(30)$-values] or that of 1-ethyl-4-methoxycarbonylpyridinium iodide [Z-values; see text] can be used as a quantitative measure of solvent polarity. The colors of the betaine show directly the spectacular sensitivity of the absorption to the nature of the solvent.

$$Z \text{ or } E_T \text{ (kcal/mol)} = 28590 / \lambda_{max}\text{, nm}$$

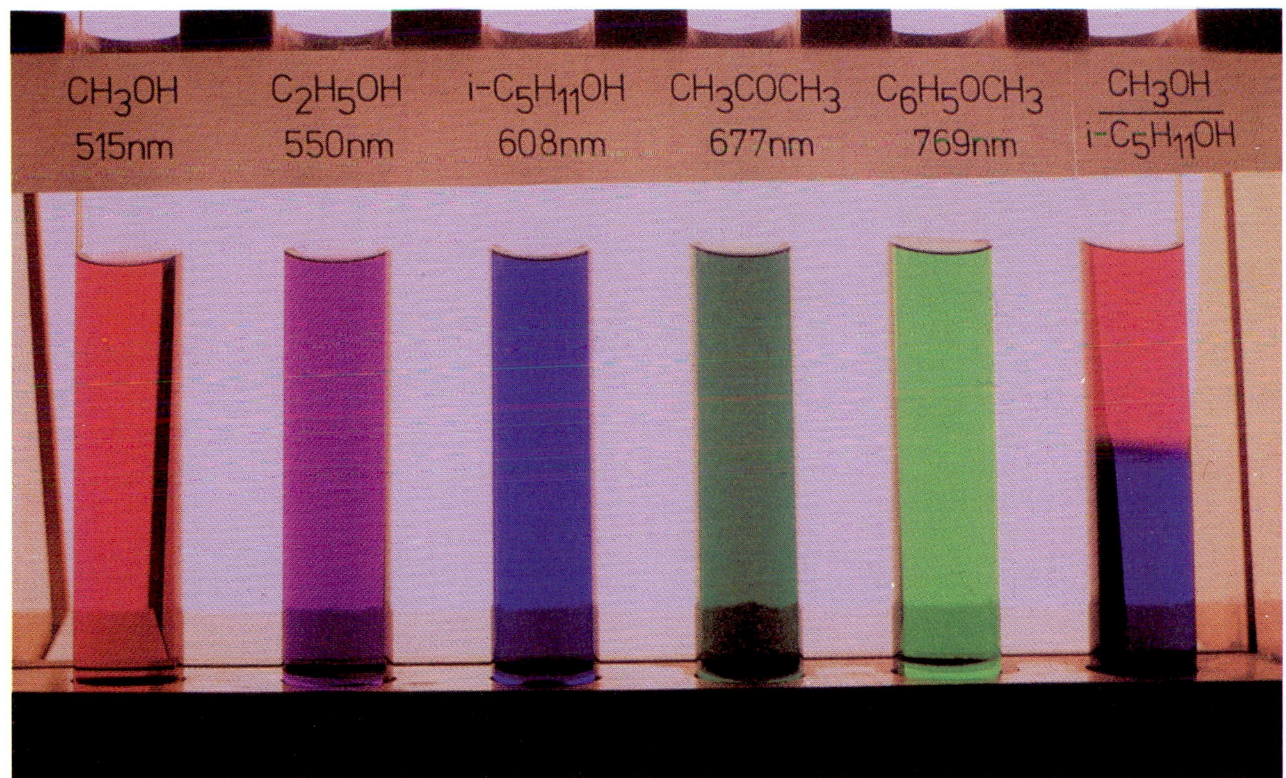

Prof. C. Reichardt, Universität Marburg

ESSAY 5 DNA Models

In a living cell, the deoxyribonucleic acid is thoroughly folded up (''condensed'') so it can fit into the tiny cell nucleus. Folding and unfolding of the DNA are a natural part of the life cycle of the cell. A single chemical cross-link between the DNA ''bases'' (purines or pyrimidines, see Chapter 33) can lead to a considerable change in the macroscopic shape of a DNA molecule. Such changes can be induced by light, either short wavelength ultraviolet (UV), which links thymine as a dimer, or longer wavelength UV, which can link pyrimidines via natural products called psoralens. The changes in DNA shape can be inferred from crystal structures of psoralen-linked bases and are illustrated here. Note how bent the psoralen-linked DNA might be, and imagine the cell struggling with this distortion! The distorted DNA can be repaired by enzymes that eventually replace the lesion, but perhaps not before some critical transcription step has been misplaced in time or missed altogether. Defective repair mechanisms may be responsible for some genetic diseases such as Xeroderma pigmentosum, Bloom's syndrome, and Cockayne's syndrome.

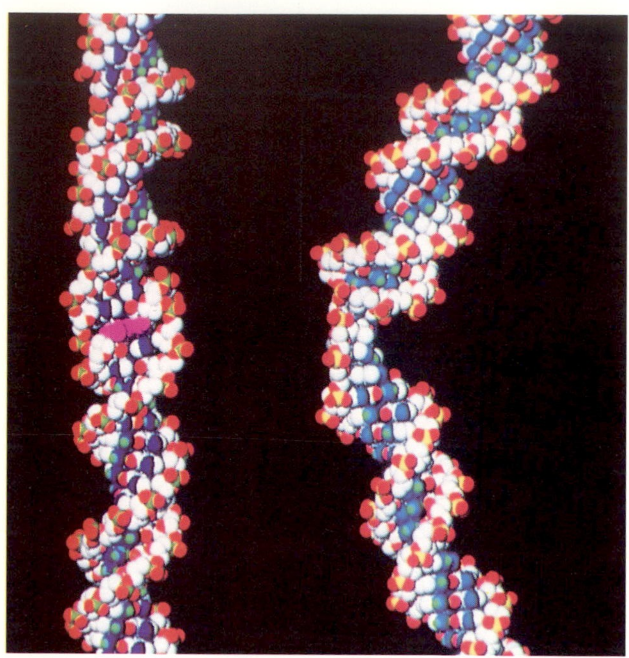

Prof. S. H. Kim, University of California, Berkeley

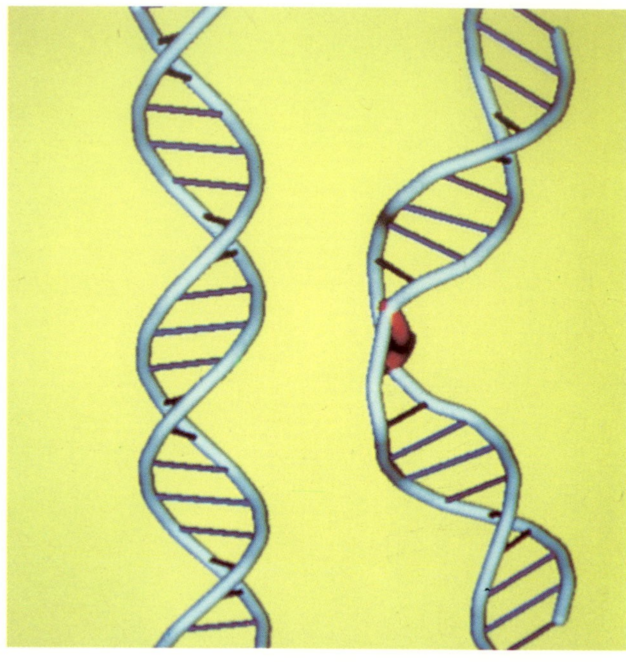

Prof. S. H. Kim, University of California, Berkeley

Prof. P. A. Bartlett, University of California, Berkeley

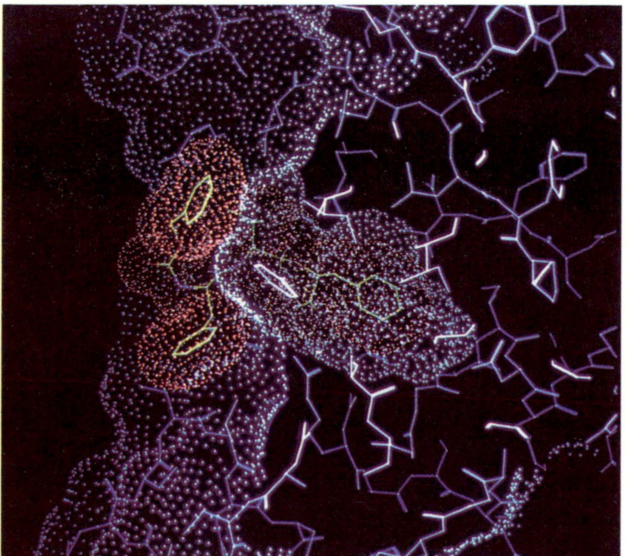

Prof. P. A. Bartlett, University of California, Berkeley

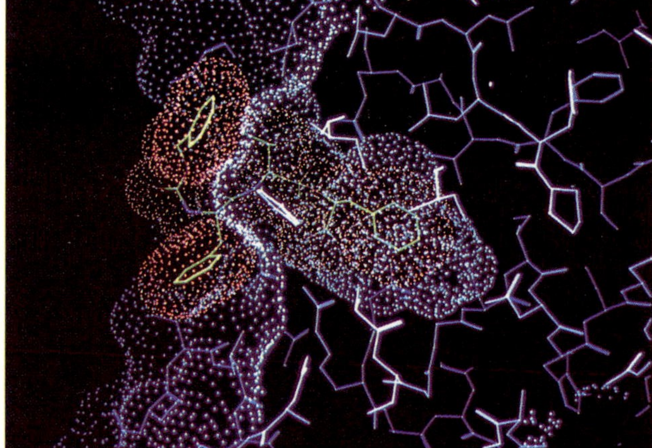

Illustrated opposite are space-filling models and simplified "ribbon" models of a normal and a psoralen-linked 48-base-pair DNA oligomer. Formulas for the 8-methoxypsoralen and 8-methoxypsoralen linked to one or two thymines are shown. Note that all of the connections to the cyclobutane rings are cis. The crystal structure of a psoralen linked to one thymine has the cis-syn arrangement. It is thought that the second thymine ring is linked to the psoralen in the same way.

Two important points to remember are, first, that psoralens are *natural* products and yet are not completely benign in their action. Some 50 different psoralens have been isolated from fruits, plants, and fungi. However, irradiation of unwanted skin cells after psoralen-treatment can be used to destroy such cells. Second, the ozone layer that has been depleted by chlorofluorocarbons (CFCs, see Chapters 6 [problem 12] and 8) is important in filtering out the UV that promotes the formation of cross-linked DNA.

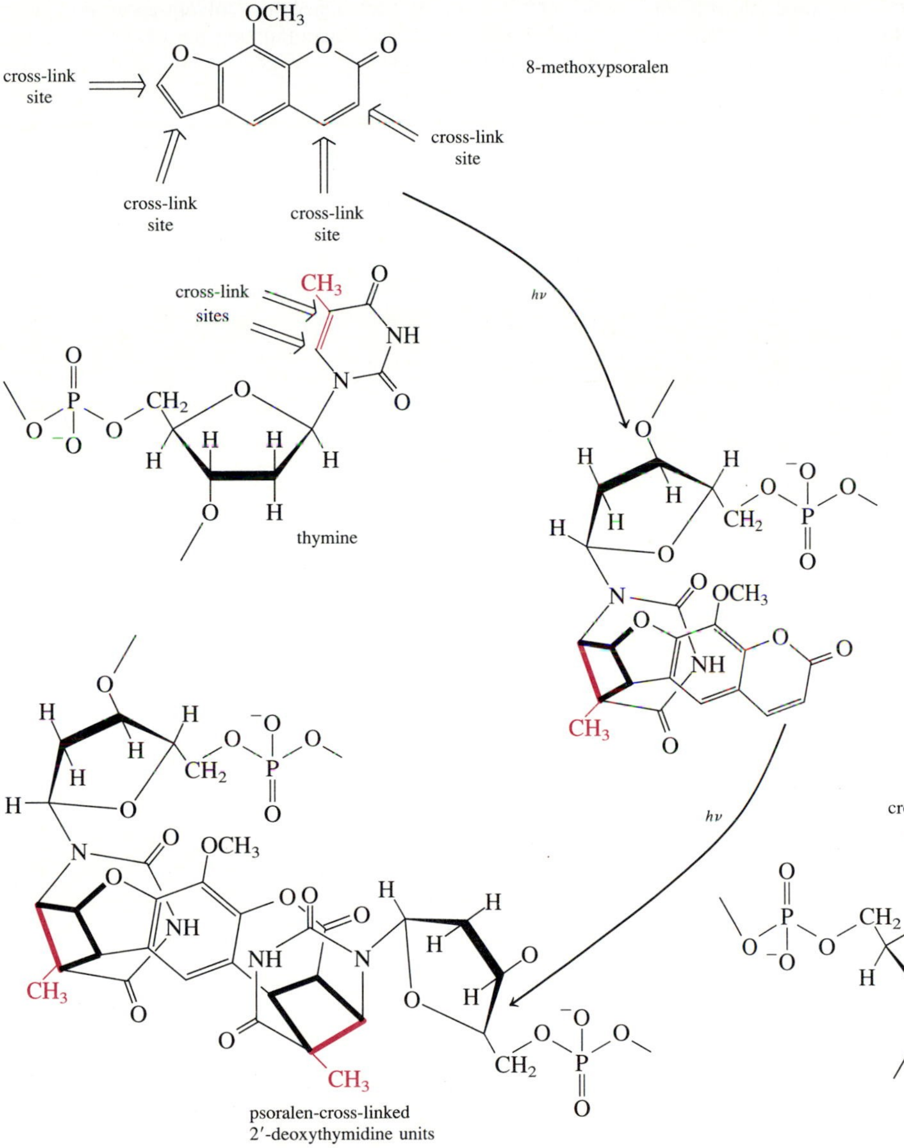

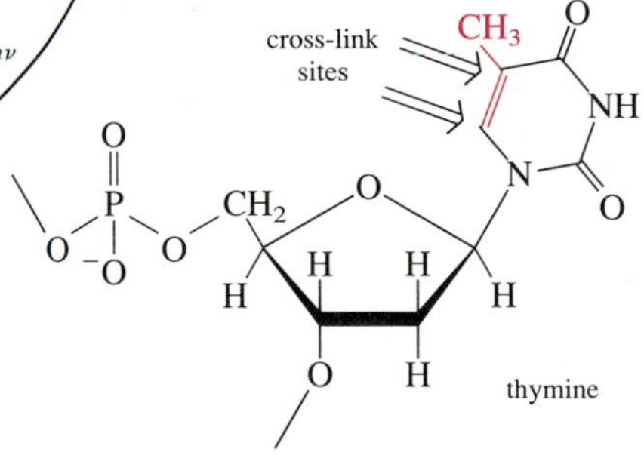

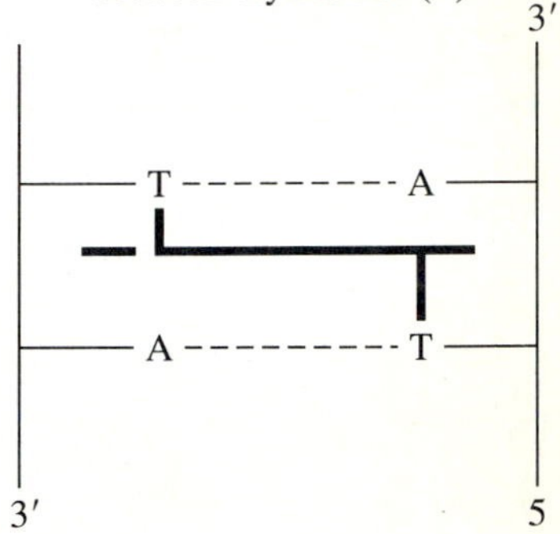

ESSAY 6 Enzymatic Targets Are Shapely

Docking a boat requires professional knowledge of currents, wind effects, and the motional properties of the elements needing to be brought together, that is, the boat and the dock. Docking one molecule into a site on another molecule, by manipulating the molecules on video displays with the techniques of molecular graphics, requires a chemist's knowledge of the size and compressibility of groups of a protein enzyme (for a substrate molecule) (see Chapter 29). The success of docking is an expression of the degree of molecular recognition (see Chapter 33). Docking a possible drug molecule, a surrogate for the true substrate, into a receptor site is one method of testing the drug candidate for its possible effectiveness. A protagonist, or more simply, an agonist, will activate the receptor. An antagonist will combine and block the receptor from combining with an agonist or activator.

(continues)

When the crystal structure of the receptor protein is known, we can carry out such computer docking experiments without the need for synthesizing every candidate molecule. The molecules on the screen can be altered in a fraction of the time that actual synthesis would require. However, ultimately we must use our laboratory skills to prepare the best choices. The inhibition of a receptor protein or enzyme can have therapeutic value, such as in the inhibition of bacterial cell-wall synthesis by penicillin antibiotics.

The hydrolysis of esters and amides usually proceeds through a tetrahedral intermediate, in which a species such as hydroxide ion has added to the carbonyl group of an ester or an amide (see Section 19.6). Look at the formulas for Cbz-Phe-Val-Phe, a substrate for the enzyme zinc carboxypeptidase A, the intermediate in the zinc enzyme-catalyzed hydrolysis of Cbz-Phe-Val-Phe, and the peptide analogue, Cbz-Phe-Val-PO_2^--Phe. The Cbz-Phe-Val-PO_2^--Phe is a surrogate for the intermediate or the transition state to the intermediate in hydrolysis. The inhibitor shown, Cbz-Phe-Val-PO_2^--Phe, forms with carboxypeptidase A the strongest enzyme-inhibitor complex yet discovered, the complex having a dissociation constant estimated as 11 femtomolar, or 11×10^{-15} M. The red dots show the van der Waals surface of the inhibitor; the blue dots mark the "solvent-accessible" surface of the protein. Imagine manipulating the molecules on the screen, turning them first one way, and then the other, as if you held them in your hand.

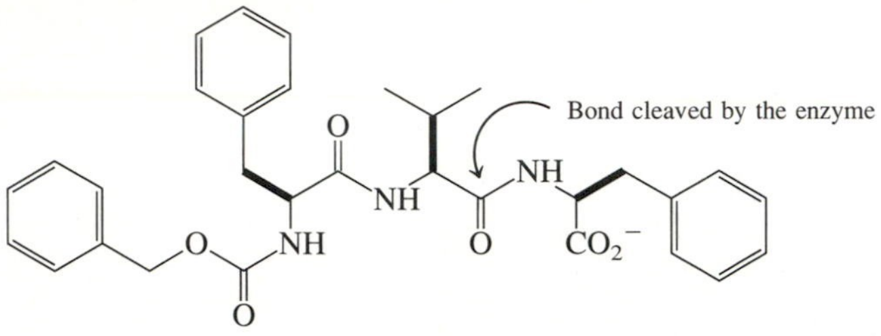

Bond cleaved by the enzyme

O NH NH NH O O CO_2^- O

Cbz = carbobenzyloxy

Cbz-Phe-Val-Phe, a substrate for carboxypeptidase A

O NH NH O NH HO O^- CO_2^- $Glu_{270}COOH$ Zn^+ Arg_{127} O

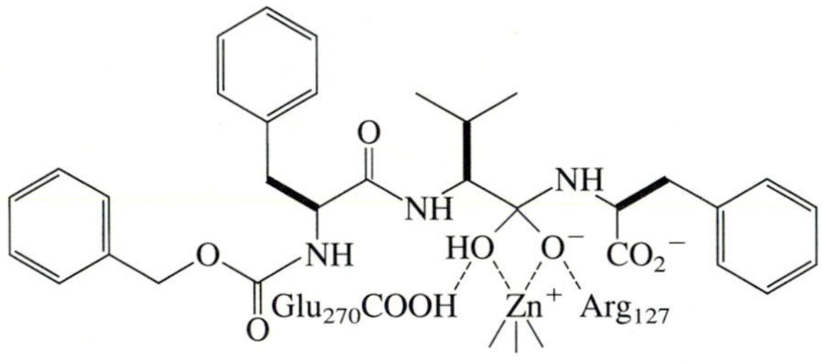

Intermediate in hydrolysis of Cbz-Phe-Val-Phe

O NH P O O NH O O^- O CO_2^- O $Glu_{270}COOH$ Zn^+ Arg_{127}

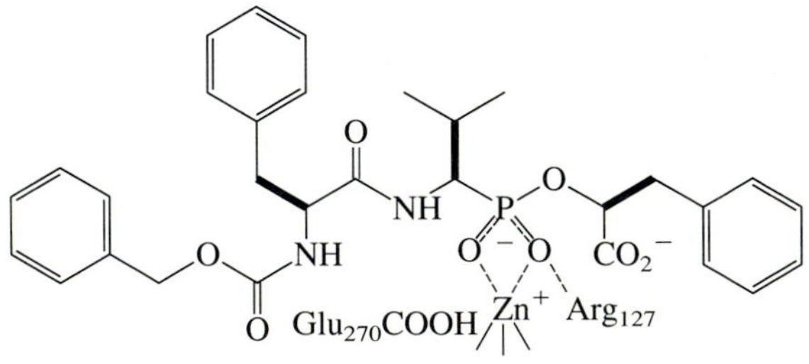

Phosphonate analog of tetrahedral intermediate, a potent inhibitor of carboxypeptidase A

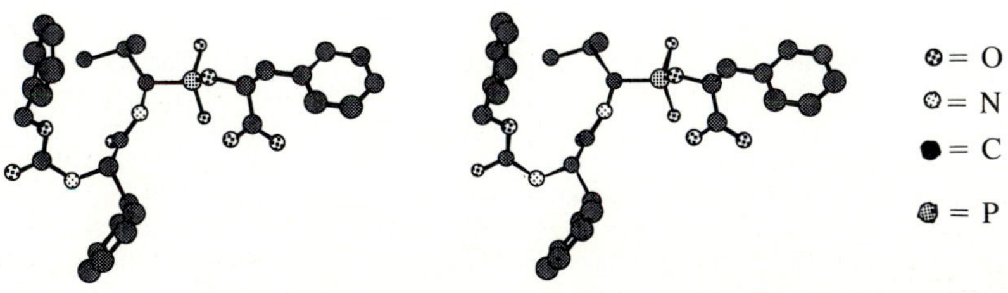

= O
= N
= C
= P

A 3-D drawing of the phosphonate analog shown in the molecular graphics representation of the enzyme active site.

ESSAY 7 Chemical Bonds Hold Up Tall Trees

Lignin is a polymer that constitutes about 25% of the weight of wood. It is therefore one of the most common natural products. To produce white paper, we must remove the dark-colored lignin. Chemists and biologists are searching for ways to remove the lignin that are less dangerous for the environment than previous methods. Lignin is also a potential resource that is used relatively little; the flavoring material vanillin is one product derived from lignin. The nature of the polymer is of great interest since the great strength of wood, shown in the stability, enormous height, and weight of tall trees (see photograph), depends upon the lignin matrix surrounding the cellulose fibrils. Lignin is formed by a free-radical process initiated by free radicals formed from enzymes from alkylaromatic precursors (see Chapter 30). However, the reaction occurs away from enzyme active sites and the product is formed in a nonstereospecific manner and is thus optically inactive. The cellulose fibrils are held together by hydrogen bonds. The hemicelluloses bind to the cellulose by hydrogen bonds, but are covalently linked to the lignin polymer. The great strength of multiple hydrogen bonds is shown here; multiple weak bonds are the essence of molecular recognition (Chapter 33).

glycoside
hemicellulose
biaryl link
tetrahydrofuran
alkyl aryl ether
biaryl ether
HO
OH
CH_3O
OCH_3

Partial structure of lignin, showing some of the groups and bonding arrangements present in the optically inactive polymer. The polymer forms by a nonstereospecific free radical reaction, which forms many different types of linkages between the reacting moieties (see Chapter 30)

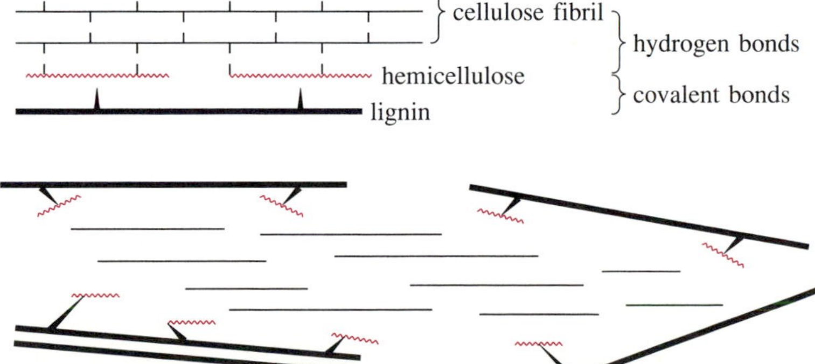

Cellulose fibrils embedded in lignin matrix linked to hemicellulose chains. Hemicelluloses are 1,4-linked sugars of moderate molecular weight, in which the most prevalent sugars are the pentose, xylose and arabinose and the hexoses, glucose and mannose (see Chapter 28)

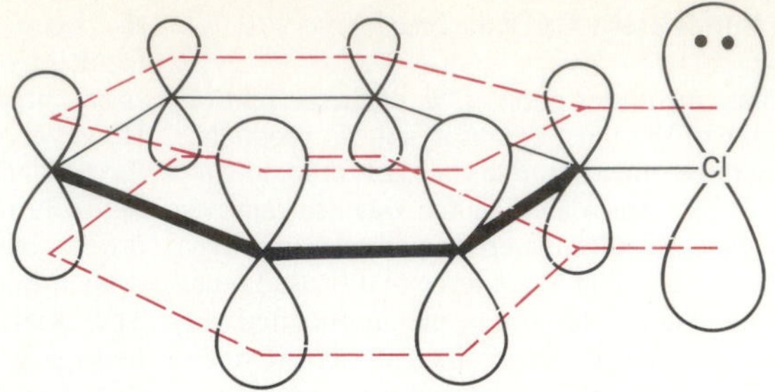

FIGURE 21.9 Conjugation of a chlorine lone pair with the benzene π-system. Actually, a chlorine 3p-orbital is involved, and the additional node is omitted for clarity.

Cl

1.75 D

Chlorobenzene has a dipole moment of 1.75 D in the gas phase. The direction of the dipole is undoubtedly also from carbon to chlorine. The magnitude of the dipole moment of chlorobenzene is smaller than that of methyl chloride for two reasons. The carbon-chlorine bond in methyl chloride may be represented approximately as C_{sp^3}—Cl_p. The bond in chlorobenzene is approximately C_{sp^2}—Cl_p. The higher s-character of the benzene orbital makes it more electronegative than an sp^3-orbital; hence, the electronegativity difference with the more electronegative chlorine orbital is reduced. The second contribution to the reduced dipole moment in chlorobenzene results from conjugation of one of the chlorine lone pairs with the benzene π-system, illustrated in Figure 21.9. The lone pair is actually part of the π-system and can be represented in terms of resonance structures by

$$\left[\text{C}_6\text{H}_5\text{—}\ddot{\text{Cl}}\text{:} \longleftrightarrow \text{=}\overset{+}{\text{Cl}}\text{:} \longleftrightarrow \text{=}\overset{+}{\text{Cl}}\text{:} \longleftrightarrow \text{=}\overset{+}{\text{Cl}}\text{:} \right]$$

The effect of conjugation is small; the ionic structures contribute only a slight amount to the overall electronic structure. This slight conjugation, however, is equivalent to a dipole moment for the π-system, μ_π, oriented in the opposite direction from that associated with the carbon-chlorine σ-bond, μ_σ. The net dipole moment is the vector sum and is less than that of μ_π alone.

Cl

μ_σ

μ_π

μ_{tot}

Dipole moments of some other benzene derivatives in the gas phase are summarized in Table 21.2. The dipole moments of multiply substituted benzenes are generally close to the vector sum of the constituent dipoles. p-Dichlorobenzene has a net dipole moment of zero because the two component carbon-chlorine dipoles oppose and cancel each other.

Cl— —Cl

net $\mu = 1.75 + (-1.75) = 0$

TABLE 21.2 Dipole Moments of Substituted Benzenes

Compound	μ, D (gas phase)	Compound	μ, D (gas phase)
C_6H_6	0	p-$C_6H_4Cl_2$	0
C_6H_5F	1.63	o-$CH_3C_6H_4Cl$	1.57
C_6H_5Cl	1.75	p-$CH_3C_6H_4Cl$	2.21
C_6H_5Br	1.72	o-$CH_3C_6H_4F$	1.35
C_6H_5I	1.71	m-$CH_3C_6H_4F$	1.85
$C_6H_5CH_3$	0.37	p-$CH_3C_6H_4F$	2.01
$C_6H_5NO_2$	4.28	m-$ClC_6H_4NO_2$	3.72
o-$C_6H_4Cl_2$	2.52	p-$ClC_6H_4NO_2$	2.81
m-$C_6H_4Cl_2$	1.68		

Because of the geometry of the hexagonal benzene ring, *ortho* and *meta* vector sums are given simply as

$$\mu = (\mu_1^2 + \mu_2^2 \pm \mu_1\mu_2)^{1/2} \qquad \begin{matrix} + \textit{ ortho} \\ - \textit{ meta} \end{matrix}$$

This equation generally is quite satisfactory for *meta* groups but frequently inadequate for *ortho* groups. *Ortho* groups are so close to each other that electronic effects are mutually perturbed. For example, this equation applied to *o*- and *m*-dichlorobenzene gives

$$\mu_o = [1.75^2 + 1.75^2 + (1.75)(1.75)]^{1/2} = 3.03 \text{ D}$$
$$\mu_m = [1.75^2 + 1.75^2 - (1.75)(1.75)]^{1/2} = 1.75 \text{ D}$$

The *meta* result is close to the experimental value of 1.68 D, but the calculated *ortho* value is substantially higher than the experimental value of 2.52 D.

Toluene has a small but distinct dipole moment of 0.37 D. We note from the data in Table 21.2 that the dipole moment of *p*-chlorotoluene is approximately that of the sum of the dipole moments of toluene and chlorobenzene; hence, both component dipoles are operating in the same direction.

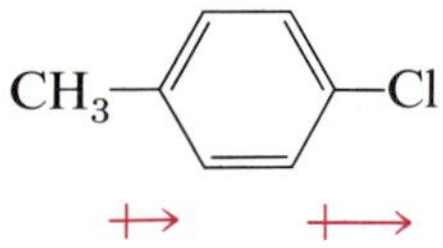

CH_3—C_6H_4—Cl

+→ 0.37 D   +→ 1.75 D

0.37 + 1.75 = 2.12 D   (experimental μ = 2.21 D)

The dipole moment of toluene results in part from the character of the C_{methyl}—C_{ring} bond. This bond can be described approximately as C_{sp^3}—C_{sp^2}. The sp^2-orbital is more electronegative than the sp^3-orbital and produces an electronic displacement corresponding to the direction of the dipole moment indicated for toluene.

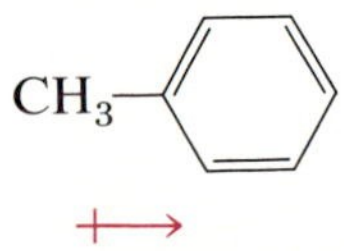

CH_3—C_6H_5

+→

The same approach applied to nitrobenzene derivatives shows that the direction of the dipole in nitrobenzene is away from the benzene ring.

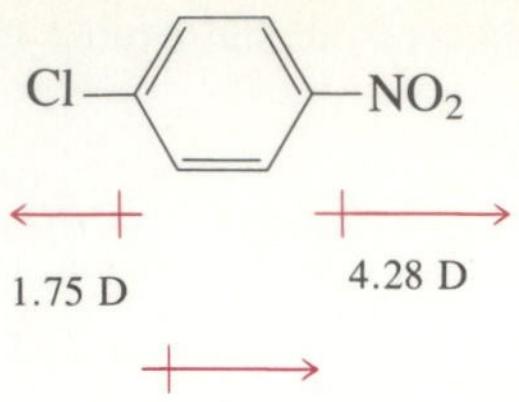

Cl—⟨benzene ring⟩—NO_2

1.75 D 4.28 D

$\mu_{net} = 4.28 - 1.75 = 2.53$ D (experimental $\mu = 2.81$ D)

The direction thus derived for the dipole moment in nitrobenzene is what we would have expected from the electronic structure of the nitro group.

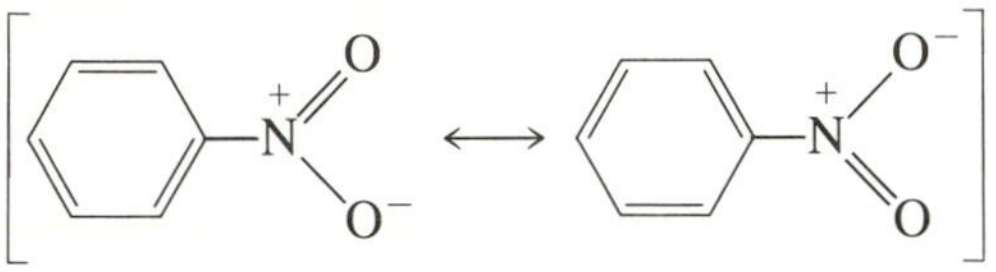

The relatively high magnitude of μ for nitrobenzene also follows from the formal charges required in the Lewis structures for the nitro group.

EXERCISE 21.7 Calculate the dipole moment expected for *p*-fluorotoluene by vector addition and compare with the experimental result in Table 21.2.

21.5 Side-Chain Reactions

A. Free Radical Halogenation

At ordinary temperatures benzene itself does not undergo the type of free radical chlorination typical of alkanes. The bond-dissociation energy of the phenyl-hydrogen bond is rather high ($DH° = 111$ kcal mole^{-1}), undoubtedly because the bond involved is C_{sp^2}—H_s and has extra *s*-character. Consequently, the hydrogen transfer reaction is endothermic and has been observed only at high temperature.

$$C_6H_5\text{—}H + Cl\cdot \rightleftharpoons C_6H_5\cdot + HCl \qquad \Delta H° = +8 \text{ kcal mole}^{-1}$$

Instead, chlorine atoms tend to add to the ring with the ultimate formation of a hexachlorocyclohexane.

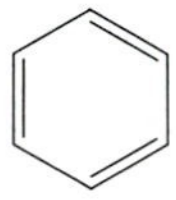

$$\text{benzene} + Cl_2 \xrightarrow{h\nu} C_6H_6Cl_6$$

benzene hexachloride
1,2,3,4,5,6-hexachlorocyclohexane

Eight geometric isomers are possible. The so-called γ-isomer (gammexane, lindane) has insecticidal properties and constitutes 18% of the mixture.

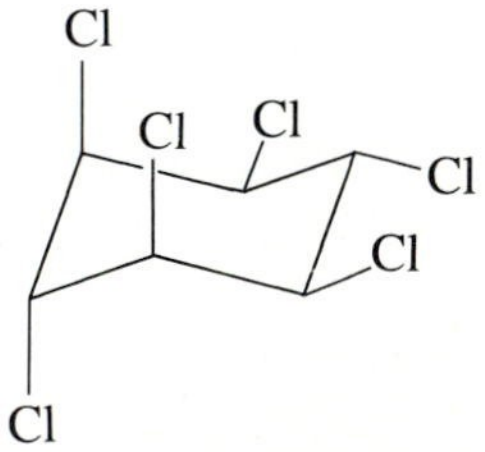

γ-benzene hexachloride

Reaction of the benzene hexachlorides with hot alcoholic potassium hydroxide gives 1,2,4-trichlorobenzene.

$$C_6H_6Cl_6 + 3\ C_2H_5O^- \xrightarrow{\Delta} \text{1,2,4-trichlorobenzene (Cl, Cl, Cl)} + 3\ C_2H_5OH + 3\ Cl^-$$

In contrast, toluene undergoes smooth free radical chlorination on the methyl group to give benzyl chloride. Benzyl chloride undergoes further halogenation to give benzal chloride and benzotrichloride.

$$C_6H_5CH_3 \xrightarrow[h\nu]{Cl_2} C_6H_5CH_2Cl \xrightarrow[h\nu]{Cl_2} C_6H_5CHCl_2 \xrightarrow[h\nu]{Cl_2} C_6H_5CCl_3$$

benzyl chloride — benzal chloride — benzotrichloride

The extent of chlorination may be controlled by monitoring the amount of chlorine used.

The reaction of toluene with chlorine atoms occurs exclusively at the methyl group because of the low bond-dissociation energy of the benzyl-hydrogen bond (DH° = 88 kcal mole^{-1}).

$$C_6H_5CH_3 + Cl\cdot \longrightarrow C_6H_5CH_2\cdot + HCl \qquad \Delta H^\circ = -16 \text{ kcal mole}^{-1}$$

benzyl radical

The benzyl radical is especially stable for the same reason the allyl radical is stabilized (Section 20.1.D). Delocalization of the odd electron into the ring spreads out and diffuses the free radical character of the molecule. This conjugation can be represented by orbital overlap between the carbon $2p$-orbital containing the odd electron and the ring π-system, as in Figure 21.10.

Alternatively, the conjugated system can be represented by resonance structures.

$$\left[C_6H_5CH_2\cdot \longleftrightarrow CH_2{=}C_6H_5\cdot\,(ortho) \longleftrightarrow CH_2{=}C_6H_5\cdot\,(para) \longleftrightarrow CH_2{=}C_6H_5\cdot\,(ortho') \right]$$

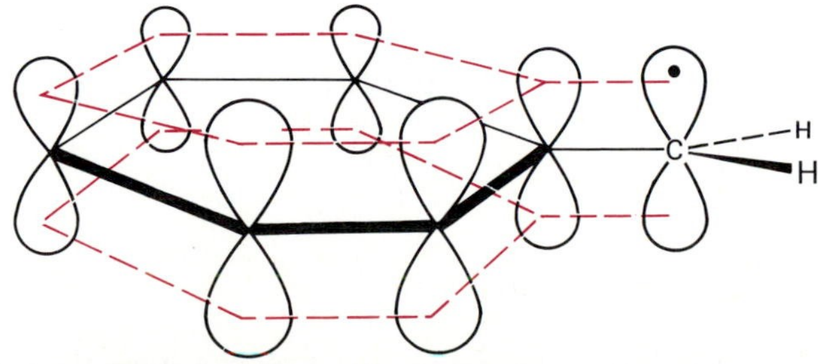

FIGURE 21.10 Delocalization of the benzyl radical.

In the next step of the chain halogenation reaction, benzyl radical reacts with chlorine to regenerate a chlorine atom, which then continues the chain.

$$C_6H_5CH_2\cdot + Cl_2 \longrightarrow C_6H_5CH_2Cl + Cl\cdot$$

Note that benzyl radical reacts exclusively at the exocyclic position. The *ortho* and *para* positions do have odd-electron character, but reaction at these positions produces a chloride that does not have the aromatic stability of a benzene ring. The course of the reaction is determined by the substantial difference in the thermodynamics of reaction.

$$C_6H_5CH_2\cdot + Cl_2 \longrightarrow Cl\cdot + C_6H_5CH_2Cl \qquad \Delta H^\circ = -15 \text{ kcal mole}^{-1}$$

$$C_6H_5CH_2\cdot + Cl_2 \longrightarrow Cl\cdot + \text{4-chloro-1-methylene-2,5-cyclohexadiene (CH}_2\text{; H, Cl)} \qquad \Delta H^\circ \cong +23 \text{ kcal mole}^{-1}$$

The free radical chain bromination of toluene is exactly analogous and is a suitable route to benzyl bromide. With xylene the two methyl groups undergo successive halogenation.

$$o\text{-}C_6H_4(CH_3)_2 + 2\,Br_2 \xrightarrow[125^\circ C]{h\nu} o\text{-}C_6H_4(CH_2Br)_2$$

(48–53%)
α,α′-dibromo-*o*-xylene

$$o\text{-}C_6H_4(CH_3)_2 + 4\,Br_2 \xrightarrow[175^\circ C]{h\nu} o\text{-}C_6H_4(CHBr_2)_2$$

(74–80%)
α,α,α′,α′-tetrabromo-*o*-xylene

With the higher alkylbenzenes chlorination is limited in its synthetic utility because reaction along the alkyl chain occurs in addition to reaction at the benzylic position.

$$C_6H_5CH_2CH_3 + Cl_2 \xrightarrow{h\nu} C_6H_5CHClCH_3 + C_6H_5CH_2CH_2Cl$$

(56%)
α-chloroethylbenzene
1-chloro-1-phenylethane

(44%)
β-chloroethylbenzene
1-chloro-2-phenylethane

However, bromination occurs exclusively at the benzylic position.

$$C_6H_5CH_2CH_3 + Br_2 \xrightarrow{h\nu} C_6H_5CHBrCH_3$$

α-bromoethylbenzene
1-bromo-1-phenylethane

This difference in behavior again reflects the greater reactivity of chlorine atoms compared to bromine; recall that bromine generally is a more selective reagent than chlorine (Section 6.3).

EXERCISE 21.8 Show how cumene can be converted to 2-phenylpropene.

EXERCISE 21.9 Write a chain reaction summary for the reaction of bromine with toluene.

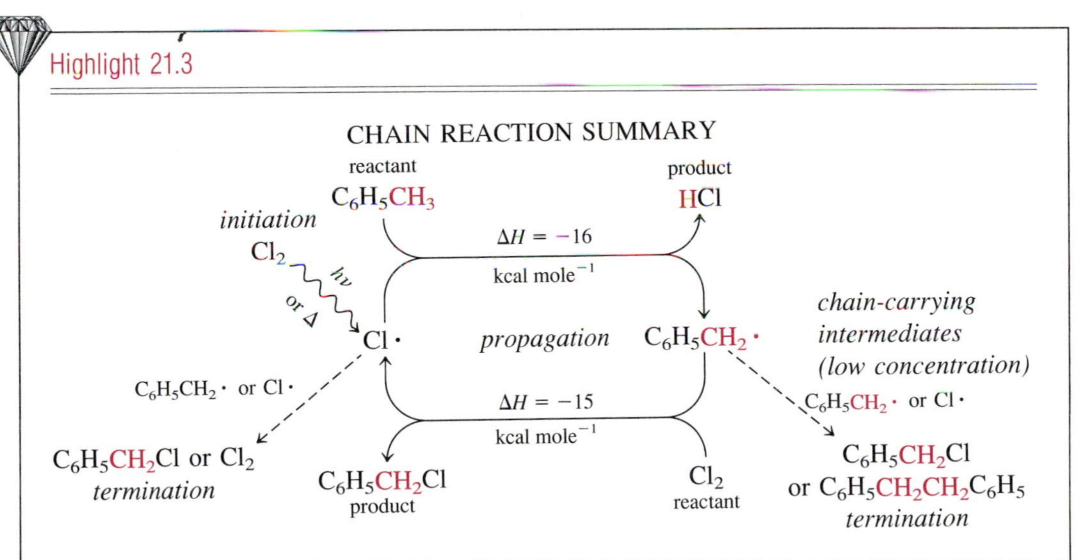

B. Benzylic Displacement and Carbocation Reactions

The reactions of phenylalkyl systems are more or less comparable to those of analogous alkyl systems—halides undergo displacements, eliminations, formation of Grignard reagents, and so on. However, when the halogen is α to a benzene ring, the compounds are especially reactive. Benzyl halides are generally at least 100 times as reactive as ethyl halides in S_N2 displacement reactions. This high reactivity is attributed to conjugation of the ring π-electrons in the transition state (Figure 21.11). Recall that the transition state for displacements on allyl halides are similarly stabilized, as shown in Highlight 20.1. Accordingly, such displacement reactions are straightforward and facile.

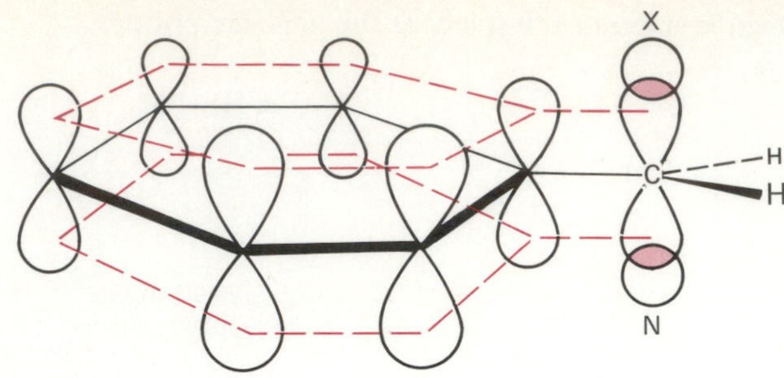

FIGURE 21.11 Transition state for S_N2 reaction with a benzyl halide showing the conjugation of the reacting center with the benzene π-system.

$$p\text{-}CH_3OC_6H_4CH_2Cl \xrightarrow[\text{acetone}]{\text{NaCN}} p\text{-}CH_3OC_6H_4CH_2CN$$

(74–81%)
p-anisylacetonitrile

$$p\text{-}O_2NC_6H_4CH_2Cl \xrightarrow[CH_3COOH]{NaOCCH_3 \,(\text{C=O})} p\text{-}O_2NC_6H_4CH_2OCCH_3 \,(\text{C=O}) \xrightarrow{NaOH} p\text{-}O_2NC_6H_4CH_2OH$$

p-nitrobenzyl chloride

(64–71%)
p-nitrobenzyl alcohol

Benzylic compounds also react rapidly by the S_N1 mechanism because of the relative stability of the benzyl cation.

$$C_6H_5CH_2Cl \longrightarrow Cl^- + \left[C_6H_5\overset{+}{C}H_2 \longleftrightarrow \text{(+ at ortho)} \longleftrightarrow \text{(+ at para)} \longleftrightarrow \text{(+ at ortho)} \right]$$

In fact, the gas-phase enthalpy of ionization of benzyl chloride is more comparable to that of secondary or tertiary alkyl chlorides than to that of primary alkyl chlorides.

		$\Delta H°$, *kcal mole*$^{-1}$
$C_6H_5CH_2Cl \rightleftharpoons$	$C_6H_5CH_2^+ + Cl^-$	154
$C_2H_5Cl \rightleftharpoons$	$C_2H_5^+ + Cl^-$	193
$(CH_3)_2CHCl \rightleftharpoons$	$(CH_3)_2CH^+ + Cl^-$	172
$(CH_3)_3CCl \rightleftharpoons$	$(CH_3)_3C^+ + Cl^-$	157

When the carbocation center is conjugated with two or three benzene rings, the positive charge is distributed to a still greater extent. For example, triphenylmethyl cation has ten resonance structures in which the charge is spread to six *ortho* and three *para* positions. Consequently, triphenylmethyl chloride ionizes readily and shows exceptional reactivity.

A liquid sulfur dioxide solution is colored yellow and conducts electricity because of the triphenylmethyl cations and chloride ions present.

$$(C_6H_5)_3CCl \overset{SO_2}{\rightleftharpoons} (C_6H_5)_3C^+ + Cl^-$$

(yellow color)

$$K = \frac{[(C_6H_5)_3C^+][Cl^-]}{[(C_6H_5)_3CCl]} = 4 \times 10^{-5}\ M$$

Similarly, triphenylmethanol is converted into substantial amounts of triphenylmethyl cation in strong aqueous sulfuric acid.

$$(C_6H_5)_3COH + H^+ \rightleftharpoons (C_6H_5)_3C^+ + H_2O$$

$$K = \frac{[(C_6H_5)_3C^+]}{[(C_6H_5)_3COH][H^+]} = 2 \times 10^{-7}\ M^{-1}$$

To give some idea of relative magnitudes of reactivity, benzyl chloride undergoes S_N1-type reactions much more slowly than *t*-butyl chloride, diphenylmethyl chloride is 10^1 to 10^3 times faster than *t*-butyl chloride, and triphenylmethyl chloride is 10^6 to 10^7 times more reactive than *t*-butyl chloride.

Order of S_N1 reactivity

$$(C_6H_5)_3CCl > (C_6H_5)_2CHCl > (CH_3)_3CCl > C_6H_5CH_2Cl$$

In fact, the rate of S_N1 reaction of triphenylmethyl chloride with ethanol is comparable to the rate at which the solid triphenylmethyl chloride dissolves.

The most effective conjugation between the carbocation center and the benzene π-electrons in triphenylmethyl cation requires that the whole molecule be coplanar. In this type of structure, however, the *ortho* hydrogens of the phenyl groups would be only about 0.5 Å apart; the resulting steric repulsion forces the rings to tilt apart. The actual structure of triphenylmethyl cation is that of a three-bladed propeller. This twisting of the phenyl groups does somewhat diminish the magnitude of conjugation between the central carbon and each ring, but the effect is not large.

The high reactivity of benzylic compounds in displacement and carbocation reactions is also seen in reactions of benzyl alcohols.

CH_2OH, CH_3 → ($SOCl_2$, benzene / one drop pyridine) → CH_2Cl, CH_3

o-methylbenzyl alcohol → (75–89%) α-chloro-*o*-xylene

$CHOHCH_3$, Cl → ($KHSO_4$, Δ) → $CH{=}CH_2$, Cl

1-(*m*-chlorophenyl)ethanol → (80–82%) *m*-chlorostyrene

EXERCISE 21.10 Write the resonance structures for *m*- and *p*-methylbenzyl cations and predict which of these isomeric cations is the more stable.

C. Oxidation

Like allylic alcohols, benzylic alcohols may be oxidized to corresponding aldehydes or ketones (Section 20.3.A), using manganese dioxide, a reagent that does not bring about the oxidation of normal alcohols.

$$C_6H_5CHOHCH(CH_3)_2 \xrightarrow[\text{benzene}, \Delta]{MnO_2} C_6H_5COCH(CH_3)_2$$

(77%)
isobutyrophenone

The benzene ring is rather stable to oxidizing agents, and under appropriate conditions side-chain alkyl groups are oxidized instead. Sodium dichromate in aqueous sulfuric acid or acetic acid is a common laboratory reagent for this purpose, but aqueous nitric acid and potassium permanganate have also been used.

$$p\text{-}CH_3C_6H_4NO_2 \xrightarrow[H_2SO_4, \Delta]{Na_2Cr_2O_7} p\text{-}HOOCC_6H_4NO_2$$

(82–86%)
p-nitrobenzoic acid

$$o\text{-}C_6H_4(CH_3)_2 \xrightarrow[155^\circ C]{\text{aq. } HNO_3} o\text{-}CH_3C_6H_4COOH$$

(53–55%)
o-toluic acid

The detailed reaction mechanisms by which these oxidations occur are rather complex. They involve numerous intermediates including chromate and permanganate esters, but they also appear to involve an intermediate benzyl cation.

$$C_6H_5CH_3 \xrightarrow{[O]} C_6H_5CH_2^+$$

As we have seen, this carbocation is relatively stable because of conjugation of the positive charge with the benzene ring. Reaction with water yields benzyl alcohol, which can oxidize further. Larger side chains can also be oxidized completely so long as there is one benzylic hydrogen for the initial oxidation. Cleavage reactions of larger side chains probably involve the formation of an intermediate alkene.

$$\text{PhCH}_2\text{CH}_2\text{CH}_3 \xrightarrow{[O]} [\text{Ph}\overset{+}{\text{C}}\text{HCH}_2\text{CH}_3] \xrightarrow{-\text{H}^+} [\text{PhCH}{=}\text{CHCH}_3] \xrightarrow{[O]} \underset{\text{benzoic acid}}{\text{PhCOOH}}$$

The more extensive oxidation required in these reactions often results in lower yields so that they are not as useful for laboratory preparations as they are for structural identification. When there is no benzylic hydrogen, the side chain resists oxidation. For example, vigorous conditions are required for the oxidation of *t*-butylbenzene, and the product is trimethylacetic acid, the product of oxidation of the benzene ring.

$$\text{C}_6\text{H}_5\text{C(CH}_3)_3 \xrightarrow{\text{KMnO}_4} (\text{CH}_3)_3\text{CCOOH}$$

pivalic acid
trimethylacetic acid

Oxidation of side-chain methyl groups is an important industrial route to aromatic carboxylic acids. The most important oxidizing agent for such reactions is air.

$$\text{C}_6\text{H}_5\text{CH}_3 \xrightarrow[\Delta]{\text{O}_2,\ \text{Co(OAc)}_3,\ \text{Mn(OAc)}_2} \text{C}_6\text{H}_5\text{COOH} + \text{H}_2\text{O}$$

An important industrial reaction of this general type is the oxidation of *p*-xylene to the dicarboxylic acid (Section 27.6).

EXERCISE 21.11 Suggest two different two-step ways by which toluene can be converted into benzyl alcohol.

D. Acidity of Alkylbenzenes

Benzyl anion is stabilized by delocalization of the negative charge into the benzene ring.

$$\left[\text{C}_6\text{H}_5\text{CH}_2^- \longleftrightarrow \text{CH}_2 \longleftrightarrow \text{CH}_2 \longleftrightarrow \text{CH}_2 \right]$$

As a result, toluene is more acidic than the alkanes. Its pK_a is about 41 compared to a value of about 50 for ethane. Toluene is still a very weak acid and is not significantly converted to the anion even with $NaNH_2$ in liquid ammonia. As additional benzene rings are added, however, the acidity increases markedly. Some relevant pK_a data are summarized in Table 21.3.

TABLE 21.3 Acidity of Some Hydrocarbons

Hydrocarbon	Conjugate Base	pK_a
ethane	$CH_3CH_2^-$	~50
benzene	$C_6H_5^-$	43
toluene	$C_6H_5CH_2^-$	41
diphenylmethane	$(C_6H_5)_2CH^-$	33
triphenylmethane	$(C_6H_5)_3C^-$	31

Di- and triphenylmethane are sufficiently acidic to be converted significantly to the corresponding carbanions with sodium amide. The resulting anions react as typical nucleophiles in alkylation reactions.

$$(C_6H_5)_2CH_2 \xrightarrow[\text{liq. } NH_3]{NaNH_2} \xrightarrow[(C_2H_5)_2O]{CH_3(CH_2)_2CH_2Br} (C_6H_5)_2CH(CH_2)_3CH_3$$

(92%)
1,1-diphenylpentane

21.6 Reduction

A. Catalytic Hydrogenation

Benzene rings are substantially more resistant to catalytic hydrogenation than alkenes or alkynes. In molecules that contain both a double bond and a benzene ring, the double bond can be preferentially hydrogenated without difficulty.

Ph–CH=CH–Ph $\xrightarrow[\text{EtOH}]{H_2,\ Pt}$ Ph–CH₂CH₂–Ph

trans-stilbene   1,2-diphenylethane

Hydrogenation of the benzene ring occurs under more vigorous conditions and yields the corresponding cyclohexane. It is generally impractical to stop the reaction at an intermediate stage, since cyclohexadienes and cyclohexenes hydrogenate more readily than benzenes. Dialkylbenzenes tend to give predominantly the *cis*-dialkylcyclohexane, although the exact stereochemistry of the reduction depends on the reaction conditions and catalysts used. Platinum or palladium catalysts may be used at temperatures near 100 °C; acetic acid is a common solvent. Nevertheless, reactions under these conditions are often inconveniently slow, and ruthenium or rhodium on carbon is often more successful for hydrogenation of aromatic rings.

o-xylene (CH_3, CH_3) $\xrightarrow[\text{100°C, 1000 psi}]{H_2,\ 5\%\ Rh/C}$ cis-1,2-dimethylcyclohexane + trans-1,2-dimethylcyclohexane

about 9 : 1

With aromatic aldehydes and ketones the functional group undergoes reduction faster than the ring.

$$\text{2-bromo-5-methoxybenzaldehyde (Br, CHO, } OCH_3) \xrightarrow[C_2H_5OH]{H_2\text{-}PtO_2} \text{2-bromo-5-methoxybenzyl alcohol (Br, } CH_2OH, OCH_3)$$

2-bromo-5-methoxybenzaldehyde   (83%) 2-bromo-5-methoxybenzyl alcohol

B. Hydrogenolysis of Benzylic Groups

Benzylic halides can be converted to hydrocarbons in the same ways that suffice for converting alkyl halides to alkanes—formation and hydrolysis of a Grignard reagent or by reduction with lithium aluminum hydride. Benzylic alcohols may be converted into the corresponding halide or sulfonate ester and thence into the hydrocarbon.

CH_2OH (p-cyclopropyl) $\xrightarrow{SOCl_2}$ CH_2Cl (p-cyclopropyl) $\xrightarrow[\text{ether}]{Mg}$ $\xrightarrow{H_2O}$ CH_3 (p-cyclopropyl)

p-cyclopropyltoluene

CH_2Br (benzene) $\xrightarrow{LiAlH_4}$ CH_3 (benzene)

benzyl bromide — (78%) toluene

These reactions differ from comparable reactions of normal alkyl halides only in that they occur a little more rapidly since the reactants are benzylic (Section 21.5.B).

A benzylic alcohol may be reduced directly to the corresponding hydrocarbon by treatment with hydrogen in the presence of palladium and a small amount of perchloric acid.

$$R_1-C(OH)(C_6H_5)-R_2 \xrightarrow[HClO_4]{H_2/Pd} R_1CHR_2(C_6H_5) + H_2O$$

Since the carbonyl group in aromatic aldehydes and ketones is usually hydrogenated more rapidly than the benzene ring, the carbonyl group in such compounds can be *completely removed* by the use of two equivalents of hydrogen.

HOOC (keto compound) $\xrightarrow{H_2\text{-}Pd/C}$ HOOC (product)

(84%)

This type of process, in which hydrogen breaks a single bond, is known as **hydrogenolysis.** Another example is

$$CH_3-C_6H_4-CH(OH)-C_6H_4-CH_3 \xrightarrow[HClO_4]{H_2,\ Pd/C} CH_3-C_6H_4-CH_2-C_6H_4-CH_3$$

di-(*p*-tolyl)methanol — di-(*p*-tolyl)methane

EXERCISE 21.12 Outline a synthesis of phenylcyclohexane, starting with benzene and cyclohexanone.

C. Birch Reduction

Aromatic rings can be reduced by alkali metals in a mixture of liquid ammonia and alcohol. The product of this reduction is an unconjugated cyclohexadiene.

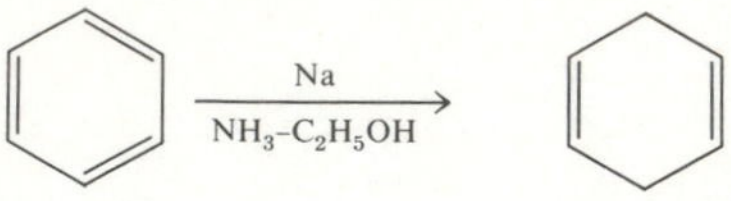

$$\xrightarrow[NH_3\text{-}C_2H_5OH]{Na}$$

1,4-cyclohexadiene

Recall that a similar reduction is used to prepare trans alkenes from alkynes (Section 12.6.A). A solution of sodium in liquid ammonia contains solvated electrons, which add to a benzene ring to give a radical anion. Benzene and alkylbenzenes are not readily reduced, and the equilibrium lies far to the left.

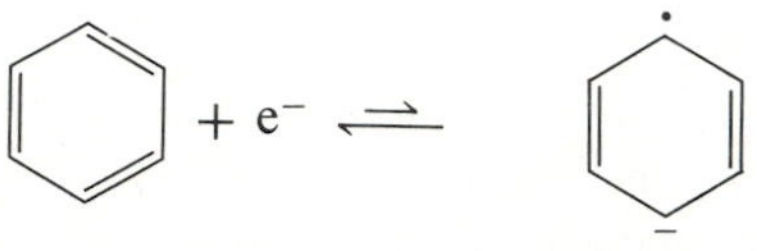

$$+ e^- \rightleftharpoons$$

benzene radical anion

Note that the radical anion has seven electrons in the benzene π-system. Many resonance structures can be written for this species; some are shown below.

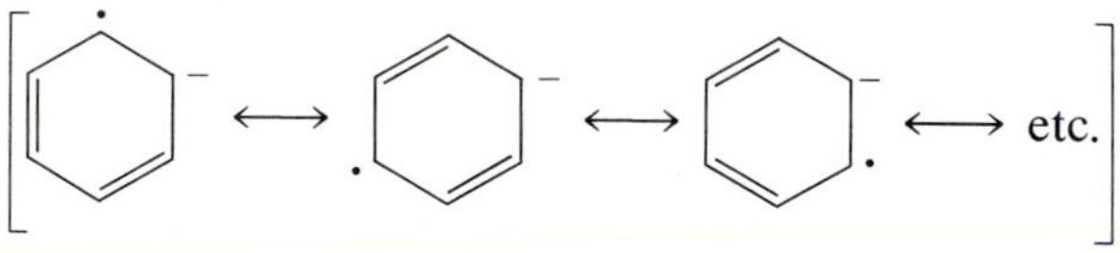

$$\longleftrightarrow \quad \longleftrightarrow \quad \longleftrightarrow \text{ etc.}$$

Nevertheless, benzene radical anion is less stable than benzene, and the ion reacts readily with proton donors. Ammonia itself is too weakly acidic to react, but ethanol is a sufficiently strong acid to protonate the radical anion. The resulting cyclohexadienyl radical immediately reacts with another solvated electron to form the corresponding cyclohexadienyl anion.

$$+ C_2H_5OH \longrightarrow \quad \text{H H} + C_2H_5O^- \quad \xrightarrow{e^-} \quad \text{H H}$$

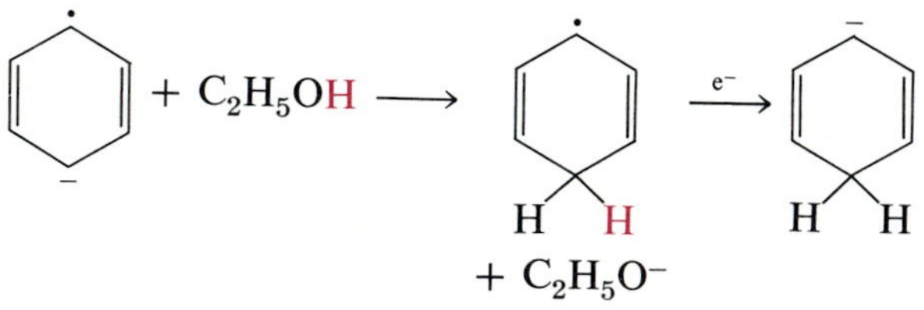

This anion is a strong base and reacts immediately with ethanol to give 1,4-cyclohexadiene and ethoxide ion.

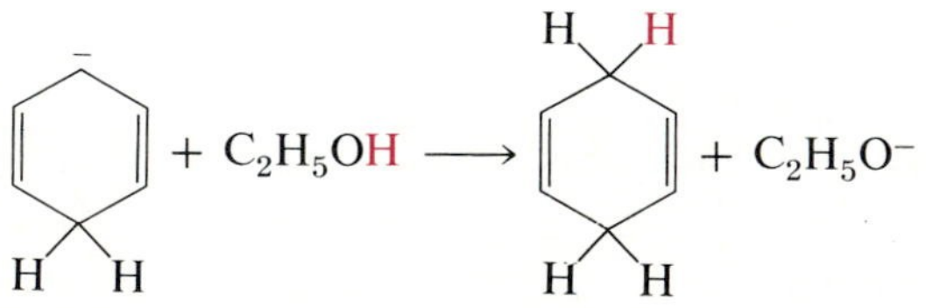

$$\text{H H} + C_2H_5OH \longrightarrow \text{H H H H} + C_2H_5O^-$$

Cyclohexadienyl anion is a conjugated carbanion of the allylic type, and the negative charge is correspondingly distributed over several carbons, as indicated by the resonance structures

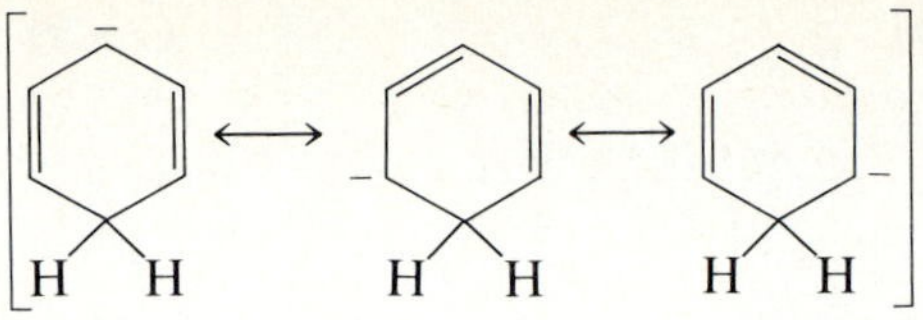

Protonation at the central carbon is much faster than at the end carbons of the conjugated chain. This result is quite general, even though the product of protonation at the terminal carbon of the conjugated chain produces a conjugated diene. The reason is not readily apparent, although various more or less sophisticated explanations have been given for this unusual effect.

Since the product contains isolated double bonds, no further reduction takes place, and the cyclohexadiene may be isolated in good yield. On prolonged contact with base, the carbanion is re-formed, allowing eventual isomerization of the unconjugated diene to the conjugated isomer, which is rapidly reduced to the monoene.

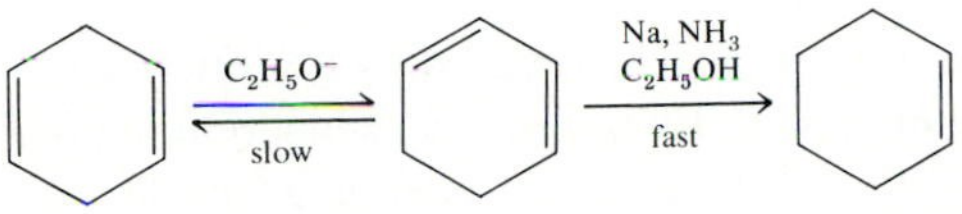

$C_2H_5O^-$ / slow ; Na, NH_3, C_2H_5OH / fast

Thus, by a proper choice of solvent and temperature one may reduce the benzene ring to either the 1,4-cyclohexadiene or the cyclohexene.

With substituted benzenes a single product is often formed in good yield.

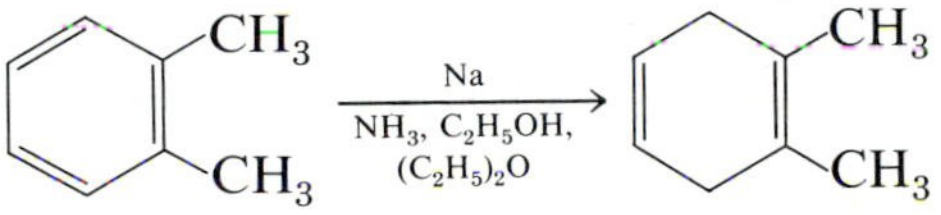

CH_3, CH_3 — Na / NH_3, C_2H_5OH, $(C_2H_5)_2O$ → CH_3, CH_3

(77–92%)
1,2-dimethyl-1,4-cyclohexadiene

Sodium is added in pieces to a mixture of liquid ammonia, ether, ethanol, and *o*-xylene cooled in a dry ice bath. The ammonia is allowed to evaporate, water is added, and the washed and dried organic layer is distilled.

Birch reduction is particularly useful with anisole and alkylanisoles. Addition of hydrogen always occurs in such a way that an enol ether is produced. Hydrolysis occurs readily (Section 14.6.B) to give a β,γ-unsaturated ketone. Under the acidic conditions of the hydrolysis, the double bond moves into conjugation with the carbonyl group (Section 20.3.A).

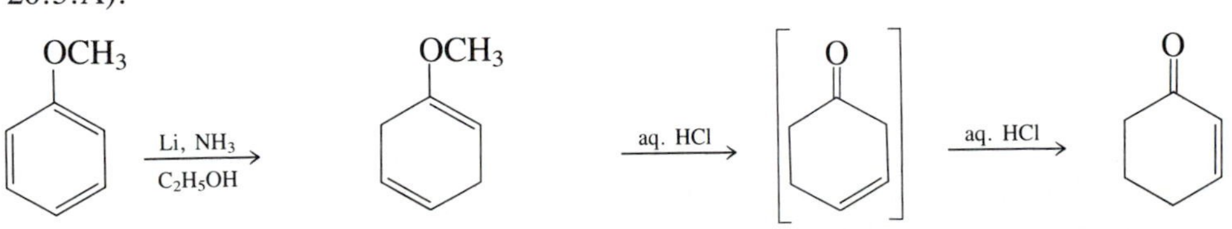

OCH_3 — Li, NH_3 / C_2H_5OH → OCH_3 — aq. HCl → O — aq. HCl → O

anisole; (84%) 1-methoxycyclohexa-1,4-diene; 3-cyclohexenone; 2-cyclohexenone

In contrast to the Birch reductions of toluene and anisole, which provide 2-substituted 1,4-cyclohexadienes, reduction of benzoic acid gives the completely unconjugated product.

COOH

1. Na, C_2H_5OH, NH_3
2. H_3O^+

COOH

(89–95%)
2,5-cyclohexadiene-1-carboxylic acid

Double bonds conjugated to the aromatic ring can also be reduced by alkali metals in liquid ammonia.

CH_3
C=CH_2

Li, NH_3

$CH(CH_3)_2$

(80%)

Although the benzene ring can also be reduced by lithium in ammonia, it is not as reactive as the conjugated double bond. Thus, selective reduction is possible, as in the foregoing example. Of course, the Birch reduction cannot generally be applied to systems that contain other easily reducible functions such as halogens, nitro groups, or carbonyl functions.

EXERCISE 21.13 What reaction products, if any, would you expect in the reactions of *cis*-2-butene, 2-phenylpropene, and 4-phenyl-1-butene with H_2-Pd/C in C_2H_5OH containing a small amount of perchloric acid?

21.7 Aromatic Transition States

Several types of reaction have transition states in which bonds being made or broken create a cyclic system of interacting orbitals with six electrons that have enhanced stability suggestive of aromatic character. These reactions are part of a group known collectively as **pericyclic reactions.** An example is the Diels-Alder reaction, in which a diene reacts with an alkene to give a cyclohexene derivative (Section 20.5). The transition state involves a cycle of six orbitals and six electrons that resembles the cyclic π-system of benzene.

‡

cf.

There is an apparent difference between the two cyclic systems. In benzene, the cyclic conjugation consists of π-overlap of six p-orbitals. In the Diels-Alder transition state, two of the interactions involve σ-overlap of two pairs of p-orbitals (Figure 20.12). Nevertheless, at the transition state this σ-overlap produces relatively weak bonds—more like π-bonds than normal σ-bonds—and the cyclic conjugation is effectively like that in benzene. In Figure 20.12 note that the p-orbitals that are involved in σ-overlap can still engage in π-overlap with their p-orbital neighbors.

The four-electron component of a [4 + 2] cycloaddition reaction need not be only a diene. An example of an alternative four-electron component is ozone. Its reaction with an alkene to form a molozonide (page 284) is also a [4 + 2] cycloaddition reaction that involves an aromatic six-electron transition state.

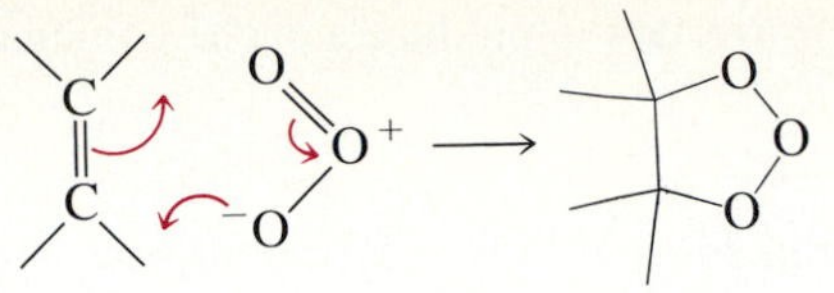

Another type of pericyclic reaction is the **Cope rearrangement,** a thermal rearrangement reaction of 1,5-dienes.

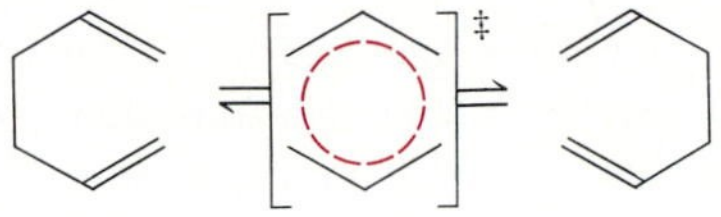

As shown by the foregoing example, the product is also a 1,5-diene. In some cases, such as with 1,5-hexatriene itself, the product is the same as the reactant, and it is not apparent that reaction has occurred upon heating. With suitably substituted or labeled dienes, however, the rearrangement is easily detected.

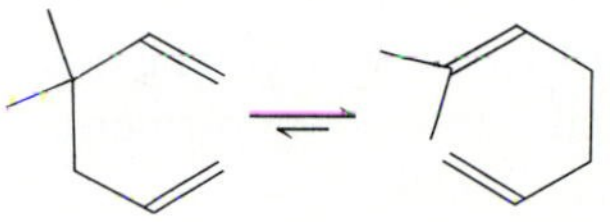

In this example the product contains more highly substituted double bonds and is predominant at equilibrium.

The central bond in 1,5-hexadiene or biallyl is rather weak; DH° is 58 kcal mole^{-1} compared to a normal carbon-carbon DH° of 82 kcal mole^{-1} for the central bond in hexane. As this bond begins to break, however, a new carbon-carbon bond forms at the opposite ends of the two allyl π-systems that are starting to be produced. An orbital diagram of the transition state is given in Figure 21.12. A cyclic system of six interacting p-orbitals is involved much as in the conjugating system of benzene. Two of the p-orbital interactions are of the σ- rather than π-type in a manner comparable to the transition state of the Diels-Alder reaction. As in that case, the combination of relatively weak σ-overlap

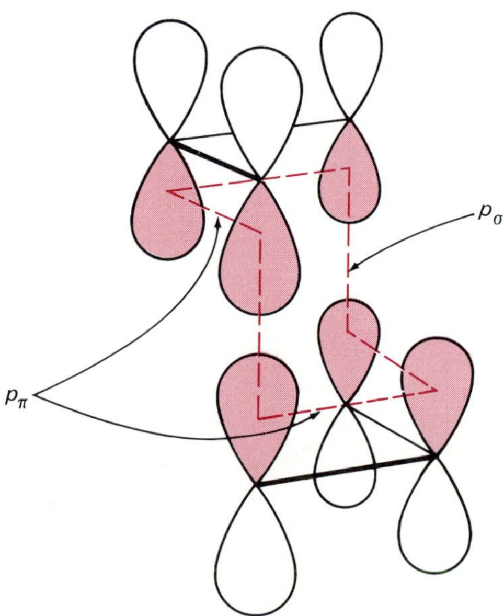

FIGURE 21.12 Electronic structure of the transition state for a Cope rearrangement.

combines with the other π-overlaps to produce a cyclic conjugation entirely analogous to benzene.

EXERCISE 21.14 Although *trans*-1,2-divinylcyclopropane is a relatively stable compound, the *cis*-isomer rearranges rapidly even at 0 °C. Suggest a reason for the high reactivity of this isomer and give the structure of the rearrangement product.

Reactions similar to the Cope rearrangement occur for compounds with atoms other than carbon in the cyclic conjugating chain. An important example is the **Claisen rearrangement,** a rearrangement reaction of allyl vinyl ethers.

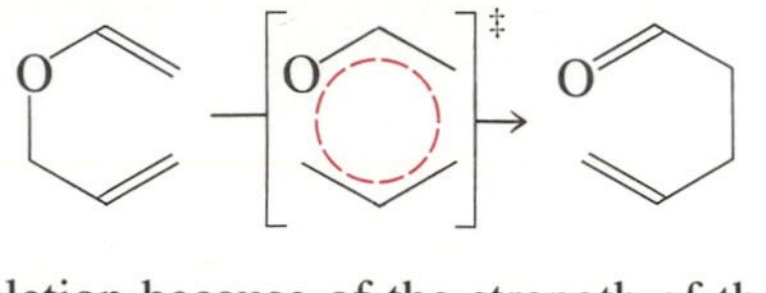

The reaction goes to completion because of the strength of the carbonyl bond produced. Vinyl ethers may be prepared by an exchange reaction of commercially available ethyl vinyl ether with an alcohol and catalyzed by mercuric ion, recalling the acid-catalyzed hydrolysis discussed in Section 14.6.B.

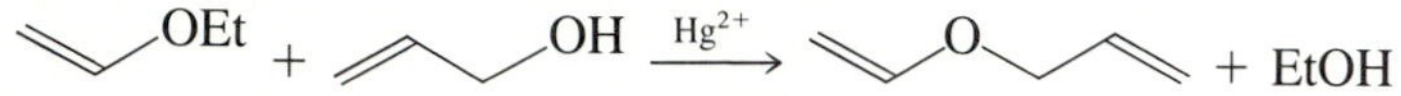

Hence, the reaction provides an efficient method for adding a two-carbon aldehyde group to an allylic unit.

EXERCISE 21.15 (*R*)-2-cyclohexenol is treated with ethyl vinyl ether and the product is heated to produce an aldehyde. Give the structure and configuration of the product.

The Cope and Claisen rearrangements are two versions of the same general reaction. In both cases a bond between one end of two three-atom fragments is being broken while a new bond at the opposite end is being formed. These reactions are part of a class of such reactions called **sigmatropic rearrangements.** In the Cope and Claisen rearrangements

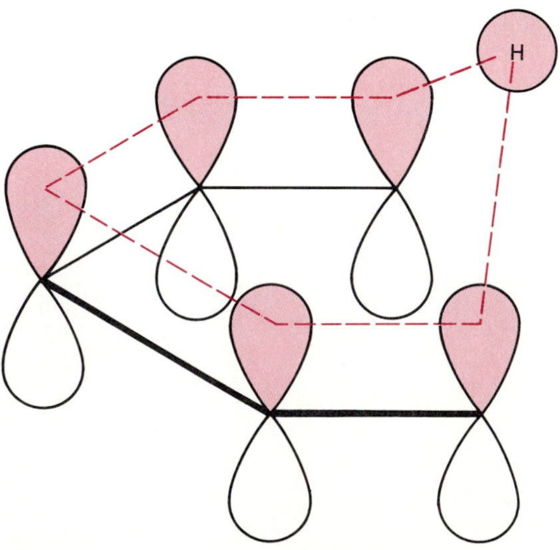

FIGURE 21.13 Orbital interactions for a [1.5] sigmatropic rearrangement.

the two rearranging fragments involved are both three atoms in length; hence, both reactions are sigmatropic rearrangements *of order [3.3]*.

Another sigmatropic rearrangement that involves a benzene-like transition state is a 1,5-hydrogen shift, a [1.5] sigmatropic rearrangement. An example of this reaction is given by the thermal rearrangement of 5-methyl-1,3-hexadiene to 2-methyl-2,4-hexadiene.

H ⇌ H

An orbital model of the transition state shows its relationship to the cyclic conjugation in benzene (Figure 21.13).

EXERCISE 21.16 (a) The rearrangement of 5-methyl-1,3-hexadiene to 2-methyl-2,4-hexadiene actually results in an equilibrium mixture of the two compounds. Which compound predominates at equilibrium? What is the transition state for the reverse reaction?

(b) 5-Methylcyclopentadiene rearranges on mild warming to a mixture that contains 1-methylcyclopentadiene and 2-methylcyclopentadiene. Show how the rearrangements involved are of the [1.5] sigmatropic type.

ASIDE

The persistence of chlorofluorocarbons (CFCs; see Chapters 6 [problem 12] and 8) is environmentally harmful, since sunlight acts on them to produce chlorine atoms that decrease the ozone concentration in the upper atmosphere. The persistence of the benzene ring in chemical reactions is a favorable circumstance for chemists since many useful benzene derivatives can be prepared. Show how the stability reflected in the persistence of the ring can be measured using hydrogenation, carefully defining all terms. Write out the reaction of chlorine and toluene in the presence of light. Is the benzene ring present in the products?

PROBLEMS

1. Write structures corresponding to each of the following names.

a. *m*-fluoroanisole
b. 2,4,6-tribromobenzoic acid
c. 2,4-dinitrotoluene
d. α-bromomesitylene
e. *m*-divinylbenzene
f. *p*-cyanophenylacetylene
g. *o*-diisopropylbenzene
h. 2-bromo-6-chloroaniline

2. Give an acceptable name for each of the following structures.

a. Cl–C_6H_4–CH(CCl_3)–C_6H_4–Cl (DDT)

b. $CH_2CH_2CH_3$ / Br (para-substituted benzene)

c. $CH(CH_3)_2$ / Br / I (substituted benzene)

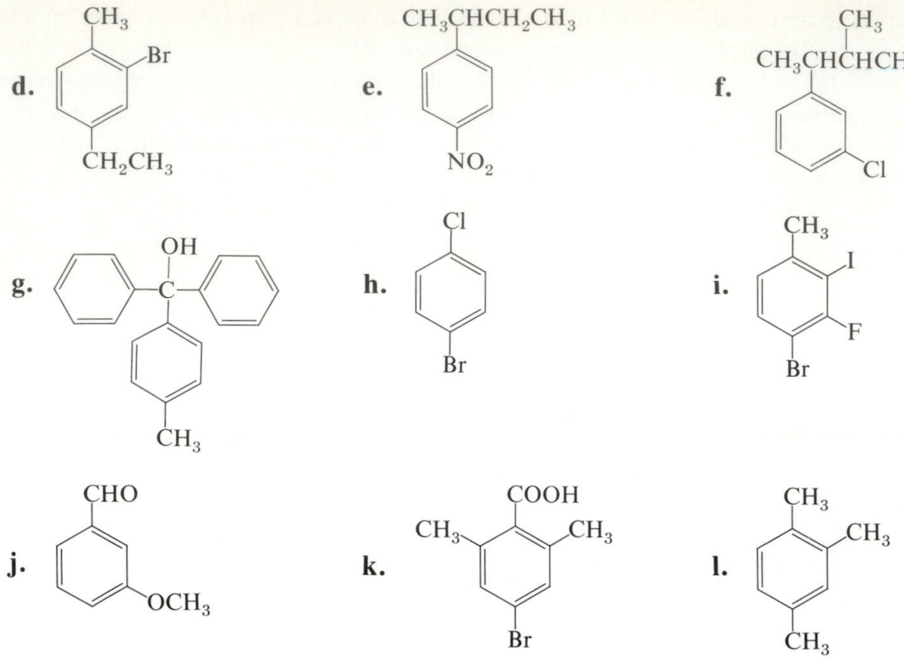

3. In Kekulé's day, one puzzling aspect of his dynamic theory for benzene was provided by 1,2-dimethylbenzene. According to his theory, there should be two distinct such compounds, one with a double bond between the two methyl-substituted carbons and one with a single bond in this position.

CH_3 CH_3 CH_3 CH_3

Only a single 1,2-dimethylbenzene is known, however.

a. Does Ladenburg's formula solve this problem?

b. Explain with modern resonance theory.

4. When passed through a hot tube, acetylene gives fair amounts of benzene. What is $\Delta H°$ for the reaction?

$$3\ HC{\equiv}CH \longrightarrow$$

The entropy change for this reaction is $\Delta S° = -79.7$ eu. How do you explain the negative sign of this entropy change? Calculate $\Delta G°$ for the reaction at 25 °C. Where does the equilibrium lie at room temperature? This reaction does not occur spontaneously at room temperature. Can you give a reason?

5. a. A common method for estimating the empirical resonance energy of benzene is to take the heat of hydrogenation of one Kekulé resonance structure as three times that of cyclohexene. What value of the empirical resonance energy does this procedure yield? Note how the exact value of the empirical resonance energy depends so markedly on the model used for a hypothetical system.

b. The heat of hydrogenation of cyclooctene to cyclooctane is -23.3 kcal mole^{-1}. That for 1,3,5,7-cyclooctatetraene is -100.9 kcal mole^{-1}. Use the procedure in (a) to calculate an empirical resonance energy for cyclooctatetraene. How do you interpret this result?

c. Another method for estimating empirical resonance energies makes use of Appendix III, Average Bond Energies. In this table $E(\text{C—H}) = 99$, $E(\text{C—C}) = 83$, $E(\text{C=C}) = 146$ kcal mole^{-1}. Calculate the total bond energy of a hypothetical cyclohexatriene. This energy is the so-called heat of atomization, the heat required to dissociate a molecule into all of its constituent separated atoms. For benzene, this heat is actually $\Delta H°_{\text{atom}} = +1318$ kcal mole^{-1}. What value for the empirical resonance energy results?

6. Consider the possible free radical chain chlorination of benzene.

$$C_6H_6 + Cl\cdot \longrightarrow C_6H_5\cdot + HCl$$

$$C_6H_5\cdot + Cl_2 \longrightarrow C_6H_5Cl + Cl\cdot$$

From the data in Appendix I calculate $\Delta H°$ for each reaction. What conditions would you recommend for accomplishing this reaction?

7. The CMR spectra of the three isomeric diethylbenzenes are shown below. Which is which?

a.

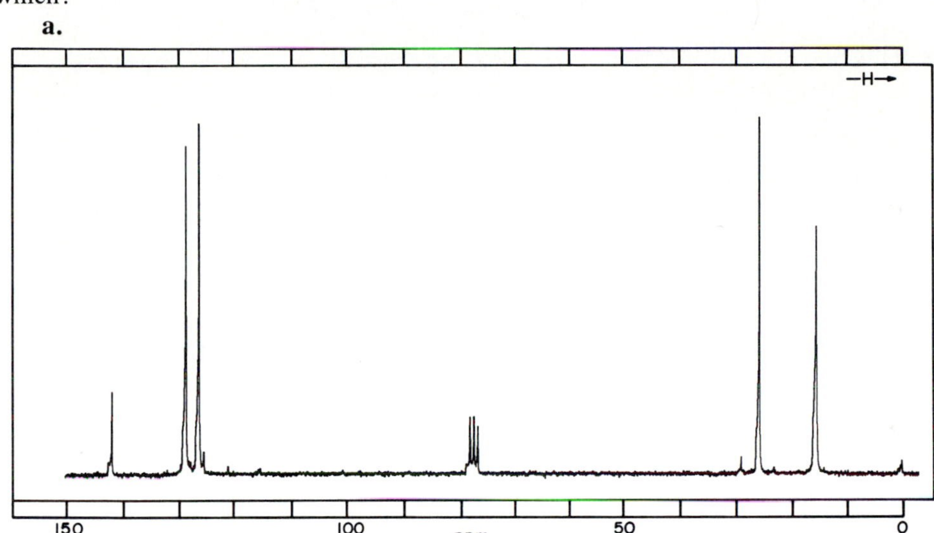

b.

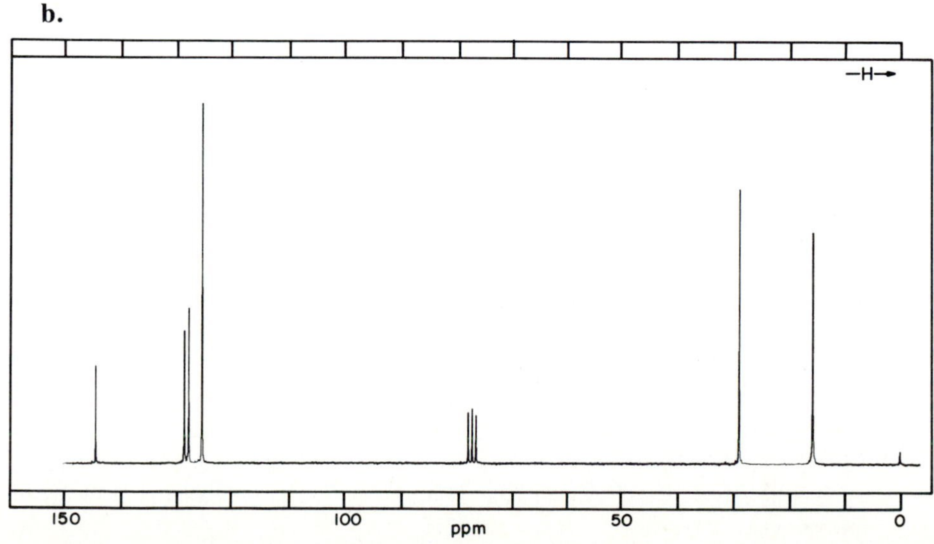

c.

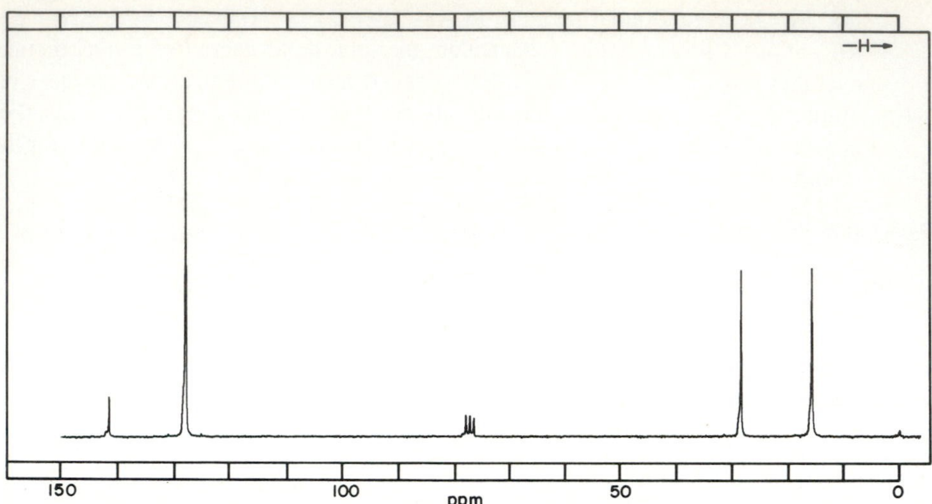

8. Give the principal product of the following reactions or reaction sequences.

a. $C_6H_5CH_2CH_3 \xrightarrow[h\nu]{Br_2} \xrightarrow[ether]{Mg} \xrightarrow{CH_2\text{—}CH_2 \text{ (O)}} \xrightarrow{H^+}$

b. $p\text{-}CH_3C_6H_4CH_3 \xrightarrow[C_2H_5OH]{Na,\ NH_3}$

c. $C_6H_5CH(CH_3)_2 \xrightarrow[h\nu]{Br_2\ (1\ mole)} \xrightarrow[\Delta]{CH_3OH}$

d. $C_6H_5CH_3 \xrightarrow[H_2SO_4,\ \Delta]{Na_2Cr_2O_7} \xrightarrow[H^-,\ \Delta]{CH_3OH} \xrightarrow[ether]{CH_3MgI}$

e. $p\text{-}CH_3CHOHC_6H_4CH_2CH_2OH \xrightarrow[HClO_4]{H_2\text{–}Pd/C}$

f. $p\text{-}CH_3CHOHC_6H_4CH_2CH_2OH \xrightarrow{MnO_2}$

g. $p\text{-}CH_3C_6H_4OCH_3 \xrightarrow[C_2H_5OH]{Li\text{—}NH_3} \xrightarrow{H_3O^+}$

9. Show how one may synthesize each of the following compounds, starting with benzene, toluene, xylene, or ethylbenzene.

a. $C_6H_5CH(CH_2CH_3)_2$

b. 1-methyl-1,4-cyclohexadiene (CH_3)

c. $C_6H_5\text{—}C{\equiv}CH$

d. $C_6H_5\text{—}CH_2CH_2\text{—}C_6H_5$

e. $p\text{-}CH_2DC_6H_4CH_3$

f. $C_6H_5CH_2C{\equiv}CH$

10. a. Write out the steps of the free radical chain bromination of toluene to give benzyl bromide.

b. From ΔH_f° and DH° values listed in Appendices I and II, calculate ΔH° for the reactions in part (a).

c. Compare these values with those for ethane and the tertiary position of 2-methylpropane. How feasible are these brominations?

11. Write structures for the eight possible benzene hexachlorides. Which one is capable of optical isomerism? Which one is slowest to eliminate HCl by the E2 mechanism?

12. *o*-Phthalyl alcohol, 1,2-bis(hydroxymethyl)benzene, on treatment with acid, gives the corresponding cyclic ether.

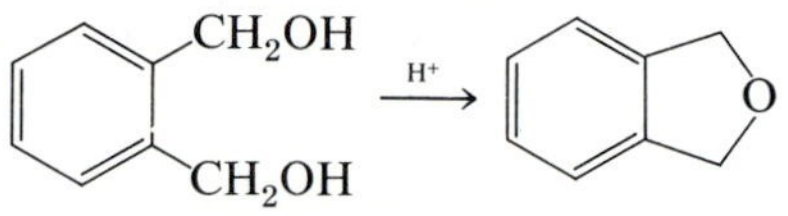

CH_2OH $\xrightarrow{H^+}$ O

CH_2OH

Give a reasonable mechanism for this reaction.

13. Allyl acetate is treated with the strong base lithium diisopropylamide. The resulting enolate solution is kept at room temperature for 2 hr and then quenched with dilute acid. The product is found to be 4-pentenoic acid. Propose a reaction mechanism.

14. Heating the deuterated hydrocarbon shown produces via a Cope rearrangement an equilibrium mixture in which the deuterium is also on the bridge position. Of the two methylene proton positions, however, only a single deuterated isomer is produced. Which is it?

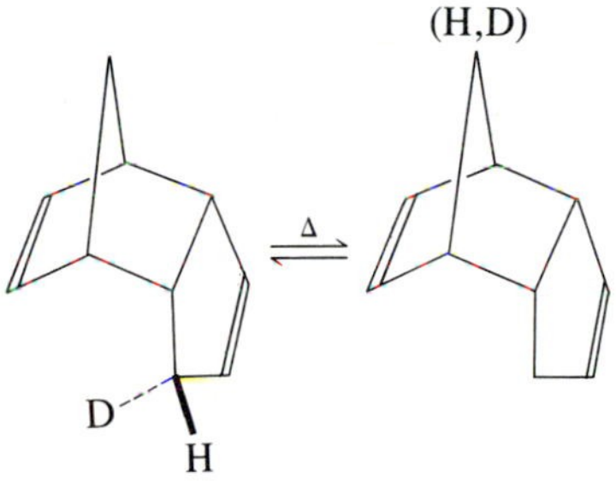

(H,D)

$\overset{\Delta}{\rightleftharpoons}$

D H

15. The **ene reaction** is a reaction of alkenes that is closely analogous to a Diels-Alder reaction. An example of an ene reaction is

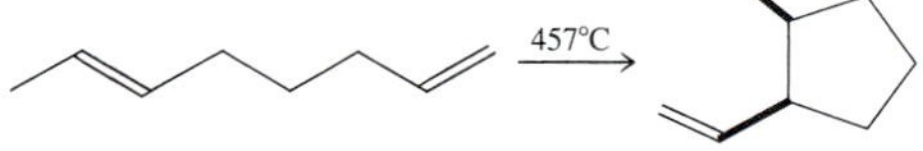

$\xrightarrow{457°C}$

Give the reaction mechanism and show how it involves an aromatic transition state.

16. Show the products of thermal [1.5] sigmatropic rearrangements of the following compounds.

a. 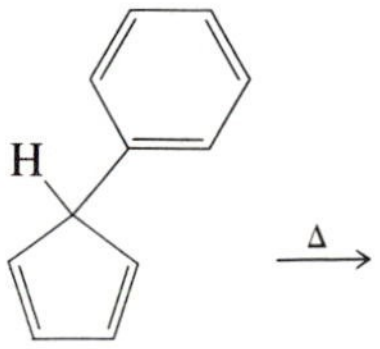 H $\xrightarrow{\Delta}$ **b.** 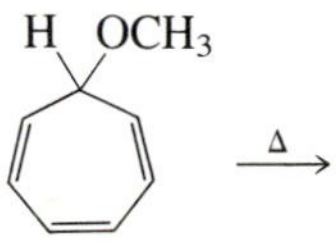 H OCH_3 $\xrightarrow{\Delta}$ **c.** $\xrightarrow{220°C}$

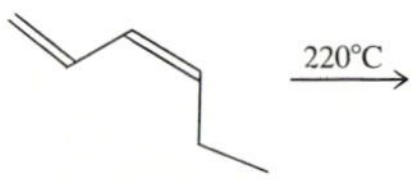

17. On page 636 we showed that the reaction of ozone with an alkene to give a molozonide is an example of a [4 + 2] cycloaddition reaction that proceeds through an aromatic six-electron transition state. In an ozonolysis reaction, the initially formed molozonide undergoes a rapid rearrangement to the normal ozonide.

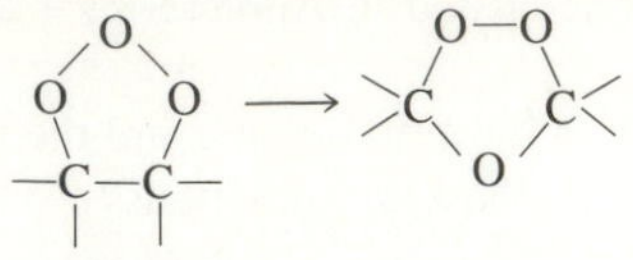

molozonide → ozonide

Propose a two-step reaction mechanism, in which both steps involve aromatic six-electron transition states analogous to that involved in formation of the molozonide.

18. Show that one could distinguish among *o*-xylene, *m*-xylene, and *p*-xylene by carefully examining the number of isomers produced by mononitration. (This method was Körner's absolute method for distinguishing the isomers of a benzene derivative.)

CHAPTER 22

MOLECULAR ORBITAL THEORY

MOST of our discussions of electronic structure have made use of resonance structures, although we have used molecular orbital concepts from time to time. In Section 2.7 we considered how bonds arise from the overlap of atomic orbitals. We have learned that the overlap of **two** atomic orbitals gives rise to **two** molecular orbitals (MOs), one bonding and one antibonding. When the orbital overlap occurs in a σ-fashion—along the bond axis—the bonding and antibonding MOs are symbolized as σ and σ^*, respectively. When the orbitals are *p*-orbitals overlapping in a π-fashion, as in double bonds (Section 11.1), the resulting bonding and antibonding MOs are π and π^*, respectively. When more than two orbitals overlap mutually, a more complex pattern of bonding and antibonding MOs results in which the MOs extend over more than two atoms. Such multicenter or *delocalized* molecular orbitals are particularly important in π-electron systems because they are involved in the unique properties of conjugated molecules. The nature of these MOs will be discussed in this chapter.

22.1 Molecular Orbital Description of Allyl and Butadiene

The resonance description of allylic conjugation involves alternative descriptions of bonding by pairs of electrons in normal Lewis structures.

$$\left[H_2C{=}CH{-}\overset{*}{C}H_2 \longleftrightarrow H_2\overset{*}{C}{-}CH{=}CH_2 \right]$$

In the foregoing structures, the asterisk represents a positive or negative charge or an odd electron. Most of the electrons bond the skeleton of the compound and are not involved in

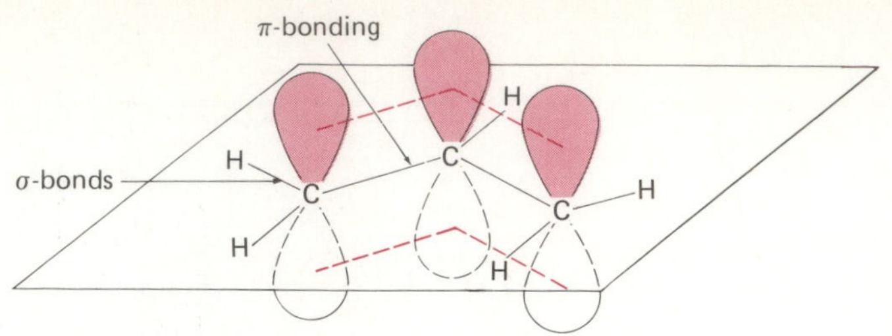

FIGURE 22.1 σ - and π-bonds in allyl systems.

the resonance stabilization or conjugation. In the molecular orbital picture these electrons form the σ-bonding framework of the molecule. The π-system consists of p-orbitals overlapping to form π-bonds above and below the plane of the atoms that define the allylic system (Figure 22.1).

The involvement of three p-orbitals in this manner clearly gives greater bonding, and it is not difficult to understand the stabilization that such orbital overlap bestows on allyl cation. However, according to the Pauli principle, only two electrons of opposite spin can be associated with any single orbital. What are we to do with the third and fourth π-electrons of allyl radical and anion?

Two p_z-orbitals overlapping, as in ethylene, generate ***two*** molecular orbitals, a bonding π-MO and an antibonding π^*-MO. Three p_z-orbitals overlapping, as in allyl, generate ***three*** different molecular orbitals, each having its own energy (Figure 22.2a; see also Figure 2.8). One orbital is most bonding and is occupied by two electrons of opposite spin in allyl cation. The third electron of allyl radical must be put into the second molecular orbital. The fourth electron of allyl anion can also be put into this second molecular orbital. The third molecular orbital has high energy and is not involved in bonding in allylic compounds. It is a high-lying π^* MO. These relationships are shown in Figure 22.2b.

The π-molecular orbitals can be regarded as having molecular orbital quantum numbers or their equivalent in nodes. When p-orbitals overlap as in allyl, they do so in such a way as to generate one molecular orbital having no nodes, a second having one node, and a third having two nodes.

> Since these molecular orbitals are made up of p-orbitals overlapping in a π-fashion, all of the molecular orbitals have one other node, the nodal plane of the component p-orbitals. This plane is illustrated in Figure 22.1. The nodes referred to in the preceding paragraph are nodes in addition to this nodal plane.

These molecular orbitals for allyl are shown in Figure 22.3. Recall that when functions of the same sign overlap, electron density is found in the overlap region between the nuclei, and bonding results. Electron density is zero at a node. The overlap of two wave functions of opposite sign creates a node in the overlap region and signifies a region devoid of electron density. The absence of such electron density to counter nuclear repulsion produces **antibonding.** Hence, the first allyl π-molecular orbital, π_1, has no nodes and is completely bonding. In π_2 there is a node going through the middle carbon. The two remaining p-orbital wave functions are so far apart that overlap is small and this molecular orbital is approximately **nonbonding.** The highest molecular orbital, π_3, has two nodes and is antibonding. In general, the greater the number of nodes, the higher the energy of an orbital and the lower the stability. The greatest stability (lowest energy) results when electrons are associated as far as possible with the most bonding molecular orbitals.

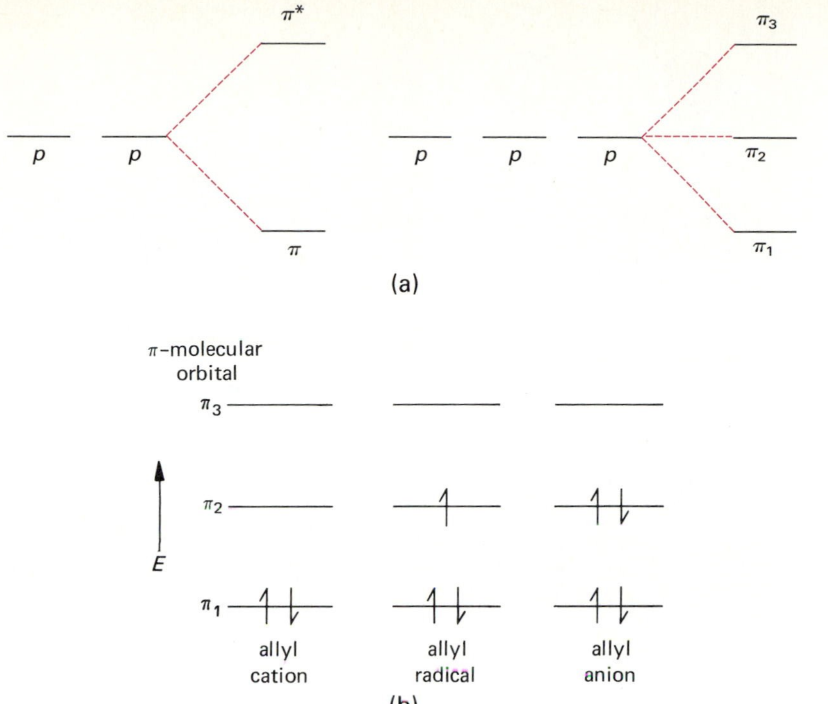

FIGURE 22.2 (a) Splitting of three orbitals into three π-molecular orbitals; (b) π-molecular orbital energies in allyl.

These molecular orbitals can be described analytically by the mathematical functions

$$\pi_1 = \frac{1}{2}p_1 + \frac{\sqrt{2}}{2}p_2 + \frac{1}{2}p_3$$

$$\pi_2 = \frac{\sqrt{2}}{2}p_1 - \frac{\sqrt{2}}{2}p_3$$

$$\pi_3 = \frac{1}{2}p_1 - \frac{\sqrt{2}}{2}p_2 + \frac{1}{2}p_3$$

in which, p_1, p_2, and p_3 are the mathematical functions for the three p-atomic orbitals. Note that the node at the middle carbon of π_2 in Figure 22.3 means simply that the coefficient of p_2 in this molecular orbital is zero. It is instructive to compare these molecular orbitals with the standing waves of other linear systems, such as a vibrating violin string or the sound waves in a pipe organ, as in Figure 22.4. The lowest-energy wave has no nodes and is either positive throughout or negative throughout. The next lowest wave has one node, and the third has two nodes. Note the close resemblance to the π-molecular orbitals of allyl. The correspondence reaffirms the common properties of all waves. Moreover, the smooth continuous nature of wave functions allows us to understand why the end coefficients in π_1 and π_3 of allyl are smaller than the middle coefficient and why the middle coefficient is zero in π_2.

The π-system of 1,3-butadiene consists of four overlapping p-orbitals, which generate four molecular orbitals. These four MOs are shown schematically in Figure 22.5. The lowest-energy orbital, π_1, has no nodes and is bonding between C-1 and C-2, between C-2 and C-3, and between C-3 and C-4. The second MO, π_2, has one node. Therefore, this orbital is bonding between C-1 and C-2 and between C-3 and C-4 but is antibonding between the center carbons. The four electrons associated with the π-system are in molec-

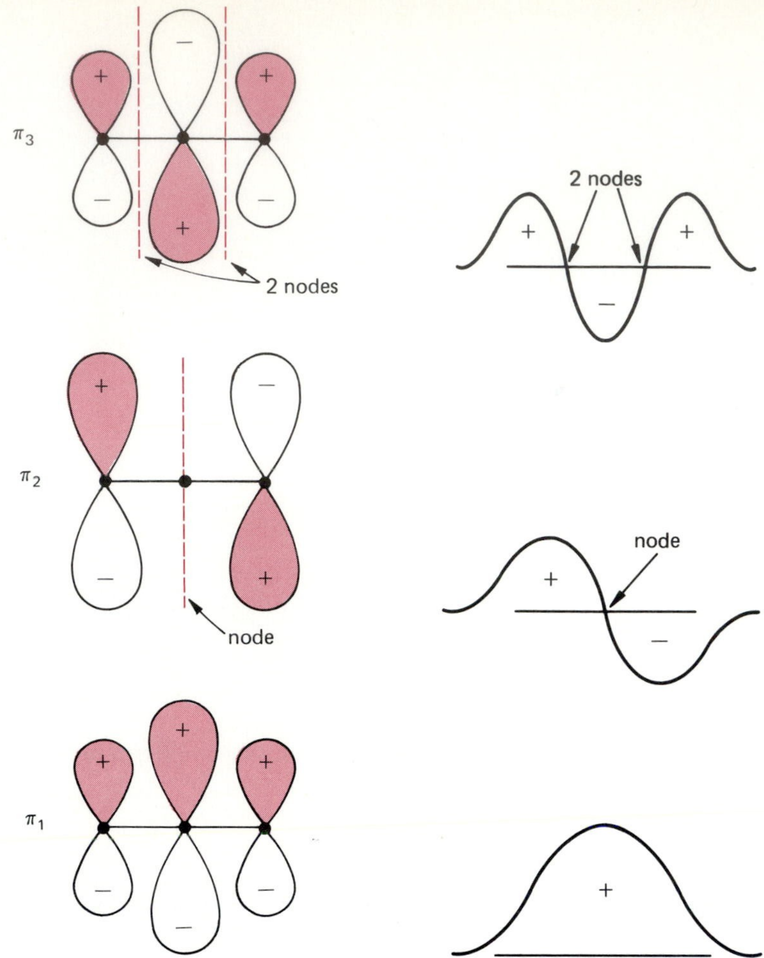

FIGURE 22.3 π-molecular orbitals of allyl. **FIGURE 22.4** Standing waves of a linear system.

ular orbitals π_1 and π_2 (Figure 22.6). Therefore, the π-bonding between C-2 and C-3 produced by π_1 is partially offset by the antibonding nature of π_2 in this region of the molecule. The antibonding character of π_2 does not quite cancel the bonding character of π_1, because the bonding and antibonding overlaps differ in magnitude. However, there is little net π-bonding between the center carbons, and the net electronic structure is given to a reasonable approximation by the normal Lewis structure

$$H_2C::CH:CH::CH_2$$

In this kind of analysis note that the nature of unoccupied MOs is not relevant; because they have no electrons in them, they do not contribute to the energy of the molecule or to its properties.

Note that the π-MOs of butadiene are analogous to those of allyl. The total number of MOs is equal to the number of p-orbitals involved in π-bonding. The number of nodes form a regular progression and they are placed so as to retain the symmetry of the molecule.

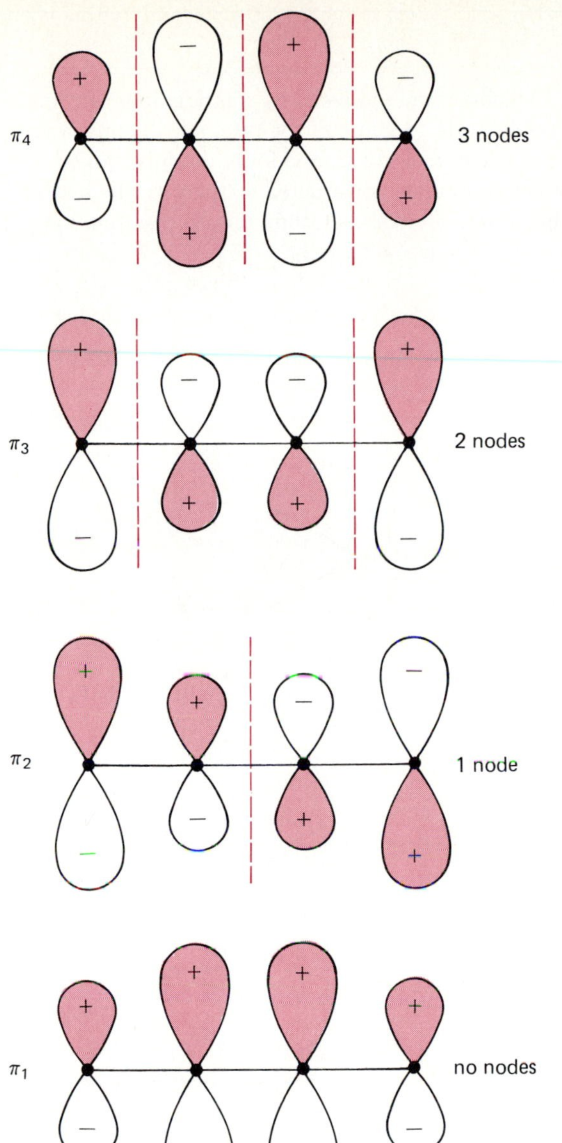

FIGURE 22.5 π-molecular orbitals of 1,3-butadiene.

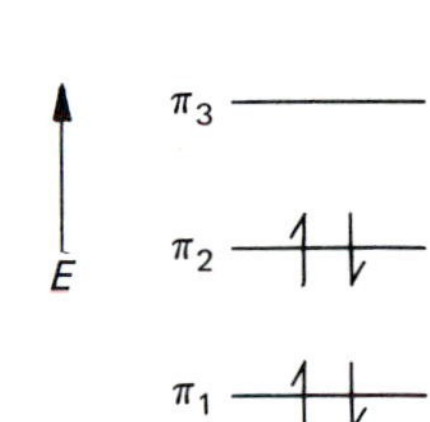

FIGURE 22.6 Relative energies of π-molecular orbitals of 1,3-butadiene.

EXERCISE Pentadienyl anion, $CH_2{=}CH{-}CH{=}CH{-}CH_2^-$, has five π-molecular orbitals with 0 to 4 nodes. Sketch the five MOs in order of their expected energy. How many π-electrons are involved and which of the π-MOs are occupied?

EXERCISE 22.2 How many π-MOs does 1,3,5-hexatriene have? Sketch these MOs and occupy appropriately with electrons analogous to Figure 22.6. Consider the bonding and antibonding characteristics of the occupied MOs and compare with a Lewis structure for hexatriene.

Highlight 22.1

π-orbitals for ethylene, allyl, and 1,3-butadiene are produced by combinations of two, three, and four C_{2p} atomic orbitals, respectively. Because of the principle of conservation of orbitals, these combinations give rise to two, three, and four molecular orbitals, respectively. In each case, the molecular orbitals are constructed mathematically by all of the possible ± combinations of the atomic orbitals that obey certain quantum mechanical rules. For this reason, some of the orbitals have one or more nodes, in addition to the normal p-orbital node that is in the plane of the molecule. In general, the more nodes an orbital has, the higher is its energy. Note that the amplitudes of some of the p-orbitals vary somewhat from this simple scheme.

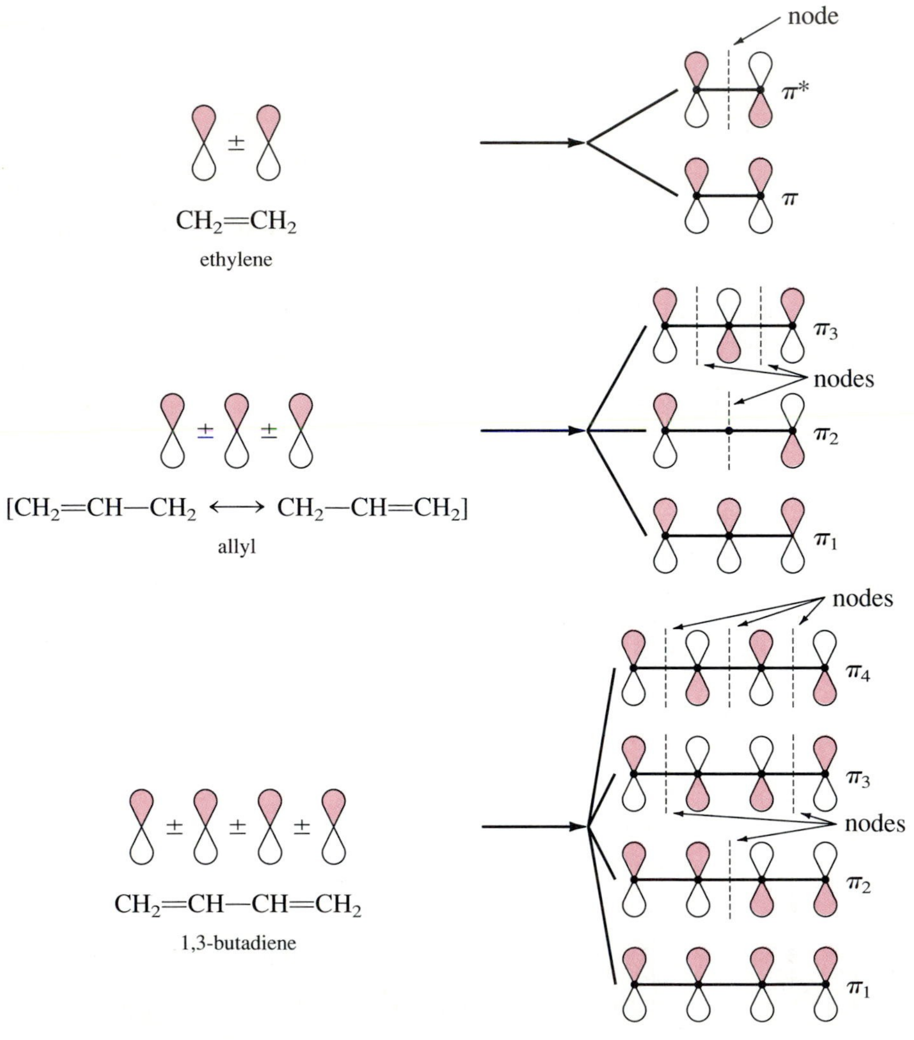

node

π^*

π

$CH_2{=}CH_2$

ethylene

π_3

nodes

π_2

π_1

$[CH_2{=}CH{-}CH_2 \longleftrightarrow CH_2{-}CH{=}CH_2]$

allyl

nodes

π_4

π_3

nodes

π_2

π_1

$CH_2{=}CH{-}CH{=}CH_2$

1,3-butadiene

Highlight 22.2

The π-molecular orbitals of allyl and 1,3-butadiene are filled by addition of electrons. For allyl, the addition of two, three, or four electrons gives the allyl cation, allyl radical, and allyl anion, respectively. In accord with the Pauli exclusion principle, no more than two electrons can be placed in an orbital. For allyl cation the two electrons fill π_1. For allyl radical the third electron must be placed in π_2. Allyl anion, with four electrons, has both π_1 and π_2 filled. Similarly, 1,3-butadiene, with four electrons, has both π_1 and π_2 filled, whereas the corresponding radical cation, which has only three electrons, has a half-filled π_2 shell.

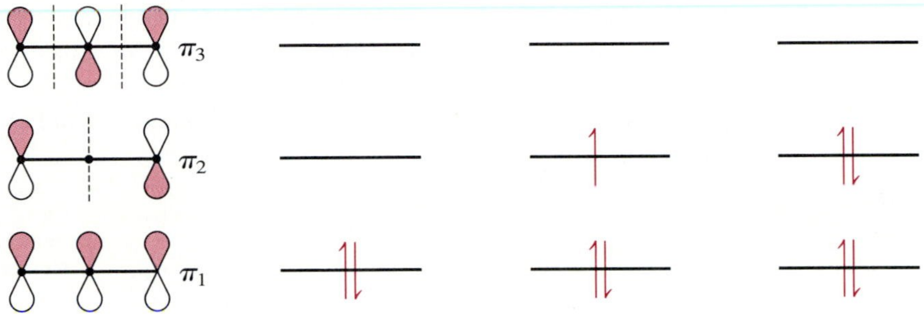

	allyl cation	allyl radical	allyl anion
π_3	—	—	—
π_2	—	↑	↑↓
π_1	↑↓	↑↓	↑↓

$$\left[\mathrm{CH_2{=}CH{-}\overset{+}{C}H_2} \longleftrightarrow \mathrm{\overset{+}{C}H_2{-}CH{=}CH_2}\right]$$

allyl cation

$$\left[\mathrm{CH_2{=}CH{-}\dot{C}H_2} \longleftrightarrow \mathrm{\dot{C}H_2{-}CH{=}CH_2}\right]$$

allyl radical

$$\left[\mathrm{CH_2{=}CH{-}\overset{-}{C}H_2} \longleftrightarrow \mathrm{\overset{-}{C}H_2{-}CH{=}CH_2}\right]$$

allyl anion

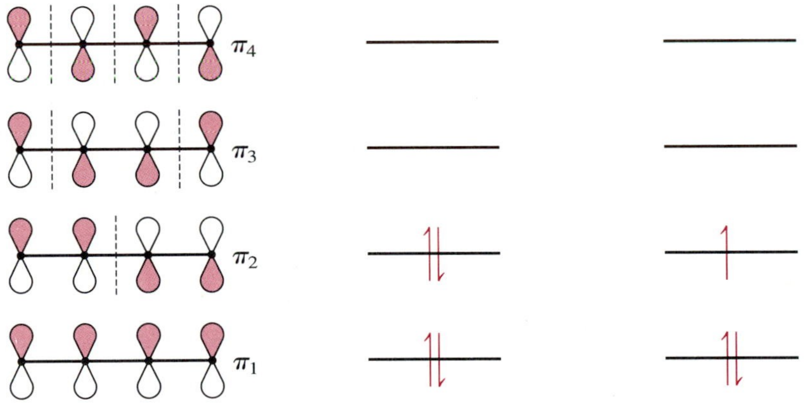

	1,3-butadiene	1,3-butadiene radical cation
π_4	—	—
π_3	—	—
π_2	↑↓	↑
π_1	↑↓	↑↓

$$\mathrm{CH_2{=}CH{-}CH{=}CH_2}$$

1,3-butadiene

$$\mathrm{CH_2{=}CH{-}\overset{+}{C}H{-}CH_2}$$

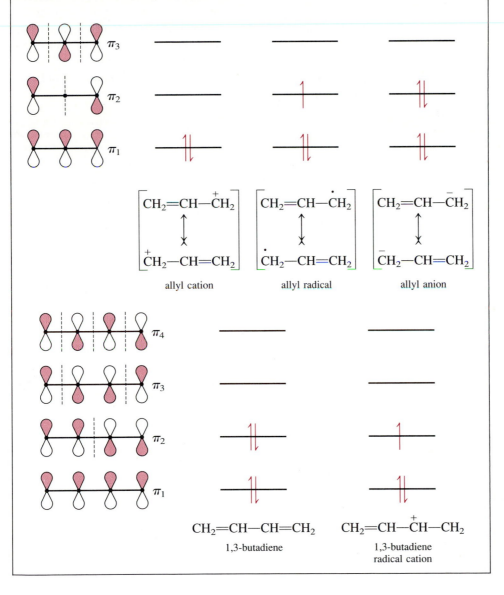

1,3-butadiene radical cation

22.2 Molecular Orbital Theory of Benzene

The π-system of benzene is made up of six overlapping *p*-orbitals on carbon, which therefore give rise to six π-molecular orbitals. The lowest, most stable molecular orbital, π_1, has no nodes and consists of all six *p*-orbitals overlapping around the ring. The next two molecular orbitals, π_2 and π_3, are not as bonding as π_1 and have higher energy. Each has one node. Note the important difference from allyl and butadiene in which only one π-MO has one node. The difference is that the π-overlap in these systems occurs linearly and can be represented as a single dimension. The cyclic overlap in benzene requires a two-dimensional representation in which nodes occur as pairs. The nodes in π_2 and π_3 are at right angles to each other. These two molecular orbitals have identical energies and are therefore said to be **degenerate.** The three molecular orbitals designated as π_1, π_2, and π_3 are the occupied π-molecular orbitals in benzene; one pair of electrons can be put in each to accommodate all six π-electrons of benzene. The three remaining π-MOs have less bonding and higher energy; they are not occupied by electrons. The relative energies of all six π-molecular orbitals and the molecular orbitals themselves are represented in Figure 22.7.

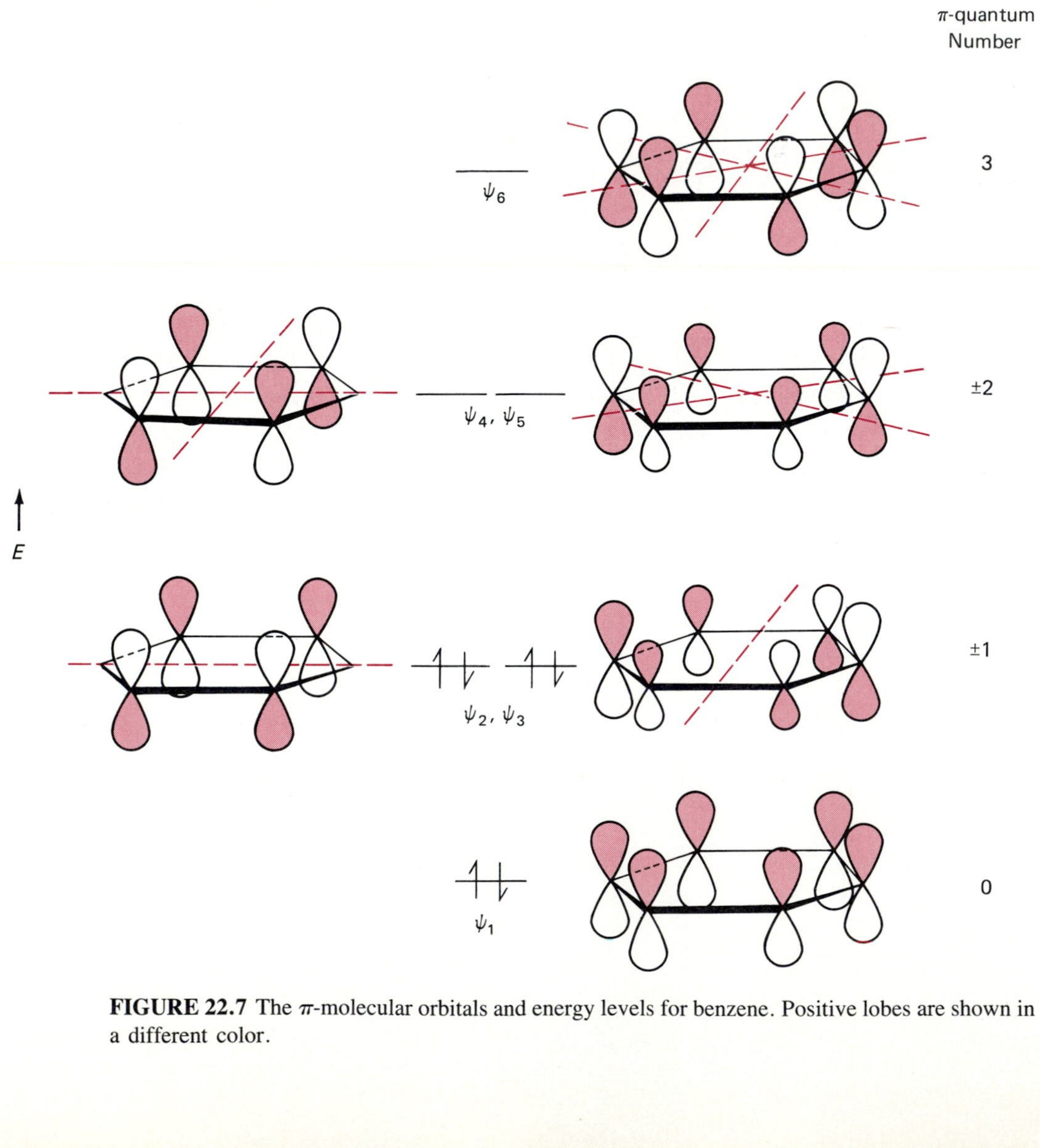

FIGURE 22.7 The π-molecular orbitals and energy levels for benzene. Positive lobes are shown in a different color.

A characteristic feature of this molecular orbital pattern is that the lowest-lying molecular orbital is a single molecular orbital; thereafter the molecular orbitals occur in pairs of equal energy until only one highest-lying level is left. These molecular orbital levels can be identified by quantum numbers, 0, ±1, ±2, and so on. Each quantum number represents a **shell** of orbitals, much as we have 2*s*- and 2*p*-shells in atomic structure. In benzene the ±1 shell is filled, and we can attribute the stability of benzene to this filled-shell structure in much the same way as the noble gases (helium, neon, argon, and so on) have stability associated with filled atomic orbital shells.

It is important to distinguish between the π-electronic system of benzene as symbolized commonly by a set of six *p*-orbitals overlapping as in Figure 21.1 and the molecular orbitals as symbolized in Figure 22.7. Allowing the six *p*-orbitals to overlap in a cyclic fashion generates the MOs of Figure 22.7. According to the Pauli principle (Section 2.5), no two electrons can have the same quantum numbers. The six π-electrons of benzene can be divided into two groups of three based on electronic spin (a quantum number, $\pm\frac{1}{2}$), but each electron in the set of three must then belong to a different orbital. For benzene we have seen that such orbitals are the π-molecular orbitals characterized by quantum numbers of 0, +1, and −1.

22.3 Aromaticity

A. Cyclooctatetraene: The Hückel 4*n* + 2 Rule

Cyclooctatetraene is a well-known hydrocarbon; it is a liquid, b.p. 152 °C, that shows all of the chemistry typical of conjugated polyenes. It polymerizes on exposure to light and air and reacts readily with acids, halogens, and other reagents. In other words, it shows none of the "aromatic" stabilization associated with benzene. If cyclooctatetraene had the structure of a planar regular octagon analogous to the hexagon of benzene, we could write two resonance structures of the benzene Kekulé type.

We would therefore anticipate a significant amount of resonance energy for such a structure. Why, then, is cyclooctatetraene not an "aromatic" compound? The π-molecular orbital energy-level pattern is shown in Figure 22.8. Six of the eight π-electrons are put

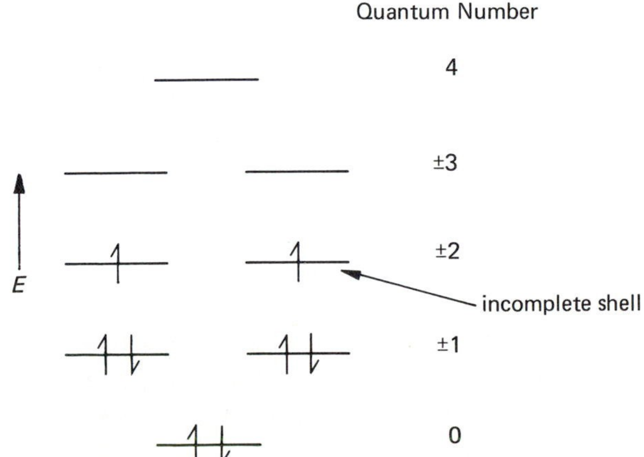

FIGURE 22.8 The π-molecular orbital energy level pattern for a planar octagonal cyclooctatetraene.

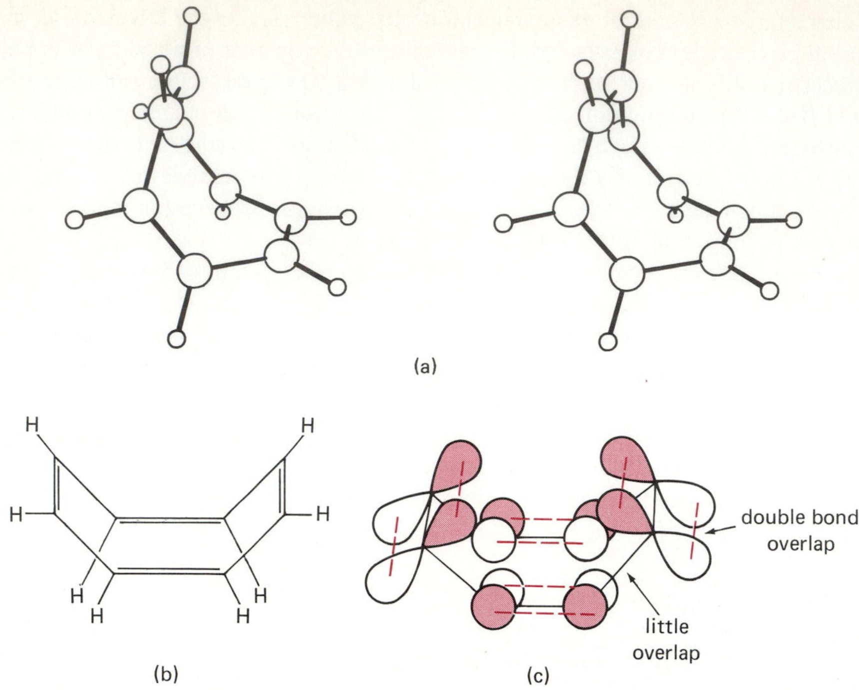

FIGURE 22.9 Cyclooctatetraene: (a) stereo representation; (b) Kekulé structure; (c) π-orbital structure.

into the three lowest molecular orbital levels, but the one pair left is not enough to fill the next shell. Thus, planar, octagonal cyclooctatetraene has an incomplete orbital shell and would therefore not be expected to have the special stability characteristic of benzene.

> Note that the last two electrons in Figure 22.8 are placed with the same spin, one in each of the degenerate orbitals. This arrangement is a consequence of Hund's rule, just as in atomic structure. Two electrons of the same spin are prevented from close approach by the Pauli principle—two electrons with the same quantum numbers cannot occupy the same region of space. Two electrons of opposite spin stay apart only because of electrostatic repulsion; hence, such a system has higher net energy and is less stable than one in which the electrons have the same spin.

In fact, the structure of cyclooctatetraene is in keeping with this analysis. The molecule is tub-shaped and has bond lengths characteristic of alternating single and double bonds (Figure 22.9).

As a result of the tub shape, the π-orbitals of adjacent double bonds are twisted with respect to one another and overlap is greatly reduced. That is, looking down any single bond, the π-orbitals are almost at right angles to each other. In short, because of the instability associated with an incomplete orbital shell in the planar octagonal geometry, cyclooctatetraene prefers a nonplanar structure in which the alternating double bonds are effectively not conjugated with each other!

Incomplete atomic orbital shells are associated with a relative ease of gaining or losing electrons to form an ion having a filled-shell electronic configuration; for example, lithium atom easily loses an electron and fluorine atom easily gains one.

$$\text{Li}\cdot \longrightarrow \text{Li}^+ + e^-$$

$$:\ddot{\underset{..}{\text{F}}}\cdot + e^- \longrightarrow :\ddot{\underset{..}{\text{F}}}:^-$$

The same behavior is seen with incomplete molecular orbital shells. Cyclooctatetraene reacts readily with alkali metals in ether solvents to form alkali metal salts of cyclooctatetraene dianion.

$$+ 2\text{ K} \xrightarrow{\text{THF}} \quad (2-) \quad + 2\text{ K}^+$$

cyclooctatetraene dianion

The dianion has the planar structure of a regular octagon with carbon-carbon bond distances of 1.4 Å, quite similar to the carbon-carbon bond distances in benzene! Cyclooctatetraene dianion has ten π-electrons, just enough to fill the molecular orbital shell with π-quantum numbers of ± 2. This reaction of cyclooctatetraene provides a remarkable demonstration of the usefulness of simple molecular orbital concepts. Cyclooctatetraene is not an aromatic system; in fact, a planar octagonal cyclooctatetraene could even be described as **antiaromatic.** However, the dianion, with two more π-electrons, is definitely an aromatic system.

The foregoing discussion illustrates a general principle. For all cyclic π-electronic systems, successive molecular orbitals above the lowest level can be characterized by quantum numbers of $\pm n$, where n is an integer.

> The absolute value of the quantum number n indicates the number of nodal planes that bisect the ring. Alternatively, and equivalently, we can consider the quantum number to represent the angular momentum of an electron circling round the ring. The lowest level then corresponds to an electron having zero angular momentum. Thereafter, the momentum can be represented clockwise or counterclockwise about the ring; hence, above zero the quantum numbers come as $\pm$integer pairs.

To summarize, a filled orbital shell corresponds to a relatively stable electronic configuration. Examples in atomic orbitals are the filled $1s$-shell of helium and the filled $2p$-shell of neon. Similarly, filled π-molecular orbital shells give the stability associated with ''aromatic'' systems and bestow that stabilization commonly known as aromatic character or aromaticity. It takes two electrons of opposite spin to fill the lowest π-molecular orbital level for which $n = 0$. Thereafter, four electrons are required to give a filled π-molecular orbital shell. That is, filled shells are associated with a total of $4n + 2$ electrons, or two ($n = 0$), six ($n = 1$), ten ($n = 2$), fourteen ($n = 3$), and so on, electrons. This rule is known as the **Hückel 4*n* + 2 rule** after Erich Hückel, the German theoretical chemist who first developed the rule in the mid-1930s.

Many examples of compounds are now known to which the Hückel rule can be applied. The results are truly remarkable for such a simple rule; a vast amount of experimental chemistry can be summarized by the generalization that *those monocyclic π systems with $4n + 2$ electrons show relative stability compared to acyclic analogs*. Furthermore, those monocyclic systems with other than $4n + 2$ electrons appear to be destabilized relative to acyclic analogs and can be said to have ''antiaromatic'' character. In succeeding sections we will summarize some of the experimental evidence for systems with several different values of n.

EXERCISE 22.3 Construct a molecular model of cyclooctatetraene. Place the model in the ''tub'' conformation described on page 654 and note the lack of overlap between adjacent double-bond π-orbitals.

B. Two-Electron Systems

One two-electron cyclic π-system is obviously ethylene, a well known and relatively stable compound. However, another cyclic π-system with two electrons is cyclopropenyl cation, a rather stable carbocation.

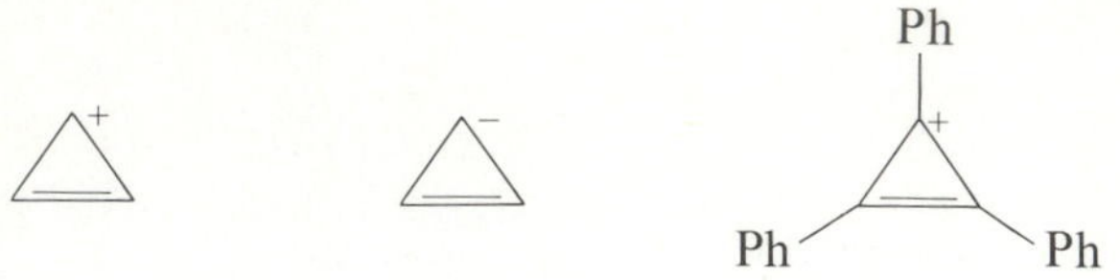

cyclopropenyl cation

cyclopropenyl anion

Ph Ph Ph

triphenylcyclopropenyl cation

Triphenylcyclopropenyl cation is such a stable carbocation that many of its salts can be isolated and stored in bottles. On the other hand, the cyclopropenyl anion is unknown. The acidity of the methylene group in the known hydrocarbon cyclopropene has been deduced from several experiments to be much less than that of alkanes.

The generalization that cyclic two-electron systems show relative stability makes its appearance in some subtle ways. For example, compounds containing the cyclopropenone ring system have unusually high dipole moments; this result is explained on the basis that the dipolar resonance structure contributes more to the resonance hybrid of cyclopropenone because it embodies the "aromatic" cyclopropenyl cation.

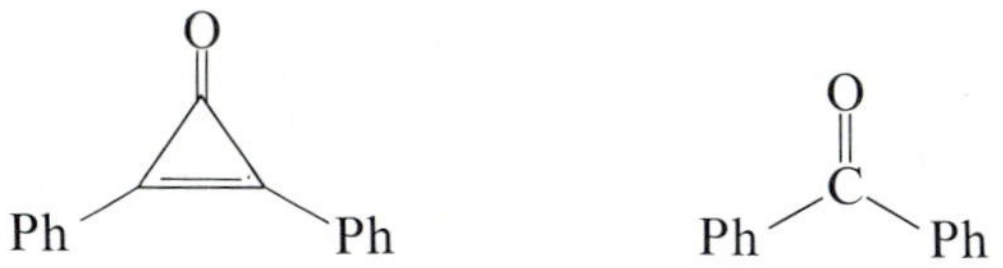

O Ph Ph

$\mu = 5.08$ D
diphenylcyclopropenone

O C Ph Ph

$\mu = 2.97$ D
benzophenone

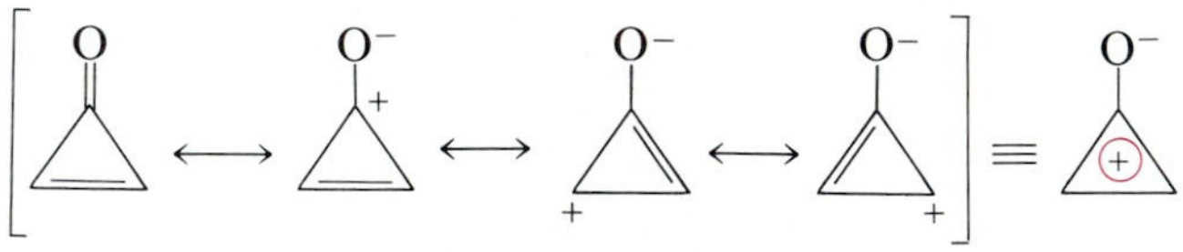

As with the structure of cyclooctatetraene dianion on page 655, this example illustrates the use of an inscribed circle to represent an aromatic cycle of $4n + 2$ electrons.

C. Six-Electron Systems

Cyclobutadiene has a cyclic π-system with four electrons and does not fit the $4n + 2$ rule; accordingly, we would expect it to have antiaromatic character. Cyclobutadiene is a known but very reactive hydrocarbon. It can be captured only at very low temperatures. Under most conditions it has but a fleeting existence and yields only dimeric products. This same reactivity is characteristic of various substituted derivatives.

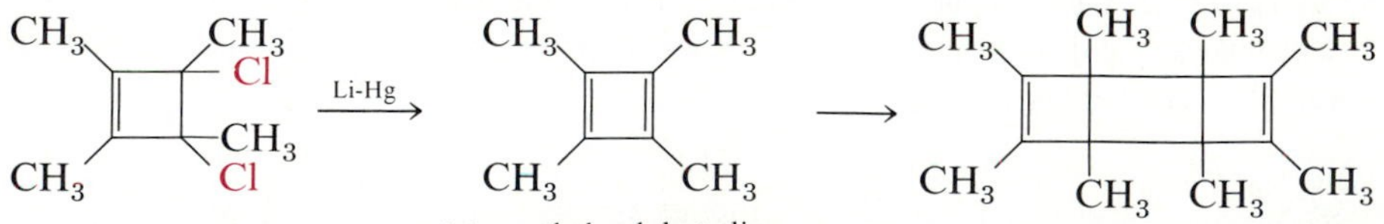

CH_3 CH_3 Cl CH_3 Cl CH_3 $\xrightarrow{\text{Li-Hg}}$ CH_3 CH_3 CH_3 CH_3 $\longrightarrow$ CH_3 CH_3 CH_3 CH_3 CH_3 CH_3 CH_3 CH_3

tetramethylcyclobutadiene

Cyclic π-systems with six electrons fit the $4n + 2$ rule, and the most important such cycle is, of course, benzene. Some other six-electron cycles are ions. Cyclopentadienyl anion is a rather stable carbanion whose conjugate acid, cyclopentadiene, is an unusually acidic hydrocarbon with a pK_a of 16. Notice that such a value is far lower than that of triphenylmethane ($pK_a = 31$). In fact, cyclopentadiene is comparable in acidity to water and the alcohols.

$$\text{cyclopentadiene} \underset{}{\overset{pK_a = 16}{\rightleftharpoons}} \text{cyclopentadienyl anion} + H^+$$

By contrast, cycloheptatriene is a nonacidic hydrocarbon; it appears to be less acidic than the open-chain heptatriene.

cycloheptatriene

$CH_3CH{=}CHCH{=}CHCH{=}CH_2$

heptatriene

The cycloheptatrienyl anion has seven equivalent resonance structures of the type shown and would be expected to have a well-distributed negative charge. But it also has an incomplete molecular orbital shell with its eight π-electrons, and this status confers anti-aromatic character (Figure 22.10).

cycloheptatrienyl anion

Figure 22.10 also shows that the situation is reversed for the corresponding cations. Cyclopentadienyl cation is highly reactive and difficult to prepare. It has only four π-electrons. On the other hand, cycloheptatrienyl cation has six π-electrons and is a remarkably stable carbocation. It is readily prepared by oxidation of cycloheptatriene and many of its salts are stable crystalline compounds.

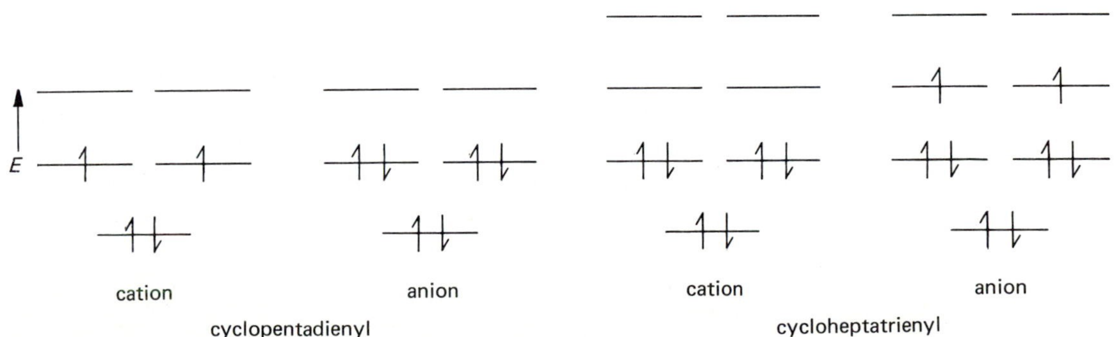

FIGURE 22.10 Molecular orbital energy levels for cyclopentadienyl and cycloheptatrienyl ions showing filled molecular orbital shells for six π-electrons.

cyclopentadienyl cation

cycloheptatrienyl cation

EXERCISE 22.4 Of the two ketones shown below, one is highly reactive and one is unusually stable. Which is which?

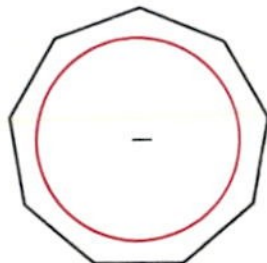

D. Ten-Electron Systems

The Hückel $4n + 2$ rule says that monocyclic π-electron systems with ten electrons will have filled π-molecular orbital shells with the highest occupied molecular orbital having quantum numbers of ± 2. We have encountered one such system in cyclooctatetraene dianion. A related system is cyclononatetraenyl anion. This anion is a known system and gives evidence of having a planar nonagon structure, despite the high angle strain in such a ring system.

cyclononatetraenyl anion

A neutral ten-π-electron hydrocarbon homologous to benzene would be cyclodecapentaene. The planar all-cis structure has highly strained bond angles. The alternative structure with two trans double bonds cannot achieve planarity because of interaction between the two interior hydrogens. As a result, cyclodecapentaene does not have the expected aromatic stability, but is instead a highly reactive hydrocarbon.

all-*cis* or all-(Z)-cyclodecapentaene

(Z,Z,E,Z,E)-cyclodecapentaene

An unusual hydrocarbon has been prepared in which the two interior hydrogens of cyclodecapentaene have been replaced by a bridging methylene group. This hydrocarbon cannot have a completely coplanar π-system, but enough cyclic overlap occurs to give the compound significant aromatic character.

bicyclo[4.4.1]undeca-1,3,5,7,9-pentaene

Highlight 22.3

A mnemonic for ordering the energy levels in cyclic [*n*]-annulenes can be drawn by inscribing the appropriate polygon in a circle with one vertex down. Each vertex corresponds to the energy level of a π-molecular orbital. The center of the circle corresponds to zero; electrons in levels below zero are **bonding** (energy has been given up to form the bond). Electrons in levels above zero are **antibonding.** Two stable ions are illustrated along with benzene.

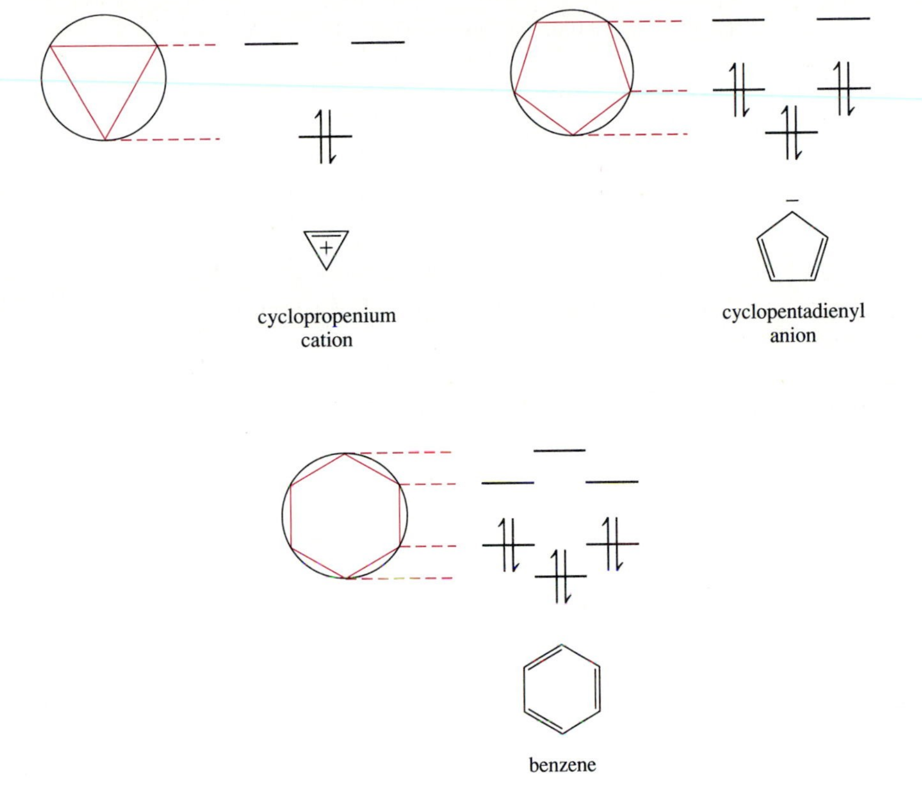

E. Larger Cyclic π-Systems

Cyclobutadiene, benzene, and cyclooctatetraene are the first three members of a family of monocyclic $(CH)_n$ compounds known as **annulenes.** Cyclobutadiene is [4]annulene, benzene is [6]annulene, and cyclooctatetraene is [8]annulene. Other fully conjugated cyclic polyenes are named in an analogous fashion.

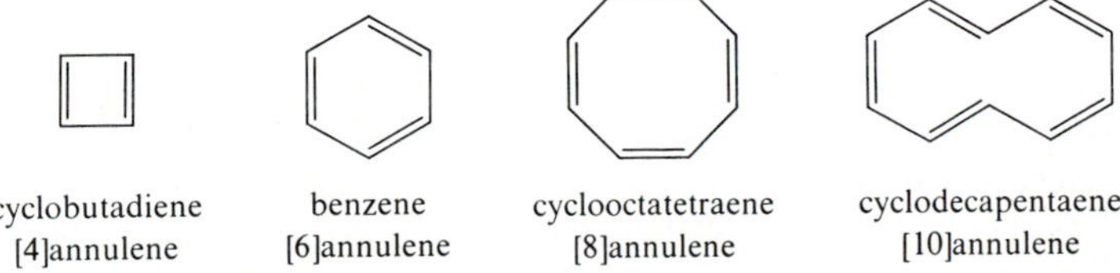

A number of larger annulenes are known that further confirm the generality of the $4n + 2$ rule. Cyclododecahexaene, [12]annulene, is a polyolefinic compound that reacts with alkali metals to give a dianion that, with 14 electrons, follows the $4n + 2$ rule. The interesting hydrocarbon depicted in Figure 22.11 has a 14-electron π-system and is a

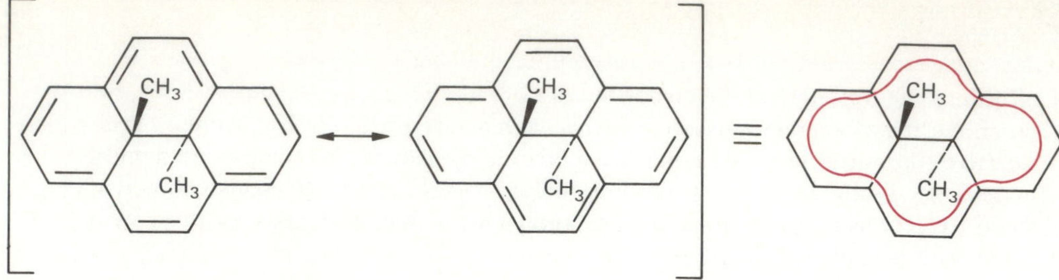

FIGURE 22.11 *trans*-10b,10c-Dimethyl-10b,10c-dihydropyrene.

stable molecule that has the properties of an aromatic system. The periphery indicated on the right of Figure 22.11 shows a [14]annulene that follows the $4n + 2$ rule.

Cyclohexadecaoctaene, [16]annulene, with 16 π-electrons, does not fit the $4n + 2$ rule. It has polyolefinic behavior, but reacts with alkali metals to form the aromatic cyclic dianion with 18 π-electrons.

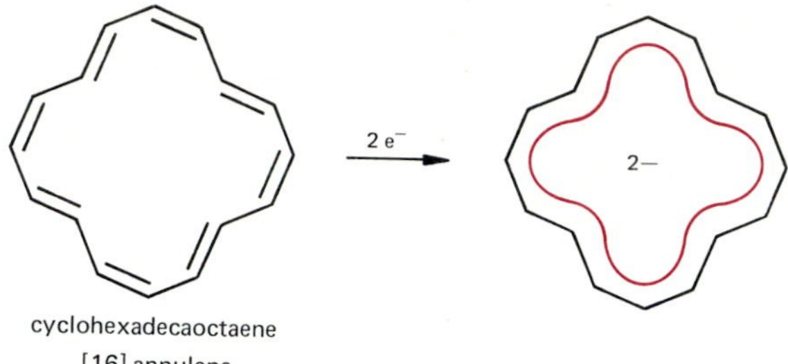

The corresponding neutral 18-π-electron hydrocarbon, cyclooctadecanonaene, [18]annulene, has been synthesized as a relatively stable brown-red compound. X-ray structure analysis suggests that the bonds have equal length.

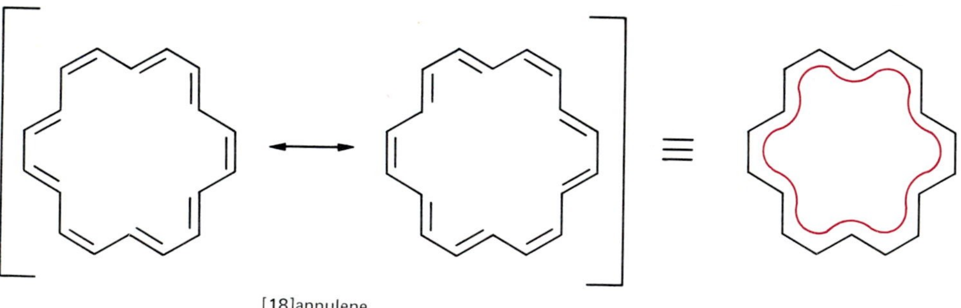

EXERCISE 22.5 Which of the following systems is expected to show aromaticity in the Hückel sense?

(a) cycloeicosadecaene
(b) cyclodoeicosaundecaene
(c) cyclooctatetraene radical cation (which is seen in the mass spectrum of cyclooctatetraene)
(d) cyclobutadiene dianion
(e) [26]annulene
(f) cycloundecapentaenyl cation.

22.4 Hückel Transition States

In the last chapter (Section 21.7) we discussed a number of reactions that have cyclic six-electron transition states with aromatic character. Examples are the Diels-Alder reaction and the Claisen rearrangement. We know that such transition states have Hückel aromaticity because they involve $4n + 2$ electrons in a cycle. We recognize that all of the orbital interactions around the ring may not be equal as they are in benzene. Especially the p_σ-types of overlap in the transition states may not be energetically equivalent to the p_π-overlaps. Such inequality of interactions around the ring has the effect of breaking the degeneracy of the levels with nodes. Nevertheless, although these levels are no longer degenerate, they are still relatively close in energy and still form MO pairs or *shells* as shown in Figure 22.12. These shells are still filled with $4n + 2$ electrons and the Hückel concept still applies.

Many pericyclic reactions involve six electrons, but the Hückel rule is also fulfilled with two, ten, etc., electrons. An important pericyclic reaction with two electrons is the rearrangement of carbocations (Section 10.6.C). Figure 22.13 shows the pericyclic nature of such a transition state. As in the related case of cyclopropenyl (Section 22.3.B) the cation with a total of two electrons has a filled shell and relative stability. The corresponding transition state for a carbanion involves four electrons and an unfilled shell (Figure 22.13c). Accordingly, carbocation rearrangements are common while carbanion rearrangements are not.

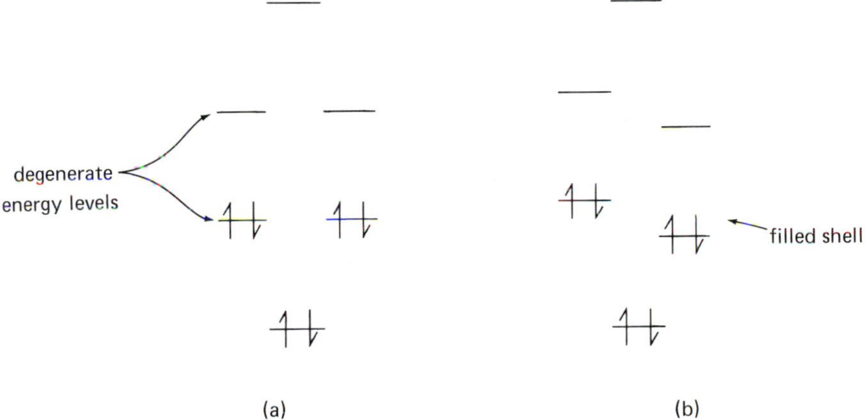

FIGURE 22.12 Comparison of MO energy levels for (a) benzene and (b) a six-electron cyclic transition state.

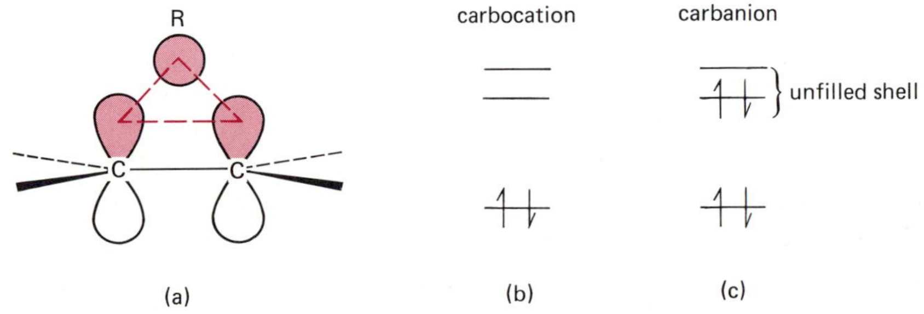

FIGURE 22.13 (a) A 1,2-rearrangement involves a three-center pericyclic transition state. (b) The carbocation case has two electrons and a stable filled shell MO pattern. (c) The carbanion case with four electrons and an unfilled shell is unstable.

In the same manner it now becomes clear why concerted [2 + 2] and [4 + 4] cycloaddition reactions (page 595) do not generally occur. The corresponding transition states involve four and eight electrons, respectively, and the electronic structures have unfilled shells. On the other hand, a number of cycloaddition reactions are known that involve 10 electrons. An example is the [8 + 2] cycloaddition reaction of heptafulvene with dimethyl acetylenedicarboxylate.

$COOCH_3$ $COOCH_3$ $COOCH_3$ $COOCH_3$

heptafulvene

dimethyl acetylenedicarboxylate

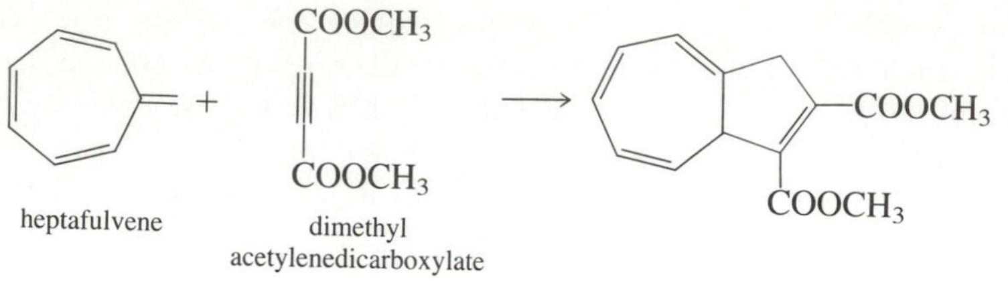

EXERCISE 22.6 (a) Sketch the expected MO energy level pattern expected for the pericyclic MO systems of the transitions states for a [2 + 2] and [4 + 4] cycloaddition and show the unfilled shells. (b) Sketch the transition state for the cycloaddition reaction of heptafulvene with dimethyl acetylenedicarboxylate and show that 10 electrons are involved in a pericyclic fashion.

22.5 Möbius Transition States

A. Electrocyclic Reactions

cis-1,3,5-Hexatriene undergoes a facile transformation on heating to give 1,3-cyclohexadiene. This type of isomerization is known as an **electrocyclic** reaction. The reaction can be perceived as proceeding through a cyclic six-membered transition state.

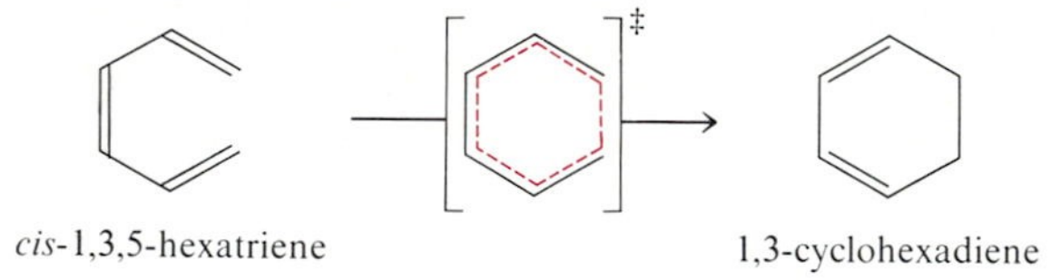

cis-1,3,5-hexatriene

1,3-cyclohexadiene

Moreover, if we follow the electrons involved by the conventional symbolism of a curved arrow for each pair of electrons, we find that three arrows are required. That is, six electrons participate in the transformation, a hint of the influence of the $4n + 2$ rule.

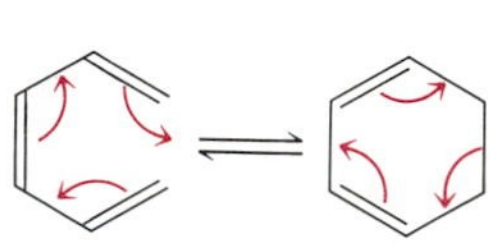

There is a complication, however, with regard to stereochemistry. In the open chain hexatriene, best π-overlap of the double bonds is achieved when all six carbons and eight hydrogens lie in the same plane—including both terminal CH_2 groups. In the cyclohexadiene, however, the H—C—H planes of the two methylene groups must be approximately perpendicular to the six-membered ring. That is, in the transformation from hexatriene to cyclohexadiene, the terminal methylene groups must rotate out of coplanarity.

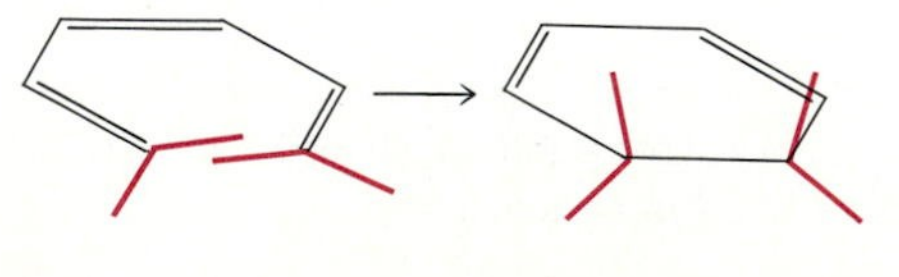

In principle, these rotations can take two possible modes. The two methylene groups can both rotate in the same sense when viewed from the same direction (**conrotatory motion**) or in the opposite sense (**disrotatory motion**).

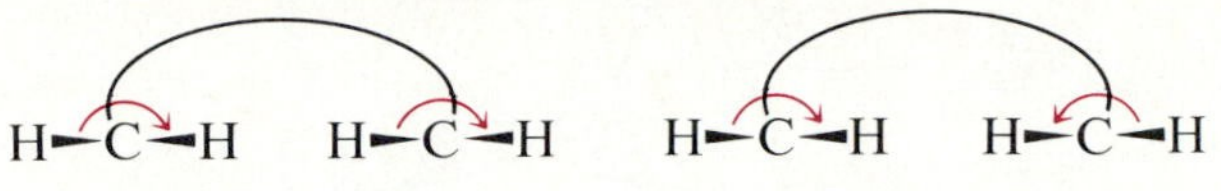

H—C—H H—C—H H—C—H H—C—H

conrotatory motion disrotatory motion

In the simple case of hexatriene itself, these alternative modes of rotation cannot be distinguished, but substituted compounds would lead to different isomers. Consider the (*E,Z,E*)-1,6-dimethyl compound as an example. If the end groups rotate in opposite directions (disrotation), the product is the *cis*-dimethylcyclohexadiene. If they rotate together (conrotation), the product is the *trans*-dimethylcyclohexadiene.

CH_3 H H CH_3 —disrotatory→ CH_3 H CH_3 H

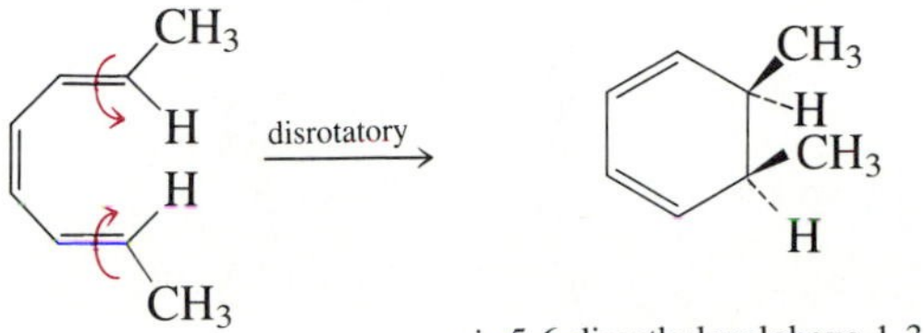

cis-5,6-dimethylcyclohexa-1,3-diene

CH_3 H H CH_3 —conrotatory→ CH_3 H H CH_3

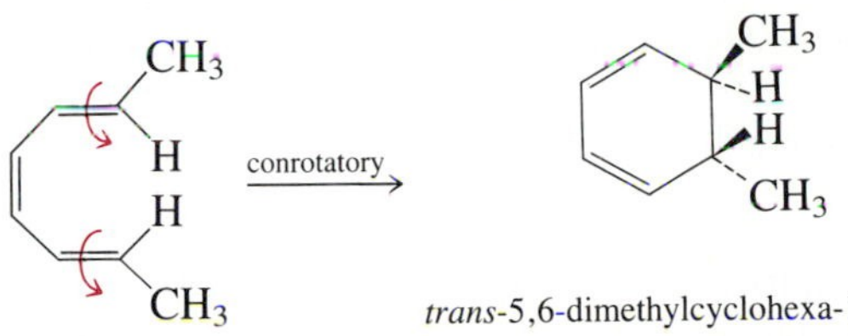

trans-5,6-dimethylcyclohexa-1,3-diene

The reaction involves the conversion of the two terminal *p*-orbitals from π-bonding to σ-bonding. At this point it is important to examine the signs of the orbital wave functions to determine whether the orbital overlaps involved are bonding (positive) or antibonding (negative) (Figure 22.14). In the starting hexatriene the *p*-orbitals have signs assigned to the wave functions to provide the most positive overlaps and lead to a set of π-molecular orbitals that describe the electronic structure. Note that for clarity only the overlap (dashed line) at the top is shown. Only the lowest occupied orbital of a set is illustrated. In disrotatory motion both terminal CH_2 groups rotate so that the positive lobes interact to give positive overlap throughout, exactly as in the related benzene system included for comparison in Figure 22.14. Even though the orbitals in the transition state for electrocyclic reaction are not aligned exactly as in benzene, the all-important overlap characteristics are the same in both; that is, the orbital overlaps are all positive around the ring in the orbital that is shown.

This result is to be contrasted with the pattern for conrotatory motion. In this case the positive lobe of one terminal *p*-orbital starts to overlap with the negative lobe of the other terminal *p*-orbital. The disrotatory transition state clearly more closely resembles orbital interactions in benzene and, indeed, the electrocyclic reactions of hexatrienes are always disrotatory. The product of the dimethyl case shown is exclusively the *cis*-dimethylcyclohexadiene, even though this product is thermodynamically less stable than the alternative trans structure.

The substantial difference in activation energies for the two cases is directly related to the stabilization energies of aromatic systems having cyclic π-systems with $4n + 2$ electrons. The transition state for disrotatory ring closure has a molecular orbital energy pattern as shown in Figure 22.15a and has a filled-shell electronic structure for six electrons.

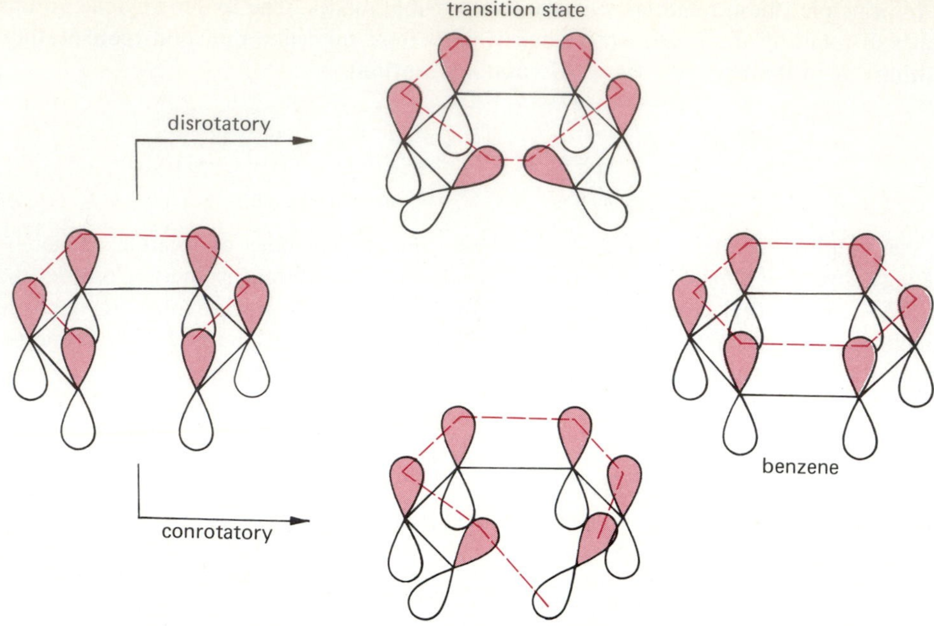

FIGURE 22.14 Benzene-like orbital interactions involved in disrotatory and conrotatory ring closure of 1,3,5-hexatriene for the lowest occupied orbital. Colored lobes represent positive wave functions, uncolored lobes represent negative wave functions.

The negative overlap required for the conrotatory ring closure gives rise to an entirely different pattern of molecular orbital energies. The negative overlap is equivalent to a node; hence, *there cannot be any molecular orbital with zero nodes*. Instead, we find a pair of molecular orbitals of similar energy with one node each, a higher pair with two nodes, and so on. This pattern is illustrated in Figure 22.15b. Six electrons leave the second shell unfilled, a condition that represents relative instability.

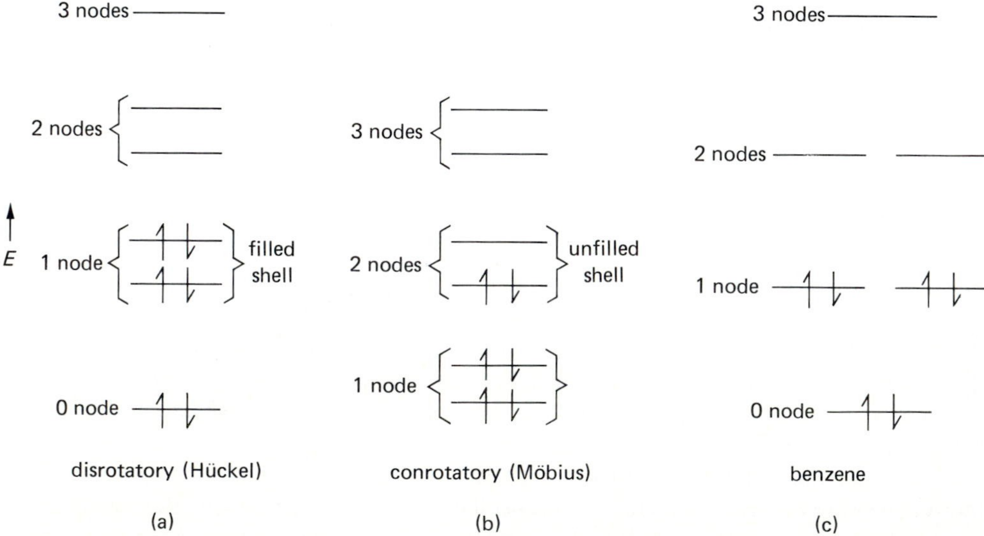

FIGURE 22.15 Energy level diagrams for alternative transition states for ring closure of hexatriene compared to that for benzene.

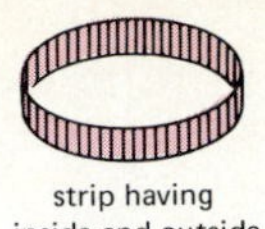

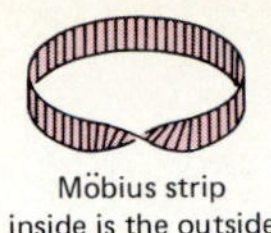

FIGURE 22.16 Illustrating the surfaces of a Möbius strip.

It is convenient to have names for these two possible patterns of molecular orbital levels. The pattern for disrotatory closure is a Hückel **molecular orbital system** and gives filled molecular orbital shells with $4n + 2$ electrons. The conrotatory pattern in Figure 22.15b is frequently referred to as a **Möbius molecular orbital system,** and has the important characteristic of giving filled molecular orbital shells with $4n$ electrons.

> This name derives from the topology of a *Möbius strip*. A Möbius strip is formed by taking a circular band, cutting in one place, giving one twist and rejoining the cut. The resulting strip has no inside or outside! Both are joined in one continuous manner (Figure 22.16). In a Hückel molecular orbital system, the *p*-orbitals are set up with a positive "top" and a negative "bottom." In the Möbius system, the negative overlap joins the "top" and the "bottom" in a manner that resembles the joining of the inside and outside of a Möbius strip.
>
> Note that this point also emphasizes the difference between setting up atomic orbitals, such as *p*-orbitals, to overlap in a given fashion (basis functions) and the set of molecular orbitals that results from such overlaps. If we start with n interacting atomic orbitals, we must end up with n molecular orbitals. The energies of the molecular orbitals depend on how the starting atomic orbitals overlap. We may summarize this discussion as follows: A set of *p*-orbitals overlapping in a cyclic manner with zero (or an even number of) negative overlap(s) gives rise to a Hückel pattern of molecular orbital energy levels to which quantum numbers can be assigned as 0, ± 1, ± 2, and so on. Cyclic interaction of a set of *p*-orbitals with one (or an odd number of) negative overlap(s) gives rise to a set of molecular orbitals having the Möbius pattern of energy levels to which quantum numbers can be assigned as ± 1, ± 2, and so on.

Let us now apply these principles to the corresponding electrocyclic ring closure of 1,3,5,7-octatetraenes to 1,3,5-cyclooctatrienes. In contrast to the (*E*,*Z*,*E*)-dimethylhexatriene case discussed previously, the (*E*,*Z*,*Z*,*E*)-dimethyloctatetraene compound shown gives, as the first product of thermal electrocyclic reaction, exclusively the *trans*-dimethylcyclooctatriene, the product of *conrotatory* motion!

CH_3 CH_3
(*E*,*Z*,*Z*,*E*)-deca-2,4,6,8-tetraene

disrotatory →
CH_3 H CH_3 H
cis-7,8-dimethylcycloocta-1,3,5-triene

conrotatory −10°C →
CH_3 H H CH_3
trans-7,8-dimethylcycloocta-1,3,5-triene
exclusive product

To see why this system changes so dramatically from the hexatriene case, we again look at the orbital overlaps involved (Figure 22.17). In disrotatory motion the overlaps involved are again all positive and give rise to a Hückel pattern of molecular orbital energy levels. But the eight electrons involved in this case do not fit the $4n + 2$ rule. The

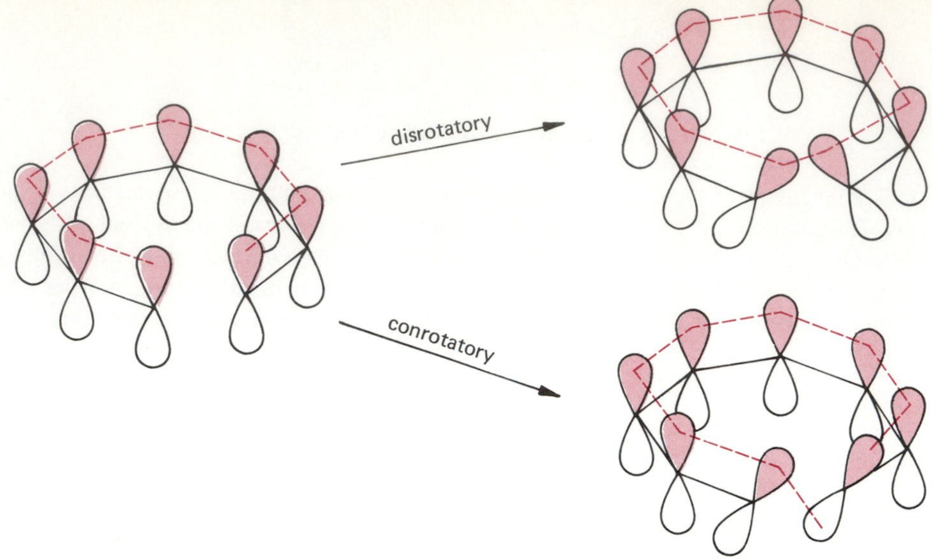

FIGURE 22.17 Orbital overlaps involved in cyclization of octatetraene.

result is the instability associated with an unfilled orbital shell. On the other hand, the conrotatory transition state gives rise to a Möbius pattern of molecular orbital levels. The eight electrons fill the first two shells and have the stability associated with filled orbital shells (Figure 22.18); that is, conrotatory ring closure of octatetraene involves a transition state that has Möbius aromatic character of the transition state for disrotatory ring closure.

This result may be generalized. *Those thermal electrocyclic reactions that involve 4n + 2 electrons react with disrotatory motion* so that the orbitals involved can overlap in the Hückel sense. *Those thermal electrocyclic reactions that involve 4n electrons react with conrotatory motion* so that the orbitals involved can overlap in the Möbius sense. These generalizations hold whether the reaction involved is that of ring closure or ring opening (principle of microscopic reversibility).

The thermal ring opening of cyclobutenes provides a further example. On heating, *cis*-3,4-dimethylcyclobutene is smoothly converted to (*E*,*Z*)-hexa-2,4-diene.

H_3C CH_3 H H $\xrightarrow{\Delta}$ CH_3 H CH_3 H

cis-3,4-dimethyl-cyclobutene

(*E*,*Z*)-hexa-2,4-diene

The reaction involves a four-electron cycle. The filled-shell molecular orbital system of the transition state thus requires Möbius overlap and conrotatory motion (Figure 22.19). Disrotatory ring opening would give a Hückel cyclic system, which, with four electrons, would be antiaromatic (Figure 22.19).

EXERCISE 22.7 Consider the conrotatory and disrotatory ring closures of butadiene to cyclobutene. Are the orbital interactions at the transition state the same as for ring opening?

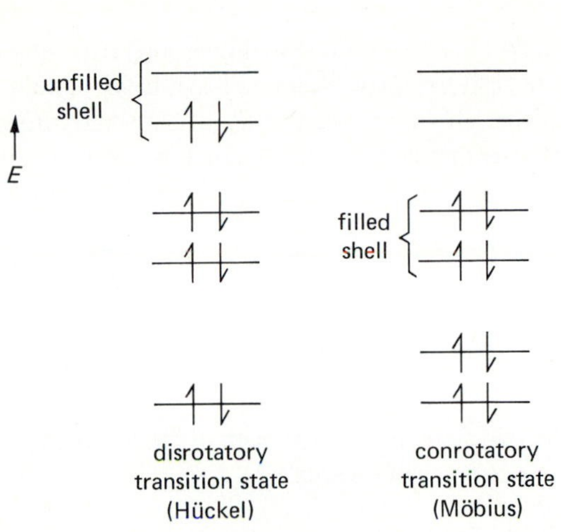

FIGURE 22.18 Energy level pattern of molecular orbitals for cyclization of octatetraene.

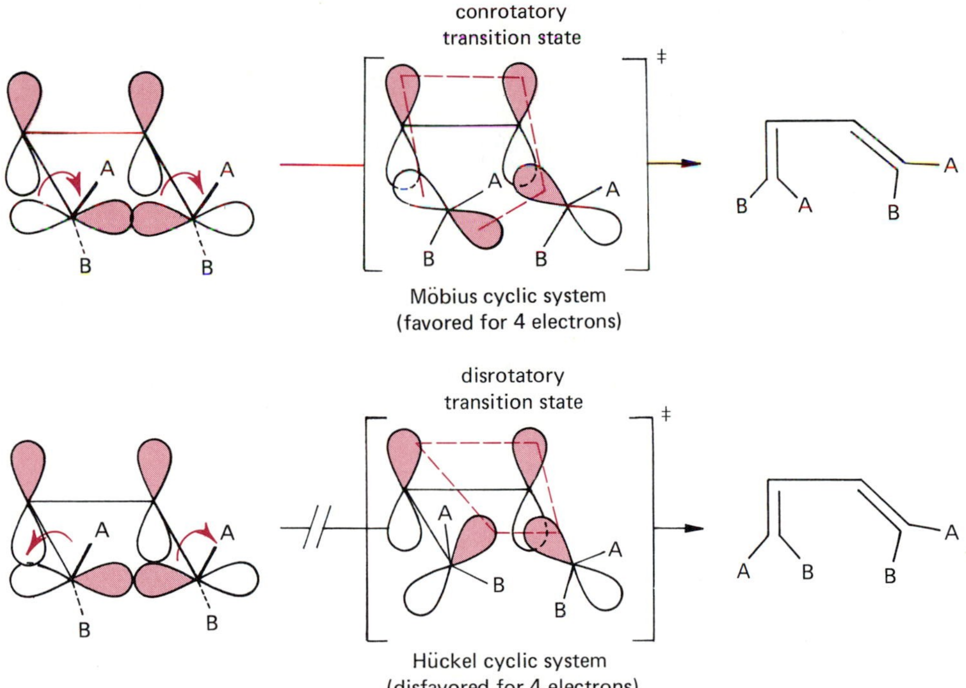

FIGURE 22.19 Orbital interactions for electrocyclic ring openings of cyclobutene.

The foregoing considerations lead to the following generalizations for thermal electrocyclic reactions.

$4n$ electrons (4, 8, 12, etc.)	conrotatory motion
$4n + 2$ electrons (2, 6, 10, 14, etc.)	disrotatory motion

The same generalizations were derived originally on the basis of symmetry properties of molecular orbitals and form part of the **Woodward-Hoffmann rules** for the stereochemistry of pericyclic reactions. We should also recognize that the changes in bond lengths and bond angles require energy, so that pericyclic reactions have an activation energy.

EXERCISE 22.8 The theoretically aromatic hydrocarbon [10]annulene has been the goal of numerous synthetic efforts. One attempted synthesis should have produced the Z,E,Z,Z,E stereoisomer. However, an isomer of 9,10-dihydronaphthalene was obtained instead. What was the stereochemistry of this product, cis or trans?

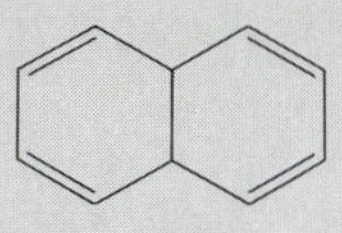

9,10-dihydronaphthalene

EXERCISE 22.9 Of the two stereoisomers of each of the following compounds, which is expected to be the more thermally stable?

(a) bicyclo[4.2.0]octa-2,4-diene
(b) bicyclo[4.2.0]oct-7-ene

EXERCISE 22.10 In each of the following thermal electrocyclic reactions, predict whether the product is cis or trans.

(a) (*E,E*)-hepta-2,5-dien-4-yl cation to 4,5-dimethylcyclopent-2-en-1-yl cation
(b) *cis*-bicyclo[5.2.0]nona-2,8-diene to bicyclo[4.3.0]nona-2,4-diene
(c) (*E,E,E*)-5-phenylnona-2,5,7-trien-4-yl anion to 1-phenyl-4,5-dimethylcyclohepta-2,6-dien-1-yl anion

B. Cycloaddition Reactions

Many examples are known of [$p + q$] cycloaddition reactions in which the number of pericyclic electrons equals 6, 10, 14, etc. These reactions are readily understood in terms of Hückel aromatic transition states. In this context, a remarkable exception would appear to be reaction of heptafulvalene with tetracyanoethylene, a [14 + 2] cycloaddition that does not fit the Hückel rule.

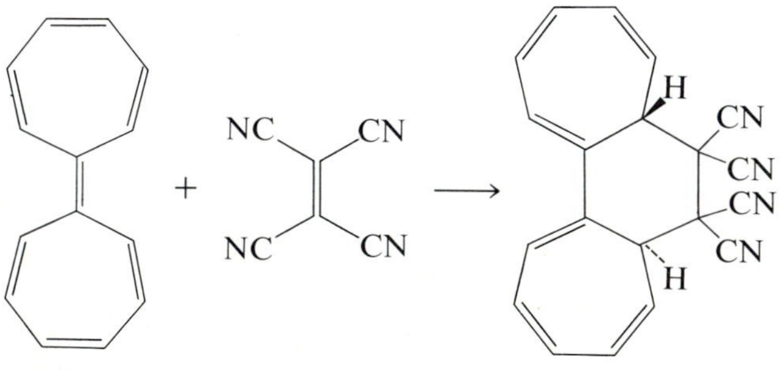

NC CN
NC CN
H
CN
CN
CN
CN
H

heptafulvalene tetracyanoethylene

However, note that the product is the result of anti addition. The corresponding transition state (Figure 22.20) involves a negative overlap that would correspond to a Möbius cyclic electronic system, a favorable transition state for a 16-electron cyclic system!

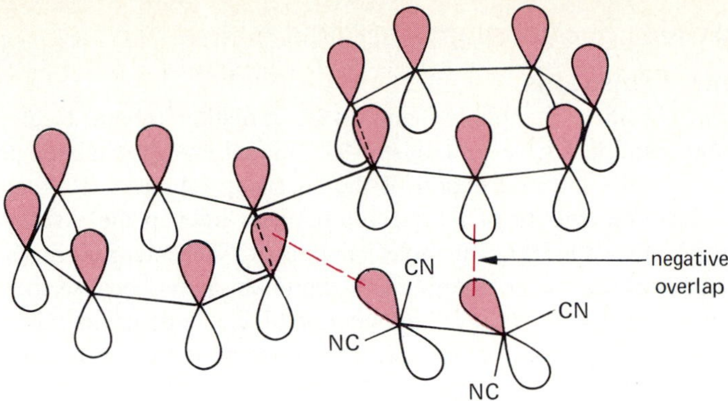

FIGURE 22.20 Orbital interactions for anti addition of tetracyanoethylene to heptafulvalene.

Addition to the same side of a π-system is called **suprafacial** and is symbolized with a subscript s; addition to opposite sides of a π-system is called **antarafacial** and is symbolized with a subscript a. Hence, the normal Diels-Alder reaction is an example of a $({}_4\pi_s + {}_2\pi_s)$ cycloaddition. The reaction of heptafulvalene with tetracyanoethylene is an example of a $({}_{14}\pi_a + {}_2\pi_s)$ cycloaddition. In general, $({}_p\pi_s + {}_q\pi_s)$ cycloadditions are thermally "allowed" when $p + q = 4n + 2$, whereas $({}_p\pi_a + {}_q\pi_s)$ thermal cycloadditions, a somewhat rarer breed, are thermally allowed when $p + q = 4n$. These generalizations also constitute part of the Woodward-Hoffmann rules for pericyclic reactions (page 668).

EXERCISE 22.11 For each of the following cycloaddition reactions determine whether the reaction is thermally allowed for the stereochemistry shown.

(a) 5,5-dimethylcyclopentadiene (CH_3, CH_3) + $^{-}CH_2$–$C(C_6H_5)$=CH_2 ⟶ product (CH_3, CH_3; $^{-}$, C_6H_5)

(b) (N) + MeOOCC≡CCOOMe ⟶ product (N, H, H, MeOOC, COOMe)

(c) + $C(CN)_2$=$C(CN)_2$ ⟶ product (CN, CN, CN, CN)

(d) + EtOOCC≡CCOOEt ⟶ product (EtOOC, COOEt, H, H)

22.6 Ultraviolet Spectroscopy

A. Electronic Transitions

A molecule can absorb a quantum of microwave radiation (about 1 cal $mole^{-1}$) and change from one rotational state to another. Vibrational energy changes are associated with light quanta in the infrared region of the spectrum (about 3–10 kcal $mole^{-1}$). A change in the electronic energy of a molecule requires light in the visible (40–70 kcal $mole^{-1}$) or ultraviolet (70–300 kcal $mole^{-1}$) regions. The energies required for such **electronic transitions** are of the magnitude of bond strengths because the electrons involved are valence electrons. That is, the energy of light quanta in this region of the electromagnetic spectrum is sufficient to **excite** an electron from a bonding to an antibonding state.

The resulting **excited electronic states,** in contrast to the **ground electronic state,** are often difficult to describe by resonance symbolism. In simple diatomic molecules an excited state can sometimes be described rather simply. For example, one excited state of LiF can be described by the process

$$\underset{\text{ground state}}{Li^+F^-} \xrightarrow{h\nu} \underset{\text{excited state}}{Li\cdot F\cdot}$$

Absorption of a photon is accompanied by a shift in electron density from fluorine to lithium, and the resulting excited state resembles two atoms held in close proximity. We shall see that such **charge-transfer transitions** are quite significant in organic molecules.

The excited states of polyatomic molecules are not usually described so simply. Fortunately, molecular orbital concepts can often be applied in a relatively simple and straightforward way. For example, the electronic transition of methane involves the excitation of an electron from a bonding molecular orbital, σ, to the corresponding antibonding molecular orbital, σ^*, as illustrated in Figure 22.21.

Recall that the bonding molecular orbital between two atoms is formed by the positive overlap of two hybrid orbitals and is symbolized as in Figure 22.22. The corresponding antibonding molecular orbital, σ^*, is produced by the negative overlap of the hybrid orbitals. This negative overlap produces an additional node between the nuclei and reduces the electron density that is so essential for covalent bonding. In the excited state, the electron in σ^* partially cancels the bonding provided by the remaining electron in σ; hence, the energy required for excitation is of the order of magnitude of bond strengths.

The bonding molecular orbitals of methane, and of alkanes generally, are relatively low in energy. Excitation of an electron requires light of high energy with a wavelength about 150 nm (1500 Å) or less. Light in this region is strongly absorbed by the oxygen in air, and spectroscopic measurements of such compounds require special instruments in which air is completely excluded. This region of the light spectrum is called the vacuum ultraviolet and is unimportant in routine organic laboratory studies.

Wavelengths of light above about 200 nm are not absorbed by air, and it is this region that is most important for organic chemists. The range of about 200–400 nm is called the ultraviolet; the visible region of the spectrum ranges from wavelengths of about 400 nm (violet light) to about 750 nm (red light). The energy of such light is insufficient to affect most σ -bonds, but it is in the range of π-electron energies, especially for conjugated systems. That is, ultraviolet-visible spectroscopy is an important spectroscopic tool for the study of conjugated multiple bonds. The π-molecular orbitals of such conjugated systems extend over several atoms. The highest occupied or least bonding of such molecular orbitals already have at least one node. Electronic excitation generally involves the transition of an electron to a molecular orbital having an additional node, and, as a general rule, the more nodes an electron has in a wave function, the less energy it takes to add another node.

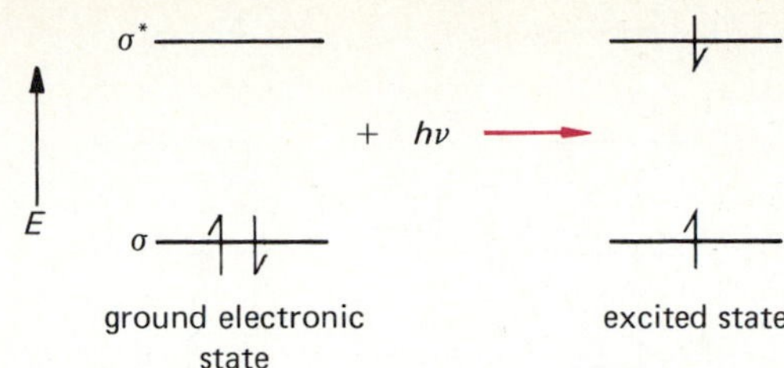

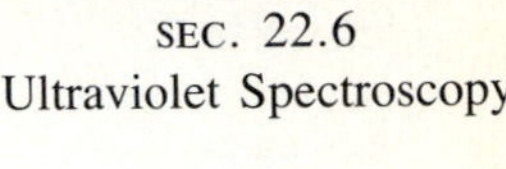

FIGURE 22.21 Ground and excited states.

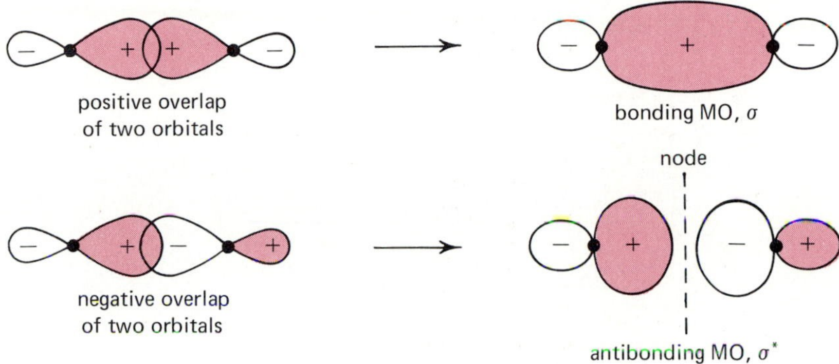

FIGURE 22.22 Bonding and antibonding molecular orbitals.

B. $\pi \rightarrow \pi^*$ Transitions

Absorption of light that produces excitation of an electron from a bonding π- to an antibonding π^*-molecular orbital is referred to as a $\pi \rightarrow \pi^*$ transition. For example, 1,3-butadiene has an intense absorption band at 217 nm (usually written as λ_{max} 217 nm; that is, lambda max = 217 nanometers) that results from the excitation of an electron from π_2 to π_3 (Figure 22.23). Recall that π_2 has one node and π_3 has two (Figure 22.5). 1,3,5-Hexatriene absorbs at longer wavelength, $\lambda_{max} = 258$ nm. It takes less energy to excite an electron from π_3, the highest occupied π-molecular orbital of hexatriene, which has two nodes, to π_4, which has three nodes. *The longer the chain of conjugation, the longer the wavelength of the absorption band.* For example, the lowest energy $\pi \rightarrow \pi^*$ transition of 1,3,5,7-octatetraene occurs at the still longer wavelength of 304 nm, whereas ethylene itself absorbs in the vacuum ultraviolet at 175 nm.

> Compounds generally have many excited electronic states, but organic chemists are mostly concerned with the lowest or more stable states, since these are the states that are accessible with the energies of ultraviolet and visible light. Many of these states can be described in terms of electron transitions that involve other than just the highest occupied and lowest vacant molecular orbitals. For example, other electronic states of butadiene arise from the electronic transition $\pi_1 \rightarrow \pi_3$ or $\pi_1 \rightarrow \pi_4$, but such transitions occur in the vacuum ultraviolet. However, in other compounds several absorption bands can occur close together. An example is benzene, discussed in Section 22.6.F.

The highly conjugated hydrocarbon *trans*-β-carotene, with eleven double bonds in conjugation, has two intense long-wavelength absorptions in alkane solution at 483 nm and 453 nm. These absorptions are in the visible region of the spectrum and correspond to blue to blue-green light. Since light of this color is absorbed by the compound, β-carotene appears yellow to orange in solution. For further discussion of color and colored compounds, see Section 36.2.

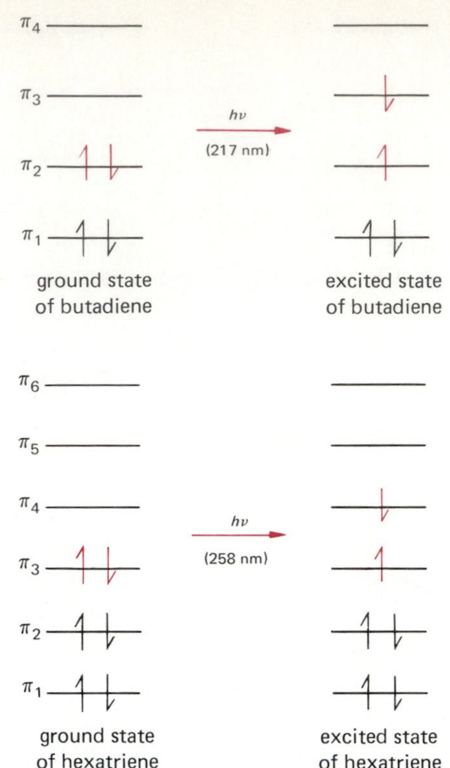

FIGURE 22.23 Electronic excitation of butadiene, $CH_2{=}CH{-}CH{=}CH_2$, and 1,3,5-hexatriene, $CH_2{=}CH{-}CH{=}CH{-}CH{=}CH_2$.

cis-β-Carotene has two absorption peaks at essentially the same wavelengths but with weaker intensities. This result is quite general; π-systems that are prevented from achieving coplanarity show significant changes from coplanar analogs, particularly in absorption intensities. In *cis*-β-carotene the two groups on the same side of the double bond sterically interfere with each other.

trans-β-carotene

H H

cis-β-carotene

α,β-Unsaturated aldehydes and ketones also have high-intensity absorptions resulting from the transition of an electron from π_2 to π_3. These $\pi \rightarrow \pi^*$ transitions occur at almost exactly the same wavelength as those for the corresponding dienes.

$CH_2{=}CHCH{=}CH_2$ λ_{max} 217 nm $CH_2{=}CHCHO$ λ_{max} 218 nm

As in the case of polyenes, the wavelength of the light absorbed by unsaturated carbonyl compounds increases as the chain of conjugation increases. The effect is illustrated in Table 22.1.

TABLE 22.1 Spectra of Some Polyene Aldehydes

Aldehyde	λ_{max}, nm
$CH_3CH{=}CHCHO$	220
$CH_3CH{=}CHCH{=}CHCHO$	270
$CH_3(CH{=}CH)_3CHO$	312
$CH_3(CH{=}CH)_4CHO$	343
$CH_3(CH{=}CH)_5CHO$	370
$CH_3(CH{=}CH)_6CHO$	393
$CH_3(CH{=}CH)_7CHO$	415

EXERCISE 22.12 Using MO energy level diagrams like those of Figure 22.2, indicate the electronic transitions for the longest wavelength bands of allyl and pentadienyl anions. Which ion is more apt to absorb in the visible region of the spectrum?

C. $n \rightarrow \pi^*$ Transitions

Carbonyl groups have another characteristic absorption that is associated with the lone-pair electrons on oxygen. Since these electrons are bound to only a single atom, they are not held as tightly as σ-electrons, and they can also be excited to π^*-molecular orbitals. The process results in a so-called $n \rightarrow \pi^*$ transition and usually occurs at relatively long wavelength.

	$CH_3C(=O)CH_3$	$CH_3C(=O)CH{=}CH_2$
$n \rightarrow \pi^*$:	λ_{max} 270 nm	λ_{max} 324 nm
$\pi \rightarrow \pi^*$:	187 nm	219 nm

The π-system of methyl vinyl ketone is more extended than that of acetone and less energy is required for the excitation in the former case. This difference is illustrated in Figure 22.24. Because of the more extensive π-system of conjugated double bonds of methyl vinyl ketone compared to acetone, both the $n \rightarrow \pi^*$ and $\pi \rightarrow \pi^*$ transitions of methyl vinyl ketone occur at longer wavelength (lower energy).

One important distinguishing characteristic of $n \rightarrow \pi^*$ transitions results from the critical feature that the lone-pair electrons tend to be concentrated in a different region of space from the π-electrons (Figure 22.25). Although $n \rightarrow \pi^*$ transitions often occur at lower energy (longer wavelength) than $\pi \rightarrow \pi^*$ transitions, they are less probable. A given quantum of $n \rightarrow \pi^*$ light must encounter many more molecules before it is absorbed than is the case for $\pi \rightarrow \pi^*$ light quanta. This difference shows up experimentally in an absorption spectrum as an intensity difference; absorptions due to $\pi \rightarrow \pi^*$ electronic transitions are generally much more intense (''strong absorption'') than those due to $n \rightarrow \pi^*$ transitions (''weak absorption''). The difference in intensity is two to three orders of magnitude.

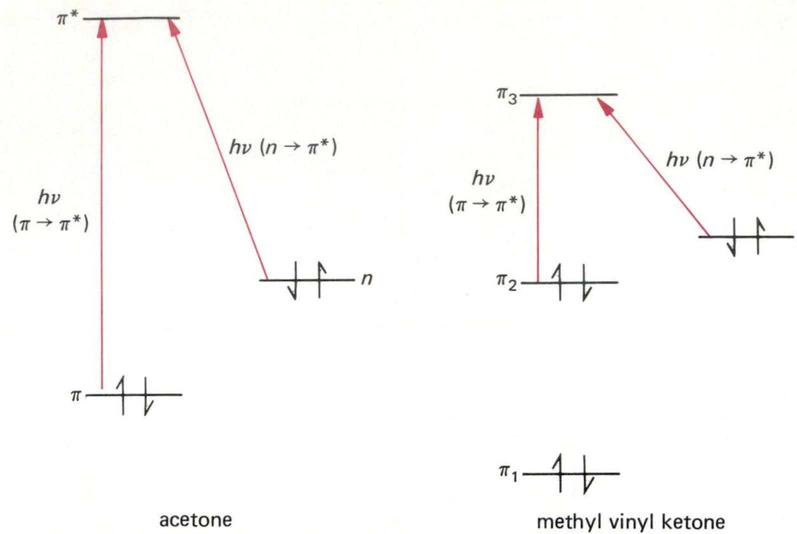

FIGURE 22.24 Illustrating $\pi \rightarrow \pi^*$ and $n \rightarrow \pi^*$ transitions in two ketones.

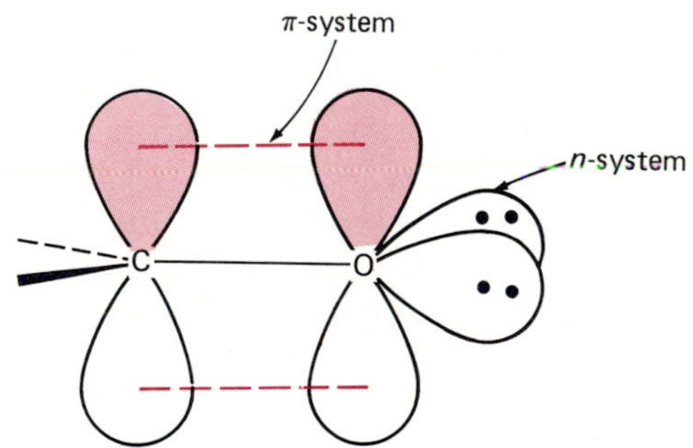

FIGURE 22.25 Lone pairs (*n*-system) and π-system of a carbonyl group.

The intensity is expressed as an **extinction coefficient** ϵ. The amount of light absorbed depends on the extinction coefficient and the number of molecules in the light path. The number of molecules depends on the concentration of the solution and the path length of the absorption cell. The amount of light that passes through a solution (transmittance) is given by Beer's law

$$\log \frac{I_0}{I} = \epsilon c d$$

where I_0 is the intensity of the light before it encounters the cell, I is the intensity of the light emerging from the cell, c is the concentration in moles per liter, and d is the path length in centimeters.

As an example, the spectrum of two concentrations of mesityl oxide, $(CH_3)_2C{=}CHCOCH_3$, in the same 1-cm cell (a common path length) is shown in Figure 22.26. A highly dilute solution is used for the $\pi \rightarrow \pi^*$ absorption at 235 nm. The extinction coefficient for this transition is calculated as

$$\epsilon = \frac{\log I_0/I}{cd} = \frac{1.18}{(9.37 \times 10^{-5})(1)} = 12{,}600 \text{ L mole}^{-1} \text{ cm}^{-1}$$

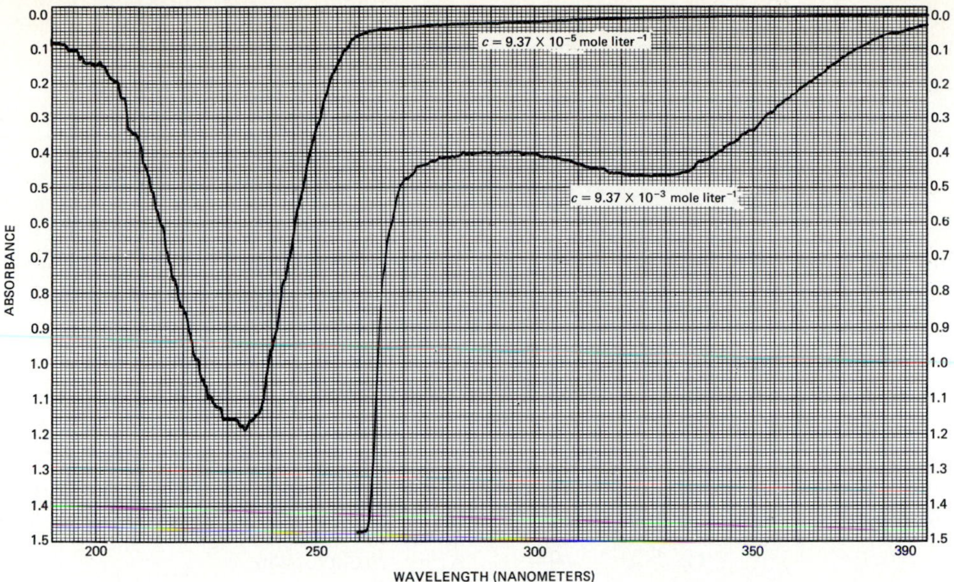

FIGURE 22.26 Ultraviolet absorption spectra of mesityl oxide, $(CH_3)_2C{=}CHCOCH_3$.

In this dilute solution the absorption due to the $n \rightarrow \pi^*$ transition is so weak it is barely discernible. A more concentrated solution gives greater absorption and, from the second curve in Figure 22.26, we may calculate ϵ for this transition at 326 nm to be

$$\epsilon = \frac{\log I_0/I}{cd} = \frac{0.47}{(9.37 \times 10^{-3})(1)} = 50$$

(Note that the units of ϵ are usually omitted.)

This concentration is so high, however, that the $\pi \rightarrow \pi^*$ transition absorbs light essentially completely at its wavelength. The ratio of the two extinction coefficients, 12,600/50 = 252, is typical for unsaturated carbonyl compounds. In general the $\pi \rightarrow \pi^*$ transitions have ϵ of about 10^4, whereas ϵ for $n \rightarrow \pi^*$ transitions are about 10–100.

EXERCISE 22.13 The ultraviolet spectrum of a solution of 0.00731 g of crotonic acid, $CH_3CH{=}CHCOOH$, in 10 mL of methanol was measured in a 1-cm cell. Although the $\pi \rightarrow \pi^*$ transition was off scale, the $n \rightarrow \pi^*$ transition at 250 nm showed an absorbance of 0.77. A 1-mL portion of this solution was diluted to 100 mL and the spectrum was recorded again. The $\pi \rightarrow \pi^*$ transition was seen clearly at 200 nm with an absorbance of 0.86. Calculate the extinction coefficients for the two transitions.

D. Charge-Transfer Transitions

Iodine is a dark violet solid, forms a violet vapor, and yields violet solutions in nonpolar solvents such as *n*-hexane or CCl_4. The solutions in ethanol, ether, and benzene are brown, but no chemical reaction has occurred. The spectrum of a solution of benzene and iodine in hexane (λ_{max} 540 nm) still shows absorptions due to benzene (Section 22.6.F) and iodine, along with a new absorption associated with a weak complex of benzene and iodine. The fairly strong ultraviolet absorption maximum of the complex is at 292 nm (ϵ 16,000). The brown color is explained by the absorption of the iodine component of the complex (λ_{max} 500 nm) which absorbs at shorter wavelengths than pure iodine. Robert S. Mulliken, an American chemical physicist who received the Nobel prize for molecular orbital theory, explained that absorption of light caused the complex to undergo a **charge-transfer transition,** described by the following sequence.

$$\text{C}_6\text{H}_6 + I_2 \rightleftharpoons \text{C}_6\text{H}_6\cdots I_2 \xrightarrow{h\nu} \text{C}_6\text{H}_6^{+}\cdots I_2^{-}$$

ground state   excited state

benzene-iodine complex

In the benzene-iodine complex, the benzene is an **electron donor** and the iodine is an **electron acceptor.** The complex is stabilized by a resonance contribution from a form in which an electron has been transferred from the benzene to the iodine. Light produces an excited state in which the primary structure is $[\text{benzene}]^+ [I_2]^-$, with a resonance contribution from the neutral form. The orbital levels involved in the transition are the π_3-level of the arene ring and the σ^*-orbital of the iodine (Figure 22.27). The new ultraviolet absorption is due to the charge-transfer transition. *A charge-transfer complex (**DA**) forms between an electron donor (**D**) and an electron acceptor (**A**) and exhibits a special light absorption not present in the spectrum of either D or A.*

$$[D\cdot^{+} A\cdot^{-} \longleftrightarrow D : A] \qquad \text{excited state}$$

$$[D : A \longleftrightarrow D\cdot^{+} A\cdot^{-}] \qquad \text{ground state}$$

major resonance form   *minor* resonance form

Electron-supplying substituents on the ring stabilize the cation radical in the excited state; mesitylene-iodine complex absorbs at 332 nm, a change of 11.8 kcal mole^{-1} in transition energy. The change per methyl group (3.9 kcal mole^{-1}) is comparable to that found for the effect of methyl groups on $\pi \rightarrow \pi^*$ electronic transitions (Section 22.6.E). There is a linear relationship between the energy needed for the charge-transfer transition and the ionization potential of the arene. Recall that the ionization potential is the energy required to remove an electron from a molecule. A similar relationship can be found for the charge-transfer transitions of different acceptors with a particular donor. The reduction potential is used as measure of the electron affinity of an acceptor. Additional examples of charge-transfer light absorptions are given in Chapter 30.

Another important type of charge-transfer complex is composed of a donor anion and an acceptor cation. The ground state complex is an ion pair that can interact with solvents in a manner dependent on their polarity. The Franck-Condon principle (light absorption occurs faster than nuclear motion) leads to large changes in the position of the charge-transfer absorption. A simplified explanation takes into account (a) that the charged ground state organizes the solvent molecules and (b) that the light pays the energy costs of the excitation. These include (1) payment for the initial solvent stabilization energy, (2) payment for the charge-transfer transition, and (3) payment for organizing the solvent in the absence of charges as if the charges were there. The last is effectively a destabilization of the system.

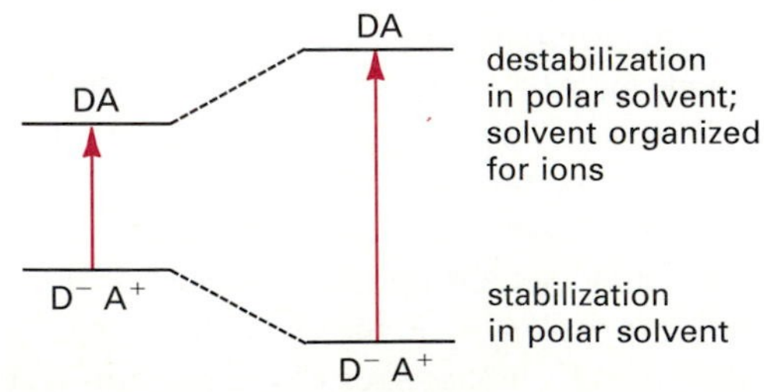

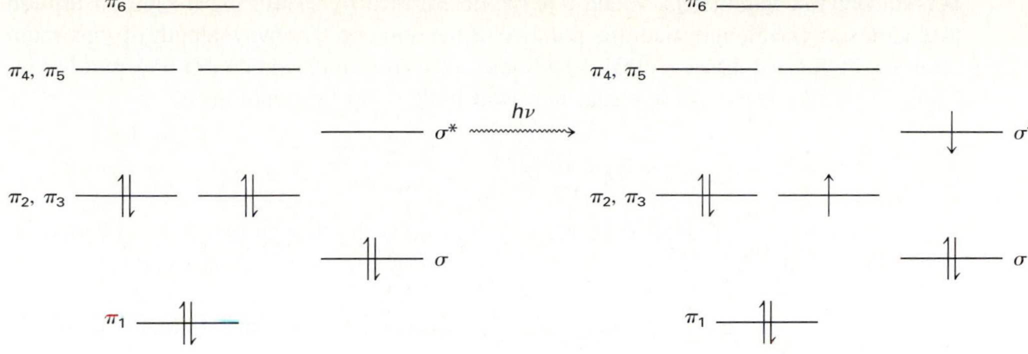

FIGURE 22.27 Orbitals involved in the benzene-iodine charge-transfer transition.

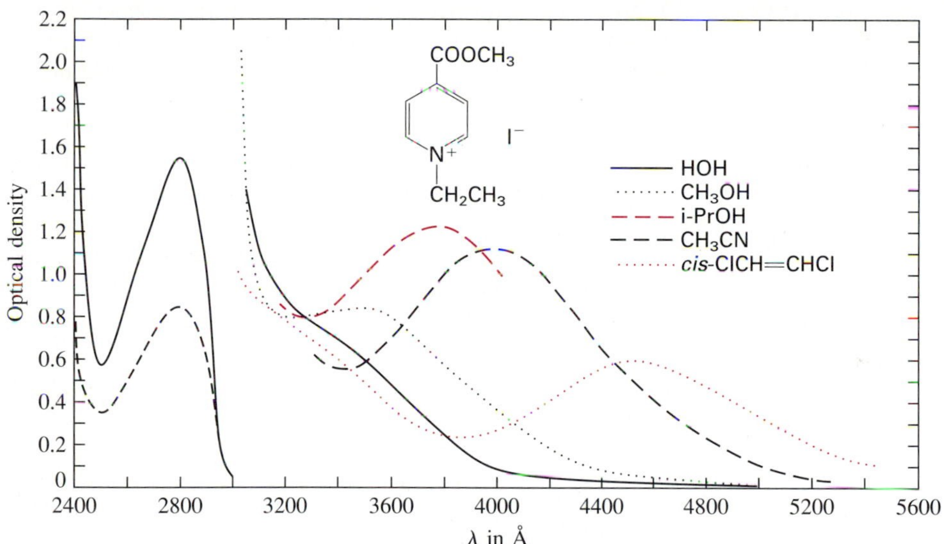

FIGURE 22.28 The variation in the position of the charge transfer absorption band for 1-ethyl-4-methoxycarbonylpyridinium iodide in solvents of different polarity.

The changes found for 1-ethyl-4-methoxycarbonylpyridinium iodide are so large (Figure 22.28) that they have been put to use as **empirical parameters of solvent polarity.**

O OCH3 C N+ CH2CH3 I⁻ —hν→ O OCH3 C N CH2CH3 I·

ground state excited state

Solvent polarity parameters, Z and E_T, are determined by measuring a light absorption that varies in wavelength with the polarity of the solvent. The wavelength of maximum light absorption for either 1-ethyl-4-carbomethoxypyridinium iodide (Z) or a pyridinium phenolate betaine (E_T) is converted into kcal mole^{-1} by the relationship

$$Z\ (\text{kcal mole}^{-1}) = -\frac{28590}{\lambda_{max}(\text{nm})}$$

Some Z- and the closely related E_T-values are given with other properties for a variety of useful solvents in Table 22.2. The great advantages of a spectroscopic method for estimating solvent polarity are speed, wide applicability (can usually be measured for a wide variety of pure and mixed solvents), and cost-effectiveness. A visual estimate of solvent polarity can often be made in moments just from the color of a solution.

Highlight 22.4

Solvent molecules interact with reactant molecules through hydrogen bonds, electrostatic interactions, and van der Waals interactions. The sum of these interactions is called solvent polarity. Solvent polarity is given by either a macroscopic parameter, D, the dielectric constant, or the microscopic parameters, Z or E_T. All can be measured experimentally.

TABLE 22.2 Empirical Polarity Parameters and Other Properties of Solvents

No.	Solvent	m.p., °C	b.p., °C	d[a]	visc[b] (η), cp	d.m.[c] (μ), D	diel[d] (D)	r.i.[e] (n_D^{20})	$E_T(30)$, kcal mole^{-1}	Z kcal mole^{-1}
1	Water	0.0	100.0	1.000^{4}	1.00	1.8	78.5	1.3330	63.1	94.6
2	Glycerol	18.2	290	1.26	1412	—	42.5	1.4746	57.0	86.5
3	Formamide	2.5	210	1.13	3.76	3.4	109.5	1.4475	56.6	83.3
4	1,2-Ethanediol	−13.0	198	1.11	26.1^{15}	2.0	37.7	1.4318	56.3	85.1
5	Methanol	−97.8	65	0.791	0.551	2.87	32.6	1.3286	55.5	83.6
6	N-Methylformamide	−3.8	180.5	1.01^{15}	1.99^{15}	3.8	182.4	1.4310	54.1	—
7	Ethanol	−117	78.3	0.789	1.08^{25}	1.7	24.3	1.3611	51.9	79.6
8	Acetic Acid	16.6	118.5	1.05	1.31	1–1.5	6.2	1.3721	51.2	79.2
9	Acetonitrile	−43.8	81.6	0.782	0.375^{15}	3.5	37.5	1.3442	46.0	71.3
10	Dimethylformamide (DMF)	−61	153	0.949	0.924	3.8	36.7	1.4269^{25}	43.8	68.4
11	Acetone	−95.3	56.2	0.790	0.337^{15}	2.7	20.7	1.3588	42.2	65.5
12	Chloroform	−63.5	61.2	1.50^{15}	0.596^{15}	1.1	4.7	1.4433^{25}	39.1	63.2
13	Ethyl acetate	−83.6	77.1	0.900	0.473^{15}	1.85	6.0	1.3724	38.1	—
14	Tetrahydrofuran	−65	65.4	0.889	0.55	1.7	7.4	1.4050	37.4	—
15	Diethyl ether	−116	34.6	0.713	0.247^{15}	1.25	4.2	1.3526	34.6	—
16	Benzene	5.5	80.1	0.879	0.649	0.0	2.3	1.5011	34.5	—
17	n-Hexane	−95	68.8	0.659^{20}	0.313	0.0	1.9	1.3749	30.9	—

[a] g cm^{-3} at 20 °C, except as noted.
[b] Viscosity in centipoises (cp) at 20° C, unless otherwise indicated by superscript.
[c] Dipole moment in Debyes.
[d] Dielectric constant at 25 °C.
[e] Refractive index.

Further discussion of charge-transfer complexes is beyond the scope of this book. Note that charge-separation lies at the heart of photosynthesis, and that excitation to a complex occurs at much lower energies than simple ionization. The idea has been used to create the record of the image in certain types of photocopying machines. An electric field promotes the necessary further separation of positive and negative charge after the initial photoexcitation.

E. Alkyl Substituents

We saw in Section 22.6.B that 1,3-butadiene has a $\pi \rightarrow \pi^*$ transition at 217 nm. Alkyl-substituted butadienes have the same π-system, but their absorption spectra vary significantly.

	λ_{max}, nm
$CH_2{=}CHCH{=}CH_2$	217
$CH_2{=}C(CH_3)CH{=}CH_2$	220
$CH_3CH{=}CHCH{=}CH_2$	223.5
$CH_2{=}C(CH_3){-}C(CH_3){=}CH_2$	226
$CH_3CH{=}CHCH{=}CHCH_3$	227

Each methyl group increases the wavelength of the absorption peak by 3–7 nm. A similar effect shows up with unsaturated carbonyl compounds.

	λ_{max}, nm ($\pi \rightarrow \pi^*$)
$CH_2{=}CHCOCH_3$	219
$CH_3CH{=}CHCOCH_3$	224
$(CH_3)_2C{=}CHCOCH_3$	235

This effect arises from the overlap of σ-orbitals in the alkyl substituent with the π-system. The resulting **hyperconjugation** is symbolized in Figure 22.29. The term was previously introduced (page 196) in connection with the stabilization of carbocations by alkyl groups.

Hyperconjugation has a smaller effect on electronic spectra than conjugation. Adding a methyl group to butadiene has little more than 10% of the effect of adding another vinyl group.

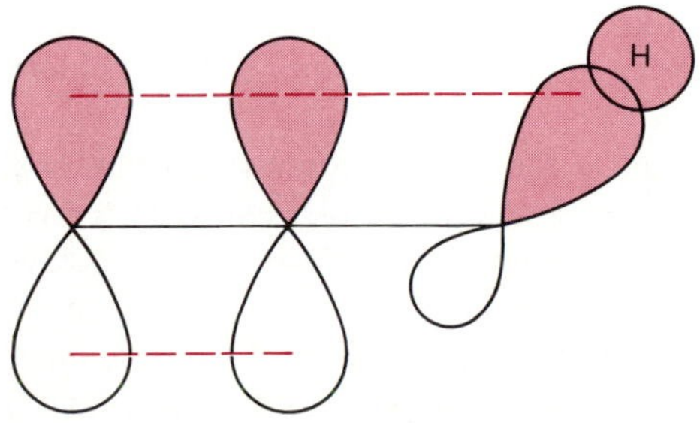

FIGURE 22.29 Hyperconjugation between a carbon-hydrogen σ-bond and the π-system of a double bond.

F. Benzene

The π-system of benzene has two highest occupied molecular orbitals, π_2 and π_3, and two lowest vacant molecular orbitals, π_4 and π_5 (Figure 22.7). We might expect to see four kinds of $\pi \rightarrow \pi^*$ transitions: $\pi_2 \rightarrow \pi_4$, $\pi_2 \rightarrow \pi_5$, $\pi_3 \rightarrow \pi_4$, and $\pi_3 \rightarrow \pi_5$. These four transitions all correspond to the same energy, and for this type of situation there is a breakdown in our simple picture of an electronic excitation as involving the transition from one molecular orbital to another. Several low-lying excited states of benzene exist that we would have to describe as various composites of the four simple transitions described above. An adequate treatment of the ultraviolet spectrum of benzene requires a rather complex quantum-mechanical discussion, which we will not develop. The longest-wavelength absorption of benzene gives a series of sharp bands centered at 255 nm with $\epsilon = 230$, a relatively low intensity for a $\pi \rightarrow \pi^*$ transition. This low value results from the high symmetry of benzene, which gives this absorption a relatively low probability. Such an absorption is called **symmetry forbidden.**

A vinyl group attached to a benzene ring constitutes a conjugated system. Styrene has two principal absorption bands: λ_{max} 244 nm with $\epsilon = 12{,}000$ and λ_{max} 282 nm with $\epsilon = 450$. The more intense band is a polyene type of $\pi \rightarrow \pi^*$ transition, whereas the less intense band corresponds to a substituted benzene. 1,2-Diphenylethylene (stilbene) allows another comparison of cis and trans isomers similar to the carotene case discussed in Section 22.6.B.

Ph(H)C=C(H)Ph	Ph(H)C=C(H)Ph
trans-stilbene	*cis*-stilbene
$\lambda_{max} = 295$ nm	$\lambda_{max} = 280$ nm
$\epsilon = 27{,}000$	$\epsilon = 13{,}500$

trans-Stilbene has no significant steric interactions. The compound has an extended coplanar π-system. In *cis*-stilbene, however, the two phenyl groups are on the same side of the double bond and sterically interfere with each other. The rings cannot both be coplanar with the double bond, and π-conjugation is not as effective as it is in the trans isomer. The result is a small change in λ_{max} but a large decrease in the extinction coefficient.

G. Other Functional Groups

Alcohols and ethers do not have conjugated π-systems and are transparent in the normal ultraviolet and visible regions. Ethanol and ether are common solvents for recording ultraviolet spectra. Sulfides, however, have relatively intense absorption at about 210 nm with a weaker band at about 230 nm. These absorptions are probably associated with transition of a lone-pair electron on sulfur to a sulfur 3*d*-orbital.

The carbonyl group in carboxylic acid derivatives is significantly different from that in ketones. Alkanoic acids have a low-intensity band about 200–210 nm, anhydrides absorb at somewhat longer wavelength, and the absorption of acid chlorides at still longer wavelengths, at about 235 nm.

Simple acetylenes absorb in the vacuum ultraviolet. Conjugated triple bonds show the type of absorption in the accessible ultraviolet expected for extended π-systems. The C≡N group of nitriles also absorbs at short wavelength, below 160 nm.

The simple alkyl fluorides and chlorides have no absorption maxima in the normal ultraviolet region. Alkyl bromides and iodides, however, do have λ_{max} in the region between 200–210 nm and 250–260 nm, respectively. These absorptions are attributed to transition of a lone-pair electron to an antibonding σ^*-orbital. Carbon-bromine and carbon-iodine bonds are sufficiently weak that the corresponding σ^*-orbitals have low enough energy to give transition energies in this ultraviolet region.

H. Photochemical Reactions

An excited state has more electronic energy than the ground state, and such states are generally rather short lived. The excess energy is generally dissipated within times that depend upon the nature of the excited state, the environment of the molecule, and the pathways open to the state. These times can range from 10^{-12} sec (psec) to more than 10^{-6} sec (μsec). One important way in which this energy is lost is by conversion of the electronic energy to vibrational and rotational energy. That is, the energy of moving electrons is converted in part to that of moving nuclei. Such energy, in turn, may simply be distributed as translational energy to other colliding molecules, in which case the net result has been the conversion of light to heat. The mechanisms, rates, and consequences of excited state energy dissipation are taken up in Section 36.3.

Alternatively, the vibrational energy may suffice to cause rearrangements or to break bonds. We saw one example of bond breaking in the light-initiated chlorination of alkanes (Section 6.3).

$$Cl_2 \xrightarrow{h\nu} 2\ Cl\cdot$$

This reaction could have been expressed as follows.

$$Cl_2 \xrightarrow{h\nu} Cl_2^* \longrightarrow 2\ Cl\cdot$$

In this representation, Cl_2* refers to an electronically excited state of Cl_2. The light promotes an electron to a chlorine-chlorine σ*-orbital. An electron in an antibonding orbital produces a weaker bond than when such an orbital is vacant and generally gives rise to a lower bond-dissociation energy.

Many examples of different types of photochemical reactions are known for organic compounds, but we will discuss only one at this point (for further examples see Section 36.3). In the electronically excited state of an alkene the double bond is generally weaker than in the ground state, and cis-trans isomerization is more facile.

$$\underset{\textit{trans}\text{-stilbene}}{(Ph)(H)C{=}C(H)(Ph)} \xrightarrow{h\nu} \underset{\textit{cis}\text{-stilbene}}{(Ph)(H)C{=}C(Ph)(H)}$$

This type of photochemical reaction is a critical step in the pathway that converts light received by cells in the retina of the eye into a nerve signal for the visual cortex, that part of the brain concerned with vision. Vitamin A_1 (retinol) is an alcohol that is oxidized enzymatically to vitamin A aldehyde, retinal. The 11-cis form of the aldehyde reacts with a lysine amino group of the protein, opsin, to produce an imine (a Schiff base, Section 14.6.C). This imine is the light-sensitive protein rhodopsin, found in discoid structures within the rod cells of the retina. The imine is present in the immonium form ($RC{=}NH^+R'$) which absorbs at longer wavelengths than the imine. Rhodopsin absorbs at 500 nm, forming excited rhodopsin Rh*. Within several psec at room temperature, the cis double bond is converted into the trans form, which then rapidly decays to the ground state. The energy stored in the excited state is converted into conformational changes in the protein. The changes can be followed because each intermediate has a different absorption maximum. The final conformational isomer, metarhodopsin II, λ_{max} 380 nm, is produced within 3 msec and catalyzes the exchange of the nucleotide guanosine triphosphate GTP for guanosine diphosphate GDP (see Chapter 33) within a G-protein complex. Many such exchanges (~500) occur before phosphorylation of metarhodopsin II by an enzyme stops the process. In sum, vision depends on light converting rhodopsin into a catalyst (metarhodopsin II), which produces many molecules of a biochemically active GTP-containing G-protein. The light signal is thus amplified, and is further amplified by additional biochemical reactions.

Phosphorylated metarhodopsin II is hydrolyzed to opsin and *trans*-retinal. The trans aldehyde is converted to the cis form by an enzyme, retinal isomerase, and the cycle starts anew. A wavelength of 500 nm corresponds to the blue-green region of the light spectrum and suggests why the rods are so sensitive to light of this color. Only a few light quanta are required to give a visual response to the dark-adapted eye. Bright light causes temporary impairment of vision because it depletes the rhodopsin and time is required for the protein to be reconstructed via the retinal isomerase cycle.

CHO

CHO

cis-retinal

trans-retinal

EXERCISE 22.14 A 500-nm light quantum accomplishes cis-trans isomerism of retinal. To what energy does light of wavelength 500 nm correspond?

22.7 Perturbational MO Approach to Reactivity

Consider what happens to the molecular orbitals of two reactants at the start of a reaction as the wave functions just begin to interact and overlap. Many of the MOs of one reactant will interact with MOs of the second reactant to produce a complex array of MOs of the combined system, but these interactions can be dissected into two fundamental types. One is the interaction of a filled MO of reactant A with a filled MO of reactant B. As illustrated in Figure 22.30a, bonding and antibonding interactions cause a splitting of the two energy levels involved, but because both levels are doubly occupied there is little *net* change in energy. The upper level actually increases somewhat more than the lower level decreases so that there is some net increase in energy. This slight energy increase amounts to a net repulsion that contributes to the total energy of activation required for reaction. The second case involves the interaction of a filled MO of reactant A with a vacant MO of reactant B. The levels again split as before but since only occupied MOs contribute to the energy of a molecule, the result is now a net *stabilization* (Figure 22.30b).

The magnitude of the energy splitting between orbitals depends primarily on their energy difference and on the amplitude of the wave functions at the point of interaction (their overlap). Two wave functions of greatly different energy interact only little. Similarly, only if both wave functions have significant amplitude in the region of interaction

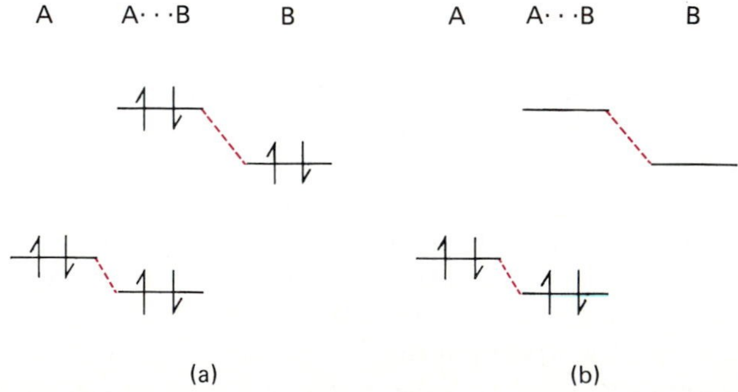

FIGURE 22.30 Two types of orbital interactions between reactants A and B.

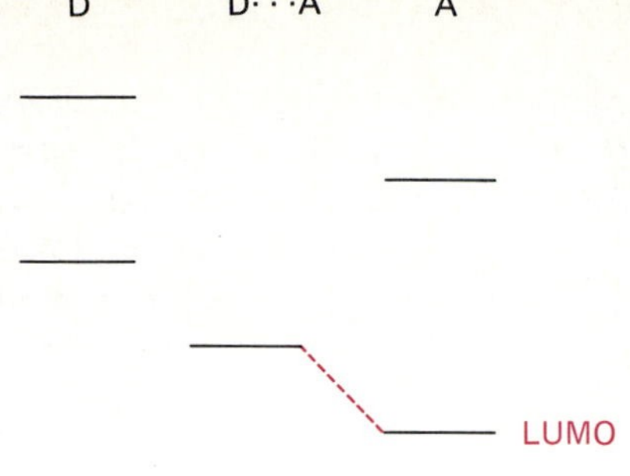

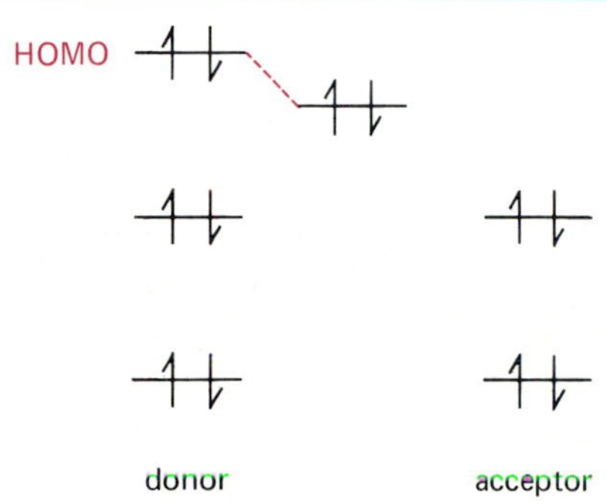

FIGURE 22.31 HOMO-LUMO interaction between donor (D) and acceptor (A) reactants.

will such interaction be important. For these reasons, the most important interactions at the beginning of a reaction are those involving the **highest occupied molecular orbital (HOMO)** of one reactant with the **lowest unoccupied molecular orbital (LUMO)** of the other reactant. These two types of MOs are also referred to as the **frontier orbitals.**

The most common case is that in which one reactant has the higher HOMO and the other reactant has the lower LUMO. A high HOMO implies weakly held electrons that can be easily lost or donated; hence, the reactant with the higher HOMO is called the **donor.** Similarly, a low-lying LUMO is one that can readily accept additional electrons. Accordingly, the reagent with the lower LUMO is called the **acceptor.** The corresponding HOMO-LUMO interaction between a donor and acceptor (Figure 22.31) can be dominating and determine the course of reaction. This interaction is directly analogous to that in charge-transfer complexes discussed in Section 22.6.D.

In an S_N2 reaction the donor is the entering nucleophile whose HOMO is a lone-pair orbital. The acceptor is the organic substrate whose LUMO for the reaction is a σ^*-orbital involving the leaving group. The LUMO of methyl fluoride, for example, is illustrated in Figure 22.32. Attack by the nucleophile at the carbon-fluorine bond involves interaction at a node with consequent little net overlap (Figure 22.33). Reaction at the rear of the carbon-fluorine bond, however, involves interaction with the large lobe at that point. Overlap with the HOMO of the nucleophile is substantial and attack at this point is favored. Moreover, the new lower energy level that results from bonding interaction of the HOMO and LUMO in Figure 22.31 belongs to a combined MO that includes LUMO character of the acceptor. Since this new combined MO is occupied by two electrons that formerly resided in an MO (the HOMO) exclusively on the donor, the net result is transfer of electron density from the HOMO of the donor to the LUMO of the acceptor. Putting electron density into an MO that has carbon-fluorine antibonding character weakens the bond to fluorine, whereas the positive overlap provides bonding character to the incoming

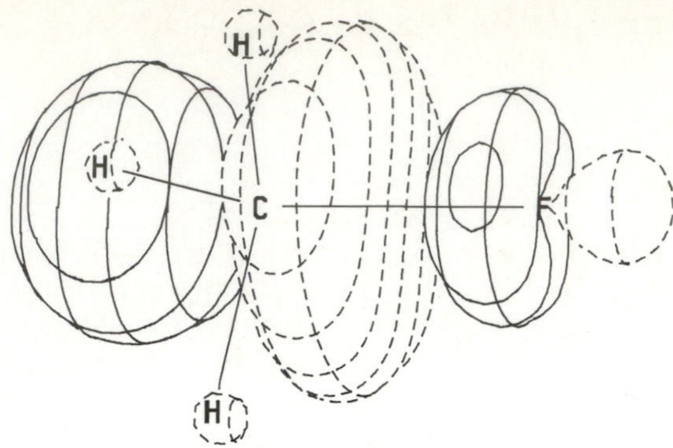

FIGURE 22.32 The LUMO of methyl fluoride is essentially a carbon-fluorine σ^*-orbital. [Reproduced with permission from W. L. Jorgensen and L. Salem, *The Organic Chemist's Book of Orbitals,* Academic Press, New York, 1973.]

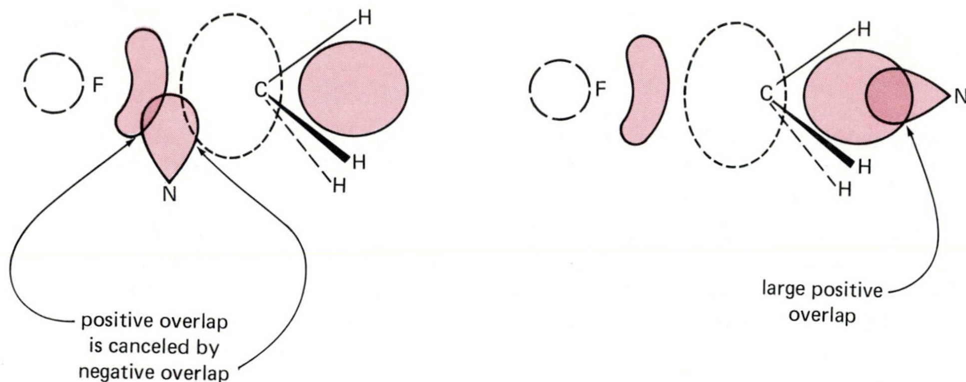

FIGURE 22.33 HOMO-LUMO interaction is weak for S_N2 attack at the carbon-fluorine bond (a) but strong at the rear of the bond (b).

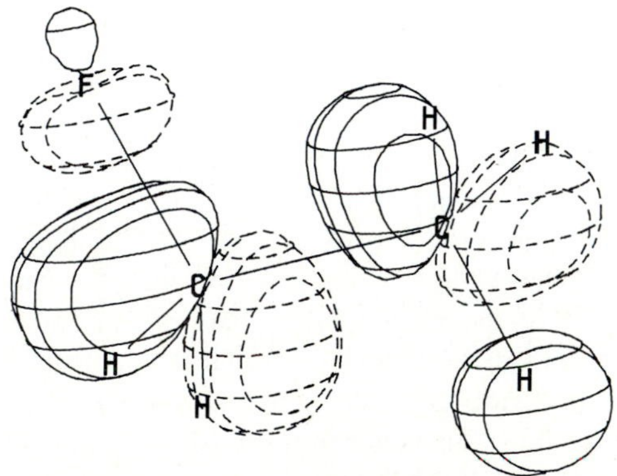

FIGURE 22.34 The LUMO of ethyl fluoride encompasses the entire molecule, including the methyl group. [Reproduced with permission from W. L. Jorgensen and L. Salem, *The Organic Chemist's Book of Orbitals,* Academic Press, New York, 1973.]

nucleophile. In short, this HOMO-LUMO interaction shows us a displacement reaction with inversion of configuration in the making.

EXERCISE 22.15 Figure 22.34 shows the LUMO of ethyl fluoride in a staggered configuration. Locate two regions of maximum overlap with an attacking nucleophile and compare with the expected mechanisms for S_N2 and E2 reactions. For the E2 reaction show how putting electron density into the LUMO starts to form the double bond as well as weakening the bonds to the leaving fluoride and proton.

EXERCISE 22.16 Examine the π-MOs of allyl cation (Figure 22.3) and deduce whether reaction with a nucleophile occurs at an end or central carbon.

Further discussion of the frontier orbital approach is beyond the scope of the present course but is an active subject of research in organic chemistry.

ASIDE

A node is the point at which a periodic curve crosses the zero line. A lode is a rich vein of metallic ore, which can be gold or silver. A chemical exercise that is highly rewarding is that of finding the nodes in a molecular orbital. Write a set of six π-molecular orbitals derived from six *p*-orbitals arranged linearly in which there are zero to five nodes. What compound does this describe? Make the set cyclic; how many nodes can you now write for the orbitals? Show the distribution of six electrons in the levels corresponding to the orbitals.

PROBLEMS

1. a. The heat of hydrogenation of cyclooctene to cyclooctane is -23.3 kcal mole^{-1}. That for cyclooctatetraene is -100.9 kcal mole^{-1}. Use these data to calculate an empirical resonance energy for cyclooctatetraene. How do you interpret this result compared to the resonance energy of benzene?

b. Use the table of average bond energies in Appendix III to estimate the atomization energy of cyclooctatetraene. The experimental value is $\Delta H^\circ_{atom} = +1713$ kcal mole^{-1}. What value does this method give for the empirical resonance energy of cyclooctatetraene? Compare this result with that in part (a).

c. Thermochemical measurements on [18]annulene give a derived heat of atomization, $\Delta H^\circ_{atom} = +3890$ kcal mole^{-1}. From the method in part (b), calculate the corresponding empirical resonance energy. How does this result compare with expectations from the $4n + 2$ rule?

2. Which of the following hydrocarbons is expected to be the most acidic? Why?

CH_3 CH_3 CH_2 CH_3 CH_2

3. Determine the number of π-electrons involved in each of the following cyclic systems and determine whether each is aromatic according to the Hückel rule.

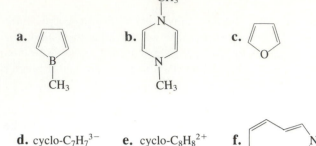

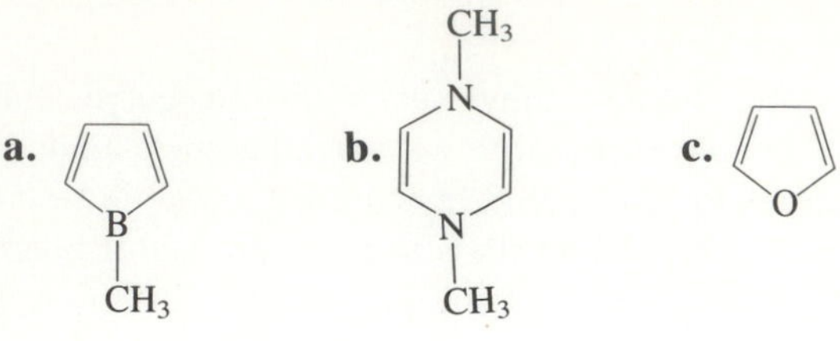

d. cyclo-$C_7H_7^{3-}$ **e.** cyclo-$C_8H_8^{2+}$ **f.** NH

4. Which *two* of the following species are *not* aromatic?

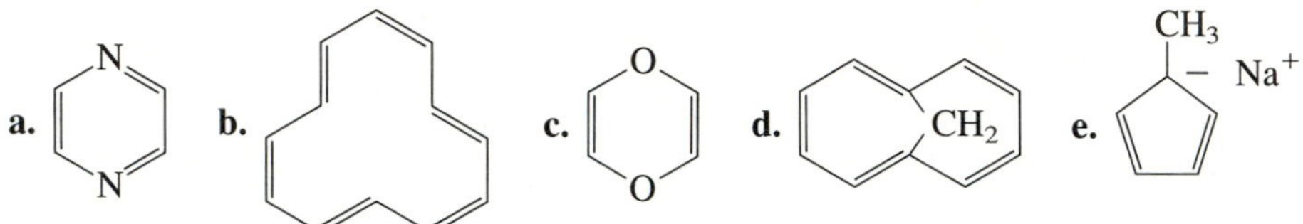

5. Figure 22.8 gives the MO energy level pattern for planar [8]annulene. Sketch the MOs corresponding to each energy.

6. Which of the following represents the highest occupied molecular orbital (HOMO) of the pentadienyl cation, $CH_2{=}CH{-}CH{=}CH{-}CH_2^+$?

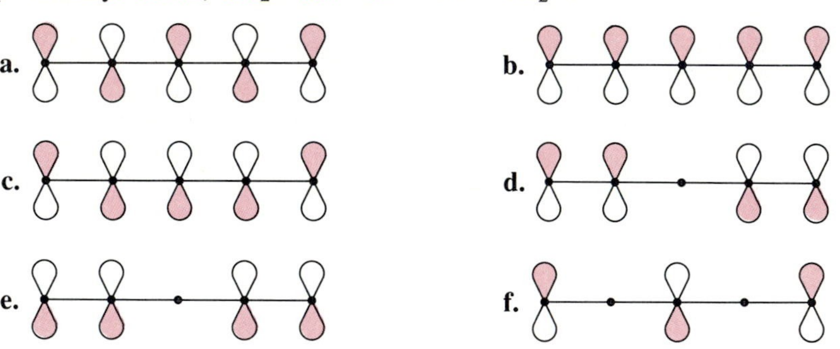

7. Each of the following cycloaddition reactions can be written with a pericyclic transition state. Which ones follow the Hückel $4n + 2$ rule?

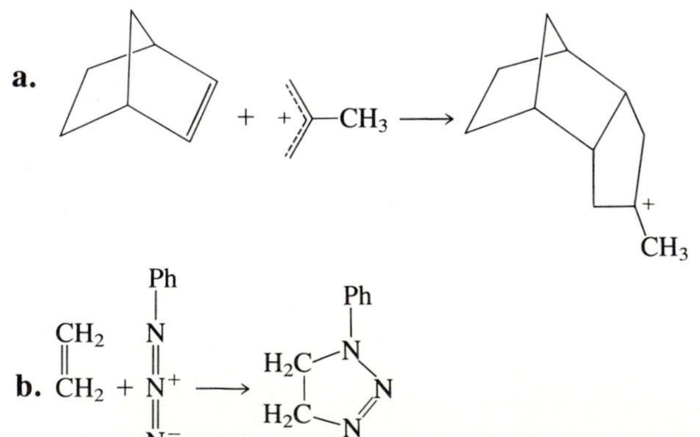

c. + COOMe ⟶ COOMe

d. + O ⟶ O

8. The following known reaction does not occur with the corresponding seven-membered ring analog. Explain.

+ PhLi ⟶ Ph Li$^+$

+ PhLi ⟶/⟶

9. Consider the cycloaddition reaction of allyl cation with ethylene to give cyclopentyl cation and with butadiene to give a cycloheptenyl cation.

a. Which reaction is preferred from considerations of HOMO-LUMO interactions?

b. Which transition state has Hückel aromaticity?

10. Alkyl bromides and iodides are normally stored in the dark or in dark bottles. On exposure to light they slowly turn brown or violet, respectively. Give an explanation for this phenomenon based on a reasonable photochemical mechanism.

11. Indicate which of the following compounds would be suitable as solvents for recording normal ultraviolet spectra of substrates and briefly explain your choices: methanol; perfluoropropane; 1-chlorobutane; ethyl ether; ethyl iodide; methylene bromide; methyl butyl sulfide; benzene; cyclohexane; acetonitrile

12. Rank the following polyenes in terms of the *longest wavelength* absorption in their ultraviolet spectra?

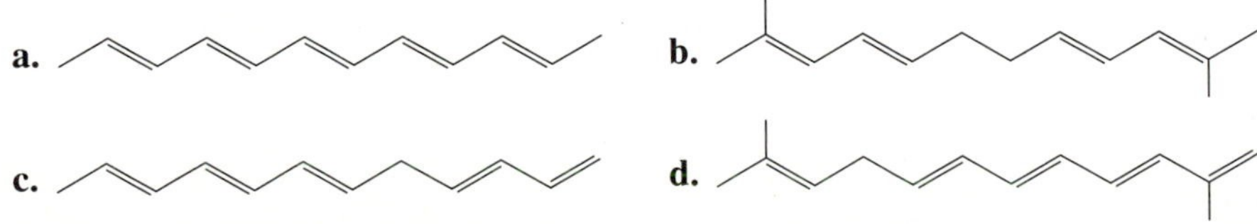

13. Both the Z-value and the dielectric constant of water (Table 22.2) are higher than those of methanol. (a) Explain this in terms of the structure of each of the solvents.

(b) Would you expect *t*-butyl chloride to react more rapidly in water or in methanol? Explain your answer, using an energy versus reaction coordinate diagram.

14. A number of simple conjugated polyenes, H(CH=CH)$_n$H, are now known up to $n = 10$; λ_{max} in nanometers corresponding to values of n are as follows: 2, 217; 3, 268; 4, 304; 5, 334; 6, 364; 7, 390; 8, 410; 10, 447. A crude model of such a conjugated π-system is that of an electron in a box having the dimensions of the π-system. A quantum-mechanical treatment of such a model suggests that $1/\lambda_{max}$ should be approximately a linear function of $1/n$. Test this prediction with the data given and try to interpolate to find λ_{max} for the missing polyene with $n = 9$.

15. The ultraviolet spectrum of 3,6,6-trimethylcyclohex-2-en-1-one is shown below. The concentration is 1.486×10^{-5} g mL^{-1} in ethanol. Calculate ϵ and determine λ_{max}.

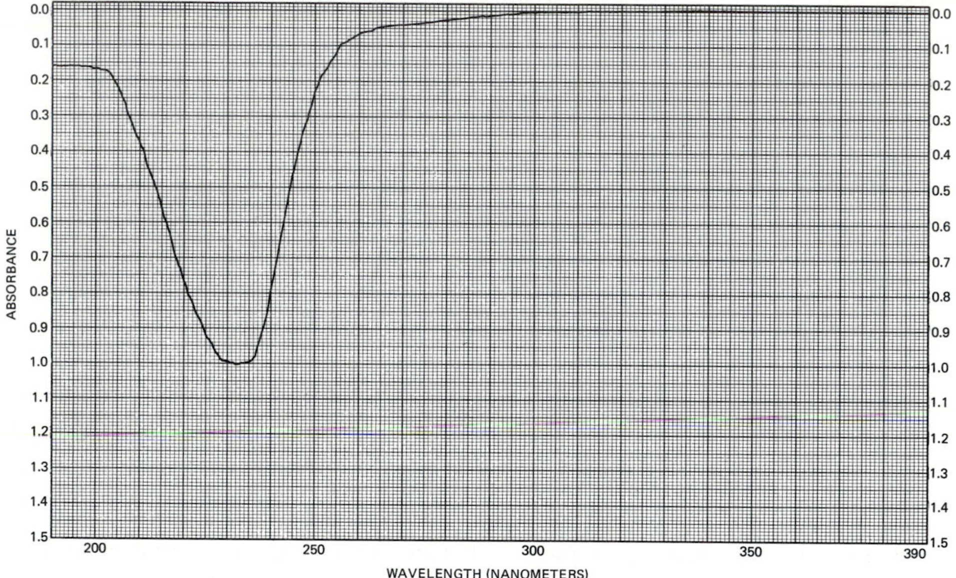

16. The effect of a methyl group on a molecular orbital depends on the magnitude of the wave function at the point of attachment. Explain why the effect of a methyl group on the UV spectrum of butadiene is greater at the 1-position than the 2-position.

17. A violet solution of iodine in tetrachloromethane was added to toluene. The solution instantly turned brown, but on removal of the solvent, no additional materials could be detected. What happened? Explain in detail the nature of the color.

CHAPTER 23

ELECTROPHILIC AROMATIC SUBSTITUTION

In Chapter 21 we encountered aromatic compounds for the first time and learned something of the special properties of the aromatic ring. In this chapter we will look at the most important reaction of the aromatic ring—substitution of one electrophile (usually a proton) by another electron-deficient species. The reaction applies to a number of different electrophilic reagents and provides an important route to many substituted aromatic compounds. Moreover, when applied to benzene systems already containing one or more substituents, the reaction shows specificity and reactivity effects that are readily rationalized by theory. Thus, this chapter provides an especially integrated combination of theory and synthesis.

23.1 Halogenation

Alkenes react rapidly with bromine even at low temperatures to give the product of *addition* of bromine.

$$CH_2{=}CH_2 + Br_2 \longrightarrow BrCH_2CH_2Br \qquad \Delta H^\circ = -29.2 \text{ kcal mole}^{-1}$$

The reaction is highly exothermic because two carbon-bromine bonds are substantially more stable than a bromine-bromine bond and the second bond of a double bond. The corresponding addition reaction of benzene is slightly endothermic.

$$C_6H_6 + Br_2 \longrightarrow C_6H_6Br_2 \qquad \Delta H^\circ \cong 2 \text{ kcal mole}^{-1}$$

Such an addition reaction destroys the cyclic π-electronic system of benzene. Note that the difference in ΔH° for the two cases is approximately the resonance energy of benzene.

Benzene does react with bromine, but the reaction requires the use of appropriate Lewis acids such as ferric bromide. The product of the reaction is the result of *substitution* rather than addition.

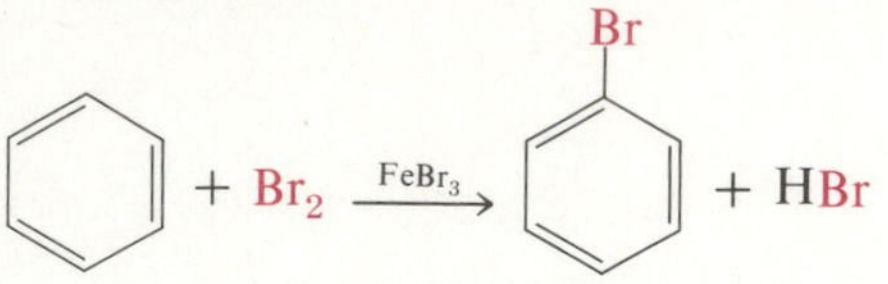

$$C_6H_6 + Br_2 \xrightarrow{FeBr_3} C_6H_5Br + HBr \qquad \Delta H^\circ = -10.8 \text{ kcal mole}^{-1}$$

Sixty grams of bromine is added slowly to a mixture of 33 g of benzene and 2 g of iron filings. The mixture is warmed until the red vapors of bromine are no longer visible, about one half hour. Water is added, and the washed and dried organic layer is distilled to give 40 g of bromobenzene, b.p. 156 °C.

In this procedure, the iron reacts rapidly with bromine to give ferric bromide. Anhydrous ferric halides are Lewis acids and react avidly with bases such as water.

$$FeX_3 + H_2O \rightleftharpoons H_2O : FeX_3 \longrightarrow \text{higher hydrates}$$

Anhydrous ferric halides are difficult to keep pure and are frequently made from the elements as needed, as in the foregoing procedure.

It is the Lewis-acid character of ferric salts that allows them to function as catalysts in this reaction. Aluminum halides are used frequently for the same purpose. Recall that the first step in the bromination of an alkene is a displacement by the alkene as a nucleophile on bromine with bromide ion as a leaving group (Section 11.6.B).

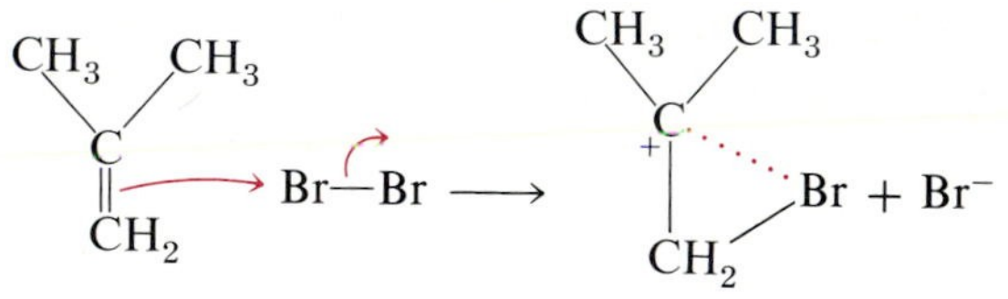

$$(CH_3)_2C{=}CH_2 + Br{-}Br \longrightarrow (CH_3)_2\overset{+}{C}\cdots Br{-}CH_2 + Br^-$$

In nonpolar solvents, the leaving bromide ion requires additional solvation by bromine (page 568).

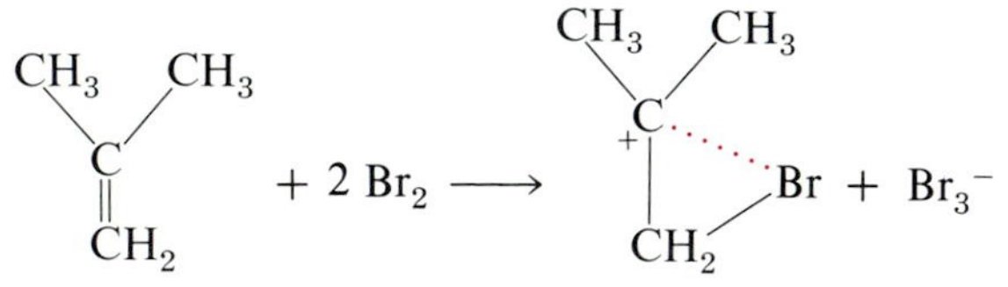

$$(CH_3)_2C{=}CH_2 + 2\,Br_2 \longrightarrow (CH_3)_2\overset{+}{C}\cdots Br{-}CH_2 + Br_3^-$$

In this case a bromine molecule serves as a mild Lewis acid to help pull bromide ion from bromine. This type of ''pull'' is provided more powerfully by a stronger Lewis acid such as ferric bromide.

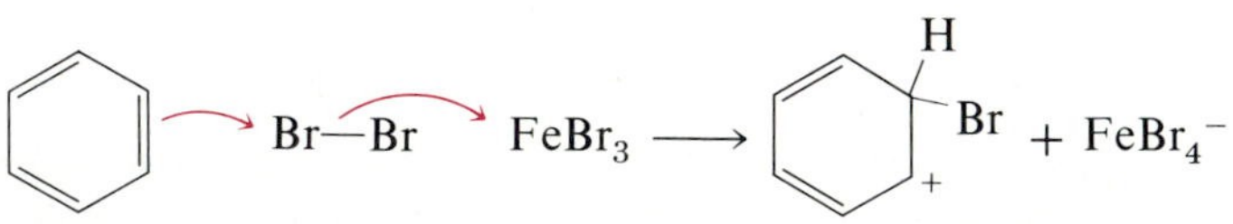

$$C_6H_6 \; Br{-}Br \; FeBr_3 \longrightarrow C_6H_6Br^+ + FeBr_4^-$$

Benzene is a much weaker nucleophilic reagent than a simple alkene and requires a more electrophilic reagent for reaction.

The intermediate in the bromination of benzene is a conjugated carbocation. Its structure may be expressed by three Lewis structures

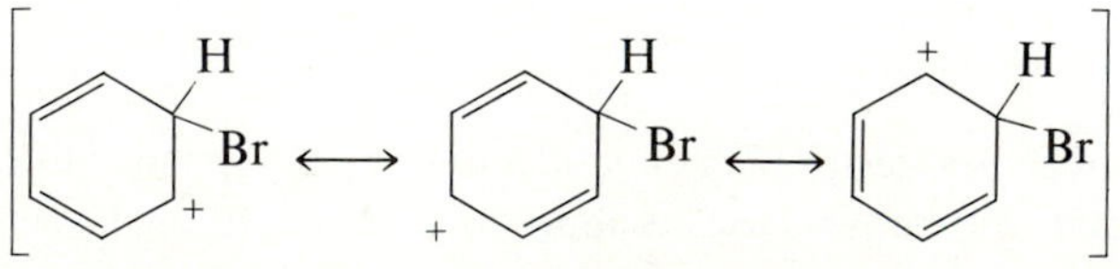

$$\left[C_6H_6Br^+ \longleftrightarrow C_6H_6Br^+ \longleftrightarrow C_6H_6Br^+ \right]$$

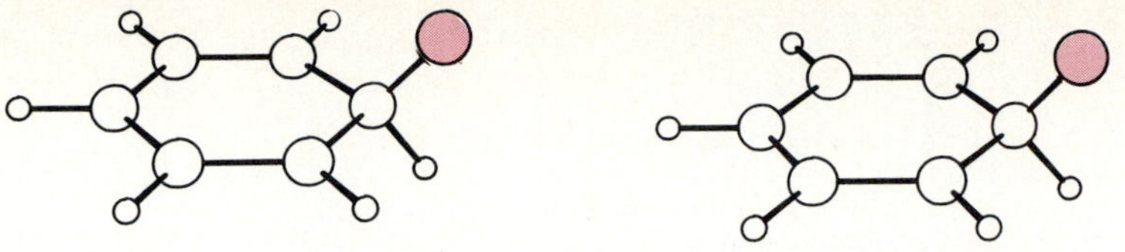

FIGURE 23.1 Stereo representation of the intermediate in the bromination of benzene.

The resulting structure is that of an approximately tetrahedral carbon attached to a planar pentadienyl cation, as shown by the stereo representation in Figure 23.1.

This resonance-stabilized pentadienyl cation is often symbolized by using a dotted line to indicate that the positive charge is delocalized over the three positions indicated in the foregoing resonance structures.

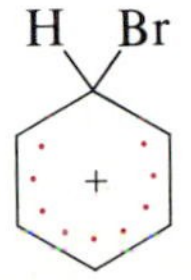

Again, however, this symbol conveys no accounting of electrons. The student is urged to use Lewis structures exclusively at this stage in order to understand more fully the electron displacements that occur in reactions.

Carbocations can generally react with a nucleophilic reagent, rearrange, or lose a proton. Reaction of the pentadienyl cation intermediate with a nucleophile would give a product without the benzene ring resonance. Consequently, this type of reaction is rarely observed in electrophilic aromatic reactions. Rearrangements are significant only in some special cases to be discussed later. The only important reaction of our bromination intermediate is loss of a proton to restore the cyclic π-system and yield the substitution product.

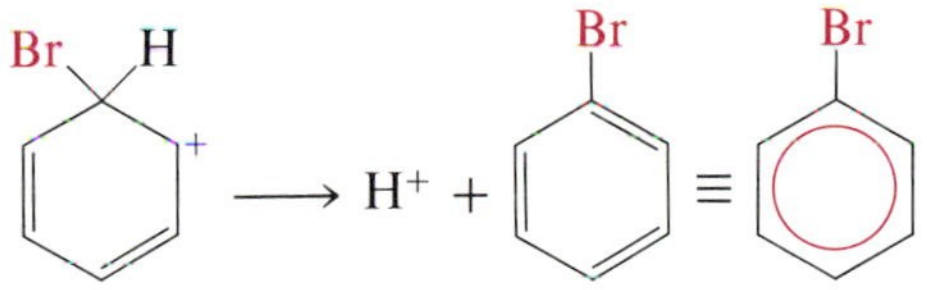

The overall reaction sequence is as follows.

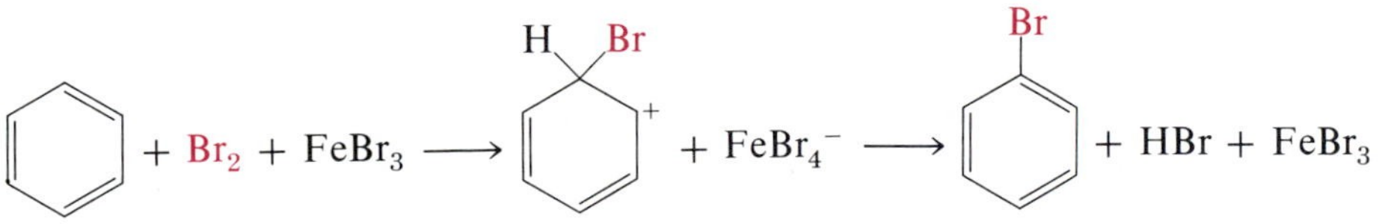

The first step is rate determining as indicated by the energy profile shown in Figure 23.2. The experimental evidence for this reaction mechanism comes from many studies of chemical kinetics, isotope effects, and structural effects.

The reaction mechanism for bromination of benzene is general for other electrophilic aromatic substitutions as well. Reaction occurs with an electron-deficient (electrophilic) species to give a pentadienyl cation intermediate, which loses a proton to give the substituted benzene product.

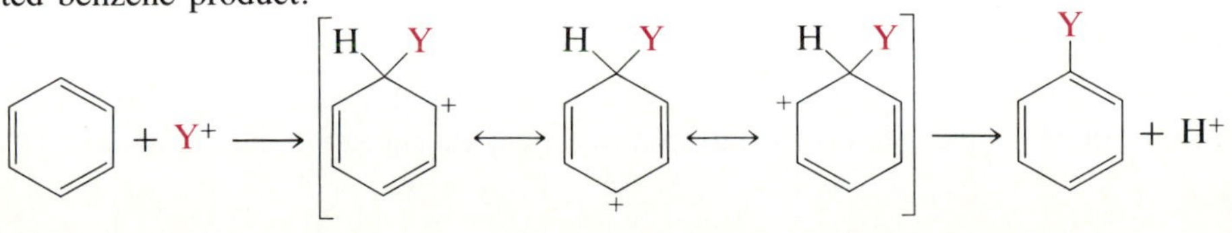

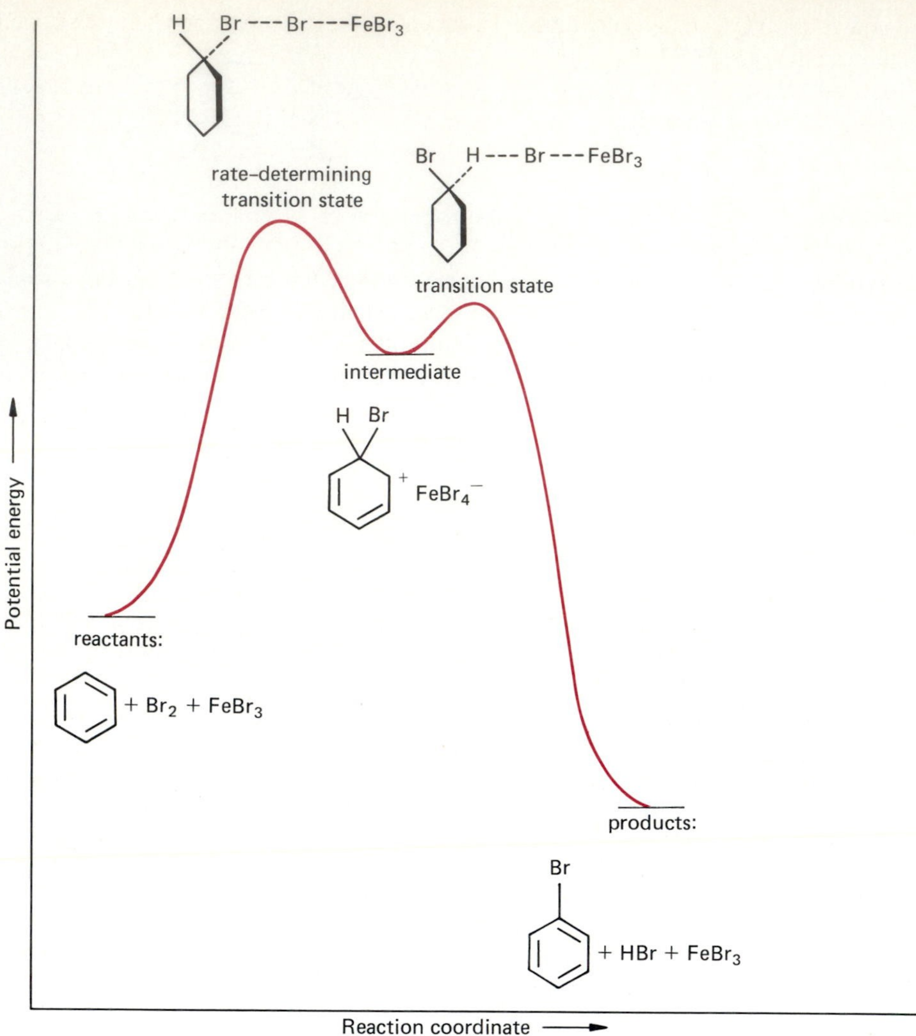

FIGURE 23.2 Reaction profile for bromination of benzene.

Chlorination is directly analogous to bromination.

$$C_6H_6 + Cl_2 \xrightarrow{AlCl_3} C_6H_5Cl + HCl$$

$$p\text{-}CH_3C_6H_4CH_3 + Cl_2 \xrightarrow{CH_3COOH} \text{2-chloro-1,4-dimethylbenzene} + HCl$$

The last example allows an important comparison of these electrophilic halogenations and the free radical halogenations of methylbenzenes that we encountered in Section 21.5.A.

Recall that under free radical conditions substitution occurs in the side chain, principally at the benzylic position.

Iodobenzene can be prepared by using iodine and an oxidizing agent under acidic conditions. Suitable oxidizing agents include nitric acid or arsenic acid (H_3AsO_4).

$$C_6H_6 + I_2 + HNO_3 \xrightarrow{50°C} NO + NO_2 + \underset{(86\text{–}87\%)}{C_6H_5I}$$

The reaction involves a normal aromatic electrophilic substitution by iodonium ion obtained by oxidation

$$I_2 \xrightarrow{-2\,e^-} 2\,I^+$$

EXERCISE 23.1 Write the three important resonance structures for the intermediate obtained by reaction of *p*-xylene with iodonium ion.

Highlight 23.1

An **electrophile** is a species (usually positively charged) that has affinity for the electron pairs in **nucleophiles.** For reaction to occur, weak nucleophiles require strong electrophiles. Lewis acids are used to strengthen electrophiles. Electrophilic substitution (E^+) of aromatic compounds is useful for certain types of conversions, as illustrated by the transformation of benzene into bromobenzene. Ferric bromide is a Lewis acid that increases the electrophilicity of bromine.

Br
Br$_2$
FeBr$_3$

Br—Br FeBr$_3$ ⟶ H Br + + FeBr$_4^-$ ⟶

Lewis acid
increases electrophilicity of bromine

pentadienyl carbocation
stabilized by delocalization

Br
+ FeBr$_3$ + HBr

23.2 Protonation

Benzene is an extremely weak base, much weaker than an alkene. Benzene is only slightly protonated in concentrated sulfuric acid, whereas isobutylene is significantly protonated even in sulfuric acid containing water. Some protonation does occur, however, and the amount can be significant if substituents are present. For example, hexamethylbenzene is 50% protonated in 90% aqueous sulfuric acid.

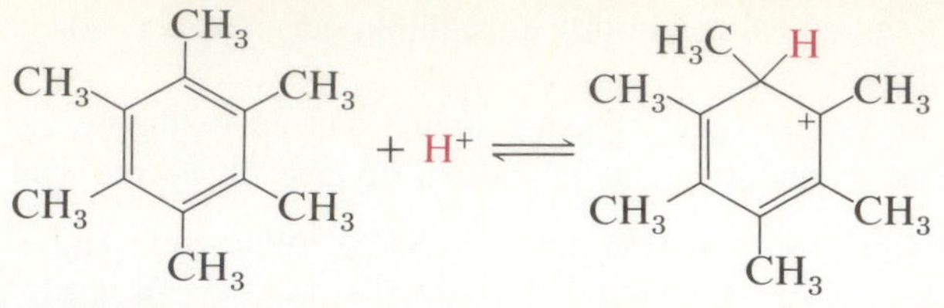

Protonation of benzene can be detected by hydrogen isotope exchange reactions in acid. If benzene is stirred for several days at room temperature with 80% aqueous sulfuric acid containing deuterium or tritium, the isotope distributes between the benzene and the aqueous acid.

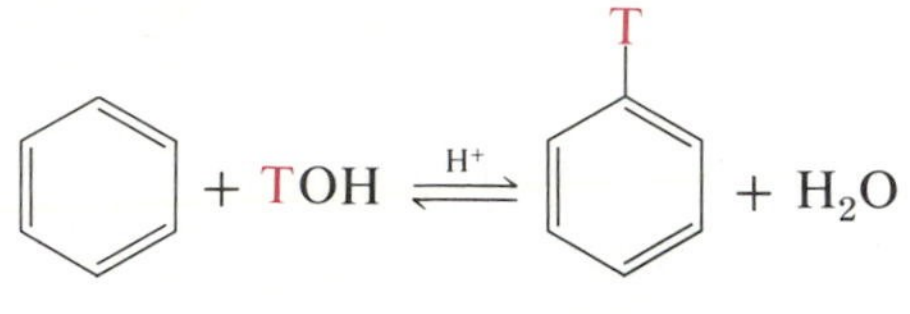

benzene-*t*

Tritium is normally used as a **radioactive tracer isotope.** It is typically used in a ratio of less than 1 ppm of ordinary hydrogen. Therefore, in an exchange process such as this, it is unlikely that a given molecule will have more than one tritium bound to it. The radioactivity of tritium can be measured by a sensitive instrument called a **liquid scintillation counter**; hence, tritium incorporation can be precisely measured by using only a small amount of the isotope. On the other hand, deuterium is used as a **macroscopic isotope.** Incorporation is monitored by less sensitive analytical techniques, such as NMR or mass spectrometry. The exchange reaction will give mixtures of deuterated benzenes containing varying numbers of deuterium atoms attached to the ring. The amount of deuterium incorporation will depend on the relative amounts of ^1H and ^2H isotopes in the hydrogen "pool." If a large excess of D_2SO_4 and D_2O is used, benzene-d_6, C_6D_6, can be obtained.

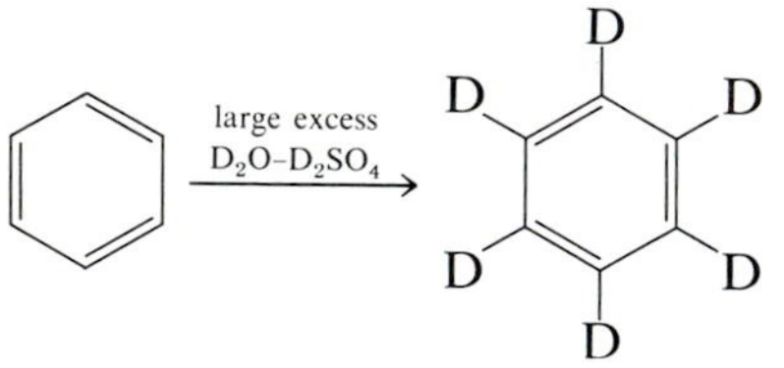

benzene-d_6

The exchange reaction is a simple type of electrophilic aromatic substitution reaction in which the electrophilic reagent is D^+.

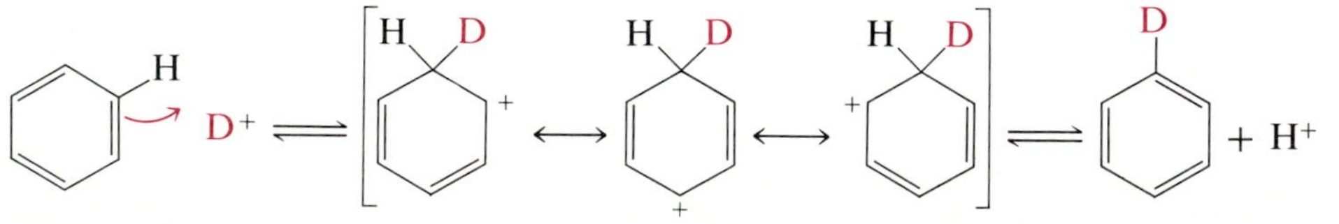

The intermediate pentadienyl cation undergoes only one significant reaction—loss of a proton (or a deuteron). This reaction, which regenerates the aromatic π-system, is much faster than its reaction with water. With alkyl cations, reaction with a nucleophilic species is a much more important reaction, because elimination of a proton from such carbocations does not have the formation of an aromatic ring as an additional driving force.

EXERCISE 23.2 What product would you expect from treatment of benzene with excess D_2O-D_2SO_4 containing some tritiated water?

23.3 Nitration

The reaction of alkenes with nitric acid is not a generally useful reaction. Addition of nitric acid to the double bond is accompanied by more or less oxidation. However, benzene is quite stable to most oxidizing agents, and its reaction with nitric acid is an important organic reaction. Actually, the nitrating reagent generally used is a mixture of concentrated nitric acid and sulfuric acid.

$$C_6H_6 + HNO_3 \xrightarrow[50°C]{H_2SO_4} C_6H_5NO_2 + H_2O$$

nitrobenzene

> To a flask containing 65 g of benzene is added a mixture of 110 mL of conc. H_2SO_4 and 85 mL of conc. HNO_3. The acid mixture is added in portions so that the temperature does not exceed 50 °C. After all of the acid has been added, the reaction mixture is cooled and the oily nitrobenzene layer is separated, washed, and distilled. The yield of pure product is 85–88 g (83–86%).

The nitro group is an important functional group in aromatic chemistry because it may be converted into many other functional groups. The nitration reaction thus provides a route to many substituted aromatic compounds. The chemistry of the nitro group will be detailed in Section 25.1. Many properties of the nitro group can be interpreted on the basis of a resonance hybrid of two Lewis structures.

$$\left[R{-}\overset{+}{N}(=O){-}O^- \longleftrightarrow R{-}\overset{+}{N}(-O^-){=}O \right] \equiv R{-}\overset{+}{N}(\cdots O^{\frac{1}{2}-})(\cdots O^{\frac{1}{2}-})$$

In these structures the O—N—O system is seen to have an allylic anion type of π-system.

In a mixture of nitric and sulfuric acids, an equilibrium is established in which many species are present. One of these species is the nitronium ion, NO_2^+, which has been detected by spectroscopic methods. In the mixture of acids, it is produced by a process in which sulfuric acid functions as an acid and nitric acid functions as a base.

$$H_2SO_4 + HONO_2 \rightleftharpoons H_2\overset{+}{O}NO_2 + HSO_4^-$$

$$H_2\overset{+}{O}NO_2 + H_2SO_4 \rightleftharpoons H_3O^+ + NO_2^+ + HSO_4^-$$

$$2\ H_2SO_4 + HONO_2 \rightleftharpoons H_3O^+ + NO_2^+ + 2\ HSO_4^-$$

The structure of nitronium ion is known from spectroscopic measurements. It is related to the isoelectronic compound carbon dioxide. The molecule is linear and is a powerful electrophilic reagent.

$$O{=}\overset{+}{N}{=}O$$

nitronium ion

> Nitronium perchlorate, $NO_2^+\ ClO_4^-$, is a white crystalline solid prepared by the reaction of $HClO_4$ with dinitrogen pentoxide, N_2O_5. Although the reagent is a violent oxidizing agent towards most organic compounds, nitrations may be carried out in nitrobenzene solution.

Nitronium ion reacts directly with benzene to give a pentadienyl cation intermediate.

$$C_6H_6 + O{=}N^+{=}O \longrightarrow \left[\text{arenium ion (H, } N^+(=O)O^-)\right] \xrightarrow{-H^+} C_6H_5\text{—}NO_2 + H^+$$

Note that reaction occurs on nitrogen rather than oxygen.

> Reaction at oxygen gives a nitrite compound, R—O—NO. Nitrites are unstable under such strongly acidic conditions and decompose to products containing carbon-oxygen bonds. These oxidation products react further to give highly colored polymeric compounds. The formation of more or less tarry by-products is a usual side reaction in most aromatic nitration reactions.

Aromatic nitro compounds are important intermediates for the synthesis of other aromatic derivatives. The most important reaction of the nitro group is reduction. We shall return to this subject in Section 24.6.C.

EXERCISE 23.3 Write equations showing all of the steps involved in the nitration of *p*-xylene. Show all contributing resonance structures for the intermediate carbocation. From a consideration of these resonance structures, can you suggest why *p*-xylene undergoes nitration more rapidly than benzene?

23.4 Friedel-Crafts Reactions

A. Acylations

The electrophile in electrophilic aromatic substitution can also be a carbocation. Such reactions are called **Friedel-Crafts reactions.** The most useful version of the reaction is Friedel-Crafts **acylation** in which the entering electrophile is an acyl group, RCO—, derived from a carboxylic acid derivative, usually an acyl halide or anhydride. The carbonyl group in such acid derivatives is sufficiently basic that formation of a complex occurs with strong Lewis acids such as aluminum chloride.

$$RCOX + AlCl_3 \rightleftharpoons \left[R\text{—}\overset{\overset{+}{O}\text{—}\overset{-}{Al}Cl_3}{\overset{\|}{C}}\text{—}X \longleftrightarrow R\text{—}\overset{O\text{—}\overset{-}{Al}Cl_3}{\overset{|}{\underset{+}{C}}}\text{—}X \right]$$

> Aluminum chloride, $AlCl_3$, can be prepared by the direct reaction of aluminum with chlorine or hydrogen chloride. Anhydrous aluminum chloride is available as a white powder that fumes in air and has the strong odor of HCl from reaction with atmospheric moisture. It can be sublimed and is very soluble in many organic solvents such as nitrobenzene, somewhat soluble in carbon tetrachloride, and slightly soluble in benzene, in which it exists as a dimer, Al_2Cl_6. Anhydrous aluminum chloride reacts vigorously with water with evolution of HCl. It is a strong Lewis acid that forms complexes with most oxygen-containing compounds. For laboratory use it is kept in tightly sealed bottles, and the fine powder is handled in air as little as possible, preferably in a hood.

The carbocation character of a carbonyl carbon is greatly enhanced by coordination to aluminum chloride, and in many cases the complex itself is sufficiently electrophilic to react with aromatic rings. In other cases, the complex exists in equilibrium with a small amount of the corresponding **acylium** ion, which is an even more powerful electrophile.

$$R{-}\overset{\overset{+}{O}\overset{-}{A}lCl_3}{\overset{\|}{C}}{-}X \rightleftharpoons \left[R{-}C{\equiv}\overset{+}{O}: \longleftrightarrow R{-}\overset{+}{C}{=}\ddot{O}\right] + AlXCl_3^-$$

acylium ion

$$C_6H_6 + R{-}C{\equiv}O^+ \longrightarrow [C_6H_6{-}COR]^+ \text{ (H, CR=O on the same ring carbon)}$$

$$[C_6H_6{-}COR]^+ \longrightarrow H^+ + C_6H_5COR$$

As shown in the foregoing equations, the mechanism for reaction of the acylium ion with benzene is completely analogous to that of other electrophilic reagents. The final product is an aromatic ketone whose carbonyl group is sufficiently basic to be complexed completely by aluminum chloride.

$$C_6H_5COR + AlCl_3 \rightleftharpoons C_6H_5{-}\underset{}{C}(R){=}\overset{+}{O}{-}\overset{-}{A}lCl_3$$

This complex is the actual reaction product. The work-up procedure involves treatment with water or dilute hydrochloric acid to decompose the complex and dissolve the aluminum salts. The liberated ketone remains in the organic layer and is isolated by crystallization or distillation. Because it complexes with the product, aluminum chloride must be used in equimolar amounts. Furthermore, the complexed ketone is resistant to further reaction so that high yields of pure product are readily available by this reaction. Friedel-Crafts acylation is an important and useful reaction in aromatic chemistry. An example is the acylation of benzene with acetyl chloride.

$$C_6H_6 + CH_3COCl + AlCl_3 \longrightarrow C_6H_5COCH_3 \cdot AlCl_3 + HCl$$

$$C_6H_5COCH_3 \cdot AlCl_3 + H_2O \longrightarrow C_6H_5COCH_3 + \text{aq. } AlCl_3$$

acetophenone

To a cooled mixture of 40 g of anhydrous aluminum chloride in 88 g of dry benzene, 29 g of acetyl chloride is added slowly with stirring or shaking. The HCl evolved is absorbed in a suitable trap. When the addition is complete, the mixture is warmed to 50 °C for 1 hr. After cooling, ice and water are added, and the benzene layer is washed, dried, and distilled. The product acetophenone is distilled, b.p. 201 °C, in a yield of 27 g.

EXERCISE 23.4 Outline multistep syntheses for the synthesis from benzene and other necessary organic reagents of each of the following compounds.

(a) 1-phenylpropane (b) 2-phenyl-2-propanol (c) styrene

B. Alkylations

Benzene undergoes Friedel-Crafts **alkylation** when treated with an alkyl halide and a Lewis-acid catalyst such as $FeBr_3$ or $AlCl_3$. An example is the reaction of benzene with *t*-butyl chloride to give *t*-butylbenzene.

$$C_6H_6 + (CH_3)_3CCl \xrightarrow{FeCl_3} C_6H_5C(CH_3)_3 + HCl$$

In this Friedel-Crafts alkylation the attacking electrophile is the *t*-butyl cation, which is produced in the reaction of *t*-butyl chloride with $FeCl_3$. In the absence of other nucleophiles, this electrophilic species reacts with the aromatic ring.

$$(CH_3)_3CCl + FeCl_3 \rightleftharpoons (CH_3)_3C^+ \; FeCl_4^-$$

$$C_6H_6 + (CH_3)_3C^+ \longrightarrow [C_6H_6C(CH_3)_3]^+$$

$$[C_6H_6C(CH_3)_3]^+ \longrightarrow C_6H_5C(CH_3)_3 + H^+$$

Friedel-Crafts alkylation has two important limitations that severely restrict its usefulness and render the reaction generally less valuable than acylation. As we shall see in Section 23.6, alkylbenzenes are generally *more* reactive in electrophilic substitution reactions than is benzene itself. Hence, Friedel-Crafts alkylation tends to give overalkylation, so that dialkyl and higher alkylated by-products are formed.

$$C_6H_6 \xrightarrow[(CH_3)_3CCl]{FeCl_3} C_6H_5C(CH_3)_3 \xrightarrow[\substack{(CH_3)_3CCl \\ \text{(faster)}}]{FeCl_3} C_6H_4[C(CH_3)_3]_2$$

The only practical way of controlling such additional reactions is to keep benzene in large excess. This approach is practical with benzene itself, since it is an inexpensive compound, but it is impractical with most substituted benzenes, which are more expensive.

Another important limitation of Friedel-Crafts alkylations relates to an alternative reaction of many carbocations, particularly in the absence of reactive nucleophiles, namely, rearrangement to isomeric carbocations. Isopropyl chloride or bromide react normally with aluminum chloride and benzene to give isopropylbenzene.

$$C_6H_6 + (CH_3)_2CHCl \xrightarrow{AlCl_3} C_6H_5CH(CH_3)_2 + HCl$$

However, 1-chloropropane also gives isopropylbenzene under these conditions. Rearrangement to the secondary carbocation is essentially complete.

$$CH_3\overset{H}{\overset{|}{C}}H{-}CH_2{-}Cl\cdots AlCl_3 \longrightarrow CH_3\overset{+}{C}HCH_3\ AlCl_4^- \longrightarrow C_6H_5CH(CH_3)_2$$

Primary alkyl halides are less reactive than secondary or tertiary halides, and higher temperatures are normally required. Under some conditions, the rearrangement of primary systems is only partial. Under these conditions, a displacement reaction by benzene on the alkyl halide coordinated with the Lewis acid competes with carbocation rearrangement. It should be emphasized, however, that at least some rearrangement always occurs with suitable primary systems and such rearrangement greatly limits the utility of this reaction.

Friedel-Crafts alkylations can also be accomplished with alcohols and a catalyst such as aluminum chloride or boron trifluoride. The reaction has the same limitations as the alkyl halide reactions in requiring a large excess of benzene and in giving rearrangement products in suitable cases. In addition, one reaction product is water, which coordinates with Lewis acids. Thus, with alcohols a stoichiometric amount of Lewis acid is required.

OH / OH $\xrightarrow[\text{benzene, }0°C]{BF_3}$ Ph (67%) 2-phenylpentane + Ph (33%) 3-phenylpentane $+ H_2O:BF_3$

In this example, the 2:1 ratio of 2-phenylpentane and 3-phenylpentane is exactly equal to the statistical mixture of isomeric secondary carbocations. Thus, equilibration of the isomeric pentyl cations is rapid compared to their rate of reaction with benzene.

+ $\rightleftharpoons$ + $\rightleftharpoons$ +

EXERCISE 23.5 Explain why reaction of 3-methyl-2-butanol with boron fluoride and benzene gives mostly 2-methyl-2-phenylbutane and no appreciable amount of 2-methyl-3-phenylbutane. Suggest a synthesis of the latter hydrocarbon making use of a Friedel-Crafts acylation in the synthetic sequence.

Alkylation reactions can also be accomplished with alkenes. Typical catalysts used, HF/BF_3 and $HCl/AlCl_3$, generate carbocations in the usual way.

$$(CH_3)_2C{=}CH_2 + C_6H_6 \xrightarrow{HF,\ BF_3} C_6H_5C(CH_3)_3$$

$$(CH_3)_2C{=}CH_2 + HF + BF_3 \rightleftharpoons (CH_3)_3C^+\ BF_4^-$$

This reaction is used industrially to prepare alkylbenzenes, but it is not an important laboratory reaction.

A reaction that is closely related to Friedel-Crafts alkylation is **chloromethylation,** the reaction of aromatic rings with formaldehyde, hydrogen chloride, and a Lewis acid such as zinc chloride.

$$C_6H_6 + CH_2O + HCl \xrightarrow[60°C]{ZnCl_2} C_6H_5CH_2Cl + H_2O$$

(79%)

The reaction is an electrophilic aromatic substitution, probably by the oxonium ion formed by coordination of the formaldehyde with the Lewis acid.

$$[H_2C{=}\overset{+}{O}{-}\overset{-}{Z}nCl_2 \longleftrightarrow H_2\overset{+}{C}{-}O{-}\overset{-}{Z}nCl_2]$$

The resulting reagent will react with aromatic rings that are at least as reactive as benzene. The product of electrophilic aromatic substitution by the coordinated aldehyde is the corresponding alcohol, but this alcohol is benzylic, and in the presence of $ZnCl_2$-HCl it is converted rapidly to the corresponding chloride.

$$p\text{-}CH_3C_6H_4CH_3 + CH_2{=}\overset{+}{O}{-}\overset{-}{Z}nCl_2 \longrightarrow \left[2,5\text{-}(CH_3)_2C_6H_3CH_2O{-}ZnCl + HCl \right] \xrightarrow{HCl} 2,5\text{-}(CH_3)_2C_6H_3CH_2Cl$$

> The chloromethylation reaction must be conducted in an efficient hood and with extreme care. Under these reaction conditions bis(chloromethyl)ether, $ClCH_2OCH_2Cl$, is produced. This compound is a potent carcinogen and should be avoided whenever possible. Alkylation of DNA by reactive halides can produce short- and long-term toxic effects such as teratogenicity (effects on embryo development), mutagenicity (changes in genetic messages), carcinogenicity (cancer-causing effects), and effects on development.

An important side reaction in chloromethylation reactions is reaction of the product, which is a reactive alkyl halide, with the starting aromatic compound. For example, chloromethylation of benzene gives some diphenylmethane, which arises from reaction of the initial product, benzyl chloride, with benzene.

$$C_6H_6 \longrightarrow C_6H_5CH_2Cl \xrightarrow[C_6H_6]{ZnCl_2} C_6H_5{-}CH_2{-}C_6H_5$$

diphenylmethane

Of course, by simply using excess benzene and adjusting the reaction conditions appropriately, this side reaction can be made a practical method for the synthesis of diphenylmethane. A related reaction is the Friedel-Crafts alkylation of benzene with carbon tetrachloride.

$$C_6H_6 + CCl_4 \xrightarrow[\Delta]{AlCl_3} (C_6H_5)_3CCl$$

(84–86%)
triphenylmethyl chloride

Triphenylmethyl chloride or ''trityl'' chloride, is a colorless crystalline solid, m.p. 111–112 °C, that forms the starting point of a fascinating chapter of organic chemistry. In experiments reported in 1900, Moses Gomberg treated triphenylmethyl chloride with finely divided silver in an inert atmosphere to obtain a white solid hydrocarbon formulated as hexaphenylethane. In organic solvents this hydrocarbon gives yellow solutions that rapidly absorb oxygen from the atmosphere. These solutions contain a relatively stable free radical, triphenylmethyl.

$$(C_6H_5)_3C{-}C(C_6H_5)_3 \rightleftharpoons 2\,(C_6H_5)_3C\cdot \xrightarrow{O_2} \underset{\text{bis-triphenylmethyl peroxide}}{(C_6H_5)_3COOC(C_6H_5)_3}$$

The stability of triphenylmethyl radical stems from π-conjugation in the radical and steric hindrance in the dimer. Equilibrium studies showed the central bond in the dimer to have a strength of only 11 kcal mole^{-1}. A remarkable epilogue to this story was provided by recent structural studies based primarily on NMR evidence that show that the hydrocarbon considered to be hexaphenylethane for the better part of a century does not have this structure at all, but is instead the product of dimerization at one *para* position.

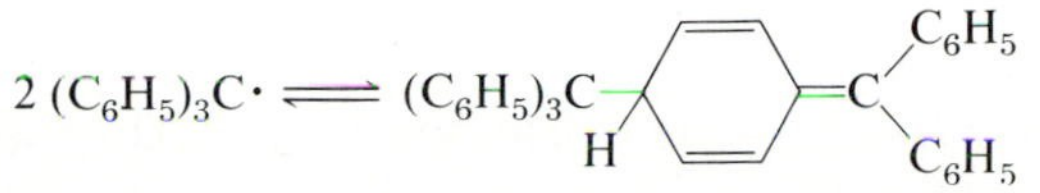

$$2\,(C_6H_5)_3C\cdot \rightleftharpoons (C_6H_5)_3C(H){-}C_6H_5{=}C(C_6H_5)_2$$

In other words, hexaphenylethane is so congested that it is less stable than the isomer shown despite the loss of the resonance energy of one benzene ring in this structure! However, a true hexaphenylethane molecule is formed if steric hindrance is increased around all the *para* positions with *t*-butyl groups at all the *meta* positions. Hexakis-(3,5-di-*t*-butylphenyl) ethane (Figure 23.3) has a central C—C distance of 1.64 Å.

EXERCISE 23.6 Write the equations illustrating the reaction of benzene with the following reagents.

(a) butanoyl chloride, $AlCl_3$
(b) 2-methylpropene, HF, BF_3
(c) 2-butanol, H_2SO_4
(d) formaldehyde, HCl, $ZnCl_2$

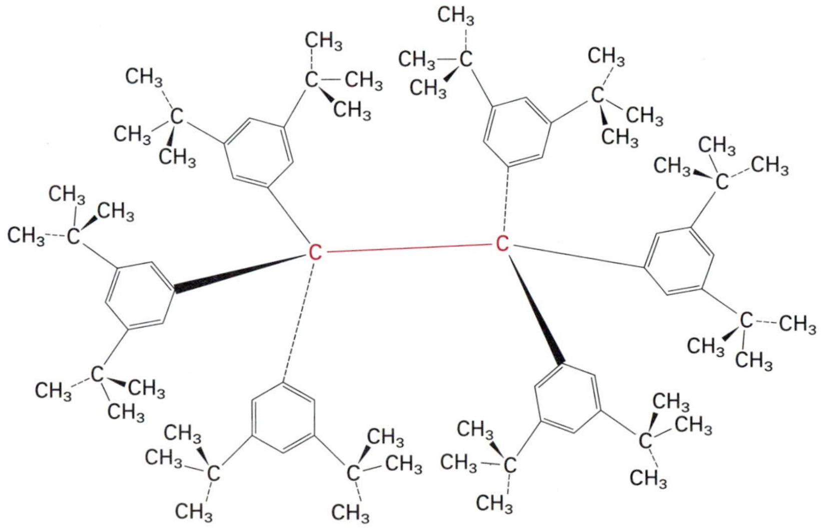

hexakis-(3,5-di-*t*-butylphenyl) ethane

FIGURE 23.3 The formula for hexakis-(3,5-di-*t*-butylphenyl) ethane

Highlight 23.2

Some electrophiles that are useful for substitution on aromatic rings are nitronium ion, acylium ion, carbocations, and oxymethyl carbocation zinc comples. These electrophiles are generated as follows.

$$\text{HO—NO}_2 + \text{H}^+ \rightleftharpoons \text{H}_2\text{O}^+\text{—NO}_2 \rightleftharpoons \underset{\text{nitronium ion}}{\text{NO}_2^{\ +}} + \text{H}_2\text{O}$$

$$\text{R(Cl)C}{=}\overset{+}{\text{O}}\text{—}\overset{-}{\text{AlCl}_3} \rightleftharpoons \text{AlCl}_4^{\ -} + \underset{\text{acylium ion}}{\left[\text{R—}\overset{+}{\text{C}}{=}\text{O} \longleftrightarrow \text{R—C}{\equiv}\overset{+}{\text{O}}\right]}$$

$$\text{R—Cl} + \text{FeCl}_3 \rightleftharpoons \underset{\text{carbocation}}{\text{R}^+} + \text{FeCl}_4^{\ -}$$

$$\text{H}_2\text{C}{=}\text{O} + \text{ZnCl}_2 \rightleftharpoons \underset{\text{oxymethyl carbocation zinc complex}}{\left[\text{H}_2\text{C}{=}\overset{+}{\text{O}}\text{—}\overset{-}{\text{Zn}}\text{Cl}_2 \longleftrightarrow \text{H}_2\overset{+}{\text{C}}\text{—O—}\overset{-}{\text{Zn}}\text{Cl}_2\right]}$$

23.5 Orientation in Electrophilic Aromatic Substitution

Benzene can give only a single monosubstituted product in electrophilic aromatic substitution. However, substitution on a compound that already has a group attached to the ring can give three products. The two substituents in a disubstituted benzene can be arranged *ortho, meta,* or *para* with respect to each other. These three isomers are generally not formed in equal amounts. The product distribution in such cases is affected by the substituent already present on the ring. With some groups further substitution gives mainly the *ortho-* and *para*-disubstituted products. Examples are seen in the products formed from bromination of bromobenzene or anisole or in the nitration of toluene.

Bromobenzene $\xrightarrow[\text{FeBr}_3]{\text{Br}_2}$ *p*-dibromobenzene (87%) *para* + *o*-dibromobenzene (13%) *ortho* + *m*-dibromobenzene (0.1%) *meta*

Anisole (OCH_3) $\xrightarrow[\text{CH}_3\text{COOH}]{\text{Br}_2}$ *p*-bromoanisole (96%) *para* + *o*-bromoanisole (4%) *ortho*

CH_3–C_6H_5 $\xrightarrow[30°C]{HNO_3,\ H_2SO_4}$ o-$CH_3C_6H_4NO_2$ + p-$CH_3C_6H_4NO_2$ + m-$CH_3C_6H_4NO_2$

(62%) *ortho* (33%) *para* (5%) *meta*

On the other hand, further substitution on some substituted benzenes gives essentially all *meta*-disubstituted product, with little or none of the *ortho* and *para* products. Examples are bromination and nitration of nitrobenzene and nitration of methyl benzoate.

$C_6H_5NO_2$ $\xrightarrow[aq.\ H_2SO_4]{Br_2,\ Ag_2SO_4}$ m-$BrC_6H_4NO_2$

(76% yield)
(essentially no
ortho or *para*)

$C_6H_5NO_2$ $\xrightarrow[100°C]{fuming\ HNO_3,\ H_2SO_4}$ m-$C_6H_4(NO_2)_2$ + o-$C_6H_4(NO_2)_2$ + p-$C_6H_4(NO_2)_2$

(93%) *meta* (6%) *ortho* (1%) *para*

$C_6H_5COOCH_3$ $\xrightarrow[10°C]{HNO_3,\ H_2SO_4}$ m-$O_2NC_6H_4COOCH_3$

(81–85% yield)

As the foregoing examples suggest, substituent groups can be divided into two categories, those that are ***ortho,para* directors** and those that are ***meta* directors.** Bromo, methyl, and methoxy groups are *ortho,para* directors and nitro and ester groups are *meta* directors. Note that *ortho* and *para* are produced together, although the *ortho*/*para* ratio may vary with different groups and under different reaction conditions with the same group.

Substituent groups may also be characterized with respect to their effect on the rate of further substitution reactions. Some substituents cause the aromatic ring to be less reactive than benzene itself; these groups are said to be **deactivating.** An example is the nitro group, which is a powerful deactivating group. Nitrobenzene is much less reactive than benzene, as shown by the conditions required for nitration of benzene (page 695) and nitrobenzene (above). Nitrobenzene is unreactive enough to be used as a solvent for nitrations with nitronium perchlorate. When two deactivating groups are attached to the ring, even more drastic conditions are necessary for further substitution. For example, *m*-dinitrobenzene may be converted into 1,3,5-trinitrobenzene in 45% yield by heating

60 g of the dinitrobenzene with 1 kg of fuming sulfuric acid and 0.5 kg of fuming nitric acid at 100 °C for 5 days.

> Commercial concentrated sulfuric acid is 98% H_2SO_4. Fuming sulfuric acid contains additional dissolved sulfur trioxide (Section 26.5.B). Concentrated nitric acid is a solution approximately 70% by weight of HNO_3 in water. The solution is colorless but becomes yellow by photochemical decomposition to yield NO_2. The red "fuming" nitric acid contains additional dissolved NO_2.

Carbonyl groups are also deactivating groups. Thus, the products of Friedel-Crafts acylation are less reactive than the starting material. Consequently, it is quite easy to achieve monosubstitution in such reactions. In fact, Friedel-Crafts acylation is generally not applicable at all to aromatic rings that contain a strongly deactivating group. For example, nitrobenzene does not react under Friedel-Crafts acylation conditions and is even used as a solvent for such reactions!

Other groups are **activating**; electrophilic substitution on rings containing these groups is more rapid than with benzene. As we mentioned in our discussion of Friedel-Crafts alkylation, alkyl groups are activating groups, and hence alkylbenzenes are more reactive than benzene itself. This effect has an important consequence, since it means that the product of Friedel-Crafts alkylation is *more reactive* than the starting material. Thus, it is difficult to avoid the formation of overalkylation by-products.

The activating effect of alkyl groups can also be seen in the conditions required for nitration of mesitylene, which may be compared with the conditions used for the nitration of benzene itself (page 695).

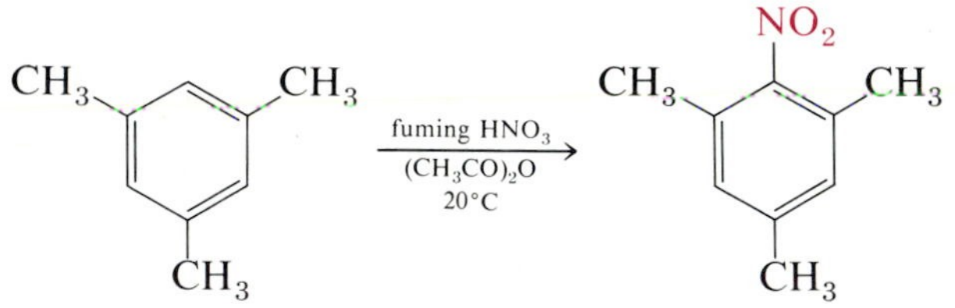

CH₃ CH₃ CH₃ — fuming HNO_3, $(CH_3CO)_2O$, 20°C → NO_2, CH₃ CH₃ CH₃

mesitylene — (74–76%) nitromesitylene

Mesitylene, with its three activating groups, is so reactive that it undergoes halogenation even without the normal Lewis-acid catalyst (see Section 23.1).

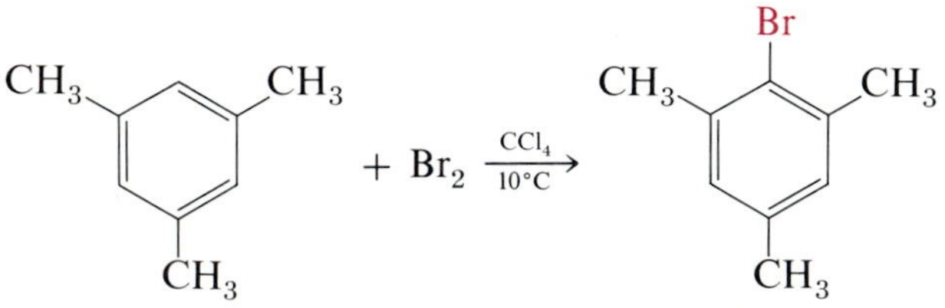

mesitylene + Br_2 — CCl_4, 10°C → Br (79–82%) bromomesitylene

In such a case the reaction can be considered to be a displacement reaction on halogen with the ring acting as a nucleophile.

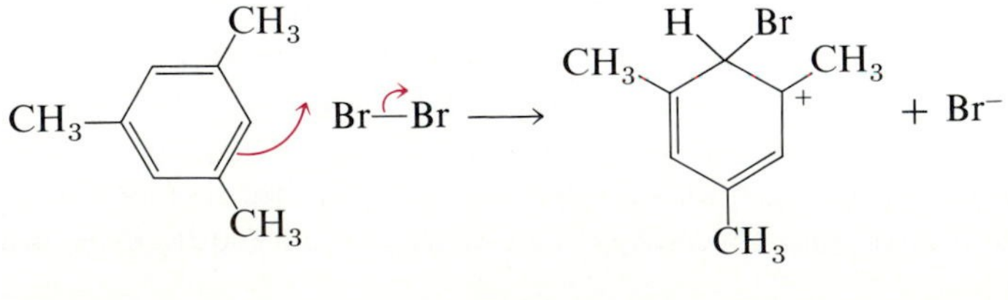

$$\text{mesitylene} + Br{-}Br \longrightarrow \text{H, Br arenium ion (+)} + Br^-$$

Mesitylene is especially reactive because the intermediate produced is highly stabilized; all three of the usual resonance structures correspond to tertiary carbocations:

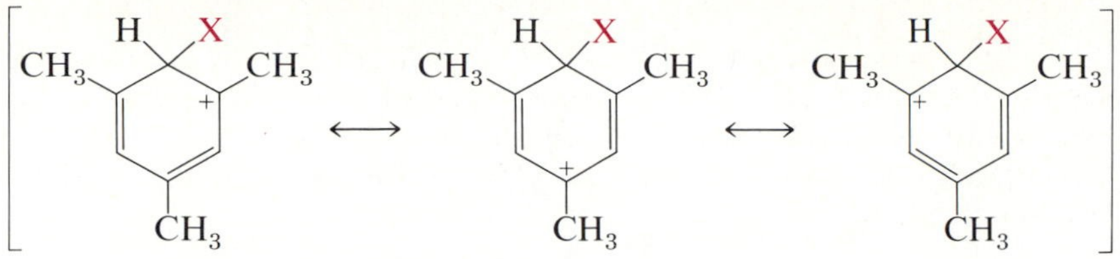

We will elaborate on the foregoing rationale for the activating effect of alkyl groups in Sections 23.6 and 23.7. For the present, however, it is useful to note that substituent groups may be grouped into three different classes with regard to whether they are activating or deactivating and whether they are *ortho,para* directing or *meta* directing.

1. ***Ortho,para directing and activating.*** Functional groups in this category include R (alkyl), NH_2, NR_2, and NHCOR (amino, alkylamino, and amide), OH, OR, and OCOR (hydroxy, alkoxy, and ester).
2. ***Ortho,para directing and deactivating.*** The most important functional groups in this category are the halogens, F, Cl, Br, and I.
3. ***Meta directing and deactivating.*** This group includes NO_2 (nitro), SO_3H (sulfonic acid), and all carbonyl compounds: COOH, COOR, CHO, and COR (carboxylic acids, esters, aldehydes, and ketones).

Note that all activating groups are *ortho,para* directors and all *meta* directors are deactivating. These generalizations derive from many experimental observations and form a set of empirical and useful rules. However, these rules are also subject to a consistent and satisfying interpretation by the modern theory of organic chemistry. This theory has its basis in the electron-donating and electron-attracting character of different functional groups, as discussed in the next section.

EXERCISE 23.7 Make a six-by-four matrix. On one side put the aromatic compounds toluene, acetophenone, bromobenzene, anisole, nitrobenzene, and acetanilide. On the other side put the four reactions bromination, nitration, Friedel-Crafts alkylation and Friedel-Crafts acylation. Each intersection of the matrix represents a reaction. What products (if any) are formed in each reaction? Which reactions occur more rapidly than the analogous reaction on benzene itself?

23.6 Theory of Orientation in Electrophilic Aromatic Substitution

In Section 23.1 we learned that the mechanism of electrophilic aromatic substitution involves combination of a positive or electrophilic species with a pair of π-electrons of the benzene ring to form an intermediate having a pentadienyl cation structure.

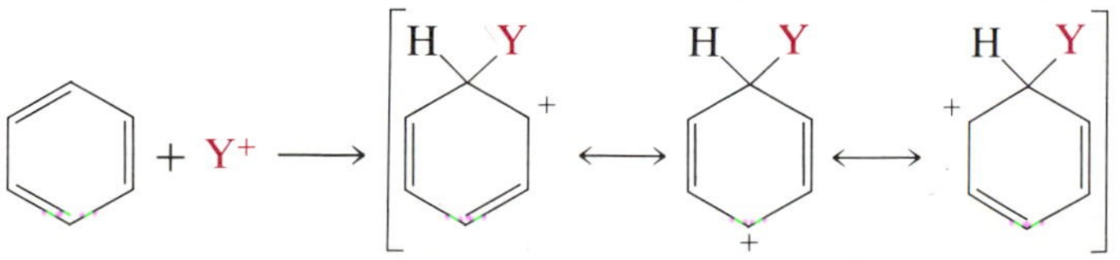

$$C_6H_6 + Y^+ \longrightarrow \left[\cdots \longleftrightarrow \cdots \longleftrightarrow \cdots \right]$$

The transition state has much of the character of the pentadienyl cation intermediate to which it leads. Those factors that affect the relative energy or stability of the intermediate

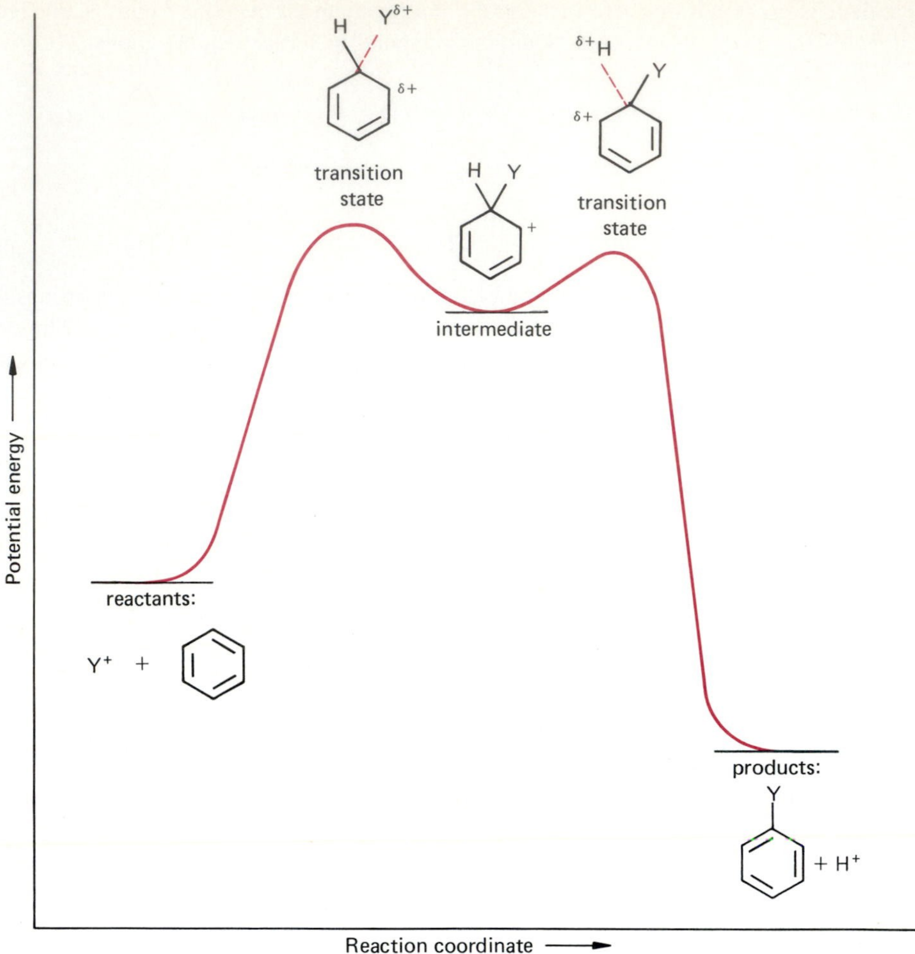

FIGURE 23.4 Energy profile for electrophilic substitution on benzene.

also affect to a lesser but substantial degree the relative energy or stability of the transition state. An energy profile for substitution on benzene is illustrated in Figure 23.4. The modern electronic theory of orientation in electrophilic aromatic substitution involves an assessment of the effect of a substituent on the relative energies of the pentadienyl cation-like transition state for reaction at different possible positions.

For example, reaction at the *ortho* position of toluene gives rise to a transition state that resembles the intermediate

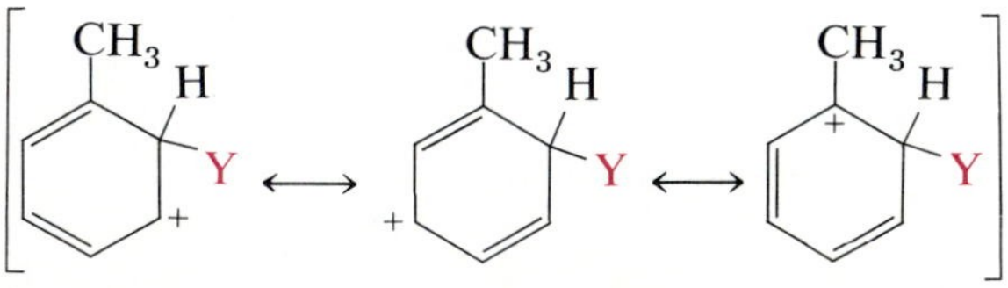

Two of the structures are those of secondary carbocations, but the third corresponds to a more stable tertiary carbocation. As a result, this intermediate and *hence also the transition state that leads to it* are more stable—have lower energy—than the corresponding intermediate and transition state for benzene in which all three resonance structures are those of secondary carbocations. The *ortho* position of toluene is therefore expected to be more reactive than a single position of benzene.

> This argument must be put on a per-hydrogen basis. Without specific orientation preferences, statistics alone would give a reactivity ratio for benzene/*ortho*/*meta*/*para* of 6:2:2:1.

Reaction at the *meta* position gives rise to the following resonance structures.

All three structures are those of secondary carbocations. Each structure is stabilized slightly by the C_{methyl}—C_{ring} dipole.

Correspondingly, the *meta* position of toluene is expected to be somewhat more reactive than a benzene position but not nearly as reactive as the *ortho* position.

Finally, we apply this approach to the *para* position to generate the resonance structures

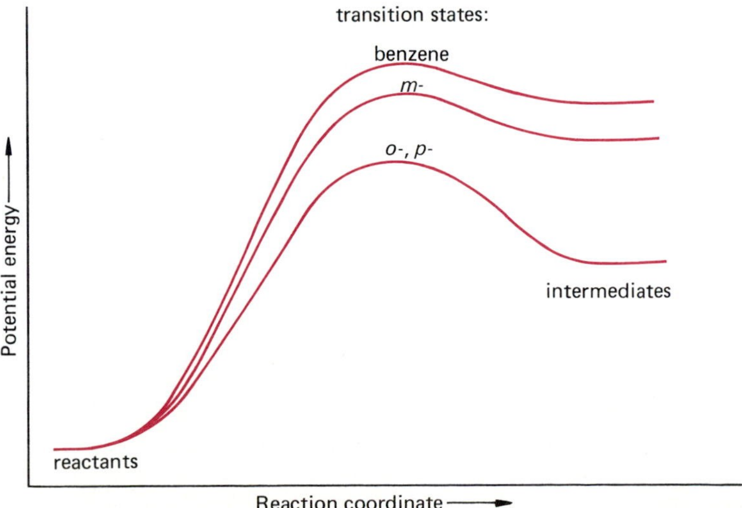

FIGURE 23.5 Energy profile for reaction at *ortho, para,* and *meta* positions of toluene compared to benzene.

Here again we find two secondary carbocation structures and one tertiary carbocation. The overall energy of the transition state is comparable to that for *ortho* substitution. Indeed, this approach does not distinguish between preference for *ortho* relative to *para* substitution, but does indicate why substituents divide into the two broad groups of *ortho,para* and *meta* directors.

The resulting energy profile for reaction at toluene is compared with that for benzene in Figure 23.5. The alternative pathways differ less in the transition state than in the intermediate. In the transition state, only a partial positive charge has to be distributed; in the intermediate, a full positive charge is developed along with a fully formed C—Y bond. We usually examine the formulas for the intermediates because it is easier to write the formulas with full rather than partial bonds. The same argument applies to the positive charge within the benzene ring in the transition state. The net result is that of predominant *ortho,para* orientation; although the *meta* position is more reactive than a single benzene position, the *ortho* and *para* positions are even more so.

We next apply this approach to the corresponding reaction at the *ortho, para,* and *meta* positions of nitrobenzene and derive three sets of resonance structures for the intermediates (and transition states) involved.

ortho

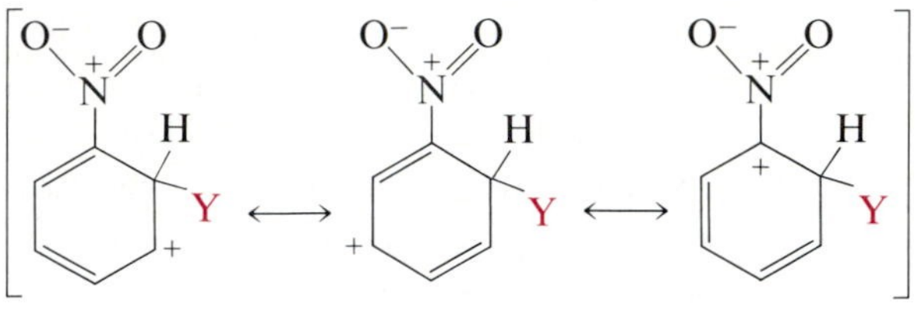

O⁻ O N H Y ⟷ O⁻ O N H Y ⟷ O⁻ O N H Y

para

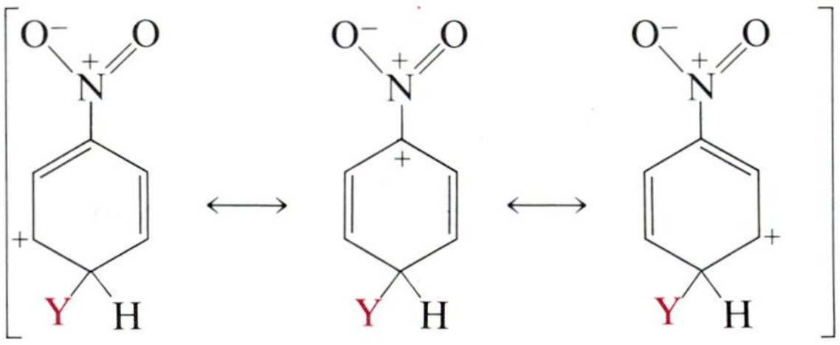

O⁻ O N Y H ⟷ O⁻ O N Y H ⟷ O⁻ O N Y H

meta

O⁻ O N H Y ⟷ O⁻ O N H Y ⟷ O⁻ O N H Y

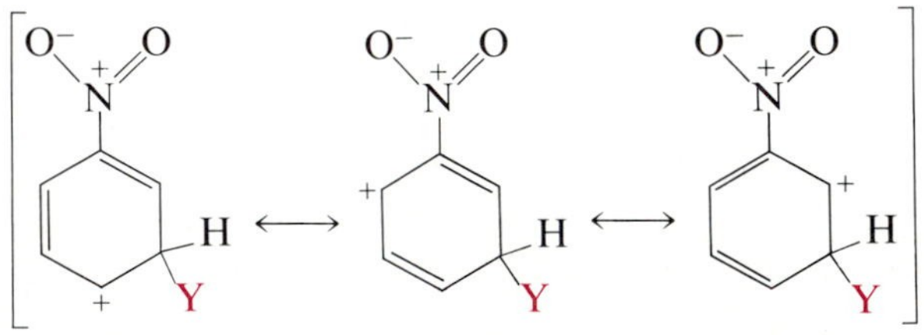

All of these structures involve the electrostatic repulsion of the carbocation charge with the strong dipole of the nitro group.

$$(+) \qquad \overset{\longleftrightarrow}{C\text{—}NO_2}$$

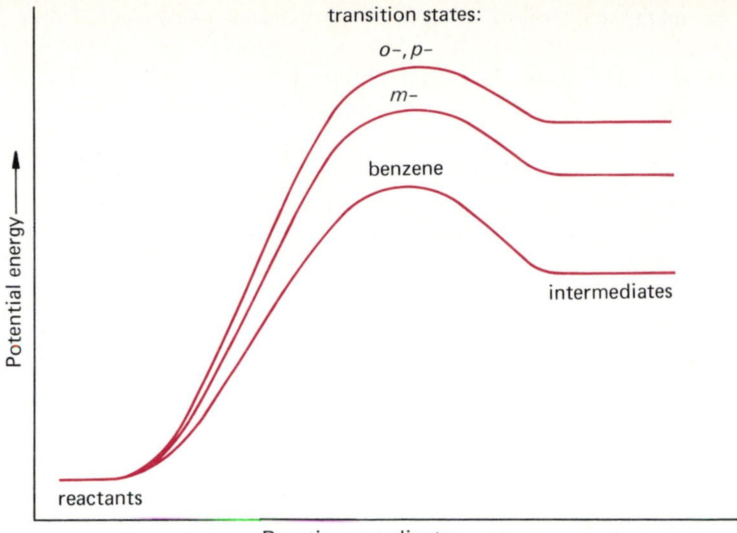

FIGURE 23.6 The intermediate derived from *meta* reaction of nitrobenzene is formed less readily than that from attack at benzene but more readily than that from reaction at the *ortho* or *para* positions.

That is, every one of these structures is substantially less stable than the corresponding structure for reaction at benzene; hence, all positions in nitrobenzene are expected to be deactivated relative to benzene. For reaction at the *ortho* and *para* positions, however, one structure is that of a carbocation right next to the positive nitrogen of the nitro group. This structure in each case is of such high energy compared to the other structures, in which the positive charges are separated by one or more atoms, that it contributes very little to the overall resonance hybrid. The *meta* reaction involves only structures in which the positive charges are separated; thus, although the transition state for *meta* substitution is of higher energy than for reaction at benzene, it is of lower energy than those for reaction at the *ortho* or *para* positions. We can phrase this result another way: the *meta* reaction is deactivated less than *ortho* or *para* reaction. The corresponding reaction profiles are summarized in Figure 23.6.

These principles apply generally to other types of substituents. For anisole, we may write the same sets of three resonance structures.

ortho

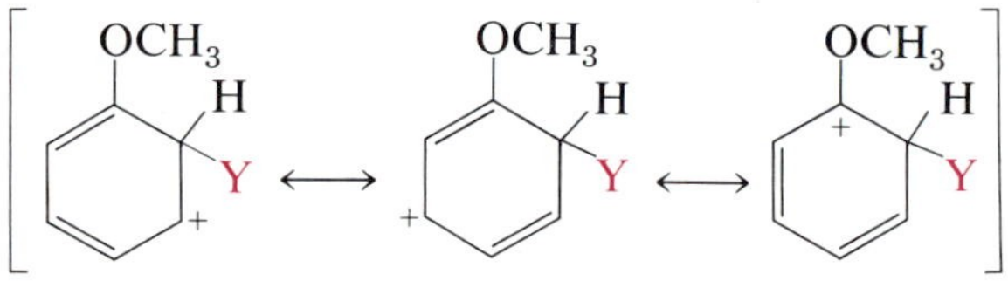

para

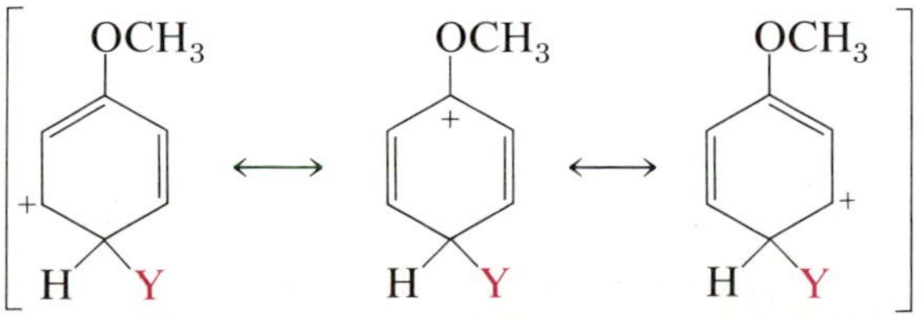

meta

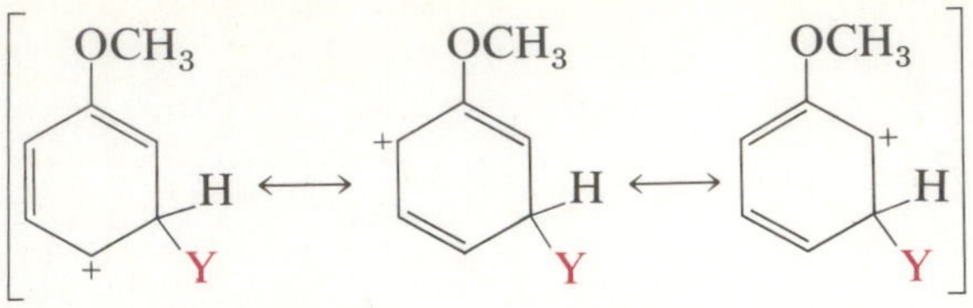

All of these structures are expected to be somewhat destabilized by the electrostatic interaction with the C—O dipole, but reaction at the *ortho* and *para* positions also corresponds to oxonium ions.

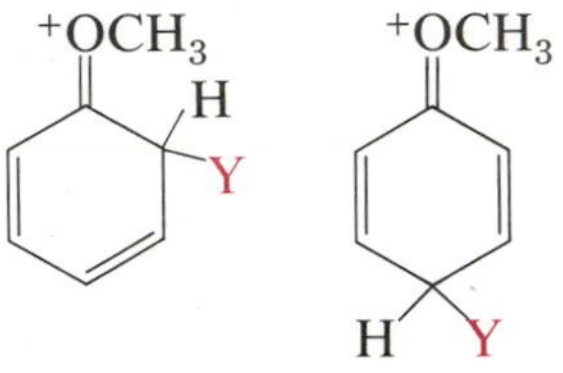

These additional structures greatly stabilize the intermediates and the transition states leading to them. Similar structures are involved in acid-catalyzed reactions of carbonyl compounds.

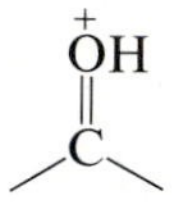

The oxonium ion structures so dominate the system that the *ortho* and *para* positions of anisole are highly activated compared to benzene. We shall see in Chapter 30 that electrophilic substitution reactions at these positions in phenols and phenyl ethers and esters are accomplished under rather mild conditions. On the other hand, reaction at the *meta* positions is expected to be somewhat less facile than in benzene.

Reactions at the *ortho* and *para* positions of aromatic amines involve related immonium ion structures, for example,

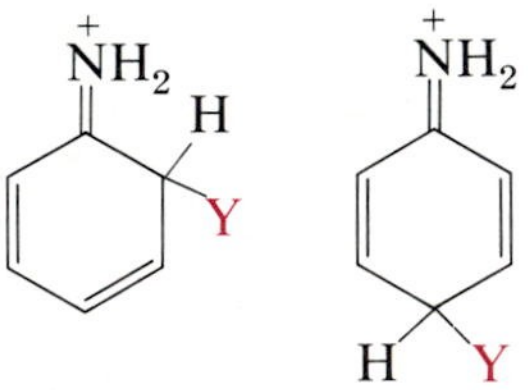

Consequently these positions are also highly activated relative to benzene.

Let us now apply the procedure to a halobenzene. Reaction at the *meta* position gives the three structures

meta

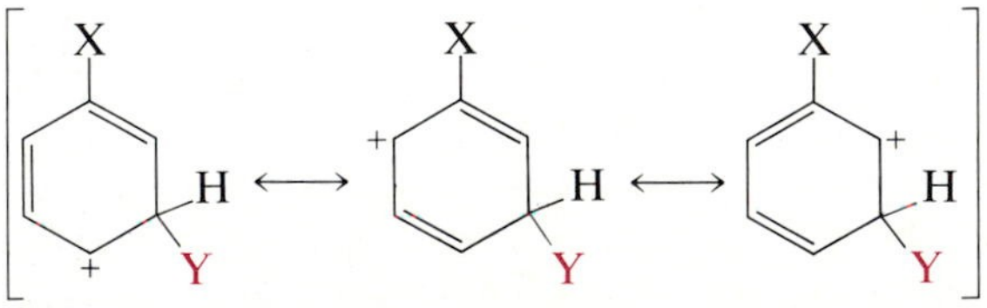

All three structures are strongly destabilized by electrostatic interaction of the positive charge with the carbon-halogen dipole.

(+) $\overset{\longleftrightarrow}{C{-}X}$

Accordingly, the *meta* position in the halobenzene is strongly deactivated relative to benzene. Reactions at the *ortho* and *para* positions involve similar carbocation structures destabilized by interaction with the carbon-halogen dipole.

ortho

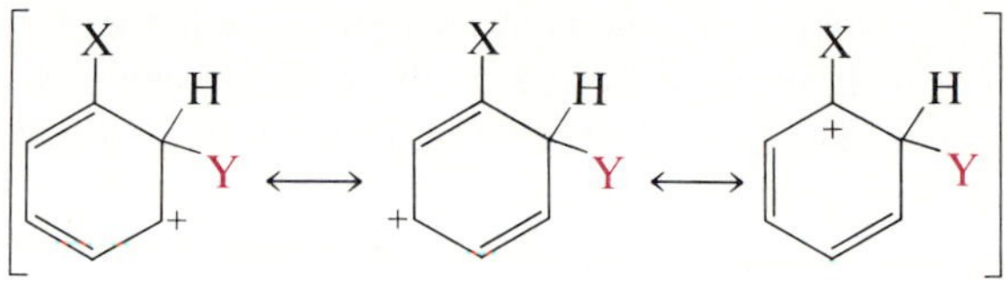

para

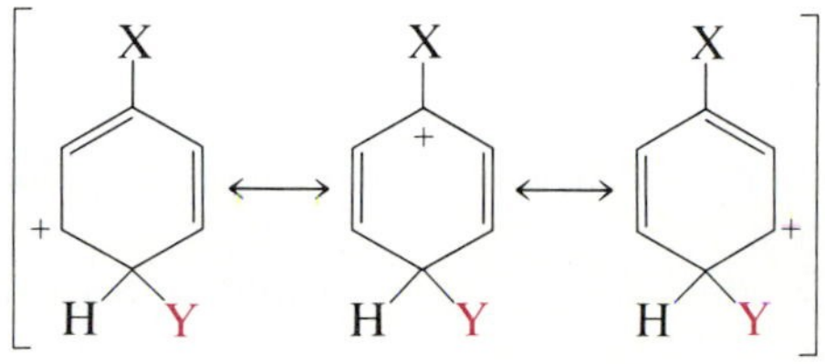

In both cases, however, one structure is that of an α-halocarbocation in which interaction with a halogen lone pair is possible to give the halonium ion structures.

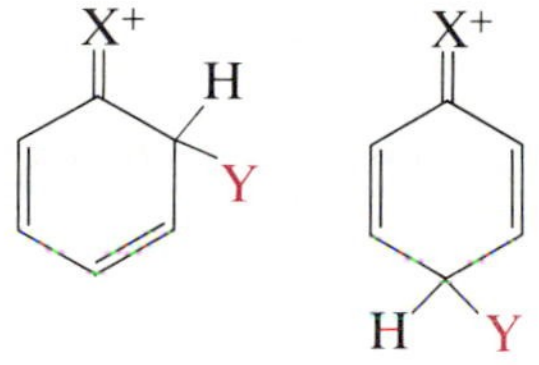

Such halonium ion structures are not nearly as stable as related oxonium and immonium ions. In practice, the additional contribution of such structures does not compensate for the deactivating effect of the carbon-halogen dipoles on the other structures, but it does make reaction at the *ortho* and *para* positions far more facile than at the *meta* position.

EXERCISE 23.8 (a) Draw energy profiles for reactions at the *ortho, para,* and *meta* positions of anisole compared to benzene. (b) Draw an energy profile for reaction of a halobenzene compared to benzene. (c) Use the theory developed in this section to predict the orientation specificity and reactivity relative to benzene of vinyl ($CH{=}CH_2$) and formyl (CHO) groups.

Highlight 23.3

The first substituent can influence the position of entry of a second substituent. **Electron-donating groups** activate the ring and are ***ortho,para* directing.** Alkyl groups activate by hyperconjugation; other groups activate by conjugation. The stronger the electron donation, the more reactive the aromatic ring. Higher electron density favors the nucleophilic activity of the ring towards electrophiles. Halogen groups withdraw electrons inductively but supply electrons through π-conjugation. The **halogens** are **deactivating,** but ***ortho,para* directing. Electron-withdrawing groups** deactivate the ring, and are ***meta* directing.**

23.7 Quantitative Reactivities: Partial Rate Factors

Nitration reactions have been studied extensively for many aromatic compounds, and relative reactivities at different positions have been determined. Furthermore, by studying the reaction of a mixture of benzene and some other compound, it is often possible to determine the quantitative reactivities of various positions relative to a benzene position. These statistically corrected relative reactivities are known as partial rate factors.

For example, the reaction of equimolar amounts of toluene and benzene with a small amount of nitric acid in acetic anhydride at 30 °C gives one part of nitrobenzene to 27 parts of nitrotoluenes. The nitrotoluenes formed are 58.1% *ortho,* 3.7% *meta,* and 38.2% *para*. The partial rate factors, f_i, are calculated as follows.

$$\begin{aligned} f_o &= (\text{fraction } o)\left(\frac{\text{no. benzene positions}}{\text{no. } o \text{ positions}}\right)\left(\frac{\text{toluene reactivity}}{\text{benzene reactivity}}\right) \\ &= (0.581)(\tfrac{6}{2})(27) = 47 \\ f_m &= (0.037)(\tfrac{6}{2})(27) = 3 \\ f_p &= (0.382)(\tfrac{6}{1})(27) = 62 \end{aligned}$$

Note that the *meta* position is more reactive than a benzene position, as predicted by the theory developed in Section 23.6.

Partial rate factors for nitration of several substituted benzenes are summarized in Table 23.1. Some effects are clearly apparent in these results. For example, a *t*-butyl group has much the same effect as a methyl group in the *meta* and *para* positions, but at the *ortho* position *t*-butylbenzene is much less reactive than toluene. The difference is clearly to be attributed to steric hindrance caused by the bulky *t*-butyl group. At the distant *meta* and *para* positions the size of the alkyl group has little effect.

TABLE 23.1 Partial Rate Factors for Nitration

Substituent	ortho	meta	para
CH_3	47	3	62
$C(CH_3)_3$	5.5	4	75
CH_2Cl	0.29	0.14	0.95
F	0.06	—	0.8
Cl	0.03	0.0009	0.14
Br	0.03	0.0011	0.11
I	0.2	0.01	0.6
NO_2	1.8×10^{-6}	28×10^{-6}	0.2×10^{-6}
$COOC_2H_5$	1×10^{-4}	2.5×10^{-4}	0.24×10^{-4}

The halobenzenes follow the theory outlined in Section 23.6. All positions are less reactive than benzene, but the *meta* positions are more strongly deactivated than *ortho* and *para*. The chloromethyl group is of special interest since the stabilizing effect of an alkyl group is decreased by the deactivating effect of the carbon-chlorine dipole. The result is a net *ortho,para* orientation with a little net deactivation.

But quantitative data for many electrophilic substitution reactions are sparse. The amounts formed of some isomers are so minute as to defy detection even by modern gas chromatography analytical methods. One approach to obtaining quantitative reactivity results for all positions in a given molecule, even when they differ greatly in reactivity, has been to study the simplest possible electrophilic aromatic substitution reaction, the replacement of one hydrogen isotope by another. Examples of this reaction were given in Section 23.2.

In principle, it is possible to prepare a variety of specifically labeled aromatic compounds and to study quantitatively the rate of loss of the hydrogen isotope under a consistent set of acidic conditions. A comparison of the rates of replacement of deuterium by hydrogen (protodedeuteration) of specifically deuterated anisoles with the corresponding rate for deuteriobenzene in aqueous perchloric acid gives the results displayed in Table 23.2. These results demonstrate the high reactivity of the *ortho* and *para* positions compared to benzene and the lower reactivity in the *meta* position. These same relative rates are expected to correspond approximately to nitration as well and imply that nitration of anisole gives only a few parts per million of *m*-nitro product. This minute quantity is extremely difficult to detect directly in the product mixture.

TABLE 23.2 Relative Rates of Protodedeuteriation in Aqueous Perchloric Acid

Benzene-D	2-D-anisole (OCH_3, D ortho)	3-D-anisole (OCH_3, D meta)	4-D-anisole (OCH_3, D para)
1	6×10^4	0.3	2×10^4

EXERCISE 23.9 Using the data in Table 23.1, calculate the percent composition of the three chloronitrobenzenes formed by nitration of chlorobenzene.

23.8 Effects of Multiple Substituents

The relative rates of replacement of tritium by hydrogen (protodetritiation) in trifluoroacetic acid for toluene and the dimethylbenzenes compared to benzene are summarized in Table 23.3. The energy effects of two methyl groups are approximately additive compared to the effect of one methyl group in toluene. For example, the 3-position in *o*-dimethylbenzene is *ortho* to one methyl and *meta* to the other. The predicted reactivity is therefore (219)(6.1) = 1340, which agrees exactly with the experimental reactivity.

TABLE 23.3 Relative Rates of Protodetritiation in Trifluoroacetic Acid

Compound	Position	Rate (predicted)
Toluene (CH_3)	ortho	219
	meta	6.1
	para	445
o-Dimethylbenzene (CH_3, CH_3)	3-position	1340 (1340)
	4-position	1900 (2720)
p-Dimethylbenzene (CH_3, CH_3)	2-position	1460 (1340)
m-Dimethylbenzene (CH_3, CH_3)	2-position	36600 (48000)
	4-position	78800 (97500)

The product of the two partial rate factors is taken because it is the activation energy quantities that are additive

$$\Delta G^\ddagger \text{ (3-position in } o\text{-dimethylbenzene)} = \Delta G^\ddagger(o\text{-}) + \Delta G^\ddagger(m\text{-})$$

The energies are related to the logarithms of the rate constants.

$$RT \ln k(\text{3-position}) = RT \ln (o\text{-}) + RT \ln (m\text{-})$$

$$\log f(\text{3-position}) = \log f_o + \log f_m = \log f_o f_m$$

$$f(\text{3-position}) = f_o f_m$$

The relative reactivities of the toluene positions were used as partial rate factors to derive the predicted reactivities of the dimethylbenzenes given in parentheses in Table 23.3. The approximate agreement can be generalized to electrophilic substitution reactions of polysubstituted benzenes. That is, the net orientation effects of two or more substituents can be predicted approximately by examining the effects of each substituent separately. If all substituents orient preferentially to the same positions, such positions are strongly preferred. For example, nitration of the following disubstituted benzenes gives the percentage of nitration at each position as indicated.

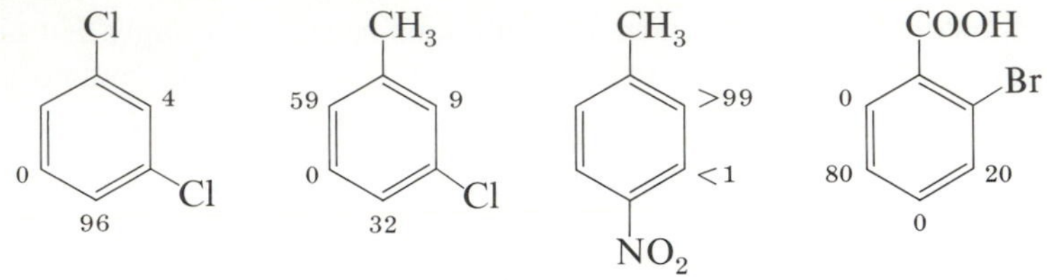

In *m*-chlorotoluene, the 5-position is *meta* both to chlorine and to methyl, and no significant reaction occurs at this position. The other positions are all *ortho* or *para* to both groups, and a disagreeable mixture results. In *p*-nitrotoluene, however, the highly favored 2-position is *ortho* to the *ortho,para*-directing methyl and *meta* to the *meta*-directing nitro.

If the groups already present have conflicting orientation preferences, it is helpful to divide substituents into three classes:

1. Strongly activating *ortho,para* directors, such as OR and NR_2.
2. Alkyl groups and halogens.
3. All *meta* directors.

If two substituents belong to different classes, the orientation effect of the superior class dominates. The following nitration results are examples.

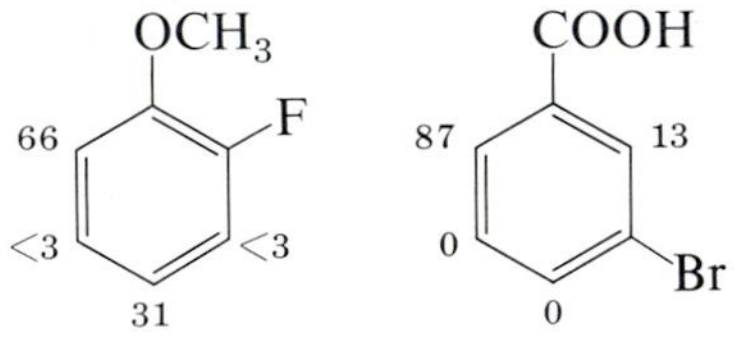

Note that the effects of all *ortho,para* directors dominate over *meta* directors.

Finally, if both substituents are in the same class, all bets are off and horrible mixtures can be anticipated. The following nitration results are examples.

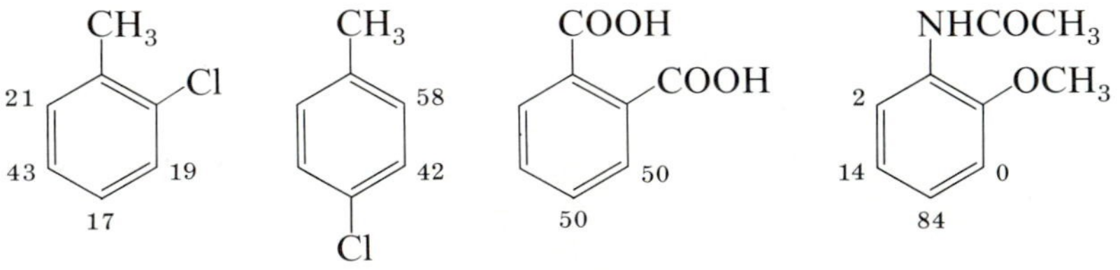

In our subsequent studies of the reactions of functional groups on benzene rings, we shall see that many syntheses can be accomplished by aromatic substitution reactions combined with functional group transformations. In such sequences, the order in which reactions are accomplished is of great importance because of the orientation preferences of different groups. One example will demonstrate this point.

$$p\text{-}O_2NC_6H_4CH_3 \xrightarrow[H_2SO_4]{HNO_3} 2,4\text{-}(O_2N)_2C_6H_3CH_3 \xrightarrow[H_2SO_4]{Na_2Cr_2O_7} 2,4\text{-}(O_2N)_2C_6H_3COOH$$

$$p\text{-}O_2NC_6H_4CH_3 \xrightarrow[H_2SO_4]{Na_2Cr_2O_7} p\text{-}O_2NC_6H_4COOH \xrightarrow[H_2SO_4]{HNO_3} 2,4\text{-}(O_2N)_2C_6H_3COOH + 3,4\text{-}(O_2N)_2C_6H_3COOH$$

mixture

The first route is clearly to be preferred as a preparation of 2,4-dinitrobenzoic acid.

EXERCISE 23.10 What major product or products are expected in nitration of (a) *p*-nitroanisole, (b) *m*-nitrobenzoic acid, (c) *p*-bromochlorobenzene, and (d) *m*-chlorotoluene?

23.9 Synthetic Utility of Electrophilic Aromatic Substitution

In this chapter we have seen how some important functional groups can be introduced directly into benzene and many of its derivatives by electrophilic substitution reactions. Important examples of such synthetically useful reactions are halogenation, nitration, and Friedel-Crafts acylation. In Chapter 26 we will learn of an additional useful reaction, sulfonation. Each of these functional groups can serve as a substrate for additional electrophilic substitution reactions, or the group can be converted to other functional groups. Halogens can be converted via lithium or Grignard reagents to a variety of functional groups. These and other reactions of aromatic halides will be discussed in Chapter 30. Nitro compounds can be reduced to amines, which in turn can be transformed to many different groups as detailed in Chapter 25. Aromatic ketones can participate in the usual reactions of carbonyl groups. One important example is reduction of the carbonyl group to a methylene group by either the Wolff-Kishner or the Clemmensen method (Section 14.8.D).

$$C_6H_5CO(CH_2)_4CH_3 \xrightarrow{H_2NNH_2} \xrightarrow[\Delta]{KOH} C_6H_5CH_2(CH_2)_4CH_3$$

(82%)
hexylbenzene

$$C_6H_5CO(CH_2)_{16}CH_3 \xrightarrow[\Delta]{Zn/Hg,\ HCl} C_6H_5CH_2(CH_2)_{16}CH_3$$

(77%)
octadecylbenzene

Since Friedel-Crafts *acylation* is generally a clean, high-yield reaction, the combination of acylation and reduction is generally to be preferred to Friedel-Crafts *alkylation*.

In our discussion of electrophilic substitution reactions we have considered the effects on orientation and reactivity of other substituents already on the benzene ring. We now need to consider how electrophilic substitution may be used in a practical sense to prepare polysubstituted benzenes. We want especially to consider how the process may be used in some cases to provide practical syntheses of pure compounds.

Remember that the goal of any chemical synthesis is generally to prepare **one pure compound** for some purpose. Therefore, whenever possible, one must use reactions that do not give mixtures of isomers. When there is no known method that provides only one isomer, a synthesis may still be acceptable if the desired isomer is produced in substantial amounts (hopefully as the *major* product) and if it may be separated in some way from the unwanted isomers.

Some electrophilic substitution reactions fit the first criterion; that is, one of the possible isomers is produced almost exclusively. Substitution on *meta*-orienting compounds usually falls into this category. Thus the following substitution reactions are good preparative reactions.

CHO → (HNO_3, H_2SO_4) → CHO, NO_2 (75–84%)

NO_2 + Br_2 → ($FeBr_3$) → NO_2, Br (60–75%)

Recall that the Friedel-Crafts acylation reaction often *does not work when the ring already contains a* meta-*directing group*. Thus *m*-nitroacetophenone may be prepared by nitration of acetophenone but not by Friedel-Crafts acylation of nitrobenzene.

acetophenone (O=C–CH_3) → (HNO_3, H_2SO_4) → O=C–CH_3, NO_2 (55%)

NO_2 → (CH_3COCl, $AlCl_3$) → no reaction

In many of these reactions, a few percent of the *ortho* and *para* isomers are produced. However, if the major isomer is crystalline, as is usually the case, it may easily be purified by recrystallization.

When the substituent already in the ring is an *ortho,para* director, mixtures invariably result, as we have seen in previous sections. In such cases direct electrophilic substitution is less satisfactory as a synthetic method. However, some benzene derivatives may still be obtained in this manner, particularly the *para* isomers. Because of its symmetrical nature, the *para* isomer usually has a significantly higher melting point than the *ortho* or *meta* isomer. Some representative data are summarized in Table 23.4. Recall that a higher melting point represents a more stable crystal lattice and lower solubility. Consequently the higher melting *para* isomer may often be crystallized from the mixture of *ortho* and *para* products of direct substitution. It is generally not possible to isolate the *ortho* isomer in a pure state by this technique.

OH OH

+ Br_2 $\xrightarrow{CS_2}$

Br

(80–84%)
m.p. 66°C

Cl Cl

+ Br_2 $\xrightarrow{FeBr_3}$

Br

(60%)
m.p. 68°C

Br Br

$\xrightarrow[H_2SO_4]{HNO_3}$

NO_2

(72%)
m.p. 127°C

TABLE 23.4 Melting Points of Disubstituted Benzenes

	Melting Point, °C		
Substituents	*ortho*	*meta*	*para*
Br, Br	7	−7	87
Cl, Cl	−17	−25	53
Br, Cl	−12	−22	68
CH_3, Br	−26	−40	29
CH_3, NO_2	−10	16	55
Br, NO_2	43	56	127
Cl, NO_2	35	46	84
Br, COOH	150	155	255
Cl, COOH	142	158	243
OH, Br	6	33	66

Another useful generalization is that the acylating agent obtained by coordination of aluminum chloride with an acyl halide behaves as a rather bulky reagent. Consequently, Friedel-Crafts acylation reactions tend to give almost completely *para* products, which are usually easy to separate from the small amounts of other isomers.

$$C_6H_5CH_3 + C_6H_5COCl \xrightarrow{AlCl_3} o\text{-}CH_3C_6H_4COC_6H_5 + m\text{-}CH_3C_6H_4COC_6H_5 + p\text{-}CH_3C_6H_4COC_6H_5$$

(9%) *o*-methylbenzophenone; (1%) *m*-methylbenzophenone; (90%) *p*-methylbenzophenone

On the other hand, Friedel-Crafts alkylations tend to be rather nonspecific. Often the orientations appear to be quite unusual. For example, under mild conditions with aluminum chloride in acetonitrile, isopropylation of toluene gives predominantly the expected *ortho* and *para* products, but there is also a substantial amount of *meta* product.

$$C_6H_5CH_3 \longrightarrow o\text{-}CH_3C_6H_4CH(CH_3)_2 + m\text{-}CH_3C_6H_4CH(CH_3)_2 + p\text{-}CH_3C_6H_4CH(CH_3)_2$$

(63%) (12%) (25%)

This unusual behavior is due to rearrangements that alkylbenzenes undergo under the conditions of Friedel-Crafts alkylation. For example, if the foregoing mixture of isopropyltoluenes is treated under vigorous conditions with $AlCl_3$ and HCl, the product is exclusively the *meta* isomer. Although we shall not go into these rearrangements in detail, the student should be aware that Friedel-Crafts alkylations are often complicated and difficult to predict. Thus the reaction has less general utility as a synthetic method.

Since *ortho* and *para* isomers usually have closely similar boiling points, fractional distillation is usually not a satisfactory method for separation of such isomer mixtures, but

TABLE 23.5 Boiling Points of Disubstituted Benzenes

	Boiling Point, °C		
Substituents	*ortho*	*meta*	*para*
Br, Br	225	218	219
Cl, Cl	181	173	174
Br, Cl	204	196	196
CH_3, Br	182	184	184
CH_3, Cl	159	162	162
Br, NO_2	258	265	256
Cl, NO_2	246	236	242
CH_3, NO_2	220	233	238
NO_2, NO_2	319	291	299
OCH_3, NO_2	277	258	274

there are exceptions to this generalization. Some representative data collected in Table 23.5 show that *o*- and *p*-nitrotoluenes differ sufficiently in boiling point to be separable by fractional distillation. On the other hand, the melting points of the bromotoluenes are too low for effective crystallization, and their boiling points are too close for simple fractionation; hence, the bromination of toluene is *not* a satisfactory route to any of the bromotoluenes.

In summary, direct electrophilic substitution is a useful synthetic method as such if only one isomer is produced or if the mixture can be conveniently separated by physical means. To predict whether such a reaction will be useful, the chemist must consider both the mechanism of the reaction—that is, what the isomer distribution is expected to be—and the probable physical properties of the expected products. We shall see in future chapters that the utility of electrophilic substitution may be extended by modification and interrelation of functional groups and by a technique in which one or more positions on the ring are temporarily deactivated or blocked.

EXERCISE 23.11 In this chapter we have learned how Br and $COCH_3$ groups can be introduced into a number of aromatic compounds by bromination and Friedel-Crafts acylation, respectively. Review the transformations you have already learned of ArBr to $ArCH_2CH{=}CH_2$, ArCOOH, $ArCR_2OH$ and of $ArCOCH_3$ to ArCOOH, $ArCH_2CH_3$, $ArC(CH_3)_2OH$. How is each of these transformations affected by the presence in the aryl group of each of the following functions: CHO, COOH, Br, NO_2?

EXERCISE 23.12 From the partial rate factors in Table 23.1 calculate the percent *ortho*, *meta*, and *para* nitration products from nitrobenzene and ethyl benzoate.

ASIDE

The German chemist Mitscherlich first prepared nitrobenzene from benzene in 1834. Why does one use a mixture of nitric and sulfuric acids for this transformation? Write out all steps in the reaction and draw an energy versus reaction coordinate diagram. Is the benzene ring present in the intermediate? Does the benzene ring persist at all stages of an electrophilic reaction?

PROBLEMS

1. Benzene can be iodinated with iodine and an oxidizing agent such as nitric acid or hydrogen peroxide. The actual electrophilic reagent in this reaction is probably IOH_2^+, which may be regarded as I^+ bound to a water molecule. Write a balanced equation for the generation of this intermediate from I_2 and H_2O_2. Include this as part of an overall mechanism for the reaction of I_2 and H_2O_2 with benzene to give iodobenzene, C_6H_5I.

2. a. The chloromethylation reaction of benzene with formaldehyde

$$C_6H_6 + CH_2O + HCl \xrightarrow{ZnCl_2} \underset{\text{benzyl chloride}}{C_6H_5CH_2Cl} + H_2O$$

could involve as the principal electrophilic reagent either $CH_2{=}\overset{+}{O}{-}\overset{-}{Zn}Cl_2$ or $^+CH_2Cl$. Write complete reaction mechanisms using both intermediates. Note that under these reaction conditions, benzyl alcohol, $C_6H_5CH_2OH$, reacts rapidly with $ZnCl_2$ and HCl (Lucas reagent) to give benzyl chloride.

b. In such chloromethylation reactions a carcinogenic agent, bis(chloromethyl) ether, $ClCH_2OCH_2Cl$, is produced as a by-product. Write a plausible mechanism for formation of this compound from formaldehyde, HCl, and $ZnCl_2$, showing each intermediate involved.

3. Benzene can be mercurated to give phenylmercuric acetate, $C_6H_5HgOOCCH_3$, with mercuric acetate in acetic acid containing some perchloric acid as an acid catalyst. The electrophilic reagent involved is probably $^+HgOOCCH_3$. Write a complete reaction mechanism.

4. Biphenyl, C_6H_5-C_6H_5, may be considered as a benzene with a phenyl substituent. Show why this hydrocarbon is expected to direct to the *ortho,para* positions, using resonance structures.

5. Use resonance structures to show why the COOH group in benzoic acid is a *meta* director.

6. Indicate the principal mononitration product or products expected from each of the following compounds.

a. 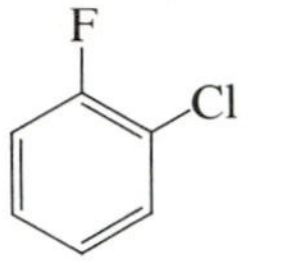 F, Cl (2-chlorofluorobenzene)
b. 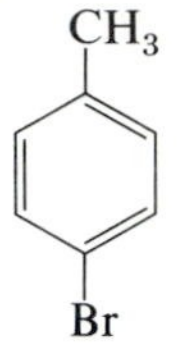 CH_3, Br (4-bromotoluene)
c. 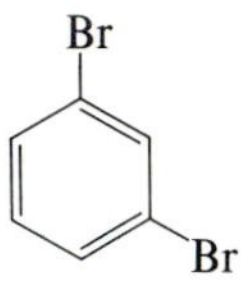 Br, Br (1,3-dibromobenzene)
d.  NH_2, CH_3 (4-methylaniline)
e. 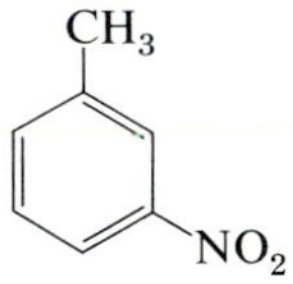 CH_3, NO_2 (3-nitrotoluene)
f. 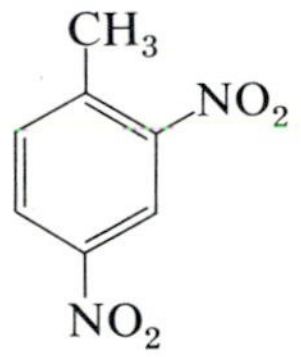 CH_3, NO_2, NO_2 (2,4-dinitrotoluene)
g.  $COOCH_3$, CH_3 (methyl 2-methylbenzoate)
h. 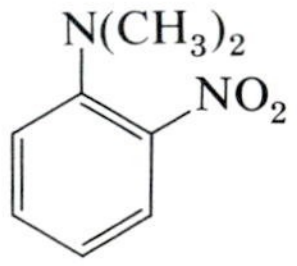 $N(CH_3)_2$, NO_2 (N,N-dimethyl-2-nitroaniline)
i. 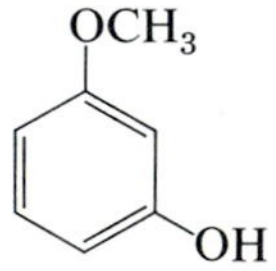 OCH_3, OH (3-methoxyphenol)
j. 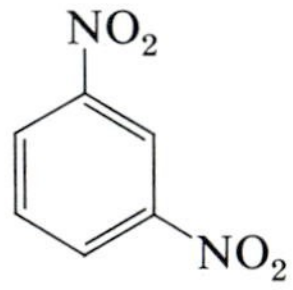 NO_2, NO_2 (1,3-dinitrobenzene)

7. a. Toluene is 605 times as reactive as benzene toward bromination in aqueous acetic acid. The bromotoluenes produced are 32.9% *ortho*, 0.3% *meta*, and 66.8% *para*. Calculate the partial rate factors.

b. The partial rate factors for chlorination of toluene are *ortho*, 620; *meta*, 5.0; *para*, 820. Calculate the isomer distribution in chlorination of *m*-xylene (*m*-dimethylbenzene). The experimental result is 77% 4-, 23% 2-, and 0% 5-chloro substitution.

c. The partial rate factors for chlorination of chlorobenzene are *ortho*, 0.1; *meta*, 0.002; *para*, 0.41. Calculate the isomer distribution in chlorination of *p*-chlorotoluene (the experimental result is 77% 2,4-dichlorotoluene and 23% 3,4-dichlorotoluene).

8. Which of the following compounds can probably be prepared in a pure state from benzene by using two successive electrophilic substitution reactions? For each compound, write out the reaction sequence and describe how the intermediates and products would be purified.

a. (Cl, Br) b. (COCH$_3$, COCH$_3$) c. (NO$_2$, COCH$_3$)

d. (Cl, COCH$_2$CH$_3$) e. (Cl, NO$_2$) f. (C(CH$_3$)$_3$, NO$_2$)

9. Which of the following compounds can probably be prepared in a pure state by electrophilic substitution on a disubstituted benzene? Outline the method in each case.

a. (OCH$_3$, NO$_2$, NO$_2$) b. (COOH, NO$_2$, COOH) c. (NO$_2$, O$_2$N, NO$_2$)

d. (NO$_2$, NO$_2$, NO$_2$) e. (Cl, Cl, NO$_2$) f. (OH, Br, Br)

g. (CH$_3$, Cl, CH$_2$CH$_3$) h. (OCH$_3$, O$_2$N, CH$_3$) i. (NO$_2$, Br, COOH)

j. (OCH$_3$, NO$_2$, CHO)

10. Toluene is *ortho,para* directing, whereas trifluoromethylbenzene, $C_6H_5CF_3$, is *meta* directing. Explain.

11. An interesting variant of Friedel-Crafts acylation is the Gatterman-Koch aldehyde synthesis, the reaction of an aromatic hydrocarbon with carbon monoxide and hydrogen chloride in the presence of a Lewis acid such as aluminum chloride. The reaction is equivalent to a Friedel-Crafts acylation with formyl chloride, HCOCl. Reaction with toluene gives primarily *p*-tolualdehyde (*p*-methylbenzaldehyde). Write a reasonable reaction mechanism.

12. **a.** The reaction of benzene with isobutyl alcohol and BF_3 gives primarily *t*-butylbenzene. Explain.

b. The reaction of 2-butanol and BF_3 with benzene at 0 °C gives 2-phenylbutane in good yield. When 2-butanol-2-*d*, $CH_3CDOHCH_2CH_3$, is used, a mixture of deuterated compounds is obtained that includes major amounts of

$$CH_3\underset{\displaystyle C_6H_5}{\underset{|}{C}}DCH_2CH_3 \quad \text{and} \quad CH_3\underset{\displaystyle C_6H_5}{\underset{|}{C}}HCHDCH_3$$

Explain.

c. By contrast, the reaction of $CD_3CHOHCH_3$ with benzene and BF_3 gives the following compound in good yield.

$$CD_3\underset{\displaystyle C_6H_5}{\underset{|}{C}}HCH_3$$

with no deuterium scrambling. How do you account for this difference?

d. 2-Propanol-1-d_3, $CD_3CHOHCH_3$, has a stereocenter and significant optical activity. According to the mechanism of the alkylation reaction, what do you expect for the steric course of the reaction of this optically active alcohol with benzene and BF_3?

13. When a solution of *p*-di-(3-pentyl)benzene in benzene is treated with aluminum chloride at 25 °C, a rapid transfer of a pentyl group occurs to give monopentylbenzene. The reaction product is approximately one part of 2-phenylpentane and two parts of 3-phenylpentane. Account for these results with a reasonable reaction mechanism.

14. Show how all three nitrobenzoic acids can be prepared from toluene.

15. On heating with aqueous sulfuric acid, styrene reacts to form a dimer in good yield.

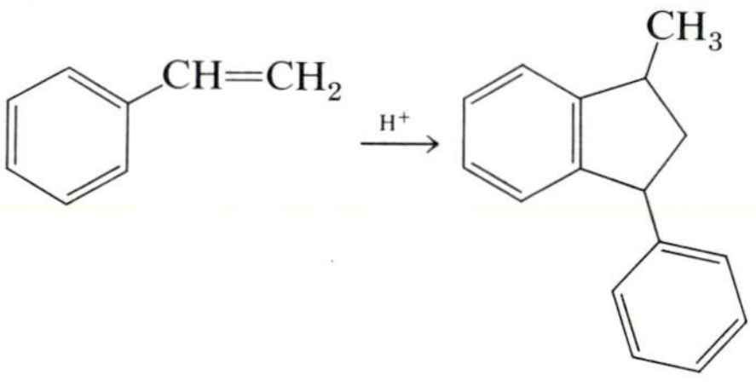

$$C_6H_5CH{=}CH_2 \xrightarrow{H^+} \text{1-methyl-3-phenylindane}$$

(77–81%)

Write a reasonable mechanism, showing all intermediates involved.

16. Show how each of the following compounds can be prepared from benzene or toluene in a practical manner.

a. 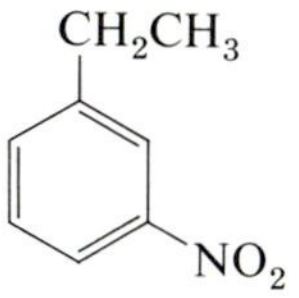 $C_6H_4(CH_2CH_3)(NO_2)$ (meta)

b. $C_6H_5CH(CH_2CH_3)_2$

c. $C_6H_4(CH_2CH_3)(NO_2)$ (para)

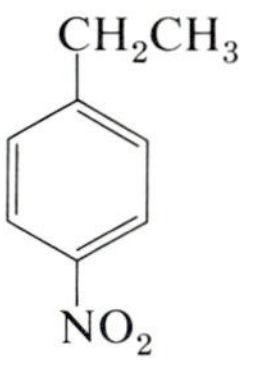

d. 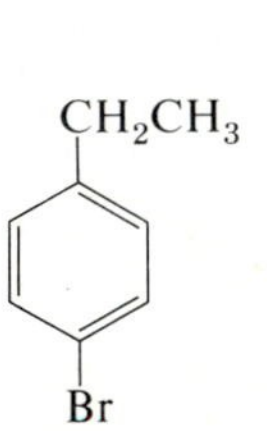 $C_6H_4(CH_2CH_3)(Br)$ (para)

e. 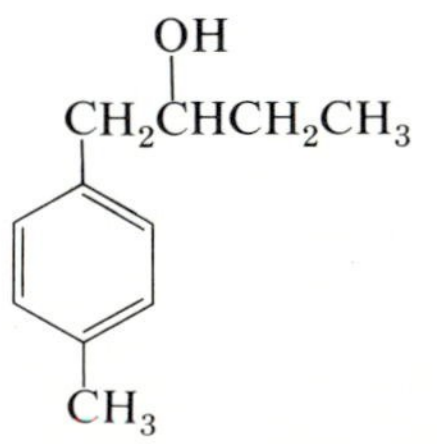 $p\text{-}CH_3C_6H_4CH_2\underset{|}{\overset{OH}{C}}HCH_2CH_3$

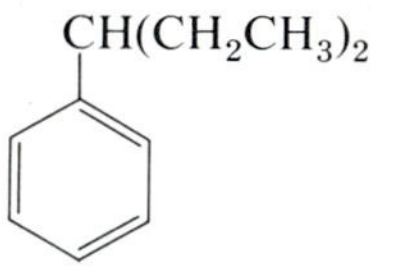

f.  $C_6H_4(COOCH_3)(Br)$ (meta)

17. The solvolysis reaction of 2-chloro-2-phenylpropane in aqueous acetone is an S_N1 carbocation process that yields 2-phenyl-2-propanol as the principal product.

a. Write out the mechanism of this reaction, showing any intermediates involved.

b. The rate of reaction depends markedly on substituents in the phenyl group. The order of reactivity given by *para* substituents is

$$CH_3O > CH_3 > H > NO_2$$

Explain, using resonance structures.

18. The dissociation of triarylmethyl chlorides into ions in liquid sulfur dioxide solution has been studied quantitatively, and a number of dissociation constants have been measured for the equilibrium

$$Ar_3CCl \overset{K}{\rightleftharpoons} Ar_3C^+ + Cl^-$$

Rank the following compounds in order of increasing K and explain. (m-ClC_6H_4)CCl, (p-$O_2NC_6H_4$)$_3$CCl, (m-$CH_3C_6H_4$)$_3$CCl, (p-$CH_3C_6H_4$)$_3$CCl, (p-$CH_3OC_6H_4$)$_3$CCl, $(C_6H_5)_3CCl$

19. a. Solvolysis of 2-methyl-2-phenylpropyl tosylate in acetic acid gives primarily a mixture of 2-methyl-1-phenyl-2-propyl acetate and 2-methyl-1-phenylpropene. Write a reasonable reaction mechanism.

b. Solvolyses of 3-phenyl-2-butyl tosylates in acetic acid also give mixtures of alkenes and esters. The ester product of solvolysis of the optically active 2S,3R diastereomer is racemic [equal amounts of (2*S*,3*R*)-3-phenyl-2-butyl acetate and (2*R*,3*S*)-3-phenyl-2-butyl acetate], whereas that from the 2R,3R tosylate is the optically active ester (2*R*,3*R*)-3-phenyl-2-butyl acetate. Provide a reasonable explanation.

CHAPTER 24

AMINES

24.1 Structure

Amines are compounds in which one or more alkyl or aryl groups are attached to nitrogen. They may be considered to be the organic relatives of ammonia in the same way that alcohols and ethers are related to water.

H_2O	ROH	R_2O
water	alcohols	ethers

$$\underset{\text{ammonia}}{NH_3} \qquad \underbrace{RNH_2 \quad R_2NH \quad R_3N}_{\text{amines}}$$

Amines are classified as **primary, secondary,** or **tertiary** according to the number of alkyl or aryl groups joined to the nitrogen. Note that these descriptive adjectives are used here to denote the *degree of substitution* on nitrogen, not the nature of the substituent groups. In secondary and tertiary amines the alkyl or aryl groups may be the same or different.

Some Primary Amines

CH_3NH_2 $\quad (CH_3)_2CHNH_2$ $\quad (CH_3)_3CNH_2$

Some Secondary Amines

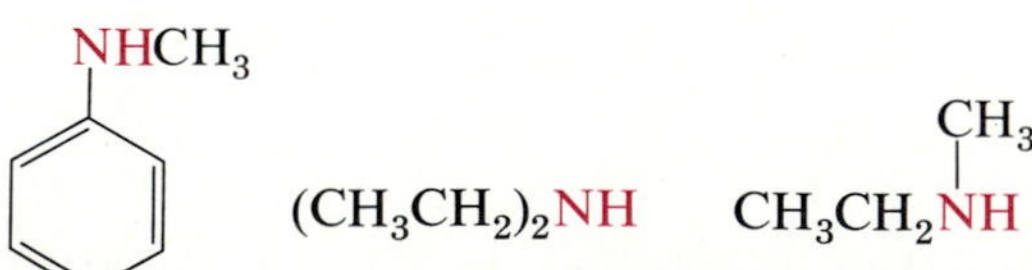

Some Tertiary Amines

$(CH_3)_3N$ $\quad$ $C_6H_5N(CH_3)_2$ $\quad$ $C_6H_{11}N(CH_3)_2$

Quaternary ammonium compounds are related to simple inorganic ammonium salts. Again the four groups joined to nitrogen in the ammonium ion may be the same or different.

$(CH_3)_4\overset{+}{N}\ Cl^-$ $\quad$ $(CH_3CH_2)_2\overset{+}{N}(CH_3)_2\ Br^-$ $\quad$ $CH_3CH_2CH_2\overset{+}{N}(CH_3)_3\ OH^-$ $\quad$ $C_6H_5\overset{+}{N}(CH_3)_3\,Br^-$

Recall that ammonia has a pyramidal shape. The nitrogen-hydrogen bond length is 1.008 Å, and the H—N—H bond angle is 107.3°. The hybridization of nitrogen is approximately sp^3. It forms three approximately sp^3-s σ-bonds to hydrogen and has a **nonbonding electron pair** that occupies the other approximately sp^3-orbital. Amines have similar structures, as shown in Figure 24.1.

ammonia

Bond Length, Å		Bond Angle, deg	
NH	1.011	HNH	105.9
CN	1.474	HNC	112.9

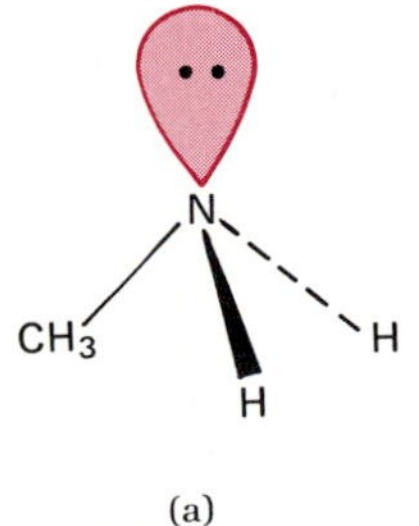

(a)

Bond Length, Å		Bond Angle, deg	
CN	1.47	CNC	108

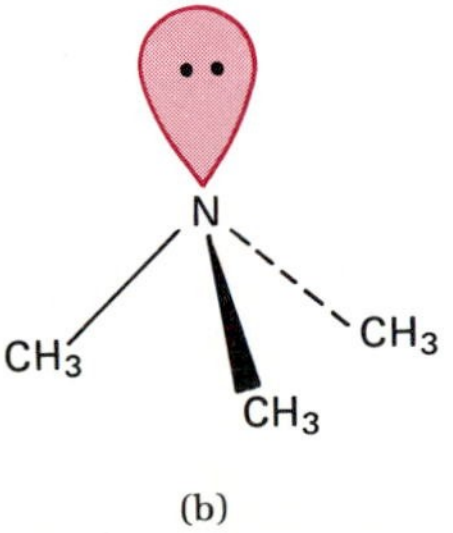

(b)

FIGURE 24.1 Simple amine structures: (a) methylamine; (b) trimethylamine.

Consider the following progression of bond lengths (in Å).

$CH_3—CH_3$	1.531	$H—CH_3$	1.085
$CH_3—NH_2$	1.474	$H—NH_2$	1.012
$CH_3—OH$	1.427	$H—OH$	0.957
$CH_3—F$	1.385	$H—F$	0.917

As we proceed along the first row of the periodic table, the increasing nuclear charge causes the electron orbitals to shrink and results in shorter bonds.

The nonbonding electron pair is important in the chemistry of amines, since it is responsible for the typical basic and nucleophilic properties of these compounds. Amines that have an aryl group attached to nitrogen are characterized by somewhat larger H—N—H and H—N—C angles; that is, the nitrogen is more nearly planar than in alkylamines. We will discuss the reason for this difference in Section 24.4.

Because of the pyramidal geometry, an amine with three different groups joined to nitrogen is chiral (alternatively, amines may be regarded as approximately tetrahedral with the nonbonding pair being the fourth "group").

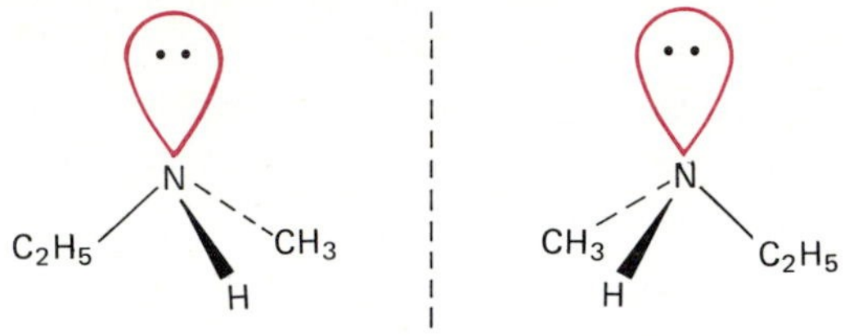

C_2H_5 N CH_3 H | CH_3 N C_2H_5 H

Recall that enantiomeric carbon compounds may be separated and that the individual enantiomers are quite stable because it is necessary to break and reform bonds to interconvert them. In contrast, the two enantiomers of a chiral amine are readily interconvertible by a process known as **nitrogen inversion.** For simple amines the activation energy required for inversion is rather small, on the order of 6 kcal mole^{-1}. In the planar transition state for inversion the nitrogen has sp^2-hybridization with the lone pair in the p_z-orbital.

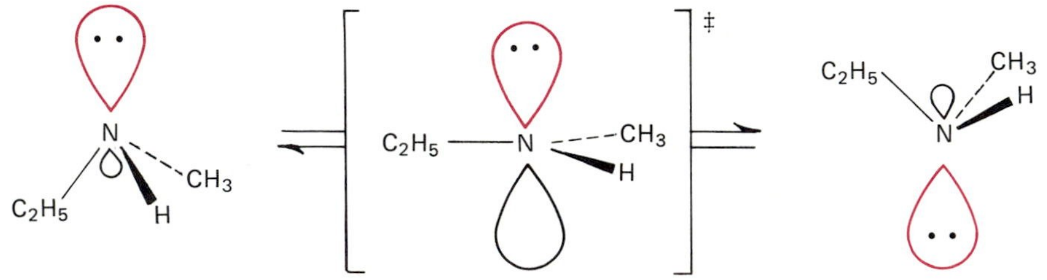

N CH_3 C_2H_5 H ⇌ [C_2H_5—N CH_3 H]‡ ⇌ C_2H_5 CH_3 H N

For quaternary ammonium compounds such inversion is not possible and chiral ions may be separated into enantiomers that are relatively stable.

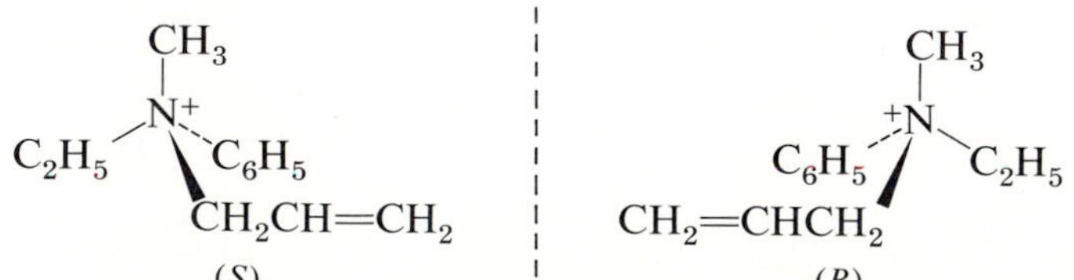

CH_3 N^+ C_2H_5 C_6H_5 $CH_2CH{=}CH_2$ (*S*) | CH_3 ^+N C_6H_5 C_2H_5 $CH_2{=}CHCH_2$ (*R*)

EXERCISE 24.1 The optically active allylethylmethylphenylammonium halides racemize slowly in solution. The rate of racemization is temperature dependent and is faster for the iodide than for the bromide. Propose a mechanism for the racemization.

$$C_2H_5-\overset{\overset{CH_3}{|}}{N^+}(C_6H_5)(CH_2CH{=}CH_2)\ X^- \rightleftharpoons C_6H_5-\overset{\overset{CH_3}{|}}{N^+}(C_2H_5)(CH_2{=}CHCH_2)\ X^-$$

24.2 Nomenclature

Like most other classes of organic compounds, amines have been named in several ways. Simple amines are usually referred to by common names, which are derived by using the suffix **-amine,** preceded by the name or names of the alkyl groups. The names are written as one word.

CH_3NH_2 methylamine

$(CH_3CH_2)_2NH$ diethylamine

$(CH_3CH_2CH_2)_3N$ tri-*n*-propylamine

$CH_3CH_2N(H)CH_3$ ethylmethylamine

$CH_3CH_2N(CH_2CH_3)CH_3$ diethylmethylamine

$C_3H_5-N(CH_3)CH_2CH_3$ cyclopropylethylmethylamine

Under the IUPAC rules amines are named as derivatives of a parent hydrocarbon by using the prefix **amino-** to designate the group NH_2.

CH_3NH_2 aminomethane

$CH_3CH_2CH(CH_3)NH_2$ 2-aminobutane

$(CH_3)_2CHCH_2CH(NH_2)CH_2CH_3$ 4-amino-2-methylhexane

In this system secondary and tertiary amines are named by using a compound prefix that includes the names of all but the largest alkyl group.

$CH_3CH_2N(CH_3)_2$ dimethylaminoethane

$(CH_3)_2CHCH_2CH(CH_3)N(CH_2CH_3)_2$ 2-(diethylamino)-4-methylpentane

$CH_3CH_2CH(CH_3)N(CH_3)CH_2CH_3$ 2-(ethylmethylamino)butane

$CH_3CH_2CH(CH_3)CH_2CH_2N(H)CH_3$ 1-(methylamino)-3-methylpentane

The simplest arylamine is **aniline.** This well-entrenched common name has the official sanction of the IUPAC. Simple derivatives are named as substituted anilines.

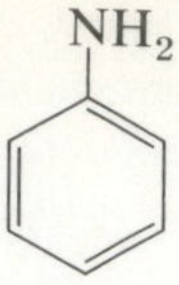

NH_2 — aniline

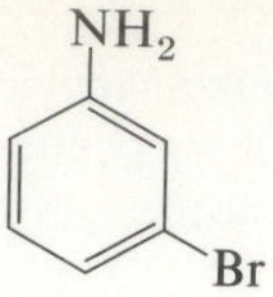

NH_2, Br — *m*-bromoaniline

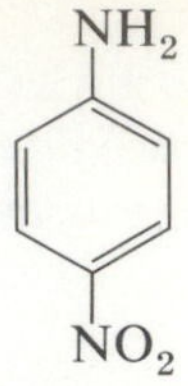

NH_2, NO_2 — *p*-nitroaniline

$N(CH_3)_2$ — N,N-dimethylaniline

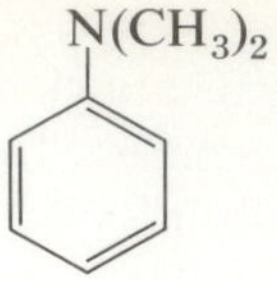

When it is necessary to name a compound containing the amino group as a derivative of some other function, the prefix **amino-** is employed.

$CH_3NHCH_2CH_2CH_2COOH$
4-(methylamino)butanoic acid

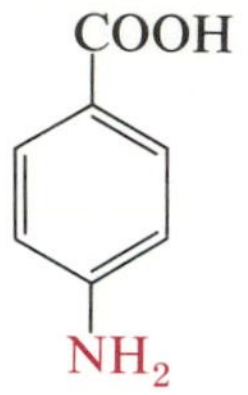

COOH, NH_2
p-aminobenzoic acid

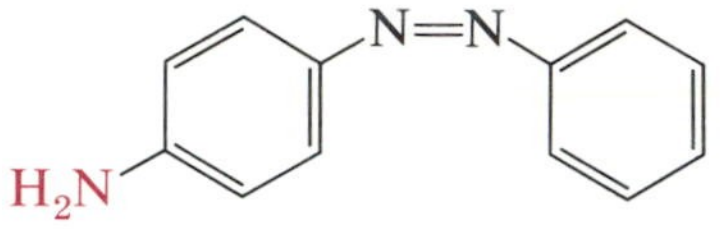

H_2N, N=N
p-aminoazobenzene
[now: *cis*-1-(*p*-aminophenyl)-2-phenyldiazene]

The ''azo'' designation for the group —N=N— may also be called **diazene** (see Chapter 25). The compound shown here, *p*-aminoazobenzene, is *cis*-1-(*p*-aminophenyl)-2-phenyldiazene.

Several aromatic amines have trivial names that have received IUPAC sanction. Some of the more important examples are shown below.

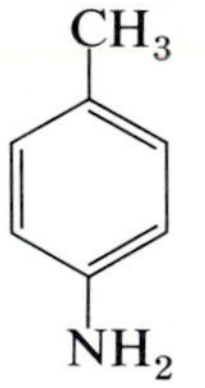

CH_3, NH_2
p-toluidine
p-aminotoluene

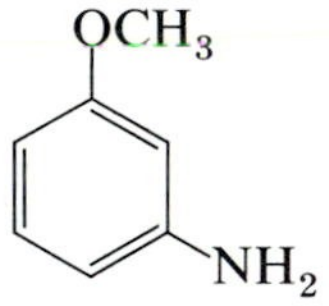

OCH_3, NH_2
m-anisidine
m-methoxyaniline

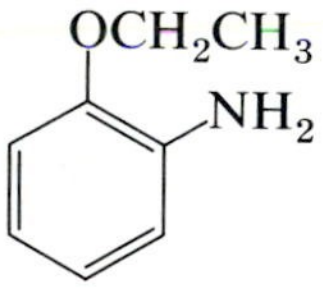

OCH_2CH_3, NH_2
o-phenetidine
o-ethoxyaniline

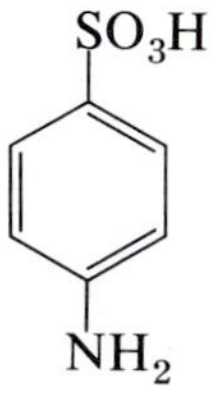

SO_3H, NH_2
sulfanilic acid
p-aminobenzenesulfonic acid

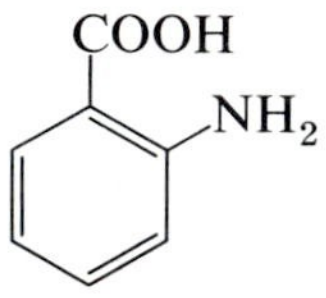

COOH, NH_2
anthranilic acid
o-aminobenzoic acid

Chemical Abstracts has adopted another system for naming amines that is more rational than either the common or IUPAC system and will probably gain universal acceptance for the nomenclature of these compounds. In this system amines are named in the same manner as are alcohols (Section 10.2). The name of the alkane is modified by replacing the final **-e** by the suffix **-amine.**

$CH_3CH_2CH_2NH_2$
propanamine

$(CH_3)_2CHCH_2CH_2CH_2CH_2NH_2$
5-methylhexanamine

$(CH_3)_2CHNH_2$
2-propanamine

For secondary and tertiary amines the parent alkane is taken to be the alkyl group with the longest chain. If two alkyl groups are "tied" by this criterion, the parent alkane is the one with the greater number of substituents. The remaining alkyl groups are named as substituents by using the prefix **N**- to indicate that they are attached to nitrogen.

$(CH_3CH_2)_2NH$	$CH_3CH_2CH_2NHCH_2CH(CH_3)_2$	$(CH_3CH_2CH_2)_2NCH_3$
N-ethylethanamine	2-methyl-N-propylpropanamine	N-methyl-N-propylpropanamine

The *Chemical Abstracts* name for aniline is benzenamine; derivatives are named accordingly.

$N(CH_3)_2$ — N,N-dimethylbenzenamine

CH_3 / NH_2 — 4-methylbenzenamine

OCH_3 / NH_2 — 3-methoxybenzenamine

In this book we shall use common names for simple amines such as methylamine, triethylamine, and di-*n*-propylamine. Because it is so widely used in the chemical literature, we shall retain the name aniline for the simplest aromatic amine and name derivatives as substituted anilines. For more complex amines, we shall use the *Chemical Abstracts* system.

EXERCISE 24.2 Write the *Chemical Abstracts* names for all of the amines depicted on pages 727 and 728.

24.3 Physical Properties and Spectra

A. Physical Properties

The melting points, boiling points, and densities of some simple amines are collected in Table 24.1. As with other classes of compounds, certain trends are evident in the properties. All three properties increase with molecular weight as a consequence of the greater intermolecular attraction with the larger members in the series.

> An individual molecule has a particular dipole moment in the gas phase. In the liquid or solid phase, the dipoles can interact with one another in a way determined by the magnitude of the moment and the distance between the dipoles as well as other interactions such as hydrogen-bonding or van der Waals interactions. The distance in turn depends on molecular geometry and compromises among the various other interactions. The structure and boiling point of a liquid depend on the average energy of the various interactions. The structure and melting point of a solid depend on more specific interactions, and a substance can have several different crystal forms with different melting points. The melting and boiling points reflect the energy required to overcome the sum of multiple intermolecular interactions.

Like alcohols, the lower amines show the effect of hydrogen bonding (Section 10.3). Since nitrogen is not as electronegative as oxygen, the N—H—N hydrogen bond is not as strong as the analogous O—H—O bond. Thus, primary amines have boiling points that are intermediate between those of alkanes and alcohols of comparable molecular weight (Figure 24.2), just as ammonia, b.p. −33 °C, is intermediate between methane, b.p. −161 °C, and water, b.p. 100 °C.

Hydrogen bonding is more important with primary than with secondary amines and is

TABLE 24.1 Physical Properties of Amines

	Molecular Weight	Melting Point, °C	Boiling Point, °C	Density
Primary Amines				
CH_3NH_2	31	−94	−6.3	0.6628
$CH_3CH_2NH_2$	45	−81	16.6	0.6829
$CH_3CH_2CH_2NH_2$	59	−83	47.8	0.7173
$CH_3CH_2CH_2CH_2NH_2$	73	−49	77.8	0.7414
Secondary Amines				
$(CH_3)_2NH$	45	−93	7.4	0.6804
$(CH_3CH_2)_2NH$	73	−48	56.3	0.7056
$(CH_3CH_2CH_2)_2NH$	101	−40	110	0.7400
$(CH_3CH_2CH_2CH_2)_2NH$	129	−60	159	0.7670
Tertiary Amines				
$(CH_3)_3N$	59	−117	2.9	0.6356
$(CH_3CH_2)_3N$	101	−114	89.3	0.7256
$(CH_3CH_2CH_2)_3N$	143	−94	155	0.7558
$(CH_3CH_2CH_2CH_2)_3N$	185		213	0.7771

not possible at all with tertiary amines. Thus, a primary amine always boils higher than a secondary or tertiary amine of the same molecular weight (Figure 24.2).

B. Infrared Spectra

The characteristic infrared absorptions of amines are associated with the nitrogen-hydrogen bonds. Typical bands are summarized in Table 24.2. For diagnostic purposes, the C—N absorptions are not very useful because these bands occur in a spectral region that

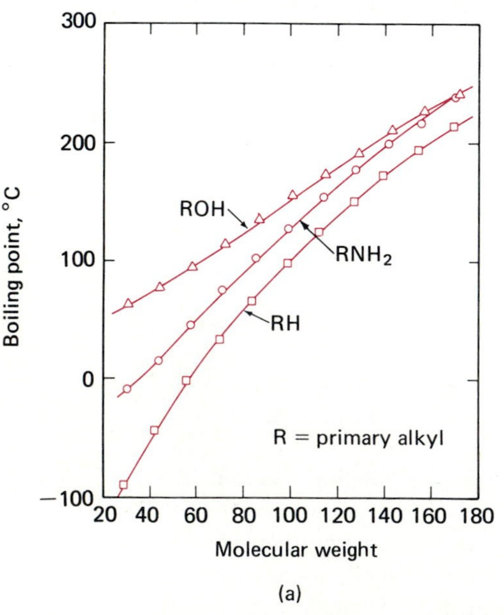

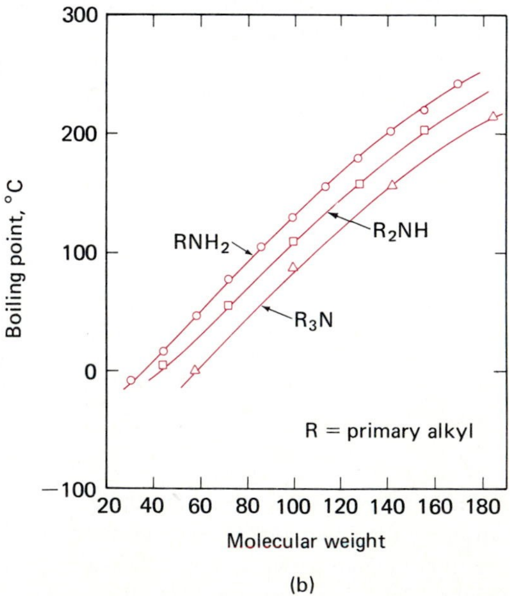

FIGURE 24.2 Boiling points: (a) alkanes, alcohols, and primary amines; (b) primary, secondary, and tertiary amines.

TABLE 24.2 Infrared Spectra of Amines

Frequency, cm^{-1}	Intensity	Assignment	Compound Type
3500, 3400 (doublet)	weak	N—H stretching	primary
3310–3350	very weak	N—H stretching	secondary
1580–1650	medium to strong	N—H bending	primary
666–909	medium to strong	N—H wagging	primary, secondary

normally also contains many bands for other types of compounds (1020–1250 cm^{-1}). Particularly useful absorptions are the weak N—H stretching bands of primary amines, the N—H bending mode of primary amines, and the N—H wagging mode for primary and secondary amines. The N—H stretch of secondary amines is so weak that it is often not observed. Infrared spectroscopy is not useful in diagnosing the presence of a tertiary amino group. The spectrum of *n*-hexylamine is shown in Figure 24.3.

C. Nuclear Magnetic Resonance Spectra

Since nitrogen is more electronegative than carbon, the protons near the amino group are deshielded. The downfield shifts are not as pronounced as in the case of alcohols and ethers (Section 13.12). As with alcohols and ethers, the exact chemical shift is dependent upon whether the protons are part of a CH_3, a CH_2, or a CH group.

	CH_3NR_2	$R'CH_2NR_2$	R'_2CHNR_2
δ, ppm:	2.2	2.4	2.8

Protons β to nitrogen are affected to a much smaller extent; they are normally seen in the range $\delta = 1.1$–1.7 ppm.

Protons bound directly to the nitrogen in primary and secondary amines may resonate anywhere in the region from $\delta = 0.6$ ppm to $\delta = 3.0$ ppm. The exact resonance position is dependent on the purity of the sample, the nature of the solvent, the concentration, and the temperature at which the measurement is made. Coupling of the type H—C—N—H is usually not observed because of proton exchange. The spectrum of di-*n*-propylamine is shown in Figure 24.4.

The CMR spectra of amines resemble those of corresponding alcohols, except that a carbon directly bonded to nitrogen does not experience as a great a downfield shift as one bonded to oxygen. Representative data are shown in Table 24.3 (see also Table 13.3, page 337).

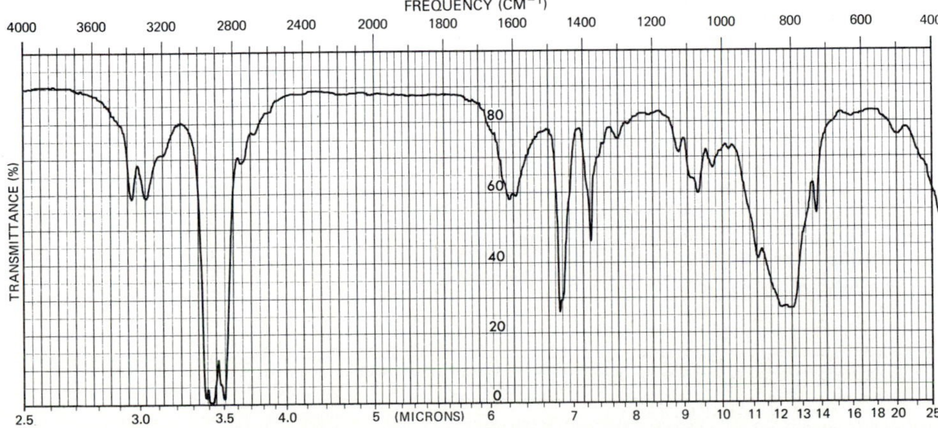

FIGURE 24.3 Infrared spectrum of *n*-hexylamine.

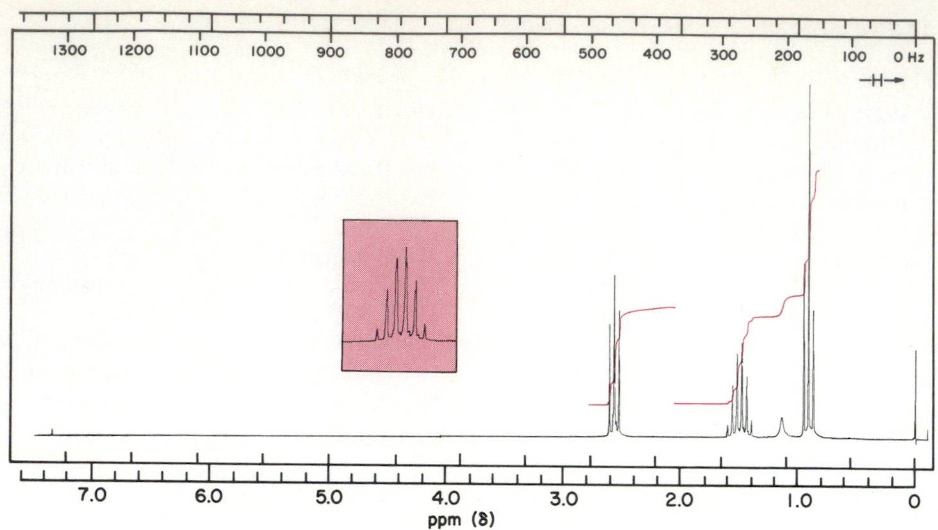

FIGURE 24.4 NMR spectrum of $(CH_3CH_2CH_2)_2NH$.

TABLE 24.3 CMR Chemical Shifts for Some Amines

	C-1	C-2	C-3	C-4	C-5
methylamine	28.3				
ethylamine	36.9	19.0			
propylamine	44.5	27.3	11.2		
butylamine	42.3	36.7	20.4	14.0	
pentylamine	42.5	34.0	29.7	23.0	14.3

EXERCISE 24.3 An unknown compound is suspected to be an amine because of its weakly basic properties. The infrared spectrum shows no absorption in the 3300–3500 cm^{-1} or 600–950 cm^{-1} regions. The CMR spectrum has bands at δ 17.0, 20.6, 39.0 and 48.4. The NMR spectrum is shown below.

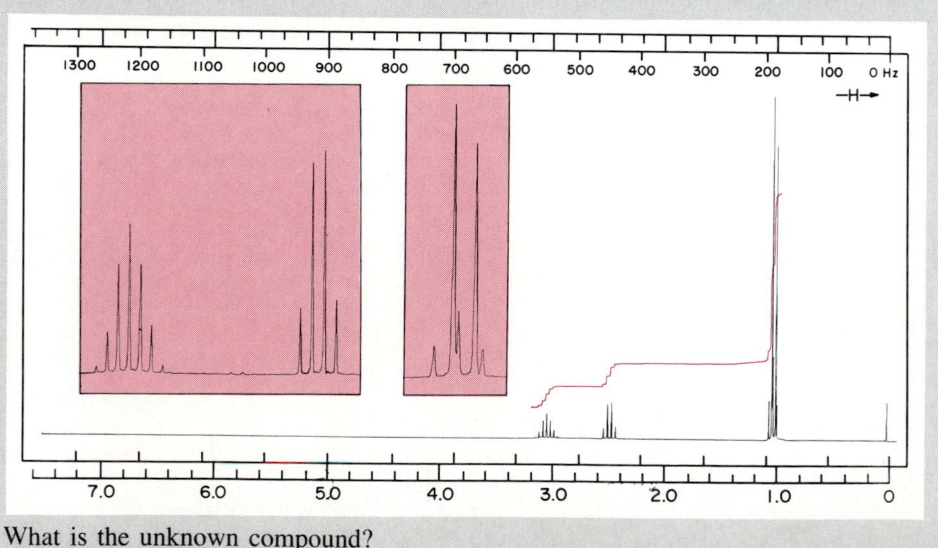

What is the unknown compound?

24.4 Basicity

Because of the presence of a nonbonding electron pair on nitrogen, amines are Lewis bases just like alcohols and ethers (Sections 10.6.B and 19.5). Nitrogen is not as electronegative as oxygen, and amines have a greater tendency to react with a proton than alcohols. Looking at it another way, alkyloxonium ions are more acidic than alkylammonium ions.

$$\underset{\text{less basic}}{CH_3OH} + H^+ \rightleftharpoons \underset{\text{more acidic}}{CH_3\overset{+}{O}H_2}$$

$$\underset{\text{more basic}}{CH_3NH_2} + H^+ \rightleftharpoons \underset{\text{less acidic}}{CH_3\overset{+}{N}H_3}$$

Since amines are much more basic than water, aqueous solutions of amines have basic properties.

$$RNH_2 + H_2O \overset{K_b}{\rightleftharpoons} RNH_3^+ + OH^- \qquad (24\text{-}1)$$

When comparing the base strengths of amines, it is convenient to refer to the dissociation constant of the corresponding ammonium ion. This equilibrium constant, like other dissociation constants, is called K_a (Section 4.5).

$$RNH_3^+ + H_2O \overset{K_a}{\rightleftharpoons} RNH_2 + H_3O^+$$

$$K_a = \frac{[RNH_2][H_3O^+]}{[RNH_3^+]}$$

As usual, the concentration of water is not included in the equilibrium expression because it is present in large excess and is essentially constant (Section 18.4.A).

The pK_as for some typical ammonium ions are collected in Table 24.4, together with the pK_a for ammonium ion for reference. Notice that the simple alkylammonium ions all have pK_as in the range 10–11 and are therefore slightly less acidic than NH_4^+ itself. In other words, amines are only slightly more basic than NH_3.

TABLE 24.4 Acidity of Some Alkylammonium Ions

Conjugate Acid	pK_a, 25°C
NH_4^+	9.24
$CH_3NH_3^+$	10.62
$CH_3CH_2NH_3^+$	10.64
$(CH_3)_3CNH_3^+$	10.68
$(CH_3)_2NH_2^+$	10.73
$(CH_3CH_2)_2NH_2^+$	10.94
$(CH_3)_3NH^+$	9.79
$(CH_3CH_2)_3NH^+$	10.75

It is important to distinguish between K_a for the dissociation of NH_4^+ and K_a for NH_3 itself and not to confuse them. Ammonia itself is an extremely weak acid; the pK_a for NH_3 is about 34. Consequently, the conjugate base of ammonia, NH_2^-, is an exceedingly strong base. Analogous anions derived by deprotonation of amines are known and are useful reagents for some organic reactions. Because amines are such feeble acids, powerful bases are needed for deprotonation; alkyllithium compounds are commonly used.

$$\underset{\substack{\text{dipropylamine} \\ pK_a \cong 40}}{(CH_3CH_2CH_2)_2NH} + \underset{\text{butyllithium}}{n\text{-}C_4H_9Li} \longrightarrow \underset{\substack{\text{lithium} \\ \text{dipropylamide}}}{(CH_3CH_2CH_2)_2N^- \, Li^+} + \underset{\substack{\text{butane} \\ pK_a \cong 50}}{C_4H_{10}}$$

EXERCISE 24.4 In some older books, one will find the basicity of amines discussed in terms of K_b, the equilibrium constant for reaction (24-1); K_b is given by the expression

$$K_b = \frac{[RNH_3^+][OH^-]}{[RNH_2]}$$

Using the relationship for the dissociation of water, $K_w = [H_3O^+][OH^-] = 10^{-14}\ M^2$, show that $pK_a + pK_b = 14$. What is pK_b for NH_2^-?

In aqueous solution arylamines are substantially less basic than alkylamines. Correspondingly, the acidity of anilinium ion is substantially greater than that of alkylammonium ions.

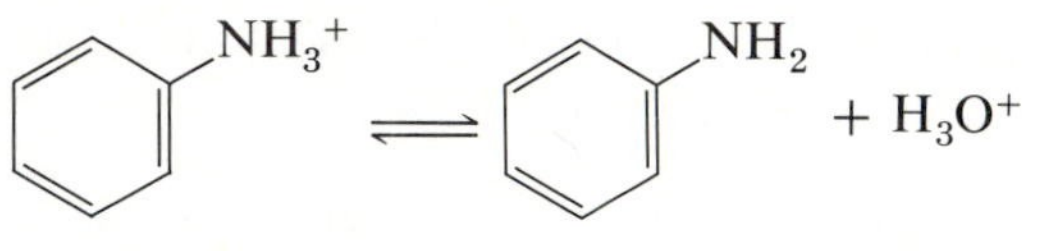

$$C_6H_5NH_3^+ \rightleftharpoons C_6H_5NH_2 + H_3O^+$$

$$K_a = 2.5 \times 10^{-5}\ M \qquad pK_a = 4.60$$

$$(CH_3)_2CHNH_3^+ \rightleftharpoons (CH_3)_2CHNH_2 + H_3O^+$$

$$K_a = 2.5 \times 10^{-12}\ M \qquad pK_a = 11.60$$

Aliphatic amines have basicity comparable to dilute solutions of sodium hydroxide; the basicity of aniline is comparable to that of sodium acetate.

The reduced basicity of aniline compared to aliphatic amines may be attributed in part to the electron-attracting inductive effect of a phenyl group; for example phenylacetic acid ($pK_a = 4.31$) is more acidic than acetic acid ($pK_a = 4.6$). However, this effect is small compared to the effect of delocalization of the nitrogen lone pair into the benzene ring.

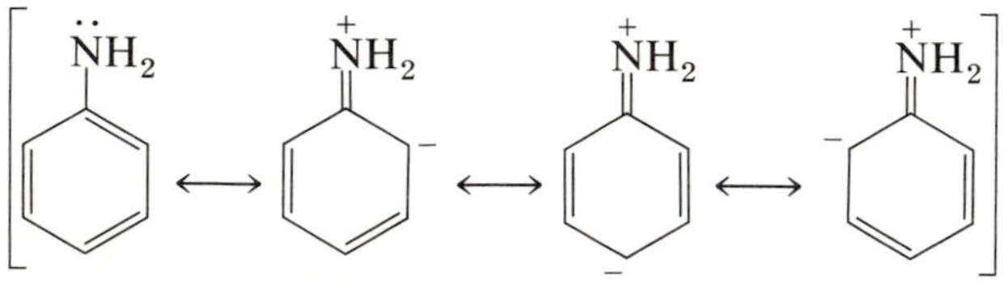

[resonance structures of aniline: $\ddot{N}H_2$ ⟷ $\overset{+}{N}H_2$ (ortho −) ⟷ $\overset{+}{N}H_2$ (para −) ⟷ $\overset{+}{N}H_2$ (ortho −)]

This delocalization renders the lone pair less accessible for bonding. Alternatively and equivalently, this delocalization effect can be expressed as a resonance stabilization of the amine that is not present in the ammonium ion. This energy effect is illustrated in Figure 24.5. The resonance energy of conjugation results in displacement of the protonation equilibrium toward the amine.

Ammonia itself and amines generally have a pyramidal structure (Section 24.1); the H—N—H bond angle in ammonia is 107.3°. The most effective conjugation of the nitrogen lone pair with the benzene ring would be obtained for a lone pair in a *p*-orbital parallel to the *p*-orbitals of the aromatic π-system. However, lone pairs are generally more stable in orbitals having some *s*-character. In the case of aniline, an energy compromise is

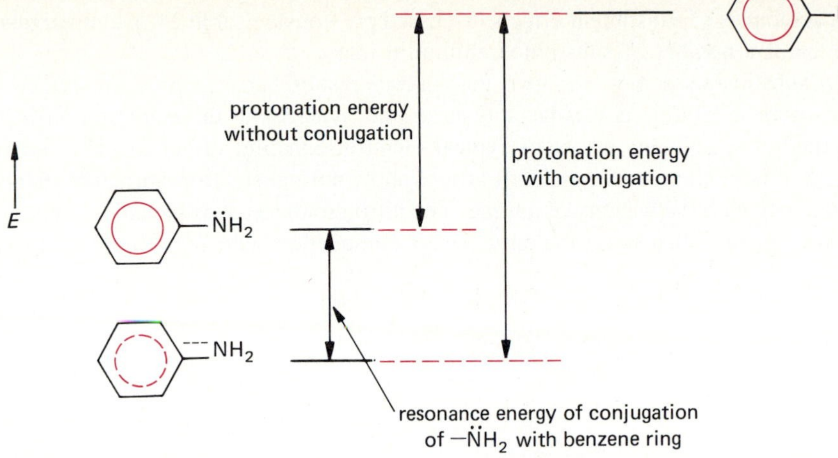

FIGURE 24.5 Conjugation with the phenyl ring decreases the basicity of the amino group in aniline.

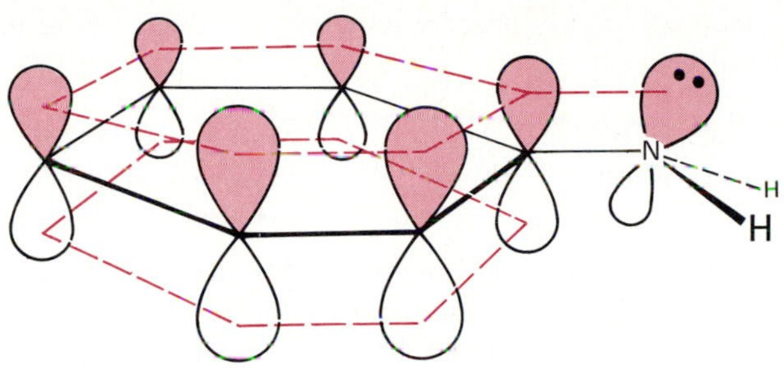

FIGURE 24.6 The partially pyramidal amino group in aniline can still conjugate with the phenyl π-system.

reached in which the lone-pair orbital has more p-character than in ammonia but in which the orbital retains some s-character. As a result, the NH_2 group in aniline is still pyramidal but with a larger H—N—H angle (113.9°) than in ammonia. The H—N—H plane intersects the plane of the benzene ring at an angle of 39.4°. The orbital structure of aniline is represented in Figure 24.6.

TABLE 24.5 pK_as of Anilinium Ions

	pK_a, 25°C		
Substituent	*ortho*	*meta*	*para*
H	4.60	4.60	4.60
benzoyl			2.17
bromo	2.53	3.58	3.86
chloro	2.65	3.52	3.98
cyano	0.95	2.75	1.74
fluoro	3.20	3.57	4.65
iodo	2.60	3.60	3.78
methoxy	4.52	4.23	5.34
methyl	4.44	4.72	5.10
nitro	−0.26	2.47	1.00
trifluoromethyl		3.20	2.75

Substituents on the aniline ring affect basicity in ways that are generally interpretable with the principles of substituent effects discussed previously. Table 24.5 summarizes the pK_a values of a number of substituted anilinium ions.

Ortho substituents sometimes give unexpected results because of steric effects; for example, *o*-methylaniline is less basic than aniline, whereas in the *meta* and *para* positions a methyl substituent exerts its typical electron-donating effect to give enhanced basicity. Bromo, chloro, iodo, and CF_3 groups show normal electron-attracting inductive effects that decrease the basicity of aniline. The nitro group has an especially potent effect in the *para* position that is attributed to direct conjugation with the amino group.

NH_2 … N^+ (^-O, O) $\longleftrightarrow$ $\overset{+}{N}H_2$ … N^+ (^-O, O^-)

Because of their basic properties, amines form salts with acids. Since these salts are ionic compounds they are usually water soluble even in cases where the corresponding amine is insoluble in water.

$$CH_3(CH_2)_9NH_2 + HCl \longrightarrow CH_3(CH_2)_9\overset{+}{N}H_3\ Cl^-$$

n-decylamine (insoluble in H_2O) — *n*-decylammonium chloride (soluble in H_2O)

Even though aromatic amines are only one millionth as basic as alkylamines (see Tables 24.4 and 24.5), they are still protonated even in dilute acidic solutions. For example, aniline is essentially completely protonated in 0.1 M HCl solution (pH = 1). Hence, although aniline is only slightly soluble in water, it dissolves completely in dilute hydrohalic and sulfuric acids. The nitroanilines are less basic but also dissolve in strong acids. 2,4-Dinitroanilinium ion has pK = −4.4; this amine is soluble only in rather concentrated acids.

EXERCISE 24.5 Using the data in Tables 24.4 and 24.5, calculate equilibrium constants for the following equilibria.

(a) $CH_3CH_2NH_3^+ + C_6H_5NH_2 \rightleftharpoons CH_3CH_2NH_2 + C_6H_5NH_3^+$

(b) $p\text{-}CF_3C_6H_4NH_3^+ + C_6H_5NH_2 \rightleftharpoons p\text{-}CF_3C_6H_4NH_2 + C_6H_5NH_3^+$

The basicity of amines provides a convenient method for separating amines from neutral organic compounds. For example, a mixture of *n*-decylamine (b.p. 221 °C) and dodecane (b.p. 216 °C) is difficult to separate by fractional distillation. The two compounds may be separated easily by *extracting* the mixture with sufficient 10% aqueous hydrochloric acid to convert all of the amine into the ammonium salt. The alkane, being insoluble in water, is unaffected by this treatment, and the ammonium salt dissolves in the water layer. The layers may be separated by use of a separatory funnel to give the pure alkane. A strong base such as sodium hydroxide is then added to the aqueous solution to neutralize the ammonium salt and liberate the free amine. The water-insoluble amine forms a second layer that can be separated.

$$CH_3(CH_2)_{10}CH_3 + CH_3(CH_2)_9NH_2$$

$\downarrow$ HCl, H_2O

organic layer: $CH_3(CH_2)_{10}CH_3$

water layer: $CH_3(CH_2)_9\overset{+}{N}H_3\ Cl^-$

$\downarrow$ NaOH

organic layer: $CH_3(CH_2)_9NH_2$

water layer: $Na^+\ Cl^-$

Amines also form salts with carboxylic acids. Again the salts are ionic and are often water soluble.

$$CH_3\overset{O}{\overset{\|}{C}}OH + CH_3NH_2 \longrightarrow CH_3\overset{O}{\overset{\|}{C}}O^-\ H_3\overset{+}{N}CH_3$$

acetic acid; methylamine; methylammonium acetate

This salt-forming reaction is often used as a method for **resolving** racemic mixtures of organic acids.

> The student should review the basic principles of stereochemistry in Chapter 7. **Resolution** is the term used to describe the separation of two enantiomers from each other.

Consider racemic α-hydroxypropionic acid (lactic acid). Recall that the two enantiomers have identical physical properties and cannot be separated by crystallization or distillation techniques. The mixture will react with methylamine to give racemic methylammonium lactate, the enantiomers of which also cannot be separated by physical methods.

$$\left.\begin{matrix} COOH \\ H\!-\!C\!-\!OH \\ CH_3 \\ \text{(}R\text{)-lactic acid} \\ + \\ COOH \\ HO\!-\!C\!-\!H \\ CH_3 \\ \text{(}S\text{)-lactic acid} \end{matrix}\right\} + CH_3NH_2 \longrightarrow \left\{\begin{matrix} CO_2^-\ CH_3\overset{+}{N}H_3 \\ H\!-\!C\!-\!OH \\ CH_3 \\ \text{methylammonium (}R\text{)-lactate} \\ + \\ CO_2^-\ CH_3\overset{+}{N}H_3 \\ HO\!-\!C\!-\!H \\ CH_3 \\ \text{methylammonium (}S\text{)-lactate} \end{matrix}\right.$$

However, consider the situation when one enantiomer of a chiral amine is used to form the salt.

$$\left.\begin{matrix} \begin{matrix} COOH \\ H-C-OH \\ CH_3 \end{matrix} \\ (R)\text{-lactic acid} \\ + \\ \begin{matrix} COOH \\ HO-C-H \\ CH_3 \end{matrix} \\ (S)\text{-lactic acid} \end{matrix}\right\} + \begin{matrix} NH_2 \\ C_6H_5-C-CH_3 \\ H \end{matrix} \longrightarrow \left\{\begin{matrix} \begin{matrix} NH_3^+ \\ C_6H_5-C-CH_3 \\ H \end{matrix} \; \begin{matrix} CO_2^- \\ H-C-OH \\ CH_3 \end{matrix} \\ (S)\text{-1-phenylethylammonium } (R)\text{-lactate} \\ \begin{matrix} NH_3^+ \\ C_6H_5-C-CH_3 \\ H \end{matrix} \; \begin{matrix} CO_2^- \\ HO-C-H \\ CH_3 \end{matrix} \\ (S)\text{-1-phenylethylammonium } (S)\text{-lactate} \end{matrix}\right.$$

(S)-1-phenylethylamine

The two salts are now diastereomeric rather than enantiomeric, and they have different physical properties. For example, the *S,R* salt may be more soluble in some solvents than the *S,S* salt. Because of this difference in solubility, the two salts can be separated by fractional crystallization. Each of the diastereomeric salts can then be treated with a strong acid such as hydrochloric or sulfuric acid to liberate the free carboxylic acid. Acidification of the *S,R* salt gives enantiomerically pure (*R*)-lactic acid, whereas similar treatment of the *S,S* salt gives pure (*S*)-lactic acid.

$$\left.\begin{matrix} (S)\text{-acid} \\ + \\ (R)\text{-acid} \end{matrix}\right\} + (S)\text{-amine} \longrightarrow \left\{\begin{matrix} (S,S)\text{-salt} \\ + \\ (S,R)\text{-salt} \end{matrix}\right. \xrightarrow{\text{separate}} \begin{matrix} (S,S)\text{-salt} \\ \downarrow \text{HCl} \\ (S)\text{-acid} \\ + \\ (S)\text{-ammonium chloride} \end{matrix} \quad + \quad \begin{matrix} (S,R)\text{-salt} \\ \downarrow \text{HCl} \\ (R)\text{-acid} \\ + \\ (S)\text{-ammonium chloride} \end{matrix}$$

Of course, in order to use this technique for resolution, suitable optically active amines must be available. Fortunately, a number of such compounds are readily available and relatively inexpensive. A particularly useful source of such resolving agents is the class of naturally occurring amines called alkaloids, which occur in nature in only one enantiomeric form. Examples are strychnine and brucine (Section 36.5.C). Another frequently used resolving agent is 1-phenyl-2-propanamine (amphetamine). Although not a natural product, synthetic amphetamine is readily available in both enantiomeric forms.

$$C_6H_5-CH_2-\overset{NH_2}{\underset{H}{C}}-CH_3$$

amphetamine

EXERCISE 24.6 The method outlined for the resolution of racemic lactic acid is sometimes known as the *method of diastereomeric salts*. The same principle may be used, even when actual salts are not employed. For example, show how you could resolve racemic 2-octanol if you had enantiomerically homogeneous (*S*)-1-methoxy-1-phenylacetic acid.

24.5 Quaternary Ammonium Compounds

A. Tertiary Amines as Nucleophiles

Recall that there is a correlation between Lewis basicity and the nucleophilicity of a species (mentioned briefly in Section 9.5). Amines are more basic than alcohols or ethers, and they are also more nucleophilic. For example, a mixture of diethyl ether and methyl iodide does not react under ordinary conditions, but triethylamine and methyl iodide react violently at room temperature. If the reaction is carried out in a solvent to moderate its vigor, the product, which is a tetraalkylammonium iodide, may be obtained in good yield.

$$(C_2H_5)_2O + CH_3I \xrightarrow{25°C} \text{no reaction}$$

$$(C_2H_5)_3N + CH_3I \xrightarrow[\text{ether}]{25°C} (C_2H_5)_3\overset{+}{N}CH_3\ I^-$$

methyltriethylammonium iodide

Such compounds, which have four alkyl groups replacing the four hydrogens of the ammonium ion, are called **quaternary ammonium compounds.** Since they are ionic, they are generally water soluble and have fairly high melting points. They often decompose at the melting point.

$(CH_3)_4N^+\ Cl^-$	$(CH_3CH_2CH_2)_4N^+\ Br^-$
tetramethylammonium chloride m.p. 420°C	tetrapropylammonium bromide m.p. 252°C

Quaternary ammonium compounds are important as intermediates in some reactions that we shall encounter and also have been important in nature almost from the beginning of life. For example, acetylcholine is an important neurotransmitter found in organisms over almost the whole evolutionary scale.

$$CH_3\overset{\overset{\displaystyle O}{\|}}{C}OCH_2CH_2\overset{+}{N}(CH_3)_3\ Br^-$$

acetylcholine bromide

It occurs in nematodes (hookworm), arthropods (spiders), molluscs (shellfish), and vertebrates (fish to mammals). Choline esters of diacylglyceryl phosphates are important phospholipids.

Neurotransmitters are molecules released in groups of 1000–10,000 on the passage of a nerve signal to a nerve ending. Neurotransmitters include acetylcholine, adrenaline, and the amino acids glutamic acid and γ-aminobutyric acid (Chapter 29). After release, the molecules diffuse across a 100–200 Å space to a cell containing receptors specific for particular neurotransmitters. At mammalian muscles, groups of the nicotinic acetylcholine receptor combine with the acetylcholine released by the nerve. A channel for ions is opened and activation and contraction of the muscle fiber follow. The toxic alkaloid, *d*-tubocurarine, a quaternary ammonium compound, blocks the nicotinic acetylcholine receptor, an effect that is medicinally useful in relaxing muscles during surgery.

d-tubacararine

Quaternary ammonium hydroxides are as basic as alkali hydroxides. They decompose on heating (Hofmann degradation; Section 24.7.E) and find use as basic catalysts in organic systems.

Highlight 24.1

Amines (primary, RNH_2, secondary, R_2NH, and tertiary, R_3N) are derivatives of ammonia. The lone pair on nitrogen (the "fourth substituent") is nucleophilic toward electrophilic agents such as protons and alkyl halides.

$$\text{amines} + H^+ \rightleftharpoons RNH_3^+,\ R_2NH_2^+,\ R_3NH^+ \qquad \text{(reversible reaction)}$$

$$\text{amines} + \mathbf{R}X \rightleftharpoons RNH_2\mathbf{R}^+,\ R_2NH\mathbf{R}^+,\ R_3N\mathbf{R}^+$$

The protonated species can dissociate to form RNH**R** or R_2N**R** and react further with **R**X. The lack of specificity diminishes the synthetic utility of the reaction. Quaternary ammonium ions ($R_3N\mathbf{R}^+$) are generally stable.

B. Phase-Transfer Catalysis

In Section 11.6.F we learned that chloroform reacts with strong bases to form dichlorocarbene, which can then add to double bonds to give dichlorocyclopropanes. If a solution of cyclohexene in chloroform is stirred with 50% aqueous sodium hydroxide, only small yields of the cyclopropane are formed. The hydroxide ion stays in the aqueous phase, and the only reaction that occurs is at the interface between the organic and aqueous phases. However, if a small amount of benzyltriethylammonium chloride is added to the heterogeneous mixture, rapid reaction occurs and 7,7-dichlorobicyclo[4.1.0]heptane is isolated in 72% yield.

$$\text{cyclohexene} + CHCl_3 \xrightarrow[\text{50\% aq. NaOH}]{C_6H_5CH_2\overset{+}{N}Et_3\ Cl^-} \text{7,7-dichlorobicyclo[4.1.0]heptane}$$

(72%)

To understand what has happened, we need to recognize that although the quaternary ammonium compound is a salt soluble in water, it also has a large organic group and has solubility in organic solvents *as an ion pair*. The quaternary ammonium chloride is used because of its availability and convenience in handling. In the presence of a large excess of hydroxide ion in the aqueous solution, benzyltriethylammonium hydroxide ion pairs are formed. Some benzyltriethylammonium hydroxide ion pairs are transferred into the chloroform layer. A combination of hydrophobic binding produced by exclusion of the large, relatively hydrophobic organic cation from the water and favorable van der Waals interactions in the chloroform compensate for the loss of hydroxide ion solvation by water. Hydroxide ion is especially reactive in this medium because of the diminished hydrogen bonding. Reaction of hydroxide ion with chloroform produces chloride ion and dichlorocarbene in the chloroform solution, in which there also is a high concentration of cyclohexene. The chloride ion associates with the benzyltriethylammonium ion to form an ion pair and remains dissolved.

$$C_6H_5CH_2NEt_3^+(aq) + OH^-(aq) \rightleftharpoons C_6H_5CH_2NEt_3^+\ OH^-(CHCl_3)$$

$$C_6H_5CH_2NEt_3^+\ OH^- + CHCl_3 \longrightarrow H_2O + C_6H_5CH_2NEt_3^+\ CCl_3^-$$

$$C_6H_5CH_2NEt_3^+\ CCl_3^- \longrightarrow :CCl_2 + C_6H_5CH_2NEt_3^+\ Cl^-$$

The benzyltriethylammonium chloride ion pairs diffuse into the aqueous phase, where the ammonium ion can again pick up a hydroxide ion and begin the cycle anew. The key to the procedure is the solubility of the quaternary ammonium salt in both water and organic solvents.

$$C_6H_5CH_2N(CH_2CH_3)_3^+\ OH^-$$

soluble in both water and organic solvents

The catalysis effected by this technique is called **phase-transfer catalysis** and can be applied to a number of different types of reaction. The general procedure is to use concentrated solutions with an aqueous and an organic phase. The quaternary ammonium salt used need only have organic groups that are sufficiently large to provide solubility in organic solvents. Among the ones commonly used are tetrabutylammonium, methyltrioctylammonium, and hexadecyltrimethylammonium salts. Some additional examples of applications of phase-transfer catalysis are given below.

$$CH_3(CH_2)_7CH{=}CH_2 \text{ (benzene soln.)} \xrightarrow[\text{aq. KMnO}_4,\ 40\text{–}50\,°\text{C}]{(CH_3(CH_2)_6CH_2)_3\overset{+}{N}CH_3\ Cl^-} CH_3(CH_2)_7COOH \quad (91\%)$$

$$C_6H_5CH_2COCH_3 + CH_3(CH_2)_3Br \xrightarrow[\text{50\% aq. NaOH}]{C_6H_5CH_2\overset{+}{N}Et_3\ OH^-} CH_3COCH(C_6H_5)CH_2CH_2CH_2CH_3 \quad (90\%)$$

$$CH_3(CH_2)_9Br \xrightarrow[\text{aq. NaSCN},\ 100\,°\text{C}]{(n\text{-}C_6H_{13})_3\overset{+}{N}CH_3\ Cl^-} CH_3(CH_2)_9SCN \quad (100\%)$$

Note that the examples include alkylation, oxidation, and displacement reactions, that the anions are not restricted to hydroxide ion and that various temperatures can be used.

EXERCISE 24.7 For each of the three foregoing examples of phase-transfer catalysis, what species are in the aqueous phase? In the organic phase? Which species are passing from one phase to another?

24.6 Synthesis

A. Direct Alkylation of Ammonia or Other Amines

In Section 9.2 it was mentioned that ammonia reacts with primary alkyl halides by the S_N2 mechanism to give alkylammonium halides. In principle, this type of displacement reaction might be used as a way of synthesizing primary amines.

$$CH_3CH_2Br + NH_3 \longrightarrow CH_3CH_2NH_3^+\ Br^- \xrightarrow{NaOH} CH_3CH_2NH_2$$

In practice, this method is not very useful because of the side reactions that occur. The product alkylammonium ion is fairly acidic and can transfer a proton to a molecule of ammonia that has not yet reacted to give the primary amine and the ammonium ion. Since the primary amine is also nucleophilic, it can undergo further reaction giving a secondary amine. By similar equilibria and further alkylation, the tertiary amine and even the quaternary ammonium compound can be formed. The actual result is a complex mixture even when equivalent molar amounts of ammonia and alkyl halide are used.

$$RBr + NH_3 \longrightarrow RNH_3^+\ Br^-$$
$$RNH_3^+ + NH_3 \rightleftharpoons RNH_2 + NH_4^+$$
$$RNH_2 + RBr \longrightarrow R_2NH_2^+\ Br^-$$
$$R_2NH_2^+ + NH_3 \rightleftharpoons R_2NH + NH_4^+$$
$$R_2NH + RBr \longrightarrow R_3NH^+\ Br^-$$
$$R_3NH^+ + NH_3 \rightleftharpoons R_3N + NH_4^+$$
$$R_3N + RBr \longrightarrow R_4N^+\ Br^-$$

The "overalkylation" can be suppressed by using a large excess of ammonia or the amine being alkylated. This ploy is only practical in cases where the amine is relatively inexpensive and sufficiently volatile that the unreacted excess can be easily removed. An example is the preparation of *n*-butylamine by the reaction of *n*-butyl bromide with ammonia.

$$CH_3CH_2CH_2CH_2Br + NH_3 \longrightarrow \xrightarrow{NaOH} \underset{(47\%)}{CH_3CH_2CH_2CH_2NH_2}$$

> A solution of 300 g of NH_3 (20 moles) in 8 L of 90% aqueous ethanol is prepared. *n*-Butyl bromide is added slowly until 1507 g (11 moles) has been added. The reaction mixture is stirred at 25 °C for 48 hr and then made basic with aqueous NaOH. Fractional distillation of the organic layer gives 388 g of *n*-butylamine (47%) along with some di-*n*-butylamine and tri-*n*-butylamine, which have higher boiling points.

Secondary and tertiary amines can also be prepared this way, but the yields are again often low due to overalkylation. Also, if the amine is not readily available or is expensive, it is undesirable to use it in excess. In many cases where a pure primary, secondary, or tertiary amine is desired, direct alkylation is not a practical synthetic method. Several indirect methods have been devised to accomplish this purpose, and we shall study some of them in later parts of this section.

We saw in Section 24.4 that aromatic amines are much less basic than alkylamines. They are also less nucleophilic, and their reactions with alkyl halides require somewhat more vigorous conditions. Since these amines are less reactive nucleophiles, it is easier to achieve monoalkylation, as illustrated with the following synthesis of N-benzylaniline.

$$C_6H_5NH_2 \xrightarrow[\text{aq. NaHCO}_3,\ 90\text{–}95\,^\circ\text{C}]{C_6H_5CH_2Cl} C_6H_5NHCH_2C_6H_5$$

(85–87%)
N-benzylaniline

The sodium bicarbonate serves to neutralize the HCl that is produced in the reaction.

B. Indirect Alkylation: The Gabriel Synthesis

Pure primary amines can be prepared in good yield by a method called the **Gabriel synthesis.** This method involves the alkylation of a ''protected'' form of ammonia. The compound phthalimide (Section 27.6.C) is prepared from ammonia and the dicarboxylic acid phthalic acid. Imides have acidic properties because the negative charge of the conjugate base is delocalized over both oxygens and the nitrogen. The pK_a of phthalimide is 8.3. In aqueous basic solution the compound is converted almost completely into the anion.

$$\text{phthalimide (NH)} + OH^- \longrightarrow [\text{N}^- \longleftrightarrow \text{O}^-\text{ resonance form}] + H_2O$$

phthalimide

The phthalimide anion has nucleophilic properties and can enter into displacement reactions with alkyl halides. Reaction could in principle take place on either oxygen or nitrogen, but since nitrogen is more nucleophilic, it occurs mostly on nitrogen. Further alkylation cannot occur because there are no acidic protons. The product is an N-alkylphthalimide, and hydrolysis gives the amine and phthalic acid.

$$\text{phthalimide N}^- + CH_3CH_2CH_2CH_2Br \longrightarrow$$

$$\text{phthalimide-}NCH_2CH_2CH_2CH_3 + Br^- \xrightarrow[H_2O]{NaOH}$$

$$C_6H_4(CO_2^-)_2 + H_2NCH_2CH_2CH_2CH_3$$

The best solvent for the alkylation appears to be dimethylformamide $HCON(CH_3)_2$. The Gabriel synthesis is frequently used in the preparation of α-amino carboxylic acids, and we shall encounter it again in that context in Chapter 29.

EXERCISE 24.8 Explain why the Gabriel synthesis *cannot* be used to prepare each of the following amines.

(a) neopentylamine (b) *t*-butylamine (c) di-*n*-propylamine

C. Reduction of Nitro Compounds

Nitro compounds undergo ready reduction to yield primary amines. Because aromatic nitro compounds of a wide variety are available from nitration of aromatic compounds (Chapter 23), this method constitutes the most general synthesis of aromatic amines. Reduction can be accomplished by catalytic hydrogenation or by the use of chemical reducing agents in acidic solution.

CH_3, NO_2, $CH(CH_3)_2$ $\xrightarrow[\text{pressure, }\Delta]{H_2/Ni}$ CH_3, NH_2, $CH(CH_3)_2$

(87–90%)
2-methyl-5-isopropylaniline

CH_3, NO_2, NO_2 $\xrightarrow[C_2H_5OH,\ HCl,\ \Delta]{Fe}$ CH_3, NH_2, NH_2

(74%)
2,4-diaminotoluene

Many chemical reducing agents have been used for the conversion of aromatic nitro groups to amines. Among the most common are metals and acid, usually iron or zinc and dilute hydrochloric acid. Stannous chloride, $SnCl_2$, and hydrochloric acid are an especially useful combination when other reducible groups, such as carbonyl groups, are present.

NO_2, CHO $\xrightarrow[\text{conc. HCl}]{SnCl_2}$ [NH_3^+, CHO] $\xrightarrow{NaOH}$ NH_2, CHO

m-nitrobenzaldehyde *m*-aminobenzaldehyde

Reduction can be applied to unsymmetrical dinitro compounds as well, and selective conversions are sometimes possible.

NO_2, NO_2 $\xrightarrow[CH_3OH,\ \Delta]{NaSH}$ NH_2, NO_2

(79–85%)

m-dinitrobenzene *m*-nitroaniline

EXERCISE 24.9 We saw in this section that some aromatic amines are readily synthesized from aromatic compounds by the two-step sequence of (1) nitration and (2) reduction of the NO_2 group to NH_2. Show how this method can be applied to the synthesis of 2,4-dimethylaniline. Explain why the method is not applicable for the synthesis of *m*-methoxyaniline or *o*-aminobenzaldehyde.

D. Reduction of Nitriles

We saw in Section 12.6.A that nitriles are reduced by hydrogen and a catalyst or by lithium aluminum hydride in an ether solvent to give primary amines.

$$RC{\equiv}N \xrightarrow[\text{or LiAlH}_4]{H_2/\text{cat.}} RCH_2NH_2$$

In the catalytic hydrogenation procedure, secondary amines are often produced as by-products. The initially produced imine can disproportionate by reaction with some of the primary amine already produced in the reduction to give a new imine. Hydrogenation of this imine gives the secondary amine.

$$\begin{aligned}
&(1)\ RCN + H_2 \longrightarrow RCH{=}NH \\
&(2)\ RCH{=}NH + H_2 \longrightarrow RCH_2NH_2 \\
&(3)\ RCH{=}NH + RCH_2NH_2 \rightleftharpoons RCH{=}NCH_2R + NH_3 \\
&(4)\ RCH{=}NCH_2R + H_2 \longrightarrow RCH_2NHCH_2R
\end{aligned}$$

This side reaction may be suppressed by carrying out the hydrogenation in the presence of excess NH_3, which forces equilibrium (3) to the left.

$$RCN \xrightarrow[\text{EtOH,NH}_3]{H_2\text{-Pd/C}} RCH_2NH_2$$

Secondary amine formation can also be minimized by carrying out the reaction in acetic anhydride as solvent. The primary amine produced is rapidly converted into the amide.

$$C_6H_5{-}CH_2C{\equiv}N \xrightarrow[\text{Ac}_2\text{O}]{H_2\text{-Ni}} C_6H_5{-}CH_2CH_2NH\overset{\overset{\displaystyle O}{\|}}{C}CH_3$$

(97%)

The amine can then be obtained by hydrolysis of the amide. Since nitriles are easily available by several methods, many primary amines can be synthesized by this procedure.

$$\text{cyclopentanone} + CN^- \xrightarrow{H^+} \text{1-hydroxycyclopentanecarbonitrile (HO, CN)} \xrightarrow{\text{LiAlH}_4} \text{1-(aminomethyl)cyclopentanol (HO, CH}_2\text{NH}_2)$$

Notice that cyanide ion, CN^-, is a synthon for the group CH_2NH_2.

EXERCISE 24.10 Write equations showing how the following conversions can be carried out.

(a) $(CH_3)_2CHCH_2CH_2Br \longrightarrow (CH_3)_2CHCH_2CH_2CH_2NH_2$

(b) $(CH_3)_2CHCHO \longrightarrow (CH_3)_2CH\overset{\overset{\displaystyle OH}{|}}{C}HCH_2NH_2$

E. Reduction of Oximes

Aldoximes and ketoximes, which are prepared from aldehydes or ketones by reaction with hydroxylamine (Section 14.6.C), can be reduced to primary amines. Since oximes are easily produced in high yield, this is a useful synthetic method.

$$CH_3CH_2CH_2\overset{\overset{\displaystyle O}{\|}}{C}CH_3 + H_2NOH \longrightarrow CH_3CH_2CH_2\overset{\overset{\displaystyle NOH}{\|}}{C}CH_3 \xrightarrow[C_2H_5OH]{H_2\text{-}Ni} CH_3CH_2CH_2\overset{\overset{\displaystyle NH_2}{|}}{C}HCH_3$$

(85%)

2,2-Diphenylcyclohexanone + H_2NOH $\longrightarrow$ 2,2-diphenylcyclohexanone oxime $\xrightarrow[ether]{LiAlH_4}$ 2,2-diphenylcyclohexylamine

(80%)

F. Reduction of Imines: Reductive Amination

Ammonia and primary amines condense with aldehydes and ketones to give imines (Section 14.6.C). In the case of ammonia, the imines are unstable and cannot be isolated. However, if a mixture of a carbonyl compound and ammonia is treated with hydrogen and a suitable hydrogenation catalyst, the C=N bond of the unstable imine is reduced and an amine results. The process is often called "reductive amination."

$$C_6H_5CHO + NH_3 \rightleftharpoons [C_6H_5CH{=}NH] \xrightarrow{H_2\text{-}Ni} C_6H_5CH_2NH_2$$

benzaldehyde — (89%) benzylamine

A significant side reaction complicates the reductive amination method. As the primary amine begins to build up, it can condense with the starting aldehyde to give a different imine. Reduction of this imine gives a secondary amine.

$$C_6H_5CH_2NH_2 + C_6H_5CHO \rightleftharpoons [C_6H_5CH_2N{=}CHC_6H_5] \xrightarrow{H_2\text{-}Ni} C_6H_5CH_2\overset{\overset{\displaystyle H}{|}}{N}CH_2C_6H_5$$

dibenzylamine

This side reaction can be minimized by using a large excess of ammonia in the reaction medium. On the other hand, it may actually be exploited and used as a method for the synthesis of secondary amines, as shown by the following reaction. This example also demonstrates that ketones can be used as well as aldehydes.

$$HOCH_2CH_2NH_2 + CH_3\overset{\overset{\displaystyle O}{\|}}{C}CH_3 \xrightarrow[C_2H_5OH]{H_2\text{-}Pt} HOCH_2CH_2NH\overset{\overset{\displaystyle CH_3}{|}}{C}HCH_3$$

(95%)

One version of reductive amination, which is frequently employed for the synthesis of tertiary amines where at least one of the alkyl groups is methyl, is the **Eschweiler-Clarke** reaction. Instead of hydrogen, the reducing agent is formic acid, which is oxidized to carbon dioxide.

$$\text{2-}C_6H_5\text{-piperidine (N–H)} + CH_2{=}O + HCOOH \xrightarrow{100°C} \text{2-}C_6H_5\text{-1-}CH_3\text{-piperidine} + CO_2$$

(94%)

$$CH_3\overset{CH_3}{\underset{CH_3}{C}}NH_2 + CH_2{=}O + HCOOH \xrightarrow{100°C} CH_3\overset{CH_3}{\underset{CH_3}{C}}N(CH_3)_2$$

(95%)
t-butyldimethylamine

As shown by the two foregoing examples, the reaction proceeds in excellent yield. The intermediate that is reduced is an **immonium ion,** and the reduction can be visualized as follows.

$$R_2NH + CH_2{=}O \overset{H^+}{\rightleftharpoons} R_2N{-}CH_2OH$$

$$R_2N{-}CH_2OH + HCOOH \rightleftharpoons R_2\overset{+}{N}{=}CH_2 + HCO_2^- + H_2O$$

$$R_2\overset{+}{N}{=}CH_2 + H{-}\overset{O}{\overset{\|}{C}}{-}O^- \longrightarrow R_2N{-}CH_3 + CO_2$$

An earlier version of this reaction, called the **Leukart** reaction, gives lower yields, but is more general. It can be used to prepare primary, secondary, or tertiary amines. In this method, the ketone is heated with a formate salt or a formamide at 180–200 °C.

$$C_6H_5\overset{O}{\overset{\|}{C}}CH_3 + HCO_2^-\ NH_4^+ \xrightarrow{180°C} C_6H_5\overset{NH_2}{\overset{|}{C}}HCH_3$$

(60–66%)

$$\text{cyclooctanone} + H\overset{O}{\overset{\|}{C}}N(CH_3)_2 \xrightarrow{200°C} \text{cyclooctyl-}N(CH_3)_2$$

(75%)
(dimethylamino)cyclooctane

EXERCISE 24.11 What are the products of each of the following reactions?

(a) 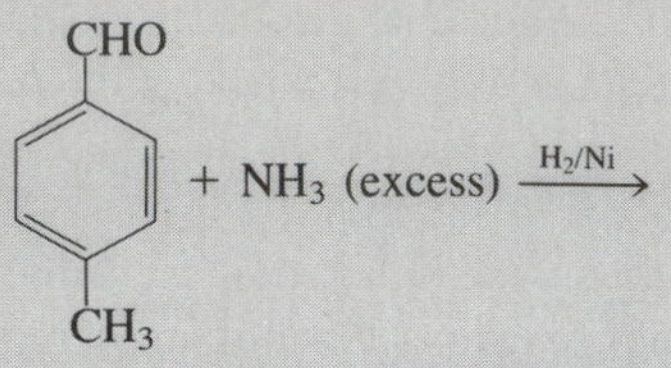 p-$CH_3C_6H_4CHO$ + NH_3 (excess) $\xrightarrow{H_2/Ni}$

(b) 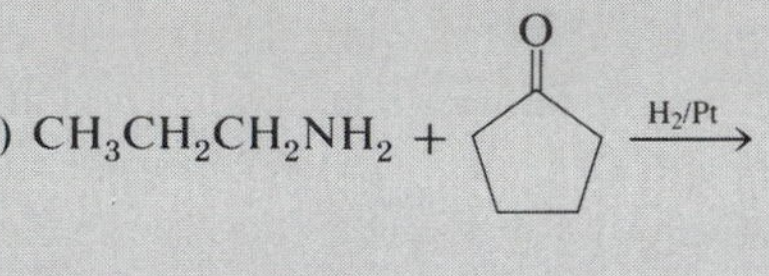 $CH_3CH_2CH_2NH_2$ + cyclopentanone $\xrightarrow{H_2/Pt}$

(c) $C_6H_5CH_2\overset{O}{\overset{\|}{C}}CH_3 + H\overset{O}{\overset{\|}{C}}N(CH_3)_2 \xrightarrow{200°C}$

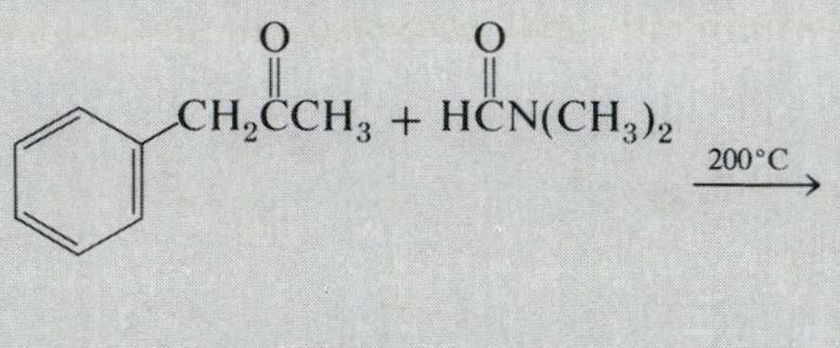

(d) 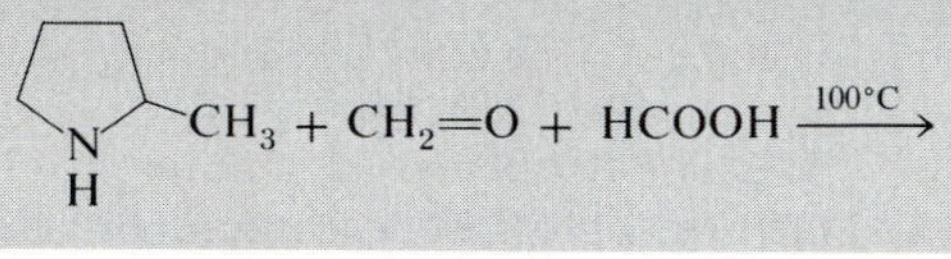 2-methylpyrrolidine (CH_3 on ring, N–H) $+ CH_2{=}O + HCOOH \xrightarrow{100°C}$

G. Reduction of Amides

Amides are reduced by lithium aluminum hydride in refluxing ether to give amines (Section 19.8). The reduction is unusual in that a C=O group is reduced to CH_2. Yields are generally good.

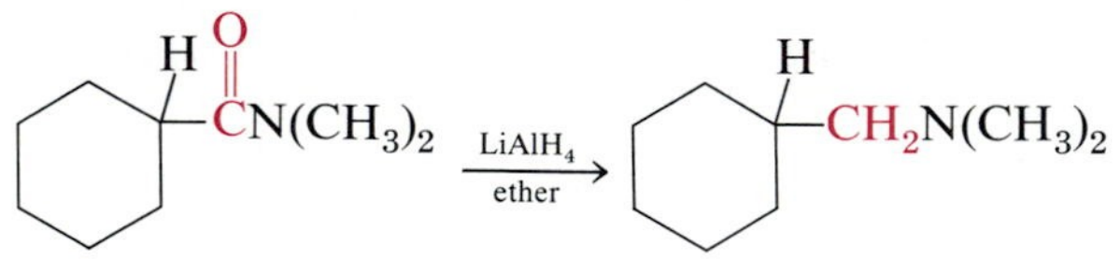

$$C_6H_{11}\overset{O}{\overset{\|}{C}}N(CH_3)_2 \xrightarrow[\text{ether}]{LiAlH_4} C_6H_{11}CH_2N(CH_3)_2$$

N,N-dimethylcyclohexanecarboxamide → (88%) N,N-dimethylcyclohexylmethanamine

Diborane, B_2H_6, can also be used as the reducing agent.

$$(CH_3)_3C\overset{O}{\overset{\|}{C}}N(CH_3)_2 \xrightarrow[THF]{B_2H_6} (CH_3)_3CCH_2N(CH_3)_2$$

N,N,2,2-tetramethylpropanamide → (79%) N,N,2,2-tetramethylpropanamine

The method also serves as a method to prepare primary or secondary amines, depending on the structure of the amide used.

$$R\overset{O}{\overset{\|}{C}}NH_2 \xrightarrow{LiAlH_4} RCH_2NH_2$$

$$R\overset{O}{\overset{\|}{C}}NHR' \xrightarrow{LiAlH_4} RCH_2NHR'$$

EXERCISE 24.12 The two-step sequence of (1) acylation of an amine and (2) reduction of the resulting amide to an amine constitutes a method for the indirect addition of an alkyl group to nitrogen. Show how this sequence of reactions can be used to accomplish the following transformations.

(a) $CH_2CH_2CH_2CH_2NH_2 \longrightarrow CH_3CH_2CH_2CH_2NHCH_2CH(CH_3)_2$

(b) $(CH_3)_2CHNH_2 \longrightarrow (CH_3)_2CHN(CH_2CH_3)CH_2CH_2CH_2CH_3$

H. Reduction of Azides

Organic azides are readily prepared by displacement reactions from azide ion, N_3^-, and alkyl halides. The azido group is similar in electronegativity to the bromo group, but cannot be displaced as easily. Hydrogen azide, HN_3, is a weak acid (b.p. 37 °C, pK_a 4.68) and a powerful vasodilator, known to cause headaches in those who inhale some. Compounds containing a high azido content can explode; phenyl azide (azidobenzene) can be distilled at 73.5 °C at 23 mm, but decomposes or explodes if distillation is attempted under atmospheric pressure. Azides have a highly characteristic infrared band between 2140–2240 cm^{-1}. They are readily reduced to primary amines by hydrogen or hydrazine over platinum.

$$CH_3(CH_2)_8CH_2\mathbf{I} \xrightarrow[(CH_3)_2SO]{NaN_3} CH_3(CH_2)_8CH_2N_3 \xrightarrow[\text{Pd/CH}_3\text{OH, reflux, 2 hr}]{NH_2NH_2} CH_3(CH_2)_8CH_2NH_2$$

1-azidodecane; (71%) 1-aminodecane

Another facile method for the conversion of azides into primary amines is known as the **Staudinger reaction** and involves the use of triphenylphosphine. The reaction is suitable for the preparation of amines with a variety of other functional groups such as nitro, epoxy, hydroxy, cyano, and ethoxycarbonyl.

$$NCCH_2CH_2CH_2CH_2N_3 + Ph_3P \xrightarrow[\text{THF}]{H_2O} NCCH_2CH_2CH_2CH_2NH_2 + Ph_3PO \quad (-N_2)$$

5-azidopentanenitrile; (82%) 5-aminopentanenitrile

$$NCCH_2(CH_2)_4C{\equiv}CCH_2CH_2N_3 + Ph_3P \xrightarrow[\text{THF}]{H_2O} NCCH_2(CH_2)_4C{\equiv}CCH_2CH_2NH_2 + Ph_3PO \quad (-N_2)$$

10-azido-7-decynenitrile; (92%) 10-amino-7-decynenitrile

The Staudinger reaction proceeds by way of an intermediate called an iminophosphorane, $RN{=}PPh_3$. The formation of the iminophosphorane occurs via a detectable intermediate which may be $RN{=}N{-}N{=}PPh_3$.

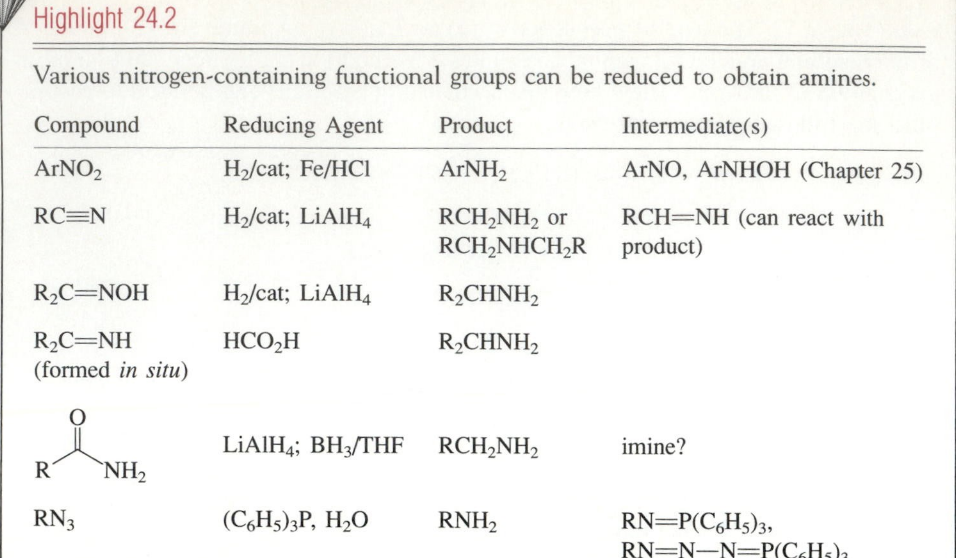

Highlight 24.2

Various nitrogen-containing functional groups can be reduced to obtain amines.

Compound	Reducing Agent	Product	Intermediate(s)
$ArNO_2$	H_2/cat; Fe/HCl	$ArNH_2$	ArNO, ArNHOH (Chapter 25)
$RC{\equiv}N$	H_2/cat; $LiAlH_4$	RCH_2NH_2 or RCH_2NHCH_2R	$RCH{=}NH$ (can react with product)
$R_2C{=}NOH$	H_2/cat; $LiAlH_4$	R_2CHNH_2	
$R_2C{=}NH$ (formed *in situ*)	HCO_2H	R_2CHNH_2	
$RC(=O)NH_2$	$LiAlH_4$; BH_3/THF	RCH_2NH_2	imine?
RN_3	$(C_6H_5)_3P$, H_2O	RNH_2	$RN{=}P(C_6H_5)_3$, $RN{=}N{-}N{=}P(C_6H_5)_3$

I. Preparation from Carboxylic Acids: The Hofmann, Curtius, and Schmidt Rearrangements

The Hofmann, Curtius, and Schmidt rearrangements all accomplish the same overall process—conversion of a carboxylic acid to a primary amine with loss of the carboxy carbon of the acid.

$$R{-}\overset{O}{\overset{\|}{C}}OH \longrightarrow R{-}NH_2 + CO_2$$

Examples of the three reactions are shown below.

Hofmann

$$CH_3(CH_2)_7CH_2\,CONH_2 + Cl_2 + OH^- \longrightarrow CH_3(CH_2)_7CH_2\,NH_2$$

decanamide → nonanamine (66%)

Curtius

$$\text{c-}C_3H_5{-}\overset{O}{\overset{\|}{C}}Cl + Na^+N_3^- \longrightarrow \text{c-}C_3H_5{-}\overset{O}{\overset{\|}{C}}N_3 \xrightarrow[H_2O]{\Delta} \text{c-}C_3H_5{-}NH_2$$

cyclopropanecarbonyl chloride → cyclopropanecarbonyl azide → cyclopropanamine (60%)

Schmidt

$$m\text{-}ClC_6H_4COOH + NaN_3 \xrightarrow[2.\ NaOH]{1.\ H_2SO_4,\ CHCl_3} m\text{-}ClC_6H_4NH_2$$

m-chlorobenzoic acid → *m*-chloroaniline (75%)

Although the Hofmann rearrangement begins with an amide, the Curtius reaction with an acyl azide, and the Schmidt reaction with an acid, the three reactions are related mechanistically. In each case, the crucial intermediate is probably the same, an **acyl nitrene.**

The probable mechanism for the Hofmann rearrangement is outlined below. The first two steps are simply the base-promoted halogenation of the amide, which is mechanistically related to the base-promoted halogenation of ketones (Section 15.1.D).

$$\text{(1)}\quad R\overset{O}{\overset{\|}{C}}NH_2 + OH^- \rightleftharpoons R\overset{O}{\overset{\|}{C}}NH^- + H_2O$$

$$\text{(2)}\quad R\overset{O}{\overset{\|}{C}}NH^- + Cl_2 \rightleftharpoons R\overset{O}{\overset{\|}{C}}NHCl + Cl^-$$

The N-chloroamide is more acidic than the starting amide and also reacts with base to give the corresponding anion. This intermediate loses chloride ion to give a highly reactive intermediate called a **nitrene.** Nitrenes are neutral molecules in which the nitrogen has only six electrons; they are structurally similar to carbenes (Section 11.6.F).

$$\text{(3)}\quad R\overset{O}{\overset{\|}{C}}\ddot{N}HCl + OH^- \rightleftharpoons R\overset{O}{\overset{\|}{C}}\ddot{\underset{..}{N}}^{-}Cl + H_2O$$

$$\text{(4)}\quad R\overset{O}{\overset{\|}{C}}\ddot{\underset{..}{N}}^{-}Cl \rightleftharpoons \underset{\text{an acyl nitrene}}{R\overset{O}{\overset{\|}{C}}-\ddot{\underset{..}{N}}} + Cl^-$$

Acyl nitrenes undergo a rapid rearrangement to give compounds called **isocyanates,** which are similar to allenes (Section 20.2.C) or ketenes (Section 20.3.C).

$$\text{(5)}\quad R-\overset{O}{\overset{\|}{C}}-\ddot{N}: \longrightarrow \underset{\text{an isocyanate}}{R-\ddot{N}=C=O}$$

Like ketenes, isocyanates react rapidly with water; the products in this case are **carbamic acids,** which are thermally unstable. Decarboxylation of the carbamic acid occurs to give the amine and carbon dioxide.

$$\text{(6)}\quad \underset{\text{an alkyl isocyanate}}{R-N=C=O} + H_2O \longrightarrow \underset{\text{a carbamic acid}}{RNH\overset{O}{\overset{\|}{C}}OH}$$

$$\text{(7)}\quad RNH\overset{O}{\overset{\|}{C}}OH \longrightarrow RNH_2 + CO_2$$

If the Hofmann rearrangement is carried out in alcohol solution rather than in water, steps 6 and 7 are not possible. Instead, the isocyanate adds the alcohol to give the ester of the carbamic acid, which is stable and may be isolated.

$$\text{(6')}\quad R-N=C=O + CH_3OH \longrightarrow \underset{\text{a methyl carbamate}}{RNH\overset{O}{\overset{\|}{C}}OCH_3}$$

In the Curtius reaction, the acyl azide loses nitrogen upon heating to give the acyl nitrene directly. The remaining steps 5–7 are the same. Acyl azides are potentially explosive, and the decomposition is therefore somewhat hazardous to carry out.

$$R-\overset{O}{\overset{\|}{C}}-\ddot{\underset{..}{N}}-\overset{+}{N}\equiv N: \xrightarrow{\Delta} R-\overset{O}{\overset{\|}{C}}-\ddot{N}: + N_2$$

The Schmidt reaction also proceeds by way of the acyl azide, which is formed by reaction of the carboxylic acid with hydrazoic acid, HN_3, under the acidic conditions of the reaction.

$$R-\overset{O}{\overset{\|}{C}}-OH + HN_3 \xrightarrow{H_2SO_4} R-\overset{O}{\overset{\|}{C}}-N_3 + H_2O$$

EXERCISE 24.13 Write equations showing the preparation of 3-methylpentanamine and *o*-bromoaniline by the Hofmann, Curtius, and Schmidt rearrangements.

24.7 Reactions

Certain important reactions of amines that have already been presented will not be discussed further here. These are the reactions with protons (Section 24.4) and with alkyl halides (Sections 24.5 and 24.6.A).

A. Formation of Amides

Recall that ammonia and primary and secondary amines react with acyl halides and acid anhydrides to give amides (Section 19.7.B). If an acyl halide is used, the hydrohalic acid produced will neutralize an additional equivalent of amine.

$$CH_2{=}CH\overset{O}{\overset{\|}{C}}Cl + 2\,CH_3NH_2 \longrightarrow CH_2{=}CH\overset{O}{\overset{\|}{C}}NHCH_3 + CH_3NH_3^+\,Cl^-$$

Aromatic amines can also be converted to amides with acyl chlorides or anhydrides. We shall find that this is frequently a useful procedure because the amide group is less strongly activating than the amino group in electrophilic aromatic substitution reactions. Thus the high reactivity of the amine can be moderated by conversion into an amide. For this **moderating** purpose the acetyl group is used most often and is usually introduced with acetic anhydride. Acetanilides can also be prepared by direct heating of aniline or a substituted aniline with acetic acid, but this process is slower.

$$C_6H_5NH_2 \xrightarrow[-H_2O,\ \Delta]{CH_3COOH} C_6H_5NHCOCH_3$$

acetanilide

Acetanilide is a colorless crystalline solid, m.p. 114 °C, that behaves as a neutral compound under normal conditions. The basic character of aniline is reduced by the acetyl group. The conjugate acids of amides have pK_as of about 0 to +1; that is, they are extensively protonated in 10% sulfuric acid.

$$RNH\overset{^+OH}{\overset{\|}{C}}CH_3 \overset{K_a}{\rightleftharpoons} H_3O^+ + RNHCOCH_3 \qquad K_a \cong 0.1\text{–}1\ M$$

Acetanilide is a somewhat weaker base, just as aniline is a weaker base than aliphatic amines; the pK_a of the conjugate acid of acetanilide is about −1 to −2. It is protonated by strong sulfuric acid solutions. Acetanilide is also a weak acid with a pK_a estimated to be about 15. It is not appreciably soluble in dilute aqueous alkali hydroxides and requires more basic conditions to form the conjugate anion.

$$p\text{-}C_2H_5OC_6H_4NH_2 \xrightarrow[CH_3COONa]{\text{dil. HCl}} \xrightarrow{(CH_3CO)_2O} p\text{-}C_2H_5OC_6H_4NHCOCH_3$$

p-phenetidine
p-ethoxyaniline

phenacetin
p-ethoxyacetanilide

Phenacetin has been used as an analgesic and in mixture with aspirin and caffeine as formerly popular over-the-counter analgesic pills. It has since been removed from such compositions because of not uncommon addiction accompanied by kidney damage.

The mixture of HCl and sodium acetate creates a buffered medium that keeps the amine in solution as the ammonium salt in equilibrium with a small amount of free amine. The free amine reacts rapidly with acetic anhydride to form the amide in high yield and in a pure state. Acetic anhydride hydrolyzes slowly under these conditions. The amide can be hydrolyzed back to the amine by heating with alcoholic HCl.

B. Reactions with Nitrous Acid

Amines undergo interesting reactions with nitrous acid, HNO_2. The reaction products depend on whether the amine is primary, secondary, or tertiary, and whether it is aromatic or aliphatic.

Secondary amines, either aromatic or aliphatic, give N-nitroso compounds, also known as **nitrosamines.** Aliphatic nitrosamines are often colored.

$$C_5H_{10}NH + HNO_2 \longrightarrow C_5H_{10}N\text{—}NO + H_2O$$

piperidine

N-nitrosopiperidine
(yellow)

$$C_6H_5NHCH_3 \xrightarrow[HCl]{NaNO_2} C_6H_5N(CH_3)NO$$

N-nitroso-N-methylaniline

The reaction mechanism can be thought of in terms of the following steps.

$$(1)\ HO\text{—}N{=}O + H^+ \rightleftharpoons H_2\overset{+}{O}\text{—}N{=}O$$

$$(2)\ H_2\overset{+}{O}\text{—}N{=}O \rightleftharpoons H_2O + {}^+\ddot{N}{=}\ddot{O}$$

$$(3)\ R_2\ddot{N}H + NO^+ \rightleftharpoons R_2\overset{+}{N}(H)\text{—}N{=}O$$

$$\text{(4)}\quad R_2\overset{+}{N}(N{=}O)(H) \rightleftharpoons R_2N{-}N{=}O + H^+$$

Since tertiary amines have no proton on nitrogen, step 4 is blocked. Thus no overall reaction occurs by this pathway. However, if the tertiary amine is aromatic, an alternative reaction path is available.

$$C_6H_5N(CH_3)_2 \xrightarrow[8°C]{NaNO_2,\ HCl} p\text{-}ON{-}C_6H_4{-}N(CH_3)_2$$

(80–89%)
p-nitrosodimethylaniline

The mild electrophile NO^+ (the **nitrosonium ion**) attacks the highly reactive aromatic ring to yield the beautiful green *p*-nitroso-N,N-dimethylaniline. The electrophilic aromatic substitution occurs only at the *para* position (for other reactions, see Section 24.7.D).

With primary amines, the reaction with nitrous acid proceeds beyond the nitrosamine stage and a **diazonium compound** is produced.

$$\text{(5)}\quad R{-}\ddot{N}H{-}N{=}O + H^+ \rightleftharpoons \left[R{-}\ddot{N}H{-}N{=}\overset{+}{O}H \longleftrightarrow R{-}\overset{+}{N}H{=}N{-}OH\right]$$

$$\text{(6)}\quad R{-}\overset{+}{N}H{=}N{-}OH \rightleftharpoons R{-}N{=}N{-}OH + H^+$$

$$\text{(7)}\quad R{-}\ddot{N}{=}\ddot{N}{-}OH + H^+ \rightleftharpoons R{-}\ddot{N}{=}\ddot{N}{-}OH_2^+$$

$$\text{(8)}\quad R{-}\ddot{N}{=}\ddot{N}{-}OH_2^+ \rightleftharpoons H_2O + R{-}\overset{+}{N}{\equiv}\ddot{N}$$

alkanediazonium cation

Alkanediazonium compounds are exceedingly unstable and decompose, even at low temperatures, to give nitrogen and various products and intermediates including carbocations. An example is the reaction of *n*-butylamine with nitrous acid, generated *in situ* from sodium nitrite and aqueous hydrochloric acid.

$$CH_3(CH_2)_3NH_2 \xrightarrow[H_2O,\ 25°C]{NaNO_2,\ HCl} \underset{(25\%)}{CH_3(CH_2)_3OH} + \underset{(13\%)}{CH_3CH_2CH(OH)CH_3} + \underset{(5\%)}{CH_3(CH_2)_3Cl} +$$

$$\underset{(3\%)}{CH_3CH_2CHClCH_3} + \underset{(26\%)}{CH_3CH_2CH{=}CH_2} + \underset{(3\%)}{\textit{cis-}CH_3CH{=}CHCH_3} + \underset{(7\%)}{\textit{trans-}CH_3CH{=}CHCH_3}$$

Because the reaction with nitrous acid gives complex mixtures of products, it is not a generally useful one with aliphatic amines.

The diazonium ions produced by the reaction of primary aromatic amines are more stable than alkanediazonium ions. Aqueous solutions of these **arenediazonium ions** are stable at ice-bath temperatures.

$$C_6H_5NH_2 \xrightarrow[\text{aq. HCl, } 5\,^\circ\text{C}]{\text{NaNO}_2} C_6H_5N^+{\equiv}N$$

benzenediazonium cation

The overall process of converting a primary aromatic amine into an arenediazonium salt is called **diazotization.** As we shall see in Chapter 25, diazotization provides access to a wide variety of substituted aromatic compounds.

The mechanism of diazotization formulated on pages 753–54 is probably oversimplified. In solution nitrous acid is in equilibrium with several other species, such as dinitrogen trioxide, the anhydride of nitrous acid.

$$2\ \text{HONO} \rightleftharpoons H_2O + N_2O_3$$

With halide ions, the equilibria contain nitrosyl halides, which are the mixed anhydrides of nitrous acid and hydrohalic acids.

$$\text{HONO} + \text{HX} \rightleftharpoons \text{ONX} + H_2O$$

The actual nitrosating agent in many cases is probably N_2O_3, although in solutions containing halide ion the corresponding nitrosyl halide may also play a role. Nitrous acid itself is an intermediate oxidation state of nitrogen and disproportionates to nitric oxide and nitric acid.

$$3\ \text{HONO} \longrightarrow 2\ \text{NO} + H_3O^+ + NO_3^-$$

The rate of this reaction is temperature-dependent and is also strongly dependent on the concentration of nitrous acid—the rate is proportional to $[\text{HONO}]^4$. Consequently, nitrous acid solutions are usually kept cold and dilute and are used immediately.

EXERCISE 24.14 Write equations showing the reaction product(s) expected from treatment of each of the following amines with an aqueous solution of $NaNO_2$ and HCl at 0 °C.

(a) *n*-propylamine
(b) di-*n*-propylamine
(c) tri-*n*-propylamine
(d) aniline
(e) N-ethylaniline
(f) N,N-diethylaniline

Highlight 24.3

The reactions of amine with the electrophile $N{=}O^+$ are diverse and important.

Primary alkanamine

$$R{-}\ddot{N}H_2 + \overset{+}{N}{=}O \longrightarrow R{-}\overset{+}{N}H_2{-}N{=}O \underset{+H^+}{\overset{-H^+}{\rightleftharpoons}} R{-}\ddot{N}H{-}N{=}\overset{+}{O}{-}H \longrightarrow$$

$$R{-}\overset{+}{N}H{=}N{-}O{-}H \underset{+H^+}{\overset{-H^+}{\rightleftharpoons}} R{-}\ddot{N}{=}N{-}\overset{+}{O}H_2 \longrightarrow R{-}\overset{+}{N}{\equiv}N{:} + H_2O$$

alkanediazonium ion
(unstable, decomposes to carbocations)

Primary arenamine

$$Ar{-}\ddot{N}H_2 + \overset{+}{N}{=}O \longrightarrow \longrightarrow \longrightarrow R{-}\overset{+}{N}{\equiv}N{:}$$

arenediazonium ion
(fairly stable useful chemical reactant)

Secondary N-alkylalkanamine

$$R{-}\ddot{N}(R)H + \overset{+}{N}{=}O \longrightarrow R{-}\overset{+}{N}(R)H{-}N{=}O \underset{+H^+}{\overset{-H^+}{\rightleftharpoons}} R{-}\ddot{N}(R){-}N{=}O$$

N-nitroso-*sec*-amine

Tertiary N,N-dialkylarenamine

$$C_6H_5{-}\ddot{N}R_2 + \overset{+}{N}{=}O \longrightarrow \longrightarrow O{=}N{-}C_6H_4{-}\ddot{N}R_2$$

p-nitroso-N,N-dialkylarenamine

C. Oxidation

As expected for good electron donors, amines are oxidized with ease. With primary amines, oxidation is complicated by the variety of reaction paths that are available. Few useful oxidation reactions are known for this class. Secondary amines are easily oxidized to hydroxylamines. Again yields are generally poor due to over oxidation.

$$R_2NH + H_2O_2 \longrightarrow R_2NOH + H_2O$$

Tertiary amines are oxidized cleanly to tertiary amine oxides. Useful oxidants are H_2O_2 or organic peroxyacids, RCO_3H.

$$C_6H_{11}{-}CH_2N(CH_3)_2 + H_2O_2 \longrightarrow C_6H_{11}{-}CH_2\overset{+}{N}(O^-)(CH_3)_2$$

(90%)

A mixture of 49 g of N,N-dimethylcyclohexylmethanamine, 45 mL of methanol, and 120 g of 30% aqueous H_2O_2 is kept at room temperature for 36 hr. The excess H_2O_2 is destroyed by the addition of a small amount of colloidal platinum. The solution is then filtered and evaporated to obtain the crude amine oxide in greater than 90% yield.

Amine oxides fall into the class of organic compounds for which no completely uncharged Lewis structure may be written. The Lewis electron-dot representation of trimethylamine oxide shows that both the oxygen and the nitrogen have octet configurations and that they bear (−) and (+) formal charges, respectively.

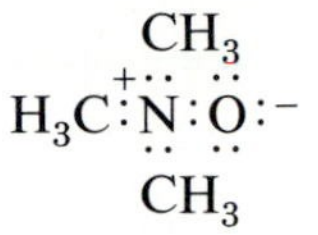

$$\begin{array}{c} CH_3 \\ H_3C:\overset{+}{N}:\ddot{O}:^- \\ CH_3 \end{array}$$

trimethylamine oxide

Aromatic amines are readily oxidized by a variety of oxidizing agents as well as by air. As a result, the oxidation of other functional groups cannot usually be carried out as satisfactorily if amino groups are also present.

The nature of amine oxidations is demonstrated by oxidation of *p*-bis(dimethylamino)benzene, which gives a relatively stable *radical cation* called Wurster's blue.

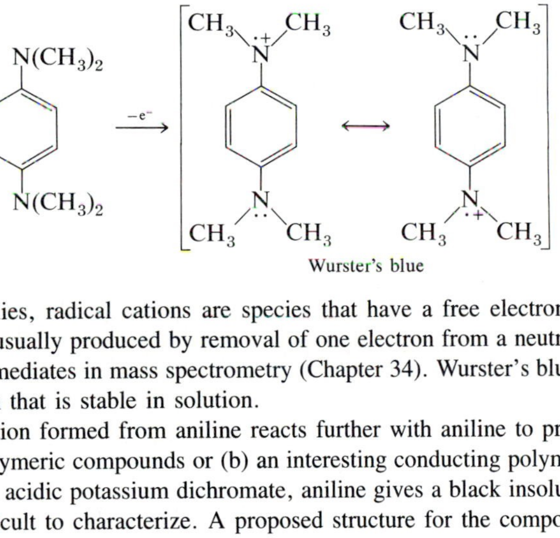

$$p\text{-}(CH_3)_2N{-}C_6H_4{-}N(CH_3)_2 \xrightarrow{-e^-} \left[(CH_3)_2\overset{\cdot +}{N}{-}C_6H_4{-}\ddot{N}(CH_3)_2 \longleftrightarrow (CH_3)_2\ddot{N}{-}C_6H_4{-}\overset{\cdot +}{N}(CH_3)_2 \right]$$

Wurster's blue

As the name implies, radical cations are species that have a free electron and a positive charge. They are usually produced by removal of one electron from a neutral molecule and are common intermediates in mass spectrometry (Chapter 34). Wurster's blue is an example of a radical cation that is stable in solution.

The radical cation formed from aniline reacts further with aniline to produce either (a) highly colored polymeric compounds or (b) an interesting conducting polymer, polyaniline. On treatment with acidic potassium dichromate, aniline gives a black insoluble dye, aniline black, that is difficult to characterize. A proposed structure for the compound is

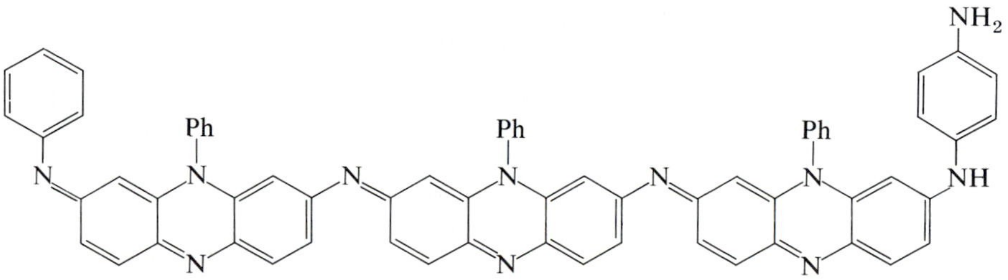

Oxidation of aniline with potassium persulfate in 1.2 M hydrochloric acid yields polyaniline, a linear polymer that is conducting in the appropriate oxidation state. Conducting polymers represent an exciting recently developed field of organic chemistry (see Chapter 36).

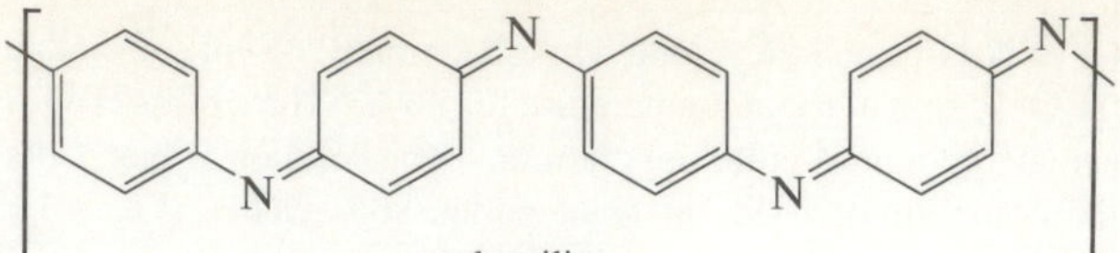

polyaniline

Still other oxidation conditions give *p*-benzoquinone, a compound we will consider in detail in Chapter 30.

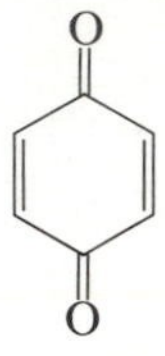

p-benzoquinone

D. Electrophilic Aromatic Substitution

Aromatic amines are highly activated toward substitution in the ring by electrophilic reagents. Reaction with such amines generally occurs under rather mild conditions. For example, halogenation is so facile that all unsubstituted *ortho* and *para* positions become substituted.

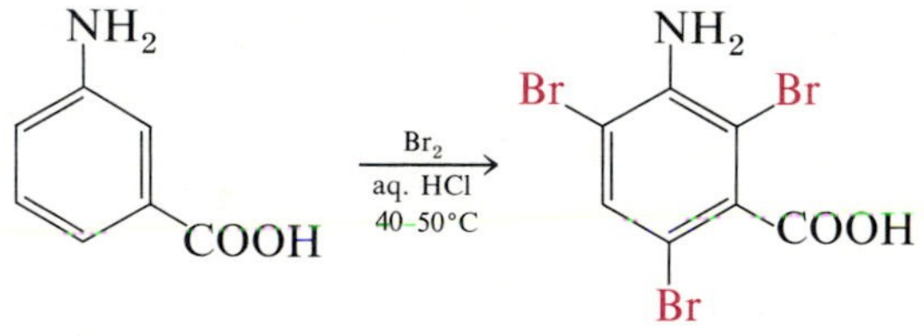

Br_2, aq. HCl, 40–50°C

m-aminobenzoic acid → 3-amino-2,4,6-tribromobenzoic acid

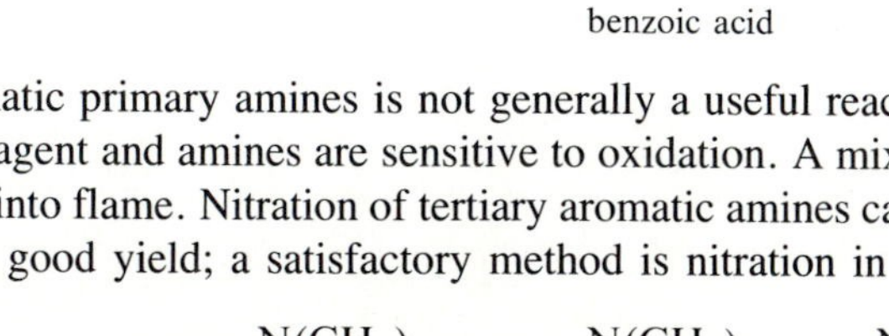

Cl_2, aq. HCl

anthranilic acid → (69–78%) 2-amino-3,5-dichlorobenzoic acid

Nitration of aromatic primary amines is not generally a useful reaction because nitric acid is an oxidizing agent and amines are sensitive to oxidation. A mixture of aniline and nitric acid can burst into flame. Nitration of tertiary aromatic amines can be accomplished conveniently and in good yield; a satisfactory method is nitration in acetic acid.

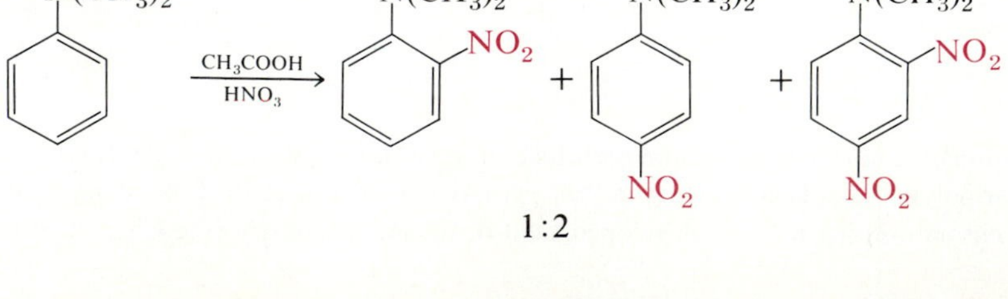

$C_6H_5N(CH_3)_2 \xrightarrow[HNO_3]{CH_3COOH}$ o-$O_2NC_6H_4N(CH_3)_2$ + p-$O_2NC_6H_4N(CH_3)_2$ + 2,4-$(O_2N)_2C_6H_3N(CH_3)_2$

1:2

The *ortho*/*para* ratio is 1:2, and the amount of 2,4-dinitro-N,N-dimethylaniline depends on the reaction conditions.

In general, electrophilic aromatic substitution reactions can be applied to tertiary aromatic amines. Furthermore, the dialkylamino group is such a strongly activating substituent that rather mild reaction conditions can be used.

$N(CH_3)_2$, NO_2 $\xrightarrow[\text{room temp.}]{Br_2,\ CH_3COOH/NaOOCCH_3}$ $N(CH_3)_2$, NO_2, Br

In this example, no additional Lewis-acid catalyst is required, even with a deactivating nitro group in the ring.

Friedel-Crafts acylations can also be accomplished under mild conditions.

$N(CH_3)_2$ $\xrightarrow[20°C]{COCl_2}$ $N(CH_3)_2$, COCl $\xrightarrow[ZnCl_2]{C_6H_5N(CH_3)_2}$ $(CH_3)_2N$—C(=O)—$N(CH_3)_2$

p,p'-bis-(dimethylamino)benzophenone
Michler's ketone
(a dye intermediate)

Aromatic tertiary amines can undergo a useful variant of Friedel-Crafts acylation, the **Vilsmeier reaction,** the reaction of an aromatic compound with dimethylformamide and phosphorus oxychloride to produce an aldehyde. The reaction requires a rather activated aromatic ring and is unsuccessful with aromatic compounds of only moderate reactivity, such as benzene, toluene, and chlorobenzene.

$N(CH_3)_2$ + $HCON(CH_3)_2$ $\xrightarrow[2.\ H_2O]{1.\ POCl_3}$ $N(CH_3)_2$, CHO

dimethylaniline; dimethylformamide (DMF); (80–84%) *p*-dimethylaminobenzaldehyde

The electrophilic reagent in the Vilsmeier reaction is a chloroimmonium ion, which is formed in the following manner.

$$Cl{-}P(=O)(Cl){-}Cl + H{-}C(=O){-}N(CH_3)_2 \rightleftharpoons Cl_2P(=O)O{-}CH{=}\overset{+}{N}(CH_3)_2\ Cl^- \rightleftharpoons$$

$$Cl_2P(=O)O{-}CH(Cl){-}N(CH_3)_2 \rightleftharpoons Cl{-}CH{=}\overset{+}{N}(CH_3)_2\ Cl_2PO_2^-$$

This electrophilic species reacts only with aromatic rings that contain highly activating groups such as OH and NR_2. The initial product, an α-chloroamine, hydrolyzes rapidly during work-up to afford the aldehyde.

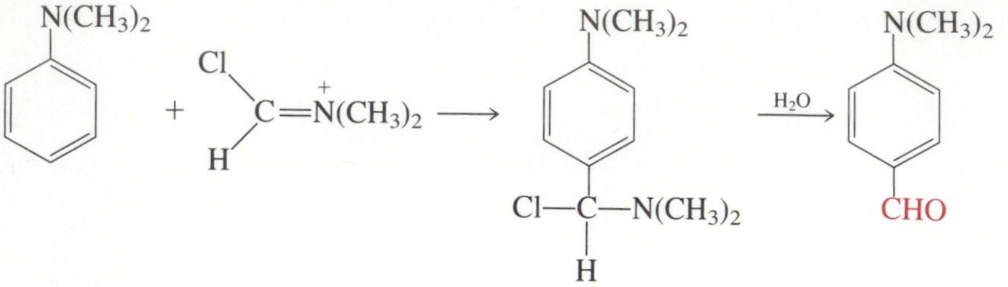

$$C_6H_5N(CH_3)_2 + ClCH{=}\overset{+}{N}(CH_3)_2 \longrightarrow p\text{-}(CH_3)_2NC_6H_4\text{—}CHCl\text{—}N(CH_3)_2 \xrightarrow{H_2O} p\text{-}(CH_3)_2NC_6H_4CHO$$

The Vilsmeier reaction has been an important industrial process, particularly for formylation of reactive heterocyclic compounds. However, large quantities of by-product phosphorus compounds are produced, and the disposal of these waste materials presents an ever-increasing problem. Thus, there is a trend away from using this method.

The reactivity of aromatic amines to electrophilic reagents is moderated when the amino group is converted into an amide. The nitrogen lone pair of the amine helps to stabilize the developing positive charge in the electrophilic substitution transition state; in the amide, the pair is partially tied up by conjugation with the carbonyl group. Amide groups are still *ortho,para* directors, but they are not nearly as activating as amino groups and electrophilic reactions on amide derivatives are readily controlled. Another way of understanding the moderating effect of the acyl group on nitrogen is by consideration of the resonance structures for the intermediate that results from electrophilic attack *para* to the amide group.

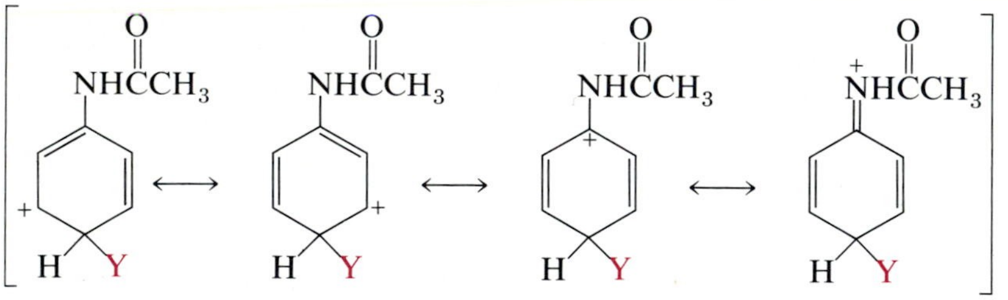

$$\left[\text{NHCCH}_3 \text{ (O)}, \text{H Y} \longleftrightarrow \text{NHCCH}_3 \longleftrightarrow \text{NHCCH}_3 \longleftrightarrow \overset{+}{\text{N}}\text{HCCH}_3\right]$$

The fact that the developing positive charge can be placed on the nitrogen still causes attack at this position (or at the *ortho* position) to be more favorable than attack at the *meta* position. However, this resonance structure is not quite as important as it is in the intermediate resulting from attack at the *para* position of the corresponding amine because the positive nitrogen is now adjacent to the positive carbon of the carbonyl group.

Acetanilides are widely used as substrates for electrophilic aromatic substitution reactions. Several examples are given below.

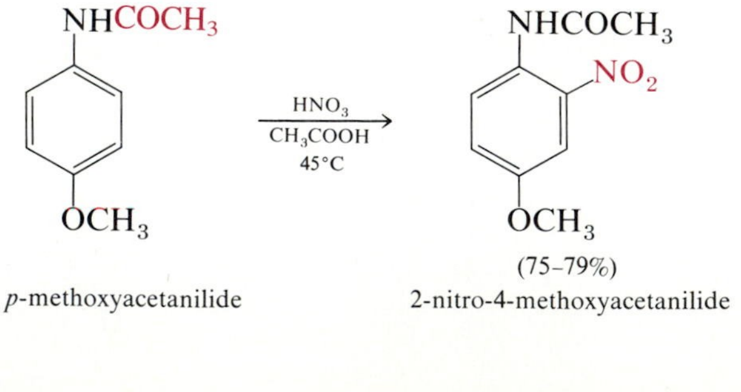

$$p\text{-}CH_3OC_6H_4NHCOCH_3 \xrightarrow[\text{45°C}]{HNO_3,\ CH_3COOH} \text{2-}NO_2\text{-4-}CH_3OC_6H_3NHCOCH_3$$

p-methoxyacetanilide → (75–79%) 2-nitro-4-methoxyacetanilide

$$p\text{-}CH_3C_6H_4NH_2 \xrightarrow[\Delta]{CH_3COOH} p\text{-}CH_3C_6H_4NHCOCH_3 \xrightarrow[CH_3COOH,\ 55°C]{Br_2} \text{2-Br-4-}CH_3C_6H_3NHCOCH_3 \xrightarrow[\text{alcohol},\ \Delta]{HCl} \xrightarrow{NaOH} \text{2-Br-4-}CH_3C_6H_3NH_2$$

(60–67%)
2-bromo-4-methylaniline

$$C_6H_5NHCOCH_3 \longrightarrow o\text{-}O_2NC_6H_4NHCOCH_3 + p\text{-}O_2NC_6H_4NHCOCH_3$$

	ortho	para
90% aq. HNO_3, −20°C	(23%)	(77%)
HNO_3, Ac_2O, 20°C	(68%)	(30%)

The last examples show that the *ortho/para* ratio can sometimes be altered by choosing the proper reaction conditions.

EXERCISE 24.15 Write equations showing the reactions of aniline, *m*-nitroaniline, and *p*-methoxyaniline with the following reagents or series of reagents.

(a) Br_2, water
(b) (i) acetic anhydride, pyridine; (ii) nitric acid, acetic acid; (iii) HCl, ethanol, heat; (iv) 1 N aqueous NaOH (work-up)

E. Elimination of the Amino Group: The Cope and Hofmann Elimination Reactions

Simple amines undergo neither base-catalyzed nor acid-catalyzed elimination reactions. In the former case, the leaving group would be NH_2^-, which is the conjugate base of a very weak acid, ammonia ($pK_a = 34$). In the latter case, the leaving group NH_3 is still not a very good one, since its conjugate acid, the ammonium ion, NH_4^+, has a relatively low acidity ($pK_a = 9.4$).

$$B:^- + H{-}\overset{|}{\underset{|}{C}}{-}\overset{|}{\underset{|}{C}}{-}NH_2 \not\longrightarrow BH + \rangle C{=}C\langle + {}^-{:}NH_2$$

$$B: + H{-}\overset{|}{\underset{|}{C}}{-}\overset{|}{\underset{|}{C}}{-}\overset{+}{N}H_3 \not\longrightarrow BH^+ + \rangle C{=}C\langle + {:}NH_3$$

However, quaternary ammonium hydroxides do undergo elimination upon being heated.

$$CH_3CH_2\overset{+}{N}(CH_2CH_3)_3\ OH^- \xrightarrow{\Delta} CH_2{=}CH_2 + N(CH_2CH_3)_3 + H_2O$$

tetraethylammonium hydroxide — triethylamine

Elimination proceeds by the E2 mechanism with hydroxide ion as the attacking base.

$$HO^- + H{-}CH_2{-}CH_2{-}\overset{+}{N}R_3 \longrightarrow \left[\overset{\delta-}{HO}\cdots H\cdots CH_2 \text{=} CH_2\cdots \overset{\delta+}{N}R_3\right]^{\ddagger} \longrightarrow HOH + CH_2{=}CH_2 + NR_3$$

The elimination reaction itself is the final step in a process known as **Hofmann degradation.** In this process a primary, secondary, or tertiary amine is first treated with enough methyl iodide to convert it into the quaternary ammonium iodide. The iodide is then replaced by hydroxide by treatment with silver oxide and water. The elimination reaction to give the alkene is effected by heating the dry quaternary ammonium hydroxide at 100 °C or higher. In the process, the carbon-nitrogen bond is broken and an amine and an alkene are produced.

$$C_6H_{11}NH_2 \xrightarrow[K_2CO_3]{CH_3I} C_6H_{11}\overset{+}{N}(CH_3)_3\ I^- \xrightarrow[H_2O]{Ag_2O} C_6H_{11}\overset{+}{N}(CH_3)_3\ OH^- \xrightarrow{120°C} \text{cyclohexene} + N(CH_3)_3 + H_2O$$

If the amine is cyclic, then the product is an amino alkene.

$$\text{N-methylpiperidine} \xrightarrow{CH_3I} \text{N,N-dimethylpiperidinium iodide} \xrightarrow[H_2O]{Ag_2O} \text{N,N-dimethylpiperidinium } OH^- \xrightarrow{\Delta} \text{N,N-dimethylpent-4-en-1-amine} + H_2O$$

N-methyl-piperidine; N,N-dimethyl-piperidinium iodide; N,N-dimethyl-pent-4-en-1-amine

The process can be repeated with the initial amino alkene to yield a diene, liberating the nitrogen as trimethylamine **(Hofmann exhaustive methylation).**

$$CH_2{=}CHCH_2CH_2CH_2N(CH_3)_2 \xrightarrow{CH_3I} \xrightarrow[H_2O]{Ag_2O} \xrightarrow{\Delta} CH_2{=}CHCH_2CH{=}CH_2 + N(CH_3)_3$$

1,4-pentadiene

A side reaction that may occur is S_N2 displacement. This is seldom a significant reaction except in cases where there are no β-hydrogens.

$$(CH_3)_3\overset{+}{N}{-}CH_3\ \ OH^- \xrightarrow[\Delta]{H_2O} (CH_3)_3N + CH_3OH$$

When the quaternary ammonium hydroxide has two or more different β-hydrogens, more than one alkene may be formed in the elimination. Unlike normal bimolecular eliminations with alkyl halides, the Hofmann elimination gives predominantly the less highly substituted alkene, as comparison of the following examples shows.

$$CH_3CH_2\underset{}{\overset{Br}{CH}}CH_3 \xrightarrow[EtOH,\ \Delta]{NaOEt} \underset{(81\%)}{CH_3CH{=}CHCH_3} + \underset{(19\%)}{CH_3CH_2CH{=}CH_2}$$

$$CH_3CH_2\overset{\overset{+}{N}(CH_3)_3}{CH}CH_3\ \ OH^- \longrightarrow \underset{(5\%)}{CH_3CH{=}CHCH_3} + \underset{(95\%)}{CH_3CH_2CH{=}CH_2}$$

This type of behavior is also seen when the ammonium compound contains two different alkyl groups that may be lost as the alkene.

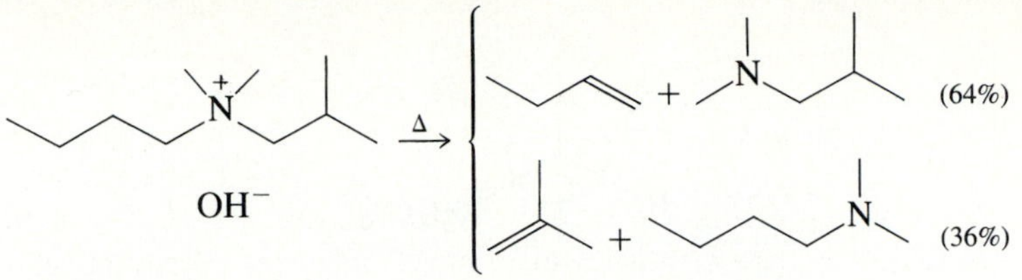

(64%)

(36%)

For ammonium hydroxides having only simple alkyl groups, such as the foregoing examples, the mode of elimination may be generalized as follows (Hofmann rule): "In the decomposition of quaternary ammonium hydroxides, the hydrogen is lost most easily from CH_3, next from RCH_2, and least easily from R_2CH." The direction of elimination in the Hofmann reaction is probably governed mostly by steric factors. The generality of the rule is shown by the following additional examples.

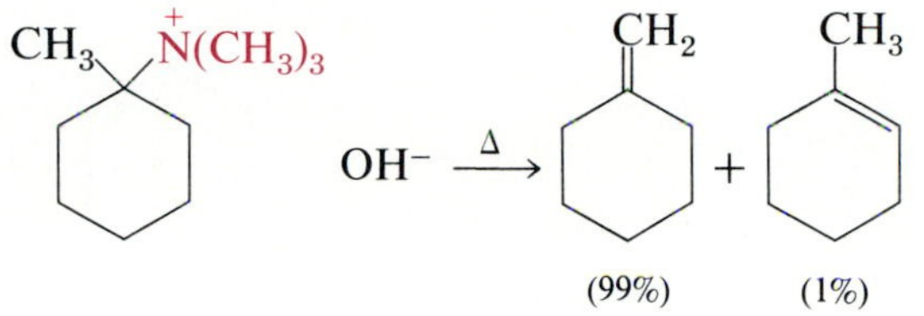

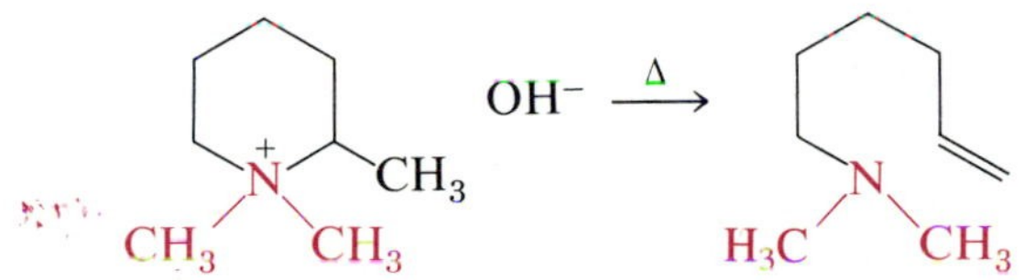

(99%) (1%)

One way in which steric effects contribute to this preference for the less substituted olefin is by their effect on the populations of different conformations. For example, in the 2-butyl case, the most stable conformation may be represented by a Newman projection of the C-2—C-3 bond as follows.

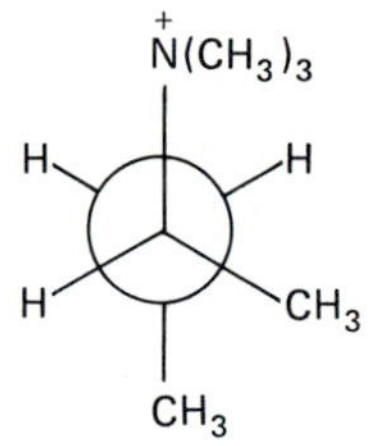

However, this conformation has no anti hydrogen at C-3. Anti elimination can only occur in one of the two conformations in which the trimethylammonium group is anti to a hydrogen.

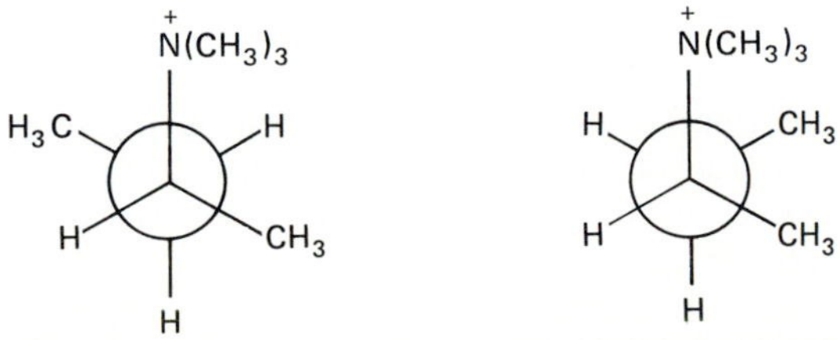

Both of these conformations have a methyl group gauche to the bulky trimethylammonium group, a group comparable in size to *t*-butyl. Hence, the populations of these conformations are small. On the other hand, all conformations with respect to the C-1—C-2 bond have an anti hydrogen.

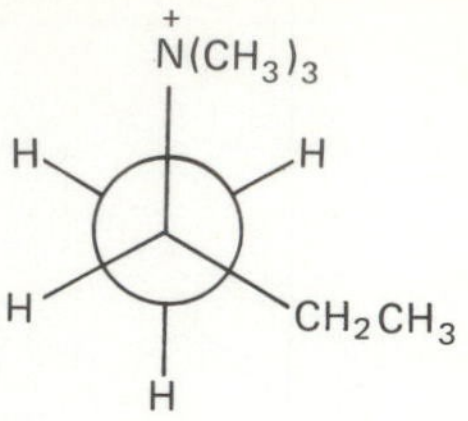

$$\overset{+}{N}(CH_3)_3$$

H H H CH_2CH_3 H

Although removal of a proton from C-3 would be faster because a more stable disubstituted ethylene results (Section 11.4), the population of conformations with anti hydrogen at C-3 is so small that the inherently slower reaction at C-1 dominates. An interesting example is the elimination reaction of the following cyclohexane derivative.

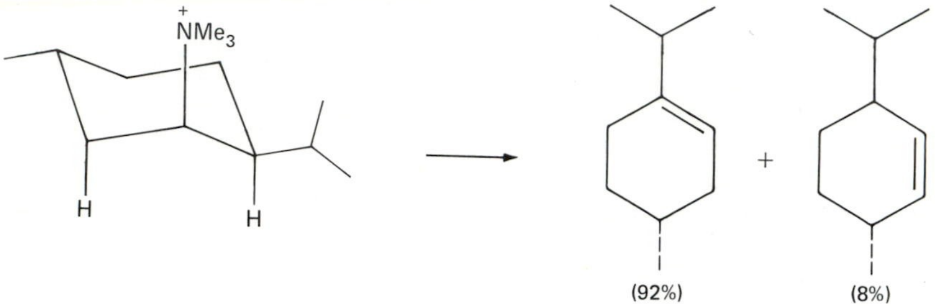

$\overset{+}{N}Me_3$ H H → (92%) + (8%)

This compound has a fixed conformation with two anti hydrogens, one secondary and one tertiary. Reaction at the tertiary hydrogen is faster and gives the more highly substituted olefin.

When electron-withdrawing groups are attached to one of the β-carbons, the Hofmann rule is not followed.

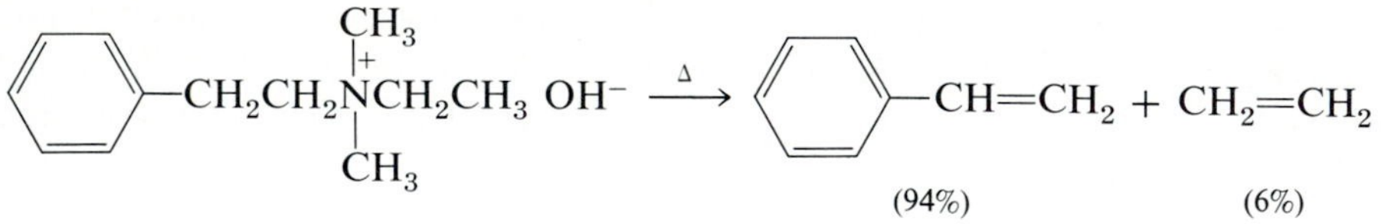

$$C_6H_5CH_2CH_2\overset{+}{N}(CH_3)_2CH_2CH_3\ OH^- \xrightarrow{\Delta} C_6H_5CH{=}CH_2 + CH_2{=}CH_2$$

(94%) (6%)

Elimination of the amino group may also be brought about by the thermal elimination of amine oxides (Section 24.7.C). These compounds undergo elimination when heated to 150–200 °C, provided that there is at least one hydrogen β to the nitrogen. The reaction is called the **Cope elimination** and is a useful alternative to the Hofmann degradation as a method for removing nitrogen from a compound. It is also useful as a preparative method for certain alkenes.

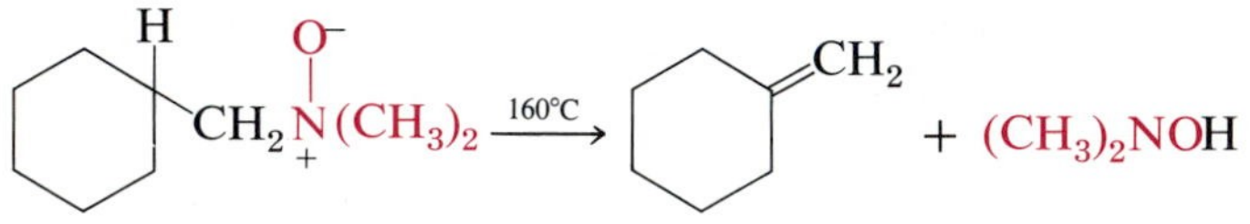

$$C_6H_{11}CH_2\overset{+}{N}(O^-)(CH_3)_2 \xrightarrow{160°C} C_6H_{10}{=}CH_2 + (CH_3)_2NOH$$

Crude N,N-dimethylcyclohexylmethanamine oxide (about 50 g) is placed in a flask that has been evacuated to a pressure of about 10 torr. The liquefied amine oxide is heated at 160 °C for 2 hr. Water is added, and the alkene layer is separated and distilled to obtain 30 g of methylenecyclohexane (98%).

The transformation of an amine oxide into an alkene proceeds by a sort of internal E2 process in which the oxide oxygen acts as the attacking base, abstracting the β-proton in a concerted reaction.

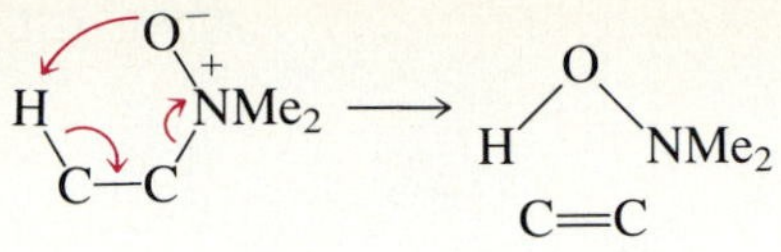

$$H-C-C(-O^-)(-\overset{+}{N}Me_2) \longrightarrow H-O-NMe_2 \quad C=C$$

This mechanism is supported by experiments that clearly show the elimination to be **syn,** meaning that the groups depart from the same side.

EXERCISE 24.16 Write equations showing the product(s) expected from each of the following reactions.

(a) N,N-dimethyl-2-pentanamine + methyl iodide; silver hydroxide; heat
(b) N,N-dimethyl-1-octanamine + methyl iodide; silver hydroxide; heat
(c) triethylamine + hydrogen peroxide; heat
(d) (1*R*,2*S*)-1-deuterio-N,N,2-trimethyl-1-butanamine + hydrogen peroxide; heat

24.8 Enamines and Immonium Ions

Enamines are compounds in which an amino group is attached directly to a carbon-carbon double bond. They are the nitrogen analogs of enols.

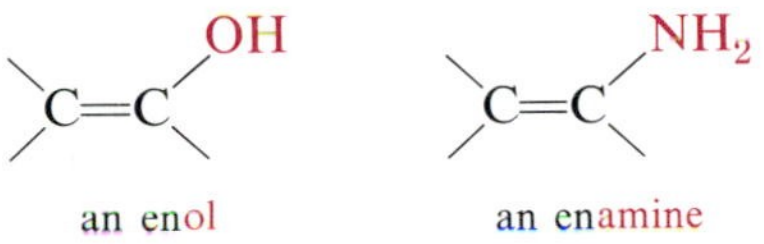

$$>C=C<^{OH} \qquad >C=C<^{NH_2}$$

an enol   an enamine

Like enols, enamines are generally unstable and undergo rapid conversion into the imine isomer.

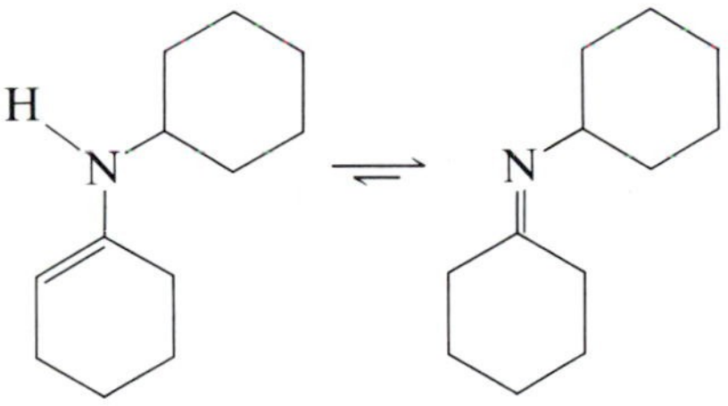

When the nitrogen of an enamine is tertiary, such isomerization cannot occur, and the enamine can be isolated and handled.

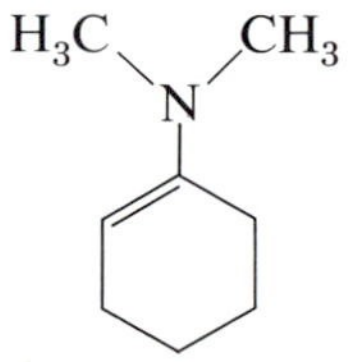

N,N-dimethyl-1-cyclohexenamine

Tertiary enamines are prepared by reaction of a secondary amine with an aldehyde or ketone. Water must be removed as it is formed in order to shift the equilibrium to the enamine product. Cyclic secondary amines are commonly used.

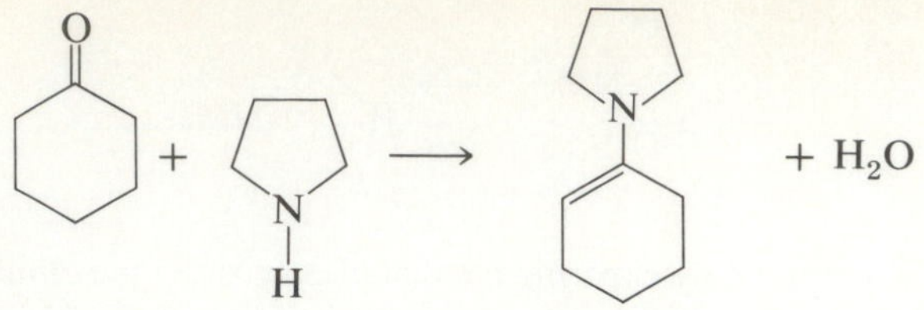

cyclohexanone + pyrrolidine ⟶ N-(1-cyclohexenyl)pyrrolidine + H_2O

pyrrolidine

N-(1-cyclo-
hexenyl)pyrrolidine

The mechanism for enamine formation is similar to the mechanism for formation of an imine with the loss of the proton in the final step being similar to that in enol formation; the reaction is subject to both acid and base catalysis (Sections 14.6.C and 15.1.A).

(1) 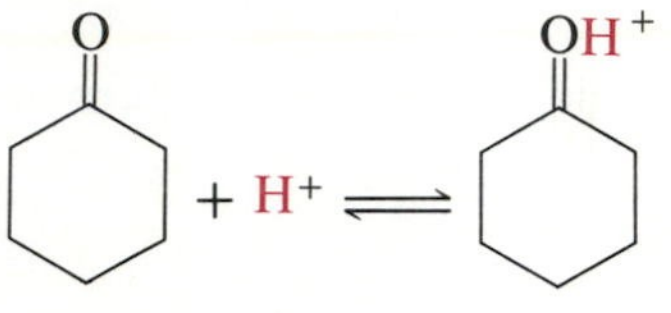 cyclohexanone + H^+ ⇌ protonated cyclohexanone (OH^+)

(2) 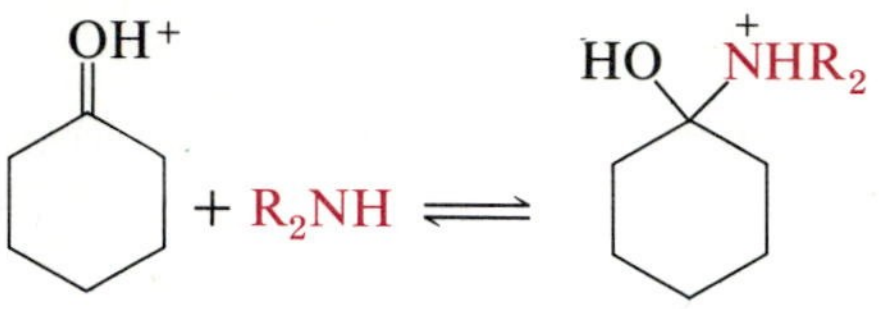 protonated cyclohexanone (OH^+) + R_2NH ⇌ 1-HO, 1-$\overset{+}{N}HR_2$ cyclohexane

(3) 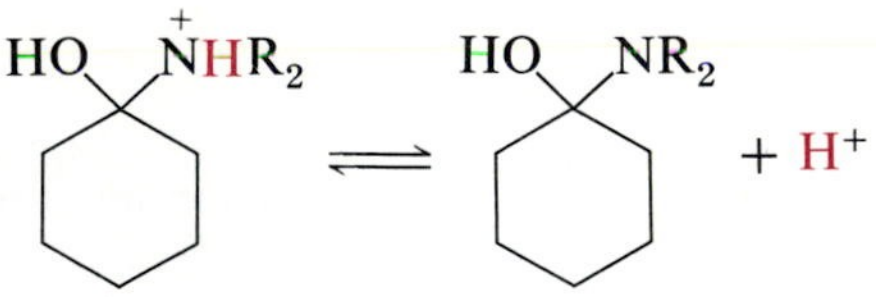 1-HO, 1-$\overset{+}{N}HR_2$ cyclohexane ⇌ 1-HO, 1-NR_2 cyclohexane + H^+

(4) 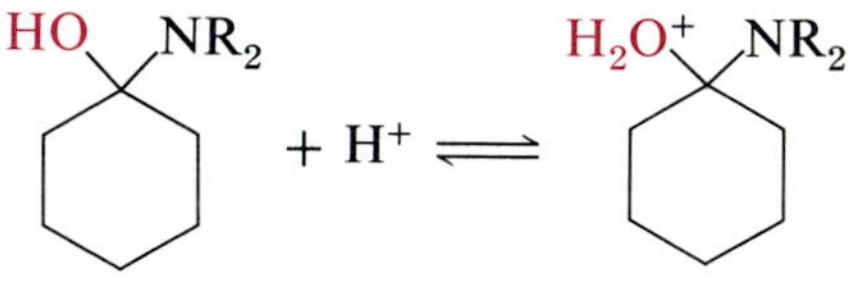 1-HO, 1-NR_2 cyclohexane + H^+ ⇌ 1-H_2O^+, 1-NR_2 cyclohexane

(5) 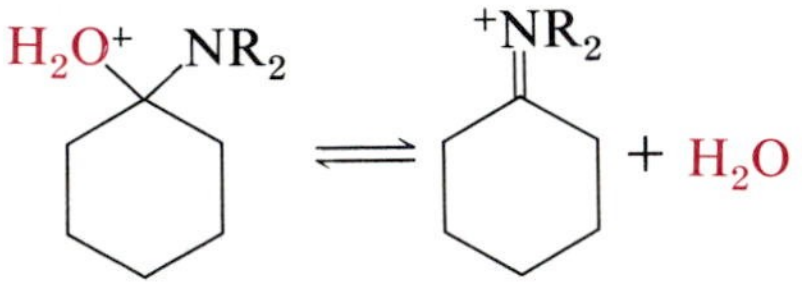 1-H_2O^+, 1-NR_2 cyclohexane ⇌ cyclohexylidene iminium ($^+NR_2$) + H_2O

(6) 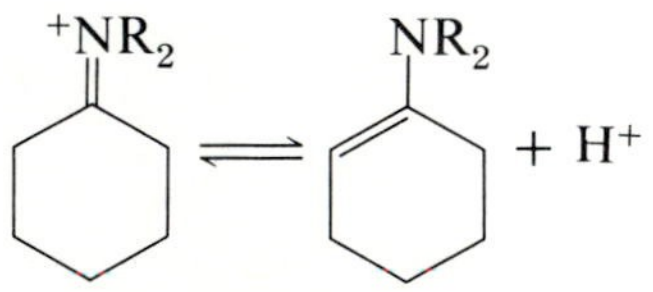 cyclohexylidene iminium ($^+NR_2$) ⇌ 1-NR_2 cyclohexene + H^+

The products are sensitive to aqueous acid and revert to the carbonyl compound and the amine in dilute acid.

$$\text{1-pyrrolidinylcyclohexene} + H_3O^+ \longrightarrow \text{cyclohexanone} + \text{pyrrolidinium } (H_2N^+)$$

Enamines are useful intermediates in some reactions because the β-carbon of the double bond has nucleophilic character, as shown in the following resonance structures.

$$\left[:NR_2 \text{ (enamine)} \longleftrightarrow {}^+NR_2 \text{ (iminium, carbanion)} \right]$$

Reaction occurs rapidly with reactive alkyl halides to give alkylated immonium compounds, which undergo facile hydrolysis to give the alkylated ketone.

$$\text{1-pyrrolidinylcyclohexene} + CH_2{=}CHCH_2Br \longrightarrow \left[\text{immonium } Br^- \right] \xrightarrow{H_2O} \text{2-allylcyclohexanone} + \text{pyrrolidinium } (\overset{+}{N}H_2)\ Br^-$$

2-allylcyclohexanone

Enamines can also be prepared by the dehydrogenation of tertiary amines. Dehydrogenation is accomplished by oxidation to an immonium ion with mercuric acetate, followed by deprotonation of the immonium ion with sodium bicarbonate.

$$\text{quinolizidine} + Hg(OAc)_2 \longrightarrow \text{immonium } {}^-OAc \xrightarrow{NaHCO_3} \text{enamine}$$

Immonium ions are intermediates in an important reaction for formation of carbon-carbon bonds, the **Mannich reaction.** In this reaction, a secondary amine, a ketone, or other carbonyl compound that can undergo easy enolization, and an aldehyde (often formaldehyde) combine to give a β-amino ketone.

$$C_6H_5\overset{O}{\overset{\|}{C}}CH_3 + CH_2O + (CH_3)_2NH_2^+\ Cl^- \longrightarrow C_6H_5\overset{O}{\overset{\|}{C}}CH_2CH_2N(CH_3)_2$$

acetophenone

β-(N,N-dimethylamino)propiophenone

A mixture of 60 g of acetophenone, 52.7 g of dimethylamine hydrochloride, 19.8 g of paraformaldehyde, 1 mL of concentrated hydrochloric acid, and 80 mL of 95% ethanol is heated on a steam bath for 2 hr. The warm yellow solution is diluted with 400 mL of acetone and cooled in an ice bath. The product separates as large crystals, 72–77 g (68–72%), m.p. 138–141 °C.

The Mannich reaction involves an intermediate immonium ion, which reacts with the enol form of the ketone.

$$CH_2{=}O + R_2\overset{+}{N}H_2 \longrightarrow CH_2{=}\overset{+}{N}R_2 + H_2O$$

$$R'C(\ddot{O}H){=}CH_2 + CH_2{=}\overset{+}{N}R_2 \longrightarrow R'C({=}\overset{+}{O}H){-}CH_2{-}CH_2{-}\ddot{N}R_2$$

We shall see that chemistry such as this is important in connection with methods for preparation of heterocyclic compounds (Chapter 32).

EXERCISE 24.17 Show how the Mannich reaction can be used to prepare 4-methyl-5-(N,N-dimethylamino)-3-pentanone.

The conversion of the amino acid serine (Chapter 29) to the keto acid pyruvic acid involves an elimination of water in a reaction catalyzed by a pyridoxal-phosphate-dependent enzyme. Enamines are the critical intermediates in the sequence that

serine

pyridoxal phosphate (vitamin B_6)

$-H_2O$

enamine

immonium ion

H_2O, $-NH_3$

pyruvate ion

enamine

immonium ion

immonium ion

begins with formation of an imine from the amino group of serine and the aldehyde of the cofactor, pyridoxal phosphate. The first-formed imine rearranges to an isomeric imine, which now undergoes an elimination reaction to form an enamine. The enamine reacts with water to yield the enamine derived from the carbonyl compound, pyruvic acid. Hydrolysis of the enamine leads to pyruvic acid, a central compound in metabolism. The sequence shown opposite illustrates many of the points we have studied about the reaction of amines and carbonyl groups. Pyridoxal phosphate is a cofactor for many important biochemical reactions. Humans require vitamin B_6, which is the name assigned to the corresponding alcohol, *pyridoxine*. Although the vitamin is nontoxic, unlike vitamin A (see page 681), a sensory neuropathy develops on a dose of 2000 mg daily and withdrawal symptoms have been noted after 200 mg daily. Pyridoxal phosphate is produced efficiently in the human body from pyridoxine.

ASIDE

The salt of Ammon (L. *sal ammoniac*) or, more likely, Amun, was first collected from the dung of camels near the temple of the god Amun in Thebes. From there the pharaohs controlled the Middle Kingdom of ancient Egypt about 1750 B.C. The salt is ammonium chloride, and alkali released a pungent gas that was thus called ammonia. Amines derive their name from ammonia. How many isomers are there of the amines with the formula $C_4H_{11}N$? Name them and write one synthesis for each. Rank them in order of rate of reaction with methyl iodide and give the product or products of the reaction.

PROBLEMS

1. Name the following compounds. For amines, use the convention enunciated at the end of Section 24.2.

a. $CH_3CH_2CH(CH_3)CH_2NH_2$

b. $(CH_3)_3N$

c. $CH_2{=}CHCH_2N(CH_3)(CH_2CH_3)$

d. 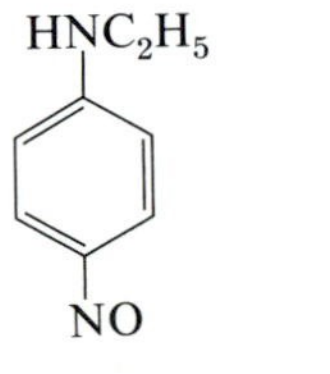 p-$BrC_6H_4N^+(CH_3)_3\ Cl^-$ (benzene ring bearing $^+N(CH_3)_3\ Cl^-$ and, para, Br)

e. 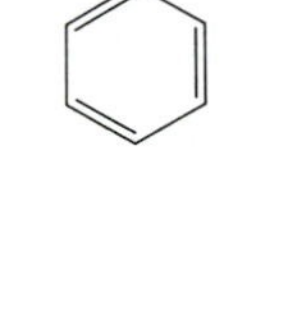 $C_6H_5N(NO)C_2H_5$ (benzene ring bearing N with ON and C_2H_5)

f. 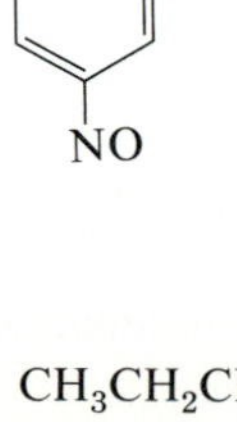 p-$ONC_6H_4NHC_2H_5$ (benzene ring bearing HNC_2H_5 and, para, NO)

g. $CH_3CH_2\overset{+}{N}(CH_3)_3\ I^-$

h. $CH_3CH_2CH_2\overset{+}{N}(O^-)(CH_3)_2$

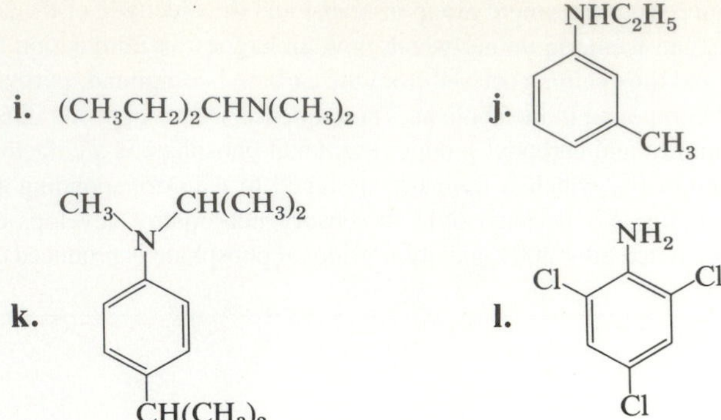

2. The NMR and IR spectra of an unknown compound are shown below. The CMR spectrum shows four resonances, with δ 8.2, 24.3, 35.0, and 49.0 ppm. Propose a structure for the compound.

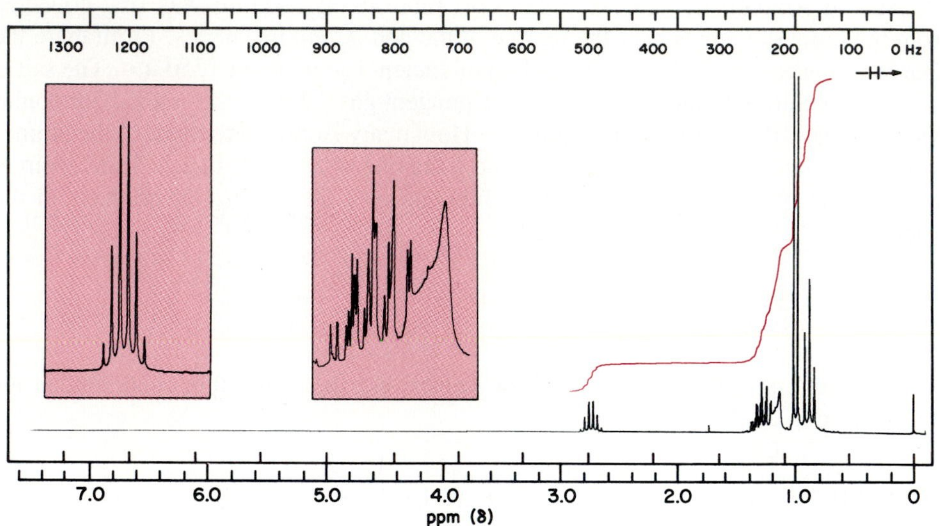

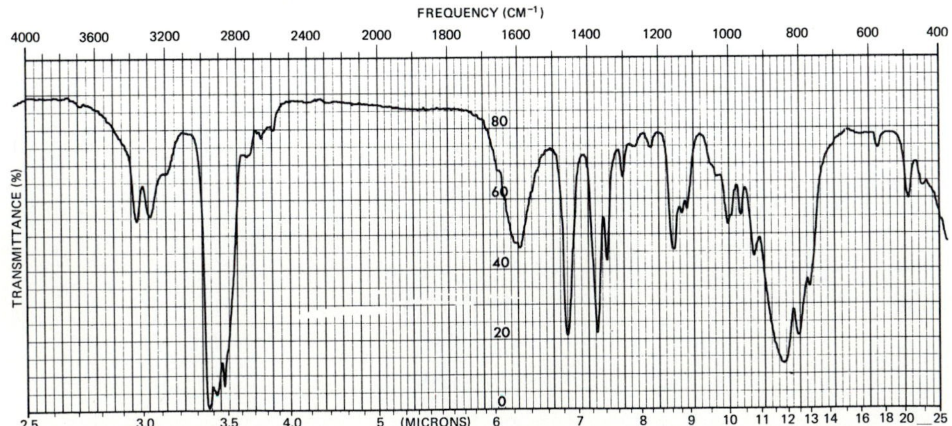

3. The NMR and IR spectra of an unknown compound are shown opposite. Propose a structure for the compound.

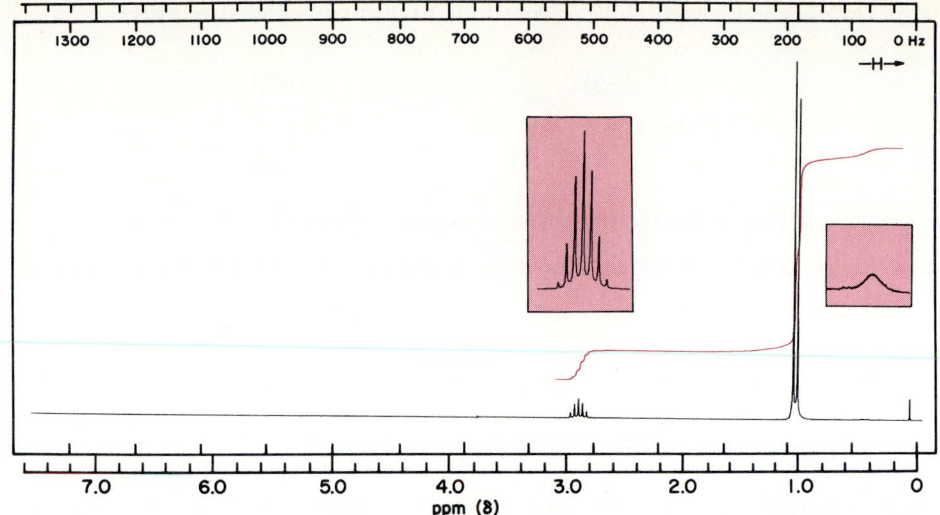

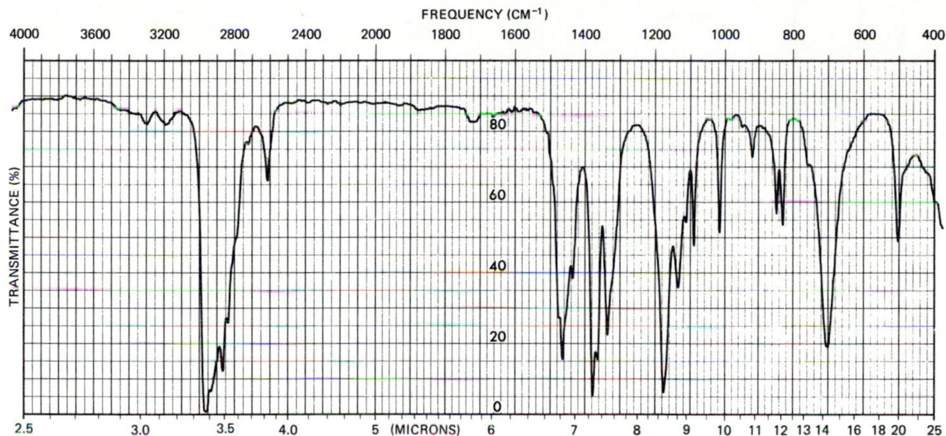

4. Although the inversion barrier for trimethylamine is only 6 kcal mole^{-1}, that for the heterocyclic tertiary amine N-methylaziridine is about 19 kcal mole^{-1}. Propose an explanation.

$$\begin{array}{l} CH_2 \\ | \quad \rangle N{-}CH_3 \\ CH_2 \end{array}$$

N-methylaziridine

5. Consider a solution of methylamine in water.

a. At what pH are the CH_3NH_2 and $CH_3NH_3^+$ concentrations exactly equal?

b. Calculate the $[CH_3NH_2]/[CH_3NH_3^+]$ ratio at pH 6, 8, 10, and 12.

6. Consider the reaction of methylamine with acetic acid.

$$CH_3NH_2 + CH_3COOH \overset{K}{\rightleftharpoons} CH_3NH_3^+ + CH_3CO_2^-$$

a. Using the data in Tables 18.3 and 24.4, calculate K.

b. At what pH does $[CH_3CO_2^-] = [CH_3NH_3^+]$?

7. a. Propose a method for separating a mixture of cyclohexanecarboxylic acid, tributylamine, and decane.

b. Alcohols react with phthalic anhydride to give monophthalate esters.

ROH + [phthalic anhydride] ⟶ [benzene ring with COOH and COOR]

Suggest a method for the resolution of racemic 2-octanol.

8. What is (are) the principal organic product(s) of each of the following reactions?

a. p-$CH_3C_6H_4NHCOCH_3$ $\xrightarrow{HNO_3,\ CH_3COOH}$

b. p-$CH_3C_6H_4NO_2$ $\xrightarrow[\Delta]{SnCl_2,\ HCl}$

c. p-$H_2NC_6H_4NHCOCH_3$ $\xrightarrow{Cl_2,\ aq.\ HCl}$

d. m-$O_2NC_6H_4COOH$ $\xrightarrow{H_2/Pt}$

e. $C_6H_5COCH_2CH_3$ $\xrightarrow[(CH_3)_2NH,\ H^+]{CH_2{=}O}$

f. $(CH_3)_2C_6H_2(NO_2)_2$ $\xrightarrow[CH_3OH]{NaSH}$

9. Suggest a sequence of reactions involving the Mannich condensation and the Hofmann elimination that can be used to convert acetone into methyl vinyl ketone.

10. How will each of the following compounds behave with aqueous nitrous acid?

a. $C_6H_5NH_2$ **b.** $C_6H_5NHCH_3$ **c.** $C_6H_5NHCOCH_3$

d. $C_6H_5N(CH_3)_2$ **e.** $C_6H_5CH_2NH_2$ **f.** $C_6H_5NHNH_2$

g. $C_6H_5CH_2NHCH_3$ **h.** $C_6H_5CH_2N(CH_3)_2$ **i.** $C_6H_5CH_2NHCOCH_3$

11. The dipole moment of *p*-(N,N-dimethylamino)benzonitrile, 6.60 D, is substantially greater than the sum of the dipole moments of N,N-dimethylaniline, 1.57 D, and benzonitrile, 3.93 D. Explain.

12. Write out the mechanism for bromination of N,N-dimethylaniline in the *para* position with Br_2 and show why this compound is so much more reactive than benzene.

13. Although *o*-methylaniline ($pK_a = 4.44$) is a somewhat weaker base than aniline ($pK_a = 4.60$), *o*-methyl-N,N-dimethylaniline ($pK_a = 6.11$) is a much stronger base than N,N-dimethylaniline ($pK_a = 5.15$). Give a rational explanation.

14. In each of the following pairs of compounds, which is the more basic in aqueous solution? Give a brief explanation.

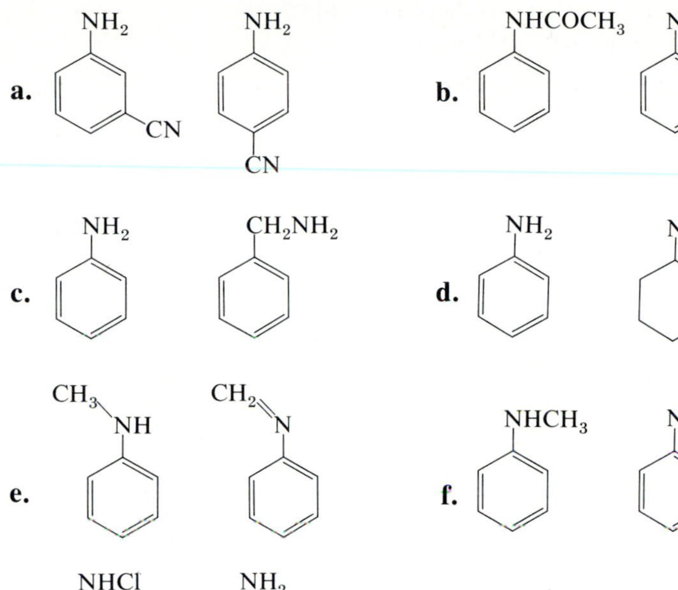

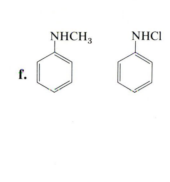

15. Outline a synthesis of each of the following compounds from alcohols containing five or fewer carbon atoms.

a. $CH_3(CH_2)_4NH_2$ **b.** $CH_3(CH_2)_4N(CH_3)_2$

c. $CH_3CH_2CH_2N(CH_3)CH_2CH_2CH_3$ **d.** $(CH_3)_2CHCH_2CH_2NHCH_2CH_3$

16. Propose a method for the stereospecific conversion of
a. (*R*)-2-octanol into (*S*)-2-octylamine
b. (*R*)-2-octanol into (*R*)-2-octylamine

17. Propose a synthesis for each of the following compounds.

a. $H_2NCH(CH_3)CH(OH)CH_2CH_3$ (mixture of diastereomers)

b. H_2N, H, CH_3 —C—C— H_3C, H, OH (indicated diastereomer only)

c. 1-(HO)(CH$_2$NH$_2$)cyclohexane: HO, CH_2NH_2 on cyclohexane ring

d. 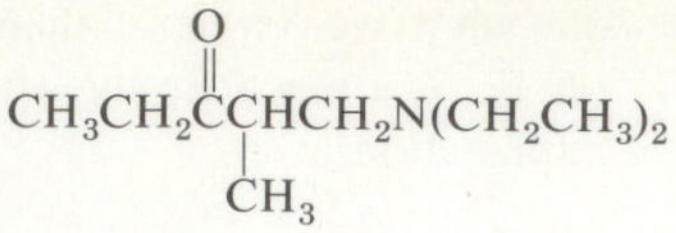 $CH_3CH_2\overset{O}{\overset{\|}{C}}\underset{CH_3}{\underset{|}{C}}HCH_2N(CH_2CH_3)_2$

e. 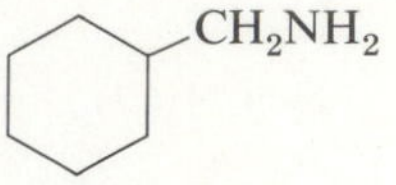 (cyclohexyl)CH_2NH_2 (*Hint:* See Section 21.6)

f. 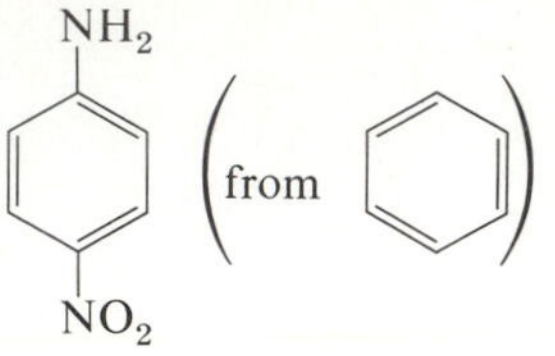 p-nitroaniline (NH_2, NO_2) (from benzene)

g. 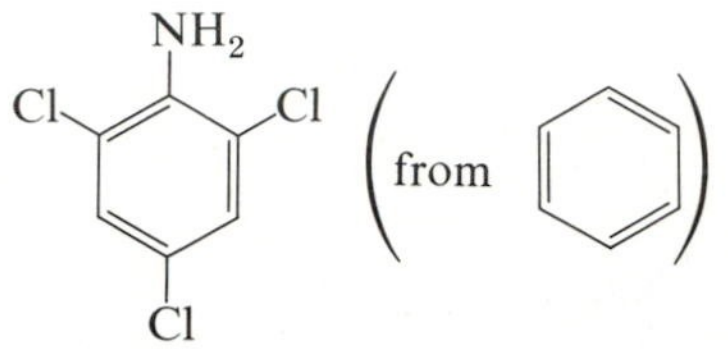 2,4,6-trichloroaniline (NH_2, Cl, Cl, Cl) (from benzene)

18. Diphenylamine, $(C_6H_5)_2NH$, is a rather weak base; the pK_a of the conjugate acid, 0.79, shows that diphenylamine is about 10^{-4} as basic as aniline. Give a reasonable explanation.

19. The trimethylanilinium cation, $C_6H_5N(CH_3)_3{}^+$ is prepared by an S_N2 reaction of dimethylaniline with a methyl halide or sulfonate. The compound undergoes a number of electrophilic substitution reactions such as nitration. Write the resonance structures involved for the intermediate produced by reaction at the *meta* and *para* positions and determine whether the trimethylammonium group is activating or deactivating, and *ortho,para* or *meta* directing.

20. Show how to accomplish each of the following conversions.

a. $CH_2{=}CHCO_2C_2H_5 \longrightarrow H_2NCH_2CH_2CH_2CH_2N(CH_3)_2$

b. 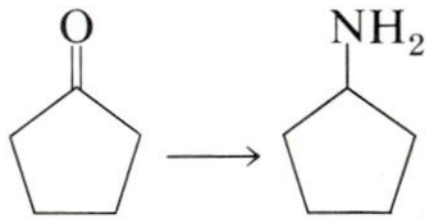 cyclopentanone (O) $\longrightarrow$ cyclopentylamine (NH_2)

c. 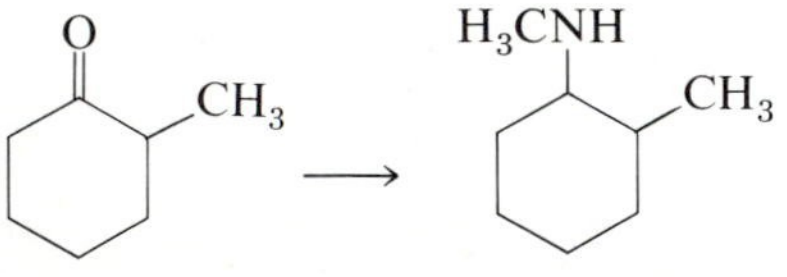 2-methylcyclohexanone (O, CH_3) $\longrightarrow$ (H_3CNH, CH_3)

d. 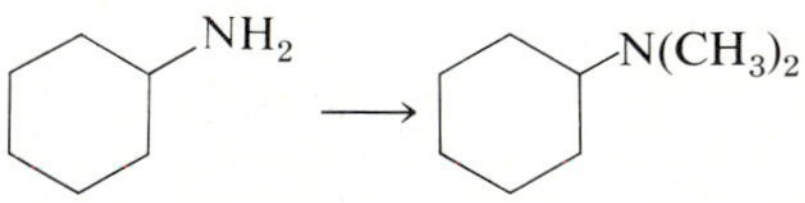 (cyclohexyl)NH_2 $\longrightarrow$ (cyclohexyl)$N(CH_3)_2$

e. 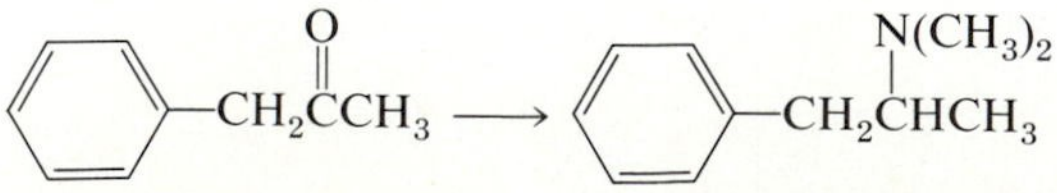 $C_6H_5CH_2\overset{O}{\overset{\|}{C}}CH_3 \longrightarrow C_6H_5CH_2\underset{}{\overset{N(CH_3)_2}{\overset{|}{C}}}HCH_3$

21. What is the expected product when piperidine is subjected to each of the following sets of reactions?

a. piperidine $\xrightarrow{CH_3\overset{O}{\overset{\|}{C}}Cl}$ $\xrightarrow{LiAlH_4}$

b. piperidine $\xrightarrow[(-H_2O)]{\text{cyclohexanone}}$

c. piperidine $\xrightarrow{HNO_2}$

d. piperidine $\xrightarrow[CH_2{=}O,\ HCl]{C_6H_5COCH_3}$

e. piperidine $\xrightarrow[Na_2CO_3]{2CH_3I}$ $\xrightarrow{Ag_2O}$ $\xrightarrow{\Delta}$

f. piperidine $\xrightarrow{H_2O_2}$

22. *cis*-2-Butene is subjected to the following sequence of reactions.

$$\text{cis-}CH_3CH{=}CHCH_3 \xrightarrow[2.\ H_2O_2,\ OH^-]{1.\ B_2D_6} CH_3CHD\overset{OH}{\overset{|}{C}}HCH_3 \xrightarrow{C_6H_5SO_2Cl}$$

$$CH_3CHD\overset{OSO_2C_6H_5}{\overset{|}{C}}HCH_3 \xrightarrow[\text{(page 743)}]{\text{Gabriel synthesis}} CH_3CHD\overset{NH_2}{\overset{|}{C}}HCH_3 \xrightarrow[HCO_2H]{CH_2O} CH_3CHD\overset{N(CH_3)_2}{\overset{|}{C}}HCH_3 \xrightarrow{H_2O_2}$$

$$CH_3CHD\overset{^-O-\overset{+}{N}(CH_3)_2}{\overset{|}{C}}HCH_3 \xrightarrow{150°C} \text{1-butene} + \textit{cis}\text{-2-butene} + \textit{trans}\text{-2-butene}$$

Two of the butene isomers produced in the pyrolysis contain one atom of deuterium per molecule and the other isomer contains only hydrogen. Which isomer contains no deuterium? Explain.

23. Predict the major product in each of the following elimination reactions.

a. $CH_3CH_2\overset{+}{N}(CH_3)_2CH_2CH(CH_3)_2\ OH^- \xrightarrow{\Delta}$

b. $(CH_3)_2CH\overset{^+N(CH_3)_3}{\overset{|}{C}}HCH_3\ OH^- \xrightarrow{\Delta}$

c. $(CH_3CH_2)_3\overset{+}{N}CH_2CH_2\overset{O}{\overset{\|}{C}}CH_3\ OH^- \xrightarrow{\Delta}$

d. bicyclic ammonium ion (bridgehead H; CH_3 substituents on both rings; $\overset{+}{N}$–CH_3) $OH^- \xrightarrow{\Delta}$

e. cyclohexane bearing $^-O-\overset{+}{N}(CH_3)_2$ and adjacent CH_3

24. Show how to accomplish each of the following conversions.

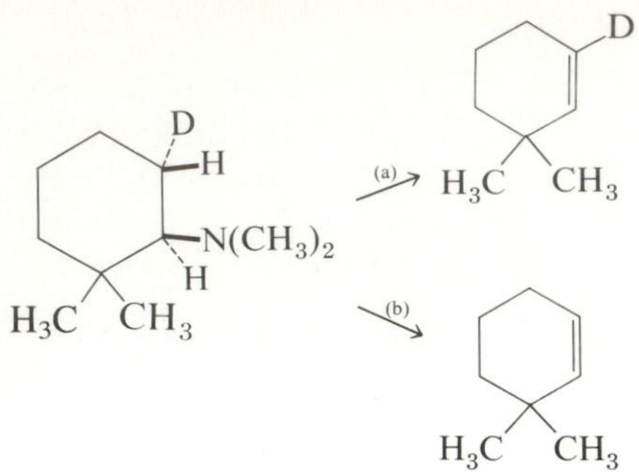

25. Reaction of (cyclopentylmethyl)amine with aqueous nitrous acid gives a mixture of two alcohols and three olefins. Deduce their structures using a reasonable reaction mechanism.

26. Show how each of the following reactions can be carried out using phase-transfer catalysis.

a. $n\text{-}C_{10}H_{21}Br + NaOOCCH_3 \longrightarrow n\text{-}C_{10}H_{21}OOCCH_3$

b. $3,4\text{-}(CH_3)_2C_6H_3COCH_3 + NaOD + D_2O \longrightarrow 3,4\text{-}(CH_3)_2C_6H_3COCD_3$

c. $(CH_3)_2CH(CH_2)_4OH + (CH_3)_2SO_4 \longrightarrow (CH_3)_2CH(CH_2)_4OCH_3$

d. $(CH_3)_2CHCHO + C_6H_5CH_2Cl \longrightarrow C_6H_5CH_2C(CH_3)_2CHO$

e. $C_6H_5CH{=}CH_2 + CHCl_3 \longrightarrow$ C_6H_5–(2,2-dichlorocyclopropyl)

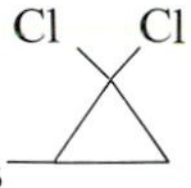

27. Which member of each of the following pairs of substituted ammonium ions is the more acidic? Explain briefly.

a. $ClCH_2CH_2NH_3^+$; $CH_3CH_2CH_2NH_3^+$
b. $CH_3ONH_3^+$; $CH_3NH_3^+$
c. $CH_3CONH_3^+$; $CH_3CH_2NH_3^+$
d. $CH_2{=}NH_2^+$; $CH_3NH_3^+$
e. $CH_3OOCCH_2NH_3^+$; $CH_3CH_2CH_2NH_3^+$
f. $CH_2{=}CHNH_3^+$; $CH_3CH_2NH_3^+$

28. Granatine, $C_9H_{17}N$, is an alkaloid that occurs in pomegranate. Two stages of the Hofmann exhaustive methylation (page 762) remove the nitrogen and yield a mixture of cyclooctadienes identified by catalytic hydrogenation to cyclooctane. The ultraviolet spectrum of the mixture shows the absence of the conjugated diene, 1,3-cyclooctadiene. Deduce the structure of granatine.

29. N-Chloroacetanilide is converted to a mixture of 32% *o*-chloroacetanilide and 68% *p*-chloroacetanilide in the presence of HCl. The use of ^{36}Cl-enriched HCl finds the isotopic Cl incorporated into the product. The reaction of acetanilide with chlorine under the same reaction conditions gives the same product composition. Write a reaction mechanism for the rearrangement of N-chloroacetanilide (Orton rearrangement) to account for these facts.

30. Treatment of (1-hydroxycyclohexyl)methylamine with aqueous nitrous acid gives a single product in good yield. The product shows a strong infrared absorption at

$1705\ cm^{-1}$ and has a four-line CMR spectrum; δ 23.5, 29.7, 42.7, and 211.7. What is the structure of this product? Propose a mechanism for its formation.

31. 1-Bromodecane reacts with sodium azide in ethanol to form A, a compound with a boiling point similar to that of the bromide. Reaction of A with triphenylphosphine yields B, which reacts with water to yield C, which has an equivalent weight of 157 by titration with hydrochloric acid. Reduction of A with hydrazine or hydrogenation also gives C. Write the reactions for the sequence, identifying A, B, and C and naming each product.

32. Following are five synthetic methods, each of which can be used to prepare some kinds of amines.

Method **A**: Gabriel synthesis: phthalimide (N—H) $\xrightarrow[\text{2. } H_2SO_4,\ H_2O,\ 100\ °C]{\text{1. KOH, R—Br}}$ RNH_2

Method **B**: Formation and reduction of a nitrile: R—Cl $\xrightarrow[\text{2. } H_2/Ni]{\text{1. NaCN, EtOH, reflux}}$ RCH_2NH_2

Method **C**: Preparation and reduction of an amide: R′NHR″ $\xrightarrow[\text{2. } LiAlH_4]{\text{1. RCOCl}}$ $RCH_2N(R'')R'$

Method **D**: Preparation and reduction of an oxime: R′COR″ $\xrightarrow[\text{2. } LiAlH_4]{\text{1. } H_2NOH}$ $R'CH(NH_2)R''$

Method **E**: Curtius rearrangement: RCOCl $\xrightarrow[\text{2. } H_2O\text{, heat}]{\text{1. } NaN_3,\ CH_3CN\text{, heat}}$ RNH_2

Fill in the following table with Y (yes) if the method can be used to prepare the given amine and N (no) if the method cannot be used for this amine.

	Method **A**	Method **B**	Method **C**	Method **D**	Method **E**
a. $(CH_3)_3CCH_2NH_2$	______	______	______	______	______
b. cyclohexyl–$CH_2N(CH_3)_2$	______	______	______	______	______
c. $(CH_3)_2CHCH_2CH_2NH_2$	______	______	______	______	______
d. 3-chlorophenyl–NH_2 (Cl, NH_2)	______	______	______	______	______
e. $(CH_3)_2CHNHCH_2CH_3$	______	______	______	______	______

CHAPTER 25

OTHER NITROGEN FUNCTIONS

The amino group is the most important functional group of nitrogen. Nevertheless, a number of other nitrogen-containing functional groups are known, and some have been mentioned from time to time in this text. Several of these other nitrogen functional groups are listed in Table 25.1 and will be discussed in this chapter. We shall see that they vary in importance to organic chemistry. Some of these functional groups have special importance when they are attached to aromatic rings. An example is the diazonium group, a particularly significant aromatic functional group that is useful for the preparation of a wide variety of other compounds.

25.1 Nitro Compounds

Nitroalkanes are relatively rare, although a few of the simpler ones are commercially available. Examples are nitromethane (which is used as a high-power fuel in racing engines), nitroethane, and 2-nitropropane.

$$\left[CH_3\overset{+}{N}(-O^-)(=O) \longleftrightarrow CH_3\overset{+}{N}(=O)(-O^-) \right] \equiv CH_3NO_2 \qquad CH_3CH_2NO_2 \qquad (CH_3)_2CHNO_2$$

nitromethane — nitroethane — 2-nitropropane

Aromatic nitro compounds are much more common because they are easily prepared by the electrophilic nitration of aromatic compounds (Section 23.3).

A. Nitroalkanes

Nitroalkanes are prepared industrially by the free radical nitration of alkanes (see problem 13, page 121).

$$CH_4 + HNO_3 \xrightarrow{400°C} CH_3NO_2 + H_2O$$

TABLE 25.1 Functional Groups Containing Nitrogen

Structure	Name	Example	
$R{-}NO_2$	nitro	$C_6H_5NO_2$	nitrobenzene
$R{-}NCO$	isocyanate	C_6H_5NCO	phenyl isocyanate
$R{-}NHCOOR'$	urethane, carbamate	$C_6H_5NHCOOCH_3$	methyl N-phenylcarbamate
$R{-}NHCONH{-}R'$	urea	H_2NCONH_2	urea
$R{-}N_3$	azide	$CH_3CH_2N_3$	ethyl azide
$R{-}N{=}N{-}R'$	azo	$C_6H_5N{=}NC_6H_5$	azobenzene
$R{-}\overset{+}{N}(O^-){=}N{-}R$	azoxy	$C_6H_5\overset{+}{N}(O^-){=}NC_6H_5$	azoxybenzene
$R{-}NHNH_2$	hydrazine, diazine	$C_6H_5NHNH_2$	phenylhydrazine
$R_2C{=}N_2$	diazo	$CH_2{=}N_2$	diazomethane
$R{-}N_2{}^+$	diazonium	$C_6H_5N_2{}^+\ Cl^-$	benzenediazonium chloride

Some nitro compounds can be prepared in the laboratory by the displacement of alkyl halides with nitrite ion. Since nitrite is an ambident anion, some alkyl nitrite is usually produced as a by-product (Section 9.4).

$$CH_3(CH_2)_5CH(I)CH_3 + NaNO_2 \longrightarrow \underset{(58\%)}{CH_3(CH_2)_5CH(NO_2)CH_3} + \underset{(30\%)}{CH_3(CH_2)_5CH(ONO)CH_3}$$

Isoamyl nitrite, $(CH_3)_2CHCH_2CH_2ONO$, and in fact, many organic nitrate esters, especially glyceryl trinitrate, $O_2NOCH_2CH(ONO_2)CH_2ONO_2$ (''nitroglycerin''), are used to counteract the effects of angina, heart muscle pain. The volatility of isoamyl nitrite allows rapid delivery via the nose. The mechanism of biological action involves the formation in blood vessels of nitric oxide, NO, the *natural agent* that leads to relaxation of blood vessel walls (vasodilation) and thus promotes blood flow. Although such compounds are not particularly toxic, the biological effects warrant caution when carrying out syntheses in which the relatively volatile nitrites could be by-products. Nitroglycerin in pure form is a highly explosive liquid that is converted to the safer dynamite by adsorption on an inert solid.

Yields of nitroalkane are higher when silver nitrite is used, but this added economy is tempered by the cost of the silver salt.

$$CH_3(CH_2)_6CH_2I + AgNO_2 \longrightarrow \underset{(83\%)}{CH_3(CH_2)_6CH_2NO_2} + \underset{(11\%)}{CH_3(CH_2)_6CH_2ONO}$$

The most striking chemical property of nitroalkanes is their acidity. The pK_a of nitromethane is 10.2, that of nitroethane is 8.5, and that of 2-nitropropane is 7.8. 2-Nitropropane is so acidic that it is extensively ionized in neutral solution (pH = 7). Like carboxylic acids and ketones, nitro compounds owe their acidity to the fact that the corresponding carbanion is conjugated to electronegative oxygens.

$$CH_3\overset{+}{N}(=O)O^- \longrightarrow H^+ + \left[:\overset{-}{C}H_2{-}\overset{+}{N}(=O)O^- \longleftrightarrow :\overset{-}{C}H_2{-}\overset{+}{N}(O^-){=}O \longleftrightarrow CH_2{=}\overset{+}{N}(O^-)O^- \right]$$

methylnitronate ion

The anions derived from nitroalkanes are nucleophilic and enter into typical nucleophilic reactions. One particularly useful reaction is the **Henry reaction,** which is analogous to the aldol addition reaction of aldehyde and ketone enolates (Section 15.2).

$$CH_3(CH_2)_7CHO + CH_3NO_2 \xrightarrow[\text{EtOH}]{\text{NaOH}} CH_3(CH_2)_7CH(OH)CH_2NO_2$$

(80%)
1-nitro-2-decanol

Since nitro compounds are so acidic, only weakly basic catalysts are required. In the case of aromatic aldehydes, dehydration of the initial β-hydroxy nitro compound usually results.

$$C_6H_5\text{—}CHO + CH_3NO_2 \xrightarrow[25°C]{n\text{-}C_5H_{11}NH_2} C_6H_5\text{—}CH{=}CHNO_2$$

(75%)
β-nitrostyrene

Another general reaction of nitro compounds is reduction to the corresponding amine (Section 24.6.C). Reduction of β-hydroxy nitroalkanes produced by the Henry reaction provides a convenient method for preparing 1,2-amino alcohols (compare page 745).

$$HO\text{—}CH_2CH_2C(CH_3)_2\text{—}NO_2 \xrightarrow[H_2O,\ H_2SO_4]{Fe,\ FeSO_4} HO\text{—}CH_2CH_2C(CH_3)_2\text{—}NH_3^+\ HSO_4^- \xrightarrow{NaOH} HO\text{—}CH_2CH_2C(CH_3)_2\text{—}NH_2$$

EXERCISE 25.1 Show how 2-pentanone can be converted into each of the following compounds.

(a) 2-methylpentanamine (b) 2-hydroxy-2-methylpentanamine

B. Nitroarenes

Nitrobenzene and related nitro compounds are generally high-boiling liquids. Nitrobenzene is a pale yellow oil, b.p. 210–211 °C, having a characteristic odor of almonds. It was used at one time in shoe polish, but its use for this purpose has now been discontinued because it is readily absorbed through the skin and is poisonous.

2,4,6-Trinitrotoluene, TNT, is an important explosive. It is relatively insensitive to shock and is used with a detonator. It melts at 81 °C, and it can be poured as the melt into containers for the manufacture of high explosive devices. 1,3,5-Trinitrobenzene is less sensitive than TNT to shock and has more explosive power, but is more difficult to prepare. Direct introduction of the third nitro group into toluene is assisted by the methyl group. Small amounts of 1,3,5-trinitrobenzene are prepared by oxidation of TNT to trinitrobenzoic acid.

$$2,4,6\text{-}(O_2N)_3C_6H_2CH_3 \xrightarrow[H_2SO_4,\ 45\text{–}55°C]{Na_2Cr_2O_7} 2,4,6\text{-}(O_2N)_3C_6H_2COOH$$

2,4,6-trinitrotoluene

(57–69%)
2,4,6-trinitrobenzoic acid

2,4,6-Trinitrobenzoic acid is a strong acid ($pK_a = 0.7$) whose anion decomposes on heating to give carbon dioxide and a phenyl anion that is stabilized by the electron-attracting inductive effect of the three nitro groups.

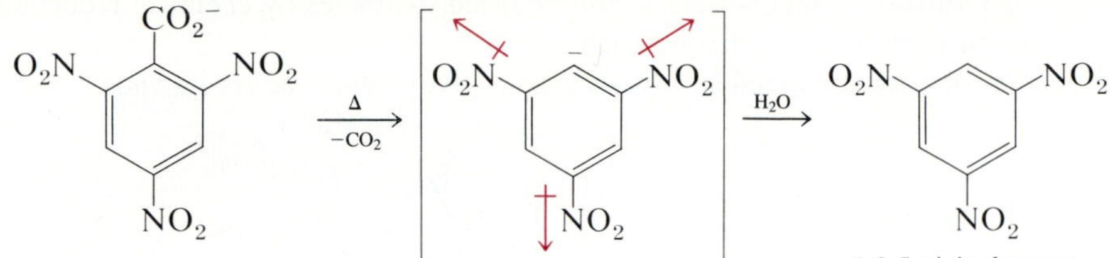

1,3,5-trinitrobenzene

EXERCISE 25.2 The protons attached to the methyl carbon in 2,4,6-trinitrotoluene (TNT) are fairly acidic. For example, a mixture of benzaldehyde and TNT react with the aid of a secondary amine catalyst to give an adduct that is analogous to the product of the Henry reaction.

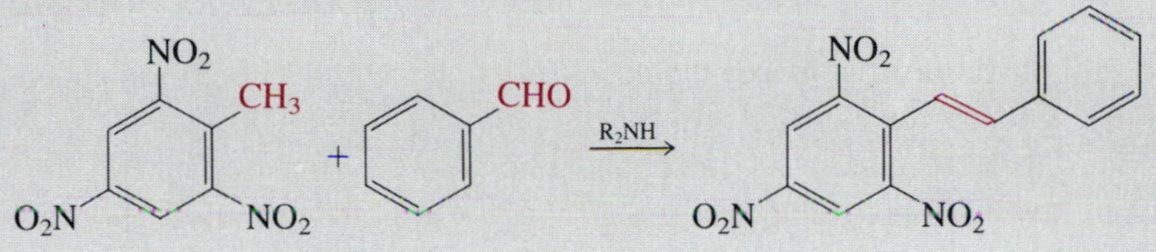

Suggest a reason for the acidity of these protons.

C. Reactions of Nitroarenes

The nitro group is relatively stable to many reagents. It is generally inert to acids and most electrophilic reagents; hence, it may be present in a ring when reactions with such reagents are used. The nitro group is also stable to most oxidizing agents, but it reacts with Grignard reagents and other strongly basic compounds such as lithium aluminum hydride. The most important reaction of the nitro group in aromatic compounds is reduction, but the reduction product depends on the reaction conditions used. Catalytic hydrogenation and reduction in acidic media yield the corresponding amine (Section 24.6.C).

$$ArNO_2 \longrightarrow ArNH_2$$

Reduction of the nitro group actually proceeds in a series of two-electron steps. In acid the intermediate compounds cannot be isolated, but are reduced rapidly in turn.

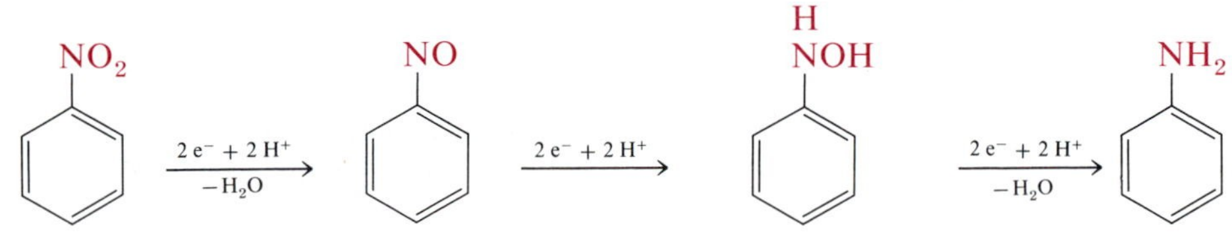

nitrobenzene $\xrightarrow[-H_2O]{2e^- + 2H^+}$ nitrosobenzene $\xrightarrow{2e^- + 2H^+}$ phenylhydroxylamine $\xrightarrow[-H_2O]{2e^- + 2H^+}$ aniline

In neutral media a higher reduction potential is required, and reduction is readily stopped at the hydroxylamine stage.

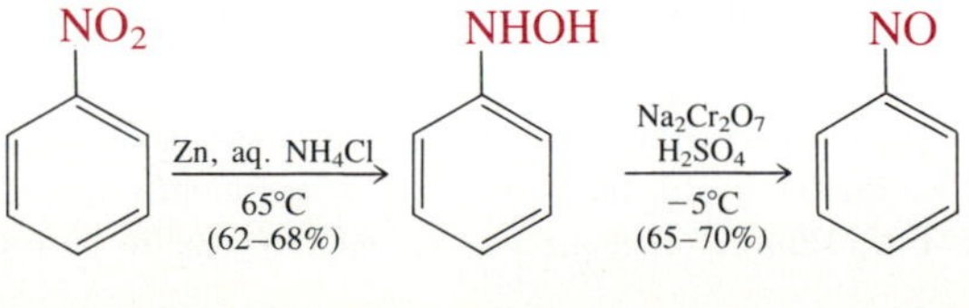

$C_6H_5NO_2 \xrightarrow[\substack{65°C \\ (62\text{–}68\%)}]{Zn,\ aq.\ NH_4Cl} C_6H_5NHOH \xrightarrow[\substack{-5°C \\ (65\text{–}70\%)}]{Na_2Cr_2O_7,\ H_2SO_4} C_6H_5NO$

Aromatic hydroxylamines are relatively unimportant compounds. Phenylhydroxylamine is a water-soluble, crystalline solid, m.p. 82 °C, that deteriorates in storage. It may be oxidized to the bright green nitrosobenzene as shown in the foregoing equation. Hydroxylamines and nitroso compounds are readily reduced to amines by chemical reduction in acidic solution or by catalytic hydrogenation.

Reduction of nitro compounds in basic media gives binuclear compounds.

NO_2 (nitrobenzene) — As_2O_3, aq. NaOH, Δ → Ph–N=N⁺(O⁻)–Ph (85%) azoxybenzene

— Zn, alc. NaOH, Δ → Ph–N=N–Ph (84–86%) azobenzene

— H_2NNH_2, Ru/C, alc. KOH → Ph–NHNH–Ph (80%) hydrazobenzene

All of these compounds are reduced to aniline under acidic conditions. They may also be interconverted by the following reactions.

Ph–N=N⁺(O⁻)–Ph ⇌ [$(C_2H_5O)_3P$ →; ← H_2O_2, CH_3COOH, 65°C] Ph–N=N–Ph ⇌ [Pd/C, H_2NNH_2 →; ← air or NaOBr] Ph–NHNH–Ph

These binuclear compounds may best be considered to arise by condensation reactions between the intermediates formed during reduction.

Ph–N=O + Ph–N(OH)H → [Ph–N(O⁻)–N(OH)–Ph] $\xrightarrow{-OH^-}$ Ph–N⁺(O⁻)=N–Ph

In fact, azoxybenzene can be prepared by the base-catalyzed condensation of phenylhydroxylamine with nitrosobenzene. The azoxy function is the least important functional group among these compounds. Azobenzene is a bright orange-red solid. Although azobenzene itself has only limited significance, the azo linkage is an important component of azo dyes (Section 36.2). Although the term ''azo'' is well-entrenched, diazene should be used where possible to emphasize the presence of *two* nitrogens in a *double* bond. Note that an unsymmetrical diazene can form two different oxides, which can be differentiated by nitrogen number, as in 1-phenyl-2-*p*-tolyldiazene 1-oxide.

Many azo compounds show cis-trans isomerism. The trans isomer is generally the more stable, and the activation energy for the conversion is sufficiently low that the cis isomer is generally not seen. For example, azobenzene can be converted in part to the cis isomer by photolysis, but the activation energy required to convert back to *trans*-azobenzene is only 23–25 kcal $mole^{-1}$ in various solvents. This reaction has a half-life on the order of hours at room temperature.

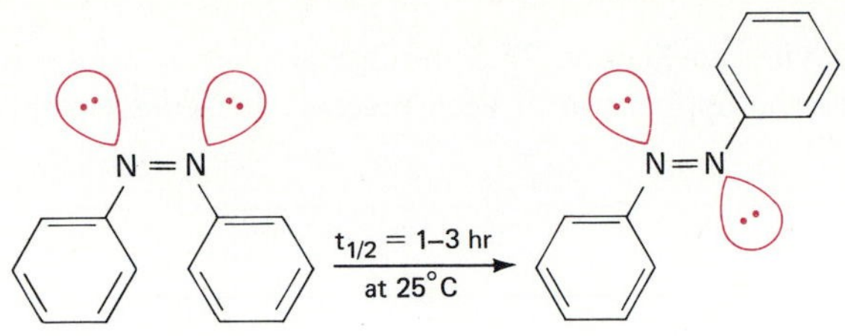

cis-azobenzene $\xrightarrow[\text{at } 25^\circ C]{t_{1/2} = 1\text{–}3 \text{ hr}}$ trans-azobenzene

The color of azobenzenes is due to an $n \rightarrow \pi^*$ transition (Section 22.6.C). Absorption occurs at longer wavelengths than for carbonyl $n \rightarrow \pi^*$ transitions due to the "splitting" of the non-bonding electron levels. Combination of the nonbonding *n*-orbitals forms two new levels, one much higher (n_-) and one much lower (n_+). Both levels have two electrons; exciting an electron from the upper level takes relatively little energy. In other words, light absorption occurs at relatively low energies and long wavelengths.

Hydrazobenzene or 1,2-diphenylhydrazine is a colorless solid that oxidizes in air to azobenzene at a moderate rate. It is significant principally because of a rearrangement that it undergoes in strongly acidic solution, the **benzidine rearrangement.**

$$C_6H_5-NHNH-C_6H_5 \xrightarrow[\Delta]{H^+} H_2N-C_6H_4-C_6H_4-NH_2$$

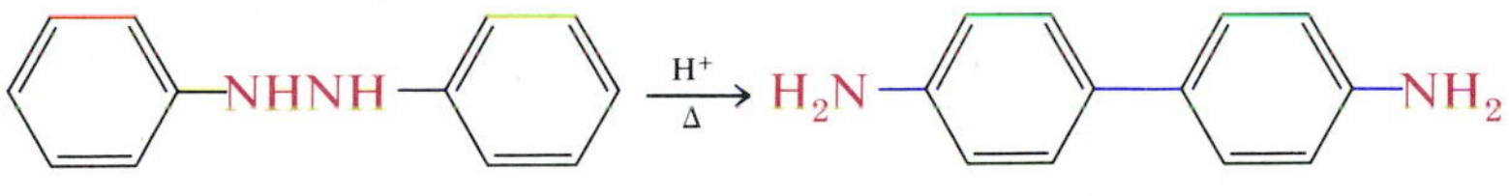

hydrazobenzene → 4,4′-diaminobiphenyl (benzidine)

This remarkable reaction involves the mono- or diprotonated salt in which bonding occurs between the *para* positions as the nitrogen-nitrogen bond is broken.

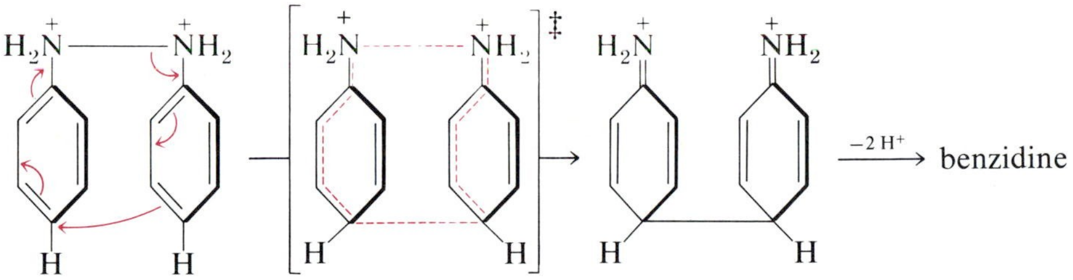

$$\overset{+}{H_2N}-\overset{+}{NH_2} \longrightarrow [\ldots]^{\ddagger} \longrightarrow \ldots \xrightarrow{-2\,H^+} \text{benzidine}$$

transition state

The benzidine rearrangement is an example of a [5.5] sigmatropic rearrangement and involves a 10-electron Hückel transition state. Benzidine has had important uses as an intermediate in dye manufacture, but the compound has been found to be carcinogenic.

Azoxybenzene, azobenzene, and hydrazobenzene are all conveniently reduced to aniline with sodium hydrosulfite.

$$C_6H_5-N{=}N-C_6H_5 + 2\,Na_2S_2O_4 + 2\,H_2O \longrightarrow 2\,C_6H_5NH_2 + 4\,NaHSO_3$$

Sodium hydrosulfite, or sodium dithionite, is a useful reagent in neutral or alkaline solution. The dithionite dianion is in equilibrium with the radical anion, $\cdot SO_2^-$, which can be detected by ESR, electron spin resonance, and which is the actual reducing agent. In the presence of a catalyst, sodium dithionite (as Fieser's solution) can be used to remove small amounts of oxygen from a gas stream. In acidic medium it decomposes with the liberation of sulfur. It is especially useful in the reductive cleavage of the azo groups in azo dyes. These dyes are generally water-soluble, and reduction is accomplished by adding sodium hydrosulfite to an aqueous solution until the color of the dye has been discharged.

EXERCISE 25.3 Review in Section 23.3 the preparation of *o*-nitrotoluene. Write the equations showing the application of each reaction presented in this section with *o*-nitrotoluene.

Highlight 25.1

Nitro derivatives are especially important in syntheses of aromatic compounds, primarily through conversion to amines and other reduced compounds by reduction.

$$C_6H_5NO_2 \xrightarrow{[H]} C_6H_5NHOH \xrightarrow{[O]} C_6H_5N{=}O$$

$$C_6H_5NO_2 \xrightarrow{[H]} [C_6H_5NHOH + C_6H_5N{=}O] \longrightarrow C_6H_5N{=}\overset{+}{N}(O^-)C_6H_5 \longrightarrow C_6H_5N{=}NC_6H_5 \longrightarrow C_6H_5NH{-}NHC_6H_5$$

$$C_6H_5NO_2 \xrightarrow{[H]} [C_6H_5N{=}O] \longrightarrow [C_6H_5NHOH] \longrightarrow C_6H_5NH_2$$

Highlight 25.2

The nitro group is strongly electron-withdrawing and stabilizes carbanions more effectively than the carbonyl group. The pK_a of acetone is 19, but the pK_a of nitromethane is 10.2.

$$CH_3{-}C({=}O){-}CH_3 \underset{+H^+}{\overset{-H^+}{\rightleftharpoons}} \left[CH_2{=}C(O^-){-}CH_3 \longleftrightarrow {}^-CH_2{-}C({=}O){-}CH_3\right] \underset{+H^+}{\overset{-H^+}{\rightleftharpoons}} CH_2{=}C(OH){-}CH_3$$

enol

$$CH_3{-}\overset{+}{N}({=}O){-}O^- \underset{+H^+}{\overset{-H^+}{\rightleftharpoons}} \left[CH_2{=}\overset{+}{N}(O^-){-}O^- \longleftrightarrow {}^-CH_2{-}\overset{+}{N}({=}O){-}O^-\right] \underset{+H^+}{\overset{-H^+}{\rightleftharpoons}} CH_2{=}\overset{+}{N}(OH){-}O^-$$

nitronic acid

25.2 Isocyanates, Carbamates, and Ureas

We have encountered alkyl isocyanates previously as intermediates in the Hofmann rearrangement of amides (Section 24.6.I). They can also be prepared by displacement of alkyl halides with cyanate ion. This ion is ambident and reacts preferentially at the nitrogen end.

$$CH_3CH_2CH_2CH_2Br + Na^+ \ NCO^- \longrightarrow \underset{n\text{-butyl isocyanate}}{CH_3CH_2CH_2CH_2NCO}$$

Isocyanates react with water to give N-alkyl carbamic acids, which are unstable and spontaneously lose carbon dioxide to give the corresponding amine. The reaction proceeds via the **zwitterion** or **internal salt** (Ger., *Zwitter*, hybrid), $RNH_2{}^+COO^-$, followed by decarboxylation. Rather than the anion formed in the case described in Section 25.1.B, the stable product is an even more favorable neutral amine.

$$R{-}NCO + H_2O \longrightarrow [R{-}NHCOOH] \longrightarrow RNH_2 + CO_2$$

Isocyanates give carbamate esters with alcohols and ureas with amines.

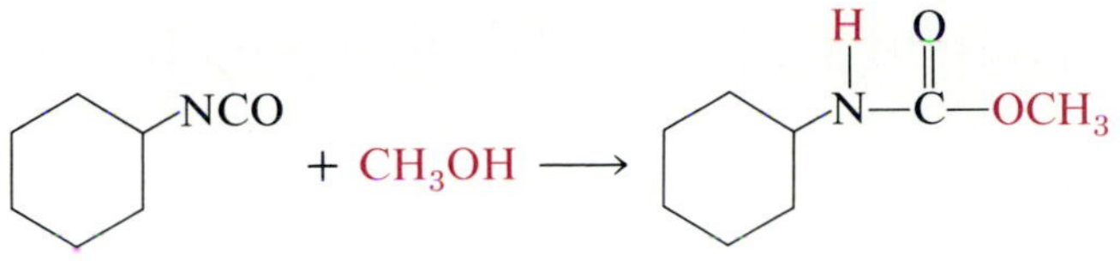

$$C_6H_{11}{-}NCO + CH_3OH \longrightarrow C_6H_{11}{-}\overset{H}{N}{-}\overset{O}{\overset{\|}{C}}{-}OCH_3$$

methyl N-cyclohexylcarbamate

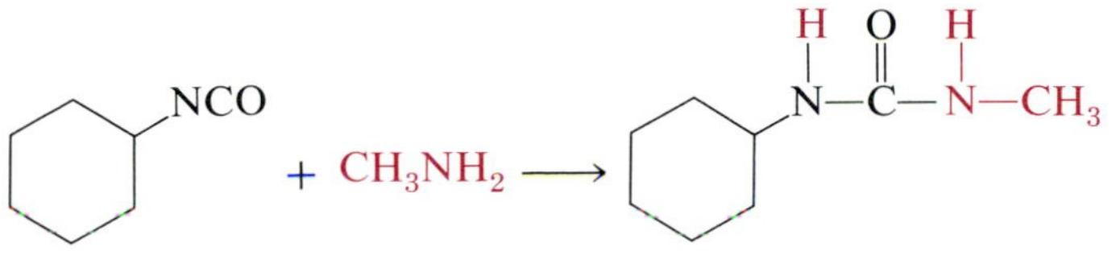

$$C_6H_{11}{-}NCO + CH_3NH_2 \longrightarrow C_6H_{11}{-}\overset{H}{N}{-}\overset{O}{\overset{\|}{C}}{-}\overset{H}{N}{-}CH_3$$

N-methyl-N′-cyclohexylurea

Carbamate esters are also called urethanes. An important class of commercial polymers is the polyurethanes, which are formed from an aromatic diisocyanate and a diol. One type of diol used is actually a low molecular weight copolymer made from ethylene glycol and adipic acid. When this polymer, which has free hydroxy end groups, is mixed with the diisocyanate, a larger polymer is produced.

$$HOCH_2CH_2OH + HOOC(CH_2)_4COOH \longrightarrow$$

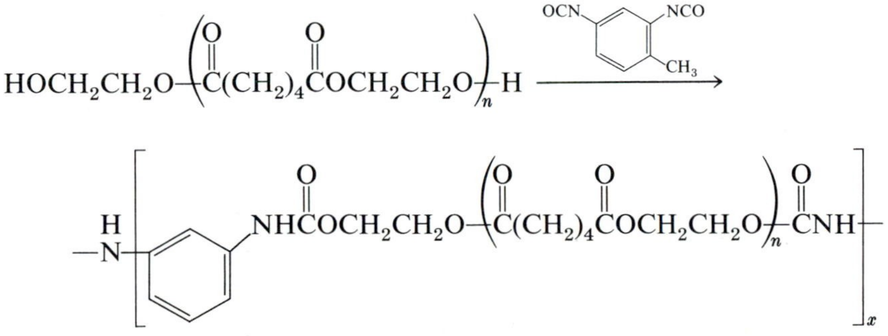

$$HOCH_2CH_2O{\left(\overset{O}{\overset{\|}{C}}(CH_2)_4\overset{O}{\overset{\|}{C}}OCH_2CH_2O\right)_n}H \xrightarrow{\text{OCN-C}_6\text{H}_3(\text{CH}_3)\text{-NCO}}$$

$$\left[{-}\overset{H}{N}{-}C_6H_3(CH_3){-}NH\overset{O}{\overset{\|}{C}}OCH_2CH_2O{\left(\overset{O}{\overset{\|}{C}}(CH_2)_4\overset{O}{\overset{\|}{C}}OCH_2CH_2O\right)_n}\overset{O}{\overset{\|}{C}}NH{-}\right]_x$$

a polyurethane

In the manufacturing process, a little water is mixed in with the diol. Some of the diisocyanate reacts with water to give an aromatic diamine and carbon dioxide. The carbon dioxide forms bubbles that are trapped in the bulk of the polymer as it solidifies. The result is a spongy product called polyurethane foam.

$$\text{OCN}-C_6H_3(\text{CH}_3)-\text{NCO} + 2\,H_2O \longrightarrow H_2N-C_6H_3(\text{CH}_3)-NH_2 + 2\,CO_2$$

25.3 Azides

Organic azides are compounds with the general formula RN_3 and were introduced in Section 24.6.H. They are related to the inorganic acid, hydrazoic acid, HN_3. Azide ion, N_3^-, is a resonance hybrid of the following important dipolar structures.

$$\left[:\text{N}{\equiv}\overset{+}{\text{N}}{-}\ddot{\underset{..}{\text{N}}}:^{2-} \longleftrightarrow :\overset{-}{\underset{..}{\text{N}}}{=}\overset{+}{\text{N}}{=}\overset{-}{\underset{..}{\text{N}}}: \longleftrightarrow {}^{2-}:\ddot{\underset{..}{\text{N}}}{-}\overset{+}{\text{N}}{\equiv}\text{N}:\right]$$

This anion is relatively nonbasic for anionic nitrogen (the pK_a of HN_3 is 4.68) and is a good nucleophile. Alkyl azides are best prepared by nucleophilic displacement on alkyl halides.

$$CH_3CH_2CH_2CH_2Br + N_3^- \xrightarrow[H_2O \atop (90\%)]{CH_3OH} CH_3CH_2CH_2CH_2N_3 + Br^-$$

A mixture of 34.5 g of NaN_3, 68.5 g of *n*-butyl bromide, 70 mL of water, and 25 mL of methanol is refluxed for 24 hr. The *n*-butyl azide separates as an oily layer. It is dried and distilled behind a safety barricade to obtain 40 g of pure *n*-butyl azide (90%).

Azide ion is also sufficiently nucleophilic to open the epoxide ring. β-Hydroxyalkyl azides may be prepared in this way.

$$CH_3\overset{\;O\;}{CH{-}CH_2} + N_3^- \xrightarrow[25°C,\ 24\ hr]{H_2O} CH_3\overset{OH}{\overset{|}{C}}HCH_2N_3 \quad (70\%)$$

Acyl azides may be prepared from acyl halides and azide ion (Section 24.6.I).

$$R\overset{O}{\overset{\|}{C}}Cl + N_3^- \longrightarrow R\overset{O}{\overset{\|}{C}}N_3 + Cl^-$$

Alkyl azides are reduced by lithium aluminum hydride or by catalytic hydrogenation to give the corresponding amines (Section 24.6.H). The two-step process of (1) displacement of halide ion by azide ion, and (2) reduction of the resulting azide provides a convenient synthesis of pure primary amines. Also, since azide ion is relatively nonbasic yet still highly nucleophilic, its substitution/elimination ratio is high, even with secondary and β-branched alkyl halides.

$$CH_3CH_2CH_2CH(Br)CH_3 + Na^+\,N_3^- \xrightarrow[H_2O]{C_2H_5OH} \underset{\text{2-pentyl azide}}{CH_3CH_2CH_2CH(N_3)CH_3} \xrightarrow{LiAlH_4} \underset{\text{2-pentanamine}}{CH_3CH_2CH_2CH(NH_2)CH_3}$$

Both alkyl and acyl azides are thermally unstable and lose nitrogen on heating. In some cases, particularly when the nitrogen content of the molecule is higher than about 25%, the decomposition can occur with explosive violence. Decomposition of alkyl azides gives a complex mixture of products. Acyl azides decompose to the acyl nitrene, which rearranges to an isocyanate (Schmidt and Curtius rearrangements; Section 24.6.I).

$$R-\overset{\displaystyle O}{\overset{\|}{C}}-N_3 \longrightarrow N_2 + R-\overset{\displaystyle O}{\overset{\|}{C}}-\ddot{\underset{\cdot\cdot}{N}} \longrightarrow R-N=C=O$$

an isocyanate

EXERCISE 25.4 Compare the Lewis structures of an isocyanate and an azide. What is the expected product of the reaction of benzyl chloride with sodium cyanate in methanol solution?

25.4 Diazo Compounds

Diazo compounds have the general formula $R_2C{=}N_2$. The electronic structure of diazomethane, the simplest diazo compound, shows that the carbon has nucleophilic properties.

$$\left[{}^{-}\ddot{C}H_2-\overset{+}{N}\equiv N: \longleftrightarrow CH_2=\overset{+}{N}=\underset{\cdot\cdot}{N}:^{-}\right]$$

diazomethane

We have encountered diazomethane previously as a reagent for converting carboxylic acids into methyl esters (Section 18.7.A).

$$R\overset{\displaystyle O}{\overset{\|}{C}}OH + CH_2N_2 \longrightarrow R\overset{\displaystyle O}{\overset{\|}{C}}OCH_3 + N_2$$

Diazomethane is prepared by treating an N-methyl-N-nitrosoamide with concentrated potassium hydroxide solution.

$$R\overset{\displaystyle O}{\overset{\|}{C}}N(CH_3)(N{=}O) + OH^- \longrightarrow CH_2N_2 + RCO_2^- + H_2O$$

The preparation is carried out in a two-phase mixture consisting of ether and aqueous KOH. The diazomethane dissolves in the ether as it is formed and it is generally used as an ether solution.

Other diazo compounds are also known. α-Diazo ketones and α-diazo esters are relatively stable since the carbonyl group can delocalize the carbanionic electron pair.

$$\left[R-\overset{\displaystyle O}{\overset{\|}{C}}-\ddot{C}H-\overset{+}{N}\equiv N: \longleftrightarrow R-\overset{\displaystyle O^-}{\overset{|}{C}}=CH-\overset{+}{N}\equiv N: \longleftrightarrow R-\overset{\displaystyle O}{\overset{\|}{C}}-CH=\overset{+}{N}=\underset{\cdot\cdot}{N}:^{-}\right]$$

This type of diazo compound is conveniently prepared by the reaction of diazomethane with an acyl halide.

$$C_6H_5\overset{\displaystyle O}{\overset{\|}{C}}Cl + 2\,CH_2N_2 \longrightarrow C_6H_5\overset{\displaystyle O}{\overset{\|}{C}}CHN_2 + CH_3Cl + N_2$$

Excess diazomethane must be used to react with the HCl that is produced in the reaction. If only one equivalent of diazomethane is used, a chloromethyl ketone results.

$$C_6H_5CH_2COCl + CH_2N_2 \longrightarrow [C_6H_5CH_2COCHN_2 + HCl] \longrightarrow C_6H_5CH_2COCH_2Cl + N_2$$

(83–85%)

The reaction of diazomethane with acyl halides is another reaction that shows the nucleophilic nature of the carbon in this compound. The mechanism of the reaction can be visualized as follows.

$$(1)\ R-C(=O)-Cl + \overset{-}{:}CH_2-\overset{+}{N}\equiv N: \rightleftharpoons R-C(O^-)(Cl)-CH_2-\overset{+}{N}\equiv N:$$

$$(2)\ R-C(O^-)(Cl)-CH_2-\overset{+}{N}\equiv N: \rightleftharpoons R-C(=O)-CH_2-\overset{+}{N}\equiv N: + Cl^-$$

$$(3)\ R-C(=O)-CH_2-\overset{+}{N}\equiv N: \rightleftharpoons R-C(=O)-CHN_2 + H^+$$

If there is no excess diazomethane to react with the proton liberated in step 3, the chloride ion displaces nitrogen to give the chloro ketone.

$$(4)\ Cl^- + R-C(=O)-CH_2-\overset{+}{N}\equiv N: \longrightarrow R-C(=O)-CH_2Cl + :N\equiv N:$$

Like azides, diazo compounds lose nitrogen either thermally or when irradiated with ultraviolet light. The decomposition is catalyzed by transition metals such as copper or rhodium. The initial product is a carbene (Section 11.6.F), which then reacts further. In the case of α-diazo ketones, the resulting acylcarbene rearranges to give a ketene (see the Curtius rearrangement; Section 24.6.I).

$$C_6H_5-C(=N_2)-C(=O)-C_6H_5 \xrightarrow[-N_2]{110°C} [C_6H_5-\ddot{C}-C(=O)-C_6H_5] \longrightarrow (C_6H_5)_2C=C=O$$

(65%)
diphenylketene

The carbenes derived from some diazo compounds may be trapped by reaction with an alkene; the products are cyclopropane derivatives.

$$\text{cyclohexene} + N_2CHCOOC_2H_5 \xrightarrow[Cu]{\Delta} \text{bicyclo[4.1.0]heptane-7-}(H)(COOC_2H_5) + N_2$$

ethyl 7-bicyclo[4.1.0]-heptanecarboxylate

EXERCISE 25.5 Show how propanoic acid can be converted into the following compounds by routes involving the use of diazomethane.

(a) methyl propanoate (b) 1-diazo-2-butanone (c) 1-chloro-2-butanone
(d) ethylketene (a reactive ketene that rapidly dimerizes)

25.5 Diazonium Salts

In Section 24.7.B we found that aromatic amines react with nitrous acid in aqueous solution to give solutions of arenediazonium salts, which are moderately stable if kept cold.

$$\text{C}_6\text{H}_5\text{NH}_2 \xrightarrow[\text{H}_2\text{O, 0°C}]{\text{NaNO}_2\text{, HCl}} \text{C}_6\text{H}_5\text{N}_2^+\ \text{Cl}^-$$

aniline → benzenediazonium chloride

These unstable compounds comprise an important class of synthetic intermediates. In a sense, they are the "Grignard reagents" of aromatic chemistry, since they can be used in the synthesis of such a wide variety of aromatic compounds. In this section we shall discuss the chemistry of this useful group of compounds.

A. Acid-Base Equilibria of Arenediazonium Ions

In acid solution arenediazonium salts have the diazonium ion structure with a linear C—N—N bond system. The diazonium ion has a π-system that can conjugate with the aromatic π-system (Figure 25.1). This conjugation is responsible in part for the relative stability of these compounds. Recall that aliphatic diazonium ions are not at all stable and generally react immediately upon formation (Section 24.7.B; Highlight 24.3).

Arenediazonium ions behave as dibasic acids. The two steps in the equilibria are represented as

$$\text{Ar—}\overset{+}{\text{N}}\text{≡N} + 2\,\text{H}_2\text{O} \overset{K_1}{\rightleftharpoons} \underset{\text{arenediazohydroxide}}{\text{Ar—N=NOH}} + \text{H}_3\text{O}^+$$

$$\text{Ar—N=NOH} + \text{H}_2\text{O} \overset{K_2}{\rightleftharpoons} \underset{\text{arenediazotate ion}}{\text{Ar—N=N—O}^-} + \text{H}_3\text{O}^+$$

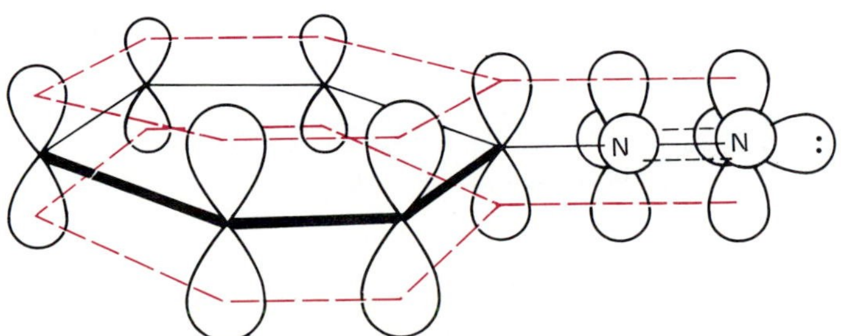

FIGURE 25.1 Orbital structure of benzenediazonium cation.

The diazonium ions represent an unusual class of dibasic acids in that $K_2 \gg K_1$. That is, the arenediazohydroxide is present only in small amount. For the phenyl group, equal concentrations of benzenediazonium ion and benzenediazotate are present at a pH of 11.9. Even in neutral solutions with pH = 7, the diazonium ions are generally the most predominant among the species present.

Benzenediazotate ion exists in syn and anti forms, like other compounds containing carbon-nitrogen and nitrogen-nitrogen double bonds. The anti form is the more stable, but the less stable syn isomer is that formed first by reaction of the diazonium cation with hydroxide ion.

$$C_6H_5N_2^+ + 2\,OH^- \xrightarrow{\text{fast}} \textit{syn}\text{-}C_6H_5N{=}NO^- \xrightarrow{\text{slow}} \textit{anti}\text{-}C_6H_5N{=}NO^-$$

syn-benzenediazotate ion   *anti*-benzenediazotate ion

Alkanediazonium ions react by direct nucleophilic displacement of nitrogen (S_N2 mechanism), formation of carbocations (S_N1 mechanism, with attendant rearrangements), and elimination by both the E1 and E2 mechanisms (Section 24.7.B). None of these pathways is readily available to arenediazonium ions. The most likely reaction, formation of an aryl cation, is limited by the high energy of these species. In phenyl cation, the empty orbital has approximately sp^2-hybridization and cannot conjugate with the π-electronic system (Figure 25.2). For example, the enthalpy of formation in the gas phase of phenyl cation from chlorobenzene is about the same as that for the ionization of vinyl chloride and is almost as high as the enthalpy of formation of methyl cation from methyl chloride.

$$\begin{array}{rll} C_6H_5Cl \longrightarrow & C_6H_5^+ + Cl^- & \Delta H^\circ = 224 \text{ kcal mole}^{-1} \\ CH_2{=}CHCl \longrightarrow & C_2H_3^+ + Cl^- & \Delta H^\circ = 225 \text{ kcal mole}^{-1} \\ CH_3Cl \longrightarrow & CH_3^+ + Cl^- & \Delta H^\circ = 228 \text{ kcal mole}^{-1} \end{array}$$

Nevertheless, many reactions of diazonium ions are the reactions expected if aryl cations were intermediates. In many preparative methods, however, the aryl cation is not free; it is combined as a complex with a metal, often copper. Other reactions involve free radical intermediates. All of the following types of compounds can be prepared by appropriate reactions of the arenediazonium ions, ArN_2^+.

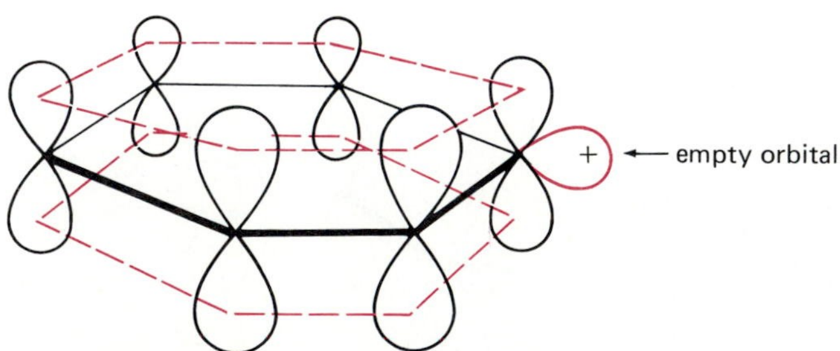

FIGURE 25.2 Orbital diagram of phenyl cation.

ArOH	ArCN	Ar—Ar′
ArI	ArF	$ArNHNH_2$
ArSH	$ArNO_2$	ArN_3
ArCl	ArH	ArN=NAr′
ArBr	Ar—Ar	

B. Thermal Decomposition of Diazonium Salts: Formation of ArOH, ArI, and ArSH

Aqueous solutions of arenediazonium ions are not stable. Nitrogen gas is evolved slowly in the cold and rapidly on heating. The net reaction is that of hydrolysis.

2-bromo-4-methylaniline $\xrightarrow[5°C]{NaNO_2,\ aq.\ H_2SO_4}$ [N_2^+, Br, CH_3 diazonium salt] $\xrightarrow[H_2O]{\Delta}$ 2-bromo-4-methylphenol (80–92%)

m-aminobenzaldehyde dimethyl acetal (NH_2, $CH(OCH_3)_2$) $\xrightarrow[5°C]{NaNO_2,\ aq.\ H_2SO_4}\ \xrightarrow{\Delta}$ m-hydroxybenzaldehyde (OH, CHO) $+ CH_3OH + N_2$

m-aminobenzaldehyde dimethyl acetal

m-hydroxybenzaldehyde

Hydrolysis of the diazonium ion appears to be an S_N1 type of process involving the aryl cation.

$$ArN_2^+ \longrightarrow N_2 + Ar^+$$

The aryl cation forms despite its high energy because of the great stability of nitrogen. That is, the formation of N_2 is a powerful driving force for the decomposition of diazonium ions. The aryl cation intermediate is highly reactive and reacts rapidly with water to form the corresponding phenol.

$$Ar^+ + H_2O \longrightarrow Ar{-}OH_2^+ \rightleftharpoons ArOH + H^+$$

However, the aryl cation also reacts with other nucleophiles that may be present—such as halide ion. If HCl is used in the diazotization, some chloroarene is also produced.

$$Ar^+ + Cl^- \longrightarrow ArCl$$

For this reason sulfuric acid is normally used as the acid in diazotizations in which the diazonium salt is to be thermally decomposed. Bisulfate ion is a much poorer nucleophile than chloride ion and does not compete well with water for the aryl cation.

When aqueous solutions of diazonium ions containing chloride ion are allowed to decompose, the rate of reaction is independent of the chloride ion concentration, but the amount of chloroarene formed is proportional to $[Cl^-]$. Thus, the rate-determining step does not depend on chloride ion, but the product-determining steps do. This result is interpretable by the scheme

$$Ar{-}N_2^+ \xrightarrow{\text{slow}} N_2 + Ar^+ \begin{cases} \xrightarrow{H_2O} ArOH + H_3O^+ \\ \xrightarrow{Cl^-} ArCl \end{cases}$$

The competition of chloride ion with water for the intermediate aryl cation is usually inadequate for this method to be a successful preparation of chloroarenes; the Sandmeyer reaction (Section 25.5.C) is generally better.

Highly nucleophilic anions can compete successfully with water for the intermediate aryl cation and lead to satisfactory syntheses. A useful example is the preparation of aryl iodides by treatment of arenediazonium compounds with aqueous potassium iodide.

$$C_6H_5NH_2 \xrightarrow[7^\circ C]{NaNO_2,\ aq.\ HCl} C_6H_5N_2^+ \xrightarrow[25^\circ C]{KI} C_6H_5I$$

(74–76%)

In this case HCl is used for the diazotization, and the resulting solution contains Cl^-. Nevertheless, the more nucleophilic I^- dominates the reaction. The diazotization can also be accomplished with hydriodic acid, but this acid is far more expensive than HCl.

Reaction of aqueous diazonium salts with HS^- or with metal polysulfides has been used for preparation of thiophenols, but violent reactions and explosions have been reported and the method is not recommended. An alternative route involves potassium ethyl xanthate, $KSCSOC_2H_5$, which is available commercially from the reaction of potassium ethoxide with carbon disulfide.

$$m\text{-}CH_3C_6H_4NH_2 \xrightarrow[0-4^\circ C]{NaNO_2,\ aq.\ HCl} [m\text{-}CH_3C_6H_4N_2^+] \xrightarrow[40-45^\circ C]{KSC(=S)OC_2H_5} m\text{-}CH_3C_6H_4SC(=S)OC_2H_5 \xrightarrow{1.\ aq.\ KOH,\ \Delta\quad 2.\ H^+} m\text{-}CH_3C_6H_4SH$$

m-toluidine; *m*-tolyl ethyl xanthate; (63–75%) *m*-methylthiophenol *m*-thiocresol

This reaction involves a variation of the hydrolysis reaction above. Reaction of the diazonium ion with the ethyl xanthate ion gives first the diazoxanthate.

$$Ar{-}N{=}N{-}SC(=S)OC_2H_5$$

This intermediate decomposes with liberation of nitrogen by an ion-pair or radical mechanism. Even this method is hazardous because the intermediate diazoxanthate can detonate and should be allowed to decompose as formed. The use of traces of nickel—even a nichrome stirrer—has been recommended to facilitate the controlled decomposition.

EXERCISE 25.6 Write equations showing how *p*-xylene may be converted into 2,5-dimethylphenol and into 2,5-dimethyl-1-iodobenzene.

C. The Sandmeyer Reaction: Preparation of ArCl, ArBr, and ArCN

Decomposition of diazonium salts is catalyzed by cuprous salts. In laboratory practice, the cold diazonium solution is added dropwise to a hot suspension of cuprous bromide, chloride, or cyanide to give the corresponding aromatic product in a method known as the **Sandmeyer reaction.** The process is the only practical way to obtain certain aryl halides uncontaminated by isomers.

$$\text{}\xrightarrow[\text{aq. HBr, }10^\circ\text{C}]{NaNO_2}\xrightarrow[\text{HBr}]{\text{CuBr}}$$

o-chloroaniline → (89–95%) *o*-bromochlorobenzene

$$\xrightarrow[\text{aq. HCl, }0^\circ\text{C}]{NaNO_2}\xrightarrow[0^\circ\text{C}]{\text{CuCl}}$$

p-toluidine → (70–79%) *p*-chlorotoluene

Note in these cases that the hydrohalic acid is used to correspond to the halogen introduced; the use of HCl with CuBr would give a mixture of chloro and bromo product.

The Sandmeyer reaction is also a preferred method for the preparation of most aryl cyanides.

$$\xrightarrow[\text{aq. HCl, }5^\circ\text{C}]{NaNO_2}\xrightarrow[0\text{–}5^\circ\text{C}]{\text{CuCN}}$$

p-toluidine → (64–70%) *p*-tolunitrile

Note that the decomposition with CuCN occurs even in the cold.

> Cupric chloride is normally obtained as a blue-green hydrate, $CuCl_2 \cdot 2H_2O$. This color is characteristic of many cupric salts. Cupric chloride and bromide are readily soluble in water. Cuprous bromide and chloride are white, insoluble powders prepared by reducing an aqueous solution of cupric sulfate and sodium bromide or chloride with sodium bisulfite. The cuprous halide precipitates as a white powder, which is filtered and used directly. On standing in air the white cuprous salts darken by oxidation. Cuprous cyanide is prepared by treating an aqueous suspension of cuprous chloride with sodium cyanide. Cuprous cyanide is also insoluble in water, but dissolves in excess sodium cyanide with formation of a complex, $Cu(CN)_2^-$. This solution is used directly in the Sandmeyer reaction.

The aromatic nitriles prepared by the Sandmeyer reaction can, of course, be hydrolyzed to carboxylic acids, reduced to benzylamines, treated with Grignard reagents to produce ketones, and so on. Consequently, the diazonium salts provide an entry to a host of aromatic compounds.

EXERCISE 25.7 Write equations showing the conversion of benzene into each of the following compounds by multistep procedures involving the Sandmeyer reaction as one step.

(a) chlorobenzene (b) bromobenzene (c) benzoic acid
(d) benzylamine (e) benzamide

D. Preparation of Fluoro- and Nitroarenes

Some diazonium salts are fairly stable and can be isolated and handled. One such salt is the tetrafluoroborate. This salt is prepared by diazotization with sodium nitrite and tetrafluoroboric acid. The diazonium tetrafluoroborate usually precipitates and is filtered.

The isolated diazonium tetrafluoroborate salts are useful in two significant reactions. In one reaction a suspension of the salt in aqueous sodium nitrite is treated with copper powder. Nitrogen is evolved, and the corresponding nitro compound is produced.

NH_2 / NO_2 $\xrightarrow[\text{aq. } HBF_4,\ 0°C]{NaNO_2}$ $N_2^+\ BF_4^-$ / NO_2 $\xrightarrow[\text{Cu powder}]{\text{aq. } NaNO_2}$ NO_2 / NO_2

p-nitroaniline — *p*-nitrobenzenediazonium fluoborate — (67–82%) *p*-dinitrobenzene

This example is similar to the Sandmeyer reaction, which was discussed in the previous section. Note that in this case copper powder, rather than a cuprous salt, is used to bring about decomposition of the diazonium ion. This variant is sometimes called the **Gatterman reaction.** The Gatterman method can also be used to prepare aryl halides, but it is not as useful for this purpose as is the Sandmeyer reaction.

The isolated diazonium tetrafluoroborate can be decomposed thermally either as the dry salt or in an inert solvent such as THF to provide a satisfactory preparation of aryl fluorides (**Schiemann reaction**).

NH_2 / CH_3 $\xrightarrow[HF,\ BF_3,\ 0°C]{NaNO_2}$ $N_2^+\ BF_4^-$ / CH_3 $\xrightarrow{\Delta}$ F / CH_3 $+ N_2 + BF_3$

m-toluidine — (76–84%) *m*-methylbenzenediazonium fluoborate — *m*-fluorotoluene

An improved procedure makes use of hexafluorophosphoric acid, HPF_6. The corresponding diazonium hexafluorophosphates are less soluble than the tetrafluoroborates and are obtained in generally higher yield. The dry salt is thermally decomposed to form the aryl fluoride.

NH_2 / Br $\xrightarrow[\text{aq. HCl},\ HPF_6]{NaNO_2}$ $N_2^+\ PF_6^-$ / Br $\xrightarrow{165\,°C}$ F / Br $+ N_2 + PF_5$

o-bromoaniline — *o*-bromobenzenediazonium hexafluorophosphate — (76–78%) *o*-bromofluorobenzene

EXERCISE 25.8 Outline methods for the synthesis of all three isomeric dinitrobenzenes.

E. Replacement of the Diazonium Group by Hydrogen

The diazonium group may also be replaced by hydrogen. This reaction allows use of the amino group to direct the orientation of an electrophilic aromatic substitution reaction, after which the amino group is removed. The most generally useful reagent for the reaction is hypophosphorous acid, H_3PO_2.

$$\text{3-aminobenzoic acid} \xrightarrow[H_2O]{Br_2} \text{2,4,6-tribromo-3-aminobenzoic acid} \xrightarrow[-5^\circ C]{NaNO_2,\ aq.\ H_2SO_4} \xrightarrow[-5^\circ C]{H_3PO_2} \text{2,4,6-tribromobenzoic acid}$$

Hypophosphorous acid is a low-melting (m.p. 26.5 °C), crystalline compound having a structure with two phosphorus-hydrogen bonds.

$$H{-}\overset{\displaystyle H}{\underset{\displaystyle OH}{P}}{=}O$$

Salts of the acid are prepared by treating white phosphorus with alkali or alkaline earth hydroxides. The free acid can be liberated from the water-soluble calcium salt with sulfuric acid. Aqueous solutions are available commercially and can be used directly. The monobasic acid has $pK_a = 1.2$ and is a powerful reducing agent.

In the foregoing example, the amino group is used to direct the facile introduction of three bromines (Section 24.7.D). Since NH_2 is a powerful *ortho,para*-directing group, reaction occurs at the positions indicated. Having served its function of activating the ring and directing the incoming bromines to specific positions, the amino group is diazotized and the diazonium group is replaced by hydrogen. The yield of tribromobenzoic acid is 70–80% from *m*-aminobenzoic acid.

Some diamines can be diazotized at both amino groups and subsequent reaction of both diazonium groups can be accomplished. An example is provided by the following reaction of a substituted benzidine.

$$\text{4,4'-diamino-3,3'-dimethylbiphenyl} \xrightarrow[aq.\ HCl]{NaNO_2} \xrightarrow[25^\circ C]{H_3PO_2} \text{3,3'-dimethylbiphenyl}$$

EXERCISE 25.9 Write equations showing how benzene can be converted into 1,3,5-trichlorobenzene.

F. Arylation Reactions

Arylation provides a convenient preparation of unsymmetrical biaryls in which the diazonium group is replaced by an aromatic ring. Diazotization is carried out in the usual way except that a minimum of water is used. The solution is made basic, and the resulting concentrated aqueous solution is stirred at 0 °C with a liquid arene.

$$p\text{-}CH_3C_6H_4NH_2 \xrightarrow[\text{aq. HCl}]{NaNO_2} \xrightarrow[\text{benzene, 5°C}]{NaOH} CH_3\text{-}C_6H_4\text{-}C_6H_5$$

(21%)
4-methylbiphenyl

The reaction is called the **Gomberg-Bachmann reaction.** Yields are generally low, but the starting materials are often readily available, and there are few other methods for the synthesis of such biaryls (Section 31.2.A).

The mechanism of the Gomberg-Bachmann reaction involves free radical intermediates. In basic solution the diazonium salt is in equilibrium with the covalent diazohydroxide (Section 25.5.A), a species that can react with a second diazonium ion to form the relatively nonpolar bis-(diazenyl) ether.

$$\text{Ar—N}\equiv\text{N}^+ + \text{Ar—N=N—OH} \rightarrow \text{Ar—N=N—O—N=N—Ar}$$

The latter is extracted into the organic phase, and undergoes homolytic fission to an aryl radical, a nitrogen molecule, and an aryldiazenyloxy radical. The aryl radical adds to the arene, giving an adduct radical. The adduct radical donates a hydrogen atom to the oxy radical, forming the biaryl and the diazohydroxide. The diazohydroxide can return to the aqueous phase and there react with another diazonium ion. The overall result is substitution of a hydrogen atom of the arene by the aryl radical corresponding to the beginning arylamine. The driving force for simultaneous rupture of two bonds is the formation of the highly stable nitrogen molecule.

$$\text{Ar—N=N—O—N=N—Ar} \longrightarrow \text{Ar}\cdot + \text{N}\equiv\text{N} + \cdot\text{O—N=N—Ar}$$

$$C_6H_6 + \text{Ar}\cdot \longrightarrow C_6H_6\text{Ar}\cdot \xrightarrow{\cdot\text{O—N=N—Ar}} C_6H_5\text{—Ar} + \text{H—O—N=N—Ar}$$

EXERCISE 25.10 There are at least two ways in which the aryldiazeneoxyl radical can react with the adduct formed from the phenyl radical and benzene. Write each pathway and try to decide which is more likely.

Although the mechanism involves intermediate free radicals, it does not appear to be a radical chain reaction. The reaction is somewhat different than most free radical reactions in that the concentration of radicals becomes rather high. The typical low yields result from the many alternative reactions that are available to the intermediate radicals.

Because the aromatic substitution involves radicals, the orientation rules for electrophilic agents do not hold; almost all substituents tend to give *ortho* and *para* orientation, and mixtures of products are common. In using this method to prepare an unsymmetrical biaryl, it is best to start with the substituents on the diazonium ring and to keep the ring to be added as simple as possible.

$$C_6H_5N_2^+ \xrightarrow[\text{toluene}]{\text{NaOH}} C_6H_5\text{—}C_6H_4CH_3$$

(66% *o*-, 19% *m*-, 14% *p*-)

The same reaction can usually be carried out in better yield by using an N-nitrosoamide. This intermediate is prepared in straightforward fashion from the amine and is heated in an aromatic solvent. The N-nitrosoamide rearranges to a diazo ester, which forms the same aryl radical intermediate involved in the Gomberg-Bachmann reaction.

$$m\text{-}NO_2C_6H_4NH_2 \xrightarrow[\text{AcOH, }\Delta]{(CH_3CO)_2O} [m\text{-}NO_2C_6H_4NHCOCH_3] \xrightarrow[CH_3COOH]{N_2O_3} m\text{-}NO_2C_6H_4N(NO)COCH_3 \xrightarrow[\text{benzene, }\Delta]{NaOOCCH_3} m\text{-}NO_2C_6H_4\text{—}C_6H_5$$

N-nitroso-*m*-nitro-acetanilide

(56–60%)
3-nitrobiphenyl

The rearrangement of the intermediate nitrosoamide can be regarded as an intramolecular transesterification.

$$\left[Ar\text{—}N(N{=}O)\text{—}C(=O)\text{—}CH_3 \longleftrightarrow Ar\text{—}\overset{+}{N}(=N\text{—}O^-)\text{—}C(=O)\text{—}CH_3\right] \longrightarrow \left[Ar\text{—}\overset{+}{N}(=N\text{—}O)\text{—}C(O^-)\text{—}CH_3\right] \longrightarrow$$

$$Ar\text{—}N{=}N\text{—}O\text{—}\overset{O}{\overset{\|}{C}}CH_3 \longrightarrow Ar\cdot + N_2 + \cdot O\overset{O}{\overset{\|}{C}}CH_3$$

EXERCISE 25.11 4-Ethylbiphenyl can be prepared by the Gomberg-Bachmann reaction of 4-ethylaniline with benzene or of aniline with ethylbenzene. Write the equations for these two syntheses and explain why one combination is superior to the other.

G. Diazonium Ions as Electrophiles: Azo Compounds

The diazonium cation bears a resemblance to some other species that are known as intermediates in electrophilic aromatic substitution reactions.

$$Ar\text{—}\overset{+}{N}{\equiv}N: \qquad R\text{—}C{\equiv}\overset{+}{O}: \qquad \overset{+}{O}{\equiv}N:$$

Arenediazonium ions *can* react as electrophilic reagents in aromatic substitutions, but they are such mild reagents that only highly activated rings can be used. In practice, such reactions are limited primarily to aromatic amines and phenols.

Benzenediazonium ion reacts with excess aniline in acidic medium to give the aromatic substitution product *p*-aminoazobenzene.

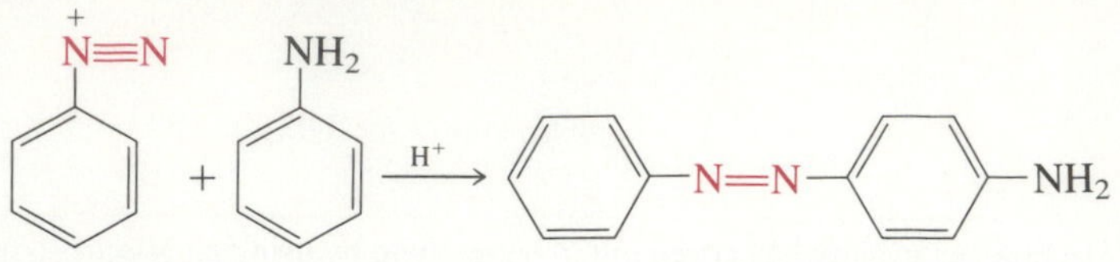

p-aminoazobenzene

The mechanism of the foregoing reaction actually involves initial attack of the benzenediazonium ion on the nitrogen of aniline, to give diazoaminobenzene, a derivative of the unstable inorganic compound triazene, HN=N—NH_2.

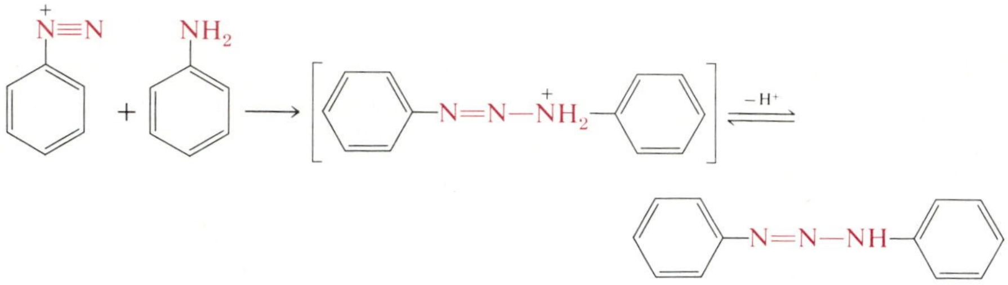

(82–85%)
diazoaminobenzene
1,3-diphenyltriazene

Diazoaminobenzene reacts with aniline to give the aromatic substitution product *p*-aminoazobenzene.

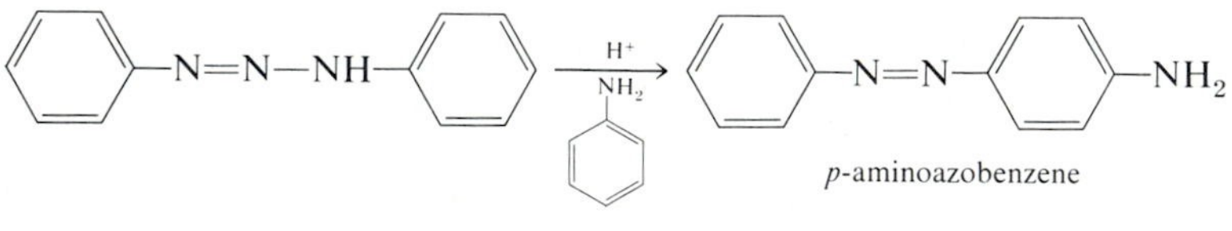

p-aminoazobenzene

This reaction is usually assumed to involve a reversal of diazoaminobenzene formation followed by the slower reaction of benzenediazonium cation with the *para* position of aniline.

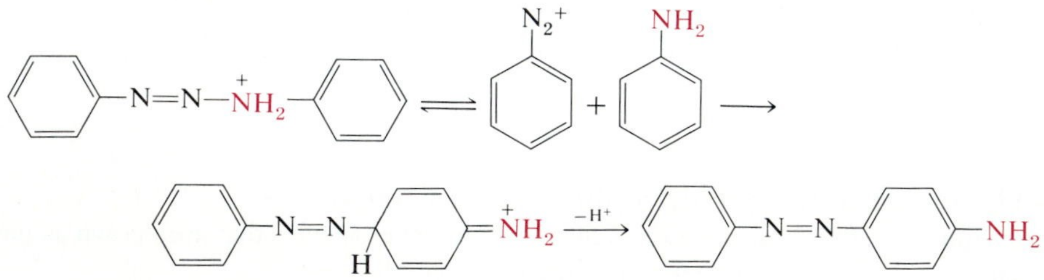

This mechanism incorporates the frequently encountered distinction between kinetically and thermodynamically controlled reactions and explains many features of the reaction. However, the reaction also produces a variety of minor by-products that suggest a more complex reaction mechanism.

Electrophilic substitution of aniline derivatives with diazonium ions is an important method of synthesis of colored substances that are used as dyes.

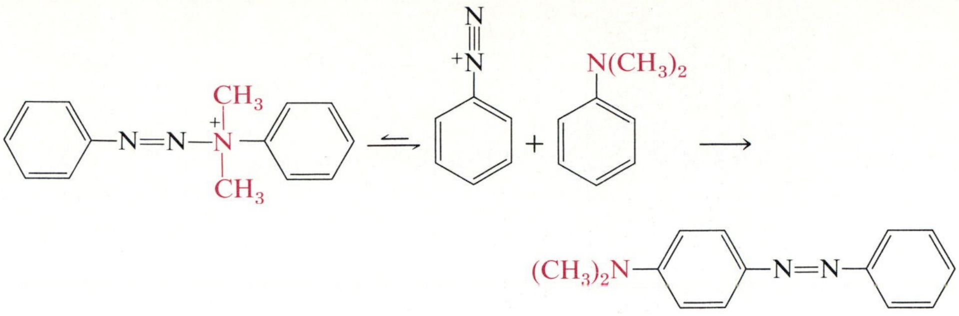

p-dimethylaminoazobenzene
butter yellow

Butter yellow was used at one time as a yellow food coloring, but as a suspected carcinogen it is no longer used for this purpose. Substituted azoarenes form an important class of dyes (Section 36.3). Several are also useful as indicators in the laboratory. Methyl orange, p-dimethylaminoazobenzene-p'-sulfonic acid, is prepared from diazotized sulfanilic acid and N,N-dimethylaniline.

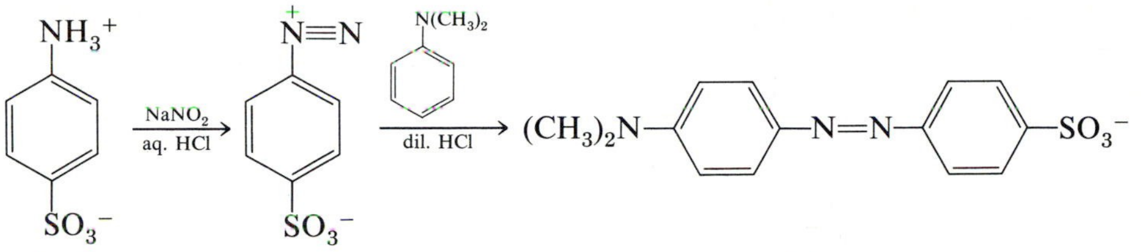

sulfanilic acid

methyl orange

The product is isolated as the sodium salt by salting out with sodium chloride. Note that these so-called ''coupling reactions'' of diazonium salts occur almost exclusively at the *para* position. Reaction occurs generally at the *ortho* position only when the *para* position is blocked.

The azo group has nitrogen lone pairs and is expected to show basic properties. However, each of these lone pairs is in an approximately sp^2-orbital and is less basic than an amino lone pair in which the orbital has less s-character. Furthermore, the adjacent nitrogen further reduces the basicity. As a result the azo group in azobenzene itself is a rather weak base; the pK_a of the protonated compound is -2.5.

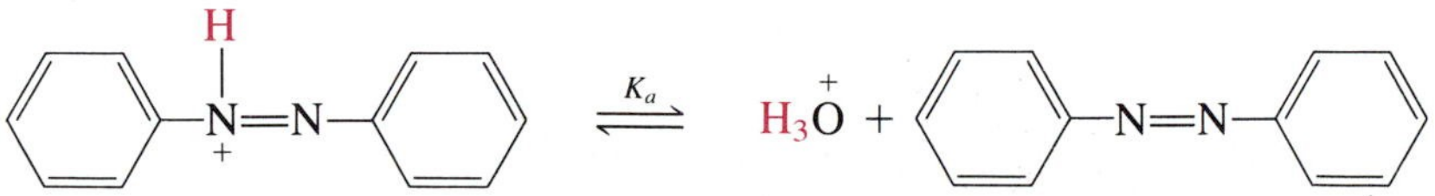

$K_a = 300$ M

Azobenzene itself is protonated only in rather strong acid. Methyl orange has a pK_a of 3.5; this value refers to the protonated azo group, not to the dimethylamino or sulfonic acid groups. At pH above 3.5, methyl orange is in the yellow azo form. At pH lower than 3.5 it is present in the red protonated form.

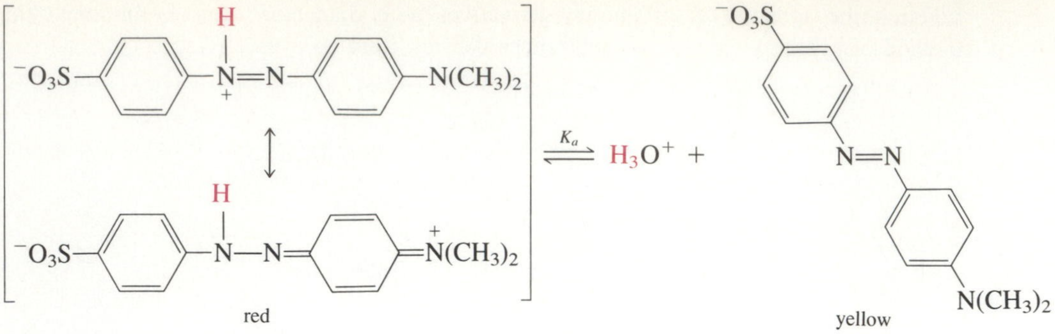

$$\left[{}^{-}O_3S-C_6H_4-\overset{H}{\underset{+}{N}}=N-C_6H_4-N(CH_3)_2 \longleftrightarrow {}^{-}O_3S-C_6H_4-\overset{H}{N}-N=C_6H_4=\overset{+}{N}(CH_3)_2 \right] \overset{K_a}{\rightleftharpoons} H_3O^+ + {}^{-}O_3S-C_6H_4-N=N-C_6H_4-N(CH_3)_2$$

red — yellow

$$K_a = 3 \times 10^{-4}\ M$$

Note that the azo-protonated form of methyl orange is stabilized by the *p*-$N(CH_3)_2$ group. This stabilization renders the protonated form less acidic than protonated azobenzene.

H. Synthetic Utility of Arenediazonium Salts

We saw in Chapter 23 that the nitro group can be introduced with ease into a wide variety of aromatic compounds. We have also seen that the nitro group can be reduced to the amino group by several reliable methods (Section 24.6.C) and that the resulting aromatic amines can be converted into arenediazonium salts (Section 24.7.B). In this section we have learned how the diazonium group can be replaced by OH, I, SH, Cl, Br, F, CN, NO_2, H, and Ar. Thus, *any of these functions can be introduced into an aromatic ring at any position that can be nitrated.*

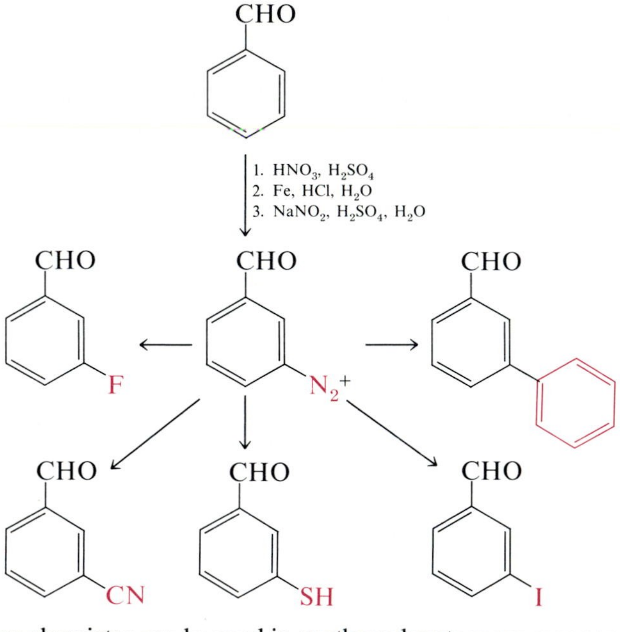

CHO (benzaldehyde)

1. HNO_3, H_2SO_4
2. Fe, HCl, H_2O
3. $NaNO_2$, H_2SO_4, H_2O

CHO, N_2^+ (meta) → CHO/F; CHO/phenyl; CHO/CN; CHO/SH; CHO/I

But diazonium chemistry can be used in another advantageous manner for the synthesis of aromatic compounds. Suppose, for example, that it is desired to convert radiolabelled benzoic acid into radiolabelled *p*-nitrobenzoic acid.

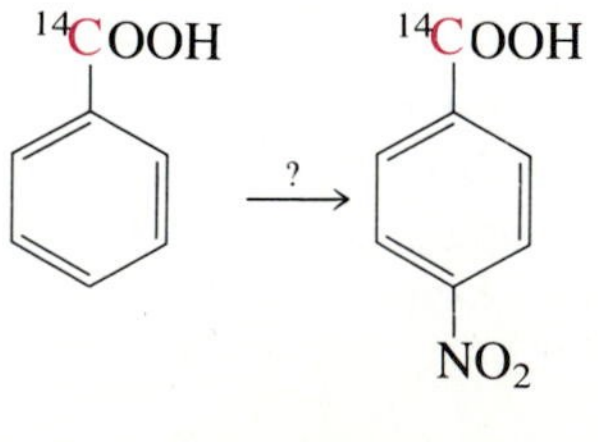

$$C_6H_5{}^{14}COOH \xrightarrow{?} p\text{-}O_2N{-}C_6H_4{}^{14}COOH$$

We are unable to accomplish this conversion by direct nitration because the COOH group is a strong *meta* director (Section 23.5). However, in *m*-acetamidobenzoic acid, which we can prepare in a straightforward synthesis from benzoic acid, the *ortho,para*-directing $NHCOCH_3$ group overcomes the *meta*-directing COOH group, and nitration occurs primarily *para* to COOH. The amide protecting group is then removed by hydrolysis, the resulting amine is diazotized, and the diazonium salt is treated with H_3PO_2.

$$C_6H_5{}^{14}COOH \xrightarrow[H_2SO_4]{HNO_3} m\text{-}O_2NC_6H_4{}^{14}COOH \xrightarrow[H_2O]{Fe,\ HCl} m\text{-}H_2NC_6H_4{}^{14}COOH \xrightarrow{(CH_3CO)_2O} m\text{-}CH_3CONHC_6H_4{}^{14}COOH \xrightarrow[H_2SO_4]{HNO_3}$$

$$\text{3-NHCOCH}_3\text{-4-NO}_2C_6H_3{}^{14}COOH \xrightarrow[H_2O]{NaOH} \text{3-NH}_2\text{-4-NO}_2C_6H_3{}^{14}COOH \xrightarrow[H_2O]{NaNO_2,\ HCl} \text{3-N}_2^+\text{-4-NO}_2C_6H_3{}^{14}COOH \xrightarrow{H_3PO_2} p\text{-}O_2NC_6H_4{}^{14}COOH$$

With these examples, we see the central position that the versatile diazonium function holds in aromatic synthesis.

EXERCISE 25.12 Write equations showing how the strategy just introduced may be used in the conversion of *p*-toluic acid into 2-chloro-4-methylbenzoic acid.

Highlight 25.3

The diazonium ion appears in many useful compounds, either as such, or as an important contributing resonance form.

$$[R-\overset{-}{N}-\overset{+}{N}\equiv N \longleftrightarrow R-N=\overset{+}{N}=N^-]$$

alkyl (or aryl) azide

$$[R-\overset{-}{C}H-\overset{+}{N}\equiv N \longleftrightarrow R-CH=\overset{+}{N}=N^-]$$

diazoalkane

$$R-\overset{+}{N}\equiv N \qquad\qquad Ar-\overset{+}{N}\equiv N$$

alkanediazonium ion	arenediazonium ion
unstable, yields R^+, or undergoes S_N2 or E2 reactions	*moderately stable,* undergoes S_N1 (unstable Ar^+) or S_N2 type reactions, or "coupling reactions" in which a diazene derivative is formed

$$[\overset{-}{O}-\overset{+}{N}\equiv N \longleftrightarrow O=\overset{+}{N}=N^-]$$

nitrous oxide
"laughing gas," *stable,*
used as anesthetic and aerosol propellant

ASIDE

The path to discovery is ill-marked; chance favors those who are prepared to pay attention. Several chemists had converted aniline to phenol with nitrous acid, but in 1858, Griess found that treatment of 4-amino-2,6-dinitrophenol (see Chapter 30) with nitrous acid in ethanol gave the corresponding diazonium compound, which, luckily, separated as very explosive yellow leaflets. Analysis showed its composition to be that of a diazo compound (hence the name), although its structure is actually that of a diazonium oxide, a zwitterion. He realized the generality of the reaction and made many arenediazonium ions. Write out all the steps in the conversion for the formation of phenol from aniline using nitrous acid and those for the formation of the zwitterion mentioned above. How would you replace the diazonium ion with a fluoro, chloro, or cyano group?

PROBLEMS

1. What is the principal product obtained from *p*-toluenediazonium cation with each of the following reagents?

a. I^-
b. CuCN
c. OH^- (cold)
d. H_2O (hot)
e. CuBr
f. $NaNO_2$, copper powder
g. aq. NaOH, benzene, 5 °C
h. (i) $NaBF_4$; (ii) heat
i. H_3PO_2
j. CuCl
k. N,N-diethylaniline
l. (i) HPF_6; (ii) heat

2. Each of the following compounds is a significant dye intermediate. Give a practical laboratory preparation for each starting with benzene or toluene.

a. 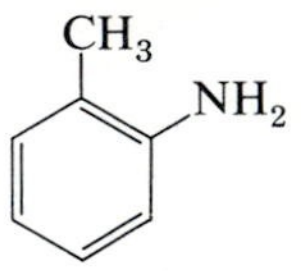 CH_3, NH_2
b. 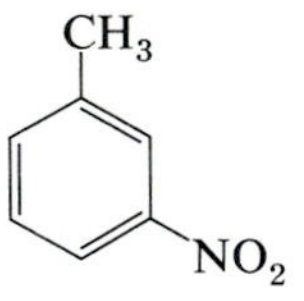 CH_3, NO_2
c. 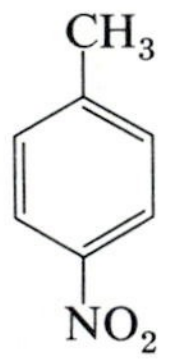 CH_3, NO_2
d. 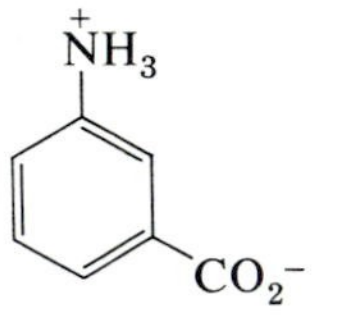 $\overset{+}{N}H_3$, CO_2^-
e. 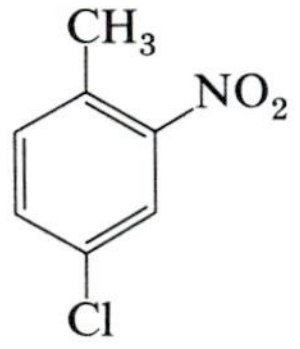 CH_3, NO_2, Cl
f. 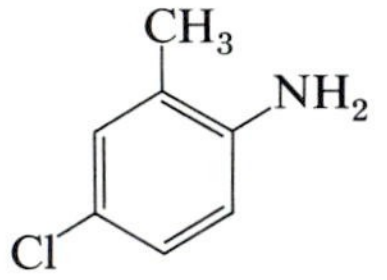 CH_3, NH_2, Cl
g. 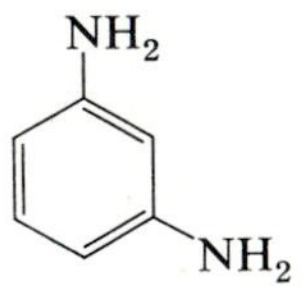 NH_2, NH_2
h. $N(CH_3)_2$, NH_2
i. 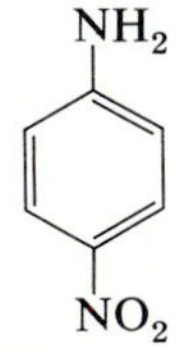 NH_2, NO_2

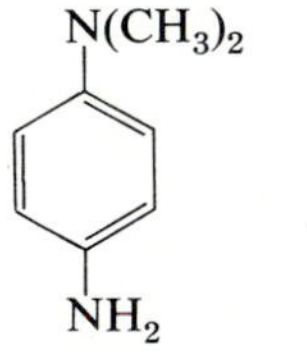

3. Show how each of the following conversions can be accomplished in a practical manner.

a. 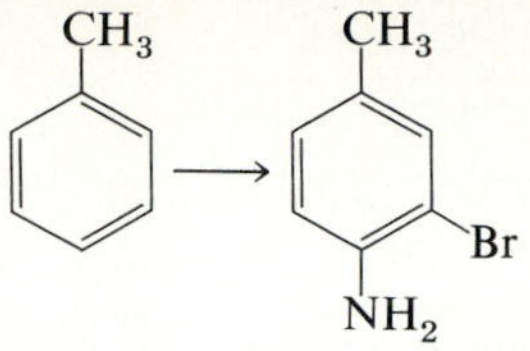 CH_3 (benzene) ⟶ CH_3, Br, NH_2 (benzene)

b. 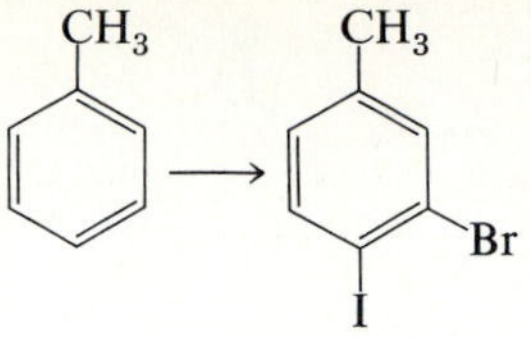 CH_3 (benzene) ⟶ CH_3, Br, I (benzene)

c. 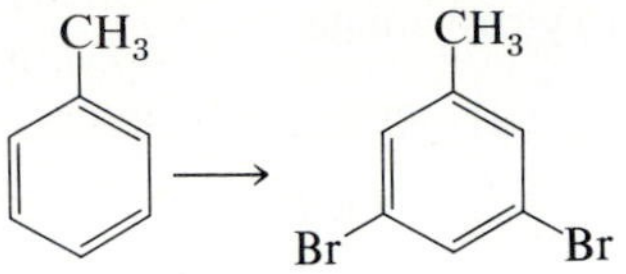 CH_3 (benzene) ⟶ CH_3, Br, Br (benzene)

d. 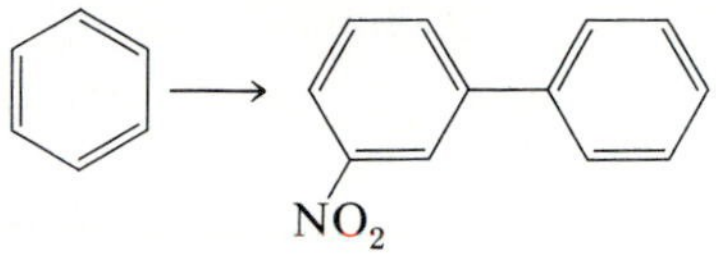 (benzene) ⟶ NO_2 (biphenyl)

e. 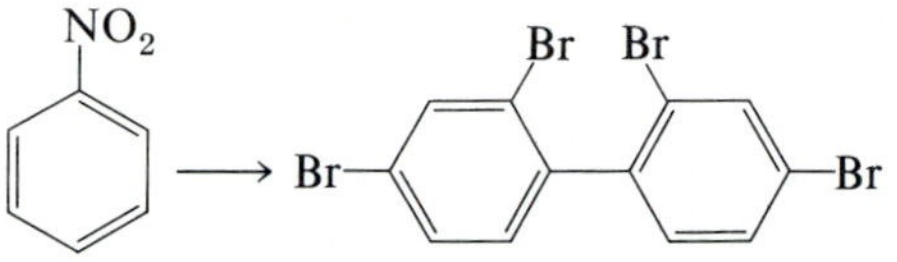 NO_2 (benzene) ⟶ Br, Br Br, Br (biphenyl)

f. 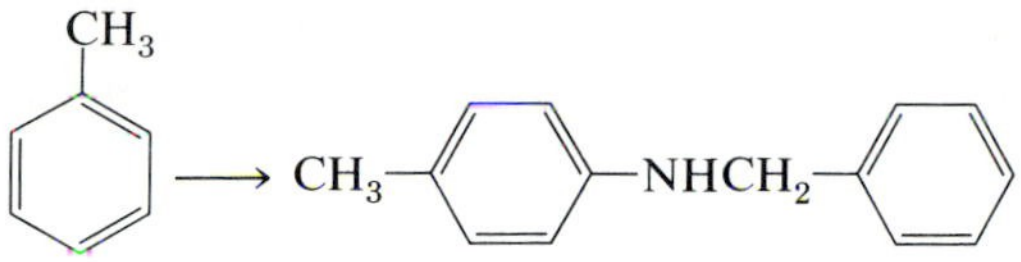 CH_3 (benzene) ⟶ CH_3–(benzene)–$NHCH_2$–(benzene)

g. 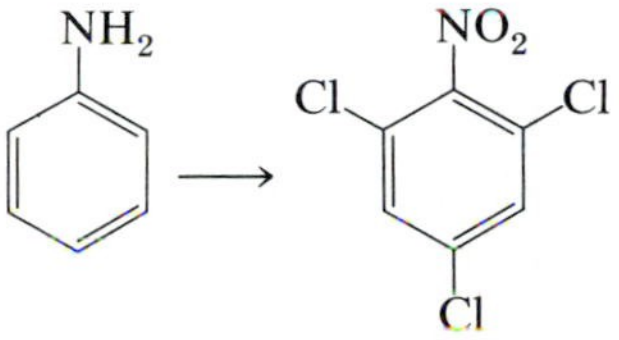 NH_2 (benzene) ⟶ NO_2, Cl, Cl, Cl (benzene)

h. 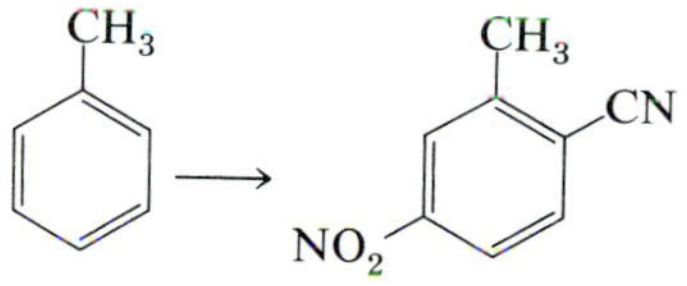 CH_3 (benzene) ⟶ CH_3, CN, NO_2 (benzene)

4. A small amount of methyl orange is added to a solution containing equimolar amounts of acetic acid and sodium acetate. Is this solution yellow or red?

5. A diazonium salt prepared from *p*-nitroaniline, when decomposed in nitrobenzene, gives a 69% yield of 4,4′-dinitrobiphenyl. That is, reaction of the aryl radical formed from the diazonium salt occurs primarily at the *para* position of the nitrobenzene. Give a reasonable explanation of this orientation behavior.

6. Give the principal reduction product from *m*-nitrotoluene under each of the following conditions.

a. Zn, alcoholic NaOH
b. Pt/H_2
c. Zn, aq. NH_4Cl
d. $SnCl_2$, HCl
e. H_2NNH_2, Ru/C, alcoholic KOH
f. As_2O_3, aq. NaOH

7. When an α-diazo ketone is irradiated or heated in aqueous solution, the product obtained is a carboxylic acid.

$$R\overset{O}{\overset{\|}{C}}CHN_2 \xrightarrow[H_2O]{h\nu \text{ or } \Delta} RCH_2COOH$$

a. Propose a mechanism for the transformation (**Wolff rearrangement**).
b. Predict the product when diazoacetone is irradiated in methanol solution.

8. When cyclohexanone is treated with diazomethane, a mixture of cycloheptanone and methylenecyclohexane oxide is produced. Propose a mechanism.

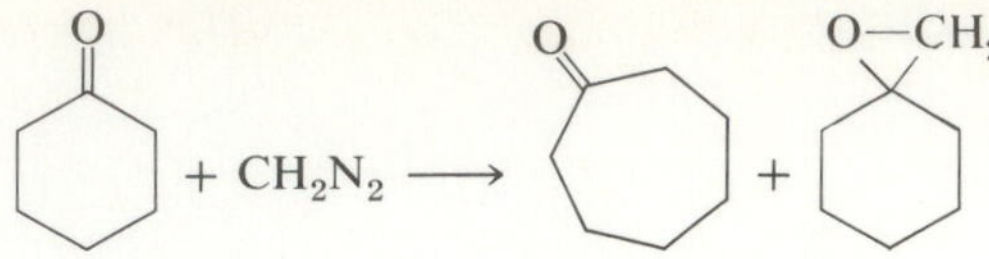

cyclohexanone + CH_2N_2 ⟶ cycloheptanone + spiro-epoxide (O—CH_2)

9. When cyclohexanecarboxamide is treated with bromine and sodium methoxide in methanol, the product obtained is methyl N-cyclohexylcarbamate.

$$\text{C}_6\text{H}_{11}\overset{\text{O}}{\overset{\|}{\text{C}}}\text{NH}_2 + \text{Br}_2 \xrightarrow[\text{CH}_3\text{OH}]{\text{CH}_3\text{O}^-} \text{C}_6\text{H}_{11}\overset{\text{H}}{\overset{|}{\text{N}}}\text{—}\overset{\text{O}}{\overset{\|}{\text{C}}}\text{OCH}_3$$

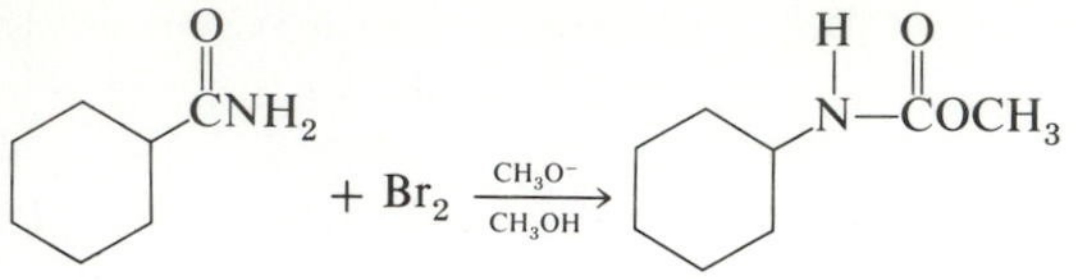

Rationalize with a plausible mechanism.

10. When ethyl N-cyclohexylcarbamate is refluxed with 1 M KOH in methanol for 100 hr, the only product obtained is the methyl ester in 95% yield.

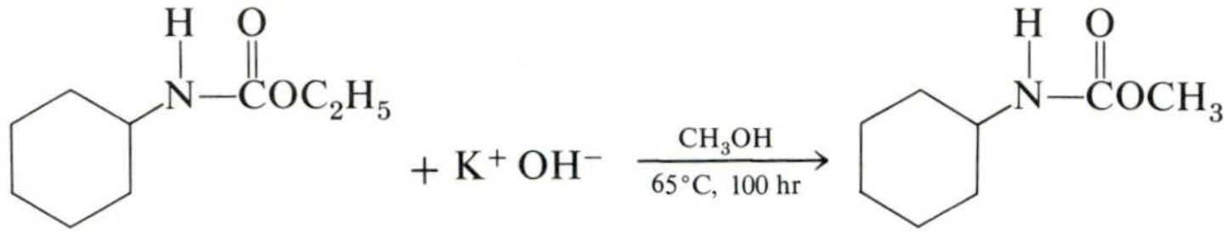

$$\text{C}_6\text{H}_{11}\overset{\text{H}}{\overset{|}{\text{N}}}\text{—}\overset{\text{O}}{\overset{\|}{\text{C}}}\text{OC}_2\text{H}_5 + \text{K}^+\,\text{OH}^- \xrightarrow[65\,^\circ\text{C, 100 hr}]{\text{CH}_3\text{OH}} \text{C}_6\text{H}_{11}\overset{\text{H}}{\overset{|}{\text{N}}}\text{—}\overset{\text{O}}{\overset{\|}{\text{C}}}\text{OCH}_3 \quad (95\%)$$

Explain why no cyclohexylamine is produced.

11. A laboratory technician attempted to prepare *p*-bromocumene by diazotizing *p*-aminocumene with a mixture of aqueous sodium nitrite and hydrochloric acid at 0 °C followed by reaction with hot cuprous bromide, but a mixture of products was obtained. What was the nature of the mixture?

12. When *p*-bromoaniline was diazotized with sodium nitrite and hydrochloric acid, and the mixture was allowed to decompose at room temperature, the *p*-bromophenol formed as principal product was found to be contaminated by some *p*-chlorophenol. Explain.

13. Show how each of the following transformations can be accomplished. Write the structures of intermediates, if appropriate.

a. 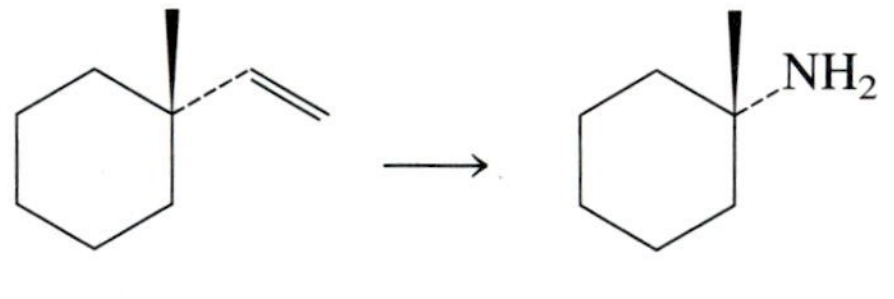 1-methyl-1-vinylcyclohexane ⟶ 1-methylcyclohexylamine (NH_2)

b. 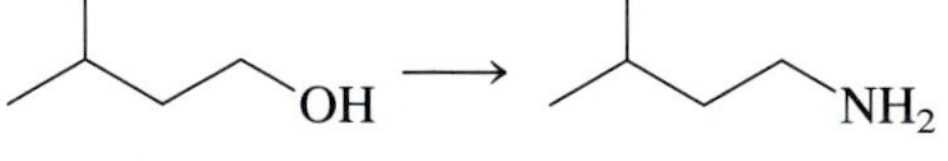 3-methylbutanol (OH) ⟶ 3-methylbutylamine (NH_2)

c. 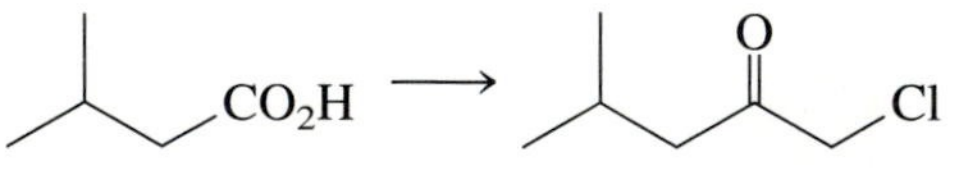 3-methylbutanoic acid (CO_2H) ⟶ 1-chloro-4-methyl-2-pentanone (Cl)

d. 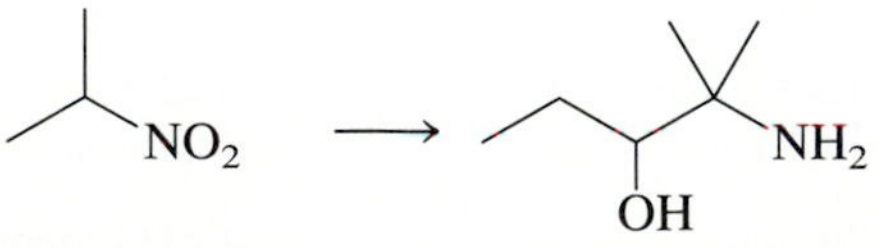 2-nitropropane (NO_2) ⟶ 1-amino-2,2-dimethyl... amino alcohol (NH_2, OH)

14. Suggest a mechanism for the following rearrangement reaction (the **Tiffeneau rearrangement**).

$$\text{1-(aminomethyl)cyclohexanol} \xrightarrow[\text{(59–65\%)}]{\text{NaNO}_2\text{, HCl, H}_2\text{O, 25 °C}} \text{cycloheptanone}$$

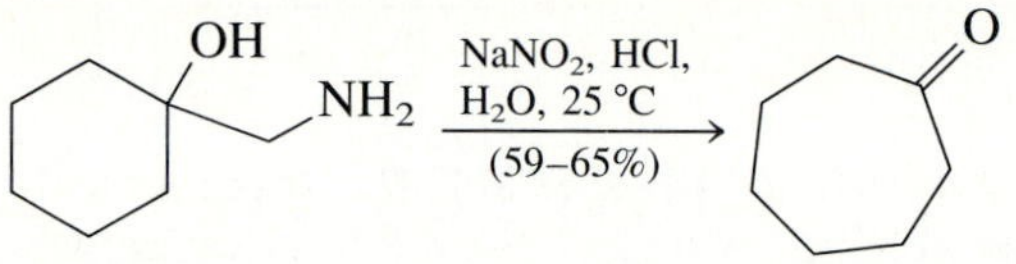

15. The substrate for the Tiffeneau rearrangement in problem 14 was prepared from cyclohexanone by the two-step sequence: (1) Henry reaction with nitromethane, (2) reduction of the nitro group with $LiAlH_4$. Discuss the possible use of this protocol as a general method for the conversion of one ketone into another by the insertion of a CH_2 group into one of the carbonyl-α-carbon bonds. For which of the following substrates would this process prove troublesome? Explain.

a. 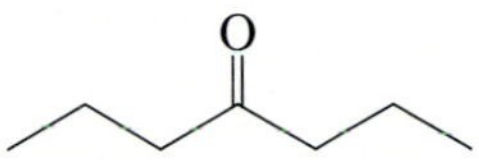 (3-hexanone, O)

b. 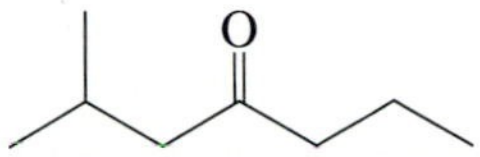 (2-methyl-4-heptanone, O)

c. 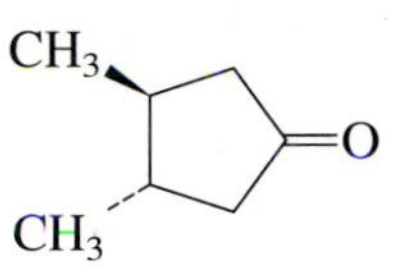 (3,4-dimethylcyclopentanone: CH_3, CH_3, O)

d. 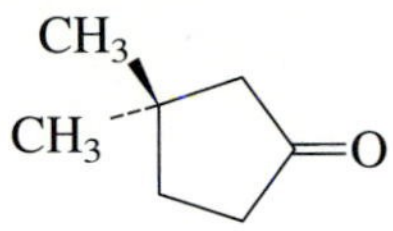 (3,3-dimethylcyclopentanone: CH_3, CH_3, O)

CHAPTER 26

SULFUR, PHOSPHORUS, AND SILICON COMPOUNDS

SEVERAL interesting functional groups contain sulfur or phosphorus as a central element. Although these classes of compounds are not as common as those containing oxygen and nitrogen, some of them have particular significance in biochemistry. In addition, sulfur-, phosphorus-, and silicon-containing compounds have important uses as synthetic intermediates. Sulfur, phosphorus, and silicon are in the same columns of the periodic table as oxygen, nitrogen, and carbon, respectively. Thus, we expect to find similarities in the chemistry of analogous sulfur and oxygen functions. Similarly, we expect comparable chemistry of analogous nitrogen and phosphorus compounds. Finally, we expect that organosilicon compounds will show some of the typical chemistry of their carbon analogs. To some extent this correspondence is observed. However, the third-period elements are both less electronegative and more polarizable than their second-period relatives and these differences result in significant quantitative differences in chemistry. Furthermore, sulfur and phosphorus have higher oxidation states available and can form some compounds that have no counterparts in oxygen and nitrogen chemistry. We have already encountered some of these functional groups (for example, sulfonate esters, phosphorus ylides). In this chapter we shall take an abbreviated look at the characteristic organic chemistry of sulfur, phosphorus, and silicon.

26.1 Thiols and Sulfides

Thiols, RSH, and sulfides, R_2S, bear an obvious relationship to alcohols and ethers. In the IUPAC system the alkane name is combined with the suffix -**thiol** in the same way that alcohols are named as alkanols. One difference is that the final -**e** of the alkane name is retained in naming thiols.

$$\underset{\text{ethanethiol}}{CH_3CH_2SH} \qquad \underset{\text{2-butanethiol}}{CH_3CH_2\overset{\overset{\displaystyle CH_3}{|}}{C}HSH}$$

A common name is that of alkyl mercaptan, analogous to alkyl alcohol, but the mercaptan nomenclature is falling into disuse in favor of IUPAC systematic names. However, **mercapto** is still used for a thiol group substituent, e.g., 2-mercaptoethanol for $HSCH_2CH_2OH$.

Sulfides are commonly named in a manner analogous to the common nomenclature of ethers. The two alkyl group names are followed by the word **sulfide.**

$$\underset{\text{dimethyl sulfide}}{CH_3SCH_3} \qquad \underset{\text{ethyl propyl sulfide}}{CH_3CH_2SCH_2CH_2CH_3}$$

In the IUPAC system sulfides are named as alkylthioalkanes. The prefix **alkylthio-** is analogous to **alkoxy-** and refers to a group RS—. As with ethers, the larger of the two alkyl groups is taken as the stem.

$$\underset{\text{2-methyl-1-(methylthio)butane}}{CH_3SCH_2\overset{\overset{\displaystyle CH_3}{|}}{C}HCH_2CH_3} \qquad \underset{\text{1-isopropylthiobutane}}{CH_3\overset{\overset{\displaystyle CH_3}{|}}{C}HSCH_2CH_2CH_2CH_3}$$

The IUPAC system for naming sulfides is only used in practice for complex structures that are not conveniently named as dialkyl sulfides.

The principal structural differences between methanethiol and methanol are that the carbon-sulfur bond is about 0.4 Å longer than the carbon-oxygen bond and the C—S—H angle is relatively more acute than the C—O—H angle. In thiols, as in H_2S itself, sulfur uses orbitals that are rich in *p*-character for bonding. The barrier to rotation about the carbon-sulfur bond is identical to that about the carbon-oxygen bond in methanol, 1.1 kcal mole^{-1}.

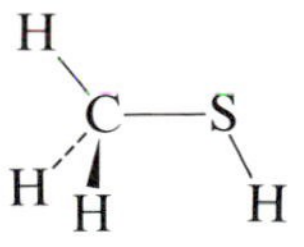

Bond Distances, Å		Bond Angles, deg	
C—H	1.10	H—C—H	110.2
C—S	1.82	H—C—S	108
S—H	1.33	C—S—H	100.3

A similar structure is found for dimethyl sulfide. Again, the C—S—C angle is relatively small (98.9°), corresponding to carbon-sulfur bonds in which sulfur uses a high percentage of its 3*p*-orbitals.

Thiols have boiling points that are almost normal for their molecular weight; they generally boil somewhat higher than the corresponding chlorides. For example, ethanethiol has b.p. 37 °C compared to ethyl chloride, b.p. 13 °C. Thiols are stronger acids than alcohols, just as H_2S is a stronger acid than water. The pK_a of ethanethiol, 10.60, indicates that the compound is completely converted to its anion by hydroxide ion.

$$\underset{pK_a\ 10.6}{C_2H_5SH} + OH^- \rightleftharpoons C_2H_5S^- + \underset{pK_a\ 15.7}{H_2O}$$

Although thiols are more acidic than alcohols, sulfur is less electronegative than oxygen. Hence, thiols have lower dipole moments (CH_3SH, $\mu = 1.26$ D; CH_3OH, $\mu = 1.71$ D) and hydrogen bonding between thiol molecules is much weaker than for alcohols. However, hydrogen-bonding from the acidic SH protons to water oxygen is significant, and the thiols have some water solubility.

The most impressive property of volatile thiols is their odor. Their intensely disagreeable odors discourage use as laboratory reagents. The thiols of the defensive odor used by the skunk are a mixture of 3-methylbutane-1-thiol, *trans*-2-butene-1-thiol and *trans*-2-buten-1-yl methyl disulfide. Methanethiol gives the urine of most persons a distinctive odor after eating asparagus although there is some genetic variability in the ability to detect the odor. The nose is more sensitive than any laboratory instrument in detecting ethanethiol; one part in 50 billion parts of air can be detected. Small amounts of isopentyl thiols are included in fuel gas, which is otherwise almost odorless, to warn users about leaks. The lower molecular weight sulfides have similarly repugnant odors. A small increase in molecular weight can have a substantial effect on odor; an important component of roasted coffee aroma is 2-furylmethanethiol (furfuryl thiol), presumably formed from pentoses and sulfur-containing amino acids. (The conversion of pentoses to furfural, 2-furaldehyde, is described in detail in Section 32.3.B.)

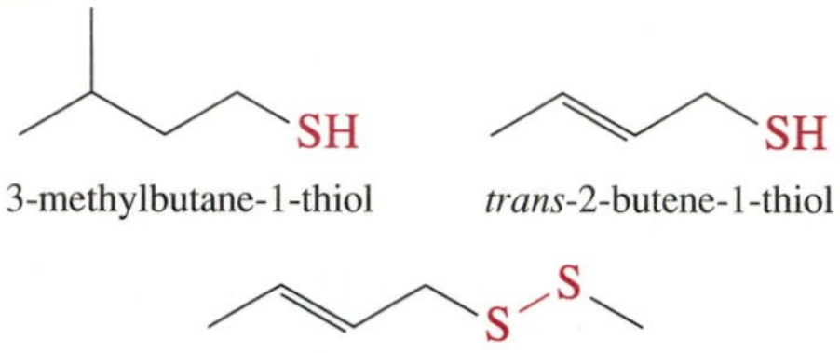

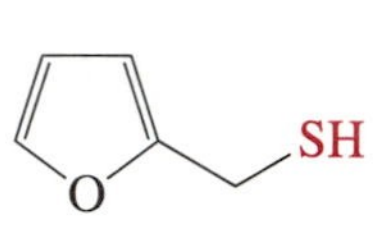

SH
3-methylbutane-1-thiol

SH
trans-2-butene-1-thiol

S S
trans-2-buten-1-yl methyl disulfide

Skunk

O SH
2-furylmethanethiol

Roast coffee

Thioaldehydes and thioketones are known, but are relatively rare. The lower molecular weight compounds are red liquids with intensely obnoxious odors. Although they may be prepared in a monomeric form, they rapidly polymerize to give cyclic trimers.

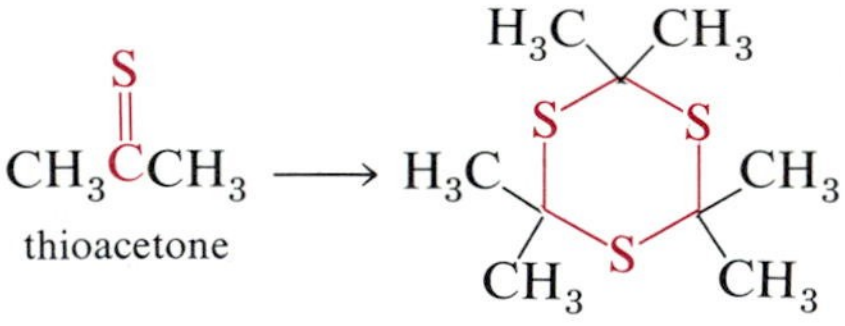

$$CH_3\overset{\overset{S}{\|}}{C}CH_3 \longrightarrow$$ (cyclic trimer: 2,2,4,4,6,6-hexamethyl-1,3,5-trithiane; H_3C, CH_3 groups; S atoms in ring)

thioacetone

Thioacids, RCOSH, and thioesters, RCOSR′, are more common.

$$CH_3\overset{\overset{O}{\|}}{C}SH \qquad CH_3CH_2\overset{\overset{O}{\|}}{C}SC(CH_3)_3$$

t-butyl thiopropanoate

Note that the stable form of thioacids is the one with C═O and S—H groups and not the alternative structure with C═S and O—H groups. A particularly important example of a thioester is acetyl coenzyme A, acetyl coA, an important intermediate used by nature in the biosynthesis of numerous organic compounds (page 541).

EXERCISE 26.1 Write the structure corresponding to each of the following names.

(a) ethyl isopropyl sulfide (b) butyl mercaptan (c) 3-methylthiooctane
(d) diphenyl sulfide (e) 3-pentanethiol (f) thioacetaldehyde
(g) ethyl thioacetate

Thiols can be prepared from alkyl halides by displacement with hydrosulfide ion, HS^-, in ethanol solution.

$$CH_3(CH_2)_{16}CH_2I + Na^+ SH^- \xrightarrow{C_2H_5OH} CH_3(CH_2)_{16}CH_2SH + Na^+ I^-$$

1-octadecanethiol

In preparing thiols by this method it is necessary to employ a large excess of hydrosulfide because of the equilibrium

$$HS^- + RSH \rightleftharpoons H_2S + RS^-$$

The thiol anion produced by this equilibrium is itself a good nucleophile and can react with the alkyl halide to give the corresponding sulfide.

$$RS^- + RBr \longrightarrow RSR + Br^-$$

The use of a large excess of hydrosulfide makes its reaction with the alkyl halide more probable and maximizes the yield of thiol.

For this reason HS^- has been almost exclusively replaced by thiourea for such displacement reactions.

$$H_2N{-}\overset{\overset{\displaystyle S}{\|}}{C}{-}NH_2$$

thiourea

Thiourea, a commercially available solid, m.p. 178 °C, is soluble in water and alcohols. The sulfur is nucleophilic and readily takes part in S_N2 reactions on alkyl halides. The product salt is readily hydrolyzed to the alkanethiol.

$$\underset{n\text{-dodecyl bromide}}{CH_3(CH_2)_{11}Br} + H_2N{-}\overset{\overset{\displaystyle S}{\|}}{C}{-}NH_2 \xrightarrow[\Delta]{95\%\ ethanol} CH_3(CH_2)_{11}S{-}\overset{\overset{\displaystyle \overset{+}{N}H_2}{\|}}{C}NH_2\ Br^-$$

$$CH_3(CH_2)_{11}S{-}\overset{\overset{\displaystyle \overset{+}{N}H_2}{\|}}{C}NH_2\ Br^- \xrightarrow[\Delta]{aq.\ NaOH} \underset{\substack{(79\text{–}83\%)\\ \text{1-dodecanethiol}}}{CH_3(CH_2)_{11}SH} + H_2N\overset{\overset{\displaystyle O}{\|}}{C}NH_2$$

This route avoids formation of sulfides. The other product of the hydrolysis is urea, H_2NCONH_2 (Section 25.2).

Thiols can also be prepared by reaction of Grignard reagents with sulfur.

$$(CH_3)_3CMgBr + S_8 \longrightarrow (CH_3)_3CSMgBr \xrightarrow{HCl} \underset{\substack{t\text{-butyl mercaptan}\\ \text{2-methyl-2-propanethiol}}}{(CH_3)_3CSH} + MgBrCl$$

Both symmetrical and unsymmetrical sulfides can be prepared by S_N2 reactions of alkylthio anions with alkyl halides or sulfonates.

$$CH_3CH_2SH + OH^- \xrightarrow{H_2O} CH_3CH_2S^-$$

$$CH_3CH_2S^- + (CH_3)_2CHCH_2Br \longrightarrow \underset{\substack{(95\%)\\ \text{ethyl isobutyl sulfide}}}{(CH_3)_2CHCH_2SCH_2CH_3}$$

This general method for preparing dialkyl sulfides is directly analogous to the Williamson ether synthesis (Section 10.9).

EXERCISE 26.2 Show how the following conversions can be accomplished.

(a) $(CH_3)_3CCH_2Br \rightarrow (CH_3)_3CCH_2SH$
(b) $CH_3CH_2CH_2OH \rightarrow (CH_3CH_2CH_2)_2S$

26.3 Reactions of Thiols and Sulfides

One of the similarities between thiol chemistry and alcohol chemistry is that both classes of compounds are weak acids. However, just as HCl is a stronger acid in aqueous solution than HF, RSH compounds are substantially more acidic than their ROH counterparts. The pK_a of ethanethiol, 10.6, tells us that the compound is half-ionized at pH 10.6, which corresponds roughly to the pH of a 5% sodium carbonate solution. Thus, unlike alcohols, thiols may be converted essentially quantitatively into the corresponding anions in aqueous solution. The relative acidities of alcohols and thiols are explained in Section 26.1.

CH_3CH_2SH	CH_3CH_2OH
ethanethiol	ethanol
$pK_a = 10.6$	$pK_a = 15.7$

Thiols are readily oxidized to disulfides. The disulfide bond is weak and is easily reduced to give the thiol.

$$2\ RSH \underset{\text{reduction}}{\overset{\text{oxidation}}{\rightleftharpoons}} RS{-}SR$$

a disulfide

Mild oxidizing agents such as iodine suffice for the oxidation. The reaction with iodine occurs by way of a sulfenyl iodide, RSI. A closely related reaction is oxidation with diazenedicarboxylic acid bis-N,N-dimethylamide, which is used for biological thiols such as the tripeptide, glutathione (see below). This method avoids the reaction of iodine with reactive aromatic amino acids or aromatic amines (see Section 23.5). Formation of the sulfenyl iodide involves nucleophilic displacement of iodide by thiolate attack *on iodine*. Formation of the disulfide occurs via nucleophilic displacement of iodide by attack of thiolate *on sulfur*. The thiolate adds to the diazene double bond ("displacement" of an electron pair) to form a sulfenylhydrazide, followed by a thiolate attack *on sulfur* to yield a disulfide and a hydrazide.

$$2\ CH_3CH_2SH + I_2 \longrightarrow CH_3CH_2SSCH_2CH_3 + 2HI$$

diethyl disulfide

$$RSH \underset{+H^+}{\overset{-H^+}{\rightleftharpoons}} RS^- \quad I{-}I \longrightarrow RSI + I^-$$

alkyl sulfenyl iodide

$$RS^- \quad RSI \longrightarrow RSSR + I^-$$

dialkyl disulfide

$$(CH_3)_2NC(=O)N{=}NC(=O)N(CH_3)_2 \xrightarrow{RS^-,\ H^+} (CH_3)_2NC(=O)N(SR){-}N(H)C(=O)N(CH_3)_2 \xrightarrow{H^+}$$

alkyl sulfenylhydrazide

$$(CH_3)_2N-C(=\overset{+}{O}H)-N(-S-R)-NH-C(=O)-N(CH_3)_2 \;\; {}^{-}SR \longrightarrow (CH_3)_2N-C(OH)=N-NH-C(=O)-N(CH_3)_2 + R-S-S-R$$

dialkyl disulfide

$$\xrightarrow[-H^+]{+H^+} \text{ tautomerism}$$

$$(CH_3)_2N-C(=O)-NH-NH-C(=O)-N(CH_3)_2$$

hydrazide

A commonly used reducing agent for regeneration of the thiol is lithium in liquid ammonia. Other agents for the conversion of disulfides into thiols include sodium borohydride and, for compounds containing groups sensitive to borohydride or lithium in ammonia, triphenylphosphine in acid solution.

$$CH_3CH_2SSCH_2CH_3 \xrightarrow[NH_3]{Li} \xrightarrow[\text{(work-up)}]{H_3O^+} 2\ CH_3CH_2SH$$

$$Ph_3P \;\; R-S-S-R \rightleftharpoons R-S-\overset{+}{P}Ph_3 \;\; {}^{-}S-R \xrightarrow{H_3O^+} O{=}PPh_3 + 2\ HS-R$$

The balance between thiols and disulfides is especially important in biological systems. The interiors of cells contain a high proportion of thiols, while the disulfide link is one of the bonds that defines the structures of many proteins and hormones. Extracellular proteins generally have disulfide bonds rather than thiols. The most common thiol in biological systems is the tripeptide thiol γ-glutamylcysteinylglycine or GSH. (See Chapter 29 for discussion of amino acids and peptides.) The weakness of the RS—H bond makes GSH a good hydrogen donor for free radicals. Such free radicals are generally detrimental to the cell and might be produced by ionizing radiation. Certain thiols are used as radioprotective agents during treatment of cancers by radiation.

$$GSH + R\cdot \longrightarrow GS\cdot + RH$$

$$GS\cdot + GS\cdot \longrightarrow GSSG$$

$$GSSG \xrightarrow[\textit{glutathione disulfide reductase}]{NADPH} 2\ GSH$$

$G = \gamma$-glutamylcysteinylglycine

An interesting natural product is thioctic acid, which contains a cyclic disulfide link. Thioctic acid is normally linked to a protein through an amide bond and forms an acetylthiol ester as an intermediate in the formation of acetyl coA. There is some evidence that thioctic acid is useful in the treatment of poisoning by α-amanitin, the most toxic component of *Amanita phalloides* and other *Amanita* mushrooms.

The toxin, α-amanitin, inhibits ribonucleic acid polymerase II, and is insidiously slow and devastating in its action. The enzyme is involved in messenger ribonucleic acid (mRNA) synthesis, and so is critical to the ongoing performance of the cell. The messenger RNA is a transcription of the deoxyribonucleic acid (DNA) code and provides the instructions (code) that are translated into a protein amino acid sequence at the ribosome (see Chapter 33).

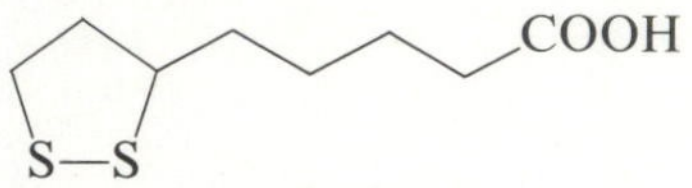

COOH

S—S

thioctic acid

The sulfur-sulfur bond of disulfides is susceptible to cleavage by nucleophiles as noted above for triphenylphosphine and borohydride.

$$N:^- + RS—SR \longrightarrow N—SR + RS:^-$$

A synthetically useful version of this reaction is found in the thiolation of ketone and ester enolates.

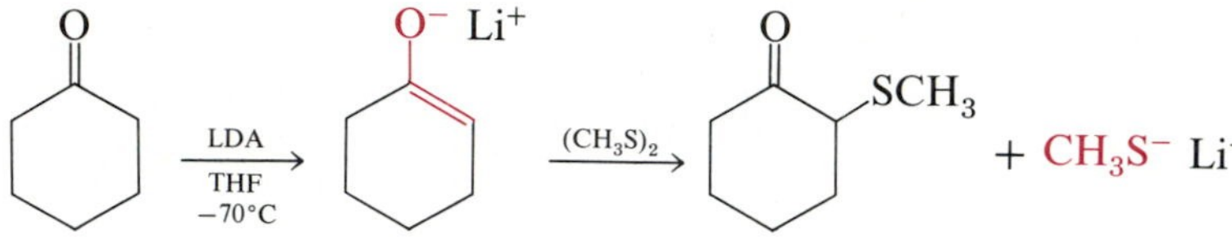

$$\xrightarrow[\text{THF, } -70°C]{\text{LDA}} O^-\ Li^+ \xrightarrow{(CH_3S)_2} SCH_3 + CH_3S^-\ Li^+$$

Strong oxidants such as potassium permanganate or hot nitric acid oxidize thiols or disulfides to the corresponding sulfonic acids. If a mixture of chlorine and nitric acid is used as the oxidant, the corresponding sulfonyl chloride is obtained. Sulfonyl chlorides have the same relationship to sulfonic acids as acyl halides have to carboxylic acids. Thus they are hydrolyzed to sulfonic acids by water.

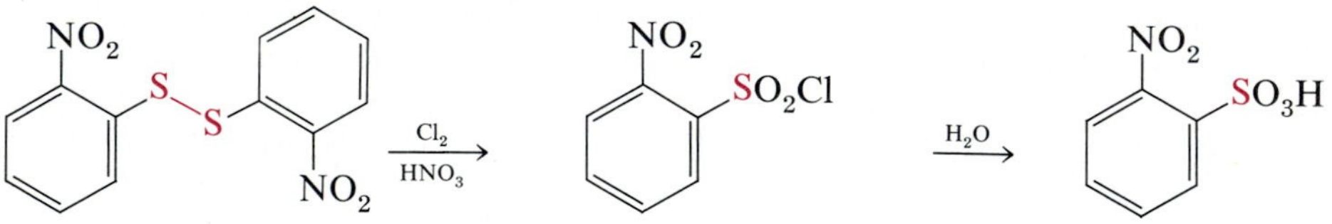

$$\xrightarrow[HNO_3]{Cl_2} \qquad \xrightarrow{H_2O}$$

NO_2, S, S, NO_2 — NO_2, SO_2Cl — NO_2, SO_3H

84%

di-*o*-nitrophenyl disulfide — *o*-nitrobenzenesulfonyl chloride — *o*-nitrobenzenesulfonic acid

Sulfides are also easily oxidized. The initial oxidation product is a **sulfoxide.** Further oxidation of the sulfoxide yields a **sulfone.**

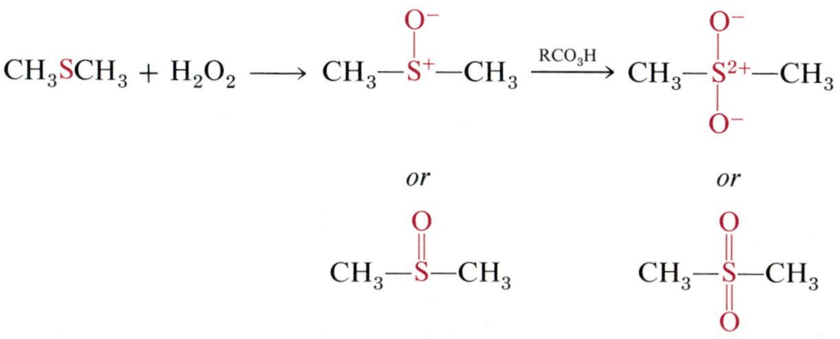

$$CH_3SCH_3 + H_2O_2 \longrightarrow CH_3—\overset{O^-}{\underset{}{S^+}}—CH_3 \xrightarrow{RCO_3H} CH_3—\overset{O^-}{\underset{O^-}{S^{2+}}}—CH_3$$

or *or*

$$CH_3—\overset{O}{\overset{\|}{S}}—CH_3 \qquad CH_3—\overset{O}{\overset{\|}{\underset{\underset{O}{\|}}{S}}}—CH_3$$

dimethyl sulfoxide (DMSO) — dimethyl sulfone

An especially convenient oxidant, which converts sulfides to sulfoxides without the danger of over-oxidation to the sulfone, is sodium periodate, $NaIO_4$.

$$C_6H_5\text{—}S\text{—}CH_3 + NaIO_4 \xrightarrow[0°C]{H_2O} C_6H_5\text{—}\overset{O^-}{\overset{|}{S^+}}\text{—}CH_3$$

methyl phenyl sulfide — (99%) methyl phenyl sulfoxide

To 210 mL of a 0.5 M solution of $NaIO_4$ at 0 °C is added 12.4 g of methyl phenyl sulfide. The mixture is stirred at ice-bath temperature overnight, the precipitated $NaIO_3$ is removed by filtration, and the filtrate is extracted with chloroform. Evaporation of solvent gives a crude product, which is distilled to obtain 13.9 g (99%) of methyl phenyl sulfoxide, m.p 29–30 °C.

Sulfur compounds such as sulfoxides and sulfones are frequently represented for convenience as having sulfur-oxygen double bonds and an expanded octet around sulfur. However, the sulfur-oxygen bond is not a double bond in the same sense as carbon-carbon or carbon-oxygen double bonds in which the "second bond" is viewed as arising from π-overlap of atomic *p*-orbitals. Because sulfur has *d*-orbitals of rather low energy, it has long been speculated that sulfur-oxygen "double" bonds result from overlap of an oxygen *p*-orbital with a sulfur *d*-orbital.

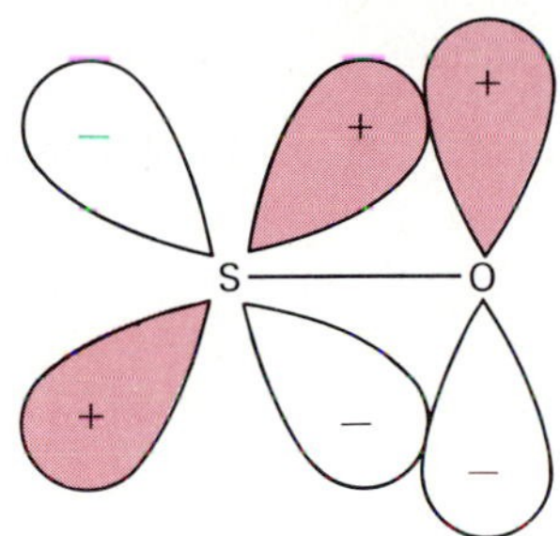

Some such bonding may be involved, but classical coulombic attraction of the partially positive sulfur for the partially negative oxygen provides much of the bonding. It also appears that the high polarizability of the sulfur valence electrons is involved.

We have previously encountered dimethyl sulfoxide as an important solvent in the class of polar aprotic solvents (Section 9.5.B). It is a relatively inexpensive colorless liquid that is miscible with water in all proportions. It is prepared industrially by the NO_2-catalyzed air-oxidation of dimethyl sulfide, a by-product produced in tonnage quantities in the sulfite pulping process for paper manufacture.

$$CH_3SCH_3 + \tfrac{1}{2}\,O_2 \xrightarrow{NO_2} CH_3\overset{\overset{O}{\|}}{S}CH_3$$

dimethyl sulfide — dimethyl sulfoxide (DMSO)

Dimethyl sulfoxide owes its utility to the fact that it readily dissolves many inorganic salts as well as most organic compounds. It is an excellent solvent for reactions such as displacements on alkyl halides. However, the potent solvent properties of DMSO result in a particular hazard that is associated with its use. The compound diffuses through the skin almost instantaneously and carries any solutes along with it. For this reason, one should always wear impermeable gloves when working with DMSO solutions.

Note that sulfoxides are similar in electronic structure to amine oxides (Section 24.7.C). Like amine oxides, sulfoxides undergo thermal elimination to give alkenes (Section 24.7.E). The process is a syn elimination and can sometimes occur under fairly mild conditions.

$$\text{(96\%)}$$

The sulfur-containing by-product in this reaction is a **sulfenic acid.** Recall that a sulfenyl iodide was formed in the reaction of a thiol with iodine. Sulfenic acids are unstable and undergo disproportionation to the corresponding **sulfinic acid** and disulfide, which are the actual by-products of a sulfoxide elimination.

$$3\ CH_3SOH \longrightarrow CH_3SSCH_3 + CH_3SO_2H + H_2O$$

methanesulfenic acid — dimethyl disulfide — methanesulfinic acid

The elimination of sulfoxides can also occur under biological conditions. Garlic, *Allium sativum,* contains S-allylcysteine S-oxide or alliin, a colorless, odorless solid. The enzyme

allinase

alliin — allylsulfenic acid

allicin

allylsulfenic acid — thioacrolein (thiopropenal)

(*Z*)-ajoene — (*E*)-ajoene

allinase present in garlic catalyzes an elimination that produces allylsulfenic acid. One molecule of the sulfenic acid can react as an S-nucleophile toward a second sulfenic acid to form allyl allylthiosulfinate, allicin. The latter can also undergo β-elimination to yield allylsulfenic acid and thioacrolein, which dimerizes. Other reactions also occur to produce a rather complex mixture. Garlic has some reputation as a folk medicine for prevention of stroke, coronary thrombosis, and atherosclerosis. Significant components of the mixture formed from allicin are (*E*)- and (*Z*)-ajoenes, for which a mechanism of formation can be written. These seem to prevent aggregation of platelets in blood, and therefore might decrease the occurrence of events associated with clotting. However, the odor of the compounds is so overpowering that their use is not likely to become general.

EXERCISE 26.3 Using reactions learned in this section, show how the following transformation can be accomplished.

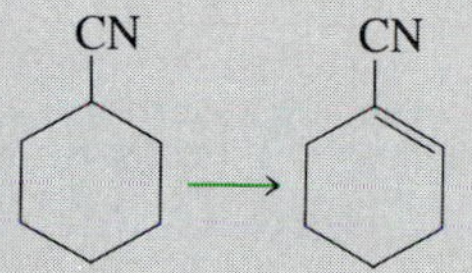

CN-cyclohexane $\longrightarrow$ 1-cyanocyclohexene (CN)

The sulfur in an dialkyl sulfide is nucleophilic. Sulfides react readily with alkyl halides by the normal S_N2 mechanism to produce **trialkylsulfonium salts,** which are usually hygroscopic solids.

$$(CH_3)_2\ddot{S}: + CH_3{-}I \longrightarrow (CH_3)_2\ddot{S}^+{-}CH_3\ I^-$$

trimethylsulfonium iodide

Like other S_N2 displacements, the reaction works best with primary halides.

$$(CH_3)_2CHCH_2Br + (CH_3)_2S \longrightarrow (CH_3)_2CHCH_2\overset{+}{S}(CH_3)_2\ Br^-$$

isobutyldimethylsulfonium bromide

When trialkylsulfonium salts are heated the reaction reverses. Halide ion acts as the nucleophile, and the dialkyl sulfide is the leaving group. The driving force for reaction is removal of the volatile products.

$$Br:^- + CH_3{-}\overset{+}{S}(CH_3)_2 \xrightarrow{\Delta} CH_3Br\uparrow + (CH_3)_2S\uparrow$$

Nature makes extensive use of this S_N2 reaction. The compound S-adenosylmethionine is a methylating agent in biochemical S_N2 reactions, which are catalyzed by appropriate enzymes. It can be regarded as the body's equivalent of methyl iodide.

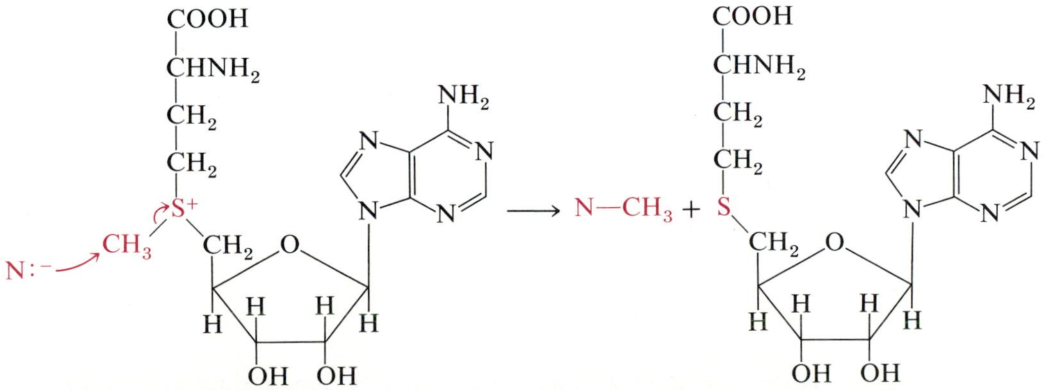

N:⁻ + S-adenosylmethionine [COOH–CHNH$_2$–CH$_2$–CH$_2$–S^+(CH$_3$)–CH$_2$-ribose(OH, OH)-adenine (NH$_2$)] $\longrightarrow$ N—CH_3 + S-adenosylhomocysteine [COOH–CHNH$_2$–CH$_2$–CH$_2$–S–CH$_2$-ribose(OH, OH)-adenine (NH$_2$)]

S-adenosylmethionine

The nucleophiles subjected to methylation by S-adenosylmethionine (SAM) include phosphatidylethanolamine to form phosphatidylcholine (Section 19.11.D), lysine to form ϵ-N-methyllysine (Chapter 29), proteins, ribonucleic acid (RNA), and deoxyribonucleic acid (DNA).

Like 1,2- and 1,3-diols, the analogous 1,2- and 1,3-dithiols react with aldehydes and ketones under conditions of acid catalysis to give cyclic thioacetals.

$$CH_3CHO + HS(CH_2)_3SH \xrightarrow{BF_3}$$ (2-methyl-1,3-dithiane: six-membered ring with S, S and CH_3)

1,3-propanedithiol — 2-methyl-1,3-dithiane

EXERCISE 26.4 Write a mechanism for the reaction of acetaldehyde and 1,3-propanedithiol to give 2-methyl-1,3-dithiane.

The thioacetals that are prepared in this way using 1,3-propanedithiol are useful as synthetic reagents (Section 26.8). The carbon-sulfur bond can be reductively cleaved by certain reagents, the most common of which is Raney nickel. The products of the reaction are the hydrocarbons formed by hydrogenolysis of each carbon-sulfur bond.

$$R{-}S{-}R' + H_2 \xrightarrow{\text{Raney Ni}} RH + R'H$$

Desulfurization of thioacetals provides a method for net deoxygenation of aldehydes and ketones and is complementary to the Wolff-Kishner and Clemmensen deoxygenations (Section 14.8.D).

(2-oxocyclohexane, O, $COOC_2H_5$) $+ CH_3SH \xrightarrow{BF_3}$ (cyclohexane with CH_3S, SCH_3, $COOC_2H_5$) $\xrightarrow{\text{Raney Ni}}$ (cyclohexane, $COOC_2H_5$)

(75%)

EXERCISE 26.5 Write equations illustrating the following reaction sequences beginning with 1-butanethiol.

(a) (i) NaOH, CH_3Br; (ii) H_2O_2; (iii) 150 °C (b) (i) I_2, KI; (ii) Li, NH_3
(c) (i) NaOH, CH_3CH_2I; (ii) CH_3I
(d) (i) NaOH, CH_3CH_2Br; (ii) $NaIO_4$, H_2O, 0 °C

Highlight 26.1

Thiols, especially in the form of the thiolate anion, are excellent nucleophiles.

$$RSH \longrightarrow RS^-$$

Among the wide variety of substrates that yield useful products are the following.

$$R{-}S^- + R{-}Br \longrightarrow R{-}S{-}R + Br^-$$

dialkyl sulfide

$$R{-}S^- + I{-}I \longrightarrow R{-}S{-}I + I^-$$

alkylsulfenyl iodide

$$R{-}S^- + R'S{-}SR'' \longrightarrow R{-}S{-}SR' + R''S^-$$

dialkyl disulfide

alkylsulfenylhydrazine

dithiane

$$+ \ ^-O{-}\bar{B}F_3 \xrightarrow{H^+} H_2O + BF_3$$

26.4 Sulfate Esters

Sulfuric acid, H_2SO_4, is a strong dibasic inorganic acid with $pK_1 \cong -5$ and $pK_2 = 1.99$.

$$H_2SO_4 \rightleftharpoons H^+ + HSO_4^- \qquad K_a \cong 1.4 \times 10^5 \text{ M}$$

$$HSO_4^- \rightleftharpoons H^+ + SO_4^{2-} \qquad K_a = 1.0 \times 10^{-2} \text{ M}$$

The acidity of sulfuric acid in dilute aqueous solution corresponds to $K_1 \cong 1.4 \times 10^5$ M, which means that the acid is completely dissociated. However, its effective acidity or ''protonating power'' increases markedly in highly concentrated solutions. In order to quantify the protonating power of concentrated solutions, a property known as an ''acidity function'' has been defined. The acidity function is a property of a given medium that provides a quantitative measure of the proton donating ability of the medium. The best known acidity function is the Hammett acidity function, which is derived from a series of weak bases that are protonated only in exceedingly ''acidic'' media. For a given medium the ratio of protonated and unprotonated forms of the indicator base is measured, usually spectrophotometrically. The Hammett acidity function, H_0, is defined in terms of the pK of the indicator base and the negative logarithm of the ratio of protonated and unprotonated species.

$$H_0 = pK(BH^+) - \log\frac{[BH^+]}{[B]}$$

In dilute aqueous acid solutions H_0 is equal to the pH of the solution. Hammett acidity functions for some sulfuric acid solutions are listed in Table 26.1. Note that the effective acidity of sulfuric acid increases by $10^{1.59}$ or 39-fold in going from 30% to 50% sulfuric acid and by a factor of $10^{3.11}$ or 1288-fold in going from 70% to 90% sulfuric acid.

TABLE 26.1 H_0 for Sulfuric Acid–Water Mixtures

% H_2SO_4	H_0
5	−0.02
10	−0.43
30	−1.82
50	−3.41
70	−5.92
90	−9.03
95	−9.73
98	−10.27
99	−10.57
100	−11.94

Both mono- and diesters of sulfuric acid are known. Like sulfuric acid itself, the sulfuric acid esters are often considered as resonance hybrids involving an expanded sulfur octet.

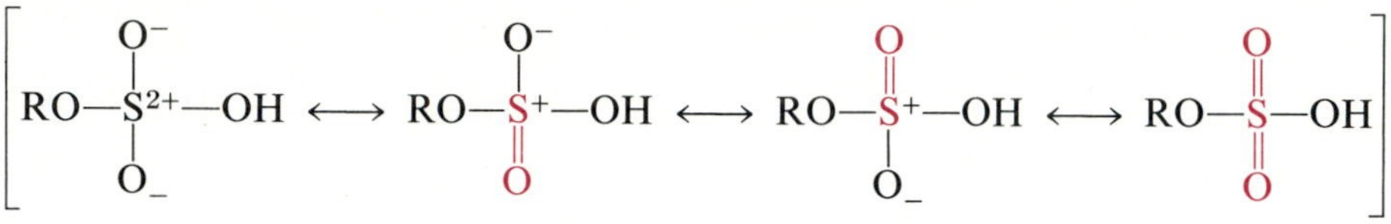

$$\left[RO{-}\overset{O^-}{\underset{O^-}{S^{2+}}}{-}OH \longleftrightarrow RO{-}\overset{O^-}{\underset{\|O}{S^+}}{-}OH \longleftrightarrow RO{-}\overset{O\|}{\underset{O^-}{S^+}}{-}OH \longleftrightarrow RO{-}\overset{O\|}{\underset{\|O}{S}}{-}OH \right]$$

For convenience, we shall only use the Kekulé structure represented as having two S=O bonds.

Diesters of sulfuric acid are named by combining the alkyl group name(s) with the word ''sulfate'' just as though they were salts of sulfuric acid.

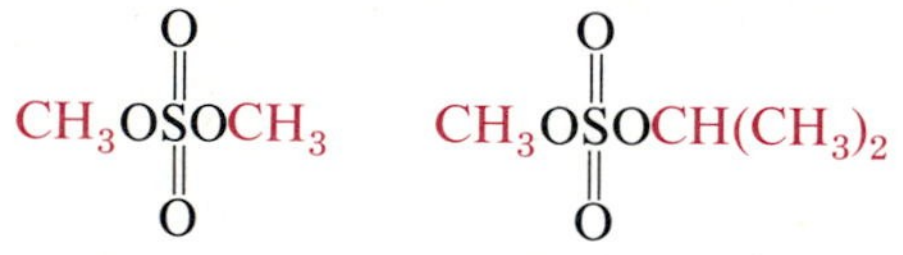

$$CH_3O\overset{O}{\underset{O}{S}}OCH_3 \qquad CH_3O\overset{O}{\underset{O}{S}}OCH(CH_3)_2$$

dimethyl sulfate methyl isopropyl sulfate

Monoesters are named as alkylsulfuric acids.

$$CH_3\,O\overset{O}{\underset{O}{S}}OH \qquad CH_3CH_2\,O\overset{O}{\underset{O}{S}}OH$$

methylsulfuric acid ethylsulfuric acid

Dialkyl sulfates are highly polar compounds and generally have rather high boiling points. Their water solubility is surprisingly low (Table 26.2). Alkylsulfuric acids are approximately as acidic as sulfuric acid itself.

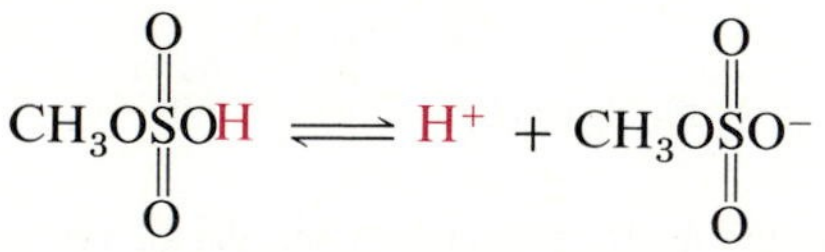

$$CH_3O\overset{O}{\underset{O}{S}}OH \rightleftharpoons H^+ + CH_3O\overset{O}{\underset{O}{S}}O^-$$

They readily form inorganic salts, which are named as metal alkyl sulfates.

$$CH_3OSO_3^-\ Na^+ \qquad (CH_3CH_2OSO_3^-)_2\ Ba^{2+}$$

sodium methyl sulfate barium ethyl sulfate

TABLE 26.2 Physical Properties of Dialkyl Sulfates

	Melting Point, °C	Boiling Point, °C	Solubility in H_2O, g/100 ml
$CH_3OSO_2OCH_3$	−27	188	2.8
$CH_3CH_2OSO_2OCH_2CH_3$	−25	210	very low

The monoesters are rarely encountered as reagents in organic chemistry. Ethylsulfuric acid is an intermediate in the industrial hydration of ethylene to give ethanol.

$$CH_2{=}CH_2 + H_2SO_4 \xrightarrow{0°C} CH_3CH_2OSO_3H \xrightarrow{H_2O} CH_3CH_2OH + H_2SO_4$$

Dimethyl sulfate and diethyl sulfate are encountered rather more frequently as organic reagents. Both diesters are readily available, inexpensive materials. They are prepared commercially from the corresponding alcohol and sulfuric acid.

$$2\ ROH + H_2SO_4 \longrightarrow ROSO_2OR + H_2O$$

Since alkylsulfuric acids are such strong acids, the alkyl sulfate ion is a good leaving group, roughly comparable to iodide ion. Hence, dimethyl sulfate and diethyl sulfate readily enter into S_N2 displacement processes (Section 9.6).

$$(CH_3)_2CHCH_2O^- + CH_3OSO_2OCH_3 \longrightarrow (CH_3)_2CHCH_2OCH_3 + CH_3OSO_3^-$$

They are used in organic chemistry mainly for this purpose—as alkylating agents.

EXERCISE 26.6 Fluorosulfonic acid, FSO_3H, is sometimes used as a reagent in organic synthesis. Write Kekulé structures for this substance and explain why it is a much stronger acid than sulfuric acid.

26.5 Sulfonic Acids

Sulfonic acids contain the functional group SO_3H joined to carbon. They are named as alkane**sulfonic acids** or arene**sulfonic acids.**

CH_3SO_2OH (S=O above and below) — methanesulfonic acid

$(CH_3)_2CHSO_2OH$ — 2-propanesulfonic acid

CF_3SO_2OH — trifluoromethanesulfonic acid

$C_6H_5SO_3H$ — benzenesulfonic acid

$p\text{-}CH_3C_6H_4SO_3H$ — *p*-toluenesulfonic acid

Sulfonic acids are strong acids, as strong as typical inorganic acids.

$$CH_3CH_2SO_2OH \rightleftharpoons H^+ + CH_3CH_2SO_3^-$$

Because of the inductive effect of the fluorines, trifluoromethanesulfonic acid is much more acidic and, indeed, is one of the strongest acids known.

A. Alkanesulfonic Acids

Alkanesulfonic acids can be prepared by nucleophilic displacement of alkyl halides with bisulfite ion, an ambident anion. Because of the greater nucleophilicity of sulfur, alkylation occurs primarily on sulfur rather than on oxygen. The initial product is the salt of the sulfonic acid, which is converted into the sulfonic acid by treatment with strong acid.

$$(CH_3)_2CHCH_2CH_2Br + HO{-}\ddot{S}(=O){-}O^-\,Na^+ \xrightarrow{H_2O} (CH_3)_2CHCH_2CH_2SO_2O^-\,Na^+$$

(96%)
sodium 3-methylbutanesulfonate

Sodium salts of α-hydroxysulfonic acids are obtained by the addition of sodium bisulfite to aldehydes and some ketones.

$$CH_3CHO + NaHSO_3 \longrightarrow CH_3CH(OH){-}SO_2O^-\,Na^+$$

(89%)
sodium 1-hydroxyethanesulfonate

$$CH_3COCH_3 + NaHSO_3 \longrightarrow (CH_3)_2C(OH)SO_3^-\,Na^+$$

(59%)
sodium 2-hydroxy-2-propanesulfonate

Sulfonic acid esters are best prepared from sulfonyl chlorides, which are obtained from the sodium sulfonates by treatment with phosphorus pentachloride (PCl_5) or thionyl chloride ($SOCl_2$).

$$CH_3CH_2SO_2O^-\,Na^+ + PCl_5 \longrightarrow CH_3CH_2SO_2Cl + POCl_3 + NaCl$$

ethanesulfonyl chloride

$$CH_3CH_2SO_2Cl + CH_3O^-\,Na^+ \longrightarrow CH_3CH_2SO_2OCH_3 + NaCl$$

methyl ethanesulfonate

EXERCISE 26.7 Outline a synthesis of methyl 3-methylbutanesulfonate starting with 1-chloro-3-methylbutane.

As with the alkyl sulfates, the alkanesulfonates are potent alkylating agents because the sulfonate ion is a reactive leaving group. One class of alkanesulfonates in common use is the esters of methanesulfonic acid, which are prepared from methanesulfonyl chloride (''mesyl chloride''), an inexpensive commercial material. Methanesulfonates, frequently called ''mesylates,'' are used in substitution and elimination processes in the same way as alkyl halides.

$$C_6H_{11}OH + CH_3SO_2Cl \xrightarrow[0°C]{\text{pyridine}} C_6H_{11}OSO_2CH_3$$

methanesulfonyl chloride "mesyl chloride" — cyclohexyl methanesulfonate

$$C_6H_{11}OSO_2CH_3 + CH_3CO_2^-\ K^+ \xrightarrow{\text{acetone}} C_6H_{11}OCCH_3 + CH_3SO_3^-\ K^+$$

(product: $C_6H_{11}OC(=O)CH_3$)

B. Arenesulfonic Acids

Aromatic sulfonic acids are more common than the aliphatic acids because of their availability through electrophilic sulfonation reactions.

In the reactions of alkenes with sulfuric acid, the acid acts primarily as a protonating reagent to produce a carbocation that reacts with any nucleophile present (Section 11.6.C). We have seen that benzene itself undergoes reversible protonation in sulfuric acid (Section 23.2). However, to be detected, the resulting proton exchange must be followed by means of a hydrogen isotope.

$$C_6H_6 + H^+ \rightleftharpoons C_6H_6 + H^+$$

In concentrated sulfuric acid, a substantial amount of sulfur trioxide, SO_3, is present as a result of the following equilibrium.

$$2\ H_2SO_4 \rightleftharpoons SO_3 + H_3O^+ + HSO_4^-$$

Sulfur trioxide is electrophilic, and its reaction with benzene to give benzenesulfonic acid is a useful and important one. In practice, the reaction is usually carried out with fuming sulfuric acid, a solution of sulfur trioxide in sulfuric acid.

$$C_6H_6 + SO_3 \xrightarrow[40°C]{H_2SO_4} C_6H_5SO_3H$$

benzenesulfonic acid

Sulfur trioxide, SO_3, exists in several allotropic forms. The so-called α and β forms are polymers that form long fibrous needles. The monomer is a liquid, called the γ form; it is available commercially with an inhibitor which is added to prevent polymerization. Sulfur trioxide is prepared by the catalytic oxidation of sulfur dioxide with oxygen. Sulfur trioxide is the anhydride of sulfuric acid and reacts vigorously with water with evolution of much heat. The reaction with heavy water, D_2O, is used to prepare D_2SO_4. Sulfuric acid is prepared commercially by dissolving sulfur trioxide in sulfuric acid to produce "fuming sulfuric acid." Commercial fuming sulfuric acid contains 7–8% of SO_3. Dilution with water gives ordinary concentrated sulfuric acid.

Sulfonation is a general reaction, and occurs with substituted benzenes as well as with benzene itself.

Cl / fuming H_2SO_4, Δ → Cl / SO_3H

CH_3 / conc. H_2SO_4, 0°C → CH_3 / SO_3H (43%) + CH_3 / SO_3H (4%) + CH_3 / SO_3H (53%)

Note that milder reaction conditions suffice to bring about sulfonation of toluene, which is more reactive than either chlorobenzene or benzene because of the electron-donating effect of the methyl group. The isomer distribution often depends on the exact experimental conditions. For example, at 100 °C a typical product composition from toluene is 13% *ortho*, 8% *meta*, and 79% *para*. The sulfonation reaction is reversible, and the product depends on whether the reaction conditions favor kinetic or thermodynamic control. In the sulfonation of toluene at low temperature, the reaction product is the product of kinetic control; that is, the product composition reflects relative energies of transition states. At higher temperatures the reverse reaction has a significant rate, and the reaction takes on the aspects of an equilibrium.

$$\text{Ar—H} + SO_3 \rightleftharpoons \text{Ar—}SO_3H$$

important at high temperatures

The sulfonic group is a rather bulky group, and steric interaction with *ortho* substituents is significant. At equilibrium the relatively unhindered *p*-toluenesulfonic acid dominates over *o*-toluenesulfonic acid. Such steric effects are much less evident in the transition state for sulfonation, because the carbon-sulfur bond is not yet completely formed. Hence, *o*- and *p*-toluenesulfonic acids are formed at comparable rates.

According to the principle of microscopic reversibility (page 107), the back reaction must be the exact reverse of the forward reaction. The forward reaction is a reaction with sulfur trioxide to form a dipolar neutral intermediate that loses a proton to form the arenesulfonate ion.

$$\text{C}_6\text{H}_6 + SO_3 \rightleftharpoons [\text{C}_6\text{H}_6SO_3]^{\pm} \rightleftharpoons \text{C}_6\text{H}_5SO_3^- + H^+$$

Consequently, the reverse reaction involves reaction by a proton at a ring carbon of the sulfonate ion. This reverse reaction is faster for a more hindered *ortho*-substituted sulfonic acid because steric congestion effects are relieved in the dipolar intermediate.

EXERCISE 26.8 Does the product composition given above for sulfonation of toluene at 100 °C correspond to the equilibrium composition?

Reversal of sulfonation can be accomplished by heating the sulfonic acid in dilute aqueous sulfuric acid. In this way the sulfonic acid group can serve as a protecting group to direct aromatic substitution into other positions. An example of this strategy is provided by the following preparation of pure *o*-bromophenol.

$$\text{C}_6\text{H}_5OH \xrightarrow[100°C]{H_2SO_4} \text{2,4-}(SO_3H)_2\text{C}_6\text{H}_3OH \xrightarrow[H_2O]{Br_2,\ OH^-} \text{2-Br-4,6-}(SO_3H)_2\text{C}_6\text{H}_2OH \xrightarrow[100°C]{H_3O^+} o\text{-BrC}_6\text{H}_4OH$$

(40–43%)

In this example, the highly reactive aromatic ring is sulfonated at the *ortho* and *para* positions, relative to the activating hydroxy group. Bromine is then introduced at the remaining *ortho* position (which is also *meta* to the sulfonic acid groups). Finally, sulfonation is reversed to obtain the pure *o*-bromophenol.

The sulfonic acid function is deactivating and strongly *meta*-directing in electrophilic aromatic substitution reactions. This is easy to understand in light of the Lewis structure of benzenesulfonic acid, which has a doubly positive sulfur adjacent to the ring. Because of this deactivation, introduction of a second sulfonic acid group into an aromatic ring is much more difficult than introduction of the first one. Benzene can be disulfonated by the use of hot 20% fuming sulfuric acid (sulfuric acid containing 20% sulfur trioxide).

$$\text{C}_6\text{H}_6 \xrightarrow[90°C]{20\%\ \text{fuming}\ H_2SO_4} m\text{-C}_6\text{H}_4(SO_3H)_2$$

(90%)
m-benzenedisulfonic acid

As was shown by the foregoing example of sulfonation of phenol, aromatic amines and phenols undergo sulfonation readily. Aniline undergoes sulfonation first on the nitrogen; the initial product must be heated to obtain the ring-sulfonated product.

$$\text{C}_6\text{H}_5NH_2 \xrightarrow[180\text{–}190°C]{H_2SO_4} [\text{C}_6\text{H}_5NHSO_3H] \longrightarrow p\text{-}H_3N^+\text{C}_6\text{H}_4SO_3^-$$

sulfanilic acid

Sulfanilic acid contains an acidic and a basic group in the same molecule and exists in the **zwitterionic** or **internal salt** form.

Arenesulfonyl chlorides may be prepared by sulfonation of aromatic compounds with chlorosulfonic acid, $ClSO_3H$. An example is chlorosulfonation of acetanilide.

$$C_6H_5NHCOCH_3 \xrightarrow{ClSO_3H} p\text{-}CH_3CONHC_6H_4SO_2Cl$$

acetanilide — *p*-acetamidobenzenesulfonyl chloride

Arenesulfonic acids are strong acids, about as strong as hydrochloric acid. They are completely dissociated in aqueous solution, and are normally rather water soluble. Indeed, their water solubility presents problems in isolations. Consequently, the products of sulfonation reactions are usually isolated as salts. The sodium salts, like most sodium salts, are also water soluble, but they are generally not as soluble as sodium sulfate or sodium chloride. The less soluble sodium arenesulfonates can usually be ''salted out'' by saturation of the aqueous solution with sodium sulfate or sodium chloride.

$$ArSO_3^- + Na^+ \rightleftharpoons ArSO_3^-\,Na^+\downarrow$$

The sodium salts of benzenesulfonic acid that have a long alkyl side chain are detergents and surfactants (surface-active compounds). The sulfonate end is hydrophilic and the alkane end is hydrophobic; the amphiphilic combination forms micelles that can solubilize hydrophobic lipid materials in water or hydrophilic polar materials in hydrocarbons (Section 18.4.D). One can disperse a material (particle sizes 0.1 μ to 10 μ) in a liquid in which it is not soluble by using a surface-active or emulsifying agent.

At one time the alkane side chain was made by the carbocation polymerization of propylene to give a tetrameric olefin that was used to alkylate benzene; this ''alkylate'' was then sulfonated to give the product, which was widely used in many common household detergents.

$$CH_3CH{=}CH_2 \longrightarrow CH_3\overset{CH_3}{\overset{|}{C}}H(CH_2\overset{CH_3}{\overset{|}{C}}H)_2CH{=}\overset{CH_3}{\overset{|}{C}}H \xrightarrow[\text{benzene}]{AlCl_3}$$

$$CH_3\overset{CH}{\overset{|}{C}}H(CH_2\overset{CH_3}{\overset{|}{C}}H)_2CH_2\overset{CH_3}{\overset{|}{C}}H{-}C_6H_5 \longrightarrow CH_3(\overset{CH_3}{\overset{|}{C}}HCH_2)_3\overset{CH_3}{\overset{|}{C}}H{-}C_6H_4{-}SO_3^-\,Na^+$$

The widespread use of large quantities of this material caused problems in the purification of sewage effluent because the branched chains were only slowly biodegradable. The detergent industry has now completely replaced this product with one prepared from a mixture of straight-chain C_{12}–C_{15} alkanes. The hydrocarbon mixture is chlorinated and used for Friedel-Crafts alkylation of benzene. The resulting mixture of phenylalkanes is sulfonated to give a product that has the straight chain alkyl group necessary for rapid biodegradability by bacteria.

$$\begin{matrix} CH_3(CH_2)_x \\ \diagdown \\ CH{-}C_6H_4{-}SO_3^-\,Na^+ \\ \diagup \\ CH_3(CH_2)_y \end{matrix}$$

modern household detergent

p-Toluenesulfonic acid is readily available as the crystalline monohydrate, $C_7H_7SO_3^-\ H_3O^+$, m.p. 105 °C. It is prepared by salting out the product from sulfonation of toluene with concentrated hydrochloric acid. Alternatively, the barium salt is treated with the stoichiometric amount of sulfuric acid, and the insoluble barium sulfate is filtered. The filtrate is a strong acid, and concentrated solutions will dehydrate cellulose (filter paper!) just like sulfuric acid. *p*-Toluenesulfonic acid is used as an acid catalyst in many organic reactions.

EXERCISE 26.9 *p*-Chlorobenzenesulfonic acid is a relatively inexpensive, commercially available material. Suggest a way in which it can be used to prepare 2,6-dinitrochlorobenzene.

The sulfonate group in aromatic sulfonic acid can be replaced by **nucleophilic aromatic substitution** reactions. The conditions are drastic: fusion with alkali hydroxide or other salts at temperatures of 200–350 °C.

$SO_3^-\ Na^+$; CH_3 — NaOH–KOH, 230–330°C → H^+ → OH ; CH_3

sodium *p*-toluenesulfonate

(63–72%)
p-cresol

$SO_3^-\ Na^+$ — NaCN, 285–300°C → CN

sodium 1-naphthalenesulfonate

(60–70%)
1-naphthonitrile

The method is used primarily for preparing phenols and nitriles. The necessary reaction conditions are tolerated by few other functional groups; hence, the scope of the reaction is limited. Other nucleophilic aromatic substitution reactions will be considered in Section 30.3.

An unusual feature of this reaction is the fact that it takes place in a fused-salt medium. Such media are highly polar liquids composed almost wholly of ions. In general, neutral organic compounds are not soluble in such ionic media; hence, we have encountered them only rarely in organic reactions. Fused-salt media are useful in organic chemistry only when the organic compound is itself a salt.

EXERCISE 26.10 Show how toluene can be converted into 2,4-dihydroxytoluene.

As with the aliphatic sulfonic acids, one of the most important reactions of aromatic sulfonic acids is conversion to the corresponding sulfonyl chloride. This reaction is most conveniently carried out on the sodium salt by treatment with PCl_5 or $POCl_3$. The product can be distilled or crystallized from benzene. Benzenesulfonyl chloride is a high boiling liquid, b.p. 251.5 °C, and *p*-toluenesulfonyl chloride is a solid, m.p. 68 °C. Alternatively, the acid chloride may be prepared by direct sulfonation with chlorosulfonic acid as discussed on page 824.

$$C_6H_6 \xrightarrow{ClSO_3H} C_6H_5SO_2Cl \xleftarrow{PCl_5} C_6H_5SO_3H$$

p-Toluenesulfonyl chloride (often called ''tosyl'' chloride) is used to prepare *p*-toluenesulfonate esters (''tosylates'') from alcohols. The procedure involves combining the reagents with excess pyridine at room temperature. Pyridinium chloride separates from solution; the mixture is then added to dilute hydrochloric acid, and the product tosylate is filtered or extracted into ether.

$$ROH + p\text{-}CH_3C_6H_4SO_2Cl + C_5H_5N \longrightarrow ROSO_2\text{-}C_6H_4\text{-}CH_3 + C_5H_5NH^+\ Cl^-$$

an alkyl tosylate

Sulfonyl chlorides are effective in Friedel-Crafts acylations. The products of such acylations are aromatic sulfones (see page 812).

$$C_6H_5SO_2Cl \xrightarrow[C_6H_6]{AlCl_3} C_6H_5SO_2C_6H_5$$

phenyl sulfone

Reaction of sulfonyl chlorides with ammonia or amines gives the corresponding sulfonamides. Many such compounds have important medicinal use as antibacterial agents. Examples are sulfanilamide (*p*-aminobenzenesulfonamide), sulfadiazine [*p*-amino-N-(2-pyrimidyl)benzenesulfonamide], and sulfathiazole [*p*-amino-N-(2-thiazolyl)benzenesulfonamide]. These compounds are in general bacteriostatic rather than bactericidal, inhibiting the formation of folic acid, an essential vitamin, from *p*-aminobenzoic acid. They are now used in combination with an inhibitor for a reductive step in the utilization of folic acid.

$p\text{-}NH_2C_6H_4SO_2NH_2$ — sulfanilamide

$p\text{-}NH_2C_6H_4SO_2NH\text{-}(2\text{-pyrimidyl})$ — sulfadiazine

$p\text{-}NH_2C_6H_4SO_2NH\text{-}(2\text{-thiazolyl})$ — sulfathiazole

Reduction of a sulfonyl chloride with zinc and water gives a sulfinic acid (page 814).

$$p\text{-}CH_3C_6H_4SO_2Cl \xrightarrow[90°C]{Zn,\ H_2O} \xrightarrow[NaOH]{Na_2CO_3} p\text{-}CH_3C_6H_4SO_2^-\ Na^+$$

(64%)
sodium *p*-toluenesulfinate

The salts of sulfinic acids have the interesting property of being ambident nucleophiles (Section 9.4). In their reactions with alkyl halides, reaction occurs mainly on sulfur rather than oxygen to produce sulfones.

$$p\text{-}CH_3C_6H_4\overset{+}{S}(\ddot{O}^-)_2 + CH_3I \longrightarrow p\text{-}CH_3C_6H_4SO_2CH_3 + I^-$$

methyl *p*-tolyl sulfone

Under more strongly reducing conditions sulfonyl chlorides give the corresponding thiols.

EXERCISE 26.11 Sulfonation of bromobenzene gives *p*-bromobenzenesulfonic acid, which is isolated as the sodium salt. This salt is frequently called sodium brosylate, and the brosylate group is sometimes used instead of the analogous tosylate group. How can sodium brosylate be converted into brosyl chloride? Give the reaction product resulting from treatment of *p*-bromobenzenesulfonyl chloride with each of the following sets of reagents.

(a) $CH_3CH_2CH_2CH_2OH$, pyridine
(b) toluene, $AlCl_3$
(c) (i) Zn, water; (ii) Na_2CO_3; (iii) $C_6H_5CH_2Cl$
(d) Zn, H_2SO_4

EXERCISE 26.12 Sulfonium salts with three different groups, $RR'R''S^+$, are chiral and have inversion barriers that are sufficiently high (25–30 kcal $mole^{-1}$) that enantiomerically homogeneous samples can be obtained at room temperature. Which two of the following compounds are also chiral?

(a) methyl phenyl sulfide
(b) methyl phenyl sulfoxide
(c) methyl phenyl sulfone
(d) methyl benzenesulfonate
(e) methyl benzenesulfinate

Of the three achiral compounds, propose chiral analogs that incorporate isotopes.

Highlight 26.2

Oxidized sulfur groups give rise to somewhat stable anions on carbon, moderately stable anions on nitrogen, and very stable anions on oxygen. Examples are

$$CH_3{-}SO{-}CH_3 \rightleftharpoons CH_3{-}SO{-}CH_2^- \qquad pK_a \cong 35$$

dimethyl sulfoxide

$$CH_3{-}SO_2{-}CH_3 \rightleftharpoons CH_3{-}SO_2{-}CH_2^- \qquad pK_a \cong 31$$

dimethyl sulfone

$$CH_3{-}SO_2{-}NH_2 \rightleftharpoons CH_3{-}SO_2{-}NH^- \qquad pK_a \cong 12$$

methanesulfonamide

$$CH_3{-}S{-}OH \rightleftharpoons CH_3{-}S{-}O^- \qquad pK_a < 10$$

methanesulfenic acid

$$CH_3{-}SO{-}OH \rightleftharpoons CH_3{-}SO{-}O^- \qquad pK_a \cong 2$$

methanesulfinic acid

$$CH_3{-}SO_2{-}OH \rightleftharpoons CH_3{-}SO_2{-}O^- \qquad pK_a \cong -6$$

methanesulfonic acid

26.6 Phosphines and Phosphonium Salts

Phosphines are the phosphorus analogs of amines. They are named by appending the suffix **-phosphine** to the stem names of the alkyl groups attached to phosphorus.

$CH_3CH_2PH_2$	$(CH_3CH_2CH_2CH_2)_2PH$	$(C_6H_5)_3P$
ethylphosphine	di-*n*-butylphosphine	triphenylphosphine

Tertiary phosphines are the most important. They are conveniently prepared by reaction of Grignard reagents with phosphorus trichloride.

$$\text{C}_6\text{H}_{11}\text{—MgBr} + PCl_3 \xrightarrow[\Delta]{\text{ether}} (\text{C}_6\text{H}_{11})_3\text{P}$$

tricyclohexylphosphine

Phosphines are more highly pyramidal than amines, but there is considerable variation in the bond angles about phosphorus. Trimethylphosphine has a C—P—C angle of 99°, somewhat expanded from the H—P—H angle of 93° in phosphine itself. Triphenylphosphine has a C—P—C angle of 103°.

Since phosphines are analogs of amines, we expect the phosphorus lone pair to show characteristic basic properties. Phosphines do act as Lewis bases, but the base strength is strongly dependent on structure, particularly on the degree of substitution at phosphorus. The changes may depend in part on bond-angle variations between the pyramidal phosphine and tetrahedral phosphonium salt. Representative pK_a values for the protonated forms of some phosphines are collected in Table 26.3. Comparison of the pK_as in Table 26.3 with the pK_as of amines given in Table 24.4 (page 733) shows that tertiary phosphines are about 100-fold less basic than the corresponding amines. For secondary and primary phosphines the difference is much greater. Phosphine itself, PH_3, shows no basic properties whatsoever; the pK_a of PH_4^+ has been estimated to be -14! Like the analogous amines, phenylphosphines are less basic than the alkyl compounds.

Even though phosphines are considerably less basic than amines, the phosphorus is highly polarizable, and phosphines are highly nucleophilic and readily participate in S_N2 reactions.

$$(C_6H_5)_3P + C_6H_5CH_2Cl \longrightarrow (C_6H_5)_3\overset{+}{P}CH_2C_6H_5\ Cl^-$$

benzyltriphenylphosphonium chloride

TABLE 26.3 Acidity of Phosphonium Ions

R_3PH^+	pK_a
$(C_2H_5)_3PH^+$	8.69
$(n\text{-}C_4H_9)_3PH^+$	8.43
$(C_4H_9)_2PH_2^+$	4.51
$(CH_3)_2PH_2^+$	3.91
$i\text{-}C_4H_9PH_3^+$	−0.02
$n\text{-}C_8H_{17}PH_3^+$	0.43
$(C_6H_5)_3PH^+$	2.73
$(C_6H_5)_2PH_2^+$	0.03

Phosphonium salts are most important for their use as reagents in the Wittig synthesis of alkenes (Section 14.7.D).

Tertiary phosphines also act as good donor ligands toward metals and are commonly used in the preparation of organometallic complexes such as tris(phenylphosphine) rhodium(I) chloride (Section 36.1).

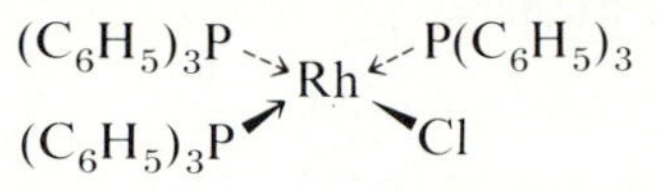

$$\begin{array}{ccc} (C_6H_5)_3P \dashrightarrow & \text{Rh} & \dashleftarrow P(C_6H_5)_3 \\ (C_6H_5)_3P \nearrow & & \searrow Cl \end{array}$$

Trialkylphosphines are readily oxidized, even by air, to the corresponding phosphine oxides.

$$(n\text{-}C_4H_9)_3P \xrightarrow{[O]} (n\text{-}C_4H_9)_3\overset{+}{P}—O^-$$

tri-*n*-butylphosphine → tri-*n*-butylphosphine oxide

EXERCISE 26.13 Review the discussion of the Wittig reaction in Section 14.7.D. Write the equations illustrating the use of benzyltriphenylphosphonium chloride to prepare several alkenes. How could the triphenylphosphine that is used in this sequence be prepared?

26.7 Phosphate and Phosphonate Esters

There are several oxyacids of phosphorus. The most common one is orthophosphoric acid, more commonly called simply phosphoric acid, H_3PO_4. When orthophosphoric acid is heated above 210 °C, it loses water with the formation of pyrophosphoric acid, which may be regarded as an anhydride of phosphoric acid. In phosphorus-oxygen derivatives, the P=O bonds are more accurately written as $P^+—O^-$, a normal Lewis structure that accounts for acidity, but for convenience are usually shown as P=O.

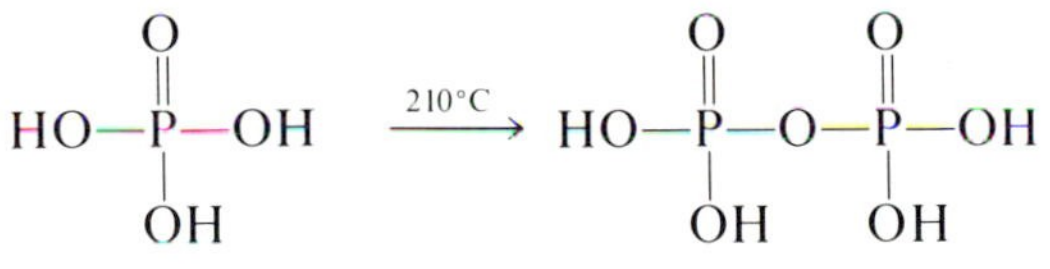

$$HO—\underset{OH}{\overset{O}{\overset{\|}{P}}}—OH \xrightarrow{210°C} HO—\underset{OH}{\overset{O}{\overset{\|}{P}}}—O—\underset{OH}{\overset{O}{\overset{\|}{P}}}—OH$$

orthophosphoric acid → pyrophosphoric acid

"Polyphosphoric acid" (PPA) is a mixture of phosphoric anhydrides that is prepared by heating H_3PO_4 with phosphorus pentoxide, P_2O_5. It consists of about 55% triphosphoric acid, the remainder being H_3PO_4 and higher polyphosphoric acids.

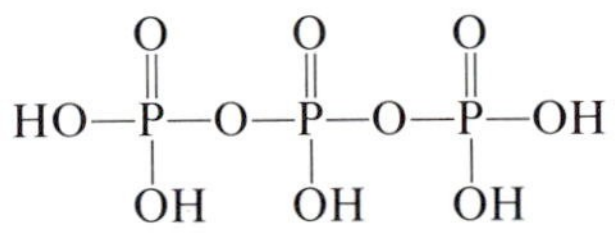

$$HO—\underset{OH}{\overset{O}{\overset{\|}{P}}}—O—\underset{OH}{\overset{O}{\overset{\|}{P}}}—O—\underset{OH}{\overset{O}{\overset{\|}{P}}}—OH$$

triphosphoric acid

Polyphosphoric acid is sometimes used as an acid catalyst in organic reactions. Derivatives of triphosphoric acid, such as adenosine triphosphate (ATP) (page 834) and guanosine triphosphate (GTP), play a central role in the control and growth of biological cells.

Orthophosphoric acid is a tribasic acid having $pK_1 = 2.15$, $pK_2 = 7.20$, and $pK_3 = 12.38$.

$$H_3PO_4 \rightleftharpoons H^+ + H_2PO_4^- \qquad K_a = 7.1 \times 10^{-3}\ M$$
$$H_2PO_4^- \rightleftharpoons H^+ + HPO_4^{2-} \qquad K_a = 6.3 \times 10^{-8}\ M$$
$$HPO_4^{2-} \rightleftharpoons H^+ + PO_4^{3-} \qquad K_a = 4.2 \times 10^{-13}\ M$$

It may form mono-, di-, and triesters.

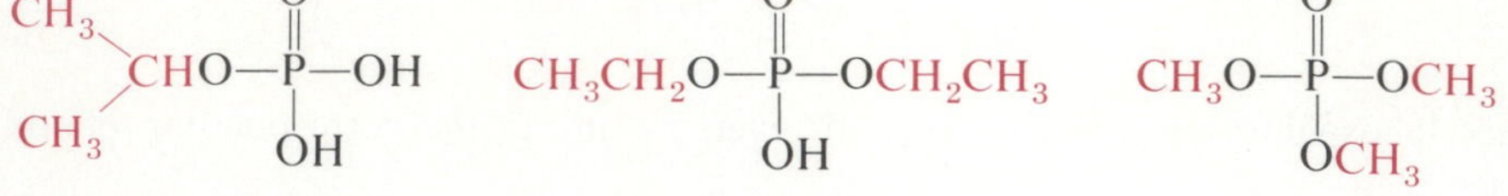

$(CH_3)_2CHO-P(=O)(OH)-OH$ isopropyl phosphate

$CH_3CH_2O-P(=O)(OH)-OCH_2CH_3$ diethyl phosphate

$CH_3O-P(=O)(OCH_3)-OCH_3$ trimethyl phosphate

The mono- and diesters still contain OH groups and have acidic properties. They are actually stronger acids than phosphoric acid itself.

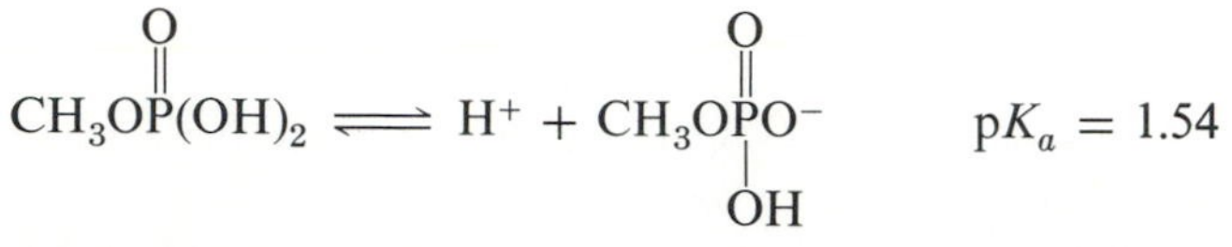

$$CH_3OP(=O)(OH)_2 \rightleftharpoons H^+ + CH_3OP(=O)(OH)O^- \qquad pK_a = 1.54$$

methyl phosphate

$$CH_3OP(=O)(OCH_3)OH \rightleftharpoons H^+ + CH_3OP(=O)(OCH_3)O^- \qquad pK_a = 1.29$$

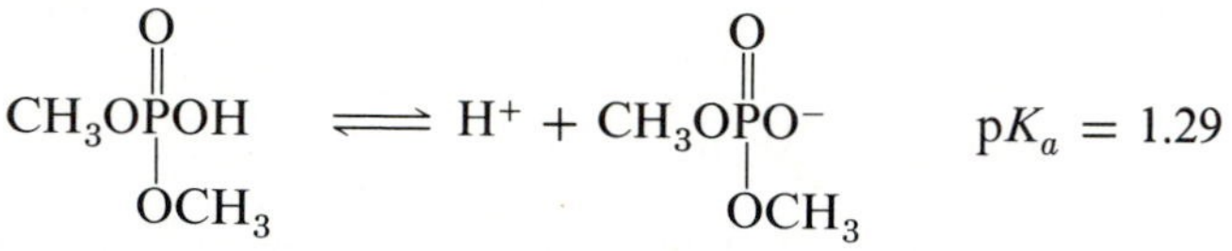

dimethyl phosphate

Analogous esters are possible for pyrophosphoric acid, but the most common are the monoesters such as cyclohexyl pyrophosphate. Tetraethyl pyrophosphate had some use as an insecticide, but is very toxic for mammals (LD_{50} for rats, about 1.1 mg kg^{-1}; LD_{50} = a dose that is lethal to 50% of the test subjects, given as mg per kg of animal).

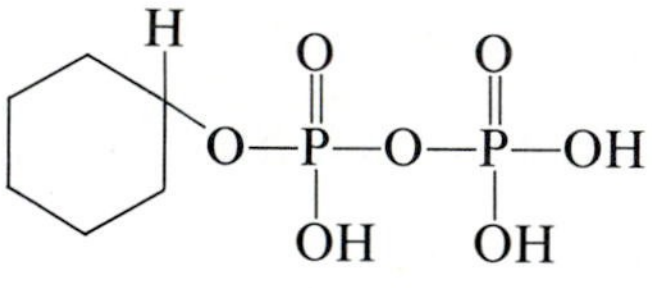

$C_6H_{11}O-P(=O)(OH)-O-P(=O)(OH)-OH$

cyclohexyl pyrophosphate

Phosphate triesters are commonly prepared from the alcohol and phosphorus oxychloride, which is the acyl halide corresponding to phosphoric acid.

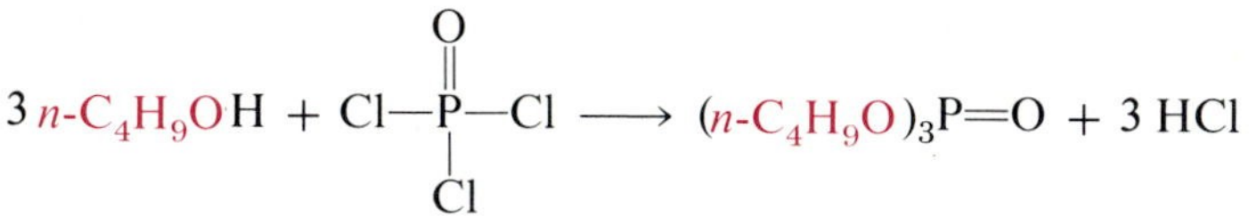

$$3\,n\text{-}C_4H_9OH + Cl-P(=O)(Cl)-Cl \longrightarrow (n\text{-}C_4H_9O)_3P{=}O + 3\,HCl$$

tributyl phosphate

Compounds prepared by the replacement of two of the chlorines of $POCl_3$ (phosphorochloridates) can be used in a similar way to prepare mixed phosphates.

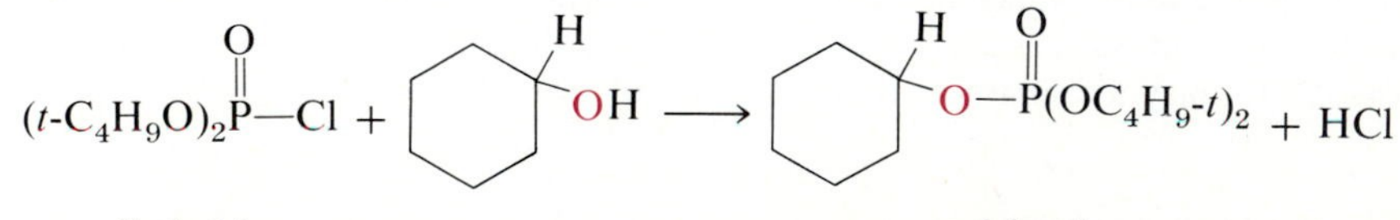

$$(t\text{-}C_4H_9O)_2P(=O)-Cl + C_6H_{11}OH \longrightarrow C_6H_{11}O-P(=O)(OC_4H_9\text{-}t)_2 + HCl$$

di-t-butyl phosphorochloridate — cyclohexyl di-t-butyl phosphate

The only reaction of phosphate esters that we shall consider here is hydrolysis. Hydrolysis may be either acid- or base-catalyzed and may involve either carbon-oxygen or phosphorus-oxygen bond rupture. Under basic conditions hydrolysis occurs mainly by an addition-elimination mechanism, similar to that involved in the hydrolysis of carboxylic acid esters.

$$CH_3O{-}\overset{\overset{O}{\|}}{\underset{\underset{OCH_3}{|}}{P}}{-}OCH_3 + OH^- \rightleftharpoons CH_3O{-}\overset{HO\quad O^-}{\underset{\underset{OCH_3}{|}}{P}}{-}OCH_3 \rightleftharpoons$$

$$CH_3O{-}\overset{\overset{OH}{|}}{\underset{\underset{O}{\|}}{P}}{-}OCH_3 + CH_3O^- \longrightarrow CH_3O{-}\overset{\overset{O}{\|}}{\underset{\underset{O^-}{|}}{P}}{-}OCH_3 + CH_3OH$$

The first alkyl group of a trialkyl phosphate is hydrolyzed most easily with the second and third groups being hydrolyzed rather more sluggishly.

Under acidic conditions carbon-oxygen bond cleavage is the predominant mode of hydrolysis, although P—O rupture is also observed in some cases. Cleavage of the carbon-oxygen bonds can occur by either the S_N2 or the S_N1 mechanism, the former being preferred with primary alkyl phosphates and the latter with tertiary systems.

$$S_N2:\ CH_3O{-}\overset{\overset{O}{\|}}{\underset{\underset{OCH_3}{|}}{P}}{-}OCH_3 + H^+ \rightleftharpoons CH_3O{-}\overset{\overset{OH}{|}}{\underset{\underset{O-CH_3}{|}}{P^+}}{-}OCH_3$$

$$CH_3O{-}\overset{\overset{OH}{|}}{\underset{\underset{O-CH_3\ \leftarrow :\ddot{O}H_2}{|}}{P^+}}{-}OCH_3 \rightleftharpoons CH_3O{-}\overset{\overset{OH}{|}}{\underset{\underset{O}{\|}}{P}}{-}OCH_3 + CH_3OH + H^+$$

$$S_N1:\ t\text{-}C_4H_9O{-}\overset{\overset{O}{\|}}{\underset{\underset{OH}{|}}{P}}{-}OH + H^+ \rightleftharpoons t\text{-}C_4H_9O{-}\overset{\overset{OH}{|}}{\underset{\underset{OH}{|}}{P^+}}{-}OH$$

$$t\text{-}C_4H_9O{-}\overset{\overset{OH}{|}}{\underset{\underset{OH}{|}}{P^+}}{-}OH \rightleftharpoons O{=}\overset{\overset{OH}{|}}{\underset{\underset{OH}{|}}{P}}{-}OH + t\text{-}C_4H_9^+ \xrightarrow{H_2O} t\text{-}C_4H_9OH + H^+$$

As a leaving group in such substitution reactions, phosphate is comparable to bromide ion. The pyrophosphate group is a somewhat better leaving group, being about 100-fold more effective than iodide ion. The pyrophosphate group is an important leaving group in nucleophilic substitution reactions that occur in nature. A number of compounds are built up by plants (**biosynthesized**) from acetic acid units. By a series of enzyme-catalyzed steps, acetic acid is transformed into the compound isopentenyl pyrophosphate, which undergoes enzyme-catalyzed isomerization to γ,γ-dimethylallyl pyrophosphate.

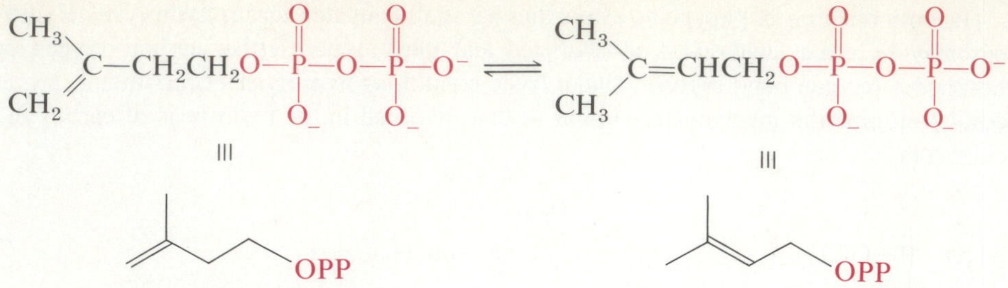

$CH_2{=}C(CH_3){-}CH_2CH_2O{-}P(=O)(O^-){-}O{-}P(=O)(O^-){-}O^- \rightleftharpoons (CH_3)_2C{=}CHCH_2O{-}P(=O)(O^-){-}O{-}P(=O)(O^-){-}O^-$

OPP — isopentenyl pyrophosphate

OPP — γ,γ-dimethylallyl pyrophosphate

Since the pyrophosphate ion is such a good leaving group, γ,γ-dimethylallyl pyrophosphate readily ionizes to give the allylic cation.

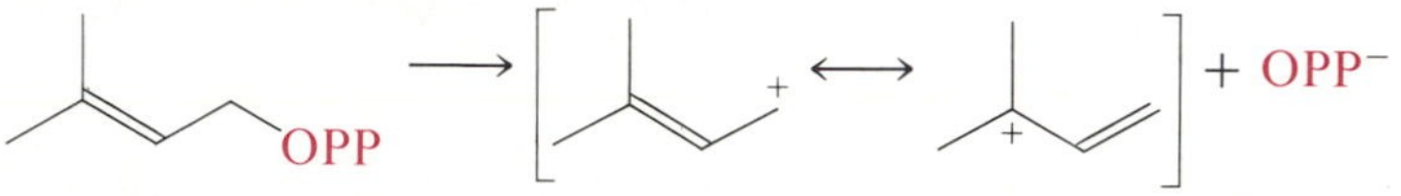

OPP ⟶ [+ ⟷ +] + OPP⁻

The dimethylallyl cation so produced reacts with isopentenyl pyrophosphate to give a new carbocation that eliminates a proton to give geranyl pyrophosphate.

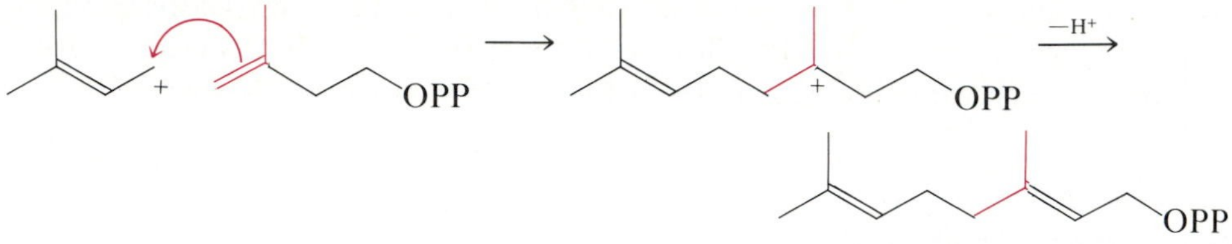

+ OPP ⟶ + OPP $\xrightarrow{-H^+}$

OPP

geranyl pyrophosphate

Although the foregoing reactions are illustrated as simple carbocation reactions, they are undoubtedly under enzyme control. Repetition of these types of reactions leads to more complex structures. Phosphate esters of carbohydrates are also important natural products (Section 28.9). Deoxyribonucleic acid (DNA) contains deoxyribose phosphate units; ribonucleic acid (RNA) includes ribose phosphate units (Section 33.5).

Phosphorous acid, H_3PO_3, is a less important oxyphosphorus acid. Trialkyl phosphites are generally prepared by treatment of phosphorus trichloride with the alcohol and pyridine.

$$PCl_3 + 3\ C_2H_5OH + 3\ C_5H_5N \longrightarrow \underset{\text{triethyl phosphite}}{(C_2H_5O)_3P} + 3\ C_5H_5NH^+\ Cl^-$$

Phosphonic acids contain the functional group PO_3H_2 attached to carbon. They are named as alkylphosphonic acids.

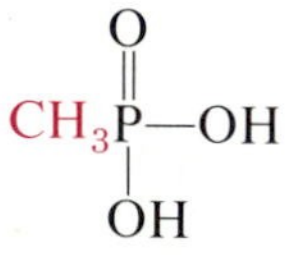

$CH_3P(=O)(OH){-}OH$ — methylphosphonic acid

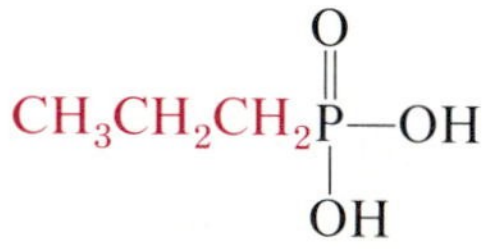

$CH_3CH_2CH_2P(=O)(OH){-}OH$ — *n*-propylphosphonic acid

The most common derivatives are the diesters.

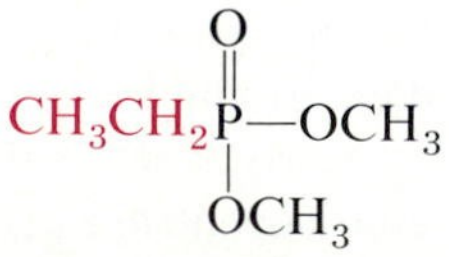

$CH_3CH_2P(=O)(OCH_3){-}OCH_3$ — dimethyl ethylphosphonate

$CH_3CH_2P(=O)(OCH_2CH_3){-}OCH_2CH_3$ — diethyl ethylphosphonate

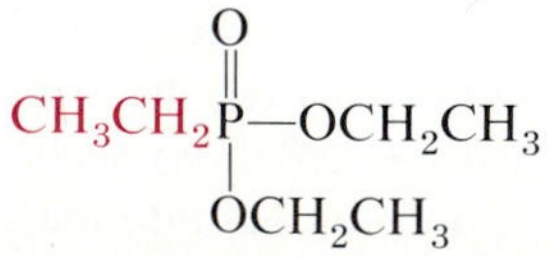

Dialkyl phosphonates are best prepared from trialkyl phosphites by a reaction known as the **Arbuzov-Michaelis reaction.** For example, when trimethyl phosphite is heated at 200 °C with a catalytic amount of methyl iodide, dimethyl methylphosphonate is produced in virtually quantitative yield.

$$\underset{\text{trimethyl phosphite}}{(CH_3O)_3P:} \xrightarrow[200\,°C]{CH_3I} \underset{\text{dimethyl methylphosphonate}}{CH_3\overset{O}{\overset{\|}{P}}(OCH_3)_2}$$

The reaction mechanism involves two successive S_N2 processes. In the first step the nucleophilic phosphorus of the trialkyl phosphite displaces iodide from methyl iodide, giving an alkyltrialkoxyphosphonium salt.

$$(1)\ (CH_3O)_3P: + CH_3{-}I \longrightarrow \underset{\text{methyltrimethoxyphosphonium iodide}}{(CH_3O)_3P^+{-}CH_3\ I^-}$$

The liberated iodide ion attacks one of the methoxy groups in a second S_N2 process, displacing the neutral dialkyl phosphonate.

$$(2)\ I:^- + CH_3{-}O{-}\underset{OCH_3}{\overset{OCH_3}{P^+}}{-}CH_3 \longrightarrow CH_3I + \underset{\text{dimethyl methylphosphonate}}{CH_3{-}\underset{OCH_3}{\overset{O}{\overset{\|}{P}}}{-}OCH_3}$$

Since the alkyl halide is regenerated in the second step, only a catalytic amount is required in order to initiate reaction.

EXERCISE 26.14 A mixture of 1.0 mole of triethyl phosphite and 0.01 mole of methyl iodide is heated at 200 °C. What is the reaction product?

The Arbuzov-Michaelis reaction has been applied to the synthesis of numerous dialkyl phosphonates. If a full equivalent of alkyl halide is used, dialkyl phosphonates having different groups attached to oxygen and phosphorus can be prepared.

$$\underset{\text{triethyl phosphite}}{(CH_3CH_2O)_3P:} + \underset{\text{ethyl bromoacetate}}{BrCH_2COOC_2H_5} \xrightarrow{200\,°C} \underset{\text{triethyl phosphonoacetate}}{C_2H_5O\overset{O}{\overset{\|}{C}}CH_2\overset{O}{\overset{\|}{P}}(OC_2H_5)_2} + CH_3CH_2Br$$

Monoalkyl phosphonates are readily obtained from the diesters by alkaline hydrolysis. Hydrolysis of the second group is more difficult.

$$\underset{\text{dimethyl butylphosphonate}}{CH_3CH_2CH_2CH_2\overset{O}{\overset{\|}{P}}(OCH_3)_2} + NaOH \xrightarrow[\Delta]{C_2H_5OH}$$

$$\underset{\text{sodium methyl butylphosphonate}}{CH_3CH_2CH_2CH_2\underset{OCH_3}{\overset{O}{\overset{\|}{P}}}O^-Na^+} \xrightarrow{\text{dil. } H_2SO_4} \underset{\text{methyl butylphosphonate}}{CH_3CH_2CH_2CH_2\underset{OCH_3}{\overset{O}{\overset{\|}{P}}}OH}$$

A phosphonic acid derivative, N-(phosphonomethyl)glycine, is used on a large scale in the form of the isopropylamine salt as the herbicide Roundup.

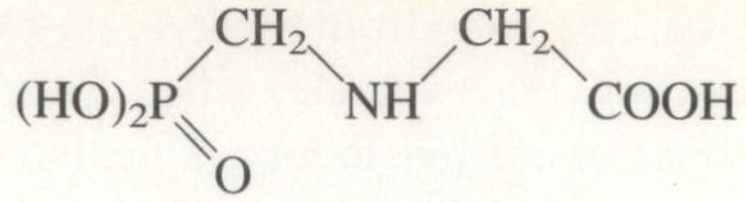

$$(HO)_2P(=O)CH_2NHCH_2COOH$$

N-(phosphonomethyl)glycine
(glyphosate)

EXERCISE 26.15 Write equations illustrating the preparation of the following phosphate and phosphonate esters using $POCl_3$ and PCl_3 as the source of phosphorus.

(a) $CH_3OP(=O)(OC_2H_5)_2$ (b) $(i\text{-}C_4H_9O_3)_3PO$

(c) $CH_3CH_2CH_2P(=O)(OC_2H_5)_2$ (d) $CH_3CH_2P(=O)(OH)OCH_2CH_3$

Phosphoric acid and polyphosphoric acids are particularly important in biochemistry, and in organic chemistry largely as insecticides or as extractants for uranium and related actinide elements. For example, 20% tributyl phosphate in kerosene extracts an isotope of element 91, protoactinium-231 (lifetime 3.28×10^5 years), as $^{231}Pa(V)$ from a solution prepared from the residues of extraction of uranium from Congo ore. Biochemical examples are

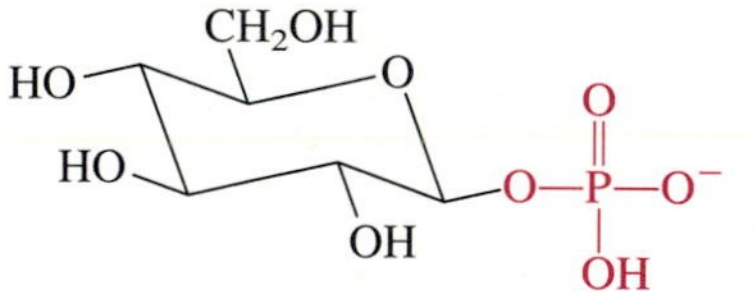

glucose 1-phosphate
(see discussion of sugar phosphates, Section 28.9)

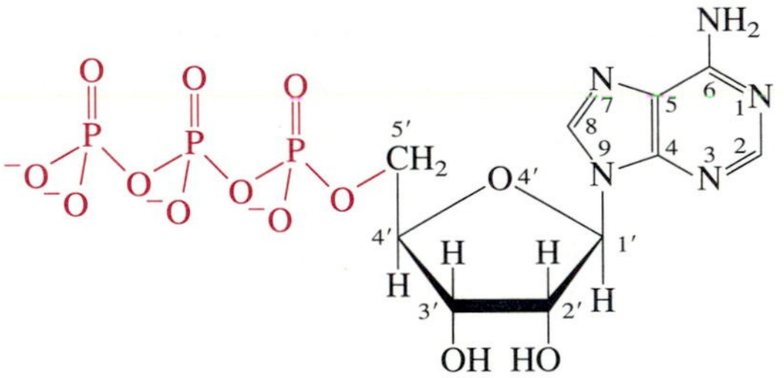

adenosine 5′-triphosphate (ATP)
(see discussion of the synthesis of DNA, Section 33.5.C)

26.8 Sulfur- and Phosphorus-Stabilized Carbanions

Sulfur- and phosphorus-containing functional groups stabilize adjacent carbanions to varying degrees, depending on the exact nature of the function. This property gives rise to carbanionic reagents that have important uses in organic synthesis.

The most well-known and useful compounds in this class are the phosphorus ylides, or Wittig reagents (Section 14.7.D). These compounds are formed by reaction of alkyl phosphonium salts with a strong base such as n-butyllithium.

$$(C_6H_5)_3\overset{+}{P}—CH_3 + n\text{-}C_4H_9Li \longrightarrow (C_6H_5)_3\overset{+}{P}—CH_2^- + n\text{-}C_4H_{10}$$

The bonding in such ylides is still a subject of controversy. At one time, it was thought that the ylide phosphorus-carbon bond could be described as a double bond resulting from overlap of the carbon p-orbital with a phosphorus d-orbital. However, quantum-mechanical calculations suggest that such p-d double bonds are relatively unimportant. As in the case of the sulfur-oxygen bond (page 813), a large part of the bonding in ylides is electrostatic—attraction of the positive phosphorus for the negative carbon. However, this cannot be the complete answer, since alkylammonium salts are not converted into ylides under conditions that suffice to form phosphorus ylides. Electrostatic bonding in the nitrogen ylide should be as important as in a phosphorus ylide.

$$R_3\overset{+}{N}—CH_2^-$$

an ammonium ylide

The higher polarizability of the phosphorus valence electrons is probably involved in stabilizing the dipolar ylide structure. Such polarization can be viewed as an induced dipole on phosphorus.

$$R_3\overset{+}{\underset{\leftarrow\!+}{P}}—CH_2^-$$

Whatever the exact explanation for the bonding in phosphorus ylides, it is quantitatively significant. Although pK_as for simple phosphonium salts have not been measured, salts that also contain a keto or ester function attached to the acidic position are substantially more acidic than 1,3-diketones and 1,3-keto esters (Section 27.7).

$$(C_6H_5)_3\overset{+}{P}—CH_2\overset{O}{\overset{\|}{C}}C_6H_5 \qquad (C_6H_5)_3\overset{+}{P}—CH_2COOC_2H_5$$

pK_a = 6.0   pK_a = 9.1

From these pK_as it may be concluded that alkyltriphenylphosphonium salts are somewhat more acidic than ketones! Since the pK_a of acetone is 19, it may be estimated that the pK_a of methyltriphenylphosphonium ion is on the order of 15-18.

Simple phosphonates may be deprotonated by strong bases such as *n*-butyllithium, and the resulting carbanions add to aldehydes and ketones. Aqueous work-up of the initial adducts affords β-hydroxyphosphonates.

$$CH_3\overset{O}{\overset{\|}{P}}(OC_2H_5)_2 + n\text{-}C_4H_9Li \longrightarrow {}^-CH_2\overset{O}{\overset{\|}{P}}(OC_2H_5)_2 \xrightarrow{CH_3CHO}$$

diethyl methylphosphonate

$$CH_3\overset{O^-}{\overset{|}{C}}HCH_2\overset{O}{\overset{\|}{P}}(OC_2H_5)_2 \xrightarrow{H_2O} CH_3\overset{OH}{\overset{|}{C}}HCH_2\overset{O}{\overset{\|}{P}}(OC_2H_5)_2$$

diethyl 2-hydroxypropylphosphonate

Phosphonates that have a carbonyl group attached to the α-carbon are more acidic, having pK_as of about 15.

$$C_2H_5OOCCH_2\overset{O}{\overset{\|}{P}}(OC_2H_5)_2$$

triethyl phosphonoacetate
estimated p$K_a \cong$ 15

Comparison of this value with the pK_as of malonic ester and acetoacetic ester (Table 27.6, page 889) shows that the dialkylphosphono group, $(RO)_2PO$, is not quite as effective as a carbonyl group at stabilizing an adjacent carbanion. Anions of such activated phosphonates add to aldehydes and ketones, and the resulting products eliminate dialkylphosphonate ion to give α,β-unsaturated esters or ketones (**Horner-Emmons** reaction).

$$(C_2H_5O)_2\overset{O}{\overset{\|}{P}}CH_2CN \xrightarrow{NaH} \xrightarrow{CH_3O-C_6H_4-CHO} CH_3O-C_6H_4-CH{=}CHCN$$

(88%)

EXERCISE 26.16 Write equations showing how trimethyl phosphite can be employed in each of the following multistep transformations.

(a) $CH_3C(=O)CH_3 \longrightarrow (CH_3)_2C(OH)CH_2P(=O)(OCH_3)_2$

(b) $CH_3CHO \longrightarrow CH_3CH{=}CHCOOCH_3$

Sulfur also stabilizes adjacent carbanions. An example is thioanisole, which can be deprotonated by *n*-butyllithium. The resulting lithium compound can be alkylated by primary alkyl halides and also reacts with aldehydes and ketones.

$$C_6H_5SCH_3 \xrightarrow[\text{THF}]{n\text{-}C_4H_9Li} C_6H_5SCH_2Li$$

$$C_6H_5SCH_2Li \xrightarrow[\text{THF, 25°C}]{CH_3(CH_2)_9I} C_6H_5SCH_2(CH_2)_9CH_3 \quad (93\%)$$

$$C_6H_5SCH_2Li \xrightarrow{CH_3CHO} C_6H_5SCH_2CH(OH)CH_3 \quad (85\%)$$

The acetal protons in dithioacetals are activated by two sulfur atoms and are correspondingly more acidic than simple sulfides. An extensively studied member of this class is 1,3-dithiane, the product of reaction of 1,3-propanedithiol and formaldehyde. The pK_a of 1,3-dithiane is less than 40, and it is readily converted into the 1,3-dithianyl anion by treatment with *n*-butyllithium.

$$CH_2{=}O + HS(CH_2)_3SH \xrightarrow{HCl} \text{1,3-dithiane} \xrightarrow{n\text{-BuLi}} \text{2-lithio-1,3-dithiane } (Li^+) + C_4H_{10}$$

1,3-dithiane

Analogous anions may be produced from other 1,3-dithianes.

The 1,3-dithianyl anions are nucleophilic and undergo addition to aldehyde and ketone carbonyl groups. The resulting thioacetals are stable to normal hydrolytic conditions, but hydrolyze easily when treated with mercuric chloride in aqueous acetonitrile.

$$\text{2-methyl-1,3-dithiane} \xrightarrow{n\text{-BuLi}} \xrightarrow[\text{2. } H_3O^+]{\text{1. } C_6H_5CHO} \text{2-methyl-2-(hydroxy(phenyl)methyl)-1,3-dithiane } (91\%) \xrightarrow[H_2O,\ CH_3CN]{HgCl_2} CH_3COCH(OH)C_6H_5 \ (90\%)$$

The overall process constitutes one method for the synthesis of α-hydroxy ketones (Section 27.4.A).

The dithiane method can also be used as a method of preparing certain β-hydroxy ketones. If the dithianyl anion is used as a nucleophile to open an epoxide ring (Section 10.11.A), hydrolysis of the resulting hydroxy dithiane affords the β-hydroxy ketone.

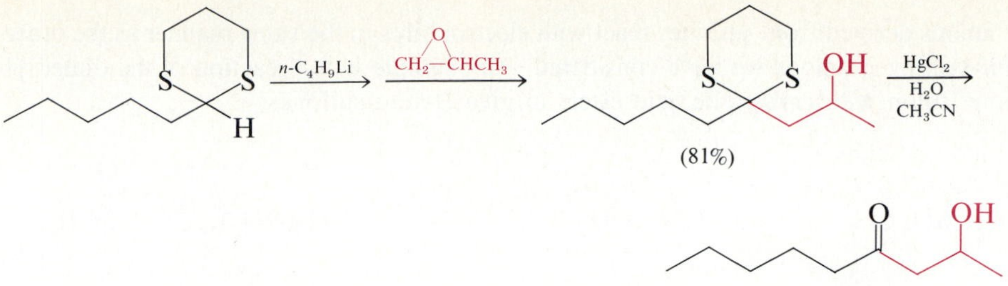

$\xrightarrow{n\text{-}C_4H_9Li}$ $\xrightarrow{CH_2\text{—}CHCH_3 \text{ (epoxide)}}$ (81%) $\xrightarrow[H_2O,\ CH_3CN]{HgCl_2}$ (83%)

2-hydroxy-4-nonanone

EXERCISE 26.17 Outline a synthesis of each of the following compounds by routes employing dithianyl anions.

(a) 4-hydroxyheptan-3-one (b) 4-hydroxyheptan-2-one

Like phosphonium salts, the analogous sulfonium salts are acidic. Because the sulfur bears a positive charge, which aids in stabilizing the conjugate base, sulfonium salts are much more acidic than simple sulfides or even dithioacetals. One way in which the enhanced acidity of sulfonium salts is shown is by base-catalyzed exchange of the acidic protons in deuterated media such as D_2O.

$$(CH_3)_3S^+\ I^- \longrightarrow (CD_3)_3S^+\ I^-$$

Although exact pK_as of such sulfonium ions have not been measured, they are probably on the order of 20. If a strong base such as *n*-butyllithium is used, the sulfonium salt can be converted completely into ylides, which are analogous to the phosphonium ylides.

$$(CH_3)_3S^+\ I^- + n\text{-}C_4H_9Li \xrightarrow[-70°C]{THF} (CH_3)_2\overset{+}{S}\text{—}\overset{-}{C}H_2 + C_4H_{10} + LiI$$

Sulfonium ylides are unstable at temperatures higher than 0 °C. However, at lower temperatures they add to aldehydes and ketones. The initial product is a zwitterion that behaves differently from the zwitterion produced by addition of a phosphonium ylide to a carbonyl compound (Section 14.7.D). In this case, the alkoxide ion acts as the nucleophile in an intramolecular S_N2 process and dimethyl sulfide is the leaving group. The product is an epoxide.

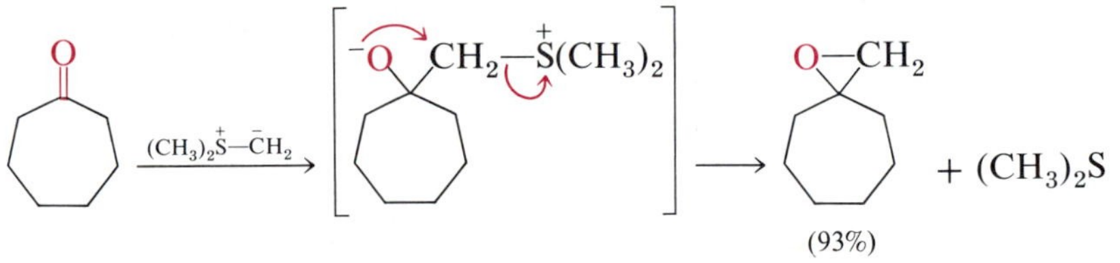

$\xrightarrow{(CH_3)_2\overset{+}{S}\text{—}\overset{-}{C}H_2}$ $\left[{}^-O\ \ CH_2\text{—}\overset{+}{S}(CH_3)_2 \right]$ $\longrightarrow$ O—CH$_2$ $+ (CH_3)_2S$

(93%)

Sulfoxides and sulfones (page 812) are also relatively acidic; the pK_as of dimethyl sulfoxide and dimethyl sulfone are 35 and 31, respectively, in DMSO solution.

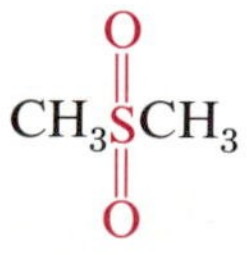

$CH_3\overset{O}{\underset{O}{S}}CH_3$	$CH_3\overset{O}{S}CH_3$
dimethyl sulfone pK_a = 31 (in DMSO)	dimethyl sulfoxide pK_a = 35 (in DMSO)

The anions derived from sulfones react with electrophiles in the same manner as the other sulfur-stabilized anions we have considered. An example is the reaction of the dimethyl sulfone anion with carboxylic acid esters to give β-keto sulfones.

$$CH_3SO_2CH_3 \xrightarrow[DMSO]{NaH} CH_3SO_2CH_2^- \xrightarrow{CH_3(CH_2)_{14}COOCH_3} CH_3(CH_2)_{14}\overset{O}{\overset{\|}{C}}CH_2\underset{\underset{O}{\|}}{\overset{\overset{O}{\|}}{S}}CH_3$$

(83%)

This reaction is analogous to the mixed Claisen condensation (Section 19.9). Note that dimethyl sulfoxide is used as a solvent even though it is also acidic. The difference of 4 pK_a units between dimethyl sulfone and DMSO suffices for synthetic selectivity.

EXERCISE 26.18 What is the structure of the product of each of the following reactions?

(a) dimethyl sulfoxide + n-butyllithium, followed by benzaldehyde
(b) trimethylsulfonium iodide + n-butyllithium, followed by benzaldehyde
(c) 2-phenyl-1,3-dithiane + n-butyllithium, followed by benzaldehyde

Highlight 26.3

Carbanions are useful in synthesis because their reactions with carbon electrophiles result in the formation of carbon-carbon bonds and permit us to build up larger organic molecules from small, readily available ones. Examples of carbanions that we have encountered previously are organometallic reagents (RLi and RMgX), the anions of acetylenes ($RC{\equiv}C^-$), and the enolate ions obtained by deprotonation of carbonyl compounds at their α-positions ($C{=}C{-}O^-$). Carbanions can also be formed with relative ease by removal of a proton from a carbon that is bonded to a polarizable atom like sulfur or phosphorus, especially if the sulfur or phosphorus has a positive charge. The resulting species are called **ylides.**

$$(CH_3)_2\overset{+}{S}{-}CH_3 + n\text{-}C_4H_9Li \longrightarrow (CH_3)_2\overset{+}{S}{-}CH_2^-$$

trimethylsulfonium ion → trimethylsulfonium ylide

$$(C_6H_5)_3\overset{+}{P}{-}CH_3 + n\text{-}C_4H_9Li \longrightarrow (C_6H_5)_3\overset{+}{P}{-}CH_2^-$$

triphenylmethylphosphonium ion → triphenylmethylphosphonium ylide

Both sulfonium and phosphonium ylides display nucleophilic properties and add readily to ketones. However, the intermediates of these two additions suffer different fates. In the sulfonium ion case the alkoxide group *displaces* $(CH_3)_2S$ to give an oxirane.

(axial attack)

$CH_3\overset{+}{S}(CH_3)CH_2^-$ + 4-tert-butylcyclohexanone ⟶

$(CH_3)_2\overset{+}{S}CH_2$–C–$O^-$ intermediate ⟶ epoxide + $(CH_3)_2S$

The analogous intermediate from addition to the phosphonium ylide *adds to phosphorus* to give an intermediate oxephosphetane, which eliminates triphenylphosphine oxide to give the product alkene. This useful process for forming alkenes is known as a **Wittig reaction.**

4-tert-butylcyclohexanone + $^-CH_2—\overset{+}{P}(C_6H_5)_3$ ⟶ O^- … $\overset{+}{P}(C_6H_5)_3$ ⟶

(equatorial attack)

O—$P(C_6H_5)_3$ ⟶ CH_2 alkene + $(C_6H_5)_3\overset{+}{P}$—O

oxephosphetane

26.9 Organosilicon Compounds: Structure and Properties

Because silicon is just below carbon in the periodic table, it is expected that there would be structural similarities in the two classes of compounds, and that analogous physical and chemical properties would be found. To some extent, such similarities exist. For example, tetramethylsilane (b.p. 26.5 °C) and neopentane (b.p. 9.4 °C) are both volatile substances in which the central atom is at the center of a perfect tetrahedron of methyl groups.

$(H_3C)_4Si$ — tetramethylsilane, b.p. 26.5°C

$(H_3C)_4C$ — neopentane, b.p. 9.4°C

The similarity in structure stems from a similarity in bonding; for silicon the orbitals used are Si_{sp^3}, while for carbon they are C_{sp^3}. However, silicon is a third-period element, and utilizes 3*s* and 3*p* orbitals, so its bonds are longer but are roughly comparable in strength to corresponding carbon bonds. For example, in tetramethylsilane the carbon-silicon bond length is 1.89 Å and the bond dissociation energy is 89 kcal mole^{-1}. Comparable values for the carbon-carbon bonds in neopentane are 1.54 Å and 87 kcal mole^{-1}.

Tetramesityldisilene (mesityl = 2,4,6-trimethylphenyl) can be prepared by dimerization of dimesitylsilylene; the Si═Si bond is 2.16 Å in length. Compounds containing a carbon-silicon double bond can be prepared in the vapor phase, and are highly reactive materials that polymerize or react with other materials present. The π-bond of a Si═C bond is weak (39 kcal mole^{-1} versus 65 kcal mole^{-1} for the π-bond of a C═C bond) for several reasons. An important factor is the "mismatch" that results from the π-overlap of Si_{3p} and C_{2p} orbitals. As shown by the following diagram, the Si_{3p} orbital has one more node than a C_{2p} orbital. Although there *is* a bonding interaction between the parallel *p*-orbitals, there is also an antibonding one. In addition, the greater length of a silicon-carbon bond, relative to a carbon-carbon bond, results in less effective π-overlap.

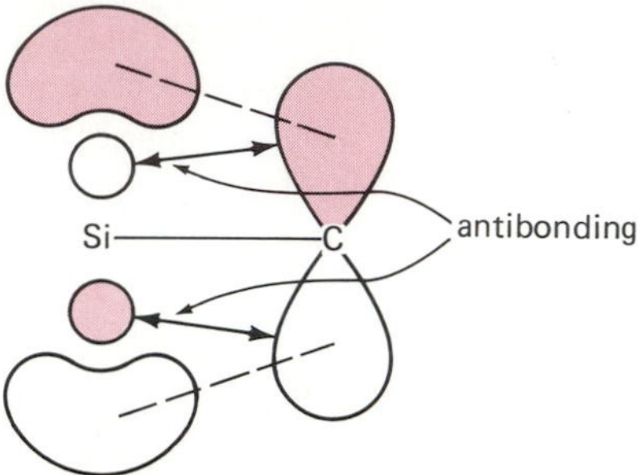

On the other hand, silicon-halogen and silicon-oxygen bonds are especially strong, relative to the analogous carbon-hetero atom bonds. Representative values are given in Table 26.4. The enhanced strength of silicon-heteroatom bonds is largely due to the fact that silicon is more electropositive than carbon (Table 8.6, page 161). As a consequence, there is a substantial coulombic component to the bond energy. In addition, the polarizability of silicon provides an effective method of stabilization of oxygen and halogen lone pairs.

$$[\text{Si—X} \longleftrightarrow \text{Si}^+ \ :\text{X}^-]$$

We shall see that the great strength of silicon-heteroatom bonds plays an important role in the chemistry of organosilicon compounds.

TABLE 26.4 Bond Lengths and Bond-Dissociation Energies for Silicon Compounds

Compound	Bond Length, Å	$DH°$, kcal mole^{-1}
$(CH_3)_3Si$—H		90
$(CH_3)_3Si$—CH_3	1.89	89
$(CH_3)_3Si$—OCH_3	1.48	127
$(CH_3)_3Si$—F	1.60	193
$(CH_3)_3Si$—Cl	2.05	113
$(CH_3)_3Si$—Br	2.21	96
$(CH_3)_3Si$—I	2.44	77

26.10 Organosilicon Compounds: Preparation

The ultimate raw materials for the preparation of silicon compounds are metallic silicon and the silicon halides. The reaction of silicon with alkyl and aryl halides, usually under the influence of copper catalysis, is an important industrial process.

$$2\ CH_3Cl + Si \xrightarrow[250°C]{Cu} (CH_3)_2SiCl_2 + \text{by-products}$$
(70–90%)

Although the foregoing reaction gives a number of by-products, the yield of dichlorodialkylsilane is good, and the reactants are inexpensive. Dichlorodimethylsilane is prepared in tonnage quantity for the production of silicone polymers, which are formed by hydrolysis.

$$(CH_3)_2SiCl_2 + H_2O \longrightarrow \left[-O-\underset{CH_3}{\overset{CH_3}{Si}}-O-\underset{CH_3}{\overset{CH_3}{Si}}- \right]_n$$

Organosilicon compounds are readily available by the reaction of Grignard reagents and organolithium compounds with silicon halides. Reaction of silicon tetrachloride with excess methylmagnesium chloride gives tetramethylsilane.

$$4\ CH_3MgCl + SiCl_4 \longrightarrow (CH_3)_4Si + 4\ MgCl_2$$

However, each substitution of chlorine by alkyl makes the next substitution more difficult, and it is easy to achieve stepwise replacement of the four halogens.

$$CH_3MgCl + SiCl_4 \longrightarrow CH_3SiCl_3 + MgCl_2$$

$$CH_3MgCl + CH_3SiCl_3 \longrightarrow (CH_3)_2SiCl_2 + MgCl_2$$

$$(CH_3)_3C\,Li + (CH_3)_2SiCl_2 \longrightarrow (CH_3)_3C\,\underset{CH_3}{\overset{CH_3}{Si}}Cl + LiCl$$

The last example shows that organosilicon compounds having two or more different alkyl or aryl groups bonded to silicon can be easily prepared.

EXERCISE 26.19 Outline syntheses of the following compounds.

(a) phenyltrimethylsilane
(b) diphenylsilyl dichloride
(c) allyltrimethylsilane
(d) vinyltrimethylsilane

26.11 Organosilicon Compounds: Reactions

A. Nucleophilic Substitution at Silicon

Nucleophilic substitution at silicon is much easier than in corresponding carbon compounds. For example, trimethylsilyl chloride reacts with alcohols at room temperature to give the corresponding ethers; an amine such as triethylamine or pyridine is usually added to neutralize the HCl produced in the reaction.

$$CH_3CH_2OH + (CH_3)_3SiCl \xrightarrow{Et_3N} CH_3CH_2OSi(CH_3)_3 + Et_3\overset{+}{N}H\ Cl^-$$

The high reactivity of the silicon-halogen bond toward nucleophilic substitution is also seen in the fact that trimethylsilyl chloride reacts with enolate ions at −78 °C to give silyl enol ethers.

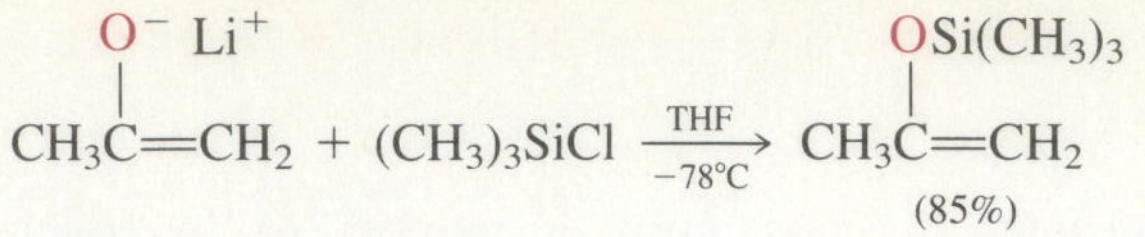

$$\underset{\displaystyle CH_3C{=}CH_2}{\overset{\displaystyle O^-\ Li^+}{|}} + (CH_3)_3SiCl \xrightarrow[-78°C]{THF} \underset{\displaystyle CH_3C{=}CH_2}{\overset{\displaystyle OSi(CH_3)_3}{|}}$$

(85%)

In principle, nucleophilic substitution at silicon could occur by either the S_N1 or S_N2 mechanism. In practice, substitution by the S_N1 mechanism is virtually unknown, not because the R_3Si^+ ion is especially unstable, but because the S_N2-Si mechanism is especially good. The two foregoing examples show that even a tertiary silyl halide is easily replaced under typical S_N2 conditions. In part, this is because the silicon-carbon bonds are much longer than carbon-carbon bonds; thus, steric hindrance effects at silicon are much smaller than at carbon. In addition, the great strength of silicon-heteroatom bonds stabilizes the transition state for nucleophilic substitution:

$$RO^{\delta-}\cdots\cdots Si(CH_3)_3 \cdots\cdots Cl^{\delta-}$$

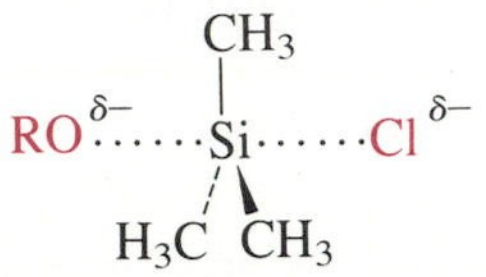

The pentacoordinate structure is probably an intermediate; for example, the carbon-fluorine bond is so strong that the product of addition of fluoride ion to trimethylsilyl fluoride is an isolable, stable species.

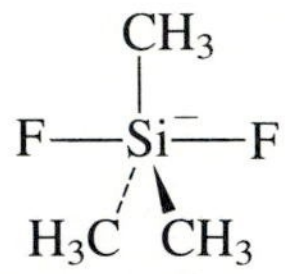

$$F{-}\overset{-}{Si}(CH_3)_3{-}F$$

difluorotrimethylsiliconate ion

The difluorotrimethylsiliconate ion has a trigonal bipyramid geometry, as shown. Its structure is analogous to that of PCl_5 (page 17).

Pentacoordinated species such as the difluorotrimethylsiliconate ion have two kinds of groups attached to silicon—the three that form the base of the bipyramid (the equatorial positions) and the two that form the apices of the bipyramid (the apical positions). The three equatorial ligands are bonded to Si by orbitals that are essentially Si_{sp^2}. The two apical ligands are bonded by weaker bonds to the two lobes of the remaining Si_{3p} orbital. Consequently, the Si-equatorial bond lengths are shorter than the Si-apical bonds (see the bond lengths in PCl_5, page 17).

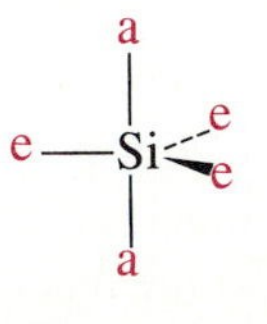

a

e—Si(e)(e)

a

a = apical positions; e = equatorial position

Species such as this are capable of a molecular rearrangement called **pseudorotation,** in which the two apical ligands switch positions with two of the three equatorial ligands. The process is illustrated by the following example, in which ligand 2 may be considered to be the "pivot" for the pseudorotation, which results in the isomerization of the apical ligands 1 and 5 to equatorial positions while the equatorial ligands 3 and 4 adopt apical positions.

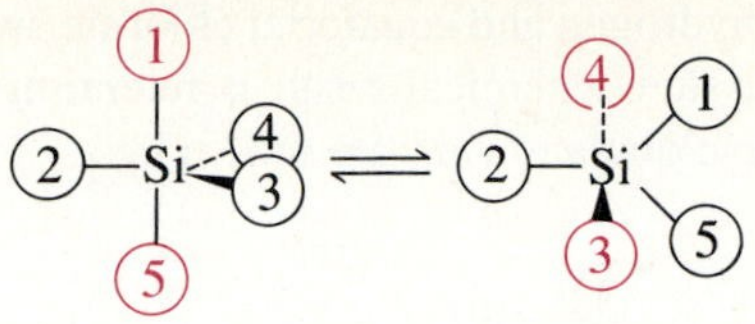

The foregoing pseudorotation is actually a form of molecular vibration in which the 2—Si—1 and 2—Si—5 angles enlarge from 90° to 120° while the 2—Si—3 and 2—Si—4 angles shrink from 120° to 90°.

EXERCISE 26.20 How many pseudorotations are required in order to change the difluorotrimethylsiliconate ion from a structure having both fluorines in apical positions into one having one apical and one equatorial fluorine?

Silicon can be a stereocenter, and the stereochemistry of nucleophilic substitution at silicon has been investigated. Most substitutions proceed with *inversion of configuration,* just as in carbon compounds. The system that has been most investigated is the methylnaphthylphenylsilyl series of compounds (Me = methyl; Ph = phenyl; Np = 1-naphthyl, see Chapter 31).

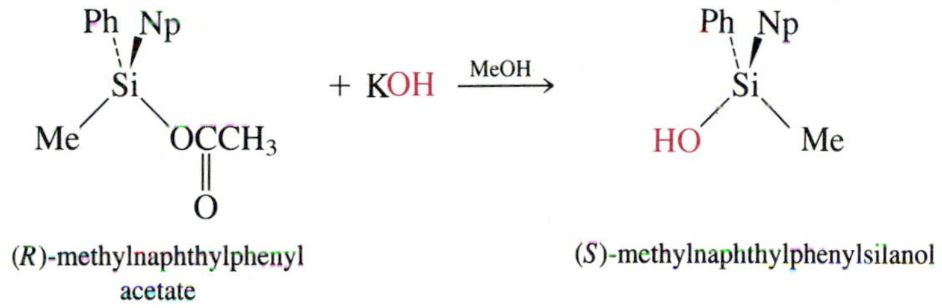

(*R*)-methylnaphthylphenyl acetate + KOH $\xrightarrow{\text{MeOH}}$ (*S*)-methylnaphthylphenylsilanol

However, nucleophilic substitution with *retention of configuration* is also known in organosilicon compounds. An example is seen in the following reaction of a cyclic silane.

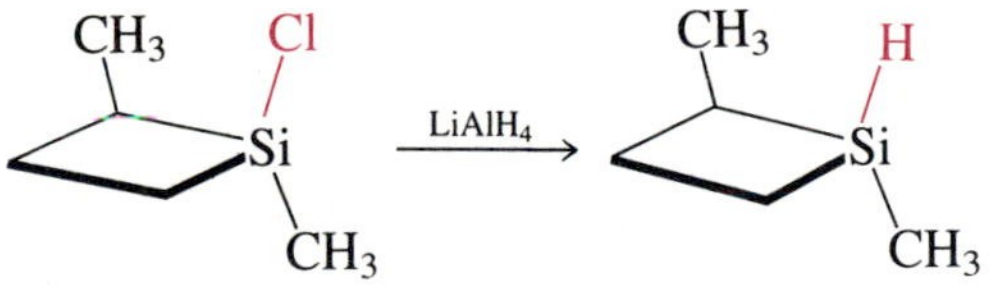

$$\xrightarrow{\text{LiAlH}_4} \qquad (26\text{-}1)$$

The mechanism for the latter kind of substitution is one in which the incoming nucleophile adds to silicon to give a stable trigonal bipyramid intermediate. On the basis of a large number of experiments, the following preference rules have been worked out for substitution reactions that involve the intermediacy of such intermediates.

1. Nucleophiles enter, and leaving groups depart, from apical positions.
2. Electronegative substituents prefer apical positions.
3. Four- and five-membered rings must span one apical and one equatorial position.

In the substitution reaction shown in equation 26-1, the first rule tells us that the incoming nucleophile (hydride) must initially occupy an apical position and rule 3 tells us that one of the ring carbons must occupy the other apical position. However, rule 1 also requires that the leaving chloride ion depart from an apical position. Thus, the initial adduct exists until pseudorotation places Cl in an apical position. The pivot for the pseudorotation is the methyl group; C-1 of silacyclobutane ring becomes equatorial and C-3 becomes apical.

Simultaneously, the apical hydrogen and equatorial chlorine switch roles. When the apical chloride ion leaves, the net stereochemical result is retention of configuration—the two methyl groups on the silacyclobutane ring are still trans.

CH_3 Cl Si CH_3 $\xrightarrow{H^-}$ CH_3 1(a) (e) Cl Si—H (a) 3(e) CH_3 (e) ⇌

CH_3 1(e) (e) H (a) Si—Cl 3(a) CH_3 (e) $\xrightarrow{-Cl^-}$ CH_3 H Si CH_3

EXERCISE 26.21 Consider a chiral silane $R^1R^2R^3SiCl$ that undergoes nucleophilic substitution by a mechanism involving *two* pseudorotation steps of an intermediate trigonal bipyramid intermediate. What is the expected stereochemistry of the process?

B. Electrophilic Cleavage of the Carbon-Silicon Bond

Organosilicon compounds owe a great deal of their special utility to the fact that the carbon-silicon bond can be heterolytically cleaved relatively easily. For example, arylsilanes react with a variety of electrophiles with net **electrophilic substitution** of the silyl group.

COOH, $Si(Me_3)_3$ + Br_2 ⟶ COOH, Br + Me_3SiBr

CH_3, $SiMe_3$ + (O, Cl) $\xrightarrow{AlCl_3}$ CH_3, O + Me_3SiCl

In both of the foregoing reactions the electrophile appears to defy the normal directing effect of the groups attached to the benzene ring; reaction occurs *meta* to the methyl group and *para* to the carboxy group. In fact, these examples illustrate a strong propensity of the silyl group to direct the incoming electrophile to the position it occupies (''ipso attack''). The same property is shown by the silicon substituent in triethylsilylbenzene, which undergoes protodesilylation 10^4 times faster than proton exchange in benzene.

$SiEt_3$ + H_2O $\xrightarrow{H^+}$ + Et_3SiOH $\qquad k_{rel} = 10^4$

$$C_6H_6 + D_2O \xrightarrow{D^+} C_6H_5D + HOD \qquad k_{rel} = 1$$

The directing effect of the silicon stems from the fact that *silicon strongly stabilizes a cationic center β to it*. This stabilization results partly from the fact that silicon is relatively electropositive. However, part of the stabilization appears to result from a type of hyperconjugation of the empty *p*-orbital of the carbocation with the relatively diffuse carbon-silicon bond (see also page 196).

The ability of silicon to stabilize a β-carbocation is manifest also in the reactions of vinylsilanes. Although normal addition reactions do occur in some cases, electrophilic substitution is not uncommon.

$$CH_3CH{=}CHSiMe_3 \xrightarrow[20°C]{Br_2} CH_3CHBrCHBrSiMe_3$$

$$PhCH{=}CHSiMe_3 \xrightarrow[CCl_4]{Br_2} PhCH{=}CHBr$$

EXERCISE 26.22 Show how each of the following transformations can be accomplished.

(a) *o*-bromotoluene to *o*-nitrotoluene
(b) (*Z*)-1-trimethylsilylpropene to (*Z*)-1-bromo-1-trimethylsilyl-1-propene

Allylsilanes undergo ready electrophilic substitution with specific rearrangement of the double bond.

$$\text{(SiMe}_3\text{, COOMe cyclohexadiene)} \xrightarrow[\text{benzene, 80 °C}]{p\text{-TsOH}} \text{(COOMe cyclohexadiene)}$$

The mechanism of this reaction involves protonation of the double bond, with concomitant loss of the silicon. The cation-stabilizing effect of silicon is graphically illustrated in this example, since protonation produces a secondary, rather than a tertiary, carbocation.

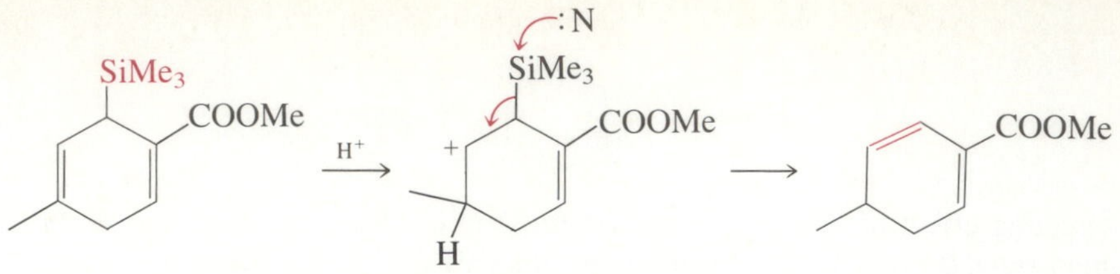

$:N$ SiMe$_3$ COOMe $\xrightarrow{H^+}$ SiMe$_3$ COOMe H $\longrightarrow$ COOMe

Other electrophiles may also enter into the reaction. Again, specific rearrangement of the double bond occurs.

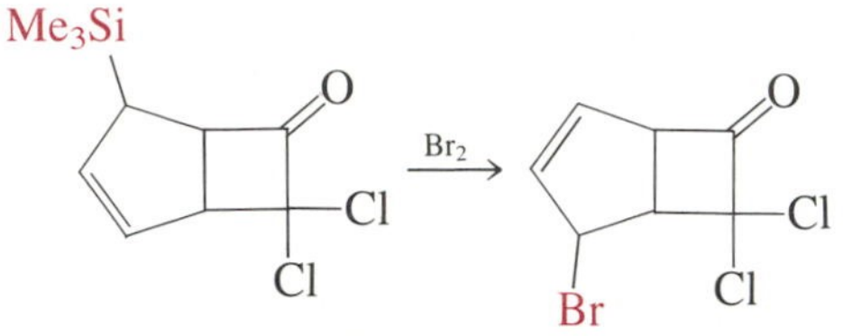

Me$_3$Si O Cl Cl $\xrightarrow{Br_2}$ O Cl Br Cl

EXERCISE 26.23 Write mechanisms for the following reactions.

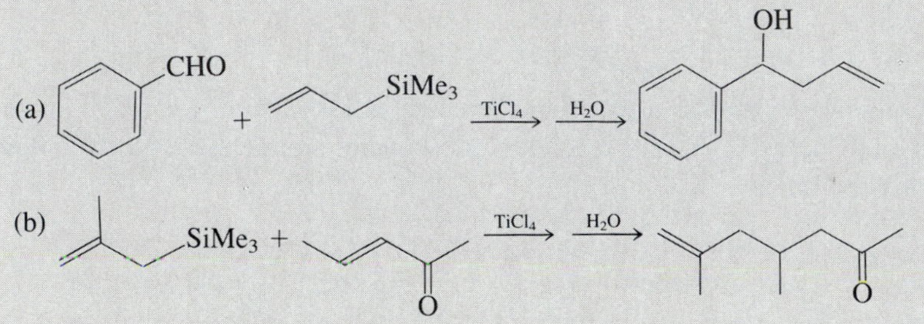

(a) CHO + SiMe$_3$ $\xrightarrow{TiCl_4}$ $\xrightarrow{H_2O}$ OH

(b) SiMe$_3$ + O $\xrightarrow{TiCl_4}$ $\xrightarrow{H_2O}$ O

C. Silyl Ethers as Protecting Groups

Because the silicon-oxygen bond is easily formed by nucleophilic substitution on a silyl halide, and because it is easily hydrolyzed, silicon finds much use in protecting groups (Section 16.4). Several reagents are in general use for this purpose. The most common is trimethylsilyl chloride, which reacts with alcohols in the presence of a weak base such as triethylamine.

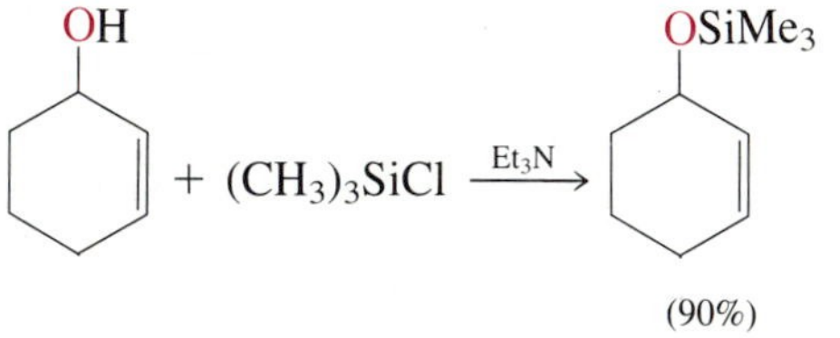

OH + $(CH_3)_3SiCl$ $\xrightarrow{Et_3N}$ OSiMe$_3$

(90%)

After desired reaction or sequence of reactions has been carried out elsewhere in the molecule, the protecting group is easily removed by the use of dilute aqueous acid.

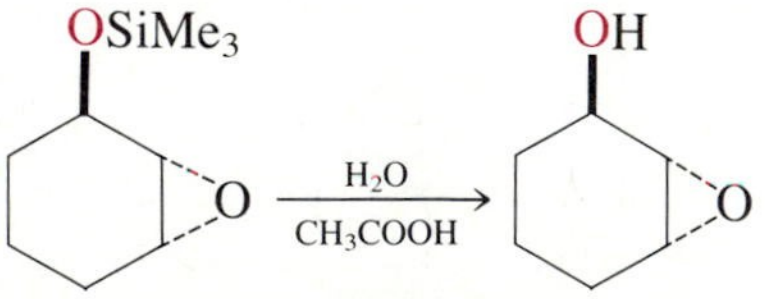

OSiMe$_3$ O $\xrightarrow[CH_3COOH]{H_2O}$ OH O

Another useful protecting group, which is more stable to some reaction conditions than the trimethylsilyl group, is *t*-butyldimethylsilyl. This group is normally introduced by treating the alcohol with a mixture of *t*-butyldimethylsilyl chloride and the heterocyclic base imidazole (Section 32.5).

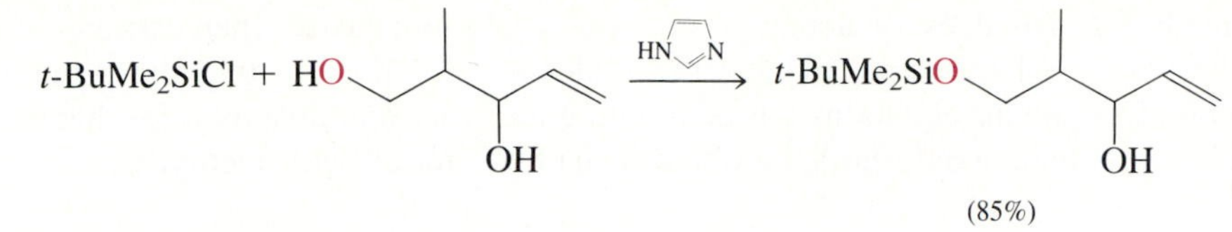

t-BuMe$_2$SiCl + HO(chain)(OH) —(imidazole, HN N)→ *t*-BuMe$_2$SiO(chain)(OH)
(85%)

This example also demonstrates that primary hydroxy groups react more rapidly than secondary or tertiary ones. In most cases, selective protection of the less substituted hydroxy may be achieved.

Like trimethylsilyl ethers, the *t*-butyldimethylsilyl ethers can sometimes be hydrolyzed by dilute aqueous acid. However, a convenient alternative is cleavage by fluoride ion. Benzyltrimethylammonium fluoride, which is soluble in organic solvents, is frequently used for this purpose. The reaction is carried out in a solvent such as THF in the presence of a small amount of water.

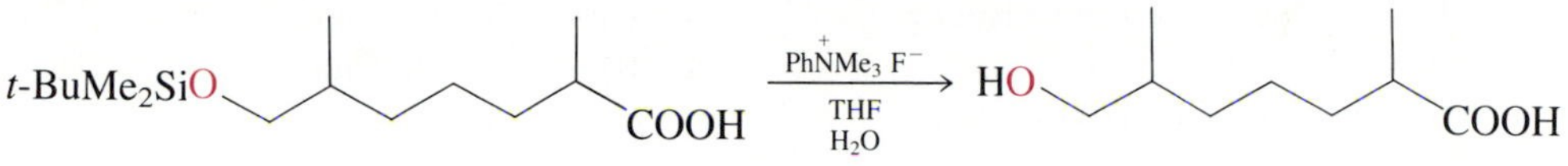

t-BuMe$_2$SiO(chain)COOH —($PhN^{+}Me_3\ F^-$, THF, H_2O)→ HO(chain)COOH

EXERCISE 26.24 Illustrate the use of silyl protecting groups in the following transformations.

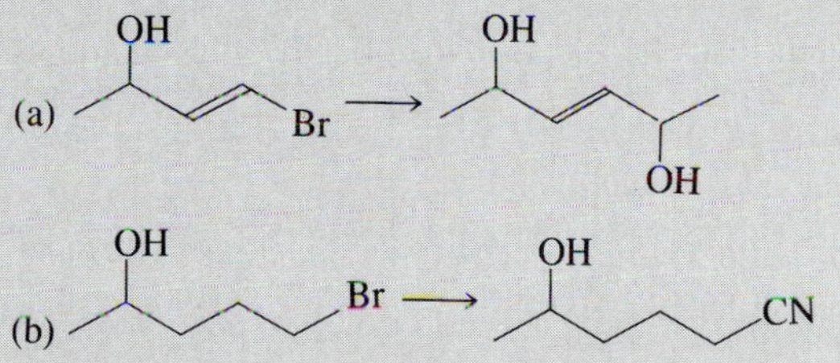

(a) OH ... Br ⟶ OH ... OH

(b) OH ... Br ⟶ OH ... CN

Explain why the protecting group is needed in each case.

ASIDE

In Genesis 19:24 it is written that God rained sulfur and fire on Sodom and Gomorrah. Sulfur in the atmosphere is still a problem as the chief component of "acid rain," while the fire may have been due to burning phosphorus. The name for silicon originates in the word for flint (L., *silex*), a stone used by cavemen to construct primitive tools. Pure silicon is now used in semiconductor "chips," the basis for modern electronic computers. Sulfur (thiols, sulfonic acids), phosphorus (phosphorus ylides), and silicon (trimethylsilyl groups) are important in organic chemistry, and two—sulfur (glutathione, cysteine; see Chapter 29) and phosphorus (nucleotide triphosphates, nucleic acids; see Chapter 33)—are central to biology.

A famous organosulfur chemist, Emmett Reid, instructed a young student at a southern U.S. university to prepare *n*-butyl sulfide from *n*-butyl bromide. The student gladly did so, neglecting to mention that he liked to work with kilos of material. At some point he lost control of the reaction, and an intermediate in the synthesis with a skunk-like odor saturated him and the building, in fact causing the temporary evacuation of the entire campus. What chemical reactions was the student trying to carry out? Write equations for the reactions and write out the mechanism. What was the intermediate? Why does a reaction go out of control more easily when carried out on a large scale? Which chemical reagents could he have used to trap the offending compound? Write the equations for the trapping reactions and give their mechanisms. Which compounds would he have obtained had he started with methyl iodide?

PROBLEMS

1. Give the structure corresponding to each of the following names.

a. ethyl neopentyl sulfide
b. isobutyl mercaptan
c. 2-methylthiocyclopentanone
d. dibutyl disulfide
e. cyclohexanethiol
f. isobutylsulfuric acid
g. diethyl sulfate
h. methyl *p*-nitrobenzenesulfonate
i. tributyl phosphate
j. diethyl ethylphosphonate
k. cyclohexyl methanesulfonate
l. triphenyl phosphite
m. allyl pyrophosphate
n. trimethylpropylsilane
o. diphenylmethylsilyl chloride

2. Give the IUPAC name corresponding to each of the following structures.

a. $(CH_3)_2CHCH_2CHCH_3$ (SH on the second CH)

b. $(CH_3)_2CHCHCH_2CH_3$ (SCH_3 on the second CH)

c. $(C_5H_9)_4P^+ Br^-$ (tetracyclopentylphosphonium, drawn as cyclopentyl in parentheses, subscript 4)

d. benzene ring with $CH(CH_3)_2$ and, para, SO_2NHCH_3

e. benzene ring with Br, ortho $COOCH_2C(CH_3)_3$, and SO_2OH para to Br

f. benzene ring with SO_2Cl and, para, NO_2

g. benzene ring with $SO_2—CH_3$ and, para, Br

h. cyclopropyl—S—S—cyclopropyl

i. $CH_3CH_2\overset{O}{\overset{\|}{P}}(OEt)_2$

j. cyclohexane ring with $SiMe_3$

3. What is the expected product of each of the following reactions?

a. $t\text{-BuO}^- K^+ + CH_3OSO_2OCH_3 \xrightarrow[\Delta]{t\text{-BuOH}}$

b. $CH_3CH_2S^+(CH_3)_2\ I^- \xrightarrow{\Delta}$

c. $(CH_3)_3CCH_2S^+(CH_2CH_3)_2\ Br^- \xrightarrow{\Delta}$

d. 1-methylcyclohexyl with $^+S(CH_2CH_3)_2$, CH_3 $Cl^- \xrightarrow{\Delta}$

e. 2-naphthalenesulfonic acid (SO_3H) $\xrightarrow[\text{fuse}]{\text{NaOH–KOH}}$

f. $C_6H_5SO_2^- + C_6H_5CH_2Cl \longrightarrow$

g. $HS(CH_2)_4SH \xrightarrow{I_2,\ KI}$

h. $HS(CH_2)_3SH +$ C_6H_5–CHO $\xrightarrow{HCl}$

i. 1,3-dithiane (S, S, H, H) $\xrightarrow{n\text{-}C_4H_9Li} \xrightarrow{CH_3\overset{O}{\overset{\|}{C}}CH_3}$

j. $(CH_3)_3S^+\ Br^- \xrightarrow{n\text{-}C_4H_9Li} \xrightarrow{C_6H_5CHO}$

4. What is the expected product of each of the following reactions?

a. $(CH_3CH_2CH_2CH_2O)_3P + \underset{\text{(trace)}}{CH_3CH_2CH_2CH_2Br} \xrightarrow{200°C}$

b. $CH_3CH_2CH_2\overset{O}{\overset{\|}{P}}(OCH_3)_2 + NaOH \xrightarrow[\Delta]{C_2H_5OH} \xrightarrow[\text{(workup)}]{H^+}$

c. $(C_2H_5O)_2\overset{O}{\overset{\|}{P}}Cl + (CH_3)_2CHCH_2CH_2OH \longrightarrow$

d. $POCl_3 + CH_3OH$ (excess) $\longrightarrow$

e. $(C_6H_5)_3P + CH_2{=}CHCH_2Br \xrightarrow{\Delta}$

f. $CH_3\overset{O}{\overset{\|}{P}}(OCH_3)_2 \xrightarrow{n\text{-}C_4H_9Li}$ $\xrightarrow{\text{cyclohexanone}}$

g. $CH_3CH_2\overset{O}{\overset{\|}{P}}(OCH_3)_2 \xrightarrow{n\text{-BuLi}} \xrightarrow{CH_3CO_2CH_3}$

5. What is the expected product of each of the following reactions?

a. 1-(trimethylsilylmethyl)cyclohexene ($SiMe_3$) $+ Br_2 \xrightarrow[25°C]{CCl_4}$

b. $CH_3CH{=}CHCH_2SiMe_3 + DCl \xrightarrow[0°C]{CH_2Cl_2}$

c. 3,5-dimethylphenyl–$SiMe_3$ (CH_3, CH_3, $SiMe_3$) $+ Cl_2 \xrightarrow{FeCl_3}$

d. $C_6H_5Br \xrightarrow[\text{ether}]{Mg} \xrightarrow[25°C]{(CH_3)_2SiCl_2}$

e. $CH_3CH{=}CHCH_2SiMe_3 +$ C_6H_5COCl $\xrightarrow[CHCl_3]{AlCl_3}$

6. To minimize the odor caused by traces of thiols it is recommended that reaction vessels be rinsed with aqueous potassium permanganate as soon as possible. Nitric acid may also be used for this purpose. The thiols are oxidized to alkanesulfonic acids, and the

sulfides are oxidized to sulfones. Write balanced equations for each of the following reactions.

a. $CH_3SH + KMnO_4 \xrightarrow{H_2O} CH_3SO_3K + MnO_2 + KOH$

b. $(CH_3)_2S + KMnO_4 \xrightarrow{H_2O} CH_3SO_2CH_3 + MnO_2 + KOH$

c. $(CH_3)_2S + HNO_3 \longrightarrow CH_3SO_2CH_3 + H_2O + NO_2$

7. The following reaction sequences are impractical. Determine what is wrong in each case.

a. $(CH_3)_3CO^- \; K^+ \xrightarrow{PCl_3} [(CH_3)_3CO]_3P \xrightarrow[200°C]{(CH_3)_3CBr} (CH_3)_3C\overset{O}{\overset{\|}{P}}(OC(CH_3)_3)_2$

b. CH_2CH_3 (benzene) $\xrightarrow{H_2SO_4}$ CH_2CH_3, SO_3H (benzene) $\xrightarrow[AlCl_3]{CH_3COCl}$ CH_2CH_3, $COCH_3$, SO_3H (benzene) $\xrightarrow[\Delta]{H^+}$ CH_2CH_3, $COCH_3$ (benzene)

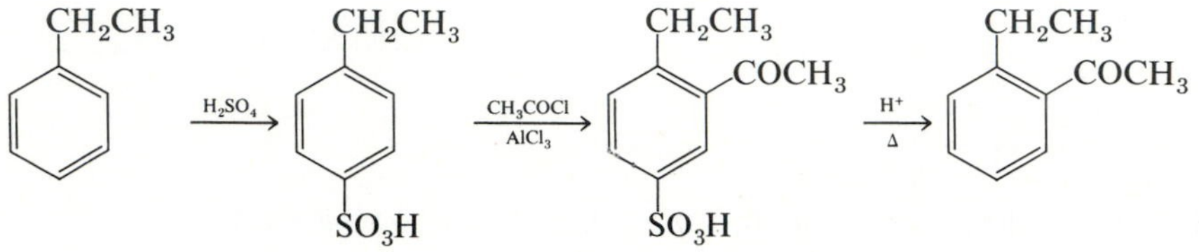

c. (benzene) $\xrightarrow[AlCl_3]{CH_3CH_2COCl}$ $COCH_2CH_3$ (benzene) $\xrightarrow[H_2SO_4]{fuming}$ $\xrightarrow{NaCl}$ $COCH_2CH_3$, $SO_3^- \; Na^+$ (benzene) $\xrightarrow[fuse]{KOH}$ $\xrightarrow{H^+}$ $COCH_2CH_3$, OH (benzene)

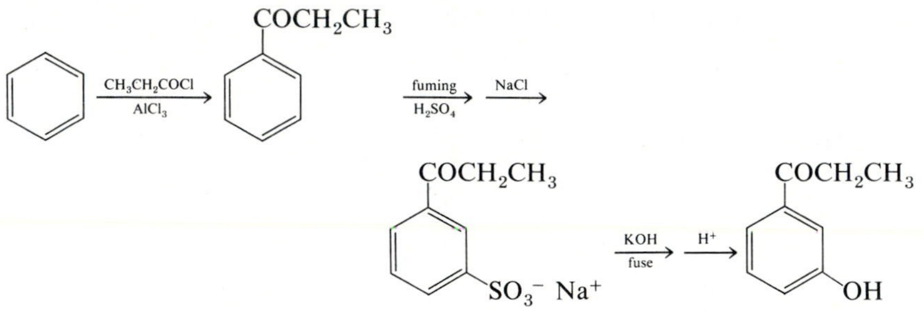

d. CH_3, Br (benzene) $\xrightarrow[2.\ Me_3SiCl]{1.\ Mg}$ CH_3, $SiMe_3$ (benzene) $\xrightarrow[FeCl_3]{Cl_2}$ CH_3, Cl, $SiMe_3$ (benzene) $\xrightarrow[AlCl_3]{CH_3COCl}$ CH_3, Cl, $C(=O)CH_3$ (benzene)

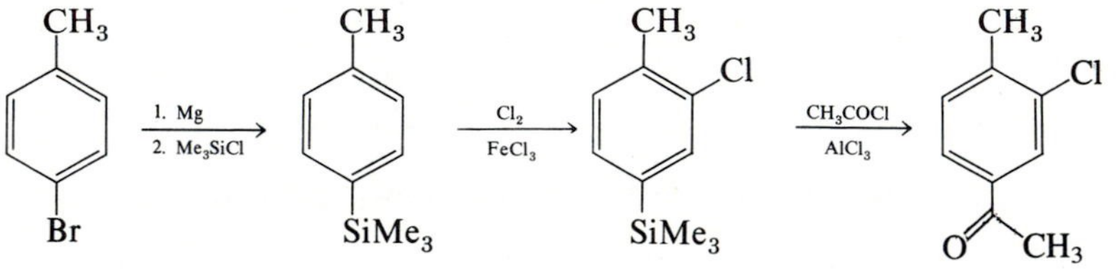

8. Show how to accomplish each of the following conversions in a practical manner.

a. $(CH_3)_3CBr \longrightarrow (CH_3)_3CSCH_3$

b. $CH_3CH_2CH_2CH_2OH \longrightarrow (CH_3CH_2CH_2CH_2)_2S_2$

c. CH_3 (benzene) $\longrightarrow$ CH_3(benzene)$-SO_2-$(benzene)CH_3

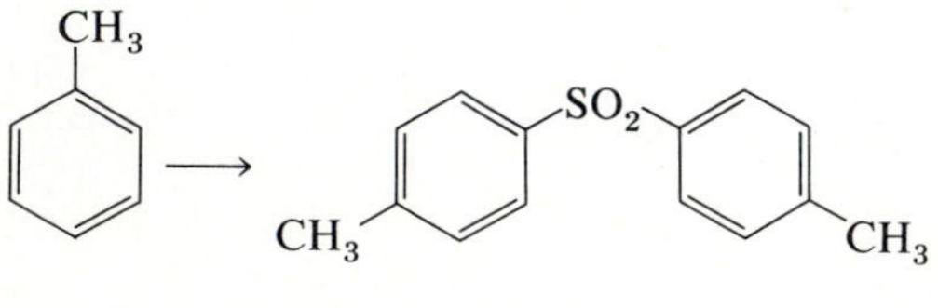

d. NH_2 (benzene) $\longrightarrow$ NO_2, SO_3H (benzene)

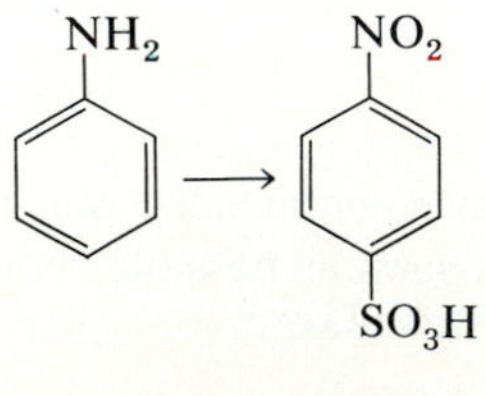

e. CH_3 (benzene) $\longrightarrow$ CH_2SO_3H (benzene)

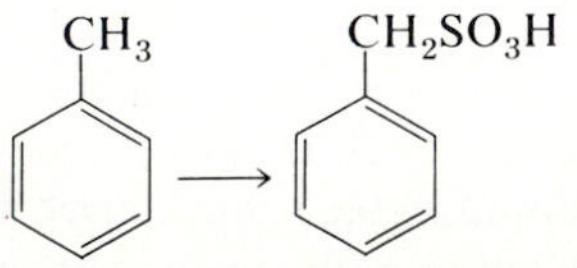

f. CH_3–C_6H_4–SO_3H ⟶ CH_3–C_6H_4–COOH

g. cyclohexanone ⟶ 1-(hydroxymethyl)cyclohexanol (OH, CH_2OH)

h. sulfolane ⟶ 2-propylsulfolane ($CH_2CH_2CH_3$)

9. How may each of the following transformations be accomplished?

a. cyclopentanol (–OH) ⟶ (cyclopentyl)$_3$P

b. benzyl alcohol (OH) ⟶ $C_6H_5CH_2P(=O)(OCH_2C_6H_5)_2$

c. benzaldehyde (CHO) ⟶ $C_6H_5C(=O)CH_2CH(OH)CH_2CH_2CH_3$

d. cyclopentanone (=O) ⟶ cyclopentylidene–CH–CN

10. Mustard gas or bis(β-chloroethyl) sulfide, $(ClCH_2CH_2)_2S$, is an oily liquid that was used extensively as a poison gas in World War I. It is a deadly vesicant that causes blindness and numerous other effects. The active agent is actually the cyclic sulfonium salt.

$$ClCH_2CH_2{-}\overset{+}{S}\langle{}^{CH_2}_{CH_2}|$$

This intermediate reacts with nucleophilic materials in the body. The formation of the cyclic sulfonium salt can be regarded as an intramolecular S_N2 reaction. Write out the reaction mechanism. What mechanism does this process suggest for the subsequent reaction of the cyclic sulfonium salt with nucleophilic reagents?

11. Thiols are used as inhibitors in free radical reactions. In such use they end up as disulfides. The bond-dissociation energy of CH_3S—H is 91 kcal mole^{-1}. Calculate $\Delta H°$ for the reaction of methane with the methylthio radical.

$$CH_4 + CH_3S\cdot \rightleftharpoons CH_3\cdot + CH_3SH$$

In which direction does the equilibrium lie? Explain how CH_3SH works as an inhibitor of radical chain reactions.

12. Compare the relative acidities of benzenesulfonic acid with benzoic acid and of benzenesulfonamide with benzamide. Predict the relative acidities of acetophenone and phenyl methyl sulfone.

13. The normal phosphorus-oxygen single bond distance is about 1.60 Å. The coordinate phosphorus-oxygen bond ($^+P—O^-$ or P═O) as in phosphorus oxychloride and in phosphates is almost 0.2 Å shorter. What does this comparison imply about the relative bond strengths of the two types of phosphorus-oxygen bonds? The rearrangement of trimethyl phosphite to dimethyl methylphosphonate is exothermic by 47 kcal mole^{-1}. What do you think might be the principal driving force for this reaction?

14. The intermediate in base-catalyzed hydrolysis of phosphate and phosphonate esters is a pentacoordinated species similar to that involved in nucleophilic substitution of some silicon compounds. The three generalizations enumerated on page 843 also apply to these intermediates. In hydrolysis of the following cyclic phosphate esters, the six-membered compound undergoes hydrolysis of both methoxy and ring alkoxy bonds, but in the five-membered compound, only ring cleavage is observed. Explain.

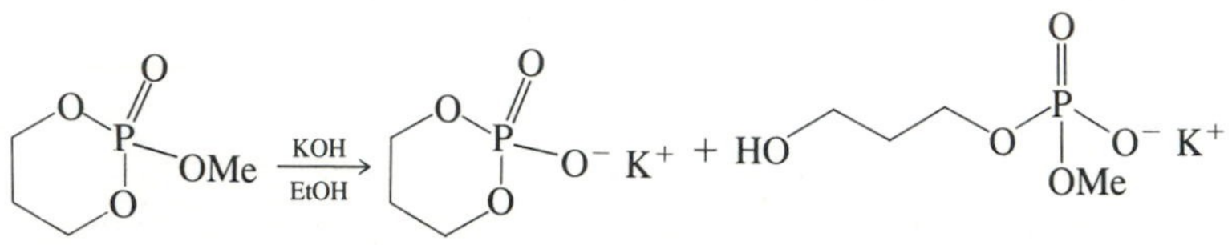

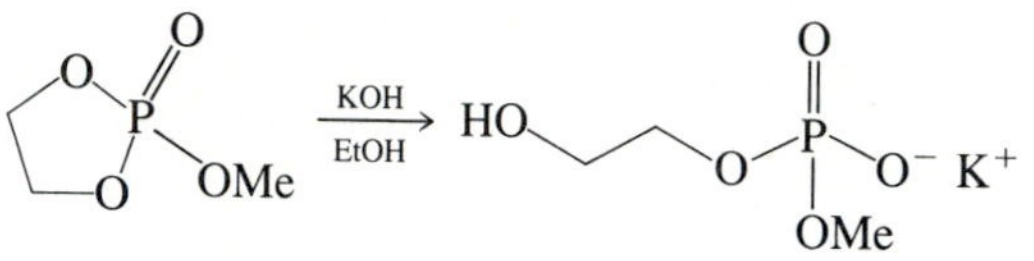

O, P, O, O, OMe —KOH, EtOH→ O, P, O, O, O$^-$ K$^+$ + HO, O, P, O, O$^-$ K$^+$, OMe

O, P, O, O, OMe —KOH, EtOH→ HO, O, P, O, O$^-$ K$^+$, OMe

15. One of the products of the reaction of sodium 3,5-dibromo-4-aminobenzenesulfonate with aqueous bromine is 2,4,6-tribromoaniline, the product of an ipso-substitution reaction. Write a reasonable reaction mechanism.

16. β-Keto sulfoxides react with aqueous acids or other acidic reagents to give α-hydroxy sulfides (**Pummerer rearrangement**).

$$C_6H_5\overset{O}{\overset{\|}{C}}CH_2\overset{O}{\overset{\|}{S}}CH_3 \xrightarrow[\text{DMSO, 25°C}]{\text{HCl, }H_2O} C_6H_5\overset{O}{\overset{\|}{C}}\underset{OH}{\underset{|}{C}}HSCH_3$$

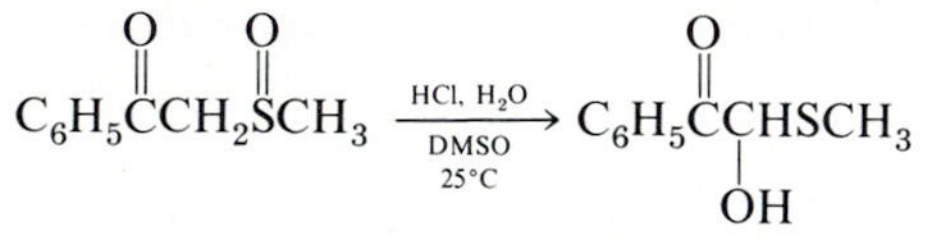

(80%)

Propose a mechanism for the Pummerer rearrangement. What product will result if such a β-keto sulfoxide is treated with acetic anhydride in pyridine?

17. Suggest a general synthetic method whereby organosulfur chemistry could be used to convert saturated ketones and esters into their α,β-unsaturated counterparts.

18. Oxidation of diphenyl disulfide with hydrogen peroxide gives a dioxide that could be formulated as either a disulfoxide or as a thiosulfonic ester.

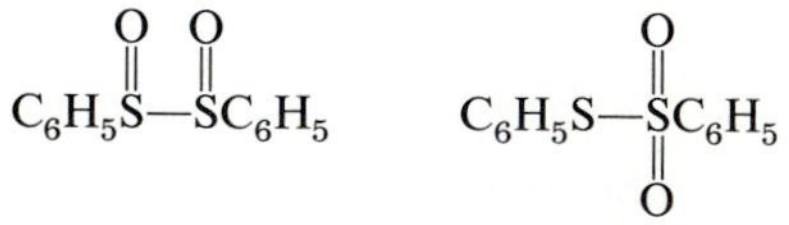

$$C_6H_5\overset{O}{\overset{\|}{S}}—\overset{O}{\overset{\|}{S}}C_6H_5 \qquad C_6H_5S—\underset{O}{\underset{\|}{\overset{O}{\overset{\|}{S}}}}C_6H_5$$

a disulfoxide — a thiosulfonic ester

Based on thermodynamic analogies available in Appendix I, which formulation do you think is correct?

19. Oxidation of trithioformaldehyde gives a mixture of two bis-sulfoxides, A and B. Further oxidation of A with H_2O_2 gives a single tris-sulfoxide, C, whereas further oxidation of B gives a mixture of tris-sulfoxides, C and D. What are the structures of the bis-sulfoxides, A and B, and the tris-sulfoxides, C and D?

trithioformaldehyde

20. Show how 1-trimethylsilyl-1,3-butadiene can be used to synthesize the following compounds.

a.

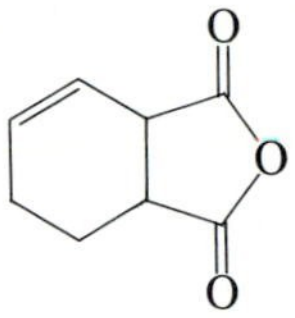

b.

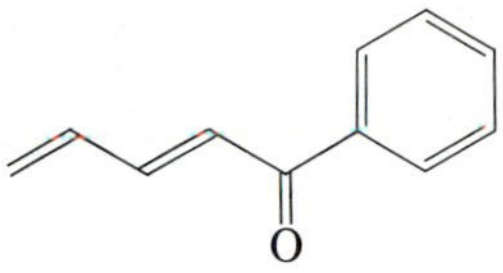

21. Treatment of α-trimethylsilylbenzyl alcohol with alcoholic potassium hydroxide gives benzyltrimethylsilane (**Brook rearrangement**). Propose a mechanism for the rearrangement.

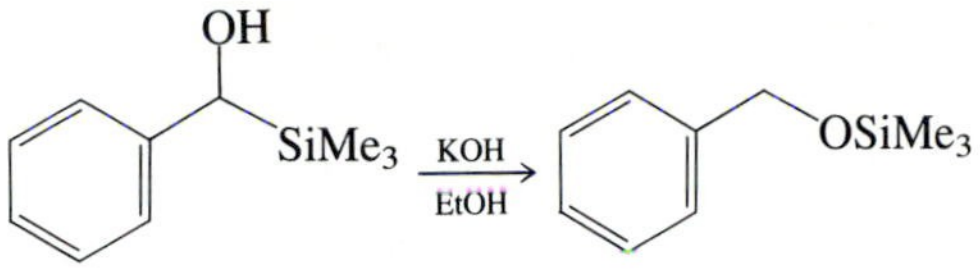

OH

$SiMe_3$ $\xrightarrow[\text{EtOH}]{\text{KOH}}$ $OSiMe_3$

Rationalize the fact that, when the Brook rearrangement is carried out with a chiral substrate in which silicon is a stereocenter, retention of configuration at silicon is observed.

22. Explain the difference between the acidities of thiols and alcohols. The pK_a for H_2O is about 16 while that for H_2S is about 7. Consider the thermodynamic cycle (1) vaporization of H_2S from aqueous solution, (2) dissociation of H_2S into $H\cdot$ and $\cdot SH$, (3) ionization of $H\cdot$ to H^+ and e^-, (4) addition of an electron to $\cdot SH$ to give SH^- (electron affinity of **thiyl radical**), solvation of H^+, solvation of thiolate (hydrosulfide) anion, SH^-. From the cycle, the heat of conversion of H_2S to H_3O^+ and SH^- can be estimated as a few kcal per mole. A similar cycle for the conversion of H_2O to H_3O^+ and OH^- leads to a larger value. The difference arises from the bond energies (H—S, 89 kcal $mole^{-1}$, O—H, 119 kcal $mole^{-1}$) and electron affinities [EA($\cdot$SH), 52 kcal $mole^{-1}$, EA($\cdot$OH), 40 kcal $mole^{-1}$] as compared with greater solvation energy of OH^- than SH^-. The difference in bond energies for RS—H and RO—H is not as great as that for HS—H and HO—H.

23. The tripeptide glutathione (page 549) is treated with iodine. What is the product and what is the mechanism of its formation?

CHAPTER 27

DIFUNCTIONAL COMPOUNDS

27.1 Introduction

To a first approximation the chemical properties of difunctional compounds are a summation of those of the individual functions. For example, cyclohex-2-en-1-one is a difunctional compound that undergoes typical alkene reactions (catalytic hydrogenation) and normal ketone reactions (reduction by lithium aluminum hydride).

H_2–Pd

LiAlH$_4$ H_2O

H OH

However, in many cases the two functional groups interact in such a way as to give the compound chemical properties that are not observed with the simple monofunctional compounds. In cyclohex-2-en-1-one the two functional groups form a conjugated system (Chapter 20), so this molecule undergoes some special reactions, such as 1,4-addition of lithium dimethylcuprate (Section 20.3.A).

$$+ \ LiCu(CH_3)_2 \longrightarrow \xrightarrow{H_3O^+}$$

CH_3

This is a special reaction of the difunctional compound because neither simple alkenes nor simple ketones react with the reagent.

In other cases the chemical properties of a difunctional compound are similar to those of a corresponding monofunctional compound in a qualitative sense but not in a quantitative sense. An example is the reaction of allyl bromide with azide ion. The reaction is a normal S_N2 replacement of a primary halide, but since the organic group is allylic, the reaction is over 50 times faster than it is with propyl bromide (Section 20.1.B).

$$CH_2{=}CHCH_2Br + N_3^- \xrightarrow{\text{faster}} CH_2{=}CHCH_2N_3 + Br^-$$

$$CH_3CH_2CH_2Br + N_3^- \xrightarrow{\text{slower}} CH_3CH_2CH_2N_3 + Br^-$$

In still other cases two functional groups in a molecule may enter into a chemical reaction with each other. An example is the intramolecular S_N2 reaction leading to a cyclic ether; another is the intramolecular reaction leading to cyclic amines (problem 13, Chapter 9).

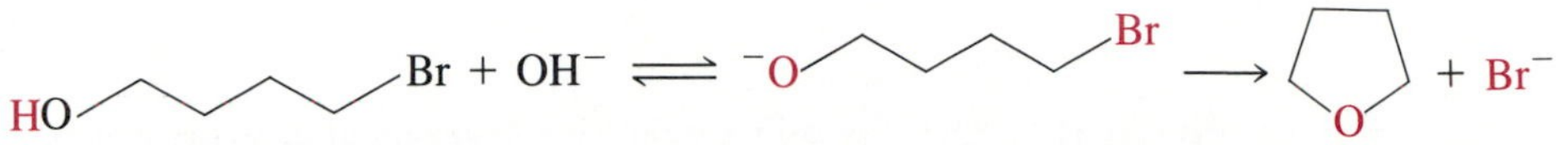

$$HO(CH_2)_4Br + OH^- \rightleftharpoons {}^-O(CH_2)_4Br \longrightarrow \text{tetrahydrofuran} + Br^-$$

As may be seen in the foregoing examples, we have already encountered the reactions of a number of difunctional compounds, mainly in the study of conjugated systems (Chapter 20). In this chapter, we shall take up a few specific types of difunctional compounds, pointing out some of the unique chemistry that results from the cooperation or interaction of the two functional groups. Specific difunctional compounds we shall consider at this time are those containing the functional groups OH and C=O: diols, diketones, dicarboxylic acids, hydroxy aldehydes, hydroxy ketones, hydroxy acids, and keto acids. The chemistry of these difunctional compounds forms a necessary foundation for our study of the chemistry of carbohydrates (Chapter 28). In Chapter 29 we shall consider another large and important class of difunctional compounds, amino acids.

27.2 Nomenclature of Difunctional Compounds

Recall that most simple monofunctional compounds are named in such a way that the ending of the name denotes the functional group: acetic **acid,** 3-pentan**ol,** cyclohexan**one,** 1-but**ene.** Alkyl halides are exceptions to this generalization, in that they are considered as derivatives of the parent alkane, for example, 2-chloroheptane. When a compound contains two like functional groups, it is generally named in the same way except that the typical group suffix is combined with **di-** to indicate the presence of two groups. Numbers are used to locate the positions of the groups on the carbon skeleton. Bis- or bis(group), twice, tris- or tris(group), three times, tetrakis- or tetrakis(group), four times, are prefixes thtat are used if di-, tri-, and tetra- would cause confusion, as in the case of complex substituents.

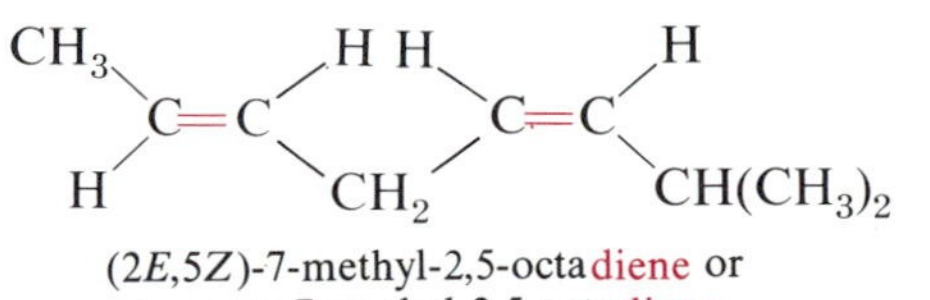

$$CH_3CH{=}CHCH_2CH{=}CHCH(CH_3)_2$$

(2*E*,5*Z*)-7-methyl-2,5-octadiene or *trans,cis*-7-methyl-2,5-octadiene

$$CH_3C{\equiv}C{-}C{\equiv}CH$$

1,3-pentadiyne

$$CH_3\overset{O}{\overset{\|}{C}}CH_2\overset{O}{\overset{\|}{C}}CH_3$$

2,4-pentanedione

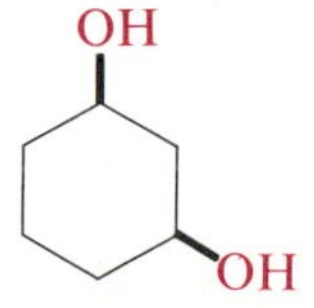

cis-1,3-cyclohexanediol

$$CH_3\overset{Br}{\overset{|}{C}}HCH_2\overset{CH_3}{\overset{|}{C}}HCH_2CH_2Br$$

1,5-dibromo-3-methylhexane

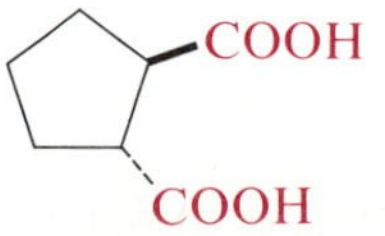

trans-1,2-cyclopentane dicarboxylic acid

TABLE 27.1 Names of Some Dicarboxylic Acids

n	Formula	Common	IUPAC
C_2	$HO\overset{O}{\overset{\|}{C}}\overset{O}{\overset{\|}{C}}OH$	oxalic acid	ethanedioic acid
C_3	$HO\overset{O}{\overset{\|}{C}}CH_2\overset{O}{\overset{\|}{C}}OH$	malonic acid	propanedioic acid
C_4	$HO\overset{O}{\overset{\|}{C}}(CH_2)_2\overset{O}{\overset{\|}{C}}OH$	succinic acid	butanedioic acid
C_5	$HO\overset{O}{\overset{\|}{C}}(CH_2)_3\overset{O}{\overset{\|}{C}}OH$	glutaric acid	pentanedioic acid
C_6	$HO\overset{O}{\overset{\|}{C}}(CH_2)_4\overset{O}{\overset{\|}{C}}OH$	adipic acid	hexanedioic acid
C_7	$HO\overset{O}{\overset{\|}{C}}(CH_2)_5\overset{O}{\overset{\|}{C}}OH$	pimelic acid	heptanedioic acid
C_8	$HO\overset{O}{\overset{\|}{C}}(CH_2)_6\overset{O}{\overset{\|}{C}}OH$	suberic acid	octanedioic acid
C_9	$HO\overset{O}{\overset{\|}{C}}(CH_2)_7\overset{O}{\overset{\|}{C}}OH$	azelaic acid	nonanedioic acid
C_{10}	$HO\overset{O}{\overset{\|}{C}}(CH_2)_8\overset{O}{\overset{\|}{C}}OH$	sebacic acid	decanedioic acid

Diols are sometimes called **glycols.** This is a trivial nomenclature widely used in the chemical industry, particularly for some of the simpler diols, which are important commercial items. For example, ethylene glycol (1,2-ethanediol) is the most widely used antifreeze additive for automobile radiators.

The aliphatic dicarboxylic acids having up to ten carbons in their chains have common names that are used extensively in the chemical literature (Table 27.1). Although these names are no longer used for indexing purposes, the student should be aware of them since they were uniformly used before about 1975. The benzenedicarboxylic acids are known as phthalic, isophthalic, and terephthalic acids. The last-named acid is a highly important industrial material; it forms one of the building blocks of the synthetic fiber known as polyester, Dacron, or Terylene.

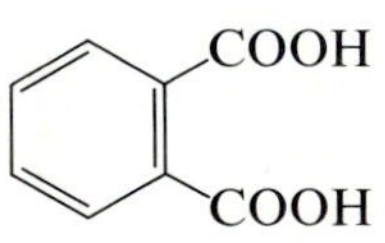

COOH COOH (phthalic acid structure)

phthalic acid
1,2-benzenedicarboxylic acid

COOH COOH (isophthalic acid structure)

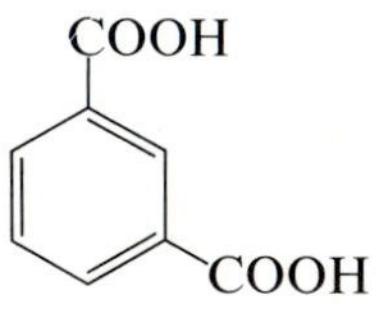

isophthalic acid
1,3-benzenedicarboxylic acid

COOH COOH (terephthalic acid structure)

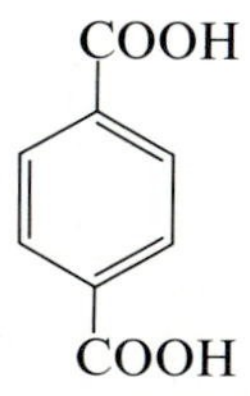

terephthalic acid
1,4-benzenedicarboxylic acid

Two unsaturated aliphatic diacids that have widely used common names are maleic and fumaric acids.

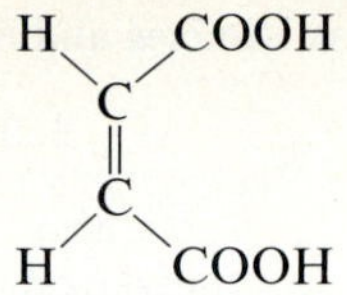

$$\underset{\substack{\text{maleic acid}\\ cis\text{-butenedioic acid}}}{\begin{matrix}\text{H} & & \text{COOH}\\ & \text{C} & \\ & \| & \\ & \text{C} & \\ \text{H} & & \text{COOH}\end{matrix}} \qquad \underset{\substack{\text{fumaric acid}\\ trans\text{-butenedioic acid}}}{\begin{matrix}\text{H} & & \text{COOH}\\ & \text{C} & \\ & \| & \\ & \text{C} & \\ \text{HOOC} & & \text{H}\end{matrix}}$$

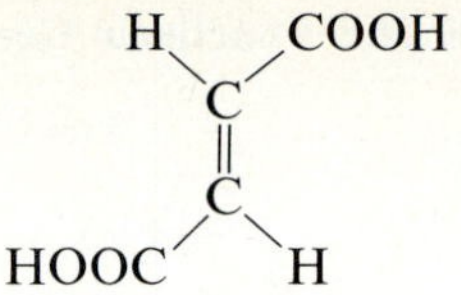

Aldehydes and functional derivatives corresponding to common diacids are frequently named as derivatives of the acids, in the same manner as is used to name simple aldehydes and functional derivatives.

$$\underset{\text{succinaldehyde}}{\overset{\text{O}\qquad\qquad\text{O}}{\text{HCCH}_2\text{CH}_2\text{CH}}} \qquad \underset{\beta,\beta\text{-dimethylglutaraldehyde}}{\text{HCCH}_2\text{C(CH}_3)_2\text{CH}_2\text{CH}}\ (\text{with } \overset{\text{O}}{\|} \text{ on each terminal C})$$

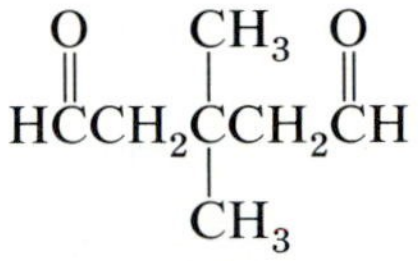

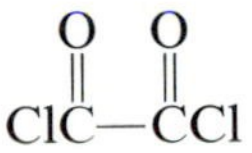

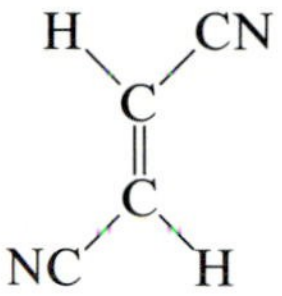

$$\underset{\text{oxalyl chloride}}{\text{ClC(=O)}-\text{C(=O)Cl}} \qquad \underset{\text{dimethyl adipate}}{\text{CH}_3\text{OC(=O)CH}_2\text{CH}_2\text{CH}_2\text{CH}_2\text{C(=O)OCH}_3} \qquad \underset{\text{fumaronitrile}}{\begin{matrix}\text{H} & & \text{CN}\\ & \text{C} & \\ & \| & \\ & \text{C} & \\ \text{NC} & & \text{H}\end{matrix}}$$

When a compound contains two different functional groups, one of the groups (the principal function) is usually expressed in the ending of the name and the other as a prefix.

$$\text{HOCH}_2\text{CH}_2\text{COOH}$$

3-hydroxypropanoic acid
β-hydroxypropionic acid

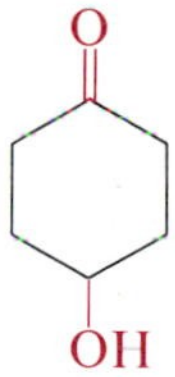

4-hydroxycyclohexanone

Alkenes and alkynes are exceptions in that the double or triple bond cannot be expressed as a prefix. For compounds containing a multiple bond and another functional group, two suffixes are used.

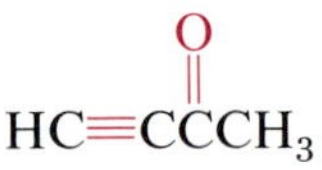

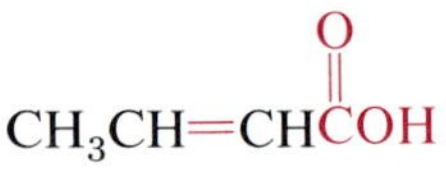

$$\underset{\text{3-buten-1-ol}}{\text{CH}_2\!=\!\text{CHCH}_2\text{CH}_2\text{OH}} \qquad \underset{\text{3-butyn-2-one}}{\text{HC}\!\equiv\!\text{C}\overset{\text{O}}{\overset{\|}{\text{C}}}\text{CH}_3} \qquad \underset{\text{2-butenoic acid}}{\text{CH}_3\text{CH}\!=\!\text{CH}\overset{\text{O}}{\overset{\|}{\text{C}}}\text{OH}}$$

In the naming of a difunctional compound, a choice must be made as to which group is the principal function. The generally accepted order is carboxylic acid, sulfonic acid, ester, acyl halide, amide, nitrile, aldehyde, ketone, alcohol, thiol, amine, alkyne, alkene. Since alkenes and alkynes cannot be designated by prefixes, they are always indicated by a second suffix, which is placed before the final suffix of any function higher in the order. Table 27.2 contains a listing of the common functions with the appropriate prefix and suffix used to designate each one.

TABLE 27.2 Functional Groups as Prefixes and Suffixes

Group	Prefix	Suffix
—COOH	carboxy-	-oic acid -carboxylic acid
$—SO_3H$	sulfo-	-sulfonic acid
—COOR	alkoxycarbonyl-	-carboxylate
—COCl	chloroformyl-	-oyl chloride -carbonyl chloride
$—CONH_2$	carbamoyl-	-amide -carboxamide
—CN	cyano-	-nitrile -carbonitrile
—CHO	formyl- oxo-	-al -carboxaldehyde -carbaldehyde
—C(=O)—	oxo- (IUPAC) keto- (common)	-one
—OH	hydroxy-	-ol
—SH	mercapto-	-thiol
$—NH_2$	amino-	-amine
—C≡C—	—	-yne
—C=C—	—	-ene
—Cl	chloro-	—

EXERCISE 27.1 Write structures of a six-carbon compound containing each pairwise combination of the following functional groups: OH, C=C, C=O (both aldehyde and ketone), COOH. Assign a name to each of your 15 structures.

27.3 Diols

A. Preparation

1,2-Diols are usually prepared from the corresponding alkene by the net addition of two hydroxy groups to the double bond (**hydroxylation**). Direct hydroxylation may be accomplished by oxidation of the alkene with $KMnO_4$ or OsO_4 (Section 11.6.E). Overall hydroxylation may be achieved by conversion of the alkene to an epoxide (Section 11.6.E), which is then hydrolyzed to the 1,2-diol (Section 10.11.A).

Direct hydroxylation with $KMnO_4$ or OsO_4 is a stereospecific process; the overall result is syn addition of the two hydroxy groups to the double bond.

$$\text{cis-cyclooctene} + H_2O_2 \xrightarrow{OsO_4} \text{cis-1,2-cyclooctanediol}$$

cis-cyclooctene *cis*-1,2-cyclooctanediol

The two-step procedure for hydroxylation of an alkene is also stereospecific, but this process produces net anti addition of the two hydroxy groups to the double bond. For example, cyclohexene reacts with peroxyacetic acid to give cyclohexene oxide. This compound undergoes acid-catalyzed ring opening by the S_N2 mechanism, resulting in inversion of configuration at one of the two carbon-oxygen bonds. Thus, the overall result is formation of the trans 1,2-diol.

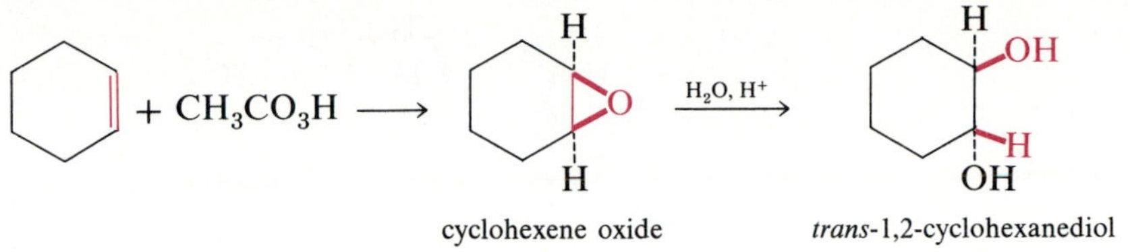

+ CH_3CO_3H ⟶ [cyclohexene oxide] $\xrightarrow{H_2O,\ H^+}$ [*trans*-1,2-cyclohexanediol]

cyclohexene oxide

trans-1,2-cyclohexanediol

Different diols result from addition to the cis or trans isomer of the alkene. For a symmetrical alkene, such as 2-butene, syn hydroxylation of the cis isomer gives a meso diol.

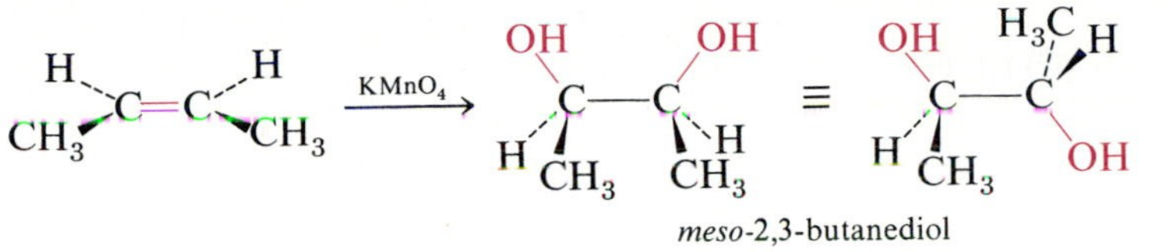

$\xrightarrow{KMnO_4}$

meso-2,3-butanediol

Syn hydroxylation of the trans isomer gives a 50:50 mixture of two enantiomeric diols. These two products arise from addition of the reagent to the two faces of the planar alkene molecule. Since the reagent is achiral, the transition states leading to the two products are enantiomeric and equal in energy. The product is therefore a racemate (Section 7.4). To distinguish this mixture of enantiomers from the meso diol, it is frequently designated as (±) or *dl* (meaning an equimolar or racemic mixture of the dextrorotatory and levorotatory enantiomers).

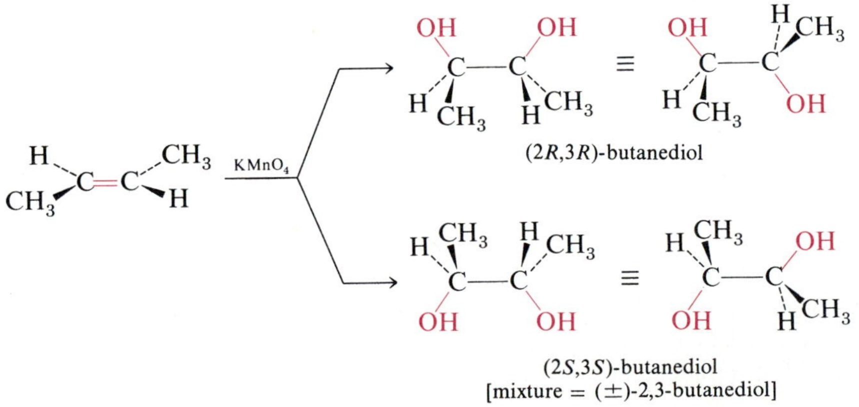

$\xrightarrow{KMnO_4}$

(2*R*,3*R*)-butanediol

(2*S*,3*S*)-butanediol
[mixture = (±)-2,3-butanediol]

For anti hydroxylation, the situation is just exactly reversed—the cis alkene gives the (±)-diol and the trans alkene affords the meso diol. When the acyclic alkene is not symmetrical, both syn and anti additions of each isomer produce racemic mixtures. The following equations illustrate the syn hydroxylation of the two isomers of 2-pentene.

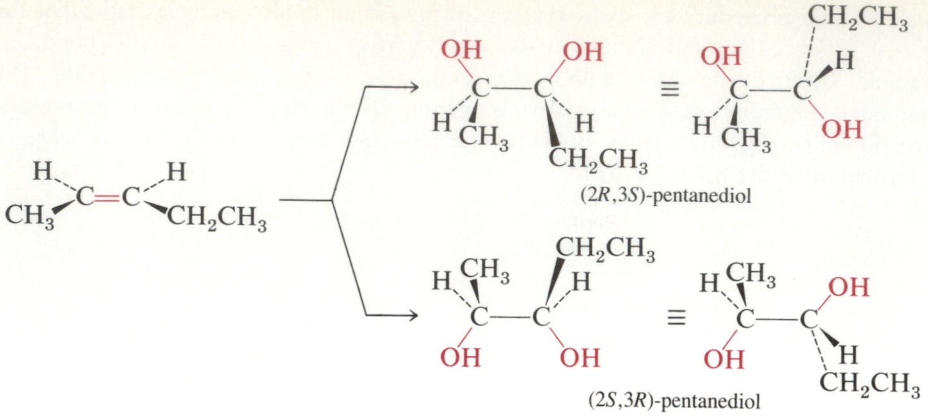

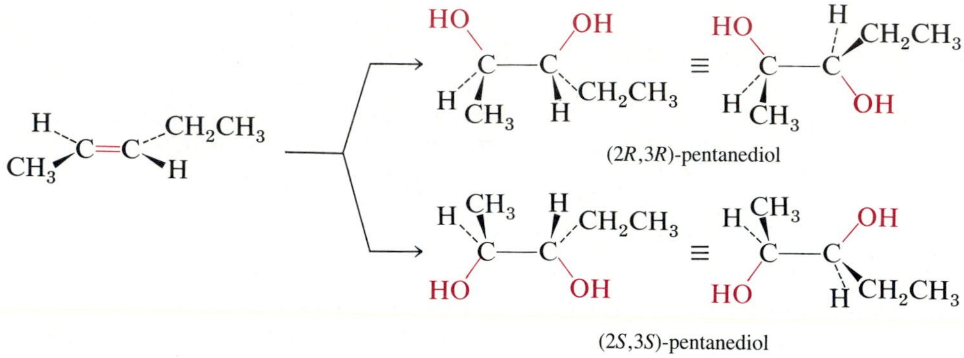

$CH_3CH{=}CHCH_2CH_3$ (cis) → (2R,3S)-pentanediol + (2S,3R)-pentanediol

(2R,3S)-pentanediol

(2S,3R)-pentanediol

$CH_3CH{=}CHCH_2CH_3$ (trans) → (2R,3R)-pentanediol + (2S,3S)-pentanediol

(2R,3R)-pentanediol

(2S,3S)-pentanediol

The 2*R*,3*S* and 2*S*,3*R* isomers of 2,3-pentanediol are called **erythro** isomers, and the 2*R*,3*R* and 2*S*,3*S* isomers are called **threo** isomers. These names derive from carbohydrate chemistry (Chapter 28) and are frequently used for other simple difunctional compounds. When a compound contains two stereocenters that have two identical attached groups and a third that differs, the isomer that would be meso if the third groups were identical is the erythro isomer. The other isomer is the threo isomer.

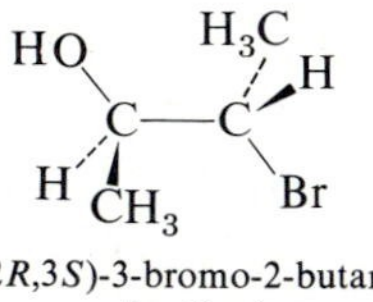

(2*R*,3*S*)-3-bromo-2-butanol (erythro)

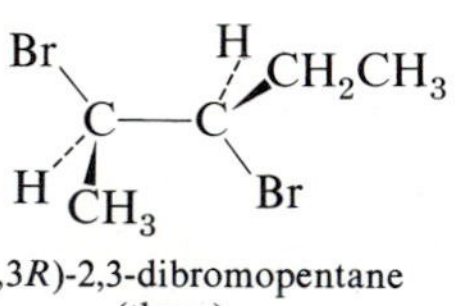

(2*R*,3*R*)-2,3-dibromopentane (threo)

The most reliable reagent for the preparation of vicinal diols by syn addition is OsO_4. However, its expense is a problem (the 1991 cost of OsO_4 was \$71.25 per gram). Fortunately, hydroxylation methods have been developed that require only a catalytic amount of OsO_4. The technique that is used in such catalytic osmylations is to include in the reaction mixture an oxidizing agent (a "co-oxidant") that does not affect alkenes but does react with the initially formed cyclic osmate to give the vicinal diol with regeneration of OsO_4. The best co-oxidants are tertiary amine oxides (page 757), and the reagent of choice is N-methylmorpholine N-oxide (NMO), which becomes reduced to the water-soluble by-product N-methylmorpholine.

$OCOCH_3$ + N-methylmorpholine N-oxide $\xrightarrow[\text{(74\%)}]{0.1 \text{ equiv } OsO_4,\ C_2H_5OH,\ H_2O}$ diol + N-methylmorpholine

To a solution of 0.5 g (2.0 mmol) of OsO_4 and 5 g (19 mmol) of N-methylmorpholine N-oxide in 20 ml of THF and 70 ml of *t*-butyl alcohol is added 5 g (19 mmol) of alkene. After 48 hr at room temperature the reaction mixture is diluted with water and extracted with ether. The ether extract is washed with cold, dilute hydrochloric acid to remove residual N-methylmorpholine and then dried over $MgSO_4$. Evaporation of the ether gives 4.2 g (74%) of diol, m.p. 152–154 °C.

An important recent development has been the discovery that the addition of chiral amines causes the OsO_4 addition to take place preferentially from one of the two enantiotopic faces of the alkene so that one of the two enantiomeric diols is formed in excess, relative to the other. The best chiral reagents are ethers of the natural product quinine (page 1255); an example is the following complex amino ether called MEQ.

MEQ

The chiral additive presumably forms a reactive complex with OsO_4. Because MEQ is derived from a natural product that exists in only one of the two possible enantiomeric forms, the active hydroxylation reagent is enantiomerically homogeneous. Therefore, the two transition states for attack on the enantiotopic faces of the alkene are diastereomeric and have different energies. Thus, hydroxylation of one of the enantiotopic faces has a lower activation energy than hydroxylation of the other enantiotopic face, and the result is that one enantiomer is formed in excess over the other.

$COOCH_3$ $\xrightarrow[t\text{-BuOH},\ H_2O]{1 \text{ mol \% OsO4},\ 2 \text{ mol \% MEQ},\ K_3Fe(CN)_6}$ (99%) + (1%)

The co-oxidant in this hydroxylation is ferric ion, which is used in the form of the complex salt potassium ferricyanide.

The foregoing reaction is an example of **asymmetric synthesis,** in which one of the two possible enantiomers is formed in greater amount than the other through the use of a chiral reagent or reaction additive. The efficiency of an asymmetric synthesis is described in terms of its **"enantiomeric excess,"** or % **ee** (page 416). In this case, the two enantiomers are formed in a ratio of 99:1, so the reaction proceeds with 98% ee (99 − 1 = 98).

EXERCISE 27.2 In Section 11.4 we learned that both cis and trans isomers are capable of existence in rings containing eight or more members. What is the structure of the diol produced by hydroxylation of *trans*-cyclooctene with $KMnO_4$? Use your molecular model kit in formulating the answer to this question. Compare your answer with the reaction given for *cis*-cyclooctene on page 858.

Reductive dimerization of ketones serves as a preparation for some symmetrical 1,2-diols. The reducing agent is generally an electropositive metal, such as sodium or magnesium; titanium (III) generated by the combination of $TiCl_4$ and magnesium amalgam can also be used. The reaction occurs by electron transfer from the metal to the ketone to produce a **ketyl, or radical anion.** Dimerization of two radical anions affords the dianion of a 1,2-diol, which is protonated in a separate step to the diol itself.

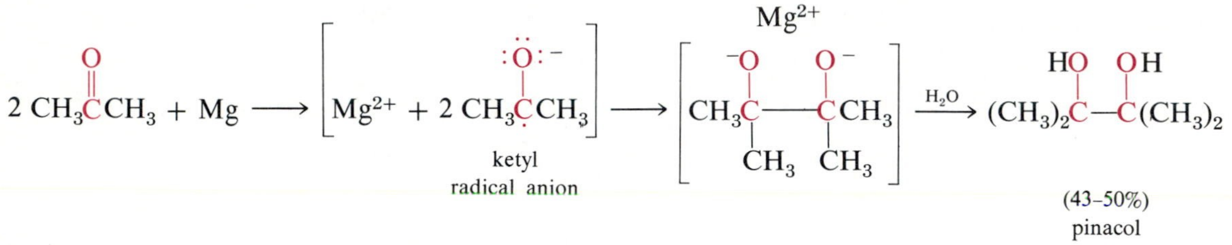

Because the diol produced from acetone has the trivial name "pinacol," this reaction is called the **pinacol reaction.**

Recently, special reducing agents have been developed that permit the effective cross coupling of two different aldehydes to give unsymmetrical 1,2-diols. An example is V^{2+}, which will bring about the coupling of two different aldehydes if one has a coordinating group elsewhere in the molecule **(Pedersen reaction).**

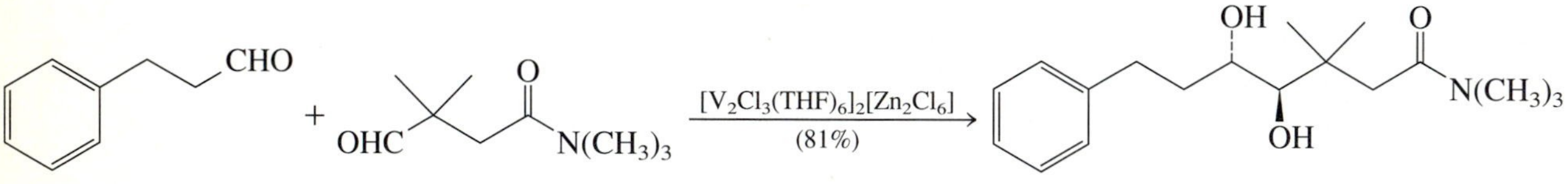

The reagent in the foregoing reaction is prepared from vanadium chloride and zinc dust. It has the interesting dimeric structure shown.

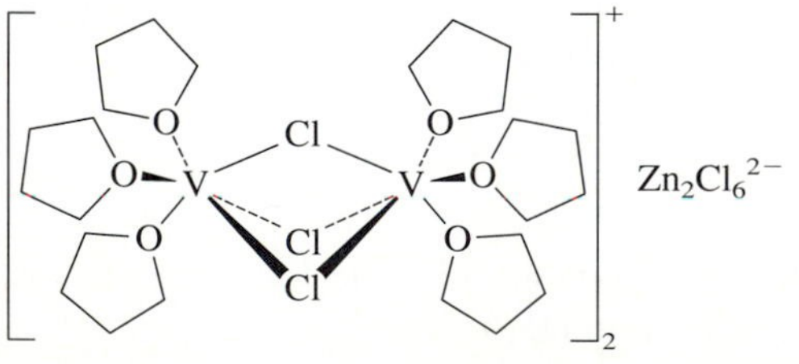

Other types of diols are generally prepared by reduction of the appropriate dicarbonyl compounds, as indicated by the following examples.

$$C_2H_5OC(=O)(CH_2)_8C(=O)OC_2H_5 \xrightarrow{LiAlH_4} HOCH_2(CH_2)_8CH_2OH$$

(75%)
1,10-decanediol

$$CH_3C(=O)(CH_2)_3C(=O)CH_3 + NaBH_4 \xrightarrow{C_2H_5OH} CH_3CH(OH)(CH_2)_3CH(OH)CH_3$$

(80%)
2,6-heptanediol

The **Sharpless asymmetric epoxidation** reaction is a valuable method for the synthesis of diols and related compounds. The reaction is specific for allylic alcohols and is carried out with titanium tetra-isopropoxide and *t*-butyl hydroperoxide in the presence of the chiral additive diethyl tartrate. The product is an epoxy alcohol that can be transformed into 1,3- and 1,2-diols by appropriate methods. Since both enantiomers of tartaric acid are commercially available, Sharpless epoxidation can be used to prepare either enantiomer of a desired product.

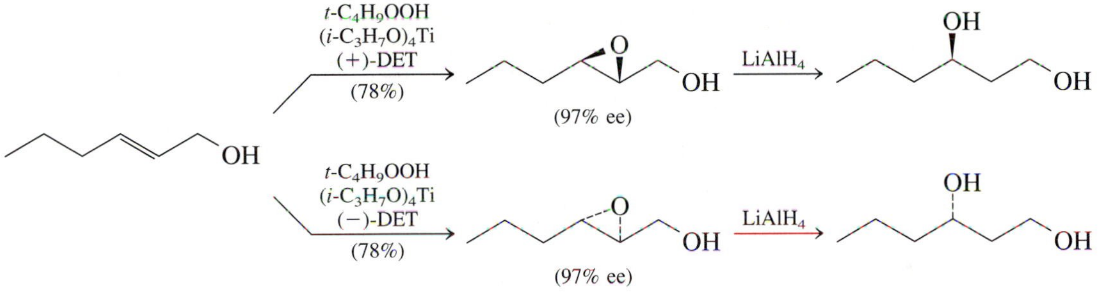

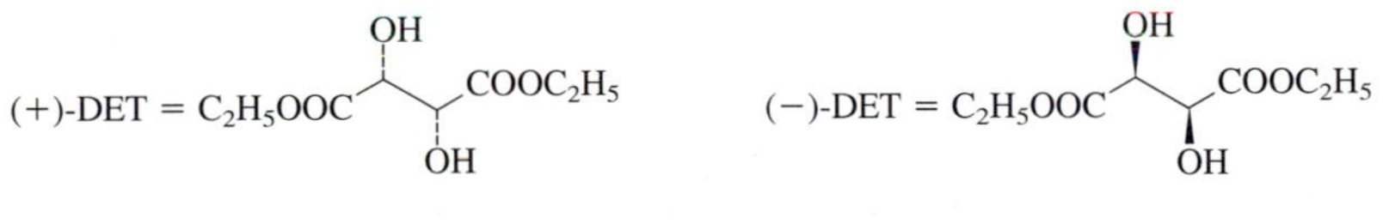

OH $\xrightarrow[(78\%)]{t\text{-}C_4H_9OOH,\ (i\text{-}C_3H_7O)_4Ti,\ (+)\text{-DET}}$ O, OH (97% ee) $\xrightarrow{LiAlH_4}$ OH, OH

OH $\xrightarrow[(78\%)]{t\text{-}C_4H_9OOH,\ (i\text{-}C_3H_7O)_4Ti,\ (-)\text{-DET}}$ O, OH (97% ee) $\xrightarrow{LiAlH_4}$ OH, OH

(+)-DET = C_2H_5OOC–CH(OH)–CH(OH)–$COOC_2H_5$ (−)-DET = C_2H_5OOC–CH(OH)–CH(OH)–$COOC_2H_5$

EXERCISE 27.3 Write equations for the overall anti hydroxylation of each of the following alkenes using the sequence of reagents (1) peroxyacetic acid and (2) aqueous sulfuric acid. Clearly illustrate the stereochemistry at each step of the process.

(a) *cis*-2-butene (b) *trans*-2-butene (c) *cis*-2-pentene
(d) *trans*-2-pentene (e) *trans*-cyclooctene

B. Reactions

One unique reaction of 1,2- and 1,3-diols is their reaction with aldehydes and ketones to form cyclic acetals (Section 14.6.B).

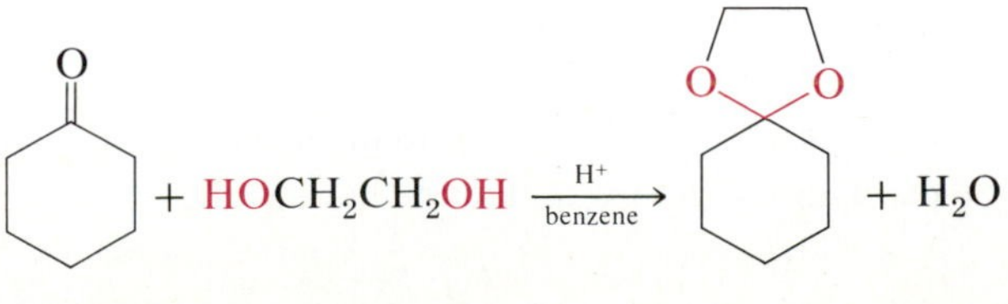

$$\text{cyclohexanone} + HOCH_2CH_2OH \xrightarrow[\text{benzene}]{H^+} \text{cyclic acetal} + H_2O$$

Diols with more than three carbons intervening between the two hydroxy functions do not generally give the reaction, because the resulting ring would be seven-membered or greater.

Dehydration of 1,2-diols under acid catalysis is frequently accompanied by skeletal rearrangement. For example, pinacol (2,3-dimethylbutane-2,3-diol) reacts with sulfuric acid to give *t*-butyl methyl ketone, which has the trivial name "pinacolone."

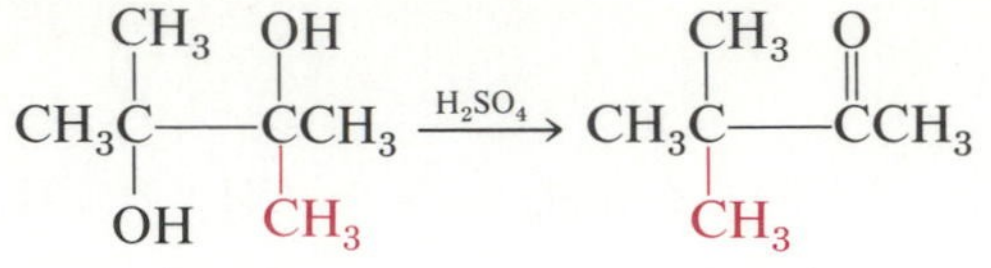

pinacol — pinacolone

The mechanism of this pinacol rearrangement involves 1,2-migration of a methyl group and its bonding electron pair from one carbinyl position to an adjacent electron-deficient center (Section 10.6.C). The driving force for the rearrangement is formation of a stable oxonium ion, the conjugate acid of a ketone.

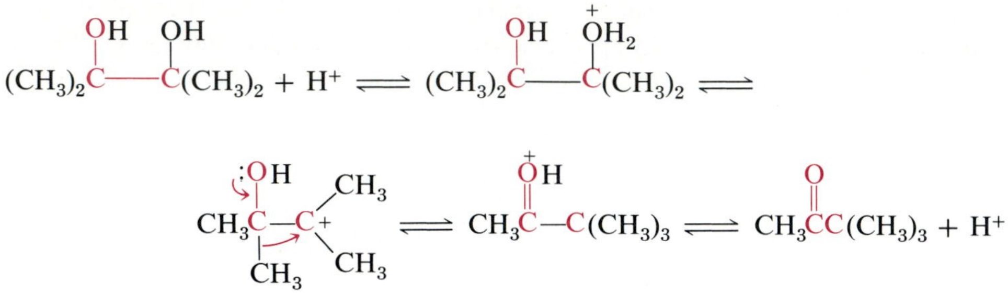

tertiary carbocation — oxonium ion — pinacolone

By combining the pinacol reaction with this acid-catalyzed rearrangement process, interesting and unusual compounds can be prepared.

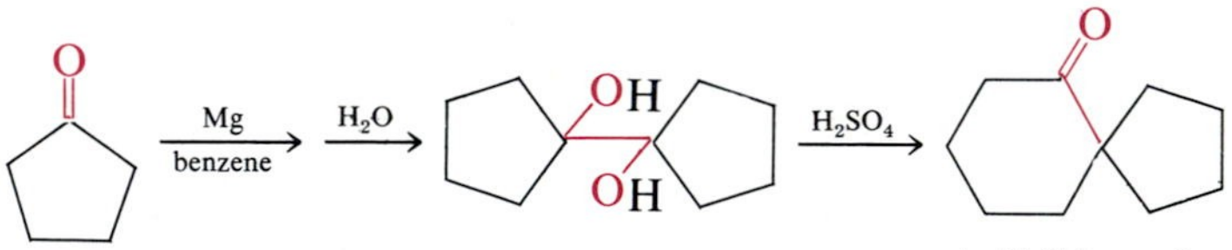

spiro[4.5]decan-6-one

Bicyclic compounds having one carbon common to both rings are **spiro** compounds. The nomenclature is based on the following scheme.

$(CH_2)_nC(CH_2)_m \equiv$ spiro[*n*.*m*]alkane

Numbering starts next to the common carbon and proceeds around the smaller ring first.

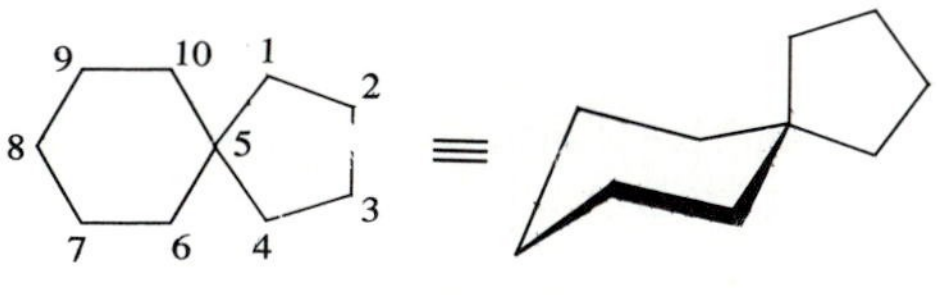

spiro[4.5]decane

Dehydration of 1,4- and 1,5-diols often leads to the formation of cyclic ethers, particularly when one of the hydroxy groups is tertiary.

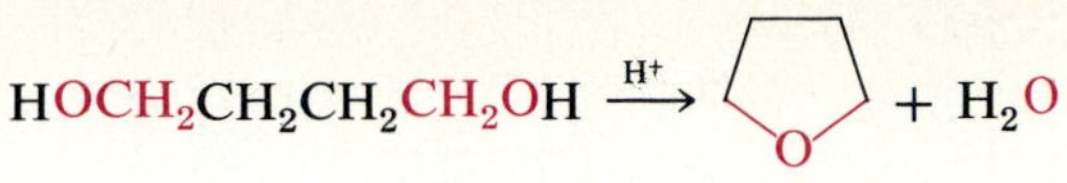

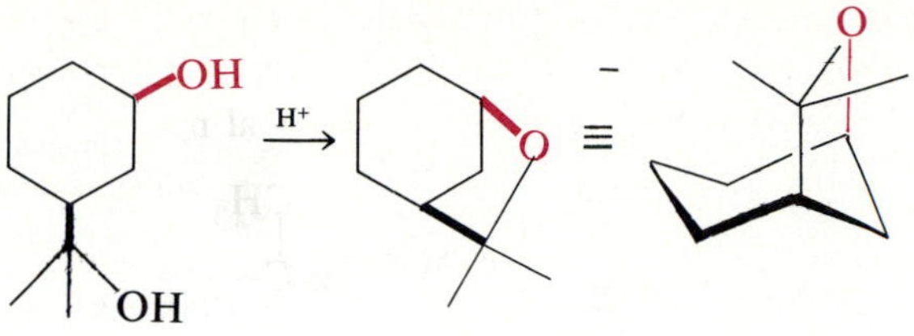

$$HOCH_2CH_2CH_2CH_2OH \xrightarrow{H^+} \text{(tetrahydrofuran)} + H_2O$$

In the first of the two foregoing examples, the reaction undoubtedly occurs by intramolecular nucleophilic displacement on the initially formed oxonium ion.

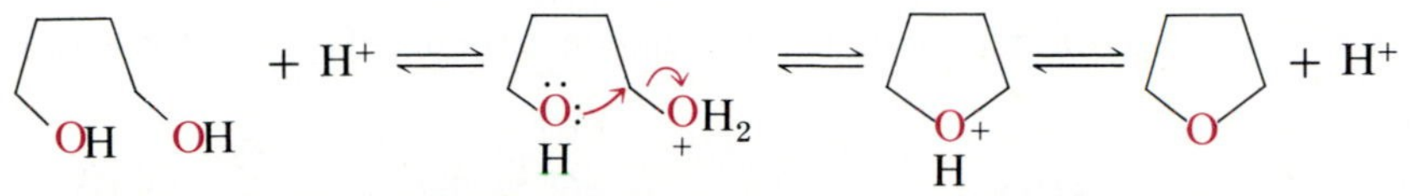

The second example probably involves the formation of a tertiary carbocation which is trapped by the secondary hydroxy group.

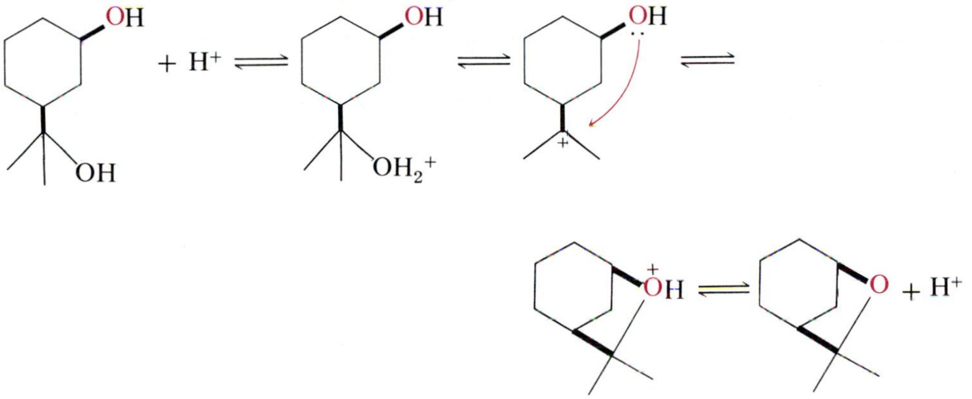

In this case the cyclization is possible because the two groups are cis; the trans analog cannot give a cyclic product.

EXERCISE 27.4 What are the principal products of the pinacol rearrangements of the following diol when Y = CH_3 and Y = NO_2?

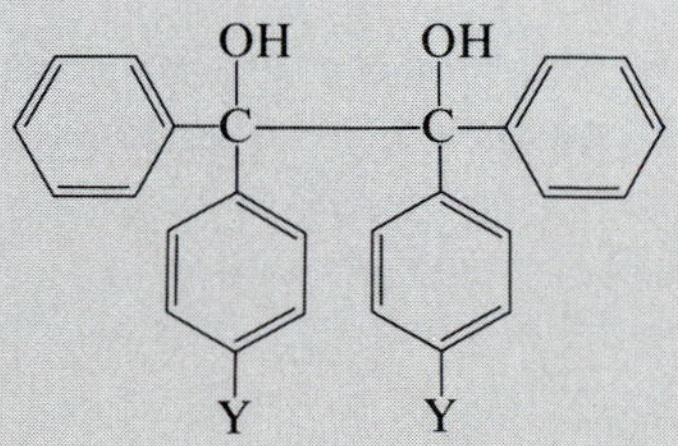

1,2-Diols undergo easy cleavage of the carbon-carbon bond joining the two hydroxy carbons when treated with periodic acid (''per-iodic'' acid, HIO_4) or lead tetraacetate. Combined with the hydroxylation process, this oxidation constitutes a method for the cleavage of alkenes that is complementary to ozonolysis (Section 11.6.E).

$$\text{norbornane}{=}CH_2 \xrightarrow[H_2O_2]{OsO_4} \text{(OH, } CH_2OH\text{)} \xrightarrow[\substack{CH_3OH\text{-}H_2O \\ (25\,°C)}]{HIO_4} \text{norbornan-2-one}{=}O + CH_2{=}O$$

(85%)
bicyclo[2.2.1]heptan-2-one
norbornanone

$$\begin{matrix} COOC_4H_9 \\ | \\ CHOH \\ | \\ CHOH \\ | \\ COOC_4H_9 \end{matrix} + Pb(OAc)_4 \xrightarrow{\text{benzene}} 2\ \begin{matrix} COOC_4H_9 \\ | \\ CHO \end{matrix} + Pb(OAc)_2 + 2\ CH_3COOH$$

(77–87%)

The periodic oxidation involves the formation of a cyclic diester of periodic acid. Decomposition of this cyclic diester yields the two carbonyl fragments and iodic acid.

$$\begin{matrix} -\overset{|}{C}-OH \\ -\underset{|}{C}-OH \end{matrix} + HIO_4 \longrightarrow \begin{matrix} -\overset{|}{C}-O \\ -\underset{|}{C}-O \end{matrix}\!\!>\!I(OH)(=O)_2 \longrightarrow \begin{matrix} -\overset{|}{C}=O \\ -\underset{|}{C}=O \end{matrix} + HIO_3$$

$$+ H_2O$$

Various procedures have been developed in which alkene hydroxylation and the diol cleavage reactions are combined into one operation. One such reaction (the **Lemieux-Johnson reaction**) involves treating an alkene with sodium periodate and a catalytic amount of osmium tetroxide.

$$\text{cyclohexene} + NaIO_4 \xrightarrow[H_2O]{OsO_4} OHC(CH_2)_4CHO + NaIO_3$$

A mixture of 15 mL of ether, 15 mL of water, 0.41 g of cyclohexene, and 0.065 g of OsO_4 is stirred at 25 °C while 2.32 g of $NaIO_4$ is added over a period of 40 min. After an additional 80 min at 25 °C, the product adipaldehyde is isolated in 77% yield.

EXERCISE 27.5 What is the product of the reaction of cyclohexanone with magnesium, followed by hot aqueous sulfuric acid?

27.4 Hydroxy Aldehydes and Ketones

A. Synthesis

α-Hydroxy ketones result from the treatment of esters with sodium in an inert solvent such as ether or benzene. Such compounds are called **acyloins,** and the reaction is called the **acyloin condensation.** The initial product of the reaction is the disodium salt of an enediol, which is hydrolyzed to give the acyloin.

$$2\ CH_3CH_2CH_2C(=O)OMe + 4\ Na \xrightarrow{\text{ether}}$$

$$CH_3CH_2CH_2C(O^-\,Na^+){=}C(O^-\,Na^+)CH_2CH_2CH_3 + 2\ NaOMe \xrightarrow{H_2O} CH_3CH_2CH_2C(=O)CH(OH)CH_2CH_2CH_3$$

an acyloin

The acyloin condensation is a useful method for the synthesis of ring compounds, particularly for medium-sized rings (8-13 members). In such cases, the reaction must be carried out under conditions of high dilution to suppress intermolecular reactions.

$$\text{(cyclic diester)}\ \begin{matrix}\text{COOCH}_3\\ \text{COOCH}_3\end{matrix} \xrightarrow[\text{xylene}]{\text{Na}} \xrightarrow{H_2O} \text{(cyclic C(=O)–CH(OH))}$$

(66%)
2-hydroxycyclodecanone

The acyloin condensation is related mechanistically to the pinacol reaction (page 862) in that electron transfer from sodium to the ester carbonyl produces an intermediate ketyl. The chief side reaction is the Claisen condensation (Section 19.9), which stems from the alkoxide ion produced as a by-product in the reaction.

Aromatic aldehydes are converted into acyloins by sodium cyanide in aqueous ethanol. The reaction is called the **benzoin condensation,** and cyanide ion is a specific catalyst.

$$2\ C_6H_5CHO \xrightarrow[H_2O,\ C_2H_5OH]{CN^-} C_6H_5\overset{O}{\overset{\|}{C}}-\overset{OH}{\overset{|}{C}H}C_6H_5$$

benzaldehyde — (92%) benzoin

The catalyst functions by first adding to the carbonyl group to form the cyanohydrin (see Section 14.7.B). The former aldehyde hydrogen is now α to a cyano group and is sufficiently acidic to be removed by a base. The resulting carbanion then adds to another molecule of aldehyde to give an intermediate cyano diol. Elimination of cyanide ion yields the acyloin and regenerates the catalyst.

$$RCHO + CN^- \rightleftharpoons R\overset{O^-}{\overset{|}{C}}HCN \underset{^-OH}{\overset{H_2O}{\rightleftharpoons}} R\overset{OH}{\overset{|}{C}}HCN \underset{H_2O}{\overset{OH^-}{\rightleftharpoons}} R\overset{OH}{\overset{|}{\underset{-}{C}}}-CN \overset{RCHO}{\rightleftharpoons}$$

$$R\overset{HO}{\overset{|}{C}}(CN)-\overset{O^-}{\overset{|}{C}}HR \underset{^-OH}{\overset{H_2O}{\rightleftharpoons}} R\overset{HO}{\overset{|}{C}}(CN)-\overset{OH}{\overset{|}{C}}HR \underset{H_2O}{\overset{^-OH}{\rightleftharpoons}} R\overset{^-O}{\overset{|}{C}}(CN)-\overset{OH}{\overset{|}{C}}HR \rightleftharpoons R\overset{O}{\overset{\|}{C}}-\overset{OH}{\overset{|}{C}}HR + CN^-$$

Although the acyloin and benzoin condensations produce the same type of product, it is important to remember that they involve *entirely different mechanisms*.

The most general synthesis of β-hydroxy aldehydes and ketones is the aldol addition reaction (Section 15.2). Recall that simple aldehydes condense to form β-hydroxy aldehydes when treated with cold aqueous base.

$$2\ CH_3CHO \xrightarrow[\substack{H_2O\\ 0°C}]{NaOH} CH_3\overset{OH}{\overset{|}{C}}HCH_2CHO$$

Under more forcing conditions, such as are necessary to accomplish the initial condensation with aldehydes of more than six carbons, the β-hydroxy aldehyde undergoes dehydration to give the α,β-unsaturated aldehyde.

$$\text{heptanal} \xrightarrow[\text{EtOH, reflux}]{NaOC_2H_5} \text{2-pentyl-2-nonenal (70\%)}$$

Mixed aldol reactions can be performed by converting a ketone completely into the lithium enolate, which is then allowed to react with an aldehyde. Hydrolysis of the initially formed alkoxide with water affords the β-hydroxy ketone, usually in good yield.

$$CH_3\overset{O}{\overset{\|}{C}}CH_3 \xrightarrow[\text{THF, } -70°C]{(i\text{-}C_3H_7)_2NLi} \xrightarrow{C_6H_5CHO} \xrightarrow{H_2O} C_6H_5\overset{OH}{\overset{|}{C}}HCH_2\overset{O}{\overset{\|}{C}}CH_3$$

(88%)
4-hydroxy-4-phenyl-2-butanone

EXERCISE 27.6 The acyloin, benzoin, and aldol addition reactions may be used as routes to diols. Show how the following diols can be prepared using one of these reactions as a key step.

(a) 1,2-diphenylethane-1,2-diol
(b) hexane-3,4-diol
(c) butane-1,3-diol
(d) hexane-2,4-diol

Discuss the stereochemical problem that arises in three of these syntheses.

B. Reactions

β-Hydroxy aldehydes and ketones undergo acid-catalyzed dehydration more easily than normal alcohols. The following examples illustrate the magnitude of the differences.

Relative Rate

$$CH_3\overset{OH}{\overset{|}{C}}HCH_2CH_2CH_3 \xrightarrow{H_2SO_4} \text{mixture of pentenes} \qquad 1$$

$$CH_3\overset{OH}{\overset{|}{C}}HCH_2\overset{O}{\overset{\|}{C}}CH_3 \xrightarrow{H_2SO_4} CH_3CH{=}CH\overset{O}{\overset{\|}{C}}CH_3 \qquad >10^5$$

Recall that the dehydration of a secondary or tertiary alcohol involves the formation of an intermediate carbocation; the rate of formation of this intermediate determines the rate of dehydration (Section 10.6.D).

$$(1)\quad -\overset{H}{\overset{|}{\underset{|}{C}}}-\overset{OH}{\overset{|}{\underset{|}{C}}}- + H^+ \overset{\text{fast}}{\rightleftharpoons} -\overset{H}{\overset{|}{\underset{|}{C}}}-\overset{\overset{+}{O}H_2}{\overset{|}{\underset{|}{C}}}-$$

$$(2)\quad -\overset{H}{\overset{|}{\underset{|}{C}}}-\overset{\overset{+}{O}H_2}{\overset{|}{\underset{|}{C}}}- \overset{\text{slow}}{\rightleftharpoons} -\overset{H}{\overset{|}{\underset{|}{C}}}-C^+\!\!< + H_2O$$

$$(3)\quad -\overset{H}{\overset{|}{\underset{|}{C}}}-C^+\!\!< \overset{\text{fast}}{\rightleftharpoons} >C{=}C< + H^+$$

β-Hydroxy ketones undergo dehydration by a different mechanism, involving the enol form of the ketone. The rate-determining step is formation of the enol. Elimination of water from the protonated enol gives a resonance-stabilized oxonium ion, which is simply the protonated form of the α,β-unsaturated ketone.

(1) —C(OH)—CH—C(=O)— $\overset{\text{slow}}{\rightleftharpoons}$ —C(OH)—C=C(OH)—

(2) —C(OH)—C=C(OH)— + H^+ $\overset{\text{fast}}{\rightleftharpoons}$ —C($\overset{+}{O}H_2$)—C=C(OH)—

(3) —C($\overset{+}{O}H_2$)—C=C(OH)— $\overset{\text{fast}}{\rightleftharpoons}$ [$>\overset{+}{C}$—C=C(OH)— $\longleftrightarrow$ >C=C—C^+(OH)— $\longleftrightarrow$ >C=C—C(=$^+$OH)—]

(4) >C=C—C(=$\overset{+}{O}H$)— $\overset{\text{fast}}{\rightleftharpoons}$ >C=C—C(=O)— + H^+

Normal alcohols do not undergo dehydration under basic conditions, as shown by the fact that *t*-butyl alcohol solutions of potassium *t*-butoxide are quite stable. However, β-hydroxy aldehydes and ketones undergo dehydration fairly easily under basic conditions. In this case the dehydration actually proceeds via the enolate ion (Section 15.1.B).

—C(OH)—CH—C(=O)— + OH^- $\rightleftharpoons$ —C(OH)—C=C(O^-)—

—C(OH)—C=C(O^-)— $\rightleftharpoons$ >C=C—C(=O)— + OH^-

In contrast to the easy dehydration of β-hydroxy carbonyl compounds, α-hydroxy ketones undergo acid-catalyzed dehydration with even more difficulty than normal alcohols. In this case, the intermediate carbocation would be destabilized by the inductive effect of the adjacent carbonyl group.

—C($\overset{+}{O}H_2$)—C(=O)— $\xrightarrow{\text{very slow}}$ [$>\overset{+}{C}$—C(=O)— $\longleftrightarrow$ $>\overset{+}{C}$—$\overset{+}{C}$(O^-)—]

An example is the preparation of 3-methylbut-3-en-2-one by heating a mixture of the α-hydroxy ketone and *p*-toluenesulfonic acid in an oil bath at 150 °C. These conditions are far more vigorous than required for dehydration of normal tertiary alcohols.

$$CH_3C(OH)(CH_3)—CCH_3(=O) + CH_3—C_6H_4—SO_3H \xrightarrow{150°C} CH_2{=}C(CH_3)CCH_3(=O)$$

(74%)
3-methylbut-3-en-2-one

EXERCISE 27.7 The following reaction represents a useful method for the preparation of α,β-unsaturated ketones.

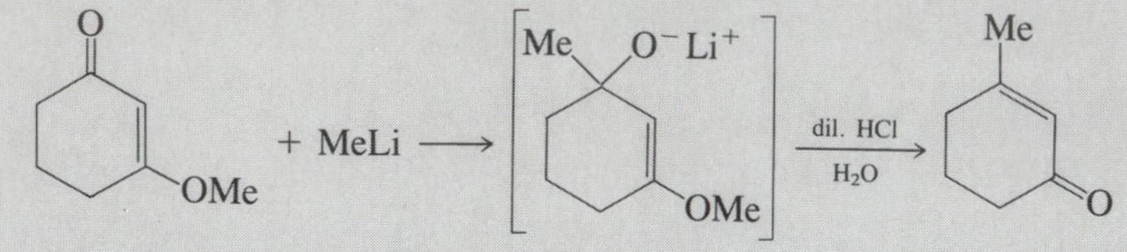
O
OMe
+ MeLi
Me
O⁻ Li⁺
OMe
dil. HCl
H₂O
Me
O

Account for the fact that the intermediate tertiary alcohol is dehydrated with great ease—more easily than a normal tertiary alcohol.

Many hydroxy aldehydes and ketones exist to some extent in a cyclic hemiacetal form (Section 14.6.B). This tendency is particularly pronounced when the ring is five- or six-membered.

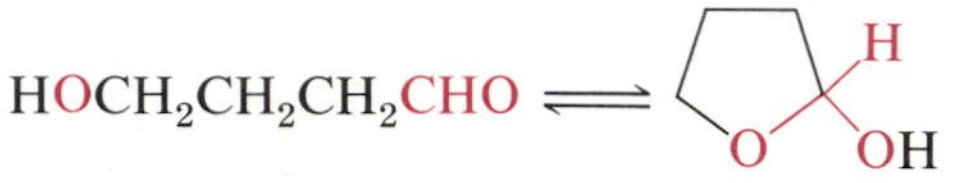
HOCH₂CH₂CH₂CHO
H
O
OH

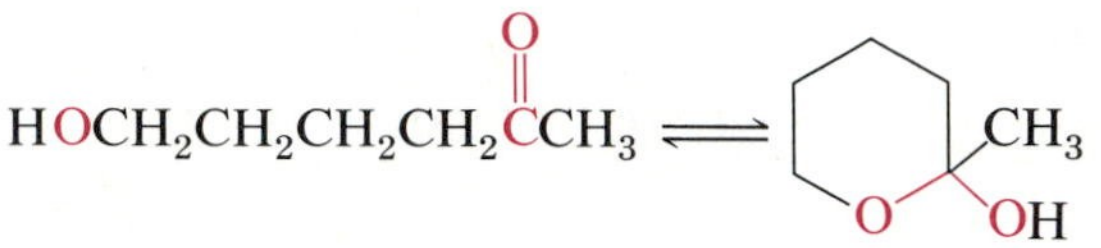
O
HOCH₂CH₂CH₂CH₂CCH₃
CH₃
O
OH

The data in Table 27.3 show that the cyclic form predominates with 4- and 5-hydroxy aldehydes. The formation of a cyclic hemiacetal from a hydroxy aldehyde or ketone is subject to acid or base catalysis, as in the formation of acetals by intermolecular reaction (Section 14.6.B). However, when five- or six-membered rings are involved, the cyclization is so facile that it occurs even under neutral conditions. Thus, any reaction that would nominally give a 4- or 5-hydroxy aldehyde or ketone will yield an equilibrium of the open-chain and ring-closed isomers.

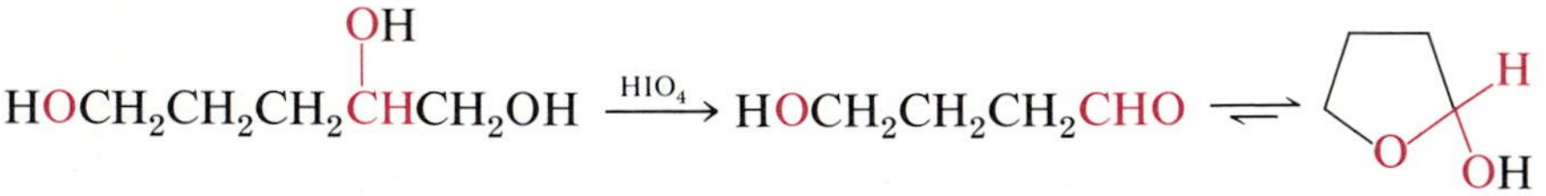
OH
HOCH₂CH₂CH₂CHCH₂OH
HIO₄
HOCH₂CH₂CH₂CHO
H
O
OH

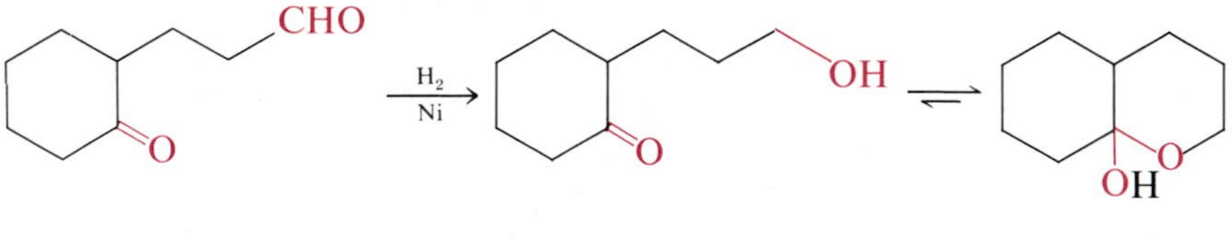
CHO
O
H₂
Ni
OH
O
O
OH

TABLE 27.3 Cyclic Hemiacetal-Hydroxyaldehyde Equilibria

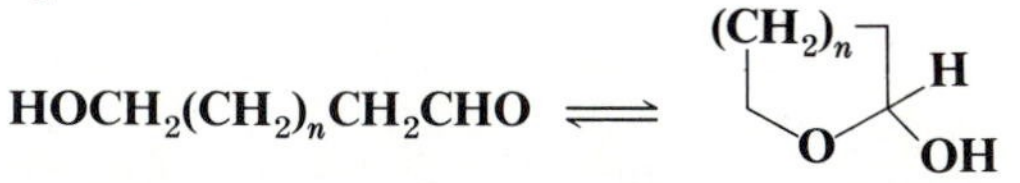
HOCH₂(CH₂)ₙCH₂CHO
(CH₂)ₙ
H
O
OH

n	Ring Size	Percent Free Aldehyde
1	5	11
2	6	6
3	7	85

Like noncyclic hemiacetals, these compounds react with alcohols under acid catalysis to give acetals.

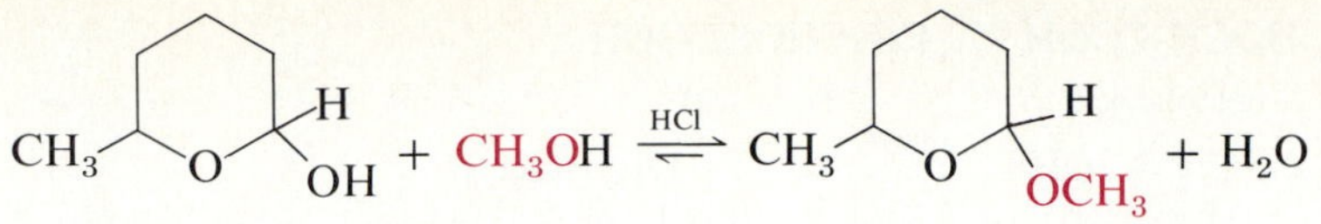

$$CH_3\text{-(2-tetrahydropyranyl, H, OH)} + CH_3OH \overset{HCl}{\rightleftharpoons} CH_3\text{-(2-tetrahydropyranyl, H, OCH}_3\text{)} + H_2O$$

Since there is usually a small amount of the open-chain hydroxy carbonyl compound in equilibrium with the cyclic hemiacetal form, solutions of such compounds can show reactions of either form, as the following examples show.

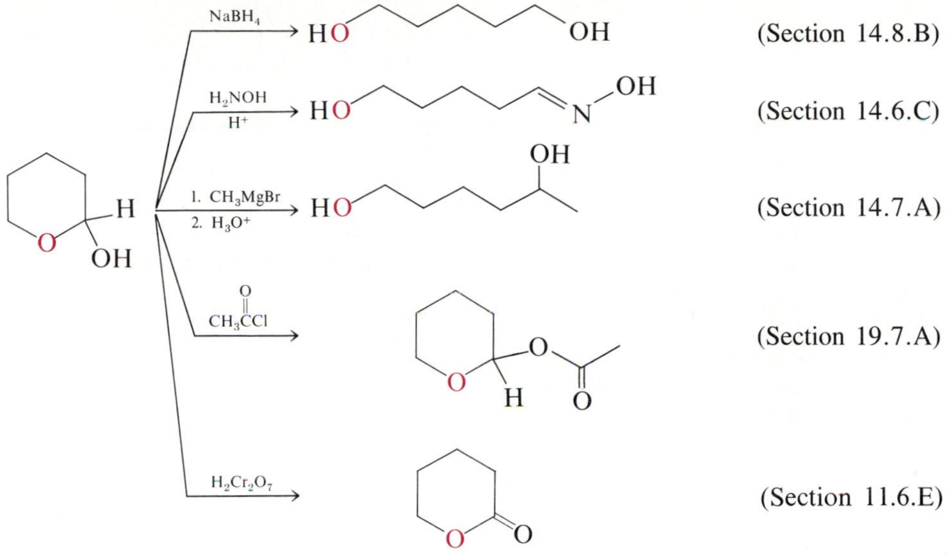

Cyclic hemiacetal (2-hydroxytetrahydropyran, H, OH):

$\xrightarrow{NaBH_4}$ HO~~~~~OH (Section 14.8.B)

$\xrightarrow[H^+]{H_2NOH}$ HO~~~~CH=N–OH (Section 14.6.C)

$\xrightarrow[2.\ H_3O^+]{1.\ CH_3MgBr}$ HO~~~~CH(OH)CH$_3$ (Section 14.7.A)

$\xrightarrow{CH_3CCl\ (O)}$ 2-acetoxytetrahydropyran (H, O–C(=O)CH$_3$) (Section 19.7.A)

$\xrightarrow{H_2Cr_2O_7}$ δ-valerolactone (ring O, C=O) (Section 11.6.E)

α-Hydroxy aldehydes and ketones, like 1,2-diols, are oxidized with C—C bond cleavage by periodic acid.

$$CH_3\overset{O}{\overset{\|}{C}}-\overset{OH}{\overset{|}{C}}HCH_3 + HIO_4 \longrightarrow CH_3COOH + CH_3CHO + HIO_3$$

$$CH_3CH_2\overset{OH}{\overset{|}{C}}HCHO + HIO_4 \longrightarrow CH_3CH_2CHO + H\overset{O}{\overset{\|}{C}}OH + HIO_3$$

The reaction constitutes a useful method for structure determination in the carbohydrate field (Chapter 28).

EXERCISE 27.8 Write mechanisms for the five reactions of the cyclic hemiacetal of 5-hydroxypentanal shown above. Clearly indicate each step.

27.5 Hydroxy Acids

A. Natural Occurrence

Many hydroxy acids are important in nature and have trivial names that are in common use. Glycolic acid is a constituent of cane-sugar juice. Lactic acid is responsible for the characteristic taste of sour milk; the odor is due to traces of volatile fatty acids. Malic acid occurs in fruit juices. Tartaric acid has been known since antiquity as the monopotassium salt (cream of tartar), which deposits in the lees of wine. Citric acid is a hydroxytricarboxylic acid that is widespread in nature and the initial component in the most important metabolic cycle, the tricarboxylic acid cycle. It is especially prevalent, as its trivial name implies, in the juice of citrus fruits. Chiral hydroxy acids are normally found in nature in the enantiomerically homogeneous, optically active form.

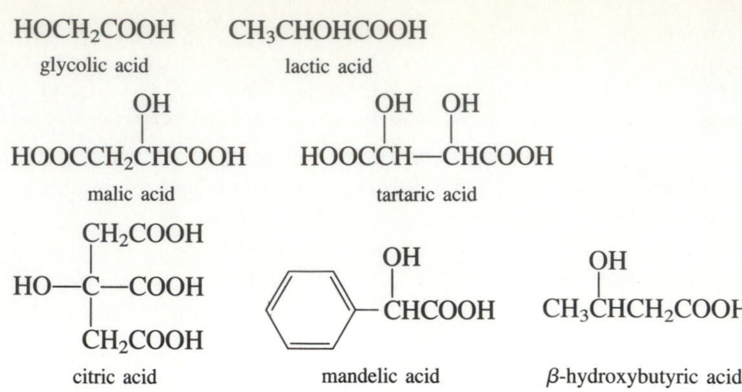

(*S*)-β-Hydroxybutyric acid forms a polymeric ester with itself that is widespread in cells and cell membranes. The natural polyester is finding increasing use as a commercial polymer because it is biodegradable. By contrast, the (*R*)-acid does not occur in nature and is *teratogenic* (page 700). Space-filling models of both are shown on the cover of this textbook.

The **tricarboxylic acid cycle** is a series of reactions starting with acetyl coA (see page 541) and oxaloacetic acid, the net effect being the oxidation of acetate to CO_2. We note below the mostly familiar organic reaction types that correspond to the biochemical reactions. The thiamine-catalyzed decarboxylation has some new aspects described in Section 27.7.C. Three of

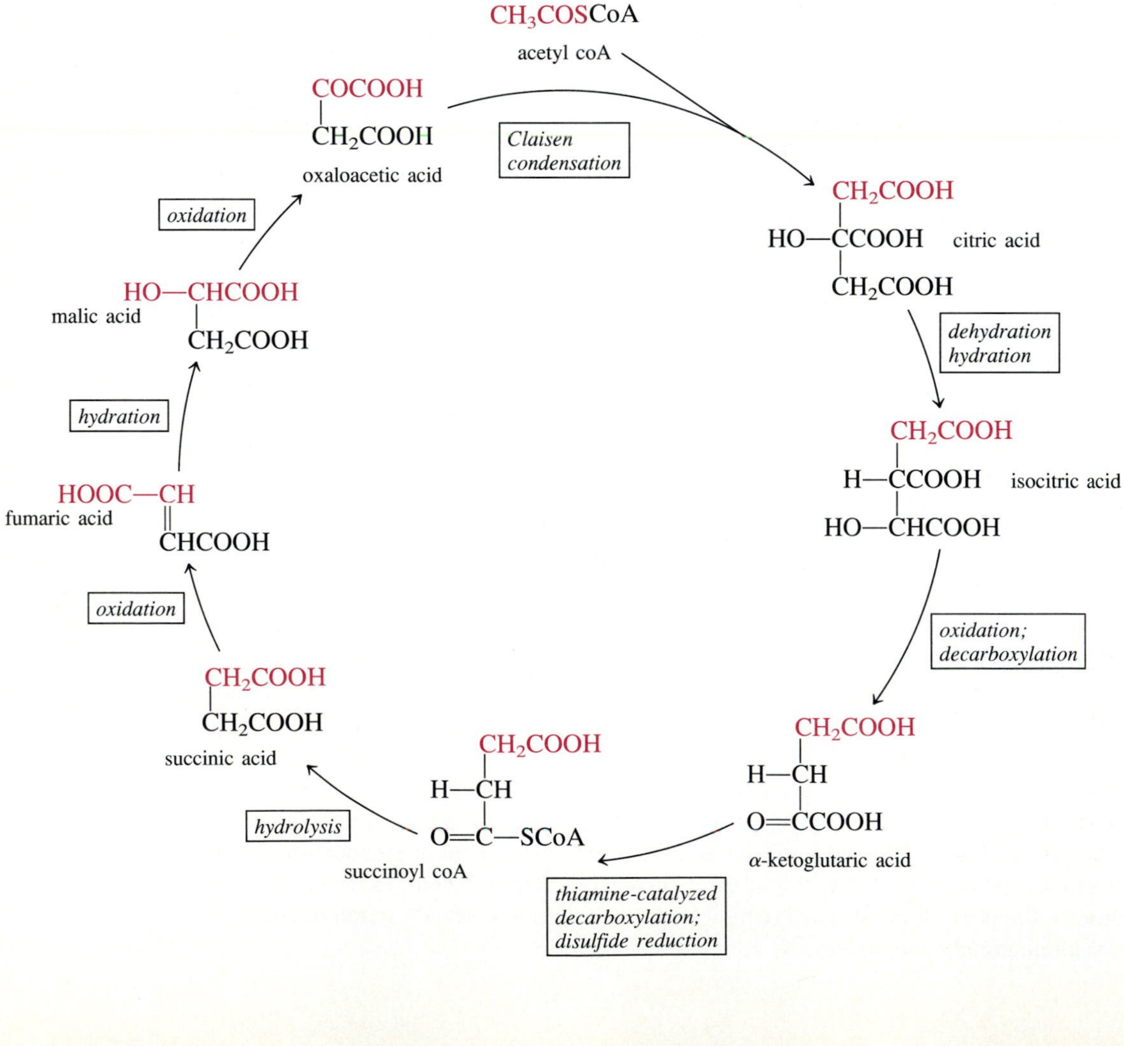

$CH_3COSCoA$
acetyl coA

$COCOOH$ | CH_2COOH
oxaloacetic acid

Claisen condensation

CH_2COOH | HO—$CCOOH$ | CH_2COOH
citric acid

dehydration hydration

CH_2COOH | H—$CCOOH$ | HO—$CHCOOH$
isocitric acid

oxidation; decarboxylation

CH_2COOH | H—CH | O=$CCOOH$
α-ketoglutaric acid

thiamine-catalyzed decarboxylation; disulfide reduction

CH_2COOH | H—CH | O=C—SCoA
succinoyl coA

hydrolysis

CH_2COOH | CH_2COOH
succinic acid

oxidation

HOOC—CH || CHCOOH
fumaric acid

hydration

HO—CHCOOH | CH_2COOH
malic acid

oxidation

the eight compounds in the cycle (citric, isocitric, and malic acids) are α-hydroxy carboxylic acids, and two are α-keto carboxylic acids (oxaloacetic and α-ketoglutaric acids). Amusingly, the first publication of one of the authors [L. Friedman and E. M. Kosower, ''Preparation of α-Ketoglutaric Acid,'' *Org. Syn.* **26,** 42–44 (1946)] described a synthetic procedure for α-ketoglutaric acid and included an acknowledgment of help from another of the authors (A. Streitwieser).

Both the (+) and (−) forms of tartaric acid are found in nature, although the (+) acid is by far the more common. Two optically inactive forms are known. **''Racemic acid,''** m.p. 206 °C, is simply a mixture of (+)- and (−)-tartaric acids. *meso*-Tartaric acid, m.p. 140 °C, is the *R*,*S* diastereomer.

HOOC COOH; C—C; H, OH; OH, H

(+)-(*R*,*R*)-tartaric acid

HOOC COOH; C—C; H, H; OH, OH

meso-tartaric acid

Tartaric acid played an important role in the development of stereochemistry. In 1848, Louis Pasteur noticed that crystals of sodium ammonium tartrate are chiral and that all of the crystals show chirality in the same sense. He proceeded to investigate 19 different tartrate salts and found that they all gave chiral crystals. On the basis of these observations, he postulated that there is a relationship between the chirality of the crystals and the fact that, in solution, the salts rotate the plane of polarized light.

However, there was a problem. The optically inactive racemic acid, obtained as a by-product in the crystallization of tartaric acid, was also known at this time. Racemic acid and tartaric acid were recognized to be isomers, and Mitscherlich had reported that crystals of sodium ammonium tartrate and sodium ammonium racemate are identical in all respects except that the tartrate gives a dextrorotatory solution whereas the racemate gives an optically inactive solution. This report could only be rationalized to Pasteur's hypothesis if the crystals of sodium ammonium racemate turned out to be achiral.

Pasteur repeated Mitscherlich's work on sodium ammonium racemate and was disappointed to discover that Mitscherlich had been correct and that crystals of the racemate salt are indeed chiral. Upon closer examination, however, he noticed that the crystals are not all chiral in the same sense. In his words, ''the hemihedral faces which in the tartrate are all turned one way are in the racemate inclined sometimes to the right and sometimes to the left.'' In short, the racemate salt gives a mixture of nonsuperimposable mirror image crystals. Using a pair of tweezers, Pasteur carefully separated the left-handed from the right-handed crystals, dissolved each in water, and measured their optical rotations. To his great excitement, he discovered that one solution was dextrorotatory and the other was levorotatory. When he converted the separated salts back to the free acids, he found that one was identical with natural (+)-tartaric acid and that the other was a new tartaric acid isomer, identical in all respects save the sign of its optical rotation. Pasteur had accomplished the first resolution—separation of a racemate into its component enantiomers.

Pasteur's work paved the way for an understanding of stereoisomerism. He made the important suggestion that since the crystals of the enantiomeric salts show handedness, the molecules themselves might also show handedness—and this before the idea of chemical bonds had even been conceived.

B. Synthesis

α-Hydroxy acids are most commonly prepared by hydrolysis of α-halo acids. Recall that α-halo acids are readily available by bromination of carboxylic acids (Section 18.7.B). Thus the two-step sequence provides a way to introduce the hydroxy group at the α-position of a carboxylic acid.

$$CH_3CH_2COOH \xrightarrow[P]{Br_2} CH_3CHBrCOOH \xrightarrow{OH^-} CH_3CHOHCO_2^-$$

β-Hydroxy acids cannot be prepared by hydrolysis of the corresponding β-halo acid, since these compounds undergo elimination in base to give the unsaturated acids.

$$CH_3CHBrCH_2COOH \xrightarrow{OH^-} CH_3CH{=}CHCO_2^-$$

α-Hydroxy acids are also generally available by hydrolysis of cyanohydrins, which result from the reaction of HCN with aldehydes or ketones (Section 14.7.B). Since the addition of HCN to a carbonyl group is reversed by the strongly basic conditions necessary to hydrolyze a nitrile to an acid, the hydrolysis is done under acidic conditions.

$$CH_3CH_2\overset{O}{\overset{\|}{C}}CH_3 \xrightarrow{HCN} CH_3CH_2\underset{OH}{\overset{CH_3}{C}}CN \xrightarrow[25°C]{\text{conc. HCl}} CH_3CH_2\underset{OH}{\overset{CH_3}{C}}COOH$$

(55%)
2-methyl-2-hydroxybutanoic acid

β-Hydroxy acids and their derivatives are available by methods analogous to the aldol addition reaction (Sections 15.2, 19.9).

OMe; LDA, THF, −78°C; CHO; H_2O; OH, O, OMe

(95%)
methyl 3-hydroxy-2-methyl-3-phenylpropanoate

A useful variant of this basic method is the **Evans asymmetric aldol reaction,** in which the enolate is formed from a chiral carbonyl compound. Because the chiral enolate is used in only one enantiomeric form, one of the stereoisomeric β-hydroxy carbonyl compounds is formed in excess, relative to the others. The product of the Evans reaction can be hydrolyzed to obtain the β-hydroxy acid with excellent enantiomeric excess.

O, O, N, O; $(C_4H_9)_2BOSO_2CF_3$, $(C_2H_5)_3N$, CH_2Cl_2; CHO; OH, O, O, N, O; LiOH, H_2O, C_2H_5OH

OH, COOH + H, N, O, O

(98% ee)

The foregoing reaction illustrates the use of a boron enolate, which is produced by reaction of a ketone or related carbonyl compound with a dialkylboron trifluoromethanesulfonate.

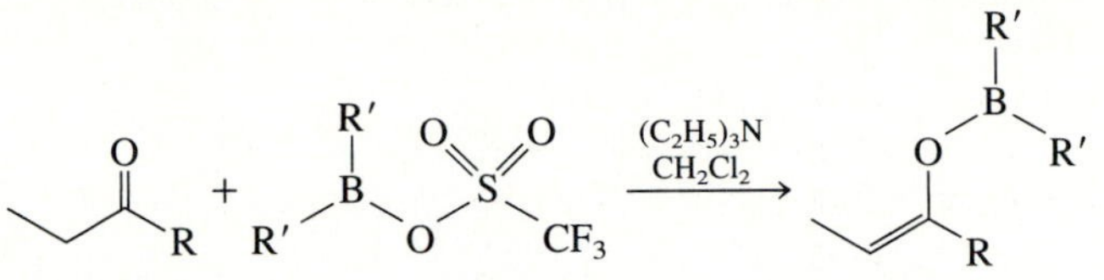

$$\text{CH}_3\text{CH}_2\text{COR} + R'_2B{-}O{-}SO_2{-}CF_3 \xrightarrow[CH_2Cl_2]{(C_2H_5)_3N} \text{CH}_3\text{CH}{=}C(R){-}O{-}BR'_2$$

Boron enolates are similar to the alkali metal enolates that were discussed in Chapter 15, except that the boron-oxygen bond is covalent, rather than ionic, as in the case of lithium, sodium, and potassium enolates. As a result, boron enolates are soluble in organic solvents like dichloromethane. In addition, boron enolates are not nearly as reactive as alkali metal enolates.

A variety of hydroxy acids are available by hydrolysis of lactones, which can be obtained by the Baeyer-Villiger oxidation of cyclic ketones (Section 14.8.A).

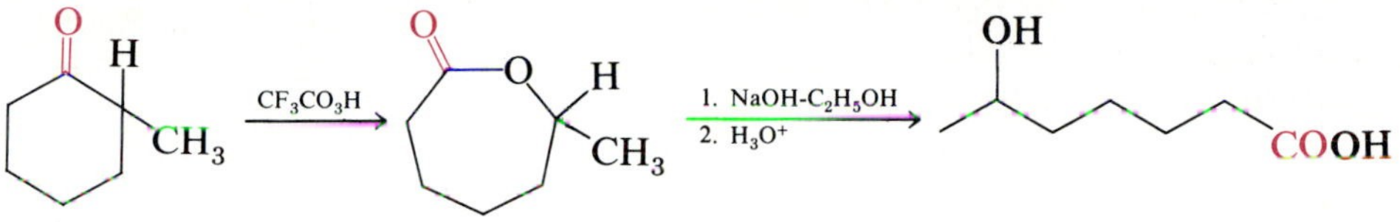

$$\text{2-methylcyclohexanone} \xrightarrow{CF_3CO_3H} \text{lactone (7-membered ring, CH}_3\text{)} \xrightarrow[2.\ H_3O^+]{1.\ NaOH\text{-}C_2H_5OH} CH_3CH(OH)CH_2CH_2CH_2CH_2COOH$$

EXERCISE 27.9 Show how 2-hydroxypentanoic acid may be prepared from (a) pentanoic acid and (b) butanal. Outline a synthesis of 3-hydroxypentanoic acid.

C. Reactions

Recall that carboxylic acids react with alcohols under acid catalysis to yield esters (Section 18.7.C).

$$\text{RCOH} + \text{R'OH} \overset{H^+}{\rightleftharpoons} \text{RCOR'} + H_2O$$

A hydroxy acid contains both of these functional groups, and thus it can undergo intramolecular esterification to yield a cyclic ester, which is called a **lactone.**

$$HOCH_2CH_2CH_2COOH \overset{H^+}{\rightleftharpoons} \text{(five-membered lactone)} + H_2O$$

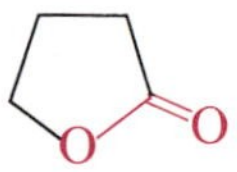

γ-hydroxybutyric acid — γ-butyrolactone

Lactonization, like normal esterification, is an equilibrium process. Only when the lactone has a five- or six-membered ring is there a substantial amount of lactone present under equilibrium conditions, as shown by the data in Table 27.4. These data also reveal that alkyl substitution on the ring increases the amount of lactone present at equilibrium.

TABLE 27.4 Hydrolytic Equilibria of Lactones

Lactone Formula	Equilibrium Composition	
	Hydroxy Acid, %	Lactone, %
(4-membered lactone)	100	0
(5-membered lactone)	27	73
(5-membered lactone, CH_3)	5	95
(5-membered lactone, CH_3, CH_3)	2	98
(6-membered lactone)	91	9
(6-membered lactone, CH_3)	79	21
(6-membered lactone, CH_3, CH_3)	75	25
(7-membered lactone)	~100	~0

Although the larger lactones do not exist to any appreciable extent in equilibrium with the free hydroxy acids, such lactones can be prepared under the proper conditions. It is necessary to treat the hydroxy acid with acid under conditions where the water formed in the reaction is removed so as to shift the unfavorable equilibrium toward the lactone. It is also necessary to operate in very dilute solution so as to minimize the intermolecular esterification reaction, which leads to a polymer.

OH, COOH $\xrightarrow[\text{benzene}]{H^+}$ (O, O) $+ H_2O$

15-hydroxypentadecanoic acid (0.007 M) — (100%) 15-hydroxypentadecanoic acid lactone

γ-Lactones and δ-lactones arise from the hydroxy acids so readily that it is often not necessary even to add acid to catalyze the intramolecular esterification; mere traces of acid in the solvent or on the glassware suffice to bring about lactonization. Thus, in any reaction that would yield a 4- or 5-hydroxy acid, the product isolated is often the corresponding lactone.

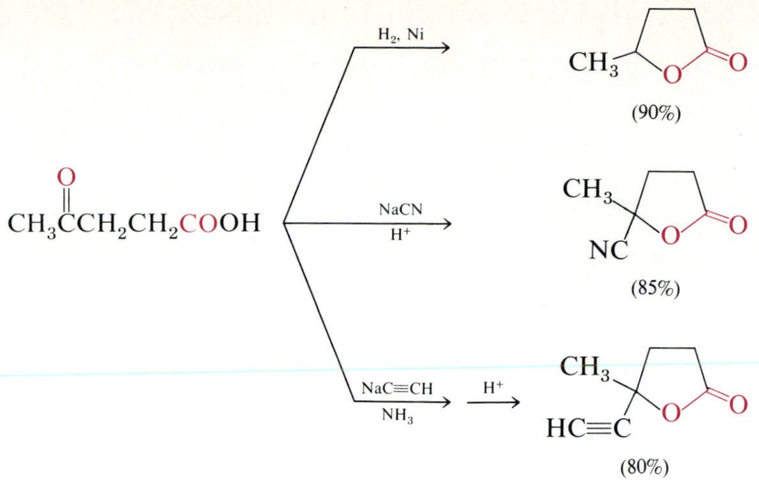

$$CH_3\overset{O}{\overset{\|}{C}}CH_2CH_2COOH \xrightarrow{H_2,\ Ni} \text{(5-methyl lactone)}\ (90\%)$$

$$\xrightarrow[H^+]{NaCN} \text{(5-cyano-5-methyl lactone)}\ (85\%)$$

$$\xrightarrow[NH_3]{NaC\equiv CH} \xrightarrow{H^+} \text{(5-ethynyl-5-methyl lactone)}\ (80\%)$$

Lactones can also result from reactions of other substituted carboxylic acids, as shown by the following examples.

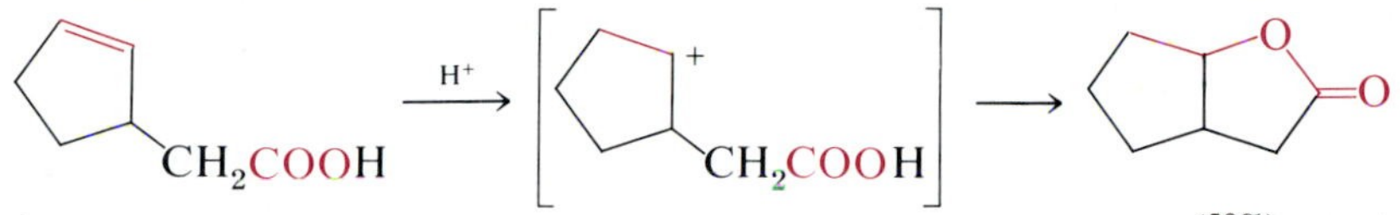

$$\text{(cyclopentenyl)}CH_2COOH \xrightarrow{H^+} \left[\text{(cyclopentyl cation)}CH_2COOH\right] \longrightarrow \text{(bicyclic lactone)}\ (50\%)$$

$$Br(CH_2)_{n}COOH \xrightarrow{NaOH} \left[\text{Br}\ \cdots\ O^-\right] \longrightarrow \text{(macrocyclic lactone)}\ (56\%)$$

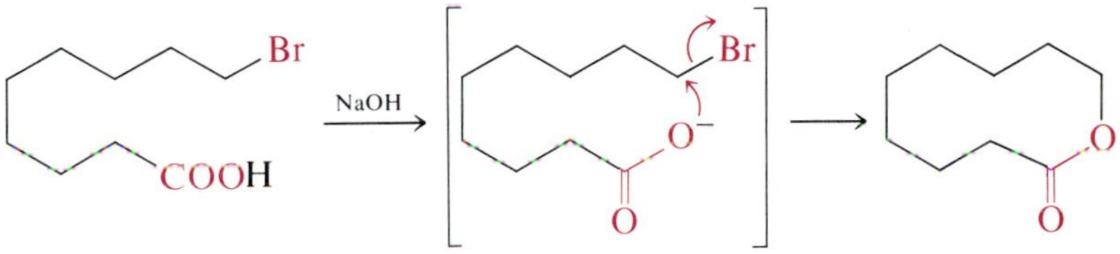

The latter reaction must be carried out under high dilution to suppress intermolecular displacement reactions.

EXERCISE 27.10 Suggest a multistep synthesis of the following lactone, starting with ethyl acetate and 1-bromo-3-methyl-2-butene.

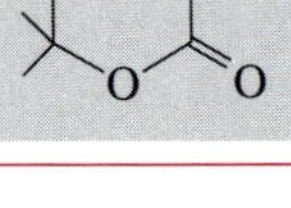

(6,6-dimethyl six-membered lactone)

Hydroxy acids that cannot form five- or six-membered rings undergo polymerization unless the reaction is carried out under high dilution conditions.

$$n\ HO(CH_2)_8\overset{O}{\overset{\|}{C}}OH \longrightarrow HO(CH_2)_8\overset{O}{\overset{\|}{C}}O(CH_2)_8\overset{O}{\overset{\|}{C}}O(CH_2)_8\overset{O}{\overset{\|}{C}}O\ldots$$

α-Hydroxy acids cannot form a stable lactone ring (three-membered), so they undergo intermolecular self-esterification under acid catalysis. However, the initial dimeric product is now a form of 5-hydroxy acid, so lactonization occurs. The product, which is a dilactone containing two molecules of the original α-hydroxy acid, is called a **lactide.**

$$2\,CH_3\overset{OH}{\overset{|}{C}}HCOOH \xrightarrow[-H_2O]{H^+} CH_3\overset{HO}{\overset{|}{C}}H\overset{O}{\overset{\|}{C}}O\underset{CH_3}{\underset{|}{C}}HCOOH \xrightarrow{-H_2O} \text{lactide}$$

lactic acid — lactide

Like β-hydroxy aldehydes and ketones, β-hydroxy acids and their derivatives undergo dehydration easily under acidic conditions. The mechanism is similar to that for dehydration of the other β-hydroxy carbonyl compounds discussed previously (Sections 20.3.A and 27.4.B). Since conjugation of a double bond with an acid or ester carbonyl group is less stabilizing than with an aldehyde or ketone carbonyl (Section 20.3), mixtures of the α,β-unsaturated and β,γ-unsaturated acids often result from dehydration of a β-hydroxy acid.

$$CH_3CH_2\overset{OH}{\underset{CH_3}{C}}CH_2COOC_2H_5 \xrightarrow[\Delta]{KHSO_4} CH_3CH_2\overset{CH_3}{\overset{|}{C}}{=}CHCOOC_2H_5 + CH_3CH{=}\overset{CH_3}{\overset{|}{C}}CH_2COOC_2H_5$$

(57%) (43%)

EXERCISE 27.11 What are the principal products when each of the following hydroxy acids is treated with acid?

(a) 2-hydroxybutanoic acid
(b) 3-hydroxybutanoic acid
(c) 4-hydroxybutanoic acid

27.6 Dicarboxylic Acids

The simple aliphatic dicarboxylic acids are fairly widespread in nature and crystallize readily from aqueous solutions. Consequently, they are easy to isolate and were among the earliest known organic compounds. Oxalic acid occurs in many plants, such as rhubarb, usually as the potassium salt. The insoluble calcium salt is found in plant cells and in some calculi, which are stony deposits found in the human body. The acid is poisonous. Succinic acid occurs in fossils, fungi, lichens, and amber. It was first isolated in 1546 from the distillate of amber. Glutaric acid occurs in sugar beets and is also found in the aqueous extract of crude wool. Adipic acid can also be isolated from sugar beets, but it is normally synthesized from cyclohexane and its derivatives, as discussed in the next section.

A. Synthesis

Several dicarboxylic acids can be prepared by methods involving the hydrolysis of nitriles. For example, malonic acid is prepared from chloroacetic acid via cyanoacetic acid. The displacement reaction and the alkaline hydrolysis are carried out in one operation, and the product is isolated in about 80% yield.

$$ClCH_2COOH + NaOH + NaCN \xrightarrow{H_2O} [NCCH_2CO_2^-\ Na^+] \xrightarrow[H_2O]{NaOH}$$

$$Na^+\ {}^-O_2CCH_2CO_2^-\ Na^+ \xrightarrow{HCl} HOOCCH_2COOH$$

(77–82%)

A similar example is the synthesis of glutaric acid by the acid-catalyzed hydrolysis of 1,3-dicyanopropane.

$$BrCH_2CH_2CH_2Br \xrightarrow[H_2O]{NaCN} NCCH_2CH_2CH_2CN \xrightarrow[H_2O]{HCl} HOOCCH_2CH_2CH_2COOH$$

(82%) (84%) glutaric acid

Succinic acid derivatives are often available by conjugate addition of cyanide to α,β-unsaturated esters (Section 20.3.B). Hydrolysis of the β-cyano acid yields the corresponding succinic acid.

$$CH_3CH{=}CHCOOC_2H_5 + NaCN \xrightarrow{H_2O} CH_3\underset{}{\overset{CN}{C}}HCH_2COOH \xrightarrow[H_2O]{Ba(OH)_2} \xrightarrow{H_3O^+} CH_3\overset{COOH}{C}HCH_2COOH$$

(66–70%)
α-methylsuccinic acid
2-methylbutanedioic acid

> EXERCISE 27.12 Show how α-phenylsuccinic acid can be prepared starting with benzaldehyde and methyl acetate.

Certain diacids are conveniently prepared by the oxidation of cyclic alkenes or ketones. This is particularly true for adipic acid derivatives because cyclohexane derivatives are generally readily available. Examples are

butadiene + acrylonitrile (CN) $\xrightarrow{120°C}$ cyclohex-3-ene-carbonitrile $\xrightarrow{KMnO_4}$ 3-cyanohexanedioic acid (HOOC, HOOC, CN)

cyclohex-3-ene-carbonitrile

3-cyanohexanedioic acid

3,5-dimethylcyclohexanone $\xrightarrow[100°C]{HNO_3}$ 2,4-dimethylhexanedioic acid (COOH, COOH, CH_3, CH_3) (60%)

3,5-dimethylcyclohexanone

2,4-dimethylhexanedioic acid

Adipic acid is manufactured on a large scale for use in making nylon and derived polymers (Section 36.4). One of several methods that are used for its synthesis is the oxidation of cyclohexane or cyclohexene.

benzene $\xrightarrow[500\ psi,\ 200°C]{H_2/Pt\text{-}Al_2O_3}$ cyclohexane $\xrightarrow[Co(OOCCH_3)_3]{O_2}$ $\xrightarrow[CH_3COOH\ cat.]{O_2}$ HOOC(CH₂)₄COOH

adipic acid

One acid that is usually considered to be an inorganic acid of carbon is carbonic acid. However, important organic derivatives are known. The diacyl chloride, phosgene, $COCl_2$, is prepared commercially by allowing CO and Cl_2 to react in the presence of a catalyst. Phosgene reacts with alcohols to give dialkyl carbonates.

$$\mathrm{COCl_2 + 2\,ROH \longrightarrow RO\overset{\overset{\displaystyle O}{\|}}{C}OR + 2\,HCl}$$

dialkyl
carbonate

The diamide of carbonic acid, urea, H_2NCONH_2, is a metabolic product that has important commercial (7,700,000 tons were produced in the United States alone in 1989) and historical significance in organic chemistry (Chapter 1).

On a weight basis, the most important dicarboxylic acid by far is terephthalic acid (4,250,000 tons compared to 820,000 tons of adipic acid were manufactured in the United States in 1989), which is prepared on an industrial scale by the air oxidation of *p*-xylene (Section 21.5.C). The oxidation of the two side chains occurs in stages, first to *p*-toluic acid, which in turn is oxidized to terephthalic acid.

CH_3 CH_3 + O_2 $\xrightarrow[\text{Co(III)}]{CH_3COOH}$ CH_3 $COOH$ + O_2 $\xrightarrow[\text{Co(III)}]{CH_3COOH}$ $COOH$ $COOH$

p-toluic acid

terephthalic
acid

The first step proceeds much more easily than the second. Indeed, the oxidation of *p*-toluic acid requires such high temperatures that oxidation of the acetic acid solvent becomes a significant cost concern and corrosion of reaction vessels is a problem.

Phthalic acid is another important industrial product. Various high-boiling esters, particularly the bis-2-ethylhexyl ester, are widely used as plasticizers. Phthalic acid is prepared commercially by the oxidation of naphthalene (Section 31.3.D) or *o*-xylene. On heating it readily loses water to produce phthalic anhydride, a compound with a characteristic odor that forms long colorless needles on sublimation.

naphthalene $\xrightarrow[360\text{–}450°C]{O_2,\ V_2O_5}$ [phthalic acid: $COOH$, $COOH$] $\longrightarrow$ phthalic anhydride

o-xylene (CH_3, CH_3) $\xrightarrow[CH_3COOH]{O_2,\ Co^{3+}}$ phthalic acid

naphthalene

o-xylene

phthalic acid

phthalic
anhydride

Phthalic anhydride is used in the manufacture of glyptal resins, highly cross-linked, infusible polyesters prepared by heating the anhydride with glycerol. Potassium hydrogen phthalate is a well-characterized compound available in pure anhydrous form. It is used as a primary standard in titrations with bases.

B. Acidity

The dicarboxylic acids are dihydric acids and are characterized by two dissociation constants, K_1 and K_2.

$$HOOC(CH_2)_nCOOH \overset{K_1}{\rightleftharpoons} HOOC(CH_2)_nCO_2^- + H^+$$

$$HOOC(CH_2)_nCO_2^- \overset{K_2}{\rightleftharpoons} {}^-O_2C(CH_2)_nCO_2^- + H^+$$

The dissociation constants for several diacids are summarized in Table 27.5. If we treat the COOH group as a substituent in acetic acid, YCH_2COOH, the higher acidity of malonic acid compared to acetic acid ($pK_a = 4.76$) indicates that the COOH group acts as an electron-attracting inductive group.

> Be careful of statistical effects in this comparison. Malonic acid has two COOH groups that can lose a proton and would be expected to have a first dissociation constant twice that of acetic acid because of this statistical effect alone.

The acid-strengthening effect of a carboxylic acid substituent is not unexpected. All carbonyl groups have this effect because of the associated dipole which provides electrostatic stabilization of the negative charge of a carboxylate anion.

electrostatic attraction

$$HOOC\sim\sim CO_2^-$$

TABLE 27.5 Acidity of Alkanedioic Acids

Acid	$K_1 \times 10^{-5}$ M	$K_2 \times 10^{-5}$ M	pK_1	pK_2
oxalic	5400	5.4	1.27	4.27
malonic	140	0.20	2.85	5.70
succinic	6.2	0.23	4.21	5.64
glutaric	4.6	0.39	4.34	5.41
adipic	3.7	0.39	4.43	5.41

On the other hand, K_2 for a dicarboxylic acid is generally less than the dissociation constant of acetic acid. The presence of a carboxylate ion substituent reduces the acidity of an acid. This effect is clearly associated with the electrostatic repulsion of two negative charges in the dicarboxylate ion.

electrostatic repulsion

$${}^-O_2C\sim\sim CO_2^-$$

> Once more, be careful of statistical effects in this comparison. The dianion has two COO^- groups that can gain a proton and the mono acid monoanion would be expected to have a dissociation constant half that of acetic acid because of this statistical effect alone.

As expected for such a phenomenon, both the acid-strengthening effect of a carboxylic acid substituent and the acid-weakening effect of a carboxylate anion diminish with distance down a chain.

> The second dissociation constant of oxalic acid seems anomalous by this comparison. Oxalate monoanion is more acidic than a neutral alkanoic acid despite the high electrostatic repulsion inherent in the oxalate dianion. This exception to the above-mentioned generalization is probably a solvation phenomenon and is associated with the high charge density on the oxalate dianion.

C. Reactions of Dicarboxylic Acids and Their Derivatives

Dicarboxylic acids undergo a variety of thermal reactions. Anhydrous oxalic acid can be sublimed by careful heating, but at higher temperatures it decomposes to carbon dioxide and formic acid. Formic acid also decomposes under these conditions to carbon monoxide and water.

Malonic acid decarboxylates smoothly at 150 °C to give acetic acid.

$$HOOCCH_2COOH \xrightarrow{150\,°C} CO_2 + CH_3COOH$$

This reaction is general for all substituted malonic acids and for β-keto acids as well. The mechanism may involve a cyclic six-center transition state similar to that discussed previously for the Diels-Alder reaction (Sections 20.5, 21.7). The initial product is an enol, which rapidly isomerizes to acetic acid.

Decarboxylation

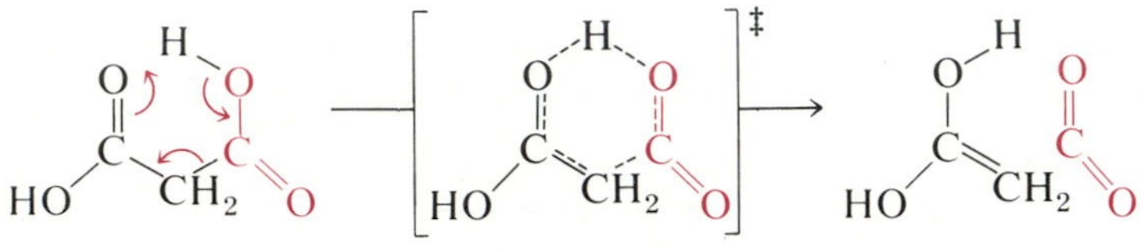

Diels–Alder Reaction

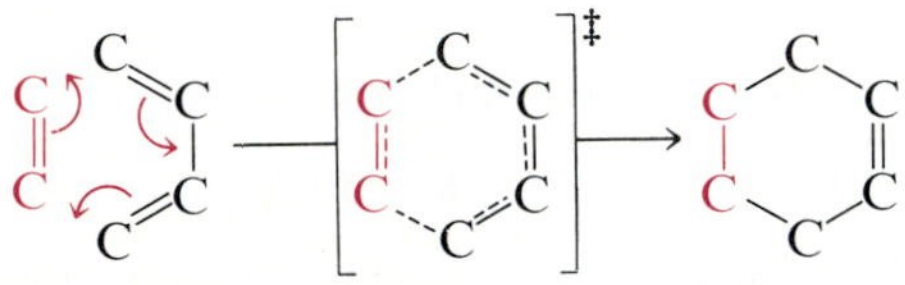

Decarboxylation of substituted malonic acids is a frequently used process for the synthesis of carboxylic acids (Section 27.7.C). Decarboxylation is usually accomplished by heating the pure diacid at 120–180 °C for several hours.

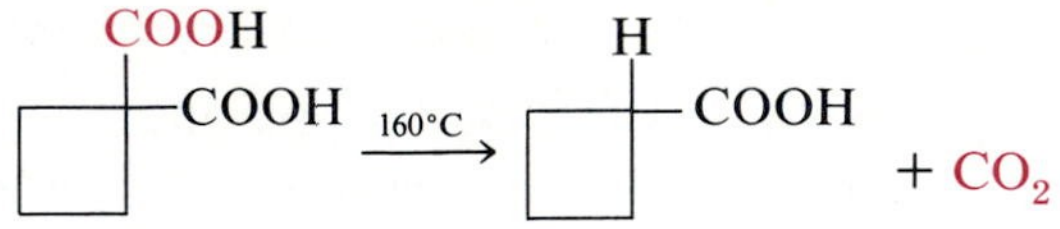

$$\text{1,1-cyclobutanedicarboxylic acid} \xrightarrow{160\,°C} \text{cyclobutanecarboxylic acid} + CO_2$$

Succinic and glutaric acids lose water on heating to give cyclic anhydrides.

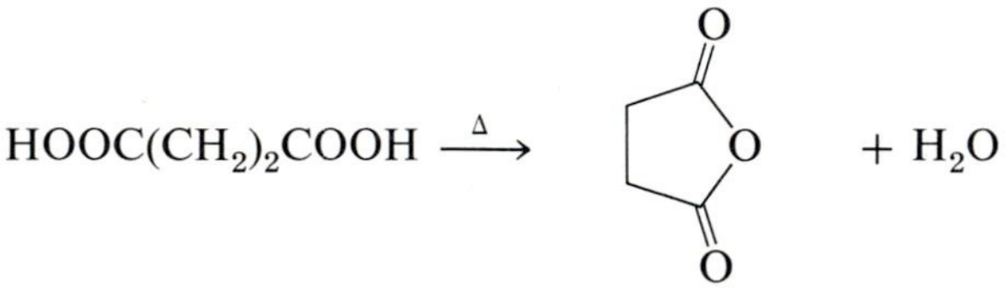

$$HOOC(CH_2)_2COOH \xrightarrow{\Delta} \text{succinic anhydride} + H_2O$$

succinic anhydride

However, the preparation of these anhydrides is best accomplished by heating with acetyl chloride or acetic anhydride. These reagents react with the carboxylic acid to form a "mixed" anhydride and either HCl or acetic acid. The mixed anhydrides undergo ring closure much more rapidly than the dicarboxylic acid.

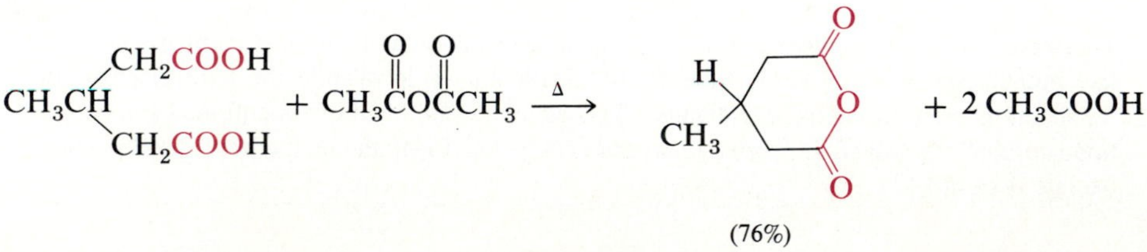

$$CH_3CH(CH_2COOH)_2 + CH_3COCOCH_3 \xrightarrow{\Delta} \text{3-methylglutaric anhydride} + 2\,CH_3COOH$$

(76%)

The easy dehydration of phthalic acid to phthalic anhydride was mentioned in Section 27.6.A. Other dehydrating agents that have been used for the formation of cyclic anhydrides are PCl_5, P_2O_5, $POCl_3$, and $SOCl_2$.

Succinic and glutaric acid and their derivatives also form cyclic **imides** with ammonia and primary amines. Five-membered ring imides form the most readily; pyrolysis of the diammonium salt often gives excellent yields.

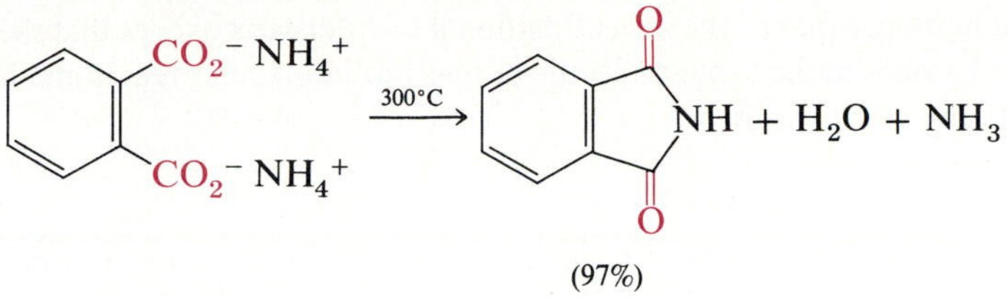

$CO_2^- NH_4^+$ / $CO_2^- NH_4^+$ $\xrightarrow{300°C}$ NH + H_2O + NH_3

ammonium phthalate — (97%) phthalimide

Six-membered ring imides form less readily; a convenient method of preparation involves pyrolysis of the monoamide of the corresponding dicarboxylic acid, as illustrated by the following example.

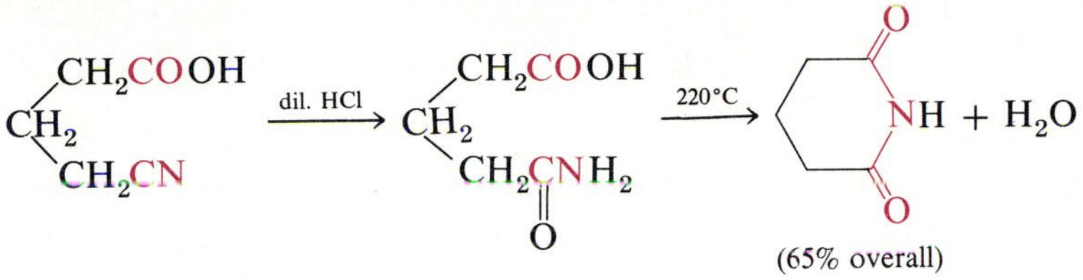

CH_2COOH / CH_2 / CH_2CN $\xrightarrow{\text{dil. HCl}}$ CH_2COOH / CH_2 / CH_2CNH_2 (C=O) $\xrightarrow{220°C}$ NH + H_2O

(65% overall) glutarimide

EXERCISE 27.13 There are four isomeric diacids having the formula $C_5H_8O_4$. Write their structures and predict the product of heating each isomer at 200 °C for 2 hr.

Adipic and pimelic acid esters undergo an intramolecular Claisen condensation (Section 19.9) known as a **Dieckmann condensation;** the products of such reactions are five- and six-membered cyclic β-keto esters.

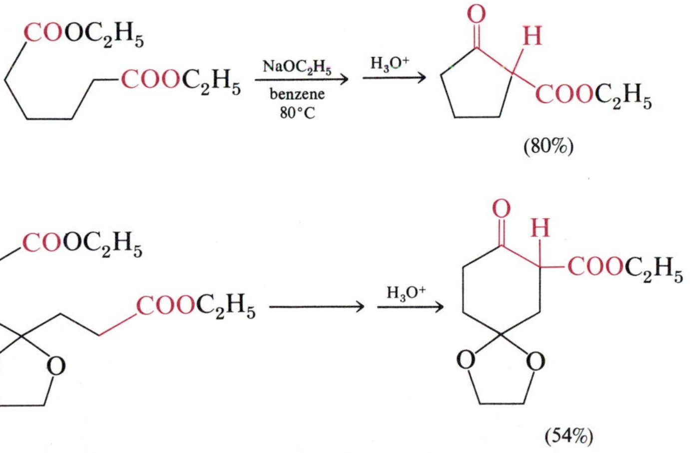

$COOC_2H_5$ / $COOC_2H_5$ $\xrightarrow[\text{benzene, 80°C}]{NaOC_2H_5}$ $\xrightarrow{H_3O^+}$ H, $COOC_2H_5$

(80%)

$COOC_2H_5$ / $COOC_2H_5$ $\longrightarrow$ $\xrightarrow{H_3O^+}$ H, $COOC_2H_5$

(54%)

The Dieckmann condensation is not satisfactory for the preparation of other sized rings.

EXERCISE 27.14 Write the stepwise mechanism for the reaction of diethyl adipate with sodium ethoxide in refluxing benzene.

27.7 Diketones, Keto Aldehydes, Keto Acids, and Keto Esters

This diverse group of difunctional compounds is best considered together because their chemistry is dominated by the interaction of two carbonyl groups in each case. As we shall see, there are interesting aspects to the chemistry of the relatively rare 1,2-compounds. The most important group of dicarbonyl compounds are the 1,3-isomers because of their importance in synthesis. Other dicarbonyl compounds show chemical behavior that is simply that of the monofunctional counterparts except that the presence of two functional groups in the same molecule allows intramolecular reactions, leading to the formation of ring compounds.

A. Synthesis

α-Diketones may be obtained by the mild oxidation of α-hydroxy ketones, which are available by the acyloin condensation (Section 27.4.A). Since the product α-diketones are also susceptible to oxidation (with cleavage of the carbonyl-carbonyl bond), especially mild oxidants must be used. Cupric acetate is particularly effective.

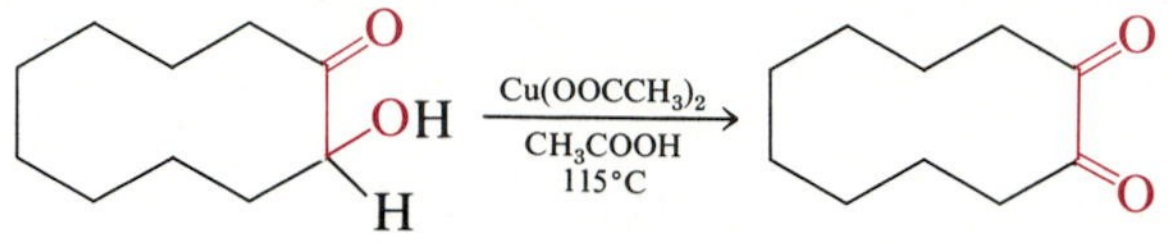

$$\xrightarrow[\text{CH}_3\text{COOH},\ 115\,^\circ\text{C}]{\text{Cu(OOCCH}_3)_2}$$

(88%)

α-Diketones and α-keto aldehydes are also available by the direct oxidation of simple ketones with selenium dioxide.

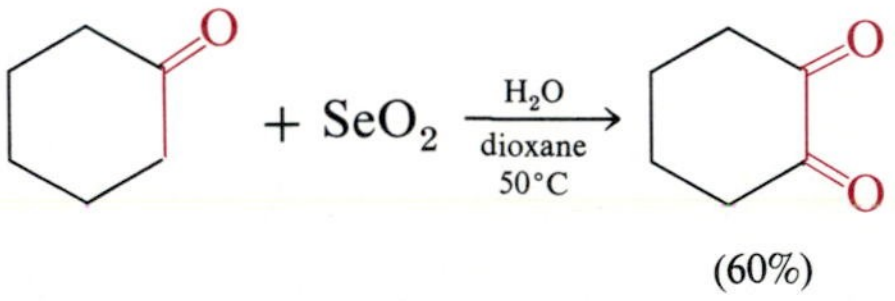

$$+ \text{SeO}_2 \xrightarrow[\text{dioxane},\ 50\,^\circ\text{C}]{\text{H}_2\text{O}}$$

(60%)

> Selenium dioxide, SeO_2, is a white, crystalline material that melts at 340 °C. It is prepared by oxidizing selenium metal with nitric acid. Although it is rather high melting, it has a substantial vapor pressure at moderate temperatures (12.5 torr at 70 °C). The yellowish green vapor has a pungent odor. In the body it is reduced to selenium metal, which may produce liver damage. Prolonged occupational exposure to selenium or SeO_2 leads to a garlic odor of breath and sweat. Selenium is an essential microconstituent of many organisms, appearing in enzymes such as glutathione peroxidase as selenocysteine.

1,3-Dicarbonyl compounds are almost uniformly prepared by some version of the Claisen condensation. In Section 19.9 we saw that esters react with base to give β-keto esters.

$$2\ \text{CH}_3\text{COOC}_2\text{H}_5 \xrightarrow{\text{C}_2\text{H}_5\text{O}^-\text{Na}^+} \xrightarrow{\text{H}_3\text{O}^+} \text{CH}_3\overset{\text{O}}{\overset{\|}{\text{C}}}\text{CH}_2\text{COOC}_2\text{H}_5$$

β-Diketones and β-keto aldehydes may be prepared by a mixed Claisen condensation using a ketone and an ester.

$$\text{CH}_3\overset{\text{O}}{\overset{\|}{\text{C}}}\text{CH}_3 + \text{CH}_3\overset{\text{O}}{\overset{\|}{\text{C}}}\text{OEt} \xrightarrow[\text{ether}]{\text{NaH}} \xrightarrow{\text{H}_3\text{O}^+} \text{CH}_3\overset{\text{O}}{\overset{\|}{\text{C}}}\text{CH}_2\overset{\text{O}}{\overset{\|}{\text{C}}}\text{CH}_3$$

(85%)

$$\text{C}_6\text{H}_5\overset{\text{O}}{\overset{\|}{\text{C}}}\text{OEt} + \text{C}_6\text{H}_5\overset{\text{O}}{\overset{\|}{\text{C}}}\text{CH}_3 \xrightarrow[\text{benzene}]{\text{NaNH}_2} \xrightarrow{\text{H}_3\text{O}^+} \text{C}_6\text{H}_5\overset{\text{O}}{\overset{\|}{\text{C}}}\text{CH}_2\overset{\text{O}}{\overset{\|}{\text{C}}}\text{C}_6\text{H}_5$$

(73%)

When ethyl formate is used in a mixed Claisen condensation, the product is a β-keto aldehyde, which exists almost entirely in the enolic form (Section 27.7.B).

$$\text{cyclohexanone} + HCOOEt \xrightarrow[\text{EtOH}]{\text{NaOEt}} \xrightarrow{H_3O^+} \text{2-(hydroxymethylene)cyclohexanone (=CHOH)} \rightleftharpoons \text{2-hydroxycyclohex-1-ene-carbaldehyde (CHO)}$$

(75%)

The mixed Claisen condensation of ketones and esters works well because ketones are considerably more acidic than are esters (Section 19.9). Thus, in the basic medium, the ketone is deprotonated to a larger extent than the ester.

$$CH_3\overset{O}{\overset{\|}{C}}CH_3 + EtO^- \rightleftharpoons CH_3\overset{O^-}{\overset{|}{C}}{=}CH_2 + EtOH \qquad K \cong 10^{-3}$$

$$CH_3\overset{O}{\overset{\|}{C}}OEt + EtO^- \rightleftharpoons CH_2{=}\overset{O^-}{\overset{|}{C}}OEt + EtOH \qquad K \cong 10^{-9}$$

Of course, once the ketone enolate is formed, it can react with another nonionized ketone molecule (aldol addition) or with the ester. However, the aldol reaction is usually thermodynamically unfavorable with ketones (page 433), and this reaction is only a minor side reaction.

$$CH_3\overset{O^-}{\overset{|}{C}}{=}CH_2 + CH_3\overset{O}{\overset{\|}{C}}CH_3 \rightleftharpoons CH_3\overset{O}{\overset{\|}{C}}CH_2\overset{O^-}{\overset{|}{C}}(CH_3)_2$$

On the other hand, the Claisen condensation is driven by the all-important final deprotonation of the acidic product. Thus the β-diketone is formed in high yield.

Cyclic β-diketones are formed by *intramolecular* Claisen condensation of 1,4- and 1,5-keto esters. The reaction is a useful method for the formation of five- and six-membered rings. This reaction is clearly analogous to the Dieckmann condensation (page 883).

$$CH_3\overset{O}{\overset{\|}{C}}CH_2CH_2CH_2CH_2COOEt \xrightarrow[\text{EtOH}]{\text{NaOEt}} \xrightarrow{H_3O^+} \text{2-acetylcyclopentanone}$$

(90%)

EXERCISE 27.15 Outline syntheses of the following dicarbonyl compounds.

(a) 3,4-hexanedione (b) 2,4-pentanedione (c) 3-oxobutanal

B. Keto-Enol Equilibria in Dicarbonyl Compounds

Simple ketones exist largely in the keto form with but a trace of the enol (vinyl alcohol) form present at equilibrium (Section 15.1.A).

$$CH_3\overset{O}{\overset{\|}{C}}CH_3 \rightleftharpoons CH_3\overset{OH}{\overset{|}{C}}{=}CH_2$$

$1.5 \times 10^{-5}\%$

In contrast, 1,2- and 1,3-dicarbonyl compounds often contain a large amount of enol form in equilibrium with the dicarbonyl form. For example, 2,4-pentanedione is a mixture of 84% dione and 16% enolic form in aqueous solution. In hexane solution the compound exists almost entirely in the enolic form.

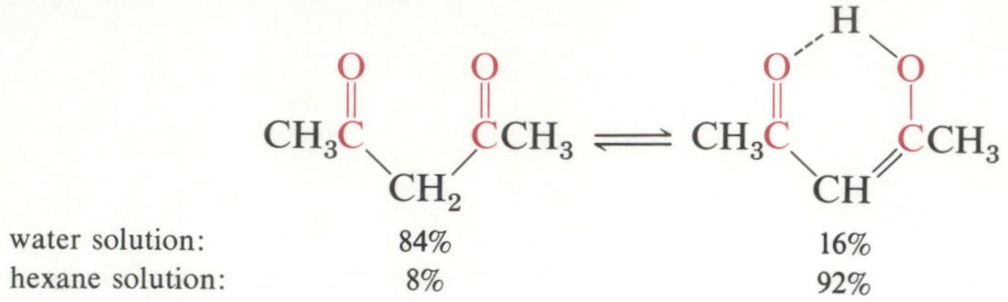

$$CH_3C(=O)CH_2C(=O)CH_3 \rightleftharpoons CH_3C(=O)CH{=}C(OH)CH_3 \quad (\text{intramolecular } O\cdots H{-}O)$$

	dione	enol
water solution:	84%	16%
hexane solution:	8%	92%

One important reason for this phenomenon is the ability of the enol to form an intramolecular hydrogen bond. Such intramolecular hydrogen bonds are especially favorable when six-membered rings are formed. The enolic form also benefits from resonance stabilization in a way not available to the dicarbonyl compound itself.

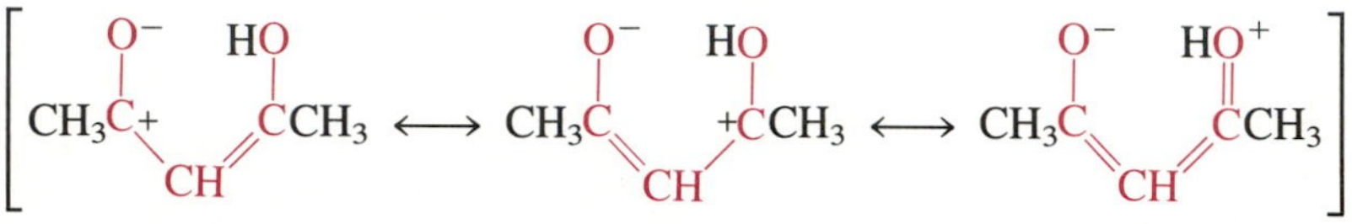

$$\left[CH_3\overset{O^-}{\overset{|}{C}}{+}\,CH{=}\overset{HO}{\overset{|}{C}}CH_3 \longleftrightarrow CH_3\overset{O^-}{\overset{|}{C}}{=}CH\,\overset{HO}{\overset{|}{+C}}CH_3 \longleftrightarrow CH_3\overset{O^-}{\overset{|}{C}}{=}CH\overset{HO^+}{\overset{\|}{C}}CH_3 \right]$$

Note that the type of delocalization shown in the foregoing example is precisely the kind that is involved in carboxylic acids.

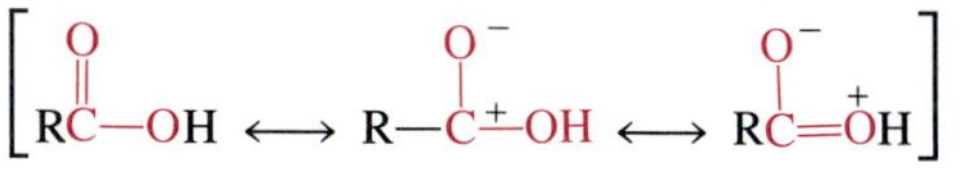

$$\left[R\overset{O}{\overset{\|}{C}}{-}OH \longleftrightarrow R{-}\overset{O^-}{\overset{|}{C}}{\overset{+}{-}}OH \longleftrightarrow R\overset{O^-}{\overset{|}{C}}{=}\overset{+}{O}H \right]$$

In the enolic form of a 1,3-diketone, the hydroxy group is conjugated *through the double bond* to the carbonyl oxygen.

Whenever two functional groups are joined to a double bond in this way, the molecule has properties similar to the corresponding compound without the double bond. This empirical concept is called the principle of **vinylogy,** and such compounds are called **vinylogs.**

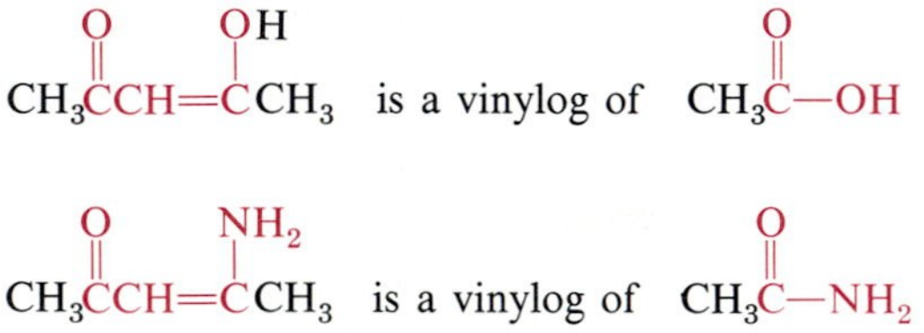

$$CH_3\overset{O}{\overset{\|}{C}}CH{=}\overset{OH}{\overset{|}{C}}CH_3 \text{ is a vinylog of } CH_3\overset{O}{\overset{\|}{C}}{-}OH$$

$$CH_3\overset{O}{\overset{\|}{C}}CH{=}\overset{NH_2}{\overset{|}{C}}CH_3 \text{ is a vinylog of } CH_3\overset{O}{\overset{\|}{C}}{-}NH_2$$

Note that the percentage of enol form at equilibrium is higher in nonpolar aprotic solvents because in such solvents the intramolecular hydrogen bond is most beneficial. In protic solvents both the dicarbonyl compound as well as the enol can hydrogen-bond to solvent molecules, and the intramolecular hydrogen bond in the enol affords no additional stabilization.

Other 1,3-dicarbonyl compounds also contain substantial amounts of enolic forms in solution. β-Keto esters are in equilibrium with significant amounts of the form in which the ketone carbonyl is enolized.

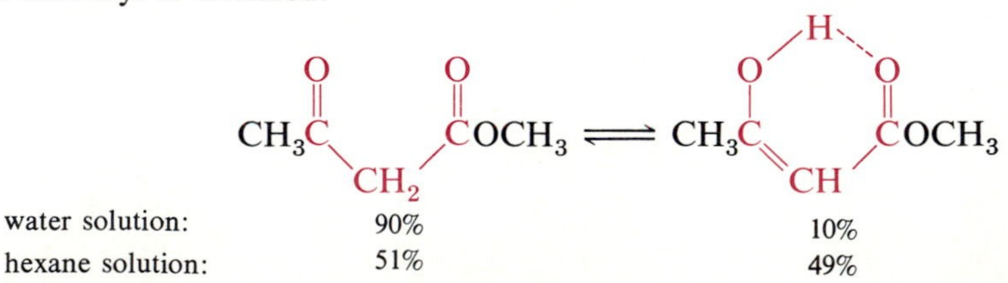

$$CH_3C(=O)CH_2C(=O)OCH_3 \rightleftharpoons CH_3C(OH){=}CHC(=O)OCH_3 \quad (\text{intramolecular } O{-}H\cdots O)$$

	keto	enol
water solution:	90%	10%
hexane solution:	51%	49%

β-Keto aldehydes exist almost entirely in the enolic form; both carbonyl groups are enolized to an appreciable extent. The two enolic forms are easily interconvertible, since only small shifts in bond distances are required.

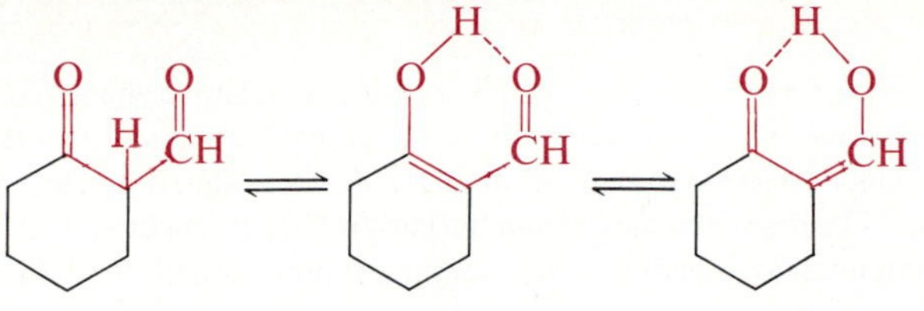

0% 76% 24%
(CCl_4 solution)

Cyclic 1,3-diketones also exist predominantly in the enolic form, even though they cannot participate in intramolecular hydrogen bonding for reasons of geometry.

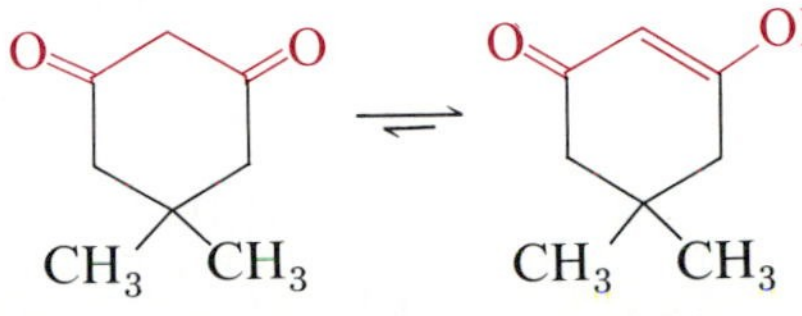

5% 95%
(water solution)

1,2-Diketones also show enhanced amounts of enol form. The main driving force for enolization in this case is relief of the electrostatic repulsion that occurs when the two electrophilic carbonyl groups are adjacent to each other.

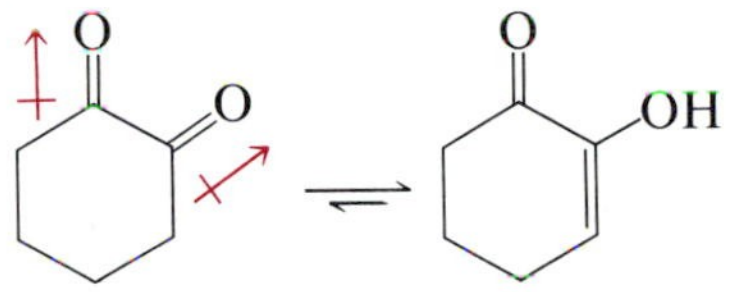

minor major

EXERCISE 27.16 Sketch the expected NMR spectrum of a CCl_4 solution of 2,4-pentanedione.

C. Decarboxylation of β-Keto Acids

β-Keto acids undergo thermal decarboxylation in the same manner as do 1,3-diacids (Section 27.6.C). In this case milder conditions suffice to bring about decarboxylation; 2-ethyl-3-oxohexanoic acid has a half-life of only 15 min at 50 °C.

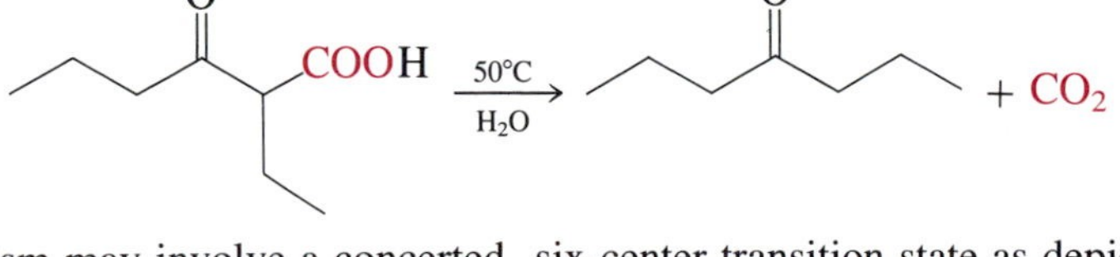

$$\xrightarrow[H_2O]{50°C} \quad + CO_2$$

The mechanism may involve a concerted, six-center transition state as depicted on page 882 for the decarboxylation of malonic acid. The initial product in the case of a β-keto acid is the enol form of the ketone. This mechanism is consistent with the resistance of bridgehead bicyclic β-keto acids to decarboxylation; the product would be a highly strained bridgehead olefin.

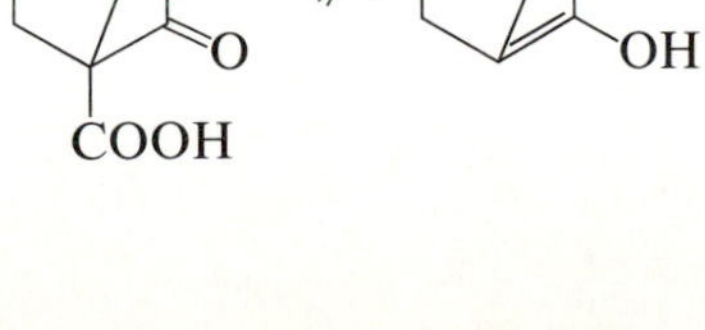

EXERCISE 27.17 Use your molecular models to confirm that bridgehead olefins such as the one just discussed are highly strained.

The decarboxylation of the β-keto acid anion requires that negative charge be accommodated by a "sink" within the molecule; formation of the enolate anion satisfies this need. Enzymatic decarboxylation of acetoacetic acid involves the conversion of the keto group to an immonium group. The negative charge from the carboxylate is "neutralized" by the positive charge in the immonium ion, leading to an enamine. Protonation of the imine and hydrolysis of the immonium ion leads to the product acetone.

$$CH_3-C(=O)-CH_2-C(=O)-O^- \longrightarrow CH_3-C(-O^-)=CH_2 + O=C=O$$

acetoacetate — acetone enolate

$$CH_3-C(=NH^+)-CH_2-C(=O)-O^- \longrightarrow CH_3-C(-NH)=CH_2 + O=C=O$$

immonium ion — enamine

Thiamine pyrophosphate, vitamin B_1, is an unusually suitable cofactor evolved by Nature to turn an α-keto acid into a β-keto acid-like intermediate. The enamine product can serve as a nucleophile towards acetaldehyde to yield acetoin. The pKa of the thiazolium ring for the 2-hydrogen is about 14, and anion formation is the enzyme-catalyzed step critical to the operation of the whole mechanism.

thiamine pyrophosphate

pyruvate

acetoin

D. 1,3-Dicarbonyl Compounds as Carbon Acids

1,3-Dicarbonyl compounds that have a hydrogen bound to the carbon between the two carbonyl groups are much stronger acids than normal aldehydes, ketones, or esters because the charge in the resulting enolate ion can be delocalized into both carbonyl groups.

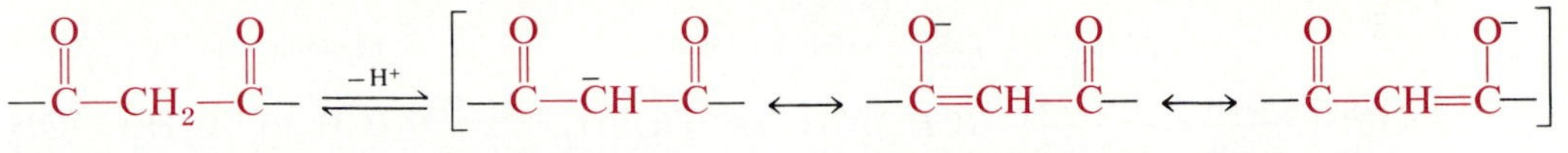

$$-\overset{\overset{O}{\|}}{C}-CH_2-\overset{\overset{O}{\|}}{C}- \xrightleftharpoons{-H^+} \left[-\overset{\overset{O}{\|}}{C}-\bar{C}H-\overset{\overset{O}{\|}}{C}- \longleftrightarrow -\overset{\overset{O^-}{|}}{C}=CH-\overset{\overset{O}{\|}}{C}- \longleftrightarrow -\overset{\overset{O}{\|}}{C}-CH=\overset{\overset{O^-}{|}}{C}-\right]$$

Some typical pK_as for such systems are contained in Table 27.6. The acidities of 1,3-dicarbonyl compounds are sufficiently high that they are converted to their conjugate bases essentially quantitatively by hydroxide ion in water or by alkoxide ion in alcoholic solvent.

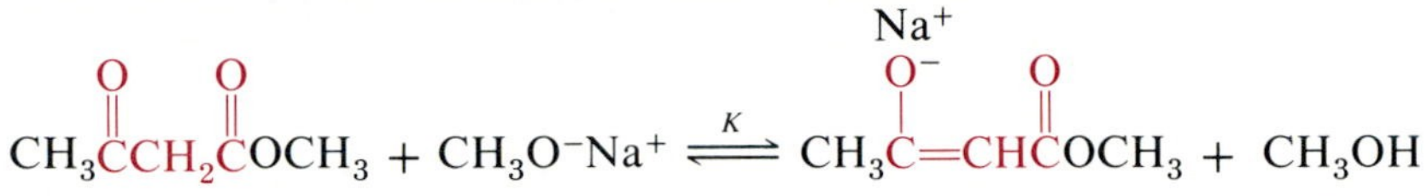

$$\underset{pK_a\ 11}{CH_3\overset{\overset{O}{\|}}{C}CH_2\overset{\overset{O}{\|}}{C}OCH_3} + CH_3O^-Na^+ \xrightleftharpoons{K} CH_3\overset{\overset{Na^+}{O^-}}{\overset{|}{C}}=CH\overset{\overset{O}{\|}}{C}OCH_3 + \underset{pK_a\ 16}{CH_3OH}$$

TABLE 27.6 Acidity of β-Dicarbonyl Compounds

Compound	pK_a
$NCCH_2\overset{\overset{O}{\|}}{C}OCH_3$	9
$CH_3\overset{\overset{O}{\|}}{C}CH_2\overset{\overset{O}{\|}}{C}CH_3$	9
$CH_3\overset{\overset{O}{\|}}{C}CH_2\overset{\overset{O}{\|}}{C}OCH_3$	11
$CH_3\overset{\overset{O}{\|}}{C}\underset{\underset{CH_3}{\vert}}{C}H\overset{\overset{O}{\|}}{C}CH_3$	11
$NCCH_2CN$	11
$CH_3O\overset{\overset{O}{\|}}{C}CH_2\overset{\overset{O}{\|}}{C}OCH_3$	13

As we shall see, these easily accessible carbanions are valuable synthetic intermediates.

EXERCISE 27.18 Using the pK_a values given in Table 27.6, estimate the equilibrium constant for the reaction of 2,4-pentanedione with sodium methoxide in methanol.

E. The Malonic Ester and Acetoacetic Ester Syntheses

The anions of 1,3-dicarbonyl compounds are nucleophiles and can take part in S_N2 displacement reactions with alkyl halides. Diethyl malonate and ethyl acetoacetate are inexpensive commercial compounds that are often alkylated in this manner. Hydrolysis of the alkylated product followed by decarboxylation of the resulting α-carboxy carboxylic acid (C=O of α-COOH is β) or β-keto acid (Sections 27.6.C, 27.7.C) provides an important general synthesis of acids and methyl ketones. The overall processes are called the **malonic ester synthesis** or the **acetoacetic ester synthesis.**

$$CH_2(COOEt)_2 \xrightarrow[EtOH]{NaOEt} \xrightarrow{n\text{-}C_4H_9Br} \underset{\substack{(80\text{–}90\%)\\ \text{diethyl } n\text{-butylmalonate}}}{n\text{-}C_4H_9CH(COOEt)_2} \xrightarrow[115°C]{\text{conc. HCl}} \underset{\text{hexanoic acid}}{n\text{-}C_4H_9CH_2COOH}$$

$$CH_2(COOEt)_2 \xrightarrow[EtOH]{NaOEt} \xrightarrow{(CH_3)_2CHI} \underset{\substack{(70\text{–}75\%)\\ \text{diethyl isopropylmalonate}}}{(CH_3)_2CHCH(COOEt)_2} \xrightarrow[115°C]{\text{conc. HCl}} \underset{\text{3-methylbutanoic acid}}{(CH_3)_2CHCH_2COOH}$$

The initially formed alkylmalonic ester can be alkylated again with the same alkyl halide or with a different one to widen the scope of the procedure.

$$CH_2(COOEt)_2 \xrightarrow[EtOH]{NaOEt} \xrightarrow{n\text{-}C_4H_9Br} n\text{-}C_4H_9CH(COOEt)_2 \xrightarrow[EtOH]{NaOEt} \xrightarrow{CH_3Br}$$

$$CH_3CH_2CH_2CH_2C(CH_3)(COOEt)_2 \xrightarrow[115°C]{\text{conc. HCl}} \underset{\text{2-methylhexanoic acid}}{CH_3CH_2CH_2CH_2CH(CH_3)COOH}$$

The overall synthetic result of alkylation and decarboxylation of malonic ester is formation of an alkyl or dialkylacetic acid.

$$CH_2(COOEt)_2 \xrightarrow{NaOEt} \xrightarrow{R'X} R'CH(COOEt)_2 \xrightarrow{NaOEt} \xrightarrow{R''X}$$

$$R'R''C(COOEt)_2 \xrightarrow[\Delta]{H^+} R'R''CHCOOH$$

A principal limitation in the synthetic sequence is that the alkylation process is an S_N2 reaction; bimolecular elimination is an expected side reaction whose importance depends on the structure of RX. If a suitable dihalide is used in the reaction, 2 moles of malonic ester can be added to both ends of a chain; alternatively, intramolecular alkylation in the second step leads to a cyclic diester.

$$CH_2(COOEt)_2 \xrightarrow[EtOH]{2\ NaOEt} \xrightarrow{Br(CH_2)_3Cl} \text{cyclobutane-1,1-}(COOEt)_2 \xrightarrow[EtOH]{KOH} \xrightarrow[\Delta]{H_3O^+} \text{cyclobutane-}CH\text{-}COOH$$

(42–44%)
cyclobutanecarboxylic acid

If ethyl acetoacetate is used as the starting material, the combination of alkylation, hydrolysis, and decarboxylation provides a synthesis of various methyl alkyl ketones.

$$CH_3COCH_2COOEt \xrightarrow[\text{2. } n\text{-BuBr}]{\text{1. NaOEt, EtOH}} \underset{(69\text{–}72\%)}{CH_3COCH(COOEt)CH_2CH_2CH_2CH_3} \xrightarrow[\text{2. } H_2SO_4,\ 25°C]{\text{1. NaOH, } H_2O} \underset{\substack{(61\%\ \text{overall})\\ \text{2-heptanone}}}{CH_3COCH_2CH_2CH_2CH_2CH_3}$$

As in the malonic ester synthesis, the starting β-keto ester can be alkylated successively with two different alkyl halides. After hydrolysis and decarboxylation, the product is a ketone that is branched at the α-carbon.

$$\text{CH}_3\text{COCH}_2\text{COOEt} \xrightarrow[\text{2. MeI}]{\text{1. NaOEt, EtOH}} \xrightarrow[\text{2. allyl bromide}]{\text{1. NaOEt, EtOH}} \text{CH}_3\text{COC(CH}_3\text{)(CH}_2\text{CH=CH}_2\text{)COOEt} \xrightarrow[\text{2. H}_2\text{O}_4\text{, 25°C}]{\text{1. NaOH, H}_2\text{O}} \text{CH}_3\text{COCH(CH}_3\text{)CH}_2\text{CH=CH}_2$$

3-methylhex-5-en-2-one

The net result is summarized below and is again subject to the usual limitations of S_N2 reactions.

$$\text{CH}_3\text{COCH}_2\text{COOEt} \xrightarrow{\text{R'X}} \xrightarrow{\text{R''X}} \xrightarrow[\Delta]{\text{H}^+} \text{CH}_3\overset{\text{O}}{\overset{\|}{\text{C}}}\text{CH}\begin{smallmatrix}\text{R'}\\ \text{R''}\end{smallmatrix}$$

Other β-diketones and β-keto esters can be alkylated in the same manner. The following examples show the tremendous utility of these reactions in building up complex organic structures.

1,3-cyclohexanedione $\xrightarrow[\text{CH}_3\text{I, H}_2\text{O}]{\text{1 equiv NaOH}}$ (70%) 2-methyl-1,3-cyclohexanedione

1,3-cyclohexanedione $\xrightarrow[\text{CH}_3\text{I, MeOH}]{\text{2 equiv NaOMe}}$ (80%) 2,2-dimethyl-1,3-cyclohexanedione

ethyl 2-oxocyclopentanecarboxylate $\xrightarrow[\text{EtOH}]{\text{NaOEt}} \xrightarrow{\text{Br(CH}_2\text{)}_3\text{Br}}$ 2-($CH_2CH_2CH_2Br$)-2-COOEt-cyclopentanone $\xrightarrow[\Delta]{\text{48\% HBr}}$ (60%) 2-(3-bromopropyl)cyclopentanone

EXERCISE 27.19 Write equations showing all of the steps involved in the preparation of the following compounds by the malonic ester and acetoacetic ester syntheses.

(a) hex-5-en-2-one
(b) 5-methylhexanoic acid
(c) 2-ethylpent-4-enoic acid
(d) 4-oxopentanoic acid

F. The Knoevenagel Condensation

Esters of malonic acid readily react with aldehydes and ketones under basic conditions to give α,β-unsaturated diesters.

$$(CH_3)_2CHCH_2CHO + CH_2(COOEt)_2 \xrightarrow[\text{benzene}, \Delta]{\text{piperidine}} (CH_3)_2CHCH_2CH{=}C(COOEt)_2$$

(78%)

$$C_6H_5CHO + CH_2(COOEt)_2 \xrightarrow[\Delta]{\text{piperidine}} C_6H_5CH{=}C(COOEt)_2$$

(80%)

The transformation is similar to the aldol reaction (Section 15.2) and the related reactions of ester enolates with aldehydes and ketones (Section 19.9). In the present case, however, only weakly basic catalysts are required, because of the high acidity of malonic esters. The reaction works best for aldehydes, although ketones sometimes can be used.

A variation of this reaction makes use of malonic acid with an amine catalyst. In this case some of the carboxylate anion is present, but since amines are weak bases, the fraction of carboxylate anion is small. The carbanion derived from the α-hydrogen is also present because the carbonyl groups of the carboxylic acid stabilize the carbanion just as they stabilize the ester carbonyls. The resulting condensation products decarboxylate on heating to give α,β-unsaturated acids. The catalyst usually employed in this reaction is the tertiary amine pyridine.

$$CH_3(CH_2)_7CHO + CH_2(COOH)_2 \xrightarrow[\Delta]{\text{pyridine}} \left[CH_3(CH_2)_7CH{=}C(COOH)_2\right] \xrightarrow{-CO_2} CH_3(CH_2)_6CH{=}CHCOOH$$

(75–85%)
dec-2-enoic acid

This version of the Knoevenagel condensation works well with all aldehydes. It can be employed with ketones, but yields are generally low.

The Wittig reaction (Section 26.8) is also an excellent method for converting an aldehyde into an α,β-unsaturated ester; the latter must be hydrolyzed if the acid is desired.

$$(C_6H_5)_3P\colon + BrCH_2COOCH_3 \longrightarrow (C_6H_5)_3\overset{+}{P}CH_2COOCH_3\ Br^- \xrightarrow{CH_3O^-}$$

$$(C_6H_5)_3\overset{+}{P}\overset{-}{C}HCOOCH_3 \xrightarrow{C_6H_5CHO} \left[\text{oxephosphetane: } O{-}P(C_6H_5)_3,\ C_6H_5,\ COOCH_3\right] \longrightarrow$$

oxephosphetane
(not isolated)

$$C_6H_5CH{=}CHCOOCH_3 + (C_6H_5)_3P{=}O$$

EXERCISE 27.20 The Knoevenagel condensation provides a synthesis of unsaturated malonic esters or of α,β-unsaturated acids. Outline syntheses of the following compounds.

(a) $CH_3CH_2CH{=}CHCOOH$

(b) $CH_3CH_2CH{=}C(COOEt)_2$

G. The Michael Addition Reaction

In Section 20.3.A, we saw that α,β-unsaturated carbonyl compounds can react with such nucleophiles as cyanide ion and Grignard reagents by either 1,2- or 1,4-addition. The 1,4-addition of a carbanion to an α,β-unsaturated carbonyl system is called a **Michael addition.** It is a common and useful reaction. For example, when a mixture of 2-cyclohexen-1-one and diethyl malonate is treated with a catalytic amount of sodium ethoxide in ethanol, the following addition reaction occurs.

$$\text{2-cyclohexen-1-one} + CH_2(COOEt)_2 \xrightarrow[\text{EtOH}]{\text{NaOEt}} \text{3-}[CH(COOEt)_2]\text{cyclohexanone}$$

(90%)

The mechanism of the Michael addition is illustrated as follows with diethyl malonate and acrolein; the product is obtained in 50% yield.

$$(1)\quad CH_2(COOEt)_2 + EtO^- \rightleftharpoons {}^-{:}CH(COOEt)_2 + EtOH$$

$$(2)\quad HC(=O)\text{—}CH\text{=}CH_2 + {}^-{:}CH(COOEt)_2 \rightleftharpoons HC(O^-)\text{=}CH\text{—}CH_2\text{—}CH(COOEt)_2$$

$$(3)\quad HC(O^-)\text{=}CHCH_2CH(COOEt)_2 + EtOH \rightleftharpoons HC(=O)CH_2CH_2CH(COOEt)_2 + EtO^-$$

A Michael addition such as this is similar to the alkylation of a carbanion by an alkyl halide—with one important exception. In the alkylation with an alkyl halide, a stoichiometric amount of base is consumed; in the Michael addition, the base functions as a catalyst. Thus, only a small amount of base need be used in Michael additions, and the process is reversible. The driving force for the reaction is the formation of a new carbon-carbon single bond at the expense of the π-bond of the unsaturated carbonyl compound; this driving force is essentially the same as that of all additions to a double bond.

Michael additions are observed between carbon acids containing an acidic proton and a variety of α,β-unsaturated carbonyl systems.

$$CH_2(COOEt)_2 + CH_2\text{=}CHC(=O)CH_3 \xrightarrow[\text{EtOH, 0°C}]{\text{NaOEt}} CH_3C(=O)CH_2CH_2CH(COOEt)_2$$

(71%)

$$CH_3C(=O)CH_2C(=O)CH_3 + CH_2\text{=}CHCN \xrightarrow[t\text{-BuOH, 25°C}]{Et_3N} CH_3C(=O)CH(CH_2CH_2CN)C(=O)CH_3$$

(77%)

$$\text{1,3-cyclohexanedione} + CH_2\text{=}CHCOOEt \xrightarrow[\text{EtOH, 0°C}]{\text{NaOEt}} \text{2-}(CH_2CH_2COOEt)\text{-1,3-cyclohexanedione}$$

If an excess of the α,β-unsaturated carbonyl component is used, it is possible to achieve dialkylation.

$$CH_2(COOEt)_2 + \underset{\text{(excess)}}{CH_2\text{=}CHC(=O)CH_3} \xrightarrow{OH^-} [CH_3C(=O)CH_2CH_2]_2C(COOEt)_2$$

(85%)

The Michael addition constitutes a useful method for the synthesis of 1,5-dicarbonyl systems. When diethyl malonate or acetoacetic ester is used as the adding group, the product can be hydrolyzed and decarboxylated to obtain the alkylated acid or ketone.

$$CH_3\overset{O}{\overset{\|}{C}}CH_2COOEt + CH_2{=}CHCOOEt \xrightarrow[EtOH]{NaOEt} CH_3\overset{O}{\overset{\|}{C}}\underset{COOEt}{\underset{|}{C}H}CH_2CH_2COOEt \xrightarrow[\Delta]{H_3O^+}$$

$$\left[CH_3\overset{O}{\overset{\|}{C}}\underset{COOH}{\underset{|}{C}H}CH_2CH_2COOH\right] \longrightarrow CH_3\overset{O}{\overset{\|}{C}}CH_2CH_2CH_2COOH + CO_2$$

EXERCISE 27.21 Make a 3 × 3 matrix with methyl acrylate, acrolein, and methyl vinyl ketone on one side and diethyl malonate, acetoacetic ester, and 2-ethoxycarbonylcyclopentanone on the other. Each of the nine intersections of the matrix represents a Michael addition reaction. Assuming that the reactants are used in a ratio of 1:1 in each case, write the structures of the nine reaction products. Now write the products expected from the hydrolysis and decarboxylation of each of the nine initial adducts.

Although the Michael addition is most successful when the carbon acid is relatively acidic, such as a 1,3-dicarbonyl compound, the reaction also occurs with simple ketones.

$$\text{2-methylcyclohexanone (CH}_3\text{, H)} + CH_2{=}CH\overset{O}{\overset{\|}{C}}C_6H_5 \xrightarrow[EtOH,\ ether]{KOEt} \text{2-methylcyclohexanone-2-}CH_2CH_2\overset{O}{\overset{\|}{C}}C_6H_5$$

(64%)

$$CH_3\overset{O}{\overset{\|}{C}}CH_2CH_3 + \underset{(excess)}{CH_2{=}CHCN} \xrightarrow[t\text{-BuOH}]{KOH} CH_3\overset{O}{\overset{\|}{C}}\underset{CH_3}{\underset{|}{C}}(CH_2CH_2CN)_2$$

A useful variant of the Michael addition occurs with methyl vinyl ketone and its derivatives. The initially formed 1,5-diketone undergoes a subsequent intramolecular aldol reaction to yield a cyclohexenone ring. The process is essentially a combination of the Michael reaction and aldol reaction and is called **Robinson annulation.**

$$\text{2-methylcyclohexanone (CH}_3\text{, H)} + CH_2{=}CH\overset{O}{\overset{\|}{C}}CH_3 \xrightarrow[EtOH]{KOH} [\text{2-methyl-2-(3-oxobutyl)cyclohexanone}] \xrightarrow{KOH^-}$$

$$[\text{bicyclic aldol, CH}_3\text{, OH, C=O}] \longrightarrow \text{bicyclic enone, CH}_3$$

(50%)

The Robinson annulation sequence has been useful in building up the carbon framework of complex natural products such as steroids (Section 36.5.B). An example of the use of the reaction as the first step in the laboratory synthesis of a steroid is given below.

$$+ \ CH_3CH_2CCH{=}CH_2 \xrightarrow[\text{EtOH}]{\text{OH}^-}$$

EXERCISE 27.22 Propose a synthesis of the following spirocyclic compound.

Highlight 27.1

The parent name of a compound containing two functional groups is chosen in the order COOH, SO_3H, COOR, RCOCl, $RCONH_2$, RCN, RCHO, RCOR', ROH, RSH, RNH_2, RC≡CR', RCH=CHR'

Although two functional groups within the same molecule can act independently, they often have a strong influence on one another in physical and chemical properties.

alkene + ketone → α,β-unsaturated ketone

Combination undergoes 1,4-addition of nucleophiles (Michael addition)

ketone + ester → β-keto ester

Combination has an acidic proton on the middle carbon and the derived enolate ion is a useful nucleophile

Highlight 27.2

The intramolecular and intermolecular reactions of two functional groups within the same molecule are interesting to contrast.

Functional Groups		Intramolecular Product	Intermolecular Product
Diols			
		Monomolecular	*Polymolecular*
HOCH$_2$OH	$\xrightarrow{-H_2O}$	CH$_2$=O formaldehyde	RO[CH$_2$O]$_n$CH$_2$OR polyformaldehyde [Delrin]
		Trimolecular (six-membered ring, three O)	
HOCH$_2$CH$_2$OH	$\xrightarrow{-H_2O}$	ethylene oxide oxirane	HO[CH$_2$CH$_2$O]$_n$CH$_2$CH$_2$OH polyethylene glycol
		Bimolecular 1,4-dioxane	
Hydroxy carboxylic acids			
HOCH$_2$CH$_2$COOH	$\xrightarrow{-H_2O}$	β-propiolactone	H[OCH$_2$CH$_2$C(=O)]$_n$OCH$_2$CH$_2$COOH polyester

ASIDE

The wheel is a device invented by humans about 5500 years ago, probably in Mesopotamia by the Sumerians. A large box mounted on two pairs of wheels coupled by axles is a cart. The interaction between the paired wheels and the box yields a transport instrument that is different in function from any of the components. Yet each component can behave independently; the box can be used for storing materials, the wheel could be taken off and used as a potter's wheel. In a parallel fashion, a molecule bearing two groups can exhibit behavior different from that of a molecule containing only one of the groups. Using the three functional groups, a hydroxy, a carboxy, and double bond, construct molecules with all possible combinations in which the groups are separated by 0, 1, 2, and 3 carbon atoms. Name the compounds and write one reaction for each compound. Identify those reactions that exhibit interaction between the groups.

PROBLEMS

1. Name the following compounds.

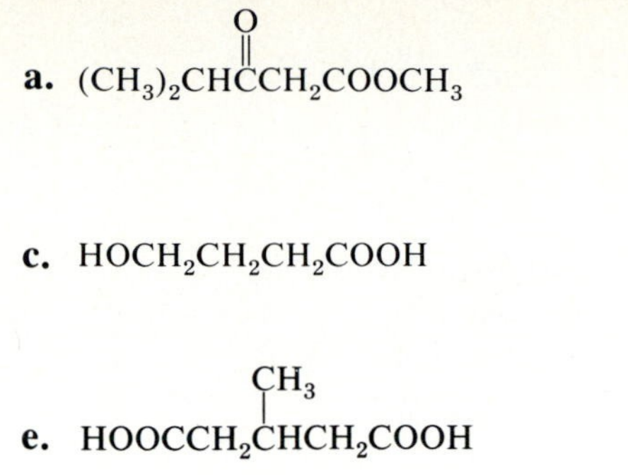

a. $(CH_3)_2CHCOCH_2COOCH_3$ **b.** $HOCH_2C(CH_3)_2CH_2OH$

c. $HOCH_2CH_2CH_2COOH$ **d.** $CH_3COCH_2CH_2CN$

e. $HOOCCH_2CH(CH_3)CH_2COOH$ **f.** $HOOCCH_2CH_2CH_2CH(CH_3)COOH$

g. $(CH_3CH_2CH_2)(CH_3)C(COOCH_3)_2$ **h.** $HOOCCH_2CH(CH_3)CH_2CHO$

i. $HOCH_2CH_2CH_2CONH_2$ **j.** $CH_2{=}CHCH_2CH_2COCH_3$

k. 2,2-dimethylcyclohexane-1,4-dione (ring with O, CH_3, CH_3, O) **l.** $CH_3COCH_2CH(CH_3)COOH$

m. $NCCH_2CH_2CH_2CN$ **n.** $CH_3CH(CH_3)CH_2COCH_2CH_2CHO$

2. Compare the stereostructure of the 1,2-cyclodecanediol produced from *cis*- and *trans*-cyclodecene by each of the following reactions.

a. (i) HCO_3H; (ii) aqueous NaOH **b.** OsO_4, H_2O_2
c. (i) aqueous Br_2; (ii) aqueous NaOH

3. Pinacol rearrangement of 1,1-diphenyl-1,2-ethanediol gives diphenylacetaldehyde and not phenylacetophenone. Explain.

$$(C_6H_5)_2C(OH)CH_2OH \xrightarrow{H^+} (C_6H_5)_2CHCHO \quad \text{(not } C_6H_5COCH_2C_6H_5\text{)}$$

4. Give the principal product(s) of each of the following reactions or reaction sequences.

a. 1-(hydroxymethyl)cyclopentanol (cyclopentane ring bearing OH and CH_2OH) $\xrightarrow{H^+}$

b. $HOCH_2C(CH_3)_2CHOHCH_2OH + HIO_4 \longrightarrow$ **c.** $HO(CH_2)_6COOH \xrightarrow{\Delta}$

d. $CH_3CH_2CHOHCOOH \xrightarrow{\Delta}$ **e.** $HOCH_2CH_2CH_2COOH \xrightarrow{\Delta}$

f. $C_6H_4(COOEt)_2$ (o-) $+ EtOOC(CH_2)_3COOEt \xrightarrow[EtOH]{EtONa} \xrightarrow[\Delta]{dil\ H^+} C_{11}H_{10}O_2$

g. H–C(D)(CH$_3$)–OH $\xrightarrow[pyridine]{C_6H_5SO_2Cl} \xrightarrow[EtONa/EtOH]{CH_2(COOEt)_2} \xrightarrow[\Delta]{dil\ H^+} C_4H_7DO_2$ optically active

h. $2\ CH_2(COOEt)_2 + C_6H_5CHO \xrightarrow[EtOH]{EtONa} \xrightarrow[\Delta]{dil\ H^+} C_{11}H_{12}O_4$

i. $C_6H_5COCH_2COOEt \xrightarrow[EtOH]{EtONa} \xrightarrow{CH_3CH_2I} \xrightarrow[EtOH]{EtONa} \xrightarrow{CH_3I} \xrightarrow[\Delta]{H^+} C_{11}H_{14}O$

j. $EtOOCCOOEt + CH_3COOEt \xrightarrow[EtOH]{EtONa}$

5. a. Write alternative Lewis structures for periodic acid, $HOIO_3$, and iodic acid, $HOIO_2$, making use of $^-O—I^+$ or O=I bonds. Be careful to count electrons and assign formal charges properly; note that iodic acid has a lone pair of electrons on iodine. Follow the changes in electron pairs symbolized in the cyclic mechanism on page 866.

b. *cis*-1,2-Cyclopentanediol is oxidized to glutaraldehyde (1,5-pentanedial) by periodic acid much more rapidly than the trans isomer. Explain.

c. Suggest a method whereby periodic acid could be used to distinguish between 1,2,3-pentanetriol and 1,2,4-pentanetriol.

6. a. 4,5-Dihydroxypentanal exists in solution largely as a cyclic hemiacetal. From the data in Table 27.3 predict which form will predominate at equilibrium.

HO–CH$_2$–CH(OH)–CH$_2$–CH$_2$–CHO ⇌ (six-membered ring hemiacetal with OH, OH) ⇌ (five-membered ring hemiacetal with OH, CH_2OH)

b. When 4,5-dihydroxypentanal is treated with silver oxide and the resulting dihydroxyhexanoic acid is treated with acid, a lactone results. What is the structure of the lactone (see Table 27.4)?

7. Propose mechanisms for the following reactions.

a. 1,3-cyclohexanedione $+ (CH_3)_2CHCH_2OH \xrightarrow[benzene]{H_2SO_4}$ 3-$OCH_2CH(CH_3)_2$-2-cyclohexenone $+ H_2O$

b. $CH_2(COOCH_3)_2 \xrightarrow[CH_3OH]{NaOCH_3} \xrightarrow{CH_2—CH_2\ (O)}$ γ-butyrolactone with $COOCH_3$ at α-carbon

c. $CH_2=CHCH_2CH_2CH(COOCH_3)_2 \xrightarrow{C_6H_5CO_3H} \xrightarrow[CH_3OH]{CH_3O^-}$ HO–cyclopentane–$C(COOCH_3)_2$

d. $p\text{-}CH_3OC_6H_4C(C_6H_5)_2CH_2OH + SOCl_2 \longrightarrow (C_6H_5)_2C{=}CHC_6H_4OCH_3\text{-}p$

8. When ethyl acetoacetate is treated with 1,3-dibromopropane and 2 moles of sodium ethoxide in ethanol, a product (A) is produced that has the formula $C_9H_{14}O_3$. Compound A has an infrared spectrum that shows only one carbonyl absorption and no OH bond. Suggest a structure for A and rationalize its formation.

9. Give the expected product for each reaction sequence.

a. 1-methylcyclohexene $\xrightarrow[H_2O_2]{OsO_4}$ $(C_7H_{14}O_2)$ $\xrightarrow{HIO_4}$ $(C_7H_{12}O_2)$

b. $CH_3CH_2\overset{O}{\overset{\|}{C}}CH_2CH_3$ $\xrightarrow[benzene]{Mg}$ $\xrightarrow{H_2O}$ $(C_{10}H_{22}O_2)$ $\xrightarrow[\Delta]{H_3O^+}$ $(C_{10}H_{20}O)$

c. furfural (2-furyl–CHO) $\xrightarrow[C_2H_5OH]{KCN}$ $(C_{10}H_8O_4)$ $\xrightarrow{Cu(OAc)_2}$ $(C_{10}H_6O_4)$ $\xrightarrow[\Delta]{conc.\ KOH}$ $\xrightarrow{H_3O^+}$ $(C_{10}H_8O_5)$
(Hint: See problem 17)

d. *trans*-2-butene $\xrightarrow{HCO_3H}$ $\xrightarrow[H_2O]{NaOH}$ $(C_4H_{10}O_2)$ $\xrightarrow[H^+]{\text{cyclopentanone}}$ $(C_9H_{16}O_2)$

10. Show how one can accomplish each of the following conversions.

a. $CH_3CH_2COOH \longrightarrow CH_3CH_2CHOHCH(CH_3)COOH$

b. $CH_3C_6H_4CHO \longrightarrow CH_3C_6H_4CH(OH)CH(OH)C_6H_4CH_3$ (para-substituted rings)

c. cyclopentanone $\longrightarrow$ dispiro lactide (6-membered ring containing two O and two C=O, each remaining carbon spiro-fused to a cyclopentane ring)

d. $CH_3CH_2COOH \longrightarrow CH_3CH_2CHOHCHOHCH_2CH_3$

e. cyclohexanone $\longrightarrow$ 2-(COOEt)cyclohexanone

f. $C_6H_5CH_2COOEt \longrightarrow C_6H_5CH(COOEt)_2$

g. $C_6H_5CH(COOEt)_2 \longrightarrow HOOCCH(C_6H_5)CH_2COOH$

11. Show how each of the following compounds can be prepared starting with diethyl malonate.

a. $(CH_3)_2CHCH_2CH_2COOH$

b. $CH_2{=}CHCH_2CH(CH_3)COOH$

c. $CH_2{=}C(CH_3)CH_2CH_2COOH$

d. O (cyclohexanone ring) CH_2COOH

e. $HOOCCH_2CH_2COOH$

f. H (cyclohexane ring) COOH

12. Show how each of the following compounds can be prepared starting with ethyl acetoacetate.

a. $CH_3\overset{O}{\overset{\|}{C}}CH_2CH_2CH_2CH(CH_3)_2$

b. $CH_3\overset{O}{\overset{\|}{C}}CH_2CH_2CH_2COOH$

c. $CH_3\overset{O}{\overset{\|}{C}}\underset{CH_2CH_3}{\underset{|}{C}H}CH_2CH_2CH_3$

d. $CH_3\overset{O}{\overset{\|}{C}}\underset{CH_3}{\underset{|}{C}H}CH_2CH_2\underset{CH_3}{\underset{|}{C}H}\overset{O}{\overset{\|}{C}}CH_3$

13. Show how each of the following compounds can be synthesized from compounds containing five or fewer carbons.

a. HOOC COOH

b. COOH COOH

c. Me_2N O COOH

d. O O

e. HOOC O

f. OH COOH

g. O O

h. O COOH

i. O

j. O

k. O COOH

l. HOOC COOH

m. O O

n. OH COOH

14. Suggest a procedure for the dialkylation of diethyl malonate with benzyl chloride using phase-transfer catalysis. What is the final product of hydrolysis of this product and heating in acid?

15. Show how the Robinson annulation can be used to prepare each compound.

a. 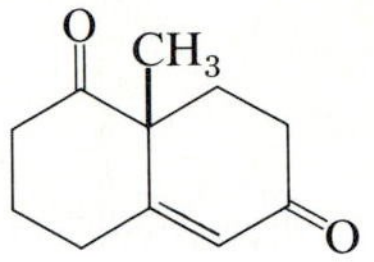 O, CH_3, O

b. 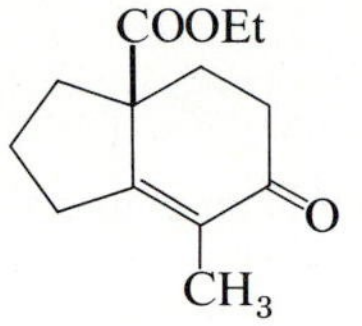 COOEt, O, CH_3

c. 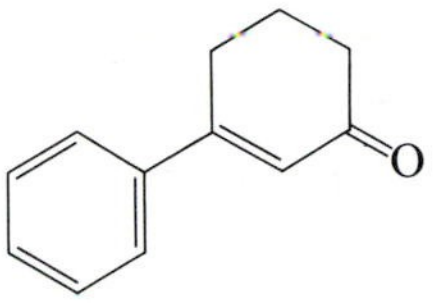 O

d. 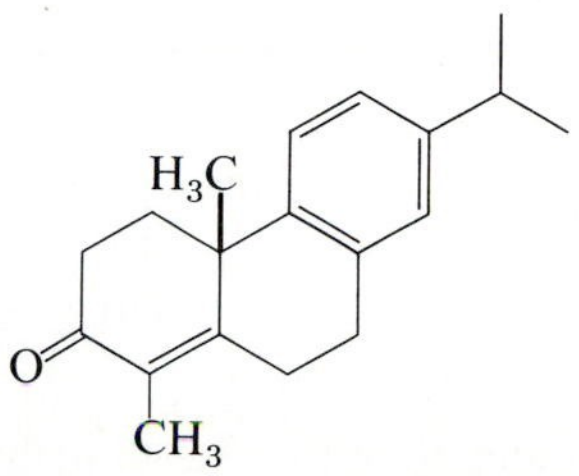 H_3C, O, CH_3

16. 1-Phenylethane-*1-d*, $C_6H_5CHDCH_3$ (B), has been prepared in optically active form. This compound is particularly interesting because its chirality is due entirely to the isotopic difference between H and D. Nevertheless, the magnitude of its rotation has the relatively high value of $[\alpha]_D \pm 0.6$. The absolute configuration of B is related to the known configuration of mandelic acid by the following sequences of reactions.

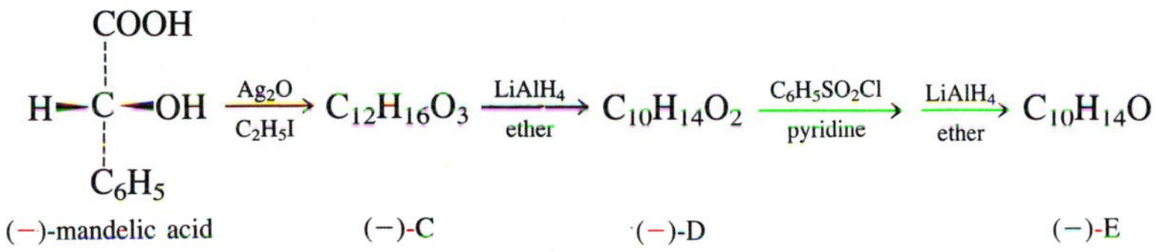

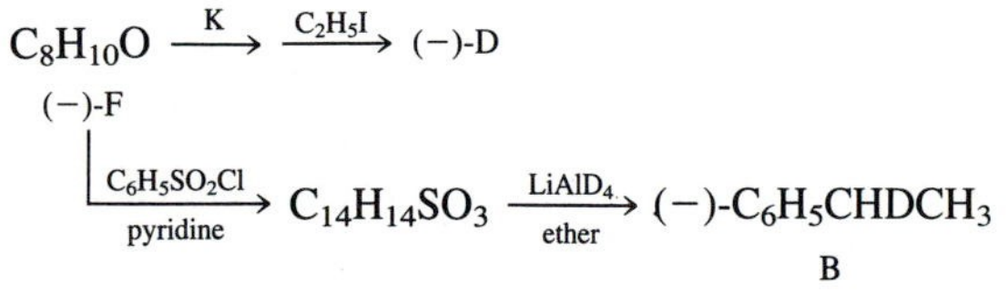

$$\underset{(-)\text{-mandelic acid}}{\text{H—C(COOH)(C}_6\text{H}_5\text{)—OH}} \xrightarrow[C_2H_5I]{Ag_2O} \underset{(-)\text{-C}}{C_{12}H_{16}O_3} \xrightarrow[\text{ether}]{LiAlH_4} \underset{(-)\text{-D}}{C_{10}H_{14}O_2} \xrightarrow[\text{pyridine}]{C_6H_5SO_2Cl} \xrightarrow[\text{ether}]{LiAlH_4} \underset{(-)\text{-E}}{C_{10}H_{14}O}$$

$$\underset{(-)\text{-F}}{C_8H_{10}O} \xrightarrow{K} \xrightarrow{C_2H_5I} (-)\text{-D}$$

$$(-)\text{-F} \xrightarrow[\text{pyridine}]{C_6H_5SO_2Cl} C_{14}H_{14}SO_3 \xrightarrow[\text{ether}]{LiAlD_4} \underset{\text{B}}{(-)\text{-}C_6H_5CHDCH_3}$$

Deduce the absolute configuration of (−)-B and the structure and configuration of each intermediate, C through F, in the sequence. Assign the proper *R*,*S* notation to each structure B through F.

17. α-Diketones undergo an interesting rearrangement reaction upon being treated with strong base.

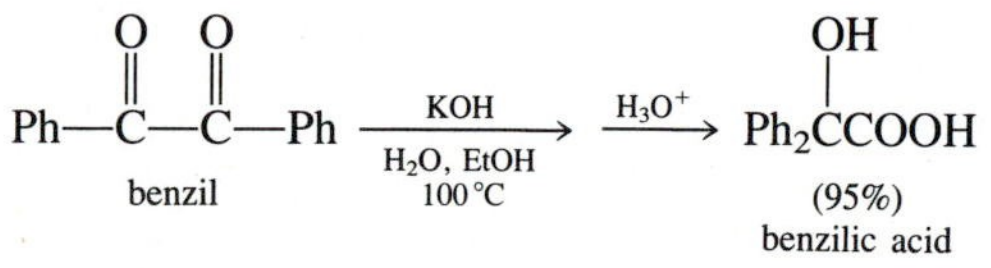

$$\underset{\text{benzil}}{\text{Ph—C(=O)—C(=O)—Ph}} \xrightarrow[\text{H}_2\text{O, EtOH, 100 °C}]{\text{KOH}} \xrightarrow{H_3O^+} \underset{\substack{(95\%)\\ \text{benzilic acid}}}{Ph_2C(OH)COOH}$$

The reaction is called the **benzilic acid rearrangement** after the trivial name of (diphenyl)hydroxyacetic acid.

a. Suggest a mechanism for the reaction.

b. Propose a mechanism for the following related transformation.

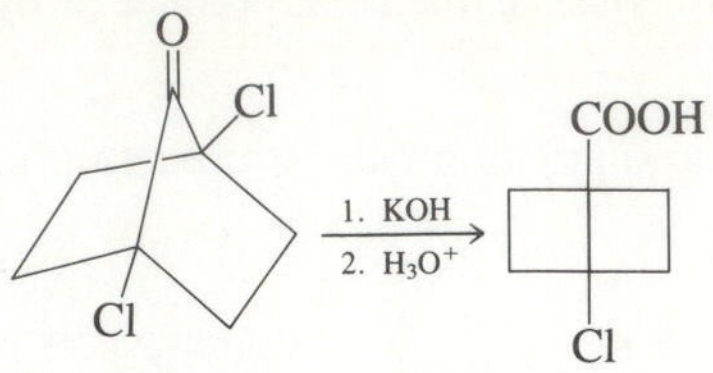

O, Cl, Cl; 1. KOH, 2. H_3O^+; COOH, Cl

c. Outline a multistep conversion of cyclohexanone into 1-hydroxycyclopentanone.

18. The following reaction is similar to the acyloin condensation. Propose a mechanism for the reaction.

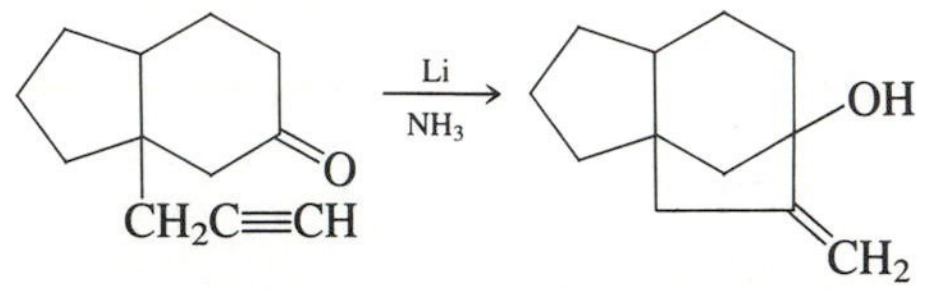

O, $CH_2C{\equiv}CH$; Li, NH_3; OH, CH_2

19. Treatment of ethyl 2,2-dimethyl-3-oxobutanoate with aqueous sodium hydroxide gives 2-methylpropanoic and acetic acids, after acidification of the basic reaction mixture.

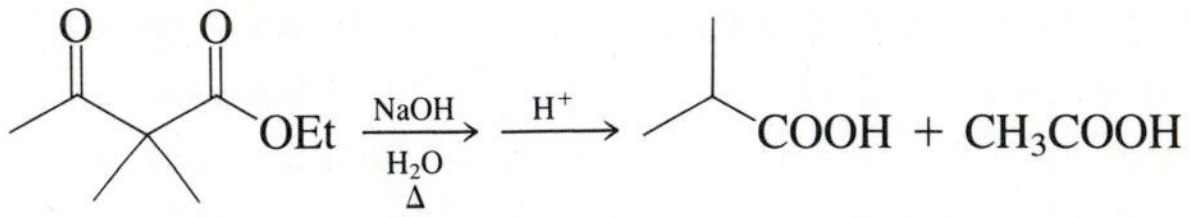

O, O, OEt; NaOH, H_2O, Δ; H^+; COOH + CH_3COOH

a. Propose a mechanism for the transformation.

b. Note the conversion of 2-methyl-2-methoxycarbonylcyclopentanone into 5-methyl-2-methoxycarbonylcyclopentanone shown below. Outline a multistep conversion of 2-methoxycarbonylcyclopentanone to 2-methyladipic acid.

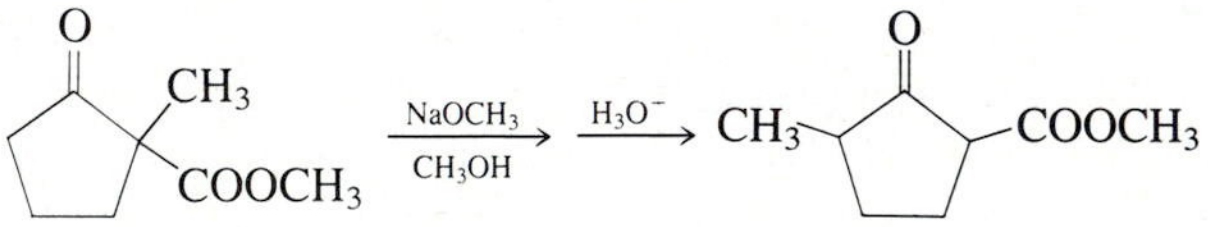

O, CH_3, $COOCH_3$; $NaOCH_3$, CH_3OH; H_3O^-; CH_3, O, $COOCH_3$

20. Present a summary in the format of Highlight 27.1 for (a) alkoxycarbonyl and carboxylic acid groups located on the same carbon, separated by two carbons, and separated by three carbons (b) two carbonyl groups connected to one another, separated by one carbon, and separated by two carbons.

21. Using the format of Highlight 27.2, tabulate the combinations expected for (a) two hydroxy groups separated by three, four, and five carbons, (b) for mercapto and carboxylic acid groups located on the same carbon atom, separated by two carbon atoms, and separated by three carbon atoms.

22. Give at least three routes to the synthesis of α,β-unsaturated acid esters. Evaluate the effectiveness of each route in terms of amounts required, time, and cost.

CHAPTER 28

CARBOHYDRATES

28.1 Introduction

The carbohydrates are an important group of naturally occurring organic compounds. They are extremely abundant in plants, comprising up to 80% of the dry weight. Especially important in the vegetable kingdom are cellulose (the chief structural material of plants), starches, pectins, and the sugars sucrose and glucose. These sugars are obtained on an industrial scale from various plant sources. More than 100,000,000 tons of sucrose is produced per year worldwide, a quantity comparable to that of sulfuric acid, 43,400,000 tons in 1989 in the United States. In higher animals, the simple sugar glucose is an essential constituent of blood and occurs in a polymeric form as glycogen in the liver and in muscle. Carbohydrate derivatives include adenosine triphosphate, which is a key material in biological energy storage and transport systems, and the nucleic acids, which control storage and transfer of genetic information in the production of proteins.

The term **carbohydrate** is used loosely to characterize the whole group of natural products related to the simple sugars. The name first arose because the simple sugars, such as glucose ($C_6H_{12}O_6$), have molecular formulas that appear to be "hydrates of carbon," that is, $C_6H_{12}O_6 = (C \cdot H_2O)_6$. Although subsequent structural investigations revealed that this simple-minded view was erroneous, the term carbohydrate has persisted.

Sugars, also called **saccharides,** are the simplest type of carbohydrate. An example is glucose, which is the cyclic hemiacetal form of one of the diastereomers of 2,3,4,5,6-pentahydroxyhexanal. As we shall see in a later section, glucose exists almost entirely in the cyclic form. In solution it is in equilibrium with a minute amount of the acyclic (i.e., noncyclic) pentahydroxyaldehyde form.

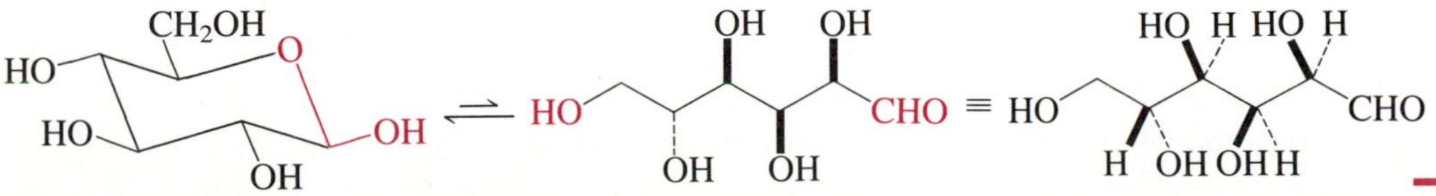

glucose

As is generally true for natural products, *the carbohydrates occur in enantiomerically homogeneous form,* and only one enantiomer is found in nature. Glucose is an example of a **monosaccharide,** a term that means that glucose is not hydrolyzable into smaller units.

Maltose is an example of a **disaccharide;** upon hydrolysis under mildly acidic conditions, maltose yields two equivalents of the monosaccharide glucose.

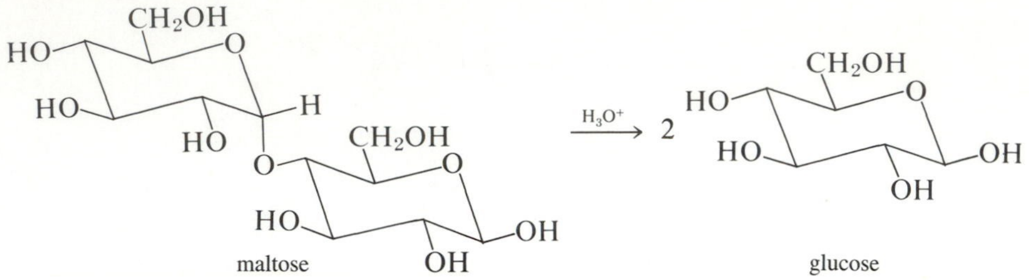

maltose $\xrightarrow{H_3O^+}$ 2 glucose

A **trisaccharide** yields three monosaccharides on hydrolysis, a **tetrasaccharide** four, and so forth. **Oligosaccharide** is a general term applied to sugar polymers containing up to eight units. **Polysaccharide** refers to polymers in which the number of subunits is greater than eight; the natural polysaccharides generally consist of 100–3000 subunits.

The monosaccharides are also characterized in terms of the number of carbons in the chain and the nature of the carbonyl group, aldehyde or ketone. Glucose, which is a six-carbon aldehyde, is a **hexose,** which specifies the number of carbons, and an **aldose,** which shows that it is an aldehyde. It is completely characterized by the general term **aldohexose.** Other aldoses are glyceraldehyde, an **aldotriose,** erythrose, an **aldotetrose,** and arabinose, an **aldopentose.** The structures of these examples are shown below as their open-chain forms.

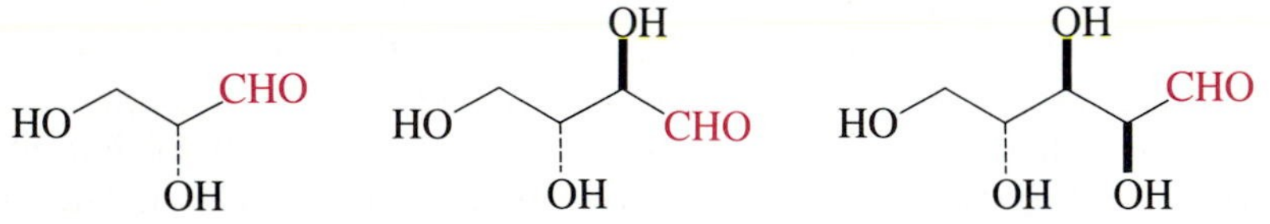

glyceraldehyde
(an aldotriose)

erythrose
(an aldotetrose)

arabinose
(an aldopentose)

Most of the naturally occurring sugars are derived from the aldoses, and the most widespread are the aldohexoses and the aldopentoses.

A few important saccharides are **ketoses,** meaning that they contain a ketone, rather than an aldehyde, carbonyl group. Fructose is an example of a **ketohexose,** a six-carbon pentahydroxy ketone. An example of a **ketopentose** is ribulose. Both compounds are shown below in open-chain form.

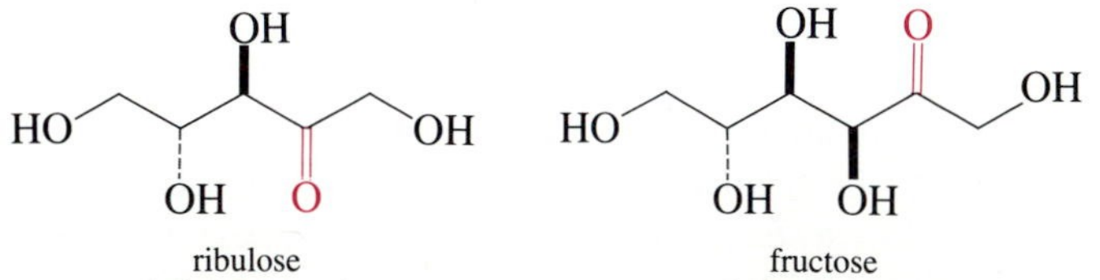

ribulose
(a ketopentose)

fructose
(a ketohexose)

EXERCISE 28.1 Reduction of the C=O group of fructose by sodium borohydride gives a mixture of two isomeric hexanehexaols. The CMR spectrum of one isomer has six absorptions, but the CMR spectrum of the other has only three. Explain.

28.2 Stereochemistry and Configurational Notation of Sugars

The simplest polyhydroxy aldehyde is the compound 2,3-dihydroxypropanal or glyceraldehyde. The molecule has one stereocenter, so there are two enantiomers (Chapter 7). The absolute configurations of the glyceraldehyde enantiomers are known, and the isomer with $[\alpha]_D = +8.7$ has the structure shown below on the left.

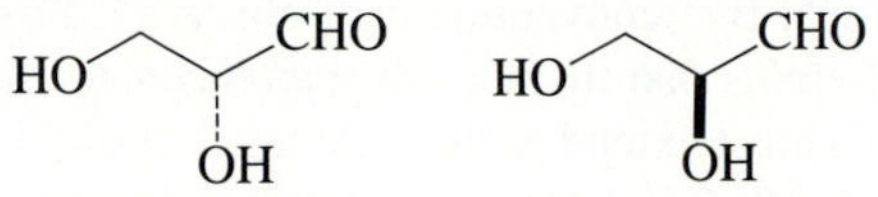

D-(+)-glyceraldehyde (*R*)-glyceraldehyde L-(−)-glyceraldehyde (*S*)-glyceraldehyde

This enantiomer can be distinguished from the other by calling it (+)-glyceraldehyde, meaning "the dextrorotatory enantiomer of glyceraldehyde." Alternatively, it can be described in a manner that specifies the absolute configuration, as (*R*)-glyceraldehyde (Section 7.3). In the carbohydrate field, it is customary to use another and older system of configurational notation, wherein this enantiomer is called D-(+)-glyceraldehyde. Under this convention, all D-sugars have the same configuration as D-(+)-glyceraldehyde at the stereocenter most distant from the carbonyl group. Sugars with the opposite configuration at this center are members of the L-family. Thus, natural glucose is D-(+)-glucose and fructose is D-(−)-fructose.

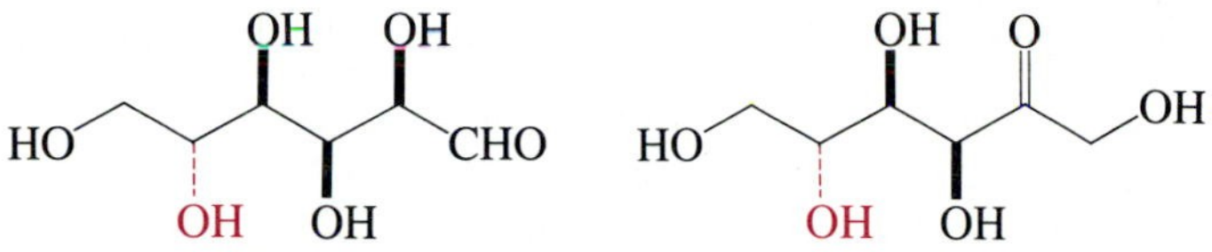

D-(+)-glucose D-(−)-fructose

The projection representations used so far in this chapter are useful and unambiguous in meaning. An alternative projection system, which is widely used in the carbohydrate field, is called the Fischer system. In a **Fischer projection** a stereocenter is represented with two of its four bonds extending horizontally, to the left and right, and with the other two extending vertically, to the top and bottom. The horizontal lines represent bonds extending forward from the stereocenter. The two vertical lines represent bonds extending back away from the plane of the page. The atom that is the stereocenter is often omitted, in which case it is symbolized by the intersection point of the two lines of a cross. A Fischer projection for (*R*)-glyceraldehyde is compared to the wedge-and-dashed-line structure (Section 7.3) and the "zigzag" structure as follows.

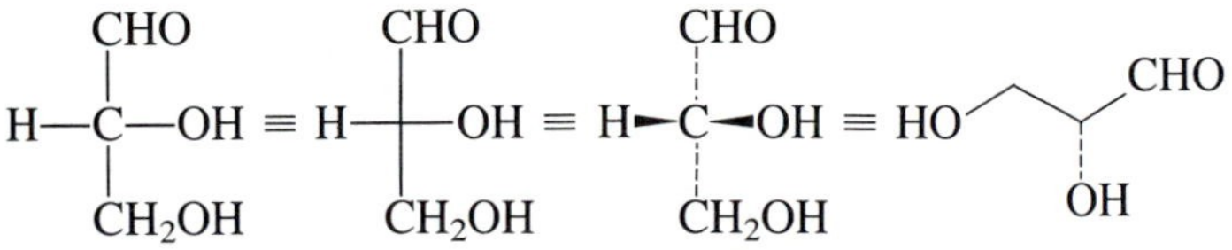

(*R*)-glyceraldehyde

It is important to remember that Fischer projections are two-dimensional projections of three-dimensional objects. For purposes of visualizing whether or not two structures are identical, these projections can be manipulated only in certain ways. In order to change one Fischer projection to another correct projection for the same enantiomer, one may interchange any *two* pairs of substituents. If only one pair of groups is interchanged, a projection for the enantiomer is generated.

$$
\begin{array}{c}
\text{CHO} \\
| \\
\text{H—C—OH} \\
| \\
\text{CH}_2\text{OH}
\end{array}
\equiv
\begin{array}{c}
\text{CH}_2\text{OH} \\
| \\
\text{HO—C—H} \\
| \\
\text{CHO}
\end{array}
\not\equiv
\begin{array}{c}
\text{CHO} \\
| \\
\text{HO—C—H} \\
| \\
\text{CH}_2\text{OH}
\end{array}
$$

Fischer projections are especially useful for compounds that have two or more stereocenters. In such cases, two conventions are observed. First, the structure must be written with the chain of atoms that includes the stereocenters drawn vertically. Second, all of the appendage bonds must extend to the right and left of this vertical representation. Fischer projections of (2*R*,3*R*)-2,3,4-trihydroxybutanal (D-erythrose) illustrate this convention.

$$
\begin{array}{c}
\text{CHO} \\
| \\
\text{H—C—OH} \\
| \\
\text{H—C—OH} \\
| \\
\text{CH}_2\text{OH}
\end{array}
\equiv
\begin{array}{c}
\text{CH}_2\text{OH} \\
| \\
\text{HO—C—H} \\
| \\
\text{HO—C—H} \\
| \\
\text{CHO}
\end{array}
$$

D-erythrose

It is important to note that the Fischer projection is simply a device for depicting the absolute configuration at the various stereocenters of a compound. It *does not* tell us anything about the preferred *conformation* of the molecule. Indeed, the Fischer projections that were just depicted for D-erythrose both correspond to the eclipsed conformation.

EXERCISE 28.2 Construct a molecular model of (2*R*,3*R*)-2,3,4-trihydroxybutanal and use it to confirm the points that were made in the foregoing paragraph.

If a compound has more than two stereocenters, the same convention is followed. The chain containing the stereocenters is placed in a vertical position with the other two bonds from each stereocenter projecting to the right and left of this chain. The following structures are Fischer projections of D-glucose and D-fructose.

$$
\begin{array}{c}
\text{CHO} \\
| \\
\text{H—C—OH} \\
| \\
\text{HO—C—H} \\
| \\
\text{H—C—OH} \\
| \\
\text{H—C—OH} \\
| \\
\text{CH}_2\text{OH}
\end{array}
\qquad
\begin{array}{c}
\text{CH}_2\text{OH} \\
| \\
\text{C=O} \\
| \\
\text{HO—C—H} \\
| \\
\text{H—C—OH} \\
| \\
\text{H—C—OH} \\
| \\
\text{CH}_2\text{OH}
\end{array}
$$

D-glucose D-fructose

One must remember to observe the conventions upon which the Fischer projections are based. Thus, the following structures, in which the chain containing the stereocenters is written in a horizontal, rather than a vertical manner, are *not valid Fischer projections*.

$$
\begin{array}{ccccccccccc}
 & & \text{H} & & \text{H} & & \text{H} & & \text{OH} & & \text{H} \\
 & & | & & | & & | & & | & & | \\
\text{HO} & — & \text{C} & — & \text{C} & — & \text{C} & — & \text{C} & — & \text{C—CHO} \\
 & & | & & | & & | & & | & & | \\
 & & \text{H} & & \text{OH} & & \text{OH} & & \text{H} & & \text{OH}
\end{array}
\qquad
\begin{array}{cccccc}
 & \text{H} & \text{H} & \text{OH} & \text{H} & \\
 & | & | & | & | & \\
\text{HOCH}_2\text{—} & + & + & + & + & \text{—CHO} \\
 & | & | & | & | & \\
 & \text{OH} & \text{OH} & \text{H} & \text{OH} &
\end{array}
$$

incorrect Fischer projections

Recall that for a compound with n stereocenters, there are 2^n possible optical isomers. Thus, there are 2 aldotrioses, 4 aldotetroses, 8 aldopentoses, and 16 aldohexoses. Half of these compounds belong to the D-family (are related to D-glyceraldehyde) and half belong to the L-family. To avoid cumbersome names, each isomer has been given a trivial name; that is, D-(+)-glucose is (2*R*,3*S*,4*R*,5*R*)-2,3,4,5,6-pentahydroxyhexanal. Fischer projections depicting the complete D-family of the aldoses in their open-chain forms are shown in Table 28.1.

The naturally occurring sugars generally belong to the D-family shown in Table 28.1, but there are an equal number of compounds with the L-configuration. Each D-sugar has an enantiomeric L-counterpart.

Recall that a molecule with two or more stereocenters is achiral if it has a plane of symmetry. Such compounds are called meso compounds (page 136). It often happens in carbohydrate chemistry that a chiral compound undergoes a chemical reaction to yield a meso product. For example, consider the reduction of the aldotetrose D-(−)-erythrose by sodium borohydride. The product is meso-1,2,3,4-butanetetraol (erythritol). The same compound would be produced by the reduction of L-(+)-erythrose.

$$\begin{array}{c} CHO \\ | \\ H\text{—}C\text{—}OH \\ | \\ H\text{—}C\text{—}OH \\ | \\ CH_2OH \end{array} \xrightarrow{NaBH_4} \begin{array}{c} CH_2OH \\ | \\ H\text{—}C\text{—}OH \\ | \\ H\text{—}C\text{—}OH \\ | \\ CH_2OH \end{array} \xleftarrow{NaBH_4} \begin{array}{c} CHO \\ | \\ HO\text{—}C\text{—}H \\ | \\ HO\text{—}C\text{—}H \\ | \\ CH_2OH \end{array}$$

D-(−)-erythrose erythritol L-(+)-erythrose

On the other hand, the aldotetroses D-(−)-threose and L-(+)-threose each yield an optically active butanetetraol on reduction.

$$\begin{array}{c} CHO \\ | \\ HO\text{—}C\text{—}H \\ | \\ H\text{—}C\text{—}OH \\ | \\ CH_2OH \end{array} \xrightarrow{NaBH_4} \begin{array}{c} CH_2OH \\ | \\ HO\text{—}C\text{—}H \\ | \\ H\text{—}C\text{—}OH \\ | \\ CH_2OH \end{array}$$

D-(−)-threose D-(+)-threitol

$$\begin{array}{c} CHO \\ | \\ H\text{—}C\text{—}OH \\ | \\ HO\text{—}C\text{—}H \\ | \\ CH_2OH \end{array} \xrightarrow{NaBH_4} \begin{array}{c} CH_2OH \\ | \\ H\text{—}C\text{—}OH \\ | \\ HO\text{—}C\text{—}H \\ | \\ CH_2OH \end{array}$$

L-(+)-threose L-(−)-threitol

The formation of a meso compound can be a powerful piece of information for use in determining the relative configuration of a compound that has two or more stereocenters. For example the fact that erythrose undergoes reduction to give a meso tetraol proves that its two stereocenters are either *R*,*R* or *S*,*S*. Conversely, since threose gives a tetraol that is optically active, it must have either the *R*,*S* or *S*,*R* configuration.

EXERCISE 28.3 Write Fischer projections for the products of $NaBH_4$ reduction of D-(+)-galactose, L-(−)-xylose, and D-(+)-mannose. Are any of these products achiral?

TABLE 28.1 The D-Family Aldoses

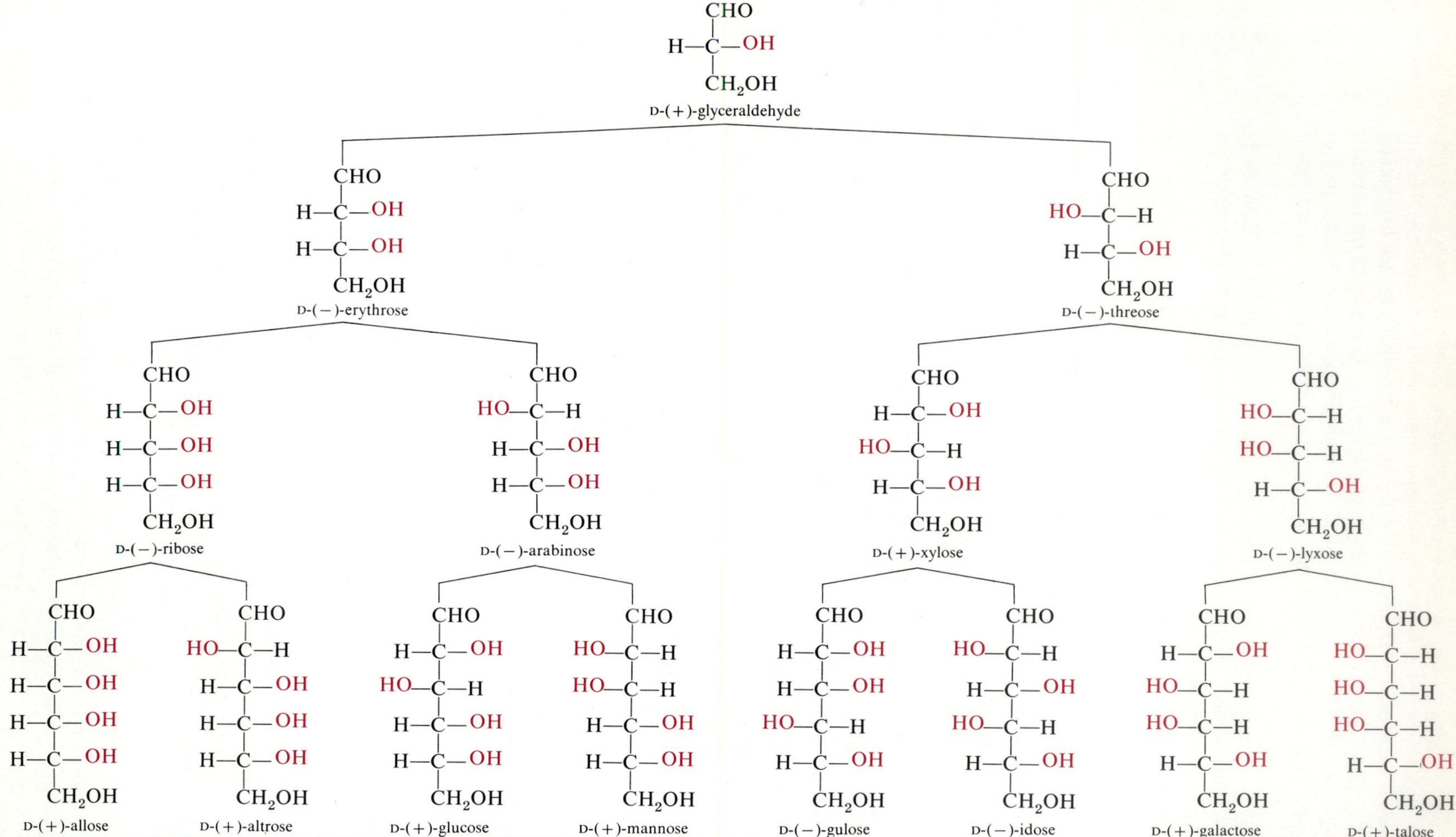

In the last chapter (Section 27.4), we learned that 4- and 5-hydroxy aldehydes and ketones exist mainly in the cyclic hemiacetal form.

$HOCH_2CH_2CH_2CHO \rightleftharpoons$ [five-membered cyclic hemiacetal: O, H, OH]

$HOCH_2CH_2CH_2CH_2CHO \rightleftharpoons$ [six-membered cyclic hemiacetal: O, H, OH]

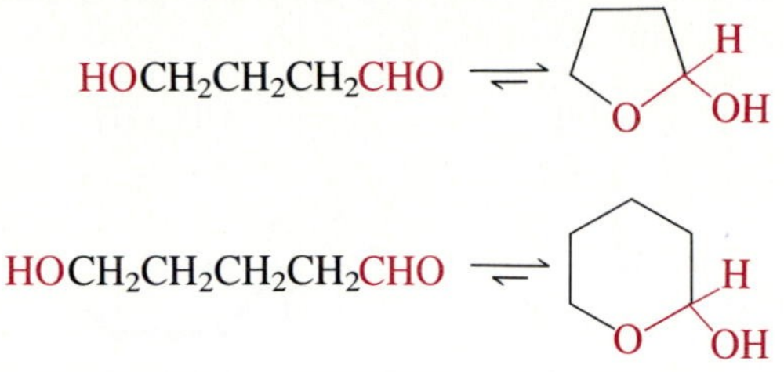

It is not surprising then that the sugars also exist in such a cyclic form. Although either the five- or six-membered hemiacetal structure is possible, almost all of the simple sugars exist in the six-membered ring form (see Table 27.3, page 870). Note that when the hemiacetal is formed, *the former aldehyde carbon becomes a stereocenter*. Thus, there are two cyclic forms of glucose.

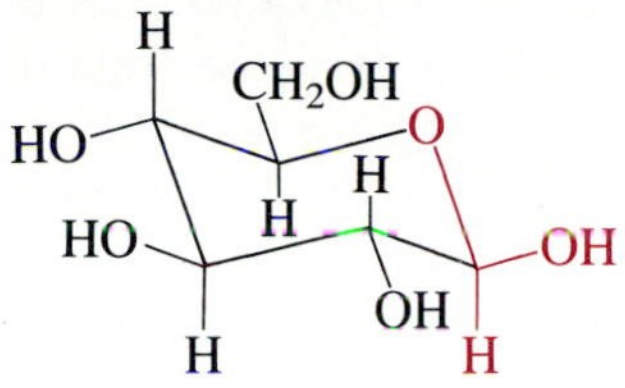

β-D-glucose 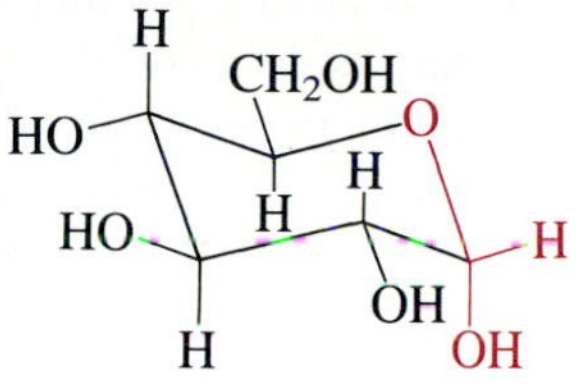 α-D-glucose

The two cyclic isomers of glucose differ only in the configuration at C-1, the hemiacetal carbon (former aldehyde carbon). Such isomers are called **anomers,** and the hemiacetal carbon is called the **anomeric** carbon. The two anomers are commonly differentiated by the Greek letters α and β; for example, α-D-glucose, β-D-glucose. For the aldohexoses, the β-anomer is the one that has the OH at C-1 and the CH_2OH at C-5 cis with respect to each other on the ring.

The Fischer projection formulas shown in Table 28.1 are a convenient way in which to represent the open-chain form of sugars. Modified Fischer projections have frequently been used to depict the cyclic hemiacetal form. For example, the two D-glucose anomers can be represented as follows.

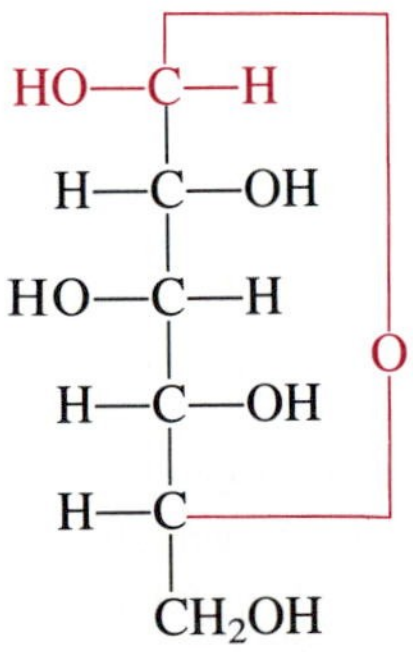

β-D-glucose 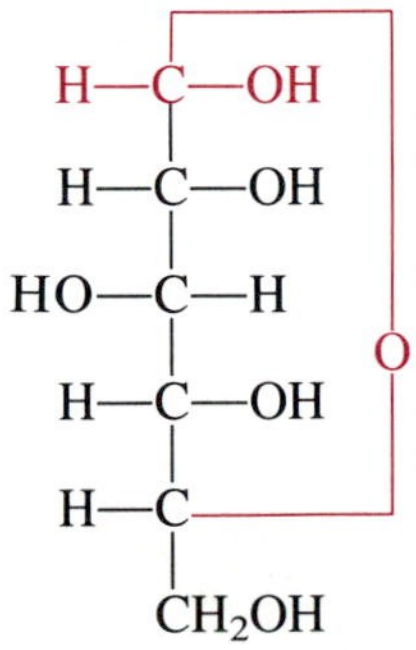 α-D-glucose

Note the convention used to represent stereochemistry at anomeric carbon. The OH at C-1 in the β-anomer is written to the left and that in the α-anomer is written to the right.

Because these modified projections lead to awkward drawings of bond lengths, which offend the sensibilities of many chemists, Haworth introduced an alternate projection formula, which is used extensively by sugar chemists. In a **Haworth projection,** the sugar ring is written as a planar hexagon with the oxygen in the upper right vertex. Substituents are indicated by straight lines through each vertex, either above or below the plane. The OH at the anomeric carbon is up in the β-anomer and down in the α-anomer. Hydrogens attached to the ring are omitted.

β-D-glucose $\qquad$ α-D-glucose

There is a simple way to convert a Fischer projection to a Haworth projection, or vice versa. The OH groups that project to the left in a Fischer projection project up in a Haworth projection. In this book, we shall use Fischer projections to depict open-chain sugars and the more accurate chair representations to depict cyclic forms.

The six-membered ring form of a sugar is called a **pyranose** from the name of the simplest heterocyclic compound containing such a ring, pyran. Thus, β-D-glucose is a pyranose form, and it may be completely described by the name β-D-glucopyranose. Although the free sugars normally do not exist as the five-membered ring form, numerous derivatives are known that have such a structure. They are called **furanoses** from the name of the parent heterocyclic compound, furan.

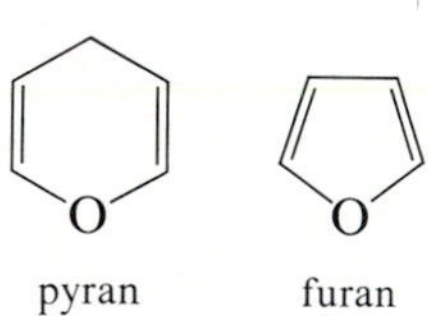

pyran $\qquad$ furan

Pure β-D-glucose has an optical rotation $[\alpha]_D = +18.7$; the α-anomer has $[\alpha]_D = +112$. Both anomers have been isolated in pure crystalline states. If either pure anomer is dissolved in water, the optical rotation of the solution gradually changes until it reaches an equilibrium value of + 52.7. This phenomenon, which was first observed in 1846, results from the interconversion of the two anomers in solution and is called **mutarotation.** At equilibrium, the solution contains 63.6% of the more stable β-anomer and 36.4% of the α-anomer.

> The phenomenon of anomerism caused considerable confusion for the early workers in carbohydrate chemistry, who believed the sugars to be acyclic compounds. After their cyclic structures were recognized, there arose the problem of how to name the two anomers for each sugar. In 1909, a system of nomenclature based purely on optical rotation was adopted. In the D series, the more dextrorotatory member of a pair of anomers is defined as the α-D-anomer and the less dextrorotatory anomer is the β-D-anomer. When reliable methods for determining the configuration at C-1 became available, it turned out that all α-anomers have the same absolute configuration at C-1.

Interconversion of the two anomers is subject to both acid and base catalysis and occurs by the normal mechanism for acetal formation and hydrolysis (Sections 14.6.B and 27.4.B). The open-chain form is probably an intermediate in the process. The mechanism of acid-catalyzed interchange of the α- and β-anomers of glucose is outlined as follows (all ring hydrogens except the one on the anomeric carbon are omitted for clarity).

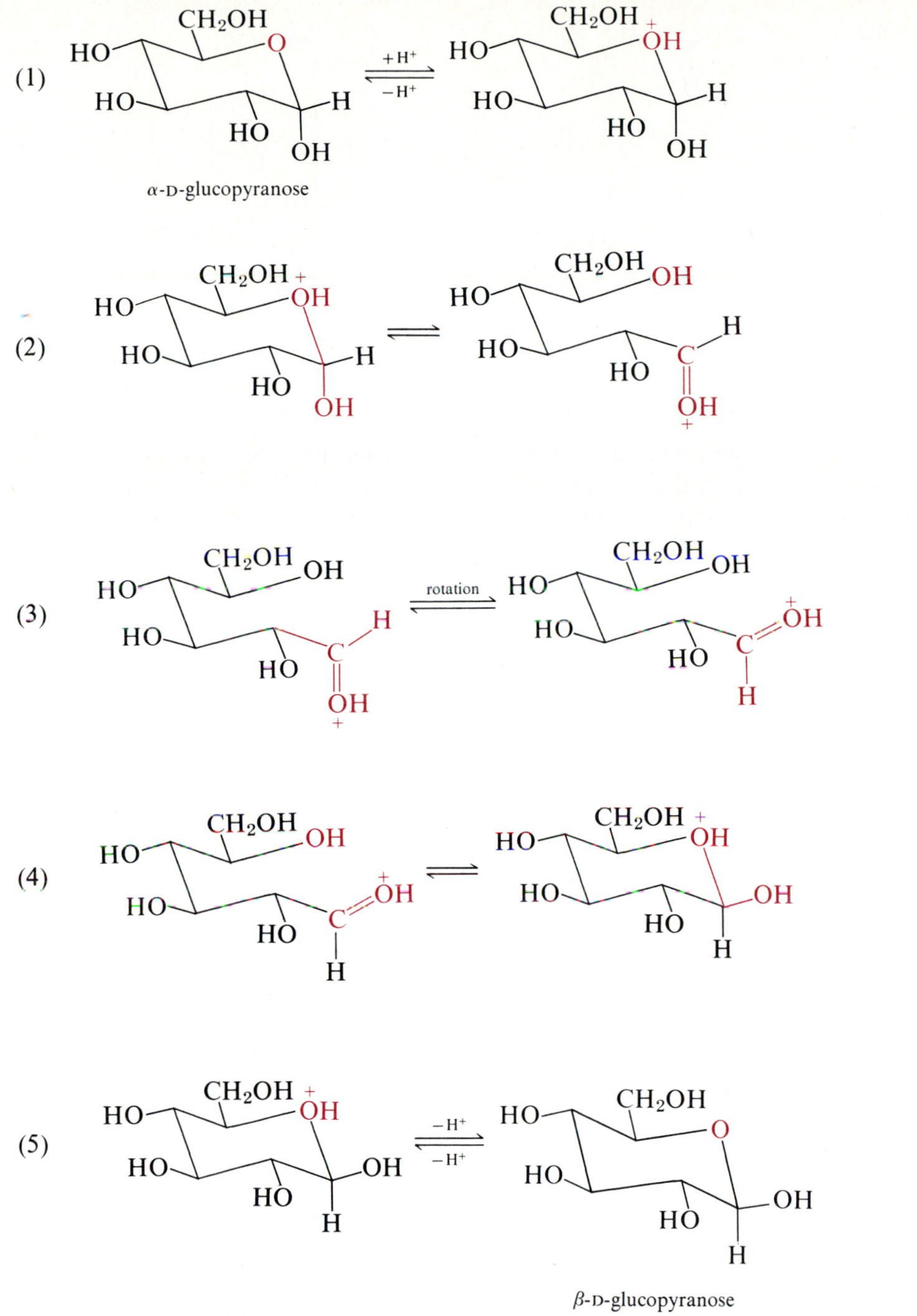

(1)

α-D-glucopyranose

(2)

(3)

(4)

(5)

β-D-glucopyranose

When an aldose is dissolved in an alcohol and the solution is treated with a mineral acid catalyst, a cyclic acetal is produced (Section 27.4.B). In carbohydrate chemistry such cyclic acetals are called **glycosides.** A glycoside derived from glucose is a **glucoside,** one derived from mannose is a **mannoside,** and so on. Like the hemiacetals, these cyclic acetals can exist in both α- and β-anomeric forms as shown below for the methyl mannosides.

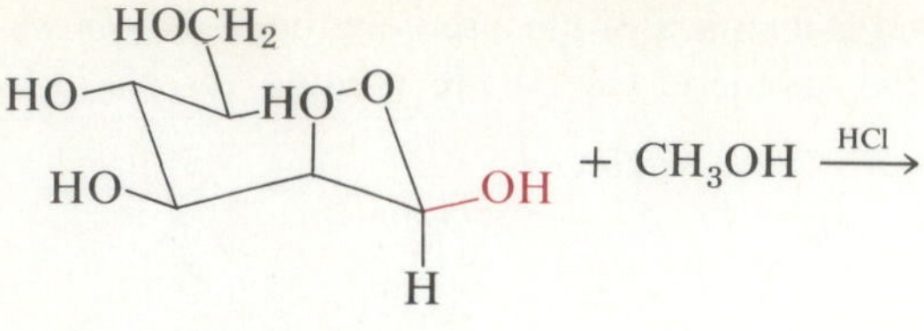

HOCH$_2$ HO HO-O HO OH H + CH_3OH $\xrightarrow{HCl}$

β-D-mannose

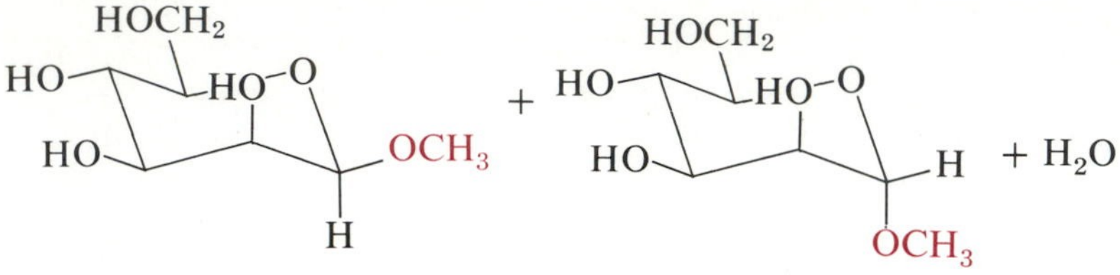

methyl β-D-mannoside
methyl β-D-mannopyranoside

+ H_2O

methyl α-D-mannoside
methyl α-D-mannopyranosid

Glycosides form only under acid catalysis; the mechanism for the formation of the methyl galactosides is outlined as follows. Recall that cyclohexene has a conformation in which four atoms of the ring are coplanar and the other two are above and below the plane, respectively (Section 11.6.B); the intermediate carbocation involved in glycoside formation probably has a similar conformation.

(1) $\underset{-H^+}{\overset{+H^+}{\rightleftharpoons}}$

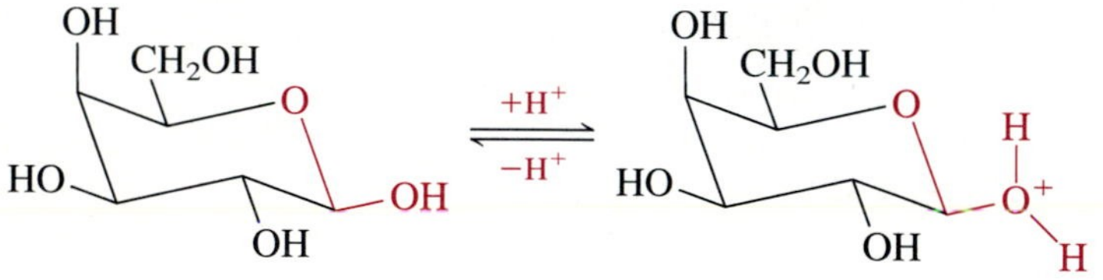

β-D-galactose

(2) $\underset{+H_2O}{\overset{-H_2O}{\rightleftharpoons}}$ ⟷

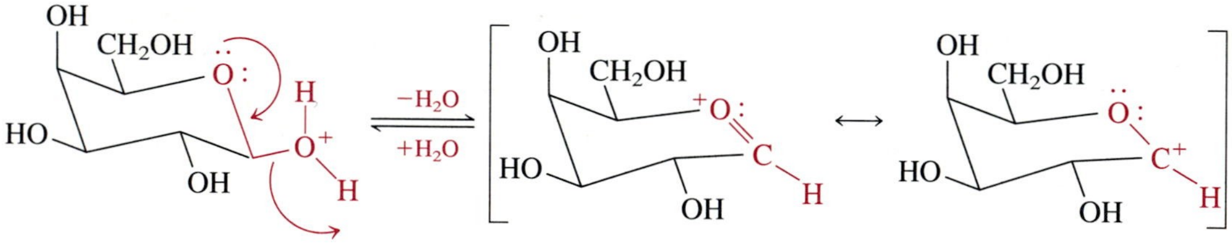

(3) $\underset{-CH_3OH}{\overset{+CH_3OH}{\rightleftharpoons}}$ $\underset{-CH_3OH}{\overset{+CH_3OH}{\rightleftharpoons}}$

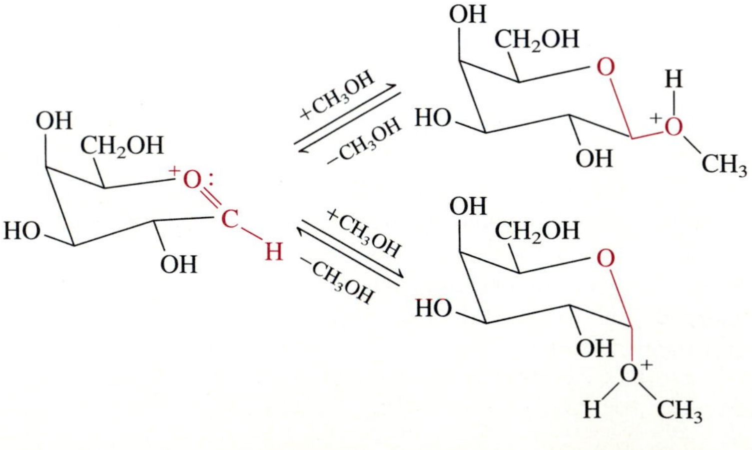

(4) 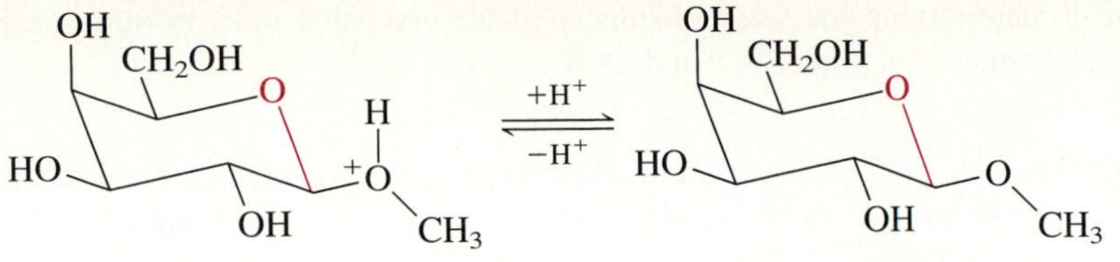 [protonated glycoside] $\underset{-H^+}{\overset{+H^+}{\rightleftharpoons}}$ methyl β-D-galactoside

methyl β-D-galactoside
methyl β-D-galactopyranoside

The formation of glycosides is a reversible process under acidic conditions. If a glycoside is treated with an acid catalyst in aqueous solution where water is present in excess, the equilibrium shifts and hydrolysis occurs. Of course, under acidic conditions a mixture of the anomeric sugars results.

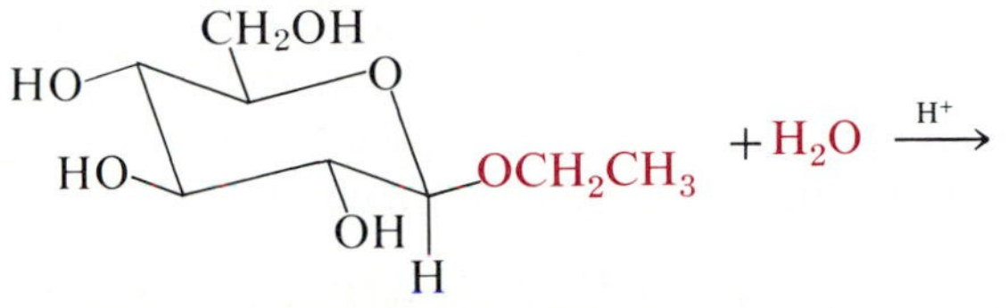

ethyl β-D-glucoside $+ H_2O \xrightarrow{H^+}$ β-D-glucose + α-D-glucose $+ C_2H_5OH$

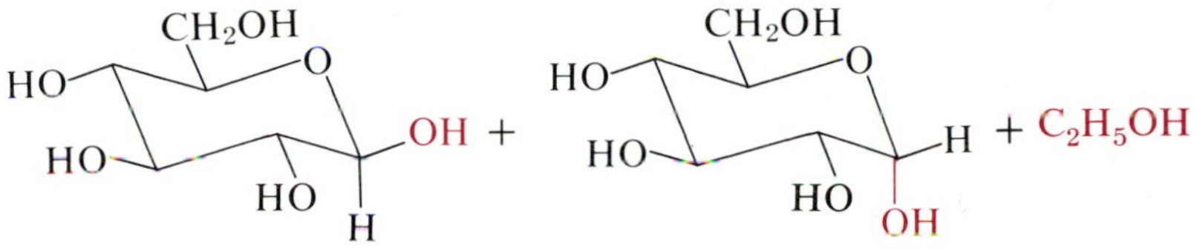

ethyl β-D-glucoside

β-D-glucose α-D-glucose

The hydrolysis of glycosides is brought about by certain **enzymes.** Enzymes are complex natural products, mainly protein in nature (Chapter 29), that function as catalysts in biological reactions. They are extremely potent catalysts, often speeding up reactions by factors as large as 10^{10}. They also show remarkable structural specificity, as shown by the present example. Methyl α-D-glucopyranoside is hydrolyzed in the presence of an enzyme, isolated from yeast, called α-glucosidase. This particular enzyme only catalyzes the hydrolysis of α-glucoside linkages; methyl β-D-glucopyranoside is unaffected by it.

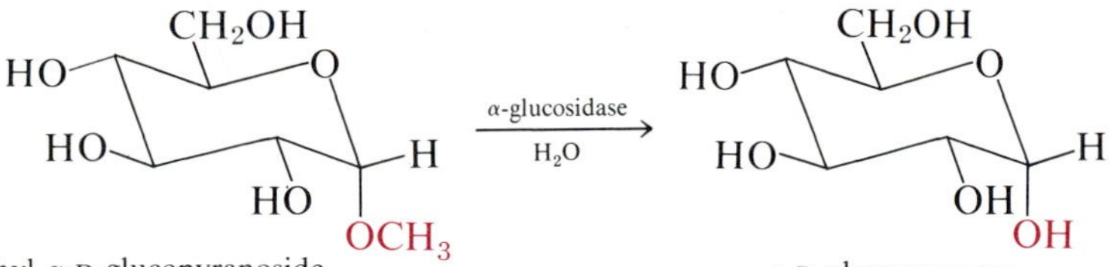

methyl α-D-glucopyranoside $\xrightarrow[H_2O]{\alpha\text{-glucosidase}}$ α-D-glucopyranose

Another enzyme, β-glucosidase, from almonds, has opposite properties; it only catalyzes the hydrolysis of β-glucosides.

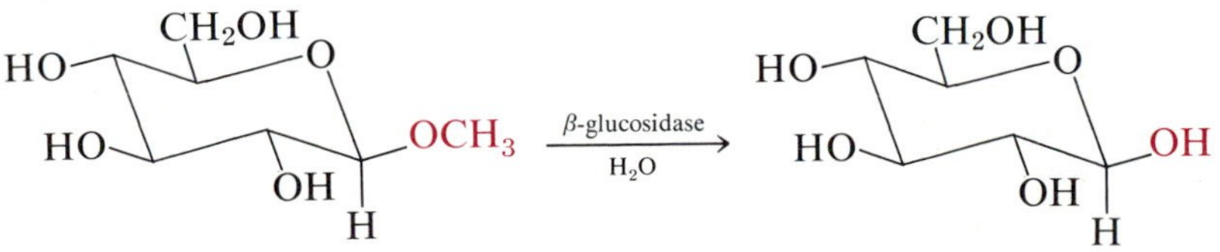

methyl β-D-glucopyranoside $\xrightarrow[H_2O]{\beta\text{-glucosidase}}$ β-D-glucopyranose

Similar enzymes are known that specifically catalyze the hydrolysis of α- and β-galactosides (α- and β-galactosidase) and other glycosidic bonds. These enzymes are useful in determining the stereochemistry of the glycoside links in oligosaccharides and polysaccharides (Sections 28.7 and 28.8).

EXERCISE 28.4 Write Haworth and chair perspective formulas for α-D-altrose. Using the chair perspective formula, write all of the steps for mutarotation to β-D-altrose.

28.4 Conformations of the Pyranoses

As has been tacitly implied in the structures used thus far in this chapter, the pyranose forms of sugars exist in a chair conformation similar to the stable conformation of cyclohexane (Section 5.6). As in cyclohexane, two alternative chair forms are possible, and the one that predominates is that one with the fewer repulsive interactions. For β-D-glucose there is a large difference between the two forms. In one form all five substituents are in equatorial positions, whereas they are all axial in the other conformation; the difference between these two conformations has been estimated to be 6 kcal $mole^{-1}$.

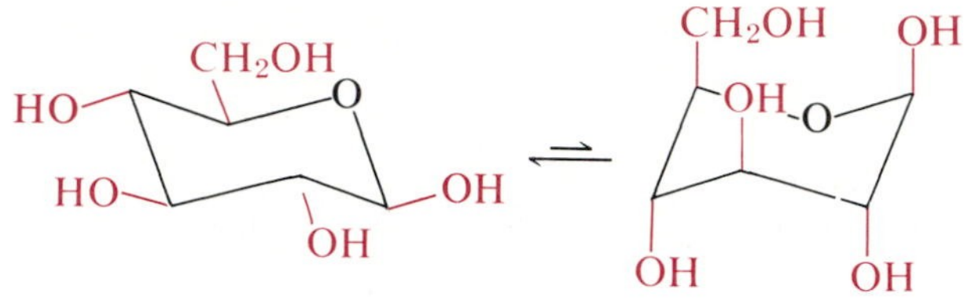

CH_2OH HO HO O OH OH ⇌ CH_2OH OH OH O OH OH

β-D-glucose

Of the eight D-aldohexoses, glucose is the only one that can have all five substituents equatorial. It may be no accident that glucose is the most abundant natural monosaccharide. A stereo structure of β-D-glucose is shown in Figure 28.1.

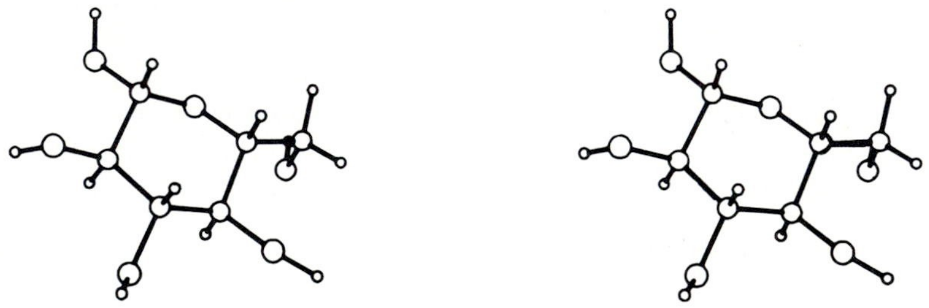

FIGURE 28.1 Stereo structure of β-D-glucose. [Reproduced with permission from *Molecular Structure and Dimensions,* International Union of Crystallography, 1972.]

The prevalence of glucose might be the consequence of prebiotic evolution (the time before any organized self-reproducing system existed), from selection during ribonucleic acid (RNA) evolution, or from selection during deoxyribonucleic acid (DNA)-protein evolution. RNA and DNA are based on pentoses and have properties well adapted to their present-day functions of storage and transfer of genetic information. Synthetic hexosenucleic acids are too rigid for these functions. The hexose glucose, which might have been formed in the prebiotic ''soup,'' was then available for transformations in various energy-generating metabolic reactions.

If one remembers that β-D-glucose has all substituents equatorial, it is easy to write conformational structures for the other aldohexoses by simply referring to Table 28.1. For example, D-allose differs from D-glucose only in the configuration at C-3. Thus, β-D-allose is

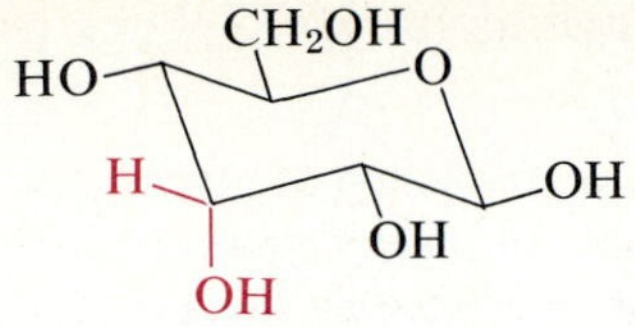

HO CH$_2$OH O H OH OH OH

β-D-allose

For most of the aldohexoses, the more stable conformation is the one with the CH$_2$OH group in an equatorial position. However, in a few cases the two conformations are nearly equal in energy and substantial amounts of both may be present at equilibrium. For α-D-idose, the stable conformation is that in which the CH$_2$OH group is axial.

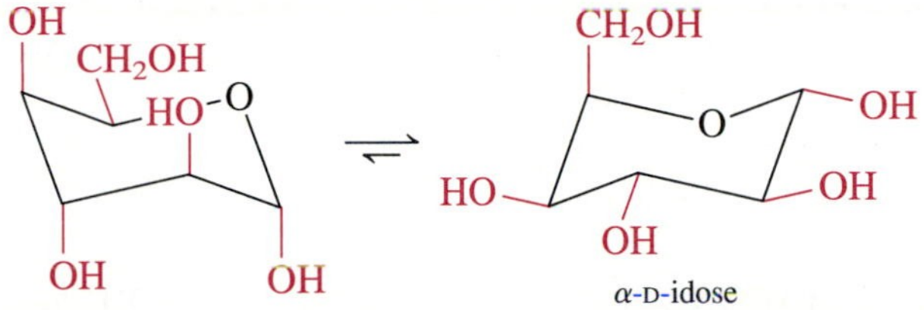

OH CH$_2$OH HO O OH OH ⇌ CH$_2$OH O OH HO OH OH

α-D-idose

EXERCISE 28.5 Write chair perspective formulas for β-D-gulose and α-D-talose.

Highlight 28.1

The eight possible aldohexoses exist largely in the pyranose hemiacetal forms. The conformations are determined mainly by the differences between axial and equatorial locations for the hydroxy and hydroxymethyl groups. The most important conformations are shown below.

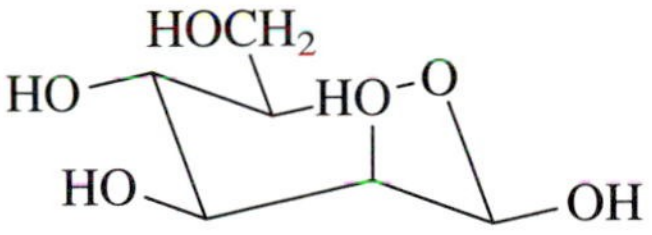

β-D-mannopyranose

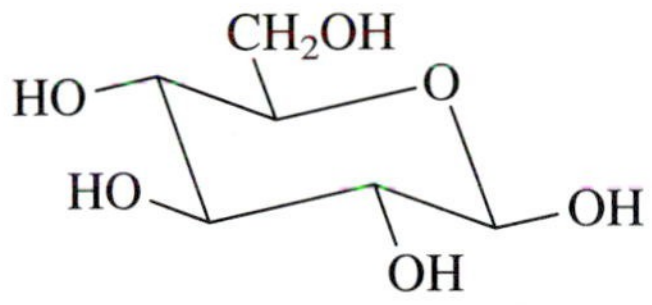

β-D-glucopyranose

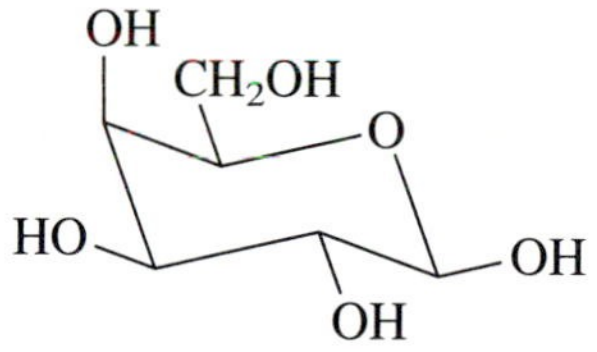

β-D-galactopyranose

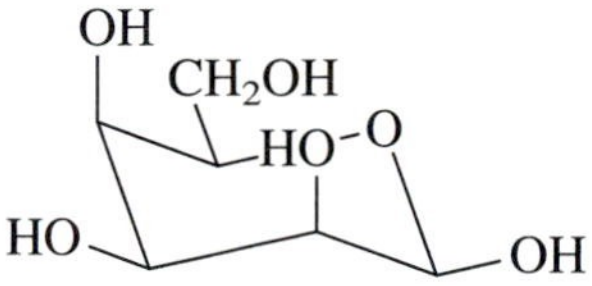

β-D-talopyranose

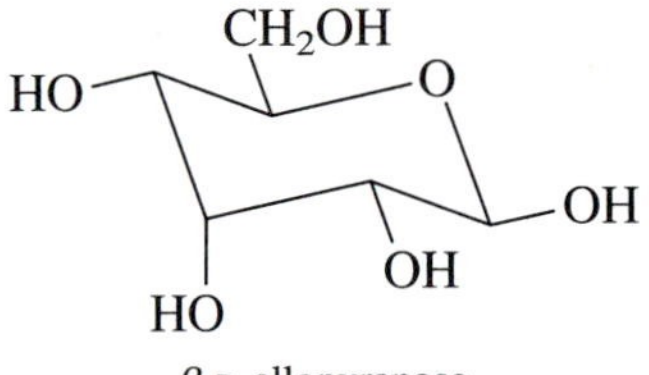

β-D-allopyranose

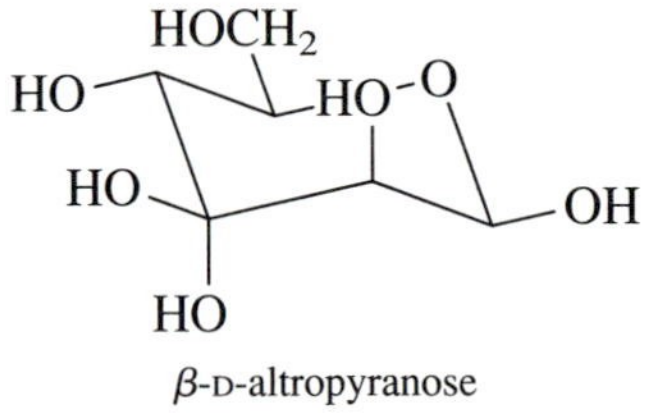

β-D-altropyranose

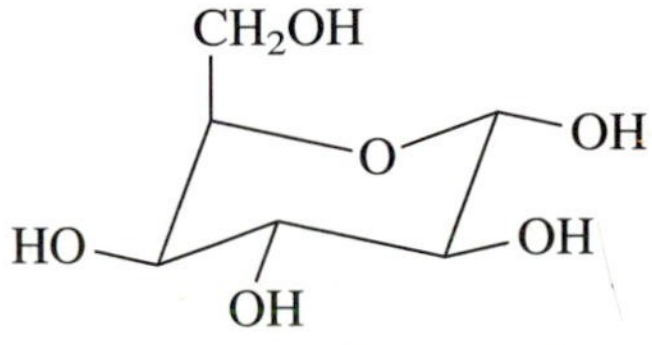

α-D-idopyranose

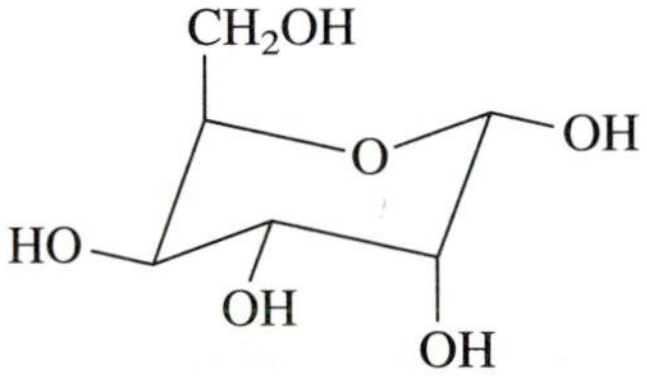

α-D-gulopyranose

28.5 Reactions of Monosaccharides

A. Ether Formation

In Section 28.3 we discussed the formation of glycosides, in which the OH group at the anomeric carbon is replaced by an alkoxy group under mildly acidic conditions. The remaining hydroxy groups are unaffected by this process because such a process would involve a primary or secondary carbocation, rather than the far more stable oxonium ion that is involved in glycoside formation.

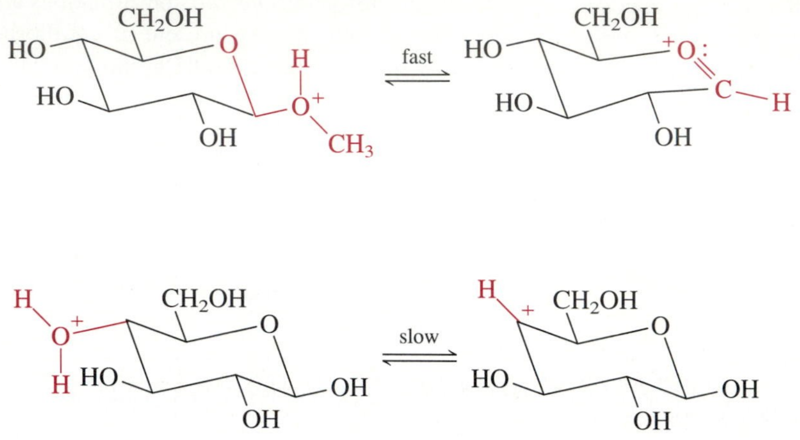

fast

slow

The other hydroxy groups can be converted into ethers by an application of the Williamson ether synthesis (Section 10.9). The most common ethers are the methyl ethers, which are prepared by treating the sugar with 30% aqueous sodium hydroxide and dimethyl sulfate, or with silver oxide and methyl iodide. Since the free aldehyde form of an aldose is not stable to strongly basic conditions, it is customary to protect the anomeric carbon by converting the sugar into the methyl glycoside. The glycoside linkage can then be cleaved by mild acid hydrolysis because the normal ether linkages are stable under these conditions.

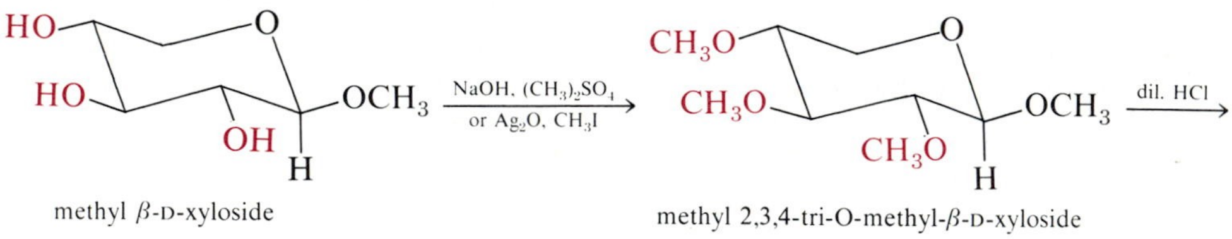

$\xrightarrow[\text{or } Ag_2O,\ CH_3I]{NaOH,\ (CH_3)_2SO_4}$ $\xrightarrow{\text{dil. HCl}}$

methyl β-D-xyloside

methyl 2,3,4-tri-O-methyl-β-D-xyloside

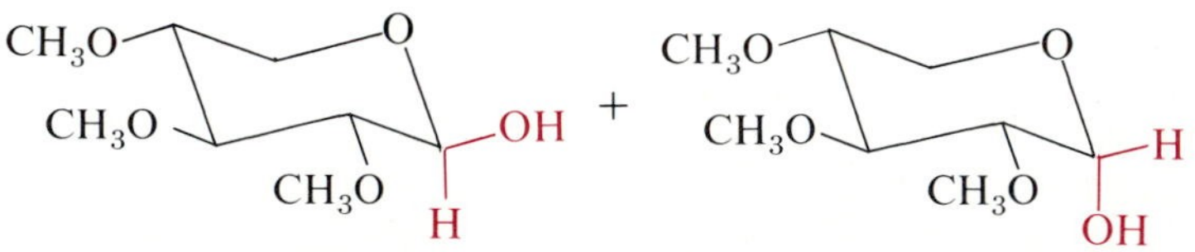

2,3,4-tri-O-methyl-β-D-xylose + 2,3,4-tri-O-methyl-α-D-xylose

Methylation can be a useful method for determining the size of the acetal ring in a glycoside. For example, oxidation of the foregoing mixture of anomeric tri-O-methylxyloses yields a 2,3,4-trimethoxyglutaric acid, thus establishing that the original methyl xyloside had the pyranose structure.

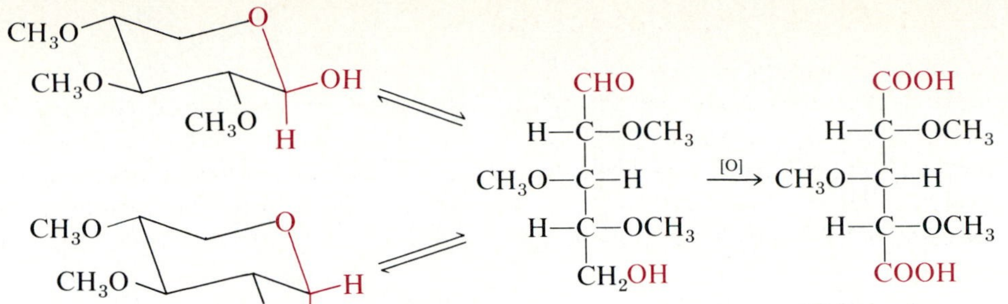

2,3,4-trimethoxyglutaric acid

EXERCISE 28.6 Using chair perspective formulas, write equations showing the reaction of methyl β-D-glucopyranoside with Ag_2O and CH_3I. What products are produced when this material is treated with HCl?

B. Formation of Cyclic Acetals

Recall that 1,2- and 1,3-diols condense with aldehydes and ketones to form cyclic acetals. If the diol is itself cyclic, the acetal forms only when the two OH groups are cis, for geometric reasons.

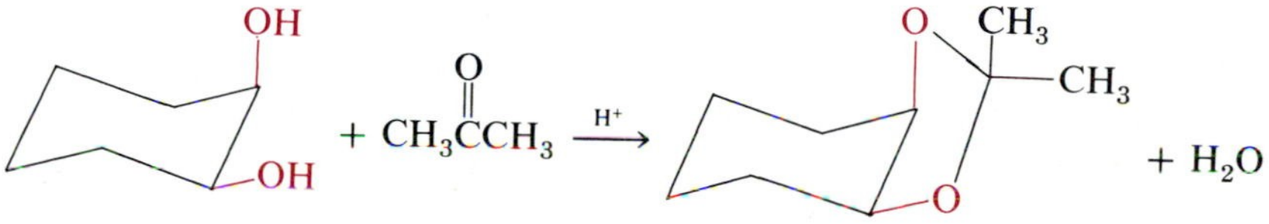

Since sugars are polyhydroxy compounds, they also undergo this reaction. The reaction is often complicated by the fact that the ring size in the product is not the same as it is in the free sugar. This complication usually occurs when the more stable pyranose form does not have cis vicinal hydroxy groups, but the furanose form does. Thus galactose reacts with acetone to give the diacetal shown because in the α-form, which is present under the acidic conditions of the reaction, there are two pairs of cis vicinal OH groups.

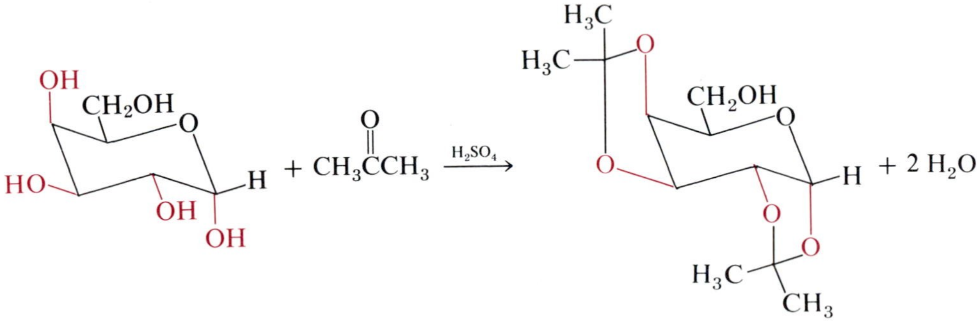

α-D-galactose

1,2:3,4-di-O-isopropylidene-α-D-galactopyranoside

Glucose, on the other hand, reacts by way of the furanose form.

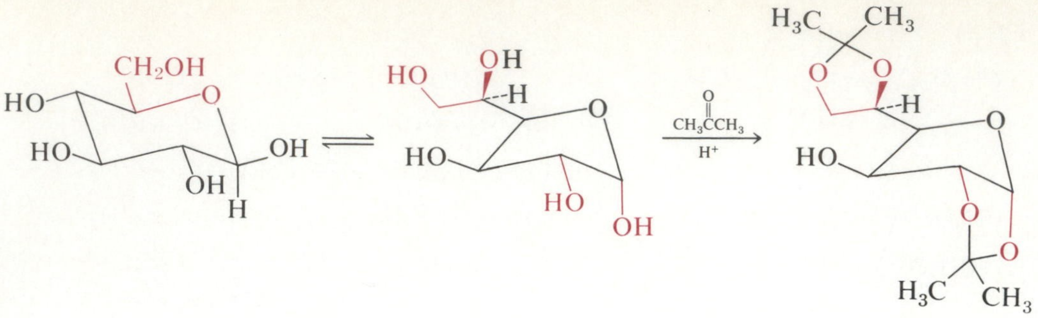

1,2:5,6-di-O-isopropylidene-α-D-glucofuranose

Similar condensations occur with aldehydes. Benzaldehyde shows a tendency to form six-membered ring acetals. Thus, benzaldehyde reacts with methyl α-D-galactoside to give the 4,6-benzylidene derivative.

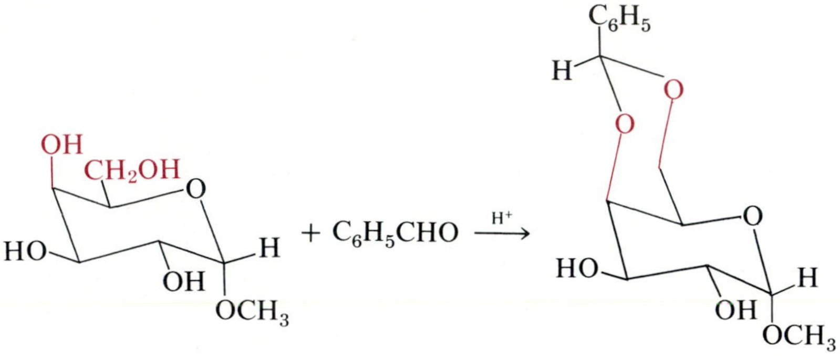

$$+ C_6H_5CHO \xrightarrow{H^+}$$

methyl 4,6-O-benzylidene-α-D-galactopyranoside

These cyclic acetals serve the useful function of protecting either two or four of the OH groups normally present in the free sugar. The acetal groups are sensitive to acid, but are relatively stable to neutral and basic conditions. Reactions can be carried out on the remaining OH groups, and the protecting groups can then be removed by mild acid hydrolysis. An example is the synthesis of 3-O-methylglucose, a feat that cannot be accomplished by selective methylation of glucose itself.

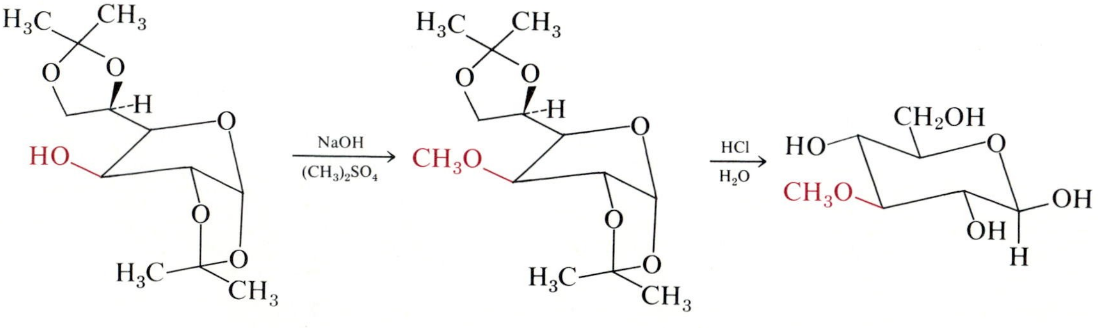

Another example is the following, which shows how D-glucose can be converted into D-allose by inversion of configuration at C-3.

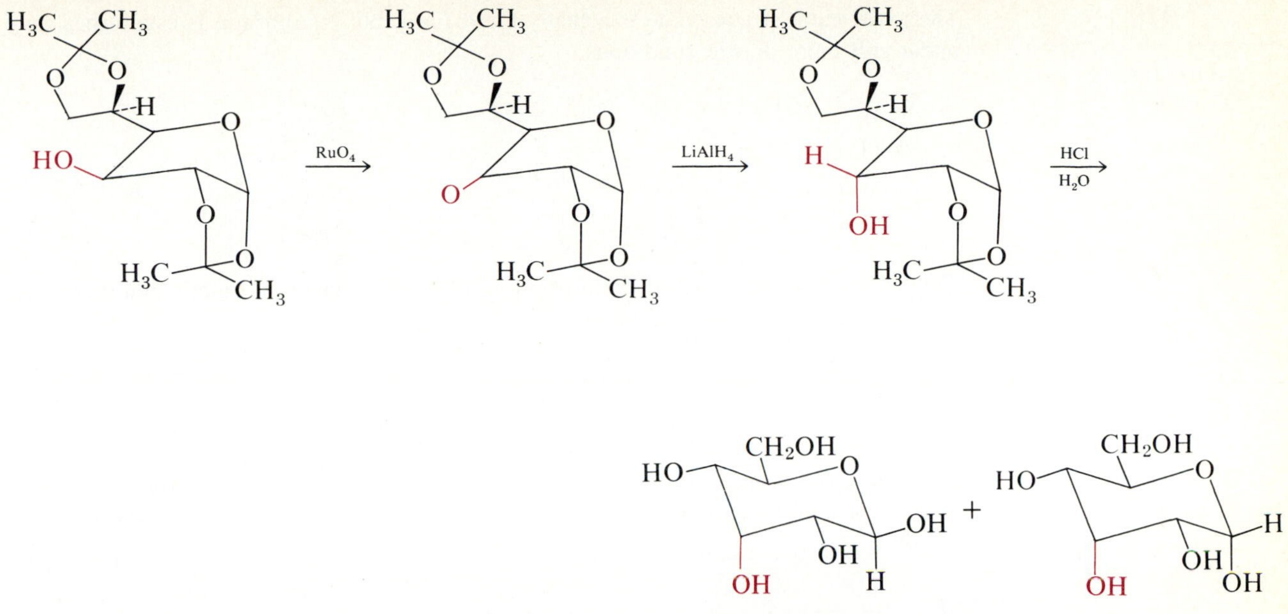

H_3C CH_3 O O H O HO O O H_3C CH_3 $\xrightarrow{RuO_4}$ H_3C CH_3 O O H O O O O H_3C CH_3 $\xrightarrow{LiAlH_4}$ H_3C CH_3 O O H O H OH O O H_3C CH_3 $\xrightarrow[H_2O]{HCl}$

CH_2OH HO O OH OH OH H + CH_2OH HO O H OH OH OH

β-D-allopyranose

α-D-allopyranose

EXERCISE 28.7 The reaction product of glucose and acetone, 1,2:5,6-di-O-isopropylidene-α-D-glucofuranose, is a valuable synthetic intermediate, as shown by the examples in this section. Suggest a way in which this intermediate might be used in a synthesis of 3-deoxyglucose (glucose lacking the hydroxy group at C-3).

C. Esterification

The hydroxy groups in sugars can be esterified by normal methods (Section 18.7.A). The most common procedure to form acetates uses acetic anhydride and a mild basic catalyst such as sodium acetate or pyridine. At low temperature, acetylation in pyridine occurs more rapidly than interconversion of the anomers; at 0 °C either α-D- or β-D-glucose gives the corresponding pentaacetate. At higher temperatures the anomers interconvert rapidly, and the β-pentaacetate is produced preferentially, since the equatorial OH of the β-anomer reacts more rapidly than does the axial OH of the α-anomer.

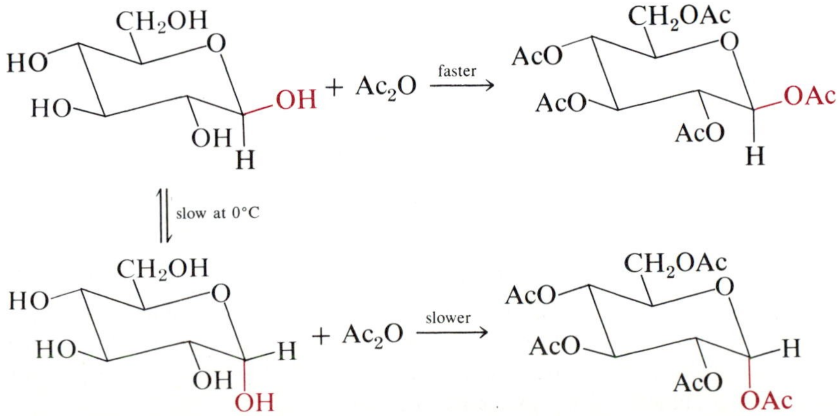

CH_2OH HO HO O OH OH H + Ac_2O $\xrightarrow{\text{faster}}$ CH_2OAc AcO AcO O OAc AcO H

$\rightleftharpoons$ slow at 0°C

CH_2OH HO HO O H OH OH + Ac_2O $\xrightarrow{\text{slower}}$ CH_2OAc AcO AcO O H AcO OAc

The more stable pentaacetate is actually the α-form, but equilibrium is established only under still more drastic conditions.

$K = 6.7$

This example provides a further illustration of the importance of kinetic and thermodynamic factors in organic reactions.

It may seem surprising that the axial α-pentaacetate is more stable than the equatorial β-pentaacetate. This equilibrium is an example of the **anomeric effect,** which frequently causes an electronegative anomeric substituent such as an alkoxy or acyloxy group to prefer the axial position. The anomeric effect is generally small and might arise from an electronic bonding effect or from minimization of dipole-dipole repulsion. Less polar solvents accentuate the effect.

Highlight 28.2

Conversion of most of the hydroxy groups of sugars to alkyl or acyl derivatives is accomplished by the usual methods, after replacing the ionizable hemiacetal hydroxy group with a methoxy group.

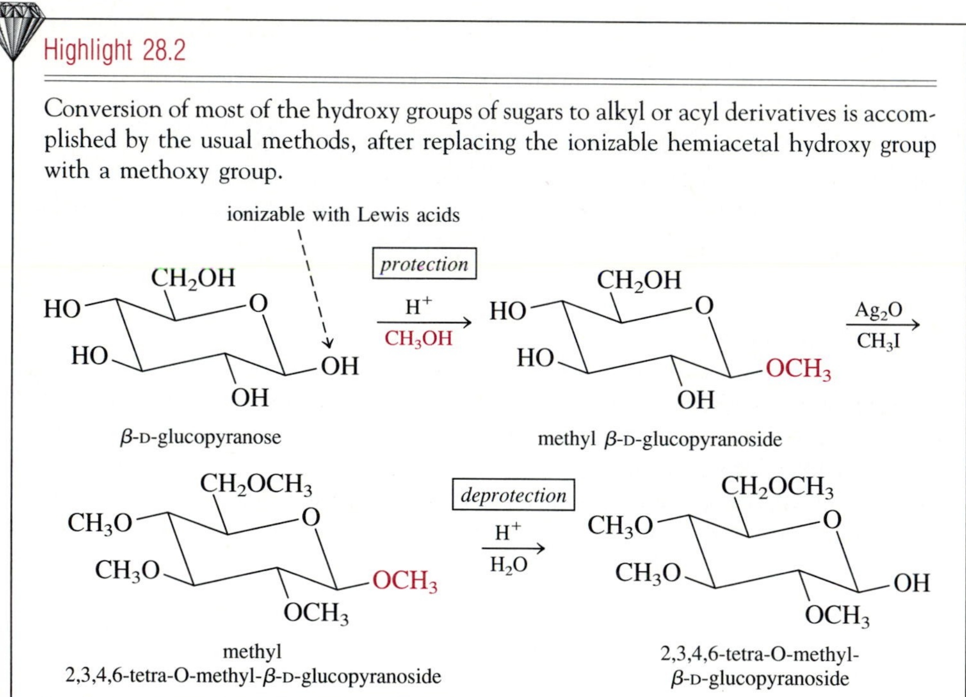

β-D-glucopyranose

methyl β-D-glucopyranoside

methyl 2,3,4,6-tetra-O-methyl-β-D-glucopyranoside

2,3,4,6-tetra-O-methyl-β-D-glucopyranoside

D. Reduction: Alditols

Monosaccharides can be reduced by various methods to the corresponding polyalcohols, which as a class are called **alditols.** Reduction of D-glucose gives D-glucitol (D-glucitol is referred to as D-sorbitol in the older literature), which also occurs in nature. The same compound is produced by the reduction of L-gulose. D-Glucitol is prepared on an industrial scale by catalytic hydrogenation of D-glucose over a nickel catalyst. The reduction probably occurs on the small amount of open-chain form that is present in equilibrium with the cyclic form. As the open-chain form is removed in this way, the equilibrium continually shifts until all of the sugar is reduced.

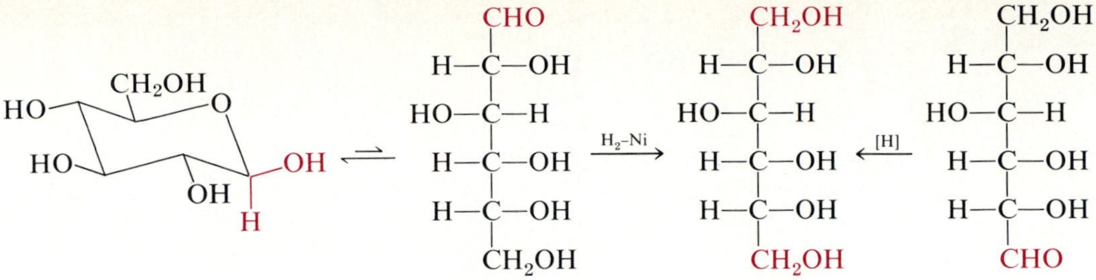

D-glucose ⇌ D-glucose (Fischer projection) $\xrightarrow{H_2\text{-Ni}}$ D-glucitol $\xleftarrow{[H]}$ L-gulose

D-Mannitol, produced by the sodium borohydride reduction of D-mannose, is widespread in nature, occurring in such varied sources as olives, marine algae, onions, and mushrooms. It is also produced, along with a little D-glucitol, in the reduction of the ketohexose D-fructose.

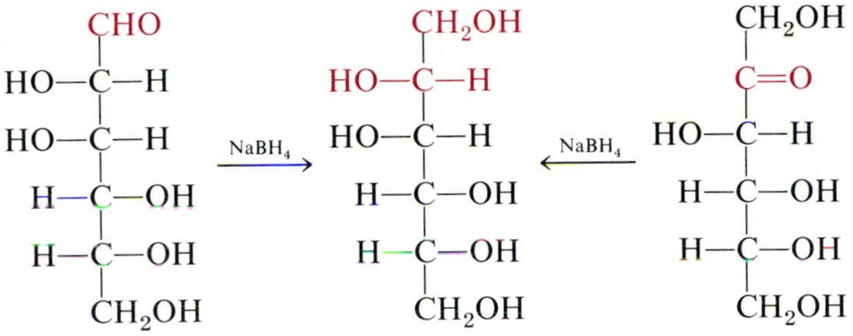

D-mannose $\xrightarrow{NaBH_4}$ D-mannitol $\xleftarrow{NaBH_4}$ D-fructose

E. Oxidation: Aldonic and Saccharic Acids

Sugars are oxidizable in several ways. In the aldoses, the most susceptible group is the aldehyde group. For preparative purposes, the most convenient method employs bromine in a buffered solution at pH 5–6. Yields of the polyhydroxy carboxylic acids (**aldonic acids**) are usually in the range 50–70%. With glucose, yields as high as 95% have been achieved.

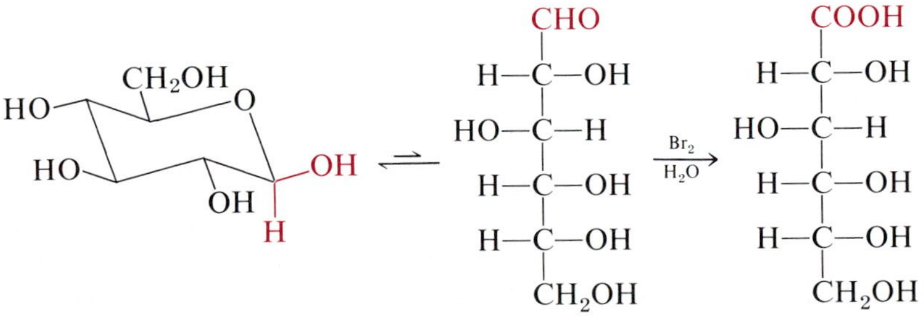

D-glucose ⇌ (Fischer projection) $\xrightarrow[H_2O]{Br_2}$ D-gluconic acid

Since they are 4-hydroxyalkanoic acids, the aldonic acids lactonize readily (Section 27.5). Although either a five- or six-membered lactone might, in principle, be formed, the more stable lactones are those containing a five-membered ring (see Table 27.4).

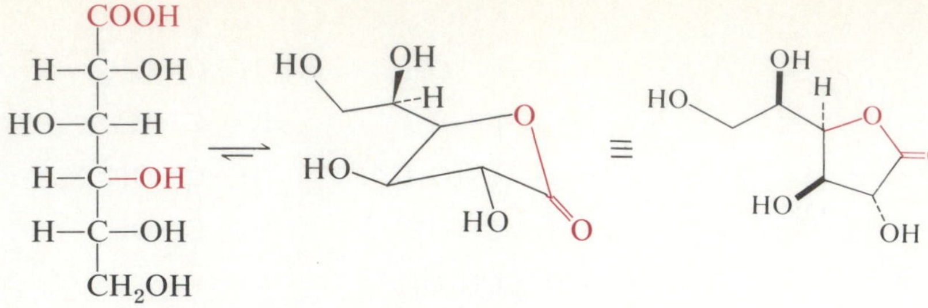

D-gluconic acid

D-gluconic acid γ-lactone

The easy oxidation of aldoses provides a basis for two qualitative tests that were widely used in the early days of sugar chemistry—Fehling's test, employing cupric ion as the oxidant, and Tollens' test, in which silver ion is the oxidant. In the Fehling reaction the presence of a potential aldehyde group is shown by the formation of cuprous oxide as a brick-red precipitate. In Tollens' test the silver ion is reduced to metallic silver, which deposits in the form of a mirror on the inside of the test tube.

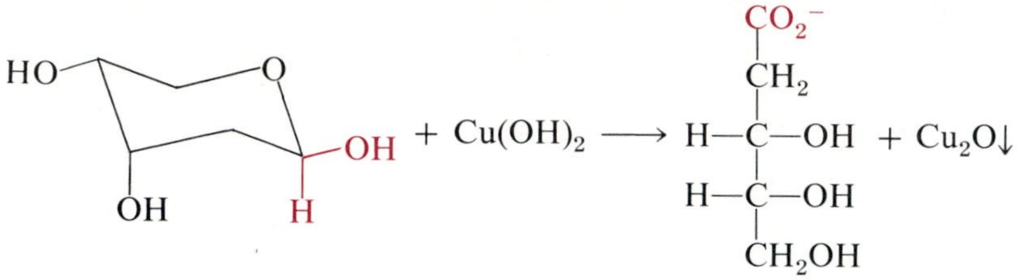

$$+ Cu(OH)_2 \longrightarrow + Cu_2O\downarrow$$

β-D-2-deoxyribose

$$+ Ag^+ \longrightarrow + Ag\downarrow$$

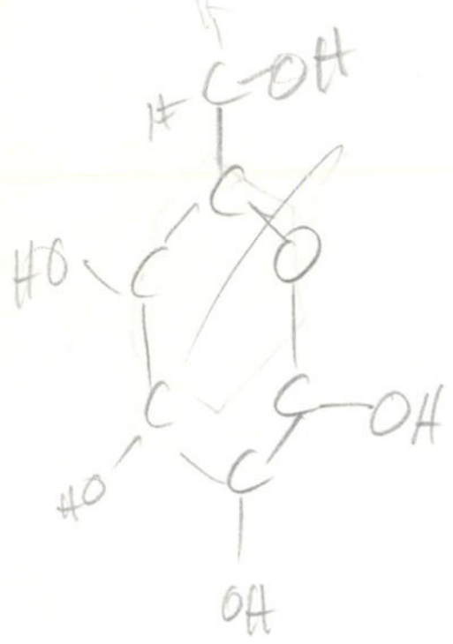

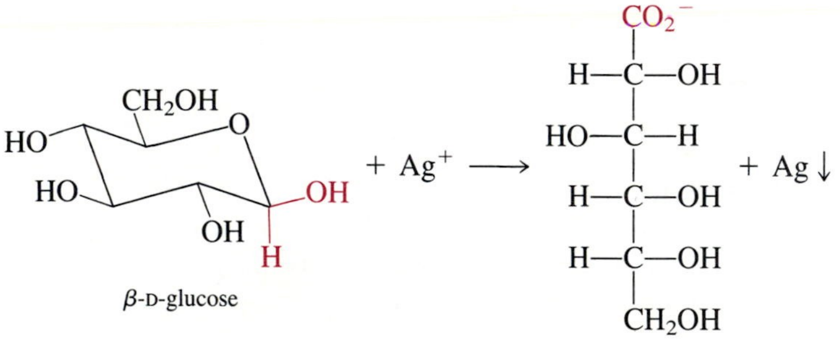

β-D-glucose

If the sugar is in the form of a glycoside, then the anomeric carbon is protected under basic conditions, and the sugar is stable to these mild oxidizing conditions.

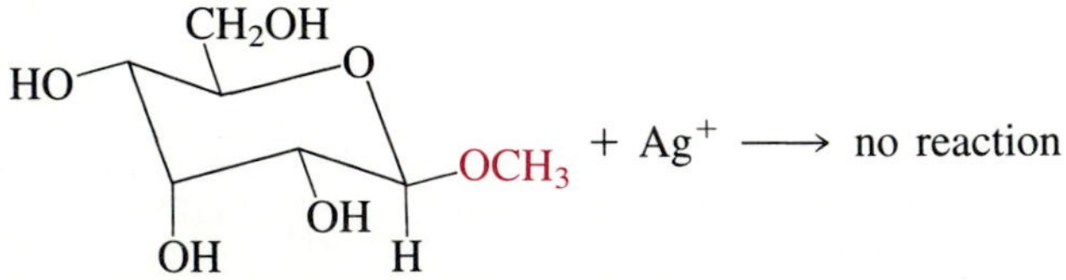

$$+ Ag^+ \longrightarrow \text{no reaction}$$

methyl β-D-allopyranoside

Such compounds are called **nonreducing sugars**; sugars that do reduce basic solutions of Cu^{2+} or Ag^+ are called **reducing sugars.**

Under more vigorous oxidizing conditions, one or more hydroxy groups can also be oxidized. The primary OH groups are attacked most readily and are generally oxidized all the way to the carboxylic acid stage. The product is a polyhydroxy dicarboxylic acid called a **saccharic acid.** A convenient oxidizing agent for the preparation of saccharic acids is aqueous nitric acid.

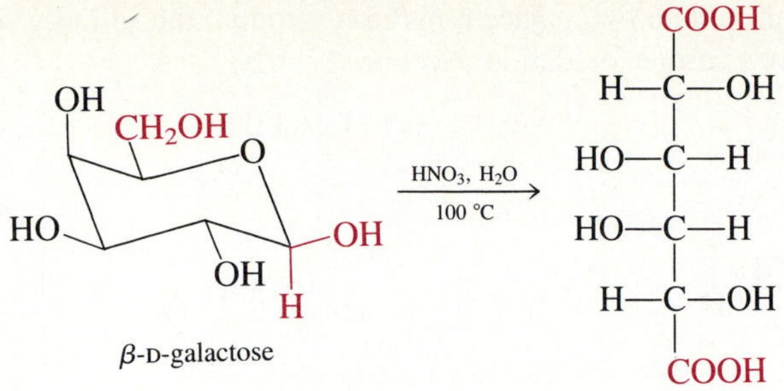

β-D-galactose

$\xrightarrow[100\ °C]{HNO_3,\ H_2O}$

mucic acid
(galactaric acid)

The saccharic acids have been useful in unraveling the puzzle of the relative configuration of the aldoses. Since the two ends of the chain in such a dicarboxylic acid are the same, meso compounds are possible, depending on the relative configuration of the stereocenters. Note, for example, that mucic acid is a meso compound and hence is optically inactive. The observation that galactose gives a meso saccharic acid automatically limits its structure to only 4 of the 16 possible aldohexoses.

Like the aldonic acids, the saccharic acids lactonize readily and are generally found to be dilactones. The 1,4:3,6-dilactone of glucaric acid, which is derived from D-glucose, is shown below.

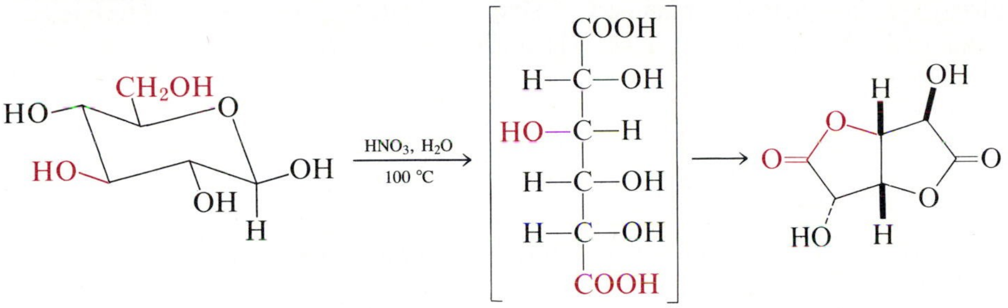

$\xrightarrow[100\ °C]{HNO_3,\ H_2O}$

EXERCISE 28.8 Write Fischer projection formulas for the aldonic and saccharic acids derived from D-galactose, D-mannose, and D-xylose. Are any of these acids achiral? What are the principal lactones expected from the three aldonic acids?

F. Oxidation by Periodic Acid

Like other vicinal diols, sugars are cleaved by periodic acid (Section 27.3.B). For example, methyl 2-deoxyribopyranoside (see D-ribose in Table 28.1) reacts with one equivalent of periodic acid to give the dialdehyde shown below.

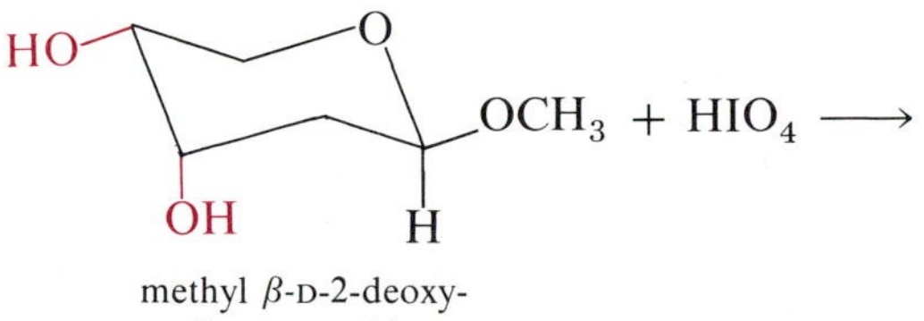

$$+ HIO_4 \longrightarrow$$

methyl β-D-2-deoxyribopyranoside

$$HIO_3 + \ldots \equiv CH_3O{-}CH(CH_2CHO)_2$$

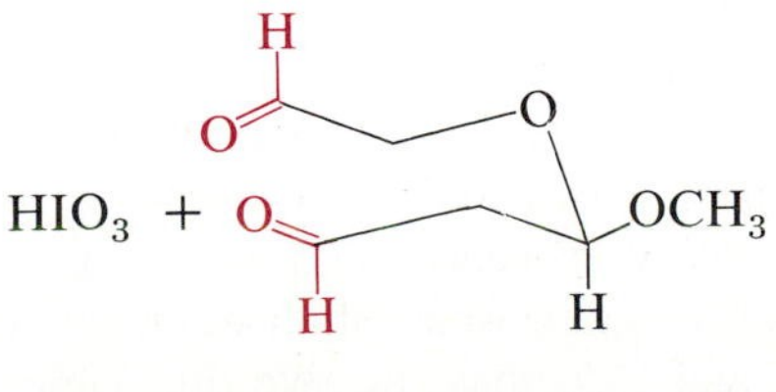

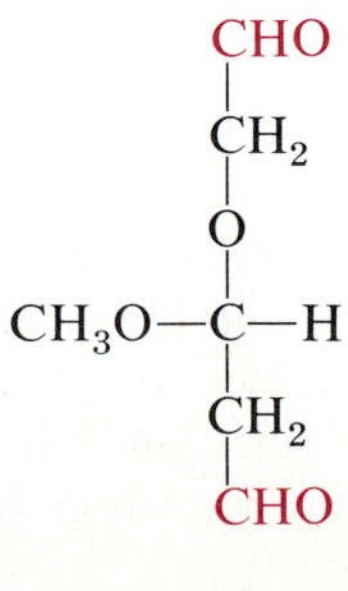

When there are more than two adjacent hydroxy groups, the initially formed α-hydroxy aldehyde undergoes further oxidation (Section 27.4.B).

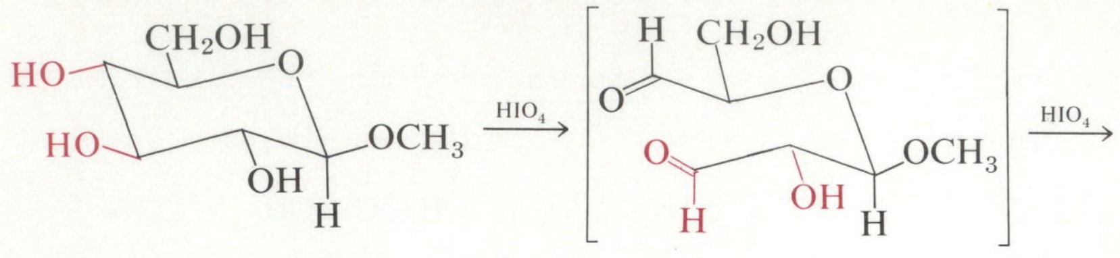

methyl β-D-glucopyranoside

$\xrightarrow{HIO_4}$ $\xrightarrow{HIO_4}$

$\equiv$

$$\begin{array}{c} CHO \\ | \\ H-C-OCH_3 \\ | \\ O \\ | \\ H-C-CH_2OH \\ | \\ CHO \end{array}$$

$+ HCOOH$

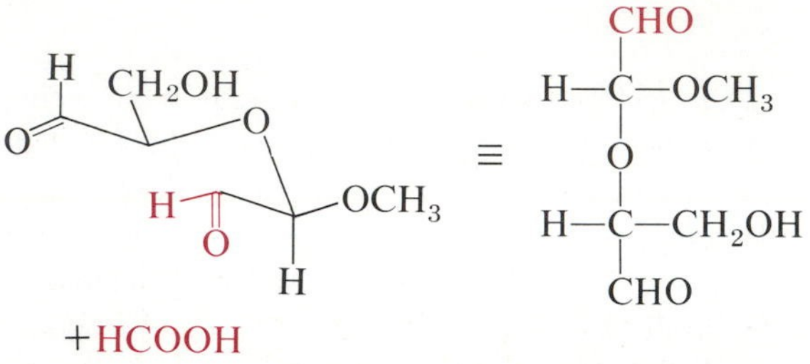

As shown in the foregoing example, each time an α-hydroxy aldehyde is cleaved, one equivalent of formic acid is produced. With the free sugars it is the open-chain form that is oxidized, and complete oxidation occurs. Glucose yields five equivalents of formic acid and one equivalent of formaldehyde, which arises from C-6.

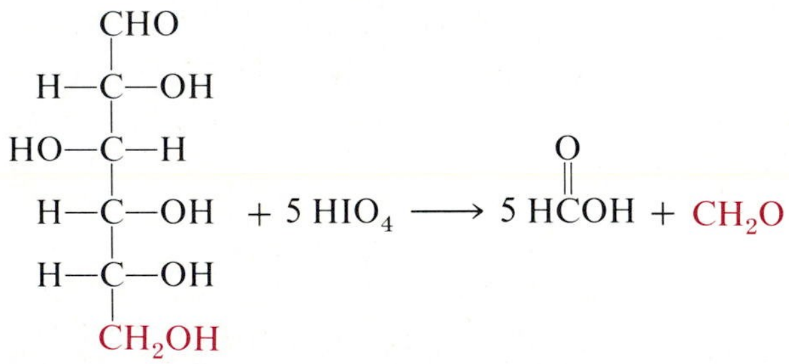

$$\begin{array}{c} CHO \\ | \\ H-C-OH \\ | \\ HO-C-H \\ | \\ H-C-OH \\ | \\ H-C-OH \\ | \\ CH_2OH \end{array} + 5\,HIO_4 \longrightarrow 5\,H\overset{O}{\overset{\|}{C}}OH + CH_2O$$

Periodate oxidation has been applied as a method for determining whether glycosides have the furanose or the pyranose structure. For example, methyl D-glucopyranoside reacts with two equivalents of the reagent and gives one equivalent of formic acid along with the dialdehyde as shown above. The corresponding furanoside also reacts with two equivalents of reagent, but it yields only one equivalent of formaldehyde because the carbon lost corresponds to C-6.

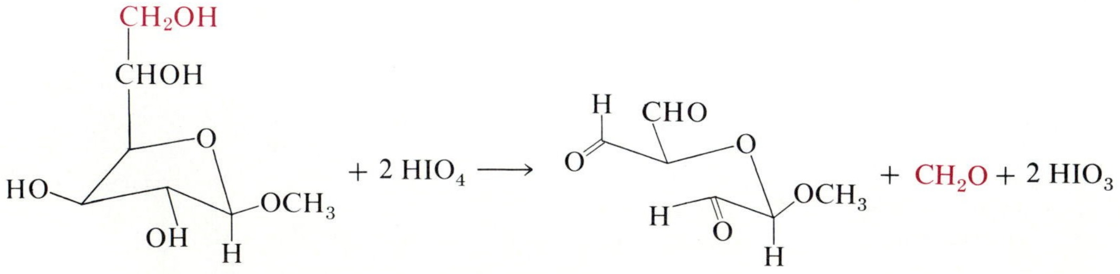

$$+ 2\,HIO_4 \longrightarrow \quad + CH_2O + 2\,HIO_3$$

Periodic acid has also been used to determine the configuration at the anomeric carbon in the pyranosides. Note that oxidation of methyl β-D-glucopyranoside yields a dialdehyde that contains two stereocenters corresponding to C-1 and C-5 in the glucoside itself. Since the methyl β-D-glucosides of all of the aldohexoses have the same absolute configuration at C-1 and C-5, they all give this same dialdehyde on oxidation. One stereocenter in the product is determined by the D-configuration, whereas the other is determined by the β-glycoside linkage.

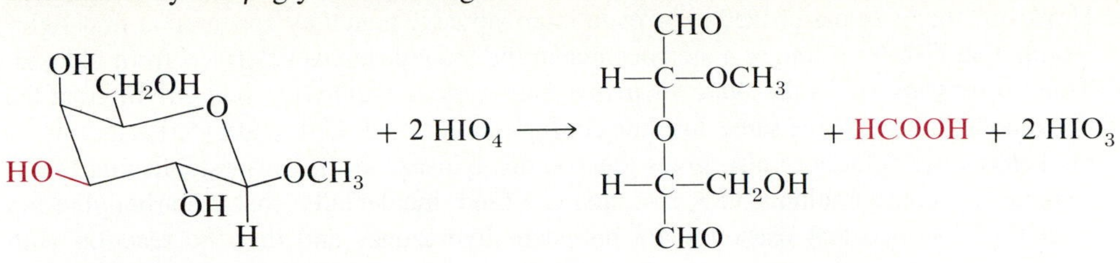

$$\text{methyl } \beta\text{-D-galactopyranoside} + 2\,HIO_4 \longrightarrow \begin{array}{c} CHO \\ | \\ H-C-OCH_3 \\ | \\ O \\ | \\ H-C-CH_2OH \\ | \\ CHO \end{array} + HCOOH + 2\,HIO_3$$

methyl β-D-galactopyranoside

EXERCISE 28.9 How many equivalents of periodic acid are required for complete oxidation of each of the following sugars? For each case, indicate how many equivalents of formaldehyde and formic acid are produced and write the structure of any dialdehyde that remains.

(a) methyl 4,6-O-benzylidene-α-D-galactopyranoside (page 918)
(b) methyl β-D-xyloside (page 916)
(c) L-(−)-threitol (page 907)

G. Phenylhydrazones and Osazones

Because of their polyhydroxy nature, sugars are rather difficult to isolate and purify. They are extremely water soluble and tend to form viscous syrups that crystallize poorly. Naturally occurring examples of such syrups are honey and molasses. These properties caused severe problems in working with sugars in the nineteenth century, before the advent of today's powerful spectroscopic methods of analysis. At that time the only way to ascertain the identity or nonidentity of two compounds was to compare melting points. In 1884 Emil Fischer introduced the use of phenylhydrazine as a reagent in sugar chemistry and opened up a new vista in the subject. Fischer found that a monosaccharide, such as glucose, will react by way of its open-chain form with phenylhydrazine in acetic acid to give a normal phenylhydrazone (Section 14.6.C). However, the initially formed phenylhydrazone reacts further with two more equivalents of phenylhydrazine to yield a derivative called an **osazone.**

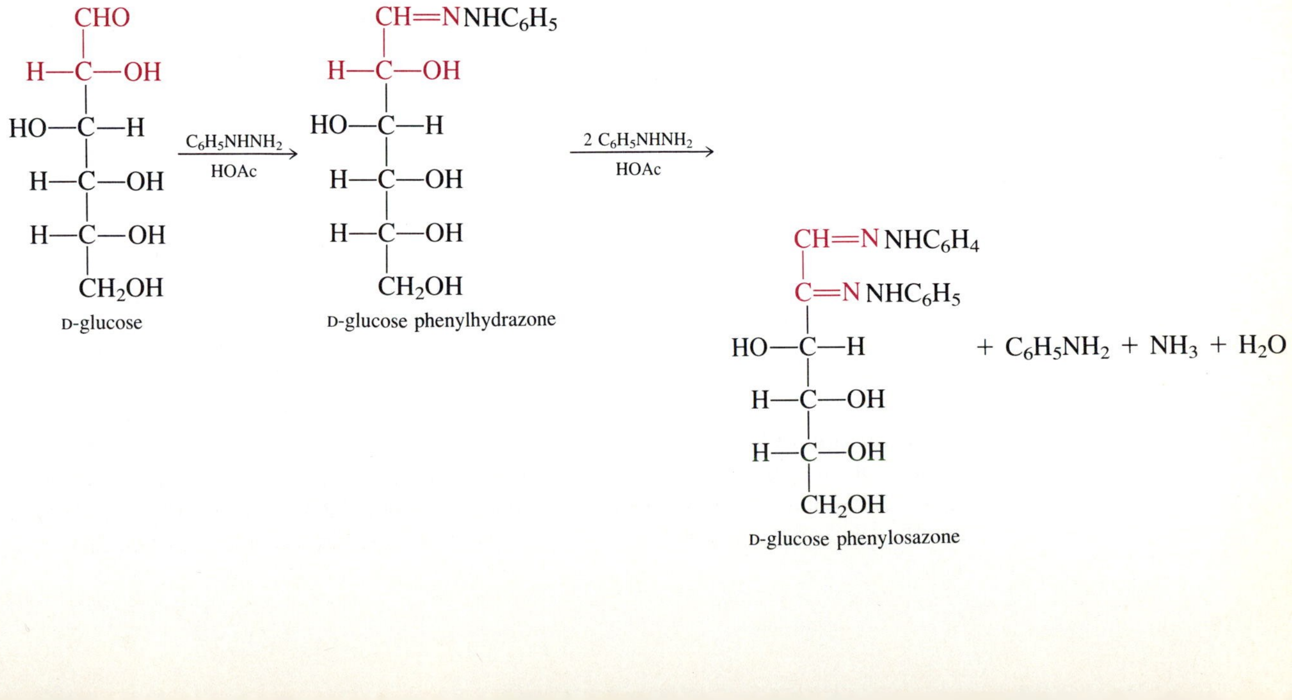

$$\begin{array}{c} CHO \\ | \\ H-C-OH \\ | \\ HO-C-H \\ | \\ H-C-OH \\ | \\ H-C-OH \\ | \\ CH_2OH \end{array} \xrightarrow[HOAc]{C_6H_5NHNH_2} \begin{array}{c} CH{=}NNHC_6H_5 \\ | \\ H-C-OH \\ | \\ HO-C-H \\ | \\ H-C-OH \\ | \\ H-C-OH \\ | \\ CH_2OH \end{array} \xrightarrow[HOAc]{2\,C_6H_5NHNH_2} \begin{array}{c} CH{=}NNHC_6H_4 \\ | \\ C{=}NNHC_6H_5 \\ | \\ HO-C-H \\ | \\ H-C-OH \\ | \\ H-C-OH \\ | \\ CH_2OH \end{array} + C_6H_5NH_2 + NH_3 + H_2O$$

D-glucose; D-glucose phenylhydrazone; D-glucose phenylosazone

The osazones produced from various sugars are bright yellow materials with characteristic crystal forms. Consequently, they were useful as derivatives for characterization. However, the osazones proved to be even more valuable than they appeared at first sight. Notice that C-2 is no longer a stereocenter in the phenylosazones derived from aldoses. Thus, *D-mannose gives the same phenylosazone as does D-glucose,* thus proving that the two aldohexoses have the same absolute configuration at C-3, C-4, and C-5. Furthermore, the ketohexose D-fructose also gives glucose phenylosazone, thereby establishing that it also has this configuration at C-3, C-4, and C-5 (and, incidentally, that its carbonyl group is at C-2). Notice that osazones are bis-phenylhydrazones and that the reaction with phenylhydrazine stops at this stage; that is, further reaction at the C-3 hydroxy group does not normally occur.

CHO
H—C—OH
HO—C—H
H—C—OH
H—C—OH
CH_2OH
D-glucose

→

$CH{=}NNHC_6H_5$
$C{=}NNHC_6H_5$
HO—C—H
H—C—OH
H—C—OH
CH_2OH
D-glucose phenylosazone

←

CHO
HO—C—H
HO—C—H
H—C—OH
H—C—OH
CH_2OH
D-mannose

↑

CH_2OH
C=O
HO—C—H
H—C—OH
H—C—OH
CH_2OH
D-fructose

EXERCISE 28.10 D-(+)-Sorbose is a ketohexose that gives the same osazone as does the aldohexose D-(−)-gulose. What is the structure of D-(+)-sorbose? What other aldohexose gives this same osazone?

H. Chain Extension: The Kiliani-Fischer Synthesis

When D-glyceraldehyde is treated with HCN, a mixture of two cyanohydrins is produced. Both have the R configuration at C-3, corresponding to the same configuration at C-2 in D-glyceraldehyde. They differ only in the configuration at C-2, the new stereocenter. Hydrolysis of the two cyanohydrins yields the same aldonic acids as are produced by the mild oxidation of the aldotetroses, D-erythrose and D-threose. Since glyceraldehyde is a chiral molecule, the transition states leading to the two cyanohydrins are diastereomeric rather than enantiomeric, and the two products are not produced in equal amounts (see Sections 7.8 and 14.7.C).

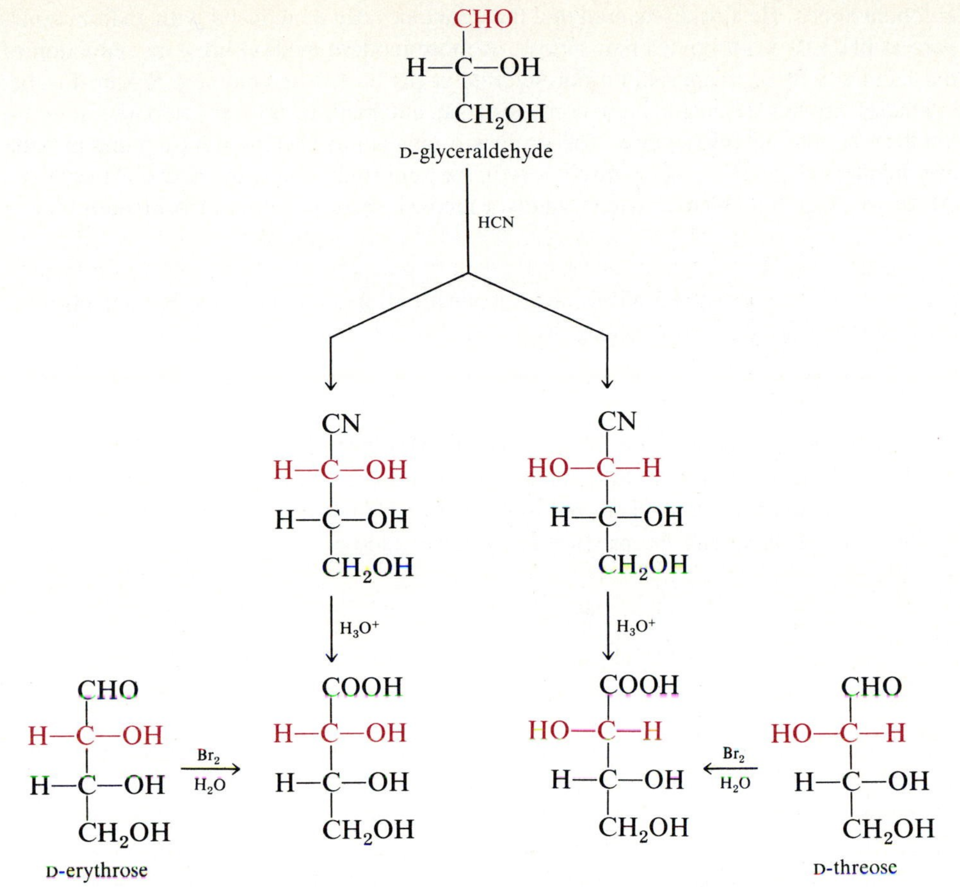

D-glyceraldehyde

HCN

H_3O^+ H_3O^+

Br_2 / H_2O Br_2 / H_2O

D-erythrose D-threose

The cyanohydrin chain-lengthening procedure has been applied extensively in sugar chemistry and has come to be known as the **Kiliani-Fischer synthesis.** Fischer discovered that the aldonic acids produced by hydrolysis of the cyanohydrins lactonize on heating to

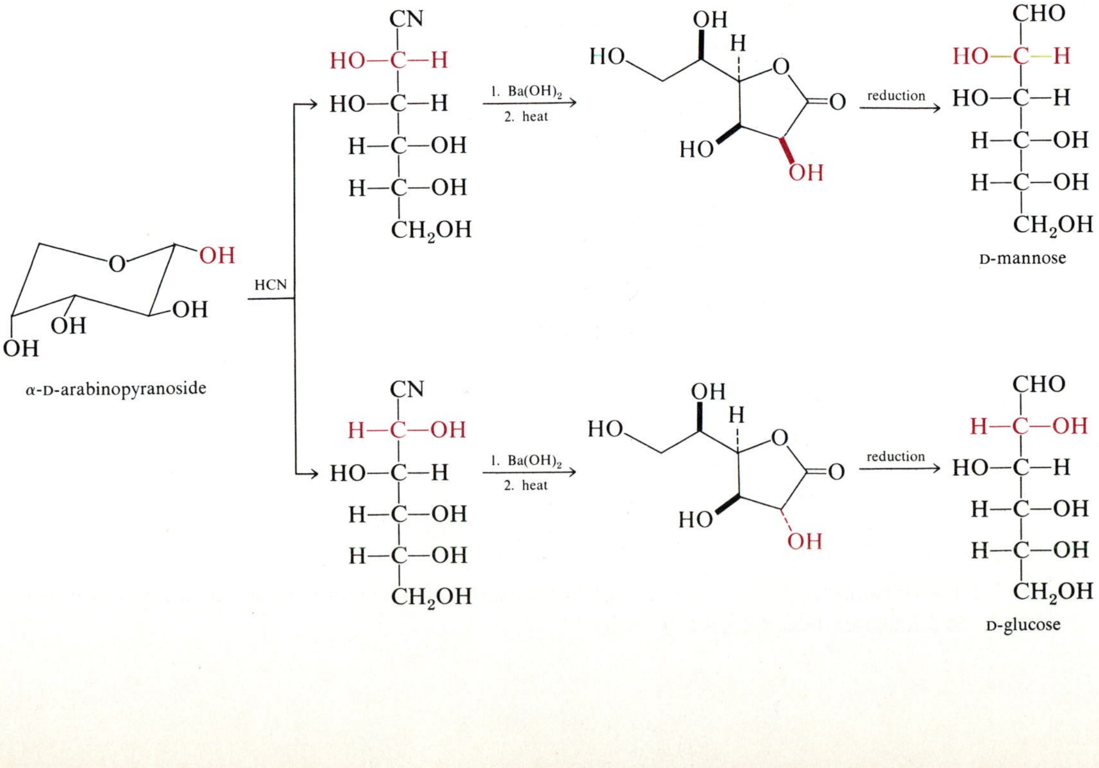

1. $Ba(OH)_2$ 2. heat — reduction

D-mannose

HCN

α-D-arabinopyranoside

1. $Ba(OH)_2$ 2. heat — reduction

D-glucose

aldonolactones. He also discovered that these lactones can be reduced with sodium amalgam at pH 3.0–3.5 to give a new aldose. A more modern method involves reduction of the lactone with aqueous sodium borohydride at pH 3–4. The complete Kiliani-Fischer synthesis provides a method for converting an aldopentose into an aldohexose or an aldohexose into an aldoheptose. The synthesis always provides two diastereomers, usually in unequal amounts, which differ only in their configuration at the new C-2 (old C-1). An example is D-arabinose, which yields a mixture of D-glucose and D-mannose.

EXERCISE 28.11 What two aldohexoses result from application of the Kiliani-Fischer synthesis to (a) D-(−)-ribose and (b) D-(+)-xylose?

I. Chain Shortening: The Ruff and Wohl Degradations

In 1896 Ruff discovered that the calcium salts of aldonic acids are oxidized by hydrogen peroxide, the reaction being catalyzed by ferric salts. The oxidation occurs with cleavage of the C-1,C-2 bond and the product is the lower aldose.

$$\underset{\displaystyle |}{\overset{\displaystyle CO_2^-\ \tfrac{1}{2}Ca^{2+}}{\underset{|}{CHOH}}} \xrightarrow[Fe^{3+}]{H_2O_2} \underset{|}{CHO} + CO_2$$

Since aldohexoses may be readily oxidized to aldonic acids by bromine water (Section 28.5.E), the two-stage process provides a way of converting an aldohexose into an aldopentose; it is called the **Ruff degradation.** Although yields are not high, the Ruff degradation has been a useful technique for the synthesis of certain aldopentoses. Two aldohexoses that differ only in configuration at C-2 yield the same aldopentose.

D-glucose (CHO; H—C—OH; HO—C—H; H—C—OH; H—C—OH; CH_2OH) $\xrightarrow[H_2O]{Br_2}$ $\xrightarrow{Ca(OH)_2}$ calcium D-gluconate ($CO_2^-\ \tfrac{1}{2}Ca^{2+}$; H—C—OH; HO—C—H; H—C—OH; H—C—OH; CH_2OH) $\xrightarrow[Fe^{3+}]{H_2O_2}$

D-mannose (CHO; HO—C—H; HO—C—H; H—C—OH; H—C—OH; CH_2OH) $\xrightarrow[H_2O]{Br_2}$ $\xrightarrow{Ca(OH)_2}$ calcium D-mannonate ($CO_2^-\ \tfrac{1}{2}Ca^{2+}$; HO—C—H; HO—C—H; H—C—OH; H—C—OH; CH_2OH) $\xrightarrow[Fe^{3+}]{H_2O_2}$

→ D-arabinose (CHO; HO—C—H; H—C—OH; H—C—OH; CH_2OH) (40–50%)

Unfortunately, the process is not very useful for the conversion of aldopentoses into aldotetroses because of low yields.

Another process, called the **Wohl degradation,** accomplishes the same overall conversion, shortening the aldose chain by the removal of C-1. The Wohl degradation is essentially the reverse of the Kiliani-Fischer synthesis. The aldose is first converted into its oxime by treatment with hydroxylamine (Section 14.6.C). When the resulting polyhydroxy oxime is heated with acetic anhydride and sodium acetate, all of the hydroxy groups are acetylated and the oxime group is dehydrated to a cyano group. The product is the acetate ester of a cyanohydrin. The ester groups are removed by treatment with base. Under the basic conditions of hydrolysis, the cyanohydrin is decomposed to the corresponding aldehyde. Again, the process does not give especially high yields, but it is applicable to pentoses as well as to hexoses.

$$\begin{array}{c} CHO \\ H-C-OH \\ H-C-OH \\ H-C-OH \\ CH_2OH \\ \text{D-ribose} \end{array} \xrightarrow{H_2NOH} \begin{array}{c} CH{=}NOH \\ H-C-OH \\ H-C-OH \\ H-C-OH \\ CH_2OH \end{array} \xrightarrow[NaOAc]{Ac_2O} \begin{array}{c} C{\equiv}N \\ H-C-OAc \\ H-C-OAc \\ H-C-OAc \\ CH_2OAc \end{array} \xrightarrow[CHCl_3]{NaOCH_3}$$

$$\left[\begin{array}{c} CN \\ H-C-OH \\ H-C-OH \\ H-C-OH \\ CH_2OH \end{array}\right] \xrightarrow{NaOCH_3} \begin{array}{c} CHO \\ H-C-OH \\ H-C-OH \\ CH_2OH \\ \text{D-erythrose} \end{array} + NaCN + CH_3OH$$

EXERCISE 28.12 What tetrose results from application of the Wohl degradation to D-(−)-lyxose? Which other pentose also affords this tetrose?

28.6 Relative Stereochemistry of the Monosaccharides: The Fischer Proof

In the late nineteenth century organic chemists were faced with a puzzle regarding the structures of the monosaccharides. A number of compounds had been isolated that were known to have the same formula and that had the same connectivity. That is, the available evidence showed that glucose, galactose, and mannose were all 2,3,4,5,6-pentahydroxyhexanals. The Le Bel-van't Hoff theory of stereoisomerism provided an explanation for this phenomenon. The challenging question was, which relative arrangement of the four stereocenters corresponds to glucose, which to mannose, and so on?

The challenge was taken up by Emil Fischer, who succeeded in establishing the correct stereostructures for D-glucose, D-mannose, D-fructose, and D-arabinose in 1891. The structure proof consists of an elegant series of logical deductions and has come to be known as "the Fischer proof." We will present a modernized version of the proof here because it typifies the method that has been used to establish the structures of all the sugars.

At the outset Fischer realized that he could establish the *relative configuration* of the various stereocenters in a sugar, but that he had no way to determine the *absolute configu-*

ration of any of the compounds. In order to understand this distinction, consider the four aldotetroses D- and L-threose and D- and L-erythrose. All four compounds are oxidized by nitric acid to saccharic acids. The enantiomeric D- and L-threoses give enantiomeric D- and L-2,3-dihydroxysuccinic acids, which are called D-tartaric acid and L-tartaric acid. However, both D- and L-erythrose are oxidized by nitric acid to a saccharic acid that is an optically inactive diacid, *meso*-tartaric acid. *meso*-Tartaric acid owes its optical inactivity to an internal symmetry plane, and hence the relative configuration of its two stereocenters is fixed. It follows that one of the erythroses has the *R*,*R* configuration and the other has the *S*,*S* configuration. However, there was no way to tell which is which.

CHO
HO—C—H
H—C—OH
CH_2OH
D-threose

$\xrightarrow{HNO_3}$

COOH
HO—C—H
H—C—OH
COOH
D-tartaric acid

CHO
H—C—OH
HO—C—H
CH_2OH
L-threose

$\xrightarrow{HNO_3}$

COOH
H—C—OH
HO—C—H
COOH
L-tartaric acid

CHO
H—C—OH
H—C—OH
CH_2OH
D-erythrose

HNO_3

CHO
HO—C—H
HO—C—H
CH_2OH
L-erythrose

HNO_3

COOH
H—C—OH
- - - - - - - - - symmetry plane
H—C—OH
COOH
meso-tartaric acid

The question of the absolute configuration was not settled until 1954, when Bijvoet determined the absolute configuration of a salt of D-tartaric acid by an X-ray crystallographic technique known as anomalous dispersion.

1. Fischer started by arbitrarily choosing what is now called the *R* configuration for the configuration at C-5 in D-glucose. He had a 50% chance of being correct in this assignment, but it has no bearing on the rest of the proof because the other centers were to be determined relative to C-5. In 1954, Bijvoet's work showed that Fischer had actually made the correct choice in an absolute sense. The structures of the eight aldohexoses having the *R* configuration at C-5 are shown in Table 28.2 and are designated **1** through **8**.

TABLE 28.2 The (5*R*)-Aldohexoses

1	2	3	4
CHO	CHO	CHO	CHO
H—C—OH	HO—C—H	H—C—OH	HO—C—H
H—C—OH	H—C—OH	HO—C—H	HO—C—H
H—C—OH	H—C—OH	H—C—OH	H—C—OH
H—C—OH	H—C—OH	H—C—OH	H—C—OH
CH_2OH	CH_2OH	CH_2OH	CH_2OH

5	6	7	8
CHO	CHO	CHO	CHO
H—C—OH	HO—C—H	H—C—OH	HO—C—H
H—C—OH	H—C—OH	HO—C—H	HO—C—H
HO—C—H	HO—C—H	HO—C—H	HO—C—H
H—C—OH	H—C—OH	H—C—OH	H—C—OH
CH_2OH	CH_2OH	CH_2OH	CH_2OH

2. It was known that glucose and mannose give the same osazone (Section 28.5.G). Therefore the two compounds have the same configuration at C-3, C-4, and C-5; they differ only at C-2. The two compounds must be **1** and **2**, **3** and **4**, **5** and **6**, or **7** and **8**.

3. *Both* D-glucose and D-mannose are oxidized by nitric acid to *optically active* saccharic acids (Section 28.5.E). The aldohexoses with structures **1** and **7** would give meso saccharic acids. Therefore D-glucose and D-mannose must be either **3** and **4** or **5** and **6**. (Remember that the two compounds differ only at C-2, so eliminating **1** also eliminates **2**, and eliminating **7** also eliminates **8**.)

4. Kiliani-Fischer chain extension of the aldopentose D-arabinose yields both D-glucose and D-mannose (Section 28.5.H). Therefore, D-arabinose has the same configuration at its C-2, C-3, and C-4 as D-glucose and D-mannose at C-3, C-4, and C-5, respectively. D-Arabinose must be either **9** or **10**.

```
     CHO
      |
  HO—C—H
      |
   H—C—OH    Kiliani-Fischer
      |      ──────────────→  3 and 4
   H—C—OH
      |
     CH2OH
       9
or
     CHO
      |
   H—C—OH
      |
  HO—C—H     Kiliani-Fischer
      |      ──────────────→  5 and 6
   H—C—OH
      |
     CH2OH
      10
```

However, oxidation of D-arabinose gives an *optically active* diacid. The saccharic acid derived from aldopentose **10** would be meso, so D-arabinose must be **9**, and D-glucose and D-mannose must be **3** and **4**.

$$\begin{array}{c} CHO \\ HO-C-H \\ H-C-OH \\ H-C-OH \\ CH_2OH \end{array} \xrightarrow{HNO_3} \begin{array}{c} COOH \\ HO-C-H \\ H-C-OH \\ H-C-OH \\ COOH \end{array}$$

9, D-arabinose — optically active

or

$$\begin{array}{c} CHO \\ H-C-OH \\ HO-C-H \\ H-C-OH \\ CH_2OH \end{array} \xrightarrow{HNO_3} \begin{array}{c} COOH \\ H-C-OH \\ \text{--}HO-C-H\text{---} \\ H-C-OH \\ COOH \end{array} \quad \text{symmetry plane}$$

10 — meso

5. Fischer developed a method that allowed him to interchange the two ends of an aldose chain. The method is fairly involved, and we need not go into the chemical details here. However, consider the results when the method is applied to the aldohexoses that have structures **3** and **4**.

$$\begin{array}{c} CHO \\ H-C-OH \\ HO-C-H \\ H-C-OH \\ H-C-OH \\ CH_2OH \end{array} \dashrightarrow \begin{array}{c} CH_2OH \\ H-C-OH \\ HO-C-H \\ H-C-OH \\ H-C-OH \\ CHO \end{array} \equiv \begin{array}{c} CHO \\ HO-C-H \\ HO-C-H \\ H-C-OH \\ HO-C-H \\ CH_2OH \end{array}$$

3 — rotate structure 180°

$$\begin{array}{c} CHO \\ HO-C-H \\ HO-C-H \\ H-C-OH \\ H-C-OH \\ CH_2OH \end{array} \dashrightarrow \begin{array}{c} CH_2OH \\ HO-C-H \\ HO-C-H \\ H-C-OH \\ H-C-OH \\ CHO \end{array} \equiv \begin{array}{c} CHO \\ HO-C-H \\ HO-C-H \\ H-C-OH \\ H-C-OH \\ CH_2OH \end{array}$$

4 — rotate structure 180°

When C-1 and C-6 are interchanged, compound **3** gives a *different aldohexose*. However, when the same operation is performed on compound **4**, the final product is the same as the starting material. Fischer applied his method to D-glucose and discovered that a new aldohexose was produced, which he named L-gulose. The proof was complete; D-glucose must have structure **3** and D-mannose must have structure **4**!

EXERCISE 28.13 Referring to Table 28.1, choose any aldohexose except glucose or mannose. Work through the reactions involved in the Fischer proof and write the structures of the compounds that would be involved if glucose has the structure you have chosen. Which experiments show that the chosen aldohexose is not glucose?

28.7 Oligosaccharides

Oligosaccharides (Gk., *oligos,* a few) are polysaccharides that yield from two to eight monosaccharide units upon hydrolysis. The most common are the disaccharides, which are dimers composed of two monosaccharides. The two monosaccharides can be the same or different. Disaccharides are joined by a glycoside linkage from the OH group of one monosaccharide to the anomeric carbon of the other.

A simple example of a disaccharide is maltose, which is produced by the enzymatic hydrolysis of starch. Maltose contains two D-glucose units, both in the pyranose form. The C-4 hydroxy group of one glucose is bound by an α-glycoside bond to the anomeric carbon of the other unit. The crystalline disaccharide is the β-anomer, which mutarotates in solution to a mixture of the α- and β-forms.

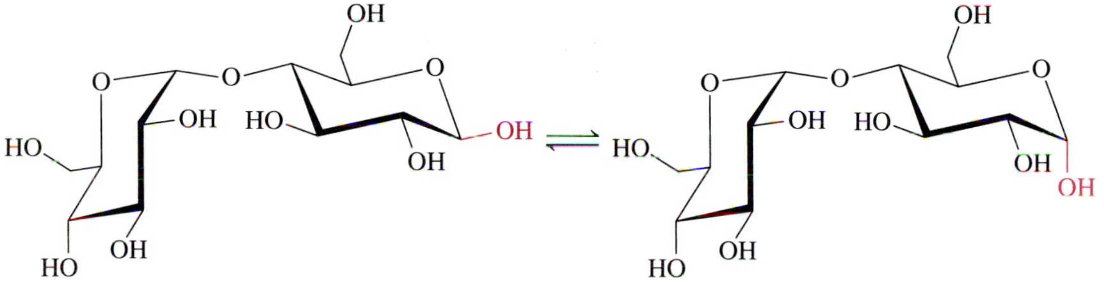

4-O-(α-D-glucopyranosyl)-β-D-glucopyranose
β-maltose

4-O-(α-D-glucopyranosyl)-α-D-glucopyranose
α-maltose

In maltose, one of the glucose units has its aldehyde carbon firmly bound in the glycosidic linkage to the other unit. However, the carbonyl group of the second ring is in the hemiacetal form, and it may therefore undergo normal carbonyl reactions, just as the monosaccharides do. Thus maltose is oxidized by Tollens' reagent and by Fehling's solution and is a reducing sugar. Disaccharides undergo most of the same reactions as do the monosaccharides. For example, maltose is oxidized by bromine water to maltobionic acid.

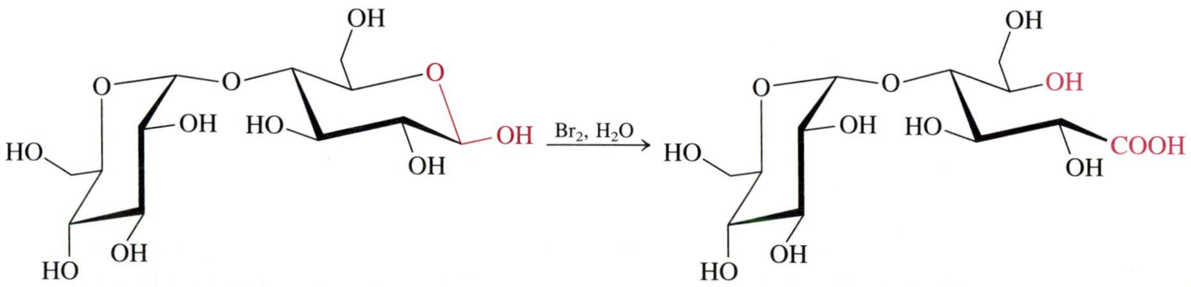

Br_2, H_2O

β-maltose

maltobionic acid

(a)

(b)

FIGURE 28.2 β-Cellobiose, 4-O-(β-D-glucopyranosyl)-β-D-glucopyranose: (a) conventional structure; (b) stereo structure. [Part (b) reproduced with permission from *Molecular Structure and Dimensions,* International Union of Crystallography, 1972.]

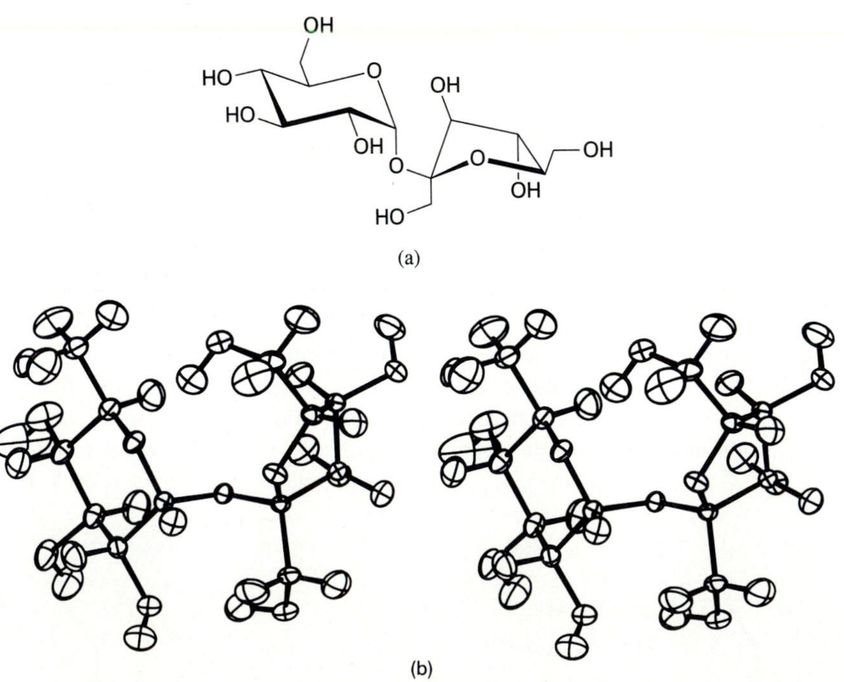

FIGURE 28.3 Sucrose, α-D-glucopyranosyl-β-D-fructofuranoside or β-D-furanosyl-α-D-glucopyranoside: (a) conventional structure; (b) stereo structure. The stereo structure also illustrates the way in which modern x-ray structure determinations are presented. The noncircular shapes of atoms, called ''thermal ellipsoids,'' represent thermal motions of the atoms in the crystal. [Part (b) reproduced with permission from G. M. Brown and H. A. Levy, *Acta Cryst.*, **B29**, 790 (1973).]

Cellobiose is a disaccharide that is obtained by the partial hydrolysis of cellulose. It is isomeric with maltose and contains a β-glycosidic linkage. The structure of β-cellobiose is shown in Figure 28.2.

Lactose is an example of a disaccharide in which the two monosaccharide units are different. It constitutes about 5% by weight of mammalian milk. It is produced commercially from whey, which is obtained as a by-product in the manufacture of cheese. Evaporation of the whey at temperatures below 95 °C causes the less soluble α-anomer to precipitate. Hydrolysis of lactose affords 1 equivalent of glucose and 1 equivalent of galactose. The galactose unit is bound in the glycoside form and both rings are pyranoses. The hydrolysis of lactose is catalyzed by an enzyme called β-galactosidase, which is specific for the hydrolysis of β-galactoside links, highlighted in the following structure.

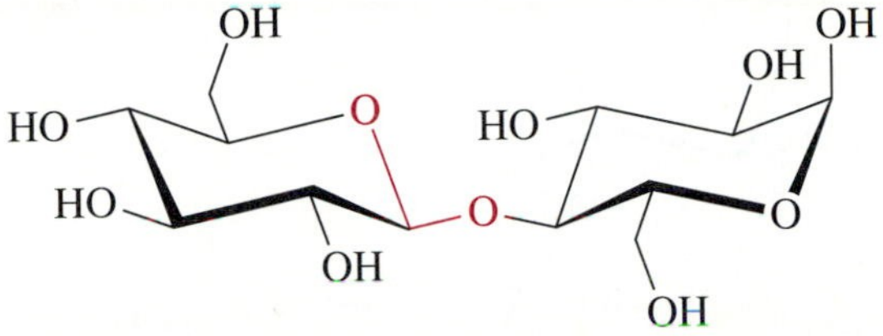

α-lactose

Sucrose is one of the most widespread sugars in nature. It is produced commercially from sugar cane and sugar beets. It is a disaccharide composed of one D-glucose and one D-fructose unit, which are joined by an acetal linkage between the two anomeric carbons. The glucose unit is in the pyranose form and the fructose is in the furanose form. The structure is shown in Figure 28.3a. Another representation, which shows the hydrogen bonding in the crystal, is given below. As highlighted in this structure, both anomeric cartons are bound in the acetal form and sucrose is a nonreducing sugar.

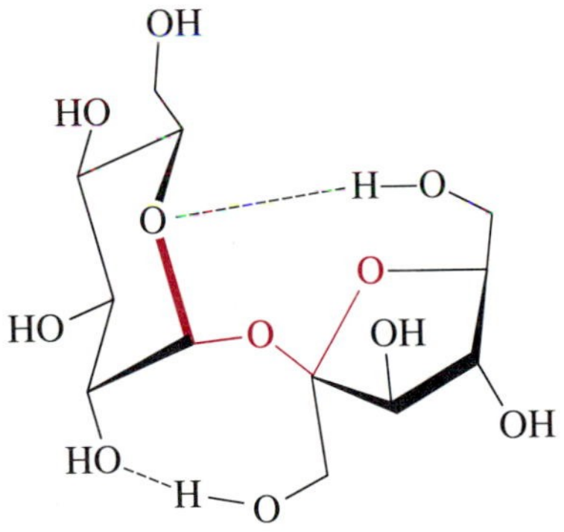

Acidic hydrolysis of sucrose yields an equimolar mixture of D-glucose and D-fructose. Sucrose itself is dextrorotatory, having an optical rotation $[\alpha]_D = +66$. D-Glucose is also dextrorotatory (the equilibrium mixture of α- and β-anomers has $[a]_D = +52.5$), but D-fructose is strongly levorotatory (the equilibrium mixture of fructose isomers has $[\alpha]_D = -92.4$). In the early days of carbohydrate chemistry D-glucose was known as "dextrose" and D-fructose was called "levulose," terms that were derived from the signs of rotation of the two monosaccharides.

In the process of hydrolysis the dextrorotatory sucrose solution becomes levorotatory because an equimolar mixture of D-glucose and D-fructose has $[\alpha]_D = -20$. This commonly encountered mixture is called "invert sugar" from the inversion in the sign of rotation that occurs during its formation. A number of organisms, including honeybees, have enzymes that catalyze the hydrolysis of sucrose. These enzymes are usually called **invertases** and are specific for the β-D-fructofuranoside linkage. Honey is largely a mixture of D-glucose, D-fructose, and sucrose. It has been shown that the equilibrium mixture resulting from hydrolysis of sucrose contains 32% β-D-glucopyranose, 18% α-D-glucopyranose, 34% β-D-fructopyranose, and 16% β-D-fructofuranose. Note that although glucose exists only in pyranose forms, fructose exists as a mixture of pyranose and furanose forms.

Raffinose is an example of a trisaccharide. It is a minor constituent in sugar beets (0.01–0.02%) and is obtained as a by-product in the isolation of sucrose from this source. Raffinose is nonreducing and on hydrolysis yields one equivalent each of D-galactose, D-glucose, and D-fructose. If the hydrolysis is catalyzed by the enzyme α-galactosidase, the products are galactose and sucrose.

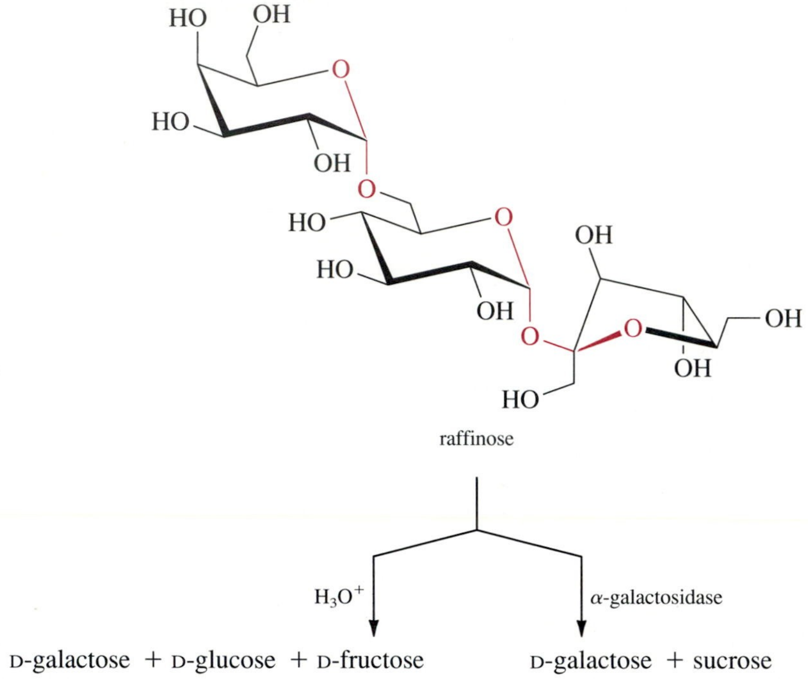

raffinose

H_3O^+ → D-galactose + D-glucose + D-fructose

α-galactosidase → D-galactose + sucrose

EXERCISE 28.14 Write the structures of the species present in the equilibrium mixture that results from hydrolysis of sucrose.

28.8 Polysaccharides

Polysaccharides differ from the oligosaccharides only in the number of monosaccharide units that make up the molecule. The majority of the natural polysaccharides contain from 80 to 100 units, but some materials have much larger molecular weights; cellulose, for example, has an average of about 3000 glucose units per molecule. Polysaccharides can have a linear structure in which the individual monosaccharides are joined one to the other by glycosidic bonds, or they can be branched. A branched polysaccharide has a linear backbone, but additional OH groups on some of the monosaccharide units are involved in glycosidic bonding to another chain of sugars. A few cyclic polysaccharides are also known. The three types are illustrated schematically in Figure 28.4.

Cellulose is probably the single most abundant organic compound on the earth. It is the chief structural component of plant cells. For example, it comprises 10–20% of the dry weight of leaves, about 50% of the weight of tree wood and bark, and about 90% of the weight of cotton fibers, from which pure cellulose is most easily obtained.

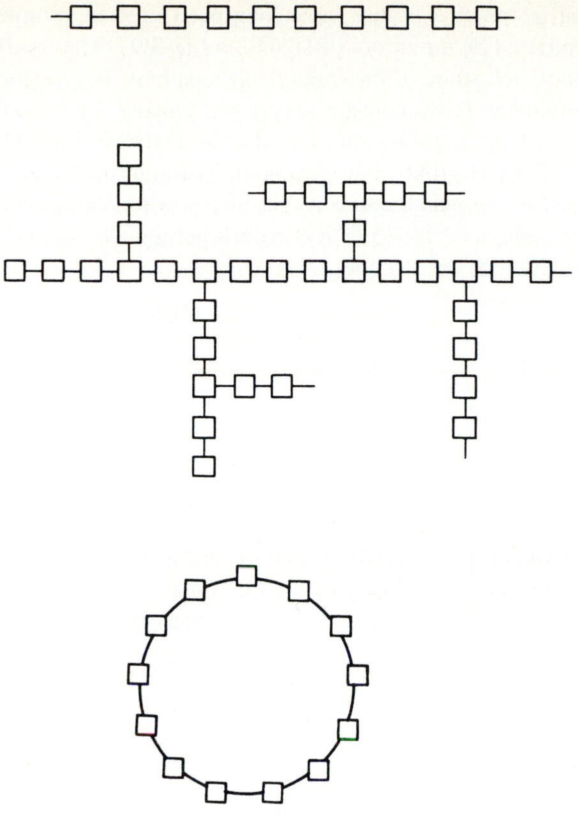

linear polysaccharide

branched polysaccharide

cyclic polysaccharide

FIGURE 28.4. Types of polysaccharides

Structurally, cellulose is a polymer of D-glucose in which the individual units are linked by β-glucoside bonds from the anomeric carbon of one unit to the C-4 hydroxy of the next unit. It may be hydrolyzed by 40% aqueous hydrochloric acid to give D-glucose in 95% yield. Partial hydrolysis, which can be brought about by enzymatic methods, yields the disaccharide cellobiose (Section 28.7). It is a linear polysaccharide, the isolated form containing an average of 3000 units per chain, corresponding to an average molecular weight of about 500,000. Some degradation occurs during the isolation; the actual "native cellulose" as it exists in plants may contain as many as 10,000–15,000 glucose units per chain, corresponding to a molecular weight of 1.6–2.4 million. The strength of wood derives principally from hydrogen bonds of the hydroxy groups of one chain to hydroxy groups of neighboring chains.

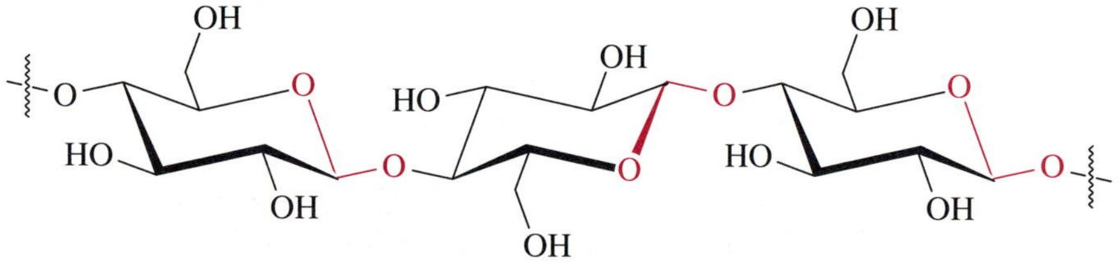

cellulose

Although the higher animals do not have enzymes that can catalyze the degradation of cellulose to glucose, such enzymes (**cellulases**) are common in microorganisms. Cellulases produced by the microflora that reside in the digestive tracts of herbivorous animals permit these animals to utilize cellulose as a food source.

Various chemically modified forms of cellulose have long been used in commercial applications. Cellulose can be nitrated by a mixture of HNO_3 and H_2SO_4. The product is a partially degraded cellulose in which some of the free OH groups have been converted into nitrate esters. The average number of nitrate ester groups per glucose unit is variable and depends on the composition of the nitrating mixture and the reaction time. Highly nitrated cellulose, in which 2.5–2.7 OH groups per glucose unit are nitrated, has explosive properties and has been used in the manufacture of blasting powder. Nitrated cellulose possessing a lower nitrogen content (2.1–2.5 ONO_2 groups per glucose unit) is used in the preparation of plastics (celluloid) and lacquers.

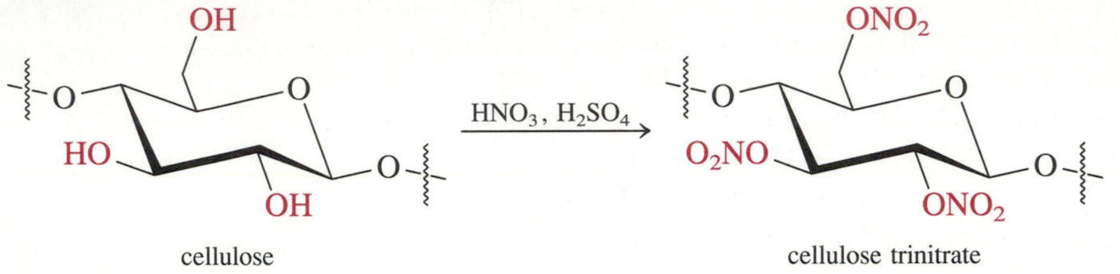

cellulose → cellulose trinitrate

Starch is the second most abundant polysaccharide and occurs in both the vegetable and animal kingdoms. It is the chief source of carbohydrate for humans and is therefore of considerable economic importance. The polysaccharide is deposited in the plant in the

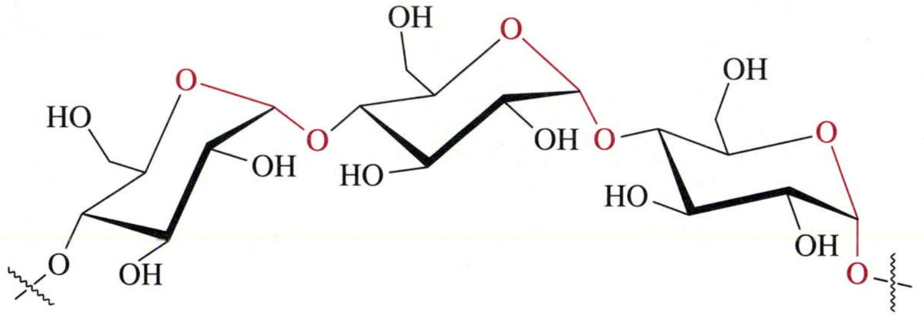

amylose

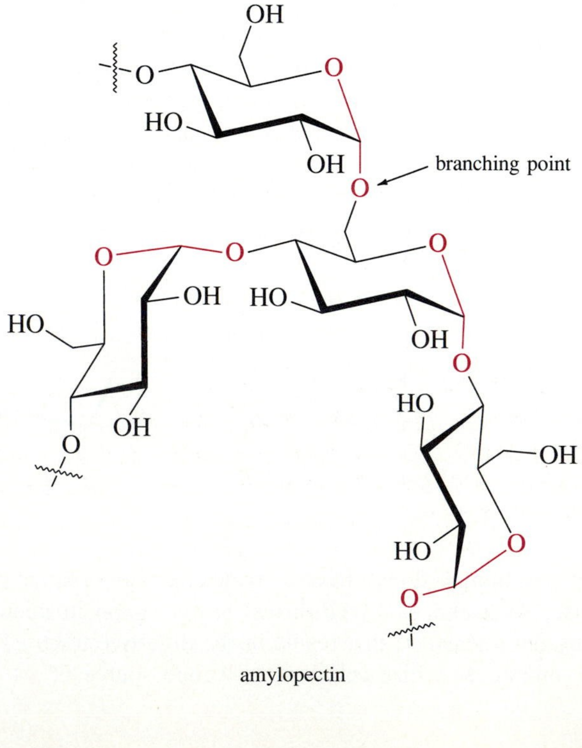

amylopectin

form of small insoluble particles called starch granules. Like cellulose, the term starch is a general one; there is a considerable variety in the nature of the starch molecules produced by a given plant. Natural starch may be separated into two gross fractions, called **amylose** and **amylopectin.**

Like cellulose, starch yields only D-glucose on hydrolysis. Although amylose appears to be essentially unbranched, amylopectin has a highly branched structure. Both types of starch have high molecular weights, corresponding to many thousands of glucose units per molecule. The main chain consists of D-glucose units bound through the C-4 OH group as in cellulose but with the glucoside bond having the α-configuration. In the branched form, the branches appear to be to the C-6 OH group. The anomeric bonds are highlighted in the structure of amylopectin. Partial hydrolysis of starch yields the disaccharide maltose (Section 28.7).

Glycogen is a polysaccharide that is structurally similar to starch. It is the form in which animals store glucose for further use. It is found in most tissues, but the best source is liver or muscle. Glycogen has a structure similar to that of amylopectin but is more highly branched.

28.9 Sugar Phosphates

The sugar phosphates are a class of carbohydrates that is particularly important in living systems. The chemistry of phosphate esters was discussed in Section 26.7. Sugar phosphates are intermediates in many metabolic processes, such as the degradation of glycogen to lactic acid in muscle (glycolysis), the conversion of sugars to alcohol in fermentation, and the biosynthesis of carbohydrates in plants by the process of photosynthesis. They are also the chain-building constituents of ribonucleic and deoxyribonucleic acids (RNA and DNA), which are of importance in the transfer of genetic information.

Typical sugar phosphates, which are known to be involved in the biosynthesis and biodegradation of the polysaccharides glycogen and starch, are α-D-glucopyranosyl phosphate and D-glucose 6-phosphate.

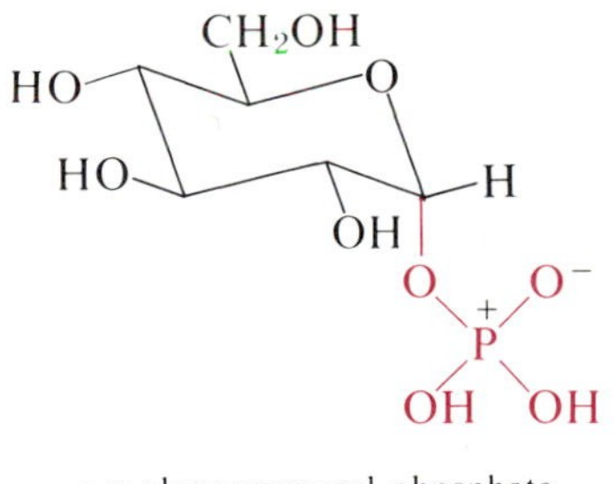

α-D-glucopyranosyl phosphate

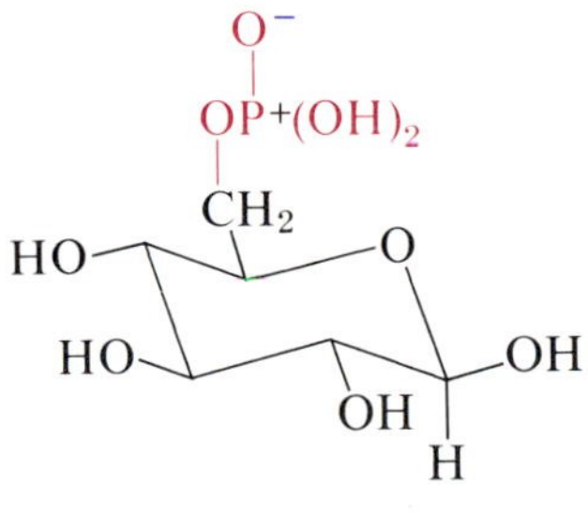

D-glucose 6-phosphate

These polysaccharides are synthesized in organisms by an enzyme-catalyzed process in which glucose units are added in a stepwise fashion onto the growing polysaccharide chain. The reactive glucose units are α-D-glucopyranosyl nucleotide diphosphates. These can be either uridine for UDP-glucose or adenine for ADP-glucose, depending on organism and tissue. The nucleotide diphosphates are produced from glucose 1-phosphate through an enzyme-catalyzed reaction with either uridine triphosphate (UTP) or adenosine triphosphate (ATP). In this form the anomeric carbon is ''activated'' toward nucleophilic substitution processes, displacement, or ionization (nucleotide diphosphate ion is a much better leaving group than hydroxide ion). The overall stereochemistry of the reaction is retention ($\alpha \rightarrow \alpha$). The enzyme must intervene in a way that promotes carbocation formation or nucleophilic displacement with an enzyme group. The enzyme group can either stabilize the carbocation as a type of polar solvent or form a reactive covalently bonded intermediate.

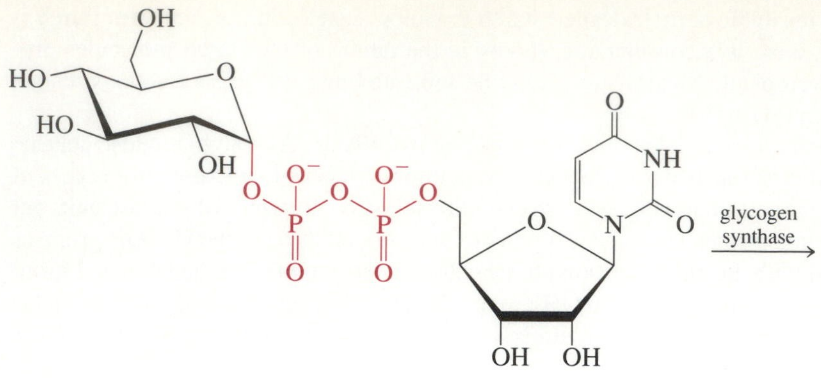

HO OH O HO OH O O⁻ P O O O⁻ P O O O O NH N O O OH OH glycogen synthase

uridine diphosphate glucose

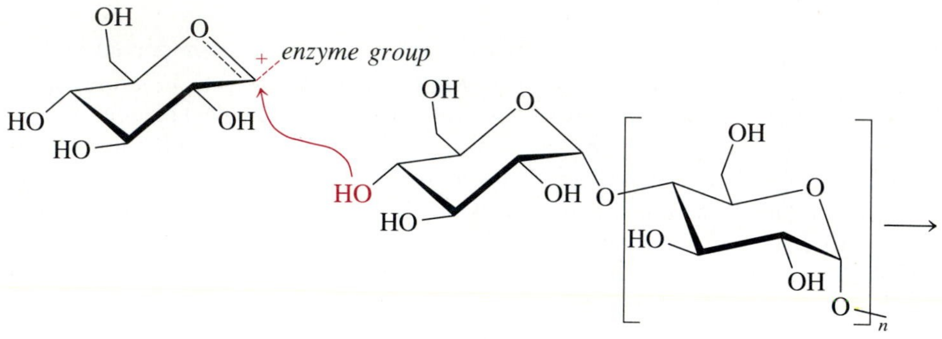

OH O + enzyme group HO HO OH OH O HO HO OH O OH O HO OH O n

glycogen

OH O HO O OH O HO HO OH OH O OH O HO OH O n

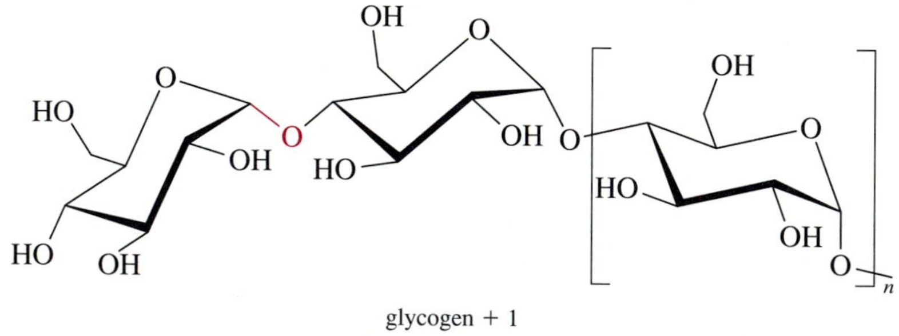

glycogen + 1

The process whereby the organism degrades, or depolymerizes, the polysaccharide directly to glucose 1-phosphate is different from the synthesis and involves a somewhat complex enzyme system called phosphorylase *b*. Similar enzymes catalyze the formation and cleavage of the 1 → 6 glycosidic bond by way of D-glucose 6-phosphate.

α-D-Glucopyranosyl uridine diphosphate (UDP-glucose) is also involved in the formation of sucrose.

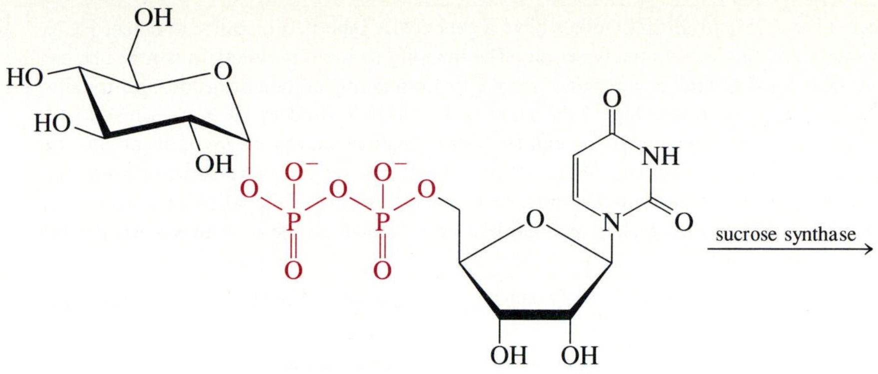

uridine diphosphate glucose

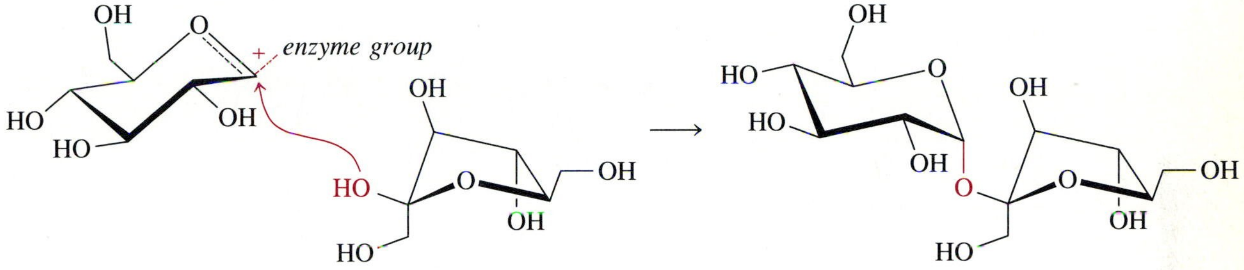

fructose

sucrose

28.10 Natural Glycosides

Sugars are often found to occur in organisms in the form of glycosides. Hydrolysis of a glycoside yields the sugar (the **glycon**) and the alkyl or aryl group to which it is bound (the **aglycon**). There are many types of glycosides, and we shall only give a few examples here.

Amygdalin was one of the first glycosides to be discovered. It occurs in bitter almonds and is a glycoside formed from the disaccharide gentiobiose and the cyanohydrin of benzaldehyde. Almonds contain an enzyme that catalyzes the conversion of amygdalin to HCN, benzaldehyde, and two molecules of D-glucose.

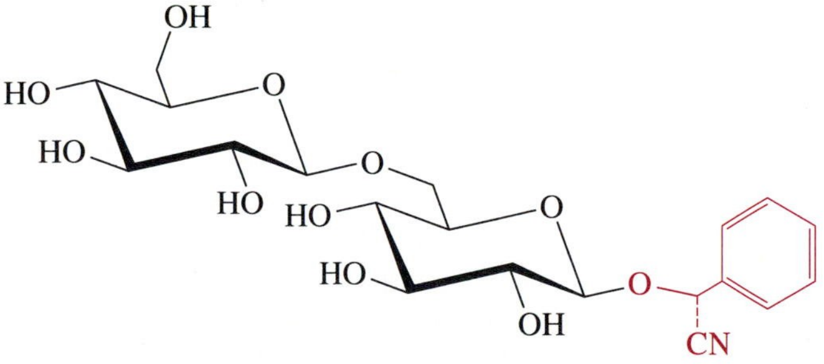

amygdalin
(R)-α-[(6-O-β-D-Glucopyranosyl-β-D-glucopyranosyl)oxy]phenylacetonitrile

Amygdalin (''laetrile'') is not effective as a cancer treatment and has chronic toxicity as a result of hydrolysis of the glycoside linkage, with the release of HCN.

> Between 15 and 25% of all individuals have a genetically inherited inability to detect HCN, which smells like almonds to many people. The inability to smell is called **anosmia;** one can become anosmic in general as a result of a cold, or from taking certain drugs or, specifically, through a genetic anosmia. One of the authors was overexposed to HCN as a result of a genetic anosmia. Anyone who works with HCN or cyanide should be aware of his or her own limitations in this regard, but **DO NOT** attempt a trial. If you carry out a benzoin condensation (Section 27.4.A) properly in the hood, you should be aware of the HCN odor. Curiously enough, benzaldehyde also has an almond-like odor for which there is no known genetic anosmia.

Another natural glycoside is peonin, which is responsible for the color of the dark red peony (Section 36.2).

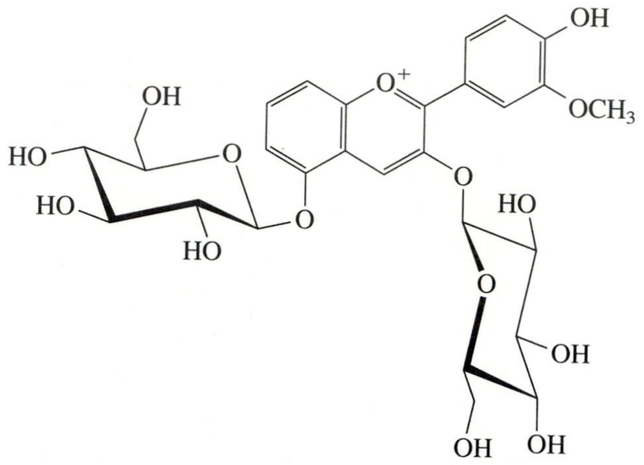

peonin

A number of naturally occurring antibiotics contain sugars bound as glycosides. The glycosyl groups often have unusual structures. An example is erythromycin A, a widely used antibiotic. The aglycon is a 14-membered lactone containing five hydroxy groups, two of which are bound to the rare monosaccharides cladinose and desosamine.

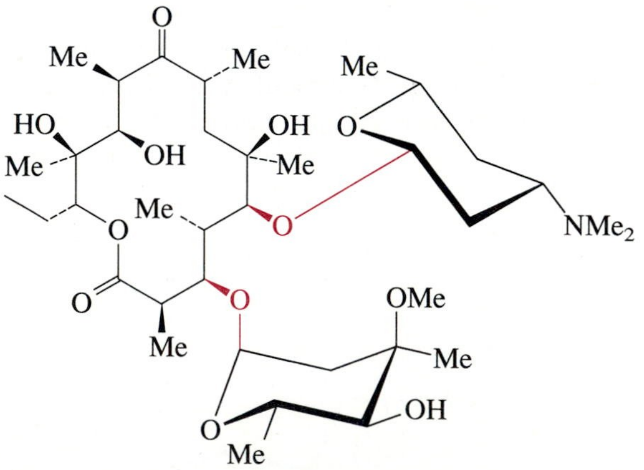

erythromycin A

28.11 Aminosaccharides and Poly(aminosaccharides)

Replacement of an OH group in a glycoside by an NH_2 group leads to an aminosugar. A poly(β-(1 $\rightarrow$ 4)-2-N-acetylamino-2-deoxyglucose) is about as abundant as starch in the animal kingdom; the polymer is called **chitin.** The chemical structure is the same as that of cellulose with the N-acetylamino group replacing a hydroxyl group. The water-insoluble material is the main constituent of the structural cells of fungi and the exoskeletons of marine invertebrates such as shrimp as well as those of the insects and other Arthropoda. Acid hydrolysis removes some of the N-acetyl groups and changes the solubility and other characteristics of the polymer. The raw material was regarded as waste but now can be converted into a number of useful, and biodegradable, products.

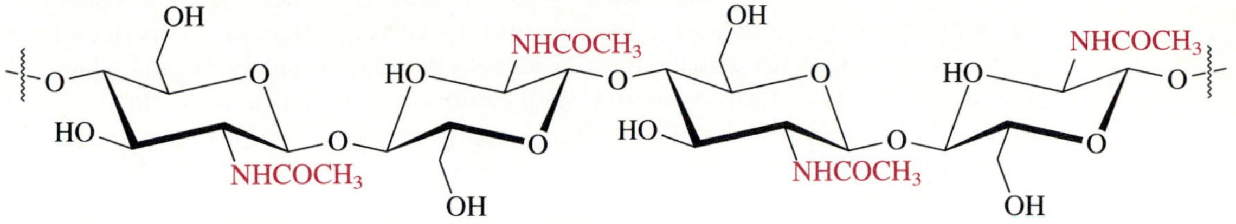

chitin

Oligomers and polymers based on aminosugars occur in cell walls of bacteria (muramic acid, an ester of lactic acid) and higher organisms (N-acetylneuraminic acid, a sialic acid).

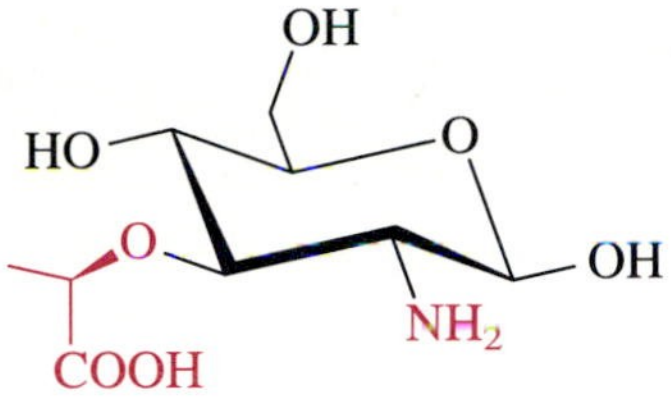

muramic acid

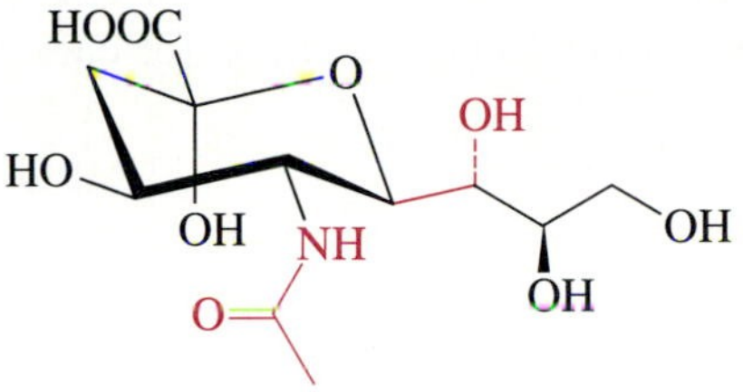

N-acetylneuraminic acid

Antigens that contain N-2-acetylamino-2-deoxyglucose and/or -galactose are present on the outer cell membranes of red blood cells and are responsible for the main blood groups (A, AB, B, and O). These aminosugars are often referred to as N-acetylglucosamine or N-acetylgalactosamine, the lack of a hydroxy group at the 2-position being understood. The degradation of gangliosides (polysaccharides present in brain cells) fails because of the lack of a hexosamidinase in Tay-Sachs disease, which occurs among Ashkenazi Jews and is invariably fatal to the carriers by the age of 3.

Aminosugars are common as components of antibiotics. We have cited erythromycin A (page 942). Paromomycin is an ambecide with an interesting structure.

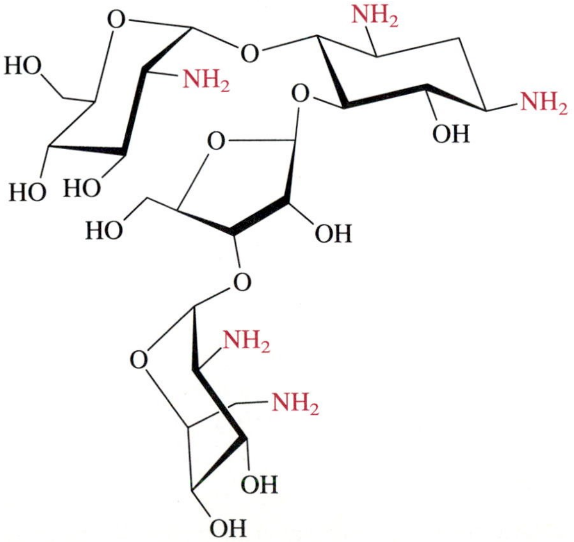

paromomycin

ASIDE

Some 12,000 years ago, an artist depicted a nubile young woman in the act of gathering honey from a wild beehive near present-day Valencia, Spain. The human craving for sweet things has not diminished. With the aid of synthetic organic chemistry materials that are hundreds of times sweeter than sucrose, glucose, or fructose have been made to satisfy the demand for a sweet taste without the calories contained in the natural sugars. If we wish to improve upon nature, we should first understand what is needed. In this case we should know the molecular structures and how sugars interact with their protein receptors. Activation of these receptors initiates a complex chain of events that eventually produce the sensation of a sweet taste. We can begin with the molecular structures of the simple sugars. Draw and identify Fischer projection formulas for all eight aldohexoses. Is fructose an aldohexose? Do these open-chain structures represent the forms in which the sugars are found? From what we learned about difunctional compounds, we might expect cyclic forms, called pyranoses, to be present. Draw and identify these forms for the aldohexoses and evaluate the stability of the different conformations for each compound.

PROBLEMS

1. Assign *R* and *S* notations to all of the aldoses in Table 28.1.

2. Construct a "family tree," similar to Table 28.1, that contains the structures for all of the D-ketoses having six or fewer carbons. Identify which aldoses and ketoses will give the same osazones.

3. Using Table 28.1, identify all of the aldoses that give meso saccharic acids on oxidation by nitric acid.

4. 5-Hydroxyheptanal exists in two cyclic hemiacetal forms. Write three-dimensional structures for the two compounds. Which is more stable? Write a mechanism for interconversion of the two forms under conditions of acid catalysis and base catalysis.

5. a. Draw three-dimensional projection structures for the two conformations of β-D-xylopyranose. Predict which conformation predominates in solution.
b. Answer part (a) for α-D-arabinopyranose.

6. Suggest a method for the synthesis of 6-deoxygalactose from galactose. (*Hint*: See page 917.)

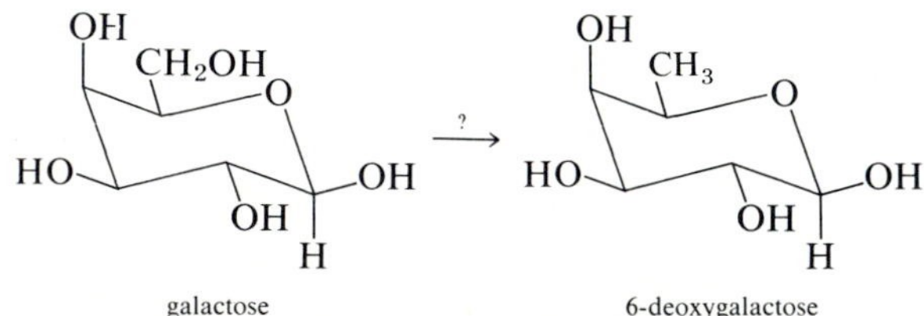

7. a. Write equations that show the application of the Kiliani-Fischer synthesis to each of the D-aldotetroses. Which aldopentoses are obtained from D-threose and which from D-erythrose?
b. Answer part (a) for the aldopentoses.

8. Write equations that show the application of the Ruff degradation to each of the D-aldohexoses. Which aldopentoses are obtained from each aldohexose?

9. Under the proper conditions D-glucose reacts with benzaldehyde to give 2,4-O-benzylidene-D-glucose.

$$\text{D-glucose} + C_6H_5CHO \xrightarrow{H^+} \text{2,4-O-benzylidene-D-glucose}$$

2,4-O-benzylidene-D-glucose

This compound is reduced to 2,4-O-benzylidene-D-glucitol, which reacts with periodic acid to give the benzylidene derivative of an aldopentose. Hydrolysis of the latter compound gives the aldopentose. What are the structure and name of the aldopentose?

10. Complete acid- or base-catalyzed hydrolysis of one class of nucleic acids yields a D-aldopentose, A, phosphoric acid, and several purine and pyrimidine bases. Nitric acid oxidation of A yields a meso diacid, B. Treatment of A with hydroxylamine forms the oxime, C, which upon treatment with acetic anhydride is converted into an acetylated cyanohydrin, D. Hydrolysis of compound D gives an aldotetrose, E, which is oxidized by nitric acid to a meso diacid, F. What are the structures of compounds A through F?

11. A substance, G, with the formula $C_5H_{10}O_5$ is reduced by $NaBH_4$ to an optically inactive product, H ($C_5H_{12}O_5$). Wohl degradation of G gives I ($C_4H_8O_4$). Compound I is oxidized by HNO_3 to an optically active dicarboxylic acid, J ($C_4H_8O_6$). What are the structures of compounds G through J, assuming that they are straight-chain sugars that belong to the D-glyceraldehyde family?

12. A disaccharide, K, $C_{11}H_{20}O_{10}$, is hydrolyzable by α-glucosidase, yielding D-glucose and a D-pentose. The disaccharide does not reduce Fehling's solution. Methylation of K with dimethyl sulfate in NaOH yields a heptamethyl ether, L, which upon acid hydrolysis yields 2,3,4,6-tetra-O-methyl-D-glucose and a pentose tri-O-methyl ether, M. Oxidation of M by bromine water yields 2,3,4-tri-O-methyl-D-ribonic acid. Assign structures to compounds K through M.

13. A naturally occurring compound, N, has the formula $C_7H_{14}O_6$. It is nonreducing and does not mutarotate. Compound N is hydrolyzed by aqueous HCl to compound O, $C_6H_{12}O_6$, a reducing sugar. Oxidation of O with dilute HNO_3 gives an optically inactive diacid, P ($C_6H_{10}O_8$). Ruff degradation of O gives a new reducing sugar, Q ($C_5H_{10}O_5$), which is oxidized by dilute HNO_3 to an optically active diacid, R ($C_5H_8O_7$). Compound N is treated successively with NaOH and dimethyl sulfate, aqueous HCl, and hot nitric acid. From the product mixture, one may isolate α,β-dimethoxysuccinic acid and α-methoxymalonic acid.

$$HOOC{-}CH(OCH_3){-}CH(OCH_3){-}COOH \qquad HOOC{-}CH(OCH_3){-}COOH$$

α,β-dimethoxysuccinic acid — α-methoxymalonic acid

a. Give structures for compounds N through R.
b. What structural ambiguity exists, if any?

14. An aldopentose, S, is oxidized to a diacid, T, which is optically active. Compound S is also degraded to an aldotetrose, U, which undergoes oxidation to an optically inactive diacid, V. Assuming that S has the D-configuration (4*R*), what are the structures of S through V?

15. Aldohexose W is reduced by sodium borohydride ($NaBH_4$) to an optically inactive alditol, X. Ruff degradation of W gives an aldopentose, Y, which is oxidized by nitric acid to an optically active saccharic acid, Z. What are compounds W through Z, assuming them to be D-sugars?

16. Oxidation of aldohexose A′ by nitric acid gives an optically active saccharic acid, B′. Ruff degradation of A′ gives an aldopentose, C′, which yields an optically inactive diacid, D′, on nitric acid oxidation. When compound A′ is subjected to a series of reactions that exchange C-1 and C-6, the same aldohexose is obtained. Assuming them to be D-sugars, what are compounds A′ through C′?

17. The optical rotations for the α- and β-anomers of D-mannose are $[\alpha]_D = +29.3$ and $[\alpha]_D = -17.0$, respectively. In water solution each form mutarotates to an equilibrium value of $[\alpha]_D = +14.2$. Calculate the percentage of each anomer present at equilibrium.

18. The disaccharide melibiose is hydrolyzed by dilute acid to a mixture of D-glucose and D-galactose. Melibiose is a reducing sugar and is oxidized by bromine water to melibionic acid, which is methylated by sodium hydroxide and dimethyl sulfate to octa-O-methylmelibionic acid. Hydrolysis of the latter gives a tetra-O-methylgluconic acid, E′, and a tetra-O-methylgalactose, F′. Compound E′ is oxidized by nitric acid to tetra-O-methylglucaric acid. Compound F′ is also obtained by the acidic hydrolysis of methyl 2,3,4,6-tetra-O-methylgalactopyranoside. Melibiose is hydrolyzed by an α-galactosidase from almonds. What is the structure of melibiose?

19. The trisaccharide gentianose is hydrolyzed by acid to 2 equivalents of D-glucose and 1 of D-fructose. Partial acid hydrolysis yields D-fructose and gentiobiose (page 941). The enzymes of almond emulsion cleave gentianose into D-glucose and sucrose. What is the structure of gentianose?

20. Write Haworth projections for the following saccharides.

a. α-D-galactopyranose **b.** methyl β-D-mannoside
c. α-maltose **d.** β-cellobiose

21. 1,2:5,6-Di-O-isopropylidene-α-D-glucofuranose reacts with aqueous acetic acid to give 1,2-O-isopropylidene-α-D-glucofuranose. That is, the 5,6-acetal is hydrolyzed selectively. This selective hydrolysis of the 5,6-acetal in compounds of this series is normal. Making use of this general reaction, along with other reactions discussed in this chapter, outline methods for the conversion of D-glucose into the following products.

a. D-(+)-xylose **b.** *meso*-2,4-dihydroxyglutaric acid **c.** L-(+)-ribose

22. It is not possible to convert glyceraldehyde into its acetal with acetone by the direct reaction of the two compounds in the presence of acid. However, D-mannitol (page 921) reacts smoothly with acetone in the presence of zinc chloride to form a bis-acetal, 1,2:5,6-di-O-isopropylidene-D-mannitol. How might this observation be used to achieve the indirect synthesis of 2,3-O-isopropylidene-D-glyceraldehyde?

23. D-Glucose, D-mannose, and D-fructose all react with three equivalents of phenylhydrazine to yield the same yellow crystalline osazone. Write the structure of the osazone and show with structural formulas what implications this result has for the stereochemistry of the sugars. Write out the mechanism of formation of osazone.

24. A compound, G′, with the formula $C_5H_{10}O_5$ is optically active and does not react with $AgNO_3$ to give a carboxylic acid. Reduction of G′ with $NaBH_4$ gives *only one* product, H′ ($C_5H_{12}O_5$), which is optically active. What are compounds G′ and H′, assuming that they are straight-chain sugars that belong to the D-glyceraldehyde family?

25. An aldopentose, I′, is oxidized by nitric acid to give an *optically active* saccharic acid, J′. Treatment of I′ with hydroxylamine gives an oxime that reacts with acetic anhydride to give an acetylated cyanohydrin, K′. Basic hydrolysis of K′ gives an aldotetrose, L′, which is oxidized by nitric acid to an *optically active* dicarboxylic acid, M′. What are the structures of compounds I′ through M′? What are the names of compounds I′ and K′?

26. Which of the following glycosides does *not* have an α-anomeric linkage.

a. OH OH HO O H HO O HO O HO OH OH

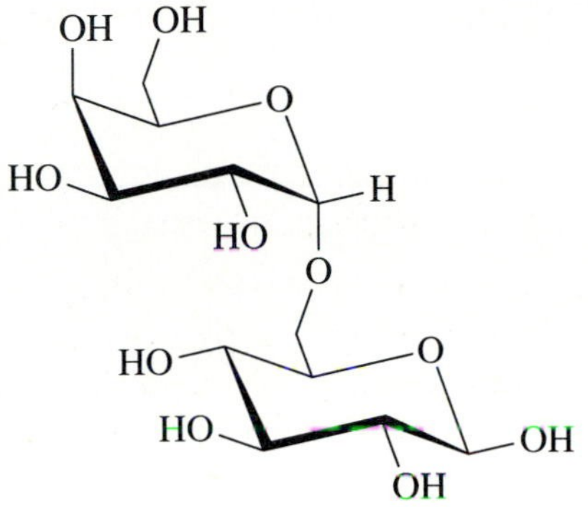

b. OH O HO HO O CN

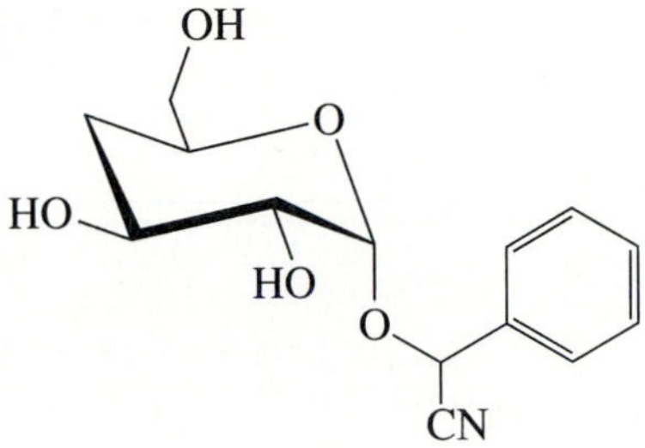

c. OH OH O HO OH HO OH O OH O OH

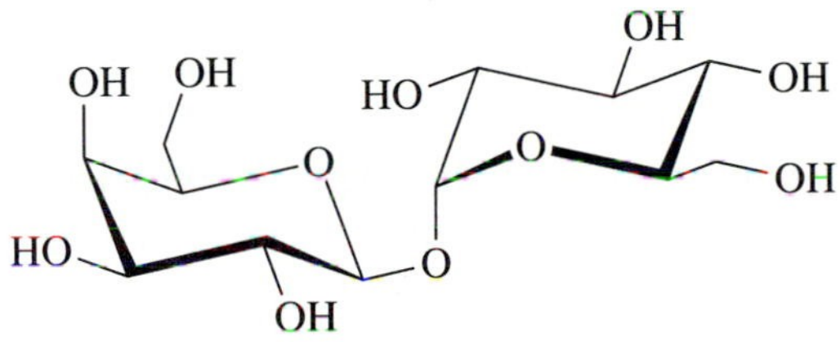

d. OH OH O HO OCH$_3$ OH

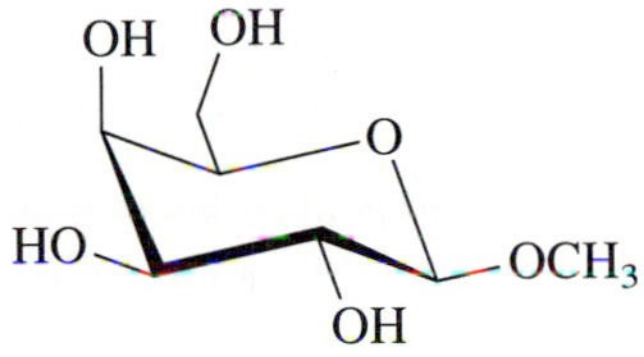

27. Write a Fischer projection formula for each of the following compounds.

a. OH OCH$_3$ HO CHO OH N(CH$_3$)$_2$

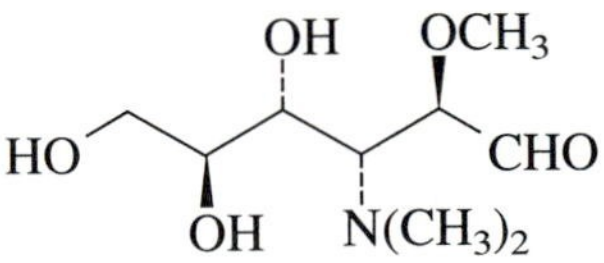

b. CH$_2$OH O OH OH OH OH

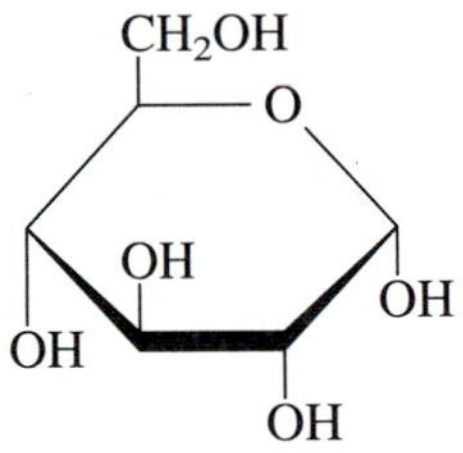

c. CH$_3$ CH$_3$ Me$_2$N O OH OCH$_3$

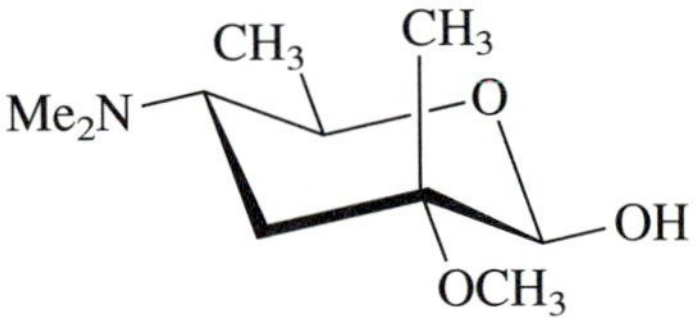

CHAPTER 29

AMINO ACIDS, PEPTIDES, AND PROTEINS

29.1 Introduction

Amino acids constitute a particularly important class of difunctional compounds because they are the building blocks from which proteins are constructed. Since the two functional groups in an amino acid are, respectively, basic and acidic, the compounds are amphoteric and actually exist as **zwitterions** or **inner salts.** For example, glycine, the simplest amino acid, exists mostly in the form shown, rather than as aminoacetic acid.

$$\underset{\text{zwitterion form}}{\underset{\text{glycine}}{H_3\overset{+}{N}CH_2CO_2^-}} \rightleftharpoons \underset{\text{amino acid form}}{\underset{\text{glycine}}{H_2NCH_2COOH}}$$

An amide bond is produced by the elimination of water from two amino acids. This linkage between two amino acid units is also called a **peptide bond,** and the resulting compounds are called **peptides.** (Peptides were originally fragments of protein produced by partial degradation with the enzyme, pepsin.) For example, the peptide formed from two molecules of glycine is glycylglycine, a **dipeptide.** Like glycine, it is amphoteric and exists as a zwitterion.

$$\underset{\text{a dipeptide}}{\underset{\text{glycylglycine}}{H_3\overset{+}{N}CH_2\overset{\overset{\displaystyle O}{\|}}{C}\text{—}NHCH_2CO_2^-}}$$

peptide bond (C—N)

As amino acid units are added, the prefix to peptide changes to reflect the number of units; a **tripeptide** contains three amino acid units, a **tetrapeptide** four, and so on.

$$H_3\overset{+}{N}CH_2\overset{O}{\overset{\|}{C}}NHCH_2\overset{O}{\overset{\|}{C}}NHCH_2CO_2^-$$

glycylglycylglycine
a tripeptide

The addition of more units or ''parts'' to the chain ultimately produces a polymer molecule (Gk., *poly,* many, *meros,* part). Such polymers are called, as a class, **polypeptides. Proteins** (Gk., *proteos,* prime) are special types of polypeptides composed primarily of about 20 different specific amino acids. They are large molecules, with molecular weights from 6000 to more than 1,000,000 (from about 50 to more than 8000 amino acids per molecule).

In this chapter we will review the individual amino acids, especially those important in nature, and their chemical and physical properties. In the discussion of the peptides and polypeptides that follows, we shall learn of the importance of the strength of the amide linkage and the dominating role played by its conformational tendencies.

29.2 Structure, Nomenclature, and Physical Properties of Amino Acids

Most of the important natural amino acids are α-amino acids; that is, the amino group occurs at the position adjacent to the carboxy function.

$$CH_3\overset{\overset{+}{N}H_3}{\overset{|}{C}}HCO_2^- \qquad H_3\overset{+}{N}CH_2CH_2CO_2^-$$

alanine, an α-amino acid — β-alanine, a β-amino acid

The important natural amino acids are listed in Table 29.1, along with both three-letter and one-letter codes that are conventionally used as abbreviations. A sequence written in the three-letter code is easier to use for chemical purposes, but the one-letter code is the only one compact enough to compare the thousands of amino acid sequences now known. The structures are all written in the amino acid form, rather than as zwitterions since alternative zwitterionic structures are possible for some.

TABLE 29.1 Amino Acids

Amino Acid Formula R—CH(NH$_2$)—COOH	Name	Abbreviation	Single Letter Code
H—CH(NH$_2$)—COOH	glycine	Gly	G
CH$_3$—CH(NH$_2$)—COOH	alanine	Ala	A
CH$_3$—CH(CH$_3$)—CH(NH$_2$)—COOH	valine	Val	V
CH$_3$—CH(CH$_3$)—CH$_2$—CH(NH$_2$)—COOH	leucine	Leu	L
CH$_3$CH$_2$—CH(CH$_3$)—CH(NH$_2$)—COOH	isoleucine	Ile	I
CH$_3$S—CH$_2$CH$_2$—CH(NH$_2$)—COOH	methionine	Met	M

continues

Amino Acid Formula $R-CH(NH_2)-COOH$	Name	Abbreviation	Single Letter Code
CH_2—CH_2—CH_2—NH—CH—COOH (ring: CH_2, CH_2, CH_2, NH, CH)	proline	Pro	P
$C_6H_5-CH_2-CH(NH_2)-COOH$ (benzene ring)	phenylalanine	Phe	F
(indole ring, NH)$-CH_2-CH(NH_2)-COOH$	tryptophan	Trp	W
$HOCH_2-CH(NH_2)-COOH$	serine	Ser	S
$CH_3-CH(OH)-CH(NH_2)-COOH$	threonine	Thr	T
$HSCH_2-CH(NH_2)-COOH$	cysteine	Cys	C
$HO-C_6H_4-CH_2-CH(NH_2)-COOH$ (benzene ring)	tyrosine	Tyr	Y
$H_2N-C(=O)-CH_2-CH(NH_2)-COOH$	asparagine	Asn	N
$H_2N-C(=O)-CH_2-CH_2-CH(NH_2)-COOH$	glutamine	Gln	Q
$HO-C(=O)-CH_2-CH(NH_2)-COOH$	aspartic acid	Asp	D
$HO-C(=O)-CH_2-CH_2-CH(NH_2)-COOH$	glutamic acid	Glu	E
$H_2N-CH_2-CH_2-CH_2-CH_2-CH(NH_2)-COOH$	lysine	Lys	K
$H_2N-C(=NH)-NH-CH_2-CH_2-CH_2-CH(NH_2)-COOH$	arginine	Arg	R
(imidazole ring: N, NH)$-CH_2-CH(NH_2)-COOH$	histidine	His	H

EXERCISE 29.1 Commit to memory the names, structures, and abbreviations of the 20 common amino acids in Table 29.1.

The inner-salt nature of the amino acids results in physical properties that are somewhat different from the properties normally found in organic compounds. Zwitterions are highly polar substances for which intermolecular electrostatic attractions lead to rather strong crystal lattice structures. Consequently, melting points are generally high. Most amino acids decompose instead of melting, and it is customary to record decomposition points (Table 29.2). In general, decomposition points are dependent on the rate of heating of the sample and are not reliable physical properties. Most of the amino acids are only sparingly soluble in water, again as a consequence of the strong intermolecular forces acting in the crystal lattice. Exceptions are glycine, alanine, proline, lysine, and arginine, which are all quite soluble in water.

TABLE 29.2 Physical Properties of Amino Acids

Amino Acid	Decomposition Point, °C	Water Solubility, g/100 ml H_2O at 25	$[\alpha]_D^{25}$	pK_1	pK_2	pK_3
glycine	233	25		2.35	9.78	
alanine	297	16.7	+8.5	2.35	9.87	
valine	315	8.9	+13.9	2.29	9.72	
leucine	293	2.4	−10.8	2.33	9.74	
isoleucine	284	4.1	+11.3	2.32	9.76	
methionine	280	3.4	−8.2	2.17	9.27	
proline	220	162	−85.0	1.95	10.64	
phenylalanine	283	3.0	−35.1	2.58	9.24	
tryptophan	289	1.1	−31.5	2.43	9.44	
serine	228	5.0	−6.8	2.19	9.44	
threonine	225	very	−28.3	2.09	9.10	
cysteine			+6.5	1.86	8.35	10.34
tyrosine	342	0.04	−10.6	2.20	9.11	10.07
asparagine	234	3.5	−5.4	2.02	8.80	
glutamine	185	3.7	+6.1	2.17	9.13	
aspartic acid	270	0.54	+25.0	1.99	3.90	10.00
glutamic acid	247	0.86	+31.4	2.13	4.32	9.95
lysine	225	very	+14.6	2.16	9.20	10.80
arginine	244	15	+12.5	1.82	8.99	13.20
histidine	287	4.2	−39.7	1.81	6.05	9.15

With the exception of glycine, all of the common amino acids are chiral molecules. The naturally occurring compounds all have the same absolute configuration at the stereocenter. As with carbohydrates, it is traditional to use the D and L nomenclature with amino acids. Natural amino acids belong to the L-series (Figure 29.1). The stereo structure of L-proline is shown in Figure 29.2. Optical rotations for the natural L-amino acids are given in Table 29.2.

EXERCISE 29.2 Assign *R* and *S* stereochemical descriptors to L-alanine, L-serine, and L-cysteine (see Section 7.3).

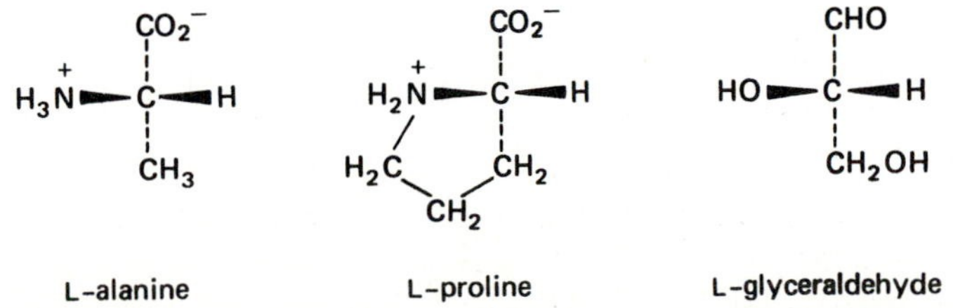

L-alanine L-proline L-glyceraldehyde

FIGURE 29.1 The relationship of L-alanine and L-proline to L-glyceraldehyde.

FIGURE 29.2 Stereo structure of L-proline. [Reproduced with permission from *Molecular Structure and Dimensions,* International Union of Crystallography, 1972.]

29.3 Acid-Base Properties of Amino Acids

Amino acids show both acidic and basic properties and are therefore **amphoteric.** In acidic solution, the amino acid is completely protonated and exists as the conjugate acid.

$$H_3\overset{+}{N}CH_2CO_2^- + H^+ \longrightarrow H_3\overset{+}{N}CH_2COOH$$

The titration curve for glycine hydrochloride is shown in Figure 29.3. The salt behaves as a typical diprotic acid.

$$H_3\overset{+}{N}CH_2COOH \overset{K_1}{\rightleftharpoons} H^+ + H_3\overset{+}{N}CH_2CO_2^-$$

$$H_3\overset{+}{N}CH_2CO_2^- \overset{K_2}{\rightleftharpoons} H^+ + H_2NCH_2CO_2^-$$

$$K_1 = \frac{[H^+][H_3\overset{+}{N}CH_2CO_2^-]}{[H_3\overset{+}{N}CH_2COOH]}$$

$$K_2 = \frac{[H^+][H_2NCH_2CO_2^-]}{[H_3\overset{+}{N}CH_2CO_2^-]}$$

When the hydrochloride has been half neutralized, $[H_3NCH_2COOH] = [H_3NCH_2CO_2^-]$. The pH of the solution at this point is equal to pK_1. This first dissociation constant refers to ionization of the COOH, which is the more acidic of the two acidic

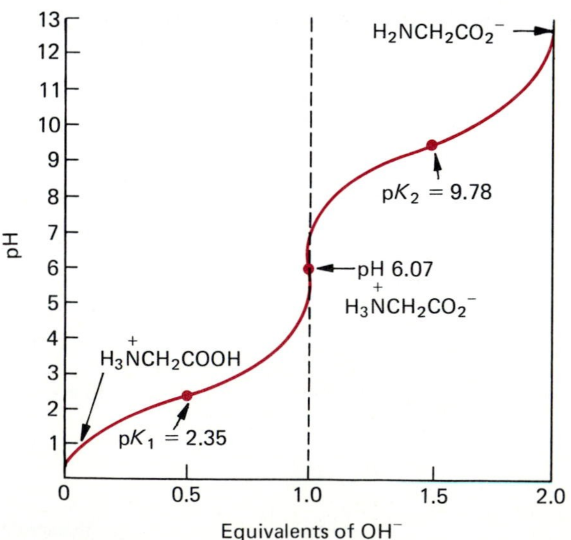

FIGURE 29.3 Titration curve for glycine hydrochloride.

groups in the dibasic acid. Note that glycine is substantially more acidic than acetic acid, which has $pK_a = 4.76$, because of the large inductive effect of the NH_3^+ group (see Sections 10.4 and 18.4).

After one equivalent of base has been added, the chief species in solution is the zwitterionic form of the amino acid itself. The pH of the solution at this point is simply the pH of a solution of the amino acid in pure water. This pH, the **isoelectric point,** produces the minimum solubility for the amino acid by maximizing intermolecular electrostatic and hydrogen-bonding interactions.

Addition of a further half-equivalent of base corresponds to half-neutralization of the acid $H_3\overset{+}{N}CH_2CO_2^-$. At this point $[H_3\overset{+}{N}CH_2CO_2^-] = [H_2NCH_2CO_2^-]$, and the pH of the solution is equal to pK_2, the dissociation constant for the protonated amino group. Note that pK_2 for glycine, 9.78, is slightly lower than that for the conjugate acid of methylamine, which has pK_a 10.4.

$$H_3\overset{+}{N}CH_2CO_2^- \rightleftharpoons H^+ + H_2NCH_2CO_2^- \qquad pK_a = 9.8$$

$$H_3\overset{+}{N}CH_3 \rightleftharpoons H^+ + H_2NCH_3 \qquad pK_a = 10.4$$

Thus, the ammonium group of glycine is slightly more acidic than the methylammonium ion.

This result may be surprising at first sight because the carboxylate anion is expected to stabilize the ammonium ion by electrostatic attraction. The difference undoubtedly results from solvation effects. As indicated in Figure 29.4, a zwitterion in which the two charges are close together is less efficiently solvated by solvent dipoles than two ions far apart.

FIGURE 29.4 Solvation effects on a zwitterion and related species.

EXERCISE 29.3 What is the principal organic species present in an aqueous solution of glycine at (a) pH 2, (b) pH 4, (c) pH 8, and (d) pH 11?

As shown in Table 29.2, most of the amino acids show similar values of pK_1 and pK_2. Aspartic and glutamic acids each have an additional carboxy group, with pK_as of 3.90 and 4.32, respectively.

$pK_2 = 3.90$ → $HOOCCH_2\overset{\overset{+}{N}H_3}{|}CHCOOH$; $\overset{+}{N}H_3$ ← $pK_3 = 10.00$; COOH ← $pK_1 = 1.99$

aspartic acid

$pK_2 = 4.32$ → $HOOCCH_2CH_2\overset{\overset{+}{N}H_3}{|}CHCOOH$; $\overset{+}{N}H_3$ ← $pK_3 = 9.95$; COOH ← $pK_1 = 2.13$

glutamic acid

Lysine has two amino groups with pK_as of 9.20 and 10.8. The more basic group is probably the one more remote from the carboxy group. Consequently, the principal form of lysine is probably the zwitterion in which the terminal amino group is protonated.

$$H_3\overset{+}{N}CH_2CH_2CH_2CH_2\underset{}{\overset{NH_2}{\overset{|}{C}}}HCO_2^-$$

lysine

Arginine contains the strongly basic guanadino group, corresponding to pK_a 13.2. It exists in the following zwitterionic form.

$$H_2N\overset{\overset{+}{N}H_2}{\overset{\|}{C}}NHCH_2CH_2CH_2\overset{NH_2}{\overset{|}{C}}HCO_2^-$$

arginine

Guanidines are compounds of the general formula

$$RNH\overset{NH}{\overset{\|}{C}}NH_2$$

Protonation of the guanidino group on the imino nitrogen results in a cation that is highly resonance stabilized. Guanidines are among the strongest organic bases.

$$RNH\overset{NH}{\overset{\|}{C}}NH_2 + H^+ \rightleftharpoons \left[RNH{-}\overset{\overset{+}{N}H_2}{\overset{\|}{C}}{-}NH_2 \longleftrightarrow R\overset{+}{N}H{=}\overset{NH_2}{\overset{|}{C}}{-}NH_2 \longleftrightarrow RNH{-}\overset{NH_2}{\overset{|}{C}}{=}\overset{+}{N}H_2\right]$$

The high pK_3 for arginine shows that the guanidino group is half protonated even at pH 13.2.

Tyrosine and histidine also contain other titratable groups, corresponding to pK_as of 10.07 and 6.05, respectively. These ionization constants refer to the phenolic hydroxy in tyrosine and the imidazole ring in histidine. The pK_a of about 10 is a normal value for a phenol (Section 30.5.B). We shall discuss imidazole in Chapter 32. The third titratable group in cysteine is the SH, which has a pK_a of 10.34, a normal value for a thiol (Section 26.1).

EXERCISE 29.4 Sketch the expected general appearance of the titration curve for histidine.

29.4 Occurrence of Amino Acids

The genetic code normally provides instructions for the incorporation of 20 amino acids, those that are listed in Table 29.1. However, many other ''rare'' amino acids are found in nature. At least one of these, selenocysteine, is coded for by a DNA codon.

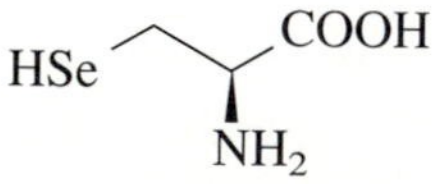

HSe–CH$_2$–CH(NH$_2$)–COOH

selenocysteine

Some amino acids are produced by modification of amino acid elements already present in proteins. Examples are the conversion of glutamic acid to carboxyglutamic acid (see vitamin K mechanism, page 1037), the modification of histidine to **diphth**amide (the target component for **diphth**eria toxin in the protein biosynthesis elongation factor, EF2), and the transformation of proline to (*R*)-4-hydroxyproline (an essential component of the structural protein collagen).

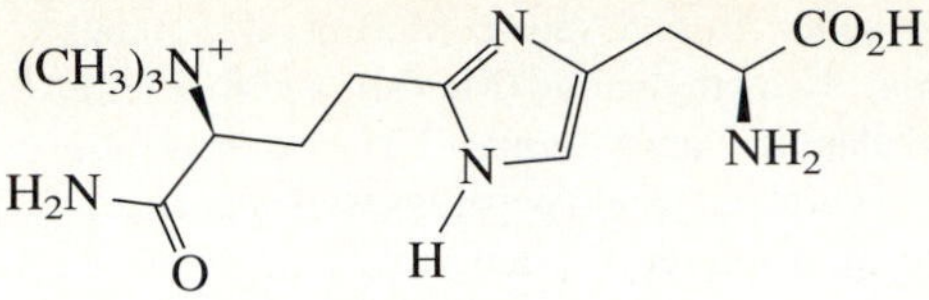

diphthamic acid

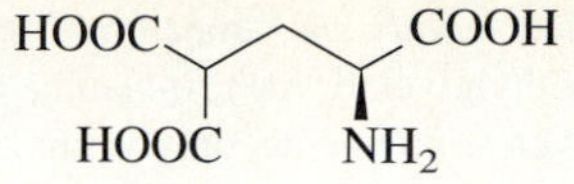

γ-carboxyglutamic acid

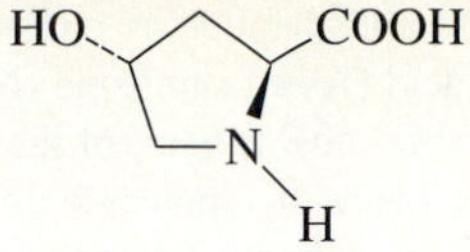

(*R*)-4-hydroxyproline

Other rare amino acids are formed in the biochemical reactions of metabolism or have roles as chemical messengers or control agents. Examples are citrulline and ornithine, which are involved in the reaction cycle that produces urea, γ-aminobutyric acid (GABA), an inhibitory neurotransmitter, and statine, a component of a naturally occurring pentapeptide that inhibits the action of the enzyme pepsin (page 983).

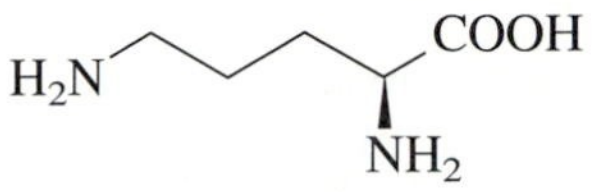

ornithine

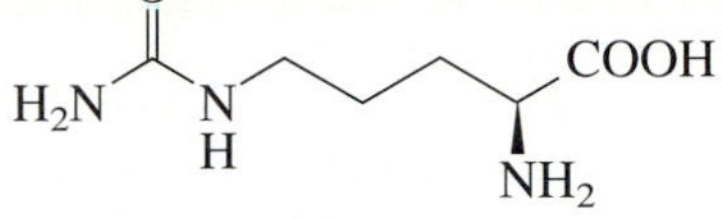

citruilline

γ-aminobutyric acid

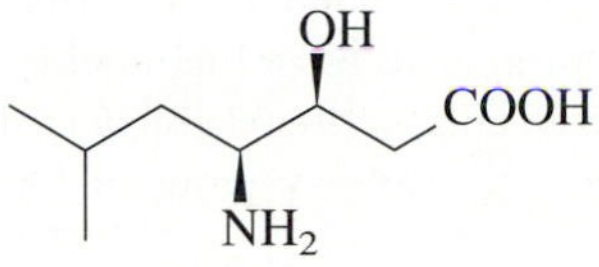

statine

Unusual amino acids have also been found as components of natural antibiotics and proteins. Examples of this class are α-aminoadipic acid, formed in the biosynthesis of penicillin, and L-(+)-MeBmt, which is a component of the powerful immunosuppressant ciclosporin.

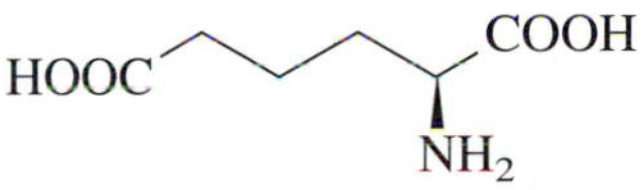

α-aminoadipic acid

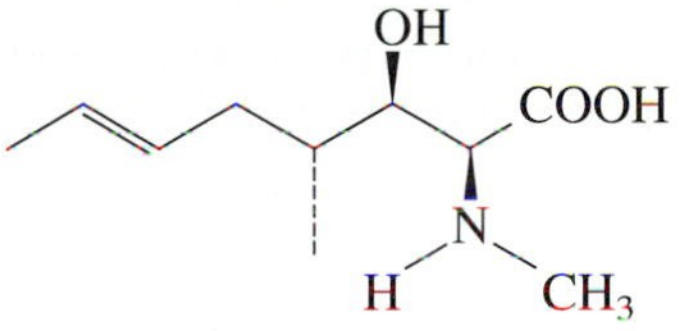

L-(+)-MeBmt

Ciclosporin is a fungal metabolite that was isolated in the 1960s. The natural product was found to have remarkable immunosuppressive properties and was therefore investigated as a possible drug to use in connection with organ transplantation. In 1973, pure ciclosporin was isolated in crystalline form, which permitted its structure to be elucidated by x-ray crystal analysis. The compound turned out to be a cyclic peptide containing 11 amino acids.

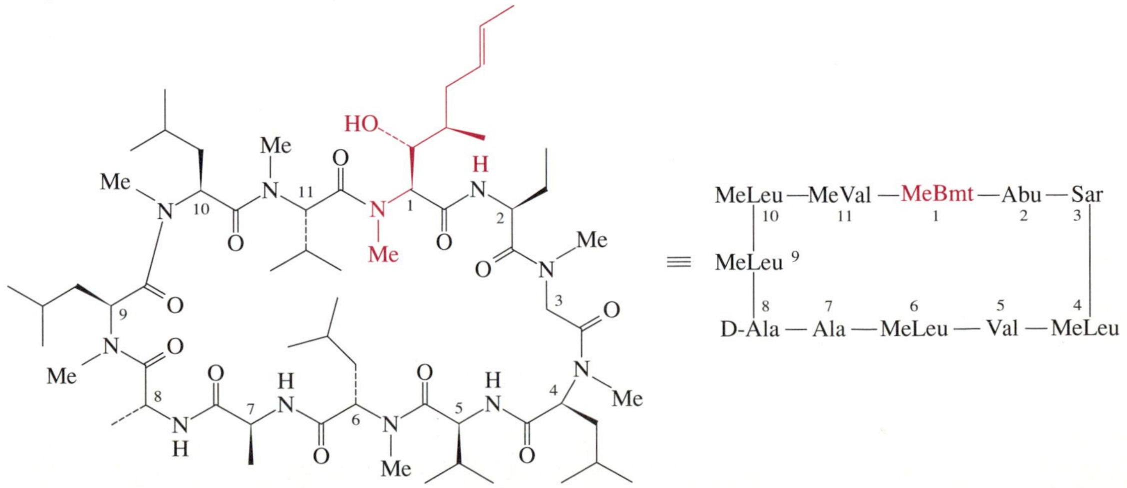

MeLeu(10)—MeVal(11)—MeBmt(1)—Abu(2)—Sar(3)—MeLeu(4)—Val(5)—MeLeu(6)—Ala(7)—D-Ala(8)—MeLeu(9)

ciclosporin

Note that this remarkable peptide is constructed *mainly* of "unnatural" amino acids. In addition to MeBmt, it contains (S)-α-aminobutyric acid (Abu), sarcosine (N-methylglycine, Sar), D-alanine (D-Ala), N-methylvaline (MeVal), and four units of N-methylleucine (MeLeu). In fact, of the eleven amino acids, only valine-5 and alanine-7 are "normal." The extensive use of unnatural amino acids in this molecule presumably evolved so as to confuse the natural enzymes **(peptidases)** that normally catalyze the hydrolysis of peptide bonds and would therefore destroy ciclosporin as a foreign peptide.

Ciclosporin was first marketed in 1983 and rapidly became the drug of choice for use in connection with solid organ (heart, kidney, liver, lung) and bone marrow transplantation. It functions to suppress the natural immune system of the donor recipient and therefore prevents rejection of the transplanted organ or marrow. The availability of ciclosporin has resulted in a rapid increase over the last decade in the number of organ transplants and also in the life expectancy of the recipients of these organs.

Highlight 29.1

The two functional groups of amino acids are, respectively, weakly basic and weakly acidic, and this **amphoteric** nature dominates their chemical and physical properties. Although an amino acid is neutral, it exists mainly in the form with CO_2^- and NH_3^+. Because of this dipolar nature, the molecules form very strong crystal lattices. As a result, amino acids have high melting points. Most amino acids are soluble in acidic or basic media and highly insoluble at the **isoelectric point,** which is the arithmetic mean of pK_a for the carboxy group, COOH, and pK_a for the protonated amine, NH_3^+.

29.5 Synthesis of Amino Acids

A. Commercial Availability

All of the common amino acids are available from chemical suppliers in optically active form. Table 29.3 lists the prices per 100 g of the amino acids quoted by various suppliers in 1990. These prices reflect several factors.

TABLE 29.3 Prices of Amino Acids

	Price per 100 g, $		
Amino Acid	L-Enantiomer	D-Enantiomer	Racemate
glycine	—	—	1.81
alanine	27.60	186.80	6.30
valine	25.95	164.40	10.40
leucine	17.30	270.20	46.80
isoleucine	57.90	21640.00	73.60[a]
methionine	22.00	103.60	5.20
proline	31.30	1519.00	371.00
phenylalanine	25.90	124.00	21.55
tryptophan	48.80	182.40	51.70
serine	35.35	191.80	16.50
threonine	53.75	148.40	32.50
cysteine	25.90	3055.00[b]	178.60
tyrosine	16.40	722.00	58.40
asparagine	13.50[c]	31.40[c]	20.00[c]
glutamine	20.90	1201.50	—
aspartic acid	6.15	106.40	5.95
glutamic acid	4.85	168.90	23.95[c]
lysine	5.70[b]	626.80[b]	34.80[b]
arginine	12.60	1096.00[b]	158.80[b,c]
histidine	25.20	306.00	102.00

[a]Mixture of diastereomers. [b]Hydrochloride salt. [c]Monohydrate.

All of the racemic amino acids are synthetic and are prepared commercially by methods to be outlined later in this section. The prices of the synthetic amino acids reflect both the ease of synthesis and the demand for the various compounds. Note that one of the more expensive racemic amino acids is the cyclic compound proline, which cannot be easily prepared by the standard methods that serve for the other amino acids.

Some of the available L-amino acids are isolated from natural sources; this is generally true when the price is lower than that for the racemate. The relatively low price of glutamic acid is a consequence of the fact that monosodium glutamate (MSG) is widely used as flavor enhancer in food preparation. The L-amino acid is prepared by a fermentation process in tonnage quantities, and its low price reflects this volume. Similarly, the L-aspartic acid is prepared on a large scale for the synthetic dipeptide sweetener, Aspartame (L-aspartyl-L-phenylalanyl methyl ester), and it has a low price. Some of the commercially available L-amino acids and all of the D-enantiomers are prepared by resolution of the synthetic racemates. Their high costs result from the additional expenses incurred in the resolution process (see Section 29.4.F).

B. Amino Acids from α-Halo Acids

α-Halo acids are available by the halogenation of carboxylic acids (Section 18.7.B). Recall that the direct alkylation of ammonia or an amine is not generally a satisfactory method for preparing amines owing to the overalkylation problem (Section 24.6.A). The reaction is somewhat better for preparing α-amino acids because the amino group in the product amino acids is less basic (by about 0.8 pK_a unit) than the amine itself. Thus the second alkylation reaction is now slower than the first. A number of α-amino acids can be prepared in this way.

$$CH_3CH(Br)COOH + NH_4^+\ OH^- \xrightarrow[25^\circ C,\ 4\ days]{H_2O} CH_3CH(\overset{+}{N}H_3)CO_2^- + NH_4^+\ Br^-$$

> α-Bromopropionic acid (153 g) is added to 5.8 L of concentrated aqueous ammonia and the resulting solution is kept at room temperature for 4 days. The solution is evaporated to dryness and extracted with warm absolute ethanol to remove ammonium bromide. The amino acid, 50 g (56%), is obtained as a white crystalline mass.

> EXERCISE 29.5 Write equations for the syntheses of phenylalanine, valine, and leucine, starting with the corresponding carboxylic acids. What special problems arise in the application of this method for the synthesis of serine or tyrosine?

C. Alkylation of N-Substituted Aminomalonic Esters

An especially useful general method for the synthesis of α-amino acids involves a variation of the malonic ester synthesis (Section 27.7.D). Diethyl malonate can be monobrominated to yield a bromide that enters into the S_N2 reaction with the potassium salt of phthalimide to give N-phthalimidomalonic ester.

$$CH_2(COOEt)_2 \xrightarrow[CCl_4]{Br_2} CHBr(COOEt)_2 + C_6H_4(CO)_2N^-\ K^+ \longrightarrow C_6H_4(CO)_2N{-}CH(COOEt)_2$$

potassium phthalimidate — N-phthalimidomalonic ester

The ester can be alkylated by a variety of alkyl halides or α,β-unsaturated carbonyl compounds. Vigorous acid hydrolysis causes hydrolysis of both ester groups and the phthalimido group and decarboxylation of the resulting malonic acid. The product is a racemic α-amino acid.

$$\text{PhthN}CH(COOEt)_2 \xrightarrow{\text{base}} \xrightarrow{RX} \text{PhthN}-\overset{R}{\overset{|}{C}}(COOEt)_2 \xrightarrow[\Delta]{H_3O^+} R\overset{\overset{+}{N}H_3}{\overset{|}{C}}HCO_2^-$$

Representative examples of this procedure are the synthesis of methionine and glutamic acid.

$$\text{PhthN}CH(COOEt)_2 \xrightarrow{\text{NaOEt}} \xrightarrow{CH_3SCH_2CH_2Cl} \xrightarrow[\Delta]{H_3O^+} CH_3SCH_2CH_2\overset{\overset{+}{N}H_3}{\overset{|}{C}}HCO_2^-$$

(50% overall)
methionine

$$\text{PhthN}CH(COOEt)_2 \xrightarrow[\text{NaOEt}]{CH_2{=}CHCOOEt} \xrightarrow[\Delta]{H_3O^+} HOOCCH_2CH_2\overset{\overset{+}{N}H_3}{\overset{|}{C}}HCO_2^-$$

(75% overall)
glutamic acid

EXERCISE 29.6 Write the equations illustrating the synthesis of aspartic acid, phenylalanine, and valine by the N-phthalimidomalonic ester method.

Other procedures similar to the foregoing are also useful. The best method utilizes the N-acetamido rather than the N-phthalimido derivative. The starting material is readily prepared from malonic ester. Treatment of the diester with nitrous acid gives a nitroso derivative, which rearranges to the oxime. Hydrogenation of the oxime in acetic anhydride solution gives acetamidomalonic ester (Section 24.6.E).

$$CH_2(COOEt)_2 + HNO_2 \longrightarrow [O{=}N{-}CH(COOEt)_2] \longrightarrow HON{=}C(COOEt)_2 \xrightarrow[Ac_2O]{H_2\text{-}Pt} CH_3\overset{O}{\overset{\|}{C}}NHCH(COOEt)_2$$

acetamidomalonic ester

The acetamidomalonic ester is alkylated, and the resulting product is hydrolyzed and decarboxylated to obtain the amino acid.

$$CH_3\overset{O}{\overset{\|}{C}}NHCH(COOEt)_2 \xrightarrow{\text{NaOEt}} \xrightarrow{\text{4-(chloromethyl)imidazole}} \xrightarrow[\Delta]{H_3O^+} \text{(imidazol-4-yl)}CH_2\overset{\overset{+}{N}H_3}{\overset{|}{C}}HCO_2^-$$

(35% overall)
histidine

$$CH_3\overset{O}{\overset{\|}{C}}NHCH(COOEt)_2 \xrightarrow{\text{NaOEt}} \xrightarrow{(CH_3)_2CHCH_2Br} \xrightarrow[\Delta]{H_3O^+} (CH_3)_2CHCH_2\overset{\overset{+}{N}H_3}{\overset{|}{C}}HCO_2^-$$

(51% overall)
leucine

EXERCISE 29.7 Write the equations illustrating the synthesis of serine, tyrosine, and valine by the N-acetamidomalonic ester method.

D. Strecker Synthesis

Another method of some generality for the preparation of α-amino acids is the hydrolysis of α-amino nitriles, which are available by the treatment of aldehydes with ammonia and HCN (**Strecker synthesis**).

$$RCHO + NH_3 + HCN \longrightarrow R\overset{NH_2}{\overset{|}{C}}HCN \xrightarrow{H_3O^+} R\overset{\overset{+}{N}H_3}{\overset{|}{C}}HCO_2^-$$

The mechanism of formation of the α-amino nitrile probably involves the addition of HCN to the imine, which is formed by condensation of the aldehyde with ammonia.

$$RCHO + NH_3 \rightleftharpoons H_2O + RCH{=}NH \xrightarrow{HCN} R\overset{NH_2}{\overset{|}{C}}HCN$$

An example of the application of the Strecker synthesis is the following preparation of phenylalanine.

$$C_6H_5CH_2CHO + NH_3 + HCN \longrightarrow C_6H_5CH_2\overset{NH_2}{\overset{|}{C}}HCN \xrightarrow[H_2O,\ \Delta]{NaOH} \xrightarrow{H_3O^+} C_6H_5CH_2\overset{\overset{+}{N}H_3}{\overset{|}{C}}HCO_2^-$$

(74%)
phenylalanine

EXERCISE 29.8 Show how you could prepare tyrosine, specifically labelled with ^{14}C in the C-1 (carboxy) position (Na ^{14}CN is available). What problem arises in application of the Strecker synthesis for the preparation of lysine?

E. Miscellaneous Methods

The foregoing methods are of general applicability for the synthesis of the simpler amino acids, either natural or unnatural. Some of the more complicated structures must be prepared in other ways. For example, the heterocyclic amino acid proline has been synthesized by the following route.

$$\text{(phthalimido)}NCH(COOEt)_2 \xrightarrow{NaOEt} \xrightarrow{Br(CH_2)_3Br} \text{(phthalimido)}N\overset{COOEt}{\underset{COOEt}{C}}CH_2CH_2CH_2Br \xrightarrow[EtOH]{NaOH}$$

$$\left[BrCH_2CH_2CH_2C(NH_2)(CO_2^-)_2 \right] \longrightarrow \left[\text{pyrrolidine-2,2-}(CO_2^-)_2 \right] \xrightarrow{H_3O^+} \text{prolinate } (\overset{+}{N}H_2,\ H,\ CO_2^-)$$

(70%)

The basic amino acid lysine has been prepared in a variety of ways. One interesting method involves application of the Schmidt reaction (Section 24.6.I) to 2-oxocyclohexanecarboxylic acid. The product is a cyclic amido acid that can be hydrolyzed to the amino dicarboxylic acid.

$$\text{2-oxocyclohexanecarboxylic acid (COOH)} \xrightarrow[H_2SO_4]{HN_3} \text{cyclic amido acid (COOH)} \xrightarrow{H_3O^+} HOOC(CH_2)_4\overset{\overset{+}{N}H_3}{CH}CO_2^-$$

A second application of the Schmidt reaction yields lysine. Fortunately, only the carboxy group that is not α to the amino group reacts; in fact, α-amino acids fail to react at all in the Schmidt reaction.

$$HOOC(CH_2)_4\overset{\overset{+}{N}H_3}{CH}CO_2^- \xrightarrow[H_2SO_4]{HN_3} H_3\overset{+}{N}(CH_2)_4\overset{NH_2}{CH}CO_2^- \quad (74\%)$$

The Schmidt reaction, introduced in Section 24.6.I as a reaction of carboxylic acids, also can be applied to ketones. It is a general method for the conversion of ketones to amides.

$$\text{cyclohexanone} \xrightarrow[H_2SO_4]{HN_3} \text{caprolactam (O, H, N)}$$

A probable mechanism for the conversion is shown below.

$$RCR(=O) \underset{}{\overset{H^+}{\rightleftharpoons}} RCR(=\overset{+}{O}H) \overset{N_3^-}{\rightleftharpoons} R{-}\underset{R}{\overset{OH}{C}}{-}\ddot{N}{-}\overset{+}{N}{\equiv}N: \longrightarrow R{-}\underset{R}{\overset{OH}{C}}{-}\ddot{N} + N_2 \longrightarrow$$

$$\left[R{-}\overset{\overset{+}{O}H}{C}{-}\ddot{N}R \longleftrightarrow R{-}\overset{OH}{C}{=}NR\right] \rightleftharpoons R{-}\overset{O}{C}{-}NHR$$

EXERCISE 29.9 Show how the Schmidt reaction could be used to prepare the natural amino acid 2-aminoadipic acid, a precursor of the important antibiotics penicillin and cephalosporin.

F. Resolution

Amino acids that are synthesized by the methods outlined in the preceding sections are obtained as racemates. It is usually desirable to have one of the two enantiomers, frequently the L-enantiomer. The trend towards chiral drugs (Chapter 16, p. 444) requires chiral synthesis or resolution. For this reason a good deal of attention has been paid to the problem of resolving racemic amino acids.

One method that may be used for the resolution of amino acids involves converting them into diastereomeric salts (Section 24.4). The amino group is usually converted into an amide so that the material is not amphoteric. For example, alanine reacts with benzoyl chloride in aqueous base to give N-benzoylalanine, which is a typical acid.

$$CH_3\overset{\overset{+}{N}H_3}{\overset{|}{C}H}CO_2^- + C_6H_5\overset{O}{\overset{\|}{C}}Cl \xrightarrow[H_2O]{OH^-} C_6H_5\overset{O}{\overset{\|}{C}}NH\overset{}{\underset{CH_3CHCOOH}{|}}$$

benzoyl chloride → N-benzoylalanine

The racemic N-benzoylalanine is resolved in the normal way (Section 24.4) with brucine or strychnine. If brucine is used, it is the brucine salt of D-alanine that is less soluble. If strychnine is used, the strychnine salt of L-alanine crystallizes. Acidification of the salts yields the D- and L-enantiomers of N-benzoylalanine. Basic hydrolysis then affords the pure enantiomeric amino acids. The process is outlined schematically as follows.

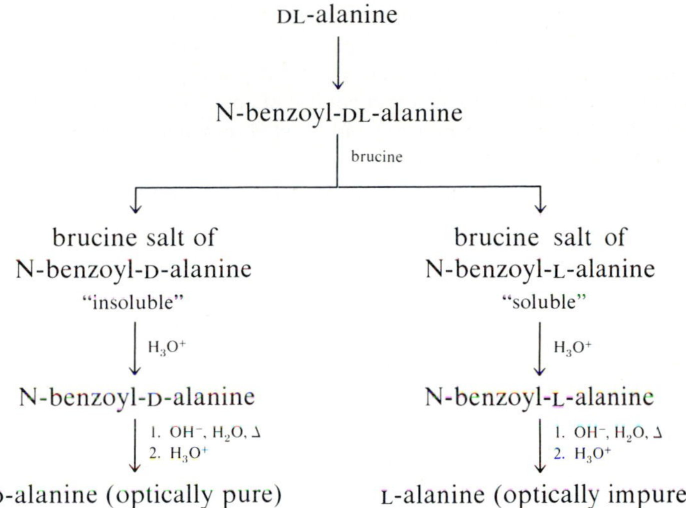

The enantiomer that forms the less soluble salt is usually obtained in an optically pure state. Since the other enantiomer is usually isolated by evaporation of the solution, it will be optically impure because some of the less soluble salt invariably remains in solution. In the case given above, the impure N-benzoyl-L-alanine may be treated with strychnine to give the insoluble strychnine salt. In this way both enantiomers may be obtained in an optically pure state.

In spite of its simplicity, the **method of diastereomeric salts** suffers from several serious drawbacks. The less soluble diastereomeric salt is usually contaminated with the other salt, and several tedious recrystallizations may be required in order to purify it. These repetitive crystallizations are wasteful of both time and material, which may often be quite valuable. There is no way to predict which chiral base will give well-defined crystals with a given amino acid or which enantiomer of the amino acid will form the less soluble salt.

Various biological procedures are much more useful for the routine large-scale resolution of amino acids. The success of biological resolution stems from the fact that organisms are generally capable of utilizing only one enantiomer of a racemic substance. Thus, if a racemic amino acid is fed to an animal or microorganism, one enantiomer is consumed. The unreacted enantiomer may then be isolated from the culture medium in the case of microorganisms or from the urine of the animal. Since L-enantiomers are utilized by almost all organisms, this method is useful for preparing optically pure D-enantiomers.

In practice, the procedure of using the whole animal for resolution is of only limited value. A more useful adaptation of the basic principle employs the use of crude enzyme preparations that catalyze some reaction on only one enantiomer. An example is the resolution of DL-leucine by *hog renal acylase,* an enzyme isolated from hog kidneys. The enzyme functions as a catalyst for the hydrolysis of amide linkages and is specific for amides of L-amino acids. For resolution, the racemic amino acid is first converted into the N-acetyl derivative, which is then incubated with a small amount of the crude enzyme preparation. The enzyme catalyzes hydrolysis of N-acetyl-L-leucine to the amino acid, leaving N-acetyl-D-leucine unchanged. The two enantiomers are easily separable, since one is acidic and the other is amphoteric.

$$\text{DL-leucine} \xrightarrow{Ac_2O} \text{N-acetyl-DL-leucine} \xrightarrow[\text{acylase}]{\text{hog renal}} \begin{cases} \text{L-leucine} \\ + \\ \text{N-acetyl-D-leucine} \end{cases}$$

A suspension of 17.3 g of N-acetyl-DL-leucine in 1 L of water is adjusted to pH 7.0 with ammonium hydroxide solution, and 0.012 g of hog renal acylase powder is added. The mixture is agitated at 38 °C for 24 hr. The mixture is acidified with 10 mL of acetic acid, filtered, and evaporated under vacuum to a volume of about 50 mL. Upon addition of ethanol, L-leucine crystallizes. The semipure amino acid is recrystallized from ethanol-water to give 5 g (80%) of optically pure L-leucine.

The filtrates from the foregoing process are acidified to pH 2 with HCl and chilled, whereupon N-acetyl-D-leucine crystallizes. One recrystallization from water gives 7 g (80%) of optically pure product. It may be hydrolyzed by refluxing with 2 N HCl to obtain pure D-leucine.

Highlight 29.2

Amino acids can be synthesized in a variety of ways, including (a) treatment of an α-bromo acid with ammonia; (b) alkylation of acetamido- or phthalimidomalonic acid; and (c) reaction of an aldehyde with ammonia and HCN **(Strecker synthesis).**

R–CH(Br)–COOH + NH_3 → R–CH(NH_3^+)–CO_2^-

AcHN–CH(COOEt)$_2$ $\xrightarrow{\text{1. base, RX; 2. } H_3O^+}$ R–CH(NH_3^+)–CO_2^-

Phthalimido–N–CH(COOEt)$_2$ $\xrightarrow{\text{1. base, RX; 2. } H_3O^+}$ R–CH(NH_3^+)–CO_2^-

RCHO $\xrightarrow{NH_3,\ HCN}$ R–CH(NH_2)–CN $\xrightarrow{H_3O^+}$ R–CH(NH_3^+)–CO_2^-

Because only one form of a chiral molecule such as an amino acid is biologically active, it is necessary to separate the two enantiomers when an amino acid is prepared by one of these methods. This process is called **resolution.** Resolution can be accomplished in two ways. One method of resolution involves the temporary conversion of the racemic amino acid into a mixture of diastereomers, which can be separated because they have different physical properties, such as solubility. The other method takes advantage of the fact that the two enantiomers have different chemical reactivity toward chiral, enantiomerically homogeneous reagents such as enzymes. A common method of resolving racemic amino acids is reaction of the enantiomeric amides with an appropriate enzyme that catalyzes the hydrolysis of only one of the two enantiomers.

$$\underset{R\text{ enantiomer}}{R\text{-}CH(NHCOCH_3)\text{-}COOH} + \underset{S\text{ enantiomer}}{R\text{-}CH(NHCOCH_3)\text{-}COOH} \xrightarrow{\textit{hog renal acylase}} \underset{R\text{ enantiomer, recovered}}{R\text{-}CH(NHCOCH_3)\text{-}COOH} + \underset{S\text{ enantiomer of product amino acid}}{R\text{-}CH(NH_3^+)\text{-}CO_2^-}$$

29.6 Reactions of Amino Acids

A. Esterification

The carboxy group of an amino acid can be esterified in the normal way. Methyl, ethyl, and benzyl esters are employed extensively as intermediates in the synthesis of peptides (Section 29.6). The methyl and ethyl esters are normally prepared by treating a suspension of the amino acid in the appropriate alcohol with anhydrous hydrogen chloride. The amino acid ester is isolated as the crystalline hydrochloride salt.

$$C_6H_5CH_2CH(\overset{+}{N}H_3)CO_2^- \xrightarrow[CH_3OH]{HCl} C_6H_5CH_2CH(\overset{+}{N}H_3\ Cl^-)COOCH_3$$

(90%)
phenylalanine methyl ester hydrochloride

Benzyl esters are often prepared using benzenesulfonic acid as the catalyst. The water produced in the reaction is removed by azeotropic distillation, thus avoiding the use of a large excess of benzyl alcohol.

$$H_3\overset{+}{N}CH_2CO_2^- + C_6H_5CH_2OH \xrightarrow{C_6H_5SO_3H} H_3\overset{+}{N}CH_2COOCH_2C_6H_5\ \ C_6H_5SO_3^-$$

(90%)
glycine benzyl ester benzenesulfonate

As we shall see later, the benzyl esters are especially useful derivatives because they can be converted back to acids by nonhydrolytic methods. For example, glycine benzyl ester reacts with hydrogen in the presence of palladium to give glycine and toluene (Section 21.6.B).

$$H_3\overset{+}{N}CH_2COOCH_2C_6H_5\ Cl^- \xrightarrow{H_2\text{-}Pd} H_3\overset{+}{N}CH_2COOH\ Cl^- + C_6H_5CH_3$$

B. Amide Formation

Acylation of the amino group in amino acids is best carried out under basic conditions, so that a substantial concentration of the free amino form is present. A typical procedure calls for treatment of a mixture of the amino acid and benzoyl chloride with concentrated aqueous sodium hydroxide. At the end of the reaction, it is necessary to acidify the aqueous solution to obtain the acidic product.

$$(CH_3)_2CH\overset{\overset{+}{N}H_3}{\overset{|}{C}}HCO_2^- + C_6H_5\overset{O}{\overset{\|}{C}}Cl \xrightarrow[\text{2 hr, 4°C}]{OH^-,\ H_2O} \xrightarrow{HCl} (CH_3)_2CH\overset{NH\overset{O}{\overset{\|}{C}}C_6H_5}{\overset{|}{C}}HCOOH$$

(80%)
N-benzoylvaline

Amides may also be prepared by reaction with acetic anhydride.

$$\text{(imidazol-4-yl)}-CH_2\overset{\overset{+}{N}H_3}{\overset{|}{C}}HCO_2^- + Ac_2O \xrightarrow[\text{2 hr}]{100°C} \text{(imidazol-4-yl)}-CH_2\overset{NH\overset{O}{\overset{\|}{C}}CH_3}{\overset{|}{C}}HCOOH$$

(80%)
N-acetylhistidine

EXERCISE 29.10 Write equations illustrating the conversion of phenylalanine into N-acetylphenylalanine methyl ester.

C. Ninhydrin Reaction

When an aqueous solution of an α-amino acid is treated with triketohydrindene hydrate (ninhydrin), a purple color is produced.

$$\text{ninhydrin hydrate (C(OH)}_2\text{)} \underset{\pm H_2O}{\rightleftharpoons} \text{ninhydrin} \qquad R-\overset{NH_3^+}{\overset{|}{C}}H-COO^- \rightleftharpoons R-\overset{NH_2}{\overset{|}{C}}H-COOH$$

$$\text{ninhydrin} + R-\overset{NH_2}{\overset{|}{C}}H-COOH \longrightarrow \text{(indandione)}=N-CH(R)-C(=O)O^- \xrightarrow{-CO_2}$$

ninhydrin

$$\text{(2-oxido-indenone)}-N=CH-R \xrightarrow[H_2O]{-RCHO} \text{ninhydrin} + NH_2-\text{(2-oxido-indenone)} \longrightarrow$$

$$\text{(indandione)}=N-\text{(3-hydroxy-indenone)}$$

purple, λ_{max} 570 nm

The reaction mechanism is straightforward, as illustrated above. The amino group of the amino acid adds to the highly reactive carbonyl group of the triketo form of ninhydrin. The resulting imine undergoes decarboxylation to give a different imine (a Schiff base, see page 398), which is hydrolyzed to an amine and an aldehyde. The amine reacts with a second ninhydrin (in the triketo form) yielding a purple imine. The formation of the purple imine is essentially quantitative and highly reliable, so much so that the overall reaction serves as the basic chemical detection method for micromole quantities of amino acids in **amino acid analysis.** Proline reacts to give an intermediate with a yellow color, with a structure probably stabilized by the contribution of an ylide form.

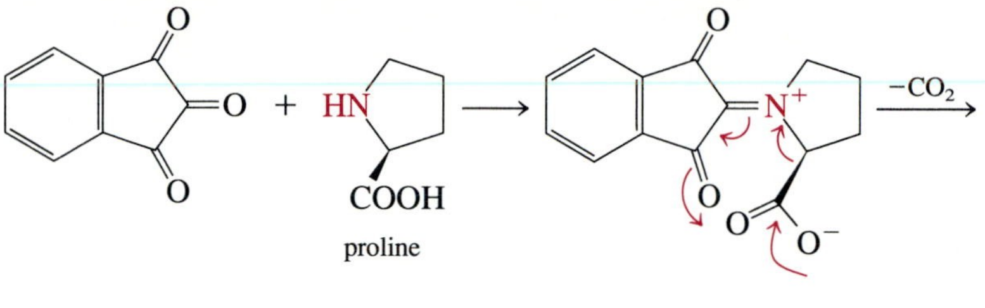

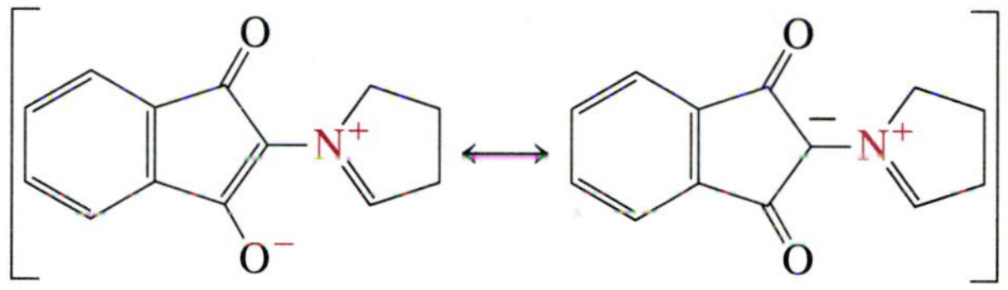

O O + HN COOH proline → O N⁺ O O O⁻ $\xrightarrow{-CO_2}$

[O N⁺ O⁻ ⟷ O N⁺ O]

yellow, λ_{max} 440 nm

29.7 Peptides

A. Structure and Nomenclature

Peptides, also called polypeptides, are amino acid polymers containing from 2 to about 50 individual units. The individual amino acids are connected by amide linkages from the amino group of one unit to the carboxy group of another. Unless a polypeptide is cyclic, it will contain a free NH_3^+ group (the N-terminal end) and a free CO_2^- group (the C-terminal end).

N-terminal amino acid C-terminal amino acid

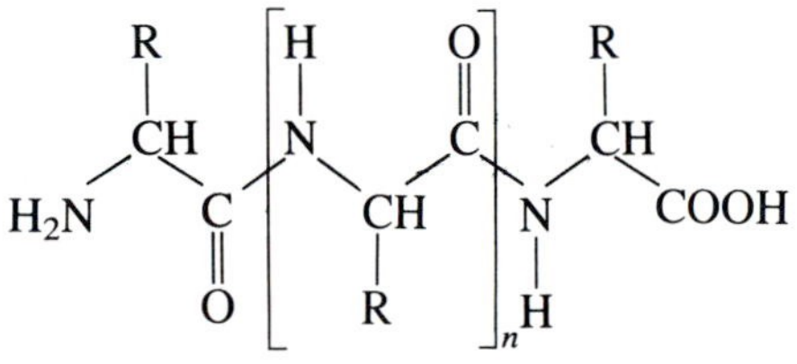

H_2N–CH(R)–C(=O)–[NH–CH(R)–C(=O)]$_n$–NH–CH(R)–COOH

a polypeptide

By convention, peptide structures are always written with the N-terminal unit on the left and the C-terminal unit on the right. They are named by prefixing the name of the C-terminal unit with the group names of the other amino acids, beginning with the N-terminal unit. Since the names tend to become rather unintelligible, a shorthand notational system is used employing either the three-letter or the one-letter codes given in Table 29.1.

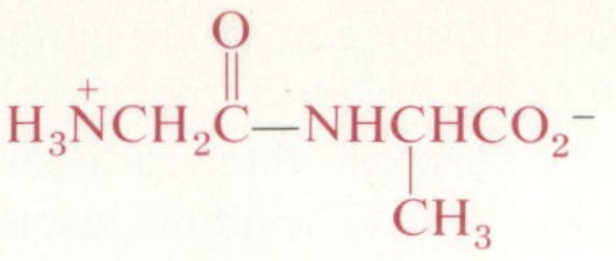

$$H_3\overset{+}{N}CH_2\overset{O}{\overset{\|}{C}}{-}NH\underset{CH_3}{\underset{|}{C}}HCO_2^-$$

glycylalanine
Gly-Ala (GA)

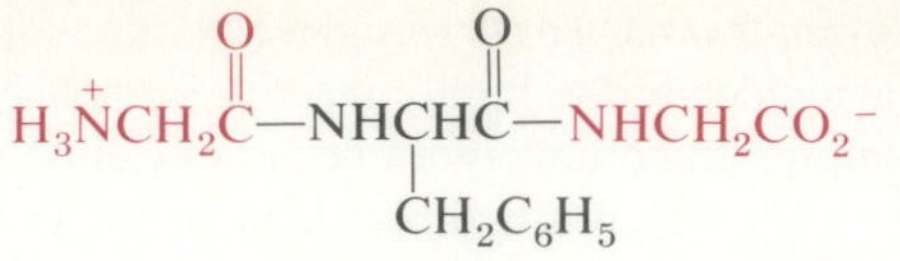

$$H_3\overset{+}{N}CH_2\overset{O}{\overset{\|}{C}}{-}NH\underset{CH_2C_6H_5}{\underset{|}{C}}H\overset{O}{\overset{\|}{C}}{-}NHCH_2CO_2^-$$

glycylphenylalanylglycine
Gly-Phe-Gly (GFG)

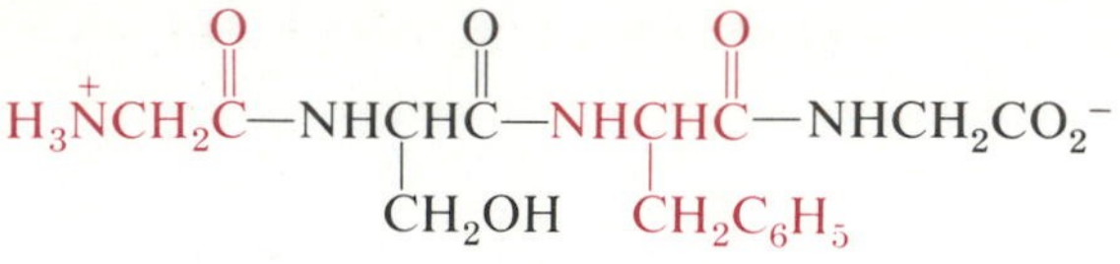

$$H_3\overset{+}{N}CH_2\overset{O}{\overset{\|}{C}}{-}NH\underset{CH_2OH}{\underset{|}{C}}H\overset{O}{\overset{\|}{C}}{-}NH\underset{CH_2C_6H_5}{\underset{|}{C}}H\overset{O}{\overset{\|}{C}}{-}NHCH_2CO_2^-$$

glycylserylphenylalanylglycine
Gly-Ser-Phe-Gly (GSFG)

The stereo structure of Gly-Phe-Gly (GFG) is shown in Figure 29.5.

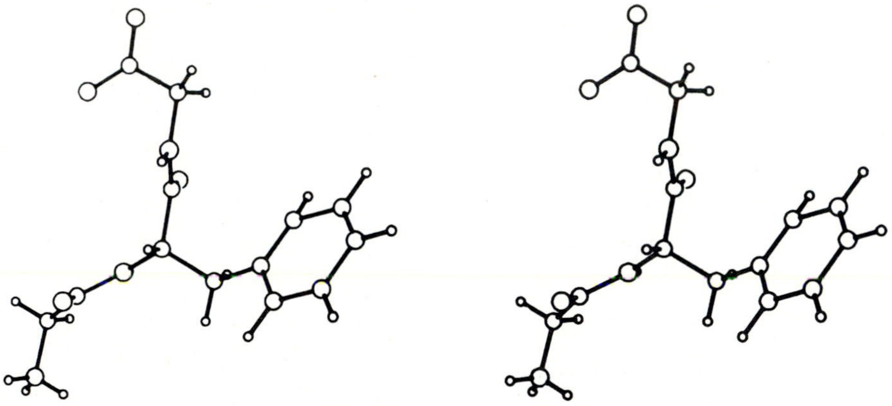

FIGURE 29.5 Stereo structure of Gly-Phe-Gly (GFG). [Reproduced with permission from *Molecular Structure and Dimensions,* International Union of Crystallography, 1972.]

Peptides are formed by partial hydrolysis of proteins, which are amino acid polymers of much higher molecular weight (more than 50 amino acid units). Upon hydrolysis of a protein, some amide linkages are broken and a complex mixture of peptides results. Complete hydrolysis gives a mixture of amino acids. Many peptides are important natural products. Neuroactive peptides are of especial interest, and some are listed in Table 29.4. An example is the nonapeptide bradykinin, which is generated from a precursor by the enzyme *kallikrein* in blood plasma under the influence of certain stimuli; part of the enzyme cascade that leads to kallikrein is related to the clotting cascade. Bradykinin has extraordinarily high pharmacological activity as the most potent **autacoid** (Gk., *autos,* self, and *akos,* medicinal agent), about ten times more potent than histamine as a vasodilator, causing a fall in blood pressure.

Arg-Pro-Pro-Gly-Phe-Ser-Pro-Phe-Arg (RPPGFSPFR)
bradykinin

The central feature of the polypeptide chain is the succession of amide linkages. Recall from our previous study (Chapter 19) that the carbon-nitrogen bond in an amide has a high degree of "double bond character" that has been attributed to delocalization of the nitro-

TABLE 29.4 Some Neuroactive Peptides

Name (no. of residues)	Amino Acid Sequence[a]
carnosine (2)	Ala-His (AH)
thyrotropin-releasing hormone (TRH) (3)	pGlu-His-ProNH$_2$ (pEHPNH$_2$)
Met-enkephalin (5)	Tyr-Gly-Gly-Phe-Met (YGGFM)
Leu-enkephalin (5)	Tyr-Gly-Gly-Phe-Leu (YGGFL)
angiotensin II (8)	Asp-Arg-Val-Tyr-Ile-His-Pro-PheNH$_2$ (DRVYIHPFNH$_2$)
cholecystokinin-like peptide (9)	Asp-Tyr(SO$_3$H)-Met-Gly-Trp-Met-Asp-PheNH$_2$ (DY(SO$_3$H)MGWMDPNH$_2$)
oxytocin (9)	Ile-Tyr-Cys / Gln-Asn-Cys-Pro-Leu-GlyNH$_2$ (IYC / QNCPLGNH$_2$)
vasopressin (9)	Phe-Tyr-Cys / Gln-Asn-Cys-Pro-Arg-GlyNH$_2$ (FYC / QNCPRGNH$_2$)
luteinizing-hormone-releasing hormone (LHRH) (10)	pGlu-His-Trp-Ser-Tyr-Gly-Leu-Arg-Pro-GlyNH$_2$ (pEHWSYGLRPGNH$_2$)
neurotensin (13)	pGlu-Leu-Tyr-Glu-Asn-Lys-Pro-Arg-Arg-Pro-Tyr-Ile-Leu (pELYGNKPRRPYIL)
bombesin (14)	pGlu-Gln-Arg-Leu-Gly-Asn-Gln-Trp-Ala-Val-Gly-His-Leu-MetNH$_2$ (pEQRLGNQWAVGHLMNH$_2$)
somatostatin (14)	Ala-Gly-Cys-Lys-Asn-Phe-Phe-Trp / Cys-Ser-Thr-Phe-Thr-Lys (AGCKNFFW / CSTFTK)
vasoactive intestinal polypeptide (VIP) (28)	His-Ser-Asp-Ala-Val-Phe-Thr-Asp-Asn-Tyr-Thr-Arg-Leu-Arg / Asn-Leu-Ile-Ser-Asn-Leu-Tyr-Lys-Lys-Val-Ala-Met-Gln-Lys / NH$_2$ (HSDAVFTDNYTRLRKQMAVKKYLNSILNNH$_2$)
β-endorphin (31)	Tyr-Gly-Gly-Phe-Met-Thr-Ser-Glu-Lys-Ser-Gln-Thr-Pro-Leu / Lys-His-Ala-Asn-Lys-Val-Ile-Ala-Asn-Lys-Phe-Leu-Thr-Val / Lys-Gly-Gln (YGGFMTSEKSQTPLVTLFKNAIVKNAHKKGQ)
ACTH (corticotropin) (39)	Ser-Tyr-Ser-Met-Glu-His-Phe-Arg-Tyr-Gly-Lys-Pro-Val-Gly / Glu-Ala-Gly-Asp-Pro-Tyr-Val-Lys-Val-Pro-Arg-Arg-Lys-Lys / Asp-Glu-Leu-Ala-Glu-Ala-Phe-Pro-Leu-Glu-PheNH$_2$ (SYSMEHFRYGKPVGKKRRPVKVYPDGAEDELAEAFPLEFNH$_2$)

[a]pE-pyroglutamic acid = ; NH$_2$ at end = amide; C—C bond in line is peptide; not in line is S—S, i.e., linked by a disulfide bond (nicotinic acetylcholine receptor α-subunit has adjacent C—C, which are *also* linked by S—S); SO$_3$H is O-sulfate.

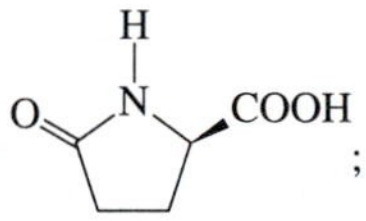

gen lone pair into the carbonyl group and that reduces the basicity of the nitrogen and causes restricted rotation about the carbon-nitrogen bond.

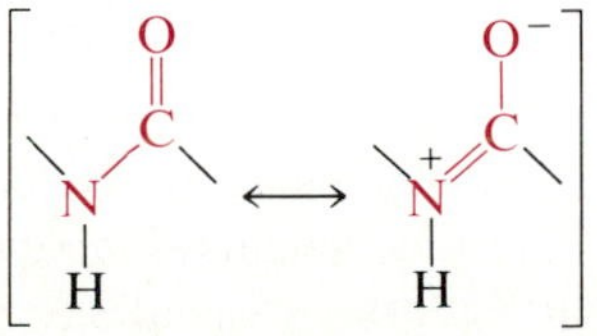

The restricted rotation has an important effect on the three-dimensional structure of proteins, as we shall see later.

The only other type of covalent bond between amino acids in proteins and peptides is the disulfide linkage between two cysteine units.

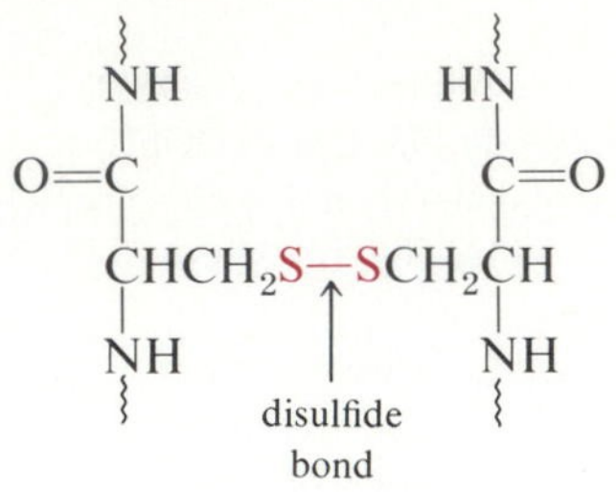

```
   ⌇                  ⌇
   NH                 HN
   |                  |
 O=C                  C=O
   |                  |
   CHCH2S—SCH2CH
   |     ↑            |
   NH    |            NH
   ⌇   disulfide      ⌇
         bond
```

Recall that disulfides, R—S—S—R, are formed by the mild oxidation of thiols (page 810). The disulfide linkage is easily reduced to regenerate the thiols. The tripeptide thiol, glutathione (γ-Glu-Cys-Gly, GSH), is a ubiquitous component of most cells on all levels of evolution. Note that this tripeptide, γ-ECG, is commonly abbreviated as GSH, but generally in a context that makes confusion with Gly-Ser-His unlikely. In addition to participation in a number of reactions and as one component of leukotrienes (page 548), the thiol group of glutathione serves as free-radical trapping agent (page 811). The disulfide, GSSG, is formed as a result of this intervention; in the cell, GSSG is reduced to GSH by the enzyme glutathione disulfide reductase and the cofactor dihydronicotinamide adenine dinucleotide phosphate [NADPH].

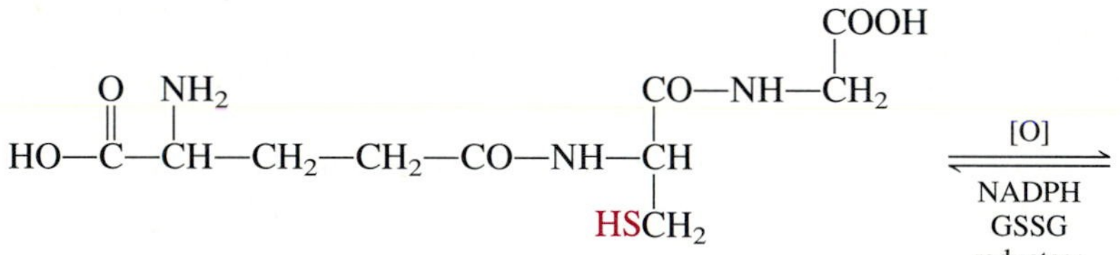

```
                                             COOH
                                             |
   O  NH2                           CO—NH—CH2
   ‖  |                             |                     [O]
HO—C—CH—CH2—CH2—CO—NH—CH                           ⇌
                                    |                   NADPH
                                  HSCH2                  GSSG
                                                       reductase
```

glutathione, γ-Glu-Cys-Gly (γ-ECG) [GSH]

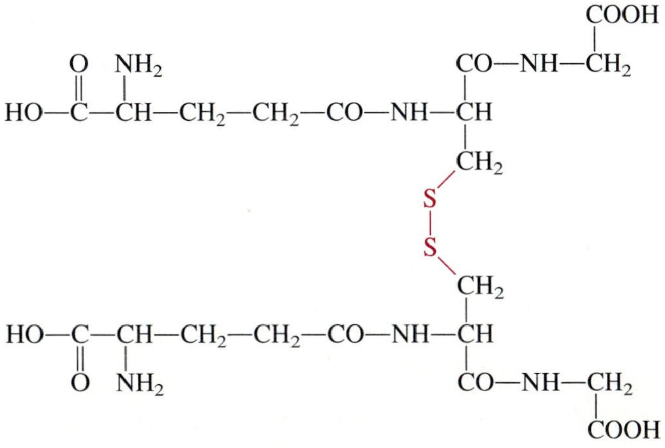

```
                                             COOH
                                             |
   O  NH2                           CO—NH—CH2
   ‖  |                             |
HO—C—CH—CH2—CH2—CO—NH—CH
                                    |
                                    CH2
                                   /
                                  S
                                  |
                                  S
                                   \
                                    CH2
                                    |
HO—C—CH—CH2—CH2—CO—NH—CH
   ‖  |                             |
   O  NH2                           CO—NH—CH2
                                             |
                                             COOH
```

glutathione disulfide [GSSG]

When such a disulfide bond occurs between two cysteine residues in the same chain, a ''loop'' results, as in the posterior pituitary hormone oxytocin (Table 29.4). If the cysteine units are in different chains, the disulfide link may bind the two chains together, as in the A and B chains of insulin (Figure 29.6).

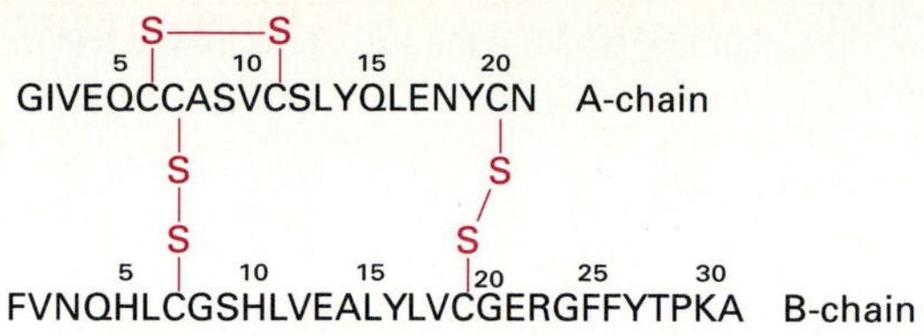

GIVEQCCASVCSLYQLENYCN A-chain

FVNQHLCGSHLVEALYLVCGERGFFYTPKA B-chain

FIGURE 29.6 Amino acid sequence and disulfide bridges of bovine insulin. The N-terminal units are at the left and the C-terminal units are at the right. Both C-terminal units occur as amides, $CONH_2$.

Like the simpler amino acids, peptides are amphoteric compounds, since they usually still contain a free α-amino and a free α-carboxy group; they exist as zwitterions. The pK_as for the two functions in a few simple peptides are listed in Table 29.5. Also included are the isoelectric points, pH_I, the pH at which the peptide is least soluble in aqueous solution.

TABLE 29.4 pK_a Values for Some Peptides

Peptide	pK_1 COOH	pK_2 NH_3	Isoelectric Point, pH_I
Gly-Gly	3.14	8.25	5.70
Gly-Ala	3.15	8.23	5.69
Ala-Gly	3.17	8.18	5.68
Gly-Gly-Gly	3.23	8.09	5.66
Ala-Ala-Ala-Ala	3.42	7.94	5.68

EXERCISE 29.11 Write the structure of the tetrapeptide Val-Phe-Ser-Leu (VFSL).

B. Synthesis of Peptides

The simplest method for the synthesis of peptides is the polymerization of an amino acid. The resulting **homopolymer** is a mixture of peptides of variable chain length. Such homopolymers are not found in nature, but the synthetic ones have been useful in understanding some of the physical and spectral properties of proteins.

$$H_3\overset{+}{N}CH_2CO_2^- \longrightarrow H_3\overset{+}{N}CH_2\overset{O}{\overset{\|}{C}}\left(NHCH_2\overset{O}{\overset{\|}{C}}\right)_n NHCH_2CO_2^-$$

polyglycine

The first product formed when two amino acids condense is a dipeptide. The terminal amino and carboxy groups are now situated so that they can interact to form a six-membered ring diamide. We have seen previously that intramolecular reactions to form five- and six-membered rings are frequently much faster than their intermolecular analogs (Sections 15.2, 27.5.C). Thus, when glycine is heated, the cyclic dimer 2,5-diketopiperazine is produced.

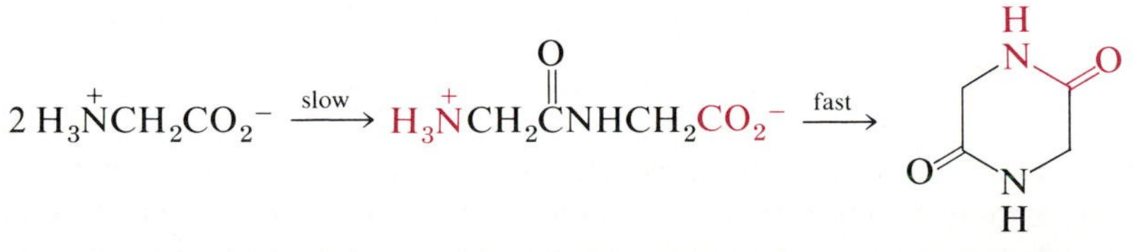

$$2\,H_3\overset{+}{N}CH_2CO_2^- \xrightarrow{\text{slow}} H_3\overset{+}{N}CH_2\overset{O}{\overset{\|}{C}}NHCH_2CO_2^- \xrightarrow{\text{fast}}$$

2,5-diketopiperazine
"glycine anhydride"

Piperazine is a heterocyclic diamine, which is numbered as shown. It is the nitrogen analog of 1,4-dioxane (Highlight 27.2 and page 215).

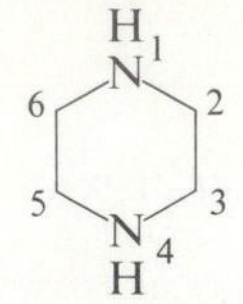

piperazine

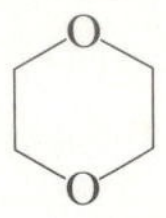

1,4-dioxane

EXERCISE 29.12 In 1914 Maillard reported a study of the polymerization of glycine. The amino acid was heated in glycerol solution. The main product of the reaction was found to be 2,5-diketopiperazine. A polypeptide fraction was produced in low yield. The predominant peptides in this fraction were the even peptides tetraglycine and hexaglycine. Explain.

Hydrolysis of one of the amide bonds in a 2,5-diketopiperazine is one method for preparing simple dipeptides.

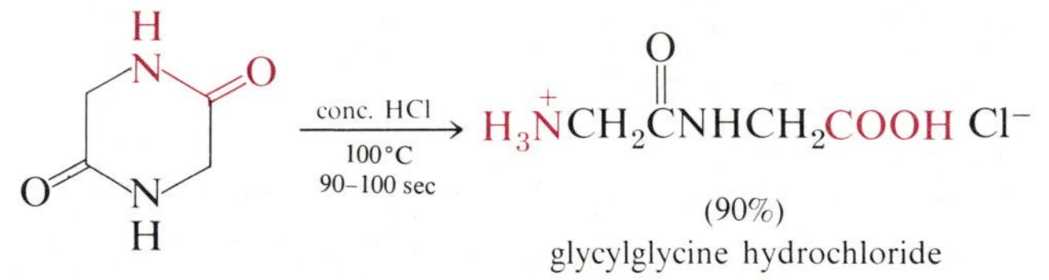

$$\xrightarrow[\text{100°C, 90–100 sec}]{\text{conc. HCl}} H_3\overset{+}{N}CH_2\overset{O}{\overset{\|}{C}}NHCH_2COOH\ Cl^-$$

(90%)
glycylglycine hydrochloride

The rational synthesis of peptides is a challenging task that has only been solved in the past few decades. In order to illustrate the difficulty, consider the synthesis of the simple dipeptide glycylalanine (Gly-Ala) from glycine and alanine. The problem is to form an amide linkage between the carboxy group of glycine and the amino group of alanine.

$$H_3\overset{+}{N}CH_2CO_2^- + H_3\overset{+}{N}\underset{}{\overset{CH_3}{\overset{|}{C}}}HCO_2^- \dashrightarrow H_3\overset{+}{N}CH_2\overset{O}{\overset{\|}{C}}NH\overset{CH_3}{\overset{|}{C}}HCO_2^-$$

The normal method for converting a carboxylic acid into an amide is to activate the carboxy group by converting it to an acyl halide and then to add the amine.

$$RCOOH \xrightarrow{SOCl_2} R\overset{O}{\overset{\|}{C}}Cl \xrightarrow{R'NH_2} R\overset{O}{\overset{\|}{C}}NHR'$$

But an amino acid cannot be converted into an acyl halide; polymerization would result. Another possibility would be the direct formation of the amide link by treatment of a mixture of the two amino acids with some dehydrating agent to remove the water produced. However, such a direct approach will give a mixture of four different dipeptides. Furthermore, each of these dipeptides can react further to give higher peptides.

$$\text{glycine} + \text{alanine} \xrightarrow{-H_2O} \text{Gly-Gly} + \text{Ala-Ala} + \text{Gly-Ala} + \text{Ala-Gly}$$

An additional complication arises for amino acids that have other reactive functional groups.

The general method that has been developed to avoid these difficulties involves the use of **protecting groups** (Section 16.4). Protecting groups have been developed for both the amino and carboxy groups, as well as for the other groups that occur in the side chains of the various amino acids. A suitable protecting group must fulfill several criteria.

1. The protecting group must be easy to introduce into the molecule.
2. It must protect the functional group under conditions of amide formation.
3. It must be removable under conditions that leave the newly created amide link intact.

Carboxy groups are normally protected by conversion into the methyl, ethyl, or benzyl ester. Since esters are hydrolyzed more easily than amides, the protecting group can be removed by alkaline hydrolysis.

$$\sim\overset{\overset{\text{O}}{\|}}{\text{C}}\text{NH}\underset{\underset{\text{R}}{|}}{\text{C}}\text{HCOOCH}_3 \xrightarrow[\text{H}_2\text{O}]{\text{OH}^-} \xrightarrow{\text{H}_3\text{O}^+} \sim\overset{\overset{\text{O}}{\|}}{\text{C}}\text{NH}\underset{\underset{\text{R}}{|}}{\text{C}}\text{HCOOH} + \text{CH}_3\text{OH}$$

Benzyl esters can be cleaved by hydrogenolysis (Section 21.6.B).

$$\sim\overset{\overset{\text{O}}{\|}}{\text{C}}\text{NH}\underset{\underset{\text{R}}{|}}{\text{C}}\text{HCOOCH}_2\text{—C}_6\text{H}_5 \xrightarrow{\text{H}_2\text{–Pd/C}} \sim\overset{\overset{\text{O}}{\|}}{\text{C}}\text{NH}\underset{\underset{\text{R}}{|}}{\text{C}}\text{HCOOH} + \text{CH}_3\text{—C}_6\text{H}_5$$

Of the many amino protecting groups that have been developed, we shall discuss only two, the benzyloxycarbonyl (**"carbobenzoxy," Cbz**) and the ***t*-butoxycarbonyl (Boc)** groups. The benzyloxycarbonyl group is introduced by treating the amino acid with benzyl chloroformate in alkaline solution.

$$\overset{+}{\text{H}_3\text{N}}\text{CH}_2\text{CO}_2^- + \text{C}_6\text{H}_5\text{—CH}_2\text{O}\overset{\overset{\text{O}}{\|}}{\text{C}}\text{Cl} \xrightarrow[\text{H}_2\text{O, 5°C, 30 min}]{\text{NaOH}} \xrightarrow{\text{H}_3\text{O}^+} \text{C}_6\text{H}_5\text{—CH}_2\text{O}\overset{\overset{\text{O}}{\|}}{\text{C}}\text{NHCH}_2\text{COOH}$$

glycine — benzyl chloroformate — (70–80%) benzyloxycarbonylglycine Cbz-Gly

> Benzyl chloroformate is the half benzyl ester, half acyl chloride of carbonic acid. It is prepared by treating benzyl alcohol with phosgene.
>
> $$\text{C}_6\text{H}_5\text{—CH}_2\text{OH} + \text{COCl}_2 \longrightarrow \text{C}_6\text{H}_5\text{—CH}_2\text{O}\overset{\overset{\text{O}}{\|}}{\text{C}}\text{Cl}$$
>
> (95–99%)

The new carbon-nitrogen linkage in a benzyloxycarbonyl amino acid is part of a carbamate grouping (Section 24.6.I). Like amides, carbamates hydrolyze with difficulty. However, the benzyl-oxygen bond can be cleaved by catalytic hydrogenolysis, yielding the unstable carbamic acid, which undergoes decarboxylation (Section 24.6.I).

$$\text{C}_6\text{H}_5\text{—CH}_2\text{O}\overset{\overset{\text{O}}{\|}}{\text{C}}\text{NH}\underset{\underset{\text{R}}{|}}{\text{C}}\text{HCOOH} \xrightarrow{\text{H}_2\text{–Pd}}$$

$$\text{C}_6\text{H}_5\text{—CH}_3 + \left[\text{HO}\overset{\overset{\text{O}}{\|}}{\text{C}}\text{NH}\underset{\underset{\text{R}}{|}}{\text{C}}\text{HCOOH}\right] \longrightarrow \text{CO}_2 + \overset{+}{\text{H}_3\text{N}}\underset{\underset{\text{R}}{|}}{\text{C}}\text{HCO}_2^-$$

The *t*-butoxycarbonyl group is introduced by treating the amino acid with *t*-butoxycarbonyloximinophenylacetonitrile ("Boc-On"). The latter compound may be purchased or prepared via the oximinophenylacetonitrile from phenylacetonitrile and methyl nitrite, followed by conversion to the chloroformyl derivative with phosgene and reaction with *t*-butyl alcohol in the presence of base.

t-butoxycarbonyloximino-phenylacetonitrile "Boc-On"

(100%)
t-butoxycarbonylproline
Boc-Pro

The *t*-butoxycarbonyl group is removed by treating the protected amino acid or peptide with anhydrous acid, such as trifluoroacetic acid or hydrogen chloride in acetic acid.

$$t\text{-BuOC(=O)NHCH(R)COOH} \xrightarrow{\text{HCl}} \text{H}_3\overset{+}{\text{N}}\text{CH(R)COOH Cl}^-$$

The initial reaction is cleavage of the alkyl-oxygen bond to give the relatively stable *t*-butyl cation and a carbamic acid. The resulting carbamic acid then decarboxylates, giving the amine.

$$(\text{CH}_3)_3\text{COC}(=\overset{+}{\text{O}}\text{H})\text{NHR} \longrightarrow (\text{CH}_3)_3\text{C}^+ + \text{HOC(=O)NHR}$$

$$\text{HOC(=O)NHR} \longrightarrow \text{CO}_2 + \text{H}_2\text{NR}$$

The most generally useful coupling reagent is **dicyclohexylcarbodiimide (DCC),** a commercially available reagent that is prepared from cyclohexylamine and carbon disulfide by the route indicated below.

$$2\ \text{C}_6\text{H}_{11}\text{–NH}_2 + \text{CS}_2 \longrightarrow \text{C}_6\text{H}_{11}\text{–NHC(=S)NH–C}_6\text{H}_{11} \xrightarrow{\text{HgO}} \text{C}_6\text{H}_{11}\text{–N=C=N–C}_6\text{H}_{11}$$

(86%)
dicyclohexylcarbodiimide
DCC

Dicyclohexylcarbodiimide is an effective catalyst for condensation of carboxylic acids with alcohols and amines. It functions by activating the free carboxy group of the N-protected amino acid. An equimolar mixture of a carboxylic acid, an amine, and DCC results in formation of the corresponding amide and the highly insoluble N,N′-dicyclohexylurea.

$$\text{RCOOH} + \text{R'NH}_2 + \text{C}_6\text{H}_{11}\text{N=C=NC}_6\text{H}_{11} \longrightarrow \text{RC(=O)NHR'} + \text{C}_6\text{H}_{11}\text{NHC(=O)NHC}_6\text{H}_{11}$$

N,N′-dicyclohexylurea

The probable mechanism for the DCC coupling reaction is outlined as follows. Addition of the carboxylic acid to the diimide gives the ester of isourea, an O-acylisourea.

$$\mathrm{R\overset{O}{\overset{\|}{C}}OH + R'N{=}C{=}NR' \longrightarrow R{-}\overset{O}{\overset{\|}{C}}{-}O{-}\overset{NHR'}{\overset{|}{C}}{=}NR'}$$

an O-acylisourea

The intermediate O-acylisourea is an *activated carboxylic acid derivative* similar in reactivity to an anhydride or an acyl halide. Nucleophilic substitution by the amine yields the amide and the dialkylurea.

$$\mathrm{R\overset{O}{\overset{\|}{C}}{-}O{-}\overset{NHR'}{\overset{|}{C}}{=}NR' + R''NH_2 \rightleftharpoons \left[R\overset{OH}{\overset{|}{\underset{NHR''}{\underset{|}{C}}}}{-}O{-}\overset{NHR'}{\overset{|}{C}}{=}NR'\right] \rightleftharpoons}$$

$$\mathrm{\left[R\overset{O^-}{\overset{|}{\underset{NHR''}{\underset{|}{C}}}}{-}O{-}\overset{NHR'}{\overset{|}{C}}{=}\overset{+}{N}HR'\right] \longrightarrow R\overset{O}{\overset{\|}{C}}NHR'' + R'NH\overset{O}{\overset{\|}{C}}NHR'}$$

An example of the synthesis of a dipeptide utilizing this method is the synthesis of threonylalanine (Thr-Ala) from benzyloxycarbonylthreonine and alanine benzyl ester.

$$\mathrm{\underset{\text{N-protected threonine}}{CbzNH\underset{\underset{CH_3}{|}}{\underset{CHOH}{\underset{|}{C}}}HCOOH} + \underset{\text{C-protected alanine}}{H_2N\underset{CH_3}{\underset{|}{C}}HCOOCH_2C_6H_5} \xrightarrow{DCC} \underset{\text{Cbz-Thr-Ala-CH}_2\text{C}_6\text{H}_5}{CbzNH\underset{\underset{CH_3}{|}}{\underset{CHOH}{\underset{|}{C}}}H\overset{O}{\overset{\|}{C}}NH\underset{CH_3}{\underset{|}{C}}HCOOCH_2C_6H_5} \xrightarrow[HOAc]{H_2\text{-}Pd/C}}$$

$$\mathrm{\underset{\text{Thr-Ala}}{H_3\overset{+}{N}\underset{\underset{CH_3}{|}}{\underset{CHOH}{\underset{|}{C}}}H\overset{O}{\overset{\|}{C}}NH\underset{CH_3}{\underset{|}{C}}HCO_2^-} + 2\ C_6H_5CH_3 + CO_2}$$

Thus far we have discussed peptide synthesis only with amino acids containing no other reactive groups. When there is another functional group present in the molecule, it too must be protected until after the peptide has been formed. Typical protecting groups are benzyloxycarbonyl for the second amino group in lysine and benzyl for the sulfur in cysteine.

$$\mathrm{C_6H_5{-}CH_2O\overset{O}{\overset{\|}{C}}NH(CH_2)_4\overset{NH_3^+}{\overset{|}{C}}HCO_2^-}$$

ε-benzyloxycarbonyllysine

$$\mathrm{C_6H_5{-}CH_2SCH_2\overset{NH_3^+}{\overset{|}{C}}HCO_2^-}$$

S-benzylcysteine

Both protecting groups are removable by cleavage with anhydrous acids such as hydrogen bromide in acetic acid. The second carboxy group in aspartic acid or glutamic acid is usually protected as a methyl or benzyl ester.

A development that has revolutionized peptide synthesis is the **solid phase technique** introduced by R. B. Merrifield of Rockefeller University. In the Merrifield method the peptide or protein is synthesized throughout a swollen cross-linked polymer network that is insoluble and can be recovered by filtration. The polymer used is polystyrene (Section 36.4) in which some of the benzene rings are substituted by —CH_2Cl groups. The polystyrene used is cross-linked with about 1% of divinylbenzenes. The particle sizes range from 20 to 70 μ in diameter.

$$\cdots CH_2-CH(C_6H_5)-CH_2-CH(C_6H_4CH_2Cl)-CH_2-CH(C_6H_5)-CH_2-CH(C_6H_5)-CH_2-CH(C_6H_4CH_2Cl)\cdots$$

chloromethylated polystyrene

Typically, about one out of every 10–100 phenyl groups is chloromethylated.

The C-terminal amino acid of the desired peptide is bound to the polymer by shaking a solution of the N-protected amino acid salt in an organic solvent such as DMF with the insoluble polymer. The product is an amino acid ester in which the alkoxy group of the ester is the polymer itself.

$$t\text{-BuOC(=O)NHCH(R)CO}^- + ClCH_2\text{—}\boxed{\text{polymer}} \longrightarrow t\text{-BuOC(=O)NHCH(R)C(=O)OCH}_2\text{—}\boxed{\text{polymer}} + Cl^-$$

polymer-bound, N-protected amino acid

Excess reagents are removed by filtration, the insoluble polymer-bound amino ester is washed and the Boc group is removed by treatment with acid.

$$t\text{-BuOC(=O)NHCH(R)C(=O)OCH}_2\text{—}\boxed{\text{polymer}} \xrightarrow{H^+} \xrightarrow{Et_3N} H_2NCH(R)C(=O)OCH_2\text{—}\boxed{\text{polymer}}$$

polymer-bound amino acid

A solution of an N-protected amino acid is then added with DCC, and the heterogeneous mixture is shaken until coupling is complete.

$$t\text{-BuOC(=O)NHCH(R')COOH} + H_2NCH(R)C(=O)OCH_2\text{—}\boxed{\text{polymer}} \xrightarrow{DCC}$$

N-protected amino acid　　polymer-bound amino acid

$$t\text{-BuOC(=O)NHCH(R')C(=O)NHCH(R)C(=O)OCH}_2\text{—}\boxed{\text{polymer}}$$

N-protected, polymer-bound dipeptide

The polymer, now bound to an N-protected dipeptide, is again filtered and washed, and a strong anhydrous acid, usually trifluoroacetic acid, is added to remove the protecting group.

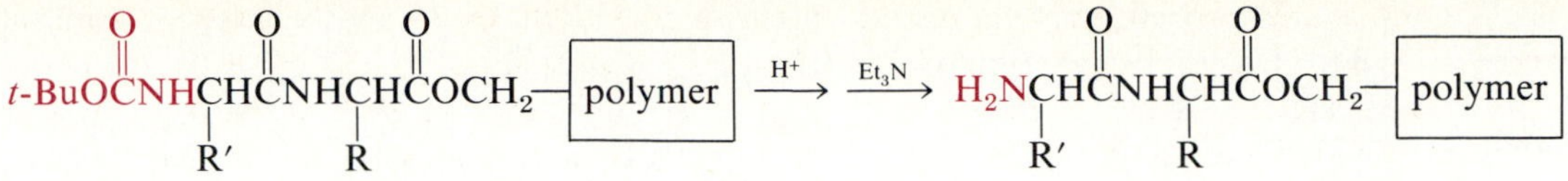

$$t\text{-BuOCNHCHCNHCHCOCH}_2\text{—[polymer]} \xrightarrow{H^+} \xrightarrow{Et_3N} H_2NCHCNHCHCOCH_2\text{—[polymer]}$$

(with C=O on each carbonyl carbon; side chains R′ and R)

polymer-bound dipeptide

The process can be repeated to add the third amino acid, and so on. At the end of the synthesis, the peptide is removed from the resin by treatment with anhydrous hydrogen fluoride. At the same time, all side-chain protecting groups are also removed. This final cleavage step does not affect the amide linkages of the peptide chain. The synthetic peptide is then purified by a suitable chromatographic method.

The great advantages of the solid-phase technique are the ease of operation and the high overall yield. Since the growing peptide chain is bound to the highly insoluble polystyrene resin, no mechanical losses are entailed in the intermediate isolation and purification stages. Furthermore, since the method involves the repetitive use of a small number of similar operations, the synthesis is easily automated. Almost all synthetic peptides are now made by the solid-phase technique.

EXERCISE 29.13 Write out all of the steps in a rational synthesis of the pentapeptide Ala-Val-Phe-Ala-Ala (AVFAA). As N-protecting groups use Cbz and for coupling use DCC. Assuming a yield of 95% in each step in your synthesis, what is the overall yield, based on the starting amino acid?

Peptides are important, mostly in their use as biological agents. The two-chain 51-amino acid polypeptide insulin is a central element in the control of glucose balance and metabolism in mammals. Individuals who lack insulin have high blood sugar and suffer from the disease diabetes. Human insulin is favored for diminishing blood glucose in diabetic patients with a minimum of side effects. Although biological systems are used for the commercial preparation, an alternative is to convert porcine (pig) insulin into human insulin by chemical means. The complete formula for bovine insulin is given in Figure 29.6. The chemical steps are illustrated here with partial formulas using three-letter codes for the amino acids.

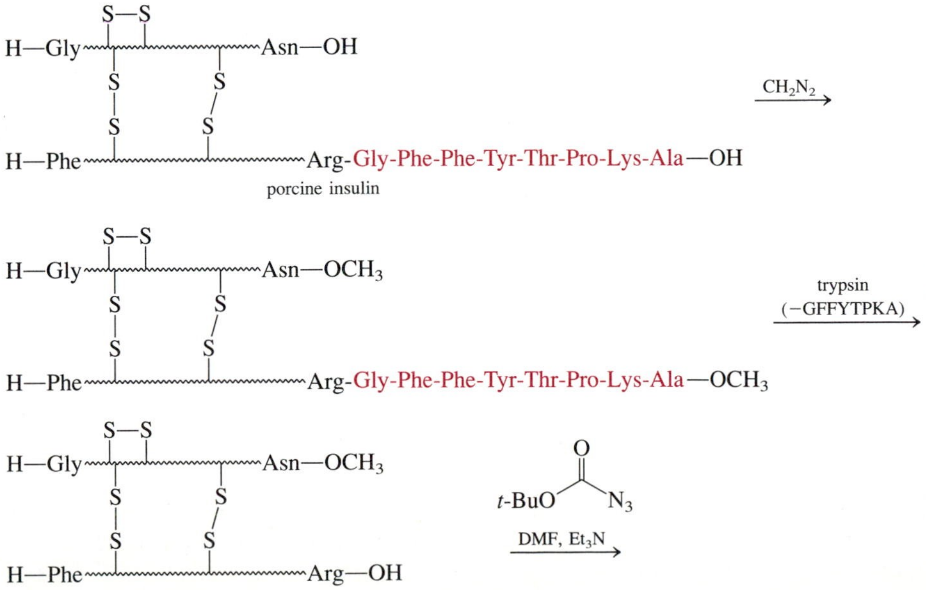

S—S

H—Gly~~~~Asn—OH

S S

S S

H—Phe~~~~Arg-Gly-Phe-Phe-Tyr-Thr-Pro-Lys-Ala—OH

porcine insulin

$\xrightarrow{CH_2N_2}$

S—S

H—Gly~~~~Asn—OCH_3

S S

S S

H—Phe~~~~Arg-Gly-Phe-Phe-Tyr-Thr-Pro-Lys-Ala—OCH_3

$\xrightarrow[\text{(−GFFYTPKA)}]{\text{trypsin}}$

S—S

H—Gly~~~~Asn—OCH_3

S S

S S

H—Phe~~~~Arg—OH

t-BuOC(=O)N_3

$\xrightarrow{DMF,\ Et_3N}$

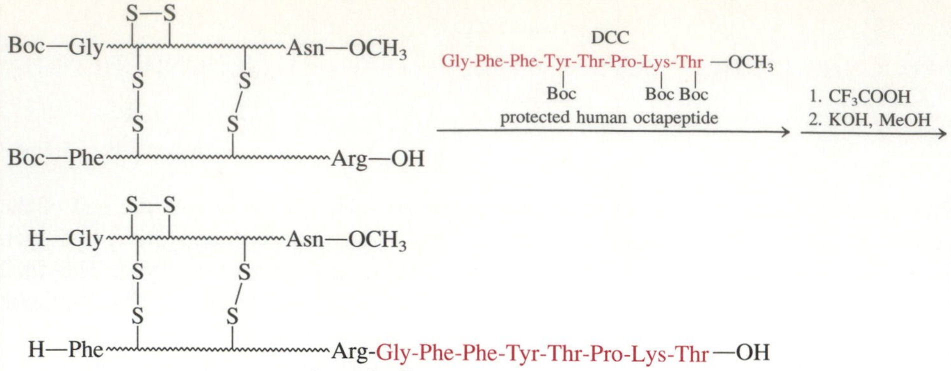

Boc—Gly~~~Asn—OCH_3 / Boc—Phe~~~Arg—OH (S—S linked) + DCC, Gly-Phe-Phe-Tyr-Thr-Pro-Lys-Thr —OCH_3 (Boc, Boc Boc) protected human octapeptide → 1. CF_3COOH 2. KOH, MeOH →

H—Gly~~~Asn—OCH_3 / H—Phe~~~Arg-Gly-Phe-Phe-Tyr-Thr-Pro-Lys-Thr—OH

human insulin

The use of artificial sweeteners diminishes sugar intake and therefore helps patients control blood sugar. They have also come into widespread use as sugar replacements in products such as soft drinks. One of the earliest artificial sweeteners was saccharin, which is still in use. Another was sodium cyclamate, which has largely been abandoned because of concerns about its possible carcinogenicity. The dipeptide L-aspartyl-L-phenylalanine methyl ester is also a very sweet substance. This dipeptide is marketed under the trade name Aspartame. Saccharin, sodium cyclamate, and Aspartame are 500, 30, and 160 times as sweet as sucrose in dilute aqueous solutions. Although it is clear that the sensation of sweetness results from the interaction of substances with certain cellular receptors, it is not clear from the dissimilar structures of saccharin, cyclamate, sucrose, and Aspartame why all should elicit such similar biological responses. Consequently, much research is currently being devoted to understanding the biochemical mechanism of sweetness.

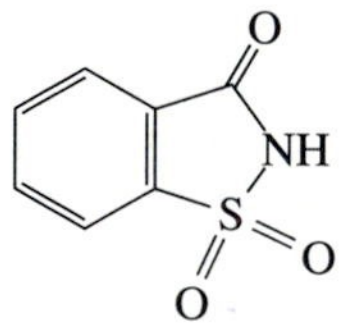

saccharin

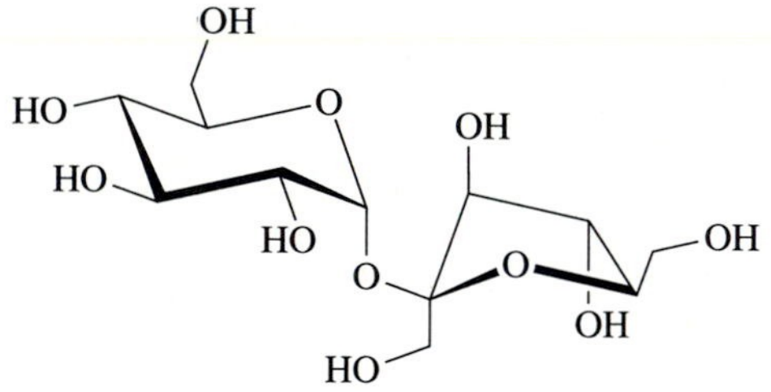

sucrose

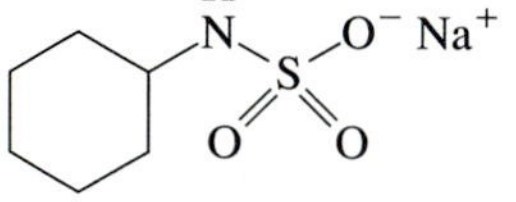

sodium cyclamate

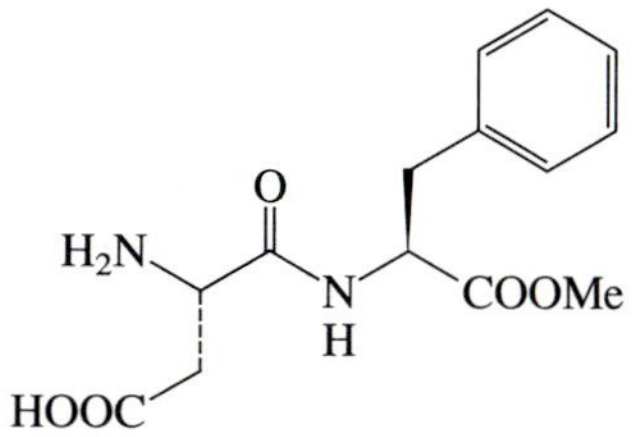

Aspartame

Highlight 29.3

The synthesis of a peptide presents a challenge in reaction selectivity because each of the reactants has both an amino and a carboxy group. In the formation of a peptide bond between two different amino acids, there are four possible combinations; the problem is to find a way that only one of these very similar reactions will occur.

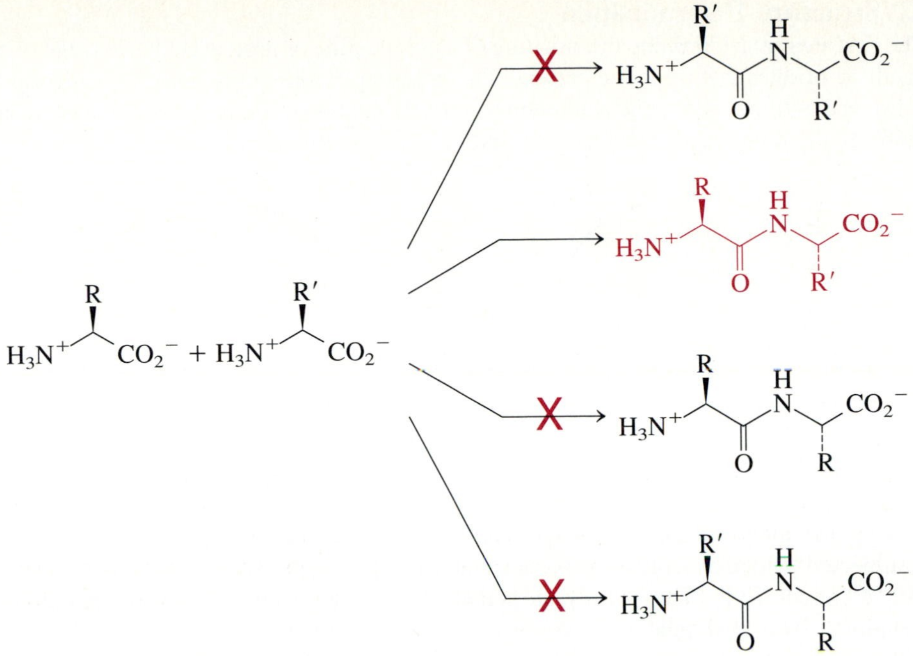

This is accomplished through the use of protecting groups. One of the reactants is converted into an amino derivative, usually the *t*-butoxycarbonyl (Boc) derivative. The other is used as an ester. The two reactants are then coupled and the protecting groups are removed to obtain the desired peptide.

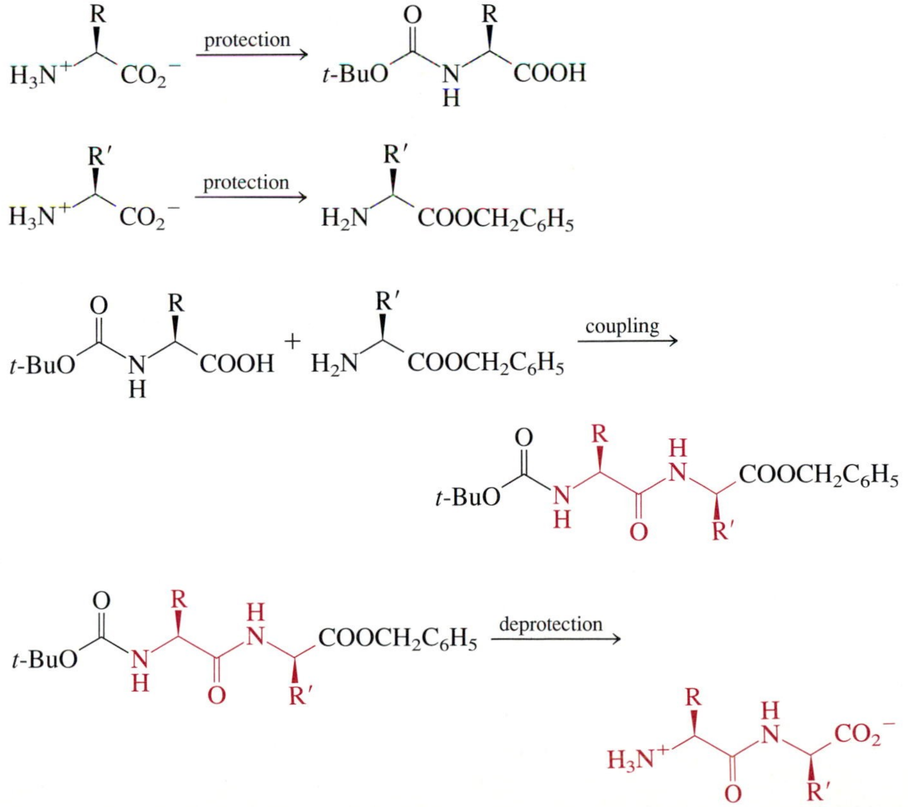

C. Structure Determination

The first step in determining the structure of a polypeptide or protein is the cleavage of any disulfide bridges that might be present. This reaction is commonly done by oxidizing the substance with peroxyformic acid, which converts the two cysteine units into cysteic acid units. If the compound contains no disulfide bridges, this step is not necessary.

$$\text{S—S} \xrightarrow{HCO_3H} SO_3H \quad SO_3H$$

The next analytical step is to determine the total amino acid composition. The material is subjected to total hydrolysis by some suitable method, typically heating with 6 N HCl at 112 °C for 24–72 hr. The hydrolyzate is then purified and analyzed by a chromatographic technique. The analytical method currently in use employs a commercial instrument called an **amino acid analyzer.** The mixture of amino acids is chromatographed on an ion exchange column with an aqueous buffer solution as eluent. The effluent from the column is automatically mixed with ninhydrin solution, and the presence of an amino acid is indicated by the typical violet color produced in the reaction (Section 29.5.C). The effluent is monitored at appropriate wavelengths with a spectrophotometer, and the absorbance is plotted by a recorder as a function of time. By comparing the chromatogram of an unknown mixture with that of a mixture of known composition, the analyst can arrive at a quantitative analysis of the mixture. The chromatogram for a standard mixture of amino acids is shown at the right in Figure 29.7. The left curve is a chromatogram of hydrolyzed bradykinin (page 966).

Other reagents are growing in use. Fluorescamine or *o*-phthalaldehyde yield fluorescent products by postcolumn derivatization (reaction with reagents after separation); phenyl isothiocyanate [by a modification of the Edman degradation described below] gives rise to phenylthiocarbamyl derivatives detectable at 254 nm and suitable for rapid separation and detection. A combination of labeling with fluorescein isothiocyanate (the strong fluorescence of fluorescein is noted in Section 30.7.B) and detection of laser-induced fluorescence after separation by capillary electrophoresis has allowed detection of less than 10,000 molecules of alanine.

There are two methods available for identifying the amino acid unit that occupies the N-terminal position in the polypeptide chain. The first is called the **Sanger method.** The NH_2 group in amino acids and peptides reacts with 2,4-dinitrofluorobenzene to form yellow 2,4-dinitrophenyl (DNP) derivatives. The reaction, illustrated for glycine, is an example of aromatic nucleophilic substitution (see Section 30.3.A).

$$2,4\text{-}(O_2N)_2C_6H_3\text{—F} + H_2NCH_2CO_2^- \longrightarrow \xrightarrow{H^+} 2,4\text{-}(O_2N)_2C_6H_3\text{—NHCH}_2\text{COOH}$$

2,4-dinitrofluorobenzene

N-(2,4-dinitrophenyl)glycine
(yellow)

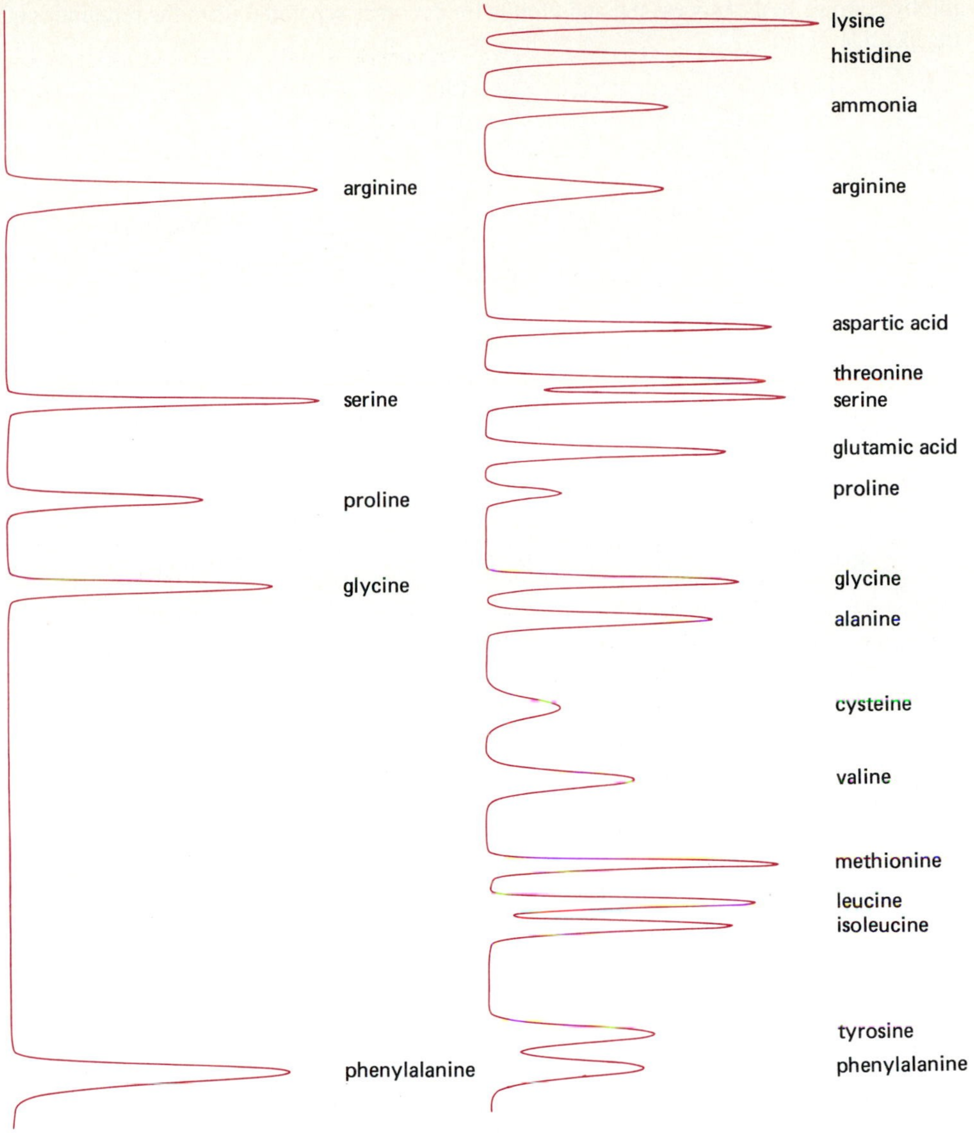

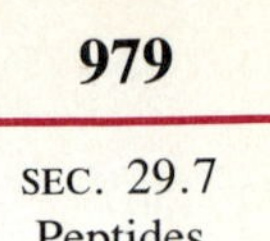

FIGURE 29.7 Amino acid analyzer traces. The right curve is an equimolar mixture. The left curve is the analysis of a sample of hydrolyzed bradykinin.

If the Sanger reaction is carried out on a peptide, the only α-amino group that is available for the reaction is the free group on the N-terminal end. Total hydrolysis of the DNP-labeled peptide then gives a mixture of amino acids, only one of which is labeled with the DNP function on the α-amino group. By knowing which amino acid bears the label, the investigator can deduce which amino acid is at the N-terminal end of the peptide.

The other technique for N-terminal analysis, which is actually more useful, is called the **Edman degradation.** In the Edman degradation, the peptide is allowed to react with phenyl isothiocyanate, $C_6H_5N{=}C{=}S$. The terminal NH_2 group reacts to form the phenylthiocarbamoyl derivative of the peptide. The labeled peptide is then treated with anhydrous HCl in an organic solvent. Although these conditions do not hydrolyze the amide linkages, the labeled amino acid undergoes a cyclization reaction, giving a phenylthiohydantoin. In the process the end group also becomes separated from the remainder of the peptide chain.

```
   Ph                          Ph                      Ph
   |                           |                       |
   N=C=S                       NH                      N
                               |                     /   \
phenyl isothiocyanate          C=S               O=C       C=S
   +                           |                   |       |
   NH2                         NH                R—CH——————NH
   |                           |          HCl
R—CH            ——→         R—CH          ——→    phenylthiohydantoin
   |                           |          H2O
   C=O                         C=O                     +
   |                           |
   NH                          NH                      NH2
   |                           |                       |
R′—CH                       R′—CH                  R′—CH
   |                           |                       |
   C=O                         C=O                     C=O
   |                           |                       |
   NH                          NH                      NH
   |                           |                       |
R″—CH                       R″—CH                  R″—CH
   |                           |                       |
   C=O                         C=O                     C=O
   ~                           ~                       ~

 peptide              phenylthiocarbamoyl-        shorter peptide
                            peptide
```

The substituted phenylthiohydantoin produced can be identified chromatographically by comparing it with known materials. Furthermore, the degraded peptide can be isolated and subjected to another cycle of the Edman degradation to identify the new N-terminal unit. The process has been automated and has been used to identify the first 60 amino acids in whale myoglobin, a protein that contains 153 amino acids in its chain, and the first 40 amino acids in each of the five subunits of the nicotinic acetylcholine receptor.

EXERCISE 29.14 A tetrapeptide is subjected to total amino acid analysis and found to contain serine, valine, alanine, and glycine. Edman degradation gives a tripeptide and the N-phenylthiohydantoin of valine. A second Edman degradation on the tripeptide gives a dipeptide and the N-phenylthiohydantoin of serine. The dipeptide is analyzed by the Sanger method and N-(2,4-dinitrophenyl)alanine is isolated. What is the structure of the original tetrapeptide? Write the equations illustrating the entire degradation sequence.

Highlight 29.4

The most fundamental element of the structure of a peptide is its linear sequence of amino acid components, and the determination of peptide structure is called **sequencing.** One way that peptide sequence can be determined is by **Edman degradation,** wherein the N-terminal amino acid is removed by successive treatment with phenyl isothiocyanate and hydrochloric acid. A second Edman degradation reveals the identity of the second amino acid in the chain, and so on.

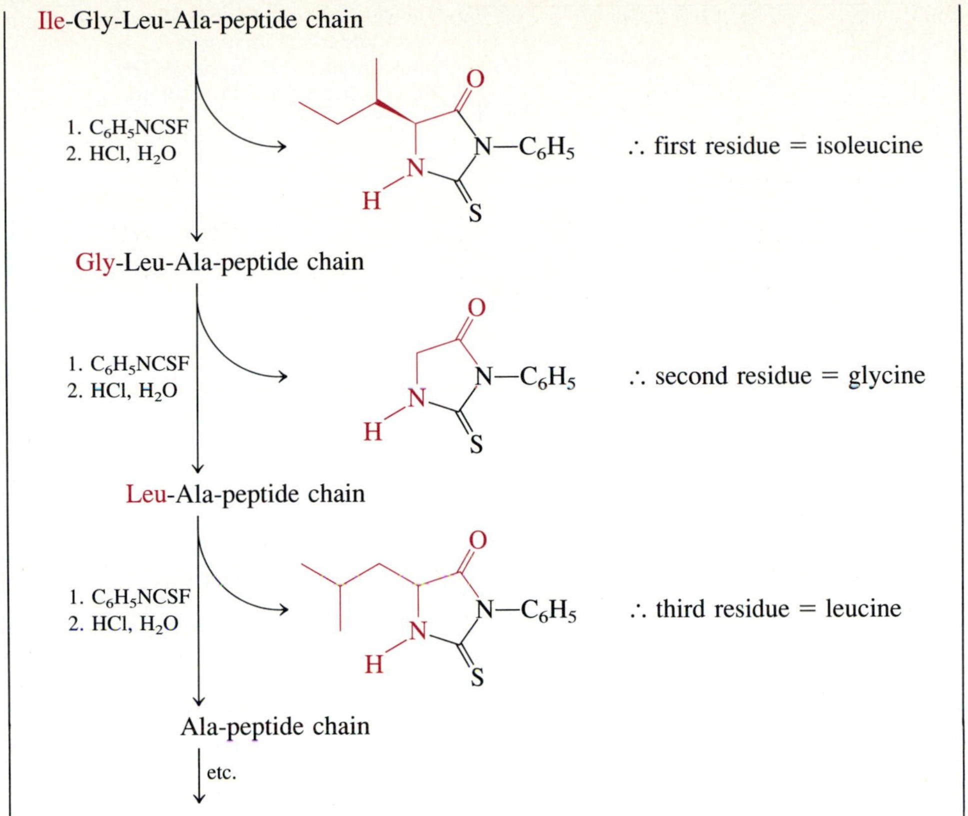

Ile-Gly-Leu-Ala-peptide chain

1. C_6H_5NCS
2. HCl, H_2O

$N—C_6H_5$ ∴ first residue = isoleucine

Gly-Leu-Ala-peptide chain

1. C_6H_5NCS
2. HCl, H_2O

$N—C_6H_5$ ∴ second residue = glycine

Leu-Ala-peptide chain

1. C_6H_5NCS
2. HCl, H_2O

$N—C_6H_5$ ∴ third residue = leucine

Ala-peptide chain

etc.

Because of its repetitive nature, the Edman degradation is amenable to automation, and commercial instruments are available that can be used to sequence small amounts of relatively large peptides.

The C-terminal amino acid can be identified by hydrolyzing with the enzyme carboxypeptidase, which specifically catalyzes the hydrolysis of the C-terminal amide link in a peptide or protein chain.

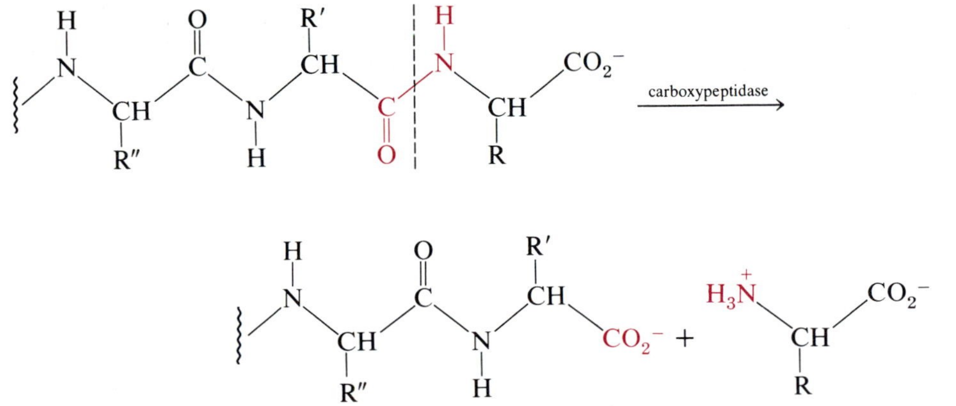

carboxypeptidase

CO_2^- + $H_3\overset{+}{N}$ CO_2^-

Thus, when the material is incubated with carboxypeptidase, the first free amino acid to appear in solution is the one that occupies the C-terminal position. Of course, once that amino acid has been removed from the chain, the enzyme continues to function and goes to work on the next residue, and so on. Eventually, the entire peptide or protein will be hydrolyzed to the constituent amino acids. By measuring the rate of appearance of amino

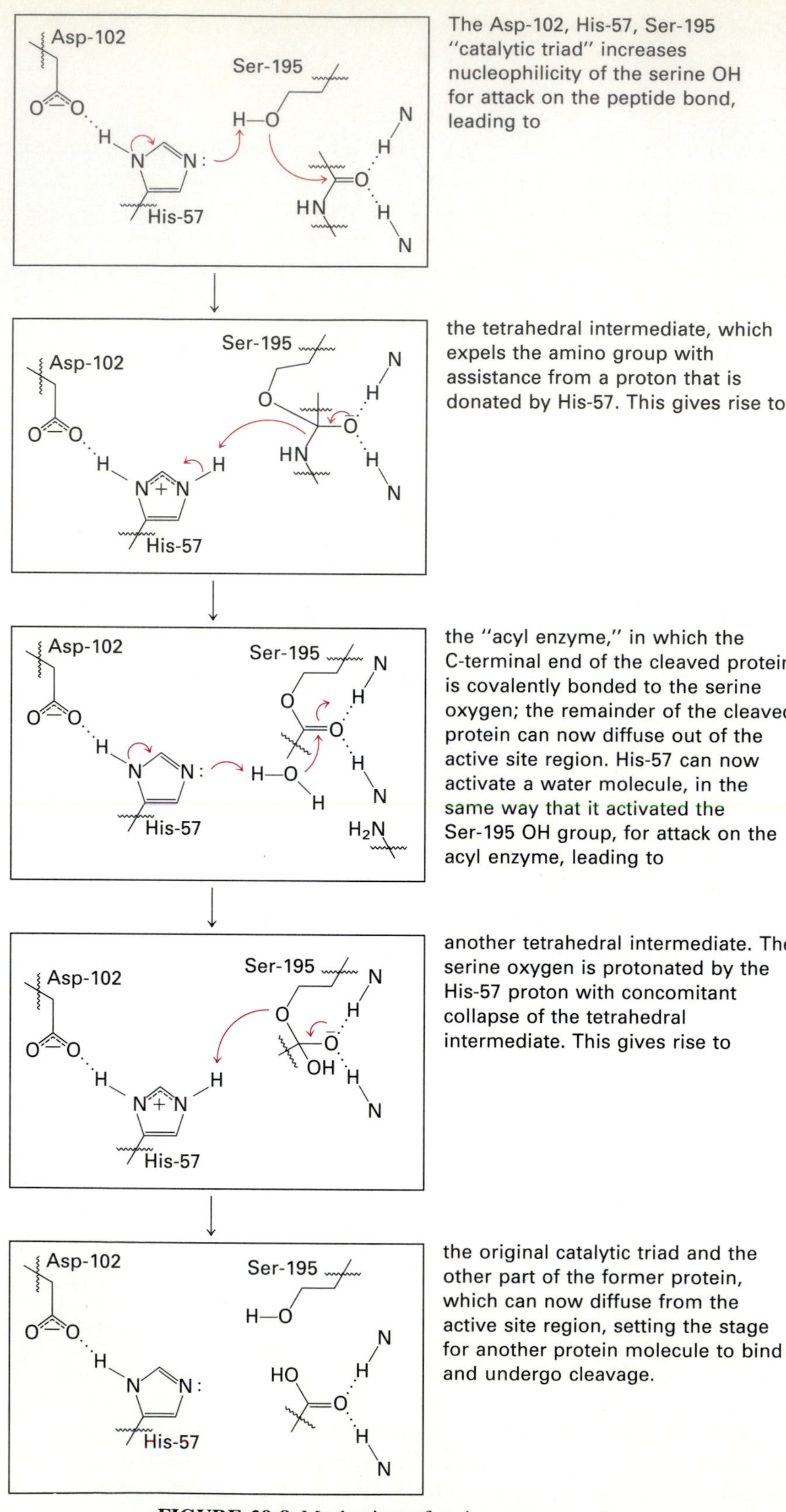

Asp-102
Ser-195
His-57
The Asp-102, His-57, Ser-195 "catalytic triad" increases nucleophilicity of the serine OH for attack on the peptide bond, leading to
the tetrahedral intermediate, which expels the amino group with assistance from a proton that is donated by His-57. This gives rise to
the "acyl enzyme," in which the C-terminal end of the cleaved protein is covalently bonded to the serine oxygen; the remainder of the cleaved protein can now diffuse out of the active site region. His-57 can now activate a water molecule, in the same way that it activated the Ser-195 OH group, for attack on the acyl enzyme, leading to
another tetrahedral intermediate. The serine oxygen is protonated by the His-57 proton with concomitant collapse of the tetrahedral intermediate. This gives rise to
the original catalytic triad and the other part of the former protein, which can now diffuse from the active site region, setting the stage for another protein molecule to bind and undergo cleavage.

FIGURE 29.8 Mechanism of serine protease action.

acids in the hydrolyzate, one may identify the C-terminal unit. In favorable cases the first three or four units can be identified in this way.

EXERCISE 29.15 A tripeptide having the empirical composition (PheGlySer, FGS) is subjected to the action of carboxypeptidase. The first free amino acid to appear in solution is phenylalanine. When the tripeptide is subjected to Edman degradation, the N-phenylthiohydantoin of glycine is obtained. What is the structure of the tripeptide?

Several methods are available to fragment a polypeptide or protein chain into smaller peptides. The most useful method is enzymatic hydrolysis. There are several classes of enzymes called **proteases** that catalyze hydrolysis of the peptide chain, usually at specific positions. One such class is that of the **serine proteases,** so named because there is a serine at the **active site** to which the cleaved acyl group becomes bound through the **serine hydroxy group.** One member of the class, **trypsin,** which occurs in the intestines of mammals, causes cleavage of peptide bonds only when the carbonyl group is part of a lysine or arginine unit. In a similar way, another serine protease, **chymotrypsin,** also an intestinal enzyme, catalyzes hydrolysis of the peptide bond formed by the carbonyl group of hydrophobic **residues** such as phenylalanine, tryptophan, and tyrosine. Note that the term *residue* can be used in place of *unit* for an *amino acid fragment* that is part of a polypeptide. A schematic mechanism for serine protease action is shown in Figure 29.8.

Pepsin, an enzyme derived from the stomach, has maximum activity at pH 1.0. This enzyme belongs to the **carboxy protease** class, in which the critical groups at the active site are two **aspartic carboxylic groups.** A related enzyme in the same class is rennin (chymosin), which is important in cheese-making. Pepsin is less specific than the serine proteases already described, causing hydrolysis between two generally hydrophobic residues such as phenylalanine, tryptophan, tyrosine, or leucine. A glutamic acid residue may be present on the carbonyl side of the peptide bond that is cleaved. Recall that peptides were originally fragments of protein produced by partial degradation with pepsin. The name was coined by Schwann in 1825 (Gk., *pepsis,* digestion) for the active principle of gastric juice.

The **thiol proteases,** in which a thiol group at the active site becomes acylated, are another class of enzymes useful for protein degradation. Papain (from papaya fruit), ficin (from figs), bromelain (from pineapple), and actinidin (from the Chinese gooseberry now called the Kiwi) are all thiol proteases. **Papain** hydrolyzes polypeptides (of at least seven residues) having a hydrophobic residue at the position once removed on the N-terminal side from the unit at which cleavage occurs.

One of the ways in which the body regulates blood pressure is through a complex series of biochemical steps starting with a protein named angiotensinogen. This protein, which circulates in the bloodstream, is cleaved by the protease renin to a decapeptide, angiotensin I. This substance is cleaved in turn by another protease called angiotensin-converting enzyme (ACE) to the octapeptide angiotensin II. Angiotensin II binds to cellular receptors in the blood vessels, resulting in vasoconstriction (contraction of the vessels) and a concomitant elevation of blood pressure.

1 5 10
Asp-Arg-Val-Tyr-Ile-His-Pro-Phe-His-Leu-Val-Ile-His-protein $\xrightarrow{\text{renin}}$
angiotensinogen

1 5 10
Asp-Arg-Val-Tyr-Ile-His-Pro-Phe-His-Leu $\xrightarrow{\text{ACE}}$
angiotensin I

1 5
Asp-Arg-Val-Tyr-Ile-His-Pro-Phe
angiotensin II

Pharmaceutical chemists have been able to capitalize on this bioregulatory system to invent useful drugs for the treatment of hypertension (elevated blood pressure). The first such drugs were compounds called ACE inhibitors, which bind strongly to the active site region of the angiotensin-converting enzyme and prevent its action on angiotensinogen. Two widely used ACE inhibitors are captropril and enalapril.

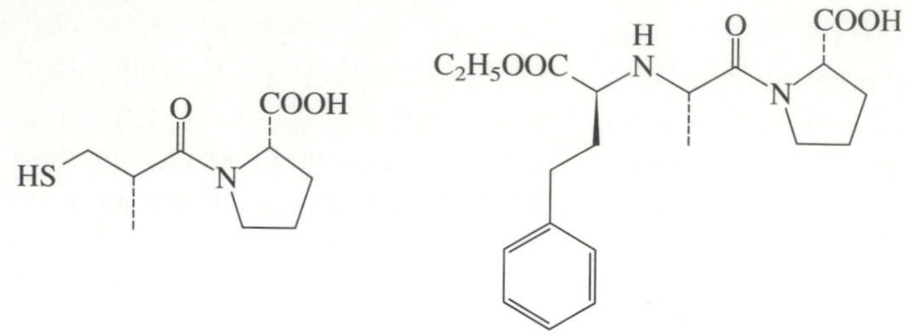

captropril

enalapril

Much current research is directed at discovering similar inhibitors of renin, since such drugs would be expected to have a similar antihypertensive effect but might have a different side-effect profile.

Another useful method for selective cleavage of polypeptide chains employs **cyanogen bromide,** BrCN. This reagent cleaves the chain only at the carbonyl group of methionine units; the methionine is converted into a C-terminal homoserine lactone unit.

$\xrightarrow{BrCN}$ + CH_3SCN

homoserine lactone unit

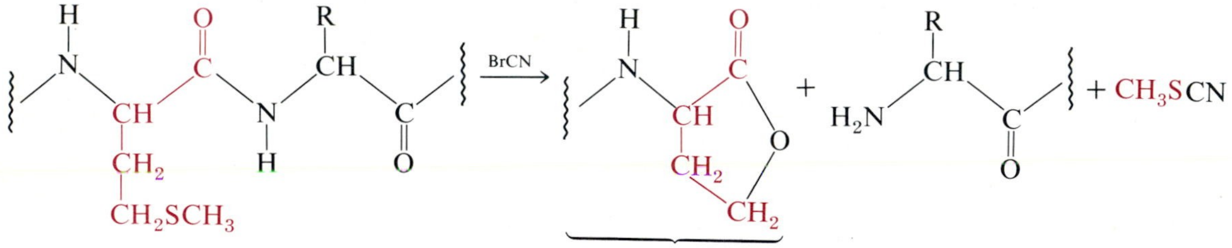

Partial degradation of the polypeptide chain, using one of the aforementioned methods, is an important step in determining the proper amino acid sequence of the molecule. Usually the purified polypeptide or protein is first incubated with trypsin, the most selective protease. The resulting mixture of peptide fragments is chromatographed, and the pure fragments are isolated. The peptides produced will usually contain from 2 to about 20 amino acid units. If the polypeptide chain is very long and if it contains relatively few lysine and arginine units, much larger fragments may be produced. The purified fragments are then analyzed for total amino acid content and subjected to repetitive Edman degradation to determine their structures.

The process is then repeated using a different cleavage method, usually cyanogen bromide. This second set of peptide fragments is then analyzed and sequenced. The various peptide blocks from the two degradation methods are then fitted together to produce a structure that unequivocally satisfies both sets of data.

As an example of the reasoning employed, consider a hypothetical eicosapeptide (20 amino acid units) having the amino acid composition $G_2A_4L_4F_3WK_2M_2SR$ (Gly_2Ala_4-$Leu_4Phe_3TrpLys_2Met_2SerArg$). End-group analysis shows that the polypeptide has alanine at the N-terminus (Sanger method) and phenylalanine at the C-terminus (carboxypeptidase). The material is hydrolyzed with trypsin to give four fragments: a tripeptide, two pentapeptides, and a heptapeptide. The four peptide fragments are each sequenced by repetitive Edman degradation and found to have the following structures.

I Trp-Phe-Arg
II Ala-Leu-Gly-Met-Lys
III Leu-Gly-Leu-Leu-Phe
IV Ala-Ala-Ser-Met-Ala-Phe-Lys

At this point the investigator knows that fragment III must correspond to the last five amino acids in the chain because trypsin does not cleave a chain at a phenylalanine carbonyl. Furthermore, fragment II or IV must correspond to the N-terminal end, but it is not possible with this information alone to write a unique complete sequence.

The intact polypeptide is then cleaved with cyanogen bromide, and the fragments are isolated, purified, and sequenced as before. Three fragments are produced, having the structures

V Ala-Leu-Gly-Met
VI Ala-Phe-Lys-Leu-Gly-Leu-Leu-Phe
VII Lys-Trp-Phe-Arg-Ala-Ala-Ser-Met

The four fragments in the first degradation and the three fragments in the second are then ordered in an overlapping way to arrive at an unambiguous structure.

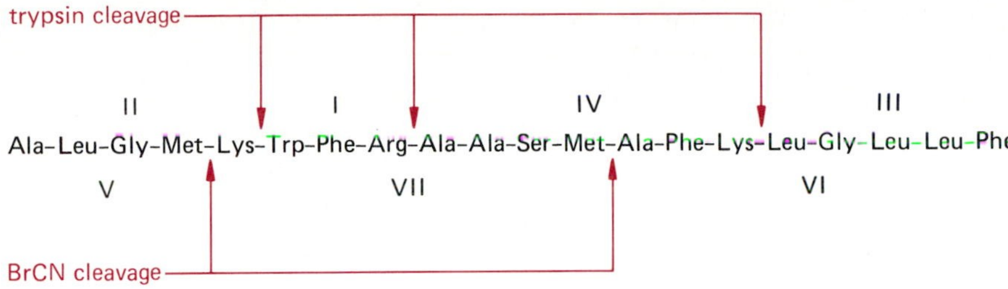

In practice, identification of the complete sequence of a complicated polypeptide or protein is rarely as simple as this example, and the actual process is usually tedious and time consuming. It often happens that almost the entire sequence is elucidated, but the exact positions of a few amino acids remain doubtful. The general process is still being improved, and new methods for routine sequencing are being developed. One technique that offers promise for sequencing relatively small polypeptides on microgram quantities is mass spectrometry (Chapter 34). However, most protein sequences are now deduced from DNA sequences and the genetic code (Chapter 33). The identification of the DNA corresponding to a particular protein is very much facilitated by using polynucleotide probes. These are made in an automated synthesizer on the basis of short amino acid sequences from the desired protein using the classical methods outlined here. Multiple polynucleotide probes are needed due to the redundancy in the genetic code.

The first major structure determination was that of bovine insulin (pages 968–69) by Sanger in 1953. The next significant accomplishment in this area was the sequencing of adrenocorticotropin (39 amino acid units), the hormone produced in the anterior pituitary gland that stimulates the adrenal cortex. Using the classical techniques, such large proteins as bovine chymotrypsinogen (245 amino acid units) and glyceraldehyde 3-phosphate dehydrogenase (333 amino acid units) have been sequenced. The new techniques of molecular biology have provided DNA and, thus, polypeptide sequences of a size and variety beyond the possibilities of the classical approach. Nevertheless, there are polypeptides and proteins for which a genetic message does not exist. For example, an enzyme can be inactivated with a ''suicide'' or irreversible substrate containing a radioactive or fluorescent substituent called a **''label.''** Sequencing of the polypeptide fragment containing the label is the only way to discover the nature of the labeled site and, thus, the active site of the enzyme.

EXERCISE 29.16 What is the structure of a pentapeptide that gives Gly-Ala, Leu-Phe, Leu-Leu, and Ala-Leu upon partial hydrolysis and the N-phenylthiohydantoin of glycine upon Edman degradation? What products will be obtained from incubation of this pentapeptide with pepsin?

Highlight 29.5

The amino acid sequence of a protein can be determined by a process of partial degradation of the chain to give a mixture of peptides that are individually sequenced by methods such as Edman degradation N-terminal analysis and carboxypeptidase C-terminal group analysis. If the sequences of enough small fragments are known, there will often be only one way in which they can be fitted together in an overlapping fashion to give the full protein sequence. An example is summarized below for determination of the sequence of an eicosapeptide that has the empirical formula Gly_2,Ala_4,Leu_4,Phe_3,Trp,Lys_2,Met_2,Ser,Arg (G_2,A_4,L_4,F_3,T,K_2,M_2,S,R). Chain degradation by reaction with trypsin and cyanogen bromide gave four and three peptides, respectively.

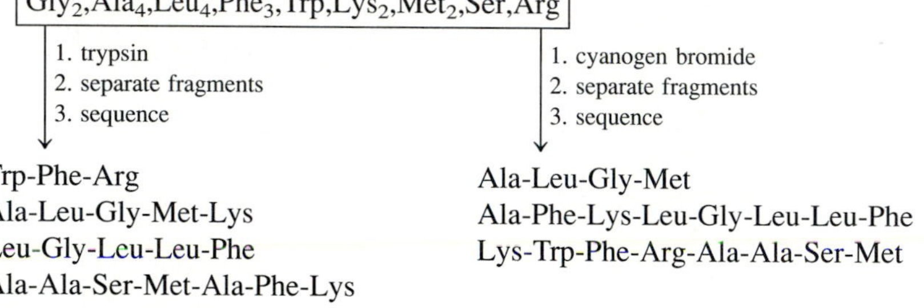

These seven fragments can only be assembled in one overlapping fashion to give a linear sequence of 20 amino acids.

Ala-Leu-Gly-Met
Ala-Leu-Gly-Met-Lys Ala-Ala-Ser-Met-Ala-Phe-Lys
Lys-Trp-Phe-Arg-Ala-Ala-Ser-Met Leu-Gly-Leu-Leu-Phe
Trp-Phe-Arg Ala-Phe-Lys-Leu-Gly-Leu-Leu-Phe

Therefore, the structure of the eicosapeptide must be

Ala-Leu-Gly-Met-Lys-Trp-Phe-Arg-Ala-Ala-Ser-Met-Ala-Phe-Lys-Leu-Gly-Leu-Leu-Phe

Note that the two chain degradations gave more information than was needed to solve the problem, as only four (in color) of the seven peptide structures are sufficient to specify the full structure of the protein.

29.8 Proteins

Proteins are natural polymers, composed of more than about 50 amino acid units. Insulin (51 units) is a polypeptide *hormone,* whereas bovine pancreatic trypsin inhibitor (BPTI) is a small protein with 56 residues. The remarkable diversity of structure and function to be found in proteins underlies the enormous complexity of life. The connection between this diverse expression of molecular properties and the properties of the individual units can be formulated in two general questions.

First, can we understand **protein folding?** Proteins can be folded or combined into wildly different shapes, as we will discuss.

Second, can we understand **protein function?** There are protein enzymes that catalyze esterification, alkene hydration, oxidation and reduction, decarboxylation, Claisen condensations, and many other reactions. There are carriers or transport proteins and ion channels; there are contractile proteins and antibodies. We have given several explicit examples of the chemical transformations associated with several enzymatic reactions (see pages 542 and 547–49). Only a few general ideas can be given in this book.

A. Molecular Shape

Proteins serve several important biological functions. On the one hand, they serve as structural material. The structural proteins tend to be **fibrous** in nature. That is, the long polypeptide chains are lined up more or less parallel to each other and are bonded one to another by hydrogen bonds. Depending on the actual three-dimensional structure of the individual protein molecule and its interaction with other similar molecules, a variety of structural forms may result. Examples are protective tissues such as hair, skin, nails, and claws (α- and β-keratins), connective tissues such as tendon (collagen), and the contractile material of muscle (myosin). Fibrous proteins are usually insoluble in water.

Proteins also have important roles as biological catalysts and regulators. They are responsible for catalyzing and regulating biochemical reactions and for the transport of various materials throughout an organism. The catalytic proteins (**enzymes**) and transport proteins tend to be **globular** in nature. In such a compound, the polypeptide chain is folded around itself so as to give the entire molecule a rounded shape. Each globular protein has its own characteristic geometry, which is a result of interactions between different sites on the chain. The intrachain interactions may be of five types: disulfide bridging, hydrogen bonding, dipolar interactions, charge-charge attractions or repulsions, or van der Waals attraction.

Sometimes each molecule of a globular protein consists of a single long polypeptide chain twisted about and folded back upon itself. In other cases the molecule is composed of several subunits. Each subunit is a single polypeptide chain that has adopted its own unique three-dimensional geometry. Several of the subunits are then bonded together by secondary forces (hydrogen bonding and van der Waals attraction) to give the total globular unit. Although they are highly complex molecules of relatively large molecular weight, globular proteins have specific molecular shapes as a result of various intrachain interactions, to be discussed later in this section. In some cases, the stable three-dimensional structure of the molecule is such that the surface contains a high percentage of amino acids having polar groups. In such a case, the globular protein is water soluble and exists in the cytoplasm or some other aqueous environment. The surfaces of other globular proteins are covered with amino acids having nonpolar side chains—such structures are found for globular proteins that exist embedded in intercellular membrane structures.

Globular proteins often carry a nonprotein molecule (the **prosthetic group**) as a part of their structure. The prosthetic group may be covalently bonded to the polypeptide chain, or it may be held in place by other forces.

B. Factors That Influence Molecular Shape

As we saw in the previous section, proteins are amino acid polymers containing more than about 50 individual units per chain. The backbone of the protein chain is the repeating unit

$$\text{—NH}\underset{}{\overset{\displaystyle \text{R} \atop |}{\text{C}}}\text{HCO—}$$

The linear amino acid sequence of a peptide or protein is referred to as its **primary structure.**

Portions of the polypeptide (protein) chain can adopt certain arrangements, either extended or helical. The nature of these arrangements is defined as the **secondary structure** of the protein. One important feature of polypeptide structure is the relationship between successive amino acid residues. This is usually expressed in terms of the dihedral angles, ϕ and ψ (Figure 29.9). The two amide bonds in the structure, $—C_1(=O_1)N_1H_1—CH(R)—C_2(=O_2)N_2H_2—$, are both planar. Each amide bond can rotate with respect to the bond to the α-carbon between them. The dihedral angle ϕ is defined as 0° for the first amide when the $C_1(=O_1)—N_1$ bond is syn to the $C_\alpha—C_2=O_2$ bond. The dihedral angle ψ is defined as 0° for the second amide when the $C_2(=O_2)—N_2$ bond is syn to the $N_1H_1—C_\alpha$ bond. These dihedral angles are shown for both 0° and 180° (ϕ,ψ) values in Figure 29.9. A plot of ψ against ϕ allows one to identify relationships that are energetically permissible. **Ramachandran maps** give an overview of the energetically accessible conformational arrangements by identifying which values of ψ are allowed for particular value of ϕ, and vice versa. The maps are not shown and further discussion will be found in more advanced books.

The most important secondary structures found in proteins are extended chains, reverse-turns, 3_{10}-helices, and α-helices. The **extended chains** form combinations so as to maximize interchain hydrogen bonding as well as van der Waals interactions. The combinations are called β-sheets which may be parallel, antiparallel, or twisted. **Reverse turns** contain a hydrogen bond and are very common. The most common helical structure is the right-handed ***α*-helix** (see below), a structure in which intrachain hydrogen bonds are favorable and the dihedral angles ϕ and ψ do not give rise to sterically unfavorable arrangements. The **3_{10}-helix** is less common but observed in many structures.

There are two kinds of secondary bonds that may exist between two different polypeptide chains, or between different regions of the same chain. Disulfide bridges between cysteine units in separate chains result in cross-linking of the two chains. An example is seen in insulin (Figure 29.6), in which the A and B chains are bonded together by two disulfide links. When the two cysteine units are in the same chain, as in oxytocin or the A chain of insulin (Table 29.4), disulfide bridging results in loops in the chain.

Hydrogen bonding is another type of secondary bonding that may occur between two different chains or between different regions of the same chain. Although hydrogen bonds are inherently weak (about 5 kcal mole^{-1} per hydrogen bond), a polypeptide chain contains many C=O and N—H groups that can engage in such bonding. The total amount of bonding that results from many small interactions is substantial and plays an important role in the actual shape or conformation of the molecule. Reciprocal hydrogen bonding can occur between the C=O and N—H groups of different chains and thus bind them together. Intrachain hydrogen bonding causes the chain to fold back on itself in some specific fashion.

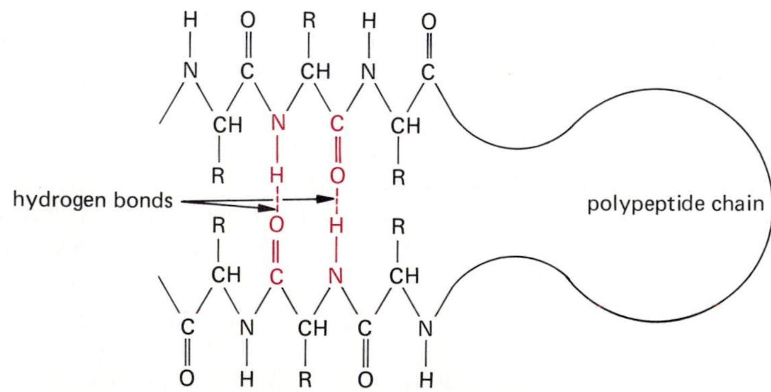

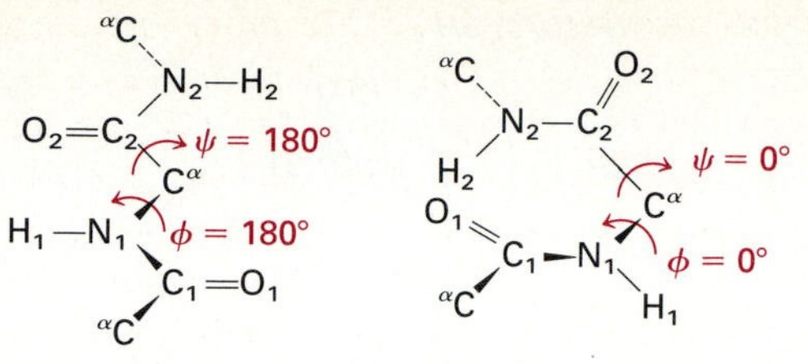

$$\psi = 180^\circ,\ \phi = 180^\circ \qquad \psi = 0^\circ,\ \phi = 0^\circ$$

FIGURE 29.9 Dihedral angles ψ and ϕ for polypeptides.

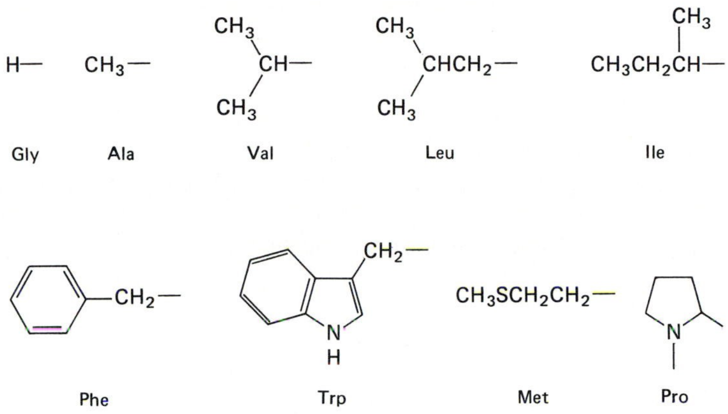

H— (Gly), CH_3— (Ala), $(CH_3)_2CH$— (Val), $(CH_3)_2CHCH_2$— (Leu), $CH_3CH_2CH(CH_3)$— (Ile), $C_6H_5CH_2$— (Phe), indolyl-CH_2— (Trp), $CH_3SCH_2CH_2$— (Met), Pro

FIGURE 29.10 Nonpolar (hydrophobic) side chains.

The combined effect of disulfide bridges and hydrogen bonds is to give the protein a preferred conformation that is referred to as its **secondary structure.**

The other important factor that governs the final molecular shape of a protein is the polar or nonpolar nature of the side-chain groups of the amino acids that constitute the molecule. Some of the side-chain groups that project from the polypeptide backbone are nonpolar or **hydrophobic** (Figure 29.10). In globular proteins, these nonpolar groups are found to be about equally distributed between the interior and the surface of the molecule.

Other side chains are polar and can hydrogen bond to water molecules. Since the globular proteins exist mainly in aqueous solutions, the polar side chains are found mainly on the outer surface of the molecule. The polar or **hydrophilic** side chains are listed in Figure 29.11. Some are neutral, and others bear either a negative or a positive charge at neutral pH.

The problem of protein folding is to understand the combined effects of all of the relatively small interactions that we have discussed. Which hydrogen bond arrangement is most favorable for the particular polypeptide? How do the various amino acid side chains interact with each other? What are the interactions of the various groups with the external environment? The various secondary structural elements (helices, extended chains) interact with one another in particular ways to form an overall molecular structure. For the combination of several elements, we can form **supersecondary structures,** which can combine into **domains.** Finally, these domains are combined into an overall structure for the protein, which is the **tertiary structure** of the compound. The association of proteins into more complex structures (hemoglobins are composed of four polypeptides or subunits; nicotinic acetylcholine receptor, of five subunits) produces the **quaternary structure.**

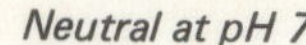

Neutral at pH 7

$HOCH_2—$	$CH_3CH(OH)—$	$HSCH_2—$	$H_2NC(=O)CH_2—$	$H_2NC(=O)CH_2CH_2—$
Ser	Thr	Cys	Asn	Gln

(imidazolyl)$CH_2—$	$HO—C_6H_4—CH_2—$
His	Tyr

Positively Charged at pH 7

$H_3\overset{+}{N}CH_2CH_2CH_2CH_2—$	$H_2N—C(=\overset{+}{N}H_2)NHCH_2CH_2CH_2—$
Lys	Arg

Negatively Charged at pH 7

$^{-}OC(=O)CH_2—$	$^{-}OC(=O)CH_2CH_2—$
Asp	Glu

FIGURE 29.11 Polar (hydrophilic) side chains.

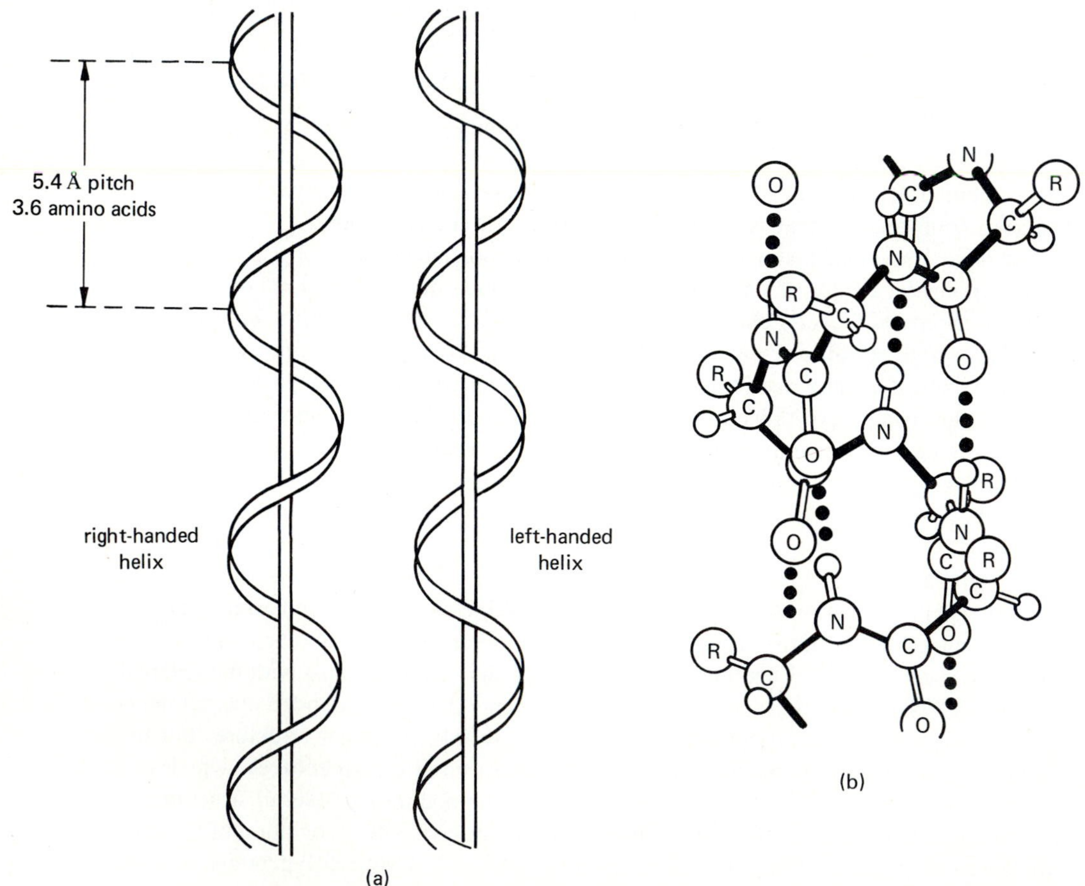

FIGURE 29.12 (a) Right-handed and left-handed helix. Note that ordinary screws are right-handed helices. (b) Diagram of a peptide α-helix.

A project to decipher the human genome has been started; it is expected that hundreds of proteins without a name or function will be discovered. The primary tool for studying these proteins is by discovering homology in sequence to known proteins. Homology in this connection implies that a sufficient similarity in sequence implies similarity in physical or chemical function. Thorough understanding of protein folding may also aid in the identification of such proteins.

C. Structure of the Fibrous Proteins

The most important type of conformation found in fibrous proteins is the **α-helix.** In this structure the polypeptide chain coils about itself in a spiral manner. The spiral or helix is held together by intrachain hydrogen bonding. The α-helix is right-handed and has a pitch of 5.4 Å or 3.6 amino acid units (Figure 29.12). Although a right-handed α-helix can

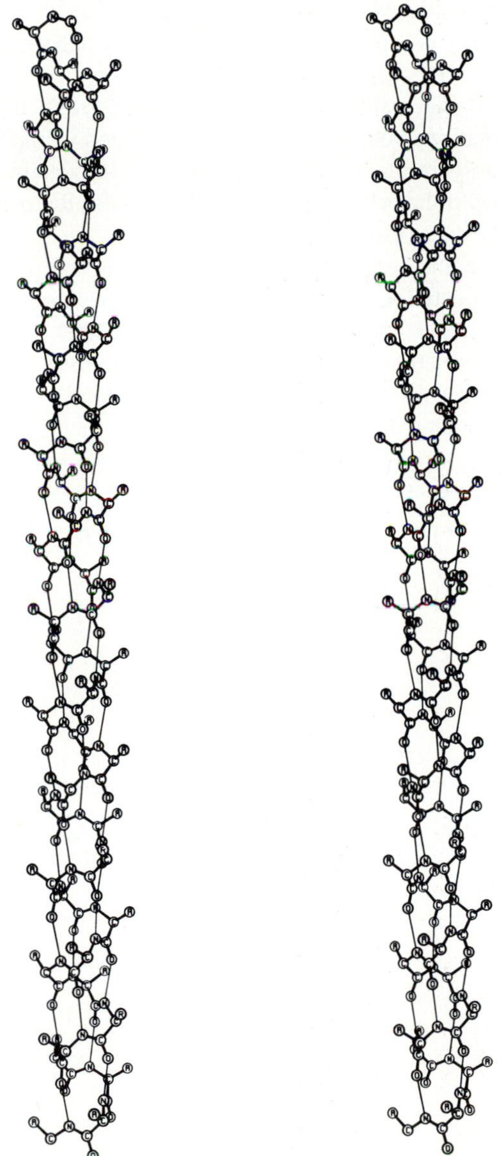

FIGURE 29.13 Stereo representation of polyalanine. [Courtesy of C. K. Johnson, Oak Ridge National Laboratory.]

form from either D- or L-amino acids (but not from DL), the right-handed version is more stable with the natural L-amino acids. A dramatic demonstration of the α-helix is shown by the stereo representation of polyalanine in Figure 29.13. However, collagen, the most important structural protein, is a right-handed superhelix made of extended left-handed helices. The structure is a repeating triplet, Gly-X-Y, for which the most frequent is Gly-Pro-Hyp (hydroxyproline).

Not all polypeptide chains can form a stable α-helix. The stability of the coil is governed by the nature of the side-chain groups and their sequence along the chain. Polyalanine, where the side chains are small and uncharged, forms a stable α-helix. However, polylysine does not. At pH 7 the terminal amino groups in the lysine side chains are all protonated. Electrostatic repulsion between the neighboring ammonium groups disrupts the regular coil and forces polylysine to adopt a **random coil** conformation. At pH 12 the lysine amino groups are uncharged, and the material spontaneously adopts the α-helical structure. In a similar way, polyglutamic acid exists as a random coil at pH 7, where the terminal carboxy groups are ionized, and as an α-helix at pH 2, where they are uncharged.

Proline is particularly interesting. Since the α-amino group in proline is part of a five-membered ring, rotation about the carbon-nitrogen bond is impossible. Furthermore, the amide nitrogen in polyproline has no hydrogens, and intrachain hydrogen bonding is not possible. Wherever proline occurs in a polypeptide chain, the α-helix is disrupted and a "kink" or "bend" results (Figure 29.14).

In some cases, such as the keratins of hair and wool, several α-helices coil about one another to produce a **superhelix.** In other cases, the helices are lined up parallel to one another and are held together by intercoil hydrogen bonding.

Another type of conformation found in the fibrous proteins is the β- or **pleated-sheet** structure of β-keratin (silk). In the β-structure, the polypeptide chains are extended in a "linear" or zigzag arrangement. Neighboring chains are bonded together by reciprocal

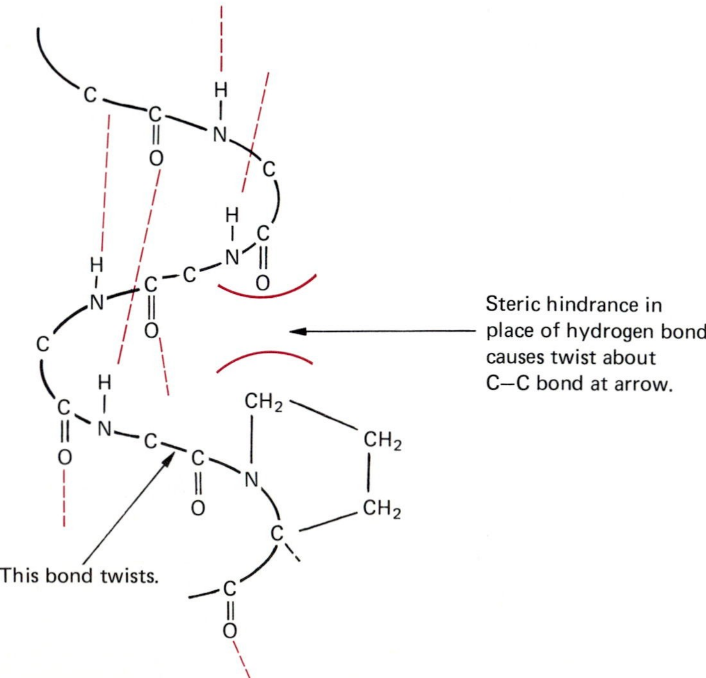
Steric hindrance in place of hydrogen bond causes twist about C—C bond at arrow.
This bond twists.

FIGURE 29.14 Showing the origin of a kink in an α-helix at proline. The proline unit in a peptide chain has no N—H for hydrogen bonding.

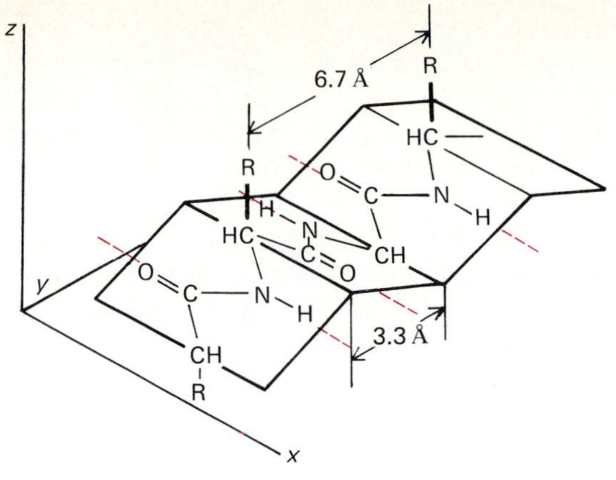

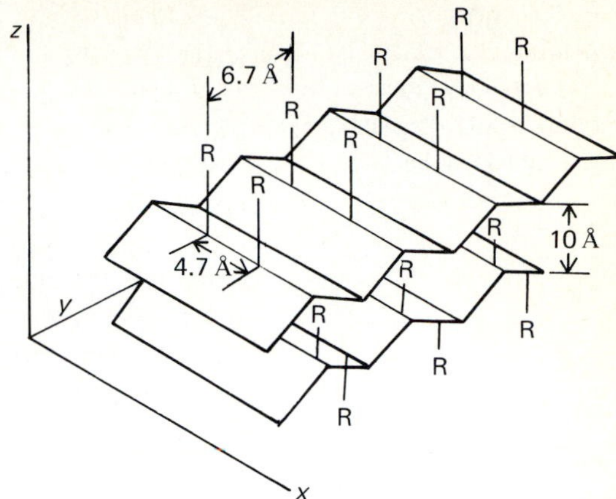

FIGURE 29.15 Schematic diagrams of pleated-sheet structure of polypeptides. The peptide bonds lie in the plane of the pleated sheet; the side chains lie above and below the sheet alternately. The polypeptide chains are held together by interchain hydrogen bonds, shown as dotted lines.

interchain hydrogen bonding. The result is a structure resembling a pleated sheet (Figure 29.15). Side-chain groups extend alternately above and below the general plane of the sheet. The pleated-sheet structure results in the side-chain groups being fairly close together. For this reason side chains that are bulky or have like charges disrupt the arrangement. In the β-keratin of silk fibroin 86% of the amino acid residues are glycine, alanine, and serine, all of which have small side chains.

D. Structure of the Globular Proteins

Globular proteins are designed by nature either to be soluble in the aqueous body fluids or in the intercellular membrane structures. They often must have a unique structure that creates an active site where the catalytic or transport function of the protein is carried out. The specific coiling that produces the proper geometry of the protein results from a delicate interplay of all the forces we have discussed up until now. Some folding is stabilized by disulfide bridges. In globular proteins that exist in the aqueous medium of the cytoplasm, the molecule tends to orient itself so that the nonpolar side chains lie inside the bulk of the structure where they attract each other by van der Waals forces. The polar side chains tend to be on the surface of the molecule where they can hydrogen bond to the solvent molecules and confer the necessary water solubility. Further coiling and compacting of the structure result from intrachain hydrogen bonds between the amide linkages inside the bulk of the molecule. Some segments of the polypeptide chain might have the typical α-helical structure, and others might be random coil. In other cases the chain might fold back on itself in the β- or pleated-sheet fashion. A schematic representation of a globular protein is shown in Figure 29.16.

If the protein contains a prosthetic group, that group will be imbedded at some point within the overall three-dimensional structure of the protein, either covalently bonded to the polypeptide chain or simply held by secondary forces. An example of a prosthetic group is heme, which is found in hemoglobin and myoglobin (Figure 29.17). In these proteins, both of which are oxygen carriers—myoglobin in muscle and hemoglobin in the bloodstream—the function of the prosthetic group is to bind an oxygen molecule. In both cases the polypeptide chain folds in such a way as to leave a hydrophobic "pocket" into which the heme just fits. The heme pocket is equipped with a histidine situated in such a way that its imidazole nitrogen can act as a fifth ligand for the ferrous ion in the center of

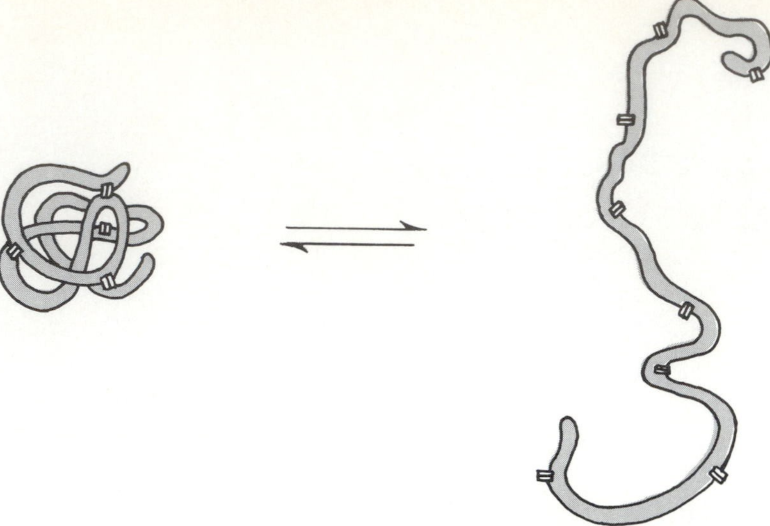

FIGURE 29.16 Schematic diagram of a globular protein with intrachain bonds (hydrogen bonds, van der Waals forces, and so on), showing reversible denaturation to random coil chain. [Adapted with permission from S. J. Baum, *Introduction to Organic and Biological Chemistry, 2nd ed,* Macmillan Publishing Co., Inc., 1978.]

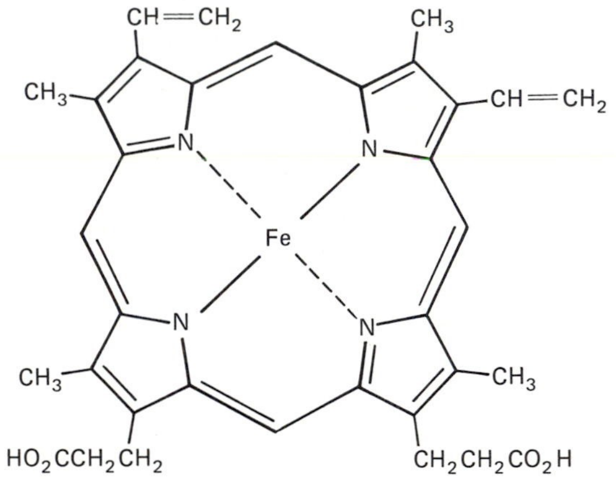

FIGURE 29.17 Hemin, the prosthetic group of hemoglobin and myoglobin.

the heme molecule. The prosthetic group is further held in its pocket by hydrogen bonding between the two propionic acid side chains and other appropriate side chains within the pocket.

The stereo representation of myoglobin in Figure 29.18 shows only the backbone of the polypeptide chain and the heme; substituent groups have been deleted for clarity. Note how the globular protein coils up on itself. There are several α-helical regions in the chain. An extensive one is seen at the top of the molecule and is viewed almost end-on in this representation. The imidazole ''fifth ligand'' (not shown) is just above the heme.

Under proper conditions the delicate three-dimensional structure of globular proteins may be disrupted. This process is called **denaturation.** Denaturation commonly occurs when the protein is subjected to extremes in temperature or pH. It is usually attended by a dramatic decrease in the water solubility of the protein. An example is the coagulation that results when skim milk is heated or acidified (denaturation of lactalbumin). A similar process is involved in the hardening of the white and the yolk of an egg upon heating.

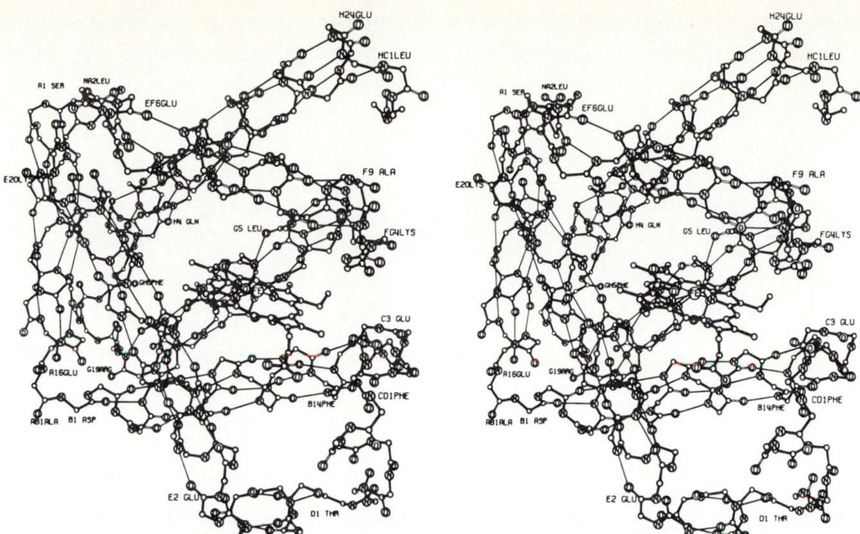

FIGURE 29.18 Stereo representation of myoglobin (side-chain substituents are not shown). [Courtesy of C. K. Johnson, Oak Ridge National Laboratory.]

Until fairly recently it was believed that denaturation was an irreversible process. It now appears, however, that in some cases the process is reversible. The reverse process is called **renaturation.** Many cases are now known in which a soluble denatured protein reverts to its natural folded geometry when the pH and temperature are adjusted back to the point where the native protein is stable. The three-dimensional structure of a protein seems to be a natural consequence of its primary structure; the unique conformation of each protein is simply a stable structure that the molecule can achieve under biological conditions.

E. Biological Functions of Proteins and Polypeptides

Although a complete discussion of the biological function of proteins is beyond the scope of this book, we shall give an overview of the topic here. In addition, we will take a brief look at the relationship between structure and function of one simple polypeptide.

As discussed previously, one important function of the proteins is structural. We have already mentioned α-keratin, the important structural component of skin, hair, feathers, and nails. Collagen is the material that forms the basis of the connective tissues—tendon, bone, and cartilage. Fibroin is the silk of spider webs and of cocoons. The hard exoskeletons of insects are composed of chitin, poly(N-acetylamino)glucose (page 942) and a modified (''tanned'') protein, sclerotin. Muscle tissue contains two important structural proteins—myosin is the stationary component and actin is the contractile component.

The regulatory proteins serve an immense variety of purposes. We have already encountered carboxypeptidase, pepsin, and trypsin, **enzymes** that catalyze the hydrolysis of polypeptide chains (Section 29.6.C). Hemoglobin and myoglobin (Figure 29.18) are heme proteins that **transport** oxygen, the former in the bloodstream and the latter in muscle tissue.

Cytochrome P-450 is a physiologically and chemically interesting protein that is responsible for various oxidation reactions. The protein contains the heme prosthetic group (see Figure 29.17). However, in cytochrome P-450 the iron differs from that in hemin in that it is bonded to another monovalent ligand (e.g., chloride) and is therefore in the +3 (III) oxidation state. This Fe(III) acquires an oxygen atom from O_2 to give the active oxidant, which is written as Fe═O. In this complex the iron is formally in the +5 (V) oxidation state. Cytochrome P-450 is a **monoxygenase;** it oxidizes alkanes to alcohols, arene and alkenes to the corresponding oxides, amines to amine oxides, and sulfides to sulfoxides.

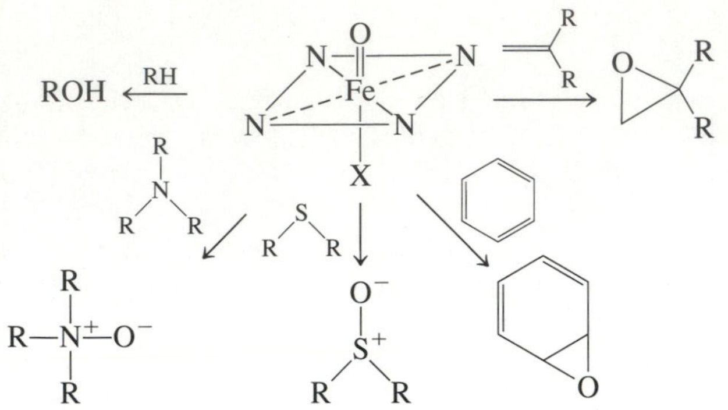

Cytochrome P-450 is an important biological defense system that is used by vertebrates and invertebrates to detoxify foreign substances such as alkaloids and terpenes from plants. Examples of the remarkable oxidation reactions brought about by cytochrome P-450 are shown below.

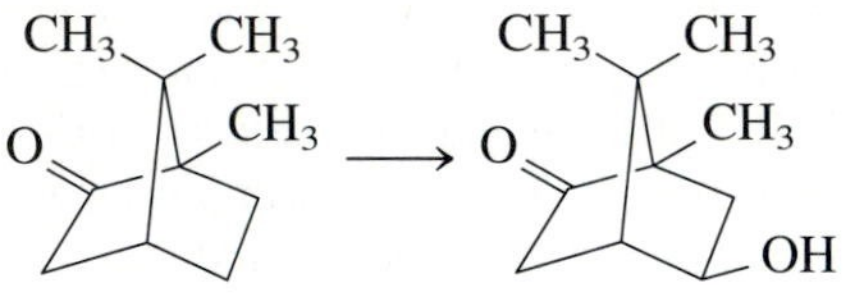

camphor

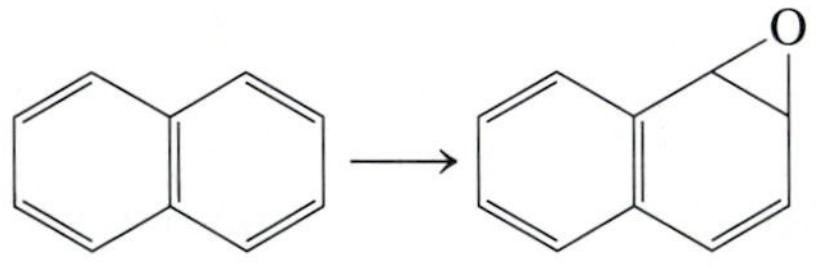

naphthalene

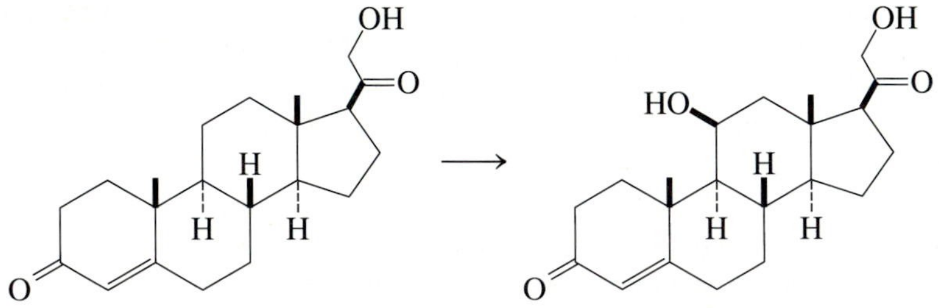

corticosterone

Some proteins are used for **storage.** Ovalbumin is employed by nature as a food reservoir in egg white. Casein plays a similar role in mammalian milk. The body manufactures antibodies as **defensive substances**—these materials form insoluble complexes with foreign substances that invade the bloodstream. Finally, some polypeptides are **hormones.** Examples are insulin (Figure 29.6), which regulates the metabolism of glucose, and β-endorphin (Table 29.4), a polypeptide with morphine-like activity that appears to be a natural pain reliever.

We shall now have a brief look at a simple polypeptide and see how one can analyze its structure and explain its biological action. Melittin is a 26-amino acid peptide that is

$$H_2N\text{—Gly-Ile-Gly-Ala-}\overset{5}{\text{Val}}\text{-Leu-Lys-Val-Leu-}\overset{10}{\text{Thr}}\text{-Thr-Gly-Leu-}$$

$$\text{Pro-}\overset{15}{\text{Ala}}\text{-Leu-Ile-Ser-Trp-}\overset{20}{\text{Ile}}\text{-Lys-Arg-Lys-Arg-}\overset{25}{\text{Gln}}\text{-Gln—}CONH_2$$

FIGURE 29.19 Amino acid sequence of melittin, the principal toxin of bee venom.

responsible for some of the the toxicity of bee venom. It has several biological effects, including a powerful ability to lyse red blood cells (erythrocytes). **Lysis** means the dissolution of the cell membrane accompanied by the release of the internal contents, most notably the protein, hemoglobin. Melittin also induces a voltage-dependent ion conductance across lipid bilayers, forms micelles with phospholipids, and causes the fusion of bilayers and membranes. At present, it appears that these properties reflect somewhat different aspects of melittin behavior.

The primary structure of melittin is shown in Figure 29.19. The secondary structure varies with ionic strength, concentration, and solvent. Melittin is quite soluble in water (>250 mg/mL). In dilute solution, the monomer exists as an extended chain or random coil; in more concentrated solution, tetramers form with the polypeptide chain mostly in an α-helical conformation. The crystal structure of melittin also shows a tetramer made of α-helical monomers. At concentrations of 1 mg/mL, monomeric melittin binds rapidly to red blood cell membranes and causes the release of hemoglobin.

Essentially the whole melittin molecule is necessary for lytic activity; peptides made of the first 20 amino acids (N-terminal or NH_2 end) or the 6 C-terminal amino acids are inactive. However, the simpler "peptide I" (NH_2-LLQSLLSLLQSLLSLLLQWLKRKR-$QQNH_2$ in single-letter code), also an α-helix, is even more lytic than melittin. To obtain an idea about the physical properties that are responsible for the lytic effect, we can look at the distribution of polar and nonpolar groups around the α-helix.

The melittin α-helix is diagrammed in the form of an "Edmundson helical wheel" in Figure 29.20. In this depiction, the amino acids of the chain are shown at the places they occupy on the periphery of the helix when the molecule is viewed along the axis of helix from the NH_2 end. In an α-helix there are 3.6 amino acid units per turn. Thus, in an Edmundson helical depiction of a polypeptide, each amino acid is displaced exactly 100° from its two immediate neighbors along the chain. In the Edmundson projection of melittin it is seen that one side of the molecule contains amino acids that have hydrophilic side-chain groups, whereas the other side contains amino acids with lipophilic side chains. (Glycine and alanine are considered as hydrophilic because the small protruding groups, H and CH_3, do not significantly shield the polar peptide backbone from an aqueous environment.) The result of this analysis is that the melittin molecule is revealed as an **amphiphilic molecule**—one which "likes" (Gk., *philos,* love) both polar and nonpolar media.

The binding of the polypeptide to a membrane or bilayer probably involves the α-helix parallel to the bilayer plane. The polar side of the helix can bind negatively charged lipids, which distorts the surrounding bilayer. A sufficiently distorted bilayer might become permeable to water. With ordinary phospholipids, the hydrophobic side of the peptide will bind and distort the reasonably ordered arrangement in the bilayer. Recall the ordered and disordered phospholipids shown in Section 19.11.D.

The aggregation of melittin within the bilayer or cell membrane is important to know but difficult to measure. At a suitable concentration, micelles are formed, showing that aggregation can occur.

EXERCISE 29.17 Make an Edmundson helical wheel projection of peptide I and compare its properties to those of melittin.

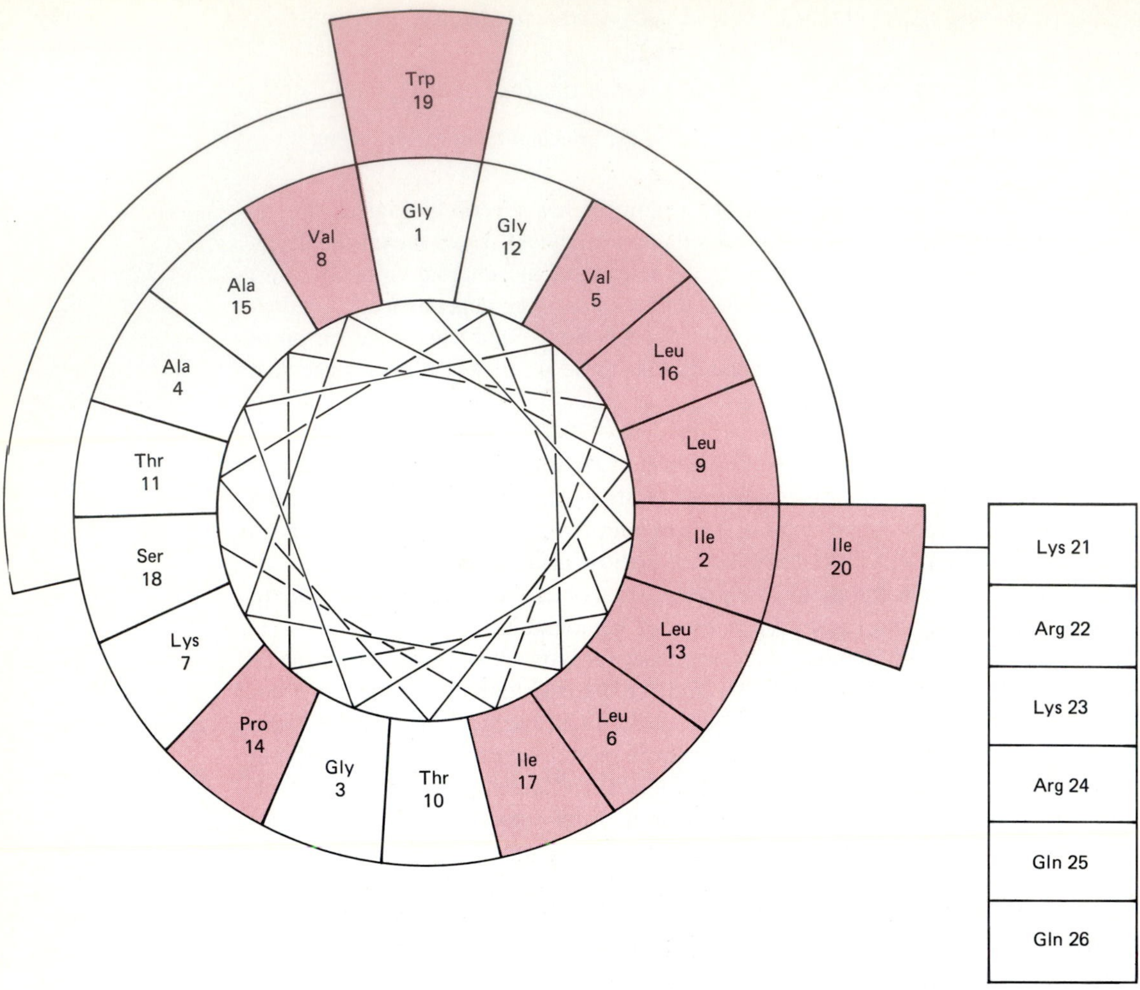

FIGURE 29.20 Edmundson helical wheel representation of melittin. Hydrophobic residues are shown in color.

ASIDE

Glue is used to cause two surfaces to adhere to one another and was used to fix inlays in wooden boxes found in Egyptian tombs dating from 3000 B.C. The glue was no doubt prepared by evaporating the hot water extract of animal hides or bones. This is mainly collagen, from which the French chemist Braconnot in 1820 obtained glycine through hydrolysis with hot dilute sulfuric acid. The substance was sweet and was at first named by the American chemist Horsford as glycocoll (sweet; Gk., *glucus,* glue, Gk., *kola*), a name later modified to glycine. DNA contains the information (using the "genetic code"; Chapter 33) for the incorporation of 20 amino acids into proteins. The L-configuration is the usual one found for amino acids in proteins, although D-amino acids occur in many natural products. Write the structure, the three-letter abbreviation, and the one-letter code with the correct configuration for each of the 20 natural amino acids. Give a synthesis for two of the amino acids. If the synthesis yields racemic product, show how you would separate the L- and D-isomers. If each step requires 3 hr, and your time is evaluated at \$3.00/hr, calculate the cost for the products, beginning with 100 g of starting material and assuming that all materials are free and a 75% yield in each step.

PROBLEMS

1. For each of the following compounds write the structure of the principal ionic species present in aqueous solution at pH 2, 7, and 12.

a. isoleucine
b. aspartic acid
c. lysine
d. glycylglycine (Gly-Gly, GG)
e. lysylglycine (Lys-Gly, KG)
f. alanylaspartylvaline (Ala-Asp-Val, ADV)
g. γ-glutamylcysteinylglycine (γ-Glu-Cys-Gly, γ-ECG, GSH)

2. Show how the isoelectric point of an amino acid can be computed from pK_1 and pK_2.

3. The pK_as for β-alanine and 4-aminobutanoic acid are shown below. Compare these values with the pK_as for the α-amino acids in Table 29.2 and explain the differences.

$H_3\overset{+}{N}CH_2CH_2CO_2^-$
β-alanine
$pK_1 = 3.55$
$pK_2 = 10.24$

$H_3\overset{+}{N}CH_2CH_2CH_2CO_2^-$
4-aminobutanoic acid
$pK_1 = 4.03$
$pK_2 = 10.56$

What are the isoelectric points for these two amino acids?

4. The dipeptide Gly-Asp has three known pK_a values: 2.81, 4.45, and 8.60. Associate each pK_a with the appropriate functional group in the structure of this peptide. Give a practical synthesis of this peptide starting with the amino acids.

5. Propose syntheses for the following amino acids.

a. $CH_3CH_2CH_2CH_2\underset{}{CH}(\overset{+}{N}H_3)CO_2^-$
b. $C_6H_5CH(\overset{+}{N}H_3)CO_2^-$
c. $(CH_3)_3CCH(\overset{+}{N}H_3)CO_2^-$
d. $CH_3CH_2C(\overset{+}{N}H_3)(CH_3)CO_2^-$
e. 1-($H_3\overset{+}{N}$)cyclohexane-1-CO_2^-
f. piperidine-2-CO_2^- ($\overset{+}{N}H_2$)

6. The following isotopically labeled amino acids are desired for biochemical research. Show how each may be prepared. The only acceptable sources of ^{14}C are $Ba^{14}CO_3$ and $Na^{14}CN$. The ^{14}C-labeled atom is marked with an asterisk in each case. Deuterated compounds may be prepared using D_2O, $LiAlD_4$, or D_2.

a. $CH_3CH(\overset{+}{N}H_3)\text{—}^*CO_2^-$
b. $CH_3\text{—}^*CH(\overset{+}{N}H_3)CO_2^-$
c. $CD_3CH(\overset{+}{N}H_3)CO_2^-$
d. $C_6H_5CH_2CD(\overset{+}{N}H_3)CO_2^-$
e. $^*CH_3SCH_2CH_2CH(\overset{+}{N}H_3)CO_2^-$
f. $(CD_3)_2CHCH(\overset{+}{N}H_3)CO_2^-$

g. $HOOC\overset{*}{C}H_2\underset{}{C}H(\overset{+}{N}H_3)CO_2^-$ **h.** $H_3\overset{+}{N}CD_2CH_2CH_2CH_2CH(NH_2)CO_2^-$

7. Glycine undergoes acid-catalyzed esterification more slowly than does propanoic acid. Explain.

8. Explain why the benzoyl group cannot be used as a N-protecting group for peptide synthesis.

9. In 1903 Emil Fischer introduced a rational method for the stepwise construction of peptides. The process, known as the α-haloacyl halide method, is outlined below in a synthesis of glycylglycine (GG).

$$BrCH_2\overset{O}{\overset{\|}{C}}Br + H_2NCH_2CO_2^- \longrightarrow BrCH_2\overset{O}{\overset{\|}{C}}NHCH_2CO_2^- \xrightarrow[\Delta]{NH_3} H_3\overset{+}{N}CH_2\overset{O}{\overset{\|}{C}}NHCH_2CO_2^-$$

a. Show how the α-haloacyl halide method can be used to synthesize glycyl-L-alanine and glycylglycyl-L-alanine.
b. Which α-haloacyl halides would be used to add alanyl or valyl units?
c. If the method is applied to L-alanine using the acyl halides in part (b), what will the products be?
d. What would be the chief problem in applying the α-haloacyl halide method to a fairly complex polypeptide such as Ala-Val-Phe-Ala-Ala (AVFAA)?

10. Write out all the steps in a synthesis of the hexapeptide Gly-Ala-Pro-Ala-Ala-Val (GAPAAV). As N-protecting groups use either the benzyloxycarbonyl or *t*-butoxycarbonyl groups. For coupling use DCC.

11. Propose a synthesis of the pentapeptide Ala-Lys-Glu-Gly-Gly (AKEGG). Note that the terminal amino group of lysine and the terminal carboxy group of glutamic acid must be protected.

12. Propose a synthesis of the decapeptide Ala-Val-Phe-Ala-Ala-Ala-Val-Phe-Ala-Ala (AVFAAAVFAA).

13. Pyroglutamic acid, pyroGlu, is a cyclic lactam obtained by heating glutamic acid.

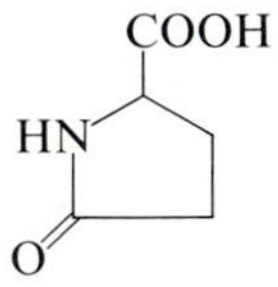

COOH
HN
O

pyroGlu

This derivative of proline occurs in an important tripeptide, thyrotropin-releasing hormone (TRH), pyroglutamylhystidylprolineamide, which occurs in brain tissue. It also occurs in the anterior lobe of the pituitary gland where it stimulates the secretion of several other hormones. Write out the structure of TRH. A sensitive assay method has been developed that makes use of synthetic hormone. Propose a synthesis of TRH from pyroglutamic acid and other required reagents.

14. The cyanogen bromide method for cleavage of peptide chains involves reaction of the nucleophilic sulfur of a methionine unit with the carbon of BrCN. Write out the complete reaction mechanism.

15. Ribonuclease A is a 124-amino acid enzyme that catalyzes the hydrolysis of the phosphate backbone of ribonucleic acid. The protein has the following primary structure:

H_2N-Lys-Glu-Thr-Ala-Ala-Ala-Lys-Phe-Glu-Arg-Glu-His-Met-Asp-Ser-Ser-Thr-Ser-Ala-Ala-Ser-Ser-Ser-Asn-Tyr-Cys-Asn-Glu-Met-Met-Lys-Ser-Arg-Asn-Leu-Thr-Lys-Asp-Arg-Cys-Lys-Pro-Val-Asn-Thr-Phe-Val-His-Glu-Ser-Leu-Ala-Asp-Val-Glu-Ala-Val-Cys-Ser-Glu-Lys-Asn-Val-Ala-Cys-Lys-Asn-Gly-Glu-Thr-Asn-Cys-Tyr-Glu-Ser-Tyr-Ser-Thr-Met-Ser-Ile-Thr-Asp-Cys-Arg-Glu-Thr-Gly-Ser-Ser-Lys-Tyr-Pro-Asn-Cys-Ala-Tyr-Lys-Thr-Thr-Glu-Ala-Asn-Lys-His-Ile-Ile-Val-Ala-Cys-Glu-Gly-Asn-Pro-Tyr-Val-Pro-Val-His-Phe-Asp-Ala-Ser-Val-COOH
(H_2NKETAAAKFEREHMDSSTSAASSSNYCNEMMKSRNLTKDRCKPVNTFVHES-LADVEAVCSEKNVACKNGETNCYESYSTMSITDCRETGSSKYPNCAYKTTEANK-HIIVACEGNPYVPVHFDASVCOOH)

There are disulfide bridges between the Cys units 26–84, 40–95, 58–110 and 65–72. What polypeptide fragments will be produced when ribonuclease A is partially hydrolyzed with (a) trypsin, (b) chymotrypsin, or (c) cyanogen bromide? Assume that the disulfide bonds are cleaved prior to hydrolysis.

16. Gastrins are heptadecapeptide (17 amino acid units) hormones that stimulate the secretion of gastric acid in the stomach of mammals. Feline gastrin has the empirical amino acid composition ($Ala_2AspGly_2Glu_5LeuMetPheProTrp_2Tyr$). The peptide was digested with chymotrypsin and four peptide fragments were isolated. The four fragments were sequenced and found to have the following structures.

I Glu-Gly-Pro-Trp (EGPW)
II Gly-Trp (GW)
III Met-Asp-Phe (MDF)
IV Leu-Glu-Glu-Glu-Glu-Ala-Ala-Tyr (LEEEEAAY)

End-group analysis revealed that the N-terminal unit is Glu and the C-terminal unit is Phe. What two structures for feline gastrin are compatible with the foregoing evidence?

17. Porcine pancreatic secretory trypsin inhibitor I is a protein containing 56 amino acid units. Acidic hydrolysis, followed by amino acid analysis, gave the following empirical composition: $D_4T_6S_6E_7P_5G_4AV_4C_6I_3L_2Y_2K_4R_2$ ($Asp_4Thr_6Ser_6Glu_7Pro_5Gly_4AlaVal_4$-$Cys_6Ile_3Leu_2Tyr_2Lys_4Arg_2$). (*Note*: Complete hydrolysis does not distinguish Gln from Glu or Asn from Asp.) After cleavage of disulfide bridges, the protein was digested with trypsin. Nine fragments were isolated and purified by chromatography. The nine fragments were each sequenced by repetitive Edman degradation. Eight of the fragments were found to have the following structures.

T-1 Lys (K)
T-2 Arg (R)
T-3 Ser-Gly-Pro-Cys (SGPC)
T-4 Thr-Ser-Pro-Gln-Arg (TSPQR)
T-5 Gln-Thr-Pro-Val-Leu-Ile-Gln-Lys (QTPVLIQK)
T-6 Ser-Asn-Glu-Cys-Val-Leu-Cys-Ser-Glu-Asn-Lys (SNECVLCSENK)
T-7 Ile-Tyr-Asn-Pro-Val-Cys-Gly-Thr-Asp-Gly-Ile-Thr-Tyr (IYNPVCGTDGITY)
T-8 Glu-Ala-Thr-Cys-Thr-Ser-Glu-Val-Ser-Gly-Cys-Pro-Lys (EATCTSEVSGCPK)

The ninth fragment contained 24 amino acid units and had the empirical composition ($Asp_4Thr_2Ser_2Glu_2Pro_2Gly_2Val_2Cys_3Ile_2LeuTyr_2Lys$). Seven cycles of Edman degradation showed that the N-terminal end of fragment T-9 had the composition

T-9 Ile-Tyr-Asn-Pro-Val-Cys-Gly . . .

Edman degradation of the intact protein showed the N-terminal unit to be Thr. The C-terminal residue was shown to be Cys.

The protein was then digested with chymotrypsin and three peptide fragments were isolated. The three chymotryptic fragments were each subjected to total hydrolysis and analyzed for amino acid composition. They were also subjected to three cycles of Edman degradation to identify the N-terminal sequence and incubated with carboxypeptidase to identify the C-terminal unit. The partial structures of the three fragments were found to be

Ch-1 Thr-Ser-Pro(Thr$_2$Ser$_2$Glu$_3$ProGlyAlaValCys$_2$IleLysArg)Tyr
Ch-2 Asn-Pro-Val(AspThr$_2$Gly$_2$CysIle)Tyr
Ch-3 Ser-Asn-Glu(AspThrSer$_2$Glu$_3$Pro$_2$GlyVal$_2$Cys$_2$IleLeu$_2$Lys$_3$Arg)Cys

The intact protein was then treated with methyl isothiocyanate. This reagent modifies the lysine side chains so that they are not cleaved by trypsin. The modified protein was digested with trypsin and three fragments were isolated. The three fragments were isolated, hydrolyzed, and analyzed and shown to have the following empirical compositions.

*T-1 (ThrSerProArgGlu)
*T-2 (ThrSerGlu$_2$Pro$_2$GlyValCysIleLeuLys)
*T-3 (Asp$_4$Thr$_4$Ser$_4$Glu$_4$Pro$_2$Gly$_3$AlaVal$_3$Cys$_5$IleLeuTyr$_2$Lys$_3$Arg)

From the data, what is the primary structure of the protein?

18. Label and illustrate the dihedral angles in the tripeptide, glutathione (γ-Glu-Cys-Gly).

19. Name and illustrate the various types of secondary structures found in proteins.

20. N-acetyl-L-cysteine is widely used as a mucolytic agent. Such an agent causes mucus to flow more freely, and is important in treatment of respiratory diseases. Devise a synthesis for the compound and draw its titration curve.

21. Give the most likely products of reaction of the enzyme cytochrome P-450 and the following compounds. Write the mechanism of each transformation.

a. hexane
b. tri-*n*-propylamine
c. cyclohexene
d. di-*n*-butyl sulfide

22. Devise a synthesis for as many as possible of the following rare amino acids.

a. selenocysteine
b. γ-carboxyglutamic acid
c. (*R*)-4-hydroxyproline
d. ornithine
e. citrulline
f. γ-aminobutyric acid
g. α-aminoadipic acid

23. *t*-Butoxycarbonyloximinophenylacetonitrile (''Boc-On,'' page 972) is prepared via the oximinophenylacetonitrile from phenylacetonitrile and methyl nitrite, followed by conversion to the chloroformyl derivative with phosgene and reaction with *t*-butyl alcohol in the presence of base. Write out each step, deducing the nature of the reagents needed and the mechanism of the reaction. Deduce how the methyl nitrite is generated and introduced into the reaction mixture. *Hints*: (1) Check the boiling points of alkyl nitrites. (2) Consider the analogy between an N=O bond and a C=O bond.

24. Write out the mechanisms for all steps shown on pages 975–76 for the conversion of porcine insulin into human insulin, including the step involving the enzyme trypsin as the reagent. *Hint*: Trypsin and chymotrypsin cleave amides by similar mechanisms but with different specificities.

25. Following is the structure of an octapeptide, with its amino and carboxy groups written in their non-ionized forms.

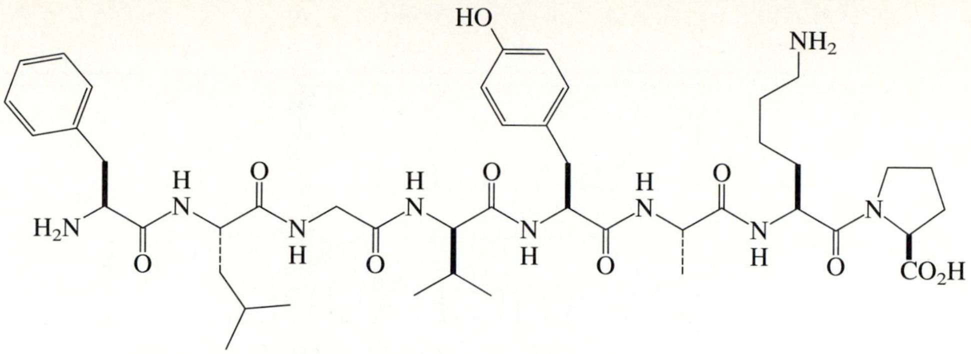

Answer the following questions about this octapeptide.

a. What is its name using both the three-letter and one-letter codes?

b. All of the amino acids except one have the natural *S* configuration. What is the common name of the one *R* amino acid?

c. What is the name of the "C-terminal" amino acid?

d. What is the name of the first amino acid to be released by an Edman degradation?

e. At physiological pH (approximately 7.0) is the net charge of this octapeptide −1, 0, or +1? Explain by writing the names of the amino acid units that are ionized at pH 7 and indicating whether each contributes a positive or a negative charge to the octapeptide.

CHAPTER 30

AROMATIC HALIDES, PHENOLS, PHENYL ETHERS, AND QUINONES

30.1 Introduction

When halogen is in the side chain of an alkyl-substituted benzene, the chemistry is essentially the same as for any alkyl halide. As we saw in Section 21.5, enhanced reactivity is seen in the displacement and carbocation reactions of benzylic halides, since the phenyl group stabilizes either an S_N2 transition state or a cationic center to which it is directly attached. Side-chain halides are prepared by methods analogous to those used for the preparation of normal alkyl halides, mainly from corresponding alcohols (Section 10.6.B). Recall that free radical halogenation is especially useful for preparation of benzylic halides (Section 21.5.A).

When halogen is attached directly to the benzene ring, there is some special chemistry, in regard to both preparations and reactions, that is different from the normal chemistry of alkyl halides. We shall take up these unique reactions pertaining to aryl halides in Sections 30.2 and 30.3. Since several of the important reactions of aryl halides give rise to phenols, it is convenient to study that important family of organic compounds next, and we shall do so in Sections 30.4–30.7. Finally, we shall take up the quinones in Section 30.8, since the chemistry of this class of compounds is intimately associated with the chemistry of phenols.

30.2 Preparation of Halobenzenes

One important preparation of ring halides is by electrophilic aromatic substitution (Section 23.1). The active reagent in these reactions is an actual or incipient halonium ion. For suitably activated rings the halogen alone can be used as the reagent.

mesitylene + Br_2 $\xrightarrow[10°C]{CCl_4}$ bromomesitylene (79–82%)

For less reactive rings a Lewis acid such as a ferric salt is used to catalyze the reaction.

$C_6H_5COCl + Cl_2 + FeCl_3 \xrightarrow{35°C} m\text{-}ClC_6H_4COCl + HCl + FeCl_3$

In general, direct electrophilic aromatic substitution is only used as a method for synthesis of halobenzenes in cases where the reaction gives essentially a single product. This is usually the case when the benzene ring contains a *meta*-directing group or when the structure of the starting material is such that only a single product can be produced. Thus bromination of *p*-nitrotoluene gives only one product.

$p\text{-}CH_3C_6H_4NO_2 \xrightarrow[75\text{–}80°C]{Br_2,\ Fe}$ 2-bromo-4-nitrotoluene

When the ring contains a single *ortho,para*-directing group, a mixture of products generally results. If the group is very large, as in the case of *t*-butyl, the amount of *ortho* product is small, and the *para*-disubstituted product is obtained in good yield.

$C_6H_5C(CH_3)_3 \xrightarrow[\text{pyridine}]{Br_2}$ $p\text{-}BrC_6H_4C(CH_3)_3$ (94%)

However, in most such cases direct halogenation gives a mixture of isomeric products that is difficult to separate. For example, bromination of toluene gives a mixture consisting of 65% *p*-bromotoluene and 35% *o*-bromotoluene.

$C_6H_5CH_3 \xrightarrow[25°C]{Br_2,\ Fe}$ $p\text{-}BrC_6H_4CH_3$ (65%) + $o\text{-}BrC_6H_4CH_3$ (35%)

In situations like this, syntheses involving arenediazonium salts are successful for the preparation of pure materials (Section 24.5). For example, although the *ortho* and *para* isomers of bromotoluene cannot easily be separated by distillation, the isomeric nitrotoluenes are readily separated by this method (see Table 23.5, page 718). Since toluene is an inexpensive starting material, direct nitration followed by fractional distillation represents an economical method for the preparation of both the *ortho* and *para* isomers.

$$\text{C}_6\text{H}_5\text{CH}_3 \xrightarrow{\text{HNO}_3,\ \text{H}_2\text{SO}_4} p\text{-CH}_3\text{C}_6\text{H}_4\text{NO}_2 + o\text{-CH}_3\text{C}_6\text{H}_4\text{NO}_2$$

(55%) b.p. 238°C; (45%) b.p. 220°C

Both isomers may be converted into the corresponding bromotoluenes by reduction of nitro to amino (Section 24.6.C), followed by a Sandmeyer reaction (Section 25.5.C). For example,

$$p\text{-CH}_3\text{C}_6\text{H}_4\text{NO}_2 \xrightarrow{\text{Fe, HCl}} p\text{-CH}_3\text{C}_6\text{H}_4\text{NH}_2 \xrightarrow[\text{2. CuBr}]{\text{1. NaNO}_2,\ \text{HBr}} p\text{-CH}_3\text{C}_6\text{H}_4\text{Br}$$

p-aminotoluene

The foregoing example illustrates the indirect replacement of a nitro group by halogen via the diazonium group. However, arenediazonium chemistry can also be utilized in another way. For example, bromination of *p*-acetamidotoluene followed by hydrolysis of the protecting group affords 4-amino-3-bromotoluene in good purity, since the acetamido function is a more powerful *ortho,para*-directing group than is methyl (Section 23.8). Diazotization of this product, followed by reduction of the arenediazonium salt with hypophosphorous acid, affords *m*-bromotoluene.

$$p\text{-CH}_3\text{C}_6\text{H}_4\text{NHCOCH}_3 \longrightarrow \text{CH}_3\text{C}_6\text{H}_3(\text{Br})\text{NHCOCH}_3 \longrightarrow \text{CH}_3\text{C}_6\text{H}_3(\text{Br})\text{NH}_2 \longrightarrow m\text{-CH}_3\text{C}_6\text{H}_4\text{Br}$$

EXERCISE 30.1 Write equations illustrating the preparation of the three chlorobromobenzenes starting with benzene.

30.3 Reactions of Halobenzenes

A. Nucleophilic Aromatic Substitution: The Addition-Elimination Mechanism

The halogen of an aryl halide can be replaced by other nucleophiles. However, such substitution reactions do *not* occur by the S_N2 mechanism. Like vinyl halides (Section 12.7), aryl halides cannot achieve the geometry necessary for a backside displacement; the ring shields the rear of the carbon-halogen bond.

N:$^-$ → (benzene)—Cl

Instead, nucleophilic substitution occurs by two other mechanisms, **addition-elimination** and **elimination-addition.**

Aryl halides that have electron-attracting groups in positions *ortho* and *para* to the halogen undergo substitution under rather mild conditions. The most effective groups for favoring this type of substitution are nitro and carbonyl.

$$o\text{-nitrochlorobenzene} \xrightarrow[130°C]{\text{aq. } Na_2CO_3} o\text{-nitrophenol}$$

o-nitrochlorobenzene — o-nitrophenol

$$p\text{-nitrochlorobenzene} \xrightarrow[CH_3OH]{CH_3O^-} p\text{-nitroanisole}$$

p-nitrochlorobenzene — p-nitroanisole

If a ring has two nitro groups in favorable positions, this type of displacement reaction is quite facile. As we will see below, favorable means that the charge of the nucleophile can be delocalized.

$$\text{2,4-dinitrochlorobenzene} \xrightarrow[\Delta]{NH_2NH_2,\ EtOH} \text{2,4-dinitrophenylhydrazine}$$

2,4-dinitrophenylhydrazine

2,4-Dinitrophenylhydrazine is a common reagent used for preparing the corresponding 2,4-dinitrophenylhydrazone derivatives of aldehydes and ketones (Section 14.6.C). These derivatives are usually crystalline compounds with well-defined melting points and are useful for characterizing aldehydes and ketones; the 2,4-dinitrophenylhydrazones are commonly abbreviated as DNPs.

The mechanism of these substitution reactions involves two steps, an addition followed by an elimination. It is analogous to the nucleophilic addition-elimination mechanism so prevalent in the chemistry of carboxylic acids and derivatives of carboxylic acids (Chapter 19).

$$R-\overset{O}{\overset{\|}{C}}-X + N:^- \rightleftharpoons \left[R-\overset{O^-}{\overset{|}{\underset{N}{\underset{|}{C}}}}-X\right] \rightleftharpoons R-\overset{O}{\overset{\|}{C}}-N + X:^-$$

In the first step the attacking nucleophile adds to the benzene ring to give a resonance-stabilized pentadienyl anion.

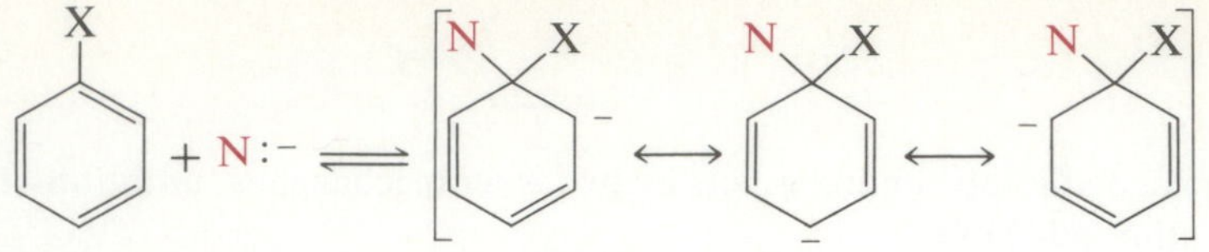

The pentadienyl anion can eject the nucleophile, regenerating the reactants, or it can eject halide ion, giving the substitution product.

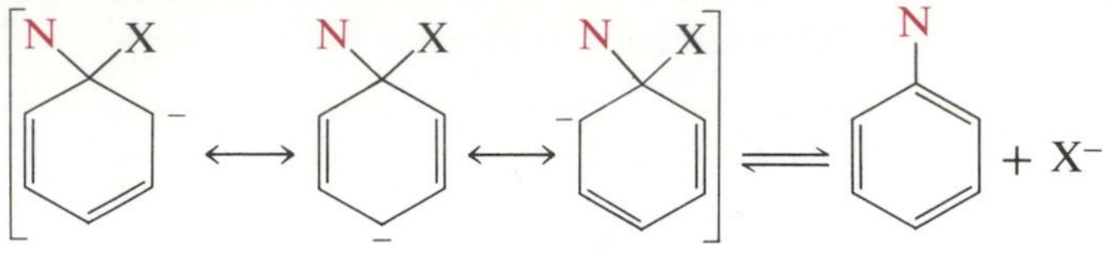

However, even a conjugated pentadienyl anion is not sufficiently stable for this mechanism to operate with such simple aryl halides as chlorobenzene or *o*-bromotoluene. Electron-attracting groups provide further resonance stabilization of the anion, thus lowering its energy enough for it to be formed as a reaction intermediate. As the foregoing resonance structures show, the nitro or carbonyl groups are most effective when they are *ortho* or *para* to the leaving group.

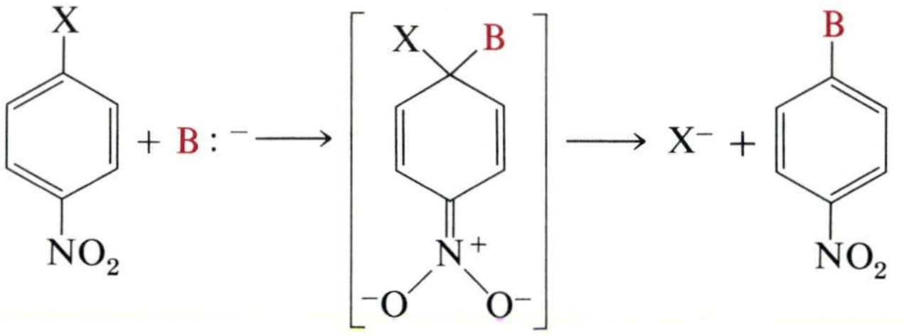

EXERCISE 30.2 Write equations illustrating the conversion of chlorobenzene to *p*-methoxyaniline.

B. Nucleophilic Aromatic Substitution: The Elimination-Addition Mechanism

After the foregoing discussion it might seem surprising that one commercial preparation of phenol involves heating chlorobenzene itself with aqueous sodium hydroxide.

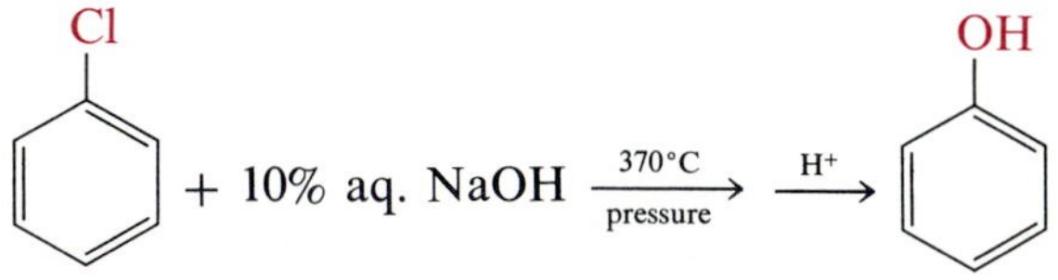

However, this reaction is not a simple displacement of chloride by hydroxide. *o*-Chlorotoluene, for example, gives not only *o*-methylphenol in this reaction but also *m*-methylphenol.

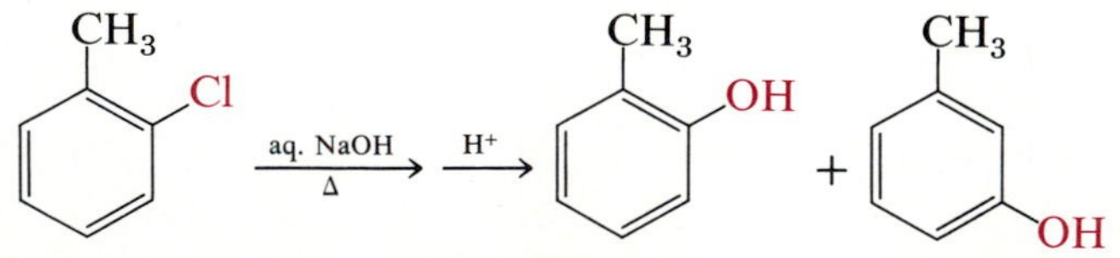

An analogous reaction occurs under milder conditions with amide ion in liquid ammonia.

$$p\text{-}CH_3C_6H_4Cl \xrightarrow[NH_3]{NH_2^-} p\text{-}CH_3C_6H_4NH_2 + m\text{-}CH_3C_6H_4NH_2$$

$$o\text{-}CH_3C_6H_4Cl \xrightarrow[NH_3]{NH_2^-} o\text{-}CH_3C_6H_4NH_2 + m\text{-}CH_3C_6H_4NH_2$$

The remarkable feature of these reactions is that the entering group substitutes not only at the position of the displaced halide but also at the ring position adjacent to the original halide. Even iodobenzene shows this behavior, as has been demonstrated using ^{14}C-labeled materials.

$$^{14}C \longrightarrow C_6H_5I^{*} \xrightarrow[NH_3]{NH_2^-} C_6H_5NH_2^{*} + C_6H_5NH_2^{*}$$

These results are rationalized by the involvement as a reactive intermediate or the product of an elimination reaction—dehydrobenzene or ''benzyne.''

$$C_6H_4(X)(H) \xrightarrow{\text{base}} [C_6H_4]$$

"benzyne"

The detailed mechanism involves a series of steps. Benzene itself is a weak acid, but its pK_a of 43 corresponds to a much higher acidity than the alkanes (pK_as $\cong$ 50). The close proximity of an electronegative halogen renders an adjacent hydrogen sufficiently acidic that it is removed by a strong base such as NH_2^-.

$$C_6H_4(I)(H) + NH_2^- \rightleftharpoons C_6H_4I^- + NH_3$$

The intermediate iodophenyl anion can itself pick up a proton to regenerate the original iodobenzene, or it can lose iodide ion.

$$C_6H_4I^- \longrightarrow I^- + C_6H_4$$

The driving force for this reaction is the formation of a stable halide ion. The ''benzyne'' generated is a very reactive intermediate. The ''triple bond'' in benzyne is highly strained. Recall that the two carbons in acetylene are *sp*-hybridized and that the H—C—C angles are 180°. That is, the two carbons of an alkyne as well as the two atoms directly attached to the triple bond comprise a linear array. For geometric reasons this is impossible when the triple bond is in a small ring. In fact, the smallest stable cycloalkyne is cyclooctyne, and even it is less stable than acyclic alkynes. Nevertheless, benzyne does form as a transient reaction intermediate. It has even been detected spectroscopically by using special techniques. The electronic structure of benzyne can be visualized readily as a distorted acetylene. A triple bond has two π-bonds, as shown in Figure 30.1. One π-bond is constructed from *p*-orbitals perpendicular to the plane of the paper (dotted line in Figure 30.1); the other π-bond derives from overlap of *p*-orbitals in the plane of the page, as illustrated. When the triple bond is distorted from linearity, one π-bond is essentially unchanged, but the other π-bond now involves hybrid orbitals directed away from each other with consequent reduced overlap. Reduced overlap means a weaker and more reactive bond. Benzyne is related in this sense to the distorted acetylenes. The two orbitals shown in Figure 30.1 provide inefficient overlap and a weak, reactive bond. The resulting strained triple bond reacts readily with any available nucleophilic reagent at either end of the triple bond.

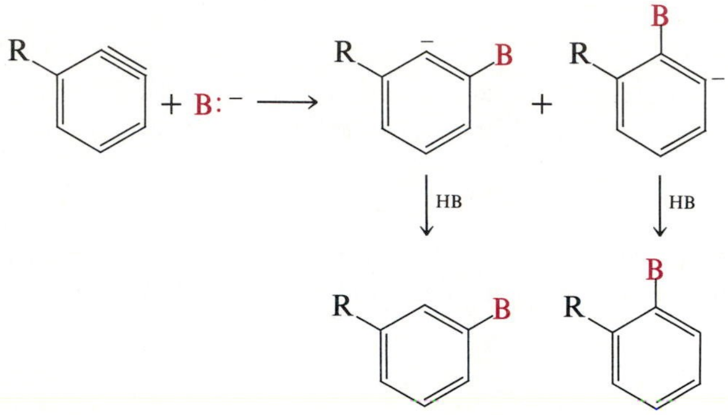

R + B:$^-$ ⟶ R, B$^-$ + R, B$^-$
HB ↓ HB ↓
R, B   R, B

Benzyne can also be generated by decomposition of the diazonium salt produced by diazotization of anthranilic acid. The diazonium compound is an inner salt and is insoluble. The dry salt will detonate and must be kept moist and handled with care. The

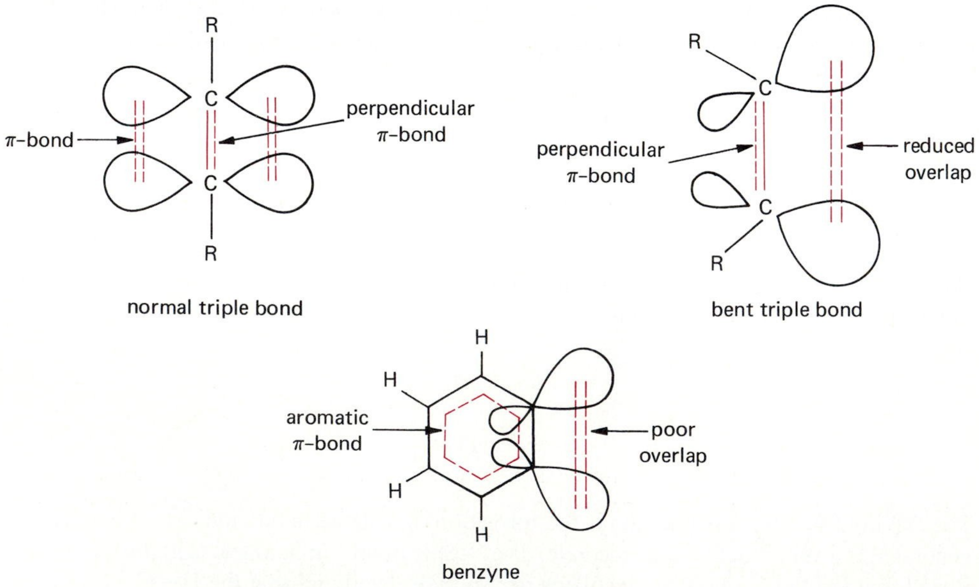

FIGURE 30.1 Electronic structure of benzyne.

controlled decomposition in ethylene chloride provides the unusual strained hydrocarbon, biphenylene.

NH_2 COOH $\xrightarrow[\text{THF}]{C_5H_{11}ONO,\ H^+}$ N_2^+ CO_2^- $\xrightarrow[\Delta]{ClCH_2CH_2Cl}$ [benzyne] $\longrightarrow$ biphenylene

anthranilic acid — benzyne — (21–30%) biphenylene

In this preparation, no nucleophilic reagent is present to react with the benzyne, and therefore dimerization occurs to a significant extent.

EXERCISE 30.3 What products are produced by treatment of *o*-, *m*-, and *p*-bromotoluene with KNH_2 in liquid NH_3?

C. Metallation

Aryl bromides form Grignard reagents in a normal fashion, and these derivatives undergo all the usual reactions of Grignard reagents; they react with aldehydes and ketones, CO_2, D_2O, and so on.

Br $\xrightarrow[\text{ether}]{Mg}$ MgBr $\xrightarrow{CO_2}$ $\xrightarrow{H^+}$ COOH

However, aryl chlorides do not react with magnesium in ether. Consequently a bromochlorobenzene can be converted into a chloro-Grignard reagent.

Br, Cl $\xrightarrow[\text{ether}]{Mg}$ MgBr, Cl $\xrightarrow{CH_3CHO}$ $\xrightarrow{H^+}$ $CHOHCH_3$, Cl

m-chlorophenylmagnesium bromide — (82–88%) 1-(*m*-chlorophenyl)ethanol

Grignard reagents *can* be produced from aryl chlorides by using tetrahydrofuran (THF) as the solvent.

Cl + Mg $\xrightarrow{\text{THF}}$ MgCl

Of course, the formation of the Grignard reagent is successful only if no functional group is present that will react with such reagents: examples of such reactive groups are NO_2, NO, COR, SO_3R, CN, OH, and NH_2.

Aryllithium reagents can be prepared from lithium metal and the aryl chloride or bromide.

$$C_6H_5Cl + 2\ Li \xrightarrow{\text{ether}} C_6H_5Li + LiCl$$

The lithium reagents generally undergo the same reactions as the Grignard reagents. Furthermore, aryllithiums can be prepared by **transmetallation** of an aryl bromide or iodide with an alkyllithium.

$$C_6H_5Br + C_4H_9Li \xrightarrow{\text{ether}} C_6H_5Li + C_4H_9Br$$

(85%)

With aryl bromides or iodides the reaction is rapid, even at low temperature. The reaction can be regarded as a displacement reaction on halogen to form the lithium salt of a more stable anion.

$$ArX + R^- \ Li^+ \longrightarrow Ar^- \ Li^+ + RX$$

Transmetallation is most successful with aromatic bromides and iodides; aryl chlorides do not react as cleanly or rapidly.

When a mixture of an alkyl halide and an aryl halide is treated with sodium in ether, coupling occurs to give the arylalkane. The reaction, known as the **Wurtz-Fittig** reaction, is sometimes an excellent preparative method.

$$C_6H_5Br + CH_3(CH_2)_3CH_2Br + 2\ Na \xrightarrow[20°C]{\text{ether}} C_6H_5(CH_2)_4CH_3 + 2\ NaBr$$

(65–70%)

In this reaction the sodium reacts first with the aryl bromide to form the arylsodium which then reacts with the alkyl bromide by the S_N2 mechanism.

$$ArBr + 2\ Na \longrightarrow ArNa + NaBr$$
$$ArNa + RX \longrightarrow ArR + NaX$$

Under the proper conditions, the yields are quite good. Like all organometallic reactions, the Wurtz-Fittig reaction is only applicable to compounds not having highly reactive functional groups (hydroxy, carbonyl, nitro, and so on).

A reaction that is superficially related to the Wurtz-Fittig reaction is the **Ullmann** reaction, in which two molecules of an aryl halide are coupled by heating with copper powder. The product is a **biaryl,** a compound in which two benzene rings are joined together.

$$o\text{-}ClC_6H_4NO_2 \xrightarrow[225°C]{Cu} \text{2,2'-dinitrobiphenyl}$$

2,2′-dinitrobiphenyl

The chemistry of biaryls will be detailed in a subsequent chapter (Chapter 31). The Ullmann reaction works well with chlorides, bromides, and iodides and is facilitated by electron-attracting groups such as NO_2 and CN functions. The reaction involves the formation of an arylcopper intermediate that undergoes a free-radical-like coupling, probably while still coordinated to copper. A dramatic acceleration in rate is achieved by ultrasonic activation.

$$\text{2-iodonitrobenzene} \xrightarrow[\text{20 kHz, 60–64 °C}]{\text{Cu,)))}} \text{2,2'-dinitrobiphenyl } (24\text{–}81\%)$$

A method of coupling aryl halides that is complementary to the Ullmann reaction involves formation of the diarylcuprate (Sections 8.8 and 20.3.A), which is oxidized by oxygen at low temperature.

$$2\ C_6H_5Li \xrightarrow[\text{THF, 0 °C}]{\text{CuBr}} (C_6H_5)_2CuLi \xrightarrow[\text{THF, −70 °C}]{O_2} C_6H_5{-}C_6H_5$$

lithium diphenylcuprate — biphenyl (75%)

The cuprate method avoids the high temperatures of the Ullmann reaction, but it is only applicable with arenes lacking functional groups that react with the aryllithium intermediate. Ultrasonic coupling with copper is then preferred.

Aryl iodides and aryl bromides react with vinyl derivatives such as methyl acrylate to yield β-aryl-α,β-unsaturated esters in one step carried out under relatively mild conditions using palladium catalysts. Substituents such as the nitro group do not interfere with the reaction. Although palladium acetate is a suitable catalyst, it is common to add triphenylphosphine. The general transformation can be carried out with heterocyclic, benzyl, and vinyl halides as well as aryl halides and is called the **Heck** reaction (see Chapter 35 for a literature search). The reaction with aryl halides involves a cycle of steps beginning with the formation of a palladoarene from the insertion of the palladium into the aryl-iodine bond. The mechanism will not be further described here.

$$C_6H_5{-}I + CH_2{=}CH{-}COOCH_3 \xrightarrow[\text{(C}_4\text{H}_9\text{)}_3\text{N, 100 °C, 1 hr}]{Pd(OAc)_2} C_6H_5{-}CH{=}CH{-}COOCH_3$$

methyl acrylate — (81%) methyl (*E*)-cinnamate

$$3\text{-}HO{-}C_6H_4{-}I + CH_2{=}CH{-}COOCH_3 \xrightarrow[\text{(C}_2\text{H}_5\text{)}_3\text{N, CH}_3\text{CN, 5 hr, 100 °C}]{Pd(OAc)_2} 3\text{-}HO{-}C_6H_4{-}CH{=}CH{-}COOCH_3$$

methyl acrylate — (95%) methyl (*E*)-3-hydroxycinnamate

$$Br{-}C_6H_4{-}I + CH_2{=}CH{-}COOCH_3 \xrightarrow[\text{(C}_2\text{H}_5\text{)}_3\text{N, CH}_3\text{CN, 5.5 hr, 100 °C}]{Pd(OAc)_2} Br{-}C_6H_4{-}CH{=}CH{-}COOCH_3$$

methyl acrylate — (68%) methyl (*E*)-4-bromocinnamate

EXERCISE 30.4 Show how *p*-bromotoluene can be converted into *p*-toluic acid (page 616), 4,4′-dimethylbiphenyl, and 4-*n*-propyltoluene.

30.4 Nomenclature of Phenols and Phenyl Ethers

Compounds having a hydroxy group directly attached to a benzene ring are called **phenols.** The term phenol is also used for the parent compound, hydroxybenzene. Hydroxybenzene may be regarded as an enol (Section 15.1), as implied by the name phenol, from **ph**enyl + **enol.** However, unlike simple ketones, which are far more stable than their corresponding enols, the analogous equilibrium for phenol lies far on the side of the enol form. The reason for this difference is the resonance energy of the aromatic ring, which provides an important stabilization of the enol form.

$$CH_3-\overset{O}{\overset{\|}{C}}-CH_3 \rightleftharpoons CH_3-\overset{OH}{\overset{|}{C}}=CH_2 \qquad \Delta H^\circ \cong +14\ \text{kcal mol}^{-1}$$

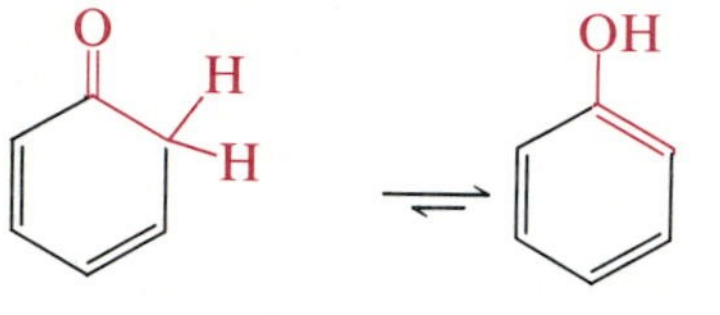

$$\Delta H^\circ \cong -16\ \text{kcal mol}^{-1}$$

cyclohexa-2,4-dien-1-one ⇌ phenol

Since the functional group occurs as a suffix in phen**ol,** many compounds containing an aromatic hydroxy group are named as derivatives of the parent compound phenol, as illustrated by the following IUPAC names. Although *Chemical Abstracts* utilizes the IUPAC-approved name phenol for the parent compound, substituted phenols are indexed as derivatives of **benzenol.** The *Chemical Abstracts* names are given in brackets for the following compounds. We will not use the name benzenol in this text.

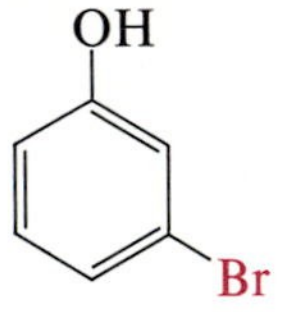

m-bromophenol [3-bromobenzenol]

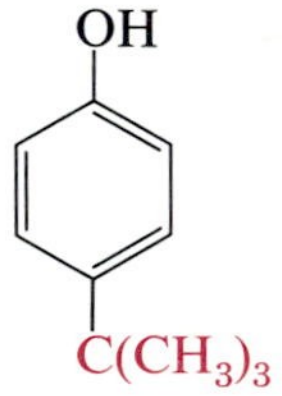

p-*t*-butylphenol [4-(1,1-dimethylethyl)benzenol]

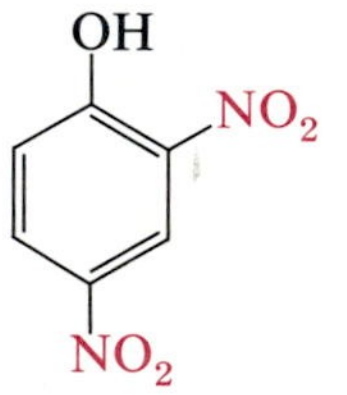

2,4-dinitrophenol [2,4-dinitrobenzenol]

Suffix groups such as sulfonic acid and carboxylic acid take priority, and when these groups are present, the hydroxy group is used as a modifying prefix.

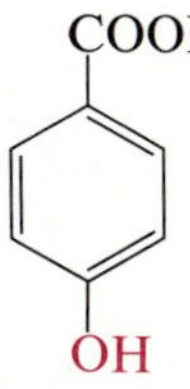

p-hydroxybenzoic acid

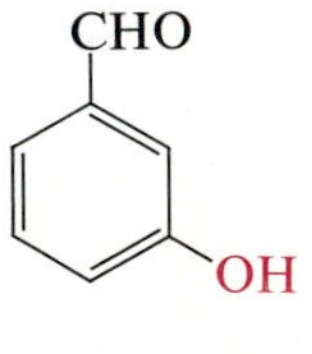

m-hydroxybenzaldehyde

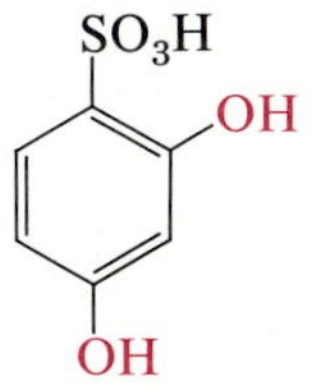

2,4-dihydroxybenzenesulfonic acid

Phenyl ethers are named in the IUPAC system as alkoxyarenes, although the ''ether'' nomenclature is used for some compounds.

$OC(CH_3)_3$ — *t*-butoxybenzene

$CH_3C_6H_4OCH_2C_6H_5$ — *p*-benzyloxytoluene

$C_6H_5OC_6H_5$ — phenyl ether

Phenols and their ethers are widespread in nature, and, as is usual for such compounds, trivial names abound. Many of these names are in such common use that they should be learned.

OCH_3 — anisole, methoxybenzene

OC_2H_5 — phenetole, ethoxybenzene

CH_3, OH — *o*-cresol, *o*-methylphenol

OH, OH — catechol, *o*-dihydroxybenzene

OH, OH — resorcinol, *m*-dihydroxybenzene

OH, OH — hydroquinone, *p*-dihydroxybenzene

OH, O_2N, NO_2, NO_2 — picric acid, 2,4,6-trinitrophenol

$COOH$, OH — salicylic acid, *o*-hydroxybenzoic acid

OH, OCH_3 — guaiacol, *o*-methoxyphenol

Note that compounds with more than one hydroxy group are named with the hydroxy prefix. Terms such as "phen-diol" or "benzene-triol" are not used.

More complex examples of naturally occurring phenols are urushiols and lignins.

OH, OH, $(CH_2)_7CH{=}CH(CH_2)_5CH_3$

a urushiol

Urushiols are the active constituents of the allergenic oils of poison ivy, sumac, and oak. They are C-3 alkylated catechols in which the side chain can be saturated or can contain up to three double bonds. In poison ivy and poison sumac the side chain contains 15 carbons; in poison oak it contains 17 carbons.

Lignins are complex natural polymers that occur together with cellulose in the "woody" part of plants such as shrubs and trees. Softwoods contain 26–32% of lignin and hardwoods contain about 17–25%. Lignin is second in abundance to cellulose as a renewable organic resource. Because lignins are high molecular weight polymers, their exact structures are not known. The links between cellulose and lignin contribute to the strength of the wood by acting as **cross-links** between the long cellulose polymer molecules. They are composed of three basic building blocks: coniferyl alcohol, sinapyl alcohol, and *p*-coumaryl alcohol.

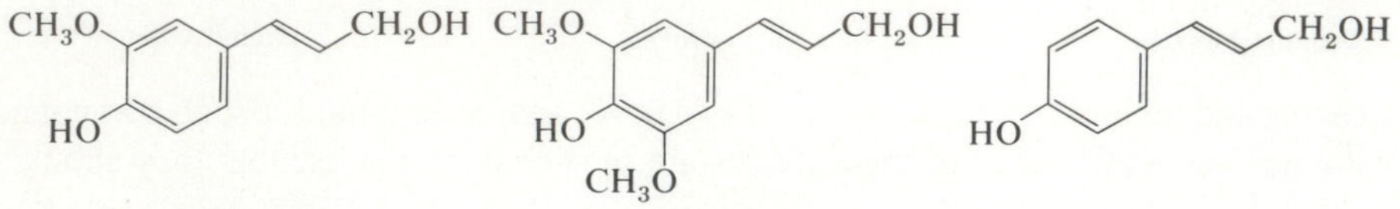

coniferyl alcohol sinapyl alcohol *p*-coumaryl alcohol

Different plants apparently have lignins of different composition. Guaiacyl lignins are derived mainly from coniferyl alcohol. Guaiacyl-syringyl lignins are also derived principally from coniferyl alcohol, as well as some sinapyl alcohol. *p*-Coumaryl alcohol is a minor constituent (1–5%) of all lignins. The lignin polymers contain a variety of types of linkages between the phenolic monomers and to cellulose units in wood. Most of the lignin must be removed to avoid colored products made from the cellulose fibers of wood. Cleaving the bond to cellulose or oxidizing the lignin components is important in the preparation of white paper such as newsprint in the paper industry. Using chlorine dioxide, ClO_2, as the oxidant in place of Cl_2 diminishes the proportion of chlorine substituted into the aromatic rings. Enzymatic cleavage of the lignin-cellulose bond by bacteria is under intensive study. Decreasing the ecological impact of the processes involved in the removal of lignin is an important field of current research.

30.5 Preparation and Properties of Phenols and Phenyl Ethers

Phenols are generally crystalline compounds with distinctive odors. Phenol itself melts at 40.9 °C, but is often found to be semiliquid because of the presence of water, which lowers the melting point; a mixture of 8% water and 92% phenol is liquid at room temperature. Although only 6.7 g dissolve in 100 mL of cold water, phenol is totally miscible with hot water. The lower alkylphenols are sparingly soluble in water; for example, *o*-cresol dissolves to the extent of 2.5 g per 100 mL of water at 25 °C. Phenol and the cresols are widely used in commercial disinfectants. Phenol turns pink on exposure to air because of oxidation. The sensitivity of phenols to air oxidation is enhanced by the presence of more than one hydroxy group and by alkali. The oxidation of phenols to quinones will be discussed in Section 30.8.B. Phenols are sufficiently acidic that they are caustic toward flesh and are poisonous.

The lower phenyl ethers are liquids; for example, anisole boils at 154 °C. Unlike phenols, the ethers are essentially insoluble in water. They lack the hydroxy group of phenol, which can hydrogen bond to water oxygens. The ether oxygens have relatively low basicity and form only weak hydrogen bonds to water hydrogens. The low basicity of the oxygens of phenyl ethers compared to aliphatic ethers stems from conjugation of a lone pair with the aromatic ring. The same phenomenon is responsible for the reduced basicity of aromatic amines compared to aliphatic amines.

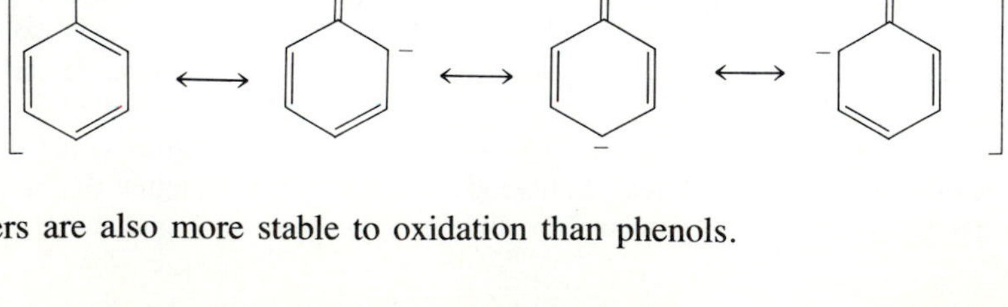

Aryl ethers are also more stable to oxidation than phenols.

A. Preparation of Phenols

All of the important preparations of phenols involve reactions that have already been discussed. Fusion of arenesulfonic acids with alkali hydroxide is an excellent method in cases where sulfonation gives a good yield of sulfonic acid and no base-sensitive functional group is present (Section 26.5).

$$ArSO_3^- \xrightarrow[\Delta]{NaOH} ArO^- \xrightarrow{H^+} ArOH$$

The hydrolysis of haloarenes with alkali at high temperature is a commercial preparation, but is not suitable for general laboratory use because of the formation of mixtures of isomeric phenols via benzyne intermediates (Section 30.3.B).

$$ArCl \xrightarrow[\Delta]{NaOH} ArO^- \xrightarrow{H^+} ArOH$$

On the other hand, compounds in which the halogen is *ortho* or *para* to strongly electron-attracting groups, such as nitro groups, undergo hydrolysis by the addition-elimination mechanism to give phenols in good yield (Section 30.3.A). The hydrolysis of arenediazonium salts is a good route to many phenols and has been discussed in Section 25.5.B.

$$ArN_2^+ \xrightarrow[\Delta]{H_2O} ArOH + H_3O^+ + N_2$$

An important industrial preparation of phenol involves the oxidation of cumene, an inexpensive hydrocarbon that can be prepared by alkylation of benzene (Section 23.4.B).

$$C_6H_6 + CH_3CH{=}CH_2 \xrightarrow{H_2SO_4} C_6H_5CH(CH_3)_2$$

cumene

Cumene is oxidized using air to obtain cumene hydroperoxide.

$$C_6H_5CH(CH_3)_2 + O_2 \longrightarrow C_6H_5C(CH_3)_2OOH$$

cumene hydroperoxide

The oxidation is a typical free radical chain process. It is especially facile because the intermediate cumyl radical is tertiary and benzylic.

$$C_6H_5CH(CH_3)_2 + R\cdot \longrightarrow C_6H_5\dot{C}(CH_3)_2 + RH$$

$$C_6H_5\dot{C}(CH_3)_2 + O_2 \longrightarrow C_6H_5C(CH_3)_2OO\cdot$$

$$C_6H_5C(CH_3)_2OO\cdot + C_6H_5CH(CH_3)_2 \longrightarrow C_6H_5C(CH_3)_2OOH + C_6H_5\dot{C}(CH_3)_2$$

propagation steps

The cumene hydroperoxide is treated with sulfuric acid to obtain phenol and acetone.

$$C_6H_5C(CH_3)_2OOH \xrightarrow{H_2SO_4} C_6H_5OH + CH_3\overset{O}{\overset{\|}{C}}CH_3$$

This interesting fragmentation is related to the Baeyer-Villiger oxidation (page 411).

EXERCISE 30.5 Write a mechanism for the reaction of cumene hydroperoxide with sulfuric acid.

The cumene process is a good example of the economics of industrial organic synthesis. It owes its economic feasibility in part to the fact that *two* important commercial products are formed. Note that in the overall process from benzene and propene to phenol and acetone the only reagent that is consumed is oxygen. In fact, the overall process amounts to a circuitous oxidation of both starting hydrocarbons. The sulfuric acid that is used both in the alkylation and in decomposition of the hydroperoxide is recycled.

EXERCISE 30.6 Write equations illustrating four ways by which benzene can be converted into phenol.

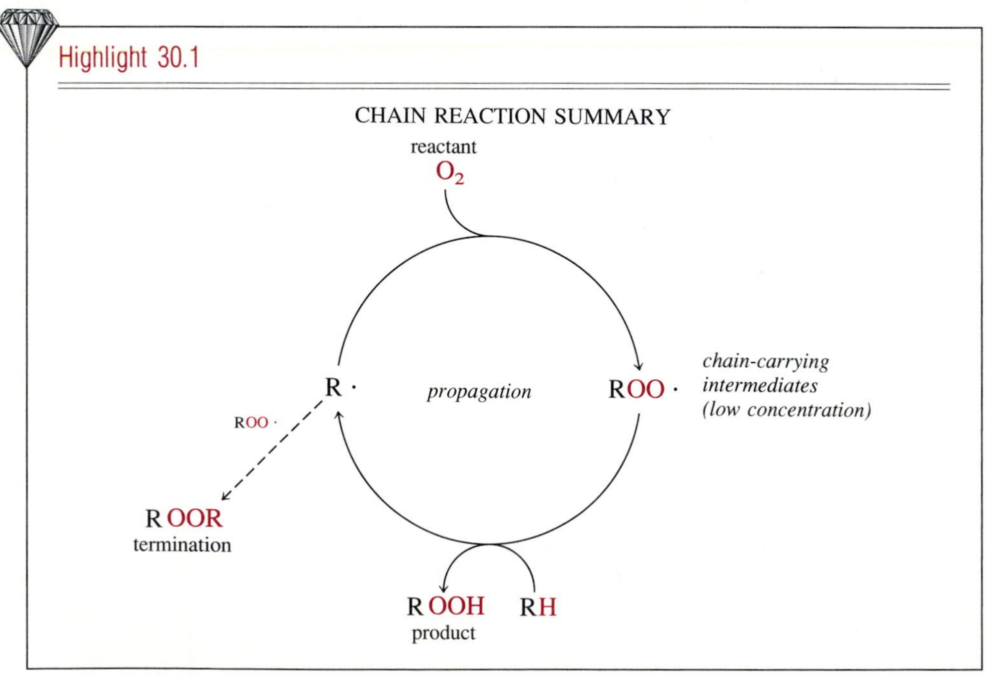

B. Acidity of Phenols

Phenol has pK_a = 10.00. The pK_as of some substituted phenols are summarized in Table 30.1; most values are in the range from 8 to 10.

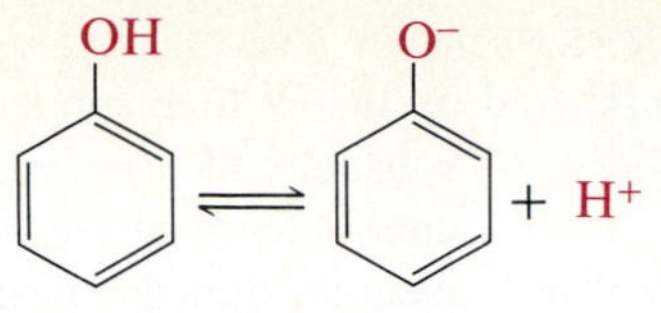

$$K_a = 1.0 \times 10^{-10}\ M \qquad pK_a = 10.00$$

Phenols are generally several orders of magnitude less acidic than carboxylic acids, but are far more acidic than alcohols. Recall that the value of the pK_a corresponds to that pH at which the conjugate acid and base are in equal concentrations. Thus, pK_as of 8–10 imply that phenols will dissolve in dilute alkali hydroxide solutions (pH 12–14) and water-insoluble phenols will not dissolve in aqueous sodium bicarbonate (pH ≅ 6–7). Carboxylic acids dissolve in aqueous bicarbonate.

TABLE 30.1 Acidities of Phenols

	pK_a (25 °C)		
Substituent	*ortho*	*meta*	*para*
H	10.00	10.00	10.00
methyl	10.29	10.09	10.26
fluoro	8.81	9.28	9.81
chloro	8.48	9.02	9.38
bromo	8.42	8.87	9.26
iodo	8.46	8.88	9.20
methoxy	9.98	9.65	10.21
methylthio		9.53	9.53
cyano			7.95
nitro	7.22	8.39	7.15

Table 30.1 shows the expected effect of substitution on acidity. Electron-donating groups on the ring are acid weakening. Halogens increase acidity, and strongly electron-attracting groups such as cyano and nitro have pronounced acid-strengthening effects. The stability of the phenolate anion is increased since the negative charge in the anion can be delocalized onto the oxygen or nitrogen of the substituent.

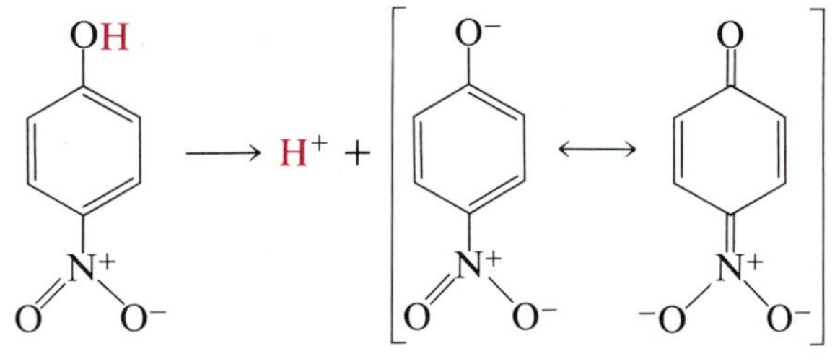

Note that a nitro group is more acid strengthening when it is *ortho* or *para* to the hydroxy group. Dinitrophenols are comparable to carboxylic acids in acidity; for example, the pK_a of 2,4-dinitrophenol is 4.09. Picric acid, 2,4,6-trinitrophenol, has pK_a = 0.25 and is a rather strong acid, comparable to trifluoroacetic acid.

EXERCISE 30.7 Using graph paper, plot the pK_as of phenols from Table 30.1 versus the pK_as of the corresponding anilinium ions from Table 24.5. Explain the result.

EXERCISE 30.8 Calculate the pH of a 0.1 M solution of phenol in water. What is the pH of a solution containing 0.1 M phenol and 0.1 M sodium phenolate?

C. Preparation of Phenyl Ethers

Alkyl phenyl ethers can be prepared by the Williamson synthesis—the S_N2 reaction of phenoxide ions with alkyl halides. As is usually the case with S_N2 reactions, this preparation works best for primary halides and is least successful with tertiary halides. The reaction can be carried out in water, acetone, dimethylformamide, or even alcohol. Because of the great difference in acidities of alcohols and phenols, the following equilibrium lies far to the left.

$$ArO^- + ROH \rightleftharpoons ArOH + RO^-$$

Thus, reaction of the alkyl halide with alkoxide ion is not an important side reaction. An example of the preparation of an aryl ethers is the formation of 1-bromo-3-phenoxypropane from phenol and 1,3-dibromopropane.

$$C_6H_5OH + BrCH_2CH_2CH_2Br \xrightarrow[\Delta]{\text{aq. NaOH}} C_6H_5OCH_2CH_2CH_2Br$$

(84–85%)
1-bromo-3-phenoxypropane

For the preparation of aryl methyl ethers, dimethyl sulfate is especially convenient.

$$C_6H_5OH \xrightarrow[(CH_3)_2SO_4]{\text{NaOH}} C_6H_5OCH_3$$

(72–75%)

Ethers can also be prepared by nucleophilic aromatic substitution in suitable cases (see Section 30.3.A).

$$\text{2,4-}(NO_2)_2C_6H_3Cl \xrightarrow[CH_3OH]{CH_3O^-} \text{2,4-}(NO_2)_2C_6H_3OCH_3$$

2,4-dinitroanisole

EXERCISE 30.9 Write an equation illustrating the synthesis of allyl phenyl ether, $C_6H_5OCH_2CH{=}CH_2$.

30.6 Reactions of Phenolate Ions

Phenolate ions can be considered as enolate ions, and many of the reactions of phenolate ions point up this relationship. It is convenient to distinguish such reactions from those of the conjugate acids, the phenols. Many reactions of phenols resemble those of the corresponding ethers enough to be considered together.

A. Halogenation

The reaction of an aqueous solution of phenol with bromine gives a precipitate of 2,4,4,6-tetrabromocyclohexa-2,5-dienone. This precipitate is normally washed with aqueous sodium bisulfite to generate 2,4,6-tribromophenol. The reactive form of phenol in this process is the phenolate ion. As successive bromines are introduced into the ring, the products are progressively more acidic, and a greater fraction of the phenol is present in the phenolate form. Thus each bromophenol is attacked more rapidly than the previous one, until the product is no longer a phenol. The overall process is similar to the bromination of a ketone under basic conditions (Section 15.1.D).

Br$_2$ → −H$^+$ ⇌ Br$_2$ → −H$^+$ ⇌ Br$_2$ → −H$^+$ ⇌ Br$_2$ → precipitates NaHSO$_3$ →

Corresponding reactions occur with chlorine and iodine and with other phenols. The net reaction is halogenation of all available *ortho* and *para* positions. Halogenation of phenol is possible under acid conditions, and the incorporation of successive halogens can be controlled (Section 30.7.B). Here also we see an analogy to the acid-catalyzed, as well as the base-catalyzed, halogenation of carbonyl compounds (Section 15.1.D).

B. Addition to Aldehydes

A characteristic reaction of enolate ions from aldehydes and ketones is the condensation with other carbonyl groups as in the aldol addition reaction (Section 15.2). A similar reaction occurs with phenolate ions. Phenol reacts with formaldehyde in the presence of dilute alkali to give a mixture of *o*- and *p*-hydroxybenzyl alcohols.

CH_2O → CH_2O^- + CH_2O^- ⇌ CH_2OH CH_2OH

The reaction is difficult to control because of further condensations that lead to a polymeric product.

Under proper conditions the final product is a dark, brittle, cross-linked polymer known as Bakelite, one of the oldest commercial plastics. Such polymers belong to a general class called phenol-formaldehyde resins.

EXERCISE 30.10 Draw the structure of a portion of the resin derived from phenol and formaldehyde.

C. Kolbe Synthesis

The reaction of carbanions with carbon dioxide to give carboxylate salts (Section 18.6.B) has its counterpart in the reaction of phenolate ions with CO_2.

$$\text{resorcinol} + CO_2 \xrightarrow{\text{aq. } KHCO_3} \xrightarrow{H^+} \text{2,4-dihydroxybenzoic acid (57–60\%)}$$

resorcinol

(57–60%)
2,4-dihydroxybenzoic acid

The reaction of phenols with carbon dioxide under basic conditions is called the **Kolbe synthesis.** The product depends on the manner in which the reaction is carried out. Carbonation of sodium phenolate at relatively low temperature gives sodium salicylate. However, the reaction is reversible, and best yields are obtained only if carried out under pressure.

$$C_6H_5O^- Na^+ + CO_2 \underset{}{\overset{150\ ^\circ C}{\rightleftharpoons}} o\text{-}HOC_6H_4CO_2^- Na^+$$

sodium salicylate
sodium *o*-hydroxybenzoate

Under more severe conditions isomerization to the more stable *para* isomer occurs. Thus potassium salicylate smoothly isomerizes to the *para* isomer at 240 °C.

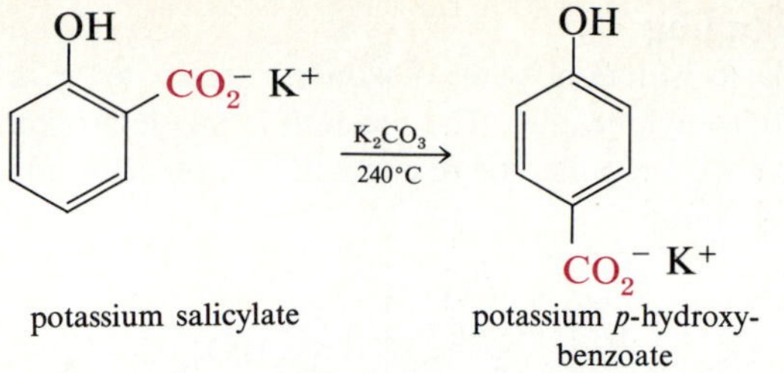

OH CO_2^- K^+ → (K_2CO_3, 240°C) OH, CO_2^- K^+

potassium salicylate — potassium *p*-hydroxybenzoate

Although the Kolbe synthesis may be written simply as the enolate condensation of phenolate ion with carbon dioxide it is clear that coordination phenomena are involved. However, these mechanistic details are not yet fully understood.

D. Reimer-Tiemann Reaction

The **Reimer-Tiemann** reaction is the reaction of a phenol with chloroform in basic solution to give a hydroxybenzaldehyde. Reaction occurs primarily in an *ortho* position unless both are blocked. The reaction mechanism involves the prior formation of dichlorocarbene by the reaction of chloroform with alkali (page 287).

$$CHCl_3 + OH^- \rightleftharpoons {}^-{:}CCl_3 \longrightarrow {:}CCl_2$$

dichlorocarbene

The dichlorocarbene then reacts with the phenolate ion to give a dichloromethyl compound, which rapidly hydrolyzes.

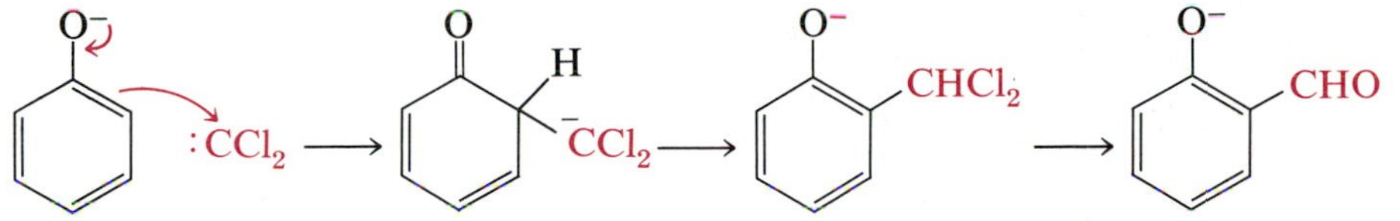

O^- + $:CCl_2$ → O, H, $\bar{C}Cl_2$ → O^-, $CHCl_2$ → O^-, CHO

The final hydrolysis reaction is facilitated by the phenoxide ion in the following way.

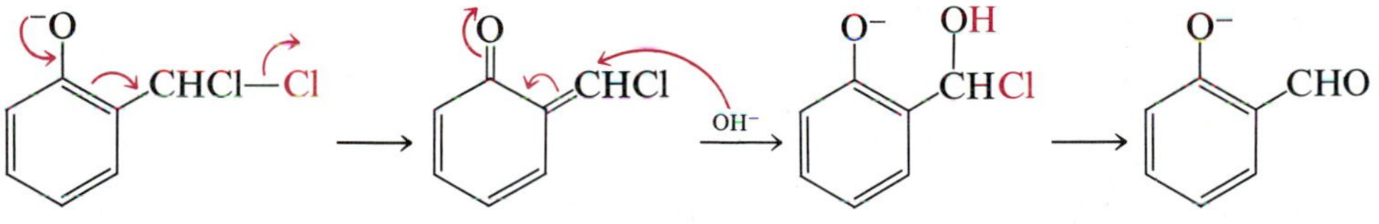

^-O, CHCl—Cl → O, CHCl → (OH^-) O^-, OH, CHCl → O^-, CHO

The essential correctness of the overall mechanism is revealed by an interesting by-product of the Reimer-Tiemann reaction on *p*-cresol.

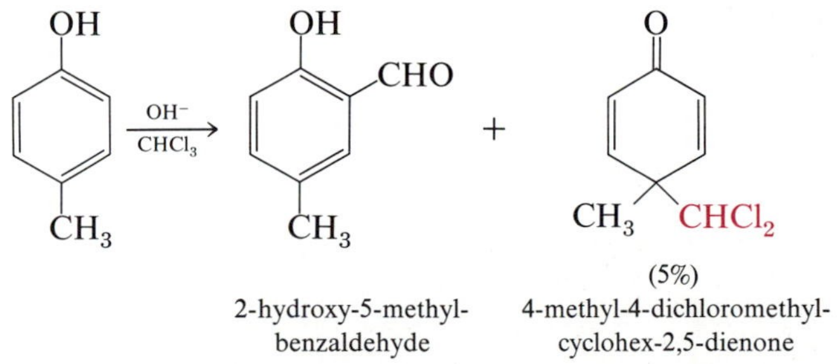

OH, CH_3 → (OH^-, $CHCl_3$) OH, CHO, CH_3 + O, CH_3, $CHCl_2$

2-hydroxy-5-methylbenzaldehyde — (5%) 4-methyl-4-dichloromethylcyclohex-2,5-dienone

EXERCISE 30.11 In the foregoing reaction explain why the dichloromethyl group is hydrolyzed to form an aldehyde group in the principal product but remains as a dichloromethyl group in the minor product.

E. Diazonium Coupling

Phenols react in basic solution with diazonium salts to give the corresponding arylazophenols (aryldiazenylphenols). The reaction is an electrophilic aromatic substitution reaction by a weak electrophile, the diazonium ion, on an aromatic ring that is highly activated by the oxide anion.

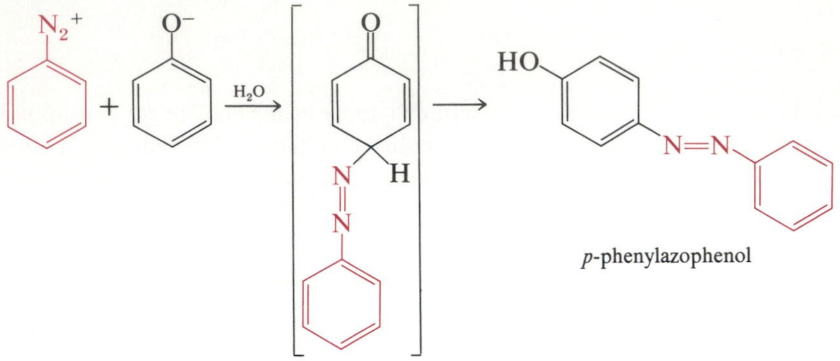

N_2^+ O^- H_2O O N H N HO N=N

p-phenylazophenol

The product is almost exclusively the *para* isomer; the *ortho* isomer is formed to the extent of only 1%. Arylazophenols constitute an important class of azo dyes. Further examples are discussed in Section 36.2.

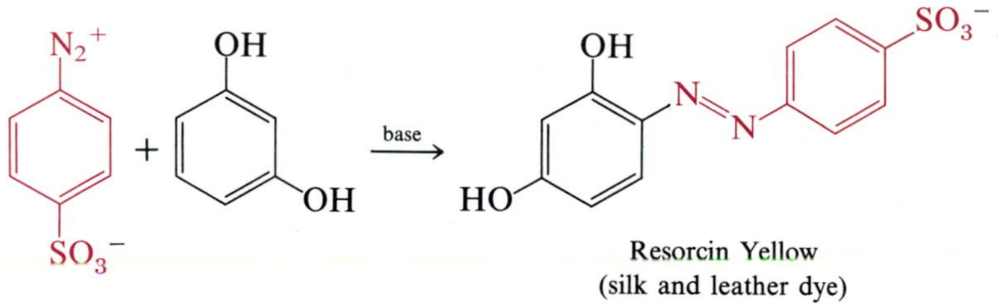

N_2^+ OH OH base OH N N SO_3^- HO SO_3^-

Resorcin Yellow
(silk and leather dye)

EXERCISE 30.12 Write equations showing the reactions (if any) of *o*-cresol with the following reagents.

(a) aqueous $NaHCO_3$
(b) (i) Br_2-H_2O; (ii) $NaHSO_3$
(c) KOH, $CHCl_3$
(d) NaOH, CO_2, 150 °C
(e) aqueous NaOH; $C_6H_5N_2^+\ Cl^-$

30.7 Reactions of Phenols and Phenyl Ethers

A. Esterification

Phenols can be converted to esters but *not* generally by direct reaction with carboxylic acids. Although the esterification equilibrium is exothermic for alcohols, it is slightly endothermic for phenols, as shown by the following gas phase enthalpies of reaction.

$$C_6H_5OH + CH_3COOH \rightleftharpoons C_6H_5OOCCH_3 + H_2O \qquad \Delta H^\circ = +1.5 \text{ kcal mole}^{-1}$$

$$C_2H_5OH + CH_3COOH \rightleftharpoons C_2H_5OOCCH_3 + H_2O \qquad \Delta H^\circ = -4.6 \text{ kcal mole}^{-1}$$

Aryl esters *can* be prepared by allowing the phenol to react with an acid chloride or anhydride under basic or acid catalysis.

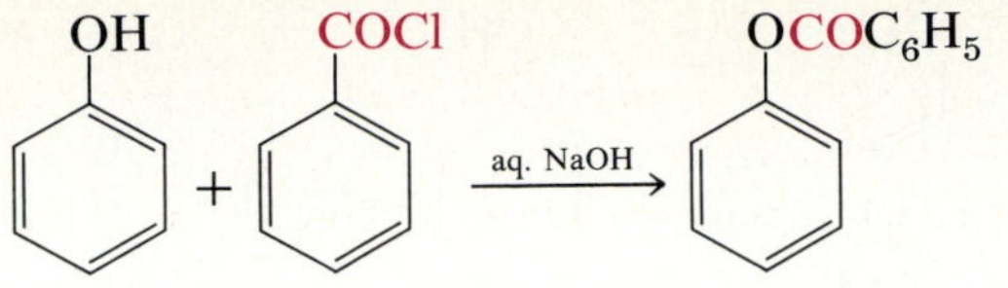

OH + COCl —aq. NaOH→ $OCOC_6H_5$

phenyl benzoate

One of the best-known aromatic acetates is acetylsalicylic acid, or aspirin, which is prepared by the acetylation of salicylic acid.

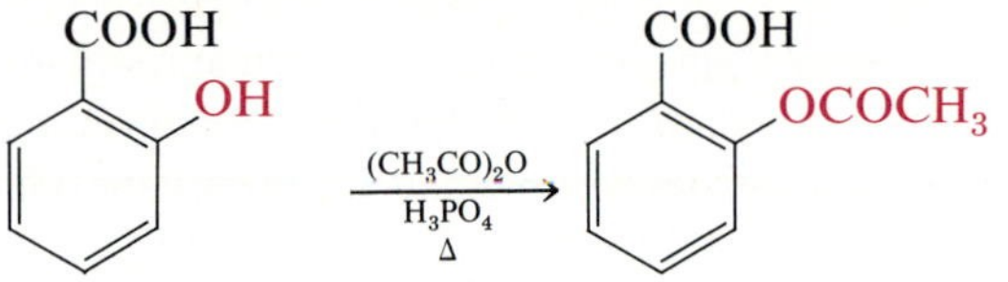

COOH, OH —$(CH_3CO)_2O$, H_3PO_4, Δ→ COOH, $OCOCH_3$

salicylic acid
o-hydroxybenzoic acid

aspirin
acetylsalicylic acid
o-acetoxybenzoic acid

Aspirin is widely used, primarily for its analgesic effect but also as an antipyretic and antirheumatic. The reactive acetyl group in acetylsalicylic acid can undergo transesterification reactions (Section 19.7.A). Acetylation of a serine hydroxyl group in the enzyme cyclooxygenase (see Section 19.11.C) inhibits the enzyme. This prevents formation of the prostaglandins (page 547) that stimulate the nerves that produce the sensation of pain.

Aspirin is not so innocuous a drug as one might imagine from its widespread use and ready availability. Repeated use may cause gastrointestinal bleeding, and large doses can provoke a host of reactions including vomiting, diarrhea, vertigo, and hallucinations. The average dose is 0.3-1 g; single doses of 10–30 g can be fatal.

B. Electrophilic Substitutions on Phenols and Phenyl Ethers

In acidic solutions electrophilic substitutions occur on the nonionized phenol. Such substitutions are still rather facile because of the activating nature of the hydroxy group, which is a strong *ortho,para* director. Reaction at one of these positions gives an intermediate cation that is essentially a protonated ketone.

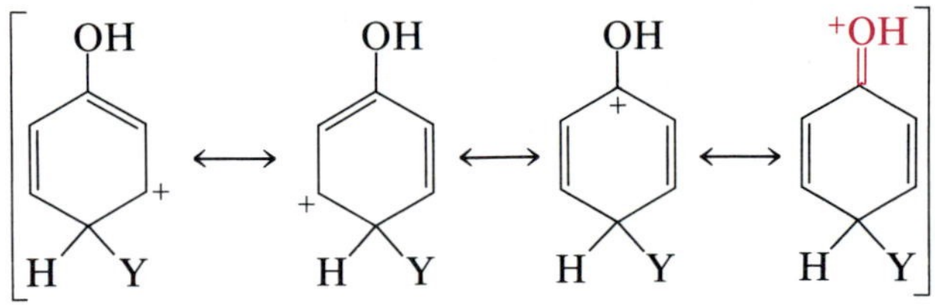

[OH, +, H Y ⟷ OH, +, H Y ⟷ OH, +, H Y ⟷ $^+$OH, H Y]

The last structure shows the role of an oxygen lone pair in stabilizing the intermediate and the transition state leading to it. Exactly the same stabilization occurs in the intermediate for electrophilic substitution of aromatic ethers so that alkoxy groups are also powerful *ortho,para* directors. Consequently, for many electrophilic aromatic substitution reactions, phenols and ethers can be considered together. The principal difference between the two groups of compounds lies in the greater water solubility of the phenols. Many electrophilic reactions of phenols can be carried out in aqueous solutions.

The phenol ring is sufficiently reactive that reaction occurs readily even with such feeble electrophiles as nitrous acid. Nitrosation is usually carried out in aqueous solution or in acetic acid; the principal product is the *p*-nitrosophenol.

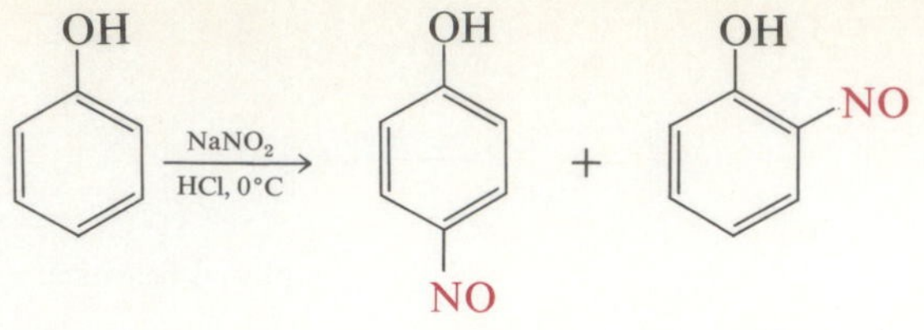

OH, NaNO$_2$ / HCl, 0°C → OH (para NO) + OH (ortho NO)

15 : 1

p-nitrosophenol *o*-nitrosophenol

Phenol is nitrated by dilute aqueous nitric acid, even at room temperature. Nitration of phenol also yields large amounts of tarry by-products produced by oxidation of the ring. Nevertheless, nitration is a satisfactory method for preparing both *o*- and *p*-nitrophenol because the isomers can be readily separated and purified.

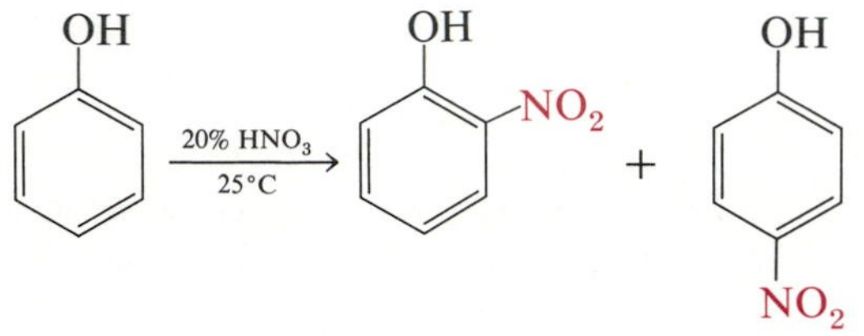

OH, 20% HNO$_3$ / 25°C → OH (ortho NO$_2$) + OH (para NO$_2$)

(30–40%) *o*-nitrophenol (15%) *p*-nitrophenol

o-Nitrophenol has lower solubility and higher volatility because of the **chelation** or intramolecular hydrogen bonding between the hydroxy group and the nitro group. Chelation (Gk., *chele,* claw) refers to formation of a ring by coordination. In this case, a cyclic O—H···O bond forms between the OH and oxygen of the nitro group.

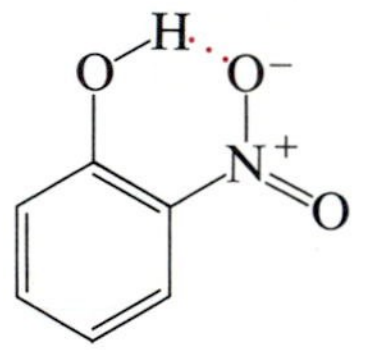

Because the acceptor hydrogen of *o*-nitrophenol is involved in chelation, it is not available for hydrogen bonding to solvent water molecules. The resulting lower solubility and higher volatility are such that *o*-nitrophenol can be steam-distilled from the reaction mixture. The *o*- and *p*-nitrophenols are also available by hydrolysis of *o*- and *p*-chloronitrobenzenes (Section 30.3.A).

2,4-Dinitrophenol can be prepared from phenol by using somewhat stronger nitric acid than is used for mononitration. However, a more convenient preparation of this phenol involves dinitration of chlorobenzene and hydrolysis of the resulting 2,4-dinitrochlorobenzene (page 1007).

Picric acid, 2,4,6-trinitrophenol, is prepared by treating phenol with concentrated sulfuric acid at 100 °C, followed by nitric acid, first at 0 °C, then at higher temperature. The first reaction that occurs in this sequence is disulfonation of the ring to give 4-hydroxybenzene-1,3-disulfonic acid. This substance is nitrated at the remaining *ortho* position by cold nitric acid. At higher temperature, sulfonic acid groups are replaced by nitro groups.

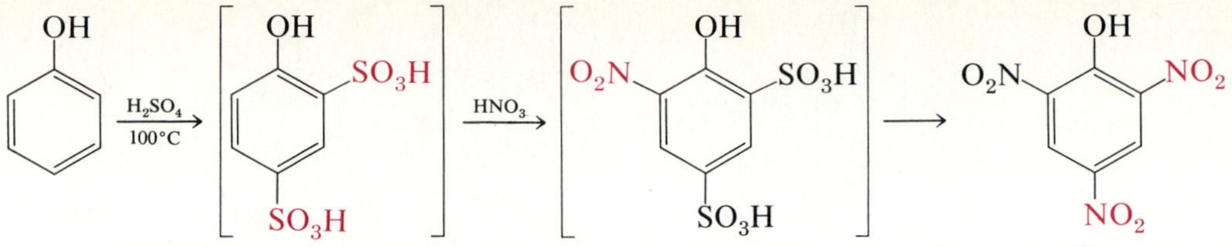

4-hydroxybenzene-1,3-disulfonic acid

(90%) picric acid

The reaction in which a sulfonic acid group is replaced by a nitro group is not uncommon in electrophilic aromatic substitutions but occurs more often as a side reaction rather than the main reaction. The mechanism is exactly the same as for substitution of a proton, except that a different cation is lost.

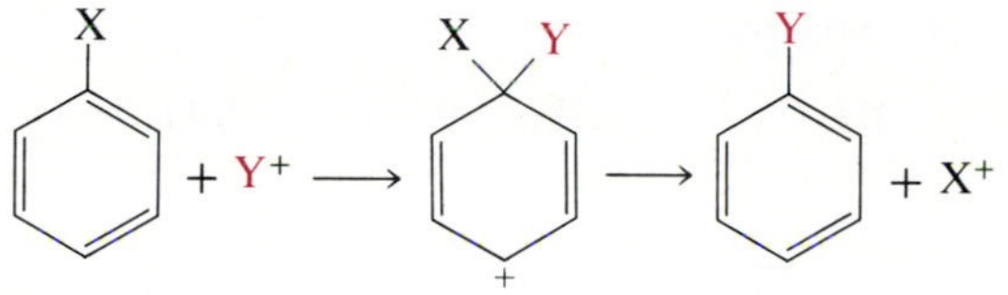

The reaction is best when strong *ortho,para*-directing groups such as OR and NR_2 are present and with functions that form relatively stable electrophilic molecules. The sulfonic acid group is prone to such replacement because it is lost as a neutral molecule, SO_3.

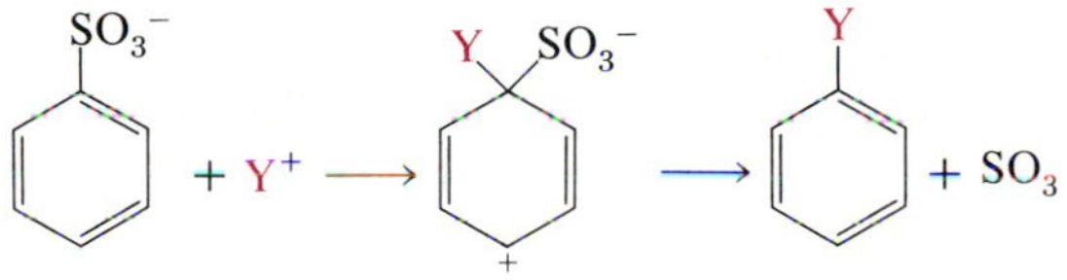

We have previously seen examples of such *ipso*-substitutions in some reactions of arylsilicon compounds (pages 844–46).

Picric acid forms yellow crystals, m.p. 123 °C and is an intensely bitter substance (Gk., *pikros,* bitter). It explodes at temperatures above 300 °C and was once used as a synthetic dye. It is important in explosives as certain salts detonate easily. At one time, picric acid was useful because of the charge-transfer complexes it forms with many compounds, especially with polycyclic aromatic hydrocarbons and their derivatives (Sections 30.8.D and 22.6.D). Such picric acid complexes are called picrates. They can be crystallized and are useful for purification purposes. On treatment with base, the picric acid component is converted to the picrate ion, which does not form complexes; thus the other component of the complex is readily recovered.

Anisole is readily nitrated to give a mixture of *o*- and *p*-nitroanisole. These compounds are also available from the corresponding chloronitrobenzenes by substitution with methoxide ion (Section 30.3.A).

Monosulfonation of phenol gives an equimolar mixture of *ortho* and *para* substitution products if the reaction is carried out at room temperature, but predominantly *p*-hydroxybenzenesulfonic acid at about 100 °C. The behavior is typical for sulfonation reactions and is due to the easy reversibility of sulfonation (Section 26.5).

OH (phenol) $\xrightarrow{98\%\ H_2SO_4}$ OH, SO_3H (ortho) + OH, SO_3H (para)

20°C	(49%)	(51%)
100°C	(10%)	(90%)

With more concentrated sulfuric acid the disulfonic acid is formed. This product can be isolated as the sodium salt or can be used directly for further reactions as in the preparation of picric acid (page 1027).

We saw in Section 30.6.A that halogenation of phenol in neutral solution involves reaction of the phenolate ion rather than the phenol itself. In acidic solution phenolate ion is suppressed, and the free phenol is involved in electrophilic halogenation. By a proper choice of reaction conditions one, two, or three halogens can be introduced into the available *ortho* and *para* positions.

p-bromophenol (mostly) $\xleftarrow[CCl_4]{Br_2}$ phenol $\xrightarrow[30°C]{Br_2,\ HBr}$ 2,4-dibromophenol (87%)

Anisole behaves in an analogous manner.

Anisole functions as an excellent substrate for Friedel-Crafts acylation. Mild reaction conditions suffice because the alkoxy group is highly activating.

OCH_3 (anisole) + succinic anhydride $\xrightarrow[C_6H_5NO_2,\ 0\text{–}5°C]{AlCl_3,\ CHCl_2CHCl_2}$ CH_3O–C_6H_4–CO–CH_2CH_2–COOH

succinic anhydride

(85%)
β-(*p*-methoxybenzoyl)-propionic acid

Other alkoxybenzenes also give good results in Friedel-Crafts acylation, provided the alkyl group is primary.

Direct Friedel-Crafts acylation of phenol is generally unsatisfactory. However, there are some exceptions. For example, treatment of phenol with acetic acid and boron trifluoride affords *p*-hydroxyacetophenone in excellent yield. Almost none of the *ortho* isomer is produced.

OH (phenol) + CH_3COOH $\xrightarrow{BF_3}$ OH, $COCH_3$ (para)

(95%)

Many other phenol acylations are known, but most proceed in low yield, partly because of competing esterification of the hydroxy group.

Phenols undergo a special Friedel-Crafts acylation with phthalic anhydride and sulfuric acid or zinc chloride. In this case two molecules of phenol condense with one molecule of phthalic anhydride to give triarylmethane derivatives known as **phthaleins.**

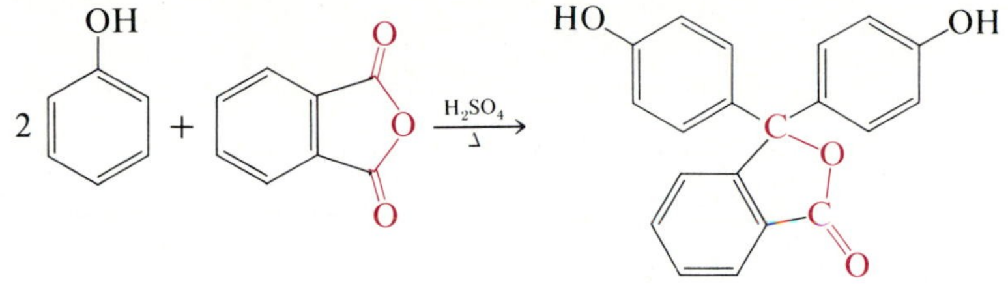

phenolphthalein
(colorless lactone form)

The phthaleins are an important class of indicators and dyes. For example, phenolphthalein has the colorless lactone structure shown in solutions below pH 8.5. Above pH 9 two protons are lost to form an intensely colored red dianion.

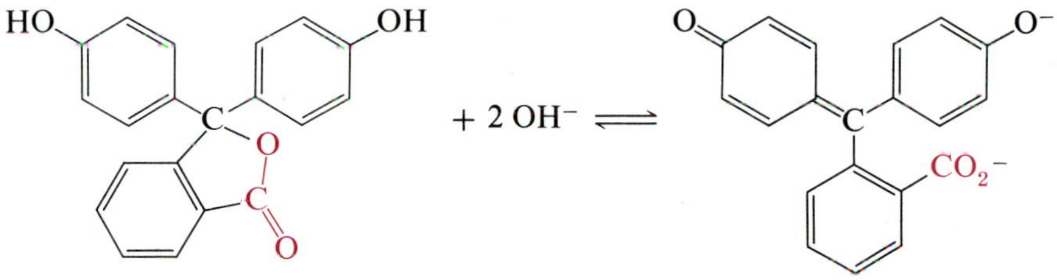

colorless

red

The color is due to a $\pi \rightarrow \pi^*$ electronic transition in the extended π-system of the ion, arising from absorption in the green and blue portions of the visible spectrum.

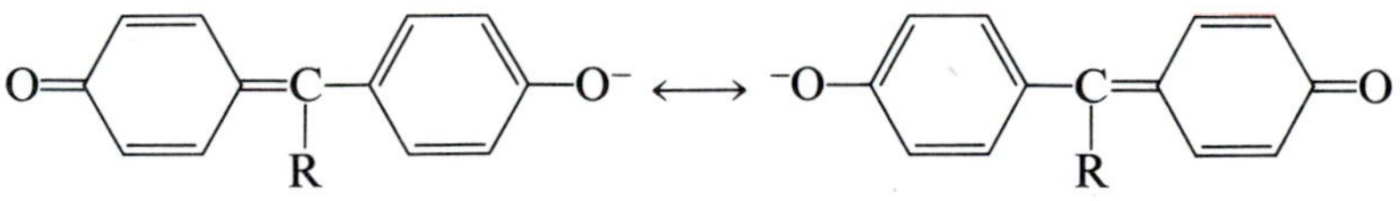

In general, the longer the conjugated system, the lower the energy of an electronic transition (Chapter 22).

Phenolphthalein is used medicinally as a laxative and is the principal active ingredient in some proprietary preparations sold as laxative agents. Its effect was discovered by accident during a Hungarian study to determine its effectiveness as a color-forming identifying agent for artificial wines. The laxative action is associated with an outflow of water into the intestine and may be due to inhibition of Na^+-K^+-ATPase. The latter catalyzes the exchange of intracellular sodium for extracellular potassium with the hydrolysis of adenosine triphosphate (ATP).

The condensation of resorcinol and phthalic anhydride gives an intensely fluorescent dye, fluorescein.

2 resorcinol + phthalic anhydride $\xrightarrow[\Delta]{H_2SO_4}$ fluorescein

resorcinol fluorescein

The yellowish green fluorescence of fluorescein is detectable even in extremely dilute solutions and has been used for tracing the course of underground rivers. Fluorescein also finds use in ophthalmology; a minute amount added to the eye provides enough fluorescence to assist the visual fitting of contact lenses under UV illumination.

Phenyl esters (Section 30.7.A) undergo a Lewis-acid-catalyzed rearrangement that amounts to intramolecular Friedel-Crafts acylation. The reaction is known as the **Fries rearrangement** and is carried out by heating the ester with aluminum chloride, often in the absence of solvent.

$OCOCH_2CH_3$ $\xrightarrow[140–150°C]{AlCl_3}$ $\xrightarrow{\text{dil. HCl}}$ (OH, $COCH_2CH_3$) + (OH, $COCH_2CH_3$)

phenyl propionate; (32–35%) *o*-hydroxypropiophenone; (45–50%) *p*-hydroxypropiophenone

The two products can be conveniently separated by fractional distillation.

EXERCISE 30.13 Why do the two foregoing hydroxypropiophenones have different volatilities? Which is expected to be the more volatile?

EXERCISE 30.14 Write equations showing the reactions, if any, of *o*-cresol with the following reagents.

(a) acetic acid, H_2SO_4
(b) acetic anhydride, H_2SO_4
(c) $NaNO_2$, HCl, 0 °C
(d) HNO_3, H_2O, 25 °C
(e) conc. H_2SO_4, 100 °C
(f) Br_2, CCl_4, 25 °C
(g) acetic acid, BF_3
(h) (i) acetic anhydride, H_2SO_4; (ii) $AlCl_3$, 150 °C
(i) phthalic anhydride, H_2SO_4, heat

C. Reactions of Ethers

Alkyl aryl ethers are cleaved by acids just as are dialkyl ethers. Hydrobromic or hydriodic acid is commonly used.

(OCH_3, OH) $\xrightarrow[\Delta]{HBr}$ (OH, OH) + CH_3Br

The reaction mechanism is the same as for aliphatic ethers; the protonated ether undergoes S_N1 or S_N2 cleavage. Because the phenyl group is not susceptible to either S_N1 or S_N2 reaction, cleavage of the aliphatic carbon-oxygen bond always occurs.

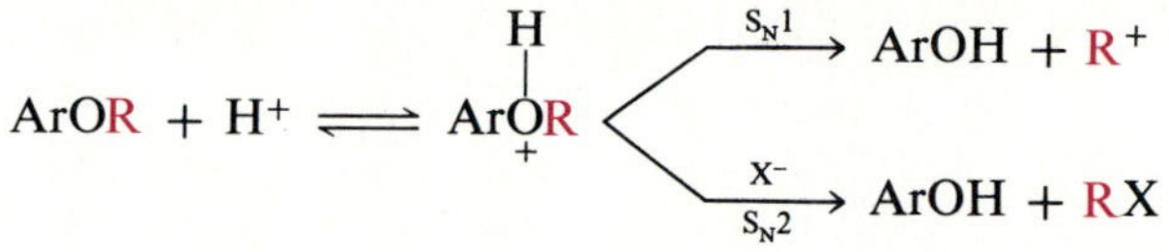

$$\text{ArOR} + \text{H}^+ \rightleftharpoons \text{Ar}\overset{\text{H}}{\underset{+}{\text{O}}}\text{R} \begin{cases} \xrightarrow{S_N1} \text{ArOH} + \text{R}^+ \\ \xrightarrow[S_N2]{X^-} \text{ArOH} + \text{RX} \end{cases}$$

When R is a tertiary alkyl group, the ether cleavage is especially facile; cleavage occurs by the S_N1 mechanism.

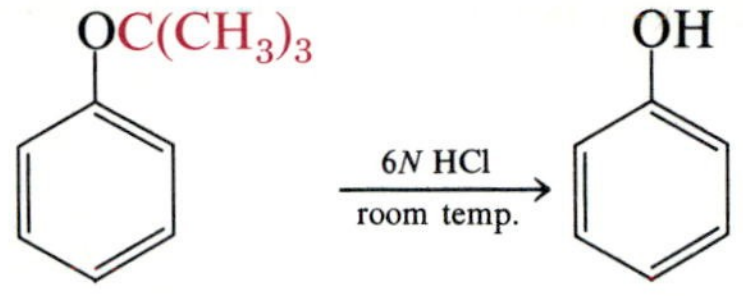

$$C_6H_5OC(CH_3)_3 \xrightarrow[\text{room temp.}]{6N\ \text{HCl}} C_6H_5OH$$

t-butyl phenyl ether

Allyl aryl ethers, like allyl vinyl ethers, undergo the Claisen rearrangement reaction (Section 21.7). The reaction requires heating to about 200 °C and results in apparent migration of the allyl group to the *ortho* position on the benzene ring.

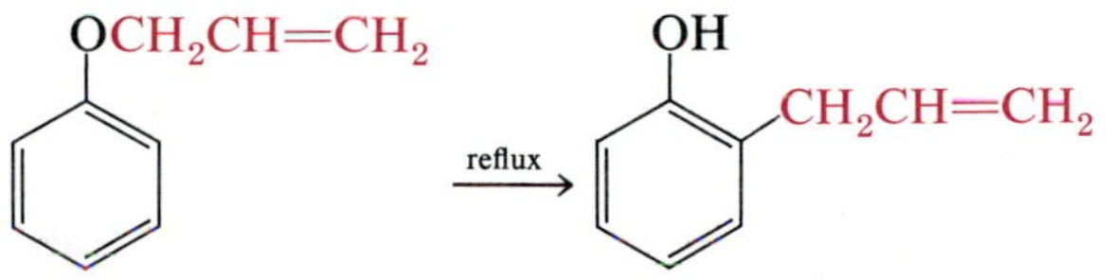

$$C_6H_5OCH_2CH{=}CH_2 \xrightarrow{\text{reflux}} o\text{-}HOC_6H_4CH_2CH{=}CH_2$$

allyl phenyl ether — (73%) *o*-allylphenol

If both *ortho* positions are occupied by substituents, rearrangement occurs to the *para* position.

The reaction mechanism involves the concerted formation of a carbon-carbon bond between the *ortho* carbon and the terminal position of the allyl group as the carbon-oxygen bond is broken.

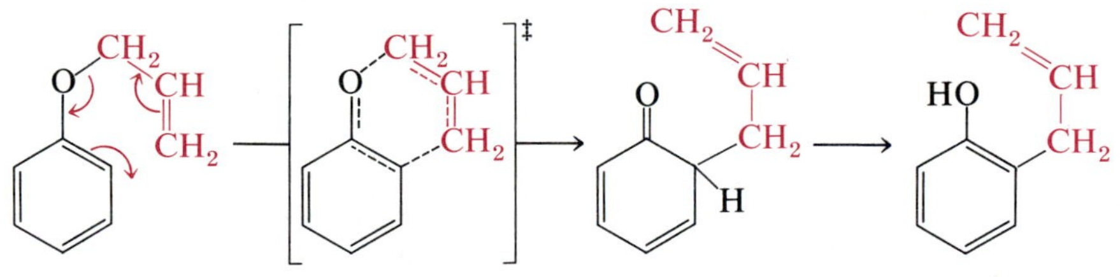

[Mechanism scheme: allyl phenyl ether (O–CH_2–CH=CH_2) → [transition state]‡ → cyclohexadienone with H and $CH_2CH{=}CH_2$ at ortho carbon → *o*-allylphenol (HO, CH_2–CH=CH_2)]

The reaction is of the pericyclic type with an aromatic six-electron transition state.

Note that the γ-carbon of the allyl group becomes attached to the benzene ring. When the ether contains an unsymmetrical allyl group the allylic rearrangement is apparent.

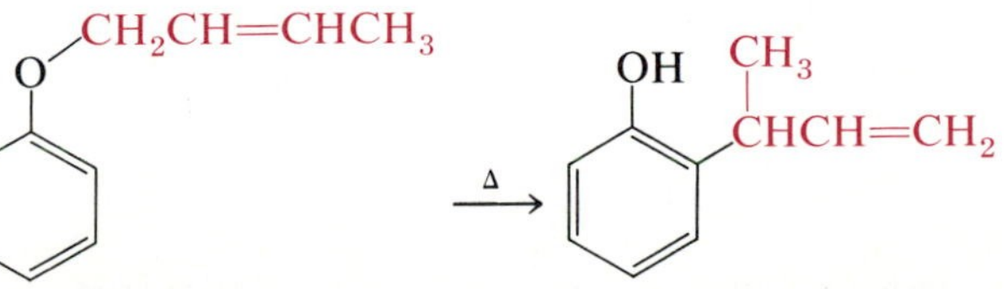

$$C_6H_5OCH_2CH{=}CHCH_3 \xrightarrow{\Delta} o\text{-}HOC_6H_4CH(CH_3)CH{=}CH_2$$

o-(α-methylallyl)phenol

If no *ortho* hydrogen is available, enolization to the phenol cannot occur, and a second rearrangement occurs to the *para* position.

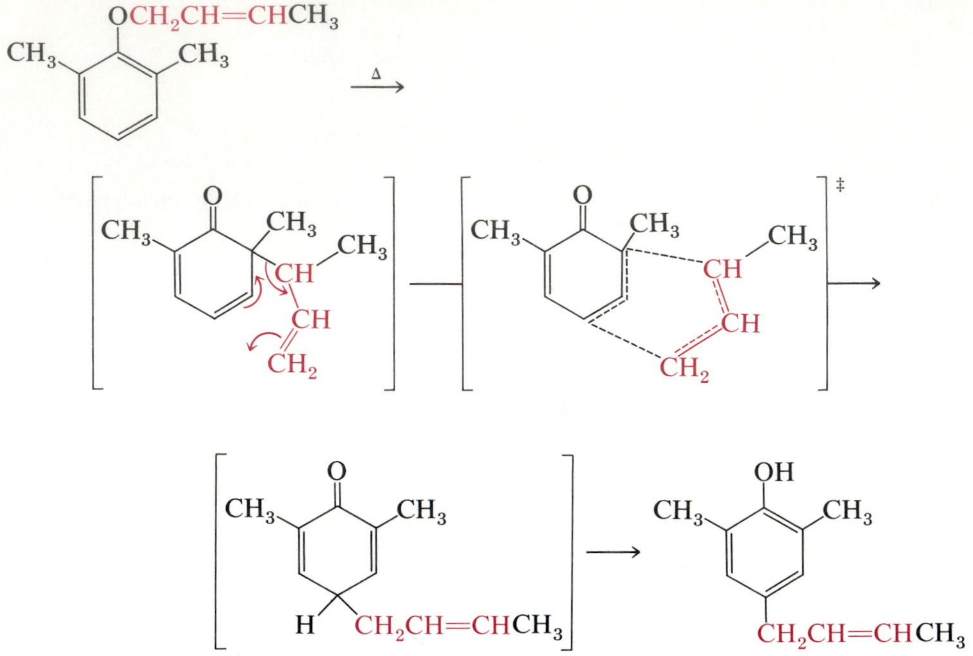

$OCH_2CH{=}CHCH_3$, CH_3, CH_3 $\xrightarrow{\Delta}$ [CH_3, O, CH_3, CH, CH, CH_2, CH_3] — [CH_3, O, CH_3, CH, CH, CH_2, CH_3]‡ → [CH_3, O, CH_3, H, $CH_2CH{=}CHCH_3$] → OH, CH_3, CH_3, $CH_2CH{=}CHCH_3$

2,6-dimethyl-4-(but-2-enyl)phenol

Note that two successive allylic rearrangements restore the original orientation of the allylic group. The second rearrangement is an example of a Cope rearrangement (Section 21.7).

EXERCISE 30.15 Write the equations showing the following reaction sequences starting with *p*-cresol.

(a) (i) NaOH, $CH_3CH{=}CHCH_2Br$; (ii) 200 °C
(b) (i) NaOH, $CH_3OSO_3CH_3$; (ii) HNO_3, Ac_2O, 10 °C; (iii) HBr, 100 °C.

30.8 Quinones

A. Nomenclature

Quinones are cyclohexadiendiones, but they are named as derivatives of aromatic systems: benzoquinones are derived from benzene, toluquinones from toluene, naphthoquinones from naphthalene, and so on. ''Quinone'' is used both as a generic term and as a common name for *p*-benzoquinone.

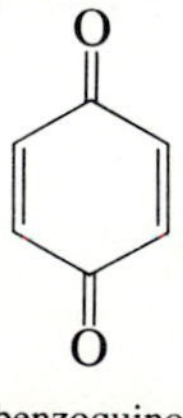

p-benzoquinone or quinone 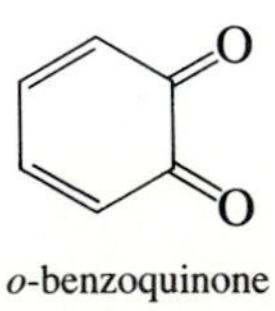 — *o*-benzoquinone 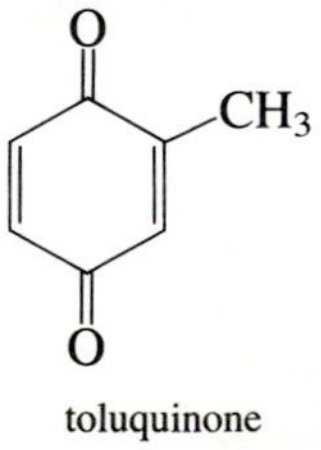 — toluquinone

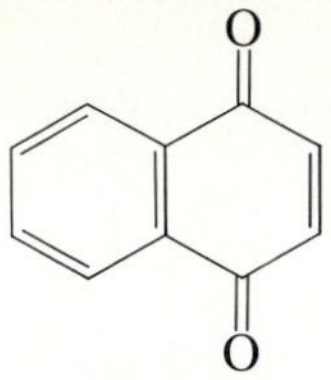

1,4-naphthoquinone 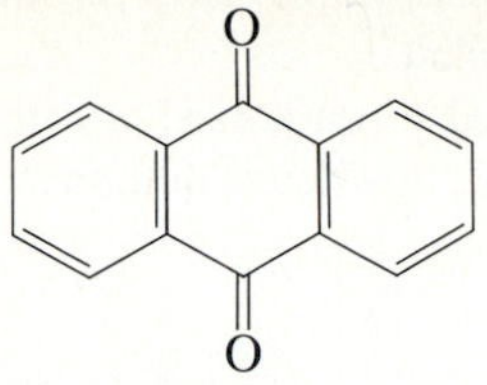 9,10-anthraquinone 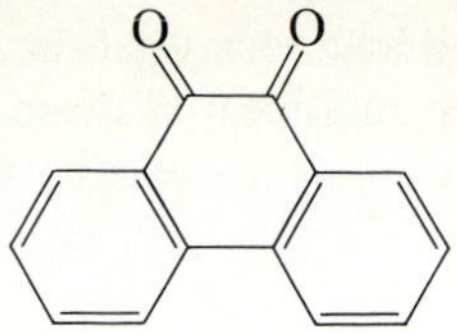 9,10-phenanthraquinone

Many quinones, especially hydroxyquinones, occur in nature. Some examples are the antibiotics fumigatin and phthiocol.

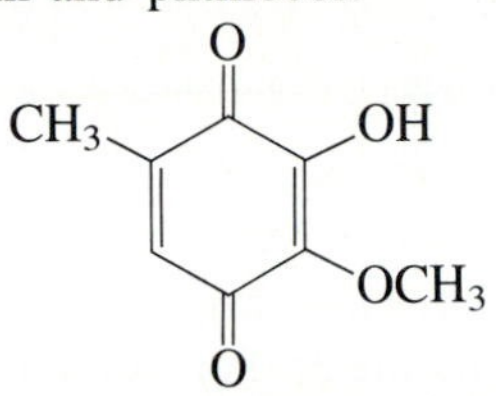

fumigatin
3-hydroxy-2-methoxy-
5-methyl-1,4-benzoquinone

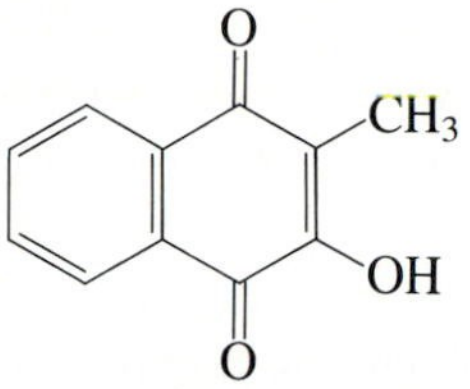

phthiocol
2-hydroxy-3-methyl-
1,4-naphthoquinone

Hydroxynaphthoquinones and hydroxyanthraquinones are also common, either free or bound to glucose. Many natural pigments have quinone structures.

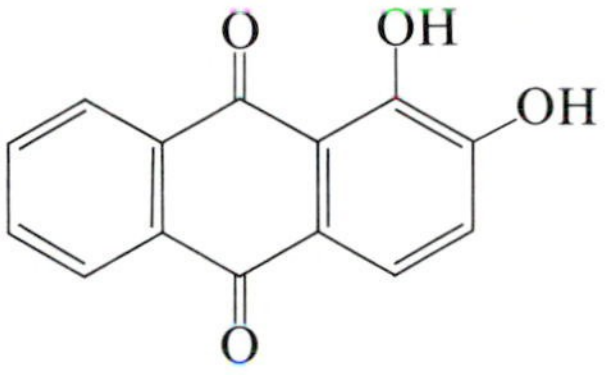

alizarin
(madder root)

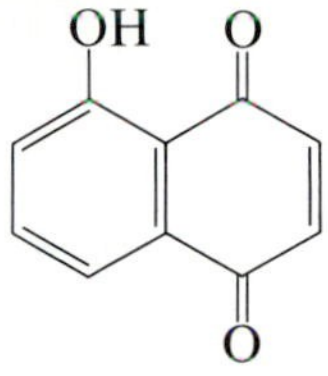

juglone
(walnut shells)

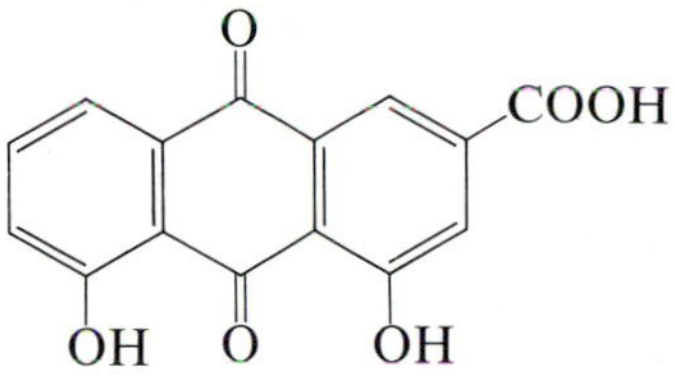

rhein
(rhubarb)

Quinone structures are frequently associated with color and the following structural units are referred to as ''quinoid'' structures.

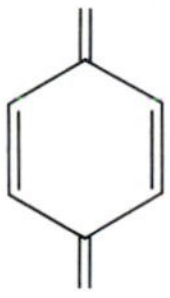

and

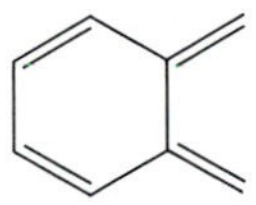

B. Preparation

The only important method for the preparation of quinones is oxidation of phenols and aromatic amino compounds. Substituted phenols or aniline derivatives can be used with some oxidizing agents. For example, *p*-benzoquinone can be prepared by oxidation of benzene or aniline with a variety of oxidizing agents, but the usual laboratory preparation involves the oxidation of hydroquinone.

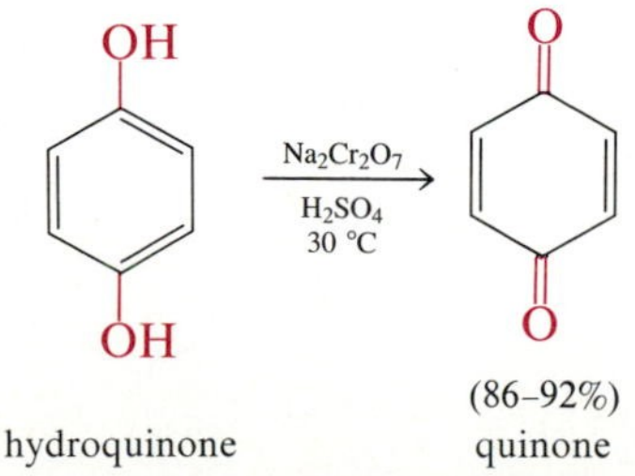

$$\xrightarrow[\substack{H_2SO_4 \\ 30\ ^\circ C}]{Na_2Cr_2O_7}$$

hydroquinone

(86–92%)
quinone

p-Benzoquinone forms yellow crystals, m.p. 115.7 °C, that are slightly soluble in water and can be sublimed or steam-distilled.

Aminophenols are easily oxidized to quinones, and this route constitutes one of the best methods for the preparation of substituted quinones.

NH_2 Cl OH $\xrightarrow[H_2SO_4,\ <30\ °C]{Na_2Cr_2O_7}$ O Cl O

2-chloro-1,4-benzoquinone

In addition to dichromic acid, $H_2Cr_2O_7$, oxidizing agents that are used for the preparation of quinones are ferric ion, dinitrogen tetroxide (N_2O_4), and sodium chlorate-vanadium pentoxide. Many other oxidizing agents have also been used, and the best one for any given compound must be determined by experiment. For example, the preparation of tetrachloro-*p*-benzoquinone (chloranil) makes advantageous use of nitric acid.

OH $\xrightarrow[\text{conc. HCl}]{Cl_2}$ $\xrightarrow[80\text{–}85\,°C]{HNO_3}$ O Cl Cl Cl Cl O

(60%)
2,3,5,6-tetrachloro-1,4-benzoquinone
chloranil

Chloranil is commercially available and has several uses in organic chemistry that we will encounter later (Sections 30.8.D and 30.3.B).

The oxidation of *o*-dihydroxybenzenes to *o*-quinones can be carried out with silver oxide in ether.

OH OH $\xrightarrow[Na_2SO_4]{Ag_2O,\ \text{ether}}$ O O

catechol — *o*-benzoquinone

o-Benzoquinone forms red crystals that are water-sensitive; anhydrous sodium sulfate is used in its preparation to remove the water formed in the oxidation.

In many cases phenyl ethers and esters undergo oxidation to the corresponding quinone with loss of the alkyl or acyl group. An example is the oxidation of 2,6-di-*t*-butyl-4-methoxyphenyl propionate by ceric ammonium nitrate.

O
‖
$OCCH_2CH_3$
$(CH_3)_3C$ $C(CH_3)_3$
OCH_3

$\xrightarrow{(NH_4)_2Ce(NO_3)_6}$

O
$(CH_3)_3C$ $C(CH_3)_3$
O

$+\ CH_3CH_2COOH$

C. Reduction-Oxidation Equilibria

Quinones, which are readily produced by oxidation of 1,2- and 1,4-dihydroxybenzenes, are easily reduced, forming again the dihydroxy derivatives. This reduction can be carried out chemically.

aq. $Na_2S_2O_4$
ether

2-methyl-1,4-naphthoquinone

(95%)
2-methyl-1,4-dihydroxy-naphthalene

However, the most important aspect of this redox system is that it is electrochemically reversible.

$+ 2\,H^+ + 2\,e^- \rightleftharpoons$

The electrical potential of this cell is given by the Nernst equation (30-1)

$$E = E^\circ + \frac{2.303\,RT}{n\mathcal{F}} \log \frac{[\text{quinone}][H^+]^2}{[\text{hydroquinone}]} \qquad (30\text{-}1)$$

in which $\mathcal{F}$ is the Faraday. At 25 °C equation (30-1) may be written as (30-2), in which the electrical potential is given in volts.

$$E^{25\,°C} = E^\circ - 0.059\ \text{pH} + 0.0296 \log \frac{[\text{quinone}]}{[\text{hydroquinone}]} \qquad (30\text{-}2)$$

The standard potential E° is that given at unit hydrogen ion concentration and equal concentrations of quinone and hydroquinone. Some values of E° are listed in Table 30.2. The more positive the value of the potential, the more readily the quinone is reduced. Note that electron-donating groups such as methyl and hydroxy stabilize the quinone form relative to the hydroquinone and result in lowering the reduction potential; electron-attracting groups such as halogen have the opposite effect.

TABLE 30.2 Reduction Potentials of Quinones

Quinone	Reduction Potential E°, volts (25 °C)
1,4-benzoquinone	0.699
2-methyl-1,4-benzoquinone	0.645
2-hydroxy-1,4-benzoquinone	0.59
2-bromo-1,4-benzoquinone	0.715
2-chloro-1,4-benzoquinone	0.713
1,2-benzoquinone	0.78
1,4-naphthoquinone	0.47
1,2-naphthoquinone	0.56
9,10-anthraquinone	0.13
9,10-phenanthraquinone	0.44

The reduction potentials in Table 30.2 allow one to see a clear parallel between redox phenomena in organic compounds and those observed with inorganic species. Recall that oxidation corresponds to the loss of electrons, and reduction to the gain of electrons.

$$Fe^{2+} \underset{\text{reduction}}{\overset{\text{oxidation}}{\rightleftharpoons}} Fe^{3+} + e^-$$

The more electron-rich a species is, the easier is its oxidation and the more difficult is its reduction. In Table 30.2 we see that the electron-attracting substituents chloro and bromo do indeed cause the quinone to be reduced more easily (more positive reduction potential). Similarly, the electron-donating substituents hydroxy and methyl cause the quinone to be reduced less easily.

The reduction of quinone occurs in two one-electron steps. The product of the first step is a radical anion that can be detected in dilute solution by the technique of electron spin resonance spectroscopy.

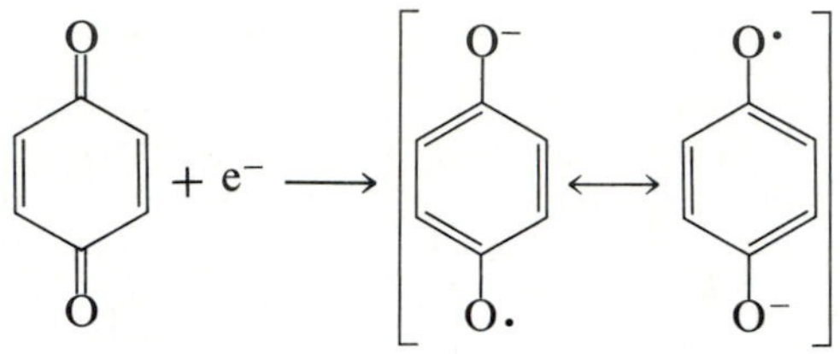

Electron spin resonance, ESR, is closely related to nuclear magnetic resonance. Electrons, like protons, have spin, and in molecules with an odd number of electrons (radicals) the resulting net electronic spin is aligned with or against an applied magnetic field. With commercial magnets the energy difference between the two states is in the microwave region of electromagnetic radiation. The resulting ESR spectra have been extremely useful for detecting small concentrations of radicals, and the details of the spectra provide important information about the electronic structures of radicals. These details, however, are beyond the scope of an introductory textbook.

The same radical anions are produced by one-electron oxidation of hydroquinone dianions.

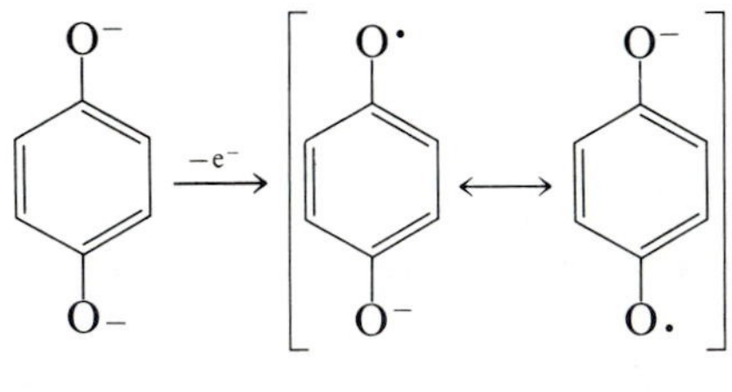

Phenoxide ions are also subject to one-electron oxidation to give the corresponding neutral phenoxyl radicals.

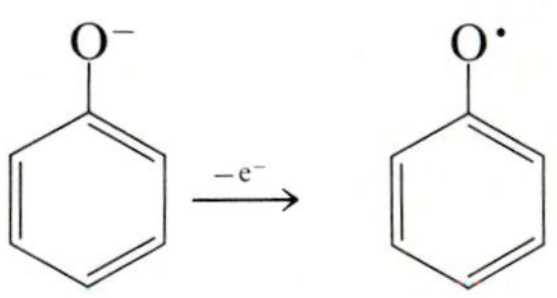

phenoxyl radical

Such radicals are involved in many of the reactions of phenols, including reactions of naturally occurring phenols.

Vitamin E, α-tocopherol, is a phenol that is widespread in plant materials. Among the several functions it may have in animals, the most important seems to be as a radical scavenger for peroxyl radicals formed from polyunsaturated acyl groups in cell membranes. Reactive free radicals appear to be damaging to membranes or other molecules present in biological systems, and vitamin E appears to protect against such damage. Free radicals have also been blamed for the aging process, but the whole phenomenon is probably too complex to have such a simple and unique explanation.

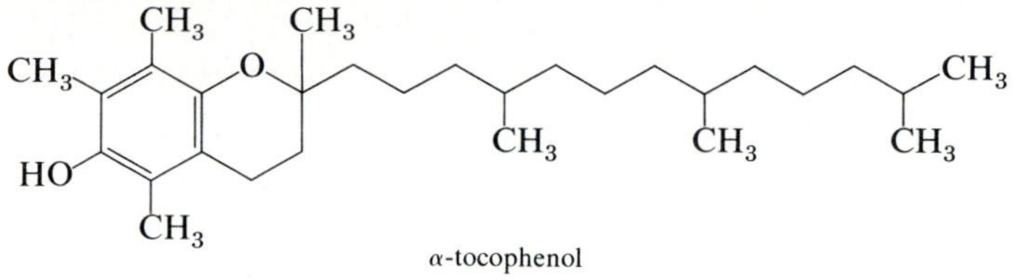

α-tocophenol
vitamin E

Quinone-hydroquinone redox systems have a number of important uses. Hydroquinone itself, for example, is an important photographic developer.

Silver bromide crystals that have become photoactivated by exposure to light are reduced by hydroquinone. The photoactivation involves transfer of an electron to a trap in the small AgBr crystal from a light absorbing substance on the surface, a **sensitizer.** The absorption maximum of the sensitizer determines the wavelength range for the film containing the silver bromide. Only the photoactivated silver bromide is reduced readily to black silver metal; the hydroquinone is oxidized to *p*-benzoquinone. The residual silver bromide is then removed by ''hypo,'' sodium thiosulfate, which forms a soluble complex with silver cation. The result is a black image where the silver bromide emulsion was exposed to light. Some developer formulas include *p*-methylaminophenol, usually as the sulfate (Elon, Metol), which is also oxidized to *p*-benzoquinone.

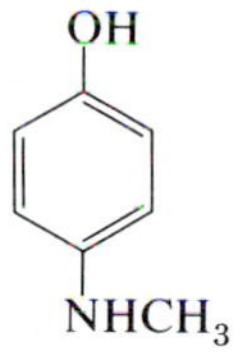

p-methylaminophenol

The oxidation-reduction reactions of hydroquinone and quinone derivatives play an important role in physiological redox processes.

Vitamin K is actually many vitamins; for example, K_1, K_2, K_3, and so on. They are all related to 1,4-naphthoquinone or compounds that are oxidized to it. For example, vitamin $K_{2(30)}$ is

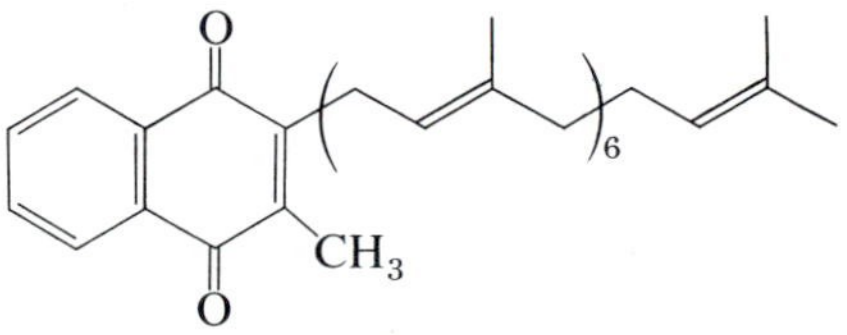

Others vary in the length of the side chain. The K vitamins are present in blood as coagulation factors. They function as cofactors for an enzyme that carboxylates (opposite of decarboxylate) glutamic acid side chains in proteins. The resulting γ-carboxyglutamic acid groups are probably important in chelation of calcium ion. A plausible mechanism for the carboxylation reaction is shown below. It addresses some unusual features of the overall reaction. The reduced form of the cofactor is required, Oxygen is needed and a 2,3-epoxide of the quinone is an intermediate product. In addition, carboxylation of an alkanoic acid at an ''unactivated''

position requires a mechanism for deprotonating the position next to the carboxylate of the glutamyl group, using only the reduced vitamin K and an enzyme.

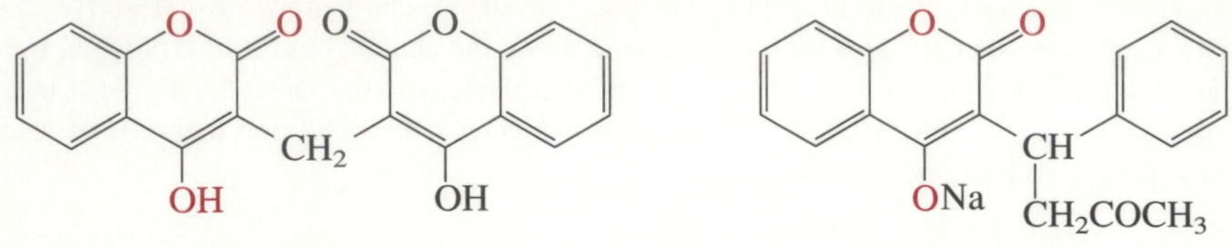

dicumarol

warfarin sodium

The modified enzymes contain substantial numbers of γ-carboxyglutamyl groups. The process of coagulation involves a cascade of proteolytic enzyme reactions, in which an enzyme is activated in order to activate the next enzyme, and so on. The carboxylation reaction is blocked by dicumarol, a compound first isolated from spoiled sweet clover as the causative agent of bleeding in cattle. A related compound, warfarin, was synthesized and used widely as a rodenticide. Warfarin blocks the reaction that reduces the epoxide to the diol. At first warfarin was considered too toxic for human use, but the failure of an attempted suicide using the substance led to its introduction as an anticoagulant under the name coumadin. Although coumadin has been widely used, the compound can nevertheless be toxic to humans.

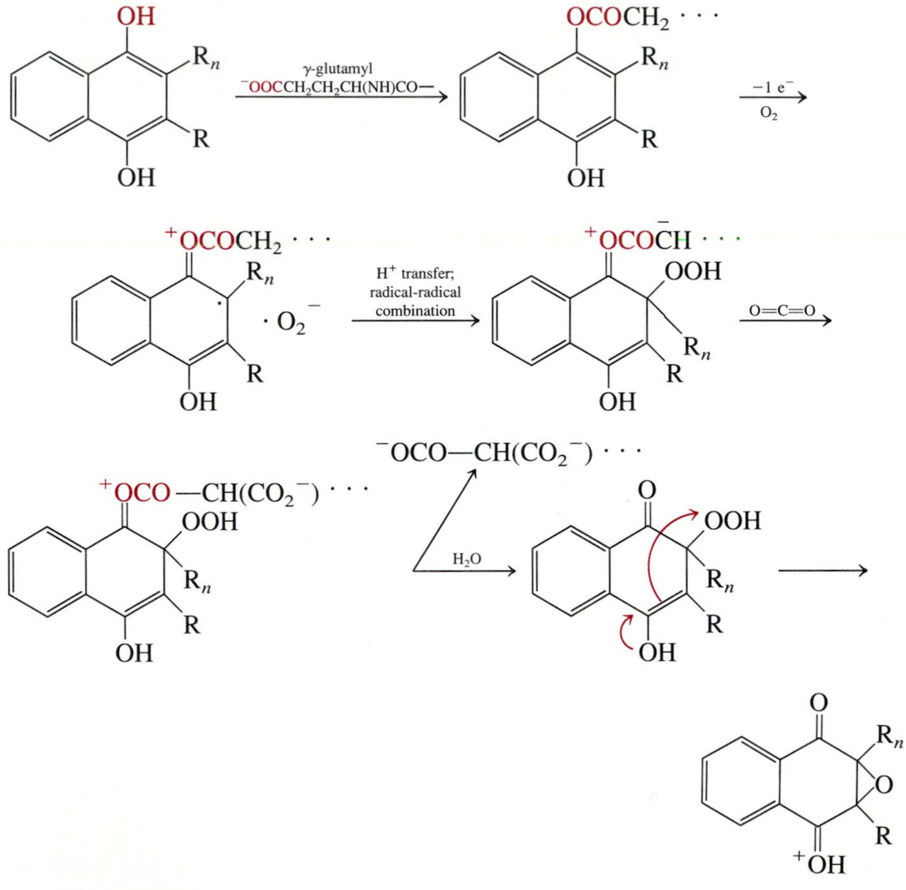

γ-glutamyl $^{-}OOCCH_2CH_2CH(NH)CO-$

$-1\ e^{-}$, O_2

H^+ transfer; radical-radical combination

O=C=O

H_2O

A related series of compounds is coenzyme Q, which occurs in many kinds of cells with $n = 6$, 8, or 10 ($n = 10$ in mammalian cells); indeed, when first discovered, it was called **ubiquinone** because it was so ubiquitous in cells. Coenzyme Q is involved in electron-transport systems, and the long isoprenoid chain is undoubtedly designed to promote solubility in the phospholipid bilayers of cell or mitochondrial membranes.

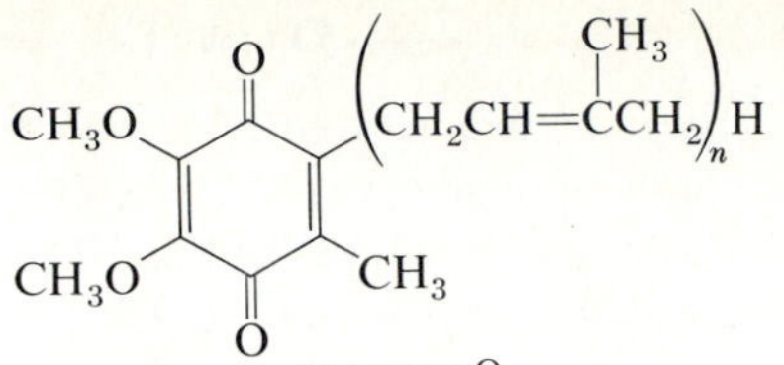
CH3
O
CH3O
CH2CH=CCH2
H
n
CH3O
CH3
O

coenzyme Q

EXERCISE 30.16 Using the reduction potentials in Table 30.2 predict the direction of equilibrium for the following reaction.

$$\left.\begin{array}{c}\text{1,4-benzoquinone}\\ +\\ \text{2-chlorohydroquinone}\end{array}\right\} \rightleftharpoons \left\{\begin{array}{c}\text{hydroquinone}\\ +\\ \text{2-chloro-1,4-benzoquinone}\end{array}\right.$$

What voltage would a cell generate that has equal concentrations of the four compounds?

D. Charge-Transfer Complexes

An equimolar mixture of *p*-benzoquinone and hydroquinone forms a dark green crystalline molecular complex, ''quinhydrone,'' having a definite melting point of 171 °C. This material dissolves in hot water, and the solution is largely dissociated into its components.

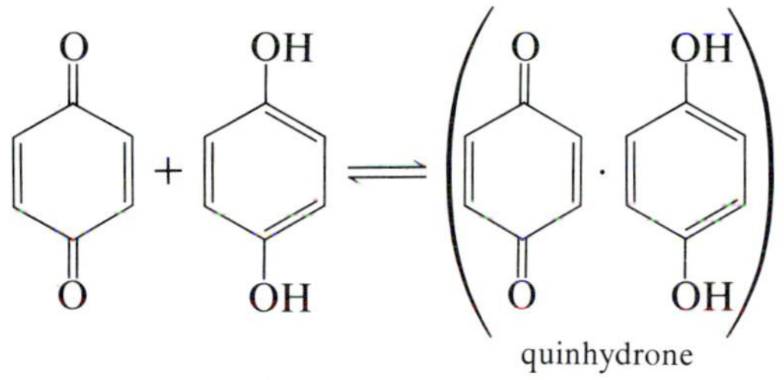
O
OH
O
OH
O
OH
O
OH

quinhydrone

The buffered solution has been used as a standard reference electrode.

The structure of the crystals consists of alternating molecules of quinone and hydroquinone with the rings parallel to each other (Figure 30.2). This complex is another example of a **charge-transfer** complex (Section 22.6.D). Recall that such complexes are characterized by one component that is electron-rich (the donor) and another component that is strongly electron attracting (the acceptor); hence, they are also known as **donor-acceptor complexes.** In resonance language the complexes are characterized as a hybrid of two resonance structures.

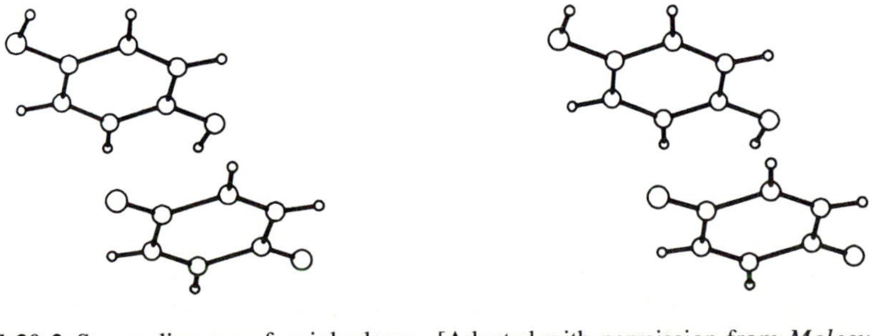

FIGURE 30.2 Stereo diagram of quinhydrone. [Adapted with permission from *Molecular Structure and Dimensions,* International Union of Crystallography, 1972.]

$$[\text{D}:\text{A} \longleftrightarrow \overset{+}{\text{D}}\cdot \overset{-}{\text{A}}\cdot]$$

The second structure, the ''charge-transfer structure,'' makes only a small contribution to the total electronic structure in the ground state. This second structure provides some of the bonding that holds the two components together (ca 10%), but, in any case, with all the various forces taken together, the bond strength involved is only a few kcal $mole^{-1}$.

In molecular orbital theory, the donors have a high-lying HOMO in which electrons are held rather loosely; that is, this highest occupied molecular orbital has a low ionization potential. The acceptors have a relatively low-lying LUMO or lowest unoccupied molecular orbital; in fact, common acceptors frequently form radical anions readily on one-electron reduction in which the electron enters the LUMO. In the complex there is some overlap of the HOMO of the donor with the LUMO of the acceptor that results in transfer of some electron density from donor to acceptor. The amount of charge transferred is small and corresponds typically to a small fraction ($\cong$0.05) of an electron. This molecular orbital approach is entirely equivalent to the resonance interpretation. A charge-transfer complex, such as those discussed in Section 22.6.D, is characterized by a special light absorption called a charge-transfer transition.

Donors and acceptors can be classified by the nature of the donor or acceptor orbitals. We have described the spectroscopic properties of benzene:iodine complexes (Section 22.6.D). In that case, benzene is a π-donor and iodine is a σ-acceptor. Ether and ethanol are n-donors, where n signifies nonbonding electrons. Benzene derivatives with electron-donating groups such as OH, OCH_3, $N(CH_3)_2$, CH_3, and so on, are π-donors and form charge-transfer complexes. Picric acid is a π-acceptor as are 1,3,5-trinitrobenzene and quinones. Especially potent are quinones with additional electron-attracting groups; chloranil (tetrachloro-*p*-benzoquinone) is an important example. The structure of the complex formed from hexamethylbenzene and chloranil is shown in Figure 30.3; this complex has a bond strength of about 5 kcal $mole^{-1}$. Compounds with several CN groups are also used as acceptors. Some examples are tetracyanoethylene and 2,3-dicyano-1,4-benzoquinone. Perhaps the most widely used is a quinone analogue, tetracyanoquinodimethane (see Highlight 30.2).

Charge-transfer complexes are intensely colored if the charge-transfer transition is in the visible region. Charge-transfer interactions are now recognized as being significant in solid-state structures in which the other interactions (dipole-dipole, van der Waals, hydrogen bonding) are weaker. Furthermore, many reactions and reaction mechanisms are now recognized to involve charge-transfer phenomena, for example the formation of the Grignard reagent (page 166). However, a detailed treatment of such phenomena must be deferred to advanced organic chemistry texts.

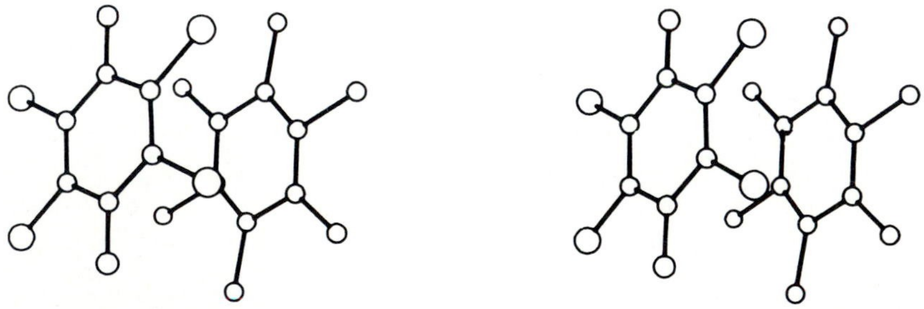

FIGURE 30.3 Stereo diagram of hexamethylbenzene-chloranil complex. [Adapted with permission from *Molecular Structure and Dimensions,* International Union of Crystallography, 1972.]

E. Reactions of Quinones

Quinones are α,β-unsaturated carbonyl compounds and show double bond reactions typical of such structures. One significant reaction is addition of hydrogen chloride.

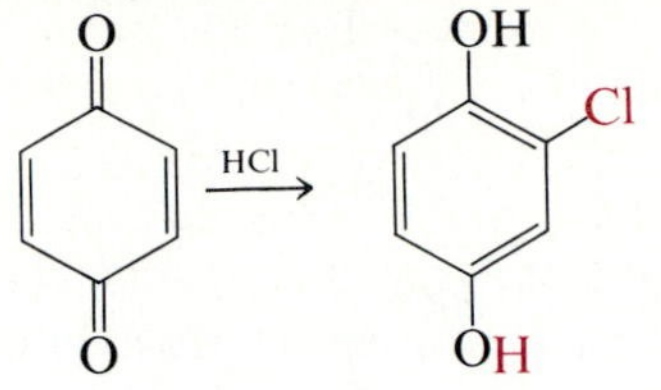

HCl

chlorohydroquinone

This reaction is simply an acid-catalyzed conjugate addition (see Section 20.3.A).

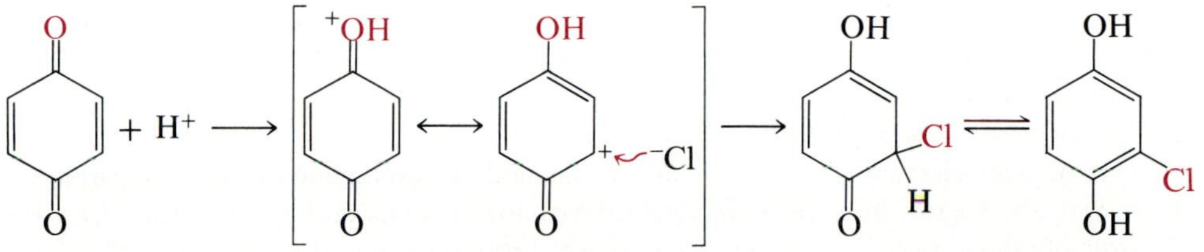

Amines add readily. 1,4-Benzoquinone reacts with aniline to give 2,5-dianilino-1,4-benzoquinone and hydroquinone.

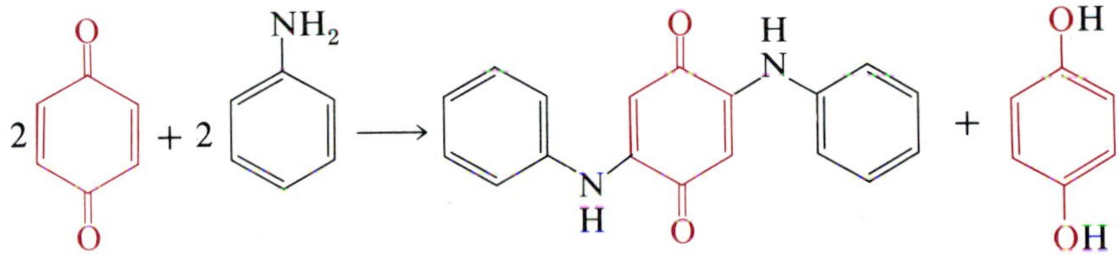

This reaction provides an interesting contrast to the addition of HCl, where the product is the chlorohydroquinone. In this case the group entering the hydroquinone ring is electron donating. Thus the initial product of conjugate addition of aniline to 1,4-benzoquinone, 2-anilinohydroquinone, is rapidly oxidized by an equivalent of 1,4-benzoquinone, which is reduced to hydroquinone.

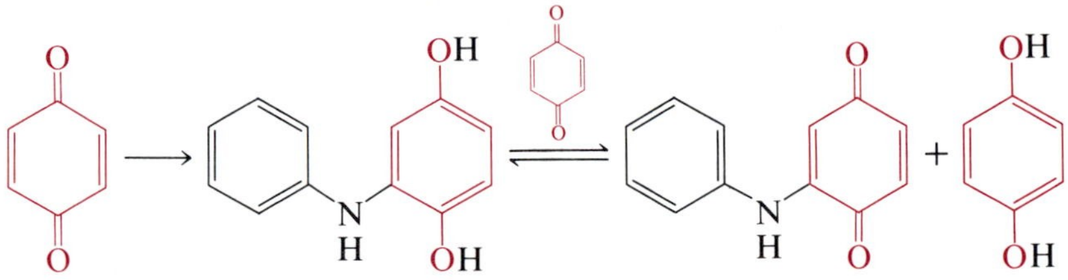

This equilibrium lies far to the right because of the reduction potentials of the two quinones. The 2-anilino-1,4-benzoquinone then undergoes a second conjugate addition resulting in the formation of 2,5-dianilinohydroquinone, which is similarly oxidized to produce the isolated product.

Quinones also function as potent dienophiles in the Diels-Alder reaction (Section 20.5). An example is the reaction of 1,4-benzoquinone with butadiene, which occurs in acetic acid solution at room temperature. The product may be isolated or it may be treated with HCl, whereupon rearrangement to the more stable hydroquinone form occurs.

$\xrightarrow[25\,°C]{CH_3COOH}$ (87%) $\xrightarrow[CH_3COOH]{HCl}$

EXERCISE 30.17 Write equations illustrating the reaction of 2,3-dimethyl-1,4-benzoquinone with each of the following reagents.

(a) HCl (b) CH_3NH_2 (c) 2-methyl-1,3-butadiene (25 °C).

Highlight 30.2

Quinones are electron acceptors. Related compounds are quinodimethanes. The parent hydrocarbon easily dimerizes to [2.2]paracyclophane, a compound in which the crystal structure shows that the benzene rings are bent. Tetracyanoquinodimethane (TCNQ) is one of the most widely used electron acceptors for preparing charge-transfer complexes.

1,4-benzoquinone

1,4-quinodimethane

tetracyano-1,4-quinodimethane (TCNQ)

[2.2]paracyclophane

ASIDE

"Real vanilla, with its complex veils of aroma and jiggling flavors, . . ." wrote one popular and serious analyst of the role of the senses in human behavior. Vanilla (little pod, Sp., *vainilla*) is a tropical climbing orchid first cultivated by the Indians of Old Mexico. The fruits contain vanilla beans and yield vanillin on ethanol extraction of the fermented and dried pods. A glucosidase hydrolyzes vanillin glucoside (see Chapter 28) during fermentation. Vanillin is produced on an industrial scale by synthesis to meet the large demand. The compound is 3-methoxy-4-hydroxybenzaldehyde, both a phenol and a phenol ether. What are the formulas of unsubstituted phenol ("phenolic smell"), unsubstituted benzaldehyde ("almond smell"), and unsubstituted anisole ("fragrant aromatic")? How does one prepare phenol from benzene industrially? Name and draw the orbital structure for the highly reactive intermediate formed in this synthesis. Show how the highly reactive intermediate can be generated by simple means in the laboratory. Write a synthesis for anisole from phenol. Can you convert a 1,2-dihydroxybenzene into a dihydroxybenzaldehyde? Which isomers would you obtain? Would this be a suitable starting point for the preparation of vanillin? If not, why not?

PROBLEMS

1. Write structures for each of the following names.

a. *m*-cresol
b. benzyne
c. 3-chloro-1,2-benzoquinone
d. *o*-methoxyphenol
e. picric acid
f. benzyl phenyl ether
g. 3-(*o*-hydroxyphenyl)pentanoic acid
h. *p*-isobutylphenol
i. 2-methoxy-1,4-naphthoquinone
j. 2,5-dichloro-1,4-benzoquinone

2. When 2,4,6-trinitroanisole is treated with methoxide ion in methanol, a red anion having the composition $C_8H_8O_8N_3^-$ is produced. Such anions are called Meisenheimer complexes after the chemist who first suggested the correct structure. What structure do you think he suggested? One of Meisenheimer's experiments compared the product of reaction of 2,4,6-trinitroanisole and ethoxide ion with the product of 2,4,6-trinitrophenyl ethyl ether and methoxide ion. What do you think he found?

3. The reaction of chlorobenzene with hot aqueous sodium hydroxide actually goes in part by way of a benzyne intermediate and in part by the addition-elimination mechanism. Reaction of chlorobenzene labeled with ^{14}C at the 1-position with 4 *M* NaOH at 340 °C gives phenol in which 58% of the ^{14}C remains at the 1-position and 42% is at the 2-position. Calculate the fraction of reaction proceeding by each of the two mechanisms.

4. Give the principal product of the following reactions or reaction sequences:

a. 3-bromo-5-chlorotoluene (ring with CH_3, Cl, Br) $\xrightarrow[\text{ether}]{\text{excess Mg}}$ $\xrightarrow{\text{DCl}}$

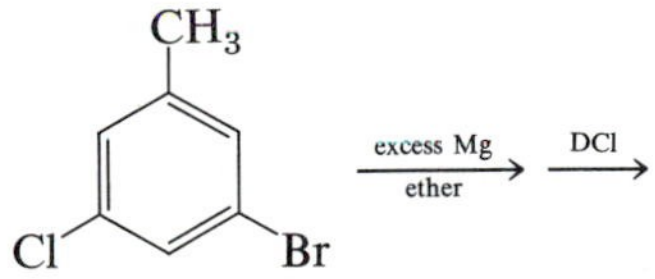

b. chlorobenzene (ring with Cl) $\xrightarrow[\text{fuming } H_2SO_4]{\text{fuming } HNO_3 \text{ (2 moles)}}$ $\xrightarrow{NH_3}$

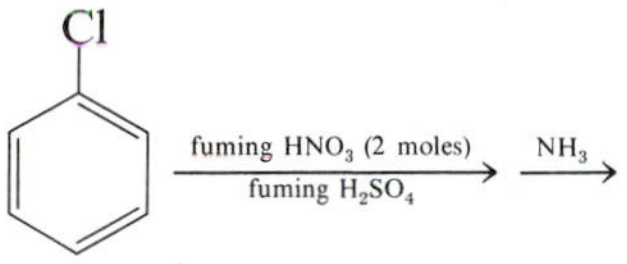

c. (ring with NO_2, Cl, Cl) $+ \text{NaOCH}_3 \xrightarrow[\Delta]{CH_3OH}$

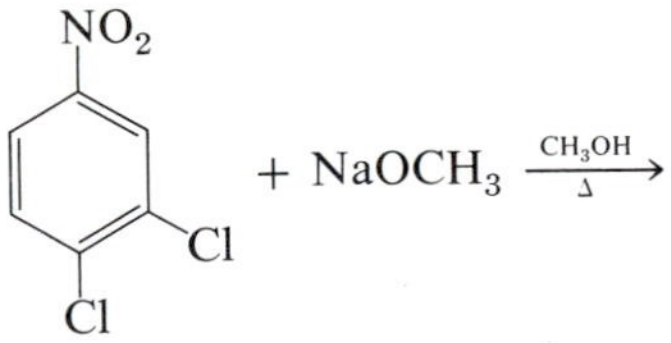

d. (ring with CH_2CH_3) $\xrightarrow[h\nu]{Br_2}$ $\xrightarrow[\text{ether}]{\text{Mg}}$ $\xrightarrow{CH_2\text{—}CH_2 \text{ (epoxide, O)}}$ $\xrightarrow{H^+}$

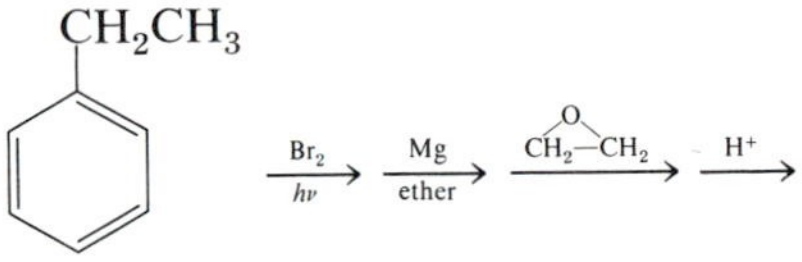

5. Write the principal reaction product or products, if any, when *o*-cresol is subjected to the following conditions.

a. $(CH_3O)_2SO_2$, NaOH
b. $Na_2S_2O_4$
c. $SnCl_2$, HCl
d. $Na_2Cr_2O_7$, H_2SO_4
e. $C_6H_5N_2^+$, aqueous NaOH
f. CH_3COOH, H_2SO_4, heat
g. aqueous NH_3
h. 98% H_2SO_4, 25 °C
i. $KMnO_4$, heat
j. cold, dilute $KMnO_4$
k. bromine water
l. $(CH_3CO)_2O$
m. $CHCl_3$, aqueous NaOH
n. HONO

o. $LiAlH_4$
p. HNO_3 (2 moles) in CH_3COOH
q. HBr, heat
r. CO_2, K_2CO_3, 240 °C
s. bromine (1 mole) in CCl_4

6. Write the principal reaction product or products, if any, when 2-methylanisole is subjected to the following conditions.

a. $Na_2S_2O_4$
b. conc. H_2SO_4
c. HBr, heat
d. bromine in CH_3COOH
e. HNO_3 (2 moles) in CH_3COOH
f. $LiAlH_4$
g. $(CH_3CO)_2O$
h. CH_3COCl, $ZnCl_2$
i. phthalic anhydride, $C_6H_5NO_2$, $AlCl_3$, 0 °C
j. Na, liquid NH_3

7. Each of the following phenol or quinone derivatives has the common or trivial name shown and is a compound of some significance. Provide the IUPAC name and show how each can be synthesized from the indicated starting material.

a. $CH_2CH_2NH_2$ / CH_3O / OCH_3 / OCH_3

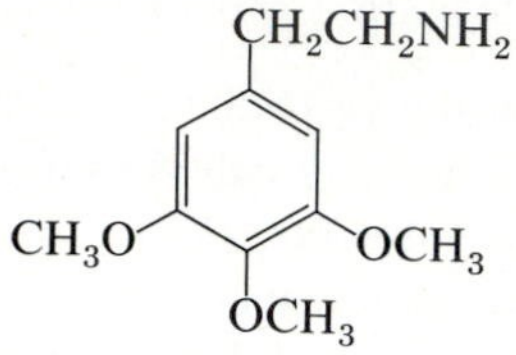

Mescaline is the active ingredient in peyote (mescal buttons) and is used as a psychotomimetic (that is, mimics psychosis) drug. Show how it can be prepared from gallic acid, 3,4,5-trihydroxybenzoic acid.

b. OH / CH_3 / O_2N / $CHCH_2CH_3$ / NO_2

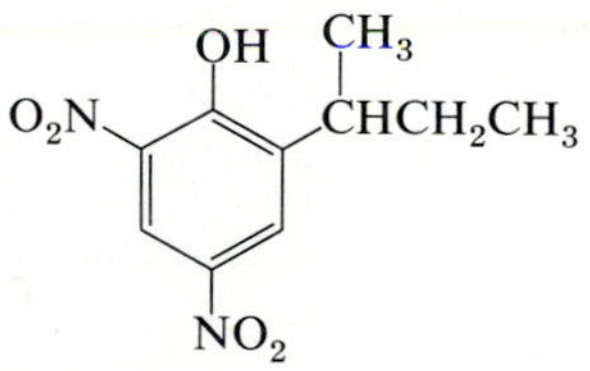

The ester of this phenol with 3-methylbut-2-enoic acid is **binapacryl,** which is used as a fungicide and mitocide. Show how the phenolic portion may be prepared from phenol.

c. O_2N—C_6H_4—N=N—C_6H_3(COOH)—OH

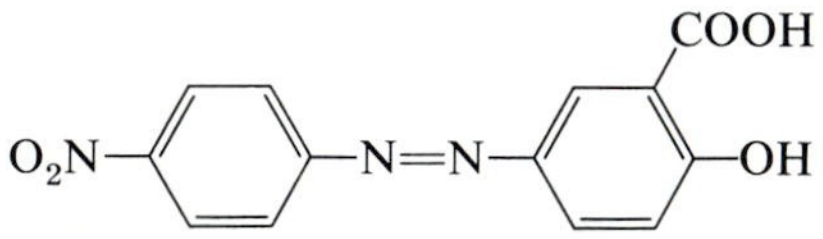

Alizarine yellow R is used as an indicator in alkaline solutions. The color changes from yellow to red over the pH range from 10 to 12. Starting from aniline, outline a synthesis of this dye.

d. $CON(C_2H_5)_2$ / OC_2H_5 / OH

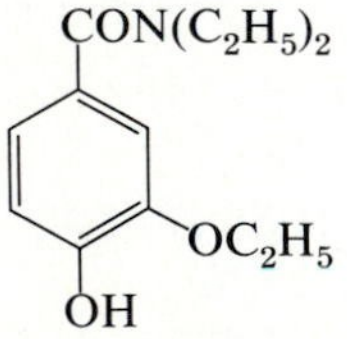

Anacardiol is a drug that acts as a stimulant on the central nervous system. Propose a synthesis, beginning with catechol.

e. 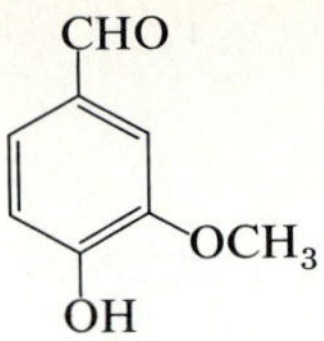 [structure: benzene ring with CHO, OCH_3, OH]

Vanillin occurs naturally in vanilla and other plant materials and is used as a flavoring agent. Indicate how it can be prepared from guaiacol.

f. 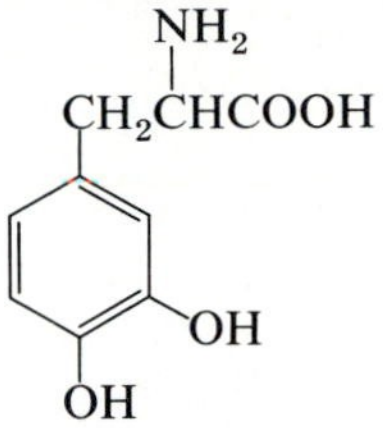 [structure: $CH_2CH(NH_2)COOH$ on benzene ring with OH, OH]

This material (L-**dopa**) is found in some beans and is also used in a treatment of Parkinson's disease. Show how the racemate can be prepared from catechol.

g. 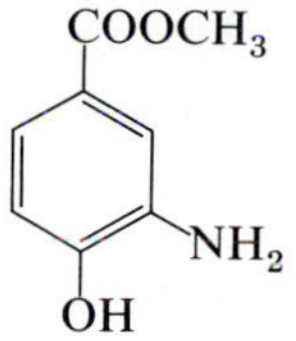 [structure: $COOCH_3$ on benzene ring with NH_2, OH]

Orthocaine is used as a surface anesthetic. Suggest a synthesis from phenol.

8. Show how each of the following conversions can be accomplished.

a. 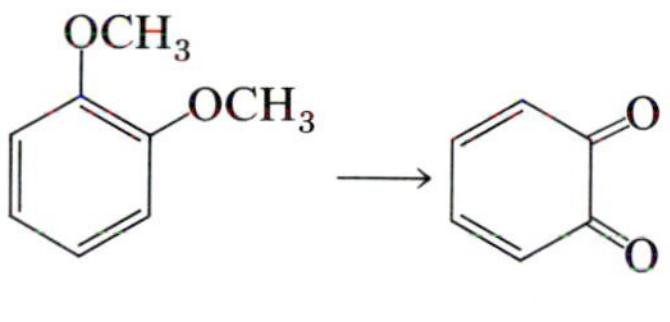 OCH_3, OCH_3 (1,2-dimethoxybenzene) ⟶ o-benzoquinone (O, O)

b. 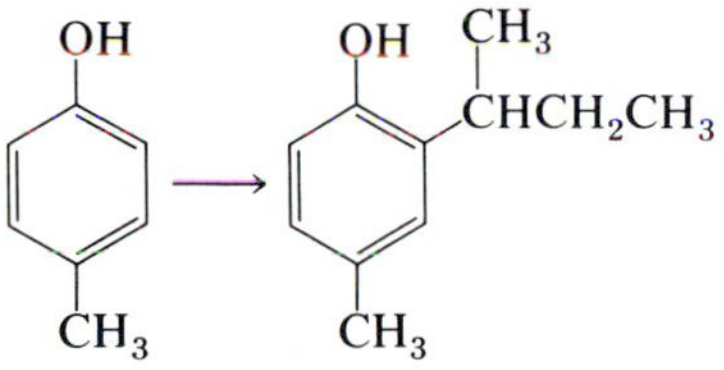 OH, CH_3 (p-cresol) ⟶ OH, $CH(CH_3)CH_2CH_3$, CH_3

c. 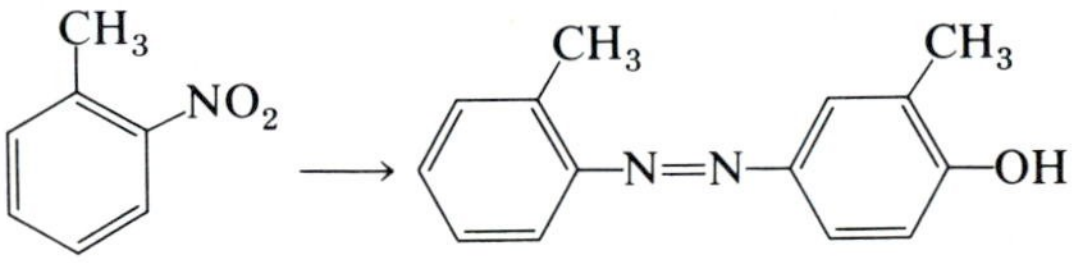 CH_3, NO_2 ⟶ CH_3-C$_6$H$_4$—N=N—C$_6$H$_3$(CH_3)—OH

d. 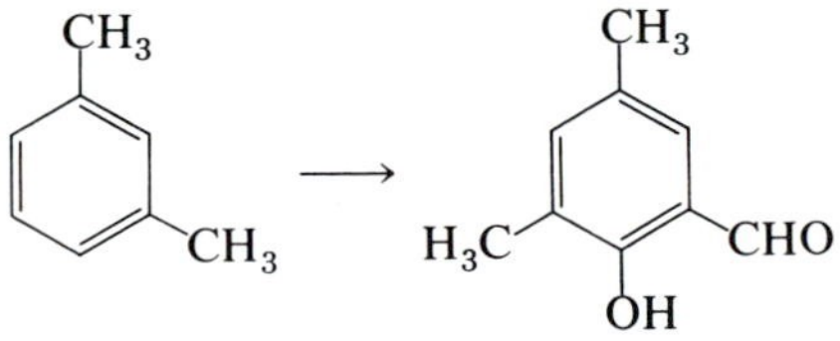 CH_3, CH_3 (m-xylene) ⟶ CH_3, H_3C, CHO, OH

e. 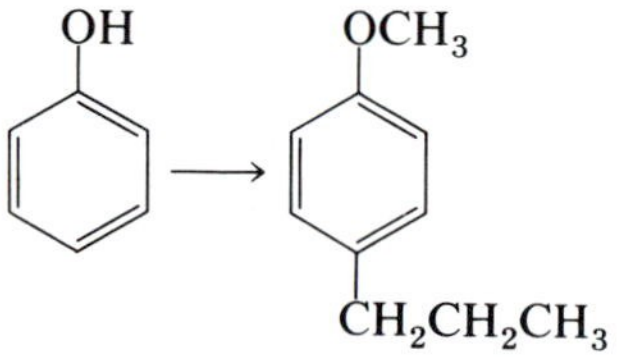 OH (phenol) ⟶ OCH_3, $CH_2CH_2CH_3$

9. What is the principal organic product of each of the following sequences?

a. $C_6H_5CH{=}CHCH_2Cl \xrightarrow{C_6H_5O^-} \xrightarrow{200°C}$

b. 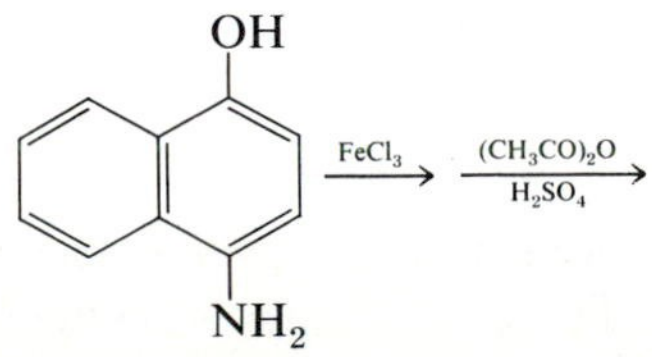 [structure: naphthalene with OH and NH_2] $\xrightarrow{FeCl_3} \xrightarrow[H_2SO_4]{(CH_3CO)_2O}$

c. $(CH_3)_2C{=}CHCH_2OCH{=}CH_2 \xrightarrow{\Delta} \xrightarrow{CH_2{=}P(C_6H_5)_3} \xrightarrow{\Delta}$

d. $C_6H_5NH_2 \xrightarrow[HCl]{NaNO_2} \xrightarrow[aq.\ NaOH]{o\text{-}CH_3OC_6H_4OH}$

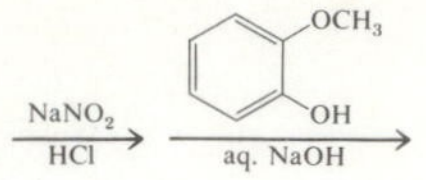

e. 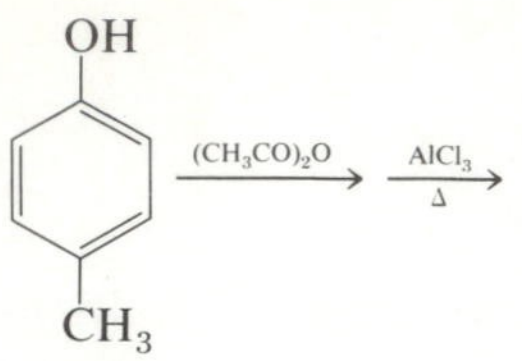 $p\text{-}CH_3C_6H_4OH \xrightarrow{(CH_3CO)_2O} \xrightarrow[\Delta]{AlCl_3}$

10. 2,6-Dichlorophenol is present in some ticks and is thought to be a sex pheromone. Devise a practical synthesis from phenol.

11. In Table 23.2 (page 713) the relative rates of protodedeuteriation of *o*-, *m*-, and *p*-anisole-*d* are summarized. Show how each of these deuteriated anisoles can be prepared uncontaminated by the other isomers.

12. Another component of urushiol, the active constituent of the irritating oil of poison ivy, is 3-pentadecyl-1,2-dihydroxybenzene. Synthesize this compound from catechol (be careful in handling the product!).

13. When *n*-butyl benzenesulfonate is heated with an ethanolic solution of potassium benzyloxide, the product is a mixture of *n*-butyl ethyl ether and *n*-butyl benzyl ether. However, if the isomeric salt, potassium *p*-methylphenolate, is used, the product is almost exclusively *n*-butyl *p*-methylphenyl ether. Explain.

14. Acetanilide is oxidized in the body by oxygen and a hydroxylase enzyme to *p*-hydroxyacetanilide. Show how this compound can be synthesized from phenol.

15. 2,2-Bis-(*p*-hydroxyphenyl)propane or "Bisphenol A," used commercially in the manufacture of epoxy resins and as a fungicide, is prepared by the reaction of phenol with acetone in acid.

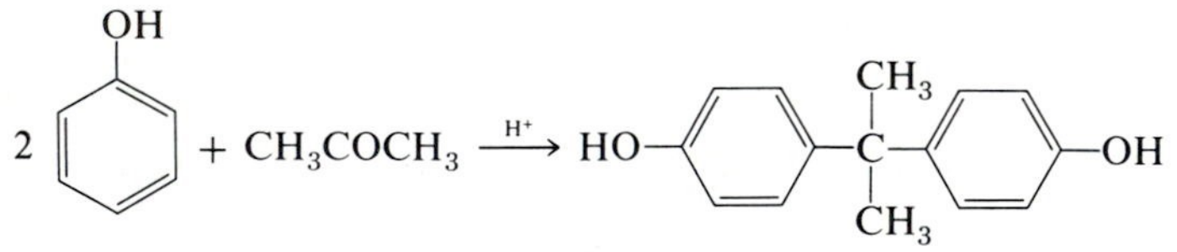

$$2\ C_6H_5OH + CH_3COCH_3 \xrightarrow{H^+} HO{-}C_6H_4{-}C(CH_3)_2{-}C_6H_4{-}OH$$

Bisphenol A

Write out the mechanism of this reaction, showing all intermediates involved.

16. The sulfonation of *p*-cymene (1-methyl-4-isopropylbenzene) gives the 2-sulfonic acid. Is this the expected orientation? Explain. Use this fact to synthesize carvacrol, 2-methyl-5-isopropylphenol, from *p*-cymene. Carvacrol is found in the essential oils from thyme, marjoram, and summer savory. It has a pleasant thymol-like odor.

17. Tri-*o*-cresyl phosphate, $(o\text{-}CH_3C_6H_4O)_3PO$, is used as a gasoline additive. Suggest a preparation.

18. Rank each group in order of increasing acidity.

a. phenol, 3-acetylphenol, 4-acetylphenol

b. *p*-dimethylaminomethylphenol, *p*-dimethylaminophenol, trimethyl-(*p*-hydroxyphenyl)ammonium ion

c. 2-hydroxy-1,4-benzoquinone, 2,5-dimethoxyphenol, 4-hydroxy-1,2-benzoquinone

19. A student attempted the following synthesis of *o*-methoxybenzyl alcohol from *o*-cresol, but got almost none of the desired product. What went wrong?

CH_3, OH → ($Na_2Cr_2O_7$, H_2SO_4) → COOH, OH → (excess CH_2N_2, ether) → $COOCH_3$, OCH_3 → ($LiAlH_4$, ether) → (H^+) → CH_2OH, OCH_3

20. Hydrolysis of 2,4,6-triaminobenzoic acid by refluxing with dilute NaOH gives 1,3,5-trihydroxybenzene (phloroglucinol).

COOH, H_2N, NH_2, NH_2 → (OH^-, Δ) → (H^+) → HO, OH, OH

Write a reasonable mechanism for this reaction. (*Hint*: The reaction involves the nonaromatic keto forms.)

21. *o*-Phenylazophenol is readily separable from the *para* isomer by steam distillation. Give a reasonable explanation for the greater volatility of the *ortho* isomer.

22. a. Write a reasonable mechanism for the sulfuric acid-catalyzed condensation of phenol with phthalic acid. Be sure to show all intermediates.

b. Phenol does not form diphenyl ether with sulfuric acid, yet the condensation of resorcinol with phthalic anhydride to give fluorescein includes the formation of an ether link from two phenolic hydroxy groups. Give a reasonable explanation.

23. Equimolar mixtures of *p*-benzoquinone with hydroquinone and of 2-chloro-1,4-benzoquinone with chlorohydroquinone in the same buffer solution are contained in separate beakers. The beakers are connected by a salt bridge, and the potential difference between them is measured. What is this potential difference? Which beaker constitutes the negative end (cathode) of this battery?

24. Allyl chloride labeled with ^{14}C is allowed to react with the anion from 2-methyl-6-allylphenol to form the corresponding ether. When this ether is heated, the Claisen rearrangement product 2-methyl-4,6-diallylphenol is formed. More than half, but not all, of the ^{14}C is found in the allyl group in the 4-position. Explain.

25. Diazotization of 2,4-dinitroaniline in aqueous solution is accompanied by some conversion to phenols in which a nitro group is replaced by a hydroxy group.

N_2^+, NO_2, NO_2 → (H_2O) → N_2^+, OH, NO_2 + N_2^+, NO_2, OH

Suggest a mechanism for this reaction.

26. In the chlorination of aniline in aqueous solution some chlorophenol is produced. Suggest a mechanism for this substitution.

27. Using chlorine dioxide, ClO_2, as the oxidant for the lignin in water in place of Cl_2 diminishes the proportion of chlorine substituted into the aromatic rings of the components shown on page 1016. Write Lewis structures for the important resonance forms of ClO_2, and deduce why the oxygen might attack aromatic rings in preference to chlorine.

28. Using the Heck reaction, design syntheses of (a) *p*-nitrocinnamic acid from aniline, (b) 2-hydroxycinnamic acid from phenol, and (c) 4-chlorocinnamic acid from benzene

29. A solution of sodium dithionite ($Na_2S_2O_4$) in water is added little by little to 1,4-benzoquinone. Describe in detail what you would observe, write the mechanism of each step of the reactions that you have deduced, and give the nature of the final product.

CHAPTER 31

POLYCYCLIC AROMATIC HYDROCARBONS

31.1 Nomenclature

Polycyclic aromatic hydrocarbons can be dissected into two broad classes: the biaryls and the condensed benzenoid hydrocarbons. The latter class is by far the larger and more important group.

The biaryls are benzenoid compounds in which two rings are linked together by a single bond. The parent system of this class is biphenyl. In numbering the ring positions, the rings are considered to be joined at the 1-position, and the two rings are distinguished by the use of primes.

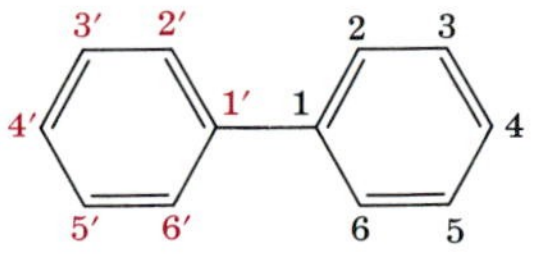

biphenyl

Simple derivatives can be named by use of *ortho, meta, para* nomenclature.

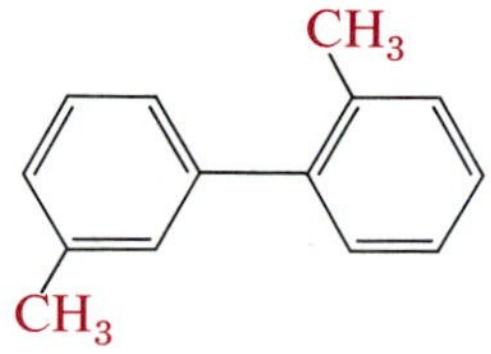

CH_3

CH_3

o,m′-dimethylbiphenyl
2,3′-dimethylbiphenyl

More complex compounds are named using numbers. Again, substituents in one ring are designated by the use of primes.

2′,6′-dichloro-6-nitrobiphenyl-3-carboxylic acid

The condensed benzenoid compounds are characterized by two or more benzene rings fused or superimposed together at *ortho* positions in such a way that each pair of rings shares two carbons. The simplest members of this group are naphthalene, with two rings, and anthracene and phenanthrene, with three rings. In the IUPAC system all carbons that may bear a substituent are numbered. Carbons that are part of a ring junction are denoted by a lowercase a or b following the number of the immediately preceding carbon. The numbering systems for naphthalene, anthracene, and phenanthrene are

naphthalene   anthracene   phenanthrene

Derivatives are named using these numbering systems.

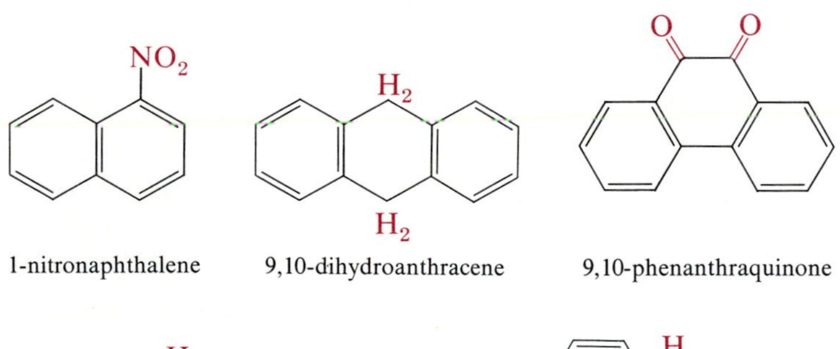

1-nitronaphthalene   9,10-dihydroanthracene   9,10-phenanthraquinone

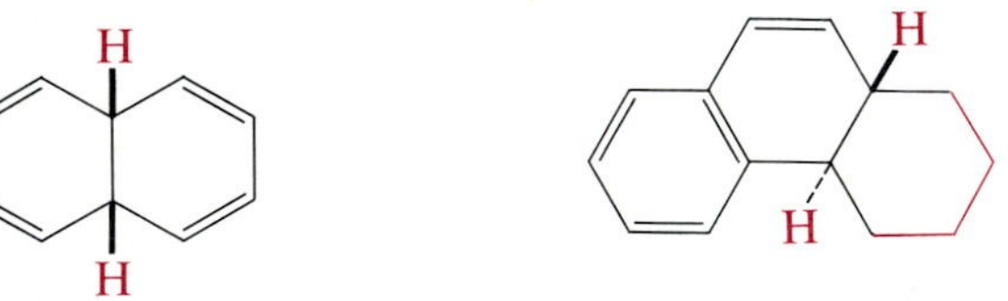

cis-4a,8a-dihydronaphthalene   *trans*-1,2,3,4,4a,10a- hexahydrophenanthrene

EXERCISE 31.1 Name the following compounds.

(a)   (b)   (c)

A. Synthesis

Biphenyl is prepared commercially by the pyrolysis of benzene.

$$2\ C_6H_6 \xrightarrow{\Delta} C_6H_5{-}C_6H_5 + H_2$$

It is a colorless crystalline solid with a melting point of 70 °C. Substituted biphenyls are prepared by electrophilic aromatic substitution reactions on the parent hydrocarbon (Section 31.2.C) or from benzene derivatives using reactions we have already studied. One of the most useful methods is the benzidine rearrangement (page 783).

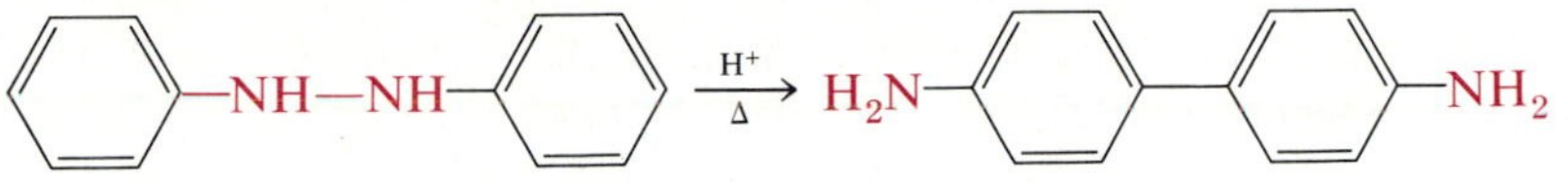

$$C_6H_5{-}NH{-}NH{-}C_6H_5 \xrightarrow[\Delta]{H^+} H_2N{-}C_6H_4{-}C_6H_4{-}NH_2$$

hydrazobenzene — benzidine, 4,4′-diaminobiphenyl

The amino groups in benzidine can be converted to many other functional groups by way of the bis-diazonium salt. Thus, benzidine and substituted benzidines are useful intermediates for the preparation of a variety of symmetrical biphenyls.

[Scheme: 2,2′-dimethyl-5,5′-dimethoxybenzidine (with CH_3O, OCH_3, H_2N, NH_2, H_3C, CH_3 substituents) $\xrightarrow[HCl]{NaNO_2}$ ("tetrazotization") bis-diazonium salt ($^+N_2$, N_2^+) $\xrightarrow{H_3PO_2}$ product]

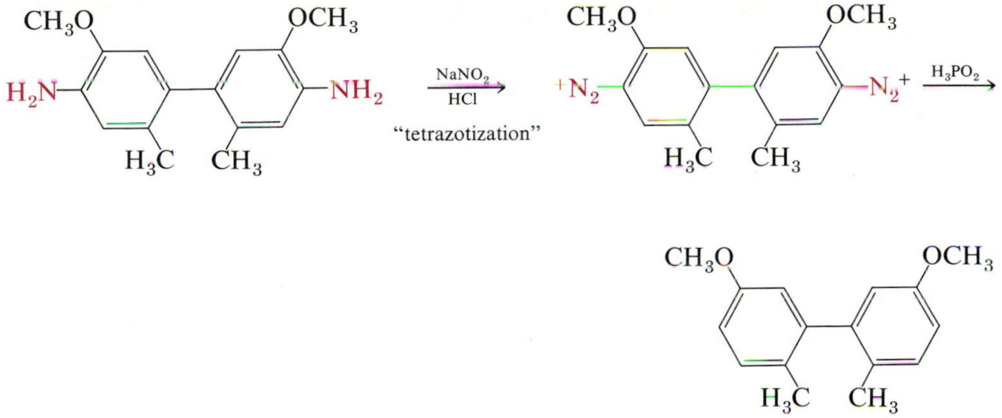

(80%)
2,2′-dimethyl-5,5′-dimethoxybiphenyl

The Ullmann reaction (Section 30.3.B) is also useful for the preparation of symmetrically substituted biphenyls, as is the oxidation of certain lithium diarylcuprates (Section 30.3.B). The Gomberg-Bachmann reaction (Section 25.5.F) is suitable for the preparation of some unsymmetrical biphenyls.

EXERCISE 31.2 Propose syntheses of (a) *m,m′*-dimethylbiphenyl and (b) *m*-methylbiphenyl beginning with *o*-nitrotoluene.

B. Structure

In the crystal both benzene rings of biphenyl lie in the same plane. However, in solution and in the vapor phase the two rings are twisted with respect to each other by an angle of about 45° (Figure 31.1). This twisting is the result of steric interactions between the 2,2′ and 6,6′ pairs of hydrogens (Figure 31.2). The magnitude of these repulsions is relatively small, only a few kcal mole^{-1}, and in the crystal is less than the stabilization obtained by stacking biphenyls together in coplanar arrays. Of course, these crystal-packing forces do

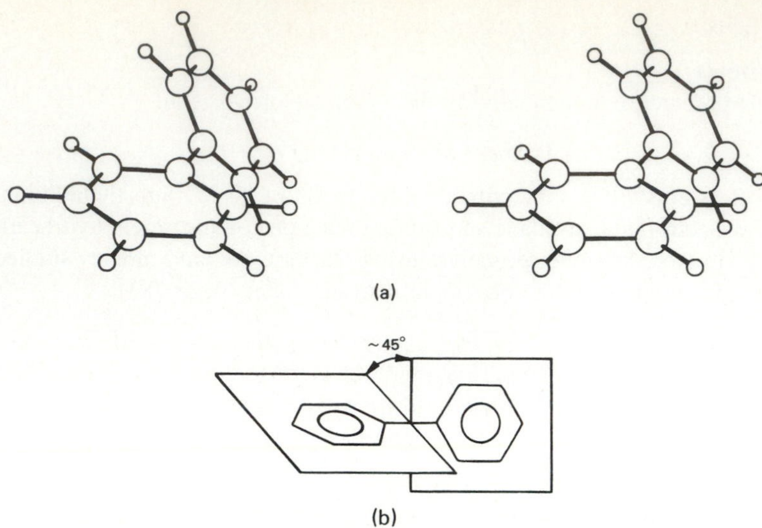

FIGURE 31.1 Structure of biphenyl in the solution or vapor phase: (a) stereo representation; (b) perspective diagram.

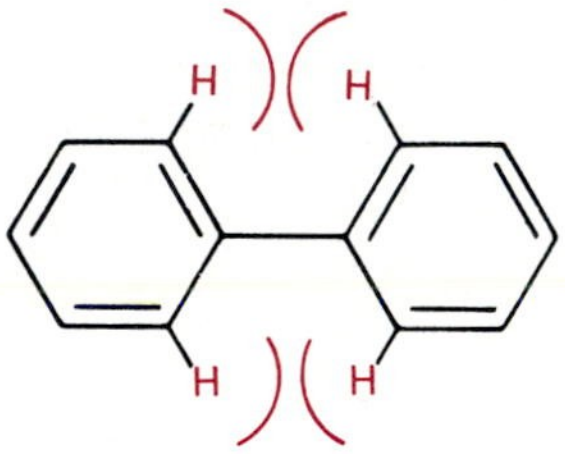

FIGURE 31.2 Steric interactions between *ortho*-hydrogens in biphenyl.

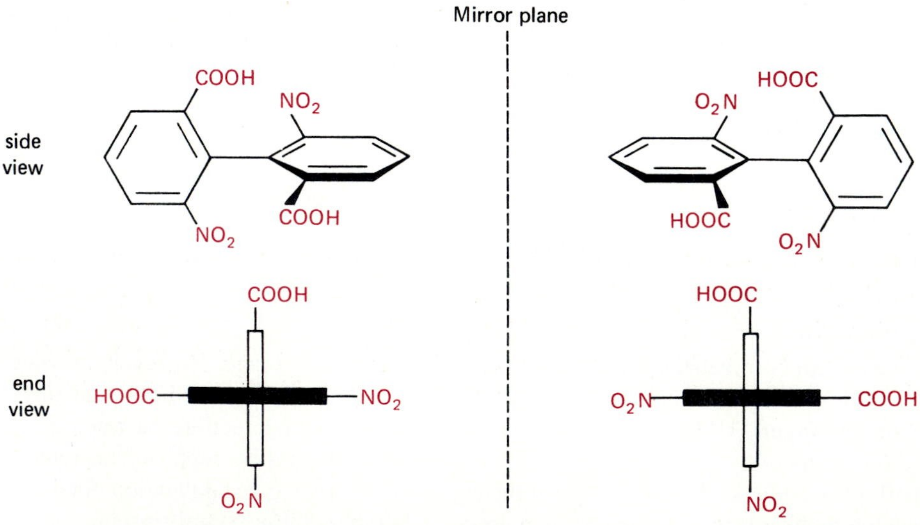

FIGURE 31.3 Enantiomers of 6,6′-dinitrobiphenyl-2,2′-dicarboxylic acid.

not exist in the vapor phase, and the twisting of the rings provides greater separation of the hydrogens.

These repulsion effects are enhanced by *ortho* substituents larger than hydrogen. When the groups are sufficiently large, rotation of the phenyl rings with respect to each other is hindered or prevented. For example, 6,6′-dinitrobiphenyl-2,2′-dicarboxylic acid can be resolved into its enantiomers and each enantiomer is stable indefinitely (Figure 31.3). The nitro and carboxylic acid groups are so bulky that they cannot pass by each other, and rotation about the bond joining the two rings is prevented.

If the bulky nitro groups are replaced by the smaller fluorine atoms, the resulting compound, 6,6′-difluorobiphenyl-2,2′-dicarboxylic acid, can still be obtained in optically active form. However, the compound racemizes readily; that is, the enantiomers are readily interconverted. The racemization process involves squeezing the fluorines past the adjacent COOH groups via a planar transition state.

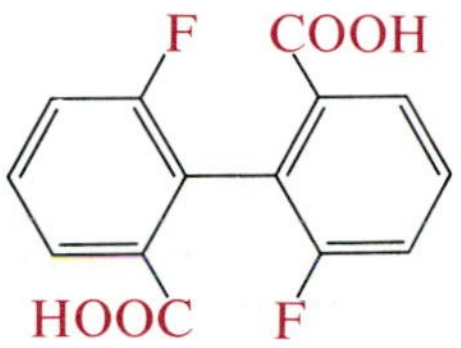

F COOH

HOOC F

6,6′-difluorobiphenyl-2,2′-dicarboxylic acid

This transition state is congested and requires the bending of bonds. The process takes energy and is measurably slow. On the other hand, all attempts to resolve biphenyl-2,2′-dicarboxylic acid (diphenic acid) have failed. The process of slipping a small hydrogen past the carboxylic acid group is so facile that racemization of enantiomers occurs rapidly.

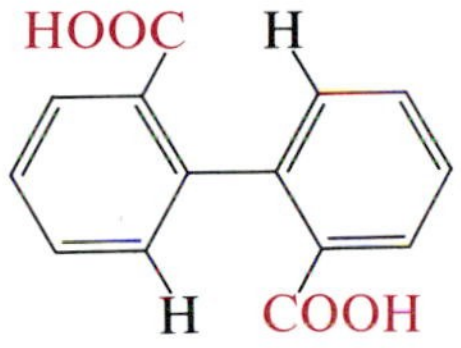

HOOC H

H COOH

diphenic acid

EXERCISE 31.3 (a) Construct a molecular model of a biphenyl having different substituents in the 2,2′- and 6,6′-positions. Demonstrate that the molecule is chiral so long as the two rings are not coplanar. (b) Assign *R* and *S* designations to the enantiomers in Figure 31.3. (*Hint*: See pages 574–75.)

C. Reactions

Biphenyl undergoes electrophilic aromatic substitution more readily than benzene; a phenyl substituent is activating and is an *ortho,para* director. Nitration in acetic anhydride solution gives primarily 2-nitrobiphenyl, but most other substitution reactions give primarily *para* orientation. Bromination, for example, gives almost wholly 4-bromobiphenyl, and excess reagent leads readily to 4,4′-dibromobiphenyl. Typical partial rate factors are

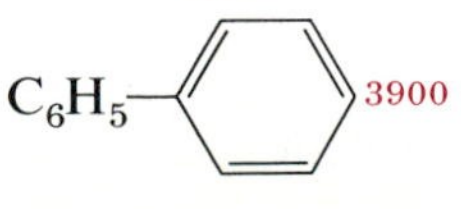

130

C_6H_5 3900

for Br_2 in 50% aqueous acetic acid

Friedel-Crafts acylation with acetyl chloride and $AlCl_3$ yields 4-acetyl- or 4,4′-diacetylbiphenyl depending on the conditions.

In general, 4-substituted and 4,4′-disubstituted biphenyls can be prepared by electrophilic substitution reactions of biphenyl. Other derivatives are constructed from benzene compounds by way of the syntheses described in Section 31.2.A.

EXERCISE 31.4 Write the resonance structures for the intermediate cation resulting from attack of Br^+ at the *para* position and at the *meta* position of biphenyl. Explain why a phenyl substituent is *ortho,para* directing.

D. Related Compounds

The terphenyls have three benzene rings linked together. All three possible isomers, *ortho, meta,* and *para,* are known. Note how the greater symmetry of the *para* isomer confers a much higher melting point.

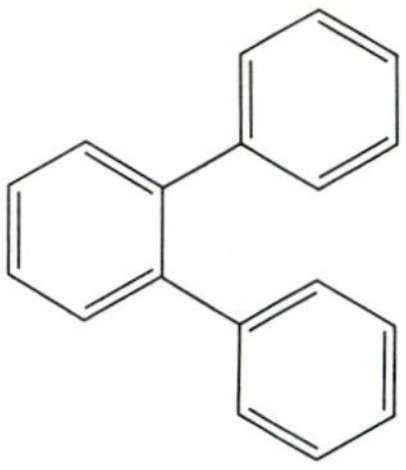

o-terphenyl
m.p. 57°C

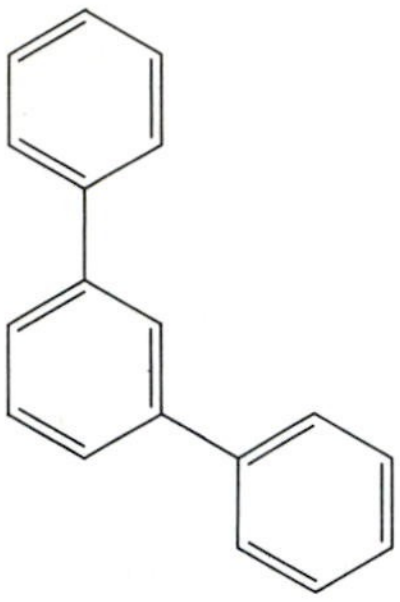

m-terphenyl
m.p. 87°C

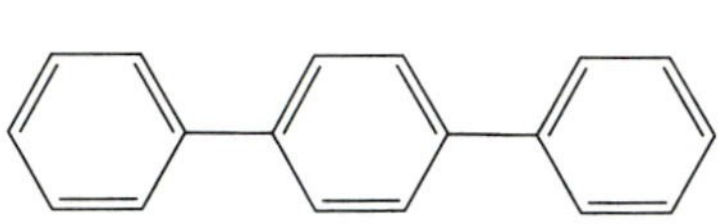

p-terphenyl
m.p. 171°C

Many of the higher polyphenyls are known, especially for the *para* isomers. *p*-Quaterphenyl has four phenyl groups linked and melts at 320 °C. *p*-Sexiphenyl is thought to melt at about 600 °C. These compounds are generally such insoluble materials that they are difficult to work with. Poly-*p*-phenylene can be converted into a conducting material and is of considerable interest as an organic conductor.

Fluorene is a biphenyl in which two *ortho* positions are linked by a methylene group. It is obtained commercially from coal tar.

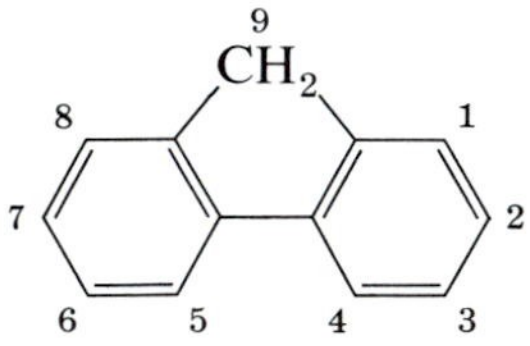

9
CH_2
8 1
7 2
6 5 4 3

fluorene

The 2- and 7-positions correspond to the *para* positions of biphenyl and are, accordingly, the most reactive positions in electrophilic aromatic substitution reactions.

CH_2 $\xrightarrow[80°C]{HNO_3,\ CH_3COOH}$ CH_2 —NO_2

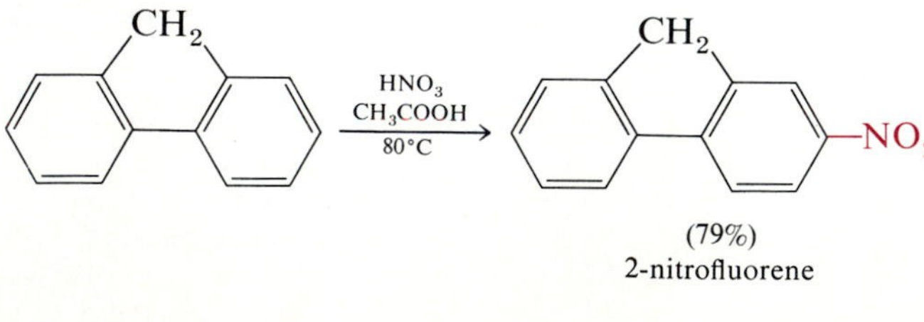

(79%)
2-nitrofluorene

The methylene group is an important center for other reactions. Oxidation gives the corresponding yellow ketone, fluorenone.

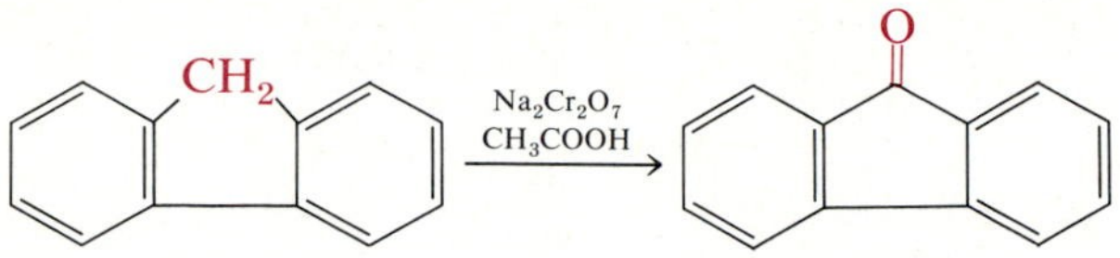

CH$_2$ $\xrightarrow{Na_2Cr_2O_7,\ CH_3COOH}$ O

fluorenone

One of the especially interesting aspects of the chemistry of fluorene is its relatively high acidity. The pK_a value of 23 puts the methylene group of this hydrocarbon in the same range as ketones and esters. Alkali metal salts can be prepared by melting with potassium hydroxide or by treatment with butyllithium.

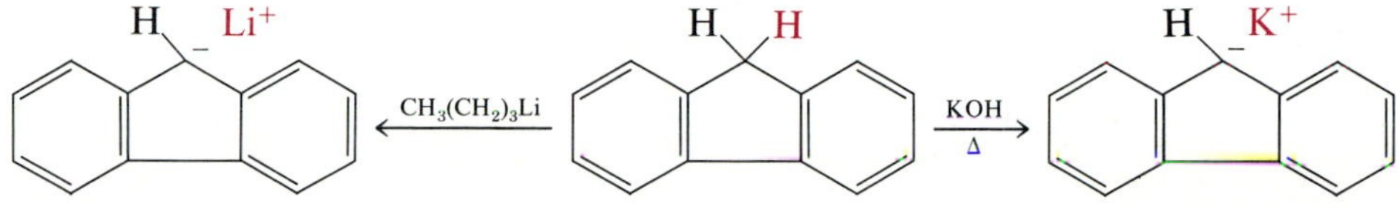

H Li$^+$ $\xleftarrow{CH_3(CH_2)_3Li}$ H H $\xrightarrow[\Delta]{KOH}$ H K$^+$

9-fluorenyllithium 9-fluorenylpotassium

The reason for this remarkably high acidity is related to the central five-membered ring structure. Cyclopentadiene is a highly acidic hydrocarbon with a pK_a of 16—an acidity comparable to that of water and alcohols (Section 22.3.C). The six-electron aromatic character of this electronic system stabilizes the anion relative to the hydrocarbon. If one or both double bonds in cyclopentadiene are replaced by benzene rings, the corresponding anion has reduced stability relative to its conjugate acid because the delocalization of negative charge disrupts the benzene conjugation.

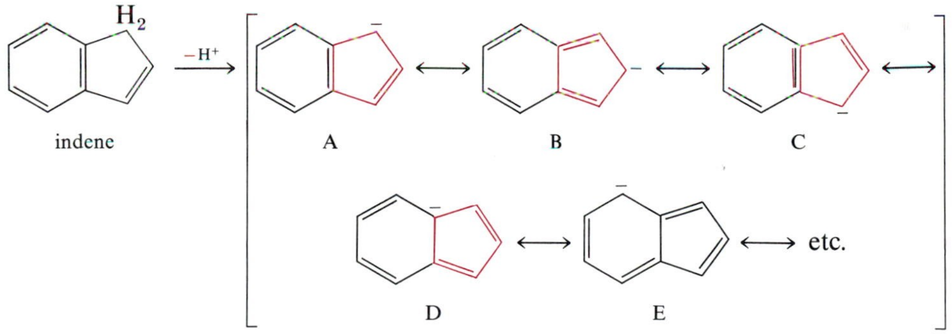

H$_2$ $\xrightarrow{-H^+}$

indene A B C

D E etc.

Indene, unlike cyclopentadiene, has a benzene ring. Structures A and C of indenyl anion also have benzene rings, but the other structures, B, D, E, and so on, have no benzene rings and are expected to be much less stable. The same principles apply to fluorene and fluorenyl anion. The corresponding pK_a values are summarized in Table 31.1 and compared with several other hydrocarbons for reference.

EXERCISE 31.5 Suggest syntheses of the following compounds.

(a) *p*-quaterphenyl (b) fluorene-9-carboxylic acid
(c) fluoren-9-ylacetic acid

TABLE 31.1 Acidities of Some Hydrocarbons

Formula	Name	pK_a
	cyclopentadiene	16
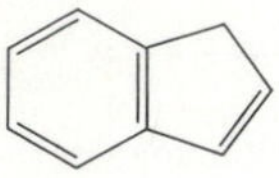	indene	20
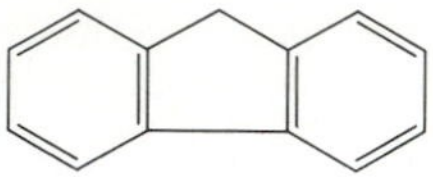	fluorene	23
$(C_6H_5)_3CH$		31
$(C_6H_5)_2CH_2$		33
$C_6H_5CH_3$		41

31.3 Naphthalene

A. Structure and Occurrence

Naphthalene is a colorless crystalline hydrocarbon, m.p. 80 °C. It sublimes readily and is isolated in quantity from coal tar.

> Coal tar is obtained from the conversion of bituminous coal to coke. The coal is heated in the absence of oxygen, giving gas and a condensate that can be redistilled. The low-boiling fraction contains benzene, toluene, and xylenes. A fraction distilling at 195–230 °C, called naphthalene oil, yields crude naphthalene on cooling. The higher-boiling coal tar is a black odoriferous complex mixture containing many polycyclic hydrocarbons and heterocyclic compounds.

Naphthalene is the parent hydrocarbon of the series of fused benzene polycyclic structures. X-ray analysis shows it to have the structure shown in Figure 31.4. The bonds are not all of the same length, but are close to the benzene value of 1.397 Å. Naphthalene can be considered to be resonance hybrid of three Kekulé structures.

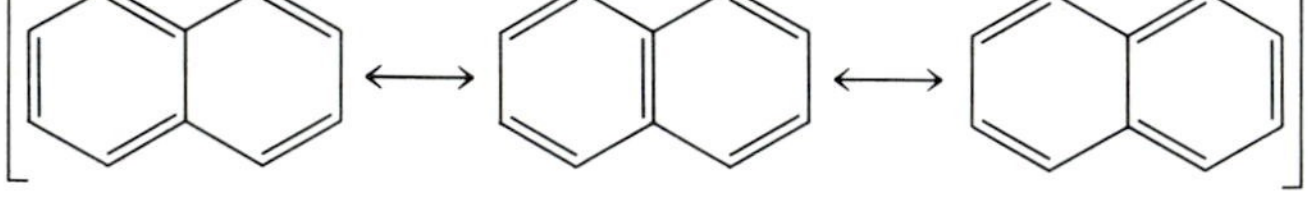

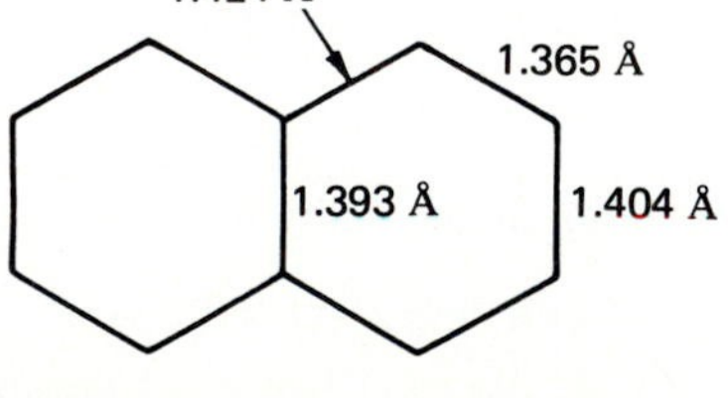

1.424 Å
1.365 Å
1.393 Å
1.404 Å

FIGURE 31.4 Structure of naphthalene.

Accordingly, it has an empirical resonance energy of about 60 kcal mole^{-1}, a value greater than that of benzene (Section 21.1.B).

Substituted naphthalenes are named using the numbering system given in Section 31.1. Monosubstituted naphthalenes are often named using α- and β-nomenclature for the 1- and 2-positions, respectively; for example,

CH_3 OH

α-methylnaphthalene
1-methylnaphthalene

β-naphthol
2-naphthol

Reduced naphthalenes are widespread in nature, particularly in terpenes and steroids (Section 36.6). The fully reduced form, decahydronaphthalene, has the trivial name decalin. Two diastereomeric forms are possible, in which the hydrogens at the ring juncture carbons are either cis or trans. The decalins can also be named using the systematic nomenclature for bicyclic compounds (pages 597–98).

H H H H

cis-bicyclo[4.4.0]decane
(*cis*-decalin)

trans-bicyclo[4.4.0]decane
(*trans*-decalin)

In both *cis*- and *trans*-decalin the two cyclohexane rings are each in the chair conformation. In *trans*-decalin the two chair cyclohexanes are fused together in such a way that each ring comprises two equatorial substituents on the other one (Figure 31.5). In *cis*-decalin the cyclohexane chairs are joined together as equatorial and axial substituents (Figure 31.6).

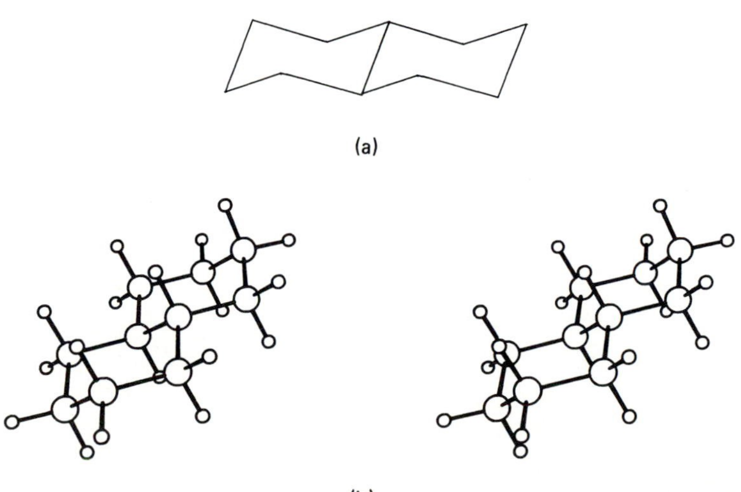

FIGURE 31.5 *trans*-Decalin: (a) conventional representation; (b) stereo structure.

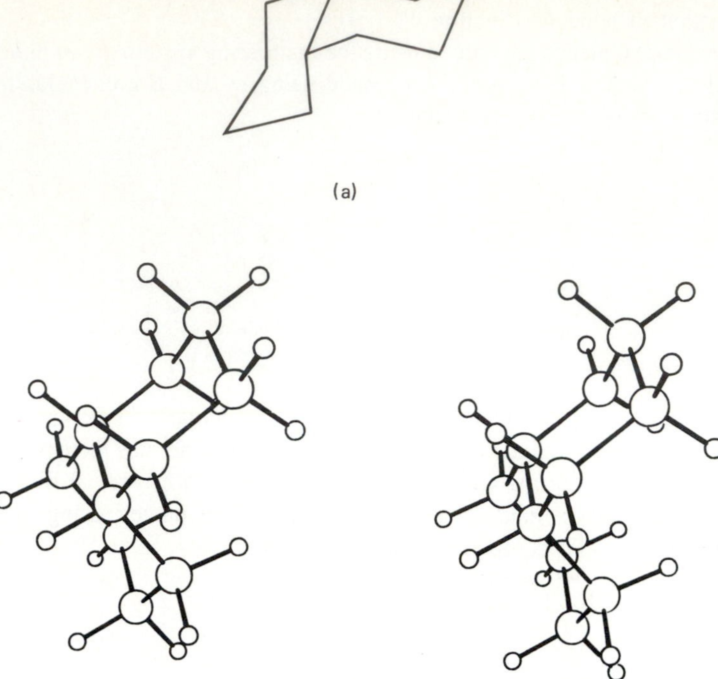

FIGURE 31.6 *cis*-Decalin: (a) conventional representation; (b) stereo structure.

EXERCISE 31.6 From the resonance structures on page 1056, determine the fractional double bond character of all of the different carbon-carbon bonds in naphthalene. A pure C_{sp^2}—C_{sp^2} single bond is expected to have a length of about 1.50 Å. Using this value and the bond lengths of ethylene and benzene, draw a smooth curve for bond length as a function of double bond character. Calculate the bond lengths expected for the different bonds in naphthalene using this curve and compare with the experimental values in Figure 31.4.

B. Synthesis

The naphthalene ring system can be prepared from suitable benzene derivatives by making use of a general method for building up a second ring that starts with a Friedel-Crafts acylation using a cyclic anhydride. A typical sequence is illustrated by the following example.

CH_3 (toluene) + succinic anhydride $\xrightarrow{AlCl_3}$ CH_3–C_6H_4–CO–CH_2CH_2–COOH (4-*p*-tolyl-4-oxobutanoic acid) $\xrightarrow[\Delta]{H_2NNH_2,\ \text{glycol},\ KOH}$

toluene   succinic anhydride   4-*p*-tolyl-4-oxo-butanoic acid

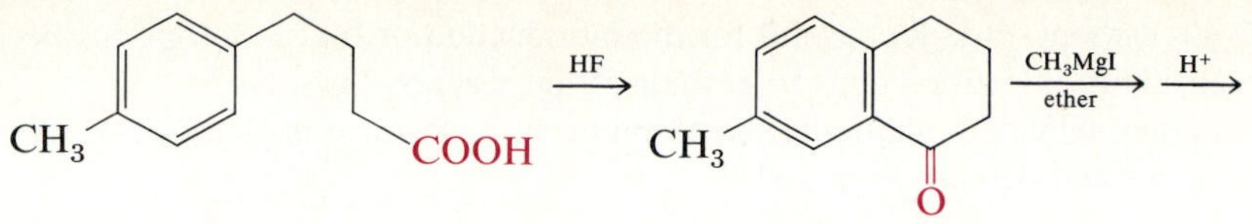

CH_3-C$_6$H$_4$-CH$_2$CH$_2$CH$_2$COOH $\xrightarrow{HF}$ (1-oxo-7-methyl-1,2,3,4-tetrahydronaphthalene) $\xrightarrow[\text{ether}]{CH_3MgI}$ $\xrightarrow{H^+}$

4-*p*-tolylbutanoic acid

1-oxo-7-methyl-1,2,3,4-tetrahydronaphthalene

$\xrightarrow[\Delta]{H^+}$ $\xrightarrow{\text{S or Se, }\Delta}$

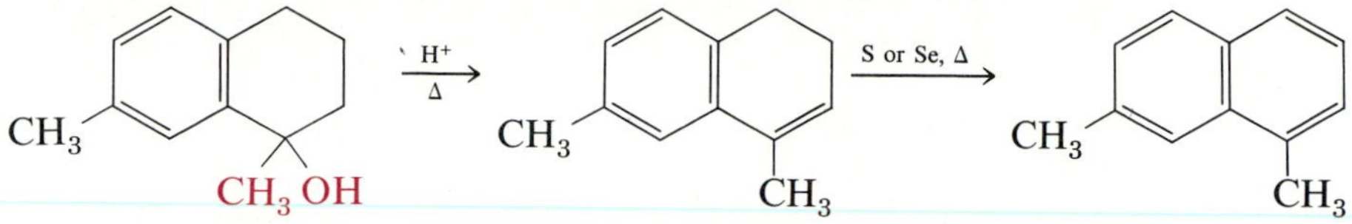

1,7-dimethyl-1,2,3,4-tetrahydro-1-naphthol

4,6-dimethyl-1,2-dihydronaphthalene

1,7-dimethylnaphthalene

Note that cyclic anhydrides function normally as acylating reagents in Friedel-Crafts reactions and that the resulting keto acids can be reduced to the corresponding alkanoic acids by Wolff-Kishner or Clemmensen conditions (Section 14.8.D). β- and γ-Arylalkanoic acids readily undergo intramolecular Friedel-Crafts acylation reactions to generate the corresponding five- or six-membered ring ketone. Commonly used reagents are $AlCl_3$ with the acid chloride and sulfuric acid, polyphosphoric acid, or liquid hydrogen fluoride with the free acid.

> The example shown makes use of liquid HF, a convenient reagent for this purpose if special precautions are taken. Anhydrous HF is a low-boiling liquid, b.p. 19 °C, available in cylinders. It is highly corrosive to glass and tissue and must be handled with due caution. The liquid is an excellent solvent for oxygen-containing organic compounds (hydrogen bonding). It does not attack polyethylene or Teflon, and these polymers make suitable reaction vessels. Because of the etching of glass windows, it is generally best to use one specific hood in a laboratory for HF reactions. The vapors should not be inhaled, and the material causes severe burns on contact with skin. For the intramolecular Friedel-Crafts reaction the carboxylic acid is weighed into a polyethylene beaker and—in an efficient hood—liquid HF is added from an inverted tank previously cooled to 5 °C (use polyethylene or rubber gloves). The mixture is stirred and the HF is allowed to evaporate over the course of several hours. The residue is mixed with aqueous Na_2CO_3 and extracted with benzene. The product is obtained by distillation or crystallization. Yields are typically 70–90%.

The last step in the foregoing synthesis of 1,7-dimethylnaphthalene illustrates the **aromatization** of a hydroaromatic compound. The driving force is formation of the stable aromatic ring. When sulfur or selenium is used for aromatization, it is concomitantly reduced to H_2S or H_2Se, respectively. Good ventilation is required for these toxic gases. Palladium metal can also be used to catalyze the aromatization. In this case, which amounts to the reverse of catalytic hydrogenation, hydrogen is eliminated.

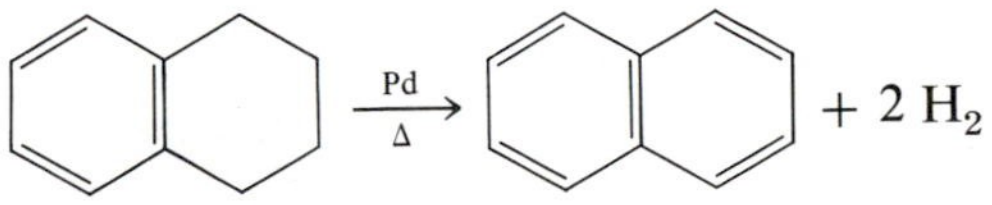

$$\text{tetralin} \xrightarrow[\Delta]{Pd} \text{naphthalene} + 2\,H_2$$

naphthalene

Tetrachloro-*p*-benzoquinone, chloranil, can also be used as a reagent for such dehydrogenation reactions. Although partially hydrogenated benzenes may be aromatized in this way, the reaction is primarily used for the synthesis of polycyclic aromatic compounds. Note that the preceding reaction sequence is adaptable to the synthesis of many different naphthalene hydrocarbons. It is less useful for the introduction of functional groups because of the sensitivity of most groups to several of the reactions involved.

Another important way of building up the second ring makes use of the Diels-Alder reaction of quinones and dienes (page 1041).

EXERCISE 31.7 Outline syntheses of (a) 1-propyl-7-methoxynaphthalene and (b) 2,3-dimethyl-1,4-naphthoquinone.

C. Electrophilic Substitution

Naphthalene undergoes a number of the usual electrophilic aromatic substitution reactions such as nitration, halogenation, sulfonation, and Friedel-Crafts acylation. The 1-position is the more reactive.

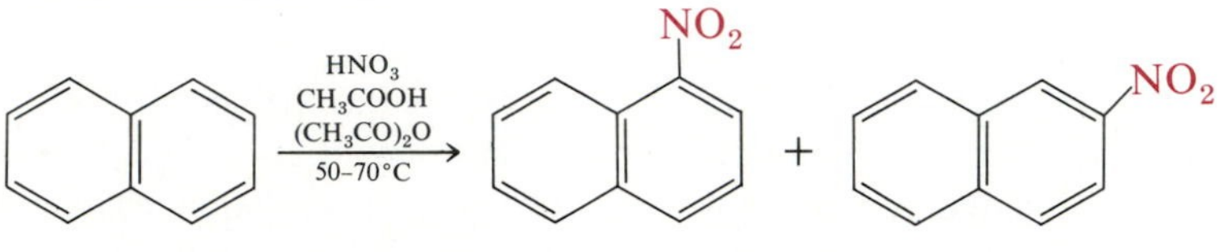

NO_2 NO_2

HNO_3 CH_3COOH $(CH_3CO)_2O$ 50–70°C

+

10:1

1-nitronaphthalene 2-nitronaphthalene

The reason for the generally greater reactivity of the 1-position can be seen by examination of the resonance structures for the two transition states or the intermediates resulting from them.

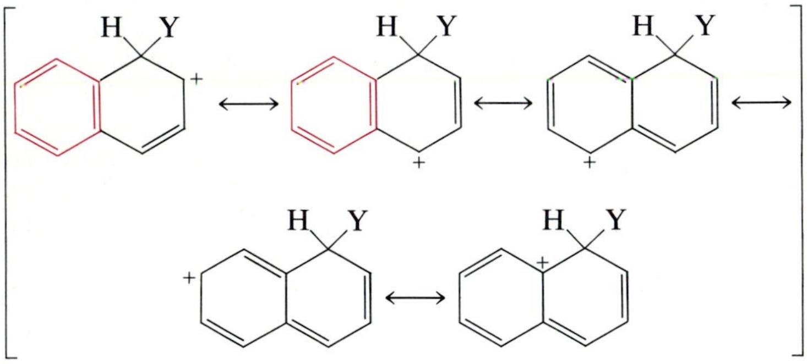

H Y H Y H Y + + + H Y H Y + +

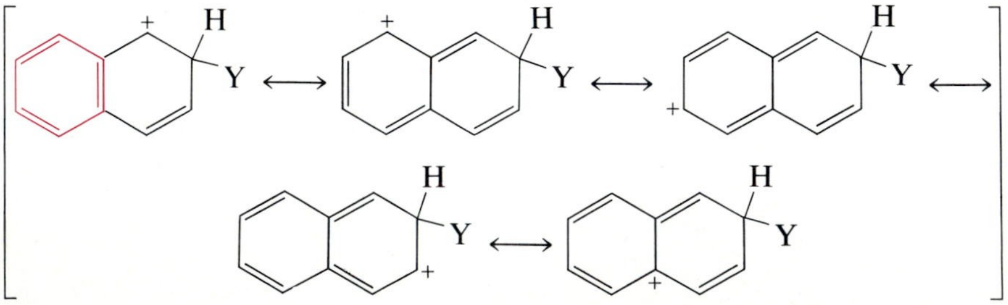

+ H + H H Y Y Y + H H Y Y + +

In both cases the positive charge can be distributed to five different positions, but these carbocation structures are not equivalent in energy. In the α-case the first two structures still have an intact benzene ring and are consequently much more stable than the remaining three structures. The first two structures contribute much more to the overall resonance hybrid. In the β-case, however, only the first structure has an intact benzene ring; the resulting resonance hybrid has higher energy than in the α-case.

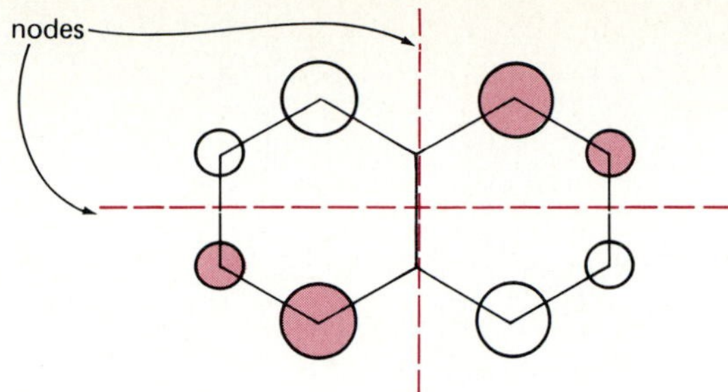

FIGURE 31.7 HOMO of naphthalene.

The same conclusion results from application of molecular orbital theory. The frontier orbital (HOMO) of naphthalene is shown in Figure 31.7. Note that the position closer to a node generally has a small magnitude of wave function than a position farther from a node. The α-position has greater wave function magnitude than the β-position and is generally more reactive.

In the nitration reaction the small amount of 2-nitronaphthalene formed is readily removed by recrystallization; hence, the nitration reaction is a satisfactory route to 1-nitronaphthalene. More vigorous nitration conditions give mixtures of 1,5- and 1,8-dinitronaphthalenes. Since the nitro group is a deactivating group, the second nitro group enters the other ring.

Bromination is also an excellent reaction and gives almost pure 1-bromonaphthalene.

Br

Br_2
CCl_4
Δ

(72–75%)
1-bromonaphthalene

Sulfonation under mild conditions gives the 1-sulfonic acid. However, at higher temperature naphthalene-2-sulfonic acid results. This pattern is the same phenomenon of kinetic versus thermodynamic control that we have seen previously for sulfonations (Section 26.5). The 1-position is the more reactive, but 1-naphthalenesulfonic acid is more hindered and less stable than the C-2 acid because the bulky sulfonic acid group is within the van der Waals radius of the C-8 hydrogen.

steric interaction

less interaction

H
SO_3H
H
SO_3H
H

naphthalene-1-sulfonic acid

naphthalene-2-sulfonic acid

Under conditions where the sulfonation reaction is reversible, the C-2 acid is the dominant product.

SO_3H

(96%)

100% H_2SO_4
<80 °C

H_2SO_4
>160 °C

95% H_2SO_4
165 °C

SO_3H

(85%)

Friedel-Crafts acylation reactions also frequently give mixtures. In general, use of $AlCl_3$ with CS_2 as solvent gives predominantly the α-product, whereas the use of nitrobenzene generally leads to the β-isomer. Separation of the isomers can be difficult or impractical. These generalizations are only approximate. The reaction products depend on the reaction conditions and the concentrations of reagents. These reactions are not simple, and the nature of the rate-determining step can differ for α- and β-reactions.

COC_6H_5

C_6H_5COCl
$AlCl_3$
CS_2

(81%)
1-benzoylnaphthalene

CH_3COCl
$AlCl_3$
$C_6H_5NO_2$

$COCH_3$

(90%)
2-acetylnaphthalene

$AlCl_3$
$C_6H_5NO_2$

COOH + COOH

(36%) (47%)
can be separated by crystallization

EXERCISE 31.8 Suggest two methods each by which naphthalene can be converted efficiently to (a) 1-benzylnaphthalene and (b) naphthalene-2-carboxylic acid.

D. Oxidation and Reduction of Naphthalene

With various oxidizing agents naphthalene is converted to 1,4-naphthoquinone, but the yields are frequently rather poor.

CrO_3
CH_3COOH
10–15 °C

(18–22%)

More vigorous oxidation results in loss of one ring; one commercial synthesis of phthalic anhydride involves catalytic oxidation of naphthalene.

air
V_2O_5
Δ

phthalic anhydride

Birch reduction (Section 21.6.C) of naphthalene yields 1,4-dihydronaphthalene. Note that in this product there is an isolated double bond that is not reduced further.

Na
liq. NH_3
alcohol

1,4-dihydronaphthalene

Catalytic hydrogenation gives either 1,2,3,4-tetrahydronaphthalene (tetralin) or decahydronaphthalene (decalin) depending on catalyst or conditions.

H_2/Rh-C or H_2/Pt-C, pressure, Δ

H_2/Ni or H_2/Pd-C, pressure, Δ

decalin

tetralin

cis-Decalin (Figure 31.6) is the predominant product of complete hydrogenation. Tetralin and the decalins are high-boiling liquids that find some use as solvents.

EXERCISE 31.9 Suggest efficient syntheses from naphthalene of (a) 1,2-bis(hydroxymethyl)benzene and (b) 2,3-epoxy-1,2,3,4-tetrahydronaphthalene.

E. Substituted Naphthalenes

Functional groups on a naphthalene ring behave more or less as their benzenoid analogs. For example, nitro groups can be reduced to amines, and bromides can be converted to Grignard or lithium reagents. An especially useful reaction is the fusion of the sulfonic acids with sodium or potassium hydroxide.

$SO_3^-\ Na^+$

NaOH, Δ; H^+

OH

2-naphthol

Since both naphthalenesulfonic acids are available by sulfonation under different conditions (pages 1061–62), this reaction provides a route to either α-naphthol or β-naphthol.

In the further electrophilic substitution reactions of monosubstituted naphthalenes some simple generalizations can be made.

1. *Meta*-directing substituents in either the 1- or 2-position generally cause substitution at the 5- and 8-positions, the α-positions of the other ring.

SO_3H $\xrightarrow{HNO_3,\ H_2SO_4}$ NO_2 SO_3H + SO_3H NO_2

2-naphthalenesulfonic acid

8-nitronaphthalene-2-sulfonic acid

5-nitronaphthalene-2-sulfonic acid

2. *Ortho,para*-directing groups in the 1-position cause substitution principally at the 4-position, but also occasionally at the 2-position as well.

$NHCOCH_3$ $\xrightarrow{HNO_3,\ CH_3COOH}$ $NHCOCH_3$ NO_2 + $NHCOCH_3$ NO_2

2-nitro-1-acetylamino-naphthalene

4-nitro-1-acetylamino-naphthalene

OCH_3 $\xrightarrow{HNO_3}$ OCH_3 NO_2

4-nitro-1-methoxy-naphthalene

3. *Ortho,para*-directing groups in the 2-position generally cause substitution at the 1-position.

OH $\xrightarrow{N_2^+\text{-}C_6H_4\text{-}SO_3^-,\ \text{aq. NaOH},\ 5°C}$ N=N SO_3^- OH

Orange II

Exceptions to these generalizations are not uncommon, especially in Friedel-Crafts acylations and sulfonation.

CH_3 —(96% H_2SO_4, Δ)→ HO_3S … CH_3

(80%)
6-methylnaphthalene-2-sulfonic acid

CH_3 + O O O —($AlCl_3$, $C_6H_5NO_2$)→ HOOC … CH_3, O

(60–79%)

One of the important reactions in naphthalene chemistry, the **Bucherer** reaction, involves the interconversion of naphthols and naphthylamines and does not apply generally in benzene chemistry.

OH —(aq. $(NH_4)_2SO_3$, 150°C, pressure)→ NH_2

2-naphthol 2-naphthylamine

2-Naphthol is readily available from 2-naphthalenesulfonic acid; hence, the Bucherer reaction provides a simple route to 2-naphthylamine which, in turn, can be converted to many other functions via the diazonium ion.

2-Naphthylamine is a powdery solid that at one time was widely used as an important intermediate in dye chemistry. This amine is carcinogenic and is no longer used.

The reaction is reversible and also provides a hydrolytic route from amine to naphthol.

NH_2 —(aq. $NaHSO_3$, Δ)→ OH

The use of aniline or substituted anilines in the reaction with 2-hydroxynaphthalene-6-sulfonic acid leads to 2-N-arylaminonaphthalene-6-sulfonic acids, which are useful fluorescent dyes.

^-O_3S … OH + H_2N … X —($NaHSO_3$, H_2O, reflux)→

^-O_3S … $\overset{H}{N}$ … X

X=H, CH_3, Cl

The sulfite or bisulfite ion is essential in this reaction. The amine and naphthol are in equilibrium with a small amount of the imine or keto form, an α,β-unsaturated system that undergoes conjugate addition by bisulfite ion much as in the formation of bisulfite addition compounds of aldehydes and ketones (page 820). The bisulfite adduct then undergoes the exchange reactions expected for a ketone or an imine.

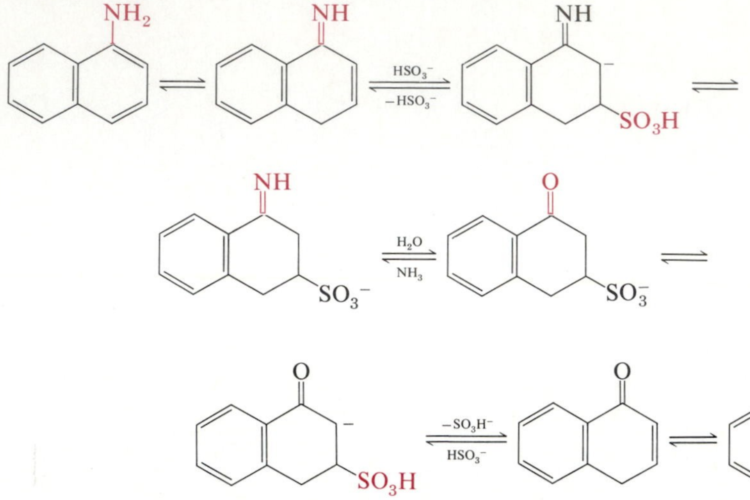

NH_2 NH HSO_3^- $-HSO_3^-$ NH SO_3H

NH SO_3^- H_2O NH_3 O SO_3^-

O SO_3H $-SO_3H^-$ HSO_3^- O OH

Polyhydroxybenzene derivatives such as 1,3,5,-trihydroxybenzene (phloroglucinol) often react as triketones; 9-hydroxyanthracene is largely in the keto form (Section 31.4.C).

EXERCISE 31.10 Write the most stable resonance structure for nitration of 2-methoxynaphthalene at each of the seven free positions. Explain why nitration occurs primarily at C-1.

31.4 Anthracene and Phenanthrene

A. Structure and Stability

The isomeric tricyclic benzenoid hydrocarbons differ significantly in thermodynamic stability; the linear system, anthracene, is almost 6 kcal mole^{-1} less stable than the angular system, phenanthrene.

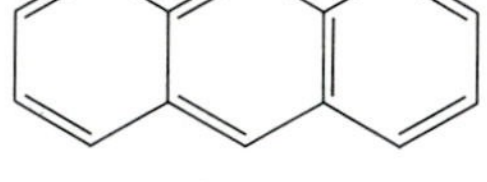

$$\Delta H_f^\circ = +55.2 \text{ kcal mole}^{-1}$$

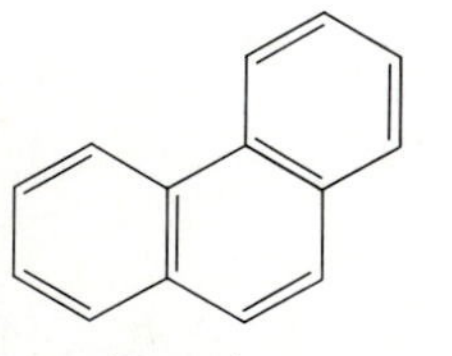

anthracene

$$\Delta H_f^\circ = +49.5 \text{ kcal mole}^{-1}$$

phenanthrene

The empirical resonance energies show a corresponding change; one set of values is 84 kcal mole^{-1} for anthracene and 91 kcal mole^{-1} for phenanthrene. The empirical resonance energy of benzene calculated in the same way is 36 kcal mole^{-1}. The resonance energies of anthracene and phenanthrene are not much more than that of two benzene rings; that is, the third ring contributes relatively little additional resonance stabilization. We shall see that this characteristic is reflected in the reactivities of these hydrocarbons.

B. Preparation of Anthracenes and Phenanthrenes

Anthracene and phenanthrene are both available from coal tar in grades that are suitable for most reactions. Commercial material requires extensive further treatment to obtain the pure hydrocarbons. When pure, anthracene (m.p. 216 °C) is colorless and exhibits a beautiful blue fluorescence. This fluorescence is diminished or altered by impurities in the commercial material. Phenanthrene also is a colorless crystalline solid (m.p. 101 °C), but it does not fluoresce.

Both ring systems can be built up from simpler compounds. Anthracene derivatives can be prepared from phthalic anhydride and benzene compounds.

$\xrightarrow[\Delta]{AlCl_3}$ COOH $\xrightarrow[\Delta]{H_2SO_4}$

benzoylbenzoic acid

anthraquinone

Note that intramolecular Friedel-Crafts occurs readily in this case even though the ring already has a carbonyl group.

Anthraquinones can be reduced directly to anthracene by several reducing agents, including sodium borohydride and boron trifluoride etherate.

NaBH$_4$
Et$_2\overset{+}{O}$:$\overset{-}{B}F_3$
diglyme

(73%)

The phenanthrene ring system can be built up from naphthalene.

HOOC

Zn-Hg
aq. HCl
toluene
Δ

HOOC

SnCl$_4$

β-(2-naphthoyl)propionic acid
4-oxo-4-(2-naphthyl)butanoic acid

(91%)
4-(2-naphthyl)-
butanoic acid

(88%)
4-oxo-1,2,3,4-
tetrahydro-
phenanthrene

Note that cyclization occurs exclusively at the 1-position of naphthalene (Section 31.3.E). A similar sequence starting from the 1-substituted naphthalene also gives the phenanthrene ring system.

COOH Zn–Hg aq. HCl Δ → COOH $SnCl_4$ →

β-(1-naphthoyl)propionic acid
4-oxo-4-(1-naphthyl)butanoic acid

(70%)
4-(1-naphthyl)butanoic acid

(92–94%)
1-oxo-1,2,3,4-tetrahydrophenanthrene

The cyclic ketones can both be converted to phenanthrene by successive reduction, dehydration, and dehydrogenation.

$LiAlH_4$ / ether → ; H^+ → ; S or Se, Δ or chloranil →

Many substituted phenanthrenes can be synthesized by variations of this general sequence.

EXERCISE 31.11 Show how the two β-naphthoylpropionic acids (page 1062) can be used to prepare 1-methylphenanthrene or 4-methylphenanthrene.

C. Reactions

Anthracene and phenanthrene undergo ready oxidation to the corresponding quinones.

$NaClO_3$, V_2O_5, CH_3COOH, H_2SO_4, Δ →

(88–91%)
9,10-anthraquinone

CrO_3, H_2SO_4, Δ →

(44–48%)
9,10-phenthraquinone

Anthraquinone can be partially reduced to give anthrone.

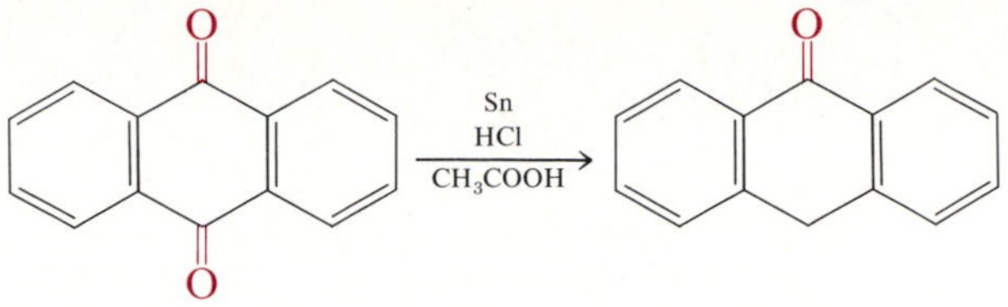

Sn
HCl
CH_3COOH

(82%)
anthrone

Anthrone is the keto form of 9-anthranol; both isomers can be isolated, but anthrone is the stable form.

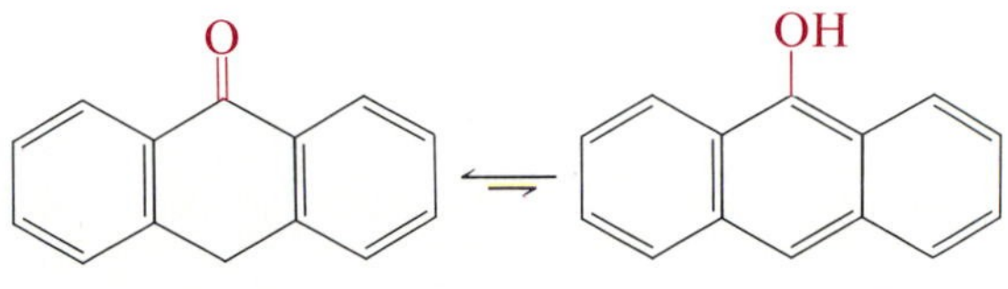

OH

anthrone

9-anthranol

Anthracene (and naphthalene) is reduced by sodium metal in tetrahydrofuran to the radical anion, a species stabilized by charge delocalization. Both anthracene and phenanthrene can be reduced readily to dihydro compounds.

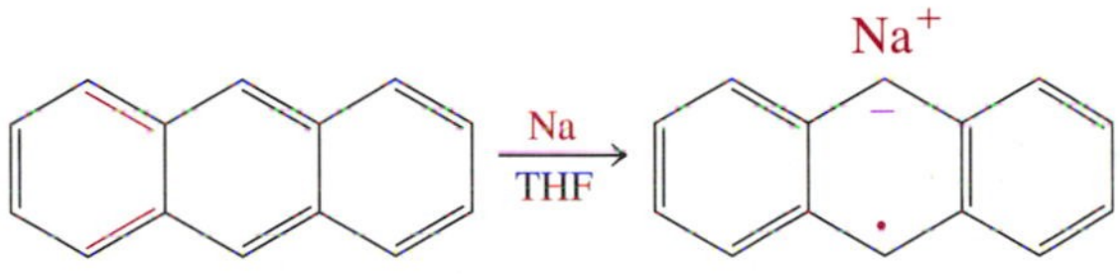

Na^+

Na
THF

sodium anthracenide

Na
EtOH
Δ

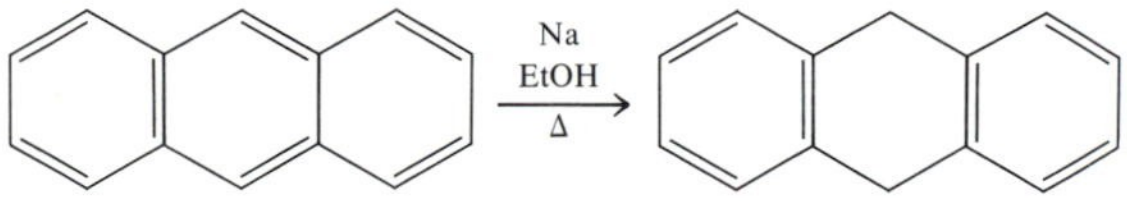

(75–79%)
9,10-dihydroanthracene

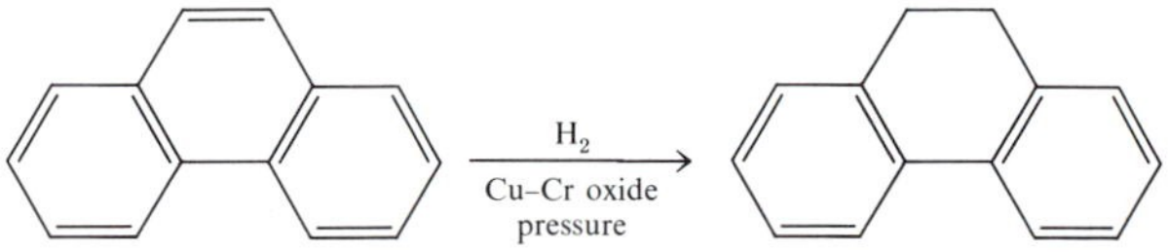

H_2
Cu-Cr oxide
pressure

(70–77%)
9,10-dihydrophenanthrene

These reactions show the distinctive reactivity of the 9,10-positions of both compounds, a reactivity inherent in the low resonance stabilization contributed by the third benzene ring (page 1066). This reactivity is also demonstrated by the ability of anthracene to undergo Diels-Alder reactions as a diene. The reaction with maleic anhydride is reversible, but the equilibrium favors the adduct.

+ xylene Δ
maleic
anhydride
(99%)

A novel reaction of this type is with benzyne (pages 1009–10) to give the unusual hydrocarbon triptycene.

F Br Mg THF 60°C F MgBr −MgBrF

(28%)
triptycene

Electrophilic aromatic substitution reactions with anthracene and phenanthrene occur most readily in the 9-position and frequently give disubstituted products.

Br_2 CCl_4 Br Br

(83–88%)
9,10-dibromoanthracene

Br_2 CCl_4 Δ Br

(90–94%)
9-bromophenanthrene

Because of the reactivity of polybenzenoid aromatic hydrocarbons, special conditions must frequently be established for individual reactions. A detailed discussion of this chemistry is beyond the scope of this book.

EXERCISE 31.12 Using resonance structures, suggest why both anthracene and phenanthrene undergo electrophilic attack primarily at the 9-position.

31.5 Higher Polybenzenoid Hydrocarbons

A large number of polybenzenoid hydrocarbons are known, and some are relatively important. Some multiring systems, with their established common names and their numbering systems, are shown.

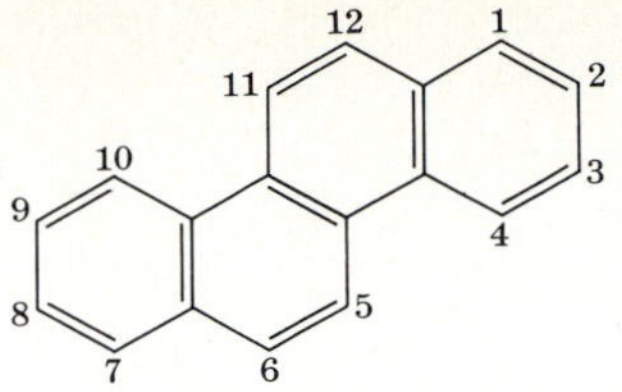

chrysene

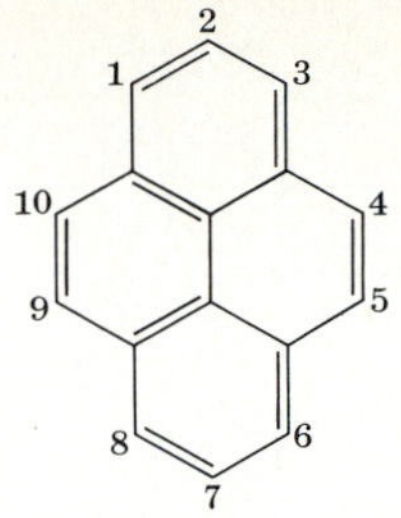

pyrene

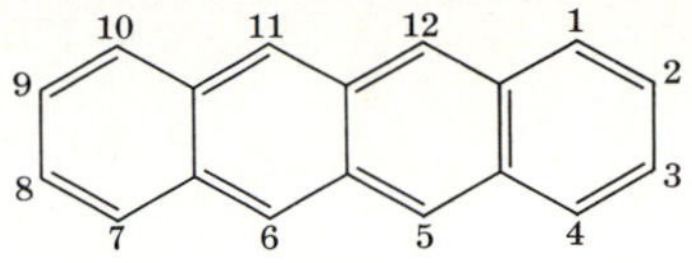

tetracene

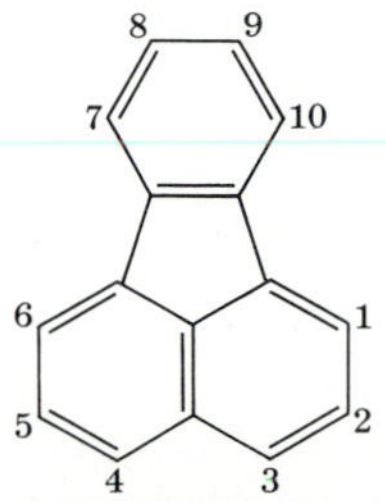

fluoranthene

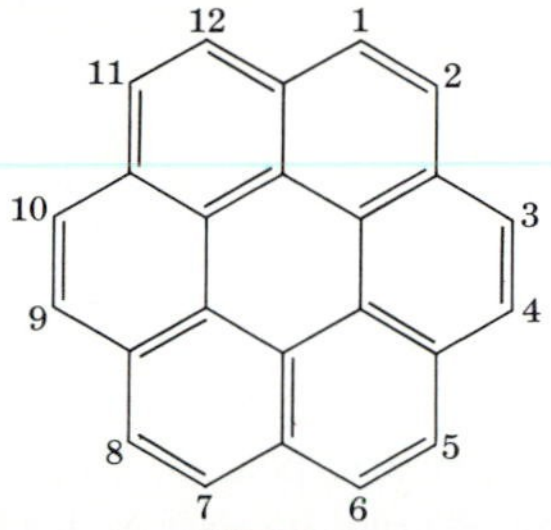

coronene

Some of these hydrocarbons are available from coal tar; others are prepared from simpler systems by building up rings in the manner shown in the preceding section for anthracene and phenanthrene.

Tetracene is an orange compound that shows much of the chemistry of anthracene. Oxidation and reduction occur readily at the 5- and 12-positions, and the hydrocarbon reacts readily as a Diels-Alder diene. Higher linear **acenes** are known: pentacene, with five fused benzene rings in a row, is blue; hexacene is green; and heptacene is a deep greenish black. The higher linear acenes are reactive, air sensitive, and difficult to obtain pure.

Chrysene is similar to phenanthrene in its reactions: it can be oxidized to the 5,6-quinone.

Pyrene is among the most important of these hydrocarbons. Pyrene undergoes the usual electrophilic aromatic substitution reactions such as halogenation, nitration, Friedel-Crafts acylation, and so on. These reactions occur exclusively at the 1-position.

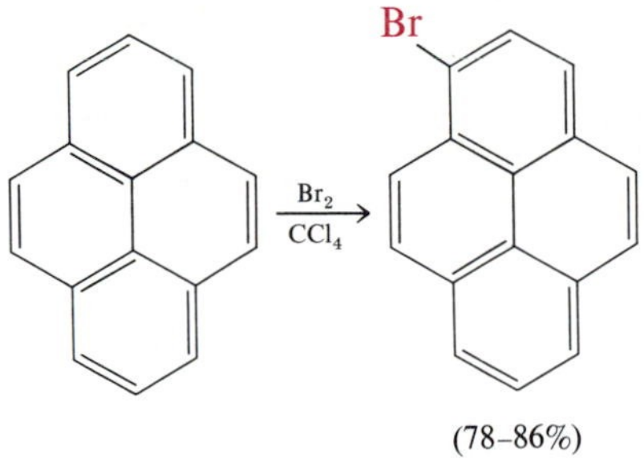

$\xrightarrow[CCl_4]{Br_2}$

(78–86%)
1-bromopyrene

> Two numbering systems have been used for pyrene, and care must be taken in reading the literature, particularly the older literature, to establish which system has been used. The system shown here is the accepted IUPAC numbering, but even today references will be found with the older nomenclature.

Polycyclic systems much larger than coronene are known. The large polycyclic hydrocarbons have low solubility, and few are significant in organic chemistry. Their properties start to approach those of graphite, an allotrope of carbon that consists of infinite planes of benzene rings with the planes separated by 3.4 Å. This distance is usually taken as the total width of the π-electronic system of benzene.

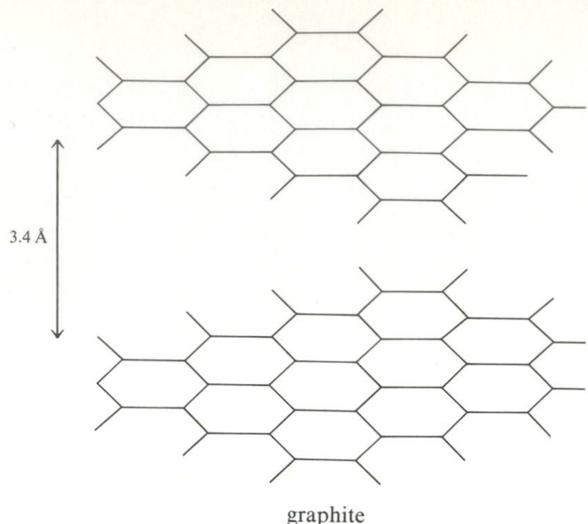

graphite

A strong mass spectrometric peak at $C_{60}{}^{+}$ (mass 720) is observed for carbon vapor. Strong resistive heating of graphite rods under helium yields graphitic soot, from which a third form of carbon, a stable C_{60} compound, can be extracted with benzene and chromatographically purified on silica gel. (The other forms of carbon are diamond and graphite.) The structure suggested is a truncated icosahedron, also called **buckminsterfullerene** for its resemblance to the geodesic domes designed by the American inventor Buckminster Fuller. There are 20 **six**-membered rings and 12 **five**-membered rings.

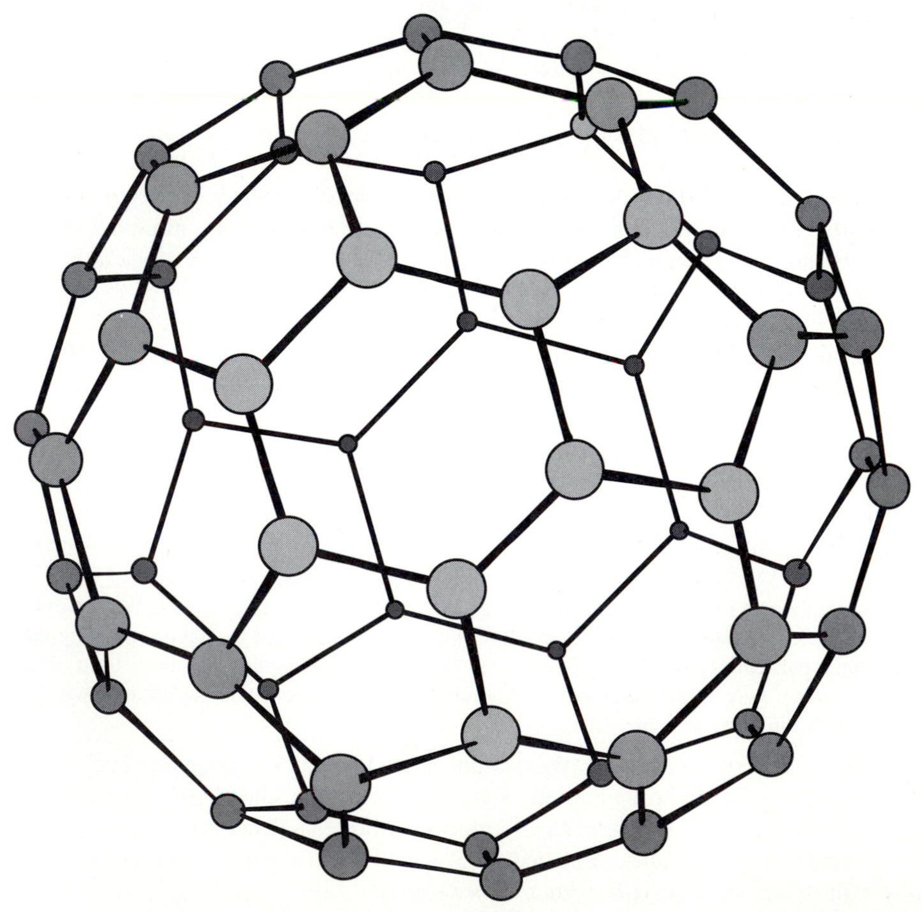

C_{60} is a remarkable three-dimensional π-system. Both molecular orbital and *ab initio* calculations indicate that C_{60} should be stable, and experiment bears out this conclusion. The compound has a *single* ^{13}C NMR peak at 143.2 ppm, four main IR peaks at 1428.5, 1182.4, 576.4, and 527.4 cm^{-1}, and the main ultraviolet and visible absorption peaks for a hexane solution at 540, 328, 256, and 211 nm. Birch reduction of C_{60} with lithium, liquid ammonia, and *t*-butyl alcohol (Section 21.6) yields several reduced derivatives, among them $C_{60}H_{36}$, which is easily dehydrogenated to C_{60} by DDQ (2,3-dichloro-5, 6-dicyano-1,4-benzoquinone). Osmium tetroxide (Section 11.6.E) in benzene containing 4-*t*-butylpyridine reacts with C_{60} to form an osmate ester in 75% yield. The stability of C_{60} is illustrated by thermal decomposition of the osmate ester to C_{60} by attempted sublimation under vacuum.

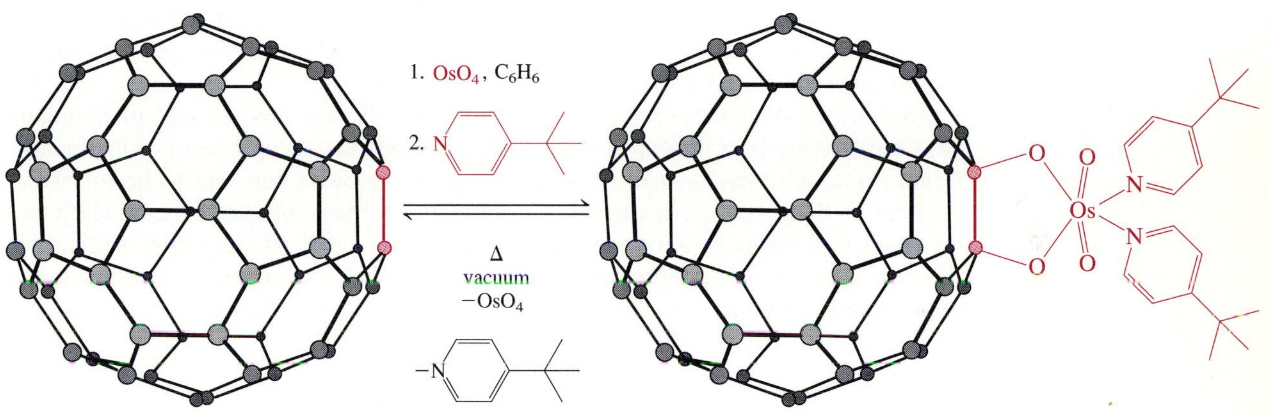

C_{60} can be reduced *reversibly* to a monoanion at −0.33 V, to a dianion at −0.73 V, and to a trianion at −1.25 V (potentials relative to a silver/silver chloride electrode). No electrochemical oxidation could be achieved up to +1.50 V, and the ionization potential of C_{60} is between 7.5 and 7.7 eV, unusually high for such an extended π-system. In 1991, films containing both C_{60} and the salt $K^+C_{60}^-$ were shown to be the first organic three-dimensional conducting solid. The chemical reactions show that the "outside" can be transformed. In addition, preparation of C_{60} from graphite rods containing metal salts leads to $MC_{60}{}^{n+}$ species, suggesting that an "inside" chemistry could be developed.

$$C_{60} \underset{-0.33\ V}{\overset{e^-}{\rightleftharpoons}} C_{60}{}^{-} \underset{-0.73\ V}{\overset{e^-}{\rightleftharpoons}} C_{60}{}^{2-} \underset{-1.25\ V}{\overset{e^-}{\rightleftharpoons}} C_{60}{}^{3-}$$

A number of polycyclic aromatic hydrocarbons are named as **benz**- or **benzo**- derivatives of simpler systems. The position of fusion of the benz-ring is represented by a lowercase italic letter that designates the side around the periphery of the parent system used for the fusion. For example, the sides of anthracene are lettered starting with side *a* between positions 1 and 2.

In this way the following hydrocarbons are derived.

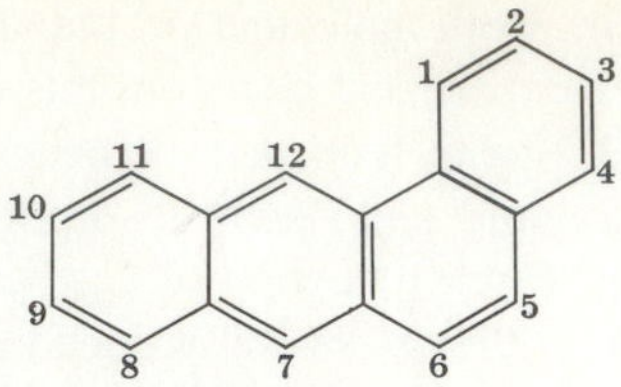

benz[*a*]anthracene

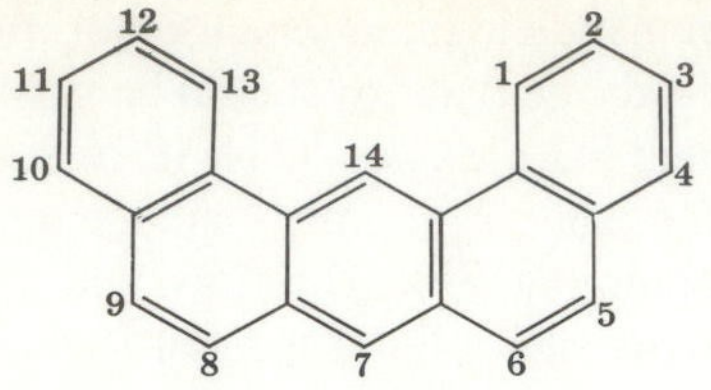

dibenz[*a,j*]anthracene

Some polycyclic aromatic hydrocarbons are highly carcinogenic compounds. Minute amounts painted on the skin of mice will produce skin tumors (epithelioma) in the course of a few months. Some of the most potent of the carcinogenic hydrocarbons are dibenz[*a,h*]anthracene (but not dibenz[*a,c*]anthracene), benzo[*a*]pyrene (but not benzo[*e*]pyrene), dibenzo[*a,i*]pyrene, and benzo[*b*]fluoranthene. These compounds occur in coal tar and in soot. A high incidence of scrotal cancer in chimney sweeps was noticed in England as early as 1775. All of these carcinogenic hydrocarbons have been detected in minute quantity in tobacco smoke.

The way in which these polycyclic aromatic hydrocarbons produce malignant tumors has been actively investigated for several decades, and the chemical part of the mechanism is now fairly well understood. The actual carcinogens turn out to be metabolic products of the polycyclic hydrocarbons. After the hydrocarbon enters a cell within the organism, it is epoxidized by a oxidase enzyme (see Section 29.7.E). This oxidation is normal for rendering the hydrocarbon more water soluble. The epoxides are converted to sugar or glutathione derivatives by nucleophilic ring opening, yielding substances that are more easily eliminated from the organism in normal detoxification reactions. An example of epoxide formation is that for benzo[*a*]pyrene, which gives rise to the highly carcinogenic diol epoxide shown.

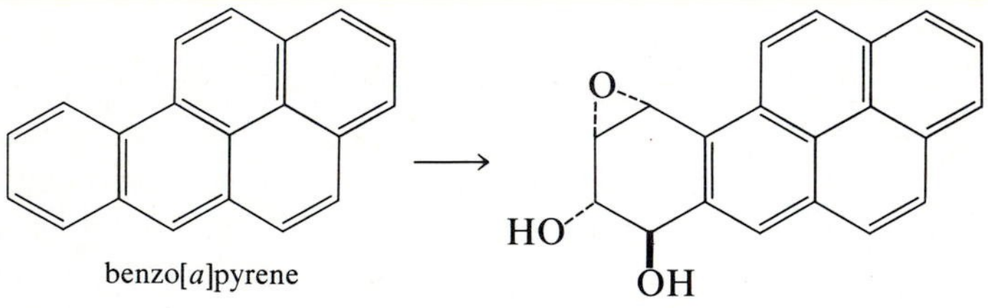

benzo[*a*]pyrene

Less reactive diol epoxides are more carcinogenic because they are not consumed in the detoxification reactions. Instead they alkylate cellular DNA, causing mutations and an eventual loss of the cell's ability to undergo controlled replication.

Benzo[*c*]phenanthrene reveals a further interesting aspect of the structural consequences of adding rings to certain aromatic systems.

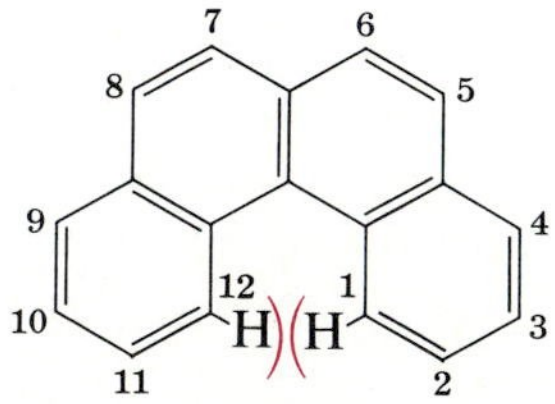

benzo[*c*]phenanthrene

The hydrogen atoms at the 1- and 12-positions interact significantly, and the molecule is forced to twist somewhat from coplanarity. The compound with two additional benzene rings cannot be planar; the material is the spirally fused hydrocarbon hexahelicene.

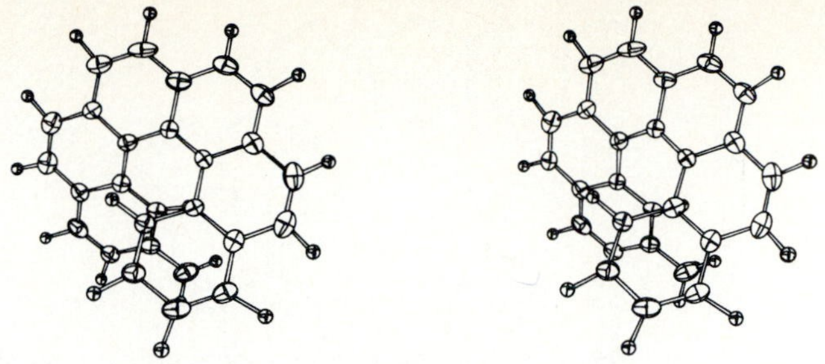

FIGURE 31.8 Stereo structure of 2-methylhexahelicene. [Reproduced with permission from K. N. Trueblood et al., *Acta Cryst.*, **B29:**223 (1973).]

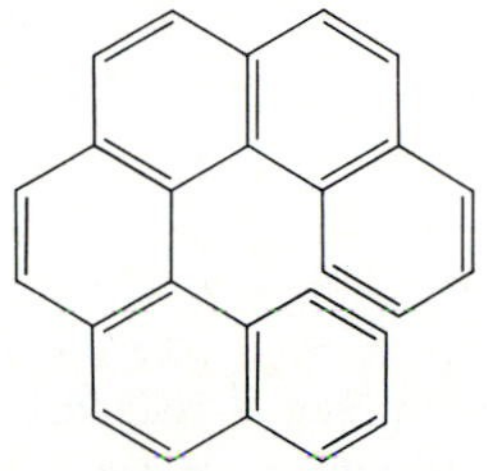

hexahelicene

If this molecule were planar, two sets of CH groups would have to exist in the same space. In practice the hydrocarbon adopts a spiral structure that is also chiral. The enantiomers of this hydrocarbon have been obtained and have enormous optical rotations, $[\alpha]_D$ 3700. The spiral structure has been demonstrated experimentally by the x-ray structure determination of 2-methylhexahelicene as shown in the stereo plot in Figure 31.8.

EXERCISE 31.13 Write the structures of the hydrocarbons named on page 1074.

Highlight 31.1

Polycyclic aromatic structures can be built up in at least two ways. First, aryl rings can be substituted for hydrogen on arenes. Second, arene rings can be fused onto other arenes. Biphenyl, 2,2′-dihydroxy-1,1′-binaphthyl, and 9,9′-bianthracenyl are examples of the first type.

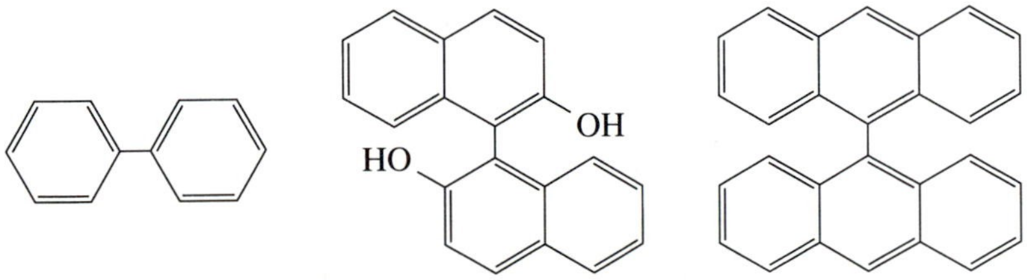

OH
HO

biphenyl   2,2′-dihydroxy-1,1′-binaphthyl   9,9′-bianthracenyl

Arene rings can be fused in a linear fashion, as in the series benzene, naphthalene, anthracene, tetracene, pentacene, and so on, or in a nonlinear fashion, resulting in phenanthrene, chrysene, coronene, the helicenes (which must be nonplanar and helical) and a recently synthesized molecule called kekulene after the scientist who proposed the correct structure for benzene.

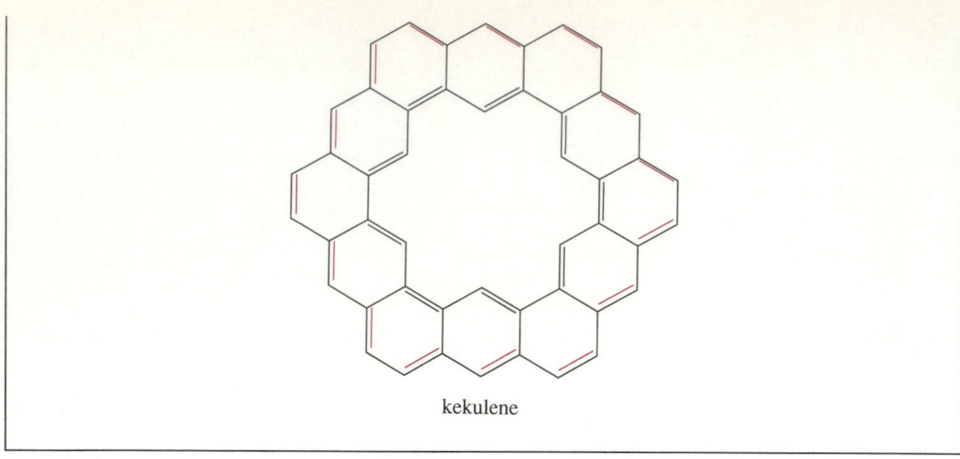

kekulene

ASIDE

"The *art* of designing tilings and patterns is clearly extremely old and well developed. By contrast the *science* of tilings and patterns, by which we mean the study of their mathematical properties, is comparatively recent and many parts of the subject remain unexplored" commented Grünbaum and Shepherd in a book on the subject. Although we are only concerned with the simplest application of these ideas to chemistry, we can experiment with chemical tiles to create examples that reflect the richness of chemistry.

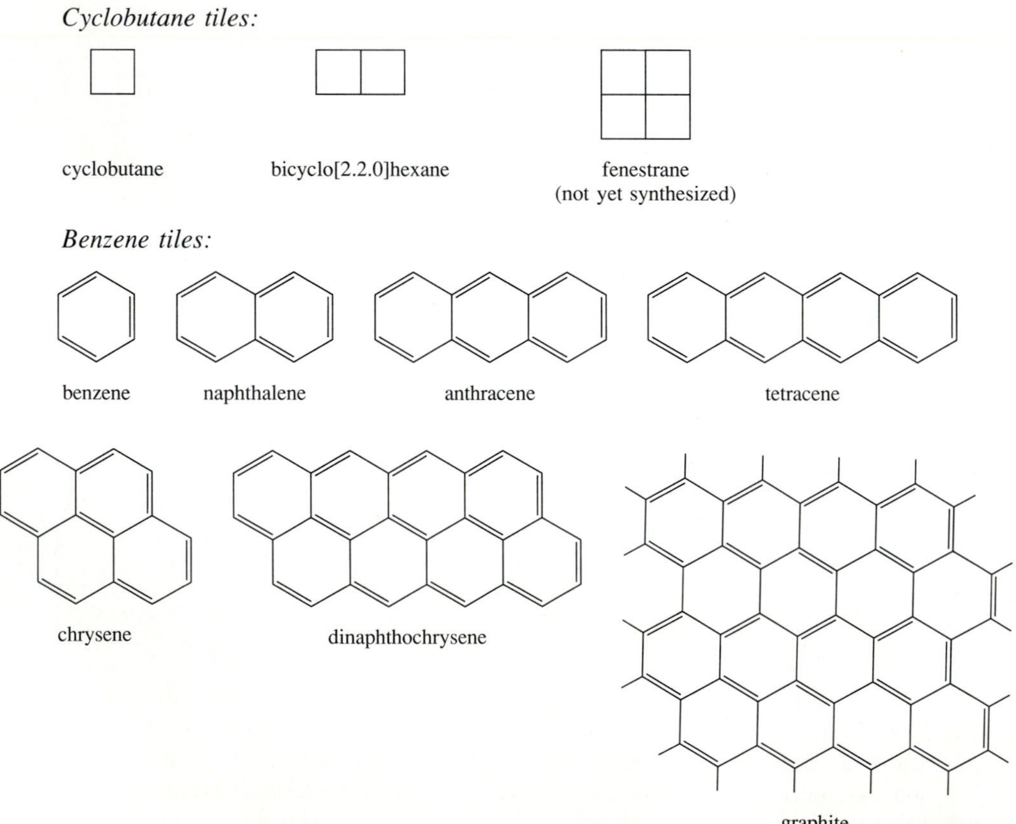

Join the benzene tiles at a single point, and name the compound the formula represents. Is the compound planar? What molecules can be created by combining benzene tiles (B) with cyclohexane tiles (C) at two points? Try the combinations (a) 1B and 1C, (b) 2C, (c) 2B, (d) 2B + C. Name the compounds. Is the combination of 2C unique? What aspect of molecular structure does this approach omit? Count the number of monobromo derivatives that can be made with the compounds derived from 1B, 2B, 3B, and 4B. Identify which would be formed by evaluating which isomers would be favored. How many different 4B compounds can you write?

PROBLEMS

1. Name each of the following compounds and show a practical synthesis starting with a suitable benzene derivative.

a.

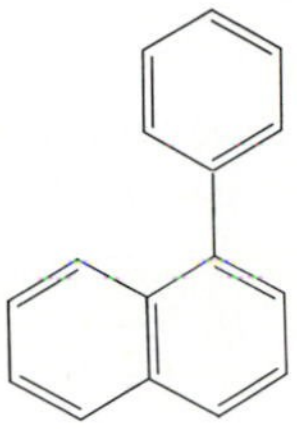

b. 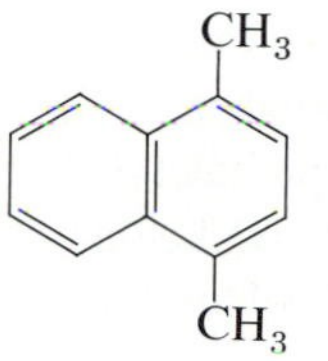 CH_3 CH_3

c. 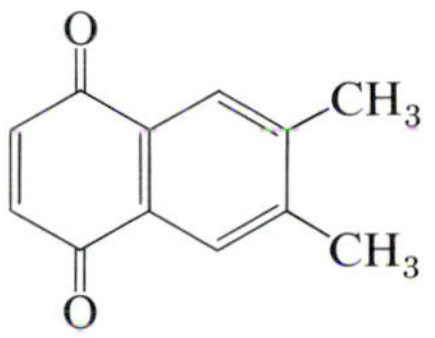 O CH_3 CH_3 O

d. 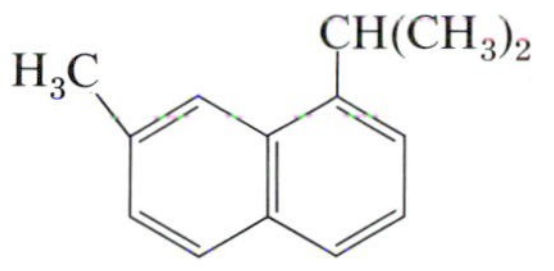 H_3C $CH(CH_3)_2$

e. 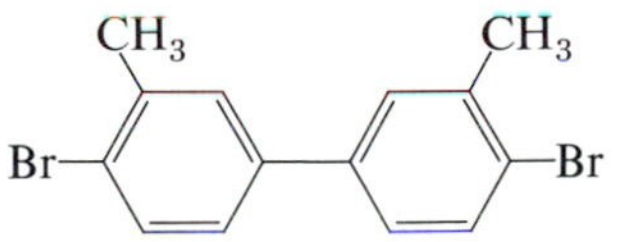 CH_3 CH_3 Br— —Br

f. 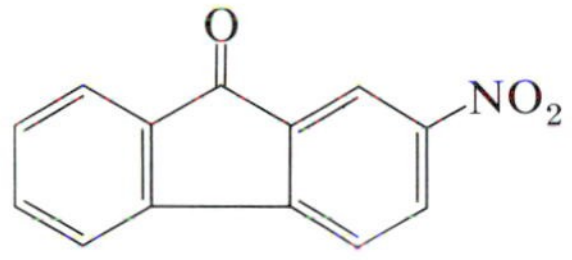 O NO_2

g. 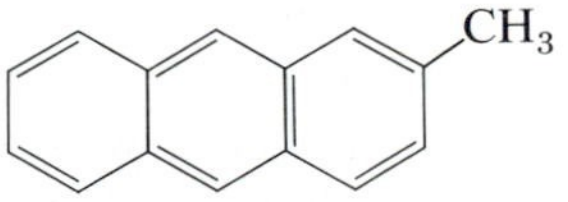 CH_3

h.

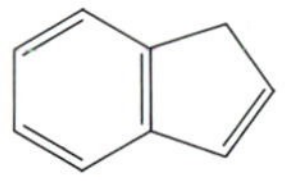

2. 6,6′-Dinitrobiphenyl-2,2′-dicarboxylic acid can be prepared by the Ullmann reaction on 2-iodo-3-nitrobenzoic acid. Give a reasonable preparation of this compound from available materials.

3. Substitution reactions of 2-methylnaphthalene with bulky electrophilic reagents tend to occur at the 6-position. Explain why this position is preferred to the sterically equivalent 7-position.

4. Write out the mechanism for the conversion of 2-naphthol to 2-naphthylamine showing every intermediate involved in the Bucherer reaction.

5. The heat of formation of naphthalene, ΔH_f°, is 36.1 kcal mole^{-1}; ΔH_f° for *trans*-decalin is −43.5 kcal mole^{-1}.

- **a.** Calculate the heat of hydrogenation of naphthalene to *trans*-decalin.
- **b.** Using the heat of hydrogenation of cyclohexene as a comparison standard, estimate the heat of hydrogenation of naphthalene in the absence of any conjugation stabilization.

c. Compare (a) and (b) to derive the corresponding empirical resonance energy of naphthalene.

6. Cadinene, $C_{15}H_{24}$, is a sesquiterpene (Section 36.6.A) occurring in the essential oils of junipers and cedars. Dehydrogenation gives the naphthalene hydrocarbon cadalene.

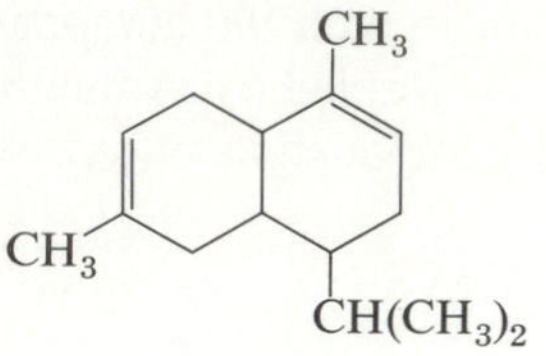

CH_3 CH_3 $CH(CH_3)_2$
cadinene

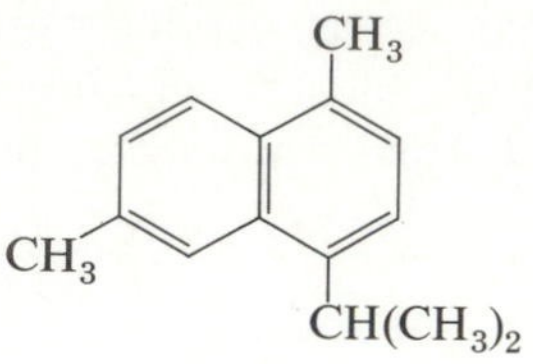

CH_3 CH_3 $CH(CH_3)_2$
cadalene

a. What is the IUPAC name for cadalene?
b. Give a rational synthesis of cadalene from toluene and any necessary aliphatic compounds.

7. Provide a practical synthesis of each of the following compounds from naphthalene:
a. 2-bromonaphthalene
b. 1-methylnaphthalene
c. 1-isopropylnaphthalene
d. 1-naphthyl propyl ketone
e. 2-phenylnaphthalene
f. 1,2-naphthoquinone
g. 1-naphthoic acid
h. naphthalene-1-*d*

8. Allyl β-naphthyl ether undergoes the Claisen rearrangement to give exclusively 1-allyl-2-naphthol. Give a reasonable explanation for the decided preference of this reaction over the alternative reaction to 3-allyl-2-naphthol.

9. The difference in empirical resonance energies of anthracene and phenanthrene can be accounted for on the basis of resonance structures. There are four Kekulé structures for anthracene and five for phenanthrene.
a. Write out both sets of resonance structures for anthracene and phenanthrene.
b. For each of the five different carbon-carbon bonds in anthracene, compare the number of resonance structures in which each is single or double and determine the fraction of double bond character (bond order). Compare with the bond lengths predicted using the curve you constructed for Exercise 31.6 with the experimental values determined by x-ray crystal structure techniques as

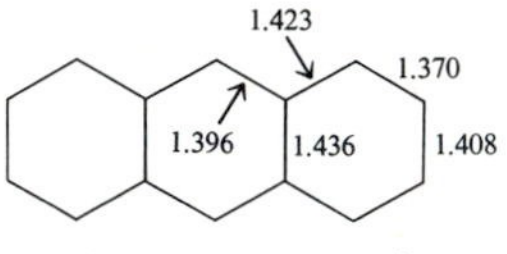

1.423 1.370 1.396 1.436 1.408

bond distances in Å

10. Give the expected dominant product or products in mononitration of each of the following compounds.

a. CH_3

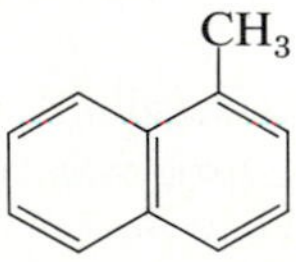

b. CN

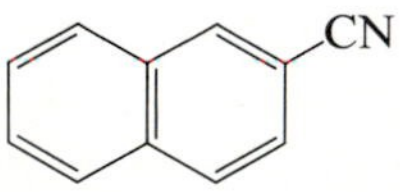

c. 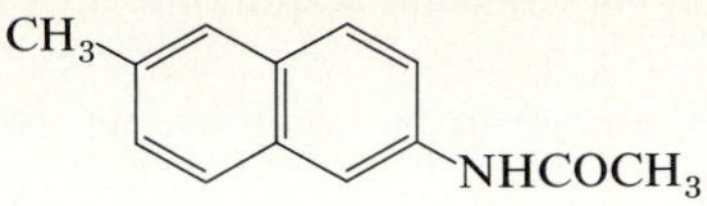 CH_3 … $NHCOCH_3$

d. 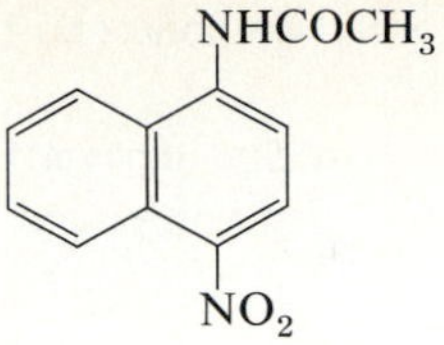 $NHCOCH_3$ … NO_2

e. 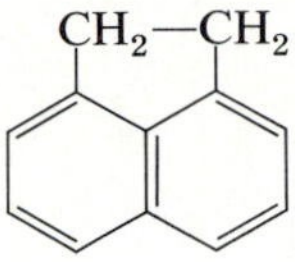 CH_2—CH_2

f.

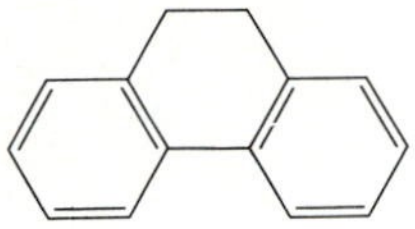

g. 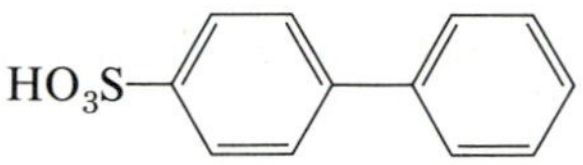 HO_3S—

h.

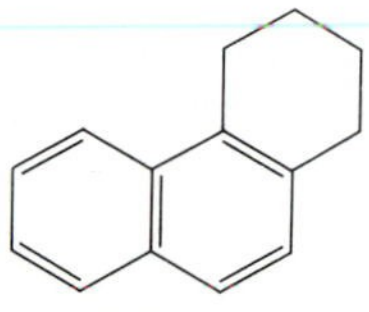

i. 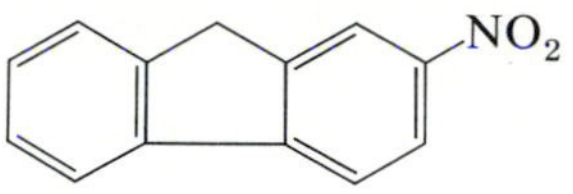 NO_2

j.

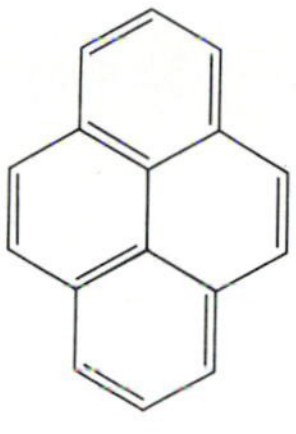

k. 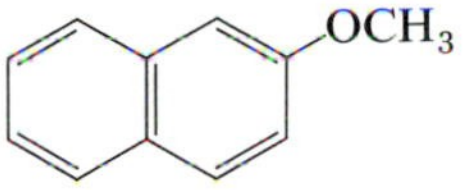 OCH_3

l.

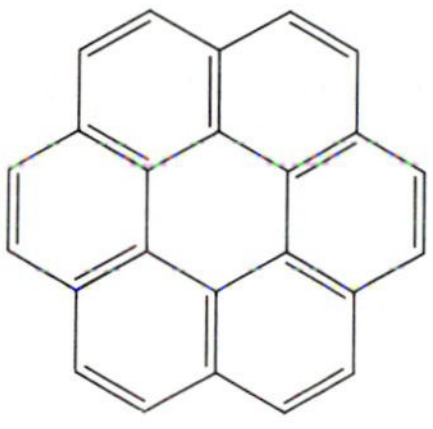

11. Acetylation of phenanthrene with acetyl chloride and $AlCl_3$ in nitrobenzene gives primarily 3-acetylphenanthrene. 2-Acetylphenanthrene is best prepared by Friedel-Crafts acetylation of 9,10-dihydrophenanthrene (note that this hydrocarbon is a biphenyl compound and the 2-position corresponds to the *para*-position of biphenyl) followed by dehydrogenation with Pd/C. Show how to prepare each of the following phenanthrene derivatives.

a. 2- and 3-phenanthrenecarboxylic acid
b. 2- and 3-aminophenanthrene
c. 2- and 3-bromophenanthrene
d. phenanthrene-2-*d* and phenanthrene-3-*d*

12. a. Starting from naphthalene or either of the monomethylnaphthalenes show how to prepare all five possible methylphenanthrenes. (*Note*: Some of these are more difficult than others.)

b. α-Methylsuccinic anhydride reacts with naphthalene and $AlCl_3$ in nitrobenzene to give about equal amounts of 4-oxo-4-(1-naphthyl)-2-methylbutanoic acid and 4-oxo-4-(2-naphthyl)-2-methylbutanoic acid. These acids can be separated and used as starting materials for problem (a). Which of the methylphenanthrenes can be prepared in this way?

13. Show how anthraquinone can be prepared from 1,4-naphthoquinone.

14. The following methyl derivatives have been shown to be carcinogenic. Supply an adequate name for each compound.

a. CH_3 **b.** H_3C CH_3 CH_3 CH_3 **c.** H_3C CH_3

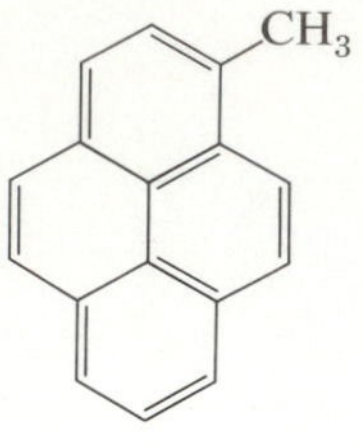

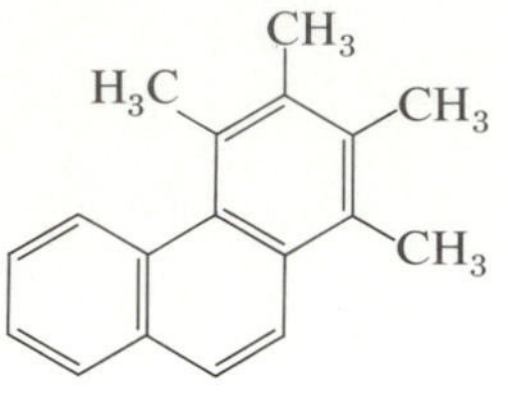

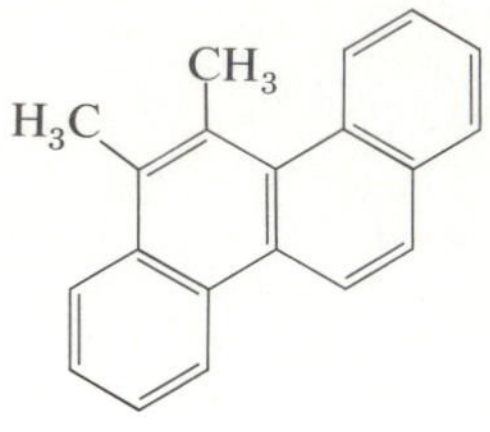

15. The acidity of fluorene is sufficiently high that it will undergo condensation reactions as do esters in alcoholic sodium ethoxide. Show how such condensation reactions can be utilized for the preparation of the following compounds.

a. fluorene-9-carboxylic acid

b. 9-methylfluorene-9-carboxylic acid

c. 9-benzoylfluorene

d. fluorene-9-carboxaldehyde

16. Suggest a procedure using phase-transfer catalysis for the alkylation of fluorene with *n*-butyl bromide.

17. a. Write a reasonable mechanism for the following reaction showing all intermediates involved.

Cl $\xrightarrow[NH_3]{NH_2^-}$ NH_2

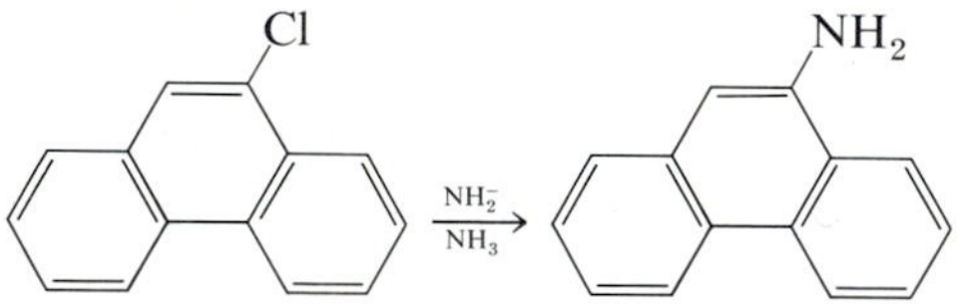

b. On the basis of this mechanism, what would be the course of reaction for 2-methyl-9-chlorophenanthrene?

18. Give a reasonable mechanism for the following reaction, showing all intermediates involved.

O O $\xrightarrow[\Delta]{H^+}$ OH OH

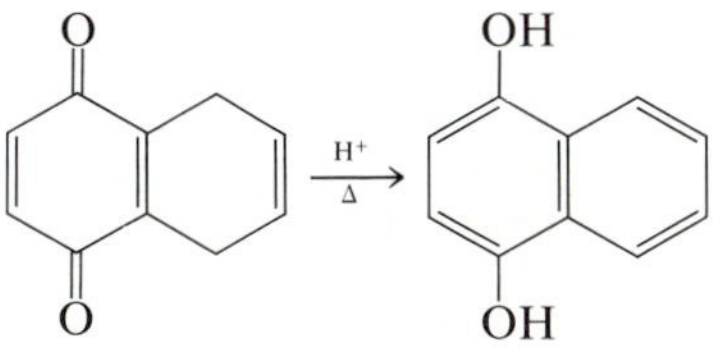

19. 2,6-Naphthoquinone is reduced more readily than 1,2-naphthoquinone. Explain.

20. Steganone is a naturally occurring biphenyl, which was shown by x-ray analysis to have the following structure.

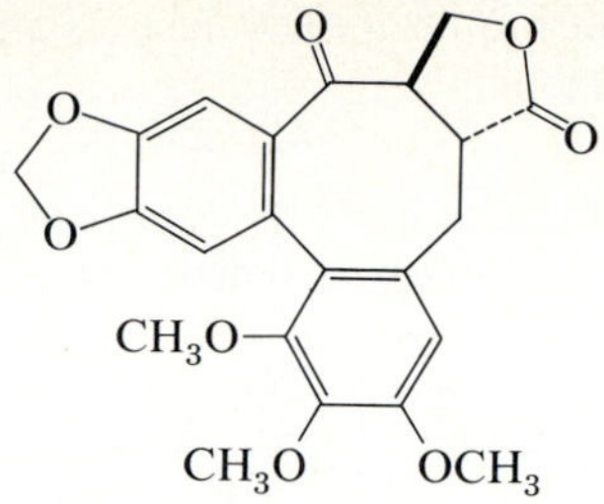

CH_3O CH_3O OCH_3

steganone

A synthesis of steganone was carried out, and a product was obtained that was different from steganone. This isomeric product, named isosteganone, was shown to have the same gross structure as steganone and was also shown to have the lactone ring fused trans to the eight-membered ring, as in steganone. What is the nature of the difference between steganone and isosteganone?

21. Triptycene (page 1070) is a triarylmethane formally similar to triphenylmethane. Triphenylmethane is a relatively acidic hydrocarbon with a pK_a of 31, whereas the pK_a of triptycene is at least 10 units higher. Explain.

22. How many stereoisomers exist for 1,3-bis-(2-bromo-6-methylphenyl)benzene? Write their structures. Which are chiral?

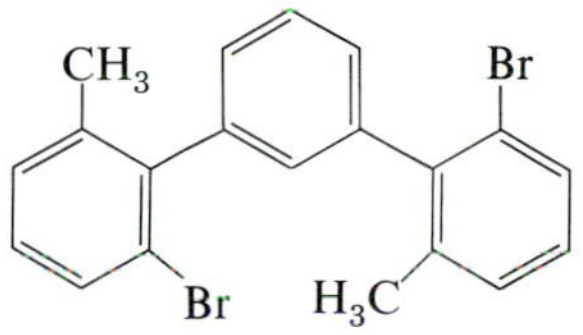

CH_3 Br Br H_3C

23. The perinaphthenyl cation is a relatively stable carbocation in which the positive charge can be distributed to six equivalent positions. Write resonance structures showing this equivalency. Use the stability of the perinaphthenyl cation to explain why the 1-position of pyrene is exceptionally reactive in electrophilic substitution reactions.

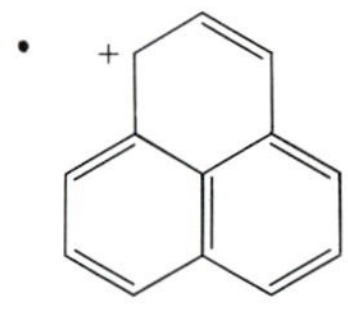

+

perinaphthenyl cation

24. We saw on page 1063 that vanadium pentoxide-catalyzed air oxidation of naphthalene gives phthalic anhydride. Application of the same reaction to 1-naphthylamine also gives phthalic anhydride. However, 1-nitronaphthalene gives 3-nitrophthalic anhydride. Explain.

25. Sketch the geometry of each molecule shown in Highlight 31.1. From their geometries, compare the chemical reactivity of biphenyl with benzene, 2,2′-dihydroxy-1,1′-binaphthyl with 2-naphthol, and 9,9′-dianthracenyl with anthracene. Can one expect any special properties for the 2,2′-dihydroxy-1,1′-binaphthyl?

26. Write at least three resonance forms for kekulene (Highlight 31.1). Predict the ^{1}H and ^{13}C NMR spectra for the compound.

27. Given the following HOMO of anthracene, does electrophilic substitution occur preferentially at the C-1, C-2, or C-9? (*Hint*: Compare Figure 31.7.)

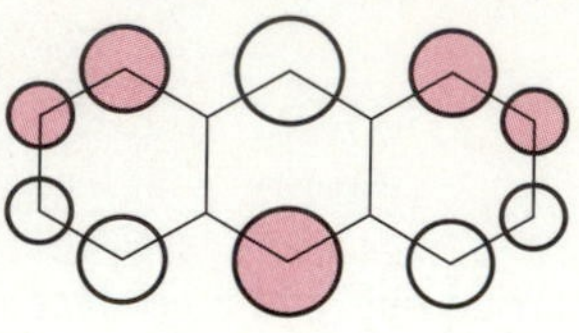

CHAPTER 32

HETEROCYCLIC COMPOUNDS

32.1 Introduction

Heterocycles are cyclic compounds in which one or more ring atoms are not carbon (that is, **heteroatoms;** Gk., *heteros*, other, different). Although heterocyclic compounds are known that incorporate many different elements into cyclic structures (for example, N, O, S, B, Al, Si, P, Sn, As, Cu), we shall consider only some of the more common systems in which the heteroatom is N, O, or S.

Heterocycles are conveniently grouped into two classes, nonaromatic and aromatic. The nonaromatic compounds have physical and chemical properties that are typical of the particular heteroatom. Thus, tetrahydrofuran and 1,4-dioxane are typical ethers, while 1,3,5-trioxane behaves as an acetal.

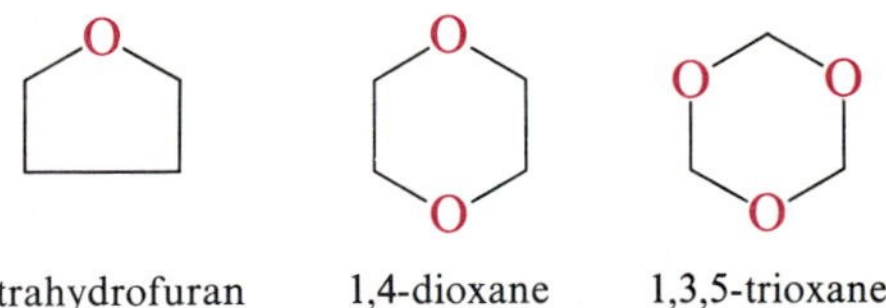

tetrahydrofuran 1,4-dioxane 1,3,5-trioxane

Pyrrolidine and piperidine are typical secondary amines and the bicyclic compound quinuclidine is a tertiary amine.

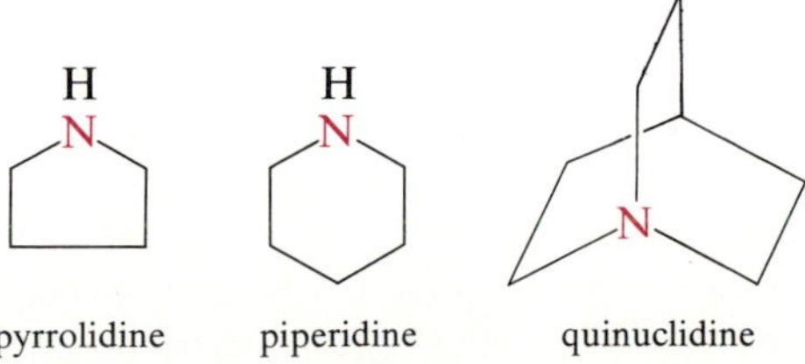

pyrrolidine piperidine quinuclidine

Since the chemistry of these compounds parallels the chemistry of their acyclic relatives, we shall treat them here only briefly.

The aromatic heterocycles include such compounds as pyridine, where nitrogen replaces one of the CH groups in benzene, and pyrrole, in which the aromatic sextet is supplied by the four electrons of the two double bonds and the lone pair on nitrogen.

pyridine   pyrrole

Other aromatic heterocycles contain more than one heteroatom, and still others contain fused aromatic rings. Examples that we will treat in more detail later include oxazole, indole, and purine.

oxazole   indole   purine

The nomenclature of these heterocyclic series is a vast sea of special names for individual ring systems and trivial names for individual compounds. In the course of developing the chemistry of some important groups of compounds we will treat the associated nomenclature. There is only one naming scheme common to all of these compounds, and it, unfortunately, is used only in cases where alternative nomenclature based on special names is awkward. This scheme is based on the corresponding hydrocarbon. The compound formed by replacing a carbon by a heteroatom is named by an appropriate prefix: **aza** for nitrogen, **oxa** for oxygen, and **thia** for sulfur. For example, the following heterocycles are considered as derivatives of bicyclo[2.2.1]heptane and bicyclo[2.2.0]hexane, respectively.

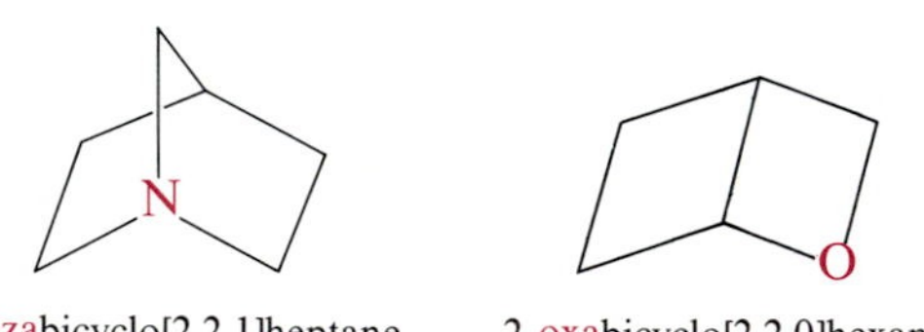

1-azabicyclo[2.2.1]heptane   2-oxabicyclo[2.2.0]hexane

For saturated, monocyclic heterocycles not containing nitrogen the ring size is designated by a suffix. For three-membered heterocycles the suffix is **-irane**; for four-membered compounds it is **-etane**; for five-membered materials **-olane**; and for six-membered heterocycles the suffix is **-ane.** Note that this system is *not used* with nitrogen-containing rings. In addition, most of the simple heterocycles have common names that are in such general use that the systematic names are rarely used. Some examples of this nomenclature are as follows.

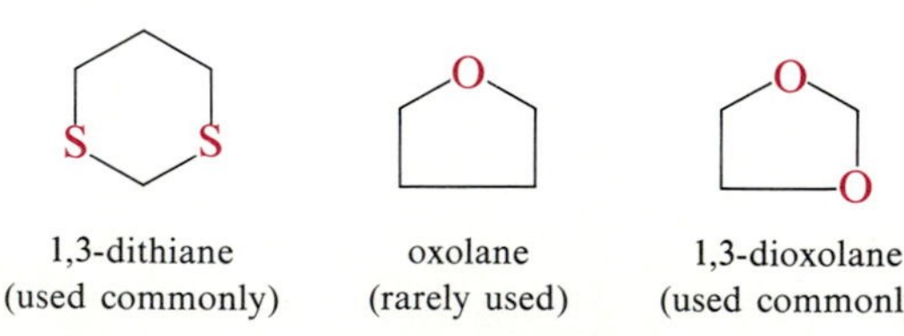

1,3-dithiane (used commonly)   oxolane (rarely used)   1,3-dioxolane (used commonly)

The commonly used names for monocyclic rings with a single heteroatom will be discussed in the next section.

32.2 Nonaromatic Heterocycles

A. Nomenclature

Names in common use of some fully saturated heterocycles containing only one heteroatom are shown below.

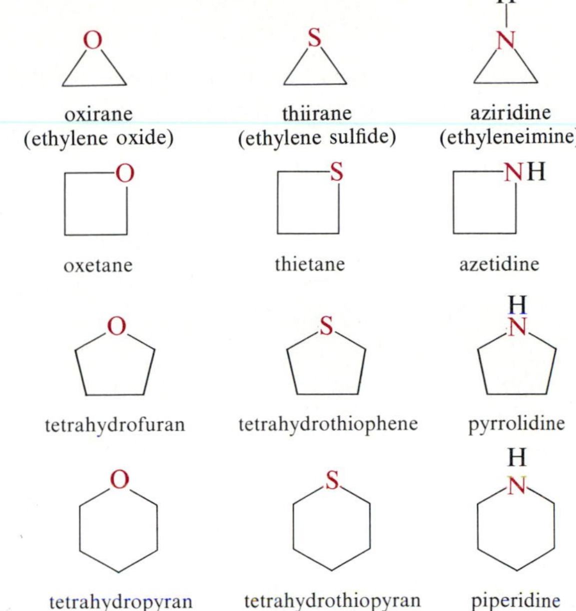

In naming substituted derivatives, the ring is numbered beginning with the heteroatom.

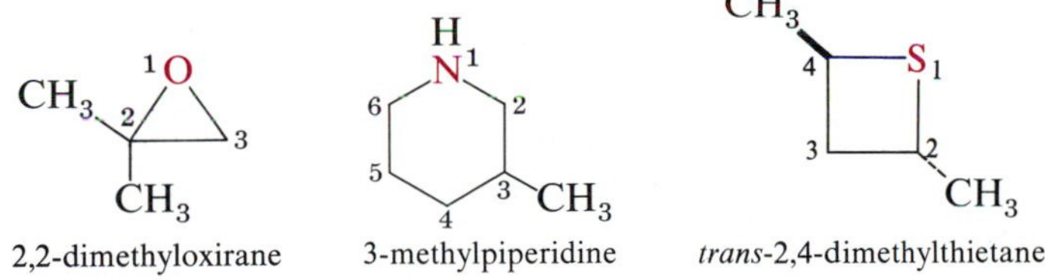

B. Three-Membered Rings

Oxiranes have been discussed previously (Sections 10.11.A and 11.6.E.). Recall that the two most general syntheses are the oxidation of alkenes with peroxyacids and the base-promoted cyclization of halohydrins (page 276).

$$CH_3CH_2CH_2\overset{CH_3}{\overset{|}{C}}{=}CH_2 + C_6H_5COOOH \longrightarrow$$

2-methyl-2-propyloxirane

$$C_6H_5CH{=}CHC_6H_5 + Br_2 \xrightarrow{H_2O} C_6H_5CHBr{-}CH(OH)C_6H_5 \xrightarrow{NaOH}$$

trans-2,3-diphenyloxirane

Aziridines are most commonly prepared by related cyclization reactions. A classical method consists of converting a β-amino alcohol into a β-amino hydrogen sulfate, which is cyclized by treatment with a strong base.

$$C_6H_5CH(OH)CH_2NH_2 \xrightarrow{H_2SO_4} C_6H_5CH(OSO_3H)CH_2NH_3^+ \xrightarrow[\Delta]{NaOH} \left[C_6H_5CH(OSO_3^-)CH_2\ddot{N}H_2\right] \longrightarrow C_6H_5\text{-aziridine}$$

(90%)
2-phenylaziridine

They may also be prepared by cyclization of β-haloalkylamines and their derivatives. An example is the conversion of an alkene into an aziridine via the iodo isocyanate and iodo carbamate.

cyclohexene + INCO ⟶ (iodo isocyanate: NCO, I) $\xrightarrow{CH_3OH}$ (iodo carbamate: $NHCOCH_3$ with C=O, I) $\xrightarrow[EtOH]{KOH}$ [aziridine $NCOCH_3$ (C=O)] $\xrightarrow[EtOH]{KOH}$ aziridine NH

cyclohexeneimine

EXERCISE 32.1 (a) Write mechanisms for each step in the foregoing synthesis of cyclohexeneimine from cyclohexene. (b) What is the structure, including stereochemistry, of the product that will be produced if this reaction sequence is applied to *trans*-2-butene?

Thiiranes are most conveniently prepared from the corresponding oxiranes. An especially useful method involves treating the epoxide with sodium thiocyanate. The ensuing reaction is formulated as follows.

cyclohexene oxide + $^-$SCN ⟶ [S—C≡N, O$^-$] ⟶ [cyclic S, O, =N$^-$] ⟶

[S$^-$, O—C≡N] ⟶ cyclohexene sulfide (S) + NCO^- (32-1)

cyclohexene sulfide

EXERCISE 32.2 What product results from the reaction of (2*R*,3*R*)-2,3-dimethyloxirane with sodium thiocyanate?

The most striking chemical property of the three-membered heterocycles is their extraordinary reactivity, which has its origin in the relief of ring strain that occurs when the ring is cleaved. Recall that oxirane is much more reactive than normal ethers and undergoes ring opening by dilute acid or by base (Section 10.11.A).

$$\text{oxirane} + H_2O \xrightarrow[\text{or } OH^-]{H^+} HOCH_2CH_2OH$$

Similar reactivity is observed with aziridines and with thiiranes.

$$\text{2-phenylaziridine} + 2\ HCl \longrightarrow C_6H_5CH(Cl)CH_2NH_3^+\ Cl^-$$

EXERCISE 32.3 From the data in Appendix I calculate and compare $\Delta H°$ for the following two reactions.

$$CH_3OCH_3 \longrightarrow \text{oxirane} + H_2$$

$$CH_3SCH_3 \longrightarrow \text{thiirane} + H_2$$

How can you rationalize the result?

C. Four-Membered Rings

The four-membered-ring heterocycles—oxetane, azetidine, and thietane—are rarer, mainly because of the difficulty of preparing four-membered rings (Section 9.9). In some favorable cases the rings can be formed by direct ring closure, but yields in such reactions are often low.

$$BrCH_2C(CH_3)_2CH_2NHCH_3 \xrightarrow[100\ °C]{50\%\ KOH} \text{1,3,3-trimethylazetidine}$$

(80%)
1,3,3-trimethylazetidine

$$ClCH_2CH_2CH_2Cl + Na_2S \xrightarrow{C_2H_5OH} \text{thietane}$$

(20–30%)
thietane

Certain four-membered-ring heterocycles may be synthesized by the [2 + 2] cycloaddition of two double bonds. Examples are the formation of β-lactones and β-lactams by the reactions of ketenes with aldehydes and imines, respectively.

$$CH_2{=}O + CH_2{=}C{=}O \xrightarrow[10\ ^\circ C]{ZnCl_2} \text{(2-oxetanone)}$$

(88%)
β-propiolactone

$$(C_6H_5)_2C{=}C{=}O + C_6H_5CH{=}NC_6H_5 \longrightarrow$$

(72%)
1,3,3,4-tetraphenylazetidin-2-one

These cycloaddition reactions do not proceed by way of concerted paths. Because a concerted transition state for a [2 + 2] cycloaddition would involve only four π-electrons, such a transition state is not aromatic and does not benefit from the special stabilization that characterizes [4 + 2] cycloaddition transition states, such as the Diels-Alder reaction (Section 21.7). Instead, stepwise mechanisms are involved.

$$(C_6H_5)_2C{=}C{=}O + C_6H_5CH{=}\ddot{N}C_6H_5 \longrightarrow (C_6H_5)_2C{=}C(O^-){-}\overset{+}{N}(C_6H_5){=}CHC_6H_5 \longrightarrow (C_6H_5)_2C{-}C{=}O \;/\; C_6H_5C(H){-}NC_6H_6$$

EXERCISE 32.4 Write a stepwise mechanism for the zinc chloride-catalyzed reaction of ketene with formaldehyde to give β-propiolactone (2-oxetanone). Rationalize the observed mode of addition; that is, why does the reaction not produce 3-oxetanone?

Like the three-membered ring analogs, oxetanes, azetidines, and thietanes are susceptible to acid-catalyzed ring-opening reactions.

$$\text{oxetane} + C_2H_5OH \xrightarrow[25\ ^\circ C]{\text{trace } H_2SO_4} C_2H_5OCH_2CH_2CH_2OH$$

(58%)

They are also more reactive than their open-chain relatives toward nucleophiles but are much less reactive than the analogous three-membered ring compounds. Note the strenuous conditions required for the ring-opening of oxetane in the following example.

$$\text{oxetane} + C_6H_5CH_2S^-\ Na^+ \xrightarrow[\substack{100\ ^\circ C\\ 6\ hr}]{H_2O} C_6H_5CH_2SCH_2CH_2CH_2OH$$

(63%)
3-benzylthio-1-propanol

D. Five- and Six-Membered Rings

One source of the saturated five-membered-ring heterocycles is reduction of available aromatic compounds derived from furan and pyrrole.

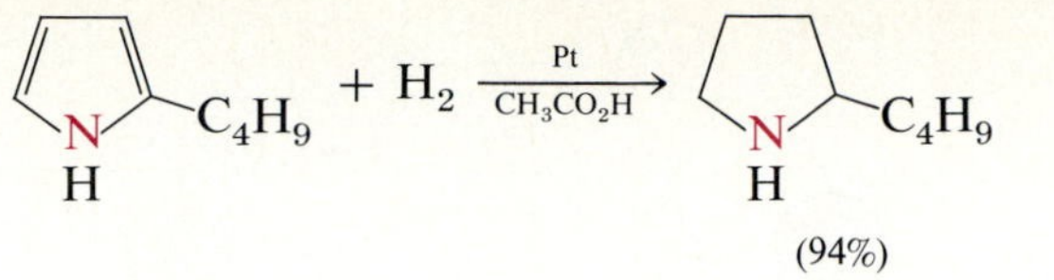

$$\text{2-butylpyrrole} + H_2 \xrightarrow[CH_3CO_2H]{Pt} \text{2-butylpyrrolidine (94\%)}$$

2-butylpyrrole   2-butylpyrrolidine

Many piperidine derivatives can be prepared by hydrogenation of the corresponding pyridine.

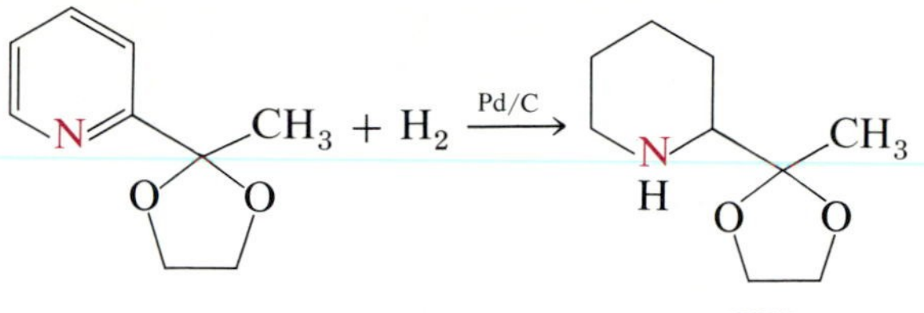

$$+ H_2 \xrightarrow{Pd/C} \quad (77\%)$$

Aside from reduction of aromatic heterocycles, the main synthetic route to the five- and six-membered-ring saturated compounds is by ring closure of suitable difunctional compounds.

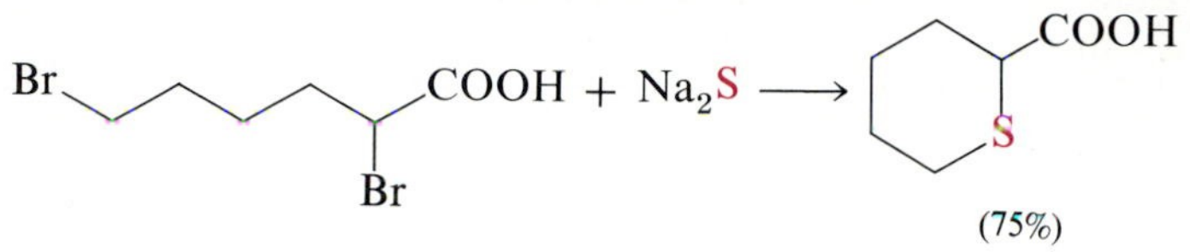

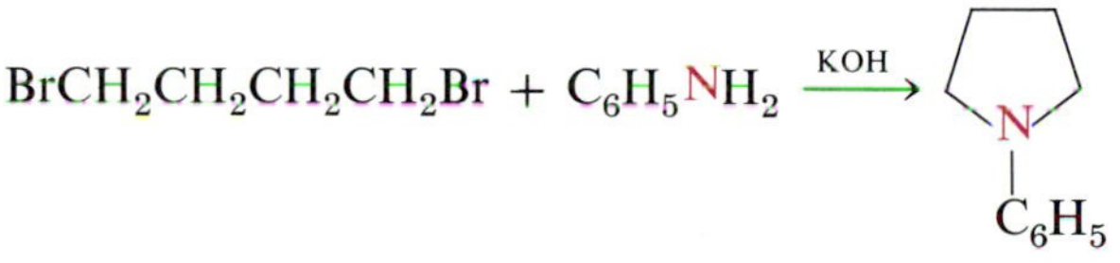

$$+ Na_2S \longrightarrow \quad (75\%)$$

$$BrCH_2CH_2CH_2CH_2Br + C_6H_5NH_2 \xrightarrow{KOH} \quad (100\%)$$

32.3 Furan, Pyrrole, and Thiophene

A. Structure and Properties

The structures of furan, pyrrole, and thiophene would suggest that they have highly reactive diene character.

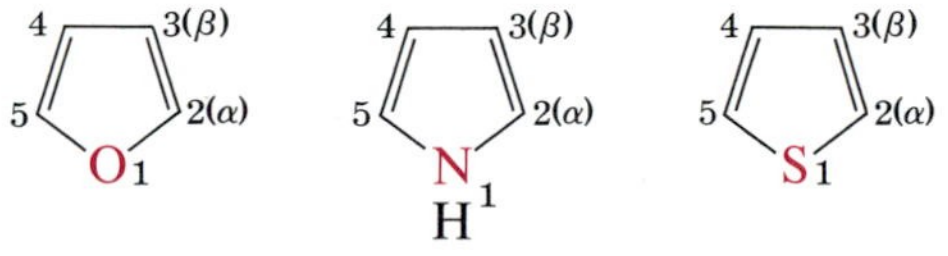

furan   pyrrole   thiophene

However, like benzene, they have many chemical properties that are not typical of dienes. They undergo substitution rather than addition reactions, and they show the effect of a ring current in their NMR spectra. In short, these heterocycles have characteristics associated with aromaticity.

From an orbital point of view, pyrrole has a planar pentagonal structure in which the four carbons and the nitrogen have sp^2-hybridization. Each ring atom forms two sp^2—sp^2 σ-bonds to its neighboring ring atoms, and each forms one sp^2—s σ-bond to a hydrogen.

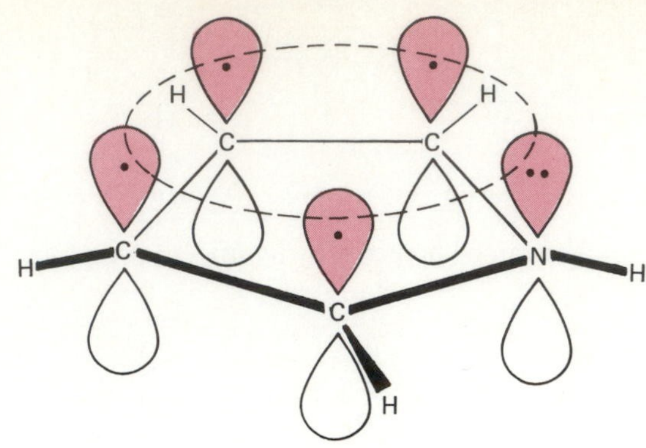

FIGURE 32.1 Orbital structure of pyrrole.

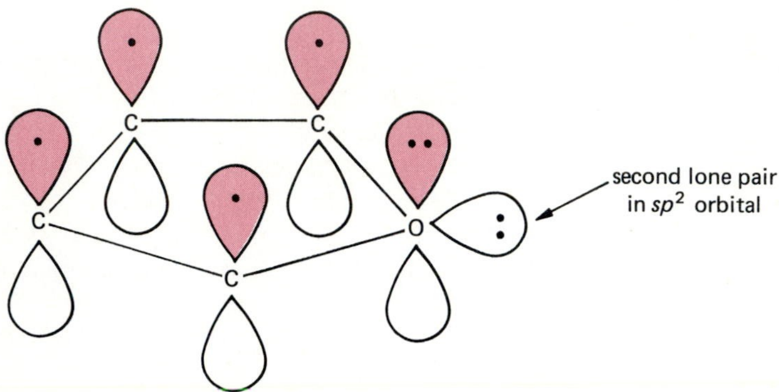

FIGURE 32.2 Orbital structure of furan.

The remaining p-orbitals on each ring atom overlap to form a π-molecular system in which the three lowest molecular orbitals are bonding. The six π-electrons (one for each carbon and two for nitrogen) fill the three bonding orbitals and give the molecule its aromatic character. Pyrrole (Figure 32.1) is isoelectronic with cyclopentadienyl anion, an unusually stable carbanion that also has a cyclic π-electronic system with six electrons (Section 22.3.C).

Furan and thiophene have similar structures. In these cases the second lone pair on the heteroatom can be considered to occupy an sp^2-orbital that is perpendicular to the π-system of the ring (Figure 32.2).

The aromatic character of these heterocycles can also be expressed by using resonance structures, which show that a pair of electrons from the heteroatom is delocalized around the ring.

This delocalization of the lone-pair electrons away from the heteroatom can be inferred from the dipole moments of these aromatic heterocycles and their nonaromatic counterparts.

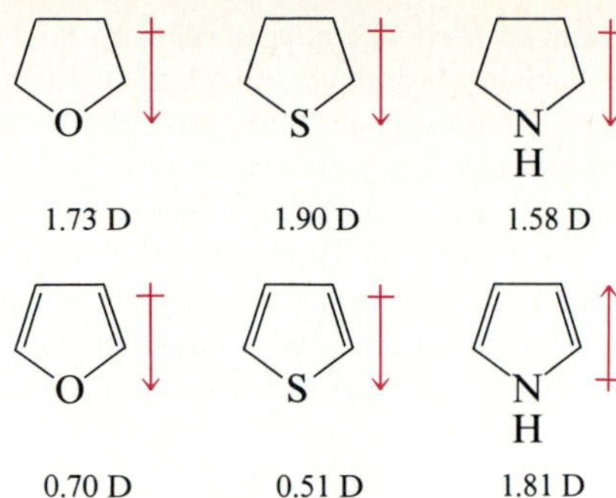

In the saturated compounds, the heteroatom is at the negative end of the dipole. In the aromatic heterocycles the dipole moment associated with the π-system opposes the σ-moment. As a result the net dipole moment of furan and thiophene is reduced. In pyrrole the π-moment is larger than the σ-moment so that the direction of the net dipole moment is actually reversed from its saturated counterpart!

Empirical resonance energies for furan, pyrrole, and thiophene can be computed from the heats of combustion for the compounds. In all cases, there is a substantial stabilization energy, although of considerably smaller magnitude than that for benzene.

Although pyrrole is an amine, it is an extremely nonbasic one because the nitrogen lone pair is involved in the aromatic sextet and is therefore less available for bonding to a proton. The pK_a of its conjugate acid is -4.4. In fact, this pK_a corresponds to a conjugate acid in which protonation has occurred predominantly on carbon rather than on nitrogen.

$$\text{pyrrole} + H^+ \rightleftharpoons \text{2H-pyrrolium ion}$$

p$K_a = -4.4$

EXERCISE 32.5 Using resonance theory, explain why pyrrole protonates on carbon, rather than on nitrogen.

Pyrrole compounds occur widely in living systems. One of the more important pyrrole compounds is the porphyrin hemin, the prosthetic group of hemoglobin and myoglobin (see Figure 29.17). A number of simple alkylpyrroles have played an important role in the elucidation of the porphyrin structures. Drastic reduction of hemin gives a complex mixture from which the four pyrroles—hemopyrrole, cryptopyrrole, phyllopyrrole, and opsopyrrole—have been isolated.

hemopyrrole — cryptopyrrole — phyllopyrrole — opsopyrrole

The red blood cell contains a high concentration of hemoglobin, which transports oxygen in an organism; 1 g of hemoglobin absorbs 1.35 mL of oxygen at STP, corresponding to exactly one molecule of O_2 per atom of iron. The oxygen binds to the hemoglobin molecule as a ligand of the iron, and the binding constant varies with the partial pressure of oxygen. In the lungs, where the partial pressure of oxygen is high, hemoglobin binds oxygen. In the tissues served by the bloodstream the oxyhemoglobin dissociates back into O_2 and hemoglobin. Some 25% of the red blood cell membrane is an anion exchange protein that catalyses the exchange of HCO_3^- for Cl^-. Some of the HCO_3^- is bound to hemoglobin as a carbamate, —$NHCOO^-$. After return to the lungs, the enzyme carbonic anhydrase facilitates the formation of carbon dioxide, which is then exchanged for additional oxygen in the lungs. Carbon monoxide is a poison because it is a stronger ligand for the iron of hemoglobin than oxygen.

The porphyrins are derivatives of porphine, a tetrapyrrole heterocycle, and occur as metal complexes in the active sites of a number of enzymes. The porphine nucleus contains a conjugating system of 18 π-centers, indicated by the color in the structure shown. This system obeys the $4n + 2$ rule and is therefore an aromatic cycle.

porphine

EXERCISE 32.6 One approach to calculating empirical resonance energies is to compare the heats of hydrogenation of furan, pyrrole, and thiophene to tetrahydrofuran, pyrrolidine, and tetrahydrothiophene, respectively, with the heat of hydrogenation of cyclopentadiene to cyclopentane. Apply this method with the data in Appendix I. Rationalize the results using resonance structures.

Highlight 32.1

The three important five-membered ring heterocycles furan, thiophene, and pyrrole are isoelectronic with the cyclopentadienyl anion. The "aromatic sextet" is comprised of one *p* electron from each carbon and two from the oxygen, nitrogen, or sulfur. Furan and thiophene have an additional pair of electrons in a sp^2 orbital that lies in the plane of the molecule, perpendicular to the aromatic π system.

furan thiophene pyrrole

Because its lone pair is involved in the aromatic sextet, pyrrole does not have the basic properties typical of amines. That is, protonation is accompanied by destruction of the aromatic sextet.

B. Synthesis

Pyrrole is prepared commercially by the fractional distillation of coal tar or by passing a mixture of furan, ammonia and steam over a catalyst at 400 °C.

Thiophene is prepared industrially by passing a mixture of butane, butenes, or butadiene and sulfur through a reactor heated to 600 °C at a rate such that the contact time is about 1 sec.

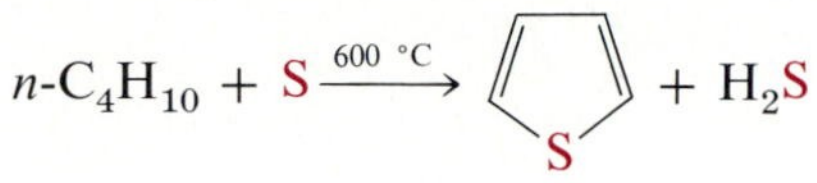

$$n\text{-}C_4H_{10} + S \xrightarrow{600\ ^\circ C} \text{thiophene} + H_2S$$

Furan, 2-furaldehyde (furfural), 2-furylmethanol, and 2-furoic acid are all inexpensive commercial items.

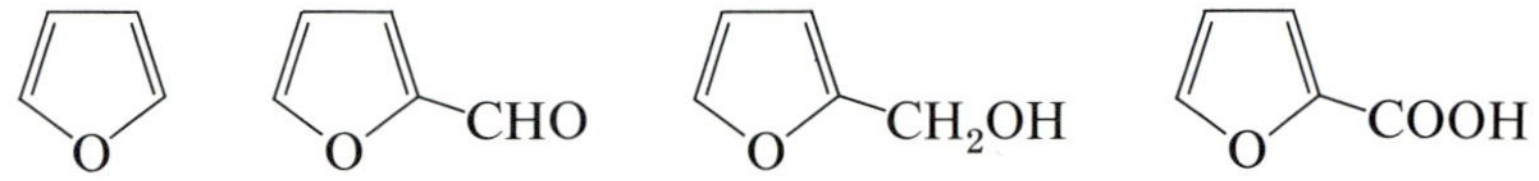

furan — 2-furaldehyde furfural (CHO) — 2-furylmethanol (CH_2OH) — 2-furoic acid (COOH)

The ultimate source of these heterocycles is furfural, which is obtained industrially by the acid hydrolysis of the polysaccharides of oat hulls, corn cobs, or straw. These polysaccharides are built up from pentose units. Dehydration of the pentose may be formulated as follows.

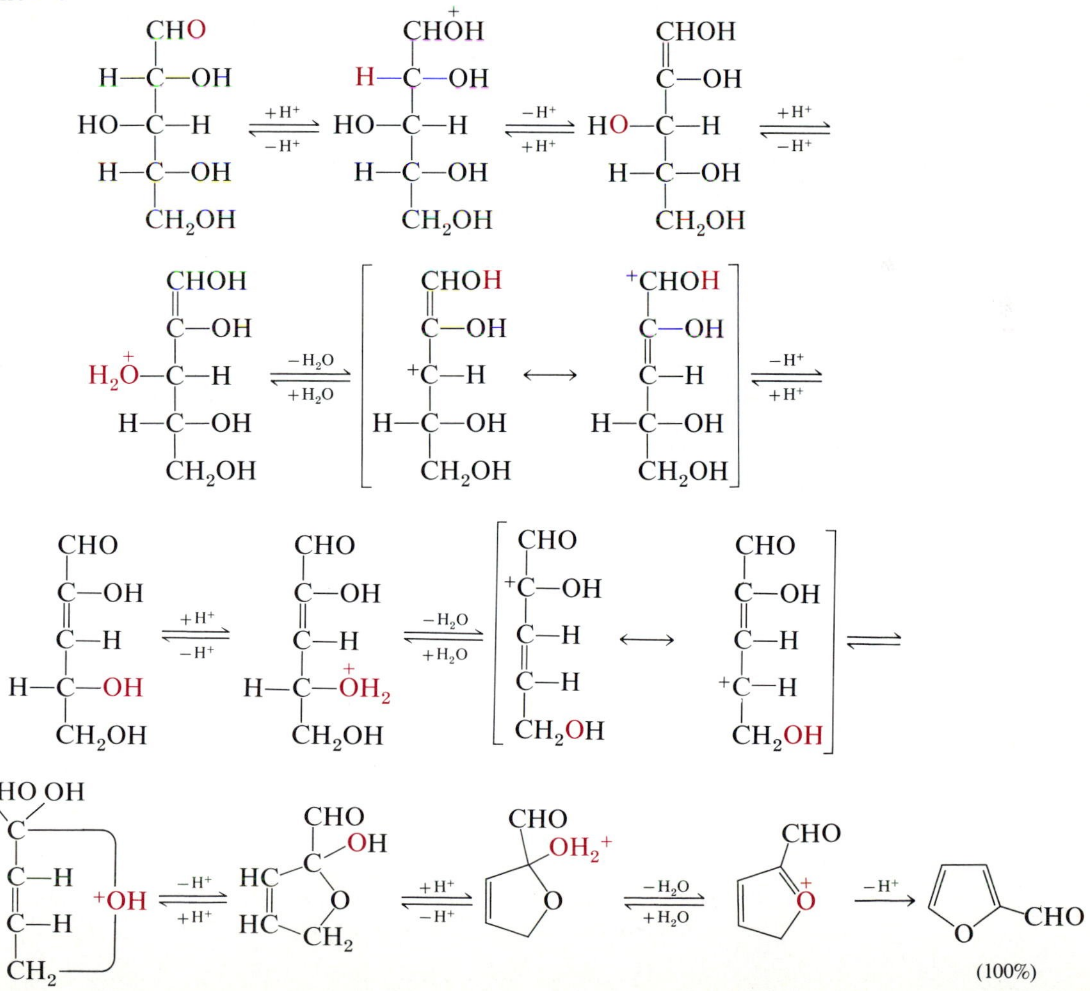

$$\text{pentose} \underset{-H^+}{\overset{+H^+}{\rightleftharpoons}} \underset{-H^+}{\overset{+H^+}{\rightleftharpoons}} \cdots \underset{+H_2O}{\overset{-H_2O}{\rightleftharpoons}} \cdots \xrightarrow{-H^+} \text{furfural (100\%)}$$

(100%)

In the foregoing mechanism, note that each cationic intermediate is an oxonium ion; simple carbocations are not involved at any point.

Substituted furans, pyrroles, and thiophenes can be prepared by electrophilic substitution on one of the available materials discussed or by a variety of cyclization reactions. The most general is the **Paal-Knorr** synthesis, in which a 1,4-dicarbonyl compound is heated with a dehydrating agent, ammonia, or an inorganic sulfide to produce the furan, pyrrole, or thiophene, respectively.

2,5-hexanedione $\xrightarrow[100\ °C]{P_2O_5}$ 2,5-dimethylfuran

2,5-hexanedione $\xrightarrow[\Delta]{P_2S_5}$ 2,5-dimethylthiophene

2,5-hexanedione $\xrightarrow[\Delta]{NH_3}$ 2,5-dimethylpyrrole

EXERCISE 32.7 Write reasonable mechanisms for the formation of 2,5-dimethylfuran and 2,5-dimethylpyrrole from 2,5-hexanedione in the foregoing reactions.

Another general method for the synthesis of substituted pyrroles is the **Knorr pyrrole synthesis,** the condensation of an α-amino ketone with a β-keto ester. The method is illustrated by the following synthesis of diethyl 3,5-dimethylpyrrole-2,4-dicarboxylate.

EtOOC, NH_2, O + COOEt, O $\xrightarrow{CH_3COOH}$ H_3C, COOEt, EtOOC, N, H, CH_3

(57–64%)

The probable mechanism of the Knorr synthesis is as follows.

O, EtOOC, $\ddot{N}H_2$, COOEt, ^+O, H $\rightleftharpoons$ O, EtOOC, $\overset{+}{N}$, H_2, COOEt, OH $\underset{H^+}{\overset{-H^+}{\rightleftharpoons}}$

O, EtOOC, N, H, COOEt, OH $\underset{-H^+}{\overset{+H^+}{\rightleftharpoons}}$ O, EtOOC, $\ddot{N}$, H, COOEt, $\overset{+}{O}H_2$ $\underset{+H_2O}{\overset{-H_2O}{\rightleftharpoons}}$

O, EtOOC, $\overset{+}{N}$, H, COOEt, H $\underset{+H^+}{\overset{-H^+}{\rightleftharpoons}}$ O, EtOOC, N, H, COOEt $\underset{-H^+}{\overset{+H^+}{\rightleftharpoons}}$

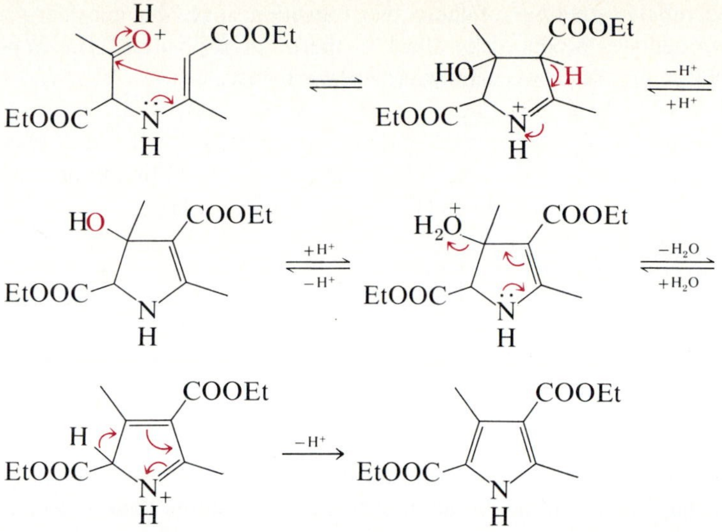

Notice how each individual step involves either an oxonium ion or an ammonium ion. Again, no unstabilized carbocation is involved.

EXERCISE 32.8 Propose syntheses of the following heterocycles from acyclic starting materials.

(a) 2-methyl-3,5-diethylthiophene

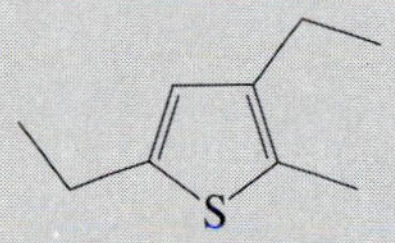

(b)

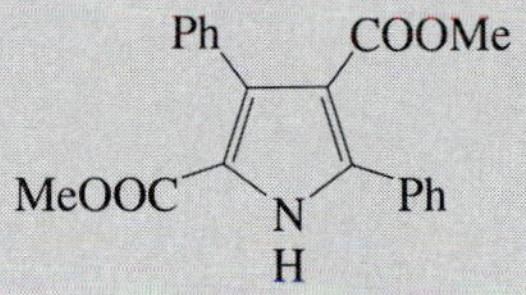

C. Reactions

The most typical reaction of furan, pyrrole, and thiophene is electrophilic substitution. All three heterocycles are much more reactive than benzene, the reactivity order being pyrrole > furan > thiophene > > benzene. To give some idea of the magnitude of this reactivity order, partial rate factors (reactivities relative to benzene) for tritium exchange with trifluoroacetic acid (page 713) for thiophene are as follows.

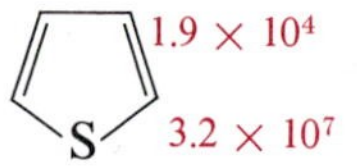

1.9×10^4

3.2×10^7

Because of this high reactivity, even mild electrophiles suffice to cause electrophilic substitution in these heterocycles. Substitution occurs predominantly at the α-position (C-2).

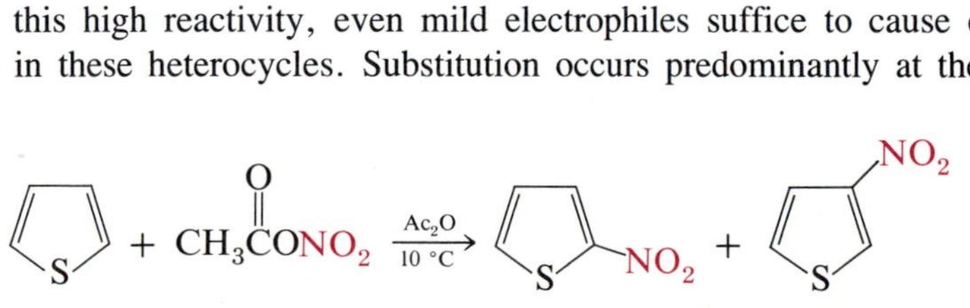

$$\text{thiophene} + CH_3\overset{O}{\overset{\|}{C}}ONO_2 \xrightarrow[10\ °C]{Ac_2O} \text{2-nitrothiophene} + \text{3-nitrothiophene}$$

acetyl nitrate — 2-nitrothiophene (70%) — 3-nitrothiophene (5%)

This orientation is understandable in terms of the mechanism of electrophilic aromatic substitution. The α/β ratio is determined by the relative energies of the transition states leading to the two isomers. As in the case of substituted benzenes (Section 23.6), we can estimate the relative energies of these two transition states by considering the actual reaction intermediates produced by attack at the α- or β-position. The important resonance structures for these two cations are shown below.

The most important of these structures are the two with the positive charge on sulfur because, in these two sulfonium cation structures, all atoms have octets of electrons. Nevertheless, as the sets of resonance structures show, the charge on the cation resulting from attack at the α-position is more extensively delocalized than that for the cation resulting from attack at the β-position. The following examples further demonstrate the generality of α-attack.

$$\text{furan} + CH_3\overset{O}{\overset{\|}{C}}O\overset{O}{\overset{\|}{C}}CH_3 \xrightarrow{BF_3} \text{2-furyl}-COCH_3$$

(75–92%)
2-acetylfuran

$$\text{pyrrole} + CH_3\overset{O}{\overset{\|}{C}}O\overset{O}{\overset{\|}{C}}CH_3 \longrightarrow \text{2-pyrrolyl}-COCH_3$$

(60%)
2-acetylpyrrole

$$\text{thiophene} + I_2 \xrightarrow[\text{benzene}]{HgO} \text{2-thienyl}-I$$

(75%)
2-iodothiophene

In the last example, note that 2-iodothiophene is the sole product of iodination, even though the reaction is carried out in benzene as solvent; that is, thiophene is so much more reactive than benzene that no significant amount of iodobenzene is formed.

Hydrolytic ring opening is a typical reaction of furans. In essence, the reaction is the reverse of the Paal-Knorr synthesis. Careful hydrolysis of furans can lead to the corresponding 1,4-dicarbonyl compounds in good yield.

$$\text{2,5-dimethylfuran} \xrightarrow[CH_3COOH]{H_2O,\ H_2SO_4} CH_3\overset{O}{\overset{\|}{C}}CH_2CH_2\overset{O}{\overset{\|}{C}}CH_3$$

(90%)

EXERCISE 32.9 Write a stepwise mechanism for the foregoing reaction.

Pyrroles are polymerized by even dilute acids, probably by a mechanism such as the following. Electrochemical oxidation yields conducting films of polypyrrole.

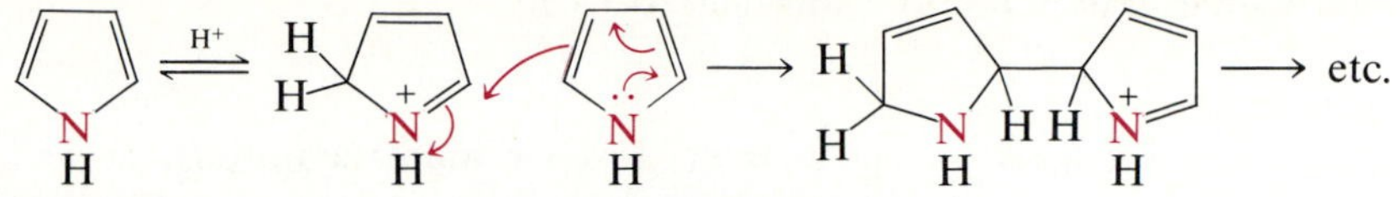

Thiophenes are more stable and do not undergo hydrolysis.

EXERCISE 32.10 Give reasonable syntheses, starting with the unsubstituted heterocycle in each case, of the following compounds.

(a) 1-(2-pyrryl)-1-propanone (b) furan-2-*d* (c) 2-(chloromethyl)thiophene

32.4 Condensed Furans, Pyrroles, and Thiophenes

A. Structure and Nomenclature

Benzofuran, indole, and benzothiophene are related to the monocyclic heterocycles in the same way that naphthalene is a benzo derivative of benzene. Carbazole is a dibenzopyrrole, and is analogous to anthracene. In benzofuran, indole, and benzothiophene, the rings are numbered beginning with the heteroatom; carbazole is numbered in a manner analogous to anthracene.

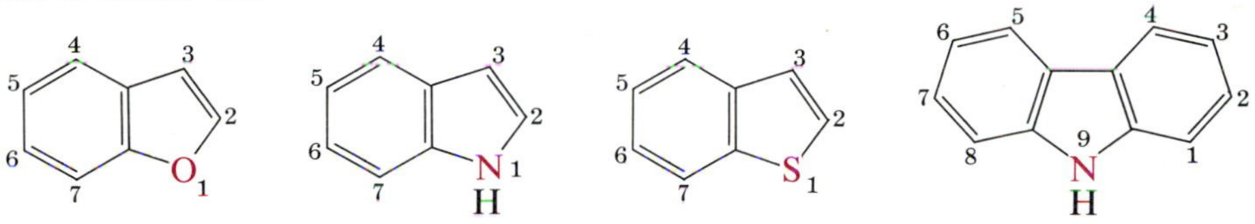

benzofuran indole benzothiophene carbazole

Of the four systems, indoles are by far the most important. Many natural products include indole structures (see Section 36.6.C).

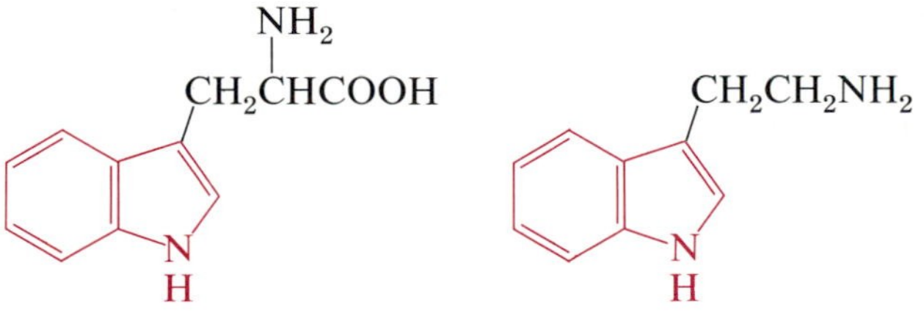

tryptophan

tryptamine

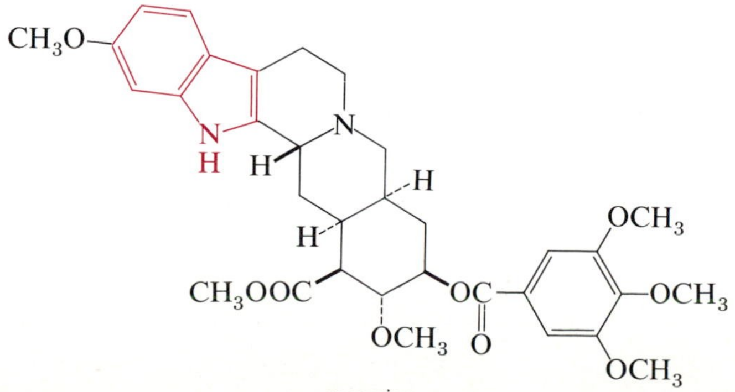

reserpine

From a chemical standpoint the chief effect of fusing the benzene ring onto the simple heterocycle is to increase the stability and to change the preferred orientation in electrophilic substitution from C-2 to C-3 (Section 32.4.C).

B. Synthesis

The most general synthesis of indoles is the **Fischer indole synthesis,** in which the phenylhydrazone of an aldehyde or ketone is treated with a catalyst such as BF_3, $ZnCl_2$, or polyphosphoric acid (PPA).

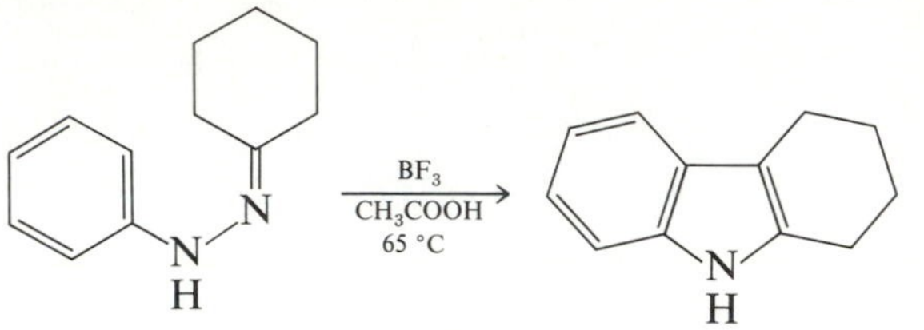

BF_3, CH_3COOH, 65 °C

(93%)
1,2,3,4-tetrahydrocarbazole

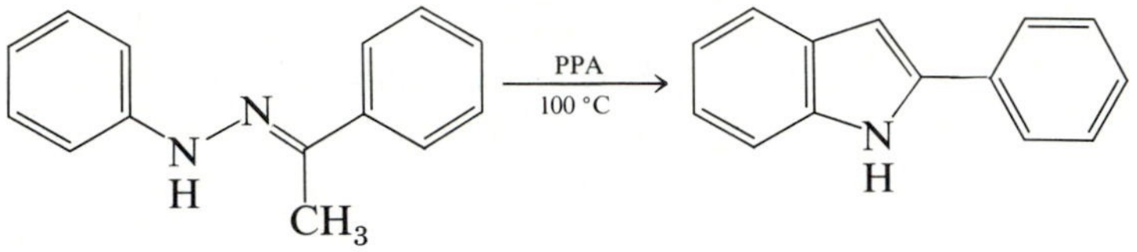

PPA, 100 °C

CH_3

(73%)
2-phenylindole

The mechanism of the Fischer synthesis has been the subject of much study. The available evidence is in accord with a pathway involving a benzidine-like rearrangement (Section 25.1.C).

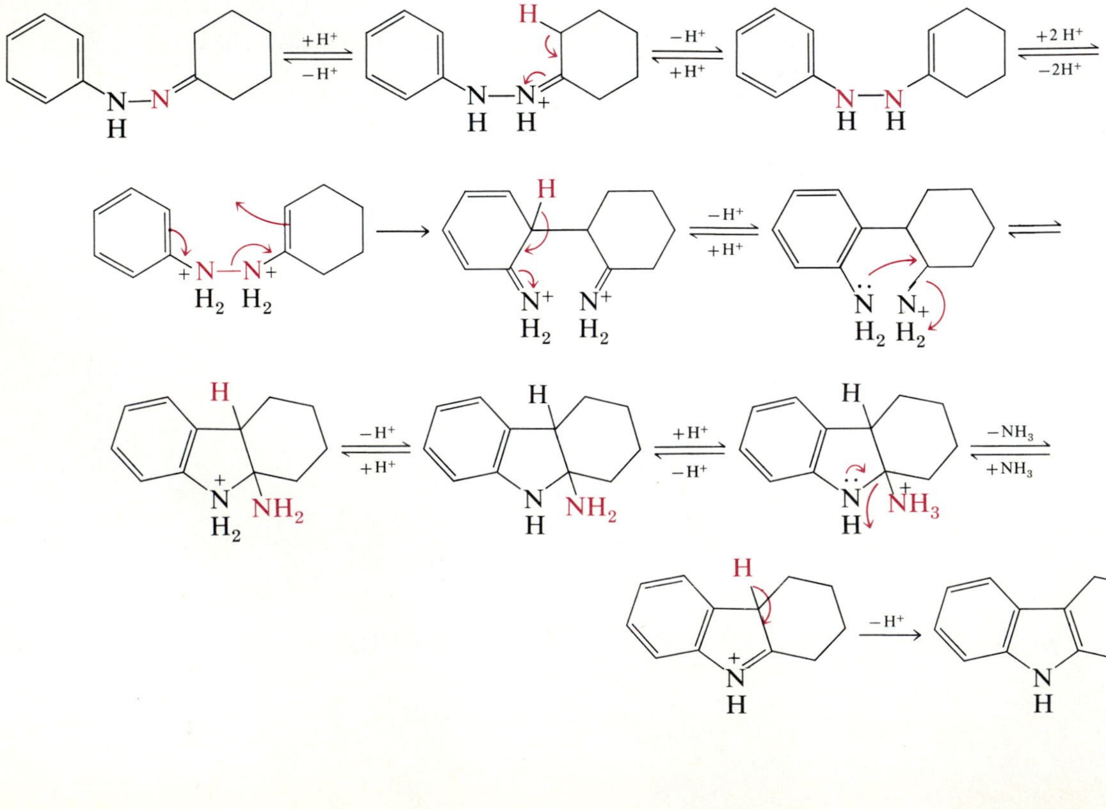

$+H^+$ / $-H^+$; $-H^+$ / $+H^+$; $+2\,H^+$ / $-2H^+$

$-H^+$ / $+H^+$

$-H^+$ / $+H^+$; $+H^+$ / $-H^+$; $-NH_3$ / $+NH_3$

NH_2 ; NH_2 ; $\overset{+}{N}H_3$

$-H^+$

The reaction fails with the phenylhydrazone of acetaldehyde and thus cannot be used to prepare indole itself. However, the phenylhydrazone of pyruvic acid does react to yield indole-2-carboxylic acid, which can be decarboxylated to give indole.

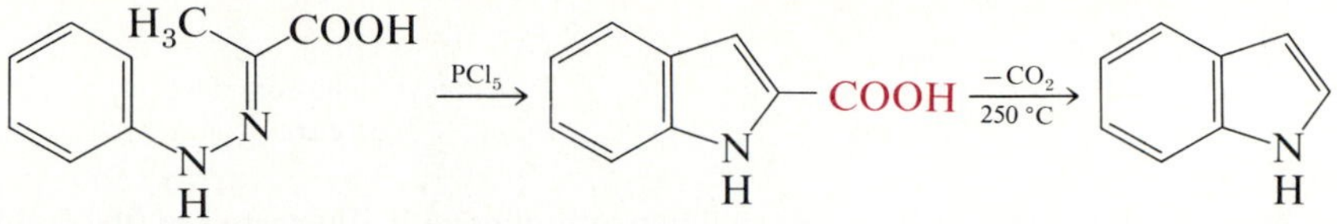

H_3C COOH N N H $\xrightarrow{PCl_5}$ COOH N H $\xrightarrow[250\ °C]{-CO_2}$ N H

EXERCISE 32.11 Show how each of the following compounds can be prepared by the Fischer indole synthesis.
(a) 2,5-dimethylindole (b) 4,6-dimethoxyindole

What problems arise in preparation of the following compounds by this method?
(c) 2-ethylindole (d) 2,4-dimethylindole

C. Reactions

All three condensed heterocycles undergo electrophilic substitution in the heterocyclic ring rather than in the benzene ring. However, each is markedly less reactive than the corresponding monocyclic heterocycle. Some partial rate factors for protodetritiation with trifluoroacetic acid (page 713) are available for benzothiophene and benzofuran.

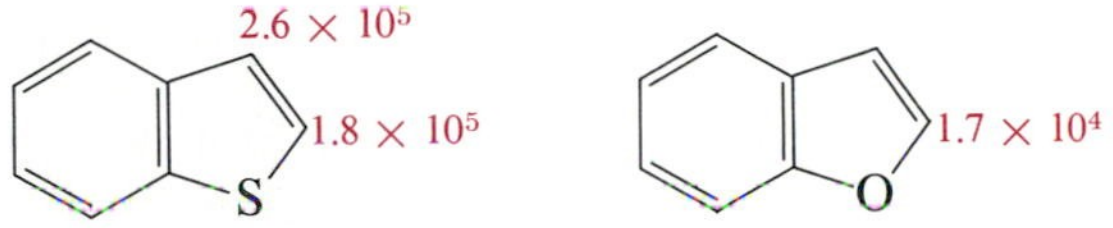

2.6×10^5 1.8×10^5 S 1.7×10^4 O

These values are at least two orders of magnitude smaller than that for the α-position of thiophene (page 1095).

The preferred orientation in electrophilic substitution reactions in these compounds can be summarized as follows.

1. In benzofuran the most reactive position is C-2.
2. In benzothiophene C-2 and C-3 have comparable reactivities, with C-3 being somewhat the more reactive.
3. In indole the most reactive position is C-3.

The way in which these generalizations apply in practice is illustrated with the following specific examples.

Electrophilic substitution in benzofuran occurs predominantly at C-2, just as in furan itself.

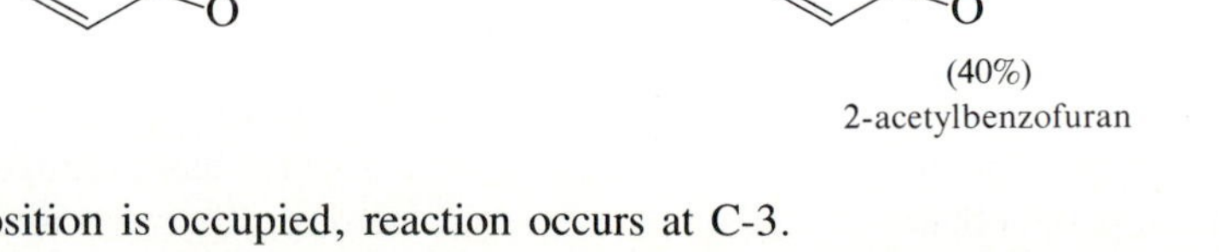

O + $CH_3\overset{O}{\overset{\|}{C}}O\overset{O}{\overset{\|}{C}}CH_3$ $\xrightarrow{SnCl_4}$ O $-COCH_3$

(40%)
2-acetylbenzofuran

If the 2-position is occupied, reaction occurs at C-3.

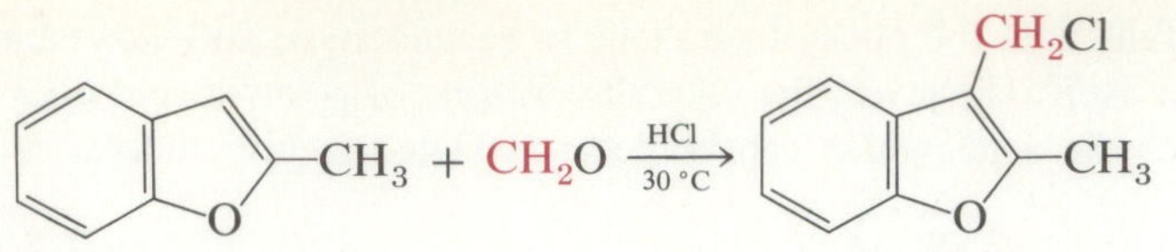

2-methylbenzofuran $-CH_3$ + CH_2O $\xrightarrow[30\ °C]{HCl}$ CH_2Cl, CH_3

(70%)
2-methyl-3-chloromethyl-benzofuran

The preferred reaction at C-3 in indole and benzothiophene is illustrated by the following reactions.

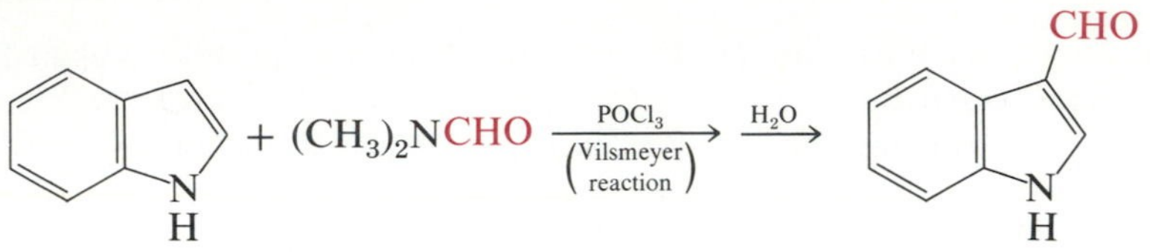

indole + $(CH_3)_2NCHO$ $\xrightarrow[\text{(Vilsmeyer reaction)}]{POCl_3}$ $\xrightarrow{H_2O}$ CHO

(97%)
indole-3-carboxaldehyde

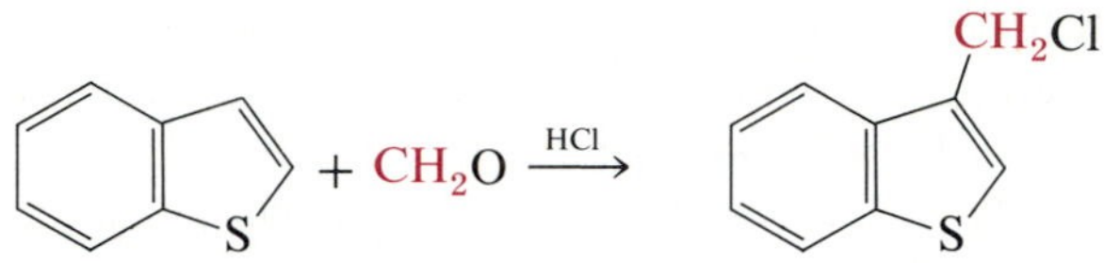

benzothiophene + CH_2O $\xrightarrow{HCl}$ CH_2Cl

(56%)
3-chloromethylbenzothiophene

With benzothiophene other isomers are usually produced as well, but are not always detected or isolated.

These orientation specificities can be rationalized by considering the intermediate ions produced by attack at C-2 and C-3. Reaction at C-2 gives a carbocation in which the charge is distributed to the benzene ring and to the heteroatom; however, the structure with the charge on the heteroatom no longer has a benzene ring. In contrast, reaction at C-3 does not permit effective distribution of charge around the benzene ring, but the electron pair on the heteroatom is utilized efficiently without disruption of the benzene resonance.

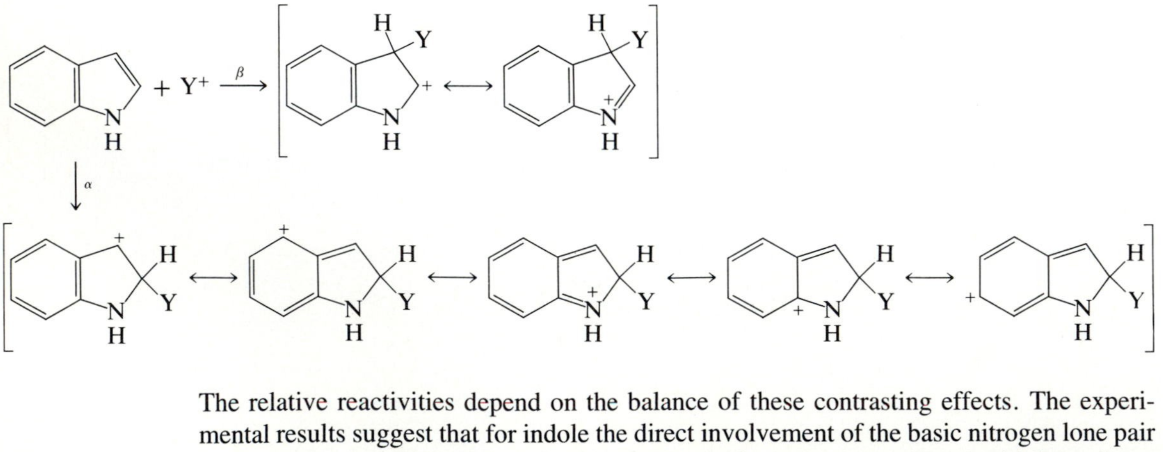

indole + Y^+ $\xrightarrow{\beta}$ [H, Y, N, H, + $\longleftrightarrow$ H, Y, +N, H]

$\downarrow \alpha$

[+ H, Y, N, H $\longleftrightarrow$ + H, Y, N, H $\longleftrightarrow$ H, Y, +N, H $\longleftrightarrow$ H, Y, +N, H $\longleftrightarrow$ + H, Y, N, H]

The relative reactivities depend on the balance of these contrasting effects. The experimental results suggest that for indole the direct involvement of the basic nitrogen lone pair is much more important than conjugation with the benzene ring, whereas with benzofuran the oxygen lone pair is less basic and the involvement of the benzene ring now dominates. In the case of benzothiophene the two effects are roughly comparable in magnitude.

EXERCISE 32.12 From the principles and examples developed in this section, work out whether the 2- or 3-position of carbazole is the more reactive in electrophilic substitution.

32.5 Azoles

A. Structure and Nomenclature

Azoles are five-membered aromatic heterocycles containing two nitrogens, one nitrogen and one oxygen, or one nitrogen and one sulfur. They are named and numbered as shown below. They may be considered as aza analogs of furan, pyrrole, and thiophene in the same way that pyridine is an aza analog of benzene (see Section 32.6).

furan　　oxazole (3-azafuran)　　benzene　　pyridine (azabenzene)

oxazole	imidazole	thiazole	isoxazole	pyrazole	isothiazole
b.p. 70°C	b.p. 263°C	b.p. 117°C	b.p. 95°C	b.p. 188°C	b.p. 113°C
	m.p. 90°C			m.p. 70°C	

From a molecular orbital standpoint the azoles are similar to the simpler aromatic heterocycles. For example, in imidazole each carbon and nitrogen can be considered to be sp^2-hybridized. One nitrogen makes two sp^2—sp^2 σ-bonds to carbon and one sp^2—s σ-bond to hydrogen. The other nitrogen has its lone pair in the third sp^2-orbital. The π-molecular orbital system is made up from the p_z-orbitals from each ring atom (Figure 32.3). Six π-electrons (one from each carbon and from one nitrogen, two from the other nitrogen) complete the aromatic shell.

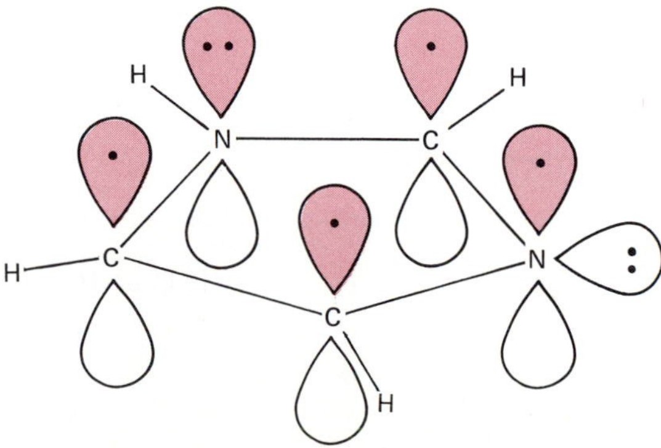

FIGURE 32.3 Orbital structure of imidazole.

> The role of the carboxylate as a charge stabilizer in the serine proteinase mechanism (Figure 29.8, page 982) is made clear by this analysis. The carboxylate is hydrogen-bonded to a σ-bonded hydrogen on the first imidazole nitrogen. The p-electrons on that nitrogen participate in the π-system, redistributing the positive charge acquired through formation of a σ-bond to a proton with the sp^2-nonbonding pair on the second nitrogen.

An examination of the physical properties of the simple azoles reveals that imidazole and pyrazole have anomalously high boiling points. They are also the only simple azoles that are solids at room temperature. These properties clearly result from intermolecular hydrogen bonding. With imidazole the hydrogen bonding is of a linear polymer, whereas pyrazole seems to exist largely as dimers.

Like pyridine (pK_a 5.2; Section 32.6.A), thiazole (pK_a 2.4), pyrazole (pK_a 2.5), and oxazole (pK_a 0.8) are weak bases. As in pyridine, the nitrogen lone pair is in an sp^2-orbital. Recall that greater s-character of lone-pair electrons is associated with heightened stability and lower basicity (Section 12.4). A similar trend is seen with nitrogen acids (Table 32.1).

The higher s-character of the pyridine lone pair compared to aliphatic amines is sufficient to account for a decrease in basicity of several powers of ten. In pyrazole, thiazole, and oxazole the basicity of the nitrogen lone pair is further reduced by the presence of the other heteroatom.

pK_a 2.4 pK_a 0.8 pK_a 2.5

TABLE 32.1 Hybridization and Acidity

Orbital	Carbon Acid	pK_a	Nitrogen Acid	pK_a
sp	HC≡CH	25	$CH_3C{\equiv}\overset{+}{N}H$	−10
sp^2	$CH_2{=}CH_2$	44	(pyridinium) $N^+{-}H$	5
sp^3	$CH_3{-}CH_3$	50	$(CH_3)_3\overset{+}{N}H$	10

In marked contrast to these results, imidazole seems to be abnormally basic for a compound with sp^2-hybridized nitrogen (pK_a 7.0). The enhanced basicity of imidazole is presumably due to the symmetry of the conjugate acid and the consequent resonance stabilization.

pK_a 7.0

Its pK_a of 7.0 means that imidazole is half protonated in neutral water. As a result the basicity of imidazole plays an important role in biological processes. The imidazole ring in the amino acid histidine is often involved as a proton acceptor in the active site of enzymes (Section 29.6.C, Figure 29.8).

histidine

The thiazole ring is also important in nature. It occurs, for example, in vitamin B_1, thiamine, a coenzyme required for the oxidative decarboxylation of α-keto acids as illustrated in Section 27.7.C.

thiamine

A tetrahydrothiazole also appears in the skeleton of penicillin, one of the first broad-spectrum antibiotics.

benzylpenicillin

Highlight 32.2

In imidazole the aromatic sextet is composed of one p-electron from each carbon, one p-electron from one of the nitrogens, and two p-electrons from the other nitrogen. This leaves one nitrogen with a second lone pair of electrons, which is in an sp^2-orbital that lies in the plane of the molecule.

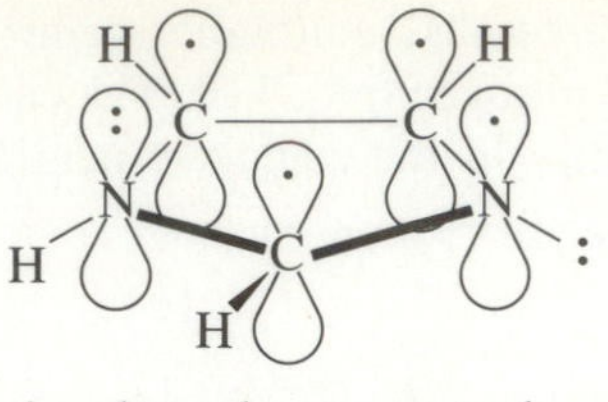

Because imidazole has a nonbonding electron pair that is not part of the aromatic sextet, it can be protonated like other amines. The imidazolium ion is a resonance hybrid of two equivalent structures and its pK_a is 7.0.

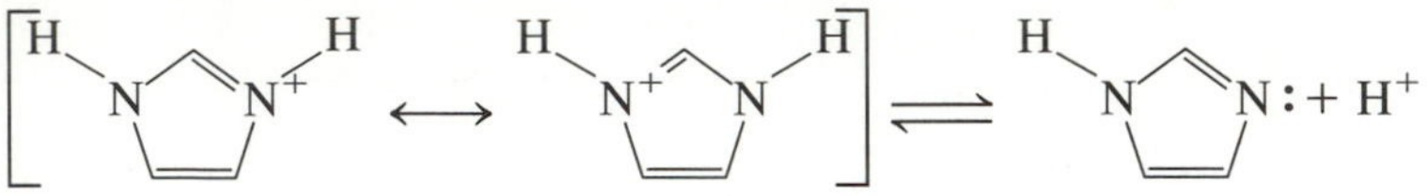

imidazolium ion imidazole

Because the imidazolium ion has a pK_a of 7, imidazole is exactly one-half protonated in neutral medium and can act as either an acid or a base. It is no accident that Nature chose the amino acid histidine to use in important proteolytic enzymes like the serine proteases (Figure 29.8).

B. Synthesis

Pyrazoles and isoxazoles can be synthesized by the reaction of hydrazine or hydroxylamine with 1,3-dicarbonyl compounds or the equivalent.

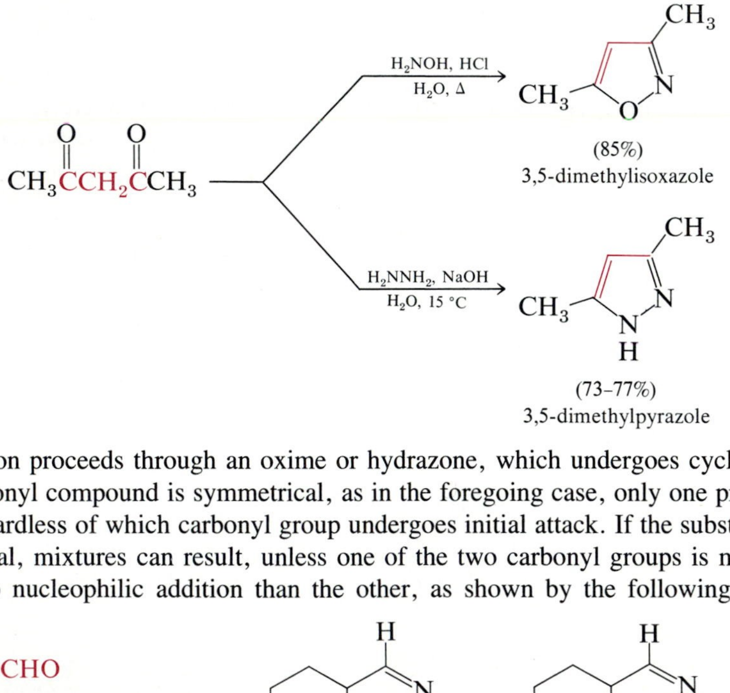

$CH_3COCH_2COCH_3$ — H_2NOH, HCl / H_2O, Δ → (85%) 3,5-dimethylisoxazole

$CH_3COCH_2COCH_3$ — H_2NNH_2, NaOH / H_2O, 15 °C → (73–77%) 3,5-dimethylpyrazole

The reaction proceeds through an oxime or hydrazone, which undergoes cyclization. If the dicarbonyl compound is symmetrical, as in the foregoing case, only one product can result, regardless of which carbonyl group undergoes initial attack. If the substrate is not symmetrical, mixtures can result, unless one of the two carbonyl groups is much more reactive to nucleophilic addition than the other, as shown by the following example.

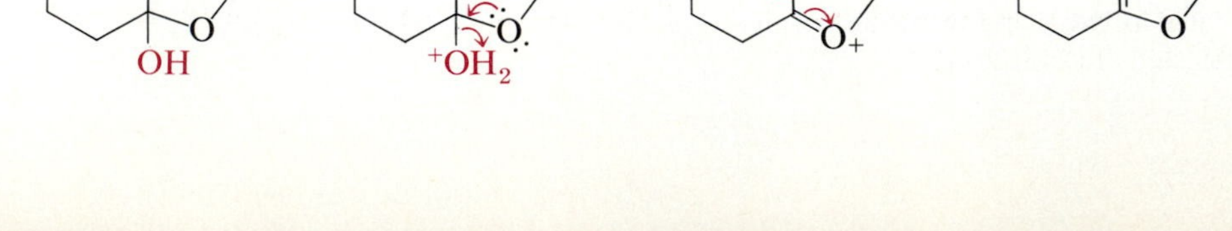

CHO + H_2NOH → ⇌ → ; H^+ ⇌ ; $-H_2O$ → ; $-H^+$ →

If a substituted hydrazine is used, a 1-substituted pyrazole results.

EXERCISE 32.13 What product is obtained when benzoylacetophenone (1,3-diphenyl-1,3-propanedione) is treated with phenylhydrazine and aqueous HCl?

An alternative synthesis of isoxazoles involves the cycloaddition of a nitrile oxide to an acetylene.

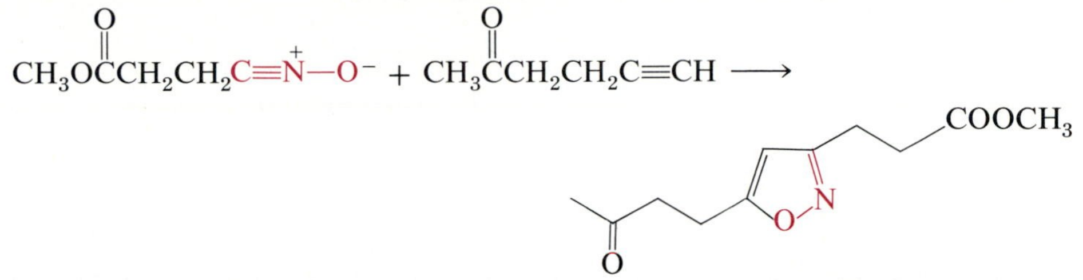

$$CH_3OCCH_2CH_2C{\equiv}\overset{+}{N}{-}O^- + CH_3CCH_2CH_2C{\equiv}CH \longrightarrow$$

(50%)

Nitrile oxides are unstable compounds generated *in situ* by the dehydrochlorination of hydroxamic acid chlorides, which are prepared by chlorination of aldoximes.

$$C_6H_5{-}CH{=}NOH \xrightarrow{Cl_2} C_6H_5{-}C(Cl){=}NOH \xrightarrow{NaOH} C_6H_5C{\equiv}\overset{+}{N}{-}O^-$$

An alternative preparation (**Mukaiyama method**) involves dehydration of nitroalkanes.

$$CH_3CH_2\overset{+}{N}(=O)O^- + C_6H_5NCO \longrightarrow CH_3C{\equiv}\overset{+}{N}{-}O^-$$

The reaction of nitrile oxides with alkynes and alkenes is called a **1,3-dipolar cycloaddition.** It is an example of a large class of such cycloaddition reactions that involve six-electron, Hückel transition states and are fully analogous to the Diels-Alder reaction (Sections 20.5, 21.7, 22.4).

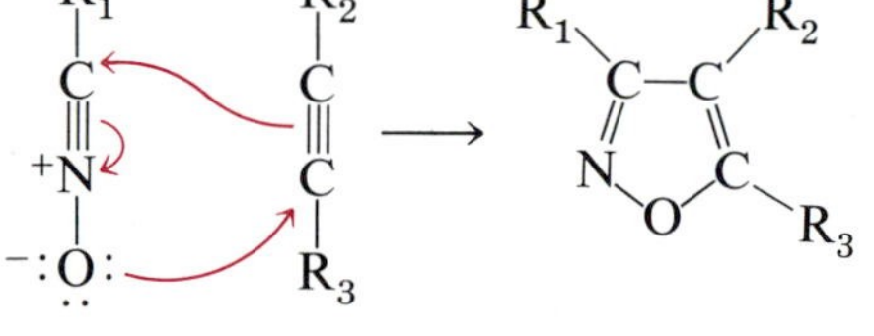

EXERCISE 32.14 (a) Draw the transition state for the preceding example and show the six pericyclic electrons.
(b) Outline a synthesis of 3-ethyl-5-isopropylisoxazole, starting with organic compounds containing five or fewer carbons.

Pyrazoles may also be prepared by 1,3-dipolar cycloaddition, this time between diazomethane and an acetylene.

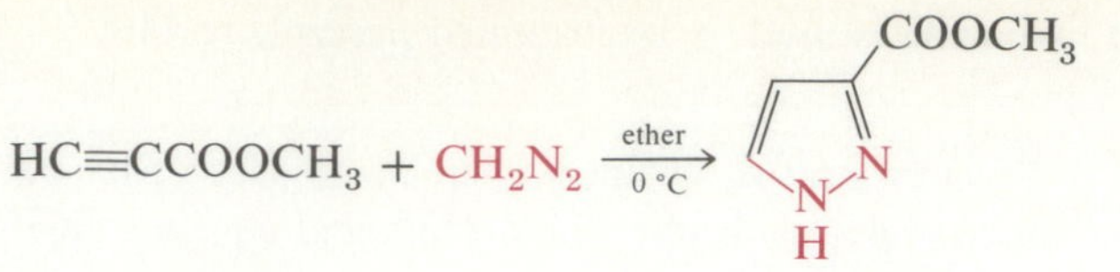

$$HC{\equiv}CCOOCH_3 + CH_2N_2 \xrightarrow[0\,^\circ C]{ether}$$ methyl pyrazole-3-carboxylate (80%)

(80%)
methyl pyrazole-3-carboxylate

The reaction is formulated in a completely analogous manner.

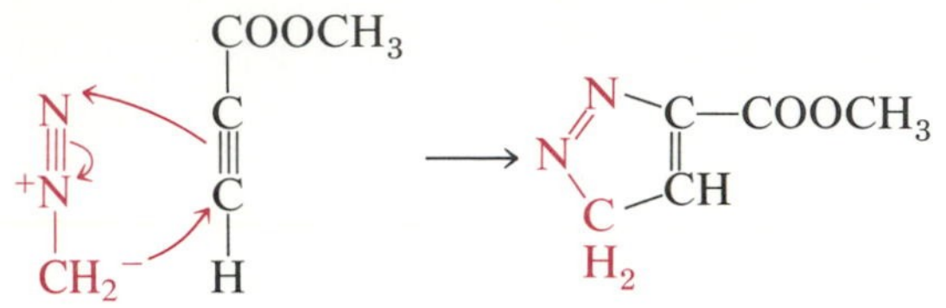

The initially formed product isomerizes to the more stable aromatic system. 3-Substituted pyrazoles bearing a proton on nitrogen exist in equilibrium with the 5-isomers. Such prototropic equilibria (Section 15.1.A) are generally slow enough so that both isomers are seen in the NMR and CMR spectra. However, the isomerization is usually fast enough that it is not practical to isolate the individual 3- and 5-isomers.

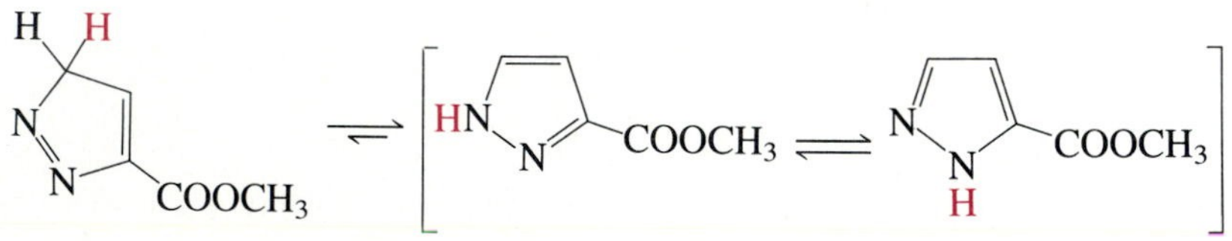

The most general synthesis of the 1,3-azoles is the dehydration of 1,4-dicarbonyl compounds, a form of Paal-Knorr cyclization.

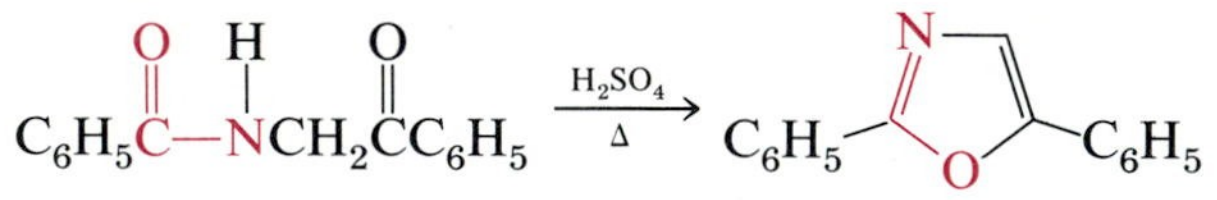

$$C_6H_5\overset{O}{\overset{\|}{C}}-\overset{H}{\overset{|}{N}}CH_2\overset{O}{\overset{\|}{C}}C_6H_5 \xrightarrow[\Delta]{H_2SO_4}$$

2,5-diphenyloxazole

$$C_6H_5\overset{O}{\overset{\|}{C}}-\overset{H}{\overset{|}{N}}\underset{C_6H_5}{\underset{|}{C}H}\overset{O}{\overset{\|}{C}}C_6H_5 \xrightarrow[\substack{HOAc\\120\,^\circ C}]{NH_4{}^+OAc^-}$$

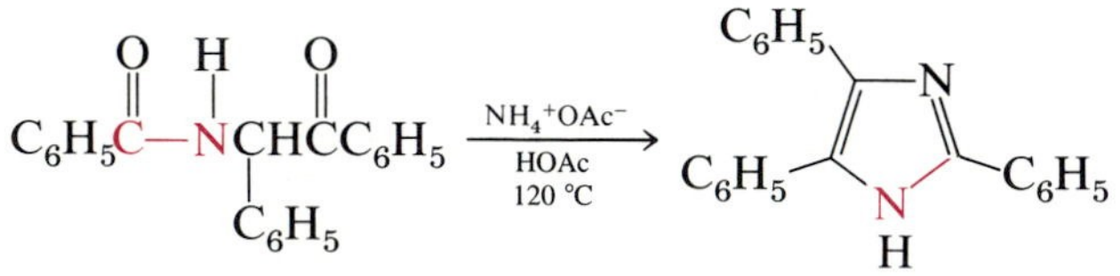

(93%)
2,4,5-triphenylimidazole

$$CH_3\overset{O}{\overset{\|}{C}}CH_2NH\overset{O}{\overset{\|}{C}}CH_3 + P_2S_5 \xrightarrow{120\,^\circ C}$$

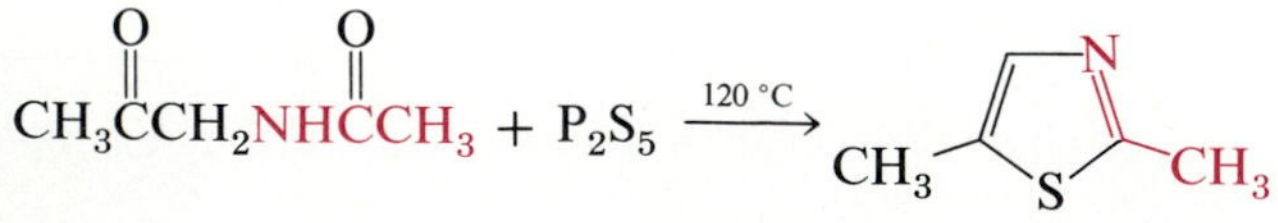

2, 5-dimethylthiazole

EXERCISE 32.15 Write reasonable reaction mechanisms for the foregoing preparations of 2,5-diphenyloxazole and 2,4,5-triphenylimidazole.

C. Reactions

The azoles are markedly less reactive toward electrophilic agents than furan, pyrrole, and thiophene. The reduced reactivity is due to the electronegative azole nitrogen. For the 1,2-azoles the reactivity order is as follows.

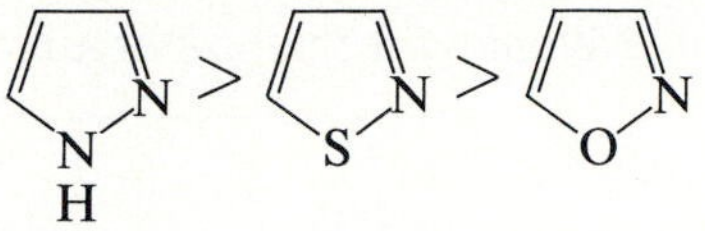

Electrophilic substitution takes place exclusively at C-4, whether or not other substituents are at C-3 and C-5.

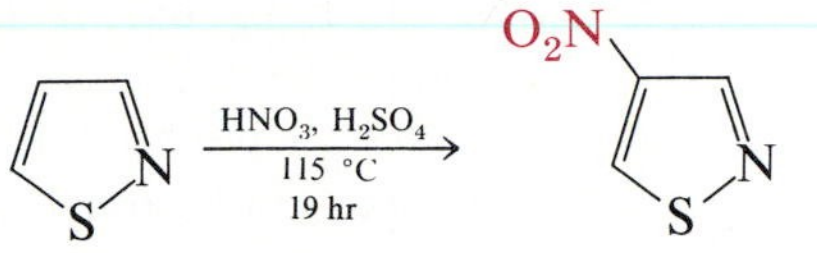

HNO_3, H_2SO_4, 115 °C, 19 hr

(97%)
4-nitroisothiazole

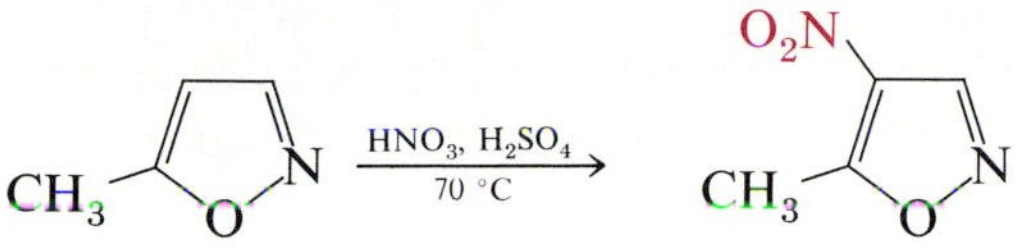

HNO_3, H_2SO_4, 70 °C

5-methylisoxazole → 4-nitro-5-methylisoxazole

EXERCISE 32.16 We mentioned in Section 32.3.C that the positions in thiophene α to the sulfur are much more reactive than the β-positions toward electrophilic substitution. Explain why the 5-position of isothiazole (α to the sulfur) is *less* reactive than the 4-position (β to the sulfur).

For the 1,3-azoles a similar reactivity order is found.

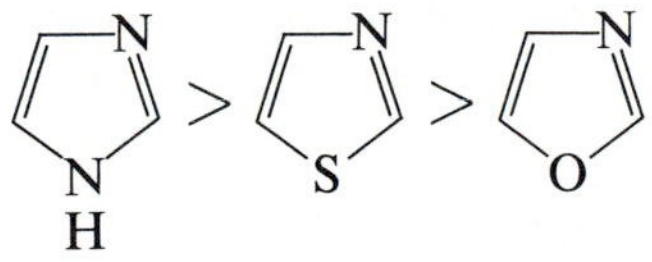

For imidazoles, which have been studied most extensively, substitution occurs preferentially at C-4 (equivalent to C-5 by *prototropy* as shown) rather than at C-2.

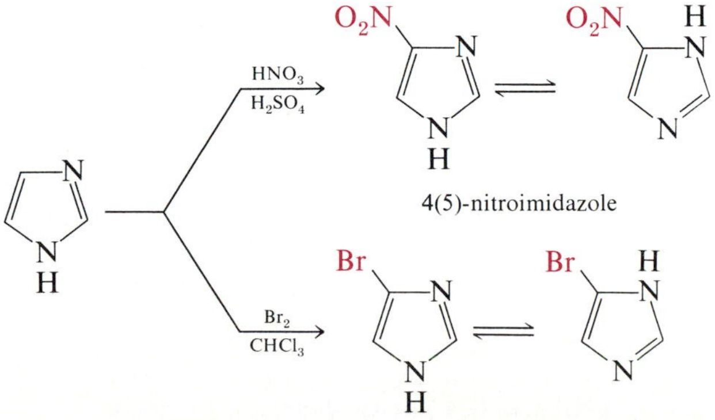

HNO_3, H_2SO_4

4(5)-nitroimidazole

Br_2, $CHCl_3$

4(5)-bromoimidazole

When both C-4 and C-5 are blocked, substitution occurs at C-2.

EXERCISE 32.17 Show how each of the following conversions can be accomplished.

(a) 1,4-dimethylimidazole to 5-bromo-1,4-dimethylimidazole
(b) 4-methylimidazole to 2-bromo-4-methyl-5-nitroimidazole

32.6 Pyridine

A. Structure and Physical Properties

Pyridine is an analog of benzene in which one of the CH units is replaced by nitrogen (Figure 32.4). The nitrogen lone pair is located in an sp^2-hybrid orbital that is perpendicular to the π-system of the ring. The effect on the basicity of the nitrogen (pK_a 5.2) has been discussed in Section 32.5.A. Various values have been deduced for the empirical resonance energy of pyridine, but it would appear to be roughly comparable to that of benzene. The resonance stabilization is shown by the two equivalent Kekulé structures and the three zwitterionic forms with negative charge on nitrogen.

The surplus negative charge on nitrogen is manifest in the dipole moment of pyridine, which is substantially greater than that of piperidine, the nonaromatic analog. That is, the π-moment is in the same direction as the σ-moment and the net moment is additive.

2.26 D　　1.17 D

As the charged resonance structures and the dipole moment show, the ring in pyridine is relatively electron-deficient, a feature that is reflected in many of the reactions of pyridine (Section 32.6.C).

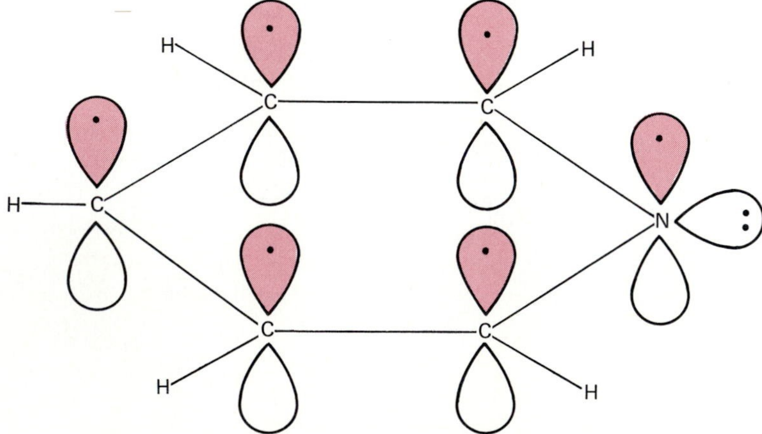

FIGURE 32.4 Orbital structure of pyridine.

Most alkylpyridines have trivial names that are in common use. The most important are the picolines (methylpyridines) and lutidines (dimethylpyridines). Collidine is the common name of 2,4,6-trimethylpyridine.

pyridine
b.p. 115 °C

α-picoline
b.p. 128 °C

β-picoline
b.p. 144 °C

γ-picoline
b.p. 144 °C

2,6-lutidine
b.p. 144 °C

2,5-lutidine
b.p. 157 °C

2,4-lutidine
b.p. 157 °C

2,3-lutidine
b.p. 163 °C

sym-collidine
b.p. 170 °C

Highlight 32.3

In pyridine the aromatic sextet is provided by one *p*-electron from each carbon and one *p*-electron from nitrogen. This leaves the nitrogen with a lone pair of electrons in an sp^2-orbital that lies in the plane of the molecule, perpendicular to the aromatic π-system.

Because it has a nonbonding electron pair that is not involved in the aromatic sextet, pyridine acts as a normal amine. However, because the basic electron pair is in an sp^2-orbital, the pK_a of pyridininium ion is only 5.2.

$$C_5H_5NH^+ \rightleftharpoons C_5H_5N + H^+$$

Thus, pyridine is one-half protonated at pH 5.2 and only about 2% protonated at pH 7.

B. Synthesis

Pyridine itself and most of the simpler alkylpyridines are available from coal tar distillates. Several syntheses are available for deriving substituted pyridines from other compounds.

The most general technique for constructing the ring is the **Hantzsch pyridine synthesis.** Although numerous variations are known, the simplest consists of the condensation of a β-keto ester with an aldehyde and ammonia. The product is a 1,4-dihydropyridine, which is subsequently aromatized by oxidation.

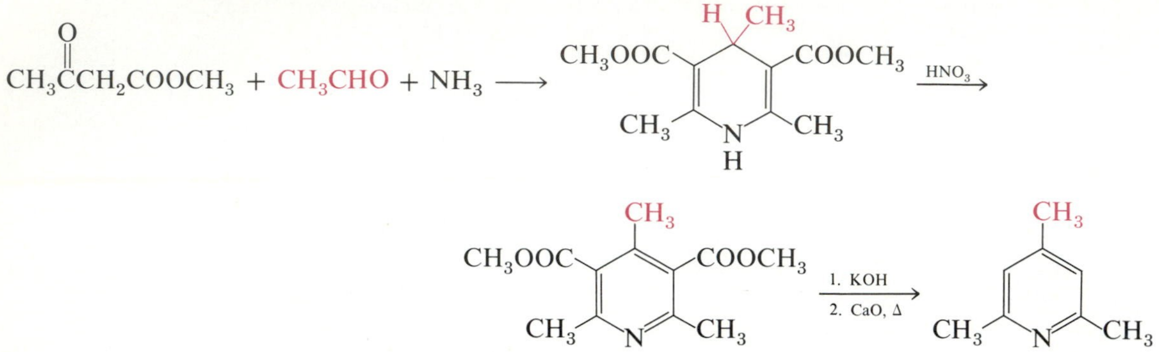

$$CH_3\overset{O}{\overset{\|}{C}}CH_2COOCH_3 + CH_3CHO + NH_3 \longrightarrow \text{(1,4-dihydropyridine: 4-H, 4-}CH_3\text{; 3,5-}COOCH_3\text{; 2,6-}CH_3\text{; N-H)} \xrightarrow{HNO_3}$$

$$\text{(pyridine: 4-}CH_3\text{; 3,5-}COOCH_3\text{; 2,6-}CH_3) \xrightarrow[\text{2. CaO, }\Delta]{\text{1. KOH}} \text{(2,4,6-trimethylpyridine)}$$

A reasonable mechanism for the Hantzsch reaction is outlined as follows. The first step is probably a Knoevenagel condensation of the aldehyde (Section 27.7.E) with the β-keto ester.

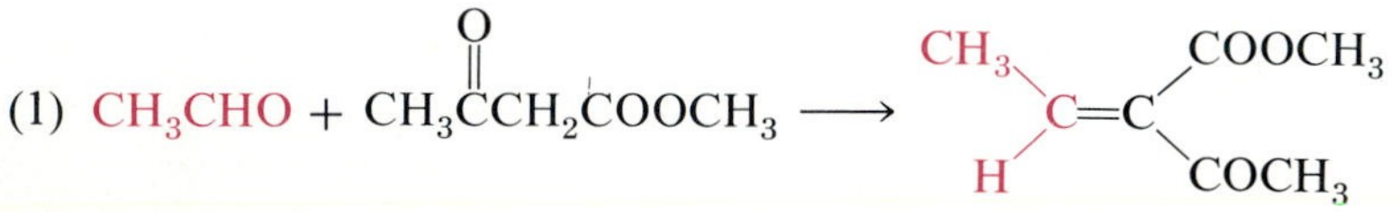

$$(1)\quad CH_3CHO + CH_3\overset{O}{\overset{\|}{C}}CH_2COOCH_3 \longrightarrow \underset{H}{\overset{CH_3}{>}}C{=}C\underset{COCH_3}{\overset{COOCH_3}{<}}$$

A part of the β-keto ester also condenses with ammonia to form an enamine.

$$(2)\quad CH_3\overset{O}{\overset{\|}{C}}CH_2COOCH_3 + NH_3 \longrightarrow CH_3\overset{NH_2}{\overset{|}{C}}{=}CHCOOCH_3 + H_2O$$

The unsaturated keto ester produced in step 1 then undergoes a condensation with the enamine produced in step 2.

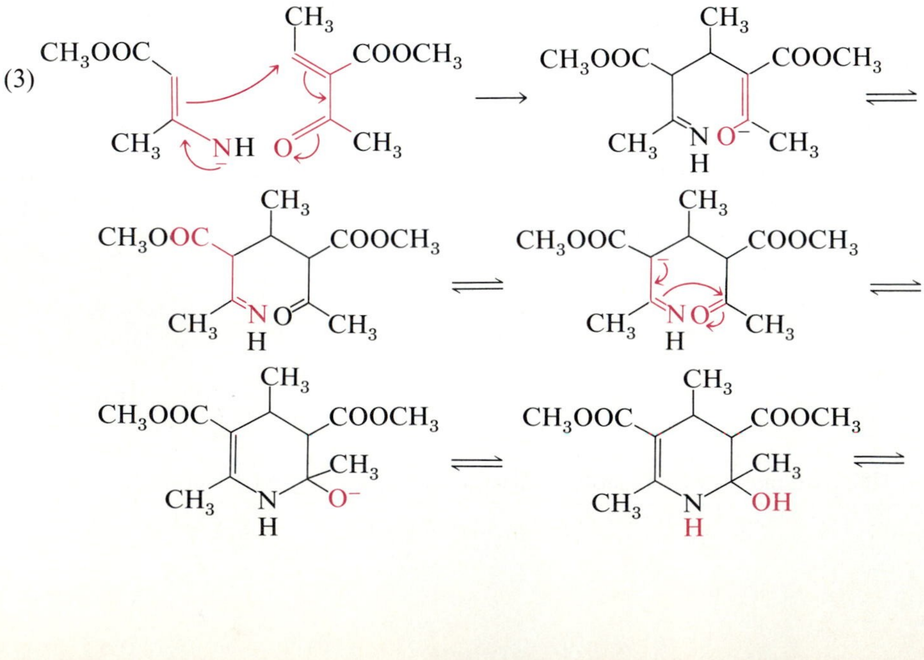

(3) [Mechanism scheme: the enamine (CH_3OOC, CH_3, NH) adds to the unsaturated keto ester (CH_3, $COOCH_3$, CH_3, O) $\longrightarrow$ ring intermediate with CH_3OOC, CH_3, $COOCH_3$, $\overset{+}{N}$H, O^-, CH_3 $\rightleftharpoons$ iminium/ketone intermediate $\rightleftharpoons$ N attacks C=O $\rightleftharpoons$ cyclic alkoxide (CH_3OOC, CH_3, $COOCH_3$, CH_3, N–H, O^-) $\rightleftharpoons$ cyclic hemiaminal (CH_3OOC, CH_3, $COOCH_3$, CH_3, N–H, OH) $\rightleftharpoons$]

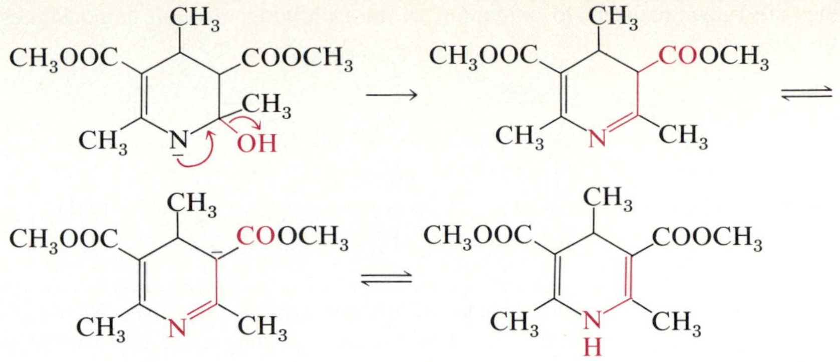

CH$_3$OOC / COOCH$_3$ / CH$_3$ / CH$_3$ / N / OH → CH$_3$OOC / COOCH$_3$ / CH$_3$ / N / CH$_3$ ⇌

CH$_3$OOC / COOCH$_3$ / CH$_3$ / N / CH$_3$ ⇌ CH$_3$OOC / COOCH$_3$ / CH$_3$ / N–H / CH$_3$

EXERCISE 32.18 The Hantzsch pyridine synthesis is usually carried out with ammonia in alcohol solution, mildly basic conditions. The carbanion intermediates in the reaction are enolate ions with the negative charge delocalized onto oxygen or nitrogen. The foregoing reaction sequence involves a number of protonations and deprotonations. What is the base for deprotonations? Where do the protons come from in the protonation steps? Trace the course of each protonation and deprotonation involved, and write resonance structures to show the stabilization of negative charge in each carbanion intermediate.

C. Reactions

The nitrogen lone pair has basic and nucleophilic properties, although both are diminished by the hybridization effect. Pyridines form salts with acids and are widely used as catalysts and "acid scavengers" in reactions where strong acids are produced.

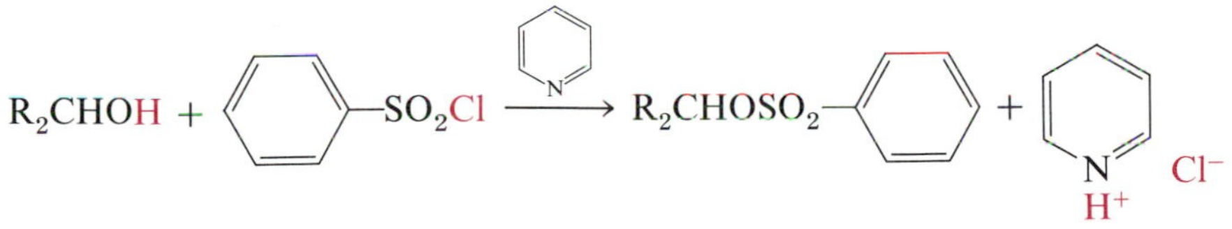

$$R_2CHOH + C_6H_5SO_2Cl \xrightarrow{\text{pyridine}} R_2CHOSO_2C_6H_5 + C_5H_5NH^+\ Cl^-$$

pyridinium chloride

The nitrogen can be alkylated by primary alkyl halides, leading to N-alkylpyridinium salts.

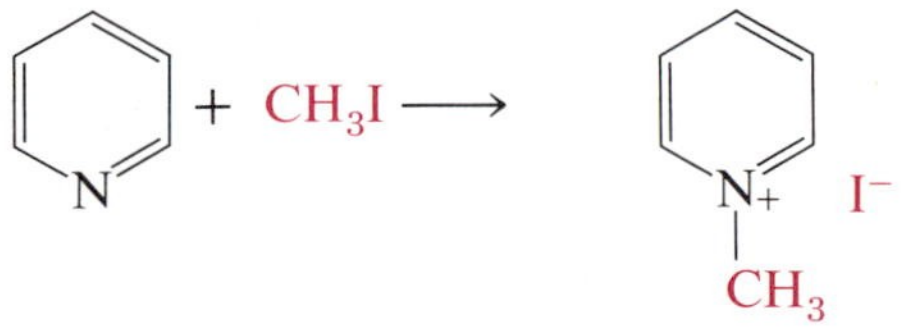

$$C_5H_5N + CH_3I \longrightarrow C_5H_5N^+CH_3\ I^-$$

N-methylpyridinium iodide

Recall that a 1-methyl-4-methoxycarbonylpyridinium iodide absorbs light to undergo a charge-transfer transition, in a solvent-sensitive process that underlies the solvent polarity parameters, Z- and E_T-values (Section 22.6.D). Pyridinium salts, especially those formed from the good donor iodide ion, are often colored as a result of charge-transfer transitions.

Pyridines are rather resistant to oxidation, as the following reaction demonstrates.

quinoline $\xrightarrow[\Delta]{KMnO_4}$ quinolinic acid (65–70%) $\xrightarrow{\Delta}$ nicotinic acid (niacin)

The foregoing reaction provides a route to β-substituted pyridine derivatives. Nicotinic acid is present in minute amounts in all living cells. The corresponding amide, niacinamide, is an essential B vitamin. The enzyme cofactors, nicotinamide adenine dinucleotide (NAD) and its phosphate (NADP), function through reduction of the pyridine ring to the 1,4-dihydropyridine derivative or reoxidation of the latter to the pyridine as illustrated for alcohol dehydrogenase. From the equation, the pH will control the position of the equilibrium. The preferred description for the mechanism involves "hydride" transfer but a definitive description has proven elusive.

$$\text{CONH}_2\text{-pyridinium-}N^+\text{-R} + CH_3—CH_2OH \xrightleftharpoons{\text{alcohol dehydrogenase}} \text{1,4-dihydro (H, H) CONH}_2\text{-pyridine-N-R} + CH_3—CH{=}O + H^+$$

R = ribose-diphosphate-ribose-adenine

Nicotinic acid is also produced by oxidation of nicotine, an alkaloid present to the extent of 2–8% in the dried leaves of *Nicotiana tabacum*. Nicotine is used as an agricultural insecticide, but is also toxic to humans; fatal doses can be absorbed through the skin, and might be as low as 25–40 mg.

nicotine $\xrightarrow{HNO_3}$ nicotinic acid (COOH)

Because of its resistance to oxidation, pyridine can even be used as a solvent for chromium trioxide oxidations (**Sarrett procedure,** Section 14.8.A). However, under the proper conditions the nitrogen can be oxidized to the N-oxide, as are other tertiary amines (Section 24.7.C).

$$\text{3-methylpyridine} + 30\%\ H_2O_2 \xrightarrow[\text{70 °C, 24 hr}]{CH_3COOH} \text{3-methylpyridine N-oxide } (N^+—O^-)$$

(75%)
3-methylpyridine N-oxide

As we shall see, pyridine N-oxides are important synthetic intermediates.

EXERCISE 32.19 For pyridine N-oxide, there are four resonance structures in which all atoms have octet electronic configurations. Write these structures, paying careful attention to formal charges. Be sure to show each nonbonding electron pair as a pair of dots. What do these resonance structures suggest about orientation in the reactions of pyridine N-oxide with electrophiles?

Pyridine is resistant to electrophilic aromatic substitution conditions, not only because of the electron-deficient ring but also because under the acidic conditions of such reactions the nitrogen is protonated or complexed with a Lewis acid. In general, pyridine is less reactive in such reactions than trimethylanilinium ion.

$\overset{+}{N}(CH_3)_3$ — fuming HNO_3, fuming H_2SO_4 → $^+N(CH_3)_3$, NO_2

N — fuming HNO_3, fuming H_2SO_4 → no reaction

Substitution is achieved only under the most drastic conditions; reaction occurs at C-3, and yields are often poor.

N — KNO_3, HNO_3, H_2SO_4, Fe, 300 °C → NO_2, N

(22%)
3-nitropyridine

N — 20% SO_3, H_2SO_4, H_2SO_4, 230 °C, 24 hr → SO_3H, N

(71%)
pyridine-3-sulfonic acid

Alkyl and amino groups activate the ring toward electrophilic substitution. In the alkylpyridines the ring nitrogen directing influence predominates (C-3 or C-5 attack) regardless of the position of alkylation.

CH_3, N — 20% SO_3, H_2SO_4, $HgSO_4$, 230 °C, 16 hr → CH_3, SO_3H, N

β-picoline

5-methylpyridine-3-sulfonic acid

Amino groups either free or acylated govern the position of further substitution (*ortho* or *para* to the amino).

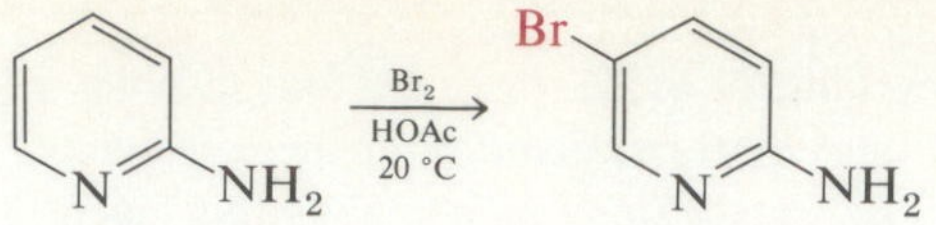

2-aminopyridine $\xrightarrow[\text{HOAc, 20 °C}]{Br_2}$ 5-bromo-2-aminopyridine (90%)

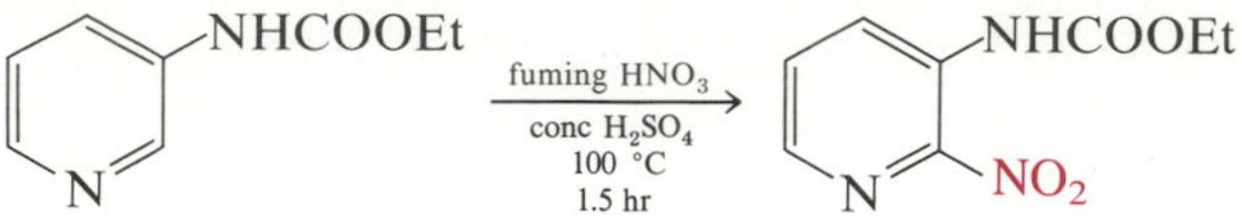

ethyl N-(3-pyridyl)-carbamate $\xrightarrow[\text{conc } H_2SO_4,\ 100\ °C,\ 1.5\ hr]{\text{fuming } HNO_3}$ ethyl N-(2-nitro-3-pyridyl)-carbamate (61%)

The predominant 3-substitution in pyridine is explainable in terms of the resonance structures of the intermediate ions, and the corresponding transition states, produced by electrophilic attack at the three positions.

Attack at C-2

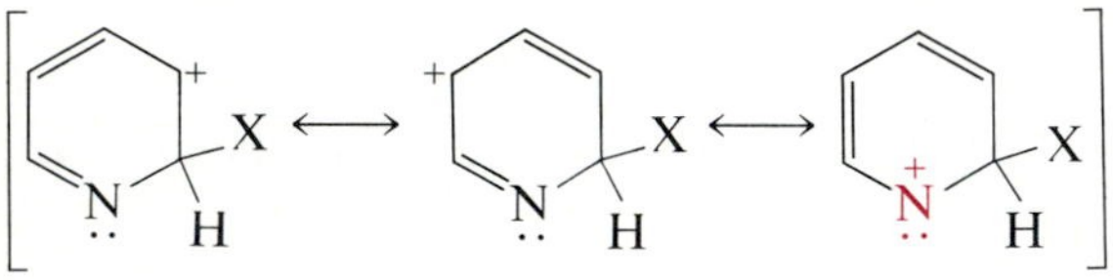

Attack at C-4

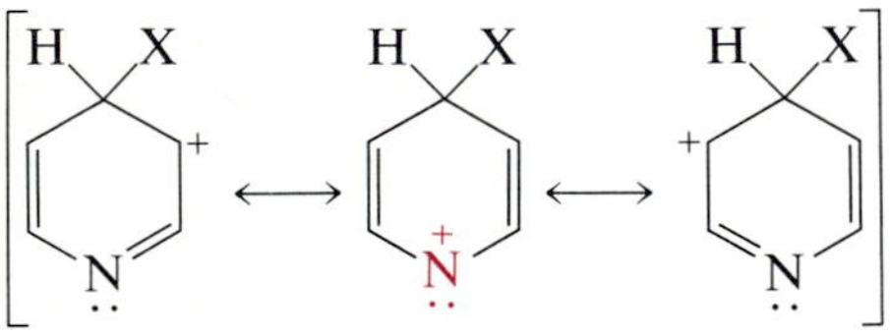

Attack at C-3

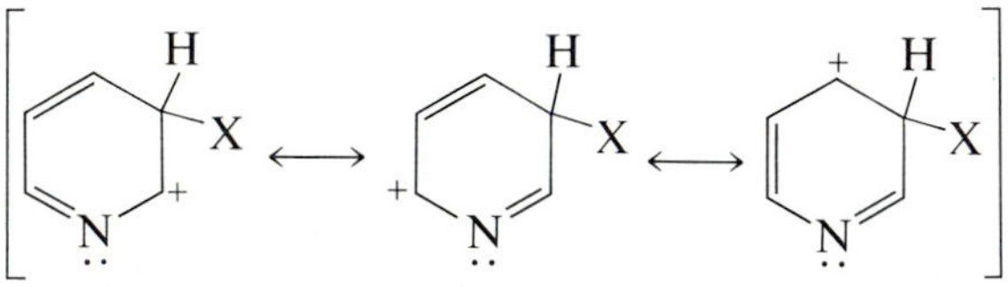

Compared with the ion produced from benzene, all three ions from pyridine are destabilized by the inductive effect of the nitrogen, especially if it is protonated or coordinated with a Lewis acid. However, the situation is much worse when attack is at C-2 or C-4 than at C-3. In the two former cases one of the structures of the intermediate ion has the positive charge on an electron-deficient nitrogen. Thus the situation in pyridine is similar to that in nitrobenzene. Electrophilic attack is retarded at all positions, but especially at C-2 and C-4.

Pyridine N-oxides undergo electrophilic substitution somewhat more readily. Reaction generally occurs at C-4.

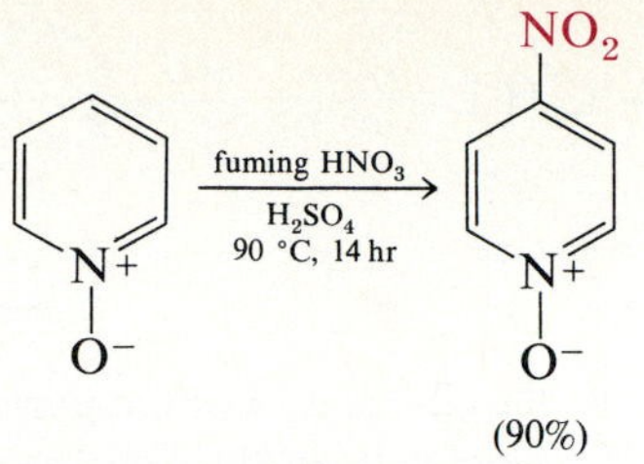

fuming HNO_3, H_2SO_4, 90 °C, 14 hr → (90%)

The N-oxide can often be used as an ''activated'' form of the pyridine. Treatment of the substituted N-oxide with PCl_3 removes the oxygen.

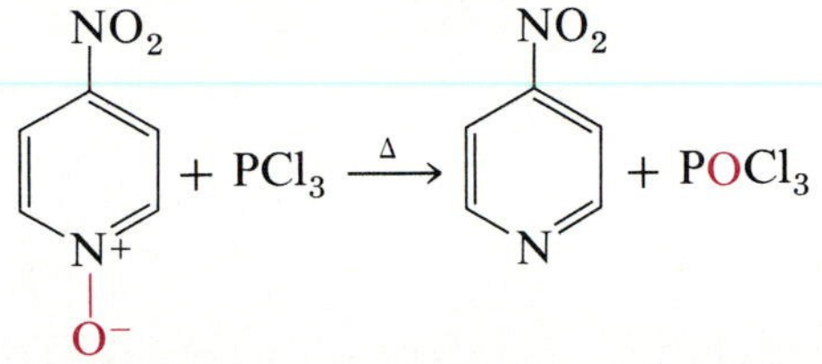

$$+ PCl_3 \xrightarrow{\Delta} + POCl_3$$

EXERCISE 32.20 Write the resonance structures for the intermediate cations produced by reaction of pyridine N-oxide with NO_2^+ at C-2, C-3, and C-4 and use these to explain the observed orientation in the reaction.

The electron-deficient nature of the pyridine ring is also manifest in the ease with which pyridines undergo **nucleophilic substitution.** A particularly useful and unusual example is the synthesis of aminopyridines by the reaction of a pyridine with an alkali metal amide (**Chichibabin reaction**).

$$+ NaNH_2 \xrightarrow[\text{reflux}]{C_6H_5N(CH_3)_2} \text{NH}^- \text{Na}^+ \xrightarrow{H_2O} NH_2$$

(70–80%)
2-aminopyridine

The reaction is initiated by attack by the nucleophile at C-2 or C-6. Attack occurs at these positions because the negative charge can be delocalized onto the ring nitrogen.

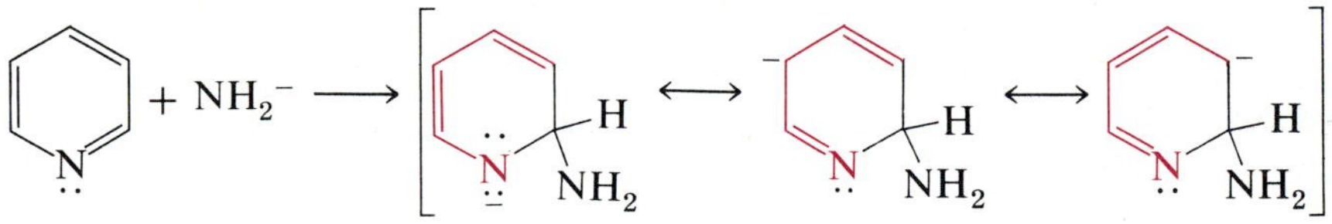

$$+ NH_2^- \longrightarrow$$

The second step is elimination of hydride ion, which reacts with the aminopyridine to give H_2. The driving force for the elimination of hydride ion is, of course, the formation of the aromatic cycle.

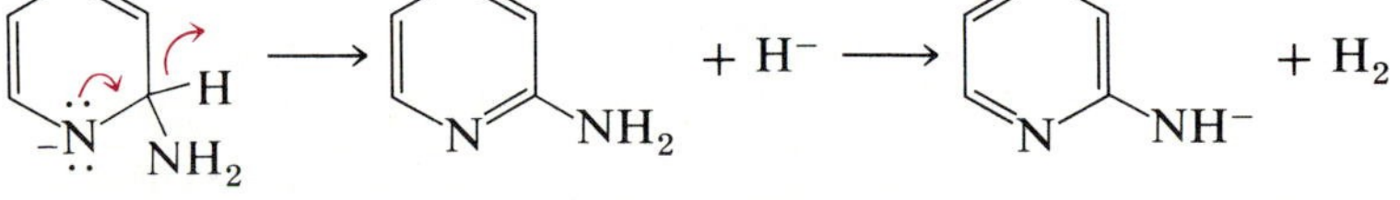

$$\longrightarrow NH_2 \quad + H^- \longrightarrow NH^- \quad + H_2$$

Chichibabin-like reactions are also observed with organolithium compounds.

$$\text{pyridine} + C_6H_5Li \xrightarrow[\text{110 °C, 8 hr}]{\text{toluene}} \xrightarrow{H_2O} \text{2-}C_6H_5\text{-pyridine}$$

(50%)
2-phenylpyridine

Attempted diazotization of the 2- and 4-aminopyridines yields the 2- and 4-hydroxypyridines, which exist completely in the keto form (for example, α-pyridone, γ-pyridone). Presumably the intermediate diazonium ion undergoes nucleophilic substitution by water.

$$\text{2-}NH_2\text{-pyridine} \xrightarrow[H_3O^+]{HNO_2} \text{2-}OH\text{-pyridine} \rightleftharpoons \alpha\text{-pyridone (N–H, C=O)}$$

α-pyridone

$$\text{4-}NH_2\text{-pyridine} \xrightarrow[H_3O^+]{HNO_2} \text{4-}OH\text{-pyridine} \rightleftharpoons \gamma\text{-pyridone (N–H, C=O)}$$

γ-pyridone

Even though the simple pyridones exist in the isomeric form with hydrogen attached to nitrogen (amide form), they still have extensive aromatic character, understandable in terms of an important dipolar resonance structure.

$$\left[\ \text{N(R)–C=O} \longleftrightarrow \overset{+}{N}\text{(R)=C–}O^-\ \right]$$

EXERCISE 32.21 Suggest a synthesis of 6-phenyl-2-pyridone, starting with pyridine.

Another feature of the pyridine ring is of interest. The methyl groups in α- and γ-picoline are comparable in acidity to those in methyl ketones and readily undergo base-catalyzed reactions. Similar reactions are seen with other pyridines that have alkyl groups at C-2 or C-4.

$$\text{4-}CH_3\text{-3-}CH_2CH_3\text{-pyridine} \xrightarrow{NaNH_2} \text{4-}\overset{-}{C}H_2\text{-3-}CH_2CH_3\text{-pyridine} \xrightarrow{CH_3I} \text{4-}CH_2CH_3\text{-3-}CH_2CH_3\text{-pyridine}$$

(80%)
3,4-diethylpyridine

The enhanced acidity at these positions is again attributed to delocalization of negative charge in the intermediate anion into the ring and especially onto the nitrogen.

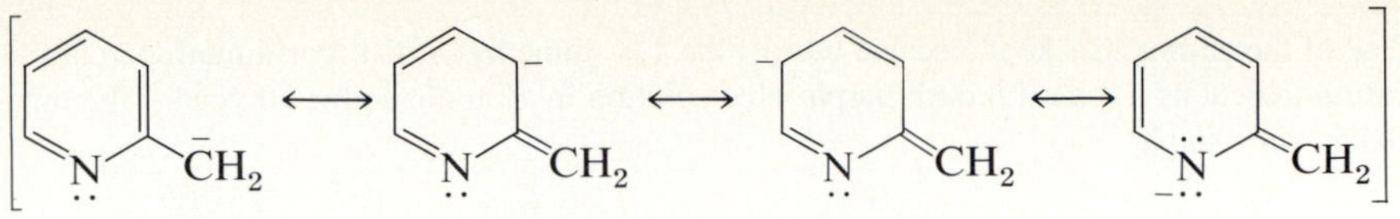

This side-chain acidity is enhanced in the N-alkylpyridinium compounds.

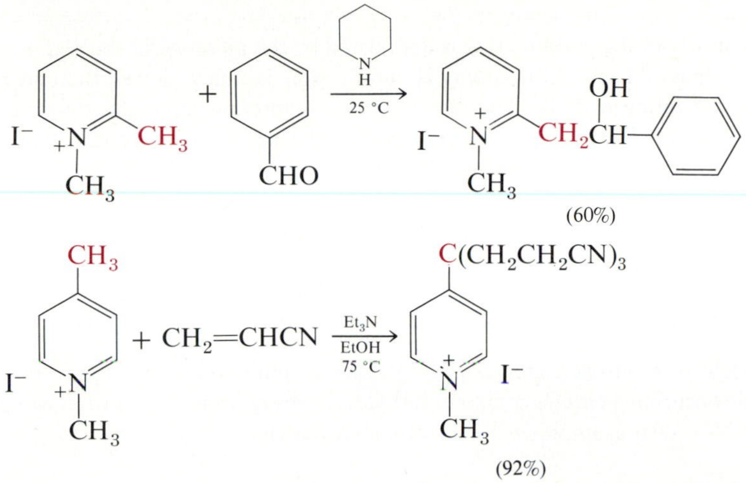

(60%)

(92%)

EXERCISE 32.22 When 2,3-dimethylpyridine N-oxide is dissolved in methanol-*d* containing sodium methoxide, three protons are exchanged by deuterium. What is the structure of the product? Explain with the use of resonance structures.

There is an interesting consequence of the increased electron affinity of the pyridinium ring on N-alkylation. Metals and other one-electron reducing agents add an electron to the pyridinium ring to form a pyridinyl free radical. With an electron-withdrawing group in the 4-position, the radicals are stable enough to isolate in pure form and can be distilled in the absence of oxygen on a vacuum line. A widely used bipyridinium compound with the common name of paraquat is reduced by the photosynthetic system of plants to a cation radical. The cation radical reacts with oxygen to form superoxide ion and hydrogen peroxide. These agents destroy the chloroplast that contains the photosynthetic apparatus, thus causing the death of the plant. Many classes of weeds can take up the dication and paraquat is widely used as a weedkiller.

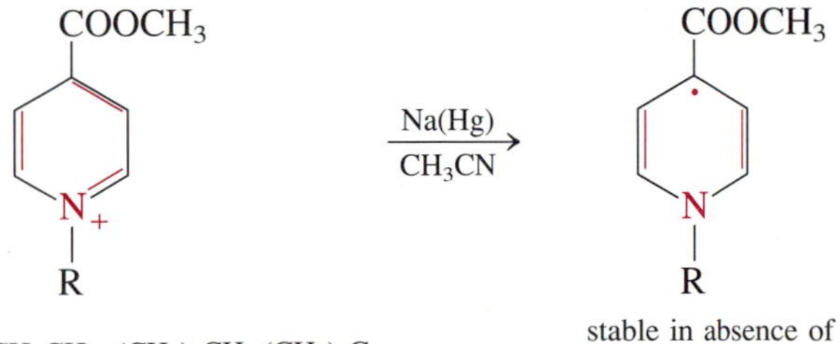

$\xrightarrow[CH_3CN]{Na(Hg)}$

R = CH_3, CH_3CH_2, $(CH_3)_2CH$, $(CH_3)_3C$

stable in absence of O_2
distills as emerald-green liquid

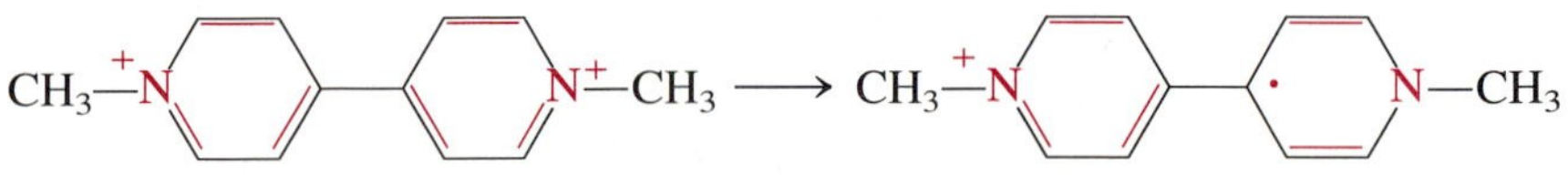

1,1′-dimethyl-4,4′-bipyridinium dication

paraquat dication; methylviologen dication

1,1′-dimethyl-4,4′-bipyridinium cation radical

paraquat cation radical; methylviologen cation radical

One of the authors has kept a sealed tube of the 1,1′-dimethyl-4,4′-bipyridinium perchlorate cation radical as a beautiful dark purple-blue solution in acetonitrile for 30 years. The material is very stable.

Highlight 32.4

The chemistry of the pyridine ring is dominated by the presence of the electronegative nitrogen atom. Electrophilic aromatic substitution is much slower than in benzene because the nitrogen destabilizes the carbocation intermediates and transition states leading to them. The major product of electrophilic aromatic substitution is that resulting from attack of the electrophile at C-3.

[pyridine + E^+ → resonance structures of the C-3 attack intermediate, each with H and E at C-3 and + on the ring]

Conversely, the nitrogen greatly facilitates nucleophilic addition, a type of reaction that only occurs on benzene compounds that have strongly electron-withdrawing groups such as NO_2. An example is the Chichibabin reaction.

[pyridine + C_6H_5Li → N-Li, 2-H, 2-C_6H_5 dihydropyridine → 2-C_6H_5-pyridine + LiH]

Another manifestation of the importance of the ring nitrogen is the fact that methyl groups at C-2 and C-4 are comparable in acidity to those of methyl ketones.

[2-methylpyridine (CH_3) + KOH ⇌ resonance structures with CH_2^- and $\bar{N}$ / $=CH_2$]

32.7 Quinoline and Isoquinoline

A. Structure and Nomenclature

Quinoline and isoquinoline are benzopyridines. The two compounds are both numbered in the same manner as naphthalene (Section 31.3.A) in such a way that nitrogen gets the smallest possible number.

[numbered structures of quinoline (N at 1) and isoquinoline (N at 2)]

quinoline
b.p. 238 °C
pK_a 4.8

isoquinoline
b.p. 243 °C
pK_a 5.4

The orbital structures of quinoline and isoquinoline are related to those of pyridine (Section 32.6.A) and naphthalene (Section 31.3.A). Both are weak bases, with pK_as comparable to that of pyridine. Alkaloids containing the quinoline and isoquinoline skeleton are widespread in the plant kingdom (Section 36.6.C).

B. Synthesis

The most general method for synthesizing quinolines is the **Skraup reaction,** in which aniline or a substituted aniline is treated with glycerol, sulfuric acid, and an oxidizing agent such as As_2O_5, ferric salts, or the nitro compound corresponding to the amine used.

$$C_6H_5NH_2 + HOCH_2CH(OH)CH_2OH \xrightarrow[As_2O_5,\ \Delta]{H_2SO_4} \text{quinoline}$$

(84–91%)

The mechanism of the Skraup reaction probably involves initial dehydration of the glycerol to give acrolein, which undergoes a 1,4-addition by the aniline. The resulting β-phenylaminopropionaldehyde is then cyclized to a dihydroquinoline, which is finally oxidized to give the product.

$$(1)\quad HOCH_2CH(OH)CH_2OH \xrightarrow{H^+} CH_2{=}CHCHO$$

$$(2)\quad C_6H_5NH_2 + CH_2{=}CHCHO \longrightarrow C_6H_5NHCH_2CH_2CHO$$

$$(3)\quad C_6H_5NHCH_2CH_2CHO \longrightarrow \text{1,2-dihydroquinoline}$$

$$(4)\quad \text{1,2-dihydroquinoline} \xrightarrow{[O]} \text{quinoline}$$

Similar results are obtained if an α,β-unsaturated ketone or aldehyde is substituted for the glycerol.

$$C_6H_5NH_2 + CH_2{=}CHCCH_3(=O) \xrightarrow[ZnCl_2]{FeCl_3} \text{4-methylquinoline}$$

(73%)
lepidine

If a saturated aldehyde is used, an initial aldol condensation occurs to give an α,β-unsaturated aldehyde that adds an aniline as in the Skraup reaction (**Döbner-Miller reaction**).

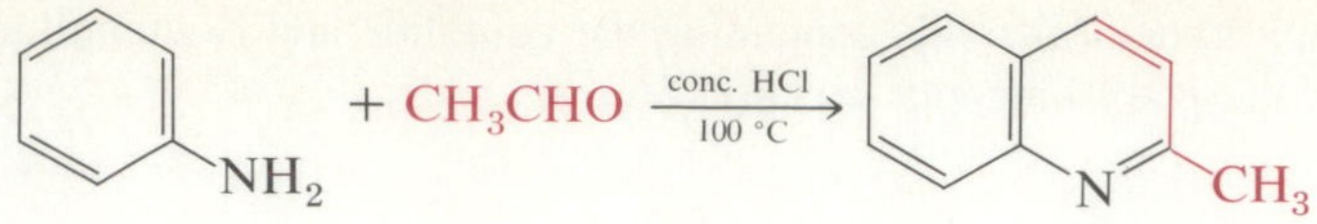

$$C_6H_5NH_2 + CH_3CHO \xrightarrow[100\ ^\circ C]{\text{conc. HCl}} \text{quinaldine (2-methylquinoline)}$$

(32%)
quinaldine

In some of these cases an oxidizing agent is not included; in these cases unsaturated reaction intermediates probably serve as oxidizing agents, but this point has not been established. The Skraup synthesis is extremely versatile; almost any desired quinoline may be prepared by using the proper combination of aniline and aldehyde, so long as the reagents will survive the hot acid conditions.

EXERCISE 32.23 Show how the Skraup synthesis can be used to prepare 6-methoxy-8-nitroquinoline. What problems do you think might exist in the application of this method for the synthesis of 5-methoxyquinoline?

A second general preparation of quinolines is the **Friedländer synthesis.** In this method an *o*-aminobenzaldehyde is condensed with a ketone.

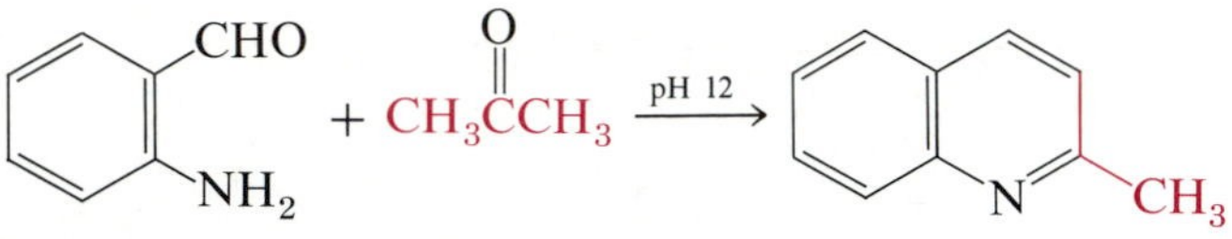

$$o\text{-}H_2NC_6H_4CHO + CH_3\overset{O}{\overset{\|}{C}}CH_3 \xrightarrow{\text{pH } 12} \text{2-methylquinoline}$$

(85%)

The Friedländer synthesis probably involves the following reaction steps.

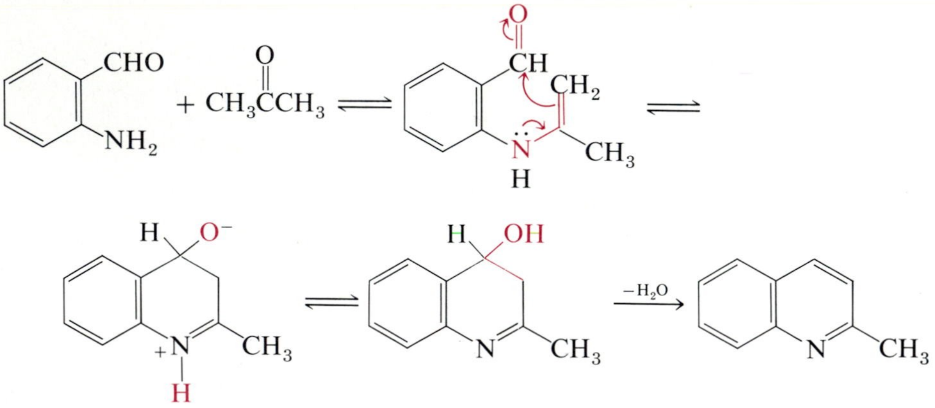

$$o\text{-}H_2NC_6H_4CHO + CH_3\overset{O}{\overset{\|}{C}}CH_3 \rightleftharpoons o\text{-}(O{=}CH)C_6H_4\ddot{N}H\text{–}C(CH_3){=}CH_2 \rightleftharpoons$$

$$\text{(H, O}^-\text{; }{+}N\text{–H, CH}_3\text{)} \rightleftharpoons \text{(H, OH; N, CH}_3\text{)} \xrightarrow{-H_2O} \text{2-methylquinoline}$$

Although substituted *o*-aminobenzaldehydes are not readily available, the parent compound is, and the reaction occurs smoothly with a variety of aldehydes and ketones. It constitutes a good method for the synthesis of quinolines substituted in the pyridine ring.

EXERCISE 32.24 Show how the Friedländer synthesis can be used to prepare the following quinoline.

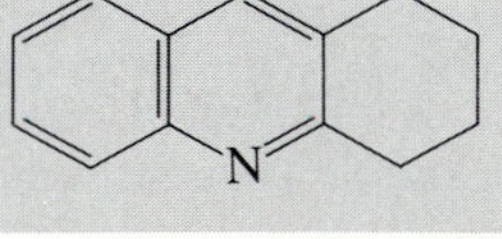

N

Isoquinolines are most easily prepared by a reaction known as the **Bischler-Napieralski synthesis.** An acyl derivative of a β-phenylethylamine is treated with a dehydrating agent to give a dihydroisoquinoline, which is dehydrogenated to the isoquinoline.

$C_6H_5CH_2CH_2NHCOCH_3 \xrightarrow[205\ °C]{P_2O_5}$ 1-methyl-3,4-dihydroisoquinoline (83%) $\xrightarrow[190\ °C]{Pd}$ 1-methylisoquinoline (83%)

(83%)

(83%)
1-methylisoquinoline

EXERCISE 32.25 *p*-Methoxybenzaldehyde is subjected to the following sequence of reactions.

$p\text{-MeOC}_6H_4CHO \xrightarrow[OH^-]{CH_3NO_2} \underset{C_9H_9NO_3}{A} \xrightarrow{H_2,\ Pd/C} \underset{C_9H_{13}NO}{B} \xrightarrow[pyridine]{C_6H_5COCl}$

$C \xrightarrow{P_2O_5,\ 200\ °C} D \xrightarrow{Pd,\ 200\ °C} \underset{C_{16}H_{13}NO}{E}$

What are the structures of compounds A through E?

C. Reactions

Quinoline and isoquinoline are considerably more reactive than pyridine in electrophilic substitution reactions. For reactions carried out in strongly acidic solution, reaction occurs on the protonated form, and substitution occurs in the benzene ring at C-5 and C-8.

quinoline $\xrightarrow[0\ °C,\ 30\ min]{HNO_3,\ H_2SO_4}$ 5-nitroquinoline + 8-nitroquinoline

(52%)
5-nitroquinoline

(48%)
8-nitroquinoline

isoquinoline $\xrightarrow[0\ °C,\ 30\ min]{HNO_3,\ H_2SO_4}$ 5-nitroisoquinoline + 8-nitroisoquinoline

(90%)
5-nitroisoquinoline

(10%)
8-nitroisoquinoline

As with pyridine N-oxide, quinoline N-oxide undergoes nitration easily; reaction occurs at C-4.

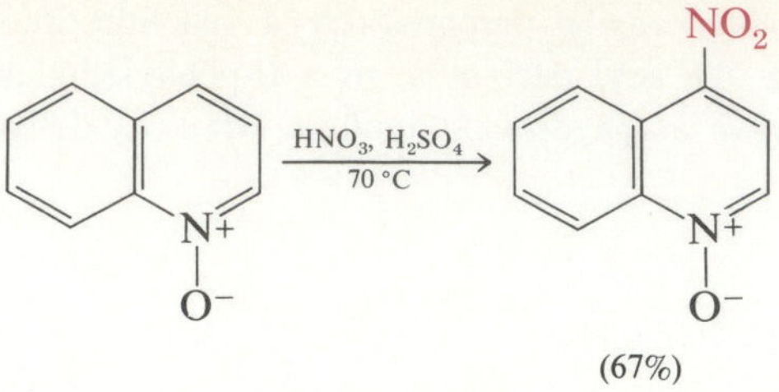

quinoline N-oxide $\xrightarrow{HNO_3,\ H_2SO_4,\ 70\ °C}$ 4-nitroquinoline N-oxide (NO_2, N^+, O^-)

(67%)

Both quinoline and isoquinoline readily undergo nucleophilic substitution reactions of the Chichibabin type.

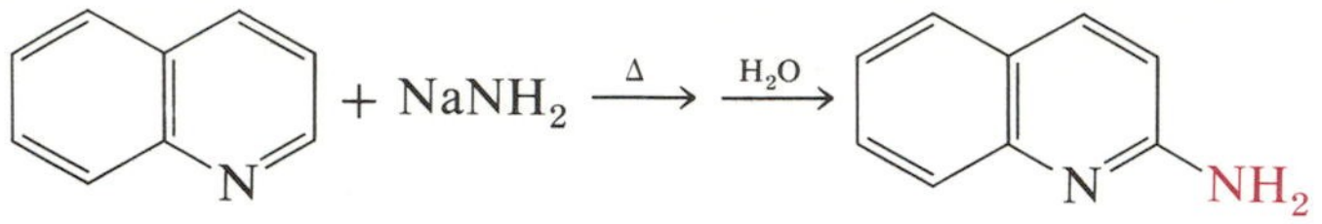

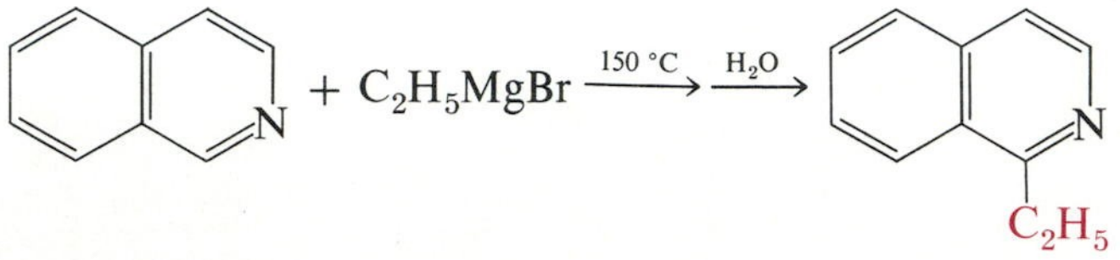

quinoline + $NaNH_2$ $\xrightarrow{\Delta}$ $\xrightarrow{H_2O}$ 2-aminoquinoline (NH_2)

isoquinoline + C_2H_5MgBr $\xrightarrow{150\ °C}$ $\xrightarrow{H_2O}$ 1-ethylisoquinoline (C_2H_5)

Like 2- and 4-alkylpyridines, 2- and 4-alkylquinolines and 1-alkylisoquinolines have α-hydrogens that are sufficiently acidic to enter into base-catalyzed reactions.

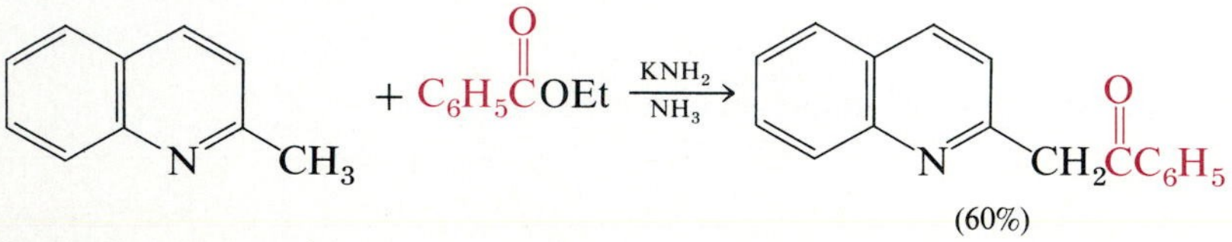

2-methylquinoline (CH_3) + $C_6H_5\overset{O}{\overset{\|}{C}}OEt$ $\xrightarrow[NH_3]{KNH_2}$ quinolin-2-yl–$CH_2\overset{O}{\overset{\|}{C}}C_6H_5$

(60%)

EXERCISE 32.26 When 1,3-dimethylisoquinoline is treated with $NaOCH_3$ in CH_3OD, the protons of the C-1 methyl group are exchanged much more rapidly than those of the C-3 methyl group. Explain.

EXERCISE 32.27 What is the expected product from quinoline in each of the following reactions:

(a) C_6H_5Li (b) CH_3I, CH_3CN (c) 2 Br_2, Ag^+, H_2SO_4
(d) 30% H_2O_2, acetic acid, 70 °C (e) $KMnO_4$, heat

32.8 Diazines

A. Structure and Occurrence

In this section, we shall take a brief look at another class of heterocycles, the diazines. The three isomeric diazabenzenes are called pyridazine, pyrimidine, and pyrazine.

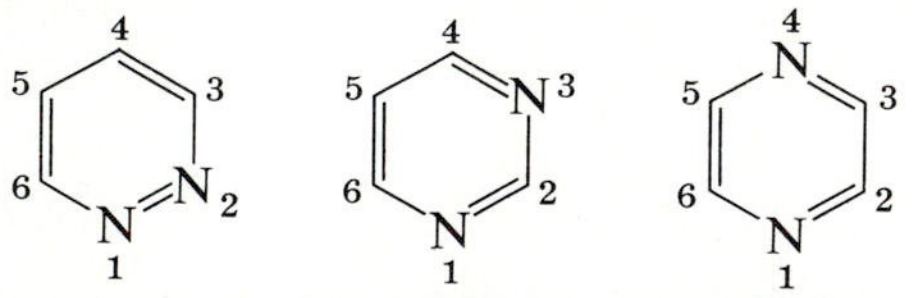

pyridazine	pyrimidine	pyrazine
b.p. 208 °C	b.p. 134 °C	b.p. 118 °C
pK_a 2.3	pK_a 1.3	pK_a 0.7

In addition to these three diazines, the bicyclic tetraaza compound, purine, is an important heterocyclic system.

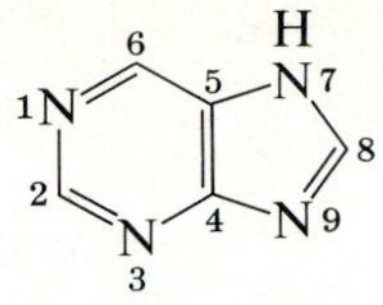

purine
m.p. 217 °C
pK_a 2.3

These ring systems, particularly those of pyrimidine and purine, occur commonly in natural products. The pyrimidines cytosine, thymine, and uracil are especially important because they are components of nucleic acids, as are the purine derivatives adenine and guanine (Chapter 33).

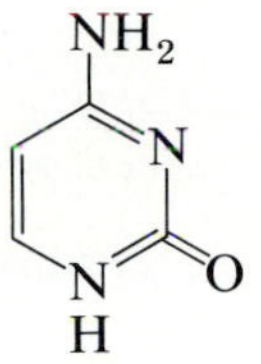

cytosine 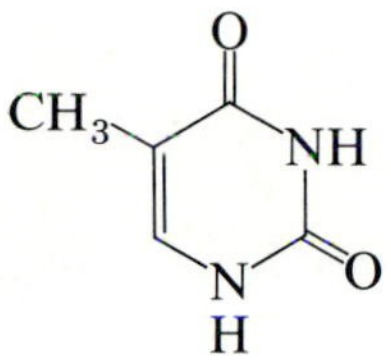
thymine 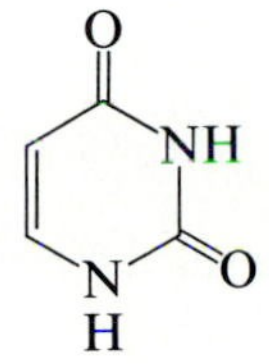
uracil

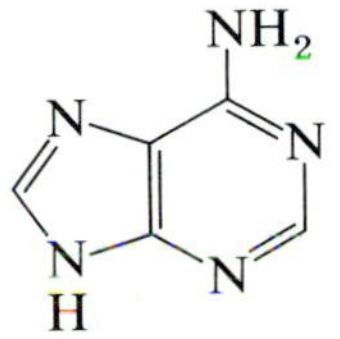

adenine
guanine 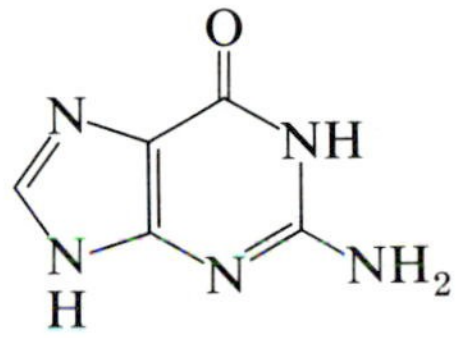

The purine nucleus also occurs in such compounds as caffeine (coffee and tea) and theobromine (cacao beans).

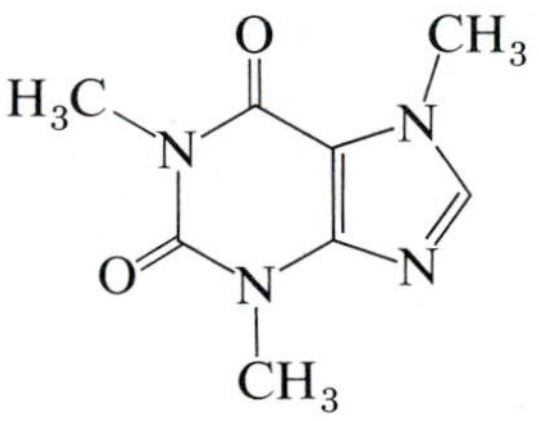

caffeine
theobromine 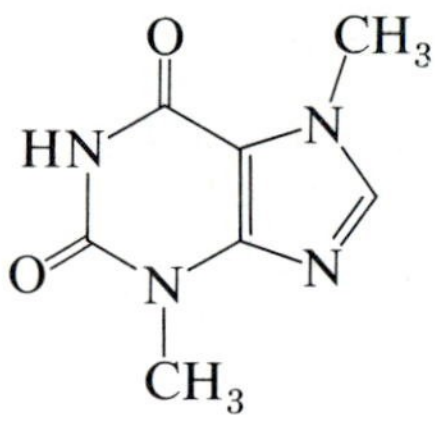

B. Synthesis

Pyridazines are prepared by the reaction of hydrazine with 1,4-dicarbonyl compounds.

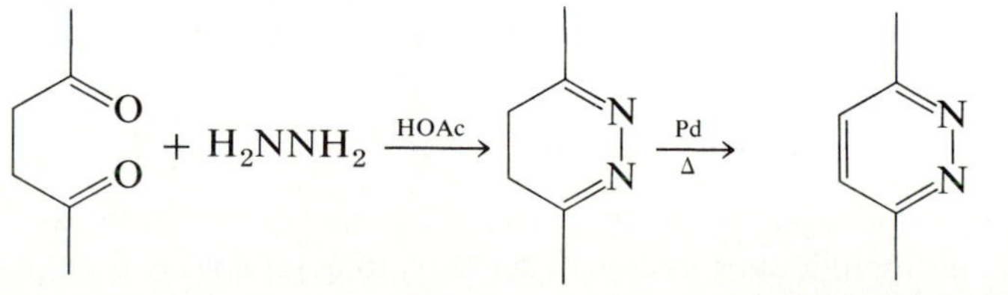

$$+ H_2NNH_2 \xrightarrow{HOAc} \xrightarrow[\Delta]{Pd}$$

3,6-dimethylpyridazine

Pyrimidines may be most easily prepared by condensation between 1,3-dicarbonyl compounds and urea or a related substance.

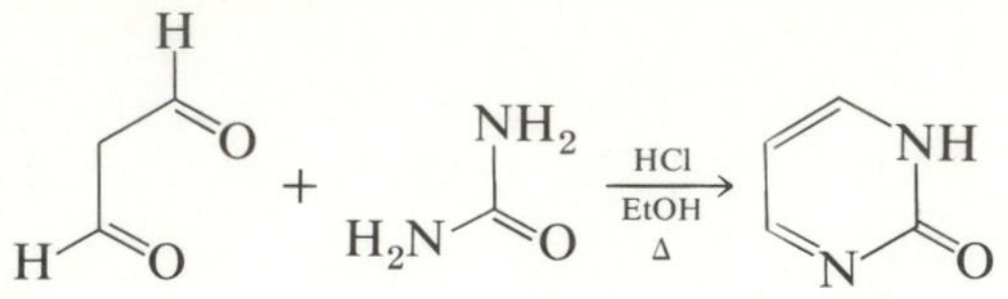

(73%)
2-pyrimidone

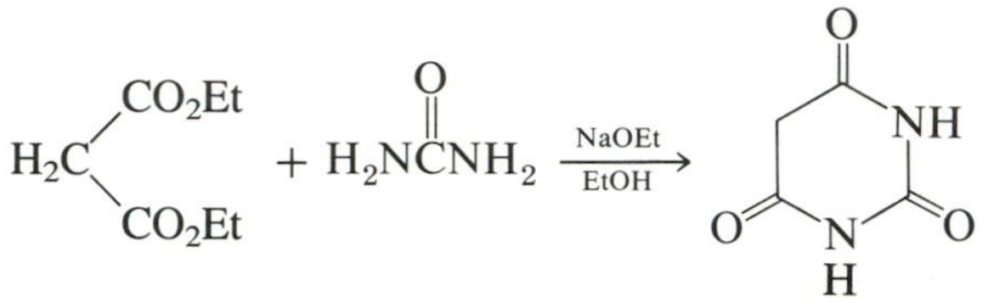

(72–78%)
barbituric acid

Note that C-2 oxygenated pyrimidines, like C-2 oxygenated pyridines, exist in the keto form.

Pyrazines result from the dimerization of α-amino carbonyl compounds. The initial dihydropyrazines can be oxidized to obtain the pyrazine.

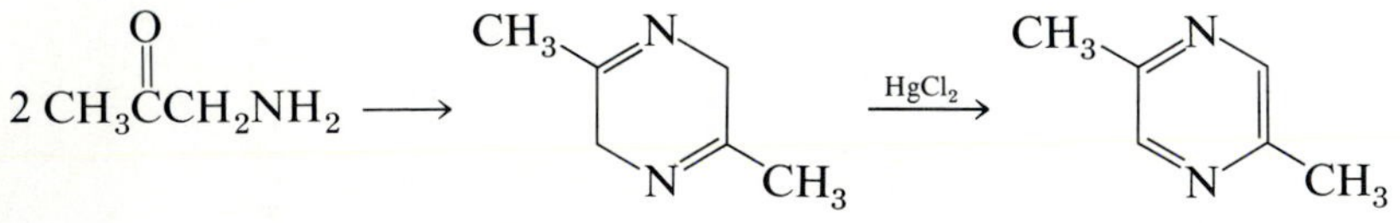

2,5-dimethylpyrazine

Pyrazines are also obtained from the condensation of 1,2-diamines with 1,2-dicarbonyl compounds. When 1,2-diaminobenzene (*o*-phenylenediamine) is used, the product is a benzopyrazine (quinoxaline). The reaction has been used as a diagnostic test for such 1,2-dicarbonyl compounds.

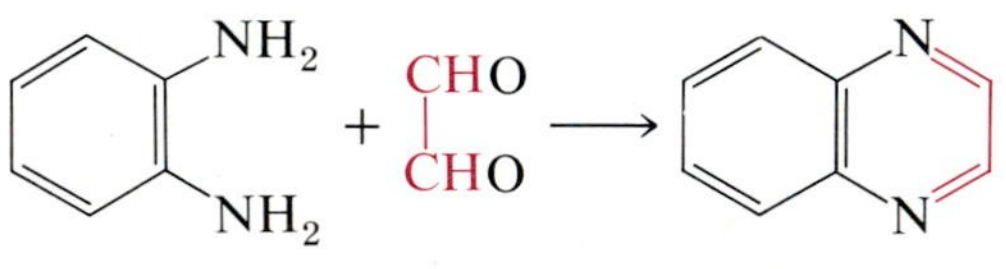

(85–90%)
quinoxaline

EXERCISE 32.28 Outline syntheses of each of the following compounds.

(a) 3,6-diphenylpyridazine
(b) 4,6-diphenyl-2-pyrimidone
(c) 2,5-diphenylpyrazine

C. Reactions

Because of the second nitrogen in the ring system, the diazines are even less reactive than pyridine toward electrophilic substitution. When activating groups are present on the ring, such substitutions may occur.

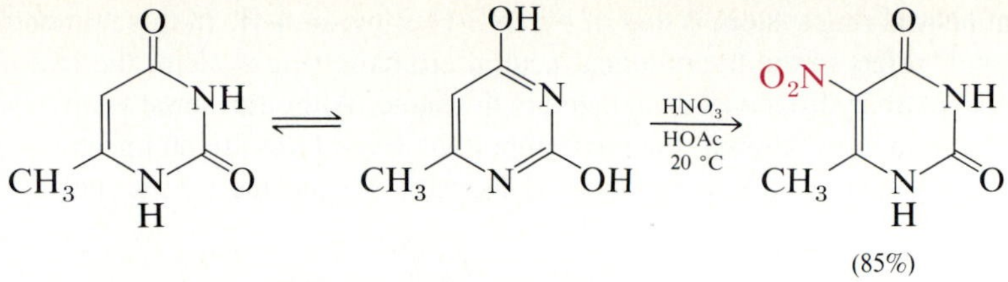

(85%)

As with quinoline and isoquinoline, attack on the benzodiazines occurs in the benzene ring.

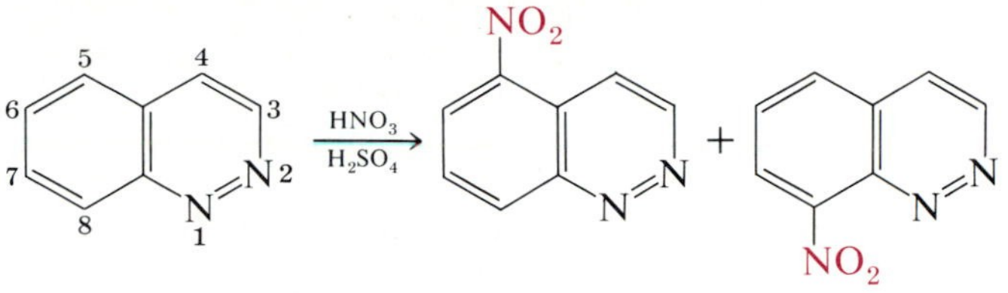

cinnoline 5-nitrocinnoline 8-nitrocinnoline

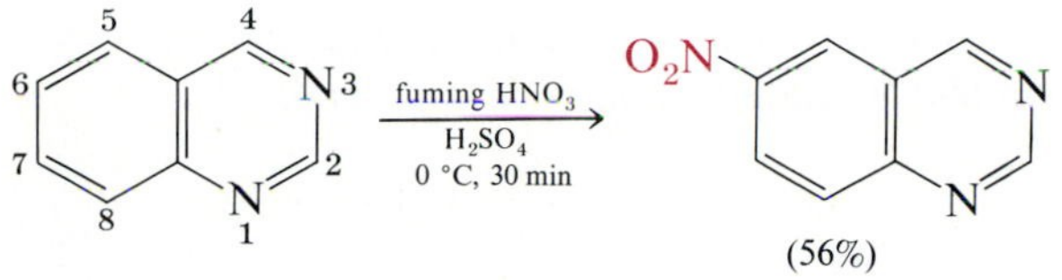

(56%)

quinazoline 6-nitroquinazoline

Many other reactions of the diazines and their benzo derivatives are similar to those observed with pyridine, quinoline, and isoquinoline. The following reactions illustrate some of these similarities.

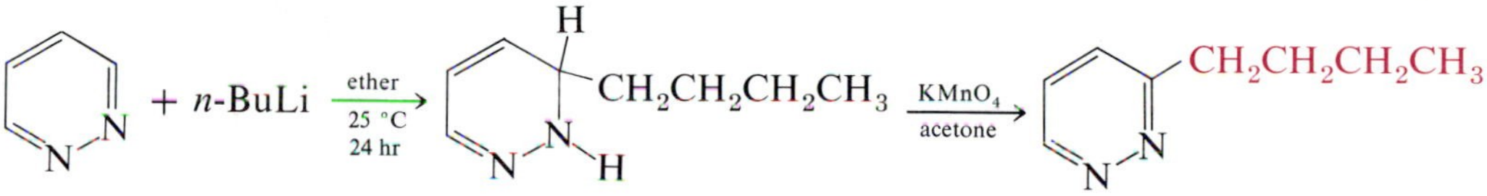

(66%)

(77%)
3-butylpyridazine

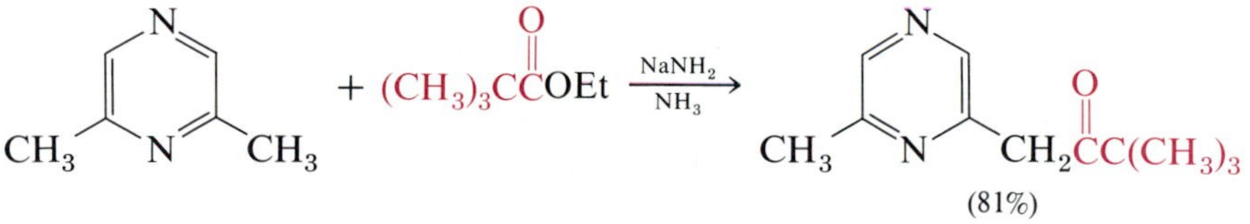

(81%)

EXERCISE 32.29 2,5-Dimethylpyrimidine reacts with benzaldehyde and alcoholic KOH to give an adduct, $C_{13}H_{12}N_2$. What is the structure of this material? Write a mechanism rationalizing its formation.

32.9 Pyrones and Pyrylium Salts

Pyrylium cations are isoelectronic with pyridines. The pyrylium cation ring system is an oxonium salt with benzenoid resonance.

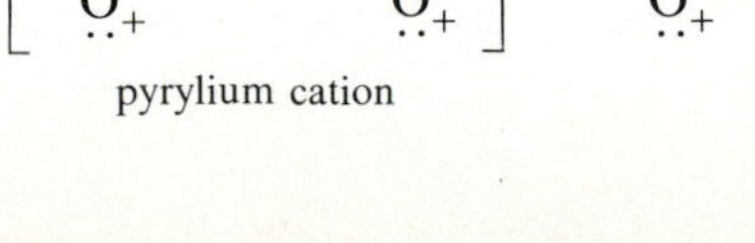

pyrylium cation

The parent neutral ring system is that of pyran-2-H or pyran-4-H. In this nomenclature the term ''pyran'' refers to the hypothetical neutral aromatic ring system; the real molecules must have an extra hydrogen as designated in the name. Although some simple derivatives of the parent pyrans are known, the most important derivatives are an unsaturated ketone, 1,4-pyrone (γ-pyrone), and an unsaturated lactone, 1,2-pyrone (α-pyrone).

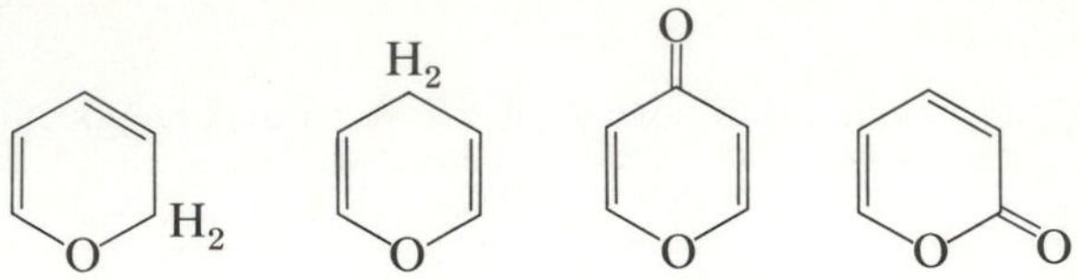

pyran-2-H pyran-4-H 1,4-pyrone γ-pyrone 1,2-pyrone α-pyrone

Pyrone and pyrylium salt structures are widespread in nature (Section 36.3.A).

A. Pyrones

Treatment of malic acid with fuming sulfuric acid produces 1,2-pyrone-5-carboxylic acid, which on heating to 650 °C decarboxylates to α-pyrone.

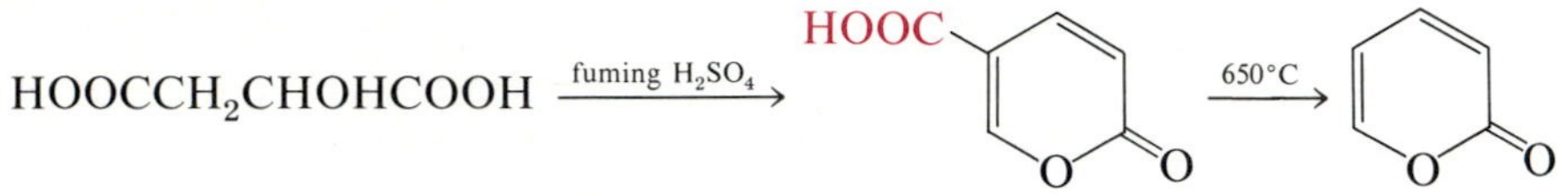

$HOOCCH_2CHOHCOOH \xrightarrow{\text{fuming } H_2SO_4}$ (HOOC) $\xrightarrow{650°C}$

malic acid (60%) coumalic acid (66–70%) α-pyrone

γ-Pyrones are available by intramolecular cyclization of 1,3,5-triketones.

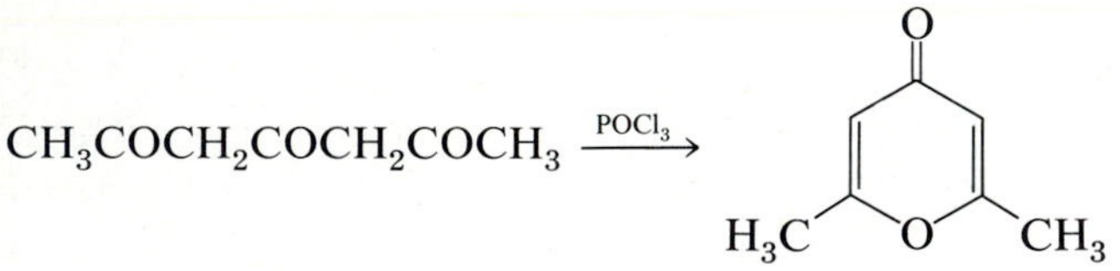

$CH_3COCH_2COCH_2COCH_3 \xrightarrow{POCl_3}$

2,6-dimethyl-1,4-pyrone

The required triketones are prepared as needed by condensation of an ester with the terminal carbon of the dianion of a β-diketone.

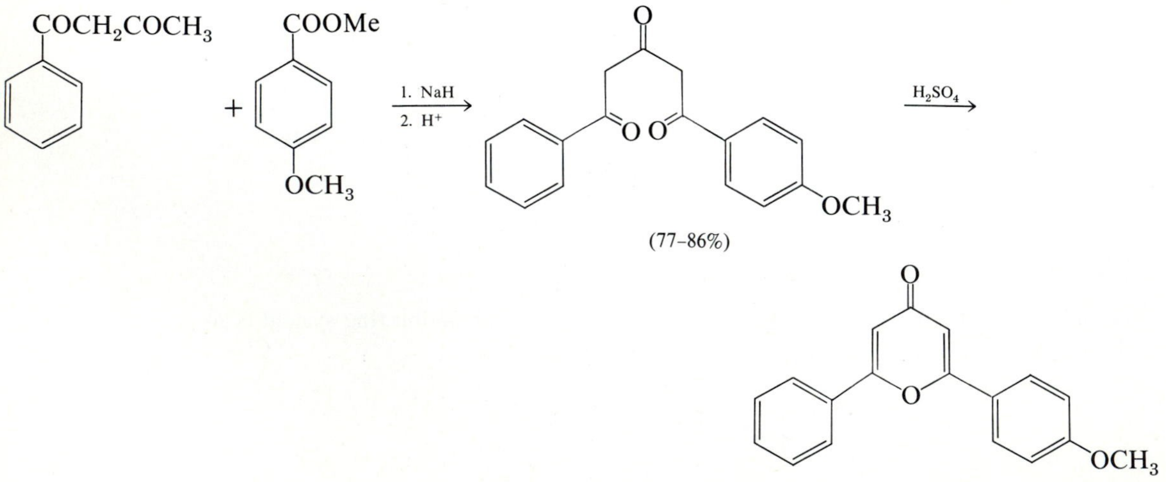

$COCH_2COCH_3$ + $COOMe$ / OCH_3 $\xrightarrow{1.\ NaH;\ 2.\ H^+}$ (77–86%) $\xrightarrow{H_2SO_4}$ (88–98%)

α-Pyrones are dienes and partake readily in Diels-Alder reactions with appropriate dienophiles. The intermediate product decarboxylates thermally, providing a synthesis of substituted benzenes.

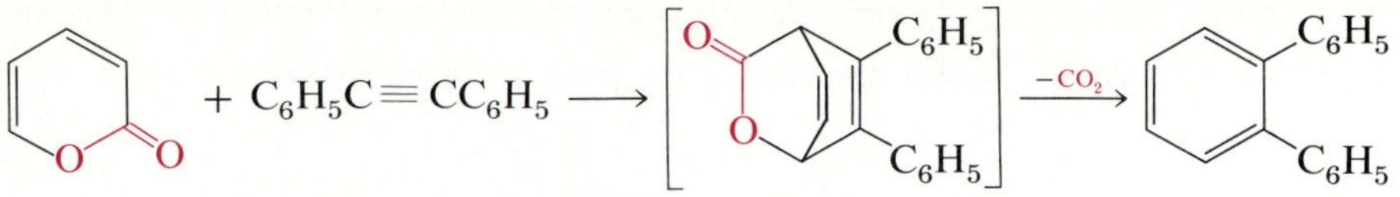

$$\text{α-pyrone} + C_6H_5C{\equiv}CC_6H_5 \longrightarrow [\text{bicyclic lactone adduct}] \xrightarrow{-CO_2} o\text{-}C_6H_4(C_6H_5)_2$$

The pyrones react with ammonia and primary amines under mild conditions to give the corresponding pyridones.

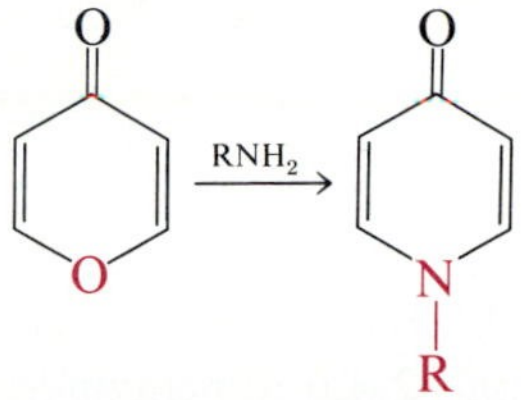

$$\text{4-pyrone} \xrightarrow{RNH_2} \text{N-R-4-pyridone}$$

Pyrones are relatively basic and salts can frequently be isolated. The pK_a of 2,6-dimethyl-1,4-pyrone is 0.4; that is, this pyrone is about as basic as trifluoroacetate ion. The conjugate acid of a pyrone can be considered as a hydroxypyrylium salt.

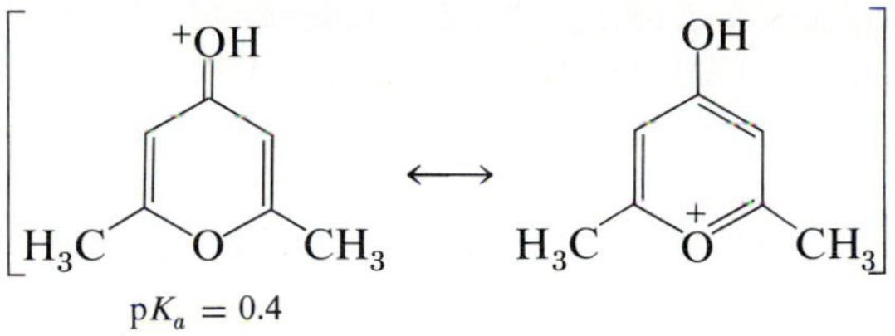

$$\left[\text{2,6-dimethyl-4-pyrone-}{}^{+}OH \longleftrightarrow \text{4-hydroxy-2,6-dimethylpyrylium}\right]$$

pK_a = 0.4

EXERCISE 32.30 The first step in the reaction of malic acid with fuming sulfuric acid is decarboxylation to give 3-oxopropanoic acid. Write a reasonable mechanism for the subsequent formation of coumalic acid.

B. Pyrylium Salts

Pyrylium salts can be prepared by treatment of pyrones with Grignard reagents. The intermediate tertiary alcohols are not isolated, but are converted directly with acid to the pyrylium cations.

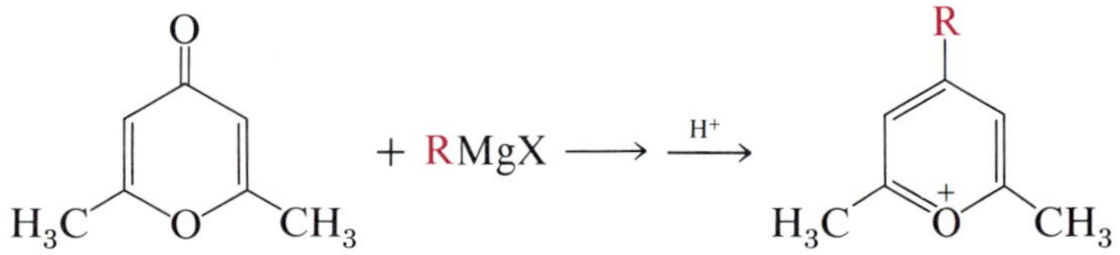

$$\text{2,6-dimethyl-4-pyrone} + RMgX \longrightarrow \xrightarrow{H^+} \text{4-R-2,6-dimethylpyrylium}$$

Alternatively, they are prepared by acid-catalyzed condensations of α,β-unsaturated ketones.

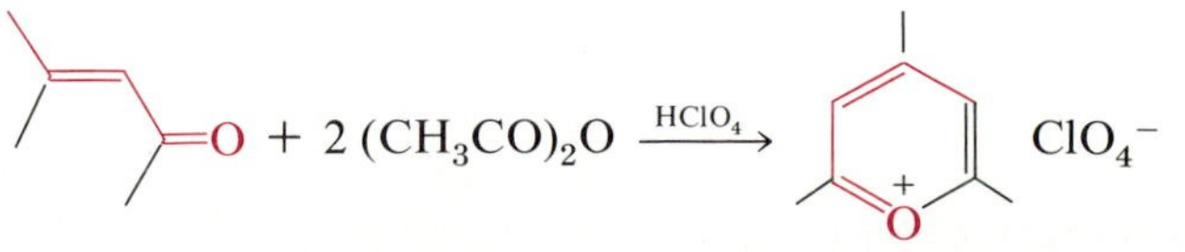

$$\text{mesityl oxide}{=}O + 2\,(CH_3CO)_2O \xrightarrow{HClO_4} \text{2,4,6-trimethylpyrylium}\ ClO_4^-$$

(54–56%)

EXERCISE 32.31 Write a reasonable reaction mechanism for the foregoing reaction. Suggest a method for the preparation of 2,4-dimethyl-6-phenylpyrylium perchlorate.

Pyrylium salts can be useful reagents. They react with nucleophilic reagents at the α-position.

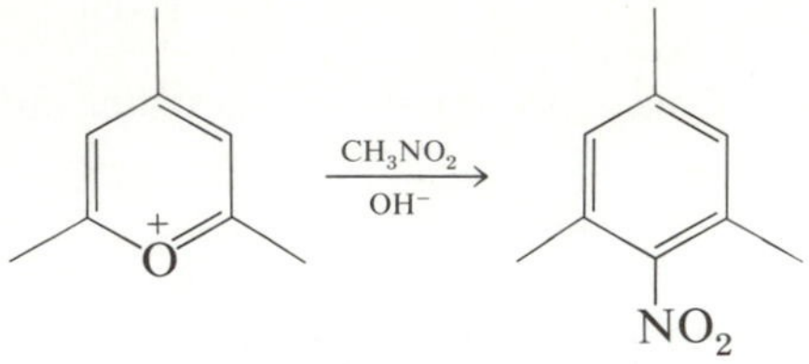

CH_3NO_2, OH^- → NO_2

EXERCISE 32.32 Write a reasonable reaction mechanism for the foregoing synthesis of 2,4,6-trimethylnitrobenzene from 2,4,6-trimethylpyrylium perchlorate. Suggest a method for the conversion of 2,6-dimethyl-1,4-pyrone into 2,6-dimethyl-4-phenylnitrobenzene.

EXERCISE 32.33 Suggest a synthesis for the compound, 2,4,6-triphenylpyridinium 2′,6′-phenolate betaine, used in the measurement of the solvent polarity parameters, the E_T-values (see Section 22.6.D). A betaine is an internal salt (Section 14.7.D), a permanent zwitterion, a generalization of the name for the compound, $(CH_3)_3N^+CH_2COO^-$, betaine, isolated from sugar beets, *Beta vulgaris*. The charge-transfer transition (Section 22.6.D) is

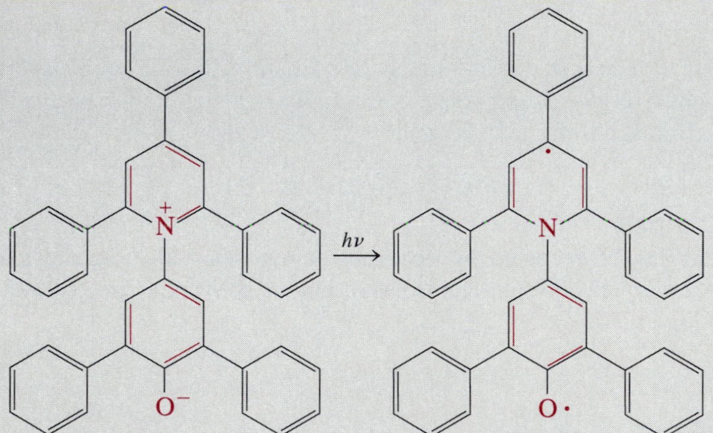

$h\nu$ →

pyridinium phenolate betaine

diradical excited state

Highlight 32.5

A table of stem suffixes illustrates the wide variety of heterocyclic compounds. The small differences between for nitrogen-containing heterocyclics and others are shown. Numbering starts at oxygen (**oxa-**), sulfur (**thia-**), or nitrogen (**aza-**) in decreasing order of preference, and proceeds in a way such that heteroatoms are assigned lowest possible numbers. (Final **a** is omitted next to a vowel). To illustrate, a three-membered ring containing two nitrogens and one double bond (the maximum) is diazirine; a saturated six-membered ring with sulfur and oxygen is oxathiane.

Stem Suffixes for Heterocyclics

	Rings with Nitrogen			Rings Without Nitrogen		
Ring	Maximum Unsaturation	One Double Bond	Saturated	Maximum Unsaturation	One Double Bond	One Saturated
3	-irine	—	-iridine	-irene	—	-irane
4	-ete	-etine	-etidine	-ete	-etene	-etane
5	-ole	-oline	-olidine	-ole	-olene	-olane
6	-ine	—	—	-in	—	-ane
7	-epine	—	—	-epin	—	-epane
8	-ocine	—	—	-ocin	—	-ocane
9	-onine	—	—	-onin	—	-onane
10	-ecine	—	—	-ecin	—	-ecane

ASIDE

Harusame ya — The spring rain
ko no ma ni miyuru — Between the trees is seen
umi no michi — A path to the sea

This classical Japanese haiku by Otsuji consists of 17 syllables, divided 5, 7, 5, and contains a reference to nature. It refers to a particular event and is happening now.

Other atoms in
a shining circle so small
the flowers take form

Our haiku, composed especially for this edition, refers to heterocyclic compounds, a class that may constitute half of all known compounds, and the changes in properties caused by the introduction of a hetero atom. Draw cyclic hydrocarbons (3–6 carbons) with one, two, or three double bonds and replace one, two, or three carbon atoms with nitrogen, oxygen, and sulfur. Name the compounds (see Highlight 32.5) and give either a synthesis or a reaction for each compound.

PROBLEMS

1. Name each of the following compounds.

a. 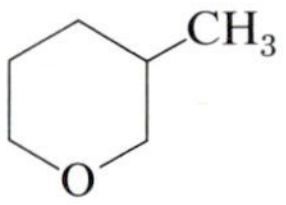 CH_3 O

b.  O N H

c. 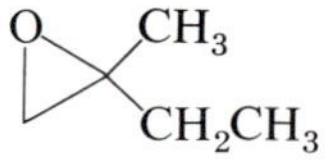 O CH_3 CH_2CH_3

d. 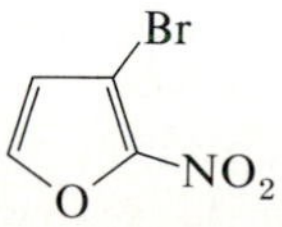 Br O NO_2

e. 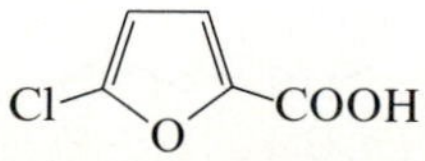 Cl O COOH

f. 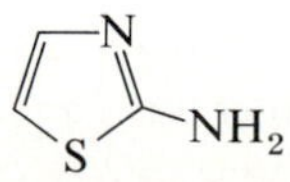 N S NH_2

g. H_3C, COOH, N, O

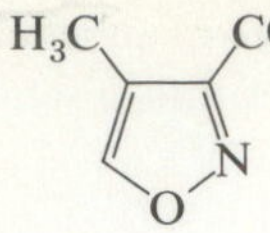

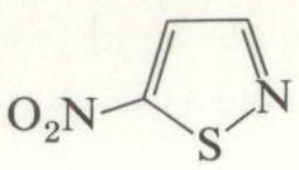

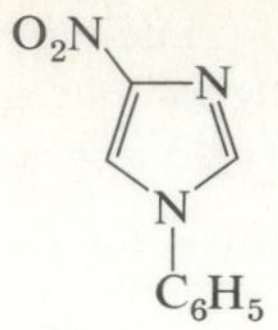

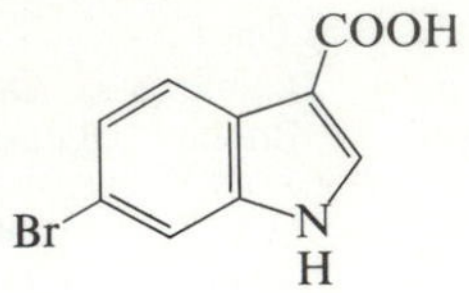

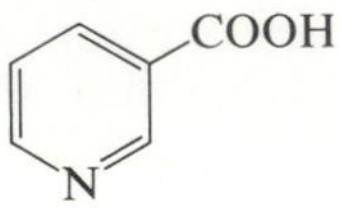

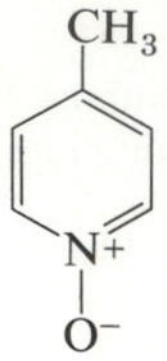

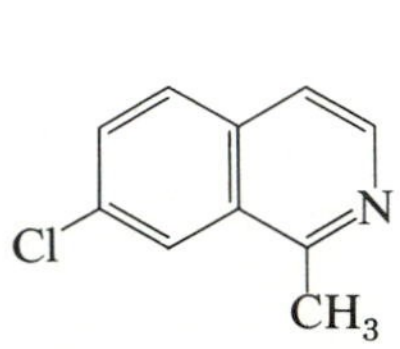

h. O_2N, S, N

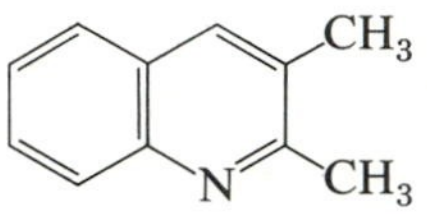

i. O_2N, N, N, C_6H_5

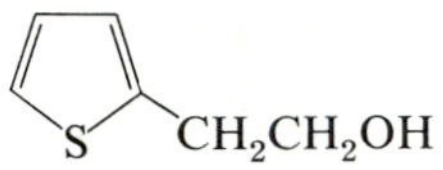

j. COOH, Br, N, H

k. COOH, N

l. CH_3, N+, O−

m. Cl, N, CH_3

n. CH_3, N, CH_3

o. CH_3, N, N, NH_2

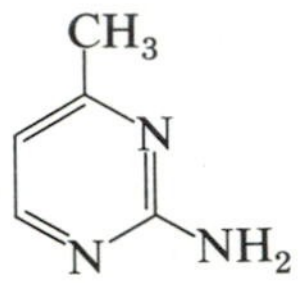

p. 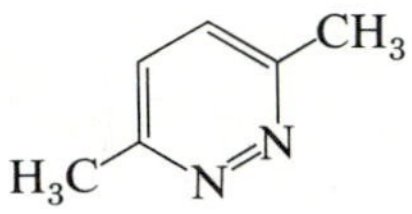 CH_3, H_3C, N=N

q. 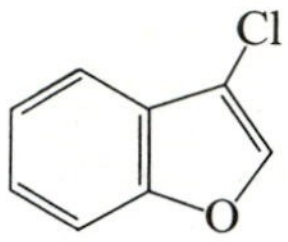 Cl, O

r. S, CH_2CH_2OH

s. O, O, OCH_3

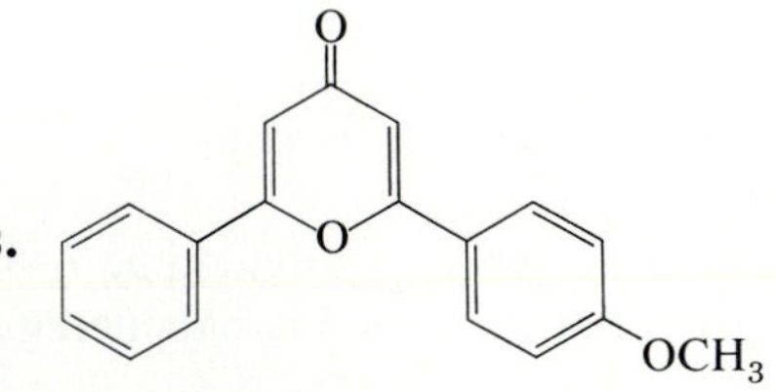

t. 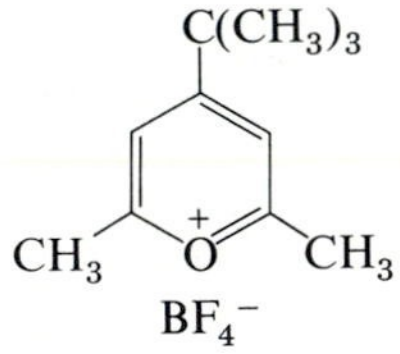 $C(CH_3)_3$, CH_3, $\overset{+}{O}$, CH_3, BF_4^-

2. Write a structure for each compound.
   a. 1,2-diphenylaziridine
   b. 2,5-dihydrofuran
   c. 1-methyl-2-pyridone
   d. 8-bromoisoquinoline
   e. 7-methyl-6-aminopurine
   f. 2-aminopurine
   g. (3-indolyl)acetic acid
   h. 4-nitroquinoline-1-oxide
   i. 4-chlorothiophene-2-carboxylic acid
   j. 2-methyl-5-phenylpyrazine
   k. 5-nitroquinoline-2-carboxylic acid
   l. 2-nitrothiazole
   m. 3-cyanoisoxazole
   n. 4,6-dimethyl-1,2-pyrone

3. Outline a synthesis for each of the following compounds.

a. 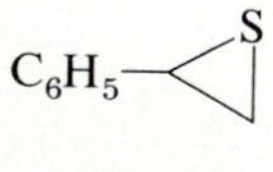 C_6H_5, S

b. H, O, CH_3, C_6H_5, H

c. H, N

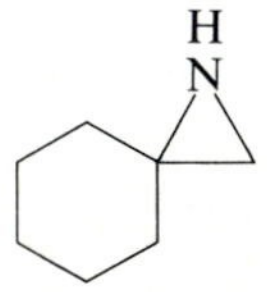

d. 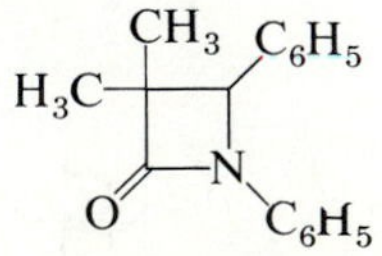 CH_3, C_6H_5, H_3C, N, O, C_6H_5

e. 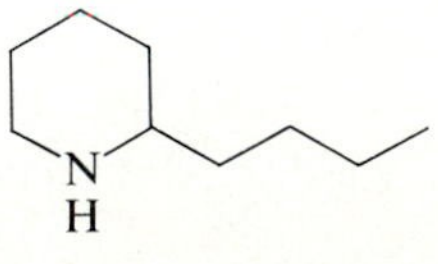 N, H

f. 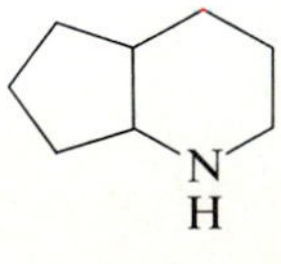 N, H

4. Outline a synthesis for each of the following compounds, starting from non-heterocyclic precursors.

a. 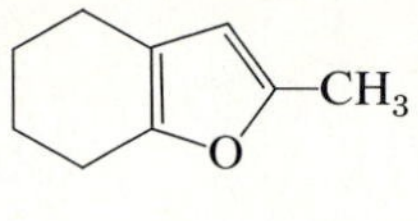 CH_3

b. 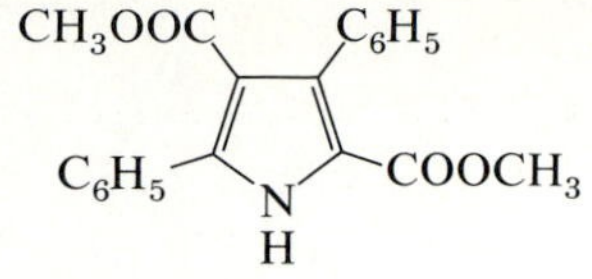 CH_3OOC C_6H_5 C_6H_5 $COOCH_3$ N H

c. CH_3 Cl N H

d. CH_3 N H CH_3O

e. CH_3 CH_2CH_3 N H

f. C_6H_5 C_6H_5 N O

g. CH_3 C_6H_5 N N H

h. CH_3OOC C_6H_5 CH_3OOC N O

i. H_3C $COOCH_3$ N N H

j. N C_6H_5 CH_3 N H

k. CH_3 N CH_3

l. N

m. CH_3 N

n. CH_3 N CH_2CH_3

o. N

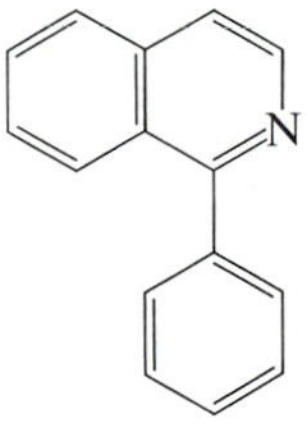

p. 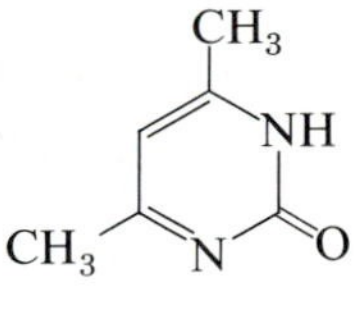 CH_3 NH CH_3 N O

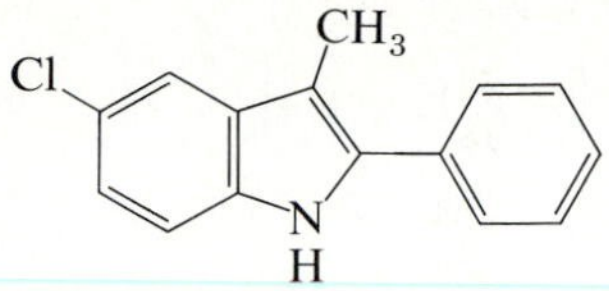

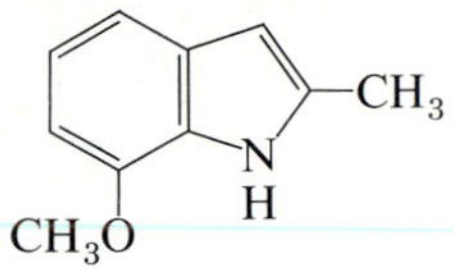

q. 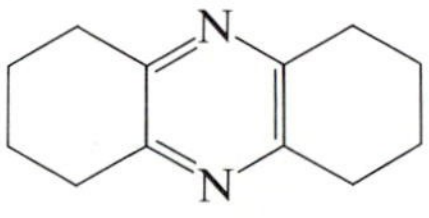 N N

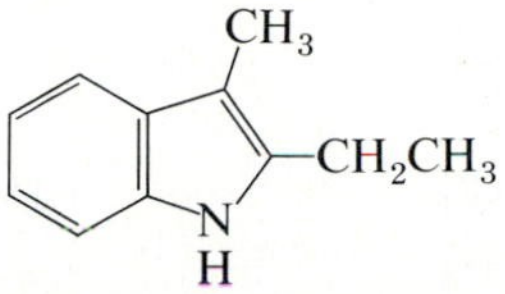

r. 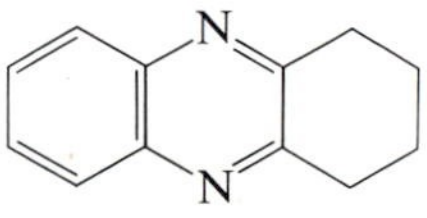 N N

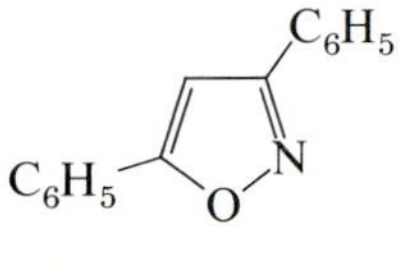

s. 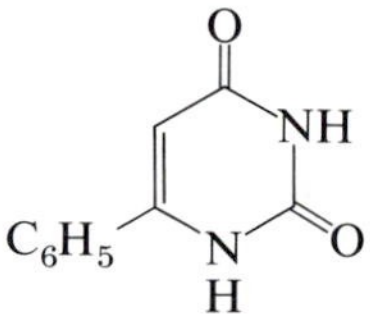 O NH C_6H_5 N O H

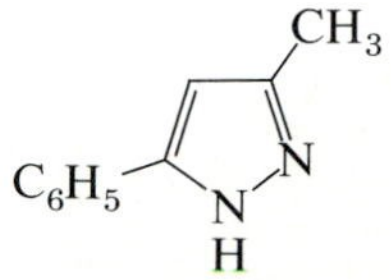

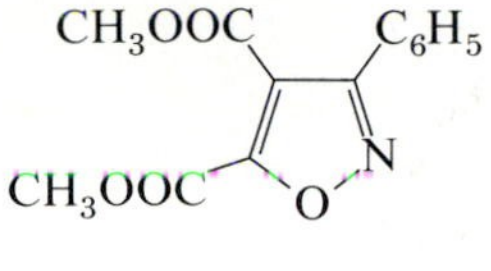

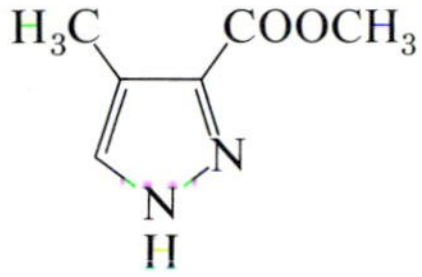

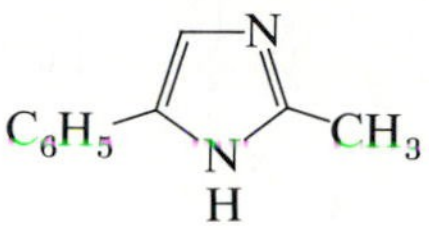

t. 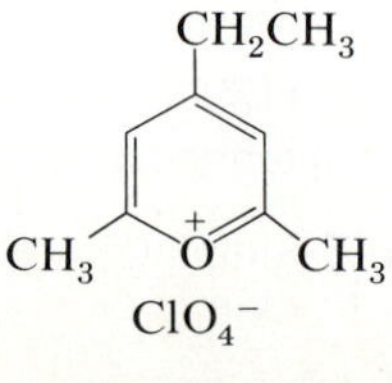 CH_2CH_3 CH_3 $\overset{+}{O}$ CH_3 ClO_4^-

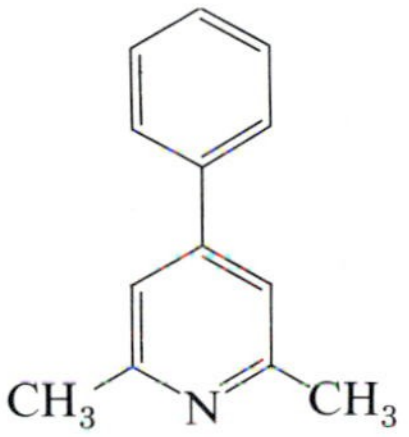

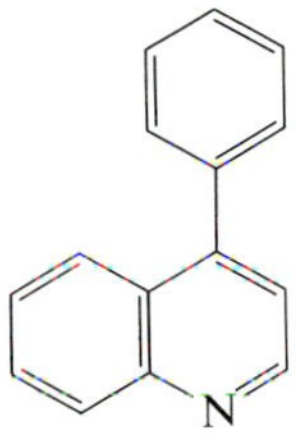

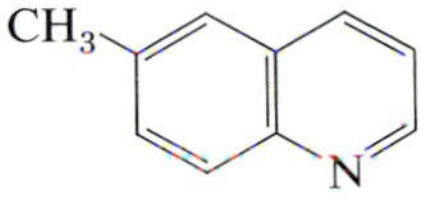

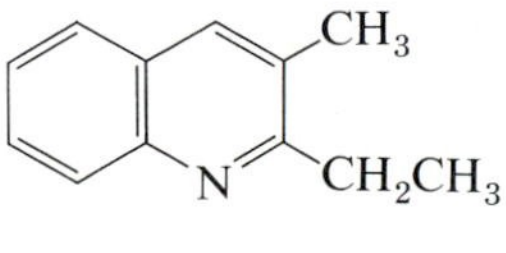

u. 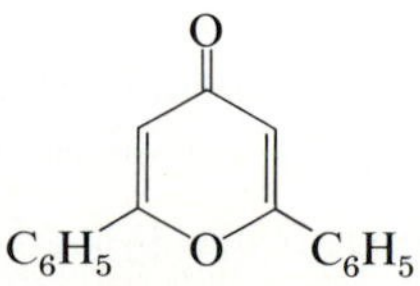 O C_6H_5 O C_6H_5

5. Outline a synthesis for each of the following compounds from the corresponding unsubstituted or alkyl-substituted heterocyclic system.

a.

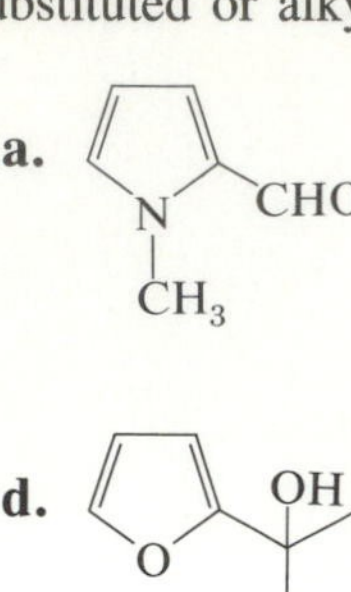

b.

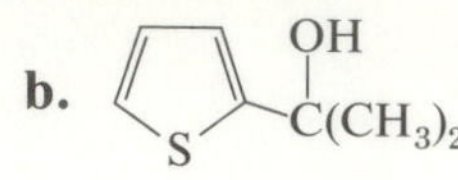

c. C_6H_5 (N)

d.

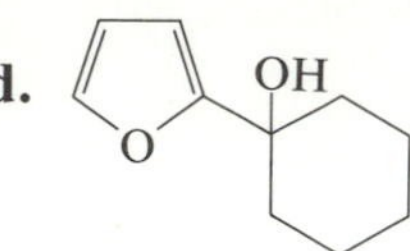

e. (O, O)

f. D (O)

g. Br (N, H)

h. NO_2 (N)

i. NO_2, CH_3 (N)

j. $NHCH_3$ (N)

k. O, CH_2CH_3 (N)

l. NO_2, CH_3 (N)

m. CH=CH (N, O)

n. NO_2 (N, N)

6. Write a reasonable mechanism that explains the following reaction.

$$\text{2-pyridyl}-CH{=}CH_2 + CH_2(COOCH_3)_2 \xrightarrow[CH_3OH]{NaOCH_3} \text{2-pyridyl}-CH_2CH_2CH(COOCH_3)_2$$

Would you expect the analogous reaction to occur with 3-vinylpyridine?

7. *o*-Aminobenzaldehyde is a useful starting material in the Friedländer synthesis of quinolines.

a. Propose a synthesis of *o*-aminobenzaldehyde.

b. Use *o*-aminobenzaldehyde in the synthesis of the following compounds.

(N) (COOEt, CH_3, N) (N)

c. The mechanism of the Friedländer synthesis given on page 1120 was abbreviated. Write out the complete mechanism, showing all of the intermediates involved.

8. The pyridine ring is so inert that Friedel-Crafts reactions fail completely. Suggest a method to synthesize phenyl 3-pyridyl ketone.

9. Predict the major product from each of the following reactions.

a. 2-nitro-5-methylthiophene $\xrightarrow[H_2SO_4]{HNO_3}$

b. 3-nitrothiophene $\xrightarrow[HOAc]{Br_2}$

c. 3-methylthiophene $\xrightarrow[H_2SO_4]{HNO_3}$

d. 3-methylindole $\xrightarrow[HOAc]{Br_2}$

e. 2,3-dimethylpyridine + benzaldehyde $\xrightarrow{KOH}$

f. 1-methylimidazole $\xrightarrow[H_2SO_4]{HNO_3}$

g. $CH_3C_6H_4NHN{=}C(CH_3)_2$ (meta) $\xrightarrow[\Delta]{PPA}$

10. Write a reasonable mechanism for the following reaction.

2-chloropyridine N-oxide $\xrightarrow[CH_3OH, \text{ reflux}]{NaOCH_3}$ 2-methoxypyridine N-oxide

11. N-Alkylated α-pyridones may be prepared by the oxidation of N-alkylpyridinium salts in basic medium; the usual oxidant is Fe^{3+}, in the form of potassium ferricyanide.

4-($COOCH_3$)-1-methylpyridinium $\xrightarrow[KOH, H_2O]{K_3Fe(CN)_6}$ 4-($COOCH_3$)-1-methyl-2-pyridone (96%)

Propose a reasonable mechanism for this reaction that accounts for the fact that hydroxide ion is necessary.

12. Pyridine N-oxide reacts with benzyl bromide to give N-benzyloxypyridinium bromide. Treatment of this salt with strong base gives benzaldehyde (92%) and pyridine. Rationalize with a reasonable mechanism.

13. Write a mechanism, showing all steps, that explains the following reaction.

$$CH_3CH_2\overset{O}{\overset{\|}{C}}CH{=}CHCl + H_2NOH \xrightarrow[\Delta]{EtOH}$$ 3-ethylisoxazole (60%) + 5-ethylisoxazole (40%)

14. Pyrrole reacts with ethylmagnesium bromide, followed by methyl iodide, to give a mixture of 2- and 3-methylpyrrole. Rationalize this result, using resonance structures where desirable.

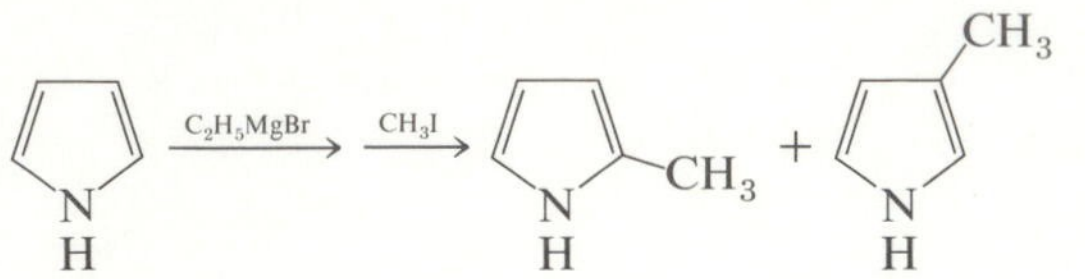

15. Treatment of 4-butyl-5-methylisoxazole with potassium *t*-butoxide, followed by mild acidic workup, provides 3-cyano-2-heptanone. Write a reasonable mechanism for this reaction. Show how the reaction can be used as one step in the conversion of cyclohexanone into 2-cyano-2-methylcyclohexanone.

16. Write a mechanism for the following reaction in which a furan is produced.

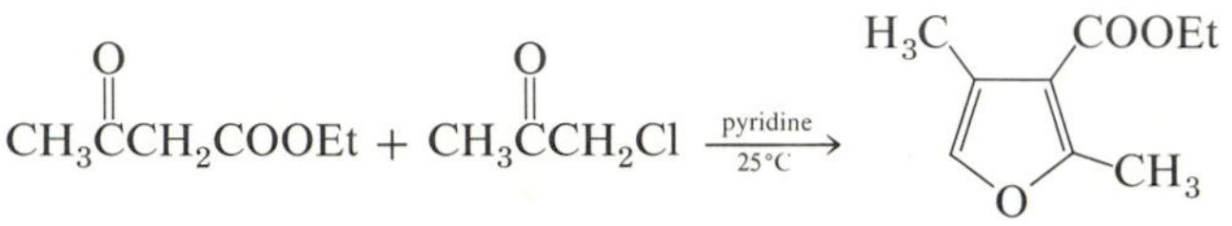

$$CH_3\overset{O}{\overset{\|}{C}}CH_2COOEt + CH_3\overset{O}{\overset{\|}{C}}CH_2Cl \xrightarrow[25^\circ C]{\text{pyridine}}$$

17. Heats of formation, ΔH_f°, for pyridine and piperidine are + 34.6 and −11.8 kcal mole^{-1}, respectively. Before these data can be used to estimate the empirical resonance energy of pyridine, we need a value for the heat of hydrogenation of a carbon-nitrogen double bond. Data for several compounds suggest a value of −21 kcal mole^{-1} for ΔH° for the reaction

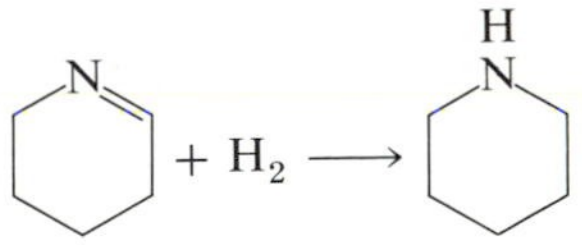

$$+ H_2 \longrightarrow$$

a. Use this information together with corresponding results for the heat of hydrogenation of cyclohexene (Appendix I) and derive an empirical resonance energy for pyridine.

b. An alternative method for calculating the empirical resonance energy is to compare the experimental heat of atomization with that obtained by use of a table of average bond energies, such as that in Appendix III. Compare the value you calculate by this method with the commonly quoted value for the resonance energy of pyridine of 23 kcal mole^{-1}. To get some insight into the source of the discrepancy, compare the calculated and observed heats of atomization of piperidine. How accurate are the results expected from the use of average bond energies?

18. Dipole moments of furan, thiophene, and pyrrole were discussed in Section 32.3.A, and the assignments of directions of the dipoles were presented. Given the following dipole moment data, deduce the directions assigned on page 1091.

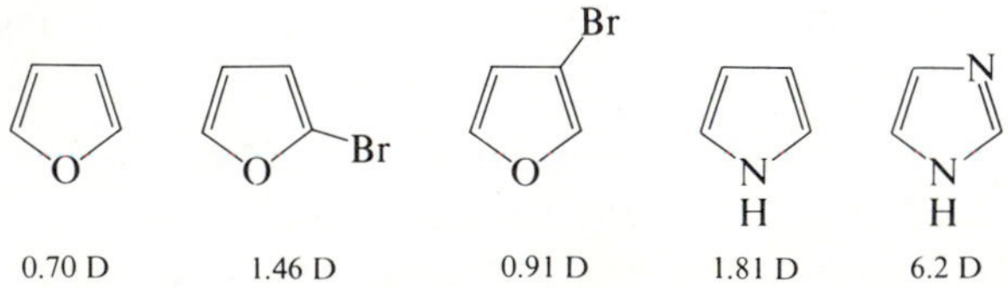

furan	2-bromofuran	3-bromofuran	pyrrole	imidazole
0.70 D	1.46 D	0.91 D	1.81 D	6.2 D

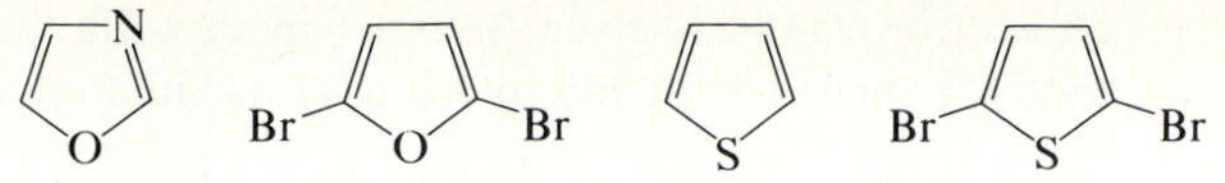

2.8 D 1.63 D 0.51 D 1.1 D

19. Given the following dipole moments, deduce the direction of the dipole moment of pyridine. Compared to the dipole moment of piperidine, is this direction reasonable?

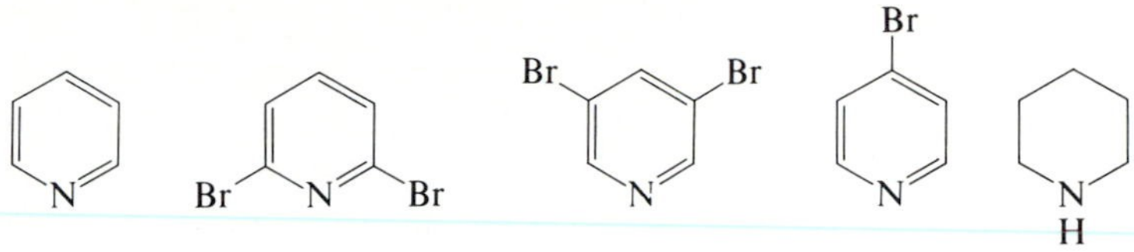

2.27 D 3.5 D 1.0 D 0.8 D 1.18 D

20. Umbelliferone is a coumarin derivative present as a glucoside in many plants. It is used commercially as a sun-screen in lotions. Show how it may be synthesized from resorcinol.

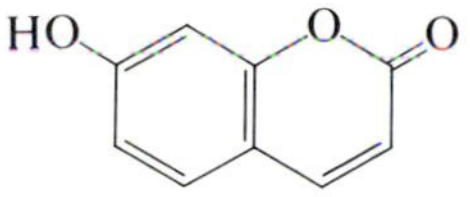

21. Papaverine, $C_{20}H_{21}NO_4$ is an alkaloid present in opium and is used as muscle relaxant. It is nonaddicting, but is classified as a narcotic. Reaction with excess hydriodic acid gives 4 moles of CH_3I and shows the presence of four CH_3O groups (Zeisel determination). Oxidation with $KMnO_4$ gives first a ketone, $C_{20}H_{19}NO_5$; continued oxidation gives a mixture from which the compounds shown below were isolated and identified. Deduce the structure of papaverine, and interpret the reactions described.

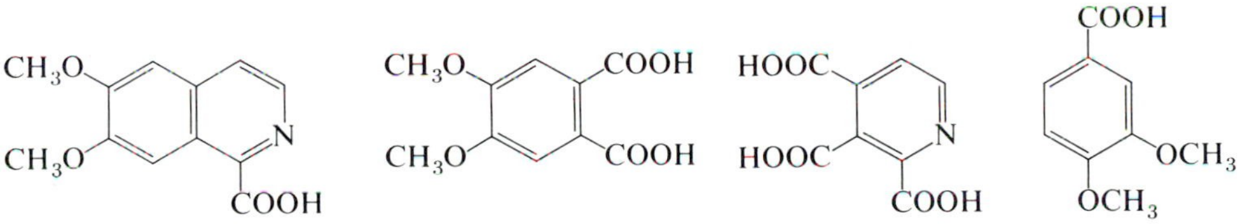

22. Write a mechanism for the base-promoted chlorination of an oxime to give a hydroxamic acid chloride (page 1105).

23. Azulene is an isomer of naphthalene that is characterized by its brilliant blue color. 4,6,8-Trimethylazulene is formed in 43–49% yield when 2,4,6-trimethylpyrylium perchlorate is treated with cyclopentadienylsodium in THF. Write a reasonable reaction mechanism for this reaction.

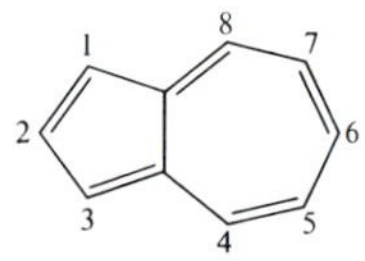

azulene

24. 1-Methyl-4-methoxycarbonylpyridinium iodide (see problem 11) is reduced by sodium amalgam in acetonitrile in the absence of oxygen to a beautiful green species that can be distilled in pure form from the reaction mixture. The product reacts with bromochloromethane to yield two products, 1-methyl-4-methoxycarbonylpyridinium bromide and a mixture of dihydropyridines, each of which contains one chlorine atom. The rate of the reaction is insensitive to solvent polarity. Describe the course of the reaction and deduce the mechanism from the facts given. (*Hint*: 1-methyl-3,5-dicyanopyridinium iodide exists mainly as the 1-methyl-3,5-dicyano-4-iodo-1,4-dihydropyridine.)

25. Rank the following three heterocyclic amines in terms of increasing basicity: pyridine, imidazole, pyrrole. Explain the relative basicities using resonance structures.

26. Using resonance structures, explain why quinoline undergoes electrophilic aromatic substitution more at C-5 and C-8 than at the other positions.

CHAPTER 33

MOLECULAR RECOGNITION: NUCLEIC ACIDS AND SOME BIOLOGICAL CATALYSTS

33.1 Introduction

The interaction between molecules is often geometrically rather specific. Even when the interactions are based primarily on mutual polarizability, as in the case of alkanes, the interactions are sufficiently precise that crystals of these compounds have definite structures. Such specificity is especially apparent when the intermolecular interactions are based on hydrogen bonding or attractions between opposite charges. An example is seen in the mutual interaction of a lone pair on a carbonyl oxygen with the proton of the OH group of another carboxylic acid. As was explained in Section 17.1, this reciprocal hydrogen bonding leads to a dimeric structure for carboxylic acids.

$$2\ R{-}C(=O)OH \rightleftharpoons R{-}C(=O\cdots H{-}O)(O{-}H\cdots O{=})C{-}R$$

In one sense, this kind of intermolecular interaction is quite specific, to the degree that carboxylic acids even exist in the dimeric form in the vapor phase under certain conditions. On the other hand, the interaction is not specific with regard to the nature of the R group of the carboxylic acid, since all carboxylic acids behave qualitatively in the same way. In a case like this, when two molecules form relatively stable complexes because of specific interactions, the molecules are said to ''recognize'' each other and the phenomenon is called **molecular recognition.**

Another example of molecular recognition that we have encountered previously is that of crown ethers for certain metal cations (Section 10.11.B). For example, 18-crown-6 "recognizes" potassium ions because its cavity is just the right size for optimum interaction of the ether lone pair electrons with the positive potassium ion.

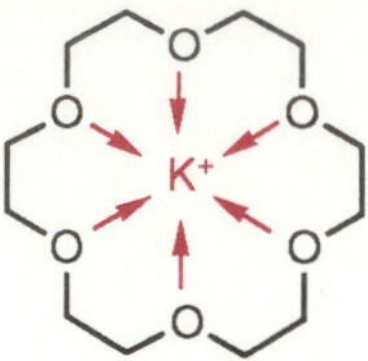

The metal cation is spherical and the interactions accordingly have little directional character, but the distance between cation and lone pairs is particularly important. As a result, 18-crown-6 preferentially binds K^+ and can "select" it from a mixture of Li^+, Na^+, K^+, Rb^+, and Cs^+. In one type of popular terminology, this kind of special intermolecular interaction is called a **host-guest** interaction. The binding sites of the host converge or turn inward, whereas those of the guest diverge or turn outward. In this case, 18-crown-6 is a "host" and potassium ion is a "guest."

When two molecules interact to form a complex, such as the carboxylic acid dimers or the crown ether–cation complexes, the process is characterized by an association constant, K, which is simply the equilibrium constant for complex formation. That is, for the reaction of A + A to give A—A, or A + B to give A—B, the association constants are

$$K_a = \frac{[\text{A—A}]}{[\text{A}]^2} \quad \text{and} \quad K_a = \frac{[\text{A—B}]}{[\text{A}][\text{B}]}$$

As with any equilibrium constant, K_a is related by the basic equation of thermodynamics to $\Delta G°$, and this quantity is referred to as the **standard free energy of binding** of the guest by the host.

$$\Delta G° = -RT \ln K_a$$

The free energy of binding is a function of the **enthalpy of binding, $\Delta H°$,** which results from the strengths of the various bonds involved in complex formation (hydrogen bonding, electrostatic interaction, van der Waals interactions, etc.), and the **entropy of binding, $\Delta S°$,** which is related to the change in the amount of order that is experienced by the two partners in the process of complex formation.

$$\Delta G° = \Delta H° - \text{T}\Delta S°$$

When two species combine to form one, the entropy change is negative and the quantity $-\text{T}\Delta S°$ is therefore positive. Thus, in order for complex formation to be favored (that is, in order for $\Delta G°$ to be negative), the enthalpy of binding must be sufficiently negative to compensate for the entropy that is lost in the process. Furthermore, the magnitude of $\Delta G°$ must be large enough to overcome any interactions that the associating molecules must sacrifice in order to form a complex with each other. These include interactions between the associating molecules and the solvent or interactions with other solute molecules. In a nonpolar or weakly interacting solvent such as cyclohexane, one polar group in a molecule is often enough to cause self-association. Thus, acetic acid forms a dimer in cyclohexane solution. However, in the more polar solvent water, the dimer is not formed because water molecules interact strongly with the monomer and the magnitude of the free energy of binding of two acetic acid molecules to each other is not sufficient to overcome these interactions. When two molecules prefer to associate with each other, even if it means the sacrifice of other interactions, the association is said to be **specific.** Such specificity in molecular recognition usually depends on there being more than one interaction between A and A or A and B. That is, two or three groups in each molecule with the appropriate geometry are generally needed for specificity.

EXERCISE 33.1 Calculate the standard free energy of binding for the following values of K_a: (a) 10 M^{-1}; (b) 5×10^3 M^{-1}; (c) 8×10^9 M^{-1}.

EXERCISE 33.2 For the association of two species to give one, the entropy is usually in the range −10 entropy units (e.u.) to −20 e.u. (The normal units of e.u. are kcal deg^{-1} $mole^{-1}$). Assuming a value of $\Delta S°$ of −16 e.u. and an enthalpy of binding of −5 kcal $mole^{-1}$, calculate the standard free energy of binding at room temperature (25 °C, 298 K) of A and B to give a complex A—B. What is the standard free energy of binding for this process at the normal physiological temperature of 37 °C?

33.2 Recognition of Guests by Synthetic Hosts

An interesting example of molecular recognition that is related to both the carboxylic acid dimers and the crown ether–metal ion complexes is seen in a series of dicarboxylic acids that have recently been synthesized. *cis,cis,cis*-1,3,5-Trimethyl-1,3,5-cyclohexanetricarboxylic acid ("Kemp's triacid") reacts with 1,3-diaminobenzene to give a polyfunctional product with the structure shown below. The geometry of this molecule is such that the two carboxy groups are just the right distance apart to form strong reciprocal hydrogen bonds, as in the case of the carboxylic acid dimers discussed previously.

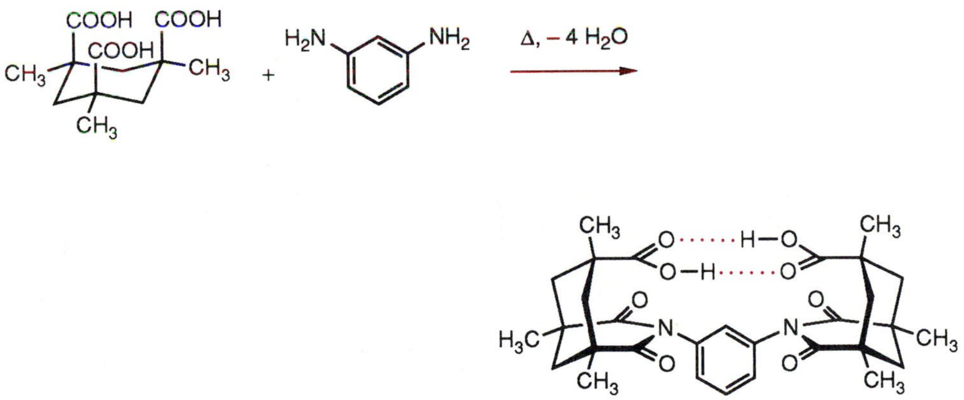

COOH COOH COOH CH_3 CH_3 CH_3 + H_2N NH_2 Δ, − 4 H_2O →

CH_3 O······H—O CH_3 O—H······O O O H_3C N N CH_3 O O CH_3 CH_3

In this compound the two carboxy groups interact with each other even more strongly than they do in a normal carboxylic acid dimer. This is because it is not necessary to bring two different molecules together in order to form the reciprocal hydrogen-bonding arrangement. Since the two functional groups are in the same molecule, it is only necessary to rotate about several C—C bonds in order to achieve the correct conformation for maximum interaction of the functional groups. The diacid is sufficiently soluble in water that its ability to complex metal ions can be measured under homogeneous conditions. The dianion strongly binds Ca^{2+} ion, with $K = 2 \times 10^5$ M^{-1} ($\Delta G° = 7.3$ kcal $mole^{-1}$), compared with the value of 7×10^3 M^{-1} ($\Delta G° = 5.3$ kcal $mole^{-1}$) for iminodiacetic acid, $HN(CH_2COOH)_2$. The diacid is thus a **receptor** for calcium ion.

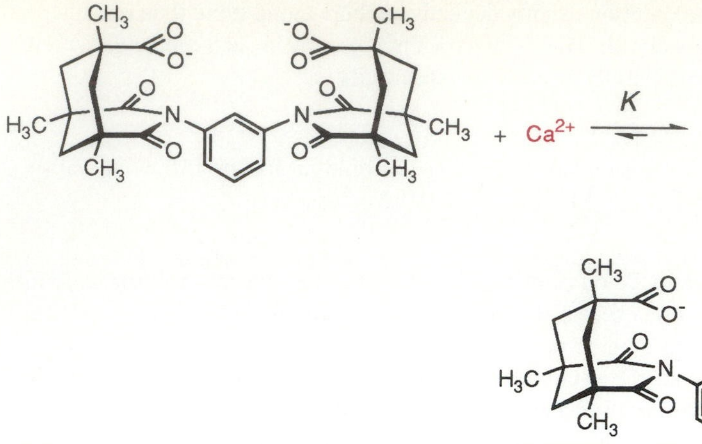

A related diacid has been prepared from Kemp's triacid and the dye ''acridine yellow,'' 2,7-dimethyl-3,6-dimethylacridine (acridine is a symmetrical dibenzopyridine). The acridine bis-(imide) derivative has a larger space between the carboxylic acid groups, and the dianion is not a particularly good host for Ca^{2+}. However, it does act as a host for the doubly protonated form of 1,4-diaza-bicyclo[2.2.2]octane, which is bound strongly (K_a is about 10^5 M^{-1}, corresponding to $\Delta G° = 6.9$ kcal mole^{-1}).

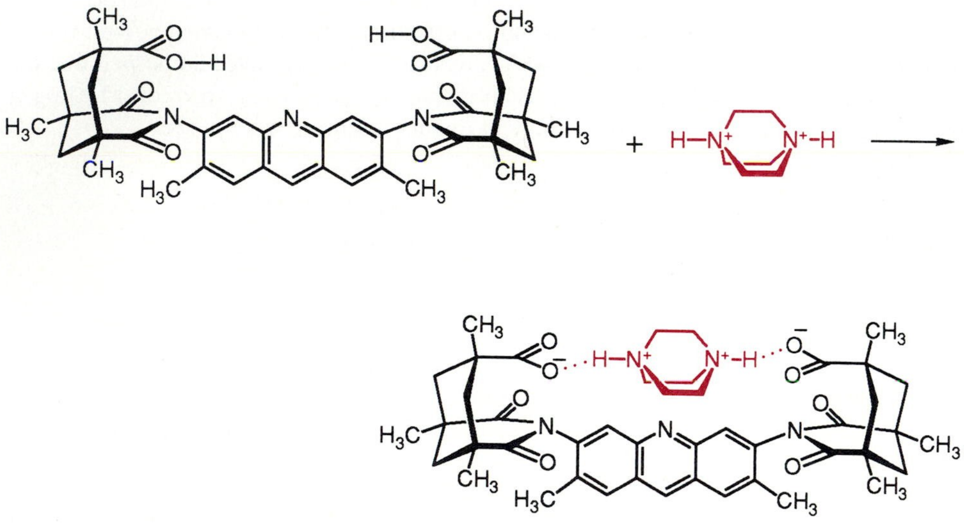

The enhanced affinity of these dicarboxylic acids for cations illustrates the important principle of **preorganization.** To understand preorganization, consider a simple molecular recognition process, complexation of potassium ion by ether molecules. Some inorganic salts, such as $KMnO_4$ and KCN, readily dissolve in hydrocarbon solvents if the polyether 18-crown-6 is added. However, these same salts are not solubilized by equivalent amounts of tetrahydrofuran (THF). This striking difference in behavior is due to the unfavorable entropy that results from organizing the THF ligands around the potassium ion. Consequently, even though the positive potassium ion could be effectively solvated by ether dipoles, the net association energy is not sufficient to overcome the lattice energy of $K^+ X^-$. On the other hand, much less entropy is lost when the six oxygens of 18-crown-6 coordinate a potassium ion because they have been **preorganized.** In a sense, this entropy has already been paid in the process of synthesizing the polyether.

However, although the six oxygens in 18-crown-6 have been preorganized for binding a potassium ion, the preorganization is not perfect. The polyether is a "floppy" molecule with many conformations available to it, whereas the complex has a rigid, nonflexible structure. In fact, it is known that uncomplexed 18-crown-6 has a structure in which the "cavity" is partially filled with inward-turned CH_2 groups.

So even with 18-crown-6, some entropy is lost in binding. An ideal host for a potassium ion would have a rigid structure with the six oxygens positioned in the free polyether precisely as they are in the complex. In such a case, complexation would be accompanied by no loss of conformational entropy and the maximum association constant would result. Much modern research has been devoted to the synthesis of better hosts to bind a variety of different guests. One example of a host for potassium ion that has been preorganized to a greater degree is the following diamino tetraether, an example of a **cryptand.**

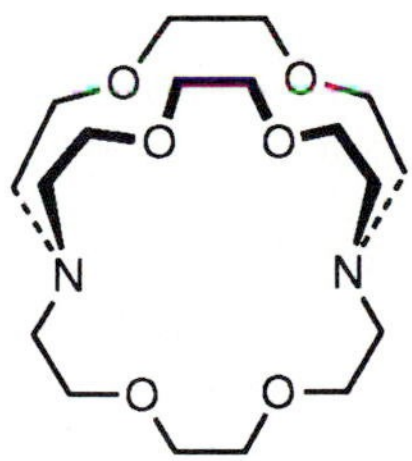

The standard free energy of binding a potassium ion by this host is 18 kcal $mole^{-1}$, compared to only 11 kcal $mole^{-1}$ for 18-crown-6.

Yet the foregoing cryptand still has some conformational flexibility. An even more rigid host is the following cyclic hexamer of *p*-methylanisole. As suggested in the drawing, the molecule exists in a conformation with each of the six benzene rings staggered so that the MeO groups project alternately up and down. The result is a barrel-shaped molecule that is a representative of a class of hosts called **spherands.** The spherand shown here has a rather small cavity and is a specific host for lithium cations. The binding free energies for complexation of Li^+, Na^+, and K^+ by this host are >23 kcal $mole^{-1}$, 19.2 kcal $mole^{-1}$, and <6 kcal $mole^{-1}$, respectively. Larger versions of this spherand show different guest specificity. For example, the analogous spherand with eight units has a much larger cavity and is selective for cesium ions.

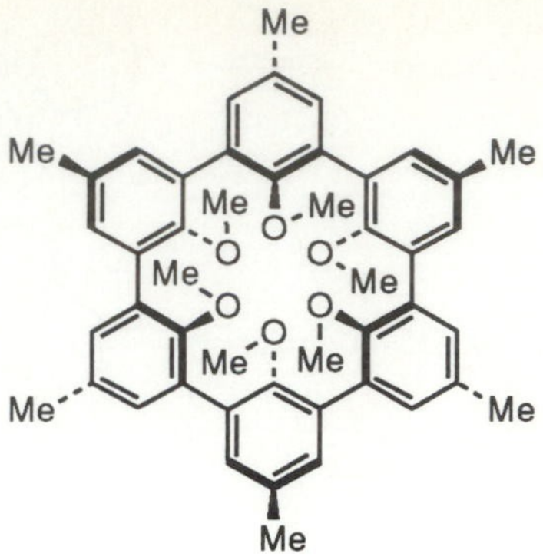

The synthesis of hosts for specific metal ions is a fascinating intellectual exercise, and it has allowed chemists to understand a great deal about the details of molecular recognition. However, it is much more than that. For example, the following tetraamide (3,4,3-LICAM) was designed by inorganic chemists as a host for Pu^{4+} ions.

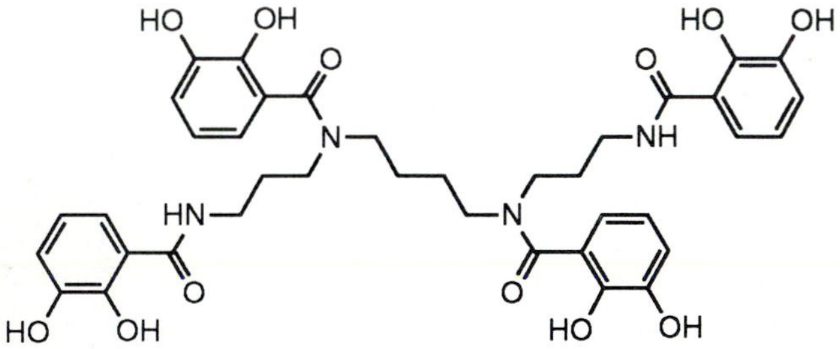

3,4,3-LICAM is an example of a useful class of ion-specific hosts that can be used for **metal decorporation therapy,** in this case for treatment of individuals who accidentally become contaminated with the exceedingly toxic element plutonium.

So far, we have only discussed hosts for cations. However, hosts for anions have also been synthesized. An example of a simple anion-binding host is the following tetraammonium ion. The geometry of this species is such that the four relatively acidic N—H groups turn inward into the cavity in just the right direction and at an appropriate distance to stabilize a chloride ion. This host shows a strong preference for Cl^- over Br^-, and it does not complex other anions at all.

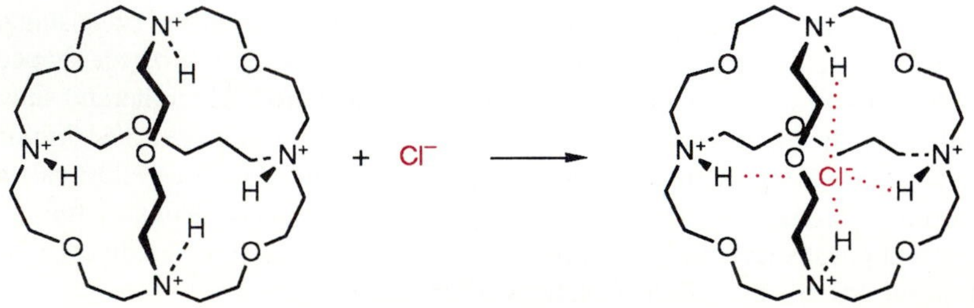

The following even simpler 24-membered hexaammonium ion strongly binds adenosine triphosphate (ATP) and even acts as a catalyst for its hydrolysis to adenosine diphosphate (for simplicity, the complex is shown with only one acidic proton on each nitrogen and the net charge of each of the phosphate groups over the phosphorus atom).

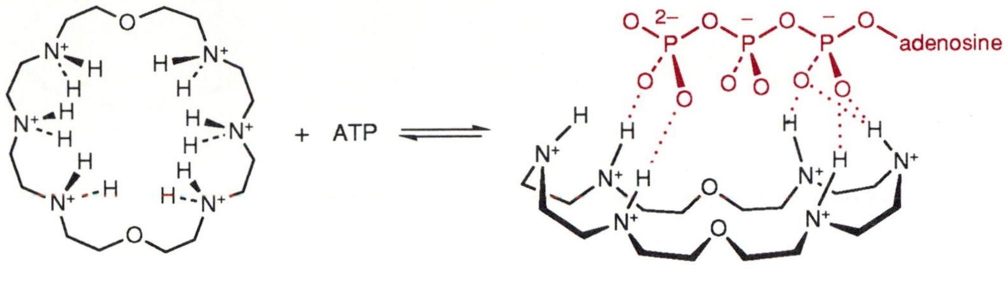

Although the actual structure of the complex of this host with ATP has not been determined, there is strong indirect evidence that it is as shown above, with five of the six ammonium ions being hydrogen bonded to four oxygens of the triphosphate. The mechanism of the hydrolysis reaction has been found to involve transfer of the terminal phosphate to the sixth ring nitrogen, giving a phosphorylated intermediate that then loses phosphate ion by reaction with a water molecule.

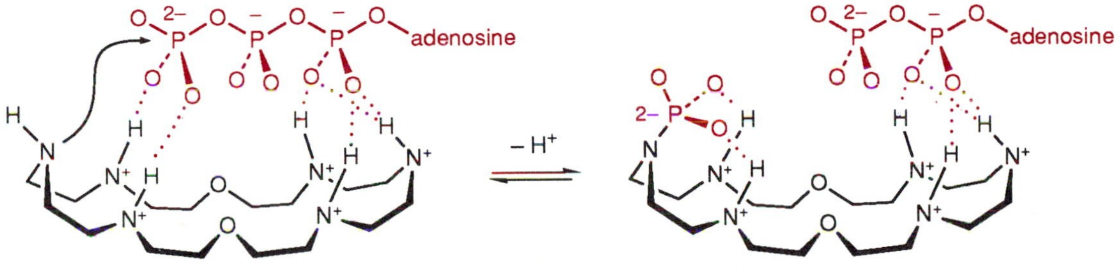

In proceeding through a receptor-bound intermediate that is hydrolyzed in a subsequent step, the mechanism resembles that discussed on page 982 for the serine proteases.

The hosts that we have discussed so far interact with their guests by polar forces, either hydrogen bonds or electrostatic interactions. However, nonpolar interactions are important components of binding in biological host-guest systems (e.g., enzymes and their substrates). Chemists have been able to imitate this kind of recognition as well. The following host has two rather different binding areas; the region at the bottom of the cavity is surrounded by the three basic nitrogen lone pairs and three polar ether functions, whereas the region at the top of the cavity is predominantly hydrocarbon in nature, being essentially the π-electron surfaces of the three benzene rings. The molecule was designed and synthesized with the anticipation that it would form specific interactions with amphiphilic molecules, compounds that have both polar and nonpolar regions. In fact, this host strongly binds methylammonium ions. The three nitrogens are ideally situated to form three hydrogen bonds to the ammonium protons, and the hydrophobic cavity is just the right size to accommodate a methyl group; other ammonium ions are not bound because their alkyl groups are too big to fit the hydrophobic binding cavity.

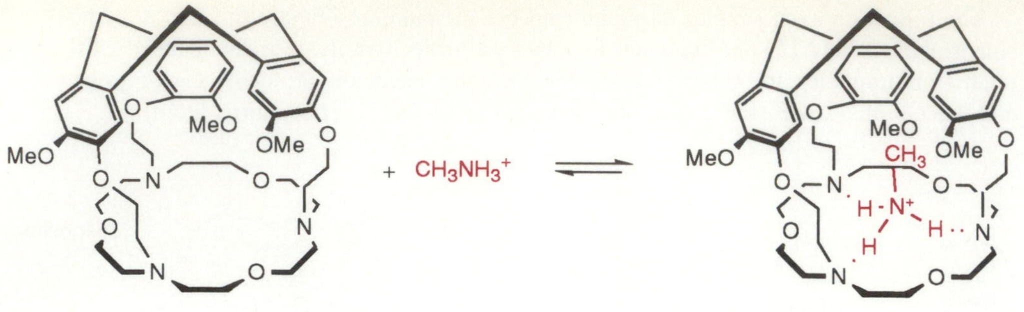

All of these synthetic hosts have one structural characteristic in common—their exterior surfaces are relatively hydrophobic, so they dissolve in organic solvents, often with a polar molecule or ion encapsulated in their polar cavities. However, Nature's hosts (e.g., enzymes, "carrier proteins") usually have hydrophilic exteriors and exist in aqueous medium. Although their binding cavities typically do contain one or more polar sites where hydrogen bonding or electrostatic attraction is important, simple van der Waals attractive forces are equally important. Chemists have also been able to model this hydrophobic binding by constructing hosts that have hydrophilic exteriors and hydrophobic interiors. With this kind of host, the goal is to solubilize a nonpolar guest in aqueous solution. Many such compounds have now been prepared. One series that has been studied extensively is typified by the following molecule. Because of the four ammonium functions, the host is soluble in aqueous solution.

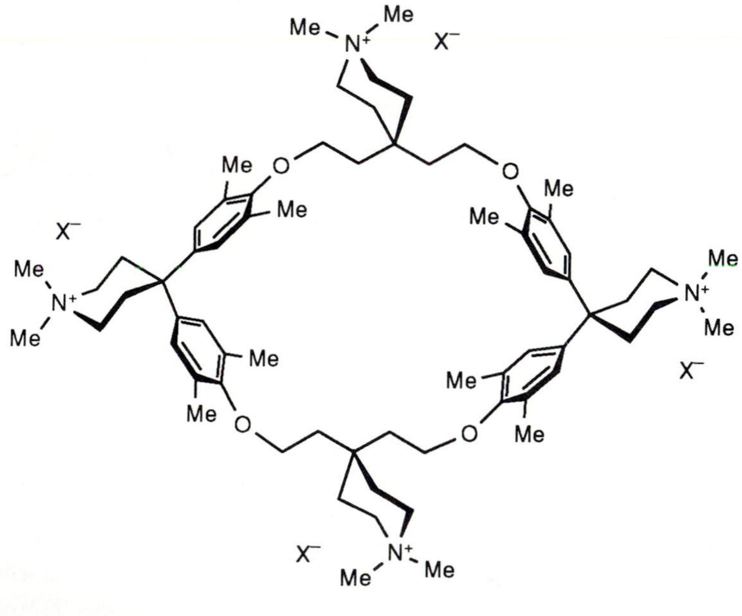

However, the four benzene rings form a cavity that can accommodate nonpolar guest molecules such as aromatic hydrocarbons. Complexes with a number of aromatic hydrocarbons have been studied. The molecule that best fits the cavity of the host is the pentacyclic compound perylene, which is shown bound in the hydrophobic pocket of the host.

With this hydrocarbon the host–guest association constant in aqueous solution is $K_a = 1.6 \times 10^7\ M^{-1}$, corresponding to a standard free energy of binding of 9.6 kcal $mole^{-1}$. As shown by the following hydrocarbon structures, the host displays specificity for guests with larger surface areas; the incremental free energy of binding is about 1.2–1.4 kcal $mole^{-1}$ per aromatic ring.

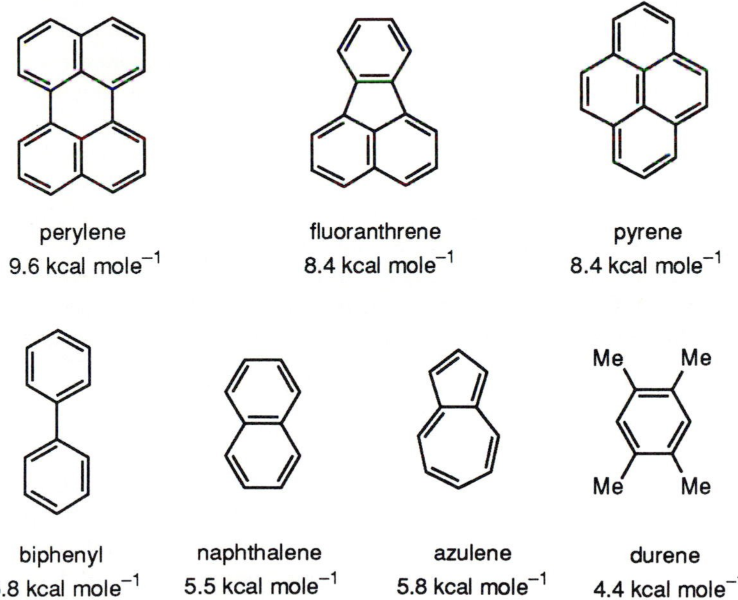

The hosts that have been discussed in this section were designed and synthesized by chemists with one goal in mind—to recognize and bind certain guest molecules. This is, in fact, the simplest function of natural host molecules. For example, some natural "ionophores" like nonactin (page 241) function by binding ions and transporting them through membranes or cell walls. Other hosts, such as myoglobin and hemoglobin, bind and transport oxygen. However, natural hosts like enzymes also cause their guests to undergo chemical reactions. With these hosts, the binding cavity must not only be well preorganized to bind the guest, but it also must contain functional groups that can enter into a chemical reaction. In addition, some natural hosts have secondary binding regions for another reactant molecule, such as a cofactor. During the last two decades, we have seen the emergence of an entire subfield of organic chemistry that is concerned with understanding the chemical principles of binding and catalysis in these natural host-guest systems. A major focus of this research community is in the design of hosts similar to those we have illustrated but also adorned with various functional groups that can react with the guest once it is bound. A further description of this topic is beyond the scope of an introductory text, but the interested student will find one or more articles on the subject in almost any current issue of the *Journal of Organic Chemistry* or the *Journal of the American Chemical Society*.

EXERCISE 33.3 A chloroform solution of the spherand discussed earlier in this section is used to extract a mixture of lithium and sodium salts from aqueous solution. Given binding free energies of 23 kcal $mole^{-1}$ for Li^+ and 19.2 kcal $mole^{-1}$ for Na^+ calculate the Li^+/Na^+ ratio that will be found in the chloroform solution.

Highlight 33.1

Molecular recognition is the process whereby two molecules interact through noncovalent forces such as hydrogen bonds, electrostatic interactions, and van der Waals attractive forces. Complexing partners are referred to as hosts and guests. The binding sites of the host converge or turn inward, whereas those of the guest diverge or turn outward. For the complex to be favored over the uncomplexed state, the binding free energy, $\Delta G°$, must be negative and of sufficient magnitude to compensate for other interactions (e.g., with solvent molecules or other solute molecules) that the interacting molecules sacrifice in forming the complex. In order for complexation to be specific for a given host-guest pair, it is generally necessary that the two interacting molecules have at least two or three points of interaction. Chemists have synthesized a wide variety of hosts. Some have nonpolar surfaces and bind polar molecules or ions in hydrophilic cavities using directed hydrogen bonds or electrostatic interactions. Others have hydrophilic exterior surfaces and accept nonpolar guests in hydrophobic binding sites. One of the important features that contributes to strong host–guest interactions is preorganization of the host for optimum interaction of its binding regions with those of the guest.

33.3 Molecular Recognition by an Enzyme

Nature makes much use of the principles of molecular recognition to do its business. This is especially true in the chemistry of proteins. We saw in Chapter 29 that proteins have functions as structural materials (e.g., collagen, the structural material of the connective tissues), as transporters of other molecules (e.g., hemoglobin and myoglobin, which carry oxygen from one place to another), and as catalysts (e.g., the serine proteases, which catalyze the hydrolysis of other peptides and proteins). All of these important jobs involve molecular recognition. In this section, we will briefly examine an enzyme that catalyzes one step of a complex set of chemical reactions that Nature uses for synthesis of many aromatic compounds.

The synthesis of the aromatic amino acids phenylalanine, tyrosine, and tryptophan in plants and microorganisms is effected via the "shikimate pathway." The following scheme shows some of the major intermediates in this biosynthetic pathway.

phosphoenolpyruvate (PEP) + erythrose 4-phosphate →

3-deoxy-D-*arabino*-heptulose 7-phosphate → (*dehydroquinate synthase*) → dehydroquinate → phenylalanine, tyrosine, tryptophan

The enzyme that catalyzes one of the last steps is inhibited by one of the most widely used herbicides, glyphosate (Section 25.7). The formation of dehydroquinate is catalyzed by an enzyme, dehydroquinate (DHQ) synthase. DHQ synthase has 362 amino acids, a divalent metal ion (probably Zn^{2+}), and a bound redox cofactor, nicotinamide adenine dinucleotide (NAD). In general, a relatively small enzyme such as DHQ synthase has just one active site and catalyzes just one type of reaction. Although there is no change in oxidation state in the conversion of 3-deoxy-D-arabinoheptulose 7-phosphate to dehydroquinate, the presence of NAD in the enzyme dehydroquinate synthase suggests that oxidation and reduction are involved in the overall transformation. A mechanism for the formation of dehydroquinate is shown in Figure 33.1. Note that a carbocyclic ring in the product has replaced the pyranose ring of the substrate for DHQ synthase.

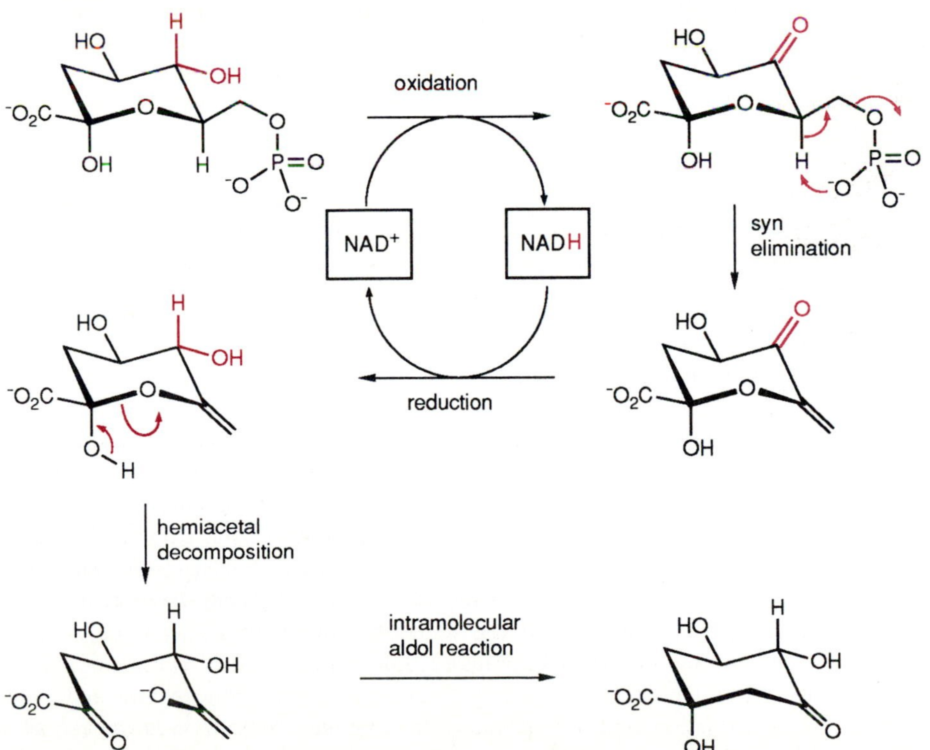

FIGURE 33.1 A detailed mechanism for the biosynthesis of dehydroquinate.

The variety of reactions that occur during the conversion of 3-deoxy-D-arabinoheptulose 7-phosphate to dehydroquinate is unusual and includes oxidation, syn elimination, reduction, hemiacetal decomposition, and an intramolecular aldol reaction. What does dehydroquinate synthase recognize in the substrate molecule? Could the multitude of reactions that occur imply that there are multiple types of recognition? The answer to this question seems to be that there is only one kind of recognition. In fact, studies with various analogs suggest that the enzyme catalyzes only two of the steps, oxidation of a secondary hydroxy group to a carbonyl and its subsequent reduction to a secondary hydroxy group.

This conclusion was based on the results of several experiments. One experiment was designed to determine the stereochemical course, syn or anti, of the elimination reaction. An analog of 3-deoxy-D-arabinoheptulose 7-phosphate was prepared that lacked the hydroxy group at C-2 and was specifically deuterated at C-6. Because the analog lacks the C-2 OH group, the hemiacetal decomposition step cannot occur. This allows an examination of the stereochemical outcome of the elimination process. The analog was treated with DHQ synthase and the product was isolated and analyzed to determine the position of the deuterium label. The results showed that the elimination is syn. If it had been anti, the positions of the deuterium and hydrogen on the double bond would have been reversed. The following two equations illustrate this stereochemical point.

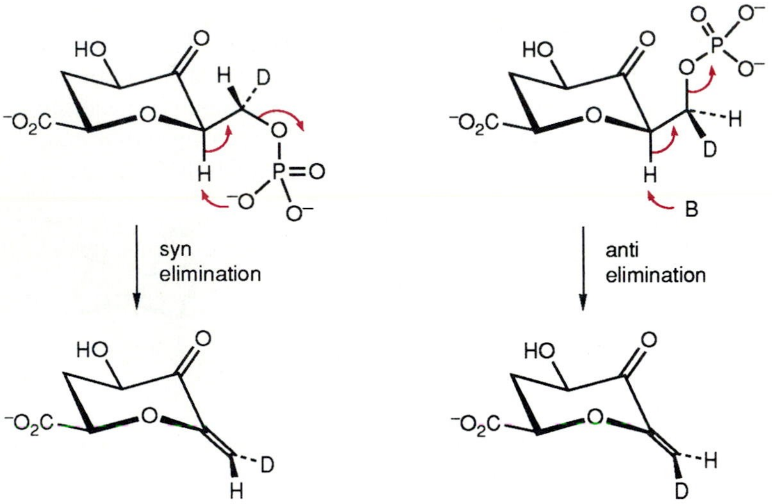

syn elimination

anti elimination

Since the elimination step turns out to be syn, it is possible that the base that removes the proton is a lone pair from the phosphate leaving group itself. The function of the enzyme, then, would be to hold the substrate in the right conformation for this elimination reaction, not to supply the base to remove a proton. Evidence that this is the case came from the behavior of several other analogs of 3-deoxy-D-arabinoheptulose 7-phosphate. The trans vinylphosphonate analog is prevented by the double bond from achieving the conformation that can undergo intramolecular syn elimination (boxed insert), but the cis vinylphosphonate analog can achieve the required geometry. When these two analogs were incubated with DHQ synthase, it was found that the trans isomer is only weakly bound by the enzyme and that it does not undergo the normal oxidation reaction. On the other hand, the cis vinylphosphonate analog is strongly bound by the enzyme and undergoes the normal oxidation reaction. Because the enzyme also catalyzes the reduction reaction, the last point must be established by carrying out the reaction in the presence of a large excess of the oxidized form of the cofactor, NAD^+. Note that since the analog is carbocyclic, the normal elimination reaction is blocked.

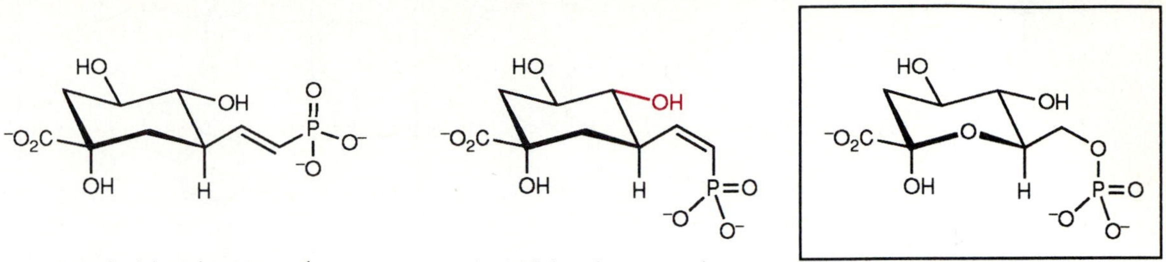

HO, OH, ^-O_2C, OH, H, O, P, O^-, ^-O

trans vinylphosphonate analog

HO, OH, ^-O_2C, OH, H, P=O, ^-O, O^-

cis vinylphosphonate analog

HO, OH, ^-O_2C, O, OH, H, O, P=O, ^-O, O^-

DHQ synthase

HO, O, ^-O_2C, OH, H, P=O, ^-O, O^-

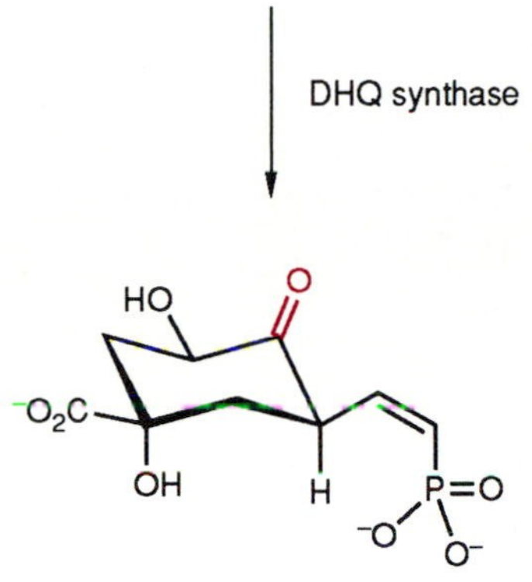

Finally, it was found that treatment of the cis vinylphosphonate analog with DHQ synthase in D_2O solution results in exchange of the proton next to the vinylphosphonate group.

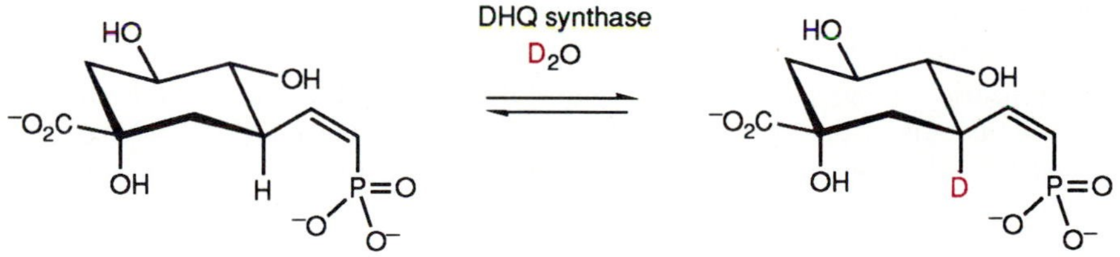

HO, OH, ^-O_2C, OH, H, P=O, ^-O, O^-

DHQ synthase
D_2O

HO, OH, ^-O_2C, OH, D, P=O, ^-O, O^-

The implication of these and other related experiments is that the function of DHQ synthase is to bind 3-deoxy-D-arabinoheptulose 7-phosphate and hold it in a conformation similar to that illustrated, with the phosphate group near the axial hydrogen that must be removed in the elimination reaction. The enzyme also binds the cofactor NAD^+, which oxidizes the C-3 hydroxy group to a carbonyl group. The acidity of the proton at C-5 is significantly increased because it is now α to a ketone carbonyl. In other words, the enzyme has set the stage for the elimination reaction, which then occurs spontaneously to give the enolate depicted in Figure 33.1.

One further experiment suggested that the final intramolecular aldol reaction also does not involve catalysis by the enzyme. This point was tested by carrying out the aldol reaction in the absence of the enzyme. The substrate was prepared as shown in the following scheme.

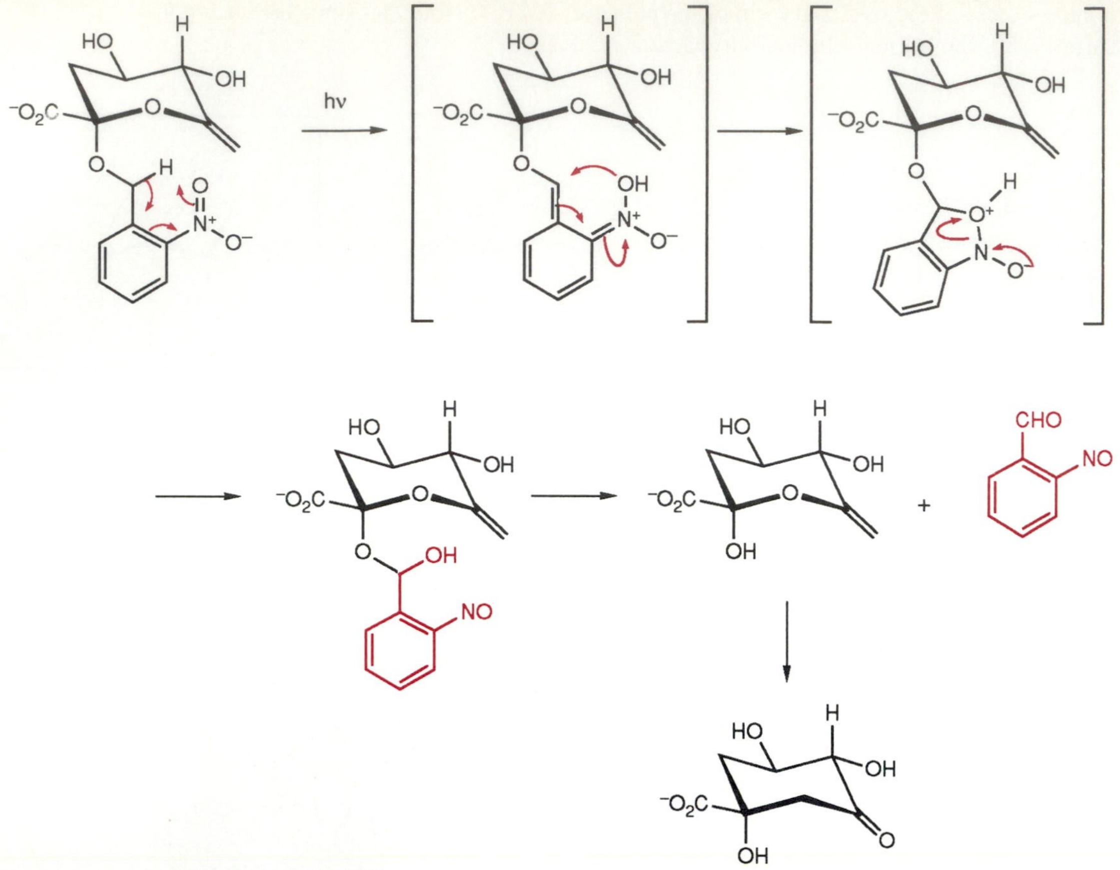

Photochemical removal of the 2-nitrobenzyl group from the 2-nitrobenzyl ether of the hemiacetal gives dehydroquinate in the absence of the DHQ synthase. Thus, hemiacetal decomposition and enolate addition to the carbonyl appear to occur spontaneously. Furthermore, the stereochemical outcome of the spontaneous aldol reaction is the same as that seen when dihydroquinate is produced enzymatically.

In this section we have taken a close look at the methods that were used to investigate some of the details of the mechanism of action of one enzyme, DHQ synthase. Similar investigations have been carried out with dozens of other enzymes. Molecular recognition often plays an important role in such research projects, as in the example of DHQ synthase, where the ability of the enzyme to recognize and bind the cis vinylphosphonate analog much better than it does the trans isomer provided an important clue to the mechanism of the elimination reaction.

33.4 Catalytically Active Antibodies

Antibodies are complex proteins that are produced by the vertebrate immune system to defend against foreign proteins. They work by recognizing the foreign molecule and forming a strong complex with it. The part of the foreign molecule that is recognized by an antibody is called an **antigenic group** or an **epitope,** and the foreign molecule itself is called an **antigen.** Cross-reactivity between an epitope that is native to the vertebrate and a similar foreign epitope can lead to autoimmune disease; the foreign epitope stimulates an antibody that recognizes the native epitope, with deleterious effects. Antibodies (Figure 33.2) are immunoglobulins (IgD, IgE, IgM, IgG, IgA) composed of two light chains and two heavy chains. Each of the chains is divided into subregions. Each light chain has a

Variable Region(V)

Hypervariable loops

Constant regions(C)

Hinge region(H)

HEAVY CHAINS

Antibody
(immunoglobulin)

(a)

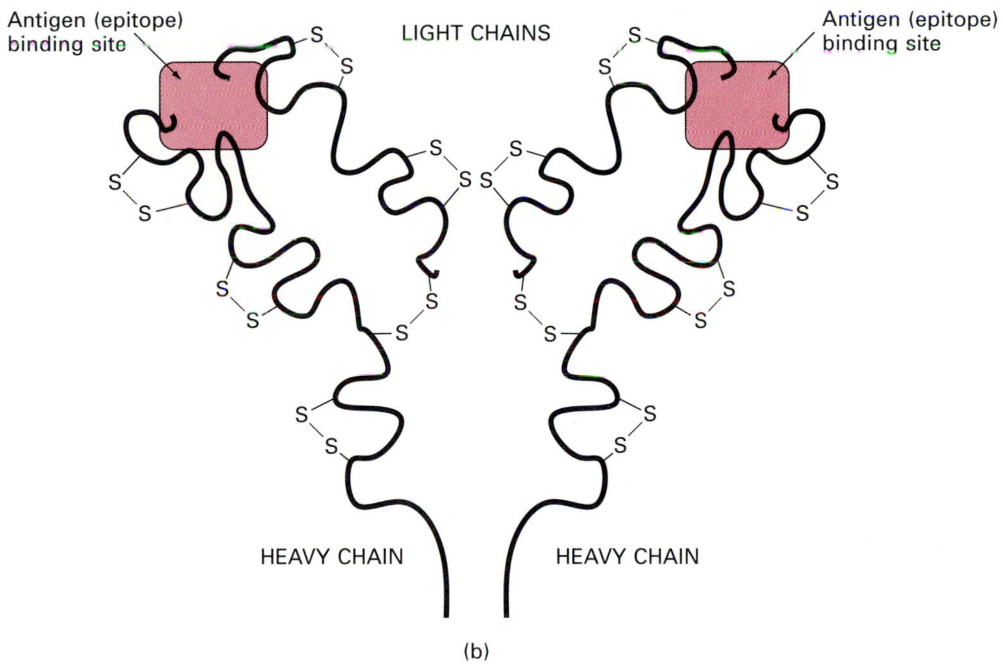

(b)

FIGURE 33.2 Two representations of an immunoglobulin antibody. (a) The relative positions of the light and heavy chains. Within each chain are variable regions (V) and constant regions (C); the heavy chain has a hinge region (H). Highly variable (''hypervariable'') regions are shown as dark bands. A number of disulfide bonds are shown to illustrate the nature of the links between the chains. Diagram (b) shows the heavy and light chains and indicates that a substantial number of disulfide bonds contribute to intrachain conformational arrangements. Most important, the antigen-bonding sites are shown to be between the hypervariable loops of both the light and heavy chains.

variable region and a constant region. Each heavy chain has a variable region, three constant regions, and a hinge region. Each variable region has three hypervariable loops at which epitope binding occurs. The light chains are bound to the heavy chains by a disulfide bond and the two heavy chains are linked to each other by three disulfide bonds.

It is estimated that there might be as many as 10^8 to 10^{12} antibodies and that each can recognize 10^{12} to 10^{14} different molecules. Antibodies are produced in quantity when a membrane-bound immunoglobulin recognizes an epitope well enough for effective binding. The formation of the biologically effective receptor–epitope complex initiates in the cell a cascade of biochemical and biological processes that multiplies the number of antibody-carrying cells by a large factor and thereby amplifies the amount of antibody present in the organism. Thus, the presence of the antigen (epitope) induces the proliferation of cells that produce appropriate antibodies, a process that is called **clonal selection.** A number of different antibodies that recognize the same antigen are generally produced. Such a mixture of different antibodies that have similar binding properties is called **polyclonal.**

It is also possible to prepare antibody samples in which every molecule has precisely the same structure. These are called **monoclonal** antibodies. Monoclonal antibodies are obtained from hybrid cells **(hybridomas)** that are generated by fusing antibody-producing cells from a spleen with myeloma cells. The fused cells grow readily in tissue culture with hypoxanthine as a nucleotide source. Each group of cells **(clone)** produces a single, structurally unique antibody. The desired clone is recognized by showing that it binds or recognizes the original antigen. After isolation and further culturing, one can produce the monoclonal antibodies in sufficient quantities to use them as reagents for other purposes.

An antibody recognizes an antigen in much the same manner as some of the other cases of molecular recognition we have discussed—a combination of reciprocal hydrogen bonding, electrostatic interactions, and van der Waals attractive forces. Antibodies are elicited to large, proteinaceous antigens much more easily than to small molecules. To prepare an antibody that recognizes a small molecule, it is coupled to a carrier protein such as bovine serum albumin (BSA) or keyhole limpet hemocyanin (KLH). The resulting modified protein causes the generation of some antibodies that recognize and bind it in the region where the small molecule (called a **hapten**) is attached.

Highlight 33.2

Structurally homogeneous (monoclonal) antibodies can be obtained in the following manner. First, the antigenic small molecule (called a hapten) is attached to a common protein such as bovine serum albumin or keyhole limpet hemocyanin. This modified protein is then used to inoculate an appropriate organism. In response to the introduction of the foreign protein antibody-producing cells are activated. These antibody-producing cells are fused with myeloma cells to produce hybridomas that are sorted by a biological technique known as cloning. Each clone produces a unique antibody against the modified protein antigen. The clones can be multiplied to obtain significant quantities of monoclonal antibodies.

Molecular recognition of an epitope requires a set of appropriate functional groups in the receptor. For example, an antibody to a substance containing a positively charged group is likely to have a negatively charged aspartate or glutamate in the binding site.

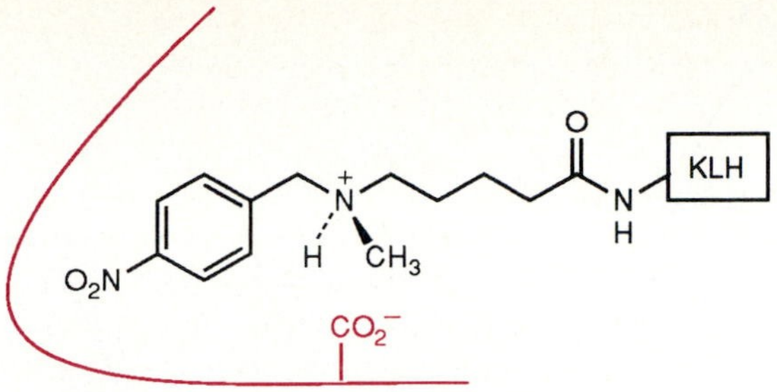

In one clever exploitation of this natural complementarity, a conjugate or KLH and the hapten 5-(N-4-nitrobenzyl-N-methyl)aminopentanoic acid was used to generate monoclonal antibodies. The purified antibodies were then used as artificial ''enzymes'' to catalyze the elimination of HF from the sterically related substrate 4′-nitrophenyl-2-fluoro-ethyl methyl ketone. In this case the antibody recognizes and binds the ketone because its molecular shape is similar to that of the hapten that was used to elicit the antibody in the first place. Once the ketone is bound, the negatively charged carboxylate group that stabilized the positive charge of the ammonium group in the antibody-antigen complex is now appropriately located to act as a base, causing elimination of HF. It was found that many of the monoclonal antibodies prepared in this way caused elimination of HF to occur at rates almost 10^5 faster than acetate ion acting on the same substrate.

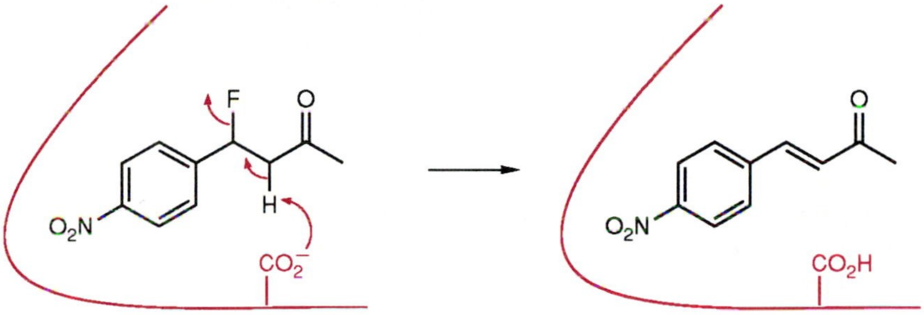

Catalytic antibodies have also been designed by taking advantage of a useful concept of enzymology, that an enzyme can catalyze a reaction by preferentially binding and stabilizing the transition structure for the reaction. The basic idea can be easily understood by considering a simple one-step transformation of reactant A to product B, proceeding through the transition structure $A^{\ddagger}$ (Figure 33.3). The rate of the reaction is directly proportional to the energy of activation, E_a, which is equal to the difference in energy of reactant A and transition structure $A^{\ddagger}$. Suppose the enzyme is constructed in such a way that it recognizes and binds the transition structure better than it does either the reactant or the product of the reaction. That is, if the free energy of binding of the transition structure ($\Delta G°_{\text{E-TS}}$) is more negative than that of the reactant ($\Delta G°_{\text{E-R}}$), the transformation will proceed more rapidly than in the absence of the enzyme.

A compound that resembles the transition structure for a reaction more than it does the reactant or product of the reaction is called a **transition state analog.** Because an enzyme recognizes and binds the transition state analog more strongly than it does the reactant, a transition state analog often acts as a **reversible inhibitor** of the enzyme. That is, the analog occupies the active site of the enzyme and prevents the substrate from being bound, thus inhibiting the reaction. The principle of transition state analogs is often used for the invention of pharmaceutical products, drugs that will interfere with a certain enzymatic process, thus achieving some desirable goal. The ACE inhibitors discussed on page 984 are transition state analogs for the hydrolysis of angiotensin I by angiotensin-converting enzyme.

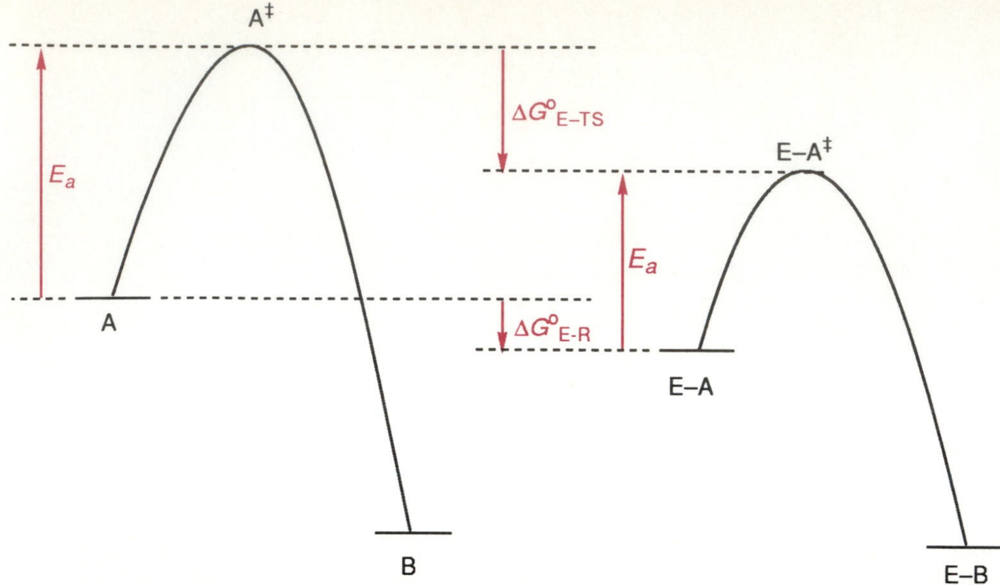

FIGURE 33.3 In the uncatalyzed transformation of A to B, the reaction is related to the energy of activation, E_a. The free energy of binding of the enzyme to the transition structure for the reaction, $\Delta G°_{\text{E-TS}}$, is larger than the free energy of binding of the enzyme to the reactant, $\Delta G°_{\text{E-R}}$. As a result, the activation energy for the reaction is lowered and the reaction is faster.

Transition state analogs can also be used as haptens to generate antibodies that are, in a sense, tailor-made enzymes for a certain reaction. A simple example is the hydrolysis of methyl 4-nitrophenyl carbonate.

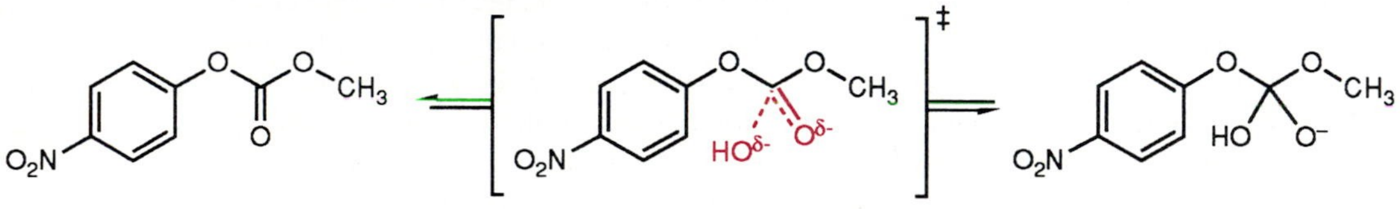

Monoclonal antibodies were elicited to a conjugate of a phosphate analog. The phosphate group is tetrahedral, like the transition structure for the hydrolysis of the carbonate. Moreover, it exists under physiological conditions as a monoanion in which the two oxygens share the negative charge (only one Lewis structure is shown below). Several of the monoclonal antibodies produced in this way were found to catalyze hydrolysis of methyl 4-nitrophenyl carbonate.

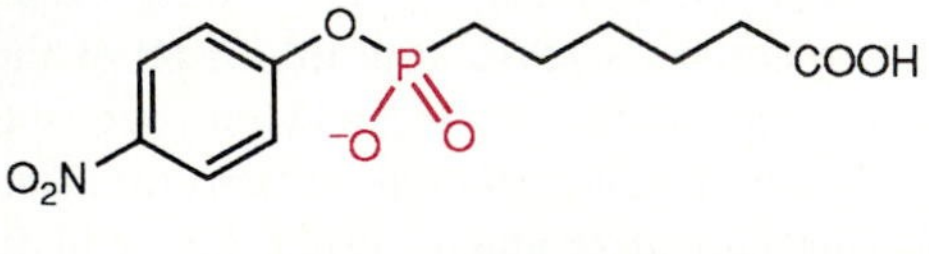

This strategy has been employed to create artificial enzymes that catalyze a large variety of organic reactions in addition to simple HF elimination and ester hydrolysis.

Examples of other reactions whose rates are accelerated by appropriately constructed catalytic antibodies are the Diels-Alder reaction (Section 20.5), the Claisen rearrangement (page 638), and the formation of Schiff bases (page 399).

Highlight 33.3

A transition state analog is a compound that structurally resembles the transition structure for a reaction more than it does the reactants or products. If the reaction is enzyme-catalyzed, it is often found that a transition state analog inhibits the normal reaction by competing with the normal substrate for the enzyme binding site. Because the inhibitor is not covalently linked to the enzyme, this kind of inhibition is referred to as reversible. When a transition state analog is used as a hapten to elicit monoclonal antibodies, the resulting antibodies often function as artificial enzymes and catalyze the reaction whose transition structure the analog mimics.

EXERCISE 33.4 Draw energy versus reaction coordinate diagrams for a hydroxide-catalyzed hydrolysis and compare this to a diagram for an antibody-catalyzed hydrolysis of a carbonate ester.

33.5 Nucleic Acids

A. Molecular Components of Nucleic Acids

Nucleic acids are natural molecules that are involved in the storage and expression of genetic information and in protein biosynthesis. As we will see, they are enormous organic molecules whose structure and function are dominated by the principles of molecular recognition.

There are two types of nucleic acid, **ribonucleic acid (RNA)** and **deoxyribonucleic acid (DNA).** In mammals, the DNA is located mainly in the nucleus of the cell. In higher plants, as much as 15% of the cellular DNA is located in the chloroplast, the photosynthesizing organelle of green plants. The RNA is located mainly in the cytoplasm, the intracellular solution. Both DNA and RNA are biopolymers in which the repeating monomer units are called **nucleotides.** A nucleotide, in turn, is composed of three parts, one unit each of phosphate, a pentose sugar, and a heterocyclic base. The combination of a sugar with a heterocyclic base is called a **nucleoside.** The heterocyclic base can be either a pyrimidine or a purine. In RNA the pentose is D-ribose, the pyrimidine can be thymine or cytosine, and the purine, adenine or guanine. The combinations are often referred to as **ribonucleotides** or **ribonucleosides.** In DNA, the pentose is 2-deoxy-D-ribose, the pyrimidine can be thymine or cytosine, and the purine, adenine or guanine. The combinations are therefore **2′-deoxyribonucleotides** or **2′-deoxyribonucleosides.** The names for the nucleosides and nucleotides are illustrated in Figures 33.4–33.6 for the RNA components, cytidylic and uridylic acids, and the DNA components, 2′-deoxyadenylic, 2′-deoxyguanylic, and 2′-deoxythymidylic acids. The regular numbering systems used in the heterocyclic moiety and the primed numbers in the sugar portion of the molecule are also shown in the figures. For each class of nucleic acid there are four nucleotide monomers. In RNA, one finds cytidylic (C), uridylic (U), adenylic (A), and guanylic (G) acids. In DNA are found 2′-deoxyadenylic (dA), 2′-deoxyguanylic (dG), 2′-deoxycytidylic (dC), and 2′-deoxythymidylic (dT) acids.

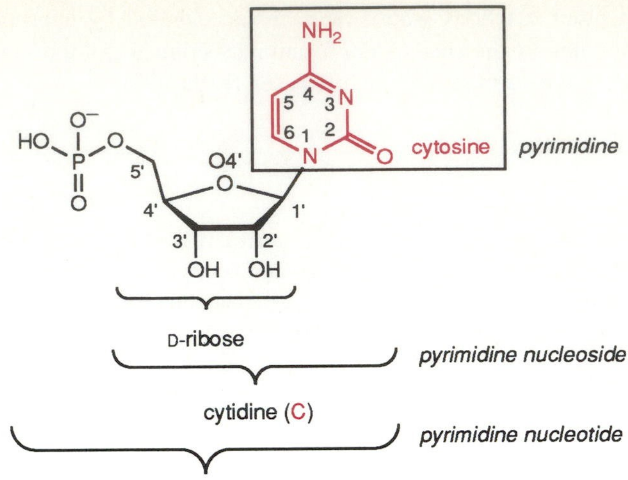

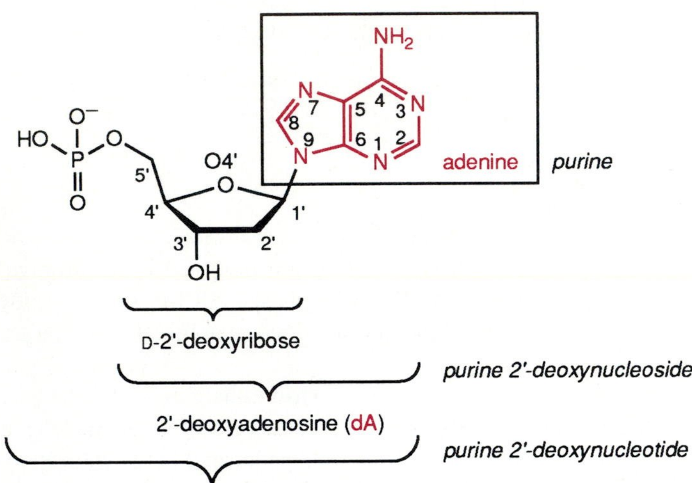

FIGURE 33.4 Cytidylic and 2′-deoxyadenylic acid building blocks for nucleic acids.

Base-catalyzed hydrolysis of a nucleotide removes the phosphate group and yields the nucleoside. The nucleosides formed from the RNA nucleotides are cytidine, uridine, adenosine, and guanosine. The nucleosides derived from the DNA nucleotides are the corresponding 2′-deoxy analogs, 2′-deoxycytidine, 2′-deoxyadenosine, 2′-deoxyguanosine, and 2′-deoxythymidine.

The nucleic acids are formed in nature by enzymatic polymerization of nucleoside triphosphates. Oligodeoxyribonucleotides are much smaller molecules that contain 15–50 nucleotide units. They can be synthesized in the laboratory by the automated execution of a sequence of reactions in a programmable **DNA synthesizer.** The nucleic acids can be extremely large molecules, some as long as a whole eukaryotic chromosome with 10^8 residues (nucleotide units) and molecular weights of more than 300 billion. The nucleic acid backbone is a copolymer of phosphoric acid and either ribose (RNA) or 2-deoxyribose (DNA) units. Linked to C-1′ of each of the pentose units is one of the four heterocycles, adenine, guanine, cytosine, and uracil (RNA) or thymine (DNA). A DNA

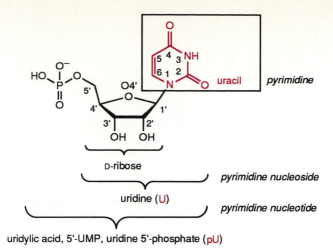

uracil
pyrimidine
D-ribose
pyrimidine nucleoside
uridine (U)
pyrimidine nucleotide
uridylic acid, 5'-UMP, uridine 5'-phosphate (pU)

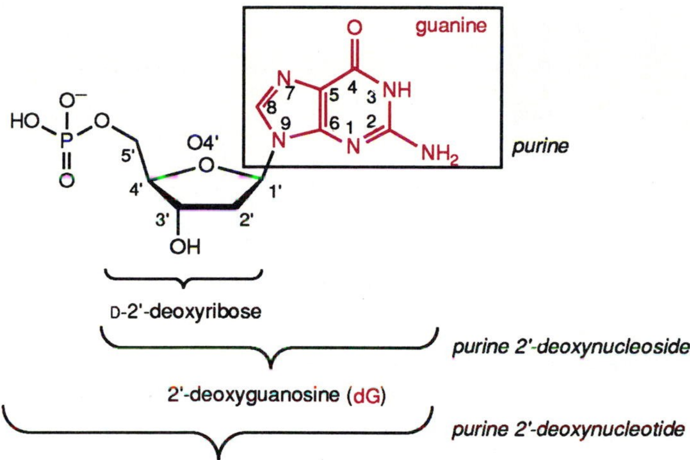

FIGURE 33.5 Uridylic and 2′-deoxyguanylic acid building blocks for nucleic acids.

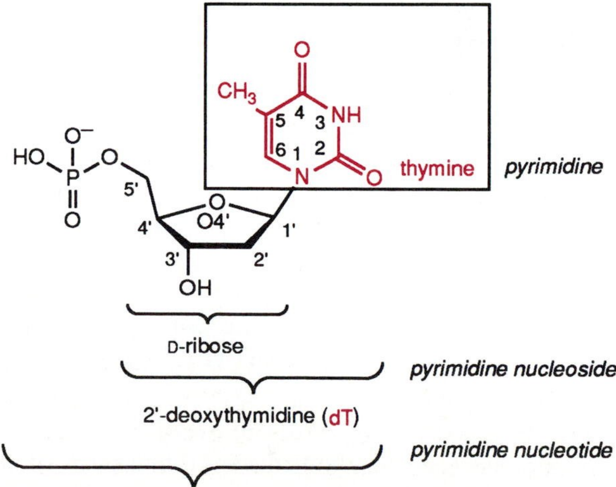

FIGURE 33.6 2′-Deoxythymidylic acid building block for nucleic acids.

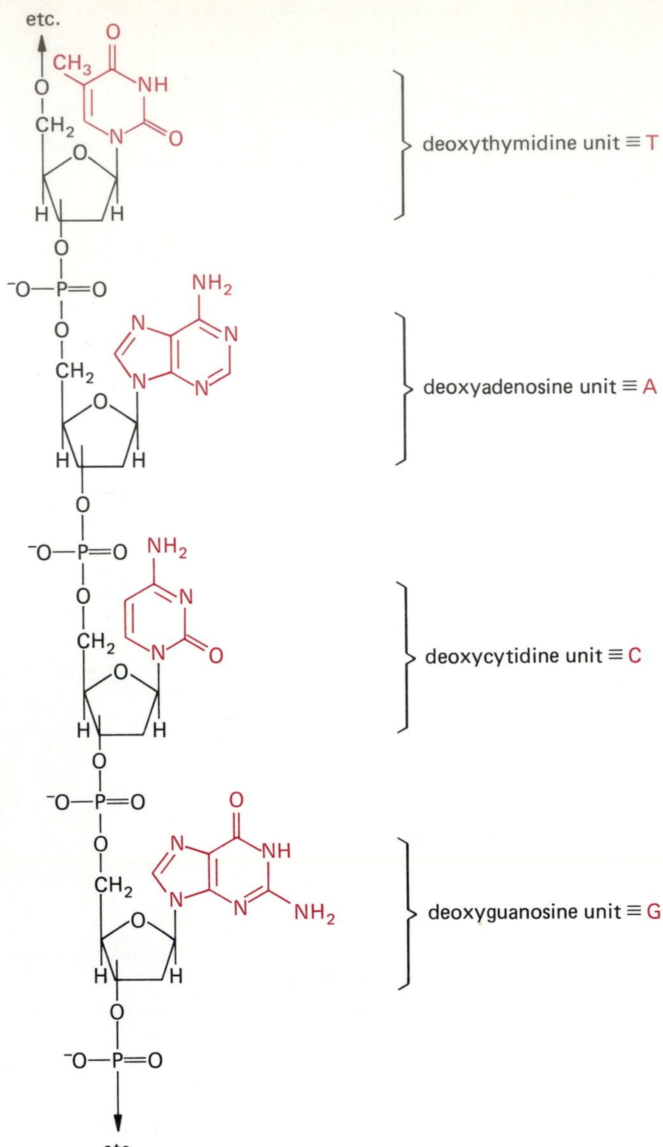

FIGURE 33.7 A portion of a deoxyribonucleic acid (DNA) molecule. The 5′-end is at the top and the 3′-end is at the bottom.

sequence is illustrated in Figure 33.7. The beginning of the chain is the phosphate bonded to C-5′ of the pentose (the 5′-end); the end of the chain is at the C-3′ of the pentose to which the phosphate is bonded in the chain (the 3′-end). As shown in Figure 33.7, each unit is labeled with a one-letter symbol to show which heterocyclic base is attached. The symbols, which are the letters of the **genetic code,** are **T** (2′-deoxythymidine), **A** (2′-deoxyadenosine), **C** (2′-deoxycytidine), and **G** (2′-deoxyguanosine).

B. Molecular Recognition in Nucleic Acids

That heredity exists has been clear to humans for thousands of years. The monk Gregor Mendel found that heredity depends upon discrete units that we now call genes and is governed by definite laws that are the basis of the science of genetics. Later, genes were associated with chromosomes, which contain both DNA and protein. Experiments showed that DNA can transform certain bacteria from one type to another, and several

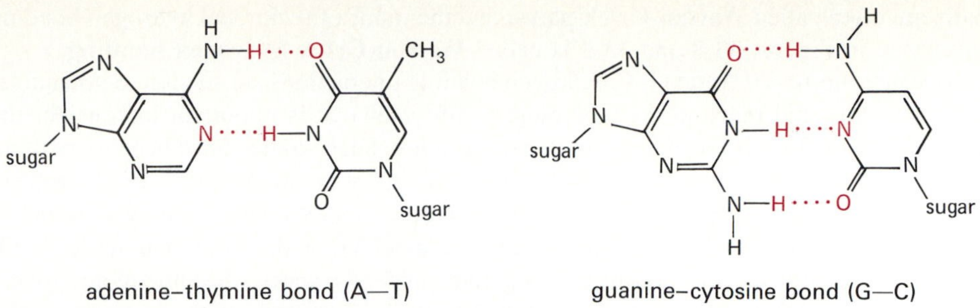

adenine–thymine bond (A—T)

guanine–cytosine bond (G—C)

FIGURE 33.8 Base pairs of DNA. The complementary hydrogen bonds responsible for the DNA double-stranded structure are shown in color. In RNA, analogous pairing exists between adenine and uracil and between guanine and cytosine.

years later J. D. Watson and F. H. C. Crick deduced the pairing scheme by which two strands of DNA recognize each other. The elementary unit of molecular recognition in DNA is the pairing interaction between adenine and thymine or between guanine and cytosine, as shown in Figure 33.8. The molecular geometry is such that adenine forms strong reciprocal hydrogen bonds to thymine (the A–T pair) and guanine forms strong reciprocal hydrogen bonds to cytosine (the G–C pair). The short stretch of a strand pair shown in Figure 33.9 illustrates the general character of the pairing. The two strands run in opposite (antiparallel) directions from the 5′-end to the 3′-end, the right-hand strand from top to bottom and the left-hand strand from the bottom to top. The A–T and G–C

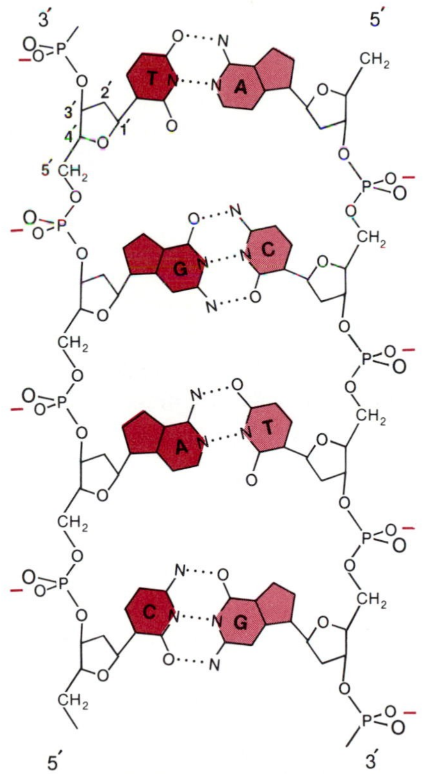

FIGURE 33.9 The DNA double helix is unrolled to show the phosphodiester backbone of the two complementary strands and the base pairs that hold them together. The backbones run in opposite (antiparallel) directions. Each base pair has one purine base, adenine (A) or guanine (G), and one pyrimidine base, thymine (T) or cytosine (C), connected by hydrogen bonds. [Illustration copyright © by Irving Geis.]

pairs are often called **Watson-Crick pairs,** and the motif of reciprocal hydrogen bonding illustrated in Figures 33.8 and 33.9 is called **Watson-Crick hydrogen bonding.**

Because the A–T(U) and G–C hydrogen bonding interactions are so vital to polynucleotide structure and function (and therefore to life itself), it is important to consider the intrinsic interactions between the parent heterocycles. Such studies have been carried out with simple heterocyclic compounds that are models for the nucleosides. In chloroform solution at 25°C, the association of 1-cyclohexyluracil and 9-ethyladenine was found to have $K_a = 103\ M^{-1}$, corresponding to $\Delta H°$ of association of -6.2 kcal $mole^{-1}$. Of course, the entropy of the association is negative, since two species become more ordered in the process; $\Delta S°$ for the process was found to be -12 entropy units (e.u.). Thus, the free energy of association for this process, $\Delta G°$, is -2.4 kcal $mole^{-1}$ at 25 °C. These values can be compared with those for self-association of 9-ethyladenine and 1-cyclohexyluracil, which were found to have K_a values of 3 and 6 M^{-1}, respectively. These self-association constants correspond to enthalpies of association, $\Delta H°$, of -4.0 and -4.3 kcal $mole^{-1}$, respectively. Because both of the self-association reactions also have negative entropies of association, about -11 e.u. in each case, the free energies of self-association of 1-cyclohexyluracil and 9-ethyladenine at 25 °C are only -0.4 and -0.7 kcal $mole^{-1}$, respectively.

1-cyclohexyluracil + 9-ethyladenine ⇌ "A–U complex"

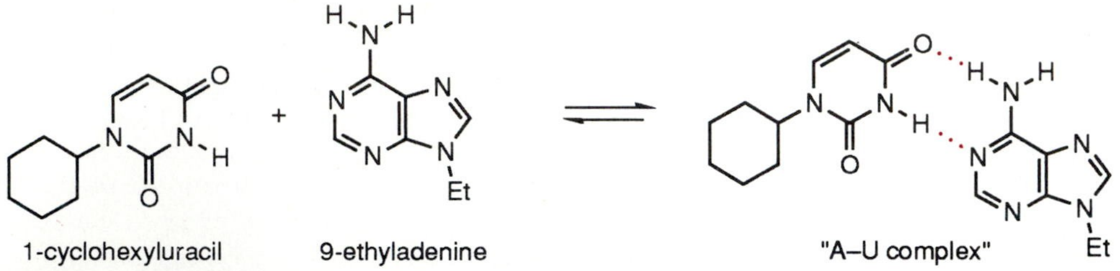

EXERCISE 33.4 Using the foregoing values of $\Delta G°$, calculate the composition at 25 °C of a chloroform solution that is nominally 0.1 M each in 1-cyclohexyluracil and 9-ethyladenine.

Similar studies have been carried out with guanosine and cytidine in dimethyl sulfoxide (DMSO) at 32 °C. In this case the thermodynamic parameters of the association were found to be $K_a = 3.7\ M^{-1}$, $\Delta H° = -5.8$ kcal $mole^{-1}$, $\Delta S° = -16$ e.u., and $\Delta G° = -0.7$ kcal $mole^{-1}$. The self-associations of cytidine with itself and of guanosine with itself were found in this study to be very small, with K_a values on the order of 0.1 M^{-1}. It is important to note that the values given here for A–U association in chloroform and G–C association in DMSO cannot be directly compared because DMSO is a polar aprotic solvent (page 186) that can form strong hydrogen bonds with monomeric cytidine and guanosine. The monomers must sacrifice these strong interactions with the solvent in

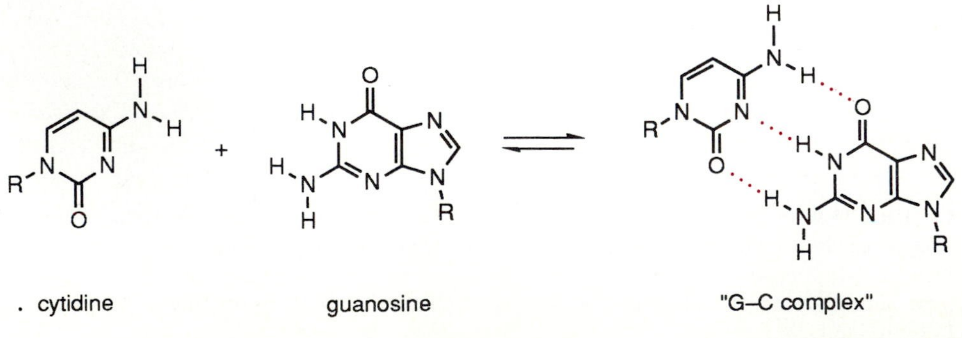

cytidine + guanosine ⇌ "G–C complex"

order to associate. Other studies indicate that, under comparable conditions, the enthalpy of association, $\Delta H°$, is about -4 kcal mole^{-1} for A–T(U) and -6 kcal mole^{-1} for G–C. These enthalpies of association correspond to approximately 2 kcal mole^{-1} per hydrogen bond, a normal value for such interactions.

Studies such as these show that there is high specificity in the recognition of the four nucleic acid bases for each other. That is, of the ten bimolecular complexes that can be formed by association of A, T(U), G, and C, only the A–T(U) and G–C complexes have substantially negative free energies of binding. However, even with the ideal combinations, the binding energies are rather small, considering the unfavorable entropy of -11 to -16 e.u., corresponding to $T\Delta S°$ of $+3.3$ to $+4.8$ kcal mole^{-1} at 25 °C. Furthermore, in aqueous solution even the monomeric nucleic acids profit from hydrogen bonding with solvent molecules. In aqueous solution, the net free energy of binding due to the reciprocal hydrogen bonds in the A–T(U) and G–C base pairs is undoubtedly less than -1 kcal mole^{-1}, and the association might even be slightly endothermic. How can such a small free energy of binding be a major factor in the structure of DNA, the very blueprint of life? There is a simple answer to this question. The free energy of binding adenine to uracil is the sum of a rather negative enthalpy of binding (about -4 kcal mole^{-1}) and an unfavorable entropy term that is of comparable magnitude at room temperature. However, consider the free energy of binding a dinucleotide AA to its complement, UU.

$$\text{A—A} \quad + \quad \text{U—U} \quad \rightleftharpoons \quad \begin{array}{c}\text{A—A}\\ \vdots \quad \vdots \\ \text{U—U}\end{array}$$

In this case, each of the A–U pairs will contribute about -4 kcal mole^{-1} to the enthalpy of binding, so the hydrogen bond forces binding A–A to U–U are about twice those that bind monomeric A to monomeric U. However, the unfavorable entropy of association is approximately the same whether the association is a pair of monomeric bases or a pair of complementary oligonucleotides. For the association of a pair of monomers, the unfavorable entropy almost exactly cancels the favorable enthalpy of association, but, the longer the oligonucleotide, the less effect the entropy term has. In a double-stranded oligonucleotide, the complementary strands are held together by many base pairs, and the many weak interactions combine to give a large total inter-strand binding energy.

Highlight 33.4

The pairings of adenine with thymine via two hydrogen bonds and of guanine with cytosine by three hydrogen bonds are the specific elements of molecular recognition underlying the structure of deoxyribonucleic acid (DNA).

DNA can occur in several forms, designated B, C, D, E, T, and Z. The usual form is the B-form, which is a double-stranded helix of two individual molecules about 20 Å apart. As has just been explained, the two chains are held together by reciprocal hydrogen bonding between pairs of bases in opposite positions in the two chains. The specific recognition of thymine by adenine and guanine by cytosine results in the two interwoven chains having exactly complementary structures. Thus, if a segment of one strand has the base sequence 5′GCAATGCC3′, the complementary strand has the base sequence 5′GGCATTGC3′, and the two chains are hydrogen bonded together as symbolized by

```
5'-end ~~G—C—A—A—T—G—C—C~~3'-end
         :  :  :  :  :  :  :  :
3'-end ~~C—G—T—T—A—C—G—G~~5'-end
```

A three-dimensional view of a stretch of double helical DNA is shown in Figure 33.10. One notable characteristic of the B-form double helix is the occurrence of alternating **wide (major)** and **narrow (minor) grooves.** These grooves are the sites of many important interactions. The wide groove is on the side away from the nitrogen atoms that are attached to the C-1′-positions of the deoxyribose. B-form DNA is composed of right-handed helices with a 34 Å pitch, the distance from one point in the helix to the same point in the next turn. There are 10 residues per turn, from which we know that the helix advances 3.4 Å per residue. It follows that a stretched-out chromosome could be 3.4×10^8 Å long or 3.4 cm! A typical cell might be 2×10^{-3} cm long, so it is clear that DNA must be folded and packed into tertiary and quaternary structures that fit within the cell.

The helical structure of B-form DNA that is shown in Figure 33.10 illustrates **base stacking,** another element of molecular recognition that plays an important role in DNA structure. Note that the flat heterocyclic bases lie in planes that are roughly parallel to each other. The pitch of the helix tells us that the average distance between these parallel planes is about 3.4 Å, exactly the same distance that separates the planes of graphite (page 1072). Furthermore, x-ray studies of crystalline nucleic acid bases have shown that they form stacks with the molecules 3.4 Å apart. The hydrophobic recognition of one nucleic acid base for another provides an attractive force that helps to bind the two strands together. In addition, it is the aromatic inter-plane distance of 3.4 Å that causes DNA to have its helical shape!

A consideration of this double-stranded, helical nature of DNA helps us to understand another peculiarity of its structure. Why is it that we do not find other pairs, for example, A–A? As shown by the following structure, two adenines can form reciprocal hydrogen bonds, and the previously discussed studies of association of 9-ethyladenine showed that the free energy of association of this model compound in chloroform solution is actually negative.

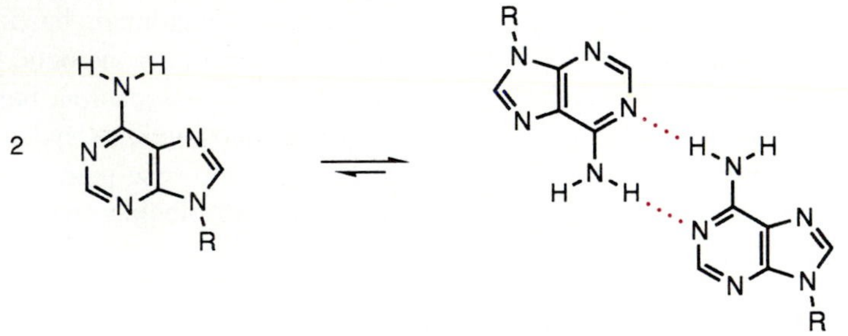

Once again, the answer is simple. The reason we do not find A–A pairs is that the total width of an A–A hydrogen-bonded dimer is too great to fit the 20-Å strand separation in double-stranded, helical DNA.

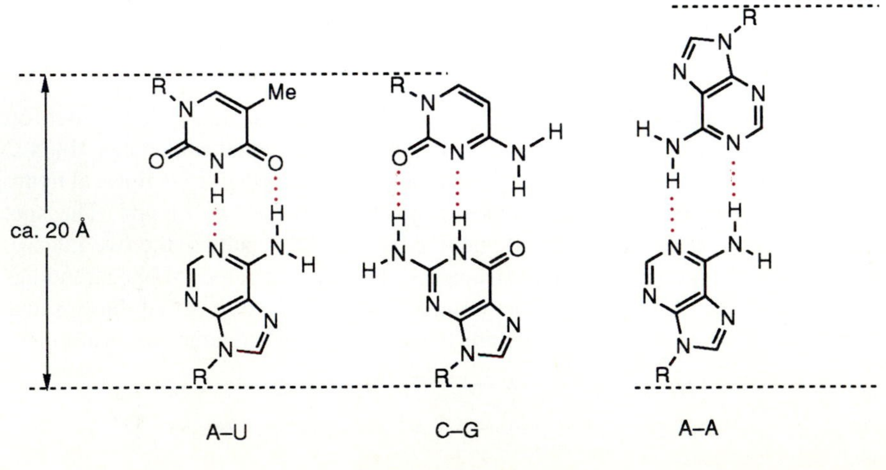

ca. 20 Å

A–U C–G A–A

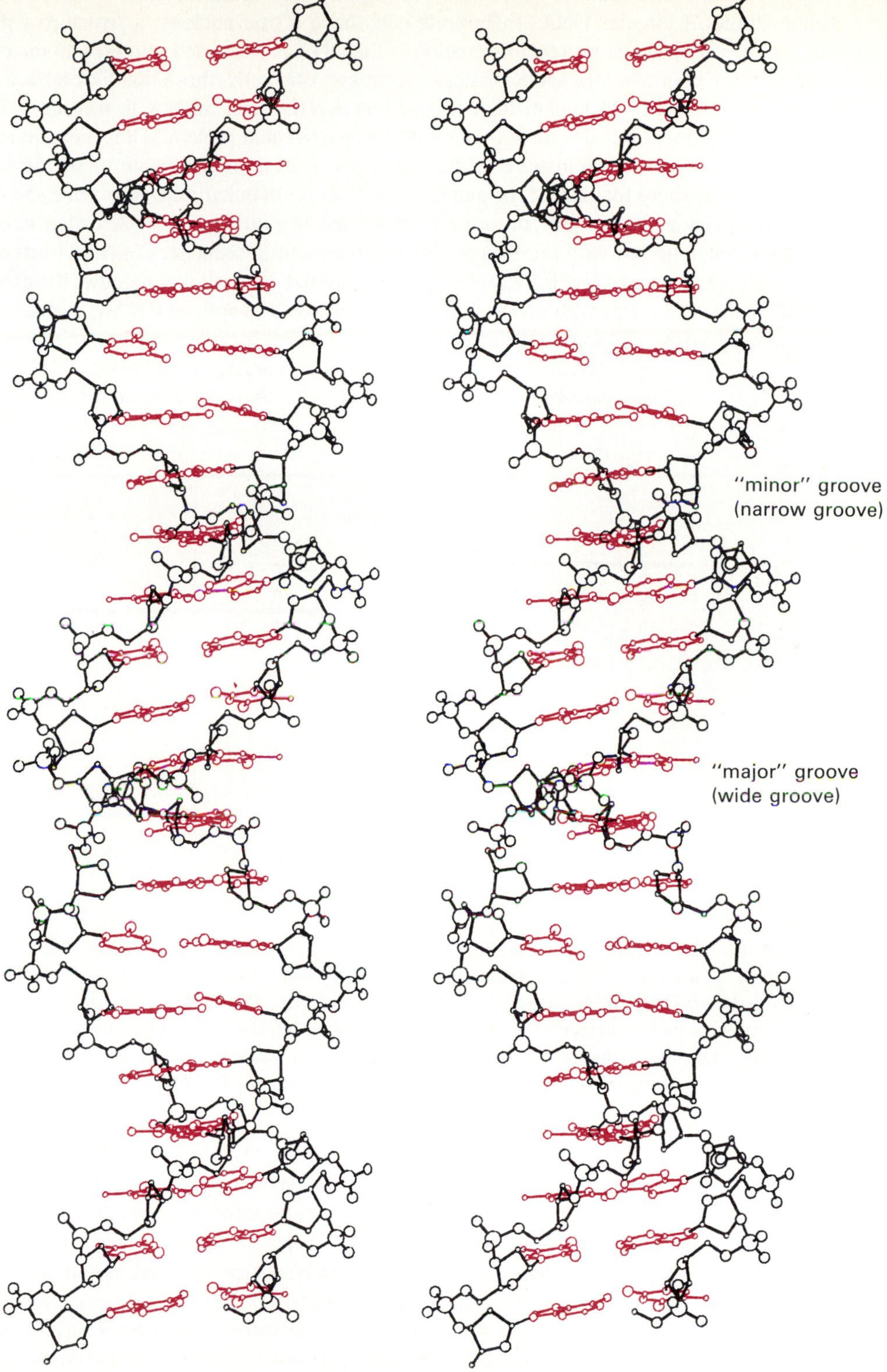

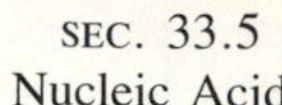

"minor" groove
(narrow groove)

"major" groove
(wide groove)

FIGURE 33.10 Stereo structure of a stretch of helical B-form DNA. The base sequence of each strand is GCGAATTCGC. [Illustration copyright © by Irving Geis.]

DNA occurs in the nuclei of all cells. In prokaryotic organisms, the ''nucleus'' is a single strand of circular DNA. Eukaryotic cells have a true nucleus, a structure with a nuclear membrane. The genetic information in the DNA is converted into protein material of the cell in a complicated series of steps. In prokaryotic cells, those that lack a nucleus, the DNA is transcribed into **mRNA** (messenger RNA). The mRNA is translated into protein at the ribosome, a complex combination of RNA and protein. The ''genetic message'' is read by **tRNA** (transfer RNA) and an enzyme system. There are many different tRNAs, one or more for each of the natural amino acids. In eukaryotic cells, an extra step is needed because the genetic message is contained in portions of DNA called **exons,** which are interspersed with sections of DNA (''intervening sequences'') called **introns.** The DNA is transcribed into RNA, which is then ''edited'' by splicing together the exons, a process that can occur with nonprotein RNA catalysts. The spliced RNA is mRNA that is translated at the ribosome into protein.

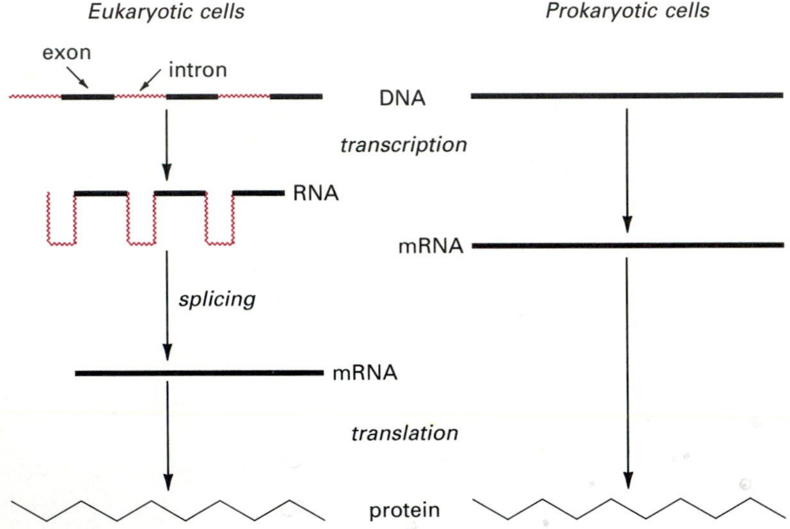

DNA exons and introns are transcribed into RNA (pre-mRNA or hnRNA, see Section 33.4.C). In the early 1980s, S. Altman and T. R. Cech found that these were excised even in the absence of protein by action of the RNA intron itself. The introns that had catalytic activity were called ribozymes. Such activity kindled an interest in an older idea that RNA preceded DNA in evolution, i.e., that there was an RNA world before there was a DNA-protein world. The origins of organic molecules and prebiotic evolution are fascinating subjects in themselves.

The DNA carries the information necessary for the exact duplication of the cell and, in fact, for the construction of the entire organism. Genetic information is duplicated in a process called DNA replication. The strands must separate, a result that can also be effected by raising the temperature. Recall that the two species involved in mutual recognition can also interact with the solvent. As the temperature rises, the rate of dissociation of the strands increases and the equilibrium shifts to a situation in which the strands have separated and the heterocyclic bases form hydrogen bonds to solvent molecules rather than to their complements on another chain. The separation of the strands is readily seen through the increase in the ultraviolet absorption of the heterocyclic bases. The **melting temperature** is that at which half of the maximum absorption increase has been reached in a plot of absorbance versus temperature. For short sequences, the process is reversible, but the cooling curve is usually not the same as the heating curve. There are three hydrogen bonds in G–C pairs as opposed to two hydrogen bonds in A–T pairs. Not surprisingly, the higher the G–C content, the higher the DNA melting temperature.

EXERCISE 33.5 One technique that has been used to gain information about the nucleic acid composition of DNA is "buoyant-density" measurement. In this experiment a concentrated solution of cesium chloride is subjected to a strong centrifugal force in a powerful centrifuge. This results in a linear density gradient. For example, the densities at the top and bottom of a 1-cm column are 1.65 g mL^{-1} and 1.76 g mL^{-1}, respectively. If a sample of DNA is present during establishment of the gradient, it collects in a narrow band at the point where its buoyant density is equal to that of the cesium chloride solution. One of the main pieces of information from buoyant density measurements is a measure of the G–C content of the DNA sample. Why do you think buoyant density is related to the G–C content?

The genetic information in the DNA specifies the amino acid sequence in proteins. The information resides in the precise sequence of purine and pyrimidine bases attached to its phosphodiester backbone. Each set of three bases specifies one amino acid, or in a few cases, an essential instruction. As noted above, the genetic code is made of letters that are organized as a **triplet code.** The principal function of DNA is as the master blueprint for the production of proteins, the essential structural materials and catalysts for all cellular reactions. The discrete unit of hereditary information, the **gene,** corresponds to the information for the primary structure of a particular protein. The **genetic code** consists of a sequence of three-letter words, each letter representing a heterocyclic base and each word, a **triad** of heterocyclic bases, specifying a particular amino acid. Although there are only 20 "natural" amino acids in proteins, the four different bases can be arranged in a total of 64 different words **(codons)** of triads. Of the 64, 61 are used to code for the various amino acids and three are "stop" messages. The full genetic code is given in Figure 33.11. For example, the sequence of bases AAA in a gene codes for phenylalanine, GAA for leucine,

	A		G		T		C	
A	AAA	Phe	AGA	Ser	ATA	Tyr	ACA	Cys
	AAG	Phe	AGG	Ser	ATG	Tyr	ACG	Cys
	AAT	Leu	AGT	Ser	ATT	stop	ACT	stop
	AAC	Leu	AGC	Ser	ATC	stop	ACC	Trp
G	GAA	Leu	GGA	Pro	GTA	His	GCA	Arg
	GAG	Leu	GGG	Pro	GTG	His	GCG	Arg
	GAT	Leu	GGT	Pro	GTT	Gln	GCT	Arg
	GAC	Leu	GGC	Pro	GTC	Gln	GCC	Arg
T	TAA	Ile	TGA	Thr	TTA	Asn	TCA	Ser
	TAG	Ile	TGG	Thr	TTG	Asn	TCG	Ser
	TAT	Ile	TGT	Thr	TTT	Lys	TCT	Arg
	TAC	Met	TGC	Thr	TTC	Lys	TCC	Arg
C	CAA	Val	CGA	Ala	CTA	Asp	CCA	Gly
	CAG	Val	CGG	Ala	CTG	Asp	CCG	Gly
	CAT	Val	CGT	Ala	CTT	Asp	CCT	Gly
	CAC	Val	CGC	Ala	CTC	Glu	CCC	Gly

FIGURE 33.11 The genetic code. The third base of each triad is less specific than the first two. In RNA, each word (an RNA **codon**) is a complement of the one shown here. For example, the RNA complement of CCC (glycine) is GGG, and that for AAA (phenylalanine) is UUU (uracil replaces thymine).

and so on. As the figure shows, there is a considerable amount of redundancy in the code, since it has 64 words and only 21 are needed. The third base in each triad is less specific than the first two; this phenomenon is referred to as "wobble" and the third base in the codon is called the "wobble base."

EXERCISE 33.6 What is the structure of the peptide coded by the following DNA sequence?

GTACCCGATTGACAGCAGAATAATAATTACGAAGAAATT

Unlike DNA, RNA molecules exist as single strands with rather irregular structures. The three general types of RNA have already been mentioned. Ribosomal RNA **(rRNA)** constitutes the major amount (90%) and serves a structural and catalytic function. Since there are thousands of different possible proteins that can be programmed by DNA, there are correspondingly thousands of different possible messenger RNA **(mRNA)** types. Only a fraction are present at any one time, with mRNA constituting 3% of the total RNA. The mRNA functions as a template for the synthesis of proteins (see scheme above). The mRNA binds many ribosomes, forming a polyribosome, so that many proteins can be synthesized at the same time. Transfer RNA **(tRNA)**, about 7% of the total RNA, has an **anticodon** that recognizes the appropriate codon on mRNA and delivers the appropriate amino acid to the growing protein chain. There is at least one transfer RNA for each of the amino acids. The principles involved in the recognition of a codon by an anticodon are the same as those that are responsible for the double-stranded structure of DNA—reciprocal hydrogen bonding within A–U and G–C pairs.

C. Determination of the Primary Structure of DNA: Sequencing

Many fascinating details of DNA and RNA chemistry have been worked out over the last quarter century. Much of the story of the genetic code and how it functions has been unraveled, but much more remains to be learned. It is not appropriate for us to go into very many of the details of the molecular biology of gene transcription and protein synthesis in this introductory chemistry text. We will, however, take a brief look at one important chemical aspect of nucleic acid chemistry, the methods by which DNA primary structures are determined. Because this process tells us the sequence of bases along the phosphodiester backbone of the molecule, it is called **sequencing.** The DNA sequences are used to obtain information about family and genetic relationships and to deduce the amino acid sequences of proteins.

Before a given specimen of DNA can be sequenced, a sample suitable for analysis must be prepared. DNA is located mostly within the nucleus of a cell and is isolated by rupturing the cell, centrifuging down the nucleus, and extracting the nuclear material. A general scheme is given in Figure 33.12. The genomic DNA is then enzymatically cut into small restriction fragments by **restriction nucleases,** enzymes that recognize particular sequences in the DNA by precisely the kind of recognition principles that we have discussed in this chapter. A number of restriction enzymes have been characterized. The specificity of a restriction enzyme depends on the length of the polynucleotide sequence that it recognizes and binds. The specificity of the restriction enzyme is an important consideration. The probability of finding a particular sequence of n bases is $1/P$ (in which $P = 4^n/2 + 4^{n/2}/2$). Thus, the chances of finding the specific sequence AACG is $1/(4^4/2 + 4^2/2) = 1/136$. Two restriction enzymes that are often used in laboratory degradations

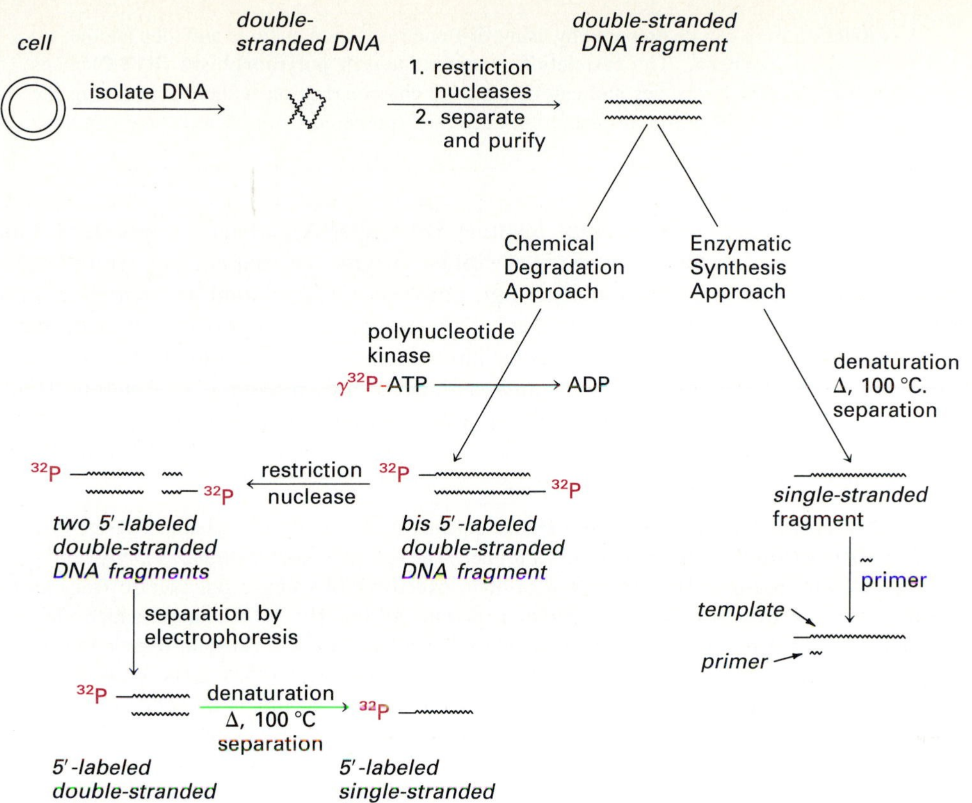

FIGURE 33.12 Preparation of single-stranded DNA fragments for further analysis, either by the chemical degradation approach or the enzymatic synthesis approach.

of DNA to smaller fragments are Eco R1 from *Escherichia coli* and Hind III from *Hemophilus influenzae*. These restriction enzymes are ''6-cutters,'' meaning that they each recognize a unique sequence of six bases and cause strand cleavage at these points.

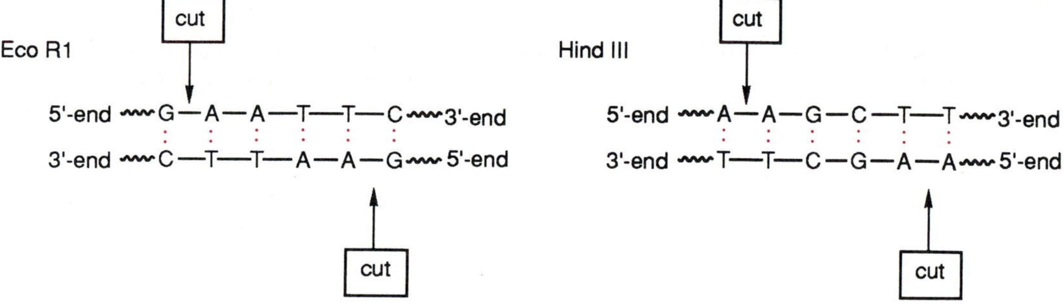

The chances of finding the specific GAATTC or AAGCTT sequences that are recognized by Eco RI and Hind III are 1/2080. A series of two or three different restriction enzyme treatments is normally sufficient to reduce fragment length to 200–300 **bp** or **base pairs,** a size that can be sequenced by the methods to be described. After degrading the DNA sample by the use of two or three different restriction enzyme treatments, a short piece of double-stranded DNA is obtained from the mixture of fragments by ''cloning,'' a procedure too complex for the present discussion.

A **restriction map** can be prepared by using first one restriction enzyme and then another on the separated fragments. The **restriction fragment length polymorphism (RFLP)** maps differ from species to species and can be used for characterization without sequencing the DNA. More than 100 different restriction nucleases, prepared from bacteria, are commercially available.

Another source for DNA is **complementary DNA (cDNA),** which is synthesized from mRNA as a template with a **reverse transcriptase** enzyme, present in certain retroviruses (RNA viruses). For amplification, the single-stranded cDNA is used as a template for a DNA polymerase, and the double-stranded cDNA is cloned or amplified by the **polymerase chain reaction (PCR).** cDNA lacks the introns present in the whole DNA and is more likely to contain uninterrupted nucleotide sequences corresponding to proteins. Thus, sequencing DNA fragments from cDNA is useful for deducing protein sequences.

A DNA polymerase can duplicate a DNA sequence from an existing chain, a mixture of 2′-deoxynucleoside triphosphates, and a primer. The resulting double-stranded DNA can be separated by heating. The simple but brilliant expedient of using a polymerase from an organism that lives at elevated temperatures permits a ''one-pot'' DNA amplification. The organism *Thermus aquaticus (Taq),* from which the polymerase was isolated, lives in the hot springs of Yellowstone National Park at temperatures above 70–75 °C. The sequence of procedures known as the **polymerase chain reaction (PCR)** consists of the following steps: Cycle 1: (1) heating to dissociate a sample of double-stranded DNA; (2) annealing to promote combination with primers (short, 15–30 base pair oligonucleotides); (3) primer extension by *Taq* polymerase; Cycle 2: (1) heating to dissociate two molecules of double-stranded DNA; (2) annealing to promote combination with primers; (3) primer extension by *Taq* polymerase; and so on for as many cycles as are necessary to make the desired number of DNA molecules. This process can be automated inexpensively. Production of more than 100,000 DNA copies per original is easily achieved in a short time (10–50 cycles in less than a day), making possible the amplification and analysis of DNA molecules that are available in very small amounts, even one or two molecules. Since the *Taq* polymerase can be used for polymerization at a relatively high temperature, the degree of **molecular recognition** that is required for the attachment of the nucleotides is higher (lower tendency to associate) and the **fidelity** of the transcription is higher.

One efficient method for determining the base sequence in DNA is the **Maxam-Gilbert method** or **chemical degradation approach** introduced in 1977. One first labels the end of the DNA chain with a radioactive isotope of phosphorus (^{32}P). Radiolabeling is accomplished by treating the DNA with an enzyme, such as polynucleotide kinase, that transfers a phosphate group from radiolabeled ATP (γ-^{32}P-ATP) to the 5′-ends of DNA molecules.

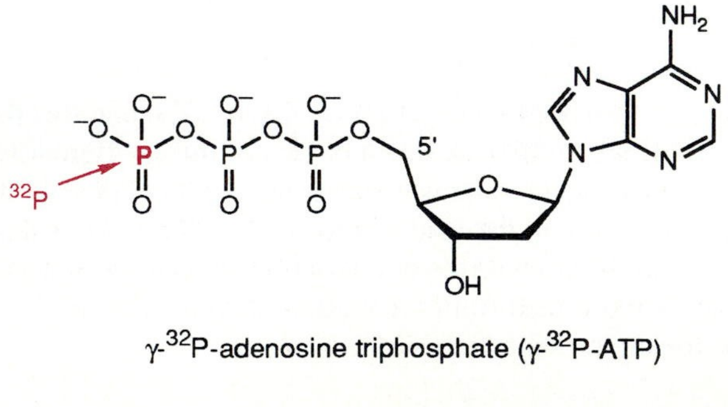

γ-^{32}P-adenosine triphosphate (γ-^{32}P-ATP)

Other enzymes can also be used to label the chains at the 3′-ends. The purpose of the ^{32}P is to render the end of the DNA chain radioactive so that fragments can be easily detected, as will be explained later.

The radiolabeled DNA is subjected to some reagent that causes base-specific cleavage of the chain. For example, treatment of the specimen with dimethyl sulfate, followed by aqueous base results in specific rupture of the imidazole rings of guanines. Further treatment with piperidine and then aqueous base eliminates the two phosphate groups, thus cleaving the phosphodiester chain at each site occupied by the hydrolyzed guanines. The chemistry of the process can be represented in the comprehensive reaction scheme shown in Figure 33.13. In the first step, N-7 of the guanine acts as a nucleophile in an S_N2 reaction with the dimethyl sulfate, giving a resonance-stabilized imidazole cation.

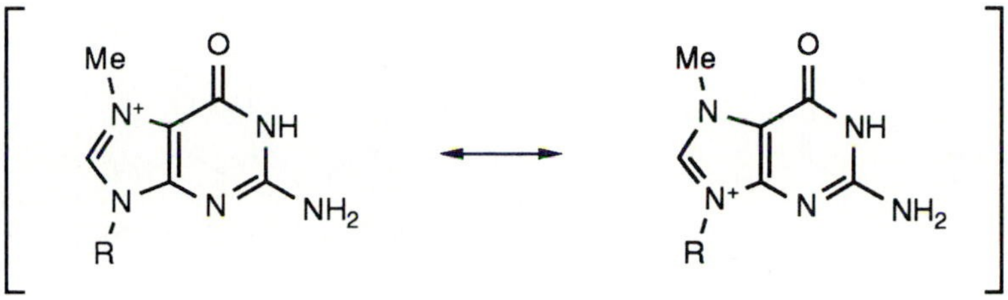

Reactivity toward nucleophiles is enhanced in the quaternized heterocyclic ring at C-8. Attack by hydroxide on the carbon of the polar C═N bond yields a carbinolamine reminiscent of the intermediate in base-catalyzed amide hydrolysis. Further reaction with hydroxide opens the five-membered ring of the purine and yields a riboside equivalent to a carbinolamine. The latter can open to a Schiff base (page 399) that is susceptible to attack by piperidine to form a piperidinium ion in which the ribose ring remains open. With the ribose ring now open, the phosphodiester chain itself is vulnerable to cleavage. The immonium ion activates the α-hydrogen towards attack by base, resulting in an E2 reaction (Sections 9.7, 11.5) with loss of the β-phosphate group. A second E2 reaction causes the loss of the γ-hydrogen and the δ-phosphate, releasing what was the 5′-end of the DNA chain. A controlled set of reactions has thus cleaved the DNA chain at the location of a specific base, in this instance, guanine.

Application of the foregoing chemical cleavage method to a DNA molecule that has been labeled with ^{32}P at the 5′-end produces a complex mixture of fragments. All of the fragments that still contain the 5′-end are radioactive. Of course, there are many fragments that do not contain the 5′-end of the molecule, but since the analytical technique only detects radioactive molecules, these are invisible and of no consequence. An important element of the technique is that cleavage is not total. Thus, a large number of fragments are produced, and most end with a base *that was next to guanine in the intact DNA*. The mixture is separated using gel electrophoresis, which depends on diffusion under the influence of a strong electric field. In this technique, the molecules travel through a gel at a rate related to the chain length—the smaller the fragment, the more rapidly it moves. The pattern in the gel is revealed by autoradiography, a sensitive technique that reveals the presence of the radioactive fragments as dark lines on the developed film. The distance of the line from the origin is inversely related to the distance of a guanine from the 5′-end of the intact DNA chain.

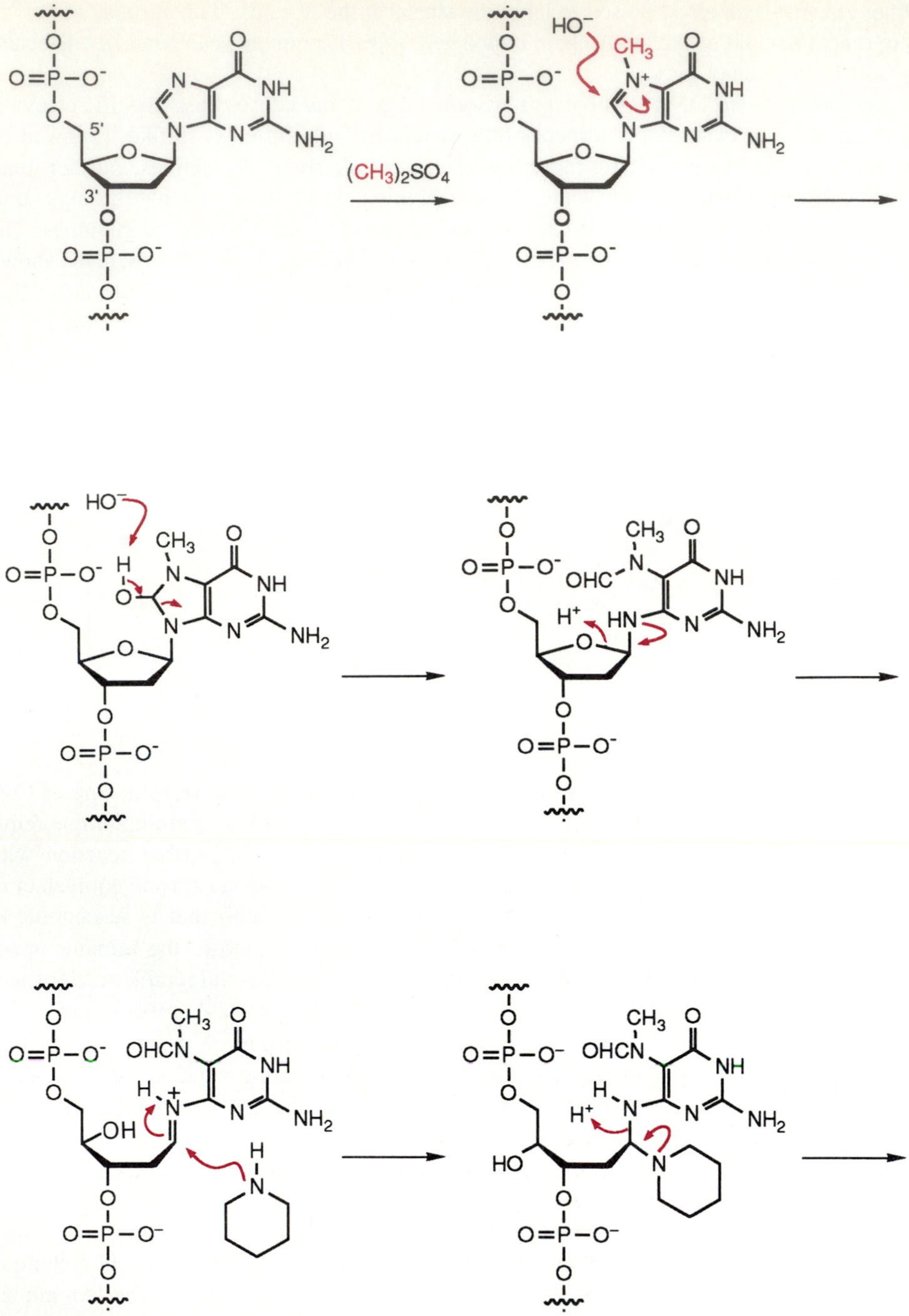

$(CH_3)_2SO_4$

FIGURE 33.13 A comprehensive reaction scheme for explaining the removal of guanine from a DNA segment.

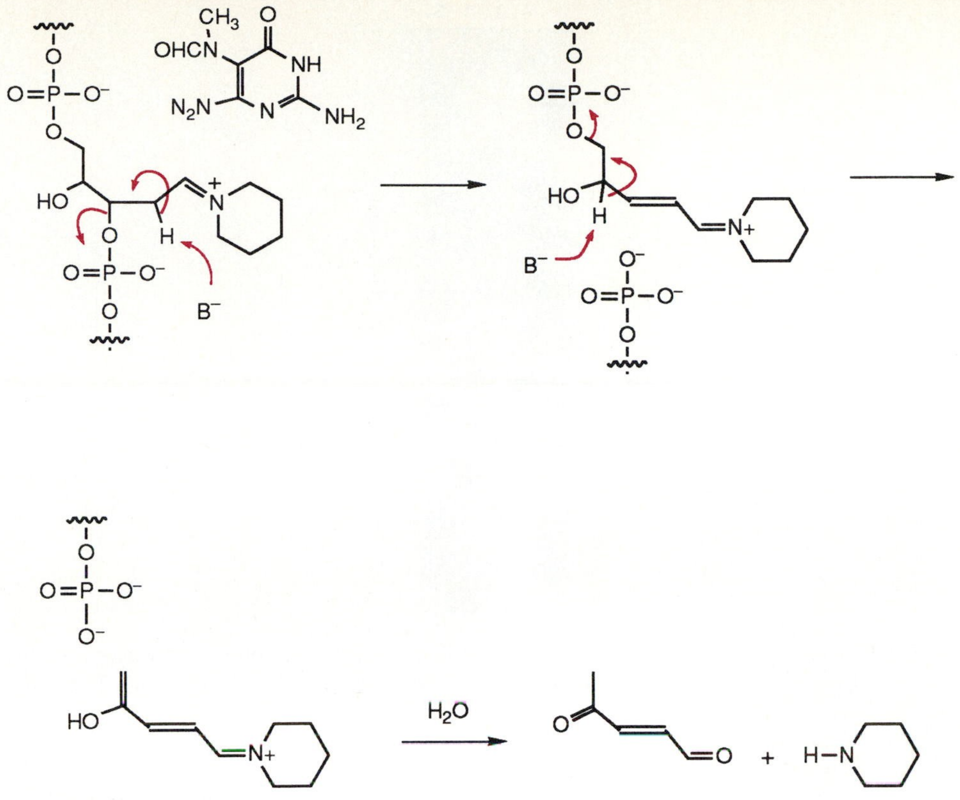

FIGURE 33.13 Continued.

Radiography is a technique in which radioactivity is used for imaging. Most people are familiar with the use of x-rays for this purpose. The x-rays are passed through the object to be examined onto a sheet of photographic film. X-rays ionize molecules to produce electrons that activate silver halide crystals for development (page 1238). The degree of exposure of the film is related to the intensity of radiation that passes through various regions of the object. In autoradiography, the substance to be detected is radioactive and produces the radiation. A mixture of radioactive substances is subjected to a thin layer chromatographic or thin layer gel electrophoretic separation. The plate carrying the separated components is then placed in contact with a photographic film for a period of time, often in the presence of an enhancer to convert more radiation into detectable photons or electrons. When the film is developed, dark spots or bands appear at points corresponding to the positions of radioactive compounds on the chromatogram or electrophoretic gel.

Chemistry such as that described has been developed for cleavage of DNA at each of the other three bases. To fully sequence a DNA molecule, one applies each of the four base-selective cleavage reactions to the radiolabeled DNA and subjects the reaction mixtures to gel electrophoresis. The autoradiographs resulting from these four analyses reveal the presence of every single base in the molecule. An example of an autoradiograph is shown in Figure 33.14. In this example, the two outer ''lanes'' were created by electrophoresis of a mixture of fragments that were created by application of the guanine-specific

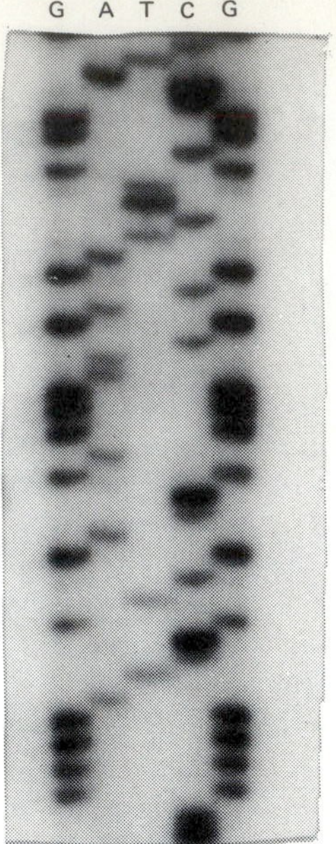

FIGURE 33.14 Autoradiograph of a sequencing gel. Reading from the bottom, the sequence of this 44-base deoxyribonucleotide is CCGGGGATCCGTCGACCGAGGGAACGACGATCTTGC-GGCCATCG.

method just described. The other three lanes were created by electrophoresis of mixtures produced by treatment of the radiolabeled deoxyribonucleotide with reagents specific for cleavage at adenine, thymine, and cytidine, respectively. The sequence is ''read'' from the bottom, by noting which lane is darkened for each position. For example, in this case, the bottom two spots are in the C-lane, the next four are in the G-lane, the seventh is in the A-lane, etc. In practice, segments of DNA up to 250 bases in length can be sequenced by this method.

EXERCISE 33.7 A second reaction that is used in the Maxam-Gilbert sequencing method involves treatment of the DNA fragment first with hydrazine (H_2NNH_2) and then with piperidine. This pair of reactions results in chain cleavage only at thymine and cytidine sites. Write a plausible mechanism that explains the selectivity of the cleavage.

A second useful method for sequencing DNA is an enzymatic synthesis approach developed by F. Sanger, S. Nicklen, and A. R. Coulson in 1977. In this method, a **primer** (a 10–30-bp oligonucleotide) is needed to initiate chain extension on a DNA template. The principal idea is that 2′,3′-dideoxynucleoside triphosphate (ddHetTP) can replace 2′-deoxynucleoside triphosphates as substrates in a polymerase reaction. The growing chain terminates when a dideoxynucleoside is attached because there is no 3′-OH group at which to continue chain growth. A small amount of ddHetTP is sufficient to produce a series of DNA chains, which can be dissociated from the template by heating, and then separated by gel electrophoresis. In a reaction mixture containing ddATP and the four nucleoside triphosphates, the chains end at every location at which A would appear in normal polymerization. Three other reaction mixtures, for ddGTP, ddTTP and ddCTP, give chains with lengths corresponding to the positions of G, T, and C, respectively. The usual polymerase I can be replaced by *Taq* polymerase for use at higher temperatures.

For reading the DNA sequence after electrophoresis, one can prepare ^{32}P- or ^{35}S-labeled chains from α-^{32}P- or α-^{35}S-labeled 2′,3′-dideoxynucleoside triphosphates (for ^{35}S, α-thiotriphosphates) and use autoradiography for four parallel lanes on the gel. Another promising approach that has recently been introduced involves the use of fluorescent-labeled 2′,3′-dideoxynucleoside triphosphates. A different fluorescent label is used for each heterocyclic base. The dNA chains produced in four different polymerase reactions with the differently labeled 2′,3′-dideoxynucleoside triphosphates are mixed and separated by gel electrophoresis. Two of a set of fluorescent triphosphates that are incorporated by the polymerase reaction are

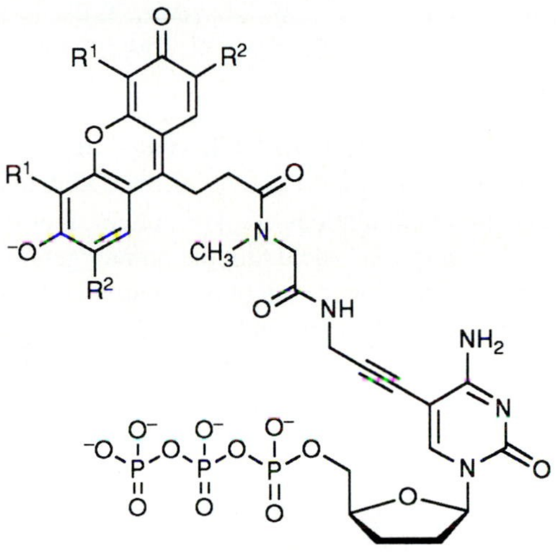

fluorescein-linked 2',3'-dideoxycytidine
5'-triphosphate (ddCTP)

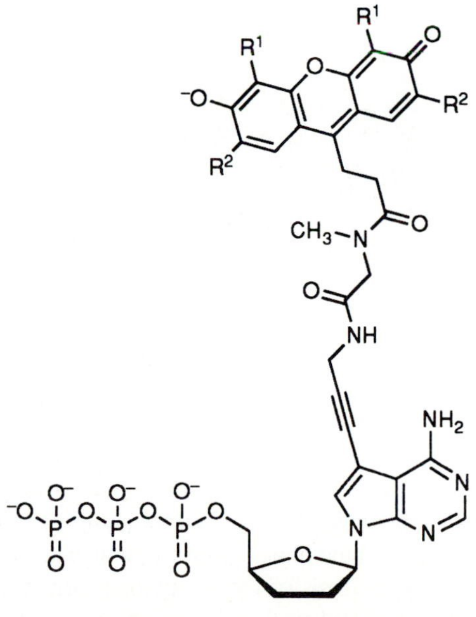

fluorescein-linked 2',3'-dideoxy-7-deazaadenosine
5'-triphosphate (ddATP)

The sequence is then read by detecting laser-induced fluorescence in a process that is easy to automate. The project to read the whole human genome depends upon lowering the costs, increasing the speed, and automating the procedures. Fluorescence labeling appears to be ideal for the purpose, although at its current state of development, it is not as sensitive as autoradiography.

EXERCISE 33.8 It is possible to prepare synthetic RNA from any combination of the four nucleotides that one desires. This synthetic RNA can then be used to induce the synthesis of various polypeptides. For example, administration of polyuridylic acid to the bacterium *Escherichia coli* causes it to synthesize polyphenylalanine. What polypeptide is produced by *E. coli* that has been administered polyadenylic acid? What is the expected empirical composition of the polypeptide produced by *E. coli* that has been administered at random 1:1 copolymer of uridylic and guanidylic acids?

D. High-Specificity Recognition and Cleavage of DNA

The restriction enzymes (Section 33.2.C) used to cleave DNA into fragments suitable for sequencing or other genetic manipulations are 4–8 cutters, that is, they recognize four to eight base pairs. For many purposes, including the human genome sequencing project, a cleavage system that recognizes a longer run of base pairs, and is therefore more specific, would be desirable. Such a reagent has recently been devised by making ingenious use of another remarkable aspect of molecular recognition.

In addition to forming base pairs, nucleotides can also form **base triplets.** Such triplets are possible because one adenine can complex two thymines simultaneously. The same situation is true for guanine, which can hydrogen bond to two cytosines at the same time. The phenomenon is known as **Hoogsteen hydrogen bonding** because it was first discovered in the x-ray crystal structure of a complex of 1-methylthymine and 9-methyladenine by K. Hoogsteen in 1963. Hoogsteen hydrogen bonding is illustrated below for the TAT base triplet. A CGC triplet is formed if a cytidine is protonated at N-3. In these triplets, the Hoogsteen hydrogen bonds are shown in color; the other hydrogen bonds are the normal Watson-Crick hydrogen bonds.

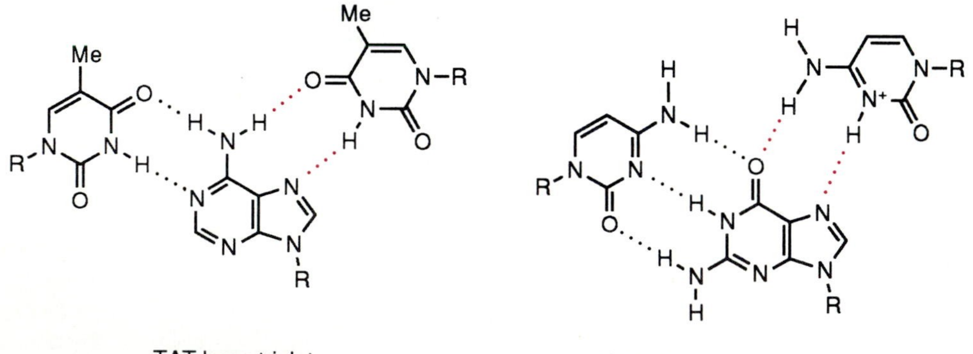

TAT base triplet

C^+GC base triplet

It was discovered in the 1950s that RNA can, under certain conditions, form a **triple helix** in which a third strand binds in the major groove of a normal double-stranded double helix. It was later found that some forms of DNA can also form such triple helices. In a triple helix, the third strand is attached to the ''middle'' strand by Hoogsteen hydrogen bonds. The triple helix phenomenon has been exploited to develop a high-specificity recognition reagent. To demonstrate the method, a 20-base oligonucleotide with the sequence AAAAAGAAAGAAGAAAAGAA was inserted by the techniques of genetic engineering into chromosome III of yeast, *Saccharomyces cerevisiae*. The complementary 20-base oligonucleotide (a "20-mer"), having the sequence TTCTTTTCTTCTT-TCTTTTT, was prepared chemically using a DNA synthesizer. The thymines at either end of the synthetic 20-mer were modified as shown below by attachment of ethylenediaminetetraacetic acid (EDTA). The EDTA units will each form a strong complex with a Fe^{2+} ion.

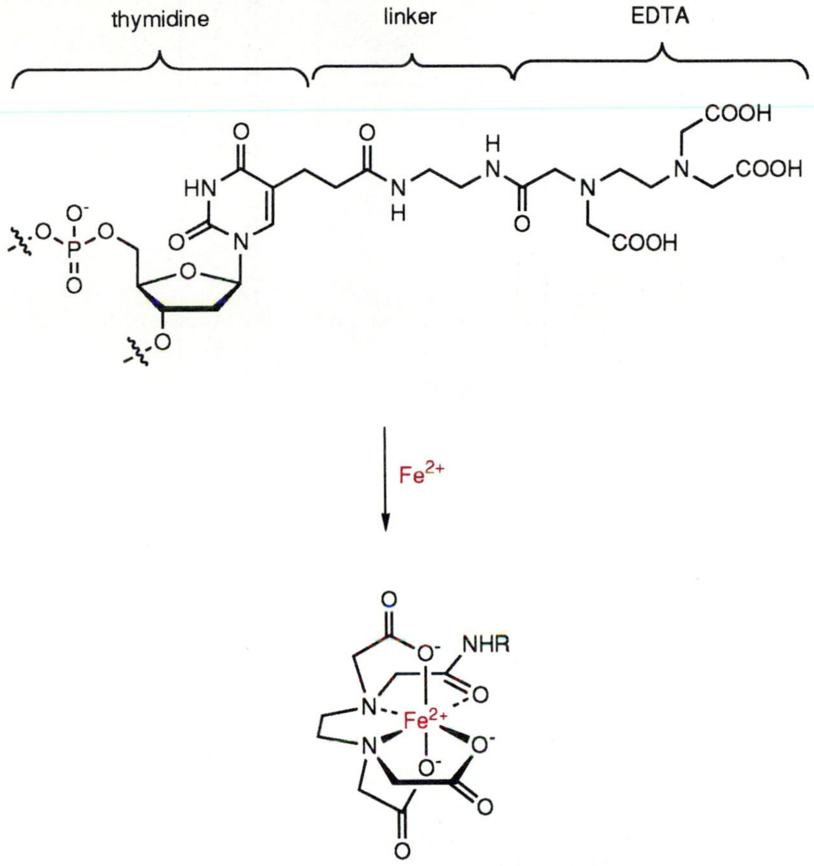

thymidine linker EDTA

$\xrightarrow{Fe^{2+}}$

When the synthetic 20-mer, bearing the EDTA molecules at each end, is added to the genetically modified yeast chromosome, it finds the AAAAAGAAAGAAGAAAAGAA sequence and binds in the major groove via Hoogsteen hydrogen bonds to form a triple helix structure. A ferrous salt is added to the solution and the two EDTA molecules form strong complexes with the Fe^{2+} ions. When dithiothreitol (DTT) is added, a reaction occurs that involves the EDTA–Fe^{2+} complex and oxygen. This reaction produces the highly reactive hydroxyl radical (HO·), a species that is so reactive that it is consumed within a short distance of the point at which it is generated. One of the many reactions that the hydroxyl radicals undergo is abstraction of a hydrogen atom from one of the positions of the deoxyribose ring (recall the sensitivity of ethers to autoxidation, Section 10.10). The end result of the reaction of this hydrogen with HO· is cleavage of the DNA chain. Because HO· has such a short lifetime, most of the radicals react within one or two nucleotide positions from the place where the T* was bound. The triple-stranded complex is illustrated in Figure 33.15.

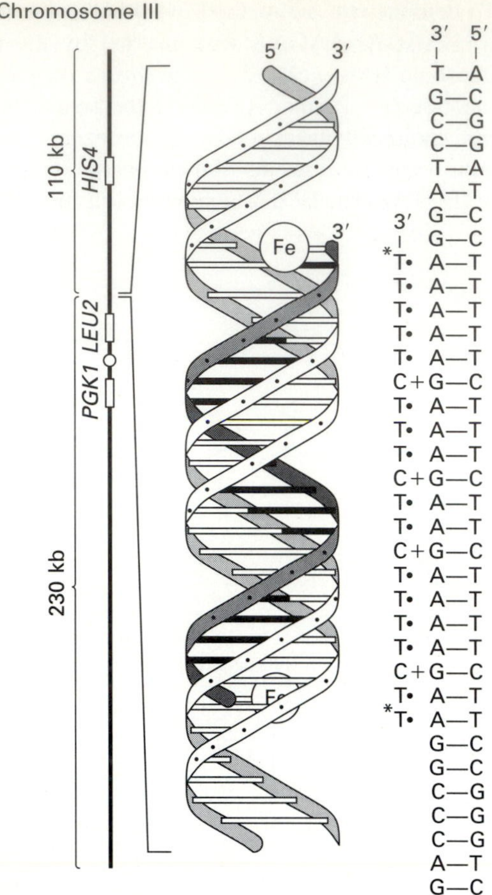

FIGURE 33.15 A segment of chromosome III of *Saccharomyces cerevisiae* containing an added target sequence AAAAAGAAAGAAGAAAAGAA. The recognizing oligonucleotide TTTTTCT-TTCTTCTTTTCTT forms a triple helix. The ·OH-generating Fe–EDTA complex units (marked Fe) are located at the ends of the oligonucleotide. Reaction with oxygen leads to formation of fragments of the expected size by cutting the double-stranded DNA in two places. The sequence is a tiny portion of the total chromosome, for which the base pair lengths on either side of the cut are indicated. [Reprinted with permission from S. A. Strobel and P. B. Dervan, *Science,* **249,** 73–75 (1990). Copyright 1990 by the AAAS.]

ASIDE

The earliest known mechanical lock dates to 2000 B.C. and comes from the Khorsabad palace near Nineveh, the destroyed capital of Assyria. The key fits, that is, is "recognized" by the lock. Recognition on a molecular scale was important in the evolution of life. After the primal explosion (the "Big Bang"), the condensation of matter, the formation of stars, nucleosynthesis, and formation of planets and molecules, it became possible for some of these molecules to associate, that is, to recognize one another. Purines and pyrimidines might have formed from HCN, and ribose 2,4-diphosphate might have formed from formaldehyde and glycolaldehyde (hydroxyacetaldehyde) phosphate, making possible the formation of ribonucleotides and then ribonucleic acids (RNA). One theory has it that self-replicating systems first evolved with RNA. The RNA polymer recognized and associated with the components for another polymer and thus made the assembly of a replicate feasible, perhaps through the enzymatic action of the RNA itself. Later, it is thought, DNA (deoxyribonucleic acid) replaced RNA for storage because of its superior intermolecular recognition ability. Name the individual nucleotides that are part of DNA and write structural formulas illustrating the intermolecular interactions. The genetic code of DNA is constructed of nucleotide triplets. Make a table showing the codes for the individual amino acids. Illustrate with structural formulas the difference between RNA and DNA.

PROBLEMS

1. Write formulas for the following compounds.

a. γ-^{32}P-adenosine triphosphate **b.** 2′-deoxyadenylic acid
c. cytidine 5′-phosphate **d.** uridine
e. guanosine diphosphate **f.** 2′-deoxythymidine 5′-phosphate

2. Give the commonly used abbreviations for the compounds listed in problem 1.

3. Name the following compounds with both formal and abbreviated names.

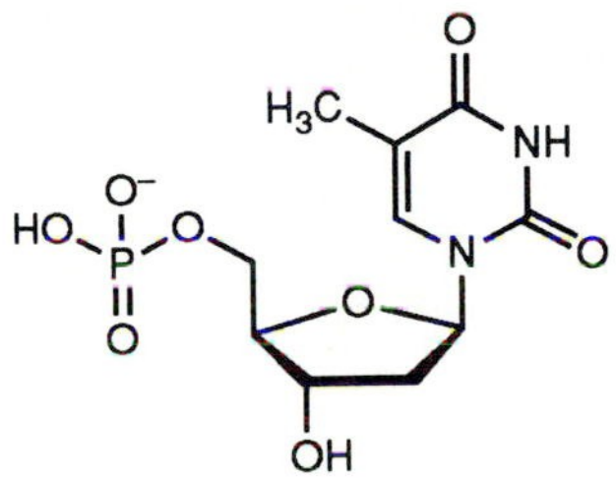

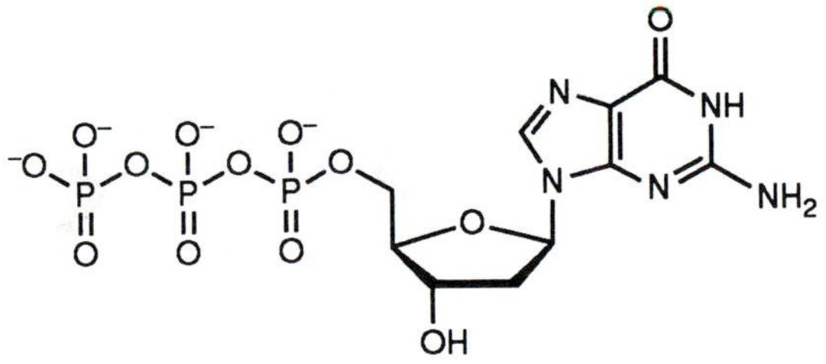

4. Spell your name with the single letter code for amino acids. For example, ''ED'' is the dipeptide Glu-Asp. Write out the DNA and mRNA sequences required for transcription and translation into the peptide sequence. How many alternate sequences would serve the same purpose?

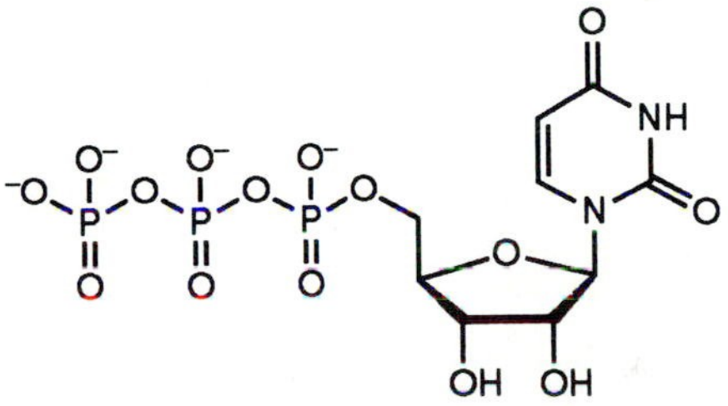

5. What is the net charge carried by DNA at neutral pH? What is the net charge carried by the tripeptide, GSH, γ-Glu-Cys-Gly? GSH is the thiol that protects many cellular constituents against free radicals, since the GS radical dimerizes rather than abstracting hydrogen from a C—H bond. Hydroxyl radicals, HO·, are among the agents generated by ionizing radiation that cause damage to DNA. GSH is not very effective in protecting DNA. Design a reagent that would be more effective. (*Hint:* Recall Coulomb's law, that opposite charges attract one another, and like charges repel.)

6. An archeologist found a bit of material in a 25,000,000-year-old bone that might have contained a very small amount of DNA. How would he or she go about amplifying the amount so that the material could be studied and sequenced? Give a step-by-step procedure.

7. Design a transition state analog that would be suitable for stimulating the production of a catalytic antibody suitable for the hydrolysis of isopropyl 4-nitrophenyl carbonate.

8. Compare the chemical degradation and enzymatic synthesis approaches to DNA sequencing. Are there reasons for choosing one in preference to the other?

9. Write out a step-by-step mechanism for the degradation at an adenine nucleotide by the chemical degradation method.

10. The structures of all new proteins that are reported in the chemical literature are collected into a computerized database. This database can be searched automatically by researchers who isolate a protein containing a certain sequence of amino acids and want to find out if that sequence has been observed in some other protein. Since there are 20 amino acids, the chances of finding a certain pentapeptide sequence are $(1/20)^5$, or 1/3,200,000. A search for the pentapeptide sequence Leu-Ile-Val-Glu-Ser (LIVES) reveals that the sequence has never been reported. However, a search for the sequence Glu-Leu-Val-Ile-Ser (ELVIS) reveals that it has been reported approximately ten times (out of only a few hundred total proteins that are included in the database). Write several DNA sequences that code for the pentapeptide ELVIS.

CHAPTER 34

MASS SPECTROMETRY

34.1 Introduction

When a beam of electrons of energy greater than the ionization energy [about 8–13 electron volts (eV) (185–300 kcal $mole^{-1}$) for most compounds] is passed through a sample of an organic compound in the vapor state, ionization of some molecules occurs. In one form of ionization, one of the valence electrons of the molecule is lost, leaving behind a **radical cation.**

$$e^- + CH_4 \longrightarrow 2\ e^- + [CH_4]^{\dot{+}}$$

In CH_4 eight valence electrons bond the four hydrogens to carbon. The symbol $[CH_4]^{\dot{+}}$ represents a structure in which seven valence electrons bond the four hydrogens to carbon. The + sign shows that the species has a net positive charge. The · signifies that the species has an odd number of electrons. If a mixture of compounds is bombarded with electrons, a mixture of radical cations differing in mass will obviously be produced.

$$\begin{array}{llll} CH_4,\ C_2H_6 & \longrightarrow & [CH_4]^{\dot{+}}, & [C_2H_6]^{\dot{+}} \\ m/z: & & 16 & 30 \end{array}$$

where m/z = mass-to-charge ratio (m is the mass of the molecular radical cation and z is the number of charges).

In practice, even when a pure substance is bombarded with electrons, a mixture of cations is produced. As will be discussed later, some of the ions initially formed break up or **fragment** into smaller ions. For example, methane can give cations with masses of 16, 15, 14, 13, and 12.

$$\begin{array}{lllllll} CH_4 & \longrightarrow & [CH_4]^{\dot{+}}, & CH_3^+, & [CH_2]^{\dot{+}}, & CH^+, & [C]^{\dot{+}} \\ m/z: & & 16 & 15 & 14 & 13 & 12 \end{array}$$

Lewis structures for these cations are

$$\left[\begin{array}{c}\mathrm{H}\\ \mathrm{H}:\ddot{\mathrm{C}}\cdot\mathrm{H}\\ \ddot{\mathrm{H}}\end{array}\right]^{+} \qquad \begin{array}{c}\mathrm{H}\\ \mathrm{H}:\ddot{\mathrm{C}}^{+}\\ \ddot{\mathrm{H}}\end{array} \qquad \left[\begin{array}{c}\mathrm{H}\\ \mathrm{H}:\ddot{\mathrm{C}}\cdot\end{array}\right]^{+} \qquad \mathrm{H}:\dot{\mathrm{C}}^{+} \qquad \left[:\dot{\mathrm{C}}\right]^{+}$$

Note that the cations having an even number of hydrogens are **radical cations,** whereas the cations having an odd number of hydrogens are normal **carbocations.**

A **mass spectrometer** is an instrument that is designed to ionize molecules, often in the gas phase, separate the ions produced on the basis of their mass-to-charge ratio, and record the relative amounts of different ions produced. A **mass spectrum** is a plot of the data obtained from the mass spectrometer. It is customary for the mass-to-charge ratio (m/z) to be plotted as the abscissa and the number of ions or relative intensity (height of each peak) to appear as the ordinate. Mass spectrometry differs from spectroscopy in that no absorption of light is involved. Nevertheless, it has been called a "spectroscopy" because the mass "spectrum" resembles other kinds of spectra.

EXERCISE 34.1 The ionization energy for methane is greater than that for ethylene. Explain.

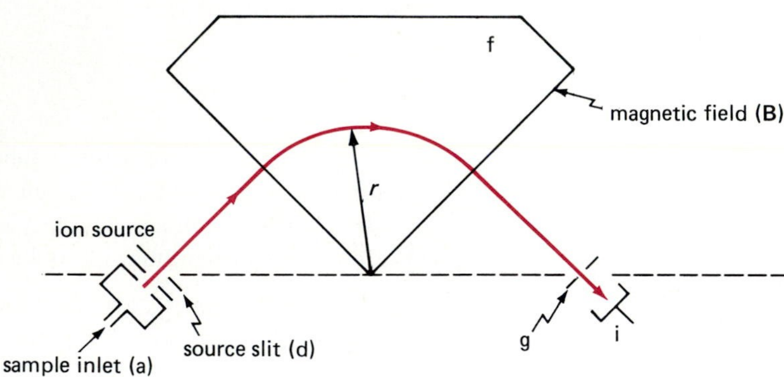

(a) schematic diagram

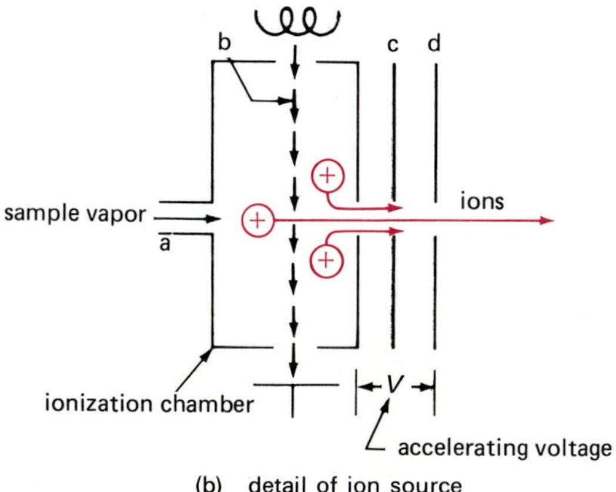

(b) detail of ion source

FIGURE 34.1 90° sector magnetic deflection mass spectrometer.

One of the most common types of mass spectrometer currently in use is the **magnetic sector mass spectrometer.** A sketch of a 90° magnetic sector instrument is shown in Figure 34.1. The sample vapor is introduced at the sample inlet a, usually at low pressure (10^{-5} to 10^{-6} torr). A low pressure is used to minimize the number of collisions between ions and nonionized molecules. Such collisions lead to reactions that produce new ions containing parts of both collision partners. Such ions are often interesting in their own right but lead to difficulties in interpretation of the data. After the sample vapor enters the ion source (see enlarged insert, Figure 34.1b), it passes through the electron beam (b) where ionization occurs. The resulting ions pass out of the ionization chamber and between two charged plates (c), which serve to focus the ion beam. There is a difference in potential of several thousand volts between the ionization chamber and the source slit (d). In this region, the ions are accelerated and pass through slit (d); after traveling a short distance they pass into the magnetic field.

The radius of the path followed by an ion of mass m in a magnetic field depends on its charge ze (e is the electronic charge) and the accelerating potential (V). The energy acquired by an ion accelerated across a potential drop is equal to its kinetic energy.

$$zeV = \tfrac{1}{2}mv^2 \tag{34-1}$$

In a uniform perpendicular magnetic field of strength **B** the ion experiences a centripetal force $\mathbf{B}zev$, where v is the velocity of the ion. Because the ion path is circular, the force on the ion is equal to mv^2/r, where r is the radius of the path followed by the ion.

$$\mathbf{B}zev = \frac{mv^2}{r} \tag{34-2}$$

$$r = \frac{mv}{ze\mathbf{B}} \tag{34-3}$$

Most of the ions are singly charged ($z = 1$). As a collection of ions of different masses enters the magnetic field region (f), each ion follows a circular path with a radius given by the foregoing equation. Ions of larger m/z follow a path of greater radius, and ions of lesser m/z follow a path of smaller radius. In the example diagrammed in Figure 34.2 the ions of $m/z = y$ are passing through the collector slit (g) and impinging upon the ion collector (i).

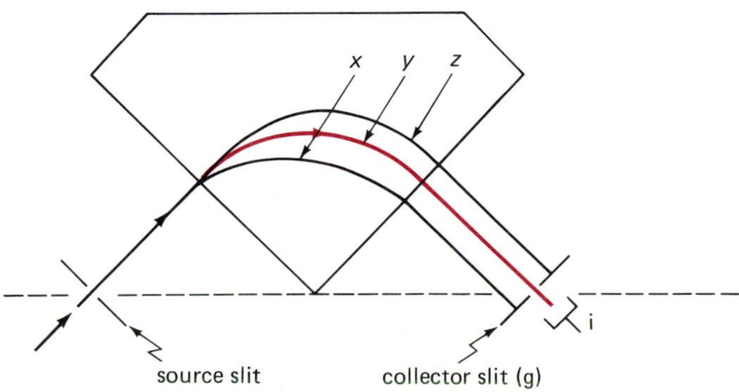

FIGURE 34.2 Ions with $m/z = y$ are focused on the collector (i) through slit (g).

Elimination of the velocity term from equations (34-1) and (34-3) gives

$$\frac{m}{z} = \frac{\mathbf{B}^2 r^2 e}{2V} \tag{34-4}$$

This relationship shows that for an ion of given mass-to-charge ratio (m/z), the radius of deflection r can be increased by decreasing **B**, the magnetic field strength. (Less deflection with lower magnetic field = larger radius of deflection.) For example, in the case diagrammed in Figure 34.2 a slight decrease in **B** will cause the radius of deflection of all of the ions to increase somewhat. In Figure 34.3 ions of $m/z = y$ no longer pass through the slit and into the collector, but ions of $m/z = x$ do.

Note that the same effect might have been obtained by increasing V slightly or by moving the collector slit slightly to the left. In actual practice this last technique is inconvenient, and scanning V has other disadvantages. Scanning of the spectrum is usually achieved by **magnetic scanning;** that is, the accelerating voltage V is kept constant while the magnetic field strength **B** is increased. As **B** is increased, ions of progressively higher m/z attain the necessary radius of deflection to pass through the collector slit (g) and into the ion collector (i).

As the ions enter the collector, they impinge upon an electron multiplier detector where a minute current is produced and amplified. The magnitude of this current is proportional to the intensity of the ion beam. The current produced is fed to a computer, which processes the data. The computer may have a stored library of thousands of spectra with which the sample spectrum can be compared. The current produced for various values of

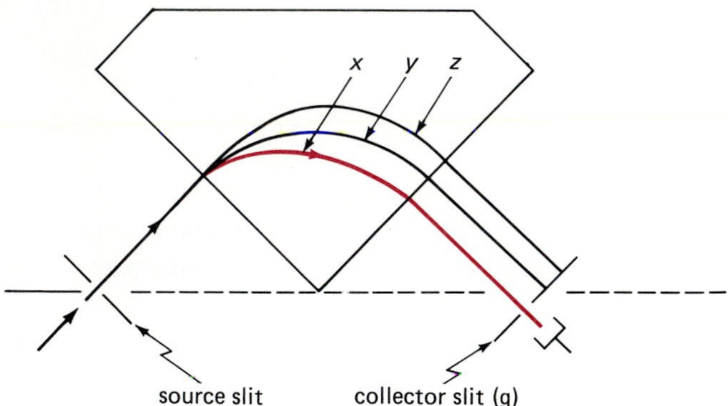

FIGURE 34.3 At lower **B**, ions with $m/z = x$ are now focused on the collector (i).

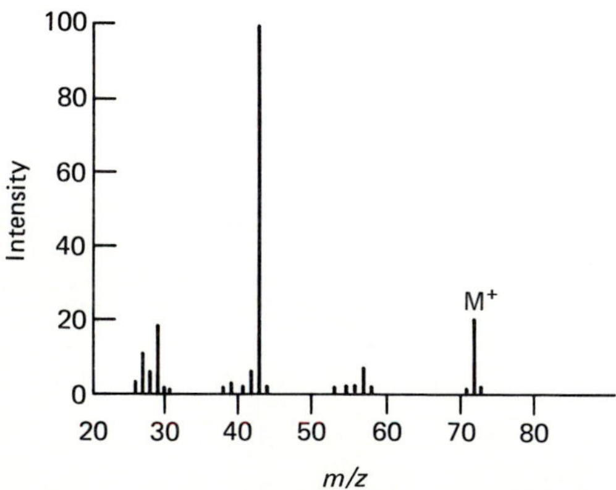

FIGURE 34.4 Mass spectrum of 2-butanone.

m/*z* is printed in a tabular manner and usually plotted as a bar graph. The most intense peak (the "base peak") is assigned the arbitrary intensity value of 100, and all other peaks are given their proportionate value. A mass spectrum recorded in this manner is shown in Figure 34.4.

EXERCISE 34.2 Note in Figure 34.4 that the most intense peak corresponds to a value of *m*/*z* of 43. To what portion of the 2-butanone molecule does such a mass correspond?

34.3 The Molecular Ion: Molecular Formula

The molecular weight of a compound is one datum that can usually be obtained by visual inspection of a mass spectrum. Although the radical cations produced by the initial electron ionization usually undergo extensive fragmentation to give cations of smaller *m*/*z* (next section), the particle of highest *m*/*z* generally (but not always) corresponds to the ionized molecule, and $m/z = M$ for this particle (called the **molecular ion** and abbreviated M^+) gives the molecular weight of the compound.

If the spectrum is measured with a "high-resolution" spectrometer, it is possible to determine a unique molecular formula for any peak in a mass spectrum, including the molecular ion. This is possible because atomic masses are not integers. For example, consider the molecules CO, N_2, and C_2H_4, all of which have a **nominal mass** of 28. The actual masses of the four atomic particles are H = 1.007825, C = 12.000000 (by definition), N = 14.003074, O = 15.994915. Therefore, the actual masses of CO, N_2 and C_2H_4 are as follows.

^{12}C	12.0000	$^{14}N_2$	28.0061	$^{12}C_2$	24.0000
^{16}O	15.9949			$^{1}H_4$	4.0314
	27.9949				28.0314

Since a high-resolution spectrometer can readily measure mass with an accuracy of better than 1 part in 100,000, and can separate masses that differ by 1 part in 10,000, the above three masses are readily distinguishable, as shown in Figure 34.5.

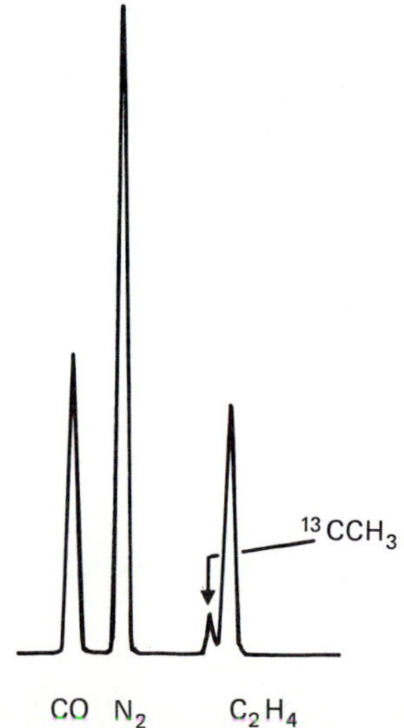

FIGURE 34.5 High-resolution mass spectrum of a mixture of ethylene, nitrogen, and carbon monoxide.

Because the mass spectrometer measures the exact m/z for each ion and because most of the elements commonly found in organic compounds have more than one naturally occurring isotope, a given peak will usually be accompanied by several isotope peaks. Table 34.1 shows the common isotopes of some of the elements.

TABLE 34.1 Natural Abundance of Common Isotopes

Element	Abundance, %			
hydrogen	99.985 ^{1}H	0.015 ^{2}H		
carbon	98.893 ^{12}C	1.107 ^{13}C		
nitrogen	99.634 ^{14}N	0.366 ^{15}N		
oxygen	99.759 ^{16}O	0.037 ^{17}O	0.204 ^{18}O	
sulfur	95.0 ^{32}S	0.76 ^{33}S	4.22 ^{34}S	0.014 ^{36}S
fluorine	100 ^{19}F			
chlorine	75.77 ^{35}Cl	24.23 ^{37}Cl		
bromine	50.69 ^{79}Br	49.31 ^{81}Br		
iodine	100 ^{127}I			

The ^{13}C abundance, 1.107%, is the content of oceanic carbonate. Organic compounds in the biosphere run around 1.08% because of isotope effects.

The ^{13}C abundance, 1.107%, is the content of oceanic carbonate. Organic compounds in the biosphere have ^{13}C contents of around 1.08% because of isotope effects.

Consider the molecular ion derived from methane. Most of the methane molecules are $^{12}C^{1}H_4$ and have the nominal mass 16. However, a few molecules are either $^{13}C^{1}H_4$ or $^{12}C^{2}H_1{}^{1}H_3$ and have the nominal mass 17. An even smaller number of molecules have both a ^{13}C and an ^{2}H or have two ^{2}H isotopes and therefore have the nominal mass 18. An exact expression for the ratio of isotopic massed $(M + 1)/M$ can be derived from probability mathematics but is rather complex. The theoretical intensities of the various isotope peaks can be looked up in special tables compiled for this purpose. However, the contributions of ^{2}H and ^{17}O to $(M + 1)/M$ are relatively small and the ratio is given to a satisfactory approximation for most compounds having few N and S atoms by equation (34-5).

$$\frac{M+1}{M} = \frac{0.01107}{0.98893}c + 0.00015h + 0.00367n + 0.00037o + 0.0080s \quad (34\text{-}5)$$

where M = intensity of the molecular ion (ions containing no heavy isotopes), $M + 1$ = intensity of the molecular ion + 1 peak (ions containing one ^{13}C, ^{2}H, ^{15}N, ^{17}O, or ^{33}S) and c, h, n, o, s = the number of carbons, hydrogens, nitrogens, oxygens, sulfurs.

Using this relationship, we can readily estimate the intensity of the $M + 1$ peak in the mass spectrum of methane.

$$\frac{M+1}{M} = 0.01119(1) + 0.00015(4) = 0.01179$$

Thus the peak at m/z 17 in the mass spectrum of methane should be approximately 1.18% as intense as the peak at m/z 16.

EXERCISE 34.3 Estimate the intensity of the $M + 1$ peak for the following compounds.

(a) decane (b) 1-decanol (c) 1-decanamine

Note that the principal contributor to the $M + 1$ peak is ^{13}C. This is partly because of the relatively large relative abundance of ^{13}C (see Table 34.1) and partly because most organic compounds contain many more carbon atoms than they do oxygens or nitrogens. In fact, a useful rule of thumb is that the $M + 1$ peak will be 1.1% for each carbon in the molecule.

A similar relationship may be derived for calculation of the intensity of the $M + 2$ peak. However, in order to obtain an exact figure, a lengthy computation is required. For most compounds the $M + 2$ peak is small. However, for compounds containing chlorine or bromine, the $M + 2$ isotopic peak is substantial. The characteristic doublets observed in the mass spectra of compounds containing chlorine and bromine are an excellent way of diagnosing for the presence of these elements, as shown in Figures 34.6 and 34.7.

One use to which isotope peaks may be put is in approximating the molecular formula of the parent ion in the mass spectrum of an unknown compound. However, one must exercise caution when applying the foregoing computations. First, the $M + 1$ peak is generally much less intense than the parent ion. Unless the parent ion is a fairly strong one, its isotope peak may be too weak to measure accurately. Second, intermolecular proton transfer reaction can give $M + 1$ peaks that are not due to isotopes. Third, the presence of a small amount of impurity with a strong peak at $M + 1$ of the sample will interfere with accurate measurement.

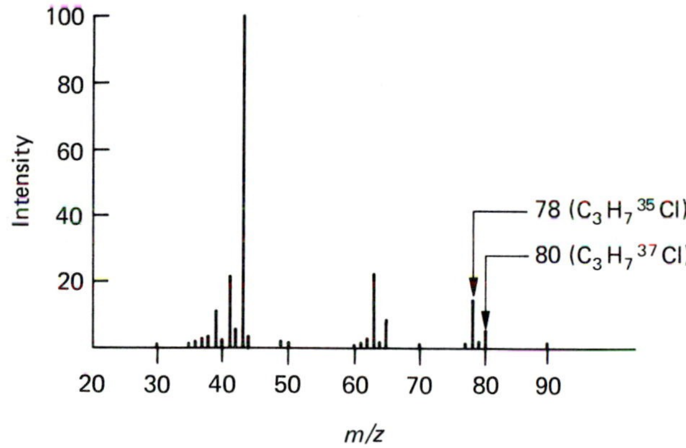

FIGURE 34.6 Mass spectrum of 2-chloropropane.

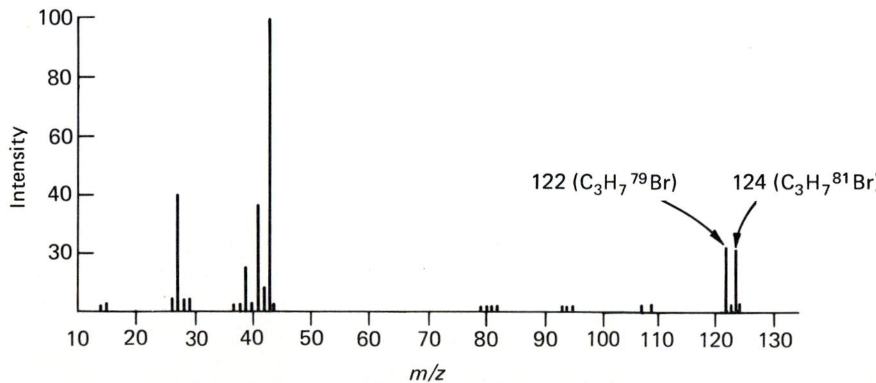

FIGURE 34.7 Mass spectrum of 1-bromopropane.

EXERCISE 34.4 Sketch the expected appearance of the mass spectrum of 1,2-dibromoethane in the region of the molecular ion(s).

34.4 Fragmentation

A. Simple Bond Cleavage

When an electron interacts with a molecule in the ionization chamber of the mass spectrometer, ionization will occur if the impinging electron transfers to the molecule an amount of energy equal to or greater than its ionization potential. The ionization potentials for several organic molecules are given in Table 34.2. When the colliding electron transfers more energy than is required for ionization, a part of the excess energy will normally be carried away by the radical cation produced in the collision. If the molecular ion gains enough surplus energy, bond cleavage (fragmentation) may occur, with the resultant formation of a new cation and a free radical. Typically, the electron beams employed in the ionization process have an energy of 50–70 eV (1150–1610 kcal $mole^{-1}$). Since this is far in excess of the typical bond energies encountered in organic compounds (50–130 kcal $mole^{-1}$), fragmentation is normally extensive.

TABLE 34.2 Ionization Energies

Compound	Ionization Energy, electron volts (eV)
benzene	9.25
aniline	7.70
acetylene	11.40
ethylene	10.52
methane	12.98
methanol	10.85
methyl chloride	11.35

Consider the case of the simplest hydrocarbon, methane. The mass spectrum of methane is shown in Figure 34.8 in bar graph form as well as tabular form. Note that the base peak (most intense peak) corresponds to the molecular ion (*m*/*z* 16). Note also the monoisotopic peak at *m*/*z* 17 (*M* + 1), which has an intensity 1.11% that of the molecular ion, within 0.07% of the intensity predicted by theory. Examination of the mass spectrum

m/*z*	Intensity
1	3.4
2	0.2
12	2.8
13	8.0
14	16.0
15	86.0
16	100.0
17	1.11

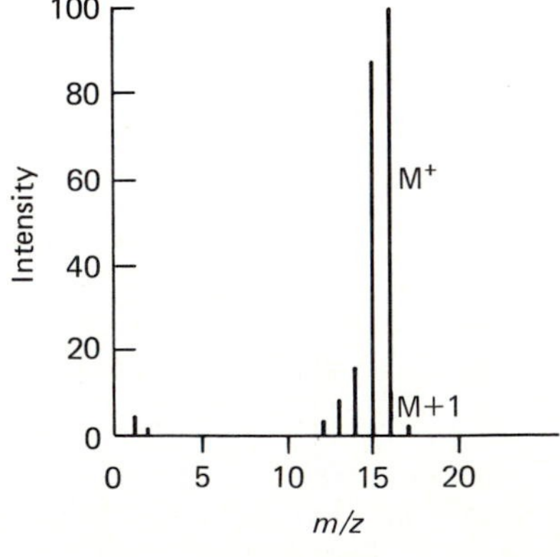

FIGURE 34.8 Mass spectrum of methane.

reveals that cations are also produced and measured that have m/z values of 15, 14, 13, 12, 2, and 1. The following modes of fragmentation can be postulated to explain these various cationic fragments. Initial ionization yields the molecular ion, with m/z 16.

$$CH_4 + e^- \longrightarrow \underset{m/z\ 16}{[CH_4]^{\dot{+}}} + 2e^-$$

Some of these ions move into the accelerating region and are passed into the magnetic field. However, since they possess a large amount of excess energy, many undergo fragmentation before leaving the ionization chamber, giving a methyl cation (m/z 15) and a hydrogen atom.

$$[CH_4]^{\dot{+}} \longrightarrow \underset{m/z\ 15}{CH_3^+} + H\cdot$$

Occasionally this cleavage occurs in such a way as to produce a methyl radical and a bare proton (m/z 1).

$$[CH_4]^{\dot{+}} \longrightarrow CH_3\cdot + \underset{m/z\ 1}{H^+}$$

The fragment CH_3^+ can be accelerated, deflected, and collected as a cation of m/z 15, or it too may undergo fragmentation, giving a hydrogen atom and a new radical cation of m/z 14.

$$CH_3^+ \longrightarrow \underset{m/z\ 14}{[CH_2]^{\dot{+}}} + H\cdot$$

Similar events give rise to fragment ions of m/z 13 and 12.

$$[CH_2]^{\dot{+}} \longrightarrow \underset{m/z\ 13}{CH^+} + H\cdot$$

$$CH^+ \longrightarrow \underset{m/z\ 12}{[C]^{\dot{+}}} + H\cdot$$

Occasionally an ion ejects an ionized hydrogen molecule, giving rise to the weak peak at m/z 2.

$$[CH_4]^{\dot{+}} \longrightarrow CH_2 + \underset{m/z\ 2}{[H_2]^{\dot{+}}}$$

More complicated alkanes give very complicated spectra, containing a large number of peaks. However, most of these fragment peaks are of low intensity. The more intense fragment peaks have m/z values of $M - 15$, $M - 29$, $M - 43$, $M - 57$, and so on, corresponding to scission of the hydrocarbon chain at various places along its length. The spectrum of *n*-dodecane, plotted in Figure 34.9, is illustrative. There is a reasonably intense molecular ion (4% of the base peak) at m/z 170. The peak at m/z 155, corresponding to loss of CH_3 ($M - 15$) is so weak as not to be noticeable. However, the peaks at m/z 141 ($M - 29$), 127 ($M - 43$), and so on, are apparent. Note that intensity decreases regularly as mass increases beyond m/z 43 (corresponding to $C_3H_7^+$). The modes of fragmentation responsible for the spectrum of *n*-dodecane are indicated in Figure 34.10.

When there is a branch point in the chain, an unusually large amount of fragmentation occurs there because a more stable carbocation results. Thus, in 2-methylpentane, loss of C_3H_7 or CH_3 is much greater than loss of C_2H_5, since the former modes give secondary carbocations, whereas the latter gives a primary carbocation.

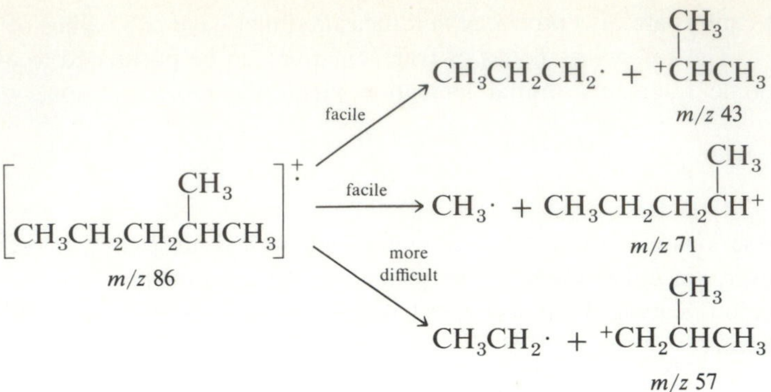

The spectrum of 2-methylpentane, plotted in Figure 34.11, illustrates this behavior.
On the other hand, the isomeric hydrocarbon 3-methylpentane can cleave in three ways

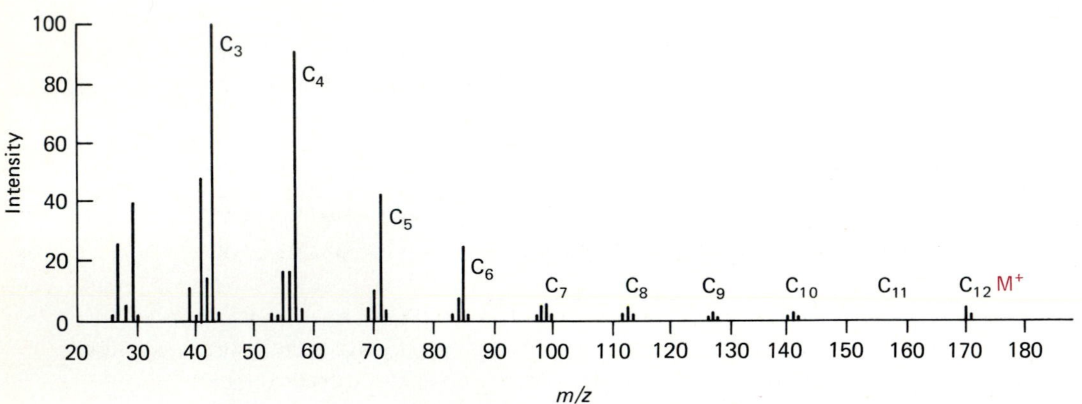

FIGURE 34.9 Mass spectrum of *n*-dodecane.

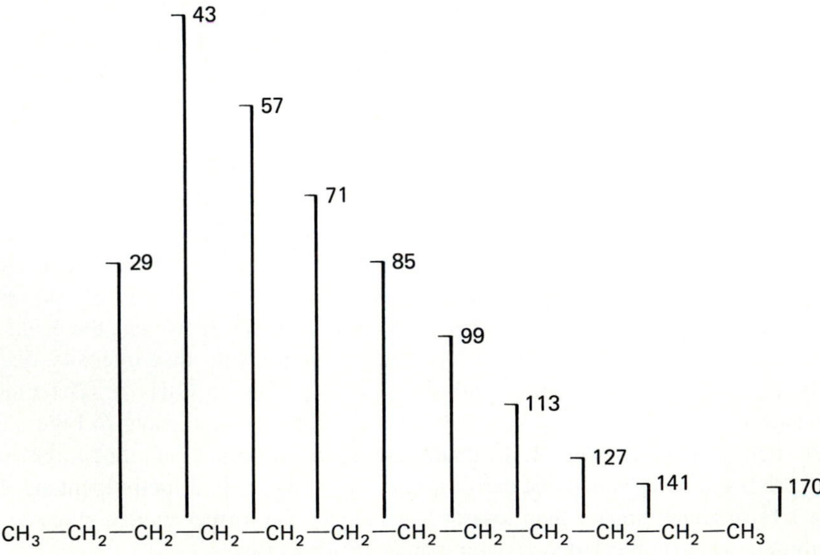

FIGURE 34.10 Fragmentation of *n*-dodecane.

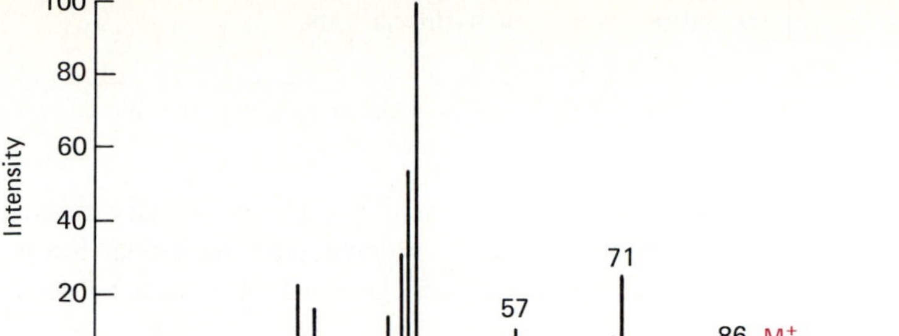

FIGURE 34.11 Mass spectrum of 2-methylpentane.

so as to give a secondary carbocation. Two of these cleavages amount to loss of C_2H_5. Correspondingly, the $M - 29$ peak in its spectrum, shown in Figure 34.12, is the most intense peak.

$$\left[\begin{array}{c}CH_3\\ CH_3CH_2\overset{|}{C}HCH_2CH_3\end{array}\right]^{+\cdot} \longrightarrow CH_3CH_2\cdot + {}^{+}\overset{\overset{CH_3}{|}}{C}HCH_2CH_3$$

m/z 86 $\qquad\qquad$ m/z 57

Note that 3-methylpentane cannot undergo a simple cleavage to give an ion with *m/z* 43. The peak in its spectrum with this value must arise by a process involving some sort of skeletal rearrangement.

The mode of fragmentation in the preceding discussion is common in mass spectrometry. A radical cation usually undergoes bond cleavage in such a manner as to give the *most stable cationic fragment*. What we know about the relative stabilities of various cations from other areas of organic chemistry may often be used to predict how fragmentation will occur in a mass spectrometer. The case of the methylpentanes is a good example of this principle. In Chapter 9 we discussed the S_N1 reactions of alkyl halides to give carbocationic intermediates and found a reactivity order tertiary > secondary > primary. From this order, and other data, we concluded that tertiary carbocations are more stable than secondary ones, which are, in turn, more stable than primary carbocations. Although these results are for solution processes and mass spectrometry measures the results of

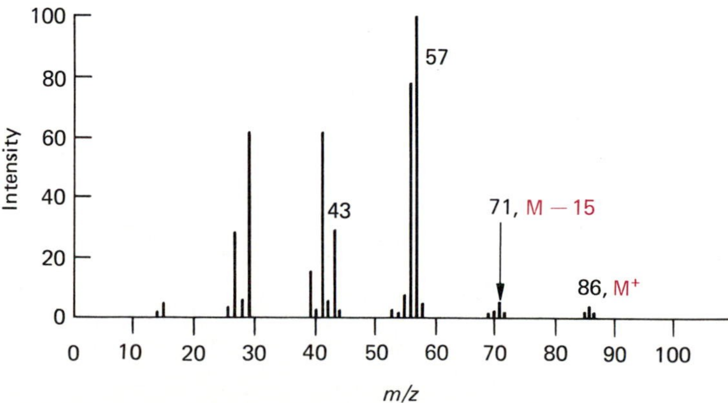

FIGURE 34.12 Mass spectrum of 3-methylpentane.

vapor phase processes, we can use our qualitative knowledge of carbocation stabilities to "interpret" the fragmentation pattern of hydrocarbons.

Some of the enthalpy data for ionization of alkyl chlorides given in Table 9.7 on page 195 were actually obtained by mass spectrometric methods.

In alkanes with a quaternary carbon, fragmentation to give tertiary carbocations is so facile that such hydrocarbons frequently give no detectable molecular ion peak. On the other hand, alkenes and aromatic hydrocarbons generally give rather intense molecular ion peaks.

EXERCISE 34.5 What are the principal fragments expected from 3,3-dimethylheptane?

Simple one-bond cleavage is also a prominent fragmentation mode in amines. Cleavage of a bond *adjacent* to a carbon-nitrogen bond gives an alkyl radical and an immonium ion. Primary amines that are not branched at the carbon attached to nitrogen show an intense fragment with *m/z* 30.

$$[CH_3CH_2CH_2NH_2]^{+\cdot} \longrightarrow CH_3CH_2\cdot + \underset{m/z\,=\,30}{CH_2{=}NH_2^{+}}$$

When the amine is branched at the nitrogen-bearing carbon, an analogous cleavage occurs, leading to a homologous immonium ion; loss of the larger group is preferred.

$$\left[\begin{matrix} CH_3 \\ | \\ CH_3CH_2CH_2CHNH_2 \end{matrix}\right]^{+\cdot} \longrightarrow CH_3CH_2CH_2\cdot + \underset{m/z\,=\,44}{CH_3CH{=}NH_2^{+}}$$

These cleavage patterns are illustrated by the spectra of isobutylamine and *t*-butylamine shown in Figures 34.13 and 34.14.

EXERCISE 34.6 Hydrocarbons and oxygen-containing compounds always have *even molecular weights,* and therefore the *m/z* of M^+ for such a compound is even. What generalization can you make about fragment ions that result from single-bond cleavage? What generalizations can be made regarding the molecular ions of monoamines and the simple fragment ions from these compounds?

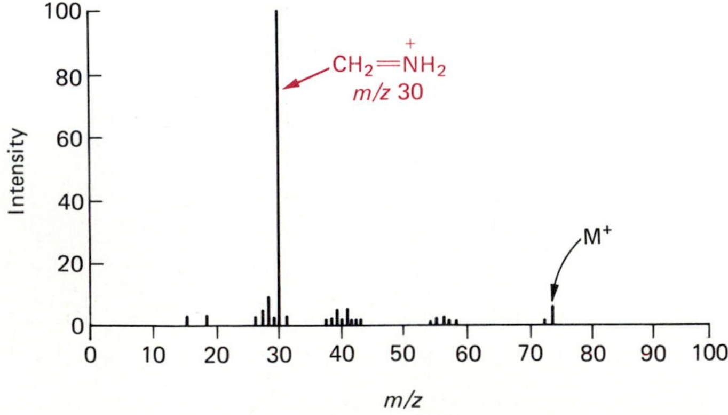

FIGURE 34.13 Mass spectrum of isobutylamine, $(CH_3)_2CHCH_2NH_2$.

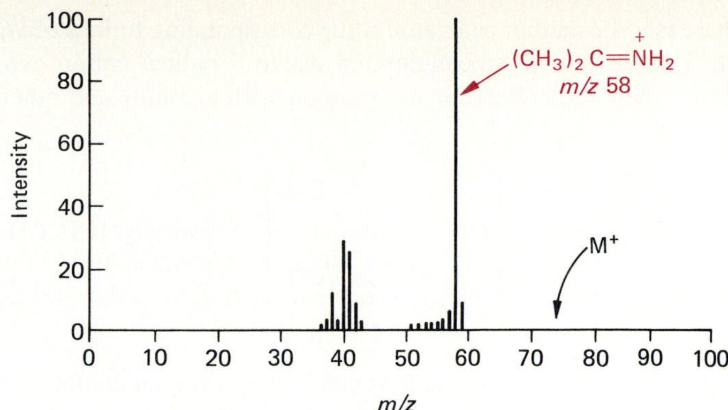

FIGURE 34.14 Mass spectrum of *t*-butylamine, $(CH_3)_3CNH_2$.

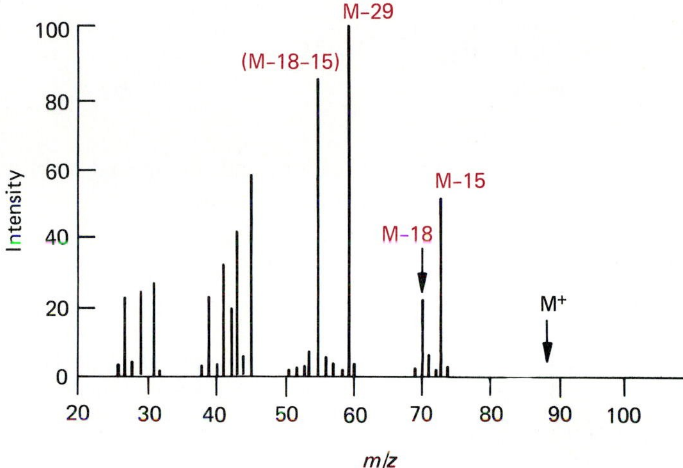

FIGURE 34.15 Mass spectrum of 2-methyl-2-butanol.

B. Two-Bond Cleavage: Elimination of a Neutral Molecule

Some compounds give extremely weak molecular ion peaks. This tends to happen when some form of fragmentation is particularly easy. Such behavior is typical of alcohols, which often give no detectable molecular ion whatsoever. The spectrum of 2-methyl-2-butanol in Figure 34.15 illustrates this phenomenon.

The molecular ion, which would appear at *m/z* 88, is not observed. Instead, sizeable peaks are observed at *m/z* values of 73 ($M - 15$) and 59 ($M - 29$), corresponding to cleavage of the radical ion so as to give stable oxonium ions.

$$\left[CH_3CH_2{-}\overset{OH}{\underset{CH_3}{C}}{-}CH_3\right]^{\dot{+}} \longrightarrow CH_3\cdot + \left[CH_3CH_2\overset{OH}{\underset{+}{C}}CH_3 \longleftrightarrow CH_3CH_2\overset{OH^+}{\overset{\|}{C}}CH_3\right] \quad m/z\ 73$$

$$\left[CH_3CH_2{-}\overset{OH}{\underset{CH_3}{C}}{-}CH_3\right]^{\dot{+}} \longrightarrow CH_3CH_2\cdot + \left[CH_3\overset{OH}{\underset{+}{C}}CH_3 \longleftrightarrow CH_3\overset{OH^+}{\overset{\|}{C}}CH_3\right] \quad m/z\ 59$$

In addition, there is a substantial peak at *m/z* 70, corresponding to loss of water from the molecular ion. This type of fragmentation, in which a radical cation expels a neutral molecule to give a new radical cation, is common with alcohols and ethers.

$$\left[CH_3CH_2-\underset{CH_3}{\overset{OH}{C}}-CH_3\right]^{\dot{+}} \xrightarrow{-H_2O} \underset{m/z\ 70}{\left[CH_3CH{=}C\begin{matrix}CH_3\\CH_3\end{matrix}\right]^{\dot{+}}} \quad \text{or} \quad \underset{m/z\ 70}{\left[CH_3CH_2\overset{CH_3}{C}{=}CH_2\right]^{\dot{+}}}$$

Of course, these new radical cations ions can undergo fragmentation of the type first discussed. The peak at *m/z* 55 probably arises from such a stepwise path.

$$\left[CH_3CH_2\underset{CH_3}{\overset{OH}{C}}-CH_3\right]^{\dot{+}} \xrightarrow{-H_2O} \underset{m/z\ 70}{\left[CH_3CH_2\overset{CH_3}{C}{=}CH_2\right]^{\dot{+}}} \xrightarrow{-CH_3\cdot} \underset{m/z\ 55}{\left[{}^{+}CH_2-\overset{CH_3}{C}{=}CH_2 \longleftrightarrow CH_2{=}\overset{CH_3}{C}-CH_2^{+}\right]}$$

The *m/z* 55 fragment is a substituted allyl cation, and the special stability of this ion (Section 20.1.A) is the reason that this fragment is so intense.

When the mass spectrum of an unknown compound does not contain a peak corresponding to the molecular ion, it is easy to be led astray in deducing the structure of the material. However, a careful examination of the mass spectrum of such a compound usually allows one to deduce that elimination of a neutral molecule has occurred and that the even peak of highest *m/z* is not the molecular ion. As an example, consider the following mass spectrum.

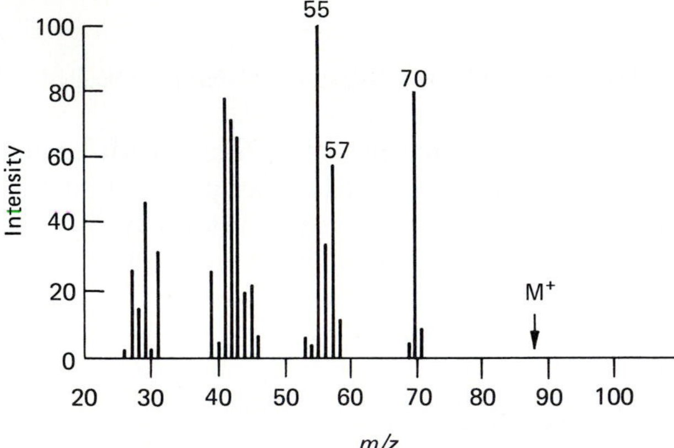

The *m/z* fragment could be considered to be the molecular ion of a compound with the molecular weight C_5H_{10}, and the *m/z* 55 fragment would then be due to loss of methyl. However, the rather intense *m/z* 57 fragment would have to correspond to loss of CH, a mechanistically unreasonable process. Thus, the internal evidence in this mass spectrum suggests that the *m/z* 70 fragment is not, in fact, a molecular ion.

In addition, the chemist usually has other evidence that might not be consistent with the obvious interpretation of a mass spectrum that does not contain a molecular ion. In the present case, for example, the infrared spectrum contains a strong absorption at 3400 cm^{-1}, strongly suggesting the presence of a hydroxy group. If the *m/z* 70 fragment *were* the molecular ion of an alcohol, then we would expect a $M - 18$ fragment with *m/z* 52. The absence of such a peak is further evidence that the spectrum does not contain a molecular ion peak.

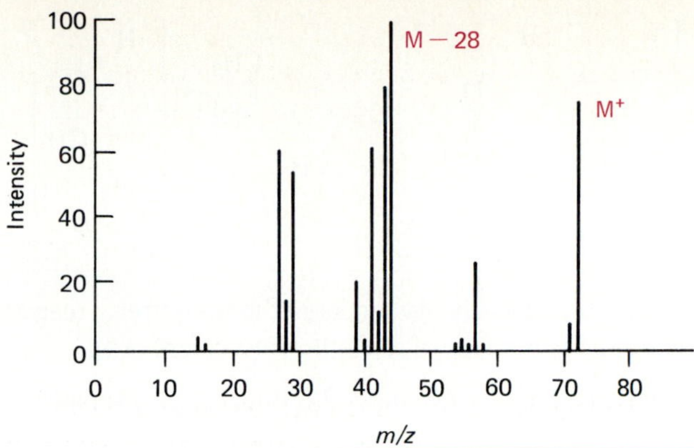

FIGURE 34.16 Mass spectrum of butyraldehyde.

There is one other type of fragmentation, also involving expulsion of a neutral molecule that we will develop here. The spectrum of butyraldehyde is plotted in Figure 34.16. The most striking thing about the spectrum is the fact that the base peak (*m*/*z* 44) is an even number. Thus it must correspond to expulsion of a molecule, rather than a radical, from the molecular ion. Extensive studies suggest that this fragment arises in the following way.

$$\left[\text{H}-CH_2-CH_2-CH_2-CH{=}O\right]^{+\cdot} \longrightarrow CH_2{=}CH_2 + \left[CH_2{=}CH-O-\text{H}\right]^{+\cdot}$$

m/*z* 72 → *m*/*z* 44

There is some evidence that suggests that this fragmentation may involve two distinct steps, transfer of a hydrogen atom to the carbonyl oxygen from the γ-carbon followed by scission of the α,β-bond.

$$\left[\text{H}-CH_2-CH_2-CH_2-CH{=}O\right]^{+\cdot} \longrightarrow \cdot CH_2-CH_2-CH_2-CH{=}O^{+}-\text{H} \longrightarrow CH_2{=}CH_2 + \left[CH_2{=}CH-\text{HO}\right]^{+\cdot}$$

This rearrangement reaction is called a **McLafferty rearrangement.** It can provide useful information concerning the structure of isomeric aldehydes and ketones. For example, 2-methylbutanal and 3-methylbutanal both undergo the rearrangement. In the former case one observes an intense peak at *m*/*z* 58, but in the latter the rearrangement peak occurs at *m*/*z* 44.

$$\left[\text{H}-CH_2-CH_2-CH(CH_3)-CH{=}O\right]^{+\cdot} \longrightarrow CH_2{=}CH_2 + \left[CH_3-CH{=}CH-O-\text{H}\right]^{+\cdot}$$

m/*z* 86 → *m*/*z* 58

$$\left[\begin{array}{c} \text{H} \\ \text{CH}_2 \quad \text{O} \\ \text{CH} \quad \text{CH} \\ \text{CH}_3 \quad \text{CH}_2 \end{array}\right]^{\dot{+}} \longleftrightarrow \begin{array}{c}\text{CH}_2 \\ \| \\ \text{CH} \\ \text{CH}_3\end{array} + \left[\begin{array}{c} \text{H} \\ \text{O} \\ \text{CH} \\ \text{CH}_2 \end{array}\right]^{\dot{+}}$$

m/z 86 *m/z* 44

EXERCISE 34.7 Ketones and esters also undergo the McLafferty rearrangement. What fragments are expected from each of the following compounds?

(a) 2-butylcyclohexanone (b) butyl 2,2-dimethylpropanoate

An additional fragmentation common to ketones is cleavage of a bond to the carbonyl group to give a cation of the oxonium ion type.

$$\left[\text{R}-\overset{\overset{\text{O}}{\|}}{\text{C}}-\text{R}'\right]^{\dot{+}} + \xrightarrow{-\text{R}'\cdot} \left[\text{R}-\text{C}\equiv\ddot{\text{O}}^{+} \longleftrightarrow \text{R}-\overset{+}{\text{C}}=\ddot{\underset{\cdot\cdot}{\text{O}}}\right]$$

EXERCISE 34.8 Write equations showing the four principal fragmentation products expected in the mass spectrum of 2-methyl-4-heptanone. There are two different McLafferty rearrangement ions and two different α-cleavages leading to oxonium ions.

34.5 Advanced Techniques

Mass spectrometers vary in price from those attached to a gas chromatograph (system \$50,000) to sophisticated double mass spectrometers (MS/MS) (about \$1,000,000). Mass spectroscopy is one of the ultimate tools of organic chemical analysis, and is effective for as little as a few femtograms (10^{-15} g) of material. Although a detailed discussion of these advanced techniques goes beyond the scope of this text, the student should be aware of the possibilities.

The general scheme for mass spectroscopic analysis can be summarized in a flow diagram.

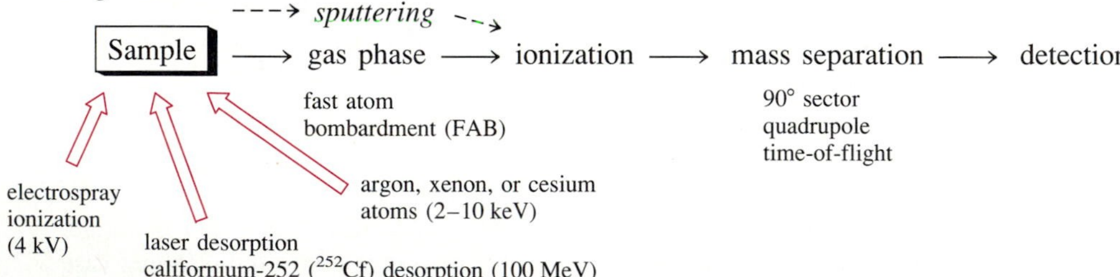

Electrospray ionization is induced by a high voltage being applied to fine droplets emanating from a stainless steel needle, then removing the solvent as, for example, from a sample of the protein carbonic anhydrase in methanol-water by a stream of nitrogen. The protein acquires a number of protons, and gives mass species with *m/z* between 700 and 1400 Da (Da = Daltons, a unit of molecular weight). A computer analysis of the multiple ions leads to a molecular weight of 29,012. It seems to be possible to measure molecular weights for proteins up to 100,000 Da. The mass separation is effected with a quadrupole mass spectrometer and the procedure can be carried out directly on the fractions obtained from high performance liquid chromatographic (HPLC) fractions. Organic molecules of considerable complexity can be investigated by mass spectrometry.

ASIDE

Bubbles of gas entrapped in meteorites older than life on Earth, or in amber 100 million years old, or in sealed jars taken from ancient tombs untouched for thousands of years are all minute samples that might be analyzed by mass spectrometry. A mass spectrometer will be the primary analytical instrument on board the Cassini-Huygens mission to Titan, the sixth moon of the sixth plant, Saturn. The purpose is to evaluate the nature of prebiotic molecules. If we are the operators for this mission, we must practice on simple but realistic samples. The atmospheric pressure on Titan must be low, but the temperature is also low. Suppose over the eons, deuterated compounds had accumulated because of their slightly lower vapor pressure. Sketch the mass spectra and mark the species expected for CH_4, CH_3D, CH_2D_2, CHD_3, and CD_4. Write out the expected fragmentation reactions.

PROBLEMS

1. **a.** Estimate the relative intensity of the peaks at m/z 112, 114, and 116 in the mass spectrum of 1,2-dichloropropane.

b. A compound shows a molecular ion at m/z 138 with a ratio of $(M + 1)/M$ of 0.111. Show how this piece of information can be used to distinguish among the three formulas $C_{10}H_{18}$, $C_8H_{10}O_2$, and $C_8H_{14}N_2$.

2. Estimate the intensity of the $M + 1$ peak, relative to the M^+ peak, for each of the following compounds.

a. dimethyl adipate
b. 1,2-diaminonaphthalene
c. 1-phenylheptane
d. *n*-hexacosane ($C_{60}H_{122}$)
e. methyl iodide
f. hexafluoroethane

3. The mass spectrum of N-propylaniline has a substantial fragment with m/z 106 ($M - 29$). Account for the fact that N-propyl-*p*-nitroaniline shows almost no $M - 29$ peak.

4. An unknown compound contains only carbon and hydrogen. Its mass spectrum is shown below. Propose a structure for the compound.

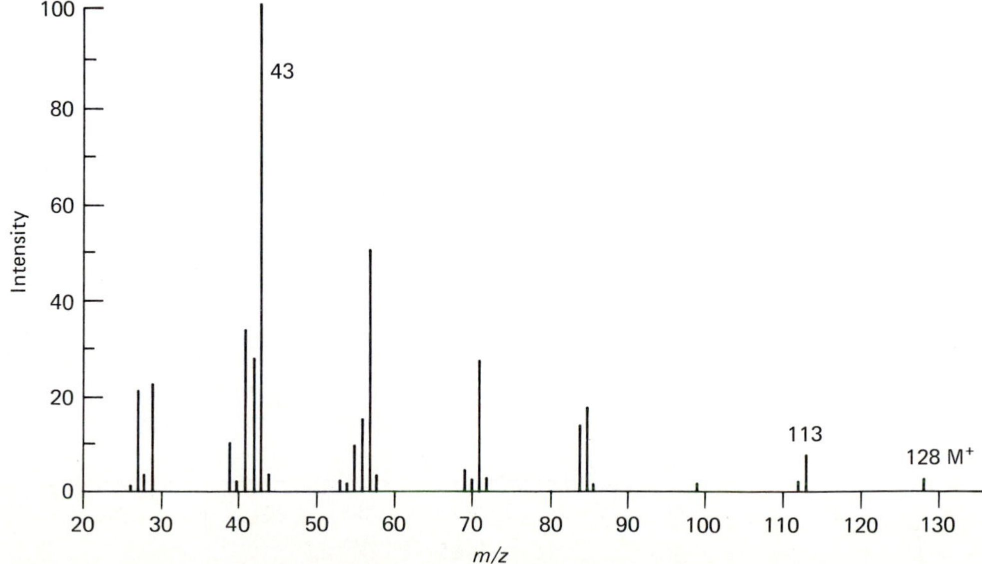

5. The following mass spectra are of 2,2-dimethylpentane, 2,3-dimethylpentane, and 2,4-dimethylpentane. Assign structures on the basis of the mass spectra.

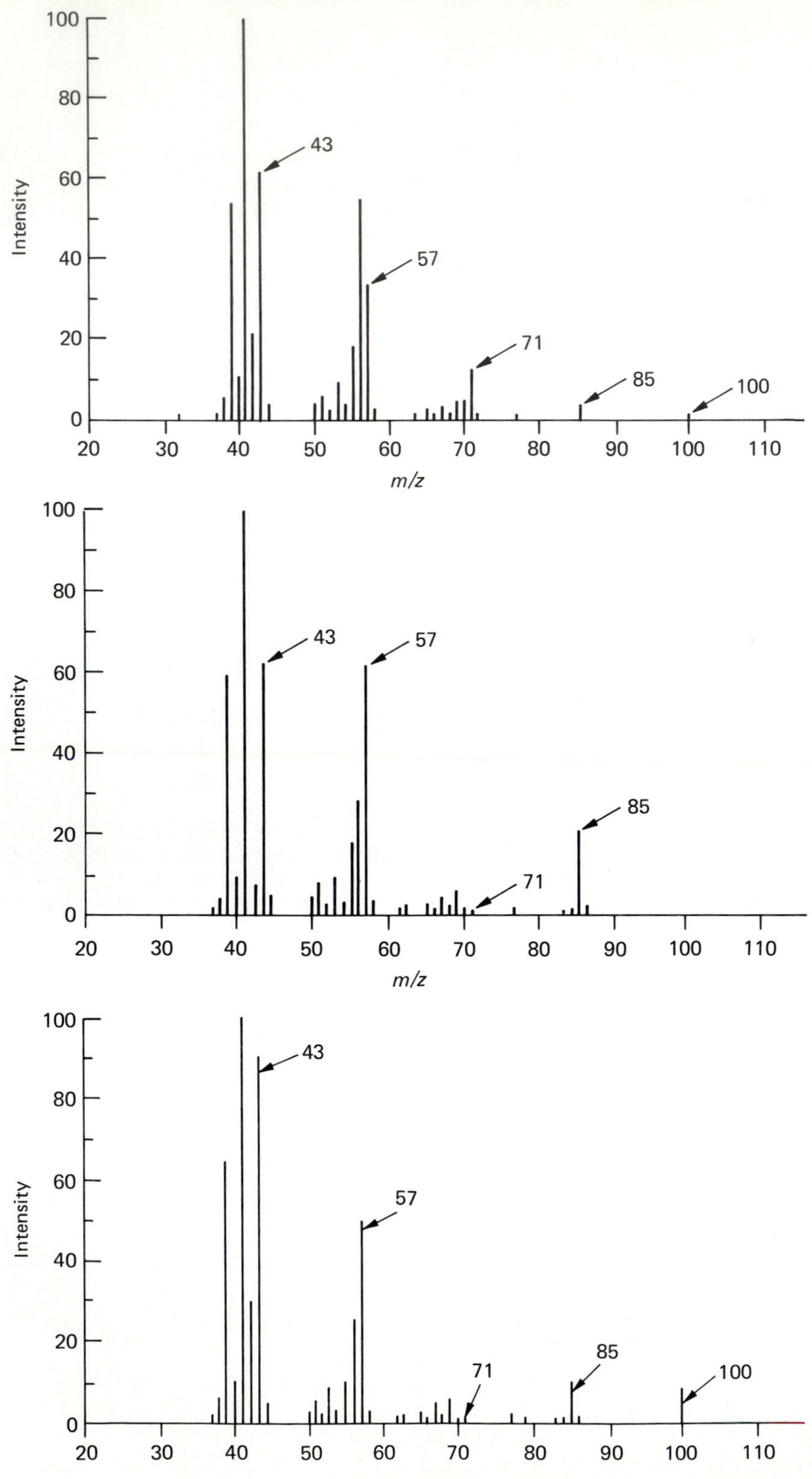

6. Identify the following compound from its IR and mass spectra.

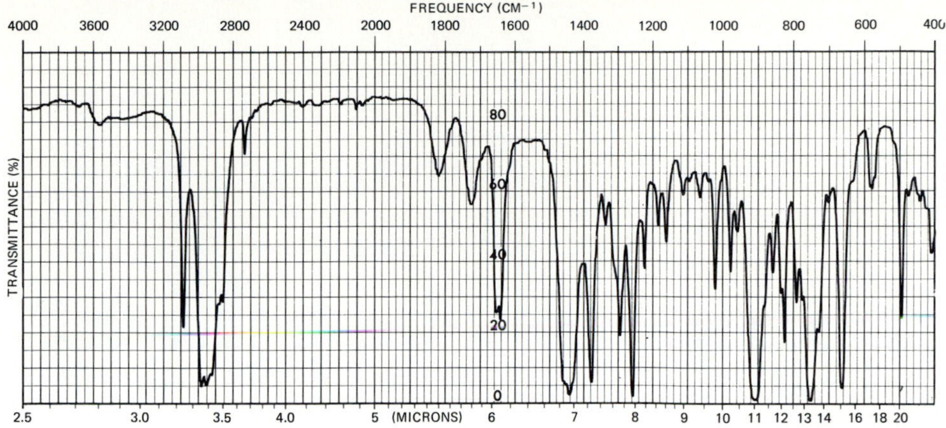

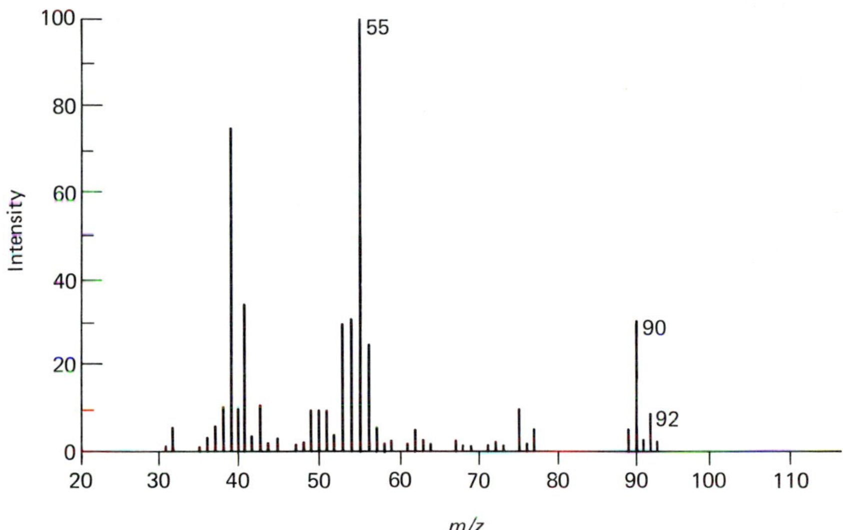

7. A compound, A, has the following properties: (i) the infrared spectrum contains a strong absorption at 1690 cm^{-1}; (ii) the ^{1}H NMR spectrum has three-proton singlets with $\delta = 1.5$ ppm and $\delta = 2.0$ ppm; (iii) the ^{13}C NMR spectrum (proton decoupled) has seven peaks; (iv) the mass spectrum shows a molecular ion with $m/z = 134$.

- **a.** What is the structure of A?
- **b.** The mass spectrum of A also has a peak with m/z 135. What is its approximate intensity (expressed as percent of the m/z 134 peak)?

8. What information can be deduced from each of the following observations relating to mass spectra? In each case data are presented as m/z (relative intensity).

- **a.** In the high region of the mass spectrum of compound B the following ions are seen: 155 (1.3), 154 (12), 136 (100).
- **b.** In the high mass region of the mass spectrum of compound C the following ions are seen: 298 (8.8), 297 (80), 282 (65), 254 (100).
- **c.** A reaction is carried out and the crude product analyzed by mass spectrometry. The high mass region of the spectrum of the reaction product has the following ions: 143 (4.1), 142 (40), 135 (2.9), 134 (36).
- **d.** A marine natural product, D, shows a mass spectrum that has the following ions in the high mass region: 294 (4), 292 (8), 290 (4).

9. Two hydrocarbons were isolated from blue-green algae. The mass spectra of the two hydrocarbons, shown below, provided a clue to their structures. Suggest structures for the two hydrocarbons.

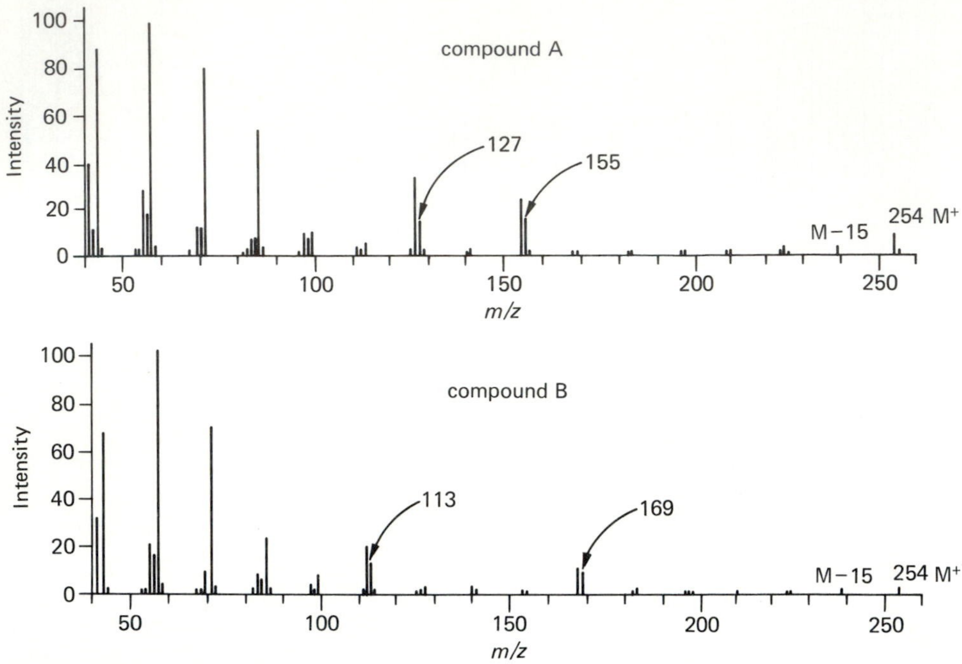

10. An unknown compound gives the following rather simple mass spectrum. What can you deduce about the structure of the compound from this spectrum?

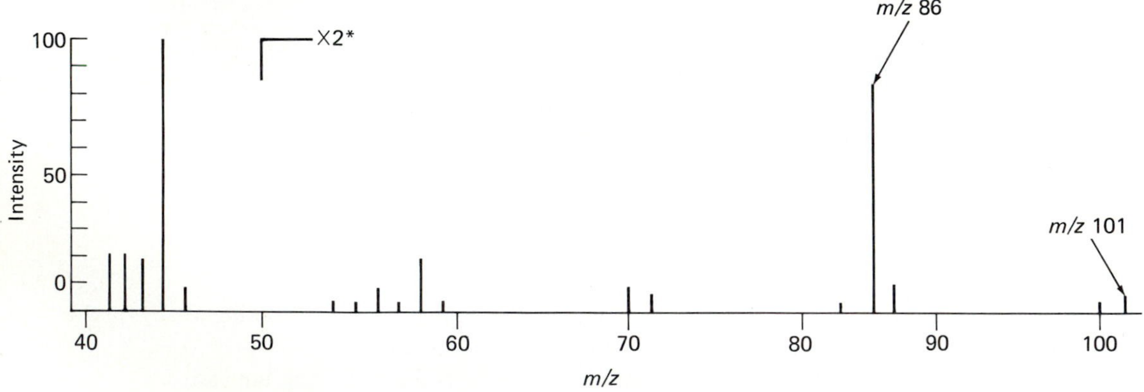

*The vertical scale is expanded by a factor of 2 above m/z 50.

The CMR spectrum of the compound shows only two resonances. Can an unambiguous structure be assigned?

11. A compound has infrared absorption at 1710 cm^{-1}. Its mass spectrum follows. Suggest a structure of the compound.

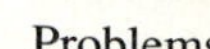

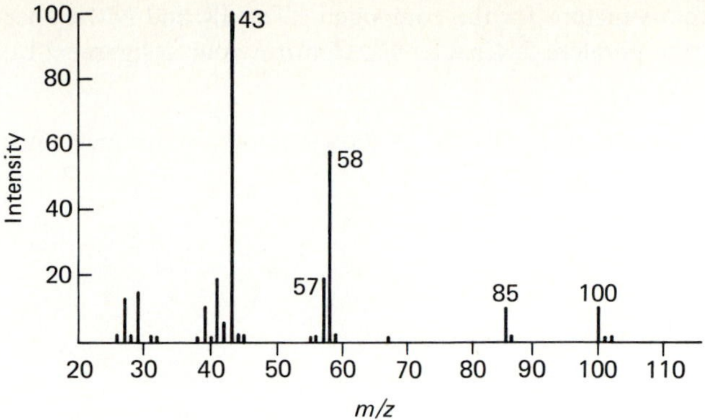

12. Propose a structure for the following compound from its mass spectrum. The IR spectrum shows a strong absorption at 1710 cm^{-1}.

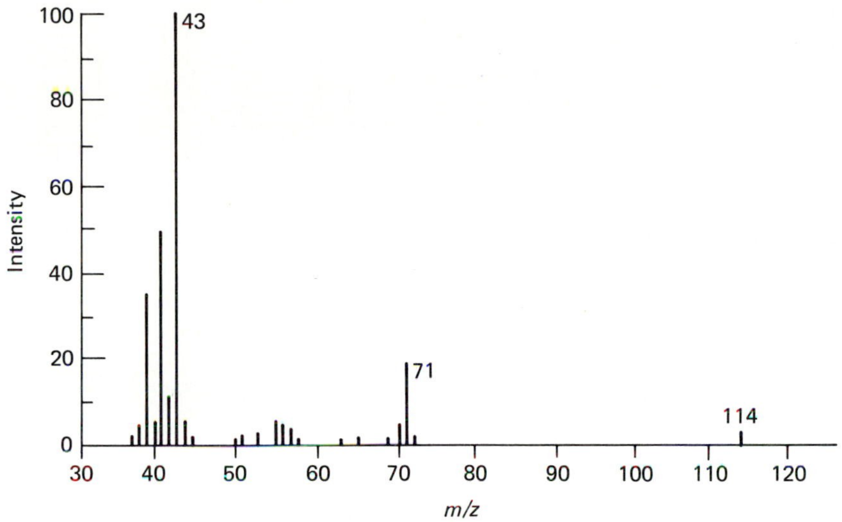

13. An unknown compound shows strong IR absorption at 3400 cm^{-1}. Its mass spectrum is shown below.

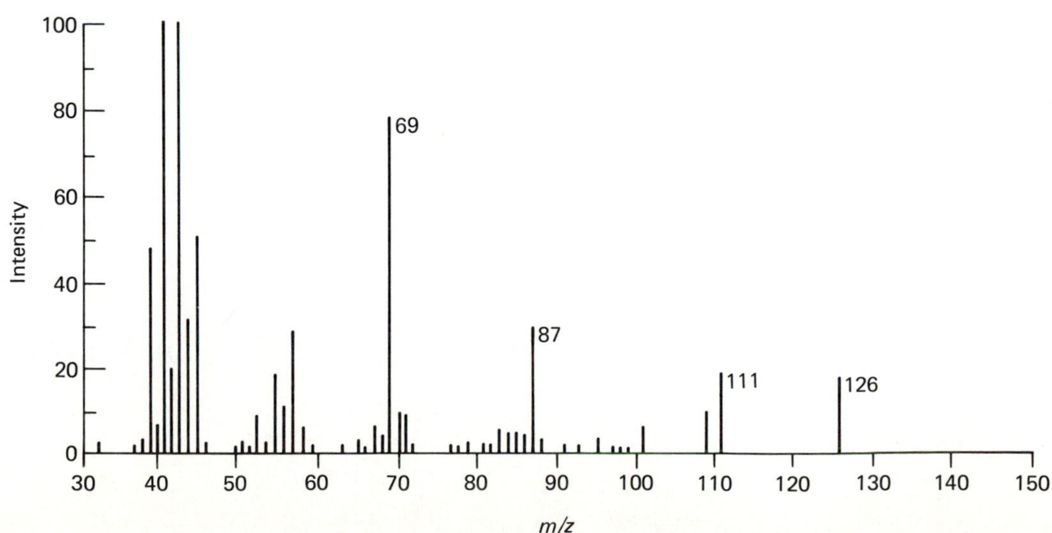

Propose a possible structure for the compound. [The IR and NMR spectra of the compound are shown in problem 5, Chapter 17. Confirm your assignment by examination of these spectra.]

14. The mass spectrum of 2-octanone is shown below. Write mechanisms showing the origin of the principal fragments.

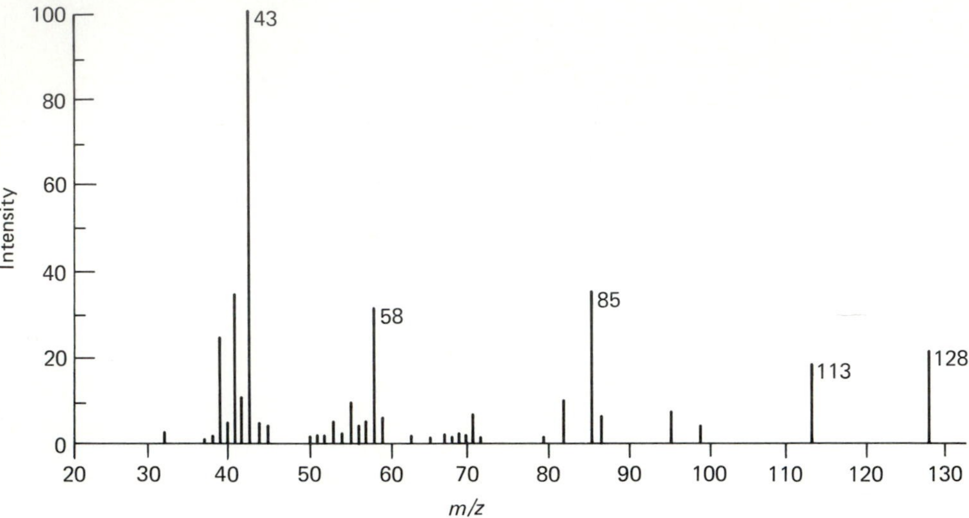

15. The following mass spectrum is given by the compound whose NMR and IR spectra are depicted in problem 2, Chapter 24.

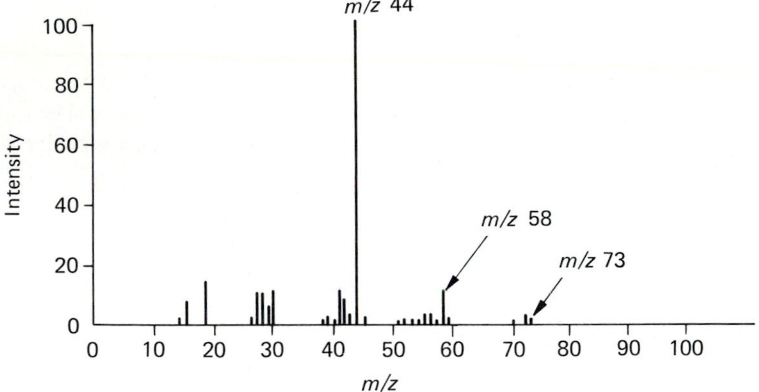

Suggest a structure for the compound.

16. The ionization potential of 2-methylbutane is 10.35 eV or 238.7 kcal mole^{-1} (1 eV = 23.06 kcal mole^{-1}); hence, ΔH_f° for 2-methylbutane cation is obtained from ΔH_f° of 2-methylbutane as 36.9 + 238.7 = 201.8 kcal mole^{-1}. From the following ΔH_f° values given for possible fragmentation products of the radical cation, calculate ΔH° for all of the possible carbon-carbon fragmentations of $[(CH_3)_2CHCH_2CH_3]^{+}$.

	ΔH_f°, kcal mole^{-1}		ΔH_f°, kcal mole^{-1}
CH_3^+	262	$CH_3\cdot$	35
$C_2H_5^+$	219	$C_2H_5\cdot$	29
$(CH_3)_2CHCH_2^+$	205	$(CH_3)_2CHCH_2\cdot$	17
$(CH_3)_2CH^+$	187	$(CH_3)_2CH\cdot$	21
$CH_3CH_2\overset{+}{C}HCH_3$	192	$CH_3CH_2\dot{C}HCH_3$	16

CHAPTER 35

THE CHEMICAL LITERATURE

35.1 Research Journals

The total knowledge of chemistry is contained in hundreds of thousands of books and journals that are known collectively as **the literature.** New knowledge is communicated to the world for the first time as a **paper** or **communication** in a **research journal.** There are perhaps 10,000 journals that publish original articles on chemical topics, but only about 50 are of general interest to most chemists. Some journals, such as the *Journal of the American Chemical Society,* publish articles in all branches of chemistry. Others, such as the *Journal of Organic Chemistry,* only publish articles dealing with a specific area. A partial listing of typical journals that would be of interest to an organic chemist, with the normal abbreviation printed in italic type, follows. The language(s) used in each journal is also indicated.

1. *Angew*andte *Chem*ie (German)
2. *Angew*andte *Chem*ie *Int*ernational *Ed*ition in *Eng*lish (English)
3. Justus Liebig's *Ann*alen der Chemie (German)
4. *Bull*etin of the *Chem*ical *Soc*iety of *Jap*an (English)
5. *Can*adian *J*ournal of *Chem*istry (English, French)
6. *Chem*ische *Ber*ichte (German)
7. *J*ournal of the *Chem*ical *Soc*iety, *Chem*ical *Commun*ications (English)
8. *Coll*ection of *Czech*oslovak *Chem*ical *Commun*ications (English)
9. *Compt*es *rend*us hebdomadaries, *Ser*ies *C* (French)
10. *Helv*etica *Chim*ica *Acta* (German, French, English)
11. *J*ournal of the *Am*erican *Chem*ical *Soc*iety (English)
12. *J*ournal of the *Chem*ical *Soc*iety, *Dalton Trans*actions (English)
13. *J*ournal of the *Chem*ical *Soc*iety, *Perkin Trans*actions (English)
14. *J*ournal of *Heterocycl*ic *Chem*istry (English)

15. *J*ournal of the *Indian Chem*ical *Soc*iety (English)
16. *J*ournal of *Med*icinal *Chem*istry (English)
17. *J*ournal of *Organometal*lic *Chem*istry (English, German, French)
18. *J*ournal of *Org*anic *Chem*istry (English)
19. *Synthesis* (English)
20. *Syn*thetic *Commun*ications (English)
21. *Tetrahedron* (English, German, French)
22. *Tetrahedron Lett*ers (English, German, French)
23. *Organometallics* (English)

An original article in a research journal can be in the form of a **full paper,** a **note,** or a **communication.** A full paper is a complete report on a research project, with full experimental details and interpretation. It is always accompanied by a short abstract, written by the authors. A note is a final report on a project of smaller scope. It includes experimental details, but has no abstract. A communication is a preliminary report on a finding of unusual significance. Communications are extremely concise, often less than 1000 words, and have little or no experimental detail. In most cases a communication will be followed later by a full paper after the project has been completed. Some journals, such as *J. Am. Chem. Soc.* and *J. Org. Chem.,* publish both papers and communications, and others, such as *Tetrahedron Lett.* and *J. Chem. Soc., Chem. Commun.,* publish only communications. Research articles are documented with references to the literature, to other research articles, and to books. The traditional form for such a **literature citation** is Author(s), *journal abbreviation,* **volume number,** page number (year). For example,

H. O. House and B. M. Trost, *J. Org. Chem.,* **30,** 2052 (1965).

However, in 1979, the American Chemical Society journals recommended a new form for literature citations. In the newly recommended form the authors' last names are given first, followed by first names and initials, followed by the *journal abbreviation,* **year,** *volume number,* and page number. For example,

House, H. O.; Trost, B. M., *J. Org. Chem.,* **1965,** *30,* 2052.

Because of the increase in punctuation required in this system, it has not gained universal favor with chemists. At the present time, both formats are in use.

If a practicing chemist is to keep abreast of the developments in his field, it is essential to peruse a number of research journals regularly as they appear. All of the journals listed above appear periodically, usually weekly, semimonthly, or monthly. Most chemists regularly scan the tables of contents of a dozen or so journals that publish articles in areas of interest to them.

35.2 Books and Review Articles

The original research journals comprise the **primary literature** of chemistry; they are the ultimate source that must be consulted for authoritative information on any subject. A second category of chemical literature is classed as **secondary literature.** The secondary literature consists of reference books and review articles in which the primary literature is collated and interpreted.

A. Handbooks

There are a number of excellent handbooks that compile data about individual organic compounds. The most extensive and most useful is the *Handbuch der Organischen Chemie,* commonly known as *Beilstein,* after its first editor. *Beilstein* is a multivolume handbook that lists all known organic compounds, together with their physical properties, methods of preparation, chemical properties, and any other available information. The main disadvantage of *Beilstein* is that it is not up to date. All of the literature through 1929 is completely covered, and most classes of compounds are covered through 1959. Further volumes covering the period to 1979 for some subjects have appeared. We shall consider the use of *Beilstein* in Section 35.5.

The Handbook of Chemistry and Physics, published by CRC Press, Inc., Boca Raton, Florida, is revised regularly. It contains a useful collection of data and a copy can be found on the desks of almost all practicing chemists. The most important table for organic chemists is ''Physical Constants of Organic Compounds,'' which occupies a major portion of the book. This table contains the name, formula, color, and several important physical properties for several thousand common organic compounds. Compounds are listed alphabetically by the IUPAC names. A similar volume is Lange's *Handbook of Chemistry,* McGraw-Hill Book Company, New York.

The Dictionary of Organic Compounds, edited by Heilbron, Cook, Bunbury, and Hey, is a five-volume handbook published by Oxford University Press, New York. It contains names, formulas, physical properties, and references for about 40,000 organic compounds. Compounds are listed alphabetically and there is no index. There are 10 supplemental volumes, which contain a large number of additional compounds.

The Merck Index: An Encyclopedia of Chemicals, Drugs, and Biologicals is published periodically by Merck and Company, Rahway, New Jersey. Over 10,000 compounds are described in the 11th edition, 1989, which is usually referred to as *The Merck Index.* It concentrates on compounds of medicinal importance, but covers most simple organic compounds, whether or not they have significant physiological properties. In addition to names and formulas, the Merck Index lists physical properties, methods of synthesis, physiological properties, and medicinal uses and also gives the generic and trade names for all compounds that are used as drugs.

The Aldrich Catalog of Chemical Compounds has become a useful secondary source for laboratory chemists. This book is primarily a catalog of the more than 16,000 compounds that may be purchased from the Aldrich Chemical Company, Inc.; it is updated annually. However, it also contains physical properties and the *Beilstein* reference for each compound. Furthermore, the entries are cross-referenced to two other reference books that contain NMR and IR spectra. Compounds are listed alphabetically by name and there is a formula index. Users should be cautioned that the nomenclature employed in the Aldrich catalog is a mixture of IUPAC, *Chemical Abstracts,* and trivial.

B. Review Articles

A review article is a survey of a single limited topic. For example, a chemist might assemble all of the information available on a topic by reading the original research articles and condense the information into a review article, frequently with a personal interpretation of the subject. There are several periodicals that specialize in publishing review articles. A few that are important to organic chemists are

1. *Chem*ical *Rev*iews (English)
2. *Chem*ical *Soc*iety *Rev*iews (English)
3. *Angew*andte *Chem*ie (German)
4. *Angew*andte *Chem*ie *Int*ernational *Ed*ition in *Eng*lish (English)
5. *Fortschr*itte der *Chem*ischen *Forsch*ung (German, English)
6. *Rev*iews of *Pure* and *Appl*ied *Chem*istry
7. *Synthesis* (English)
8. *Organomet*allic *Chem*istry *Rev*iews (English)
9. *Acc*ounts of *Chem*ical *Res*earch (English)

In addition to review journals such as these, there are a number of open-ended serial publications that are published at somewhat irregular intervals in hardbound form. These books are similar in content and format to the normal review journals. A few examples are

1. *Advances* in *Carbohydrate Chemistry*.
2. *Advances* in *Heterocyclic Chemistry*. Recent: Vol. 47.
3. *Advances* in *Photochemistry*.
4. *Progress* in *Physical Organic Chemistry*. Recent: Vol. 17.
5. *Advances* in *Physical Organic Chemistry*. Recent: Vol. 25.
6. *Organic Reactions*. Recent: Vol. 37, A. S. Kende (Ed.).
7. *Organic Reaction Mechanisms* in 24 volumes. Recent: 1988, A. C. Knipe and W. E. Watts (Eds.).
8. *Topics* in *Stereochemistry*. Recent: Vol. 19, E. L. Eliel (Ed.), 1990.
9. *Topics* in *Current Chemistry*.
10. *Advances* in *Organometallic Chemistry*.

Organic Reactions is a particularly important reference source for organic chemists. It is published approximately yearly and contains review articles on general reactions, for example, "The Wittig Reaction" and "The Clemmensen Reaction." The articles are accompanied by extensive tables of applications of the reaction. At the end of each volume there are cumulative subject and author indices.

C. Comprehensive Books and Advanced Texts

The daunting increase in the amount of primary literature has encouraged a corresponding effort to make the information accessible by means of comprehensive books and even extended multivolume textbooks. One useful series covers the *Chemistry of Functional Groups* with each volume containing comprehensive information on, for example, azido, carbonyl, and amino groups, or double and triple carbon-carbon bonds. The coverage also includes organometallic and organosilicon compounds. There were some 40 different titles in 80 volumes by the end of 1991. Some of the especially useful series are

1. S. Patai and Z. Rappoport, *The Chemistry of Functional Groups,* Wiley-Interscience (many volumes). Recent volume: *Chemistry of Sulphenic Acids, Esters and Derivatives,* 1990.
2. *Comprehensive Organic Chemistry,* 6 volumes, Pergamon, Oxford, 1979.
3. *Comprehensive Heterocyclic Chemistry,* 8 volumes, Pergamon, Oxford, 1984.
4. *Comprehensive Organometallic Chemistry,* 9 volumes, Pergamon, Oxford, 1982.
5. *Comprehensive Organic Synthesis,* 8 volumes, Pergamon, Oxford, 1991.
6. *Methoden der Organischen Chemie (Houben-Weyl),* 4th ed. (many volumes), G. Thieme, Stuttgart (German).
7. Rodd's *Chemistry of Carbon Compounds,* 5 volumes, Elsevier, Amsterdam, 1960.
8. J. March, *Advanced Organic Chemistry,* Wiley-Interscience, New York, 1985.

D. Monographs

There are a large number of excellent books available that provide in-depth surveys of specific areas. The number of such monographs is far too great to list here, and the student is referred to the card or computer catalog in his or her own library. Several examples, merely to indicate the types of topics covered, are

1. D. R. Dalton, *The Alkaloids,* Marcel Dekker, Inc., New York and Basel, 1979.
2. *Comprehensive Carbanion Chemistry,* Volume II, E. Buncel and T. Durst (Eds.), Elsevier, 1984.
3. C. Reichardt, *Solvents and Solvent Effects in Organic Chemistry,* 2nd ed., VCH, Weinheim, 1988.
4. H. Zollinger, *Color Chemistry,* VCH, Weinheim, 1987.
5. *Techniques of Chemistry,* Wiley-Interscience, New York (many volumes), various editions.
6. *Chemistry of Heterocyclic Compounds,* Wiley-Interscience, New York. Recent volumes: *Pyrroles,* Vol. 48, part 1, R. A. Jones (Ed.), 1990; *Isoxazoles,* Vol. 49, part 1, P. Grunanger and P. Vita-Finzi (Eds.), 1990.
7. M. Bodanszky, A. Bodanszky, *The Practice of Peptide Synthesis,* Springer, Heidelberg, 1984.
8. S. F. Mason, *Molecular Optical Activity and the Chiral Discriminations,* Cambridge, 1982.
9. *The Alkaloids,* Vol. 37, *Antitumor Bisindole Alkaloids from Catharanthus roseus (L.),* A. Brossi & M. Suffness (Eds.), Academic Press, 1990.

E. Books Covering Methods and Reagents

There are several useful books that are devoted to synthetic methods or to reagents used in organic reactions. *Organic Syntheses* is published by John Wiley & Sons, New York. It is a collection of procedures for the preparation of specific compounds. The work has appeared annually since 1921, the latest being Vol. 69 edited by L. A. Paquette. The procedures for each 10-year period are collected in cumulative volumes, of which seven now exist. The seventh cumulative volume was published in 1989. The procedures in *Organic Syntheses* are submitted by any chemist who wishes to do so and are then tested in the laboratory of a member of the Editorial Board. Although the methods given pertain to specific compounds, an attempt is made to include procedures that have general applicability. For this reason *Organic Syntheses* is a useful source of model procedures when the chemist wishes to carry out a new preparation. The cumulative volumes are each thoroughly indexed, and there is a collective index for the first five cumulative volumes.

Theilheimer, *Synthetic Methods of Organic Chemistry,* S. Karger Verlag, Basel, is an annual compilation of synthetic methods. It is organized by way of a system based upon types of bond formations or bond cleavages. There is an index with each volume and a cumulative index after each fifth volume.

Reagents for Organic Synthesis by L. F. Fieser and M. Fieser (Wiley, New York) is an exceedingly useful compendium of reagents and catalysts used in organic chemistry. Including the main volume, 14 volumes are now available. Volume 14 covers the literature between 1986 and 1988. The work gives information on how each reagent is prepared, commercial suppliers, and references to its uses.

35.3 Abstract Journals

Abstract journals are periodicals that publish short abstracts of articles that have appeared in the original research journals. There are currently two such publications devoted to the original chemical literature, *Chemical Abstracts* and *Referativnyl Zhurnal* (Russian). A German abstract journal, *Chemisches Zentralblatt,* ceased publication in 1970, but it is frequently useful for retrieving information published before that date.

Chemical Abstracts is published weekly by the American Chemical Society and includes abstracts in English of nearly every paper that contains chemical information, regardless of the original language. Abstracts appear from 3 to 12 months after the appearance of the original paper. The abstracts are grouped into 80 sections, of which sections 21–34 pertain to organic chemistry. Each individual abstract is preceded by the authors' names, the authors' address, the journal citation, and the language of the original article.

Although many chemists use *Chemical Abstracts* routinely to keep abreast of a broad area of chemistry, it is most useful because of its indices. From its beginning in 1907 until 1961 there were annual indices. Since 1962 there have been semiannual indices, covering the periods January–June and July–December. From 1907 until 1956 there were published additional 10-year indices. Since 1957 the cumulative indices have appeared at 5-year intervals. The most recent complete index is the *Eleventh Collective Index,* covering the period 1982–1986. The *Twelfth Collective Index* is scheduled to appear in 1992. Each annual and collective index has a subject index and an author index. Formula indices for the periods 1920–1946, 1947–1956, 1957–1961, 1962–1966, 1967–1971, 1972–1976, 1977–1981, and 1982–1986 are also available. The most useful of the indices is the formula index, which can be used to look up specific compounds. The subject index can be used to search for topics (such as "oxidation" or "kinetics") or for specific compounds by name. Although it is easier to search for a specific compound with the formula index, there is frequently merit in using the subject index. For example, the chemist may need information about derivatives of *o*-hydroxybenzoic acid in which there are additional substituents attached to the benzene ring. It would be virtually impossible to carry out such a search with the formula index.

To search *Chemical Abstracts* for information concerning a given compound, one looks up the formula of the compound in each of the collective indices and then in the semiannual indices that have appeared since the last collective index. Following each entry is an abstract number. The abstract numbers are used to locate the abstracts, and these are scanned. If it appears from an abstract that the original paper contains information of use, then this source is consulted.

For example, suppose we wish to know what has been published regarding the carcinogenic (tumor-producing) properties of the hydrocarbon benz[*a*]anthracene during the period 1967–1971.

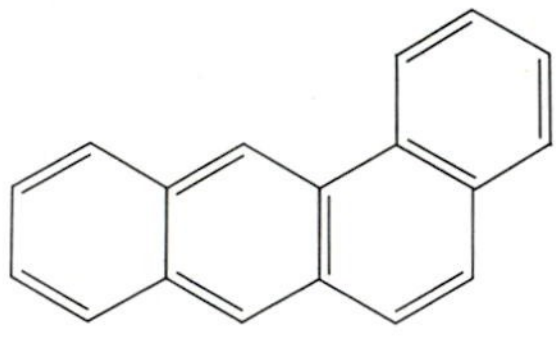

benz[*a*]anthracene

Consulting the *Eighth Collective Index,* which covers the period, we find the listing

Benz[*a*]anthracene [*56-55-3*]

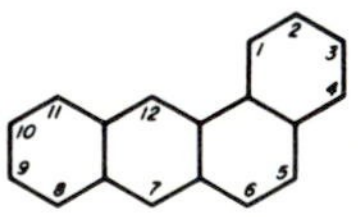

[a]

Following this listing, there are a number of indexed topics, in alphabetical order. A portion is shown.

bond length and heat of dissocn. of, **69:** 35282v
bond length in, **72:** 6380g
bond lengths and energy levels of, **75:** 5052b
bond order and localization energy of, in excited
  state, polarizability in relation to, **68:** 82565d
bromination of, mechanism of, **66:** 37049m
bud induction by, in tobacco callus culture, **74:**
  2887w
butoxyphenyl siloxanes contg., degradation of, **72:**
  133523f
carbonate dehydratase inhibition by, **67:** 115102k
carcinogenesis by, two-stage process in relation to,
  73: 33378q
carcinogenic activity of
  67: 98520a; **72:** 10910x
  activation energy of elec. conduction in relation
    to, **70:** 42062x
  mol. orbitals in relation to, **72:** 64737q
carcinogenic activity of hydrocarbons and, **72:**
  53064s
carcinogenic and menadione reductase-inducing
  activities of, **68:** 47948j
as carcinogen in human environment, **69:** 25669n
cell invasiveness in presence of, **68:** 103288h
cell protection against dimethylbenzanthracene by,
  aryl hydroxylase formation in, **71:** 122020t
cheese contamination with, from wax coating, **72:**
  109888h
chem. shifts of, mol. orbital calcn. in relation to, **73:**
  55252w

The number after each topic indicates the *Chemical Abstracts* volume number and abstract number where the information will be found. For example, the listing **67:**98520a means that abstract 98520 in volume 67 contains information on the carcinogenic activity of benz[*a*]anthracene. Going to volume 67 of the abstracts (1967), we find the following abstract.

98520a The carcinogenic activities in mice of compounds related to benz[a]anthracene. E. Boyland and P. Sims (Roy. Cancer Hosp., London). *Int. J. Cancer* **2**(5), 500–4(1967)(Eng). The carcinogenic activities of 18 aromatic hydrocarbons and their metabolic intermediates were compared after 3–10 s.c. injections of 1 mg. into C57 black mice. The monohydroxymethyl derivs. of 7,12-dimethylbenz[*a*]anthracene and some related compds. were active carcinogens, but were much less so than the parent hydrocarbon. Epoxides formed at the 5,6-bond (K-region) of chrysene, benz[*a*]anthracene, 7-methylbenz[*a*]anthracene, and dibenz[*a,h*]anthracene produced tumors when given at high dose levels, but were not as active as the parent hydrocarbons. The epoxide derived from phenanthrene was inactive. All of the compds. were prepd. by known methods with the exception of 7,12-dimethylbenz[*a*]anthracene, dibenz[*a,h*]anthracene, and chrysene which were obtained com. and 7-(diacetoxymethyl)benz[*a*]anthracene which was prepd. by heating benz[*a*]anthracene-7-carboxaldehyde under reflux with Ac_2O for 6 hrs. It sepd. from EtOH in needles, m. 196°. CTJN

If we desire more complete information, we may consult the original article, which was published in the *International Journal of Cancer Research,* volume 2, on page 500, in 1967.

In order to use *Chemical Abstracts* efficiently, it helps to have a good command of organic nomenclature. All compounds are listed as derivatives of a parent compound, for example,

$$CH_3\overset{\displaystyle Cl}{\overset{|}{C}}HCH_2CH_2COOH \equiv \text{valeric acid, 4-chloro}$$

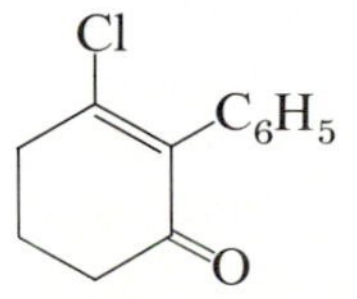

≡ 2-cyclohexen-1-one, 3-chloro, 2-phenyl

Note that *Chemical Abstracts* does not always use IUPAC nomenclature. At the beginning of each collective index, there is an extensive section dealing with the system of nomenclature used in indexing. In cases where it is not clear which name is used for indexing a particular compound, the formula index is useful. However, the formula index is much more tedious to use because one must often sift through an extensive list of isomers. However, because of the frequency with which the compilers of *Chemical Abstracts* have changed nomenclature systems in recent years, the formula index has become an indispensable aid in searching for more complex structures.

Beginning with the issuance of the volume indices covering the January-June 1972 period, the Subject Index has been issued in three parts: the Chemical Substance Index, the General Subject Index, and the Index Guide. All references to distinct, definable chemical substances are collected in the Chemical Substance Index, and all entries pertinent to any other topics (concepts, processes, organism names, diseases, reactions, generalized classes of compounds, and so on) are found in the General Subject Index. The Index Guide serves to guide the user quickly and efficiently to the proper headings in these two indices. The Index Guide can be used to find a Chemical Substance Index name for trivial, commercial, and other nonsystematically named substances. It represents a compilation of indexing cross-references, preferred index headings, synonyms, and general index notes on thousands of chemical terms and names and should be consulted before using either the Chemical Substance or General Subject Index. The Index Guide is supplemented annually to cover additions and changes that may occur within a volume indexing period.

35.4 Citation Analysis

Suppose that a scientist writes an article on a given subject. If the author has been thorough, there will be citations of articles that are most important and relevant to the subject because in preparing the article, the author has carried out a literature search for the readers. Further suppose that you find a particular point or procedure of interest and you want to know more about it. You look up the reference and read the original paper. After some thought, you decide to make use of the information for a particular experiment or analysis. You then check the literature to find out if others have carried out the same idea. You look up the specific compound you want to synthesize or analyze theoretically and it seems that no one else has done the specific experiment or analysis. Should you then go ahead and carry out the research? Suppose someone has done a similar experiment or analysis on a compound with an additional methyl group? How could you find out? If the other scientist used the same article that you found, and wrote an article in which he cited that reference, you can find it easily. Let us say that the article is one by R. F. Heck in *Accounts of Chemical Research,* volume 12, page 146, 1979, entitled ''Palladium-Catalyzed Reactions of Organic Halides with Olefins.''

A new approach to the analysis of the literature was developed by the chemist and information scientist Eugene Garfield. Note that analysis of the literature can be a full-time occupation. He reasoned that information could be traced through citations, the references to the work of others. After some early problems, he founded the Institute for Scientific Information, which has published the *Science Citation Index* since 1960. All of the citations to a particular first author are listed as shown in a sample of the 1988 listing for R. F. Heck, an American organometallic chemist who was professor at the University of Delaware. The 1979 article by Heck is found in the listing. In 1988, 20 articles cited the 1979 article.

HECK R
55 HELV CHIM ACTA 38 184
KAUPP G T CURR CHEM R 146 57 88
55 HELV CHIM ACTA 38 1541
CHUCHANI G INT J CH K 20 145 88
57 J AM CHEM SOC 79 3105
ARAKI Y CARBOHY RES 171 125 87
BONNETDE.D J ORG CHEM 53 754 88
57 J AM CHEM SOC 79 3114
BONNETDE.D J ORG CHEM 53 754 88
57 J AM CHEM SOC 79 3432
FUJIO M TETRAHEDR L 29 93 88
70 GAS WASSERFAH WASSE 111 223
JAEGGI NE ZBL BAKT B 186 494 88
88 KERNTECHNIK 53 56
CZECH J KERNTECHNIK 53 83 88
FABIAN H 53 11 88
HECK RF
60 CHEM IND 17 467
CORNELY W CHEM ZEITUN 112 191 88
60 J AM CHEM SOC 82 4438
KUKUSHKI.YN ZH OBS KH R 57 1921 87
60 2E ACT C INT CAT PAR 671
FUHRMANN E CHEM ZEITUN 112 295 88
61 J AM CHEM SOC 83 4023
CONSIGLI.G ORGANOMETAL N 7 778 88
FORD PC ADV ORGMET R 28 139 88
MAJOR A J MOL CATAL L 45 275 88
PINO P ORG SYNTH 50-9 338 88
TYLER DR PROG INORG R 36 125 88
61 J AM CHEM SOC 83 4024
FUHRMANN E CHEM ZEITUN 112 295 88
JENNER G J ORGMET CH 346 237 88
62 J AM CHEM SOC 84 2499
CUTLER AR CHEM REV R 88 1363 88
63 J AM CHEM SOC 85 651
EILBRACH.P CHEM BER 421 519 88
FUHRMANN E CHEM ZEITUN 112 295 88
63 J AM CHEM SOC 85 655
SCHUBERT U J ORGMET CH 340 101 88
63 J AM CHEM SOC 85 657
ASTRUC D CHEM REV R 88 1189 88
BRODIE NMJ INT J CH K 20 467 88
63 J AM CHEM SOC 85 1460
BERTANI R INORG CHEM 27 2809 88
IOBAL J CHEM LETT 1157 88
63 J AM CHEM SOC 85 2013
OJIMA I CHEM REV 88 1011 88
SCHORE NE R 88 1081 88
63 J AM CHEM SOC 85 2779
BRYNDZA HE CHEM REV R 88 1163 88
CORNELY W CHEM ZEITUN 112 191 88
KHAN MMT J MOL CATAL L 44 179 88
OJIMA I CHEM REV 88 1011 88
64 J ORGANOMET CHEM 2 195
FORD PC ADV ORGMET R 28 139 88
KOLOMNIK.IS USP KH R 57 729 88
64 J AM CHEM SOC 86 2580
BRODIE NMJ INT J CH K 20 467 88
DAVIS R POLYHEDRON 7 425 88
LEY SV PHI T ROY A 326 633 88
64 J AM CHEM SOC 86 2796
CALDERAZ.F GAZ CHIM IT 118 583 88
64 J AM CHEM SOC 86 2819
SCHORE NE CHEM REV R 88 1081 88
65 INORG CHEM 4 855
BRUNNER H ORGANOMETAL 7 1283 88
66 ADV ORGANOMETAL CHEM 4 243
AXE FU J AM CHEM S 110 3728 88
68 J AM CHEM SOC 90 313
BRUCE MI AUST J CHEM 41 1407 88
CALU N REV RO CHIM 33 59 88
68 J AM CHEM SOC 90 5518
ABRAMOVI.RA TETRAHEDRON R 44 3039 88
ARVIDSSO.LE J MED CHEM 31 92 88
CHEN QY J CHEM S P1 563 88
DAVIES SG 2597 87
EBERBACH W HELV CHIM A 71 404 88
GIRLING IR J CHEM S P1 1317 88
LAROCK RC TETRAHEDR L 29 905 88
NIZOVA GV B ACAD SCI 36 1920 87
PETERSON JR CAN J CHEM 66 1670 88
VICENTE J J CHEM S DA 141 88
VORBRUGG.H HETEROCYCLE R 27 2659 88

68 J AM CHEM SOC 90 5526
HECK RF ORG SYNTH 50-9 815 88
LAROCK RC TETRAHEDR L 29 905 88
VICENTE J J CHEM S DA 141 88
VORBRUGG.H HETEROCYCLE R 27 2659 88
68 J AM CHEM SOC 90 5531
CHEN QY J CHEM S P1 563 88
68 J AM CHEM SOC 90 5535
ANDERSSO.CM J ORG CHEM 53 235 88
BARTON DHR TETRAHEDRON 44 8387 88
CHEN QY J CHEM S P1 563 88
LUBELL WD J AM CHEM S 110 7447 88
VICENTE J J CHEM S DA 141 88
VORBRUGG.H HETEROCYCLE R 27 2659 88
68 J AM CHEM SOC 90 5538
SUN KS J CHEM S CH 209 88
TAMARU Y J AM CHEM S 110 3994 88
68 J AM CHEM SOC 90 5542
ABRAMOVI.RA TETRAHEDRON R 44 3039 88
NIZOVA GV B ACAD SCI 36 1920 87
SAITO K J ORGMET CH 338 265 88
SODERBER.BC J ORG CHEM 53 2925 88
VICENTE J J CHEM S DA 141 88
68 J AM CHEM SOC 90 5546
GULEVICH YV USP KH R 57 529 88
SUN KS J CHEM S CH 209 88
68 ORGANIC SYNTHESIS VI 1 373
RAO BN I J CHEM B N 27 84 88
69 ACCOUNTS CHEM RES 2 10
ABRAMOVI.RA TETRAHEDRON R 44 3039 88
BIANCHIN.C J AM CHEM S 110 6411 88
IRELAND RE ORG SYNTH 50-9 459 88
69 J AM CHEM SOC 90 5531
69 J AM CHEM SOC 90 5538
69 J AM CHEM SOC 90 5546
VICENTE J J CHEM S DA 141 88
69 J AM CHEM SOC 91 6707
ANDERSSO.CM J ORG CHEM 53 4257 88
CHEN QY J CHEM S P1 563 88
NIZOVA GV B ACAD SCI 36 1920 87
OJIMA I CHEM REV 88 1011 88
WADA A SYNTHESIS-S 771 88
71 J AM CHEM SOC 93 6896
LAROCK RC TETRAHEDR L 29 905 88
NIZOVA GV 8 ACAD SCI 36 1920 87
OJIMA I CHEM REV 88 1011 88
72 J ORG CHEM 37 2320
BENHADDO.R ORGANOMETAL 7 2435 88
FOURNET G TETRAHEDRON 44 5809 88
KOVALEVA LF ZH ORG KH 24 650 88
LEBEDEV SA J ORGMET CH 344 253 88
SAITO K " 338 265 88
72 J AM CHEM SOC 94 2712
FORD PC ADV ORGMET R 28 139 88
MILSTEIN D ACC CHEM RE R 21 428 88
OJIMA I CHEM REV 88 1011 88
VICENTE J J CHEM S DA 141 88
74 ORGANOTRANSITION MET
ASTRUC D CHEM REV R 88 1189 88
KRAFFT ME ORGANOMETAL 7 2528 88
" TETRAHEDR L 29 6421 88
MILE B ORGANOMETAL 7 1278 88
MIRALLES.J J PHYS CHEM 92 4853 88
74 ORGANOTRANSITION MET 11
MILLER DG J PHYS CHEM 92 6081 88
74 ORGANOTRANSITION MET 80
SEN A ACC CHEM RE R 21 421 88
74 ORGANOTRANSITION MET 201
VINOGRAD.MG J ORGMET CH 348 23 88
74 ORGANOTRANSITION MET 252
SCHORE NE CHEM REV R 88 1081 88
76 J ORG CHEM 41 265
HECK RF ORG SYNTH 50-9 815 88
HERSCOVI.J J ORG CHEM 52 5691 87
77 ADV CATALYSIS 26 323
BRUMBAUG.JS J AM CHEM S 110 803 88
MERKUSHE.EB SYNTHESIS-S R 923 88
MOSER WR J AM CHEM S 110 2816 88
77 ANN NY ACAD SCI 295 201
DAVIES SG J CHEM S P1 2597 87
78 J ORG CHEM 43 2941
GREINER A MAKRO CH-R 9 581 88

78 PURE APPL CHEM 50 91
BENHADDO.R ORGANOMETAL 7 2435 88
78 PURE APPL CHEM 50 691
BENHADDO.R J CHEM S CH 247 88
GIRLING IR J CHEM S P1 1317 88
KAMIGATA N B CHEM S J 61 3575 88
WANG YA J MOL CATAL 45 127 88
ZHANG ZY REACT POLYM 9 249 88
79 ACCOUNTS CHEM RES 12 145
REISCH J LIEB ANN CH 543 88
79 ACCOUNTS CHEM RES 12 146
ABRAMOVI.RA TETRAHEDRON R 44 3039 88
ANDERSSO.CM J ORG CHEM 53 235 88
BENHADDO.R J CHEM S CH 247 88
" ORGANOMETAL 7 2435 88
BOZELL JJ J AM CHEM S N 110 2655 88
BUCKLE DR J CHEM R-S 394 87
DAVIES SG J CHEM S P1 2597 87
ERNST RD CHEM REV R 88 1255 88
HEITZ W MAKROM CHE 189 119 88
KAMIGATA N B CHEM S J 61 3575 88
KARABELA.K J ORG CHEM 53 4909 88
MEINHART JD B CHEM S J 61 171 88
PRASAD JS TETRAHEDR L 29 4267 88
SMADJA W " 29 1283 88
SOLBERG J ACT CHEM B 41 712 87
VORBRUGG.H HETEROCYCLE R 27 2659 88
WADA A SYNTHESIS-S 771 88
WANG YA J MOL CATAL 45 127 88
WOLFF S SYNTHESIS-S 760 88
ZHANG ZY REACT POLYM 9 249 88
79 J AM CHEM SOC 101 5281
BURNS B TETRAHEDR L 29 4329 88
80 PLATINUM METALS REV 24 58
GULEVICH YV USP KH R 57 529 88
" ZH ORG KH 24 2126 88
81 PURE APPL CHEM 52 2323
CHEN QY J CHEM S P1 563 88
81 PURE APPL CHEM 53 2323
CHALONER PA J ORGMET CH R 337 431 87
VOBRUGG.H HETEROCYCLE R 27 2659 88
82 ORG REACT 27 345
ABELMAN MM J AM CHEM S N 110 2328 88
ABRAMOVI.RA TETRAHEDRON R 44 3039 88
ANDERSSO.CM J ORG CHEM 53 235 88
" " N 53 2112 88
" " 53 4257 88
BOZELL JJ J AM CHEM S N 110 2655 88
BRANCHAU.BP TETRAHEDR L 29 167 88
BUCKLE DR J CHEM R-S 394 87
BURNS B TETRAHEDR L 29 4325 88
DAVIES SG J CHEM S P1 2597 87
EARLEY WG TETRAHEDR L 29 3785 88
GREINER A MAKRO CH-RC 9 581 88
GRIGG R TETRAHEDRON 44 2033 88
HATANAKA Y J ORG CHEM 53 918 88
HEITZ W MAKROM CHE 189 119 88
JUNG H MAKRO CH-RC 9 373 88
KALININ VN DAN SSSR 298 119 88
KARABELA.K J ORG CHEM 53 4909 88
KASAHARA A CHEM IND L N 50 88
" " N 51 88
" " N 239 88
" " N 467 88
" " N 728 88
" SYNTHESIS-S N 704 88
KEANA JFW J ORG CHEM 53 2268 88
KIRBY AJ GAZ CHIM IT 117 667 87
KITAJIMA H NIP KAG KAI N 239 88
LAROCK RC TETRAHEDR L 29 905 88
" " 29 6399 88
LAURON H J CHEM EDUC 65 632 88
LEBEDEV SA J ORGMET CH 344 253 88
MERKUSHE.EB SYNTHESIS-S R 923 88
NAORA H B CHEM S J 61 2859 88
NEGISHI E TETRAHEDR L 29 2915 88
OCONNOR B " 29 3903 88
RADNER F J ORG CHEM 53 3548 88
REISCH J LIEB ANN CH 543 88
SAKAMOTO T HETEROCYCLE 27 257 88
" " 27 2225 88
SMADJA W TETRAHEDR L 29 1283 88
VORBRUGG.H HETEROCYCLE R 27 2659 88
WADA A SYNTHESIS-S 771 88
WANG YA J MOL CATAL 45 127 88
ZHANG Y J ORG CHEM L 53 5588 88
ZHANG ZY REACT POLYM 9 249 88

82 ORGANIC REACTIONS 27 351
ITAYA T YAKUGAKU ZA R 108 697 88
85 PALLADIUM REAGENTS 0
ABELMAN MM J AM CHEM S N 110 2328 88
ABRAMOVI.RA TETRAHEDRON R 44 3039 88
ANDERSSO.CM J ORG CHEM 53 235 88
" " N 53 2112 88
BARTON DHR PUR A CHEM 60 1549 88
" TETRAHEDR L 29 3533 88
" TETRAHEDRON 44 5661 88
CHEN QY J CHEM S P1 563 88
CRILLEY MML " 2061 88
DAVIES SG " 2597 87
EREN D J AM CHEM S 110 4356 88
GENET JP TETRAHEDRON 44 5263 88
GIRLING IR J CHEM S P1 1317 88
HAACK RA TETRAHEDR L 29 2783 88
HANZAWA Y CHEM PHARM L 36 4209 88
HATANAKA Y J ORG CHEM 53 918 88
HEITZ W MAKROM CHE 189 119 88
HOSHINO Y TETRAHEDR L 29 3983 88
IRITANI K " 29 1799 88
KARABELA.K J ORG CHEM 53 4909 88
KEINAN E PUR A CHEM 60 89 88
KIJI J CHEM LETT 957 88
KOGA T " 1141 88
KOWALSKI MH ORGANOMETAL N 7 1227 88
LAROCK RC TETRAHEDR L 29 905 88
LINDSAY OM J CHEM S P1 569 88
MIYAURA N J SYN ORG J 46 848 88
MORENOMA.M J ORG CHEM 53 5328 88
" TETRAHEDR L 29 581 88
NAIR V J ORG CHEM 53 3051 88
NEGISHI E TETRAHEDR L 29 2915 88
OCONNOR B " 29 3903 88
OKUKADO N CHEM LETT 1449 88
SAKAMOTO T HETEROCYCLE 27 257 88
" " 27 2225 88
" SYNTHESIS-S N 485 88
SCOTT WJ ACC CHEM RE R 21 47 88
SEN A " R 21 421 88
SOMEI M HETEROCYCLE 27 1585 88
TAKEDA K TETRAHEDR L 29 4105 88
TSUJI J " 29 343 88
ZHANG Y J ORG CHEM L 53 5588 88
85 PALLADIUM REAGENTS 0 18
YONEYAMA M MACROMOLEC 21 1908 88
85 PALLADIUM REAGENTS 0 19
MUZART J B S CHIM FR 731 88
85 PALLADIUM REAGENTS 0 117
VITAGLIA.A J ORGMET CH N 349 C 22 88
85 PALLADIUM REAGENTS 0 122
CARROLL WE J MOL CATAL L 44 213 88
85 PALLADIUM REAGENTS 0 179
TAKAHASH.H J ORGMET CH 350 227 88
WADA A SYNTHESIS-S 771 88
85 PALLADIUM REAGENTS 0 180
BOGER DL J ORG CHEM 53 1405 88
85 PALLADIUM REAGENTS 0 276
ANDERSSO.CM J ORG CHEM 53 4257 88
ITAYA T TETRAHEDR L 29 4129 88
" YAKUGAKU ZA R 108 697 88
85 PALLADIUM REAGENTS 0 279
SMADJA W TETRAHEDR L 29 1283 88
85 PALLADIUM REAGENTS 0 299
SAKAMOTO T CHEM.PHARM N 36 2248 88
85 PALLADIUM REAGENTS 0 316
YAMASHIT.H J ORGMET CH 356 125 88
85 PALLADIUM REAGENTS 0 322
IYODA M J CHEM S CH 65 88
85 PALLADIUM REAGENTS 0 341
KIM YJ ORGANOMETAL 7 2182 88
85 PALLADIUM REAGENTS 0 408
BANERJEE AA J CHEM S CH 1275 88
85 PALLADIUM REAGENTS 0 CH 2
CHRISOPE DR J AM CHEM S 110 230 88
85 PALLADIUM REAGENTS 0 CH 3
DUNKERTO.LV J CARB CHEM 7 49 88
85 PALLADIUM REAGENTS 0 CH 6
KAMIGATA N B CHEM S J 61 3575 88
LIEBESKI.LS PUR A CHEM 60 27 88
86 PALLADIUM REAGENTS 0
HEGEDUS LS J ORGMET CH R 343 147 88

You can now check the 20 articles to see if they have done anything similar to your proposed experiment. The *Source Index* lists the titles of the citing articles so that you can make a judgement as to how relevant the article might be before you look for it. By examining other years of the *Science Citation Index,* you can build up a fairly complete list of all those who used the reference. Some libraries have 5-year indices, which make the search faster. Of course, you can check the citations to each of the authors in your list, in case someone used them as the primary reference and did not cite the original source. In the case of an organic synthetic procedure, the experience of others might be helpful in planning the experiment. A citation search can quickly find all those who have used the procedure and have already written articles on their work. Clearly, a time gap between the original work and the appearance of the article that refers to it must exist. In the case of important discoveries, on which many are anxious to work and publish fairly quickly, 1 or 2 years might elapse. For the more usual reports, a gap of 3–6 years can be expected.

35.5 Beilstein

Beilstein's *Handbuch der Organischen Chemie* is shelved in the reference section of most chemical libraries. There have been four editions of the work and the first three are mainly useful for old references. The fourth edition (*vierte Auflage*) consists of a main series (*das Hauptwerk*) and five supplementary series (*erstes, zweites, drittes,* and *viertes Ergänzungswerk* and *Fifth Supplementary Series*). As the titles indicate, the language of publication has changed from German to English. The periods covered by the various series are

Main series	antiquity–1909
First supplement	1910–1919
Second supplement	1920–1929
Third supplement	1930–1949
Fourth supplement	1950–1959
Fifth supplement	1960–1979 (incomplete)

The main series consists of 27 volumes (*Bands*), each bound as a separate book. Each supplementary series also consists of 27 volumes, and entries in the supplements are cross-referenced to the main series. Volumes in the supplementary series are sometimes bound as more than one book, and in some cases two or more volumes are bound together.

Compounds are grouped into three major divisions, in the following manner.

Division	Volumes
Acyclische reihe Acyclic compounds	1–4
Isocyclische reihe Carbocyclic compounds	5–16
Heterocyclische reihe Heterocyclic compounds	17–27

There is a fourth minor division—carbohydrates, rubber-like compounds, and carotenoids—contained in volumes 30 and 31, which only appeared with the first extension. The contents of the various volumes are shown in Table 35.1. In addition, the table indicates which supplements had been completed by the middle of 1990. Every conceivable compound can be assigned to a system number, whether or not the compound has been reported in the literature. If one knows the method that is used by the *Beilstein* staff to decide the system number of a given compound, then one can look up the substance in this manner. Unfortunately, the procedure used in assigning system numbers is sufficiently involved that most chemists cannot readily use it. However, the system numbers are also employed for cross-referencing between the Supplements and the Main Series, and they are very useful for finding supplementary information about a given compound after it has been located in the Main Series (or *vice versa*).

Volumes 28 and 29 are a subject index (*Generalsachregister*) and a formula index (*Generalformelregister*), respectively. The most recent indices are part of the fourth sup-

TABLE 35.1 *BEILSTEIN'S HANDBUCH DER ORGANISCHEN CHEMIE*

Volume	Contents	System Numbers	Complete through
1	acyclic hydrocarbons, alcohols, ketones	1–151	1959
2	acyclic carboxylic acids	152–194	1959
3	acyclic hydroxy acids, keto acids	195–322	1959
4	acyclic sulfinic acids, amines, phosphines	323–449	1959
5	cyclic hydrocarbons	450–498	1959
6	cyclic alcohols	499–608	1959
7	cyclic ketones	609–736	1959
8	cyclic hydroxy ketones	737–890	1959
9	cyclic acids	891–1050	1949[a]
10	cyclic hydroxy acids, keto acids	1051–1504	1959
11	cyclic sulfinic and sulfonic acids	1505–1591	1959
12	cyclic amines	1592–1739	1959
13	cyclic polyamines, amino alcohols	1740–1871	1959
14	cyclic amino ketones, amino acids	1872–1928	1959
15	cyclic hydroxylamines, hydrazines	1929–2084	1959
16	cyclic azo, phosphines, organometallics	2085–2358	1959
17	heterocyclic, 1 oxygen	2359–2503	1979
18	heterocyclic, 1 oxygen	2504–2665	1979
19	heterocyclic, 2–9 oxygens	2666–3031	1979
20	heterocyclic, 1 nitrogen	3032–3102	1979
21	heterocyclic, 1 nitrogen	3103–3241	1979[b]
22	heterocyclic, 1 nitrogen	3242–3457	1959
23	heterocyclic, 2 nitrogens	3458–3554	1959
24	heterocyclic, 2 nitrogens	3555–3633	1959
25	heterocyclic, 2 nitrogens	3634–3793	1959
26	heterocyclic, 3–8 nitrogens	3794–4187	1959
27	heterocyclic, other ring systems	4188–4720	1959

[a]System numbers 891–968 are complete through 1959.
[b]System numbers 3238–3241 are complete only through 1959

plement and they cover the main series and all four supplements, that is, through 1959. The name indices are contained in 22 books divided according to volumes; the formula indices are in 20 books. One cannot rely completely on the indices, because only representative compounds are indexed. However, they are useful to obtain rapidly the approximate location of a compound in the handbook, particularly for heterocyclic compounds. The index listing gives the volume and page numbers where the compound will be found. Bold type indicates the volume number and normal type indicates the page number; supplementary series page numbers are preceded by the appropriate Roman numeral. For example, volume 28 of the second supplementary series contains the listing for indole (Ger., *Indol*)

Indol **20,** 304, I 121, II 196; **21** II 567.

Thus, we find indole listed on page 304 in volume 20 of the main series (H), on page 121 of volume 20 of the first supplementary series (E1), and on page 196 of volume 20 of the second supplementary series (EII). The final entry refers to a correction, which appeared on page 567, at the end of volume 21 of the second supplementary series. We find the same listing in volume 29 of the second supplementary series, which is the formula index, under C_8H_7N, the formula of indole. We can easily find the entries for Indol in the third + fourth (E III/IV **20/5,** 3176) and for indole in the fifth supplementary series (E V **20/7,** 5) with either the **H20,** 304 reference or the system number.

Although the *Beilstein* indices are useful, one should become familiar with the basic organizational system of the handbook if it is to be used to best advantage. In each of the first two major divisions—acyclic compounds and carbocyclic compounds—compounds are listed according to the following order of basic classes.

1. Hydrocarbons (*Kohlenwasserstoffe*), RH.
2. Hydroxy compounds (*Oxyverbindungen*), ROH.
3. Carbonyl compounds (*Oxoverbindungen*), $R_2C{=}O$.
4. Carboxylic acids (*Carbonsäuren*), RCOOH.
5. Sulfinic acids (*Sulfinsäuren*), RSO_2H.
6. Sulfonic acids (*Sulfonsäuren*), RSO_3H.
7. Selenium acids (*Seleninsäuren* and *Selenosäuren*), $RSeO_2H$ and $RSeO_3H$.
8. Amines (*Amine*), RNH_2, R_2NH, R_3N.
9. Hydroxylamines (*Hydroxylamine*), RNHOH.
10. Hydrazines (*Hydrazine*), $RNHNH_2$.
11. Azo compounds (*Azo-Verbindungen*), RN=NH.

Following these basic classes, there are a further 27 rare classes, which we do not list.

The handbook begins with acyclic hydrocarbons; the very first entry is methane, CH_4. After all of the derivatives of methane have been listed, one finds ethane, followed by its derivatives, and so on, through all the hydrocarbons having the empirical formula C_nH_{2n+2}. When all alkanes and their substitution derivatives have been listed, hydrocarbons with the formula C_nH_{2n} follow, beginning with ethylene (C_2H_4), and going on up in carbon number. Next are listed hydrocarbons with the formula C_nH_{2n-2}. In this section we find alkynes and dienes; the first entry is acetylene, C_2H_2. The following group of compounds has the general formula C_nH_{2n-4}, then C_nH_{2n-6}, and so on. Thus, within a class of compounds, such as hydrocarbons, compounds are listed in order of increasing unsaturation. The general formula for the compounds listed on a given pair of pages is printed at the top of the left-hand page.

After all hydrocarbons and their derivatives have been listed, the hydroxy compounds are listed. In the acyclic division, the first hydroxy compound is methanol, CH_3OH, which has the empirical formula $C_nH_{2n+2}O$. Following the alcohols of this formula, one finds alcohols with the formula $C_nH_{2n}O$, and so on. When all mono alcohols have been listed, the diols are listed, beginning with the $C_nH_{2n+2}O_2$ compounds. Next come the triols, tetraols, and so on. When the alcohols have been exhausted, the aldehydes and ketones are listed, and so on down the list of classes of compounds.

Polyfunctional compounds are indexed *under the class that occurs last in the listing*. For example, hydroxycarboxylic acids are indexed under carboxylic acids, amino sulfonic acids under amines, and so on. When three or more of the basic functional groups are present, the same rule applies; a hydroxy amino acid will be found under the amines.

Following each compound in the handbook, one will find its derivatives. The derivatives are of three types and are listed in the following order.

1. Functional Derivatives. These compounds are derivatives of the basic functional group and are hydrolyzable (in principle) to the parent compound. For example, dimethyl ether and methyl nitrate are both considered as functional derivatives of methyl alcohol and are indexed after it.

$$CH_3OCH_3 \xrightarrow{H_2O} 2\,CH_3OH$$

$$CH_3ONO_2 \xrightarrow{H_2O} CH_3OH + HNO_3$$

2. Substitution Derivatives. These are compounds in which a C—H has been replaced by C—X, C—NO, C—NO_2, or C—N_3. They are listed in the order

1. Halides
   (a) Fluorides, such as CH_3F.
   (b) Chlorides, such as CH_3Cl.
   (c) Bromides, such as CH_3Br.
   (d) Iodides, such as CH_3I.
2. Nitroso derivatives, such as CH_3NO.
3. Nitro derivatives, such as CH_3NO_2.
4. Azido derivatives, such as CH_3N_3.

When there is more than one of the same group attached to the basic compound, the polysubstituted compounds follow the monosubstituted compound. For example, the fluorinated methanes appear in the order CH_3F, CH_2F_2, CHF_3, CF_4. When two different substitution groups are present, the compound is listed under the group that occurs last in the foregoing list. Thus, fluorochloromethane, CH_2FCl, appears immediately after methyl chloride, CH_3Cl; in effect, CH_2FCl is considered as a substitution derivative of CH_3Cl. Likewise, chloronitromethane, $ClCH_2NO_2$, follows nitromethane in the listing. One must be careful not to confuse substitution derivatives with functional derivatives. For example, methyl hypochlorite, CH_3OCl, is listed with the functional derivatives of methyl alcohol because, in principle, it is hydrolyzable to methyl alcohol.

$$CH_3OCl \xrightarrow{H_2O} CH_3OH + HOCl$$

3. Sulfur and Selenium Compounds. These compounds are listed as replacement derivatives under the corresponding oxygen compound. For example, methyl mercaptan, CH_3SH, and dimethyl selenide, $(CH_3)_2Se$, are listed last under the derivatives of methyl alcohol. Similarly, dithioacetic acid, CH_3CS_2H, is found among the final listings that follow acetic acid.

A similar organization is followed with the carbocyclic compounds. For heterocyclic compounds, the same scheme is used, but there is an additional division into *hetero numbers*. Most practicing chemists do not use the hetero numbers but, rather, rely on the subject or formula index to locate the parent heterocycle in the handbook. One must remember that many familiar compounds not normally thought of as heterocyclic compounds indeed are. For example, succinic anhydride will be found in the third division, as a dicarbonyl derivative of the heterocycle tetrahydrofuran.

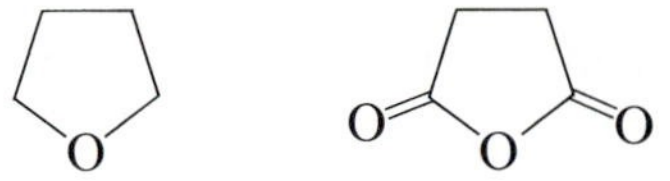

O O O O

tetrahydrofuran succinic anhydride

35.6 Computer Searches of the Literature

The explosive growth in the scientific literature has stimulated the growth of information science. The problem facing scientists is how to find the information they need. A chemist must formulate questions in the most precise manner possible, in order to obtain a reasonable amount of useful information.

Fortunately, an enormous growth in the capacity and power of computers coupled with a tremendous decrease in their cost has made several developments possible. First, it has become possible to store information efficiently in the form of various kinds of **data bases.** Second, individuals can gain access to these data bases via a **personal computer** and a **telephone connection** or **modem.** Third, the international scientific community has agreed to cooperate in the form of an international scientific and technical information network called **STN International.** The service is offered jointly by the Chemical Abstracts Service (CAS), which is a division of the American Chemical Society, the FIZ Karlsruhe, a scientific and educational organization in Karlsruhe, Germany, and the Japan Information Center of Science and Technology (JICST) in Tokyo, Japan.

Organic chemists face special problems, since they may wish to acquire information about conversions of one type of compound into another without knowing which particular compounds have already been subjected to the conversion of interest. For example, how can one oxidize a secondary alcohol to a ketone if a primary alcohol is present in the molecule at the γ-position? We can now do literature searches for **substructures.** Search programs allow you to use either an IBM PC or PC-compatible (AT or higher level) or a Macintosh (Plus, SE, or II) to search various databases. A hard disk and a modem are also required. For example, you may search the literature via these data bases on the basis of a Chemical Abstracts **registry number,** a **molecular structure** (which you draw on the computer screen), a **chemical name,** a **trade name,** a **chemical formula,** a **generic structure,** or a **molecular substructure.** You can search for reactions and you can search Beilstein.

The cost is not small, from \$50 to \$150 per connect hour, plus additional charges for printing and searching. However, with properly formulated questions, one can obtain a thorough literature search at lower cost than by the classical method of searching in the library. Academic institutions are usually eligible for considerable discounts. However, one loses the inestimable value of browsing through the literature and finding unrelated information of great usefulness.

If you have access to a literature searching program, you should compare its use to that of the regular literature search. In looking for substructures, the computer would be superior. The increased cost of maintaining a complete library has forced many institutions to curtail the number of journals to which they subscribe. Computer searches in principle make all of the world's journals accessible, if not directly, then at least via the abstracts. Other types of data bases (**CASREACT** contained about 250,000 chemical reactions in 1990) make new types of literature searches possible.

ASIDE

Chiral compounds are prevalent in nature; specific stereoisomers of medicinal agents are often the most active or even the only efficacious agent for a given purpose. Suppose we extend this idea to perfume materials, synthetic organic compounds used in the formulation of perfumes. Are the odors of enantiomers (see Chapter 7) different? Is it worthwhile to undertake a synthesis of some new compounds? We must first look at the literature. (You can actually find an answer to this question in Section 36.5.A.) Let us take a specific problem, the synthesis of *(R)*-2-butyl phenyl ether. How do you go about it? Where do you start? If possible, find a book that deals with the relationship of structure to odor, such as R. W. Moncrieff, *The Chemical Senses,* 3rd ed. (CRC Press, Boca Raton, FL, 1967). We learn that methyl phenyl ether (methoxybenzene) is strongly aromatic and fragrant, that ethyl phenyl ether is similar but less strong, and that 1-heptyl phenyl ether has the odor of opoponax, which is pleasant, mild, and distinctive. This information encourages us to think that 2-butyl phenyl ether might indeed be of interest as a perfume material. Outline, step by step, how you would go about the synthesis, beginning with the literature search and ending with the synthesis of the optically active target compound.

Problems

1. Select recent issues of several chemical journals from the current periodicals shelf of your school library. Browse through them and note the form of full papers, notes (in those journals that publish them), and communications.

2. Look up the indicated information in each of the following secondary sources.
 a. The boiling point and melting point of *p*-bromobenzonitrile in *The Handbook of Chemistry and Physics*.
 b. The toxicity of nicotine in *The Merck Index*.
 c. The method of preparation of 9-methylanthracene in *The Dictionary of Organic Compounds*.

3. a. Find a review article on the Reformatsky reaction in *Organic Reactions*.
 b. Examine a recent issue of *Chemical Reviews* and note the form of the articles and the type of subjects covered.

4. a. Find a procedure for the preparation of cyclobutanecarboxylic acid in *Organic Syntheses*.
 b. Find a leading literature reference to the use of manganese dioxide for the oxidation of allylic alcohols in *Reagents for Organic Synthesis*.

5. a. Examine a current (unbound) issue of *Chemical Abstracts*. Scan the pages containing Sections 21–34 and note the form and content of the abstracts.
 b. Using the most recent Author Index, see what papers have been published during that period by A. Streitwieser, C. H. Heathcock, and E. M. Kosower.
 c. Using the 1920–1946 Formula Index, look up 2-hydroxy-1-naphthoic acid.
 d. Using the most recent Chemical Substance Index, see what sorts of derivatives of *o*-hydroxybenzoic acid are listed.

6. Find the *Science Citation Index* in your library. Look for Barton, D. H. R. Are the initials important? Are there other Barton entries with different initials? Select a recent Barton reference and look it up. Now look at the titles of the citing articles and decide whether they are relevant to the subject. Check one or two of the citing articles and describe the connection between these articles and the Barton article.

7. Locate Beilstein's *Handbuch der Organischen Chemie* in your school library. Update Table 35.1 by determining what system numbers have now been completed through 1979. Use the Cumulative Formula Indices that are shelved with the Fourth and Fifth Supplements to look up quinoline. What is its system number?

8. Look up the following compounds in Beilstein's *Handbuch der Organischen Chemie*. Record the melting point and/or boiling point, the system number, and the page number in the main series and each supplementary series where the compound is found.

a. 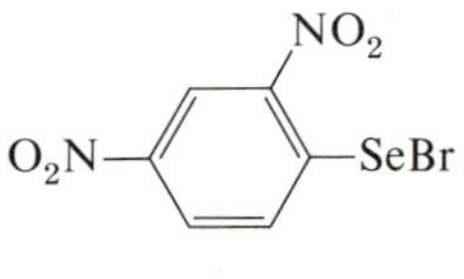 O_2N—(benzene ring with NO_2)—SeBr

b. 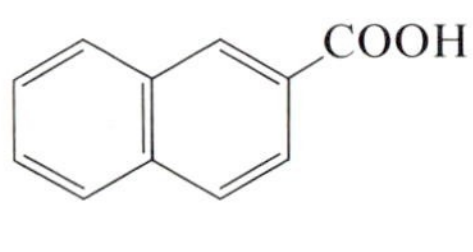 (naphthalene)—COOH

c. 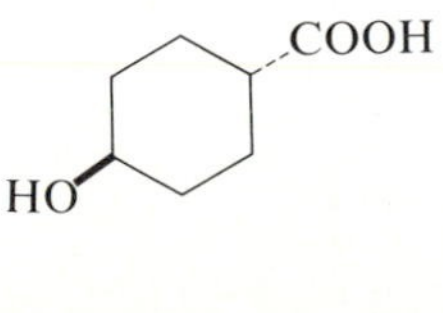 HO—(cyclohexane)—COOH

d. $ClCH_2CH_2CH_2OH$

e. (cyclopentenyl)$-(CH_2)_{12}COOCH_3$

f. $CH_2{=}C(CON_3)CH_2CON_3$

g. β-eudesmol

h. 2-chloro-4,6-dimethylpyridine

9. Do a complete literature search for each of the following compounds using *Beilstein* to cover the literature up to the time of the most recent *Beilstein* reference and the *Chemical Abstracts* formula indices to cover the literature since that time. Indicate where you looked (with time periods covered) and what references you found. If the original research journals are available in your library, scan the pertinent articles and record any physical constants (e.g., melting point, color, crystal form, spectra). List the references you find in one of the two formats discussed on page 1202, being careful to use the correct journal abbreviations as used by the American Chemical Society journals [see *Chemical Abstracts Service Source Index* (1907–1974 Cumulative)].

a. N-benzoyl-2,6-dimethylpiperidine

b. $CH_2{=}CH{-}C(CH_3)_2{-}COCH_3$

c. $CH_3CH_2CH_2CH_2CH_2CH(N_3)CH_3$

d. 3-chloro-2-methylcyclohex-2-enone

CHAPTER 36

SPECIAL TOPICS

36.1 Transition Metal Organometallic Compounds

A. Structure

Transition metal organometallic chemistry has been an area of incredible growth over the past four decades. Many such compounds are particularly important as catalysts in industrial processes, and research continues to discover new and better catalysts. The subject is at the interface between organic and inorganic chemistry. In this section we can provide only the bare essentials of an introduction to this large, growing, and important field of chemistry.

The transition metals have a partially filled *d*-orbital shell. The maximum number of electrons in a filled valence shell for an element in the fourth period of the periodic table (the potassium-krypton row) is

$$4s^2 4p^6 3d^{10} = 18$$

Thus there is a tendency for these elements to surround themselves with 18 electrons and achieve an ''inert'' rare gas electronic configuration, just as elements in the second and third periods strive to achieve an eight-electron configuration. We will see that a great many transition metal organometallic compounds have structures in which the metal is associated with either 16 or 18 electrons and that many reactions can be rationalized in terms of filling out the inert electron configuration. This ''16- or 18-electron rule'' provides a useful framework within which to organize the reactions of such compounds.

First we must take up the question of how to ''count'' the electrons in transition metal organometallic compounds. The abbreviated periodic table in Figure 36.1 shows the elements of groups 1 and 2 and of the three transition series. The numbers at the top of each group correspond to the total number of electrons in the valence shell of an element in that group, regardless of whether the electrons reside in *s*-, *p*-, or *d*-orbitals in the atom itself. The transition elements within the colored box are those to which the 16- or 18-electron rule is most applicable, for reasons that we will not discuss.

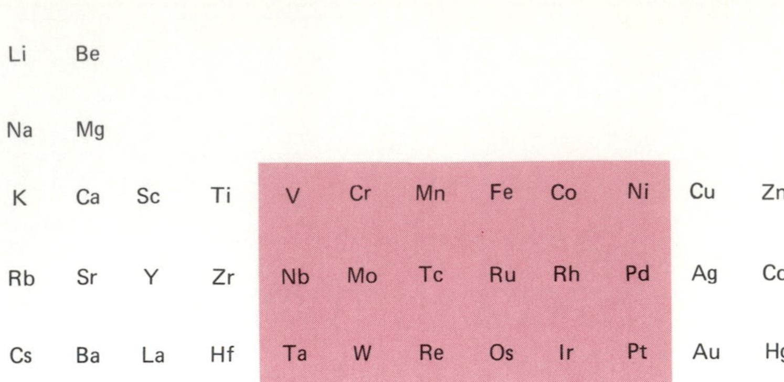

FIGURE 36.1 Periodic table of the elements, showing the number of electrons in the valence shell for each element in the three transition series. The elements within the colored box most often obey the 16–18-electron rule.

In general, electron counting in transition metal compounds works exactly as it does for Lewis structures of second and third period elements except that "donor" bonds are more common with metal systems. Recall that in such a bond two electrons are provided by a suitable donor atom or group, the **ligand.** Both electrons "belong" to the ligand for the purpose of determining formal charges, but do contribute to establishing a rare gas configuration. Common ligands of this type are $H_2\ddot{O}{}^{\times}_{\times}$, $R_2\ddot{O}{}^{\times}_{\times}$, $R_3N{}^{\times}_{\times}$, $R_3P{}^{\times}_{\times}$, $:\ddot{O}{=}C{}^{\times}_{\times}$, in which the donor electron pair is indicated by small xs.

To illustrate the electron-counting procedure, consider the compound molybdenum hexacarbonyl. Six electrons come from the valence shell of the molybdenum atom (Figure 36.1). Each carbon monoxide ligand "donates" two more electrons. Thus the total electronic configuration about the molybdenum in $Mo(CO)_6$ is 18.

$$Mo(CO)_6:\ \text{OC}\rightarrow\text{Mo}\leftarrow\text{CO (six CO ligands)}$$

molybdenum hexacarbonyl
(18 electrons)

	6	electrons on Mo
$6 \times 2 =$	12	six CO ligands
	18	total electrons

None of the 12 carbonyl electrons is counted as "belonging" to Mo; hence, Mo has only the six valence electrons of its own and is formally neutral. Note that these six electrons are not included in the symbolic representation. They would clutter up the diagram; rather, they are implied by the elemental symbol Mo.

A second example is methylmanganese pentacarbonyl. In this compound there are two types of bonds to the metal. The five carbon monoxide ligands each form a "donor" bond in the same manner as the ligands in $Mo(CO)_6$. However, the Mn—CH_3 bond may be viewed as a normal two-electron σ-bond, with one electron being contributed by carbon and one by manganese. Thus Mn has six valence electrons left that are not indicated in the structure—seven from a manganese atom (see Figure 36.1) minus one used in the bond to CH_3. The total electron count is arrived at as shown.

O
C
↓
OC⇢Mn⇠CO
OC➔ ➘CH_3
↑
C
O

methylmanganese pentacarbonyl
(18 electrons)

7	electrons on Mn
−1	Mn electron used in the Mn—CH_3 bond
5 × 2 = 10	five CO ligands
2	the Mn—CH_3 bonding pair
18	total electrons

In this compound the electrons "belonging" to Mn are the six Mn valence electrons, which are not used in bonding and are not shown on the structure, plus one-half of the electrons in the Mn—$CH_3$$\sigma$-bond. Thus the metal is formally neutral.

Another kind of common ligand in transition metal organometallic compounds is the carbon-carbon multiple bond or π-bond of an alkene or alkyne. Accordingly, these groups usually are called π-donors to distinguish them from the σ-donors, the ligands discussed above with lone-pair electrons. The electron count works in the same way. For example, consider the following rhodium compound.

H
|
Ph_3P⇢Rh ← CH_2=CH_2
Ph_3P➔ ↑
C
O

9	electrons on Rh
−1	Rh electron used in the Rh—H bond
2 × 2 = 4	two phosphine ligands
2	CO ligand
2	π-bond of ethylene
2	the Rh—H bonding pair
18	total electrons

Again, the formal charge on the metal is zero.

The bonding of ethylene to transition metals ranges from rather weak to quite strong. A strong bond could perhaps be represented as

M<(CH_2—CH_2)

It is easy to show that both representations give the same electron count.

In all of the examples discussed above the transition metal has achieved an 18-electron configuration and is said to be **coordinatively saturated.**

B. Chemical Reactions

Transition metal organometallic compounds undergo a wide variety of reactions. Fortunately, almost all known reactions, including various steps in individual reaction mechanisms, can be grouped into five or six basic classes. We consider a few of these fundamental reaction types here.

1. Lewis Acid Association-Dissociation. An example is the reaction of tetrakis(triphenylphosphine)nickel with HCl.

$$\mathrm{(Ph_3P)_4Ni} + \mathrm{HCl} \rightleftharpoons [\mathrm{(Ph_3P)_4Ni{-}H}]^+ \ \mathrm{Cl^-}$$

(18 electrons) (18 electrons)

The forward reaction is association of the metal with the Lewis acid H^+. The reverse reaction is dissociation of the Lewis acid from the metal. In Lewis acid association-dissociation the electron count of the metal is unchanged. Another example is the reaction of manganese pentacarbonyl anion with methyl bromide. The reaction may be viewed as a nucleophilic displacement of bromide ion from the alkyl bromide that proceeds by the S_N2 mechanism. However, it also corresponds to association of the manganese with the Lewis acid "CH_3^+."

$$[\mathrm{Mn(CO)_5}]^- + \mathrm{CH_3{-}Br} \longrightarrow \mathrm{(OC)_5Mn{-}CH_3} + \mathrm{Br^-}$$

(18 electrons) (18 electrons)

2. Lewis Base Association-Dissociation. This is the most common reaction in transition metal organometallic chemistry. An example is seen in the reaction of nickel tetracarbonyl with triphenylphosphine. The reaction mechanism consists of two steps. In the first, a CO ligand dissociates from the nickel, giving the intermediate nickel tricarbonyl. This species has an electron count of 16 and is **coordinatively unsaturated.** This electron-deficient intermediate reacts with triphenylphosphine by Lewis base association.

$$\mathrm{Ni(CO)_4} \longrightarrow \mathrm{Ni(CO)_3} + :\mathrm{CO}$$

(18 electrons) (16 electrons)

$$\mathrm{Ni(CO)_3} + :\mathrm{PPh_3} \longrightarrow \mathrm{Ph_3PNi(CO)_3}$$

(16 electrons) (18 electrons)

In Lewis base association-dissociation *the electron count of the metal changes by* $\pm$ *2.*

3. Oxidative Addition-Reductive Elimination. Many transition metal organometallic reactions involve the addition of a σ-bond to the metal. An example is the reaction of hydrogen with bis(triphenylphosphine)chloroiridium carbonyl. In this complex, the iridium has a 16-electron configuration. It exists in the coordinatively unsaturated form because the triphenylphosphine ligands are so large. The complex reacts with hydrogen to form two new iridium-hydrogen bonds, giving the 18-electron complex shown. The reaction is called **oxidative addition** because the formal oxidation state of the metal changes from + 1 to + 3.

$$\mathrm{(Ph_3P)_2Ir(Cl)(CO)} + \mathrm{H_2} \longrightarrow \mathrm{(Ph_3P)_2Ir(Cl)(CO)H_2}$$

(16 electrons) (18 electrons)

> The "oxidation state" of a metal in an organometallic compound can be defined in the following manner. Consider only the groups that are σ-bonded (H, R, Cl, CN, OH, etc.). The metal atom is considered to be more electropositive than any of these groups. Therefore it has a formal "oxidation number" of +1 for each such σ-bond. Donor ligands are ignored. The oxidation state of the metal is the sum of the oxidation numbers, minus one for each negative charge on the molecule and plus one for each positive charge on the complex. Some transition metal complexes contain more complex π-bonded ligands that cannot be treated in this simplified way, but they are beyond the scope of this introductory discussion.

The reverse of oxidative addition, dissociation of a hydrogen molecule with regeneration of the original iridium complex, is called **reductive elimination.** Note that the electron count of the metal changes by ± 2 in oxidative addition-reductive elimination.

4. Insertion-Deinsertion. Occasionally a donor ligand undergoes **insertion** into a σ-bond from the metal to another atom. In the process a donor bond and a σ-bond are traded for a new σ-bond. Thus the electron count of the metal decreases by two. In the reverse reaction the electron count increases by two. An example of insertion-deinsertion is the rearrangement of the 18-electron complex methylmanganese pentacarbonyl to the 16-electron complex acetylmanganese tetracarbonyl.

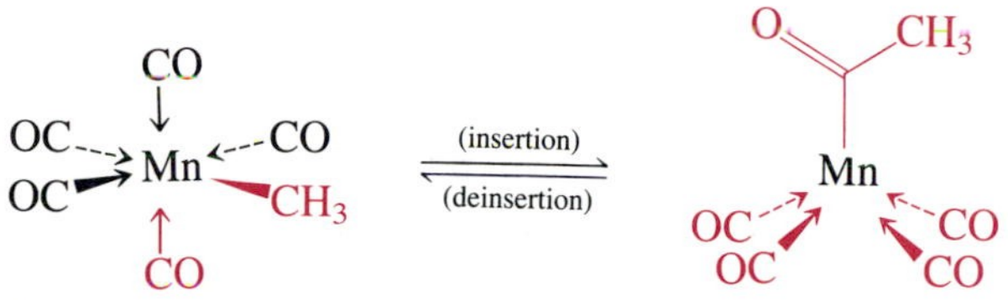

$$(CO)_5Mn{-}CH_3 \underset{\text{(deinsertion)}}{\overset{\text{(insertion)}}{\rightleftharpoons}} (CO)_4Mn{-}C(=O)CH_3$$

methylmanganese pentacarbonyl (18 electrons) — acetylmanganese tetracarbonyl (16 electrons)

Now let us briefly examine a few transition metal organometallic reactions and characterize them in terms of the foregoing fundamental transformations: association, dissociation, addition, elimination, insertion, and deinsertion. Tris(triphenylphosphine)carbonylrhodium hydride is an effective catalyst for the hydrogenation of alkenes.

$$CH_2{=}CH_2 + H_2 \xrightarrow{(Ph_3P)_3RhHCO} CH_3CH_3$$

The reaction can be understood in terms of the following stepwise mechanism (for clarity of presentation, the triphenylphosphine ligands are symbolized by L).

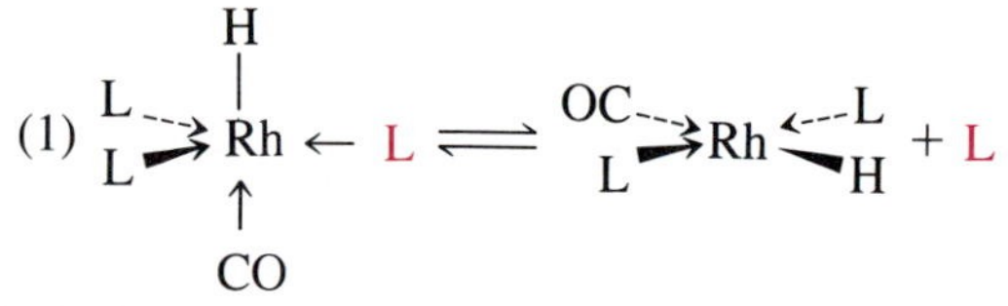

$$(1)\quad L_2Rh(H)(CO) \leftarrow L \rightleftharpoons (OC)(L)_2Rh{-}H + L$$

(18 electrons) — (16 electrons)

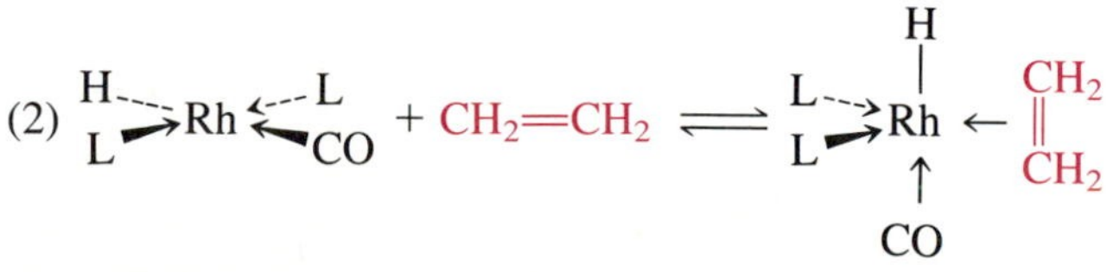

$$(2)\quad (H)(L)_2Rh(CO) + CH_2{=}CH_2 \rightleftharpoons L_2Rh(H)(CO) \leftarrow \begin{matrix} CH_2 \\ \| \\ CH_2 \end{matrix}$$

(16 electrons) — (18 electrons)

$$(3)\quad L_2Rh(H)(CO) \leftarrow CH_2{=}CH_2 \rightleftharpoons (OC)(L)_2Rh{-}CH_2CH_2{-}H$$

(18 electrons) (16 electrons)

$$(4)\quad (OC)(L)_2Rh{-}CH_2CH_3 + H_2 \rightleftharpoons (OC)(L)_2Rh(H)_2(CH_2CH_3)$$

(16 electrons) (18 electrons)

$$(5)\quad (OC)(L)_2Rh(H)_2(CH_2CH_3) \rightleftharpoons (OC)(L)_2Rh{-}H + CH_3CH_3$$

(18 electrons) (16 electrons)

The mechanism consists of the following sequence of processes.

1. Lewis base dissociation
2. Lewis base association
3. Insertion
4. Oxidative addition
5. Reductive elimination

The products of step 5 are the alkane molecule and the active catalyst, which is now free to participate again in steps 2–5 to bring about the hydrogenation of another alkene molecule. The fact that an active species can be regenerated means that the organometallic compound is a **catalyst** via this cycle of reactions.

A related catalyst, tris(triphenylphosphine)rhodium chloride, also known as Wilkinson's catalyst, also brings about hydrogenation of alkenes by a mechanism similar to that just described.

$$(Ph_3P)_2Rh(PPh_3)Cl$$

tris(triphenylphosphine) rhodium chloride
(Wilkinson's catalyst)

Wilkinson's catalyst also functions to bring about **decarbonylation** of aldehydes.

$$C_6H_5,C_6H_5\text{-cyclopropyl}(CH_3)(CHO) \xrightarrow[\Delta]{(Ph_3P)_3RhCl} C_6H_5,C_6H_5\text{-cyclopropyl}(CH_3)(H) + CO$$

The reaction mechanism can be formulated as proceeding through the following steps.

$$(1)\quad L_2Rh(L)Cl + RCHO \rightleftharpoons L_2Rh(H)(Cl)(L)(COR)$$

(16 electrons) (18 electrons)

$$(2)\quad L_2Rh(H)(Cl)(L)(COR) \rightleftharpoons L_2Rh(H)(Cl)(COR) + L$$

(18 electrons) (16 electrons)

$$(3)\quad L_2Rh(H)(Cl)(COR) \rightleftharpoons L_2Rh(H)(Cl)(R)(CO)$$

(16 electrons) (18 electrons)

$$(4)\quad L_2Rh(H)(Cl)(R)(CO) \rightleftharpoons L_2Rh(Cl)(CO) + R{-}H$$

(18 electrons) (16 electrons)

$$(5)\quad L_2Rh(Cl)(CO) + L \rightleftharpoons L_2Rh(CO)(L)Cl$$

(16 electrons) (18 electrons)

$$(6)\quad L_2Rh(CO)(L)Cl \rightleftharpoons L_2Rh(L)Cl + CO$$

(18 electrons) (16 electrons)

Like the iridium complex shown on page 1220, Wilkinson's catalyst exists in a coordinatively unsaturated 16-electron state. It can therefore undergo oxidative addition (which changes the electron count of the metal by +2) without prior dissociation of a donor ligand. Thus, the steps are

1. Oxidative addition
2. Lewis base dissociation
3. Deinsertion
4. Reductive elimination
5. Lewis base association
6. Lewis base dissociation

This mechanism is probably a little oversimplified, since there is evidence that free radical pairs might be involved in step 3. Nevertheless, the gross features are no doubt correct.

The simple structural ideas and mechanistic classifications we have introduced are but first steps toward a complete understanding of the rich chemistry of these fascinating compounds.

EXERCISE 36.1 (a) Some transition metals form neutral complexes with carbon monoxide (metal carbonyls); for example, see nickel carbonyl on page 1220. Chromium and iron also form stable carbonyls. Write the expected structures of these compounds.

(b) Other transition metals form stable metal carbonyl anions of the type $M(CO)_n^-$. Write the expected structures of the manganese and cobalt carbonyl anions and explain their stability

EXERCISE 36.2 The rate of exchange of triphenylphosphine for carbon monoxide in nickel carbonyl is inversely proportional to the pressure of carbon monoxide in contact with the reaction solution

$$Ni(CO)_4 + Ph_3P \longrightarrow Ph_3PNi(CO)_3 + CO$$

Explain.

EXERCISE 36.3 Hydroformylation of alkenes (the "oxo reaction") is an important industrial process. For example, ethylene reacts with hydrogen and carbon monoxide in the presence of hydridocobalt tetracarbonyl to give propionaldehyde.

$$CH_2{=}CH_2 + CO + H_2 \xrightarrow[100\text{-}120°C]{HCo(CO)_4} \underset{(72\%)}{CH_3CH_2CHO}$$

Propose a mechanism that is consistent with the "16- or 18-electron rule." Classify each step in your mechanism by reaction type (i.e., Lewis acid association, insertion, etc.).

EXERCISE 36.4 Write a plausible mechanism for the hydrogenation of ethylene with Wilkinson's catalyst. Wilkinson's catalyst also brings about the decarbonylation of acyl halides.

$$CH_3CH_2CH_2CH_2\overset{O}{\overset{\|}{C}}Cl \xrightarrow{(Ph_3P)_3RhCl} CH_3CH_2CH_2CH_2Cl + CO$$

Propose a mechanism for this reaction.

36.2 Colored Organic Materials

A. Color

All compounds can be excited electronically by electromagnetic radiation (Section 22.6). For most organic compounds such transitions are in the ultraviolet region of the spectrum, and such compounds are white and colorless. However, when an electronic transition is in the visible range (about 400–750 nm), the compound will appear to us as colored. The colors perceived for different wavelengths of light are summarized in Table 36.1.

Light of a given wavelength is perceived as the indicated color. However, if that wavelength is *absorbed,* we perceive the complementary color. Some compounds appear

TABLE 36.1 Color and Wavelength

Wavelength of light, nm	Color	Complementary color
400–430	violet	green-yellow
430–480	blue	yellow
480–490	green-blue	orange
490–510	blue-green	red
510–530	green	purple
530–570	yellow-green	violet
570–580	yellow	blue
580–600	orange	green-blue
600–680	red	blue-green
680–750	purple	green

to have a yellow color even though their λ_{max} are all in the ultraviolet region. In such cases a "tail" of an absorption band stretches into the visible. Since the light absorbed is violet or blue, we see the compound as the complementary color of yellow.

Intensely colored materials have absorptions in the visible region. For organic compounds, such electronic absorptions are generally $\pi \rightarrow \pi^*$ or $n \rightarrow \pi^*$ transitions and involve extended π-electronic systems. That is, color in organic compounds is generally a property of π-structure. If the absorption band is narrow or sharp, the color will appear to us as bright or brilliant and clean. A broad absorption band, or more than one band in the visible region, gives colors that we perceive as dull or "muddy."

B. Natural Colored Materials

A large number of naturally occurring organic compounds are brightly colored. Some find a role in nature because of their color. The most important to life is the porphyrin derivative, chlorophyll a, which absorbs sunlight and converts the energy by charge transfer into a form suitable for biosynthesis and oxygen production. Other examples are the colored pigments of flowers for attracting bees to transfer pollen and the pigments responsible for the camouflage of some insects and animals. Bright colors sometimes signal the unpalatability, toxicity or aggressiveness of an organism. Not all colors in nature are due to light absorption by specific compounds but to the structure of the colored material. Natural blue color is mostly due to a blue light scattering structure (spongy cells) accompanied by dark melanin to absorb other wavelengths or by a yellow pigment to produce a green appearance. Examples are the blue feathers of the bluejay and bluebird. Interference is responsible for the more complex and beautiful phenomenon of iridescence such as that found in the peacock as first noted by Isaac Newton. More often, however, specific organic compounds are involved. A few of the important classes of compounds with representative examples are summarized here.

Anthocyanins provide much of the color of the plant world. They are responsible for the red color of buds and young shoots and for the purple of autumn leaves as the green chlorophyll decomposes with the approach of winter. Their colors depend in part on the pH of their environment. For example, the blue cornflower and red rose have the same anthocyanin, cyanin. The blue color is that of the potassium salt. Anthocyanins are actually present as glycosides. Hydrolysis gives the corresponding anthocyanidins; that is, anthocyanidins are the aglycones (page 941) of anthocyanins.

Only three anthocyanidins are important: cyanidin (crimson to blue-red flowers, cherries, cranberries), pelargonidin (pelargonium, geranium), and delphinidin (delphinium, pansy, grape). These compounds have the following structures.

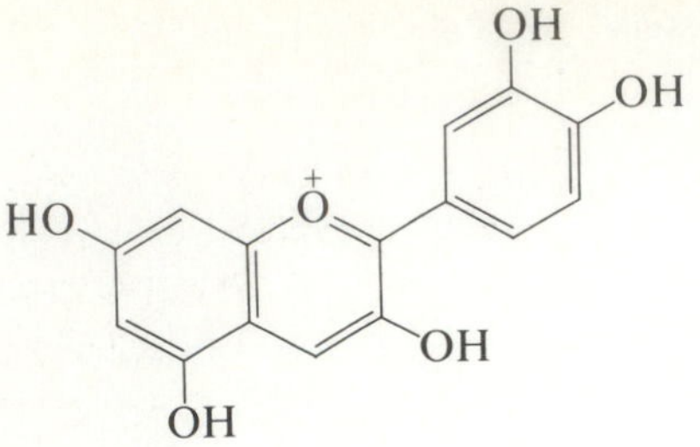

cyanidin

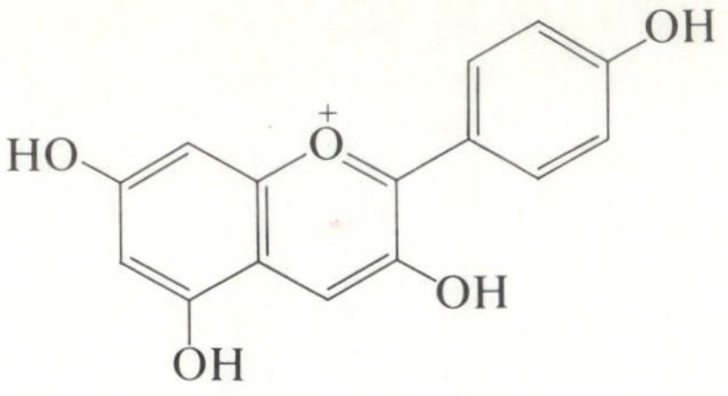

pelargonidin

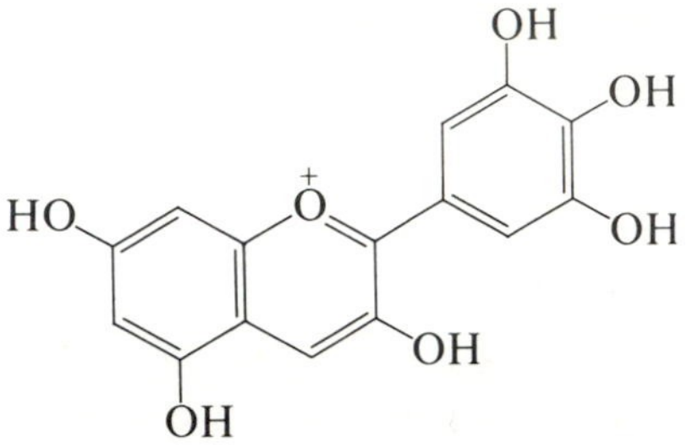

delphinidin

These pyrylium salts (Section 32.9) are generally considered as derivatives of a parent structure, flavone, a nucleus that is widespread in nature.

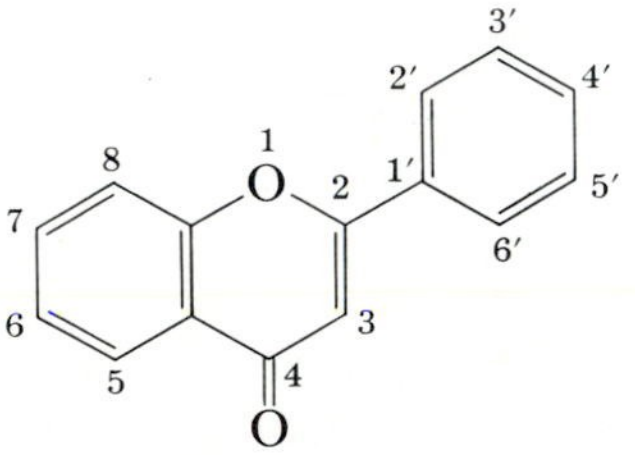

flavone

In the corresponding glycosides, the sugar units are attached at the 3- and 5-positions.

Carotenoids occur in organisms over the whole evolutionary range, from bacteria and fungi to complex plants and animals. Examples are β-carotene, which is a precursor for vitamin A, and lycopene, which occurs in tomatoes and ripe fruit. β-Carotene is manufactured on a large scale as a safe food pigment.

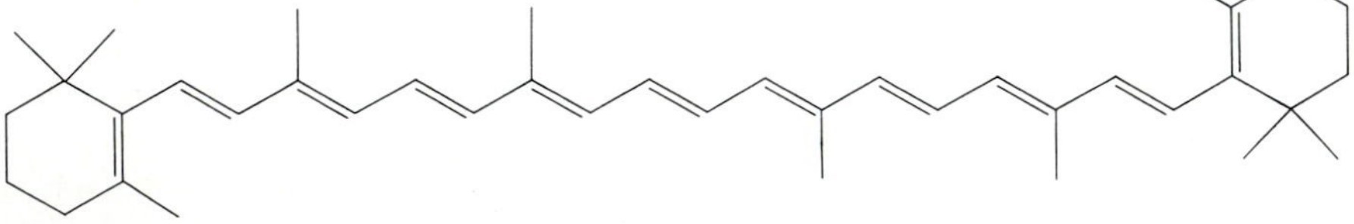

β-carotene

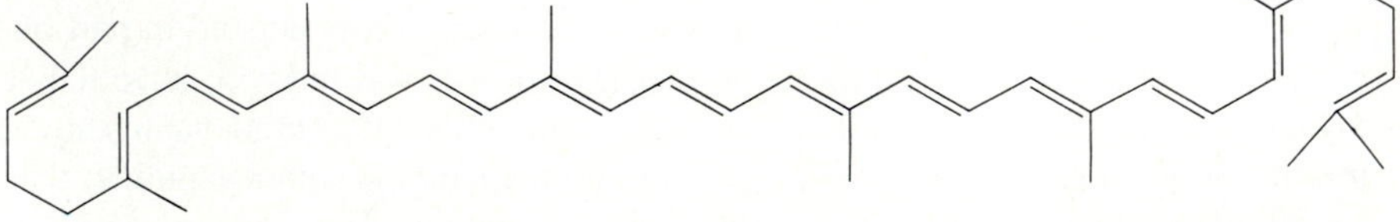

lycopene

The color of these hydrocarbons clearly comes from $\pi \rightarrow \pi^*$ transitions of the long conjugated system. Note that these compounds are terpenes (Section 36.5). Carotenoids also occur in marine biology, for example, in the skins of fish, sea stars, anemones, corals, and crustaceans, frequently in combination with proteins. Denaturation of the protein in boiling water frees the carotenoid and unmasks its color as in the red color of boiled lobster.

Naphthoquinones and **anthraquinones** occur in both animals and plants. Some examples are given in Section 30.8.A. Echinochrome is a polyhydroxynaphthoquinone that occurs as a red pigment in the sea urchin and sand dollar.

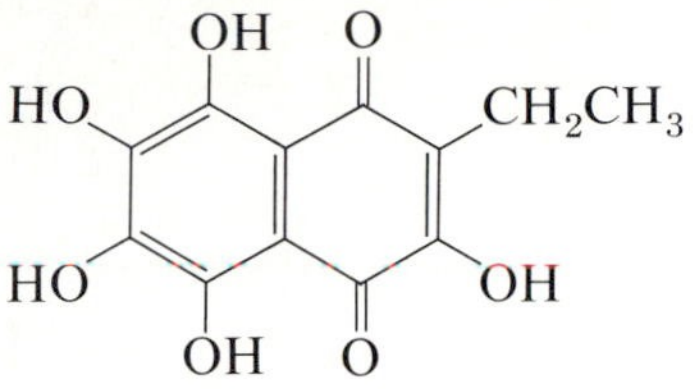

echinochrome

Cochineal is a dried female insect, *Coccus cacti* L., used for a red coloring in food products, cosmetics, and pigments. The principal constituent is carminic acid, a C-glucoside of a polyhydroxyanthraquinone, as shown below.

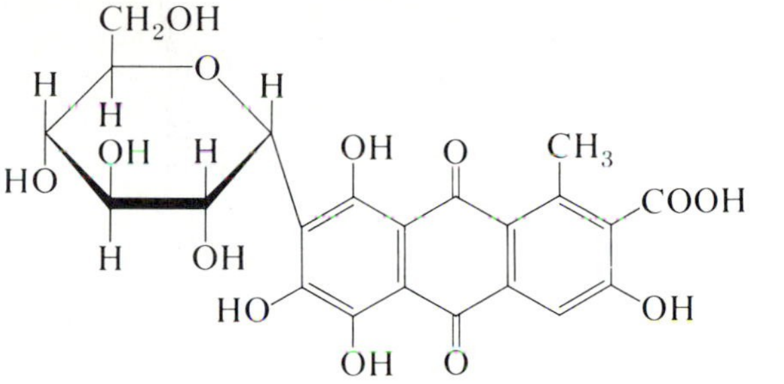

Melanins are polymeric quinoidal compounds derived from the oxidation and polymerization of 1,2-dihydroxybenzenes such as 3,4-dihydroxyphenylethylamine (dopamine) or the amino acid 3,4-dihydroxyphenylalanine (**dopa**). Melanins occur in such varied places as feathers (see page 1225), hair, eyes, and the ink of cephalopods. They occur in the skin of all humans, except albinos, and are responsible for the varied skin coloration among the races of man. Albinos and certain white animals lack the oxygenase enzyme required to convert tyrosine to dopa.

Insect eyes are made of many **ommatidia,** each containing **receptor cells** and **pigment cells** and an optical system. Granules in the pigment cells are called **ommochromes** from which the yellow **xanthommatin** can be isolated. The pigments absorb excess light and protect the rhodopsins from overexposure. Xanthopterin, a yellow pigment found in butterfly wings and in animals belongs to the **pterins,** which include folic acid (see page 826). **Porphyrins** (hemin, page 994; chlorophyll), and **indigoids** (indigotin or indigo; occurs as glycoside in many plants; used as a blue dye; see next section) are other natural pigments.

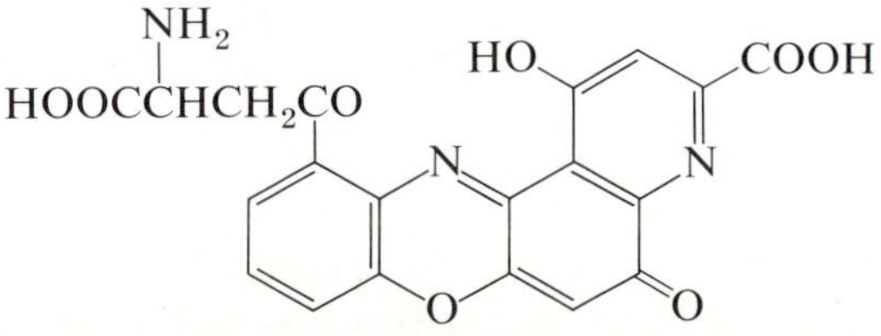

xanthommatin

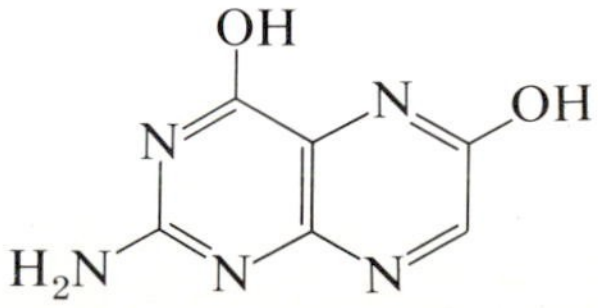

xanthopterin

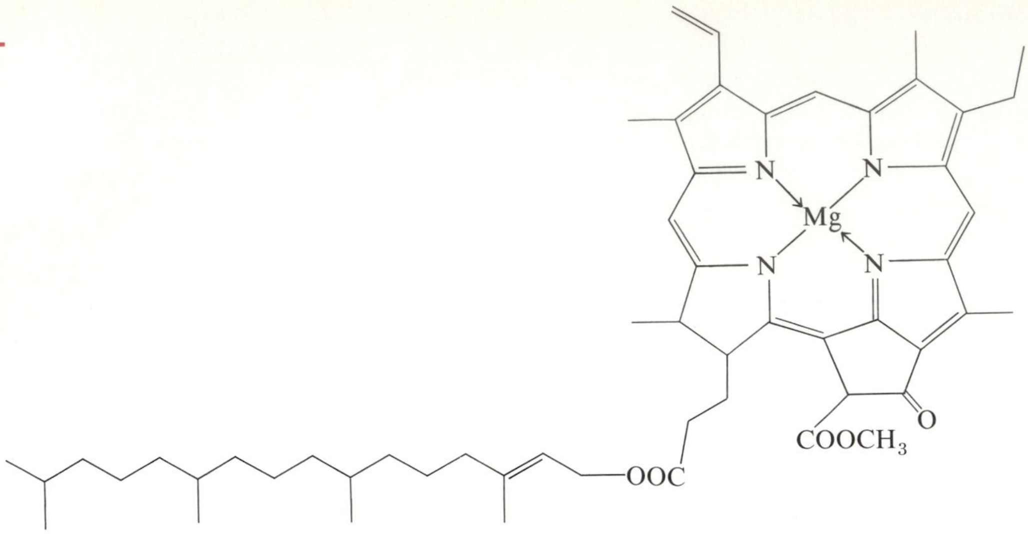

chlorophyll a

C. Dyes and Dyeing

A colorant is either a pigment or a dye. Dyes are coloring materials that are normally applied from a solution and will bind in some manner to a substrate, which might be a textile, paper, leather, hair, or other materials. Dyes can be converted to pigments after application, weakening the sharp distinction between the two classes. Ideally, dyes are fast to light and to washing. Dyes have been known to man for thousands of years. Early dyes were entirely of natural origin, but common dyes in use today are almost all synthetic. Different methods were and still are used for combining the dye with the fiber. Some of the principal categories follow.

Vat dyes are exemplified by indigo, a highly insoluble blue compound known to the ancient world. The precursor, indoxyl, or 3-hydroxyindole, is present in an Indian plant, *Indigofera tinctoria,* or the European woad, *Isatis tinctoria,* known to ancient Britain, as a glycoside and is released by enzymatic cleavage. The indoxyl is oxidized by oxygen to indigo, probably via dimerization of an intermediate free radical.

Indigo was used in Egypt before 2000 B.C. In classical times, indigo was reduced by fermentation to a colorless and soluble ''leuco'' compound. The material to be dyed was immersed in this solution and then exposed to air to reoxidize the leuco base. Indigo is now produced synthetically and is reduced to the leuco form with sodium dithionite. It can be oxidized by exposure to air or more quickly by use of an oxidizing agent such as sodium perborate. The insoluble blue pigment (λ_{max} 606 nm) so produced is ''locked'' within the fiber. The 6,6′-dibromoindigo (λ_{max} 590 nm) is Tyrian purple, prepared from mollusks (*Murex*) that were found near the ancient city of Tyre. The dye was so expensive (10,000 mollusks yielded 1 g) that only royalty could afford it, hence the expression ''born to the purple.'' It is now a relatively inexpensive dye that still finds some use.

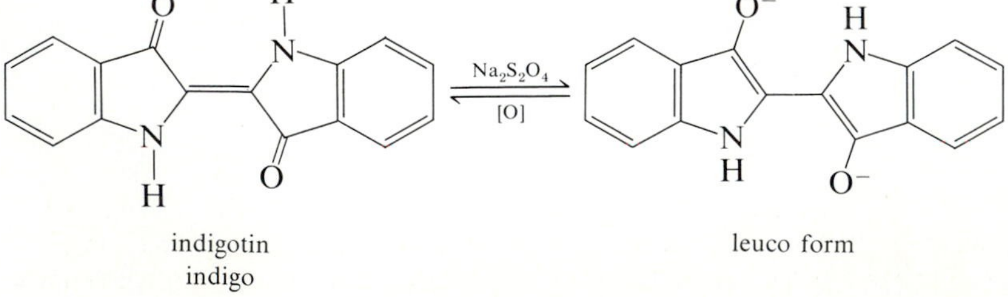

indigotin
indigo

leuco form

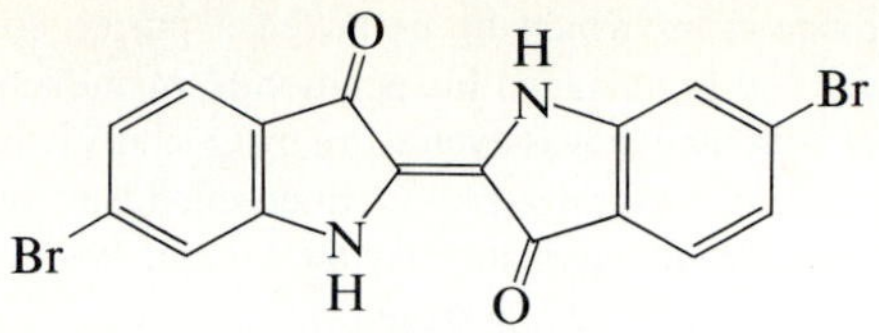

Tyrian purple
6,6′-dibromoindigo

EXERCISE 36.5 Write the formula for indoxyl and show how it could be oxidized to indigo by oxygen. [*Hint:* Protonated superoxide ion, $O_2^{-\cdot}$, would be a good hydrogen-abstracting agent.]

Certain dyes were bound to fibers only with a mordant (L. *mordere,* to bite), usually a metal salt that forms an insoluble complex or ''lake'' with the dye. The dye is applied to fiber or cloth that has been pretreated with a metal salt. An example known to the ancient world was the extract of the madder root, which was mordanted with aluminum salts to produce a color known as Turkey red. Other metal salts give different colors. The actual dye that coordinates with the metal is alizarin. Alizarin was first synthesized in 1869, and shortly thereafter synthetically manufactured material drove the natural product from the market with important economic repercussions.

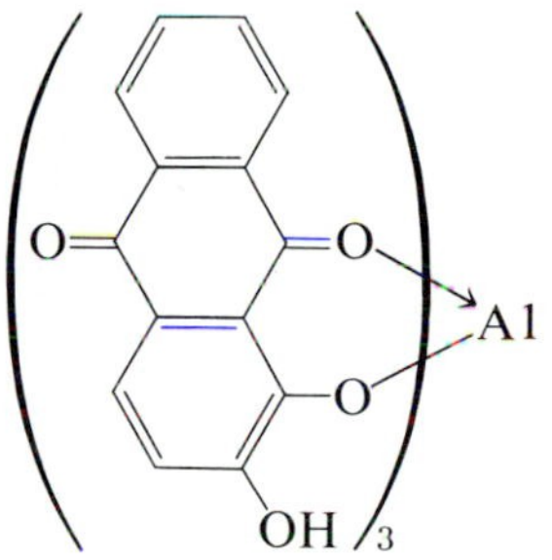

alizarin mordanted with Al^{3+}

Direct dyes can be applied to the fiber directly from an aqueous solution. This process is especially applicable to wool and silk. These fibers are proteins that incorporate both acidic and basic groups that can combine with basic and acid dyes, respectively. An example is mauve, the dye that started the modern synthetic dyestuff industry but is no longer used.

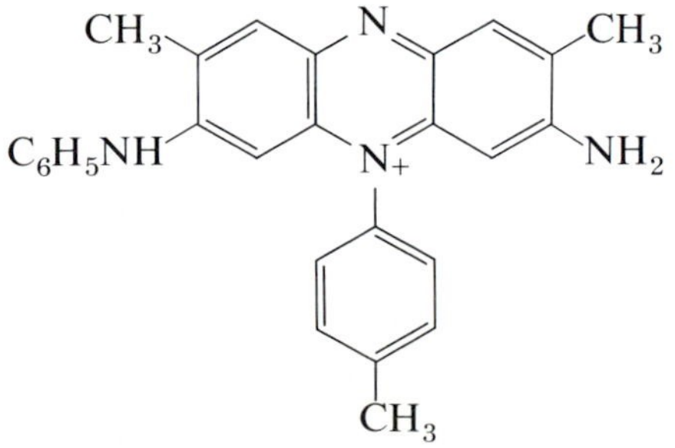

mauve

William Henry Perkin was a student at the Royal College of Chemistry when, in 1856 in his home laboratory, at the age of 18, he treated aniline sulfate with sodium dichromate and obtained a black precipitate from which he extracted a purple compound. This material showed promise as a dye, and he resigned his position to manufacture it. The product was successful and cloth dyed with mauve was even worn by Queen Victoria. Not long afterward, additional synthetic dyes were synthesized by German chemists and the synthetic dyestuff industry gradually became a German industry. By the time of World War I, almost all of the world production of synthetic dyes was German.

Perkin's success in the discovery of mauve was based on the fact that his ''aniline'' was impure. It was prepared by nitrating and reducing ''benzene'' that contained substantial amounts of toluene!

Mauve is a positively charged dye molecule that can associate with the anionic groups of a textile fiber. An example of a direct acid dye is benzopurpurin 4B, which has sulfonate ion groups that can pair with cationic centers in the fiber.

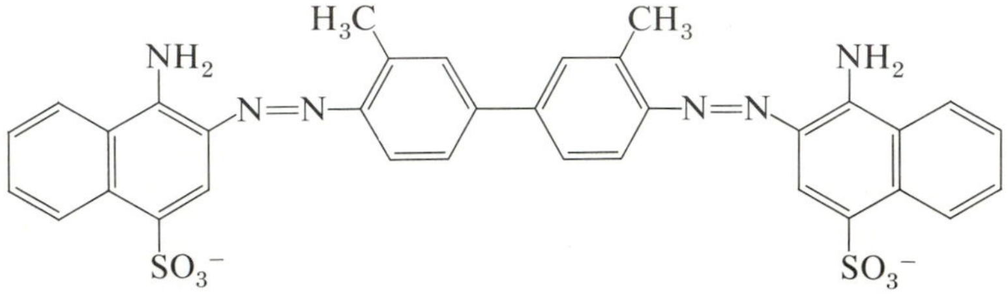

benzopurpurin 4B

Disperse dyes are used as aqueous dispersions of finely divided dyes or colloidal suspensions that form solid solutions of the dye within the fiber. They are especially useful for polyester synthetic fibers. These fibers have no acidic or basic groups for binding with direct dyes and are sensitive to hydrolysis in the strongly alkaline conditions of vat dyeing. Disperse dyes tend to have important limitations. They frequently lack fastness to washing, tend to sublime out on ironing, and are subject to fading with NO_2 or ozone in the atmosphere, a condition known as *gas fading*.

Dyes are also classified on the basis of chemical structure. These structures frequently contain a functional group that is principally involved in the $n \rightarrow \pi^*$ and $\pi \rightarrow \pi^*$ transitions that give rise to the color. Examples of such groups, called **chromophores,** are the azo group, —N=N—, the carbonyl groups in quinones, and extended chains of conjugation. Some of the principal chemical classes of dyes follow.

Azo dyes form the largest chemical class of dyestuffs. These dyes number in the thousands. They consist of a diazotized amine coupled to an amine or a phenol and have one or more azo linkages. An example of a diazo dye, a dye with two azo groups, is direct blue 2B, prepared by coupling tetrazotized benzidine with H-acid (8-amino-1-naphthol-3,6-disulfonic acid) in alkaline solution. If H-acid is coupled with a diazonium ion in dilute acid solution, reaction occurs next to the amino group. Both positions can be coupled to different diazonium salts.

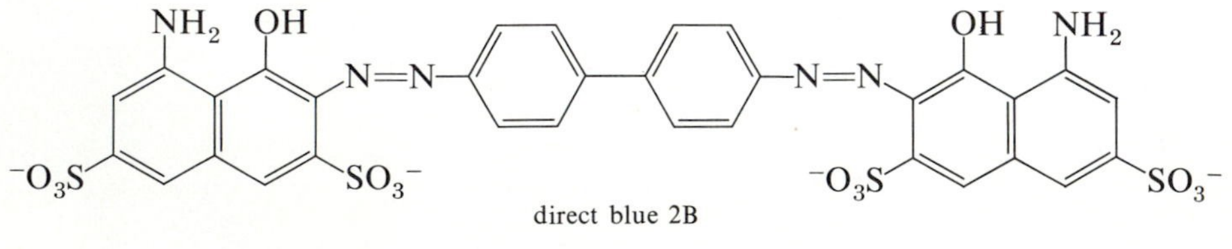

direct blue 2B

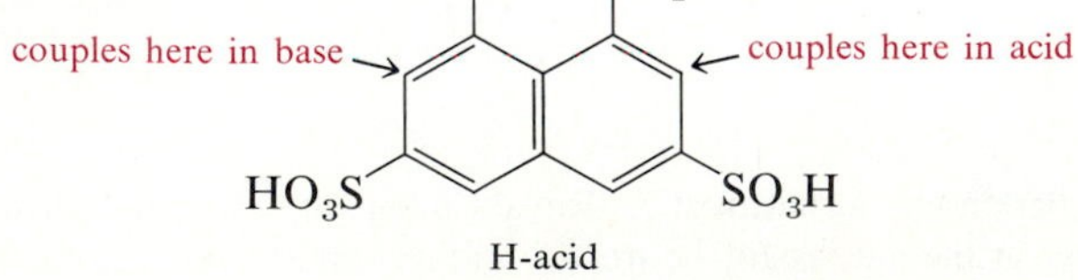

couples here in base
couples here in acid

H-acid

Triphenylmethane dyes are derivatives of triphenylmethyl cation. They are basic dyes for wool or silk or for suitably mordanted cotton. Malachite green is a typical example that is prepared by condensing benzaldehyde with dimethylaniline and oxidizing the intermediate leuco base.

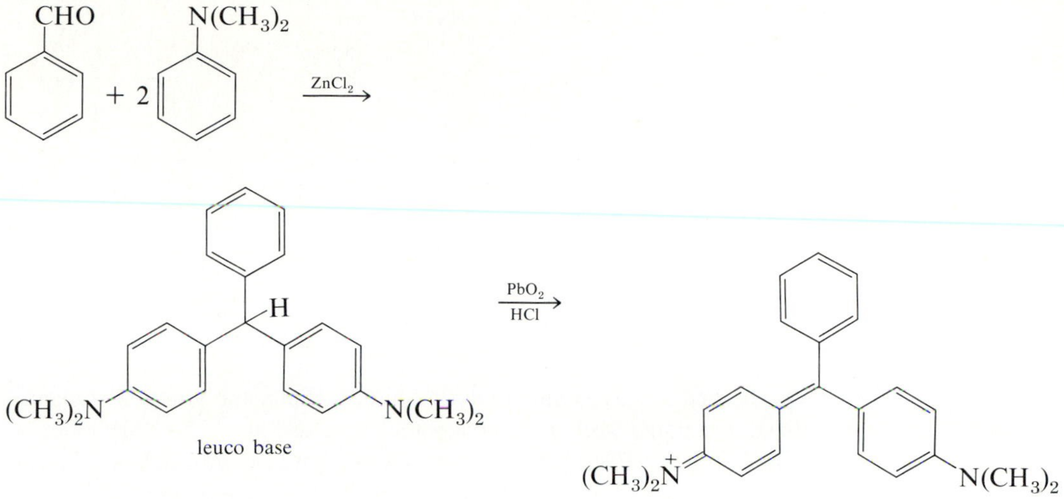

CHO + 2 $N(CH_3)_2$ $\xrightarrow{ZnCl_2}$

$(CH_3)_2N$ H $N(CH_3)_2$

leuco base

$\xrightarrow[HCl]{PbO_2}$

$(CH_3)_2\overset{+}{N}$ $N(CH_3)_2$

malachite green

Anthraquinone dyes are generally vat dyes as exemplified by alizarin. More complex examples are higher molecular weight compounds prepared by oxidizing anthraquinone derivatives under basic conditions.

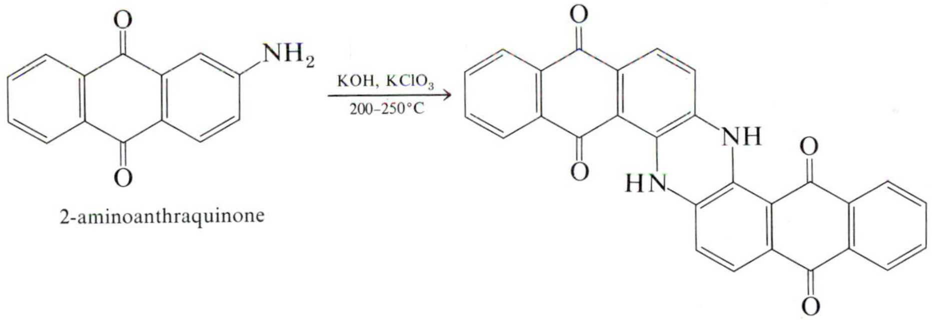

O NH_2 O

2-aminoanthraquinone

$\xrightarrow[200\text{–}250°C]{KOH,\ KClO_3}$

O O NH O HN O

indanthrone
indanthrene blue R

Azine dyes are derivatives of phenoxazine, phenothiazine, or phenazine. Mauve and aniline black (page 757) are derivatives of phenazine. Methylene blue is a thiazine derivative used as a bacteriological stain.

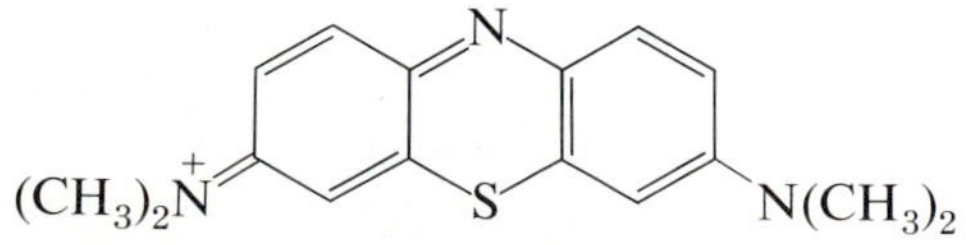

N $(CH_3)_2\overset{+}{N}$ S $N(CH_3)_2$

methylene blue

Phthalocyanines are used as pigments rather than as dyes. An important member of this class is copper phthalocyanine, a brilliant blue pigment that can be prepared by heating phthalonitrile with copper.

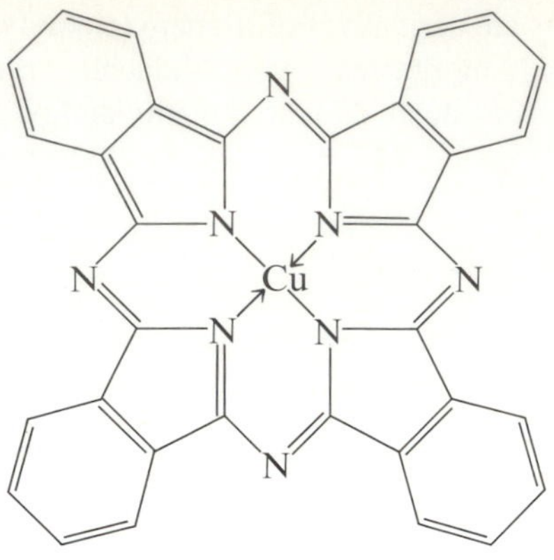

copper phthalocyanine

EXERCISE 36.6 The first azo dye found to have a direct affinity for cotton is prepared from tetrazotized benzidine and 4-aminonaphthalene-1-sulfonic acid (naphthionic acid). What is its structure? The dye has λ_{max} = 497 nm but has two bands in mineral acid solution, λ_{max} = 647 and 590 nm. What are the colors of the dye in neutral and acidic solutions?

EXERCISE 36.7 Naphthol Blue Black B is prepared by coupling H-acid with diazotized aniline in basic solution and diazotized *p*-nitroaniline in acidic solution. What is the structure of this dye?

EXERCISE 36.8 Write a reasonable mechanism for the preparation of the leuco base of malachite green.

36.3 Photochemistry

A. Electronically Excited States

Most organic molecules have an even number of electrons with all electrons paired. Within each pair, the opposing electron spins cancel, and the molecule has no net electronic spin. Such an electronic structure is called a **singlet** state. When a ground-state singlet absorbs a photon of sufficient energy, it is converted to an excited singlet state. The process is a **vertical transition**; that is, electronic excitation is so fast that the excited state has the same geometry of bond distances and bond angles as the ground state. The most stable geometry of the excited state often differs from that of the ground state with the result that the excited electronic state is often formed in an excited vibrational state as well. These relationships are illustrated in Figure 36.2. A vertical transition of the type illustrated is also known as a **Franck-Condon** transition.

Figure 36.2 is simplified in showing energy levels as a function of a single bond coordinate. For real molecules, there are many bonds and the resulting pattern of energy levels is multidimensional and complex. Furthermore, there are many excited singlet states, which may be represented collectively as S_i. The lowest excited singlet state is then represented as S_1.

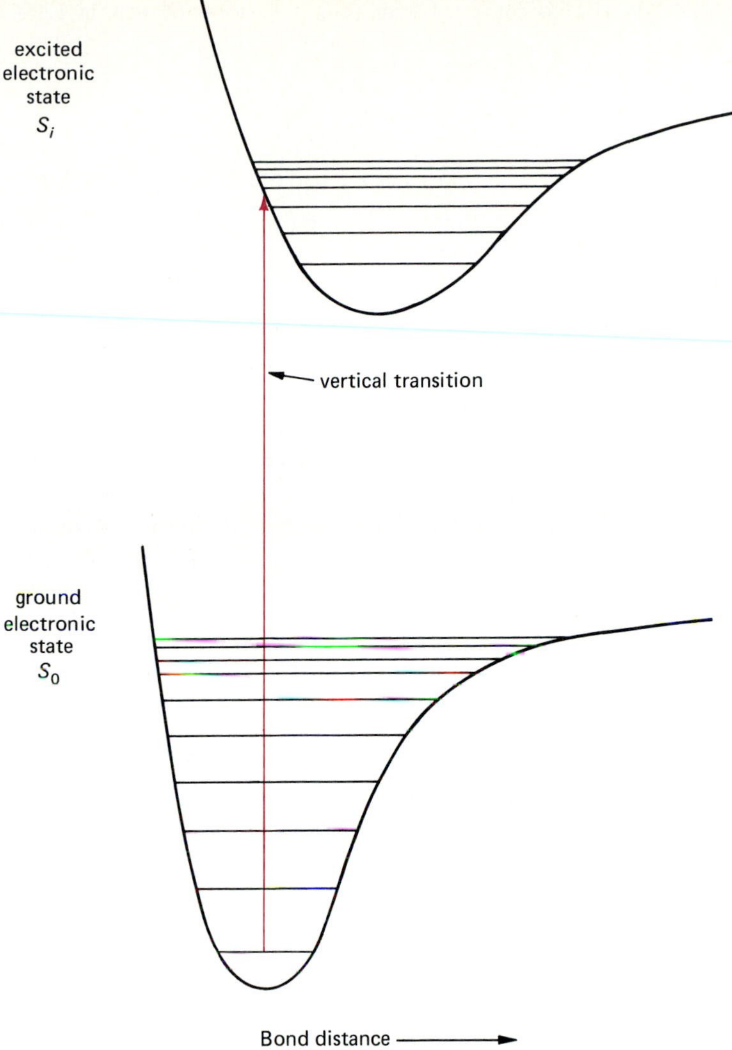

FIGURE 36.2 Vertical or Franck-Condon electronic transition.

We may now ask how the excited molecule disposes of its energy. The first formed excited state (S_1, S_2, etc.) will have a **local excess** of energy at the light-absorbing group. The first-formed state generally **redistributes** its extra vibrational energy to other bonds in the molecule and falls to an excited vibrational level of this state (**internal vibrational redistribution**). The time required for this process is very short, about 10^{-13} sec (100 femtosec or fsec, 0.1 psec), the time needed for a single vibration. In the next step, this vibrationally and electronically excited state gives up more energy in collisions with solvent and **relaxes** to the lowest vibrational level of the lowest excited state, S_1, a process that takes 2–10 psec (**vibrational relaxation**). An alternative route for loss of energy is via conversion to an excited vibrational level of a lower electronic state, a process called **internal conversion.** For example, S_2 is converted into S_1^* (* indicates vibrational excitation) in less than a psec, but the S_1 to S_0^* internal conversion can take about 10^{-11} sec, or about 10^2 vibrations, and is often slower. The internal conversion to S_0^* produces a very hot molecule, as if one had heated the material to temperatures above 1000 °C.

The dye molecule, oxazine, is converted from an S_n state to an S_1^* state in 180 fsec.

$(C_2H_5)_2N^+$ ... O ... $N(C_2H_5)_2$
ClO_4^-

oxazine 1

The **slowest organic reaction** for which there is a rate estimate is that of the racemization of amino acids which takes perhaps 10^4 years without catalysis; the **fastest organic reaction** is that of double bond photoisomerization, which can occur within tens of femtoseconds.

Usually, the first formed excited state is S_1, if the longest wavelength band is used for excitation and particularly when other excited states are much higher in energy. The lifetime of the S_1 state in its lowest vibrational level is about 10^{-8} to 10^{-7} sec. The state can lose its energy in five different ways: **emission** of **fluorescence,** radiationless **internal conversion** to a vibrationally excited ground state, **chemical reaction (photochemistry),** radiationless **intersystem crossing** to a **triplet state,** and **energy transfer** to another molecule. These alternatives are discussed below with the help of a Jablonski diagram (Figure 36.3).

1. S_1 can emit a photon and undergo an electronic transition to the ground state, a process called **fluorescence.** Because the excited state has lost energy by relaxation before fluorescence occurs, the fluorescence photon has less energy than the exciting photon. Fluorescence is light of longer wavelength than the light required for the original excitation.
2. S_1 can be transformed into an excited vibrational state of the ground state, S_0^*. This is an **internal conversion** and is a nonradiative process. The excited vibrational level again transfers energy to the solvent molecules in its environment,

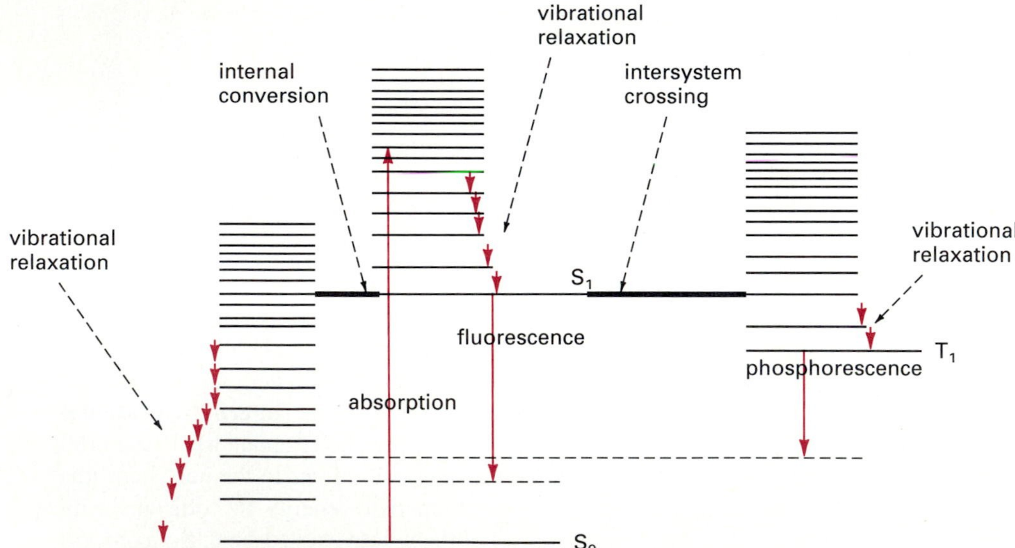

FIGURE 36.3 Jablonski diagram. Horizontal dark lines show level-level interaction of two types: internal conversion, in which an excited state, S_n can be transformed into a lower electronic state without emission of radiation, and intersystem crossing, in which a spin-paired singlet state is transformed into a spin-parallel triplet state. Vibrational relaxation proceeds level by level to the lowest state in a given manifold. Light absorption and emission processes are shown as vertical arrows. The characteristic times for the various processes are cited in the text. For simplicity, only S_1 is shown and the pathways for energy transfer and chemical reaction are omitted.

which then move or vibrate faster, i.e., become hotter. Eventually the molecule in S_0 achieves the equilibrium vibrational level distribution expected for the temperature of the environment. The net result is the conversion of the original light quantum into heat.

3. S_1 can undergo internal conversion to an excited vibrational level of the ground state of a different compound, an isomer whose lowest vibrational level of S_0 corresponds to a different geometry than our starting material. Alternatively, S_1 can react on collision with another molecule. In either case, we have achieved a **photochemical reaction.**

4. S_1 can undergo **intersystem crossing** to the triplet state, T_1. A triplet state is one in which one electron spin has been changed so that the molecule has two electrons that cannot pair. The lowest triplet state is usually of higher energy than the ground state (the rare exceptions are compounds with "ground-state triplets"). Nevertheless, T_1 is of lower energy than S_1. Electrons with the same spin tend to stay apart because of the Pauli principle. As a result, the electrostatic energy of electronic repulsion is less than in comparable singlet states. In many compounds, the switching of an electronic spin is an improbable process, and triplet states are not important in the photochemistry of such compounds. In certain other compounds, particularly $\pi \rightarrow \pi^*$ states of polycyclic aromatic hydrocarbons and $n \rightarrow \pi^*$ states of many ketones, the process of intersystem crossing is more probable. The process can take as little as 10^{-9} sec (nanosec or nsec) but is often a fraction of 10^{-6} sec (microsec). Since the lifetime of S_1 is generally in the range of 10^{-7} to 10^{-8} sec, in the case of fast intersystem crossing, almost all of the excited states intersystem cross to T_1.

5. **Energy transfer** is a process in which a second molecule accepts the excitation energy from the first. Since the process is dependent on the distance r between the molecules (on r^6), concentration is an important parameter. The transfer is like emission by the first molecule and absorption by the second. Impurities in solids can sometimes emit all of the energy absorbed.

Triplet states are fairly long-lived, with lifetimes of greater than 10^{-5} sec, and in some cases up to a second or so. One reason for such long lifetimes is that conversion to S_0 again requires switching an electronic spin. The rate at which intersystem crossing takes place depends in large part on the energy difference between the two states. The energy difference between S_1 and T_1 is generally much less than between T_1 and S_0; hence, the latter intersystem crossing is much less probable.

Triplet states themselves have four possible ways for shedding their excess energy.

1. T_1 can thermally decay to S_0 by spin inversion coupled to internal conversion to S_0^*. The net result in this case is again the conversion of a light quantum to heat.

2. T_1 can emit a photon. This process is called **phosphorescence.** As in the case of fluorescence, phosphorescence is light of longer wavelength than that of the initial exciting light. Phosphorescence is a low-probability event because of the change in electron spin required. It is generally observed only at low temperatures so that thermally activated conversion to S_0 or chemical reactions are slowed (a rigid or viscous matrix may also be effective).

3. T_1 can rearrange to the T_1 of an isomeric molecule or undergo intersystem crossing to the S_0 of an isomer. Either process results in a **photochemical isomerization.** Alternatively, T_1 can react on collision with another molecule in a photochemical reaction characteristic of a diradical. Recall that a triplet state molecule has two unpaired electrons.

4. T_1 can transfer its electronic spin to another molecule and become converted to S_0. This process is called **triplet energy transfer** and is symbolized as

$$T_1 + S'_0 \longrightarrow S_0 + T'_1$$

This reaction is actually an equilibrium governed by the usual free energy requirements, but it is usually important only when the reaction shown is exothermic.

B. Photochemical Reactions

Some photochemical reactions involve simple isomerizations. An example is the cis-trans interconversion of alkenes. In this case the excited state involves a twisted double bond (an example is Figure 11.5), and conversion to the ground state has an equal probability of going cis or trans. If a wavelength of light can be chosen at which the trans isomer absorbs and the cis does not, a trans isomer can be converted completely to the cis. In other cases a photochemical *stationary state* is achieved.

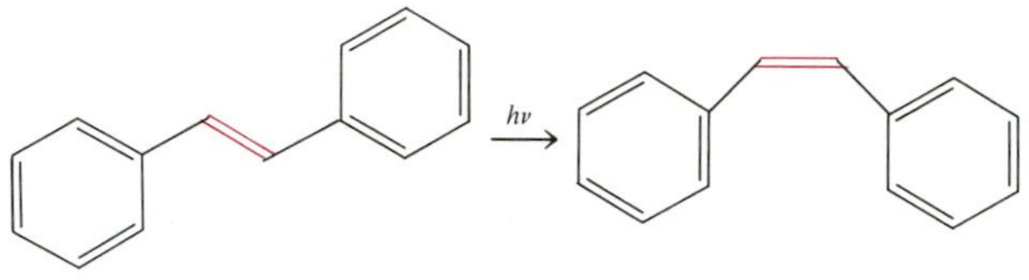

$h\nu$

trans-stilbene *cis*-stilbene

The *trans*-stilbene to *cis*-stilbene conversion is an intermediate step in the modern synthesis of phenanthrenes. Irradiation of a *trans*-stilbene in cyclohexane solution leads to a dihydrophenanthrene that is oxidized with iodine to the phenanthrene. Substituted phenanthrenes bearing groups as diverse as CH_3, CN, COOH, CF_3, CH_3O and halogen can be produced, as we illustrate with the case of 3-fluorophenanthrene.

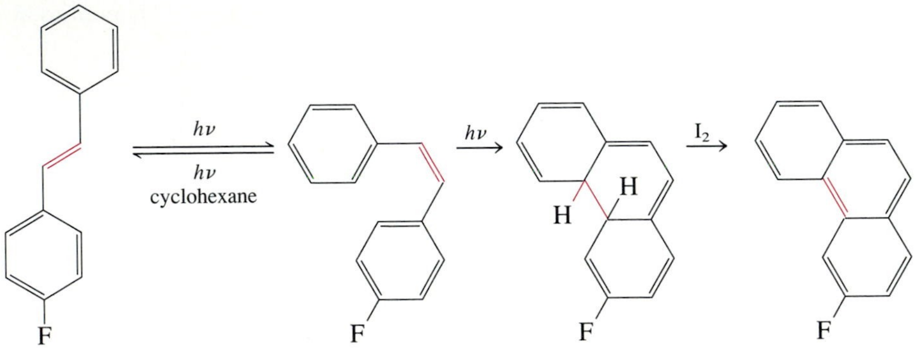

F $h\nu$ $h\nu$ cyclohexane F $h\nu$ H H F I_2 F

trans-p-fluorostilbene (76%) 3-fluorophenanthrene

The breaking of bonds is a common photochemical reaction. We saw an early example in the dissociation of chlorine to initiate free radical chain reactions (Section 6.3). This reaction is also common for ketones and is called a **Norrish type I** reaction.

$$CH_3COCH_3 \xrightarrow{h\nu} CH_3CO\cdot + CH_3\cdot$$

$$CH_3CO\cdot \longrightarrow CH_3\cdot + CO$$

$$2\,CH_3\cdot \longrightarrow CH_3CH_3$$

For cyclic ketones, the photochemical product is a **diradical** that can undergo further reactions.

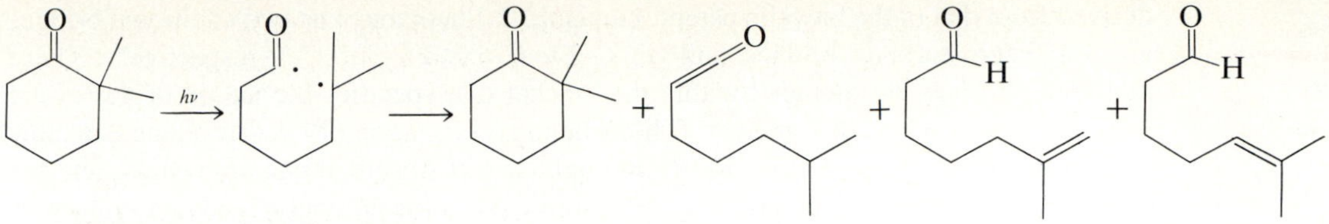

Another reaction that ketones can undergo involves intramolecular hydrogen atom transfer via a six-membered-ring transition state to form another diradical. This reaction is called a **Norrish type II** process.

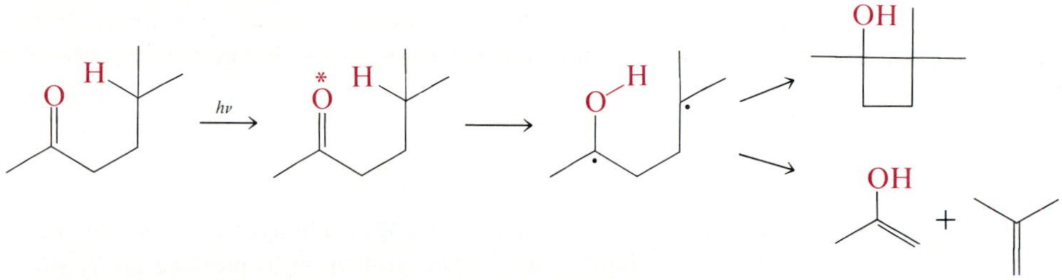

Note the use of an asterisk to indicate an excited *state*. We have previously used this symbol to refer to an antibonding orbital. The asterisk symbolism is used commonly in both contexts.

α,β-Unsaturated carbonyl compounds can undergo photochemical dimerization with the formation of a four-membered ring.

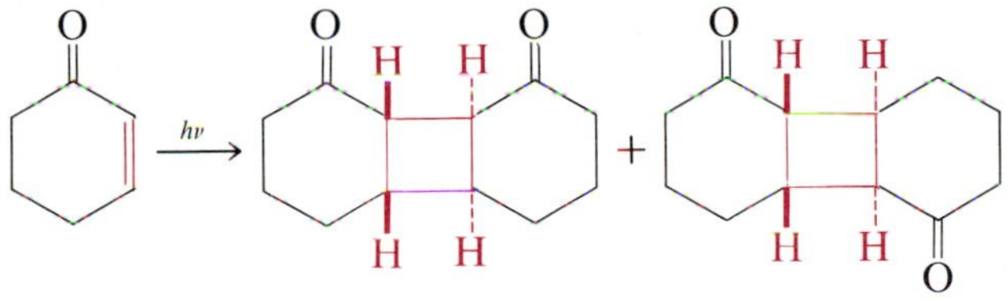

The formation of four-membered rings is also common with dienes.

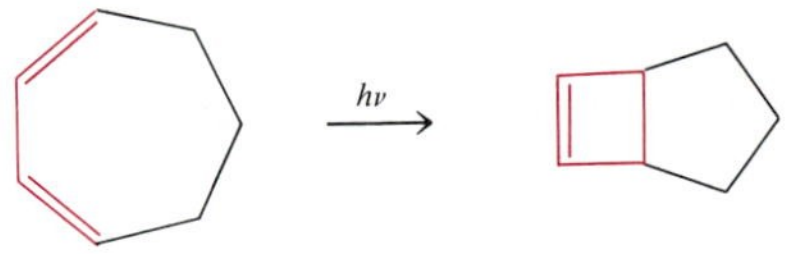

1,3-cycloheptadiene

bicyclo[3.2.0]hept-6-ene

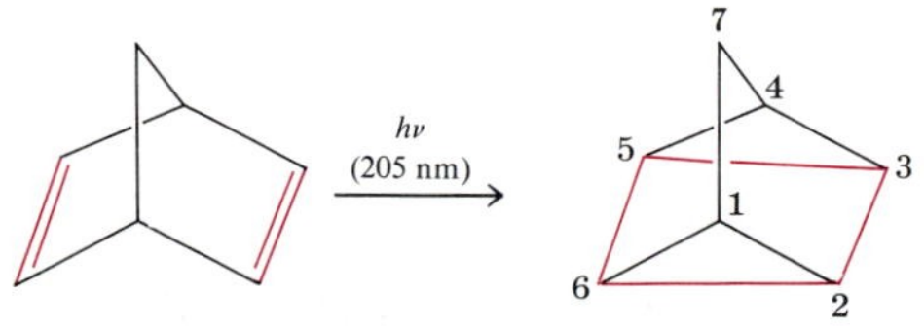

norbornadiene
bicyclo[2.2.1]-
hepta-2,5-diene

quadricyclene
tetracyclo[2.2.1.$0^{2,6}$.$0^{3,5}$]-
heptane

Quadricyclene is an unusual hydrocarbon whose trivial name derives from its tetracyclic nature. The compound is considered to be tetracyclic because four carbon-carbon bonds must be broken to obtain an acyclic system. Note how the systematic name is derived from that of the bicyclic parent, bicyclo[2.2.1]heptane. The two additional bridges are both "zero-carbon" bridges, and join C-2 to C-6 and C-3 to C-5, respectively. These additional bridges are included within the bracket that specifies the nature of the cyclic skeleton as the locants $0^{2,6}$ and $0^{3,5}$. Photochemical formation of quadricyclene is a slow process and gives a low yield. Most of the light quanta absorbed end up as heat, and the **quantum yield** is low. The quantum yield is defined as the number of product molecules divided by the number of light quanta absorbed.

The conversion becomes much more efficient when the reactive intermediate is produced by triplet energy transfer.

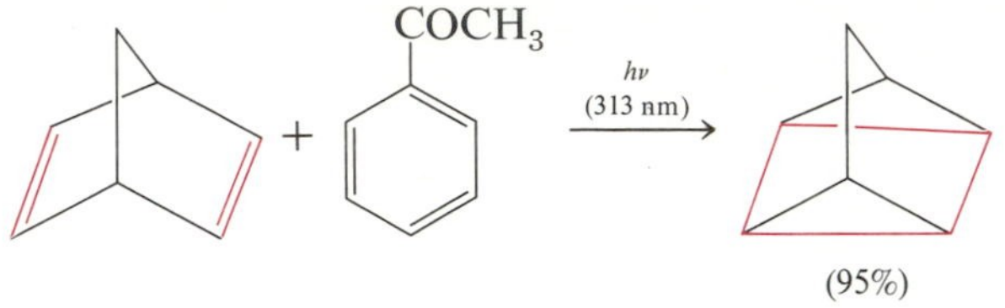

COCH$_3$ + ... $h\nu$ (313 nm) →

(95%)

Norbornadiene is transparent to light of 313 nm wavelength. Phenyl methyl ketone (acetophenone), however, absorbs this light in an $n \rightarrow \pi^*$ transition to produce an S_1 state. Intersystem crossing occurs readily to T_1 of the ketone. On colliding with a molecule of norbornadiene, the ketone T_1 gives up its triplet character and is converted back to the ground state. Norbornadiene is converted to its T_1, which undergoes the changes in geometry required to intersystem cross to the ground state of quadricyclene. In this example, the acetophenone functions as a **triplet sensitizer.**

Dye sensitizers for the silver bromide used in photographic film work in a related fashion. First, sensitizers extend the wavelength range of photography since silver bromide absorbs in the near ultraviolet. Second, the excited sensitizer can transfer energy to the silver bromide. The excitation promotes an electron from bromide ion into an excited state from which it is readily transferred to a *silver ion,* producing a *silver atom*. Such states in many solids are like those of conjugated molecules and are called conduction bands. They are found in semiconductors.

The silver atoms are catalysts for the reduction of all of the silver ions in a grain by the reducing agent in the developer (see page 1037). The electron for reducing the silver ion may also come directly from the dye by electron transfer. The grains that contain silver atoms form the latent image, which is then developed by reduction of the entire grain with developer. The organic dyes used as sensitizers are the basis for the success of the photographic industry and are an important application of organic chemistry.

The foregoing reactions have as the net result a [2 + 2] cycloaddition, a reaction that is generally not observed as a ground state concerted reaction because of the violation of the Hückel $4n + 2$ rule. The ground state rules for pericyclic reactions are frequently violated in photochemical processes.

Some photochemical reactions involve rather deep-seated rearrangements.

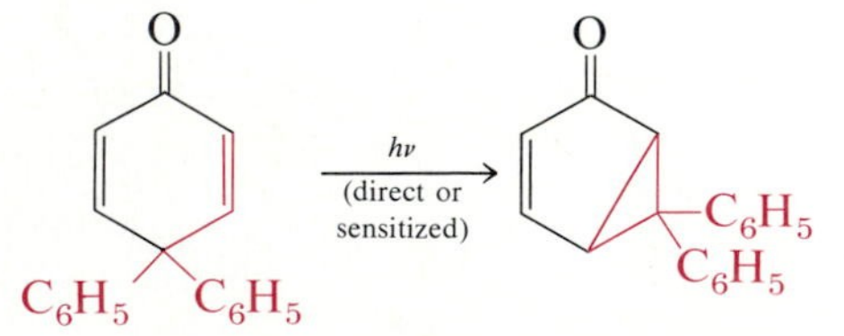

O ... C_6H_5 C_6H_5 $\xrightarrow[\text{(direct or sensitized)}]{h\nu}$ O ... C_6H_5 C_6H_5

This reaction can be rationalized by the following bond-switching process in the excited state.

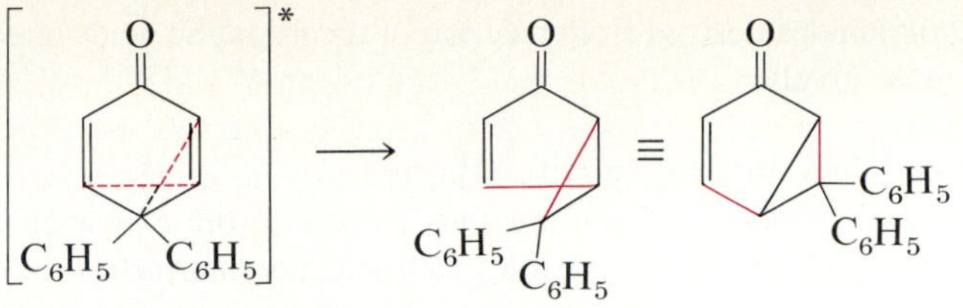

Mercury vapor is often used as a photosensitizer.

$$\mathrm{Hg} \xrightarrow[(253.7\ \mathrm{nm})]{h\nu} \mathrm{Hg}^*$$

$$\mathrm{Hg}^* + \mathrm{A} \longrightarrow \mathrm{Hg} + \mathrm{A}^*$$

Small amounts of mercury vapor suffice to make the following reactions preparatively useful.

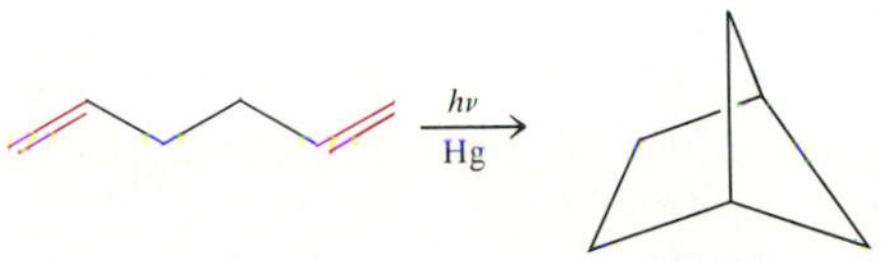

bicyclo[2.1.1]hexane

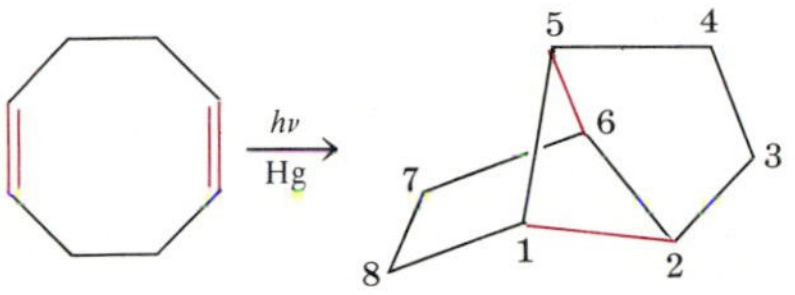

tricyclo[3.3.0.0^{2,6}]octane

EXERCISE 36.9 For the Norrish type I reaction of 2,2-dimethylcyclohexanone shown on page 1237 show how each of the products is produced.

EXERCISE 36.10 Heating (*E,Z*)-cyclonona-1,3-diene gives one stereoisomer of bicyclo[5.2.0]nona-8-ene and photolysis gives the other. Explain.

36.4 Polymer Chemistry

Polymers are large molecules that are generally built up from much smaller units or **monomers.** The *degree of polymerization* refers to the average number of monomer units per polymer molecule. We have encountered polymers frequently in our study of organic chemistry. Especially important are such natural polymers as polysaccharides (Section 28.8) and polypeptides and proteins (Section 29.7). Synthetic polymers have become important components of modern industrial society as vital *materials* of various kinds. These materials may be characterized by their behavior on heating. At high temperatures a polymer is often a viscous liquid in which molecular chains have some mobility relative to each other. At lower temperatures some regions within the polymer (crystallites) may have the regular structure characteristic of crystals. The corresponding phase transition is called the *crystalline melting point*. Alternatively, the polymer may be amorphous, a vitreous or glassy solid at low temperatures in which molecular chains or coils are effectively frozen but not in a regular pattern. Such a phase is characterized by a temperature range referred to as the *glass transition temperature*, T_g.

Synthetic polymers are also categorized by the type of polymerization reaction used. **Addition** polymers often involve the conversion of multiple bonds in monomers to bonds between the monomer units. Important examples are the polymerization of alkenes and dienes and can involve carbocation, free radical, and carbanion or organometallic intermediates. **Condensation** polymers result from a reaction of monomer units, usually accompanied by loss of a small molecule such as water or an alcohol. Examples are the polyesters and polyamides derived from reaction of dicarboxylic acids or esters with diols and diamines, respectively.

A. Carbocation Polymerization of Alkenes

In the case of alkenes, polymerization amounts merely to the exchange of π-bonds for σ-bonds and is thermodynamically feasible. Cationic polymerization is not generally a practical method for preparing useful polymers. The process is used for the dimerization and trimerization of certain alkenes and provides an excellent demonstration of the ways in which monomer units combine to give larger molecules.

As mentioned in Section 11.6.C, isobutylene is absorbed and hydrated by 60–65% aqueous sulfuric acid. Under more vigorous conditions (50% H_2SO_4 at 100 °C), the intermediate carbocation can react with alkene to form a new tertiary carbocation. Deprotonation of this new carbocation gives a mixture of alkenes known as "diisobutylenes."

$$(CH_3)_3C^+ + (CH_3)_2C{=}CH_2 \xrightarrow{\text{slow}} (CH_3)_3C{-}CH_2{-}\overset{+}{C}(CH_3)_2 \xrightarrow{-H^+}$$

$$\underset{(20\%)}{(CH_3)_3CCH{=}C(CH_3)_2} + \underset{(80\%)}{(CH_3)_3CCH_2\overset{\overset{\displaystyle CH_3}{|}}{C}{=}CH_2}$$

"diisobutylenes"

Catalytic hydrogenation of this mixture gives 2,2,4-trimethylpentane, the so-called "iso-octane" used as a standard for octane ratings of gasolines (Section 6.2). Under still more vigorous conditions isobutylene reacts with sulfuric acid to produce a mixture of trimeric alkenes, "triisobutylenes."

$$3\,(CH_3)_2C{=}CH_2 \xrightarrow{H_2SO_4} CH_3\underset{\underset{\displaystyle CH_3}{|}}{\overset{\overset{\displaystyle CH_3}{|}}{C}}CH_2\underset{\underset{\displaystyle CH_3}{|}}{\overset{\overset{\displaystyle CH_3}{|}}{C}}CH_2\overset{\overset{\displaystyle CH_3}{|}}{C}{=}CH_2 + CH_3\underset{\underset{\displaystyle CH_3}{|}}{\overset{\overset{\displaystyle CH_3}{|}}{C}}CH_2\underset{\underset{\displaystyle CH_3}{|}}{\overset{\overset{\displaystyle CH_3}{|}}{C}}CH{=}\overset{\overset{\displaystyle CH_3}{|}}{C}CH_3$$

"triisobutylenes"

Higher polymers and undesirable tars generally result from the reaction of other alkenes with strong hot acid.

In the absence of suitable alternative nucleophilic compounds to react with the carbocation intermediates, reaction with alkene is the only reaction mode possible. Reaction of isobutylene with a small amount of boron trifluoride occurs at low temperature to produce a high molecular weight polymer.

$$n\,(CH_3)_2C{=}CH_2 \xrightarrow[-200^\circ C]{BF_3} H{-}\left(CH_2{-}\underset{\underset{\displaystyle CH_3}{|}}{\overset{\overset{\displaystyle CH_3}{|}}{C}}{-}\right)_{n-1}{-}CH_2\overset{\overset{\displaystyle CH_3}{|}}{\underset{\underset{\displaystyle CH_2}{\|}}{C}}$$

(n = a large number)

Boron trifluoride does not react with alkenes in the rigorous absence of moisture. With traces of water, carbocations are produced. With isobutylene, for example, the intermediate salt, $(CH_3)_3C^+\ {}^-BF_3OH$, is produced in low concentration. The anion ${}^-BF_3OH$ has low nucleophilicity, and the *t*-butyl cation is free to react with isobutylene to start the cationic polymerization.

In some cases the carbocation will abstract a tertiary hydrogen from an alkane. A reaction of this type is used to produce ''isooctane'' directly from isobutylene and isobutane.

$$(CH_3)_3CH + (CH_3)_2C{=}CH_2 \xrightarrow[-25^\circ C]{HF} \underset{\text{2,2,4-trimethylpentane "isooctane"}}{(CH_3)_3CCH_2CH(CH_3)_2}$$

A reasonable mechanism for this alkylation reaction is

$$(CH_3)_2C{=}CH_2 \overset{HF}{\rightleftharpoons} (CH_3)_3C^+ \xrightarrow{(CH_3)_2C=CH_2} (CH_3)_3CCH_2\overset{+}{C}(CH_3)_2$$

$$(CH_3)_3CCH_2\overset{+}{C}(CH_3)_2 + (CH_3)_3CH \rightleftharpoons (CH_3)_3CCH_2CH(CH_3)_2 + (CH_3)_3C^+$$

Under these conditions the dimeric carbocation does not react with more isobutylene, but instead abstracts hydrogen from isobutane to provide more *t*-butyl cation to continue the chain of reactions. This is an example of **chain transfer,** a process that occurs generally in addition polymerization and which limits the length of polymer chains.

B. Free Radical Vinyl Polymerization

Radical polymerization is initiated by the addition of free radicals to an alkene double bond, as in the reactions discussed in Section 11.6.G. In those reactions, a reagent (HBr, Br_2, etc.) is present to react with the intermediate alkyl radical. In the absence of such a reagent the reaction of hydrocarbon radicals with alkenes can become the principal reaction to produce high molecular weight polymer chains. This reaction is an exceedingly important industrial process. Huge quantities of polyethylene (9.73 billion pounds in 1989) are made annually by this mechanism. The polymerization of ethylene requires high temperature and pressures.

$$n\ CH_2{=}CH_2 \xrightarrow[>100^\circ C;\ 15{,}000\ psi]{\text{trace } O_2 \text{ or peroxides}} Y{-}(CH_2CH_2)_n{-}Z$$

where n is a large number on the order of 1000

The end groups Y and Z depend on the initiators used and the termination reactions involved. The principal termination steps for ethylene polymerization are disproportionation and combination, as summarized in the following sequence of steps.

$$Y\cdot + CH_2{=}CH_2 \longrightarrow Y{-}CH_2{-}CH_2\cdot$$

$$YCH_2CH_2\cdot + CH_2{=}CH_2 \longrightarrow YCH_2CH_2CH_2CH_2\cdot \xrightarrow{etc.} Y(CH_2CH_2)_nCH_2CH_2\cdot$$

$$\left.\begin{matrix} Y(CH_2CH_2)_nCH_2CH_2\cdot \\ + \\ Y(CH_2CH_2)_mCH_2CH_2\cdot \end{matrix}\right\} \begin{matrix} \xrightarrow{\text{disproportionation}} Y(CH_2CH_2)_nCH_2CH_3 + Y(CH_2CH_2)_mCH{=}CH_2 \\ \\ \xrightarrow{\text{combination}} Y(CH_2CH_2)_nCH_2CH_2CH_2CH_2(CH_2CH_2)_mY \end{matrix}$$

The product of this so-called ''high-temperature polymerization'' of ethylene does not have the simple linear structure shown. Ethyl and butyl groups are known to occur along the polymethylene chain, probably because of hydrogen abstraction reactions of the following type.

$$Y(CH_2CH_2)_n{-}\underset{H}{CH}{-}CH_2{-}CH_2{-}CH_2{-}CH_2\cdot \longrightarrow Y(CH_2CH_2)_n\overset{CH_2CH_2CH_2CH_3}{\dot{C}H}$$

chain grows here

The result is a **branched-chain** polymer. **Linear polyethylene** is made by an entirely different process described later.

Vinyl chloride, tetrafluoroethylene, and styrene are other monomers converted to important polymers by free radical polymerizations. The Markovnikov addition of radicals to vinyl chloride applies with high specificity so that the product polymer has a complete head-to-tail structure.

$$Y\cdot + CH_2{=}CHCl \longrightarrow Y{-}CH_2\dot{C}HCl \longrightarrow \longrightarrow \underset{\text{polyvinyl chloride}}{Y(CH_2CHCl)_nZ}$$

Vinyl chloride is manufactured on an enormous scale, primarily for making polyvinyl chloride, PVC. The 1989 production of PVC was 8.49 billion pounds. Vinyl chloride is manufactured mostly by dehydrochlorination of 1,2-dichloroethane (ethylene dichloride). In 1974 the Occupational Safety and Health Administration concluded that vinyl chloride is a human carcinogen and set maximum limits to exposure.

Polyvinyl chloride is an extremely hard resin. In order to alter the physical properties of the polymer, low molecular weight liquids called **plasticizers** are added in the polymer formulation. Bis-2-ethylhexyl phthalate is one of the compounds added to polyvinyl chloride as a plasticizer. The resulting polymer has a tough leathery or rubber-like texture. It is used in plastic squeeze bottles, imitation leather upholstery, pipes, and so on.

Polytetrafluoroethylene or ''Teflon'' is a perfluoro polymer having great resistance to acids and organic solvents. It is used to coat ''nonstick'' frying pans and other cooking surfaces.

$$\underset{\text{Teflon}}{X{-}(CF_2CF_2)_n{-}Y}$$

The polymerization of tetrafluoroethylene was discovered through the alertness of a Du Pont chemist, Roy Plunkett. A gas cylinder which appeared to be ''empty'' was found to weigh what was expected for a full tank. The cylinder was cut open to reveal the polymer, an act that initiated an important industry.

Polystyrene is an inexpensive plastic used to manufacture many familiar household items. It is a hard, colorless, somewhat brittle material.

$$\left(\begin{matrix} C_6H_5 \\ | \\ CHCH_2 \end{matrix}\right)_n$$

In the simple formulation of polystyrene, the end groups have been omitted. This simplification is common in the symbolism of polymer chemistry. The end groups constitute a minute portion of a high molecular weight polymer, although their character can have a significant effect on the properties of the polymers.

Polymerization of styrene together with the difunctional compound, divinylbenzene, leads to a polymer containing **cross-links.** Each of the two vinyl groups participates in the growth of separate chains. Since the independent chains can grow in different directions, the resulting polymer is a three-dimensional network. Cross-linking has a large effect on physical properties because it restricts the relative mobility of polymer chains. Polystyrene, for example, is soluble in many solvents such as benzene, toluene, and carbon tetrachloride. A polymer made with only 0.1% divinylbenzene, however, no longer dissolves but only *swells*. This property is important in many uses of polystyrene-derived materials. An example is the polymer used for the Merrifield peptide syntheses (page

974). The 1989 production of polystyrene in its various forms, including copolymers, was 5.10 billion pounds.

Acrylonitrile is another important monomer manufactured in large quantity for use in synthetic fibers and polymers; its 1989 production in the United States was 1.31 million tons. It was once prepared industrially by addition of HCN to acetylene.

$$HC{\equiv}CH + HCN \xrightarrow[NH_4Cl]{CuCl} CH_2{=}CHCN$$

It is now prepared by a cheaper process that involves the catalytic oxidation of propene in the presence of ammonia.

$$2\ CH_3CH{=}CH_2 + 3\ O_2 + 2\ NH_3 \xrightarrow[\text{catalyst}]{450^\circ} 2\ CH_2{=}CHCN + 6\ H_2O$$

Free radical polymerization of acrylonitrile in aqueous solution gives a polymer that can be spun to give the textile Orlon or Acrilan.

$$CH_2{=}CHCN \longrightarrow {-}\!\left(CH_2{-}\underset{\displaystyle CN}{\underset{|}{CH}}\right)_{\!n}\!{-}$$

Orlon, Acrilan

The methyl ester of α-methylacrylic acid is also an important monomer. It is prepared from acetone by the following sequence.

$$CH_3COCH_3 + HCN \longrightarrow (CH_3)_2\overset{\displaystyle OH}{\overset{|}{C}}CN \xrightarrow{H_2SO_4} CH_2{=}\overset{\displaystyle CH_3}{\overset{|}{C}}CONH_2 \xrightarrow[H_2SO_4]{CH_3OH} CH_2{=}\overset{\displaystyle CH_3}{\overset{|}{C}}COOCH_3$$

Poly(methyl methacrylate) prepared by free radical polymerization is a stiff transparent plastic known as Lucite or Plexiglas.

$$CH_2{=}\overset{\displaystyle CH_3}{\overset{|}{C}}{-}COOCH_3 \longrightarrow {-}\!\left(CH_2{-}\overset{\displaystyle CH_3}{\overset{|}{\underset{\displaystyle COOCH_3}{\underset{|}{C}}}}{-}\right)_{\!n}\!{-}$$

poly(methyl methacrylate)
Lucite, Plexiglas

Note that with both acrylonitrile and methyl methacrylate free radical polymerization involves almost exclusively head-to-tail combination of the monomers.

C. Anionic and Organometallic Polymerization

In anionic polymerization, initiation is accomplished by addition of a nucleophile to a carbon-carbon double bond. Simple olefins are inert to most nucleophilic or basic reagents and only when the anion itself is an extremely powerful base such as *t*-butyllithium will addition to the double bond occur.

$$(CH_3)_3C^-\ Li^+ + CH_2{=}CH_2 \longrightarrow (CH_3)_3CCH_2CH_2^-\ Li^+$$

Since primary carbanions are more stable than tertiary carbanions, the reaction as shown has favorable thermodynamics. This particular reaction is of limited use because *t*-butyllithium is such a highly reactive compound and abstracts a proton from diethyl ether rapidly at room temperature.

Other kinds of organometallic intermediates provide rapid polymerization and are much more important. A particularly significant example is a catalyst (**Ziegler-Natta** catalyst) prepared from aluminum alkyls (R_3Al) and titanium tetrachloride. This catalyst polymerizes simple olefins by a mechanism that depends on the carbanion character of the carbon of the carbon-aluminum bond and the ability of the transition metal, titanium, to coordinate with the π-bonds of alkenes. The result is a rapid polymerization reaction that is used extensively with ethylene and propylene. Linear polyethylene, $(CH_2CH_2)_n$, prepared in this way, is more crystalline than high temperature polyethylene and has a higher density and melting point. The long chains of linear polyethylene can lie together in a regular manner in the solid without the defects in regularity imposed by the random branches of the high temperature polymer. Total United States production of polyethylene in 1989 was 17.83 billion pounds, of which almost one half was linear polyethylene.

Polypropylene presents an interesting further aspect of polymer stereochemistry. In addition to polymerization exclusively in a head-to-tail fashion by Ziegler-Natta catalysts, the polymer has the methyl groups entirely on one side of the zigzag backbone.

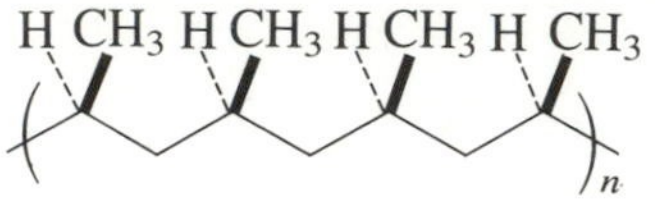

H CH_3 H CH_3 H CH_3 H CH_3

n

This isomeric form is called **isotactic.** In isotactic polypropylene there is significant steric hindrance between methyl appendages and the backbone twists regularly to give the polymer a helical structure. Note that the products of free radical polymerization have a random or **atactic** orientation of substituents. The other possible regular pattern in which substituents alternate on opposite sides of the backbone chain (**syndiotactic**) is rare. The total 1989 production of polypropylene was 7.35 billion pounds.

D. Diene Polymers

Certain tropical plants, among them the rubber tree, *Hevea brasiliensis,* exude a liquid suspension of rubber, now called a **latex.** The French word for rubber, *caoutchouc,* is a South American Indian word meaning ''weeping wood.'' The Aztecs and Mayas played games with rubber balls and made protective clothes coated with rubber that had been prepared with heat and smoke. Today, the latex is coagulated with acetic acid and converted into sheets of ''crêpe rubber.'' Natural rubber consists mostly of polyisoprene in which the double bonds are cis.

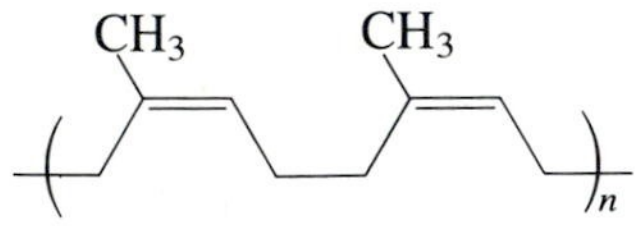

CH_3 CH_3

n

Joseph Priestley, the discoverer of oxygen, found in 1770 that caoutchouc was good for rubbing out pencil marks, hence, the name rubber.

Natural rubber becomes sticky in summer and hard in winter, so is not a useful elastomer or rubber. Modification by vulcanization was discovered by Charles Goodyear in 1839 and involves heating the raw rubber with sulfur. The process appears to involve addition of sulfur units to the double bonds with the production of cross-links between the polymer chains. Because of these cross-links, the polymer resists distortion and tends to return to its original shape.

S S

S

Elasticity has important structural requirements. If a polymer has regular repeating units, regions of the polymer may pack together by van der Waals forces in a manner similar to crystals. Such polymers are more or less crystalline and tend to be hard solids. Polymers that have flexible and irregular chains tend to be less rigid, but such a polymer is not an elastomer unless it returns to its original shape when the stress is removed. Hence, elastomers tend to have flexible chains with varying amounts of cross-linking.

Some plants produce a polyisoprene with trans double bonds.

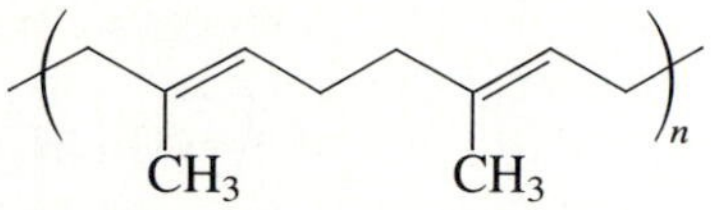

$$\left(\text{—}CH_2C(CH_3)\text{=}CHCH_2CH_2C(CH_3)\text{=}CHCH_2\text{—} \right)_n$$

This material, known as gutta-percha, is a harder and less elastomeric natural polymer than rubber.

Isoprene polymerizes under the influence of acids or Ziegler-Natta catalysts to a polyisoprene with rubber-like properties. The double bonds in this synthetic polyisoprene are both cis and trans. Synthetic rubbers, however, are derived primarily from butadiene. Buna S or GRS is a **copolymer** of 4–5 moles of butadiene to 1 mole of styrene. In this free radical polymerization, the butadiene adds by cis- and trans-1,4- and 1,2-additions. There are three repeating units in the polymer. Styrene-butadiene copolymers are the principal synthetic rubbers; 1989 production in the United States was 3.9 million pounds. In 1990, world rubber production was about 25 billion pounds, of which only 9.25 billion pounds were natural rubber.

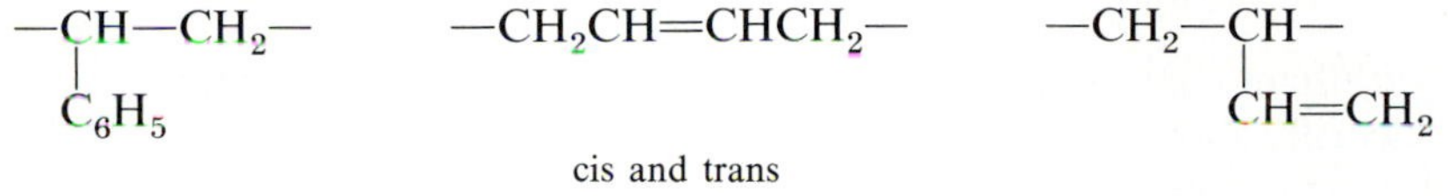

$$\text{—}CH(C_6H_5)\text{—}CH_2\text{—} \qquad \text{—}CH_2CH\text{=}CHCH_2\text{—} \qquad \text{—}CH_2\text{—}CH(CH\text{=}CH_2)\text{—}$$

cis and trans

In copolymerization the growing polymer radical has a choice of reacting with two different alkenes. The relative reactivities of the two monomer units are important in determining the composition of the resulting polymer. If one monomer is much more reactive in free radical polymerizations than the other, it will tend to form a **homopolymer** and be consumed before the second monomer starts to become incorporated. Polymers that consist of large segments of homopolymers joined in a copolymer are known as **block** copolymers.

Butadiene is also polymerized with Ziegler-Natta catalysts or by alkali metal catalysts based on alkylsodium formulations or lithium dispersions. Some of these methods are highly specific and give either cis- or trans-1,4- or 1,2-addition. A terpolymer produced from acrylonitrile, butadiene, and styrene is an inexpensive plastic known as ABS.

Neoprene is a synthetic elastomer obtained by the free radical polymerization of chloroprene, 2-chloro-1,3-butadiene. Neoprene has unique properties, such as resistance to oils, oxygen, and heat.

E. Condensation Polymers

Polymers result from the reaction of dicarboxylic acids or derivatives with diols or diamines. For example, an important **polyester** known as Dacron or Terylene is prepared by the reaction of dimethyl terephthalate and ethylene glycol. In one industrial process, the two reactants are heated together and methanol is distilled from the reactor. Polyesters are the major synthetic fiber; 1989 production was 3.59 billion pounds.

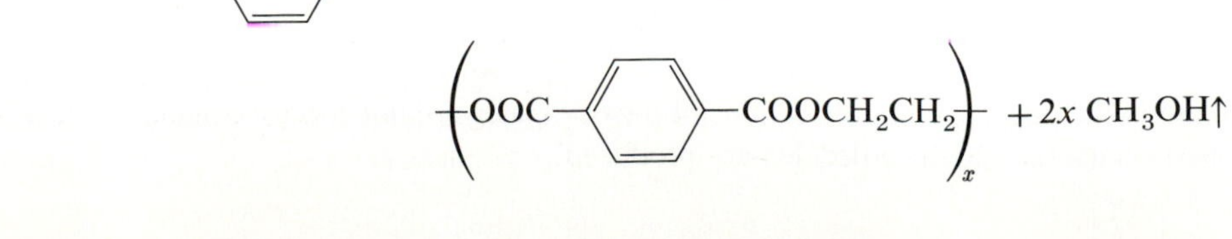

$$x\, CH_3OOC\text{—}C_6H_4\text{—}COOCH_3 + x\, HOCH_2CH_2OH \xrightarrow{\Delta} \left(OOC\text{—}C_6H_4\text{—}COOCH_2CH_2 \right)_x + 2x\, CH_3OH\uparrow$$

The best known **polyamide** is nylon 6,6, which is a copolymer formed from 1,6-hexanediamine and adipic acid. The polymer is manufactured by heating an equimolar mixture of the two monomers at 270 °C at a pressure of about 10 atm.

$$H_2N(CH_2)_6NH_2 + HOOC(CH_2)_4COOH \xrightarrow[10\ atm]{270°C} \left(NH(CH_2)_6NH\overset{O}{\overset{\|}{C}}(CH_2)_4\overset{O}{\overset{\|}{C}} \right)_n + H_2O$$

nylon 6,6

Another form of nylon is nylon 6, which can be produced by polymerization of the amino acid 6-aminohexanoic acid. The corresponding lactam, caprolactam, is produced industrially from cyclohexane and is the usual starting material for the polymer. Total 1989 production of various forms of nylon was 2.74 billion pounds.

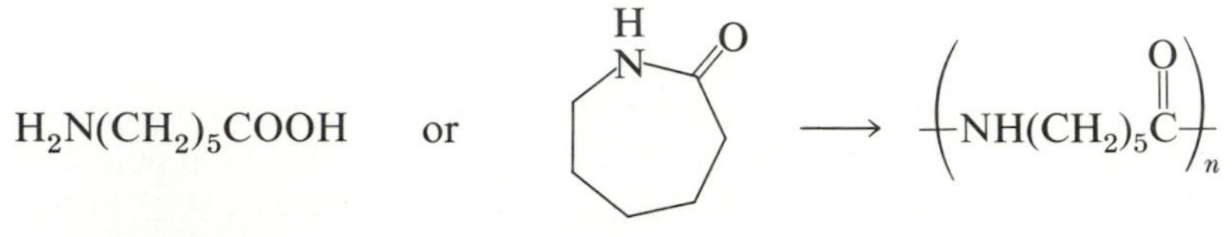

$$H_2N(CH_2)_5COOH \quad \text{or} \quad \text{caprolactam} \longrightarrow \left(NH(CH_2)_5\overset{O}{\overset{\|}{C}} \right)_n$$

6-aminohexanoic acid — caprolactam — nylon 6

Polyurethanes are formed from an aromatic diisocyanate and a diol. One type of diol used is actually a low molecular weight copolymer made from ethylene glycol and adipic acid. When this polymer, which has free hydroxy end groups, is mixed with the diisocyanate, a larger polymer is produced.

$$HOCH_2CH_2OH + HOOC(CH_2)_4COOH \longrightarrow$$

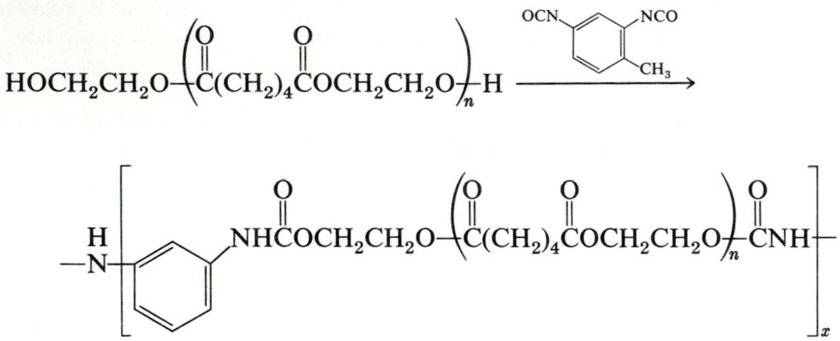

$$HOCH_2CH_2O\left(\overset{O}{\overset{\|}{C}}(CH_2)_4\overset{O}{\overset{\|}{C}}OCH_2CH_2O\right)_nH \xrightarrow{\text{OCN, NCO, } CH_3 \text{ (toluene diisocyanate)}}$$

$$\left[-\overset{H}{N}-C_6H_3-NH\overset{O}{\overset{\|}{C}}OCH_2CH_2O\left(\overset{O}{\overset{\|}{C}}(CH_2)_4\overset{O}{\overset{\|}{C}}OCH_2CH_2O\right)_n\overset{O}{\overset{\|}{C}}NH- \right]_x$$

a polyurethane

In the manufacturing process, a calculated amount of water is mixed with the diol. Some of the diisocyanate reacts with the water to give an aromatic diamine and carbon dioxide. The carbon dioxide forms bubbles that are trapped in the bulk of the polymer as it solidifies. The result is a spongy product called **polyurethane foam.** Diamines react with diisocyanates to give **polyureas.**

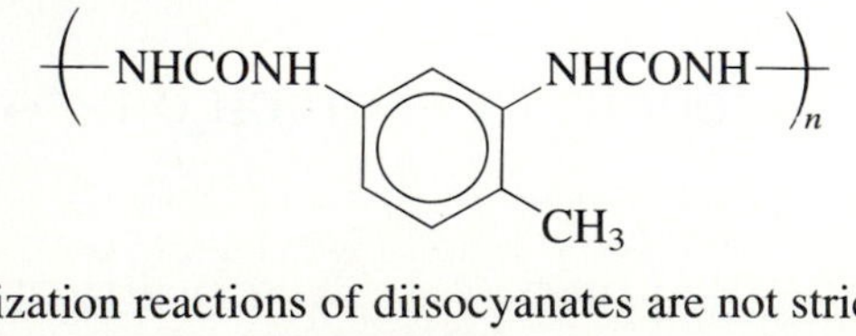

$$\left(NHCONH-C_6H_3(CH_3)-NHCONH \right)_n$$

Note that the polymerization reactions of diisocyanates are not strictly condensation reactions because no small molecules are produced.

EXERCISE 36.11 Vinyl acetate undergoes free radical polymerization. What is the expected structure of the polymer? This polymer can be hydrolyzed to a useful water-soluble polyalcohol. Why is this polyalcohol not produced directly by polymerization of a monomer?

EXERCISE 36.12 When styrene is copolymerized with maleic anhydride under free radical conditions the growing polymer can have two types of radical ends. What are they? The copolymerization of two monomers A and B is characterized by two copolymerization ratios, r_A and r_B, in which r_A is defined as the rate constant for reaction of an A radical end with monomer A divided by the rate constant for reaction with monomer B. If both r_A and r_B equal unity, a random copolymer results. For copolymerization of styrene with maleic anhydride both copolymerization ratios are very small. What does this imply about the structure of the copolymer?

EXERCISE 36.13 Pyromellitic anhydride is the dianhydride of benzene-1,2,4,5-tetracarboxylic acid. It reacts with *p*-phenylenediamine (*p*-diaminobenzene) to form a strong condensation polymer stable at high temperatures. What is its structure?

36.5 Natural Products: Terpenes, Steroids, and Alkaloids

Metabolism is the collection of chemical processes by which an organism creates and maintains its substance and obtains energy in order to grow and function. Almost all of these chemical processes involve organic compounds and reactions and naturally fall under the purview of the organic chemist. The metabolic processes of various organisms have many common elements, in sugar metabolism, in the tricarboxylic acid cycle, in the use of adenosine triphosphates, as might be anticipated among **primary metabolites** and **primary metabolic processes** arising from a common genetic code. Nevertheless, there is another more species-specific group of **secondary metabolites** and **secondary metabolic processes** that are amazingly varied and complex. The study of these processes is the subject of an entire discipline of science—biochemistry. Many of the end products of metabolism are readily isolable organic compounds and have historical importance in organic chemistry. These compounds are grouped together under the broad heading of **natural products.**

There are many different classes of naturally occurring compounds, and some, such as fats (Section 19.11.B), carbohydrates (Chapter 28), proteins (Chapter 29), and nucleic acids (Section 33.5), have already been encountered in this book. These natural products, together with a relatively small number of related substances, occur in almost all organisms and are central to their structure and operation.

Secondary metabolites are not necessarily of secondary importance to the organism, but their distribution in nature tends to be much more species-dependent. Examples of such secondary metabolites are the terpenes, the steroids, and the alkaloids. Because of the central role these natural products have played in the development of organic chemistry, we briefly examine their structures at this point.

A. Terpenes

The **terpenes** are a class of organic compounds that are the most abundant components of the **essential oils** of many plants and flowers. Essential oils are obtained by distilling the plants with water; the oil that separates from the distillate usually has an odor highly characteristic of and identified with the plant of origin. In the days of alchemists this procedure was common. The resulting mixture of organic compounds was thought to be the essence of the plant, hence the term essential oil. Terpenes are synthesized by organisms (''biosynthesized'') from acetic acid by way of the important biochemical intermediate isopentenyl pyrophosphate. Terpene structures may generally be dissected into several ''isoprene units.''

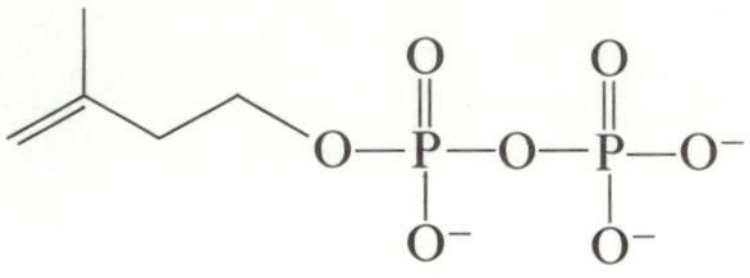

isopentenyl pyrophosphate 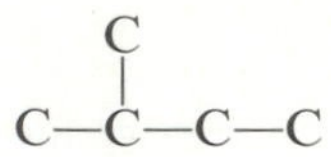   an isoprene unit

Compounds derived from a single isoprene unit are rare in nature. However, compounds composed of two isoprene units are common. These materials are called **monoterpenes.** The simplest acyclic example is geraniol, a constituent of the oil of geranium. A related monoterpene is citronellal, which is responsible for the characteristic aroma of lemon oil.

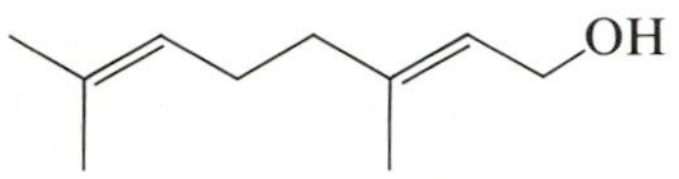

geraniol 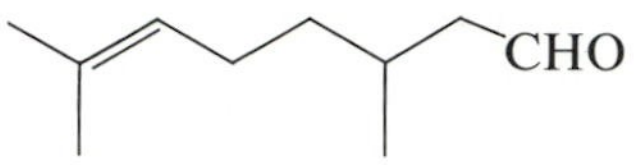   citronellal

Menthol (peppermint oil derived from *Mentha piperita*), β-pinene (turpentine, a liquid obtained by distillation of the wood of the loblolly pine tree, *Pinus taeda*), and camphor (obtained from the wood and leaves of the camphor laurel tree, *Cinnamomum camphora*) are examples of cyclic monoterpenes.

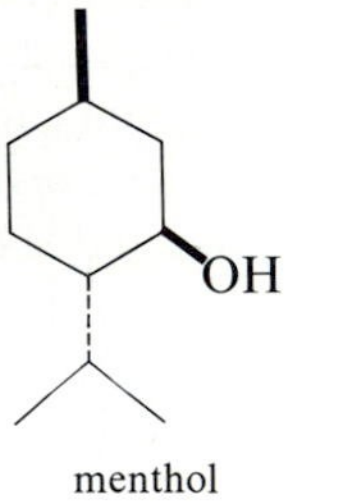

menthol 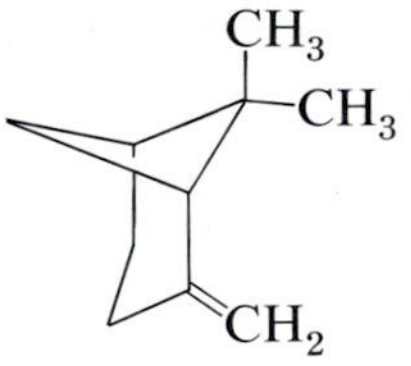 β-pinene 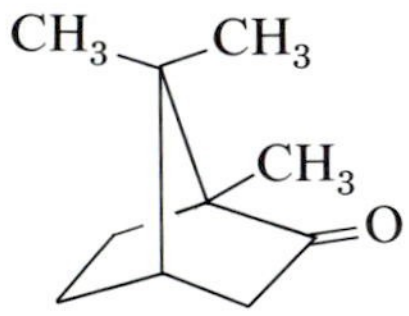   camphor

The odor of a compound can be not only an attractive characteristic but also of considerable commercial importance for perfumes and soaps. Recall that the chiral nature of a drug can be of medical significance. Some enantiomers differ considerably in odor as shown in the following examples.

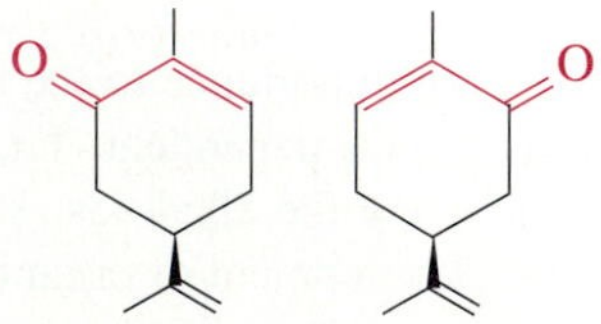

R(−) spearmint   carvone   *S*(+) caraway

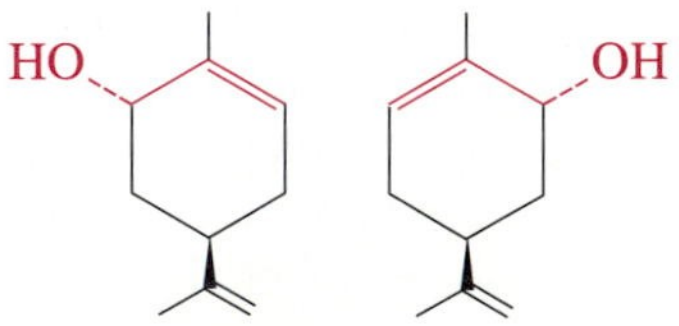

R(−) minty   carveol   *S*(+) musty

Sesquiterpenes are C_{15} compounds that are composed of three isoprene units. Farnesol can be regarded as the parent acyclic alcohol. The alcohol is easily formed from the corresponding pyrophosphate, a key intermediate in C_{15} biosynthesis, and farnesol occurs in the essential oils of at least 35 species including rose, acacia, and cyclamen. It has the characteristic odor of lily of the valley and is used in perfumery. A simple monocyclic sesquiterpene is bisabolene, which is found in the oils of bergamot and myrrh (from shrubs of the genus *Commiphora* found in southern Arabia).

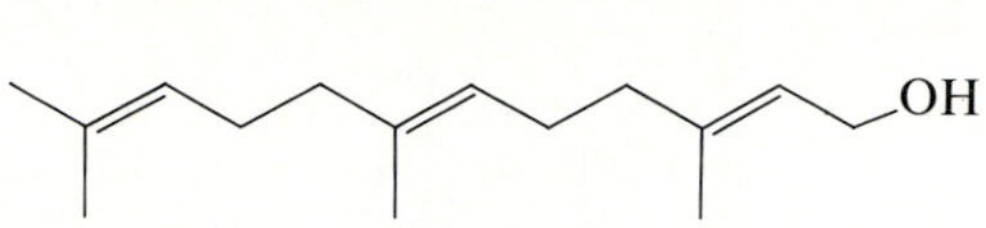

OH

farnesol

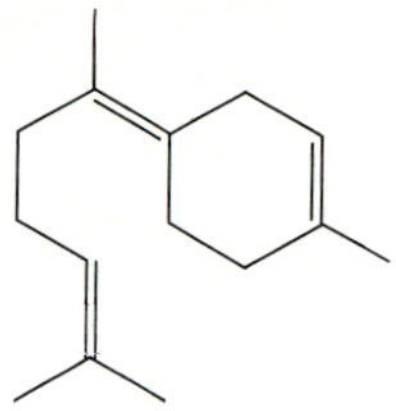

bisabolene

Most terpenes have cyclic or polycyclic structures. A tremendous variety of fascinating structures are known. Examples are nootkatone (aroma of grapefruit), β-santalol (sandalwood oil), guaiol (guaiacum wood), and copaene (copaiba balsam oil).

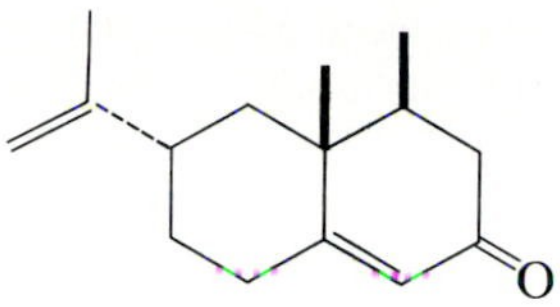

O

nootkatone

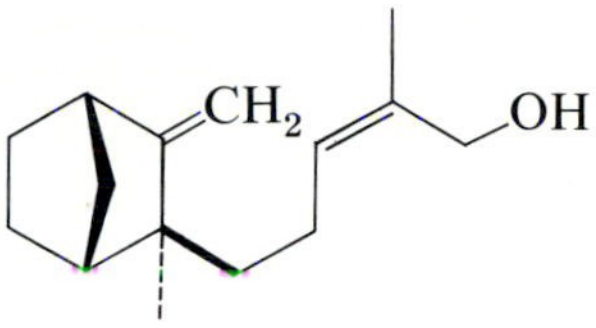

CH_2 OH

β-santalol

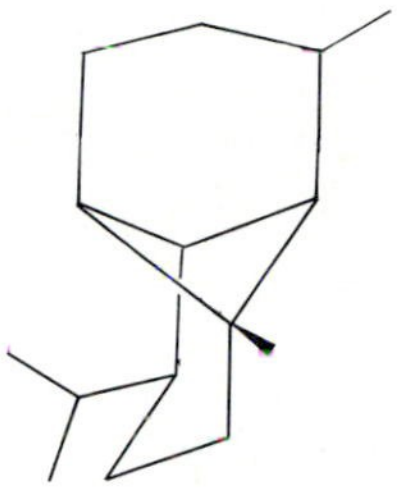

copaene

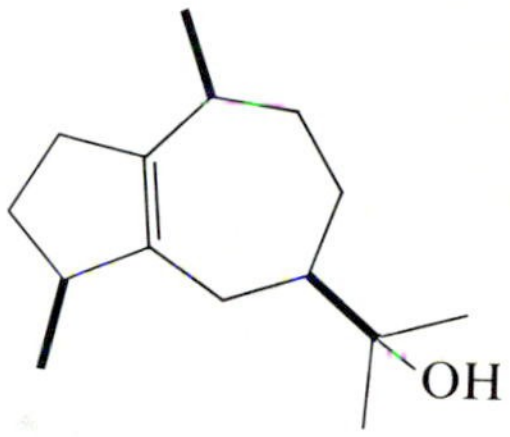

OH

guaiol

Diterpenes, C_{20} compounds composed of four isoprene units, include the important retinal (vitamin A aldehyde) which combines with the protein opsin to form the active pigment of the retina of the eye (see page 681) and abietic acid, a component of rosin, the nonvolatile exudate of coniferous (cone-bearing, mostly evergreen) trees.

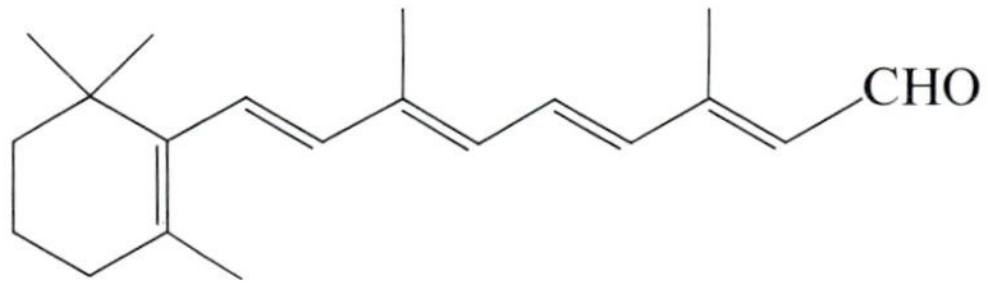

CHO

retinal

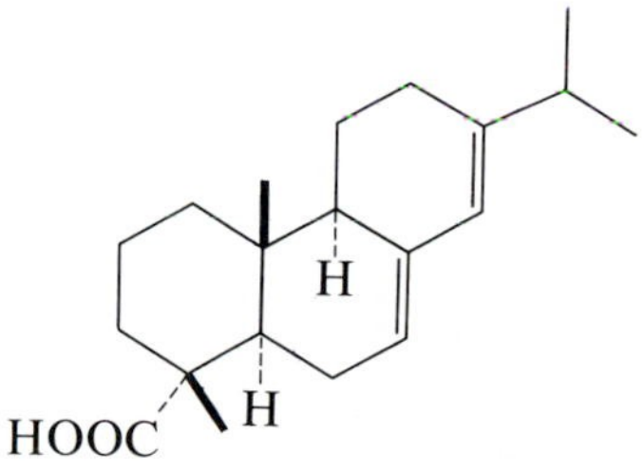

H

HOOC H

abietic acid

Terpenes having 25 carbons (sesterterpenes) are rare. However, the C_{30} compounds, or **triterpenes,** are common. An interesting example is squalene, a high-boiling viscous oil that is found in large quantities in shark liver oil. It may be isolated in smaller amounts from olive oil, wheat germ oil, rice bran oil, and yeast, and it is an intermediate in the biosynthesis of steroids.

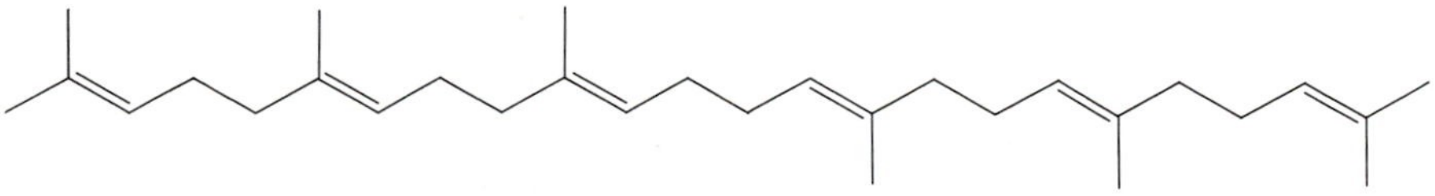

squalene

Other triterpenes are polycyclic, such as β-amyrin, a major constituent of the resin of the Manila elemi tree, and cyclolaudenol, a component of the neutral fraction of opium.

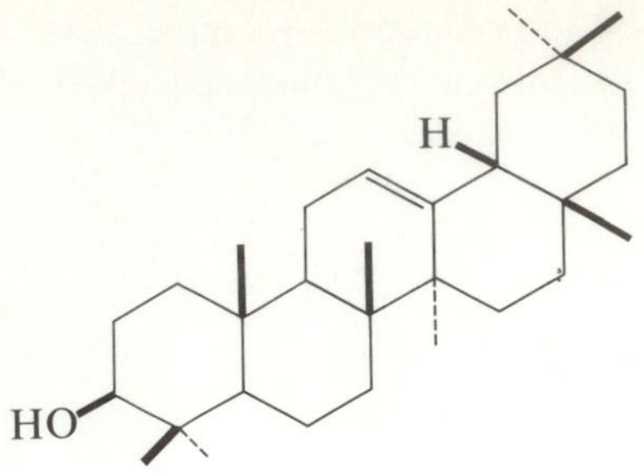

HO H H HO

β-amyrin

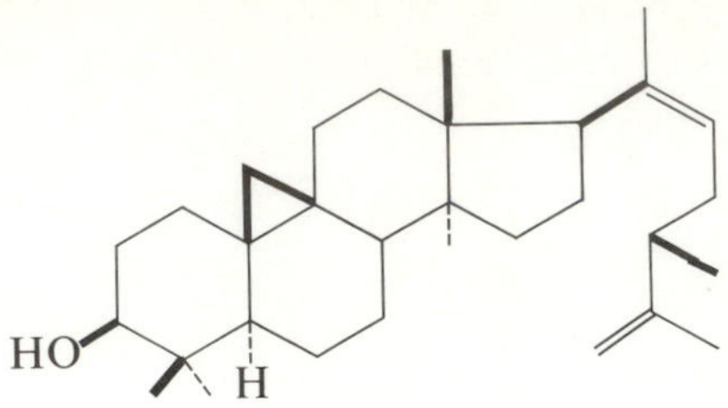

cyclolaudenol

Natural rubber and gutta-percha (pages 1244–45) are polyterpenes, being made up of a large number of isoprene units.

In the structural formulas used thus far for terpenes, note that the appendage methyl groups, both in side chains and attached to a cyclic nucleus, are indicated only as lines. This convention is widely used in depicting terpene and steroid structures. Of course, the convention of using a bold line for a methyl substituent that projects upward from the general plane of the ring and a dashed line for a methyl substituent that projects downward from the general plane is still followed.

The elucidation of the structures of the terpenes has provided a fascinating and important chapter in organic chemistry that really started only about a half century ago. Early terpene research led to the recognition of skeletal rearrangements that were among the first examples of carbocation rearrangements. A particularly important example is the camphene hydrochloride-isobornyl chloride rearrangement, which we can recognize as a simple 1,2-alkyl rearrangement.

HCl → Cl ⇌ Cl^- + →

camphene
(many essential oils)

camphene hydrochloride

+ Cl^- ≡ Cl^- + → H Cl

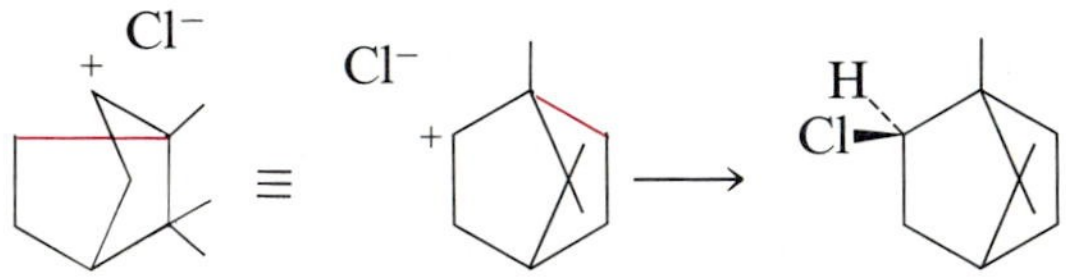

isobornyl chloride

Similar rearrangements are widespread in terpene chemistry. A further example is the rearrangement of longifolene to longifolene hydrochloride. The bond that migrates is shown in color.

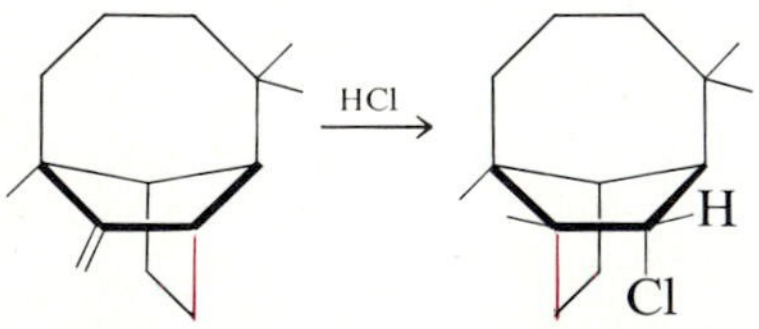

HCl → H Cl

longifolene

longifolene
hydrochloride

The stereo structure of longifolene hydrochloride is shown in Figure 36.4.

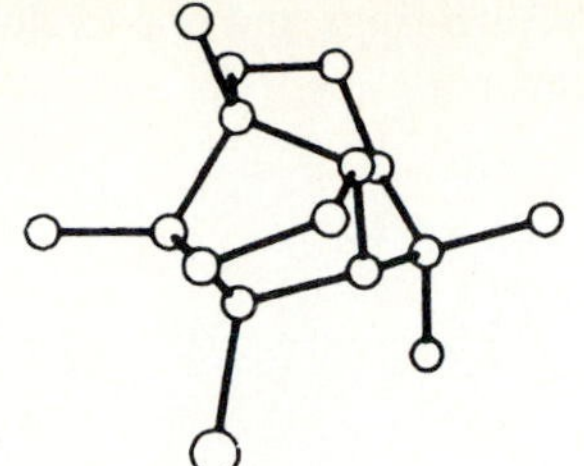
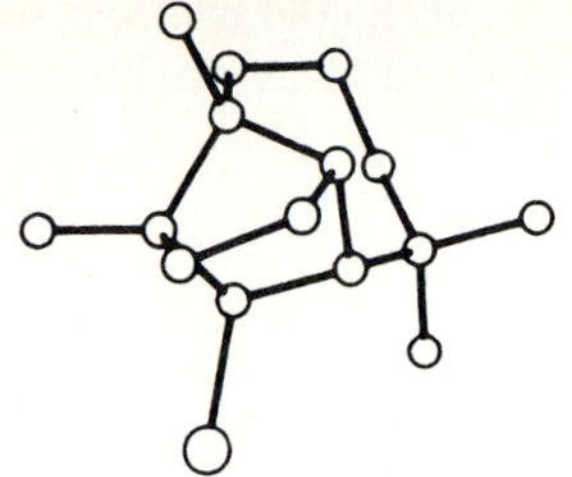

FIGURE 36.4 Stereo structure of longifolene hydrochloride. Note that hydrogens are not shown. [Reproduced with permission from *Molecular Structure and Dimensions,* International Union of Crystallography, 1972.]

Synthetic routes to the simpler terpenes are now available, but many of the more complex polycyclic terpenes provide synthetic challenges that intrigue present-day synthetic research chemists.

EXERCISE 36.14 Locate the isoprene units in menthol, β-pinene, camphor, β-santalol, copaene, guaiol, abietic acid, β-amyrin, and longifolene. Note that nootkatone cannot be dissected into three isoprene units. In fact, it has been shown that this sesquiterpene is biosynthesized by a route that involves a 1,2-alkyl rearrangement.

B. Steroids

Steroids are tetracyclic natural products that are related to the terpenes in that they are biosynthesized by a similar route. An important example is cholesterol, the major component of human gall stones (Gk., *chole,* bile).

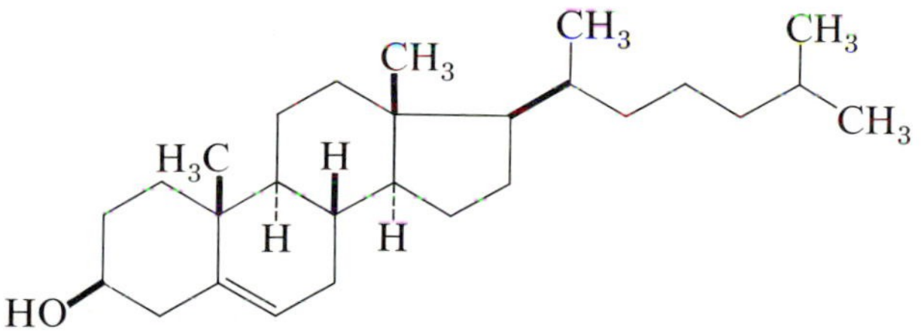

cholesterol

Actually, cholesterol is present in some amount in all normal animal tissues, but it is concentrated in the brain and in the spinal cord. The total amount present in a 180 lb person is 240 g, about 0.5 lb! It is present partly as the free alcohol and partly esterified with fatty acids.

The structure of cholesterol illustrates the basic steroid skeleton, which is that of a hydrogenated 1,2-cyclopentenophenanthrene having two methyl substituents at C-10 and C-13 and an additional side chain at C-17. The stereochemistry at the various stereocenters is almost invariably that shown, and in subsequent examples we will not indicate stereochemistry unless it differs from the usual.

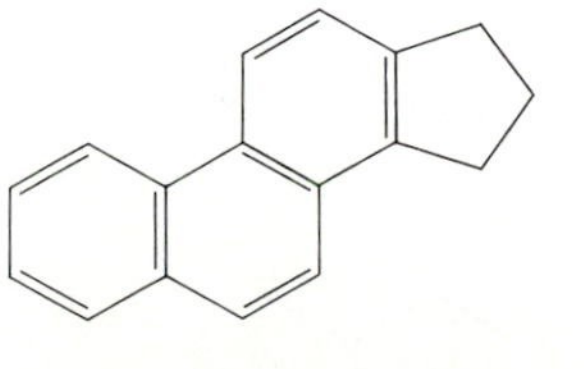

1,2-cyclopentenophenanthrene

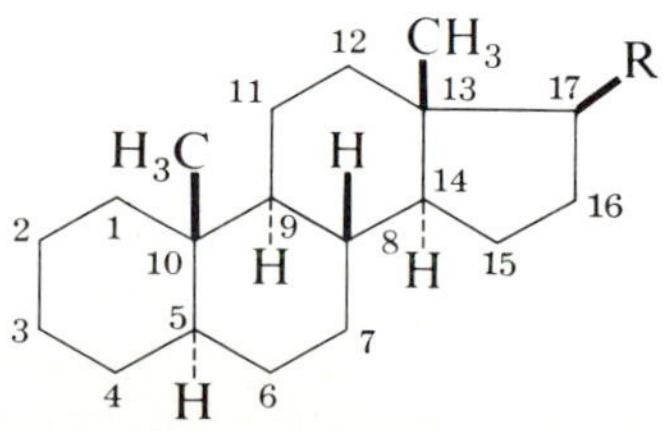

general steroid ring structure

The tetracyclic steroid nucleus consists of three cyclohexane rings and one cyclopentane ring, each fused to its neighboring ring in a trans manner.

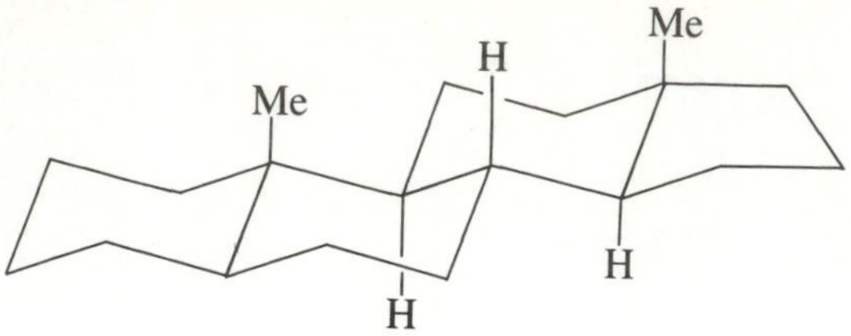

Other steroids are also common constituents of animal tissues and play important roles in normal biological process. Cholic acid, deoxycholic acid, and chenodeoxycholic acid occur in the bile duct.

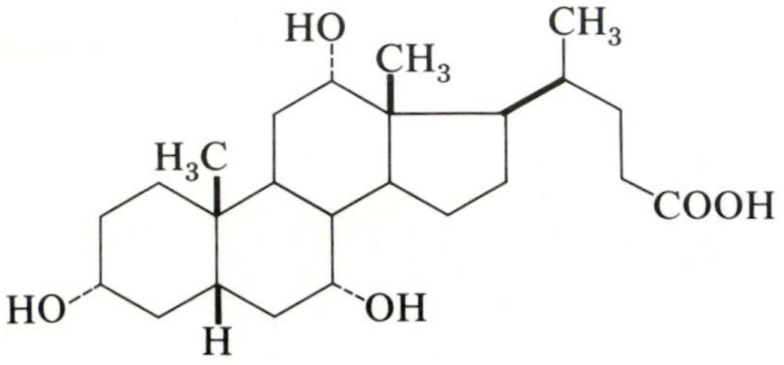

cholic acid

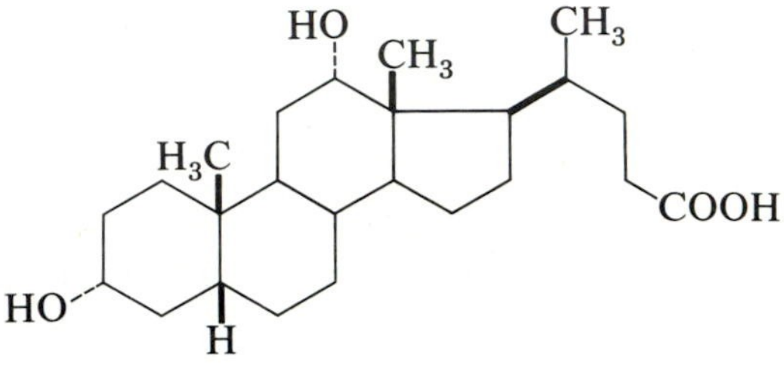

deoxycholic acid

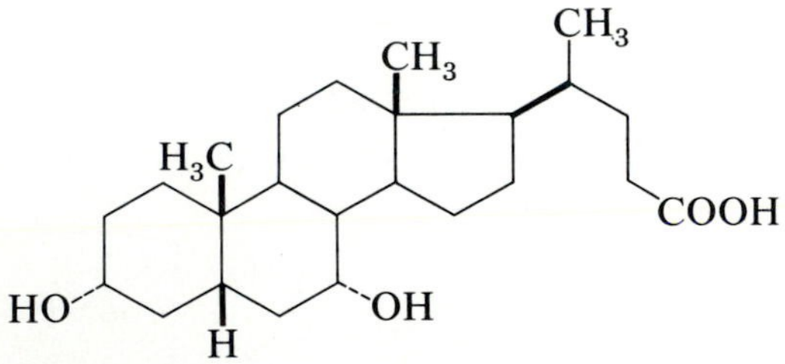

chenodeoxycholic acid

The bile acids exist as amides of the amino acid glycine, H_2NCH_2COOH, or the aminosulfonic acid taurine, $H_2NCH_2CH_2SO_3H$. The sodium salts of the bile acids form micelles (Section 18.4.D), which solubilize lipids in the intestinal tract and thus promote the absorption of fats.

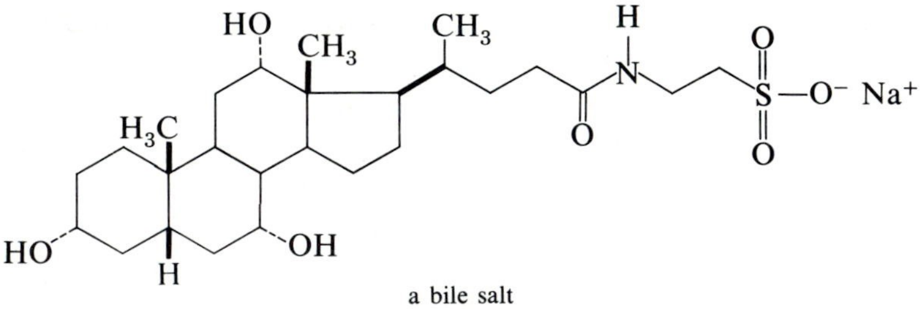

a bile salt

Estrone, progesterone, testosterone, and androsterone are steroid sex hormones.

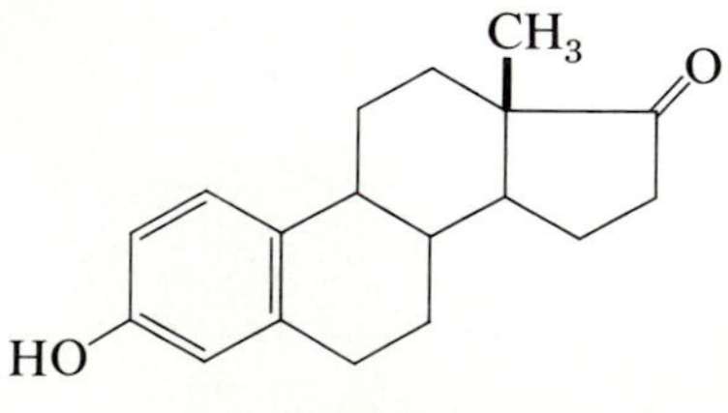

estrone

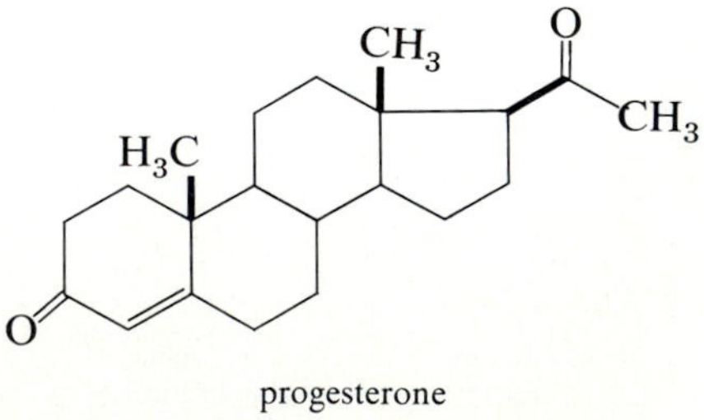

progesterone

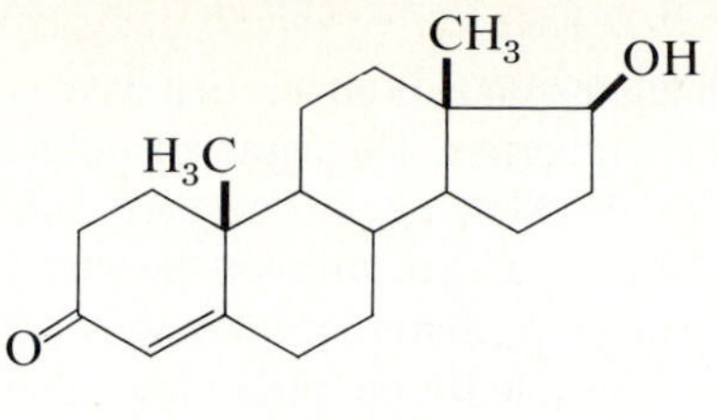

testosterone

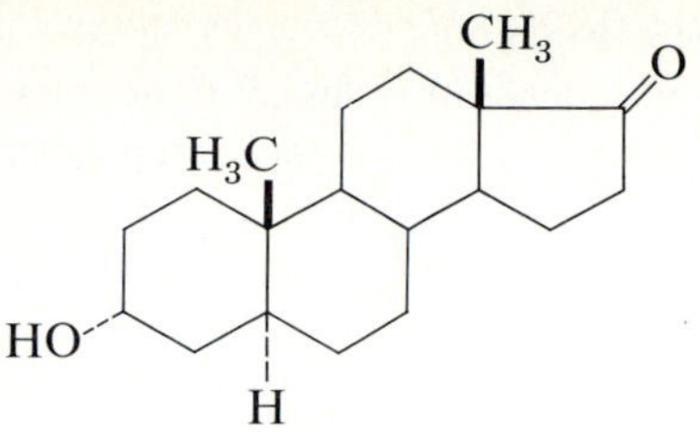

androsterone

Estrone is an example of an **estrogen,** or female sex hormone. Estrogens are secreted by the ovary and are responsible for the typical female sexual characteristics. Progesterone is another type of female sex hormone. It is also produced in the ovary and is the progestational hormone of the placenta and corpus luteus. Testosterone and androsterone are **androgens,** or male sex hormones. They are produced in the testes and are responsible for the typical male sexual characteristics.

One of the most dramatic achievements of synthetic organic chemistry, and one that has already had profound impact on the history and mores of human societies, has been the development of ''the pill.'' Actually, there are a number of different oral contraceptives in use. They are mainly synthetic steroids that interfere in some way with the normal estrus or progestational cycle in the female. One example is norethindrone, also known by the trade name Norlutin.

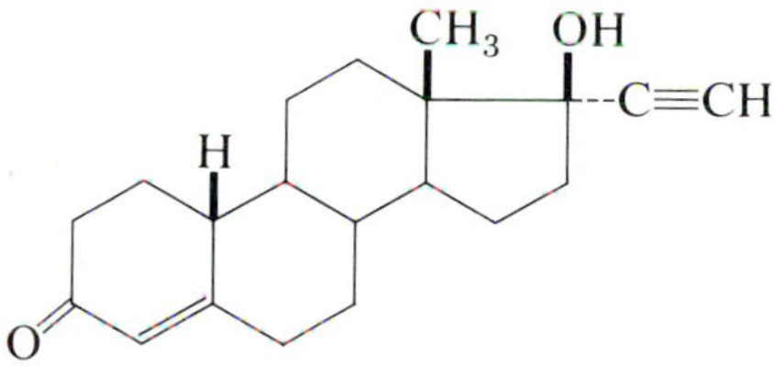

norethindrone
Norlutin

Steroids are widespread in the plant kingdom as well as in animals. One example is digitalis, a preparation made from the dried seeds and leaves of the purple foxglove. Historically, digitalis was used as a poison and as a medicine in heart therapy. The active agents in digitalis are **cardiac glycosides,** complex molecules built up from a steroid and several carbohydrates. Hydrolysis of digitoxin, one of the cardiac glycosides from digitalis, yields the steroid digitoxigenin.

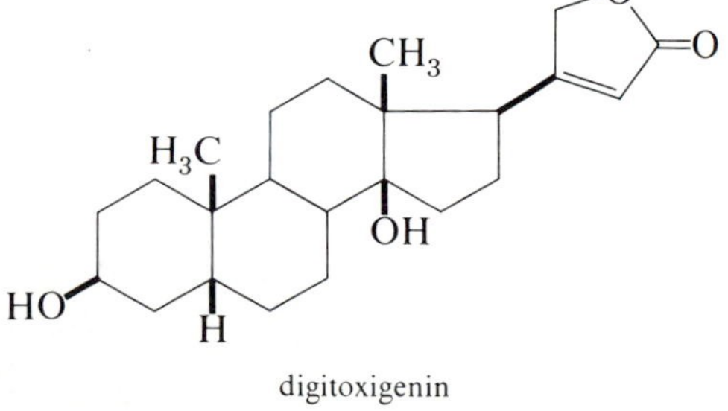

digitoxigenin

EXERCISE 36.15 On page 1252 is shown a three-dimensional perspective of the basic steroid nucleus. Write similar perspective drawings for the male sex hormone androsterone and the bile acid cholic acid.

C. Alkaloids

Alkaloids constitute a class of basic, nitrogen-containing plant products that have complex structures and possess significant pharmacological properties. The name alkaloid, or "alkali-like," was first proposed by the pharmacist W. Meissner in the early nineteenth century before anything was known about the chemical structures of the compounds.

Morphine was the first alkaloid isolated in a pure state; Sertürner accomplished this in 1805. The compound occurs in poppies and is responsible for the physiological effect of opium.

morphine

Other members of the morphine family are the O-methyl derivative codeine and the diacetyl derivative heroin.

codeine

heroin

The stereo structure of codeine hydrobromide is shown in Figure 36.5.

Another common family of rather simple alkaloids is related to phenylethylamine. An example is mescaline, which occurs in several species of cactus. It is the active principle of mescal buttons, which were once used by some American Indians in religious rites. It has more recently gained notoriety as an illegal hallucinogen. However, studies have

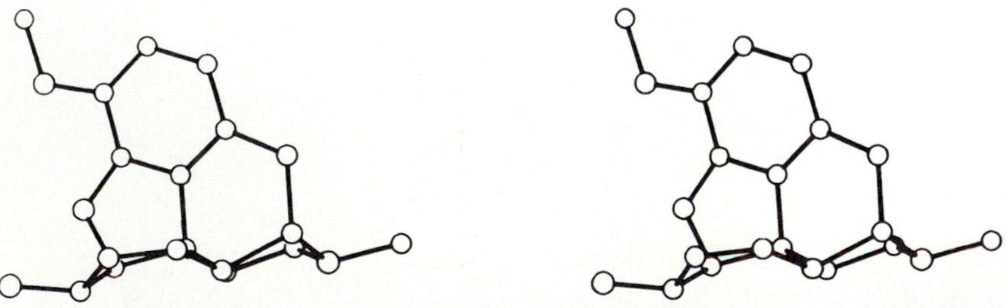

FIGURE 36.5 Stereo structure of codeine hydrobromide. Hydrogens are not shown. [Reproduced with permission from *Molecular Structure and Dimensions,* International Union of Crystallography, 1972.]

shown that virtually all ''mescaline'' in street sales is actually LSD (lysergic acid diethylamide), which is an even more potent hallucinogen.

CH_3O $CH_2CH_2NH_2$ CH_3O OCH_3

mescaline

O ‖ $CN(C_2H_5)_2$ H—N H N CH_3

lysergic acid diethylamide
"LSD"

Note that both of these hallucinogens contain a β-phenylethylamine grouping, as does amphetamine (page 738).

Another representative alkaloid is the tropane alkaloid cocaine, which is used as an external local anesthetic. It has become a substance of abuse. The precise mechanism of euphoria induction is not yet known.

CH_3 N CO_2CH_3 H O ‖ OCC_6H_5 H

cocaine

Quinine is an alkaloid from cinchona bark, which is still one of the major antimalarial agents. Resistance to quinine by malaria parasites is a growing danger.

H HO H CH_3O N N

quinine

Nicotine is the chief alkaloid of the tobacco plant.

H N N CH_3

nicotine

Strychnine and brucine are intricate constructions by nature, being heptacyclic alkaloids. These have been used as rodent poisons. They also find use as resolving agents in organic chemistry (page 738), since they are inexpensive and optically active and form well-defined salts with a variety of organic acids.

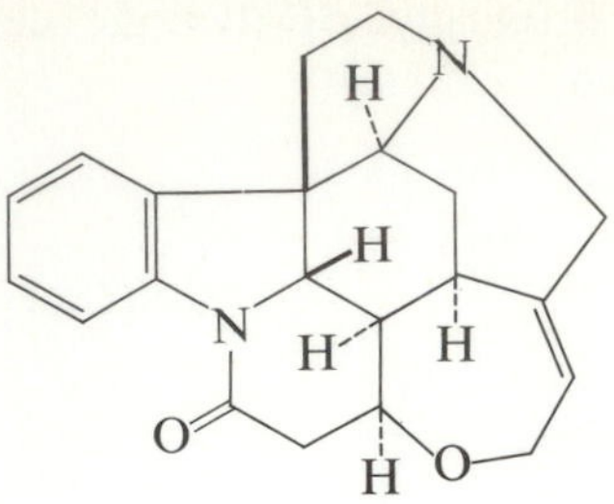

strychnine

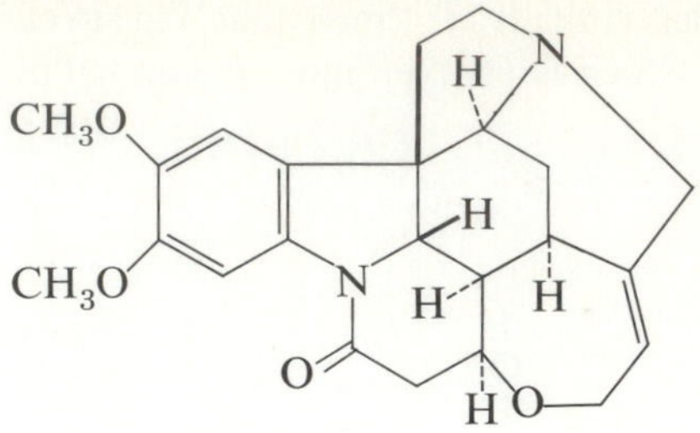

brucine

Other interesting structures are coniine, the toxic principle of poison hemlock, *Conium maculatum,* the fatal potion taken by the Greek philosopher Socrates in 399 B.C.; sparteine, a constituent of black lupin beans; and lycopodine, a fascinating tetracyclic component of the club moss.

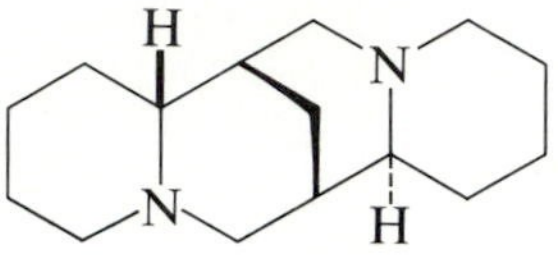

sparteine

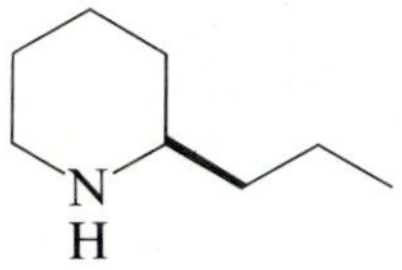

coniine

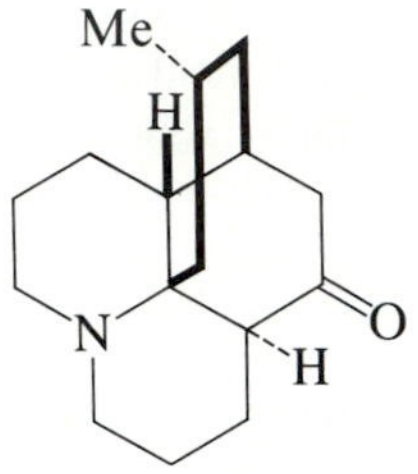

lycopodine

The examples cited show that alkaloids typically have potent physiological properties. In fact, this characteristic was partly responsible for the fact that the alkaloids were among the first organic compounds to be isolated in a pure state. Another reason for their early recognition as discrete chemical entities is the fact that they are rather easy to obtain from complex plant material because of their basic nature.

APPENDIX I

HEATS OF FORMATION

ΔH_f° (gas, 25°C), kcal mole^{-1}

Alkanes

methane	—17.9	2,2-dimethylpropane	—40.3
ethane	—20.2	hexane	—39.9
propane	—24.8	2-methylpentane	—41.8
butane	—30.4	3-methylpentane	—41.1
2-methylpropane	—32.4	2,2-dimethylbutane	—44.5
pentane	—35.1	2,3-dimethylbutane	—42.6
2-methylbutane	—36.9		

Cycloalkanes

cyclopropane	12.7	methylcyclopentane	—25.3
cyclobutane	6.8	methylcyclohexane	—37.0
cyclopentane	—18.4	ethylcyclohexane	—41.0
cyclohexane	—29.5	1,1-dimethylcyclohexane	—43.2
cycloheptane	—28.2	*cis*-1,2-dimethylcyclohexane	—41.1
cyclooctane	—29.7	*trans*-1,2-dimethylcyclohexane	—43.0
cyclononane	—31.7	*cis*-1,3-dimethylcyclohexane	—44.1
cyclodecane	—36.9	*trans*-1,3-dimethylcyclohexane	—42.2
cubane	148.7	*cis*-1,4-dimethylcyclohexane	—42.2
		trans-1,4-dimethylcyclohexane	—44.1

Alkenes

ethylene	12.5	2-methyl-1-butene	— 8.6
propene	4.9	2-methyl-2-butene	—10.1
1-butene	— 0.2	cyclobutene	37.5
cis-2-butene	— 1.9	cyclopentene	8.2
trans-2-butene	— 3.0	cyclohexene	— 1.1
2-methylpropene	— 4.3	1-methylcyclohexene	—10.3
1-pentene	— 5.3	cycloheptene	— 2.2
cis-2-pentene	— 7.0	cyclooctene	— 6.5
trans-2-pentene	— 7.9		

Alkynes and Polyenes

acetylene	54.3	*cis*-1,3-pentadiene	19.1
propyne	44.4	*trans*-1,3-pentadiene	18.1
1-butyne	39.5	1,4-pentadiene	25.3
2-butyne	34.7	2-methyl-1,3-butadiene	18.1
allene	45.6	cyclopentadiene	31.9
1,2-butadiene	38.8	1,3-cyclohexadiene	25.4
1,3-butadiene	26.1	1,3,5,7-cyclooctatetraene	71.1
1,2-pentadiene	33.6		

Aromatic Hydrocarbons

benzene	19.8	styrene	35.3
toluene	12.0	naphthalene	36.1
o-xylene	4.6	1,2,3,4-tetrahydronaphthalene	7.3
m-xylene	4.1	anthracene	55.2
p-xylene	4.3	9,10-dihydroanthracene	38.2
ethylbenzene	7.1	phenanthrene	49.5

Alcohols

methanol	—48.1	*t*-butyl alcohol	—74.7
ethanol	—56.2	cyclopentanol	—58.0
allyl alcohol	—29.6	cyclohexanol	—68.4
1-propanol	—61.2	benzyl alcohol	—24.0
2-propanol	—65.1	ethylene glycol	—93.9

Ethers

dimethyl ether	—44.0	1,1-dimethoxyethane	— 93.3
ethylene oxide	—12.6	2,2-dimethoxypropane	—101.9
tetrahydrofuran	—44.0	anisole	— 17.3
diethyl ether	—60.3		

Aldehydes and Ketones

formaldehyde	—26.0	butanal	—49.0
acetaldehyde	—39.7	cyclopentanone	—46.0
propionaldehyde	—45.5	cyclohexanone	—54.0
acetone	—51.9	benzaldehyde	— 8.8
2-butenal	—24.0		

Other Oxygen Compounds

formic acid	— 90.6	benzoic acid	— 70.1
acetic acid	—103.3	acetic anhydride	—137.1
vinyl acetate	— 75.5	furan	— 8.3
methyl acetate	−97.9	phenol	— 23.0
ethyl acetate	—106.3		

Nitrogen Compounds

methylamine	— 5.5	pyridine	34.6
dimethylamine	— 4.7	piperidine	—11.8
trimethylamine	— 5.7	aniline	20.8
ethylamine	—11.4	benzonitrile	51.5
acrylonitrile	44.1	dimethylformamide	—45.8
acetonitrile	17.6	acetanilide	—30.8
propionitrile	12.1	methyl nitrite	—15.8
pyrrole	25.9	nitromethane	—17.9
pyrrolidine	— 0.8	glycine	—93.7

Halogen Compounds

methyl chloride	—20.6	bromobenzene	25.2
methylene chloride	—23.0	chlorobenzene	12.2
chloroform	—24.6	acetyl chloride	—58.4
carbon tetrachloride	—25.2	methyl fluoride	−56.8
vinyl chloride	8.6	methyl bromide	— 9.1
ethyl chloride	—26.1	methyl iodide	3.4
n-propyl chloride	—31.0	ethyl bromide	—15.2
isopropyl chloride	—33.6	benzyl chloride	4.5

Sulfur Compounds

methanethiol	−5.4	thiirane	19.7
ethanethiol	−11.0	dimethyl sulfoxide	−36.1
dimethyl sulfide	−8.9	dimethyl sulfone	−89.1
dimethyl disulfide	−5.6	thiophene	27.6
thiophenol	26.9	tetrahydrothiophene	−8.1

Inorganic Compounds

CO_2	−94.05	NH_3	—10.9
H_2O	—57.80	CO	—26.42
H_2S	−4.8	H_2NNH_2	22.7
SO_2	−71.0	O_3	34.0
HCl	—22.1	NO_2	7.9
Br_2	7.4	HF	—65.0
HBr	— 8.7	HNO_3	—32.1
I_2	14.9	HNO_2	—18.4
HI	6.3	H_2O_2	—32.53
H_2O_2	−32.5	NO	21.6
		HCN	31.2

Atoms and Radicals

H	52.1	$CH_3\cdot$	35
Li	38.4	$C_2H_5\cdot$	29
		$CH_3CH_2CH_2\cdot$	24
C	170.9	$(CH_3)_2CH\cdot$	21
N	113.0	$(CH_3)_3C\cdot$	12
O	59.6	$CH_2{=}CH\cdot$	72
		$HC{\equiv}C\cdot$	135
F	18.9	$CH_2{=}CHCH_2\cdot$	39
Cl	28.9	$C_6H_5CH_2\cdot$	48
Br	26.7	$C_6H_5\cdot$	79
I	25.5	$CH_3CO\cdot$	−6
S	65.7	$CH_3CO_2\cdot$	−50
Na	25.8	$CH_3O\cdot$	4
HO	9.4	$C_2H_5O\cdot$	−4
H_2N	44	$HCO\cdot$	9
CN	104	$HOOC\cdot$	−53
SH	34	$CH_3SO_2\cdot$	−61
		$CH_3COCH_2\cdot$	−12
		cyclopropyl·	67
		$CH_3OOC\cdot$	−40

Cations and Anions

CH_3^+	262	H^-	34.7
$C_2H_5^+$	219	HO^-	−32.7
$(CH_3)_2CH^+$	187	F^-	−59.5
$(CH_3)_3C^+$	163	Cl^-	−54.5
$CH_2{=}CHCH_2^+$	225	Br^-	−50.9
$C_6H_5CH_2^+$	214	I^-	−45.1
H^+	365.7	CN^-	16
		$CH_3CO_2^-$	−120.5
		NO_2^-	−27.1

APPENDIX II

BOND-DISSOCIATION ENERGIES

***DH*°, kcal mole^{-1} for A—B Bonds**

	A	(52.1) B : H	(18.9) F	(28.9) Cl	(26.7) Br	(25.5) I	(9.4) OH	(44) NH_2	(35) Me	(29) Et	(21) *i*-Pr	(12) *t*-Bu	(79) Ph	(104) CN
(35)	methyl	105	110	85	71	57	93	85	90	86	86	84	102	122
(29)	ethyl	101	111	83	71	56	95	85	89	88	87	85	101	121
(24)	propyl	101	110	84	71	56	95	85	89	88	86	85	101	120
(21)	isopropyl	98	109	84	72	56.5	96	85	89	87	85	82	99	119
(12)	*t*-butyl	96	113	84	70	55	96	85	87	95	82	77	99	
(79)	phenyl	111	126	96	80.5	65	111	102	102	100	99	96	115	131
(48)	benzyl	88		72	58	48	81	71	76	75	74	73	90	
(39)	allyl	86		68	54	41	78		74	70	70	67		
(−6)	acetyl	86	119	81	66	49	107		81	79	77	75	93.5	
(−4)	ethoxy	104					44		83	85			101	
(72)	vinyl	112		92	80				102	101	100	95	105	132
(52.1)	H	104.2	135.8	103.2	87.5	71.3	119	107	105	101	98	96	111	125

Numbers in parentheses are the heats of formation, ΔH_f°, for the corresponding atom or radical.

APPENDIX III

AVERAGE BOND ENERGIES

Average Bond Energies, kcal mole^{-1}

H	C	N	O	F	Si	S	Cl	Br	I	
104	99	93	111	135	76	83	103	87	71	H
	83[a]	73[b]	86[c]	116[d]	72	65	81	68	52	C
		39	53[e]	65			46			N
			47	45	108		52	48	56	O
				37	135					F
					53		91	74	56	Si
						60	61	52		S
							58			Cl
								46		Br
									36	I

[a] C=C 146, C≡C 200.

[b] C=N 147, C≡N 213.

[c] C=O 176 (aldehydes), 179 (ketones).

[d] In CF_4.

[e] In nitrites and nitrates.

APPENDIX IV

ACID DISSOCIATION CONSTANTS

Acidities of Inorganic Acids at 25°C

Name	Formula	pK_a
ammonia		34[a]
ammonium ion	NH_4^+	9.24
boric acid	H_3BO_3	9.24
carbon dioxide	CO_2	6.35[b]
cyanic acid	HOCN	3.46
hydrazinium ion	$H_2N\overset{+}{N}H_3$	7.94
hydrazoic acid	HN_3	4.68
hydriodic acid	HI	−5.2
hydrobromic acid	HBr	−4.7
hydrochloric acid	HCl	−2.2
hydrocyanic acid	HCN	9.22
hydrofluoric acid	HF	3.18
hydrogen peroxide	H_2O_2	11.65
hydrogen selenide	H_2Se	3.89 (11.0)[c]
hydrogen sulfide	H_2S	6.97 (12.9)[c]
hydroxylammonium ion	$H_3\overset{+}{N}OH$	5.95
hypobromous acid	HOBr	8.6
hypochlorous acid	HOCl	7.53
hypophosphorus acid	H_3PO_2	1.2
nitric acid	$HONO_2$	−1.3
nitrous acid	HONO	3.23
periodic acid	H_3IO_5	1.55 (8.27)[c]
phosphoric acid	$(HO)_3PO$	2.15 (7.20, 12.38)[c]
sulfuric acid	$(HO)_2SO_2$	≈ −5.2 (1.99)[c]
sulfurous acid	$(HO)_2SO$	1.8 (7.2)[c]
thiocyanic acid	HCNS	−1.9

[a] In liquid ammonia.

[b] For the equilibrium $CO_2(aq) = H^+(aq) + HCO_3^-(aq)$.

[c] Second and third acidity constants in parentheses.

Acidities of Organic Acids at 25°C

Acid	pK_a	Acid	pK_a
$CH_3\overset{O}{\overset{\|}{\underset{+}{N}}}OH$	−11.9	$(CH_3)_2C{=}\overset{H}{\underset{+}{N}}OH$	−1.9
$C_6H_5\overset{O}{\overset{\|}{\underset{+}{N}}}OH$	−11.3	CH_3SO_3H	≈−1.2
$C_6H_5C{\equiv}\overset{+}{N}H$	−10.5	$CH_3\overset{OH}{\overset{\|}{C}}{=}NH_2^+$	≈0
$CH_3C{\equiv}\overset{+}{N}H$	−10.1	$CH_3(CH_2)_3\overset{+}{P}H_3$	0
$CH_3\overset{H}{\overset{\|}{C}}{=}\overset{+}{O}H$	≈−8	$(CH_3)_2\overset{+}{S}OH$	0
$C_6H_5\overset{OH}{\overset{\|}{C}}{=}\underset{+}{O}H$	−7.3	CF_3COOH	0.2
$(CH_3)_2C{=}\overset{+}{O}H$	−7.2	O_2N, NO_2, NO_2 ring with OH (2,4,6-trinitrophenol)	0.25
$C_6H_5\overset{H}{\overset{\|}{C}}{=}\overset{+}{O}H$	−7.1	pyridine ring $\overset{+}{N}$–OH	0.79
$CH_3\overset{+}{S}H_2$	−6.8	$(C_6H_5)_2\overset{+}{N}H_2$	0.8
$C_6H_5\overset{+}{O}H_2$	−6.7	$O_2N{-}C_6H_4{-}\overset{+}{N}H_3$	1.00
$C_6H_5\overset{H}{O}CH_3$	≈−6.5	$C_6H_5\overset{+}{N}H{=}\overset{OH}{\overset{\|}{C}}C_6H_5$	2.17
$CH_3\overset{OC_2H_5}{\overset{\|}{C}}{=}\underset{+}{O}H$	−6.5	$CH_3\overset{+}{P}H_3$	≈2.5
$C_6H_5N{=}\overset{OH}{\overset{\|}{\underset{+}{N}}}C_6H_5$	−6.45	$O_2N{-}C_6H_4{-}COOH$	3.42
$C_6H_5\overset{^+OH}{\overset{\|}{C}}CH_3$	−6.2	$CH_2(NO_2)_2$	3.57
$CH_3\overset{^+OH}{\overset{\|}{C}}OH$	−6.1	$(CH_3)_2\overset{+}{P}H_2$	3.91
$(CH_3)_2\overset{+}{S}H$	−5.4	$O_2N{-}C_6H_3(NO_2){-}OH$ (NO_2 ortho to OH)	4.09
$(CH_3)_3C\overset{+}{O}H_2$	−3.8	$C_6H_5\overset{+}{N}H_3$	4.60
$(CH_3CH_2)_2\overset{+}{O}H$	−3.6	$(CH_3)_3\overset{+}{N}OH$	4.7
$(CH_3)_2CH\overset{+}{O}H_2$	−3.2	CH_3COOH	4.74
$C_6H_5N{=}\overset{H}{\underset{+}{N}}C_6H_5$	−2.9		
$C_2H_5\overset{+}{O}H_2$	−2.4		
$CH_3\overset{+}{O}H_2$	−2.2		
$C_6H_5\overset{OH}{\overset{\|}{C}}{=}\overset{+}{N}H_2$	−2.0		

Acidities of Organic Acids at 25°C (continued)

Acid	pK_a	Acid	pK_a
N^+ H	5.29		16.0
$(CH_3CO)_3CH$	5.85	$C_6H_5COCH_3$	16
$\overset{+}{N}H$, N H	7.0	$(CH_3)_3COH$	18
O_2N—⟨ ⟩—OH	7.15	CH_3COCH_3	20
C_6H_5SH	7.8		20
$(CH_3)_3\overset{+}{P}H$	8.65		23
$(CH_3CO)_2CH_2$	9	$CH_3SO_2CH_3$	23
$(CH_3)_3\overset{+}{N}H$	9.79	$CH_3COOC_2H_5$	24.5
C_6H_5OH	10.00	$HC{\equiv}CH$	≈25
CH_3NO_2	10.21	CH_3CN	≈25
CH_3CH_2SH	10.60	$(C_6H_5)_3CH$	31.5
$CH_3\overset{+}{N}H_3$	10.62	$(C_6H_5)_2CH_2$	34
$(CH_3)_2\overset{+}{N}H_2$	10.73	$C_2H_5NH_2$	≈35
$CH_3COCH_2COOC_2H_5$	11	$C_6H_5CH_3$	41
$CH_2(CN)_2$	11.2	—H	43
CF_3CH_2OH	12.4	$CH_2{=}CH_2$	44
$CH_2(COOC_2H_5)_2$	13.3		46
$(CH_3SO_2)_2CH_2$	14	CH_4	≈49
CH_3OH	15.5	C_2H_6	≈50
$(CH_3)_2CHCHO$	15.5		≈52
C_2H_5OH	15.9		

APPENDIX V

PROTON CHEMICAL SHIFTS

Proton Chemical Shifts, δ, ppm, for C—H

Y	CH_3Y	CH_3—C—Y	CH_3—C—C—Y	R—CH_2—Y	RCH_2—C—Y	R_2CH—Y
H	0.23	0.9	0.9	0.9	1.3	1.3
CH=CH_2	1.71	1.0		2.0		1.7
C≡CH	1.80	1.2	1.0	2.1	1.5	2.6
C_6H_5	2.35	1.3	1.0	2.6	1.7	2.9
F	4.27	1.2		4.4		
Cl	3.06	1.5	1.1	3.5	1.8	4.1
Br	2.69	1.7	1.1	3.4	1.9	4.2
I	2.16	1.9	1.0	3.2	1.9	4.2
OH	3.39	1.2	0.9	3.5	1.5	3.9
OR	3.24	1.2	1.1	3.3	1.6	3.6
OAc	3.67	1.3	1.1	4.0	1.6	4.9
CHO	2.18	1.1	1.0	2.4	1.7	2.4
$COCH_3$	2.09	1.1	0.9	2.4	1.6	2.5
COOH	2.08	1.2	1.0	2.3	1.7	2.6
NH_2	2.47	1.1	0.9	2.7	1.4	3.1
$NHCOCH_3$	2.71	1.1	1.0	3.2	1.6	4.0
SH	2.00	1.3	1.0	2.5	1.6	3.2
CN	1.98	1.4	1.1	2.3	1.7	2.7
NO_2	4.29	1.6	1.0	4.3	2.0	4.4

Proton Chemical Shifts for Y—H

Group Type	δ, ppm
ROH	0.5–5.5
ArOH	4–8
RCOOH	10–13
R_2C=NOH	7.4–10.2
RSH	0.9–2.5
ArSH	3–4
RSO_3H	11–12
RNH_2, R_2NH	0.4–3.5
$ArNH_2$, ArRNH	2.9–4.8
$RCONH_2$	5.0–6.5
RCONHR	6.0–8.2
RCONHAr	7.8–9.4

APPENDIX VI

SYMBOLS AND ABBREVIATIONS

Å — Ångstrom unit (10^{-8} cm)
Ac — acetyl group, $CH_3CO—$
Ar — aryl radical
$[\alpha]$ — specific optical activity
aq. — aqueous
Boc — *t*-butoxycarbonyl group, $(CH_3)_3COCO—$
n-Bu — *n*-butyl group, $CH_3CH_2CH_2CH_2—$
t-Bu — *t*-butyl group, $(CH_3)_3C—$
CMR — ^{13}C magnetic resonance
Cbz — benzyloxycarbonyl group, $C_6H_5CH_2OCO—$
D — Debye (10^{-18} esu cm); measure of dipole moment
DCC — dicyclohexylcarbodiimide, $C_6H_{11}N{=}C{=}NC_6H_{11}$
δ — chemical shift downfield from TMS, given as ppm
Δ — symbol for heat supplied to a reaction
$\Delta G°$ — standard Gibbs free energy of reaction
$\Delta G^‡$ — Gibbs free energy of activation
$\Delta H°$ — standard enthalpy of reaction
$\Delta H_f°$ — enthalpy of formation from standard states
$\Delta H^‡$ — enthalpy of activation
$\Delta S°$ — standard entropy of reaction
$\Delta S^‡$ — entropy of activation
$DH°$ — bond dissociation energy
DIBAL — diisobutylaluminum hydride, $[(CH_3)_2CHCH_2]_2AlH$
diglyme — di-2-methoxyethyl ether, $(CH_3OCH_2CH_2)_2O$
DMF — dimethylformamide, $(CH_3)_2NCHO$
DMSO — dimethyl sulfoxide, $(CH_3)_2SO$
DNP — 2,4-dinitrophenyl group, 2,4-$(O_2N)_2C_6H_3—$ *or* 2,4-dinitrophenylhydrazone, 2,4-$(O_2N)_2C_6H_3NHN{=}$
E — *entgegen*, opposite sides in (*E,Z*) nomenclature of alkenes
E1 — unimolecular elimination reaction mechanism
E2 — bimolecular elimination reaction mechanism
EA — electron affinity
Et — ethyl group, $CH_3CH_2—$
eu — entropy units, cal deg^{-1} $mole^{-1}$
f_i — partial rate factor at position *i*
glyme — 1,2-dimethoxyethane, $CH_3OCH_2CH_2OCH_3$
H — magnetic field

Symbol	Meaning
HMPT	hexamethylphosphoric triamide, $[(CH_3)_2N]_3PO$
HOMO	highest occupied molecular orbital
$h\nu$	symbol for light
Hz	Hertz (sec^{-1} or cycles per second)
IP	ionization potential
IR	infrared
J	coupling constant, usually in Hz
k	rate constant for reaction
K	equilibrium constant for reaction
K_a	acid dissociation constant
LDA	lithium diisopropylamide, $LiN[CH(CH_3)_2]_2$
LUMO	lowest unoccupied molecular orbital
Me	methyl group, CH_3—
m/z	mass-to-charge ratio in mass spectrometry
MHz	megaHertz $\equiv 10^6$ Hz
μ	dipole moment
NMR	nuclear magnetic resonance
NR	no reaction; also indicated by $\not\longrightarrow$
Ph	phenyl radical, C_6H_5—
pH	measure of acidity $\equiv -\log [H^+]$
pK_a	measure of acid strength $\equiv -\log K_a$
PMR	proton magnetic resonance
PPA	polyphosphoric acid
ψ	wave function or orbital
R	alkyl or cycloalkyl group
(R,S)	designation of stereochemical configuration
S_N1	unimolecular nucleophilic substitution mechanism
S_N2	bimolecular nucleophilic substitution mechanism
THF	tetrahydrofuran, $\overline{CH_2CH_2CH_2CH_2O}$
TMS	tetramethylsilane, $(CH_3)_4Si$
Ts	tosyl or *p*-toluenesulfonyl group, *p*-$CH_3C_6H_4SO_2$—
UV	ultraviolet
X	halogen group
xs	excess
Z	*zusammen*, same side in (E,Z) nomenclature of alkenes
	symbol for flow of electron pair

INDEX

C

H

J

K

L

N

O

U

V

W

X

Y

Z

Periodic Table of the Elements

1 H 1.008																	2 He 4.003
3 Li 6.94	4 Be 9.01											5 B 10.81	6 C 12.011	7 N 14.01	8 O 16.00	9 F 19.00	10 Ne 20.18
11 Na 22.99	12 Mg 24.31											13 Al 26.98	14 Si 28.09	15 P 30.97	16 S 32.06	17 Cl 35.45	18 Ar 39.95
19 K 39.10	20 Ca 40.08	21 Sc 44.96	22 Ti 47.90	23 V 50.94	24 Cr 52.00	25 Mn 54.94	26 Fe 55.85	27 Co 58.93	28 Ni 58.71	29 Cu 63.55	30 Zn 65.37	31 Ga 69.72	32 Ge 72.59	33 As 74.92	34 Se 78.96	35 Br 79.90	36 Kr 83.80
37 Rb 85.47	38 Sr 87.62	39 Y 88.91	40 Zr 91.22	41 Nb 92.91	42 Mo 95.94	43 Tc 98.91	44 Ru 101.07	45 Rh 102.91	46 Pd 106.4	47 Ag 107.87	48 Cd 112.40	49 In 114.82	50 Sn 118.69	51 Sb 121.75	52 Te 127.60	53 I 126.90	54 Xe 131.30
55 Cs 132.91	56 Ba 137.34	57 La 138.91	72 Hf 178.49	73 Ta 180.95	74 W 183.85	75 Re 186.2	76 Os 190.2	77 Ir 192.2	78 Pt 195.09	79 Au 196.97	80 Hg 200.59	81 Tl 204.37	82 Pb 207.19	83 Bi 208.98	84 Po (209)	85 At (210)	86 Rn (222)
87 Fr (223)	88 Ra 226.03	89 Ac (227)	104 Unq* (261)	105 Unp* (262)	106 Unh* (263)	107 Uns* (262)	108 Uno* (265)	109 Una* (266)									

Lanthanides	58 Ce 140.12	59 Pr 140.91	60 Nd 144.24	61 Pm (145)	62 Sm 150.35	63 Eu 151.96	64 Gd 157.25	65 Tb 158.93	66 Dy 162.50	67 Ho 164.93	68 Er 167.26	69 Tm 168.93	70 Yb 173.04	71 Lu 174.97
Actinides	90 Th 232.04	91 Pa (231)	92 U 238.03	93 Np (237)	94 Pu (244)	95 Am (243)	96 Cm (247)	97 Bk (249)	98 Cf (249)	99 Es (254)	100 Fm (257)	101 Md (258)	102 No (259)	103 Lr (260)

*Symbol (and name) provisional.

Numbers in parentheses: available radioactive isotope of longest half-life.